FOUNDATIONS IN
MICROBIOLOGY

FOUNDATIONS IN
MICROBIOLOGY

Kathleen Park
TALARO

Barry
CHESS

Mc
Graw
Hill
Education

FOUNDATIONS IN MICROBIOLOGY, TENTH EDITION

1 2 3 4 5 6 7 8 9 0 LWI 21 20 19 18 17

ISBN 978–1–259–70521-2
MHID 1–259–70521–8

Chief Product Officer, SVP Products & Markets: *G. Scott Virkler*
Vice President, General Manager, Products & Markets: *Marty Lange*
Vice President, Content Production & Technology Services: *Betsy Whalen*
Managing Director: *Lynn Breithaupt*
Brand Manager: *Marija Magner*
Director of Development: *Rose M. Koos*
Product Developer: *Mandy Clark*
Digital Product Analyst: *John J. Theobald*
Marketing Manager: *Jessica Cannavo*
Director, Content Production: *Linda Avenarius*
Program Manager: *Angie FitzPatrick*
Content Project Managers: *(core): Jayne Klein; (assessment): Brent dela Cruz*
Senior Buyer: *Laura Fuller*
Designer: *Tara McDermott*
Cover Image: © *National Institute of Allergy and Infectious Diseases*
Content Licensing Specialists: *(Image): Carrie K. Burger; (Text): Lorraine Buczek*
Compositor: *Aptara®, Inc.*
Typeface: *STIX Mathjax*
Printer: *LSC Communications*

Library of Congress Cataloging-in-Publication Data

Names: Talaro, Kathleen P., author. | Chess, Barry, author.
 Title: Foundations in microbiology / Kathleen Park Talaro, Pasadena City
 College, Barry Chess, Pasadena City College.
 Description: Tenth edition. | New York, NY : McGraw-Hill Education, [2018]
 Identifiers: LCCN 2016040028| ISBN 9781259705212 (alk. paper) | ISBN
 1259705218 (alk. paper)
 Subjects: LCSH: Microbiology. | Medical microbiology.
 Classification: LCC QR41.2 .T35 2018 | DDC 616.9/041—dc23 LC record available at
 https://lccn.loc.gov/2016040028

2013041001

www.mhhe.com

Brief Contents

About the Authors

Kathleen Park Talaro is a microbiologist, educator, author, and artist. She has been nurturing her love of microbiology since her youth growing up on an Idaho farm where she was first fascinated by tiny creatures she could just barely see swimming in a pond. This interest in the microbial world led to a biology major at Idaho State University, where she worked as a teaching assistant and scientific illustrator for one of her professors. This was the beginning of an avocation that she continues today—that of lending her artistic hand to interpretation of scientific concepts. She continued her education at Arizona State University, Occidental College, California Institute of Technology, and California State University.

She taught microbiology and major's biology courses at Pasadena City College for 30 years, during which time she developed new curricula and refined laboratory experiments. She has been an author of, and contributor to, several publications of the William C. Brown Company and McGraw-Hill Publishers since the early 1980s, first illustrating and writing for laboratory manuals and later developing this textbook. She has also served as a co-author with Kelly Cowan on the first two editions of *Microbiology: A Systems Approach.*

Kathy continues to make microbiology a major focus of her life and is passionate about conveying the significance and practical knowledge of the subject to students, colleagues, family, friends, and practically anyone who shows interest. In addition to her writing and illustration, she keeps current by attending conferences and participating in the American Society for Microbiology and its undergraduate educational programs. She is gratified by the many supportive notes and letters she has received over the years from devotees of microbiology and users of her book.

She lives in Altadena, California, with husband Dave Bedrosian and son David. Whenever she can, she visits her family in Idaho. In her spare time, she enjoys photography, reading true crime books, music, crossword puzzles, and playing with her rescued kitties.

Barry Chess has been teaching microbiology at Pasadena City College for 20 years. He received his Bachelor's and Master's degrees from the California State University and did postgraduate work at the University of California, where his research focused on the expression of eukaryotic genes involved in the development of muscle and bone.

At Pasadena City College, Barry developed a new course in human genetics and helped to institute a biotechnology program. He regularly teaches courses in microbiology, general biology, and genetics, and works with students completing independent research projects in biology and microbiology. Over the past several years, Barry's interests have begun to focus on innovative methods of teaching that increase student success. He has written cases for the National Center for Case Study Teaching in Science and given talks at national meetings on the effectiveness of case studies in the classroom. His laboratory manual, *Laboratory Applications in Microbiology: A Case Study Approach,* is currently in its third edition. He feels very fortunate to be collaborating with Kathy Talaro, with whom he has worked in the classroom for more than a decade, on this tenth edition. Barry is a member of the American Society for Microbiology and the American Association for the Advancement of Science and regularly attends meetings in his fields of interest, both to keep current of changes in the discipline and to exchange teaching and learning strategies with others in the field.

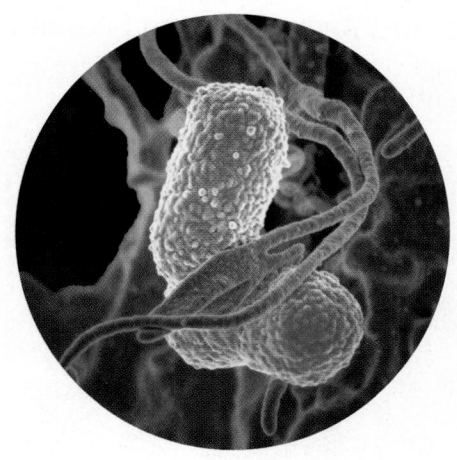

A major intent of this textbook has always been to promote an understanding of microbes and their intimate involvement in the lives of humans, but our other aim is to stimulate an appreciation that goes far beyond that. We want you to be awed by these tiniest creatures and the tremendous impact they have on all of the earth's natural activities. We hope you are inspired enough to embrace that knowledge throughout your lives.

Required=Results

©Getty Images/iStockphoto

McGraw-Hill Connect®
Learn Without Limits

Connect is a teaching and learning platform that is proven to deliver better results for students and instructors.

Connect empowers students by continually adapting to deliver precisely what they need, when they need it, and how they need it, so your class time is more engaging and effective.

73% of instructors who use Connect require it; instructor satisfaction increases by 28% when Connect is required.

Connect's Impact on Retention Rates, Pass Rates, and Average Exam Scores

70.1% 89.9% 72.5% 85.2% 71.0% 80.1%

Retention Rates Course Pass Rates Average Exam Scores

without Connect with Connect

Using **Connect** improves retention rates by **19.8%**, passing rates by **12.7%**, and exam scores by **9.1%**.

Analytics

Connect Insight®

Connect Insight is Connect's new one-of-a-kind visual analytics dashboard that provides at-a-glance information regarding student performance, which is immediately actionable. By presenting assignment, assessment, and topical performance results together with a time metric that is easily visible for aggregate or individual results, Connect Insight gives the user the ability to take a just-in-time approach to teaching and learning, which was never before available. Connect Insight presents data that helps instructors improve class performance in a way that is efficient and effective.

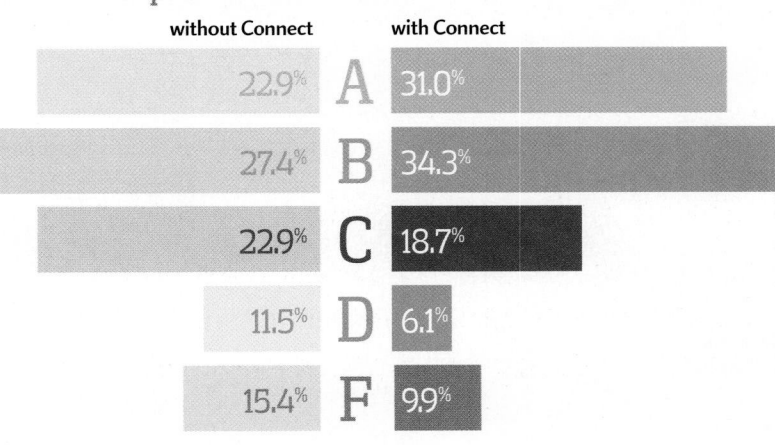

Impact on Final Course Grade Distribution

	without Connect	with Connect
A	22.9%	31.0%
B	27.4%	34.3%
C	22.9%	18.7%
D	11.5%	6.1%
F	15.4%	9.9%

Adaptive

THE **ADAPTIVE** READING EXPERIENCE DESIGNED TO TRANSFORM THE WAY STUDENTS READ

More students earn **A's** and **B's** when they use McGraw-Hill Education **Adaptive** products.

SmartBook®

Proven to help students improve grades and study more efficiently, SmartBook contains the same content within the print book, but actively tailors that content to the needs of the individual. SmartBook's adaptive technology provides precise, personalized instruction on what the student should do next, guiding the student to master and remember key concepts, targeting gaps in knowledge and offering customized feedback, and driving the student toward comprehension and retention of the subject matter. Available on tablets, SmartBook puts learning at the student's fingertips—anywhere, anytime.

Over **8 billion questions** have been answered, making McGraw-Hill Education products more intelligent, reliable, and precise.

STUDENTS WANT

Mc Graw Hill Education **SMARTBOOK®**

95% of students reported **SmartBook** to be a more effective way of reading material.

100% of students want to use the Practice Quiz feature available within **SmartBook** to help them study.

100% of students reported having reliable access to off-campus wifi.

90% of students say they would purchase **SmartBook** over print alone.

95% of students reported that **SmartBook** would impact their study skills in a positive way.

Mc Graw Hill Education

*Findings based on 2015 focus group results administered by McGraw-Hill Education

www.mheducation.com

Connecting Instructors to Students—
Digital Tools for Your Success

Save time with auto-graded assessments. Gather powerful performance data.

McGraw-Hill Connect for Prescott's Microbiology provides online presentation, assignment, and assessment solutions, connecting your students with the tools and resources they'll need to achieve success.

Homework and Assessment

With **Connect for Talaro's Foundations in Microbiology,** you can deliver auto-graded assignments, quizzes, and tests online. Choose from a robust set of interactive questions and activities using high-quality art from the textbook and animations. Assignable content is available for every Learning Outcome in the book and is categorized according to the **ASM Curriculum Guidelines.** As an instructor, you can edit existing questions and author entirely new ones.

Significant faculty demand for content at higher Bloom's levels led us to examine assessment quality and consistency of our Connect content, to develop a scientific approach to systemically increase critical-thinking levels, and develop balanced digital assessments that promote student learning. The increased challenge at higher Bloom's levels will help the student grow intellectually and be better prepared to contribute to society.

Instructor Resources

Customize your lecture with tools such as PowerPoint® presentations, animations, and editable art from the textbook. An instructor's manual for the text saves you time in developing your course.

Learn more at connect.mheducation.com.

x

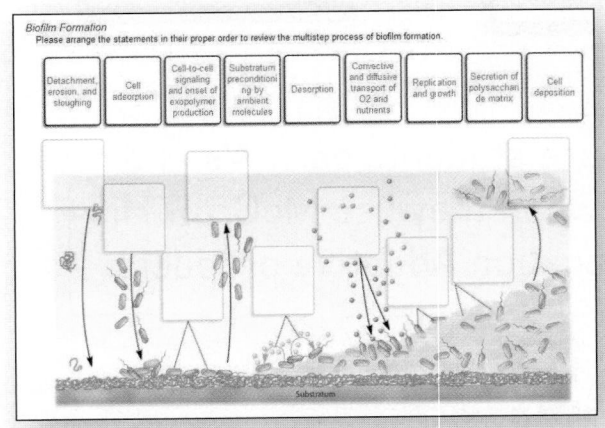

Biofilm Formation
Please arrange the statements in their proper order to review the multistep process of biofilm formation.

Detailed Reports

Track individual student performance—by question, by assignment, or in relation to the class overall—with detailed grade reports. Integrate grade reports easily with your Learning Management Systems (LMS).

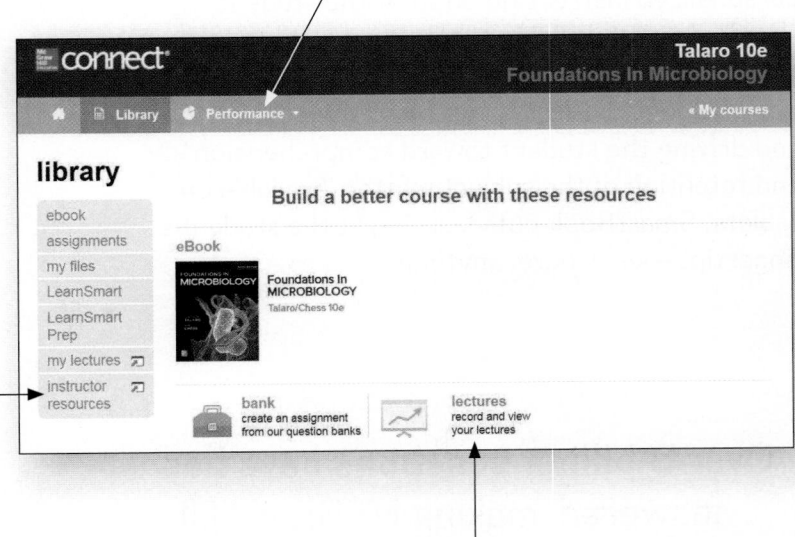

Lecture Capture

McGraw-Hill Tegrity® records and distributes your class lecture with just a click of a button. Students can view anytime, anywhere via computer or mobile device. Indexed as you record, students can use keywords to find exactly what they want to study.

Integrated and Adaptive Laboratory Tools

LearnSmart Labs® is an adaptive simulated lab experience that brings meaningful scientific exploration to students. Through a series of adaptive questions, LearnSmart Labs identifies a student's knowledge gaps and provides resources to quickly and efficiently close those gaps. Once students have mastered the necessary basic skills and concepts, they engage in a highly realistic simulated lab experience that allows for mistakes and the execution of the scientific method.

LearnSmart® Prep is an adaptive learning tool that prepares students for college-level work in Microbiology. LearnSmart Prep individually identifies concepts the student does not fully understand and provides learning resources to teach essential concepts so he or she enters the classroom prepared. Data-driven reports highlight areas where students are struggling, helping to accurately identify weak areas.

The **Profile** of a Student Success Learning Tool

Art and organization of content make this book unique

Carefully crafting a textbook to be a truly useful learning tool for students takes time and dedication. Every line of text and every piece of art in this book is scrutinized for instructional usefulness, placement, and pedagogy, and then reexamined with each revision. In this tenth edition, the authors have gone through the book page by page, with more depth than ever before, to make sure it maintains its instructional quality, fantastic art program, relevant and current material, and engaging, user-friendly writing style. Since the first edition, the goals of this book have been to explain complex topics clearly and vividly, and to present the material in a straightforward way that students can understand. The tenth edition continues to meet these goals with the most digitally integrated, up-to-date, and pedagogically important revision yet.

Like a great masterpiece hanging in a museum, *Foundations in Microbiology* is not only beautiful but also tells a story, composed of many pieces. A great textbook must be carefully constructed to place art where it makes the most sense in the flow of the narrative; create process figures that break down complex processes into their simplest parts; provide explanations at the correct level for the student audience; and offer pedagogical tools that help all types of learners. Many textbook authors write the narrative of their book and call it a day. It is the rare author team, indeed, that examines each page and makes changes based on what will help the students the most, so that when the pieces come together, the result is an expertly crafted learning tool—a story of the microbial world.

Kathy Talaro introduces new art to a revision by carefully sketching out what she envisions in precise detail, with accompanying instructions to the illustrator. The result is accurate, beautifully rendered art that helps difficult concepts come to life.

The **Structure** of a Student Success Learning Tool

Chapter-Opening Case Studies

Each chapter opens with a Case Study Part 1, which helps the students appreciate and understand how microbiology impacts their lives. Appropriate line art, micrographs, and quotes have been added to the chapter-opening page to help the students pull together the big picture and grasp the relevance of the material they're about to learn. The questions that directly follow Parts 1 and 2 of the Case Study challenge students to begin to think critically about relevant text references that will help them answer the questions as they work through the chapter. The Case Study Perspective wraps up the case and can be found on the Connect website.

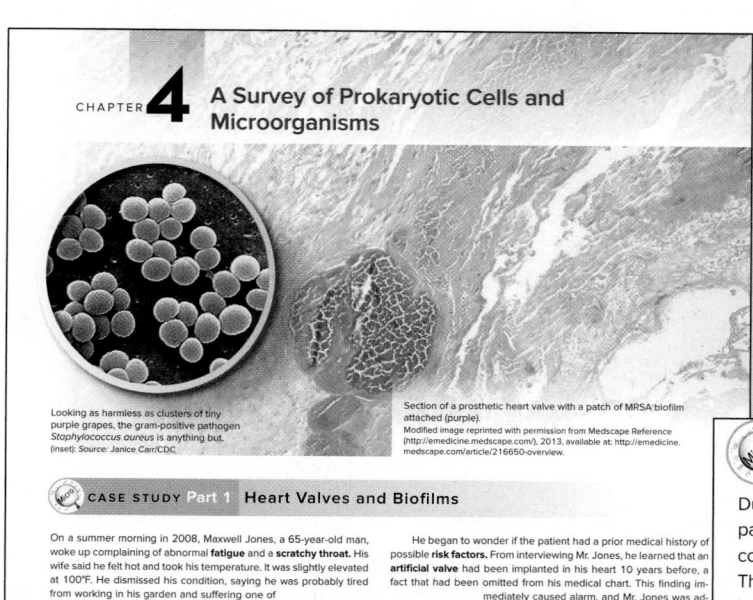

CHAPTER **4**

A Survey of Prokaryotic Cells and Microorganisms

Looking as harmless as clusters of tiny purple grapes, the gram-positive pathogen *Staphylococcus aureus* is anything but. (Inset) *Source: Janice Carr/CDC*

Section of a prosthetic heart valve with a patch of MRSA biofilm attached (purple).
Modified image reprinted with permission from Medscape Reference (http://emedicine.medscape.com/), 2013, available at: http://emedicine.medscape.com/article/216650-overview.

CASE STUDY Part 1 Heart Valves and Biofilms

On a summer morning in 2008, Maxwell Jones, a 65-year-old man, woke up complaining of abnormal **fatigue** and a **scratchy throat.** His wife said he felt hot and took his temperature. It was slightly elevated at 100°F. He dismissed his condition, saying he was probably tired from working in his garden and suffering one of his regular allergy attacks. Over the next few days, his list of symptoms grew. He lost his appetite, his **joints** and muscles were **sore,** and he woke up wringing wet from **night sweats. He** continued to have a **fever,** and his wife was worried over how pale he looked. She insisted he see a physician, who performed a physical and took a **throat culture.** Mr. Jones was sent home with instructions to take **oral penicillin** and acetaminophen (Tylenol), and to come back in a week.

At the next appointment the patient reported that he still had some of the same symptoms, including the fever, and that now he had begun to have **headaches, rapid breathing,** and **coughing. The** physician recorded a **rapid heart rate** and slight **heart murmur.** When the lab report indicated that the throat culture was **negative** for bacterial pathogens, he had to look for other causes.

"At least 65% of chronic infections are caused by microbial biofilms."

He began to wonder if the patient had a prior medical history of possible **risk factors.** From interviewing Mr. Jones, he learned that an **artificial valve** had been implanted in his heart 10 years before, a fact that had been omitted from his medical chart. This finding immediately caused alarm, and Mr. Jones was admitted to the intensive care unit and placed on a mixture of **intravenous antibiotics.** Tests for **blood cultures** and a **white blood cell count** were ordered as backup. By that evening, Mr. Jones had become confused and lost consciousness. He was rushed to the operating room but died during open heart surgery.

■ *What appear to be the most important facts in this case?*

■ *Explain why Mr. Jones's throat culture was negative for infection.*

To continue the Case Study, go to Case Study Part 2 at the end of the chapter.

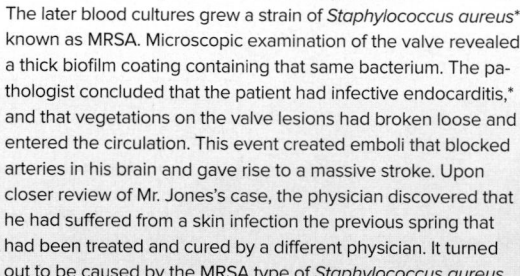

CASE STUDY Part 2

During an autopsy of Mr. Jones's body, the pathologist observed that the prosthetic valve was covered with small patches he called vegetations. The later blood cultures grew a strain of *Staphylococcus aureus** known as MRSA. Microscopic examination of the valve revealed a thick biofilm coating containing that same bacterium. The pathologist concluded that the patient had infective endocarditis,* and that vegetations on the valve lesions had broken loose and entered the circulation. This event created emboli that blocked arteries in his brain and gave rise to a massive stroke. Upon closer review of Mr. Jones's case, the physician discovered that he had suffered from a skin infection the previous spring that had been treated and cured by a different physician. It turned out to be caused by the MRSA type of *Staphylococcus aureus.*

Most bacteria can form structured multicellular communities, or biofilms, on objects in a moist environment. This is even true of bacterial pathogens in the body. The CDC estimates that at least 65% of chronic infections are caused by microbial biofilms. In this case, the MRSA bacteria in the patient's skin infection must have entered the circulation and colonized the artificial valve over several weeks to months. Most cases of chronic endocarditis are caused by biofilms on valves. When the biofilm grows into larger vegetations, portions of it break loose into the circulation. These infect the blood and are spread into organs, causing fever and other signs and symptoms, including the ones that were fatal. MRSA is an emerging pathogen that started as a problem in the hospital but is now prominent in nonhospital settings as well.

■ *What does the acronym MRSA mean, and what is its significance?*

The **Art** of a Student Success Learning Tool

Author's experience and talent transforms difficult concepts

Truly instructional artwork has always been a hallmark feature of *Foundations in Microbiology*. Kathy Talaro's experiences as a teacher, microbiologist, and illustrator have given her a unique perspective and the ability to transform abstract concepts into scientifically accurate and educational illustrations. Powerful artwork that paints a conceptual picture for students is more important than ever for today's visual learners. *Foundations in Microbiology's* art program combines vivid colors, multidimensionality, and self-contained narrative to help students study the challenging concepts of microbiology.

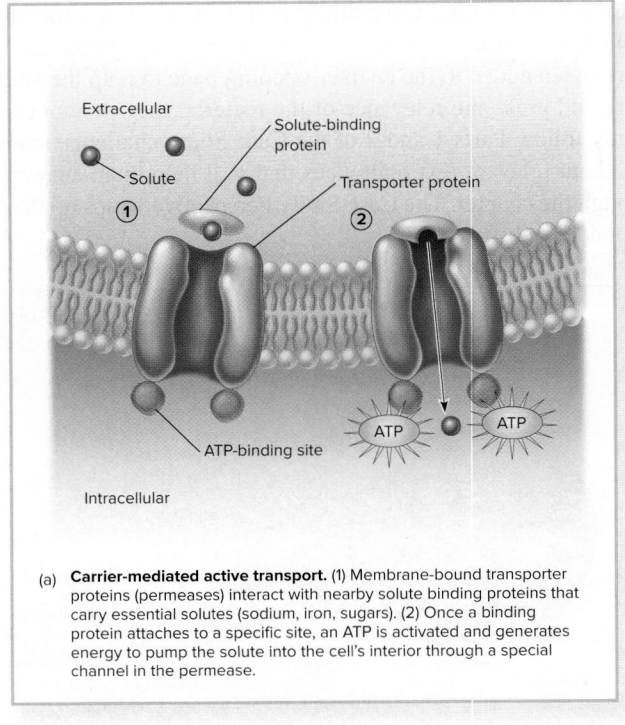

(a) **Carrier-mediated active transport.** (1) Membrane-bound transporter proteins (permeases) interact with nearby solute binding proteins that carry essential solutes (sodium, iron, sugars). (2) Once a binding protein attaches to a specific site, an ATP is activated and generates energy to pump the solute into the cell's interior through a special channel in the permease.

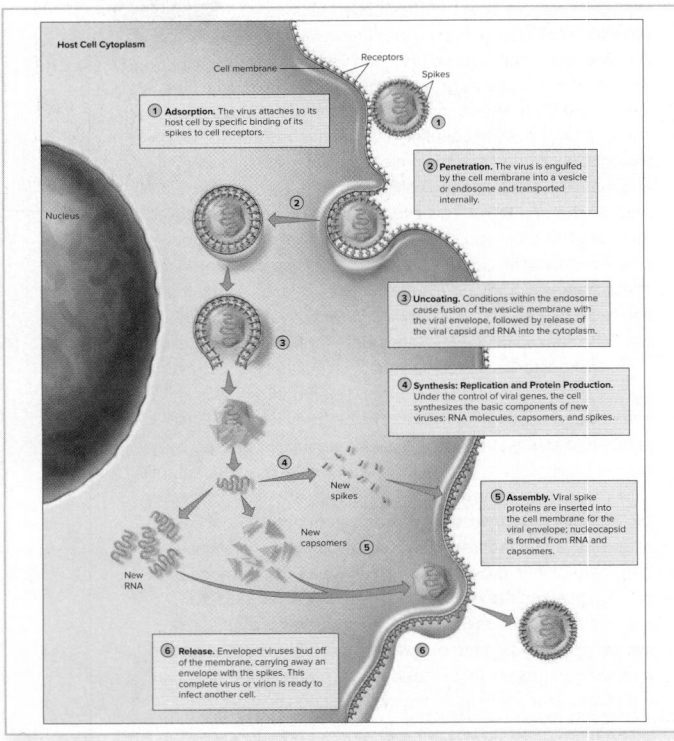

Process Figures

Many difficult microbiological concepts are best portrayed by breaking them down into stages that students will find easy to follow. These process figures show each step clearly numbered within a yellow circle and correlated to accompanying narrative to benefit all types of learners. A distinctive process icon precedes the figure number. The accompanying legend provides additional explanation.

Real clinical photos help students visualize

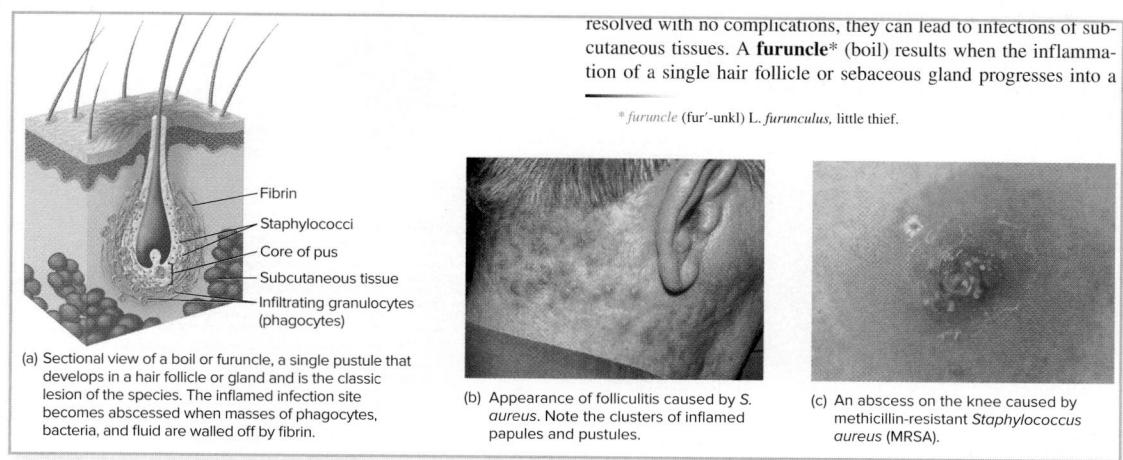

resolved with no complications, they can lead to infections of subcutaneous tissues. A **furuncle*** (boil) results when the inflammation of a single hair follicle or sebaceous gland progresses into a

* *furuncle* (fur'-unkl) L. *furunculus,* little thief.

- Fibrin
- Staphylococci
- Core of pus
- Subcutaneous tissue
- Infiltrating granulocytes (phagocytes)

(a) Sectional view of a boil or furuncle, a single pustule that develops in a hair follicle or gland and is the classic lesion of the species. The inflamed infection site becomes abscessed when masses of phagocytes, bacteria, and fluid are walled off by fibrin.

(b) Appearance of folliculitis caused by *S. aureus.* Note the clusters of inflamed papules and pustules.

(c) An abscess on the knee caused by methicillin-resistant *Staphylococcus aureus* (MRSA).

Clinical Photos

Color photos of individuals affected by disease provide students with a real-life, clinical view of how microorganisms manifest themselves in the human body.

Glycoprotein spikes

- Matrix protein
- Nucleocapsid

(a)

(b)

Combination Figures

Line drawings combined with photos give students two perspectives: the realism of photos and the explanatory clarity of illustrations. The authors chose this method of presentation often to help students comprehend difficult concepts.

The **Purpose** of a Student Success Learning Tool

Secret World of Microbes

The living world abounds with incredible, fascinating microbes that have yet to be discovered or completely understood. This feature enriches our coverage of the latest research discoveries and applications in the field of microbiology. Almost like reading a mystery novel, *The Secret World of Microbes* reveals little-known and surprising facts about this hidden realm.

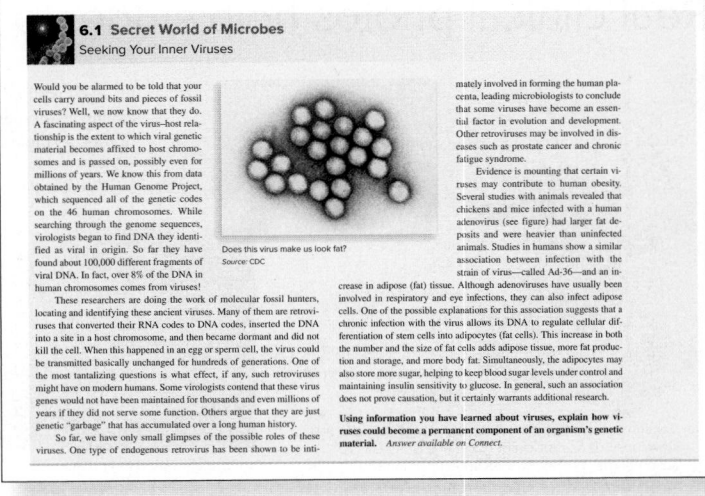

6.1 Secret World of Microbes
Seeking Your Inner Viruses

Would you be alarmed to be told that your cells carry around bits and pieces of fossil viruses? Well, we now know that they do. A fascinating aspect of the virus–host relationship is the extent to which viral genetic material becomes affixed to host chromosomes and is passed on, possibly even for millions of years. We know this from data obtained by the Human Genome Project, which sequenced all of the genetic codes on the 46 human chromosomes. While searching through the genome sequences, virologists began to find DNA they identified as viral in origin. So far they have found about 100,000 different fragments of viral DNA. In fact, over 8% of the DNA in human chromosomes comes from viruses!

These researchers are doing the work of molecular fossil hunters, locating and identifying these ancient viruses. Many of them are retroviruses that converted their RNA codes to DNA codes, inserted the DNA into a site in a host chromosome, and then became dormant and did not kill the cell. When this happened in an egg or sperm cell, the virus could be transmitted basically unchanged for hundreds of generations. One of the most tantalizing questions is what effect, if any, such retroviruses might have on modern humans. Some virologists contend that these virus genes would not have been maintained for thousands and even millions of years if they did not serve some function. Others argue that they are just genetic "garbage" that has accumulated over a long human history.

So far, we have only small glimpses of the possible roles of these viruses. One type of endogenous retrovirus has been shown to be inti-

mately involved in forming the human placenta, leading microbiologists to conclude that some viruses have become an essential factor in evolution and development. Other retroviruses may be involved in diseases such as prostate cancer and chronic fatigue syndrome.

Evidence is mounting that certain viruses may contribute to human obesity. Several studies with animals revealed that chickens and mice infected with a human adenovirus (see figure) had larger fat deposits and were heavier than uninfected animals. Studies in humans show a similar association between infection with the strain of virus—called Ad-36—and an increase in adipose (fat) tissue. Although adenoviruses have usually been involved in respiratory and eye infections, they can also infect adipose cells. One of the possible explanations for this association suggests that a chronic infection with the virus alters its DNA to regulate cellular differentiation of stem cells into adipocytes (fat cells). This increase in both the number and the size of fat cells adds adipose tissue, more fat production and storage, and more body fat. Simultaneously, the adipocytes may also store more sugar, helping to keep blood sugar levels under control and maintaining insulin sensitivity to glucose. In general, such an association does not prove causation, but it certainly warrants additional research.

Using information you have learned about viruses, explain how viruses could become a permanent component of an organism's genetic material. *Answer available on Connect.*

Does this virus make us look fat?
Source: CDC

Check Your Progress and Learning Outcomes

Check Your Progress SECTION 13.1

1. Describe the significant relationships that humans have with microbes.
2. Explain what is meant by *microbiota* and *microbiome* and summarize their importance to humans.
3. Differentiate between contamination, colonization, infection, and disease, and explain some possible outcomes in each.
4. How are infectious diseases different from other diseases?
5. Outline the general body areas that are sterile and those regions that harbor normal resident microbiota.
6. Differentiate betwee[n]
7. Explain the factors [that]
 newborn intestine an[d]

Learning Outcomes and Check Your Progress

Every numbered section in the book opens with Expected Learning Outcomes and closes with assessment questions (Check Your Progress). The Learning Outcomes are tightly correlated to digital material. Instructors can easily measure student learning in relation to the specific learning outcomes used in their course. You can also assign Check Your Progress questions to students through McGraw-Hill Connect.

6.1 Overview of Viruses

Expected Learning Outcomes

1. Indicate how viruses were discovered and characterized.
2. Describe the unique characteristics of viruses.
3. Discuss the origin and importance of viruses.

Pathogen Profiles

Pathogen Profiles are abbreviated snapshots of the major pathogens in each disease chapter. The pathogen is featured in a micrograph, along with a description of the microscopic morphology, identification descriptions, habitat information, and virulence factors. Artwork displays the primary infections/disease, as well as the organs and systems primarily impacted.

Pathogen Profile #2 *Streptococcus pyogenes*

Modified image reprinted with permission from Medscape Reference (http://emedicine.medscape.com/), 2013, available at: http://emedicine.medscape.com/article/228936-overview.

Microscopic Morphology Gram-positive cocci arranged in chains and pairs; very rarely motile; non-spore-forming.

Identified by Results of a catalase test are used to distinguish *Streptococcus* (negative) from *Staphylococcus* (positive). Beta-hemolysis and sensitivity to bacitracin are hallmarks of *S. pyogenes*. Rapid methods of identification use monoclonal antibodies to detect the C-carbohydrate found on the cell surface of *S. pyogenes*. Such tests provide accurate identification in as little as 10 minutes.

Habitat A fairly strict parasite, *S. pyogenes* is found in the throat, nasopharynx, and occasionally the skin of humans. From 5% to 15% of persons are asymptomatic carriers.

Virulence Factors *S. pyogenes* possesses several cell surface antigens that serve as virulence factors. C-carbohydrate helps prevent the bacterium from being dissolved by the lysozyme of the host; fimbriae on the outer surface of the cell enhance adherence of the bacterium; M-protein helps the cell resist phagocytosis while also improving adherence; and C5a protease catalyzes the cleavage of the C5a protein of the complement system, inhibiting the actions of complement. Most strains of *S. pyogenes* are covered with a capsule composed of hyaluronic acid

(HA) identical to the HA found in host cells, preventing an immune response by the host. Two different hemolysins, streptolysin O (SLO) and streptolysin S (SLS), cause damage to leukocytes, and liver and heart muscle, whereas erythrogenic toxin produces fever and the bright red rash characteristic of *S. pyogenes* disease. Invasion of the body is aided by several enzymes that digest fibrin clots (streptokinase), connective tissue (hyaluronidase), or DNA (streptodornase).

Primary Infections/Disease Local cutaneous infections include pyoderma (impetigo) or the more invasive erysipelas. Infection of the tonsils or pharyngeal mucous membranes can lead to streptococcal pharyngitis (strep throat), which, if left untreated, may lead to scarlet fever. Rarer infections include streptococcal toxic shock syndrome, *S. pyogenes* pneumonia, and necrotizing fasciitis. Long-term complications of *S. pyogenes* infections include rheumatic fever and acute glomerulonephritis.

Control and Treatment Control of *S. pyogenes* infection involves limiting contact between carriers of the bacterium and immunocompromised potential hosts. Patients should be isolated, and care must be taken when handling infectious secretions. As the bacterium shows little drug resistance, treatment is generally a simple course of penicillin.

Pathogen Profile #3 *Clostridium difficile*

Source: CDC

Microscopic Morphology Gram-positive bacilli, present singly or in short chains. Endospores are subterminal and distend the cell, altering its shape.

Identified by Gram reaction and endospore formation. *Clostridium* is differentiated from *Bacillus* as the former is typically a strict anaerobe and the latter is not. ELISA is often used to detect toxins of *C. difficile* in fecal samples.

Habitat Found in small numbers as part of the normal microbiota of the intestine.

Virulence Factors Enterotoxins that cause epithelial necrosis of the colon.

Primary Infections/Disease *Clostridium difficile* infection (CDI) refers to disease caused by the overgrowth of *C. difficile*. Symptoms may range from diarrhea to inflammation of the colon, cecal perforation, and, rarely, death. Although *C. difficile* is ordinarily present in low numbers, treatment with broad-spectrum antibiotics may disrupt the normal microbiota of the colon, leading to a *C. difficile* superinfection.

Control and Treatment Mild cases generally respond to withdrawal of the antibiotic. Severe cases are treated with oral vancomycin or metronidazole, along with probiotics or fecal microbiota transplants to restore the normal microbiota.

The **Framework** of a Student Success Learning Tool

Pedagogy created to promote active learning

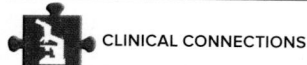

CLINICAL CONNECTIONS

An Outbreak of Fungal Meningitis

Most fungi are not invasive and do not ordinarily cause serious infections unless a patient's immune system is compromised or the fungus is accidentally introduced into sterile tissues. In 2012 we witnessed how a simple medical procedure could turn into a medical nightmare because a common, mostly harmless fungus got into the wrong place at the wrong time. It all started when a small compounding pharmacy in Massachusetts unknowingly sent out hundreds of mold-contaminated vials of medication to medical facilities for injections to control pain. These vials were sent to 23 states and used to inject the drug into the spinal columns or joints of around 14,000 patients. By the time any problems were reported, several hundred cases of infection had occurred, half of which settled in the meninges. The most drastic outcome was the deaths of 39 patients from complications of meningitis. After months of investigation, the CDC isolated a black mold, *Exserohilum rostratum*, from both the patients and the drug vials.

This mold resides in plants and soil, from which it spreads into the air and many human habitats. But it is not considered a human pathogen, and infections with it are very rare. Examination of the compounding facility uncovered negligence and poor quality controls, along with dirty preparation rooms. Mold spores were introduced during filling of the vials, and because the medication lacked preservatives, they survived and grew. The owner of the compounding pharmacy and the head pharmacist were each charged with 25 counts of second-degree murder, their trial is expected to start in late 2016.

This case drives home several important facts about fungi: (1) They can grow rapidly even in low nutrient environments; (2) just a single spore introduced into a sterile environment, whether it is a vial of medicine or the human body, can easily multiply into millions of fungal cells; and (3) even supposedly "harmless" fungi are often opportunistic, meaning that they will infect tissues "if given an opportunity." This case also emphasizes the need for zero tolerance for microbes of any kind in a drug that is being injected—such a procedure demands sterility. When you think of it, the patients were actually being inoculated in a way that assured the development of serious mycoses.

Explain how a supposedly harmless, airborne mold could get all the way into the brain and cause meningitis. *Answer available on Connect.*

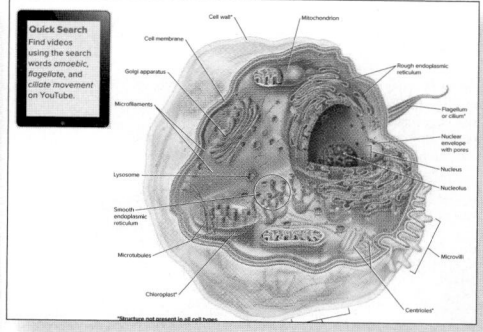

Quick Search

This feature reminds students that videos, animation, and pictorial displays that provide further information on the topic are just a "click" away using their smart-phone, tablet, or computer. This integration of learning via technology helps students become more engaged and empowered in their study of the featured topic.

Tables

This edition contains numerous illustrated tables. Horizontal contrasting lines set off each entry, making them easy to read.

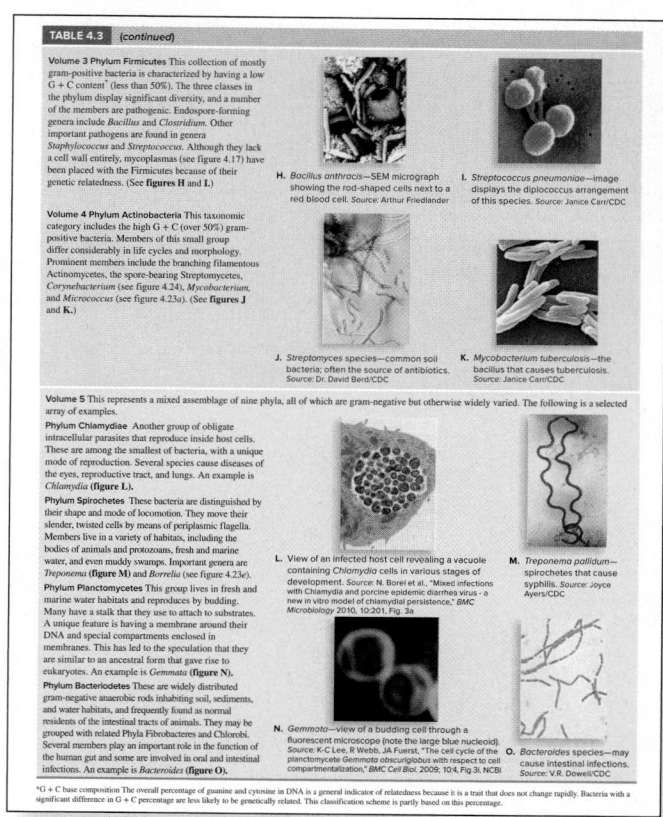

TABLE 4.3 (continued)

Volume 3 Phylum Firmicutes This collection of mostly gram-positive bacteria is characterized by having a low G + C content* (less than 50%). The three classes in the phylum display significant diversity, and a number of the members are pathogenic. Endospore-forming genera include *Bacillus* and *Clostridium*. Other important pathogens are found in genera *Staphylococcus* and *Streptococcus*. Although they lack a cell wall entirely, mycoplasmas (see figure 4.17) have been placed with the Firmicutes because of their genetic relatedness. (See **figures H** and **L**.)

H. *Bacillus anthracis*—SEM micrograph showing the rod-shaped cells next to a red blood cell. Source: Arthur Friedlander

I. *Streptococcus pneumoniae*—image displays the diplococcus arrangement of this species. Source: Janice Carr/CDC

Volume 4 Phylum Actinobacteria This taxonomic category includes the high G + C (over 50%) gram-positive bacteria. Members of this small group differ considerably in life cycles and morphology. Prominent members include the branching filamentous Actinomycetes, the spore-bearing Streptomycetes, *Corynebacterium* (see figure 4.24), *Mycobacterium*, and *Micrococcus* (see figure 4.23a). (See **figures J** and **K.**)

J. *Streptomyces* species—common soil bacteria; often the source of antibiotics. Source: Dr. David Berd/CDC

K. *Mycobacterium tuberculosis*—the bacillus that causes tuberculosis. Source: Janice Carr/CDC

Volume 5 This represents a mixed assemblage of nine phyla, all of which are gram-negative but otherwise widely varied. The following is a selected array of examples.

Phylum Chlamydiae Another group of obligate intracellular parasites that reproduce inside host cells. These are among the smallest of bacteria, with a unique mode of reproduction. Several species cause diseases of the eyes, reproductive tract, and lungs. An example is *Chlamydia* (**figure L**).

Phylum Spirochetes These bacteria are distinguished by their shape and mode of locomotion. They move their slender, twisted cells by means of periplasmic flagella. Members live in a variety of habitats, including the bodies of animals and protozoans, fresh and marine water, and even muddy swamps. Important genera are *Treponema* (**figure M**) and *Borrelia* (see figure 4.23e).

Phylum Planctomycetes This group lives in fresh and marine water habitats and reproduces by budding. Many have a stalk that they use to attach to substrates. A unique feature is having a membrane around their DNA and special compartments enclosed in membranes. This has led to the speculation that they are similar to an ancestral form that gave rise to eukaryotes. An example is *Gemmata* (**figure N**).

Phylum Bacteriodetes These gram-negative anaerobic rods inhabiting soil, sediments, and water habitats, and frequently found as normal residents of the intestinal tracts of animals. They may be grouped with related Phyla Fibrobacteres and Chlorobi. Several members play an important role in the function of the human gut and some are involved in oral and intestinal infections. An example is *Bacteroides* (**figure O**).

L. View of an infected host cell revealing a vacuole containing *Chlamydia* cells in various stages of development. Source: N. Borel et al., "Mixed infections with Chlamydia and porcine epidemic diarrhea virus - a new in vitro model of chlamydial persistence," *BMC Microbiology* 2010, 10:201, Fig. 3a

M. *Treponema pallidum*—spirochetes that cause syphilis. Source: Joyce Ayers/CDC

N. *Gemmata*—view of a budding cell through a fluorescent microscope (note the large blue nucleoid). Source: K-C Lee, R Webb, JA Fuerst, "The cell cycle of the planctomycete *Gemmata obscuriglobus* with respect to cell compartmentalization," *BMC Cell Biol*. 2009; 10:4, Fig 3i, NCBI

O. *Bacteroides* species—may cause intestinal infections. Source: V.R. Dowell/CDC

*G + C base composition The overall percentage of guanine and cytosine in DNA is a general indicator of relatedness because it is a trait that does not change rapidly. Bacteria with a significant difference in G + C percentage are less likely to be genetically related. This classification scheme is partly based on this percentage.

5. A mnemonic device to keep track of this is *LEO* says *GER:* *L*ose *E*lectrons *O*xidized; *G*ain *E*lectrons *R*educed.

Footnotes

Footnotes provide the reader with additional information about the text content.

This new feature follows the opening case study in Chapters 1-17 and 26-27. Students are provided with a descriptive pictorial guide for the main topics covered within these respective chapters.

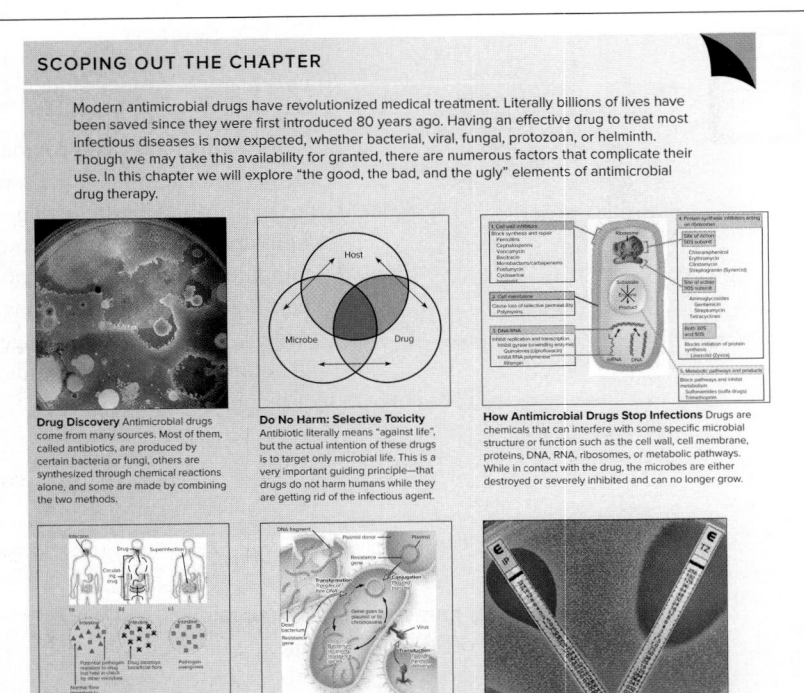

SCOPING OUT THE CHAPTER

Modern antimicrobial drugs have revolutionized medical treatment. Literally billions of lives have been saved since they were first introduced 80 years ago. Having an effective drug to treat most infectious diseases is now expected, whether bacterial, viral, fungal, protozoan, or helminth. Though we may take this availability for granted, there are numerous factors that complicate their use. In this chapter we will explore "the good, the bad, and the ugly" elements of antimicrobial drug therapy.

Drug Discovery Antimicrobial drugs come from many sources. Most of them, called antibiotics, are produced by certain bacteria or fungi, others are synthesized through chemical reactions alone, and some are made by combining the two methods.

Do No Harm: Selective Toxicity Antibiotic literally means "against life", but the actual intention of these drugs is to target only microbial life. This is a very important guiding principle—that drugs do not harm humans while they are getting rid of the infectious agent.

How Antimicrobial Drugs Stop Infections Drugs are chemicals that can interfere with some specific microbial structure or function such as the cell wall, cell membrane, proteins, DNA, RNA, ribosomes, or metabolic pathways. While in contact with the drug, the microbes are either destroyed or severely inhibited and can no longer grow.

Toxicity and Other Side Effects The promise of antimicrobial therapy is spoiled by the numerous possibilities of adverse side effects. Drugs can harm body tissues and organs, disrupt the normal microbiota that help keep a balance in the body's organs, and induce allergies and hypersensitivities.

The Global Race Against Drug Resistance Microbes are very adept at rapidly altering their physiology and genetics to adapt to drugs, making them less effective. They may develop enzyme systems that dismantle the drugs, block the drug's entrance, expel the drugs, or use an alternate pathway that bypasses the drug's effects.

Choosing an Appropriate Drug Selection of an effective drug is guided by several factors. Among the most important considerations are the nature of the infectious agent, knowing which . . . sensitive to, the possible side effec . . . and the medical condition of the p . . .

(Drug Discovery) © Kathy Park Talaro; (Choosing an Appropriate Drug) © Copyright AB BIODISK 2008. Re-printed with permission of AB BIODISK

SCOPING OUT THE CHAPTER

We don't need to take a course in ornithology to be able to recognize the structure of a bird's wing and describe how it functions. And even when they're too far away for us to see them clearly, we can instantly know that a group of animals flying in a "V" formation is a flock of birds, not butterflies or bats. And though they are about as different as two animals could be, we all understand intuitively that a hummingbird and a turkey are related and should be grouped together. In this chapter we will gain the same type of familiarity with bacterial cells, by studying their structure, function, and evolutionary history.

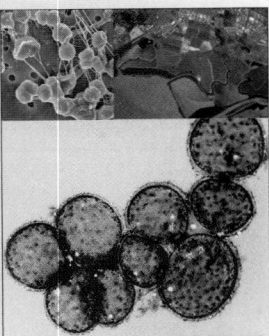

Cellular Structure The chapter opens with a discussion of what makes up a prokaryotic cell. Beginning with external structures like flagella, moving to the fluid-mosaic barrier of the cell membrane, and finally to internal structures like inclusion bodies. Understanding the anatomy of a cell is key to understanding its biology.

Bacterial Shapes and Arrangements The cell wall and cytoskeleton of a bacterial cell are responsible for its shape, and correctly recognizing the shape of a cell is one of the first steps in determining its identity. The grouping of individual cells into more complex arrangements reveals a great deal about the manner in which a cell multiplies.

Classification and Unusual Bacteria The chapter continues by explaining the ways in which prokaryotic cells may be organized based on their evolutionary relationships. Finally, we introduce several examples of novel bacteria that thrive in boiling water (top left), extraordinarily high concentrations of salt (top right), or are so big that they threaten to redefine what it means to be a bacterium (bottom).

(left: top): Source: Louisa Howard/Dartmouth Electron Microscope Facility; (left: bottom): © Kwangshin Kim/Science Source; (middle: top-left): Source: Janice Carr/CDC; (middle: top-right): Source: Joyce Ayers/CDC; (middle: bottom): Source: Jeff Hageman, M.H.S./Janice Carr/CDC; (right: top-left): Source: Maryland Astrobiology Consortium, NASA and STScI; (right: top-right): Source: NASA Johnson Space Center/ISS007EB738/ (http://eol.jsc.nasa.gov); (right: bottom): © Heide N. Schulz/Max Planck Institue for Marine Microbiology

The Planning of a Student Success Learning Tool

Pedagogy designed for varied learning styles

The end-of-chapter material for the tenth edition has been carefully planned and updated to promote active learning and provide review for different learning styles and levels of Bloom's Taxonomy. Questions are divided into two levels:

Level I. Knowledge and Comprehension
Level II. Application, Analysis, Evaluation, and Synthesis

The consistent layout of each chapter allows students to develop a learning strategy and gain confidence in their ability to master the concepts, leading to success in the class!

Case Study Review

These questions provide a quick check of concepts covered by the Case Study and allow instructors to assess students on the case study material.

Writing Challenge

Writing Challenge questions are suggested as a writing experience. Students are asked to compose a one- or two-paragraph response using the factual information learned in the chapter.

Case Study Review

1. Which of these is/are an example(s) of neglected tropical pro[...]
diseases?
 a. hookworm d. a and b
 b. Chagas disease e. b and c
 c. leishmaniasis f. all of these

Writing Challenge

For each question, compose a one- or two-paragraph answer[...]
Check Your Progress questions can also be used for writing[...]

1. Describe the anatomy and functions of each of the majo[...]
 organelles.
2. Trace the synthesis of cell products, their processing, a[...]
 packaging through the organelle network.
3. a. What is the reproductive potential of molds in term[...]

Chapter Summary with Key Terms

A brief outline of the chapter's main concepts is provided for students, with important terms highlighted. Key terms are also included in the glossary at the end of the book.

Chapter Summary with Key Terms

5.1 **The History of Eukaryotes**
5.2 **Form and Function of the Eukaryotic Cell: External Structures**
 A. The exterior configuration of eukaryotic cells is complex and displays numerous structures not found in prokaryotic cells. Biologists have accumulated much evidence that eukaryotic cells evolved through **endosymbiosis** between early prokaryotic cells.
 B. Major external structural features include: appendages (**cilia, flagella**), **glycocalyx**, cell wall, and cytoplasmic (or cell) membrane.
5.3 **Form and Function of the Eukaryotic Cell: Internal Structures**
 A. The internal structure of eukaryotic cells is compartmentalized into individual organelles.
 B. Major **organelles** and internal structural features include: **nucleus, nucleolus, endoplasmic reticulum, Golgi complex, mitochondria, chloroplasts), ribosomes, cytoskeleton (microfilaments, microtubules).**
5.4 **Eukaryotic-Prokaryotic Comparisons and Taxonomy of**

Multiple-Choice Questions

Select the correct answer from the answers provided. For questions with[...]
statement.

1. Both flagella and cilia are found primarily in
 a. algae c. fungi
 b. protozoa d. both b and c
2. Features of the nuclear envelope include
 a. ribosomes

Multiple-Choice Questions

Students can assess their knowledge of basic concepts by answering these questions and looking up the correct answers in appendix D. In addition, SmartBook allows for students to quiz themselves interactively using these questions.

End-of-Chapter Questions

Questions are divided into two levels.

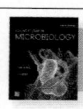

Level I. Knowledge and Comprehension

These questions require a working knowledge of[...]
and the ability to recall and understand the inform[...]

Level II. Application, Analysis, Evaluation, and Synthesis

These problems go beyond just restating facts and require higher levels of understanding and an ability to interpret, problem solve, transfer knowledge to new situations, create models, and predict outcomes.

The Innovation of a Student Success Learning Tool

Concept Mapping

An Introduction to Concept Mapping can be found on Connect.

 Concept Mapping

On Connect you can find an Introduction to Concept Mapping that provides guidance for working with concept maps along with concept-mapping activities for this chapter.

Critical Thinking

Using the facts and concepts they just studied, students must reason and problem solve to answer these specially developed questions. Questions do not have a single correct answer and thus open doors to discussion and application.

 Critical Thinking

Critical thinking is the ability to reason and solve problems using facts and concepts. These questions can be approached from a number of angles, and in most cases, they do not have a single correct answer.

1. Explain the ways that mitochondria resemble rickettsias and chloroplasts resemble cyanobacteria.

2. Give the common name of a eukaryotic microbe that is unicellular, walled, nonphotosynthetic, nonmotile, and bud-forming.

3. How are the eukaryotic ribosomes and cell membranes different from those of prokaryotes?

4. What general type of multicellular parasite is composed primarily of thin sacs of reproductive organs?

5. a. Name two parasites that are transmitted in the cyst form.
 b. How must a non-cyst-forming pathogenic protozoan be transmitted? Why?

6. Explain what factors could cause opportunistic mycoses to be a growing medical problem.

7. a. How are bacterial endospores and cysts of protozoa alike?
 b. How do they differ?

8. For what reasons would a eukaryotic cell evolve an endoplasmic reticulum and a Golgi apparatus?

9. Can you think of a simple test to determine if a child is suffering from pinworms? Hint: Clear adhesive tape is involved.

Visual Challenge

Visual Challenge questions take images and concepts learned in other chapters and ask students to apply that knowledge to concepts covered in the current chapter.

Visual Challenge

1. What term is used to describe a single species exhibiting both cell types shown below, and which types of organisms would most likely have this trait?

The Revision of a Student Success Learning Tool

Changes to Foundations in Microbiology, Tenth Edition

Overall Changes:

- A new feature "Scoping Out The Chapter" has been placed after the opening case studies. This page will give readers a descriptive pictorial guide for the main topics covered in chapters 1–17 and 26–27.
- Ten chapters (6, 7, 13, 19, 21, 22, 24, 25, 26, and 27) contain new case studies chosen for their relevance to major themes in the chapter.
- Approximately 175 new and replacement photographs have been included in the revision.
- Numerous images and figures have been revised and corrected.
- Clinical Connections boxes and side notes have a tinted screen added to set them off from the regular text. Several new Clinical Corrections boxes have been added.
- Coverage of diseases, statistics, and graphic data has been updated.
- Most chapters contain new links and quick searches for exploring topics on the internet.
- Special effort has been directed towards clarifying terms, wording, and definitions to improve understanding of more difficult concepts.

Chapter-Specific Changes:

Chapter 1

- The chapter opens with a new case study featuring microorganisms living in extreme habitats
- Epidemiology statistics have been updated throughout the chapter
- New information on the spread of chikungunya virus and Zika virus has been added
- Information concerning the ongoing pertussis epidemic has been updated
- Information on the link between microorganisms and chronic disease has been updated
- The topic of microbial evolution and classification has been updated

Chapter 2

- Discussions of the manner in which electron shells are filled and the importance of valence electrons to the formation of covalent bonds have been clarified
- The section on polymeric biomolecules (DNA, RNA, lipids, proteins, starches) has been clarified

Chapter 3

- New photos have been added to illustrate differences in resolution between light microscopes and electron microscopes
- New photos have been added to the discussions of fluorescence microscopy, electron microscopy, and selective and differential media

Chapter 4

- A new discussion and figure concerning bacterial microcompartments has been added.
- New photographs for a hyperthermophile and bacterial inclusion bodies

Chapter 5

- The case study concerning neglected tropical diseases (NTDs) has been updated to include the awarding of the 2015 Nobel Prize in Physiology or Medicine to scientists working in this area
- New tables summarize the function of structures within the eukaryotic cell
- New photos of the nucleus and mitochondria emphasize the importance of these organelles
- Update on *Pseudogymnoascus destructans,* the fungus responsible for white nose syndrome in bats

Chapter 6

- The chapter opens with a new case study focused on highly pathogenic avian influenza.
- The role of Adenovirus Ad-36 in weight gain and regulation of blood sugar levels has been updated.
- New photomicrograph of an Ebola virus budding from an infected cell

Chapter 7

- A new case study "A Creature of Habitat" describes the serious problem of cystic fibrosis and its connection with recurring *Pseudomonas* infections.
- New photographs for satellitism and an anaerobic growth chamber
- New information on biofilm formation

Chapter 8

- Addition of coenzymes to table on cofactors.
- Clarification of how the term fermentation is used under different contexts

Chapter 9

- Improved figure showing input by regulatory RNA
- Revised table on types of mutations.
- Updated box on regulatory, noncoding RNA, and riboswitches

- Improved consistency of figures for conjugation and transduction

Chapter 10

- Added details of newer DNA sequencing technologies
- Updated box on the human genome
- Revised tables on genetically-engineered animals
- New graph on genetically engineered crops
- Revised and updated Clinical Connections covering gene therapy
- The term DNA fingerprinting has been replaced with DNA profiling
- Figure on standardized DNA profiling has been revised
- Reorganized section on different uses of DNA profiling
- A note describing the gene editing technology of CRSPR has been added.

Chapter 11

- Updated case study on an outbreak of hepatitis C in a colonscopy clinic
- Integrated historical aspects of microbial control into main text and removed Making Connections box 11.1.
- New Clinical Connections box discusses the sterilization of reusable medical devices
- Revised the box on use of triclosan, including new FDA ruling

Chapter 12

- Integrated Making connections 12.2 on discovery of drugs into main chapter
- Added a new figure on the chemical synthesis of penicillin drugs
- Included new categories of antibacterial and antiviral drugs
- Updated drug resistance box and added a new figure showing carbapenem-resistant enterobacteriaceae (CRE)

Chapter 13

- New case study "Fatal Filaments from Far Away Africa" that covers the Ebola epidemic in Africa and its spread to the United States.
- Introduced new information on the importance of the microbiome to general human physiology
- Coverage of the relationship of the placental microbiome to infant development and the development of the intestinal microbiome in newborns.
- New surveillance figures for HIV infection, pertussis, and Ebola fever.
- Updated figure on healthcare associated infections (HAI); replacing use of nosocomial infections with the more commonly used HAI

The **Effort** of a Student Success Learning Tool

- New visual challenge figures to differentiate among different epidemiological patterns for diseases

Chapter 14
- Added new information on the hygiene hypothesis
- Clarified figure on the actions of complement
- Removed discussion of fever from Clinical Connections box and integrated it into text

Chapter 15
- Reorganized the order of introduction of T cell and B cell actions and functions; T cells now are covered first, followed by B cells.
- Revised figure 15.1 to align with new order of coverage.
- Added side note to focus on the functions of T regulatory cells with new information on biologic drugs based on this type of T cell
- Updated the list of monoclonal antibody-based drugs and currently-approved vaccine schedules.
- Coverage of the breast microbiome and the role breast milk has in the development of the immune systems of infants.

Chapter 16
- Revised allergen count figure
- New photographs of atopic and contact dermatitis
- New photograph of blood typing
- New photograph of rheumatoid arthritis
- Illustration of child with velocardiofacial (DiGeorge) syndrome

Chapter 17
- Updated box on point-of-care testing
- New example of the direct fluorescent antibody test
- Replacement figure for rapid identification testing
- New examples of serological test results

Chapter 18
- New electron photomicrograph of methicillin-resistant *Staphylococcus aureus* has been added
- New photos of erysipelas and limb necrosis due to meningococcemia
- Updated recommendations for treatment of bacterial infections
- Updated statistics on the prevalence of sexually transmitted diseases
- The discussion of meningococcemia and meningitis has been clarified

Chapter 19
- The chapter opens with a new case study concerning *Listeria monocytogenes*

- New photomicrographs of *Bacillus anthracis*, *Corynebacterium diphtheriae*, and fluorescently labeled *Mycobacterium tuberculosis* have been added
- New photographs for myonecrosis, erysipeloid, the Mantoux skin test for tuberculosis, paucibacillary leprosy, multibacillary leprosy, fish tank granuloma, and actinomycosis
- Expanded and updated discussion of the use of fecal microbiota transplants as a treatment of *C-difficile* infection
- New electron micrograph of *Mycobacterium tuberculosis*, updated worldwide statistics for tuberculosis, and updated treatment recommendations for both active and latent tuberculosis
- Updated classification of leprosy to match WHO standards

Chapter 20
- New photomicrograph of *Pseudomonas aeruginosa* and new photo of cutaneous *Pseudomonas* infection
- Updated treatment recommendations for *Pseudomonas* infection, *Brucellosis,* and *Tularemia*
- New information on pertactin-deficient strains of *Bordetella pertussis*
- Updated discussion of *E. coli* pathotypes
- New section on Carbepenem-resistant Enterobacteriaceae infections
- New section on naming conventions in *Salmonella*

Chapter 21
- Chapter opens with a new case study on Q fever and live cell transplantation
- New photographs of *Coxiella burnetti, Treponema pallidum, Borrelia burgdorferi, Vibrio cholera, Campylobacter jejuni, Orienta tsutsugamushi*, and *lxodes scapularis*
- Updated statistics on syphilis
- New treatment recommendations for cholera
- New photos of dental caries and oral bacteria

Chapter 22
- Case study has been updated to include the latest facts concerning the fungal meningitis outbreak connected to the New England Compounding Center
- Updates on antifungal drugs and epidemiological statistics
- New photographs of cutaneous blastomycosis, *Tinea pedis, Aspergillus,* and aspergillosis
- Reclassification of zygomycosis as mucormycosis

Chapter 23
- Updated drug recommendations for parasitic diseases
- New discussion on genetically engineered mosquitoes resistant to *Plasmodium sp.*
- New feature on Carlos Chagas and his importance to the field of parasitology
- The latest information about phase 3/4 trials of malaria vaccine RTS,S

Chapter 24
- The chapter begins with a new case study concerning unusual varicella zoster virus transmission
- New photos of herpes simplex type 1, neonatal herpes, and lymphocytes infected with Epstein-Barr virus
- Updated recommendations for treatment of neonatal herpes
- Update on treatment and prevention of HPV

Chapter 25
- New case study on measles and subacute sclerosing panencephalitis
- Updates include information on the Ebola outbreak of 2014–2016, the ongoing Zika virus outbreak, and widespread outbreaks of chikungunya virus
- Updated information on influenza vaccines and new chemotherapeutic treatments for influenza
- New information on the measles outbreak of 2015 along with discussion and references to online documentaries about vaccine skepticism
- Distribution maps for *Aedes* mosquitoes, the vector of dengue, chikungunya, and Zika viruses
- Feature on the *Aedes* mosquito
- Information about the recently approved vaccine to prevent dengue fever
- Updates on treatment strategies for HIV, including the use of pre-exposure prophylaxis (PrEP)

Chapter 26
- New case study on drinking water contamination as a result of harmful algae blooms
- New photos of *Rhizobium* root nodules and mycorrhizae
- New discussion concerning fracking as a potential contaminant of groundwater

Chapter 27
- New case study concerning three separate outbreaks of food poisoning

Acknowledgments

This edition marks the 24th anniversary of the first publication of Foundations in 1993. Looking back over the previous nine editions, the authors are struck by the extensive discoveries and new developments in the science of microbiology that are reflected in the changing content and character of this book. This 10th edition is no exception. The one thing that has remained constant and unchanging over these years is the outstanding collaboration we enjoy with the editorial and production staff at McGraw-Hill Education. This time around, we have been fortunate to have the able assistance and expertise of product developer Mandy Clark, keeping us on track and providing much needed moral support. We also appreciate the insights and contributions of brand manager Marija Magner and marketing manager Jessica Cannavo. Our project manager Jayne Klein has been an experienced and knowledgeable guide through the intricacies of a digital-style revision.

Other valued members of our team who have been instrumental in developing the text's visual elements are Carrie Burger, the content licensing specialist, Danny Meldung at Photo Affairs, and the designer Tara McDermott, who has produced another striking book and cover design. Some of the unsung heroes of authors are the readers who must sift through the text with a fine-tooth comb, checking for errors, grammatical usage, and consistency in style. This tedious job fell this time to copy editor Wendy Nelson. After poring over 800 plus pages of text in a few months, she may feel like she has taken a crash course in microbiology.

It takes about a year and a half to complete a textbook revision—a process that involves editing manuscript, writing new text, illustration, research, and much more. During this time, the entire text and art program are inspected at least six times by the authors and team members. Even with the keenest eyes and spell checks, some typos, errors, oversights, and other mistakes may end up on the printed page. If you find any of these or wish to make other comments, feel free to contact the publisher, sales representative, or authors (ktalaro@aol.com and bxchess@Pasadena.edu.)

We hope that you enjoy your explorations in the microbial world and that this fascinating science will leave a lasting impression on you.

—*Kathy Talaro and Barry Chess*

How to Maximize Your Learning Curve

Most of you are probably taking this course as a prerequisite to nursing, dental hygiene, medicine, pharmacy, optometry, physician assistant, or other health science programs. Because you are preparing for professions that involve interactions with patients, you will be concerned with infection control and precautions, which in turn requires you to think about microbes and how to manage them. This means you must not only be knowledgeable about the characteristics of bacteria, viruses, and other microbes, and their physiology and primary niches in the world, but you must also have a grasp of disease transmission, the infectious process, disinfection procedures, and drug treatments. You will need to understand how the immune system interacts with microorganisms and the effects of immunization. All of these areas bring their own vocabulary and language—much of it new to you—and mastering it will require time, motivation, and preparation. A valid question students often ask is: "How can I learn this information to increase my success in the course as well as retain it for the future?"

Right from the first, you need to be guided by how your instructor has organized your course. Because there is more information than could be covered in one semester or quarter, your instructor will select what he or she wants to emphasize and will construct reading assignments and a study outline that corresponds to lectures and discussion sessions. Many instructors have a detailed syllabus or study guide that directs the class to specific content areas and vocabulary words. Others may have their own website to distribute assignments and even sample exams. Whatever materials are provided, this should be your primary guide in preparing to study.

The next consideration involves your own learning style and what works best for you. To be successful, you must commit essential concepts and terminology to memory. A list of how we retain information called the "pyramid of learning" has been proposed by Edgar Dale: We remember about 10% of what we read; 20% of what we hear; 50% of what we see and hear; 70% of what we discuss with others; 80% of what we experience personally; and 95% of what we teach to someone else.

There are clearly many ways to go about assimilating information. Mainly, you will want to focus on more than just reading alone to gather the most important points from a chapter. Try to incorporate writing, drawing simple diagrams, and discussion or study with others. You must attend lecture and laboratory sessions to listen to your instructors or teaching assistants explain the material. You can rewrite the notes you've taken during lecture, or outline them to organize the main points. This begins the process of laying down memory. You should go over concepts with others—perhaps a tutor or study group—and even take on the role of the teacher-presenter part of the time. With these kinds of interactions, you will move beyond simple rote memorization of words and will come to *understand* the ideas and be able to apply them later.

A way to assess your understanding and level of learning is to test yourself. You may use the exam questions in the text, on the Connect website, or make up your own. LearnSmart, available within the Connect site, is an excellent way to map your own, individualized learning program. It helps to track what you know, pinpoint what you don't know, and creates personalized questions based on your progress.

Another big factor in learning is the frequency of studying. It is far more effective to spend an hour or so each day for two weeks than a marathon cramming session on one weekend. If you approach the subject in small bites and remain connected with the terminology and topics, over time it will become yours and you will find that the pieces begin to fit together. Just remember that repetition and experience are the most effective ways to acquire knowledge.

In the final analysis, the process of learning comes down to self-motivation and attitude. There is a big difference between forcing yourself to memorize something to get by and really wanting to know and understand it. Therein is the key to most success and achievement, no matter what your final goals. And though it is true that mastering the subject matter in this textbook requires time and effort, millions of students will affirm how worthwhile such knowledge has been in their professions and everyday life.

Contents

CHAPTER 5

A Survey of Eukaryotic Cells and Microorganisms 124

CHAPTER 6

An Introduction to Viruses 160

CHAPTER 7

Microbial Nutrition, Ecology, and Growth 188

CHAPTER 8

An Introduction to Microbial Metabolism: The Chemical Crossroads of Life 222

CHAPTER 9

An Introduction to Microbial Genetics 260

CHAPTER 10

Genetic Engineering: A Revolution in Molecular Biology 298

CHAPTER 11

Physical and Chemical Agents for Microbial Control 327

CHAPTER 12

Drugs, Microbes, Host—The Elements of Chemotherapy 360

CHAPTER 13

Microbe-Human Interactions: Infection, Disease, and Epidemiology 397

CHAPTER 14

An Introduction to Host Defenses and Innate Immunities 437

The Main Themes of Microbiology

Streams of methane bubbles rise from the seafloor, the product of microscopic organisms adapted to the extreme environment of the deep ocean.
Source: NOAA Okeanos Explorer Program, 2013 ROV Shakedown and Field Trials in the U.S. Atlantic Canyons

A microbial ecologist carries a water sample from subglacial Lake Whillans, 640 kilometers from the South Pole.
© JT Thomas

CASE STUDY Part 1 Microbes Find A Way

A frozen white wasteland, a toxic soup, a lunar landscape. Three fairly common descriptions of an environment so harsh—cold, toxic, or lacking nutrients—that no life can survive. Lake Whillans, a small, shallow lake trapped beneath half a mile of ice, certainly fits that description. Located 640 kilometers from the South Pole, Lake Whillans is completely encased in ice and sits at a slant, pressed against the side of a hill far below the icy surface. As heat from the core of the Earth melts the bottom of the Antarctic ice sheet, a few milliliters of liquid water are added to the lake each year.

Subglacial lakes like Lake Whillans were discovered only in the late 1990s when ice-penetrating radar and satellite measurements allowed researchers to see through the dense ice sheets that cover the polar regions of the planet. The next phase of the project was—as has been the case as long as humans have been exploring their environments—to determine what, if anything, lived in the newly discovered area. Although the immediate instinct would be to drill through the ice and sample the water in the lake, **microbial ecologists** realized that sampling Lake Whillans was not terribly different from performing surgery on a human patient; **aseptic techniques** would have to be followed so that external microbes were not allowed to contaminate the lake. Drilling equipment was **sterilized** using a combination of ultraviolet light and hydrogen peroxide, the same techniques routinely used in hospitals and laboratories, and the water used to bore through the ice was filtered to remove even the smallest microorganisms. When the drill penetrated the last of the ice, it entered the lake, which at −0.5°C was several degrees warmer than the Antarctic surface.

Over the next few days, until the drilling hole froze shut, scientists and graduate students collected 30 liters of water and several

> "It's the first time we've gotten a real insight into what organisms might live beneath the Antarctic continent."
>
> **David Pearce, a microbiologist at Northumbria University, UK**

sediment samples from the lake. Study of those samples, which continues today, reveals that Lake Whillans hosts a vibrant ecosystem. **DNA analysis** revealed nearly 4,000 different **microbial species,** and each milliliter of lake water contained more than 130,000 cells, comparable to what one finds in the deepest oceans. The biggest difference between life in Lake Whillans and ecosystems found on the surface of the planet is the lack of sunlight. In terrestrial lakes **photosynthetic microorganisms** use the energy in sunlight to convert dissolved carbon dioxide into sugars. Because sunlight can't penetrate the half mile of ice covering Lake Whillans, many of the microbes in the lake derive energy from the oxidation of iron, sulfur, or ammonium compounds, a strategy used by some deep-sea bacteria. If it turns out, as many scientists believe, that the microorganisms in Lake Whillans supply minerals and nutrients to the surrounding ocean, then this small, dark, cold, invisible lake may have a tremendous effect on the ecosystem surrounding it. Not bad for a frozen wasteland.

- *One of the environmental pressures microorganisms from Lake Whillans had to adapt to was the ability to grow in very cold temperatures. What were several other environmental challenges these microbes faced?*

- *What fields of microbiology were used to initially study these microbes, and what fields could be involved in the further study of the isolated cells?*

To continue this Case Study, go to Case Study Part 2 at the end of the chapter.

SCOPING OUT THE CHAPTER

Microbiology is sometimes regarded as an esoteric science, concerned primarily with keeping the milk fresh and getting everyone to wash their hands. In fact, microbiology encompasses a number of interrelated disciplines, a rich history, and more organisms than any other branch of biology. In this chapter you'll get a quick tour of the field.

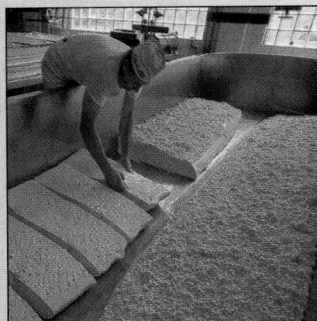

The Fields of Microbiology
Not every microbiologist works in a lab. Making cheese, wine and bread all rely on a knowledge of microbiology.

The Power of Microbiology
The vats here contain algae that are producing oil through photosynthesis, a potentially endless source of clean energy.

The Evolution of Microorganisms
All life on Earth has evolved from simple microbial cells. Evolutionary trees display the relationship between simpler and more complex organisms.

The Organisms of Microbiology
The vast majority of microorganisms pose no risk to humans. Some however, like certain strains of the bacterium *Staphylococcus aureus,* not only cause severe disease, but are resistant to almost all the drugs commonly used to fight them.

The History of Microbiology
The period from 1850 to 1950 is sometimes known as the **Golden Age of Microbiology,** when people like Louis Pasteur invented the very science of microbiology itself. The recent advent of powerful molecular biology techniques has caused many scientists to view the current era as a second Golden Age.

(Fields of Microbiology): © Joe Munroe/Science Source; (Power of Microbiology): *Source:* Christopher Botnick/NOAA; (Organisms of Microbiology): *Source:* Matthew J. Arduino, DRPH/Janice Haney Carr/CDC; (History of Microbiology): © Dea Picture Library/Getty Images

1.1 The Scope of Microbiology

 Expected Learning Outcomes

1. Define *microbiology* and *microorganisms,* and identify the major organisms included in the science of microbiology.

2. Name and define the primary fields included in microbiological studies.

As we observe the natural world, teeming with life, we cannot help but be struck by its beauty and complexity. But for every feature that is visible to the naked eye, there are millions of other features that are concealed from our sight by their small size. This alternate microscopic universe is populated by a vast microbial menagerie that is equally beautiful and complex. To sum up the presence of microbes in one word, they are **ubiquitous.*** They are found in all

* *ubiquitous* (yoo-bik′-wih-tis) L. *ubique,* everywhere and *ous,* having. Being, or seeming to be, everywhere at the same time.

natural habitats and most of those that have been created by humans. As scientists continue to explore remote and unusual environments, the one kind of entity they always find is microbes. These exist deep beneath the polar ice caps, in the ocean to a depth of 7 miles, in hot springs and thermal vents, in toxic waste dumps, and even in the clouds.

Microbiology is a specialized area of biology that deals with tiny life forms that are not readily observed without magnification, which is to say they are **microscopic.*** These microscopic organisms are collectively referred to as **microorganisms, microbes,*** or several other terms, depending upon the purpose. Some people call them "germs" or "bugs" in reference to their role in infection and disease, but those terms have other biological meanings and perhaps place undue emphasis on the disagreeable reputation of microorganisms. The major groups of microorganisms included in this study are **bacteria, viruses, fungi, protozoa, algae,** and **helminths** (parasitic worms). As we will see in subsequent chapters, each group exhibits a distinct collection of biological characteristics. The nature of microorganisms makes them both easy and difficult to study. Easy, because they reproduce so rapidly and can usually be grown in large numbers in the laboratory. Difficult, because we can't observe or analyze them without special techniques, especially the use of microscopes (see chapter 3).

Microbiology is one of the largest and most complex of the biological sciences because it integrates subject matter from many diverse disciplines. Microbiologists study every aspect of microbes—their genetics, their physiology, characteristics that may be harmful or beneficial, the ways they interact with the environment, the ways they interact with other organisms, and their uses in industry and agriculture.

See **table 1.1** for an overview of some fields and occupations that involve basic study or applications in microbiology. Each major discipline in microbiology contains numerous subdivisions or specialties that deal with a specific subject area or field (table 1.1). In fact, many areas of this science have become so specialized that it is not uncommon for a microbiologist to spend an entire career concentrating on a single group or type of microbe, biochemical process, or disease.

The specialty professions of microbiology include:

- geomicrobiologists, who focus on the roles of microbes in the development of earth's crust (table 1.1B);
- marine microbiologists, who study the oceans and its smallest inhabitants;
- medical technologists, who do the tests that help diagnose pathogenic microbes and diseases associated with them;
- nurse epidemiologists, who analyze the occurrence of infectious diseases in hospitals; and
- astrobiologists, who study the possibilities of organisms in space (see the Case Study for chapter 2).

Studies in microbiology have led to greater understanding of many general biological principles. For example, the study of microorganisms established universal concepts concerning the chemistry of life (see chapters 2 and 8), systems of inheritance (see

chapter 9), and the global cycles of nutrients, minerals, and gases (see chapter 26).

1.2 General Characteristics of Microorganisms and Their Roles in the Earth's Environments

 Expected Learning Outcomes

3. Describe the basic characteristics of prokaryotic cells and eukaryotic cells and their evolutionary origins.

4. State several ways that microbes are involved in the earth's ecosystems.

5. Describe the cellular makeup of microorganisms and their size range, and indicate how viruses differ from cellular microbes.

The Origins and Dominance of Microorganisms

For billions of years, microbes have extensively shaped the development of the earth's habitats and influenced the evolution of other life forms. It is understandable that scientists searching for life on other planets first look for signs of microorganisms.

The fossil record uncovered in ancient rocks and sediments points to bacteria-like cells having existed on earth for at least 3.5 billion years (**figure 1.1**). Early microorganisms of this type dominated the earth's life forms for the first 2 billion years. These ancient cells were small and simple, and lacked specialized internal structures to carry out their functions. It is apparent that genetic material of these cells was not bound into a separate compartment called a nucleus or "karyon." The term assigned to cells and microbes of this type is **prokaryotic,*** meaning "before the nucleus." About 1.8 billion years ago, there appeared in the fossil record a more complex cell, which had developed a nucleus and various specialized internal structures called **organelles.*** These types of cells and organisms are defined as **eukaryotic*** in reference to their "true" nucleus. **Figure 1.2** compares the two cell types and includes some examples of viruses for comparison. In chapter 5 we will learn more about the origins of eukaryotic cells—they didn't arise suddenly out of nowhere; they evolved over millennia from prokaryotic cells through an intriguing process called endosymbiosis. The early eukaryotes, probably similar to algae and protozoa, started lines of evolution that eventually gave rise to fungi, plants, and multicellular animals such as worms and insects. You can see from figure 1.1 how long that took! The bacteria preceded even the earliest animals by about 3 billion years. This is a good indication that humans are not likely to, nor should we try to, eliminate microorganisms from our environment. Having existed for eons, they are absolutely essential for maintaining the planet's life-giving characteristics.

* *microscopic* (my″-kroh-skaw′-pik) Gr. *mikros*, small, and *scopein*, to see.
* *microbe* (my′-krohb) Gr. *mikros*, small, and *bios*, life.

* *prokaryotic* (proh″-kar-ee-ah′-tik) Gr. *pro*, before, and *karyon*, nucleus. Sometimes spelled procaryotic.
* *organelles* (or-gan′-elz) Gr. *organa*, tool, and *ella*, little.
* *eukaryotic* (yoo″-kar-ee-ah′-tik) Gr. *eu*, true or good, and *karyon*, nucleus. Sometimes spelled eucaryotic.

TABLE 1.1 A Sampling of Fields and Occupations in Microbiology

A. Public Health Microbiology and Epidemiology

These branches monitor and control the spread of diseases in communities. Some of the institutions charged with this task are the U.S. Public Health Service (USPHS) and the Centers for Disease Control and Prevention (CDC). The CDC collects information and statistics on diseases from around the United States and publishes it in a newsletter, *The Morbidity and Mortality Weekly Report* (see chapter 13).

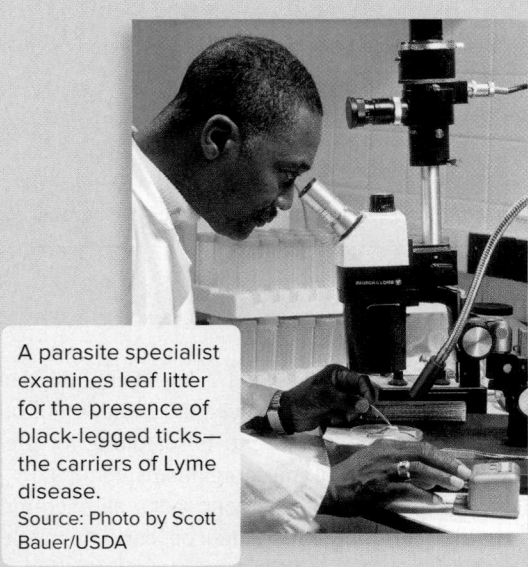

A parasite specialist examines leaf litter for the presence of black-legged ticks—the carriers of Lyme disease.
Source: Photo by Scott Bauer/USDA

B. Environmental Microbiology

This field encompasses the study of microorganisms and their ecological relationships in such natural habitats as soil and water.

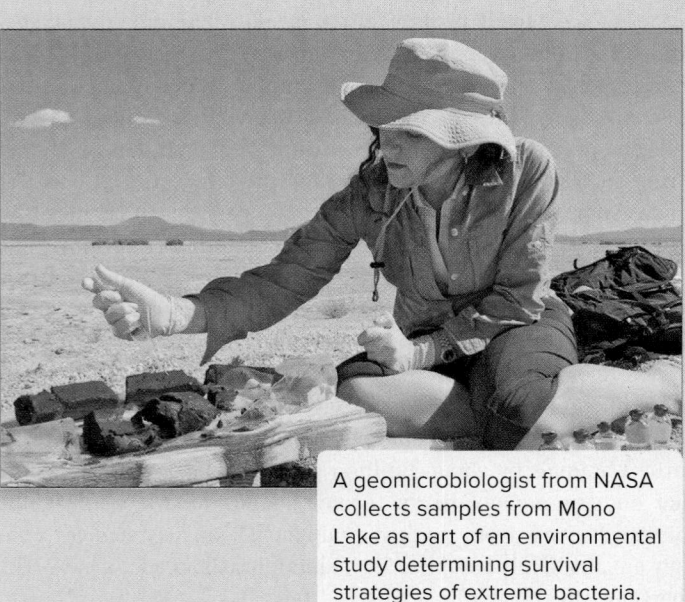

A geomicrobiologist from NASA collects samples from Mono Lake as part of an environmental study determining survival strategies of extreme bacteria.
© Henry Bortman

C. Biotechnology and Industrial Microbiology

This branch is defined by any process that harnesses the actions of living things to derive a desired product, ranging from beer to stem cells. It includes industrial microbiology, which uses microbes to produce and harvest large quantities of such substances as vaccines, vitamins, drugs, and enzymes (see chapters 10 and 27).

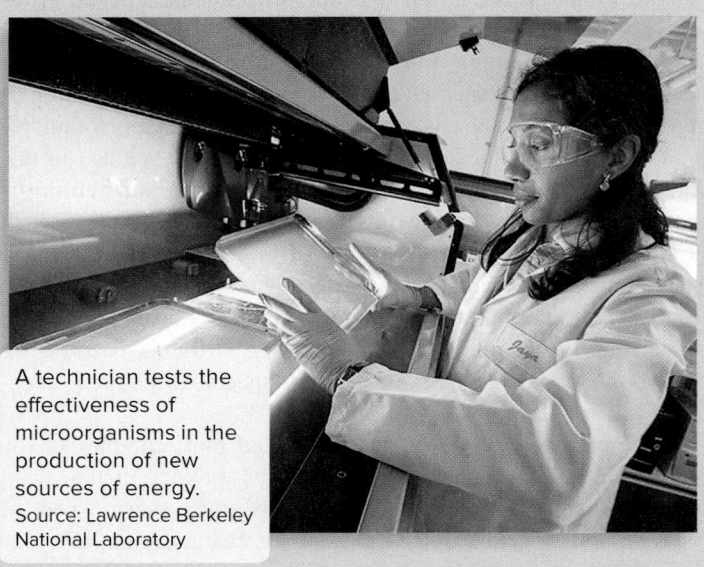

A technician tests the effectiveness of microorganisms in the production of new sources of energy.
Source: Lawrence Berkeley National Laboratory

D. Immunology

This branch studies the complex web of protective substances and reactions caused by invading microbes and other harmful entities. It includes such diverse areas as blood testing, vaccination, and allergy (see chapters 15, 16, and 17).

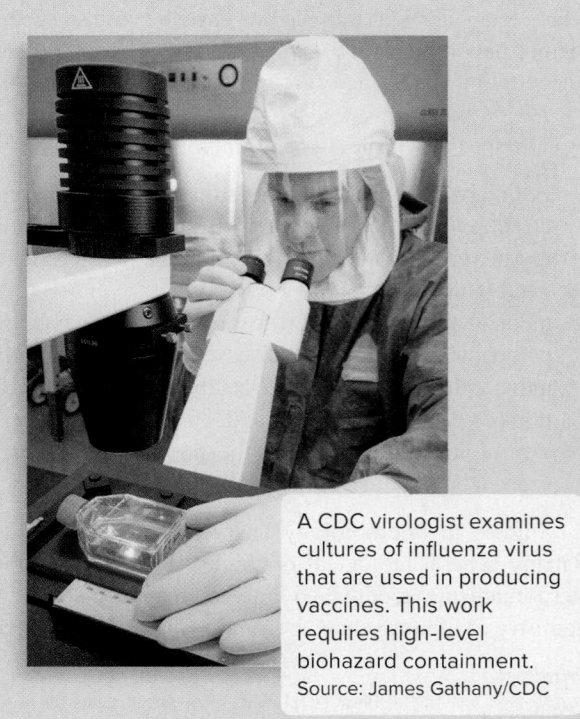

A CDC virologist examines cultures of influenza virus that are used in producing vaccines. This work requires high-level biohazard containment.
Source: James Gathany/CDC

E. Genetic Engineering and Recombinant DNA Technology

These interrelated fields involve deliberate alterations of the genetic makeup of organisms to create novel microbes, plants, and animals with unique behavior and physiology. This is a rapidly expanding field that often complements biotechnology (see chapter 10).

A bacteriologist from the U.S. Department of Energy checks cultures of genetically modified bacteria for growth.
Source: Biological and Environmental Research Information System, Oak Ridge National Laboratory. Sponsored by the U.S. Department of Energy Biological and Environmental Research Program.

F. Agricultural Microbiology

This branch is concerned with the relationships between microbes and domesticated plants and animals. Plant specialists focus on plant diseases, soil fertility, and nutritional interactions. Animal specialists work with infectious diseases and other interactions between animals and microorganisms.

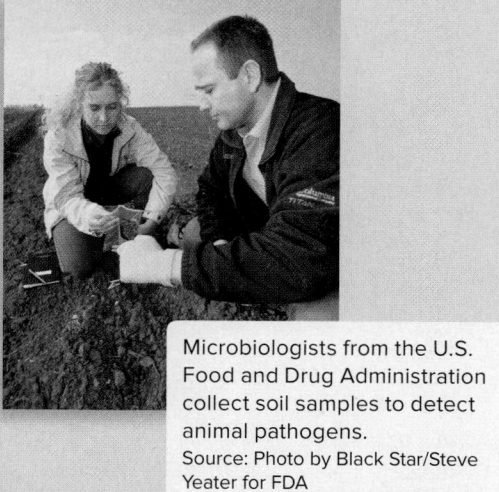

Microbiologists from the U.S. Food and Drug Administration collect soil samples to detect animal pathogens.
Source: Photo by Black Star/Steve Yeater for FDA

G. Food Microbiologists

These scientists are concerned with the impact of microbes on the food supply, including such areas as food spoilage, food-borne diseases, and production.

A U.S. Department of Agriculture technician observes tests for the presence of *Escherichia coli* in foods.
Source: Photo by Keith Weller/USDA

H. Branches of Microbiology

Branch	Chapter	Involved in the Study of:
Bacteriology	4	The bacteria—small single-celled prokaryotic organisms
Mycology	5, 22	The fungi, a group of eukaryotes that includes both microscopic eukaryotes (molds and yeasts) and larger organisms (mushrooms, puffballs)
Protozoology	5, 23	The protozoa—animal-like and mostly single-celled eukaryotes
Virology	6, 24, 25	Viruses—minute, noncellular particles that parasitize cells
Parasitology	5, 23	Parasitism and parasitic organisms—traditionally including pathogenic protozoa, helminth worms, and certain insects
Phycology or Algology	5	Simple photosynthetic eukaryotes, the algae, ranging from single-celled forms to large seaweeds
Morphology	4, 5, 6	The detailed structure of microorganisms
Physiology	7, 8	Microbial function (metabolism) at the cellular and molecular levels
Taxonomy	1, 4, 5, 17	Classification, naming, and identification of microorganisms
Microbial Genetics, Molecular Biology	9, 10	The function of genetic material and biochemical reactions that make up a cell's metabolism
Microbial Ecology	7, 26	Interrelationships between microbes and the environment; the roles of microorganisms in the nutrient cycles and natural ecosystems

A medical microbiologist tests specimens for evidence of antibodies to the human immunodeficiency virus (HIV).
Source: James Gathany/CDC

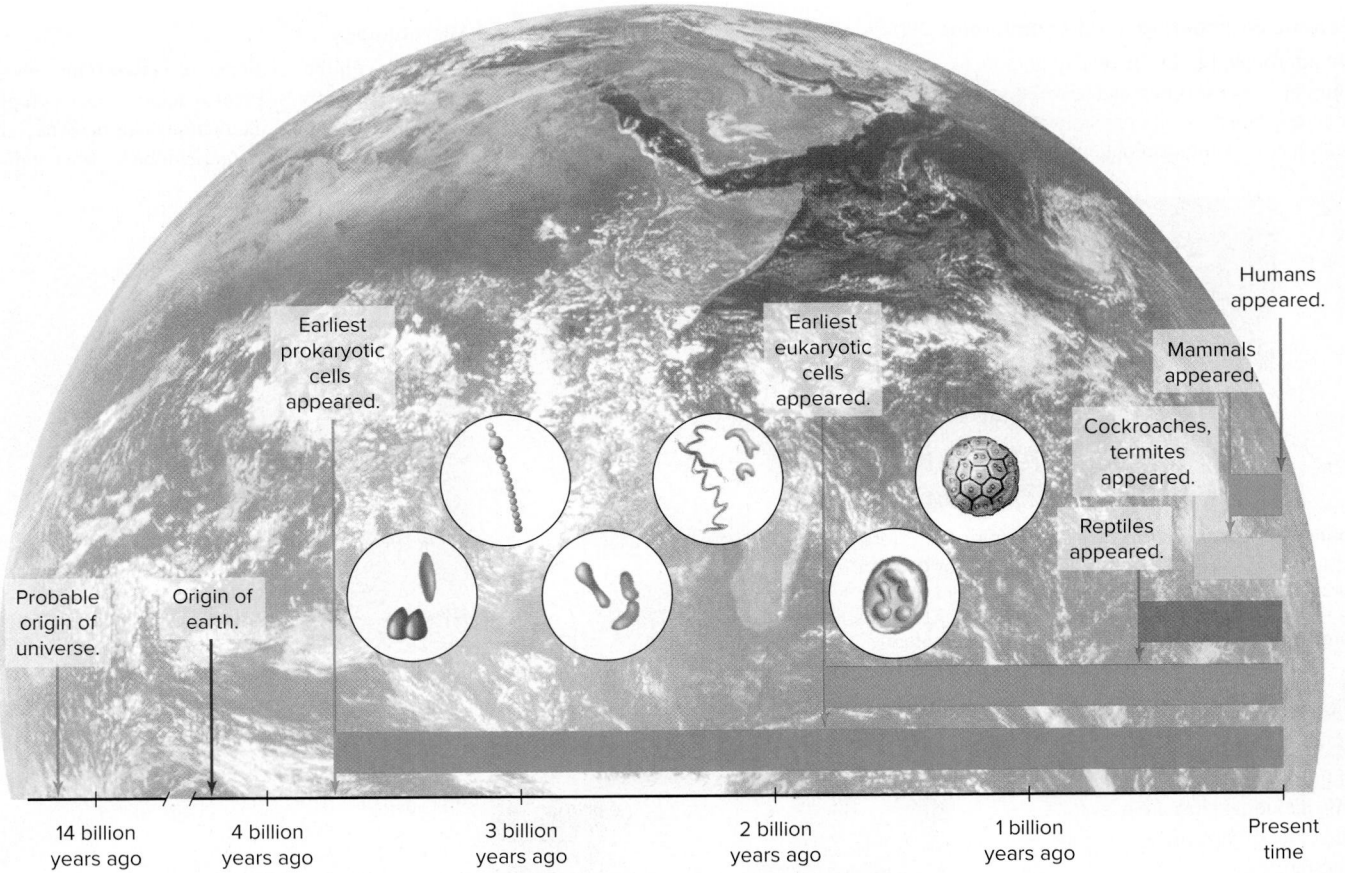

Figure 1.1 **Evolutionary time line.** The first simple prokaryotes appeared on earth approximately 3.5 billion years ago, and the first eukaryotes arose about 2 billion years ago. Although these appearances seem abrupt, hundreds of millions of years of earth's history passed while they were evolving to these stages. The fossil record for these periods is incomplete because many of the earliest microbes were too delicate to fossilize. *Source:* NASA

The Cellular Organization of Microorganisms

As a general rule, prokaryotic cells are smaller than eukaryotic cells, and in addition to lacking a nucleus, they lack organelles, which are structures in cells bound by one or more membranes. Examples of organelles include the mitochondria and Golgi

complexes, and several others, which perform specific functions such as transport, feeding, energy release and use, and synthesis. Prokaryotes perform similar functions, but they lack dedicated organelles to carry them out (figure 1.2).

The body plan of most microorganisms consists of a single cell or clusters of cells (**figure 1.3**). All prokaryotes are microorganisms,

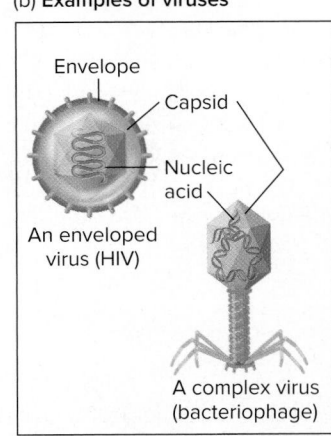

Figure 1.2 **Basic structure of cells and viruses.** **(a)** Comparison of a prokaryotic cell and a eukaryotic cell. **(b)** Two examples of viruses. These cell types and viruses are discussed in more detail in chapters 4, 5, and 6.

Reproductive spores

Bacteria: *Mycobacterium tuberculosis,* a rod-shaped cell (15,500x).

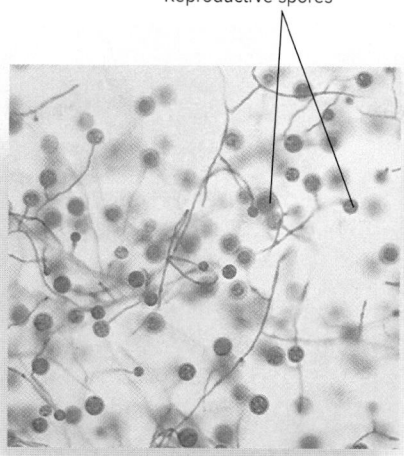

Fungi: *Histoplasma capsulatum,* with lollipop-like reproductive structures (750x). This agent is the cause of Ohio Valley fever.

Algae: desmids, *Spirogyra* filament, and diatoms (golden cells) (500x).

A single virus particle

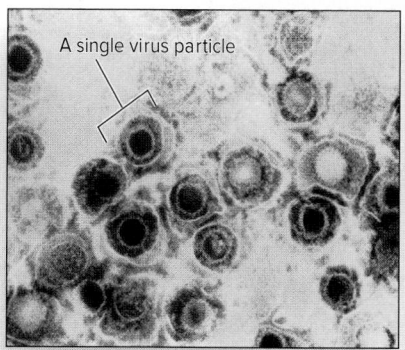

Virus: *Herpes simplex,* the cause of cold sores (100,000x).

Protozoa: A protozoan, *Oxytricha trifallax* bearing tufts of cilia that function like tiny legs (3,500x).

Helminths: Roundworms of *Trichinella spiralis* coiled in the muscle of a host (250x). This worm causes trichinellosis.

Figure 1.3 The six basic types of microorganisms. Organisms are not shown at the same magnifications; approximate magnification is provided. To see these microorganisms arrayed more accurately to scale, look for them in figure 1.4. (bacteria): *Source:* Janice Carr/CDC; (fungi): *Source:* Dr. Libero Ajello/CDC; (algae): © Charles Krebs Photography; (virus): *Source:* CDC; (protozoa): *Source:* National Human Genome Research Institute; (helminths): *Source:* CDC

and they include the bacteria and archaea (see figure 1.14). Only some of the eukaryotes are microorganisms: primarily algae, protozoa, molds and yeasts (types of fungi), and certain animals such as worms and arthropods. Not all members of these last two groups are microscopic, but certain members are still included in the study of microbiology because worms can be involved in infections and may require a microscope to identify them. Some arthropods such as fleas and ticks may also be carriers of infectious diseases. Additional coverage on cell types and microorganisms appears in chapters 4 and 5.

composed essentially of a small amount of hereditary material wrapped up in a protein covering. Some biologists refer to viruses as parasitic particles; others consider them to be very primitive organisms. Despite this slight disagreement, the impact of viruses is undeniable. Not only are they the most common microbes on earth, but they invade their hosts' cells and can inflict serious damage and death.

An adenovirus

Where Do the Viruses Fit?

Viruses are considered one type of microbe because they are microscopic and can cause infections and disease, but they are not cells. They are small particles that exist at a level of complexity somewhere between large molecules and cells (see figure 1.4). Viruses are much simpler than cells; they are

Microbial Dimensions: How Small Is Small?

When we say that microbes are too small to be seen with the unaided eye, what sorts of dimensions are we talking about? This concept is best visualized by comparing microbial groups with some organisms of the macroscopic world and also with the molecules and atoms of the molecular world (**figure 1.4**). The dimensions

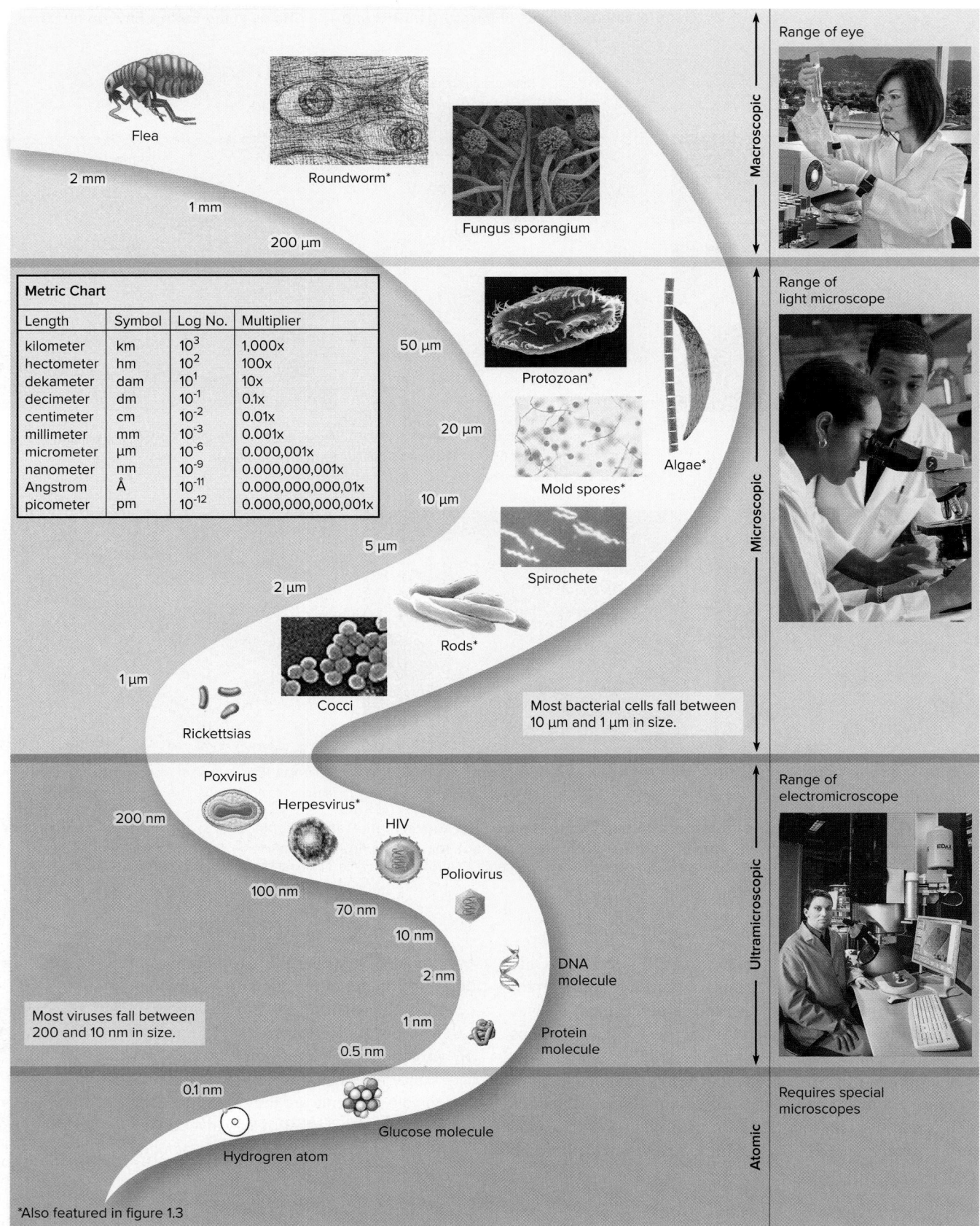

Metric Chart

Length	Symbol	Log No.	Multiplier
kilometer	km	10^3	1,000x
hectometer	hm	10^2	100x
dekameter	dam	10^1	10x
decimeter	dm	10^{-1}	0.1x
centimeter	cm	10^{-2}	0.01x
millimeter	mm	10^{-3}	0.001x
micrometer	μm	10^{-6}	0.000,001x
nanometer	nm	10^{-9}	0.000,000,001x
Angstrom	Å	10^{-11}	0.000,000,000,01x
picometer	pm	10^{-12}	0.000,000,000,001x

Range of eye

Macroscopic

Flea
2 mm
1 mm
Roundworm*
200 μm
Fungus sporangium

Range of light microscope

Microscopic

50 μm
Protozoan*
Algae*
20 μm
Mold spores*
10 μm
Spirochete
5 μm
2 μm
Rods*
1 μm
Cocci
Rickettsias

Most bacterial cells fall between 10 μm and 1 μm in size.

Range of electromicroscope

Ultramicroscopic

Poxvirus
Herpesvirus*
200 nm
HIV
100 nm
Poliovirus
70 nm
10 nm
2 nm
DNA molecule
1 nm
0.5 nm
Protein molecule

Most viruses fall between 200 and 10 nm in size.

0.1 nm

Requires special microscopes

Atomic

Glucose molecule
Hydrogren atom

*Also featured in figure 1.3

Figure 1.4 The sizes of the smallest organisms and objects. Even though they are all very small, they still display extensive variations in size. This illustration organizes the common measurements used in microbiology along with examples of organisms or items that fall into these measurement ranges. The scale includes macroscopic, microscopic, ultramicroscopic, and atomic dimensions. Most microbes we study measure somewhere between 100 micrometers (μm) and 10 nanometers (nm) overall. The examples are more or less to scale within a size zone but not between size zones. (roundworm): *Source:* CDC; (fungus): © Dennis Kunkel Microscopy, Inc./Phototake; (protozoan): *Source:* National Human Genome Research Institute; (algae): © Charles Krebs Photography; (mold spores): Dr. Libero Ajello/CDC; (spirochete): *Source:* CDC; (rods, cocci): *Source:* Janice Carr/CDC; (herpesvirus): *Source:* CDC; (range of eye): *Source:* Berkeley Lab - Roy Kaltschmidt, photographer; (range of light microscope): *Source:* Rhoda Baer (photographer)/National Cancer Institute; (range of electromicroscope): *Source:* Pacific Northwest National Laboratory

of macroscopic organisms are usually given in centimeters (cm) and meters (m), whereas those of most microorganisms fall within the range of micrometers (μm) and, sometimes, nanometers (nm) and millimeters (mm). The size range of most microbes extends from the smallest viruses, measuring around 10 nm and actually not much bigger than a large molecule, to protozoans measuring 3 to 4 mm and visible with the naked eye.

Microbial Involvement in Energy and Nutrient Flow

The microbes in all natural environments have lived and evolved there for billions of years. We do not yet know all of their roles, but it is likely they are vital components of the structure and function of these ecosystems.

Microbes are deeply involved in the flow of energy and food through the earth's ecosystems.[1] Most people are aware that plants carry out **photosynthesis,** which is the light-fueled conversion of carbon dioxide to organic material, accompanied by the formation of oxygen. But microorganisms were photosynthesizing long before the first plants appeared. In fact, they were responsible for changing the atmosphere of the earth from one without oxygen to one with oxygen. Today photosynthetic microorganisms (including algae) account for more than 50% of the earth's photosynthesis, contributing the majority of the oxygen to the atmosphere **(figure 1.5a).**

Another process that helps keep the earth in balance is the process of biological **decomposition** and nutrient recycling. Decomposition involves the breakdown of dead matter and wastes into simple compounds that can be directed back into the natural cycles of living things **(figure 1.5b).** If it were not for multitudes of bacteria and fungi, many chemical elements would become locked up and unavailable to organisms. In the long-term scheme of things, microorganisms are the main forces that drive the structure and content of the soil, water, and atmosphere. For example:

- Earth's temperature is regulated by "greenhouse gases," such as carbon dioxide and methane, that create an insulation layer in the atmosphere and help retain heat. A significant proportion of these gases is produced by microbes living in the environment and in the digestive tracts of animals.
- Recent estimates propose that, based on weight and numbers, up to 50% of all organisms exist within and beneath the earth's crust in soil, rocks, and even the frozen Antarctic **(figure 1.5c).** It is increasingly evident that this enormous underground community of microbes is a major force in weathering, mineral extraction, and soil formation.
- Bacteria and fungi live in complex associations with plants. They assist the plants in obtaining nutrients and water and may protect them against disease. Microbes form similar interrelationships with animals, notably as residents of numerous bodily sites.

1. Ecosystems are communities of living organisms and their surrounding environment.

(a)

(b)

(c)

Figure 1.5 **A microscopic wonderland. (a)** A summer pond is heavily laden with surface scum that reveals several different types of green algae called desmids (600×). **(b)** A rotting tomato being invaded by a fuzzy forest of mold. The fungus is *Botrytis,* a common decomposer of tomatoes and grapes (250×). **(c)** Even a dry lake in Antarctica, one of the coldest places on earth (−35°C), can harbor microbes under its icy sheet. Here we see a red cyanobacterium, *Nostoc* (3,000×), that has probably been frozen in suspended animation there for 3,000 years. Like the example discussed in the chapter-opening case study, this environment may serve as a model for what may one day be discovered on other planets. (a): *Source:* Photo by Lynn Betts, USDA Natural Resources Conservation Service; (a, inset) © Stephen Sharnoff/National Geographic Creative; (b & b, inset): © Kathy Park Talaro; (c) © Peter Doran/University of Illinois, Chicago; (c, inset) Image courtesy of the Priscu Research Group, Montana State University, Bozeman

 Check Your Progress SECTIONS 1.1–1.2

1. Define what is meant by the term *microorganism* and outline the important contributions microorganisms make to the earth's ecosystems.
2. Describe five different ways in which humans exploit microorganisms for our benefit.
3. Identify the groups of microorganisms included in the scope of microbiology, and explain the criteria for including these groups in the field of microbiology.
4. Observe figure 1.3 and place the microbes pictured there in a size ranking, going from smallest to largest. Use the magnification as your gauge.
5. Construct a table that displays all microbial groups based on what kind of cells they have or do not have.
6. Explain this statement: Microorganisms—we need to live with them because we can't live without them.

1.3 Human Use of Microorganisms

Expected Learning Outcome

6. Discuss the ways microorganisms can be used to create solutions for environmental problems and industrial products.

The incredible diversity and versatility seen in microbes make them excellent candidates for solving human problems. By accident or choice, humans have been using microorganisms for thousands of years to improve life and even to further human progress. Yeasts, a type of microscopic fungi, cause bread to rise and ferment sugar to make alcoholic beverages. Historical records show that households in ancient Egypt kept moldy loaves of bread to apply directly to wounds and lesions, which was probably the first use of penicillin! The manipulation of microorganisms to make products in an industrial setting is called **biotechnology.** * One newer application of this process uses farmed algae to extract a form of oil (biodiesel) to be used in place of petroleum products (**figure 1.6***a*).

Genetic engineering is a newer area of biotechnology that manipulates the genetics of microbes, plants, and animals for the purpose of creating new products and genetically modified organisms. One powerful technique for designing new organisms is termed **recombinant DNA.** This technology makes it possible to deliberately alter DNA[2] and to switch genetic material from one organism to another. Bacteria and fungi were some of the first organisms to be genetically engineered, because their relatively simple genetic material is readily manipulated in the laboratory. Recombinant DNA technology has unlimited potential in terms of medical, industrial, and agricultural uses. Microbes can be engineered to synthesize desirable proteins such as drugs, hormones, and enzymes (see table 1.1*C*).

Among the genetically unique organisms that have been designed by bioengineers are bacteria that contain a natural pesticide, yeasts that produce human hormones, pigs that produce hemoglobin, and plants that are resistant to disease (see table 1.1*E*). The

(a)

(b)

Figure 1.6 **Microbes at work. (a)** A scientist from the National Oceanic and Atmospheric Agency (NOAA) demonstrates a series of biodiesel reactors that culture single-celled algae (inset 750×) as a source of oil. This new "green" renewable energy source looks very promising. **(b)** Biotechnology meets bioremediation. Scientists at Pacific Northwest National Laboratories (PNNL) test the capacity of two newly discovered bacteria—*Shewanella* (green) and *Synechococcus* (yellow) (1,000×)—to reduce and detoxify radioactive waste. The process, carried out in large bioreactors, could speed the cleanup of hazardous nuclear waste deposits. (a): *Source:* Christopher Botnick/NOAA; (a, inset): © Yuuji Tsukii, Protist Information Server; (b & b, inset): *Source:* Pacific Northwest National Laboratory

techniques have also paved the way for characterizing human genetic material and diseases.

Another way of tapping into the unlimited potential of microorganisms is the relatively new science of **bioremediation.** * This process introduces microbes into the environment to restore stability or to clean up toxic pollutants. Bioremediation is required to control

* *biotechnology* (by′-oh-tek-nol″-oh-gee) The use of microbes or their products in the commercial or industrial realm.
2. DNA, or deoxyribonucleic acid, the chemical substance that comprises the genetic material of organisms.

* *bioremediation* (by′-oh-ree-mee-dee-ay″-shun) *bios,* life; *re,* again; *mederi,* to heal. The use of biological agents to remedy environmental problems.

the massive levels of pollution that result from human activities. Microbes have a surprising capacity to break down chemicals that would be harmful to other organisms. Agencies and companies have developed microbes to handle oil spills and detoxify sites contaminated with heavy metals, pesticides, and even radioactive wastes **(figure 1.6b).** The solid waste disposal industry is focusing on methods for degrading the tons of garbage in landfills, especially plastics and paper products. One form of bioremediation that has been in use for some time is the treatment of water and sewage. With clean freshwater supplies dwindling worldwide, it will become even more important to find ways to reclaim polluted water.

1.4 Microbial Roles in Infectious Diseases

 Expected Learning Outcomes

7. Review the roles of microorganisms as parasites and pathogens that cause infection and disease.

8. Define what is meant by emerging and reemerging diseases.

It is important to remember that the large majority of microorganisms are relatively harmless, have quantifiable benefits to humans and the environment, and in many cases are essential to life as we know it. They are free living, and derive everything they need to survive from the surrounding environment. Much of the time they form cohesive communities with other organisms, sharing habitat and nutrients. Examples include the natural partnerships that are found in symbioses and biofilms.[3]

Some microbes have adapted to a non-free-living lifestyle called parasitism. A **parasite** lives in or on the body of a larger organism called the **host** and derives most of its sustenance from that host. A parasite's actions generally damage the host through infection and disease. Another term that can be used to specify this type of microbe is **pathogen**.*

Humanity is plagued by nearly 2,000 different pathogens that can cause various types of disease. Infectious diseases still devastate human populations worldwide, despite significant strides in understanding and treating them. The most recent estimates from the World Health Organization (WHO) point to around 10 billion infections of all types across the world every year. There are more infections than people because many people acquire more than one infection. Infectious diseases are also among the most common causes of death in much of humanity, and they still kill a significant percentage of the U.S. population. The worldwide death toll from infections is about 13 million people per year, and 90% of the deaths are caused by just six infectious agents. **Table 1.2** illustrates the toll of some common infectious diseases, while **figure 1.7** compares the death rate among four groups from around the world that differ significantly in socioeconomic levels. It is quite evident which world inhabitants suffer the most from infectious diseases. **Table 1.3** displays the number of people affected by what are commonly known as neglected tropical diseases (NTDs), a collection of conditions that thrive among the worlds

3. A biofilm is a complex network of microbes and their secretions that form in most natural environments, discussed further in chapter 4.

* *pathogen* (path'-oh-jen) Gr. *pathos,* disease, and *gennan,* to produce. Disease-causing agents.

TABLE 1.2	Worldwide Morbidity and Mortality of Common Diseases*	
Disease	**New Cases per Year**	**Deaths per Year**
Influenza	3–5 million cases of severe illness	250,000–500,000
Typhoid fever	21,000,000	200,000
Measles	20,000,000	145,700
HIV/AIDS	2,300,000	1,500,000
Dengue fever	96,000,000	22,000
Shigellosis	80,000,000–165,000,000	600,000
Viral hepatitis (A and B)	380,000,000	1,280,000
Malaria	200,000,000	627,000
Tuberculosis	9,000,000**	1,500,000

*Estimates from most recent CDC and World Health Organization statistics.

**As many as 2,000,000,000 people are believed to carry the tuberculosis bacterium, most as long-term carriers.

Source: Data from World Health Organization

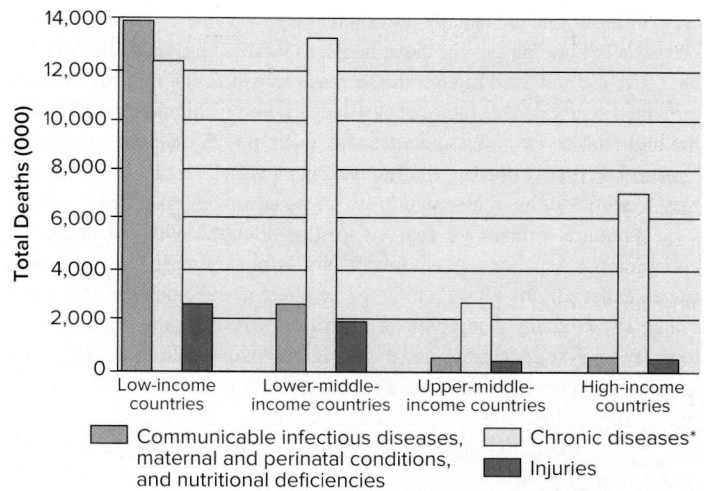

Figure 1.7 Chart comparing the major causes of death, by world socioeconomic levels. The relationship between income and rate of death from infectious diseases is most evident in this comparison. What other correlations can we make from these data?

TABLE 1.3	Neglected Tropical Diseases
Disease	**Number of Cases**
Ascariasis	1,000,000,000
Hookworm infection	700,000,000
Onchocerciasis (river blindness)	26,000,000
Lymphatic filariasis	120,000,000
Schistosomiasis	240,000,000
Trachoma	41,000,000
Trichuriasis	800,000,000

1.1 MAKING CONNECTIONS

The Changing Spectacle of Infectious Diseases

The middle of the last century was a time of great confidence in science and medicine. The introduction of antibiotics in the 1940s and a lengthening list of vaccines for preventing numerous diseases caused many medical experts to declare victory. For a short time there was a sense that infectious diseases were going to be completely manageable. But this optimistic viewpoint was on a collision course with a vast army of very tiny invaders that are everywhere—namely, microorganisms. Because humans are constantly interacting with microbes, we serve as a handy incubator for infectious diseases, both those newly recognized and older ones previously identified.

Altogether, government agencies are keeping track of more than 100 emerging and reemerging infectious diseases. The newer or emerging diseases usually erupt suddenly with no warning (Middle East respiratory syndrome). Older or reemerging diseases demonstrate just how difficult it is to eradicate microbes and the diseases they cause, even though we are very aware of them and often have drugs and vaccines to combat them. Only smallpox has been completely eliminated, although we are very close to eradicating polio. In fact, we continue to experience epidemics of childhood diseases that are usually preventable with vaccines. Starting in 2010, pertussis (whooping cough) cases began to increase dramatically over the next 2 years, and even though the increase subsided, the number of cases remained nearly double their pre-2010 levels. Some communities recorded the highest rates of pertussis in decades, even though children were being vaccinated. It now appears that the vaccine's effectiveness wears off and may require additional inoculations to create adequate protection.

What are some of the reasons for this dilemma with our microbial cohabitants? A major contributing factor is our increased mobility and travel, especially by air—an infected person can travel around the world before showing any symptoms of infection, carrying the infectious agent to many far-flung locations and exposing populations along the way, who in turn can infect their contacts. A second factor is the spread of diseases by vectors, living organisms such as fleas, ticks, or mosquitoes. Emerging viruses like chikungunya, dengue, and Zika are all spread by the *Aedes* mosquito, which is so aggressive it routinely follows people indoors to partake of a blood meal (see figure).

Other significant effects involve our expanding population and global food-growing practices. As we continue to encroach into new territory and wild habitats, there is potential for contact with emerging pathogens, as we saw with Ebola fever, Lyme disease, and hantavirus pulmonary syndrome. Our agricultural practices can unearth microbes that were lying dormant or hidden. A bacterium carried in the intestine of domestic cattle, *Escherichia coli* O157:H7, the agent of a serious kidney disease, has been associated with hundreds of thousands of infections from food and water contami-

nated with cattle feces. Much mass-produced fresh food can also travel around the world, infecting people along the way. Several large outbreaks of salmonellosis, shigellosis, and listeriosis have been traced to contaminated dairy and poultry products, and vegetables. In some areas of the world, farmers knowingly or unknowingly use fecally contaminated water and fertilizer on fresh produce that will be on your table the next week.

Then there is the matter of the incredible resistance of microbes. You know about their capacity to live where no other living things could—volcanoes, salt lakes, radioactive waste pits—so it has not been a surprise to discover how readily they can adapt to drugs we use to treat them. The emergence of drug-resistant "superbugs" has become a massive problem in medicine. Some forms of *Staphylococcus aureus* (MRSA) and *Mycobacterium tuberculosis* are resistant to so many drugs that there are few, and sometimes no, treatment choices left.

As hard as we may try to manage microbes, we keep coming up against a potent reality, a sentiment summed up by the renowned microbiologist Louis Pasteur over 130 years ago when he declared: "Microbes will have the last word."

Why is it unlikely that infectious diseases can ever be completely eradicated? *Answer available on Connect.*

The *Aedes aegypticus* mosquito is the vector for several emerging viral diseases. In this female mosquito, feeding on her photographer, blood can clearly be seen within the fascicle (feeding apparatus) and filling the distended abdomen of the mosquito. Because this species is found throughout the Americas, it is thought to be only a matter of time before the Zika, dengue, and chikungunya viruses are well established in the United States.

Source: James Gathany/Prof. Frank Hadley Collins, Dir., Cntr. for Global Health and Infectious Diseases, Univ. of Notre Dame/CDC

poorest populations and receive far too little attention. Most NTDs are easily treatable with drugs or preventable with vaccines.

Those hardest hit are residents in countries where access to adequate medical care is lacking. One-third of the earth's 7 billion inhabitants live on less than $1 per day, are malnourished, are not fully immunized, and have no access to drugs. Take the case of malaria, caused by a microorganism transmitted by mosquitoes, which kills 1 to 2 million people every year worldwide. Currently the most effective way for citizens of developing countries to avoid infection is to sleep under a bed net, because the mosquitoes are

most active in the evening. Yet even this inexpensive solution is beyond the reach of people in many developing countries who cannot afford the $3 to $5 for nets to protect their family. Fortunately for many countries in the malaria zone, several international organizations have collaborated to provide special insecticide-treated nets that can help lower the rate of infections.

Emerging and Reemerging Diseases Among the more significant factors in the overall picture of infectious diseases are emerging and reemerging diseases (**1.1 Making Connections**). **Emerging**

diseases are newly identified conditions that are being reported in increasing numbers. Since 1980 at least 87 novel infectious agents have arisen within the human population. Some of them have been associated with a specific location (Ebola fever virus), whereas others have become pandemics, meaning they spread across continents (human immunodeficiency virus—HIV). A number of them cause **zoonoses,*** which are infectious diseases native to animals that can be transmitted to humans. One recent example is chikungunya virus spread by mosquitoes to humans and other mammals. This virus traveled from the Caribbean to Florida in 2014. It is unclear how fast the virus will spread throughout the United States, as conditions become less favorable to the life cycle of mosquitoes as one moves north.

Reemerging diseases are older, well-known diseases that are increasing in occurrence for reasons outlined in 1.1 Making Connections. Among the most common reemerging infectious diseases are tuberculosis (TB), influenza, malaria, cholera, and hepatitis B. Tuberculosis, which has been known since ancient times, still causes 8 million new infections and kills 1 million to 2 million people every year. As you will see, numerous factors play a part in the tenaciousness of infectious diseases, but fundamental to all of them is the formidable capacity of microbes to adapt to alterations in the individual, community, and environment.

CLINICAL CONNECTIONS

Are Microbes a Hidden Cause of Diseases?

One of the most eye-opening discoveries has been that many diseases once considered noninfectious probably do involve microbial infection. Most scientists expect that, in time, a majority of chronic conditions will be be linked to microbial agents. The most famous of these is gastric ulcers, now known to be caused by a bacterium called *Helicobacter* (see chapter 21). But there are more. Diseases as disparate as type 1 diabetes, obsessive compulsive disorder, and coronary artery disease have been linked to chronic infections with microorganisms. Even the microbiome, the collection of microorganisms we all carry with us even when healthy, has been shown to have a much greater effect on our health than was previously thought. Recent studies have linked changes in the microbiome population to metabolic syndrome, a collection of health conditions including high cholesterol, hypertension, high blood sugar levels, and excess fat, all of which can raise the risk of heart disease, stroke, and diabetes. It seems that the golden age of microbiological discovery, during which all of the "obvious" diseases were characterized and cures or preventions were devised for them, should more accurately be referred to as the *first* golden age. We're now discovering the roles of microorganisms in hidden but slowly destructive diseases. These include female infertility caused by *Chlamydia* infection and malignancies such as liver cancer (hepatitis viruses) and cervical cancer (human papillomavirus). In fact, epidemiologists analyzing statistics on world cancer have estimated that one in six cancers can be associated with an infectious agent.

Another important development in infectious disease trends is the increasing number of patients with weakened defenses who are kept alive for extended periods. We are becoming more susceptible to infectious disease precisely because of advances in medicine.

* *zoonoses* (zoh″-uh-noh′-seez) Gr. *zoion,* animal and *nosos,* disease. Any disease indigenous to animals transmissible to humans.

People are living longer. Sicker people are staying alive much longer than in the past. Older and sicker people have heightened susceptibility to what we might call "garden-variety" microbes.

Why would chronic infections be more likely to be associated with diseases like cancer? *Answer available on Connect.*

 Check Your Progress SECTIONS 1.3–1.4

7. Describe several ways the beneficial qualities of microbes greatly outweigh microbes' roles as infectious agents.

8. Look up in the index some of the diseases shown in table 1.2 and determine which ones could be prevented by vaccines or cured with drugs. Are there other ways (besides vaccines) to prevent any of these?

9. Distinguish between emerging and reemerging infectious diseases and explain what factors contribute to their development.

1.5 The Historical Foundations of Microbiology

 Expected Learning Outcomes

9. Outline the major events in the history of microbiology, including the major contributors to the early development of microscopy, medical advances, aseptic techniques, and the germ theory of disease.

10. Explain the main features of the scientific method, and differentiate between inductive and deductive reasoning and between hypothesis and theory.

If not for the extensive interest, curiosity, and devotion of thousands of microbiologists over the last 300 years, we would know little about the microscopic realm that surrounds us. Many of the discoveries in this science have resulted from the prior work of men and women who toiled long hours in dimly lit laboratories with the crudest of tools. Each additional insight, whether large or small, has added to our current knowledge of life forms and processes. This section summarizes the prominent discoveries made in the past 300 years: microscopy, the rise of the scientific method, and the development of medical microbiology, including the germ theory and the origins of modern microbiological techniques. The table "Significant Events in Microbiology," found in Online Appendix 2, summarizes some of the pivotal events in microbiology from its earliest beginnings to the present.

The Development of the Microscope: "Seeing Is Believing"

It is likely that from the very earliest history, humans noticed that when certain foods spoiled, they became inedible or caused illness and yet other "spoiled" foods did no harm and even had enhanced flavor. Even several centuries ago, there was already a sense that diseases such as the black plague and smallpox were caused by some sort of transmissible matter. But the causes of such phenomena were obscure because the technology to study them was lacking. Consequently, they remained cloaked in mystery and regarded with superstition—a trend that led even well-educated scientists to believe in spontaneous generation (**1.2 Making Connections**).

1.2 MAKING CONNECTIONS

The Fall of Superstition and the Rise of Microbiology

For thousands of years, people believed that certain living things arose from vital forces present in nonliving or decomposing matter. This ancient idea, known as **spontaneous generation,** was continually reinforced as people observed that meat left out in the open soon "produced" maggots, that mushrooms appeared on rotting wood, that rats and mice emerged from piles of litter, and that other magical phenomena occurred. Though some of these early ideas seem quaint and ridiculous in light of modern knowledge, we must remember that, at the time, mysteries in life were accepted and the scientific method was not widely practiced.

Even after single-celled organisms were discovered during the mid-1600s, the idea of spontaneous generation persisted. Some scientists assumed that microscopic beings were an early stage in the development of more complex ones.

Over the subsequent 200 years, scientists waged an experimental battle over the two hypotheses that could explain the origin of simple life forms. Some tenaciously clung to the idea of **abiogenesis*** which embraced spontaneous generation. On the other side were advocates of **biogenesis,*** saying that living things arise only from others of their same kind. There were serious proponents on both sides, and each side put forth what appeared on the surface to be plausible explanations of why their evidence was more correct. Gradually the abiogenesis hypothesis was abandoned as convincing evidence for biogenesis continued to mount. The following series of experiments were among the most important in finally tipping the balance.

Some of the important variables to be considered in testing the hypotheses were the effects of nutrients, air, and heat, and the presence of preexisting life forms in the environment. One of the first people to test the spontaneous generation theory was **Francesco Redi** of Italy. He conducted a simple experiment in which he placed meat in a jar and covered the jar with fine gauze. Flies gathering at the jar were blocked from entering and thus laid their eggs on the outside of the gauze. The maggots subsequently developed without access to the meat, indicating that maggots were the offspring of flies and did not arise from some "vital force" in the meat. This and related experiments laid to rest the idea that more complex animals such as insects and mice developed through abiogenesis, but it did not convince many scientists of the day that simpler organisms could not arise in that way.

The Frenchman **Louis Jablot** reasoned that even microscopic organisms must have parents, and his experiments with infusions (dried hay steeped in water) supported that hypothesis. He divided an infusion that had been boiled to destroy any living things into two containers: a heated container that was closed to the air and a heated container that was freely open to the air. Only the open vessel developed microorganisms, which he presumed had entered in air laden with dust. Regrettably, the validation of biogenesis was temporarily set back by **John Needham,** an Englishman who did similar experiments using mutton gravy. His results were in conflict with Jablot's because both his heated and unheated test containers teemed with microbes. Unfortunately, his experiments were done before the realization that heat-resistant microbes are not usually

killed by mere boiling. Apparently Jablot had been lucky; his infusions were sterile.

Then, in the mid-1800s, the acclaimed microbiologist **Louis Pasteur** entered the arena. He had recently been studying the roles of microorganisms in the fermentation of beer and wine, and it was clear to him that these processes were brought about by the activities of microbes introduced into the beverage from air, fruits, and grains. The methods he used to discount abiogenesis were simple yet brilliant.

Pasteur's Experiment

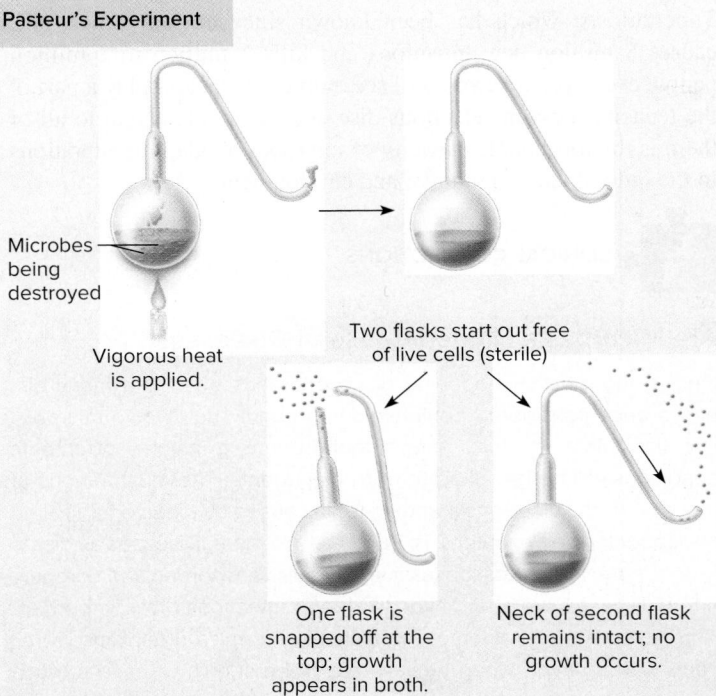

Microbes being destroyed

Vigorous heat is applied.

Two flasks start out free of live cells (sterile)

One flask is snapped off at the top; growth appears in broth.

Neck of second flask remains intact; no growth occurs.

To further clarify that air and dust were the source of microbes, Pasteur filled flasks with broth and fashioned their openings into elongate, swan-neck-shaped tubes. The flasks' openings were freely open to the air but were curved so that gravity would cause any airborne dust particles to deposit in the lower part of the necks. He heated the flasks to sterilize the broth and then incubated them. As long as the flask remained intact, the broth remained sterile, but if the neck was broken off so that dust fell directly down into the container, microbial growth immediately commenced.

Pasteur summed up his findings, "For I have kept from them, and am still keeping from them, that one thing which is above the power of man to make; I have kept from them the germs that float in the air, I have kept from them life."

What type of microorganisms were likely responsible for the misleading results of John Needham's experiment and were absent in Jablot's and Pasteur's experiments? *Answer available on Connect.*

* *abiogenesis* (ah-bee″-oh-jen′-uh-sis) L. *a,* without, *bios,* life, and *genesis,* beginning.
* *biogenesis* (by-oh-gen′-uh-sis) to begin with life.

Figure 1.8 An oil painting of Antonie van Leeuwenhoek (1632–1723) sitting in his laboratory. How astonishing to realize that this curious and original man was the first human being on earth ever to glimpse the "cavorting wee beasties" (his description) of the microbial world! © Bettmann/Corbis

(a)

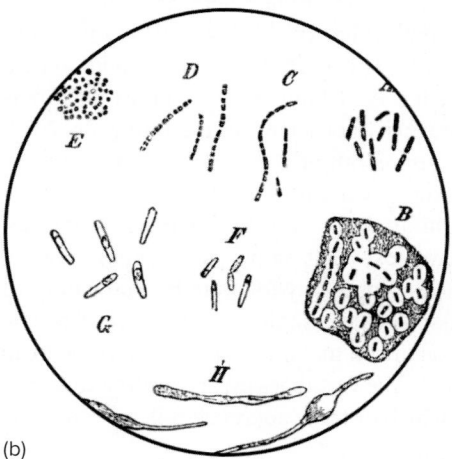

(b)

Figure 1.9 Leeuwenhoek's microscope. (a) A brass replica of a Leeuwenhoek microscope and how it is held (inset). **(b)** Early illustrations of bacterial cells from a sample of milk, magnified about 300×. These drawings closely resemble those Leeuwenhoek made of his animalcules. (a & a, inset): © Kathy Park Talaro; (b): *Source:* Trouessart, E.L., *Microbes Ferments and Moulds.* © 1895 D. Appleton and Company.

True awareness of the widespread distribution of microorganisms and some of their characteristics was finally made possible by the development of the first microscopes. These devices revealed microbes as discrete entities sharing many of the characteristics of larger, visible plants and animals. Several early scientists fashioned magnifying lenses and microscopes, but these lacked the optical clarity needed for examining bacteria and other small, single-celled organisms. The most careful and exacting observations awaited the simple single-lens microscope hand-fashioned by **Antonie van Leeuwenhoek,** a Dutch linen merchant and self-made microbiologist **(figure 1.8).**

During the late 1600s in Holland, Leeuwenhoek used his early lenses to examine the thread patterns of the draperies and upholstery he sold in his shop. Between customers, he retired to the workbench in the back of his shop, grinding glass lenses to ever-finer specifications. He could see with increasing clarity, but after a few years he became interested in things other than thread counts. He took rainwater from a clay pot, smeared it on his specimen holder, and peered at it through his finest lens. He found "animals appearing to me ten thousand times more than those which may be perceived in the water with the naked eye."

He didn't stop there. He scraped plaque from his teeth, and from the teeth of volunteers who had never cleaned their teeth in their lives, and took a close look at that. He recorded: "In the said matter there were many very little living animalcules, very prettily a-moving. . . . Moreover, the other animalcules were in such enormous numbers, that all the water . . . seemed to be alive." Leeuwenhoek started sending his observations to the Royal Society of London, and eventually he was recognized as a scientist of great merit.

Quick Search

Search for "Through van Leeuwenhoek's Eyes: Microbiology in a Nutshell" on YouTube to watch a video inspired by Leeuwenhoek's world

Leeuwenhoek constructed more than 250 small, powerful microscopes that could magnify up to 300 times **(figure 1.9).** Considering that he had no formal training in science and that he was the first person ever to faithfully record this strange new world, his descriptions of bacteria and protozoa (which he called "animalcules") were astute and precise. Because of Leeuwenhoek's extraordinary contributions to microbiology, he is sometimes considered the father of bacteriology and protozoology.

From the time of Leeuwenhoek, microscopes became more complex and improved, with the addition of refined lenses, a condenser, finer focusing devices, and built-in light sources. The prototype of the modern compound microscope, in use from about the mid-1800s, was capable of magnifications of 1,000 times or more, largely because it had two sets of lenses for magnification. Even our modern laboratory microscopes are not greatly different in basic structure and function from those early microscopes. The technical characteristics of microscopes and microscopy are a major focus of chapter 3.

The Scientific Method and the Search for Knowledge

The research that led to acceptance of biogenesis provides us with one example of the early development of science-based thought. Over the next few sections we will glimpse other important milestones in the development of the scientific method, such as vaccination, germ theory, asepsis, and Koch's postulates. The impact of science is so pervasive that you may not realize how much of our everyday life is built upon applications of the scientific method. Vaccines, antibiotics, space travel, computers, medical diagnosis, and DNA testing exist primarily because of the work of thousands of scientists doing objective observations and collecting evidence that is measurable, can be expressed quantitatively, and is subject to critical analysis.

The information obtained through the scientific method is explanatory and predictive. It aims to explain how and why phenomena occur and to predict what is expected to happen under known conditions. Two very recent events dramatically contrast the differences between scientific versus nonscientific reasoning. In 2012, NASA sent the *Curiosity* spacecraft on a 225-million-mile journey to Mars and landed it at a very precise time and place—an incredible achievement that brought to bear vast knowledge of physics, engineering, and mathematics. That same year, rumors started about doomsday predictions supposedly found in interpretations of an ancient Mayan calendar. A surprisingly large number of people all over the world were convinced that the world would come to an end on December 21, 2012. The fact that you are reading this textbook absolutely verifies which of these approaches is more reliable and predictive.

How do scientists apply the scientific method? In the **deductive reasoning** approach, a scientist uses general observations of some phenomenon to develop a set of facts to explain that phenomenon—that is, they *deduce* the facts that can account for what they have observed. This early explanation is considered a **hypothesis,** and however tentative it may start out, it is still based on scientific thought rather than subjective beliefs that come from superstition or myth. A valid hypothesis will allow for experimentation and testing and can be shown to be false. An example of a workable hypothesis based on deduction might be the speculation that a disease such as hemophilia is an inheritable condition. This would pave the way for specific experiments that test for the influence of genetics. A nonworkable hypothesis would be that hemophilia is caused by a curse placed on the royal family of England. Because supernatural beliefs offer no basis for experimentation and must be taken on faith, they can never be subjected to the rigors of the scientific method.

With **inductive reasoning,** one applies specific observations to develop a general explanation. This method is often used in the early phases of evaluation and can formulate a generalization to be tested deductively. In the previous example, induction might begin with the observation of a family in which several people have hemophilia, and this may lead to the general idea that it is inheritable.

A lengthy process of experimentation, analysis, and testing eventually leads to conclusions that either support or refute the hypothesis. If experiments do not uphold the hypothesis—that is, if it is found to be flawed—the hypothesis or some part of it is reconsidered. This does not mean the results are invalid; it means the hypothesis may require reworking or additional tests. Eventually it is either discarded or modified to fit the results of the experiment. If the hypothesis is supported by the experiment, it is not (or should not be) immediately accepted as fact. It then must be tested and retested. Indeed, this is an important guideline in the acceptance of a hypothesis. The results of the experiment must be published and repeated by other investigators.

In time, as each hypothesis is supported by a growing body of data and survives rigorous scrutiny, it moves to the next level of acceptance—the **theory.** A theory is a collection of statements, propositions, or concepts that explains or accounts for a natural event. A theory is not the result of a single experiment repeated over and over again but is an entire body of ideas that expresses or interprets many aspects of a phenomenon. When an unsupported idea is dismissed as being "just a theory," this is an incorrect use of the term as far as science is concerned. A theory is far from a weak notion or wild guess. It is a viable explanation that has stood the test of time and has yet to be disproved by serious scientific inquiries. Often theories develop and progress through decades of research and are added to and modified by new findings. At some point, evidence of the accuracy and predictability of a theory is so compelling that the next level of confidence is reached, and the theory becomes a **law,** or principle. For example, the **germ theory of disease** has been so thoroughly tested that it has clearly passed into the realm of law.

Science and its hypotheses and theories must progress along with technology. As advances in instrumentation allow new, more detailed views of living phenomena, old theories may be reexamined and altered and new ones proposed. Scientific knowledge is accumulative, and it must have built-in flexibility to accommodate new findings. It is for these reasons that scientists do not take a stance that theories or even laws are absolutely proved.

Figure 1.10 provides a summary of the scientific method in action using Edward Jenner's monumental discovery of vaccines. What is remarkable about Jenner's work is that he was the first to use scientific thought to construct a rigorous experimental model to inoculate people against disease, and he carried it through to its completion. It is also remarkable that he did this knowing nothing about viruses or even microbes. He worked out the concept of safely conferring artificial immunity long before there was any understanding of the immune system (see 15.2 Making Connections).

The Development of Medical Microbiology

Early experiments on the sources of microorganisms led to the profound realization that microbes are everywhere: Not only are air and

Observations/ information gathering

1. Dr. Jenner observed that cows had a form of pox similar to smallpox. Jenner also noted that milkmaids acquired cowpox only on the hands, and they appeared to be immune to smallpox.

Formation of Jenner's hypothesis

2. Jenner deduced that the cowpox was closely related to smallpox and could possibly be used on patients to provide protection similar to that of the milkmaids he had seen.

Testing the hypothesis, experiment I

3. Jenner took scrapings from cowpox blisters on the hand of a milkmaid and inoculated them into a boy who had not had smallpox. He developed minor symptoms but remained healthy.

Testing the hypothesis, experiment II

4. After a few weeks, the child was exposed twice to the pus from an active smallpox lesion. He did not acquire smallpox and appeared to have immune protection.

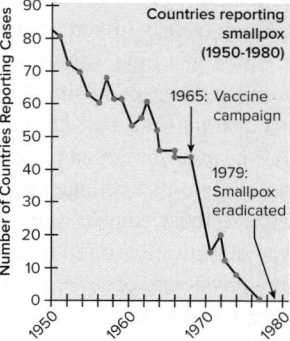

Reproducibility of results

5. Jenner went on to inoculate 23 other test subjects with cowpox. For the first time, he used lesions from one child to inoculate another. All subjects remained protected from smallpox.

Publishing of results; other medical testing

6. Jenner wrote a paper detailing his experiment. He called his technique *vaccination*, from the Latin *vacca* for cow. Other local English physicians began to vaccinate patients with some success.

Vaccination theory becomes widespread.

7. Over the next 100 years vaccination was brought to the rest of the world through local programs. Scientists used Jenner's methods to develop vaccines for other pathogens. The theory of artificial immunity became well established.

Smallpox is eradicated from the world.

8. A massive vaccination campaign aimed to reduce cases and to stamp out the disease completely. Billions of doses given over a decade reduced smallpox to zero. The last cases occurred in 1977, and in 1979 the disease was declared eradicated.

Figure 1.10 Edward Jenner and the saga of the smallpox vaccine. Jenner's work documents the first attempt based on the scientific method to control an infectious disease—smallpox. This disease was characterized by raised skin blisters called pox, and it often caused severe damage to organs. Throughout its long history this deadly disease decimated many populations worldwide, until 1977 when the last case was reported.

dust full of them, but the entire surface of the earth, its waters, and all objects are inhabited by them. This discovery led to immediate applications in medicine. Thus the seeds of medical microbiology were sown in the middle to latter half of the nineteenth century with the introduction of the first practical vaccine, the germ theory of disease, and the resulting use of sterile, aseptic, and pure culture techniques.

Jenner and the Introduction of Vaccination

We saw in figure 1.10 how the English physician and scientist **Edward Jenner** modeled the scientific method. His experiments ultimately gave rise to the first viable method to control smallpox by inoculating patients with a closely related disease agent. His contributions to medical science so changed the practice of medicine that he is considered the "Father of Immunology." It is often said of Jenner that his discovery saved more lives than any other in history. His work marked the beginning of an era of great scientific achievement—one that produced some of the most far-reaching developments in microbiology and medicine.

The Discovery of Spores and Sterilization

Following Pasteur's inventive work with infusions (1.2 Making Connections), it was not long before English physicist **John Tyndall**

provided the initial evidence that some of the microbes in dust and air have very high heat resistance and that particularly vigorous treatment is required to destroy them. Later, the discovery and detailed description of heat-resistant bacterial endospores by **Ferdinand Cohn,** a German botanist, clarified why heat would sometimes fail to completely eliminate all microorganisms. The modern sense of the word **sterile,** meaning completely free of all infectious agents, including endospores, viruses, and prions, had its beginnings here (see chapter 11). The capacity to sterilize objects and materials is an absolutely essential part of microbiology, medicine, dentistry, and some industries.

The Development of Aseptic Techniques

From earliest history, humans experienced a vague sense that "unseen forces" or "poisonous vapors" emanating from decomposing matter could cause disease. As the study of microbiology became more scientific and the invisible was made visible, the fear of such mysterious vapors was replaced by the knowledge, and sometimes even the fear, of "germs." About 130 years ago, the first studies by **Robert Koch** clearly linked a microscopic organism with a specific disease. Since that time, microbiologists have conducted a continuous search for disease-causing agents.

At the same time that abiogenesis was being hotly debated, a few budding microbiologists began to suspect that microorganisms could cause not only spoilage and decay but also infectious diseases. It occurred to these rugged individualists that even the human body itself was a source of infection. **Dr. Oliver Wendell Holmes,** an American physician, observed that mothers who gave birth at home experienced fewer infections than did mothers who gave birth in the hospital, and the Hungarian **Dr. Ignaz Semmelweis** showed quite clearly that women became infected in the maternity ward after examinations by physicians coming directly from the autopsy room.

The English surgeon **Joseph Lister** took notice of these observations and was the first to introduce **aseptic* techniques** aimed at reducing microbes in a medical setting and preventing wound infections. Lister's concept of asepsis was much more limited than our modern precautions. It mainly involved disinfecting the hands and the air with strong antiseptic chemicals, such as phenol, prior to surgery. It is hard for us to believe, but as recently as the late 1800s surgeons wore street clothes in the operating room and had little idea that hand washing was important. Lister's techniques and the application of heat for sterilization became the bases for microbial control by physical and chemical methods, which are still in use today.

The Discovery of Pathogens and the Germ Theory of Disease

Two ingenious founders of microbiology, **Louis Pasteur** of France **(figure 1.11)** and **Robert Koch** of Germany **(figure 1.12),** introduced techniques that are still used today. Pasteur made enormous contributions to our understanding of the roles of microorganisms in many aspects of medicine and industry. He developed two vaccines

* *aseptic* (ay-sep′-tik) Gr. *a,* no, and *sepsis,* decay or infection. These techniques are aimed at reducing pathogens and do not necessarily sterilize.

Figure 1.11 Photograph of Louis Pasteur (1822–1895), the father of microbiology. Few microbiologists can match the scope and impact of Pasteur's contributions to the science of microbiology.
© Dea Picture Library/Getty Images

Figure 1.12 Photograph of Robert Koch looking through a microscope with colleague Richard Pfeiffer looking on. Robert Koch won the Nobel Prize for Physiology or Medicine in 1905 for his work on *Mycobacterium tuberculosis*. Richard Pfeiffer discovered *Haemophilus influenzae* and was a pioneer in typhoid vaccination.
© Bettmann/Corbis

(rabies and anthrax) and clarified the actions of microbes in wine and beer fermentation. He invented pasteurization and completed some of the first studies showing that human diseases could arise from infection. These studies, supported by the work of other scientists, became known as the **germ theory of disease.** Pasteur's contemporary, Koch, established *Koch's postulates,* a series of proofs that verified the germ theory and could establish whether an organism was pathogenic and which disease it caused (see chapter 13). Around 1875 Koch used this experimental system to show that anthrax is caused by a bacterium called *Bacillus anthracis.* So useful were his postulates that the causative agents of 20 other diseases were discovered

between 1875 and 1900, and even today they serve as a basic premise for establishing a link between pathogens and diseases.

Numerous exciting technologies emerged from Koch's prolific and probing laboratory work. During this golden age of microbiology, he realized that study of the microbial world would require separating microbes from each other and growing them in culture. It is not an overstatement to say that he and his colleagues invented many of the techniques that are described in chapter 3: inoculation, isolation, media, maintenance of pure cultures, and preparation of specimens for microscopic examination. Other highlights in this era of discovery are presented in later chapters on microbial control (see chapter 11) and vaccination (see chapter 15).

 Check Your Progress SECTION 1.5

10. Outline the most significant discoveries and events in microscopy, culture techniques, and other methods of handling or controlling microbes.
11. Differentiate between a hypothesis and a theory. If someone says a scientific explanation is "only a theory," what do they really mean?
12. Is the germ theory of disease actually a law, and why?
13. Why was the abandonment of the spontaneous generation theory so significant?

1.6 Taxonomy: Organizing, Classifying, and Naming Microorganisms

 Expected Learning Outcomes

11. Define *taxonomy* and its supporting terms *classification*, *nomenclature,* and *identification*.
12. Explain how the levels of a taxonomic scheme relate to each other. Give the names of the levels, and place them in a hierarchy.
13. Describe the goals of nomenclature and how the binomial system is structured. Know how to correctly write a scientific name.

Students just beginning their microbiology studies are often dismayed by the seemingly endless array of new, unusual, and sometimes confusing names for microorganisms. Learning microbial **nomenclature*** is very much like learning a new language, and occasionally its demands may be a bit overwhelming. But paying attention to proper microbial names is just like following a baseball game or a theater production: You cannot tell the players apart without a program! Your understanding and appreciation of microorganisms will be greatly improved by learning a few general rules about how they are named.

The formal system for organizing, classifying, and naming living things is **taxonomy.*** This science originated more than

250 years ago when **Carl von Linné** (also known as Linnaeus; 1701–1778), a Swedish botanist, laid down the basic rules for taxonomic categories, or **taxa.** Von Linné realized early on that a system for recognizing and defining the properties of living things would prevent chaos in scientific studies by providing each organism with a unique name and an exact "slot" in which to catalog it. This classification would then serve as a means for future identification of that same organism and permit workers in many biological fields to know if they were indeed discussing the same organism. The von Linné system has served well in categorizing the 2 million or more different types of organisms that have been discovered since that time.

The primary concerns of taxonomy are classification, nomenclature, and identification. These three areas are interrelated and play a vital role in keeping a dynamic inventory of the extensive array of living things. **Classification** is an orderly arrangement of organisms into groups that indicate evolutionary relationships and history. **Nomenclature** is the system of assigning names to the various taxonomic rankings of each microbial species. **Identification** is the process of using the specific characteristics and capabilities of an organism to determine its exact identity and placement in taxonomy. A survey of some general methods of identification appears in chapter 3.

The Levels of Classification

The main taxa, or groups, in a classification scheme are organized into several descending ranks called a **hierarchy.** It begins with **domain,** which is a giant, all-inclusive category based on a unique cell type, and ends with **species,*** the smallest and most specific taxon. All the members of a domain share only one or few general characteristics, whereas members of a species are essentially the same kind of organism—that is, they share the majority of their characteristics. The order of taxa between the top and bottom levels is, in descending order: **domain, kingdom, phylum*** or **division,**[4] **class, order, family, genus,*** and **species.** Thus, each domain may be subdivided into a series of kingdoms, each kingdom is made up of several phyla, each phylum contains several classes, and so on. Because taxonomic schemes are to some extent artificial, certain groups of organisms cannot be arranged in an exact eight-part taxonomic scheme. In those cases, additional levels can be imposed immediately above (super) or below (sub) a taxon, giving us such categories as superphylum and subclass.

To illustrate how this hierarchy works, we compare the taxonomic breakdowns of a human and a common pond protozoan **(figure 1.13).** Humans and protozoa belong to the same domain (Eukarya) but are placed in different kingdoms. To emphasize just how broad the category kingdom is, ponder the fact that humans belong to the same kingdom as jellyfish. Of the several

* *nomenclature* (noh′-men-klay″-chur) L. *nomen,* name, and *clare,* to call. A system of naming.
* *taxonomy* (tacks-on″-uh-mee) Gr. *taxis,* arrangement, and *nomos,* name.

* *species* (spee′-sheez) L. *specere,* kind. In biology, this term is always in the plural form.
* *phylum* (fy′-lum) pl. phyla (fye′-luh) Gr. *phylon,* race.
4. The term *phylum* is used for protozoa, animals, bacteria, and fungi. Division is for algae and plants.
* *genus* (jee′-nus) pl. genera (jen′-er-uh) L. birth, kind.

Domain: Eukarya (All eukaryotic organisms)

Kingdom: Animalia

Sea squirt Lemur Sea star

Phylum: Chordata

Class: Mammalia

Order: Primates

Family: Hominoidea

Genus: *Homo*

Species: *sapiens*
Scientific name: *Homo sapiens*

(a)

Domain: Eukarya (All eukaryotic organisms)

Kingdom: Protista
Includes
protozoa
and algae

Phylum: Ciliophora
Protozoa with cilia
Covered by flexible
pellicle
Contain two types
of nuclei

Class: Oligohymenophora
Single, rapidly
swimming cells
Regular rows of cilia
Distinct ciliated oral
groove

Order: Peniculida
Uniform dense cilia
dispersed over cell
Oral cilia are peniculae
Trichocysts in outer
membrane

Family: Parameciidae
Cells round to elongate
Rotate while swimming
Deep oral grooves

Genus: *Paramecium*
Ovoid, cigar- and foot-shaped cells

Species: *caudatum*
Cells elongate, cylindrical
Blunt at one end and tapered to a
point at the other
Scientific name: *Paramecium caudatum*

(b)

Figure 1.13 **Sample taxonomy.** Two organisms belonging to Domain Eukarya, traced through their taxonomic series. **(a)** Modern humans, *Homo sapiens*. **(b)** A common protozoan, *Paramecium caudatum*.

phyla within this kingdom, humans are in the Phylum Chordata, but even a phylum is rather all-inclusive, considering that humans share it with other vertebrates as well as with creatures called sea squirts. The next level, Class Mammalia, narrows the field considerably by grouping only those vertebrates that have hair and suckle their young. Humans belong to the Order Primates, a group that also includes apes, monkeys, and lemurs. Next comes the Family Hominoidea, containing only humans and apes. The final levels are our genus, *Homo* (all races of modern and ancient humans), and our species, *sapiens* (meaning wise). Notice that for both the human and the protozoan, the

categories become less inclusive and the individual members more closely related and similar in overall appearance. Other examples of classification schemes are provided in sections of chapters 4 and 5 and in several later chapters. A superior source for the taxonomic breakdown of microbes is Wikipedia. Go there to search the scientific name of any species, and its taxonomy will be shown in a box on the upper-right portion of the first page.

We need to remember that all taxonomic hierarchies are based on the judgment of scientists with certain expertise in a particular group of organisms and that not all other experts may agree with

the system being used. Consequently, no taxa are permanent to any degree; they are constantly being revised and refined as new information becomes available or new viewpoints become prevalent. Our primary aim in introducing taxonomy is to present an organizational tool that helps you keep track of the various microbial groups and recognize their major categories. For the most part our emphasis will remain on the higher-level taxa (phylum, class) and genus and species.

Assigning Scientific Names

Many larger organisms are known by a common name suggested by certain dominant features. For example, a bird species may be called a yellow-bellied sapsucker, or a flowering plant, a sunflower. Some species of microorganisms (especially pathogens) are also sometimes designated by informal names, such as the gonococcus (*Neisseria gonorrhoeae*) or the TB bacillus (*Mycobacterium tuberculosis*), but this is not the usual practice. If we were to adopt common names such as the "little yellow coccus" (for *Micrococcus luteus**) or the "club-shaped diphtheria bacterium" (for *Corynebacterium diphtheriae**), the terminology would become even more cumbersome and challenging than scientific names. Even worse, common names are notorious for varying from region to region, even within the same country. A decided advantage of standardized nomenclature is that it provides a universal language that enables scientists from all countries on the earth to freely exchange information.

The **scientific name,** also known as the **specific epithet,** is assigned by using a **binomial** (two-name) **system** of nomenclature. The scientific name is always a combination of the generic (genus) name followed by the species name. The generic part of the scientific name is capitalized, and the species part begins with a lowercase letter. Both should be italicized (or underlined if italics are not available), as follows:

Histoplasma capsulatum or <u>**Histoplasma capsulatum**</u>

Because other taxonomic levels are not italicized and consist of only one word, one can always recognize a scientific name. An organism's scientific name is sometimes abbreviated to save space, as in *H. capsulatum,* but only if the genus name has already been stated. The source for nomenclature is usually Latin or Greek. If other languages such as English or French are used, the endings of these words are revised to have Latin endings. In general, the name first applied to a species will be the one that takes precedence over all others.

An international group oversees the naming of every new organism discovered, making sure that standard procedures have been followed and that there is not already an earlier name for the organism or another organism with that same name. The inspiration for names is extremely varied and often rather imaginative. Some species have been named in honor of a microbiologist who originally discovered the microbe or who has made outstanding contributions to the field. Other names may designate a characteristic of the microbe (shape, color), a location where it was found, or a disease or symptom that it causes. Some examples of scientific names, pronunciations, and origins are:

1. ***Histoplasma capsulatum*** (hiss″-toh-plaz′-mah cap″-soo-lay′-tum) Gr. *histo,* tissue, *plasm,* to form, and L. *capsula,* small sheath. A fungus that causes Ohio Valley fever.

2. ***Trichinella spiralis*** (trik′-ih-nel′-uh spy-ral′-iss) Gr. *trichos,* hair, *ella,* little, and L. *spira,* coiled. The nematode worm that causes the food-borne infection, trichinellosis.

3. ***Shewanella oneidensis*** (shee″-wan-el′-uh oh-ny″-den′-siss). Named for British bacteriologist J. M. Shewan and Lake Oneida, New York, where it was discovered. This is a remarkable species that can bioremediate radioactive metals in contaminated waste sites.

4. ***Bordetella pertussis*** (bor″-duh-tel′-uh pur-tuss′-iss) After Jules Bordet, a Belgian microbiologist who discovered this bacterium, and L. *per,* severe, and *tussis,* cough. This is the cause of pertussis, or whooping cough.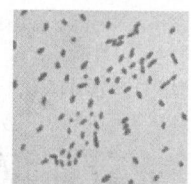

1. *Source:* Dr. Libero Ajello/CDC; **2.** *Source:* CDC; **3.** *Source:* Rizlan Bencheikh and Bruce Arey, Environmental Molecular Sciences Laboratory, DOE Pacific Northwest National Laboratory; **4.** *Source:* CDC

When you encounter the name of a microorganism in the chapters ahead, it is helpful to take the time to sound it out one syllable at a time and repeat until it seems familiar. You are much more likely to remember the names that way—and they will become part of the new language you will be learning.

Quick Search

Search the web using the phrase "Bacterial Pathogen Pronunciation Station" for help in correctly pronouncing some common scientific names.

✓ *Check Your Progress* SECTION 1.6

14. Differentiate between taxonomy, classification, and nomenclature.
15. What is the basis for a phylogenetic system of classification?
16. Explain the binomial system of nomenclature and give the correct order of taxa, going from most general to most specific.
17. Explain some of the benefits of using scientific names for organisms.

* *micrococcus luteus* (my″-kroh-kok′-us loo′-tee-us) Gr. *micros,* small, and *kokkus,* berry; L. *luteus,* yellow.

* *corynebacterium diphtheriae* (kor-eye″-nee-bak-ter′-ee-yum dif′-theer-ee-eye) Gr. *coryne,* club, *bacterion,* little rod, and *diphtheriae,* the causative agent of the disease diphtheria.

1.7 The Origin and Evolution of Microorganisms

 Expected Learning Outcomes

14. Discuss the fundamentals of evolution, evidence used to verify evolutionary trends, and the use of evolutionary theory in the study of organisms.

15. Explain the concepts behind the organization of the two main trees of life, and indicate where the major groups of microorganisms fall on these trees.

16. Explain the bases for classification, taxonomy, and nomenclature.

17. Recall the order of taxa and the system of notation used in creating scientific names.

All Life Is Related and Connected Through Evolution

As we indicated earlier, *taxonomy,* the classification of biological species, is a system used to organize all of the forms of life. In biology today, there are different methods for deciding on taxonomic categories, but they all rely on the history and relatedness of organisms. The natural relatedness between groups of living things is called their **phylogeny.** Biologists can apply their knowledge of phylogenetic relationships to develop a system of taxonomy.

To understand how organisms originate, we must understand some fundamentals of **evolution.** You have no doubt heard comments that dismiss evolution as "only a theory" as though there remain significant problems with its acceptance. But you have also learned that a scientific theory is a highly documented and well-established concept. The body of knowledge that has accumulated over hundreds of years regarding the process of evolution is so significant that scientists from all disciplines consider evolution to be a fact. It is an important theme that underlies all of biology, including microbiology. Put simply, the scientific principle of evolution states that living things change gradually through billions of years and that these changes result in various types of structural and functional adaptations through many generations. The process of evolution is selective: Those changes that most favor the survival of a particular organism or group of organisms tend to be retained, and those that are less beneficial to survival tend to be lost. The great naturalist Charles Darwin labeled this process natural selection. We do not have the space here to present a detailed analysis of evolutionary theories, but the occurrence of evolution is supported by a tremendous amount of evidence from the fossil record and from the study of **morphology** (structure), **physiology** (function), and **genetics** (inheritance). Evolution accounts for the millions of different species on the earth and their adaptation to its many and diverse habitats.

Evolution is founded on two premises: (1) that all new species originate from preexisting species through inheritance of traits, and (2) that closely related organisms have similar features because they evolved from common ancestral forms. Usually evolution progresses toward greater complexity, and

evolutionary stages range from simple, less evolved forms that are close to an ancestral organism to more complex, evolved forms that have advanced beyond the ancestral forms. Regardless of their evolutionary history, all species presently residing on the earth are modern, but some have arisen more recently in evolutionary history than others.

Traditionally we present the history of life, or phylogeny, in the form of branching trees that are designed to show the origins of various life forms (**figures 1.14** and **1.15**). At the base are the oldest ancestral forms (somewhat like roots), and the trunk indicates the progression of major lines that emerge through selection. Branches split off the main trunk as further selection and modification occurs. With this arrangement, more closely related organisms appear nearer each other on the tree. Any tree of life is nothing more than a system for classifying organisms, and the characteristics used for the process are specified by the person or group creating the tree. Classifying organisms alphabetically (aardvark, acinetobacter, aloe, antelope) or by size (giant redwood, blue whale . . .) are equally valid strategies. However, the most scientifically useful classification schemes group organisms according to their biological characteristics, using the tree to display the evolutionary relationships between organisms.

Systems for Presenting a Universal Tree of Life

The earliest classification schemes assigned every living thing to either the Plant Kingdom or the Animal Kingdom, even though most organisms were a poor fit for either. As knowledge grew concerning cellular structure, means of acquiring nutrients, and how organisms moved about, Robert Whittaker developed a five-kingdom system (figure 1.14) in the 1960s that remained the standard for many years. Individual members of the five kingdoms—the **monera, fungi, protists, plants,** and **animals**—were easily distinguished from one another using techniques common to most laboratories during the mid-twentieth century.

By the late twentieth century, the manner in which organisms were studied was undergoing a dramatic change. The techniques of molecular biology allowed scientists to examine the structure and function of individual genes and to clarify the relationships between organisms based on similarities in their genetic material, rather than simply shared characteristics. Using these new molecular methods, Carl Woese and George Fox proposed a classification system that some have likened to a shrub of life rather than a tree (figure 1.15). In this system, organisms are most broadly classified as belonging to one of three domains. The members of the **Domain Bacteria** have prokaryotic cells and are what most people think of as traditional bacterial species. The **Domain Eukarya** contains all of the organisms that display a eukaryotic cell structure, and includes the Kingdom Protista, Kingdom Fungi, Kingdom Plantae, and Kingdom Animalia. Members of the **Domain Archaea** also possess prokaryotic cells, but these cells are so distinct from those found in the Bacteria and the Eukarya that they have been placed in a domain of their own. Most members of the Domain Archaea are characterized by their ability to live in extreme environments or produce novel metabolic byproducts. Some species thrive in water

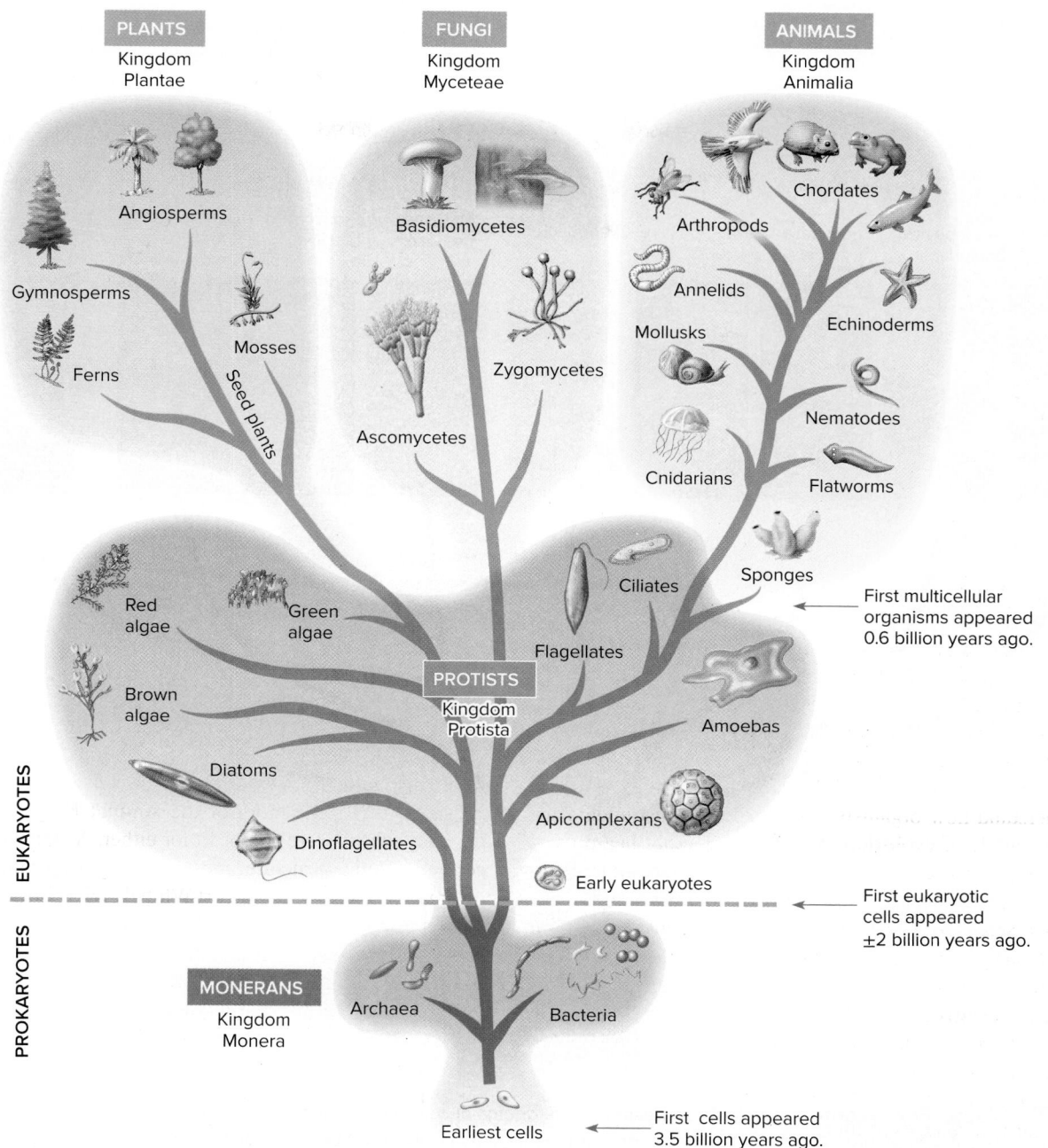

Figure 1.14 **Traditional Whittaker system of classification.** In this system, kingdoms are based on cell structure and type, the nature of body organization, and nutritional type. Bacteria and Archaea (monerans) have prokaryotic cells and are unicellular. Protists have eukaryotic cells and are simple unicellular and colonial organisms. They can be photosynthetic (algae), or they can feed on other organisms (protozoa). Fungi are eukaryotic cells with unicellular or multicellular bodies; they have cell walls and are not photosynthetic. Plants have eukaryotic cells, are multicellular, have cell walls, and are photosynthetic. Animals have eukaryotic cells, are multicellular, do not have cell walls, and derive nutrients from other organisms.

near undersea volcanic vents that has been heated to near boiling or in concentrations of salt as high as 35%, while others produce methane gas as a byproduct of their metabolism.

Although the Woese-Fox system represents the most accurate relationship between organisms, which classification scheme is ultimately used does not affect the presentation of most microbes because we tend to discuss them at the genus and species level. But be aware that biological taxonomy—and more important, our understanding of the early evolution of organisms—is in a period of transition. It is essential that any methods of classification reflect our current understanding and can be altered as new information is uncovered (see figure 4.27). Also, please note that viruses are *not* included in any of the classification or evolutionary schemes, because they are not cells and their position cannot be given with any confidence. Their special taxonomy is discussed in chapter 6.

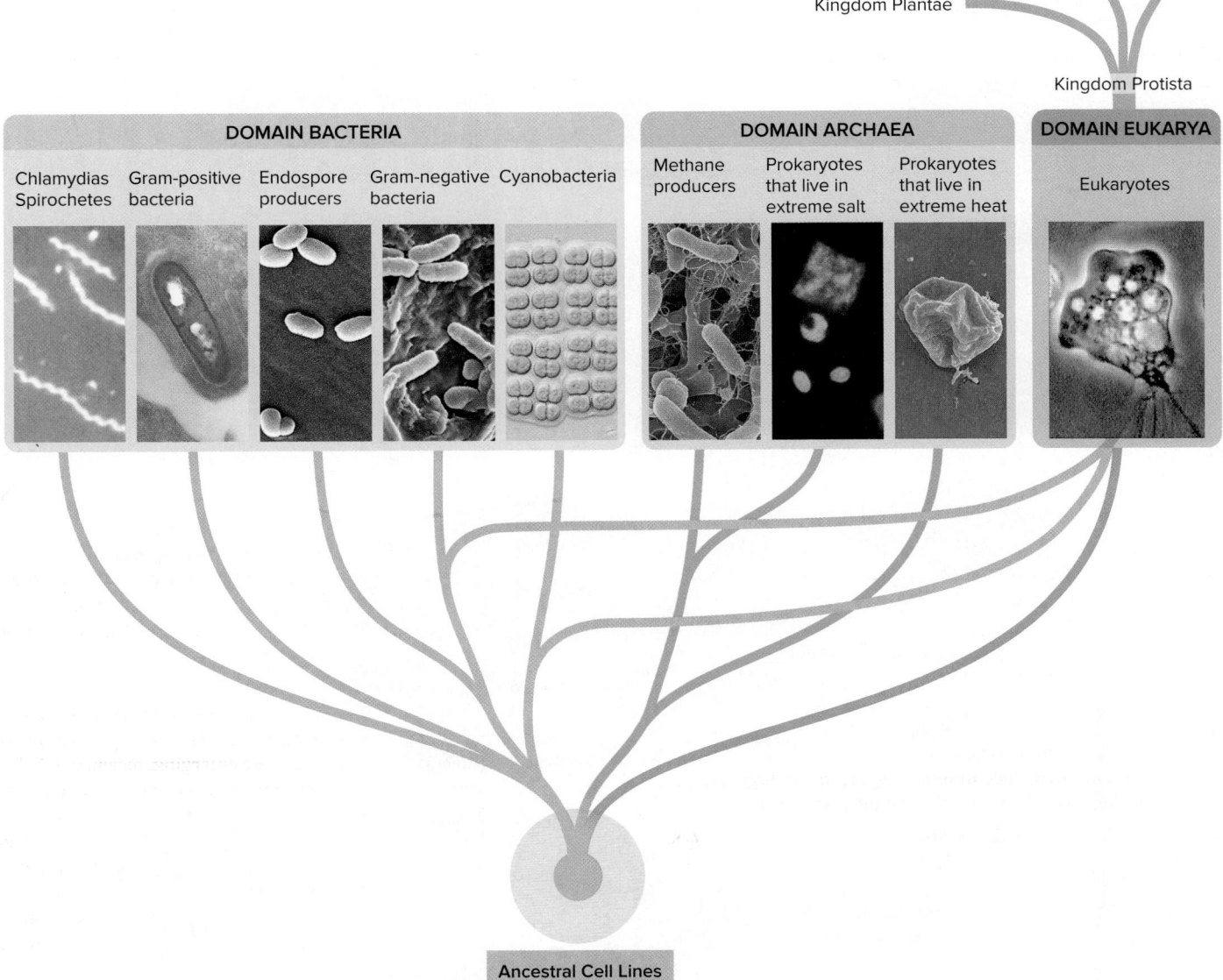

Kingdom Kingdom
Fungi Animalia

Kingdom Plantae

Kingdom Protista

DOMAIN BACTERIA

Chlamydias Gram-positive Endospore Gram-negative Cyanobacteria
Spirochetes bacteria producers bacteria

DOMAIN ARCHAEA

Methane Prokaryotes Prokaryotes
producers that live in that live in
extreme salt extreme heat

DOMAIN EUKARYA

Eukaryotes

Ancestral Cell Lines
(first living cells)

Figure 1.15 Woese-Fox system. A system for representing the origins of cell lines and major taxonomic groups as proposed by Carl Woese and colleagues. They propose three distinct cell lines placed in superkingdoms called domains. We know little about the earliest cells, called progenotes, except that they were prokaryotic. Lines of these early primitive cells gave rise to the Domains Bacteria and Archaea. Molecular evidence indicates that the Archaea started the branch that would eventually become the Domain Eukarya. Notice that two early bacterial lines contributed to the evolution of the Eukarya. Selected representatives of the Domains are included. The traditional eukaryotic kingdoms are still present with this system (see figure 1.14). Further details of classification systems are covered in chapters 4 and 5. (1): *Source:* CDC; (2): *Source:* Dr. Balasubr Swaminathan/Peggy Hayes/CDC; (3): *Source:* Laura Rose/CDC; (4): *Source:* Janice Carr/CDC; (5): Courtesy Jason Oyadomari; (6): © Dr. Adrian Hetzer; (7): © Dr. Mike Dyall-Smith, University of Melbourne; (8): *Source:* From Stand Genomic Sci. 2011 July 1; 4(3): 381-392 doi: 10.4056/sigs.2014648; (9): *Source:* CDC-DPDx

✓ *Check Your Progress* SECTION 1.7

18. Define what is meant by the term *evolution*.

19. Explain how the process of evolution is responsible for the millions of different species on the earth and their adaptation to its many and diverse habitats. Cite examples in your answer.

20. Looking at the tree of life (figure 1.14), determine which kingdom or kingdoms humans are most closely related to.

21. Archaea are often found living in extreme conditions of heat, salt, and acidity, which are similar to those found in early earth. Speculate on the origin of life, especially as it relates to the archaea.

22. Compare the domain system with the five-kingdom system. Does the newer system change the basic idea of prokaryotes and eukaryotes? What is the third cell type?

 CASE STUDY Part 2

If Lake Whillans is a frozen wasteland, then Pitch Lake is a toxic soup. Located in Trinidad and Tobago, Pitch Lake is the world's largest asphalt lake—a hot, toxic pool of oily tar, teeming with poisonous hydrocarbons and carbon dioxide—and another environment seemingly too hostile for life to exist. So it was surprising when scientists studying tiny water droplets trapped in the oil discovered vibrant communities of microorganisms, many of which were adapted to life in extreme conditions. Furthermore, chemical analysis of the oil surrounding the water droplets indicated that the microorganisms inside were degrading the oil, using it as a source of energy and producing methane gas as an end product, a process known as methanogenesis.

- *The organisms found in Pitch Lake degrade oil as a way to supply themselves with energy. How could humans use this same biodegradation to benefit our own environment?*

- *Methane is the most predominant component of natural gas. How could methanogenesis be of value to humans?*

These topics are further explored in chapter 7.

To conclude this Case Study, go to Connect.

Quick Search

To view videos of expeditions to Lake Whillans or Pitch Lake, search the Internet for the videos "Antarctica's Secret Garden" and "Pitch lake, are we alone?"

 Chapter Summary with Key Terms

1.1 The Scope of Microbiology
- A. **Microbiology** is the study of **bacteria, viruses, fungi, protozoa,** and **algae,** which are collectively called **microorganisms,** or microbes. In general, microorganisms are **microscopic** and, unlike **macroscopic** organisms, which are readily visible, they require magnification to be adequately observed or studied.
- B. The science of microbiology is diverse and has branched out into many subsciences and applications. Important subsciences include **immunology, epidemiology,** and public health, food, dairy, aquatic, and industrial microbiology.

1.2 General Characteristics of Microorganisms and Their Roles in the Earth's Environments

Microbes live in most of the world's habitats and are indispensable for normal, balanced life on earth. They play many roles in the functioning of the earth's ecosystems.
- A. Microbes are **ubiquitous.**
- B. There are many kinds of relationships between microorganisms and humans; most are beneficial, but a few are harmful.
- C. Microbes are involved in nutrient production and energy flow. Algae and certain bacteria trap the sun's energy to produce food through **photosynthesis.**
- D. Other microbes are responsible for the breakdown and recycling of nutrients through **decomposition.** Microbes are essential to the maintenance of the air, soil, and water.
- E. Microbial cells are either the small, relatively simple, nonnucleated **prokaryotic** variety or the larger, more complex **eukaryotic** type that contain a **nucleus** and **organelles.**
- F. **Viruses** are microbes but not cells. They are smaller in size and infect their prokaryotic or eukaryotic hosts in order to reproduce themselves.
- G. **Parasites** and **pathogens** are microorganisms that invade the bodies of **hosts,** often causing damage through infection and disease.

1.3 Human Use of Microorganisms

Microbes have been called upon to solve environmental, agricultural, and medical problems.
- A. **Biotechnology** applies the power of microbes to the manufacture of industrial products, foods, and drugs.
- B. Microbes form the basis of **genetic engineering** and **recombinant DNA** technology, which alter genetic material to produce new products and modified life forms.
- C. In **bioremediation,** microbes are used to clean up pollutants and wastes in natural environments.

1.4 Microbial Roles in Infectious Diseases
- A. Nearly 2,000 microbes are **pathogens** that cause infectious diseases that result in high levels of mortality and morbidity (illness). Many infections are **emerging,** meaning that they are newly identified pathogens gaining greater prominence. Many older diseases are reemerging.
- B. Some diseases previously thought to be noninfectious may involve microbial infections (e.g., *Helicobacter,* causing gastric ulcers, and coxsackieviruses, causing diabetes).
- C. An increasing number of individuals have weak immune systems, which makes them more susceptible to infectious diseases.

1.5 The Historical Foundations of Microbiology
- A. Microbiology as a science is about 300 years old. Hundreds of contributors have provided discoveries and knowledge to enrich our understanding.
- B. With his simple microscope, van Leeuwenhoek discovered organisms he called animalcules. As a consequence of his findings and the rise of the **scientific method,** the notion of **spontaneous generation,** or **abiogenesis,** was eventually abandoned for **biogenesis.** The scientific method develops rational **hypotheses** and **theories** that can be tested. Theories that withstand repeated scrutiny become **laws** in time.
- C. Early microbiology blossomed with the conceptual developments of **sterilization, aseptic techniques,** and the **germ theory of disease.** Prominent scientists from this period include Robert Koch, Louis Pasteur, and Joseph Lister.

1.6 Taxonomy: Organizing, Classifying, and Naming Microorganisms
- A. **Taxonomy** is a hierarchical scheme for the **classification, identification,** and **nomenclature** of organisms, which are grouped in categories called **taxa,** based on features ranging from general to specific.

B. Starting with the broadest category, the taxa are **domain, kingdom, phylum** (or **division**), **class, order, family, genus,** and **species.** Organisms are assigned **binomial scientific names** consisting of their genus and species names.

1.7 The Origin and Evolution of Microorganisms

A. All life on earth **evolved** from simple cells appearing in ancient oceans about 3.5 billion years ago.

B. Evolutionary change occurs when the environment places pressure on organisms that selects for the survival of those with more adaptive inheritable traits.

C. All new species are the products of preexisting species, and their ancestry may be traced by examining fossils, morphology, physiology, genetics, and other scientific forms of investigation.

D. The records of **phylogeny** are displayed in the form of a tree of life that shows organisms' relatedness.

E. One classification scheme is based on a five-kingdom organization developed by Whittaker that includes:
1. Kingdom Prokaryotae (Monera), containing eubacteria and archaea;
2. Kingdom Protista, containing primitive unicellular microbes such as algae and protozoa;
3. Kingdom Myceteae, containing the fungi;
4. Kingdom Animalia, containing animals; and
5. Kingdom Plantae, containing plants.

F. A newer classification scheme for living things is based on the genetic structure of their ribosomes. This Woese-Fox system recognizes three **Domains: Archaea,** simple prokaryotes that often live in extreme environments; **Bacteria,** typical prokaryotes; and **Eukarya,** all types of eukaryotic organisms.

Level I. Knowledge and Comprehension

These questions require a working knowledge of the concepts in the chapter and the ability to recall and understand the information you have studied.

Multiple-Choice Questions

Select the correct answer from the answers provided. For questions with blanks, choose the combination of answers that most accurately completes the statement.

1. Which of the following is not considered a microorganism?
 a. archaeon
 b. bacterium
 c. protozoan
 d. mushroom

2. An area of microbiology that is concerned with the occurrence of disease in human populations is
 a. immunology
 b. parasitology
 c. epidemiology
 d. bioremediation

3. Which process involves the deliberate alteration of an organism's genetic material?
 a. bioremediation
 b. biotechnology
 c. decomposition
 d. recombinant DNA

4. A prominent difference between prokaryotic and eukaryotic cells is the
 a. larger size of prokaryotes
 b. lack of pigmentation in eukaryotes
 c. presence of a nucleus in eukaryotes
 d. presence of a cell wall in prokaryotes

5. Which of the following parts was absent from Leeuwenhoek's microscopes?
 a. focusing screw
 b. lens
 c. specimen holder
 d. condenser

6. Abiogenesis refers to the
 a. spontaneous generation of organisms from nonliving matter
 b. development of life forms from preexisting life forms
 c. development of aseptic technique
 d. germ theory of disease

7. A hypothesis can be defined as
 a. a belief based on knowledge
 b. knowledge based on belief
 c. a scientific explanation that is subject to testing
 d. a theory that has been thoroughly tested

8. Which early microbiologist was most responsible for developing standard microbiology laboratory techniques?
 a. Louis Pasteur
 b. Robert Koch
 c. Carl von Linné
 d. John Tyndall

9. Which scientist is most responsible for finally laying the theory of spontaneous generation to rest?
 a. Joseph Lister
 b. Robert Koch
 c. Francesco Redi
 d. Louis Pasteur

10. When a hypothesis has been thoroughly supported by long-term study and data, it is considered
 a. a law
 b. a speculation
 c. a theory
 d. proved

11. Which is the correct order of the taxonomic categories, going from most specific to most general?
 a. domain, kingdom, phylum, class, order, family, genus, species
 b. division, domain, kingdom, class, family, genus, species
 c. species, genus, family, order, class, phylum, kingdom, domain
 d. species, family, class, order, phylum, kingdom

12. By definition, organisms in the same _____ are more closely related than are those in the same _____.
 a. order, family
 b. class, phylum
 c. family, genus
 d. phylum, division

13. Which of the following are prokaryotic?
 a. bacteria
 b. archaea
 c. protists
 d. both a and b

14. Order the following items by size, using numbers: 1 = smallest and 8 = largest.
 _____ human immunodeficiency virus HIV
 _____ protozoan
 _____ rickettsia
 _____ protein
 _____ worm
 _____ coccus-shaped bacterium
 _____ spirochete
 _____ atom

15. Which of the following is not an emerging infectious disease?
 a. avian influenza
 b. Lyme disease
 c. common cold
 d. West Nile fever

16. How would you categorize a virus?
 a. as prokaryotic
 b. as eukaryotic
 c. as an archaeon
 d. none of these choices
 Explain your choice for question 16.

Case Study Review

1. Many of the bacteria in Lake Whillans derive energy from the oxidation of chemical compounds. What might have made this necessary?
 a. exceedingly low temperatures
 b. a lack of sunlight
 c. the size of the bacterial populations they found
 d. a lack of fresh water

2. How was the number of different species in Lake Whillans determined?
 a. counting individual cells
 b. analyzing the DNA recovered from the samples
 c. growing the cells and then classifying them according to their structural characteristics
 d. measuring the concentration of cells in each milliliter of lake water

3. Do you think scientists working at Pitch Lake were at a great risk of infection from the organisms growing in the lake? Explain.

Writing Challenge

For each question, compose a one- or two-paragraph answer that includes the factual information needed to completely address the question. Check Your Progress questions can also be used for writing-challenge exercises.

1. What does it mean to say microbes are ubiquitous?

2. What is meant by diversity with respect to organisms?

3. What events, discoveries, or inventions were probably the most significant in the development of microbiology and why?

4. Explain how microbiologists use the scientific method to develop theories and explanations for microbial phenomena.

5. Explain how microbes are classified into groups according to evolutionary relationships, provided with standard scientific names, and identified by specific characteristics.

6. a. What are some of the sources for "new" infectious diseases?
 b. Comment on the sensational ways in which some tabloid media portray the dangers of infectious diseases.

Concept Mapping

On Connect you can find an Introduction to Concept Mapping that provides guidance for working with concept maps along with concept-mapping activities for this chapter.

Level II. Application, Analysis, Evaluation, and Synthesis

These problems go beyond just restating facts and require higher levels of understanding and an ability to interpret, problem solve, transfer knowledge to new situations, create models, and predict outcomes.

Critical Thinking

Critical thinking is the ability to reason and solve problems using facts and concepts. These questions can be approached from a number of angles, and in most cases they do not have a single correct answer.

1. What do you suppose the world would be like if there were cures for all infectious diseases and a means to destroy all microbes? What characteristics of microbes would prevent this from happening?

2. How would you describe the types of scientific reasoning in the various experiments for supporting and denying spontaneous generation?

3. Give the technical name of a microbiologist who researches or works with protozoa in a termite's gut, bacteria that live in volcanoes, the tapeworms of dogs, molds that cause food poisoning, emerging viral diseases, the metabolism of bacteria that live in acid swamps, the classification of *Paramecium*.

4. Name the six most common infectious agents on earth and the four microbes that cause the highest rates of cancer.

5. What is the ultimate way that microbes will, as Pasteur said, have the "last word"?

6. Can you develop a scientific hypothesis and means of testing the cause of stomach ulcers? (Are they caused by an infection? by too much acid? by a genetic disorder?)

7. Construct the scientific name of a newly discovered species of bacterium using your name or the name of a pet, a place, or a unique characteristic. Be sure to use proper notation and endings.

Visual Challenge

Use the following pattern to construct an outline of the scientific reasoning that was involved in developing the germ theory of disease similar to figure 1.10.

Observations

↓

Hypothesis

↓

Testing the Hypothesis

↓

Establishment of Theory

↓

Accepted Principle

STUDY TIP: Quickly create a customized practice quiz for this chapter by going to your Connect SmartBook "My Reports" section. Don't have SmartBook and want to learn more? Go to **whysmartbook.com.**

A highly magnified view of a 4-billion-year-old Martian meteorite uncovers tiny rods that resemble earthly prokaryotes, but are they?
Source: Photo by David McKay/NASA

Brown streaks radiating down the slopes of the Hale Crater are flowing streams of water—the first indisputable evidence that liquid water exists on Mars.
Source: JPL-Caltech/Univ. of Arizona/NASA

 CASE STUDY Part 1 Following the Water and Carbon on Mars

Throughout history, humans have gazed at the night sky and pondered the possible existence of other life forms in the universe. Telescopes have allowed closer glimpses of celestial bodies, but they have never been able to show the details needed to verify the presence of **life.** As technology advanced in the 1970s, the National Aeronautics and Space Administration **(NASA)** began a series of voyages to directly explore our own solar system. Astronomers knew from sophisticated **chemical** and **physical** analyses that most of the planets in our solar system had far too extreme conditions to **support life** as we know it, with the possible exception of Mars. Mars has a profile somewhat similar to earth's and is one of the closer planets, orbiting an average of 225 million miles away from earth. It is known to be very dry and cold (the temperature range is 80°F to −225°F), but there was a reasonable possibility that at least **simple organisms** could have developed there.

> "The only life forms that could possibly survive the extreme conditions of the planet in the past or present are microorganisms."
>
> **Kathy Park Talaro**

Starting in 1976, a series of space explorations was launched to find evidence that could point to signs of life on Mars. From these studies sprang a new science—**astrobiology**—which applies principles from biology, chemistry, and geology to investigate the possibilities of extraterrestrial life. One conclusion of NASA scientists that has driven many of their searches is that the only life forms that could possibly survive the extreme conditions of the planet in the past or present are microorganisms.

The significant questions being tackled by astrobiologists include these: What are the basic requirements of life? What would signs of life look like? As is commonly done for many ideas of a biological nature, they turned to **chemistry** for answers. One of their assumptions was that because some of the same chemical substances on earth would likely exist on Mars, Martian life forms would have a similar profile, at least chemically. Attention was turned to substances that are part of all life on earth: mainly **water** and **carbon-based chemicals.**

The main goals of one of the first Mars space vehicles, *Viking Explorer*, were to sample for **water, carbon dioxide,** and **organic compounds,** and to **culture microbes** from Martian soil. The results were mostly negative or inconclusive, except for one—that the atmosphere contained high levels of carbon dioxide gas. Not surprisingly, the samples yielded no Martian microbes. A discovery in 1996 of tiny rod-shaped objects in a 4-billion-year-old **Martian meteorite** from the Antarctic has been, and continues to be, the subject of much scientific scrutiny. This tantalizing image bears a resemblance to prokaryotic cells, but there is still a controversy about whether these are **ancient microbes** or geologic artifacts.

- *What chemical properties of water make it essential for living processes?*

- *What characteristics of carbon make it the central building block of life?*

To continue the Case Study, go to Case Study Part 2 at the end of the chapter.

SCOPING OUT THE CHAPTER

It's not surprising that a search for life on other planets always begins by deciding whether conditions for life are possible; after all, life is chemistry. Without a basic understanding of the rules of chemistry, the study of biology is a hopeless endeavor. Why chemical bonds form, how proteins assume a specific shape, and the chemical characteristics of macromolecules important to living systems are just a few of the topics we touch on in this chapter.

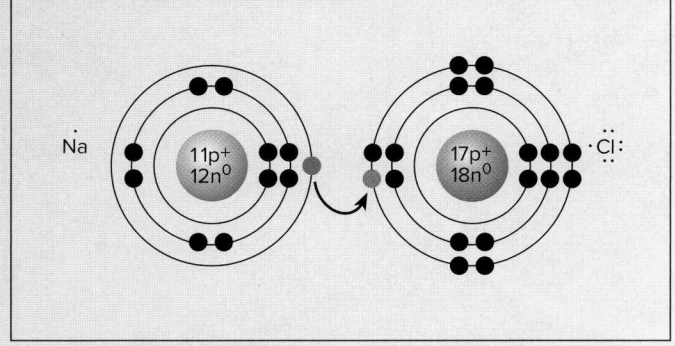

Atomic Structure
All matter is composed of atoms, and the number of protons, neutrons, and electrons ultimately determines the characteristics of each element.

Chemical Properties
Atoms react with one another, making or breaking chemical bonds, to create complex molecules and compounds.

Carbohydrates
Simple carbohydrates are used as a source of energy, while more complex carbohydrates provide structure to the walls of plants, algae, and bacterial cells.

Lipids
An important energy source because of their high energy to weight ratio, lipids serve as a major component of cell membranes.

Proteins
The three-dimensional shape of proteins is essential to their function, whether as structural parts of the cell or as enzymes, catalyzing biochemical reactions.

Nucleic acids
The genetic information of cells and viruses is carried in the nucleic acids DNA and RNA.

(proteins): Image from the RCSB PDB (www.rcsb.org) of PDB ID 1R4Q (Fraser, M.E., Fujinaga, M., Cherney, M.M., Melton-Celsa, A.R., Twiddy, E.M., O'Brien, A.D., James, M.N.G. (2004) Structure of shiga toxin type 2 (Stx2) from Escherichia coli O157:H7.)

2.1 Atoms: Fundamental Building Blocks of All Matter in the Universe

 Expected Learning Outcomes

1. Describe the properties of atoms and identify the relationships of the particles they contain.

2. Characterize elements and their isotopes.

3. Explain the differences between atomic number, mass number, and atomic weight.

4. List the major elements associated with life.

5. Describe electron orbitals and energy shells and how they are filled.

The universe is composed of an infinite variety of substances existing in gaseous, liquid, and solid states. All such tangible materials that occupy space and have mass are called **matter.**

Figure 2.1 **Models of atomic structure.** Three-dimensional models of hydrogen and carbon that approximate their actual structure. The atomic nucleus is surrounded by electrons that travel within orbitals and occur in energy levels called shells. **(a1)** Hydrogen has just one orbital and one shell. **(a2)** Carbon has four orbitals and two shells. The outermost orbitals are most accurately portrayed as sets of lobe-shaped pairs rather than circles or spheres. **(b1, 2)** Simple models of hydrogen and carbon are not strictly accurate representations of atomic structure but are designed as a quick reference to the numbers and arrangements of shells and electrons, and the numbers of protons and neutrons in the nucleus. (Not to accurate scale.)

The organization of matter—whether air, rocks, or bacteria—begins with individual building blocks called atoms. An **atom** is defined as the smallest particle that cannot be subdivided into smaller substances without losing its properties. Even in a science dealing with very small things, an atom's minute size is striking; for example, an oxygen atom is only 0.0000000013 mm (0.0013 nm) in diameter, and 1 million of them in a cluster would barely be visible to the naked eye.

Although scientists have not directly observed the detailed structure of an atom, the exact composition of atoms has been well established by extensive physical analysis using sophisticated instruments. In general, an atom derives its properties from a combination of subatomic particles called **protons** (p$^+$), which are positively charged; **neutrons** (n^0), which have no charge (are neutral); and **electrons** (e$^-$), which are negatively charged. Protons and neutrons are similar to one another in both mass and volume, while each is about 2,000 times as heavy as an electron. The protons and neutrons make up a central core, or atomic **nucleus**[1], that is surrounded by 1 or more electrons **(figure 2.1).** The nucleus makes up the larger mass (weight) of the atom, whereas the electron region, sometimes called the "electron cloud," accounts for the greater volume. To get a perspective on proportions, consider this: If an atom

were the size of a football stadium, the nucleus would be about the size of a marble! The stability of atomic structure is largely maintained by

- the mutual attraction of the protons and electrons (opposite charges attract each other), and
- the exact balance of proton number and electron number, which causes the opposing charges to cancel each other out.

At least in theory, then, isolated intact atoms do not carry a charge.

Different Types of Atoms: Elements and Their Properties

All atoms share the same fundamental structure. All protons are identical, all neutrons are identical, and all electrons are identical. But when these subatomic particles come together in specific, varied combinations, unique types of atoms called **elements** result. Each element is a pure substance that has a characteristic atomic structure and predictable chemical behavior. To date, 118 elements have been described. Ninety-four of them are naturally occurring, and the rest were artificially produced by manipulating the particles in the nucleus. By convention, an element is assigned a distinctive name with an abbreviated shorthand symbol. The elements essential to life, often called bioelements, are outlined in

1. Be careful not to confuse the nucleus of an atom with the nucleus of a cell.

TABLE 2.1	Chemical Characteristics of Major Elements of Life			
Element	Atomic Symbol*	Atomic Number	Atomic Mass**	Ionized Forms (see section 2.3)
Calcium	Ca	20	40.08	Ca^{2+}
Carbon	C	6	12.01	CO_3^{2-}
Carbon˙	C-14	6	14.0	
Chlorine	Cl	17	35.45	Cl^-
Cobalt	Co	27	58.93	Co^{2+}, Co^{3+}
Copper	Cu	29	63.54	Cu^+, Cu^{2+}
Hydrogen	H	1	1.00	H^+
Hydrogen˙	H-3	1	3.01	
Iodine	I	53	126.9	I^-
Iodine˙	I-131	53	131.0	
Iron	Fe	26	55.84	Fe^{2+}, Fe^{3+}
Magnesium	Mg	12	24.30	Mg^{2+}
Manganese	Mn	25	54.93	Mn^{2+}, Mn^{3+}
Nitrogen	N	7	14.0	NO_3^- (nitrate)
Oxygen	O	8	15.99	
Phosphorus	P	15	31.97	PO_4^{3-} (phosphate)
Phosphorus˙	P-32	15	32	
Potassium	K	19	39.10	K^+
Sodium	Na	11	22.99	Na^+
Sulfur	S	16	32.06	SO_4^{-2} (sulfate)
Zinc	Zn	30	65.38	Zn^{2+}

*Based on the Latin name of the element. The first letter is always capitalized; if there is a second letter, it is always lowercased.

**The atomic mass or weight is equal to the average mass number for the isotopes of that element.

˙An isotope of the element.

table 2.1. In some form or another, they make up the matter of all earth's organisms and viruses. Microorganisms have a particularly important relationship with most of these elements. This is due to the roles microbes play in the cycling processes that maintain the elements in forms usable by other organisms.

The vast majority of life forms require only about 20 of the 94 naturally occurring elements. The other 70 or so elements are not critical to life, and a number of them, such as arsenic and uranium, can be highly toxic to cells. As we will see in chapter 7, microbes have some incredible "survival skills" that enable them to occupy extreme habitats containing large amounts of these elements. The word that describes such a microbe is *extremophile.* * For additional information on the functions of elements and the involvement of microbes in their recycling, look ahead to table 7.1.

Quick Search

Find Carl Sagan's engaging overview of "The Chemical Elements" on YouTube.

* *An extremophile* (ex-tree′-moh-fyl″) is a microbe that can live in very severe conditions that would be harmful to other organisms.

The Major Elements of Life and Their Primary Characteristics

The unique properties of each element result from the numbers of protons, neutrons, and electrons it contains, and each element can be identified by certain physical measurements.

Each element is assigned an **atomic number (AN)** based on the number of protons it has. The atomic number is a valuable measurement because an element's proton number does not vary, and knowing it automatically tells you the usual number of electrons (recall that a neutral atom has an equal number of protons and electrons). Another useful measurement is the **mass number (MN),** equal to the number of protons and neutrons. If one knows the mass number and the atomic number, it is possible to determine the number of neutrons by subtraction. Hydrogen is a unique element because its common form has only one proton, one electron, and no neutron, making it the only element with the same atomic and mass number.

Isotopes are variant forms of the same element that differ in the number of neutrons and thus have different mass numbers. These multiple forms occur naturally in certain proportions. Carbon, for example, exists primarily as carbon 12 with 6 neutrons (MN = 12); but a small amount (about 1%) consists of carbon 13 with 7 neutrons and carbon 14 with 8 neutrons. Although isotopes have virtually the same chemical properties, some of them have unstable nuclei that spontaneously release energy in the form of radiation. Such *radioactive isotopes* play a role in a number of research and medical applications. Because they emit detectable energy, they can be used to trace the position of key atoms or molecules in chemical reactions, they are tools in diagnosis and treatment, and they are even applied in sterilization procedures (see chapter 11).

Clarifying Mass, Weight, and Related Terms

Mass refers to the quantity of matter that an atomic particle contains. The proton and neutron have almost exactly the same mass, which is about 1.66×10^{-24} grams, a unit of measurement known as a Dalton (Da) or unified atomic mass unit (U). All elements can be measured in these units. The terms *mass* and *weight* are often used interchangeably in biology, even though they apply to two different but related aspects of matter. Weight is a measurement of the gravitational pull on the mass of a particle, atom, or object. Consequently, it is possible for something with the same mass to have different weights. For example, an astronaut on the earth (normal gravity) would weigh more than the same astronaut on the moon (weak gravity). Atomic weight has been the traditional usage for biologists, because most chemical reactions and biological activities occur within the normal gravitational conditions on earth. This permits use of atomic weight as a standard of comparison. You will also see the terms *formula weight* and *molecular weight* used interchangeably, and they are indeed synonyms. They both mean the sum of atomic weights of all atoms in a molecule.

Another important measurement of an element is its **atomic mass** or **atomic weight.** This is given as the average mass numbers of all isotopic forms (table 2.1). You will notice that this number may not come out even, because most elements have several isotopes and differing proportions of them.

Figure 2.2 Examples of biologically important atoms. Featured element boxes contain information on symbol, atomic number, atomic mass, and electron shell patterns. Simple models show how the shells are filled by electrons as the atomic numbers increase. Chemists depict elements in shorthand form (red Lewis structures) that indicate only the valence electrons, because these are the electrons involved in chemical bonds. In the background is a partial display of the periodic table of elements showing the position of these elements in the periodic table.

Electron Orbitals and Shells

The structure of an atom can be envisioned as a central nucleus surrounded by a cloud of electrons that constantly spin at exceedingly high speed within confined pathways surrounding the nucleus (see figure 2.1). The pathways taken by electrons are called **orbitals,** which are not actual objects or exact locations but represent volumes of three-dimensional space in which an electron is likely to be found. The position of an electron at any one instant is dictated by its energy level or **shell,** proceeding from the lower-level energy electrons nearest the nucleus to the higher-energy electrons in the outer shells.

Electrons fill the orbitals and shells in *pairs,* starting with the shell nearest the nucleus:

- The first shell contains one orbital and a maximum of 2 electrons.
- The second shell has four orbitals and up to 8 electrons.

- The third shell with nine orbitals can hold up to 18 electrons.
- The fourth shell with 16 orbitals contains up to 32 electrons.

The number of orbitals and shells and how completely they are filled depend on the numbers of electrons, so that each element will have a unique pattern. For example:

- Helium (AN = 2) has only a filled first shell of 2 electrons.
- Oxygen (AN = 8) has a filled first shell and a partially filled second shell of 6 electrons.
- Magnesium (AN = 12) has a filled first shell, a filled second shell, and a third shell that has only one orbital, so is nearly empty.

As we will see, the chemical properties of an element are controlled mainly by the distribution of electrons in the outermost shell. **Figure 2.2** presents various simplified models of atomic structure and electron maps, superimposed over a partial display of the periodic table of elements.

 Check Your Progress SECTION 2.1

1. How are the concepts of an atom and an element related? What causes elements to differ?
2. What are the subatomic particles and how do they contribute to the structure and character of atoms?
3. How are mass number and atomic number derived? What is the atomic mass or weight?
4. What is distinctive about isotopes of elements, and why are they important?
5. Why is an isolated atom neutral?
6. Describe the concept of the atomic nucleus, electron orbitals, and shells.

2.2 Bonds and Molecules

 Expected Learning Outcomes

6. Explain how elements make chemical bonds to form molecules and compounds.
7. State the relationship among an atom, a molecule, and a compound.
8. Identify the differences between covalent, ionic, and hydrogen bonds.
9. Summarize the concepts of valence, polarity, and diatomic elements.
10. Describe ionization and distinguish between anions and cations.
11. Compare oxidation and reduction and their effects.

Most elements do not exist naturally in pure, uncombined form but are bound together as molecules and compounds. A **molecule*** is a distinct chemical substance that results from the combination of two or more atoms. Molecules come in a great variety. A few of them such as oxygen (O_2) consist of the same element (see "The Nature of Diatomic Elements"). But most molecules—for example, carbon dioxide (CO_2) and water (H_2O)—contain two or more different elements and are more appropriately termed **compounds.** So compounds are one major type of molecule. Other examples of compounds are biological molecules such as proteins, sugars, and fats.

The Nature of Diatomic Elements

You will notice that hydrogen, oxygen, nitrogen, chlorine, and iodine are often shown in notation with a subscript 2—H_2 or O_2. These elements are *diatomic* (two atoms), meaning that in their pure elemental state, they exist in pairs, rather than as a single atom. In every case, these elements are more stable as a pair of atoms joined by a covalent bond than as a single atom. The reason for this

* *molecule* (mol´-ih-kyool) L. *molecula,* little mass.

phenomenon can be explained by their valences. The electrons in the outer shell are configured so as to complete a full outer shell for both atoms when they bind. You can see this for yourself by observing figure 2.4. It is interesting that most of the diatomic elements are gases.

When atoms bind together in molecules, they lose the properties of the atom and take on the properties of the combined substance. In the same way that an atom has an atomic weight, a molecule has a **formula mass** or **molecular weight**[2] (MW), which is calculated from the sum of all of the atomic masses of the atoms it contains.

Chemical bonds of molecules are created when two or more atoms share, donate (lose), or accept (gain) electrons **(figure 2.3).** This capacity for making bonds, termed **valence,*** is determined by the number of electrons that an atom has to lose or share with other atoms during bond formation. The valence electrons determine the degree of reactivity and the types of bonds an element can make. Elements with a filled outer orbital are relatively nonreactive because they have no extra electrons to share with or donate to other atoms. For example, helium has one filled shell, with no tendency either to give up electrons or to take them from other elements, making it a stable, inert (nonreactive) gas. Elements with partially filled outer orbitals are less stable and are more apt to form some sort of bond.

2. Biologists prefer to use this term.
 * *valence* (vay´-lents) L. *valentia,* strength. The binding qualities of an atom dictated by the number of electrons in its outermost shell.

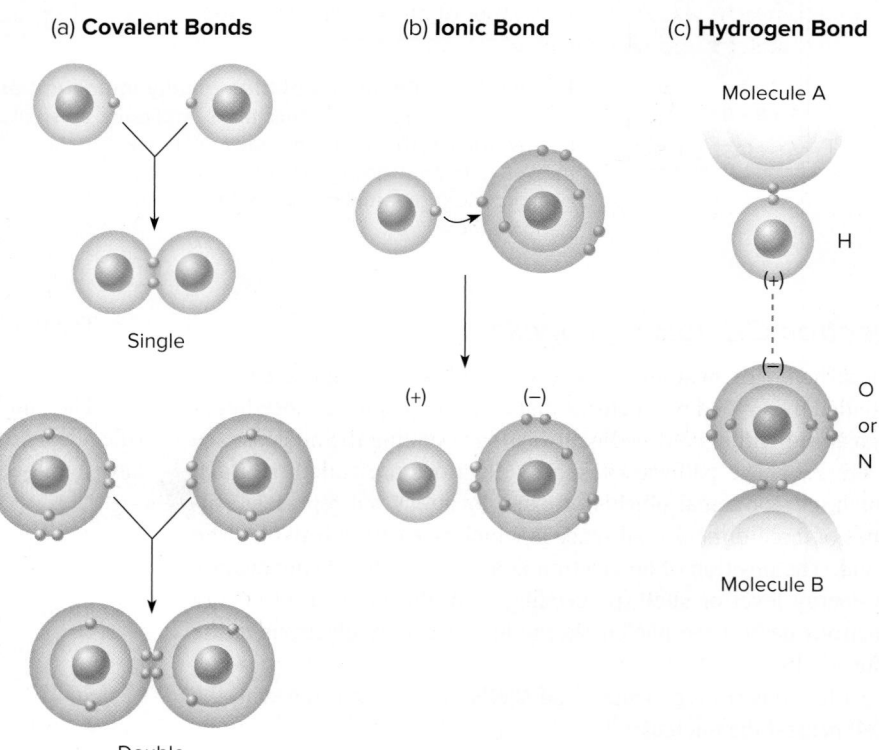

(a) **Covalent Bonds**

Single

Double

(b) **Ionic Bond**

(+) (−)

(c) **Hydrogen Bond**

Molecule A

H
(+)

(−)

O
or
N

Molecule B

Figure 2.3 General representation of three types of bonding. (a) Covalent bonds, both single and double. **(b)** Ionic bond. **(c)** Hydrogen bond. Note that hydrogen bonds are represented in models and formulas by dotted lines, as shown in (c).

Figure 2.4 Examples of molecules with covalent bonding. (a) A hydrogen molecule is formed when two hydrogen atoms share their electrons and form a single bond. **(b)** In a double bond, the outer orbitals of two oxygen atoms overlap and permit the sharing of 4 electrons (one pair from each) and the saturation of the outer orbital for both. **(c)** Simple, three-dimensional, and working models of methane. Note that carbon has 4 electrons to share and hydrogens each have one, thereby completing the shells for all atoms in the compound and creating 4 single bonds. Note that (a) and (b) also show the formation of diatomic molecules.

Molecular oxygen (O_2)

Double bond

Methane (CH_4)

Many chemical reactions are based on the tendency of atoms with unfilled outer shells to gain greater stability by achieving, or at least approximating, a filled outer shell. For example, an atom (such as oxygen) that can accept two additional electrons will bond readily with atoms (such as hydrogen) that can share or donate electrons. We explore some additional examples of the basic types of bonding in the following section.

In addition to reactivity, the number of electrons in the outer shell also dictates the number of chemical bonds an atom can make. For instance, hydrogen can bind with one other atom, oxygen can bind with up to two other atoms, and carbon can bind with four.

Covalent Bonds: Molecules with Shared Electrons

Covalent (cooperative valence) **bonds** form between atoms with valences that suit them to sharing electrons rather than to donating or receiving them. A simple example is hydrogen gas (H_2), which consists of two hydrogen atoms. A hydrogen atom has only a single electron, but when two of them combine, each will bring its electron to orbit about both nuclei, thereby approaching a filled orbital (2 electrons) for both atoms and thus creating a single covalent bond **(figure 2.4a).** Covalent bonding also occurs in oxygen gas (O_2), but with a difference. Because each atom has 2 electrons to share in this molecule, the combination creates two pairs of shared electrons, also known as a double covalent bond **(figure 2.4b).** The majority of the molecules associated with living things are composed of single and double covalent bonds between the most common biological elements (carbon, hydrogen, oxygen, nitrogen, sulfur, and phosphorus), which are discussed in more depth in chapter 7. A slightly more complex pattern of covalent bonding is shown for methane gas (CH_4) in **figure 2.4c.**

Polarity in Molecules

Electronegativity refers to the tendency of an atom or molecule to attract electrons. When atoms of different electronegativity form covalent bonds, the electrons will not be shared equally and will be pulled toward the more electronegative atom. This force causes one end of a molecule to assume a partial negative charge and the other end to assume a partial positive charge. A molecule with such an unequal distribution of charges is termed **polar** and shows polarity—meaning it has positive and negative poles.

Observe the water molecule shown in **figure 2.5** and note that, because the oxygen atom is larger and has more protons than the hydrogen atoms, it will have a stronger attraction for the shared electrons than the hydrogen atoms have. Because the electrons will spend more time near the oxygen, it will express a partial negative charge. The electrons are less attracted to the hydrogen orbitals, causing the positive charge of the hydrogen's single proton to

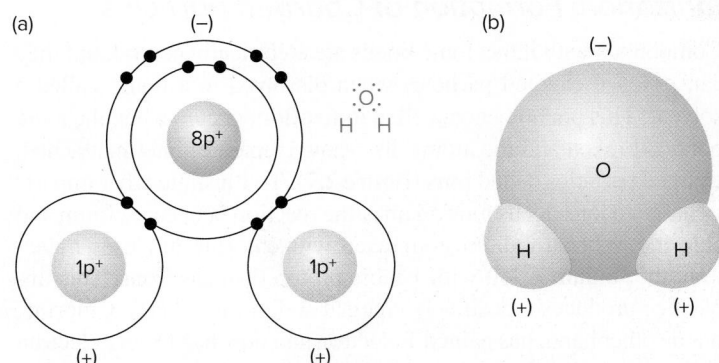

Figure 2.5 Polar molecule. (a) A simple model and **(b)** a three-dimensional model of a water molecule indicate the polarity, or unequal distribution, of electrical charge, which is caused by the pull of the shared electrons toward the oxygen side of the molecule.

dominate. The polar nature of water plays an important role in a number of biological reactions, which are discussed later. Polarity is a significant property of many large molecules in living systems and greatly influences both their reactivity and their structure.

When covalent bonds are formed between atoms that have the same or similar electronegativity, the electrons are shared equally between the two atoms. Because of this balanced distribution, no part of the molecule has a greater attraction for the electrons. This sort of electrically neutral molecule is termed **nonpolar.** Examples of nonpolar molecules are oxygen, methane (see figure 2.4), and lipids (see figure 2.18).

Ionic Bonds: Electron Transfer Among Atoms

In reactions that form **ionic bonds,** electrons are transferred completely from one atom to another and are not shared. These reactions invariably occur between atoms with complementary valences. This means that one atom has an unfilled outer shell that can readily accept electrons and the other atom has an unfilled outer shell that will readily give up electrons. A striking example is the reaction that occurs between sodium (Na) and chlorine (Cl). Elemental sodium is a soft, lustrous metal so reactive that it can burn flesh, and molecular chlorine is a very poisonous yellow gas. But when the two are combined, they form sodium chloride[4] (NaCl)—the familiar nontoxic table salt—a compound with properties quite different from either parent element **(figure 2.6).**

How does this transformation occur? Sodium has 11 electrons (2 in shell one, 8 in shell two, and only 1 in shell three), so it is 7 short of having a complete outer shell. Chlorine has 17 electrons (2 in shell one, 8 in shell two, and 7 in shell three), making it 1 short of a complete outer shell. When these two atoms come together, the sodium atom will readily donate its single valence electron and the chlorine atom will avidly receive it. The reaction is slightly more involved than a single sodium atom's combining with a single chloride atom, but the details do not negate the fundamental reaction as described here. The outcome of this reaction is a solid crystal complex that interlinks millions of sodium and chloride atoms (figure 2.6c, d). The binding of Na^+ and Cl^- exists in three dimensions. Each Na is surrounded by 6 Cls and vice versa. Their charges balance out, and no single molecule of NaCl is present.

Ionization: Formation of Charged Particles

Compounds with intact ionic bonds are electrically neutral, but they can produce charged particles when dissolved in a liquid called a solvent. This phenomenon, called **ionization,** occurs when the ionic bond is broken and the atoms dissociate (separate) into unattached, charged particles called **ions (figure 2.7).** To illustrate what imparts a charge to ions, let us look again at the reaction between sodium and chlorine. When a sodium atom reacts with chlorine and loses 1 electron, the sodium is left with 1 more proton than electrons. This imbalance produces a positively charged sodium ion (Na^+). Chlorine, on the other hand, has gained 1 electron and now has 1 more electron than protons, producing a negatively charged ion (Cl^-). Positively

(a)

Sodium atom (Na) Chlorine atom (Cl)

(b)

(c)

Sodium Chloride

(d)

Figure 2.6 Ionic bonding between sodium and chlorine.
(a) During the reaction, sodium loses its single outer orbital electron to chlorine, thereby filling chlorine's outer shell. **(b)** Simple model of ionic bonding. **(c)** Sodium and chloride ions form large molecules, or crystals, in which the two atoms alternate in a definite, regular, geometric pattern. **(d)** Note the cubic nature of NaCl crystals at the macroscopic level.
© Kathy Park Talaro

charged ions are termed **cations,*** and negatively charged ions are termed **anions.*** Substances such as salts, acids, and bases that release ions when dissolved in water are termed **electrolytes** because their charges enable them to conduct an electrical current.

Owing to the general rule that particles of like charge repel each other and those of opposite charge attract each other, we can

4. In general, when a salt is formed, the ending of the name of the negatively charged ion is changed to *-ide.*

* *cation* (kat'-eye-on) A positively charged ion that migrates toward the negative pole, or cathode, of an electrical field.
* *anion* (an'-eye-on) A negatively charged ion that migrates toward the positive pole, or anode.

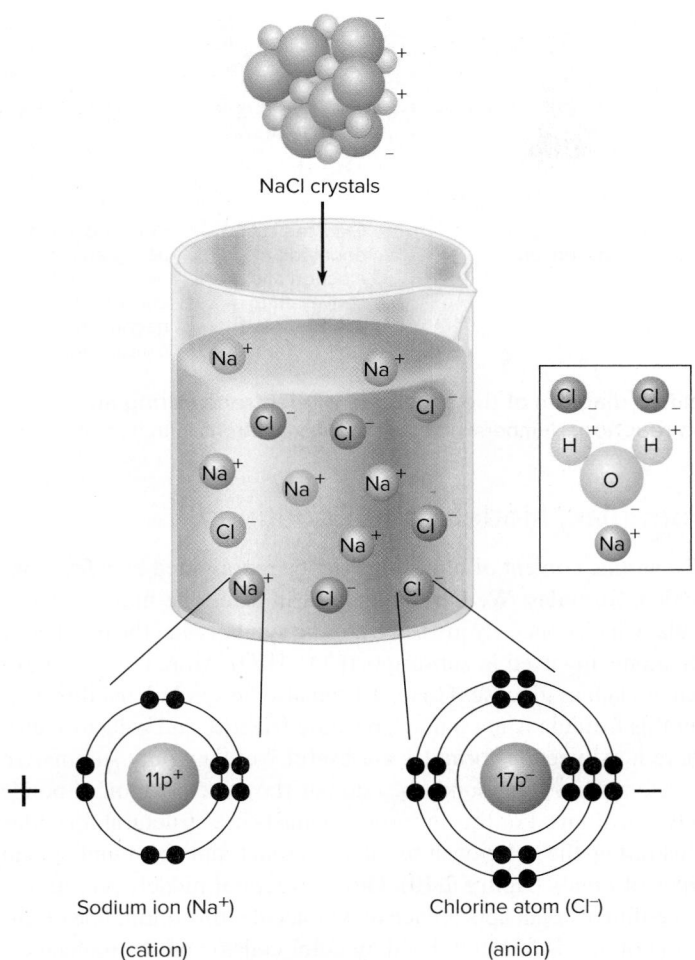

Figure 2.7 **Ionization.** When NaCl in the crystalline form is added to water, the ions are released from the crystal as separate charged particles (cations and anions) into solution. (See also figure 2.12.) In this solution, Cl⁻ ions are attracted to the hydrogen component of water, and Na⁺ ions are attracted to the oxygen (box).

Figure 2.8 **Hydrogen bonding in water.** Because of the polarity of water molecules, the negatively charged oxygen end of one water molecule is weakly attracted to the positively charged hydrogen end of an adjacent water molecule.

expect ions to interact electrostatically with other ions and polar molecules. Such interactions are important in many cellular chemical reactions, in the formation of solutions, and in the reactions microorganisms have with dyes.

Hydrogen Bonding

Some types of bonding do not involve sharing, losing, or gaining electrons but instead are due to attractive forces between nearby molecules or atoms. One such bond is a **hydrogen bond,** a weak electrostatic force that forms between a hydrogen covalently bonded to one molecule and an oxygen or nitrogen atom on the same molecule or on a different molecule. This phenomenon occurs because a hydrogen atom in a covalent bond tends to be positively charged. Thus, it can attract a nearby negatively charged atom and form an easily disrupted bridge with it. This type of bonding is usually represented in molecular models with a dotted line. A simple example of hydrogen bonding occurs between water molecules (**figure 2.8**). More extensive hydrogen bonding is partly responsible for the structure and stability of proteins and nucleic acids, as you will see later on.

Weak molecular interactions similar to hydrogen bonds that play major roles in the shape and function of biological molecules are **van der Waals forces.** The basis for these interactions is also an attraction of two regions on atoms or molecules that are opposite in charge, but van der Waals forces can occur between nearly any types of molecules and not just those containing hydrogen, oxygen, and nitrogen. These forces come into play whenever the electrons in molecules move about their orbits and become unevenly distributed. This unevenness leads to short-term "sticky spots" in the molecule—some positively charged and others negatively charged. When such regions are located close together, their opposite charges pull them together. These forces can hold even large molecules together because of the cumulative effects of numerous sites of interaction. They not only function between molecules but may occur within different regions of the same large molecule. Van der Waals forces are a significant factor in protein folding and stability (see figure 2.23, steps 3 and 4).

Electron Transfer and Oxidation-Reduction Reactions

The metabolic work of cells, such as synthesis, movement, and digestion, revolves around energy exchanges and transfers. The management of energy in cells is almost exclusively dependent on chemical rather than physical reactions, because most cells are far too delicate to operate with heat, radiation, and other more potent forms of energy. The outer-shell electrons are readily portable and easily manipulated sources of energy. It is in fact the movement of electrons from molecule to molecule that accounts for most energy exchanges in cells. Fundamentally, then, cells must have a supply of atoms that can gain or lose electrons if they are to carry out life processes.

The phenomenon in which electrons are transferred from one atom or molecule to another is termed an **oxidation-reduction** (shortened to **redox**) **reaction.** The term *oxidation* was originally adopted for reactions involving the addition of oxygen, but this is no longer the case. In current usage, *oxidation* includes any reaction that causes an atom to lose electrons. Because all redox reactions occur in pairs, it follows that *reduction* is the result of a different atom gaining

these same electrons. Keep in mind that because electrons are being added during reduction, the atom that receives them will become more negative; and that is the meaning of *reduction* in this context. It does not imply that an atom is getting smaller. In fact, reduction often results in a greater complexity of the molecule.

To analyze the phenomenon, let us again review the production of NaCl but from a different standpoint. When these two atoms, called the redox pair, react to form sodium chloride, a sodium atom gives up an electron to a chlorine atom. During this reaction, sodium is oxidized because it loses an electron, and chlorine is reduced because it gains an electron **(figure 2.9).** With this system, an atom such as sodium that can donate electrons and thereby reduce another atom is a **reducing agent.** An atom that can receive extra electrons and thereby oxidize another molecule is an **oxidizing agent.** You may find this concept easier to keep straight if you think of redox agents as partners: The reducing partner gives its electrons away and is oxidized; the oxidizing partner receives the electrons and is reduced.[5]

Redox reactions are essential to many of the biochemical processes discussed in chapter 8. In cellular metabolism, electrons are frequently transferred from one molecule to another as described here. In other reactions, oxidation and reduction occur with the transfer of a hydrogen atom (a proton and an electron) from one compound to another.

Figure 2.9 Simplified diagram of the exchange of electrons during an oxidation-reduction reaction. Numbers indicate the total electrons in that shell.

Reducing agent can donate an electron.

Oxidizing agent can accept an electron.

Oxidized cation donated an electron and converted to a positively charged ion.

Reduced anion accepted the electron and converted to a negatively charged ion.

 Check Your Progress SECTION 2.2

7. Explain how the concepts of molecules and compounds are related.
8. Distinguish between the general reactions in covalent, ionic, and hydrogen bonds.
9. Which kinds of elements tend to make covalent bonds?
10. Distinguish between a single and a double bond.
11. Define polarity and explain what causes it.
12. Which kinds of elements tend to make ionic bonds?
13. Differentiate between an anion and a cation, using examples.
14. Differentiate between oxidation and reduction, and between an oxidizing agent and a reducing agent, using examples.

2.3 Chemical Reactions, Solutions, and pH

 Expected Learning Outcomes

12. Classify different forms of chemical shorthand and types of reactions.

13. Explain solutes, solvents, and hydration.

14. Differentiate between hydrophilic and hydrophobic.

15. Describe the pH scale and how it was derived; define *acid, base,* and *neutral levels.*

5. A mnemonic device to keep track of this is *LEO* says *GER: Lose Electrons Oxidized; Gain Electrons Reduced.*

Formulas, Models, and Equations

The atomic content of molecules can be represented by a few convenient formulas. We have already been using the molecular formula, which concisely gives the atomic symbols and the number of the atoms involved in subscripts (CO_2, H_2O). More complex molecules such as glucose ($C_6H_{12}O_6$) can also be symbolized this way, but this formula is not unique, because fructose and galactose also share it. Molecular formulas are useful, but they only summarize the atoms in a compound; they do not show the position of bonds between atoms. For this purpose, chemists use structural formulas illustrating the relationships of the atoms and the number and types of bonds **(figure 2.10).** Other structural models present the three-dimensional appearance of a molecule, illustrating the orientation of atoms (differentiated by color codes) and the molecule's overall shape **(figure 2.11).** Many complex molecules such as proteins are now represented by computer-generated images (see figure 2.23, step 4).

Molecules, including those in cells, are constantly involved in **chemical reactions,** leading to changes in the composition of the matter they contain. These changes generally involve the breaking and making of bonds and the rearrangement of atoms. The chemical substances that start a reaction and that are changed by the reaction are called the **reactants.** The substances that result from the reaction are called the **products.** Keep in mind that all of the matter in any reaction is retained in some form, and the same types and numbers of atoms going into the reaction will be present in the products. Chemists and biologists use shorthand to summarize the content of a reaction by means of a **chemical equation.** In an equation, the reactant(s) are on the left of an arrow and the product(s) are on the right. The number of atoms of each element must be balanced on either side of the arrow. Note that the numbers of reactants and products are indicated by a coefficient in front of the formula (no coefficient means 1). We have already reviewed the reaction with sodium and chloride, which would be shown with this equation:

$$2Na + Cl_2 \rightarrow 2NaCl$$

Most equations do not give the details or even exact order of the reaction but are meant to keep the expression a simple overview of the process being shown. Some of the common reactions in organisms are syntheses, decompositions, and exchanges.

| (a) Molecular formulas | H_2 | O_2 | H_2O | CO_2 | CH_4 |

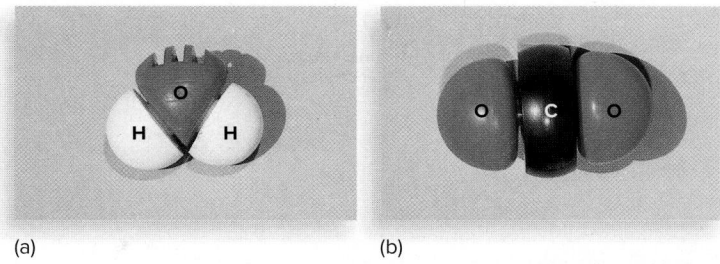

(a) (b)

Figure 2.11 **Three-dimensional, or space-filling, models of (a) water, and (b) carbon dioxide.** The red atoms are oxygen, the white ones hydrogen, and the black ones carbon. (a and b): © John W. Hole

Figure 2.10 **Comparison of molecular and structural formulas. (a)** Molecular formulas provide a brief summary of the elements in a compound. **(b1)** Structural formulas clarify the exact relationships of the atoms in the molecule, depicting single bonds by a single line and double bonds by two lines. **(b2)** In structural formulas of organic compounds, cyclic or ringed compounds may be completely labeled, or **(b3)** they may be presented in a shorthand form in which carbons are assumed to be at the angles and attached to hydrogens. See figure 2.16 for structural formulas of three sugars with the same molecular formula, $C_6H_{12}O_6$.

During **exchange reactions,** the reactants trade portions between each other and release products that are combinations of the two:

$$AB + XY \rightleftharpoons AY + XB$$

This type of reaction occurs between an acid and a base when they react to form water and a salt:

$$HCl + NaOH \rightarrow NaCl + H_2O$$

The reactions in biological systems can be reversible, meaning that reactants and products can be converted back and forth. These reversible reactions are symbolized with a double arrow, each pointing in opposite directions, as in the exchange reaction shown earlier. Whether a reaction is reversible depends on the proportions of these compounds, the difference in energy state of the reactants and products, and the presence of **catalysts** (substances that increase the rate of a reaction). Additional reactants coming from another reaction can also be indicated by arrows that enter or leave at the main arrow:

$$X + Y \xrightarrow{\quad CD \searrow C \quad} XYD$$

In a **synthesis* reaction,** the reactants bond together in a manner that produces an entirely new molecule (reactant A plus reactant B yields product AB). An example is the production of sulfur dioxide, a by-product of burning sulfur fuels and an important component of smog:

$$S + O_2 \rightarrow SO_2$$

Some synthesis reactions are not such simple combinations. When water is synthesized, for example, the reaction does not really involve one oxygen atom combining with two hydrogen atoms, because elemental oxygen exists as O_2 and elemental hydrogen exists as H_2. A more accurate equation for this reaction is:

$$2H_2 + O_2 \rightarrow 2H_2O$$

In **decomposition reactions,** the bonds on a single reactant molecule are permanently broken to release two or more product molecules. A simple example can be shown for the common chemical hydrogen peroxide:

$$2H_2O_2 \rightarrow 2H_2O + O_2$$

Solutions: Homogeneous Mixtures of Molecules

A **solution** is a mixture of one or more substances called **solutes** uniformly dispersed in a dissolving medium called a **solvent.** An important characteristic of a solution is that the solute cannot be separated by filtration or ordinary settling. The solute can be gaseous, liquid, or solid, and the solvent is usually a liquid. Examples of solutions are salt or sugar dissolved in water and iodine dissolved in alcohol. In general, a solvent will dissolve a solute only if it has similar electrical characteristics as indicated by the rule of solubility, expressed simply as "like dissolves like." For example, water is a polar molecule and will readily dissolve an ionic solute such as NaCl, yet a nonpolar solvent such as benzene will not dissolve NaCl.

Water is the most common solvent in natural systems, having several characteristics that suit it to this role. The polarity of the water molecule causes it to form hydrogen bonds with other water molecules, but it can also interact readily with charged or polar molecules.

When ionic solutes such as NaCl crystals are added to water, the Na^+ and Cl^- are released into solution. Dissolution occurs because Na^+ is attracted to the negative pole of the water molecule and Cl^- is attracted to the positive poles. In this way, they are drawn

*synthesis (sin′-thuh-sis) Gr. synthesis, putting together.

Figure 2.12 Hydration spheres formed around ions in solution. In this example, a sodium cation attracts the negatively charged region of water molecules, and a chloride anion attracts the positively charged region of water molecules. In both cases, the ions become surrounded by spherical layers of specific numbers and arrangements of water molecules.

away from the crystal separately into solution. As it leaves, each ion becomes **hydrated,** which means that it is surrounded by a sphere of water molecules **(figure 2.12).** Molecules such as salt or sugar that attract water to their surface are termed **hydrophilic.*** Nonpolar molecules, such as benzene, that repel water are considered **hydrophobic.*** A third class of molecules, such as the phospholipids in cell membranes, are considered **amphipathic*** because they have both hydrophilic and hydrophobic properties.

Because most biological activities take place in aqueous (water-based) solutions, the concentration of these solutions can be very important (see chapter 7). The **concentration** of a solution expresses the amount of solute dissolved in a certain amount of solvent. It can be figured by percentage or molarity. Percentage is the ratio of solute in solution expressed as some combination of weight or volume.

A common way to calculate concentration by percentage is to use the weight of the solute, measured in grams (g), dissolved in a specified volume of solvent, measured in milliliters (ml). For example, dissolving 3 g of NaCl in the amount of water to produce 100 ml of solution is a 3% solution; dissolving 30 g in water up to

* *hydrophilic* (hy-droh-fil′-ik) Gr. *hydros,* water, and *philos,* to love.
* *hydrophobic* (hy-droh-fob′-ik) Gr. *phobos,* fear.
* *amphipathic* (am′-fy-path′-ik) Gr. *amphi,* both.

100 ml of solution produces a 30% solution; and dissolving 3 g in 1,000 ml (1 liter) produces a 0.3% solution.

A frequent way to express concentration of biological solutions is by its molar concentration, or *molarity* (M). A standard molar solution is obtained by dissolving 1 *mole,* defined as the molecular weight of the compound in grams, in 1 L (1,000 ml) of solution. To make a 1 M solution of sodium chloride, we would dissolve 58 g of NaCl to give 1 L of solution; a 0.1 M solution would require 5.8 g of NaCl in 1 L of solution.

Acidity, Alkalinity, and the pH Scale

Another factor with far-reaching impact on living things is the concentration of acidic or basic solutions in their environment. To understand how solutions become acidic or basic, we must look again at the behavior of water molecules. Hydrogens and oxygen tend to remain bonded by covalent bonds, but under certain conditions, a water molecule may dissociate. This occurs when a single hydrogen atom breaks away as an ionic H^+, or **hydrogen ion,** leaving the remainder of the molecule in the form of an OH^-, or **hydroxide ion.** The H^+ ion is positively charged because it is essentially a hydrogen that has lost its electron; the OH^- is negatively charged because it remains in possession of that electron. Ionization of water is constantly occurring, but in pure water containing no other ions, H^+ and OH^- are produced in equal amounts, and the solution remains neutral. By one definition, a solution is considered **acidic** when one of its components (an acid) releases excess hydrogen ions.[6] A solution is **basic** when a component (a base) releases excess hydroxide ions, so that there is no longer a balance between the two ions. Another term used interchangeably with basic is **alkaline.**

To measure the acid and base concentrations of solutions, scientists use the **pH** scale, a graduated numerical scale that ranges from 0 (the most acidic) to 14 (the most basic). This scale is a useful standard for rating the relative acid or base content of a substance. Use **figure 2.13** to familiarize yourself with the pH readings of some common substances. It is not an arbitrary scale but actually a mathematical derivation based on the negative logarithm (reviewed in "Exponents" found in Connect) of the concentration of H^+ ions in moles per liter (symbolized as $[H^+]$) in a solution, represented as:

$$pH = -\log[H^+]$$

6. Actually, it forms a hydronium ion (H_3O^+), but for simplicity's sake we will use the notation H^+.

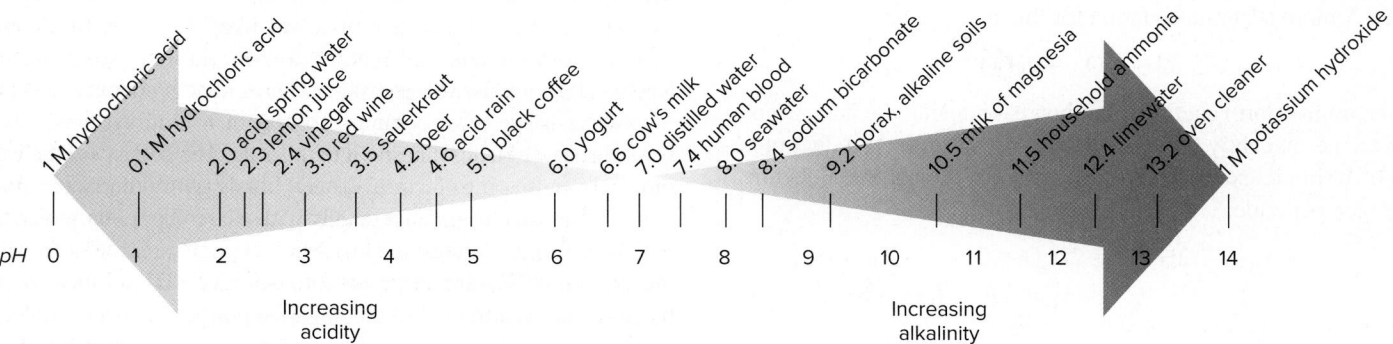

Figure 2.13 The pH scale. Shown are the relative degrees of acidity and alkalinity and the approximate pH readings for various substances.

TABLE 2.2	Hydrogen Ion and Hydroxide Ion Concentrations at a Given pH		
Moles/L of Hydrogen Ions	Logarithm	pH	Moles/L of OH⁻
1.0	10^0	0	10^{-14}
0.1	10^{-1}	1	10^{-13}
0.01	10^{-2}	2	10^{-12}
0.001	10^{-3}	3	10^{-11}
0.0001	10^{-4}	4	10^{-10}
0.00001	10^{-5}	5	10^{-9}
0.000001	10^{-6}	6	10^{-8}
0.0000001	10^{-7}	7	10^{-7}
0.00000001	10^{-8}	8	10^{-6}
0.000000001	10^{-9}	9	10^{-5}
0.0000000001	10^{-10}	10	10^{-4}
0.00000000001	10^{-11}	11	10^{-3}
0.000000000001	10^{-12}	12	10^{-2}
0.0000000000001	10^{-13}	13	10^{-1}
0.00000000000001	10^{-14}	14	10^0

Acidic solutions have a greater concentration of H⁺ than OH⁻, starting with: pH 0, which contains 1.0 moles H⁺/liter. Each of the subsequent whole-number readings in the scale reduces the [H⁺] by tenfold:

- pH 1 contains [0.1 moles H⁺/L].
- pH 2 contains [0.01 moles H⁺/L].
- Continuing in the same manner up to pH 14, which contains [0.00000000000001 moles H⁺/L].

These same concentrations can be represented more manageably by exponents:

- pH 2 has an [H⁺] of 10^{-2} moles.
- pH 14 has an [H⁺] of 10^{-14} moles **(table 2.2).**

It is evident that the pH units are derived from the exponent itself. Even though the basis for the pH scale is [H⁺], it is important to note that as the [H⁺] in a solution decreases, the [OH⁻] increases in direct proportion. At midpoint—pH 7 or **neutrality**—the concentrations are exactly equal and neither predominates, this being the pH of pure water previously mentioned.

In summary, the pH scale can be used to rate or determine the degree of acidity or alkalinity of a solution. On this scale, a pH below 7 is acidic, and the lower the pH, the greater the acidity; a pH above 7 is basic, and the higher the pH, the greater the alkalinity. Incidentally, although pHs are given here in even whole numbers, more often pH readings are given in decimal form; for example, pH 4.5 or 6.8 (acidic) and pH 7.4 or 10.2 (basic). Because of the damaging effects of very concentrated acids or bases, most cells operate best under neutral, weakly acidic, or weakly basic conditions (see chapter 7).

Aqueous solutions containing both acids and bases may be involved in **neutralization** reactions, which give rise to water and other neutral by-products. For example, when equal molar solutions of hydrochloric acid (HCl) and sodium hydroxide (NaOH, a base) are mixed, the reaction proceeds as follows:

$$HCl + NaOH \rightarrow H_2O + NaCl$$

Here the acid and base ionize to H⁺ and OH⁻ ions, which form water, and other ions, Na⁺ and Cl⁻, which form sodium chloride. Any product other than water that arises when acids and bases react is called a salt. Many of the organic acids (such as lactic and succinic acids) that function in **metabolism*** are available as the acid and the salt form (such as lactate, succinate), depending on the conditions in the cell (see chapter 8).

 Check Your Progress SECTION 2.3

15. Review the types of chemical reactions and the general ways they can be expressed in equations.
16. Define *solution*, *solvent*, and *solute*.
17. What properties of water make it an effective biological solvent, and how does a molecule like NaCl become dissolved in it?
18. What is molarity? Tell how to make a 1 M solution of $Mg_3(PO_4)_2$ and a 0.1 M solution of $CaSO_4$.
19. What determines whether a substance is an acid or a base? Briefly outline the pH scale.

2.4 The Chemistry of Carbon and Organic Compounds

 Expected Learning Outcomes

16. Describe the chemistry of carbon and the difference between inorganic and organic compounds.
17. Identify functional groups and give some examples.
18. Define *macromolecule, polymer,* and *monomer.*

So far our main focus has been on the characteristics of atoms, ions, and small, simple substances that play diverse roles in the structure and function of living things. These substances are often lumped together in a category called **inorganic chemicals.** A chemical is usually inorganic if it does not contain *both* carbon and hydrogen. Examples of inorganic chemicals include H_2O, O_2, NaCl (sodium chloride), $Mg_3(PO_4)_2$ (magnesium phosphate), $CaCO_3$ (calcium carbonate), and CO_2 (carbon dioxide). In reality, however, most of the chemical reactions and structures of living things occur at the level of more complex molecules, termed **organic chemicals.** The minimum requirement for a compound to be considered organic is that it contains a basic framework of carbon bonded to hydrogens. Organic molecules vary in complexity from the simplest, methane (CH_4; see figure 2.4c), which has a molecular weight of 16, to certain antibody molecules (produced by an immune reaction) that have a molecular weight of nearly 1 million and are among the most complex molecules on earth. Most organic chemicals in cells contain other elements such as oxygen, nitrogen, and phosphorus in addition to the carbon and hydrogen.

The role of carbon as the fundamental element of life can best be understood if we look at its chemistry and bonding patterns. The valence of carbon makes it an ideal atomic building block to form the backbone of organic molecules; it has 4 electrons in its outer orbital to be shared with other atoms (including other carbons) through

* *metabolism* (muh-tab′-oh-lizm) A general term referring to the totality of chemical and physical processes occurring in the cell.

(a)

Linear

Branched

Ringed

(b)

Figure 2.14 The versatility of bonding in carbon. In most compounds, each carbon makes a total of four bonds. **(a)** Single, double, and triple bonds can be made with other carbons, oxygen, and nitrogen; single bonds are made with hydrogen. Simple electron models show how the electrons are shared in these bonds. **(b)** Multiple bonding of carbons can give rise to long chains, branched compounds, and ringed compounds, many of which are extraordinarily large and complex.

covalent bonding. As a result, it can form stable chains containing thousands of carbon atoms and still has bonding sites available for forming covalent bonds with numerous other atoms. Carbon forms bonds that are linear, branched, or ringed, and it can form four single bonds, two double bonds, or one triple bond **(figure 2.14).** Several unique kinds of molecules are made of pure carbon alone. You may be familiar with diamonds and graphite. A relatively new discovery, fullerene, is made by carbon atoms double-bonding with each other

to produce a hollow sphere. Look ahead to the discussion of nanoparticles in **Clinical Connections** to see a type of this molecule, which is considered one of the most common molecules in the universe, and one that may have many earthly uses.

Functional Groups of Organic Compounds

One important advantage of carbon's serving as the molecular skeleton for living things is that it is free to bind with a variety of special molecular groups or accessory molecules called **functional groups.** Functional groups help define the chemical class of certain groups of organic compounds and confer unique reactive properties on the whole molecule **(table 2.3).** Because each type of functional group behaves in a distinctive manner, reactions of an organic

TABLE 2.3	Representative Functional Groups and Organic Compounds That Contain Them	
Formula of Functional Group	Name	Can Be Found In
R* — O — H	Hydroxyl	Alcohols, carbohydrates
R — C(=O)(OH)	Carboxyl	Fatty acids, proteins, organic acids
R — C(H)(H) — NH₂	Amino	Proteins, nucleic acids
R — C(=O) — O — R	Ester	Lipids
R — C(H)(H) — SH	Sulfhydryl	Cysteine (amino acid), proteins
R — C(=O) — H	Carbonyl, terminal end	Aldehydes, polysaccharides
R — C(=O) — C —	Carbonyl, internal	Ketones, polysaccharides
R — O — P(=O)(OH) — OH	Phosphate	DNA, RNA, ATP

*The R designation on a molecule is shorthand for remainder, and its placement in a formula indicates that what is attached at that site varies from one compound to another.

TABLE 2.4	Macromolecules and Their Functions	
Macromolecule	**Description/Basic Structure**	**Examples/Notes**
Carbohydrates		
Monosaccharides	3- to 7-carbon sugars	Glucose, fructose / Sugars involved in metabolic reactions; building block of disaccharides and polysaccharides
Disaccharides	Two monosaccharides	Maltose (malt sugar) / Composed of two glucoses; an important breakdown product of starch
		Lactose (milk sugar) / Composed of glucose and galactose
		Sucrose (table sugar) / Composed of glucose and fructose
Polysaccharides	Chains of monosaccharides	Starch, cellulose, glycogen / Cell wall, food storage
Lipids		
Triglycerides	Fatty acids + glycerol	Fats, oils / Major component of cell membranes; storage
Phospholipids	Fatty acids + glycerol + phosphate	Membranes
Waxes	Fatty acids, alcohols	Mycolic acid / Cell wall of mycobacteria
Steroids	Ringed structure	Cholesterol, ergosterol / Membranes of eukaryotes and some bacteria
Proteins		
Polypeptides	Amino acids in a chain bound by peptide bonds	Enzymes; part of cell membrane, cell wall, ribosomes, antibodies / Metabolic reactions; structural components
Nucleic acids	Nucleotides, composed of pentose sugar + phosphate + nitrogenous base	Purines: adenine, guanine; Pyrimidines: cytosine, thymine, uracil
Deoxyribonucleic acid (DNA)	Contains deoxyribose sugar and thymine, not uracil	Chromosomes; genetic material of viruses / Inheritance
Ribonucleic acid (RNA)	Contains ribose sugar and uracil, not thymine	Ribosomes; mRNA, tRNA / Expression of genetic traits
Adenosine triphosphate (ATP)	Contains adenine, ribose sugar, and 3 phosphate groups	A high-energy compound that gives off energy to power reactions in cells

compound can be predicted by knowing the kind of functional group or groups it carries. Many synthesis, decomposition, and transfer reactions rely upon functional groups such as R—OH or R—NH$_2$. The R designation on a molecule is shorthand for remainder, and its placement in a formula indicates that the group attached at that site varies from one compound to another.

Organic Macromolecules: Superstructures of Life

The compounds of life fall into the realm of **biochemistry.** Biochemicals are organic compounds produced by (or that are components of) living things, and they include four main families: carbohydrates, lipids, proteins, and nucleic acids **(table 2.4).** The compounds in these groups are assembled from smaller molecular subunits, or building blocks, and because they are often very large compounds, they are termed **macromolecules.** All macromolecules except lipids are formed by polymerization, a process in which repeating subunits termed **monomers*** are bound into chains of various lengths termed **polymers.*** For example, amino acids (monomers), when arranged in a chain, form proteins (polymers). The large size and complex, three-dimensional shapes of macromolecules enable them to function as structural components, molecular messengers, energy sources, enzymes (biochemical

catalysts), nutrient stores, and sources of genetic information. In section 2.5 and in later chapters, we consider numerous concepts relating to the roles of macromolecules in cells. Table 2.4 will also be a useful reference when you study metabolism in chapter 8.

 Check Your Progress SECTION 2.4

20. What atoms must be present in a molecule for it to be considered organic?
21. Name several inorganic compounds.
22. What characteristics of carbon make it ideal for the formation of organic compounds?
23. What are functional groups?
24. Differentiate between a monomer and a polymer. How are polymers formed?

2.5 Molecules of Life: Carbohydrates

 Expected Learning Outcomes

19. Define *carbohydrate* and know the functional groups that characterize carbohydrates.
20. Distinguish among mono-, di-, and polysaccharides, and describe how their bonds are made.
21. Discuss the functions of carbohydrates in cells.

* *monomer* (mahn′-oh-mur) Gr. *mono,* one, and *meros,* part.
* *polymer* (pahl′-ee-mur) Gr. *poly,* many; also the root for polysaccharide, polypeptide, and polynucleotide.

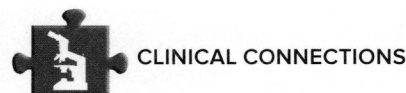

CLINICAL CONNECTIONS

Nanoparticles—Molecule-Sized "Smart" Pills

In the near future, it may be possible to treat an infection or cancer by taking a single tiny dose of medicine by nose, mouth, or skin. This is just one of the revolutionary possibilities being explored by the science of **chemotherapy,** defined as the use of chemical compounds (drugs) to treat or prevent diseases. Traditionally drugs have been swallowed in pill form or injected, but often these methods leave a great deal to be desired. For example, a drug being swallowed may be absorbed in unpredictable amounts, and it may not reliably reach the site of action; or it may not be specific to the target cells and may cause harm to healthy tissues.

For several years pharmaceutical researchers have been exploring alternate ways to administer drugs using tiny delivery systems called **nanoparticles.*** This idea comes from the science of **nanotechnology,** defined as the study and application of molecule-sized devices or structures (look ahead to 3.1 Making Connections). Of several molecule-sized systems under development, two that look very promising are liposomes and fullerenes. The structure of these two types of nanoparticles suits them well for carrying drugs to a precise location and controlling their release. In both cases, they display a spherical, hollow structure that provides spaces for holding drugs, and both have surfaces that could be manipulated to bind to a specific organ or cell type.

Liposomes are lipid-based molecules with the lipids arranged in a bilayer similar to a cell membrane **(figure *a*).** They are soluble in water, could carry a variety of drugs in their core, and readily fuse with the membrane of the cell they are targeting, immediately delivering their contents directly into the cytoplasm. Note also that liposomes can also display molecules that improve recognition and entry of the drug. Liposomes are being considered for treating tumors and viral infections, delivering antibiotics to infected tissues and organs, vaccine delivery, and even some types of genetic therapies.

Fullerenes are an unusual group of molecules that consist only of carbon double-bonded to other carbons, forming a lattice network that can assume a number of configurations. The most adaptable configuration is the spheroid form called a buckyball,* composed of 60 carbons **(figure *b*).** These molecules can also have drugs loaded into their cores and be directed toward specific cells or organs, but they are not broken down in the body and can remain in place significantly longer than traditional drug carriers. They are being tested as a means of treating cancers with radioactive isotopes and delivering antiviral drugs into infected cells.

Nanomedicine is one of the most rapidly growing areas of research. What other applications could grow out of this use of such tiny delivery systems? *Answer available on Connect.*

**nano* for nanometer, which is one-billionth of a meter in length, about 1/80,000 the diameter of a human hair.

*This term and fullerene were chosen in honor of Buckminster Fuller, who designed the geodesic dome, an architectural structure that is constructed on the same plan as these tiny molecules.

Protective layer against immune destruction

Homing peptide

DNA

Lipid-soluble drug in bilayer

Drug crystallized in aqueous fluid

Lipid bilayer

(a) Diagram of a liposome is shown carrying drugs and displaying specific surface molecules to help in homing to the target and drug delivery.

(b) Structure of a buckyball indicates its "soccer ball" appearance of hexagonal carbons. Courtesy of Pete Marenfeld/NOAO/AURA/NSF

The term **carbohydrate** originates from the way that most members of this chemical class resemble combinations of carbon and water. Carbohydrates can be generally represented by the formula $(CH_2O)_n$, in which *n* indicates the number of units that combine to make the finished carbohydrate. The basic structure of a simple carbohydrate monomer is a backbone of carbon bound to two or more hydroxyl groups. Because they also have either an aldehyde or a ketone group, they are often designated as *polyhydroxy* **aldehydes** or **ketones (figure 2.15).** In simple terms, a sugar such as glucose is an aldehyde with a terminal carbonyl group bonded to a

Figure 2.15 Common classes of carbohydrates. (a) Major saccharide groups, named for the number of sugar units each contains. **(b)** Three hexoses with the same molecular formula ($C_6H_{12}O_6$) and different structural formulas. Both linear and ring models are given. The linear form emphasizes aldehyde and ketone groups, although in solution the sugars exist in the ring form. Note that the carbons are numbered in red so as to keep track of reactions within and between monosaccharides.

hydrogen and another carbon. Fructose sugar is a ketone with a carbonyl group bonded between two carbons.

Carbohydrates exist in a great variety of configurations. The common term sugar (**saccharide***) refers to a simple carbohydrate, such as a monosaccharide or a disaccharide, that has a sweet taste. A **monosaccharide** is a simple polyhydroxy aldehyde or ketone molecule containing from 3 to 7 carbons; a **disaccharide** is a combination of two monosaccharides; and a **polysaccharide** is a polymer of five or more monosaccharides bound in linear or branched chain patterns (figure 2.15). Monosaccharides and disaccharides are specified by combining a prefix that describes some characteristic of the sugar with the suffix -**ose.** For example, **hexoses** are composed of 6 carbons, and **pentoses** contain 5 carbons. **Glucose** (Greek for sweet) is the most common and universally important hexose; **fructose** is named for fruit (one of its sources); and xylose, a pentose, derives its name from the Greek word for wood. Disaccharides are named similarly: **lactose** (Latin for milk) is an important component of milk; **maltose** means malt sugar; and **sucrose** (French for sugar) is common table sugar or cane sugar.

The Nature of Carbohydrate Bonds

The subunits of disaccharides and polysaccharides are linked by means of **glycosidic bonds,** in which carbons (each is assigned a number) on adjacent sugar units are bonded to the same oxygen atom like links in a chain **(figure 2.16).** For example:

- Maltose is formed when the number 1 carbon on a glucose bonds to the oxygen on the number 4 carbon on a second glucose.
- Sucrose is formed when glucose and fructose bind oxygen between their number 1 and number 2 carbons.
- Lactose is formed when glucose and galactose connect by their number 1 and number 4 carbons.

This bond is formed by one carbon giving up its OH group and the other (the one contributing the oxygen to the bond) losing the H from its OH group. Because a water molecule is produced, this reaction is known as **dehydration synthesis,** a process common to most polymerization reactions (see proteins, section 2.7). Three polysaccharides (starch, cellulose, and glycogen) are structurally and biochemically distinct, even though all are polymers of the same monosaccharide—glucose. The basis for their differences lies primarily in the exact way the glucose molecules are bound together, which greatly affects the characteristics of the end product **(figure 2.17).** The

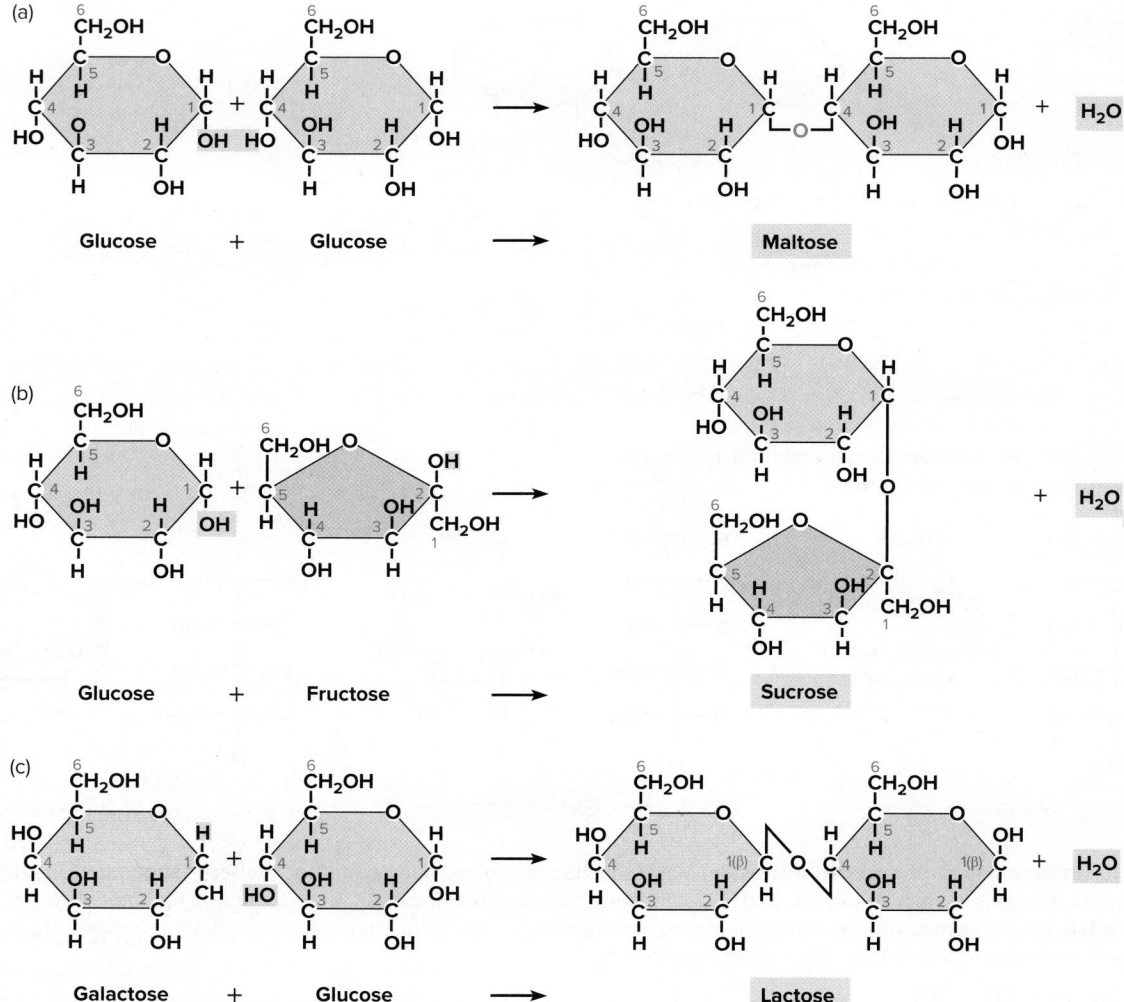

Figure 2.16 Glycosidic bond in three common disaccharides. (a) Formation of the 1,4 bond between two α glucoses to produce maltose. **(b)** Formation of the 1,2 bond between glucose and fructose to produce sucrose. **(c)** A 1,4 bond between a galactose and glucose produces lactose. Note that all three dehydration synthesis reactions release water.

(a) Cellulose **(b) Starch**

Figure 2.17 Polysaccharides. (a) Cellulose is composed of β-glucose bonded in 1,4 bonds that produce linear, lengthy chains of polysaccharides that are H-bonded along their length. This accounts for fibrous support structures of plants. **(b)** Starch is also composed of glucose polymers, in this case α glucose. The main structure is amylose bonded in a 1,4 pattern, with side branches of amylopectin bonded by 1,6 bonds. The entire molecule is compact and granular.

synthesis and breakage of each type of bond require a specialized catalyst called an enzyme (see chapter 8).

The Functions of Carbohydrates in Cells

Carbohydrates are the most abundant biological molecules in nature. They play numerous roles in cell structure, adhesion, and metabolism.

Polysaccharides typically contribute to structural support and protection and serve as nutrient and energy stores. The cell walls in plants and many microscopic algae derive their strength and rigidity from **cellulose,** a long, fibrous polymer (figure 2.17*a*). Because of this role, cellulose is probably one of the most common organic substances on the earth, yet it is digestible only by certain bacteria, fungi, and protozoa that produce the enzyme cellulase. These microbes, called decomposers, play an essential role in breaking down and recycling plant materials (see figure 7.2). Some bacteria secrete slime layers of a glucose polymer called *dextran*. This substance causes a sticky layer to develop on teeth that leads to plaque, described later in chapter 21.

Other structural polysaccharides can be conjugated (chemically bonded) to amino acids, nitrogen bases, lipids, or proteins. **Agar,** a polysaccharide indispensable for preparing solid culture media, is a natural component of certain seaweeds. It is a complex polymer of galactose and sulfur-containing carbohydrates. **Chitin** (ky-tun), a polymer of glucosamine (a sugar with an amino functional group), is a major compound in the cell walls of fungi and the exoskeletons of insects. **Peptidoglycan*** is one special class of compounds in which polysaccharides (glycans) are linked to peptide fragments (a short chain of amino acids). This molecule provides the main source of structural support to the bacterial cell wall. The cell wall of gram-negative bacteria also contains **lipopolysaccharide,** a complex of lipid and polysaccharide responsible for symptoms such as fever and shock (see chapters 4 and 13).

The outer surface of many cells has a delicate "sugar coating" composed of polysaccharides bound in various ways to proteins (the combination is called mucoprotein or glycoprotein). This structure, called the **glycocalyx,*** functions in attachment to other cells or as a site for *receptors*—surface molecules that receive and respond to external stimuli. Small sugar molecules account for the differences in human blood types, and carbohydrates are a component of large protein molecules called antibodies. Some viruses have glycoproteins on their surface for binding to and invading their host cells.

Polysaccharides are usually stored by cells in the form of glucose polymers, such as **starch** (figure 2.17*b*) or **glycogen,** that are readily tapped as a source of energy and other metabolic needs. Because a water molecule is required for breaking the bond between two glucose molecules, digestion is also termed **hydrolysis.*** Starch is the primary storage food of green plants, microscopic algae, and some fungi; glycogen (animal starch) is a stored carbohydrate for animals and certain groups of bacteria and protozoa.

2.6 Molecules of Life: Lipids

 Expected Learning Outcomes

22. Define *lipid, triglyceride, phospholipid, fatty acid,* and *cholesterol.*

23. Describe how an ester bond is formed.

24. Discuss major functions of lipids in cells.

The term **lipid,** derived from the Greek word *lipos,* meaning fat, is not a chemical designation but an operational term for a variety of substances that are not soluble in polar solvents such as water (recall that oil and water do not mix) but will dissolve in nonpolar solvents such as benzene and chloroform. This property occurs because the substances we call lipids contain relatively long or complex C—H (hydrocarbon) chains that are nonpolar and thus hydrophobic. The main groups of compounds classified as lipids are triglycerides, phospholipids, steroids, and waxes.

Important storage lipids are the **triglycerides,** a category that includes fats and oils. Triglycerides are composed of a single molecule of glycerol bound to three **fatty acids (figure 2.18). Glycerol** is a 3-carbon alcohol[7] with three OH groups that serve as binding sites. Fatty acids are long-chain unbranched hydrocarbon molecules with a carboxyl group (COOH) at one end that is free to bind to the glycerol. The bond that forms between the —OH group and the —COOH is defined as an **ester bond** (figure 2.18*a*). The hydrocarbon portion of a fatty acid can vary in length from 4 to 24 carbons and, depending on the fat, it may be *saturated* or *unsaturated*. A saturated fatty acid has all of the carbons in the chain bonded to hydrogens with single bonds. Fatty acids having at least one carbon—carbon double bond are considered unsaturated (figure 2.18*b*). Fats that contain such fatty acids are described with these terms as well. The structure of fatty acids is what gives fats and oils (liquid fats) their greasy, insoluble nature. In general, solid fats (such as beef tallow) are more saturated, and oils (or liquid fats) are more unsaturated.

In most cells, triglycerides are stored in long-term concentrated form as droplets or globules. When the ester linkage is acted on by digestive enzymes called lipases, the fatty acids and glycerol are freed to be used in metabolism. Fatty acids are a superior source of energy, yielding twice as much per gram as other storage molecules (starch). Soaps are K[+] or Na[+] salts of fatty acids whose qualities make them excellent grease removers and cleaners (see chapter 11).

Membrane Lipids

The **phospholipids** serve as a major structural component of cell membranes. Although phospholipids also contain glycerol and fatty acids, they differ significantly from triglycerides. Phospholipids contain only two fatty acids attached to the glycerol, and the third glycerol binding site holds a phosphate group. The phosphate is in turn bonded to an alcohol, which varies from one phospholipid to another **(figure 2.19*a*).** This class of lipids has a hydrophilic region from the charge on the phosphoric acid–alcohol "head" of the molecule and a hydrophobic region that corresponds to the

* *peptidoglycan* (pep-tih-doh-gly´-kan).
* *glycocalyx* (gly˝-koh-kay´-lix) Gr. *glycos,* sweet and *calyx,* covering.

7. Alcohols are hydrocarbons containing an —OH functional group.

Figure 2.18 Synthesis and structure of a triglyceride. (a) Because a water molecule is released at each ester bond, this is another form of dehydration synthesis. The jagged lines and R symbol represent the hydrocarbon chains of the fatty acids, which are commonly very long. **(b)** Structural models of two kinds of fatty acids. (1) A saturated fatty acid has long, straight chains that readily pack together. (2) An unsaturated fatty acid has at least one double bond and cannot pack densely due to the bends in the long chain caused by the double bonds. **(c)** Three-dimensional models of triglycerides containing saturated (1) and unsaturated fatty acids (2). Which of these two fats is an oil and which one is solid?

long, uncharged "tail" (formed by the fatty acids). When exposed to an aqueous solution, the charged heads are attracted to the water phase, and the nonpolar tails are repelled from the water phase **(figure 2.19b).** This property causes lipids to naturally assume single and double layers (bilayers), which contribute to their biological significance in membranes. When two single layers of polar lipids come together to form a double layer, the outer hydrophilic face of each single layer will orient itself toward the solution, and the hydrophobic portions will become immersed in the core of the bilayer. The structure of lipid bilayers confers characteristics on membranes such as selective permeability and fluid nature.

Miscellaneous Lipids

Steroids are complex ringed compounds commonly found in cell membranes and animal hormones. The best known of these is the sterol (meaning a steroid with an OH group) called **cholesterol** **(figure 2.20).** Cholesterol reinforces the structure of the cell mem-

brane in animal cells and in an unusual group of cell-wall-deficient bacteria called the mycoplasmas (see chapter 4).

The cell membranes of fungi also contain a unique sterol, called ergosterol. *Prostaglandins* are fatty acid derivatives found in trace amounts that function in inflammatory and allergic reactions, blood clotting, and smooth muscle contraction. Chemically, a *wax* is an ester formed between a long-chain alcohol and a saturated fatty acid. The resulting material is typically pliable and soft when warmed but hard and water-resistant when cold (paraffin, for example). Among living things, fur, feathers, fruits, leaves, human skin, and insect exoskeletons are naturally waterproofed with a coating of wax. Bacteria that cause tuberculosis and leprosy produce a wax (wax D) that contributes to their pathogenicity.

The Structure and Functions of Membranes

The word **membrane** appears frequently in descriptions of cells in chapters 4 and 5. The word itself can be used to indicate a lining or

Variable alcohol group

Phosphate

Charged head

Glycerol

Tail

Double bond creates a kink.

Fatty acids

(a)

Polar lipid molecule

Polar head

Nonpolar tails

Water

1. Phospholipids in single layer

Water — — Water

2. Phospholipid bilayer

(b)

Figure 2.19 Phospholipids—membrane molecules.
(a) A model of a single molecule of a phospholipid. The phosphate-alcohol head lends a charge to one end of the molecule; its long, trailing hydrocarbon chain is uncharged. (b) Phospholipids in water-based solutions become arranged (1) in single layers called micelles, with the charged head oriented toward the water phase and the hydrophobic nonpolar tail buried away from the water phase, or (2) in double-layered systems with the hydrophobic tails sandwiched between two hydrophilic layers.

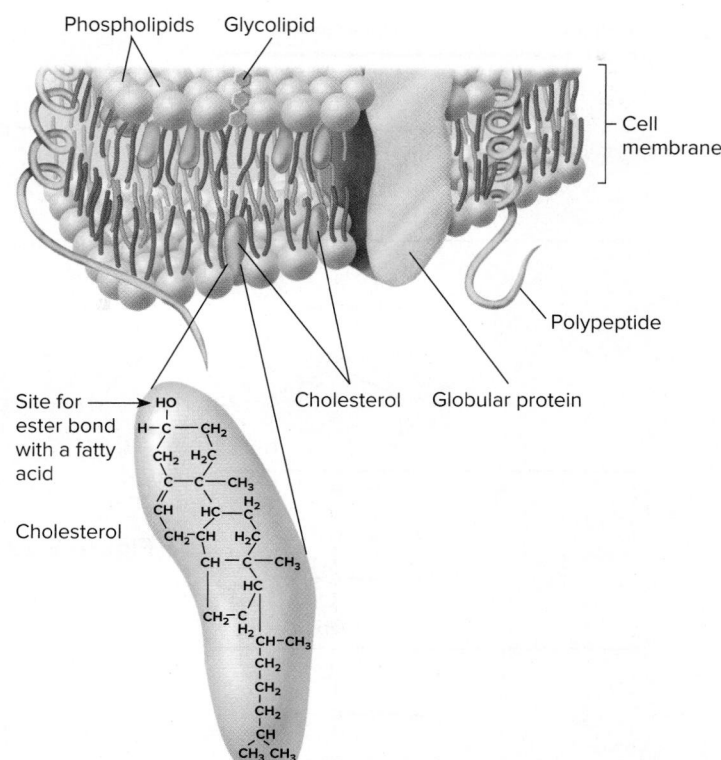

Phospholipids Glycolipid

Cell membrane

Polypeptide

Site for ester bond with a fatty acid

Cholesterol

Cholesterol Globular protein

Figure 2.20 Cutaway view of a membrane with its bilayer of lipids. The primary lipid is phospholipid; however, cholesterol is inserted in some membranes. Other structures are protein and glycolipid molecules. Cholesterol can become esterified with fatty acids at its OH^- group, imparting a polar quality similar to that of phospholipids.

covering including such multicellular structures as the mucous membranes of the body. At the cellular level, however, a membrane is a thin sheet of molecules composed of phospholipids and sterols (averaging about 40% of membrane content) and proteins (averaging about 60%). All cells have a membrane that completely encases the cytoplasm. Membranes are also components of eukaryotic organelles such as nuclei, mitochondria, and chloroplasts, and they appear in internal pockets of certain prokaryotic cells. Even some viruses, which are not cells, can have a membranous envelope.

A standard model of membrane structure has the lipids forming a continuous bilayer. The polar lipid heads face toward the aqueous phases, and the nonpolar tails orient toward the center of the membrane. Embedded at numerous sites in this bilayer are various-sized globular proteins (figure 2.20). Some proteins are situated only at the surface; others extend fully through the entire membrane.

Membranes are dynamic and constantly changing because the lipid phase is in motion and many proteins can migrate freely about. This fluidity is essential to such activities as engulfment of food and discharge or secretion by cells. The structure of the lipid phase provides an impenetrable barrier that accounts for the selective permeability and transport of molecules. Membrane

proteins function in receiving molecular signals (receptors), in binding and transporting nutrients, and as enzymes, topics to be discussed in chapters 7 and 8.

Check Your Progress SECTIONS 2.5–2.6

25. What is the structure of carbohydrates and glycosidic bonds?
26. Differentiate between mono-, di-, and polysaccharides, and give examples of each.
27. What are some of the functions of polysaccharides in cells?
28. Draw simple structural molecules of triglycerides and phospholipids to compare their differences and similarities. What is an ester bond?

2.7 Molecules of Life: Proteins

Expected Learning Outcomes

25. Describe the structures of peptides and polypeptides and how their bonds form.
26. Characterize the four levels of protein structure and describe the pattern of folding.
27. Summarize some of the essential functions of proteins.

Amino Acid	Structural Formula
Alanine	
Valine	
Cysteine	
Phenylalanine	
Tyrosine	

Figure 2.21 **Structural formulas of selected amino acids.** The basic structure common to all amino acids is shown in blue type; and the variable group, or R group, is placed in a colored box. Note the variations in structure of this reactive component.

Figure 2.22 **The formation of peptide bonds in a tetrapeptide.**

The predominant organic molecules in cells are **proteins,** a fitting term adopted from the Greek word *proteios,* meaning first or prime. To a large extent, the structure, behavior, and unique qualities of each living thing are a consequence of the proteins they contain. To best explain the origin of the special properties and versatility of proteins, we must examine their general structure.

The building blocks of proteins are **amino acids,** which exist in 20 different naturally occurring forms. **Figure 2.21** provides examples of several common amino acids. See Appendix A for a table that shows all 20. Various combinations of these amino acids account for the nearly infinite variety of proteins. Amino acids have a basic skeleton consisting of a carbon (called the α

carbon) linked to an amino group (NH_2), a carboxyl group (COOH), a hydrogen atom (H), and a variable R group. The variations among the amino acids occur at the R group, which is different in each amino acid and imparts the unique characteristics to the molecule and to the proteins that contain it. A covalent bond called a **peptide bond** forms between the amino group on one amino acid and the carboxyl group on another amino acid. As a result of peptide bond formation, it is possible to produce molecules varying in length from two amino acids to chains containing thousands of them.

Various terms are used to denote the nature of compounds containing peptide bonds. **Peptide*** usually refers to a molecule composed of short chains of amino acids, such as a dipeptide (two amino acids), a tripeptide (three), and a tetrapeptide (four) **(figure 2.22).** A **polypeptide** contains an unspecified number of amino acids but usually has more than 20 and is often a smaller subunit of a protein. A protein is the largest of this class of compounds and usually contains a minimum of 50 amino acids. It is common for the terms *polypeptide* and *protein* to be used interchangeably, though not all polypeptides are large enough to be considered proteins. In chapter 9 we will see that protein synthesis is not just a random connection of amino acids; it is directed by information provided in DNA.

Protein Structure and Diversity

The reason proteins are so varied and specific is that they do not function in the form of a simple straight chain of amino acids (called the primary structure). A protein has a natural tendency to assume more complex levels of organization, called the secondary, tertiary, and quaternary structures **(figure 2.23).**

The **primary (1°) structure** of a protein is simply the sequence of each amino acids in the polypeptide chain. Proteins vary extensively in the exact order, type, and number of amino acids, and it is this quality that gives rise to the unlimited diversity in protein form and function.

* *peptide* (pep′-tyd) Gr. *pepsis,* digestion.

1 The primary structure is a series of amino acids bound in a chain. Amino acids display small charged functional groups (red symbols).

Amino acids

Primary structure

2 The secondary structure develops when CO— and NH— groups on adjacent amino acids form hydrogen bonds. This action folds the chain into local configurations called the α helix and β-pleated sheet. Most proteins have both types of secondary structures.

β-pleated sheet

α helix

Secondary structure

Detail of hydrogen bond

3 The tertiary structure forms when portions of the secondary structure further interact by forming covalent disulfide bonds and additional interactions. From this emerges a stable three-dimensional molecule. Depending on the protein, this may be the final functional state.

Disulfide bond

S═S

Tertiary structure

Projected three-dimensional shape (note grooves and projections)

4 The quaternary structure exists only in proteins that consist of more than one polypeptide chain. The structure of the Shiga toxin, produced by *Shigella* bacteria, is a combination of separate polypeptides (shown in different colors).

Quaternary structure

Process Figure 2.23 **Formation of structural levels in a protein.** Red dashed lines indicate H bonds.
2.23(4): Image from the RCSB PDB (www.rcsb.org) of PDB ID 1R4Q (Fraser, M.E., Fujinaga, M., Cherney, M.M., Melton-Celsa, A.R., Twiddy, E.M., O'Brien, A.D., James, M.N.G. (2004) Structure of shiga toxin type 2 (Stx2) from Escherinchia coli O157:H7.)

A polypeptide does not remain in its primary state, but instead spontaneously arranges itself into a higher level of complexity called its **secondary (2°) structure.** The secondary structure arises from numerous hydrogen bonds occurring between the C═O and N—H groups of amino acids located near one another in the primary sequence. This bonding causes the whole chain to coil or fold into regular patterns. The coiled spiral form is called the α helix, and the folded, accordion form is called the β-pleated sheet. Polypeptides ordinarily will contain both types of configurations.

Once a chain has assumed the secondary structure, it goes on to form yet another level of folding and compacting—the **tertiary (3°) structure.** This structure arises through additional intrachain[8] forces and bonds between various parts of the α helix and β-pleated sheets. The chief actions in creating the tertiary structure are additional hydrogen bonds between charged functional groups, van der Waals forces between various parts of the polypeptide, and covalent disulfide bonds. The disulfide bonds occur between sulfur atoms on the amino acid **cysteine,*** and these bonds confer a high degree of stability to the overall protein structure. The result is a complex, three-dimensional protein that is now the completed functional state in many cases.

The most complex proteins assume a **quaternary (4°) structure,** in which two or more polypeptides interact to form a large, multiunit protein. The polypeptide units form loose associations based on weak van der Waals and other forces. The polypeptides in proteins with quaternary structure can be the same or different. The arrangement of these individual polypeptides tends to be symmetrical and will dictate the exact form of the finished protein (figure 2.23, step 4).

The most important outcome of bonding and folding is that each different type of protein develops a unique shape, and its surface displays a distinctive pattern of pockets and bumps. As a result, a protein can react only with molecules that complement or fit its particular surface features. Such a degree of specificity can provide the functional diversity required for many thousands of different cellular activities. **Enzymes** serve as the catalysts for all chemical reactions in cells, and nearly every reaction requires a different enzyme (see chapter 8). **Antibodies** are complex glycoproteins with specific regions of attachment for bacteria, viruses, and other microorganisms. Certain bacterial toxins (poisonous proteins) react with only one specific organ or tissue. Proteins embedded in the cell membrane have reactive sites restricted to a certain nutrient. Some proteins function as *receptors* to receive stimuli from the environment.

The functional, three-dimensional form of a protein is termed the *native state,* and if it is disrupted by some means, the protein is said to be *denatured.* Such agents as heat, acid, alcohol, and some disinfectants disrupt (and thus denature) the stabilizing intrachain bonds and cause the molecule to become nonfunctional, as described in chapter 11.

✔ *Check Your Progress* SECTION 2.7

29. Describe the basic structure of an amino acid and the formation of a peptide bond.

30. Differentiate between a peptide, a polypeptide, and a protein.

31. Explain what causes the various levels of structure of a protein molecule.

32. What functions do proteins perform in a cell?

2.8 Nucleic Acids: A Program for Genetics

 Expected Learning Outcomes

28. Identify a nucleic acid and differentiate between DNA and RNA.

29. Describe the structures of nucleotides and list the nitrogen bases.

30. Explain how the DNA code may be copied, and describe the basic functions of RNA.

The nucleic acids, **deoxyribonucleic acid (DNA)*** and **ribonucleic acid (RNA)*,** were originally isolated from the cell nucleus. Shortly thereafter, they were also found in other parts of nucleated cells, in cells with no nuclei (bacteria), and in viruses. The universal occurrence of nucleic acids in all known cells and viruses emphasizes their important roles as informational molecules. DNA, the master instruction manual of cells, contains a coded genetic program with detailed and specific instructions for each organism's heredity. It transfers the details of its program to RNA, "helper" molecules responsible for carrying out DNA's instructions and translating the DNA program into proteins that can perform life functions. For now, let us briefly consider the structure and some functions of DNA, RNA, and a close relative, adenosine triphosphate (ATP).

Both nucleic acids are polymers of repeating units called **nucleotides,*** each of which is composed of three smaller units: a **nitrogenous base,** a pentose (5-carbon) sugar, and a *phosphate* **(figure 2.24a).** The nitrogenous base is a cyclic compound that comes in two forms: *purines* (two rings) and *pyrimidines* (one ring). There are two types of purines—**adenine (A)** and **guanine (G)**—and three types of pyrimidines—**thymine (T), cytosine (C), and uracil (U) (figure 2.25).** A characteristic that differentiates DNA from RNA is that DNA contains all of the nitrogenous bases except uracil, and RNA contains all of the nitrogenous bases except thymine. The nitrogenous base is covalently bonded to the sugar *ribose* in RNA and *deoxyribose* (because it has one fewer oxygen than ribose) in DNA. Phosphate (PO_4^{3-}), a derivative of phosphoric acid (H_3PO_4), provides the final covalent bridge that connects sugars in series. Thus, the backbone of a nucleic acid strand is a chain of alternating phosphate-sugar-phosphate-sugar molecules, and the nitrogenous bases branch off the side of this backbone **(figure 2.24b, c).**

The Double Helix of DNA

DNA is a huge molecule formed by two very long polynucleotide (meaning that it consists of many nucleotides joined together) strands linked along their length by hydrogen bonds between complementary pairs of nitrogenous bases. The term *complementary* means that the nitrogenous bases pair in matched sets according to this

Quick Search
Look for the video "Assembly of DNA" on YouTube.

8. **Intra**chain means within the chain; **inter**chain would be between two chains.

 * *cysteine* (sis′-tuh-yeen) Gr. *Kystis,* sac. An amino acid first found in urine stones.

 * *deoxyribonucleic* (dee-ox″-ee-ry″-boh-noo-klay′-ik).

 * *ribonucleic* (ry″-boh-noo-klay′-ik) It is easy to see why the abbreviations are used!

 * *nucleotide* (noo′-klee-oh-tyd) From *nucleus* and *acid.*

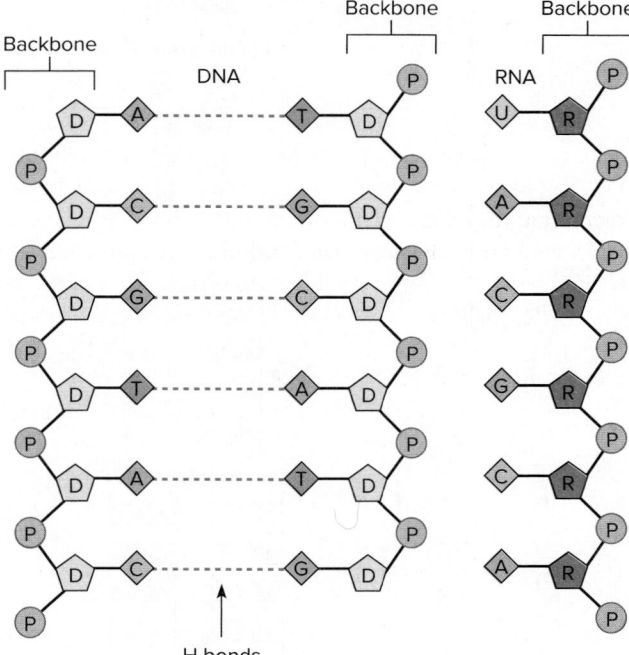

(a) A nucleotide, composed of a phosphate, a pentose sugar, and a nitrogenous base (either A,T,C,G, or U) is the monomer of both DNA and RNA.

(b) In DNA, the polymer is composed of alternating deoxyribose (D) and phosphate (P) with nitrogenous bases (A,T,C,G) attached to the deoxyribose. DNA almost always exists in pairs of strands, oriented so that the bases are paired across the central axis of the molecule.

(c) In RNA, the polymer is composed of alternating ribose (R) and phosphate (P) attached to nitrogenous bases (A,U,C,G), but it is usually a single strand.

Figure 2.24 **The general structure of nucleic acids.**

(a) **Pentose sugars**

(b) **Purine bases**

(c) **Pyrimidine bases**

Figure 2.25 **The sugars and nitrogenous bases that make up DNA and RNA. (a)** DNA contains deoxyribose, and RNA contains ribose. **(b)** A and G purine bases are found in both DNA and RNA. **(c)** Pyrimidine bases are found in both DNA and RNA, but T is found only in DNA, and U is found only in RNA.

pattern: Adenine ordinarily pairs with thymine, and cytosine with guanine. The bases are attracted in this way because each pair has oxygen, nitrogen, and hydrogen atoms positioned precisely to favor the formation of hydrogen bonds between them **(figure 2.26).**

For ease in understanding the structure of DNA, it is sometimes compared to a ladder, with the sugar-phosphate backbone representing the rails and the paired nitrogenous bases representing the steps. The flat ladder is useful for understanding basic components and orientation, but in reality DNA exists in a three-dimensional arrangement called a *double helix*. A better analogy may be a spiral staircase. In this model, the two strands (helixes or helices) coil together, with the sugar-phosphate forming outer ribbons, and the paired bases sandwiched between them (figure 2.26). As is true of protein, the structure of DNA is intimately related to its function. DNA molecules are usually extremely long, a feature that satisfies a requirement for storing genetic information in the sequence of base pairs the molecule contains. The hydrogen bonds between pairs can be disrupted when DNA is being copied, and the fixed complementary base pairing is essential to maintain the genetic code.

Making New DNA: Passing on the Genetic Message

The biological properties of cells and viruses are ultimately programmed by a master code composed of nucleic acids. This code is in the form of DNA in all cells and many viruses; a number of viruses are based on RNA alone. Regardless of the exact genetic makeup, both cells and viruses can continue to exist only if they can duplicate their genetic material and pass it on to subsequent generations. **Figure 2.27** summarizes the main steps in this process in cells.

During its division cycle, the cell has a mechanism for making a copy of its DNA by **replication,*** using the original strand as a pattern. Note that replication is guided by the double-stranded nature of DNA and the precise pairing of bases that create the master code. Replication requires the separation of the double strand into two single strands by an enzyme that helps to split the

* *replication* (reh″-plih-kay′-shun) A process that makes an exact copy.

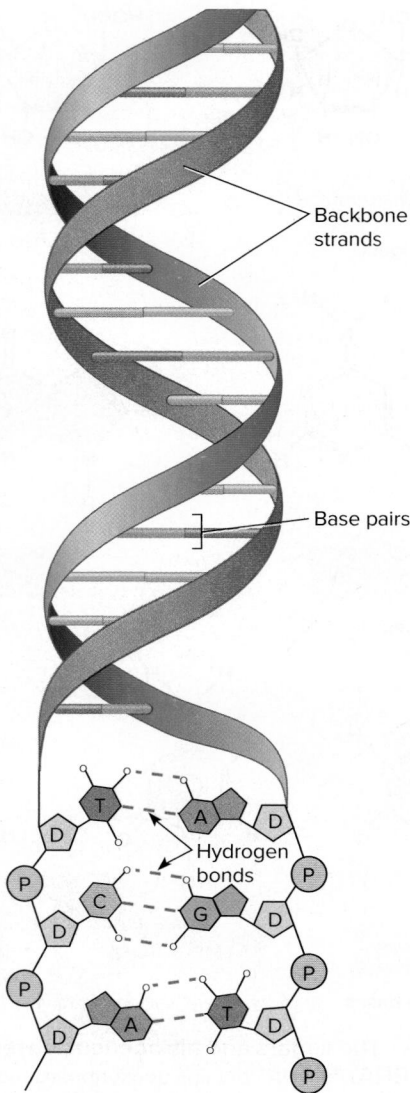

Figure 2.26 A structural representation of the double helix of DNA. At the bottom are the details of hydrogen bonds between the nitrogenous bases of the two strands.

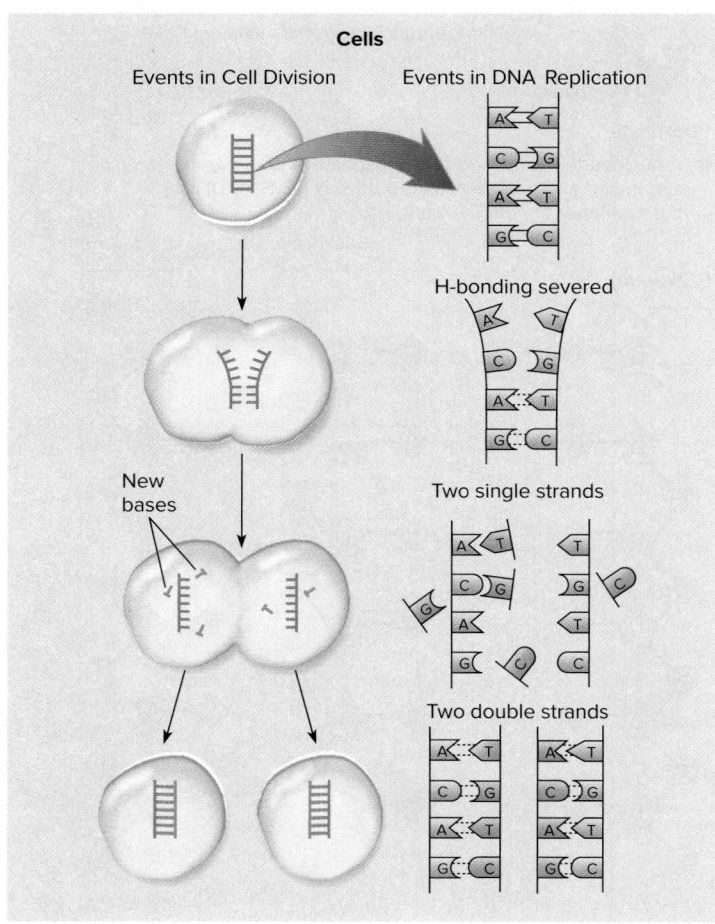

Figure 2.27 Simplified view of DNA replication in cells. The DNA in the cell's chromosome must be duplicated as the cell is dividing. This duplication is accomplished through the separation of the double DNA strand into two single strands. New strands are then synthesized using the original strands as guides to assemble the correct new complementary bases.

hydrogen bonds along the length of the molecule. This event exposes the base code and makes it available for copying. Complementary nucleotides are used to synthesize new strands of DNA that adhere to the pairing requirements of A—T and C—G. The end result is two separate double strands with the same order of bases as the original molecule. With this type of replication, each new double strand contains one of the original single strands from the starting DNA.

RNA: Organizers of Protein Synthesis

Like DNA, RNA is a polynucleotide that consists of a long chain of nucleotides. However, RNA is a single strand containing ribose sugar instead of deoxyribose and uracil instead of thymine (see figure 2.24). Several functional types of RNA are formed using the DNA template through a replication-like process. Three major types of RNA are important for protein synthesis. Messenger RNA

(mRNA) is a copy of a gene from DNA that provides instructions for the order of amino acids; transfer RNA (tRNA) is a carrier that delivers the correct amino acids during protein synthesis; and ribosomal RNA (rRNA) is a major component of ribosomes, which are the sites of protein synthesis. More information on these important processes is presented in chapter 9.

ATP: The Energy Molecule of Cells

A relative of RNA involved in an entirely different cell activity is **adenosine triphosphate (ATP).** ATP is a nucleotide containing adenine, ribose, and three phosphates rather than just one **(figure 2.28).** It belongs to a category of high-energy compounds (also including guanosine triphosphate, GTP) that give off energy when the bond is broken between the second and third (outermost) phosphate. The presence of these high-energy bonds makes it possible for ATP to release and store energy for cellular chemical reactions. Breakage of the bond of the terminal phosphate releases energy to do cellular work and also generates adenosine diphosphate (ADP). ADP can be converted back to ATP when the

(a)

(b)

Figure 2.28 Model of an ATP molecule, the chemical form of energy transfer in cells. (a) Structural formula: The wavy lines connecting the phosphates represent bonds that release large amounts of energy when broken. **(b)** Ball-and-stick model shows the arrangement of atoms in three dimensions.

third phosphate is restored, thereby serving as an energy depot. Carriers for oxidation-reduction activities (nicotinamide adenine dinucleotide [NAD], for instance) are also derivatives of nucleotides (see chapter 8).

 Check Your Progress SECTION 2.8

33. Describe a nucleotide and a polynucleotide, and compare and contrast the general structure of DNA and RNA.
34. Name the two purines and the three pyrimidines.
35. What are the functions of RNA?
36. What is ATP, and how does it function in cells?

 CASE STUDY **Part 2**

Needing to move from speculation to scientific facts, NASA started a series of missions in 2002 that could perform more detailed investigations. The most recent projects were the *Phoenix* probe in 2008 and the *Curiosity* probe in 2012. Each spacecraft was equipped with a one-ton roving science lab to analyze the Martian surface geology and soil. An important component of rover design was to explore more thoroughly whether conditions conducive to life exist on Mars now or did in the past. Rather than trying to isolate microorganisms, the Mars rover's instruments serve like remote senses—the mouth, nose, and eyes to "taste, sniff, and observe" the Martian soil and ice by measuring and assaying pH, salts, and the basic chemical makeup of its geology and atmosphere. The rover's robotic arms are able to excavate soil, rock, and ice. Other equipment can break up and vaporize rocks and perform detailed chemical analyses on the contents.

Initial results show that the soil is salty and slightly acidic. Past NASA probes had previously detected significant deposits of ice lying just below the surface. Then in September 2015 the *Mars Reconnaissance Orbiter* registered signs of liquid water and photographed streaks of water flowing downhill on craters during the warm season. This liquid is very salty, containing dissolved mixtures of common minerals.

A significant amount of methane also turned up in the samples, but more complex organic compounds have not yet been evident. Methane is formed by certain primitive prokaryotes but can also be the result of geologic activity. *Curiosity** is designed to monitor the methane content and possibly determine if it is from a biological or nonbiological source. A future mission to Mars is planned that will return samples to earth for more detailed analysis.

Quick Search
Discover other curious unknown features of Mars by googling "The 7 Biggest Mysteries of Mars."

- *Why is inorganic carbon such as CO_2 not by itself an indicator of life?*

- *For what reasons are complex organic carbon compounds considered strong evidence of living activities?*

To conclude this Case Study, go to Connect.

*To learn more about the findings of the *Curiosity* rover, go to http://mars.jpl.nasa.gov/msl/.

 Chapter Summary with Key Terms

2.1 Atoms: Fundamental Building Blocks of All Matter in the Universe
 A. Atomic Structure and Elements
 1. All **matter** in the universe is composed of minute particles called **atoms**—the simplest form of matter not divisible into a simpler substance by chemical means.

Atoms are composed of smaller particles called **protons, neutrons,** and **electrons.**
 2. a. Protons are positively (+) charged, neutrons are without charge, and electrons are negatively (−) charged.
 b. Protons and neutrons form the **nucleus** of the atom.
 c. Electrons orbit the nucleus in energy shells.

3. Atoms that differ in numbers of the protons, neutrons, and electrons are elements. **Elements** can be described by **mass number (MN),** equal to the number of protons and neutrons it has, and **atomic number (AN),** the number of protons in the nucleus, and each is known by a distinct name and symbol. Elements may exist in variant forms called **isotopes.** The **atomic mass or atomic weight** is equal to the average of the isotope mass numbers.

2.2 Bonds and Molecules
A. Atoms interact to form **chemical bonds** and **molecules.** If the atoms combining to make a molecule are different elements, then the substance is termed a **compound.**
 1. The type of bond is dictated by the electron makeup of the outer orbitals (**valence**) of the atoms. Bond types include:
 a. **Covalent bonds,** with shared electrons. The molecule is formed by one atom sharing its electrons with another atom; the balance of charge will be **polar** if unequal or **nonpolar** if equally shared/electrically neutral.
 b. **Ionic bonds,** where one atom transfers its electron(s) to another atom that can come closer to filling up its outer orbital. Dissociation of these compounds leads to the formation of charged **cations** and **anions.**
 c. **Hydrogen bonds** involve weak attractive forces between hydrogen (+ charge) and nearby oxygens and nitrogens (− charge).
 d. **Van der Waals forces** are also weak interactions between polarized zones of molecules such as proteins.
B. An **oxidation** is a loss of electrons and a **reduction** is a gain electrons. A substance that causes an oxidation by taking electrons is called an **oxidizing agent,** and a substance that causes a reduction by giving electrons is called a **reducing agent.**

2.3 Chemical Reactions, Solutions, and pH
A. Chemicals termed **reactants** can interact in a way to form different compounds termed **products.** Examples of reactions are synthesis and decomposition.
B. A **solution** is a combination of a solid, liquid, or gaseous chemical (the **solute**) dissolved in a liquid medium (the **solvent**). Water is the most common solvent in natural systems.
C. Ionization of water leads to the release of hydrogen ions (H^+) and hydroxyl (OH^-) ions. The **pH** scale expresses the concentration of H^+ such that a pH of less than 7.0 is considered **acidic,** and a pH of more than that, indicating fewer H^+, is considered **basic (alkaline).**

2.4 The Chemistry of Carbon and Organic Compounds
A. **Biochemistry** is the study of molecules that are found in living things. Most of these substances are **organic** compounds, which consist of carbon and hydrogen covalently bonded in various combinations. Some are **inorganic** compounds are composed of elements other than C and H.
B. Functional groups are small accessory molecules bonded to many organic compounds that impart unique characteristics.
C. **Macromolecules** are very large organic compounds and are generally assembled from single units called **monomers** by

polymerization. Molecules of life fall into basic categories of carbohydrates, lipids, proteins, and nucleic acids.

2.5 Molecules of Life: Carbohydrates
A. **Carbohydrates** are composed of carbon, hydrogen, and oxygen and contain aldehyde or ketone groups.
 1. **Monosaccharides** such as glucose are the simplest carbohydrates with 3 to 7 carbons; these are the monomers of carbohydrates.
 2. **Disaccharides** such as lactose consist of two monosaccharides joined by **glycosidic bonds.**
 3. **Polysaccharides** such as **starch** and **peptidoglycan** are chains of five or more monosaccharides.

2.6 Molecules of Life: Lipids
A. **Lipids** contain long hydrocarbon chains and are not soluble in polar solvents such as water due to their nonpolar, hydrophobic character. Common components of fats are **fatty acids,** elongate molecules with a carboxylic acid group. Examples are **triglycerides, phospholipids,** sterols, and waxes.

2.7 Molecules of Life: Proteins
A. **Proteins** are highly complex macromolecules that are crucial in most, if not all, life processes.
 1. **Amino acids** are the basic building blocks of proteins. They all share a basic structure of an amino group, a carboxyl group, an R group, and hydrogen bonded to a carbon atom. There are 20 different R groups, which define the basic set of 20 amino acids, found in all life forms.
 2. A **peptide** is a short chain of amino acids bound by **peptide bonds:** a **protein** contains at least 50 amino acids.
 3. The structure of a protein is very important to its function. This is described by the **primary structure** (the chain of amino acids), the **secondary structure** (formation of α helixes and β-sheets due to hydrogen bonding within the chain), **tertiary structure** (cross-links, especially disulfide bonds, between secondary structures), and **quaternary structure** (formation of multisubunit proteins). The incredible variation in shapes is the basis for the diverse roles proteins play as **enzymes, antibodies,** receptors, and structural components.

2.8 The Nucleic Acids: A Program for Genetics
A. **Nucleotides** are the building blocks of nucleic acids. They are composed of a **nitrogen base,** a pentose sugar, and a phosphate. Nitrogen bases are ringed compounds: **adenine (A), guanine (G), cytosine (C), thymine (T),** and **uracil (U).** Pentose sugars may be deoxyribose or ribose, depending on the type of nucleic acid.
B. **Deoxyribonucleic acid (DNA)** is a polymer of nucleotides that occurs as a double-stranded helix with hydrogen bonding between pairs of bases on the helices. It has all of the bases except uracil, and the pentose sugar is deoxyribose. DNA is the master code for a cell's life processes and must be transmitted to the offspring through **replication.**
C. **Ribonucleic acid (RNA)** is a polymer of nucleotides where the sugar is ribose and **uracil** is the replacement base for thymine. It is almost always found single-stranded and is used to express the DNA code into proteins.
D. **Adenosine triphosphate (ATP)** is a nucleotide with 3 phosphates that serves as the energy molecule in cells.

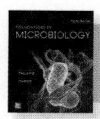

Level I. Knowledge and Comprehension

These questions require a working knowledge of the concepts in the chapter and the ability to recall and understand the information you have studied.

? Multiple-Choice Questions

Select the correct answer from the answers provided. For questions with blanks, choose the combination of answers that most accurately completes the statement.

1. The smallest unit of matter with unique characteristics is
 a. an electron
 b. a molecule
 c. an atom
 d. a proton

2. The _____ charge of a proton is exactly balanced by the _____ charge of a (an) _____.
 a. negative, positive, electron
 b. positive, neutral, neutron
 c. positive, negative, electron
 d. neutral, negative, electron

3. Electrons move around the nucleus of an atom in pathways called
 a. shells
 b. orbitals
 c. circles
 d. rings

4. Which parts of an element do not vary in number?
 a. electrons
 b. neutrons
 c. protons
 d. All of these vary.

5. If a substance contains two or more elements of different types, it is considered
 a. a compound
 b. a monomer
 c. a molecule
 d. organic

6. Bonds in which atoms share electrons are defined as _____ bonds.
 a. hydrogen
 b. ionic
 c. double
 d. covalent

7. A hydrogen bond can form between _____ adjacent to each other.
 a. two hydrogen atoms
 b. two oxygen atoms
 c. a hydrogen atom and an oxygen atom
 d. negative charges

8. An atom that can donate electrons during a reaction is called
 a. an oxidizing agent
 b. a reducing agent
 c. an ionic agent
 d. an electrolyte

9. In a solution of NaCl and water, NaCl is the _____ and water is the _____.
 a. acid, base
 b. base, acid
 c. solute, solvent
 d. solvent, solute

10. A solution with a pH of 2 _____ than a solution with a pH of 8.
 a. has less H^+
 b. has more H^+
 c. has more OH^-
 d. is less concentrated

11. Fructose is a type of
 a. disaccharide
 b. monosaccharide
 c. polysaccharide
 d. amino acid

12. Bond formation in polysaccharides and polypeptides is accompanied by the removal of a
 a. hydrogen atom
 b. hydroxyl ion
 c. carbon atom
 d. water molecule

13. The monomer unit of polysaccharides such as starch and cellulose is
 a. fructose
 b. glucose
 c. ribose
 d. lactose

14. A phospholipid contains
 a. three fatty acids bound to glycerol
 b. three fatty acids, a glycerol, and a phosphate
 c. two fatty acids and a phosphate bound to glycerol
 d. three cholesterol molecules bound to glycerol

15. Proteins are synthesized by linking amino acids with _____ bonds.
 a. disulfide
 b. glycosidic
 c. peptide
 d. ester

16. The amino acid that accounts for disulfide bonds in the tertiary structure of proteins is
 a. tyrosine
 b. glycine
 c. cysteine
 d. serine

17. DNA is a hereditary molecule that is composed of
 a. deoxyribose, phosphate, and nitrogen bases
 b. deoxyribose, a pentose, and nucleic acids
 c. sugar, proteins, and thymine
 d. adenine, phosphate, and ribose

18. What is meant by the term *DNA replication?*
 a. synthesis of nucleotides
 b. cell division
 c. interpretation of the genetic code
 d. the exact copying of the DNA code into two new molecules

19. Proteins can function as
 a. enzymes
 b. receptors

 c. antibodies
 d. all of these

20. RNA plays an important role in what biological process?
 a. replication
 b. protein synthesis
 c. lipid metabolism
 d. water transport

 Case Study Review

1. Which of the following has not been a major objective of Mars exploration?
 a. to analyze elements in soil
 b. to discover fossils
 c. to detect organic compounds
 d. to detect CO_2

2. What was a significant result of the Mars *Phoenix* project?
 a. culturing bacteria
 b. verifying water
 c. finding organic matter
 d. discovering precious metals

3. What other elements present in earth microorganisms would one expect to find on Mars?

 Writing Challenge

For each question, compose a one- or two-paragraph answer that includes the factual information needed to completely address the question. Check Your Progress questions can also be used for writing-challenge exercises.

1. Explain this statement: "All compounds are molecules, but not all molecules are compounds." Give an example.

2. What causes atoms to form chemical bonds? Why do some elements not bond readily?

3. Why are some covalent molecules polar and others nonpolar?

4. Explain why some elements are diatomic.

5. Exactly what causes the charges to form on atoms in ionic bonds?

6. Why are hydrogen bonds relatively weak?

7. What kind of substances will be expected to be hydrophilic and hydrophobic, and what makes them so?

8. How can a neutral salt be formed from acids and bases?

9. a. What characteristic of phospholipids makes them essential components of cell membranes?
 b. How are saturated and unsaturated fatty acids different?
 c. Why is the hydrophilic end of phospholipids attracted to water?

10. What makes the amino acids distinctive, and how many of them are there?

11. Why is DNA called a double helix?

12. Describe what occurs in a dehydration synthesis reaction.

13. How is our understanding of microbiology enhanced by a knowledge of chemistry?

Concept Mapping

On Connect you can find an Introduction to Concept Mapping that provides guidance for working with concept maps along with concept-mapping activities for this chapter.

Level II. Application, Analysis, Evaluation, and Synthesis

These problems go beyond just restating facts and require higher levels of understanding and an ability to interpret, problem solve, transfer knowledge to new situations, create models, and predict outcomes.

Critical Thinking

Critical thinking is the ability to reason and solve problems using facts and concepts. These questions can be approached from a number of angles, and in most cases, they do not have a single correct answer.

1. The "octet rule" in chemistry helps predict the types of bonds that atoms will form. In general, an atom will be most stable if it fills its outer shell of 8 electrons. Atoms with fewer than 4 valence electrons tend to donate electrons and those with more than 4 valence electrons tend to accept additional electrons; those with exactly 4 can do both. Using this rule, determine what category each of the following elements falls into: N, S, C, P, O, H, Ca, Fe, and Mg. (You will need to work out the valence of the atoms.)

2. Predict the kinds of bonds that occur in ammonium (NH_3), phosphate (PO_4), disulfide (S—S), and magnesium chloride ($MgCl_2$). (Use simple models, such as those in figure 2.4.)

3. Work out the following problems:
 a. Will an H bond form between H_3C—CH═O and H_2O? Draw a simple figure to support your answer.
 b. Draw the following molecules and determine which are polar: Cl_2, NH_3, CH_4.
 c. What is the pH of a solution with a concentration of 0.00001 moles/ml (M) of H^+?
 d. What is the pH of a solution with a concentration of 0.00001 moles/ml (M) of OH^-?
 e. Draw the atomic structure of magnesium and predict what kinds of bonds it will make.
 f. What kind of ion would you expect magnesium to make on the basis of its valence?

4. Distinguish between polar and ionic compounds.

5. Is galactose an aldehyde or a ketone sugar?

6. a. How many water molecules are released when a triglyceride is formed?
 b. How many peptide bonds are in a hexapeptide?

7. Observe figure 2.26 and explain why adenine forms hydrogen bonds only with thymine and why cytosine forms them only with guanine.

8. For a and b, explain your choice.
 a. Is butter an example of a saturated or an unsaturated fat?
 b. Is olive oil an example of a saturated or an unsaturated fat?
 c. Explain why sterols like cholesterol can add "stiffness" to membranes that contain them.

Visual Challenge

1. Figures 1 and 2 are both highly magnified views of biological substances. Using figure 2.17 as your basis for comparison, speculate which molecules are shown and give the reasons for them having the microscopic appearance we see here.

(1) (200X)
© Mediscan/Alamy Stock Photo

(2)
Courtesy of Professor Robert J. Linhardt, Rensselaer Polytechnic Institute

STUDY TIP: Quickly create a customized practice quiz for this chapter by going to your Connect SmartBook "My Reports" section. Don't have SmartBook and want to learn more? Go to **whysmartbook.com**.

3

Tools of the Laboratory
Methods of Studying Microorganisms

Meninges

Cerebrospinal fluid

Nasal cavity

Initial infection site

Palate

A clinical "picture of meningitis" would show meningococci (arrow) being engulfed by white blood cells in a sample of spinal fluid. These tiny cells can devastate a healthy person in a few hours. (photo): *Source:* CDC

CASE STUDY Part 1 Battling a Brain Infection

One Saturday evening in 2007, a 50-year-old woman named Kay Peterson began to suffer **flulike symptoms,** with **fever,** aching joints, sore throat, and a headache. Feeling miserable but not terribly concerned, she took some ibuprofen and went to bed. By the following morning, she began to feel increasingly ill and was unstable on her feet, **confused,** and complaining of **light-headedness.** Realizing this was more than just the flu, her husband rushed her immediately to the nearest emergency room.

An initial examination showed that most of Kay's vital signs were normal. Conditions that may provide some clues were rapid pulse and respiration, an **inflamed throat,** and a **stiff neck.** A **chest X ray** revealed no sign of **pneumonia,** and a blood test indicated an elevated **white blood cell** count. To rule out a possible brain infection, a puncture of the spinal canal was performed. As it turned out, the **cerebrospinal fluid (CSF)** the technician extracted appeared normal, microscopically and macroscopically.

Within an hour, Kay began to drift in and out of consciousness and was extremely lethargic. At one point, the medical team could

> "Intervention was truly a matter of life or death."

not find a pulse and noticed **dark brown spots** developing on her legs. When her condition appeared to be deteriorating rapidly, she was immediately taken to the intensive care unit and placed on **intravenous antibiotics.** One of the emergency room doctors later reported, "Her medical situation was so critical that our intervention was truly a matter of life or death."

Because Kay's symptoms pointed to a possible infection of the **central nervous system,** a second spinal puncture was performed. This time the spinal fluid looked cloudy. A **Gram stain** was performed right away, and cultures were started.

- *What are signs and symptoms of disease? Give examples from the case that appear to be the most diagnostically significant.*

- *What is the importance of a Gram stain in diagnosis of an infection like this?*

To continue the Case Study, go to Case Study Part 2 at the end of the chapter.

SCOPING OUT THE CHAPTER

Studying microorganisms in the laboratory is very different from studying elephants, trees, or insects. Microorganisms are invisible to the naked eye, are found everywhere, and typically must be raised in conditions quite different from their natural surroundings. In this chapter you'll get a brief overview of some general laboratory techniques carried out by microbiologists, oftentimes referred to as the six "I's". Procedures start at the central hub of specimen collection and flow from there to inoculation, incubation, and so on. But not all steps are always performed, nor do they necessarily proceed exactly in this order. Some investigators go right from sampling to microscopic inspection or from sampling to DNA testing. Others may require only inoculation and incubation on special media or test systems.

Identification
One goal of these procedures is to attach a name or identity to the microbe, using information gathered from inspection and investigation. Identification is accomplished through the use of keys, charts, and computer programs that analyze the data and arrive at a final conclusion.

Inoculation
The sample is placed into a container of medium that will support its growth. The medium may be solid or liquid, and held in tubes, plates, flasks, and even eggs. The delivery tool is usually a loop, needle, swab, or syringe.

Information Gathering
Additional tests for microbial function and characteristics may include inoculations into specialized media that determine biochemical traits, immunological testing, and genetic typing. Such tests will provide specific information unique to a certain microbe.

Specimen Collection
Microbiologists begin by sampling the object of their interest. It could be nearly any thing or place on earth. Very common sources are body fluids, foods, water, soil, plants, and animals, but even places like icebergs, volcanoes, and rocks can be sampled.

Incubation
Inoculated media are placed in a controlled environment (incubator) to promote growth. During the hours or days of this process, a culture develops as the visible growth of the microbes in the container of medium.

Inspection
Cultures are observed for the macroscopic appearance of growth characteristics. Cultures are examined under the microscope for basic details such as cell type and shape. This may be enhanced through staining and use of special microscopes.

Isolation
Some inoculation techniques can separate microbes to create isolated colonies that each contain a single type of microbe. This is invaluable for identifying the exact species of microbes in the sample, and it paves the way for making pure cultures.

3.1 Methods of Microbial Investigation

 Expected Learning Outcomes

1. Explain what unique characteristics of microorganisms make them challenging subjects for study.
2. Briefly outline the processes and purposes of the six types of procedures that are used in handling, maintaining, and studying microorganisms.

Biologists studying large organisms such as animals and plants can, for the most part, immediately see and differentiate their experimental subjects from the surrounding environment and from one another. In fact, they can use their senses of sight, smell, hearing, and even touch to detect and evaluate identifying characteristics and to keep track of growth and developmental changes.

Because microbiologists cannot rely as much as other scientists on senses other than sight, they are confronted by some unique problems. First, most habitats (such as the soil and the human mouth) harbor microbes in complex associations. It is often necessary to separate the organisms from one another so they can be identified and studied. Second, to maintain and keep track of such small research subjects, microbiologists usually will need to grow them under artificial conditions. A third difficulty in working with microbes is that they are not visible to the naked eye. This, coupled with their wide distribution, means that undesirable ones can be inconspicuously introduced into an experiment, where they may cause misleading results.

To deal with the challenges of their tiny and sometimes elusive targets, microbiologists have developed several types of procedures for investigating and characterizing microorganisms. These techniques can be summed up succinctly as the **six "I's": inoculation, incubation, isolation, inspection, information gathering,** and **identification (table 3.1).** Depending on the purposes of the particular investigator, these may be performed in several combinations and orders, but together they serve as the major tools of the microbiologist's trade. As novice microbiologists, most of you will be learning some basic microscope, inoculation, culturing, and identification techniques. Many professional researchers use more advanced investigation and identification techniques that may not even require growth or absolute isolation of the microbe in culture (see 3.1 Secret World of Microbes). If you examine the example from the **Case Study** in chapter 1, you will notice that the researchers used only some of the 6 "I's," whereas the case study in this chapter includes all of them in some form.

The first four sections of this chapter cover some of the essential concepts that revolve around the six "I's," but not necessarily in the exact order presented in table 3.1. Microscopes are so important to microbiological inquiry that we start out with the subject of microscopes, magnification, and staining techniques. This is followed by culturing procedures and media.

> **Quick Search**
> Watch "A Tour of the Microbiology Lab" on YouTube to see the daily work performed there.

 Check Your Progress SECTION 3.1

1. Name the notable features of microorganisms that have created a need for the specialized tools of microbiology.
2. In one sentence each, define what is involved in each of the six "I's."

TABLE 3.1	An Overview of Microbiology Techniques	
Technique	**Process Involves**	**Purpose and Outcome**
Inoculation	Placing a sample into a container of medium that supplies nutrients for growth and is the first stage in culturing	To increase visibility; makes it possible to handle and manage microbes in an artificial environment and begin to analyze what the sample may contain
Incubation	Exposing the inoculated medium to optimal growth conditions, generally for a few hours to days	To promote multiplication and produce the actual culture. An increase in microbe numbers will provide the higher quantities needed for further testing.
Isolation	Methods for separating individual microbes and achieving isolated colonies that can be readily distinguished from one another macroscopically*	To make additional cultures from single colonies to ensure they are pure; that is, containing only a single species of microbe for further observation and testing
Inspection	Observing cultures macroscopically for appearance of growth and microscopically for appearance of cells	To analyze initial characteristics of microbes in samples. Stains of cells may reveal information on cell type and morphology.
Information gathering	Testing of cultures using procedures that analyze biochemical and enzyme characteristics, immunologic reactions, drug sensitivity, and genetic makeup	To provide specific data and generate an overall profile of the microbes. These test results and descriptions will become key determinants in the last category, identification.
Identification	Analysis of collected data to help support a final determination of the types of microbes present in the original sample; this is accomplished by a variety of schemes	This lays the groundwork for further research into the nature and roles of these microbes; it can also provide numerous applications in infection diagnosis, food safety, biotechnology, and microbial ecology.

*Observable to the unaided or naked eye. This term usually applies to the cultural level of study.

3.2 The Microscope: Window on an Invisible Realm

 Expected Learning Outcomes

3. Describe the basic plan of an optical microscope, and differentiate between magnification and resolution.

4. Explain how the images are formed, along with the role of light and the different powers of lenses.

5. Indicate how the resolving power is determined and how resolution affects image visibility.

6. Differentiate between the major types of optical microscopes, their illumination sources, image appearance, and uses.

7. Describe the operating features of electron microscopes and how they differ from optical microscopes in illumination source, magnification, resolution, and image appearance.

8. Differentiate between transmission and scanning electron microsopes in image formation and appearance.

Imagine **Leeuwenhoek's** excitement and wonder when he first viewed a drop of rainwater and glimpsed an amazing microscopic world teeming with unearthly creatures. Beginning microbiology students still experience this sensation, and even experienced microbiologists remember their first view. The microbial existence is indeed another world, but it would remain largely uncharted without an essential tool: the microscope. Your efforts in exploring microbes will be more meaningful if you understand some essentials of **microscopy*** and specimen preparation.

Magnification and Microscope Design

The two key characteristics of a reliable microscope are **magnification,*** the ability to make objects appear enlarged, and **resolving power,** the ability to show detail.

A discovery by early microscopists that spurred the advancement of microbiology was that a clear glass sphere could act as a lens to magnify small objects. Magnification in most microscopes results from a complex interaction between visible light waves and the curvature of the lens. When a beam or ray of light transmitted through air strikes and passes through the convex surface of glass, it experiences some degree of **refraction,*** defined as the bending or change in the angle of the light ray as it passes through a medium such as a lens. The greater the difference in the composition of the two substances the light passes between, the more pronounced is the refraction. When an object is placed a certain distance from the spherical lens and illuminated with light, an optical replica, or **image,** of it is formed by the refracted light. Depending upon the size and curvature of the lens, the image appears enlarged to a particular degree, which is called its power of magnification and is usually identified with a number combined with × (read "times"). This behavior of light is evident if one looks through an everyday object such as a glass ball or a magnifying glass (**figure 3.1**). It is basic to the function of all

Figure 3.1 Effects of magnification. Demonstration of the magnification and image-forming capacity of clear glass "lenses." Given a proper source of illumination, a magnifying glass can deliver 2 to 10 times magnification. © Kathy Park Talaro

optical, or light, microscopes, though many of them have additional features that define, refine, and increase the size of the image.

The first microscopes were **simple,** meaning they contained just a single magnifying lens and a few working parts. Examples of this type of microscope are a magnifying glass, a hand lens, and Leeuwenhoek's basic little tool shown earlier in figure 1.8a. Among the refinements that led to the development of today's **compound** (two-lens) **microscope** were the addition of a second magnifying lens system, a lamp in the base to give off visible light and **illuminate*** the specimen, and a special lens called the **condenser** that converges or focuses the rays of light to a single point on the object. The fundamental parts of a modern compound light microscope are illustrated in **figure 3.2a.**

Principles of Light Microscopy

To be most effective, a microscope should provide adequate magnification, resolution, and clarity of image. Magnification of the object or specimen by a compound microscope occurs in two phases. The first lens in this system (the one closest to the specimen) is the objective lens, and the second (the one closest to the eye) is the ocular lens, or eyepiece (**figure 3.2b**). The objective forms the initial image of the specimen, called the **real image.** When the real image is projected to the plane of the eyepiece, the ocular lens magnifies it to produce a second image, the **virtual image.** The virtual image is the one that will be received by the eye and converted to a retinal and visual image. The magnifying power

* *microscopy* (mye-kraw'-skuh-pee) Gr. The science that studies microscope techniques.
* *magnification* (mag'-nih-fih-kay"-shun) L. *magnus,* great, and *ficere,* to make.
* *refraction* (ree-frakt', ree-frak'-shun) L. *refringere,* to break apart.

* *illuminate* (ill-oo'-mih-nayt) L. *illuminatus,* to light up.

of the objective alone usually ranges from 4× to 100×, and the power of the ocular alone ranges from 10× to 20×. The total power of magnification of the final image formed by the combined lenses is a product of the separate powers of the two lenses:

Power of Objective ×	Usual Power = of Ocular	Total Magnification
4× scanning objective	10× =	40×
10× low-power objective	10× =	100×
40× high dry objective	10× =	400×
100× oil immersion objective	10× =	1,000×

(a)

(b)

Figure 3.2 **(a) The main parts of a microscope.** This is a compound light microscope with two oculars (binocular) for viewing. Such microscopes usually have four objective lenses, a mechanical stage to move the specimen, a condenser, an iris diaphragm, and a built-in light source (lamp). **(b) Light pathway and image formation in light microscopes.** Light formed by the lamp is directed through the lamp filter and into the opening of the iris diaphragm; the condenser gathers the light rays and focuses them into a single point on the specimen; an enlarged image of the specimen—the real image—is next formed by the objective lens. This image is not seen but is projected into the ocular lens, which forms a final level of magnification called the virtual image. This image is observed in the field of view by the eye and perceived by the brain. Note that the lenses reverse the image and flip it upside down. (a) Courtesy of Nikon Instruments Inc., Melville, New York, USA, www.nikoninstruments.com

Microscopes are equipped with a nosepiece holding three or more objectives that can be rotated into position as needed. The power of the ocular usually remains constant for a given microscope. Depending on the power of the ocular, the total magnification of standard light microscopes can vary from 40× with the lowest-power objective (called the scanning objective) to 2,000×[1] with the highest-power objective (the oil immersion objective).

Resolution: Distinguishing Magnified Objects Clearly In addition to magnification, a microscope must also have adequate **resolution,** or **resolving power.** Resolution defines the capacity of an optical system to distinguish two adjacent objects or points from one another. For example, at a distance of 25 cm (10 in), the lens in the human eye can resolve two small objects as separate points just as long as the two objects are no closer than 0.2 mm apart. The eye examination given by optometrists is in fact a test of the resolving power of the human eye for various-size letters read at a distance of 20 feet. Because microorganisms are extremely small and usually very close together, they will not be seen with clarity or any degree of detail unless the microscope's lenses can resolve them.

A simple equation in the form of a fraction expresses the main mathematical factors that influence the expression of resolving power.

$$\text{Resolving power (RP)} = \frac{\text{Wavelength of light in nm}}{2 \times \text{Numerical aperture of objective lens}}$$

From this equation, it is evident that the resolving power is a function of the wavelength of light that forms the image, along with certain characteristics of the objective. The light source for optical microscopes consists of a band of colored wavelengths in the visible spectrum. The shortest visible wavelengths are in the violet-blue portion of the spectrum (400 nm), and the longest are in the red portion (750 nm). Because the wavelength must pass between the objects that are being resolved, shorter wavelengths (in the 400–500 nm range) will provide better resolution **(figure 3.3).**

The other factor influencing resolution is the **numerical aperture (NA),** a mathematical constant derived from the physical structure of the lens. This number represents the angle of light produced by refraction and is a measure of the quantity of light gathered by the lens. Each objective has a fixed numerical aperture value ranging from 0.1 in the lowest-power lens to approximately 1.25 in the highest-power (oil immersion) lens. Lenses with higher NAs provide better resolving power because they increase the angle of refraction and widen the cone of light entering the lens. For the oil immersion lens to arrive at its maximum resolving capacity, a drop of oil must be inserted between the tip of the lens and the specimen on the glass slide. Because immersion oil has the same optical qualities as glass, it prevents refractive loss that normally occurs as peripheral light passes from the slide into the air; this property effectively increases the numerical aperture **(figure 3.4).** There is an absolute limitation to resolution in optical microscopes, which can be demonstrated by calculating

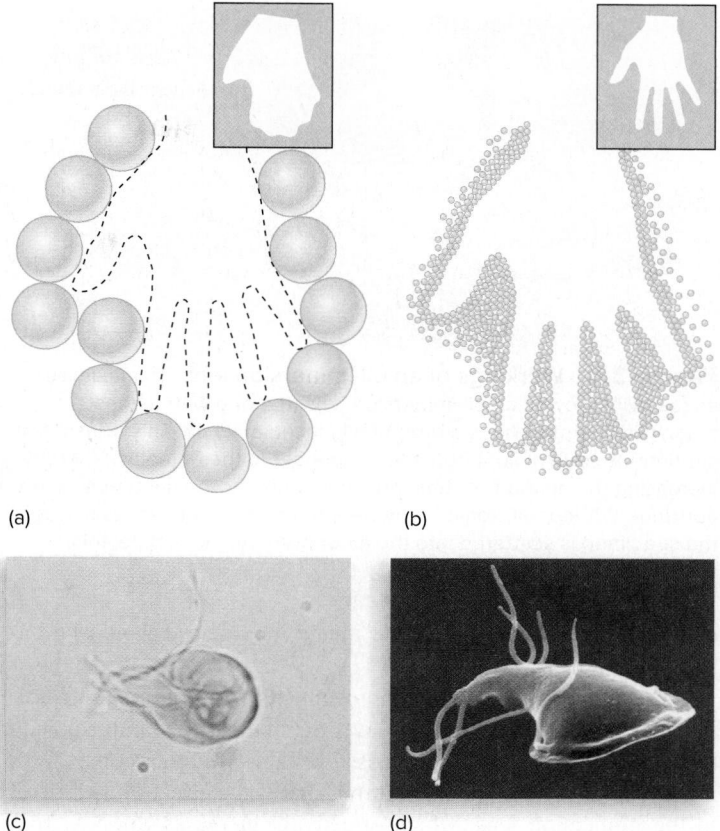

(a) (b)

(c) (d)

Figure 3.3 Effect of wavelength on resolution. A simple model demonstrates how the wavelength influences the resolving power of a microscope. Here an outline of a hand represents the object being illuminated, and two different sizes of beads represent the wavelengths of light. **(a)** The longer waves are too large to penetrate between the finer spaces, and they produce a fuzzy, undetailed image. **(b)** Shorter waves are small enough to enter small spaces and produce a much more detailed image that is recognizable as a hand. **(c)** The longer waves used by the light microscope produce an indistinct image of the protozoan parasite *Giardia*. **(d)** The much shorter waves of electrons used by the scanning electron microscope produces an image of the same organism with far greater detail. (c): *Source:* Dr. Mae Melvin/CDC; (d): *Source:* Dr. Stan Erlandsen; Dr. Dennis Feely/CDC

the resolution of the oil immersion lens using a blue-green wavelength of light:

$$\text{RP} = \frac{500 \text{ nm}}{2 \times 1.25}$$
$$= 200 \text{ nm (or } 0.2 \ \mu\text{m)}$$

In practical terms, this calculation means that the oil immersion lens can resolve any cell or cell part as long as it is at least 0.2 μm in diameter and that it can resolve two adjacent objects as long as they are no closer than 0.2 μm **(figure 3.5).** In general, organisms that are 0.5 μm or more in diameter are readily seen. This includes fungi and protozoa and some of their internal structures, and most bacteria. However, a few bacteria and most viruses are far too small to be resolved by the optical microscope and require electron microscopy. The factor that most limits the clarity of a microscope's image is its resolving power. Even if a light microscope were designed to magnify several thousand times, its resolving power

1. Some microscopes are equipped with 20× oculars or special annuli that can double their magnification.

Figure 3.4 **Workings of an oil immersion lens.** To maximize its resolving power, an oil immersion lens (the one with highest magnification) must have a drop of oil placed at its tip. This transmits a continuous cone of light from the condenser to the objective, thereby increasing the amount of light and, consequently, the numerical aperture. Without oil, some of the peripheral light that passes through the specimen is scattered into the air or onto the glass slide; this scattering decreases resolution.

could not be increased, and the image it produced would be enlarged but blurry.

Despite this limit, small improvements to resolution are possible. One is to place a blue filter over the microscope lamp, keeping the wavelength at the shortest possible value. Another comes from maximizing the numerical aperture. That is the effect of adding oil to the immersion lens, which effectively increases the NA from 1.25 to 1.4 and improves the resolving power to 0.17 μm.

Variations on the Optical Microscope

Optical microscopes that use visible light can be described by the nature of their **field of view,** or **field,** which is the circular area observed through the ocular lens. With special adaptations in lenses, condensers, and light sources, four special types of microscopes can be described: bright-field, dark-field, phase-contrast, and interference. A fifth type of optical microscope, the fluorescence microscope, uses ultraviolet radiation as the illuminating source, and a sixth, the confocal microscope, uses a laser beam. Each of these microscopes is adapted for viewing specimens in a particular way, as described in the next sections and summarized in **table 3.2.** Refer back to figure 1.4 for size comparisons of microbes and molecules.

Bright-Field Microscopy

The bright-field microscope is the most widely used type of light microscope. A bright-field microscope forms its image when light is transmitted through the specimen. The specimen, being denser and more opaque than its surroundings, absorbs some of this light, and the rest of the light is transmitted directly up through the ocular into the field. As a result, the specimen will produce an image that is darker than the surrounding brightly illuminated field. The bright-field microscope is a multipurpose instrument that can be used for both live, unstained material and preserved, stained material. The bright-field image is compared with that of other microscopes in table 3.2.

Dark-Field Microscopy

A bright-field microscope can be adapted as a dark-field microscope by adding a special disc called a *stop* to the condenser. The stop blocks all light from entering the objective lens except peripheral light that is reflected off the sides of the specimen itself. The resulting image is brightly illuminated specimens surrounded by a dark (black) field (table 3.2). The most effective use of dark-field microscopy is to visualize living cells that would be distorted by

Figure 3.5 **Effect of resolution on image visibility.** Comparison of objects that would not be resolvable versus those that would be resolvable under oil immersion at 1,000X magnification. Note that in addition to differentiating two adjacent things, good resolution also means being able to observe an object with clarity.

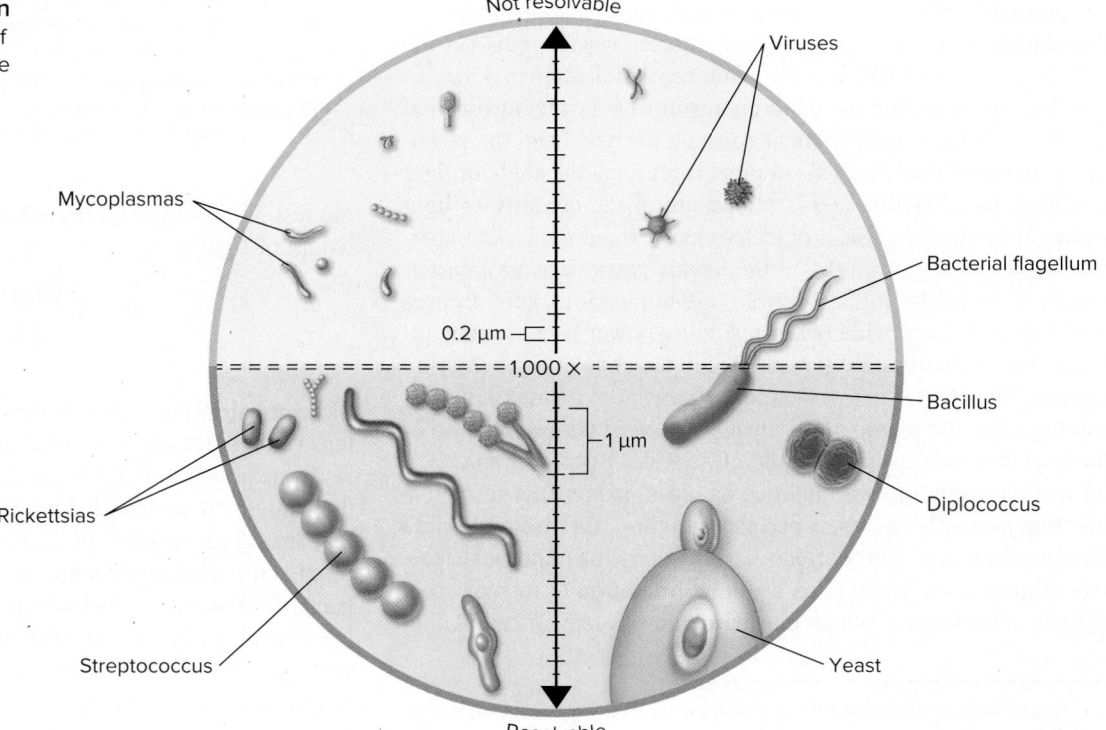

TABLE 3.2 Comparisons of Types of Microscopy

A. Comparative Images from Optical Microscopes

The subject here is yeast cells examined at 1,000× with four types of microscopes. Notice the differences in the appearance of the field and the degree of detail shown in the four methods.

I. Microscopes with visible light illumination
Maximum effective magnification = 1,000× to 2,000×*.
Maximum resolution = 0.2 μm.

Bright-field microscope	Dark-field microscope	Phase-contrast microscope	Differential interference contrast microscope
			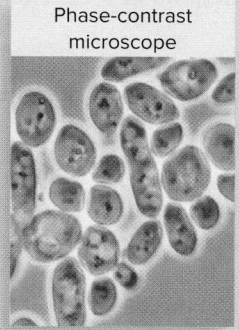
Common multipurpose microscope for live and preserved stained specimens; specimen is dark, field is white; provides fair cellular detail.	Best for observing live, unstained specimens; specimen is bright, field is black; provides outline of specimen with reduced internal cellular detail.	Used for live specimens; specimen is contrasted against gray background; excellent for internal cellular detail.	Provides very detailed, highly contrasting, three-dimensional images of live specimens.

B. Modifications of Optical Microscopes with Specialized Functions

II. Microscopes with ultraviolet or laser beam illumination
Maximum effective magnification = 1,000× to 2,000×*.
Maximum resolution = 0.2 μm**.

Fluorescent Microscope

Cells of Yersinia pestis, the bacterium responsible for plague, viewed at 200X. Ultraviolet light is used to illuminate the specimen. Antibodies labeled with fluorescent dyes emit visible light only if the antibody recognizes and binds to the cell. The specificity of this technique makes it a superior diagnostic tool.

Confocal Microscope

Two views of Paramecium 1,500X, stained by fluorescent dyes, and scanned by a laser beam, form multiple images that are combined into a three-dimensional image. Confocal microscopes may utilize visible or ultraviolet light. Note the fine detail of cilia and organelles.

C. Electron Microscopes Image Objects with a Beam of Electrons

III. Microscopes use a beam of electrons
Maximum effective magnification TEM = 1,000,000×.
Maximum effective magnification SEM = 650,000×.
Maximum resolution TEM = 0.5 nm.
Maximum resolution SEM = 10 nm.

Transmission electron microscope (TEM)	Colorized transmission electron microscope image	Scanning electron microscope (SEM)
Influenza virus particle viewed with a transmission electron microscope (TEM). This type of microscope provides the best resolution and can be used to view the internal structure of cells and viruses.	Colorized version of the same TEM image. Electron microscopes produce only black and white images, but these are often artificially colored, with the aid of a computer.	Artificially colorized scanning electron microscope (SEM) image of Middle East respiratory syndrome Coronavirus (MERS CoV) viral particles (colored yellow) on the surface of a eukaryotic cell (colored blue). The electron beam scans over the surface of the cell to produce a three-dimensional image.

*The 2,000× maximum is achieved through the use of a 20× ocular or 2× annulus.
**Three chemists recently developed a super-resolved fluorescent microscope that can resolve to 50 nm (0.05 μm).

(A1-3): © Prof. Dr. Heribert Cypionka, www.microbiological-garden.net; (A4): *Source:* Pawel Zienowicz; (B1): *Source:* Dr. Sherif Zaki; Dr. Kathi Tatti; Elizabeth White/CDC; (B2): © Anne Fleury; (C1-2): *Source:* Erskine. L. Palmer, Ph.D.; M. L. Martin/Frederick Murphy/CDC; (C3): *Source:* National Institute of Allergy and Infectious Diseases (NIAID)/CDC

drying or heat, or cannot be stained with the usual methods. It can outline the organism's shape and permit rapid recognition of swimming cells that may appear in fresh specimens, but it does not reveal fine internal details.

Phase-Contrast and Interference Microscopy

If similar objects made of clear glass, ice, and transparent plastic are immersed in the same container of water, an observer would have difficulty telling them apart because they have similar optical properties. In the same way, internal components of a live, unstained cell also lack sufficient contrast to be distinguished readily. The phase-contrast microscope contains devices that transform the subtle changes in light waves passing through the specimen into differences in light intensity. For example, thicker cell parts such as organelles alter the pathway of light to a greater extent than thinner regions such as the cytoplasm, creating contrast. The amount of internal detail visible by this method is greater than by either bright-field or dark-field methods (table 3.2). The phase-contrast microscope is most useful for observing intracellular structures such as bacterial spores, granules, and organelles, as well as the locomotor structures of eukaryotic cells.

Like the phase-contrast microscope, the **differential interference contrast (DIC)** microscope provides a detailed view of unstained, live specimens by manipulating the light. But this microscope has additional refinements that can add contrasting colors to the image. DIC microscopes produce extremely well-defined images that are often vividly colored and appear three-dimensional (table 3.2).

Scanning Confocal Microscopes

Simple optical microscopes may be unable to form a clear image at higher magnifications, because samples are often too thick for conventional lenses to focus all levels of cells simultaneously. This is especially true of larger cells with complex internal structures. Instead of using traditional illumination *scanning confocal microscopes* use a laser beam of light to scan various depths in the specimen and deliver a sharp image focusing on just a single plane. This produces an image that is a high-definition "slice" of a sample. A computer is then used to assemble the slices into a high-resolution, three-dimensional image. It is most often used on fluorescently stained specimens, but it can also be used to visualize live unstained cells and tissues (table 3.2).

Fluorescence Microscopy

The fluorescence microscope is a specially modified compound microscope furnished with an ultraviolet (UV) radiation source and filters that protect the viewer's eye from injury by these dangerous rays. The name for this type of microscopy is based on the use of certain dyes (acridine, fluorescein) and minerals that show **fluorescence.** This means that the dyes give off visible light when bombarded by shorter ultraviolet rays. For an image to be formed, the specimen must first be coated or placed in contact with a source of fluorescence. Subsequent illumination by ultraviolet radiation causes the specimen to emit visible light, producing an intense blue, yellow, orange, or red image against a black field.

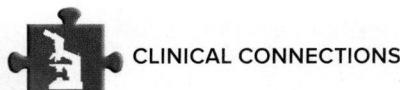 **CLINICAL CONNECTIONS**

Fluorescent Techniques in Identification

Fluorescence microscopy has its most useful applications in diagnosing infections caused by specific bacteria, protozoans, and viruses. A staining technique with fluorescent dyes is commonly used to detect *Mycobacterium tuberculosis* (the agent of tuberculosis) in patients' specimens (see figure 19.19). In a number of diagnostic procedures, fluorescent dyes are bound to specific antibodies—immunological proteins that recognize and tightly bind to a specific cells. These *fluorescent antibodies* can be used to identify the causative agents in such diseases as syphilis, chlamydia, trichomoniasis, herpes, and influenza. **Figure 3.6** displays the results of a fluorescent antibody technique used to highlight the spirochete of syphilis. A related technology uses fluorescent nucleic acid probes to differentiate between live and dead cells in mixtures or detect uncultured cells (see 3.1 Secret World of Microbes). A fluorescence microscope can be handy for locating specific microbes in complex mixtures because only those cells targeted by the technique will fluoresce.

Explain why fluorescent microscopic techniques can produce more specific results than other light microscopes. *Answer available on Connect.*

Figure 3.6 **Fluorescent staining of a sample of *Treponema pallidum,* the causative agent of syphilis.** The specific fluorescent antibodies that attach only to the pathogen and fluoresce a bright green make it a valuable diagnostic stain. Note the spiral nature of the cells (1,000X). *Source:* Russell/CDC

Electron Microscopy

If conventional light microscopes are our windows on the microscopic world, then the electron microscope (EM) is our window on the tiniest details of that world. Unlike light microscopes, the electron microscope forms an image with a beam of electrons that can be made to travel in wavelike patterns when accelerated to high speeds. These waves are 100,000 times shorter than the waves of visible light. Knowing that resolving power is a function of wavelength makes it

| | | Electron |
Characteristic	Light or Optical	(Transmission)
Useful magnification	2,000×*	1,000,000× or more
Maximum resolution	200 nm (0.2 μm)	0.5 nm
Image produced by	Light rays	Electron beam
Image focused by	Glass objective lens	Electromagnetic objective lenses
Image viewed through	Glass ocular lens	Fluorescent screen
Specimen placed on	Glass slide	Copper mesh
Specimen may be alive	Yes	Usually not
Specimen requires special stains or treatment	Depends on technique	Yes
Shows natural color	Yes	No

TABLE 3.3 Comparison of Light Microscopes and Electron Microscopes

*This maximum requires a 20× ocular.

evident that the RP fraction and number will become very small. Indeed, it is possible to resolve atoms with an electron microscope, even though the practical resolution for most biological applications is approximately 0.5 nm. The degree of resolution allows magnification to be extremely high—usually between 5,000× and 1,000,000× for biological specimens and up to 5,000,000× in some applications. Its capacity for magnification and resolution makes the EM an invaluable tool for studying the finest structure of cells and viruses. If not for electron microscopes, our understanding of biological structure and function would still be in its early theoretical stages.

In fundamental ways, the electron microscope is similar to the optical microscope. It employs components analogous to, but not necessarily the same as, those in light microscopy **(table 3.3)**. For instance, it magnifies in stages by means of two lens systems, and it has a condensing lens, a specimen holder, and a focusing apparatus. Otherwise the two types have numerous differences (table 3.3). An electron gun aims its beam through a vacuum to ring-shaped electromagnets that focus this beam on the specimen. Specimens must be pretreated with chemicals or dyes to increase contrast and usually cannot be observed in a live state. The enlarged image is displayed on a viewing screen or photographed for further study rather than being observed directly through an eyepiece. Because images produced by electrons lack color, electron micrographs (a micrograph is a photograph of a microscopic object) are always shades of black, gray, and white. The color-enhanced micrographs seen in this and other textbooks have computer-added color.

Two general forms of EM are the **transmission electron microscope (TEM)** and the **scanning electron microscope (SEM)** (see table 3.2). Transmission electron microscopes are the method of choice for viewing the detailed structure of cells and viruses. This microscope produces its image by transmitting electrons through the specimen. Because electrons cannot readily penetrate thick preparations, the specimen must be stained or coated with metals that will increase image contrast and sectioned into extremely thin slices (20–100 nm thick). The electrons passing through the specimen travel to the fluorescent screen and display a pattern or image. The darker and lighter areas of the image correspond to more and less dense parts on the specimen **(figure 3.7a)**. The TEM can also be used to produce negative images and shadow casts of whole microbes (see figure 6.3).

The scanning electron microscope provides some of the most dramatic and realistic images in existence. This instrument can create an extremely detailed three-dimensional view of all things biological—from dental plaque to tapeworm heads. To produce its images, the SEM bombards the surface of a whole, metal-coated specimen with electrons while scanning back and forth over it. A shower of electrons deflected from the surface is picked up with great accuracy by a sophisticated detector, and the electron pattern is displayed as an image on a monitor screen. The contours of the specimens resolved with scanning electron micrography are very revealing and often surprising. Areas that look smooth and flat with the light microscope display intriguing surface features with the SEM **(figure 3.7b)**. Improved technology has continued to refine electron microscopes and to develop variations on the basic plan.

Quick Search

Search the Internet for "25 Amazing Electron Microscope Images" for a unique visual experience.

(a)

(b)

Figure 3.7 Micrographs from two types of electron microscopes. (a) Colorized TEM image of the H1N1 influenza virus, displaying its internal contents. This was a new strain of virus first observed in 2009. **(b)** Colorized SEM image of two methicillin-resistant *Staphylococcus aureus* (MRSA) bacteria in the process of being phagocytized by a human neutrophil.
(a, b): Source: National Institute of Allergy and Infectious Diseases (NIAID)/CDC

3.1 MAKING CONNECTIONS

The Evolution in Resolution: Probing Microscopes

In the past, chemists, physicists, and biologists had to rely on indirect methods to gain information on the structures of the smallest molecules. But technological advances have created a new generation of microscopes that "see" atomic structure by actually feeling it. *Scanning probe microscopes* operate with a minute needle tapered to a tip that can be as narrow as a single atom! This probe scans over the exposed surface of a material and records an image of its outer texture. These revolutionary microscopes have such profound resolution that they have the potential to image single atoms and to magnify 100 million times. The scanning tunneling microscope (STM) was the first of these microscopes. It uses a tungsten probe that hovers near the surface of an object and follows its topography while simultaneously giving off an electrical signal of its pathway, which is then imaged on a screen. The STM is used primarily for detecting defects on the surfaces of electrical conductors and computer chips, but it has also provided the first incredible close-up views of DNA, the genetic material.

Another variety, the atomic force microscope (AFM), gently forces a diamond and metal probe down onto the surface of a specimen like a needle on a record. As it moves along the surface, any deflection of the metal probe is detected by a sensitive device that relays the information to an imager. The AFM is very useful in viewing the detailed functions of biological molecules such as antibodies and enzymes.

The latest versions of these microscopes have increased the resolving power to around 0.5 Å, which allowed technicians to image a pair of electrons! Such powerful tools for observing and positioning atoms have spawned a field called *nanotechnology*—the science of the "small." Scientists in this area use physics, chemistry, biology, and engineering to manipulate small molecules and atoms.

A section of graphene recently analyzed by the atomic force microscope, which clearly shows the carbon atoms that make up its structure. This image was so detailed that the bonds between the carbon atoms could be visualized and measured for the first time.

© IBM Research-Zurich

Working at these dimensions, researchers are currently creating tiny molecular tools to miniaturize computers and other electronic devices. A great deal of new research is also being devoted to the development of nanoparticles to deliver drugs, analyze DNA, and treat disease (look back at Clinical Connections, "Nanoparticles—Molecule Sized 'Smart' Pills," in chapter 2).

Looking back at figure 2.1, name some other chemical features that may become visible using these high-power microscopes. *Answer available on Connect.*

Some inventive relatives of the EM are the scanning probe microscope and atomic force microscope (**3.1 Making Connections**).

 Check Your Progress SECTION 3.2

3. Differentiate between the concepts of magnification, refraction, and resolution and explain how they contribute to the clarity of an image.
4. Briefly explain how an image is made and magnified.
5. On the basis of the formula for resolving power, explain why a smaller RP value is preferred to a larger one and explain what it means in practical terms if the resolving power is 1.0 μm.
6. What adjustments can improve a microscope's resolution?
7. Compare bright-field, dark-field, phase-contrast, confocal, and fluorescence microscopy as to field appearance, specimen appearance, light source, and uses.
8. Compare and contrast the optical compound microscope with the electron microscope.
9. Why is the resolution so superior in the electron microscope?
10. Compare the way that the image is formed in the TEM and the SEM.

3.3 Preparing Specimens for Optical Microscopes

 Expected Learning Outcomes

9. Explain the basic differences between fresh and fixed preparations for microscopy and how they are used.
10. Define dyes and describe the basic chemistry behind the process of staining.
11. Differentiate between negative and positive staining, giving examples.
12. Distinguish between simple, differential, and structural stains, including their applications.
13. Describe the process of Gram staining and how its results can aid the identification process.

A specimen for optical microscopy is generally prepared by mounting a sample on a suitable glass slide that sits on the stage between the condenser and the objective lens. The manner in which a slide

specimen, or mount is prepared depends upon: (1) the condition of the specimen, either in a living or preserved state; (2) the aims of the examiner, whether to observe overall structure, identify the microorganisms, or see movement; and (3) the type of microscopy available, whether it is bright-field, dark-field, phase-contrast, or fluorescence.

Fresh, Living Preparations

Live samples of microorganisms are used to prepare **wet mounts** so that they can be observed as near to their natural state as possible. The cells are suspended in a suitable fluid (water, broth, saline) that temporarily maintains viability and provides space and a medium for locomotion. A wet mount consists of a drop or two of the culture placed on a slide and overlaid with a cover glass. Although this preparation is quick and easy to make, it has certain disadvantages. The cover glass can damage larger cells, and the slide is very susceptible to drying and can contaminate the handler's fingers.

A more satisfactory alternative is the **hanging drop slide** (below) made with a special concave (depression) slide, an adhesive or sealant, and a coverslip from which a small drop of sample is suspended. These types of short-term mounts provide a true assessment of the size, shape, arrangement, color, and motility of cells. Even greater cellular detail can be observed with phase-contrast or interference microscopy.

Coverslip | Hanging drop containing specimen | Vaseline | Depression slide

Fixed, Stained Smears

A more permanent mount for long-term study can be obtained by preparing fixed, stained specimens. The smear technique, developed by Robert Koch more than 100 years ago, consists of spreading onto a slide a thin film made from a liquid suspension of cells and air-drying it. Next, the air-dried **smear** is usually heated gently by a process called **heat fixation** that simultaneously kills the specimen and secures it to the slide. Fixation also preserves various cellular components in a natural state with minimal distortion. Some types of fixation are performed with chemicals such as alcohol and formalin.

Like images on undeveloped photographic film, the unstained cells of a fixed smear are quite indistinct, no matter how great the magnification or how fine the resolving power of the microscope. The process of "developing" a smear to create contrast and make inconspicuous features stand out requires staining. Staining is any procedure that applies colored chemicals called dyes to specimens. Dyes impart a color to cells or cell parts by becoming affixed to them through a chemical reaction. In general, they are classified as **basic** (cationic) **dyes,** which have a positive charge, or **acidic** (anionic) **dyes,** which have a negative charge.

TABLE 3.4	Comparison of Positive and Negative Stains	
	Positive Staining	**Negative Staining**
Appearance of cell	Colored by dye	Clear and colorless
Background	Not stained (generally white)	Stained (dark gray or black)
Dyes employed	Basic dyes: Crystal violet Methylene blue Safranin Malachite green	Acidic dyes: Nigrosin India ink
Subtypes of stains	Several types: Simple stain Differential stains Gram stain Acid-fast stain Spore stain Structural stains One type of capsule stain Flagella stain Spore stain Granules stain Nucleic acid stain	Few types: Capsule Spore Negative stain of *Bacillus anthracis*, showing its capsule

(bottom): *Source:* Larry Stauffer, Oregon State Public Health Laboratory/CDC

Negative Versus Positive Staining

Two basic types of staining technique are used, depending upon how a dye reacts with the specimen (summarized in **table 3.4**). Most procedures involve a **positive stain,** in which the dye actually sticks to cells and gives them color. A **negative stain,** on the other hand, is just the reverse (like a photographic negative). The dye does not color the specimens but dries around its outer boundary, forming a silhouette. Nigrosin (a blue-black dye) and India ink (a black suspension of carbon particles) are the dyes most commonly used for negative staining. The cells themselves do not stain because these dyes are negatively charged and are repelled by the negatively charged surface of the cells. The value of negative staining is its relative simplicity and the reduced shrinkage or distortion of cells, as the smear is not heat fixed. A quick assessment can thus be made regarding cellular size, shape, and arrangement. Negative staining is also used to accentuate the capsule that surrounds certain bacteria and yeasts **(figure 3.8.).**

Staining Reactions of Dyes

Because many microbial cells lack contrast, it is necessary to use dyes to observe their detailed structure and identify them. Dyes are

colored compounds carrying double-bonded groups such as C=O, C=N, and N=N. When these groups are illuminated, they emit specific colors. Most dyes form ions when dissolved in a compatible solvent. The color-bearing ion, termed a **chromophore,** has a charge that attracts it to certain cell parts bearing the opposite charge.

Basic dyes carry a positively charged chromophore and are attracted to negatively charged cell components (nucleic acids and proteins). Because bacteria contain large amounts of negatively charged substances, they stain readily with basic dyes such as methylene blue, crystal violet, fuchsin, and safranin.

The chromophores of acidic dyes have a negative charge and therefore bind to areas of a cell carrying a positive charge. One example is eosin, a red dye used to stain blood cells. Because bacterial cells have numerous acidic substances and carry a slightly negative charge on their surface, they tend to repel acidic dyes.

(a) **Simple and Negative Stains**
Use one dye to observe cells

Methylene blue stain of *Corynebacterium* (1,000×)

Negative stain of spirilla bacteria made with nigrosin (500×)

(b) **Differential Stains**
Use two dyes to distinguish between cell types

Gram stain
Purple cells are gram-positive.
Red cells are gram-negative (1,000×).

Acid-fast stain
Red cells are acid-fast.
Blue cells are non-acid-fast (750×).

Spore stain, showing spores (green) and vegetative cells (red) (1,000×)

(c) **Structural Stains**
Special stains used to enhance cell details

Capsule stain of rod-shaped bacterium (1,000×).

Flagellar stain of *Proteus vulgaris.*
Note the fine fringe of flagella (1,500×)

Fluorescent stains of bacterial chromosome (green) and cell membrane (red) (1,500×)

Figure 3.8 Types of microbiological stains.
(a) Simple stains and negative stains. **(b)** Differential stains: Gram, acid-fast, and spore. **(c)** Structural stains: capsule, flagellar, and chromosomal. The spore stain (on right) is one method that fits both differential and structural categories.
(a1): © Harold J. Benson; (a2): Courtesy of *Tuatara: Journal of the Biological Society of Victoria University of Wellington.* J. E. Sheridan, Jan Steel and M. N. Loper. http://nzetc.victoria.ac.nz/tm/scholarly/Bio18Tuat03-fig-Bio18Tuat03_104a.html; (b1 left): Source: Bobby Strong/CDC; (b1 right): Source: CDC; (b2): © Richard J. Green/Science Source; (b3): © Steven R. Spilatro, Department of Biology, Marietta College, Marietta, OH; (c1): © Steven P. Lynch; (c2): © David Fankhauser; (c3): From I.A. Berlatzky, A. Rouvinski, S. Ben-Yehuda, "Spatial organization of a replicating bacterial chromosome," *PNAS,* 105(37):14136–14140, Fig. 4A. © 2008 by The National Academy of Sciences of the USA

Acidic dyes such as nigrosin and India ink can still be used successfully on bacteria using the negative stain method (table 3.4). With this technique, the dye settles around the cell and creates a dark background.

Simple Versus Differential Staining Positive staining methods are classified as simple, differential, or structural (see figure 3.8). **Simple stains** require only a single dye and an uncomplicated procedure, while **differential stains** use two different-colored dyes, called the *primary dye* and the *counterstain,* to distinguish between cell types or parts. These staining techniques tend to be more complex, sometimes requiring additional chemical reagents to produce the desired reaction.

Most simple staining techniques take advantage of the ready binding of bacterial cells to dyes like malachite green, crystal violet, basic fuchsin, and safranin. Simple stains cause all cells in a smear to appear more or less the same color, regardless of type, but they can still reveal characteristics such as shape, size, and arrangement. A simple stain with methylene blue is often used to stain granules in bacteria such as *Corynebacterium* (figure 3.8a), which can be a factor in identification.

Types of Differential Stains An effective differential stain uses dyes of contrasting color to clearly emphasize differences between two cell types or cell parts. Common combinations are red and purple, red and green, or pink and blue. Differential stains can also pinpoint other characteristics, such as the size, shape, and arrangement of cells. Typical examples include Gram, acid-fast, and endospore stains. Some staining techniques (spore, capsule) fall into more than one category.

Gram staining is a 130-year-old method named for its developer, **Hans Christian Gram.** Even today, it is an important diagnostic staining technique for bacteria. It permits ready differentiation of major categories based upon the color reaction of the cells: **gram-positive,** which are stained purple, and **gram-negative,** which are stained red (figure 3.8b). This difference in staining quality is due to structural variations found in the cell walls of bacteria (look ahead to chapter 4). The Gram stain is the basis of several important bacteriologic topics, including bacterial taxonomy, cell wall structure, identification, and drug therapy. As we will see in the second part of the case file for this chapter, it is a critical step in diagnosing meningitis. Gram staining is discussed in greater detail in **3.2 Making Connections.**

> **Quick Search**
>
> Search the Internet for "Microbiology Bytes: The Gram Stain" to see a demonstration of this technique.

The **acid-fast stain** is another commonly used differential stain. This stain originated as a specific method to detect *Mycobacterium tuberculosis* in specimens. These bacterial cells have compounds in their outer wall that hold fast (tightly) to the dye (carbol fuchsin) used in the acid-fast stain, even when washed with a solution containing a combination of acid and alcohol (figure 3.8b). After staining, acid-fast bacteria appear pink, while non-acid-fast bacteria are blue in color. This stain is used for other medically important mycobacteria such as the Hansen's disease (leprosy) bacillus and for *Nocardia,* an agent of lung or skin infections (see chapter 19).

The **endospore stain** (spore stain) is similar to the acid-fast method in that a dye is forced by heat into resistant survival cells called spores or endospores formed in response to adverse conditions (see chapter 4). This stain is designed to distinguish between spores and the vegetative cells that make them (figure 3.8b). Of significance in medical microbiology are the gram-positive, spore-forming members of the genus *Bacillus* (the cause of anthrax) and *Clostridium* (the cause of botulism and tetanus)—dramatic diseases of universal fascination that we consider in later chapters.

Structural stains are used to emphasize special cell parts such as capsules, endospores, and flagella that are not revealed by conventional staining methods. Capsule staining is a method of observing the microbial capsule, an unstructured protective layer surrounding the cells of some bacteria and fungi. Because the capsule does not react with most stains, it is often negatively stained with India ink, or it may be demonstrated by special positive stains. The fact that not all microbes exhibit capsules is a useful feature for identifying pathogens. One example is *Cryptococcus,* which causes a serious fungal meningitis in AIDS patients (figure 3.8c).

Flagellar staining is a method of revealing flagella, the tiny, slender filaments used by bacteria for locomotion. Because the width of bacterial flagella lies beyond the resolving power of the light microscope, in order to be seen, they must be enlarged by depositing a coating on the outside of the filament and then staining it. The presence, number, and arrangement of flagella can be helpful in identification of bacteria (figure 4.4).

 Check Your Progress SECTION 3.3

11. List the various staining methods, and briefly characterize each.
12. Explain what happens in positive staining to cause the reaction in the cell.
13. Explain what happens in negative staining that causes the final result.
14. For a stain to be considered differential, what must it do?
15. Outline the main differential stains and how they are used.

3.4 Additional Features of the Six "I's"

 Expected Learning Outcomes

14. Define *inoculation, media,* and *culture,* and describe sampling methods and instruments, and what aspects must be controlled.

15. Describe three basic techniques for isolation, including tools, media, incubation, and outcome.

16. Explain what an isolated colony is and indicate how it forms.

17. Differentiate between a pure culture, subculture, mixed culture, and contaminated culture. Define *contaminant.*

18. What kinds of data are collected during information gathering?

19. Describe some of the processes involved in identifying microbes from samples.

3.2 MAKING CONNECTIONS

The Gram Stain: A Grand Stain

In 1884, Hans Christian Gram discovered a staining technique that could be used to make bacteria in infectious specimens more visible. His technique consisted of timed, sequential applications of crystal violet (the primary dye), Gram's iodine (IKI, the mordant), an alcohol rinse (decolorizer), and a contrasting counterstain—usually, the red dye, safranin. This color choice provides differentiation between bacteria that stain purple, called gram-positive, and those that stain red, called gram-negative.

Although these staining reactions involve an attraction of the cell to a charged dye, it is important to note that the terms *gram-positive* and *gram-negative* are used to indicate not the electrical charge of cells or dyes but whether or not a cell retains the primary dye-iodine complex after decolorization. There is nothing specific in the reaction of gram-positive cells to the primary dye or in the reaction of gram-negative cells to the counterstain. The different results in the Gram stain are due to differences in the structure of the cell wall and how it reacts to the series of reagents applied to the cells.

In the first step, crystal violet stains cells in a smear all the same purple color. The second and key differentiating step is the mordant—Gram's iodine. The mordant causes the dye to form large crystals that get trapped by the meshwork of the cell wall. Because this layer in gram-positive cells is thicker, the entrapment of the dye is far more extensive in them than in gram-negative cells. Application of alcohol in the third step dissolves lipids in the outer membrane of the gram-negative cells, which removes the dye from them. By contrast, the crystals of dye tightly embedded in the gram-positive bacteria are relatively inaccessible and resistant to removal. Because gram-negative bacteria are colorless after decolorization, their presence is demonstrated by applying the counterstain safranin in the final step.

This staining method remains an important basis for bacterial classification and identification. It permits differentiation of four major

	Microscopic Appearance of Cell		Chemical Reaction in Cell (very magnified view)	
Step	Gram (+)	Gram (−)	Gram (+)	Gram (−)
1 Crystal violet (primary dye)			Both cell walls stain with the dye.	
2 Gram's iodine (mordant)			Dye crystals trapped in cell	No effect of iodine
3 Alcohol (decolorizer)			Crystals remain in cell.	Outer wall is weakened; cell loses dye.
4 Safranin (red dye counterstain)			Red dye has no effect.	Red dye stains the colorless cell.

categories based upon color reaction and shape: gram-positive rods, gram-positive cocci, gram-negative rods, and gram-negative cocci (see table 4.4). The Gram stain can also be a practical aid in diagnosing infection and in guiding drug treatment. For example, Gram staining a fresh urine or throat specimen can point to a preliminary cause of infection, and in some cases it is possible to begin drug therapy on the basis of this stain. Even in this day of elaborate and expensive medical technology, the Gram stain remains an important and unbeatable first tool in diagnosis.

Could the Gram stain be used to diagnose the flu? Why or why not?
Answer available on Connect.

Inoculation, Growth, and Identification of Cultures

To cultivate, or **culture,** * microorganisms, one introduces a tiny sample (the inoculum) into a container of **medium** * (pl. media), which provides an environment in which they multiply. This process is called **inoculation.** * The observable growth that later appears in or on the medium is known as a culture. The nature of the sample being cultured depends on the objectives of the analysis. Clinical specimens for determining the cause of an infectious disease are obtained from body fluids (blood, cerebrospinal fluid), discharges (sputum, urine, feces), or diseased tissue. Samples subject to microbiological analysis can include nearly any natural material. Some common ones are soil, water, sewage, foods, air, and inanimate objects. The important concept of media will be covered in more detail in section 3.5.

A past assumption has been that most microbes could be teased out of samples and cultured, given the proper media. But this view has had to be greatly revised (see **3.1 Secret World of Microbes**).

* *culture* (kul′-chur) Gr. *cultus,* to tend or cultivate. It can be used as a verb or a noun.
* *medium* (mee′-dee-um) pl. media; L. middle.
* *inoculation* (in-ok″-yoo-lay′-shun) L. *in,* and *oculus,* eye.

3.1 Secret World of Microbes
The Uncultured

For some time, microbiologists have had abundant evidence that culture-based methods are unable to identify many kinds of bacteria. This was first confirmed by environmental researchers, who came to believe that only about 1% (and in some environments, it was 0.001%) of microbes present in lakes, soil, and saltwater environments could be grown in laboratories by the usual methods and that, therefore, these were unknown and un-studied. These microbes are termed **viable but nonculturable,** or **VBNC.**

Microbiologists have spent several decades concocting recipes for media and having great success in growing many kinds of bacteria from many kinds of environments. Just identifying and studying those kept them very occupied. But the rise of specific, nonculturing tools based on genetic testing have led many researchers to identify microbes by analyzing their DNA alone. When these methods were used to sample various environments, they revealed a vast "jungle" of new species that had never before been cultured. These were the techniques that were used to study the bacteria isolated from Lake Whillans (see Case Study in chapter 1). This discovery seemed reasonable, because it may not yet be possible to exactly re-create the many correct media and conditions to grow such organisms in the lab.

Medical microbiologists have also missed a large proportion of microbes that are normal residents of the body. Stanford University scientists applied these techniques to subgingival plaque harvested from one of their own mouths. They used small, known fragments of DNA or probes that can highlight microbes in specimens. Oral biologists had previously recovered about 500 bacterial strains from this site; the Stanford scientists found 30 species that had never before been cultured or described. This discovery energized the medical world and spurred the use of non-culture-based methods to find VBNCs in the human body.

The new realization that our bodies are hosts to a wide variety of unknown microbes has several implications. As evolutionary microbiologist Paul Ewald has asked, "What are all those microbes doing in there?"

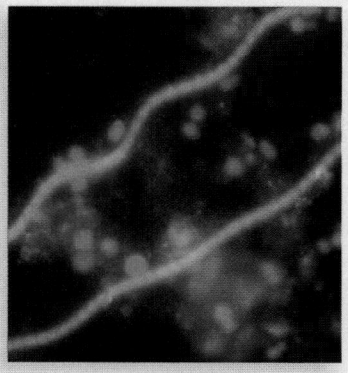

A fluorescent micrograph of a biofilm of *Vibrio cholerae* (orange cells) taken from a water sample in India. After an attempt to culture these bacteria over 495 days, a few cells enlarged, but most remained inactive. This pathogen appears to remain dormant in water, where it can be a source of cholera infection for people drinking the water.
© Dr. Rita Colwell & Dr. Munirul Alam

He points out that many oral microbes previously assumed to be innocuous are now associated with cancer and heart disease. Many of the diseases that we currently think of as noninfectious will likely be found to have an infectious cause once we continue to look for VBNCs.

In chapter 13 we will explore some of the findings from the Human Microbiome Project, which recently did a comprehensive study of the microbial content of several body sites. One important finding was the numerous roles many residents have in developing and maintaining the immune system.

Do you think that all of the VBNCs are really not culturable? Suggest some other possibilities. *Answer available on Connect.*

Some Special Requirements of Culturing

Inherent in these practices are the implementation of **sterile, aseptic,** and **pure culture**[2] techniques. Contamination during inoculation is a constant problem, so sterile techniques (media, transfer equipment) help ensure that only microbes that came from the sample are present. Another concern is the possible release of infectious agents from cultures into the environment, which is prevented by aseptic techniques. Many of these techniques revolve around keeping species in the pure culture form for further study, identification, or biotechnology applications.

Isolation Techniques

Certain **isolation** techniques are based on the concept that if an individual bacterial cell is separated from other cells and provided adequate space on a nutrient surface, it will grow into a discrete mound of cells called a **colony (figure 3.9).** Because it arises from a single cell or cluster of cells, an isolated colony consists of just one species. Proper isolation requires that a small number of cells be inoculated into a relatively large volume or over an expansive area of medium. It generally requires the following materials: a medium that has a relatively firm surface (see the description of agar in section 3.5) contained in a clear, flat covered plate called a *Petri dish,* and inoculating tools.

In the **streak plate method,** a small droplet of sample is spread with a tool called an **inoculating loop** over the surface of the medium according to a pattern that gradually thins out the sample and separates the cells spatially over several sections of the plate **(figure 3.10a, b).** Because of its ease and effectiveness, the streak plate is the method of choice for most applications.

In the **loop dilution,** or **pour plate, technique,** the sample is inoculated, also with a loop, into a series of cooled but still liquid agar tubes so as to dilute the number of cells in each successive tube in the series **(figure 3.10c, d).** Inoculated tubes are then plated out (poured) into sterile Petri dishes and are allowed to solidify

2. *Sterile* means the complete absence of viable microbes; *aseptic* refers to prevention of infection; *pure culture* refers to growth of a single species of microbe.

Mixture of cells in sample

Parent cells

Separation of cells by spreading or dilution on agar medium

Incubation

Growth increases the number of cells.

Microscopic view
Cellular level

Microbes become visible as isolated colonies containing millions of cells.

Macroscopic view
Colony level

Figure 3.9 Isolation technique. Stages in the formation of an isolated colony, showing the microscopic events and the macroscopic result. Separation techniques such as streaking can be used to isolate single cells. After numerous cell divisions, a macroscopic mound of cells, or a colony, will be formed. This is a relatively simple yet successful way to separate different types of bacteria in a mixed sample.

(harden). The number of cells per volume is so decreased that cells have ample space to form separate colonies in the second or third plate. One difference between this and the streak plate method is that in this technique, some of the colonies will develop deep in the medium itself and not just on the surface.

With the **spread plate technique,** a small volume of liquid from a diluted sample is pipetted onto the surface of the medium and spread around evenly by a sterile spreading tool (sometimes called a "hockey stick"). As with the streak plate, cells are spread over separate areas on the surface so that they can form individual colonies **(figure 3.10e, f).**

Once a container of medium has been inoculated, it is **incubated** in a temperature-controlled chamber (incubator) to encourage microbial growth. Although microbes have adapted to growth at temperatures ranging from freezing to boiling, the usual temperatures used in laboratory propagation fall between 20°C and 40°C. Incubators can also control the content of atmospheric gases such as oxygen and carbon dioxide that may be required for the growth of certain microbes. During the incubation period (ranging from a few hours to several weeks), the microbe multiplies and produces a culture with macroscopically observable growth. Microbial growth in a liquid medium materializes as cloudiness, sediment, a surface scum, or colored pigment. Growth on solid media may take the form of a spreading mat or separate colonies.

In some ways, culturing microbes is analogous to gardening. Cultures are formed by "seeding" tiny plots (media) with microbial cells. Extreme care is taken to exclude weeds (contaminants). A pure culture is a container of medium that grows only a single known species or type of microorganism **(figure 3.11c).** This type of culture is most frequently used for laboratory study, because it

allows the precise examination and control of one microorganism by itself. (Instead of the term *pure culture,* some microbiologists prefer the synonymous term *axenic culture.*) A standard method for preparing a pure culture is based on the **subculture** technique for making a second-level culture. A tiny sample of cells from a well-isolated colony is transferred into a separate container of media and incubated.

A **mixed culture (figure 3.11a)** is a container that holds two or more easily differentiated species of microorganisms, not unlike a garden plot containing both carrots and onions. A **contaminated culture (figure 3.11b)** has had **contaminants** (unwanted microbes of uncertain identity) introduced into it, like weeds into a garden. Because contaminants have the potential for disrupting experiments and tests, special procedures have been developed to control them, as you will no doubt witness in your own laboratory.

Identification Techniques

How does one determine what sorts of microorganisms have been isolated in cultures? Certainly microscopic appearance can be valuable in differentiating the smaller, simpler prokaryotic cells from the larger, more complex eukaryotic cells. Appearance can often be used to identify eukaryotic microorganisms to the level of genus or species because of their more distinctive appearance.

Bacteria are generally not as readily identifiable by these methods because very different species may appear quite similar. For them, we must include other techniques, some of which characterize their cellular metabolism. These methods, called **biochemical tests,** can determine fundamental chemical characteristics such as nutrient requirements, products given off during growth, presence of enzymes, and mechanisms for deriving energy. **Figure 3.12a** provides an example of a multiple-test, miniaturized system for obtaining biochemical characteristics. These tests are discussed in more depth in chapter 17 and in chapters that cover identification of pathogens.

By compiling macroscopic and microscopic traits together with the results of biochemical and physiological testing a complete picture of the microbe can be developed. A traditional pathway in bacterial identification uses flowcharts, also known as keys, that apply the results of tests to a selection process **(figure 3.12b).** The top of the key provides a general category from which to begin a series of branching points, usually into two pathways at each new junction. The points of separation are based on having a positive or negative result for each test. Following the pathway that fits the characteristics leads to an end point where a name for an organism is given. This process of "keying out" the organism can simplify the identification process. In some cases identification requires testing the isolated culture against known antibodies (immunologic testing) or inoculating a suitable laboratory animal. More and more, diagnostic tools are available that analyze genetic characteristics and can detect microbes based on their DNA or RNA (see chapters 10 and 17).

Note: This method works best if the spreading tool (usually an **inoculating loop**) is resterilized (flamed) after each of steps 1–3.

Loop containing sample

1 2 3 4

(a) **Steps in a Streak Plate;** this one is a four-part or quadrant streak.

(b)

Loop containing sample

1 2 3

1 2 3

(c) **Steps in Loop Dilution;** also called a pour plate or serial dilution

(d)

"Hockey stick"

1 2

(e) **Steps in a Spread Plate**

(f)

Figure 3.10 Methods for isolating bacteria. (a) Steps in a quadrant streak plate and **(b)** resulting isolated colonies of bacteria. **(c)** Steps in the loop dilution method and **(d)** the appearance of plate 3. **(e)** Spread plate and **(f)** its result. Techniques in (a) and (c) use a loopful of culture, whereas **(e)** starts with a pre-diluted sample. (b-f): © Kathy Park Talaro

Figure 3.11 Various states of cultures. (a) A mixed culture of *Micrococcus luteus* and *Escherichia coli* can be readily differentiated by their colors. **(b)** This plate of *Serratia marcescens* was overexposed to room air and has developed a large white colony. Because this intruder is not desirable and not identified, the culture is now contaminated. **(c)** Tubes containing pure cultures of *E. coli* (white), *M. luteus* (yellow), and *S. marcescens* (red) made by subculturing isolated colonies. (a-c): © Kathy Park Talaro

(a) (b) (c)

Figure 3.12 Rapid mini test system and identification key for bacterial isolates. (a) A test system for common gram-negative rods uses 20 small wells of media to be inoculated with a pure culture and incubated. A certain combination of positive and negative reactions (tabulated on the data sheet seen in the photograph) allows identification of an unknown isolate. **(b)** A sample of a key that uses test data (positive or negative reactions) to guide identification of genera and species related to *N. meningitidis*. This chart could be used to identify the isolate in the Case Study for this chapter.
(a): © John Watney/Science Source

(a)

Scheme for Differentiating Gram-Negative Cocci and Coccobacilli

(b)

 Check Your Progress SECTION 3.4

16. Clarify what is involved in inoculation, growth, and contamination.

17. Name two ways that pure, mixed, and contaminated cultures are similar and two ways that they differ from each other.

18. Explain what is involved in isolating microorganisms and why it may be necessary to do this.

19. Compare and contrast three common laboratory techniques for separating bacteria in a mixed sample.

20. Describe how an isolated colony forms.

21. Explain why an isolated colony and a pure culture are not the same thing.

3.5 Media: The Foundations of Culturing

 Expected Learning Outcomes

20. Explain the importance of media for culturing microbes in the laboratory.

21. Name the three general categories of media, based on their inherent properties and uses.

22. Compare and contrast liquid, solid, and semisolid media, giving examples.

23. Analyze chemically defined and complex media, describing their basic differences and content.

24. Describe functional media; list several different categories, and explain what characterizes each type of functional media.

25. Identify the qualities of enriched, selective, and differential media; use examples to explain their content and purposes.

26. Explain what it means to say that microorganisms are not culturable.

27. Describe live media and the circumstances that require it.

A major stimulus to the rise of microbiology in the late 1800s was the development of techniques for growing microbes out of their natural habitats and in pure form in the laboratory. This milestone enabled the close examination of a microbe and its morphology, physiology, and genetics. It was evident from the very first that for successful cultivation, the microorganisms being cultured had to be provided with all of their required nutrients in an artificial medium.

Some microbes require only a very few simple inorganic compounds for growth; others need a complex list of specific inorganic and organic compounds. This tremendous diversity is evident in the types of media that can be prepared. At least 500 different types of media are used in culturing and identifying microorganisms. Culture media are contained in test tubes, flasks, or Petri dishes, and they are inoculated by such tools as loops, needles, pipettes, and swabs. Media are extremely varied in nutrient content and consistency, and can be specially formulated for a particular purpose.

For an experiment to be properly controlled, sterile technique is necessary. This means that the inoculation must start with a sterile medium and inoculating tools with sterile tips must be used. Measures must be taken to prevent introduction of nonsterile materials, such as room air and fingers, directly into the media.

Types of Media

Most media discussed here are designed for bacteria and fungi, though algae and some protozoa can be propagated in media. Viruses can only be cultivated in live host cells. Media fall into three general categories based on their properties: physical state, chemical composition, and functional type **(table 3.5).**

(a)

(b)

Figure 3.13 Sample liquid media. (a) Liquid media tend to flow freely when the container is tilted. Urea broth is used to show a biochemical reaction in which the enzyme urease digests urea and releases ammonium. This raises the pH of the solution and causes the dye to become increasingly pink. Left: uninoculated broth, pH 7; middle: growth with no change; right: positive, pH 8.0. **(b)** Presence-absence broth is for detecting the presence of fecal bacteria in water samples. It contains lactose and bromcresol purple dye. As some fecal bacteria ferment lactose, they release acidic substances. This lowers the pH and changes the dye from purple to yellow (right is *Escherichia coli*). Nonfermenters such as *Pseudomonas* (left) grow but do not change the pH (purple color indicates neutral pH).
(a and b): © Kathy Park Talaro

Physical States of Media

Liquid media are defined as water-based solutions that do not solidify at temperatures above freezing and that tend to flow freely when the container is tilted **(figure 3.13).** These media, termed **broths,** milks, or infusions, are made by dissolving various solutes in distilled water. Growth occurs throughout the container and can then present a dispersed, cloudy, or flaky appearance. A common laboratory medium, *nutrient broth,* contains beef extract and peptone dissolved in water. Other examples of liquid media are methylene blue milk and litmus milk that contain whole milk and dyes. Fluid thioglycollate is a slightly viscous broth used for determining patterns of growth in oxygen (see figure 7.11).

At ordinary room temperature, **semisolid media** exhibit a clotlike consistency **(figure 3.14)** because they contain an amount of solidifying agent (agar or gelatin) that thickens them but does not produce a firm substrate. Semisolid media are used to determine the motility of bacteria and to

TABLE 3.5	Three Categories of Media Classification	
Physical State (Medium's Normal Consistency)	**Chemical Composition (Type of Chemicals Medium Contains)**	**Functional Type (Purpose of Medium)***
1. Liquid 2. Semisolid 3. Solid (can be converted to liquid) 4. Solid (cannot be liquefied)	1. Synthetic (chemically defined) 2. Nonsynthetic (complex; not chemically defined)	1. General purpose 2. Enriched 3. Selective 4. Differential 5. Anaerobic growth 6. Specimen transport 7. Assay 8. Enumeration

*Some media can serve more than one function. For example, a medium such as brain-heart infusion is general purpose and enriched; mannitol salt agar is both selective and differential; and blood agar is both enriched and differential.

(a)

1 2 3 4

(b)

Figure 3.14 Sample semisolid media. (a) Semisolid media have more body than liquid media but are softer than solid media. They do not flow freely and have a soft, clotlike consistency. **(b)** Sulfur indole motility (SIM) medium. When this medium is stabbed with an inoculum, the appearance of growth around the stab line can be used to determine nonmotility (2) or motility (3). Tube 1 is a control that has not been inoculated, and tube 4 shows a black precipitate that occurs around the stab when H_2S gas has been produced. (a and b): © Kathy Park Talaro

localize a reaction at a specific site. Motility test medium and sulfur indole motility (SIM) medium both contain a small amount (0.3% to 0.5%) of agar. In both cases, the medium is stabbed carefully in the center with an inoculating needle and later observed for the pattern of growth around the stab line. In addition to motility, SIM can test for physiological characteristics used in identification (hydrogen sulfide production and indole reaction).

Solid media provide a firm surface on which cells can form discrete colonies (see figure 3.10.) and are advantageous for isolating and culturing bacteria and fungi. They come in two forms: liquefiable and nonliquefiable. Liquefiable solid media, sometimes called reversible solid media, contain a solidifying agent that changes its physical properties in response to temperature. By far the most widely used and effective of these agents is **agar,**[3] a polysaccharide isolated from the red alga *Gelidium.* The benefits of agar

are numerous. It is solid at room temperature, and it melts (liquefies) at the boiling temperature of water (100°C or 212°F). Once liquefied, agar does not resolidify until it cools to 42°C (108°F), so it can be inoculated and poured in liquid form at temperatures (45°C to 50°C) that will not harm the microbes or the handler (body temperature is about 37°C or 98.6°F). Agar is flexible and moldable, and it provides a basic matrix to hold moisture and nutrients. Another useful property is that it is not readily digestible and thus not a nutrient for most microorganisms.

Any medium containing 1% to 5% agar usually has the word *agar* in its name. *Nutrient agar* is a common one. Like nutrient broth, it contains beef extract and peptone, as well as 1.5% agar by weight. Many of the examples covered in the section on functional categories of media contain agar.

Although gelatin is not nearly as satisfactory as agar, it will create a reasonably solid surface in concentrations of 10% to 15%. The main drawback for gelatin is that it can be digested by microbes and will melt at room and warmer temperatures, leaving a liquid. Agar and gelatin media are illustrated in **figure 3.15.**

Nonliquefiable solid media do not melt. They include materials such as rice grains (used to grow fungi), cooked meat media (good for anaerobes), and egg or serum media that are permanently coagulated or hardened by moist heat.

Chemical Content of Media

Media with a **chemically defined** composition are termed **synthetic.** Such media contain pure chemical nutrients that vary little from one source to another and have a molecular content specified by means of an exact formula. Synthetic media come in many forms. Some media, such as minimal media for fungi, contain nothing more than a few salts and amino acids dissolved in water. Others contain dozens of precisely measured ingredients **(table 3.6A).** Such standardized and reproducible media are most useful in research and cell culture. But they can only be used when the exact nutritional needs of the test organisms are known. For example, a medium developed to grow the parasitic protozoan *Leishmania* required 75 different chemicals.

If even one component of a given medium is not chemically definable, the medium is a **nonsynthetic, or complex,**[4] **medium (table 3.6B).** The composition of this type of medium cannot be described by an exact chemical formula. Substances that can make it nonsynthetic are extracts from animal or plant tissues, including ground-up cells and secretions. Other examples are blood, serum, meat extracts, infusions, milk, soybean digests, and peptone. Peptone is a partially digested protein, rich in amino acids, that is often used as a carbon and nitrogen source. Nutrient broth, blood agar, and MacConkey agar, though different in function and appearance, are all complex nonsynthetic media. They present a rich mixture of nutrients for microbes with complex nutritional needs.

Tables 3.6A and 3.6B provide a practical comparison of the two categories, based on media used to grow *Staphylococcus aureus.* Every substance in medium A is known to a very precise

3. This material was first employed by Dr. Hesse (see the Significant Events in Microbiology Table found at http://www.mhhe.com/talaro9, 1881).

4. *Complex* means that the medium has large molecules such as proteins, polysaccharides, lipids, and other chemicals that can vary greatly in exact composition.

(a)

(b)

Figure 3.15 Solid media that are reversible to liquids.
(a) Media containing 1%–5% agar are solid and do not move when containers are tilted or inverted. Because they are reversibly solid, they can be liquefied with heat, poured into a different container, and resolidified. **(b)** Nutrient gelatin contains enough gelatin (12%) to take on a solid consistency. The top tube shows it as a solid. The bottom tube indicates what happens when it is warmed or when microbial enzymes digest the gelatin and liquefy it. (b): © Kathy Park Talaro

degree. The dominant substances in medium B are macromolecules that contain unknown (but potentially required) nutrients. Both A and B will satisfactorily grow the bacteria. (Which one would you rather make?)

Media to Suit Every Function

Microbiologists have many types of media at their disposal, with new ones being devised all the time. Depending upon what is

TABLE 3.6A	Chemically Defined Synthetic Medium for Growth and Maintenance of Pathogenic *Staphylococcus aureus*	
0.25 Grams Each of These Amino Acids	**0.5 Grams Each of These Amino Acids**	**0.12 Grams Each of These Amino Acids**
Cystine	Arginine	Aspartic acid
Histidine	Glycine	Glutamic acid
Leucine	Isoleucine	
Phenylalanine	Lysine	
Proline	Methionine	
Tryptophan	Serine	
Tyrosine	Threonine	
	Valine	

Additional ingredients

0.005 mole nicotinamide ⎤
0.005 mole thiamine ⎥
0.005 mole pyridoxine ⎬—Vitamins
0.5 micrograms biotin ⎦

1.25 grams magnesium sulfate ⎤
1.25 grams dipotassium hydrogen phosphate ⎥
1.25 grams sodium chloride ⎬—Salts
0.125 grams iron chloride ⎦

Ingredients are dissolved in 1,000 milliliters of distilled water and buffered to a final pH of 7.0.

TABLE 3.6B	Brain-Heart Infusion Broth: A Complex, Nonsynthetic Medium for Growth and Maintenance of Pathogenic *Staphylococcus aureus*

27.5 grams brain-heart extract, peptone extract
2 grams glucose
5 grams sodium chloride
2.5 grams disodium hydrogen phosphate
Ingredients are dissolved in 1,000 milliliters of distilled water and buffered to a final pH of 7.0.

added, a microbiologist can fine-tune a medium for nearly any purpose. Microbiologists have always been aware of microbes that could not be cultivated artificially, and now we can detect a single bacterium in its natural habitat without cultivation (see 3.1 Secret World of Microbes). But even with this new technology, it is highly unlikely that microbiologists will abandon culturing, simply because it provides a constant source of microbes for detailed study, research, and diagnosis.

General-purpose media are designed to grow a broad spectrum of microbes that do not have special growth requirements. As a rule, these media are nonsynthetic (complex) and contain a mixture of nutrients that could support the growth of a variety of bacteria and fungi. Examples include nutrient agar and broth, brain-heart infusion, and trypticase soy agar (TSA). TSA is a complex medium that contains partially digested milk protein (casein), soybean digest, NaCl, and agar.

TABLE 3.7	Examples of Selective Media, Agents, and Functions	
Medium	Selective Agent	Used For
Mannitol salt agar	7.5% NaCl	Isolation of *Staphylococcus aureus* from infectious material
Enterococcus faecalis broth	Sodium azide, tetrazolium	Isolation of fecal enterococci
Phenylethanol agar (PEA)	Phenylethanol chloride	Isolation of staphylococci and streptococci
Tomato juice agar	Tomato juice, acid	Isolation of lactobacilli from saliva
MacConkey agar (MAC)	Bile, crystal violet	Isolation of gram-negative enterics
Eosin-methylene blue agar (EMB)	Bile, dyes	Isolation of coliform bacteria in specimens
Salmonella/Shigella (SS) agar	Bile, citrate, brilliant green	Isolation of *Salmonella* and *Shigella*
Sabouraud's agar (SAB)	pH of 5.6 (acid) inhibits bacteria	Isolation of fungi

An **enriched medium** contains complex organic substances such as blood, serum, hemoglobin, or special **growth factors** that must be provided to certain species in order for them to grow. Growth factors are organic compounds such as vitamins or amino acids that microbes cannot synthesize themselves. Bacteria that require growth factors and complex nutrients are termed **fastidious.** Blood agar, which is made by adding sterile animal blood (usually from sheep) to a sterile agar base (**figure 3.16***a*) is widely employed to grow fastidious streptococci and other pathogens. Pathogenic *Neisseria* and *Yersinia* species are often subultured on chocolate agar, which is made by heating blood agar and does not contain chocolate—it just has that appearance (**figure 3.16***b*).

Selective and Differential Media

Some of the cleverest and most inventive media recipes belong to the categories of selective and differential media. These media are designed for special microbial groups, and they have extensive

Growth of *Streptococcus pyogenes*
showing beta hemolysis

(a) (b)

Figure 3.16 Examples of enriched media. (a) A blood agar plate growing the pathogenic bacterium *Streptococcus pyogenes,* which is the cause of strep throat and scarlet fever. Note that this medium can also differentiate among colonies by the types of hemolysis they display. Observe the colonies with clearly defined zones (beta hemolysis) and others with less defined zones (alpha hemolysis). **(b)** A culture of *Yersinia pestis* on chocolate agar, which gets its brownish color from cooked blood and does not produce hemolysis. It is used to grow fastidious pathogens like this notorious agent of bubonic plague. (a): *Source:* Richard R. Facklam, Ph.D/CDC; (b): © Kathy Park Talaro

applications in isolation and identification. They can permit, in a single step, the preliminary identification of a genus or even a species.

A **selective medium (table 3.7)** contains one or more agents that inhibit the growth of a certain microbe or microbes (call them A, B, and C) but not another (D). This difference favors, or *selects,* microbe D and allows it to grow by itself. Selective media are very important in primary isolation of a specific type of microorganism from samples containing mixtures of different species—for example, feces, saliva, skin, water, and soil. They hasten isolation by suppressing the unwanted background organisms and allowing growth of the desired ones.

Bile salts, a component of feces, inhibit most gram-positive bacteria while permitting many gram-negative rods to grow. Media for isolating intestinal pathogens (EMB agar, MacConkey agar, Hektoen enteric [HE] agar) contain bile salts as a selective agent. Dyes such as methylene blue and crystal violet also inhibit certain gram-positive bacteria. The opposite effect can be achieved by using a different selective agent. Mannitol salt agar (MSA) contains a high concentration of NaCl (7.5%) that is quite inhibitory to most human pathogens, but not to the gram-positive bacteria in the genus *Staphylococcus,* which grow well and consequently can be amplified in mixed samples. Other agents that have selective properties are antimicrobial drugs and acid. Some selective media contain strongly inhibitory agents to favor the growth of a pathogen that would otherwise be overlooked because of its low numbers in a specimen. Selenite and brilliant green dye are used in media to isolate *Salmonella* from feces, and sodium azide is used to isolate enterococci from water and food.

Differential media grow several types of microorganisms but are designed to bring out visible differences among those microorganisms. Differences show up as variations in colony size or color, in media color changes, or in the formation of gas bubbles and precipitates (**table 3.8**). These variations come from the types of chemicals contained in the media and the ways that microbes react to them. For example, if microbe X metabolizes a certain substance not used by organism Y, then X will cause a visible change in the medium and Y will not. The simplest differential media show two reaction types such as a color change in some colonies but not in others. Several newer forms of differential media contain artificial substrates called chromogens that release a wide variety of colors, each tied to a specific microbe. **Figure 3.17***b* shows the result from

TABLE 3.8	Examples of Differential Media	
Medium	Substances That Facilitate Differentiation	Differentiates
Blood agar (BAP)	Intact red blood cells	Types of RBC damage (hemolysis)
Mannitol salt agar (MSA)	Mannitol and phenol red	Species of *Staphylococcus*
Hektoen enteric (HE) agar*	Bromthymol blue, acid fuchsin, sucrose, salicin, thiosulfate, ferric ammonium citrate, and bile	*Salmonella, Shigella,* and other lactose nonfermenters from fermenters; H_2S reactions are also observable
MacConkey agar (MAC)	Lactose, neutral red	Bacteria that ferment lactose (lowering the pH) from those that do not
Eosin-methylene blue (EMB)	Lactose, eosin, methylene blue	Same as MacConkey agar
Urea broth	Urea, phenol red	Bacteria that hydrolyze urea to ammonia and increase the pH
Sulfur indole motility (SIM)	Thiosulfate, iron	H_2S gas producers; motility; indole formation
Triple-sugar iron agar (TSIA)	Triple sugars, iron, and phenol red dye	Fermentation of sugars, H_2S production
XLD agar	Lysine, xylose, iron, thiosulfate, phenol red	Can differentiate *Enterobacter, Escherichia, Proteus, Providencia, Salmonella,* and *Shigella*

*Contains dyes and bile to inhibit gram-positive bacteria.

(a)

(b)

Figure 3.17 Media that differentiate multiple characteristics of bacteria. (a) Triple sugar iron agar (TSIA) inoculated on the surface and stabbed into the thicker region at the bottom (butt). This medium contains three sugars, phenol red dye to indicate pH changes (bright yellow is acid, various shades of red, basic), and iron salt to show H_2S gas production. Reactions are (1) no growth; (2) growth with no acid production (sugars not used); (3) acid production in the butt only; (4) acid production in all areas of the medium; (5) acid and H_2S production in butt (black precipitate). **(b)** A medium developed for culturing and identifying the most common urinary pathogens. CHROMagar Orientation™ uses color-forming reactions to distinguish at least seven species and permits rapid identification and treatment. (In the example, the bacteria were streaked so as to spell their own names. Which bacterium was probably used to write the name at the top? **(c)** MacConkey agar is used to differentiate gram-negative bacteria (see text). (a–c): © K. Talaro

(c)

Control tube

Gas bubble

Outline of Durham tube

Cloudiness indicating growth

Figure 3.18 Carbohydrate fermentation in broths. This medium is designed to show fermentation (acid production) using phenol red broth and gas formation by means of a small, inverted Durham tube for collecting gas bubbles. The tube on the left is an uninoculated negative control; the center tube is positive for acid (yellow) and gas (open space); the tube on the right shows growth but neither acid nor gas.
© Harold J. Benson

fermentation media contain sugars that can be fermented (converted to acids) and a pH indicator to show this reaction (see figure 3.18a). Media for other biochemical reactions that provide the basis for identifying bacteria and fungi are presented in several later chapters.

Transport media are used to maintain and preserve specimens that have to be held for a period of time before clinical analysis or to sustain delicate species that die rapidly if not held under stable conditions. Stuart's and Amie's transport media contain buffers and absorbants to prevent cell destruction but will not support growth. **Assay media** are used by technologists to test the effectiveness of antimicrobial drugs (see chapter 12) and by drug manufacturers to assess the effect of disinfectants, antiseptics, cosmetics, and preservatives on the growth of microorganisms. Enumeration media are used by industrial and environmental microbiologists to count the numbers of organisms in milk, water, food, soil, and other samples.

a urine culture containing six different bacteria. Other chromogenic agar is available for identifying *Staphylococcus, Listeria,* and pathogenic yeasts.

Dyes are effective differential agents because many of them are pH indicators that change color in response to the production of an acid or a base. For example, MacConkey agar contains neutral red, a dye that is yellow when neutral and pink or red when acidic **(figure 3.17c).** A common intestinal bacterium such as *Escherichia coli* that produces acid when it metabolizes the lactose in the medium develops red to pink colonies, and one such as *Salmonella* that does not produce acid remains its natural color (off-white). Some fermentation broths also contain phenol red to indicate changes in pH associated with breakdown of sugars into acids **(figure 3.18)**

A single medium can be classified in more than one category depending on the ingredients it contains. MacConkey and EMB media **(figure 3.19),** for example, appear in table 3.7 (selective media) and table 3.8 (differential media). Blood agar is both enriched and differential. Media companies have developed selective-differential media for numerous common pathogens, which makes it possible to identify them with a single streak plate. We will see examples of some of these media in later chapters.

Miscellaneous Media

A **reducing medium** contains a substance (thioglycollic acid or cystine) that absorbs oxygen or slows the penetration of oxygen in a medium, thus reducing its availability. Reducing media are important for growing anaerobic bacteria or for determining oxygen requirements of isolates (described in chapter 7). Carbohydrate

Figure 3.19 Media with both selective and differential properties. EMB agar, commonly used to detect enteric bacteria in food and dairy products, acts both selectively and differentially at the same time. Gram-negative bacteria that do not ferment the sugar lactose will grow on the medium, producing clear or slightly colored growth (*Salmonella typhimurium,* left and *Pseudomonas aeruginosa,* top). Gram-negative bacteria that do ferment lactose will produce much darker growth, sometimes with a green metallic sheen (*Escherichia coli,* right). The growth of gram-positive bacteria (*Staphylococcus aureus,* bottom) is almost completely inhibited (i.e., selected against). © McGraw-Hill Education/Lisa Burgess, photographer

A number of significant microbial groups (viruses, rickettsias, and a few bacteria) will only grow on live cells or animals. These obligate parasites have unique requirements that must be provided by living animals such as rabbits, guinea pigs, mice, chickens, and the early life stages (embryos) of birds. Such animals can be an indispensable aid for studying, growing, and identifying microorganisms. Animal inoculation is an essential step in testing the effects of drugs and the effectiveness of vaccines before they are administered to humans. Animals are an important source of antibodies, antisera, antitoxins, and other immune products that can be used in therapy or testing.

Check Your Progress SECTION 3.5

22. Describe the main purposes of media, and compare the three categories based on physical state, chemical composition, and usage.

23. Differentiate among the ingredients and functions of enriched, selective, and differential media.

24. Explain the two principal functions of dyes in media.

25. Why are some bacteria difficult to grow in the laboratory? Relate this to what you know so far about the nutrients that are added to media.

26. What conditions are necessary to cultivate viruses in the laboratory?

* *Neisseria meningitidis* (ny-serr′ee-uh men′-in-jih′-tih-dis) From the German physician, *Neisser,* and Gr. *meninx,* a membrane.
* *meningitis* (men′-in-jy′-tis) Gr. *meninx,* a membrane, and *itis,* an inflammation. Specifically, an inflammation of the meninges, the membranes that surround the brain.
* *septicemia* (sep′-tih-see′-mee-uh) Gr. *septikos,* to make putrid, and *haima,* blood. The presence of infectious agents or their toxins in the blood.

CASE STUDY Part 2

A Gram stain is one of the key tests for getting quick feedback on the kind of microbes that might be present in a sample. It is routine in meningitis because it can differentiate among several types of bacteria and detect certain other infectious agents, but it will not detect viruses. Usually within moments a lab technician can tell if there are bacterial cells and can observe their Gram reaction and shape.

Lab results from the Gram stain and culture were conclusive: Kay Peterson was infected by the meningococcus, *Neisseria meningitidis,** which is an agent of both meningitis* and septicemia.* The microscopic examination yielded the classic appearance of tiny pairs of red cocci (diplococci) and white blood cells carrying the same cocci inside. A blood culture also showed growth, indicating that the bacteria had entered her bloodstream. Plates of agar inoculated with the CSF grew typical off-white, smooth isolated colonies that tested out as *N. meningitidis.* Kay remained on the antibiotic regimen for 10 days and was released, fortunately without long-term damage.

■ *What kinds of media would be used to culture and identify this microbe?*

■ *What are some other potential microbes that could have caused this infection?*

For more information on the nature of this agent and its disease, see chapter 18 and log on to *www.cdc.gov/meningitis/index.html.*

To conclude this Case Study, go to Connect.

Chapter Summary with Key Terms

3.1 Methods of Microbial Investigation

A. Microbiology as a science is very dependent on a number of specialized laboratory techniques.

1. Initially, a specimen must be collected from a source, whether environmental or patient.

2. **Inoculation** of a **medium** with the specimen is the first step in **culturing.**

3. **Incubation** of the medium with the microbes under the right conditions creates a **culture** with visible growth.

4. **Isolation** of the microbes in the sample into discrete, separate **colonies** is one desired goal.

5. **Inspection** begins with macroscopic characteristics of the culture and continues with microscopic analysis.

6. **Information gathering** involves acquiring additional data from physiological, serological, and genetic tests.

7. **Identification** correlates the key characteristics that can pinpoint the actual species of microbe.

3.2 The Microscope: Window on an Invisible Realm

A. Optical, or light, microscopy depends on lenses that **refract** light rays, drawing the rays to a focus to produce a magnified image.

1. A simple microscope consists of a single magnifying lens, whereas a compound microscope relies on two lenses: the ocular lens and the objective lens.

2. The total power of magnification is calculated as the product of the ocular and objective magnifying powers.

3. **Resolution,** or the **resolving power,** is a measure of a microscope's capacity to make clear images of very small objects. Resolution is improved with shorter wavelengths of illumination and with a higher **numerical aperture** of the lens. Light microscopes are limited by resolution to magnifications around 2,000×.

4. Modifications in the lighting or the lens system give rise to the bright-field, dark-field, phase-contrast, interference, fluorescence, and confocal microscopes.

B. Electron microscopy depends on electromagnets that serve as lenses to focus electron beams. A **transmission electron microscope (TEM)** projects the electrons through prepared sections of the specimen, providing detailed structural images of cells, cell parts, and viruses. A **scanning electron microscope (SEM)** is more like dark-field microscopy, bouncing the electrons off the surface of the specimen to detectors.

3.3 Preparing Specimens for Optical Microscopes

A. Specimen preparation in optical microscopy varies according to the specimen, the purpose of the inspection, and the type of microscope being used.
 1. Wet mounts and hanging drop mounts permit examination of the characteristics of live cells, such as motility, shape, and arrangement.
 2. Fixed mounts are made by drying and heating a film of the specimen called a smear. This is then stained using dyes to permit visualization of cells or cell parts.
B. Staining uses either **basic** (cationic) **dyes** with positive charges or **acidic** (anionic) **dyes** with negative charges. The surfaces of microbes are negatively charged and attract basic dyes. This is the basis of **positive staining. In negative staining,** the microbe repels the dye and it stains the background. Dyes may be used alone and in combination.
 1. **Simple stains** use just one dye and highlight cell morphology.
 2. **Differential stains** require a primary dye and a contrasting counterstain in order to distinguish cell types or parts. Important differential stains include the **Gram stain, acid-fast stain,** and the endospore stain.
 3. **Structural stains** are designed to bring out distinctive characteristics. Examples include capsule stains and flagellar stains.

3.4 Additional Features of the Six "I's"

A. Inoculation of media followed by incubation produces visible growth in the form of **cultures.**
B. Techniques with solid media in Petri dishes provide a means of separating individual microbes by producing **isolated colonies.**

1. With the **streak plate,** a loop is used to thin out the sample over the surface of the medium.
2. With the **loop dilution** (pour plate), a series of tubes of media is used to dilute the sample.
3. The **spread plate** method evenly distributes a tiny drop of inoculum with a special stick.
4. Isolated colonies can be **subcultured** for further testing at this point. The goal is a **pure culture,** in most cases, or a **mixed culture. Contaminated cultures** can ruin correct analysis and study.

3.5 Media: The Foundations of Culturing

A. Artificial media allow the growth and isolation of microorganisms in the laboratory and can be classified by their physical state, chemical composition, and functional types. The nutritional requirements of microorganisms in the laboratory may be simple or complex.
B. Physical types of media include those that are **liquid,** such as **broths** and milk, those that are **semisolid,** and those that are solid. Solid media may be liquefiable, containing a solidifying agent such as **agar** or gelatin.
C. Chemical composition of a medium may be completely **chemically defined,** thus **synthetic. Nonsynthetic,** or **complex,** media contain ingredients that are not completely definable.
D. Functional types of media serve different purposes, often allowing biochemical tests to be performed at the same time. Types include **general-purpose, enriched, selective,** and **differential media.**
 1. Enriched media contain **growth factors** required by microbes.
 2. Selective media permit the growth of desired microbes while inhibiting unwanted ones.
 3. Differential media bring out visible variations in microbial growth.
 4. Others include anaerobic (reducing), assay, and enumeration media. Transport media are important for conveying certain clinical specimens to the laboratory.
E. In certain instances, microorganisms have to be grown in cell cultures or host animals.

Level I. Knowledge and Comprehension

These questions require a working knowledge of the concepts in the chapter and the ability to recall and understand the information you have studied.

 ## Multiple-Choice Questions

Select the correct answer from the answers provided. For questions with blanks, choose the combination of answers that most accurately completes the statement.

1. Which of the following is not one of the six "I's"?
 a. inspection
 b. identification
 c. illumination
 d. incubation
 e. inoculation

2. The term *culture* refers to the _____ growth of microorganisms in _____.
 a. rapid, an incubator
 b. macroscopic, media

 c. microscopic, the body
 d. artificial, colonies

3. A mixed culture is
 a. the same as a contaminated culture
 b. one that has been adequately stirred
 c. one that contains two or more known species
 d. a pond sample containing algae of a single species

4. Agar is superior to gelatin as a solidifying agent because agar
 a. does not melt at room temperature
 b. solidifies at 75°C
 c. is not usually decomposed by microorganisms
 d. both a and c

5. The process that most accounts for magnification is
 a. a condenser c. illumination
 b. refraction of light rays d. resolution

6. A subculture is a
 a. colony growing beneath the media surface
 b. culture made from a contaminant
 c. culture made in an embryo
 d. culture made from an isolated colony

7. Resolution is _____ with a longer wavelength of light.
 a. improved c. not changed
 b. worsened d. not possible

8. A real image is produced by the
 a. ocular c. condenser
 b. objective d. eye

9. A microscope that has a total magnification of 1,500× when using the oil immersion objective has an ocular of what power?
 a. 150× c. 15×
 b. 1.5× d. 30×

10. The specimen for an electron microscope is always
 a. stained with dyes c. killed
 b. sliced into thin sections d. viewed directly

11. Motility is best observed with a
 a. hanging drop preparation c. streak plate
 b. negative stain d. flagellar stain

12. Bacteria tend to stain more readily with cationic (positively charged) dyes because bacteria
 a. contain large amounts of alkaline substances c. are neutral
 b. contain large amounts of acidic substances d. have thick cell walls

13. The primary difference between a TEM and SEM is in
 a. magnification capability
 b. colored versus black-and-white images
 c. preparation of the specimen
 d. type of lenses

14. A fastidious organism must be grown on what type of medium?
 a. general-purpose medium c. synthetic medium
 b. differential medium d. enriched medium

15. What type of medium is used to maintain and preserve specimens before clinical analysis?
 a. selective medium c. enriched medium
 b. transport medium d. differential medium

16. Which of the following is NOT an optical microscope?
 a. dark-field c. atomic force
 b. confocal d. fluorescent

17. **Multiple Matching.** For each type of medium, select all descriptions that fit. For media that fit more than one description, briefly explain why this is the case.
 ____ mannitol salt agar a. selective medium
 ____ chocolate agar b. differential medium
 ____ MacConkey agar c. chemically defined (synthetic) medium
 ____ nutrient broth
 ____ brain-heart infusion broth d. enriched medium
 ____ Sabouraud's agar e. general-purpose medium
 ____ triple-sugar iron agar f. complex medium
 ____ SIM medium g. transport medium

Case Study Review

1. Which of the following techniques from the six "I's" was not used in the diagnosis?
 a. incubation
 b. inoculation
 c. inspection
 d. All of them were used.

2. Which of these would not be a *sign* of meningitis?
 a. brown blotches on the skin
 b. stiff neck
 c. cloudy spinal fluid
 d. gram-negative cocci in specimens

3. Go to figure 3.12 and trace what information was used to "key out" and identify the bacterial species that caused meningitis in the Case Study.

Writing Challenge

For each question, compose a one- or two-paragraph answer that includes the factual information needed to completely address the question. Check Your Progress questions can also be used for writing-challenge exercises.

1. a. When buying a microscope, what features are most important to check for?
 b. What is probably true of a $20 microscope that claims to magnify 1,000×?

2. How can one obtain 2,000× magnification with a 100× objective?

3. Differentiate between microscopic and macroscopic methods of observing microorganisms, citing a specific example of each method.

4. Describe the steps of the Gram stain, and explain how it can be an important diagnostic tool for infections.

5. Describe the steps you would take to isolate, cultivate, and identify a microbial pathogen from a urine sample.

6. Trace the pathway of light from its source to the eye, explaining what happens as it passes through the major parts of the microscope.

7. Evaluate the following preparations in terms of providing information on microbial size, shape, motility, and differentiation: spore stain, negative stain, simple stain, hanging drop slide, and Gram stain.

Concept Mapping

On Connect you can find an Introduction to Concept Mapping that provides guidance for working with concept maps along with concept-mapping activities for this chapter.

Level II. Application, Analysis, Evaluation, and Synthesis

These problems go beyond just restating facts and require higher levels of understanding and an ability to interpret, problem solve, transfer knowledge to new situations, create models, and predict outcomes.

Critical Thinking

Critical thinking is the ability to reason and solve problems using facts and concepts. These questions can be approached from a number of angles, and in most cases, they do not have a single correct answer.

1. A certain medium has the following composition:

Glucose	15 g
Yeast extract	5 g
Peptone	5 g
KH_2PO_4	2 g
Distilled water	1,000 ml

 a. Tell what chemical category this medium belongs to, and explain why this is true.
 b. How could you convert *Staphylococcus* medium (table 3.6A) into a nonsynthetic medium?

2. a. Name four categories that blood agar fits.
 b. Name four differential reactions that TSIA shows.
 c. Observe figure 3.14. Suggest what causes the difference in growth pattern between nonmotile and motile bacteria.

 d. Explain what a medium that is both selective and differential does, using figure 3.19.

3. a. What kind of medium might you make to selectively grow a bacterium that lives in the ocean?
 b. One that lives in the human stomach?
 c. Why are intestinal bacteria able to grow on media containing bile?

4. Go back to section 1.2 and observe the six micrographs in figure 1.3. See if you can tell what kind of microscope was used to make the photograph, based on magnification and appearance.

5. Since the dyes for the Gram stain are not specific to the bacteria they stain, what mistakes in the procedure could cause an incorrect result?

Visual Challenge

1. Examine figure 3.10*a, b* (shown here). If you performed the quadrant streak plate method using a broth culture that had 10× fewer cells than the broth used in the culture shown, which quadrant might you expect to yield isolated colonies? Explain your answer.

2. Look at Scoping Out the Chapter at the start of section 3.1. Using this figure, along with the information in the rest of the chapter, identify precisely which of the six "I's" were used in the Case Study at the start of this chapter? Which ones were used in the Case Study at the start of chapter 1? Were any used in the Case Study at the start of chapter 2?

Loop containing sample

(a) **Steps in a Streak Plate;** this one is a four-part or quadrant streak.

(b)

© Kathy Park Talaro

STUDY TIP: Quickly create a customized practice quiz for this chapter by going to your Connect SmartBook "My Reports" section. Don't have SmartBook and want to learn more? Go to **whysmartbook.com**.

A Survey of Prokaryotic Cells and Microorganisms

Looking as harmless as clusters of tiny purple grapes, the gram-positive pathogen *Staphylococcus aureus* is anything but. (inset): *Source:* Janice Carr/CDC

Section of a prosthetic heart valve with a patch of MRSA biofilm attached (purple).
Modified image reprinted with permission from Medscape Reference (http://emedicine.medscape.com/), 2013, available at: http://emedicine.medscape.com/article/216650-overview.

CASE STUDY Part 1 Heart Valves and Biofilms

On a summer morning in 2008, Maxwell Jones, a 65-year-old man, woke up complaining of abnormal **fatigue** and a **scratchy throat.** His wife said he felt hot and took his temperature. It was slightly elevated at 100°F. He dismissed his condition, saying he was probably tired from working in his garden and suffering one of his regular allergy attacks. Over the next few days, his list of symptoms grew. He lost his appetite, his **joints** and **muscles** were **sore,** and he woke up wringing wet from **night sweats.** He continued to have a **fever,** and his wife was worried over how pale he looked. She insisted he see a physician, who performed a physical and took a **throat culture.** Mr. Jones was sent home with instructions to take **oral penicillin** and acetaminophen (Tylenol), and to come back in a week.

At the next appointment the patient reported that he still had some of the same symptoms, including the fever, and that now he had begun to have **headaches, rapid breathing,** and **coughing.** The physician recorded a **rapid heart rate** and slight **heart murmur.** When the lab report indicated that the throat culture was **negative** for bacterial pathogens, he had to look for other causes.

> "At least 65% of chronic infections are caused by microbial biofilms."

He began to wonder if the patient had a prior medical history of possible **risk factors.** From interviewing Mr. Jones, he learned that an **artificial valve** had been implanted in his heart 10 years before, a fact that had been omitted from his medical chart. This finding immediately caused alarm, and Mr. Jones was admitted to the intensive care unit and placed on a mixture of **intravenous antibiotics.** Tests for **blood cultures** and a **white blood cell count** were ordered as backup. By that evening, Mr. Jones had become confused and lost consciousness. He was rushed to the operating room but died during open heart surgery.

- *What appear to be the most important facts in this case?*
- *Explain why Mr. Jones's throat culture was negative for infection.*

To continue the Case Study, go to Case Study Part 2 at the end of the chapter.

SCOPING OUT THE CHAPTER

We don't need to take a course in ornithology to be able to recognize the structure of a bird's wing and describe how it functions. And even when they're too far away for us to see them clearly, we can instantly know that a group of animals flying in a "V" formation is a flock of birds, not butterflies or bats. And though they are about as different as two animals could be, we all understand intuitively that a hummingbird and a turkey are related and should be grouped together. In this chapter we will gain the same type of familiarity with bacterial cells, by studying their structure, function, and evolutionary history.

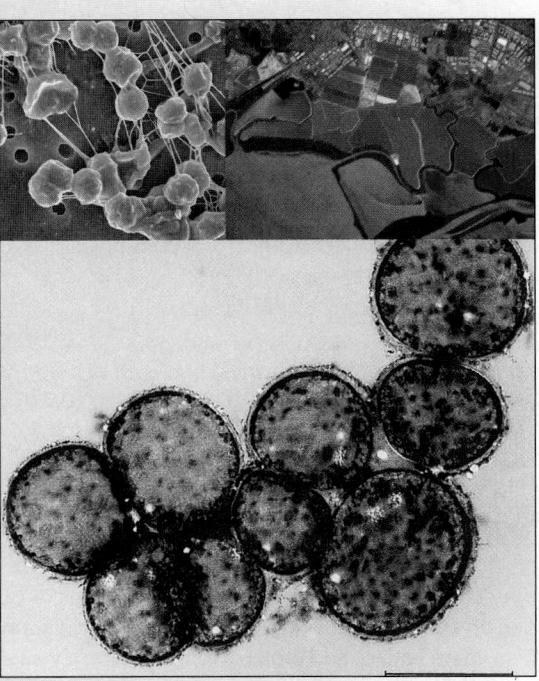

Cellular Structure The chapter opens with a discussion of what makes up a prokaryotic cell. Beginning with external structures like flagella, moving to the fluid-mosaic barrier of the cell membrane, and finally to internal structures like inclusion bodies. Understanding the anatomy of a cell is key to understanding its biology.

Bacterial Shapes and Arrangements The cell wall and cytoskeleton of a bacterial cell are responsible for its shape, and correctly recognizing the shape of a cell is one of the first steps in determining its identity. The grouping of individual cells into more complex arrangements reveals a great deal about the manner in which a cell multiplies.

Classification and Unusual Bacteria The chapter continues by explaining the ways in which prokaryotic cells may be organized based on their evolutionary relationships. Finally, we introduce several examples of novel bacteria that thrive in boiling water (top left), extraordinarily high concentrations of salt (top right), or are so big that they threaten to redefine what it means to be a bacterium (bottom).

(left: top): *Source:* Louisa Howard/Dartmouth Electron Microscope Facility; (left: bottom): © Kwangshin Kim/Science Source; (middle: top-left): *Source:* Janice Carr/CDC; (middle: top-right): *Source:* Joyce Ayers/CDC; (middle: bottom): *Source:* Jeff Hageman, M.H.S./Janice Carr/CDC; (right: top-left): *Source:* Maryland Astrobiology Consortium, NASA and STScI; (right: top-right): *Source:* NASA Johnson Space Center/ISS007E8738/ (http://eol.jsc.nasa.gov); (right: bottom): © Heide N. Schulz/Max Planck Institue for Marine Microbiology

4.1 Basic Characteristics of Cells and Life Forms

 Expected Learning Outcomes

1. Describe the fundamental characteristics of cells.

2. Identify the primary properties that define life and living things.

There is a universal biological truth that the basic unit of life is the cell, whether the organism is a bacterium whose entire body is just a single cell or an elephant made up of trillions of cells. Regardless of their origins, all cells share a few common characteristics. They tend to assume cubical, spherical, or cylindrical shapes, and have a **cell membrane** that encases an internal matrix called the **cytoplasm.** All cells have one or more chromosomes containing DNA. They all also possess ribosomes for protein synthesis, and exhibit highly complex chemical reactions. As we learned in chapter 1, all cells

discovered thus far are classified into one of two fundamentally different groups: the small, seemingly simple **prokaryotic cells** and the larger, structurally more complicated **eukaryotic cells.**

Eukaryotic cells are found in animals, plants, fungi, and protists. They contain a number of complex internal parts called **organelles** that perform useful functions for the cell. By convention, organelles are defined as cell components enclosed by membranes that carry out specific activities involving metabolism, nutrition, and synthesis. Organelles also partition the eukaryotic cell into smaller compartments. The most visible organelle is the nucleus, a roughly spherical mass surrounded by a double membrane that contains the DNA of the cell. Other organelles include the Golgi apparatus, endoplasmic reticulum, vacuoles, and mitochondria (all discussed in chapter 5).

Prokaryotic cells are found only in the bacteria and archaea. Sometimes it may seem that prokaryotes are the microbial "have-nots," because, for the sake of comparison, they are described by what they lack. They have no nucleus or other organelles. This apparent simplicity is misleading, because the fine structure of prokaryotes can be complex. Overall, prokaryotic cells can engage in nearly every activity that eukaryotic cells can, and many can function in ways that eukaryotes cannot.

We start this chapter with a description of the characteristics that impart the essence of life to cells, followed by a look at prokaryotic cell anatomy and a survey of major groups of prokaryotes. In chapter 5 we will do a similar survey of the eukaryotic world.

After you have studied the cells as presented in this chapter and chapter 5, refer to table 5.2, which summarizes the major differences between prokaryotic and eukaryotic cells.

What Is Life?

Biologists have long debated over universal characteristics we see in organisms that are solid indicators of life or of being alive. What does a cell do that sets it apart from a nonliving entity? One of the first things that often comes to mind is the ability to move or to grow. Unfortunately, taken individually, these are not life signs. After all, inanimate objects can move and crystals can grow. There is probably no single property that we can hold up as the ultimate indicator of life. In fact, defining life requires a whole collection of behaviors and properties that even the simplest organisms will have. First and foremost on this list would be a self-contained and highly organized unit to carry out the activities of life, namely a **cell.** It is the cell that is the staging area for these life-supporting phenomena: heredity, reproduction, growth, development, metabolism, responsiveness, and transport **(figure 4.1).**

- **Heredity** is the transmission of an organism's genome* to the next generation by chromosomes, which carry DNA, the molecular blueprint of life.
- **Reproduction** involves the generation of offspring necessary to continue a species's line of evolution. All organisms manifest asexual reproduction, with one cell simply dividing into two new cells by fission or mitosis. Many organisms also display sexual reproduction, involving the union of sex cells from two parents.

* *genome* (gee′-nohm) The complete set of chromosomes and genes of an organism.

Figure 4.1 **A summary of the major life-defining properties.** The cell is central to the collective activities that are synonymous with being alive.

- **Growth** has two general meanings in microbiology. In the usual sense, it means an increase in size of a population through reproduction. In another case, it refers to the enlargement of a single organism during maturation.
- **Development** includes all changes over the life span of an organism that complete the full expression of its genome.
- **Metabolism** encompasses the thousands of chemical reactions that all cells need to function. Generally, these reactions either synthesize new cell components or release energy that drives cellular activities. Both types of metabolism are supported and regulated by unique molecules of life called **enzymes.**
- **Responsiveness** is the capacity of cells to interact with external factors through irritability, communication, or movement. Cells demonstrate irritability by receiving and reacting to stimuli such as light and chemicals from their environment. They may also communicate with other cells by sending or receiving signals. Some cells show self-propulsion or motility using specialized locomotor structures.
- **Transport** is a system for controlling the flow of materials. This includes carrying nutrients and water from the external environment into the cell's interior. Without these raw materials, metabolism would cease. Cells also secrete substances, or expel waste, in the reverse direction. Movement of this type is the work of the cell membrane, which functions as the gatekeeper for cellular activities. Together, all of these life properties provide microorganisms with the capacity to adapt and evolve.

One of the main reasons viruses are generally considered nonliving is that they are not cellular. Even if one argues that they are tiny units somewhat like cells, and that they do show some traits of life such as heredity, development, and evolution, they do not display these characteristics independent of their living host cell. Outside of their host, they lack most other features of life we just described, and are inactive and inert. We will return to this discussion about viruses in chapter 6.

 Check Your Progress SECTION 4.1

1. Outline the primary indicators of life.
2. Name several general characteristics that could be used to define the prokaryotes.

4.2 Prokaryotic Profiles: The Bacteria and Archaea

 Expected Learning Outcomes

3. Describe the generalized anatomy of bacterial cells.
4. Distinguish among the types of external cell appendages.
5. Describe the structure and position of bacterial flagella and axial filaments, and their attachment patterns.
6. Explain how flagella influence motility and motile behavior.
7. Discuss the structure and functions of pili and fimbriae.
8. Define *glycocalyx,* and describe its different forms and functions.

In chapter 1 we introduced the two major types of prokaryotic cells: the bacteria and the archaea. Being the first type of cells on earth, prokaryotic cells have an evolutionary history that dates back over 3.5 billion years. At present we cannot know the precise nature of these first cells, but many microbiologists speculate that they would be similar to present-day archaea living on sulfur compounds in deep, hot oceanic vents (see figure 4.33), extreme habitats they continue to dominate 3.5 billion years later. Having evolved over the same time frame as archaea, bacteria share a similar ancient history. Biologists have estimated that these two domains together still account for at least half of the total mass of life forms on earth. The fact that these organisms have endured for so long in such a variety of habitats indicates cellular structure and function that are incredibly versatile and adaptable. The general structural plan of a prokaryotic cell can be represented with this flowchart:

Structures that are essential to the functions of all prokaryotic cells are a cell membrane, cytoplasm, ribosomes, and one, or occasionally more, chromosomes. The majority of bacteria also have a cell wall and some form of surface coating or glycocalyx. Specific structures that are found in some, but not all, bacteria are flagella, pili, fimbriae, capsules, slime layers, inclusions, an actin cytoskeleton, and endospores.

The Structure of a Generalized Bacterial Cell

Prokaryotic cells appear featureless and two-dimensional when viewed with an ordinary microscope, but this is only because of their small size. Higher magnification provides increased insight into their intricate and often complex structure (see figures 4.18 and 4.28 in later sections). The descriptions of prokaryotic structure, except where otherwise noted, refer to the **bacteria,** a category of prokaryotes with peptidoglycan in their cell walls. **Figure 4.2** shows a three-dimensional illustration of a generalized (rod-shaped) bacterial cell with most of the structures from the flowchart. As we survey the principal anatomical features of this cell, we begin with the outer cell structures and proceed to the internal contents. In sections 4.6 and 4.7 we will cover prokaryotic taxonomy, discuss the differences between bacteria and archaea, and introduce the major characteristics of archaea.

Cell Extensions and Surface Structures

Bacteria often bear accessory **appendages** sprouting from their surfaces. Appendages can be divided into two major groups: those

Figure 4.2 Structure of a typical bacterial cell. Cutaway view of a rod-shaped bacterium, showing major structural features. Note that not all components are found in all cells.

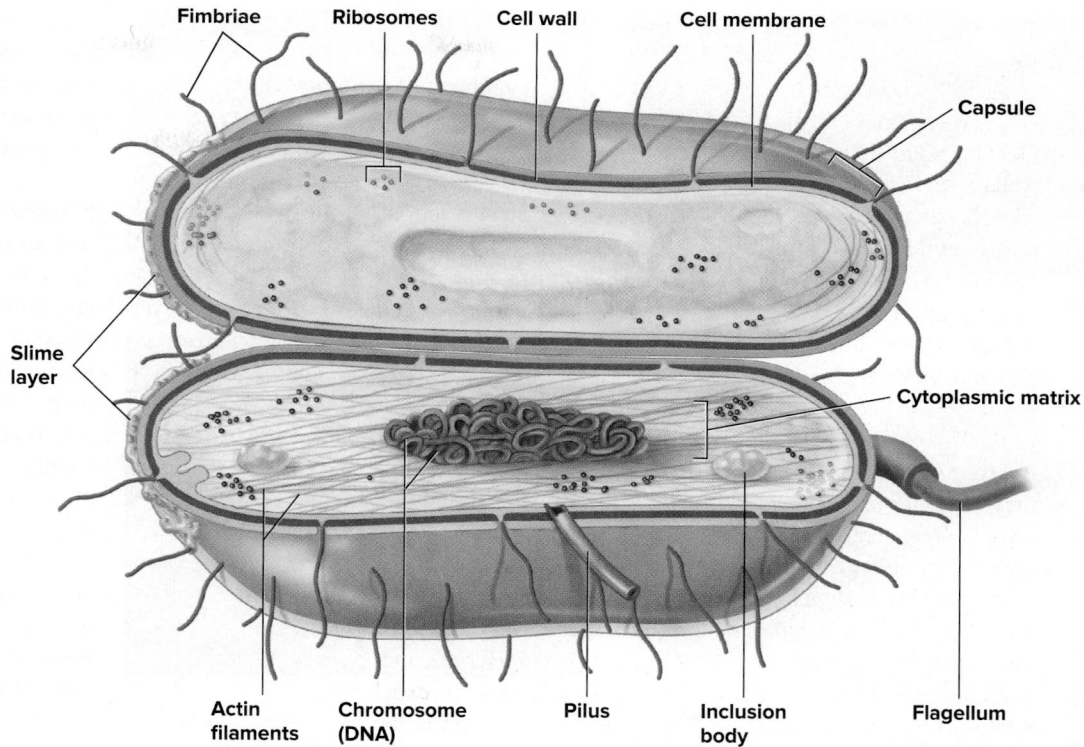

that provide motility (flagella and axial filaments) and those that provide attachment sites or channels (fimbriae and pili).

Flagella—Bacterial Propellers

The prokaryotic **flagellum*** is an appendage of phenomenal construction and is certainly unique in the biological world. Flagella provide the power of motility or self-propulsion. This allows a cell to swim freely through an aqueous habitat. The bacterial flagellum when viewed under high magnification displays three distinct parts: the filament, the hook (sheath), and the basal body **(figure 4.3).** The **filament** is a helical structure composed of a protein called flagellin. It is approximately 20 nm in diameter and varies from 1 to 70 μm in length. It is inserted into a curved, tubular hook. The hook is anchored to the cell by the **basal body,** a stack of rings firmly anchored through the cell wall to the cell membrane. The hook and its filament are free to rotate 360°—like a tiny propeller. This is in contrast to the flagella of eukaryotic cells, which undulate back and forth.

* *flagellum* (flah-jel′-em) pl. flagella; L., a whip.

Figure 4.3 Details of the flagellar basal body and its position in the cell wall. The hook, rings, and rod function together as a tiny device that rotates the filament 360°. **(a)** Structure in gram-negative cells. **(b)** Structure in gram-positive cells.

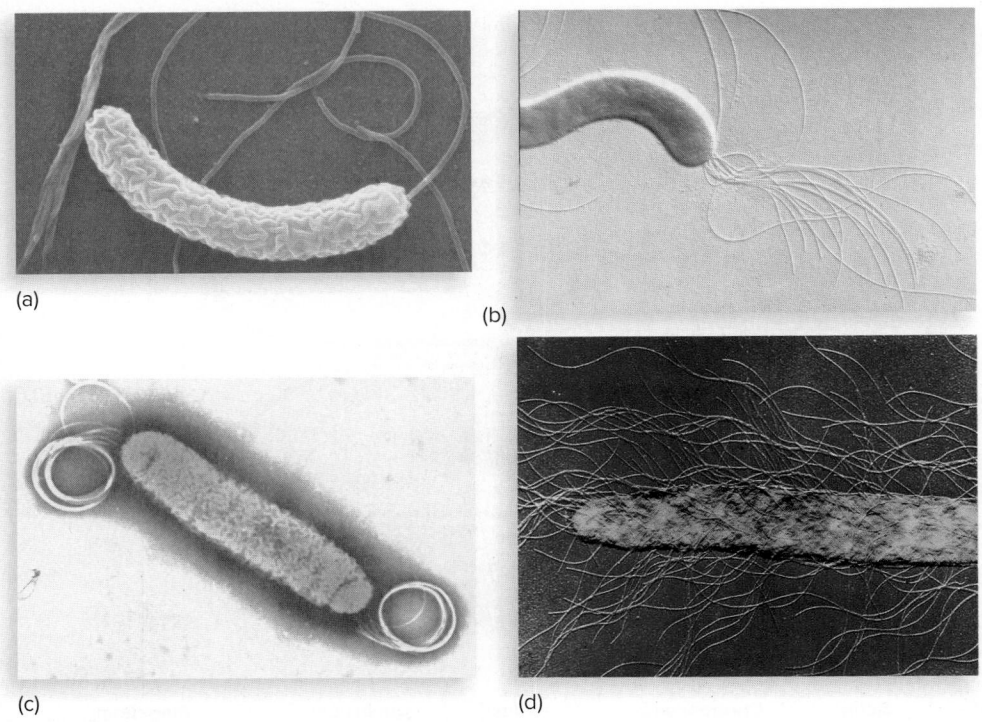

(a)

(b)

(c)

(d)

Figure 4.4 Electron micrographs depicting types of flagellar arrangements.
(a) Monotrichous flagellum on the pathogen *Vibrio cholerae* (10,000×). Note its highly textured glycocalyx. **(b)** Lophotrichous flagella on *Spirillum serpens,* a widespread aquatic bacterium (9,000×). **(c)** Unusual flagella on *Aquaspirillum* are amphitrichous and coil up into tight loops (7,500×). **(d)** *Proteus mirabilis* exhibits peritrichous flagella (10,000×).
(a): *Source:* Louisa Howard/Dartmouth Electron Microscope Facility; (b): Image courtesy Indigo® Instruments, www.indigo.com; (c): From: W.J. Strength and N.R. Krieg, "Flagellar activity in an aquatic bacterium," *Can. J. Microbiol.* May 1971, 17:1133-1137, Fig. 5. Reproduced by permission of the National Research Council of Canada; (d): © PTP/Phototake

microscopically with a hanging drop slide. A truly motile cell will flit, dart, or wobble around the field, making some progress, whereas one that is nonmotile jiggles about in one place but makes no progress.

Flagellar Responses Flagella often function as more than just a locomotor device. They are sensory appendages that can detect and respond to environmental signals. When the signal is of a chemical nature, the behavior is called **chemotaxis.*** Positive chemotaxis is movement of a cell in the direction of a favorable chemical stimulus (usually a nutrient); negative chemotaxis is movement away from a repellent (potentially harmful) compound.

The flagellum can guide bacteria in a certain direction because the system for detecting chemicals is linked to the mechanisms that drive the flagellum. Located in the cell membrane are clusters of receptors[1] that bind specific molecules coming from the immediate environment. The attachment of sufficient numbers of these molecules transmits signals to the flagellum and sets it into rotary motion. If several flagella are present, they become aligned and rotate as a group **(figure 4.5)**. As a flagellum rotates counterclockwise, the cell itself swims in a smooth linear direction toward the stimulus;

One can generalize that all spirilla, about half of the bacilli, and a small number of cocci are flagellated (these bacterial shapes are shown in figure 4.23). Flagella vary both in number and arrangement according to two general patterns: (1) In a **polar** arrangement, the flagella are attached at one or both ends of the cell. Three subtypes of this pattern are: **monotrichous,*** the only group with a single flagellum; **lophotrichous,*** posessing small bunches or tufts of flagella emerging from the same site; and **amphitrichous,*** posessing flagella at both poles of the cell. (2) In a **peritrichous*** arrangement, flagella are dispersed randomly over the surface of the cell **(figure 4.4).**

The presence of motility is one piece of information used in the laboratory identification of various groups of bacteria. Special stains or electron microscope preparations must be used to see the arrangement, because flagella are too minute to be seen in live preparations with a light microscope. Often it is sufficient to know simply whether a bacterial species is motile. One way to detect motility is to stab a tiny mass of cells into a soft (semisolid) medium in a test tube (see figure 3.14). Growth spreading rapidly through the entire medium is indicative of motility. Alternatively, cells can be observed

* *chemotaxis* (ke″-moh-tak′-sis) Gr. *chemo,* chemicals, and *taxis,* an ordering or arrangement.
1. Cell surface molecules that bind specifically with other molecules.

(a) Runs (b) Tumbles

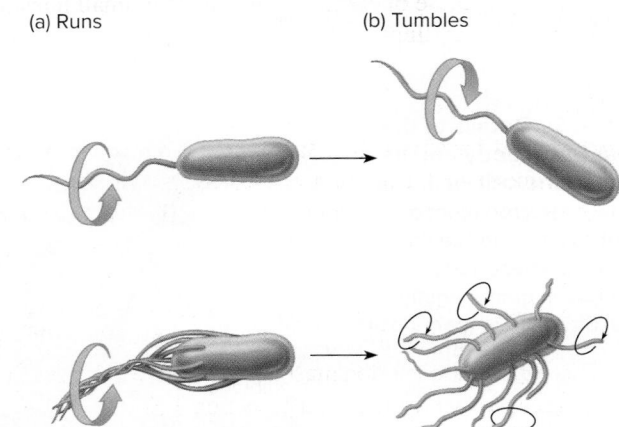

Figure 4.5 The operation of flagella and the mode of locomotion in bacteria with polar and peritrichous flagella.
(a) Runs. When a polar flagellum rotates in a counterclockwise direction, the cell swims forward in runs. In peritrichous forms, all flagella sweep toward one end of the cell and rotate as a single group. **(b)** Tumbles. When the flagellum reverses direction and rotates clockwise, the cell stops and tumbles. Peritrichous flagella also reverse direction and cause the cell to lose coordination and stop.

* *monotrichous* (mah″-noh-trik′-us) Gr. *mono,* one, and *tricho,* hair.
* *lophotrichous* (lo″-foh-trik′-us) Gr. *lopho,* tuft or ridge.
* *amphitrichous* (am″-fee-trik′-us) Gr. *amphi,* on both sides.
* *peritrichous* (per″-ee-trik′-us) Gr. *peri,* around.

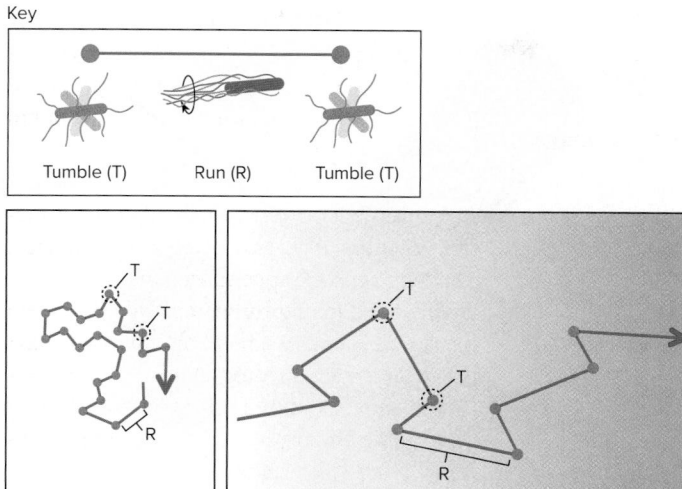

Key

Tumble (T) Run (R) Tumble (T)

(a) **No attractant or repellent** (b) **Gradient of attractant concentration**

Figure 4.6 **Chemotaxis in bacteria.** **(a)** A cell moves via a random series of short runs and tumbles when there is no attractant or repellent. **(b)** The cell spends more time on runs as it gets closer to an attractant.

this action is called a *run*. Runs are interrupted at various intervals by *tumbles* caused by the flagellum reversing its direction. This makes the cell stop and change its course. It is believed that attractant molecules inhibit tumbles, increase runs, and permit progress toward the stimulus **(figure 4.6).** Repellents cause numerous tumbles, allowing the bacterium to redirect itself away from the stimulus. Some photosynthetic bacteria exhibit *phototaxis,* a type of movement in response to light rather than chemicals.

A few pathogenic bacteria use their flagella to invade the surface of mucous membranes during infections. *Helicobacter pylori,* the agent of gastric ulcers, bores through the stomach lining, and *Vibrio cholerae,* the cause of cholera, penetrates the small intestine with the help of its flagellum.

Periplasmic Flagella

Corkscrew-shaped bacteria called **spirochetes*** show a wormlike or serpentine mode of locomotion caused by two or more long, coiled threads, the periplasmic flagella or **axial filaments.** A periplasmic flagellum is a type of internal flagellum that is enclosed in the space between the outer sheath and the cell wall peptidoglycan **(figure 4.7).** The filaments curl closely around the spirochete coils yet are free to contract and impart a twisting or flexing motion to the cell. This form of locomotion must be seen in live spirochetes to be truly appreciated (see Quick Search, section 4.5).

Nonflagellar Appendages: Fimbriae and Pili

The structures termed **fimbria*** and **pili*** both refer to common bacterial surface appendages that are involved in interactions with

* *spirochete* (spy'-roh-keet) Gr. *speira,* coil, and *chaite,* hair.
* *fimbria* (fim'-bree-ah) pl. fimbriae; L., a fringe.
* *pilus* (py'-lus) pl. pili; L., hair.

(a)

(b)

PF PC OS

Outer sheath (OS)

Protoplasmic cylinder (PC)

Periplasmic flagella (PF)

Peptidoglycan

Cell membrane

(c)

Figure 4.7 **The orientation of periplasmic flagella on the spirochete cell.** **(a)** Section through *Borrelia burgdorferi,* the spirochete of Lyme disease. The micrograph has been colorized to indicate the periplasmic flagella (yellow), the outer sheath (blue), and the protoplasmic cylinder (pink). **(b)** Longitudinal view with major cell parts labeled. **(c)** Cross section, indicating the position of the flagella with respect to other cell parts. Contraction of the filaments imparts a spinning and undulating pattern of locomotion. (a): Courtesy of Dr. Misha Kudryashev and Dr. Friedrich Frischknecht, Mol Microbiol, 2009 Mar; 71(6):1415-34

other cells but do not provide locomotion, except for some specialized pili. Some microbiologists use the terms interchangeably, but we will refer to the smaller common appendages as fimbriae and the ones with specialized functions as pili.

Fimbriae are small, bristlelike fibers emerging from the surface of many types of bacterial cells **(figure 4.8).** Their exact composition varies, but most of them contain protein. Fimbriae have an inherent tendency to stick to each other and to surfaces. They may be responsible for the mutual clinging of cells that leads to biofilms and other thick aggregates of cells on the surface of liquids and for the microbial colonization of inanimate solids such as rocks and glass **(4.1 Making Connections).** Some pathogens can colonize and infect host tissues because of a tight adhesion between their fimbriae and epithelial cells (figure 4.8*b*). For example, *Escherichia coli* colonizes the intestine and begins its invasion

(a)

(b)

Figure 4.8 Form and function of bacterial fimbriae.
(a) Several cells of pathogenic *Escherichia coli* covered with numerous fibers called fimbriae (30,000×). Note also the dark blue masses, which are chromosomes. **(b)** A row of *E. coli* cells tightly adheres by their fimbriae to the surface of an intestinal cell (12,000×). This is how the bacterium clings and gains access to the inside of cells during an infection. (G = glycocalyx). (a): © Eye of Science/Science Source; (b): From: D.R. Lloyd, S. Knutton, A. McNeish, "Identification of a New Fimbrial Structure in Enterotoxigenic *Escherichia coli* (ETEC) Serotype O148:H28 Which Adheres to Human Intestinal Mucosa: a Potentially New Human ETEC Colonization Factor," *Infection and Immunity,* January 1987, 55(1):86-92. Fig. 4A. © ASM

of tissues by this means. Mutant forms of these pathogens that lack fimbriae are unable to cause infections.

Pili come in several varieties. A versatile form called the Type IV pilus is found only in gram-negative bacteria. This structure is a flexible tube made of the protein, pilin. Bacteria with pili participate in a mating process termed **conjugation,**[2] which uses the pilus (also called a sex pilus) as a connector for transferring DNA from a donor cell to a recipient **(figure 4.9).** This role of pili is discussed further in chapter 9.

Type IV pili are a major contributor to the infectiousness of *Neisseria gonorrhoeae,* the agent of gonorrhea, by providing a mechanism for binding to the epithelial cells of the reproductive tract. Type IV pili found in *Pseudomonas* bacteria carry out a

Figure 4.9 Three bacteria in the process of conjugating.
Clearly evident are the sex pili forming mutual conjugation bridges between a donor (upper cell) and two recipients (two lower cells). Fimbriae can also be seen on the donor cell. © L. Caro/SPL/Science Source

remarkable sort of twitching motility. Cells can slide in jerking movements over moist surfaces by extending and then retracting pili in a repetitive motion. Microbiologists have found that this mechanism is every bit as complex and regulated as the flagellar mode of locomotion. This remarkable motility may be observed on YouTube by searching for "type IV pili twitching motility."

The Bacterial Surface Coating, or Glycocalyx

The bacterial cell surface is frequently exposed to severe environmental conditions. The **glycocalyx** develops as a coating of macromolecules to protect the cell and, in some cases, help it adhere to its environment. Glycocalyces differ among bacteria in thickness, organization, and chemical composition. Some bacteria are covered with a loose shield called a **slime layer** that protects them from dehydration and loss of nutrients, as well as serving in adhesion **(figure 4.10a).** Other bacteria produce **capsules** composed of repeating polysaccharide units, of protein, or

(a) (b)

Figure 4.10 Types of glycocalyces seen through cutaway views of cells. (a) The slime layer is a loose structure that is easily washed off. **(b)** The capsule is a thick, structured layer that is not readily removed.

2. Although the term *mating* is sometimes used for this process, it is not a form of reproduction.

4.1 MAKING CONNECTIONS

Biofilms—The Glue of Life

Microbes rarely live in isolation. More often they cling together in complex masses called **biofilms.** The formation of these living layers is a universal phenomenon that all of us have observed. Consider the scum that builds up in toilet bowls and shower stalls in a short time if they are not cleaned; or the algae that collect on the walls of swimming pools; and, more intimately, the constant deposition of plaque on teeth.

Fossils from ancient deposits tell us that microbes have been making biofilms for billions of years. It is through this process that they have colonized most habitats on earth and created stable communities that provide access to nutrients and other essential factors. Biofilms are often cooperative associations among several microbial groups (bacteria, fungi, algae, and protozoa) as well as plants and animals.

Substrates favorable to biofilm development have a moist, thin layer of organic material such as polysaccharides or glycoproteins deposited on their exposed surface. This has a sticky texture that attracts primary colonists, usually bacteria. These early cells attach and begin to multiply on the surface. Here, they lay down a glycocalyx consisting of fimbriae, pili, slime layers, or capsules. These substances bind the cells to the substrate and begin the development of the biofilm matrix.

As the biofilm evolves, it undergoes specific adaptations to the habitat in which it forms. In many cases the earliest colonists contribute nutrients and create microhabitats for other microbes to attach and grow into the film, forming complete communities. The biofilm varies in thickness and complexity, depending upon where it occurs and how long it keeps developing. Complexity ranges from single cell layers to thick microbial mats with dozens of dynamic interactive layers.

Biofilms are a profoundly important force in the development of terrestrial and aquatic environments. They dwell permanently in bedrock and the earth's sediments, where they play essential roles in recycling elements, leaching minerals, and forming soil. Biofilms associated with plant roots promote the mutual exchange of nutrients between the microbes and roots. As we will learn in chapter 7, the microbes in these communities have "conversations" using a special chemical vocabulary, which greatly influences the actions of the biofilm.

Researchers are currently analyzing the effects of biofilms created by our own microbiota—microbes that live naturally on the body. These associations are common on the skin and in the oral cavity and large intestine. In these locations, microbes signal each other as well as human cells in ways that shape the conditions there.

Biofilms accumulate on damaged tissues (such as heart valves), hard tissues (teeth), and foreign materials (catheters, intrauterine devices [IUDs], artificial hip joints). Microbial biofilms play a key role in the majority of infections and diseases—for example, cholera, urinary tract infections, sexually transmitted infections, endocarditis, middle ear infections, and pneumonia. New evidence indicates that bacteria turn on different genes when they are in a biofilm than when they are "free-floating," or planktonic. This altered gene expression gives the bacteria a different set of characteristics, often making them impervious to antibiotics and disinfectants. For additional discussions of biofilms, see chapters 7 and 12.

> **Quick Search**
> Search YouTube for "3D Architecture of Microbial Cities" to watch an animation about biofilms.

Describe some possible benefits of having biofilms form in or on the human body. *Answer available on Connect.*

1. Attachment of free planktonic cells to surface

2–3. Growth of cells, synthesis of extracellular matrix, stable colonization of surface

4. Maturation of biofilm formation of thick layered deposits containing sessile cells

5. Mature biofilm completes cycle by dispersing planktonic cells, which swim free to start the process again in new habitats

(a)

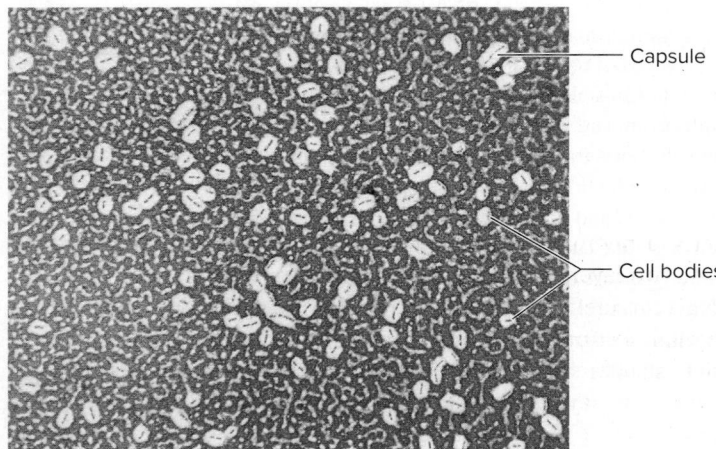

Capsule

Cell bodies

(b)

Figure 4.11 Appearance of encapsulated bacteria.
(a) Close-up view of colonies of *Bacillus* species with and without capsules. Even at the macroscopic level, the moist, slimy character of the capsule is evident. **(b)** Special staining reveals the microscopic appearance of a large, well-developed capsule (the clear "halo" around the cells) of an unidentified, rod-shaped bacterium (1,000×).
(a): © Kathy Park Talaro; (b): © Steven P. Lynch

of both. A capsule is bound more tightly to the cell than a slime layer is, and it has a thicker, gummy consistency that gives a prominently sticky (mucoid) character to the colonies of most encapsulated bacteria **(figure 4.11a).**

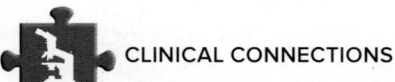

CLINICAL CONNECTIONS

The Glycocalyx and Infections

Capsules are formed by many pathogenic bacteria, such as *Streptococcus pneumoniae* (a cause of pneumonia, an infection of the lungs), *Haemophilus influenzae* (one cause of meningitis), and *Bacillus anthracis* (the cause of anthrax). Encapsulated bacterial cells generally have greater pathogenicity because capsules protect the bacteria against white blood cells called phagocytes. Phagocytes are a natural body

Biofilm matrix

Cells of biofilm

Fibrous surface of catheter

Figure 4.12 Magnified view of a biofilm. Scanning electron micrograph of a biofilm formed by a gram-negative bacterium, *Alcaligenes,* on a central venous catheter. Note the thick matrix of the biofilm caught in the fine fibers of the catheter.
Source: Janice Haney Carr/CDC

defense that can engulf and destroy foreign cells, which helps to prevent infection. A capsular coating blocks the phagocytes from attaching to, engulfing, and killing bacteria. By escaping phagocytosis, the bacteria are free to multiply and infect body tissues. Encapsulated bacteria that mutate to nonencapsulated forms usually lose their pathogenicity.

Other types of glycocalyces can be important in formation of biofilm infections (see 4.1 Making Connections). The thick, white plaque that forms on teeth comes in part from the surface slimes produced by certain streptococci in the oral cavity. This slime protects them from being dislodged from the teeth and provides a niche for persistent colonization that can lead to dental disease. The glycocalyx of some bacteria is so highly adherent that it is responsible for harmful biofilms developing on nonliving materials implanted in the body, such as plastic catheters, intrauterine devices, and metal pacemakers **(figure 4.12).**

Biofilms on heart valves can give rise to infections in other areas of the body. Explain how this happens by looking at the figure in 4.1 Making Connections. *Answer available on Connect.*

Check Your Progress SECTION 4.2

3. What other microbial groups besides bacteria have prokaryotic cells?
4. Describe the structure of a flagellum and how it operates. What are the four main types of flagellar arrangement?
5. How does the flagellum dictate the behavior of a motile bacterium? Differentiate between flagella and periplasmic flagella.
6. Differentiate between the structure and functions of pili and fimbriae.
7. Explain the position of the glycocalyx.
8. How do slime layers and capsules differ in structure and functions?
9. How can a capsule or slime layer be detected even at the level of a colony?
10. Explain how the bacterial glycocalyx and certain surface appendages contribute to biofilm formation.

4.3 The Cell Envelope: The Outer Boundary Layer of Bacteria

Expected Learning Outcomes

9. Explain the concept of the cell envelope, and describe its structure.

10. Outline the structure and functions of cell walls, and explain the role of peptidoglycan.

11. Contrast the major structure of gram-positive and gram-negative cell walls.

12. Summarize how gram-positive and gram-negative cells differ in their reactions.

13. Relate the characteristics of other types of cell walls and wall-free cells.

14. Describe the structure of the cell membrane, and explain several of its major roles in bacterial cells.

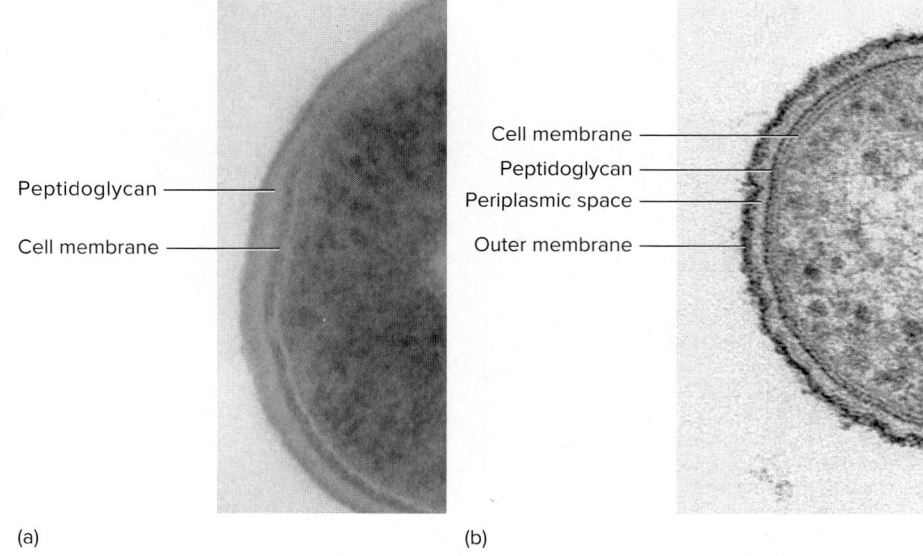

(a) (b)

Figure 4.13 Comparative views of the envelopes of gram-positive and gram-negative cells. (a) A section through a gram-positive cell wall/membrane with an interpretation of the main layers visible (92,000×). **(b)** A section through a gram-negative cell wall/membrane with an interpretation of its three sandwich-style layers (90,000×).
(a): Courtesy of Newcastle University Electron Microscopy Unit; (b): *Source:* Cynthia Goldsmith and Melissa Bell/CDC

The majority of bacteria have a chemically complex external covering, termed the cell envelope, that encloses the cytoplasm. It is composed of two main layers: the cell wall and the cell membrane. These layers are stacked together and often tightly bound into a unit like the outer husk and casings of a coconut. Although each envelope layer performs a distinct action, together they act as a single unit required for a cell's normal function and integrity.

Basic Types of Cell Envelopes

Long before the detailed anatomy of bacteria was even remotely known, a Danish physician named **Hans Christian Gram** developed a staining technique, the **Gram stain,**[3] which is commonly used to delineate two different groups of bacteria known as the gram-positive bacteria and the gram-negative bacteria. You may wish to review this stain, described in 3.2 Making Connections. Because the Gram stain does not reveal the reasons these two groups stain differently, we must turn to the electron microscope and to biochemical analysis of their structure.

The extent of the differences between gram-positive and gram-negative bacteria is evident in the physical appearance of their cell envelopes **(figure 4.13)**. In **gram-positive** cells, a microscopic section resembles an open-faced sandwich with two layers: the thick cell wall, composed primarily of peptidoglycan (defined in the next section), and the cell membrane. A similar section of a gram-negative cell envelope shows a complete sandwich with three layers: an outer membrane, a thin peptidoglycan layer, and the cell membrane. The reactions of the Gram stain are the result of these basic differences. The thick peptidoglycan of the gram-positive cell traps the crystal violet–mordant complex and makes it inaccessible to the decolorizer, leaving the cells purple. Gram-negative cell walls are thinner, and the crystal violet is relatively easy to remove with the decolorizer. In addition, the alcohol dissolves the outer membrane of the cell, which increases this loss of dye. These factors create a colorless cell that will stain with a red counterstain. **Table 4.1** provides a summary of the major similarities and differences between the wall types.

3. This text follows the American Society for Microbiology style, which calls for capitalization of the terms *Gram stain* and *Gram staining* and lowercase treatment of gram-negative and gram-positive, except in headings.

TABLE 4.1	Comparison of Gram-Positive and Gram-Negative Cell Walls	
Characteristic	**Gram-Positive**	**Gram-Negative**
Number of major layers	One	Two
Chemical composition	Peptidoglycan Teichoic acid Lipoteichoic acid Mycolic acids and polysaccharides*	Lipopolysaccharide (LPS) Lipoprotein Peptidoglycan Porin proteins
Overall thickness	Thicker (20–80 nm)	Thinner (8–11 nm)
Outer membrane	No	Yes
Periplasmic space	Narrow	Extensive
Permeability to molecules	More penetrable	Less penetrable

*In some cells.

(a) The peptidoglycan of a cell wall is a huge, three-dimensional latticework that is actually one giant molecule to surround and support the cell.

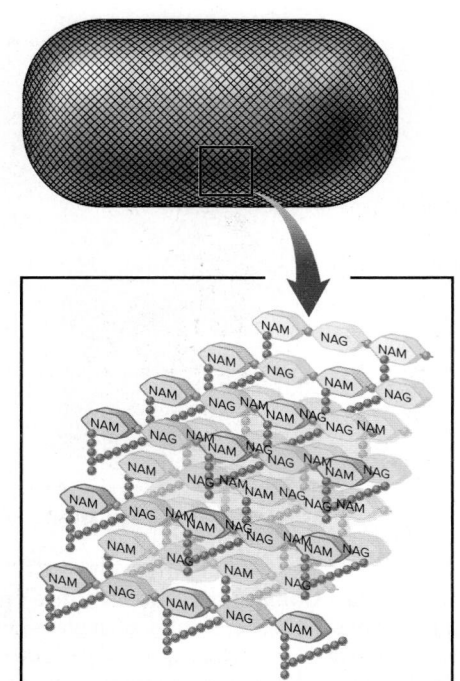

(b) This shows the molecular pattern of peptidoglycan. It has alternating glycans (NAG and NAM) bound together in long strands. The NAG stands for *N*-acetyl glucosamine, and the NAM stands for *N*-acetyl muramic acid. Adjacent muramic acid molecules on parallel chains are bound by a cross-linkage of peptides (brown spheres).

(c) An enlarged view of the links between the NAM molecules. Tetrapeptide chains branching off the muramic acids connect by amino acid interbridges. The amino acids in the interbridge can vary or may be lacking entirely. It is this linkage that provides rigid yet flexible support to the cell.

Figure 4.14 **Structure of peptidoglycan component of cell walls.**

Structure of Cell Walls

The cell wall accounts for a number of important bacterial characteristics. In general, it helps determine the shape of a bacterium, and it also provides the kind of strong structural support necessary to keep a bacterium intact despite constant changes in environmental conditions. The cell walls of most bacteria gain their relative strength and stability from a unique macromolecule called **peptidoglycan (PG).** This compound is composed of a repeating framework of long **glycan*** chains cross-linked by short peptide fragments **(figure 4.14).** The amount and exact composition of peptidoglycan vary among the major bacterial groups.

Because many bacteria live in aqueous habitats with a low solute concentration, they are constantly absorbing excess water by osmosis. Were it not for the structural support of the peptidoglycan in the cell wall, they would rupture from internal pressure. Several types of drugs used to treat infection (penicillin, cephalosporins) are effective because they target the peptide cross-links in the peptidoglycan, thereby disrupting its integrity. With their cell walls incomplete or missing, such cells have very little protection from **lysis.*** Lysozyme, an enzyme contained in tears and saliva, provides a natural defense against certain bacteria by hydrolyzing the bonds in the glycan chains and causing the wall to break down. Chapters 11 and 12 provide more information on the actions of antimicrobial chemical agents and drugs.

The Gram-Positive Cell Wall

The bulk of the gram-positive cell wall is a thick, homogeneous sheath of peptidoglycan ranging from 20 to 80 nm in thickness. It also contains tightly bound acidic polysaccharides, including teichoic acid directly attached to the peptidoglycan and lipoteichoic acid **(figure 4.15).** Cell wall **teichoic acid** is a polymer of ribitol or glycerol and phosphate embedded in the peptidoglycan sheath. Lipoteichoic acid is similar in structure but is attached to the lipids in the plasma membrane. These molecules appear to function in cell wall maintenance, enlargement during cell division, and binding of some pathogens to tissues. The cell wall of gram-positive bacteria is loosely adherent to the cell membrane, but at their junction lies a small compartment called the **periplasmic* space.** This is analogous to the larger space in gram-negatives, but it has different functions. It is a site for temporary storage of enzymes that have been released by the cell membrane. More recent studies have found that this space serves as the major site for peptidoglycan synthesis.

The Gram-Negative Cell Wall

The gram-negative cell wall is more complex in morphology because it is composed of an **outer membrane (OM)** and a thinner

* *glycan* (gly′-kan) Gr., sugar. These are large polymers of simple sugars.

* *lysis* (ly′-sis) Gr., to loosen. A process of cell destruction, as occurs in bursting.
* *periplasmic* (per″-ih-plaz′-mik) Gr. *peri,* around, and *plastos,* the fluid substances of a cell.

Gram-Positive

Wall teichoic acid

Lipoteichoic acid

Envelope

Gram-Negative

Lipopolysaccharides

Porin proteins

Phospholipids

Outer membrane layer

Peptidoglycan

Periplasmic space

Cell membrane

Lipoproteins

Membrane proteins

Periplasmic space

Membrane protein

Key

Peptidoglycan

Teichoic acid

Phospholipid

Membrane proteins

Lipopolysaccharide

Porin

Lipoprotein

Figure 4.15 **A comparison of the detailed structure of gram-positive and gram-negative cell envelopes and walls.**

shell of peptidoglycan (see figures 4.13 and 4.15). The outer membrane is somewhat similar in construction to the cell membrane, except that it contains specialized types of **lipopolysaccharides (LPS)** and **lipoproteins.** Lipopolysaccharides consist of lipid molecules bound to polysaccharides. The lipids form the matrix of the top layer of the OM, and the polysaccharide strands project from the lipid surface. The lipid portion may become toxic when it is released during infections. Its role as an endotoxin is described in chapter 13. The polysaccharides give rise to the somatic (O) antigen in gram-negative pathogens and can be used in identification. They may also function as receptors and interfere with host defenses.

Two types of proteins are located in the OM. The **porins** are inserted in the upper layer of the outer membrane. They have some regulatory control over molecules entering and leaving the cell. Many qualities of the selective permeability of gram-negative bacteria to bile, disinfectants, and drugs are due to the porins. Some structural proteins are also embedded in the upper layer of the OM. The bottom layer of the outer membrane is similar to the cell membrane in its overall structure and is composed of phospholipids and lipoproteins.

The bottom layer of the gram-negative wall is a single, thin (1–3 nm) sheet of peptidoglycan. Although it acts as a somewhat rigid protective structure, as previously described, its thinness gives gram-negative bacteria a relatively greater flexibility and sensitivity to lysis. There is a well-developed **periplasmic space** above and below the peptidoglycan. This space in gram-negative bacteria is a site of many metabolic reactions related to synthesis and transport of proteins, actions of enyzmes, and energy release.

Nontypical Cell Walls

Several bacterial groups lack the cell wall structure of gram-positive or gram-negative bacteria, and some bacteria have no cell wall at all. Although these exceptional forms can stain positive or negative in the Gram stain, examination of their fine structure and chemistry shows that they do not fit the descriptions for typical gram-negative or gram-positive cells. For example, the cells of *Mycobacterium* and *Nocardia* contain peptidoglycan and stain gram-positive, but the bulk of their cell wall is composed of unique types of lipids. One of these is a very-long-chain fatty acid called *mycolic acid,* or cord factor, that contributes to the pathogenicity of this group. The thick, waxy nature imparted to the cell wall by these lipids is also responsible for a high degree of resistance to certain chemicals and dyes. Such resistance is the basis for the **acid-fast stain** used to diagnose tuberculosis and leprosy. In this stain, hot carbol fuchsin dye becomes tenaciously attached (is held fast) to these cells so that an acid-alcohol solution will not remove the dye.

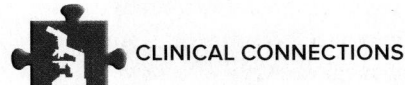 **CLINICAL CONNECTIONS**

The Cell Wall and Infections

Variations in cell wall anatomy contribute to several differences between the two cell types besides staining reactions. The outer membrane contributes an extra barrier in gram-negative bacteria that makes them more impervious to some antimicrobic chemicals such as dyes and disinfectants, so they are generally more difficult to

inhibit or kill than are gram-positive bacteria. One exception is for alcohol-based compounds, which can dissolve the lipids in the outer membrane and disturb its integrity. Treating infections caused by gram-negative bacteria often involves different drugs than gram-positive infections, because of the special requirement that the drugs must cross the outer membrane.

The cell wall or its parts can interact with human tissues and contribute to disease. The lipopolysaccharides have been referred to as *endotoxins* because they stimulate fever and shock reactions in gram-negative infections such as meningitis and typhoid fever. Proteins attached to the outer portion of the cell wall of several gram-positive species, including *Corynebacterium diphtheriae* (the agent of diphtheria) and *Streptococcus pyogenes* (the cause of strep throat), also have toxic properties. The lipids in the cell walls of certain *Mycobacterium* species are harmful to human cells as well. Because most macromolecules in the cell walls are foreign to humans, they stimulate antibody production by the immune system (see chapter 15).

Explain why penicillin can be an effective drug to treat gram-positive bacterial infections but is ineffective against most gram-negative bacteria. *Answer available on Connect.*

Mycoplasmas and Other Cell-Wall-Deficient Bacteria

Mycoplasmas are bacteria that naturally lack a cell wall. Although other bacteria require an intact cell wall to prevent the bursting of the cell, the mycoplasma cell membrane contains sterols that make it resistant to lysis. These extremely tiny bacteria range from 0.1 to 0.5 μm in size. They range in shape from filamentous to coccus or doughnut-shaped. This property of extreme variations in shape is a type of **pleomorphism.*** They can be grown on artificial media, although added sterols are required for the cell membranes of some species. Mycoplasmas are found in many habitats, including plants, soil, and animals. The most important medical species is *Mycoplasma pneumoniae* **(figure 4.16),** which adheres to the epi-

* *pleomorphism* (plee″-oh-mor′-fizm) Gr. *pleon,* more, and *morph,* form or shape. The tendency for cells of the same species to vary to some extent in shape and size.

Figure 4.16 Scanning electron micrograph of *Mycoplasma pneumoniae* (10,000×). Cells exhibit extreme variation in shape due to the lack of a cell wall. Courtesy of Dr. Heinrich Lünsdorf, HZI

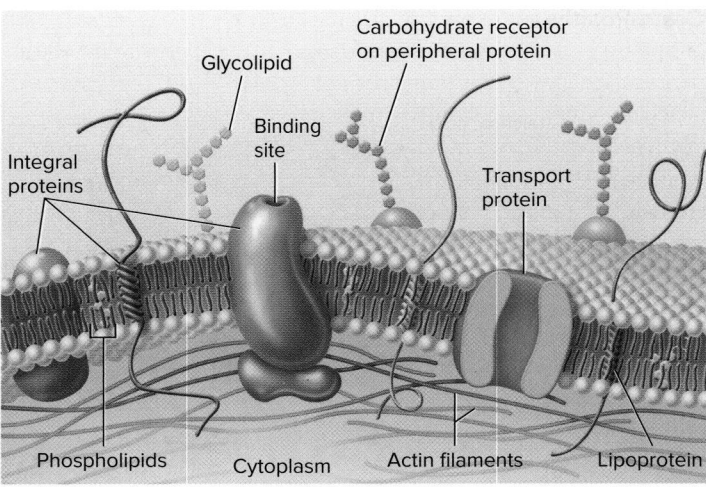

Figure 4.17 Cell membrane structure. A generalized version of the fluid mosaic model of a cell membrane indicates a bilayer of lipids with proteins embedded to some degree in the lipid matrix. Molecules attached to the proteins provide surface features for cell responsiveness and binding. This structure explains many characteristics of membranes, including flexibility, permeability, specificity, and transport.

thelial cells in the lung and causes an atypical form of pneumonia (sometimes referred to as walking pneumonia) in humans.

Some bacteria that ordinarily have a cell wall can lose it during part of their life cycle. These wall-deficient forms are referred to as **L forms** or L-phase variants (for the Lister Institute, where they were discovered). L forms arise naturally from a mutation in the wall-forming genes, or they can be induced artificially by treatment with a chemical such as lysozyme or penicillin that disrupts the cell wall. When a gram-positive cell loses its cell wall completely, it becomes a **protoplast,*** a fragile cell bounded only by a membrane that is highly susceptible to lysis. A gram-negative cells that lose the peptidoglycan layer of its cell wall generally retains its outer membrane, leaving a less fragile but nevertheless weakened **spheroplast.*** Evidence points to a role for L forms in certain chronic infections.

Cell Membrane Structure

A layer in the cell envelope appearing just beneath the cell wall is the **cell,** or **cytoplasmic, membrane,** a very thin (5–10 nm), flexible sheet structured as one complete sheath around the cytoplasm. In general composition, it is a lipid bilayer with proteins embedded to varying degrees **(figure 4.17).** This configuration, first proposed by S. J. Singer and G. L. Nicolson, is called the **fluid mosaic model.** The model describes a membrane as a continuous bilayer formed by lipids with the polar heads oriented toward the outside and the nonpolar tails toward the center of the membrane. Embedded at numerous sites in this bilayer are proteins of various sizes. Peripheral proteins are situated only at the surface and are readily removed. Integral proteins extend fully through the entire membrane and are often fixed in place. The configurations of the inner and outer sides of the membrane can be quite different because of the variations in protein shape and position.

* *protoplast* (proh′-toh-plast) Gr. *proto,* first, and *plastos,* formed.
* *spheroplast* (sfer′-oh-plast) Gr. *sphaira,* sphere.

Membranes are dynamic and constantly changing because the lipid phase is in motion, which allows some proteins to migrate to specific locations where they are needed. This fluidity is essential to such activities as cell enlargement and discharge or secretion by cells. The structure of the lipid phase provides an impenetrable barrier to many substances. This property accounts for the selective permeability and capacity to regulate transport of molecules. Bacterial cell membranes contain primarily phospholipids (making up about 30%–40% of the membrane mass) and proteins (contributing 60%–70%). Major exceptions to this description are the membranes of mycoplasmas, which contain high amounts of sterols—rigid lipids that stabilize and reinforce the membrane—and the membranes of archaea, which contain unique branched hydrocarbons rather than fatty acids. Look back at figure 2.20 to observe eukaryotic membrane structure, which is very similar to bacteria, except for the presence of cholesterol in the eukaryotic variety.

Although bacteria lack complex internal membranous organelles, some members develop stacked layers of internal membranes that carry out physiological processes related to energy and synthesis. These membranes are usually an outgrowth of the cell membrane extending into the cytoplasm, which can increase the membrane surface area available for these reactions. Examples of cells with internal membranes are cyanobacteria, whose thylakoid membranes are the sites of photosynthesis (see figure 4.28c). Even eukaryotic mitochondria, considered bacterial in origin, function by means of a series of internal membranes.

Functions of the Cell Membrane

Because bacteria have none of the eukaryotic organelles, the cell membrane provides a site for energy reactions, nutrient processing, and synthesis. The most important function of the cell membrane, however, is to act as a barrier between the inside and outside of the cell, regulating *transport*—the passage of nutrients into the cell and the discharge of wastes. Although water and small uncharged molecules can diffuse across the membrane unaided, the membrane is a *selectively permeable* structure with special carrier mechanisms for passage of most molecules (see chapter 7). The glycocalyx and cell wall can bar the passage of large molecules, but they are not the primary transport apparatus. The cell membrane is also involved in *secretion,* or the release of metabolic products into the extracellular environment.

The membranes of bacteria are an important site for a number of metabolic activities. For example, most enzymes that handle the energy reactions of respiration reside in the cell membrane (see chapter 8). Enzyme structures located in the cell membrane also help synthesize structural macromolecules to be incorporated into the cell envelope and appendages. Other products (enzymes and toxins) are secreted by the membrane into the extracellular environment.

 Check Your Progress SECTION 4.3

11. Compare the basic structure of the cell envelopes of gram-positive and gram-negative bacteria.

12. Explain the function of peptidoglycan and give a simple description of its structure.

13. What happens to a cell when its peptidoglycan is disrupted or removed?

14. How does the precise structure of the cell walls differ between gram-positive and gram-negative bacteria?

15. What other properties besides staining are different in gram-positive and gram-negative bacteria?

16. What is the periplasmic space, and what is its function?

17. Describe the medical impact of the cell walls of gram-negative and gram-positive bacteria.

18. Describe the structure of the cell membrane, and explain why it is considered selectively permeable.

19. List five essential functions that the cell membrane performs in bacteria.

4.4 Bacterial Internal Structure

Expected Learning Outcomes

15. List the contents of the cell cytoplasm.

16. Describe features of the bacterial chromosome and plasmids.

17. Characterize the bacterial ribosomes and cytoskeleton.

18. Describe inclusion bodies and granules, and explain their importance to cells.

19. Describe the life cycle of endospore-forming bacteria, including the formation and germination of endospores.

20. Discuss the resistance and significance of endospores.

Contents of the Cell Cytoplasm

The cell membrane surrounds a complex solution referred to as **cytoplasm,** or cytoplasmic matrix. This chemical "pool" is a prominent site for many of the cell's biochemical and synthetic activities. Its major component is water (70%–80%), which serves as a solvent for a complex mixture of nutrients including sugars, amino acids, and other organic molecules and salts. The components of this pool serve as building blocks for cell synthesis or as sources of energy. The cytoplasm also holds larger, discrete bodies such as the chromosome, ribosomes, granules, and actin strands.

Bacterial Chromosomes and Plasmids: The Sources of Genetic Information

The hereditary material of most bacteria exists in the form of a single circular strand of DNA designated as the **bacterial chromosome,** although a few bacteria have multiple or linear chromosomes. By definition, bacteria do not have a true nucleus. Their DNA is not enclosed by a nuclear membrane but instead is aggregated in a central area of the cell called the **nucleoid.** The chromosome is actually an extremely long molecule of DNA that is tightly coiled to fit inside the cell compartment. Arranged along its length are genetic units (genes) that carry information required for bacterial

Figure 4.18 Bacterial cells stained to highlight their chromosomes. The individual cocci of *Deinococcus radiodurans* contain one or more prominent bodies that are the chromosomes or nucleoids (1,200×). Some cells are in the process of dividing. Notice how much of the cell's space the chromosomes (light blue bodies) occupy. Courtesy of Michael J. Daly

Ribosome (70S)

Large subunit (50S) Small subunit (30S)

Figure 4.19 A model of a prokaryotic ribosome, showing the small (30S) and large (50S) subunits, both separate and joined.

maintenance and growth. When exposed to special stains or observed with an electron microscope, chromosomes have a granular or fibrous appearance **(figure 4.18).**

Although the bacterial chromosome contains all the genetic material a cell needs to survive, many bacteria contain additional genetic elements called **plasmids.** Plasmids are generally small, circular pieces of DNA that exist independently within the cytoplasm, although at times they may become integrated into the much larger bacterial chromosome. During bacterial reproduction, they are duplicated and passed on to offspring. Although plasmids are not essential for survival, they often carry genes that confer favorable traits on the bacterial cell, such as toxin production or resistance to certain antibiotics (see chapter 9). Because they can be readily manipulated in the laboratory and transferred from one bacterial cell to another, plasmids are an important component of modern genetic engineering techniques.

Ribosomes: Sites of Protein Synthesis

Every bacterial cell contains tens of thousands of **ribosomes,** which are made of RNA and protein. When viewed even by very high magnification, ribosomes show up as fine, spherical specks dispersed throughout the cytoplasm that often occur in chains (polysomes). Many are also attached to the cell membrane. Chemically, a ribosome is a combination of a special type of RNA called ribosomal RNA, or rRNA (about 60%), and protein (40%).

One method of characterizing ribosomes is by S, or Svedberg,[4] units, which rate the molecular sizes of various cell parts that have been spun down and separated by molecular weight and shape in a centrifuge. Heavier, more compact structures sediment faster and are assigned a higher S rating. Combining this method of analysis

with high-resolution electron micrography has revealed that the prokaryotic ribosome, which has an overall rating of 70S, is actually composed of two smaller subunits **(figure 4.19).** They fit together to form a miniature "factory" where protein synthesis occurs. We examine the more detailed functions of ribosomes in chapter 9.

Inclusions: Structures Within the Cell

Although prokaryotic cells lack organelles, they do possess compartmentalized structures within the cytoplasm that perform a variety of functions. For example, most bacteria are exposed to severe shifts in the availability of food. During periods of nutrient abundance, some can compensate by storing nutrients as **inclusion bodies,** or **inclusions,** of varying size, number, and content. As the environmental source of these nutrients becomes depleted, the bacterial cell can mobilize its own storehouse as required. Some inclusion bodies contain condensed, energy-rich organic substances, such as glycogen and poly β-hydroxybutyrate (PHB), within special single-layered membranes **(figure 4.20a).** Gas vesicles are a unique type of inclusion found in some aquatic bacteria that provide buoyancy and flotation. Other inclusions, also called **granules,** contain crystals of inorganic compounds and are not enclosed by membranes. Sulfur granules of photosynthetic bacteria and polyphosphate granules of *Corynebacterium* and *Mycobacterium* are of this type. The latter are an important source of building blocks for nucleic acid and ATP synthesis. They have been termed **metachromatic granules** because they stain a contrasting color (red, purple) in the presence of methylene blue dye.

A second type of inclusion does far more than simply store nutrients for later use. **Microcompartments** are polyhedral (many-sided) packets formed by proteins bound together in compact units **(figure 4.20b).** The microcompartment encloses one or more enzymes along with the reactants needed for a specific metabolic process. The best-understood microcompartment is the carboxysome, found in many cyanobacteria and other photosynthetic bacteria. It is the storage site of the enzyme responsible for CO_2 fixation, the process of converting CO_2 into organic compounds like sugar.

4. Named in honor of T. Svedberg, the Swedish chemist who developed the ultracentrifuge in 1926.

(a) (b) (c)

Figure 4.20 Bacterial inclusion bodies. (a) Large, dark particles of polyhydroxybutyrate are deposited in an insoluble, concentrated form that provides an ample, long-term supply of that nutrient (32,500×). **(b)** Rendering of a bacterial microcompartment. The outer protein shell (blue) encloses the enzymes (green) required for a specific metabolic process. **(c)** A section through *Aquaspirillum* reveals a chain of tiny iron magnets (magnetosomes = MP). These unusual bacteria use these inclusions to orient within their habitat (123,000×). (a): © Kwangshin Kim/Science Source; (c): © D. Balkwill and D. Maratea

Perhaps the most remarkable cell granule is involved neither in nutrition nor in metabolism but rather in navigation. Magnetotactic bacteria contain crystalline particles of iron oxide (magnetosomes) that have magnetic properties **(figure 4.20c).** These granules occur in a variety of bacteria living in oceans and swamps. Their primary function is to orient the cells in the earth's magnetic field, somewhat like a compass. It is thought that magnetosomes direct these bacteria into locations with favorable oxygen levels or nutrient-rich sediments.

The Bacterial Cytoskeleton

Like eukaryotes, prokaryotic cells contain a cytoskeleton.[5] Protein polymers running throughout the cell play a crucial role in defining cell shape, along with aiding the processes of cell division, partitioning of cellular structures like DNA and microcompartments, and cell motility. Some cytoskeletal proteins are analogous to the actin, tubulin, and intermediate filament proteins found in eukaryotes, while others are unique to bacterial cells. Several of these cytoskeleton systems have been studied, but the best-understood is made of actin filaments that curl within the body of the cell and help maintain its shape during cell enlargement **(figure 4.21).**

What Exactly Is a Spore?

The word *spore* can have more than one usage in microbiology. It is a generic term that refers to any of the tiny compact cells that are produced by vegetative or reproductive structures of microorganisms. Spores can be quite variable in origin, form, and function. The bacterial type discussed next is more properly called an endospore, because it is produced inside a cell. It functions in *survival,* not in reproduction. In contrast, the fungi produce many different types of spores, which serve primarily as a means of reproduction (see chapter 5).

Actin filaments

Figure 4.21 Bacterial cytoskeleton of *Bacillus*. Fluorescent stain of actin fibers appears as fine helical ribbons wound inside the cell. The inset provides a three-dimensional interpretation of its structure. © Rut Carballido-Lopez/I.N.R.A. Jouy-en-Josas, Laboratoire de Génétique Microbienne

Bacterial Endospores: An Extremely Resistant Life Form

Ample evidence indicates that the anatomy of bacteria helps them adjust rather well to adverse habitats. But of all microbial structures, nothing can compare to the bacterial **endospore** (or simply spore) for withstanding hostile conditions and facilitating survival. Endospores are dormant bodies produced by *Bacillus, Clostridium,* and several other bacterial genera. These bacteria have a two-phase life cycle that shifts between a **vegetative cell (figure 4.22,** step 1) and an endospore (figure 4.22, step 8). The vegetative cell is the metabolically active and growing phase. When exposed to certain environmental signals, it forms an endospore by a process termed **sporulation.** The spore exists in an inert, resting condition that is capable of extreme resistance and very long-term survival.

5. An intracellular framework of fibers and tubules that bind and support eukaryotic cells.

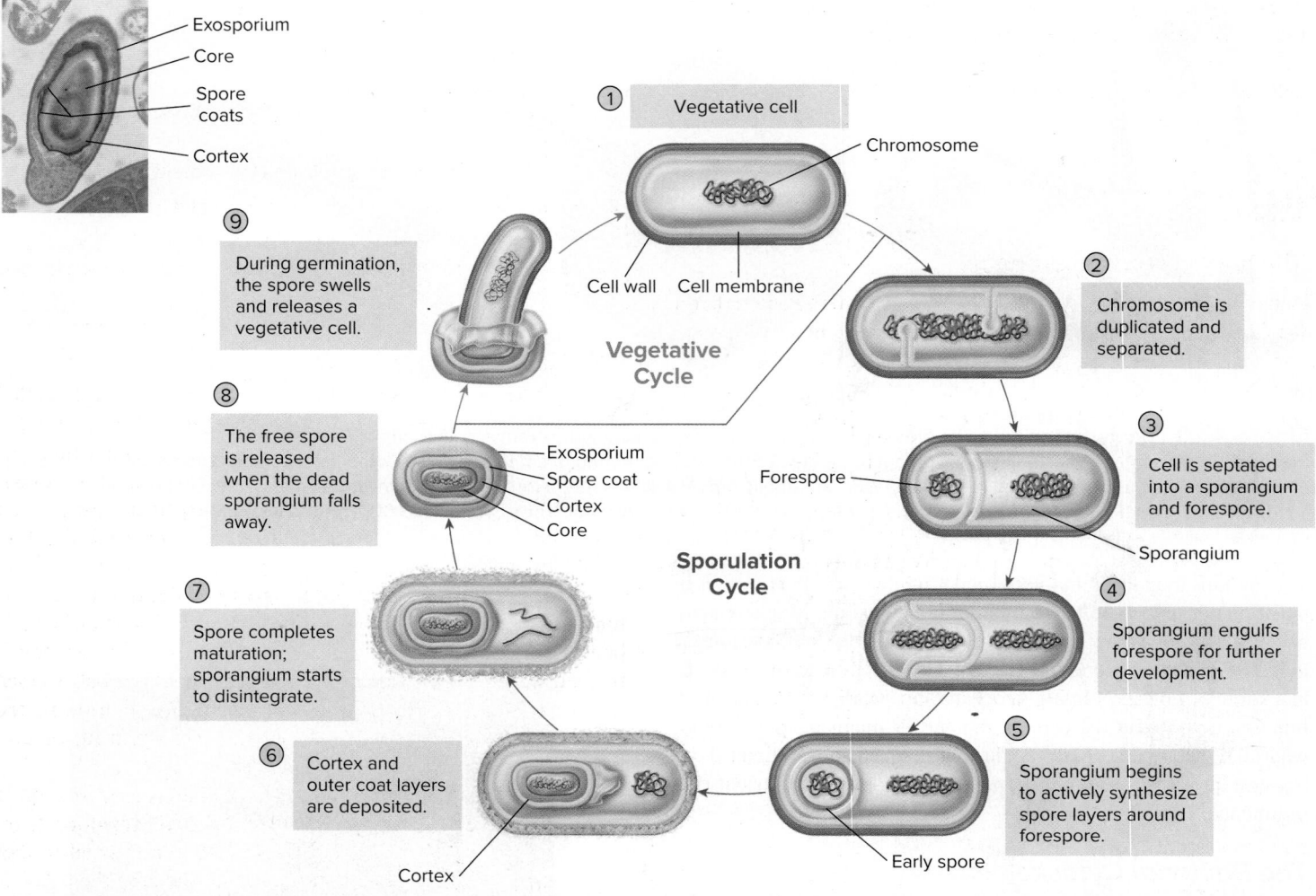

Exosporium
Core
Spore coats
Cortex

① Vegetative cell

Chromosome

Cell wall Cell membrane

Vegetative Cycle

⑨ During germination, the spore swells and releases a vegetative cell.

② Chromosome is duplicated and separated.

③ Cell is septated into a sporangium and forespore.

Forespore

Sporangium

Sporulation Cycle

⑧ The free spore is released when the dead sporangium falls away.

Exosporium
Spore coat
Cortex
Core

④ Sporangium engulfs forespore for further development.

⑦ Spore completes maturation; sporangium starts to disintegrate.

⑤ Sporangium begins to actively synthesize spore layers around forespore.

⑥ Cortex and outer coat layers are deposited.

Cortex

Early spore

Process Figure 4.22 A typical sporulation cycle from the active vegetative cell to release and germination. This process takes, on average, about 10 hours. Inset is a high magnification (10,000×) cross section of a single spore showing the dense protective layers that surround the core with its chromosome. (photo): SJ Jones, CJ Paredes, B Tracy, N Cheng, R Sillers, RS Senger, ET Papoutsakis, "The transcriptional program underlying the physiology of clostridial sporulation," *Genome Biol.,* 2008. 9:R114

Endospore Formation and Resistance

The major stimulus for endospore formation is the depletion of nutrients, especially amino acids. Once this stimulus has been received by the vegetative cell, it converts to a committed sporulating cell called a **sporangium.** Complete transformation of a vegetative cell into a sporangium and then into an endospore requires 6 to 12 hours in most spore-forming species. Figure 4.22 illustrates some major physical and chemical events in this process.

Bacterial endospores are the hardiest of all life forms, capable of withstanding extremes in heat, drying, freezing, radiation, and chemicals that would readily kill ordinary cells. Their survival under such harsh conditions is due to several factors. The heat resistance of spores has been linked to their high content of calcium and *dipicolinic acid,* although the exact role of these chemicals is not yet clear. We know, for instance, that heat destroys cells by inactivating proteins and DNA and that this process requires a certain amount of water in the protoplasm. Because the deposition of calcium dipicolinate in the endospore removes water and leaves the endospore very dehydrated, it is less vulnerable to the effects of

heat. It is also metabolically inactive and highly resistant to damage from further drying. The thick, impervious cortex and spore coats provide additional protection against radiation and chemicals.

The longevity of bacterial spores verges on immortality! One record describes the isolation of viable endospores from a fossilized bee that was 25 million years old. More recently, microbiologists unearthed a viable endospore from a 250-million-year-old salt crystal. Initial analysis of this ancient microbe indicates it is a species of *Bacillus* that is genetically different from known species.

The Germination of Endospores

After lying in a state of inactivity, endospores can be revitalized when favorable conditions arise. The breaking of dormancy, or **germination,** happens in the presence of water and a specific germination agent. Once initiated, it proceeds to completion quite rapidly (1½ hours). Although the specific germination agent varies among species, it is generally a small organic molecule such as an amino acid or an inorganic salt. This agent stimulates the formation of hydrolytic (digestive) enzymes by the endospore membranes.

These enzymes digest the cortex and expose the core to water. As the core rehydrates and takes up nutrients, it begins to grow out of the endospore coats. In time it reverts to a fully active vegetative cell, resuming the vegetative cycle (figure 4.22).

 CLINICAL CONNECTIONS

The Impact of Endospores

Although the majority of spore-forming bacteria are relatively harmless, several bacterial pathogens are sporeformers. In fact, some aspects of the diseases they cause are related to the persistence and resistance of their spores. Spores of *Bacillus anthracis,* the agent of anthrax, are classified as a high-level bioterrorism agent. The genus *Clostridium* includes even more pathogens, including *C. tetani,* the cause of tetanus (lockjaw), and *C. perfringens,* the cause of gas gangrene. When the spores of these species are embedded in a wound that contains dead tissue, they can germinate, grow, and release potent toxins. Another toxin-forming species, *C. botulinum,* is the agent of botulism, a deadly form of food poisoning. These diseases are discussed further in chapter 19.

Because they inhabit the soil and dust, endospores are a constant intruder where sterility and cleanliness are important. They resist ordinary cleaning methods that use boiling water, soaps, and disinfectants, and they frequently contaminate cultures and media. Hospitals and clinics must take precautions to guard against the potential harmful effects of endospores in wounds. Endospore destruction is a particular concern of the food-canning industry. Several endospore-forming species cause food spoilage or poisoning. Ordinary boiling (100°C) will usually not destroy such spores, so canning is carried out in pressurized steam at 120°C for 20 to 30 minutes. Such rigorous conditions will ensure that the food is sterile and free from viable bacteria.

***Clostridium difficile* is an emerging pathogen that causes thousands of intestinal infections in hospital patients every year. What qualities of this bacterium make it such a serious concern?** *Answer available on Connect.*

 Check Your Progress SECTION 4.4

20. Compare the functions of the bacterial chromosome (nucleoid) and plasmids.
21. What is unique about the structure of bacterial ribosomes, what is their function, and where are they located?
22. Compare and contrast the structure and function of inclusions and granules.
23. What are metachromatic granules, and what do they contain?
24. Describe the way endospores are formed, their structure, and their importance in the life cycle.
25. Explain why an endospore is not considered a reproductive body.
26. Why are spores so difficult to destroy?

4.5 Bacterial Shapes, Arrangements, and Sizes

 Expected Learning Outcomes

21. Describe the shapes of bacteria and their possible variants.
22. Identify several arrangements of bacteria and how they are formed.
23. Outline the size ranges among bacteria and in comparison to other organisms.

For the most part, bacteria function as independent single-celled, or unicellular, organisms. Although it is true that an individual bacterial cell can live attached to others in colonies or other such groupings, each one is fully capable of carrying out all necessary life activities, such as reproduction, metabolism, and nutrient processing unlike the more specialized cells of a multicellular organism.

Bacteria exhibit considerable variety in shape, size, and colonial arrangement. It is convenient to describe most bacteria by one of three general shapes as dictated by the configuration of the cell wall **(figure 4.23).** If the cell is spherical or ball-shaped, the bacterium is described as a **coccus.*** Cocci can be perfect spheres, but they also can exist as oval, bean-shaped, or even pointed variants.

A cell that is cylindrical (longer than wide) is termed a rod, or **bacillus.*** There is also a genus named *Bacillus* named for its rod shape. As might be expected, rods are also quite varied in their actual form. Depending on the bacterial species, they can be blocky, spindle-shaped, round-ended, long and threadlike (filamentous), or even clubbed or drumstick-shaped. When a rod is short and plump, it is called a **coccobacillus;** if it is gently curved, it is a **vibrio.***

A bacterium having the shape of a curviform or spiral-shaped cylinder is called a **spirillum,*** a rigid helix, twisted twice or more along its axis (like a corkscrew). Another spiral cell mentioned earlier in conjunction with periplasmic flagella is the **spirochete,** a more flexible form that resembles a spring. Refer to **table 4.2** for a comparison of other features of the two helical bacterial forms. Because bacterial cells look two-dimensional and flat with traditional staining and microscope techniques, they are best observed using a scanning electron microscope to emphasize their striking three-dimensional forms (figure 4.23).

Quick Search To appreciate their differences, look for "Spirochetes That Cause Lyme Disease" and "Motile Spirilla" on YouTube.

* *coccus* (kok′-us) pl. cocci (kok′-seye) Gr. *Kokkos,* berry.
* *bacillus* (bah-sil′-lus) pl. bacilli (bah-sil′-eye) L. bacill, small staff or rod.
* *vibrio* (vib′-ree-oh) L. vibrare, to shake.
* *spirillum* (spy-ril′-em) pl. spirilla; L. spira, a coil.

Key to Micrographs

(a) *Staphylococcus aureus* (10,000×) (b) *Legionella pneumophila* (6,500×) (c) *Vibrio cholerae* (13,000×) (d) *Aquaspirillum* (7,500×)
(e) *Borrelia burgdorferi* (10,000×) (f) *Streptomyces* species (1,000×)

Figure 4.23 Common bacterial shapes. Drawings show examples of shape variations for cocci, rods, vibrios, spirilla, spirochetes, and branching filaments. Below each shape is a micrograph of a representative example. (a): *Source:* Janice Carr/CDC; (b): *Source:* Janice Carr/CDC; (c): © BSIP SA/Alamy Stock Photo; (d): *Source:* Photo by De Wood. Digital colorization by Chris Pooley; (e): *Source:* Janice Haney Carr/CDC; (f): *Source:* Dr. David Berd/CDC

TABLE 4.2 Comparison of the Two Spiral-Shaped Bacteria

	Overall Appearance Helical or Spiral Forms	Mode of Locomotion	Number of Helical Turns	Gram Reaction (Cell Wall Type)	Examples of Important Types
Spirilla	Rigid helix Spirilla	Polar flagella; cells swim by rotating around like corkscrews; do not flex; have one to several flagella; can be in tufts	Varies from 1 to 20	Gram-negative	Most are nonpathogenic. One species, *Spirillum minor*, causes rat bite fever.
Spirochetes	Flexible helix Spirochete	Periplasmic flagella within sheath; cells flex; can swim by rotation or by creeping on surfaces; have 2 to 100 periplasmic flagella	Varies from 3 to 70	Gram-negative	*Treponema pallidum*, cause of syphilis; *Borrelia* and *Leptospira*, important pathogens

Figure 4.24 Pleomorphism and other morphological features of *Corynebacterium*. Cells are irregular in shape and size (800×). This genus typically exhibits a palisades arrangement, with cells in parallel array (inset). Close examination will also reveal darkly stained metachromatic granules. *Source:* Dr. P.B. Smith/CDC

It is common for cells of a single species to show **pleomorphism*** **(figure 4.24).** This is due to individual variations in cell wall structure caused by nutritional or slight hereditary differences. For example, although the cells of *Corynebacterium diphtheriae* are generally considered rod-shaped, in culture they display club-shaped, swollen, curved, filamentous, and coccoid variations. Pleomorphism reaches an extreme in the mycoplasmas, which entirely lack cell walls and thus display extreme variations in shape (see figure 4.16).

Bacterial cells can also be categorized according to **arrangement,** or style of grouping. The main factors influencing the arrangement of a particular cell type are its pattern of division and how the cells remain attached afterward. The greatest variety in arrangement occurs in cocci **(figure 4.25).** They may exist as singles, in pairs **(diplococci*),** in **tetrads** (groups of four), in irregular clusters (both **staphylococci*** and **micrococci**), or in chains of a few to hundreds of cells **(streptococci).** An even more complex grouping is a cubical packet of 8, 16, or more cells called a **sarcina.*** These different coccal groupings are the result of the division of a coccus in a single plane, in two perpendicular planes, or in several intersecting planes; after division, the resultant daughter cells remain attached.

Bacilli are less varied in arrangement because they divide only in the transverse plane (perpendicular to the axis). They occur either as single cells, as a pair of cells with their ends attached

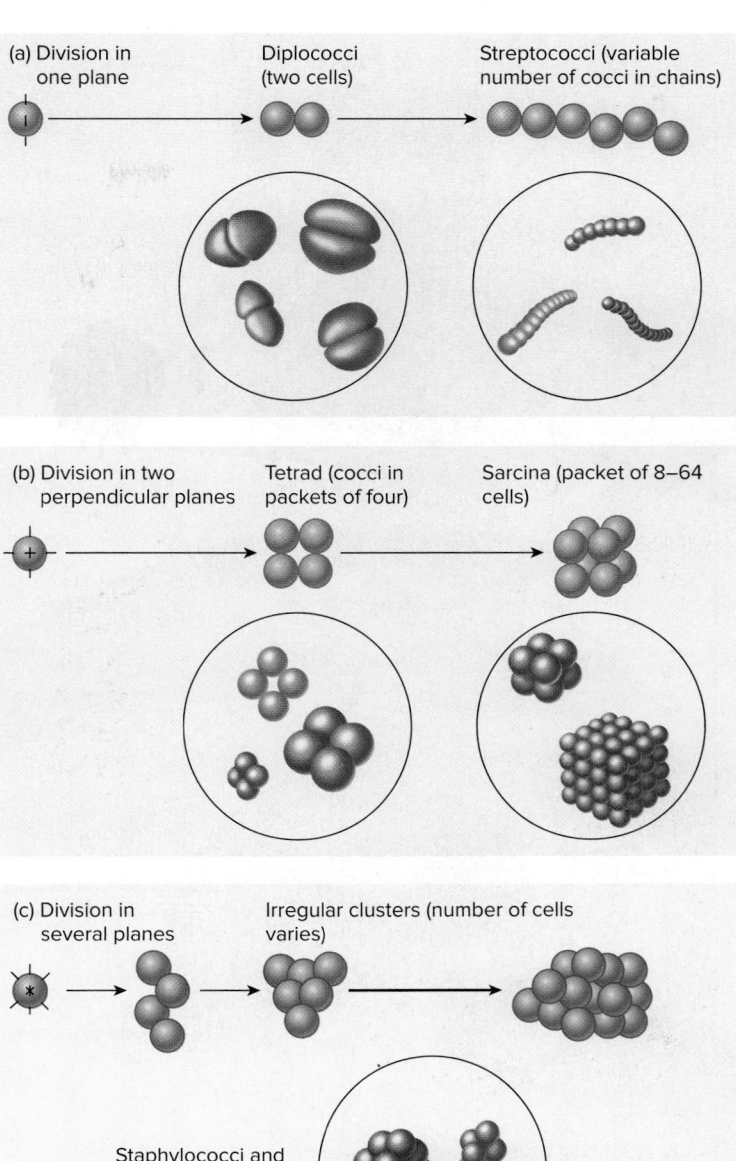

Figure 4.25 Arrangement of cocci resulting from different planes of cell division. (a) Division in one plane produces diplococci and streptococci. **(b)** Division in two planes at right angles produces tetrads and packets. **(c)** Division in several planes produces irregular clusters.

(diplobacilli), or as a chain of several cells **(streptobacilli).** A **palisades*** arrangement, typical of the corynebacteria, is formed when the cells of a chain remain partially attached by a small hinge region at the ends. The cells tend to fold (snap) back upon each other, forming a row of cells oriented side by side (see figure 4.24). The reaction can be compared

Quick Search
For a fascinating comparison of the size of microscopic structures, search for "Size and Scale" on YouTube.

* *pleomorphism* (plee″-oh-mor′-fizm) Gr. Pleon, more, and morph, form or shape.
* *diplococci* (dih-plo-kok-seye) Gr. diplo, double.
* *staphylococci* (staf″-ih-loh-kok′-seye) Gr. staphyle, a bunch of grapes.
* *sarcina* (sar′-sin-uh) L. sarcina, a packet.

* *palisades* (pal′-ih-saydz) L. pale, a stake. A fence made of a row of stakes.

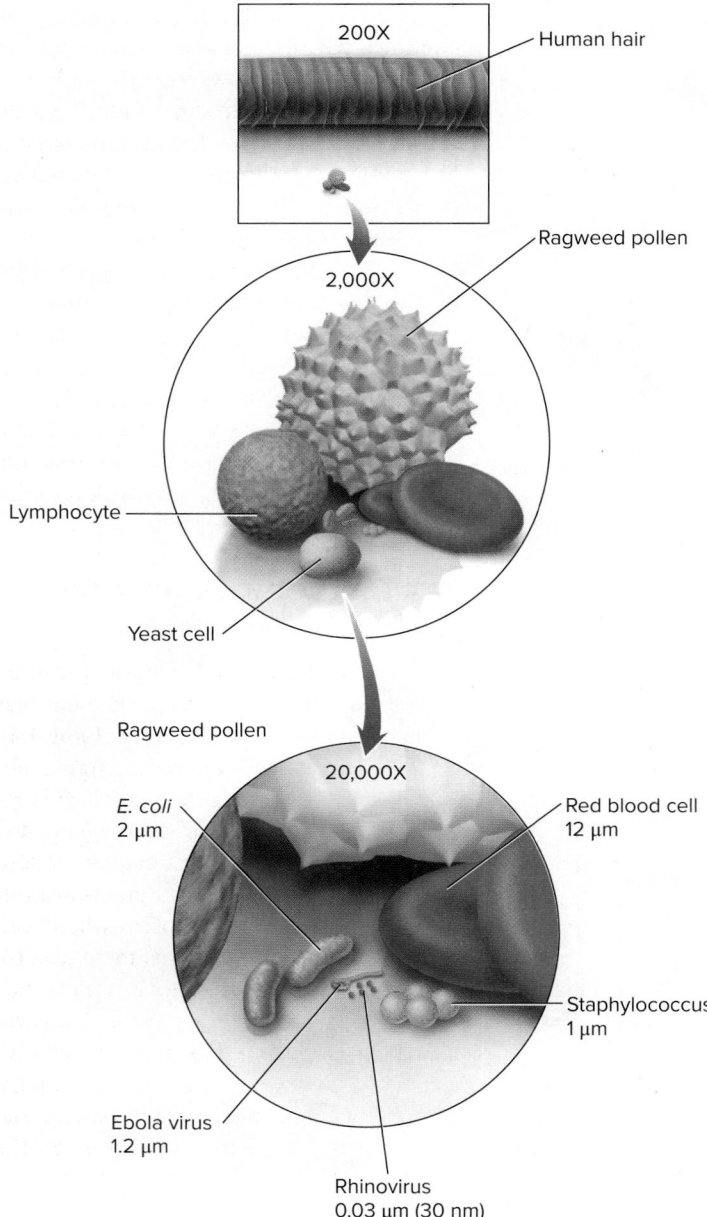

200X — Human hair

2,000X

Ragweed pollen

Lymphocyte

Yeast cell

Ragweed pollen

20,000X

E. coli
2 μm

Red blood cell
12 μm

Staphylococcus
1 μm

Ebola virus
1.2 μm

Rhinovirus
0.03 μm (30 nm)

Figure 4.26 **The dimensions of bacteria.** The sizes of bacteria range from those just barely visible with light microscopy (0.2 μm) to those measuring a thousand times that size. Cocci measure anywhere from 0.5 to 3.0 μm in diameter; bacilli range from 0.2 to 2.0 μm in diameter and from 0.5 to 20 μm in length. Note the range of sizes as compared with eukaryotic cells and viruses. Comparisons are given as average sizes.

to the behavior of boxcars on a jackknifed train, and the result looks superficially like an irregular picket fence. Spirilla are occasionally found in short chains, but spirochetes rarely remain attached after division. Comparative sizes of typical cells are presented in **figure 4.26.**

 Check Your Progress SECTION 4.5

27. Classify bacteria according to their basic shapes.
28. How are spirochetes and spirilla different?

29. What are vibrios and coccobacilli?
30. Describe pleomorphism, and give examples of bacteria with this trait.
31. What is the difference between the use of the shape bacillus and the name *Bacillus*? *Staphylococcus* and staphylococcus?
32. Rank the size ranges of bacteria according to shape, and compare bacterial size with viruses and eukaryotic cells.

4.6 Classification Systems of Prokaryotic Domains: Archaea and Bacteria

 Expected Learning Outcomes

24. Describe the purposes of classification and taxonomy in the study of prokaryotes.

25. Summarize characteristics used to classify bacteria.

26. Outline a basic system of bacterial taxonomy.

27. Explain the species and subspecies levels for bacteria.

Classification systems serve both practical and academic purposes. They aid in differentiating and identifying unknown species in medical and applied microbiology. They are also useful in organizing bacteria and as a means of studying their relationships and evolutionary origins. Since classification was first initiated around 200 years ago, several thousand species of bacteria and archaea have been identified, named, and catalogued.

Tracing the origins of and evolutionary relationships among bacteria has not been an easy task. As a rule, tiny, relatively soft organisms do not form fossils very readily. Several times since the 1960s, however, scientists have discovered billion-year-old fossils of prokaryotes that look very much like modern bacteria (see figure 4.28c).

One of the questions that has plagued taxonomists is, What characteristics are reliable indicators of closeness in ancestry? Early bacteriologists found it convenient to classify bacteria according to shape, variations in arrangement, growth characteristics, and habitat. However, as more species were discovered and as techniques for studying their biochemistry were developed, it soon became clear that similarities in cell shape, arrangement, and staining reactions do not automatically indicate relatedness. Even though gram-negative rods look alike, there are hundreds of different species with highly significant differences in biochemistry and genetics. If we attempted to classify them on the basis of Gram stain and shape alone, we could not assign them to a more specific level than Phylum! As a result, the newer classification schemes are incorporating more specific genetic and molecular data to increase accuracy.

In general, the methods a microbiologist uses to identify bacteria to the level of genus and species fall into the categories of morphology (microscopic and macroscopic), bacterial physiology or biochemistry, serological analysis, and genetic techniques (see chapter 17 and **Appendix table C.1**). Data from a cross section of such tests can produce a unique profile of each species. Many identification systems are automated and computerized to process data

Figure 4.27 A master classification scheme based on ribosomal analysis. This pattern displays the major phyla within Domains Bacteria and Archaea. Branches indicate the origins of each group, and their relative positions estimate ancestral relationships. Major groups of Domain Eukarya are shown here for comparison.

and provide a "best fit" identification. However, not all methods are used on all bacteria. A few bacteria can be identified by a special technology that analyzes only the kind of fatty acids they contain; in contrast, some are identifiable by a Gram stain and a few physiological tests; others may require a diverse spectrum of morphological, biochemical, and genetic analyses.

Bacterial Taxonomy: A Work in Progress

There is no single official system for classifying the prokaryotes. Indeed, most plans are in a state of flux as new information and methods of analysis become available. One widely used reference is *Bergey's Manual of Systematic Bacteriology,* a major resource that covers all known prokaryotes. In the past, the classification scheme in this guide has been based mostly on characteristics such as Gram stain and metabolic reactions. This is called a **phenetic,** or phenotypic, method of classification.

The second edition of this large collection now includes genetic information that clarifies the **phylogenetic** (evolutionary) history and relationships of the thousands of known species **(figure 4.27).** This change has redefined our presentation of their classification, so that prokaryotes are now placed in five major subgroups and 25 different phyla instead of two domains split into four divisions. The Domains Archaea and Bacteria based on genetic characteristics have been retained, but the bacteria of clinical importance are no longer as closely aligned, and the 250 or so species that cause disease in humans are found within seven or eight

of the revised phyla. The grouping by Gram reaction remains significant but primarily at the lower taxonomic levels. This being an introduction to the subject, we will follow a pragmatic organization that simplifies the presentation and avoids increasingly complex and often conflicting taxonomy.

The major categories of this revised taxonomic scheme are presented in **table 4.3,** which provides a brief survey of characteristics of the major taxonomic groups, along with examples of representative members (A–O). For a complete version of the major taxonomic groups and genera, go to **Appendix table C.2.**

A Diagnostic Scheme for Medical Use

Many medical microbiologists prefer an informal working system that outlines the major families and genera **(table 4.4).** This scheme uses the phenotypic qualities of bacteria in identification. It is restricted to bacterial disease agents and depends less on nomenclature. It also divides the bacteria into gram-positive, gram-negative, and those without cell walls and then subgroups them according to cell shape, arrangement, and certain physiological traits such as oxygen usage: *Aerobic* bacteria use oxygen in metabolism; *anaerobic* bacteria do not use oxygen in metabolism; and facultative bacteria may or may not use oxygen. Further tests not listed on the table would be required to separate closely related genera and species. Many of these are included in later chapters on specific bacterial groups.

Species and Subspecies in Bacteria

Among most organisms, the **species** level is a distinct, readily defined, and natural taxonomic category. In animals, for instance, a species is a distinct type of organism that can produce viable offspring only when it mates with others of its own kind. This definition does not work for bacteria primarily because they do not exhibit a typical mode of sexual reproduction. They can accept DNA from unrelated forms, and they can also alter their genetic makeup by a variety of mechanisms. Thus, it is necessary to use a modified definition for a bacterial species. Theoretically, it is a collection of bacterial cells, all of which share an overall similar pattern of traits, in contrast to other groups whose pattern differs significantly. Although the boundaries that separate two closely related species in a genus can be somewhat arbitrary, this definition still serves as a method to separate the bacteria into various kinds that can be cultured and studied. As additional information on bacterial genomes is discovered, it may be possible to define species according to specific combinations of genetic codes found only in a particular

TABLE 4.3　Overview of a Classification System for Prokaryotes Based on *Bergey's Manual,* 2nd Edition (See Appendix Table C.2 for a full outline.)

Volume 1 Domain Archaea Includes prokaryotes with unusual morphology, ecology, and modes of nutrition. Many of the known members are adapted to a habitat with extremes of temperature (hyperthermophiles), salt (halophiles), acidity (acidophiles), pressure, or lack of oxygen (anaerobic), and sometimes combinations of these. The domain has great ecological importance due to its actions in biogeochemical cycling. More coverage can be found in section 4.7.

Phylum Crenarchaeota Members depend on sulfur for growth and may inhabit hot and acidic sulfur pools and vents. Some species have adapted to cold habitats. See examples in **figures A** and 4.33.

Phylum Nanoarchaeota A newly discovered group of extremely small archaea found in salt mines and caves (see 4.1 Secret World of Microbes).

Phylum Euryarchaeota The largest and best-studied group of archaea; members' adaptations include methane production (methanogens), high saline or salt (halobacteria or halophiles), high temperature (thermophiles), and reduction of sulfur compounds. An example is *Methanocaldococcus* (**figure B**).

A. Unidentified thermophilic crenarchaeotes isolated from a deep-sea thermal vent at 160°F (71°C). *Source:* West Coast and Polar Regions Undersea Research Center, UAF/NOAA

B. *Methanocaldococcus*—a hyperthermophile inhabiting hydrothermal vents at near-boiling temperature. *Source:* Maryland Astrobiology Consortium, NASA and STScI

Volume 1 Domain Bacteria I This domain contains deeply branching and photosynthetic bacteria that are considered the oldest evolutionary lines still in existence. It contains a wide diversity of prominent bacteria, including photosynthesizers (cyanobacteria), thermophiles, radiation-resistant bacteria, halophiles, and sulfur metabolizers.

Phylum Aquificae This group contains small thermophilic rods that inhabit underwater volcanoes.

Phylum Thermotogae A phylum similar to Aquificae; includes thermophilic halophiles that live in deep-sea vents. An example is *Desulfurobacterium* (**figure C**).

Phylum Chlorobi Also known as green sulfur bacteria, this phylum consists of anaerobic bacteria that live in the muddy layers of lakes and ponds, where they photosynthesize and metabolize sulfides.

Phylum Deinococcus-Thermus A small phylum of extremophiles whose members range from being highly radiation- and desiccation-resistant cocci that live in soil and fresh water to rods that live in very hot aquatic habitats, such as hot springs and pools. *Deinococcus* can be seen in figure 4.18.

Phylum Cyanobacteria Very ubiquitous photosynthetic bacteria found in aquatic habitats, soil, and often associated with plants, fungi, and other organisms; blue-green, green, yellow, red, and orange in color. An example is *Gloeotrichia* (**figure D**; see section 4.7 for more coverage).

C. *Desulfurobacterium*—a tiny anaerobic, halophilic rod that reduces sulfur compounds. *Source:* From M. Goker et al., "Complete genome sequence of the thermophilic sulfur-reducer *Desulfurobacterium thermolithotrophum* type strain (BSAᵀ) from a deep-sea hydrothermal vent," *SIGS,* 2011, 5:3, Fig. 2

D. *Gloeotrichia*—a cyanobacterium that forms colonies of radiating filaments that float in lakes and ponds. © Barry H. Rosen, Ph. D/USGS

Volume 2 Domain Bacteria II This domain, along with the phyla that are covered in volumes 3, 4, and 5, contains bacteria that have the greatest medical impact.

Phylum Proteobacteria This phylum contains five classes representing an extremely varied cross section of over 2,000 identified species of bacteria. Though all share the characteristic of a gram-negative cell wall, the group contains bacteria with a wide range of adaptations, shapes, habitats, and ecology. Medically significant members include the obligately parasitic rickettsias; *Neisseria;* enterics such as *Escherichia coli* and *Salmonella; Vibrio* species; and other bacteria that live in the intestinal tracts of animals. The myxobacteria have the unique characteristic of being the only bacteria that can aggregate into multicellular structures (see figure 4.30; see also **figures E, F,** and **G** for examples).

E. A *Rickettsia rickettsii* cell (cause of Rocky Mountain spotted fever) being engulfed by a host cell. *Source:* Dr. Edwin P Ewing. Jr/CDC

F. *Escherichia coli* O157:H7—a food-borne pathogen. *Source:* Janice Carr/CDC

G. A vibrio-shaped thermophile, *Desulfovibrio,* living in a pond biofilm. *Source:* Pacific Northwest National Laboratory

TABLE 4.3 *(continued)*

Volume 3 Phylum Firmicutes This collection of mostly gram-positive bacteria is characterized by having a low G + C content* (less than 50%). The three classes in the phylum display significant diversity, and a number of the members are pathogenic. Endospore-forming genera include *Bacillus* and *Clostridium.* Other important pathogens are found in genera *Staphylococcus* and *Streptococcus.* Although they lack a cell wall entirely, mycoplasmas (see figure 4.17) have been placed with the Firmicutes because of their genetic relatedness. (See **figures H** and **I.**)

H. *Bacillus anthracis*—SEM micrograph showing the rod-shaped cells next to a red blood cell. *Source:* Arthur Friedlander

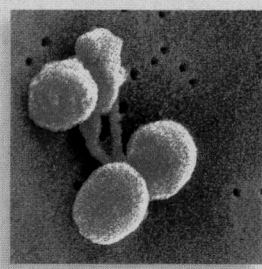

I. *Streptococcus pneumoniae*—image displays the diplococcus arrangement of this species. *Source:* Janice Carr/CDC

Volume 4 Phylum Actinobacteria This taxonomic category includes the high G + C (over 50%) gram-positive bacteria. Members of this small group differ considerably in life cycles and morphology. Prominent members include the branching filamentous Actinomycetes, the spore-bearing Streptomycetes, *Corynebacterium* (see figure 4.24), *Mycobacterium,* and *Micrococcus* (see figure 4.23*a*). (See **figures J** and **K.**)

J. *Streptomyces* species—common soil bacteria; often the source of antibiotics. *Source:* Dr. David Berd/CDC

K. *Mycobacterium tuberculosis*—the bacillus that causes tuberculosis. *Source:* Janice Carr/CDC

Volume 5 This represents a mixed assemblage of nine phyla, all of which are gram-negative but otherwise widely varied. The following is a selected array of examples.

Phylum Chlamydiae Another group of obligate intracellular parasites that reproduce inside host cells. These are among the smallest of bacteria, with a unique mode of reproduction. Several species cause diseases of the eyes, reproductive tract, and lungs. An example is *Chlamydia* (**figure L**).

Phylum Spirochetes These bacteria are distinguished by their shape and mode of locomotion. They move their slender, twisted cells by means of periplasmic flagella. Members live in a variety of habitats, including the bodies of animals and protozoans, fresh and marine water, and even muddy swamps. Important genera are *Treponema* (**figure M**) and *Borrelia* (see figure 4.23*e*).

Phylum Planctomycetes This group lives in fresh and marine water habitats and reproduces by budding. Many have a stalk that they use to attach to substrates. A unique feature is having a membrane around their DNA and special compartments enclosed in membranes. This has led to the speculation that they are similar to an ancestral form that gave rise to eukaryotes. An example is *Gemmata* (**figure N**).

Phylum Bacteriodetes These are widely distributed gram-negative anaerobic rods inhabiting soil, sediments, and water habitats, and frequently found as normal residents of the intestinal tracts of animals. They may be grouped with related Phyla Fibrobacteres and Chlorobi. Several members play an important role in the function of the human gut and some are involved in oral and intestinal infections. An example is *Bacteroides* (**figure O**).

L. View of an infected host cell revealing a vacuole containing *Chlamydia* cells in various stages of development. *Source:* N. Borel et al., "Mixed infections with Chlamydia and porcine epidemic diarrhea virus - a new in vitro model of chlamydial persistence," *BMC Microbiology* 2010, 10:201, Fig. 3a

M. *Treponema pallidum*—spirochetes that cause syphilis. *Source:* Joyce Ayers/CDC

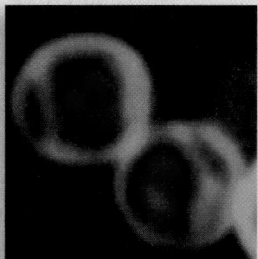

N. *Gemmata*—view of a budding cell through a fluorescent microscope (note the large blue nucleoid). *Source:* K-C Lee, R Webb, JA Fuerst, "The cell cycle of the planctomycete *Gemmata obscuriglobus* with respect to cell compartmentalization," *BMC Cell Biol.* 2009; 10:4, Fig 3i. NCBI

O. *Bacteroides* species—may cause intestinal infections. *Source:* V.R. Dowell/CDC

*G + C base composition The overall percentage of guanine and cytosine in DNA is a general indicator of relatedness because it is a trait that does not change rapidly. Bacteria with a significant difference in G + C percentage are less likely to be genetically related. This classification scheme is partly based on this percentage.

TABLE 4.4 Important Families and Genera of Pathogenic Bacteria, with Notes on Some Diseases*

I. Bacteria with gram-positive cell wall structure (Phyla Firmicutes and Actinobacteria)

A. Cocci in clusters or packets that are aerobic or facultative

Family Staphylococcaceae: *Staphylococcus* (members cause boils, skin infections)

B. Cocci in pairs and chains that are facultative

Family Streptococcaceae: *Streptococcus* (species cause strep throat, dental caries)

C. Anaerobic cocci in pairs, tetrads, irregular clusters

Family Peptococcaceae: *Peptococcus, Peptostreptococcus* (involved in wound infections)

D. Spore-forming rods

Family Bacillaceae: *Bacillus* (anthrax), *Clostridium* (tetanus, gas gangrene, botulism)

E. Non-sporeforming rods

Family Lactobacillaceae: *Lactobacillus, Listeria* (food infection), *Erysipelothrix* (erysipeloid)

Family Propionibacteriaceae: *Propionibacterium* (involved in acne)

Family Corynebacteriaceae: *Corynebacterium* (diphtheria)

Family Mycobacteriaceae: *Mycobacterium* (tuberculosis, leprosy)

Family Nocardiaceae: *Nocardia* (lung abscesses)

Family Actinomycetaceae: *Actinomyces* (dental infections)

Family Streptomycetaceae: *Streptomyces* (important source of antibiotics)

II. Bacteria with gram-negative cell wall structure (Phyla Proteobacteria, Bacteriodetes, Fusobacterium, Spirochaetes, Chlamydiae)

F. Aerobic cocci

Family Neisseraceae: *Neisseria* (gonorrhea, meningitis)

G. Aerobic coccobacilli

Family Moraxellaceae: *Moraxella, Acinetobacter*

H. Anaerobic cocci

Family Veillonellaceae: *Veillonella* (dental disease)

I. Aerobic rods

Family Pseudomonadaceae: *Pseudomonas* (pneumonia, burn infections)

Miscellaneous rods (different families): *Brucella* (undulant fever), *Bordetella* (whooping cough), *Francisella* (tularemia), *Coxiella* (Q fever) *Legionella* (Legionnaires' disease)

J. Facultative rods and vibrios

Family Enterobacteriaceae: *Escherichia, Edwardsiella, Citrobacter, Salmonella* (typhoid fever), *Shigella* (dysentery), *Klebsiella, Enterobacter, Serratia, Proteus, Yersinia* (one species causes plague)

Facultative rods and vibrios (*continued*)

Family Vibronaceae: *Vibrio* (cholera, food infection)

Family Campylobacteraceae: *Campylobacter* (enteritis)

Family Helicobacteraceae: *Helicobacter* (ulcers)

Miscellaneous genera: *Flavobacterium, Haemophilus* (meningitis), *Pasteurella* (bite infections), *Streptobacillus*

K. Anaerobic rods

Family Bacteroidaceae: *Bacteroides, Fusobacterium* (anaerobic wound and dental infections)

L. Helical and curviform bacteria

Family Spirochaetaceae: *Treponema* (syphilis), *Borrelia* (Lyme disease), *Leptospira* (kidney infection)

M. Obligate and facultative intracellular bacteria

Family Rickettsiaceae: *Rickettsia* (Rocky Mountain spotted fever)

Family Anaplasmataceae: *Ehrlichia* (human ehrlichosis)

Family Chlamydiaceae: *Chlamydia* (sexually transmitted infection)

Family Bartonellaceae: *Bartonella* (trench fever, cat scratch disease)

III. Bacteria with no cell walls (Class Mollicutes)

Family Mycoplasmataceae: *Mycoplasma* (pneumonia), *Ureaplasma* (urinary infection)

*Details of pathogens and diseases in later chapters.

isolate. The criterion currently used by most taxonomists dictates that organisms within a single species share at least 97% of the nucleotide sequence of their 16S ribosomal RNA gene. If two organisms share less than 97% of the sequence of this gene, they may be classified as two distinct species.

Because the individual members of given species can show variations, we must also define levels within species (subspecies) called **strains** and **types.** A strain or variety of bacteria is a culture derived from a single parent that differs in structure or metabolism from other cultures of that species (also called biovars or morphovars). For example, there are pigmented and nonpigmented strains of *Serratia marcescens* and flagellated and nonflagellated strains of *Pseudomonas fluorescens.* A **type** is a subspecies that can show differences in antigenic makeup (serotype or serovar), in susceptibility to bacterial viruses (phage type), and in pathogenicity (pathotype).

 Check Your Progress SECTION 4.6

33. What general characteristics are used to classify bacteria?
34. What are the most useful characteristics for categorizing bacteria into phyla?
35. Explain some of the ways species level in bacteria is defined, and name at least three ways bacteria are classified below the species level.

4.7 Survey of Prokaryotic Groups with Unusual Characteristics

 Expected Learning Outcomes

28. Differentiate various groups of photosynthetic bacteria.
29. Characterize the types of obligate intracellular bacteria.
30. Summarize the basic characteristics of archaea.
31. Compare Domain Archaea with Domains Bacteria and Eukarya.
32. Explain archaeal adaptations that place them in the category of extremophiles.

So far in our introduction to the bacterial world, we have covered a rich variety of members with exceptional modes of living and behaving. And yet we are still in the early phases of discovery. Investigations of unexplored areas of the earth continue to surprise and instruct us about the adaptations and importance of these remarkable organisms. Doing justice to such incredible diversity would require more space than we can set aside in an introductory textbook, but in this section we have selected some of the more prominent and unusual groups to present. In this mini-survey, we consider some medically important groups and some representatives of bacteria living free in the environment that are ecologically important. Many of the bacteria mentioned here do not have the morphology typical of bacteria discussed previously,

and in a few cases, they are vividly different **(4.1 Secret World of Microbes).**

Free-Living Nonpathogenic Bacteria

Photosynthetic Bacteria

The nutrition of many bacteria is heterotrophic; this means that the bacteria feed primarily off nutrients from other organisms. Photosynthetic bacteria, however, are independent cells that contain special light-trapping pigments and can use the energy of sunlight to synthesize all required nutrients from simple inorganic compounds. The two general types of photosynthetic bacteria are those that produce oxygen during photosynthesis and those that produce some other substance, such as sulfur granules or sulfates.

Cyanobacteria: Microbial Marvels

Cyanobacteria, gram-negative phototrophic* bacteria in the Phylum Cyanobacteria, are among the most dominant microorganisms on earth, with a long list of significant contributions to the planet. They are ancient in origin, probably having existed in some form for around 3 billion years. We can still find remnants of these early cells in fossil biofilms called stromatolites **(figure 4.28*b* 1,2).** Many evolutionary microbiologists theorize that, because cyanobacteria were the very first photosynthesizers, they were responsible for converting the atmosphere from anaerobic to aerobic through their production of oxygen, which made it possible for the evolution of aerobic eukaryotes. Other contributions they have made to eukaryotic evolution include the chloroplasts of algae and plants. When chloroplasts are subjected to genetic analysis, it turns out that they are close relatives of some of the oldest groups of cyanobacteria, and probably arose through endosymbiosis (see chapter 5). So the green color of a forest or a field of corn actually comes from these tiny bacteria that became a part of photosynthetic eukaryotes billions of years ago. The earth would be an entirely different place if it were not for these microorganisms.

Cyanobacteria are very diverse in distribution and morphology. They flourish in every type of aquatic environment, whether ocean, hot springs, or Antarctic ice, and they can also survive in a wide range of terrestrial habitats. Because of this widespread distribution, they are huge contributors to worldwide productivity. Two species of cyanobacteria alone—*Prochlorococcus* and *Trichodesmium*—account for 30% to 40% of biomass formation and 50% of oxygen production through photosynthesis in the oceans. They are also one of a few groups of microbes that can convert nitrogen gas (N_2) into ammonium (NH_4^+) that can be used by plants, making them essential players in the nitrogen cycle.

Their structure and arrangement are varied, from coccoid, rodlike, and spiral-shaped cells, with arrangements in filaments, clusters, and sheets, usually surrounded by a gelatinous sheath **(figure 4.28*a*, 2).** Although their primary photosynthetic pigments include green chlorophyll *b* and the bluish pigment phycocyanin, they may take on shades of yellow, orange, and red derived from other pigments.

literally "to feed with light."

4.1 Secret World of Microbes
Redefining Bacterial Size

Microbiologists keep being reminded how far we are from having a complete assessment of the bacterial world, mostly because the world is so large and bacteria are so small. There seem to be frequent reports of exceptional bacteria discovered in places like the deep-ocean volcanoes or Antarctic ice. Among the most remarkable are giant and dwarf bacteria.

Big Bacteria Break Records

In 1985 biologists discovered a new bacterium living in the intestine of surgeonfish that at the time was a candidate for the *Guinness Book of World Records*. The large cells, named *Epulopiscium fishelsoni* ("guest at a banquet of fish"), measure around 100 µm in length, although some specimens were as large as 300 µm. This record was broken when a marine microbiologist discovered an even larger species of bacteria living in ocean sediments near the African country of Namibia. These gigantic cocci are arranged in strands that look like pearls and contain hundreds of golden sulfur granules, inspiring their name, *Thiomargarita namibia* ("sulfur pearl of Namibia") (see photo). The size of the individual cells ranges from 100 up to 750 µm (0.75 mm), and many are large enough to see with the naked eye. By way of comparison, if the average bacterium were the size of a mouse, *Thiomargarita* would be as large as a blue whale!

These bacteria are found in such large numbers in the sediments that it is thought that they are essential to the ecological cycling of H_2S and other substances in this region, converting them to less toxic substances.

Miniature Microbes—Are They for Real?

At the other extreme, microbiologists are being asked to reevaluate the lower limits of bacterial size. Up until now it has been generally accepted that the smallest cells on the planet are some form of mycoplasma with dimensions of 0.2 to 0.3 µm, which is right at the limit of resolution with light microscopes. A new controversy is brewing over the discovery of tiny cells that look like dwarf bacteria but are 10 times smaller than mycoplasmas and 100 times smaller than the average bacterial cell. These minute cells have been given the name **nanobacteria** or **nanobes** (Gr. *nanos*, one-billionth).

Nanobacteria-like forms were first isolated from blood and serum samples. The tiny cells appear to grow in culture, have cell walls, and contain protein and nucleic acids, but their size range is only from 0.05 to

200 µm

Thiomargarita namibia—giant cocci. Approximately how many times larger is Thiomargarita than the nanobe shown on the right? © Heide N. Schulz/Max Planck Institute for Marine Microbiology

Three-dimensional micrograph of an ARMAN measuring about 0.25 µm across. These are fully functioning cells with a tiny nucleoid (yellow) and a small number of ribosomes (blue dots). © Banfield & Comolli

0.2 µm. Similar nanobes have been extracted by minerologists studying deposits in the ocean at temperatures up to 170°C. Many geologists are convinced that these nanobes are real cells, that they are probably similar to the first microbes on earth, and that they play a strategic role in the development of the earth's crust.

Microbiologists have been more skeptical. Experts have postulated that the minimum cell size to contain a functioning genome and reproductive and synthetic machinery is approximately 0.14 µm, which means that some of the nanobes could be artifacts or bits of larger cells that have broken free. It seems that the archaea may be the ultimate definers of smallness in prokaryotes. Microbiologists exploring acid mines in California have discovered ultrasmall archaea living in pink biofilms. These extremophiles, called ARMAN, are at the lower limits of size, between 100 and 400 nm. They are real cells, complete with a tiny genome and a reduced number of ribosomes, and are perhaps the first indisputable nanobes. Additional studies are needed to test this curious question of nanobes and possibly to answer some questions about the origins of life on earth and elsewhere in the universe.

What are some of the adaptations that a giant *Thiomargarita* would require for its survival? *Answer available on Connect.*

Within an individual cell are found extensive specialized membranes called **thylakoids,** the locations of pigments and the sites of photosynthesis (**figure 4.28c**). They also have gas vesicles that keep the cells suspended high in the water column to facilitate photosynthesis. Cyanobacteria periodically overgrow in aquatic environments to form blooms that are harmful to fish and other inhabitants, and some produce toxins that

Quick Search
Search the web for "cyanobacterium *Gloeocapsa,* which went to outer space and survived a year and a half."

could cause illness if ingested, but they are generally not medically important.

Green and Purple Sulfur Bacteria

The green and purple bacteria are also photosynthetic and contain pigments. They differ from the cyanobacteria in having a different type of chlorophyll called *bacteriochlorophyll* and by not giving off oxygen as a product of photosynthesis. They live in sulfur springs, freshwater lakes, and swamps that are deep enough for the anaerobic conditions they require yet where their pigment can still absorb wavelengths of light (**figure 4.29**). These bacteria are named for

(a1)

(a2)

Stromatolite

(b1)

10 μm

(b2)

Ribosomes

Nucleoid

Thylakoids

(c)

Figure 4.28 Cyanobacterial characteristics. (a) Striking colors and arrangements of cells. (1) *Trichodesmium* is one of the dominant planktonic forms on earth. (2) *Merismopedia* displays a slimy sheath around its regular packets of cells. **(b)** Ancient signs of cyanobacteria. (1) A sectioned stromatolite displays layers of a cyanobacterial biofilm laid down over a billion years ago. (2) A microfossil of a billion-year-old filament found in Siberian fossils. **(c)** An electron micrograph of a single cell (10,000×) reveals layers of thylakoids that are the sites of photosynthesis. (a1): © Angelicque E. White; (a2): Courtesy of Jason Oyadomari; (b1): © Richard Behl, California State University, Long Beach; (b2): © J. William Schopf/UCLA; (c): Michelle Liberton, Jotham R. Austin, II, R. Howard Berg, and Himadri B. Pakrasi, "Unique Thylakoid Membrane Architecture of a Unicellular N2-Fixing Cyanobacterium Revealed by Electron Tomography" *Plant Physiology*, vol. 155 no. 4, p. 1656-1666, Fig 2G © American Society of Plant Biologists

their predominant colors, but they can also develop brown, pink, purple, blue, and orange coloration. Both groups utilize sulfur compounds (H_2S, S) in their metabolism.

Gliding, Fruiting Bacteria

The gliding bacteria are a mixed collection of gram-negative bacteria in Phylum Proteobacteria that live in water and soil. The name

Figure 4.29 Photosynthetic bacteria. Purple-colored masses in a fall pond contain a concentrated bloom of purple sulfur bacteria (inset 1,500×). The pond also harbors a mixed population of algae (green clumps). © Kathy Park Talaro; (inset): © David J. Patterson

is derived from the tendency of members to glide over moist surfaces. The gliding property evidently involves rotation of filaments or fibers just under the outer membrane of the cell wall. They do not have flagella. Several morphological forms exist, including slender rods, long filaments, cocci, and some miniature, tree-shaped fruiting bodies. Probably the most intriguing and exceptional members of this group are the slime bacteria, or myxobacteria. What sets the myxobacteria apart from other bacteria is the complexity and advancement of their life cycle. During this cycle, the vegetative cells respond to chemotactic signals by swarming together and differentiating into a many-celled, colored structure called the fruiting body, making it one of the few bacteria that can generate a multicellular body form **(figure 4.30)**. The fruiting body is a survival structure that makes spores by a method very similar to that of certain fungi. These fruiting structures are often large enough to be seen with the unaided eye on tree bark and plant debris.

Unusual Forms of Medically Significant Bacteria

Most bacteria are free-living or parasitic forms that can metabolize and reproduce by independent means. Two groups of bacteria—the rickettsias and chlamydias—have adapted to living inside host cells. Because they cannot function without some essential factors from the host, they are considered **obligate intracellular parasites.**

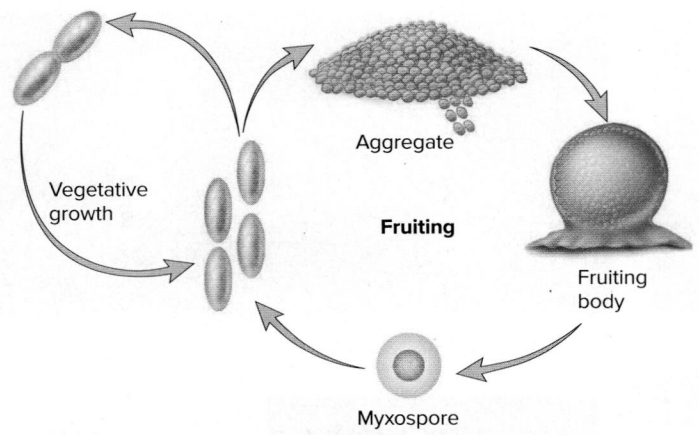

Figure 4.30 **Life cycle of *Myxococcus*.** This multicellular bacterium goes through several developmental stages in response to environmental signals.

Figure 4.31 **Transmission electron micrograph of *Rickettsia conorii* inside its host cell (100,000X).** This species causes tick-borne Mediterranean fever. *Source:* From C. Rovery, P. Brouqui, D. Raoult, "Questions on Mediterranean Spotted Fever a Century after Its Discovery," *Emerging Infectious Diseases,* 14(9), September 2008, Fig. 2. CDC

Rickettsias

Rickettsias[6] are distinctive, very tiny, gram-negative bacteria **(figure 4.31).** Although they have somewhat typical morphology, they are atypical in their life cycle and other adaptations. Most are pathogens that alternate between a mammalian host and blood-sucking arthropods,[7] such as fleas, lice, or ticks. Rickettsias cannot survive or multiply outside a host cell and cannot carry out metabolism completely on their own, so they are closely attached to their hosts. Several important human diseases are caused by rickettsias. Among these are Rocky Mountain spotted fever, caused by *Rickettsia rickettsii* (transmitted by ticks), and endemic typhus, caused by *Rickettsia typhi* (transmitted by lice). It is worth noting, in advance

6. Named for Howard Ricketts, a physician who first worked with these organisms and later lost his life to typhus.

7. An arthropod is an invertebrate with jointed legs, such as an insect, tick, or spider.

of chapter 5, that the mitochondrion, a eukaryotic organelle, is a close genetic relative to rickettsias, indicating an evolutionary link between them.

Chlamydias

Bacteria of the genera *Chlamydia* and *Chlamydophila,* termed **chlamydias,** are similar to the rickettsias in that they require host cells for growth and metabolism. But they are not closely related to them and are not transmitted by arthropods. Because of their tiny size and obligately parasitic lifestyle, they were at one time considered a type of virus (see table 4.3L). Species that carry the greatest medical impact are *Chlamydia trachomatis,* the cause of both a severe eye infection (trachoma) that can lead to blindness and one of the most common sexually transmitted diseases, and *Chlamydophila pneumoniae,* an agent in lung infections.

Diseases caused by rickettsias and chlamydias are described in more detail in the infectious disease chapters.

Archaea: The Other Prokaryotes

The discovery and characterization of novel prokaryotic cells that have unusual anatomy, physiology, and genetics changed our views of microbial taxonomy and classification (see chapter 1 and table 4.3). These single-celled, simple organisms, called **archaea,** or **archaeons,** are considered a third cell type in a separate superkingdom (the Domain Archaea). We include them in this chapter because they are prokaryotic in general structure and they do share many bacterial characteristics. But evidence is accumulating that they are more closely related to Domain Eukarya than to bacteria. For example, archaea and eukaryotes share a number of ribosomal RNA sequences that are not found in bacteria, and their protein synthesis and ribosomal subunit structures are similar. **Table 4.5** outlines selected points of comparison of the three domains.

Other ways in which the archaea differ significantly from other cell types are that certain genetic sequences are found only in their ribosomal RNA. Their cell walls are also quite different, being composed of polysaccharide or protein and lacking peptidoglycan, and some lack cell walls altogether (see figure 4.33). It is likely that the archaea are the most ancient of all life forms and have retained characteristics of the first cells that originated on the earth nearly 4 billion years ago. The early earth is thought to have contained a hot, anaerobic "soup" with sulfuric gases and salts in abundance. Many of the present-day archaea occupy the remaining habitats on the earth that have some of the same extreme conditions. For this reason they are considered **extremophiles,** or **hyperextremophiles,** meaning that they "love" the most extreme habitats on earth. Although there are also some examples of extremophilic bacteria (refer to table 4.3), they are not as widespread as archaea in ultra-extreme environments.

Metabolically the archaea exhibit nearly incredible adaptations to what would be deadly conditions for other organisms. Many of these hardy microbes are living in severe combinations of extremes as well—for instance, high acidity and high temperature, high salt and alkalinity, low temperature and high pressure. Included in this group are methane producers, hyperthermophiles, extreme halophiles, and sulfur reducers. It is worth noting that not all archaea are adapted to extremes, and many are widely distributed in more

TABLE 4.5	Comparison of Three Cellular Domains		
Characteristic	Bacteria	Archaea	Eukarya
Cell type	Prokaryotic	Prokaryotic	Eukaryotic
Chromosomes	Single, or few, circular	Single, circular	Several, linear
Types of ribosomes	70S	70S but structure is similar to 80S	80S
Unique ribosomal RNA signature sequences	+	+	+
Number of RNA sequences shared with Eukarya	One	Three	
Protein synthesis similar to Eukarya	−	+	
Presence of peptidoglycan in cell wall	+	−	−
Cell membrane lipids	Fatty acids with ester linkages	Long-chain, branched hydrocarbons with ether linkages	Fatty acids with ester linkages
Sterols in membrane	− (some exceptions)	−	+

moderate environments such as soils, oceans, and even animal intestines. By some estimates, archaea are the most common cells on the earth, and they have made major contributions to the development of the earth itself.

Members of the group called **methanogens** can convert CO_2 and H_2 into methane gas (CH_4) through unusual and complex pathways. These archaea are common inhabitants of anaerobic mud and the bottom sediments of lakes and oceans. Some are even found in the oral cavity and large intestine of humans. The gas they produce in aquatic habitats collects in swamps and may become a source of fuel. Methane may also contribute to the "greenhouse effect," which maintains the earth's temperature and can contribute to global warming.

Other types of archaea—the **extreme halophiles**—require salt to grow and may have such a high salt tolerance that they can multiply in sodium chloride solutions (36% NaCl) that would destroy most cells. They exist in the saltiest places on the earth—inland seas, salt lakes, and salt mines. They are not particularly common in the ocean because the salt content is not high enough to support them. Many of the "halobacteria" use a red pigment to synthesize ATP in the presence of light. These pigments are responsible for "red herrings," the color of the Red Sea, and the red color of salt ponds **(figure 4.32).**

Archaea adapted to growth at very low temperatures are described as **psychrophiles** (loving cold temperatures); those growing at very high temperatures are **hyperthermophiles** (loving very high temperatures). Hyperthermophiles flourish at temperatures between 80°C and 121°C (21°C above boiling) and cannot grow at 50°C. They live in volcanic waters and soils and submarine vents, and are often salt- and acid-tolerant as well **(figure 4.33).** Researchers sampling sulfur vents in the deep ocean discovered thermophilic archaea flourishing at temperatures up to 250°C—which is 150° above the temperature of boiling water! Not only were these archaea growing prolifically at this high temperature, but they were also living at 265 atmospheres of pressure. (On the earth's surface, pressure is about 1 atmosphere.) For additional discussion of the unusual adaptations of archaea, see chapter 7.

(a)

(b)

Figure 4.32 Halophiles around the world. (a) An aerial view of a salt pond at San Francisco Bay, California. The archaea that thrive in this warm, highly saline habitat produce brilliant red, pink, and orange pigments. **(b)** A sample taken from a saltern in Australia viewed by fluorescence microscopy (1,000×). Note the range of cell shapes (cocci, rods, and squares) found in this community. (a): *Source:* NASA Johnson Space Center/ISS007E8738/ (http://eol.jsc.nasa.gov); (b): © Dr. Mike Dyall-Smith, University of Melbourne

Figure 4.33 The hottest life forms on earth. A colorized transmission electron micrograph of strain 121. This member of Domain Archaea was isolated from an undersea thermal vent off the coast of Washington State. This organism thrives in habitats above the temperature of boiling water and can even survive being autoclaved.
© Derek Lovley/Kazem Kashefi/Science Source

 Check Your Progress SECTION 4.7

36. Discuss several ways in which bacteria are medically and ecologically important.

37. Name two main groups of obligate intracellular parasitic bacteria and explain why these groups can't live independently.

38. Explain the characteristics of archaea that indicate that they constitute a unique domain of living things that is neither bacterial nor eukaryotic.

39. What is meant by the terms *extremophile* and *hyperextremophile*?

40. Describe the three major archaeal lifestyles and adaptations to extreme habitats.

 Chapter Summary with Key Terms

4.1 Basic Characteristics of Cells and Life Forms
 A. All living things are composed of cells, which are complex collections of macromolecules that carry out living processes. All cells must have the minimum structure of an outer **cell membrane, cytoplasm,** a **chromosome,** and **ribosomes.**
 B. Cells can be divided into two basic types: **prokaryotes** and **eukaryotes.**
 1. Prokaryotic cells are the basic structural unit of bacteria and archaea. They lack a nucleus or organelles. They are highly successful and adaptable single-cell life forms.

 CASE STUDY Part 2

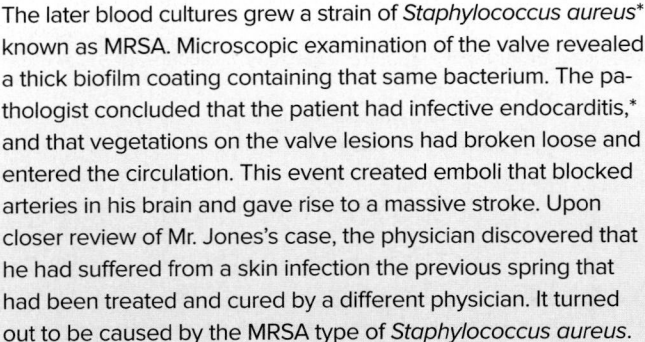

During an autopsy of Mr. Jones's body, the pathologist observed that the prosthetic valve was covered with small patches he called vegetations. The later blood cultures grew a strain of *Staphylococcus aureus** known as MRSA. Microscopic examination of the valve revealed a thick biofilm coating containing that same bacterium. The pathologist concluded that the patient had infective endocarditis,* and that vegetations on the valve lesions had broken loose and entered the circulation. This event created emboli that blocked arteries in his brain and gave rise to a massive stroke. Upon closer review of Mr. Jones's case, the physician discovered that he had suffered from a skin infection the previous spring that had been treated and cured by a different physician. It turned out to be caused by the MRSA type of *Staphylococcus aureus*.

Most bacteria can form structured multicellular communities, or biofilms, on objects in a moist environment. This is even true of bacterial pathogens in the body. The CDC estimates that at least 65% of chronic infections are caused by microbial biofilms. In this case, the MRSA bacteria in the patient's skin infection must have entered the circulation and colonized the artificial valve over several weeks to months. Most cases of chronic endocarditis are caused by biofilms on valves. When the biofilm grows into larger vegetations, portions of it break loose into the circulation. These infect the blood and are spread into organs, causing fever and other signs and symptoms, including the ones that were fatal. MRSA is an emerging pathogen that started as a problem in the hospital but is now prominent in nonhospital settings as well.

■ *What does the acronym MRSA mean, and what is its significance?*

■ *Why wasn't penicillin effective in treating the infection?*

For more background on MRSA and endocarditis, see chapter 18.

To conclude this Case Study, go to Connect.

* *Staphylococcus aureus* (staf´-uh-loh-cok´-us ar-ee-us) Gr. *staphyle,* a bunch of grapes, *kokkus,* berry, and *aurum,* golden.
* *endocarditis* (en´-doh-car-dye´-tis) Gr. endon, within, kardia, heart, and itis, an inflammation. An inflammation of the lining of the heart and its valves usually caused by infection.

 2. Eukaryotic cells contain a membrane-surrounded nucleus and a number of organelles that function in specific ways. A wide variety of organisms, from single-celled protozoans to humans, are composed of eukaryotic cells.
 3. Viruses are not generally considered living or cells, and rely on host cells to replicate.
 C. Cells show the basic essential characteristics of life. Parts of cells and macromolecules do not show these characteristics independently.
 1. The primary life indicators are heredity, reproduction, growth, metabolism, responsiveness, and transport.

4.2 Prokaryotic Profiles: The Bacteria and Archaea
 A. Prokaryotes consist of two major groups, the bacteria and the archaea. Life on earth would not be possible without them.
 B. Prokaryotic cells lack the membrane-surrounded organelles and nuclear compartment of eukaryotic cells but are still complex in their structure and function. All prokaryotes have a cell membrane, cytoplasm, ribosomes, and a chromosome.
 C. **Appendages:** Some bacteria have projections that extend from the cell.
 1. **Flagella** and internal **axial filaments** found in spirochetes are used for motility.
 2. **Fimbriae** function in adhering to the environment; pili can be involved in adhesion, movement, and genetic exchange.
 3. The **glycocalyx** may be a slime layer or a capsule.

4.3 The Cell Envelope: The Boundary Layer of Bacteria
 A. Most prokaryotes are surrounded by a protective envelope that consists of the cell wall and the cell membrane.
 B. The wall is relatively rigid due to **peptidoglycan.**
 C. Structural differences give rise to **gram-positive** and **gram-negative cells,** as differentiated by the **Gram stain.**
 1. Gram-positive bacteria contain a thick wall composed of peptidoglycan and teichoic acid in a single layer.
 2. Gram-negative bacteria have a thinner, two-layer cell wall with an **outer membrane,** a thin layer of **peptidoglycan,** and a well-developed **periplasmic space.**
 3. Wall structure gives rise to differences in staining, toxicity, and effects of drugs and disinfectants.
 4. The **cell** or **cytoplasmic membrane** is a flexible sheet that surrounds the cytoplasm. It is composed of a bilayer of lipids with attached proteins. It is a multifunctional structure involved in transport, synthesis, and energy reactions.

4.4 Bacterial Internal Structure
 The cell cytoplasm contains some or all of the following internal structures in bacteria: the **chromosome(s)** condensed in the **nucleoid; ribosomes,** which serve as the sites of protein synthesis and are 70S in size; extra genetic information in the form of **plasmids;** storage structures known as **inclusions, granules,** and **microcompartments;** a **cytoskeleton** of **bacterial actin,** which helps give the bacterium its shape; and in some bacteria an **endospore,** which is a highly resistant structure for survival, not reproduction.

4.5 Bacterial Shapes, Arrangements, and Sizes
 A. Most bacteria are unicellular and are found in a great variety of shapes, arrangements, and sizes. General shapes include **cocci, bacilli,** and helical forms such as **spirilla** and **spirochetes.** Some show great variation within the species in shape and size and are **pleomorphic.** Other variations include **coccobacilli, vibrios,** and filamentous forms.
 B. Prokaryotes divide by binary fission and do not utilize mitosis. Various **arrangements** result from cell division and are termed **diplococci, streptococci, staphylococci, tetrads,** and **sarcina** for cocci; bacilli may form pairs, chains, or **palisades.**
 C. Variant members of bacterial species are called strains and types.

4.6 Classification Systems of the Prokaryotic Domains: Archaea and Bacteria
 A. An important taxonomic system is standardized by *Bergey's Manual of Systemic Bacteriology,* which presents the prokaryotes in five major groups, as organized by volume.
 Volume 1: Domain Archaea Domain Bacteria: Deeply Branching and Phototropic Bacteria Domain Bacteria
 Volume 2: Phylum Proteobacteria
 Volume 3: Phylum Firmicutes
 Volume 4: Phylum Actinobacteria
 Volume 5: Phyla Chlamydiae, Spirochaetes, Planctomyces, Bacteriodetes, and others

4.7 Survey of Prokaryotic Groups with Unusual Characteristics
 A. Several groups of bacteria have unusual adaptations and life cycles.
 1. Medically important bacteria: **Rickettsias** and **chlamydias** are within the gram-negative group but are small **obligate intracellular parasites** that replicate within cells of the hosts they invade.
 2. Nonpathogenic bacterial groups: The majority of bacterial species are free-living and not involved in disease. Unusual groups include photosynthetic bacteria such as **cyanobacteria,** which provide oxygen to the environment, and the green and purple bacteria.
 B. **Archaea** share many characteristics with bacteria but vary in certain genetic aspects and structures such as the cell wall and ribosomes.
 1. Many are **extremophiles,** adapted to extreme environments similar to the earliest of earth's inhabitants.
 2. They are not considered medically important but are of ecological and potential economic importance.

Level I Knowledge and Comprehension

These questions require a working knowledge of the concepts in the chapter and the ability to recall and understand the information you have studied.

 Multiple-Choice Questions

Select the correct answer from the answers provided. For questions with blanks, choose the combination of answers that most accurately completes the statement.

1. Which structure is not a component of all cells?
 a. cell wall c. genetic material
 b. cell membrane d. ribosomes

2. Viruses are not considered living things because
 a. they are not cells c. they lack metabolism
 b. they cannot reproduce by themselves d. All of these are correct.

3. Which of the following is not found in all bacterial cells?
 a. cell membrane c. ribosomes
 b. a nucleoid d. actin cytoskeleton

4. The major locomotor structures in bacteria are
 a. flagella c. fimbriae
 b. pili d. cilia

5. Pili are appendages in _____ bacteria that serve as a means of
 _____.
 a. gram-positive, genetic exchange
 b. gram-positive, attachment
 c. gram-negative, genetic exchange
 d. gram-negative, protection

6. An example of a glycocalyx is
 a. a capsule c. outer membrane
 b. pili d. a cell wall

7. Which of the following is a primary bacterial cell wall function?
 a. transport c. support
 b. motility d. adhesion

8. Which of the following is present in both gram-positive and gram-negative cell walls?
 a. an outer membrane c. teichoic acid
 b. peptidoglycan d. lipopolysaccharides

9. Metachromatic granules are concentrated _____ found in _____.
 a. fat, *Mycobacterium* c. sulfur, *Thiobacillus*
 b. dipicolinic acid, *Bacillus* d. PO₄, *Corynebacterium*

10. Which of the following prokaryotes lacks cell walls?
 a. rickettsias c. *Mycobacterium*
 b. mycoplasmas d. archaea

11. The LPS layer in gram-negative cell walls releases _____ that cause _____.
 a. enzymes, pain c. techoic acid, inflammation
 b. endotoxins, fever d. phosphates, lysis

12. Bacterial endospores function in
 a. reproduction c. protein synthesis
 b. survival d. storage

13. An arrangement in packets of eight cells is described as a _____.
 a. micrococcus c. tetrad
 b. diplococcus d. sarcina

14. The major difference between a spirochete and a spirillum is
 a. presence of flagella c. the nature of motility
 b. the presence of twists d. size

15. Which phylum contains bacteria with a gram-positive cell wall?
 a. Proteobacteria c. Firmicutes
 b. Chlorobi d. Spirochetes

16. To which taxonomic group do cyanobacteria belong?
 a. Domain Archaea c. Domain Bacteria
 b. Phylum Actinobacteria d. Phylum Fusobacteria

17. Which stain is used to distinguish differences between the cell walls of medically important bacteria?
 a. simple stain c. Gram stain
 b. acridine orange stain d. negative stain

18. The first living cells on earth would most likely resemble which of these?
 a. a cyanobacterium c. a gram-positive cell
 b. an endospore former d. an archaeon

 Case Study Review

1. What is true of the condition endocarditis?
 a. It occurs in the heart muscle.
 b. It is caused by microbes growing in the internal organs.
 c. It is an infection of the heart valves and lining.
 d. It can be transmitted to others.

2. Where did the MRSA pathogen likely originate?
 a. from the artificial valve itself c. from the surgery
 b. from the patient's earlier skin infection d. from the patient's wife

3. What is a biofilm, and how did it form on the heart valve?

 Writing Challenge

For each question, compose a one- or two-paragraph answer that includes the factual information needed to completely address the question. Check Your Progress questions can also be used for writing-challenge exercises.

1. Label the parts on the bacterial cell featured here and write a brief description of its function.

2. Discuss the collection of properties that are used to define life and the prokaryotic cell structures that are involved in carrying out these life processes.

3. Describe the basic process of biofilm formation.

4. What leads microbiologists to believe the archaea are more closely related to eukaryotes than to bacteria?

5. What is required to kill endospores? How do you suppose archaeologists were able to date some spores as being thousands (or millions) of years old?

On Connect you can find an Introduction to Concept Mapping that provides guidance for working with concept maps along with concept-mapping activities for this chapter.

Level II Application, Analysis, Evaluation, and Synthesis

These problems go beyond just restating facts and require higher levels of understanding and an ability to interpret, problem solve, transfer knowledge to new situations, create models, and predict outcomes.

Critical thinking is the ability to reason and solve problems using facts and concepts. These questions can be approached from a number of angles, and in most cases, they do not have a single correct answer.

1. Using clay, demonstrate how cocci can divide in several planes and show the outcome of this division. Show how the arrangements of bacilli occur, including palisades.

2. Using a corkscrew and a spring to compare the flexibility and locomotion of spirilla and spirochetes, explain which cell type is represented by each object.

3. Under the microscope, you see a rod-shaped cell that is swimming rapidly forward.
 a. What do you automatically know about that bacterium's structure?
 b. Propose another function of flagella besides locomotion.

4. Name an acid-fast bacterium and explain what characteristics make this bacterium different from other gram-positive bacteria.

5. a. Name a bacterial group that uses chlorophyll to photosynthesize.
 b. Describe the two major groups of photosynthetic bacteria and how they are similar and different.

6. Propose a hypothesis to explain how bacteria and archaea could have, together, given rise to eukaryotes.

7. Explain or illustrate exactly what will happen to the cell wall if the synthesis of the interbridge is blocked by penicillin.

8. Ask your lab instructor to help you make a biofilm and examine it under the microscope. One possible technique is to suspend a glass slide in an aquarium for a few weeks, then carefully air-dry, fix, and Gram stain it. Observe the diversity of cell types.

9. Describe the shapes and arrangements of bacteria in figure 4.23*a*, *b*, *e*, and *f*.

Visual Challenge

1. From chapter 3, figure 3.20*b*. Which bacteria have a well-developed capsule: "*Klebsiella*" or "*S. aureus*"? Defend your answer.

© Kathy Park Talaro

2. What kinds of cells are shown here? Explain what causes the colors and appearance of the stages you see.

Courtesy of Kit Pogliano and Mark Sharp/UCSD

A Survey of Eukaryotic Cells and Microorganisms

Hookworm larvae burrowing into the intestine (20×)

Larval schistosome, a parasitic roundworm (200×)

Trypanosoma, a flagellated protozoan (1,500×)

Examples of microorganisms that cause neglected diseases, distributed across the earth's tropical and subtropical zones (bands of blue) (top): *Source:* NASA Earth Observations; (bottom: left): *Source:* CDC; (bottom: middle): *Source:* Dr. Sulzer/CDC; (bottom: right): *Source:* Dr. Myron G. Schultz/CDC

CASE STUDY Part 1 Neglected Tropical Diseases

Some of the worst suffering in the world originates from a group of ancient infectious diseases that exist primarily in the **tropical** and **subtropical** regions of Africa, India, Latin America, and Asia. Nearly 1.4 billion people worldwide, sometimes referred to as the **"bottom billion,"** experience disability or death as a result of 17 common diseases.

These **neglected tropical diseases,** or **NTDs,** have received less attention than highly publicized diseases such as **AIDS, malaria,** and **tuberculosis,** and they have been largely ignored by the medical establishment in many countries. Unfortunately, they also tend to occur in the poorest rural areas, where access to treatment is often severely limited. "Almost everyone in the bottom billion has at least one of these diseases," said Dr. Peter Hotez, a **parasitologist** and medical doctor at George Washington University. Taken together, NTDs rank second only to HIV/AIDS in their medical, social, and economic impact.

Eleven of the 17 neglected disease pathogens are **eukaryotic parasites,** either parasitic **helminth worms** or **protozoans.** The leading parasitic worm infections are **ascariasis,*** **trichiuriasis, hookworm,**

> "The impact of Avermectin and Artemisinin goes far beyond reducing the disease burden of individuals."
>
> **Hans Forssberg of the Nobel Committee**

and **schistosomiasis,** and the leading protozoan infections are **Chagas disease, African trypanosomiasis,** and **leishmaniasis.** The **chronic, progressive** actions of these infectious agents can cause debility and disfigurement. Consider Chagas disease, caused by the protozoan *Trypanosoma cruzi,* which over time **lodges in the heart and other organs,** destroys health, and shortens life. Some forms of leishmaniasis affect the skin and give rise to growths that **deform** the face and other body parts. The growth of **hookworms** causes **blood loss,** saps strength, and impairs development. Some parasites cause **blindness** and others **disfigure the limbs.** In addition to medical effects, many victims become unable to go to school or work and are harshly ostracized by their communities.

- *What subjects are studied by the science of parasitology?*
- *What can you conclude about the total number of NTD cases, given that everyone in the bottom billion has at least one infection?*

To continue the Case Study, go to Case Study Part 2 at the end of the chapter.

* Some diseases are named by adding the suffix *-iasis* to the name of the organism that causes it.

SCOPING OUT THE CHAPTER

At the heart of all eukaryotic organisms is the eukaryotic cell. An evolutionary leap from its prokaryotic predecessor, this cell is hundreds of times larger than a typical prokaryotic cell and far more complex. The development of cellular organelles allow biochemical reactions to occur more efficiently, and organisms based on eukaryotic cells are free to evolve in ways that simply aren't available to prokaryotes. This chapter provides an introduction to these incredibly complex cells.

The cellular organelles within eukaryotic cells work in concert with one another to accomplish complex tasks like synthesizing and modifying proteins.

Although the vast majority of fungi are harmless saprobes which recycle nutrients in the environment, we tend to be more interested in the few that cause disease, or are used in the production of food and drink.

Algae typically have few health risks associated with them but many industrial uses. The cell walls of the diatoms seen here are used to filter drinking water, and their photosynthetic activity is responsible for most of the oil on Earth.

Protozoans like *Giardia intestinalis* are routinely present in natural bodies of water and ingestion can cause severe disease.

Parasitic worms like this pork tapeworm (*Taenia solium*) use a combination of hooks and suckers to attach to the intestinal wall of their host. These infestations are exceeding difficult to treat because the worm's size and biology makes it more like its human host than an invading microbial organism.

5.1 The History of Eukaryotes

Expected Learning Outcomes

1. Describe the evolutionary history of eukaryotic cells.
2. Provide a substantial theory regarding how eukaryotic cells originated and how multicellularity came to be.
3. List the eukaryotic groups and their body plans.

Using evidence from ancient sediments, paleontologists estimate that the earliest eukaryotic cells first evolved around 2 billion years ago. Other intriguing evidence comes from fossilized cells discovered in shale deposits from China and Russia that are 600 million to nearly a billion years old. It is remarkable how much they resemble modern-day algae or protozoa (**figure 5.1**).

When biologists began to focus on the possible origins of these cells, they turned to an analysis of modern cells, and from these studies emerged convincing evidence that eukaryotic cells evolved from prokaryotic cells through a process of **endosymbiosis.** According to the endosymbiotic hypothesis, eukaryotic cells arose when a much larger prokaryotic cell engulfed smaller prokaryotic cells, which began to live and reproduce inside the larger cell rather than being destroyed. As the smaller cells took up permanent residence, they came to perform specialized functions for the larger cell, such as food synthesis and oxygen utilization, that enhanced the larger cell's versatility and survival. Over time the cells evolved into a single functioning entity, and the relationship became obligatory; some of the smaller cells had become **organelles,** the distinguishing feature of eukaryotic cells (**figure 5.2**). The structure of these early cells was so versatile that eukaryotic microorganisms soon spread out into available habitats and adopted greatly diverse styles of living.

(a)

(b)

Figure 5.1 Ancient eukaryotic protists discovered in fossilized rocks. (a) A cell preserved in Siberian shale deposits dates from 850 million to 950 million years ago. **(b)** A disclike cell was recovered from a Chinese rock dating 590 million to 610 million years ago. Both cells are relatively simple, with (a) showing chloroplast-like bodies that would be found in an alga. Example (b) has a cell wall with spines, very similar to the modern alga *Pediastrum.* (a and b): © Andrew H. Knoll

A larger prokaryotic cell, possibly an archaea, containing a flexible outer envelope and folded internal membranes.

A smaller prokaryotic cell, related to modern rickettsias, that can live intracellularly.

Membrane forms around the DNA, creating a primitive nucleus.

The larger cell engulfs the smaller one; smaller one survives and remains surrounded by the vacuolar membrane.

Smaller bacterium becomes a permanent resident of its host's cytoplasm; it multiplies and is passed on during cell division. It utilizes aerobic metabolism and increases energy availability for the host.

Early endoplasmic reticulum

Early mitochondria

Ancestral eukaryotic cell develops additional membrane pouches that become the endoplasmic reticulum and Golgi apparatus.

Ancestral cell

Photosynthetic bacteria (ancient cyanobacteria) are also engulfed; they develop into chloroplasts.

Chloroplast

Cell wall

Many protozoa, animals

Some groups of algae, higher plants

Figure 5.2 A possible model for evolution of eukaryotic organisms according to the endosymbiotic theory. A large archaeal cell with its flexible envelope engulfs a smaller bacterial cell. Folds in the archaeal cell membrane wrap around the chromosome to form a nuclear envelope, The engulfed bacterial cell forges a metabolic relationship with the archaea, making use of archaeal molecules and contributing energy through aerobic respiration. Evolution of a stable mutualistic existence would maintain both cell types and create the eukaryotic cell and its organelles.

| TABLE 5.1 | Eukaryotic Organisms Studied in Microbiology | | |
|---|---|---|
| **Unicellular, a Few Colonial** | **May Be Unicellular, Colonial, or Multicellular** | **Multicellular Except Reproductive Stages** |
| Protozoa | Fungi, Algae | Helminths (parasitic worms) Arthropods (animal vectors of diseases) |

Much of the evidence supporting the endosymbiotic theory concerned the similarity between the organelles of modern eukaryotic cells and the structure of bacteria. In many ways the mitochondrion of eukaryotic cells behaves as a tiny cell within a cell. It is capable of independent division, contains a circular chromosome with bacterial DNA sequences, and has ribosomes that are clearly prokaryotic. Mitochondria also have bacterial membranes and can be inhibited by drugs that affect only bacteria. In a similar vein, genetic studies indicate that chloroplasts likely originated from endosymbiotic cyanobacteria that provided their host cells with a built-in feeding mechanism. Evidence is seen in protists, such as *Paulinella chromatophora*, that harbor specialized chloroplasts with cyanobacterial origins. The endosymbiotic theory is so well accepted that bacteriologists have placed both mitochondria and chloroplasts on the family tree of bacteria (see figure 1.15). In fact, the closest relative of mitochondria are rickettsias, which are also obligate intracellular bacteria!

The first primitive eukaryotes were probably single-celled, independent microorganisms, but over time some forms began to cluster in permanent groupings called colonies. With further evolution, some of the cells within colonies became *specialized,* or adapted to perform a particular function advantageous to the whole colony, such as locomotion, feeding, or reproduction. Complex multicellular organisms evolved as individual cells in the organism lost the ability to survive apart from the intact colony. Although a multicellular organism is composed of many cells, it is more than just a disorganized assemblage of cells like a colony. Rather, it is composed of distinct groups of cells that cannot exist independently of the rest of the body. Within a multicellular organism, groupings of cells that have a specific function are termed *tissues,* and groups of tissues make up *organs.*

Looking at modern eukaryotic organisms, we find examples of many levels of cellular complexity (**table 5.1**). All protozoa, as well as numerous algae and fungi, are unicellular. Truly multicellular organisms are found only among plants and animals and some of the fungi (mushrooms) and algae (seaweeds). Only certain eukaryotes are traditionally studied by microbiologists—primarily the protozoa, the microscopic algae and fungi, and animal parasites, or helminths.

5.2 Form and Function of the Eukaryotic Cell: External Structures

 Expected Learning Outcomes

4. Describe the plan of a basic eukaryotic cell and organelles, and indicate the structures all cells possess and those found only in some groups.

5. Describe the types of eukaryotic locomotor appendages.

6. Differentiate the structure and functions of flagella and cilia, and the types of cells that possess them.

7. Define the glycocalyx for eukaryotic cells and list its basic functions.

8. Characterize the cell wall and membrane of eukaryotic cells.

The cells of eukaryotic organisms are so varied that no one member can serve as a typical example, so a composite structure of a eukaryotic cell is depicted in **figure 5.3**. No single type of microbial cell would contain all structures represented. It is evident that the organelles impart much greater complexity and compartmentalization compared to a prokaryotic cell. Differences among fungi, protozoa, algae, and animal cells are introduced in later sections.

> **Quick Search**
>
> Look up "An Interactive Tour of the Cell" at the National Science Foundation website to explore the dynamics of cell structure.

The flowchart shown here maps the structural organization of a eukaryotic cell. Compare this outline to the one found in chapter 4.

In general, eukaryotic microbial cells have a cytoplasmic membrane, nucleus, mitochondria, endoplasmic reticulum, Golgi apparatus, vacuoles, cytoskeleton, and glycocalyx. A cell wall, locomotor appendages, and chloroplasts are found only in some groups. In the following sections, we cover the microscopic structure and functions of the eukaryotic cell. As with the prokaryotes, we begin on the outside and proceed inward through the cell.

Locomotor Appendages: Cilia and Flagella

Motility allows a microorganism to locate life-sustaining nutrients and to migrate toward positive stimuli such as sunlight; it also permits avoidance of harmful substances and stimuli. Locomotion by means of flagella is common in protozoa, algae, and a few fungal and animal cells. Cilia are found only in protozoa and animal cells.

Cell wall*

Mitochondrion

Cell membrane

Rough endoplasmic reticulum

Golgi apparatus

Flagellum or cilium*

Microfilaments

Nuclear envelope with pores

Nucleus

Lysosome

Nucleolus

Smooth endoplasmic reticulum

Microvilli

Microtubules

Chloroplast*

Centrioles*

***Structure not present in all cell types**

Glycocalyx

Figure 5.3 Overview of composite eukaryotic cell. This drawing represents all structures associated with eukaryotic cells, but no microbial cell possesses all of them. See figures 5.16, 5.23, and 5.25 for examples of individual cell types.

Although they share the same name, the **flagella** of eukaryotes are much different from those of prokaryotes. The eukaryotic flagellum is thicker (by a factor of 10), has a much different construction, and is covered by an extension of the cell membrane. A flagellum is a long, sheathed cylinder containing regularly spaced hollow tubules—**microtubules**—that extend along its entire length (**figure 5.4b**). A cross section reveals nine pairs of closely attached microtubules surrounding a single central pair. This scheme, called the 9 + 2 arrangement, is a typical pattern of flagella and cilia (**figure 5.4a**). The nine pairs are linked together and anchored to the pair in the center. This architecture permits the microtubules to "walk" by sliding past each other, whipping the flagellum back and forth. Although details of this process are too complex to discuss here, it involves expenditure of energy and a coordinating mechanism in the cell membrane. Flagella can move the cell by pushing it

forward like a fishtail or by pulling it by a lashing or twirling motion (**figure 5.4c**). The placement and number of flagella can be useful in identifying flagellated protozoa and certain algae.

Cilia are very similar in overall architecture to flagella, but they are shorter and more numerous (some cells have several thousand). They are found only in certain protozoa and animal cells. In the ciliated protozoa, the cilia occur in rows over the cell surface, where they beat back and forth in regular oarlike strokes (**figure 5.5**) and provide rapid motility. The fastest ciliated protozoan can swim up to 2,500 μm/s—a meter and a half per minute! On some cells, cilia also function as feeding and filtering structures.

Quick Search
Find videos using the search words *amoebic, flagellate,* and *ciliate movement* on YouTube.

Figure 5.4 **The structure of cilia and flagella. (a)** A cross section through a protozoan cilium reveals the typical 9 + 2 arrangement of microtubules seen in both cilia and flagella. **(b)** Longitudinal section through a cilium, showing the lengthwise orientation of the microtubules and the basal body (bb) from which they arise. Note the membrane that surrounds the cilium that is an extension of the cell membrane and shows that it is indeed an organelle. **(c)** Locomotor patterns seen in flagellates. (a and b): Courtesy of Richard Allen

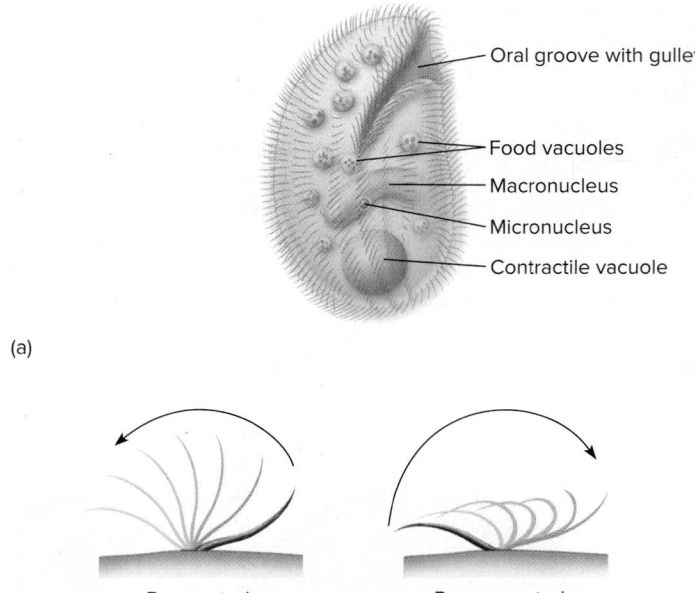

Figure 5.5 **Locomotion in ciliates. (a)** A simple ciliate, *Balantidium,* has regular rows of cilia over its surface and within its oral groove and gullet for capturing and transporting food. **(b)** Cilia beat in coordinated waves, driving the cell forward and backward. The pattern of ciliary movement is like a swimmer's arms, with a power forward stroke and a repositioning stroke. Ciliary moves are made possible by the microtubules they contain.

The Glycocalyx

Most eukaryotic microbes have a **glycocalyx,** an outermost boundary that comes into direct contact with the environment. This structure is usually composed of polysaccharides and appears as a network of fibers, a slime layer, or a capsule much like the glycocalyx of prokaryotes. From its position as the exposed cell layer, the glycocalyx serves a variety of functions. Most prominently, it promotes adherence to environmental surfaces and the development of biofilms and mats. It also serves important receptor and communication functions and offers some protection against environmental changes.

The nature of the layer beneath the glycocalyx varies among the several eukaryotic groups. Fungi and most algae have a thick, rigid cell wall surrounding a cell membrane. Protozoa, a few algae, and all animal cells lack a cell wall and are encased primarily by a cell membrane.

Form and Function of the Eukaryotic Cell: Boundary Structures

The Cell Wall

The cell walls of the fungi and algae are rigid and provide structural support and shape, but they are different in chemical composition from prokaryotic cell walls. Fungal cell walls have a thick, inner layer of polysaccharide fibers composed of chitin or cellulose and a thin outer layer of mixed glycans. The cell walls of algae are quite varied in chemical composition. Substances commonly found in various algal cell walls include various forms of sugar, such as cellulose, pectin,[1] and mannans,[2] along with minerals such as silicon dioxide and calcium carbonate.

The Cytoplasmic Membrane

The cytoplasmic (cell) membrane of eukaryotic cells is a typical bilayer of phospholipids in which protein molecules are embedded (see figures 2.20 and 4.17). In addition to phospholipids, eukaryotic membranes also contain *sterols* of various kinds. Sterols are different

1. A polysaccharide composed of galacturonic acid subunits.

2. A polymer of the sugar known as mannose.

TABLE 5.2	Function of External and Boundary Structures of the Eukaryotic Cell
Cilia and flagella	Cell movement
Glycocalyx	Adherence to surfaces; development of biofilms and mats; protection
Cell wall	Structural support and shape
Cytoplasmic membrane	Selective permeability; cell-cell interactions; adhesion to surfaces; signal transduction

from phospholipids in both structure and behavior, as you may recall from chapter 2. Their relative rigidity confers stability on eukaryotic membranes. This strengthening feature is extremely important in cells that lack a cell wall. Cytoplasmic membranes of eukaryotes are functionally similar to those of prokaryotes, serving as selectively permeable barriers in transport. Unlike prokaryotes, eukaryotes have extensive membrane-bound structures that can account for 60% to 80% of the membrane's volume. These structures, primarily proteins and carbohydrates, are responsible for interactions between cells, adhesion to surfaces, secretion of products made within the cell, and communication between the cell and its environment (signal transduction).

 Check Your Progress SECTIONS 5.1 AND 5.2

1. Briefly explain how the eukaryotic cell could have evolved from prokaryotic ones.
2. Which kingdoms of the five-kingdom system contain eukaryotic microorganisms?
3. How do unicellular, colonial, and multicellular organisms differ from each other? Give examples of each type.
4. How are flagella and cilia similar? How are they different?
5. Which eukaryotic cells have a cell wall?
6. What are the functions of the glycocalyx, cell wall, and membrane?

5.3 Form and Function of the Eukaryotic Cell: Internal Structures

 Expected Learning Outcomes

9. Describe the structure of the nucleus and its outstanding features.
10. Outline the stages in cell division and mitosis.
11. Describe the structure of the two types of endoplasmic reticulum and their functions.
12. Identify the parts of the Golgi apparatus, and explain its basic actions and uses in the cell.
13. Summarize the stages in processing by the nucleus, endoplasmic reticulum, and Golgi apparatus involved in synthesis, packaging, and export.
14. Describe the structure of a mitochondrion, and explain its importance and functions.
15. Describe the structure of chloroplasts, and explain their importance and functions.
16. Discuss features of eukaryotic ribosomes.
17. Indicate the basic structure of the cytoskeleton, and explain its main features and functions.

The Nucleus: The Control Center

The **nucleus** is a large compact body that is the most prominent organelle of eukaryotic cells. It is separated from the cell cytoplasm by an external boundary called a **nuclear envelope** or membrane. The envelope has a unique architecture. It is composed of two parallel membranes separated by a narrow space, and it is perforated with small, regularly spaced openings, or pores, at sites where the two membranes unite (**figure 5.6**). The pores are structured to serve as selective passageways for molecules to migrate between the nucleus

(a)

(b)

Figure 5.6 The nucleus. (a) Electron micrograph section of an interphase nucleus, showing its most prominent features. **(b)** Cutaway three-dimensional view of the relationships of the nuclear envelope, pores, and endoplasmic reticiulum. (a): © Don W. Fawcett/Science Source

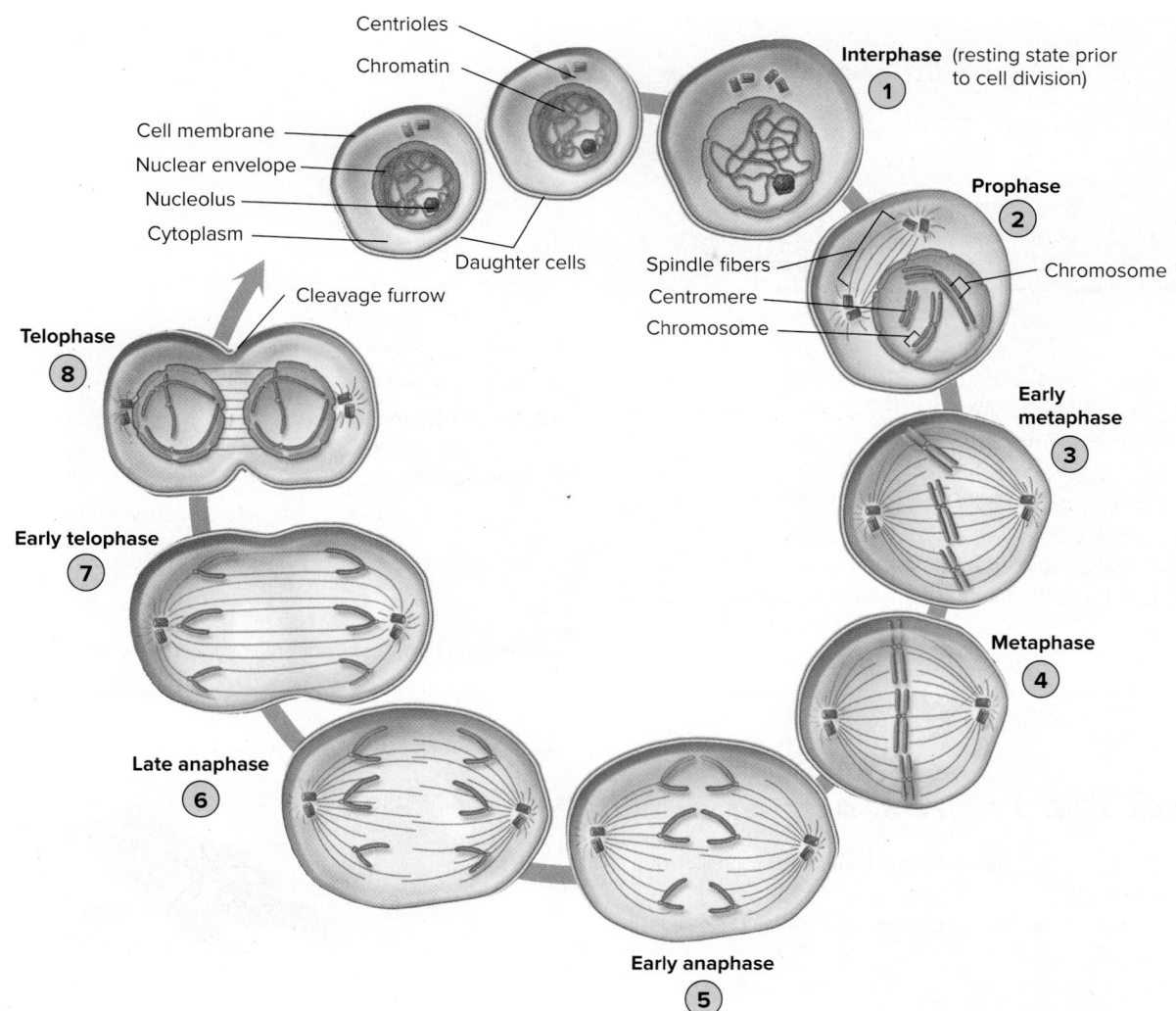

Process Figure 5.7 **Changes in the cell and nucleus during mitosis of a eukaryotic cell.** (1) Before mitosis (at interphase), chromosomes are visible only as chromatin. (2) As mitosis proceeds (early prophase), chromosomes take on a fine, threadlike appearance as they condense, and the nuclear membrane and nucleolus are temporarily disrupted. (3)–(4) By metaphase, the chromosomes are fully visible as X-shaped structures. The shape is due to duplicated chromosomes attached at a central point, the centromere. (5)–(6) Spindle fibers attach to these and facilitate the separation of individual chromosomes during anaphase. (7)–(8) Telophase completes chromosomal separation and division of the cell proper into daughter cells. Note: This mechanism is how wall-free cells divide. Cells with walls will have a different pattern.

and cytoplasm. The main body of the nucleus consists of a matrix called the nucleoplasm and a granular mass, the **nucleolus.** The nucleolus is the site for ribosomal RNA synthesis and a collection area for ribosomal subunits. The subunits are transported through the nuclear pores into the cytoplasm for final assembly into ribosomes.

A prominent feature of the nucleoplasm in stained preparations is a network of dark fibers known as **chromatin** because of its attraction for dyes. Chromatin consists primarily of DNA, along with many small proteins called **histones. Chromosomes,** the large structures within the nucleus of eukaryotic cells that contain the genetic information of the cell are composed of chromatin. In nondividing cells, the chromosomes within the nucleus are not readily visible because they are long, linear DNA molecules bound in varying degrees to histone proteins, and are far too fine to be resolved as distinct structures without extremely high magnification. During **mitosis,** however, when the duplicated chromosomes are separated equally into daughter cells, the chromosomes themselves become

readily visible as discrete bodies **(figure 5.7).** This appearance arises when the DNA becomes highly condensed by forming coils and supercoils around the histones to prevent the chromosomes from tangling as they are separated into new cells. This process is described in more detail in chapter 9.

Although we correctly view the nucleus as the primary genetic center, it does not function in isolation. As we show in the next three sections, it is closely tied to cytoplasmic organelles that perform elaborate cell functions.

Endoplasmic Reticulum: A Passageway and Production System for Eukaryotes

The **endoplasmic reticulum (ER)** is an interconnected network of membranous, hollow sacs that synthesize and transport cell substances **(figure 5.8*a*).** Originating from, and continuous with, the outer membrane of the nuclear envelope, the ER serves as a passageway for

Figure 5.8 The origin and structures of the rough endoplasmic reticulum (RER) and Golgi apparatus. (a) Schematic view of the origin of the endoplasmic reticulum from the outer membrane of the nuclear envelope. **(b)** Three-dimensional projection of the RER. **(c)** Detail of the orientation of a ribosome on the RER membrane. Proteins are collected within the RER cisternae and distributed through its network to other destinations. **(d)** The flattened layers of the Golgi apparatus receive protein-carrying transport vesicles from the RER, process their contents, and transport condensing vesicles to sites of various cellular functions.

materials between the nucleus and the cytoplasm. It also provides significant compartmentalization for numerous cell activities. The two types of endoplasmic reticulum are the **rough endoplasmic reticulum (RER)** and the **smooth endoplasmic reticulum (SER).** These two ER networks are similar in origin and are directly connected to each other, but they differ in some aspects of structure and function.

The RER consists of parallel, flattened sacs called **cisternae,*** and it appears "rough" in electron micrographs because its outer surface is studded with ribosomes **(figure 5.8b).** This architecture ties in with its role in synthesis of proteins, which are collected in the lumen **(figure 5.8c)** of the cisternae and processed for further transport to the Golgi apparatus **(figure 5.8d).** The SER lacks ribosomes and is more tubular in structure (see figures 5.3 and 5.8a). Although it communicates with the RER and also transports molecules within the cell, it has specialized functions. Most important among them

are the synthesis of nonprotein molecules such as lipids, and detoxification of metabolic waste products and other toxic substances.

Golgi Apparatus: A Packaging Machine

The **Golgi**[3] **apparatus,** also called the **Golgi complex** or body, is the site in the cell that collects proteins and and packages them for transport to their final destinations. It is a discrete organelle consisting of layers of flattened, disc-shaped sacs also called cisternae, giving an appearance of a stack of pita breads. These sacs have outer limiting membranes and cavities like those of the endoplasmic reticulum, but they do not form a continuous network (figure 5.8d). This organelle is always closely associated with the endoplasmic reticulum both in its location and function. At a site where it

* *cisternae* (siss-tur′-nee) L., *cisterna,* reservoir.

3. *golgi* (gol′-jee) Named for C. Golgi, an Italian histologist who first described the apparatus in 1898.

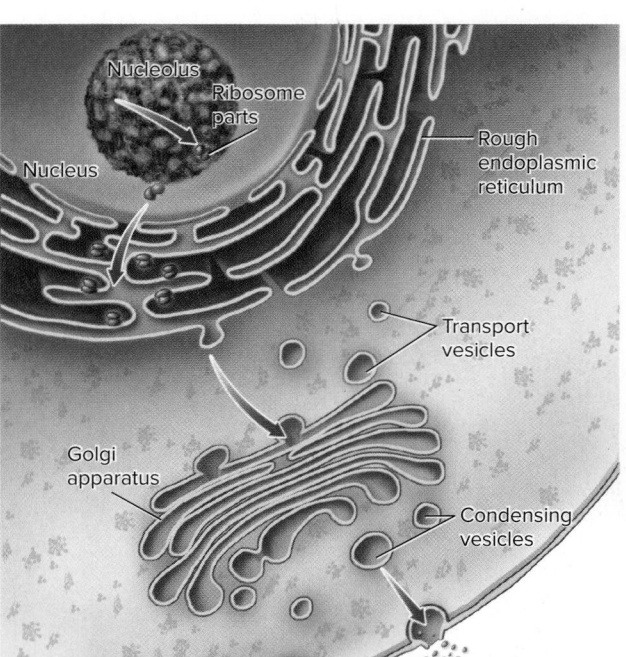

Figure 5.9 The synthesis and transport machine. Function requires cooperation and interaction among the system of organelles. The nucleolus provides a constant supply of ribosomes that travel through the nuclear pores to the RER. Once in place, ribosomes synthesize proteins that are assembled by the RER into vesicles and moved to the nearby Golgi. Finished vesicles leave the Golgi after processing and migrate to the cell membrane. From there they may be secreted (as here) or act internally.

borders on the Golgi apparatus, the endoplasmic reticulum buds off tiny membrane-bound packets of protein called **transport vesicles** that are picked up by the forming face of the Golgi apparatus. Once in the complex itself, the proteins are further modified for transport by the addition of polysaccharides and lipids. The final action of this apparatus is to pinch off finished **condensing vesicles** that will be conveyed to organelles such as lysosomes or transported outside the cell as **secretory vesicles (figure 5.9).**

Nucleus, Endoplasmic Reticulum, and Golgi Apparatus: Nature's Assembly Line

As the keeper of the eukaryotic genetic code, the nucleus ultimately governs and regulates all cell activities. But because the nucleus remains fixed in a specific cellular site, it must direct these activities through a structural and chemical network (figure 5.9). This network includes ribosomes, which originate in the nucleus, and the rough endoplasmic reticulum, which is continuously connected with the nuclear envelope.

Initially, a segment of the genetic code of DNA containing the instructions for producing a protein is copied into RNA and passed out through the nuclear pores directly to the ribosomes on the endoplasmic reticulum. Here, specific proteins are synthesized from the RNA code and deposited in the lumen (space) of the endoplasmic reticulum. Details of this process are covered in chapter 9. After being transported to the Golgi apparatus, the protein products are

chemically modified and packaged into **vesicles*** that can be used by the cell in a variety of ways. Some of the vesicles contain enzymes to digest food inside the cell; other vesicles are secreted to digest materials outside the cell; and yet others are important in the enlargement and repair of the cell wall and membrane.

A **lysosome*** is one type of vesicle originating from the Golgi apparatus that contains a variety of enzymes. Lysosomes are involved in intracellular digestion of food particles and in protection against invading microorganisms. They also participate in digestion and removal of cell debris in damaged tissue.

Other types of vesicles include **vacuoles,*** which are membrane-bound sacs containing fluids or solid particles to be digested, excreted, or stored. They are formed in phagocytic cells (certain white blood cells and protozoa) in response to food and other substances that have been engulfed. The contents of a food vacuole are digested through the merger of the vacuole with a lysosome. This merged structure is called a phagosome **(figure 5.10).** Other types of vacuoles are used in storing reserve food such as fats and glycogen. Protozoa living in freshwater habitats regulate osmotic pressure by means of contractile vacuoles, which regularly expel excess water that has diffused into the cell (discussed in section 7.3).

Mitochondria: Energy Generators of the Cell

None of the cellular activities of the genetic assembly line could proceed without a constant supply of energy, the bulk of which is generated in most eukaryotes by **mitochondria.*** When viewed with light microscopy, mitochondria appear as tiny round or elongated particles scattered throughout the cytoplasm. Higher magnification reveals that a mitochondrion consists of a smooth, continuous outer membrane that forms the external contour and an inner, folded membrane nestled neatly within the outer membrane **(figure 5.11).** The folds on the inner membrane, called **cristae,*** vary in exact structure among eukaryotic cell types. Plant, animal, and fungal mitochondria have lamellar cristae folded into shelflike layers. Those of algae and protozoa are tubular, fingerlike projections or flattened discs.

The cristae membranes hold the enzymes and electron carriers of aerobic respiration. This is an oxygen-using process that extracts chemical energy contained in nutrient molecules and stores it in the form of high-energy molecules, or ATP. More detailed functions of mitochondria are covered in chapter 8. The spaces around the cristae are filled with a chemically complex fluid called the **matrix,*** which holds ribosomes, DNA, and a pool of enzymes and other compounds involved in the metabolic cycle. Mitochondria (along with chloroplasts) are unique among organelles in that they divide independently of the cell, contain circular strands of DNA, and have prokaryote-sized 70S ribosomes. These findings have given support to the endosymbiotic theory of their evolutionary origins discussed in section 5.1.

Chloroplasts: Photosynthesis Machines

Chloroplasts are remarkable organelles found in algae and plant cells that are capable of converting the energy of sunlight into

* *vesicle* (ves′-ik-l) L. *vesios,* bladder. A small sac containing fluid.
* *lysosome* (ly′-soh-sohm) Gr. *lysis,* dissolution, and *soma,* body.
* *vacuole* (vak′-yoo-ohl) L. *vacuus,* empty. Any membranous space in the cytoplasm.
* *mitochondria* (my″-toh-kon′-dree-uh) sing. mitochondrion; Gr. *mitos,* thread, and *chondrion,* granule.
* *cristae* (kris′-te) sing. crista; L. *crista,* a comb.
* *matrix* (may′-triks) L. *mater,* mother or origin.

Figure 5.10 **The functions of lysosomes during phagocytosis.**

(a)

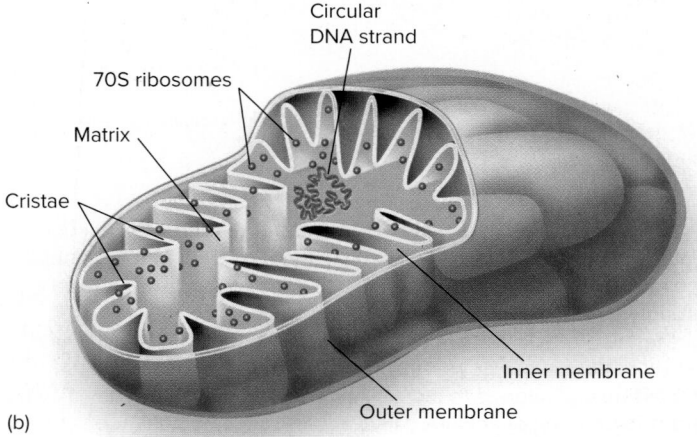

(b)

Figure 5.11 **General structure of a mitochondrion. (a)** An electron micrograph of a cross section. In most cells, mitochondria are elliptical or spherical, or occasionally filament-like. **(b)** A cutaway three-dimensional illustration details the double membrane structure and the extensive foldings of the inner membrane. (a): © Keith R. Porter/ Science Source

chemical energy through photosynthesis. The photosynthetic role of chloroplasts makes them important primary producers of organic nutrients upon which all other organisms (except certain bacteria and archaea) ultimately depend. Another important photosynthetic product of chloroplasts is oxygen gas. Although chloroplasts resemble mitochondria, chloroplasts are larger, contain special pigments, and are much more varied in shape. Be reminded that chloroplasts are thought to have originated from intracellular cyanobacteria and resemble them in many ways.

There are differences among various algal chloroplasts, but most are generally composed of two membranes, one enclosing the other. The smooth, outer membrane completely covers an inner membrane folded into small, disclike sacs called **thylakoids** that are stacked upon one another into **grana.** These structures carry the green pigment chlorophyll and sometimes additional pigments as well. Surrounding the thylakoids is a ground substance called the **stroma*** **(figure 5.12).** The role of the photosynthetic pigments is to absorb and transform solar energy into chemical energy, which is then used during reactions in the stroma to synthesize carbohydrates. We further explore some important aspects of photosynthesis in chapters 7 and 8.

Ribosomes: Protein Synthesizers

In an electron micrograph of a eukaryotic cell, ribosomes are numerous, tiny particles that give a "dotted" appearance to the cytoplasm. Ribosomes are distributed in two ways: Some are scattered freely in the cytoplasm and cytoskeleton; others are intimately associated with the rough endoplasmic reticulum, as previously described. Multiple ribosomes are often found arranged in short

* *stroma* (stroh′-mah) Gr. *stroma,* mattress or bed.

Figure 5.12 Details of an algal chloroplast. An artist's three-dimensional rendition of a chloroplast and its major features.

TABLE 5.3	Function of Internal Structures within the Eukaryotic Cell
Nucleus	Genetic center of the cell; repository of DNA; synthesis of RNA
Nucleolus	Ribosomal RNA synthesis; ribosome construction
Endoplasmic reticulum	Transport of materials; lipid synthesis
Golgi apparatus	Packaging and modification of proteins prior to secretion
Lysosomes	Intracellular digestion
Vacuoles	Temporary storage and transport; digestion (food vacuoles); regulation of osmotic pressure (water vacuoles)
Mitochondria	Energy production using the Krebs cycle, electron transport, and oxidative phosphorylation
Chloroplast	Conversion of sunlight into chemical energy through photosynthesis
Ribosomes	Protein synthesis
Cytoskeleton	Composed of microfilaments and microtubules; provides cell structure and movement; anchors organelles

chains called polyribosomes (polysomes). The basic structure of eukaryotic ribosomes is similar to that of prokaryotic ribosomes, described in chapter 4. Both are composed of large and small subunits of ribonucleoprotein (see figure 5.8). By contrast, however, the ribosomes of eukaryotes (except in the mitochondria and chloroplasts) are of the larger 80S variety that is a combination of 60S and 40S subunits. As in the prokaryotes, eukaryotic ribosomes are the staging areas for protein synthesis.

The Cytoskeleton: A Support Network

All cells share a generalized region encased by the cell membrane called the cytoplasm or, in eukaryotic cells, the **cytoplasmic matrix.**

This complex area houses the organelles and sustains major metabolic and synthetic activities. It also contains the framework of support in cells lacking walls. The matrix is interwoven by a flexible framework of molecules called the **cytoskeleton (figure 5.13).** This framework appears to have several functions, such as anchoring organelles, providing support, and permitting shape changes and movement in some cells. The two main types of cytoskeletal elements are *microfilaments* and *microtubules*.

Microfilaments are thin strands composed of the protein actin that attach to the cell membrane and form a network through the

(b)

Figure 5.13 A model of the cytoskeleton. (a) Depicted is the relationship between microtubules, microfilaments, and organelles. (Not to scale.) **(b)** The cytoskeleton of a large eukaryotic cell is highlighted by fluorescent dyes. Microtubules are stained green, microfilaments are red, and the nucleus is blue.
(b): Courtesy of Life Technologies, Carlsbad, CA

cytoplasm. Some microfilaments are responsible for movements of the cytoplasm, often made evident by the streaming of organelles around the cell in a cyclic pattern. Other microfilaments are active in *amoeboid motion,* a type of movement, typical of cells such as amoebas and phagocytes, that produces extensions of the cell membrane (pseudopods) into which the cytoplasm flows.

Microtubules are long, hollow tubes that maintain the shape of eukaryotic cells such as protozoa that lack cell walls. They may also serve as an alternative transport system for molecules. The spindle fibers that play an essential role in mitosis are actually microtubules that attach to chromosomes and separate them into daughter cells. As indicated earlier, microtubules are also responsible for the movement of cilia and flagella.

 Check Your Progress SECTION 5.3

7. In what ways does the nucleus function like the "brain" of the cell?
8. Explain how ribosomes and the nucleolus are related.
9. How does the nucleus communicate with the cytoplasm?
10. Compare and contrast the smooth ER, the rough ER, and the Golgi apparatus in structure and function.
11. Compare the structures and functions of the mitochondrion and the chloroplast.

12. What makes the mitochondrion and chloroplast unique among the organelles?
13. Describe some of the ways that organisms use lysosomes.
14. For what reasons would a cell need a "skeleton"?

5.4 Eukaryotic-Prokaryotic Comparisons and Taxonomy of Eukaryotes

 Expected Learning Outcomes

18. Compare and contrast prokaryotic cells, eukaryotic cells, and viruses.
19. Outline the basics of eukaryotic taxonomy.
20. Explain what is meant by the term *protist,* and outline the types of organisms belonging to this designation.

Now that we have introduced the major characteristics of eukaryotic cells, in this section we will summarize these characteristics and compare and contrast them with prokaryotic cells (see chapter 4), especially from the standpoints of structure, physiology, and other traits **(table 5.4)**. We have included viruses for the purposes of summarizing the biological characteristics of all microbes and to show you just

TABLE 5.4	A General Comparison of Prokaryotic and Eukaryotic Cells and Viruses			
Function or Structure	Characteristic*	Prokaryotic Cells	Eukaryotic Cells	Viruses**
Genetics	Nucleic acids	+	+	+
	True nucleus	–	+	–
	Nuclear envelope	–	+	–
	Nucleoid	+	–	–
Reproduction	Mitosis	–	+	–
	Production of sex cells	+/–	+	–
	Binary fission	+	+	–
Biosynthesis	Independent	+	+	–
	Golgi apparatus	–	+	–
	Endoplasmic reticulum	–	+	–
	Ribosomes	+***	+	–
Respiration	Enzymes	+	+	–
	Mitochondria	–	+	–
Photosynthesis	Pigments	+/–	+/–	–
	Chloroplasts	–	+/–	–
Motility/locomotor structures	Flagella	+/–***	+/–	–
	Cilia	–	+/–	–
Shape/protection	Cell membrane	+	+	–
	Cell wall	+/–***	+/–	– (have capsids instead)
	Capsule	+/–	+/–	–
Size (in general)		0.5–3 µm****	2–100 µm	< 0.2 µm

*+ means most members of the group exhibit this characteristic; – means most lack it; +/– means some members have it and some do not.
**Viruses cannot participate in metabolic or genetic activity outside their host cells.
***The prokaryotic type is structurally very different.
****Much smaller and much larger bacteria exist; see 4.1 Secret World of Microbes.

how much viruses differ from cells. We will explore this fascinating group of microbes in chapter 6.

Overview of Taxonomy

Exploring the origins of eukaryotic cells with molecular technology has significantly clarified our understanding of relationships among the organisms in Domain Eukarya. The phenetic characteristics traditionally used for placing plants, animals, and fungi into separate kingdoms are general cell type, level of organization or body plan, and nutritional type. It now appears that these criteria really do reflect accurate differences among these organisms and give rise to the same basic kingdoms, as do genetic tests using ribosomal RNA (see figure 1.15).

Because our understanding of the phylogenetic relationships is still in development, there is not yet a single official system of taxonomy for presenting all of the eukaryotes. One of the proposed alternate plans of classification is shown in **figure 5.14.** This simplified system, also based on data from ribosomal RNA analysis, retains some of the traditional kingdoms.

The one group that has created the greatest challenge in establishing reliable relationships is **Kingdom Protista**—the protists. Any simple eukaryotic cell that lacked multicellular structure or cell specialization has been placed into the Kingdom Protista generally as an alga (photosynthetic) or protozoan (nonphotosynthetic). Although exceptions arise with this classification scheme—notably multicellular algae and the photosynthetic protozoans—most eukaryotic microbes have been readily accommodated in the Kingdom Protista as traditionally presented.

Newer genetic data reveal that the organisms we call protists may be as different from each other as plants are different from animals. They are highly complex organisms that are likely to have evolved from a number of separate ancestors.

Microbiologists prefer to set up taxonomic systems that reflect true relationships based on all valid scientific data, including genotypic, molecular, morphological, physiological, and ecological. As a result, several alternative taxonomic strategies are being proposed to deal with this changing picture. The number of new kingdoms suggested for distributing the protists range from five to 25 or more.

Note that figure 5.14 compares an rRNA-based system as first outlined in chapter 1 alongside the original Kingdom Protista with its assorted phyla and divisions. One advantage of this original plan is that it continues to use most of the original taxa at the phylum, division, and lower levels, especially as they relate to medically significant microbial groups. We feel the term **protist** is still a useful shorthand reference and will continue to use it for any eukaryote that is not a fungus, animal, or plant.

In the interest of space and ease in presentation, we have adopted a system of taxonomy that emphasizes natural groupings with

Figure 5.14 **RNA-based phylogenetic tree and taxonomy of Domain Eukarya.** A comparison of two possible systems for presenting the taxonomy of major eukaryotic groups. This textbook will largely follow the simpler system on the right, using Kingdoms, Divisions, and Phyla to separate the groups of eukaryotes.

the least complexity. A final note about taxonomy: No one system is perfect or permanent. Let your instructor guide you as to which one is satisfactory for your course.

With the general structure of the eukaryotic cell in mind, let us next examine the range of adaptations that this cell type has undergone. The following sections contain a general survey of the principal eukaryotic microorganisms—fungi, algae, protozoa, and parasitic worms—while also introducing elements of their structure, life history, classification, identification, and importance. A survey of diseases associated with some of these microbes is found in chapters 22 and 23.

5.5 The Kingdom of the Fungi

 Expected Learning Outcomes

21. Describe the basic characteristics of the Kingdom Fungi in terms of general types of cells and organisms, structure, and nutrition.

22. Differentiate between characteristics of yeasts and of molds, and define fungal spores.

23. Classify types of fungal spores and explain their functions.

24. Discuss the main features of fungal classification and representative examples of each group.

25. Explain how fungi are identified.

26. Discuss the importance of fungi in ecology, agriculture, commerce, and medicine.

The position of the **fungi*** in the biological world has been debated for many years. They were originally classified with the green plants (along with algae and bacteria), and later they were placed in a group with algae and protozoa (the Protista). Even at that time, however, many microbiologists were struck by several unique qualities of fungi that warranted placement into a separate kingdom. Confirmation of their status by genetic testing eliminated any question that they belong in a kingdom of their own.

* *fungi* (fun′-jy) sing. fungus; Gr. *fungos,* mushroom.

The **Kingdom Fungi,** or **Eumycota,** is filled with organisms of great variety and complexity that have survived on earth for approximately 650 million years. About 100,000 species are known, although experts estimate a count much higher than this—perhaps as high as 1.5 million. For practical purposes, mycologists divide the fungi into two groups: the *macroscopic fungi* (mushrooms, puffballs, gill fungi) and the *microscopic fungi* (molds, yeasts). Although the majority of fungi are either unicellular or colonial, a few complex forms such as mushrooms and puffballs are considered multicellular.

Chemical traits of fungal cells include the possession of a polysaccharide, **chitin,** in their cell walls and a sterol, ergosterol, in their cell membranes. Cells of the microscopic fungi exist in two basic morphological types: hyphae and yeasts. **Hyphae*** are long, threadlike cells that make up the bodies of filamentous fungi, or **molds (figure 5.15).** A **yeast** cell is distinguished by its round to oval shape and by its mode of asexual reproduction. It grows swellings on its surface called buds, which then become separate cells **(figure 5.16a).** Some species form a **pseudohypha,*** a chain of yeasts formed when buds remain attached in a row **(figure 5.16c);** because of its manner of formation, it is not a true hypha like that of molds. Although some fungal cells exist only in a yeast form and others occur primarily as hyphae, a few, called **dimorphic,*** can take either form, depending upon growth conditions, such as changing temperature. This variability in growth form is particularly characteristic of some pathogenic molds.

* *hyphae* (hy′-fuh) pl. hyphae (hy′-fee); Gr. *hyphe,* a web.
* *pseudohypha* (soo″-doh-hy′-fuh) pl. pseudohyphae; Gr. *pseudo,* false, and *hyphe.*
* *dimorphic* (dy-mor′-fik) Gr. *di,* two, and *morphe,* form.

(a)

Septum

(b)

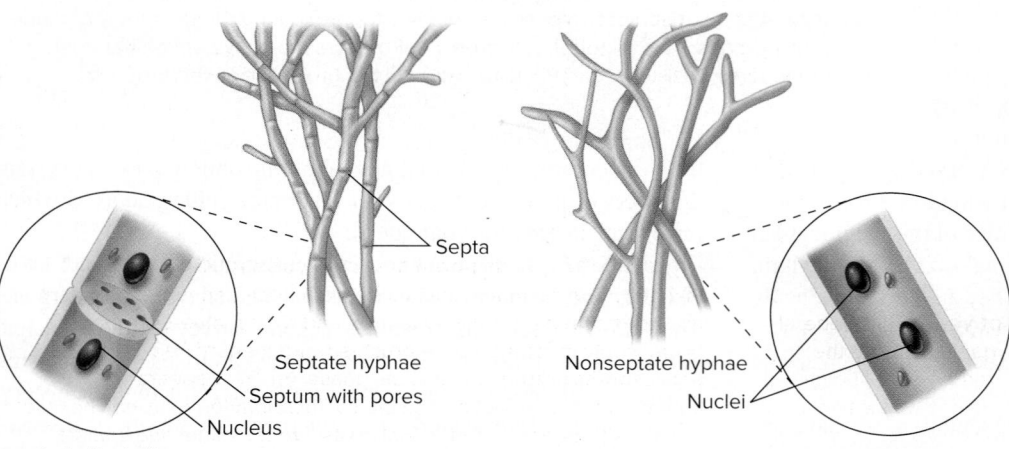

Septa

Septate hyphae
Septum with pores
Nucleus

As in *Penicillium*

(c)

Nonseptate hyphae
Nuclei

As in *Rhizopus*

Figure 5.15 Macroscopic and microscopic views of molds. (a) A container of food (left too long in the refrigerator) has developed a miniature forest of mold colonies. Note the various textures of mycelia and the array of color differences due to spores. **(b)** Close-up of hyphal structure (1,200×). **(c)** Basic structural types of hyphae. The septate hyphae develop small pores that allow communication between cells. Nonseptate hyphae lack septa and are single, long multinucleate cells.
(a): © Kathy Park Talaro; (b): © Dr. Judy A. Murphy, San Joaquin Delta College, Department of Microscopy, Stockton, CA

(a)

Fungal (Yeast) Cell

Bud — Bud scar

Ribosomes

Mitochondrion

Endoplasmic reticulum

Nucleus

Nucleolus

Cell wall

Cell membrane

Golgi apparatus

Storage vacuole

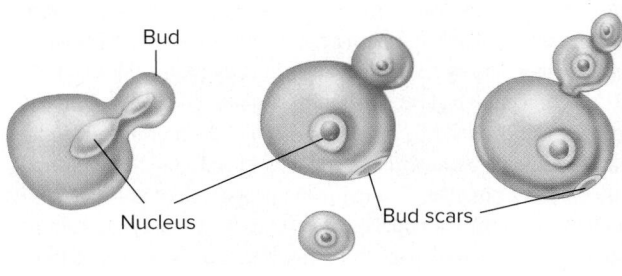

(b)

Bud

Nucleus

Bud scars

Pseudohypha

(c)

Figure 5.16 Microscopic morphology of yeasts. (a) General structure of a yeast cell, representing major organelles. Note the presence of a cell wall and lack of locomotor organelles. **(b)** Scanning electron micrograph of *Malassezia furfur,* a yeast that causes a type of superficial skin infection (25,000×). **(c)** Formation and release of yeast buds and pseudohypha (a chain of budding yeast cells). (b): *Source:* Janice Carr/CDC

Fungal Nutrition

All fungi are **heterotrophic.*** They acquire nutrients from a wide variety of organic sources or **substrates (figure 5.17).** Most fungi are **saprobes,*** meaning that they obtain these substrates from the remnants of dead plants and animals in soil or aquatic habitats. Fungi can also be parasites on the bodies of living animals or plants, although very few fungi absolutely require a living host. In general, the fungus penetrates the substrate and secretes enzymes that reduce it to small molecules that can be absorbed. Fungi have enzymes for digesting an incredible

* *heterotrophic* (het-ur-oh-tro′-fik) Gr. *hetero,* other, and *troph,* to feed. A type of nutrition that relies on an organic nutrient source.
* *saprobe* (sap′-rohb) Gr. *sapros,* rotten, and *bios,* to live. Also called saprotroph or saprophyte.

(a)

(b)

Figure 5.17 Nutritional sources (substrates) for fungi.
(a) *Penicillium notatum* mold, a very common decomposer of citrus fruit, is known for its velvety texture and typical blue-green color. Microscopic inset shows the brush arrangement of *Penicillium* phialospores (220×), the asexual phase. **(b)** The appearance of athlete's foot, an infection caused by *Trichophyton rubrum.* The inset is a 1,000× magnification revealing its hyphae (blue) and conidia (brown). (a): © Kathy Park Talaro; (a: inset): © Natures Geometry/Alamy Stock Photo; (b): © DermNet New Zealand Trust; (b: inset): *Source:* Dr. Libero Ajello/CDC

5.1 Secret World of Microbes
Fungi: A Force of Nature

Life in the Fungal Jungle

Fungi are among the most prolific microbes on earth. So dominant are they that, by one estimate, the average topsoil contains nearly 9 tons of fungal mycelia per acre. In this habitat, they associate with a wide variety of other organisms, from bacteria to animals to plants.

The activities of fungi under natural conditions include nutritional and protective interactions with plants and algae, decomposition of organic matter and recycling of the minerals it contains, and contributions to biofilm networks that are needed to successfully colonize new habitats.

Some fungi have developed into normal residents of plants called endophytes. These fungi live inside the tissues of the plant itself and protect it against disease, presumably by secreting chemicals that deter infections by other, pathogenic fungi. Understanding the roles of endophytes promises to open up a new field of biological control to inoculate plants with harmless fungi rather than using more toxic fungicides.

Lichens are unusual hybrid organisms that form when a fungus combines with a photosynthetic microbe, either an alga or a cyanobacterium. The fungus acquires nutrients from its smaller partner and probably provides a protective haven that favors survival of the lichen body. Lichens occur in most habitats on earth, but they are especially important as early invaders of rock and in soil formation. Fungi also interact with animals in their habitats, most notably with small roundworms, called nematodes, and insects. Several species of molds prey upon or parasitize nematodes.

Is There a Fungus in the House?

The fact that fungi are so widespread also means that they frequently share human living quarters, especially in locations that provide ample moisture and nutrients. Often their presence is harmless and limited to a film of mildew on shower stalls or in other moist environments.

In some cases, depending on the amount of contamination and the type of mold, these indoor fungi can also give rise to various medical problems. Such common air contaminants as *Aspergillus, Cladosporium,* and *Stachybotrys* all have the capacity to give off airborne spores and toxins that, when inhaled, cause a whole spectrum of symptoms sometimes referred to as "sick building syndrome." The usual source of harmful fungi is the presence of chronically water-damaged walls, ceilings, and other building materials that have come to harbor these fungi. People exposed to these houses or buildings report symptoms that range from

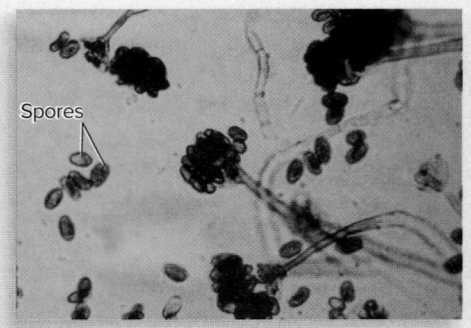

Mycelium and spores in the ascomycete *Stachybotrys.* This black mold (400×) is implicated in building contamination that leads to toxic diseases. © Neil Carlson, Department of Environmental Health and Safety

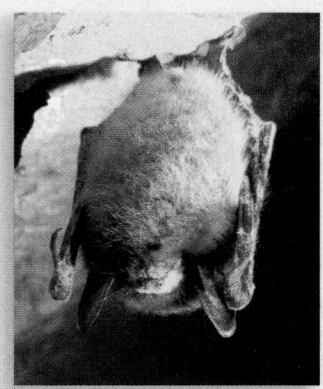

A little brown bat suffering from white nose syndrome. Note the fungus on the snout, wings, and body. *Source:* U.S. Fish and Wildlife Service

skin rash, flulike reactions, sore throat, and headaches to fatigue, diarrhea, allergies, and immune suppression.

Bat Populations Decimated by Fungal Infections

For several years wildlife personnel and animal experts have reported mass deaths of bats in caves throughout the Northeast and some parts of the southern United States. At first the cause was not clear, but workers observed a whitish growth on the muzzle of the dead bats and called the condition "white nose syndrome." Closer inspection revealed that the white growth was *Pseudogymnoascus destructans,* a newly recognized fungal pathogen that inhabits the caves where bats hibernate. The animals apparently became infected in the winter, and by spring they were already dead or sick and dying. By 2015 the infection had been noted in 25 states, and bat populations in the Northeast had declined by approximately 80%. Unfortunately, little can be done at this point, because the infection is spread from bat to bat and no viable treatment is possible. Though many people might at first not be concerned by losses in the bat population, bats are the primary predator of insects, and the agricultural damage from unsuppressed insect populations has been estimated at up to $50 billion per year in the United States.

What characteristics of fungi allow them to spread into such a wide variety of habitats? *Answer available on Connect.*

array of substances, including feathers, hair, cellulose, petroleum products, wood, even rubber. It is likely that every naturally occurring organic material on the earth can be attacked by some type of fungus.

Fungi are often found in nutritionally poor or adverse environments. Various fungal types thrive in substrates with higher amounts of salt, sugar, or acid content, at relatively high temperatures, and even in snow and glaciers.

The medical and agricultural impact of fungi is extensive. A number of species cause **mycoses** (fungal infections) in animals, and thousands of species are important plant pathogens. Fungal toxins may cause disease in humans, and airborne fungi are a frequent cause of allergies and other medical conditions (**5.1 Secret World of Microbes**).

Organization of Microscopic Fungi

The cells of most microscopic fungi grow in loose associations or colonies. The colonies of yeasts are much like those of bacteria in that they have a soft, uniform texture and appearance. The colonies of filamentous fungi are noted for the striking cottony, hairy, or velvety textures that arise from their microscopic organization and morphology. The woven, intertwining mass of hyphae that makes up the body or colony of a mold is called a **mycelium.***

* *mycelium* (my′-see-lee-yum) pl. mycelia; Gr. *mykes,* root word for fungus.

Although hyphae contain the usual eukaryotic organelles, they also have some unique organizational features. In most fungi, the hyphae are divided into segments by cross walls, or **septa,** and such hyphae are referred to as septate (see figure 5.15c). The nature of the septa varies from solid partitions with no communication between the compartments to partial walls with small pores that allow the flow of organelles and nutrients between adjacent compartments. Nonseptate hyphae consist of one long, continuous cell *not* divided into individual compartments by cross walls. With this construction, the cytoplasm and organelles move freely from one region to another, and each hyphal element can have several nuclei (see figure 5.15c).

Hyphae can also be classified according to their particular function. Vegetative hyphae (mycelia) are responsible for the visible mass of growth that appears on the surface of a food source and penetrates it to digest and absorb nutrients. During the development of a fungal colony, the vegetative hyphae give rise to structures called reproductive, or fertile, hyphae that branch off vegetative mycelium. These hyphae are responsible for the production of fungal reproductive bodies called spores, discussed next. Other specializations of hyphae are illustrated in **figure 5.18.**

Reproductive Strategies and Spore Formation

Fungi have many complex and successful reproductive strategies. Most can propagate by the simple outward growth of existing

(b) **Reproductive Hyphae**

Sporangium

Sporangiospores

Surface hyphae

Submerged hyphae

Rhizoids

Spore

Germ tube

Substrate

(a) **Vegetative Hyphae**

Hypha

(c) **Germination**

Figure 5.18 Functional types of hyphae using the mold *Rhizopus* as an example. (a) Vegetative hyphae are those surface and submerged filaments that digest, absorb, and distribute nutrients from the substrate. This species also has special anchoring structures called rhizoids. **(b)** Later, as the mold matures, it sprouts reproductive hyphae that produce asexual spores called sporangiospores. **(c)** During the asexual life cycle, the free mold spores settle on a substrate and send out germ tubes that elongate into hyphae that produce an extensive mycelium. So prolific are the fungi that a single colony of mold can easily contain 5,000 spore-bearing structures. If each of these released 2,000 single spores, and if every spore were able to germinate, we would soon find ourselves in a sea of mycelia. Most spores do not germinate, but enough are successful to keep the numbers of fungi and their spores very high in most habitats.

hyphae or by fragmentation, in which a separated piece of mycelium can generate a whole new colony. But the primary reproductive mode of fungi involves the production of various types of **spores.** Do not confuse fungal spores with the extremely resistant, nonreproductive bacterial endospores. Fungal spores function not only in multiplication but also in survival (providing genetic variation) and dissemination. Because of their compactness and relatively light weight, fungal spores are dispersed widely through the environment by air, water, and living things. Upon encountering a favorable substrate, a spore will germinate and produce a new fungus colony in a very short time (figure 5.18).

The fungi exhibit such a marked diversity in spores that they are largely classified and identified by their spores and spore-forming structures. Although there are some elaborate systems for naming and classifying spores, we present only a basic overview of the principal types. The most general subdivision is based on the way the spores arise. Asexual spores are the products of mitotic division of a single parent cell, and sexual spores are formed through a process involving the fusing of two parental nuclei followed by meiosis.

Asexual Spore Formation

On the basis of the nature of the reproductive hypha and the manner in which the spores originate, there are two subtypes of asexual spore **(figure 5.19):**

1. **Sporangiospores** (figure 5.19a) are formed by successive cleavages within a saclike head called a **sporangium,*** which is attached to a stalk, the sporangiophore. These spores are initially enclosed but are released when the sporangium ruptures. Some sporangiospores are the end result of the sexual phase of the life cycle as seen in figure 5.20.
2. **Conidia*** (conidiospores) are free spores not enclosed by a spore-bearing sac (figure 5.19b). They develop either by pinching off the tip of a special fertile hypha or by segmentation of a preexisting vegetative hypha. Conidia are the most common asexual spores, and they occur in these forms:

 arthrospore (ar′-thro-spor) Gr. *arthron,* joint. A rectangular spore formed when a septate hypha fragments at the cross walls.
 chlamydospore (klam-ih′-doh-spor) Gr. *chlamys,* cloak. A spherical conidium formed by the thickening of a hyphal cell. It is released when the surrounding hypha fractures, and it serves as a survival or resting cell.
 blastospore. A spore produced by budding from a parent cell that is a yeast or another conidium; also called a bud.
 phialospore (fy′-ah-lo-spor) Gr. *phialos,* a vessel. A conidium budded from the mouth of a vase-shaped spore-bearing cell called a phialide or *sterigma,* leaving a small collar.
 microconidium and *macroconidium.* The smaller and larger conidia formed by the same fungus under varying conditions. Microconidia are one-celled, and macroconidia have two or more cells.
 porospore. A conidium that grows out through small pores in the spore-bearing cell; some are composed of several cells.

* *sporangium* (spo-ran′-jee-um) pl. sporangia; Gr. *sporos,* seed, and *angeion,* vessel.
* *conidia* (koh-nid′-ee-uh) sing. conidium; Gr. *konidion,* a particle of dust.

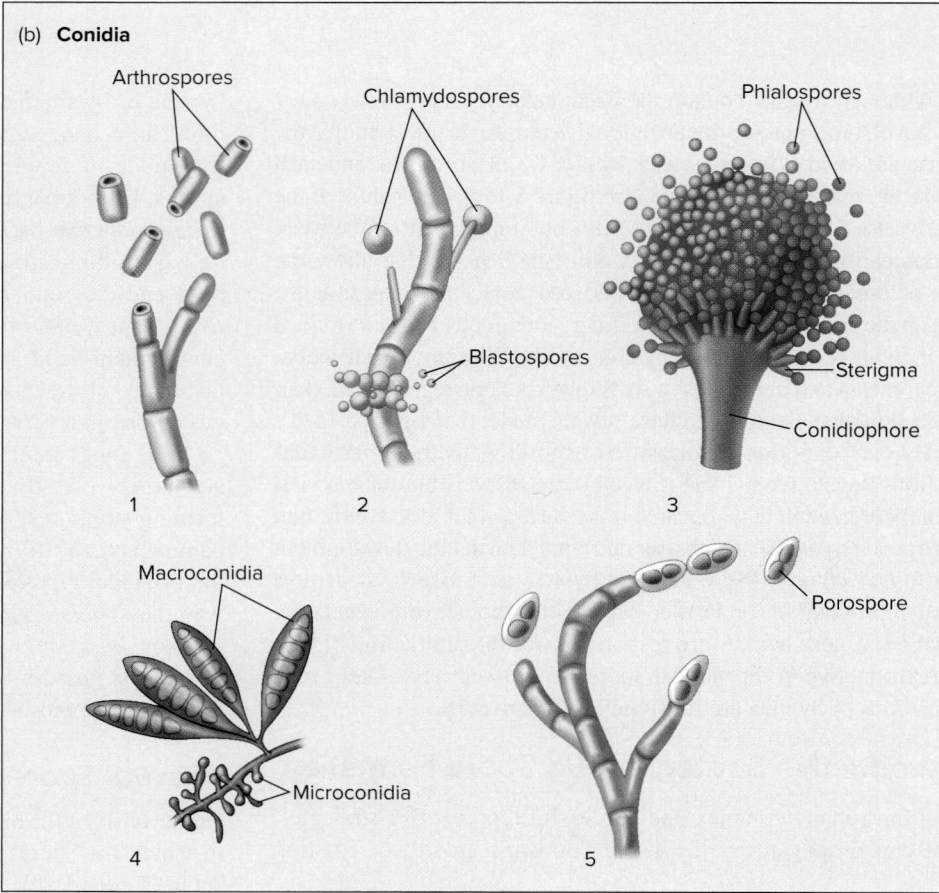

Figure 5.19 **Types of asexual mold spores. (a)** Sporangiospores: (1) *Absidia,* (2) *Syncephalastrum,* (table 5.5) **(b)** Conidia: (1) arthrospores (e.g., *Coccidioides*), (2) chlamydospores and blastospores (e.g., *Candida albicans*), (3) phialospores (e.g., *Aspergillus,* table 5.5), (4) macroconidia and microconidia (e.g., *Microsporum,* table 5.5), and (5) porospores (e.g., *Alternaria*).

Sexual Spore Formation

Because fungi can propagate successfully by producing millions of asexual spores, it is natural to wonder about the survival potential of sexual spores. The answer lies in important variations that occur when fungi of different genetic makeup combine their genetic material. As in plants and animals, a union of genes from two parents creates offspring with combinations of genes different from that of either parent. The offspring from such a union can have slight variations in form and function that are potentially advantageous to the adaptation and evolution of their species.

The majority of fungi produce sexual spores at some point. The nature of this process varies from the simple fusion of fertile hyphae of two different strains to a complex union of differentiated male and female structures and the development of special fruiting bodies. We consider the three most common sexual spores: zygospores, ascospores, and basidiospores. These spore types provide an important basis for classifying the major fungal divisions.

Zygospores* are sturdy diploid spores formed when hyphae of two opposite strains (called the plus and minus strains) fuse and create a diploid zygote that swells and becomes covered by strong, spiny walls **(figure 5.20).**

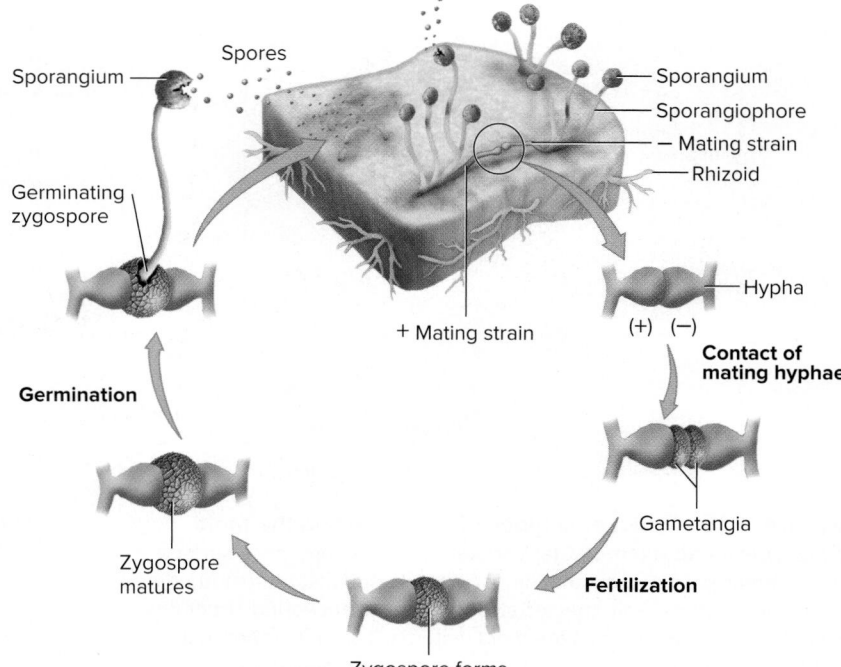

Figure 5.20 **Formation of zygospores in *Rhizopus stolonifer.*** Sexual reproduction occurs when two mating strains of hyphae grow together, fuse, and form a mature diploid zygospore. Germination of the zygospore involves production of a haploid sporangium that looks just like the asexual one shown in figure 5.19, but the individual spores contain genes from two different parents. So, in this case, the sporangiospores are the result of sexual recombination.

* *zygospores* (zy′-goh-sporz) Gr. *zygon,* yoke, to join.

142

When its wall is disrupted and moisture and nutrient conditions are suitable, the zygospore germinates and forms a mycelium that gives rise to a sporangium. Meiosis of diploid cells of the sporangium results in haploid nuclei that develop into sporangiospores. Both the sporangia and the sporangiospores that arise from sexual processes are outwardly identical to the asexual type, but because the spores arose from the union of two separate fungal parents, they are not genetically identical.

In general, haploid spores called **ascospores*** are created inside a special fungal sac, or **ascus** (pl. asci) **(figure 5.21).** Although details can vary among types of fungi, the ascus and ascospores are formed when two different strains or sexes join together to produce offspring. In many species, the male sexual organ fuses with the female sexual organ. The end result is a number of terminal cells called *dikaryons,* each containing a diploid nucleus. Through differentiation, each of these cells enlarges to form an ascus, and its diploid nucleus undergoes meiosis to form four to eight haploid nuclei that will mature into ascospores. A ripe ascus breaks open and releases the ascospores. Some species form an elaborate fruiting body to hold the asci (inset, figure 5.21).

Basidiospores* are haploid sexual spores formed on the outside of a club-shaped cell called a **basidium (figure 5.22).** In general, spore formation follows the same pattern of two mating types coming together, fusing, and forming terminal cells with diploid nuclei. Each of these cells becomes a basidium, and its nucleus produces, through

* *ascospores* (as′-koh-sporz) Gr. *ascos,* a sac.
* *basidiospores* (bah-sid′-ee-oh-sporz) Gr. *basidi,* a pedestal.

Figure 5.21 Production of ascospores in a cup fungus. Inset shows the cup-shaped fruiting body that houses the asci.

meiosis, four haploid nuclei. These nuclei are extruded through the top of the basidium, where they develop into basidiospores. Notice the location of the basidia along the gills in mushrooms, which are often dark from the spores they contain. It may be a surprise to discover that the fleshy part of a mushroom is actually a fruiting body designed to protect and help disseminate its sexual spores.

Fungal Classification

It is often difficult for microbiologists to assign logical and useful classification schemes to microorganisms that also reflect their evolutionary relationships. This difficulty is due to the fact that the organisms do not always perfectly fit the neat categories made for them, and even experts cannot always agree on the nature of the categories. The fungi are no exception, and there are several ways to classify them. For our purposes, we adopt a classification scheme with a medical mycology emphasis, in which the Kingdom Eumycota is subdivided into several phyla based upon the type of sexual reproduction, hyphal structure, and genetic profile. **Table 5.5** outlines the major phyla, their characteristics, and typical members.

Figure 5.22 The cycle of sexual spore formation in basidiomycota. A mushroom is one example to observe formation of the sexual spores. Following mating between parental hyphae underground, the mycelium gives rise to the fruiting body that we commonly identify as a mushroom. Under the cap are gills bearing club-shaped basidia which form and release basidiospores. Because they are genetically different, these basidiospores produce different parental hyphae that begin the cycle again.

Phylum I—Zygomycota (also Zygomycetes)

- Sexual spores: zygospores
- Asexual spores: mostly sporangiospores, some conidia
- Hyphae are usually nonseptate. If septate, the septa are complete
- Most species are free-living saprobes; some are animal parasites
- Can be obnoxious contaminants in the laboratory, food spoilage agents, and destructive to crops
- Examples of common molds: *Rhizopus*, a black bread mold; *Mucor; Absidia; Circinella* (**figure A**, *Syncephalastrum*, and **figure B**, *Circinella*)

A. *Syncephalastrum,* a common inhabitant of soil and plants (400×).
© Dennis Kunkel Microscopy, Inc./Phototake

B. A beautiful mold, *Circinella* is known for its curved sporangiophore (500×).
© Kathy Park Talaro

Phylum II—Ascomycota (also Ascomycetes)

- Sexual spores: most produce ascospores in asci
- Asexual spores: many types of conidia, formed at the tips of conidiophores
- Hyphae with porous septa
- Examples: This is by far the largest phylum. Members vary from macroscopic mushrooms to microscopic molds and yeasts.
 - *Penicillium* is one source of antibiotics (see figure 5.17).
 - *Aspergillus* is a common airborne mold that may be involved in respiratory infections and toxicity (**figure C**).
 - *Saccharomyces* is a yeast used in making bread and beer (**figure D**).
 - Includes many human and plant pathogens, such as *Pneumocystis (carinii) jiroveci*, a pathogen of AIDS patients.
 - *Histoplasma* is the cause of Ohio Valley fever (see figure 1.3).
 - *Trichophyton* is one cause of ringworm, which is a common name for certain fungal skin infections that often grow in a ringed pattern (see figure 22.18).
 - *Coccidioides immitis* is the cause of Valley fever; *Candida albicans* is the cause of various yeast infections; and *Stachybotrys* is a toxic mold (see 5.1 Secret World of Microbes).

C. *Aspergillus fumigatus,* displaying its flowery conidial heads (600×).
Source: CDC

D. X-ray photograph of *Saccharomyces* with a prominent nucleus (blue sphere) and bud (5,000×).
Source: Carolyn Larabell. www.cellimagelibrary.org/images/143

Phylum III—Basidiomycota (also Basidiomycetes)

Many members are familiar macroscopic forms such as mushrooms and puffballs, but this phylum also includes a number of microscopic plant pathogens called rusts and smuts that attack and destroy major crops, This has extensive impact on agriculture and global food production.

- Sexual reproduction by means of basidia and basidiospores
- Asexual spores: conidia
- Incompletely septate hyphae
- Some plant parasites and one human pathogen
- Fleshy fruiting bodies are common.
- Examples: mushrooms (**figure E**), puffballs, bracket fungi, and plant pathogens (rusts and smuts)
- The one human pathogen, the yeast *Cryptococcus neoformans,* causes an invasive systemic infection in several organs, including the skin, brain, and lungs (**figure F**).

E. *Armillaria* (honey) mushrooms, sprouting from the base of a tree, arose from a vast underground mycelium that covers 2,200 acres and stretches 3.5 miles across. © Gregory M. Filip

F. *Cryptococcus neoformans*—a yeast that causes cryptococcosis (400×). Note the capsule around each cell.
Source: Dr. Leanor Haley/CDC

Phylum IV—Chytridimycota

Members of this phylum are unusual, primitive fungi commonly called **chytrids.** * Their cellular morphology ranges from single cells to clusters and colonies. They generally do not form hyphae or yeast-type cells. The one feature that most characterizes them is the presence of special flagellated spores, called zoospores and gametes. Most chytrids are saprobic and free-living in soil, water, and decaying matter. A significant number are parasites of plants, animals, and other microbes (**figure G**), but they are not known to cause human disease. They are known to be serious frog pathogens, often responsible for destroying whole populations in some habitats.

Chytrid cells

Diatom cell

10.0 μm

G. Parasitic chytrid fungi. The filamentous diatom *Chaetoceros* is attacked by flagellated chytrids that dissolve and destroy their host (400×). © Joyce E. Longcore, University of Maine

* *chytrid* (kit′-rid) Gr. *chytridion,* little pot.

Fungi That Produce Only Asexual Spores (Imperfect)

From the beginnings of fungal classification, any fungus that lacked a sexual state was called "imperfect" and was placed in a catchall category, the Fungi Imperfecti. A species would remain classified in that category until its sexual state was described. Gradually many species were found to make sexual spores, and they were assigned to the taxonomic grouping that best fit those spores. In other cases they have been reassigned because genetic analysis indicated they belonged to an established phylum, usually to Ascomycota. This means that there is really no need for the "imperfect" taxon.

Fungal Identification and Cultivation

Fungi are identified in medical specimens by first being isolated on special types of media and then being observed macroscopically and microscopically. Examples of media for cultivating fungi are cornmeal, blood, and Sabouraud's agar. The latter medium is useful in isolating fungi from mixed samples because of its low pH, which inhibits the growth of bacteria but not of most fungi. Because the fungi are classified into general groups by the presence and type of sexual spores, it would seem logical to identify them in the same way, but sexual spores are rarely if ever demonstrated in the laboratory setting. As a result, the asexual spore-forming structures and spores are usually used to identify organisms to the level of genus and species. Other characteristics that contribute to identification are hyphal type, colony texture and pigmentation, physiological characteristics, and genetic makeup.

Fungi in Medicine, Nature, and Industry

Nearly all fungi are free-living and do not require a host to complete their life cycles. Even among those fungi that are pathogenic, most human infection occurs through accidental contact with an environmental source such as soil, water, or dust. Humans are generally quite resistant to fungal infection, except for two main types of fungal pathogens: the primary pathogens, which can infect even healthy persons, and the opportunistic pathogens, which attack persons who are already weakened in some way.

Mycoses (fungal infections) vary in the way the agent enters the body and the degree of tissue involvement **(table 5.6).** The list of opportunistic fungal pathogens has been increasing in the past few years because of newer medical techniques that keep immunocompromised patients alive. Even so-called harmless species found in the air and dust around us may be able to cause opportunistic infections in patients who already have AIDS, cancer, or diabetes.

Fungi are involved in other medical conditions besides infections. Fungal cell walls give off chemical substances that can cause allergies. The toxins produced by poisonous mushrooms can induce neurological disturbances and even death. The mold *Aspergillus flavus* synthesizes a potentially lethal poison called aflatoxin,[4]

4. From aspergillus, flavus, toxin.

TABLE 5.6 Major Fungal Infections of Humans

Degree of Tissue Involvement and Area Affected	Name of Infection	Name of Causative Fungus
Superficial (not deeply invasive)		
Outer epidermis	Tinea versicolor	*Malassezia furfur*
Epidermis, hair, and dermis can be attacked	Dermatophytosis, also called tinea or ringworm of the scalp, body, feet (athlete's foot), toenails	*Microsporum, Trichophyton*, and *Epidermophyton*
Mucous membranes, skin, nail	Candidiasis, or yeast infection	*Candida albicans* *
Systemic (deep; organism enters lungs; can invade other organs)		
Lung	Coccidioidomycosis (San Joaquin Valley fever)	*Coccidioides immitis*
	North American blastomycosis (Chicago disease)	*Blastomyces dermatitidis*
	Histoplasmosis (Ohio Valley fever)	*Histoplasma capsulatum*
	Cryptococcosis (torulosis)	*Cryptococcus neoformans*
Lung, skin	Paracoccidioidomycosis (South American blastomycosis)	*Paracoccidioides brasiliensis*

*This fungus can cause severe, invasive systemic infections in patients with cancer, AIDS, or other debilitating diseases.

which is the cause of a disease in domestic animals that have eaten grain contaminated with the mold and is also a cause of liver cancer in humans.

Fungi pose an ever-present economic hindrance to the agricultural industry. A number of species are pathogenic to field plants such as corn and grain, and fungi also rot fresh produce during shipping and storage. It has been estimated that as much as 40% of the yearly fruit crop is consumed not by humans but by fungi.

On the beneficial side, however, fungi play an essential role in decomposing organic matter and returning essential minerals to the soil. They form stable associations with plant roots (mycorrhizae) that increase the ability of the roots to absorb water and nutrients. Industry has tapped the biochemical potential of fungi to produce large quantities of antibiotics, alcohol, organic acids, and vitamins. Some fungi are eaten or used to impart flavorings to food. The yeast *Saccharomyces* produces the alcohol in beer and wine and the gas that causes bread to rise. Blue cheeses, soy sauce, and cured meats derive their unique flavors from the actions of fungi.

CLINICAL CONNECTIONS

An Outbreak of Fungal Meningitis

Most fungi are not invasive and do not ordinarily cause serious infections unless a patient's immune system is compromised or the fungus is accidentally introduced into sterile tissues. In 2012 we witnessed how a simple medical procedure could turn into a medical nightmare because a common, mostly harmless fungus got into the wrong place at the wrong time. It all started when a small compounding pharmacy in Massachusetts unknowingly sent out hundreds of mold-contaminated vials of medication to medical facilities for injections to control pain. These vials were sent to 23 states and used to inject the drug into the spinal columns or joints of around 14,000 patients. By the time any problems were reported, several hundred cases of infection had occurred, half of which settled in the meninges. The most drastic outcome was the deaths of 39 patients from complications of meningitis. After months of investigation, the CDC isolated a black mold, *Exserohilum rostratum*, from both the patients and the drug vials.

This mold resides in plants and soil, from which it spreads into the air and many human habitats. But it is not considered a human pathogen, and infections with it are very rare. Examination of the compounding facility uncovered negligence and poor quality controls, along with dirty preparation rooms. Mold spores were introduced during filling of the vials, and because the medication lacked preservatives, they survived and grew. The owner of the compounding pharmacy and the head pharmacist were each charged with 25 counts of second-degree murder, their trial is expected to start in late 2016.

This case drives home several important facts about fungi: (1) They can grow rapidly even in low nutrient environments; (2) just a single spore introduced into a sterile environment, whether it is a vial of medicine or the human body, can easily multiply into millions of fungal cells; and (3) even supposedly "harmless" fungi are often opportunistic, meaning that they will infect tissues "if given an opportunity." This case also emphasizes the need for zero tolerance for microbes of any kind in a drug that is being injected—such a procedure demands sterility. When you think of it, the patients were actually being inoculated in a way that assured the development of serious mycoses.

Explain how a supposedly harmless, airborne mold could get all the way into the brain and cause meningitis. *Answer available on Connect.*

✅ *Check Your Progress* SECTIONS 5.4 AND 5.5

15. Review the major differences and similarities between prokaryotic and eukaryotic cells.

16. Compare the structures of yeast and hyphal cells and differentiate between yeasts and molds.

17. Tell how fungi obtain nutrients, and in what habitats one would expect to find them.

18. Describe the two main types of asexual fungal spores and how they are formed. Name several types of conidia.

19. Describe the three main types of sexual spores, and construct a simple diagram to show how each is formed.

20. Explain general features of fungal classification, give examples of the four fungal phyla, and describe their structure and significance.

21. Define the term *mycosis* and explain the levels of invasion of the body by fungi.

5.6 Survey of Protists: Algae

 Expected Learning Outcomes

27. Discuss the major characteristics of algae, and explain how they are classified.

28. Describe several ways that algae are important microorganisms.

Even though the terms *algae* and *protozoa* do not have taxonomic status, they are still scientifically useful. These are terms, like *protist,* that provide a shorthand label for certain eukaryotes. Microbiologists use such general terms to reference organisms that possess a collection of predictable characteristics. For example, protozoa are considered unicellular eukaryotic protists that lack tissues and share similarities in cell structure, nutrition, life cycles, and biochemistry. They are all microorganisms, and most of them are motile. Algae are eukaryotic protists, usually unicellular or colonial, that photosynthesize with chlorophyll *a*. They lack vascular systems for transport and have simple reproductive structures.

The Algae: Photosynthetic Protists

The **algae** are a group of photosynthetic organisms most readily recognized by their larger members, such as seaweeds and kelps. In addition to being beautifully colored and diverse in appearance, they vary in length from a few micrometers to 100 meters. Algae occur in unicellular, colonial, and filamentous forms, and the larger forms can possess tissues and simple organs. Examples of algal forms are shown in **figure 5.23** and **table 5.7.** An algal cell exhibits most of the organelles (figure 5.23*a*). The most noticeable of these are the chloroplasts, which contain, in addition to the green pigment chlorophyll, a number of other pigments that create the yellow, red, and brown coloration of some groups.

Algae are widespread inhabitants of fresh and marine waters. They are one of the main components of the large floating community of microscopic organisms called **plankton.** In this capacity,

> **Quick Search**
>
> Find the video "Brink: Algae to Oil" on the Science Channel to learn about a new source of "green" products.

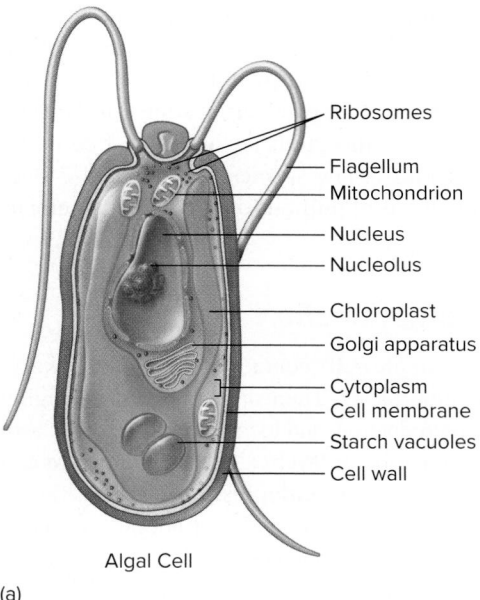

Ribosomes

Flagellum
Mitochondrion

Nucleus
Nucleolus

Chloroplast

Golgi apparatus

Cytoplasm
Cell membrane
Starch vacuoles
Cell wall

Algal Cell

(a)

Diatoms

(b)

(c)

(d)

Figure 5.23 Representative microscopic algae. (a) Structure of *Chlamydomonas,* a motile green alga, indicating major organelles. **(b)** A strew of beautiful algae called diatoms shows the intricate and varied structure of their silica cell wall. **(c)** A member of the Chlorophyta, *Oedogonium,* displaying its green filaments and reproductive structures, including eggs (clear spheres) and zygotes (dark spheres) (200×). **(d)** Spring pond with mixed algal bloom. Active photosynthesis and oxygen release are evident in surface bubbles.
(b): © Jan Hinsch/Science Source; (c, d): © Kathy Park Talaro

TABLE 5.7	Summary of Algal Characteristics			
Division/Common Name	**Organization**	**Cell Wall**	**Pigmentation**	**Ecology/Importance**
Euglenophyta (euglenids)	Mainly unicellular, motile by flagella	None, pellicle instead	Chlorophyll, carotenoids, xanthophyll	Some are close relatives of Mastigophora
Pyrrophyta (dinoflagellates)	Unicellular, dual flagella	Cellulose or atypical wall	Chlorophyll, carotenoids	Cause of "red tide"
Chrysophyta (diatoms or golden-brown algae)	Mainly unicellular, some filamentous forms, unusual form of motility	Silicon dioxide	Chlorophyll, fucoxanthin	Diatomaceous earth, major component of plankton
Phaeophyta (brown algae—kelps)	Multicellular, vascular system, holdfasts	Cellulose, system, holdfasts	Chlorophyll, carotenoids, fucoxanthin	Source of an emulsifier, alginate
Rhodophyta (red seaweeds)	Multicellular	Cellulose	Chlorophyll, carotenoids, xanthophyll, phycobilin	Source of agar and carrageenan, a food additive
Chlorophyta (green algae, grouped with plants)	Varies from unicellular, colonial, filamentous, to multicellular	Cellulose	Chlorophyll, carotenoids, xanthophyll	Precursor of higher plants

they play an essential role in aquatic food webs and contribute significantly to the oxygen content of the atmosphere through photosynthesis. One of the most prevalent groups on earth are single-celled chrysophyta called *diatoms* (figure 5.23*b*). These beautiful algae have silicate cell walls and golden pigment in their chloroplasts.

Other algal habitats include the surface of soil, rocks, and plants; and several species are even hardy enough to live in hot springs or snowbanks. A desert-adapted green alga blooms during spring rains and survives by forming resistant spores (figure 5.23*c, d*).

Animal tissues would be rather inhospitable to algae, so algae are rarely infectious. One exception, *Prototheca,* is an unusual non-photosynthetic alga associated with skin and subcutaneous infections in humans and animals.

The primary medical threat from algae is due to a type of food poisoning caused by the toxins of marine algae such as dinoflagellates. During particular seasons of the year, the overgrowth of these motile algae imparts a brilliant red color to the water, which is referred to as a "red tide" (see figure 26.16). When intertidal animals feed, their bodies accumulate toxins given off by the algae that can persist for several months. Paralytic shellfish poisoning is caused by eating exposed clams or other invertebrates. It is marked by severe neurological symptoms and can be fatal. Ciguatera is a serious intoxication caused by algal toxins that have accumulated in fish such as bass and mackerel. Cooking does not destroy the toxin, and there is no antidote.

Several episodes of a severe infection caused by *Pfiesteria piscicida,* a toxic algal form, have been reported over the past several years in the United States. The disease was first reported in fish and was later transmitted to humans. This newly identified species occurs in at least 20 forms, including spores, cysts, and amoebas, that can release potent toxins. Both fish and humans develop neurological symptoms and bloody skin lesions. The cause of the epidemic has been traced to nutrient-rich agricultural runoff water that promoted the sudden "bloom" of *Pfiesteria.* These microbes first attacked and killed millions of fish and later people whose occupations exposed them to fish and contaminated water.

5.7 Survey of Protists: Protozoa

 Expected Learning Outcomes

29. Summarize the main characteristics of protozoan form, nutrition, and locomotion.

30. Describe the general life cycle and mode of reproduction in protozoans.

31. Explain how protozoans are identified and classified.

32. Outline a classification scheme for protozoans, and provide examples of important members of each group.

33. Explain some biological properties of parasites, and list some common protozoan pathogens.

If a poll were taken to choose the most engrossing and vivid group of microorganisms, many biologists would choose the protozoa. Although their name comes from the Greek for "first animals,"

they are far from being simple, primitive organisms and they are not animals. The protozoa constitute a diverse large group (about 65,000 known species) of single-celled creatures that display startling properties when it comes to movement, feeding, and behavior. Although most members of this group are harmless, free-living inhabitants of water and soil, a few species are parasites collectively responsible for hundreds of millions of infections of humans each year.

Protozoan Form and Function

Most protozoan cells are single cells containing the major eukaryotic organelles except chloroplasts. Their organelles can be highly specialized for feeding, reproduction, and locomotion. The cytoplasm is usually divided into a clear outer layer called the **ectoplasm** and a granular inner region called the **endoplasm.** Ectoplasm is involved in locomotion, feeding, and protection. Endoplasm houses the nucleus, mitochondria, and food and contractile vacuoles. Some ciliates and flagellates[5] even have organelles that work somewhat like a primitive nervous system to coordinate movement. Because protozoa lack a cell wall, they have a certain amount of flexibility. Their outer boundary is a cell membrane that regulates the movement of food, wastes, and secretions. Cell shape can remain constant (as in most ciliates) or can change constantly (as in amoebas). Certain amoebas (foraminiferans) encase themselves in hard shells made of calcium carbonate. The size of most protozoan cells falls within the range of 3 to 300 µm. Some notable exceptions are giant amoebas and ciliates that are large enough (3–4 mm in length) to be seen swimming in pond water.

Nutritional and Habitat Range

Protozoa are heterotrophic and usually require their food in a complex organic form. Free-living species graze on live cells of bacteria and algae, and even scavenge dead plant or animal debris. Some species have special feeding structures such as oral grooves, which carry food particles into a passageway or gullet that packages the captured food into vacuoles for digestion. A remarkable feeding adaptation can be seen in the ciliate *Didinium* (see table 5.8, figure E), which can easily devour another microbe that is nearly its size. Some protozoa absorb food directly through the cell membrane. Parasitic species live on the fluids of their host, such as plasma and digestive juices, or they can actively feed on tissues.

Although protozoa have adapted to a wide range of habitats, their main limiting factor is the availability of moisture. Their predominant habitats are fresh and marine water, soil, plants, and animals. Even extremes in temperature and pH are not a barrier to their existence; hardy species are found in hot springs, ice, and habitats with low or high pH.

Styles of Locomotion

Except for one group (the Apicomplexa), protozoa are motile by means of **pseudopods** ("false foot"), **flagella,** or **cilia.** A few species have both pseudopods (also called pseudopodia) and

5. The terms *ciliate* and *flagellate* are common names of protozoan groups that move by means of cilia and flagella.

flagella. Some unusual protozoa move by a gliding or twisting movement that does not appear to involve any of these locomotor structures. Pseudopods are blunt, branched, or long and pointed, depending on the particular species. The flowing action of the pseudopods results in amoeboid motion, and pseudopods also serve as feeding structures in many amoebas.

The structure and behavior of flagella and cilia were discussed in section 5.1. Flagella vary in number from one to several, and in certain species they are attached along the length of the cell by an extension of the cytoplasmic membrane called the *undulating membrane* (see figure 5.27a). In most ciliates, the cilia are distributed over the surface of the cell in characteristic patterns. Because of the tremendous variety in ciliary arrangements and functions, ciliates are among the most diverse and awesome cells in the biological world. In certain protozoa, cilia line the oral groove and function in feeding; in others, they fuse together to form stiff props that serve as primitive rows of walking legs.

Life Cycles and Reproduction

Most protozoa are recognized by a motile feeding stage called the **trophozoite*** that requires ample food and moisture to remain active. A large number of species are also capable of entering into a dormant, resting stage called a **cyst** when conditions in the environment become unfavorable for growth and feeding. During *encystment,* the trophozoite cell rounds up into a sphere, and its ectoplasm secretes a tough, thick cuticle around the cell membrane **(figure 5.24).** Because cysts are more resistant than ordinary cells to heat, drying, and chemicals, they can survive adverse periods. They can be dispersed by air currents and may even be an important factor in the spread of

diseases such as amebic dysentery. If provided with moisture and nutrients, a cyst breaks open and releases the active trophozoite.

The life cycles of protozoans vary from simple to complex. Several protozoan groups exist only in the trophozoite state. Many alternate between a trophozoite and a cyst stage, depending on the conditions of the habitat. The life cycle of a parasitic protozoan dictates its mode of transmission to other hosts. For example, the flagellate *Trichomonas vaginalis* causes a common sexually transmitted infection. Because it does not form cysts, it is more delicate and must be transmitted by intimate contact between sexual partners. In contrast, intestinal pathogens such as *Entamoeba histolytica* and *Giardia lamblia* form cysts thatv are readily transmitted in contaminated water and foods.

All protozoa reproduce by relatively simple, asexual methods, usually mitotic cell division. Several parasitic species, including the agents of malaria and toxoplasmosis, reproduce asexually inside a host cell by multiple fission. Sexual reproduction also occurs during the life cycle of most protozoa. Ciliates participate in **conjugation,** a form of genetic exchange in which members of two different mating types fuse temporarily and exchange nuclei. This process of sexual recombination yields new and different genetic combinations that can be advantageous in evolution.

Protozoan Identification and Cultivation

Taxonomists have not escaped problems classifying protozoa, which are very diverse and frequently frustrate attempts to generalize or place them in neat groupings. We will use a simple system of four groups, based on method of motility, mode of reproduction, and stages in the life cycle.

* *trophozoite* (trof′-oh-zoh′-yte) Gr. trophonikos, to nourish, and zoon, animal.

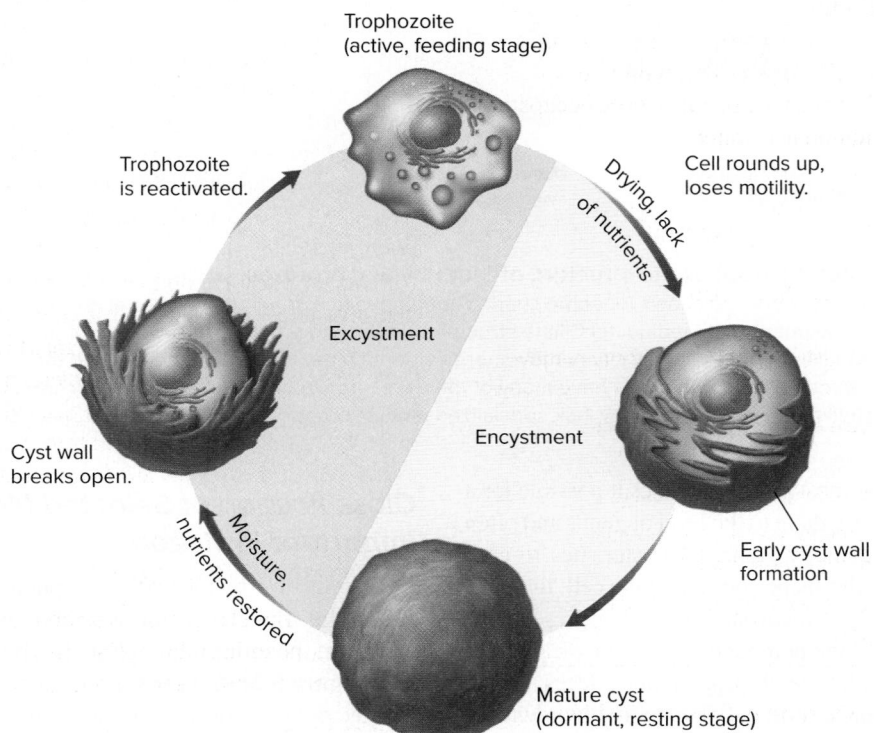

Trophozoite
(active, feeding stage)

Trophozoite is reactivated.

Drying, lack of nutrients

Cell rounds up, loses motility.

Excystment

Encystment

Cyst wall breaks open.

Early cyst wall formation

Moisture, nutrients restored

Mature cyst
(dormant, resting stage)

Figure 5.24 **The general life cycle exhibited by many protozoa.** All protozoa have a trophozoite form, but not all produce cysts.

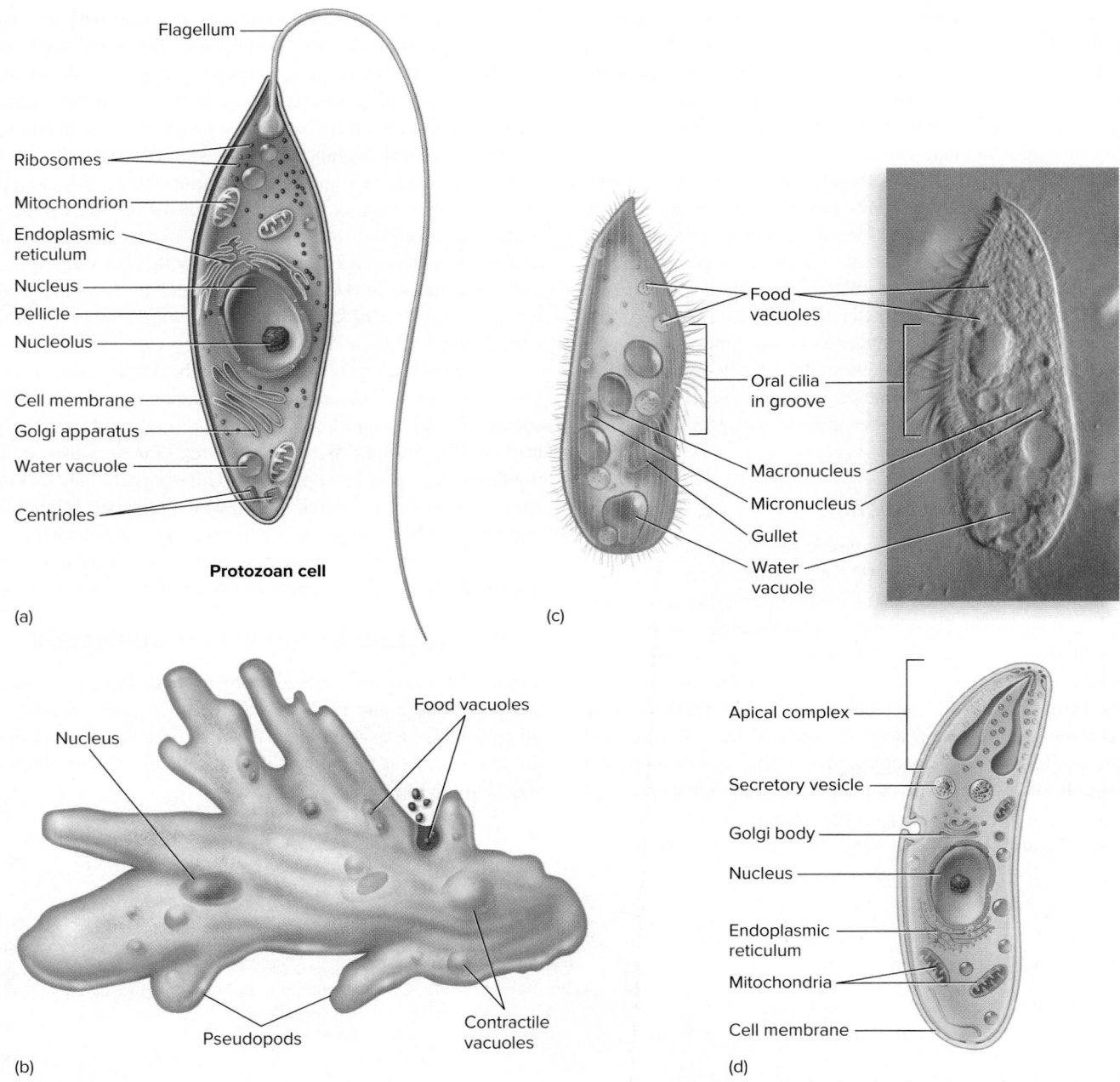

Figure 5.25 **Comparisons of the trophozoite structure of four Phyla of protozoa. (a)** General structure of a mastigophoran such as *Peranema.* The outer pellicle is flexible and allows for some shape changing in this group. **(b)** Structural appearance of a typical amoeba. Pseudopods are active in both movement and feeding. **(c)** Ciliate structure. Image and artist's interpretation of the ciliate *Blepharisma,* a common inhabitant of pond water. Note the pattern of ciliary movement evident on the left side of the cell and the details of its internal structure. **(d)** Apicomplexan trophozoite structure. Apicomplexans have many of the usual protozoan organelles but also contain an unusual body, the apical complex, involved in feeding. Unlike other protozoa, they lack specialized motility organelles. (c): © BioMedia/Science Source

The unique appearance of most protozoa makes it possible for a knowledgeable person to identify them to the level of genus and often species by microscopic morphology alone. Characteristics to consider in identification include the shape and size of the cell; the type, number, and distribution of locomotor structures; the presence of special organelles or cysts; and the number of nuclei. Medical specimens taken from blood, sputum, cerebrospinal fluid, feces, or the vagina are smeared directly onto a slide and observed with or without special stains. Occasionally protozoa are cultivated on artificial media or in laboratory animals for further identification or study.

Classification of Selected Medically Important Protozoa

Most protozoa fit the basic morphology of one of four types of cells: the flagellated mastigophorans, the pseudopod-bearing amoebas, the ciliated ciliophorans, and the nonmotile apicomplexans. **Figure 5.25** depicts various aspects of these protozoan types. The overall taxonomy we are using here classifies the protozoa into four Phyla based on motility and other cell characteristics that are outlined with examples in **table 5.8.**

TABLE 5.8 Selected Taxonomy of Protozoa, with Group Characteristics and Examples

The Mastigophora (Also Called Zoomastigophora)

- Motility is primarily by flagella alone or by both flagellar and amoeboid motion.
- Single nucleus.
- Sexual reproduction, when present, by syngamy*; division by longitudinal fission.
- Several parasitic forms lack mitochondria and Golgi apparatus.
- Most species form cysts and are free-living; the group also includes several parasites.
- Some species are found in loose aggregates or colonies, but most are solitary.
- Members include: *Trypanosoma* and *Leishmania,* important blood pathogens spread by insect vectors (see figure 5.26); *Giardia,* an intestinal parasite spread in water contaminated with feces (see **figure B**); *Trichomonas,* a parasite of the reproductive tract of humans spread by sexual contact (see figure 23.6).
- A large symbiont found in termites, *Trichonympha,* sprouts masses of flagella from its oral cavity **(figure A)**.

A. *Trichonympha*—a common symbiotic flagellate in termite intestines, displaying numerous flagella (400×).
© David J. Patterson

B. Trophozoite of *Giardia lamblia*—a widespread intestinal pathogen (2,000×).
Source: Janice Carr/CDC

The Sarcodina* (Amoebas)

- Cell form is primarily an amoeba (see figure 5.24*b*).
- Major locomotor organelles are pseudopods, although some species have flagellated reproductive states.
- Asexual reproduction by fission.
- Mostly uninucleate; usually encyst
- Most amoebas are free-living and not infectious **(figure C)**.
- *Entamoeba histolytica* is a parasite of humans **(figure D)**.

C. *Polychaos*—a giant amoeba uses several pseudopods to feed on diatoms (100×).
© David J. Patterson/MBL/Biological Discovery in Woods Hole

D. Trophozoite of *Entamoeba histolytica*—the cause of amebic dysentery (1,000×).
Source: Dr. Green/CDC

The Ciliophora (Ciliates)

- Trophozoites are motile by cilia.
- Some have cilia in tufts for feeding and attachment; most develop cysts.
- Have both macronuclei and micronuclei; division by transverse fission.
- Most have a well-developed mouth and feeding organelle.
- Show relatively advanced behavior **(figure E)**.
- The majority of ciliates are free-living and harmless.
- One pathogen is *Balantidium* **(figure F)**.

E. *Didinium* (predator, right)—preparing to attack and engulf *Paramecium* (prey, left) in an age-old ciliate struggle (1,200×). *Source:* Gregory Antipa, San Francisco State University. www.cellimagelibrary.org/ images/22781

F. *Balantidium coli*—a parasite of swine that can also cause intestinal infections in humans.
Source: Oregon Department of Health/CDC-DPDx

The Apicomplexa (Sporozoa)

- Motility is absent in most cells except male gametes.
- Life cycles are complex, with well-developed asexual and sexual stages.
- Sporozoa produce special sporelike cells called **sporozoites*** **(figures G** and **H)** following sexual reproduction, which are important in transmission of infections.
- Most form thick-walled zygotes called oocysts; entire group is parasitic.
- *Plasmodium,* the most prevalent protozoan parasite, causes 100 million to 300 million cases of malaria each year worldwide. It is an intracellular parasite with a complex cycle alternating between humans and mosquitoes (figure H).
- *Toxoplasma gondii* causes an acute infection (toxoplasmosis) in humans, which is acquired from cats and other animals (see figure 23.16).
- *Cryptosporidium* is an emerging intestinal pathogen transmitted by contaminated water (see figure G).

H. Sporozoite of *Plasmodium*—the agent of malaria, migrating through the gut of a mosquito.
Source: Ute Frevert; false color by Margaret Shear doi:10.1371/image.pbio.v03.i06.g001

G. Sporozoite of *Cryptosporidium*—a common waterborne pathogen. In this colorized electron micrograph the sporozoite is attached to the surface of the intestines.
© London School of Hygiene & Tropical Medicine/Science Source

*The fusion of haploid gametes to produce a diploid zygote.

*Some biologists prefer to combine Mastigophora and Sarcodina into the phylum Sarcomastigophora. Because the algal group Euglenophyta has flagella, it may also be included in the Mastigophora.

* *sporozoite* (spor″-oh-zoh′-yte) Gr. sporos, seed, and zoon, animal.

TABLE 5.9	Major Pathogenic Protozoa, Infections, and Primary Sources	
Protozoan/Disease		**Reservoir/Source**
Amoeboid Protozoa		
Amoebiasis: *Entamoeba histolytica*		Human/water and food
Brain infection: *Naegleria, Acanthamoeba*		Free-living in water
Ciliated Protozoa		
Balantidiosis: *Balantidium coli*		Zoonotic in pigs
Flagellated Protozoa		
Giardiasis: *Giardia lamblia*		Zoonotic/water and food
Trichomoniasis: *T. hominis, T. vaginalis*		Human
Hemoflagellates		
Trypanosomiasis: *Trypanosoma brucei, T. cruzi*		Zoonotic/vector-borne
Leishmaniasis: *Leishmania donovani, L. tropica, L. brasiliensis*		Zoonotic/vector-borne
Apicomplexan Protozoa		
Malaria: *Plasmodium vivax, P. falciparum, P. malariae*		Human/vector-borne
Toxoplasmosis: *Toxoplasma gondii*		Zoonotic/vector-borne
Cryptosporidiosis: *Cryptosporidium*		Free-living/water and food
Cyclosporiasis: *Cyclospora cayetanensis*		Water/fresh produce

Important Protozoan Pathogens

Although protozoan infections are very common, they are actually caused by only a small number of species often restricted geographically to the tropics and subtropics (**table 5.9**). In this survey, we look at examples from two protozoan groups that illustrate some of the main features of protozoan diseases.

> **Quick Search**
> Look for "What Are Malaria Protozoa" on YouTube to learn about the most dominant parasitic protozoan.

Pathogenic Flagellates: Trypanosomes

Trypanosomes are protozoa belonging to the genus *Trypanosoma*.* The two most important representatives are *T. brucei* and *T. cruzi,* species that are closely related but geographically restricted. *Trypanosoma brucei* occurs in Africa, where it causes an estimated 50,000 new cases of sleeping sickness each year. *Trypanosoma cruzi,* the cause of Chagas disease,[6] is endemic to South and Central America, where it infects several million people a year. Both species have long, crescent-shaped cells with a single flagellum that is sometimes attached to the cell body by an undulating membrane. Both occur in the blood during infection and are transmitted by blood-sucking vectors. We use *T. cruzi* to illustrate the phases of a trypanosomal life cycle and to demonstrate the complexity of parasitic relationships.

* *Trypanosoma* (try″-pan-oh-soh′-mah) Gr. *Trypanon,* borer, and *soma,* body.
6. Named for Carlos Chagas, the discoverer of *T. cruzi.*

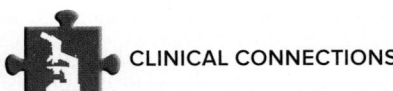

CLINICAL CONNECTIONS

The Parasitic Way of Life

Protozoa are traditionally studied along with the helminths in the science of parasitology. Although a parasite is usually defined as an organism that obtains food and other requirements at the expense of a host, the term *parasite* is also used to denote protozoan and helminth pathogens. The range of host-parasite relationships can be very broad. At one extreme are the so-called "good" parasites, which occupy their host with little harm. An example is *Trichomonas tenax,* a flagellate found in the mouth that may infect and damage the gingiva in people with poor oral hygiene. At the other extreme are parasites (*Plasmodium, Trypanosoma*) that multiply in host tissues such as the blood or brain, causing severe damage and disease. Between these two extremes are parasites of varying pathogenicity, depending on their particular adaptations.

Most human parasites go through three general stages:

- The microbe is transmitted to the human host from a source such as soil, water, food, other humans, or animals.
- The microbe invades and multiplies in the host, producing more parasites that can infect other suitable hosts.
- The microbe leaves the host in large numbers by a specific means and must enter a new host to survive.

There are numerous variations on this theme. For instance, the microbe can invade more than one host species (alternate hosts) and undergo several changes as it cycles through these hosts, such as sexual reproduction or encystment. Some microbes are spread from human to human by means of **vectors,*** defined as animals such as insects that carry diseases. Others can be spread through bodily fluids and feces.

What correlation can be made between the geographic location of a parasite and its vector? *Answer available on Connect.*

* *vector* (vek′-tur) L. vector, one who carries.

The trypanosome of Chagas disease relies on the close relationship of a warm-blooded mammal and an insect that feeds on mammalian blood. The mammalian hosts are numerous, including dogs, cats, opossums, and armadillos. The vector is the *reduviid** *bug,* an insect that is sometimes called the "kissing bug" because of its habit of biting its host at the corner of the mouth. Transmission occurs from bug to mammal and from mammal to bug but usually not from mammal to mammal, except across the placenta during pregnancy. The general phases of this cycle are presented in **figure 5.26.**

The trypanosome trophozoite multiplies in the intestinal tract of the reduviid bug and is harbored in the feces. The bug seeks a host and bites the mucous membranes, usually of the eye, nose, or lips, releasing the trypanosome in feces near the bite. Ironically, the victims themselves inadvertently contribute to the entry of the microbe by scratching the bite wound. Once in the body, the trypanosomes become established and multiply in muscle and white blood cells. Periodically these parasitized cells rupture, releasing large numbers of new trophozoites into the blood. Eventually the trypanosomes can spread to many

* *reduviid* (ree-doo′-vee-id) A member of a large family of flying insects with sucking, beaklike mouths.

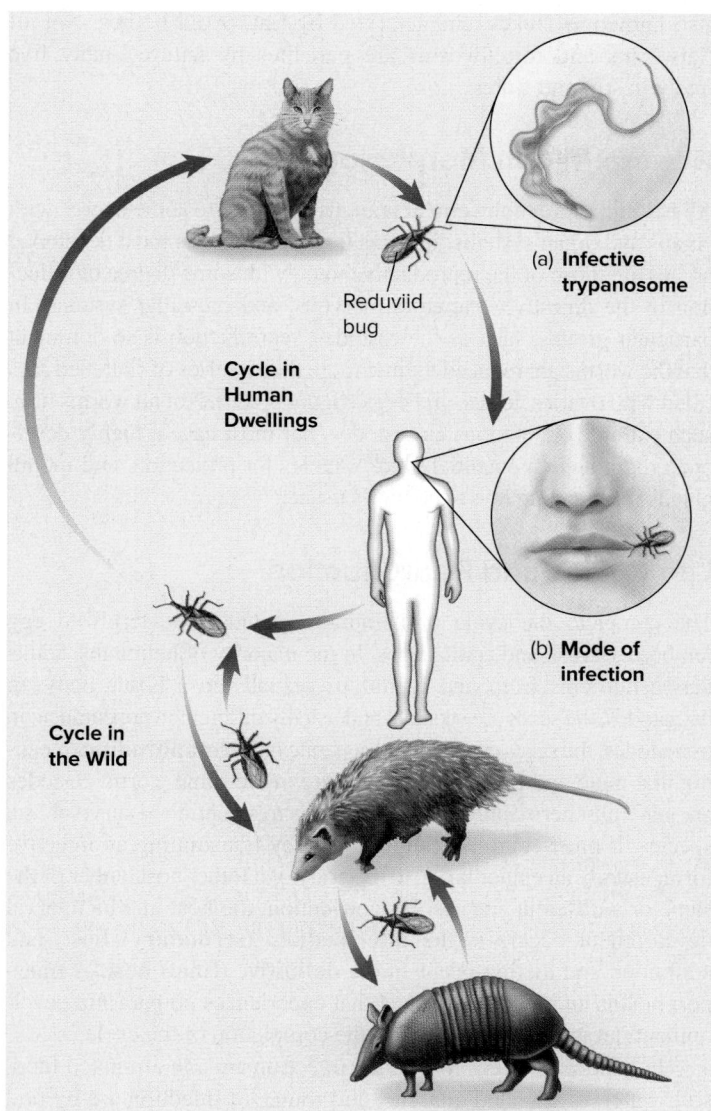

Figure 5.26 **Cycle of transmission in Chagas disease.**
Trypanosomes **(inset a)** are transmitted among domestic and wild mammalian hosts and human hosts by means of a bite from the kissing bug **(inset b)**.

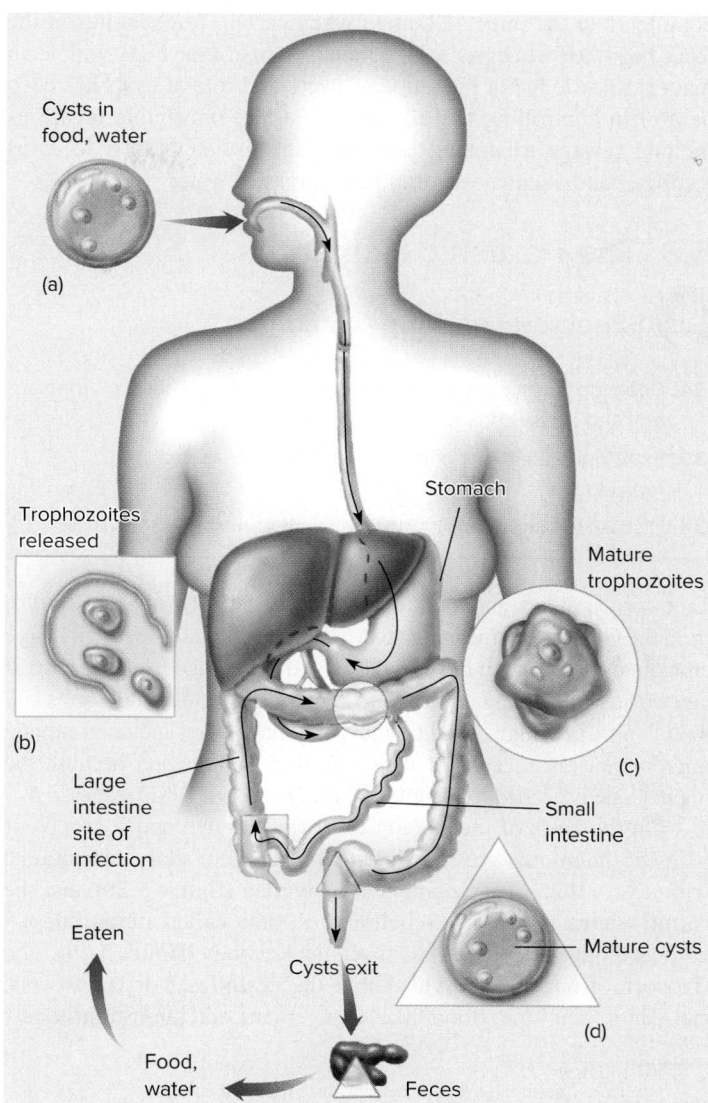

Figure 5.27 **Stages in the infection and transmission of amebic dysentery.** Arrows show the route of infection; insets show the appearance of *Entamoeba histolytica.* **(a)** Cysts are eaten. **(b)** Trophozoites (amoebas) emerge from cysts. **(c)** Trophozoites invade the large intestinal wall (see table 5.8, figure D). **(d)** Mature cysts are released in the feces and may be spread through contaminated food and water.

systems, including the lymphoid organs, heart, liver, and brain. Manifestations of the resultant disease range from mild to very severe and include fever, inflammation, and heart and brain damage. In many cases the disease has an extended course and can cause death.

Infective Amoebas: *Entamoeba*

Several species of amoebas cause disease in humans, but probably the most common disease is amoebiasis, or amebic **dysentery,*** caused by *Entamoeba histolytica.* This microbe is widely distributed in the world, from northern zones to the tropics, and is nearly always associated with humans. Amebic dysentery is the fourth most common protozoan infection in the world. This microbe has a life cycle quite different from the trypanosomes in that it does not involve multiple hosts and a blood-sucking vector. It lives part of its cycle as a trophozoite

and part as a cyst. Because the cyst is the more resistant form and can survive in water and soil for several weeks, it is the more important stage for transmission. The primary way that people become infected is by ingesting food or water contaminated with human feces.

Figure 5.27 shows the major features of the amebic dysentery cycle, starting with the ingestion of cysts. The viable, heavy-walled cyst passes through the stomach unharmed. Once inside the small intestine, the cyst germinates into a large multinucleate amoeba that subsequently divides to form small amoebas (the trophozoite stage). These trophozoites migrate to the large intestine and begin to feed and grow. From this site, they can penetrate the lining of the intestine and invade the liver, lungs, and skin. Common symptoms include gastrointestinal disturbances such as nausea, vomiting, and diarrhea, leading to weight loss and dehydration. The cycle is

* *dysentery* (dis′-en-ter″-ee) Any inflammation of the intestine accompanied by bloody stools. It can be caused by a number of factors, both microbial and nonmicrobial.

completed in the infected human when certain trophozoites in the feces begin to form cysts, which then pass out of the body with fecal matter. Knowledge of the amoebic cycle and role of cysts has been helpful in controlling the disease. Important preventive measures include sewage treatment, curtailing the use of human feces as fertilizer, and adequate sanitation of food and water.

5.8 The Parasitic Helminths

 Expected Learning Outcomes

34. Describe the major groups of helminths and their basic morphology and classification.

35. Explain the elements of helminth biology, life cycles, and reproduction.

36. Discuss the importance of the helminth parasites.

Tapeworms, flukes, and roundworms are collectively called *helminths,* from the Greek word meaning "worm." Adult animals are usually large enough to be seen with the naked eye, and they range from the longest tapeworms, measuring up to about 25 m in length, to roundworms less than 1 mm in length. Traditionally they have been included among microorganisms because of their infective abilities and because the microscope is needed to identify their tiny egg and larval stages.

On the basis of morphological form, the two major groups of parasitic helminths are the **flatworms** (Phylum Platyhelminthes), with a very thin, often segmented body plan **(figure 5.28),** and the **roundworms** (Phylum Aschelminthes, also called **nematodes**),* with an elongate, cylindrical, unsegmented body **(figure 5.29).** The flatworm group is subdivided into the **cestodes,*** or tapeworms, named for their long, ribbonlike arrangement, and the **trematodes,***

* *nematode* (neem′-ah-tohd) Gr. nemato, thread, and eidos, form.
* *cestode* (sess′-tohd) L. cestus, a belt, and ode, like.
* *trematode* (treem′-a-tohd) Gr. trema, hole. Named for the appearance of having tiny holes.

also known as flukes, characterized by flat, ovoid bodies. Not all flatworms and roundworms are parasites by nature; many live freely in soil and water.

General Worm Morphology

All helminths are multicellular animals equipped to some degree with organs and organ systems. In parasitic helminths, the most developed organs are those of the reproductive tract, with some degree of reduction in the digestive, excretory, nervous, and muscular systems. In particular groups, such as the cestodes, reproduction is so dominant that the worms are reduced to little more than a series of flattened sacs filled with ovaries, testes, and eggs (figure 5.28*a*). Not all worms have such extreme adaptations as cestodes, but most have a highly developed reproductive potential, thick cuticles for protection, and mouth glands for breaking down the host's tissue.

Life Cycles and Reproduction

The complete life cycle of helminths includes the fertilized egg (embryo), larval, and adult stages. In the majority of helminths, adults derive nutrients from and reproduce sexually in a host's body. In nematodes, the sexes are separate and usually different in appearance; in trematodes, the sexes can be either separate or **hermaphroditic,** meaning that male and female sex organs are in the same worm; cestodes are generally hermaphroditic. For a parasite's continued survival as a species, it must complete the life cycle by transmitting an infective form, usually an egg or larva, to the body of another host, either of the same or a different species. By convention, the host in which larval development occurs is the **intermediate (secondary) host,** and adulthood and mating occur in the **definitive (final) host.** A transport host is an intermediate host that experiences no parasitic development but is an essential link in the completion of the cycle.

In general, sources for human infection are contaminated food, soil, water, or infected animals, and routes of infection are by oral intake or penetration of unbroken skin. Humans are the definitive hosts for many of the parasites, and in about half the diseases, they

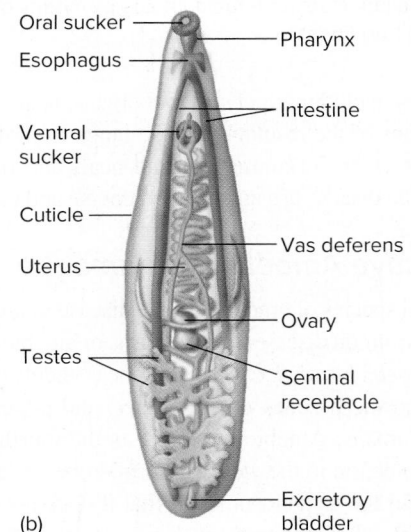

Figure 5.28 Parasitic flatworms. (a) A cestode (beef tapeworm), showing the scolex; long, tapelike body; and magnified views of immature and mature proglottids (body segments). **(b)** The structure of a trematode (liver fluke). Note the suckers that attach to host tissue and the dominance of reproductive and digestive organs.

are also the sole biological reservoir. In other cases, animals or insect vectors serve as reservoirs or are required to complete worm development. In the majority of helminth infections, the worms must leave their host to complete the entire life cycle.

Fertilized eggs are usually released to the environment and are provided with a protective shell and extra food to aid their development into larvae. Even so, most eggs and larvae are vulnerable to heat, cold, drying, and predators, and are destroyed or unable to reach a new host. To counteract this formidable mortality rate, certain worms have adapted a reproductive capacity that borders on the incredible: A single female *Ascaris*[7] can lay 200,000 eggs a day, and a large female can contain over 25 million eggs at varying stages of development! If only a tiny number of these eggs makes it to another host, the parasite will have been successful in completing its life cycle.

A Helminth Cycle: The Pinworm

To illustrate a helminth cycle in humans, we use the example of a roundworm, *Enterobius vermicularis,* the pinworm or seatworm. This worm causes a very common infestation of the large intestine (figure 5.29). Worms range from 2 to 12 mm long and have a tapered, curved cylinder shape. The condition they cause, enterobiasis, is usually a simple, uncomplicated infection that does not spread beyond the intestine.

A cycle starts when a person swallows microscopic eggs picked up from another infected person by direct contact or by touching contaminated surfaces. The eggs hatch in the intestine and then release larvae that mature into adult worms within about one month. Male and female worms mate, and the female migrates out to the anus to deposit eggs, which cause intense itchiness that is relieved by scratching. Herein lies a significant means of dispersal: Scratching contaminates the fingers, which, in turn, transfers eggs to bedclothes and other inanimate objects. This person becomes a host and a source of eggs, and can spread them to others in addition to reinfesting himself, a cycle known as the oral-fecal route of transmission. Enterobiasis occurs most often among families and in other close living situations. Its distribution is worldwide among all socioeconomic groups, but it occurs more frequently in children, presumably through their less careful attention to hygiene.

Helminth Classification and Identification

The helminths are classified according to their shape; their size; the degree of development of various organs; the presence of hooks, suckers, or other special structures; the mode of reproduction; the kinds of hosts; and the appearance of eggs and larvae. They are identified in the laboratory by microscopic detection of the adult worm or its larvae and eggs, which often have distinctive shapes or external and internal structures. Occasionally, they are cultured in order to verify all of the life stages.

Distribution and Importance of Parasitic Worms

About 50 species of helminths parasitize humans. They are distributed in all areas of the world that support human life. Some worms are restricted to a given geographic region, and many have a higher incidence in tropical areas. This knowledge must be tempered with

7. *Ascaris* is a genus of parasitic intestinal roundworms.

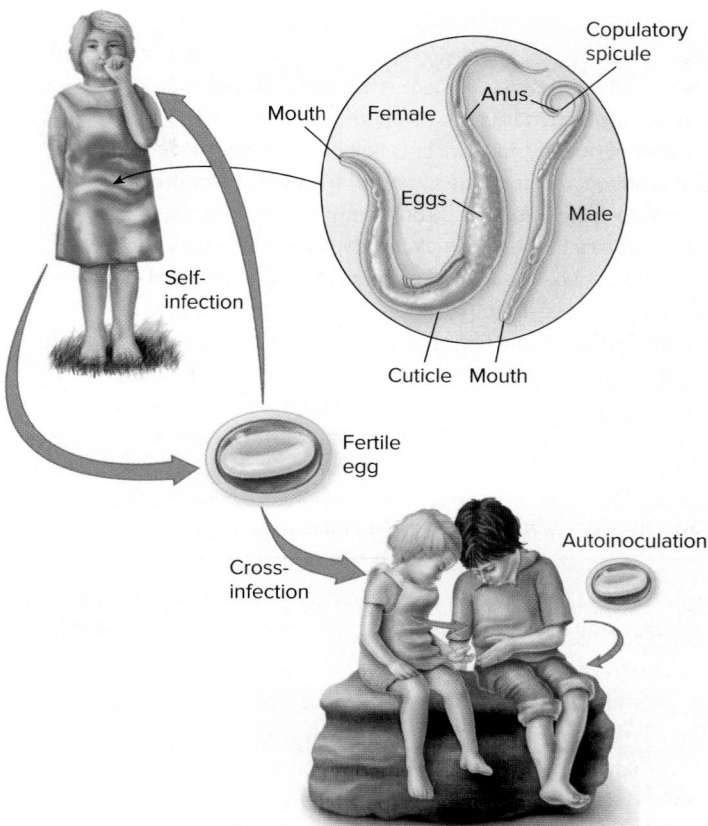

Figure 5.29 The life cycle of the pinworm, a roundworm. Eggs are the infective stage and are transmitted by unclean hands. Children frequently reinfect themselves and also pass the parasite on to others.

the realization that jet-age travel, along with human migration, is gradually changing the patterns of worm infections, especially of those species that do not require alternate hosts or special climatic conditions for development. The yearly estimate of worldwide cases numbers in the billions, and these are not confined to developing countries. A conservative estimate places 50 million helminth infections in North America alone. The primary targets are children lacking access to adequate diet and medical care.

Check Your Progress SECTIONS 5.6–5.8

22. What is a working definition of a "protist"?
23. Describe the principal characteristics of algae that separate them from protozoa.
24. How are algae important? Give examples of algae with medical importance.
25. Explain the general characteristics of the protozoan life cycle.
26. Describe the protozoan adaptations for feeding.
27. Briefly outline the characteristics of the four protozoan groups and name important pathogens in each group.
28. What characteristics set the apicomplexa apart from the other protozoan groups?
29. Discuss the adaptations of parasitic worms to their lifestyles, and explain why these adaptations are necessary or advantageous to the worms' survival.
30. Outline the basic steps in an infection cycle of a pathogenic protozoan and a helminth.

CASE STUDY Part 2

There is much more to the story of neglected diseases of poverty. Most victims first become infected as children and continue to harbor the parasite for years, even for a lifetime. The chronic nature of the infection adds enormously to the accumulated damage that occurs. Infected people may become involved in an ongoing cycle that continues to produce more parasites and increase their survival and transmission. Many protozoans produce resistant survival cells called cysts, and worms go through complex reproductive phases with eggs and larvae. Some parasites are spread by direct contact with an infected person, some burrow into the skin, and others are ingested in contaminated soil or water. A significant factor in transmission of several parasites is the involvement of arthropod vectors such as mosquitoes and flies.

For most of these diseases, there is effective drug therapy, but getting the drugs to the poorest of the poor has been difficult. Many of the countries are not only deeply impoverished,

but are involved in civil wars and other conflicts, creating upheaval in an already overburdened system. One billion of the world's poorest live on less than $1 per day, and 2.7 billion live on less than $2 per day. Most of these people are "out of sight and out of mind"—living in remote rural areas far removed from medical facilities. NTDs are not given much priority for funding because they are not high-profile diseases with noticeable, acute symptoms requiring immediate medical care. Any progress in controlling these NTDs will require a global partnership that brings together resources from many countries.

- *What is the definition of a vector, and how are vectors involved in these diseases?*

- *What are some possible solutions to the problem of neglected diseases?*

For information on NTDs and their current status, search for NTDs on the World Health Organization website.

To conclude this Case Study, go to Connect.

Chapter Summary with Key Terms

5.1 The History of Eukaryotes

5.2 Form and Function of the Eukaryotic Cell: External Structures
 A. The exterior configuration of eukaryotic cells is complex and displays numerous structures not found in prokaryotic cells. Biologists have accumulated much evidence that eukaryotic cells evolved through **endosymbiosis** between early prokaryotic cells.
 B. Major external structural features include: appendages (**cilia, flagella**), **glycocalyx,** cell wall, and cytoplasmic (or cell) membrane.

5.3 Form and Function of the Eukaryotic Cell: Internal Structures
 A. The internal structure of eukaryotic cells is compartmentalized into individual organelles.
 B. Major **organelles** and internal structural features include: **nucleus, nucleolus, endoplasmic reticulum, Golgi complex, mitochondria, chloroplasts), ribosomes, cytoskeleton (microfilaments, microtubules).**

5.4 Eukaryotic-Prokaryotic Comparisons and Taxonomy of Eukaryotes
 A. Comparisons between eukaryotic cells and prokaryotic cells show major differences in structure, size, metabolism, motility, and body form.
 B. Taxonomic groups of the Domain Eukarya are based on level of organization, body plan, cell structure, nutrition, metabolism, and certain genetic characteristics.

5.5 The Kingdom of the Fungi
 Common names of the macroscopic fungi are mushrooms, bracket fungi, and puffballs. Microscopic fungi are known as **yeasts** and **molds.**
 A. *Overall Morphology:* At the cellular (microscopic) level, fungi are typical eukaryotic cells, with thick cell walls. Yeasts are single cells that form buds and **pseudohyphae. Hyphae** are long, tubular filaments that can be septate or nonseptate and grow in a network called a **mycelium;** hyphae are characteristic of the filamentous fungi called molds.

 B. *Nutritional Mode/Distribution:* All are **heterotrophic.** The majority are harmless **saprobes** living off organic substrates such as dead animal and plant tissues. A few are **parasites,** living on the tissues of other organisms, but none is obligate. Distribution is extremely widespread in many habitats.
 C. *Reproduction:* Primarily through **spores** formed on special reproductive hyphae. In asexual reproduction, spores are formed through budding, partitioning of a hypha, or in special sporogenous structures; examples are **conidia** and **sporangiospores.** In sexual reproduction, spores are formed following fusion of male and female strains and the formation of a sexual structure; sexual spores are one basis for classification.
 D. *Major Groups:* The four main phyla among the terrestrial fungi, given with sexual spore type, are **Zygomycota (zygospores), Ascomycota (ascospores), Basidiomycota (basidiospores),** and **Chytridiomycota** (motile zoospores).
 E. *Importance:* Fungi are essential decomposers of plant and animal detritis in the environment. Economically beneficial as sources of antibiotics; used in making foods and in genetic studies. Adverse impacts include decomposition of fruits and vegetables, and human infections, or **mycoses;** some produce substances that are toxic if eaten or inhaled.

5.6 Survey of Protists: Algae
 General group that traditionally includes single-celled and colonial eukaryotic microbes that lack organization into tissues.
 A. *Overall Morphology:* Are unicellular, colonial, filamentous or larger forms such as seaweeds.
 B. *Nutritional Mode/Distribution:* Photosynthetic; freshwater and marine water habitats; main component of **plankton.**
 C. *Importance:* Provide the basis of the food web in most aquatic habitats and are major producers of oxygen. Certain algae produce neurotoxins that are harmful to humans and animals.

5.7 Survey of Protists: Protozoa
Include large single-celled organisms; a few are pathogens.

A. *Overall Morphology:* Most are unicellular; lack a cell wall. The cytoplasm is divided into ectoplasm and endoplasm. Active, feeding stage is the **trophozoite;** many convert to a resistant, dormant stage called a **cyst.** All but one group has some form of organelle for **motility.**

B. *Nutritional Mode/Distribution:* All are heterotrophic. Most are free-living in a moist habitat (water, soil); feed by engulfing other microorganisms and organic matter.

C. *Reproduction:* Asexual by binary fission and **mitosis,** budding; sexual by fusion of free-swimming gametes, conjugation.

D. *Major Groups:* Protozoa are subdivided into four groups based upon mode of locomotion and type of reproduction: **Mastigophora,** the flagellates, motile by flagella; **Sarcodina,** the amoebas, motile by **pseudopods; Ciliophora,** the ciliates, motile by cilia; **Apicomplexa,** motility not well developed; produce unique reproductive structures.

E. *Importance:* Ecologically important in food webs and decomposing organic matter. Medical significance: hundreds of millions of people are afflicted with one of the many protozoan infections (malaria, trypanosomiasis, amoebiasis). Can be spread from host to host by insect **vectors.**

5.8 The Parasitic Helminths
Includes three categories: roundworms, tapeworms, and flukes.

A. *Overall Morphology:* Animal cells; multicellular; individual organs specialized for reproduction, digestion, movement, protection, though some of these are reduced.

B. *Reproductive Mode:* Includes embryo, larval, and adult stages. Majority reproduce sexually. Sexes may be **hermaphroditic.**

C. *Epidemiology:* Developing countries in the tropics are hardest hit by helminth infections; transmitted via ingestion, vectors, and direct contact with infectious stages. They afflict billions of humans.

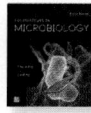

Level I. Knowledge and Comprehension

These questions require a working knowledge of the concepts in the chapter and the ability to recall and understand the information you have studied.

Multiple-Choice Questions

Select the correct answer from the answers provided. For questions with blanks, choose the combination of answers that most accurately completes the statement.

1. Both flagella and cilia are found primarily in
 a. algae
 b. protozoa
 c. fungi
 d. both b and c

2. Features of the nuclear envelope include
 a. ribosomes
 b. a double membrane structure
 c. pores that allow communication with the cytoplasm
 d. b and c
 e. all of these

3. The cell wall is usually found in which eukaryotes?
 a. fungi
 b. algae
 c. protozoa
 d. a and b

4. What is embedded in rough endoplasmic reticulum?
 a. ribosomes
 b. Golgi apparatus
 c. chromatin
 d. vesicles

5. Yeasts are _____ fungi, and molds are _____ fungi.
 a. macroscopic, microscopic
 b. unicellular, filamentous
 c. motile, nonmotile
 d. water, terrestrial

6. In general, fungi derive nutrients through
 a. photosynthesis
 b. engulfing bacteria
 c. digesting organic substrates
 d. parasitism

7. A hypha divided into compartments by cross walls is called
 a. nonseptate
 b. imperfect
 c. septate
 d. perfect

8. Algae generally contain some type of
 a. spore
 b. chlorophyll
 c. locomotor organelle
 d. toxin

9. Which algal group is most closely related to plants?
 a. diatoms
 b. Chlorophyta
 c. Euglenophyta
 d. dinoflagellates

10. Which characteristic(s) is/are not typical of protozoan cells?
 a. locomotor organelle
 b. cyst
 c. spore
 d. trophozoite

11. The protozoan trophozoite is the
 a. active feeding stage
 b. inactive dormant stage
 c. infective stage
 d. spore-forming stage

12. All mature sporozoa are
 a. parasitic
 b. nonmotile
 c. carried by an arthropod vector
 d. both a and b

13. Parasitic helminths reproduce with
 a. spores
 b. eggs and sperm
 c. mitosis
 d. cysts
 e. all of these

14. Mitochondria likely originated from
 a. archaea
 b. invaginations of the cell membrane
 c. rickettsias
 d. cyanobacteria

15. Human fungal infections involve and affect what areas of the human body?
 a. skin
 b. mucous membranes
 c. lungs
 d. all of these

16. Most helminth infections
 a. are localized to one site in the body
 b. spread through major systems of the body
 c. develop within the spleen
 d. develop within the liver

Multiple Matching. Select the description that best fits the word in the left column.

17. _____ diatom
18. _____ *Rhizopus*
19. _____ *Histoplasma*
20. _____ *Cryptococcus*
21. _____ euglenid
22. _____ dinoflagellate
23. _____ *Trichomonas*
24. _____ *Entamoeba*
25. _____ *Plasmodium*
26. _____ *Enterobius*

 a. the cause of malaria
 b. single-celled alga with silica in its cell wall
 c. fungal cause of Ohio Valley fever
 d. the cause of amebic dysentery
 e. genus of black bread mold
 f. helminth worm involved in pinworm infection
 g. motile flagellated alga with an eyespot
 h. a yeast that infects the lungs
 i. flagellated protozoan genus that causes an STD
 j. alga that causes red tides

Case Study Review

1. Which of these is/are an example(s) of neglected tropical protozoan diseases?
 a. hookworm
 b. Chagas disease
 c. leishmaniasis
 d. a and b
 e. b and c
 f. all of these

2. Which is a possible vector of a tropical eukaryotic parasite?
 a. contaminated drinking water
 b. spoiled food
 c. biting fly
 d. person with a cough

3. Provide some explanations for why the eukaryotic parasites are so widespread and successful.

Writing Challenge

For each question, compose a one- or two-paragraph answer that includes the factual information needed to completely address the question. Check Your Progress questions can also be used for writing-challenge exercises.

1. Describe the anatomy and functions of each of the major eukaryotic organelles.
2. Trace the synthesis of cell products, their processing, and their packaging through the organelle network.
3. a. What is the reproductive potential of molds in terms of spore production?
 b. How do mold spores differ from bacterial endospores?
4. a. Fill in the following summary table for defining, comparing, and contrasting eukaryotic cells.
 b. Briefly describe the manner of nutrition and body plan (unicellular, colonial, filamentous, or multicellular) for each group.
 c. Explain some ways that helminths differ from the protozoa and algae in structure and behavior.

Organelle/ Structure	Briefly Describe Functions in Cell	Commonly Present In (Check)		
		Fungi	Algae	Protozoa
Flagella				
Cilia				
Glycocalyx				
Cell wall				
Cell membrane				
Nucleus				
Mitochondria				
Chloroplasts				
Endoplasmic reticulum				
Ribosomes				
Cytoskeleton				
Lysosomes				
Microvilli				
Centrioles				

Concept Mapping

On Connect you can find an Introduction to Concept Mapping that provides guidance for working with concept maps along with concept-mapping activities for this chapter.

Level II. Application, Analysis, Evaluation, and Synthesis

These problems go beyond just restating facts and require higher levels of understanding and an ability to interpret, problem solve, transfer knowledge to new situations, create models, and predict outcomes.

Critical Thinking

Critical thinking is the ability to reason and solve problems using facts and concepts. These questions can be approached from a number of angles, and in most cases, they do not have a single correct answer.

1. Explain the ways that mitochondria resemble rickettsias and chloroplasts resemble cyanobacteria.

2. Give the common name of a eukaryotic microbe that is unicellular, walled, nonphotosynthetic, nonmotile, and bud-forming.

3. How are the eukaryotic ribosomes and cell membranes different from those of prokaryotes?

4. What general type of multicellular parasite is composed primarily of thin sacs of reproductive organs?

5. a. Name two parasites that are transmitted in the cyst form.
 b. How must a non-cyst-forming pathogenic protozoan be transmitted? Why?

6. Explain what factors could cause opportunistic mycoses to be a growing medical problem.

7. a. How are bacterial endospores and cysts of protozoa alike?
 b. How do they differ?

8. For what reasons would a eukaryotic cell evolve an endoplasmic reticulum and a Golgi apparatus?

9. Can you think of a simple test to determine if a child is suffering from pinworms? Hint: Clear adhesive tape is involved.

Visual Challenge

1. What term is used to describe a single species exhibiting both cell types shown below, and which types of organisms would most likely have this trait?

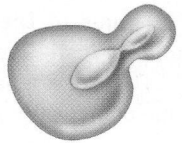

2. Label the major structures you can observe in the images in figure 5.17a, table 5.5A, B, and D, and table 5.8C.

An Introduction to Viruses

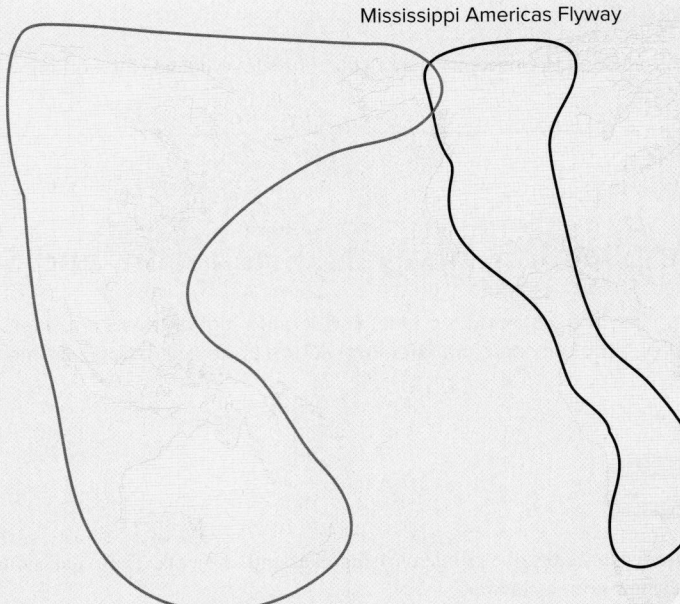

Mississippi Americas Flyway

Migratory patterns allow wild birds to spread influenza virus to domesticated flocks, sometimes creating new strains of the virus in the process.

East Asian-Australasian Flyway

Dear Customer

Due to the shortage and increased price of eggs, we will be adding $1.00 to each breakfast entrée.

Thank You

Source: Bob Nichols/USDA

A model of the influenza virus displays its hemagglutinin spikes (blue) and neuraminidase spikes (red). The green strands within are the RNA molecules of the virus. *Source:* Dan Higgins/CDC

CASE STUDY Part 1 A Killer Virus

On May 12, 2015, U.S. inspectors detected the presence of a highly **pathogenic influenza virus** circulating within a large population in Nebraska. To slow the spread of the virus, all 1.7 million members of the population were killed.

Of course, the population in which the virus was discovered was a commercial chicken flock and not humans, but there are nonetheless serious consequences involved. The 1.7 million reflected only the latest **outbreak** in an **epidemic** that had led to the culling of 48 million chickens and turkeys since the beginning of the year. For the ranchers involved, such an outbreak means reduced profits at best and, at worst, bankruptcy and loss of family farms going back generations. By the end of May the cost of eggs had risen 120%, forcing stores and restaurants across the country to raise prices. In some cases, chickens, ducks, and pigeons that were kept as pets in backyard flocks were also euthanized to prevent the spread of avian influenza.

The virus was identified as **H5N2** by researchers, an alphanumeric designation based on two types of proteins found on the viral surface. H refers to **hemagglutinin,** a protein that allows the virus to initiate infection by binding to cells of the respiratory tract. N is shorthand for **neuraminidase,** an enzyme that digests a component of mucus, allowing the virus to penetrate further into the respiratory tract. Together, these two molecules help define exactly what type of cells are susceptible to infection and how serious that infection

> "The H5N2 subtype was characterized as a highly pathogenic avian influenza."

may eventually become. With regard to birds, there are 16 H **subtypes** (H1–H16) and 9 N subtypes (N1–N9) resulting in 144 possible combinations. The H5N2 subtype moves rapidly from bird to bird and causes a serious infection, leading to its designation as a **h**ighly **p**athogenic **a**vian **i**nfluenza—HPAI H5N2—but the virus seems to move very poorly from birds to humans and from one human to another.

The standard response to an outbreak of avian influenza in domestic birds consists of five steps: (1) **Quarantine**—the movement of birds and equipment is restricted. (2) **Eradicate**—all birds within the affected flock are euthanized. (3) **Monitor**—Wild and domestic birds in a broad area surrounding the quarantine site are scrutinized for the presence of the virus. (4) **Disinfect**—The virus is destroyed in the affected location. (5) **Test**—Confirm that the poultry farm is free of avian influenza virus. Despite repeatedly applying this protocol to flocks throughout the country, the virus continued to reappear.

- *Which viral protein is most involved in determining the "host range" of the influenza virus?*

- *If the standard response to viral outbreaks outlined above is correctly followed, why does the virus keep returning?*

To continue the Case Study, go to Case Study Part 2 at the end of the chapter.

SCOPING OUT THE CHAPTER

Smaller and simpler than even the most modest prokaryotic cell, viruses are responsible for some of the most virulent diseases on Earth (Ebola fever) as well as some of the most mundane (the common cold). In this chapter we survey viruses—so small they cannot be seen without an electron microscope, so simple that most scientists don't consider them to be alive, and yet each of us has been infected by them many, many times.

Composed only of an inner molecule of genetic material surrounded by a protein coat, and occasionally an outer envelope, viruses are far simpler than cells.

Proteins anchored in the viral envelope guide the attachment of the virus to its host cell. The virus enters the host and redirects the metabolism of the host cell toward the production of hundreds of additional viral particles.

After replication, the newly created viruses exit the cell and commence the search for a new host.

Unlike bacterial cells, the growth of viruses in the lab requires that a living host be used as the "growth medium." Cell culture, live animals, and eggs are all commonly used to cultivate viruses.

Even simpler than viruses, prions consist exclusively of protein. They cause a number of transmissible spongiform encepahalopathies, named for the spongy appearance of brain cells infected with the pathogen.

(top: right): *Source:* National Institute of Allergy and Infectious Diseases (NIAID)/CDC; (bottom: left): © State Hygenic Laboratory at The University of Iowa; (bottom: right): *Source:* Sherif Zaki; MD; PhD; Wun-Ju Shieh; MD; PhD; MPH/CDC

6.1 Overview of Viruses

 Expected Learning Outcomes

1. Indicate how viruses were discovered and characterized.
2. Describe the unique characteristics of viruses.
3. Discuss the origin and importance of viruses.

Early Searches for the Tiniest Microbes

The discovery of the light microscope made it possible to see first-hand the agents of many bacterial, fungal, and protozoan diseases. But the techniques for observing and cultivating these relatively large microorganisms were useless for viruses. For many years the cause of viral infections such as smallpox and polio was unknown, even though it was clear that the diseases were transmitted from person to person. The French bacteriologist Louis Pasteur was

certainly on the right track when he postulated that rabies was caused by a "living thing" smaller than bacteria, and in 1884 he was able to develop the first vaccine for rabies.

The first substantial revelations about the unique characteristics of viruses occurred in the 1890s. First, Dmitri Ivanovski and Martinus Beijerinck showed that a disease in tobacco was caused by a virus (tobacco mosaic virus). In fact, it was Beijerinck who first proposed the term **virus** (L., "poison") to denote this particular group of infectious agents. This discovery was followed by Friedrich Loeffler and Paul Frosch's isolation of the virus that causes foot-and-mouth disease in cattle. These early researchers found that when infectious fluids from host organisms were passed through porcelain filters designed to trap bacteria, the filtrate remained infectious even though they could not see the infectious agent with a microscope.

Over the succeeding decades, a remarkable picture of the physical, chemical, and biological nature of viruses began to take form. Years of experimentation were required to show that viruses were noncellular particles with a definite size, shape, and chemical composition. Specialized techniques were developed to culture them in the laboratory. In time, virology grew into a multifaceted discipline that promised to provide much information on disease, genetics, and even life itself (**6.1 Making Connections**).

The Position of Viruses in the Biological Spectrum

Viruses are a unique group of biological entities known to infect every type of cell, including bacteria, algae, fungi, protozoa, plants, and animals. Although the emphasis in this chapter is on animal viruses, much credit for our knowledge must be given to experiments with bacterial and plant viruses. The exceptional and curious nature of viruses prompts numerous questions, including:

1. How did viruses originate?
2. Are they organisms; that is, are they alive?
3. What are their distinctive biological characteristics?
4. How can particles so small, simple, and seemingly insignificant be capable of causing disease and death?
5. What is the connection between viruses and cancer?

In this chapter we address these ideas and many others.

There is no universal agreement on how and when viruses originated. But there is little disagreement that they have been in existence for billions of years, probably ever since the early history of cells. One theory explains that viruses arose from loose strands of genetic material released by cells. These then developed a protective coating and a capacity to reenter a cell and use its machinery to reproduce. An alternate explanation for their origins suggests that they were once cells that have regressed to a highly parasitic existence inside other cells. Both of these viewpoints have some supportive evidence, and it is entirely possible that viruses have originated in more than one way.

In terms of numbers, viruses are considered the most abundant microbes on earth. To quote microbiologist Carl Zimmer, we live on "a planet of viruses." Based on a decade of research on the viral populations in the ocean, "virus hunters" now have

evidence that there are hundreds of thousands of distinct virus types that have never been described. Many of them are parasites of bacteria, which are also abundant in most ecosystems. By one estimate, viruses outnumber bacteria by a factor of 10. Because viruses tend to interact with the genetic material of their host cells and can carry genes from one host cell to another (chapter 9), they have played an important part in the evolution of Bacteria, Archaea, and Eukarya.

Viruses are different from their host cells in size, structure, behavior, and physiology. They are a type of **obligate intracellular parasite** that cannot multiply unless it invades a specific host cell and instructs its genetic and metabolic machinery to make and release quantities of new viruses. Because of this characteristic, viruses are capable of causing serious damage and disease. Other unique properties of viruses are summarized in **table 6.1.**

The unusual structure and behavior of viruses have led to debates about their connection to the rest of the microbial world. One viewpoint holds that viruses are unable to exist independently from the host cell, so they are not living things but are more akin to large, infectious molecules. Another viewpoint proposes that even though viruses do not exhibit most of the life processes of cells (look back at section 4.1), they can direct those processes and thus are certainly more than inert and lifeless molecules. Depending upon the circumstances, both views are defensible. The importance of this debate is more philosophical than practical, because viruses are agents of disease and must be dealt with through control, therapy, and prevention, whether we regard them as living or not. In keeping with their special position in the biological spectrum, it is best to describe viruses as *infectious particles* (rather than organisms) and as either *active* or *inactive* (rather than alive or dead).

TABLE 6.1	Properties of Viruses

- Obligate intracellular parasites of bacteria, protozoa, fungi, algae, plants, and animals
- Ultramicroscopic size, ranging from 20 nm up to 450 nm (diameter)
- Not cellular in nature; structure is very compact and economical
- Do not independently fulfill the characteristics of life
- Inactive macromolecules outside the host cell and active only inside host cells
- Basic structure consists of protein shell (capsid) surrounding nucleic acid core
- Nucleic acid of the viral genome is either DNA or RNA but not both
- Nucleic acid can be double-stranded DNA, single-stranded DNA, single-stranded RNA, or double-stranded RNA
- Molecules on virus surface impart high specificity for attachment to host cell
- Multiply by taking control of host cell's genetic material and regulating the synthesis and assembly of new viruses
- Lack enzymes for most metabolic processes
- Lack machinery for synthesizing proteins

6.1 MAKING CONNECTIONS

An Alternate View of Viruses

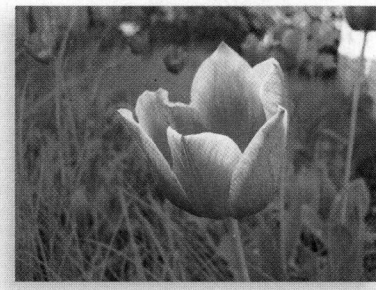
© Kathy Park Talaro

Looking at this beautiful tulip, one would never guess that its beauty derives partly from the flower's being infected with tulip mosaic virus, which alters the development of the plant cells and causes varying patterns of colors in the petals. Aside from this, the virus does not immediately destroy the plants. Despite the reputation of viruses as destructive pathogens, there is another side to viruses—that of being benign and, in some cases, even beneficial.

Research in extreme habitats in Antarctica uncovered a surprisingly rich diversity of viruses inhabiting a frigid lake. Analysis of the genetic material of the samples revealed more than 10,000 unique viruses gathered from just a small area. Other studies of the ocean have documented the presence of 10 million viruses per milliliter of water—10 times the number of cellular microbes. This hints at the dominant role viruses have in the world's ecosystems, not only because they infect and can control populations of eukaryotic and prokaryotic cells, but also because they appear to influence a number of biological activities, including photosynthesis and the cycling of nutrients.

Over the past several years, biomedical experts have been looking at viruses as vehicles to treat infections and disease. Vaccine experts have engineered new types of vaccines by combining a less harmful virus such as vaccinia or adenovirus with genetic material from a pathogen such as HIV and herpes simplex. This technique creates a vaccine that provides immunity but does not expose the person to the intact pathogen. Several of these types of vaccines are being tested.

The "harmless virus" approach is also being used to treat genetic diseases such as cystic fibrosis and sickle-cell anemia. With gene therapy, the normal gene is inserted into a virus vector, such as an adenovirus, and the patient is infected with this altered virus. It is hoped that the virus will introduce the needed gene into the cells and correct the defect. Dozens of experimental trials are under way to develop potential cures for diseases, with mixed success (look ahead to chapter 10).

An older therapy getting a second look involves use of bacteriophages to treat bacterial infections (phage therapy). The basis behind the therapy is that bacterial viruses can selectively attack and destroy their host bacteria without damaging human cells. Experiments with animals indicate that this method controls some infections as well as traditional drugs can. Potential applications include adding phage suspensions to skin grafts and to intravenous fluids to control infections.

The U.S. Department of Agriculture has recently approved bacteriophage treatments to prevent food infections. One type is designed to spray on livestock to reduce the incidence of pathogenic *Escherichia coli.* The other is intended for use on processed meats and poultry to control *Listeria monocytogenes,* another prominent food-borne pathogen.

Explain why bacterial viruses would be harmless to humans.
Answer available on Connect.

Check Your Progress SECTION 6.1

1. Describe 10 *unique* characteristics of viruses (can include structure, behavior, multiplication).
2. After consulting table 6.1, what additional facts can you state about viruses, especially as compared with cells?
3. Explain what it means to be an obligate intracellular parasite.
4. What are some other ways to describe the sort of parasitism exhibited by viruses?

6.2 The General Structure of Viruses

Expected Learning Outcomes

4. Describe the general structure and size range of viruses.
5. Distinguish among types of capsids and nucleocapsids.
6. Describe envelopes and spikes, and discuss their origins.
7. Explain the functions of capsids, nucleocapsids, envelopes, and spikes.
8. Summarize the different viral groups based on their basic structure.

Size Range

As a group, viruses are the smallest infectious agents (with some unusual exceptions to be discussed in section 6.8). Their size places them in the realm of the *ultramicroscopic.* This term means that most of them are so minute (< 0.2 μm) that an electron microscope is necessary to detect them or to examine their fine structure. They are dwarfed by their host cells: More than 2,000 bacterial viruses could fit into an average bacterial cell, and more than 50 million polioviruses could be accommodated by an average human cell. Animal viruses range in size from the small parvoviruses[1] (around 20 nm in diameter) to megaviruses and pandoraviruses[2] that are as large as small bacteria (up to 1,000 nm in width) **(figure 6.1).** Some cylindrical viruses are relatively long (800 nm in length) but so narrow in diameter (15 nm) that they still cannot be seen without the high magnification and resolution of an electron microscope. Figure 6.1 compares the sizes of several viruses with prokaryotic and eukaryotic cells and molecules.

1. DNA viruses that cause respiratory infections in humans.
2. Groups of large, complex viruses that are parasites inside cells of ocean-dwelling amoebas.

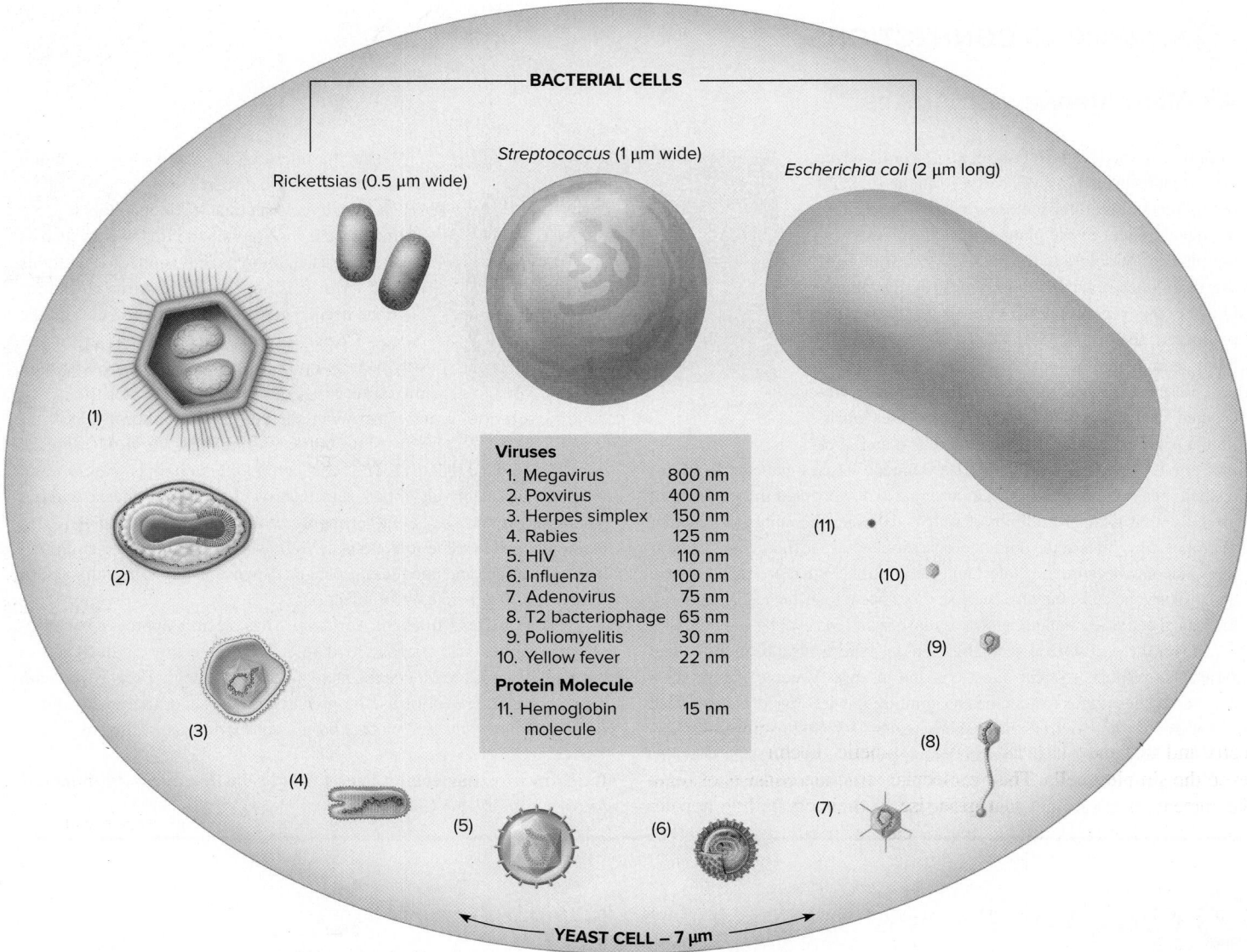

Figure 6.1 **Size comparison of viruses with a yeast cell colored blue (eukaryotic) and various bacteria (prokaryotic).** Viruses range from largest (1) to smallest (10). A molecule of a large protein (11) is included to indicate relative size of macromolecules.

Viral architecture is most readily observed through special stains in combination with electron microscopy. Negative staining uses an opaque salt to outline the shape of the virus against a dark background and to enhance textural features on the viral surface. Internal details are revealed by positive staining of specific parts of the virus such as protein or nucleic acid. The *shadowcasting* technique showers a virus preparation with a dense metallic vapor directed from a certain angle. These techniques are featured in several figures throughout this chapter.

Monster Viruses

In chapter 4 we featured discoveries of unusual types of bacteria. There are also some exceptional viruses. Probably the most outstanding examples are megaviruses **(figure 6.2)** and pandoraviruses. These are giants in the viral world, averaging about 500 to 1,000 nm in diameter, 20 to 50 times larger than an average virus and even larger than small bacteria such as rickettsias. The megaviruses were first discovered as parasites inside the cells of amoebas living in aquatic habitats. Pandoraviruses were isolated from their amoebic hosts living in sediments deep in the floor of the ocean. Both giant viruses lack cellular structure and ribosomes, metabolic enzymes, and fission-style division. They have a typical viral cycle, multiplying in the host cytoplasm in well-developed "virus factories." Although both are unusually large for viruses, the pandoraviruses have the largest capsids and genomes ever discovered in a virus, containing 2,500 genes!

Where these unusual viruses fit on the "tree of life," if anywhere at all, is still the subject of debate. Some microbiologists suggest that they may belong to a separate evolutionary line of microbes that is different in origin from other viral groups. Another possibility is that these viruses are related to ancient forms that evolved into the first cells.

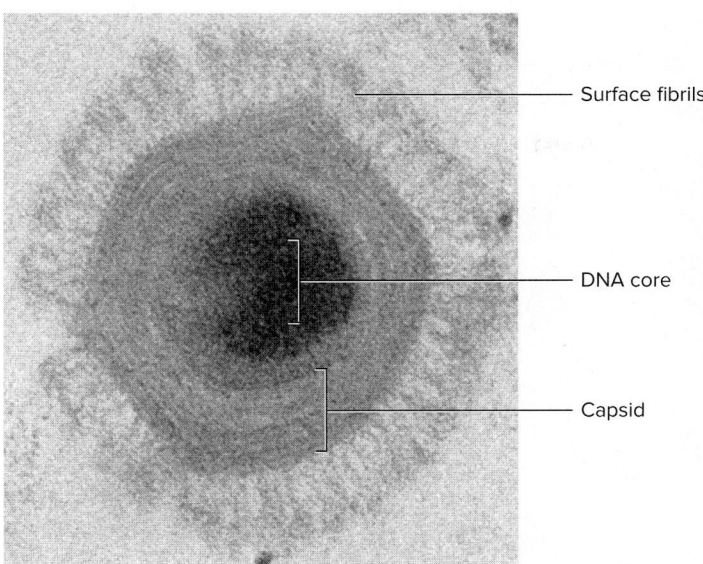

Figure 6.2 A monster virus—the megavirus. A TEM of one virus particle reveals its unique geometric shape, dark DNA core, and fine surface fibrils (4,500×). These viruses would be readily visible with an ordinary light microscope. Courtesy of C. Abergel/IGS-IMM, CNRS-AMU

Labels on figure: Surface fibrils; DNA core; Capsid

(a) (b)

Figure 6.3 The crystalline nature of viruses. (a) Light microscope magnification (1,200×) of purified poliovirus crystals. **(b)** Highly magnified (200,000×) negative stain of hundreds of polioviruses shows the tight geometric arrangements of particles. (a): *Source:* F. L. Schaffer, C. E. Schwerdt, "Crystallization of purified MEF-1 poliomyelitis virus particles," *PNAS,* 41(12): 1020-1023, 1955, fig. 2b; (b): *Source:* CDC

Viral Components: Capsids, Nucleic Acids, and Envelopes

It is important to realize that viruses bear no real resemblance to cells and that they lack any of the synthetic machinery found in even the simplest cells. Their molecular structure consists of regular, repeating molecules that give rise to their crystalline appearance. Indeed, many purified viruses can form large aggregates or crystals if subjected to special treatments **(figure 6.3).** The general plan of virus organization is very simple and compact. Viruses contain only those parts needed to invade and control a host cell: an external coating and a core containing one or more nucleic acid strands of either DNA or RNA. This pattern can be represented with a flowchart:

Virus particle
- Covering
 - Capsid
 - Envelope (not found in all viruses)
- Central core
 - Nucleic acid molecule(s) (DNA or RNA)
 - Matrix proteins Enzymes (not found in all viruses)

All viruses have a protein **capsid,*** or shell, that surrounds the nucleic acid in the central core. Together, the capsid and the nucleic acid are referred to as the **nucleocapsid.** Viruses that consist of only a nucleocapsid are considered **naked viruses (figure 6.4a).** Of

the 20 families of animal viruses, 13 are **enveloped**—that is, they possess an additional covering external to the capsid called an envelope, which is a modified section of the host's cell membrane **(figure 6.4b).** As we shall see later, the enveloped viruses also

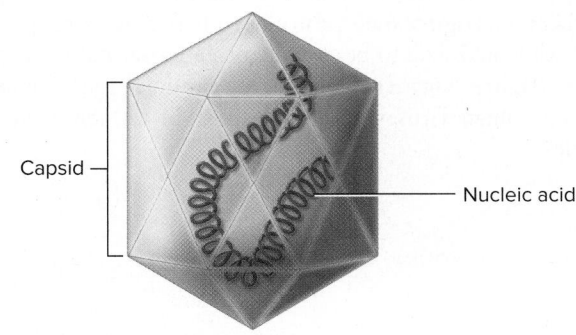

Labels: Capsid; Nucleic acid

(a) Naked Nucleocapsid Virus

Labels: Envelope with membrane; Spike; Capsid; Matrix; Nucleic acid

(b) Enveloped Virus

Figure 6.4 Generalized structure of viruses. (a) The simplest virus is a naked nucleocapsid consisting of a geometric capsid assembled around one or more nucleic acid strands. **(b)** An enveloped virus is composed of a nucleocapsid surrounded by a flexible membrane called an envelope. The envelope usually has special receptor spikes inserted into its membrane.

* *capsid* (kap′-sid) L. *capsa,* box.

differ from the naked viruses in the possible ways that they enter and leave a host cell.

The Viral Capsid: The Protective Outer Shell

When a virus particle is magnified several hundred thousand times, the capsid appears as the most prominent geometric feature (figure 6.4). In general, the capsid of any virus is constructed from a number of identical protein subunits called **capsomers.*** The capsomers can spontaneously self-assemble into the finished capsid. The shape and arrangement of the capsomers results in the production of one of two different types of capsids: helical or icosahedral.

> **Quick Search**
> Fun with viruses! Search the Internet for "paper virus models" or a "bacteriophage balloon."

The simpler **helical capsids** have rod-shaped capsomers that bind together to form a series of hollow discs resembling a bracelet. During the formation of the nucleocapsid, these discs link together and form a continuous helix into which the nucleic acid strand is coiled **(figure 6.5).** In electron micrographs, the appearance of a helical capsid varies with the type of virus. The nucleocapsids of naked helical viruses are very rigid and tightly wound into a cylinder-shaped package **(figure 6.6a).** Several viruses that infect plants are of this type **(figure 6.6b).** Enveloped helical nucleocapsids are more flexible and tend to be arranged as a looser helix within the envelope **(figure 6.6c, d).** This type of capsid is found in several enveloped human viruses, including those of influenza, measles, and rabies.

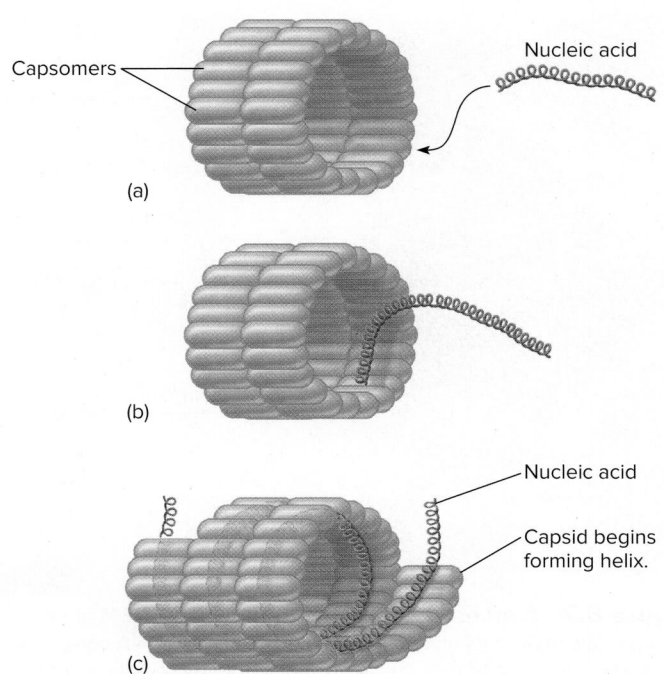

Figure 6.5 Assembly of helical nucleocapsids. (a) Capsomers assemble into hollow discs. **(b)** The nucleic acid is inserted into the center of the disc. **(c)** Elongation of the nucleocapsid progresses from both ends, as the nucleic acid is wound "within" the lengthening helix.

Several major virus families have capsids arranged in an **icosahedron***—a three-dimensional, symmetrical polygon, with 20 sides and 12 evenly spaced corners. The arrangements of the

* *capsomer* (kap′-soh-meer) L. *capsa*, box, and *mer*, part.

* *icosahedron* (eye″-koh-suh-hee′-drun) Gr. *eikosi*, twenty, and *hedra*, side. A type of polygon.

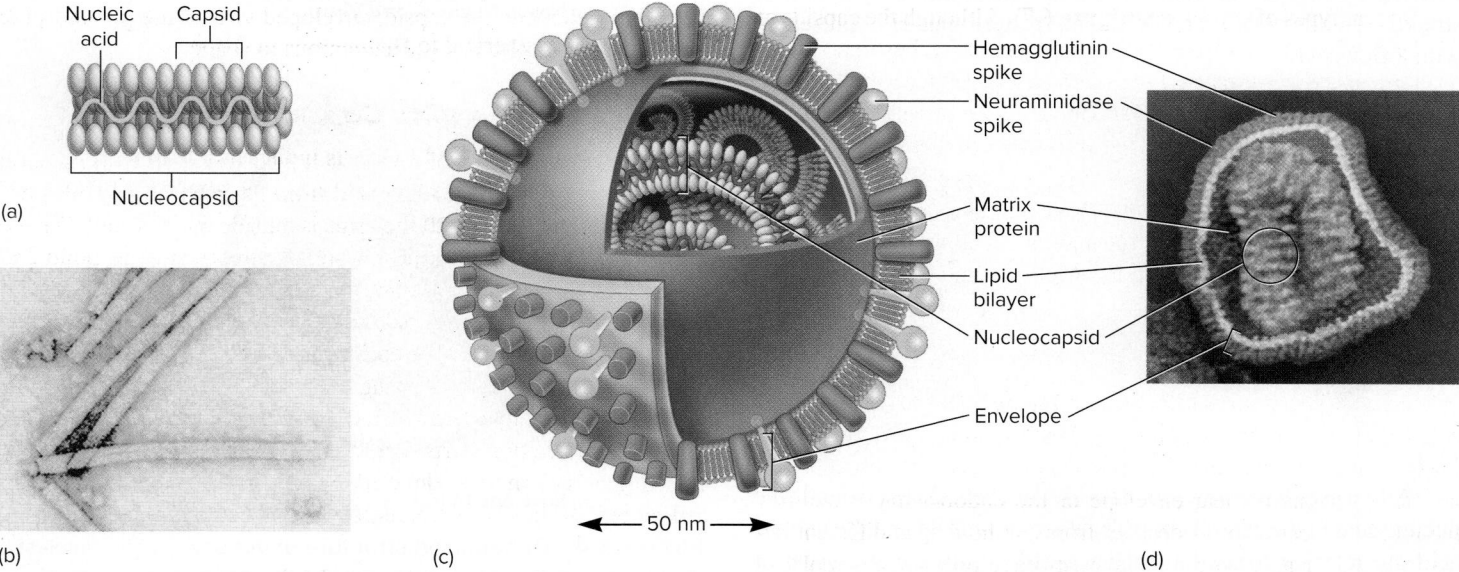

Figure 6.6 Typical variations of viruses with helical nucleocapsids (a) Schematic view and **(b)** a negative stain magnified 160,000× of tobacco mosaic virus, a naked, helical virus. Note the elongate cylindrical morphology. **(c)** Schematic view and **(d)** a colorized micrograph featuring a positive stain of the influenza virus (300,000×), an enveloped helical virus. This virus has a well-developed envelope with prominent spikes. (b): *Source:* Electron and Confocal Microscopy Laboratory, Agricultural Research Service, U. S. Department of Agriculture; (d): *Source:* Frederick Murphy/CDC

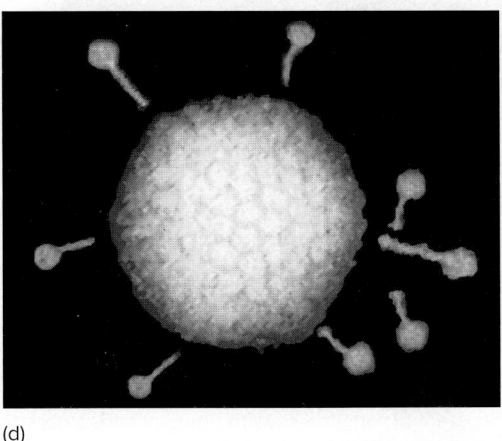

Figure 6.7 **The structure and formation of an icosahedral virus (adenovirus is the model).** **(a)** A facet or "face" of the capsid is composed of 21 identical capsomers arranged in a triangular shape. A vertex or "point" consists of five capsomers arranged with a single penton in the center. Other viruses can vary in the number, types, and arrangement of capsomers. **(b)** An assembled virus shows how the facets and vertices come together to form a shell around the nucleic acid. **(c)** A three-dimensional model (640,000×) of this virus shows fibers attached to the pentons. **(d)** A negative stain of this virus highlights its texture and fibers that have fallen off. (d): © Dr. Linda Stannard, UCT/Science Source

capsomers vary from one virus to another. The capsid of some viruses contains a single type of capsomer, whereas others may contain several types of capsomers **(figure 6.7).** Although the capsids of all icosahedral viruses have this sort of symmetry, they can vary in the number of capsomers; for example, a poliovirus has 32 capsomers and an adenovirus has 242. During assembly of the virus, the nucleic acid is packed into the center of this icosahedron, forming a nucleocapsid. Another factor that alters the appearance of icosahedral viruses is whether or not they have an outer envelope. Inspect **figure 6.8** to compare a rotavirus and its naked nucleocapsid with herpes simplex (cold sores) and its enveloped nucleocapsid.

The Viral Envelope

When enveloped viruses (mostly animal) are released from the host cell, they take with them a bit of the host's membrane system in the form of an envelope. Some viruses bud off the cell membrane; others leave via the nuclear envelope or the endoplasmic reticulum. Although the envelope is derived from the host, it is different because some or all of the regular membrane proteins are replaced with special viral proteins during the virus assembly process (see figure 6.11). Some proteins form a binding layer between the envelope and capsid of the virus, and glycoproteins (proteins bound to a carbohydrate) remain exposed on the outside of the envelope. These

protruding molecules, called **spikes** or peplomers, are essential for the attachment of viruses to the next host cell. Because the envelope is more supple than the capsid, enveloped viruses are pleomorphic and range from spherical to filamentous in shape.

Functions of the Viral Capsid/Envelope

The outermost covering of a virus is indispensable to viral function because it protects the nucleic acid from the effects of various enzymes and chemicals when the virus is outside the host cell. We see this in the capsids of enteric (intestinal) viruses such as polio and hepatitis A, which are resistant to the acid- and protein-digesting enzymes of the gastrointestinal tract. Capsids and envelopes are also responsible for helping to introduce the viral DNA or RNA into a suitable host cell, first by binding to the cell surface and then by assisting in penetration of the viral nucleic acid (discussed in later sections). In addition, parts of viral capsids and envelopes stimulate the immune system to produce antibodies that can neutralize viruses and protect the host's cells against future infections (see chapter 15).

Complex Viruses: Atypical Viruses

Two special groups of viruses, termed **complex** viruses **(figure 6.9),** are more intricate in structure than the helical, icosahedral, naked, or enveloped viruses just described. The **poxviruses** (including the agent

(a)

240–300 nm

Core membrane

Nucleic acid

200 nm

Outer envelope

Soluble protein antigens

Lateral body

(a)

Capsomers

Envelope Capsid DNA core

(b)

Figure 6.8 **Two types of icosahedral viruses, highly magnified.**
(a) Left view: A negative stain of rotaviruses with unusual capsomers that look like spokes on a wheel (150,000×); right view is a three-dimensional model of this virus. **(b)** Herpes simplex virus, a type of enveloped icosahedral virus (300,000×). (a: left): *Source:* Fred P. Williams, Jr./EPA; (a: right): © Dr. Linda Stannard, UCT/Science Source; (b): © Eye of Science/Science Source

Capsid head

Nucleic acid

Collar

Tail fibers

Sheath

Tail pins

Base plate

(b)

of smallpox) are very large DNA viruses that lack a typical capsid and are covered by a dense layer of lipoproteins and coarse fibrils on their outer surface. Some members of another group of very complex viruses, the **bacteriophages,*** have a polyhedral capsid head as well as a helical tail and fibers for attachment to the host cell. Their mode of multiplication is covered in section 6.5. **Figure 6.10** summarizes the primary morphological types found among the viruses.

Nucleic Acids: At the Core of a Virus

The sum total of the genetic information carried by an organism is known as its **genome.** So far, one biological constant is that the genome of all organisms is carried by and expressed through nucleic acids (DNA, RNA). Even viruses are no exception to this rule, but there is a significant difference. Unlike cells, which contain both DNA and RNA, **viruses contain either DNA or RNA,** *but not both,* **as the primary genetic material.** Because viruses must pack into a tiny space all of the genes necessary to instruct the host cell to make new viruses, the number of viral genes is usually quite

(c)

Figure 6.9 **Detailed structure of complex viruses.**
(a) Section through the vaccinia virus, a poxvirus, shows its internal components. **(b)** Diagram of a T4 bacteriophage of *E. coli,* (280,000×). **(c)** Photomicrograph of a bacteriophage from cyanobacteria. (c): © Bin Ni, Chisholm Lab, MIT

* *bacteriophage* (bak-teer′-ee-oh-fayj″) From *bacteria,* and Gr. *phagein,* to eat. These viruses parasitize bacteria.

A. Complex Viruses

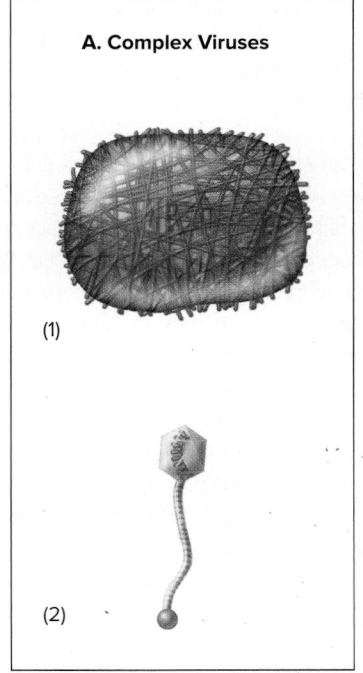

B. Enveloped Viruses

Helical Icosahedral

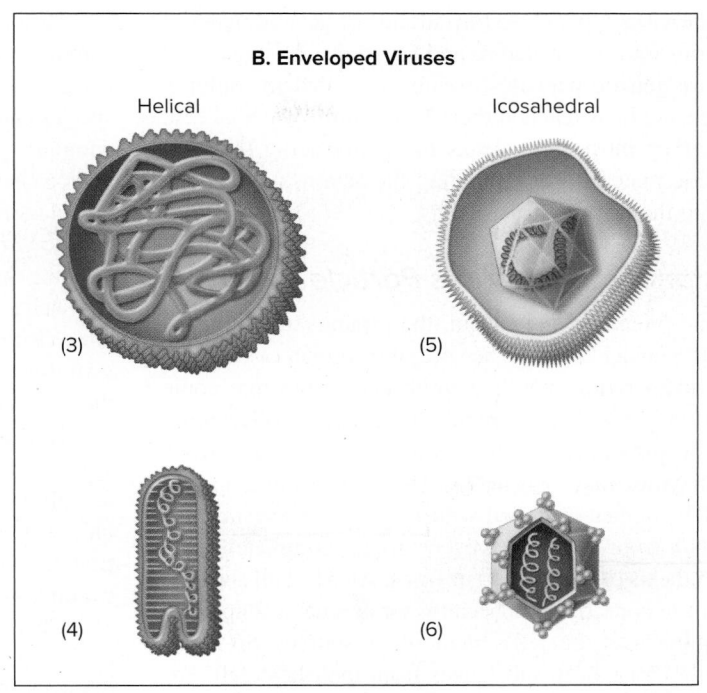

C. Nonenveloped Naked Viruses

Helical Icosahedral

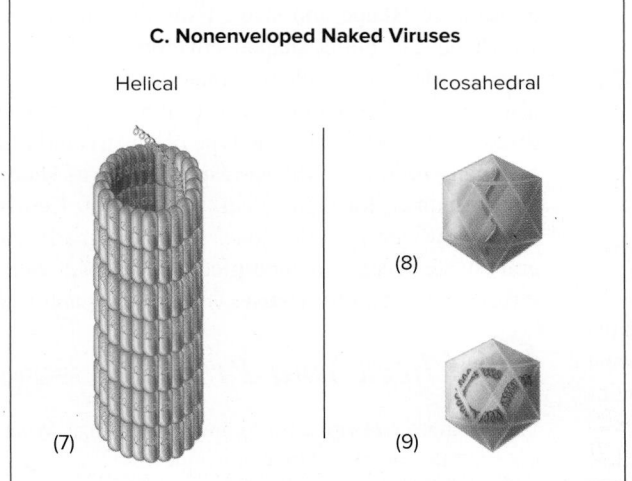

A. Complex viruses:
 (1) poxvirus, a large DNA virus
 (2) flexible-tailed bacteriophage

B. Enveloped viruses:
 With a helical nucleocapsid:
 (3) mumps virus
 (4) rhabdovirus
 With an icosahedral nucleocapsid:
 (5) herpesvirus
 (6) HIV (AIDS)

C. Naked viruses:
 Helical capsid:
 (7) plum poxvirus
 Icosahedral capsid:
 (8) poliovirus
 (9) papillomavirus

Figure 6.10 Basic types of viral morphology.

small compared with that of a cell. It varies from nine genes in human immunodeficiency virus (HIV) to as many as 2,500 genes in pandoraviruses. By comparison, the bacterium *Escherichia coli* has approximately 4,000 genes, and a human cell has around 21,000 genes. Having a larger genome allows cells to carry out the complex metabolic activity necessary for independent life. Viruses typically possess only the genes needed to invade host cells and redirect their synthetic machinery to make new viruses.

In chapter 2 you learned that DNA usually exists as a double-stranded molecule and that RNA is single-stranded. Although most viruses follow this same pattern, viruses exhibit variety in how their RNA or DNA is configured. DNA viruses can have single-stranded (ss) or double-stranded (ds) DNA; the dsDNA can be arranged linearly or in circles. RNA viruses can be double-stranded but are more often single-stranded. Notable examples are the parvoviruses,

which contain single-stranded DNA, and reoviruses (a cause of respiratory and intestinal tract infections), which contain double-stranded RNA. You will learn in chapter 9 that all proteins are made by "translating" the nucleic acid code on a single strand of RNA into an amino acid sequence. Single-stranded RNA genomes that are ready for immediate translation into proteins are called *positive-strand* RNA. RNA genomes that have to be converted into the proper form for translation are called *negative-strand* RNA.

RNA genomes may also be *segmented,* meaning that the individual genes exist in separate RNA molecules, similar to the way that the human (DNA) genome is spread across 23 separate chromosomes. The influenza virus (an orthomyxovirus) is an example of this form. Another RNA virus with some unusual features is a *retrovirus,* one of the few virus types that converts its nucleic acid from RNA to DNA inside its host cell.

Whatever the virus type, these tiny strands of genetic material carry the blueprint for viral structure and functions. In a very real sense, viruses are genetic parasites because they cannot multiply until their nucleic acid has reached their haven inside the host cell. At a minimum, they must carry genes for synthesizing the viral capsid and genetic material, for regulating the actions of the host, and for packaging the mature virus.

Other Substances in the Virus Particle

In addition to the protein of the capsid, the proteins and lipids of envelopes, and the nucleic acid of the core, viruses can contain enzymes for specific operations within their host cell. They may come with preformed enzymes that are required for viral **replication.*** Examples include **polymerases*** that synthesize DNA and RNA and replicase enzymes that copy RNA. The human immunodeficiency virus (HIV) comes equipped with *reverse transcriptase* for synthesizing DNA from RNA. However, viruses completely lack the genes for synthesis of metabolic enzymes. As we shall see, this deficiency has little consequence, because viruses have adapted to assume total control over the cell's metabolic resources. Some viruses can actually carry away substances from their host cell. For instance, arenaviruses pack along host ribosomes, and retroviruses "borrow" the host's tRNA molecules.

6.3 How Viruses Are Classified and Named

 Expected Learning Outcomes

9. Explain the classification scheme used for viruses.

10. Indicate the characteristics used in identifying and naming viruses.

Although viruses are not classified as members of the domains discussed in chapter 1, they are diverse enough to require their own classification scheme to aid in their study and identification. In an informal and general way, we have already begun classifying viruses—as animal, plant, or bacterial viruses; enveloped or naked viruses; DNA or RNA viruses; and helical or icosahedral viruses. These introductory categories are certainly useful in organization and description, but the study of specific viruses requires a more standardized method of nomenclature. For many years, the animal viruses were classified mainly on the basis of their hosts and the kind of diseases they caused. Newer systems for naming viruses also take into account the actual nature of the virus particles themselves, with only partial emphasis on host and disease. The main criteria presently used to group viruses are structure, chemical composition, and similarities in genetic makeup, which indicate evolutionary relatedness.

The International Committee on the Taxonomy of Viruses lists seven orders, 104 families, and 505 genera of viruses. Nomenclature of viruses follows these conventions: Virus families are italicized and given the suffix *-viridae,* and virus genera are likewise italicized and end in *-virus.*

Historically, some virologists had created an informal *species* naming system that mirrors the species names in higher organisms, using genus and species epithets such as *Measles morbillivirus.* The species category has created a lot of controversy within the virology community. Many scientists argue that nonorganisms such as viruses are too changeable and that fine distinctions used for deciding on species classifications will quickly disappear. All the same, over the past decade virologists have largely accepted the concept of viral species, defining them as distinct virus types that share a collection of properties such as host range, pathogenicity, and genetic makeup. The influenza viruses are an example of how variations in the receptors can give rise to different strains or subtypes, even though they are basically the same viruses in other characteristics (Case Study, this chapter).

Because the use of standardized species names are not as widely used, the genus or common English vernacular names (such as poliovirus and rabies virus) predominate in discussions of specific viruses in this text. **Table 6.2** illustrates a system of classification for important viruses and the diseases they cause. To-scale examples of each virus family are included.

Characteristics used for placement in a particular family include type of capsid, nucleic acid strand number, presence and type of envelope, overall viral size, and area of the host cell in which the virus multiplies. Some virus families are named for their microscopic appearance (shape and size). Examples include **rhabdoviruses,*** which have a bullet-shaped envelope, and **togaviruses,*** which have a cloaklike envelope. Anatomical or geographic areas have also been used in naming. For instance, **adenoviruses*** were first discovered in adenoids (one type of tonsil), and **hantaviruses** were originally isolated in the Korean Province of Hantaan. Viruses can also be named for their effects on the host. **Lentiviruses*** tend to cause slow, chronic infections. Acronyms made from blending several characteristics include **picornaviruses,*** which are tiny RNA viruses, and **hepadnaviruses** (for hepatitis and DNA).

 Check Your Progress SECTIONS 6.2 AND 6.3

5. Characterize viruses according to size range. What does it mean to say that they are ultramicroscopic?

6. Describe the general structure of viruses.

7. What is the capsid, and what is its function? How are the two types of capsids constructed?

8. What is a nucleocapsid? Give examples of viruses with the two capsid types.

9. What is an enveloped virus, and how does the envelope arise? What are spikes, how are they formed, and what is their function?

10. What are bacteriophages, and what is unique about their structure?

11. How are the poxviruses different from other animal viruses?

12. Explain the characteristics used in viral classification and the nature of viral families.

13. How are generic and common names used?

* *rhabdovirus* (rab″-doh-vy′-rus) Gr. *rhabdo,* little rod.

* *togavirus* (toh″-guh-vy′-rus) L. *toga,* covering or robe.

* *adenovirus* (ad″-uh-noh-vy′-rus) G. *aden,* gland.

* *lentivirus* (len″-tee-vy′-rus) Gr. *lente,* slow. HIV, the AIDS virus, belongs in this group.

* *picornavirus* (py-kor″-nah-vy′-rus) Sp. *pico,* small, plus RNA.

* *replication* (rep-lih-kay′-shun) L. *replicare,* to reply. To make an exact duplicate.

* *polymerase* (pol-im′-ur-ace) An enzyme that synthesizes a large molecule from smaller subunits.

Nucleic Acid Type	Family	Genus of Virus	Common Name of Genus Members	Name of Disease
DNA Viruses				
Poxviridae Chordopoxvirinae *Herpesviridae*	*Poxviridae*	*Orthopoxvirus*	Variola and vaccinia	Smallpox, cowpox
	Herpesviridae	*Simplexvirus*	Herpes simplex 1 virus (HSV-1)	Fever blister, cold sores
			Herpes simplex 2 virus (HSV-2)	Genital herpes
		Varicellovirus	Varicella zoster virus (VZV)	Chickenpox, shingles
		Cytomegalovirus	Human cytomegalovirus (CMV)	CMV infections
Adenoviridae	*Adenoviridae*	*Mastadenovirus*	Human adenoviruses	Adenovirus infection
Papillomaviridae *Polyomaviridae*	*Papillomaviridae*	*Papillomavirus*	Human papillomavirus (HPV)	Several types of warts
	Polyomaviridae	*Polyomavirus*	JC virus (JCV)	Progressive multifocal leukoencephalopathy
Hepadnaviridae *Parvoviridae* Parvovirinae	*Hepadnaviridae*	*Hepadnavirus*	Hepatitis B virus (HBV or Dane particle)	Serum hepatitis
	Parvoviridae	*Erythrovirus*	Parvovirus B19	Erythema infectiosum
RNA Viruses				
Picornaviridae	*Picornaviridae*	*Enterovirus*	Poliovirus	Poliomyelitis
			Coxsackievirus	Hand-foot-mouth disease
Caliciviridae		*Hepatovirus*	Hepatitis A virus (HAV)	Short-term hepatitis
		Rhinovirus	Human rhinovirus	Common cold, bronchitis
Togaviridae	*Caliciviridae*	*Calicivirus*	Norwalk virus	Viral diarrhea, Norwalk virus syndrome
	Togaviridae	*Alphavirus*	Eastern equine encephalitis virus	Eastern equine encephalitis (EEE)
			Western equine encephalitis virus	Western equine encephalitis (WEE)
			Yellow fever virus	Yellow fever
			St. Louis encephalitis virus	St. Louis encephalitis
Flaviviridae	*Flaviviridae*	*Rubivirus*	Rubella virus	Rubella (German measles)
		Flavivirus	Dengue fever virus	Dengue fever
			West Nile fever virus	West Nile fever
			Zika virus	Zika fever
Bunyaviridae	*Bunyaviridae*	*Bunyavirus*	Bunyamwera viruses	California encephalitis
		Hantavirus	Sin Nombre virus	Respiratory syndrome
		Phlebovirus	Rift Valley fever virus	Rift Valley fever
		Nairovirus	Crimean–Congo hemorrhagic fever (CCHF) virus	Crimean-Congo hemorrhagic fever
Reoviridae	*Filoviridae*	*Filovirus*	Ebola, Marburg virus	Ebola fever
	Reoviridae	*Coltivirus*	Colorado tick fever virus	Colorado tick fever
		Rotavirus	Human rotavirus	Rotavirus gastroenteritis
Orthomyxoviridae *Paramyxoviridae*	*Orthomyxoviridae*	*Influenzavirus*	Influenza virus, type A (Asian, Hong Kong, and swine influenza viruses)	Influenza or "flu"
	Paramyxoviridae	*Paramyxovirus*	Parainfluenza virus	Parainfluenza
			Mumps virus	Mumps
		Morbillivirus	Measles virus	Measles (red)
		Pneumovirus	Respiratory syncytial virus (RSV)	Common cold syndrome

Filoviridae

TABLE 6.2 *(continued)*

Rhabdoviridae	*Lyssavirus*	Rabies virus	Rabies (hydrophobia)	
Retroviridae	*Oncornavirus*	Human T-cell leukemia virus (HTLV)	T-cell leukemia	
	Lentivirus	HIV (human immunodeficiency viruses 1 and 2)	Acquired immunodeficiency syndrome (AIDS)	
Arenaviridae	*Arenavirus*	Lassa virus	Lassa fever	
Coronaviridae	*Coronavirus*	Infectious bronchitis virus (IBV)	Bronchitis	
		Enteric corona virus	Coronavirus enteritis	
		SARS virus	Severe acute respiratory syndrome	

Rhabdoviridae

Retroviridae

Arenaviridae

Coronaviridae

100 nm

6.4 Modes of Viral Multiplication

Expected Learning Outcomes

11. Describe the virus-host relationship.
12. Relate the stages in the multiplication cycle of animal viruses, and summarize what is happening in each stage.
13. Describe three ways in which animal viruses enter into a host cell.
14. Define *replication* as it relates to viruses.
15. Explain two ways in which animal viruses are released by a host cell.
16. Describe cytopathic effects of viruses and the possible results of persistent viral infections.

Viruses are closely associated with their hosts. In addition to providing the viral habitat, the host cell is absolutely necessary for viral multiplication, which is synonymous with infection. The way in which viruses invade a host cell is an extraordinary biological phenomenon. Viruses have been aptly described as minute parasites that seize control of the synthetic and genetic machinery of cells. The nature of this cycle profoundly affects pathogenicity, transmission, the responses of the immune defenses, and human measures to control viral infections. From these perspectives, we cannot overemphasize the importance of a working knowledge of the relationship between viruses and their host cells.

Multiplication Cycles in Animal Viruses

The general phases in the life cycle of animal viruses are **adsorption,** * **penetration, synthesis, assembly,** and **release** from the host cell. Viruses vary in the exact mechanisms of these processes, but we will use a simple animal virus to illustrate the major events **(figure 6.11).** Other examples are given in chapter 9.

Adsorption and Host Range

Invasion begins when the virus encounters a susceptible host cell and adsorbs specifically to receptor sites on the cell membrane. The membrane receptors that viruses attach to are usually glycoproteins the cell requires for its normal function. For example, the rabies virus affixes to receptors found on mammalian nerve cells, and the human immunodeficiency virus (HIV) attaches to the CD4 protein on certain white blood cells. The mode of attachment varies between the two general types of viruses. In enveloped forms such as influenza virus and HIV, glycoprotein spikes bind to the cell membrane receptors. Viruses with naked nucleocapsids (adenovirus, for example) use surface receptors on their capsids that adhere to cell membrane receptors **(figure 6.12).**

Because a virus can invade its host cell only through making an exact fit with a specific host molecule, the range of hosts it can infect in a natural setting is limited. This limitation, known as the **host range,** can vary from one virus to another. For example, hepatitis B infects only liver cells of humans; the poliovirus infects primarily intestinal and nerve cells of primates (humans, apes, and monkeys); and the rabies virus infects nerve cells of most mammals. Cells that lack compatible virus receptors are resistant to adsorption and invasion by that virus. This explains why, for example, human liver cells are not infected by the canine hepatitis virus and dog liver cells cannot host the human hepatitis A virus. It also explains why viruses usually have tissue specificities called **tropisms*** for certain cells in the body. The hepatitis B virus targets the liver, and the mumps virus targets salivary glands. Many viruses can be manipulated in the laboratory to infect cells that they do not infect naturally, thus making it possible to cultivate them.

> **Quick Search**
> Look up "Flu Attack! How a Virus Invades Your Body" on the Internet.

Penetration/Uncoating of Animal Viruses

For an animal virus to successfully infect a cell, it must penetrate the cell membrane of the host cell and deliver the viral nucleic acid into the host cell's interior. How animal viruses do this varies with the type of virus and type of host cell, but most of them enter through one of two means, fusion or endocytosis. In the case of **fusion,** the viral envelope fuses directly with the host cell membrane, so it can occur only in enveloped viruses **(figure 6.13a).** Following attachment of the virus to host cell receptors, the lipids within the adjacent membranes become rearranged so that the nucleocapsid can be translocated into the cytoplasm for the synthesis phase.

In the **endocytosis** version of penetration, the virus can be either enveloped **(figure 6.13b)** or naked **(figure 6.13c),** and it is

* *adsorption* (ad-sorp'-shun) L. ad, to, and *sorbere,* to suck. The attachment of one thing onto the surface of another.

* *tropism* (troh'-pizm) Gr. *trope,* a turn. Having a special affinity for an object or substance.

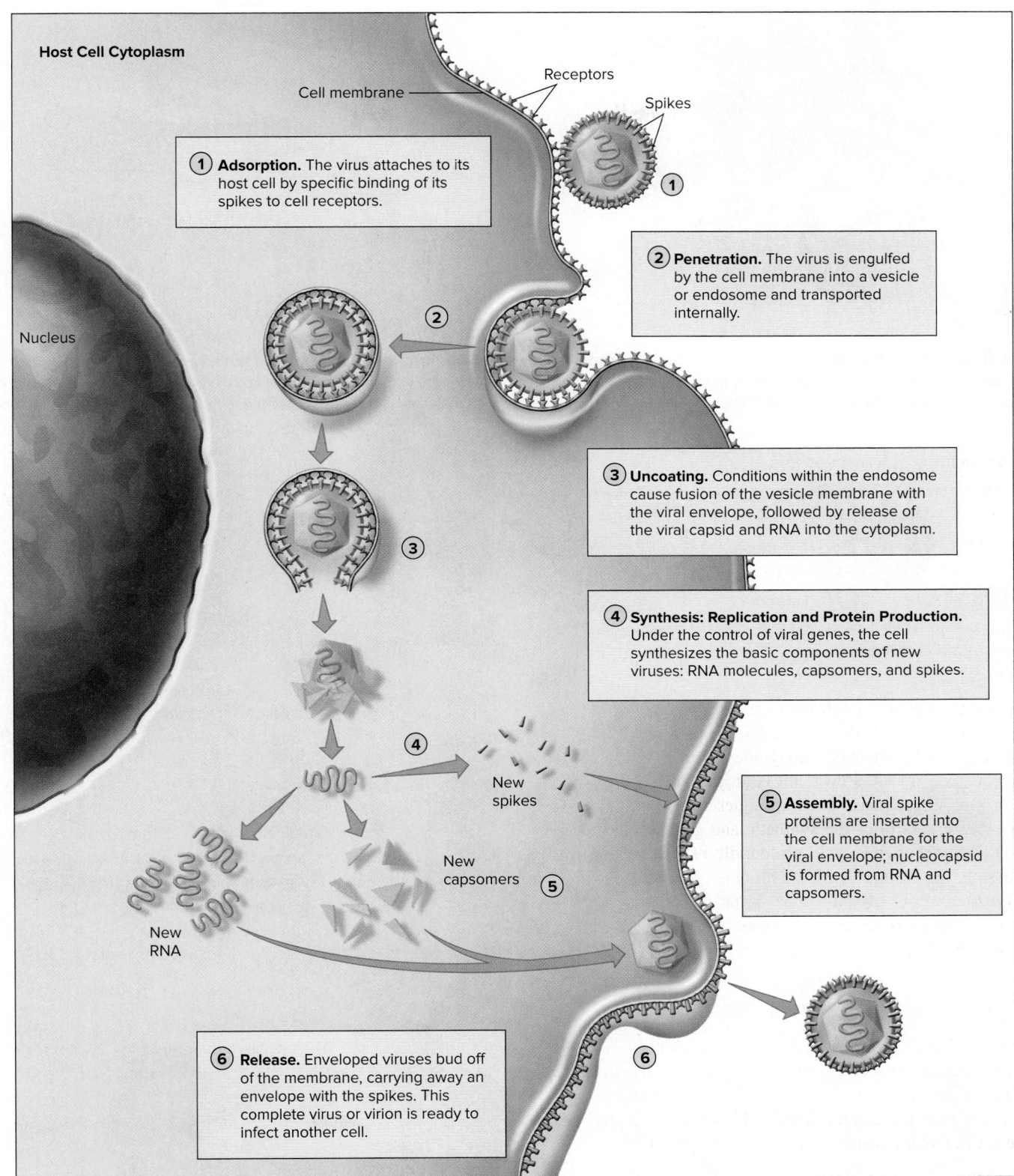

Host Cell Cytoplasm

Cell membrane

Receptors

Spikes

① **Adsorption.** The virus attaches to its host cell by specific binding of its spikes to cell receptors.

② **Penetration.** The virus is engulfed by the cell membrane into a vesicle or endosome and transported internally.

Nucleus

② ③

③ **Uncoating.** Conditions within the endosome cause fusion of the vesicle membrane with the viral envelope, followed by release of the viral capsid and RNA into the cytoplasm.

④ **Synthesis: Replication and Protein Production.** Under the control of viral genes, the cell synthesizes the basic components of new viruses: RNA molecules, capsomers, and spikes.

④

New spikes

⑤ **Assembly.** Viral spike proteins are inserted into the cell membrane for the viral envelope; nucleocapsid is formed from RNA and capsomers.

New capsomers ⑤

New RNA

⑥ **Release.** Enveloped viruses bud off of the membrane, carrying away an envelope with the spikes. This complete virus or virion is ready to infect another cell.

⑥

Process Figure 6.11 **General features in the multiplication cycle of an enveloped animal virus.** Using an RNA virus (rubella virus), the major events are outlined, although other viruses will vary in exact details of the cycle.

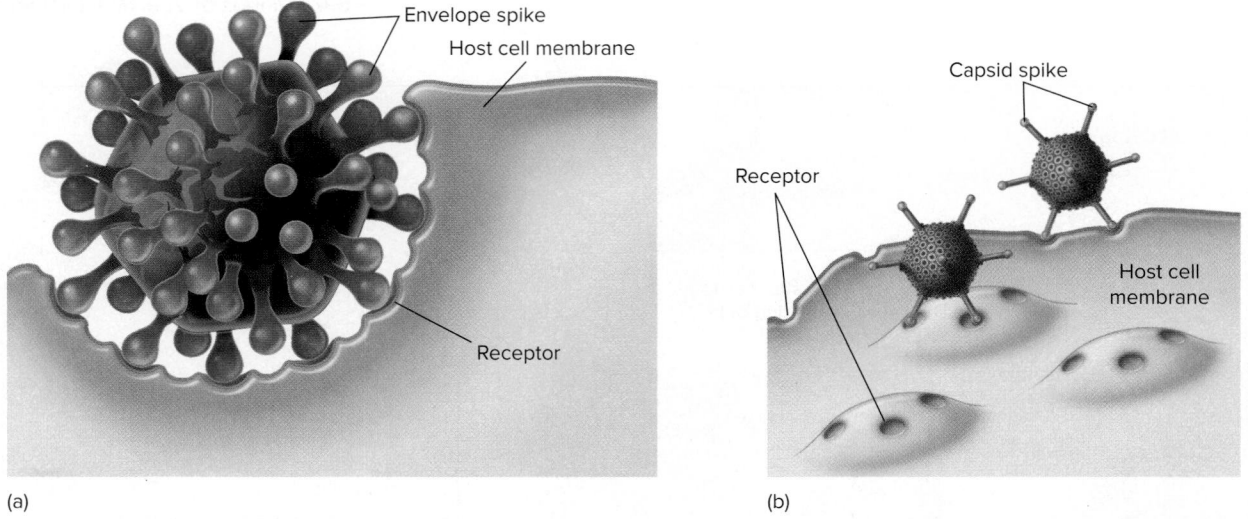

(a)

(b)

Figure 6.12 **The mode by which animal viruses adsorb to the host cell membrane. (a)** An enveloped coronavirus with prominent spikes. The configuration of the spike has a complementary fit for cell receptors. The process in which the virus lands on the cell and plugs into receptors is termed docking. **(b)** An adenovirus has a naked capsid that adheres to its host cell by nestling surface molecules on its capsid into the receptors on the host cell's membrane.

(a)

Virus spikes
Host cell membrane
Receptors
Nucleic acid (RNA)

Irreversible attachment

Membrane fusion
Receptor-spike complex

Entry of nucleocapsid

Free RNA
Uncoating of nucleic acid

(b)

Host cell membrane
Host cell receptors
Nucleic acid (DNA)

Specific attachment

Engulfment

Vesicle
Virus in vesicle

Uncoating step

Vesicle, envelope, and capsid break down

Free DNA

(c)

Capsid
Nucleic acid
Receptor

Adhesion of virus to host receptors

Engulfment into vesicle

RNA

Viral RNA is released from vesicle

Process Figure 6.13 **Modes of virus penetration.** Pictured in **(a)** is the process of fusion between the envelope of the virus and the host cell membrane, with release of the nucleocapsid into the cytoplasm. This is the mechanism for mumps and HIV viruses. Both (b) and (c) are examples of entry by endocytosis or engulfment, followed by uncoating. **(b)** In an enveloped virus such as herpesvirus, the entire virus is taken into a vesicle with subsequent release of the DNA. **(c)** A naked virus such as poliovirus is first taken into a vesicle and then releases its nucleic acid through a pore or rupture in the membrane.

(a)

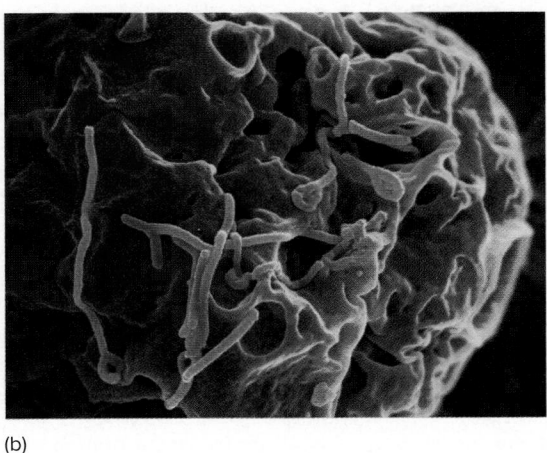

(b)

Figure 6.14 Maturation and release of enveloped viruses.
(a) As parainfluenza virus is budded off the membrane, it simultaneously picks up an envelope and spikes. **(b)** Ebola viruses leave their host cell by budding off its surface. (b): *Source:* National Institute of Allergy and Infectious Diseases (NIAID)/CDC

engulfed entirely into a vesicle called an endosome after its initial attachment. Once inside the cell, the virus is **uncoated.** This means that the vesicle membrane becomes altered so that the viral nucleocapsid or nucleic acid can be released into the cytoplasm. In some viruses the vesicle membrane fuses with the virus, and in other cases the vesicle membrane develops openings for the virus to leave. There are numerous variations in the details of this process.

Synthesis: Replication and Protein Production

The synthetic and replicative phases of animal viruses are highly regulated and extremely complex at the molecular level. Free viral nucleic acid exerts control over the host's metabolism and synthetic machinery. How this control proceeds will vary, depending on whether the virus is a DNA or an RNA virus. In general, the DNA viruses (except poxviruses) enter the host cell's nucleus and are replicated and assembled there. RNA viruses are replicated and assembled in the cytoplasm, with some exceptions.

The details of animal virus replication are discussed in chapter 9. Here we provide a brief overview of the process, using an RNA virus

as a model. Almost immediately upon entry, the viral nucleic acid alters the genetic expression of the host and instructs it to synthesize the building blocks for new viruses. For example, it was recently shown that HIV depends on the expression of 250 human genes to complete its multiplication cycle. First, the RNA of the virus becomes a message for synthesizing viral proteins (translation). Most viruses with positive-strand RNA molecules already contain the correct message for translation into proteins. Viruses with negative-strand RNA molecules must first be converted into a positive-strand message. Some viruses come equipped with the necessary enzymes for synthesis of viral components; others utilize those of the host. During the final phase, the host's replication and synthesis machinery produces new RNA, proteins for the capsid, spikes, and viral enzymes.

Assembly of Animal Viruses: Host Cell as Factory

Toward the end of the cycle—during the assembly stage—mature virus particles are constructed from the growing pool of parts. In most instances, the capsid is first laid down as an empty shell that will serve as a receptacle for the nucleic acid strand. Electron micrographs taken during this time show cells with masses of viruses, often enclosed in packets (see figure 6.15a). One important event leading to the release of enveloped viruses is the insertion of viral spikes into the host's cell membrane so they can be picked up as the virus buds off with its envelope, as discussed.

Release of Mature Viruses

To complete the cycle, assembled viruses leave their host in one of two ways. Nonenveloped and complex viruses that reach maturation in the cell nucleus or cytoplasm are released through cell **lysis** or rupturing. Enveloped viruses are liberated by **budding** or **exocytosis**[3] from the membranes of the cytoplasm, nucleus, endoplasmic reticulum, or vesicles. During this process the nucleocapsid binds to the membrane, which curves completely around it and forms a small pouch. Pinching off the pouch releases the virus with its envelope **(figure 6.14).** Budding of enveloped viruses causes them to be shed gradually, without the sudden destruction of the cell. Regardless of how the virus leaves, most active viral infections are ultimately lethal to the cell because of accumulated damage. Lethal

3. For enveloped viruses, these terms are interchangeable. They mean the release of a virus from an animal cell by enclosing it in a portion of membrane derived from the cell.

damages include a permanent shutdown of metabolism and genetic expression, destruction of cell membrane and organelles, toxicity of virus components, and release of lysosomes.

The length of a multiplication cycle, from adsorption to lysis, varies to some extent, but it is usually measured in hours. A simple virus such as poliovirus takes about 8 hours; parvovirus takes 16 to 18 hours; and more complex viruses, such as herpesviruses, require 72 hours or more.

A fully formed, extracellular virus particle that is virulent and able to establish infection in a host is called a **virion.*** The number of virions released by infected cells is variable, controlled by factors such as the size of the virus and the health of the host cell. About 3,000 to 4,000 virions are released from a single cell infected with poxviruses, whereas a poliovirus-infected cell can release over 100,000 virions. If even a small number of these virions happens to meet other susceptible cells and infect them, the potential for viral proliferation is immense.

Visible Damage to the Host Cell

The short- and long-term effects of viral infections on animal cells are well documented. **Cytopathic*** **effects** (CPEs) are defined as virus-induced damage to the cell that alters its microscopic appearance. Individual cells can become disoriented, undergo gross changes in shape or size, or develop intracellular changes **(figure 6.15a).** It is common to note **inclusion bodies,** or compacted masses of viruses or damaged cell organelles, in the nucleus and cytoplasm **(figure 6.15b).** Examination of cells and tissues for cytopathic effects has been a traditional tool for diagnosing viral infections that is usually supplemented with more specific serological and molecular methods (see chapter 17). **Table 6.3** summarizes some prominent cytopathic effects associated with specific viruses. One very common CPE is the

* *virion* (vir′-ee-on) Gr. *iso,* poison.
* *cytopathic* (sy″-toh-path′-ik) Gr. *cyto,* cell, and *pathos,* disease.

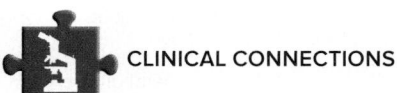

CLINICAL CONNECTIONS

Persistent Viral Infections

As a general rule, a virus infection kills its host cell, but some cells escape destruction by harboring the virus in some form. These so-called **persistent infections** can last from a few weeks to years and even for the life of the host. Take, for example, the hepatitis B virus, which produces a chronic progressive infection of the liver that may lead to cirrhosis and cancer. Several viruses remain in a **latent state,** meaning that they remain inactive or nonproductive over long periods. Examples of this are herpes simplex viruses (cold sores and genital herpes) and herpes zoster virus (chickenpox and shingles). Both viruses go into latency in nerve cells and later emerge under the influence of various stimuli to cause recurrent symptoms.

Some animal viruses enter their host cell and permanently alter its genetic material, in many cases leading to cancer. These viruses are termed *oncogenic,* and their effect on the cell is called **transformation.** A startling feature of some of these viruses is that their nucleic acid becomes integrated into the host DNA. Transformed cells generally have an increased rate of growth, alterations in chromosomes, changes in the cell's surface molecules, and the capacity to divide for an indefinite period. Mammalian viruses involved in the development of tumors are called **oncoviruses.** Some of these are DNA viruses such as papillomavirus, which is associated with cervical cancer, and herpesviruses such as Epstein-Barr virus, which is a cofactor in Burkitt's lymphoma in genetically predisposed individuals. These findings have spurred a great deal of speculation on the possible involvement of viruses in cancers whose cause is still unknown. Additional information on the connection between viruses and cancer is found in chapter 24.

Why would the integration of viral genes into a human chromosome cause cancer? *Answer available on Connect.*

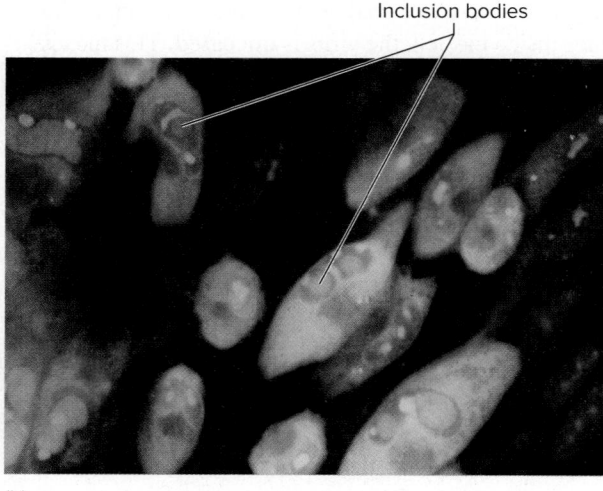

Inclusion bodies

(a)

(b)

Figure 6.15 **Cytopathic changes in cells and cell cultures infected by viruses. (a)** Human epithelial cells (400×) infected by herpes simplex virus demonstrate multinucleate giant cells; inset with a cluster of viruses from an intranuclear inclusion (100,000×). **(b)** Fluorescent-stained human cells infected with cytomegalovirus. Note the inclusion bodies (1,000×) and the loss of cohesive junctions between cells.
(a, a inset): *Source:* CDC; (b): © Massimo Battaglia, INeMM CNR, Rome Italy

TABLE 6.3	Cytopathic Changes in Selected Virus-Infected Animal Cells
Virus	**Response in Animal Cell**
Smallpox virus	Cells round up; inclusions appear in cytoplasm
Herpes simplex	Cells fuse to form multinucleated syncytia; nuclear inclusions (see figure 6.15)
Adenovirus	Clumping of cells; nuclear inclusions
Poliovirus	Cell lysis; no inclusions
Reovirus	Cell enlargement; vacuoles and inclusions in cytoplasm
Influenza virus	Cells round up; no inclusions
Rabies virus	No change in cell shape; cytoplasmic inclusions (Negri bodies)
Measles virus	Syncytia form (multinucleate)

fusion of multiple host cells into a **syncytium,** a single large cell containing multiple nuclei. These syncytia are a result of some viruses' ability to fuse together the membranes of several host cells. One virus (respiratory syncytial virus) is even named for this effect.

 Check Your Progress SECTION 6.4

14. Write a narrative that describes the stages in the multiplication of an enveloped animal virus.
15. Describe some ways animal viruses penetrate the host cell. What is uncoating?
16. Summarize the two major ways in which animal viruses leave their host cells.
17. What is meant by the term *virion?*
18. Describe several cytopathic effects of viruses. What causes the changed appearance of the host cell, and how might it be used to diagnose viral infections?
19. Explain what it means for a virus to become persistent or latent, and how these events are important.
20. Briefly describe the effects of an oncogenic virus.

6.5 The Multiplication Cycle in Bacteriophages

 Expected Learning Outcomes

17. Describe the stages in the multiplication cycle of bacteriophages.
18. Explain what is meant by *lysogeny, prophage, lysogenic induction,* and *lysogenic conversion.*
19. Compare the major stages in multiplication of animal viruses and bacteriophages.

For the sake of comparison, we now turn to multiplication in bacterial viruses—the **bacteriophages.** When Frederick Twort and Felix d'Herelle discovered these viruses in 1915, they first thought that the bacterial host cells were being eaten by some unseen parasite, which they called a bacteriophage.

So far as is known, all bacteria are parasitized by various specific viruses. Most bacteriophages (often shortened to *phage*) contain double-stranded DNA, though single-stranded DNA and RNA types exist as well. Bacteriophages are of great interest to medical microbiologists because they can make the bacteria they infect more virulent for humans.

Probably the most widely studied bacteriophages are those of the intestinal bacterium *Escherichia coli*—especially the T-even **phages** and the lambda phages. They are complex in structure, with an icosahedral capsid head containing DNA, a central tube (surrounded by a sheath), collar, base plate, tail pins, and fibers (see figure 6.9*b*). Momentarily setting aside a strictly scientific and objective tone, it is tempting to think of these extraordinary viruses as minute spacecrafts docking on an alien planet, ready to unload their genetic cargo.

Bacteriophages go through stages similar to those of the animal viruses described earlier **(figure 6.16).** First, they adsorb to host bacteria using specific receptors on the bacterial surface. Although the entire phage cannot enter the host cell, the nucleic acid is injected through a rigid tube the phage inserts through the bacterial membrane and wall **(figure 6.17).** The empty capsid remains attached to the cell surface. Entry of the nucleic acid stops host cell DNA replication and protein synthesis, and it soon prepares the cell machinery for viral replication and synthesis of viral proteins. As the host cell produces new phage parts, the parts spontaneously assemble into bacteriophages.

An average-size *Escherichia coli* cell can contain up to 200 new phage units at the end of this period. Eventually, the host cell becomes so packed with viruses that it undergoes **lysis** and splits open, thereby releasing the mature virions **(figure 6.18).** This process is hastened by viral enzymes produced late in the infection cycle that weaken the cell envelope. Upon release, the virulent phages can spread to other susceptible bacterial cells and begin a new cycle of infection.

Lysogeny: The Silent Virus Infection

The lethal effects of a virulent phage on the host cell present a dramatic view of virus-host interaction. However, not all bacteriophages go immediately into a lytic cycle. Depending upon the conditions in the bacterial host, some special DNA viruses, called **temperate* phages,** undergo adsorption and penetration into the bacterial host but are not replicated or released immediately. Instead, the viral DNA enters an inactive **prophage*** state, during which it is usually inserted into the bacterial chromosome. This viral DNA will be retained by the bacterial cell and copied during its normal cell division so that the cell's progeny will also have the phage DNA. This condition, in which the host chromosome carries bacteriophage DNA, is termed **lysogeny.*** Because viral particles are not produced, the bacterial cells carrying temperate phages do not lyse and they appear entirely normal. Later, in a process called **induction,** the prophage in a lysogenic cell will be activated and progress directly into viral replication and the lytic cycle. The lysogenic phase is depicted in the green section of figure 6.16. Lysogeny is a less deadly

* *temperate* (tem′-pur-ut) A reduction in intensity.
* *prophage* (pro′-fayj) L. pro, before, plus phage.
* *lysogeny* (ly-soj′-uhn-ee) The potential ability to produce phage.

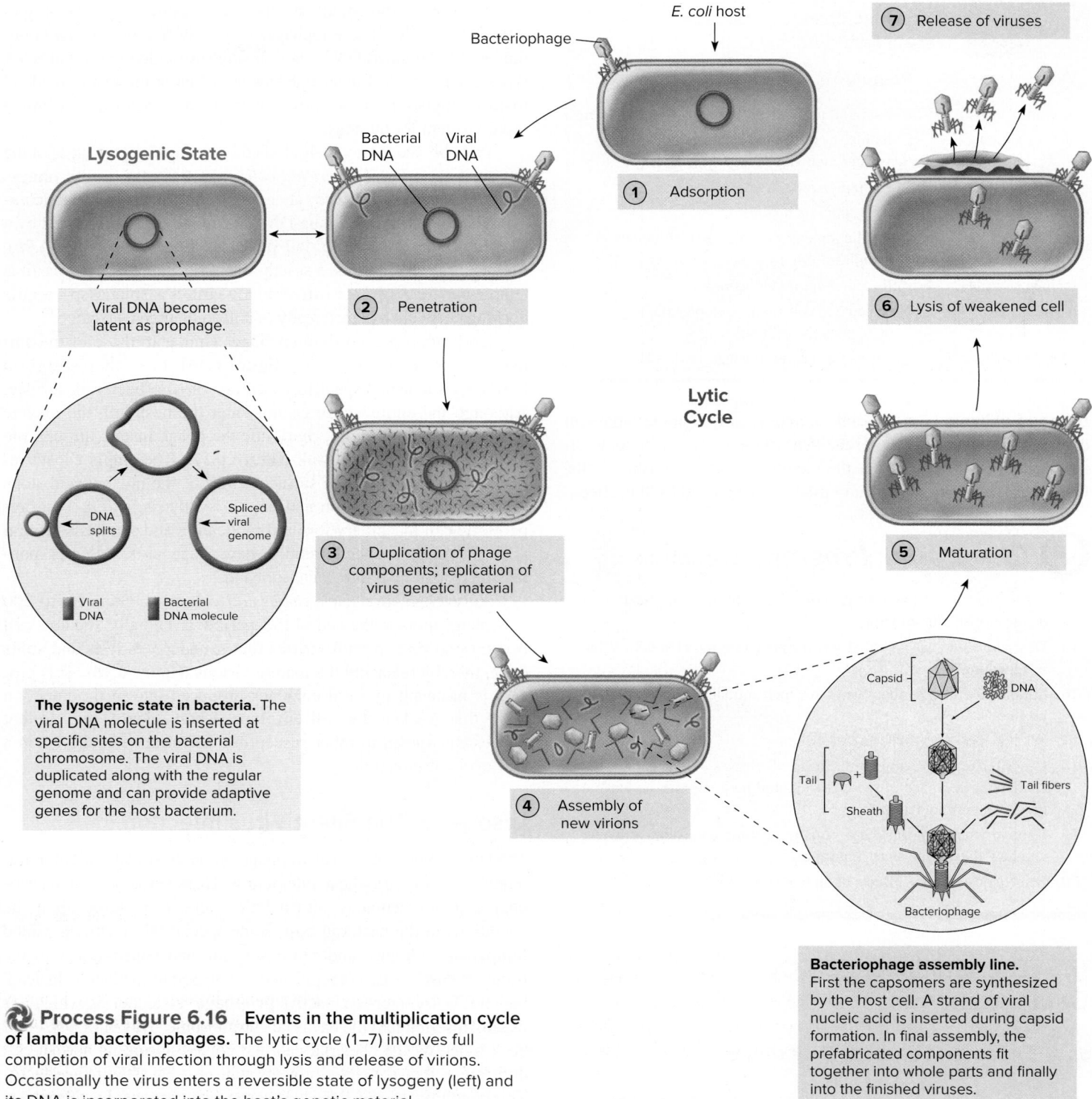

Lysogenic State

Viral DNA becomes latent as prophage.

DNA splits

Spliced viral genome

Viral DNA

Bacterial DNA molecule

The lysogenic state in bacteria. The viral DNA molecule is inserted at specific sites on the bacterial chromosome. The viral DNA is duplicated along with the regular genome and can provide adaptive genes for the host bacterium.

E. coli host

Bacteriophage

1 Adsorption

Bacterial DNA Viral DNA

2 Penetration

3 Duplication of phage components; replication of virus genetic material

4 Assembly of new virions

Lytic Cycle

7 Release of viruses

6 Lysis of weakened cell

5 Maturation

Capsid DNA

Tail Tail fibers

Sheath

Bacteriophage

Bacteriophage assembly line. First the capsomers are synthesized by the host cell. A strand of viral nucleic acid is inserted during capsid formation. In final assembly, the prefabricated components fit together into whole parts and finally into the finished viruses.

Process Figure 6.16 **Events in the multiplication cycle of lambda bacteriophages.** The lytic cycle (1–7) involves full completion of viral infection through lysis and release of virions. Occasionally the virus enters a reversible state of lysogeny (left) and its DNA is incorporated into the host's genetic material.

form of infection than the full lytic cycle and is thought to be an advancement that allows the virus to spread without killing the host.

Because of the intimate association between the genetic material of the virus and host, phages occasionally serve as transporters of bacterial genes from one bacterium to another and consequently can play a profound role in bacterial genetics. This phenomenon, called transduction, is one way that genes for toxin production and drug resistance are transferred between bacteria (see chapters 9 and 12).

Phages have profound effects on the infectiousness and virulence of pathogenic bacteria. Viral genomes often encode toxins, enzymes, and other factors that can alter the course and outcome of infections. When a bacterium acquires genes from its temperate phage, it is called **lysogenic conversion** (see figure 6.16). The phenomenon was first discovered in the 1950s in *Corynebacterium diphtheriae,* the pathogen that causes diphtheria. The diphtheria toxin responsible for the severe nature of the disease is a bacteriophage product. *C. diphtheriae*

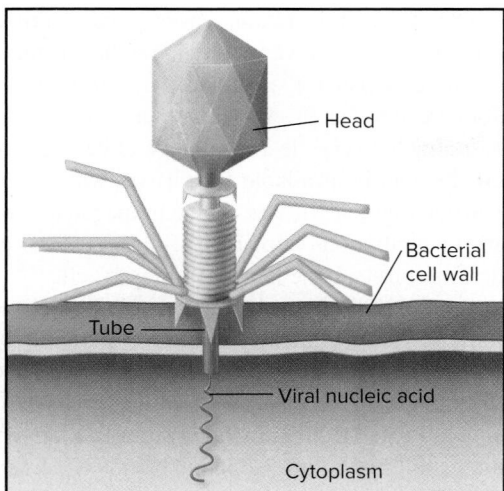

Figure 6.17 **Penetration of a bacterial cell by a bacteriophage.** After adsorption, the phage plate becomes embedded in the cell wall, and the sheath contracts, pushing the tube through the cell wall and membrane and releasing the nucleic acid into the interior of the cell.

Figure 6.18 **A weakened bacterial cell, crowded with viruses.** The cell has ruptured and released numerous virions that can then attack nearby susceptible host cells. Note the empty heads of "spent" phages lined up around the ruptured wall. © Lee D. Simon/Science Source

TABLE 6.4	Comparison of Bacteriophage and Animal Viral Multiplication	
	Bacteriophage	**Animal Virus**
Adsorption	Precise attachment of special tail fibers to cell wall	Attachment of capsid or envelope to cell surface receptors
Penetration	Injection of nucleic acid through cell wall; no uncoating of nucleic acid	Whole virus is engulfed and uncoated, or virus surface fuses with cell membrane; nucleic acid is released
Synthesis and Assembly	Occurs in cytoplasm (no nucleus present) Cessation of host synthesis Viral DNA or RNA replicated Viral components synthesized	Occurs in cytoplasm and nucleus Cessation of host synthesis Viral DNA or RNA replicated Viral components synthesized
Viral Persistence	Lysogeny	Latency, chronic infection, cancer
Release from Host Cell	Cell lyses after being weakened by viral enzymes.	Some cells lyse; enveloped viruses bud off host cell membrane.
Cell Destruction	Sudden by lysis	Sudden (lysis) or slow (budding)

cells without the phage are less pathogenic. Other bacteria that are made virulent by their prophages are *Vibrio cholerae,* the agent of cholera, and *Clostridium botulinum,* the cause of botulism.

The cycles of bacterial and animal viruses illustrate different patterns of viral multiplication, which are summarized in **table 6.4.**

 Check Your Progress SECTION 6.5

21. In simple terms, what does the viral nucleic acid do once it gets into the cell?
22. What processes are involved in bacteriophage assembly?
23. What is a prophage or temperate phage? What is lysogeny?
24. Compare and contrast the main phases in the lytic multiplication cycle in bacteriophages and animal viruses.
25. Why is penetration so different between the two groups?

6.6 Techniques in Cultivating and Identifying Animal Viruses

 Expected Learning Outcomes

20. Describe the general purposes of cultivating viruses.
21. Compare the methods and uses of cell culture, bird embryos, and live animals in growing viruses.

One problem hampering early animal virologists was their inability to propagate specific viruses routinely in pure culture and in sufficient quantities for their studies. Virtually all of the pioneering attempts at cultivation had to be performed in an organism that was the usual host for the virus. But this method had its limitations. How could researchers have ever traced the stages of viral multiplication if they had been restricted to the natural host, especially in the case of human viruses? Fortunately, systems of cultivation with broader applications were developed, including **in vitro*** cell (or tissue) culture methods and **in vivo*** inoculation of laboratory-bred animals and embryonic bird tissues. Such use of substitute host

* *in vitro* (in vee'-troh) L. *vitros,* glass. Experiments performed in test tubes or other artificial environments.
* *in vivo* (in vee'-voh) L. *vivos,* life. Experiments performed in a living body.

systems permits greater control, uniformity, and wide-scale harvesting of viruses.

The primary purposes of viral cultivation are (1) to isolate and identify viruses in clinical specimens; (2) to prepare viruses for vaccines; and (3) to do detailed research on viral structure, multiplication cycles, genetics, and effects on host cells.

Using Cell (Tissue) Culture Techniques

The most important early discovery that led to easier cultivation of viruses in the laboratory was the development of a simple and effective way to grow populations of isolated animal cells in culture. These types of in vitro cultivation systems are termed **cell culture** or tissue culture. Although these terms are used interchangeably, cell culture is probably a more accurate description. This method makes it possible to propagate most viruses. Much of the virologist's work involves developing and maintaining these cultures. Animal cell cultures are grown in sterile chambers with special media that contain the correct nutrients required by animal cells to survive. The cultured cells grow in the form of a *monolayer,* a single, confluent sheet of cells that supports viral multiplication and permits close inspection of the culture for signs of infection **(figure 6.19).**

Cultures of animal cells usually exist in the primary or continuous form. *Primary cell cultures* are prepared by placing freshly isolated animal tissue in a growth medium. Embryonic, fetal, adult, and even cancerous tissues have served as sources of primary cultures. A primary culture retains several characteristics of the original tissue from which it was derived, but this original line generally has a limited existence. Eventually it will die out or mutate into a line of cells that can grow by continuous subculture in fresh nutrient medium. One very clear advantage of cell culture is that a specific cell line can be available for viruses with a very narrow host range. Strictly human viruses can be propagated in one of several primary or continuous human cell lines, such as embryonic kidney cells, fibroblasts, bone marrow, or heart cells.

One way to detect the growth of a virus in culture is to observe degeneration and lysis of infected cells in a monolayer of a cell culture **(figure 16.19a).** The areas where virus-infected cells have been destroyed show up as clear, well-defined patches in the cell sheet called **plaques*** **(figure 6.19c).** Plaques are essentially the macroscopic manifestation of cytopathic effects (CPEs), discussed in section 6.4. This same technique is used to detect and count bacteriophages, because they also produce plaques when grown in soft agar cultures of their host cells. A plaque develops when the viruses released by an infected host cell radiate out to adjacent host cells. As new cells become infected, they die and release more viruses, and so on. As this process continues, the infection spreads gradually and symmetrically from the original point of infection, causing the macroscopic appearance of round, clear spaces that correspond to areas of lysed cells.

* *plaque* (plak) Fr. *placke,* patch or spot.

(a)

(b)

(c)

Figure 6.19 **Microscopic views of normal and infected cell cultures. (a)** Cell culture plate infected with herpes simplex virus. Clear, round spaces are regions of infection (plaques) where the host cells have been destroyed by viruses. **(b)** Closeup of a normal, uninfected region of cultured cells. **(c)** Closeup of plaques, which consist of open spaces where cells have been disrupted by viral infection.
(a): Courtesy of Marylou Gibson, Ph.D., VIRAPUR; (b and c): *Source:* CDC

Using Bird Embryos

An embryo is an early developmental stage of animals marked by rapid differentiation of cells. Birds undergo their embryonic period within the closed protective case of an egg, which makes an incubating

(a)

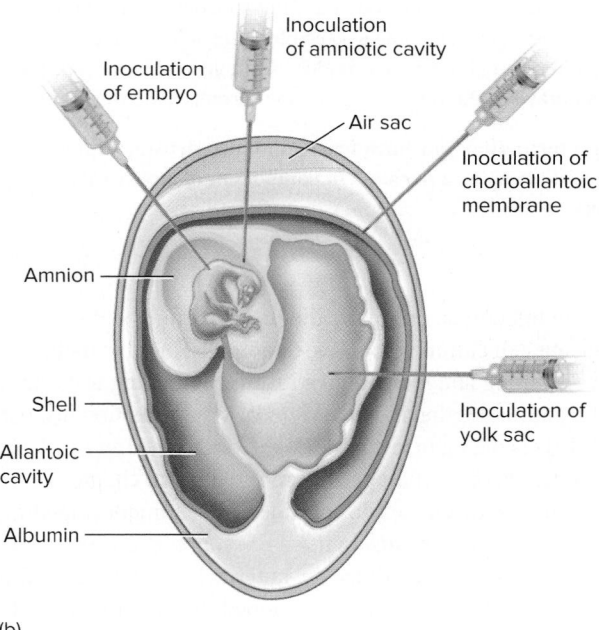

(b)

Figure 6.20 Cultivating animal viruses in a developing bird embryo. (a) A technician inoculates fertilized chicken eggs with viruses in the first stage of preparing vaccines. This process requires the highest levels of sterile and aseptic precautions. Some types of Influenza vaccine are prepared this way. **(b)** The shell is perforated using sterile techniques, and a virus preparation is injected into a site selected to grow the viruses. Targets include the allantoic cavity, a sac for embryonic waste removal; the amniotic cavity that cushions and protects the embryo; the chorioallantoic membrane, for embryonic gas exchange; the yolk sac, for the nourishment of the embryo; and the embryo itself. (a): © State Hygenic Laboratory at The University of Iowa

bird egg a nearly perfect system for viral propagation. It is an intact and self-supporting unit, complete with its own sterile environment and nourishment. Furthermore, it furnishes several embryonic tissues that readily support viral multiplication. Every year hundreds of millions of chicken embryos are inoculated to prepare influenza vaccines.

Chicken, duck, and turkey eggs are the most common choices for inoculation. The egg must be injected through the shell, necessitating rigorous sterile techniques to prevent contamination by bacteria and fungi from the air and the outer surface of the shell. The exact tissue inoculated is guided by the type of virus being cultivated and the goals of the experiment (**figure 6.20**).

Viruses multiplying in embryos may or may not cause effects visible to the naked eye. The signs of viral growth include death of the embryo, defects in embryonic development, and localized areas of damage in the membranes, resulting in discrete, opaque spots called pocks (a variant of *pox*). Embryonic fluids and tissues can be prepared for direct examination with an electron microscope. Certain viruses can also be detected by their ability to agglutinate red blood cells (form big clumps) or by their reaction with an antibody of known specificity that will affix to its corresponding virus, if it is present.

Using Live Animal Inoculation

Specially bred strains of white mice, rats, hamsters, guinea pigs, and rabbits are the usual choices for animal cultivation of viruses. Invertebrates or nonhuman primates are occasionally used as well. Because viruses can exhibit host specificity, certain animals can propagate a given virus more readily than others. Depending on the particular experiment, tests can be performed on adult, juvenile, or newborn animals. The animal is exposed to the virus by injection of a viral preparation or specimen into the brain, blood, muscle, body cavity, skin, or footpads.

A very unusual method of cultivating viruses that also became a source of controversy happened when scientists at the State University of New York were able to artificially create a virus that is virtually identical to natural poliovirus in a test tube without live cells. They used DNA nucleotides to synthesize the viral genome according to the published poliovirus sequence. They then added an enzyme that would transcribe the DNA sequence into the RNA genome characteristic of poliovirus. They ended up with viruses that mimicked the structure, infectivity, and replication cycle of the poliovirus.

6.7 Viral Infection, Detection, and Treatment

 Expected Learning Outcomes

22. Discuss the medical impact and importance of viruses.

23. Explain how animal viral infections are treated and detected.

The number of viral infections worldwide is nearly impossible to measure accurately. Certainly viruses are the most common cause of acute infections that do not result in hospitalization, especially when one considers widespread diseases such as colds, measles, chickenpox, influenza, herpes, and warts. If one also takes into account

6.1 Secret World of Microbes
Seeking Your Inner Viruses

Would you be alarmed to be told that your cells carry around bits and pieces of fossil viruses? Well, we now know that they do. A fascinating aspect of the virus–host relationship is the extent to which viral genetic material becomes affixed to host chromosomes and is passed on, possibly even for millions of years. We know this from data obtained by the Human Genome Project, which sequenced all of the genetic codes on the 46 human chromosomes. While searching through the genome sequences, virologists began to find DNA they identified as viral in origin. So far they have found about 100,000 different fragments of viral DNA. In fact, over 8% of the DNA in human chromosomes comes from viruses!

Does this virus make us look fat?
Source: CDC

These researchers are doing the work of molecular fossil hunters, locating and identifying these ancient viruses. Many of them are retroviruses that converted their RNA codes to DNA codes, inserted the DNA into a site in a host chromosome, and then became dormant and did not kill the cell. When this happened in an egg or sperm cell, the virus could be transmitted basically unchanged for hundreds of generations. One of the most tantalizing questions is what effect, if any, such retroviruses might have on modern humans. Some virologists contend that these virus genes would not have been maintained for thousands and even millions of years if they did not serve some function. Others argue that they are just genetic "garbage" that has accumulated over a long human history.

So far, we have only small glimpses of the possible roles of these viruses. One type of endogenous retrovirus has been shown to be inti-mately involved in forming the human placenta, leading microbiologists to conclude that some viruses have become an essential factor in evolution and development. Other retroviruses may be involved in diseases such as prostate cancer and chronic fatigue syndrome.

Evidence is mounting that certain viruses may contribute to human obesity. Several studies with animals revealed that chickens and mice infected with a human adenovirus (see figure) had larger fat deposits and were heavier than uninfected animals. Studies in humans show a similar association between infection with the strain of virus—called Ad-36—and an increase in adipose (fat) tissue. Although adenoviruses have usually been involved in respiratory and eye infections, they can also infect adipose cells. One of the possible explanations for this association suggests that a chronic infection with the virus allows its DNA to regulate cellular differentiation of stem cells into adipocytes (fat cells). This increase in both the number and the size of fat cells adds adipose tissue, more fat production and storage, and more body fat. Simultaneously, the adipocytes may also store more sugar, helping to keep blood sugar levels under control and maintaining insulin sensitivity to glucose. In general, such an association does not prove causation, but it certainly warrants additional research.

Using information you have learned about viruses, explain how viruses could become a permanent component of an organism's genetic material. *Answer available on Connect.*

prominent viral infections found only in certain regions of the world (dengue fever, Rift Valley fever, and yellow fever), the total could easily exceed several billion cases each year. Although most viral infections do not result in death, some, such as rabies, HIV, and Ebola, have very high mortality rates, and others (polio, hepatitis) can lead to long-term debility. Continuing research is focused on the connection of viruses to chronic afflictions of unknown cause, such as type 1 diabetes, multiple sclerosis, various cancers, and even conditions such as obesity (**6.1 Secret World of Microbes**).

Because some viral diseases can be life threatening, it is essential to have a correct diagnosis as soon as possible. Obtaining the overall clinical picture of the disease (specific signs) is often the first step in diagnosis. This may be followed by identification of the virus in clinical specimens by means of rapid tests that detect the virus or signs of cytopathic changes in cells or tissues (see CMV herpesvirus, figure 6.15). Immunofluorescence techniques or direct examination with an electron microscope are often used for this (see figure 6.8). Samples can also be screened for the presence of indicator molecules (antigens) from the virus itself. A standard procedure for many viruses is the polymerase chain reaction (PCR), which can detect and amplify minute amounts of viral nucleic acid in a sample.

In certain infections, definitive diagnosis requires cultivation of the virus using cell culture, embryos, or animals, but this method can be time-consuming and slow to give results. Screening tests can detect specific antibodies that indicate signs of virus infection in a patient's blood. This is the main test for HIV infection (see figure 17.16). Additional details of viral diagnosis are provided in chapter 17.

The nature of viruses has at times been a major impediment to effective therapy. Because viruses are not bacteria, antibiotics aimed at bacterial infections do not work for viruses. Although more antiviral drugs are being developed, most of them block virus replication by targeting the function of host cells. This can cause severe side effects. Antiviral drugs are designed to target one of the steps in the viral life cycle you learned about earlier in this chapter. Azidothymidine (AZT), a drug used to treat AIDS, targets the nucleic acid synthesis stage. A different class of HIV drugs, the protease inhibitors, disrupts the final assembly phase of the viral life cycle. Another compound that shows some potential for treating and preventing viral infections is a naturally occurring human cell product called *interferon* (see chapters 12 and 14). Vaccines that stimulate immunity are an extremely valuable tool but are available for only a limited number of viral diseases (see chapter 15).

 Check Your Progress SECTIONS 6.6 AND 6.7

26. Describe the three main techniques for cultivating viruses.

27. What are the advantages of using cell culture, and what are the disadvantages?

28. What are cell lines and monolayers, and how are plaques formed?

29. What are the primary problems in treating viral infections with drugs?

6.8 Prions and Other Nonviral Infectious Particles

 Expected Learning Outcomes

24. Describe the properties of nonviral infectious particles.

25. Discuss the importance of prions and viroids and the diseases they cause.

Prions are a group of noncellular infectious agents that are not viruses and really belong in a category all by themselves. The term *prion* is derived from the words **pr**oteinaceous **in**fectious particle to suggest its primary structure—that of a naked protein molecule. They are quite remarkable in being the only biologically active agent that lacks any sort of nucleic acid (DNA or RNA). Up until their discovery about 35 years ago, it was considered impossible that an agent lacking genetic material could ever be infectious or transmissible.

The diseases associated with prions are known as **transmissible spongiform encephalopathies (TSEs).** This description recognizes that the diseases are spread from host to host by direct contact, contaminated food, or other means. It also refers to the effects of the agent on nervous tissue, which develops a spongelike appearance due to loss of nerve and glial cells (see figure 25.28*b*). Another pathological effect observed in these diseases is the buildup of tiny protein fibrils in the brain tissue (**figures 6.21*a* and 6.21*b*).** Several forms of prion diseases are known in mammals, including scrapie in sheep, bovine spongiform encephalopathy (mad cow disease) in cattle, and wasting disease in elk, deer, and mink. These diseases have a long latent period (usually several years) before the first symptoms of brain degeneration appear. The animals lose

(a)

Prion protein fibers

(b)

① Prion proteins infect a nerve cell.

Prion proteins

② Upon contact with normal proteins, the prions are able to shift the configuration of the normal proteins, converting them to prions.

③ When this process creates large numbers of prions, they bind tightly together and form elongate chains.

④ As the chains build up, they form fibers within the cell, which interfere with the cell's function and destroy it.

Normal proteins

Process Figure 6.21 **The appearance and mechanisms of prion-based diseases. (a)** A magnified view of nerve cells infected with prion protein. Red areas indicate places in the cytoplasm where the prion fibers have deposited. **(b)** A diagram of a nerve cell to outline the stages in the conversion of normal cell proteins into prions and the formation of prion fibers. (a): *Source:* National Institute of Allergy and Infectious Diseases

coordination, have difficulty moving, and eventually progress to collapse and death.

Humans are host to similar chronic diseases, including *Creutzfeldt-Jakob Disease* (CJD), kuru, fatal familial insomnia, and others. In all of these conditions, the brain progressively deteriorates and the patient loses motor coordination, along with sensory and cognitive abilities. There is no treatment and most cases so far have been fatal. In the 1990s, a different variant of CJD that appeared in Europe was traced to people eating meat from cows infected with the bovine form of encephalopathy. Several hundred people developed the disease and a majority of them have since died. This was the first indicator that prions of animals could cause infections in humans. It sparked a crisis in the beef industry and strict controls on imported beef. Although only three cases of human disease have occurred in the United States, infected cattle have been reported.

The medical importance of these novel infectious agents has led to a great deal of research into how they function. Researchers have discovered that prion or prionlike proteins are very common in the cell membranes of plants, yeasts, and animals. One widely accepted theory suggests that the prion protein is an abnormal version of a normal protein. When the prion comes in contact with a normal protein, it can induce spontaneous abnormal folding in the normal protein. Ultimately the buildup of these converted proteins damages and kills the cell (figure 6.21*b*).

Another serious issue with prions is their extreme resistance. They are not destroyed by disinfectants, radiation, or the usual sterilization techniques. Even treatments with extreme high temperatures and concentrated chemicals are not always reliable methods to eliminate them. Additional information on prion diseases can be found in chapter 25.

Other fascinating viruslike agents in human disease are defective forms called satellite viruses that depend on other viruses for replication. Two remarkable examples are the adeno-associated virus (AAV), which can replicate only in cells infected with adenovirus, and the delta agent, a naked strand of RNA that is expressed only in the presence of the hepatitis B virus and can worsen the severity of liver damage.

Plants are also parasitized by viruslike agents called **viroids** that differ from ordinary viruses by being very small (about one-tenth the size of an average virus) and being composed of only naked strands of RNA, lacking a capsid or any other type of coating. The existence of these unusual agents is one bit of supportive evidence that viruses may have evolved from naked bits of nucleic acid. Viroids are significant pathogens in several economically important plants, including tomatoes, potatoes, cucumbers, citrus trees, and chrysanthemums.

Looking Ahead at the Chapters to Come

We have completed our survey of prokaryotes, eukaryotes, and viruses and have described the significant characteristics of these three groups. Now that we know more about these microbes, we will use chapters 7, 8, and 9 to explore how they maintain themselves, beginning with microbial nutrition (chapter 7) followed by metabolism (chapter 8), and ending with genetics (chapter 9).

Check Your Progress SECTION 6.8

30. What is the principal effect of the prion agent of Creutzfeldt-Jakob disease?

31. Explain how prions and viroids are different from viruses.

CASE STUDY Part 2

Avian influenza viruses are found worldwide in both wild and domestic birds, and are generally adapted to migratory waterfowl and shorebird species. The birds usually suffer no ill effects as a result of infection, but they do excrete the virus in their feces and through secretions from their nose, mouth, and eyes. As the birds travel along major migratory pathways from the northernmost reaches of Canada to South America, they carry the virus with them, creating opportunities for viral exchange as they commingle with domestic flocks. Recommendations made by the U.S. Department of Agriculture to reduce avian influenza outbreaks include eliminating pools of standing water that attract wild birds, reducing food sources, and installing wildlife deterrents like scarecrows, all actions meant to minimize contact between wild and domestic flocks. Unfortunately, even the best attempts at eliminating interactions between domestic and wild fowl are not completely effective.

Most migration corridors are restricted to continental landmasses; birds rarely cross oceans. However, the area between the eastern edge of Russia and western Alaska is uniquely an area where major migratory pathways cross continental boundaries. The East Asian–Australasian Flyway extends from Australia north through Asia and Russia, and east to Alaska, where it intersects the Mississippi Americas Flyway, which descends south through Canada, the United States, Mexico, and Central and South America. This overlap allows birds from Asia to come into direct contact with North American birds, establishing a path for viral transmission across continents.

The Asian connection in this case was especially interesting when a highly pathogenic avian influenza virus of Asian origin, HPAI H5N8, spread rapidly along wild-bird migratory pathways during 2014 and was found in a state along the Mississippi river in 2015. The H5N8 virus then mixed with North American avian influenza viruses, creating a new, mixed-origin virus, HPAI H5N2. This mixed-origin virus contains the Asian-origin H5 part of the virus, which is highly pathogenic to poultry. The N parts of the virus came from North American low-pathogenic avian influenza viruses.

■ *All influenza viruses contain both H and N proteins. Why is HPAI H5N2 referred to as a "mixed origin" virus?*

■ *What are "sentinel chickens"? (A trip to the Internet may be useful.)*

See chapter 25 for additional details on influenza and track the latest information on flu outbreaks at www.cdc.gov/flu/.

To conclude this Case Study, go to Connect.

Chapter Summary with Key Terms

6.1 Overview of Viruses

A. Viruses, being much smaller than bacteria, fungi, and protozoa, had to be indirectly studied until the 20th century, when they were finally seen with an electron microscope.

B. Scientists don't agree about whether viruses are living or not, but it is clear they do not fulfill all of the requirements of life. They are obligate intracellular and genetic parasites.

6.2 The General Structure of Viruses

A. Viruses are infectious particles and not cells; lack organelles and locomotion of any kind; are large, complex molecules; and can be crystalline in form. A virus particle is composed of a **nucleic acid** core (DNA or RNA, not both) surrounded by a geometric protein shell, or **capsid;** the combination is called a **nucleocapsid;** a capsid is **helical** or **icosahedral** in configuration; many are covered by a membranous **envelope** containing viral protein **spikes.** Complex viruses have additional external and internal structures.

B. *Shapes/Sizes:* Icosahedral, helical, spherical, and cylindrical shaped. Smallest infectious forms range from the largest poxvirus (0.45 mm or 450 nm) to the smallest viruses (0.02 mm or 20 nm).

C. *Nutritional and Other Requirements:* Lack enzymes for processing food or generating energy; are tied entirely to the host cell for all needs (obligate intracellular parasites).

D. Viruses are known to parasitize all types of cells, including bacteria, algae, fungi, protozoa, animals, and plants. Each viral type is limited in its host range to a single species or group, mostly due to specificity of adsorption of virus to specific host receptors.

6.3 How Viruses Are Classified and Named

A. The two major types of viruses are **DNA** and **RNA viruses**. These are further subdivided into families, depending on shape and size of capsid, presence or absence of an envelope, whether double- or single-stranded nucleic acid, antigenic similarities, and host cell.

B. Viruses are classified into orders, families, and genera. These groupings are based on virus structure, chemical composition, and genetic makeup.

6.4 Modes of Viral Multiplication

A. *Multiplication Cycle: Animal Cells*

B. The life cycle steps of an animal virus are **adsorption, penetration, synthesis** and **assembly,** and **release** from the host cell. A fully infective virus particle is a **virion.**

C. Some animal viruses cause chronic and **persistent infections.**

D. Viruses that alter host genetic material may cause *oncogenic* effects.

6.5 The Multiplication Cycle in Bacteriophages

A. **Bacteriophages** are viruses that attack bacteria. They penetrate by injecting their nucleic acid and are released as virulent phages upon **lysis** of the cell.

B. Some viruses go into a **latent,** or **lysogenic,** phase in which they integrate into the DNA of the host cell and later may be active and produce a lytic infection.

6.6 Techniques in Cultivating and Identifying Animal Viruses

A. The need for an intracellular habitat makes it necessary to grow viruses in living cells, either in **cell culture** (cultures of live, isolated host cells), in bird embryos, or in the intact host animal.

B. *Identification:* Viruses are identified by means of **cytopathic effects** (CPE) in host cells, direct examination of viruses or their components in samples, genetic analysis to detect virus nucleic acid, and growing viruses in culture.

6.7 Viral Infection, Detection, and Treatment

A. *Medical:* Viruses attach to specific target hosts or cells. They cause a variety of infectious diseases, ranging from mild respiratory illnesses (common cold) to destructive and potentially fatal conditions (rabies, AIDS). Some viruses can cause birth defects and cancer in humans and other animals.

B. *Research:* Viruses have become an invaluable tool for studying basic genetic principles. Current research is focused on the connection of viruses to afflictions of unknown causes, such as type I diabetes and multiple sclerosis.

C. Viral infections are detected by direct examination of specimens, genetic tests, testing patients' blood, and characteristic symptoms. Some viral infections can be treated with drugs that block viral replication.

6.8 Prions and Other Nonviral Infectious Particles

A. Spongiform encephalopathies are chronic persistent neurological diseases caused by **prions.**

B. Prions are infectious agents composed of a protein that can alter the structure of nerve cells.

C. Examples of neurological diseases include "mad cow disease" and Creutzfeldt-Jakob disease.

D. Other noncellular infectious agents include satellite viruses and **viroids.**

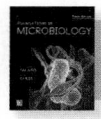

Level I. Knowledge and Comprehension

These questions require a working knowledge of the concepts in the chapter and the ability to recall and understand the information you have studied.

 Multiple-Choice Questions

Select the correct answer from the answers provided. For questions with blanks, choose the combination of answers that most accurately completes the statement.

1. A virus is a tiny infectious
 a. cell
 b. living thing
 c. particle
 d. nucleic acid

2. Viruses are known to infect
 a. plants
 b. bacteria
 c. fungi
 d. all organisms

3. The capsid is composed of protein subunits called
 a. spikes
 b. protomers
 c. virions
 d. capsomers

4. The envelope of an animal virus is derived from the _____ of its host cell.
 a. cell wall
 b. membrane
 c. glycocalyx
 d. receptors

5. The nucleic acid of a virus is
 a. DNA only
 b. RNA only
 c. both DNA and RNA
 d. either DNA or RNA

6. The general steps in a viral multiplication cycle are
 a. adsorption, penetration, synthesis, assembly, and release
 b. endocytosis, replication, assembly, and budding
 c. adsorption, duplication, assembly, and lysis
 d. endocytosis, penetration, replication, maturation, and exocytosis

7. A prophage is a/an _____ stage in the cycle of _____.
 a. latent, bacterial viruses
 b. infective, RNA viruses
 c. early, poxviruses
 d. late, enveloped viruses

8. The nucleic acid of animal viruses enters the host cell through
 a. injection
 b. fusion
 c. endocytosis
 d. b and c

9. In general, RNA viruses multiply in the cell _____, and DNA viruses multiply in the cell _____.
 a. nucleus, cytoplasm
 b. cytoplasm, nucleus
 c. vesicles, ribosomes
 d. endoplasmic reticulum, nucleolus

10. Enveloped viruses carry surface receptors called
 a. buds
 b. spikes
 c. fibers
 d. sheaths

11. Viruses that persist in the cell and cause recurrent disease are considered
 a. oncogenic
 b. cytopathic
 c. latent
 d. resistant

12. Viruses cannot be cultivated in
 a. cell culture
 b. bird embryos
 c. live mammals
 d. blood agar

13. Clear patches in cell cultures that indicate sites of virus infection are called
 a. plaques
 b. pocks
 c. colonies
 d. prions

14. Which of these is *not* a general pattern of virus morphology?
 a. enveloped, helical
 b. naked, icosahedral
 c. enveloped, icosahedral
 d. complex, helical

15. Which of these is true of prions?
 a. They are small RNA viruses.
 b. They replicate in the nucleus.
 c. They lack protein.
 d. They cause death of brain cells.

 ## Case Study Review

1. Which receptor of the influenza virus is most involved in binding to the respiratory cells?
 a. neuraminidase spikes
 b. hemagglutinin spikes
 c. the viral capsid
 d. the viral envelope

2. While H5N2 is of great concern economically, it is not considered an imminent threat to human health. This is mostly because _____.
 a. if birds are properly cooked the virus cannot spread
 b. the virus moves poorly from bird to bird
 c. the virus cause only mild disease in most birds
 d. the virus moves poorly from bird to human and from human to human

3. The standard response to an outbreak of avian influenza consists of five steps: (1) Quarantine, (2) Eradicate, (3) Monitor, (4) Disinfect, and (5) Test. Explain the importance of each step in stopping the spread of the virus.

 ## Writing Challenge

For each question, compose a one- or two-paragraph answer that includes the factual information needed to completely address the question. Check Your Progress questions can also be used for writing-challenge exercises.

1. a. What characteristics of viruses could be used to describe them as life forms?
 b. What makes them more similar to lifeless molecules?

2. HIV attacks only specific types of human cells, such as certain white blood cells and nerve cells. Can you explain why a virus can enter some types of human cells but not others?

3. a. Since viruses lack metabolic enzymes, how can they synthesize necessary components?
 b. Name some enzymes that viruses may have for invading and completing their cycles.

4. a. What dictates the host range of animal viruses?
 b. How is the influenza virus different in its host range from most other viruses?

 ## Concept Mapping

On Connect you can find an Introduction to Concept Mapping that provides guidance for working with concept maps along with concept-mapping activities for this chapter.

Level II. Application, Analysis, Evaluation, and Synthesis

These problems go beyond just restating facts and require higher levels of understanding and an ability to interpret, problem solve, transfer knowledge to new situations, create models, and predict outcomes.

Critical Thinking

Critical thinking is the ability to reason and solve problems using facts and concepts. These questions can be approached from a number of angles, and in most cases, they do not have a single correct answer.

1. Comment on the possible origin of viruses. Is it not curious that the human cell welcomes a virus in and hospitably removes its coat as if it were an old acquaintance?

2. a. If viruses that normally form envelopes were prevented from budding, would they still be infectious? Why or why not?
 b. If only the RNA of an influenza virus were injected into a cell by itself, could it cause a lytic infection?

3. The end result of most viral infections is death of the host cell.
 a. If this is the case, how can we account for such differences in the damage that viruses do? (Compare the effects of the cold virus with those of the rabies virus.)
 b. Describe the adaptations of viruses that do not immediately kill the host cell and explain what their functions may be.

4. a. Given that DNA viruses can be carried in the DNA of the host cell's chromosomes, comment on what this phenomenon means in terms of inheritance in the offspring.

 b. Discuss the connection between viruses and cancers, giving possible mechanisms for viruses that cause cancer.

5. a. If you were involved in developing an antiviral drug, what would be some important considerations? (Can a drug "kill" a virus?)
 b. How could multiplication be blocked?

6. Is there such a thing as a "good virus"? Explain why or why not. Consider both bacteriophages and viruses of eukaryotic organisms.

7. Discuss some advantages and disadvantages of bacteriophage therapy in treating bacterial infections.

8. a. Consult table 6.2, right-hand column, to determine which viral diseases you have had and which ones you have been vaccinated against.
 b. Which viruses would more likely be possible oncoviruses? Why?

9. Of the following, list those that are viral infections: cholera, rabies, plague, cold sores, whooping cough, tetanus, genital warts, gonorrhea, mumps, Rocky Mountain spotted fever, syphilis, rubella, rat bite fever.

Visual Challenge

1. Label the parts of viruses in figures 6.7*d*, 6.9*c*, and 6.10.

2. How would you describe the kind of capsid found on this virus from figure 6.2? In what ways is this virus different from other viruses?

Courtesy of C. Abergel/IGS-IMM, CNRS-AMU

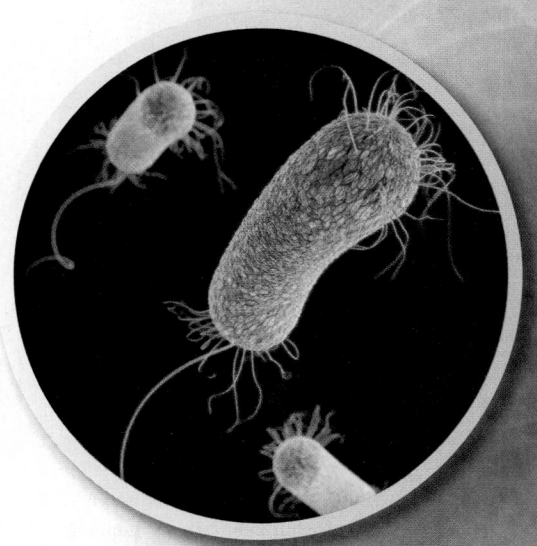

Pseudomonas aeruginosa has found its perfect niche in the lungs of CF patients. The fine whiskers, or fimbriae, help it attach to the lung tissue. *Source:* U.S. Centers for Disease Control and Prevention - Medical Illustrator/CDC

An X-ray view of the lungs of a cystic fibrosis (CF) patient with pattern of diffuse gray masses, evidence of the thick biofilms entrenched there. *Source:* National Human Genome Research Institute

CASE STUDY Part 1 A Creature of Habitat

Molly Kenner had suffered from **persistent respiratory symptoms** since she was a young child. Her family physician originally diagnosed her with **asthma,** complicated by **bronchitis** and **pneumonia,** and prescribed anti-allergy medicines, antibiotics, and steroid inhalers. This therapy worked for a few months, but eventually her symptoms of **coughing, chest pain,** and **congestion** would return. On many occasions she was ill enough to require hospitalization and a breathing machine. This cycle went on for several years. By the time she was in her twenties, her lungs showed signs of permanent damage and she often struggled to breathe.

As her quality of life continued to decline, her family and friends urged her to seek more expert medical **intervention** with a **lung specialist** at a university teaching hospital. Based on her history, a physician at the clinic ordered a **lung biopsy** and **chest X-rays.** Samples of her lung tissues grew a pure culture of ***Pseudomonas aeruginosa,*** described as gram-negative bacillus with a single flagellum and noticeable blue-green pigment. The species is extremely **flexible** in its environmental and nutritional requirements. Although its usual **habitat** is soil, water, and plant matter, it is frequently isolated from skin and other regions of the human body. It occasionally causes infections, especially in those with **weakened host defenses,** but it is not generally pathogenic to healthy individuals.

Molly's doctor immediately understood the significance of this finding, because the lungs are ordinarily supposed to be **sterile** and do not harbor a **resident microbial population.** Yet here was direct

> "Samples of her lung tissues grew a pure culture of *Pseudomonas aeruginosa.*"

evidence of a **massive infection** developing as large **nodules** and patches throughout the lungs. The combination of her history and the isolation of this particular species were powerful clues in understanding the real basis of Molly's medical condition. The specialist immediately ordered a **genetic test** that came back as positive for a disease called **cystic fibrosis (CF).** Cystic fibrosis is a **recessive disorder** caused by a **mutation** in a gene called the **CFTR** or cystic fibrosis transmembrane receptor gene. The normal, nonmutated gene codes for a protein that controls the **active transport** of **chloride** across epithelial cells. Chloride helps to maintain the liquid content of the **mucous** coating of lung membranes needed by the **natural host defenses** to maintain lung health. Because Molly inherited a **defective CFTR gene** from each parent, her cells **cannot secrete** a sufficient amount of chloride into the lung tissues. As a result, the mucus becomes thick and tenacious and creates a specialized habitat that supports the **long-term establishment** of *P. aeruginosa.*

- *What terms from the chapter would you use to describe the natural habitat, biochemical characteristics, and ecological niche of Pseudomonas aeruginosa?*

- *What sorts of adjustments would P. aeruginosa have to make in adapting successfully to the human lung?*

To continue the Case Study, go to Case Study Part 2 at the end of the chapter.

SCOPING OUT THE CHAPTER

The earth is home to an incredibly complex web of organisms that is constantly interacting with each other as well as with the environment. This network is ubiquitous and inseparable and has existed in some form for billions of years. It has shaped the very structure and function of the earth. In this chapter we are introduced to the endless ways that microbes make a living, including nutrition and energy sources, adaptations, interrelationships, and growth.

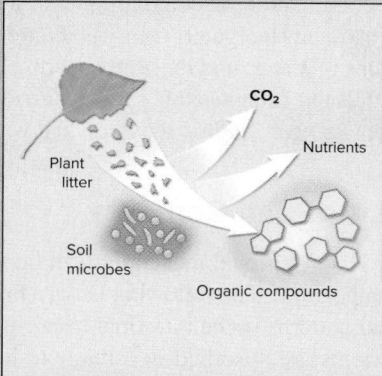

How Microbes Get Nourishment
Just like their larger counterparts, microorganisms must absorb a broad array of chemical substances from their immediate environments and convert them to their own particular forms of molecules.

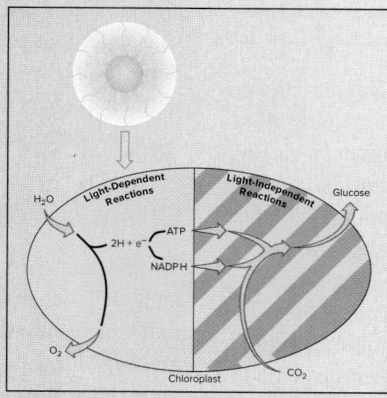

What Drives Microbial Engines?
Microbes require a usable form of energy (sunlight is one example) to power metabolic activities such as synthesis, growth, and transport.

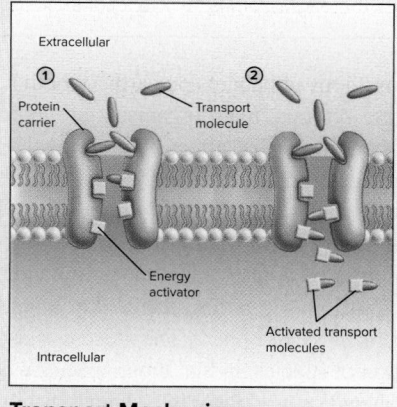

Transport Mechanisms
Many microbial activities are dependent on the movement of materials into and out of their cells. Membranes have very specialized passageways for carrying out this process.

Adaptations to Habitat and Niche
Microbes set up residence in almost every nook and cranny of the earth. Wherever they end up, they experience a range of conditions that favors their survival (for example, this little pink algae that thrives in glaciers).

Abundant Partnerships
Most microorganisms live closely with other organisms, often forming long term associations such as symbiosis, parasitism, and biofilms.

Expanding the Population
As is true of all organisms, microbes must produce offspring to maintain their lineage. A common way they can rapidly increase their numbers is for cells to undergo numerous, repeated cycles of division.

(bottom: left, all): Images courtesy of Nozomu Takeuchi

7.1 Microbial Nutrition

Expected Learning Outcomes

1. Describe the major environmental factors to which microbes must adapt for survival.
2. Define *nutrition* and *nutrients* and their subcategories based on need and quantity.
3. Differentiate between organic and inorganic nutrients.
4. Discuss the origins, types, and functions of bioelements and nutrients.

There are millions of habitats on earth, of both natural and human origin. In these settings, microorganisms are exposed to a tremendous variety of conditions that affect their survival. Environmental factors with the greatest impact on microorganisms are nutrient and energy sources, temperature, gas content, water, salt, pH, radiation, and other organisms (**figure 7.1**). Microbes survive in their habitats through the process of gradual adjustment of anatomy and physiology, a process called **adaptation.*** It is this adaptability that allows microbes to inhabit all parts of the biosphere. The process is also a major force behind the evolution of species, in that those species with greater adaptability will survive and pass on favorable characteristics to their offspring.

Microbial **nutrition** is a complex activity through which microbes acquire chemical compounds from the environment in order to sustain life. These compounds, called **nutrients,** are absorbed, assimilated, and subsequently used for cellular metabolism, growth, energy, and many other functions. With respect to their nutrition, microbes are not really so different from humans and other macroorganisms. Although bacteria living deep in a swamp on a diet of inorganic sulfur, or protozoa digesting wood in a termite's intestine, may seem to show radical adaptations, even these organisms require a constant influx of certain specific substances from their habitat.

In general, all living things have an absolute need for the **bioelements,** traditionally listed as carbon, hydrogen, oxygen, phosphorus, potassium, nitrogen, sulfur, calcium, iron, sodium, chlorine, magnesium, and a few other other elements.[1] Beyond these basic requirements, microbes show significant differences in the source, chemical form, and amount of the elements they use. Any substance, whether an element or compound, that an organism must get from a source outside its cells is called an **essential nutrient.** Once absorbed, these nutrients are processed and transformed into the chemicals of the cell.

Two categories of essential nutrients are **macronutrients** and **micronutrients.** Macronutrients are required in relatively large quantities and play principal roles in cell structure and metabolism. Examples of macronutrients are compounds such as sugars and amino acids that contain carbon, hydrogen, and oxygen. Micronutrients, or **trace elements,** such as manganese, zinc, and nickel are present in much smaller amounts and are involved in enzyme function and maintenance of protein structure. What constitutes a micronutrient can vary from one microbe to another and often must be

determined in the laboratory. This determination is made by deliberately omitting the substance in question from a growth medium to see if the microbe can grow in its absence.

Another way to categorize nutrients is according to their carbon content. Most **organic** nutrients are molecules that contain a basic framework of carbon and hydrogen. Natural organic molecules are nearly always the products of living things. They range from the simplest organic molecule, methane (CH_4), to large polymers (carbohydrates, lipids, proteins, and nucleic acids).

In contrast, an **inorganic** nutrient is composed of an element or elements lacking some combination of both carbon and hydrogen. The natural reservoirs of many inorganic compounds are mineral deposits in the soil, bodies of water, and the atmosphere. Examples include metals and their salts (magnesium sulfate, ferric nitrate, sodium phosphate), gases (oxygen, carbon dioxide), and water.

Chemical Analysis of Cell Contents

To gain insight into a cell's nutritional makeup, it can be useful to analyze its chemical composition. The following list is a brief summary of some nutritional patterns in the intestinal bacterium *Escherichia coli*. Some nutrients are absorbed in a ready-to-use form, and others must be synthesized by the cell from simple compounds:

- Water content is the highest of all components (70%).
- About 97% of the dry cell weight is composed of organic compounds.
- Proteins are the most prevalent organic compound.
- About 96% of the cell is composed of six elements (represented by CHONPS).
- Chemical elements are needed in the overall scheme of cell growth, but most of them are available to the cell as compounds and not as pure elements.
- A cell as "simple" as *E. coli* contains on the order of 5,000 different compounds, yet it needs to absorb only a few types of nutrients to synthesize this great diversity. These include $(NH_4)_2SO_4$, $FeCl_2$, $NaCl$, trace elements, glucose, KH_2PO_4, $MgSO_4$, $CaHPO_4$, and water.

Escherichia coli

Forms, Sources, and Functions of Essential Nutrients

The elements that comprise nutrients ultimately exist in an environmental inorganic reservoir of some type. These reservoirs not only serve as a source of these elements but can be replenished by the activities of organisms. Thus, elements cycle in a pattern from an inorganic form in an environmental reservoir to an organic form in organisms. Organisms may, in turn, serve as a continuing source of that element for other organisms. Eventually the element is recycled to the inorganic form, usually by the actions of microorganisms. All in all, a tremendous variety of microorganisms are involved in processing the elements. From a larger perspective, the overall cycle of life on this planet depends upon the combined

* *adaptation* L. *adaptare,* to fit. Changes in structure and function that improve an organism's survival in a given environment.

1. See critical thinking question 1. b for a useful mnemonic device to recall the essential elements.

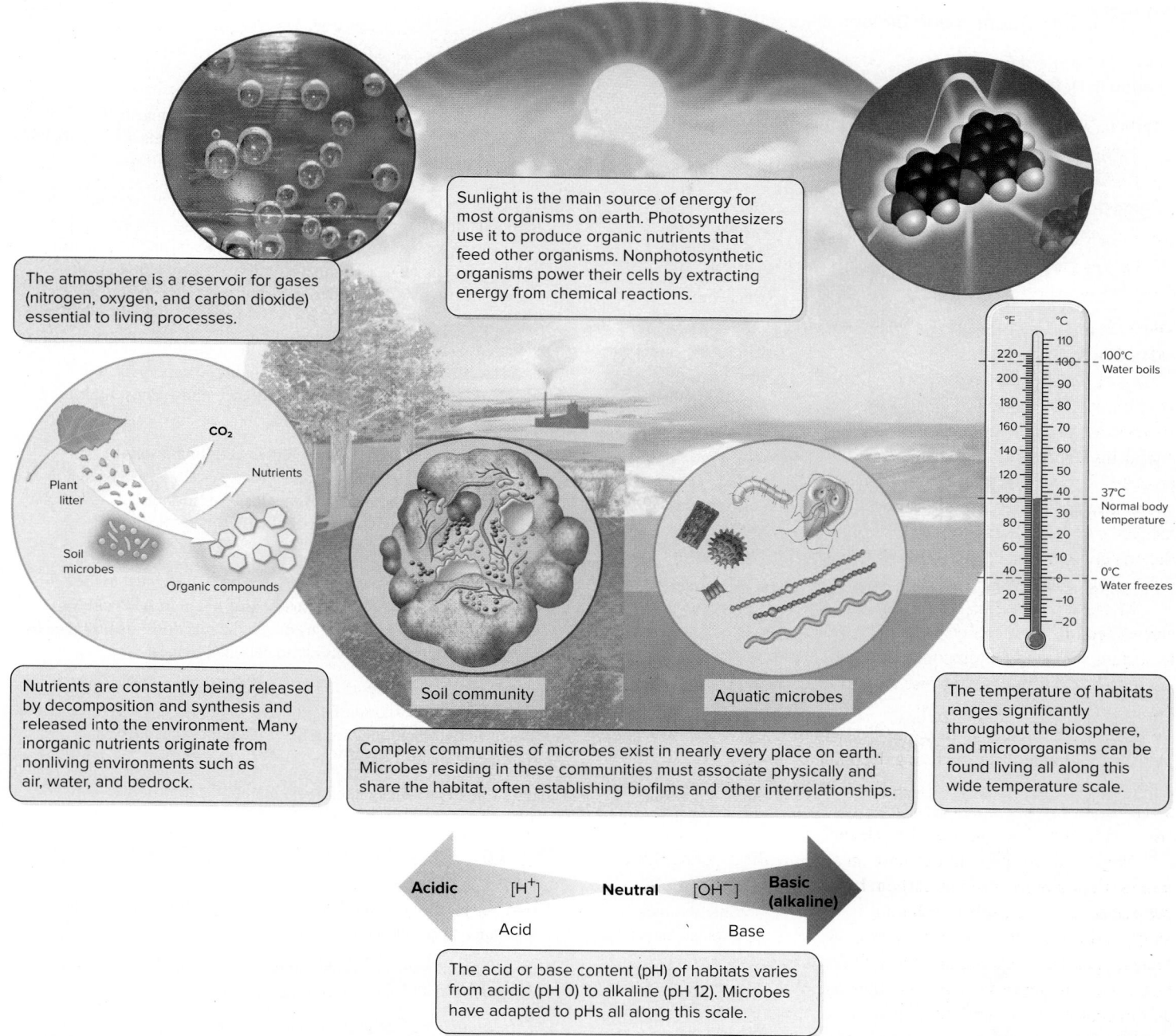

The atmosphere is a reservoir for gases (nitrogen, oxygen, and carbon dioxide) essential to living processes.

Sunlight is the main source of energy for most organisms on earth. Photosynthesizers use it to produce organic nutrients that feed other organisms. Nonphotosynthetic organisms power their cells by extracting energy from chemical reactions.

Nutrients are constantly being released by decomposition and synthesis and released into the environment. Many inorganic nutrients originate from nonliving environments such as air, water, and bedrock.

Soil community

Aquatic microbes

Complex communities of microbes exist in nearly every place on earth. Microbes residing in these communities must associate physically and share the habitat, often establishing biofilms and other interrelationships.

The temperature of habitats ranges significantly throughout the biosphere, and microorganisms can be found living all along this wide temperature scale.

Acidic [H⁺] Neutral [OH⁻] Basic (alkaline)

Acid Base

The acid or base content (pH) of habitats varies from acidic (pH 0) to alkaline (pH 12). Microbes have adapted to pHs all along this scale.

Figure 7.1 **Environmental conditions that influence microbial adaptations.** The earth's habitats provide a constant supply of energy, nutrients, and gases; maintain a certain pH and temperature; and establish communities of interactive organisms. (photo, top left): © K. Talaro

action of several interacting nutritional schemes, each performing a necessary step. In fact, as we shall see in a later chapter, the framework of microbial nutrition underlies the nutrient cycles of all life on earth.

The source of nutrients is extremely varied: Microbes such as photosynthetic bacteria obtain their nutrients entirely in inorganic form from the environment. Others require a combination of organic and inorganic nutrients. For example, parasites that invade and live on the human body derive all essential nutrients from host tissues, tissue fluids, secretions, and wastes.

Refer to **table 7.1** to see an overview of the major bioelements, compounds, their sources, and their importance to microorganisms.

Carbon-Based Nutritional Types

The element carbon is so key to the structure and metabolism of all life forms that the source of carbon defines two basic nutritional groups:

• A **heterotroph*** is an organism that must obtain its carbon in an organic form. Because most organic carbon originates from organisms, heterotrophs are nutritionally dependent on other life forms. Among the common organic molecules that can satisfy this requirement are proteins, carbohydrates, lipids, and nucleic acids. In most cases, these nutrients provide several

* *heterotroph* (het′-uhr-oh-trohf) Gr. *hetero,* other, and *troph,* to feed.

TABLE 7.1 Sources and Biological Functions of Essential Elements and Nutrients

Element/Nutrient Forms Found in Nature		Sources/Reservoirs of Compounds	Significance to Cells
Carbon	CO_2 (carbon dioxide) gas CO_3^{2-} (carbonate) Organic compounds	Air (0.036%*) Sediments/soils Living things	CO_2 is produced by respiration and used in photosynthesis; CO_3^{2-} is found in cell walls and skeletons; organic compounds are essential to the structure and function of all organisms and viruses.
Nitrogen	N_2 gas NO_3^- (nitrate) NO_2^- (nitrite) NH_3 (ammonium) Organic nitrogen (proteins, nucleic acids)	Air (79%*) Soil and water Soil and water Soil and water Organisms	Nitrogen gas is available only to certain microbes that fix it into other inorganic nitrogen compounds—nitrates, nitrites, and ammonium—the primary sources of nitrogen for algae, plants, and the majority of bacteria; animals and protozoa require organic nitrogen; all organisms use NH_3 to synthesize amino acids and nucleic acids.
Oxygen	O_2 gas Oxides H_2O	Air (20%*), a major product of photosynthesis Soil	Oxygen gas is necessary for the metabolism of nutrients by aerobes. Oxygen is a significant element in organic compounds and inorganic compounds (see water, sulfates, phosphates, nitrates, carbon dioxide in this table).
Hydrogen	H_2 gas H_2O H_2S (hydrogen sulfide) CH_4 (methane) Organic compounds	Waters, swamps, mud Volcanoes, vents Swamps organisms	Water is the most abundant compound in cells and a solvent for metabolic reactions; H_2, H_2S, and CH_4 gases are produced and used by bacteria and archaea; H^+ ions are the basis for transfers of cellular energy and help maintain the pH of cells.
Phosphorus	PO_4^{3-} (phosphate)	Rocks, mineral deposits Soil	Phosphate, a key component of DNA and RNA, is critical to the genetic makeup of cells and viruses; also found in ATP and NAD, where it takes part in numerous metabolic reactions; its presence in phospholipids provides stability to cell membranes.
Sulfur	S SO_4^{2-} (sulfate) SH (sulfhydryl)	Mineral deposits, volcanic sediments Soil	Elemental sulfur (S) is oxidized by some bacteria as an energy source; sulfur is found in vitamin B1; sulfhydryl groups are part of certain amino acids, where they form disulfide bonds that shape and stabilize proteins.
Potassium	K^+	Mineral deposits, ocean water, soil	Plays a role in protein synthesis and membrane transport
Sodium	Na^+	Same as potassium	Major participant in membrane actions; maintains osmotic pressure in cells
Calcium	Ca^+	Oceanic sediments, rocks, and soil	A component of protozoan shells (as $CaCO_3$); stabilizes cell walls; adds resistance to bacterial endospores
Magnesium	Mg^{2+}	Geologic sediments, rocks, and soil	A central atom in the chlorophyll molecule; required for function of membranes, ribosomes, and some enzymes
Chloride	Cl^-	Ocean water, salt lakes	May function in membrane transport; required by obligate halophiles to regulate osmotic pressure
Zinc	Zn^{2+}	Rocks, soil	An enzyme cofactor; regulates eukaryotic genetics
Iron	Fe^{2+}	Rocks, soil	Essential element for the structure of respiratory proteins (cytochromes)
Micronutrients: copper, cobalt, nickel, molybdenum, manganese, iodine		Geologic sediments, soil	Required in tiny amounts to serve as cofactors in specialized enzyme systems of some microbes but not all

*As a portion of the earth's atmosphere.

other elements as well. Some organic nutrients (such as **mono-saccharides** and amino acids) already exist in a form that is simple enough for absorption, but many larger molecules must be digested by the cell before absorption. Not all heterotrophs can use the same organic carbon sources. Some are restricted to a few substrates, whereas others (certain *Pseudomonas* bacteria, for example) are so versatile that they can metabolize hundreds of different substrates.

- An **autotroph*** is an organism that uses inorganic CO_2 as its carbon source. Because autotrophs have the special capacity to convert CO_2 into organic compounds, they are not nutritionally dependent on other living things. We later enlarge on the topic of nutritional types based on carbon and energy sources.

* *autotroph* (aw'-toh-trohf) Gr. *auto,* self, and *troph,* to feed.

Carbon Source Versus Carbon Function

It may be helpful to clear up any confusion about the *source* of carbon as opposed to the *intracellular function* of carbon compounds. Although a distinction is made between the form of carbon compounds that cells absorb as nutrients (inorganic or organic), the majority of carbon compounds involved in the normal structure and metabolism of all cells are organic. So, even microbes that use carbon dioxide as their primary source of carbon will convert this CO_2 into organic compounds once it gets into the cell.

Growth Factors: Essential Organic Nutrients

Many **fastidious** bacteria are not as versatile as *E.coli* and lack the genetic and metabolic mechanisms to synthesize every organic compound they need for survival. An essential nutrient such as an amino acid, nitrogenous base, or vitamin that cannot be synthesized by an organism and must be provided as a nutrient is considered a **growth factor.** For example, all cells require 20 different amino acids for proper assembly of proteins, but many cells cannot synthesize all of them. Those that must be obtained from food are called essential amino acids. A notable example of the need for growth factors occurs in *Haemophilus influenzae,* a bacterium that causes meningitis and respiratory infections in humans. It can grow only when hemin (factor X), NAD^+ (factor V), thiamine and pantothenic acid (vitamins), uracil, and cysteine are provided by another organism or a growth medium.

 Check Your Progress SECTION 7.1

1. Summarize the main environmental factors that microbes encounter and their basic characteristics.
2. Differentiate between micronutrients, macronutrients, and essential nutrients.
3. List the general functions of the essential bioelements in the cell.
4. Define growth factors and metallic ions with examples, and explain their functions in cells.

7.2 Classification of Nutritional Types

 Expected Learning Outcomes

5. Describe the main categories of nutritional types among organisms.
6. Distinguish different types of autotrophs and their energy sources.
7. Distinguish different types of heterotrophs and their energy sources.

The earth's limitless habitats and microbial adaptations are matched by an elaborate menu of microbial nutritional schemes. Fortunately, most organisms show consistent trends and can be described by a few general categories (**table 7.2**) and a few selected terms (**7.1 Making Connections**).

The main determinants of a microbe's nutritional type are its sources of carbon and energy. We saw earlier that microbes can be defined according to their carbon sources as autotrophs or heterotrophs. Another system classifies them according to their energy source as

TABLE 7.2	**Nutritional Categories of Microbes by Energy and Carbon Source**	
Category/ Carbon Source	**Energy Source**	**Example**
Autotroph/CO_2	**Nonliving Environment**	
Photoautotroph	Sunlight	Photosynthetic organisms, such as algae, plants, cyanobacteria
Chemoautotroph	Simple inorganic chemicals	Only certain bacteria and archaea, such as methanogens and deep-sea vent bacteria
Heterotroph/ Organic Carbon	**Other Organisms or Sunlight**	
Chemoheterotroph	Metabolic conversion of the nutrients from other organisms	Protozoa, fungi, many bacteria, animals
1. Saprobe	Metabolize the organic matter from dead organisms	Fungi, bacteria, some protozoa (decomposers)
2. Symbiotic microbes	Obtain organic matter from living organisms	Parasites, commensals, mutualistic microbes
Photoheterotroph	Sunlight or organic matter	Purple and green photosynthetic bacteria

phototrophs or **chemotrophs.** Microbes that photosynthesize (acquire energy from light) are phototrophs, and those that gain energy from chemical compounds are chemotrophs. The terms for carbon and energy source are often merged into a single word for convenience (as in table 7.2). The categories described here are meant to cover only the major nutritional groups and do not include unusual exceptions.

Autotrophs and Their Energy Sources

Autotrophs derive energy from one of two possible nonliving sources: sunlight and chemical reactions involving simple chemicals.

Photoautotrophs and Photosynthesis **Photoautotrophs** are photosynthetic; that is, they capture the energy of light rays and transform it into chemical energy that can be used in cell metabolism. In general, photosynthesis relies on special pigments to collect the light and uses the energy to convert CO_2 into simple organic compounds. There are two major variations in the mechanisms of photosynthesis.

Oxygenic (oxygen-producing) photosynthesis can be summed up by the equation:

$$CO_2 + H_2O \xrightarrow[\text{by chlorophyll}]{\text{Sunlight absorbed}} (CH_2O)_n + O_2$$

in which $(CH_2O)_n$ is shorthand for a carbohydrate. This type of photosynthesis occurs in plants, algae, and cyanobacteria and uses chlorophyll as the primary pigment. Carbohydrates formed by the reaction can be used by the cell to synthesize other cell components.

7.1 MAKING CONNECTIONS

A Guide to Terminology

Much of the vocabulary for describing microbial adaptations is based on some common root words. These are combined in various ways that assist in discussing the types of nutritional or ecological adaptations, as shown in the list below.

Modifier terms are also used to specify the nature of an organism's adaptations. **Obligate** or **strict** means being confined to a narrow niche or habitat, such as an obligate thermophile that requires high temperatures to grow. By contrast, **facultative*** means not being so restricted but adapting to a wider range of environmental conditions. A facultative halophile can grow with or without high salt concentrations. **Tolerance** is another term that can be used to describe the capacity to survive a range of conditions.

Using this guide, define obligate halophile, facultative psychrophile, and aerotolerant anaerobe. *Answer available on the Connect website.*

* *facultative* (fak′-uhl-tay-tiv) Gr. *facult,* capability or skill, and *tatos,* most.

Root	Meaning	Example of Use
troph-	Food, nourishment	Trophozoite—the feeding stage of protozoa
-phile	To love	Extremophile—an organism that has adapted to ("loves") extreme environments
-obe	To live	Microbe—to "live small"
hetero-	Other	Heterotroph—an organism that requires nutrients from other organisms
auto-	Self	Autotroph—an organism that "feeds by itself"
photo-	Light	Phototroph—an organism that uses light as an energy source
chemo-	Chemical	Chemotroph—an organism that uses chemicals rather than light for energy
sapro-	Rotten	Saprobe—an organism that lives on dead organic matter
halo-	Salt	Halophile—an organism that can grow in high-salt environments
thermo-	Heat	Thermophile—an organism that grows at high temperatures
psychro-	Cold	Psychrophile—an organism that grows at cold temperatures
aero-	Air (O_2)	Aerobe—an organism that uses oxygen in metabolism

This topic is covered in greater depth in chapter 8. Because these organisms are the primary producers in most ecosystems, they constitute the basis of food chains by providing nutrition for heterotrophs. This type of photosynthesis is also responsible for maintaining the level of oxygen gas in the atmosphere that is so vital to many organisms.

The other form of photosynthesis is termed **anoxygenic** (no oxygen produced). It may be summarized by the equation:

$$CO_2 + H_2S \xrightarrow[\text{by bacteriochlorophyll}]{\text{Sunlight absorbed}} (CH_2O)_n + S^0 + H_2O$$

Note that this type of photosynthesis is different in several respects. (1) It uses a unique pigment, bacteriochlorophyll; (2) its hydrogen source is hydrogen sulfide gas; (3) it gives off elemental sulfur as one product, and (4) the reactions all occur in the absence of oxygen. Common groups of photosynthetic bacteria are the purple and green sulfur bacteria that live in various aquatic habitats, often in mixtures with other photosynthetic microbes (see figure 4.32).

Chemoautotrophy—A Radical Existence Compared to common, familiar organisms, **chemoautotrophs** have adapted to the most stringent nutritional strategy on earth. All chemoautotrophs are bacteria or archaea that survive totally on inorganic substances such as minerals and gases. They require neither light nor organic nutrients in any form, and they derive energy in diverse and sometimes surprising ways. In very simple terms, they remove electrons from inorganic substrates such as hydrogen gas, hydrogen sulfide, sulfur, or iron and combine them with other inorganic substances such as carbon dioxide, oxygen, and hydrogen. These reactions release simple organic

molecules and a modest amount of energy to drive the synthetic processes of the cell. **Figure 7.2a** provides an example of an unusual bacterium that inhabits hot springs in New Zealand. Chemoautotrophs play an important part in recycling inorganic nutrients and elements. For an example of chemoautotrophy and its importance to deep-sea communities, see 7.1 Secret World of Microbes.

The Methanogen World **Methanogens*** are a unique type of chemoautotroph widely distributed in the earth's habitats. All known methanogens are archaea, and many of them are found in extreme habitats, ranging from hot springs and vents to the deepest, coldest parts of the ocean **(figure 7.2b)**. Others are common in soil, swamps, and even the intestines of humans and other animals.

Methanogens' metabolism is adapted to producing methane gas (CH_4 or "swamp gas") by reducing carbon dioxide, using hydrogen gas under anaerobic conditions, summarized as:

$$4 H_2 + CO_2 \rightarrow CH_4 + 2 H_2O$$

Researchers sampling deep underneath the seafloor have uncovered massive deposits of methanogens. The immensity of this community has led one group of scientists to estimate that it comprises nearly one-third of all life on the planet! Evidence points to its extreme age: It has been embedded in the earth's crust for billions of years. Placed under tremendous pressures, the methane it releases becomes

* *methanogen* (meth-an″oh-gen′) From *methane,* a colorless, odorless gas, and *gennan,* to produce.

(a) (b)

Figure 7.2 **Examples of chemotrophy. (a)** *Venenivibrio,* an extremophilic bacterium that lives in acidic hot springs and derives energy by combining hydrogen gas with oxygen to form water and hydrogen peroxide. **(b)** *Methanocaldococcus jannaschii* is an archaeon and methanogen that inhabits hot vents in the seafloor. It also metabolizes hydrogen gas but its product is methane gas (180,000×). (a): © Dr. Adrian Hetzer; (b): © Electron Microscope Lab, UC Berkeley

frozen into crystals. Some of it serves as a nutrient source for other extreme archaea, and some of it escapes into the ocean. From this position, it may be a significant factor in the development of the earth's climate and atmosphere. Some microbial ecologists suggest that rises in sea temperature could increase the melting of these methane deposits. Because methane is an important "greenhouse gas," this process could contribute considerably to ongoing global climate change.

Heterotrophs and Their Energy Sources

The majority of heterotrophic microorganisms are **chemoorganotrophs** that derive both carbon and energy from organic compounds. Processing these organic molecules by respiration or fermentation releases energy in the form of ATP. We will explore these topics in greater depth in chapter 8. An example of chemoheterotrophy is *aerobic respiration,* the principal energy-yielding pathway in animals, most protozoa and fungi, and aerobic bacteria. It can be simply represented by the equation:

$$\text{Glucose } [C_6H_{12}O_6] + 6O_2 \rightarrow 6CO_2 + 6H_2O + \text{Energy(ATP)}$$

Note that this reaction is complementary to photosynthesis. Here, glucose and oxygen are reactants and carbon dioxide is given off. Indeed, the earth's balance of both energy and metabolic gases greatly depends on this relationship. Chemoheterotrophic microbes are all similar in requiring an organic carbon source, but they differ in how they obtain it. This category includes a wide variety of organisms that feed on other organisms: animals, protozoa, fungi, and many types of bacteria. Examples of varied nutritional patterns include grazers, scavengers, and predators, but the dominant chemoheterotrophic microorganisms fit the description of saprobes or symbionts. **Saprobes**[2] are free-living microbes that feed primarily on organic detritis released by dead organisms, and **symbionts*** derive their organic nutri-

ents from the bodies of living organisms. These categories are not absolute, in that some saprobes can adapt to living organisms, and some symbionts can obtain nutrients from a nonliving source. Because many of these nutritional patterns are also ecological in nature, they will be given fuller coverage in section 7.5. In this section, we introduce saprobes and parasites as they relate to nutrition.

Saprobic Microorganisms The primary niche of saprobes is as decomposers of plant litter, animal matter, and dead microbes. If not for the work of decomposers, the earth would gradually fill up with organic material and the nutrients it contains would not be recycled. Most saprobes, notably bacteria and fungi, have a rigid cell wall and cannot engulf large particles of food. To compensate, they release enzymes to the extracellular environment and digest the food particles into smaller molecules that can be transported into the cell (**figure 7.3**).

Many saprobes exist strictly on dead organic matter in environmental reservoirs such as soil and water, and are unable to adapt to the body of a live host. This group includes free-living protozoa, most fungi, and a variety of bacteria. Apparently there are fewer of these microbes than was once thought, and many supposedly nonpathogenic saprobes can adapt to and invade a susceptible host. When a saprobe does infect a host, it is considered a *facultative parasite.* Because such an infection usually occurs when the host is compromised, the microbe is said to be an **opportunistic pathogen.** A notable example is *Pseudomonas aeruginosa*—its natural habitat is soil and water, yet it is a frequent cause of infections in hospital patients and compromised individuals (as seen in this chapter's Case Study).

Parasitic Microorganisms Most human infections are caused by heterotrophic microorganisms, but those having the greatest adverse impact on human health are **parasites.** By definition, a parasite is a microbe that invades the body of a host, uses it as a habitat and source of nutrients, and, in the process, harms the host to some degree. Such parasitic microbes that grow inside sterile tissues and cause damage, and even death, are also termed **pathogens.** Parasites range from viruses to helminth worms, and they can live on the body (ectoparasites), in the organs and tissues (endoparasites), or even within cells

2. Synonyms are *saprotroph* and *saprophyte.* We prefer to use the terms *saprobe* and *saprotroph* because they are more consistent with other terminology.

* *symbiont* (sim-bee′-ont) An organism that lives in association with another organism, from the term symbiosis (sim″-bye-oh′-sis) Gr, *syn,* together, and *bios,* to live. Literally, living together.

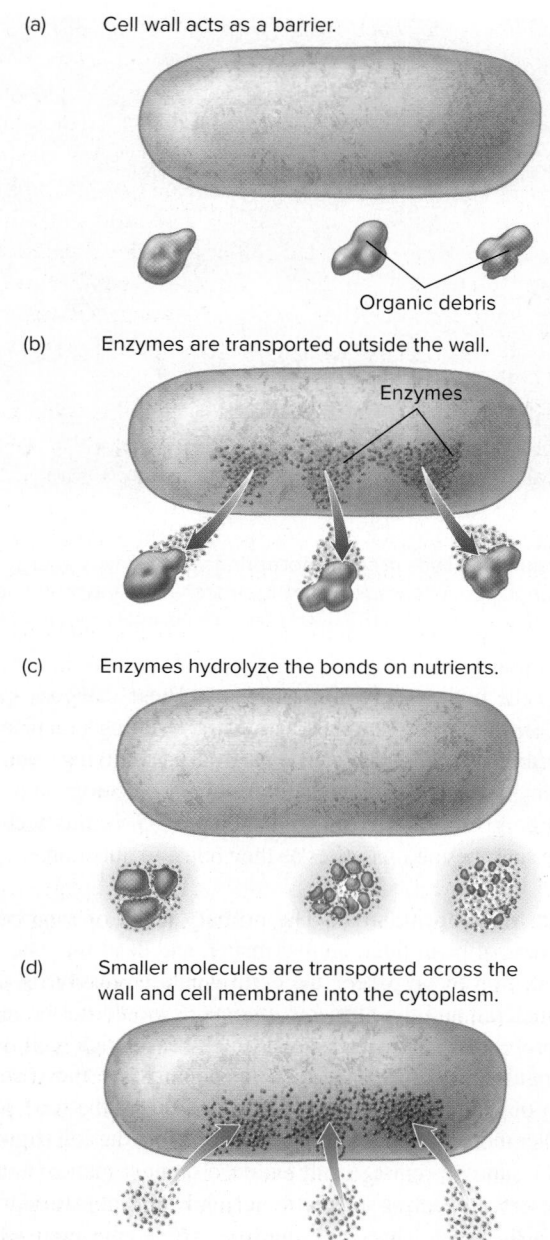

(a) Cell wall acts as a barrier.

Organic debris

(b) Enzymes are transported outside the wall.

Enzymes

(c) Enzymes hydrolyze the bonds on nutrients.

(d) Smaller molecules are transported across the wall and cell membrane into the cytoplasm.

Figure 7.3 **Extracellular digestion in a saprobe with a cell wall (bacterium or fungus). (a)** A walled cell is inflexible and cannot engulf large pieces of organic debris. **(b)** In response to a usable food substrate, the cell synthesizes enzymes that are transported across the wall into the extracellular environment. **(c)** The enzymes hydrolyze the bonds in the debris molecules. **(d)** Digestion produces molecules small enough to be transported into the cytoplasm.

(intracellular parasites, the most extreme type). The more successful parasites generally have no fatal effects and may eventually evolve to a less harmful relationship with their host (see section 7.5).

Obligate parasites (for example, the leprosy bacillus and the syphilis spirochete) are so dependent on their hosts that they are unable to grow outside of the host. Less strict parasites such as the gonococcus and pneumococcus can be cultured artificially if provided with the correct nutrients and environmental conditions in the laboratory.

 Check Your Progress SECTION 7.2

5. Compare autotrophs and heterotrophs with respect to the form of carbon-based nutrients they require.

6. Describe the nutritional strategy of two types of chemoautotrophs described in this section.

7. Contrast chemoautotrophs and chemoheterotrophs as to carbon and energy sources and other unique strategies they may have.

8. What are the main differences between saprobes, symbionts, and parasites?

7.3 Transport: Movement of Substances Across the Cell Membrane

Expected Learning Outcomes

8. Describe the basic factors in diffusion and passive transport systems.

9. Define *osmosis* and describe varying osmotic conditions.

10. Analyze adaptations microbes make in response to osmosis.

11. Describe the features of active transport and differentiate among its mechanisms.

A microorganism's habitat provides necessary substances—some abundant, others scarce—that must still be taken into the cell for nutrition. The reverse is also true: cells must transport waste materials out of the cell (and into the environment). Whatever the direction, transport occurs across the **cell membrane,** the structure specialized for this role. This is true even in organisms with cell walls (bacteria, algae, and fungi), because the cell wall is usually only a partial, nonselective barrier. In this section we discuss some important physical forces in cell transport.

Diffusion and Molecular Motion

All atoms and molecules, regardless of being in a solid, liquid, or gas, are in continuous movement. As the temperature increases, the molecular movement becomes faster due to an increase in kinetic energy. In any solution, including cytoplasm, these moving molecules cannot travel very far without having collisions with other molecules, and they will bounce off each other like millions of colliding pool balls every second **(figure 7.4).** As a result of each collision, the directions of the colliding molecules are altered and the direction of any one molecule is unpredictable and considered random. An example would be a situation in which molecules of a substance are more concentrated in one area than in another. Just by random thermal movement, the molecules will become dispersed away from an area of higher concentration to an area of lower concentration. Over time this will evenly distribute the molecules in the solution. This net movement of molecules down their concentration gradient by random thermal motion is known as **diffusion.** It can be demonstrated by a variety of simple observations. A drop of perfume released into one part of a room is soon smelled in another part, or a lump of sugar in a cup of tea spreads through the whole cup without stirring (figure 7.4).

How Molecules Diffuse in Aqueous Solutions

Figure 7.4 Diffusion of molecules in aqueous solutions.
A high concentration of sugar exists in the cube at the bottom of the liquid. An imaginary molecular view of this area shows that sugar molecules are in a constant state of motion. Those at the edge of the cube diffuse from the concentrated area into more dilute regions. As diffusion continues, the sugar will spread evenly throughout the aqueous phase, and eventually there will be no gradient. At that point, the system is said to be in equilibrium.

Diffusion is a driving force in cell activities, but its effects are greatly controlled by membranes. Two special cases are osmosis and facilitated diffusion. Both processes are considered a type of **passive transport.** This means that the cell does not expend extra energy for them to function. The inherent energy of the molecules moving down a gradient does the work of transport.

The Diffusion of Water: Osmosis

Diffusion of water through a selectively permeable membrane, a process called **osmosis,*** is a physical phenomenon that is easily demonstrated in the laboratory with nonliving materials. Simple experiments provide a model of how cells deal with various solute concentrations in aqueous solutions **(figure 7.5).** In an osmotic system, the membrane is *selectively,* or *differentially, permeable,* having passageways that allow free diffusion of water but can block certain dissolved molecules. When this type of membrane is placed between solutions of differing concentrations where the solute (protein, for example) cannot pass across, then under the laws of diffusion, water will diffuse at a faster rate from the side that has more water to the side that has less

water. As long as the concentrations of the solutions differ, one side will experience a net loss of water and the other a net gain of water until equilibrium is reached and the rate of diffusion is equalized.

Osmosis in living systems is similar to the model shown in figure 7.5. Living membranes generally block the entrance and exit of larger molecules and permit free diffusion of water. Because most cells contain and are surrounded by some sort of aqueous solution, osmosis can have far-reaching effects on cellular activities and survival. Depending on the water content of a cell as compared with its environment, a cell can gain or lose water, or it may remain unaffected. Terms that we use for describing these conditions are *isotonic, hypotonic,* and *hypertonic*[3] **(figure 7.6).**

Under **isotonic*** conditions, the environment is equal in solute concentration to the cell's internal environment; and because diffusion of water proceeds at the same rate in both directions, there is no net change in cell volume. Isotonic solutions are generally the most stable environments for cells, because they are already in an osmotic steady state with the cell. Parasites living in host tissues are most likely to be living in isotonic habitats.

Under **hypotonic*** conditions, the solute concentration of the external environment is lower than that of the cell's internal environment. Pure water provides the most hypotonic environment for cells because it has no solute. Because the net direction of osmosis is from the hypotonic solution into the cell, cells without walls swell and can burst when exposed to this condition.

Slight hypotonicity is tolerated quite well by most bacteria because of their rigid cell walls. The light flow of water into the cell keeps the cell membrane fully extended and the cytoplasm full. This is the optimum condition for the many processes occurring in and on the membrane.

A cell in a **hypertonic*** environment is exposed to a solution with higher solute concentration than its cytoplasm. Because hypertonicity will force water to diffuse out of a cell, it is said to create high *osmotic pressure* or potential. In cells with a wall, water loss causes shrinkage of the protoplast away from the wall, a condition called **plasmolysis.*** Although the whole cell does not collapse, this event can still damage and even kill many kinds of cells. The effect on cells lacking a wall is to shrink down and usually to collapse (figure 7.6). The growth-limiting effect of hypertonic solutions on microbes is the principle behind using concentrated salt and sugar solutions as preservatives for food, such as in salted hams and fish. Such solutions effectively remove moisture from the food, which inhibits the growth of microbial contaminants.

Adaptations to Osmotic Variations in the Environment

Let us now see how specific microbes have adapted osmotically to their environments. In general, isotonic conditions place little stress on cells, so survival depends on counteracting the adverse effects of hypertonic and hypotonic environments.

3. It will help you to recall these osmotic conditions if you remember that the prefixes *iso-, hypo-,* and *hyper-* refer to the environment *outside* of the cell.
 * *isotonic* (eye-soh-tahn´-ik) Gr. *iso,* same, and *tonos,* tension.
 * *hypotonic* (hy-poh-tahn´-ik) Gr. *hypo,* under, and *tonos,* tension.
 * *hypertonic* (hy-pur-tahn´-ik) Gr. *hyper,* above, and *tonos,* tension.
 * *plasmolysis* (plaz´-moh-ly´-sis)

* *osmosis* (oz-moh´-sis) Gr. *osmos,* impulse, and *osis,* a process.

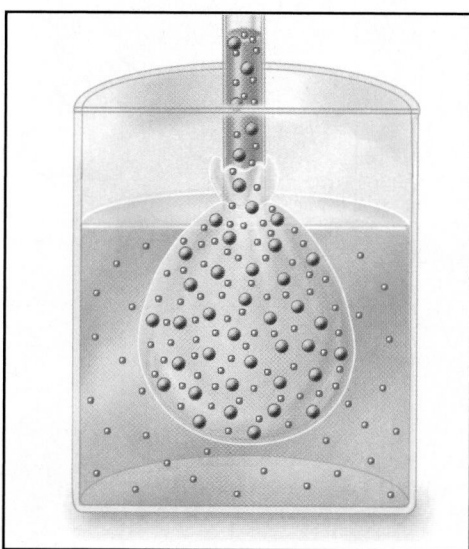

(a) Inset shows a close-up of the osmotic process. The gradient goes from the outer container (higher concentration of H$_2$O) to the sac (lower concentration of H$_2$O). Some water will diffuse the opposite direction but the net gradient favors osmosis into the sac.

(b) As the H$_2$O diffuses into the sac, the volume increases and forces the excess solution into the tube, which will rise continually.

(c) Even as the solution becomes diluted, there will still be osmosis into the sac. Equilibrium will not occur because the solutions can never become equal. (Why?)

Figure 7.5 Model system to demonstrate osmosis. Here we have a solution enclosed in a membranous sac and attached to a hollow tube. The membrane is permeable to water (solvent) but not to solute. The sac is immersed in a container of pure water and observed over time. **(a)** Inset shows a close-up of the osmotic process. The gradient goes from the outer container (higher concentration of H$_2$O) to the sac (lower concentration of H$_2$O). Some water will diffuse the opposite direction but the net gradient favors osmosis into the sac. **(b)** As the H$_2$O diffuses into the sac, the volume increases and forces the excess solution into the tube, which will rise continually. **(c)** Even as the solution becomes diluted, there will still be osmosis into the sac. Equilibrium will not occur because the solutions can never become equal. (Why?)

An alga and an amoeba living in fresh pond water are examples of cells that live in constantly hypotonic conditions. The rate of water diffusing across the cell membrane into the cytoplasm is rapid and constant, and the cells would die without a way to adapt. As with bacteria, the majority of algae have a cell wall that protects them from bursting even as the cytoplasmic membrane becomes **turgid*** from pressure. The amoeba has no cell wall to protect it, so it must expend energy to deal with the influx of water. This is accomplished with a water, or contractile, vacuole that siphons excess water out of the cell like a tiny pump.

A microbe living in a high-salt environment (hypertonic) has the opposite problem and must either restrict its loss of water to the environment or increase the salinity of its internal environment. Halobacteria living in the Great Salt Lake and the Dead Sea must absorb salt to make their cells isotonic with the environment; thus, they have a physiological need for a high-salt concentration in their habitats (see halophiles in section 7.4).

The Movement of Solutes Across Membranes

Simple diffusion works well for movement of small nonpolar molecules such as oxygen or lipid soluble molecules that readily pass

* *turgid* (ter'-jid) A condition of being swollen or congested.

through membranes. But many substances that cells require are polar and ionic chemicals with greatly reduced permeability. Simple diffusion alone would be unable to transport these substances. One way that cells have adapted to this limitation involves a process called **facilitated diffusion (figure 7.7).** This passive transport mechanism utilizes a carrier protein in the membrane that will bind a specific substance. This binding changes the conformation of the carrier proteins in a way that facilitates movement of the substance across the membrane. Once the substance is transported, the carrier protein resumes its original shape and is ready to transport again. These carrier proteins exhibit **specificity,** which means that they bind and transport only a single type of molecule. For example, a carrier protein that transports sodium will not bind glucose. A second characteristic exhibited by facilitated diffusion is **saturation.** The rate of transport of a substance is limited by the number of binding sites on the transport proteins. As the substance's concentration increases, so does the rate of transport until the concentration of the transported substance causes all of the transporters' binding sites to be occupied. Then the rate of transport reaches a steady state and cannot move faster despite further increases in the substance's concentration.

Other examples of passive protein carriers found in a wide variety of cells are *aquaporins*, also known as water channels. These openings in the membrane facilitate passive transport of water

Cells with cell walls

| Isotonic Solution | Hypotonic Solution | Hypertonic Solution |

Cell wall

Cell membrane

Cell membrane

Water concentration is equal inside and outside the cell, thus rates of diffusion are equal in both directions.

Net diffusion of water is into the cell; this swells the protoplast and pushes it tightly against the wall. Wall usually prevents cell from bursting.

Water diffuses out of the cell and shrinks the cell membrane away from the cell wall; process is known as **plasmolysis.**

Cells lacking walls

Cell membrane

Early

Late (osmolysis)

Early

Late

Rates of diffusion are equal in both directions.

Diffusion of water into the cell causes it to swell, and may burst it if no mechanism exists to remove the water.

Water diffusing out of the cell causes it to shrink and become distorted.

⟶ Direction of net water movement.

Figure 7.6 Cell responses to solutions of differing osmotic content.

molecules following an existing osmotic gradient. They appear to be involved in regulating volume and osmotic pressure.

Active Transport: Bringing in Molecules Against a Gradient

Free-living microbes exist under relatively nutrient-starved conditions and cannot rely completely on slow and rather inefficient passive transport mechanisms. To ensure a constant supply of nutrients and other required substances, microbes must capture those that are in low concentrations and actively transport them into the cell. Cells must likewise transport substances in the reverse direction to the external environment. Some microbes have such efficient active transport

Figure 7.7 Facilitated diffusion. Facilitated diffusion involves the attachment of a molecule to a specific protein carrier. Bonding of the molecule causes a conformational change in the protein that facilitates the molecule's passage across the membrane (left view). The membrane protein releases the molecule into the cell interior (right view). The cell does not have to expend energy for transport.

(a) **Carrier-mediated active transport.** (1) Membrane-bound transporter proteins (permeases) interact with nearby solute binding proteins that carry essential solutes (sodium, iron, sugars). (2) Once a binding protein attaches to a specific site, an ATP is activated and generates energy to pump the solute into the cell's interior through a special channel in the permease.

(b) **In group translocation,** (1) a specific molecule is actively captured, but on its passage through the membrane protein carrier (2) it is chemically altered or activated for use in the cell. By coupling transport with synthesis, the cell conserves energy.

(c) **Endocytosis.** With phagocytosis, solid particles are engulfed by flexible cell extensions or pseudopods (1-4) (1,000X). (5) With pinocytosis, fluids and/or dissoved substances are enclosed in vesicles by very fine protrusions called microvilli (3,000X). Oil droplets fuse with the membrane and are released directly into the cell.

Figure 7.8 Active transport. In active transport mechanisms, energy is expended (ATP) to transport the molecule across the cell membrane.

systems that an essential nutrient can be found in intracellular concentrations that are hundreds of times greater than in the habitat.

Features inherent in **active transport** systems are:

1. the transport of nutrients against the diffusion gradient or in the same direction as the natural gradient but at a rate faster than by diffusion alone,

2. the presence of specific membrane proteins (permeases and pumps; **figure 7.8a**), and

3. the expenditure of additional cellular energy in the form of ATP-driven uptake. Examples of substances transported actively are monosaccharides, amino acids, organic acids, phosphates, and metal ions.

★A

TABLE 7.3	Summary of Transport Processes in Cells			
General Process	**Nature of Transport**	**Examples**	**Description**	**Qualities**
Passive	Energy expenditure by the cell is not required. Substances exist in a gradient and move from areas of higher concentration toward areas of lower concentration in the gradient.	Diffusion osmosis	A fundamental property of atoms and molecules that exist in a state of random motion	Nonspecific Brownian movement Movement of small uncharged molecules across membranes
		Facilitated diffusion	Molecule binds to a carrier protein in membrane and is carried across to other side.	Molecule specific; transports both ways Transports sugars, amino acids, water
Active	Energy expenditure is required. Molecules need not exist in a gradient. Rate of transport is increased. Transport may occur against a concentration gradient.	Carrier-mediated active transport	Atoms or molecules are pumped into or out of the cell by specialized receptors; driven by ATP or other high-energy molecules	Transports simple sugars, amino acids, inorganic ions (Na^+, K^+)
		Group translocation	Molecule is moved across membrane and simultaneously converted to a metabolically useful substance.	Alternate system for transporting nutrients (sugars, amino acids)
		Bulk transport	Mass transport of large particles, cells, and liquids by engulfment and vesicle formation	Process is endocytosis; examples are phagocytosis and pinocytosis

Carrier-mediated active transport involves specific membrane proteins that bind both ATP and the molecules to be transported. Release of energy from ATP drives the movement of the molecule through the protein carrier. This can occur in either direction. Some bacteria transport certain sugars, amino acids, vitamins, and phosphate into the cell by this mechanism. Other bacteria can actively pump drugs out of the cell, thereby providing them resistance to the drugs. Another type of active transport pumps can rapidly carry ions such as K^+, Na^+, and H^+ across the membrane. It is this type of transport protein being defective in cystic fibrosis patients that is responsible for its basic pathology (see this chapter's Case Study).

Another type of active transport, **group translocation,** couples the transport of a nutrient with its conversion to a substance that is immediately useful inside the cell **(figure 7.8b).** This method is used by certain bacteria to transport sugars (glucose, fructose) while simultaneously adding phosphate molecules that activate them in preparation for a metabolic cycle.

Endocytosis: Eating and Drinking by Cells

Some cells can transport large molecules, particles, liquids, or even other cells across their cell membranes. Because the cell usually expends energy to carry out this movement, it is also a form of active transport. The substances transported do not pass physically through the membrane but are carried into the cell by **endocytosis.** First the cell encloses the substance in its membrane, simultaneously forming a vacuole and engulfing it **(figure 7.8c).** Amoebas and certain white blood cells ingest whole cells or large solid matter by a type of endocytosis called **phagocytosis.** Liquids, such as oils or molecules in solution, enter the cell through **pinocytosis.** The mechanisms for transport of molecules into cells are summarized in **table 7.3.**

Check Your Progress SECTION 7.3

9. Compare and contrast passive and active forms of transport, using examples of what is being transported and the requirements for each.

10. How are phagocytosis and pinocytosis similar? How are they different?

11. Explain the differences between facilitated diffusion and group translocation.

12. Compare the effects of isotonic, hypotonic, and hypertonic solutions on an amoeba and on a bacterial cell. If a cell lives in a hypotonic environment, what will occur if it is placed in a hypertonic one? Answer for the opposite case as well.

7.4 Environmental Factors That Influence Microbes

Expected Learning Outcomes

12. Differentiate between habitat and niche.

13. Describe the range of temperatures a microbe can function within.

14. Explain the adaptive temperature groups, with examples of microbes that exist in them.

15. List the major gases and describe microbial requirements for these gases.

16. Outline the adaptations of microbial groups to variations in pH.

17. Identify microbial adaptations to osmotic pressure.

7.1 Secret World of Microbes
Going to Extremes

Any extreme habitat—whether hot, cold, salty, acidic, alkaline, high-pressure, arid, oxygen-free, or toxic—is likely to harbor microorganisms with special adaptations to these conditions. Most such inhabitants are archaea and bacteria, but certain fungi, protozoans, algae, and even viruses are also capable of living in severe habitats. Microbiologists have termed such remarkable organisms **extremophiles.** Discoveries of new types of extremophiles flourishing in forbidding environments are reported rather frequently. For microbiologists these findings are often astounding but not unexpected, because they confirm our perception that the earth has been "incubating" microbes for billions of years and there are still multitudes of hidden life forms to uncover.

Extreme Temperatures

Some of the most extreme habitats are hot springs, geysers, volcanoes, and ocean vents, all of which support flourishing microbial populations. Temperatures in these regions range from 50°C (122°F) to well above the boiling point of water, with some ocean vents approaching 350°C (662°F). Many heat-adapted microbes are ancient archaea whose genetics and metabolism are extremely modified for this mode of existence. A unique ecosystem based on hydrogen sulfide–oxidizing bacteria exists in the hydrothermal vents lying along deep oceanic ridges. Other denizens of hot volcanic waters are unusual invertebrates such as the Pompeii worm, which frequents rocky hideouts around sea vents measuring 80°C (176°F) **(photo a).** Their survival is greatly dependent on symbiotic bacteria that form an insulated coat on the worm's surface.

A large part of the earth exists at cold temperatures. Microbes settle and grow throughout the Arctic and Antarctic and in the deepest parts of the ocean in near-freezing temperatures. Several species of algae and fungi thrive on the surfaces of snow and glacier ice (see figure 7.10).

More surprising still is that some bacteria and algae flourish in the sea ice of Antarctica. Although the ice appears to be completely solid, it is honeycombed by various-size pores and tunnels filled with a solution that is −20°C (−4°F). These frigid microhabitats are home to a microcosm of planktonic bacteria, algae, and predators that feed on them.

Life in Very Salty Places

The growth of most microbial cells is inhibited by high amounts of salt; for this reason, salt is a common food preservative. But there are many salt lovers, including rich communities of bacteria and algae living in oceans, salt lakes, and inland seas, some of which are saturated with salt (30%). Microbiologists recently discovered salt pockets that supported archaea living in water 100 times more concentrated than seawater.

"Conan the Bacterium"

A tiny gram-positive coccus, *Deinococcus radiodurans* **(photo b),** has been called "Conan the bacterium" because of its capacity to survive extreme drying and very high levels of radiation. This bacterium is widespread in many habitats, but its resistance to radiation sets it apart from every other known organism. The DNA and proteins of most organisms are very sensitive to gamma rays and other high-energy radiation, which is why radiation is so damaging to cells. But we now know its secret. *Deinococcus* carries around several copies of its genome, making it easily replaceable, and it contains radiation-resistant enzymes that help it rapidly repair DNA damage. This microbe is being studied for its potential in bioremediation of radioactive waste and as a model organism for Mars studies.

Brave "Old" World

When samples of ancient Antarctic ice (several million years old) were brought into the lab, they yielded up living bacteria. Some of the oldest

Microbes are exposed to a wide variety of environmental factors that affect growth and survival. Microbial ecology focuses on ways in which microorganisms deal with or adapt to such factors as heat, cold, gases, acid, radiation, osmotic and hydrostatic pressures, and even other microbes. Adaptation involves a complex adjustment in biochemistry or genetics that enables long-term survival and growth. The term biologists use to describe the totality of adaptations organisms make to their habitats is **niche.*** For most microbes, environmental factors fundamentally affect the function of metabolic enzymes. Thus, survival is largely a matter of whether the enzyme systems of microorganisms can continue to function even in a changing environment. Looking at **7.1 Secret World of Microbes,** will introduce you to the broad range of conditions that support microbial life.

Adaptations to Temperature

Microbial cells cannot control their temperature and therefore assume the ambient temperature of their natural habitats. So, to survive they must adapt to whatever temperature variations are encountered in their habitat. The range of temperatures for microbial growth can be expressed as three **cardinal temperatures. The minimum temperature**

is the lowest temperature that permits a microbe's continued growth and metabolism; below this temperature, its activities are inhibited. The **maximum temperature** is the highest temperature at which growth and metabolism can proceed. If the temperature rises slightly above maximum, growth will stop. If it continues to rise beyond that point, the enzymes and nucleic acids will eventually become permanently inactivated—a condition known as **denaturation**—and the cell will die. This is why heat works so well as an agent in microbial control. The **optimum temperature** covers a small range, intermediate between the minimum and maximum, which promotes the fastest rate of growth and metabolism. Only rarely is the optimum a single point.

Depending on their natural habitats, some microbes have a narrow cardinal range, others a broad one. Some strict parasites will not grow if the temperature varies more than a few degrees below or above the host's body temperature. For instance, rhinoviruses (one cause of the common cold) multiply successfully only in tissues that are slightly below normal body temperature (33°C to 35°C or 91°F to 95°F). Other microbes are not so limited. Strains of *Staphylococcus aureus* grow within the range of 6°C to 46°C (43°F to 114°F), and the intestinal bacterium *Enterococcus faecalis* grows within the range of 0°C to 44°C (32°F to 112°F).

Another way to express temperature adaptation is to describe whether an organism grows optimally in a cold, moderate, or hot

* *niche* (nitch) Fr. *nichier,* to nest.

isolates were unusual, very slow-growing forms, but others were similar to modern bacteria.

Another bizarre, ancient ultraextremophile was discovered 3 kilometers down in an African gold mine. The bacterium *Desulforudis* is the only occupant of this totally dark, anaerobic, hot, alkaline habitat, and it derives its energy from the radioactive decay of uranium. This is the first organism on earth known to have established its own ecosystem and to sustain itself without the contributions of other organisms.

Numerous species have carved a niche for themselves in the deepest layers of mud, swamps, and oceans, where oxygen gas and sunlight cannot penetrate. The predominant living things in the deepest part of the oceans (10,000 m or below) are pressure- and cold-loving microorganisms. Even parched zones in sand dunes and deserts harbor a hardy brand of microbes; and thriving bacterial populations can be found in petroleum, coal, and mineral deposits containing copper, zinc, gold, and uranium. Microbes that can adapt and survive under such lethal conditions are called toxophiles.

As a rule, a microbe that has adapted to an extreme habitat will not survive in a moderate one. And except for rare cases, none of the organisms living in these extremes is a pathogen, because the human body would be a rather hostile habitat for them.

Explain how extremophiles can survive such severe and adverse conditions. *Answer available on Connect.*

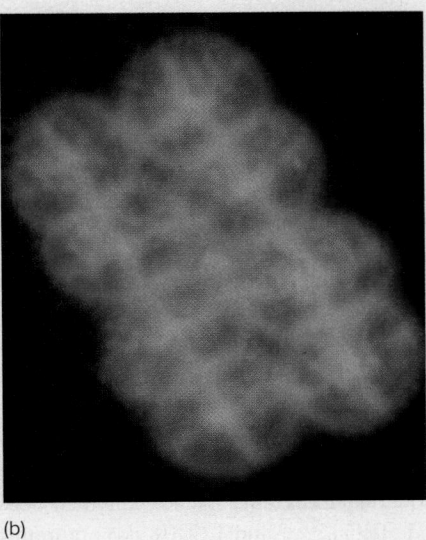

(a) The hottest animal on earth is this Pompeii worm wearing a coat of symbiotic bacteria (fine filaments) to protect against the heat. (b) A fluorescent micrograph of *Deinococcus radiodurans,* showing its packet arrangement, cell walls (red), and chromosomes (green). (a): Image contributed by M. Cottrell and C. Cary, College of Marine Studies, University of Delaware; (b): © Dea Slade

temperature range. The terms used for these ecological groups are *psychrophile, mesophile,* and *thermophile* (**figure 7.9**).

A **psychrophile*** is a microorganism (bacterium, archaea, fungus, or alga) with an optimum temperature below 15°C (59°F) but generally can grow at 0°C (32°F). It is obligate with respect to cold and generally cannot grow above 20°C (68°F). Laboratory work with true psychrophiles can be a real challenge. Inoculations have to be done in a cold room because ordinary room temperature can be lethal to these organisms. Unlike most laboratory cultures, storage in the refrigerator incubates rather than inhibits them. As one might predict, true psychrophiles are not capable of surviving in the human body and do not cause infections. Instead, they populate some of the coldest places on earth, including snowfields (**figure 7.10**), polar ice, permafrost, and the deep ocean. Recently bacteria were isolated from an undersea lake in Antarctica that were living at −15°C (5°F). True psychrophiles must

Figure 7.9 Ecological groups by temperature of adaptation. Psychrophiles can grow at or near 0°C and have an optimum below 15°C. As a group, mesophiles can grow between 10°C and 50°C, but their optima usually fall between 20°C and 40°C. One type of mesophile termed a psychrotroph is capable of growth at temperatures below 20°C. Generally speaking, thermophiles require temperatures above 45°C and grow optimally between this temperature and 80°C. Extreme thermophiles (hyperthermophiles) are bacteria and archaea with optima above 80°C. Note that the extremes of the ranges can overlap to some extent.

psychrophile (sy′-kroh-fyl) Gr. *psychros,* cold, and *philos,* to love.

(a) (b)

Figure 7.10 Red snow. (a) The surface of an Alaskan glacier provides a perfect habitat for psychrophilic photosynthetic organisms such as *Chlamydomonas nivalis*. **(b)** Microscopic views of this snow alga, which is classified as a "green" alga although a red pigment dominates at this stage of its life cycle (600x). (a and b): Images courtesy of Nozomu Takeuchi

be distinguished from *psychrotrophs* or *facultative psychrophiles* that grow slowly in cold but have an optimum temperature above 20°C (68°F). Psychrotrophs such as *Staphylococcus aureus* and *Listeria monocytogenes* are a concern because they can grow in refrigerated food and cause food-borne illness.

The majority of medically significant microorganisms are **mesophiles,*** organisms that grow at intermediate temperatures. Although an individual species can grow between the extremes of 10°C and 50°C (50°F and 122°F), the optimum growth temperatures (optima) of most mesophiles fall into the range of 20°C to 40°C (68°F to 104°F). Organisms in this group inhabit animals and plants as well as soil and water in temperate, subtropical, and tropical regions. Most human pathogens have optima somewhere between 30°C and 40°C (human body temperature is 37°C or 98°F). *Thermoduric* microbes, which can survive short exposure to high temperatures but are normally mesophiles, are common contaminants of heated or pasteurized foods (see chapter 11). Examples include heat-resistant cysts such as *Giardia* or sporeformers such as *Bacillus* and *Clostridium*.

A **thermophile*** is a microbe that grows optimally at temperatures greater than 45°C (113°F). Such heat-loving microbes live in soil and water associated with volcanic activity, in compost piles, and in habitats directly exposed to the sun. Thermophiles generally range in growth temperatures from 45°C to 80°C (113°F to 176°F). Most eukaryotic forms cannot survive above 60°C (140°F), but a few bacteria and archaea, called **hyperthermophiles,** are adapted to life in extreme temperatures. They can grow at between 80°C and 121°C (250°F), currently thought to be the highest temperature limit endured by enzymes and cell structures. Strict thermophiles are so heat tolerant that researchers can use a heat-sterilizing device to isolate them from non-thermophiles in culture.

Currently biotechnology companies are intensely interested in thermal microorganisms. One of the most profitable discoveries so far was a strict thermophile *Thermus aquaticus,* which produces an enzyme that can make copies of DNA even at high temperatures. This enzyme—Taq polymerase—is now an essential component of the polymerase chain reaction or PCR, a process used in many areas of medicine, forensics, and biotechnology (see chapter 10).

Gas Requirements

The atmospheric gases that most influence microbial growth are oxygen (O_2) and carbon dioxide (CO_2). Oxygen comprises about 20% of the atmosphere and CO_2 about 0.04%. Oxygen gas has a great impact on microbial adaptation. It is an important respiratory gas, and it is also a powerful oxidizing agent that exists in many toxic forms. In general, microbes fall into one of three categories: those that use oxygen and can detoxify it; those that can neither use oxygen nor detoxify it; and those that do not use oxygen but can detoxify it.

How Microbes Process Oxygen

As oxygen gas enters into cellular reactions, it can be transformed into several toxic products. Singlet oxygen (1O_2) is an extremely reactive molecule produced by both living and nonliving processes. Notably, it is one of the substances produced by phagocytes to kill invading bacteria (see chapter 14). The buildup of singlet oxygen and the oxidation of membrane lipids and other molecules can damage and destroy a cell. The highly reactive superoxide ion (O_2^-), peroxide (H_2O_2), and hydroxyl radicals (OH) are other destructive metabolic by-products of oxygen. To survive these toxic oxygen products, many microorganisms have developed enzymes capable of scavenging and neutralizing these chemicals. The complete conversion of superoxide ion into less harmful oxygen gas involves a two-step process and at least two enzymes:

Step 1. $2O_2^- + 2H^+ \xrightarrow{\text{Superoxide dismutase}} H_2O_2 \text{ (hydrogen peroxide)} + O_2$

Step 2. $2H_2O_2 \xrightarrow{\text{Catalase}} 2H_2O + O_2$

In this series of reactions essential for aerobic organisms, the superoxide ion is first converted to hydrogen peroxide and normal oxygen by the action of an enzyme called superoxide dismutase. Because hydrogen peroxide is also toxic to cells (it is used as a disinfectant and antiseptic), it must be degraded by an enzyme—either catalase or peroxidase—into water and oxygen. If a microbe is not capable of dealing with toxic oxygen by these or similar mechanisms, it will be restricted to habitats free of oxygen.

With respect to oxygen requirements, several general categories are recognized. An **aerobe*** (aerobic organism) can use gaseous oxygen in its metabolism and possesses the enzymes needed to process toxic oxygen products. An organism that cannot grow without oxygen is an **obligate aerobe.** Most fungi and protozoa, as well as many bacteria (genera *Micrococcus* and *Bacillus*), have strict requirements for oxygen in their metabolism.

A **facultative anaerobe** is an aerobe that does not require oxygen for its metabolism and is capable of growth in its absence. This type of organism metabolizes by aerobic respiration when oxygen is present, but in its absence, it adopts an anaerobic mode of metabolism such as fermentation. Facultative anaerobes usually possess catalase and superoxide dismutase. A number of bacterial pathogens

* *mesophiles* (mez'-oh-fylz) Gr. *mesos*, middle.
* *thermophile* (thur'-moh-fyl) Gr. *therme*, heat.

* *aerobe* (air'-ohb) Although the prefix means air, it is used in the sense of oxygen.

fall into this group. This includes gram-negative intestinal bacteria and staphylococci. A **microaerophile*** does not grow at normal atmospheric concentrations of oxygen but requires a small amount of it (1%–15%) in metabolism. Most organisms in this category live in a habitat such as soil, water, or the human body that provides small amounts of oxygen but is not directly exposed to the atmosphere.

A true **anaerobe** (anaerobic microorganism) lacks the metabolic enzyme systems for using oxygen gas in respiration. Because **strict**, or **obligate, anaerobes** also lack the enzymes for processing toxic oxygen, they cannot tolerate any free oxygen in the immediate environment and will die if exposed to it. Strict anaerobes live in highly reduced habitats, such as deep muds, lakes, oceans, and soil.

Determining the oxygen requirements of a microbe from a biochemical standpoint can be a very time-consuming process. An initial clarification comes from cultures made with reducing media that contain an oxygen-removing chemical such as thioglycollate. The location of growth in a tube of fluid thioglycollate medium is a fair indicator of an organism's adaptation to oxygen use **(figure 7.11)**. Growing strictly anaerobic bacteria usually requires special media, methods of incubation, and handling chambers that exclude oxygen. **Figure 7.12** shows special systems for handling and growing anaerobes.

Even though human cells use oxygen, and oxygen is found in the blood and tissues, some body sites present anaerobic pockets or microhabitats where colonization or infection can occur. Dental caries are partly due to the complex actions of aerobic and anaerobic bacteria in plaque. Most gingival infections consist of similar mixtures of oral bacteria that have invaded damaged gum tissues. Another common site for anaerobic infections is the large intestine, a

relatively oxygen-free habitat that harbors a rich assortment of strictly anaerobic bacteria. Anaerobic infections can occur following abdominal surgery and traumatic injuries (gas gangrene and tetanus).

Aerotolerant anaerobes do not utilize oxygen gas but can survive and grow in its presence. These anaerobes are not harmed by oxygen, and some of them possess alternate mechanisms for breaking down peroxide and superoxide. For instance, lactobacilli, which are common residents of the intestine, inactivate these compounds with manganese ions.

Although microbes require some carbon dioxide in their metabolism, **capnophiles** grow best at higher CO_2 concentrations (3%–10%) than are normally present in the atmosphere (0.033%). This becomes important in the initial isolation of some pathogens from clinical specimens, notably *Neisseria* (gonorrhea, meningitis),

(a)

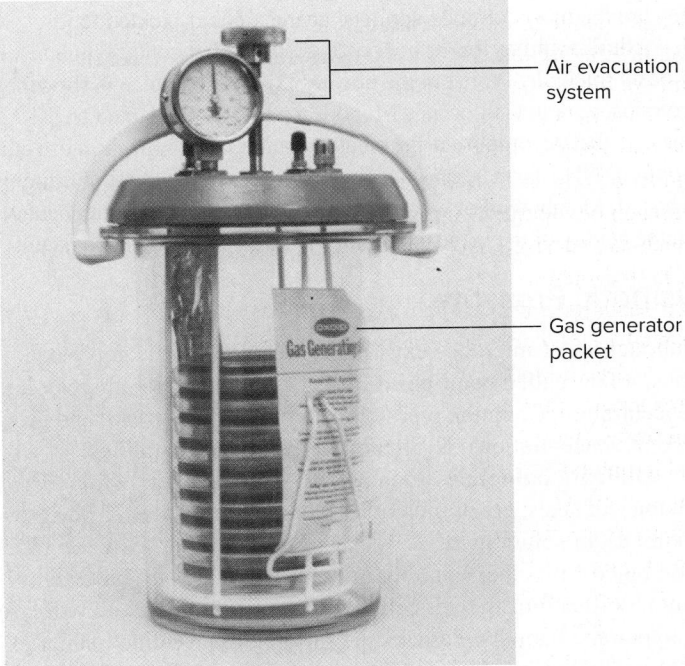

Air evacuation system

Gas generator packet

(b)

Figure 7.12 Culturing techniques for anaerobes. **(a)** An anaerobic environmental chamber is equipped with ports for handling strict anaerobes without exposing them to air. It also has provisions for incubation and inspection in a completely O_2-free system. **(b)** An anaerobic jar, equipped to evacuate air and be refilled with oxygen-free gases such as nitrogen or carbon dioxide. (a): *Source:* Keith Weller, USDA/ARS; (b): *Source:* Dr. Gilda Jones/CDC

* *microaerophile* (myk″-roh-air′-oh-fyl)

Figure 7.11 Use of thioglycollate broth to demonstrate oxygen requirements. Thioglycollate is a reducing medium that can establish a gradation in oxygen content. Oxygen concentration is highest at the top of the tube and absent in the deeper regions. When a series of tubes is inoculated with bacteria that differ in O_2 requirements, the relative position of growth provides some indication of their adaptations to oxygen use. Tube 1 (far left): aerobic (*Pseudomonas aeruginosa*); Tube 2: facultative (*Staphylococcus aureus*); Tube 3: facultative (*Escherichia coli*); Tube 4: obligate anaerobe (*Clostridium butyricum*). © Terese M. Barta, Ph.D.

Brucella (undulant fever), and *Streptococcus pneumoniae*. Incubation is carried out in a CO_2 incubator that provides the correct range of CO_2 (figure 7.12*b*). Keep in mind that CO_2 is an essential nutrient for autotrophs, which use it to synthesize organic compounds.

Effects of pH

Microbial growth and survival are also influenced by the pH of the habitat. The pH was defined in chapter 2 as the degree of acidity or alkalinity of a solution. It is expressed by the pH scale, a series of numbers ranging from 0 to 14, pH 7 being neither acidic nor alkaline. As the pH value decreases toward 0, the acidity increases, and as the pH increases toward 14, the alkalinity increases. The majority of organisms live or grow in habitats between pH 6 and pH 8 because stronger acids and bases can be highly damaging to enzymes and other cellular substances.

Although most microorganisms living in soil, fresh water, or the bodies of plants and animals are **neutrophiles,** living within the range of pH 5.5 to 8, some can manage exposure to pH extremes. Highly acidic or alkaline habitats such as acidic bogs and streams or alkaline soils and ponds can provide numerous habitats for specialized microbial communities. Obligate **acidophiles** include *Euglena mutabilis*, an alga that grows in acid pools between 0 and 1.0 pH, and *Thermoplasma,* an archaea that lives in hot coal piles at a pH of 1 to 2 and will lyse if exposed to pH 7. A few species of algae, archaea, and bacteria can survive at a pH near that of concentrated hydrochloric acid near a pH of 0. Not only do they require such a low pH for growth, but a few species of bacteria actually help maintain the low pH by releasing strong acid. Because many molds and yeasts tolerate moderate acidity, they are the most common spoilage agents of acid-pickled foods.

Alkalinophiles live in hot pools and soils that contain high levels of basic minerals. Probably the upward limit of tolerance is shown by proteobacteria in California's Mono Lake that have adapted to pH 12. Bacteria that decompose urine create alkaline conditions, because ammonium (NH_4^+) can be produced when urea (a component of urine) is digested. Metabolism of urea is one way that *Proteus* spp. can neutralize the acidity of the urine to colonize and infect the urinary system.

Osmotic Pressure

Although most microbes exist under hypotonic or isotonic conditions, a few, called **osmophiles,** live in habitats with a high solute concentration. Common types of osmophiles that survive and grow in high concentrations of salt are referred to as **halophiles.***

Obligate halophiles such as *Halobacterium* and *Halococcus* inhabit salt lakes, ponds, and other hypersaline habitats. They grow optimally in solutions of 25% NaCl but require at least 9% NaCl (combined with other salts) for growth. These archaea have significant modifications in their cell walls and membranes, and will lyse in hypotonic habitats. Perhaps the most extreme saline habitat exists in a shallow pond in frigid Antarctica. The salt concentration of Don Juan Pond tests at 33% $CaCl_2$. For a time researchers thought this site would be a good model for duplicating the lifeless conditions on Mars, until they discovered an active biofilm community of cyanobacteria and diatoms residing there!

Some microbes adapt to wide concentrations in solutes. These are considered **osmotolerant** and are sometimes referred to as **facultative**

* *halophiles* (hay´-loh-fylz)

halophiles. Such organisms are remarkably resistant to salt, even though they do not normally reside in high-salt environments. For example, *Staphylococcus aureus* can grow on NaCl media ranging from 0.1% up to 20%. Although it is common to use high concentrations of salt and sugar to preserve food (jellies, syrups, and brines), many bacteria and fungi thrive under these conditions and are common spoilage agents. One particularly hardy sugar-loving or *saccharophilic* yeast withstands the high sugar concentration of honey and candy.

Miscellaneous Environmental Factors

Descent into the ocean depths subjects organisms to increasing hydrostatic pressure. Deep-sea microbes called **barophiles** exist under pressures many times that of the atmosphere. Marine biologists sampling deep-sea trenches 7 miles below the surface isolated unusual eukaryotes called foraminifera that were being exposed to pressures 1,100 times greater than what would occur at the sea's surface. These microbes are so strictly adapted to high pressures that they will rupture when exposed to normal atmospheric pressure at sea level.

Because of the high water content of cytoplasm, all cells require water from their environment to sustain growth and metabolism. Water is the solvent for cell chemicals, and it is needed for enzyme function and digestion of macromolecules. A certain amount of water on the external surface of the cell is required for the diffusion of nutrients and wastes. Even in apparently dry habitats, such as sand or dry soil, the particles retain a thin layer of water usable by microorganisms. Only dormant, dehydrated cell stages (for example, spores and cysts) tolerate extreme drying because of the inactivity of their enzymes.

 Check Your Progress SECTION 7.4

13. Why are most pathogens mesophilic?
14. What are the ecological roles of psychrophiles and thermophiles?
15. Explain what it means to be an obligate intracellular parasite. Name three groups of obligate intracellular parasites.
16. Where in the body are anaerobic habitats apt to be found?
17. Where do superoxide ions and hydrogen peroxide originate? What are their toxic effects?

7.5 Ecological Associations Among Microorganisms

 Expected Learning Outcomes

18. Outline the types of associations among microorganisms and describe their basic differences.

19. Explain what occurs in symbiosis and in coevolution.

20. Differentiate among mutual, commensal, and parasitic associations, providing examples.

21. Explain syntrophy and amensalism, using examples.

22. Describe interactions between humans and their microbiota.

Two major unifying themes of this chapter are microbial adaptability and **ubiquity.** The presence of microbes in most natural settings and their interactions with other organisms in these settings have

7.2 Secret World of Microbes
A Mutual Attraction

A tremendous variety of mutualistic symbioses occur in nature. These associations gradually coevolve as participating members come to require some substance or habitat that the other members provide. In many cases they become so interdependent that they will not survive outside of this association.

Many protozoan cells receive growth factors from symbiotic bacteria and algae that are mutually nurtured by the protozoan cell. One peculiar ciliate propels itself by affixing symbiotic bacteria to its cell membrane to act as "oars." Other ciliates and amoebas harbor specialized bacteria or algae inside their cells to provide nutrients.

Mutualism Between Microbes and Animals

Microorganisms carry on mutual symbiotic relationships with animals as diverse as sponges, worms, and mammals. Complex, multipartner mutualism is especially striking in termites, which harbor 250 or more specialized microbes inside their gut. The termite's role is simply to gnaw off pieces of wood, but it cannot digest the wood. **Endosymbiotic** protozoans and bacteria do the final work of breaking down the wood particles, and all of the mutualists share in the nutrients provided (see photos).

Bacteria and protozoa are essential in the operation of the rumen (a complex, four-chambered stomach) of cud-chewing mammals. These mammals produce no enzymes of their own to break down the cellulose that is a major part of their diet, but the microbial population harbored in their rumens does. The complex food materials are digested through several stages, during which time the animal regurgitates and chews the partially digested plant matter (the cud) and occasionally burps methane produced by the microbial symbionts.

Thermal Vent Communities

Another fascinating relationship has been found in the deep hydrothermal vents in the seafloor, where geologic forces spread the crustal plates and release superheated fluid (over 350°C or 662°F) and gas. These vents are a focus of tremendous biological and geologic activity. The source of energy in this community is not the sun, because the vents are too deep for light to penetrate (2,600 m). Instead, this ecosystem is based on a massive chemoautotrophic bacterial population that oxidizes the abundant hydrogen sulfide (H_2S) gas given off by the volcanic activity there. As the bottom of the food web, these bacteria serve as the primary producers of nutrients that service a broad spectrum of specialized animals. One example, the tube worm *Riftia*, has become so specialized that it has lost its gastrointestinal tract and feeds solely from its bacterial symbionts.

Quick Search

Search "Hydrothermal Vents" or "Black Smokers" on YouTube to watch videos of these ecosystems.

Describe some of the ecological benefits of mutualism. *Answer available on Connect.*

Left: A termite displayed alongside its gut. Examples of gut microbes involved in wood digestion are protozoa (middle panel, 500x) and spirochete bacteria (right panel, 1000x). © Dr. Jared R. Leadbetter

produced an abundance of specialized adaptations and niches of incredible diversity. If we go by the constant discoveries of new organisms and new associations, it is likely that every possible combination and example of microbial association exists somewhere on earth. Even with such a great variety, we can organize these microbial alliances into six basic patterns, ranging from organisms living a totally codependent existence to free-living microbes that interact in loose and temporary unions.

Probably the best-described and best-understood interactions are **symbioses,**[4]* defined as close associations between organisms that are advantageous to at least one of the members. Recall that the members of symbiotic relationships are termed symbionts. Symbioses can be obligatory or nonobligatory, can involve animals, plants, and other microbes, and can include complex multipartner interactions. Some symbionts live internally (endosymbionts), whereas others are affixed to the outside surfaces of their partners (ectosymbionts). **Figure 7.13** provides an overview of the major categories of symbiosis with examples.

The most intimate and interdependent type of symbiosis is **mutualism,** a relationship in which all members share in the benefits of the relationship. Many mutualistic associations have developed over hundreds of millions, and even billions, of years of shared evolution, a process termed **coevolution.** Coevolving symbionts remain in very close contact and must evolve together to sustain themselves. Any changes in partner A exert a selective pressure on partner B to adapt to these changes, and vice versa. These adaptations occur at the genetic level, and in some cases genes are

4. Although it is sometimes misused as a synonym for mutualism, be aware that the term *symbiosis* does not refer only to mutually beneficial associations, as parasitism is also a form of symbiosis.
 symbiosis, symbioses (sim-bye-oh′-sis, sym-bye-oh′-seez) Gr. sym, together, and bios, to live

Figure 7.13 **Part 1—Symbiosis: A shared existence.** These relationships vary in the degree of interaction and its outcome. (a1): *Source:* Scott Bauer/USDA; (a2): © Louisa Howard - Dartmouth Electron Microscope Facility; (b1): © WaterFrame/Alamy Stock Photo; (b2): Courtesy Todd LaJeunesse; (c): Courtesy of Dr. Ralf Wagner; (d): *Source:* Nick Hill/USDA; (e): © Kathy Park Talaro; (f1): *Source:* CDC-DPDx /Image courtesy of Dr. CSBR Prasad, Vindhya Clinic and Diagnostic Lab, India; (f2): *Source:* Janice Carr/CDC; (g): *Source:* Cynthia Goldsmith/CDC; (h1): *Source:* James Gathany/CDC; (h2): *Source:* Blaine A. Mathison/CDC DPDX

Obligate Mutualism Organisms are so intimately associated that they require each other to survive. Root nodules **(a1)** have nitrogen-fixing endosymbiotic bacteria that supply the plant with usable nitrogen and provide a nurturing habitat for the bacteria **(a2 inset)**. Jellyfish **(b1)** and corals rely on endosymbiotic algae called dinoflagellates **(b2 inset)** for survival.

(a1) Root nodules on a legume; **(a2)** *Bradyhizobium* bacteria inside a nodule (3,000×)

(b1) A *Casseopeia* jellyfish gets its color and nutrition from **(b2)** dinoflagellates (400×).

Nonobligate Mutualism Organisms interact at the cellular level for mutual benefit, but they can be separated and live apart. The protozoan in **(c)** engulfs the algae but absorbs the nutrients they release and shelters them. **(d)** The plant supplies nutrients to the fungus and the fungus protects the plants against drying and insects.

(c) The ciliophoran *Euplotes* (800×) harbors unicellular green algae.

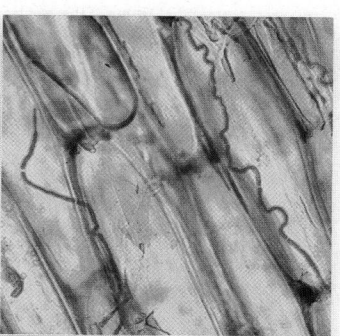

(d) Fungal hyphae (blue filaments—500×) growing in close contact with the cells of a grass leaf.

Commensalism The members have an unequal relationship. One partner is favored by the association and the other is not harmed or helped. **(e)** Tiny colonies of *Haemophilus* absorb required growth factors given off by *Staphylococcus*. **(f1 and 2)** Human commensals associated with the epidermis make a living off flakes and excretions, generally with neutral effects.

(e) Satellite *Haemophilus* colonies clustered around *Staphylococcus* (white line of growth).

(f1) *Demodex* mite lives in or around human hair follicles. (100×) and **(f2)** *Micrococcus luteus* bacteria live on the skin surface (2,000×)

Parasitism A microbe invades the sterile regions of a host and occupies its tissues and cells, causing some degree of damage. **(g)** All viruses are parasites that invade cells and take over their function. **(h)** Malaria shows multilevel parasitism. Mosquitoes **(h1)** are blood-sucking ectoparasites of humans that carry their own parasites **(h2)** that can also infect humans.

(g) Sin Nombre hantavirus, a human pathogen carried in the waste of mice (5,000×).

(h1) The malaria vector, a female *Anopheles* mosquito; **(h2)** malaria parasites (*Plasmodium*, 1,000×) from blood

Figure 7.13 **Part 2—Additional microbial adaptations.** (j): © MedicalRF.com; (k): © Paul Bertner/Getty Images RF; (l1): *Source:* Jörg Barke, Ryan F Seipke, Sabine Grüschow, Darren Heavens, Nizar Drou, Mervyn J Bibb, Rebecca JM Goss, Douglas W Yu, and Matthew I Hutchings; (l2): © Microfield Scientific Ltd/Science Source;

Syntrophy Microbes sharing a habitat feed off substances released by other organisms, **(i)** *Azotobacter* releases NH_4 that feeds *Cellulomonas*, and *Cellulomonas* degrades cellulose that feeds *Azotobacter*. **(j)** A dust mite lives in human settings and feeds off dead skin flakes.

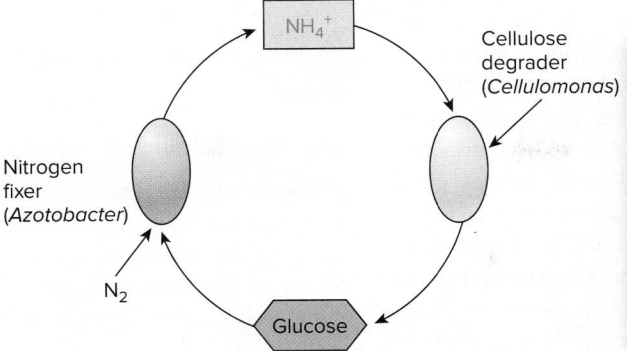

(i) Cycle of cross-feeding in two soil bacteria

(j) *Dermatophagoides* mite (100×)

Amensalism One member of an association produces a substance that harms or kills another. **(k)** Ants have complex symbiotic relationships that involve mutualism and amensalism with fungi and bacteria. **(l1 and l2)** In the amensal phase of their ecology, the ants cultivate actinomycetes to protect their habitat from microbial pests.

(k) Leaf cutting ants gathering food for a fungus garden that is their source of food.

Fungus

Zone with no growth where antibiotic inhibits undesirable fungus

Actinomycete colony that produces antibiotic

(l1) Antibiosis by an actinomycete against a pathogenic fungus; **(l2)** micrograph of actinomycete (800×)

exchanged between the partners. Over very long periods, the mutualists can become completely interdependent. We see this in *Buchnera* endosymbionts of certain aphids. These bacteria provide the aphid with needed amino acids and the aphid provides them a protective habitat. The bacteria are now so obligate that they have lost a large part of their genome and cannot survive outside the aphid. The aphids too have come to require the bacteria for survival. Be reminded that endosymbiotic bacteria likely gave rise to the mitochondria and chloroplasts of eukaryotic cells. Other well-studied examples of obligate mutualism can be found in termites, corals, root nodules, tube worms, and ants (**figure 7.13a,** and **7.2 Secret World of Microbes**).

The nondependent forms of mutualism are also referred to as **cooperation.** Here, organisms gain mutual benefit from their association, but they can survive independently outside of the partnership. Many of these cooperative relationships are coevolving to greater dependency. One example is the protozoan *Euplotes,* which harbors endosymbiotic algae in its cells (**figure 7.13c**). Although both members can survive apart from this mutualistic interaction, they appear to receive significant survival benefits from being together.

Other examples include fungi that grow within plant tissues and protect the plant from drought and insects while being harbored and fed (**figure 7.13d**). Hot-vent Pompeii worms harbor ectosymbiotic filamentous bacteria on their surfaces to protect against high temperatures and toxic metals (7.2 Secret World of Microbes).

Another common symbiotic relationship is **commensalism.*** Commensal organisms live in close contact, but only one partner benefits, while the other receives neither benefit nor harm from the union. The members of the association are called **commensals.** One form of commensalism observed in cultured microbes is termed the "satellite phenomenon," in which one species releases various growth factors that are required by a nearby species to grow. The feeder microbe grows around its partner as tiny colonies, while the passive partner is not harmed or helped (**figure 7.13e**). Many of the microbes that occupy a niche on the human body are considered commensals. They make a living by feeding off dead skin and secretions, but do not usually cause harm (**figure 7.13f1, 2**). The ongoing microbiome studies will no doubt find that many human residents are actually mutualists that contribute to human health.

The last type of symbiotic interrelationship is **parasitism,** first introduced in section 7.2. This is the one type of symbiosis that is primarily beneficial to the parasite and harmful to the host. Parasitism is widespread throughout nature. All organisms host parasites, and even some viruses have recently been found to harbor their own viral parasites! Many parasitic relationships also undergo coevolution, and it is rather common for the parasite to be completely dependent on the host.

* *commensalism* (kuh-men'-sul-izm) L. *com,* together, and *mensa,* table.

209

There are various degrees of parasitism. The most extreme form is **obligate intracellular parasitism,** which means the microbe spends all or most of its life cycle inside the host cell, from which it derives essential nutrients and other types of support. Examples include viruses (genetic and metabolic parasites) **(figure 7.13g);** *Rickettsia* and *Chlamydia* bacteria (energy parasites); and apicomplexan protozoans like *Plasmodium* (cause of malaria and hemoglobin parasites) **(figure 7.13h1, 2).** The kinds of harm that parasites do to their hosts varies from superficial damage (ringworm on the skin from a fungal infection) to significant damage (peptic ulcers) to rapid death (rabies virus). It is reasonable to think that most parasites modulate the damage they do and can even coevolve into a commensal or mutualistic existence as the host develops tolerance.

Additional interactions can be important factors in the ecology and adaptations of organisms. One of these, **syntrophy,*** or **cross-feeding,** is communal feeding between organisms sharing a habitat. In essence, products given off by one organism are usable by another **(figure 7.13i).** The organisms involved do not require this relationship for survival, but they generally benefit from it. Many syntrophic associations occur in aquatic habitats and soils within biofilms and are related to nutrient and bioelement recycling. A simple example involves a pair of free-living soil bacteria that share their metabolic products cyclically **(figure 7.13j).** *Cellulomonas* uses its enzymes to digest plant cellulose into glucose, but it cannot fix nitrogen. *Azotobacter* fixes nitrogen gas from the air and releases ammonium, but it does not digest cellulose. The glucose is a source of carbon for *Azotobacter*, and the ammonium supplies *Cellulomonas* with usable nitrogen. Even more complex interactions occur in the cycling of elements such as sulfur, phosphorus, and nitrogen.

Amensalism* describes an action of one microbe that causes an adverse effect in another microbe. It usually involves antagonism or competition and occurs in a community where microbes are sharing space and nutrient sources. Some microbes effectively compete by using up a vital nutrient to grow faster and dominate the habitat. Other microbes may release specific inhibitory chemicals into the surrounding environment. This is the description of **antibiosis**—the release of natural chemicals or **antibiotics**—that inhibit or kill microbes. Many fungi and bacteria are adapted to this survival strategy. An intriguing example can be found in the complex symbiosis of certain ants that actively cultivate specialized gardens of fungi as a source of food. To protect the fungi from other microbes, the ants also cultivate species of filamentous bacteria called actinomycetes that produce antibiotics. The ants spread the actinomycetes through their gardens to protect the fungi from invasion by parasites. **Figure 7.13k, l** features some aspects of this relationship.

Biofilms—A Microbial Conversation

Among the most common and prolific associations of microbes are **biofilms,** first described in 4.1 Making Connections. Biofilms result when organisms attach to a substrate by some form of extracellular matrix that binds them together in complex organized layers. Biofilms are so prevalent that they dominate the structure of most natural

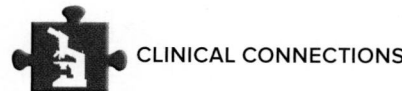

CLINICAL CONNECTIONS

Our Microbial Passengers

Microbial interrelationships obviously affect humans in many ways, but those having the most impact are microbes that naturally live on or in the body and parasites that attack the body. The normal residents, called **microbiota,** are so abundant that, cell-for-cell, they outnumber us by 3 to 10 times.* From this standpoint, one could argue that we are more microbial than human. Obviously, a community of microbes this large is bound to have enormous effects on our physiology, immunity, and genetics.

Resident microbes come from nearly all groups and include viruses, archaea, bacteria, fungi, and protozoa that have adapted to specialized niches throughout our surfaces. Certainly some are commensals that make a living on the body without harm or benefit, but a greater number are probably mutualistic, providing some form of nutrition or protection. For example, lactobacilli growing in the vagina maintain an acidic pH that can protect the female reproductive tract from infections (see photo), and other species are involved in digestive processes and the function of the intestine. The mutualist *Bacteroides thetaiotaomicron* processes complex food molecules and even regulates the expression of the human genome. The stabilizing effects of bacteria on the intestinal environment are the basis behind probiotic supplements and foods. Results of hundreds of new studies are rapidly accumulating information on the complex interrelationships we have with our microbiota.

Most of the time our 100 trillion microbial passengers do more good than harm, but if certain types are displaced from their niche or are allowed to enter the sterile tissues, they become opportunistic pathogens and can cause infections. Microbes living in biofilms can create mixed infections, meaning that several microbes acting together invade the tissues and cause damage. This is the pattern in dental caries, periodontal disease, and gas gangrene.

Stained vaginal smear reveals a large pink epithelial cell harboring gram-positive rods of *Lactobacillus,* an important member of the normal microbiota that helps protect against infection. *Source:* Dr. Mike Miller/CDC

Explain the role of broad-spectrum antibiotics in disrupting the microbial symbionts of humans and the possible problems this can cause. *Answer available on Connect.*

* *syntrophy* (sin′-trof-ee) Gr, *sym,* together, and *troph,* to feed.
* *amensalism* (ae-men′-sul-izm) Gr, *a,* not, and L. *mensa,* table.

* On average, the human body contains 10^{13} cells (30 trillion) and 10^{14} microbes (100 trillion). These figures are estimates, which is why it is more accurate to provide a range.

environments on earth. This tendency of microbes to form biofilm communities is an ancient and effective adaptive strategy. Not only do biofilms favor microbial persistence in habitats, but they also offer greater access to life-sustaining conditions. In a sense, these living networks operate as "superorganisms" that influence such microbial activities as adaptation to a particular habitat, content of soil and water, nutrient cycling, and even the course of infections.

It is generally accepted that the individual cells in the bodies of multicellular organisms such as animals and plants have the capacity to produce, receive, and react to chemical signals such as hormones made by other cells. For many years biologists regarded most single-celled microbes as simple individuals without comparable properties, other than to cling together in colonies. But these assumptions have turned out to be incorrect. It is now evident that microbes show a well-developed capacity to communicate and cooperate in the formation and function of biofilms. This is especially true of bacteria, although fungi and other microorganisms can participate in these activities.

One idea to explain the development and behavior of biofilms is termed **quorum sensing.** This process occurs in several stages, including self-monitoring of cell density, secretion of chemical signals, and genetic activation **(figure 7.14)**. Early in biofilm formation, free-floating or swimming microbes—often described as **planktonic**—are attracted to a surface and come to rest or settle down. Settling stimulates the cells to secrete a slimy or adhesive matrix, usually made of polysaccharide, that binds them to the substrate. Once attached, cells begin to release **inducer* molecules** that accumulate as the cell population grows. By this means, they can monitor the size of their own population. In time, a critical number of cells, termed a **quorum,*** accumulates and this ensures that there will be sufficient quantities of inducer molecules. These inducer molecules enter biofilm cells and stimulate specific genes on their chromosomes to begin expression.

The nature of this expression varies, but generally it allows the biofilm to react as a unit. For example, by coordinating the expression of genes that code for proteins, the biofilm can simultaneously produce large quantities of a digestive enzyme or toxin. This regulation of expression accounts for several observations made about microbial activities. For instance, it explains how saprobic microbes in soil and water rapidly break down complex substrates and how some pathogens release their toxins into the tissues simultaneously.

The effect of quorum sensing has also given greater insight into how pathogens invade their hosts and produce large quantities of substances that damage host defenses. It has been well studied in a number of pathogenic bacteria. Infections by *Pseudomonas aeruginosa* give rise to tenacious lung biofilms in cystic fibrosis patients (see the Case Study for this chapter) and some types of pneumonia. *Staphylococcus aureus* commonly forms biofilms on

Quick Search

Watch "How Biofilms Form" and "Quorum Sensing" on YouTube.

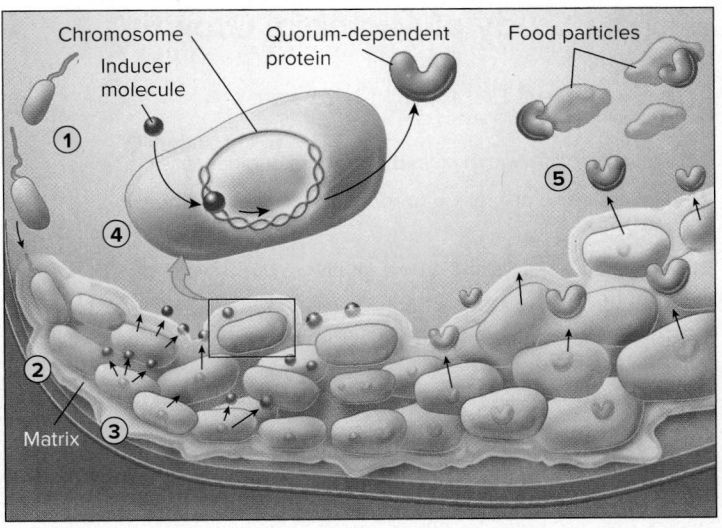

① Free-swimming cells lose their motility and settle down onto a surface or substrate.

② Cells synthesize an adhesive matrix that holds them tightly to the substrate.

③ When biofilm grows to a certain density (quorum), the cells release inducer molecules that can coordinate a response.

④ Enlargement of one cell to show genetic induction. Inducer molecule stimulates expression of a particular gene and synthesis of a protein product, such as an enzyme.

⑤ Cells secrete their enzymes in unison to digest food particles.

Process Figure 7.14 Stages in biofilm formation, quorum sensing, induction and expression, shown in progression from left to right.

inanimate medical devices and in wounds. *Streptococcus* species are the initial colonists that create dental plaque on tooth surfaces.

Although the best-studied biofilms involve just a single type of microorganism, most biofilms observed in nature are polymicrobial. In fact, many symbiotic and cooperative relationships are based on complex communication patterns among coexisting organisms. As each organism in the biofilm carries out its specific niche, signaling among the members sustains the overall partnership. Biofilms are known to be a rich ground for genetic transfers among neighboring cells that involve conjugation and transformation (see chapter 9). As our knowledge of biofilm patterns grows, it will likely lead to greater understanding of their involvement in infections and their contributions to disinfectant and drug resistance (see chapter 12).

✓ *Check Your Progress* SECTION 7.5

18. Explain the process of coevolution and how it influences the development of microorganisms.

19. Define *symbiosis* and differentiate among mutualism, commensalism, syntrophy, parasitism, and amensalism, using examples.

20. Outline quorum sensing by using a flow diagram to organize the stages.

21. Relate several advantages to communication within a biofilm.

* *inducer* (in-doos′-ur) L. *inducere,* to lead. Something that brings about or causes an effect.

* *quorum* (kwor′-uhm) L. *qui,* who. Indicates the minimum number of group members necessary to conduct business.

7.6 The Study of Microbial Growth

 Expected Learning Outcomes

23. Define *growth* and explain the process of binary fission.
24. Describe the process of population growth and how it is measured.
25. Explain the stages in the population growth curve and its practical importance.
26. Relate the techniques and tools used to detect and count cells in culture.

When microbes are provided with nutrients and the required environmental factors, they become metabolically active and grow. Growth takes place on two levels. On one level, a cell synthesizes new cell components and increases its size; on the other level, the number of cells in the population increases. This capacity for multiplication, increasing the size of the population by cell division, has tremendous importance in microbial control, infectious disease, and biotechnology. In the following sections, we focus primarily on the characteristics of bacterial growth that are generally representative of single-celled microorganisms.

The Basis of Population Growth: Binary Fission and the Bacterial Cell Cycle

The division of a bacterial cell occurs mainly through **binary,** or **transverse, fission.**[5] The term *binary* means that one cell becomes two, and *transverse* refers to the division plane forming across the width of the cell. During the bacterial cell division cycle, the parent cell enlarges, duplicates its chromosome, and forms a central transverse septum that divides the cell into two daughter cells. This process is repeated at intervals by each new daughter cell in turn; and with each successive round of division, the population increases. The stages in this continuous process are shown in greater detail in **figures 7.15** and **7.16.**

Quick Search
Search the Internet for the video "A Nice Introduction to Bacterial Growth."

The Rate of Population Growth

The time required for a complete fission cycle—from parent cell to two new daughter cells—is called the **generation,** or **doubling,**

5. Research with bacteria that are not culturable indicates that not all cells divide this way. Some bud and others divide by multiple fission.

(1) **A parent cell at the beginning of the cell cycle**
What cannot be seen is the synthesis and activity gearing up for cell division.

(2) **Chromosome replication and cell enlargement**
The parent cell duplicates the chromosome and synthesizes new structures that enlarge the cell in preparation for the daughter cells.

(3) **Chromosome division and septation**
The chromosomes affix to the cytoskeleton and are separated into the forming cells. The cell lays down a septum that begins to wall off the new cells. Other components (ribosomes) are equally distributed to the developing cells.

(4) **Completion of cell compartments**
The septum is synthesized completely through the center, and the cell membrane patches itself so that there are two separate cell chambers.

(5) **End of cell division cycle**
Daughter cells are now independent units. Some species will separate completely as shown here, while others will remain attached, forming chains or pairs, for example.

 Cell wall　 Cell membrane　○ Chromosome 1　○ Chromosome 2　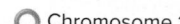 Cytoskeleton　• Ribosomes

Process Figure 7.15　**Stages in the bacterial cell cycle and binary fission of a rod-shaped bacterium.**

(b)

Number of cells	1	2	4	8	16	32
Number of generations		1	2	3	4	5
Exponential value		2^1 (2×1)	2^2 (2×2)	2^3 (2×2×2)	2^4 (2×2×2×2)	2^5 (2×2×2×2×2)

(a)

Figure 7.16 **The mathematics of population growth. (a)** Starting with a single cell, if each product of reproduction goes on to divide by binary fission, the population doubles with each new cell division or generation. This process can be represented by logarithms (2 raised to an exponent) or by simple numbers. **(b)** Plotting the logarithm of the cells produces a straight line indicative of exponential growth, whereas plotting the cell numbers arithmetically gives a curved slope. *Note that the left scale is logarithmic and the right scale is arithmetic.

time. The term *generation* has a meaning similar to that for humans. It is the period between an individual's birth and the time of producing offspring. In bacteria, each new fission cycle or generation increases the population by a factor of 2, or doubles it. Thus, the initial parent stage consists of 1 cell, the first generation consists of 2 cells, the second 4, the third 8, then 16, 32, 64, and so on. As long as the environment remains favorable, this doubling effect can continue at a constant rate. With the passing of each generation, the population will double, over and over again.

The length of the generation time is a measure of the growth rate of an organism. Compared with the growth rates of most other living things, bacteria are notoriously rapid. The average generation time is 30 to 60 minutes under optimum conditions. The shortest generation times average 5 to 10 minutes, and longest generation times require days. For example, *Mycobacterium leprae,* the cause of Hansen's disease, has a generation time of 10 to 30 days—as long as in some animals. Most pathogens have relatively short doubling times. Cultures of *Salmonella enteritidis* and *Staphylococcus aureus,* bacteria that cause food-borne illness, double in 20 to 30 minutes, which explains why leaving food at room temperature even for a short period has caused many a person to be suddenly stricken with an attack of food-borne disease. In a few hours, a population of these bacteria can easily grow from a small number of cells to several million.

Figure 7.16 shows some quantitative characteristics of growth: (1) The cell population size can be represented by the number 2 with an exponent (2^1, 2^2, 2^3, 2^4); (2) the exponent increases by one in each generation; and (3) the number of the exponent also gives the number of the generation. This growth pattern is termed **exponential.** Because these populations often contain very large numbers of cells, it is useful to express them by means of exponents or logarithms (see "Exponents," Online Appendix 3). The data from a growing bacterial population are graphed by plotting the number of cells as a function of time. The cell number can be represented logarithmically or arithmetically.

Plotting the logarithm number over time provides a straight line indicative of exponential growth. Plotting the data arithmetically gives a constantly curved slope. In general, logarithmic graphs are preferred because an accurate cell number is easier to read, especially during early growth phases.

Predicting the number of cells that will arise during a long growth period (yielding millions of cells) is based on a relatively simple concept. One could use the method of addition 2 + 2 = 4; 4 + 4 = 8; 8 + 8 = 16; 16 + 16 = 32, and so on, or a method of multiplication (for example, $2^5 = 2 \times 2 \times 2 \times 2 \times 2$), but it is easy to see that for 20 or 30 generations, this calculation could be very tedious. An easier way to calculate the size of a population over time is to use an equation such as:

$$N_f = (N_i)2^n$$

In this equation, N_f is the total number of cells in the population at some point in the growth phase, N_i is the starting number, the exponent n denotes the generation number, and 2^n represents the number of cells in that generation. If we know any two of the values, the other values can be calculated. Let us use the example of *Staphylococcus aureus* to calculate how many cells (N_f) will be present in an egg salad sandwich after it sits in a warm car for 4 hours. We will assume that N_i is 10 (number of cells deposited in the sandwich while it was being prepared). To derive n, we need to divide 4 hours (240 minutes) by the generation time (we will use 20 minutes). This calculation comes out to 12, so 2^n is equal to 2^{12}. Using a calculator or table,[6] we find that 2^{12} is 4,096.

Final number (N_f) = 10 × 4,096
= 40,960 bacterial cells in the sandwich

This same equation, with modifications, is used to determine the generation time, a more complex calculation that requires

6. See "Exponents" found at http://www.mhhe.com/talaro9 for a table with powers of 2.

knowing the number of cells at the beginning and end of a growth period. Such data are obtained through actual testing by a method discussed in the following section.

Determinants of Population Growth

In reality, a population of bacteria does not maintain its potential growth rate and does not double endlessly, because in most systems numerous factors prevent the cells from continuously dividing at their maximum rate. Quantitative laboratory studies indicate that a population typically displays a predictable pattern, or **growth curve,** over time. The method traditionally used to observe the population growth pattern is a viable count technique, in which the live cells in a culture are sampled, grown, and counted during a growth period, as described in the next section.

The Viable Plate Count: Batch Culture Method

A growing population is established by inoculating a flask containing a known quantity of sterile liquid medium with a few cells of a pure culture. The flask is incubated at that bacterium's optimum temperature and timed. The population size at any point in the growth cycle is quantified by removing a tiny measured sample of the culture from the growth chamber and plating it out on a solid medium to develop isolated colonies. This procedure is repeated at evenly spaced intervals (i.e., every hour for 24 hours) **(figure 7.17).**

Evaluating the samples involves a common and important principle in microbiology: One colony on a plate represents one cell or colony-forming unit (CFU) from the original sample. Because the CFU of some bacteria is actually composed of several cells (consider the clustered arrangement of *Staphylococcus,* for instance), using a colony count can underestimate the exact population size to an extent. This is not a serious problem, because in such bacteria the CFU is the smallest unit of colony formation and dispersal. Multiplication of the number of colonies in a single sample by the container's volume gives a fair estimate of the total population size (number of cells) at any given point. The growth curve is determined by graphing the number for each sample in sequence for the whole incubation period (see figure 7.18).

Because of the scarcity of cells in the early stages of growth, some samples can give a zero reading even if there are viable cells in the culture. The sampling itself can remove enough viable cells to alter the tabulations, but because the purpose is to compare relative trends in growth, these factors do not significantly change the overall pattern.

Stages in the Normal Growth Curve

The system of batch culturing just described is *closed,* meaning that nutrients and space are finite and there is no mechanism for the removal of waste products. Data from an entire growth period of 3 to 4 days typically produce a curve with a series of phases termed the lag phase, the exponential growth (log) phase, the stationary phase, and the death phase **(figure 7.18).**

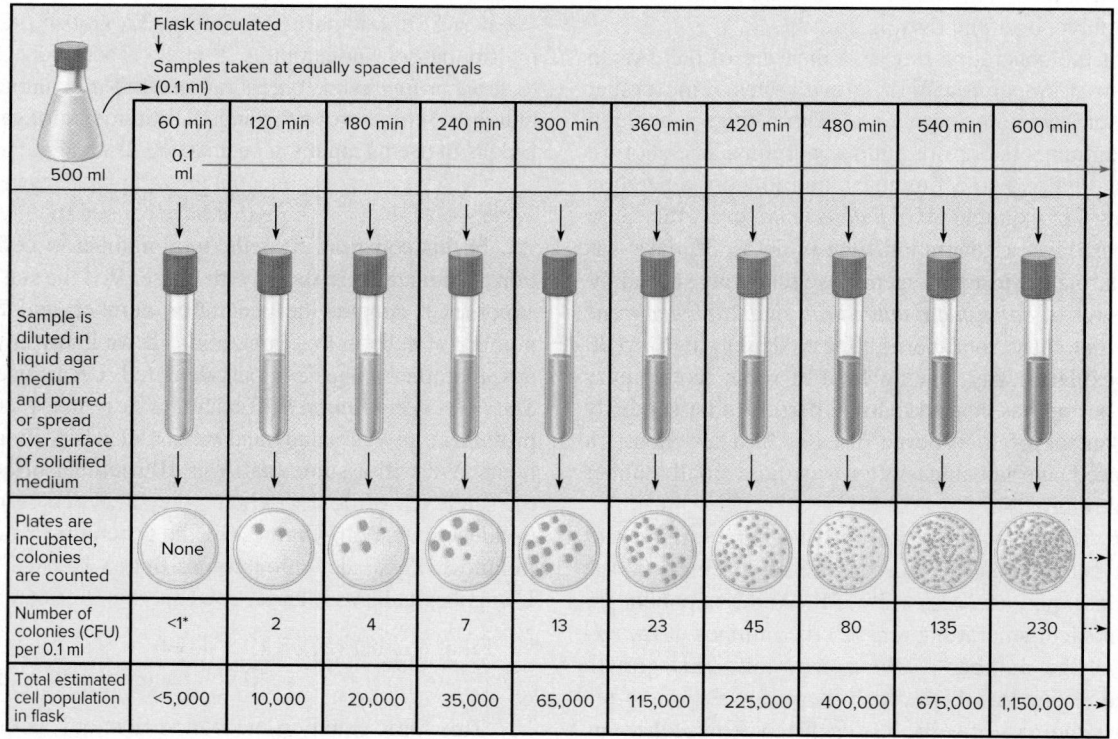

Flask inoculated ↓ Samples taken at equally spaced intervals (0.1 ml)	60 min 0.1 ml	120 min	180 min	240 min	300 min	360 min	420 min	480 min	540 min	600 min
Plates are incubated, colonies are counted	None									
Number of colonies (CFU) per 0.1 ml	<1*	2	4	7	13	23	45	80	135	230
Total estimated cell population in flask	<5,000	10,000	20,000	35,000	65,000	115,000	225,000	400,000	675,000	1,150,000

*The culture is probably in lag period with no new cells added.

Figure 7.17 The general steps in a viable plate count. 1. Place a few cells into a sterile liquid medium. **2.** Incubate this culture over a period of several hours. **3.** Sample the broth at regular intervals during incubation. **4.** Plate each sample onto solid media. **5.** Count the number of colonies present after incubation.

Figure 7.18 **The growth curve in a bacterial culture.** On this graph, the number of viable cells expressed as a logarithm (log) is plotted against time. See text for discussion of the various phases. Note that with a generation time of 30 minutes, the population has risen from 10 (10^1) cells to 1,000,000,000 (10^9) cells in only 16 hours. Even though they are not counted, dead cells are indicated to show stages in the cycle where death affects the appearance of the curve.

The **lag phase** is an early "flat" period on the graph when the population appears not to be growing or is growing at less than the exponential rate. Growth lags primarily because

1. the newly inoculated cells require a period of adjustment, enlargement, and synthesis of DNA, enzymes, and ribosomes;
2. the binary fission cycle has not reached its maximum rate; and
3. the population of cells is so sparse or dilute that the sampling misses them.

The length of the lag period varies from one population to another, depending on the condition of the microbes and medium. It is important to note that even though the population of cells is not increasing (growing), individual cells are metabolically active as they synthesize new components and prepare to divide.

The cells reach the maximum rate of cell division during the **exponential growth (logarithmic or log) phase,** a period during which the curve increases geometrically. This phase will continue as long as cells have adequate nutrients and the environment is favorable. During this phase, the population realizes its potential generation time, and growth is balanced and genetically coordinated.

At the **stationary growth phase,** the population reaches a size limit in which some cells divide more slowly or stop dividing entirely and may even have died. The curve levels off because the rate of cell inhibition or death balances out the rate of multiplication. The number of viable cells has reached maximum and remains relatively constant during this period. The decline in the growth rate is caused by several factors. A common reason is the depletion of nutrients and oxygen. Another is that the increased cell density often causes an accumulation of organic acids and other toxic biochemicals.

As the limiting factors intensify, the population begins to show a decline, seen as a downward slope on the curve. This stage has been called the **death phase,** but what is actually happening is more complex. Some cells go into dormancy and remain viable but do not grow. Some cells enter a starvation mode that helps them resist the lack of

nutrients and other factors. Depending upon the species, some microbes could remain in these states for long periods. It is also apparent that many cells go through a programmed cell death and lyse during this phase, which accounts for the overall reduction in population size. Under these conditions, it is possible for persistent members of the culture to survive by using the nutrients released by the dead cells.

Practical Importance of the Growth Curve The tendency for populations to exhibit phases of rapid growth, slow growth, and death has important implications in microbial control, infection, food microbiology, and culture technology. Antimicrobial agents such as heat and disinfectants rapidly accelerate the death phase in all populations, but microbes in the exponential growth phase are more vulnerable to these agents than are those that have entered the stationary phase. In general, actively growing cells are more vulnerable to conditions that disrupt cell metabolism and binary fission.

Growth patterns in microorganisms can account for the stages of infection (see chapter 13). A person shedding bacteria in the early and middle stages of an infection is more likely to spread it to others than is a person in the late stages. The course of an infection is also influenced by the relatively faster rate of multiplication of the microbe, which can overwhelm the slower growth rate of the host's own cellular defenses.

Understanding the stages of cell growth is crucial for work with cultures. Sometimes a culture that has reached the stationary phase is incubated under the mistaken impression that enough nutrients are present for the culture to continue to multiply. In most cases it is unwise to incubate a culture beyond the stationary phase, because doing so will reduce the number of active cells and the culture could die out completely. It is also preferable to do stains (an exception is the spore stain) and motility tests on young cultures, because the cells will show their natural size and correct reaction, and motile cells will have functioning flagella.

For certain research or industrial applications, closed-batch culturing with its four phases is inefficient. The alternative is an

Figure 7.19 **Turbidity measurements as indicators of population density and amount of growth. (a)** Holding a broth to the light is a practical method of checking for gross differences in cloudiness (turbidity). The broth on the left is transparent, indicating little or no growth; the broth on the right is cloudy and opaque, indicating heavy growth. **(b)** The eye is not sensitive enough to pick up fine degrees in turbidity; more sensitive measurements can be made with a spectrophotometer. A culture placed into the machine is exposed to a beam of light. A detector picks up the amount of light transmitted through the tube and registers this on a screen as absorbance. In **(b1)**, very little light is being scattered and the absorbance reads low, indicating few cells are present. In **(b2)**, more light is scattered, and the absorbance reads higher, indicating significant numbers of cells and growth. (a): © K. Talaro

automatic growth chamber called the **chemostat,** or continuous culture system. This device can admit a steady stream of new nutrients and siphon off used media and old bacterial cells, thereby stabilizing the growth rate and cell number. The chemostat is very similar to industrial fermentors used to produce vitamins and antibiotics. It has the advantage of maintaining the culture in a biochemically active state and preventing it from entering the death phase.

Other Methods of Analyzing Population Growth

Microbiologists have developed several alternative ways of analyzing bacterial growth qualitatively and quantitatively. One of the simplest methods for estimating the size of a population is through turbidometry. This technique relies on the simple observation that a tube of clear nutrient solution becomes cloudy, or **turbid,** as microbes grow in it. In general, a greater turbidity indicates a higher microbial count. Differences in turbidity can be measured by means of sensitive instruments such as a spectrophotometer, which measures the ability of cells to scatter light as it passes through a liquid culture, a phenomenon called absorbance. The higher the absorbance measurement, the more growth or cell mass there is in the culture, and vice versa **(figure 7.19*b*).**

Quantifying and Analyzing Cultures

Turbidity readings are useful for evaluating relative amounts of growth, but if a more quantitative evaluation is required, the viable colony count or some other enumeration (counting) procedure is necessary. The **direct,** or **total, cell count** involves counting the number of cells in a sample microscopically **(figure 7.20).** This technique, very similar to that used in blood cell counts, employs a special microscope slide (cytometer) calibrated to accept a tiny sample that is spread over a premeasured grid. The cell count from a cytometer can be used to estimate the total number of cells in a larger sample (for instance, of milk or water). One inherent inaccuracy in this method as well as in spectrophotometry is that no distinction can be made between dead and live cells, both of which

Figure 7.20 **Direct microscopic count of bacteria.** A small sample is placed on the grid under a cover glass. Individual cells, both living and dead, are counted. This number can be used to calculate the total count of a sample.

are included in the count. Additional procedures that can assay the microbial content of water are the most probable number (MPN) and membrane filtration techniques discussed later in the book.

Counting can be automated by sensitive devices such as the *Coulter counter,* which electronically scans a culture as it passes through a tiny pipette. As each cell flows by, it is detected and registered on an electronic sensor. A **flow cytometer** is a device that works on a similar principle. It is widely used for counting, sorting, sizing, and identifying cells in suspension **(figure 7.21*a*).** For most applications, the cells to be analyzed have to be stained with different fluorescent dyes. They are streamed one at a time through a flow chamber and illuminated by a laser that causes each one to fluoresce in a specific color. As each color signal is detected, it is compiled in graph form **(figure 7.21*b*).** Final separation is accomplished by magnets that pull the cells into collection chambers according to their charge. Some of the more common applications include blood cell counts and blood cell separation and identification; analysis of cancer cells; and differentiation of gram-negative and gram-positive bacteria. Even live and dead cells can be differentiated with this method.

Although flow cytometry can be used to count bacteria in natural samples without the need for culturing them, it still requires

(a)

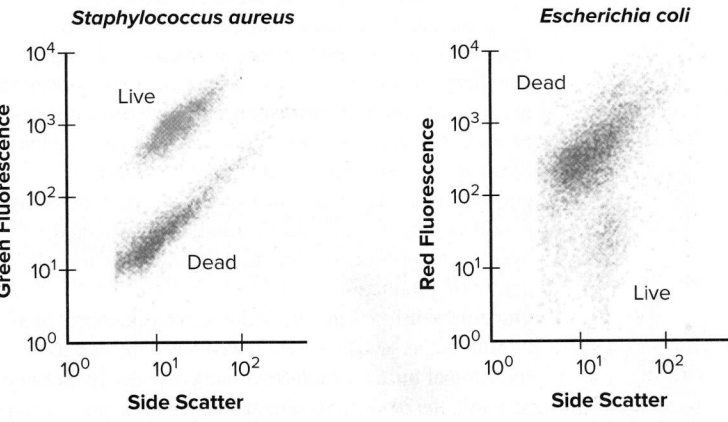

(b)

Figure 7.21 Design and functions of a flow cytometer. (a) The main body of the device is a sheath and flow chamber, which holds fluid and cells. As the labeled cells pass through it, they are lasered and detected by the color they emit. The cells are then diverted into collection chambers. **(b)** A graphic readout analyzes the pattern of cells detected. The result here reveals live and dead bacterial cells stained with red and green fluorescent colors. Each dot on the graph represents one bacterial cell. To see an animation of this process go to http://www.helmholtz-muenchen.de/imi/zs/e-prinzip.html.

fluorescent labeling of the cells, which is not always possible. In this event, there are newer DNA-based techniques (for instance, real-time PCR) that can detect and differentiate microbes that have not been isolated, cultured, or identified.

Check Your Progress SECTION 7.6

22. What is microbial growth? Explain the stages in the bacterial cell cycle and binary fission.
23. Why is growth called exponential? What causes the pattern of the growth curve?
24. Contrast two different methods of detecting and counting microbial populations, and discuss their advantages and disadvantages.
25. Explain the relationship between colony counts and colony-forming units. Why can one use the number of colonies as an index of population size?

CASE STUDY Part 2

Cystic fibrosis is the most common autosomal recessive disease reported in the United States, affecting approximately one person in 2,500. Most cases are diagnosed within the first three years of life, although a number of cases are not diagnosed until adulthood, as happened with Molly Kenner. Currently there are about 30,000 children and adults living with some manifestations of CF. Although pulmonary symptoms are a dominant feature, CF also effects any organ or system that requires open chloride channels to function normally. This would include the gastrointestinal tract, liver, pancreas, and reproductive tract.

The exact pathology of CF is still being studied, but it appears that when lung epithelial cells cannot secrete chloride into the extracellular spaces, the osmosis of water into the lung mucus is also greatly reduced. This drying effect causes thick plugs of mucus to collect within the alveoli and along the smaller bronchioles. As a consequence, the natural defenses, primarily cilia and white blood cells, are unable to trap bacteria and remove them from the lungs. Inhaled bacteria that remain in this environment can become firmly entrenched.

This situation sets the scene for *Pseudomonas aeruginosa,* the bacterial pathogen that causes the greatest morbidity and mortality in CF. Around 80% of CF patients will eventually become colonized by this species, which stimulates a constant state of inflammation that damages the airways and alveoli. A significant factor in the life cycle of *P. aeruginosa* is its formation of biofilms through quorum sensing. The most common strain found in CF patients is the "mucoid" variety, which produces thick, sticky polysaccharide slime layers that help it adhere to the lung epithelium. Once established, it undergoes physiological changes that increase its virulence. The biofilm is extremely difficult to treat and eradicate.

■ *Explain the possible mechanisms by which Pseudomonas aeruginosa invades and creates biofilms through quorum sensing.*

■ *What are some major difficulties in treating cystic fibrosis as it relates to this infection?*

To conclude this Case Study, go to Connect.

 Chapter Summary with Key Terms

7.1 Microbial Nutrition
Environmental factors that have the greatest impact on microbes are nutrients, temperature, pH, amount of available water, atmospheric gases, light, pressure, and other organisms. **Nutrition** consists of taking in chemical substances (nutrients) and assimilating and extracting energy from them.
 A. Substances required for survival are **essential nutrients**—usually containing the elements (C, H, N, O, P, S, Na, Cl, K, Ca, Fe, Mg). Essential nutrients are considered **macronutrients** (required in larger amounts) or **micronutrients** (trace elements required in smaller amounts—Zn, Mn, Cu). Nutrients are classed as either **inorganic** or **organic.**
 B. A **growth factor** is an organic nutrient (amino acid and vitamin) that cannot be synthesized and must be provided.

7.2 Classification of Nutritional Types
 A. An **autotroph** depends on carbon dioxide for its carbon needs. An autotroph that derives energy from light is a **photoautotroph;** one that extracts energy from inorganic substances is a **chemoautotroph. Methanogens** are chemoautotrophs that produce methane.
 B. A **heterotroph or chemoorganotroph** acquires carbon from organic molecules originating from the bodies of dead or living organisms. A **saprobe** is a decomposer that feeds primarily upon dead organic matter, and a **parasite** feeds from a live host and usually causes harm. Parasites that cause damage or death are called **pathogens.**

7.3 Transport: Movement of Substances Across the Cell Membrane
 A. A microbial cell must take in nutrients from its surroundings by transporting them across the cell membrane.
 B. Passive transport involves the natural movement of substances down a concentration gradient and requires no additional energy **(diffusion).** A form of passive transport that can move specific substances is **facilitated diffusion.**
 C. **Osmosis** is diffusion of water through a selectively permeable membrane.
 D. Osmotic changes that affect cells are **hypotonic** solutions, which contain less solute than the cell, and **hypertonic** solutions, which contain more solute than the cell. **Isotonic** solutions have the same solute concentration as the inside of the cell. A **halophile** thrives in hypertonic surroundings, and an **obligate halophile** requires a salt concentration of at least 15% but grows optimally in 25%.
 E. In **active transport,** substances are taken into the cell by a process that consumes energy. In **group translocation,** molecules are altered during transport.
 F. **Phagocytosis** and **pinocytosis** are forms of active transport in which bulk quantities of solid and fluid material are taken into the cell.

7.4 Environmental Factors That Influence Microbes
Every organism adapts to a particular habitat and niche.
 A. *Temperature:* An organism exhibits **optimum, minimum,** and **maximum temperatures.** Organisms that cannot grow above 20°C but thrive below 15°C and continue to grow even at 0°C are known as **psychrophiles. Mesophiles** grow from 10°C to 50°C, having temperature optima from 20°C to 40°C. The growth range of **thermophiles** is 45°C to 80°C. Extreme thermophiles may grow to temperatures of 250°C.
 B. *Oxygen Requirements:* The ecological need for free oxygen (O_2) is based on whether a cell can handle toxic by-products such as superoxide and peroxide.
 1. **Aerobes** grow in normal atmospheric oxygen and have enzymes to handle toxic oxygen by-products. An aerobic organism capable of living without oxygen if necessary is a **facultative anaerobe.** An aerobe that requires a small amount of oxygen but does not grow under anaerobic conditions is a **microaerophile.**
 2. **Strict (obligate) anaerobes** do not use free oxygen and cannot produce enzymes to dismantle reactive oxides. They are actually damaged or killed by oxygen. An **aerotolerant anaerobe** cannot use oxygen for respiration yet is not injured by it.
 C. *Effects of pH:* Acidity and alkalinity affect the activity and integrity of enzymes and the structural components of a cell. Optimum pH for most microbes ranges from 6 to 8. **Acidophiles** prefer a pH below 7, **alkalinophiles** prefer a pH above 7, and **neutrophils** grow best around pH 7.
 D. *Other Environmental Factors:* Radiation and barometric pressure affect microbial growth. A **barophile** is adapted to life under high pressure (bottom dwellers in the ocean, for example).

7.5 Ecological Associations Among Microorganisms
 A. *Microbial Interactions:* Microbes coexist in varied relationships in nature.
 1. Types of **symbiosis** are **mutualism,** a reciprocal, obligatory, and beneficial relationship between two organisms, and **commensalism,** an organism receiving benefits from another without harming the other organism in the relationship. **Parasitism** occurs between a host and an infectious agent; the host is harmed in the interaction.
 2. **Syntrophism** is a beneficial but not obligatory coexistence in which organisms cross-feed. **Amensalism** entails competition, inhibition, and injury directed against an opposing organism. A special case is antibiotic production.
 3. **Quorum sensing** is an interaction among members of a biofilm that results in a coordinated reaction, such as secretion of **inducer molecules** and enzymes. By behaving as a unit, the biofilm remains stable and organized. It can occupy and exploit a wide variety of habitats.

7.6 The Study of Microbial Growth
 A. During the bacterial cell cycle, the parent cell splits to form a pair of daughter cells by means of **binary,** or **transverse, fission.**
 B. It involves replication of the chromosome and the equal division of other cell components.
 C. The duration of each division is called the **generation,** or **doubling, time.** A population theoretically doubles with each generation, so the growth rate is **exponential,** and each cycle increases in geometric progression.
 D. A **growth curve** is a graphic representation of a closed population over time, using an estimate of live cells called a viable count. The initial flat period of the curve is called the **lag phase,** followed by the **exponential growth phase,** in

which viable cells increase in logarithmic progression. Adverse environmental conditions create a plateau, or **stationary growth phase.** In the **death phase,** nutrient depletion and waste buildup cause increased cell death.

E. Cell numbers can be counted by a viable plate count, microscope counting chamber, Coulter counter, or, **flow cytometer.** Cell growth can also be determined by turbidometry and a total cell count.

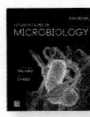

Level I. Knowledge and Comprehension

These questions require a working knowledge of the concepts in the chapter and the ability to recall and understand the information you have studied.

 Multiple-Choice Questions

Select the correct answer from the answers provided. For questions with blanks, choose the combination of answers that most accurately completes the statement.

1. An organic nutrient essential to an organism's metabolism that cannot be synthesized itself is termed a/an
 a. trace element c. growth factor
 b. micronutrient d. mineral

2. The source of the necessary elements of life is
 a. an inorganic environmental reservoir
 b. the sun
 c. rocks
 d. the air

3. An organism that can synthesize all its required organic components from CO_2 using energy from the sun is a
 a. photoautotroph c. chemoautotroph
 b. photoheterotroph d. chemoheterotroph

4. An obligate halophile requires high
 a. pH c. salt
 b. temperature d. pressure

5. Chemoautotrophs can survive on _____ alone.
 a. minerals c. minerals and CO_2
 b. CO_2 d. methane

6. Which of the following substances are required by *all* organisms?
 a. organic nutrients c. growth factors
 b. inorganic nutrients d. oxygen gas

7. A pathogen would most accurately be described as a
 a. parasite c. saprobe
 b. commensal d. symbiont

8. Which of the following is true of passive transport?
 a. it requires a gradient c. it includes endocytosis
 b. it uses the cell wall d. it only moves water

9. A cell exposed to a hypertonic environment will _____ by osmosis.
 a. gain water c. neither gain nor lose water
 b. lose water d. burst

10. Active transport of a substance across a membrane requires
 a. a gradient c. water
 b. the expenditure of ATP d. diffusion

11. Psychrophiles would be expected to grow
 a. in hot springs c. at refrigeration temperatures
 b. on the human body d. at low pH

12. Which of the following is *not* involved in quorum sensing?
 a. accumulation of inducer molecules
 b. a certain level of biofilm density
 c. self-monitoring
 d. release of genetic material

13. Superoxide ion is toxic to strict anaerobes because they lack
 a. catalase c. dismutase
 b. peroxidase d. oxidase

14. The time required for a cell to undergo binary fission is called the
 a. exponential growth rate c. generation time
 b. growth curve d. lag period

15. In a viable plate count, each _____ represents a _____ from the sample population.
 a. cell, colony c. hour, generation
 b. colony, cell d. cell, generation

16. The stage in population growth with the highest rate of cell division is the
 a. stationary phase c. exponential phase
 b. lag phase d. enlargement phase

 Case Study Review

1. What dysfunction tied to cystic fibrosis increases the establishment of *Pseudomonas aeruginosa* in the lungs?
 a. increased chloride transport
 b. decreased chloride transport
 c. lack of mucus
 d. pneumonia

2. Select a factor or factors that favor the invasion and growth of *Pseudomonas aeruginosa* in the lungs.
 a. fimbriae
 b. mucoid glycocalyx
 c. blockage of host defenses.
 d. b and c
 e. a, b, and c

3. Describe how active transport and osmosis play a part in the normal lung compared to that of cystic fibrosis.

Writing Challenge

For each question (except 7), compose a one- or two-paragraph answer that includes the factual information needed to completely address the question. Check Your Progress questions can also be used for writing-challenge exercises.

1. Look at the following diagrams and predict in which direction osmosis will take place. Use arrows to show the net direction of osmosis. Is one of these microbes a halophile? Which one?

2. Describe the basic metabolism of methanogens.

3. Discuss extreme environments and types of extremophiles that live there.

4. Differentiate between mutualism and syntrophy with examples.

5. a. What biochemical events in quorum sensing ensure that a biofilm can function as a unit?
 b. Explain how biofilm self-monitoring would favor infections.

6. Explain what is happening to the population at points A, B, C, and D in the following diagram.

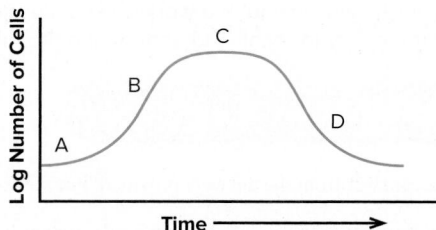

7. Fill in the following table:

Source of Carbon	Usual Source of Energy	Example
Photoautotroph		
Photoheterotroph		
Chemoautotroph		
Chemoheterotroph		
Saprobe		
Parasite		

Concept Mapping

On Connect you can find an Introduction to Concept Mapping that provides guidance for working with concept maps and a sample concept map for this chapter.

Level II. Application, Analysis, Evaluation, and Synthesis

These problems go beyond just restating facts and require higher levels of understanding and an ability to interpret, problem solve, transfer knowledge to new situations, create models, and predict outcomes.

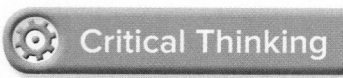
Critical Thinking

Critical thinking is the ability to reason and solve problems using facts and concepts. These questions can be approached from a number of angles, and in most cases, they do not have a single correct answer.

1. a. Is there a microbe that could grow on a medium that contains only the following compounds dissolved in water: $CaCO_3$, $MgNO_3$, $FeCl_2$, $ZnSO_4$, and glucose? Defend your answer.
 b. A useful mnemonic device for keeping track of major elements required by most organisms is: **C HOPKINS Ca Fe, Na Cl Mg**ood. Read aloud, it becomes "See Hopkin's Cafe, NaCl (salt) M good!" What does each letter stand for?

2. Describe how one might determine the nutrient requirements of a microbe from Mars. If, after exhausting various nutrient schemes, it still does not grow, what other factors might one take into account?

3. Patients with ketoacidosis associated with diabetes are especially susceptible to fungal infections. Propose why this might be the case.

4. a. Explain how osmotic pressure and pH are used in preserving foods.
 b. What effects do they have on microbes?

5. Provide some suggestions for treating anaerobic infections by manipulating the gas content of the environment.

6. Could anaerobes and aerobes coexist in the same habitat? Why or why not?

7. a. If an egg salad sandwich sitting in a warm car for 4 hours develops 40,960 bacterial cells, how many more cells would result with just one more hour of incubation? Use the same criteria that were stated in the sample problem under the heading "Rate of Population Growth" in section 7.6.

b. How many would result after 10 additional hours of incubation?
c. What do your answers tell you about using clean techniques in food preparation and storage (other than aesthetic considerations)?

8. The noted microbial ecologist Martin W. Beijerinck has stated: "Everything is everywhere, the environment selects." Use this concept to explain what ultimately determines whether a microorganism can live in a certain habitat.

9. Describe the similarities and differences between a mutualistic endosymbiont and a parasitic endosymbiont.

1. Place appropriate points on the axes and draw and label separate graphs that fit a psychrotroph and an obligate thermophile.

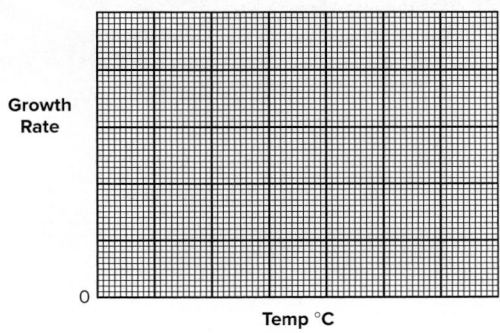

2. Enter points on the axes and graph the probable results for an acidophile, a neutrophile, and an alkalinophile.

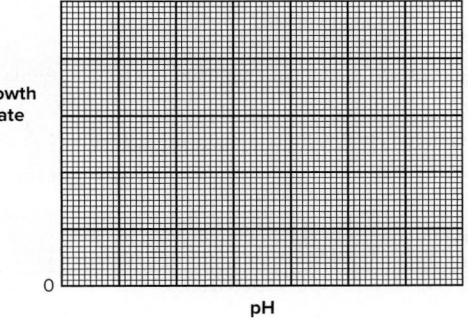

An Introduction to Microbial Metabolism:
The Chemical Crossroads of Life

Bubbles of carbon dioxide rise from a vat of fermenting brew.
© Paul Bock/Alamy Stock Photo

Clostridium botulinum—a hidden pathogen.
Source: CDC

Saccharomyces yeast—one workhorse of fermentation.
© Science Photo Library/Getty Images RF

CASE STUDY Part 1 Recipe for a Toxic Brew

This Case Study provides insight into the practical importance of microbial biochemistry to everyday life. It brings together elements of **fermentation** and **alcohol formation** and introduces the concept of anaerobic respiration by a microbe that synthesizes a toxin with deadly potential. To help understand the circumstances of this Case Study, we will revisit several concepts from chapter 7, including nutrition, temperature, pH, and metabolism in the absence of oxygen gas.

In October 2011 an inmate at Utah State Prison was brought to the infirmary with **vision** and **swallowing difficulties, muscular weakness,** and **vomiting.** The man was hospitalized for medical evaluation. When he admitted to having recently consumed an illegal, **prison-made wine,** his doctors believed they were dealing with a case of food poisoning. A number of other inmates were interviewed about drinking this substance, and 12 of them confessed that they had also consumed it about **30 hours earlier.** When some of these prisoners started having similar symptoms, medical personnel concluded that the wine was probably the source of the attack and hospitalized the other sick inmates. Specimens from the **wine, stool, stomach,** and **bloodstream** were sent for analysis. Final results were delayed because specialized testing done by the Centers for Disease Control and Prevention takes several days.

Preparation of the illicit homemade wine called "pruno" is rather common in prisons. This alcohol is usually made from a

> "The rank smell and taste of pruno often cause prisoners to gag, but they swallow it anyway."

mixture of crushed fruit, juice, water, a source of sugar, and a starter culture from an older brew. The preparation is placed in **sealed, airtight** plastic containers and incubated in warm tap water for several days. These conditions foster **anaerobic growth** and alcoholic fermentation, which will bring final alcohol levels as high as 15%. The orange-colored liquid portion is strained off and consumed. The rank smell and taste of pruno often cause prisoners to gag, but they swallow it anyway. Unfortunately their desire for alcohol overcomes the beverage's foul qualities.

Investigators continued to trace the possible events that led to this outbreak. Examination of the materials used to brew the wine yielded bits of a baked **potato,** which is a departure from the usual "recipe." One of the inmates admitted that he had stored the potato in a **sealed container** for several days before using it. Other batches of pruno that had been made at the same time did not contain potato and did not cause illness.

- *What substances and conditions are required for fermentation, and what are its usual products?*

- *What possible signs, if any, might have indicated that the prison brew was tainted?*

To continue the Case Study, go to Case Study Part 2 at the end of the chapter.

SCOPING OUT THE CHAPTER

Metabolism: The chemical traffic of cell metabolism can be compared to a busy highway going in all directions, with on ramps and off ramps, intersections, side streets, green lights, and stop signs, all producing a constant flurry of change—but in the case of metabolism the changes are to the biochemical structure of cells.

Enzymes: Chemical Gatekeepers
Like traffic signals and signs, enzymes ensure the correct biochemical reactions by providing specific sites for these changes to take place.

Metabolic Pathways—Guiding the Flow of Reactions Down the Right Track
These exist in two basic patterns

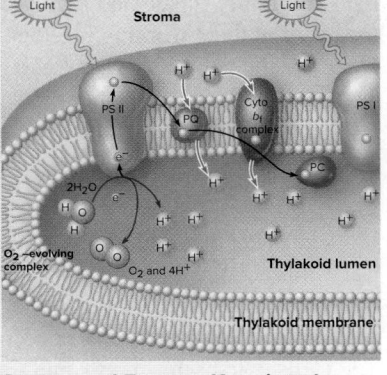

Sources of Energy: Keeping the Metabolic Machine Powered
Photosynthetic organisms operate by collecting light energy and converting it into cellular energy.

Anabolic pathways synthesize and build up cellular components and structures.

Catabolic pathways break down fuels and release their potential energy for cellular work.

The chemical reactions of life are fueled by changes to molecular bonds that can release energy.

8.1 An Introduction to Metabolism and Enzymes

 Expected Learning Outcomes

1. Define *metabolism* and differentiate its two types.
2. Describe the basic functions of enzymes in cells.
3. Outline the prominent characteristics of enzymes.
4. Explain how enzymes lower the energy required for a reaction to occur.
5. Discuss enzyme structure, and interactions between enzymes and substrates.
6. Describe the types of enzyme functions and methods of naming them.
7. Summarize key features of enzyme regulation.

Metabolism, from the Greek term *metaballein,* meaning change, pertains to all chemical workings of cells. It is a shorthand term for encapsulating almost any activity or behavior of an organism, from the general to the specific, including growth, synthesis, transport, digestion, energy release and consumption, and movement. Although metabolism entails thousands of different reactions, most of them can be placed into one of two general categories. In the case of **catabolism,** larger molecules are degraded or broken down into smaller molecules, usually with the release of energy. In the case of **anabolism,** also called biosynthesis, larger molecules are built from smaller ones, which results in the formation of cell structures. It is usually driven by energy derived from catabolism. Even though they have somewhat opposite effects, these two forms of metabolism are tightly linked and highly complementary. As we will see, the result of them working together is that all metabolism is coordinated for the cell to carry out essential life functions **(figure 8.1).** The major metabolic functions in cells and the sections of the chapter that cover them are listed here:

1. Enzymes serve as catalysts for both catabolic and anabolic activities. Because of their importance in all areas of microbiology, the remainder of section 8.1 is devoted to characteristics of enzymes.
2. Cells acquire energy from an external source (outside the cell) such as nutrients or the sun. Discussion of this idea began in chapter 7 and continues in sections 8.2 and 8.6.
3. Cells extract the energy and transfer it to a high-energy molecule, adenosine triphosphate (ATP), a universal compound for managing energy. This is accomplished by several different types of catabolic pathways, covered in sections 8.3 and 8.4.
4. Cells use ATP to power anabolic processes. Simple building blocks are bonded into larger molecules and cell structures,

thereby creating greater complexity through the input of this energy. Anabolic pathways are covered in section 8.5.

We suggest you review relevant sections in chapter 2 covering organic molecules and biochemistry.

Enzymes: Catalyzing the Chemical Reactions of Life

A microbial cell could be viewed as a microscopic factory, complete with basic building materials, a source of energy, and a genetic "blueprint" for running its extensive network of metabolic reactions. But the chemical reactions of life would never proceed without a special class of proteins called **enzymes.** Enzymes are a remarkable example of **catalysts,** chemicals that increase the rate of a chemical reaction without becoming part of the products or being consumed in the reaction. Do not make the mistake of thinking that an enzyme creates a reaction. Because of the free energy inherent in molecules, a reaction could occur spontaneously at some point even without an enzyme but at a very slow rate. A study of the enzyme urease shows that it increases the rate of the breakdown of urea by a factor of 100 trillion as compared to an uncatalyzed reaction. Because most uncatalyzed metabolic reactions do not occur fast enough to sustain cell processes, enzymes are indispensable to life. Other major qualities of enzymes are summarized in **table 8.1.**

How Do Enzymes Work?

We have said that an enzyme speeds up the rate of a metabolic reaction, but just how does it do this? During a chemical reaction, reactants are converted to products by bond formation or breakage. A certain amount of energy is required to initiate every such reaction,

Figure 8.1 **Simplified model of metabolism.** Chemical reactions that occur in cells fall into two major categories. Catabolism involves the breakdown of complex organic molecules to extract energy and form simpler end products. Anabolism uses the energy to synthesize necessary macromolecules and cell structures from simple precursors, increasing chemical complexity. The two systems interact in a continuous cycle that maintains the metabolic balance inside cells.

8.1 MAKING CONNECTIONS

Enzymes as Biochemical Levers

An analogy will allow us to envision the relationship of enzymes to the energy of activation. A large boulder sitting precariously on a cliff's edge contains a great deal of potential energy, but it will not fall to the ground and release this energy unless something disturbs it. It might eventually be disturbed spontaneously as the cliff erodes below it, but that could take a very long time. Maneuvering it with a crowbar (to overcome the resistance of the boulder's weight) will cause it to tumble down freely by its own momentum. If we relate this analogy to chemicals, the reactants are the boulder on the cliff, the activation energy is the energy needed to move the boulder, the enzyme is the crowbar, and the product is the boulder at the bottom of the cliff. Like the crowbar, an enzyme permits a reaction to occur rapidly by getting it over the energy "hump." Although this analogy concretely illustrates the concept of the energy of activation, it is imperfect in that the enzyme, unlike the person with the crowbar, does not actually add any energy to the system.

This phenomenon can be represented by plotting the energy of the reaction against the direction of the reaction. The curve of reaction is shown in the accompanying graph. Reactants must overcome the energy of activation (E_{act}) for the reaction to proceed. When a catalyst is present, it lowers the E_{act}. It does this by providing the reactants with an alternate pathway, requiring less energy, by which they can progress to the final state. Note that the products reach the same final energy state with or without an enzyme.

What is the actual physical process performed by the enzyme that speeds up a reaction? *Answer available on Connect.*

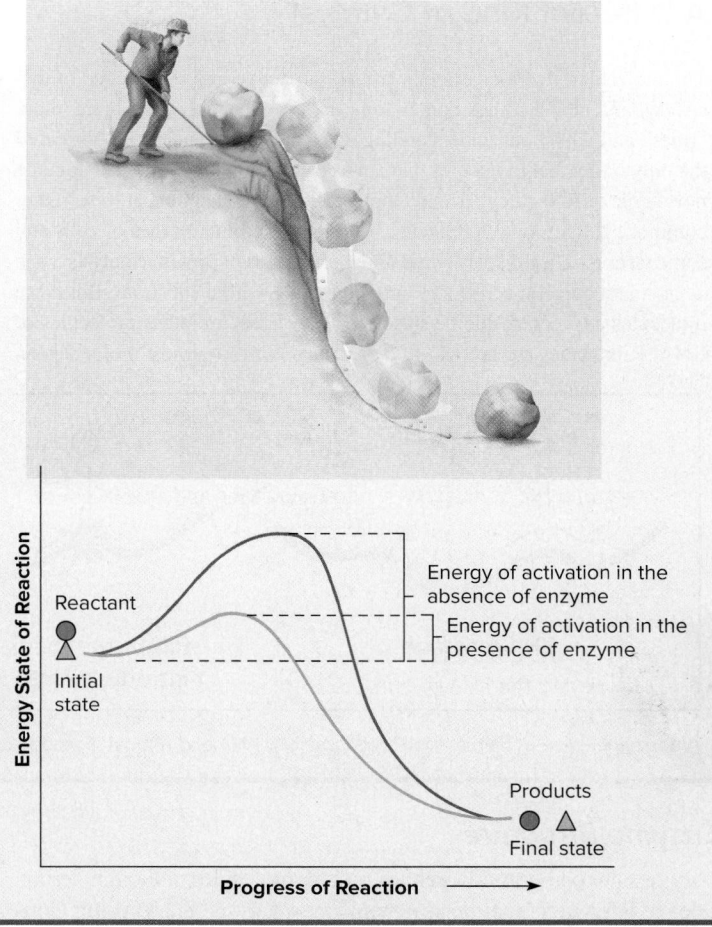

<table>
<tr><td colspan="2">TABLE 8.1 Checklist of Enzyme Characteristics</td></tr>
</table>

- Composed mostly of protein; may require nonprotein cofactors
- Act as organic catalysts to speed up the rate of cellular reactions
- Lower the activation energy required for a chemical reaction to proceed (see 8.1 Making Connections)
- Enable metabolic reactions to proceed at a speed compatible with life
- Have unique characteristics such as shape, specificity, and function
- Provide an active site for target molecules called substrates
- Are much larger in size than their substrates
- Associate closely with substrates but do not become integrated into the reaction products
- Are not used up or permanently changed by the reaction
- Can be recycled and function in extremely low concentrations
- Are greatly affected by temperature and pH
- Can be regulated by feedback and genetic mechanisms

which limits its rate. This resistance to a reaction, which must be overcome for a reaction to proceed, is measurable and is called the **energy of activation** or activation energy (**8.1 Making Connections**). In the laboratory, overcoming this initial resistance can be achieved by

- increasing thermal energy (heating) to increase molecular velocity,

- increasing the concentration of reactants to increase the rate of molecular collisions, or
- adding a catalyst.

In most living systems, the first two alternatives are not feasible, because elevating the temperature is potentially harmful, and higher concentrations of reactants are often not practical. This leaves only the action of catalysts, and enzymes fill this need efficiently and potently.

At the molecular level, an enzyme promotes a reaction by serving as a physical site upon which the reactant molecules, called **substrates,** can be positioned for various interactions. The enzyme is much larger in size than its substrate, and it presents a uniquely shaped pocket that fits only that particular substrate. Although an enzyme binds to the substrate and participates directly in changes to the substrate, it does not become a part of the products, is not used up by the reaction, and it can function over and over again. Enzyme speed, defined as the number of substrate molecules converted per enzyme per second, is well documented. Speeds range from several million for catalase to a thousand for lactate dehydrogenase. A key to understanding the roles of enzymes in metabolism is their structure.

Quick Search

Go online to find the video "Functions of Enzymes—How Do Enzymes Work?" and observe a quick animation of enzyme reactions.

8.2 MAKING CONNECTIONS

A Different Kind of Catalyst

The molecular reactions of cells are still an active and rich source of discovery, and new findings come along on a regular basis that break older "rules" and adjust our understanding. Originally proteins were considered the only biological molecules that act as catalysts. Then biologists found a novel type of RNA they termed **ribozymes.** Ribozymes appear to be rather common, with around 500 having been identified in all types of cells and some viruses. They display some of the properties of protein catalysts, such as having a specific active site and interacting with a substrate. But these molecules are remarkable because their substrates are other molecules of RNA. Ribozymes are thought to be remnants of the earliest molecules on earth that could have served as both catalysts and genetic material. In natural systems, ribozymes are involved in self-splicing or cutting of RNA molecules during processing of the genetic code, and even ribosomes function as ribozymes during protein synthesis (see chapter 9). Further research has shown that ribozymes can be designed to handle other kinds of activities, such as inhibiting gene expression. Several companies have developed and are testing ribozyme-based therapies for treating cancer and HIV infection.

Propose some effects that ribozymes have on microorganisms that would make them suitable for therapy. *Answer available on Connect.*

A ribozyme (here a "hairpin loop" type) can clip RNA and remove sections of it.

Enzyme Structure

Enzymes are composed largely of protein molecules, although certain types of RNA can function as nonenzyme catalysts (**8.2 Making Connections**). Enzymes can be classified as simple or conjugated. Simple enzymes consist of protein alone, whereas conjugated enzymes (**figure 8.2**) contain protein and nonprotein molecules. A conjugated enzyme, sometimes referred to as a **holoenzyme,** is a combination of a protein, now called the **apoenzyme,** and one or more **cofactors (table 8.2).** Cofactors are either organic molecules, called **coenzymes,** or inorganic elements (metal ions) that many enzymes require to become functional. In some enzymes, the cofactor is loosely associated with the apoenzyme by noncovalent bonds; in others, it is linked by covalent bonds.

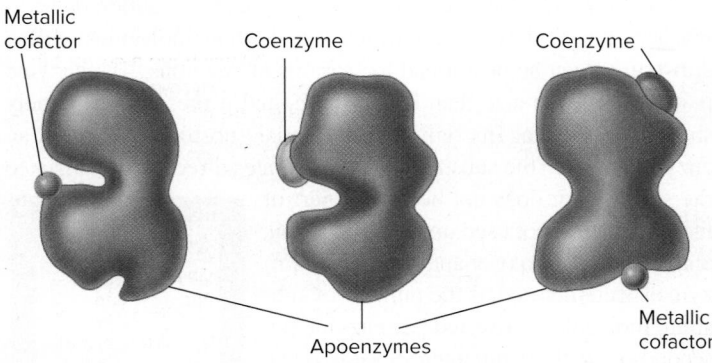

Figure 8.2 Conjugated enzyme structure. All examples have an apoenzyme (polypeptide or protein) component and one or more cofactors.

TABLE 8.2	Selected Enzymes, Catalytic Actions, and Cofactors*	
Enzyme	**Action**	**Cofactor Required**
Catalase	Breaks down hydrogen peroxide	Iron (Fe)
Oxidase	Adds electrons to oxygen	Iron, copper (Cu)
Hexokinase	Transfers phosphate to glucose	Magnesium (Mg)
Arginase	Acts on the amino acid arginine	Manganese
Nitrate reductase	Reduces nitrate to nitrite	Molybdenum (Mo)
DNA polymerase	Synthesis of DNA	Zinc (Zn) and Mg
Botulinum toxin	Hydrolyzes protein needed for vesicle transport	Zinc
Pyruvate dehydrogenase	Converts pyruvic acid to acetyl CoA and CO_2	Thiamine and magnesium
Succinate dehydrogenase	Oxidizes succinate to fumarate in the Krebs cycle	FAD (contains riboflavin)

* The first seven are metals and the last two are coenzymes containing vitamins.

Levels of Structure

(a) As the polypeptide forms intrachain bonds and folds, it assumes a three-dimensional (tertiary) state that displays an active site (AS).

(b) Because each unique polypeptide folds differently, each apoenzyme will have a differently shaped active site.

(c) More complex enzymes have a quaternary structure consisting of more than one polypeptide. New active sites may be formed by the junction of polypeptides.

Figure 8.3 How the active site and specificity of the apoenzyme arise. Review figure 2.22 for details of this process.

Apoenzymes: Specificity and the Active Site

Apoenzymes range in size from small polypeptides with about 100 amino acids and a molecular weight of 12,000 to large polypeptide conglomerates with thousands of amino acids and a molecular weight of over 1 million. Like all proteins, an apoenzyme exhibits levels of molecular complexity called the primary, secondary, tertiary, and, in larger enzymes, quaternary organization **(figure 8.3).** As we saw in chapter 2 (see figure 2.22), the first three levels of structure arise when a single polypeptide chain undergoes an automatic folding process and achieves stability by forming disulfide and other types of bonds. Folding causes the surface of the apoenzyme to acquire three-dimensional features that create the enzyme's specificity for substrates. The actual pocket where the substrate binds is called the **active site,** or **catalytic site,** and there can be from one to several such sites (figure 8.3). Because each type of enzyme has a different primary structure (type and sequence of amino acids), this gives rise to variations in folding and unique active sites.

Enzyme-Substrate Interactions

For a reaction to take place, a temporary **enzyme–substrate** union must occur at the active site. There are several explanations for the way this happens **(figure 8.4).**

Figure 8.4 Enzyme–substrate reactions. (a) When the enzyme and substrate come together, the substrate (S) will have a correct fit and position with respect to the enzyme (E). **(b)** When the ES complex is formed, it enters a transition state. During this temporary but tight interlocking union, the enzyme participates directly in breaking or making bonds. **(c)** Once the reaction is complete, the enzyme releases the products. **(d)** The induced-fit model proposes that the enzyme recognizes its substrate and adapts slightly to it so that the final binding is even more precise.

The specificity is often described as a "lock-and-key" fit in which the substrate is inserted into the active site's pocket. This is a useful analogy, but an enzyme is not a rigid lock mechanism. It is likely that the enzyme actually helps the substrate move into the active site through slight changes in its shape. This is called an *induced fit* (figure 8.4d). The bonds formed between the substrate and enzyme are weak and, of necessity, easily reversible. Once the enzyme–substrate complex has formed, appropriate reactions occur on the substrate, often with the aid of a cofactor, and a product is formed and released. The enzyme can then attach to another substrate molecule and repeat this action.

Cofactors: Supporting the Work of Enzymes

In chapter 7 you learned that microorganisms require specific metal ions called trace elements and certain organic growth factors. In many cases, the need for these substances arises from their roles as **cofactors** (table 8.2). The **metallic cofactors,** including iron, copper, magnesium, manganese, zinc, cobalt, selenium, and many others, participate in precise functions between the enzyme and its substrate. In general, metals activate enzymes, help bring the active site and substrate close together, and participate directly in chemical reactions with the enzyme–substrate complex.

Coenzymes are organic cofactors that work in conjunction with the apoenzyme to perform a necessary alteration of a substrate. The general function of a coenzyme is to remove a chemical group from one substrate molecule and add it to another substrate, thereby serving as a temporary carrier of this group **(figure 8.5).** The specific activities of coenzymes are many and varied. In section 8.3 we shall see that coenzymes carry and transfer hydrogen atoms, electrons, carbon dioxide, and amino groups. Some of the most common components of coenzymes are **vitamins,** which explains why vitamins are important to nutrition and may be required as growth factors for living things. Vitamin deficiencies prevent the complete holoenzyme from forming, which compromises both the chemical reaction and the structure or function that is dependent upon that reaction.

Classification of Enzyme Functions

Enzymes are classified and named according to characteristics such as site of action, type of action, and substrate **(8.3 Making Connections).**

Location and Regularity of Enzyme Action

Enzymes perform their tasks either inside or outside of the cell in which they were produced. After initial synthesis in the cell, **exoenzymes** are transported extracellularly, where they break down large food molecules or harmful chemicals (see discussion of hydrolysis reactions under the next heading). Examples of exoenzymes are cellulase, amylase, and penicillinase. By contrast, **endoenzymes** are retained intracellularly and function there **(figure 8.6).** Most enzymes of the metabolic pathways are of this variety.

In terms of their presence in the cell, enzymes are not all produced in equal amounts or at equal rates. Some, called **constitutive**

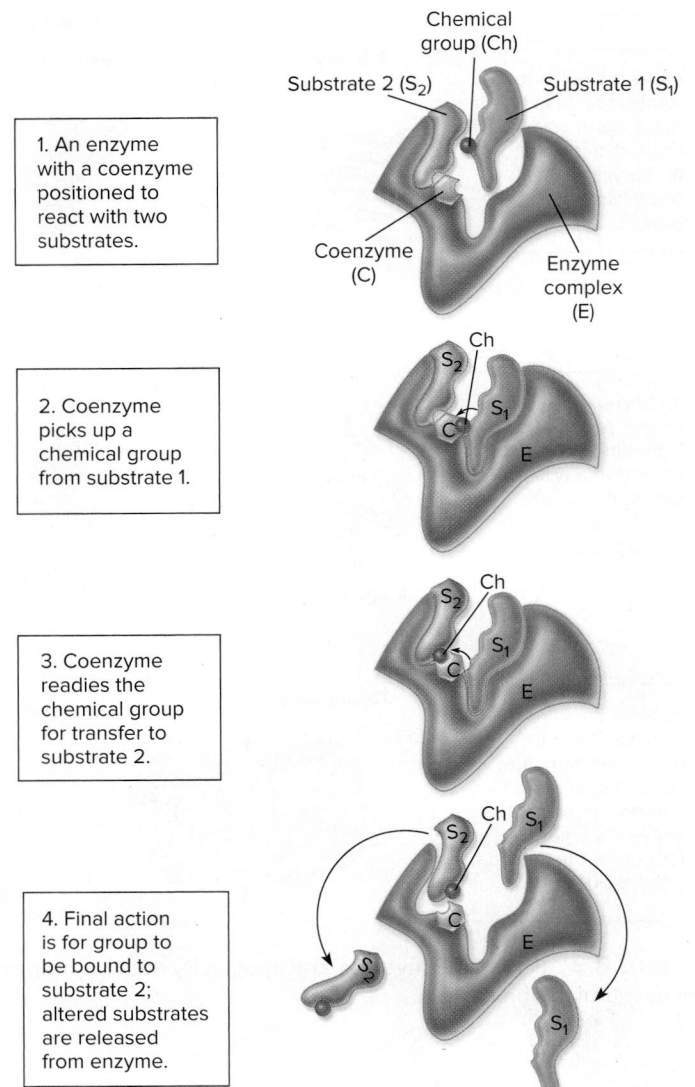

1. An enzyme with a coenzyme positioned to react with two substrates.

2. Coenzyme picks up a chemical group from substrate 1.

3. Coenzyme readies the chemical group for transfer to substrate 2.

4. Final action is for group to be bound to substrate 2; altered substrates are released from enzyme.

Figure 8.5 The carrier functions of coenzymes. Many enzymes use coenzymes to transfer chemical groups from one substrate to another.

Figure 8.6 Types of enzymes, as described by their location of action. (a) Exoenzymes are transported outside the cell to become active. **(b)** Endoenzymes remain in the cell and function there.

8.3 MAKING CONNECTIONS

The Enzyme Name Game

Most metabolic reactions require a separate and unique enzyme. Up to the present time, researchers have discovered and named over 5,000 of them. Given the complex chemistry of the cell and the extraordinary diversity of living things, many more enzymes probably remain to be discovered. A standardized system of nomenclature and classification was developed to prevent discrepancies.

In general, an enzyme name is composed of two parts: a prefix or stem word derived from a certain characteristic—usually the substrate acted upon or the type of reaction catalyzed, or both—followed by the ending *-ase.*

The system classifies the enzyme in one of these six classes, on the basis of its general biochemical action:

1. *Oxidoreductases* transfer electrons from one substrate to another, and *dehydrogenases* transfer a hydrogen from one compound to another.
2. *Transferases* transfer functional groups from one substrate to another.
3. *Hydrolases* cleave bonds on molecules with the addition of water.
4. *Lyases* add groups to or remove groups from double-bonded substrates.
5. *Isomerases* change a substrate into its isomeric form. An **isomer** is a compound that has the same molecular formula as another compound but differs in arrangement of the atoms.

6. *Synthases* are enzymes that catalyze the formation of large molecules; for example, fatty acid synthase makes one component of triglycerides.

Each enzyme is also assigned a technical name that indicates the specific reaction it catalyses. These names are useful for biochemists but are often too lengthy for routine use. More often a common name is derived from some notable feature such as the substrate or enzyme action. For example, *hydrogen peroxide oxidoreductase* is more commonly known as *catalase.* With this simpler system, an enzyme that digests a carbohydrate substrate is a *carbohydrase;* a specific carbohydrase, *amylase,* acts on starch (amylose is a major component of starch). An enzyme that hydrolyzes peptide bonds of a protein is a *proteinase, protease,* or *peptidase,* depending on the size of the protein substrate. Some fats and other lipids are digested by *lipases.* DNA is hydrolyzed by *deoxyribonuclease,* generally shortened to *DNase,* and a *polymerase* synthesizes DNA or RNA. Other examples of enzymes are presented in **table 8.A.** (See also table 8.2.)

Give the common names for enzymes that digest collagen, that make ATP, and that clot blood. *Answer available on Connect.*

TABLE 8.A	A Sampling of Enzymes, Their Substrates, and Their Reactions			
Common Name	**Systematic Name**	**Enzyme Class**	**Substrate**	**Action**
Lactase	β,D galactosidase	Hydrolase	Lactose	Breaks lactose down into glucose and galactose
Penicillinase	Beta-lactamase	Hydrolase	Penicillin	Hydrolyzes beta-lactam ring
DNA polymerase	DNA nucleotidyl-transferase	Synthase	DNA nucleosides	Synthesizes a strand of DNA using the complementary strand as a model
Lactate dehydrogenase	Same as common name	Oxidoreductase	Pyruvic acid	Catalyzes the conversion of pyruvic acid to lactic acid
Oxidase	Cytochrome oxidase	Oxidoreductase	Molecular oxygen	Catalyzes the reduction (addition of electrons and hydrogen ions) to O_2

enzymes **(figure 8.7a),** are continually present in relatively constant amounts, regardless of the amount of substrate. The enzymes involved in utilizing glucose, for example, are very important in metabolism and thus are constitutive. Other enzymes are **regulated enzymes (figure 8.7b),** the production of which is either turned on (induced) or turned off (repressed) in response to changes in concentration of the substrate or product. The level of inducible and repressible enzymes is controlled by variations in expression of the genes for these proteins, as discussed in chapter 9.

Synthesis and Hydrolysis Reactions A growing cell is in a constant state of synthesis. All such anabolic reactions require enzymes (synthases) to form covalent bonds between smaller substrate molecules. Also known as **condensation** (dehydration) **reactions,** synthesis reactions release one water molecule for each bond made **(figure 8.8a).** Catabolic reactions are also highly active in a growing cell and require their own, different set of enzymes.

For example, large substrates must be broken down or digested by enzymes into smaller molecules so they can be used. Because the breaking of bonds usually requires the input of water, digestion is also termed a **hydrolysis* reaction (figure 8.8b).**

The Sensitivity of Enzymes to Environmental Conditions

The activity of an enzyme is highly influenced by the cell's environment. In general, enzymes operate only under the natural temperature, pH, and osmotic pressure of an organism's habitat. When enzymes are subjected to departures from these normal conditions, they tend to be chemically unstable, or **labile.** Low temperatures inhibit catalysis, and high temperatures can denature the apoenzyme.

* *hydrolysis* (hy-drol′-uh-sis) Gr. *hydro,* water, and *lysis,* setting free.

(a)

or

(b)

Figure 8.7 Constitutive and regulated enzymes
(a) Constitutive enzymes are present in constant amounts in a cell. The addition of more substrate does not increase the numbers of these enzymes. **(b)** The concentration of some regulated enzymes in a cell increases or decreases in response to substrate levels.

Denaturation is a process by which the bonds that collectively maintain the native shape of the apoenzyme are disrupted. This loss causes extreme distortion of the enzyme's shape and prevents the substrate from attaching to the active site (look ahead to chapter 11). Such nonfunctional enzymes block metabolic reactions and can lead to cell death. Low or high pH and certain chemicals (heavy metals, alcohol) are also denaturing agents.

Regulation of Enzymatic Activity and Metabolic Pathways

Metabolic pathways proceed in a stepwise, highly regulated manner that maximizes the use of available nutrients and energy. The cell responds to environmental conditions by adopting those metabolic

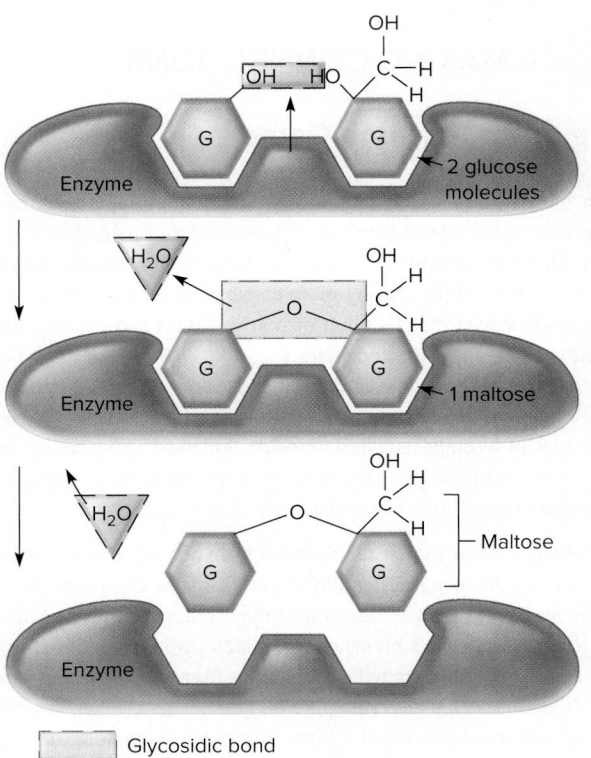

Glycosidic bond

(a) **Condensation Reaction.** Forming a glycosidic bond between two glucose molecules to generate maltose requires the removal of a water molecule.

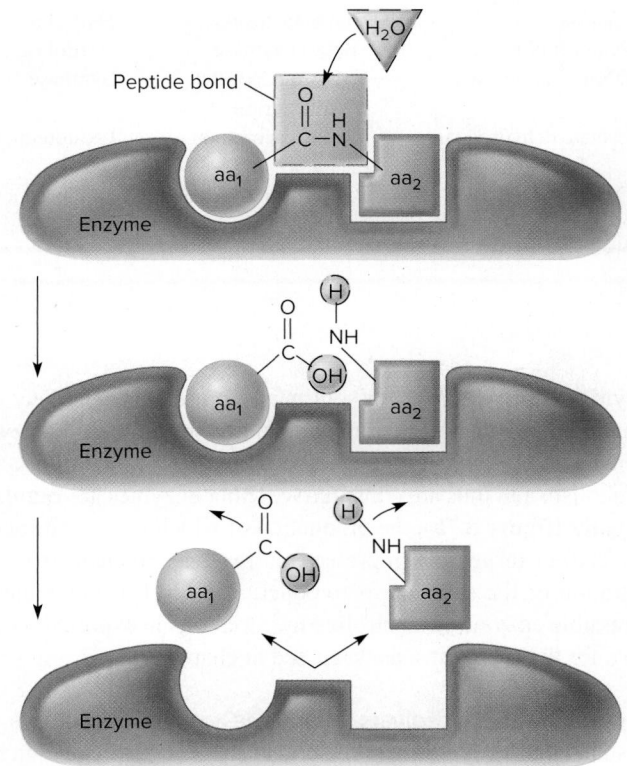

(b) **Hydrolysis Reaction.** Breaking a peptide bond between two amino acids requires a water molecule that adds OH to one amino acid and H to the other.

Figure 8.8 Examples of enzyme-catalyzed synthesis and hydrolysis reactions.

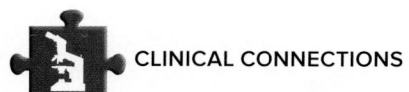

CLINICAL CONNECTIONS

The Role of Microbial Enzymes in Disease

Many pathogens secrete unique exoenzymes that help them avoid host defenses or promote their multiplication in tissues. Because these enzymes contribute to pathogenicity, they are referred to as virulence factors, or toxins in some cases (look ahead to chapter 13). *Streptococcus pyogenes* (a cause of throat and skin infections) produces a streptokinase involved in the digestion of blood clots and can assist the pathogen in invading wounds. Another exoenzyme from this bacterium is called streptolysin.[1] In mammalian hosts, streptolysin damages blood cells and tissues. It is also responsible for lysing red blood cells used in blood agar media, which can be used to identify some bacteria growing in culture (see figure 3.17).

Pseudomonas aeruginosa, a respiratory and skin pathogen, produces elastase and collagenase, which digest elastin and collagen, two proteins found in connective tissue. These increase the severity of certain lung and burn infections. Members of the genus *Clostridium* secrete toxins with significant pathological effects. Many of these toxins—for example, botulinum and tetanus toxins—have an enzymatic component that causes the symptoms of the diseases. Both of them prevent cell vesicles containing neurotransmitters from releasing their contents and thus block nerve transmission to muscles. The botulism toxin causes loss of muscle tone (flaccid paralysis); the tetanus toxin causes sustained muscle tone (rigid paralysis). *Clostridium perfringens,* an agent of gas gangrene, synthesizes lecithinase C, a lipase that profoundly damages cell membranes and accounts for the tissue death associated with this disease. Other medically important enzymes of microbes do not target tissues. Some, such as the beta-lactamases, act on portions of some penicillin and cephalosporin antibiotics and inactivate them. This makes the bacteria drug-resistant and protected from the effects of the drugs.

Even though viruses lack most features of metabolism, a few of them can express genes for enzymes. How would you describe the actions of the following viral enzymes: neuraminidase, reverse transcriptase, and integrase? *Answers available on Connect.*

1. Even though most enzyme names end in *-ase* (see 8.3 Making Connections), not all do.

reactions that most favor its growth and survival. Because enzymes are critical to these reactions, the regulation of metabolism is largely the regulation of enzymes by an elaborate system of checks and balances. Let us take a look at some general features of metabolic pathways.

Patterns of Metabolic Pathways

Metabolic reactions rarely consist of a single action or step. More often, they occur in a multistep series or pathway, with

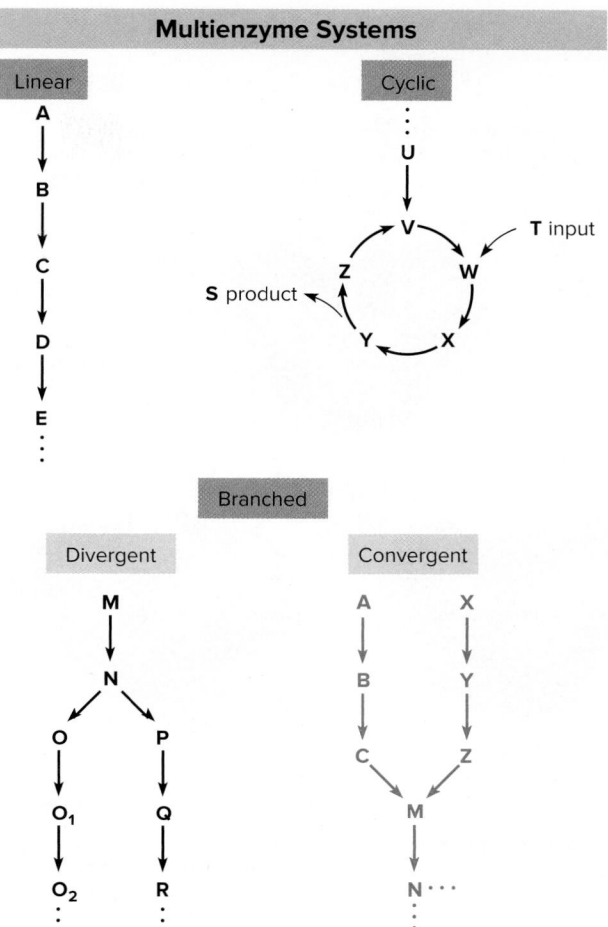

Figure 8.9 Patterns of metabolic pathways. In general, metabolic pathways consist of a linked series of individual chemical reactions that produce intermediary metabolites and lead to a final product. These pathways occur in several patterns, including linear, cyclic, and branched. Anabolic pathways involved in biosynthesis result in a more complex molecule, each step adding on a functional group, whereas catabolic pathways involve the dismantling of molecules and can generate energy. Virtually every reaction in a series involves a specific enzyme.

each step catalyzed by an enzyme. An individual reaction is shown in various ways, depending on the pathway pattern **(figure 8.9).** The product of one reaction is often the reactant (substrate) for the next, forming a linear chain of reactions. Many pathways have branches that provide alternate methods for nutrient processing. Others take a cyclic form, in which the starting molecule is regenerated to initiate another turn of the cycle. Pathways generally do not stand alone; they are interconnected and merge at many sites.

Every pathway has one or more enzyme pacemakers that set the rate of a pathway's progression. Regulation of pacemaker enzymes proceeds on two fundamental levels. Either the enzyme itself is directly inhibited or activated, or the amount of the enzyme in the system is altered (decreased or increased). When factors affect the enzyme directly, the system can be finely controlled or "tuned." When regulation is at the genetic level (enzyme synthesis), control is slower and less sensitive.

Figure 8.10 Examples of two common modes of competitive inhibition that regulate enzyme action. For a quick animation of these mechanisms, go to "Cellular Metabolism, Pathways, & Feedback" on YouTube.

Direct Controls on the Action of Enzymes

The bacterial cell has many ways of directly influencing the activity of its enzymes. It can inhibit enzyme activity by supplying a molecule that resembles the enzyme's normal substrate. The "mimic" can then occupy the enzyme's active site, preventing the actual substrate from binding there. Because the mimic cannot actually be acted on by the enzyme and a product is not released, the enzyme is effectively shut down. This is one example of **competitive inhibition,** because the mimic is competing with the substrate for the binding site **(figure 8.10).** In chapter 12 you will see that some antibiotics work on the same strategy of competing with enzymatic active sites to shut down metabolic processes.

Another form of competitive inhibition can occur with special types of enzymes that have two binding sites: the active site and another area called the **regulatory** or **allosteric* site** (figure 8.10). The action of these enzymes is controlled by the binding of molecules other than the substrate in the regulatory site. Often the regulatory molecule is the product of the enzymatic reaction itself. This provides a **negative feedback** mechanism that can slow down enzymatic activity once a certain concentration of product is produced.

Noncompetitive inhibition is a third mechanism that does not involve an inhibitor molecule competing with substrate for the active site or disable the active site. This type of inhibitor binds to the entire enzyme-substrate complex and prevents the enzyme from completing its action on the substrate.

Controls on Enzyme Synthesis

Controlling enzymes by controlling their synthesis is another effective mechanism. Because enzymes do not last indefinitely, they eventually degrade and must be replaced. This cycle works into the scheme of the cell, where replacement of enzymes can be regulated according to cell demand. The mechanisms of this system are genetic in nature; that is, they require regulation of DNA and protein synthesis, a topic we cover in more depth in chapter 9.

Enzyme repression is a means to stop further synthesis of an enzyme somewhere along its pathway. As the level of the end product from a given enzymatic reaction has built to excess, the genetic apparatus responsible for replacing these enzymes is automatically suppressed **(figure 8.11).** The response time is longer than for feedback inhibition, but its effects are more enduring.

A response that resembles the inverse of enzyme repression is **enzyme induction.** In this process, enzymes appear (are induced)

* *allosteric* (al-oh-stair′-ik) Gr. *allos,* other, and *steros,* solid. Literally, "another space."

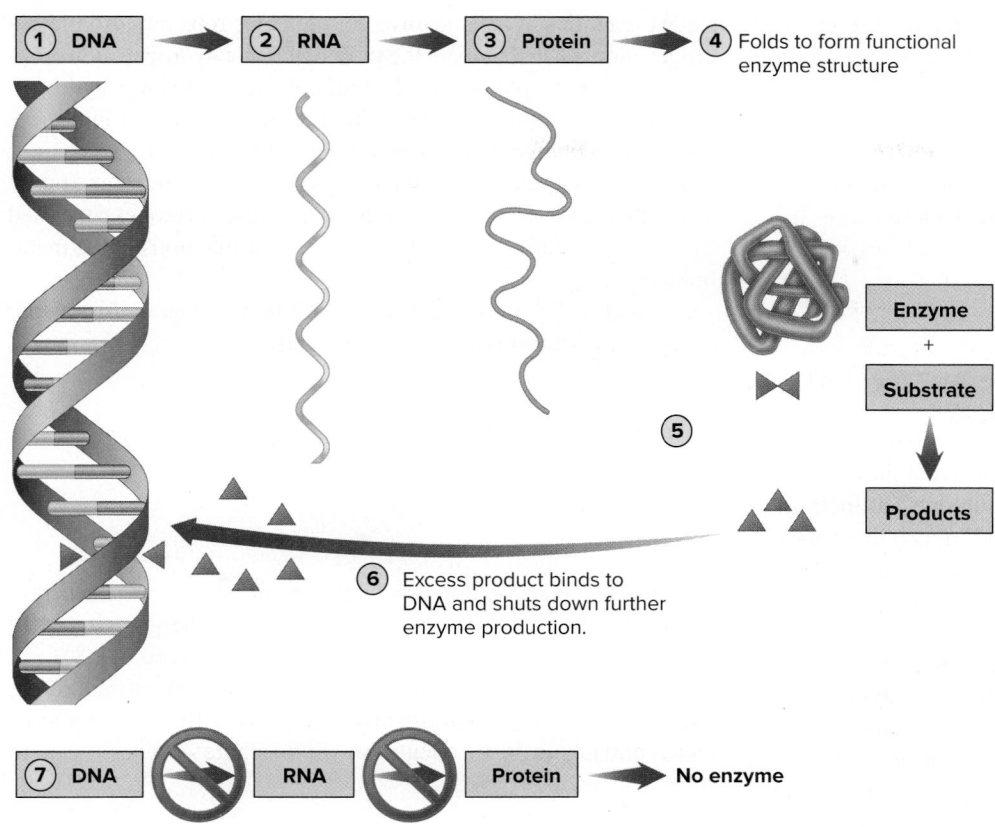

only when suitable substrates are present; that is, the synthesis of enzyme is induced by its substrate. This response enables the organism to adapt to a variety of nutrients, and it also prevents a microbe from wasting energy by making enzymes for which substrates are lacking.

 Check Your Progress SECTION 8.1

1. Differentiate between catabolism and anabolism using examples.
2. Describe 10 important biochemical properties of enzymes.
3. Describe the chemistry of enzymes, and explain how the apoenzyme forms.
4. Show diagrammatically the interaction of holoenzyme and its substrate and general products that can be formed from a reaction.
5. Differentiate among the chemical composition and functions of various cofactors. Provide examples of each type.
6. Summarize the direct and indirect controls that regulate enzymes.

8.2 The Pursuit and Utilization of Energy

 Expected Learning Outcomes

8. Relate the diverse sources of energy and the cellular management of energy.
9. Describe biological oxidation-reduction and energy transfer.
10. Discuss the structure, basic formation, and roles of ATP in cells.

Energy is defined as the capacity to do work or to cause change. In order to carry out the work of an array of metabolic processes, cells require constant input and expenditure of some form of usable energy. Energy commonly exists in these forms:

- thermal (heat) energy from molecular motion,
- radiant (wave) energy from visible light or other rays,
- electrical energy from a flow of electrons,
- mechanical energy from a physical change in position,
- atomic energy from reactions in the nucleus of an atom, and
- chemical energy present in the bonds of molecules.

Cells are far too fragile to rely in any constant way on thermal or atomic energy for cell transactions. Except for certain chemoautotrophic bacteria, the ultimate source of energy is the sun, but only phototrophs can tap this source directly. This capacity of photosynthesis to convert solar energy into chemical energy provides both a nutritional and an energy basis for all heterotrophic living things. We cover this essential form of metabolism in section 8.6. For the most part, only chemical energy and reactions can fuel cellular energetics of nonphotosynthetic organisms.

In this section, we further explore the role of energy as it is related to the use of nutrients, redox carriers, and the generation of ATP.

Cell Energetics

Cells manage energy in the form of chemical reactions that change molecules. This often involves activities such as the making or breaking of bonds and the transfer of electrons. Not all cellular

reactions have the same relation to energy. Some release energy and others require it to proceed. For example, a reaction with this equation:

$$X + Y \xrightarrow{\text{Enzyme}} Z + \text{Energy}$$

releases energy as it goes forward. This type of reaction is termed **exergonic.*** Energy of this type is considered free: it is available for doing cellular work. Energy transactions such as the following:

$$\text{Energy} + A + B \xrightarrow{\text{Enzyme}} C$$

are called **endergonic*** because they are driven forward with the addition of energy.

Summaries of metabolism might make it seem that cells "create" energy from nutrients, but they do not. What they actually do is extract chemical energy already present in nutrient fuels and apply that energy toward useful work in the cell, much like a gasoline engine releases energy as it burns fuel. The engine does not actually produce energy, but it converts the potential energy of the fuel to the free energy that drives an machine.

The basis for cell energetics lies in the chemical bonds, electron transfers, and special pathways along with their enzymes. During exergonic reactions, electrons are pulled out of atoms, thereby releasing their inherent energy. These electrons are picked up by special carriers and transferred through a series of reactions until they reach a **final electron acceptor.** As we will see in section 8.3, the production of ATP is intimately tied to this process. As ATP is formed, it is used to support endergonic reactions such as synthesis. These reactions are balanced and cyclic, so that ATP is constantly being made, used, regenerated, and so on. The actual process of electron transfer is a type of redox reaction that was first described in chapter 2. Let us revisit this topic and see how it is involved in metabolism.

Biological Oxidation and Reduction: Electron and Energy Transfer

Many chemical substances readily donate or receive electrons and participate in oxidation (the loss of electrons) or reduction (the gain of electrons). The compound that loses the electrons is **oxidized,** and the compound that receives the electrons is **reduced.** Such oxidation-reduction or **redox reactions** are common in cells and indispensable to energy transactions (look back at figure 2.9). Redox reactions always occur in pairs, with an electron donor and an electron acceptor, which constitute a *conjugate pair,* or *redox pair.* The reaction can be represented as follows:

Electron donor + Electron acceptor ⟷ Electron donor + Electron acceptor
•e⁻ •e⁻
Reduced Oxidized Oxidized Reduced

* *exergonic* (ex-er-gon´-ik) Gr. *exo,* without; and *ergon,* work. In general, catabolic reactions are exergonic.
* *endergonic* (en-der-gon´-ik) Gr. *endo,* within, and *ergon,* work. Anabolic reactions tend to be endergonic.

Redox reactions in cells involve not only electrons but also the hydrogen atoms which donate them. Recall that a hydrogen atom consists of a single proton and a single electron. When a hydrogen atom gives up its single electron to a positively charged molecule, this leaves a hydrogen ion or proton (H^+). For simplicity's sake, our discussions will emphasize the process of electron transfer. But be aware that many pathways need hydrogen ions (protons) generated by oxidation-reduction to carry out essential functions in the metabolic pathways.

One example of a typical redox reaction is seen in this overall equation associated with aerobic respiration:

$$C_6H_{12}O_6 + 6O_2 \rightarrow 6CO_2 + 6H_2O + \text{Energy}$$

Note that the electron donor is hydrogen and the electron acceptor is oxygen. The electrons are transported and energy is released through a stepwise series of reactions, covered in section 8.3.

For energy to be biologically useful, it must ultimately be captured through **phosphorylation,** a process that adds an inorganic phosphate P_i to adenosine diphosphate, converting it to ATP:

$$ADP + P_i + \text{Energy} \leftrightarrows ATP$$

Several different types of carriers exist for handling and transporting electrons and protons until they can be transferred to a final electron acceptor. Such redox reactions are a constant supplier of electrons for the final "processing plant" operating in cells—the electron transport system (look ahead to figure 8.16).

Managing Electrons in Metabolism

We learned in chapter 2 that electrons are highly reactive, negatively charged particles, and that they are transferred or shared between atoms during bond formation. These electrons do not bounce around randomly in cells. Generally, they are captured and transported by some form of carrier molecule.

Much of the work of electron capture and transport in metabolic pathways falls to molecules with enhanced affinity for electrons. They readily accept electrons from one substrate and donate them to another substrate. The most common of these carriers is the coenzyme **nicotinamide adenine dinucleotide (NAD⁺) (figure 8.12).** Other related carriers are FAD (flavin adenine dinucleotide), coenzyme A, and the compounds of the respiratory chain, all discussed in section 8.3.

NAD⁺ reduces 2 hydrogen atoms and carries 2 electrons away from this reaction, represented as NADH. The electrons are carried in the nicotinamide ring, where they can be accessed during other electron transfers (figure 8.12).

At the end point the electrons will be received by a final electron acceptor, which in aerobic metabolism is molecular oxygen. An important function of this carrier system is that every transfer of electrons from NADH to another carrier molecule automatically

replenishes NAD$^+$ for use in the numerous redox reactions in cells (see example in figure below).

Adenosine Triphosphate: Metabolic Money

ATP has been described as metabolic currency because it can be earned, spent, and exchanged. You can gain some appreciation for the enormous need for ATP if you consider that even one tiny cell of *Escherichia coli* consumes around 2.5 million ATP molecules per second for synthesis alone! As a temporary energy repository, ATP provides a connection between energy-yielding catabolism and all other cellular activities that require energy. Some clues to its energy-storing properties lie in its unique molecular structure.

The Molecular Structure of ATP ATP is a three-part molecule consisting of a nitrogen base (adenine) linked to a 5-carbon sugar

Figure 8.12 **Structure of NAD$^+$ and its carrier function. (a)** The coenzyme NAD$^+$ contains the vitamin nicotinamide (niacin) bound to two ribose phosphates and an adenine. **(b)** The site of reduction is at the nicotinamide ring portion (box), which carries 2 electrons, one at the ring nitrogen and the other on a hydrogen bound to a ring carbon.

⌐ Bond that releases energy when broken

Figure 8.13 **A model of adenosine triphosphate (ATP).** A structural view of ATP and its related components ADP and AMP. Note that the last two phosphate bonds release energy when they are cleaved.

(ribose), with a chain of three phosphate groups bonded to the ribose **(figure 8.13)**. The type, arrangement, and especially the proximity of atoms in ATP combine to form a compatible but unstable high-energy molecule. The high energy of ATP originates in the orientation of the phosphate groups, which are relatively bulky and carry negative charges. The proximity of these repelling electrostatic charges imposes a strain that is most acute on the bonds between the last two phosphate groups. When the strain on the phosphate bonds is released by removal of the terminal phosphates, free energy is released.

Breaking the bonds between the second and third phosphates of ATP yields adenosine diphosphate (ADP), which can then be converted to adenosine monophosphate (AMP). It is worthwhile noting how economically a cell manages its pool of adenosine nucleotides. AMP derivatives help form the backbone of RNA and are also a major component of certain coenzymes (NAD$^+$, FAD, and coenzyme A).

The Metabolic Roles of ATP ATP is at the center of a cell's energy cycle, regardless of whether it is a chemoheterotroph using respiration, a photosynthetic organism, or a chemoautotroph. Expenditure of ATP powers nearly all processes that are part of cellular work. This energy cycle can be depicted as shown here:

ADP + P$_i$

Respiration
Fermentation
Photosynthesis
Chemoautotrophy

Chemical work
Transport work
Mechanical work

ATP

In many instances, the energy released from the high-energy bond in ATP activates individual subunits before they are enzymatically bonded during synthesis. ATP is also used to prepare a

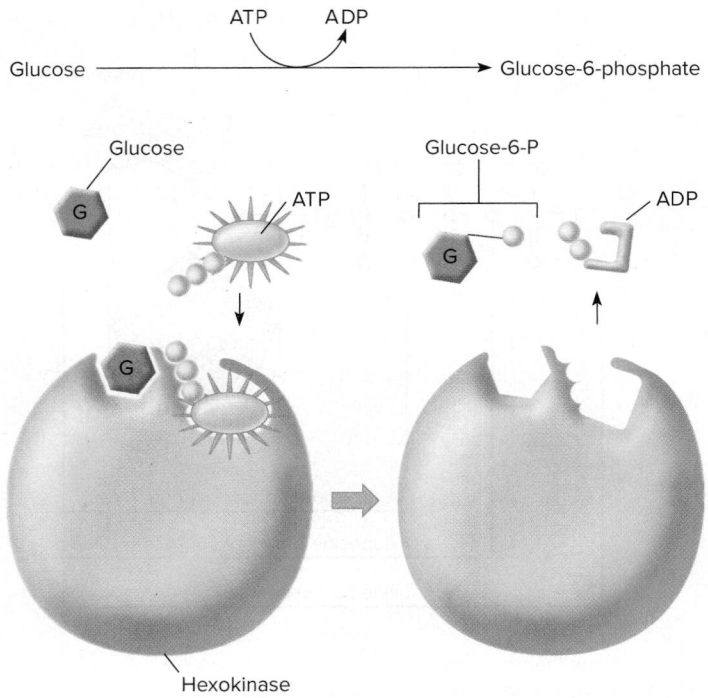

Figure 8.14 The phosphorylation of glucose by ATP. The first step in catabolizing glucose is the addition of a phosphate from ATP by an enzyme called hexokinase. This use of high-energy phosphate as an activator is a recurring feature of many metabolic pathways.

molecule for catabolism such as the phosphorylation of a 6-carbon sugar during the early stages of glycolysis **(figure 8.14).**

Because ATP needs to be constantly replenished, reactions are constantly continuing in the reverse order. The ADPs generated during phosphorylation will repeat the cycle by reacting immediately with inorganic phosphate and a high-energy source that can regenerate ATP. The energy for these high-energy phosphate bonds comes from catabolic pathways where substrates or electron carriers are harvesting energy.

Most ATPs in aerobic organisms are formed during **oxidative phosphorylation,** a series of reactions involving electron transport and an enzyme (ATP synthase) that can trap energy and store it in ATP (look ahead to figure 8.20). Some ATP molecules are formed through **substrate-level phosphorylation,** a process that all cells carry out, regardless of their metabolic type. In this mode of phosphorylation, ATP is formed by transfer of a phosphate group from a high-energy compound (substrate) directly to ADP **(figure 8.15; see also figure 8.17, step 6).**

Photosynthetic organisms have a system involving **photophosphorylation,** in which the ATP is formed by means of a light-driven

Figure 8.15 ATP formation by substrate-level phosphorylation. The substrate has been bound to a phosphate, creating a bond with high potential energy. Transfer of the high-energy phosphate to ADP gives rise to ATP and returns the substrate to its original form.

electron transport mechanism that we discuss in more detail in section 8.6.

Check Your Progress SECTION 8.2

7. Explain how oxidation of a substrate proceeds without oxygen.
8. Refer to the blue redox equation for glucose. Which substance is reduced? Which is oxidized? Which is the oxidizing agent? Which is the reducing agent?
9. In the following redox pairs, which compound is reduced and which is oxidized?
 a. NAD^+ and NADH
 b. $FADH_2$ and FAD
 c. NO_3^- and NO_2^-
10. a. Describe the roles played by ATP and NAD^+ in metabolism.
 b. What particular features of their structure lend them to these functions?
11. Name the major ways that substrate-level phosphorylation is different from oxidative phosphorylation.
12. What is meant by the concept of the "final electron acceptor"?

8.3 Pathways of Bioenergetics

Expected Learning Outcomes

11. Relate the main points of bioenergetics and energy strategies in microorganisms.
12. Describe the main catabolic pathways and their locations in aerobic respiration.
13. Define *glycolysis* and explain its input and output, its basic steps, and how it is linked to the Krebs cycle and oxidative phosphorylation.
14. Explain the basic steps in the Krebs cycle, its input and output, and how it is linked to oxidative phosphorylation.
15. Describe the components of the respiratory chain and their functions.
16. Explain the chemiosmotic mechanism of ATP formation, its relationship to electron transport, and the final step in electron transfer.
17. Summarize the results of aerobic respiration.
18. Define and describe some reactions of anaerobic respiration.

One of the scientific community's greatest achievements was deciphering the biochemical pathways of cells. Initial work with bacteria and yeasts, followed by studies with animal and plant cells, clearly demonstrated metabolic similarities and strongly supported the presence of metabolic patterns that are universal. **Bioenergetics** is a study of the mechanisms of cellular energy release, including catabolic and anabolic routes. Although these pathways are interconnected and interdependent, anabolic pathways are not simply reversals of catabolic ones. At first glance it may seem more economical to use identical pathways, but having different enzymes and divided pathways allows anabolism and catabolism to proceed simultaneously without interference.

In the following sections, we focus our discussion on the most common catabolic pathways that illustrate general principles of other pathways as well.

Catabolism: An Overview of Nutrient Breakdown and Energy Release

For most chemoheterotrophs, the primary catabolism of fuels (such as glucose) that release energy proceeds through a series of three coupled pathways:

1. **glycolysis,*** also called the Embden-Meyerhof-Parnas (EMP) pathway;[2]
2. the **Krebs cycle,** also known as the citric acid or tricarboxylic acid cycle TCA;[2] and

* *glycolysis* (gly-kol′-ih-sis) Gr. *glykys,* sweet, and *lysis,* to split.

2. The EMP pathway is named for the biochemists who first outlined its steps. Krebs is in honor of Sir Hans Krebs who, with F. A. Lipmann, delineated this pathway, an achievement for which they won the Nobel Prize in Physiology or Medicine in 1953. TCA refers to the involvement of several organic acids containing three carboxylic acid groups, citric acid being the first tricarboxylic acid formed.

3. the **respiratory chain** (electron transport and oxidative phosphorylation).

Each segment of the pathway is responsible for a specific set of actions on various products of glucose. The major interconnections of these pathways in aerobic respiration are represented in **figure 8.16,** and their reactions are summarized in **table 8.3.** In the following sections, we consider each of these pathways in greater detail.

Energy Strategies in Microorganisms

In chapter 7 we gained insight into the metabolic diversity of microbes. There are numerous strategies used by microbes to derive energy and synthesize cell components. Because microbial metabolism is so varied, we have chosen to cover elements of the most common types of metabolism, based on the three basic catabolic pathways. In previous discussions, microorganisms were

Figure 8.16 **Side-by-side comparison of the stages, reactions, and major products of pathways for aerobic respiration, anaerobic respiration, and fermentation.** Note that all begin with glycolysis but vary in use of other pathways, electron acceptors, and ATP yield.

TABLE 8.3	Metabolic Strategies Among Heterotrophic Microorganisms Growing on Glucose			
Scheme	Pathways Involved	Final Electron Acceptor	Products	Primary Pathway Found In
Aerobic respiration	Glycolysis, Krebs cycle, electron transport	O_2	ATP, CO_2, H_2O	Aerobes; facultative anaerobes
Anaerobic metabolism Fermentative	Glycolysis	Organic molecules	ATP, CO_2, ethanol, lactic acid*	Facultative, aerotolerant, strict anaerobes
Respiration	Glycolysis, Krebs cycle, electron transport	Various inorganic ions $(NO_3^-, SO_4^{2-}, CO_3^{3-})$	Highly varied, but can include CO_2, ATP, organic acids, H_2S, CH_4, N_2	Anaerobes; some facultatives and aerotolerants

* The products of microbial fermentations are extremely varied and include organic acids, alcohols, and gases.

categorized according to their requirement for oxygen gas, and this requirement is related directly to their mechanisms of energy release. As we shall see, **aerobic respiration** is a series of reactions (glycolysis, Krebs cycle, and the respiratory chain) that converts glucose to CO_2, produces H_2O, and generates energy (figure 8.16a). It relies on molecular oxygen (O_2) as the final acceptor for electrons and hydrogens, and releases a relatively large amount of ATP. Aerobic respiration is characteristic of fungi, protozoa, animals, plants, and many bacteria, and it will be the system we emphasize here.

Some facultative and aerotolerant anaerobes may use only glycolysis or similar pathways to ferment glucose. In this case, oxygen is not required, organic compounds are the final electron acceptors, and a relatively small amount of ATP is synthesized (figure 8.16c and section 8.4).

Some anaerobic microorganisms metabolize by means of **anaerobic respiration.** This system may involve the same three pathways as aerobic respiration, but it does not use molecular oxygen as the final electron acceptor (figure 8.16b). Instead, it uses other oxidized ions such as NO_3^-, SO_4^{2-}, or CO_2 for the final electron acceptor. Aspects of anaerobic respiration and fermentation are covered later in this section and the next one.

Aerobic Respiration

Aerobic respiration is a series of enzyme-catalyzed reactions in which electrons are transferred from fuel molecules such as glucose to oxygen as a final electron acceptor. This pathway is the principal energy-yielding scheme for aerobic heterotrophs, and it provides both ATP and metabolic intermediates for many other pathways in the cell, including those of protein, lipid, and carbohydrate synthesis (see figure 8.25).

Aerobic respiration in microorganisms can be summarized by this equation:

Glucose ($C_6H_{12}O_6$) + $6O_2$ + 38 ADP + 38P$_i$ → $6CO_2$ + $6H_2O$ + 38 ATP

The seeming simplicity of the equation for aerobic respiration conceals the complexity of the process. Fortunately we do not have to present all of the details to address some important concepts concerning its reactants and products, as follows:

1. the steps in the oxidation of glucose,
2. the involvement of coenzyme carriers and the final electron acceptor,
3. where and how ATP originates,

4. where carbon dioxide originates, and
5. where oxygen is required.

Glucose: The Starting Compound

Carbohydrates such as **glucose** are good fuels because these compounds are readily oxidized; that is, they are superior electron donors. The electrons they donate can be used in energy transfers. The end products of the oxidation of such organic compounds are energy-rich ATP and energy-poor carbon dioxide and water. Polysaccharides (starch, glycogen) and disaccharides (maltose, sucrose) are stored sources of glucose also available for the respiratory pathways. Although we use glucose as the main starting compound, other hexoses (fructose, galactose) and fatty acid subunits can enter the pathways of aerobic respiration as well, as we will see in section 8.5.

Glycolysis: The Starting Lineup

The process called **glycolysis (EMP)** is a pathway that converts glucose through several steps into **pyruvic acid.** Depending on the organism and the conditions, it may be only the first phase of respiration, or it may serve as the primary metabolic pathway for fermentative microbes. Glycolysis provides a way to synthesize a small amount of ATP, to release another potential source of energy—NADH, and to generate pyruvic acid, an essential intermediary metabolite. None of these reactions involve the direct input of oxygen.

Steps in the Glycolytic Pathway Glycolysis proceeds along nine linear steps. The first portion of glycolysis involves activation of glucose, which is followed by oxidation reactions of the glucose fragments, the synthesis of ATP, and the formation of pyruvic acid. Although each step of metabolism is catalyzed by a specific enzyme, we will mention only certain key ones. The following outline lists the principal steps of glycolysis. To see an overview of glycolysis, go to **figure 8.17.** Detailed illustrations appear in Appendix A.

1. First, glucose is phosphorylated at the number six carbon by an **ATP** to produce glucose-6-phosphate (see figure 8.14). This is a way of "priming" the system and preventing the glucose from being transported out of the cell.
2. Glucose-6-phosphate is converted to its isomer, fructose-6-phosphate.
3. Another ATP is spent in phosphorylating the first carbon of fructose-6-phosphate, which yields a molecule with two phosphates called fructose-1,6-diphosphate.

These steps take place in the cytoplasm of both prokaryotic and eukaryotic cells.

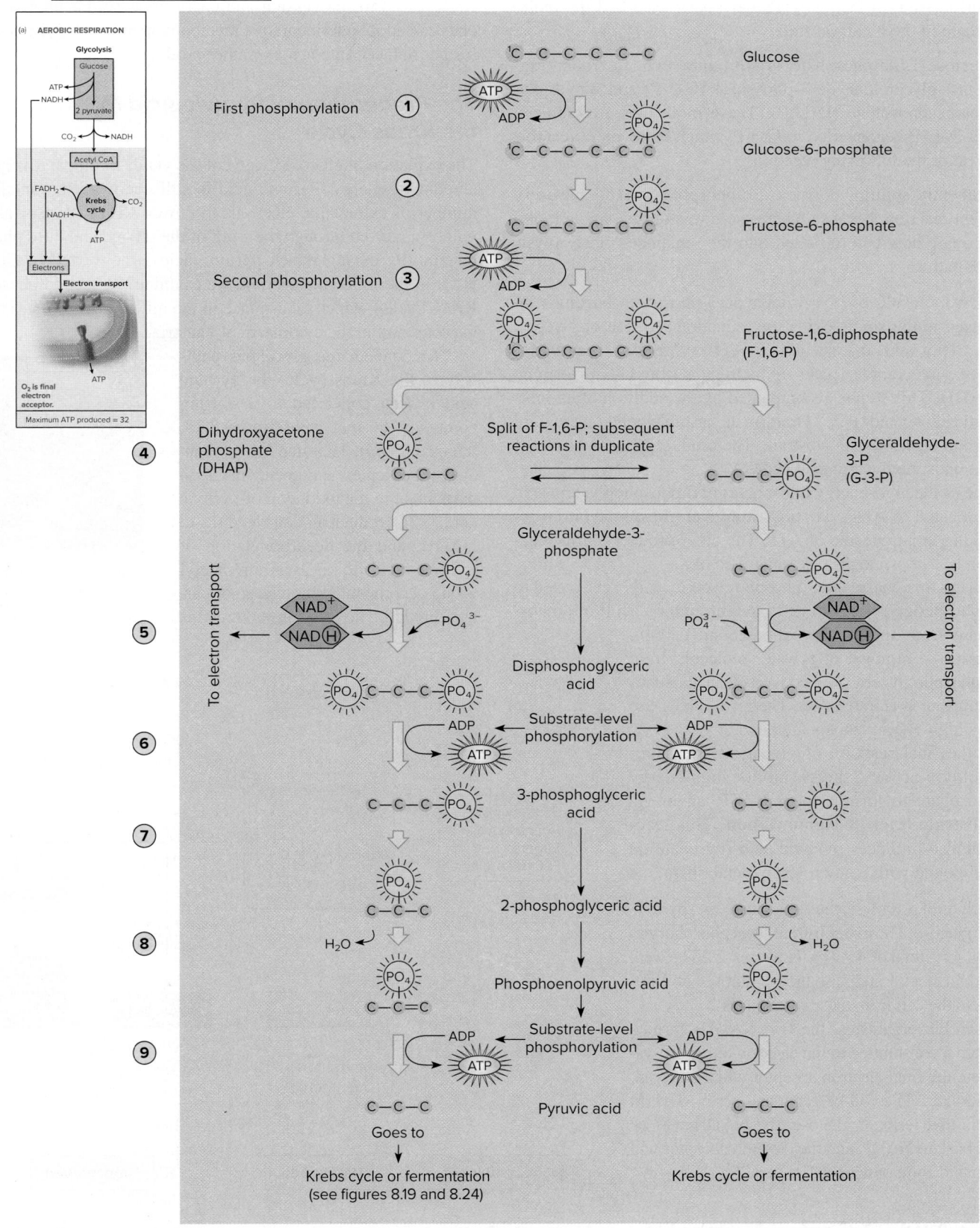

(a) AEROBIC RESPIRATION

Glycolysis

Glucose
ATP
NADH
2 pyruvate
CO_2 — NADH
Acetyl CoA
FADH₂
Krebs cycle — CO_2
NADH
ATP
Electrons
Electron transport
O_2 is final electron acceptor.
ATP

Maximum ATP produced = 32

First phosphorylation ①
ATP → ADP
PO_4

Glucose
1C—C—C—C—C—C

Glucose-6-phosphate
1C—C—C—C—C—C

②
PO_4

Fructose-6-phosphate
1C—C—C—C—C—C

Second phosphorylation ③
ATP → ADP

Fructose-1,6-diphosphate (F-1,6-P)
PO_4 PO_4
1C—C—C—C—C—C

Split of F-1,6-P; subsequent reactions in duplicate

Dihydroxyacetone phosphate (DHAP) ④
PO_4
C—C—C

Glyceraldehyde-3-P (G-3-P)
C—C—C—PO_4

Glyceraldehyde-3-phosphate

To electron transport

C—C—C—PO_4

NAD⁺
NAD(H)
← PO_4^{3-}
⑤

PO_4^{3-} →
NAD⁺
NAD(H)

To electron transport

C—C—C—PO_4

Disphosphoglyceric acid
PO_4—C—C—C—PO_4

PO_4—C—C—C—PO_4

Substrate-level phosphorylation
ADP
ATP
⑥
ADP
ATP

3-phosphoglyceric acid
C—C—C—PO_4

C—C—C—PO_4

⑦

2-phosphoglyceric acid

PO_4
C—C—C

PO_4
C—C—C

⑧
H_2O

H_2O

Phosphoenolpyruvic acid

PO_4
C—C=C

PO_4
C—C=C

Substrate-level phosphorylation
ADP
ATP
⑨
ADP
ATP

Pyruvic acid
C—C—C

C—C—C

Goes to

Krebs cycle or fermentation (see figures 8.19 and 8.24)

Goes to

Krebs cycle or fermentation

🌀 **Process Figure 8.17** **Summary of the steps in glycolysis.** The primary reaction steps are described in the accompanying text. These reactions all take place in the cytoplasm, regardless of the cell type. Inset in upper corner has a gold box to show the position of this pathway in the scheme of aerobic respiration. Only the pink carbon skeleton is shown to simplfy the dynamics of the reactions. A detailed analysis of specific chemical structures and reactions can be found in Appendix A.

Up to this point, no energy has been released, no oxidation-reduction has occurred, and, in fact, 2 ATPs have been used and the molecules remain in the 6-carbon state.

4. Fructose-1,6-diphosphate is split into two 3-carbon fragments: glyceraldehyde-3-phosphate (G-3-P) and dihydroxyacetone phosphate (DHAP). These molecules are isomers, and DHAP is converted to G-3-P, which is the more reactive form for the following reactions.

The effect of the splitting of fructose diphosphate is to double every subsequent reaction, because where there was once a single molecule, there are now two to be fed into the remainder of glycolysis and other pathways.

5. Each molecule of glyceraldehyde-3-phosphate becomes involved in the single oxidation-reduction reaction of glycolysis. This sets the scene for ATP synthesis. The **NAD⁺** coenzyme complex picks up hydrogens from G-3-P, forming **NADH.** This step is accompanied by the addition of an inorganic phosphate (PO_4^{3-}) to form an unstable bond on the third carbon of the G-3-P substrate. The product of these reactions is diphosphoglyceric acid (DPGA).

6. One of the high-energy phosphates of diphosphoglyceric acid is donated to ADP via **substrate-level phosphorylation** resulting in a molecule of ATP. The other product of this reaction is 3-phosphoglyceric acid (3-PGA).

7, 8. During this phase, the 3-phosphoglyceric acid is converted to 2-phosphoglyceric acid (2-PGA) through the shift of a phosphate from the third to the second carbon. Then, the removal of a water molecule from 2-phosphoglyceric acid converts it to phosphoenolpyruvic acid (PEPA). The result gives rise to a high-energy phosphate bond.

9. In the final reaction of glycolysis, phosphoenolpyruvic acid gives up its high-energy phosphate to form a second ATP, again via **substrate-level phosphorylation.** This reaction also produces **pyruvic acid** (pyruvate), a compound with many roles in metabolism.

The reactions of glycolysis thus give rise to two pyruvic acids per glucose. The two substrate-level phosphorylations produce a total of **4** ATPs. However, 2 ATPs were expended for steps 1 and 3, so the net number of ATPs available to the cell from these reactions is **2.**

In aerobic organisms, the two NADHs formed during step 5 are shuttled to the electron transport system, where the final electron acceptor will be oxygen, and additional ATPs will be generated per NADH. In organisms that ferment glucose, the NADH will be oxidized back to NAD⁺, and the electron acceptor will be an organic compound (see figure 8.23).

Pyruvic Acid—A Central Metabolite

Pyruvic acid occupies an important position in several pathways, and different organisms handle it in different ways **(figure 8.18).** In strictly aerobic

organisms and some anaerobes, pyruvic acid is channeled into the Krebs cycle for processing and energy release. Facultative anaerobes can adopt a fermentative metabolism, in which pyruvic acid is further reduced into acids or other products.

The Preparation of Pyruvic Acid for the Krebs Cycle

The oxidation of glucose by glycolysis yields a comparatively small amount of energy. Pyruvic acid is still energy-rich, containing a number of extractable electrons to power ATP synthesis, and this will be achieved through the work of the second and third phases of respiration, during which pyruvic acid is converted to CO_2 and H_2O. In the following section, we examine the second phase—the Krebs cycle—which takes place in the mitochondrial matrix in eukaryotes and in the cytoplasm of bacteria.

To connect the glycolysis pathway to the Krebs cycle, the pyruvic acid is first converted to a starting compound for that cycle **(figure 8.19,** tan oval at top). This step brings in an enzyme complex and coenzyme A to participate in the oxidation of pyruvic acid, the reduction of NAD⁺ to NADH, and the decarboxylation of pyruvic acid to a 2-carbon acetyl group accompanied by release of carbon dioxide. The acetyl group remains

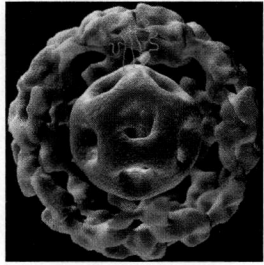

Artist's concept of pyruvate dehydrogenase *Source:* Sriram Subramaniam/National Cancer Institute (NCI)

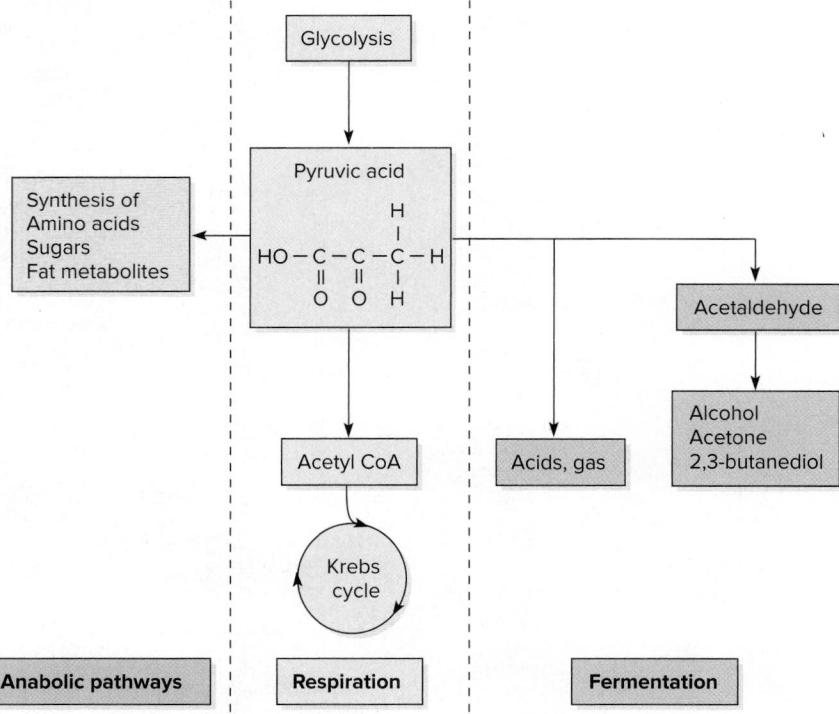

Figure 8.18 The fates of pyruvic acid (pyruvate). This metabolite is an important hub in the processing of nutrients by microbes. It may be fermented anaerobically to several end products or oxidized completely to CO_2 and H_2O through the Krebs cycle and the electron transport system. It can also serve as a source of raw material for synthesizing amino acids and carbohydrates (see figure 18.25).

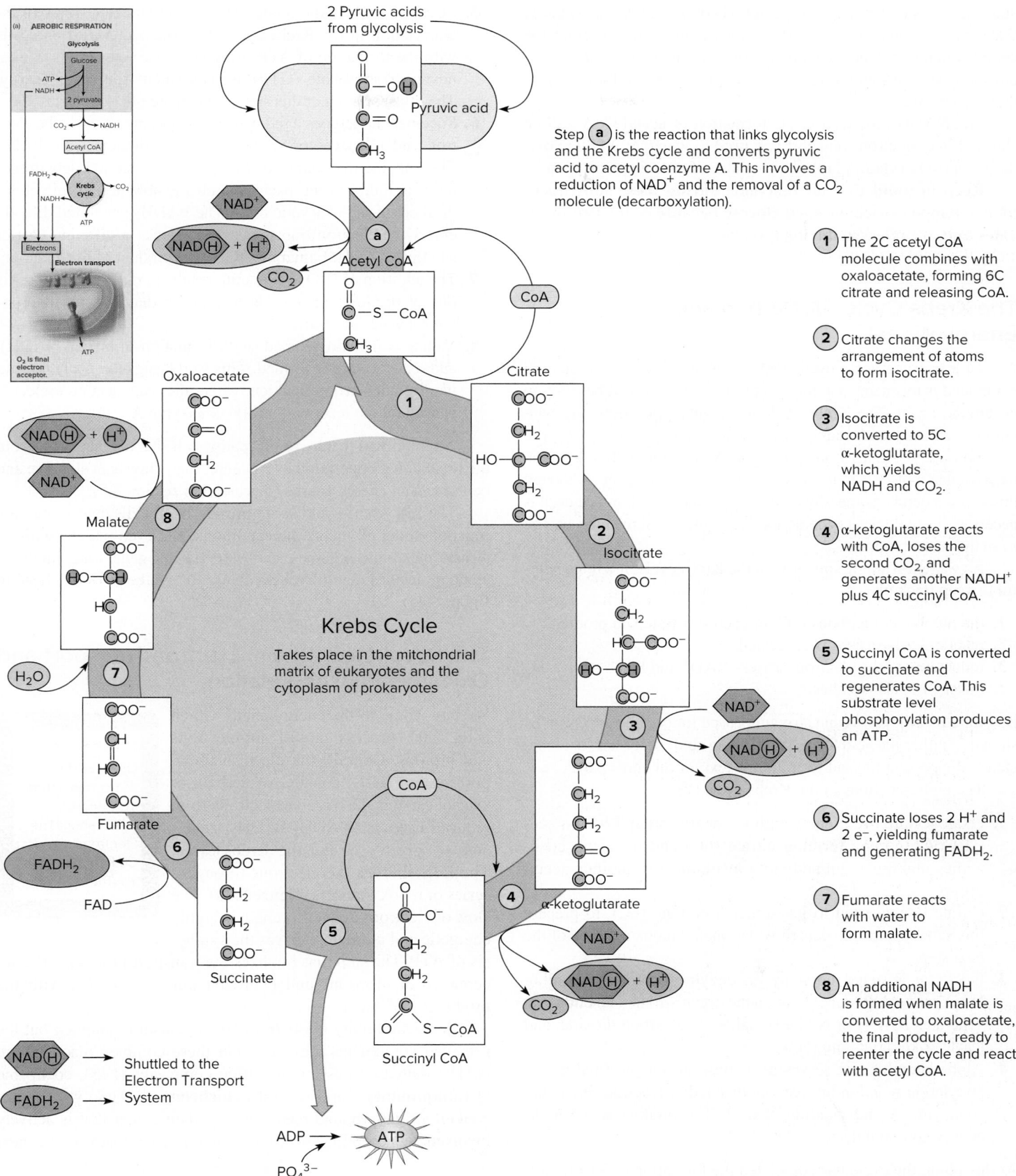

Step **a** is the reaction that links glycolysis and the Krebs cycle and converts pyruvic acid to acetyl coenzyme A. This involves a reduction of NAD^+ and the removal of a CO_2 molecule (decarboxylation).

1 The 2C acetyl CoA molecule combines with oxaloacetate, forming 6C citrate and releasing CoA.

2 Citrate changes the arrangement of atoms to form isocitrate.

3 Isocitrate is converted to 5C α-ketoglutarate, which yields NADH and CO_2.

4 α-ketoglutarate reacts with CoA, loses the second CO_2, and generates another $NADH^+$ plus 4C succinyl CoA.

5 Succinyl CoA is converted to succinate and regenerates CoA. This substrate level phosphorylation produces an ATP.

6 Succinate loses 2 H^+ and 2 e^-, yielding fumarate and generating $FADH_2$.

7 Fumarate reacts with water to form malate.

8 An additional NADH is formed when malate is converted to oxaloacetate, the final product, ready to reenter the cycle and react with acetyl CoA.

Krebs Cycle

Takes place in the mitchondrial matrix of eukaryotes and the cytoplasm of prokaryotes

NAD(H) → Shuttled to the Electron Transport

$FADH_2$ → System

Process Figure 8.19 **The reactions of a single turn of the Krebs cycle.** Each glucose will produce two turns of this pathway. The top portion (a) depicts the conversion of pyruvic acid to acetyl Co enzyme A, a necessary reaction that sets up the first step of Krebs. Note that this is an enlarged, more detailed view of the central portion of figure 8.16a.

attached to coenzyme A, forming **acetyl coenzyme A** (acetyl CoA) that feeds into the Krebs cycle. Pyruvate dehydrogenase, the enzyme complex that makes this reaction possible, is huge. In *E. coli,* it contains four vitamins and has a molecular weight of 6 million—larger than a ribosome!

The NADH produced in the formation of acetyl CoA will be shuttled into electron transport and used to generate ATP via oxidative phosphorylation in the third phase of aerobic respiration.

Keep in mind that all reactions during these subsequent phases happen twice for each glucose because of the two pyruvates that are released during glycolysis.

The Krebs Cycle—A Carbon and Energy Wheel

As you learned earlier, a cyclic pathway is one in which the starting compound is regenerated at the end of the cycle so it will be ready to go around again. The Krebs cycle has eight steps, beginning with citrate formation and ending with oxaloacetate (figure 8.19). The 2-carbon acetyl groups from coenzyme A combine with a 4-carbon oxaloacetate molecule to form citrate, a 6-carbon molecule. Although this seems a cumbersome way to dismantle such a small molecule, it is necessary in biological systems for extracting the remaining energy from the acetyl fragment.

As we take a single spin around the Krebs cycle, it will be helpful to keep track of

1. the numbers of carbons (#C) of each substrate and product,
2. reactions where CO_2 is generated,
3. reductions of the electron carriers NAD^+ and FAD, and
4. the site of ATP synthesis.

Be aware that the terms and structures used for organic acids can be shown as either the acid form (oxaloacetic acid) or its salt (oxaloacetate). We present the salt form in this text for the most part.

The eight reactions in the Krebs cycle are:

1. Oxaloacetate (4C) reacts with the acetyl group (2C) on acetyl CoA, thereby forming citrate (6C) and releasing coenzyme, making it immediately available for another acetyl group.
2. Citrate is converted to its isomer, isocitrate (6C), to prepare this substrate for the decarboxylation and redox reaction of the next step.
3. Isocitrate is acted upon by an enzyme complex including NAD^+ or NADP (depending on the organism) in a reaction that generates NADH or NADPH, splits off a carbon dioxide, and leaves α-ketoglutarate (5C).
4. Alpha-ketoglutarate serves as a substrate for the final decarboxylation reaction and yet another redox reaction, involving coenzyme A and yielding NADH. The product is the high-energy compound succinyl CoA (4C).

At this point, the cycle has completed the formation of 3 CO_2 molecules that balance out each original 3-carbon pyruvic acid released by glycolysis. The remaining steps serve not only to regenerate the oxaloacetate to start the cycle again but also to extract more energy from the intermediate compounds leading to oxaloacetate.

5. Succinyl CoA is the source of the one substrate level phosphorylation in the Krebs cycle. In most bacteria, it proceeds with the formation of ATP, although eukaryotes produce guanosine triphosphate (GTP), an equivalent source of energy. The other product at this step is succinate (4C).
6. Succinate undergoes a redox reaction, but in this case, the electron and H^+ acceptor is flavin adenine dinucleotide (FAD). The enzyme that catalyzes this reaction, succinyl dehydrogenase, is found in the bacterial cell membrane and mitochondrial crista of eukaryotic cells. The $FADH_2$ generated directly enters the electron transport system but at a different carrier site than NADH. Fumarate (4C) is the product of this reaction.
7. The addition of H_2O to fumarate results in malate (4C). This is one of the few reactions in respiration that directly incorporates a water molecule.
8. Malate is dehydrogenated (with formation of a final NADH), and oxaloacetate is formed. This step brings the cycle back to its original starting position, where the oxaloacetate molecule is available to react with acetyl coenzyme A.

An important feature of this pathway is that acetyl groups from the breakdown of certain fats can enter the pathway at step one and be used as an energy source (discussed in section 8.5).

The full Krebs cycle is not present in all cells, and it may not function under all metabolic conditions. But even in cells with alternate metabolic schemes, it is essential for generating small organic molecules that microbes require for synthesis (look ahead to figure 8.25).

The Respiratory Chain: Electron Transport and Oxidative Phosphorylation

We now come to the energy chain, which is the final "processing mill" for electrons and protons. Overall, the electron transport system (ETS) consists of a chain of special carriers that receive electrons from reduced carriers (NADH, $FADH_2$) generated by glycolysis and the Krebs cycle. The ETS shuttles the electrons through a series of redox transfers **(figure 8.20)**. The flow of electrons down this chain is highly energetic and ultimately drives the synthesis of ATP. The step that finalizes the transport process is the acceptance of electrons and hydrogen ions by oxygen, with the production of water.

Quick Search
Observe this process in action at "Cellular Respiration: The Electron Transport Chain" on YouTube.

Some variability exists from one organism to another, but the principal compounds that carry out these complex reactions are NADH dehydrogenase, iron-sulfide complexes (FeS), coenzyme Q **(ubiquinone),*** and several **cytochromes.*** The cytochromes contain a tightly bound metal atom at their center that is actively involved in accepting electrons and donating them to the next

* *ubiquinone* (yoo-bik′-wih-nohn) L. *ubique,* everywhere. A type of chemical, similar to vitamin K, that is very common in cells.
* *cytochrome* (sy′-toh-krohm) Gr. *cyto,* cell, and *kroma,* color. Cytochromes are pigmented, iron-containing molecules similar to hemoglobin.

Cutaway view of a mitochondrion

Cell membrane
Nucleus
Ribosomes
Mitochondria
Cilium

Outer membrane
Inner membrane
Matrix
Intermembrane space
Cristae
ATP synthases

① **The mitochondrion is the site of electron transport and oxidative phosphorylation in eukaryotic cells.** Its double membrane feature creates two separate compartments: the intermembrane space (brown) between the outer and inner membranes and the matrix (yellow) entirely enclosed by the inner membrane folded into cristae (blue). The electron transport proteins and the ATP synthases are stationed throughout the cristae. From here, the pathways have access to both matrix and intermembrane compartments.

② **Close up of the ETS.** NAD dehydrogenase accepts electrons from NADH and transfers them to CoQ. Electrons removed from $FADH_2$ by the FeS complex are also transferred to CoQ. These electrons are passed through the remaining carriers: the cytochrome b/c_1 complex, cytochrome c, and cytochrome a/a_3. At cytochrome a/a_3, 2 electrons combine with 2 protons (H^+) and an oxygen atom to form H_2O. Electron transport is coupled with the pumping of protons across the crista membrane into the intermembrane space by three of the carrier complexes. This process forms the basis of chemiosmosis.

② Closeup shows relative positions of cristae, compartments, ETS, and ATP synthase.

I	NADH dehydrogenase (N)
II	FeS proteins (F)
	Coenzyme Q
III	Cytochrome b
	Cytochrome c_1
	Cytochrome c
IV	Cytochrome a
	Cytochrome a_3

(a)

(b)

Crista membrane

N e⁻ e⁻ F Q e⁻ e⁻ b c₁ e⁻ c e⁻ a a₃ e⁻

H^+ ... (many throughout intermembrane space)

From glycolysis and Krebs — NAD(H) → NAD

FAD(H₂) ← From Krebs → FAD

③ ADP + P_i

$2 H^+$ $2 e^-$ $\frac{1}{2} O_2$
H O H

ATP

Matrix

F₁

ATP synthase F₀

Intermembrane space

Outer membrane

+ Charged **− Charged**

③ **(a)** Proton transport by the ETS into the intermembrane space (IMS) sets up a chemical and charge gradient between the IMS and the matrix called the proton motive force. **(b)** ATP synthase. The gradient creates an electrical potential for capturing the free energy of electron movement to drive ATP synthesis. As protons (H^+) diffuse from the intermembrane space through the F_0 portion of the ATP synthase into the matrix, the F_1 portion rotates, pulls in ADP and P_i, and releases ATP. This is an exergonic reaction that requires 3 protons for each ATP.

Figure 8.20 Electron transport, oxidative phosphorylation, the proton motive force, chemiosmosis, and ATP synthesis in the mitochondrion.

carrier in the series. The highly compartmentalized structure of the respiratory chain is an important factor in its function. Note in figure 8.20 that the electron transport carriers and enzymes are associated in complexes in the inner mitochondrial membranes in eukaryotes. The cell membrane is the equivalent structure for housing them in bacteria (**figure 8.21**).

The Carriers of Electron Transport: The Energy Cascade in the Mitochondria

The principal questions about the electron transport system are these: How are the electrons passed from one carrier to another in the series? How is this progression coupled to ATP synthesis? Where and how is oxygen utilized? Although the biochemical details of this process are somewhat complicated, the basic reactions consist of a number of redox reactions now familiar to us. In general, the carrier compounds and their associated enzymes are arranged in linear sequence and are reduced and oxidized while shuttling the electrons along to their final acceptor (**figure 8.20, part 2**).

The electron carrier complexes and associated molecules present in aerobic organisms are as follows:

1. **Complex I** consists of a huge multienzyme cluster termed NADH dehydrogenase (reductase), which receives the NADH from glycolysis and the Krebs cycle.
2. **Complex II** is made up of a series of iron-sulfur (FeS) proteins that receive electrons from $FADH_2$ produced in the sixth step of the Krebs cycle.
3. **Coenzyme Q,** or **ubiquinone,** is a mobile carrier not embedded in the membrane that can pick up electrons from both complexes I and II and donate them to complex III.
4. **Complex III** is comprised of cytochromes b and c_1, which deliver electrons to cytochrome c.
5. **Cytochrome c** is another mobile carrier that shuttles electrons between complexes III and IV.
6. **Complex IV** catalyzes the reaction between electrons, H^+, and oxygen, yielding H_2O and completing the electron transfer process.

In addition to the roles outlined here, complexes I, III, and IV have the additional function of pumping protons from the matrix into the intermembrane space to establish a charge gradient (discussed under the next heading). With each redox exchange, the energy level of the reactants is lessened. This free energy is captured and processed by **ATP synthase** complexes stationed in the cristae in close association with the ETS carriers. Each NADH that enters electron transport can potentially give rise to a maximum of 3 ATPs. Because $FADH_2$ from the Krebs cycle enters the cycle after complex I, its reactions result in a maximum yield of about 2 ATPs. This coupling of ATP synthesis to electron transport is termed **oxidative phosphorylation.**

> **Quick Search**
> Explore this concept at such sites as "Gradients: ATP Synthase" on YouTube.

The Formation of ATP and Chemiosmosis What biochemical processes are involved in coupling electron transport to the production of ATP? According to a widely accepted concept called **chemiosmosis,** during electron transport, some of the carriers actively transport protons across the cristae membrane and into the intermembrane compartment of the mitochondrion. This process sets up a concentration gradient of hydrogen ions called the **proton motive force (PMF).** The PMF consists of a difference in charge between the intermembrane compartment (+) and the matrix compartment (−) (**figure 8.20, part 3**).

Separating the charge has an effect similar to a battery, which can temporarily store potential energy. This charge will be maintained by the impermeability of the cristae membranes to protons (H^+). The only sites where protons can diffuse into the matrix compartment is at the ATP synthase complexes, which sets the stage for ATP synthesis.

ATP synthase is one of the most remarkable enzymes in the molecular world. Its structure includes two motors, an ion pump, and an enzyme that can capture energy and store it as ATP. These synthases are located throughout the membranes of the cristae in high numbers (one estimate is around 10,000 per mitochondrion). Each ATP synthase complex consists of two large units, F_0 and F_1 (figure 8.20, part 3). The F_0 portion is embedded in the membrane and can rotate like a motor to pull in protons. As protons flow through the F_0 portion by diffusion, the F_1 compartments pull in ADP and P_i. Rotation induces a change in the three-dimensional structure of ATP synthase. This adjustment creates a high-energy bond between ADP and P_i. The ATP that forms is released into the matrix, where it will serve as a constant supply for cellular reactions (figure 8.20, part 3). The enzyme is then rotated back to the start position for the next round of ATP synthesis.

An important issue is just how the ATP is delivered out of the mitochondrion and into the cytoplasm for use in cell synthesis and other activities. The mitochondrial membranes are impregnated with a number of specialized active transporters called ATP-ADP translocases that constantly transport ATP out of, and ADP into, the mitochondrion.

Bacterial ATP synthesis occurs by means of this same overall process. However, bacteria have the ETS stationed in the cell membrane, and the direction of the proton movement is from the cytoplasm to the periplasmic space between the membrane and cell wall. Bacteria may also show variations in the number and types of electron carriers (**figure 8.21**). As we will see, these differences will affect the amount of ATP produced. In both cell types, the chemiosmotic theory has been supported by tests showing that oxidative phosphorylation is blocked if the mitochondrial or bacterial cell membranes are disrupted.

The Terminal Step of Electron Transport

The terminal step, during which oxygen accepts the electrons, is catalyzed by cytochrome a/a_3, also called cytochrome oxidase. This large enzyme complex is specifically adapted to receive electrons from cytochrome c, pick up hydrogen ions from solution, and react with oxygen to form a molecule of water (figure 8.20, part 3). This reaction, though in actuality more complex, is summarized as follows:

$$2H^+ + 2e^- + \tfrac{1}{2}O_2 \rightarrow H_2O$$

Most eukaryotic aerobes have a fully functioning cytochrome system, but bacteria exhibit wide-ranging variations in this part of the

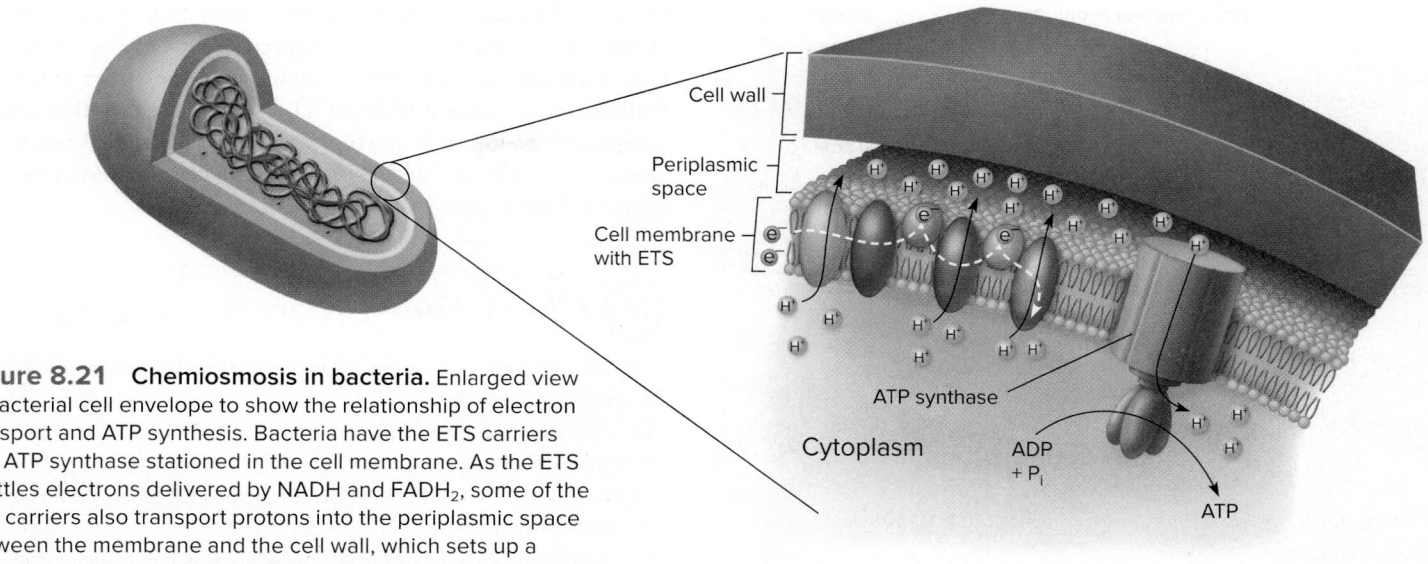

Figure 8.21 Chemiosmosis in bacteria. Enlarged view of bacterial cell envelope to show the relationship of electron transport and ATP synthesis. Bacteria have the ETS carriers and ATP synthase stationed in the cell membrane. As the ETS shuttles electrons delivered by NADH and FADH$_2$, some of the ETS carriers also transport protons into the periplasmic space between the membrane and the cell wall, which sets up a gradient similar to mitochondria. Bacteria will vary in the exact nature of the carriers in the ETS.

Figure labels: Cell wall; Periplasmic space; Cell membrane with ETS; ATP synthase; Cytoplasm; ADP + P$_i$; ATP

system. Some species lack one or more of the redox steps; others have several alternative electron transport schemes. Because many bacteria lack cytochrome oxidase, this variation can be used to differentiate among certain genera of bacteria. An oxidase detection test can be used to help identify members of the genera *Neisseria* and *Pseudomonas* and some species of *Bacillus*. Another variation in the cytochrome system is evident in certain bacteria (*Klebsiella, Enterobacter*) that can grow even in the presence of cyanide because they lack cytochrome oxidase. Cyanide will cause rapid death in humans and other eukaryotes because it blocks cytochrome oxidase, thereby completely terminating aerobic respiration, but it is harmless to these bacteria.

A potential side reaction of the respiratory chain in aerobic organisms is the incomplete reduction of oxygen to superoxide ion (O_2^-) and hydrogen peroxide (H_2O_2). As mentioned in chapter 7, these toxic oxygen products can be very damaging to cells. Aerobes have neutralizing enzymes to deal with these products, including *superoxide dismutase* and *catalase.* One exception is the genus *Streptococcus,* which can grow well in oxygen yet lacks both cytochromes and catalase. The tolerance of these organisms to oxygen can be explained by the neutralizing effects of a special peroxidase. The lack of cytochromes, catalase, and peroxidases in strict anaerobes as a rule limits their ability to process free oxygen and contributes to its toxic effects on them.

Summary of Aerobic Respiration

Reviewing the summary equation for respiration, we are now in a position to estimate the theoretical total of ATP molecules arising from a single glucose molecule through a combination of substrate-level phosphorylation and oxidation of NADH and FADH$_2$:

- There are 3 substrate-level reactions that give rise directly to ATP—2 in glycolysis and 1 in the Krebs cycle. Because these are happening **twice** for each event, they provide **6 ATPs** overall.
- There are **5 NADHs**—1 from glycolysis, 1 from pyruvic acid oxidation, and 3 from the Krebs cycle. Each NADH could

potentially power the synthesis of 3 ATPs, yielding $5 \times 3 = 15$ ATPs. Again, this happens **twice for each glucose** and could potentially generate **30 ATPs.**
- There is a single reaction in the Krebs cycle that yields FADH$_2$, which can power synthesis of 2 ATPs, and multiplied by 2 provides **4 ATPs.**
- The theoretical total for ATP synthesis could be $6 + 30 + 4 = $ **40;** however, because the first and second reactions of glycolysis use up 2 ATPs, the **final net number** is **38 ATPs.**

We emphasize that these totals are theoretical because the yields for NADH and FADH$_2$ are lower when the actual conditions in cells are taken into account. It is likely that oxidation of each NADH$^+$ provides closer to 2.5 ATPs and oxidation of FADH$_2$, about 1.75 ATPs. This will decrease the practical total to around **32 ATPs.** Another reason this total will be reduced for eukaryotic cells is that they must use active transport (and ATP) to carry the NADH from glycolysis into the mitochondrial cristae, bringing the total down to **30 ATPs** per glucose. Most prokaryotic cells probably produce even less ATP because they lack one or more of the carriers in the electron transport system or do not generate as much NADH. **Figure 8.22** charts the course of these calculations as they occur in the three phases of aerobic respiration.

No summary would be complete without considering three other key participants:

- Six carbon dioxide molecules are generated by reactions just prior to and during the Krebs cycle. These represent all of the carbons and oxygens from the original glucose molecule.
- Six oxygen molecules are consumed during the final transfer of electrons and hydrogens to cytochrome oxidase a_3.
- Six water molecules are produced in electron transport and 2 in glycolysis, but because 2 are used in the Krebs cycle, this leaves a net number of 6.

Anaerobic Respiration

Some bacteria have evolved an anaerobic respiratory system that functions like the aerobic cytochrome system, except that it utilizes

Pathways and Products **Net Output Summary****

Glycolysis — Glucose 6C	2 ATPs
	2 NADHs
Pyruvic acid 3C	2 Pyruvic acids
Transition step	2 NADHs
2C Acetyl CoA	2 CO₂s
Krebs Cycle 2 Turns	4 CO₂s
	6 NADHs
	2 ATPs
	2 FADH₂s
Electron transport and Oxidative Phosphorylation	34 ATPs
	6 H₂Os

*These compounds are processed by the electron transport system.

**This net figure accounts for total ATPs minus the 2 used in glycolysis and the doubling of the products after the split into 2 3C compounds in glycolysis.

Figure 8.22 Summary of aerobic metabolism products and ATP yields. Compare this figure with the original equation:
Glucose ($C_6H_{12}O_6$) + 6O_2 + 38 ADP + 38P_i → 6CO_2 + 6H_2O + 38 ATP.

oxygen-containing ions rather than free oxygen as the final electron acceptor. Of these, the nitrate (NO_3^-) and nitrite (NO_2^-) reduction systems are best known. The reaction in species such as *E. coli* is represented as follows:

$$\text{Nitrate reductase}$$
$$NO_3^- + NADH + H^+ \xrightarrow{\quad\quad} NO_2^- + H_2O + NAD^+$$

The enzyme nitrate reductase catalyzes the removal of oxygen from nitrate, leaving nitrite and water as products. This reaction is the basis for a physiological test used in identifying certain bacteria.

Some species of *Pseudomonas* and *Bacillus* possess enzymes that can further reduce nitrite to nitric oxide (NO), nitrous oxide (N_2O), and even nitrogen gas (N_2). This process, called **denitrification,** is a very important step in recycling nitrogen in the biosphere. Other oxygen-containing nutrients reduced anaerobically by various bacteria are carbonates and sulfates.

Most obligate anaerobes use the NADH they generate during glycolysis and the Krebs cycle to reduce some compound other than

O_2 but still follow the oxidative phosphorylation pathway. Methanogens (described in chapter 7) reduce CO_2 or CO_3^- (carbonate) to CH_4 (methane gas), and certain sulfate bacteria reduce SO_4^{2-} to S (sulfide) or hydrogen sulfide gas (H_2S). Some use metals or organic compounds as the final electron acceptor. These microbes live in subterranean habitats such as swamps and deep oceanic vents that are devoid of O_2 gas.

Check Your Progress SECTION 8.3

13. Describe the basic energy strategies of microbes and the catabolic pathways they use.
14. Compare and contrast the location of glycolysis, Krebs cycle, and electron transport in prokaryotic and eukaryotic cells.
15. Outline the basic steps in glycolysis, indicating where ATP is used and where it is given off.
16. Where does NADH originate, and what is its fate in an aerobe?
17. What is the fate of NADH in a fermentative organism?
18. Summarize the chemiosmotic theory of ATP formation.
19. How many ATPs could *theoretically* be formed from a glucose molecule carried through aerobic respiration, and how many are *probably* formed?
20. Why does the total of ATPs generated differ between bacteria and many eukaryotes?
21. Name the sources of oxygen in bacteria that use anaerobic respiration.
22. What are the final electron acceptors in aerobic, anaerobic, and fermentative metabolism?
23. Describe the reaction between electrons, H^+, and oxygen at the final step of electron transport.

8.4 The Importance of Fermentation

Expected Learning Outcomes

19. Explain what is meant by the term *fermentation* and how fermentation is used by natural biological systems and biotechnology.
20. Describe some of the processes of fermentation and the products that result.

Of all the results of pyruvate metabolism, probably the most varied is fermentation. Technically speaking, **fermentation*** is a process that involves (1) the incomplete oxidation of glucose or other carbohydrates in the absence of oxygen, (2) the oxidation of NADH to NAD^+, which is essential to the continuation of the glycolytic pathway, (3) the use of organic compounds as the terminal electron acceptors, and (4) the release of a small amount of ATP.

Over time, the term *fermentation* has acquired several looser meanings. Originally Pasteur called the microbial action of yeast

* *fermentation* (fur-men-tay′-shun) L. *fervere,* to boil, or *fermentatum,* leaven or yeast.

during wine production *ferments,* and to this day, biochemists use the term in reference to the production of ethyl alcohol by yeasts acting on glucose and other carbohydrates. Fermentation is traditionally what bacteriologists call the formation of acid, gas, and other products by the action of various bacteria on pyruvic acid. Industrial processes that produce chemicals on a massive scale through the actions of microbes are also called fermentations. Each of these usages is acceptable for one application or another.

It may seem that fermentation yielding only meager amounts of energy (2 ATPs maximum per glucose) would slow down growth. And yet the process is a common metabolic strategy among bacteria. What actually happens is that many bacteria can grow as fast as they would in the presence of oxygen. This rapid growth is made possible by an increase in the rate of glycolysis. From another standpoint, fermentation permits survival and growth in the absence of molecular oxygen and allows colonization of anaerobic environments. It also enables microorganisms with a versatile metabolism to adapt to variations in the availability of oxygen. For such facultative microbes, fermentation provides a means to function even when oxygen levels are too low for aerobic respiration or the usual electron acceptor is unavailable. An additional advantage of fermentation is that it generates a number of intermediates that can be used in **anabolic** pathways and shared within a population of microbes.

Bacteria that digest cellulose in the rumens of cattle and other ruminants are largely fermentative. After initially hydrolyzing cellulose to glucose, they ferment the glucose to organic acids, which are then absorbed as the bovine's principal energy source. Even human muscle cells can undergo a form of fermentation that permits short periods of activity after the oxygen supply in the muscle has been exhausted. Muscle cells convert pyruvic acid into lactic acid, which allows anaerobic production of ATP to proceed for a time. But this cannot go on indefinitely, and after a few minutes, the accumulated lactic acid causes muscle fatigue.

Products of Fermentation in Microorganisms

Alcoholic beverages (wine, beer, whiskey) are perhaps the most prominent among fermentation products; others are solvents (acetone, butanol), organic acids (lactic, acetic), dairy products, and many other foods. Derivatives of proteins, nucleic acids, and other organic compounds are fermented to produce vitamins, antibiotics, and even hormones such as hydrocortisone.

Fermentation products can be grouped into two general categories: alcoholic fermentation products and acidic fermentation products **(figure 8.23). Alcoholic fermentation** occurs in yeast or bacterial species that have metabolic pathways for converting pyruvic acid to ethanol. This process involves a decarboxylation of pyruvic acid to acetaldehyde, followed by a reduction of the acetaldehyde to ethanol. In oxidizing the NADH formed during glycolysis, NAD$^+$ is regenerated, thereby feeding back to and maintaining the glycolytic pathway. These processes are crucial in the production of beer and wine, though the actual techniques for arriving at the desired amount of ethanol and the prevention of unwanted side reactions are important tricks of the brewer's trade **(8.4 Making Connections).** Note that the

Figure 8.23 The fermentation systems that produce acid and alcohol. In both cases, the final electron acceptor is an organic compound. In yeasts, pyruvic acid is decarboxylated to acetaldehyde and CO_2, and the NADH given off in the glycolytic pathway reduces acetaldehyde to ethyl alcohol. In homolactic fermentative bacteria, pyruvic acid is reduced by NADH to lactic acid. Both systems regenerate NAD$^+$ to feed back into glycolysis or other cycles (pink dotted lines).

products of alcoholic fermentation include both ethanol *and* CO_2, a gas that accounts for the bubbles in champagne and beer and the rising of bread dough.

Alcohols other than ethanol can be produced during bacterial fermentation pathways. Certain clostridia produce butanol and isopropanol through a complex series of reactions. Although this process was once an important source of alcohols for industrial use, it has been largely replaced by a nonmicrobial petroleum process.

The pathways of **acidic fermentation** are extremely varied. Lactic acid bacteria ferment pyruvate in the same way that humans do—by reducing it to lactic acid. If the product of this fermentation is mainly lactic acid, as in certain species of *Streptococcus* and *Lactobacillus,* it is termed *homolactic.* The souring of milk is due largely to the production of this acid by bacteria. When glucose is fermented to a mixture of lactic acid, acetic acid, and carbon dioxide, as is the case with *Leuconostoc* and other species of *Lactobacillus,* the process is termed *heterolactic fermentation.*

Many members of the family Enterobacteriaceae (*Escherichia, Shigella,* and *Salmonella*) possess enzyme systems for converting pyruvic acid to several acids simultaneously **(figure 8.24). Mixed**

8.4 MAKING CONNECTIONS

Pasteur and the Wine-to-Vinegar Mystery

The microbiology of alcoholic fermentation was greatly clarified by Louis Pasteur after French winemakers hired him to uncover the causes of periodic spoilage in wines. Especially troublesome was the conversion of wine to vinegar and the resultant sour flavor. Up to that time, wine formation had been considered strictly a chemical process. After extensively studying beer making and wine grapes, Pasteur concluded that wine, both fine and not so fine, was the result of microbial action on the juices of the grape and that wine "disease" was caused by contaminating organisms that produced undesirable products such as acid. Although he did not know it at the time, the bacterial contaminants responsible for the acidity of the spoiled wines were likely to be *Acetobacter* or *Gluconobacter* introduced by the grapes, air, or wine-making apparatus. These common gram-negative genera further oxidized ethanol to acetic acid and are presently used in production of some types of vinegar. The following formula shows how this is accomplished:

© Kathy Park Talaro

$$H-\overset{\overset{\displaystyle H}{|}}{\underset{\underset{\displaystyle H}{|}}{C}}-\overset{\overset{\displaystyle H}{|}}{\underset{\underset{\displaystyle H}{|}}{C}}-OH \xrightarrow{O_2} H-\overset{\overset{\displaystyle H}{|}}{\underset{\underset{\displaystyle H}{|}}{C}}-C\overset{\displaystyle O}{\underset{\displaystyle OH}{}} + 2H_2O$$

Ethanol Acetic acid

Pasteur's far-reaching solution to the problem is still with us today—mild heating, or *pasteurization*, of the grape juice to destroy the contaminants, followed by inoculation of the juice with a pure yeast culture. The topic of pasteurization is explored further in chapter 11.

An analysis of wine shows that it contains, in addition to water and ethyl alcohol, thousands of other chemical compounds (see "Wine Chemistry" on Wikipedia). What does this tell you about the metabolism of yeasts during alcoholic fermentation? *Answer available on Connect.*

acid fermentation produces a combination of acetic, lactic, succinic, and formic acids, and it lowers the pH of a medium to about 4.0. *Propionibacterium* produces primarily propionic acid, which gives the characteristic flavor to Swiss cheese; gas (CO_2) from the same fermentation reaction produces the holes. Some members also further decompose formic acid completely to carbon dioxide and hydrogen gases. Because enteric bacteria commonly occupy

the intestine, this fermentative activity accounts for the accumulation of some types of gas—primarily CO_2 and H_2—in the intestine. Some bacteria reduce the organic acids and produce the neutral end product 2,3-butanediol. Several of these end products are the basis of biochemical tests described in chapter 17.

We have provided only a brief survey of fermentation products, but it is worth noting that microbes can be harnessed to synthesize

Figure 8.24 Miscellaneous products of pyruvate fermentation and some of the bacteria involved in their production.

a variety of other substances by varying the raw materials provided them. In fact, so broad is the meaning of the word *fermentation* that the large-scale industrial syntheses by microorganisms are not technically fermentations and often utilize entirely different mechanisms from those described here. They even occur aerobically, particularly in antibiotic, hormone, vitamin, and amino acid production, but they still produce a valuable commercial product.

 Check Your Progress SECTION 8.4

24. What adaptive advantages does a fermentative metabolism confer on a microbe?
25. Describe three patterns of fermentation typical of microbes.

8.5 Biosynthesis and the Crossing Pathways of Metabolism

 Expected Learning Outcomes

21. Explain how cells perform anabolic functions and what is involved in amphibolism.
22. Identify major pathways where molecules can be redirected or converted into useful cell products.
23. Briefly describe several mechanisms in biosynthesis of macromolecules.

Our discussion now turns from catabolism and energy extraction to anabolic functions and biosynthesis. In this section, we will explore this area, also called intermediary metabolism, that includes amphibolic pathways, the synthesis of simple molecules, and the synthesis of macromolecules.

The Frugality of the Cell—Waste Not, Want Not

It must be obvious by now that cells have mechanisms for careful management of carbon compounds. Rather than being dead ends, most catabolic pathways contain strategic molecular intermediates (metabolites) that can be diverted into anabolic pathways. In this way, a given molecule can serve multiple purposes, and the maximum benefit can be derived from all nutrients and metabolites of the cell pool. The property of a system to integrate catabolic and anabolic pathways to improve cell efficiency is termed **amphibolism.***

* *amphibolism* (am-fee-bol'-izm) Gr. *amphi,* two-sided, and *bole,* a throw.

At this point in the chapter, you can appreciate a more complex view of metabolism than that presented at the beginning in figure 8.1. **Figure 8.25** demonstrates the amphibolic nature of intermediary metabolism. The pathways of glucose catabolism are an especially rich "metabolic marketplace." The principal sites of amphibolic interaction occur during glycolysis (glyceraldehyde-3-phosphate and pyruvic acid) and the Krebs cycle (acetyl coenzyme A and various organic acids).

Amphibolic Sources of Cellular Building Blocks

Glyceraldehyde-3-phosphate can be diverted away from glycolysis and converted into precursors for amino acid, carbohydrate, and triglyceride (fat) synthesis. (A precursor molecule is one that gives rise to a different compound.) Earlier, we noted the numerous directions that pyruvic acid catabolism can take. In terms of synthesis, pyruvate also plays a pivotal role in providing intermediates for amino acids. In the event of an inadequate glucose supply, pyruvate serves as the starting point in glucose synthesis from various metabolic intermediates, a process called **gluconeogenesis.***

* *gluconeogenesis* (gloo″-koh-nee″-oh-gen′-uh-sis) Literally, the formation of "new" glucose.

Figure 8.25 An amphibolic view of metabolism. Intermediate compounds such as pyruvic acid and acetyl coenzyme A serve multiple functions. With comparatively small modifications, these compounds can be converted into other compounds and enter a different pathway. Note that catabolism of glucose (center) furnishes numerous intermediates for anabolic pathways that synthesize amino acids, fats, nucleic acids, and carbohydrates. These building blocks can serve in further synthesis of larger molecules to construct a wide array of cell components.

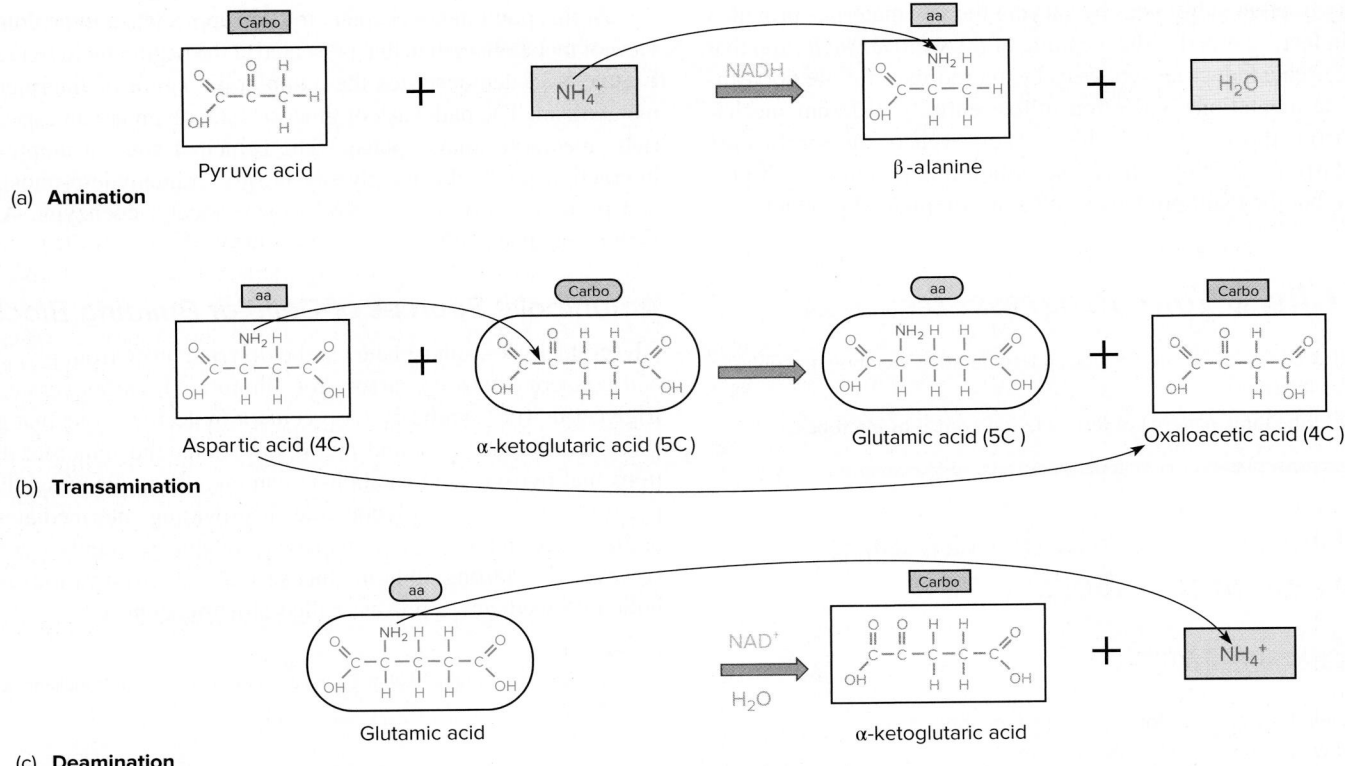

(a) **Amination**

(b) **Transamination**

(c) **Deamination**

Figure 8.26 Reactions that produce and convert amino acids. All of the reactions require energy as ATP or NADH and specialized enzymes. **(a)** Through amination (the addition of an ammonium molecule [amino group]), a carbohydrate can be converted to an amino acid. **(b)** Through transamination (transfer of an amino group from an amino acid to a carbohydrate), metabolic intermediates can be converted to amino acids that are in low supply. **(c)** Through deamination (removal of an amino group), an amino acid can be converted to a useful intermediate of carbohydrate catabolism. This is how proteins are used to derive energy. Ammonium is one waste product of deamination. Note that carbohydrate structures are in brown and amino acids are in blue.

The acetyl group that starts the Krebs cycle is another extremely versatile metabolite that can be fed into a number of synthetic pathways. This 2-carbon fragment can be converted as a single unit into one of several amino acids, or a number of these fragments can be condensed into hydrocarbon chains that are important building blocks for fatty acid and lipid synthesis. Note that the reverse is also true: Fats can be degraded to acetyl and thereby enter the Krebs cycle via acetyl coenzyme A. This aerobic process, called **beta oxidation,** releases significant energy. Oxidation of a 6-carbon fatty acid with glycerol potentially yields 50 ATPs, compared with 38 for a 6-carbon sugar.

Two carbohydrate metabolites from the Krebs cycle are oxaloacetic acid and α-ketoglutaric acid—essential intermediates in the synthesis of certain amino acids. This occurs through **amination,** the addition of an amino group to a carbon skeleton **(figure 8.26a).** A certain core group of amino acids can then be used to synthesize others. Amino acids and carbohydrates can be interchanged through **transamination** as well **(figure 8.26b).**

Pathways that synthesize the nitrogen bases (purines, pyrimidines), which are components of DNA and RNA, originate in amino acids and so can be dependent on intermediates from the Krebs cycle as well. Because the coenzymes NAD⁺, NADP, FAD, and others contain purines and pyrimidines similar to the nucleic acids, their synthetic pathways are also dependent on amino acids. During times of carbohydrate deprivation, organisms can likewise convert amino acids to intermediates of the Krebs cycle by **deamination**

(removal of an amino group) and thereby derive energy from proteins. Deamination results in the formation of nitrogen waste products such as ammonium ions or urea **(figure 8.26c).**

Formation of Macromolecules

Monosaccharides, amino acids, fatty acids, nitrogen bases, and vitamins—the building blocks that make up the various macromolecules and organelles of the cell—come from two possible sources. They can enter the cell from the outside as nutrients, or they can be synthesized through various cellular pathways. The degree to which an organism can synthesize its own building blocks (simple molecules) is determined by its genetic makeup, a factor that varies tremendously from group to group. In chapter 7 you learned that autotrophs require only CO_2 as a carbon source, a few minerals to synthesize all cell substances, and no organic nutrients. Even some heterotrophic organisms (*E. coli,* yeasts) are so metabolically complete that they can synthesize all cellular substances from minerals and one organic carbon source such as glucose. Compare this with a strict parasite lacking or having lost significant synthetic capabilities and having to obtain most essential nutrient molecules from its host.

Whatever their source, once these building blocks are added to the metabolic pool, they are available for synthesis of polymers by the cell. The details of synthesis vary among the types of macromolecules, but all of them involve the formation of bonds by specialized enzymes and the expenditure of ATP.

Carbohydrate Biosynthesis

The role of glucose in bioenergetics is so crucial that its biosynthesis is ensured by several alternative pathways. Certain structures in the cell depend on an adequate supply of glucose as well. It is the major component of the cellulose cell walls of some eukaryotes and of certain storage granules (starch, glycogen). One of the intermediaries in glycolysis, glucose-6-P, is used to form glycogen. Monosaccharides other than glucose are important in the synthesis of bacterial cell walls. Peptidoglycan contains a linked polymer of muramic acid and glucosamine. Fructose-6-P from glycolysis is used to form these two sugars. Carbohydrates (deoxyribose, ribose) are also essential building blocks in nucleic acids. Polysaccharides are the predominant components of cell surface structures such as capsules and the glycocalyx, and they are commonly found in slime layers (dextran is an example). Remember that most polymerization reactions occur via loss of a water molecule (see figure 2.16) and the input of energy.

Assembly of the Cell

The component parts of a bacterial cell are being synthesized on a continuous basis, and catabolism is also taking place, as long as nutrients are present and the cell is in a nondormant state. When anabolism produces sufficient macromolecules, and when DNA replication produces duplicate copies of the cell's genetic material, the cell undergoes binary fission. The two new cells will need twice as many ribosomes, twice as many enzymes, and so on. The cell has created these during the initial anabolic phases we have described. Before cell division, the membrane(s) and the cell wall will have increased in size to create a cell that is almost twice as big as a "newborn" cell. Once synthesized, the phospholipid bilayer components of the membranes assemble themselves spontaneously, but proteins and other components must be added to the membranes. Growth of the cell wall, accomplished by the addition and coupling of sugars and peptides, requires energy input. The catabolic processes provide constant energy for these complex building reactions.

 Check Your Progress SECTION 8.5

26. What is meant by amphibolism, and what are its most important functions?
27. Identify major ways that amino acids are processed.
28. Which macromolecules are synthesized by specialized anabolic pathways?

8.6 Photosynthesis: The Earth's Lifeline

 Expected Learning Outcomes

24. Outline the general reactions of photosynthesis and the importance it holds for the energetics of life.

25. Describe the pigment systems and how they function.

26. Describe the main events in the light-dependent reactions and the products that result.

27. Describe the main events in the light-independent reactions and the products that result.

As mentioned earlier, the ultimate source of most of the chemical energy in cells is the sun. Because this source is *directly* available only to the cells of photosynthesizers, most organisms on earth are either directly or indirectly dependent on photosynthesis, except for a few chemoautotrophs that derive their energy and nutrients solely from inorganic substrates. The other major products of photosynthesis are organic carbon compounds, which are produced from carbon dioxide through a process called **carbon fixation.** On land, green plants are the primary photosynthesizers; and in aquatic ecosystems, where 80% to 90% of all photosynthesis occurs, this role is filled by algae, cyanobacteria, and green sulfur, purple sulfur, and purple nonsulfur bacteria.

The summary equation for the main reactants and products of photosynthesis in aerobic organisms is

$$6CO_2 + 6H_2O \xrightarrow{\text{Light and pigment}} \underset{\text{Glucose}}{C_6H_{12}O_6} + 6O_2 \uparrow$$

The anatomy of photosynthetic cells is adapted to trapping sunlight, and their physiology effectively uses this solar energy to produce high-energy glucose from low-energy CO_2 and water. Photosynthetic organisms achieve this remarkable feat through a series of reactions involving light, pigment, CO_2, and water, which is used as a source for electrons.

Photosynthesis proceeds in two phases: the **light-dependent reactions,** which proceed only in the presence of light waves, and the **light-independent reactions,** which can operate without direct exposure to light, but are still reliant on the energy molecules made in the light-dependent reactions **(figure 8.27).**

Solar energy is delivered in discrete energy packets called **photons** (also called quanta) that travel as waves. The wavelengths of light operating in photosynthesis occur in the visible spectrum between 400 nanometers (violet) and 700 nanometers (red) and above. As this light strikes photosynthetic pigments, some wavelengths are absorbed, some pass through, and some are reflected. The activity that has the greatest impact on photosynthesis is the absorbance of light by photosynthetic pigments. These include the **chlorophylls,** which are green; **carotenoids,** which are yellow, orange, or red; and **phycobilins,** which are red or blue-green.[3] By far the most important of these pigments are the chlorophylls, which contain a *photocenter* that consists of a magnesium atom held in the center of a complex ringed molecule called a *porphyrin* **(figure 8.28, part 3).** As we will see, the chlorophyll molecule harvests the energy of photons and converts it to electron (chemical) energy. Accessory photosynthetic pigments such as carotenes trap light energy and shuttle it to chlorophyll, thereby functioning like antennae. These light-dependent reactions are responsible for **photophosphorylation,**

3. The color of the pigment corresponds to the wavelength of light it reflects.

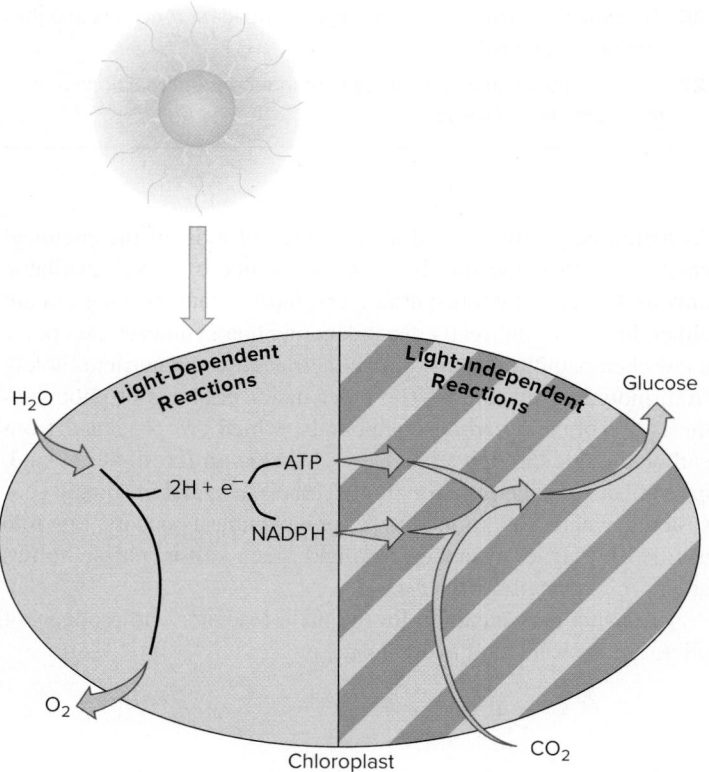

Figure 8.27 **Overview of photosynthesis.** The general reactions of photosynthesis, divided into two phases called light-dependent reactions and light-independent reactions. The light-dependent reactions involve two major events: (1) the trapping of light with a photosensitive pigment, such as chlorophyll, and converting the light energy into chemical energy in the form of ATP and NADPH; and (2) the splitting of water molecules with the release of oxygen gas. The light-independent reactions utilize ATP and NADPH produced during the light-dependent reactions to fix CO_2 into organic compounds such as glucose.

the channeling of energy extracted from light to make high-energy bonds of ATP. This sets the scene for the light-independent reactions, which use both ATP and NADPH for synthesis. During this phase, carbon atoms from CO_2 are fixed to the carbon backbones of organic molecules.

The detailed biochemistry of photosynthesis is beyond the scope of this text, but we will provide an overview of the general process as it occurs in green plants, algae and cyanobacteria. It is worth noting that many of the basic activities are biochemically similar to certain pathways of aerobic metabolism.

Quick Search

Search the Internet for videos referencing "photosynthesis light reactions."

Light-Dependent Reactions

The systems that carry the photosynthetic pigments are also the sites for the light-dependent reactions. They occur in the **thylakoid** membranes of compartments called **grana** (singular, granum) in chloroplasts **(figure 8.28, part 1)** and in thylakoid layers of the cell

membranes of cyanobacteria (see figure 4.28c). These systems exist as two separate complexes called **photosystem I (PS700)** and **photosystem II (PS680)**[4] **(figure 8.28, part 4).**

Both systems contain chlorophyll, but their chlorophylls are sensitive to different wavelengths: PS I absorbs longer wavelengths (above 680 nm) and PS II absorbs shorter wavelengths (680 nm and below), which efficiently covers the range of light to which phototrophs are exposed. Both photosystems are stationed in **antennae**—clusters of about 300 molecules of chlorophyll located in the thylakoid membranes. Antennae absorb light and convey it to **reaction centers,** where the light-dependent reactions take place (figure 8.28, part 4).

The chloroplast is composed of separate compartments similar to mitochondria. One compartment—the **stroma**—is the region surrounding the grana and thylakoids, and the other one is the space inside the thylakoids called the **lumen** (figure 8.28, part 4).

Overview of the Light-Dependent Reactions and Photophosphorylation

The thylakoids are specially situated to bring together all components of the light-dependent reactions. Most of these components are embedded in the thylakoid membranes and are accessible from both the lumen and the stroma. They include chlorophyll molecules and their photosystems (PS II and PS I)[5]; an oxygen-evolving complex; electron transport systems—consisting of a mobile electron carrier plastoquinone (PQ); cytochrome b_f complex; plastocyanin (PC), a second mobile electron carrier; ferredoxin (Fd) with nicotinamide adenine dinucleotide phosphate (NADP[+]) reductase (FNR); and ATP synthase. Several of the systems and their actions are very similar to those we previously examined in mitochondria.

The light-dependent pathway proceeds as follows:

PS II Reactions

1. When light strikes the magnesium in PS II, an electron becomes excited (is raised to a higher energy level) and is released by PS II to the first electron transport system.
2. An oxygen-evolving complex associated with PS II splits two water molecules into 4 electrons, 4 protons (H[+]), and oxygen. Because it occurs in the presence of light, this step is termed **photolysis,** and it results in **the first major product** of the light-dependent reactions—the **O_2 gas** required by all aerobic organisms, including even most photosynthesizers. In addition, the electrons released by photolysis return the PS II complex to a ground state so it is again ready to respond to light. Another important effect of photolysis is the release of protons into the lumen, which contributes to the chemiosmotic proton gradient so essential to ATP synthesis.

4. The numbers refer to the wavelength of light to which each system is most sensitive.

5. These numbers signify the order in which they were discovered, not the order of reactivity.

Process Figure 8.28 **The location and participants in the light-dependent reactions of a chloroplast. (1)** A cell of the algae, *Chlamydomonas,* with a cutaway view of a single chloroplast to show the internal structure of stroma and grana. **(2)** A granum consists of a stack of hollow membrane-bound discs called thylakoids, which house **(3)** the chlorophyll molecules, arranged in arrays (antennae) for capturing light. **(4)** The chlorophylls exist in separate systems called PS II and PS I, embedded in the thylakoid membrane. Other players in the process are an electron transport chain, an oxygen-evolving complex, a final electron acceptor ($NADP^+$), and ATP synthase. The text covers the main mechanisms of this process.

Electron Transport

1. An electron given off by PS II is picked up by the first compound of the electron transport chain—PQ—which moves it to the cytochrome b_f complex. From here, it is transported to plastocyanin and to the PS I complex.

2. An event coupled to electron transport is the delivery of protons (H^+) from the stroma side to the lumen side by the actions of the cytochrome b_f complex, which helps maintain the electrochemical gradient between the stroma and lumen.

PS I Reactions

1. When PS I is excited by light, it also releases an electron that will be transported through a different series of carriers. In addition, it picks up the electrons from PS II delivered by plastocyanin.

2. These electrons are shuttled to the final electron acceptors—ferredoxin and the FNR complex. The function of the FNR is to transfer 2 electrons (and 2 protons) to NADP$^+$, converting it to **NADPH.** This is the **second major product** of the light-dependent reactions, one which provides reducing power for the light-independent reactions.

ATP Synthase and Photophosphorylation ATP synthase complexes are distributed throughout the thylakoid membranes. They have the same structure as those of mitochondria and they, too, capture the free energy inherent in the proton motive force to synthesize ATP. As protons flow through the F_0 portion of the synthase from the lumen to the stroma, the F_1 component pulls in ADP and P_i, releasing ATP, the **third major product** of the light reactions. You may notice that the energy-associated molecules (NADPH and ATP) are released in the stroma, which situates them for immediate use in the light-independent reactions covered in the next section.

Light-Independent Reactions

The anabolic reactions of photosynthesis occur in the stroma of a chloroplast or the cytoplasm of cyanobacteria. These reactions use energy produced by the light phase to synthesize glucose by means of the Calvin cycle **(figure 8.29).**

The cycle begins at the point where CO_2 is combined with a 5-carbon acceptor molecule with two terminal phosphates called ribulose-1,5-bisphosphate (**RuBP**). This first critical step in **carbon fixation** generates a 6-carbon intermediate compound that immediately splits into two 3-carbon molecules of 3-phosphoglyceric acid (PGA). The subsequent steps use ATP and NADPH generated by the photosystems to form high-energy intermediates. First, 2 ATPs are expended to add a second phosphate to 3-PGA, producing two molecules of 1,3-bisphosphoglyceric acid (BPG). This is followed by NADPH contributing its electrons to BPG with the removal of one high-energy phosphate. These events give rise to two glyceraldehyde-3-phosphate (G3P). This molecule and its isomer dihydroxyacetone phosphate (DHAP) are key compounds in hexose synthesis leading to fructose and glucose. Some of the G3P is conveyed to the remainder of the cycle, where it participates in the regeneration of RuBP. This involves numerous steps not

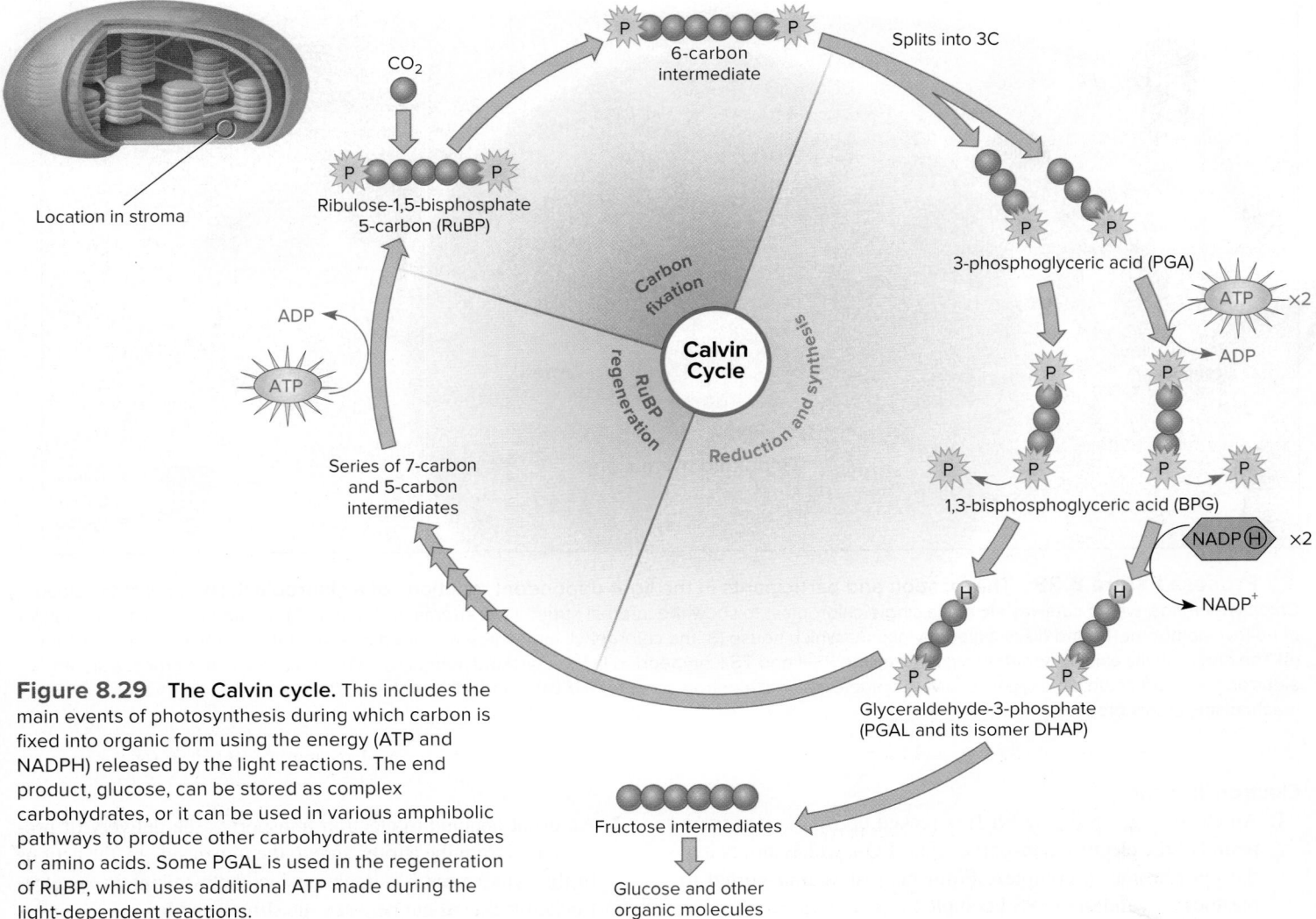

Figure 8.29 The Calvin cycle. This includes the main events of photosynthesis during which carbon is fixed into organic form using the energy (ATP and NADPH) released by the light reactions. The end product, glucose, can be stored as complex carbohydrates, or it can be used in various amphibolic pathways to produce other carbohydrate intermediates or amino acids. Some PGAL is used in the regeneration of RuBP, which uses additional ATP made during the light-dependent reactions.

depicted in figure 8.29. You may notice that this pathway is very similar to glycolysis, except that it runs in reverse (see figure 8.17).

Other Mechanisms of Photosynthesis

The **oxygenic,** or oxygen-releasing, photosynthesis described in the previous section occurs in plants, algae, and cyanobacteria and is the dominant type on the earth. Other photosynthesizers such as green and purple sulfur bacteria possess bacteriochlorophyll, which is more versatile in capturing light. They have only a cyclic photosystem I, which routes the electrons from the photocenter to the electron carriers and back to the photosystem again. This pathway generates a relatively small amount of ATP, and it may not produce NADPH. As photolithotrophs, these bacteria use H_2, H_2S, or elemental sulfur rather than H_2O as a source of electrons and reducing power. As a consequence, they are **anoxygenic** (non-oxygen-producing), and many are strict anaerobes.

 Check Your Progress SECTION 8.6

29. Indicate whether each of the following is produced during the light-independent or light-dependent reactions of photosynthesis: O_2, ATP, NADPH, and glucose.
30. During which reactions are water and CO_2 consumed, and what is the outcome?
31. What are the functions of chlorophyll and the photosystems?
32. What are the fates of the ATP and NADH in the Calvin cycle?
33. Compare oxygenic with nonoxygenic photosynthesis.

 CASE STUDY **Part 2**

Results of the tests were positive for the bacterium *Clostridium botulinum** type A, as well as its toxin, botulinum.* This finding, along with the symptoms, clearly confirmed the cases as botulism. Although the pathogen is quite sensitive to acid, the addition of water and other pH-neutral substances to the fruit-potato mixture probably maintained a low enough acid content to allow germination of *Clostridium* endospores and growth of its vegetative cells.

Botulinum neurotoxin is a complex protein molecule described as an AB toxin that consists of two functional parts (look ahead to chapter 13). It is transported in the bloodstream and carried to motor neurons in the muscles. One part of the toxin molecule (A) attaches to the membrane of the nerve cell and is transferred into the interior of the cell along with an enzyme component (B). B is a protease that hydrolyzes key secretory proteins, thereby blocking the release of a neurotransmitter (acetylcholine) from secretory vesicles. This

Molecule of botulinum toxin
Courtesy of Lacy, D.B., Tepp, W.,Cohen, A.C., DasGupta, B.R., Stevens, R.C.

blockade prevents the muscle from receiving a chemical stimulus necessary for contraction. The result is the paralysis of many large muscles, including those of the jaw, skeleton, eyes, neck, and diaphragm.

Altogether, eight prisoners developed botulism, several with severe morbidity and breathing problems. The only effective treatment for botulism is antitoxin, a serum containing antibodies that attach to the toxin and inactivate it. To be most effective, it must be given early on. Four inmates required prolonged hospitalization and continued to have residual neurological and muscular symptoms for several months, but there were no fatalities. In the past, botulism was often a death sentence because it caused respiratory failure, but modern medical care and use of mechanical ventilation have greatly improved survival. Despite its extreme toxicity, this toxin formulated as "Botox" is applied to treat a wide variety of muscular disorders, migraines, and, of course, for temporary elimination of wrinkles.

■ *Why must the antitoxin be administered in the earliest phases of botulism?*

■ *How could the inmates have prevented making such a toxic brew?*

To conclude this Case Study, go to Connect.

* *Clostridium botulinum* (klawss-trih′-dee-um bawt″-yoo-ly′-num) Gr. *kloster,* spindle, and *idion,* of a shape; L. *botulus,* sausage, and *ismus,* the result of.
* *botulinum* This toxin is one of the most potent chemical compounds on earth.

 Chapter Summary with Key Terms

8.1 An Introduction to Metabolism and Enzymes
 A. *Metabolism*
 1. **Metabolism** is the sum of biochemical activities in cells; it involves chemical changes to reactants and the release of products using well-established pathways.
 2. Metabolism is a complementary process consisting of **anabolism,** synthetic reactions that convert small molecules into large molecules, and **catabolism,** in which large molecules are degraded. Together, they generate thousands of intermediate molecular states, called metabolites, which are regulated at many levels.
 B. *Enzymes: Metabolic Catalysts*
 1. Metabolism is made possible by organic **catalysts,** or **enzymes,** that speed up reactions by lowering the

energy of activation. Enzymes are not consumed and can be reused. Each enzyme acts specifically upon its matching molecule or **substrate.**

2. *Enzyme Structure:* A conjugated enzyme consists of a protein component called the **apoenzyme** and one or more activators called **cofactors.**

3. *Enzyme Specificity:* Substrate attachment occurs in the special pocket called the **active, or catalytic, site.** In order to fit, a substrate must conform to the active site of the enzyme. This three-dimensional state is determined by the shape of the apoenzyme, making enzymes **substrate-specific.**

4. *Cofactors:* Metallic cofactors impart greater reactivity to the enzyme-substrate complex. **Coenzymes** such as **NAD⁺** (nicotinamide adenine dinucleotide) are transfer agents that pass functional groups from one substrate to another. Coenzymes usually contain **vitamins.**

5. *Enzyme Classification:* Enzyme names often consist of a prefix derived from the type of reaction or the substrate and the ending -*ase.*

6. *Microbial Enzymes and Disease:* Many pathogens secrete enzymes or toxins, which are referred to as virulence factors, that enable them to avoid host defenses.

7. *Types of Enzyme Function:* The release of water that comes with formation of new covalent bonds is a **condensation** reaction. **Hydrolysis** reactions involve addition of water to break bonds. Functional groups may be added, removed, or traded in many reactions. Coupled **redox reactions** transfer electrons and protons (H⁺) from one substrate to another.

8. *Enzyme Sensitivity:* Enzymes are **labile** (unstable) and function only within narrow operating ranges of temperature, osmotic pressure, and pH; extreme conditions can cause **denaturation.** Enzymes are frequently the targets for physical and chemical agents used in control.

C. *Regulation of Enzymatic Activity*
Regulatory controls can act on enzymes directly or on the process that gives rise to the enzymes.

1. A substance that resembles the normal substrate and can occupy the same active site is said to exert **competitive inhibition.**

2. **Allosteric inhibition** occurs when a substance binds to a regulatory site and prevents substrate binding.

3. **Noncompetitive inhibition** occurs when the regulator molecule binds the enzyme-substrate complex and keeps a product from being formed.

4. The mechanisms of **enzyme repression** and **enzyme induction** exert control at the genetic level by controlling the synthesis of key enzymes.

8.2 The Pursuit and Utilization of Energy
A. *Cell Energetics*
Energy is the capacity of a system to perform work. It is consumed in **endergonic reactions** and is released in **exergonic reactions.**

B. *A Closer Look at Biological Oxidation and Reduction*
1. Extracting energy requires a series of electron carriers arrayed in a downhill redox chain between electron donors and electron acceptors.
2. NAD⁺ is the primary carrier of electrons from one pathway to another.

3. In **substrate-level** and **oxidative phosphorylation,** energy is transferred to high-energy compounds such as ATP.

8.3 Pathways of Bioenergetics
A. *Pathways in Aerobic Respiration*
These use oxygen as a final electron acceptor and glucose as the electron and energy source. The three pathways involved are glycolysis, the Krebs cycle, and electron transport.
1. **Glycolysis** is a pathway that degrades glucose to **pyruvic acid** without requiring oxygen.
2. Pyruvic acid is an important key molecule in several catabolic and anabolic pathways.

B. *Fate of Pyruvic Acid in Krebs Cycle and Electron Transport*
1. Pyruvic acid is completely metabolized via the **Krebs cycle** and its associated electron transport chain.
2. **Acetyl coenzyme A** is the initial product of pyruvic acid processing that undergoes further oxidation and decarboxylation in the Krebs cycle and generates ATP, CO₂, and H₂O.
3. The respiratory chain carries electrons that support final energy extraction.
4. The **chemiosmotic** theory and **proton motive force** are conceptual models that explain the origin and maintenance of electropotential gradients across a membrane that leads to ATP synthesis by **ATP synthase.**
5. The final electron acceptor in aerobic respiration is oxygen. In **anaerobic respiration,** compounds such as sulfate, nitrate, or nitrite serve this function. Bacteria serve as important agents in the nitrogen cycle **(denitrification).**

8.4 The Importance of Fermentation
A. **Fermentation** is based on the glycolysis pathway in which both the electron donor and final electron acceptors are organic compounds. It is an essential source of NAD⁺ required for electron transport.
B. Fermentation enables anaerobic and facultative microbes to survive in environments devoid of oxygen. Production of alcohol, vinegar, and certain industrial solvents relies upon fermentation.

8.5 Biosynthesis and the Crossing Pathways of Metabolism
A. *Amphibolism*
The ability of a cell or system to integrate catabolic and anabolic pathways to improve efficiency is called **amphibolism.**
B. *Anabolism: Formation of Macromolecules*
Macromolecules, such as proteins, carbohydrates, and nucleic acids, are made of building blocks from two possible sources: from outside the cell (preformed) or via synthesis in one of the anabolic pathways.
C. *Versatility of Glycolysis and Krebs Cycle*
Glycolysis and the Krebs cycle are central pathways that link catabolic and anabolic pathways, allowing cells to break down different classes of molecules in order to synthesize compounds required by the cell.
1. Metabolites of these pathways double as building blocks and sources of energy. Intermediates such as pyruvic acid are convertible into amino acids through **amination.** Amino acids can be **deaminated** and used as a source of glucose and other carbohydrates (**gluconeogenesis**).
2. Two-carbon acetyl molecules from pyruvate **decarboxylation** can be used in fatty acid synthesis.

8.6 Photosynthesis: The Earth's Lifeline
Photosynthesis takes place in two stages—**light-dependent** and **light-independent reactions.**
 A. During the light-dependent reactions, **photons** of solar energy are absorbed by **chlorophyll** and other pigments in **thylakoid** membranes of **chloroplast grana** in eukaryotes and the cell membrane in cyanobacteria. Light energy is captured by two photosystems (**PS II** and **PS I**), and water is split by **photolysis.** This releases oxygen gas. The light energy harvested by photosystems becomes the chemical energy of electrons; electron transport energy is stored in NADPH and ATP; ATP is formed through **photophosphorylation** and chemiosmosis. Light energy is used to synthesize ATP and NADPH.
 B. The **Calvin cycle** (light-independent) occurs in the **stroma** of chloroplasts and uses ATP to **fix** carbon dioxide to a carrier molecule, ribulose-1,5-bisphosphate, and convert it to glucose in a multistep process.
 C. The type of photosynthesis commonly found in plants, algae, and cyanobacteria is called **oxygenic** because it liberates molecular oxygen as a by-product of the light-dependent reactions. **Anoxygenic** photosynthesis occurs in photolithotrophs, which do not produce oxygen.

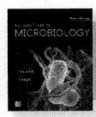

Level I. Knowledge and Comprehension

These questions require a working knowledge of the concepts in the chapter and the ability to recall and understand the information you have studied.

Multiple-Choice Questions

Select the correct answer from the answers provided. For questions with blanks, choose the combination of answers that most accurately completes the statement.

1. _____ is another term for biosynthesis.
 a. Catabolism c. Metabolism
 b. Anabolism d. Catalyst

2. Catabolism is a form of metabolism in which _____ molecules are transformed into _____ molecules.
 a. larger, smaller c. amino acid, protein
 b. smaller, larger d. glucose, starch

3. An enzyme _____ the activation energy required for a chemical reaction.
 a. increases c. lowers
 b. converts d. catalyzes

4. An enzyme
 a. becomes part of the final products
 b. is nonspecific for substrate
 c. is consumed by the reaction
 d. is heat and pH labile

5. An apoenzyme is where the _____ is located.
 a. cofactor c. redox reaction
 b. coenzyme d. active site

6. Many coenzymes contain
 a. metals c. proteins
 b. vitamins d. substrates

7. To digest cellulose in its environment, a fungus produces a/an
 a. endoenzyme c. catalase
 b. exoenzyme d. polymerase

8. Energy in biological systems is primarily
 a. electrical c. radiant
 b. chemical d. mechanical

9. Energy is carried from catabolic to anabolic reactions in the form of _____.
 a. ADP c. coenzymes
 b. high-energy ATP bonds d. inorganic phosphate

10. Exergonic reactions
 a. release potential energy c. form bonds
 b. consume energy d. occur only outside the cell

11. A reduced compound is
 a. NAD^+ c. NADH
 b. FAD d. ADP

12. Most oxidation reactions in microbial bioenergetics involve the
 a. removal of electrons c. addition of oxygen
 b. addition of electrons d. removal of oxygen

13. A product or products of glycolysis is/are
 a. ATP c. CO_2
 b. pyruvic acid d. both a and b

14. Fermentation of a glucose molecule has the potential to produce a net number of _____ ATPs.
 a. 4 b. 2 c. 38 d. 32

15. Complete oxidation of glucose in aerobic respiration by a eukaryote could potentially yield a net maximum output of _____ ATPs.
 a. 40 b. 30 c. 38 d. 2

16. The compound that enters the Krebs cycle is
 a. citric acid c. pyruvic acid
 b. oxaloacetic acid d. acetyl coenzyme A

17. The $FADH_2$ formed during the Krebs cycle enters the electron transport system at which site?
 a. NADH dehydrogenase c. complex IV
 b. complex II d. ATP synthase

18. The proton motive force is the result of
 a. ATP synthase transporting protons during ATP synthesis
 b. an electron gradient between the matrix and the intermembrane space of a mitochondrion
 c. a proton gradient between the matrix and intermembrane space of a mitochondrion
 d. a buildup of negatively charged ions

19. Photosynthetic organisms convert the energy of _____ into chemical energy.
 a. electrons c. photons
 b. protons d. hydrogen atoms

20. The Calvin cycle operates during which part of photosynthesis?
 a. only the light-dependent phase
 b. only in the dark
 c. the light-independent phase
 d. only during photosystem I

21. The oxygen produced by photosynthesis comes from:
 a. CO_2 c. H_2O
 b. NO_3 d. glucose

22. **Multiple Matching.** Match the process a, b, or c with the metabolic events in the list.
 a. glycolysis
 b. Krebs cycle
 c. electron transport/oxidative phosphorylation
 _____ H^+ and e^- are delivered to O_2 as the final acceptor.
 _____ Pyruvic acid is formed.
 _____ ATP is formed.
 _____ H_2O is produced.
 _____ CO_2 is formed.
 _____ Fructose diphosphate is split into two 3-carbon fragments.
 _____ NADH is oxidized.
 _____ ATP synthase is active.

 Case Study Review

1. What is the most likely source of the botulism bacteria that grew in the prison wine?
 a. fruit c. potato
 b. air d. sugar

2. Could other inmates develop botulism from being in contact with the ill prisoners?
 a. Yes, it can be spread in body fluids.
 b. No, a person must ingest the toxin.
 c. Yes, it can be carried by inanimate objects.
 d. No, *C. botulinum* bacteria cannot grow in human tissue.

3. Describe the conditions of nutrition, pH, temperature, oxygen, and life cycle that made it possible for both the yeast and *Clostridium* to grow and synthesize their products in the pruno mixture.

 Writing Challenge

For each question, compose a one- or two-paragraph answer that includes the factual information needed to completely address the question. Check Your Progress questions can also be used for writing-challenge exercises.

1. Discuss the relationship of
 a. anabolism to catabolism
 b. ATP to ADP
 c. glycolysis to fermentation
 d. electron transport to oxidative phosphorylation

2. Give the general name of the enzyme that
 a. converts citrate to isocitrate
 b. reduces pyruvic acid to lactic acid
 c. reduces nitrate to nitrite

3. Explain what is unique about the actions of ATP synthase.

4. Compare the general equation for aerobic metabolism (section 8.3) with the summary of aerobic metabolism (figure 8.22), verify the

final totals of reactants and products, and indicate where they occur in the pathways.

5. Describe four requirements required for fermentation to take place.

6. Is fermentation catabolic or anabolic? Explain your choice.

7. Name three electron carriers and the vitamins that are essential to their function. Explain what actions the electron carriers and their associated vitamins have in metabolic reactions.

8. Explain the similarities and differences between anaerobic respiration and fermentation.

 Concept Mapping

On Connect you can find an Introduction to Concept Mapping that provides guidance for working with concept maps and a sample concept map for this chapter.

 Level II. Application, Analysis, Evaluation, and Synthesis

These problems go beyond just restating facts and require higher levels of understanding and an ability to interpret, problem solve, transfer knowledge to new situations, create models, and predict outcomes.

Critical Thinking

Critical thinking is the ability to reason and solve problems using facts and concepts. These questions can be approached from a number of angles, and in most cases, they do not have a single correct answer.

1. Use the following graph to diagram the energetics of a chemical reaction, with and without an enzyme. Be sure to position reactants and products at appropriate points and to indicate the stages in the reaction and the energy levels.

Energy Involved in Reaction

Progress of Reaction ⟶

2. Explain what is meant by the "biochemical circle of life" that exists between phototrophs and heterotrophs.

3. Explain how it is possible for certain microbes to survive and grow in the presence of cyanide, which would kill many other organisms.

4. Suggest the advantages of having metabolic pathways staged in specific membrane or organelle locations rather than being free in the cytoplasm.

5. Two steps in glycolysis are catalyzed by allosteric enzymes. These are: (1) step 2, catabolized by phosphoglucoisomerase, and (2) step 9, catabolized by pyruvate kinase. Suggest what metabolic products might regulate these enzymes. How might one place these regulators in figure 8.17?

6. Beer production requires an early period of rapid aerobic metabolism of glucose by yeast. Given that anaerobic conditions are necessary to produce alcohol, can you explain why this step is necessary?

7. What would be the expected pHs of the matrix and intermembrane space of the mitochondrion? What about the stroma and thylakoid lumen of the chloroplast? Explain your answers.

8. At which site in the mitochondrion and chloroplast do the ATP molecules collect after being made by ATP synthases, and what is the advantage of this?

9. Explain the relationship between enzyme sensitivity and the adaptations microbes make to their environments

10. Why wouldn't antibiotics work as therapy for botulism?

Visual Challenge

1. From **chapter 7, figure 7.11** (reproduced below). Describe the likely metabolic strategies of the microbes growing in tubes 1 through 4. Specify the pathways, final electron acceptor, and end products.

© Terese M. Barta, Ph.D.

2. Look at the two figure parts (a) and (b) from figure 8.8 (reproduced below), and explain what processes are depicted and what the results of the reactions will be.

(a)

(b)

STUDY TIP: Quickly create a customized practice quiz for this chapter by going to your Connect SmartBook "My Reports" section. Don't have SmartBook and want to learn more? Go to **whysmartbook.com.**

CHAPTER **9** An Introduction to Microbial Genetics

Plasmid—a circular snippet of DNA carrying genetic treasures. *Source:* National Institutes of Health/United States Department of Health and Human Services

A mating "dance" between cells connected by conjugation pili. *Source:* © L. Caro/SPL/Science Source

Acinetobacter—tiny nonmotile gram-negative bacilli, coming to a hospital in your neighborhood? *Source:* Janice Carr/CDC

 CASE STUDY Part 1 At War with an Emerging Pathogen

A 55-year-old woman recovering from **lung transplant** surgery developed a case of pneumonia from a ventilator she needed for breathing. The bacteria isolated from her lungs were identified as **_Acinetobacter baumannii,_** a widespread resident of soil and water that has found its way into hospitals and long-term care facilities. Tests showed that all of the drugs available to treat the infection were ineffective against the pathogen—except for **colistin.** Early on, the drug appeared to be curing the infection, but after a few days it returned and the patient died of respiratory failure. Cultures taken late in the case detected a **new strain** of the infectious agent that had developed **resistance** to all known drugs, including colistin.

During the same period, several other patients in the ICU developed drug-resistant *A. baumannii* infections. This sudden outbreak of **health-care-associated infections** spurred an investigation by an infection control nurse, who took samples from multiple locations and cultured them. The **bacteria** were **isolated** from **nearly every possible source,** including medical devices, patient skin samples, bedding, hospital workers, the air, and room surfaces.

But this sort of problem with *A. baumannii* was not restricted to North America. A continent away in Iraq and Afghanistan, this same

> "Over time, so many soldiers suffered battlefield-related infections by this ubiquitous pathogen that it was given the nickname "Iraqnobacter.""

microbe became one of the most **fearsome pathogens** to infect soldiers with **battle injuries.** In the case of a 55-year-old man who sustained a grenade injury while transporting soldiers, shrapnel penetrated his thigh and fractured the femur. When X rays revealed the presence of an infection in his broken bone, he was transported to a trauma hospital. Cultures of the **wound** isolated a **gram-negative coccobacillus** that was a different strain of *A. baumannii.* Testing indicated that it was **resistant** to all drugs except for a new type of drug called **tigecycline.** After a several weeks' course of tigecycline, the infection was resolved and the patient was released. Over time, so many soldiers suffered battlefield-related infections by this ubiquitous pathogen that it was given the nickname "Iraqnobacter."

- *What are the probable origins of the pathogen, and what conditions increase its transmission in a hospital setting?*

- *What is the meaning of drug resistance from a genetic standpoint?*

To continue the Case Study, go to Case Study Part 2 at the end of the chapter.

SCOPING OUT THE CHAPTER

You probably have at least some insight into the role of DNA in the heredity of all life forms on earth. DNA has sometimes been likened to a molecular blueprint or to a chemical language, and in a sense both of these comparisons are valid. The coded nature of DNA lays out a master plan detailing how a cell will construct new parts of itself. One could also say that the DNA code spells out a complete library of information for expression of cell structure and function. In this chapter our aim is to show you what it means to say "It's in your DNA."

Chromosomes and Genes: Packages of DNA

All cells tightly wind their elongate strands of DNA into compact chromosomes. Genes bearing specific codes are located at sites along the length of the chromosomes.

DNA Replication: Copying the DNA Blueprint

The codes carried by DNA are conserved for offspring and future generations. This ensures that organisms will inherit the complete and accurate collection of genes required to survive and reproduce.

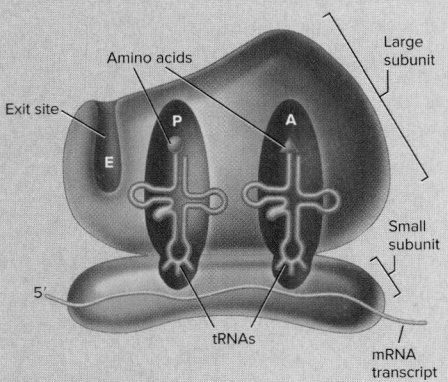

Transcription/translation: Interpreting the Blueprint

DNA acts as a model for assembling RNA copies of its codes. RNAs are transported to ribosomes to read the coded message and express it into useful cell proteins. A shorthand formula for this is: DNA→RNA→Protein.

Mutations: Changes in the DNA Code

Like any blueprint or language, mistakes can change the meaning of the DNA message, so that transcription and translation are unable to produce the intended protein, with a variety of outcomes, usually harmful.

Recombination: Transmission of DNA from One Cell to Another.

Some bacteria readily share their DNA codes with other bacteria and even with viruses, as seen in this small vignette. This results in bacteria with "new" genes and traits that help them adapt to changes in their environments.

9.1 Introduction to Genetics and Genes: Unlocking Secrets of Heredity

 Expected Learning Outcomes

1. Define *heredity, genetics, genome, gene, phenotype,* and *genotype.*
2. Compare the basic nature of genetic material in eukaryotes, prokaryotes, and viruses.
3. Explain how DNA is organized and packaged.
4. Describe the chemical structure of DNA and its significance.
5. List the nitrogen bases and explain their bonding patterns.
6. Describe the process of DNA replication as it occurs in prokaryotic cells.

Genetics is the science that studies the inheritance of biological characteristics by life forms. This subject, also known as heredity, is a wide-ranging science that examines:

- the transmission of biological properties (traits) from parent to offspring,
- the expression and variation of those traits,
- the structure and function of the genetic material, and
- how this material changes or evolves.

The study of genetics takes place on several levels **(figure 9.1).** Organismal genetics looks at the transmission and expression of genetic factors in the whole organism or cell; chromosomal genetics examines the characteristics and actions of chromosomes; and molecular genetics deals with the biochemistry of gene function. All of these levels provide insight into microbial structure, physiology, evolution, and pathogenicity. It should also be noted that the study of microbial genetics will emphasize universal themes in the genetics of all organisms, and it will increase appreciation for the astounding advances in genetic engineering covered in chapter 10.

The Nature of the Genetic Material

Deoxyribonucleic acid or **DNA** serves as the foundation for most genetic studies. In order to appreciate and understand a broad range of genetic principles, it will be essential to develop a working knowledge of DNA's characteristics. With this in mind, we will use the first section of this chapter to lay some groundwork on DNA organization, structure, and duplication. This will be followed by coverage of related concepts such as RNA and protein synthesis, genetic control, mutations, and genetic recombination in the remaining five sections.

The Levels of Structure and Function of the Genome

The **genome** is the sum total of genetic material (DNA) carried within a cell. Although most of the genome exists in the form of chromosomes, genetic material can appear in nonchromosomal sites as well **(figure 9.2).** For example, bacteria and some fungi contain tiny extra pieces of DNA **(plasmids),** and the mitochondria and chloroplasts of eukaryotes are equipped with their own functional chromosomes. Genomes of cells are composed exclusively of DNA, but viruses contain either DNA or RNA as the principal genetic material. Although the specific genome of an individual organism is unique, the general pattern of nucleic acid structure and function is similar among all organisms and viruses.

In general, a **chromosome** is a discrete cellular structure composed of a neatly packaged DNA molecule. The chromosomes of eukaryotes and bacterial cells differ in several respects. The structure of eukaryotic chromosomes consists of a DNA molecule tightly wound around histone proteins, whereas a bacterial chromosome is condensed and secured into a packet by means of a different type of protein. Eukaryotic chromosomes are located in the nucleus; they vary in number from a few to hundreds; they can occur in pairs (diploid) or singles (haploid); and they are linear in format. In contrast, most bacteria have a single, circular chromosome, although some have multiple chromosomes and a few have linear chromosomes.

All chromosomes contain a series of basic informational "packets" called genes. A **gene** can be defined from more than one

Organism level Cell level Chromosome level Molecular level

Enterobius vermicularis—a common parasitic worm

Figure 9.1 Levels of genetic study. Genetic processes can be observed at the levels of organism, cell, chromosome, and DNA sequence (molecular level). Although levels shown here are for a eukaryotic organism they would follow a similar pattern in prokaryotes.
Source: Image contributed by the Centre for Tropical Medicine and Imported Infectious Diseases, Bergen, Norway./CDC-DPDx

perspective. In classical genetics, the term refers to the fundamental unit of heredity responsible for a given trait in an organism. In the molecular and biochemical sense, it is a portion of the chromosome that provides information for a given cell function. More specifically still, it is a specific segment of DNA that contains the necessary codes to make a molecule of **protein** or **RNA.** This last definition of a gene is emphasized in this chapter.

Cells

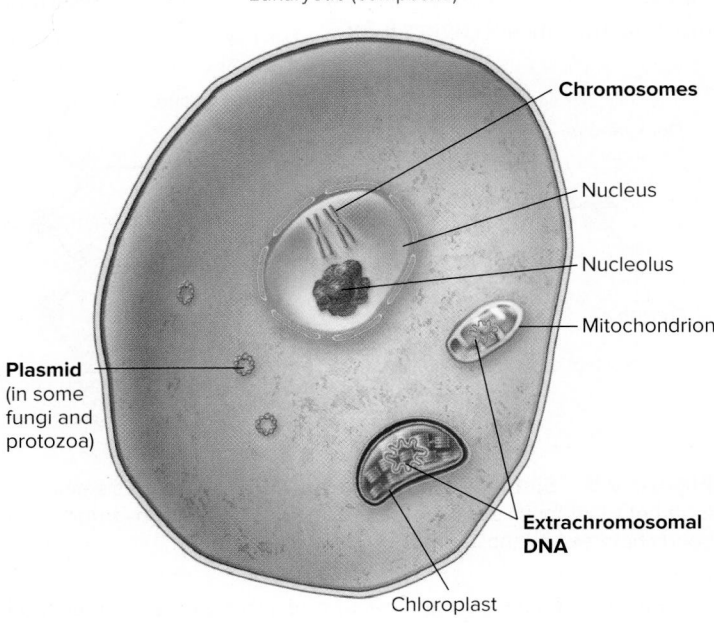

Eukaryotic (composite)

- Chromosomes
- Nucleus
- Nucleolus
- Mitochondrion
- **Plasmid** (in some fungi and protozoa)
- **Extrachromosomal DNA**
- Chloroplast

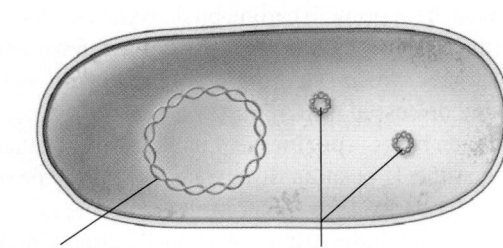

Prokaryotic

- Chromosome
- Plasmids

Viruses

DNA RNA

Figure 9.2 **The general location and forms of the genome in two cell types and selected viruses (not to scale).**

Genes fall into three basic categories: *structural genes* that code for proteins, genes that code for RNA, and *regulatory genes* that control gene expression. The collection of all of these types of genes constitutes an organism's distinctive genetic makeup, or **genotype.*** The expression of the genotype creates traits (certain structures or functions) referred to as the **phenotype.*** For example a person inherits a combination of genes (genotype) that gives a certain eye color or height (phenotype), a bacterium inherits genes that direct the formation of a flagellum or the ability to metabolize a certain substrate, and a virus has genes that dictate its capsid structure. All organisms contain more genes in their genotypes than are being seen as a phenotype at any given time. In other words, the phenotype can change depending on which genes are "turned on" or expressed.

The Size and Packaging of Genomes

Genomes vary greatly in size. Viruses have anywhere from a few to over a thousand genes; the bacterium *Escherichia coli* has a single chromosome containing 4,288 genes, and a human cell packs about five times that many into 46 chromosomes. The chromosome of *E. coli* would measure about 1 mm if unwound and stretched out linearly, and yet this fits within a cell that measures just over 1 μm across, making the stretched-out DNA 1,000 times longer than the cell **(figure 9.3).** Still, the bacterial chromosome takes up only about one-third to one-half of the cell's volume. Likewise, if the sum of all DNA contained in the 46 human chromosomes were unraveled and laid end to end, it would measure about 6 feet. How can such elongated genomes fit into the minuscule volume of a cell and, in the case of eukaryotes, into an even smaller compartment, the nucleus? The answer lies in the complex coiling of the DNA chain.

* *genotype* (jee′-noh-typ) Gr. *gennan,* to produce, and *typos,* type.
* *phenotype* (fee′-noh-typ) Gr. *phainein,* to show. The physical manifestation of gene expression.

Figure 9.3 **A disrupted *Escherichia coli* cell.** The cell has ejected its unraveled chromosome, which now appears as thin strings radiating around the outer surface (10,000×). These strings are the uncoiled DNA strand. © Dr. Jack Griffith

The Packaging of DNA

Packing the mass of DNA into the cell involves compacting the DNA molecule by means of supercoils or superhelices. In the simpler system of prokaryotes, the circular chromosome is packaged by the action of a special enzyme called a **topoisomerase** (specifically, DNA **gyrase**). This enzyme coils the chromosome into a tight bundle by introducing a reversible series of twists into the DNA molecule.

Quick Search

To see an animation of the how a bacterial chromosome is packaged, google "Cell Structure—Bacterial Chromosome Compaction."

The system in eukaryotes is more complex, with three or more levels of coiling. First, the DNA molecule of a chromosome, which is linear, is wound twice around the histone proteins, creating a chain of **nucleosomes.** The nucleosomes fold in a spiral formation upon one another. An even greater supercoiling occurs when this spiral arrangement further twists on its radius into a giant spiral with loops radiating from the outside. This extreme degree of compactness is what makes the eukaryotic chromosome visible during mitosis (**figures 9.4** and 5.6).

The condensation of DNA and protein into chromatin was once thought to function mainly as a way to fit such a long molecule into a cell. However, it is now evident that the coiling of DNA also serves to make certain segments of the genetic program either more or less available. This action provides a major control on the expression of many traits, especially in eukaryotes. Having an additional regulatory function creates an alternate mechanism to guide when and how genes are accessed. The realization that the phenotype of a cell is due to more than just the order of nucleotides in the DNA (genotype) has led to a rethinking of some of the most basic concepts of genetics (see 9.2 Making Connections).

Figure 9.4 Eukaryotic DNA interacts with a variety of proteins and small molecules. This relationship modulates the expression of many genes while at the same time allowing for the ordered condensation of DNA into a compact supermolecule.

The Structure of DNA: A Double Helix with Its Own Language

To analyze the structure of DNA at the molecular level, imagine being able to magnify a small piece of a gene about 5 million times. What you would see is one of the great marvels of life. The pieces of this molecular puzzle were finally worked out by James Watson and Francis Crick in 1953 (**9.1 Making Connections**). They determined that DNA is a giant molecule with two strands bound together into a double helix. Extensive evidence shows that the general structure of DNA is universal, except in some viruses that contain single-stranded DNA.

The basic unit of DNA structure is a **nucleotide,** composed of **phosphate, deoxyribose sugar,** and a **nitrogen base,** as shown in this simplified model (**figure 9.5**):

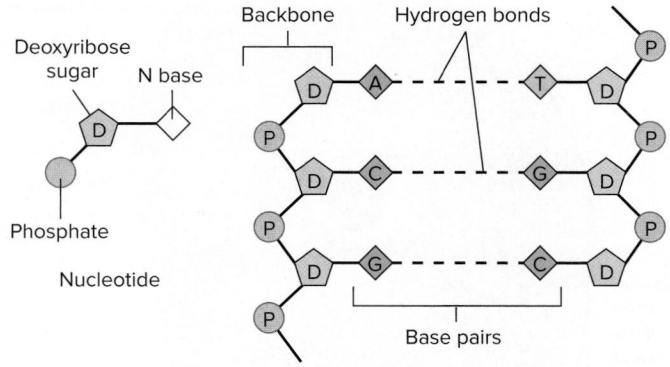

Figure 9.5 Simplified structure of DNA. Nucleotides stack together to form two opposing strands held together by hydrogen bonds between the base pairs.

Figure 9.6 lays out the structure of DNA going from general (a,b,c) to specific (d). To make the vertical sugar-phosphate backbones of the molecule, each deoxyribose sugar bonds covalently in a repeating pattern with two phosphates. One of the bonds is to the number $5'$ (read "five prime") carbon on deoxyribose, and the other is to the $3'$ (three prime) carbon. This pattern will format the direction of the strands, which run opposite to each other, as shown in figure 9.6d and discussed next.

The nitrogen bases—**purines** and **pyrimidines**—attach by covalent bonds at the $1'$ position of the sugar (figure 9.6c,d). When these nitrogen bases span the center of the molecule and bind with their complementary bases from the opposite strand, they form a double-stranded helix. The **paired bases** are held together by hydrogen bonds, which can be readily disrupted, allowing the molecule to be "unzipped" into two strands. This feature is of great importance in gaining access to the information encoded in the nitrogen base sequence. Pairing of purines and pyrimidines is not random; it is dictated by the structure of the bases and how they align to form hydrogen bonds. Thus, in DNA, the purine **adenine (A)** pairs with the pyrimidine **thymine (T),** and the purine **guanine (G)** pairs with the pyrimidine **cytosine (C).** Note that adenine forms two hydrogen bonds with thymine, and cytosine forms three hydrogen bonds with guanine. This difference will influence a number of DNA functions. Evidence also indicates that the bases are attracted to each other in this pattern because each base has a three-dimensional shape that conforms exactly to its mate. Although

9.1 MAKING CONNECTIONS

Deciphering the Structure of DNA

The search for the primary molecules of heredity was a serious focus throughout the first half of the twentieth century. At first many biologists thought that protein was the genetic material. An important milestone occurred in 1944 when Oswald Avery, Colin MacLeod, and Maclyn McCarty purified DNA and demonstrated at last that it was indeed the blueprint for life. This was followed by an avalanche of research, which continues today.

"We Have Discovered the Secret of Life!"*

One area of extreme interest concerned the molecular structure of DNA. In 1951, American biologist James Watson and English physicist Francis Crick collaborated on solving the DNA puzzle. Although they did little of the original research, they were intrigued by several findings from other scientists. It had been determined by Erwin Chargaff that any model of DNA structure would have to contain deoxyribose, phosphate, purines, and pyrimidines arranged in a way that would provide variation and a simple way of copying itself. Watson and Crick spent long hours constructing models with cardboard cutouts and kept alert for any and every bit of information that might give them an edge.

Two English biophysicists, Maurice Wilkins and Rosalind Franklin,* had been painstakingly collecting data on X-ray crystallographs of

Dr. James Watson (left) and Dr. Francis Crick (right) stand next to their model that finally explained the structure of DNA in 1953.
© A. Barrington Brown/Science Source

DNA for several years. With this technique, molecules of DNA bombarded by X rays produce a photographic image that can predict the three-dimensional structure of the molecule. After Dr. Wilkins allowed them to view their data, Watson and Crick noticed an unmistakable pattern: The molecule appeared to be a double helix. In a short time the pieces of the puzzle fell into place and a final model was assembled—a model that explained all of the qualities of DNA, including how it is copied. Although Watson and Crick were hailed for the clarity of their solution, it must be emphasized that their success was built on the considerable efforts of a number of English and American scientists. This historic discovery spawned ingenious research in all areas of molecular genetics that is still ongoing.

Quick Search
Find out more of the story by googling "The Secret of Life: Discovery of DNA Structure."

Since the discovery of the double helix in 1953, an extensive body of biochemical, microscopic, and crystallographic analysis has left little doubt that the model first proposed by Watson and Crick is correct. Scanning tunneling microscopes produce three-dimensional images of DNA that verify the helical shape and twists of DNA represented by models.

How did Watson and Crick's model of DNA accurately fit the requirements needed to predict its structure? *Answer available on Connect.*

* comment by Watson and Crick after publication of their monumental discovery.
* Unfortunately Dr. Franklin died before she could be properly recognized by receiving a Nobel Prize. Without her contributions, Watson and Crick could not have deciphered the puzzle as quickly as they did.

the base-pairing partners generally do not vary, the sequence of base pairs along the DNA molecule can assume any order, resulting in an infinite number of possible nucleotide sequences. One significant finding that has emerged from sequencing chromosomal DNA is how extremely long it can be for a single molecule. An average bacterial chromosome consists of 5 million to 6 million nucleotides; the 46 chromosomes (diploid) of humans total about 6.4 billion nucleotides.

Other important considerations of DNA structure concern the nature of the double helix itself. As indicated previously, one side of the helix runs in the opposite direction of the other, in what is called an *antiparallel arrangement*. Using the order of the bond between the carbon on deoxyribose and the phosphates, we see that one helix runs from the 5′ to 3′ direction and the other runs from the 3′ to 5′ direction (figure 9.6d). This characteristic is a significant factor in DNA synthesis, transcription, and translation. Although the DNA is constructed somewhat like a twisted ladder, it is not symmetrical. The curvature of the helix and the stepwise stacking of the nitrogen bases produce two different-size surface features, the major and minor grooves **(figure 9.7)**.

DNA Replication: Preserving the Code and Passing It On

For its long-term survival, a species must reproduce. In bacteria, this involves division of the cell by means of binary fission or budding, but it also involves the accurate duplication and separation of the genetic material into each daughter cell to ensure that each offspring receives a functional genome.

The sequence of bases along the length of DNA constitutes its "language." For this language to be preserved for hundreds of generations, the codes it contains will need to be duplicated with high fidelity. This process of duplication is called **DNA replication.** In the following example, we will show replication in bacteria, but with some exceptions it generally applies to the process as it occurs in eukaryotes and some viruses. Recall from chapter 7 that early in

binary fission, the metabolic machinery of a bacterium is programmed to initiate the duplication of the chromosome. Despite its complexity, DNA replication can be very rapid: This must be completed in a single generation time (around 20 minutes in *E. coli*).

The Significance of DNA Structure

The pairing and arrangement of nitrogen bases in DNA has two profound and essential results:

1. **Maintenance of the genetic message during reproduction.** The constancy of base-pairing guarantees that the correct order of the DNA bases will be retained during cell division. When the two strands are separated, each one provides a **template** (pattern or model) for the replication (exact copying) of a new molecule (figure 9.7). Because the sequence of one strand provides the correct pattern for its complementary strand, the codes can be duplicated with accuracy.

2. **Providing variety.** The order of bases along the length of the DNA strand provides information needed to produce RNA and protein molecules, which in turn are responsible for the **phenotype** of the cell. As we will see, changing the identity of the bases or their order in the DNA molecule can have a dramatic effect on the phenotype of the organism.

Figure 9.6 Levels of DNA structure. (a) A space-filling model of DNA shows its basic configuration as a double helix of sugar-phosphate backbones holding the nitrogen bases between them. **(b)** The twisted ladder model depicts how the complementary nitrogen bases (bars) pair with each other across the center of the molecule. **(c)** A schematic lays out the basic chemical makeup and bonding pattern of the molecule. **(d)** A more detailed view of the base pairs, hydrogen bonds, and the antiparallel arrangement of the two helices. Note that in one strand, the deoxyribose-phosphate runs in the 5′ to 3′ direction and the other side is oriented from 3′ to 5′.

It is remarkable how such a seemingly simple set of codes can account for the extreme differences among forms as diverse as a virus, a bacterium, and a human. We know that the English language, based on 26 letters, can create an infinite variety of words, but how can an apparently complex genetic language such as DNA be based on just four nitrogen base "letters"? The answer lies in the fact that the four bases of DNA are arranged into much longer sequences. An average bacterial gene may contain 1,000 nucleotides: so while DNA is made up of fewer "letters," they are arranged into much longer "words." Mathematically speaking, 1,000 nucleotides can be arranged in $4^{1,000}$ different sequences, a number so large (1.5×10^{602}) that it provides nearly endless degrees of variation.

The Overall Replication Process

What features allow the DNA molecule to be exactly duplicated, and how is its integrity retained? DNA replication requires a careful orchestration of the actions of 30 different enzymes (partial list in **table 9.1**), which separate the strands of the existing DNA molecule, copy its template, and produce two complete daughter molecules.

A simplified version of replication is shown in **figure 9.7** and includes the following:

1. uncoiling the parent DNA molecule, beginning at a predetermined point of origin;
2. unzipping the hydrogen bonds between the base pairs, thus separating the two strands and exposing the nucleotide sequence of each strand (which is normally hidden in the center of the helix) to serve as templates; and
3. synthesizing two new strands by using each single-stranded template as a pattern for adding the correct complementary nucleotides to the newly forming strand.

A critical requirement of DNA replication is that each completed daughter molecule be identical to the parent in composition, but neither one is completely new. The strand that serves as a template is an original parental DNA strand and is retained in the daughter molecule. The preservation of the parent molecule in this way, termed **semiconservative replication,** helps explain the reliability and accuracy of replication.

TABLE 9.1	Some Enzymes Involved in DNA Replication and Their Functions
Enzyme	**Function**
Helicase	Unzipping the DNA helix
Primase	Synthesizing an RNA primer
DNA polymerase III	Adding bases to the new DNA chain; proofreading the chain for mistakes
DNA polymerase I	Removing RNA primers, replacing gaps between Okazaki fragments with correct nucleotides, repairing mismatched bases
Ligase	Final binding of nicks in DNA during synthesis and repair
Gyrase	Supercoiling

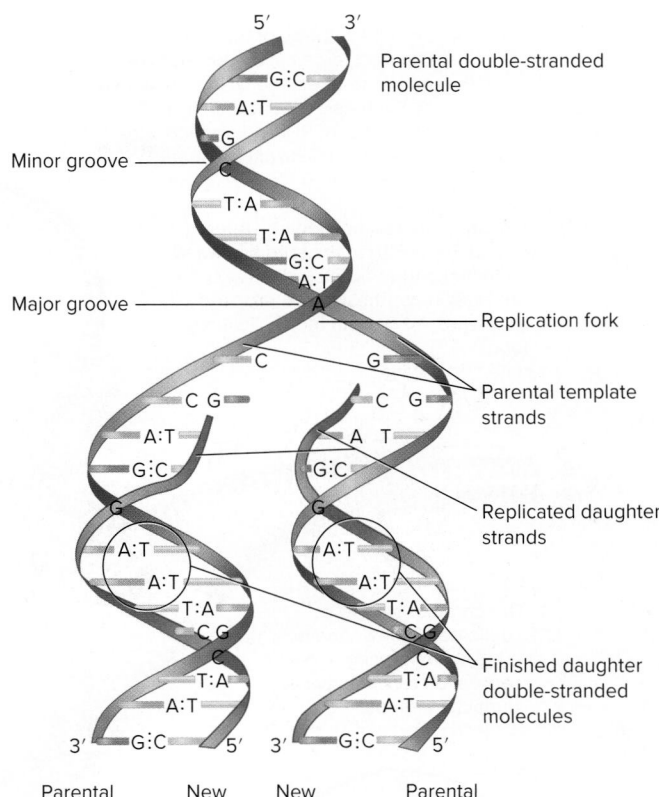

Figure 9.7 **Simplified steps to show the semiconservative replication of DNA.** The two strands of the parental double helix are unwound and separated. Each single strand will serve as a template (blue) to synthesize a new strand of DNA (red), starting at the replication fork. As synthesis proceeds, the correct nucleotides are added according to the pattern of the template. An A on the template will pair with a T on the new molecule, and a C will pair with a G. The finished daughter DNA molecules contain one strand of the newly synthesized DNA and the original template (parental) strand. This keeps the code intact because the same linear arrangement of the bases is maintained during this process. This process is covered in greater detail in figure 9.8.

Events in Replication

All chromosomes have a specific **origin of replication** site that serves as the place where replication will be initiated. It is recognized by a short sequence rich in adenine and thymine that, you will recall, are held together by only two hydrogen bonds rather than three. Because the origin of replication is AT-rich, less energy is required to separate the two strands than would be required if the origin were rich in guanine and cytosine.

The process of synthesizing a new daughter strand of DNA using the parental strand as a template is carried out by a giant enzyme complex that brings in the primary replication enzyme—**DNA polymerase III**—along with numerous accessory enzymes. The DNA polymerase III is responsible for deciphering and duplicating DNA codes, but it has some major conditions that will affect the overall process **(figure 9.8):**

- DNA polymerase cannot add nucleotides to a DNA template strand without a guide that can serve as the starting point for nucleotides to be bonded during synthesis.

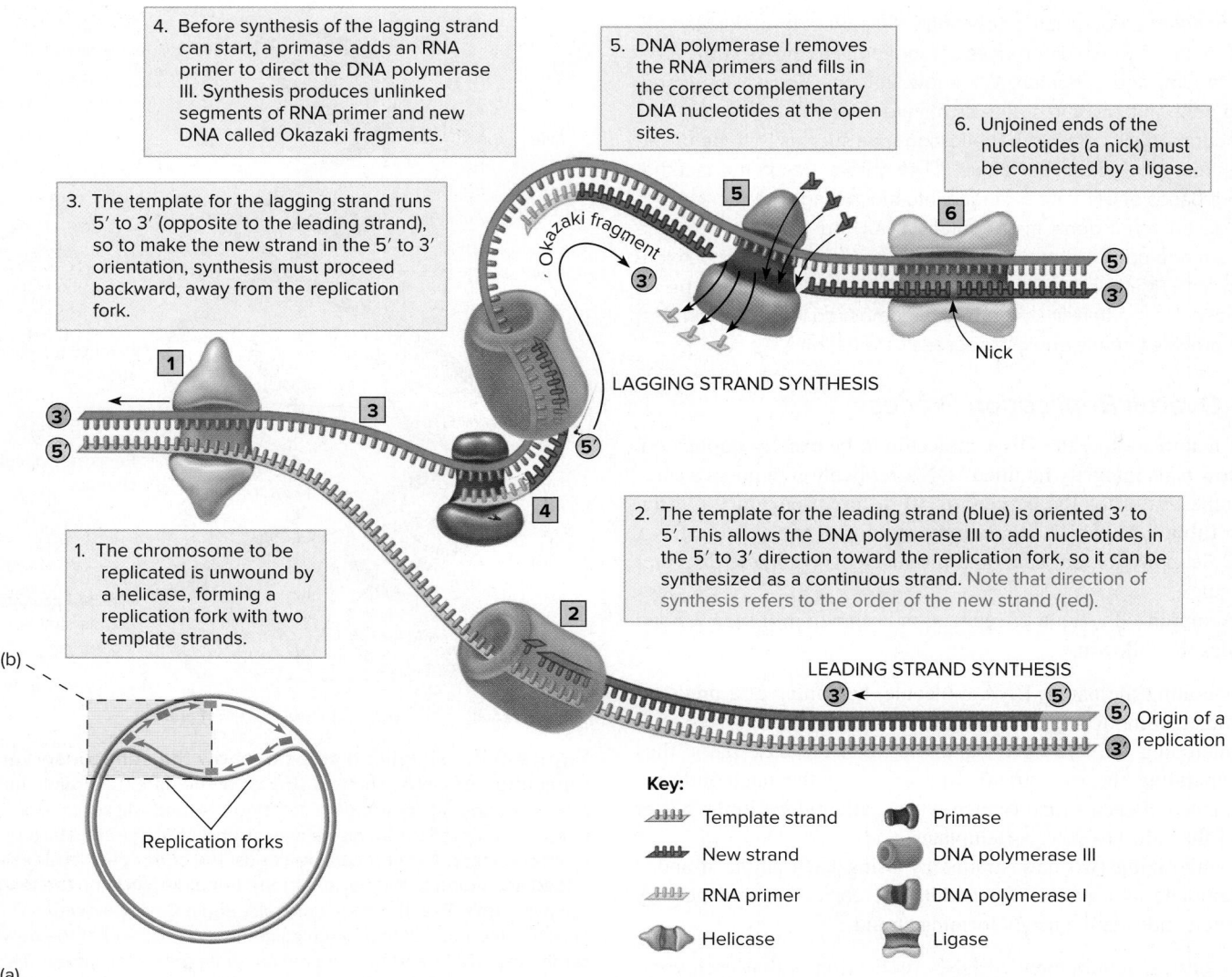

4. Before synthesis of the lagging strand can start, a primase adds an RNA primer to direct the DNA polymerase III. Synthesis produces unlinked segments of RNA primer and new DNA called Okazaki fragments.

5. DNA polymerase I removes the RNA primers and fills in the correct complementary DNA nucleotides at the open sites.

6. Unjoined ends of the nucleotides (a nick) must be connected by a ligase.

3. The template for the lagging strand runs 5′ to 3′ (opposite to the leading strand), so to make the new strand in the 5′ to 3′ orientation, synthesis must proceed backward, away from the replication fork.

Okazaki fragment

LAGGING STRAND SYNTHESIS

Nick

1. The chromosome to be replicated is unwound by a helicase, forming a replication fork with two template strands.

2. The template for the leading strand (blue) is oriented 3′ to 5′. This allows the DNA polymerase III to add nucleotides in the 5′ to 3′ direction toward the replication fork, so it can be synthesized as a continuous strand. Note that direction of synthesis refers to the order of the new strand (red).

LEADING STRAND SYNTHESIS

3′ ← 5′

5′ Origin of a
3′ replication

(b)

Replication forks

Key:

▥ Template strand		🔴 Primase	
▥ New strand		⬜ DNA polymerase III	
▥ RNA primer		🔵 DNA polymerase I	
🔵 Helicase		⬜ Ligase	

(a)

Process Figure 9.8 **The assembly line of DNA replication in a circular bacterial chromosome. (a)** A bacterial chromosome showing the overall pattern of replication. There are two replication forks where new DNA is being synthesized. **(b)** An enlarged view of the left replication fork to show the details of replication.

- Before replication can begin, a piece of RNA called a primer is inserted at the origin of replication to provide the first nucleotide with a free 3′—OH group for binding.
- DNA polymerase can **read** the template only in the 3′ to 5′ direction and can add nucleotides to the newly synthesized strand only in the 5′ to 3′ direction. Because the two template strands are oriented in opposite directions, this will necessitate an alternative type of synthesis for the template strand oriented in the 5′ to 3′ direction (see lagging strand in **figure 9.8b, part 3**).

Prior to the start of replication, enzymes called helicases (unzipping enzymes) bind to the DNA at the origin. These enzymes untwist the helix and break the hydrogen bonds holding the two strands together, resulting in two separate strands, each of which will be used as a template for the synthesis of a new strand **(figure 9.8b, step 1).**

Replication begins when RNA **primers** are synthesized by a primase at the origin of replication **(figure 9.8a).** DNA polymerase III

will use this short strand of RNA as a starting point for adding nucleotides. In the circular DNA molecule of bacteria, there are two **replication forks,** each containing its own set of replication enzymes. Each fork requires two active DNA polymerases along with several other proteins and enzymes whose main functions are to stabilize the polymerase and provide a means of removing and replacing nucleotides. The enzyme complex encircles the DNA near the replication fork and synthesizes new DNA as guided by the parental strand. As synthesis proceeds, the forks are continually opened up to expose the template for replication.

Because DNA polymerase III can only synthesize new DNA in the 5′ to 3′ direction, just one strand, called the **leading strand,** will be synthesized continuously **(figure 9.8b, step 2).** After addition of an RNA primer, a DNA polymerase complex enters and moves along the template strand in the 3′ to 5′ order toward the replication fork, adding nucleotides in the 5′ to 3′ order on the new strand.

The other template strand runs the opposite direction, which prevents the DNA polymerase from synthesizing a continuous,

unbroken strand of new DNA. This other strand, termed the **lagging strand** will be synthesized in a series of shorter fragments that will later be connected into a continuous strand (**figure 9.8b, steps 3** and **4**). It still requires primers, and each new fragment will require its own separate primer. The DNA polymerase adds nucleotides to the new strand in the 5′ to 3′ order, but it does this by going in a backward direction away from the fork. This manner of synthesis produces short fragments of DNA (100 to 1,000 bases long) called **Okazaki fragments** on the lagging strand. The lagging strand will be completed when the enzyme, **DNA polymerase I**, removes the RNA primers from the Okazaki fragments, and fills in the missing segments with the correct complementary DNA nucleotides (**figure 9.8b, step 5**). The replacement strand is not completely bonded to the finished part of the new strand, so an enzyme, **ligase,** makes these final phosphate-sugar connections (**figure 9.8b, step 6**).

The addition of nucleotides proceeds at an astonishing pace, estimated in some bacteria to be 750 bases per second at each fork! As replication proceeds, one newly synthesized strand loops down (**figure 9.9a**). When the forks have gone full circle, a termination site shuts replication down. The two circular daughter molecules remain connected briefly but are nicked and separated by a helicase (**figure 9.9b**).

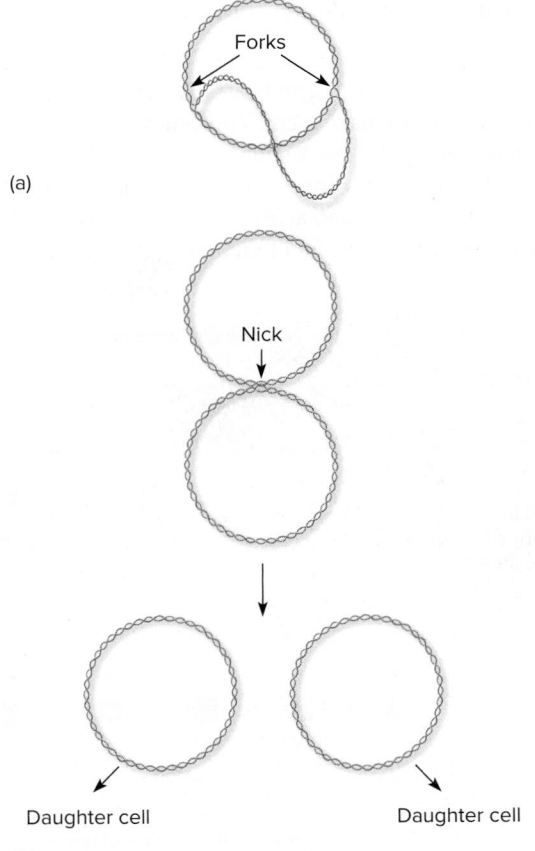

Forks

(a)

Nick

Daughter cell Daughter cell

(b)

Figure 9.9 Completion of chromosome replication in bacteria. (a) As replication proceeds, one strand loops down. **(b)** Final separation is achieved when an enzyme cuts and releases the two completed molecules. Each daughter cell receives one completed chromosome during binary fission.

Like any language, DNA is occasionally "misspelled" when an incorrect base is added to the growing chain. Studies have shown that such mistakes are made once in approximately 10^8 to 10^9 bases, but most of these are corrected. If not corrected, they are considered mutations that can lead to major cell dysfunction. Because continued cellular integrity is very dependent on accurate replication, cells have evolved their own proofreading function for DNA. DNA polymerase III, the enzyme that elongates the molecule, can also detect incorrect, unmatching bases; excise them; and replace them with the correct base. DNA polymerase I is also involved in proofreading the molecule and repairing damaged DNA.

Quick Search
Do an online search for "DNA Interactive" and explore the process of replication—copying the code.

 Check Your Progress SECTION 9.1

1. Compare the genetic material of eukaryotes, bacteria, and viruses in terms of general structure, size, and mode of replication.
2. Characterize the organization of genetic material, going from the genome down to the gene.
3. Briefly describe how DNA is packaged to fit inside a cell.
4. What are the fundamental building blocks of DNA and how are they bonded together?
5. Describe what is meant by the antiparallel arrangement of DNA.
6. Explain the synthesis of the leading and lagging strands during DNA replication and identify the enzymes required for producing a finished product.
7. Name several characteristics of DNA structure that enable it to be replicated with such great accuracy generation after generation.

9.2 Applications of the DNA Code: Transcription and Translation

 Expected Learning Outcomes

7. Present an overview of the main aspects of the flow of genetic information in cells.

8. Explain the relationship between the structure of DNA and the structure of proteins.

9. Describe the different types of RNA and their basic functions in genetic expression.

10. Relate the major events of transcription.

11. Describe the genetic code, codons, and anticodons, and how they relate to one another.

12. Recount the participants and steps in translation (protein synthesis).

13. Distinguish major points of difference between prokaryotic and eukaryotic transcription and translation.

We have explored how the genetic language of the DNA molecule is conserved through replication. Next we consider exactly how this language is expressed. Given that the sequence of bases in DNA is a genetic code, just what is the nature of this code and how is it

utilized? Although the genome is full of critical information, the DNA molecule itself cannot perform cell processes directly. Instead, its information is conveyed to RNA molecules, which carry out the instructions it contains. The concept that genetic information flows from DNA to RNA to protein has long been a central theme of molecular biology **(figure 9.10a).** It states that the master code of DNA is first used to synthesize RNA via a process called **transcription,** and the information contained in the RNA is then used to produce proteins in a process known as **translation.** The principal exceptions to this pattern are found in RNA viruses, which convert RNA to other RNA, and in retroviruses, which can convert RNA to DNA.

As with many established concepts in biology, the science of molecular genetics continues to add significant discoveries to our understanding. In addition to the RNA used to produce proteins, a wide variety of specialized RNAs act by regulating gene function **(figure 9.10b).** We will introduce some of the expanded functions for RNA in 9.3 Making Connections, but we will start with its direct participation in protein synthesis.

The Gene-Protein Connection

The Triplet Code and Its Relationship to Proteins

Several questions invariably arise concerning the relationship between genes and cell function. For instance, how does gene structure lead to the expression of traits in the individual, and what features of gene expression cause one organism to be so distinctly different from another? For answers, we must turn to the correlation between gene and protein structure. We know that each structural gene is an ordered sequence of nucleotides that codes for a protein. Because each protein is different, each gene must also differ somehow in its composition. In fact, the language of DNA exists in the order of groups of three consecutive bases designated as **triplets** or **codons,** on one DNA strand **(figure 9.11).** In short, one gene differs from another in the order and number of its triplets. An equally important part of this concept is that each triplet represents a code for a particular amino acid. When the triplet code is transcribed and translated, it dictates the type and order of amino acids in a polypeptide (protein) chain.

To summarize the key points that connect DNA and protein function:

1. DNA is a blueprint that indicates which kinds of proteins to make and how to make them. This blueprint exists in the order of triplets along the DNA strands.
2. The order of triplets directs a protein's primary structure—the order and type of amino acids in the chain—which determines its characteristic shape and function.

Figure 9.10 Summary of the flow of genetic information in cells. DNA is the ultimate storehouse and distributor of genetic information. **(a)** DNA must be deciphered into a usable cell language. It does this by transcribing its code into RNA helper molecules that translate that code into protein. **(b)** Other sections of DNA are copied to make additional types of RNA molecules that regulate genes and their products.

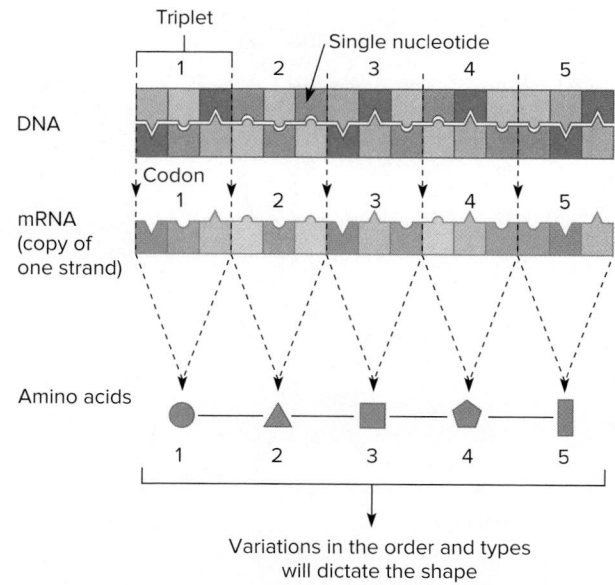

Figure 9.11 Simplified view of the DNA-protein relationship. The DNA molecule is a continuous chain of base pairs, but the sequence must be interpreted in groups of three base pairs (a triplet or codon). Each triplet transcribed into mRNA codons will translate into one amino acid; consequently, the ratio of base pairs to amino acids is 3:1.

3. Proteins contribute significantly to the phenotype by functioning as enzymes and structural molecules.

The Major Participants in Transcription and Translation

Transcription, the formation of RNA using DNA as a template, and translation, the synthesis of proteins using RNA as a template, are highly complex. A number of components participate: most prominently, messenger RNA, transfer RNA, ribosomes, several types of enzymes, and a storehouse of raw materials. After first examining each of these components, we shall see how they come together in the assembly line of the cell.

RNAs: Tools in the Cell's Assembly Line

Ribonucleic acid (RNA) is an encoded molecule like DNA, but its general structure is different in several ways:

1. It is a **single-stranded** molecule that can assume secondary and tertiary levels of complexity due to bonds within the molecule, leading to specialized forms of RNA (mRNA, tRNA, and rRNA—**figure 9.12**).
2. RNA contains **uracil,** instead of thymine, as the complementary base-pairing mate for adenine. This does not change the inherent DNA code in any way because the uracil still follows the pairing rules.
3. Although RNA, like DNA, is structured with a backbone of alternating sugar and phosphate molecules, the sugar in RNA is **ribose** rather than deoxyribose (see figure 2.24).

The many functional types of RNA range from small regulatory pieces to large structural ones. All types of RNA are formed through transcription of a DNA gene, but only mRNA is further translated into protein. See **table 9.2** for a comparison of major types of RNA and their functions. Additional forms of RNA are introduced in section 9.3.

Messenger RNA: Carrying DNA's Message

Messenger RNA (mRNA) is a transcribed version of a structural gene or genes in DNA. It is synthesized by a process similar to synthesis of the leading strand during DNA replication, and the complementary base-pairing rules ensure that the code will be faithfully copied in the mRNA transcript. The message of this transcribed strand is later read as a series of triplets called **codons (figure 9.12a),** and the length of the mRNA molecule varies from about 100 nucleotides to several thousand. The details of transcription and the function of mRNA in translation are covered later in this section.

Transfer RNA: The Key to Translation

Transfer RNAs (tRNAs) are also complementary copies of specific regions of DNA; however, they differ from mRNA. Transfer RNAs typically range from 75 to 95 nucleotides in length, and they contain sequences of bases that form hydrogen bonds with complementary sections in the same tRNA strand. At these points, the molecule bends back upon itself into several *hairpin loops,* giving the molecule a secondary *cloverleaf* structure that folds even further into a complex, three-dimensional helix **(figure 9.12b).** This compact molecule acts as a translator that converts RNA language into protein language. The bottom loop of the cloverleaf exposes a triplet, the **anticodon,** that both designates the specificity of the tRNA and complements mRNA's codons. At the opposite end of the molecule is a binding site for the amino acid that is specific for that tRNA's anticodon. For each of the 20 amino acids, there is at least one specialized type of tRNA to carry

(a) Messenger RNA (mRNA)
A short piece of messenger RNA (mRNA) illustrates the general structure of RNA: single strandedness, repeating phosphate-ribose sugar backbone attached to single nitrogen bases; use of uracil instead of thymine.

G C U	A C G	G A G	C U U	C G G	A G C	U A G
Codon 1	2	3	4	5	6	7

(b) Transfer RNA (tRNA)
Left: The tRNA strand loops back on itself to form intrachain hydrogen bonds. The result is a cloverleaf structure, shown here in simplified form. At its bottom is an anticodon that specifies the attachment of a particular amino acid at the 3' end. Right: A three-dimensional view of tRNA structure.

Figure 9.12 **Characteristics of messenger and transfer RNA.**

TABLE 9.2	Major Types of Ribonucleic Acid Involved in Protein Synthesis		
RNA Type	**Contains Codes For**	**Function In Cell**	**Translated**
Messenger (mRNA)	Sequence of amino acids in protein	Carries the DNA master code to the ribosome	Yes
Transfer (tRNA)	Specifying a given amino acid	Carries amino acids to ribosome during translation	No
Ribosomal (rRNA)	Several large structural rRNA molecules	Forms the major part of a ribosome and participates in protein synthesis	No
Primer	An RNA that can begin DNA replication	Primes DNA	No

it. Binding of an amino acid to its specific tRNA requires a specific enzyme that can correctly match each tRNA with its amino acid.

The Ribosome: A Mobile Molecular Factory for Translation

The prokaryotic (70S) ribosome is a particle composed of tightly packaged **ribosomal RNA (rRNA)** and protein. The rRNA component of the ribosome is an extensive polynucleotide molecule. It forms complex three-dimensional figures that contribute to the structure and function of ribosomes. The interactions of proteins and rRNA create the two subunits of the ribosome that engage in final translation of the genetic code (see figure 9.14). A metabolically active bacterial cell can accommodate up to 20,000 of these tiny factories—all actively engaged in reading the genetic program, taking in raw materials, and producing proteins at an impressive rate.

Transcription: The First Stage of Gene Expression

Like so many other genetic events, transcription is a marvel of efficiency. It is highly organized and guided by special codes on DNA itself. To describe it in more detail, we use the formation of mRNA in bacteria as a model. It is important to keep in mind that all types of RNA are synthesized by a similar process. The other RNAs have important functions; they just do not provide the code for the amino acids in a protein as mRNA does.

During transcription, an RNA molecule is synthesized using the codes on DNA as a guide or template. A large enzyme complex, **RNA polymerase,** is responsible for this process. This polymerase is more multipurpose than DNA polymerase, because the RNA enzyme works alone and does not require a helicase. It can bind to DNA and unwind it, as well as synthesize RNA **(figure 9.13).**

Transcription proceeds in three stages: *initiation, elongation,* and *termination.* Initiation requires the RNA polymerase to recognize a region on a gene called the **promoter region (figure 9.13, step 1).** This region consists of two sets of DNA sequences located just before the *initiation site.* The first sequence is about 35 nucleotides from the initiation site, and the second is about 10 nucleotides from it. The primary function of the promoter is to provide a position for initial binding of the RNA polymerase. A special protein molecule, the *sigma factor,* guides the RNA polymerase to the correct position on the promotor. These promoter sequences are not highly varied, and they tend to be rich in adenine and thymine

base pairs, which have one less hydrogen bond than guanine-cytosine pairs, which facilitates the separation of the DNA strands.

Prior to the first synthesis step of transcription, the RNA polymerase begins to separate the two strands of the DNA helix and forms an open "bubble" for transcription **(figure 9.13, steps 2** and **3).** This bubble serves as the space where the nucleotides of mRNA will actually be assembled. Only one strand of DNA—the **template strand**—is transcribed. This strand will have a reading frame oriented in the 3' to 5' direction recognized by the RNA polymerase, similar to replication. The mRNA made by transcription of the template strand is a message for the correct sequence of amino acids that will be linked together during translation (protein synthesis). The other strand—the **nontemplate strand**—is sometimes called the coding strand because its sequence is the same order as the mRNA (although it will have thymine instead of uracil). In some cases, this strand may serve a genetic function, but it is not transcribed. Which strand acts as the template varies from one gene to another. One important triplet appearing early in a DNA template is **TAC.** This will be transcribed into **AUG** on the mRNA, which we can see in figures 9.15 and 9.17 is the start codon, as it signals the location on the mRNA where translation starts. Note that the promotor sequences are not transcribed as part of the final mRNA molecule.

As elongation proceeds, the polymerase moves the transcription bubble forward, exposing subsequent sections of DNA. It simultaneously brings in nucleotides that are complementary to the DNA template and continues to assemble the mRNA strand. As with replication, this growing mRNA strand runs in the 5' to 3' direction. DNA that has already been transcribed rewinds back into its double helix structure **(figure 9.17, step 4).** The part of the mRNA strand that is already assembled remains attached to the enzyme complex but is kept out of the way of the processing machinery.

At termination, the polymerase recognizes a site on DNA near the end of the gene that signals the separation and release of the completed mRNA, which will next enter the ribosome for translation. As we will see in section 9.3, this mRNA delivers its message in codons that are the master code for protein synthesis. The length of the final mRNA molecule depends upon the polypeptide or protein that it encodes. In bacteria, the smallest mRNA might consist of 100 nucleotides, an average one, 600 nucleotides, and the largest, 1500.*

*What would be the corresponding numbers of amino acids in the final proteins for these three mRNAs?

Quick Search
Get an animated view of this dynamic process by searching online for "Stages of Transcription—McGraw-Hill."

1 Each gene contains a specific promoter region and a leader sequence for guiding the beginning of transcription. Next is the region of the gene that codes for a polypeptide and ends with a series of terminal sequences that stop translation.

2 DNA is unwound at the promoter by RNA polymerase. Only one strand of DNA, called the template strand, supplies the codes to be transcribed by RNA polymerase. This strand runs in the 3′ to 5′ direction.

3 The RNA polymerase moves along the DNA strand, adding complementary nucleotides as dictated by the DNA template. The mRNA strand reads in the 5′ to 3′ direction.

4 The polymerase continues transcribing until it reaches a termination site, and the mRNA transcript is released to be translated. Note that the section of the transcribed DNA is rewound into its original configuration.

Process Figure 9.13 **The major events in transcription, using mRNA as a model.**

Translation: The Second Stage of Gene Expression

In translation, all of the elements needed to synthesize a protein, from the mRNA to the tRNAs with amino acids, are brought together on the ribosomes (**figure 9.14**). The process occurs in five stages: initiation, elongation, termination, and protein folding and processing.

Initiation of Translation

In prokaryotic cells, the mRNA molecule leaves the DNA transcription site and is transported directly to ribosomes. Ribosomal subunits are specifically assembled in a way that forms sites to hold the mRNA and tRNAs. The ribosome thus recognizes these molecules and stabilizes reactions between them. The small subunit binds to the 5′ end of the mRNA and provides molecules to initiate translation. The large subunit holds the tRNAs and is actively involved in peptide bond formation by means of a specialized ribozyme, which is an RNA-based catalyst (see 8.2 Making Connections). The small subunit of the ribosome binds at a specific

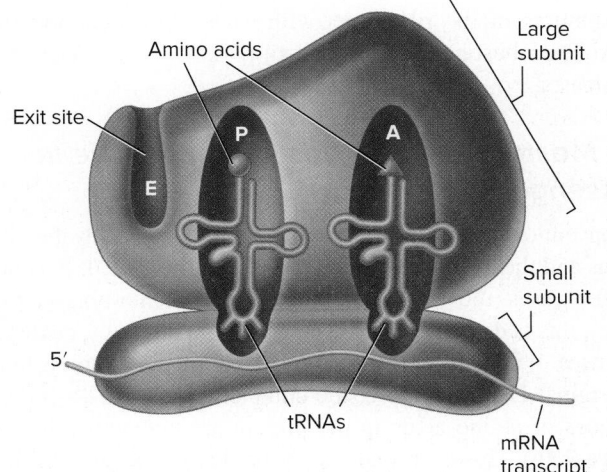

Figure 9.14 **The "players" in translation.** A ribosome serves as the stage for protein synthesis. Assembly of the small and large subunits results in specific sites for placement of the mRNA (small subunit) and two tRNAs with their amino acids (P and A sites).

		Second Base Position			
	U	**C**	**A**	**G**	
U	UUU ⎫ Phenylalanine UUC ⎭ UUA ⎫ Leucine UUG ⎭	UCU UCC Serine UCA UCG	UAU ⎫ Tyrosine UAC ⎭ UAA ⎫ STOP** UAG ⎭	UGU ⎫ Cysteine UGC ⎭ UGA STOP** UGG Tryptophan	U C A G
C	CUU CUC Leucine CUA CUG	CCU CCC Proline CCA CCG	CAU ⎫ Histidine CAC ⎭ CAA ⎫ Glutamine CAG ⎭	CGU CGC Arginine CGA CGG	U C A G
A	AUU AUC Isoleucine AUA AUG* START ,Methionine	ACU ACC Threonine ACA ACG	AAU ⎫ Asparagine AAC ⎭ AAA ⎫ Lysine AAG ⎭	AGU ⎫ Serine AGC ⎭ AGA ⎫ Arginine AGG ⎭	U C A G
G	GUU GUC Valine GUA GUG	GCU GCC Alanine GCA GCG	GAU ⎫ Aspartic acid GAC ⎭ GAA ⎫ Glutamic acid GAG ⎭	GGU GGC Glycine GGA GGG	U C A G

(First Base Position — left; Third Base Position — right)

* This codon initiates translation.
**For these codons, which give the orders to stop translation, there are no corresponding tRNAs with amino acids.

Figure 9.15 The genetic code: Codons of mRNA that specify a given amino acid. The master code for translation is provided by mRNA codons.

site on the mRNA and places the **start codon** AUG (rarely GUG) in correct alignment with the P site.

With the mRNA message in place on the assembled ribosome, the next step in translation involves entrance of tRNAs with their amino acids. The pool of cytoplasm around the region contains a complete array of tRNAs, previously charged by having the correct amino acid attached. The step in which the complementary tRNA meets with the mRNA code is guided by the two sites on the large subunit of the ribosome called the P site (left) and the A site (right).[1] Think of these sites as recessed spaces tucked within the two subunits of the ribosome, with each site accommodating a tRNA. The ribosome also has an exit or E site where used tRNAs are released.

The Master Genetic Code: The Message in Messenger RNA

By convention, the master genetic code is represented by the mRNA codons and the amino acids they specify (**figure 9.15**). Except in a very few cases (the genes of mitochondria and chloroplasts, for example), this code is universal, whether for prokaryotes, eukaryotes, or viruses. It is worth noting that once the mRNA codon is known, the original DNA sequence, the complementary tRNA code, and the types of amino acids in the protein are automatically known (**figure 9.16**). However, one cannot predict (backward) from protein structure what the exact mRNA codons are because of a factor

Figure 9.16 Interpreting the DNA code. If the DNA sequence is known, the mRNA codon can be surmised. If a codon is known, the anticodon and, finally, the amino acid sequence can be determined. The reverse is not possible (determining the exact codon or anticodon from amino acid sequence) due to the redundancy of the code.

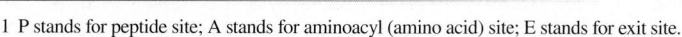

1 P stands for peptide site; A stands for aminoacyl (amino acid) site; E stands for exit site.

called **redundancy** or **degeneracy,** meaning that a particular amino acid can be coded for by more than one codon.

In figure 9.15 the mRNA codons and their corresponding amino acid specificities are given. Because there are 64 different triplet codes[2] and only 20 different amino acids, it is expected that some amino acids are represented by several codons. For example, leucine and serine can each be represented by any of six different triplets, but only tryptophan and methionine are represented by a single codon. In such codons as leucine, only the first two nucleotides are required to encode the correct amino acid, and the third nucleotide does not change its sense. This property, called *wobble,* is thought to permit some level of variation or mutation without changing the message.

The Beginning of Protein Synthesis

With mRNA serving as the guide, the stage is finally set for actual protein assembly. The correct tRNA (labeled **1** on **figure 9.17**) enters the P site and binds to the **start codon (AUG)** presented by the mRNA. Rules of pairing dictate that the **anticodon** of this tRNA will complement the mRNA codon AUG; thus, the tRNA with anticodon UAC will first occupy site P. It happens that the amino acid carried by the initiator tRNA in bacteria is *formyl methionine* ($_f$Met; see figure 9.15), though in many cases, it may not remain a permanent part of the finished protein.

Continuation and Completion of Protein Synthesis: Elongation and Termination

The mRNA comes with a specific order of codons called the *reading frame* that must be translated in that exact order. In general, the ribosome is mobile and moves along the mRNA from one codon to the next. As it moves the ribosome, it brings each succeeding mRNA codon into place and facilitates entry of the tRNA with its matching anticodon. The actual joining of the amino acids is carried out by a *ribozyme* (a catalytic RNA molecule) that is part of the large subunit of the ribosome. It forms a peptide bond between the amino acids on the adjacent tRNAs, and the polypeptide grows in length. Several steps of this process are outlined here:

- Elongation begins with the filling of the A site by a second tRNA (**figure 9.17, step 1**). The identity of this tRNA and its amino acid is dictated by the second mRNA codon.
- The entry of tRNA 2 into the A site brings the two adjacent tRNAs in favorable proximity for a peptide bond to form between the amino acids they carry. The $_f$Met is transferred from the first tRNA to amino acid 2, resulting in two coupled amino acids called a dipeptide (**figure 9.17, step 2**).
- For the next step to proceed, some room must be made on the ribosome, and the next codon in sequence must be brought into position for reading. This process is accomplished by *translocation,* the enzyme-directed shifting of the ribosome to the next position on the mRNA strand, which causes the blank tRNA (1) to be discharged from the ribosome (**figure 9.17, step 3**) at the E site.
- This simultaneously shifts the tRNA holding the dipeptide into P position. Site A is temporarily left empty. The tRNA that has

been released is now free to drift away and become recharged with an amino acid for later additions to this or another protein.

- The stage is now set for the insertion of tRNA 3 at site A as directed by the third mRNA codon (**figure 9.17, step 4**). This insertion is followed once again by peptide bond formation between the dipeptide and amino acid 3 (making a tripeptide), splitting of the peptide from tRNA 2, and translocation (**figure 9.17, step 5**).
- This action releases tRNA 2, shifts mRNA to the next position, moves tRNA 3 to position P, and opens position A for tRNA 4 (**step 6**). From this point on, peptide elongation proceeds repetitively by this same series of actions out to the end of the mRNA (**steps 7** and **8**).

The termination of protein synthesis is not simply a matter of reaching the last codon on mRNA. It is brought about by the presence of at least one special codon occurring just after the codon for the last amino acid. Termination codons—UAA, UAG, and UGA—are codons for which there is no corresponding tRNA. Also termed **nonsense** or **stop codons,** they carry a necessary message: *Stop here.* When this codon is reached, a special enzyme breaks the bond between the final tRNA and the finished polypeptide chain, releasing it from the ribosome.

Before newly made proteins can carry out their structural or enzymatic roles, they often require finishing touches. Even before the peptide chain is released from the ribosome, it begins folding upon itself to achieve its biologically active tertiary conformation. Other alterations, called *posttranslational modifications,* may be necessary. Some proteins must have the starting amino acid (formyl methionine) clipped off; proteins destined to become complex enzymes have **cofactors** added; and some join with other completed proteins to form **quaternary** levels of structure.

The operation of transcription and translation is machinelike in its precision. Protein synthesis in bacteria is both efficient and rapid. At 37°C, 12 to 17 amino acids per second are added to a growing peptide chain. An average protein consisting of about 400 amino acids requires less than half a minute for complete synthesis. Further efficiency is gained when the translation of mRNA starts while transcription is still occurring (**figure 9.18**). A single mRNA is long enough to be fed through more than one ribosome simultaneously. This permits the synthesis of hundreds of protein molecules from the same mRNA transcript arrayed along a chain of ribosomes. This **polyribosomal complex** is indeed an assembly line for mass production of proteins. Protein synthesis consumes an enormous amount of energy. Approximately 1,200 ATPs or ATP equivalents are consumed just for synthesis of an average-size protein.

Eukaryotic Transcription and Translation: Similar yet Different

Eukaryotes share many aspects of protein synthesis with prokaryotes. They show similarities in the operation of the ribosomes, the functions of start and stop codons, and the mass production of proteins by means of polyribosomes. But they do differ in significant ways. The start codon in eukaryotes is also AUG, but it codes for an alternate

Quick Search
See this in animation by searching online for "VCAC: Molecular Processes: Translation the Movie."

2 This number 64 comes from 4^3 (the 4 different nucleotides in all possible combinations of 3).

Leucine

fMet

Anticodon

mRNA

Start codon

① Entrance of tRNAs 1 and 2

Peptide bond 1

② Formation of peptide bond

Empty tRNA

P site

③ Discharge of tRNA 1 at E site

Proline

A site

④ First translocation; tRNA 2 shifts into P site; tRNA 3 enters ribosome at A

Peptide bond 2

⑤ Formation of peptide bond

Alanine

⑥ Discharge of tRNA 2; second translocation; tRNA 4 enters ribosome

Peptide bond 3

⑦ Formation of peptide bond

Stop codon

⑧ Steps 3 to 5 are repeated until stop codon is reached and protein synthesis is terminated.

Process Figure 9.17 The events in protein synthesis.

(a)

(b)

(c)

Figure 9.18 **Speeding up the protein assembly line in bacteria.** **(a)** The mRNA transcript encounters ribosomal parts immediately as it leaves the DNA. **(b)** The ribosomal factories assemble along the mRNA in a chain, each ribosome reading the message and translating it into protein. Many products will thus be well along the synthetic pathway before transcription has even terminated. **(c)** Photomicrograph of a polyribosomal complex in action. Note that the protein "tails" vary in length depending on the stage of translation (30,000×). (c) © Steven McKnight and Oscar L Miller, Department of Biology, University of Virginia

form of methionine. Another difference is that eukaryotic mRNAs code for just one protein, unlike bacterial mRNAs, which often contain information from several genes in series. See 9.2 Making Connections to discover other exceptional aspects of eukaryotic genetics.

Prokaryotic and eukaryotic cells also differ in location and structure of genes. The presence of DNA in the nucleus of eukaryotic cells means that eukaryotic transcription and translation cannot be simultaneous as it is in prokaryotes. The mRNA transcript must pass through pores in the nuclear membrane and be carried to the ribosomes in the cytoplasm or on the endoplasmic reticulum for translation.

So far we have introduced a simplified definition of a gene that fits well for the majority of prokaryotes. The pattern of their genes is colinear, meaning that their message can be directly read into proteins without any additional processing.[3] Eukaryotic genes differ significantly in this characteristic because they are not colinear. Located within their genes are one to several intragenic regions, called **introns,** that do not code for protein.[4] These introns are interspersed between coding regions, called **exons,** that will be translated into protein **(figure 9.19).** To

[3] With colinearity, an mRNA transcript can be translated directly into the correct sequence of amino acids.

[4] Introns have also been found in some prokaryotes (cyanobacteria), archaea) and even viruses.

Figure 9.19 **The split gene of eukaryotes.** Eukaryotic genes have an additional complication in their transcription/translation. Their coding sequences, or exons (E), are interrupted at intervals by segments called introns (I) that are not part of that protein's code. Introns are transcribed but not translated, which necessitates their removal by RNA splicing enzymes before translation.

illustrate this we can use words as examples. A short section of colinear prokaryotic gene might read: SAM SAW HIS NEW CAR GET HIT; a eukaryotic gene that codes for the same portion would read SAM SAW SVXF FPL HIS NEW CAR QZWVP GET HIT. The recognizable words are the exons, and the nonsense letters represent the introns.

This discontinuous genetic structure, also called a split gene, requires further processing before translation. Transcription of the entire gene with both exons and introns occurs first, producing a pre-mRNA. Next, an RNA-protein complex called a *spliceosome* recognizes the exon-intron junctions and enzymatically cuts through them in a process called RNA splicing. The action of this splicer enzyme loops the introns into lariat-shaped pieces, excises them, and joins the exons end to end. By this means, a strand of mRNA with no intron material is produced. This completed mRNA strand can then proceed to the cytoplasm to be translated.

Several different types of introns have been discovered, some of which do code for cell substances. For example, certain introns have been found to code for an enzyme called reverse transcriptase, which can convert RNA into DNA. Other introns are translated into endonucleases, enzymes that can snip DNA and allow insertions and deletions into the DNA sequence. Some introns have an innate catalytic function and can splice themselves out of an RNA transcript. A great deal of non-protein-coding DNA is proving to be vital for cell function. Human DNA has an average of 8 introns per gene, which represents a significant percentage of the DNA found in chromosomes. One role alternative splicing plays is in providing greater flexibility in creating alternative versions of genes. It is likely that future research will discover additional roles of introns in regulating gene expression **(9.2 Making Connections).**

Check Your Progress SECTION 9.2

8. How is the language of a gene expressed?
9. If a bacterial protein is 3,300 amino acids long, how many nucleotide pairs long is the gene sequence that codes for it?
10. Construct a table that compares the structure and functions of DNA and RNA.
11. Where does transcription begin?
12. What are the template and nontemplate strands of DNA?
13. Why is only one strand transcribed, and is the same strand of DNA always transcribed?
14. Distinguish between a codon and an anticodon.
15. Briefly describe the events in translation.
16. What are the functions of start and stop codons? Give examples.
17. Summarize how bacterial and eukaryotic cells differ in gene structure, transcription, and translation.
18. Discuss the roles of exons and introns.

9.3 Genetic Regulation of Protein Synthesis and Metabolism

 Expected Learning Outcomes

14. Explain the functions of operons in bacterial genetic control.
15. Describe the main features of the lactose operon and inducible operons.
16. Describe the main features of repressible operons.
17. Summarize some aspects of genetic control by RNA.

In chapter 8 we surveyed the metabolic reactions in cells and the enzymes involved in those reactions. There we learned that some enzymes are regulated and that one form of regulation occurs at the genetic level. This control mechanism ensures that genes are active only when their products are required. In this way, enzymes will be produced as they are needed and the waste of energy and materials in dead-end synthesis is prevented. A major form of gene regulation in prokaryotes is through systems called **operons.** An operon is a section of DNA that contains one or more structural genes along with a corresponding operator gene that controls transcription. Having this kind of setup will permit genes for a particular metabolic pathway to be induced or repressed in unison by the same regulatory element.

Operons are described as either inducible or repressible. The category of an operon is determined by how it is affected by the environment within the cell. Many catabolic operons are **inducible,** meaning that the operon is turned on (induced) by the substrate of the enzyme for which the structural genes code. In this way, the enzymes needed to metabolize a nutrient (lactose, for example) are produced only when that nutrient is present. **Repressible operons** often contain genes coding for anabolic enzymes, such as those used to synthesize amino acids. In the case of these operons, several genes in series are turned off (repressed) by the product synthesized by the enzyme.

The Lactose Operon: A Model for Inducible Gene Regulation in Bacteria

The best understood cell system for explaining control through genetic induction is the **lactose (*lac*) operon.** The system, first described in 1961 by François Jacob and Jacques Monod, accounts for the regulation of lactose metabolism in *Escherichia coli.* Many other operons with similar modes of action have since been identified, and together they furnish convincing evidence that a cell's metabolic actions can have great impact on gene expression.

The lactose operon has three important features **(figure 9.20):**

1. the **regulator,** composed of the gene that codes for a protein capable of repressing the operon (a **repressor**);
2. the *control locus,* composed of two areas, the promoter recognized by RNA polymerase (see figure 9.13) and the **operator,** a sequence that acts as an on/off switch for transcription; and

Revising Some Rules of Genetics

What was true for E. coli *would be true for the elephant.*
 Jacques Monod, Nobel laureate and discover of the *lac* operon

Bacteria—with their small genomes and short generation times—have been ideal models for research into molecular genetics. From this research came elegant and universal rules governing the workings of DNA. This led to the "Central Dogma of Biology," which says that DNA is the repository of genetic information, RNA is a molecule that ferries information between DNA and the ribosome, and proteins produced from the RNA codes are responsible for carrying out the structural and metabolic needs of the cell. And all of this is still accurate.

It is just that there is more to the story. Significant discoveries in eukaryotic organisms have revealed that certain aspects of their genetics have turned out to be somewhat more complex and varied. This does not change our introduction to microbial genetics for two reasons. First, the laws of genetics covered in this chapter do hold true for the vast majority of prokaryotic organisms. Second, this is an *introduction* to genetics and it is certainly fair to leave the exceptions, the unusual, and the overly complex for future courses. We would like to use this feature to introduce some of these revolutionary new findings.

The Genetic Code Is Universal with Some Exceptions

Based on early studies, it was assumed that the master genetic code was the same in all cells and viruses, as no exceptions had ever been seen. Even more convincing, eukaryotic mRNAs were appropriately translated by bacterial cells, arguing for a single genetic dictionary. But later work with mitochondria isolated from yeast and humans found a slightly different code; for example, the stop codon UGA specifies the amino acid tryptophan in mitochondria. This discovery was followed by the detection of other exceptions to the genetic code in a few prokaryotic and eukaryotic organisms. This difference in the genetic code of mitochondria is one of the more compelling pieces of evidence supporting the endosymbiotic theory (see chapter 5).

Genetic Expression Can Be Modified by Epigenetic Mechanisms

For years, geneticists have debated the effects of nature (genes) versus nurture (environment) on the development of organisms. Though both have profound influence, we now have greater understanding about the direct effects of nongenetic factors.

The term *epigenetics* is used to designate factors that effect genetic expression other than the specific sequence of nucleotides in genes. One major epigenetic affect is in the manner that DNA coils around proteins during chromatin formation. This layer of information is only now being fully appreciated and is still far from being completely understood. Epigenetic effects seem to work by increasing or decreasing the expression of certain genes by making them more or less accessible for transcription. For example, having methyl groups (—CH_3) attached to

numerous cytosine nucleotides reduces the level of transcription of a section of DNA that is methylated (**figure *a***).

To illustrate this effect, a study in mice determined that the mother's diet during gestation could greatly alter gene expression in her offspring. Genetically identical mothers (clones) were placed into two groups—one that was fed a diet high in the soy chemical genistein and the other a normal diet. The soy group had young mice with significant differences in color coat and weight. They were larger and yellower than the mice from the normal group. This was a permanent effect, because the soy group remained unchanged even when placed on a normal diet. The effect appears to be due to methylation of the embryonic genes by genistein during fetal development (**figure *b***).

90,000 Proteins . . . 21,000 Genes?

Humans have about 90,000 different proteins, and, going by our earlier understanding, we would have expected at least 90,000 different genes. Imagine the surprise most biologists felt when the Human Genome Project revealed that we have fewer than 21,000 protein-coding sequences, about 20,000 fewer than a simple corn plant.

For some time it has been known that eukaryotic genes consist of expressed sequences (exons) interspersed with intervening sequences (introns) and that the final mRNA need not contain every exon. Until our low gene count was revealed, however, no one realized how often a single gene could be used to make several different proteins. Utilizing *alternative splicing,* a cell can select which exons will be retained and spliced into a final mRNA. Like a Scrabble player deciding which lettered tiles to use to make a word, a cell can place any combination of the exons in a gene to produce many alternate mRNAs (and resulting proteins). It is currently estimated that three of every four human genes are edited in this way and that the prevalence of alternative splicing appears to increase with the complexity of an organism.

Suggest some reasons for having large parts of the genome silenced or unused. *Answer available on Connect.*

(a) Section of gene with methylation that blocks its activation

Normal cytosine Methylated cytosine

(b) One example of epigenetic effects due to methylation of bases during early development

Mice from mothers fed a high-soy diet were larger and different-colored from mice whose mothers were fed a normal diet. All mice are genetically the same (clones).
(b): © Randy L. Jirtle

3. the *structural locus,* made up of three genes, each coding for a different enzyme needed to catabolize lactose. One of the enzymes, β-galactosidase, hydrolyzes the lactose into its monosaccharides; another, permease, brings lactose across the cell membrane.

The promoter, operator, and structural components lie adjacent to one another, but the regulator can be at a distant site.

In inducible systems such as the *lac* operon, the operon is normally in an *off mode* and does not initiate enzyme synthesis when the appropriate substrate is absent **(figure 9.20, step 1).** How is the operon maintained in this mode? The key is in the repressor protein that is encoded by the regulatory gene. This relatively large molecule is **allosteric,** meaning that its activity can be altered, depending on which active site and substrate are in play. A substrate binding to one active site can distort a different active site and pre-

vent it from accepting its substrate (see figure 8.10*b*). In the absence of lactose, the repressor protein interacts with the operator and causes the operator to distort into a temporary loop configuration. This loop blocks access of the RNA polymerase to the DNA of the operator and prevents transcription. Think of the repressor as a lock on the operator, and if the operator is locked, the structural genes cannot be transcribed. Importantly, the regulator gene is located in a separate site from the operator region and is not affected by this block on the operon.

If lactose is added to the cell's environment, it triggers several events that turn the operon *on.* First, the binding of lactose to the repressor protein creates a shape change in the repressor that dislodges it from the operator segment **(figure 9.20, step 2).** This opens up the operator to allow RNA polymerase to bind to it and begin transcription. The structural genes are transcribed

① **Operon Off: No Lactose**
In the absence of lactose, a repressor protein (the product of a regulatory gene located elsewhere) attaches to the operator of the operon. This effectively blocks transcription of structural genes downstream (to its right). Suppression of transcription (and translation) prevents the unnecessary synthesis of enzymes for processing lactose.

② **Operon On: Lactose Present**
Upon entering the cell, the substrate (lactose) becomes a genetic inducer by attaching to the repressor, which is rendered inactive and falls away. The operator is no longer closed off and its DNA becomes accessible to the RNA polymerase. The RNA polymerase transcribes the structural genes, and the mRNA is translated into enzymes that can act on the lactose substrate.

Process Figure 9.20 The lactose operon in bacteria: How inducible genes are controlled by substrate.

in a single unbroken transcript coding for all three enzymes. During translation, however, each protein is synthesized separately. Because lactose is ultimately responsible for stimulating the chain of events leading to protein synthesis, it is considered an **inducer.**

As lactose is depleted, further enzyme synthesis is not necessary, so the order of events reverses. At this point, there is no longer sufficient lactose to inhibit the repressor; hence, the repressor regains its active shape for attachment to the operator. The operator is locked, and transcription of the structural genes and enzyme synthesis related to lactose are arrested.

A fine but important point about the *lac* operon is that it functions only in the absence of glucose or if the cell's energy needs are not being met by the available glucose. Glucose is the preferred carbon source because it can be used immediately in growth and does not require induction of an operon. When a supply of glucose is present, a second regulatory system ensures that the *lac* operon is inactive, regardless of lactose levels in the environment.

A Repressible Operon

Bacterial systems for synthesis of amino acids, purines and pyrimidines, and other processes work on a principle that is the reverse of the *lac* operon—that of repression. Similar factors such as repressor proteins, operators, and a series of structural genes are part of a repressible operon, but with some important differences. Unlike the *lac* operon, this operon is normally in the *on* mode and will be turned *off* only when the product of the pathway is no longer required. Excess product plays a role as a **corepressor** that slows the transcription of the operon.

A metabolically active cell that is consuming large amounts of the amino acid arginine (arg) will serve to illustrate the operation of a repressible operon. Under these conditions, the *arg* operon is set to *on,* and arginine is being actively synthesized through the action of the operon's enzymatic products (**figure 9.21, step 1**). In such an active cell, the arginine will be used immediately, and the repressor will remain inactive (unable to bind the operator) because there is too little free arginine to activate it. As the cell's metabolism begins to slow

① **Operon On: Arginine being used by cell**
A repressible operon remains on when its nutrient products (here, arginine) are in great demand by the cell. The repressor has the wrong shape to bind to the DNA operator without a corepressor, so that RNA polymerase is free to actively transcribe the genes and translation actively proceeds.

② **Operon Off: Arginine building up**
The operon is repressed when (1) arginine builds up and, serving as a corepressor, activates the repressor. (2) The activated repressor complex affixes to the operator and blocks the RNA polymerase and further transcription of genes for arginine synthesis.

Process Figure 9.21 Repressible operon: Genetic control through excess product.

9.3 MAKING CONNECTIONS

Is it Really an RNA World?

Evidence about the roles of RNAs in gene expression is accumulating rapidly. We have learned how the introns of eukaryotic mRNAs are removed by the catalytic RNA of the spliceosome complex, and the peptide bonds of proteins are catalyzed by ribosomal RNA. Based on our current understanding, it is clear that various forms of RNA dominate genetic control of transcription and translation. This type of RNA, called noncoding RNA (ncRNA) because it does not code for protein, regulates how the coded molecules are expressed.

Recently described ncRNAs in bacteria are **riboswitches**—short leader segments of mRNA that are transcribed but not expressed in a finished protein. Acting as a switch means that they can be turned on or off, depending on their folding pattern, which means they can both start or stop a step in gene expression. Some riboswitches work at the level of transcription and others alter translation, but the outcome of both is to affect protein synthesis. Riboswitches have a unique ability to sense and bind a specific molecule (ligand) such as an amino acid, vitamin, or cofactor that changes their folding pattern and action.

Riboswitches in gram-negative bacteria regulate translation at the ribosome. When the riboswitch is on, the leader sequence of mRNA forms a loop with itself that opens up the binding sequence and makes it available for the ribosome and protein synthesis. The riboswitch is turned off when a ligand binds to it and repositions the binding sequence. This makes the mRNA binding position on the ribosome inaccessible, thereby blocking translation (see figure). Like the *lac* operon, this mechanism will be reversed in the absence of the ligand, and translation can proceed. This riboswitch regulates synthesis of some vitamins.

Eukaryotes have their own versions of controls on transcription and translation. We are just beginning to understand the functions of small RNAs containing around 20 to 25 nucleotides that are found in animal and plant cells. Small RNAs have been termed "gene silencers" because they shut down final products of gene expression. Examples are microRNAs and si (short interfering) RNAs, both of which control the fate of mRNA following transcription. Both types of small RNAs attach to mRNA and keep it from being translated. In some cases, they suppress mRNA functions, and in others, they actually control the synthesis of enzymes that can destroy the mRNA. SiRNA is also known to signal the destruction of foreign RNA from viruses.

Another form of eukaryotic regulation acts at the level of the chromosomes, by means of reorganization and modification of the chromatin—a type of epigenetic effect. By fine-tuning the tightness and looseness of the chromatin structure, certain small RNAs can control which genes are accessible for function (see figure 9.4). Other small RNAs are involved in methylation of DNA, which also modulates gene availability.

Knowledge of these novel types of RNA molecules has answered some questions that came out of genome sequencing studies. Most of the DNA in organisms was found *not* to code for functional proteins. This DNA was once considered "junk," but its percentage in humans was 85%! Obviously, in bacteria as well as in humans, it is far from junk. In fact, this DNA has been preserved in the same form for many millions of years of evolution, suggesting it has a very important function. We now know that much of it codes for these important RNA regulatory molecules.

In what ways could knowledge of these regulatory RNAs inspire new types of drug therapies? *Answer available on Connect.*

down, however, the synthesized arginine is no longer used up and accumulates. The free arginine is then available to act as a corepressor by attaching to the repressor. This reaction changes the shape of the repressor, making it capable of binding to the operator and stopping transcription. Arginine will cease to be synthesized until the cell once again requires it in metabolism (**figure 9.21, step 2/2**).

Non-Operon Control Mechanisms

Analogous gene control mechanisms in eukaryotic cells are not as well understood, but it is known that gene function can be altered by intrinsic regulatory segments similar to operons. Some molecules, called transcription factors, insert on the grooves of the DNA molecule and enhance transcription of specific genes. These transcription factors can regulate gene expression in response to environmental stimuli such as nutrients, toxin levels, or even temperature. Eukaryotic genes are also regulated during growth and development, leading to the hundreds of different tissue types found in higher multicellular organisms. Considerable interest has been generated in the functions of regulatory RNAs that exert control on several levels in both prokaryotic and eukaryotic cells (**9.3 Making Connections**).

 Check Your Progress SECTION 9.3

19. What is an operon? Describe the functions of regulators, promoters, and operators.
20. Compare and contrast the *lac* operon and repressible operon systems.
21. For both figures 9.20 and 9.21, draw a simple stepwise diagram indicating the events that would reverse the operon's action.
22. At which levels of DNA regulation do small RNAs function?

9.4 Mutations: Changes in the Genetic Code

Expected Learning Outcomes

18. Define a mutation and explain how mutant forms are identified.
19. Summarize the causes and types of mutations and their effects.
20. Describe some mutation repair mechanisms.
21. Compare beneficial and detrimental effects of mutations.

As precise and predictable as the rules of genetic expression seem, permanent changes do occur in the genetic code. Indeed, genetic change is the driving force of evolution through its addition of variation to populations of organisms. Changes may become evident in altered gene expression, such as in the appearance or disappearance of anatomical or physiological traits. For example, a pigmented bacterium can lose its ability to form pigment, or a strain of the malarial parasite can develop resistance to a drug. A phenotypic change that is due to an alteration in the genotype is called a **mutation.** On a strictly molecular level, a mutation is an alteration in the nitrogen base sequence of DNA. It can involve the loss of base pairs, the addition of base pairs, or a rearrangement in the order of base pairs. Do not confuse this with genetic recombination, in which microbes transfer whole segments of genetic information between themselves.

A microorganism that exhibits a natural, nonmutated characteristic is known as a **wild type,** or wild strain. If a microorganism develops a mutation, it is called a **mutant strain.** Mutant strains can show variance in morphology, nutritional characteristics, genetic mechanisms, resistance to chemicals, temperature preference, and nearly any type of enzymatic function. Mutant strains are very useful for tracking genetic events, unraveling genetic organization, and pinpointing genetic markers.

The simplest way to detect mutant bacteria is to inoculate solid media containing differential or selective agents such as metabolic substrates or antibiotics. Growing a pure culture on a plate of media

and observing the appearance of colonies can reveal strains of a species that express genetic variations. For instance, in cultures of wild-type *E. coli* grown on differential MacConkey agar, most of the isolated colonies will indicate a positive result for lactose fermentation, but invariably there will be a small number of colonies that are negative and have lost the genes for using lactose. This colony can be readily pulled out for further study. Antibiotics are frequently added to media as selective agents for separating out strains of bacteria that are drug resistant. Figure 10.12 shows how this method works for detecting recombinant cells used in genetic engineering.

Another method for detecting and isolating mutant strains of microbes that grow in colonies is the *replica plating* technique **(figure 9.22).** This method is based on a variation between wild strains that can synthesize a nutrient, such as an amino acid, and mutant strains that cannot synthesize it. First a culture of test microbes is exposed to a chemical agent known to cause mutations and then

(a) Treatment of culture with a mutagen

(b) Use treated culture to inoculate a plate with complete growth medium and incubate. Both wild type and mutants form colonies.

(c) Replica block
Velvet surface (sterilized)
Master plate (complete medium)
Target mutant colony—position not yet known
Press replica surface onto two plates.

(c)

Replica plate (complete medium)
Replica plate (medium minus nutrient)
Incubation

(d) All strains grow
Mutant colony does not grow
No colony present at this position

Mutant colony can be located and isolated.

Figure 9.22 The replica plating method for isolating mutant bacteria. This technique was developed by Joshua Lederberg. It is based on **(a)** exposing a culture to a mutagen, **(b)** plating the culture for isolated colonies on a master plate with complete nutrients, **(c)** pressing a sterile carrier onto the surface of the colonies to pick up a tiny clump of cells and pressing it again to innoculate a plate of complete medium and then a plate of incomplete medium. **(d)** After growth, the two plates are compared. The colonies that occupy a specific position on the complete medium but are missing from that same position on the incomplete medium are mutant colonies that can be subcultured for further use.

plated on a complete medium that contains the particular nutrient. It is important for this to be done in a way that enables individual colonies to form. Colony position on the plate is retained by pressing a pickup or replica carrier onto the colonies and then using the carrier to inoculate two plates in the same orientation. These two plates contain (1) a complete medium containing the nutrient being tested for and (2) an incomplete medium that lacks it. Colonies that grow on the complete medium could be of either the wild or mutant type, but only the wild strains will grow on the incomplete medium. By comparing the two plates side by side, it is possible to locate the position of the colony on the complete medium that does not grow on the incomplete one. This mutant colony is now detectable and can be isolated.

Replica plating has been an effective technique to screen for phenotypic variants of bacteria and other microbes. It has the potential to provide information on a wide variety of metabolic characteristics. For example, it is used to detect mutants that vary in biochemical characteristics such as amino acid synthesis, substrate utilization, and sensitivity to antibiotics. A technique related to this can provide information on the rates of mutation caused by various mutagenic agents (see the Ames test, figure 9.23).

Causes of Mutations

A mutation is described as spontaneous or induced, depending upon its origin. A **spontaneous mutation** is a random change in the DNA arising from errors in replication that occur without a known cause. The frequency of spontaneous mutations has been measured for a number of organisms, and can vary considerably. Bacteria show rates ranging from high (10^{-4} or 1 mutation per base per replication out of 4,000 events) to low (10^{-11} or 1 mutation per base per replication out of 10,000,000,000). The rapid rate of bacterial reproduction allows these mutations to be observed more readily in bacteria than in most eukaryotes.

Induced mutations result from exposure to known **mutagens,** which are primarily physical or chemical agents that damage DNA and interfere with its functioning **(table 9.3).** The carefully controlled use of mutagens has proved a useful way to induce mutant strains of microorganisms for study.

TABLE 9.3	Selected Mutagenic Agents and Their Effects
Agent	**Effect**
Chemical	
Nitrous acid, bisulfite	Remove an amino group from some nitrogen bases
Ethidium bromide	Inserts between the paired bases
Acridine dyes	Cause frameshifts due to insertion between base pairs
Nitrogen base analogs	Compete with natural bases for sites on replicating DNA
Radiation	
Ionizing (gamma rays, X rays)	Form free radicals that cause single or double breaks in DNA
Ultraviolet	Causes cross-links between adjacent pyrimidines

Chemical mutagenic agents act in a variety of ways to change the DNA. Agents such as acridine dyes insert completely across the DNA helices between adjacent bases. This results in distortion of the helix and can cause frameshift mutations. Analogs[5] of the nitrogen bases (5-bromodeoxyuridine and 2-aminopurine, for example) are chemical mimics of natural bases that are incorporated into DNA during replication. Addition of these abnormal bases leads to mistakes in base-pairing. Many chemical mutagens are also carcinogens, or cancer-causing agents (see the discussion of the Ames test later in this section).

Physical agents that alter DNA are primarily types of radiation. High-energy gamma rays and X rays introduce major physical changes into DNA, which can accumulate breaks that may not be repairable. Ultraviolet (UV) radiation induces abnormal bonds between adjacent pyrimidines so that normal replication of that region is blocked. Exposure to large doses of radiation can be fatal, which is why radiation is so effective in microbial control; it can also be carcinogenic in animals. The use of radiation to control microorganisms is described further in chapter 11.

Categories of Mutations

Mutations range from large mutations, in which longer genetic sequences are gained or lost, to small ones that affect only a single base on a gene. These latter mutations, which involve addition, deletion, or substitution of single bases, are called **point mutations.**

To understand how a change in DNA influences the cell, remember that the DNA code appears in a particular order of triplets (three bases) that is transcribed into mRNA codons, each of which specifies an amino acid. A permanent alteration in the DNA that is copied faithfully into mRNA and translated can change the structure of the protein. A change in a protein can likewise change the morphology and physiology of a cell. Most mutations have a harmful effect on the cell, leading to cell dysfunction or death; these are called lethal mutations. Neutral mutations produce neither adverse nor helpful changes. A small number of mutations are beneficial in that they provide the cell with a useful change in structure or physiology.

Any change in the code that leads to placement of a different amino acid is called a **missense mutation.** A missense mutation can do one of the following:

- create a faulty, nonfunctional (or less functional) protein,
- produce a protein that functions in a different manner, or
- cause no significant alteration in protein function.

A **nonsense mutation,** on the other hand, changes a normal codon into a stop codon that does not code for an amino acid and stops the production of the protein wherever it occurs. A nonsense mutation almost always results in a nonfunctional protein. A **silent mutation** alters a base but does not change the amino acid and thus has no effect. For example, because of the redundancy of the code, ACU, ACC, ACG, and ACA all code for threonine, so a mutation that changes only the last base will not alter the sense of the message in any way. A **back-mutation** occurs when a gene that has undergone

5 An analog is a chemical structured very similarly to another chemical except for minor differences in functional groups.

mutation reverses (mutates back) to its original base composition. An **inversion** occurs when two nucleotides have switched their order, which will nearly always change the amino acid, as we see by inverting GUC (valine) to GCU (threonine).

Mutations also occur when one or more bases are inserted into or deleted from a newly synthesized DNA strand (see **table 9.4, IIB**). This type of mutation, known as a **frameshift**, is so named because the reading frame of mRNA will shift to the left or right, depending on the type of change. Frameshift mutations nearly always result in a nonfunctional protein because the codon structure will become reset from that point on, will code for a different sequence of amino acids from the original DNA, and is likely to introduce stop codons. Also note that insertion or deletion of bases in multiples of three (3, 6, 9, etc.) does not disturb the reading frame downstream from the mutation site. It can still disrupt the structure of the protein, depending on the change in the amino acid sequence. The effects of these types of mutations is modelled in **table 9.4.**

TABLE 9.4	Classification of Major Types of Mutations
	Example
I. **Wild-type (nonmutated) sequence**	THE BIG BAD CAT ATE THE FAT RED BUG
The wild-type sequence of a gene is the DNA sequence found in most organisms and is generally considered the "normal" sequence.	
II. **Categories of mutations based on type of DNA alteration**	
A. Substitution mutations	
1. Missense	THE BIG MAD CAT ATE THE FAT RED BUG
A missense mutation causes a different amino acid to be incorporated into a protein. Effects range from unnoticeable to severe, based on how the new amino acid alters protein function.	
2. Nonsense	THE BIG BAD XXX (stop)
A nonsense mutation converts a codon to a stop codon, resulting in premature termination of protein synthesis. Effects of this type of mutation are almost always severe.	
B. Inversion mutations	THE BIG ABD CAT ATE THE FAT RED BUG
	THE BIB GAD CAT ATE THE FAT RED BUG
Inversions arise when adjacent letters exchange places, which alters 1 or 2 bases, depending on the location of the inversion. This does not change the reading frame, but can result in significant changes in amino acids and protein function.	
C. Frameshift Mutations	
1. Insertion	THE BIG BAB DCA TAT ETH EFA TRE DBU G →
2. Deletion	THE BIG *ADC ATA TET HEF ATR EDB UG ←
Insertion (addition of letter) and deletion (removal of letter) mutations cause a change in the reading frame ↔ of the mRNA, resulting in a protein in which every amino acid after the mutation can be affected. Because of this, frameshift mutations almost always result in a nonfunctional protein.	

Repair of Mutations

Earlier, we indicated that DNA has a proofreading mechanism to repair mistakes in replication that might otherwise become permanent. Because mutations are potentially life-threatening, the cell has additional systems for finding and repairing DNA that has been damaged by various mutagenic agents and processes. Most ordinary DNA damage is resolved by enzymatic systems specialized for finding and fixing such defects.

DNA that has been damaged by ultraviolet radiation can be restored by photoactivation or light repair. This repair mechanism requires visible light and a light-sensitive enzyme, DNA photolyase, which can attach to sites of abnormal pyrimidine bonding and restore the original DNA structure. Ultraviolet repair mechanisms are successful only for a relatively small number of UV mutations. Mutations can be excised by a series of enzymes that remove the incorrect bases and add the correct ones. This process is known as *excision repair*. First, enzymes break the bonds between the bases and the sugar-phosphate strand at the site of the error. A different enzyme subsequently removes the defective bases one at a time, leaving a gap that will be filled in by DNA polymerase I and ligase. A repair system can also locate mismatched bases that were missed during proofreading: for example, C mistakenly paired with A, or G with T. The base must be replaced soon after the mismatch is made, or it will not be recognized by the repair enzymes.

Quick Search

See this process animated by going to the video "DNA Nucleotide Excision Repair" on YouTube.

The Ames Test

New agricultural, industrial, and medicinal chemicals are constantly being added to the environment, and exposure to them is widespread. The discovery that many such compounds are mutagenic and that up to 83% of these mutagens are linked to cancer is significant. Although animal testing has been a standard method of detecting chemicals with carcinogenic potential, a more rapid screening system called the **Ames test**[6] is also commonly used. In this inventive test, the experimental subjects are bacteria whose gene expression and mutation rate can be readily observed and monitored. The premise is that any chemical capable of mutating bacterial DNA can similarly mutate mammalian (and thus human) DNA and is therefore potentially hazardous. Many potential carcinogens (benzanthracene and aflatoxin, for example) are mutagenic agents only after being acted on by mammalian liver enzymes, so an extract of these enzymes is added to the test medium.

One indicator organism in the Ames test is a mutant strain of *Salmonella enterica*[7] that has lost the ability to synthesize the amino acid histidine, a defect highly susceptible to back-mutation because the strain also lacks DNA repair mechanisms. Mutations

6 Named for its creator, Bruce Ames.

7 *S. enterica* inhabits the intestine of poultry and causes food poisoning in humans. It is used extensively in genetic studies of bacteria.

that cause reversion to the wild strain, which is capable of synthesizing histidine, occur spontaneously at a low rate. A test agent is considered a mutagen if it enhances the rate of back-mutation beyond levels that would occur spontaneously. **Figure 9.23** compares a standard version of the test with a multiwell method that streamlines result taking. The Ames test has proved invaluable for screening an assortment of environmental and dietary chemicals for mutagenicity and carcinogenicity without resorting to animal studies.

Positive and Negative Effects of Mutations

Many mutations are not repaired. How the cell copes with them depends on the nature of the mutation and the strategies available to that organism. Mutations are passed on to the offspring of organisms during reproduction and to new viruses during replication. They become a long-term part of the gene pool. Most mutations are harmful to organisms, but some can provide adaptive advantages.

If a mutation leading to a nonfunctional protein occurs in a gene for which there is only a single copy, as in haploid or simple organisms, the cell will probably die. This happens when certain mutant strains of *E. coli* acquire mutations in the genes needed to repair damage by UV radiation. Mutations of the human genome affecting the action of a single protein (mostly enzymes) are responsible for more than 3,500 diseases.

Although most spontaneous mutations are not beneficial, a small number contribute to the success of the individual and the population by creating variant strains with alternate ways of expressing a trait. Microbes are not "aware" of this advantage and do not direct these changes; they simply respond to the environment they encounter. In the long-range view, mutations and the variations they produce are the raw materials for change in the population and, thus, for evolution.

Mutations that create variants occur frequently enough that any population contains mutant strains for a number of characteristics, but as long as the environment is stable, these mutants will never comprise more than a tiny percentage of the population. When the environment changes, however, it can become hostile for the survival of certain individuals, and only those microbes bearing protective mutations will be equipped to survive in the new environment.

Such observations can be explained by the important biological principles of natural selection and survival of the fittest. This theory states that environmental pressure selects for organisms that have greater adaptability due to beneficial mutations. They are more likely to survive, reproduce, and pass the changed genes on to succeeding generations. In time, these survivors become the dominant members of the population as long as the mutation continues to favor their survival. One of the clearest models for natural selection in action is acquired drug resistance in bacteria (discussed in chapter 12).

As we will see in section 9.5, bacteria have additional mechanisms for changing their genetic makeup by sharing genes with other bacteria through recombination.

Culture of *Salmonella* bacteria, histidine (–)

In the control setup, bacteria are plated on a histidine-free medium containing liver enzymes but lacking the test agent.

The experimental plate is prepared the same way except that it contains the test agent.

(a) **Control Plate**
Minimal medium lacking histidine and test chemical

(b) **Test Plate**
Minimal medium with test chemical and no histidine

Incubation (12 h)
Any colonies that form have back-mutated to his($^+$)

his($^+$) colonies arising from spontaneous back-mutation

his($^+$) colonies in presence of the chemical

(c) The degree of mutagenicity of the chemical agent can be calculated by comparing the number of colonies growing on the control plate with the number on the test plate. Chemicals that induce an increased incidence of back-mutation (right side) are considered carcinogens.

(d) Ames fluctuation tests using multiple wells; a blue well indicates no back-mutation occurring; yellow indicates back-mutations.

Counting the number of (+) yellow wells provides the number of back-mutations; the result on the right indicates a strong mutagen.

Figure 9.23 The Ames test. (a, b, c) In this traditional plate method, a strain of *Salmonella enterica* that cannot synthesize histidine [his (–)] is the test species. If the chemical causes an increase in mutation rate to histidine-producing cells as compared to a control, the chemical is considered mutagenic. **(d)** The newer well fluctuation test is based on the same basic scheme, however, it has greater sensitivity and is easier to evaluate. (d) Courtesy of Xenometrix AG

Check Your Progress SECTION 9.4

23. Explain what is meant by the terms *mutation* and *mutant*.

24. Describe the primary causes, types, and outcomes of mutations. Give an example of a mutation that is beneficial and one that is lethal or harmful.

25. Explain the purposes behind replica plating and the Ames test.

9.5 DNA Recombination Events

 Expected Learning Outcomes

22. Explain recombination in bacteria and what it involves.

23. Describe the main features of conjugation and its outcomes.

24. Discuss what is required for bacterial transformation and its outcomes.

25. Identify the basic processes involved in transduction and the end results.

26. Discuss transposons and their importance to microorganisms.

Genetic recombination through sexual reproduction is an important means of genetic variation in eukaryotes. Although bacteria have no exact equivalent to sexual reproduction, they exhibit a primitive means for sharing or recombining parts of their genomes. An event in which one bacterium donates DNA to another bacterium is a type of genetic transfer termed **recombination,** the end result of which is a new strain different from both the donor and the original recipient strain. In general, any organism that has acquired genes that originated in another organism is called a **recombinant.**

Recombination in bacteria depends largely on their extreme versatility in acquiring and expressing the genetic material of other bacteria and even other organisms. This capacity for genetic sharing has tremendous effects on the diversity and evolution of bacteria. Unlike spontaneous mutations, recombinations are generally beneficial. They can provide additional genes for resistance to drugs and metabolic poisons, new nutritional and metabolic schemes, increased virulence, and adaptations to changing environmental conditions.

Transmission of Genetic Material in Bacteria

DNA transfer between bacterial cells typically involves small pieces of DNA in the form of **plasmids** or chromosomal fragments. Plasmids are small, circular pieces of DNA that contain their own origin of replication and therefore can replicate independently of the bacterial chromosome. Plasmids are found in many bacteria (as well as some fungi) and typically contain a few dozen genes. Although plasmids are not necessary for bacterial survival, they often carry adaptive genes, such as those for drug resistance.

The Case Study that opens this chapter details a pathogen that has received dozens of resistance genes from other species. Chromosomal fragments that have escaped from a lysed bacterial cell are also commonly involved in the transfer of genetic information between cells. An important difference between plasmids and fragments is that, while a plasmid can be stably replicated and inherited, chromosomal fragments must integrate themselves into the bacterial chromosome in order to be replicated and eventually passed to progeny cells. The rate of natural genetic recombination varies a great deal among bacteria, but its frequency can be increased in the laboratory, where the ability to transfer genes between organisms has become an important tool of research and technology.

Depending upon the mode of transmission, the means of genetic recombination in bacteria is called conjugation, transformation, or transduction. **Conjugation** requires the attachment of two cells and the formation of a bridge that can transport DNA. **Transformation** entails the transfer of free DNA and requires no special vehicle. **Transduction** is DNA transfer mediated through the action of bacterial viruses **(table 9.5).**

Conjugation: Genetic Transmission Through Direct Contact

Conjugation is a mode of genetic recombination in which a plasmid or fragment of DNA is transferred from a donor cell to a recipient cell via a direct connection. Both gram-negative and gram-positive bacteria can conjugate, but only gram-negative cells operate with a specialized plasmid called a **fertility,** or **F factor.** This plasmid directs the synthesis of a unique **pilus,** also called a sex pilus,[8] that functions in most conjugation transfers. The recipient cell is usually a related bacterium with a recognition site on its surface for interacting with the pilus **(figure 9.24, step 1).**

A cell's role in conjugation is denoted by F^+ for the cell that has the F plasmid and by F^- for the cell that lacks it. Contact is made when a pilus grows out from the F^+ cell, attaches to the surface of the F^- cell, contracts, and draws the two cells together (figure 9.24, step 1). At the site where the pilus attaches the two cells, a mating or conjugative bridge is formed that serves as a transfer

8 This term refers to its function in the mating of bacteria, but it has nothing to do with sexual reproduction, which is absent in bacteria.

TABLE 9.5	Types of Genetic Recombination in Bacteria		
Mode	**Factors Involved**	**Direct or Indirect***	**Examples of Genes Transferred**
Conjugation	Donor cell with pilus Fertility plasmid in donor Both donor and recipient alive Bridge forms between cells to transfer DNA.	Direct	Drug resistance; resistance to metals; toxin production; enzymes; adherence molecules; degradation of toxic substances; uptake of iron
Transformation	Free donor DNA (fragment or plasmid) Live, competent recipient cell, donor usually dead	Indirect	Polysaccharide capsule; metabolic enzymes; drug resistance; unlimited with cloning techniques
Transduction	Donor is lysed bacterial cell. Defective bacteriophage is carrier of donor DNA. Live recipient cell of same species as donor	Indirect	Exotoxins; enzymes for sugar fermentation; drug resistance

*Direct means the donor and recipient are in contact during exchange; indirect means they are not.

system for the plasmid. The latest reliable information indicates that the pilus becomes part of a complex secretory system that opens a gateway between the cell walls and membranes of the two cells and serves as a passageway for the plasmid.

There are hundreds of conjugative plasmids with some variations in their properties. One of the best understood plasmids is the F factor in *E. coli,* which exhibits these patterns of transfer:

1. Because bacteria tend to conserve plasmids, the donor cell first makes a copy of the F factor, which is simultaneously transferred across the bridge **(figure 9.24, step 2).** This way, it retains the original plasmid, remains F⁺, and can continue to conjugate with other bacteria. The F⁻ cell has been changed into an F⁺ cell capable of producing a pilus and conjugating with other cells. No additional donor genes are transferred during this transfer.

2. In high-frequency recombination (Hfr) donors, the fertility factor has been integrated into the F⁺ donor chromosome.

The term *high-frequency recombination* denotes that a cell with an integrated F factor transmits its chromosomal genes at a higher frequency than other cells. This is because the F factor can direct a more comprehensive transfer of part of the donor chromosome to a recipient cell. This transfer occurs through replication of the donor chromosome along with the F factor. One strand of DNA is retained by the donor, and the other strand is transported across to the recipient cell **(figure 9.24, step 3).** The F factor may not be transferred during this process. The transfer of an entire chromosome takes about 100 minutes, but the pilus bridge between cells is ordinarily broken before this time, and rarely is the entire genome of the donor cell transferred.

Conjugation has great biomedical importance. Special **resistance (R) plasmids,** or **factors,** that bear genes for resisting antibiotics and other drugs are commonly shared among bacteria through conjugation. Transfer of R factors can confer multiple resistance to antibiotics such as tetracycline, chloramphenicol, sulfonamides,

Physical Conjugation

F factor Bacterial chromosome

F⁺

Sex pilus makes contact with F⁻ recipient cell.

F⁻

Sex pilus contracts, bringing cells together.

F⁺

F⁻

1 The pilus of donor cell (top) attaches to receptor on recipient cell and retracts to draw the two cells together. This is the mechanism for gram-negative bacteria.

F Factor Transfer

Bridge made with pilus

Donor Recipient

F⁺ F⁻

F factor (plasmid)

Chromosome

F⁺ F⁻

F factor being copied

F⁺ F⁺

F factor

2 Transfer of the F factor, or conjugative plasmid

Hfr Transfer

Pilus

Donor Recipient

Integrated F factor

Hfr cell

Chromosome

Partial copy of donor chromosome

Bridge broken Donated genes

3 High-frequency (Hfr) transfer involves transmission of chromosomal genes from a donor cell to a recipient cell. The donor chromosome is duplicated and transmitted in part to a recipient cell, where it is integrated into the chromosome.

Process Figure 9.24 Conjugation: Genetic transmission through direct contact between two cells.

and penicillin. This phenomenon is discussed further in chapter 12. Other types of R factors carry genetic codes for resistance to heavy metals (nickel and mercury) or for synthesizing virulence factors (toxins, enzymes, and adhesion molecules) that increase the pathogenicity of the bacterial strain. Conjugation studies have also provided a means to map the bacterial chromosome.

A few gram-positive bacteria (*Bacillus* and *Streptococcus,* for example) are known to recombine through conjugation. Because they lack pili, they must transport the plasmids via specialized proteins activated during adhesion of the two cells. The mechanisms involved are yet to be fully worked out.

Transformation: Capturing Free DNA from Solution

One of the cornerstone discoveries in microbial genetics was made in the late 1920s by the English biochemist Frederick Griffith working with *Streptococcus pneumoniae* and laboratory mice. The pneumococcus exists in two major strains based on the presence of the capsule, colonial morphology, and pathogenicity. Encapsulated strains bear a smooth (S) colonial appearance and are virulent; strains lacking a capsule have a rough (R) appearance and are nonvirulent. (Recall from chapter 4 that the capsule protects a bacterium from the phagocytic host defenses.) Griffith first showed that when mice were injected with a live, virulent (S) strain, they soon died (**figure 9.25*a***). Mice injected with a live, nonvirulent (R) strain remained alive and healthy (**figure 9.25*b***). Next he tried a variation on this theme. He heat-killed an S strain and injected it into mice, which remained healthy (**figure 9.25*c***). Then came the ultimate test: Griffith injected both dead S cells and live R cells into mice, with the result that the mice died from pneumococcal blood infection (**figure 9.25*d***). If killed bacterial cells do not come back to life and the nonvirulent live strain was harmless, why did the mice die? Although he did not know it at the time, Griffith had demonstrated that dead S cells, while passing through the body of the mouse, broke open and released some of their DNA (by chance, a segment containing the genes for making a capsule). A few of the live R cells subsequently picked up this loose DNA and were transformed by it into virulent, capsule-forming strains.

Later studies supported the concept that a chromosome released by a lysed cell breaks into fragments small enough to be accepted by a recipient cell and that DNA, even from a dead cell, retains its genetic code. This nonspecific acceptance by a bacterial cell of small fragments of soluble DNA from the surrounding environment is termed **transformation (figure 9.26).** Transformation is facilitated by special DNA-binding proteins on the cell wall that capture DNA from the surrounding medium. Cells that are capable of accepting genetic material through this means are termed **competent.** The new DNA is passed through a DNA uptake system in the cell wall and membrane. Within the cytoplasm, the donated strand is inserted into the chromosome of the recipient cell, now said to be *transformed.* Natural transformation is found in several groups of gram-positive and gram-negative bacterial species. In addition to genes coding for the capsule, bacteria also exchange genes for antibiotic resistance and bacteriocin synthesis in this way.

Because transformation requires no special appendages, and the donor and recipient cells do not have to be in direct contact, the process

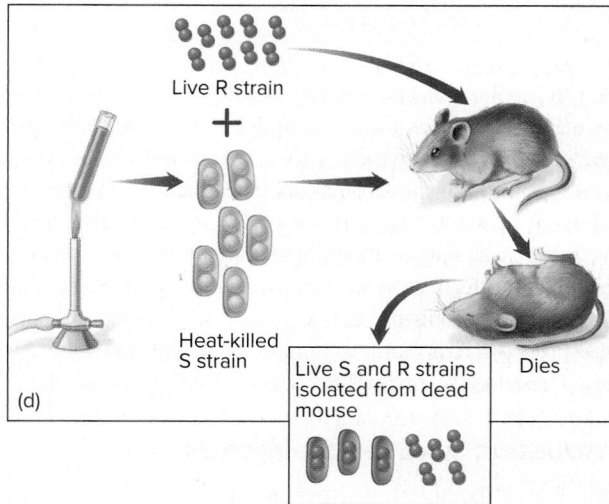

Figure 9.25 **Griffith's classic experiment in transformation.** In essence, this experiment proved that DNA released from a killed cell can be acquired by a live cell. The cell receiving this new DNA is genetically transformed—in this case, from a nonvirulent strain to a virulent one.

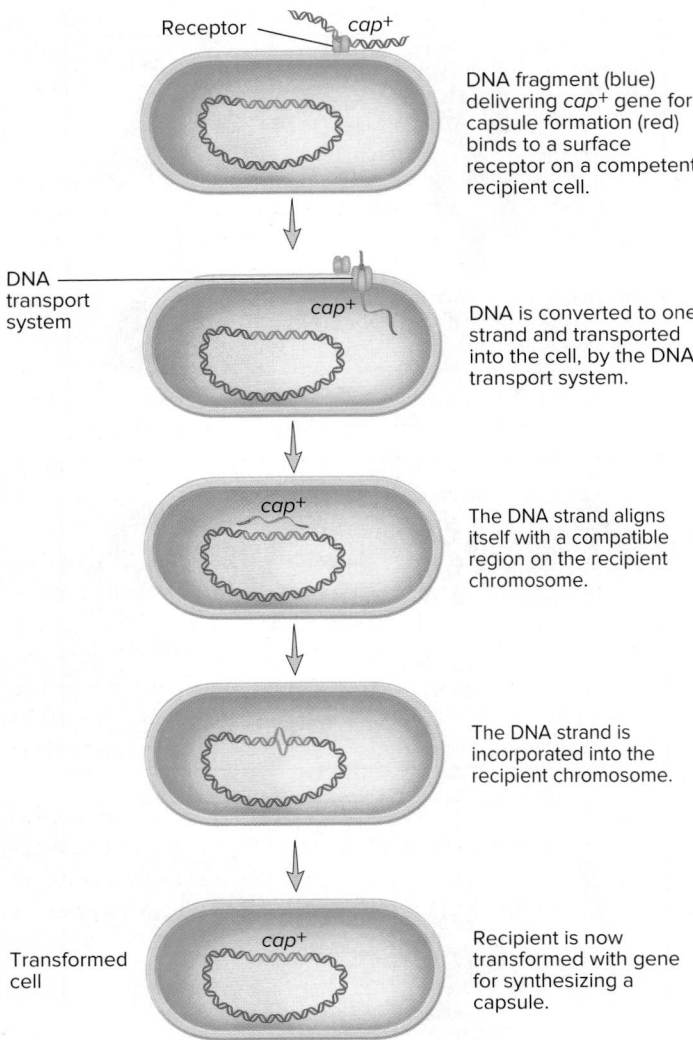

Receptor cap⁺

DNA fragment (blue) delivering *cap⁺* gene for capsule formation (red) binds to a surface receptor on a competent recipient cell.

DNA transport system cap⁺

DNA is converted to one strand and transported into the cell, by the DNA transport system.

cap⁺

The DNA strand aligns itself with a compatible region on the recipient chromosome.

The DNA strand is incorporated into the recipient chromosome.

Transformed cell cap⁺

Recipient is now transformed with gene for synthesizing a capsule.

Figure 9.26 Steps in bacterial transformation as seen in *S. pneumoniae.* This shows how genes coding for capsule formation are brought into the cell and inserted into the chromosome. *Cap⁺* indicates genes for forming a capsule.

Phage DNA
Donor (host) chromosome
(1)
Cell A (donor)

Parts of phage
Separated piece of host DNA
(2)

Newly assembled phage incorporating piece of host DNA

Lysis
(3)

DNA from donor
(4)
Cell B (recipient)

Incorporated into chromosome
(5)

Cell survives and utilizes transduced DNA

Process Figure 9.27 Generalized transduction: Genetic transfer by means of a virus carrier. (1) A phage infects cell A (the donor cell) by normal means. **(2)** During replication and assembly, a phage particle incorporates a segment of bacterial DNA by mistake. **(3)** Cell A then lyses and releases the mature phages, including the genetically altered one. **(4)** The altered phage adsorbs to and penetrates another host cell (cell B), injecting the DNA from cell A rather than viral nucleic acid. **(5)** Cell B receives this donated DNA, which recombines with its own DNA. Because the virus is defective (biologically inactive as a virus), it is unable to complete a lytic cycle. The transduced cell survives and can use this new genetic material.

is useful for applications in recombinant DNA technology. With this technique, foreign genes from a completely unrelated organism are inserted into a plasmid, which is then introduced into a competent bacterial cell through transformation. These recombinations can be carried out in a test tube, and human genes can be experimented upon and even expressed outside the human body by placing them in a microbial cell. Transformation is often used in the production of genetically engineered organisms such as yeasts, plants, and animals, and has been proposed as a future technique for curing genetic diseases in humans. These topics are covered in more detail in chapter 10.

Transduction: Viruses as Vectors

Bacteriophages (bacterial viruses) have been previously described as destructive bacterial parasites. Viruses can also serve as genetic vectors (an entity that can bring foreign DNA into a cell). The process by which a bacteriophage serves as the carrier of DNA from a donor cell to a recipient cell is **transduction.** Although it occurs naturally in a broad spectrum of bacteria, the participating bacteria

in a single transduction event must be the same species because of the specificity of viruses for host cells.

There are two versions of transduction. In *generalized transduction* (**figure 9.27**), random fragments of disintegrating host DNA are taken up by the phage during assembly. Virtually any

gene from the bacterium can be transmitted through this means. In *specialized transduction* (**figure 9.28**), a highly specific part of the host genome is regularly incorporated into the virus. This specificity is explained by the prior existence of a temperate prophage inserted in a fixed site on the bacterial chromosome. When activated, the prophage DNA separates from the bacterial chromosome, carrying a small segment of host genes with it. During a lytic cycle, these specific viral-host gene combinations are incorporated into

the viral particles and carried to another bacterial cell. Instances of transduction include the transfer of drug resistance seen in *Staphylococcus* spp. and the transmission of gene regulators in gram-negative rods such as *Escherichia* and *Salmonella*.

Transposons: "The Case of the Jumping Genes" One type of genetic transferral of great interest involves transposable elements, or **transposons.** Transposons have the distinction of shifting from one part of the genome to another and so are termed "jumping genes." When the idea of their existence in corn plants was first postulated by the geneticist Barbara McClintock, it was greeted with wide skepticism, because it had long been believed that the location of a given gene was set and that genes did not or could not move around. Now it is evident that jumping genes are widespread among prokaryotic and eukaryotic cells and viruses. Results from analysis of human DNA indicates that they make up nearly 45% of the human genome.

All transposons share the general characteristic of traveling from one location to another on the genome—from one chromosomal site to another, from a chromosome to a plasmid, or from a plasmid to a chromosome (**figure 9.29**). Because transposons can occur in plasmids, they can also be transmitted from one cell to another in bacteria and a few eukaryotes. Some transposons replicate themselves before jumping to the next location, and others simply move without replicating first.

Transposons contain DNA that codes for the enzymes needed to remove and reintegrate the transposon at another site in the genome. Flanking the coding region of the DNA are sequences called inverted repeats, which mark the point at which the transposon is

Recombination results in two possible outcomes.

Process Figure 9.28 Specialized transduction: Transfer of specific genetic material by means of a virus carrier. (1) Specialized transduction begins with a cell that contains a prophage (red). **(2)** When the virus enters a lytic cycle, it excises itself from its host cell, carrying along some host DNA. **(3)** Replication and assembly results in production of a **chimeric** virus containing viral and bacterial DNA. **(4)** Release of the recombinant virus and infection of a new host result in transfer of bacterial DNA between cells. **(5)** Recombination can occur between the new host chromosome and the virus DNA, resulting in either bacterial DNA or a combination of viral and bacterial DNA being incorporated into the bacterial chromosome.

Figure 9.29 Transposons: Shifting segments of the genome. (1) A transposon exists as a small piece of DNA integrated into the host cell chromosome (red). **(2)** The transposon may excise itself and move from one location to another in the genome, maintaining itself as a single copy per cell. **(3)** It may also replicate prior to moving, leading to an increase in the copy number and a greater effect on the genome of the host. **(4)** Finally, the transposon may jump to a plasmid, which can then be transferred to another bacterial cell.

removed or reinserted into the genome. The smallest transposons consist of only these two genetic sequences and are often referred to as *insertion elements.* A type of transposon called *retrotransposon* can transcribe DNA into RNA and then back into DNA for insertion in a new location. Another example is the *integron,* which can carry large blocks of genetic material that are especially prominent in transmitting drug resistance.

The overall effect of transposons—to scramble the genetic language—can be beneficial or adverse, depending upon such variables as where insertion occurs in a chromosome, what kinds of genes are relocated, and the type of cell involved. In bacteria, transposons are known to be involved in

- changes in traits such as colony morphology, pigmentation, and antigenic characteristics;
- replacement of damaged DNA; and
- the transfer of drug resistance in bacteria.

 Check Your Progress SECTION 9.5

26. Compare conjugation, transformation, and transduction on the basis of general method, donor, and recipient.
27. Explain the differences between general and specialized transduction, using drawings.
28. By means of a flowchart, show the possible jumps that a transposon can make. Show the involvement of viruses in its movement.

9.6 The Genetics of Animal Viruses

 Expected Learning Outcomes

27. Explain the major elements of viral genetics.
28. Compare aspects of the genetics of DNA and RNA viruses, using examples.

Viruses essentially consist of one or more pieces of DNA or RNA enclosed in a protective coating. They are genetic parasites that require access to their host cell's genetic and metabolic machinery to be replicated, transcribed, and translated, and they also have the potential for genetically changing the cells. Because they contain mainly those genes needed for the production of new viruses, the genomes of viruses tend to be very small and compact.

Viruses show extensive variety in genetic patterns. In many viruses, the nucleic acid is linear in form; in others, it is circular. The genome of most viruses exists in a single molecule, though in a few, it is segmented into several smaller molecules. Most viruses contain normal double-stranded (ds) DNA or single-stranded (ss) RNA, but other patterns exist. There are ssDNA viruses, dsRNA viruses, and retroviruses, which work backward by making dsDNA from ssRNA. In some instances, viral genes overlap one another, and in a few DNA viruses, both strands contain a translatable message.

With few exceptions, replication of the DNA molecule of DNA animal viruses occurs in the nucleus, where the cell's DNA replication machinery lies. The genome of most RNA viruses is replicated in the cytoplasm. In all viruses, viral mRNA is translated into viral proteins on host cell ribosomes using host tRNA.

Replication Strategies in Animal Viruses

Replication, Transcription, and Translation of dsDNA Viruses

Replication of dsDNA viruses is divided into phases **(figure 9.30).** During the early phase, viral DNA enters the nucleus, where several genes are transcribed into a messenger RNA. Next, the mRNA transcript then moves into the cytoplasm to be translated into viral proteins (enzymes) needed to replicate the viral DNA. This replication usually occurs in the nucleus with the host cell's own DNA polymerase, though some viruses (herpes, for example) have their own. During the late phase, other viral genes are transcribed and

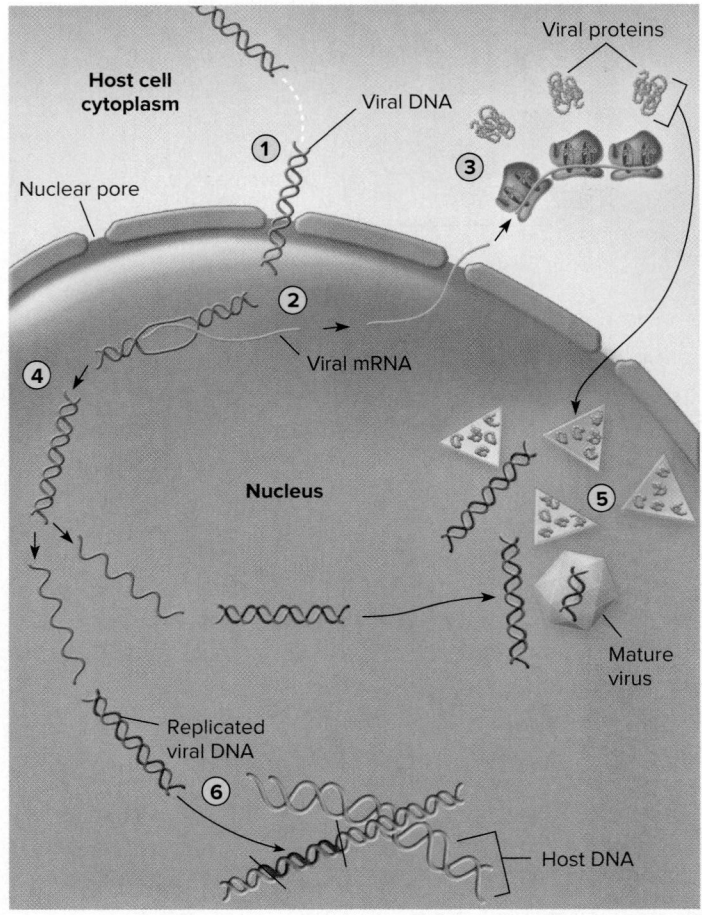

Figure 9.30 General stages in the multiplication of double-stranded DNA viruses. The virus penetrates the host cell and releases DNA, which, **(1)** enters the nucleus and **(2)** is transcribed by host cell enzymes into mRNA. **(3)** Viral mRNA leaves the nucleus and is translated into structural proteins; these proteins are transported into the nucleus. **(4)** Viral DNA is replicated repeatedly in the nucleus. **(5)** Viral DNA and proteins are assembled into a mature virus in the nucleus. **(6)** Because it is double-stranded, some viral DNA can insert itself into host DNA (latency). Some exceptions to this pattern occur in the poxviruses.

translated into proteins required to form the capsid and other structures. The new viral genomes and capsids are assembled, and the mature viruses are released by budding or cell disintegration.

Double-stranded DNA viruses can interact directly with the DNA of their host cell. In some viruses, the viral DNA becomes silently *integrated* into the host's genome by insertion at a particular site on the host genome. Persistence of viral DNA may also lead to the transformation of the host cell into a cancer cell. Several DNA viruses, including hepatitis B (HBV), the herpesviruses, and papillomaviruses (warts), are known to be initiators of cancers and are thus termed **oncogenic.** The mechanisms of oncogenic transformation involve viral genes that regulate cellular genomes and control the cell division cycle.

Replication, Transcription, and Translation of RNA Viruses

RNA viruses exhibit several differences from DNA viruses. Their genomes are smaller, they enter the host cell already in an RNA form, and the virus cycle occurs entirely in the cytoplasm for most viruses. RNA viruses can have one of the following genetic messages:

1. a positive-strand (+) genome that comes ready to be translated into proteins,
2. a negative-strand (−) genome that must be converted to a positive-strand before translation, and
3. a positive-strand (+) genome that can be converted to DNA or a dsRNA genome.

Positive-Strand Single-Stranded RNA Viruses

The ssRNA of positive-strand viruses such as polio come ready to be translated, so it is immediately translated into a large protein and cleaved into individual functional units. One of these is an RNA polymerase that initiates the replication of the viral strand. Replication of a positive-message strand happens in two steps **(figure 9.31).** First, a negative strand is synthesized using the positive strand as a template by the usual base-pairing mechanism. The resultant negative strand becomes a master template that formats the formation of new positive daughter strands. Further translation of the viral genome produces large numbers of structural proteins for final assembly and maturation of the virus.

RNA Viruses with Reverse Transcriptase: Retroviruses

A most unusual class of viruses can reverse the order of the flow of genetic information. Thus far in our discussion, all genetic entities have shown the patterns DNA → DNA, DNA → RNA, or RNA → RNA. Retroviruses, including HIV, the cause of AIDS, and HTLV I, a cause of one type of human leukemia, synthesize DNA using their RNA genome as a template. They accomplish this by means of a **reverse transcriptase** that comes packaged with each virus particle. This enzyme synthesizes a single-stranded DNA against the viral RNA template and then directs the formation of a complementary strand of this ssDNA, resulting in a double strand of viral DNA. The dsDNA strand enters the nucleus, where it can be

Figure 9.31 Replication of positive-strand, single-stranded RNA viruses. In general, these viruses do not enter the nucleus. **(1)** Penetration of viral RNA. **(2)** Because it is positive in message and single-stranded, the RNA can be directly translated on host cell ribosomes into various necessary viral proteins. **(3)** A negative genome is synthesized against the positive template to produce large numbers of positive genomes for final assembly. **(4)** The negative template is then used to synthesize a series of positive replicates. **(5)** RNA strands and proteins assemble into mature viruse released when the cell lyses.

transcribed by the usual mechanisms into new viral ssRNA and used to assemble new viral particles. When the DNA of some retroviruses becomes inserted into the host's DNA as a provirus, cells may be transformed and produce tumors. Insertion allows HIV to remain latent in an infected cell for several years.

 Check Your Progress SECTION 9.6

29. Explain why some viruses must enter the nucleus to replicate whereas others do not.
30. Explain the difference between positive-strand and negative-strand viral genomes.
31. Outline the basic steps in the replication cycles of a DS DNA virus and an RNA virus.

CASE STUDY Part 2

Cases of *A. baumannii* infections were first reported in the 1970s. At that time it was not considered a serious pathogen, and it was sensitive to most available drugs. How did it go from being an uncommon infectious agent to one that accounts for 20% of infections acquired in intensive care units? How did the number of MDR strains increase from 3% to 40%? It had not developed greater virulence.* It is actually relatively harmless for most healthy people with intact immune systems. It is an opportunist that attacks mainly immunocompromised patients with various underlying medical problems. This pathogen infects a variety of organs and systems, including respiratory, skin, urinary, and the bloodstream. Medical specialists suspect that its rapid evolution can be explained by the combination of four factors: the widespread distribution of the bacteria in the environment, the drug-rich environment of hospitals and clinics, the increase in severely compromised patients, and the pathogen's extreme genetic adaptability to challenges from new drugs.

For many years the rise of "superbugs" has become a major concern of the medical community. The working definition of this term includes any bacterial species that have developed significant resistance to several drugs that would ordinarily have been given in therapy. Two of the *A. baumannii* strains described here are defined as multiple drug resistant (MDR), in that they resist a fairly broad range of drugs. One strain fits the category of extreme drug resistance (XDR), meaning that it is resistant to all possible drugs. Both versions of this pathogen have been added to the list of emerging diseases.

- *In what ways is Acinetobacter an opportunist?*
- *Which genetic events have allowed Acinetobacter to become so resistant?*

To conclude this Case Study, go to Connect.

These topics are discussed further in chapter 12. To read more about the medical implications of this case, go to http://emedicine.medscape.com/article/236891-overview.

** virulence* (veer'-yoo-lents) Ability of a microbe to invade and damage a host.

Chapter Summary with Key Terms

9.1 Introduction to Genetics and Genes: Unlocking the Secrets of Heredity

A. Genetics is the study of heredity and can be studied at the level of the organism, genome, chromosome, gene, and DNA. Genes provide the information needed to construct proteins, which have structural or catalytic functions in the cell.

B. DNA is a long molecule in the form of a double helix. Each strand of the helix consists of a string of nucleotides.
 1. A **nucleotide** consists of a phosphate, **deoxyribose sugar,** and **nitrogen base.** The bases form **hydrogen bonds** with their counterparts on the other strand.
 2. **Adenine** base-pairs with **thymine** while **guanine** base-pairs with **cytosine.**
 3. The order of the nucleotides specifies which amino acids will be used to construct proteins during the process of translation.

C. **Replication** of DNA–making copies of the genetic program.
 1. Requires unwinding the double helix and exposing the nucleotides to serve as templates for synthesis of 2 identical molecules by **DNA polymerase.**
 2. Is **semiconservative** and proceeds in the 5′ to 3′ direction of the newly synthesized DNA.

9.2 Applications of the DNA Code: Transcription and Translation

A. DNA is used to produce RNA (**transcription**) and RNA is then used to produce protein (**translation**).

B. **RNA:** Unlike DNA, RNA is single stranded, contains **uracil** instead of thymine and **ribose** instead of deoxyribose.
 1. Major forms of RNA found in the cell include **mRNA, tRNA,** and **rRNA.**
 2. The genetic information contained in DNA is copied to produce an RNA molecule. **Codons** in the mRNA pair with **anticodons** on the tRNA to specify which amino acids to assemble on the **ribosome** during translation.

C. *Transcription and Translation:*
 1. Transcription occurs when **RNA polymerase** copies the template strand of a segment of DNA. RNA is always made in the 5′ to 3′ direction.
 2. Translation occurs when the mRNA is used to direct the synthesis of proteins on the ribosome.
 3. Codons in the mRNA pair with anticodons on the tRNA to assemble a string of amino acids. This occurs until a stop codon is reached.

D. *Eukaryotic Gene Expression:* Eukaryotic genes are composed of **exons** (expressed sequences) and **introns** (intragenic regions). The introns must be removed and the exons spliced together to create the final mRNA.

9.3 Genetic Regulation of Protein Synthesis and Metabolism

Protein synthesis is regulated through gene induction or repression, as controlled by an **operon.** Operons consist of several structural genes controlled by a common regulatory element.

A. **Inducible operons** such as the lactose operon are normally off but can be turned on by a lactose inducer.

B. **Repressible operons** are usually on but can be turned off when their end product is no longer needed.

C. Newly described types of RNA can serve in gene regulation.
 1. Micro and small interfering RNAs can block the action of other RNAs and shut down gene expression in eukaryotes.
 2. **Riboswitches** can also control RNA functions related to transcription and translation in bacteria.

9.4 Mutations: Changes in the Genetic Code
 A. Changes in the genetic code occur primarily through mutation and recombination.
 B. Inheritable changes in the genome of a microorganism are known as mutations. They usually entail an alteration in the nucleotide composition of the genetic material. Mutations may be **spontaneous** or **induced.**
 C. Mutations are detected by the use of selective media and the replica plating technique.
 D. Point mutations entail a change in one or a few bases and are categorized as **missense, nonsense, silent** mutations, and **back-mutations,** based on the effect of the change in nucleotide(s).
 E. **Frame-shift mutations** that reset the reading frame and alter downstream coding structure are caused by **deletions** and **insertions** of nucleotides.
 F. The **Ames test** measures the mutagenicity of chemicals by their ability to induce mutations in bacteria.

9.5 DNA Recombination Events
 A. **Recombination** involves the acquisition of new genetic material from an outside source.

 B. Intermicrobial transfer and genetic recombination permit gene sharing between bacteria. Major types of recombination include **conjugation, transformation,** and **transduction.**
 C. **Transposons** are DNA sequences that regularly move to different places within the genome of a cell, generating mutations and variations in chromosome structure.

9.6 The Genetics of Animal Viruses
 A. Viruses contain various forms of DNA and RNA that are transcribed and translated by the genetic machinery of their host cells to form functioning viral particles.
 1. Genomes of viruses are found in double-stranded (ds) DNA and single-stranded (ss) RNA forms. A few contain unusual ssDNA and dsRNA not found in cells.
 2. DNA viruses tend to replicate in the nucleus while RNA viruses replicate in the cytoplasm. **Retroviruses** synthesize dsDNA using ssRNA as a template.

Level I. Knowledge and Comprehension

These questions require a working knowledge of the concepts in the chapter and the ability to recall and understand the information you have studied.

 Multiple-Choice Questions

Select the correct answer from the answers provided. For questions with blanks, choose the combination of answers that most accurately completes the statement.

1. What is the smallest unit of heredity (genotype)?
 a. chromosome c. codon
 b. gene d. nucleotide

2. A nucleotide contains which of the following?
 a. 5-carbon sugar d. b and c only
 b. nitrogen base e. all of these
 c. phosphate

3. The nitrogen bases in DNA are bonded to the
 a. phosphate c. ribose
 b. deoxyribose d. hydrogen

4. DNA replication is considered semiconservative because the _____ strand will remain as half of the _____ molecule.
 a. antisense, DNA c. sense, mRNA
 b. template, final DNA d. codon, anticodon

5. In DNA, adenine is the complementary base for _____ ,and cytosine is the complement for _____
 a. guanine, thymine c. thymine, guanine
 b. uracil, guanine d. thymine, uracil

6. The base pairs are held together primarily by
 a. covalent bonds c. hydrogen bonds
 b. ionic bonds d. gyrases

7. Why must the lagging strand of DNA be replicated in short pieces?
 a. because of limited space
 b. otherwise, the helix will become distorted
 c. the DNA polymerase can synthesize in only one direction
 d. to make proofreading of code easier

8. Messenger RNA is formed by _____ of a gene on the DNA template strand.
 a. transcription c. translation
 b. replication d. transformation

9. Transfer RNA is the molecule that
 a. contributes to the structure of ribosomes
 b. adapts the genetic code to protein structure
 c. transfers the DNA code to mRNA
 d. provides the master code for amino acids

10. As a general rule, the first codon on a DNA template strand that signals the start for a protein is
 a. TAC c. ATG
 b. AUG d. UAC

11. What components are involved in transcription?
 a. sigma factor d. only b and c
 b. RNA polymerase e. all of these
 c. promotor

12. The *lac* operon is usually in the _____ position and is activated by a/an _____ molecule.
 a. on, repressor c. on, inducer
 b. off, inducer d. off, repressor

13. For mutations to have an effect on populations of microbes, they must be
 a. inheritable d. a and b
 b. permanent e. all of these
 c. beneficial

14. Which of the following characteristics is *not* true of a plasmid?
 a. It is a circular piece of DNA.
 b. It is required for normal cell function.
 c. It is found in bacteria.
 d. It can be transferred from cell to cell.

15. Which genetic material could be transmitted naturally by all three methods of intermicrobial transfer?
 a. an *E. coli* chromosome
 b. F factor
 c. drug resistance genes
 d. a and c
 e. all of these

16. Which of these features is found in eukaryotes but not bacteria?
 a. polygene mRNAs c. stop codon
 b. introns d. AUG codon

17. Which of the following is present in prokaryotes but not eukaryotes?
 a. exons
 b. polyribosomes
 c. spliceosomes
 d. simultaneous transcription and translation

18. **Multiple Matching.** Fill in the blanks with all the letters of the words below that apply.
 _____ site of protein synthesis
 _____ carries the codon
 _____ carries the anticodon
 _____ a process synonymous with mRNA synthesis
 _____ bacteriophages participate in this transfer
 _____ duplication of the DNA molecule
 _____ process in which transcribed DNA code is deciphered into a polypeptide
 _____ involves plasmids
 a. replication
 b. tRNA
 c. conjugation
 d. ribosome
 e. transduction
 f. mRNA
 g. transcription
 h. transformation
 i. translation
 j. none of these

Case Study Review

1. Which of the following is not a factor in the rise of drug-resistant *Acinetobacter baumannii*?
 a. It is very widespread in many habitats.
 b. It has acquired greater toxicity.
 c. It readily acquires genetic material from other species.
 d. The hospital environment brings together vulnerable patients and drugs.

2. Drug-resistance plasmids could be transferred in *A. baumannii* by which of the following mechanisms:
 a. conjugation pilus d. a and b
 b. bacteriophages e. a, b, and c
 c. capture by cell wall

3. Explain how it would be possible for *A. baumannii* to develop drug resistance to a drug in the body of a patient while he or she is receiving therapy.

Writing Challenge

For each question, compose a one- or two-paragraph answer that includes the factual information needed to completely address the question. Check Your Progress questions can also be used for writing-challenge exercises.

1. Explain what is meant by the following diagram.

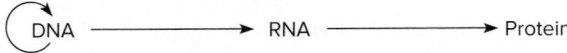

2. On paper, replicate the following segment of DNA:

 5' A T C G G C T A C G T T C A C 3'
 3' T A G C C G A T G C A A G T G 5'

 a. Show the direction of replication of the new strands and explain what causes lagging and leading strands to show different patterns of replication.
 b. Explain how this is semiconservative replication. Are the new strands identical to the original segment of DNA?

3. The following sequence represents triplets on DNA:

 TAC CAG ATA CAC TCC CCT GCG ACT

 a. Give the mRNA codons and tRNA anticodons that correspond with this sequence, and then give the sequence of amino acids in the polypeptide.
 b. Provide a different mRNA strand that can be used to synthesize this same protein.

4. Describe the actions of all of the enzymes involved in replication of a DNA molecule.

5. Compare and contrast the actions of DNA polymerase and RNA polymerase.

6. Examine the following series of words and identify what kind of a mutation occurred in each, starting with the first word in non-mutated form: BEAST FEAST BREAST BEST BEATS

 Concept Mapping

On Connect you can find an Introduction to Concept Mapping that provides guidance for working with concept maps and a sample of a concept map for this chapter.

 Level II. Application, Analysis, Evaluation, and Synthesis

These problems go beyond just restating facts and require higher levels of understanding and an ability to interpret, problem solve, transfer knowledge to new situations, create models, and predict outcomes.

 Critical Thinking

Critical thinking is the ability to reason and solve problems using facts and concepts. These questions can be approached from a number of angles, and in most cases, they do not have a single correct answer.

1. Knowing that retroviruses operate on the principle of reversing the direction of transcription from RNA to DNA, propose a drug that might possibly interfere with their replication.

2. Using the piece of DNA in writing-challenge question 3, show a deletion, an insertion, a substitution, and nonsense mutations. Which ones are frameshift mutations? Are any of your mutations nonsense? Are any missense? (Use the universal code to determine this.)

3. Why will a mistake in the RNA code alone not become a mutation?

4. The enzymes required to carry out transcription and translation are themselves produced through these same processes. Speculate which may have come first in evolution—proteins or nucleic acids—and explain your choice.

5. Why can one not reliably predict the sequence of nucleotides on mRNA or DNA by observing the amino acid sequence of proteins?

6. Activation, transcription, and translation of the *lac* operon will release which exact nutrients?

7. Explain the mechanisms by which RNA can control transcription, translation, and gene expression.

8. Explain how epigenetics is related to the expression: phenotype = genotype + environment.

9. Use the concepts of chapters, letters, a whole library, words, and books to make an analogy that fits the levels of genetic structure and function.

 Visual Challenge

1. From **figure 9.17, step 3.** Label each part of the illustration.

2. Examine **figure 8.11,** and explain which type of operon is functioning here. How would another operon differ from this?

 STUDY TIP: Quickly create a customized practice quiz for this chapter by going to your Connect SmartBook "My Reports" section. Don't have SmartBook and want to learn more? Go to **whysmartbook.com.**

10 Genetic Engineering:
A Revolution in Molecular Biology

DNA sequence readings for the normal and mutated forms of the *XIAP* gene on the X chromosome. Top is a normal reference with guanine (black G) at the site; middle is Nicholas's reading with a single mutation that substitutes an adenine (green A), and bottom is Nicholas's mother, who has a normal gene on one X chromosome and a mutated gene on the other X chromosome. (right): *Source:* Jane Ades, NHGRI

CASE STUDY Part 1 Nicholas's Story: A New Frontier in Genomic Medicine

In 2007 a desperately ill 2-year-old boy was brought to the Medical College of Wisconsin for consultation about his condition, which appeared to be a type of **inflammatory bowel disease** (IBD). The child, Nicholas Volker, had symptoms of **abcesses** and penetrating **fistulas*** in his lower gastrointestinal tract that led to severely **stunted development** and **chronic infections.**

The doctors tried everything in their arsenal to control his disease. Over time, Nicholas had hundreds of surgeries in an attempt to repair the fistulas, and he was placed on a whole range of antimicrobial and **immunosuppressive drugs.** The doctors redirected his colon with a **colostomy** and placed him on **intravenous nutrition.** After bouts of sepsis and continued deterioration of his lower intestine and abdominal cavity, they finally removed his entire colon. After each intervention, he would be relatively free of symptoms for a few months, but then he would experience a sudden setback, which was often **life-threatening.** For most of his young life he had been unable to eat solid food and so far had spent 600 days in the hospital.

Nicholas's situation became so urgent that his physicians realized they must dig deeper to find the root cause of this complex medical mystery. They agreed that he was suffering from some form of **inborn genetic disease** that caused chronic inflammation, inter-

> "Finding that mutation hidden among 6.4 billion nucleotides of the human genome seemed like an impossible task."

fered with the normal **immune protection** in his colon, and prevented healing. They started with a detailed battery of tests for immune function and compared his profile with known causes of inflammatory bowel disease. After months of analysis, they were unable to link the condition to any known defects, which meant that it was likely due to some type of rare, **unknown mutation.** Finding that mutation hidden among **6.4 billion nucleotides** of the human genome seemed like an impossible task—a one-in-a-billion gamble—that that was Nicholas's only hope.

The medical team immediately made preparations to **sequence** his genome and locate the exact genetic defect, something that had been attempted only once before on a much smaller scale. One advantage lay in the availability of the entire sequence of the **human genome** completed several years before. Its huge **database** could be used to compare genomic information, and the newer sequencing machines were becoming cheaper and faster than ever before. This search would put the emerging technology of **genomics** and **informatics** to the test and become one of the earliest uses of **genomic diagnosis.**

- *What is meant by an inborn genetic disease?*
- *What is involved in DNA sequencing?*

* *fistula* (fiss'-chew-lah) An abnormal passageway between a hollow organ and the skin.

To continue the Case Study, go to Case Study Part 2 at the end of the chapter.

SCOPING OUT THE CHAPTER

The advances in DNA science, whether medical, industrial, agricultural, or forensic, are a dominant force in many areas of our lives. None of these applications would exist if it were not for studies completed on microorganisms over the past 50 years. Simple basic research on bacteria and viruses first opened up the possibilities of modifying and manipulating DNA. Out of this came valuable tools for analyzing, cloning, cutting, splicing, hybridizing, probing, sequencing, synthesizing, and identifying DNA. Often, this knowledge has made what seemed impossible, possible. Look below for a few of the highlights.

Manipulating DNA. DNA can be removed intact from cells and altered by unzipping, cutting, splicing, and adding tracer molecules to locate and identify genes.

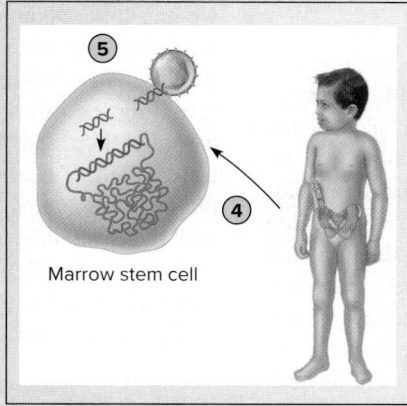

Medicine for Genetic Disorders. New technologies have the potential for correcting genetic disorders. With gene therapy, for example, recombinant microbes containing the healthy gene are inserted into cells of afflicted patients.

Techniques for Amplifying and Sequencing DNA. By imitating the processes in cells, DNA can be copied in a test tube. Then, specialized machines can read out the precise nucleotide codes of this isolated DNA.

Recombinant DNA and Cloned Genes. Selected DNA from two different organisms can be combined and introduced into a host cell, creating an entirely new genetic combination or recombinant organism.

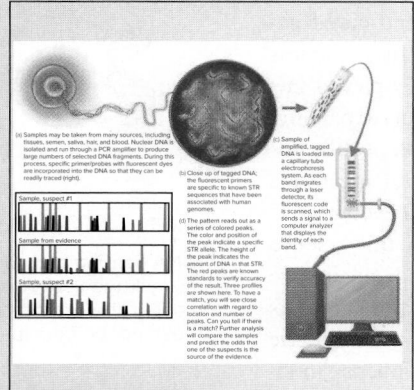

DNA Profiling and Identification. Many areas of discovery benefit from an abbreviated "picture" of DNA that displays the unique signature of a person, microbe, or unknown DNA sample. Often termed a DNA profile or fingerprint, this technique is an essential part of forensics, medicine, and research.

Genetically-modified Organisms (GMO). Animals and plants can have new genes inserted into their genomes, creating novel and highly unique forms for agricultural, medical, and research purposes.

(Recombination DNA and Cloned Genes) Jaana Tammur, Hiljar Sibul, Ene Ustav, Mart Ustav, Andres Metspalu, "A bovine papillomavirus-1 based vector restores the function of the low-density lipoprotein receptor in the receptor-deficient CHO-ldlA7 cell line," *BMC Molecular Biology* 2002, 3:5

10.1 Basic Elements and Applications of Genetic Engineering

 Expected Learning Outcomes

1. Define *genetic engineering,* and describe some of its subcategories and aims.
2. Explain the properties of DNA that lend to its manipulation in the laboratory.
3. Summarize the major methods of analyzing DNA and their results.
4. Describe the technology behind identifying, synthesizing, sequencing, and amplifying DNA.

The knowledge of DNA that has emerged since its structure was decoded 60 years ago has spawned a revolution in a wide variety of human endeavors. In many ways it has changed our view of the world, and in fact it has profoundly changed the world. It now dominates research and development in hundreds of specialties in biotechnology, medicine, forensics, diagnosis of disease, drug development and treatment, agriculture, industrial production, and ecology. Today it is possible to accomplish feats that were once only imagined. The new wave of DNA explorers have gone back in history to identify the remains of the last czar of Russia and King Richard III of England. They have extended their searches even to prehistoric eras by analyzing the genomes of the Neanderthals and a woolly mammoth. They have changed the face of agriculture by developing plants that resist disease and animals that can synthesize human hormones, and they have created entirely new strains of viruses and bacteria. Applications of this technology have diagnosed and cured genetic diseases, identified the mass victims of disasters, and become a key part of law enforcement. Because of our expanding knowledge, we are on the brink of fully understanding the complexities of life forms and their capacity to adapt and evolve.

Source: Office of Biological and Environmental Research of the U.S. Department of Energy Office of Science. *http://genomicscience.energy.gov*

The primary goals of this science are to isolate, identify, and manipulate DNA in various ways. A major part of its work falls into the category of **genetic engineering** or **bioengineering,** defined as the direct, deliberate modification of an organism's genome. It also belongs to the larger scheme of **synthetic biology** and **biotechnology,** whose major aims are the design of biological systems that can synthesize new forms of molecules, cells, organs, and even organisms.

Information on genetic engineering and its biotechnological applications is growing at such an expanding rate that some new discovery or product is disclosed almost daily. To keep this subject somewhat manageable, we present essential concepts and applications, organized under the following six topics:

- Basic techniques in manipulating and analyzing DNA
- Recombinant DNA technology
- Biochemical products of recombinant DNA technology
- Genetically modified organisms
- Genetic treatments
- Genome analysis

Tools and Techniques of DNA Technology

In chapter 9 we examined the ways in which microorganisms duplicate, change, and express their genomes. All of the intrinsic properties of DNA that we touched on hold true whether the DNA is in a bacterium or a test tube. For example, the enzyme helicase is able to unwind the two strands of the double helix just as easily inside or outside a bacterial cell. But in the laboratory we can take advantage of our knowledge of DNA chemistry to make helicase unnecessary. It turns out that when DNA is heated to just below boiling (90°C to 95°C), the two strands separate, revealing the information contained in their bases. With the nucleotides exposed, DNA can be easily identified, replicated, or transcribed. If heat-denatured DNA is then slowly cooled, complementary nucleotides will hydrogen-bond with one another, and the strands will renature or regain their familiar double-stranded form **(figure 10.1a).** As we shall see, this process is a necessary feature of the polymerase chain reaction and nucleic acid probes described later in this section.

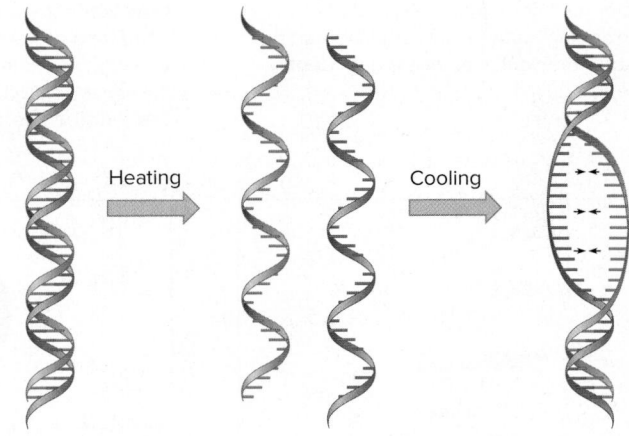

(a) **DNA heating and cooling.** DNA responds to heat by denaturing—losing its hydrogen bonding and thereby separating into its two strands. When cooled, the two strands rejoin at complementary regions. The two strands need not be from the same organisms as long as they have matching sites.

Endonuclease	EcoRI	HindIII	HaeIII
Cutting pattern	G A A T T C C T T A A G	A A G C T T T T C G A A	G G C C C C G G

(b) **Examples of endonucleases, palindromes and cutting patterns.** The first two are staggered cuts, and the third is a blunt cut.

Figure 10.1 Some properties of DNA that are an integral part of DNA manipulation. The method in **(a)** can be used to separate the DNA strands lengthwise and make them single-stranded. The method in **(b)** uses enzymes called endonucleases to make cuts across both DNA strands at specific sites called palindromes.

Enzymes for Dicing, Splicing, and Reversing Nucleic Acids

The polynucleotide strands of DNA can also be clipped crosswise at selected positions by means of **restriction endonucleases**[1] that are a natural product of bacterial cells. They recognize foreign DNA and are capable of breaking the phosphodiester bonds between adjacent nucleotides on both strands of DNA, leading to a break in the DNA strand. In the bacterial cell, these enzymes protect against the incompatible DNA from bacteriophages or plasmids. In the biotechnologist's lab, the enzymes can be used to cleave DNA at desired sites and are a must for the techniques of recombinant DNA technology.

3-D view of restriction endonuclease cutting DNA (yellow strands).
Courtesy Konrad F- Koehler, PhD

So far about 3,000 restriction endonucleases have been discovered in bacteria. Each type has a known sequence of 4 to 10 base pairs as its target, so sites of cutting can be finely controlled. These enzymes have the unique property of recognizing and clipping at sites called **palindromes**[2] **(figure 10.1b).** Palindromes are sequences of DNA bases that are identical when read from the 5′ to 3′ direction on one strand and the opposite direction on the complementary strand.

Endonucleases are named by combining the first letter of the bacterial genus, the first two letters of the species, and the endonuclease number. Thus, *Eco*RI is the first endonuclease found in *Escherichia coli's* R strain, and *Hin*dIII is the third endonuclease discovered in *Haemophilus influenzae* Type d (figure 10.1b).

Endonucleases are used in the laboratory to cut DNA into smaller pieces for further study as well as to remove and insert it during recombinant DNA techniques, described in section 10.2. Usually the enzymes make staggered symmetrical cuts that leave short, single-stranded tails called "sticky ends." Such adhesive tails will base-pair with complementary tails on other DNA fragments or plasmids. This effect makes it possible to splice genes into specific sites **(figure 10.2).** A few endonucleases such as *Hae*III make straight, blunt cuts on DNA and do not form sticky ends.

The separated segments of DNA produced by restriction endonucleases are termed **restriction fragments.** Because DNA sequences vary, even among members of the same species, differences in the cutting pattern of specific restriction endonucleases give rise to restriction fragments of differing lengths, known as restriction fragment length polymorphisms (RFLPs). RFLPs allow the direct comparison of the DNA cutting patterns of two different organisms with the same endonuclease, which is one step in preparing genes from different organisms to be spliced together (see figure 10.11).

Quick Search
Find the video "Cloning in a Plasmid Vector" on the Internet to see the process.

Action of restriction endonucleases. (1) A restriction endonuclease recognizes and cleaves DNA at the site of a specific palindromic sequence. Cleavage usually produces staggered tails called sticky ends that accept complementary tails for gene splicing. This palindrome is cut by *Aci* I. (2) The sticky ends can be used to join DNA from different organisms by cutting it with the same restriction enzyme, ensuring that both fragments have two complementary ends. This palindrome is cut by *Taq* I.

Figure 10.2 Examples of cutting patterns by restriction endonucleases and how they are used in gene splicing.

Another enzyme, called a **ligase,** is necessary to seal the sticky ends together by rejoining the phosphate-sugar bonds cut by endonucleases. Its main application is in final splicing of genes into plasmids and chromosomes.

An enzyme called **reverse transcriptase** is best known for its role in the replication of HIV and other retroviruses. It also provides geneticists with a valuable tool for converting RNA into DNA. Copies called **complementary DNA,** or **cDNA,** can be made from messenger, transfer, ribosomal, and other forms of RNA. The technique provides a valuable means of synthesizing eukaryotic genes from processed mRNA transcripts **(figure 10.3).** The bacteria often used to transcribe and translate eukaryotic genes in genetic engineering lack the cellular machinery to splice out introns (intragenic sequences) found in unprocessed eukaryotic mRNA. So the advantage with synthesized cDNA genes is that they already have the introns removed and consist only of exons and can be readily transcribed and translated by bacteria. cDNA also provides a powerful tool for converting RNA for sequencing, analyzing ribosomes and identifying RNA viruses.

Analysis of DNA

One way to produce a readable pattern of DNA fragments is through **electrophoresis (figure 10.4).** In this technique, samples are placed in compartments (wells) in a soft agar gel or similar soft medium and

1. The meaning of restriction is that the enzymes do not act upon the bacterium's own DNA; an endonuclease nicks DNA internally, not at the ends.

2. This term originated from words or phrases with letters that read the same backward or forward, as in *radar,* "Madam, I'm Adam," and "Can I attain a C?"

Figure 10.3 **Synthesizing cDNA from eukaryotic mRNA.** Bacterial cloning hosts cannot process intact eukaryotic genes. To provide a gene that will be transcribable and translatable by a bacterium, **(a)** first modified mRNA, lacking introns, is isolated from the source eukaryotic cell. **(b)** This strand is converted in a test tube to a single-stranded DNA by reverse transcriptase and double-stranded DNA by a polymerase. This copy of DNA (cDNA) can be transcribed and translated by bacteria.

Figure 10.4 **Electrophoresis setup.** Electrophoresis uses an electrical current to separate charged molecules of different sizes. The samples are loaded into small wells in a gel bathed in a buffer solution. The gel contains tiny pores that allow the passage of molecules according to their size. Current and movement go from the negative pole to the positive pole. The bands produced by the fragments are arrayed along a range from larger fragments higher up and smaller ones nearer the bottom.

subjected to an electrical current. The phosphate groups in DNA give the entire molecule an overall negative charge, which causes the DNA to move toward the positive pole in the gel. The rate of movement is based primarily on the size of the fragments. The larger fragments are pulled into the gel more slowly and remain nearer the wells, whereas the smaller fragments migrate faster and are positioned farther from the wells. The positions of DNA fragments are visualized by one of several methods **(figure 10.5).** Electrophoresis patterns can be quite distinctive and are very useful in displaying DNA fragments and comparing the degree of genetic similarities among samples in a Southern blot or DNA profile (see Visual Challenge #1).

Nucleic Acid Hybridization and Probes

Two different nucleic acids can **hybridize** by uniting at their complementary sites. All different combinations are possible: Single-stranded DNA can unite with other single-stranded DNA or RNA, and RNA can hybridize with other RNA. This knowledge has led to the development of short, single strands of 6 to 25 nucleotides called *oligonucleotides* to be used as **gene probes.** These probes consist of a segment of DNA or RNA of known sequence that can base-pair with a stretch of DNA that has a complementary sequence, if one exists in a test sample. Hybridization probes have practical value because they can detect specific nucleotide sequences in unknown samples. So that areas of hybridization can be visualized, the probes carry reporter molecules, such as fluorescent dyes, which can be visualized with ultraviolet radiation (see figure 10.6), or radioactive labels, which are revealed by placing photographic film in contact with the test reaction (figure 10.5b).

When probes hybridize with an unknown sample of DNA or RNA, they tag the precise area and degree of hybridization, and help determine the nature of nucleic acid present in a sample. In a method called the **Southern blot,**[3] DNA fragments are first separated by electrophoresis and then denatured and transferred to a special filter. A DNA probe is then incubated with the sample, and wherever this probe encounters the segment for which it is complementary, it will attach and form a hybrid. Development of the hybridization pattern will show up as one or more bands. This method is a sensitive and specific way to isolate fragments from a complex mixture and to find specific gene sequences on DNA. Much of the information obtained through Southern blotting can be achieved in more detail from another hybridization procedure called a microarray, discussed in section 10.4.

Probes have numerous applications in diagnosing the cause of an infection from a patient's specimen or identifying a culture of an unknown bacterium or virus (see chapter 17). One example is a simple and rapid method called a *hybridization test*. DNA from a test sample is isolated, denatured, placed on an absorbent filter, and combined with a microbe-specific probe. The blot is then developed and observed for areas of hybridization. Commercially available diagnostic kits are now on the market for identifying intestinal pathogens such as *E. coli, Salmonella, Campylobacter, Shigella, Clostridium difficile,* rotaviruses, and adenoviruses. Other bacterial probes exist for *Mycobacterium, Neisseria, Staphylococcus, Mycoplasma, Bordetella,*

3. Named for its developer, E. M. Southern. The "northern" blot is a similar method used to analyze RNA, whereas the "western" blot detects proteins. There is a reason scientists became scientists and not comedians.

Figure 10.5 **Separating DNA restriction fragments with electrophoresis. (a)** A restriction endonuclease is applied to a sample (using sample 3 as an example). The sample with DNA fragments is loaded into a well at the top of a gel and electrophoresed. The final position of the fragments will be visualized by staining or the method shown in (b). **(b)** A developed gel reveals the separated bands of DNA fragments. The size of a given DNA band can be determined by comparing the distance traveled to the distance traveled by a set of DNA fragments of known size called a size ladder (lane 5), measured in number of base pairs. This gel contained DNA with radioactive bases that can expose a photographic negative. *(b)* © Kathy Park Talaro

and *Chlamydia;* probes are available for herpes simplex and zoster, papilloma, hepatitis A and B, and HIV viruses. DNA probes have also been developed for human genetic markers and some types of cancer.

With another method, called *fluorescent in situ hybridization* (FISH), probes are applied to intact cells *in situ,* meaning that the DNA is left intact in the cell and probed in place rather than being extracted. The preparation is then observed microscopically for signs of genetic marker sequences on genes that have hybridized with the probe. FISH is a very effective way to locate genes on chromosomes **(figure 10.6).** *In situ* techniques can also be used to identify unknown

Figure 10.6 **Fluorescent in situ hybridization (FISH) technique.** This is a microscopic technique using fluorescently labeled probes to react with intact cells and chromosomes. The probe is mixed with test cells or chromosomes to hybridize any matching DNA sequences and observed with a UV microscope for fluorescence indicative of a particular location and type of gene. The inset displays chromosomes that have DNA from a papillomavirus integrated into chromosomes (yellow dots). Jaana Tammur, Hiljar Sibul, Ene Ustav, Mart Ustav, Andres Metspalu, "A bovine papillomavirus-1 based vector restores the function of the low-density lipoprotein receptor in the receptor-deficient CHO-ldlA7 cell line," *BMC Molecular Biology* 2002, 3:5

bacteria living in natural habitats without having to culture them, and they can be used to detect RNA in cells and tissues.

DNA Sizing, Sequencing, and Synthesizing

The relative sizes of nucleic acids are usually denoted by the number of base pairs (bp) or nucleotides they contain. For example, the palindromic sequences recognized by endonucleases are usually 4 to 10 bp in length; an average gene in *E. coli* is approximately 1,300 bp, or 1.3 kilobases (kb); and its entire genome is approximately 4,700,000 bp, 4,700 kb, or 4.7 megabases (mb). The DNA of the human mitochondrion contains 16 kb, and the Epstein-Barr virus (a cause of infectious mononucleosis) has 172 kb. Humans have approximately 6.4 billion base pairs (bbp) arrayed along 46 chromosomes. The Human Genome Project, finished in 2001, produced a map containing every base in the entire human genome (**10.1 Making Connections**).

DNA Sequencing: Determining the Exact Genetic Code One of the most powerful kinds of genetic knowledge attainable for organisms is to decipher the order and types of base pairs in their genes and genome. This process, called **DNA sequencing,** can be used on all types of DNA, including genomic, cDNA, chromosomal, plasmid, and mitochondrial DNA. Many sequencing techniques are based on Frederick Sanger's original idea, using the synthesis and analysis of a complementary strand of DNA (**figure 10.7**). Even though mostly automated techniques are used today, we will illustrate the basic principles behind this process using the Sanger type of sequencing as a model.

Because most DNA being sequenced is usually very long, it is made more manageable by cutting it into a number of shorter fragments, which are then separated. The test strands, typically several hundred nucleotides long, are denatured to expose single strands that will serve as templates to synthesize complementary strands.

(1) Original DNA to be sequenced.

(2) DNA is denatured to produce single template strand.

(3) Strand is labeled with specific primer molecule.

(4) A reaction tube contains DNA polymerase, primed single strands, all four regular nucleotides, and a small amount of the four fluorescently labeled dideoxy (dd) nucleotides that provide a visible tracer of the locations of terminal bases added by the polymerase.

(5) During replication each new strand is terminated (↓) at the point of addition of a dd nucleotide.

(6) Left: A schematic view illustrating how finished fragments will end with a labeled dideoxynucleotide. Electrophoresis separates the labeled fragments and displays them in order by size. The color of the label gives the type of base in its exact sequence. Note that this sequence is not the final product – it is the complement of the template strand of interest.

(7) Automated display of a short segment of human DNA. The computer substitutes black for G for ease in reading.

(8)

Figure 10.7 **Some basic techniques in DNA sequencing** (1-6) The sequencing of a 9 base strand using the Sanger method. The sequence in #6 is read from the bottom to the top and must be converted to its complement to get the true nucleotide order of TCGGCTAGG. (7) An example of an automated read-out. (8) Enlarged Illustration of a nanopore sequencer depicting one strand of DNA threaded through a pore to be analyzed.

10.1 MAKING CONNECTIONS

OK, the Genome's Sequenced—What's Next?

In early 2001 a news conference was held to announce that the human genome had been sequenced. The 3.2 billion base pairs (which translates to 6.4 billion nucleotides) that make up the DNA found in (nearly) every human cell had been put in the proper order. Champagne corks popped, balloons fell, bands played, reporters reported on the significance of the occasion, and . . . nothing else seemed to change for a while. Here are some important FAQs about the Human Genome Project (HGP) and what has happened since that monumental event.

Q: How big was this project?

A: The Human Genome Project was first discussed in the mid-1980s and got under way in 1990. Although the project was to have taken at least 15 years, advances in technology led to its being completed in just over 10 years at a cost of about 3 billion dollars. It was finally determined that the 6.4 billion nucleotides of DNA code for only about 21,000 genes, not the 100,000 or so that the genome was thought to contain only a few years ago. Still, this is the equivalent of a million-page book—a long read.

Q: Will we ever see any benefits from this project?

A: We already are. Sequencing the genome was only the first step. Knowing what proteins are produced in the body, what they do, and how they interact with one another is essential to understanding the workings of human physiology. By knowing the genetic signatures of different diseases, we will be able to design extraordinarily specific diagnostic tests that detect thousands of genetic diseases and their many forms. For example, the cancer drugs Iressa and Erbitux are so specific in their actions that they are only prescribed after a genetic test of the cancer has found it to be a suitable target.

Q: What have we learned so far?

A: The first HGP maps sequenced the DNA of five humans of different sexes and ethnicities. As expensive as these first whole genome maps were, improvements in the technology have made the current cost more reasonable at about $5.000 per person. Keep in mind that such an undertaking will not yet provide complete information on health, future medical problems, and other genetic information. But as more data are made available, genomic sequencing will become a viable method of diagnosis (see this chapter's Case Study).

An important research finding that emerged from worldwide analyses of DNA samples was that the genome sequence is shared among 99.9% of humans and varies by only 0.1%. But even this 0.1% amounts to at least 1 million different haplotypes (a block of DNA inherited as a unit). Studying "hap maps" has provided new data to trace the incidence of and risk for genetic diseases, the genetic makeup of modern-day populations, and the possible origins of ancestral humans.

Another significant discovery coming from comparative genome analysis was that humans have about 2.4 million polymorphisms called single nucleotide polymorphisms or SNPs, which involve variations in just a single nucleotide at the same location in a gene, meaning they are alleles. SNPs are known to affect the onset of some genetic diseases (diabetes, cancer) and to cause variations in the response to drugs, vaccines, and infectious agents. For more information, go to http://www.ornl.gov/hgmis/ or log on to the National Human Genome Research Institute.

Once the sequence of bases in an organism's genome is known, what are the next logical steps in deciphering the information it contains?
Answer available on Connect.

SNP variations at a locus of a gene. There are over 3 billion of these polymorphisms in the human genome, enough to create significant differences in human development.

Usually these fragments must be increased in number by PCR (see figure 10.8). The concepts behind the sequencing procedure involve the same enzymes and raw materials as would be present for regular DNA synthesis as performed in a cell. The fragments are bound to an RNA primer, which positions the attachment of nucleotides at its 3′ end, then a DNA polymerase adds the complementary nucleotides using the pattern of the template strand as a guide.

The reaction tube also contains all of the regular nucleotides (A, T, C, G) and a small amount of dideoxynucleotides (ddNP) that have two special features: (1) They are color-labeled with four different fluorescent dyes that can report their position in a sequence, and (2) they are missing the hydroxyl group (OH^-) bound to the 3′ carbon of deoxyribose that is required for the chemical bonding of a nucleotide to the growing DNA strand. The addition of a ddNP at any location terminates the further elongation of the strand. As the synthesis proceeds, most strands will elongate normally, except for a small percentage of fragments that will randomly have a ddNP nucleotide incorporated and be terminated there. Late in the process, all possible positions in the sequence will have a terminal ddNP nucleotide. The reaction tube will thus have a series of strands reflecting the correct sequence that can be traced by its color from the first nucleotide to the last.

During analysis, the reaction products are subjected to electrophoresis in fine capillaries that separate strands differing by only a single nucleotide in length. This arrays the fragments in order of size, and only the fragments carrying the labeled nucleotides will be readable (figure 10.7). Automatic sequencing machines scan the electrophoresis pattern and record the DNA sequence by detecting the color of each band with a sensing device. In this way, the DNA sequence of the synthesized strand is analyzed and that of the template strand can easily be inferred. Most sequencers have computers that read the sequence and display it as a graphic figure. This method of sequencing is remarkably accurate, with only about one mistake in every 1,000 bases. All of these steps have been automated to obtain the speed necessary to sequence the entire genomes of humans and other organisms. A next generation sequencer called nanopore technology is based on DNA strands being fed through tiny pores and sequenced directly without the need to synthesize complementary copies (figure 10.7, step 8). It has the potential for highly rapid sequencing of very long molecules and will probably replace many of the older sequencers.

Whole genome sequencing projects have increased as the efficiency and speed of new sequencing technologies has improved. The case study provides a striking example of how it is being used in genomic medicine to diagnose diseases. As of 2016, whole genome sequence maps of thousands of viruses, prokaryotes, and eukaryotes have been published. The eukaryotic genomes include chimpanzee, rat, mouse, a nematode worm, yeast, honeybee, cats, dogs, horses, cows, rice, and corn. One of the remarkable discoveries in this huge enterprise has been how similar the genomes of relatively unrelated organisms can be. Humans share around 99% of their genome with chimpanzees, 80% with mice, about 50% with corn, and even 30% with the roundworm *Caenorhabditis elegans*.

Although sequencing provides the ultimate genetic map, it does not automatically identify the exact genes or alleles, nor will it immediately inform us about gene function. For that we need additional information about gene expression. Analysis and storage of this massive amount of new data require specialized computers. Because the data consist of information of both a biological and mathematical content, whole new disciplines have grown up around managing these data: **genomics** (see "A Guide to the New '-Omic' Sciences") and **bioinformatics.** The job of these specialties is to analyze, compare, and classify DNA, determine protein sequences, and ultimately understand the functions of specific genes and proteins. This fact is likely to bring another golden era in biology and microbiology, that of unraveling the complexity of relationships of genes and gene regulation. Eventually it will provide a more complete understanding of such phenomena as normal cell function, disease, development, aging, and many other issues. The case study for this chapter is a powerful example of what can be possible using these new tools.

A Guide to the New "-Omic" Sciences

The ability to obtain the entire DNA sequences of organisms has spawned new terms that refer to the "total picture" of some aspect of a cell, organism, or community:

genomics The systematic study of an organism's genes and their functions.

transcriptomics Analysis of all RNA molecules in a cell. This includes mRNA, tRNA, rRNA, and the regulatory RNAs. It is an intermediate link between the activity of genes and the protein products of translation.

proteomics The study of an organism's complement of proteins (its "proteome") and functions mediated by the proteins.

metagenomics (also called "community genomics") The study of all the genomes in a particular ecological niche, as opposed to individual genomes from single species.

metabolomics The study of the complete complement of small chemicals present in a cell at any given time. Provides a snapshot of the physiological state of the cell and the end products of its metabolism.

One of the fastest growing areas of the new "-omics" is designed to survey the scope of organisms that live in various habitats. This science is called environmental genomics, or **metagenomics.** In the past it was difficult to study complex communities where millions of organisms coexist, and it has not been possible to isolate and culture most of them. Metagenomics provides microbiologists a window through which to view the diversity of such habitats and makes it possible to pull out the microbial genomes, study them, and begin to understand their interactions. This technique has been used for several years in the large-scale genetic sampling of the oceans. Other research efforts on dozens of habitats, including the human body, are yielding massive amounts of new genetic data.

Polymerase Chain Reaction: A Molecular Copy Machine for DNA Some of the techniques used to analyze DNA and RNA are limited by the small amounts of test nucleic acid available. This problem was largely solved by the invention of a simple, versatile way to amplify DNA called the **polymerase chain reaction (PCR).** This technique rapidly increases the amount of DNA in a sample without the need for making cultures or carrying out complex purification techniques. It is so sensitive that it holds the potential to detect cancer from a single cell or to diagnose an infection from a single gene copy. It is comparable to being able to pluck a single DNA "needle" out of a "haystack" of other molecules and make unlimited copies of the DNA. The rapid rate of PCR makes it possible to replicate a target DNA from a few copies to billions of copies in a few hours. It can amplify DNA fragments that consist of a few base pairs to whole genomes containing several million base pairs.

To understand the idea behind PCR, it will be instructive to review figures 9.5 and 9.6, which model DNA synthesis as it occurs naturally in cells. The PCR method uses essentially the same events, with the opening of the double strand, using the exposed strands as templates, the addition of primers, and the actions of a DNA polymerase.

Initiating the reaction requires a few specialized ingredients **(figure 10.8).** The **primers** are synthetic oligonucleotides (short DNA strands) of a known sequence of 15 to 30 bases that serve as landmarks to indicate where DNA amplification will begin. These take the place of RNA primers that would normally be provided by the cell. Depending upon the purposes and what is known about the DNA being replicated, the primers can be random, attaching to any sequence they may fit, or they may be highly specific and chosen to

(a) In cycle 1, the DNA to be amplified is denatured, primed, and replicated by a polymerase that can function at high temperatures. The two resulting strands then serve as templates for a second cycle of denaturation, priming, and synthesis.

Cycle 1

DNA Sample Segment of DNA to be amplified

Heat to 94°C

Denaturation
Strands separate.

50°C–65°C

Priming
Oligonucleotide primers attach at ends of strands to promote replication of amplicons.

Amplicons Primer Primer

Extension
Heat-stable DNA polymerase synthesizes complementary strand.

72°C

Polymerase

Cycle 1

Cycle 2

Denaturation

Completed copies New strand

Original strands

New strand

Heat to 94°C

Priming

50°C–65°C

Extension

72°C

4 copies

Cycle 2

Cycles 3, 4, . . . repeat same steps, with the copies of DNA

(b) A view of the process after 6 cycles, with 64 copies of amplified DNA. Continuing this process for 20 to 40 cycles can produce millions of identical DNA molecules.

1* fragment

2 copies

4 copies

8 copies

16 copies

32 copies

64 copies

🌀 **Process Figure 10.8 Diagram of the polymerase chain reaction. (a)** Details of the process over two cycles. **(b)** Yield after the cycle continues past the second cycle.

amplify a known gene. To keep the DNA strands separated, processing must be carried out at a relatively high temperature. This necessitates the use of heat-resistant DNA polymerases isolated from thermophilic bacteria. Examples of these unique enzymes are Taq polymerase obtained from *Thermus aquaticus* and Vent polymerase from *Thermococcus litoralis*. Enzymes isolated from these thermophilic organisms remain active at the elevated temperatures used in PCR. Another useful component of PCR is a machine called a thermal cycler that automatically initiates the cyclic temperature changes.

The PCR technique operates by repetitive cycling of three basic steps: denaturation, priming, and extension:

1. **Denaturation.** The first step involves heating target DNA to 94°C to separate it into two strands. Each strand will become a template DNA, or *amplicon*.

2. **Priming.** As the reaction is cooled to between 50°C and 65°C (depending on the exact sequence of the primers), the oligonucleotide primers base-pair with specific sites on the amplicons.

3. **Extension.** In the third phase, which proceeds at 72°C, the DNA polymerase is activated and begins to synthesize DNA from a pool of available nucleotides. Beginning at the free end of the primers on both strands, the polymerases move along the strands, which serve as templates for the synthesis of two complete double strands of DNA.

It is through cyclic repetition of these steps that DNA becomes amplified. When the DNAs formed in the first cycle are denatured, they become amplicons to be primed and extended in the second cycle. Each subsequent cycle converts the new DNAs to amplicons and doubles the number of copies. The number of cycles required to produce a million molecules is 20, but the process is usually carried out to 30 or 40 cycles. One significant advantage of this technique has been its natural adaptability to automation. A thermal cycler can perform 20 cycles on nearly 100 samples in 2 or 3 hours.

DNA amplified by PCR can be analyzed by any of the techniques discussed earlier. PCR can also be adapted to analyze RNA by initially converting an RNA sample to DNA with reverse transcriptase. This cDNA can then be amplified by PCR in the usual manner. It is by such means that ribosomal RNA and messenger RNA are readied for sequencing and RNA viruses are sequenced and identified. The PCR technique has found prominence as a powerful workhorse of molecular biology, medicine, and biotechnology. It often plays an essential role in gene mapping, the study of genetic defects and cancer, forensics, taxonomy, and evolutionary studies.

Quick Search
See how PCR looks in animation at "Polymerase Chain Reaction (PCR) from Garland Science" on YouTube.

For all of its advantages, PCR techniques can encounter problems. One concern is the introduction and amplification of nontarget DNA from the surrounding environment, such as a skin cell from the technician carrying out the PCR rather than material from the sample DNA that was supposed to be amplified. Such contamination can be minimized by using equipment and rooms dedicated for DNA analysis and maintained with the utmost degree of clean-

liness. Problems with contaminants can also be reduced by using gene-specific primers and treating samples with enzymes that can degrade the contaminating DNA before it is amplified.

 Check Your Progress SECTION 10.1

1. Define *genetic engineering* and *biotechnology*, and summarize the important purposes of these fields. Review the use of the terms *genome, chromosome, gene, DNA,* and *RNA* from chapter 9.
2. Describe the processes involved in denaturing and renaturing of DNA, and explain what is useful about this procedure.
3. Define *restriction endonuclease* and explain what it does.
4. Define *palindrome* and draw three different palindromic sequences.
5. Explain how electrophoresis works and the general way that DNA is sized.
6. How would you make a copy of DNA from an mRNA transcript, what is this molecule called, and how would it be used?
7. Briefly summarize the steps involved in DNA sequencing and describe its applications.
8. Outline the steps in the PCR technique and describe the effects of temperature change.
9. What are the functions of primer and Taq polymerase?

10.2 Recombinant DNA Technology: A New Take on Nature

 Expected Learning Outcomes

5. Explain what is involved in recombinant DNA technology.
6. Characterize the events in cloning, using an example.
7. List and discuss some protein products of recombinant DNA and their applications.

The primary intent of **recombinant DNA (rDNA) technology** is to deliberately remove genetic material from one organism and combine it with that of a different organism. Its origins can be traced to 1970, when microbiologists first began to duplicate the clever tricks bacteria do naturally with bits of extra DNA such as plasmids, transposons, and proviruses. The discovery that bacteria can readily accept, replicate, and express foreign DNA made them powerful agents for studying the genes of other organisms in isolation. The practical applications of this work were soon realized by biotechnologists. Bacteria could be genetically engineered to mass-produce substances such as hormones, enzymes, and vaccines that have been difficult to synthesize by the usual industrial methods.

Figure 10.9 provides an overview of the recombinant DNA procedure. An important objective of this technique is to form genetic **clones.** This type of cloning involves the removal of a selected

Nucleus

Human, other mammal, microorganism, or plant cell

DNA of interest is isolated or produced from mRNA (cDNA).

DNA is inserted into vector

Microbial cloning host is transformed when it picks up vector with foreign DNA.

Multiplication of cloning host to amplify gene

Transformed cell processes foreign DNA into protein

Growth of transformed cell supplies DNA for study and engineering of other organisms.

Gene therapy

Protein Production
Pharmaceuticals
Hormones
• insulin
• human growth hormone
Vaccines
• hepatitis B
Industrial chemicals

Altered Organisms with Economically Useful Traits
Transgenic plants
• pest resistance
• herbicide resistance
• improved nutritional value
Transgenic animals

Multiple Applications in Genome Research, Testing, and Therapy
Gene function
Nucleotide sequencing
Identification
Gene libraries
Gene therapy

Figure 10.9 Overview of methods and applications of genetic engineering. Practical applications of genetic engineering include the development of pharmaceuticals, genetically modified organisms, and identification techniques. See figure 10.11 for details of these steps used in recombinant DNA. *Source:* Linda Bartlett/National Cancer Institute

gene from an animal, plant, or microorganism (the genetic donor) followed by its propagation in a different host organism. Cloning requires that the desired donor gene first be selected, excised by restriction endonucleases, and isolated. The gene is next inserted into a **vector** (usually a plasmid or a virus) that will insert the DNA into a **cloning host.** The cloning host is usually a bacterium or a

yeast that can replicate the gene and transcribe and translate it into the protein product for which it codes. Next we examine the elements of gene isolation, vectors, and cloning hosts and show how they participate in a complete recombinant DNA procedure.

Technical Aspects of Recombinant DNA and Gene Cloning

The first hurdles in cloning a target gene are to locate its exact site on the genetic donor's chromosome and to isolate it. Some of the first isolated genes came from viruses, which have extremely small and manageable genomes, and later genes were isolated from bacteria and yeasts. At first the complexity of the human genome made locating specific genes very difficult. However, this process has been greatly improved by tools for mapping, identifying, and excising genes. There are many common strategies for obtaining genes in an isolated state, including these:

1. A gene can be synthesized from isolated mRNA transcripts using reverse transcriptase (cDNA).
2. A gene can be amplified from a chromosome or larger segment of isolated DNA using PCR.
3. Gene-sized segments of DNA can be synthesized with an apparatus that polymerizes nucleotides in a chosen order. This makes it possible to create genes with almost any sequence.

Although the initial phases of gene isolation and cloning can be labor-intensive, a fortunate outcome is that, once isolated, genes can be maintained in a cloning host and vector just like a microbial pure culture. Genomic libraries, which contain fragments of DNA representing all the genes in a particular species, have been created for thousands of organisms and viruses.

Characteristics of Cloning Vectors

A good recombinant vector has two indispensable qualities: It must be capable of carrying a significant piece of the donor DNA, and it must be readily accepted by the cloning host. Plasmids are excellent vectors because they are small, well characterized, and easy to manipulate, and they can be transferred into appropriate host cells through transformation. Bacteriophages also serve well because they have the natural ability to inject DNA into bacterial hosts through transduction. A modified phage vector, the *Charon*[4] phage, has had sections of its genome deleted so it can carry a fairly large segment of foreign DNA.

Today thousands of unique cloning vectors are available commercially. Although every vector has characteristics that make it ideal for a specific project, all vectors can be thought of as having three important attributes **(figure 10.10):**

1. An origin of replication (ORI) is needed somewhere on the vector so that it will be replicated by the DNA polymerase of the cloning host.
2. The vector must accept DNA of the desired size. Early plasmids were limited to an insert size of less than 10 kb of DNA, far too small for most eukaryotic genes, with their sizable introns. Newer vectors called cosmids can hold 45 kb, whereas

4. Named for the mythical boatman in Hades who carried souls across the River Styx.

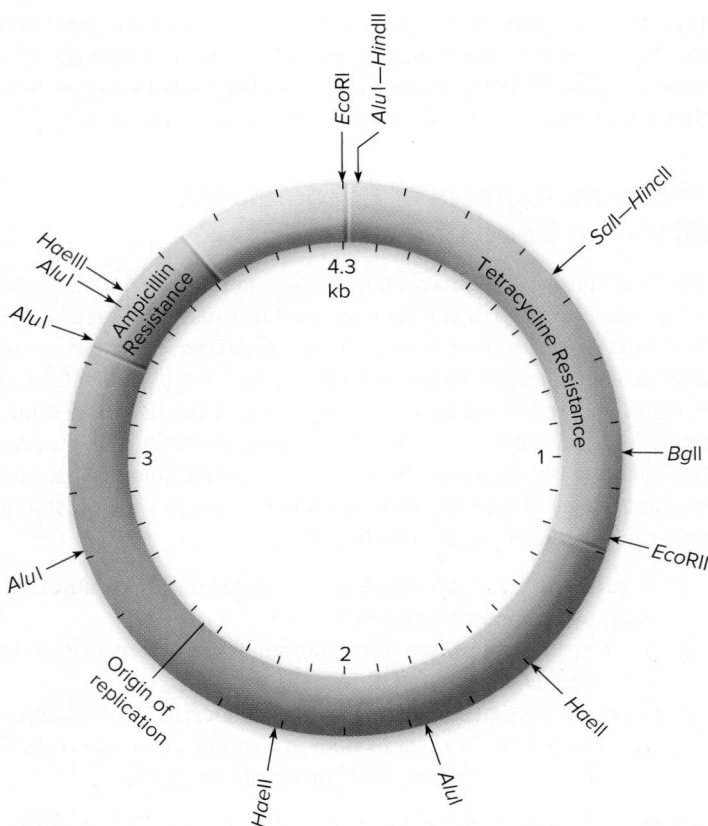

Figure 10.10 **Partial map of the pBR322 plasmid of** *Escherichia coli.* Arrows delineate sites cleaved by various restriction endonucleases. Numbers indicate the size of regions in kilobases (kb). Other important components are genes for ampicillin and tetracycline resistance and the origin of replication.

complex *bacterial artificial chromosomes* (BACs) and *yeast artificial chromosomes* (YACs) can hold as much as 300 kb and 1,000 kb, respectively.

3. Vectors typically contain a gene that confers drug resistance to their cloning host. In this way, cells can be inoculated into drug-containing media, and only those cells that harbor a plasmid will be selected for growth.

Characteristics of Cloning Hosts

The best cloning hosts possess several key characteristics **(table 10.1).** The traditional cloning host—and the one still used in basic

TABLE 10.1	Desirable Features in a Microbial Cloning Host
Rapid turnover, fast growth rate	
Can be grown in large quantities using ordinary culture methods	
Nonpathogenic	
Genome that is well delineated (mapped)	
Capable of accepting plasmid or bacteriophage vectors	
Maintains foreign genes through multiple generations	
Will secrete a high yield of proteins from expressed foreign genes	

experiments—is *Escherichia coli.* Because this bacterium was the original recombinant host, the protocols using it are well established, relatively easy, and reliable. Hundreds of specialized cloning vectors have been developed for it. The main disadvantages with this species are that the splicing of mRNA and the modification of proteins that would normally occur in the eukaryotic endoplasmic reticulum and Golgi apparatus are unavailable in this prokaryotic cloning host. One alternative host for certain industrial processes and research is the eukaryotic yeast *Saccharomyces cerevisiae,* which would be able to process and modify eukaryotic genes and products.

Certain techniques may also employ animal cell culture, and even live animals and plants to serve as cloning hosts. Although a number of recombinant proteins are produced by transformed bacterial hosts, many of the enzymes, hormones, and antibodies used in drug therapy are currently being manufactured with mammalian cell cultures as the cloning and expression hosts. One of the primary advantages to this alternative procedure is that these cell cultures can modify the proteins (adding carbohydrates, for example) so that they are biologically more active. Some forms of reproductive hormones, human growth hormone, and interferon are products of cell culture. In our coverage, we present the recombinant process as it is performed in bacteria and yeasts.

Construction of a Recombinant, Insertion into a Cloning Host, and Genetic Expression

This section emphasizes the main steps in recombinant DNA technology needed to produce a protein. A drug called alpha-2a interferon (Roferon-A) will serve as an example, but it should be remembered that the steps are essentially the same regardless of the protein being produced. This form of interferon is used to treat cancers such as hairy-cell leukemia and Kaposi sarcoma in AIDS patients (see chapter 14). The human alpha interferon gene is a DNA molecule of approximately 500 bp that codes for a polypeptide of 166 amino acids. It was originally isolated and identified from human blood cells and prepared from processed mRNA transcripts that are free of introns.

The first step in cloning the interferon gene or any desired gene is to prepare it for splicing into an *E. coli* plasmid **(figure 10.11).** This can be accomplished by digesting both the gene and the plasmid with the same restriction enzyme, resulting in complementary sticky ends on both the vector and insert DNA. In the presence of the endonuclease, the plasmid's circular molecule is nicked open, and its sticky ends are exposed. When the gene and plasmid are placed together, their free ends base-pair, and a ligase makes the final covalent seals. The resultant gene and plasmid combination is called a **recombinant.**

Following this procedure, the recombinant is introduced by transformation into the cloning host, a special laboratory strain of *E. coli* that lacks any extra plasmids that could complicate the

Quick Search
See this played out in "Steps in Cloning a Gene," a video from McGraw-Hill on YouTube.

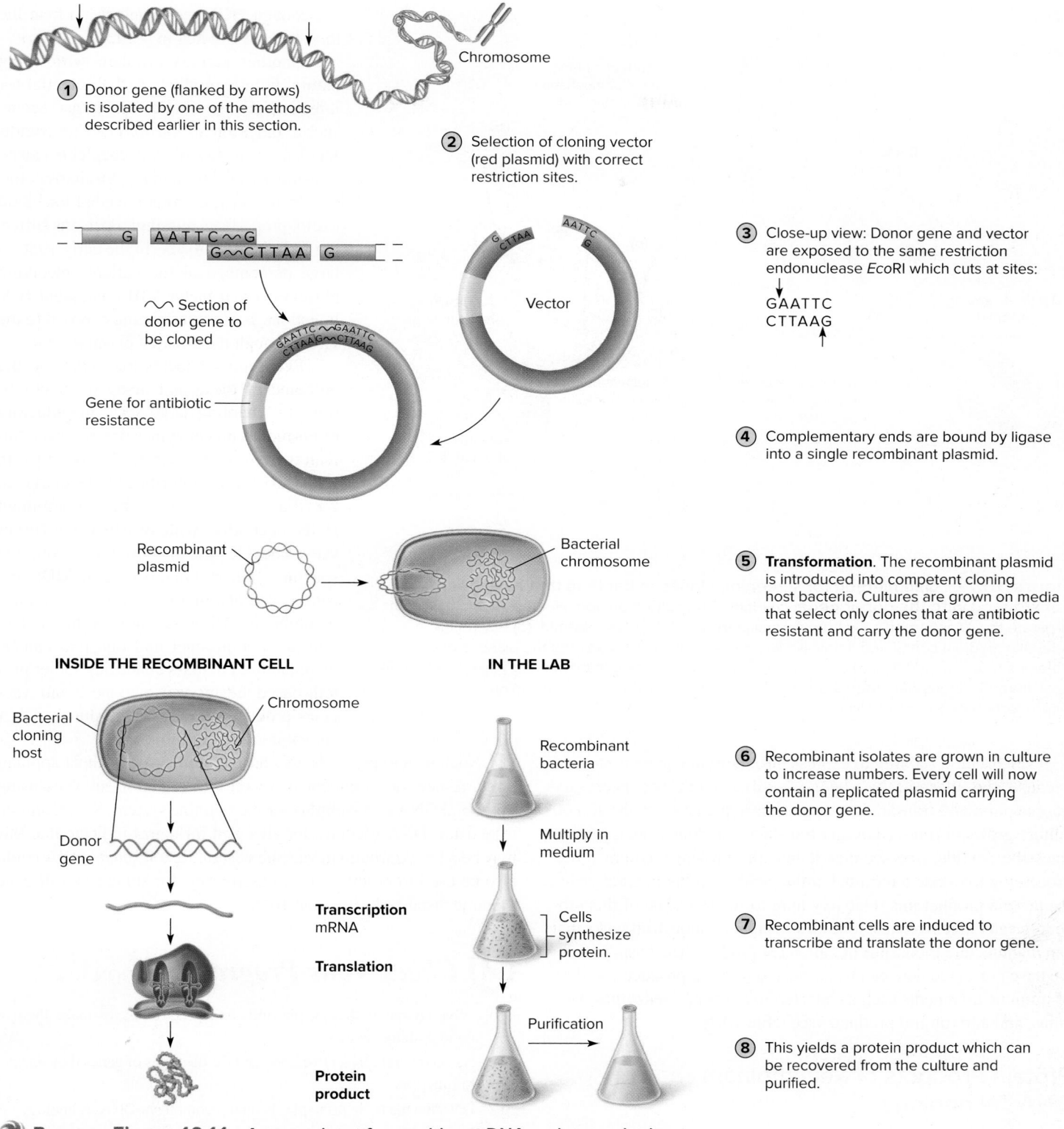

① Donor gene (flanked by arrows) is isolated by one of the methods described earlier in this section.

Chromosome

② Selection of cloning vector (red plasmid) with correct restriction sites.

~ Section of donor gene to be cloned

Vector

Gene for antibiotic resistance

③ Close-up view: Donor gene and vector are exposed to the same restriction endonuclease *Eco*RI which cuts at sites:

GAATTC
CTTAAG

④ Complementary ends are bound by ligase into a single recombinant plasmid.

Recombinant plasmid

Bacterial chromosome

⑤ **Transformation.** The recombinant plasmid is introduced into competent cloning host bacteria. Cultures are grown on media that select only clones that are antibiotic resistant and carry the donor gene.

INSIDE THE RECOMBINANT CELL

Bacterial cloning host

Chromosome

Donor gene

Transcription mRNA

Translation

Protein product

IN THE LAB

Recombinant bacteria

Multiply in medium

Cells synthesize protein.

Purification

⑥ Recombinant isolates are grown in culture to increase numbers. Every cell will now contain a replicated plasmid carrying the donor gene.

⑦ Recombinant cells are induced to transcribe and translate the donor gene.

⑧ This yields a protein product which can be recovered from the culture and purified.

Process Figure 10.11 An overview of recombinant DNA and gene cloning.

expression of the gene. Because the recombinant plasmid enters only some of the cloning host cells, it is necessary to locate these recombinant clones. Cultures are plated out on medium containing an antibiotic such as ampicillin or tetracycline, and only those clones that carry the plasmid with resistance to this antibiotic can form colonies **(figure 10.12).** These recombinant colonies are selected from the plates and cultured. As the cells multiply, the plasmid is replicated along with the cell's chromosome. In a few hours of growth, there can be billions of cells, each containing the target donor gene. Once the gene has been successfully cloned and tested, this step does not have to be repeated; the recombinant strain can be maintained in culture indefinitely.

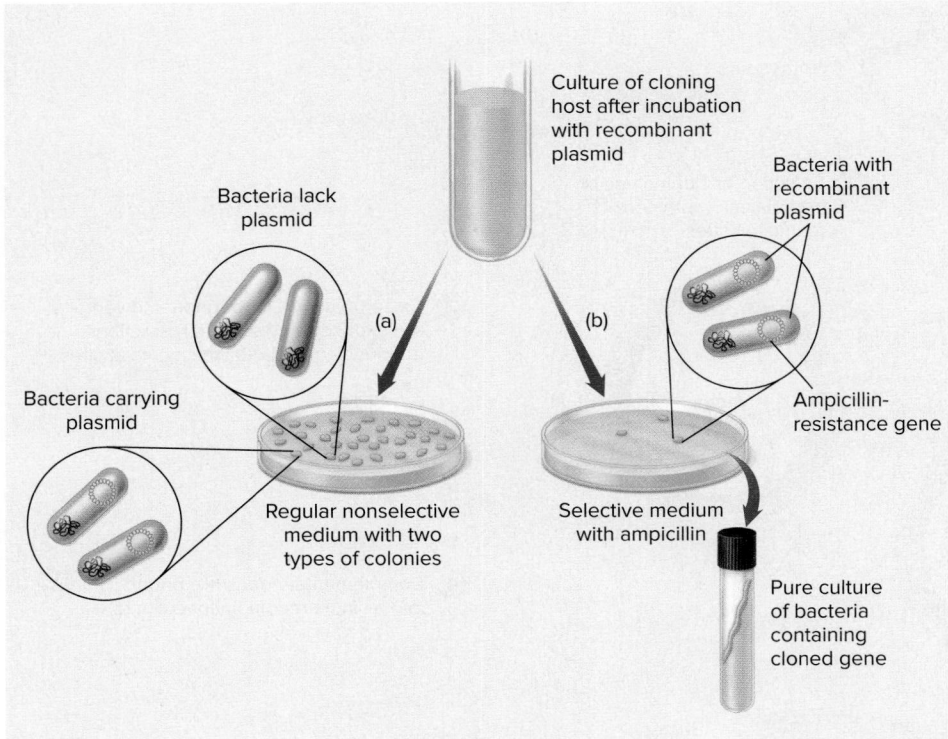

Culture of cloning
host after incubation
with recombinant
plasmid

Bacteria lack
plasmid

Bacteria with
recombinant
plasmid

(a)

(b)

Bacteria carrying
plasmid

Ampicillin-
resistance gene

Regular nonselective
medium with two
types of colonies

Selective medium
with ampicillin

Pure culture
of bacteria
containing
cloned gene

Figure 10.12 **One method for screening clones of bacteria that have been transformed with the donor gene. (a)** Plating the culture on nonselective medium will not separate the transformed cells from normal cells, which lack the plasmid. **(b)** Plating on selective medium containing ampicillin will permit only cells containing the plasmid to multiply. Colonies growing on this medium that carry the cloned gene can be used to make a culture for gene libraries, industrial production, and other processes. (Plasmids are shown disproportionate to size of cell.)

The bacteria's ability to express the eukaryotic gene is ensured because the plasmid has been modified with the necessary transcription and translation recognition sequences. As the *E. coli* culture grows, it transcribes and translates the donor gene, synthesizes the peptide, and secretes it into the growth medium. Final processing to excise a terminal amino acid from the peptide yields the protein product in a relatively pure form. The scale of this procedure can range from test tube size to gigantic industrial tanks that can manufacture thousands of gallons of product. The basic process we have presented here can be harnessed to mass-produce a variety of immune treatments such as interferon, hormones, enzymes, vaccines, and agricultural products (see table 10.2).

Protein Products of Recombinant DNA Technology

Recombinant DNA technology is used by pharmaceutical companies to manufacture medications that are difficult to procure by other means. Diseases such as diabetes and dwarfism, caused by the lack of an essential hormone, are treated by replacing the missing hormone. Insulin of animal origin was once the only form available to treat diabetes, even though such animal products can cause allergic reactions in sensitive individuals. In contrast, dwarfism cannot be treated with animal growth hormones, so originally the only source of human growth hormone (HGH) was the pituitaries of cadavers. At one time,

not enough HGH was available to treat the thousands of children in need.

Another serious problem with using natural human products is the potential for infection. For example, infectious agents such as the prion responsible for Creutzfeldt-Jakob disease, described in chapter 6, can be transmitted in this manner. Similarly, clotting factor VIII, a protein needed for blood to clot properly, is missing in persons suffering from hemophilia A. In the early 1980s a large percentage of the patients receiving plasma-derived factor VIII contracted HIV infections as a result of being exposed to the virus through the donated plasma.

Recombinant technology changed the outcome of these and many other conditions by enabling large-scale manufacture of hormones and enzymes derived from human genes. Recombinant human insulin was the first rDNA hormone to be marketed for diabetics, followed by recombinant HGH for children with dwarfism and Turner syndrome. HGH is also used to prevent the wasting syndrome that occurs in AIDS and cancer patients. In all of these applications, recombinant DNA technology has led to both a safer product and one that can be manufactured in mass quantities. Other protein-based hormones, enzymes, and vaccines produced through recombinant DNA are summarized in **table 10.2.**

Nucleic acid products also have a number of medical applications. A new development in vaccine formulation involves using microbial DNA as a stimulus for the immune system. So far, animal tests using DNA vaccines for HIV and influenza indicate that this may be a breakthrough in vaccine design. Recombinant DNA could also be used to produce DNA-based drugs for some types of gene therapy, discussed in section 10.3.

Check Your Progress SECTION 10.2

10. What characteristics of plasmids and bacteriophages make them good cloning vectors?

11. Name several types of vectors, and list the types of genes they need to function.

12. Describe the basic principles behind recombinant DNA technology.

13. Summarize the characteristics of bacteria and yeasts that make them good cloning hosts.

14. Outline the main steps in cloning a gene, including how this gene is used.

15. What is one way to determine whether a bacterial culture has received a recombinant plasmid?

16. Characterize several products that have resulted from recombinant DNA technology.

TABLE 10.2	Examples of Protein Products Made by Recombinant DNA Technology

Immune Treatments

Interferons—peptides used to treat some types of cancer, multiple sclerosis, and viral infections such as hepatitis and genital warts
Interleukins—types of cytokines that regulate the immune function of white blood cells; used in cancer treatment
Orthoclone—an immune suppressant in transplant patients
Macrophage-colony-stimulating factor—used to stimulate bone marrow activity after bone marrow grafts
Tumor necrosis factor (TNF)—used to treat cancer
Granulocyte-colony-stimulating factor (Neupogen)—developed for treating cancer patients suffering from low neutrophil counts

Hormones

Insulin—a small protein that is used to treat some forms of diabetes
Erythropoietin (EPO)—a peptide that stimulates bone marrow used to treat some forms of anemia
Tissue plasminogen activator (tPA)—can dissolve potentially dangerous blood clots
Hemoglobin A—form of artificial blood to be used in place of real blood for transfusions
Factor VIII—needed as replacement blood-clotting factor in type A hemophilia
Human growth hormone (HGH)—stimulates growth in children with dwarfism; prevents wasting syndrome

Enzymes

rH DNase (pulmozyme)—a treatment that can break down the thick lung secretions of cystic fibrosis
Antitrypsin—replacement therapy to benefit emphysema patients
PEG-SOD—a form of superoxide dismutase that minimizes damage to brain tissue after severe trauma
Lipases—added to laundry detergents

Vaccines

Vaccines for hepatitis A and B, *Haemophilus influenzae,* meningitis, influenza, rotavirus, and human papillomavirus
Experimental malaria and HIV vaccines based on recombinant surface antigens

Miscellaneous

Bovine growth hormone or bovine somatotropin (BST)—given to cows to increase milk production
Apolipoprotein—used to deter the development of fatty deposits in the arteries and to prevent strokes and heart attacks
Spider silk—a light, tough fabric for parachutes and bulletproof vest

10.3 Genetically Modified Organisms and Other Applications

 Expected Learning Outcomes

8. Define what is meant by the term *transgenic* or *genetically modified organisms.*
9. Describe the uses of genetically modified bacteria and viruses.
10. Discuss genetic modifications in plants and animals, explaining how they are carried out and giving examples of applications.
11. Explain how DNA technology can be used to treat genetic diseases.

Recombinant organisms produced through the introduction of foreign genes are called **genetically modified organisms (GMOs),** or **transgenic organisms.** Although foreign genes have been inserted into a variety of life forms including microbes, plants, and animals, the techniques used to create them owe their existence to the early work that took place in bacterial systems.

Recombinant Microbes: Modified Bacteria and Viruses

One of the first practical applications of recombinant DNA was the creation of a genetically altered strain of the bacterium *Pseudomonas syringae.* The wild strain ordinarily contains an ice formation gene that promotes ice or frost formation on moist plant surfaces. Genetic alteration of the frost gene using recombinant plasmids created a different strain that could prevent ice crystals from forming. A commercial product called Frostban was developed but never marketed due to the controversy arising from its release into the environment. In another trial, a strain of *Pseudomonas fluorescens* was engineered with the gene from a bacterium (*Bacillus thuringiensis*) that codes for an insecticide (*Bt toxin*). These recombinant bacteria are released to colonize plant roots and help destroy invading insects. All releases of recombinant microbes must be approved by the U.S. Environmental Protection Agency and are closely monitored.

Although primarily known for its ability to cause colds, the adenovirus has been bioengineered to attack and destroy cancer cells. These *oncolytic adenoviruses* have been altered so that their host range consists exclusively of cells that produce cancer-specific proteins. By simply infecting a cancerous cell and replicating, the engineered virus will eventually lyse and kill the cell. Although this

10.2 MAKING CONNECTIONS

A Moment to Think

"When we seed millions of acres of land with these plants, what happens to foraging birds, to insects, to microbes, to other animals, when they come in contact with and digest plants that are producing materials ranging from plastics to vaccines to pharmaceutical products?"

Jeremy Rifkin, Foundation on Economic Trends

"If you put a label on genetically modified food, you might as well put a skull and crossbones on it."

Norman Brasick, of Monsanto, the largest biotechnology firm selling engineered seeds

"There is a profound difference between the types of expected effects from traditional breeding and genetic engineering."

Dr. Louis Pribyl, FDA microbiologist

There is no denying that many aspects of synthetic biology are highly controversial and emotionally charged, but few would deny that this technology is here to stay. As evidenced by the graph, as of 2015, 85% to 95% of major crops are now composed of genetically modified plants. Most of these plants have been altered to either make their own insecticide or resist herbicides that kill weeds. The edible crops are already winding up in the foods we eat. With regard to animals, so far only one GMO has received approval for human consumption—the AquAdvantage salmon—a farmed Atlantic salmon that has been engineered with genes from other fish. Time will tell what kind of acceptance these fish will have in the consumer market.

What effects, if any, will these GMOs have? In many cases we do not yet know the answer, but so far we have relied on the scientists who developed this technology to test its effects and report their findings. Those who will determine the limits of genetic technology include not only scientists and politicians but voters and consumers as well. History is rife with both knee-jerk rejections to new technologies that have later proven to be invaluable (vaccinations) and complacency while harmful products were released with inadequate scrutiny (for example, Fen-Phen, Vioxx).

Ethical choices can be properly made only from a standpoint of intellectual awareness; and in this era of advertising, polling, and focus groups, people with a stake in genetic technology know that the most effective way to drum up support (both for or against) is *not* by carefully educating the public as to the uses and limits of our newfound powers but, rather, by publicizing exaggerated claims or frightening scenarios. Enlightenment is often a casualty of these advertising campaigns, and you as a student, voter, and potential consumer of bioengineered products need to realize that the truth lies somewhere between nightmare scenarios land the blissful utopia often portrayed by some proponents of biotechnology.

Throughout this chapter, a common theme has been that biotechnology, genetically modified organisms, and gene-based therapies are simply new ways of achieving long-held goals. This is, of course, a vast understatement. Society has never before had the ability to do even a fraction of the things we can do today, and at some point we must decide where we as individuals and as a society draw the line. The answer will differ for each person, based on his or her unique experiences and personal philosophy, but should certainly include a factual knowledge of the probable benefits and possible risks. We live in serious times, and these are choices requiring serious thought.

Can you think of any adverse effects might arise from eating GMO corn or fish? After all, every type of non-GMO food we eat already contains large amounts of foreign DNA from the food organism. *Answer available on Connect.*

> **Quick Search**
>
> Explore this subject by searching the Internet for "Top Ten Genetically Engineered Crops" and "Amazing Parts Created Outside the Body."

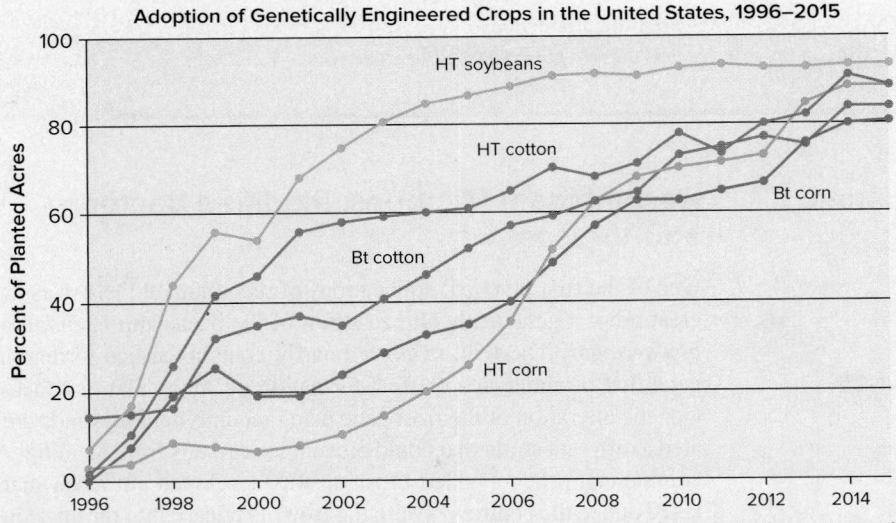

Adoption of Genetically Engineered Crops in the United States, 1996–2015

Graph tracking the percentage of crops in the United States grown from genetically engineered seeds. *Bt* is short for *Bacillus thuringiensis*, a bacterium that produces a natural insecticide, and HT indicates herbicide resistance. *Source:* United States Department of Agriculture

procedure is not yet approved for use in humans, this novel use of viral vectors speaks to the potential of genetically engineered viruses. This technology is currently used to engineer viruses to deliver genes into animals and humans (see figure 10.15).

In 2010 a milestone was reached when Craig Venter's laboratory team created an entirely new bacterial species by inserting a completely synthesized genome of artificial origin into *Mycoplasma genitalium*. The new species, named *Mycoplasma laboratorium*, contains the smallest functional genome of any bacterium, with 382 genes of around 1.1 million base pairs. So far the scale of this experiment is small—a single-celled organism—but its implications are large. The growing power of genetic engineering and synthetic biology has fostered a deeper examination of the **bioethics**[5] involved in wielding the genetics of organisms (**10.2 Making Connections**).

5. A field that relates biological issues to human conduct and moral judgment.

Recombination in Multicellular Organisms

Although the terms *genetically modified,* genetically engineered, and *transgenic organism* apply to any life form whose genome has been deliberately altered, it is often used in reference to plants and animals that possess genes that were not acquired through traditional breeding. The basic techniques covered in this chapter have given scientists a limited ability to custom-build an organism that expresses a certain trait, in a sense accelerating the pace of evolution. Because of the universal nature of the genetic code, however, scientists can go nature one step farther and combine genes from more than a single species, creating plants or animals that would never have existed otherwise. Examples of genetically modified organisms include plants with added nutritional value, animals that excrete clot-dissolving proteins in their milk, and humans who have gained the ability to produce an enzyme crucial to the immune response.

Transgenic Plants

Engineering plants to produce new traits often revolves around *Agrobacterium tumefaciens,* a plant pathogen particularly adept at transferring small amounts of DNA to plant cells. These bacteria live in soil and can invade injured plant tissues. Inside the wounded tissue, the bacteria adhere to plant cells and introduce a large plasmid termed **Ti** (tumor-inducing). This plasmid inserts into the genome of the infected plant cell and transforms it **(figure 10.13).** Over time this causes development of a tumor called *crown gall disease,* a mass of undifferentiated tissue on the stem.

The primary factor that induces this transformation is a discrete region of the Ti plasmid called T-DNA. This series of genes is integrated into a plant cell's chromosome, where it directs the synthesis of nutrients to feed the bacteria and hormones to stimulate plant growth. Even after the bacteria in a tumor are dead, the plasmid genes remain in the cell nucleus and the tumor continues to grow.

This plasmid is an effective vector for inserting foreign genes into plant genomes. The procedure involves removing the Ti plasmid, inserting a specially selected gene such as herbicide resistance into it, and returning the plasmid to *Agrobacterium.* Infection of the plant by the recombinant bacteria automatically transfers the plasmid with the foreign gene into the plant cells. Transformed cells can be grown into mature plants that express the donor gene and transmit it to their offspring. The insertion of genes with a Ti plasmid works primarily to engineer important plants such as potatoes, tomatoes, cotton, and grapes.

For plants that cannot be transformed with *Agrobacterium,* seed embryos have been successfully implanted with additional genes using gene guns—small "shotguns" that shoot the genes into plant embryos with tiny gold bullets.

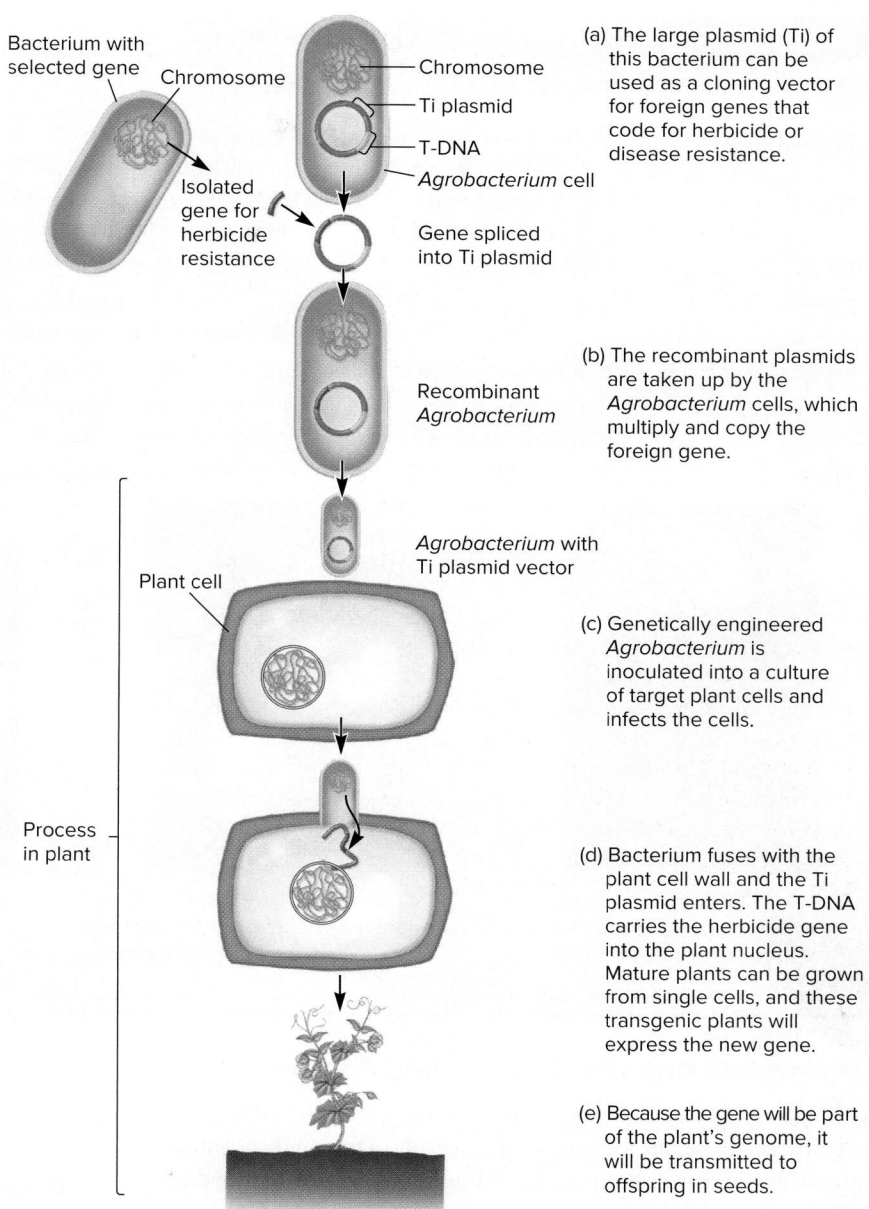

(a) The large plasmid (Ti) of this bacterium can be used as a cloning vector for foreign genes that code for herbicide or disease resistance.

(b) The recombinant plasmids are taken up by the *Agrobacterium* cells, which multiply and copy the foreign gene.

(c) Genetically engineered *Agrobacterium* is inoculated into a culture of target plant cells and infects the cells.

(d) Bacterium fuses with the plant cell wall and the Ti plasmid enters. The T-DNA carries the herbicide gene into the plant nucleus. Mature plants can be grown from single cells, and these transgenic plants will express the new gene.

(e) Because the gene will be part of the plant's genome, it will be transmitted to offspring in seeds.

Figure 10.13 Bioengineering of plants. Most techniques employ a natural tumor-producing bacterium called *Agrobacterium tumefaciens* **(a).** The example here shows the insertion of a gene to create a herbicide-resistant plant **(b-e).**

Agricultural officials had approved more than 1,200 genetically modified plants for field testing through the year 2014. Around 240 million acres of these plants were grown in the United States alone in 2012, with herbicide- and insecticide-resistant soybeans, corn, cotton, and canola making up the vast majority. The principal concern about such plants is the potential for genetic recombination with nearby wild plants, leading to the creation of "superweeds" that are resistant to herbicides and disease. This has already happened in parts of Canada, where canola plants have transferred their genes to wild mustard. Because this could alter the ecology of natural environments, the U.S. Department of Agriculture is increasing the regulation of all releases of transgenic plants. See **table 10.3** for examples of plants that have been engineered.

TABLE 10.3	Examples of Genetically Modified Plants	
Plant	**Trait(s)**	**Results**
Cherry tomatoes	High levels of anthocyanin, an antioxidant	Tomatoes were engineered with pigment gene from snapdragons, producing five times the purple pigment anthocyanin and greatly increasing the nutrition value (**figure a**).
Wheat	Resistance to head blight	Genes added to break down the cell wall of the fungus Fusarium protect wheat plants from being attacked (**figure b**).
Wheat	Improved nutrition	Engineered with genes to increase zinc and iron levels in the plant.
Golden rice	Improved nutrition	Genes added to produce beta-carotene, a precursor to vitamin A; aim is to prevent deficiencies in developing countries that lead to blindness.
Soybeans, corn, cotton	Herbicide resistance	Plants were engineered to withstand chemical herbicides.
Peanuts	Insect resistance	Genes for a biopesticide called Bt were added to kill insects that feed on the plants (**figure c**).
Banana	Vaccine production	Vaccines for hepatitis B in development
Plum	Viral resistance	Genes for resistance to plum poxvirus prevent infection and destruction of plant

(a) This purple tomato contains high levels of a pigment that has numerous health benefits.

(b) Researcher holds a genetically engineered wheat plant being tested for its resistance to a fungal infection.

(c) Peanut plant (left) genetically modified to produce its own insecticide is unharmed by larvae; unmodified plant (right) shows damage from insect larvae.

(photos): (a): © Andrew Davis and Sue Bunnewell/John Innes Centre; (b): *Source:* Jack Dykinga, USDA/ARS; (c): *Source:* Herb Pilcher, USDA/ARS

Transgenic Animals: Engineering Embryos

Animals such as mice, pigs, sheep, cattle, fish, and poultry are prominent candidates for genetic engineering. Thousands of animal strains now exist for the purposes of agriculture, industry, and research into genetic diseases.

The most effective way to insert genes into animals is to use a virus to transfect the fertilized egg or early embryo (germ-line engineering). Foreign genes can also be delivered into a fertilized egg by pulsing the egg with high voltage or by injection (**figure 10.14**). The success rate does not have to be high, because once created, a transgenic animal will pass the genes on to its offspring. An inventive newer technology known as **CRISPR** (see "CRISPR—A Power Tool for Gene Editing") is rapidly becoming the chosen method for manipulating embryonic DNA, especially to create mouse models of human disease.

The dominant animals in the genetic engineer's laboratory are transgenic forms of mice. In one pioneering experiment, fertilized mouse embryos were injected with a gene from a jellyfish that produces a green fluorescent protein. The cells of these mice fluoresce in ultraviolet radiation. One serious use has been to study the developmental stages in mice. This strategy has led to the design of animal

Several embryos recovered from female

Pronuclei

Embryos transferred to a depression slide containing culture medium

Culture medium
Oil

Holding pipette

Pronucleus
DNA to be injected
Injection pipette

As embryo is held in place, DNA is injected into one pronucleus.

Several injected embryos are placed into oviduct of receptive female.

Figure 10.14 **Steps in the creation of a transgenic animal.** Hundreds of mice and other mammals with genetic modifications have been produced this way.

models to study human genetic diseases and to use these natural systems to test new genetic therapies before they are used in humans. Animals such as sheep or goats can also be engineered to become "factories" capable of manufacturing proteins and excreting them in their milk or semen, a process often referred to (when done to produce medically useful proteins) as "pharming."

One benefit of transgenic animals is that they will automatically transcribe and translate eukaryotic genes and accurately carry out posttranslational modification of the protein. By ensuring that the protein product is produced in the milk or semen of animals, recovery and purification are made that much simpler. For

Embryonic mouse displaying green fluorescent protein
Source: National Institute of Mental Health

example, alpha-1 antitrypsin (AAT), produced in sheep, is used to treat an inherited form of emphysema. A single animal produces AAT at concentrations of 35 grams per liter of milk. A few hundred sheep could easily meet the needs of the 20,000 people a year who suffer from this debilitating disease. See **table 10.4** to examine some other examples and applications of transgenic animals.

Medical Applications of DNA Technology

One offshoot of studies with transgenic animals has been the advancement of medical treatments for human diseases that involve manipulating genetics, rather than administering drugs. The applications that show the most promise are gene therapy, which aims to replace a nonfunctional gene with its normal counterpart, and regulatory RNAs, which manage DNA expression of abnormal genes.

Gene Therapy

We have known for decades that for certain diseases, the disease phenotype is due to a missing or non-functioning protein. Often referred to as "loss of function" diseases, examples include cystic

TABLE 10.4	Selected Applications for Transgenic Animals
Animal/Product	**Application**
Production of Pharmaceuticals	
Pig/Factors VIII and IX	Treatment of human blood-clotting disorders.
Sheep/Alpha antitrypsin	Replacement protein to prevent emphysema.
Rabbit/Human interleukin-2	Stimulates the production of T cells to fight cancer.
Cow/Lactoferrin	Iron-binding protein inhibits bacterial and viral infections.
Cow/Human albumin	Used in surgery to maintain blood volume.
Goat/Tissue plasminogen activating factor	Dissolves blood clots and reduces heart damage.
Goat/Antithrombin	*Atryn* was the first drug from a genetically engineered animal approved by the FDA; used to prevent blood clots.
Research/Industry/Environment	
Anopheles mosquito malaria resistance	Mosquito embryos have new genes inserted to make them resistant to infection by *Plasmodium,* the malaria parasite. Modified mosquitoes would no longer carry and spread the parasite.
Goat/Silk fibers	Female animals secrete silk protein in their milk. Silk has many uses for constructing strong, lightweight material.
Pig/Pork meat	*Enviropig* has enzymes in its saliva to process excess phosphate in its feed, reducing pollution.
Mice, cats, pigs, monkeys/research animals, Tropical fish	Engineered with the green fluorescent protein from jellyfish that causes their cells to glow green in UV light. Animals are used to trace development in embryos and the stages of certain diseases.
Cats and dogs/Hypoallergenic	Genes that code for allergens from animal's saliva and sebaceous glands have been silenced.
Salmon/Farmed fish	Atlantic salmon engineered with the growth hormone gene from a different species of salmon to increase growth rate.
Cow/Human breast milk	Produces milk with the same composition as human breast milk.

Genetically engineered Atlantic salmon grow two times faster than regular salmon.

A mother mouse genetically modified to have mosaic-pattern fur, and her two transgenic pups with solid-pattern fur.

fibrosis, hemophilia, sickle-cell anemia, some forms of dwarfism, and muscular dystrophy. A few of these diseases can be treated by administering the correct protein—recall the use of recombinant growth hormone to treat dwarfism. But in the majority of cases, a replacement drug is not yet available. About 30 years ago, medical geneticists began to explore the possibility of restoring function by means of **gene therapy,** by inserting the normal functioning gene into the cells of afflicted patients.

The inherent potential of gene therapy would be to permanently cure the physiological dysfunction by repairing the genetic defect. There are two strategies for this therapy. In *ex vivo* therapy, the normal gene is cloned in vectors such as adenoviruses or retroviruses that have been modified to carry a corrective gene into cells of the patient. Tissues removed from the patient are incubated with these genetically modified viruses to transfect them with the normal gene. The transfected cells are then reintroduced into the patient's body by transfusion **(figure 10.15).** In contrast, the *in vivo* type of therapy skips the intermediate step of incubating excised patient tissue. Instead, the naked DNA or a virus vector is directly introduced into the patient's tissues.

① Normal gene is isolated from healthy subject.

② Gene is cloned.

③ Gene is inserted into virus vector.

④ Bone marrow sample is taken from patient with genetic defect.

⑤ Stem cells from marrow are transfected with virus (enlarged view).

⑥ Transfected cells (red dots) are infused into patient.

⑦ Patient is observed for expression of normal gene.

Marrow stem cell

Figure 10.15 Protocol for the *ex vivo* type of gene therapy in humans.

Clinical testing using various types of gene therapy is performed on human volunteers with a diagnosed genetic disorder. Over 2,000 of these trials have been or are being implemented in the United States and other countries. Most trials target cancer, single-gene defects, and infections. The most dramatic example of this type of treatment was seen in the very first gene therapy trial conducted in 1990. In this case, two girls suffering from adenosine deaminase (ADA) deficiency, a fatal immunologic disorder caused by a missing enzyme, received infusions of their own hematopoietic stem cells that had been transformed with a healthy copy of the adenosine deaminase gene.

As with most medical experimentation, some of the earlier gene therapy trials experienced setbacks. A serious problem arose with the use of the viral vectors. Even though they were supposed to be "harmless," five children infused with a recombinant mouse retrovirus developed leukemia when the vector DNA integrated at a cancer-inducing site on a chromosome. After this event, researchers worked to develop modified adeno-associated viral vectors to greatly reduce the chances of this happening.

Over the past 5 years, medical geneticists reported major successes in curing 20 children born with the devastating genetic disease termed X-1-linked SCID (severe combined deficiency disease). These children lacked a functioning immune system from birth, which researchers restored by inserting copies of the functioning gene into their stem cells using a safer recombinant virus. In another case, a team of physicians improved the vision of several children with a genetic form of progressive blindness by transferring the gene directly into the retina. Other genetic specialists have engineered T cells to treat melanoma and leukemia and have cured a form of the hemoglobin disorder called thalassemia. Other trials under way to target hemophilia and Parkinson disease appear to restore some function to patients with these conditions. A newer therapy for the treatment of lipoprotein lipase deficiency (LLD) is called alipogene tiparvovec **(Glybera).** It is based on an engineered adeno-associated virus that carries the normal gene for processing certain types of fats. When injected into the muscles of the legs, it is assimilated locally and can restore normal function to fat metabolism. So far, it is the only gene therapy approved for routine use in the United States.

Quick Search
Follow up by searching for "gene therapy" on the Internet.

Therapy with Regulatory RNAs

Tens of thousands of genetic disorders have been described. Most are due to nonfunctional or missing genetic material, but a proportion of them result from the expression of abnormal, mutated genes. These conditions, termed "gain of function" diseases, include a number of cancers, some forms of Alzheimer disease, Huntington disease, and fragile X syndrome. Since every cell in the body carries these faulty but active genes, merely replacing them is not yet an option.

For these types of diseases, genetic researchers are taking a lesson from the natural strategies used by bacteria and eukaryotic organisms for regulating gene expression (chapter 9). Recall that these

organisms use regulatory RNA or RNAi, including small interfering RNAs and microRNAs, to silence gene expression. This knowledge has driven the development of genetic-based drugs that interfere with mRNA processing and protein synthesis. Some of the most promising therapies involve the targeting of RNA virus genomes by small RNAs. They have been used to degrade HIV and respiratory syncytial viruses. They have also been tested as a treatment for macular degeneration and as a therapy to silence cancer genes.

Early clinical trials of RNAi therapeutics have been promising, with most of them dramatically lowering the rate of synthesis of the targeted protein. Because the RNAi agent can be targeted to a specific sequence of DNA, side effects are rare. The biggest obstacles involve how to deliver the RNAi drugs into deep tissues where their targets lie and the fact that these drugs would have to be taken for life.

CRISPR—A Power Tool for Gene Editing

Novel "tricks" of microbial genetics continue to be a valuable source of innovations in biotechnology. One of the newest and most powerful methods of DNA manipulation, based on a natural system found in many bacterial and archaeal cells, is termed "**c**lustered **r**egularly-**i**nterspersed **s**hort **p**alindromic **r**epeats," thankfully shortened to its acronym—**CRISPR** (pronounced "krisper"). These short segments of DNA are associated with enzymes—**CRISPR-as**sociated proteins or **Cas**—whose actions are somewhat similar to those of restriction endonucleases. For example, CRISPR/Cas combinations recognize and cut DNA at precise palindrome-like markers, which provides a survival defense against the foreign DNA of invading viruses and plasmids. A surprising feature is that CRISPR functions like an acquired immune system; it not only destroys foreign DNA sequences but also maintains a "memory" copy of each foreign DNA, which can be reactivated if this same DNA invades again.

The CRISPR/Cas system has revolutionized the process of **gene editing.** The technology is simpler, faster, and more versatile than anything else yet discovered. Researchers and biotechnologists working in almost any field can apply it as a means to remove unwanted DNA from, or add new DNA to, any genome, whether microbial, plant, animal, or human. So far most of the research has involved bacteria, yeasts, and simple animals.

Here is a brief outline of how the system works in bacteria: Bacteria are constantly accumulating foreign DNA from invading viruses and plasmids that could harm or even kill them. The genes contained in CRISPR code for the Cas nuclease enzymes that are the primary destroyers of this foreign DNA. Part of the CRISPR gene complex is a section that stores the copies of foreign DNA (called spacers). These constitute the immune memory for future encounters. Once any foreign DNA (injected by a virus, for example) has first entered the cell, it is copied onto a spacer and also transcribed into a CRISPR RNA (crRNA) molecule (the guide), which the Cas enzyme picks up and transports around the cell. When the enzyme-crRNA complex encounters any foreign DNA that matches the guide RNA, the Cas enzyme cuts the DNA at exact targeted sites, which will permanently disable the virus or disrupt the plasmid. To view a dynamic animation of this process, go to https://www.youtube.com/watch?v=2pp17E4E-O8&list=PLEbm7wJ3WiYxd19t4TIuM_HXRymY22PT1.

CRISPR/Cas has been widely used to "knock out" genes in bacteria and animals to study the effects on organism function when a particular gene has been lost. It is gaining favor as a means of creating transgenic plants and animals. A recent application of CRISPR was to insert genes into embryonic mosquitoes to make the adult mosquitoes resistant to malaria. Because of its high potential to alter genomes, the technology has met with some controversy. It has been known to cause insertional and deletional mutations and other unforeseen problems. After Chinese researchers used it on nonviable human embryos to test the possibilities of gene therapy to knock out dysfunctional genes, scientists around the world called for a moratorium to halt this sort of germ line experimentation until its safety could be ensured. Many biologists think that once these ethical issues have been worked out, CRISPR will become a source of viable treatments for genetic diseases, cancers, and even infections.

 Check Your Progress SECTION 10.3

17. Describe several uses of genetically modified bacteria and viruses in medicine or industry.
18. Explain the techniques for genetically modifying plants.
19. Why must animals usually be modified in the embryo stage to arrive at successful outcomes?
20. Describe the possible ethical or biological problems associated with eating peanuts from an engineered plant or pork from a transgenic pig.
21. What are some ethical and biological considerations when using transgenic animals to manufacture various human products?
22. Outline the uses of gene therapy and gene editing to control genetic diseases and the possible harm they may cause.

10.4 Genome Analysis: DNA Profiling and Genetic Testing

 Expected Learning Outcomes

12. Discuss the concept of the use of DNA in genome identification.

13. Describe two methods in performing a DNA analysis, its possible applications, and how it can be so precise.

14. Describe several applications of DNA profiling and microarray analysis.

A geneticist's joke begins: "How would you describe a cat?" The answer, "ACATACATACAT . . . ," is becoming more and more a reality. As we mentioned earlier, DNA technology has allowed us to accomplish many age-old goals by new and improved means. By remembering that phenotypes, whether human, bacterial, or even viral, are the result of specific sequences of DNA, we are now at a point where knowledge of DNA can be even more effective than observing phenotypes to differentiate among organisms and people. Additionally, DNA can be used to "see" the phenotype of an organism or person no longer present, as when a criminal is

identified by DNA extracted from a strand of hair left behind at the crime scene. Finally, possession of a particular sequence of DNA may indicate an increased risk of a genetic disease. Detection of a specific and unique region of DNA, or **marker,** can identify a person as being at increased risk for cancer or Alzheimer disease long before symptoms arise. The ability to detect diseases before symptoms arise is especially important for conditions such as cancer, where early treatment can make the difference between life and death. With examples like this in mind, let's look at several ways in which new DNA technology offers extraordinary glimpses into the present, past, and future.

DNA Profiling: A Unique Picture of a Genome

Although all DNA is based on a structure of nucleotides, the exact way these nucleotides are combined is unique for each organism. Such genetic variations, or polymorphisms, will vary to some extent even among close relatives (except identical twins, whose DNA comes from the same fertilized egg). A major goal of DNA technology is to discover these unique qualities, emphasize their differences, and provide a pattern for comparison and identification. **DNA profiling** (sometimes called DNA fingerprinting or typing) was developed as a **forensic*** technique in the mid-1980s by English geneticist Alex Jeffreys. The first methods of analysis used restriction enzymes to cut DNA located at specific recognition sites and to target genetic polymorphisms called RFLPs, introduced earlier in the chapter. The results were rendered into a series of bands or "barcodes" that could be used to compare several samples alongside each other. It turns out that hidden within genomes are large numbers of other types of DNA variations that can be used in identifying individuals that are even more specific and widespread than RFLP fingerprints.

Several of the methodologies discussed previously in this chapter are essential to creating a DNA profile. This includes PCR amplification for increasing the number of copies of a certain gene, electrophoresis to separate DNA fragments, and hybridization probes or primers that link to specific nucleotide sequences. The primary idea is to distinguish one isolated piece of DNA from another by comparing the genetic variations at a specific location, or **locus***. Note that DNA profiling is not the same as whole genome sequencing but instead focuses on a few selected genetic loci to produce a "snapshot"of a particular polymorphism pattern. Millions of such differences are known to exist throughout the genome. Many of these polymorphisms are located in regions of the DNA that do not code for protein and are considered normal rather than harmful variations.

The most reliable methods of DNA profiling presently available are (1) **short tandem repeats** (STRs), relatively short fragments of nuclear DNA that are covered below in some detail; (2) **mtDNA** or mitotyping to identify the DNA found in mitochondria; and (3) **single nucleotide polymorphisms** (**SNP**s) that locate variants of a single nucleotide in the population. Important features of STR analysis and mitochondrial profiling will be discussed under the next headings.

* *forensic* (for'-en-sik) The application of science to law, especially as it relates to the investigation of crime.
* *locus* (loh'-kuss) The specific site on a DNA molecule where a particular gene or sequence of nucleotides is located.

STR Profiling

The most routine methods in forensic identification are based on short tandem repeats (STRs). These are segments of DNA that consist of a short chain 2 to 8 nucleotides long. These STRs are repeated in series multiple times. Both the number of nucleotides in a sequence and the number of repeats of it vary throughout the genome. An example of this type of sequence in which 4 bases (GATA) repeat 5 times would read:

AGCT|*GATA*|*GATA*|*GATA*|*GATA*|*GATA*|GGCC

Studies have identified thousands of these repeated sequences, which provide the background data for producing a profile. Almost all of them occur in noncoding regions of the genome, but they are inherited the same way that genes are—as a pair of alleles, one from each parent. The genotype of a given allele can be the same or can be different, just as for eye color or blood type. At the heart of STR analysis is the pattern of nucleotides and number of repeats detected. This form of DNA profiling is mostly conducted by specialized machinery that is computer-driven and computer-analyzed, as outlined in **figure 10.16** and discussed below.

A sample of DNA is isolated from a person or object and tagged with a number of known STR-specific primers to ready them for PCR amplification. These primers not only complement a known sequence of DNA, they also carry fluorescent dyes for later detection. After these sequences are amplified, they are electrophoresed and scanned by a laser that detects the fluorescent pattern of the primers and thus the specific STR alleles. The resultant readout, called an electropherogram, is a series of colored peaks that coincide with the STRs from that sample (figure 10.16*d*). This pattern undergoes further analysis to indicate possible matches and a mathematical probability of the pattern being anyone other than the individual from the sample. The database contains a cross reference of the frequencies of the combinations of STRs from real populations for accuracy in comparisons.

Let us examine how this analysis would work: Imagine that a sample of blood found at a crime scene is thought to have come from a suspect and has tested positive for STR markers (alleles) A, B, C, and D. If marker A is normally present in 1 person out of 80, marker B in 1 person out of 60, marker C in 1 person out of 45, and marker D in 1 person out of 70, the probability that a single person would have all four markers is ($1/80 \times 1/60 \times 1/45 \times 1/70$) or 1 out of 15,120,000. When the suspect's result matches the evidence blood, it is a powerful indicator that the suspect was involved in the crime, especially when supported by traditional evidence, such as the suspect's having been seen near the crime scene. Adding even more STR loci can increase the probability to as high as 1 out of several billion or trillion, solidifying the match to an undeniable degree.

DNA profiling has been widely used to convict or exonerate many thousands of accused defendants in criminal cases. Besides actual fingerprints, criminals often leave other evidence at the site of a crime—hair, skin, semen, blood, saliva, and even "touch DNA" from fingertips. In fact, the technology is so reliable that many older cases for which samples are available are being reopened and solved. The FBI has established a national database (the **Combined DNA Index System** or **CODIS**) to support this

(a) Samples may be taken from many sources, including tissues, semen, saliva, hair, and blood. Nuclear DNA is isolated and run through a PCR amplifier to produce large numbers of selected DNA fragments. During this process, specific primer/probes with fluorescent dyes are incorporated into the DNA so that they can be readily traced (right).

(b) Close up of tagged DNA; the fluorescent primers are specific to known STR sequences that have been associated with human genomes.

(c) Sample of amplified, tagged DNA is loaded into a capillary tube electrophoresis system. As each band migrates through a laser detector, its fluorescent code is scanned, which sends a signal to a computer analyzer that displays the identity of each band.

(d) The pattern reads out as a series of colored peaks. The color and position of the peak indicate a specific STR allele. The height of the peak indicates the amount of DNA in that STR. The red peaks are known standards to verify accuracy of the result. Three profiles are shown here. To have a match, you will see close correlation with regard to location and number of peaks. Can you tell if there is a match? Further analysis will compare the samples and predict the odds that one of the suspects is the source of the evidence.

Sample, suspect #1

Sample from evidence

Sample, suspect #2

Figure 10.16 **(a–d) An Outline of Short Tandem Repeat (STR) Profiling.** This technique is widely used in forensics, geneology, population studies, identification of types of cancer, and genetic diseases. For excellent coverage of these techniques, go to http://nij.gov/training/pages/dna-multimedia.aspx.

effort. The CODIS technology analyzes 13 common STRs, with a 14th that determines sex. This provides a uniform source for comparisons across law enforcement and forensics. Present laws require that prisoners, felons, and other violent offenders have their CODIS profiles entered into this system to keep tabs on known criminals.

The most frequent uses of DNA profiling are in forensics, identifying missing persons, detecting genetic diseases, determining parentage, analyzing the family trees of humans and animals, tracing the lineage of organisms, and identifying microorganisms. In a stunning application of DNA profiling, forensic scientists were able to positively identify the Unibomber, Ted Kaczynski, by DNA recovered from the back of postage stamps he had licked. This same technology has also been effective in identifying the remains of astronauts from the shuttle *Columbia,* some victims of Hurricane Katrina, nearly 2,000 individuals from the World Trade Center attack, and even to verify that the person killed in the raid on a Pakistani compound in 2011 was indeed Osama Bin Laden. This sort of identification of unknown DNA always requires a sample of the known DNA

of the suspected person or close relatives who would share some of the STR alleles.

Mitotyping

Recall that only eukaryotic organisms have mitochondria and that each mitochondrion is self-sustaining and carries its own chromosome. Mitochondrial DNA (mtDNA) consists of about 16,500 base pairs that code for 37 genes. One advantage of mtDNA is that it is highly conserved in evolution and does not change greatly over long periods of time. Even with this small genome, it has a number of unique variations that are sufficiently varied to identify humans, plants, and animals to the level of species and even individuals. Another advantage of mtDNA is that it does not break down as readily as chromosomal DNA and remains intact even after being exposed to harsh conditions for thousands of years. Also, it is inherited only from the mother's egg, because sperm do not contain mitochondria. This means that children from the same mother will have the same mtDNA profile. MtDNA was integral to the identification of the Romanovs—the family of the last czar of Russia, who

were assassinated and buried in a mass grave in 1919. It has also become a tool of geneology as a way to trace the maternal lineage of families and to clarify cases of maternity

Because even fossilized or ancient DNA can remain partially intact, ancient DNA is being studied for comparative and evolutionary purposes. Anthropologists have traced the migrations of ancient people by analyzing the mitochondrial DNA found in bone fragments. Evolutionary biologists have been able to recover Neanderthal mtDNA and use it to show that these ancient primates were in a line of evolution separate from the modern human line. Biologists studying canine origins have also been able to show that the wolf is the ancestral species for modern canines.

Several large databanks are currently collecting genomic profiles on mtDNA markers for thousands of species of plants, animals, and fungi. Two of these—Genbank and Barcodes of Life—offer these sites as a way to keep track of biodiversity (the varieties of life on earth) and to provide information for researchers and commercial operations needing to identify specimens of fish, invertebrates, plants, and other organisms that cannot be identified by the usual methods.

Miscellaneous DNA analysis Single nucleotide polymorphisms, or "SNIPs," are the most specific polymorphisms because they are the most variable, occurring around 10 million times among a person's 3.2 billion base pairs. They appear as single base differences at random sites throughout the chromosomes, both in genes and noncoding DNA. SNPs are a valuable tool in disease diagnosis, population genetics, personalized medicine, and biomedical research. More and more, infectious disease laboratories are employing DNA test procedures based on SNPs to identify bacteria. Other forms of DNA typing are rapidly becoming the method of choice to identify pathogens, including *Neisseria gonorrhoeae*, *Chlamydia*, the syphilis spirochete, *Staphylococcus* MRSA, *M. tuberculosis*, and a number of viruses. It is also an essential tool for determining genetic similarities between microbes involved in disease outbreaks. For more information on SNPs, refer to 10.1 Making Connections. To see a virtual version of bacterial identification, go to http://www. hhmi.org/biointeractive/bacterial-identification-virtual-lab.

Determining Gene Activity with Microarrays

Having a DNA profile or a sequence of a genome is really only half the battle. With very few exceptions, all cells in a multicellular organism contain the same DNA, so knowing the sequence of that DNA, while certainly helpful, is of very little use when comparing two cells or tissues from the same organism. Recall that genes are expressed in response to both internal needs and external stimuli, and that although the DNA content of a cell remains constant, the mRNA (and hence protein) content at any given time provides scientists with a profile of genes currently being expressed in the cell. What truly distinguishes a liver cell from a kidney cell or a healthy cell from a diseased cell is their manner of gene expression.

Twin advances in biology and electronics have allowed biologists to view the expression of genes in any given cell using a technique called DNA *microarray analysis*. Prior to the advent of this technology, scientists were able to track the expression of at most only a few genes at a time. Given the complex interrelationship that exists between genes in a typical cell, tracking two or three genes would be of very little practical use. Microarrays are able to track the

Figure 10.17 **Gene expression analysis using microarrays.** Microarrays display large numbers of genes being expressed at the same time in the same cell. Small slides or chips are coated with fluorescently labeled genes and exposed to unknown cDNAs. A laser excites the bound cDNAs, while a detector records those spots that fluoresce. The color of each spot reveals whether the DNA on the microarray is present in the normal cell, the cancer cell, both cells, or neither. *Source:* Darryl Leja, NHGRI

expression of thousands of genes at once and are able to do so in a single efficient experiment. Microarrays consist of a "chip" (made of glass, silicon, or nylon) that carries bound sequences from tens of thousands of different known genes. A solution containing fluorescently labeled cDNA, representing all of the mRNA molecules active in a cell at a given time, is applied to the chip. The labeled cDNA is allowed to hybridize with any complementary DNA bound to the chip. Bound cDNA is then detected by exciting the fluorescent tag on the cDNA with a laser and recording the fluorescence with a detector linked to a computer. The computer is programmed to interpret this data by highlighting what mRNAs are present in the cell under a variety of conditions. Look at **figure 10.17** for an example of results comparing a normal cell with a cancer cell.

Microarrays permit the development of extraordinarily sensitive diagnostic tests that search for a specific pattern of gene expression. As an example, being able to identify a patient's cancer as one of many subtypes (rather than just, for instance, breast cancer) will allow selection of a cancer drug that appears to be most effective. Again, we see how genetic technology can greatly assist in medical decisions.

✓ *Check Your Progress* SECTION 10.4

23. Describe what a DNA profile is and how STRs and other polymorphic fragments can be used to display the unique qualities of DNA.
24. Discuss briefly how DNA profiling is being used routinely by medicine, the law, the military, and studies of human biology and history.
25. Explain the origins of mtDNA and its importance in identification.
26. Explain the difference between a DNA profile and a microarray.

CASE STUDY Part 2

In 2009, with the clock ticking down on a gravely ill child, specialists in sequencing arranged for Nicholas's genome to be analyzed. Blood samples were used to extract a significant quantity of his genomic DNA. Although a complete sequencing of his genome was considered, it would have added layers of complexity and expense, so the team decided it would be more effective to sequence only the exons— a process called *exome sequencing*—to ensure that they would be looking only at segments of DNA that code for proteins. Even this shortcut was far from easy or simple. It would require examination of 180,000 known exons. They arranged for a software specialist to design a program that could filter out "normal" exons and identify potential bits of his genome that encoded mutations that could produce abnormal proteins. The search located 16,124 possible exons of this type.

After additional passes of the DNA through the sequencer to eliminate mutations that were common and harmless, the program eventually narrowed the potentially harmful exon mutations down to about 2,000. Now the search became focused on which of these genes would be most likely involved in immune dysfunction and inflammatory bowel disease. After several weeks the specialists found a gene termed *XIAP* known to be involved in immune regulation. It was located on the X chromosome, which means that a male child will express the mutated form of it because there is no protection from an alternate allele on the Y chromosome. This gene and its protein, also termed *XIAP*, is an immune regulator with the normal function of preventing programmed cell death. If it has become mutated and unable to perform its usual function, cells will spontaneously die, leading to destruction of tissues and organs.

With further scrutiny, the nature of Nicholas's mutation became clear. He had inherited a single base substitution on the *XIAP* gene from his mother. The substitution was a guanine (G) being replaced with an adenine (A), changing the codon from TGT to TAT(see the figure in the Case Study for this chapter). This would also switch the amino acid at that position on the *XIAP* protein from cysteine to tyrosine, which would interfere with the normal three-dimensional folding of the protein. The result was such a rare alteration that it has not been found in any other known genome.

- Describe what exons are and why they would be the best choice for diagnosing a genetic disease.

- Explain how bioinformatics were applied in this case.

To conclude this Case Study, go to Connect.

Chapter Summary with Key Terms

10.1 Basic Elements and Applications of Genetic Engineering
Genetic engineering refers to the manipulation of an organism's genome and is often used in conjunction with biotechnology, the use of an organism's biochemical and metabolic pathways for industrial production of proteins.
 A. *Practical Properties of DNA*
 1. **Restriction endonucleases** are used to cut DNA at specific recognition sites, resulting in small segments of DNA that can be more easily managed in the laboratory.
 2. The enzyme **ligase** is used to covalently join DNA molecules, whereas **reverse transcriptase** is used to obtain a DNA copy of a particular mRNA.
 3. Short strands of DNA called oligonucleotides are used either as probes for specific DNA sequences or as primers for DNA polymerase.
 B. *Nucleic Acid Hybridization and Probes*
 1. The two strands of DNA can be separated (denatured) by heating. These single-stranded nucleic acids will bind, or **hybridize,** to their complementary sequence, allowing them to be used as **gene, hybridization,** or **probes.**
 a. In the Southern blot method, DNA isolated from an organism is probed to identify specific DNA sequences.
 b. **FISH** is a method of hybridizing and identifying intact DNA inside cells.
 C. *Methods Used to Size, Synthesize, and Sequence DNA*
 1. A technique called **DNA sequencing,** originally developed by Frederick Sanger, allows for the determination of the exact sequence of DNA.

 a. The ultimate map is a whole genome sequence map, which provides the exact order of all of the nucleotides that make up an entire genome of an organism. This can provide detailed information on the function of genes and can be useful in identification of genetic diseases. Sequencing is also a common method of choice to identify microbes.
 2. The **polymerase chain reaction** is a technique used to amplify a specific segment of DNA several million times in just a few hours.
 3. The Human Genome Project has completed a sequence map for humans, identifying the sequence of the approximately 21,000 genes found in our DNA.

10.2 Recombinant DNA Technology: How to Imitate Nature
 A. *Technical Aspects of Recombinant DNA Technology*
 1. **Recombinant DNA** (rDNA) methods have as their goal the transfer of DNA from one organism to another. This commonly involves inserting the DNA into a **vector** (usually a plasmid or virus) and transferring the recombinant vector to a **cloning host** (usually a bacterium or yeast) for expression.
 a. Examples of cloning vectors include plasmids, bacteriophages, and cosmids, as well as the larger bacterial and yeast artificial chromosomes (BACs and YACs).
 b. Cloning hosts are fast-growing, nonpathogenic organisms that are genetically well understood. Examples include *E. coli* and *Saccharomyces cerevisiae.*

2. **Recombinant organisms** are created by cutting both the donor DNA and the plasmid DNA with restriction endonucleases and then joining them together with ligase. The vector is then inserted into a cloning host by transformation, and those cells containing the vector are cultured and selected by their resistance to antibiotics.

3. When the appropriate recombinant cell has been identified, it is grown on an industrial scale. The cloning host expresses the gene it contains, and the resulting protein is excreted from the cell where it can be recovered from the media and purified.

B. *Protein Products of Recombinant DNA Technology*

1. In many instances, the final product of recombinant DNA technology is a protein.

2. Medical conditions such as diabetes, dwarfism, and blood-clotting disorders can be treated with recombinant insulin, human growth factor, and clotting factors produced by rDNA.

10.3 Genetically Modified Organisms and Other Applications

A. Recombinant organisms, also known as **transgenics** and **genetically-modified organisms** (GMO) include microbes, plants, and animals that have been engineered to express genes inserted from another organism.

1. Transformed bacteria produce a wide variety of drugs, vaccines, and other useful materials.

2. Modified viruses function as genetic vectors and vaccines.

3. Plants are **transfected** by a bacterium *Agrobacterium,* that can shuttle donor genes into cells.

4. Animals are genetically engineered by having selected genes inserted into an embryo.

B. *Medical Applications of DNA Technology*

1. **Gene therapy** describes a collection of techniques in which defective genes are replaced or have their expression increased or decreased. Therapy is said to be *in vivo* if cells are altered within the body and *ex vivo* if the alteration is done on cells taken from the body.

2. Several proposed treatments involve gene silencing with regulatory RNA.

10.4 Genome Analysis: DNA Profiling and Genetic Testing

A. **DNA Profiling**

1. Because the DNA sequence of an organism is unique, it can be analyzed by identifying specific genetic markers or combinations of DNA sequences unique to each individual. These are known as **polymorphisms** ("many forms") because they vary widely within a population.

2. Common polymorphisms include **short tandem repeats (STRs),** which are repeats of the same nucleotide sequence, and **single nucleotide polymorphisms (SNPs),** which are variations of a single nucleotide at a given locus on a chromosome.

3. Technologies exist for targeting these markers for use in forensic identification, disease diagnosis, anthropology, and identifying microorganisms.

4. **Mitochondrial DNA (MtDNA)** also has sequences unique to a given eukaryotic species. It is useful in tracing family history and ancient origins, and in identifying species of animals, plants, and fungi.

B. *Microarray analysis* is a method of determining which genes are actively transcribed in a cell under a variety of conditions, such as health versus disease or growth versus differentiation.

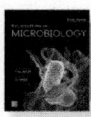

Level I. Knowledge and Comprehension

These questions require a working knowledge of the concepts in the chapter and the ability to recall and understand the information you have studied.

 Multiple-Choice Questions

Select the correct answer from the answers provided. For questions with blanks, choose the combination of answers that most accurately completes the statement.

1. Which gene is incorporated into plasmids to detect recombinant cells?
 a. restriction endonuclease
 b. virus receptors
 c. a gene for antibiotic resistance
 d. reverse transcriptase

2. Which of the following is *not* essential to carry out the polymerase chain reaction?
 a. primers
 b. DNA polymerase
 c. gel electrophoresis
 d. high temperature

3. Which of the following is *not* a part of the Sanger method to sequence DNA?
 a. dideoxynucleotides
 b. DNA polymerase
 c. electrophoresis
 d. reverse transcriptase

4. The function of ligase is to
 a. rejoin segments of DNA
 b. make longitudinal cuts in DNA
 c. synthesize cDNA
 d. break down ligaments

5. The pathogen of plant roots that is used as a cloning host is
 a. *Pseudomonas*
 b. *Agrobacterium*
 c. *Escherichia coli*
 d. *Saccharomyces cerevisiae*

6. Which of the following sequences, when combined with its complement, would be clipped by a restriction endonuclease?
 a. ATCGATCGTAGCTA
 b. AAGCTTCGAA
 c. GAATTC
 d. ACCATTGGA

7. Which DNA fragment will be closest to the top (negative pole) of an electrophoretic gel?
 a. 450 bp
 b. 3,560 bp
 c. 5 kb
 d. 1,500 bp

8. Which of the following is or are required for cloning a gene?
 a. restriction endonuclease
 b. vector
 c. cloning host
 d. a and b
 e. all of these

9. For which of the following would not require a nucleic acid probe?
 a. locating a gene on a chromosome
 b. developing a Southern blot
 c. identifying a microorganism with FISH
 d. constructing a recombinant plasmid

10. **Multiple Matching.** Match the term with its description.
 _____ nucleic acid probe
 _____ nontemplate strand
 _____ template strand
 _____ reverse transcriptase
 _____ Taq polymerase
 _____ RNAi
 _____ primer
 _____ restriction endonuclease
 a. enzyme that transcribes RNA into DNA
 b. complementary strand that blocks mRNA expression
 c. the nontranslated strand of DNA or RNA
 d. enzyme that snips DNA at palindromes
 e. oligonucleotide that initiates the PCR
 f. strand of nucleic acid that is transcribed or translated
 g. thermostable enzyme for synthesizing DNA
 h. oligonucleotide used in hybridization

 Case Study Review

1. What type of mutation caused Nicholas's disease?
 a. frameshift c. nonsense
 b. missense d. insertion

2. Which type of cells were used to extract the DNA that was sequenced?
 a. red blood cells c. white blood cells
 b. intestinal epithelium d. cheek swab

3. Lay out the genetics of Nicholas's case, including where the mutation occurred, what exact nucleotide defect was present, why this was a serious type of defect, and why he was so affected by the mutation but his mother was not.

 Writing Challenge

For each question, compose a one- or two-paragraph answer that includes the factual information needed to completely address the question. Check Your Progress questions can also be used for writing-challenge exercises.

1. Why do bacteria make restriction endonucleases?

2. What is it about the endonucleases that prevents bacteria from destroying their own DNA? Using nucleotide letters, show the kind of cut that could be made on a DNA molecule to circularize it into a plasmid.

3. Why would it be an advantage to synthesize eukaryotic genes by using reverse transcriptases to make cDNA?

4. a. Explain what hybridization is and how it is used in the Southern blot.
 b. Explain how gene probes and FISH techniques work.

5. Look at figure 10.7, and determine the correct DNA sequence for the fragment's template strand. (Be careful of orientation.)

6. a. Explain why the PCR is unlikely to amplify contaminating bacterial DNA in a sample of human DNA.

 b. Explain how PCR could be used to pick a gene out of a complex genome and amplify it.

7. The way PCR amplifies DNA is similar to the doubling in a population of growing bacteria—a single DNA strand is used to synthesize 2 DNA strands, which become 4, then 8, then 16, etc. If a complete cycle takes 3 minutes, how many strands of DNA would theoretically be present after 10 minutes? After 1 hour?

8. Explain the kinds of study involved in genomics, proteomics, and bioinformatics.

9. For what reasons would gene therapy be more effective performed on embryos rather than fully developed animals?

10. Explain the basics of DNA STR profiling for identification of unknown DNA.

 Concept Mapping

On Connect you can find an Introduction to Concept Mapping that provides guidance for working with concept maps and a sample of a concept map for this chapter.

 Level II. Application, Analysis, Evaluation, and Synthesis

These problems go beyond just restating facts and require higher levels of understanding and an ability to interpret, problem solve, transfer knowledge to new situations, create models, and predict outcomes.

⚙️ Critical Thinking

Critical thinking is the ability to reason and solve problems using facts and concepts. These questions can be approached from a number of angles, and in most cases, they do not have a single correct answer.

1. a. Give an example of a benefit of genetic engineering to society and a possible adverse outcome. Discuss.
 b. Give an example of an ecological benefit and a possible adverse side effect. Discuss.

2. a. When gene probes, DNA profiling, and sequencing make it possible to detect genetic diseases in you or one of your children, would you want you or them to be tested?
 b. How would you react to it being used as a screening method for employment or insurance?
 c. Most of us would agree to growth hormone therapy for a child with dwarfism, but what would you think about parents who want to give growth hormones to their 8-year-old son so that he will be "better at sports"?

3. Which suspect is the likely perpetrator according to the profile in figure 10.16? Explain your reasoning.

4. Trace the genetic steps in the development of a peanut plant that is resistant to insect larvae (table 10.3). (Hint: You need to use *Agrobacterium.*)

5. You are on a jury to decide whether a person committed a homicide, and you have to weigh evidence from DNA profiling. Two different sets of profiles were done: one that tested 5 markers, and one that tested 10. Both sets match the defendant's profile. Which one is more reliable and why?

6. Can you think of some reasons it would *not* be possible to re-create dinosaurs using the technology we have described in this chapter?

7. What would be some major impediments to genetically modifying human embryos with CRISPR?

8. Could actual fingerprints taken from human fingers be used to perform a DNA profile? Explain.

9. Describe the main differences between genome sequencing and DNA profiling as a means of identifying a person or microbe.

10. Itemize all of the ways that microbes have contributed to the development of genetic engineering and biotechnology.

👁 Visual Challenge

1. Below are two unrelated DNA paternity tests: one indicates the man is the father and the other shows he is not. Can you tell which is which? C/AF = Child/Alleged father in one lane combined.

Courtesy of Genelex Corporation

2. **Figure 9.25d,** shown here, shows the original transformation experiment as done by Griffith. Explain how this outcome would be accomplished with modern technology, using figure 10.11 as your guide.

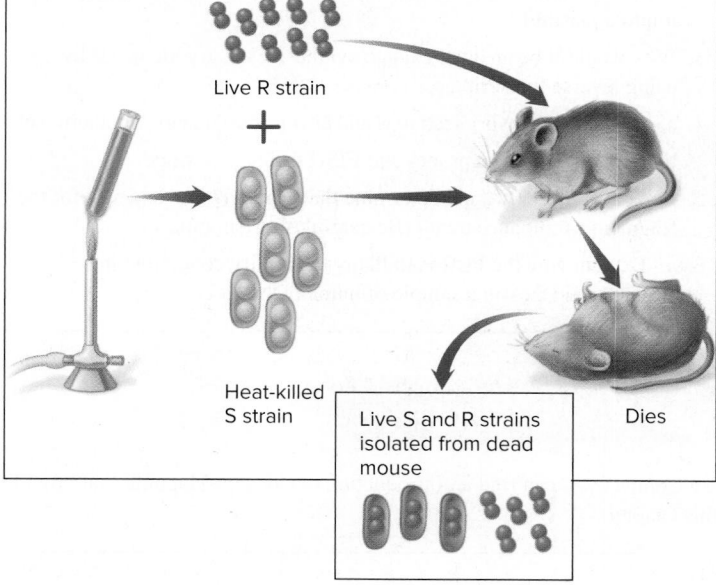

Live R strain

Heat-killed S strain

Live S and R strains isolated from dead mouse

Dies

STUDY TIP: Quickly create a customized practice quiz for this chapter by going to your Connect SmartBook "My Reports" section. Don't have SmartBook and want to learn more? Go to **whysmartbook.com.**

11 Physical and Chemical Agents for Microbial Control

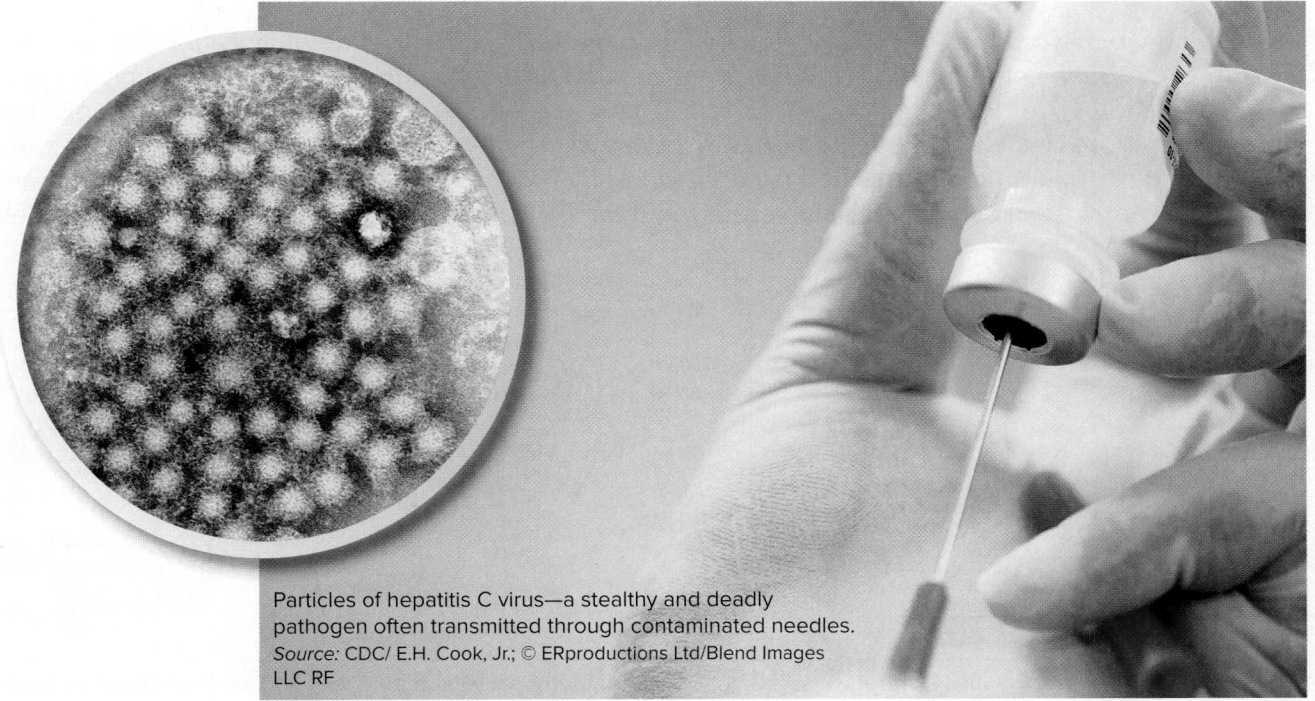

Particles of hepatitis C virus—a stealthy and deadly pathogen often transmitted through contaminated needles.
Source: CDC/ E.H. Cook, Jr.; © ERproductions Ltd/Blend Images LLC RF

CASE STUDY Part 1 The Billion Dollar Medical Error

Late in 2007 a middle-aged man was hospitalized in Las Vegas with acute symptoms of **abdominal pain, nausea,** and **vomiting.** The attending physician also noticed that his eyes and skin were **jaundiced** (yellowed) and his urine was a dark color. Laboratory tests revealed an increased level of **liver enzymes** and **antibodies** to **hepatitis C virus** (HCV), and a **nucleic acid test** confirmed that **RNA** from HCV was present in his blood. Upon questioning, the patient denied having any of the **risk factors** for this infection, such as **IV drug abuse** or **exposure to blood or blood products.** Further probing disclosed that he had recently undergone a dental procedure and two **endoscopic examinations** of his gastrointestinal tract. This case was immediately reported to the Nevada Office of Epidemiology.

Within a month, two other cases of **acute** HCV infection were diagnosed. These patients also claimed to lack the risk factors that caused this infection and they, too, had recently had endoscope procedures performed at the same clinic as the first patient. The seriousness of these findings made it necessary to request assistance from the **Centers for Disease Control and Prevention** (CDC). The connection of the three patients to the same endoscopy center made this facility the prime focus of the investigation.

> "An RN-anesthetist was observed drawing a second dose of the drug from a large, multidose vial using the same syringe that had been used to take the first dose, although he did replace the needle."

The clinic had a large, thriving practice doing **colonoscopies** and other procedures on several thousand people each year. Because this raised the potential for other patients to have become infected with the virus, the CDC required that 63,000 persons who had a procedure in this clinic be immediately notified of their possible **exposure to the virus.**

It was essential to verify whether the three patients were infected by the same **genetic strain** of HCV. When it turned out that the viruses were genetically identical, authorities were certain that the infections originated from the same **source.** Knowing that the three patients had never had any kind of contact, **epidemiologists** concluded that the clinic had to be the origin of the infection. This finding alone did not provide any clues as to what went wrong, but it had tremendous repercussions on the clinic, its patients, and the medical and insurance companies involved.

■ *How are acute and chronic infections different, and why was it so important to differentiate them in this case?*

■ *What sorts of procedures in a clinic might expose a person to bloodborne diseases?*

To continue the Case Study, go to Case Study Part 2 at the end of the chapter.

SCOPING OUT THE CHAPTER

Though we may not be aware of it, most of us go through a daily routine of activities that prevent infection and reduce contamination of the substances and objects around us. Whether it is brushing teeth, washing hands, refrigerating food, or scrubbing out the sink, we count on these simple habits to make our lives more healthful. In the medical and industrial arena, it is even more critical to use proven techniques that will get rid of unwanted microbes in body tissues, products, food, and the environment. In this chapter we will examine the numerous ways we go about managing the microbes that are all around us and on us.

Basic Goals of Antimicrobial Control
Antimicrobial means "against microbes" and includes any method for keeping undesirable, harmful microbes out of the "wrong places". Control agents are applied for sterilization, disinfection, or antisepsis.

Physical Methods: Heat that is Wet or Dry Applying increased temperature or heat is the most common method of achieving microbial death. Heat can be applied in combination with water (moist heat), or water-free (dry heat).

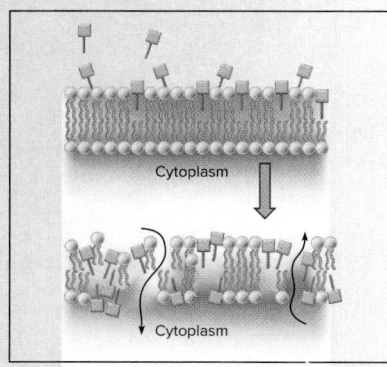

Agents of Microbial Death: How Do They Work? Both physical (energy-based) and chemical (molecule-based) agents are directed against microbial structures and functions and designed to block their reproduction and survival.

Physical Methods: Zapping with Radiation
High energy wavelengths such as gamma rays, X-rays, and even ultraviolet rays can sterilize and disinfect materials without using heat.

Chemical Methods: Molecules That Destroy Microbes. A wide range of chemicals can kill or inhibit microbes on inanimate objects or the human body. Examples are disinfectants, antiseptics, sanitizers, or degermers.

Common Chemical Categories:
Antimicrobial chemicals come in the form of liquids, gases, or solids. Classes of chemical agents include chlorine, iodine, alcohol, phenolics, peroxides, aldehydes, gases, detergents, and soaps.

(Physical Methods) © Kathy Park Talaro; (Chemical Methods) © OJO Images/SuperStock RF

11.1 Controlling Microorganisms

 Expected Learning Outcomes

1. Summarize the major categories of microbial control and its purposes.
2. Evaluate the relative resistance of different groups of infectious agents.
3. Define and differentiate among the major terms for microbial control, citing examples of each.
4. Characterize the parameters of microbial death, and describe several factors that influence the rate of death.
5. Summarize what factors influence the effectiveness of an antimicrobial agent.
6. Identify the targets of antimicrobial control agents, and explain what effects these agents have.

Much of the time in our daily existence, we take for granted tap water that is drinkable, food that is not spoiled, shelves full of products to eradicate "germs," and drugs to treat infections. Controlling our degree of exposure to potentially harmful microbes is of monumental concern to our lives, and it has been a concern even before we were aware of the germ-disease connection.

Historians have discovered that during the great plague pandemic of the Middle Ages it was commonplace to bury corpses in mass graves, burn the clothing of plague victims, and ignite aromatic woods in the houses of the sick in the belief that fumes would combat the disease. In a desperate search for some sort of protection, survivors wore peculiar garments and anointed their bodies with herbs, strong perfume, and vinegar **(figure 11.1)**. These attempts may sound foolish and antiquated, but it now appears that

Figure 11.1 Illustration of protective clothing used by doctors in the 1700s to avoid exposure to plague victims. The beaklike portion of the hood contained volatile perfumes to protect against foul odors and possibly inhaling "bad air."
© Universal History Archive/UIG/Getty Images

they may have had some benefits. Burning wood releases formaldehyde, which could have acted as a disinfectant; herbs, perfume, and vinegar contain mild antimicrobial substances. Each of these early methods, although somewhat crude, laid the foundations for microbial control methods that are still in use today.

General Considerations in Microbial Control

The methods of microbial control belong to the general category of *decontamination* procedures, in that they destroy or remove contaminants. In microbiology, contaminants are microbes present at a given place and time that are undesirable or unwanted. Most decontamination methods employ either physical agents, such as heat or radiation, or chemical agents, such as disinfectants and antiseptics. This separation is convenient, even though the categories overlap in some cases; for instance, radiation can cause damaging chemicals to form, and chemicals can generate heat. A flowchart **(figure 11.2)** summarizes the major applications and aims in microbial control.

Relative Resistance of Microbial Forms

Processes for controlling microbes generally target microorganisms that are constantly present in the external environment and on the human body and that could cause infection or spoilage. This targeted population is rarely simple or uniform; in fact, it often contains mixtures of microbes with extreme differences in resistance and harmfulness. Contaminants that can have detrimental effects if not adequately controlled include bacterial vegetative cells and endospores, fungal hyphae and spores, yeasts, protozoan trophozoites and cysts, worms, viruses, and prions. This scheme compares the general resistance of these forms to physical and chemical methods of control:

Highest resistance
> Prions: proteinaceous infectious particles (see "The Problem with Prions"); bacterial endospores, primarily from bacteria in the genera *Bacillus* and *Clostridium*.

Moderate resistance
> Protozoan cysts; some fungal sexual spores (zygospores); some viruses. In general, naked viruses are more resistant than enveloped forms.
>
> Several groups of bacteria lack endospores but have vegetative cells with increased resistance to microbial control agents. These include *Mycobacterium,* whose cell walls have thick, waxy layers that can block the entrance of many disinfectants. The cell walls of *Pseudomonas, Acinetobacter,* and other gram-negative pathogens have an outer membrane that also prevents penetration of some chemical agents. *Staphylococcus* species are among the most heat- and chemical-resistant of all bacteria due to their thick, protective peptidoglycan cell walls.

Least resistance
> Most bacterial vegetative cells; fungal spores (other than zygospores) and hyphae; enveloped viruses; yeasts; and protozoan trophozoites

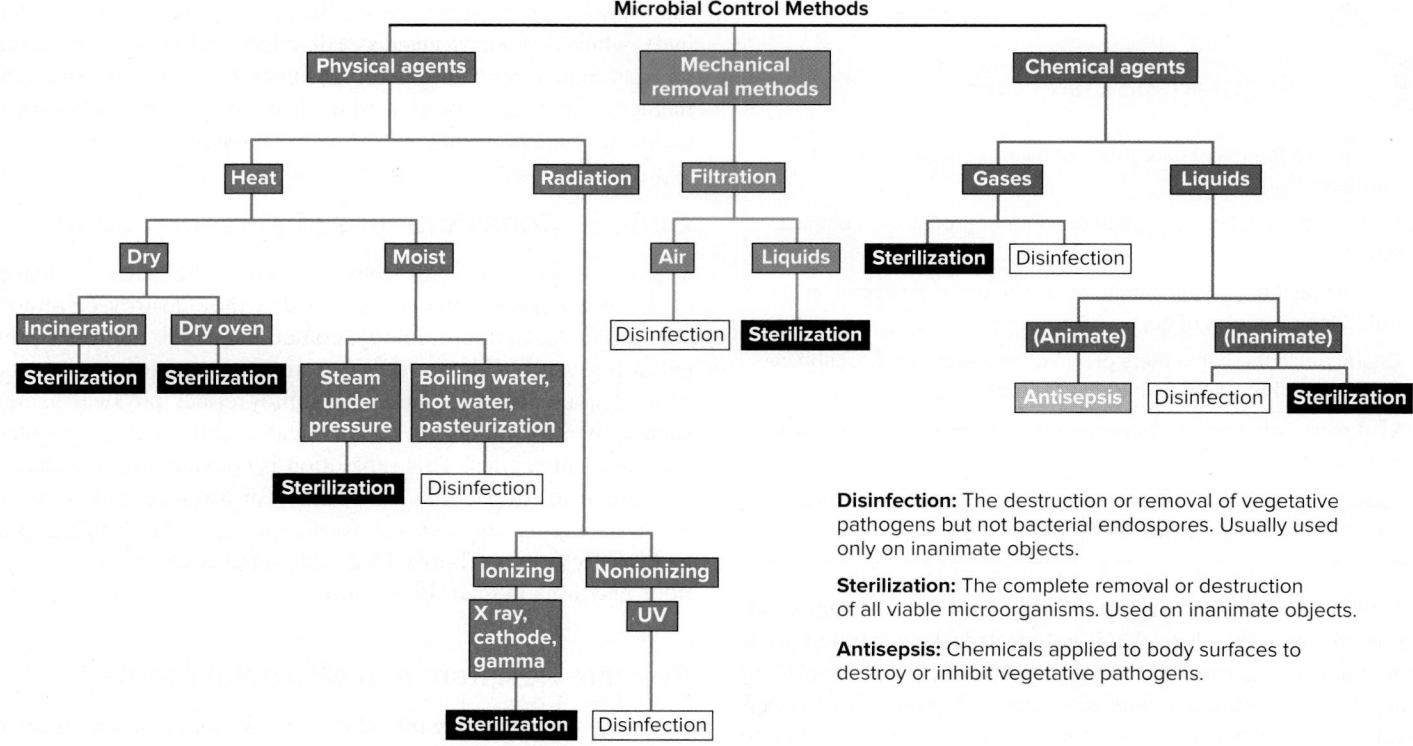

Figure 11.2 An overview of microbial control methods.

Actual comparative figures on the requirements for destroying various groups of microorganisms are shown in **table 11.1.** Bacterial endospores have traditionally been considered the most resistant microbial entities, being as much as 18 times harder to destroy than vegetative cells. Because endospores show such high resistance to microbial control and have widespread distribution, destroying them is the goal of **sterilization.** Any process that kills endospores will invariably kill all less-resistant microbial forms. Other methods of control (disinfection, antisepsis) target primarily microbes that are less hardy than endospores.

Endospore of *Bacillus* sp

S.J. Jones, C.J. Paredes, B. Tracy, N. Cheng, R. Sillers, R.S. Senger, E.T. Papoutsakis, "The transcriptional program underlying the physiology of clostridial sporulation," *Genome Biol.,* 2008. 9:R114

The Problem with Prions

The role of prions as infectious protein particles in certain progressive brain diseases (such as Creutzfeldt-Jakob disease) was first introduced in chapter 6. Not only are prions highly unusual in their actions, but they are also in a class of their own when it comes to "sterilization" procedures. This chapter defines *sterile* as the absence of all viable microbial life. But prions are not microbes or living, and most of the procedures described here are not sufficient to destroy them. Prion proteins are extraordinarily resistant to heat, chemicals, and radiation. Instruments or other objects contaminated with these unique agents must be incinerated, at very

high temperatures, or decontaminated with a combination of chemicals. The substances that have proved most effective are sodium hydroxide and hypochlorite (bleach) solutions that render them noninfectious after a significant period of exposure. When tissues, fluids, or instruments are suspected of containing prions, consultation with the CDC is recommended to ensure effective sterilization conditions. In terms of routine sterilization procedures, this chapter will treat bacterial endospores as the most resistant form of microbial life targeted by antimicrobial agents. Prions are not environmentally widespread in the way that bacterial endospores and other microbes are.

TABLE 11.1	Relative Resistance of Bacterial Endospores and Vegetative Cells to Control Agents		
Method	Endospores*	Vegetative Forms*	Relative Resistance**
Heat (moist)	120°C	80°C	1.5×
Radiation (X-ray) dosage	4,000 grays	1,000 grays	4×
Sterilizing gas (ethylene oxide)	1,200 mg/1	700 mg/1	1.7×
Sporicidal liquid (2% glutaraldehyde)	3 hours	10 min	18×

*Values are based on methods (concentration, exposure time, intensity) that are required to destroy the most resistant pathogens in each group.

**This number is a comparison of the conditions required to kill endospores versus vegetative cells and is an average of the most resistant pathogens in each group.

Terminology and Methods of Microbial Control

Through the years, a growing terminology has emerged for describing and defining measures that control microbes. To complicate matters, the everyday use of some of these terms can at times be vague or inexact. For example, occasionally one may be directed to "sterilize" or "disinfect" a patient's skin, even though this usage does not fit the technical definition of either term. To lay the groundwork for the coverage to follow, we present here a series of concepts, definitions, and usages in antimicrobial control.

Sterilization

Sterilization is a process that destroys or removes all viable microorganisms, including viruses. Any material that has been subjected to this process is said to be **sterile.** These terms should be used only in the strictest sense for methods that have been proved to sterilize. An object cannot be slightly sterile or almost sterile—it is either sterile or not sterile. Control methods that sterilize are generally reserved for inanimate objects, because sterilizing tissues within a living body would call for such harsh treatment that it would be highly dangerous and impractical. As we shall see in chapter 13, many parts of the body—the brain and most internal organs, for example—are naturally free of microbes.

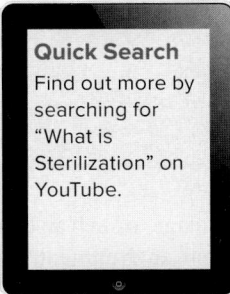

Quick Search

Find out more by searching for "What is Sterilization" on YouTube.

Sterilized products—surgical instruments, syringes, and commercially packaged foods, just to name a few—are essential to human well-being. Although most sterilization is performed with a physical agent such as heat, a few chemicals called *sterilants* can be classified as sterilizing agents because of their ability to destroy spores.

At times, sterilization is neither practicable nor necessary, and only certain groups of microbes are of concern. Some antimicrobial agents eliminate only the susceptible vegetative states of microorganisms but do not destroy the more resistant endospore and cyst stages. Keep in mind that the destruction of spores is not always a necessity, because most of the infectious diseases of humans and animals are caused by non-spore-forming microbes.

Microbicidal Agents

The root *-cide,* meaning to kill, can be combined with other terms to define an antimicrobial agent aimed at destroying a certain group of microorganisms. For example, a **bactericide** is a chemical that destroys bacteria except for those in the endospore stage. It may or may not be effective on other microbial groups. A **fungicide** is a chemical that can kill fungal spores, hyphae, and yeasts. A **virucide** is any chemical known to inactivate viruses, especially on living tissue. A **sporicidal** agent can destroy bacterial endospores, which makes it also a sterilizing agent.

Agents That Cause Microbistasis

The words *stasis* and *static* mean to stand still. They can be used in combination with various prefixes to denote a condition termed **microbistasis** in which microbes are temporarily prevented from multiplying but are not killed outright. Although killing or permanently inactivating microorganisms is the usual goal of microbial control, microbistasis can have meaningful applications. **Bacteriostatic** agents prevent the growth of bacteria on tissues or on objects in the environment, and *fungistatic* chemicals inhibit fungal growth. Materials used to control microorganisms in the body (antiseptics and drugs) often have microbistatic effects because many microbicidal compounds are highly toxic to human cells.

Germicides, Disinfection, Antisepsis

A **germicide,*** also called a *microbicide,* is any chemical agent that kills pathogenic microorganisms. A germicide can be used on inanimate (nonliving) materials or on living tissue, but it ordinarily cannot kill resistant microbial cells. Any physical or chemical agent that kills "germs" is said to have germicidal properties.

The related term **disinfection*** refers to the use of a physical process or a chemical agent—a **disinfectant**—to destroy vegetative pathogens but not bacterial endospores. It is important to note that disinfectants are normally used only on inanimate objects because, at the concentrations required to be effective, they can be toxic to living tissue. Disinfection processes can also remove the harmful products of microorganisms (toxins) from materials. Examples of disinfection include applying a diluted solution of bleach to an examination table, exposing drinking water to UV radiation, and pasteurizing milk with heat.

In modern usage, **sepsis** is defined as the growth of microorganisms in the blood and other tissues. The term **asepsis*** refers to any practice that blocks the entry of infectious agents into sterile tissues and thus prevents infection. Aseptic techniques commonly practiced in health care range from sterile methods that exclude all microbes to **antisepsis.*** With antisepsis, chemical agents called **antiseptics** are applied directly to exposed body surfaces (skin and mucous membranes), wounds, and surgical incisions to destroy or inhibit vegetative pathogens. Examples of antisepsis include preparing the skin before surgical incisions with iodine compounds, swabbing a wound with hydrogen peroxide, and ordinary hand washing with a germicidal soap.

Methods That Reduce the Numbers of Microorganisms

Although medical applications quite often require sterility, especially with regard to invasive instruments such as scalpels, clamps, dental hand tools, and the like, in other situations this high level of microbial control is often unwarranted and perhaps even unwanted. In many cases it is more important to focus on reducing the size of a microbial population, or its *microbial load.* **Sanitization*** refers to any cleansing technique that removes debris, soil, microorganisms, and toxins, and in this way reduces the potential for infection and spoilage. Soaps and detergents are the most commonly

* *germicide* (jer′-mih-syd) L. *germen,* germ, and *caedere,* to kill. *Germ* is a common term for a pathogenic microbe.
* *disinfection* (dis″-in-fek′-shun) L. *dis,* apart, and *inficere,* to corrupt.
* *asepsis* (ay-sep′-sis) Gr. *a,* no or none, and *sepsis,* decay.
* *antisepsis anti,* against.
* *sanitization* (san″ih-tih-zay′-shun) L. *sanitas,* health.

TABLE 11.2	Microbial Control Terminology	
Term	Definition	Example
Decontamination	The destruction, removal, or reduction in number of undesirable microbes	Asepsis, disinfection, sanitization, degermation, steriization
Sepsis	The growth of microorganisms in the tissues	The presence of infected wounds, blood infection
Asepsis	Techniques that prevent the entry of microorganisms into sterile tissues	Cleansing the skin with iodine prior to surgery, using sterile needles
Antiseptic	Chemicals applied to body surfaces to destroy or inhibit vegetative pathogens	Iodophors, antibacterial soap, chlorhexidine
Disinfection	Destruction of vegetative pathogens on inanimate objects	5% bleach, boiling water
Sanitization	Cleansing technique that removes microorganisms and debris from inanimate surfaces	Dishwashing, laundering clothes
Degermation	Cleansing technique that removes microorganisms and debris from living tissue	Surgical handscrub, alcohol wipes
Sterilization	The removal or destruction of all viable microbes	Autoclaving, ionizing radiation (correctly applied)

employed sanitizers. From a practical standpoint, sanitization can have an advantage over sterilization. For example, in a restaurant you could be given a sterile fork with someone else's old food on it and a sterile glass with lipstick on the rim. Given that sanitization is in general far less expensive than sterilization, its cleansing effect has obvious advantages.

With living tissues the reduction in microbial load through mechanical means is referred to as **degermation.** This process usually involves scrubbing the skin or immersing it in chemicals, or both. Microorganisms are removed both mechanically and indirectly by emulsifying the oils that coat the skin and provide a habitat for their growth. Examples of degerming procedures include the surgical handscrub, the application of alcohol wipes, and the cleansing of a wound with germicidal soap and water. **Table 11.2** summarizes several of the important terms involved in microbial control.

What Is Microbial Death?

Death is a phenomenon that involves the permanent termination of an organism's vital processes. Signs of life in complex organisms such as animals are self-evident, and death is made clear by loss of nervous function, respiration, or heartbeat. In contrast, death in microscopic organisms that are composed of just one or a few cells is hard to detect, because such organisms often reveal no conspicuous vital signs to begin with. Lethal agents (such as radiation and chemicals) do not necessarily alter the overt appearance of microbial cells. Even the loss of movement in a motile microbe cannot be used to indicate death. This situation has made it necessary to develop special qualifications that define and delineate microbial death and answer the questions: How do antimicrobial agents cause death? and Why do they vary in their effects?

The destructive effects of chemical or physical agents occur at the cellular and molecular level. As a cell is exposed to an agent such as intense heat or toxic chemicals, various cell structures break down, and the entire cell can sustain irreversible damage. At present, the most practical way to detect this damage is to determine if a microbial cell can still reproduce when exposed to a suitable environment. If the microbe has sustained metabolic or structural

damage to such an extent that it can no longer reproduce, even in an ideal environment, then it is considered nonviable. **In most circumstances of testing, the permanent loss of reproductive capacity, even under optimal growth conditions, has become the working definition of microbial death.**[1]

Factors That Affect Death Rate

The ability to define microbial death has tremendous theoretical and practical importance. It allows medicine and industry to test the conditions required to destroy microorganisms, to pinpoint the ways antimicrobial agents kill cells, and to establish standards of sterilization and disinfection in these fields. Hundreds of testing procedures have been developed for evaluating physical and chemical agents. Some examples can be found in Appendix B.

The cells of a culture show marked variation in susceptibility to a given microbicidal agent. Death of the whole population is not instantaneous but begins when a certain threshold of the microbicidal agent (some combination of time and concentration) is met. Death continues in a logarithmic manner as the time of exposure is increased **(figure 11.3)**. Because many microbicidal agents target the cell's metabolic processes, active cells (younger, rapidly dividing) tend to die more quickly than those that are less metabolically active (older, dormant). Eventually a point is reached at which survival of any cells is highly unlikely; this point is equivalent to sterilization.

The effectiveness of a particular agent is governed by several factors besides time. These additional factors influence the action of antimicrobial agents:

1. The number of microorganisms **(figure 11.3b)**. A higher load of contaminants requires more time to destroy.
2. The nature of the microorganisms in the population **(figure 11.3c)**. In most actual circumstances of disinfection and sterilization, the target population is not a single species of

1. One drawback to using this definition of microbial death is that some microbes may have entered a viable but nonculturable state, making it difficult to know if the antimicrobial agent is microbicidal or microbistatic.

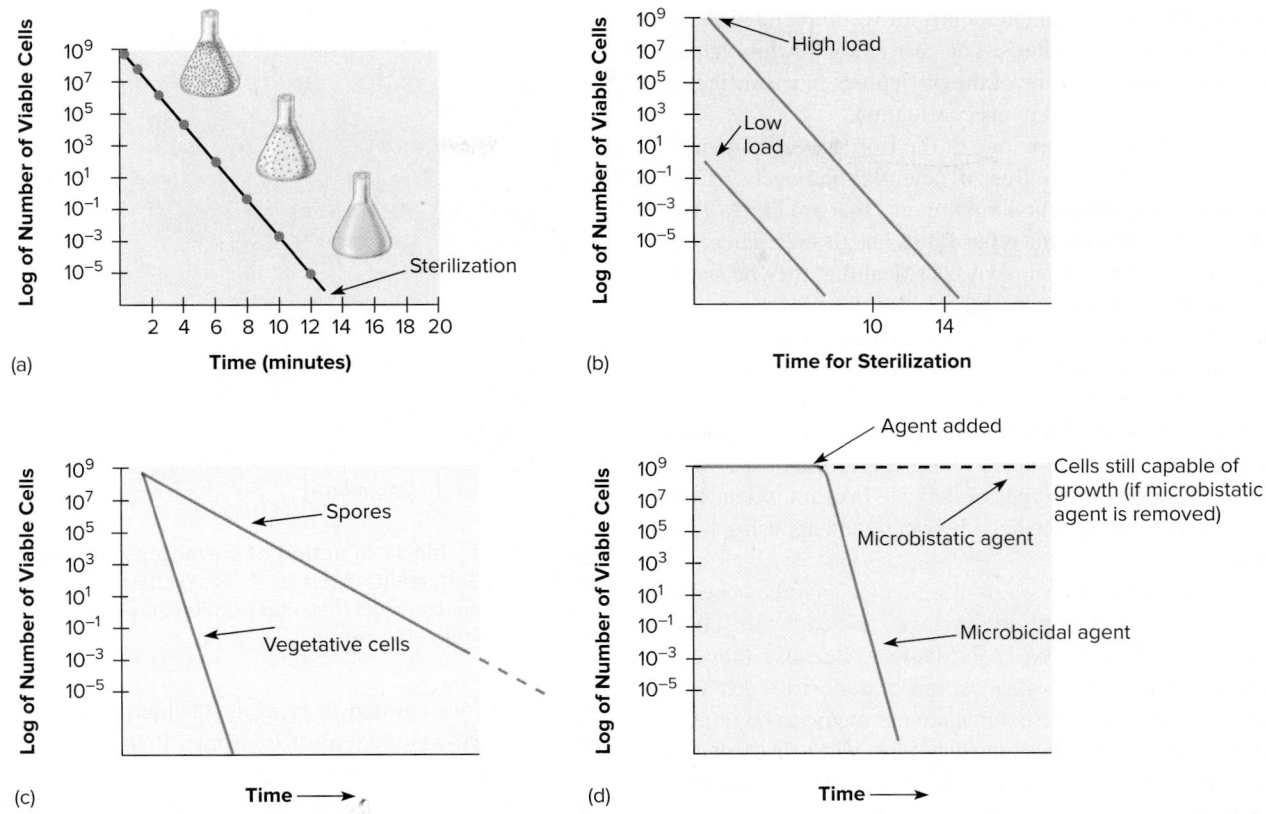

Figure 11.3 **Factors that influence the rate at which microbes are killed by antimicrobial agents. (a)** Length of exposure to the agent. During exposure to a chemical or physical agent, all cells of a microbial population, even a pure culture, do not die simultaneously. Over time, the number of viable organisms remaining in the population decreases logarithmically, giving a straight-line relationship on a graph. The point at which the number of survivors is infinitesimally small is considered sterilization. **(b)** Effect of the microbial load. **(c)** Relative resistance of spores versus vegetative forms. **(d)** Action of the agent, whether microbicidal or microbistatic.

microbe but a mixture of bacteria, fungi, spores, and viruses, presenting a broad spectrum of microbial resistance.

3. The temperature and pH of the environment.
4. The concentration (dosage, intensity) of the agent. For example, UV radiation is most effective at 260 nm and most disinfectants are more active at higher concentrations.
5. The mode of action of the agent **(figure 11.3d)**. How does it kill or inhibit microorganisms?
6. The presence of solvents, interfering organic matter, and inhibitors. Saliva, blood, and feces can inhibit the actions of disinfectants and even of heat.

Some Practical Concerns in Selecting Antimicrobial Agents

A remarkable variety of substances and objects require sterilization, disinfection, and other forms of decontamination. They range from durable materials made of glass and rubber to highly sensitive liquids such as blood plasma and even human tissues such as bone. Sometimes entire rooms with their contents and large pieces of equipment such as space vehicles and satellites may need to be sterilized. In accordance with an international agreement, all spacecraft landing on extraterrestrial bodies must have a highly reduced microbe load. This policy is to prevent the introduction of earth microbes that could potentially alter those environments or interfere with experiments being conducted by NASA. Parts of the probes are sterilized with heat, radiation, and certain chemicals (peroxide), and

then maintained in very controlled clean rooms that require extreme precautions against contamination by workers **(figure 11.4).**

There are literally hundreds of situations in health care settings alone where prevention of infection must be taken into consideration

Figure 11.4 **NASA employees work in an ultraclean room at the Jet Propulsion Lab (JPL) in Pasadena, California.** Though not sterile, the levels of microbes in the air and on surfaces are kept to a minimum. Take a tour through a JPL cleanroom at https://www.google.com/search?q=+%E2%80%9CTour+the+JPL+Clean+Room%E2%80%9D+on+Ustream&ie=utf-8&oe=utf-8. *Source:* NASA/JPL-Caltech

at all times. The CDC provides guidelines for recommended sterilization and disinfection procedures. The case study for this chapter provides a good example of some of the difficulties in ensuring that proper techniques are followed in every situation.

When choosing a procedure, one of the first questions to arise is, What will render materials free of potential pathogens? Consider, for instance, reusable dental instruments that are highly contaminated with saliva, blood, and other fluids and tissues. These are all potential sources of infection, so it is critical that they be sterilized between patients with a method that will kill endospores.

The next concern is selecting the best method of sterilization. Because they are mostly made of metal, dental instruments can be sterilized with heat, although some forms of radiation and chemical treatments could also work. Clinics generally opt for a method such as steam sterilization that is simplest, fastest, and least expensive. This would not work on other types of delicate instruments such as endoscopes, which must be sterilized between patients using heat-free techniques.

On the other hand, certain single-use medical supplies used in hospitals and clinics, such as plastic syringes and catheters, come already sterilized and packaged at the factory. Because they are usually made of heat-sensitive plastics, this is done with penetrating radiation or chemicals. These items are not meant to be reused, but they still must be decontaminated before being discarded to prevent the transmission of disease. Disposal of such medical wastes is regulated by various governmental agencies.

How Antimicrobial Agents Work: Their Modes of Action

An antimicrobial agent's adverse effect on cells is known as its **mode** (or **mechanism**) **of action.** Agents affect one or more cellular targets, inflicting damage progressively until the cell is no longer viable. Antimicrobials have a range of cellular targets, with the agents that are least selective in their targeting tending to be effective against the widest range of microbes (examples include heat and radiation). More selective agents (drugs, for example) tend to target only a single cellular component and are much more restricted in their effect on microbes.

The cellular targets of physical and chemical agents fall into four general categories: (1) cell walls, (2) cell membranes, (3) protein and nucleic acid synthesis, and (4) protein structure and function.

The Effects of Agents on the Cell Wall

The cell wall maintains the structural integrity of bacterial and fungal cells. Several types of chemical agents damage the cell wall by blocking its synthesis, digesting it, or breaking down its surface. A cell deprived of a functioning cell wall becomes fragile and is lysed very easily. Examples of this mode of action include some antimicrobial drugs (penicillins) that interfere with the synthesis of the cell wall in bacteria (described in chapter 12). Detergents and alcohol can also disrupt cell walls, especially in gram-negative bacteria.

How Agents Affect the Cell Membrane

All microorganisms have a cell membrane composed of lipids and proteins, and even some viruses have an outer membranous

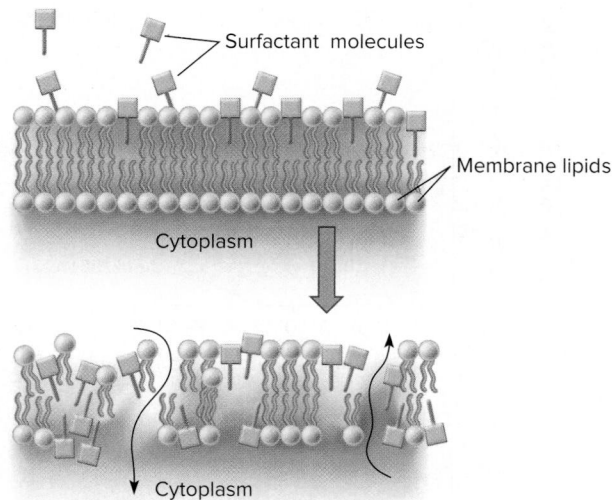

Figure 11.5 Mode of action of surfactants on the cell membrane. Surfactants inserting in the lipid bilayer disrupt it and create abnormal channels that alter permeability and cause leakage both into and out of the cell.

envelope. As we learned in previous chapters, a cell's membrane provides a two-way system of transport. If this membrane is disrupted, a cell loses its selective permeability and can neither prevent the loss of vital molecules nor bar the entry of damaging chemicals. Loss of those functions usually leads to cell death. Chemicals called **surfactants*** work as microbicidal agents because they lower the surface tension of cell membranes. Surfactants are polar molecules such as detergents with hydrophilic and hydrophobic regions that can physically bind to the lipid layer and penetrate the internal hydrophobic regions of membranes. In effect, this process "opens up" the once tight interface, leaving leaky spots that allow harmful chemicals to seep into the cell and important ions to seep out **(figure 11.5).** Alcohols exert a related effect by dissolving membrane lipids and stripping membranes away from cells.

Agents That Affect Protein and Nucleic Acid Synthesis

Microbial life depends upon an orderly and continuous supply of proteins to function as enzymes and structural molecules. As we saw in chapter 9, these proteins are synthesized on the ribosomes during translation. Any chemical that blocks this process can affect function and survival. For example, certain drugs bind to the ribosomes of bacteria in a way that stops peptide bonds from forming. In the presence of such a chemical, many bacterial cells are inhibited from forming proteins required in growth and metabolism and are thus inhibited from multiplying. Many of the drugs used to treat infections are chemicals that block protein synthesis in microbes without adversely affecting human cells. This subject is discussed in greater detail in chapter 12.

The nucleic acids are likewise necessary for the continued functioning of microbes. DNA must be regularly replicated and transcribed in growing cells, and any agent that either impedes

* *surfactant* (sir-fak′-tunt) A word derived from **surf**ace-**act**ing ag**ent.**

these processes or changes the genetic code is potentially antimicrobial. Some agents bind irreversibly to DNA, preventing both transcription and translation, whereas others are mutagenic agents. Gamma, ultraviolet, or X radiation causes mutations that result in permanent inactivation of DNA. Chemicals such as formaldehyde and ethylene oxide also interfere with DNA and RNA function.

Agents That Alter Protein Function

A microbial cell contains large quantities of proteins that function properly only if they remain in a normal three-dimensional configuration called the *native state*. The antimicrobial properties of some agents arise from their capacity to disrupt the structure of, or **denature,** proteins. In general, **denaturation** occurs when the bonds that maintain the secondary and tertiary structure of the protein are broken. Breaking these bonds will cause the protein to unfold or create random, irregular loops and coils **(figure 11.6).** One way to denature proteins is through coagulation by moist heat (think of the irreversible solidification of egg white when cooked). Chemicals such as strong organic solvents (alcohols, acids) also coagulate proteins. A few antimicrobial agents, such as metallic ions, attach to the active site of the protein and prevent it from interacting with its correct substrate. Regardless of the exact mechanism, such losses in normal protein function can promptly arrest metabolism and exert broad effects on all types of microbes.

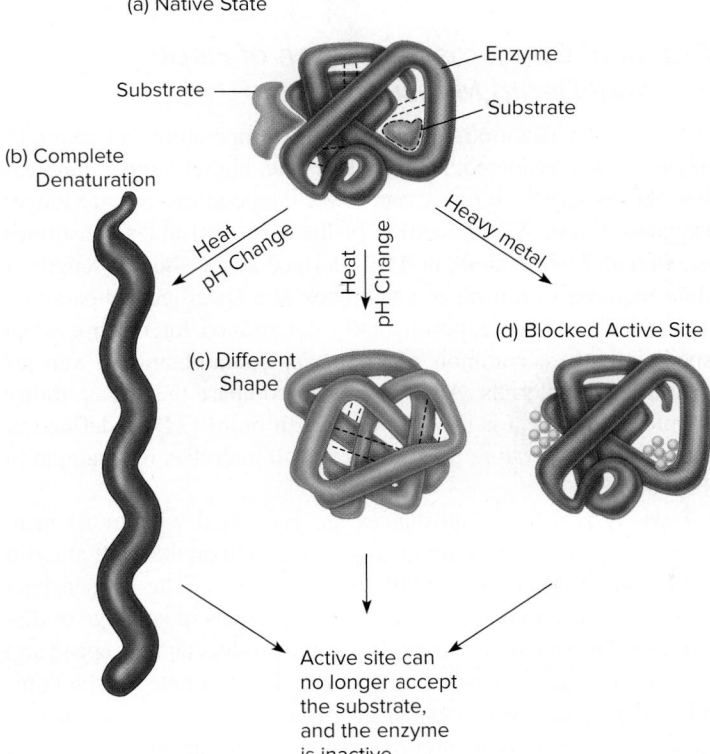

Figure 11.6 Modes of action affecting protein function.
(a) The native (functional) state is maintained by bonds that create active sites to fit the substrate. Some agents denature the protein by breaking all or some secondary and tertiary bonds. Results are **(b)** complete unfolding or **(c)** random bonding and incorrect folding. **(d)** Some agents react with functional groups on the active site and interfere with bonding.

11.2 Physical Methods of Control: Heat

Expected Learning Outcomes

7. List the major types of physical agents used in controlling microbes.
8. Describe the basic effects of temperature on microbes.
9. Define *moist* and *dry heat,* and explain their modes of action and relative effectiveness.
10. Differentiate between thermal death point and thermal death time, and provide examples of their uses.
11. Describe several moist heat methods and their chief applications for sterilization and disinfection.
12. Describe some methods using dry heat and its chief applications.
13. Explain how cold temperatures and desiccation are used in microbial control.

Microorganisms have adapted to a tremendous diversity of habitats on earth, some having severe conditions of temperature, moisture, pressure, and light. For microbes that normally withstand such extreme physical conditions, our usual attempts at control would probably have little effect. Fortunately for us, we are most interested in controlling microbes that flourish in the same environment in which humans live. The vast majority of these microbes are readily controlled by abrupt changes in their environment. Most prominent among antimicrobial physical agents is heat. Other less widely used agents include radiation and filtration. The following sections examine these and other methods, and explore their practical applications in medicine, commerce, and the home.

Effects of Temperature on Microbial Activities

A sudden departure from a microbe's temperature of adaptation is likely to have a detrimental effect on it. As a rule, temperatures that exceed the maximum growth temperature are microbicidal, whereas

temperatures below the minimum growth temperature are microbistatic. This essentially means that, for most microbes, the effects of temperatures below minimum can be reversible, whereas the effects of temperatures above maximum are not reversible and most microbes will be destroyed by them.

The two physical states of heat used in microbial control are moist and dry. **Moist heat** occurs in the form of hot water, boiling water, or steam (vaporized water). In practice, the temperature of moist heat usually ranges from 60°C to 135°C. As we shall see, the temperature of steam can be regulated by adjusting its pressure in a closed container. The expression **dry heat** denotes air with a low moisture content that has been heated by a flame or electric heating coil. In practice, the temperature of dry heat ranges from 160°C to several thousand degrees Celsius.

Mode of Action and Relative Effectiveness of Heat

Moist heat and dry heat differ in their modes of action as well as in their efficiency. Moist heat operates at lower temperatures and shorter exposure times to achieve the same effectiveness as dry heat (**table 11.3**). Although many cellular structures are damaged by moist heat, its most microbicidal effects are the coagulation and denaturation of proteins, which quickly and permanently halt cellular metabolism.

Dry heat with a moderate temperature dehydrates the cell, removing the water necessary for metabolic reactions, and it alters protein structure. However, the lack of water actually increases the stability of some protein conformations, necessitating the use of higher temperatures when dry heat is employed as a method of microbial control. At very high temperatures, dry heat oxidizes cells, burning them to ashes. This method is the one used in the laboratory when a loop is flamed or in industry when medical waste is incinerated.

Heat Resistance and Thermal Death of Spores and Vegetative Cells

Bacterial endospores exhibit the greatest resistance, and vegetative states of bacteria and fungi are the least resistant to both moist and dry heat. Destruction of spores usually requires temperatures above

TABLE 11.3	Comparison of Times and Temperatures to Achieve Sterilization with Moist and Dry Heat	
	Temperature	Time to Sterilize
Moist Heat	121°C	15 min
	125°C	10 min
	134°C	3 min
Dry Heat	121°C	600 min
	140°C	180 min
	160°C	120 min
	170°C	60 min

boiling, although resistance varies widely. The following table compares some times and temperatures required to kill sporeformers using moist versus dry heat:

Moist Heat	Temp/Time	Dry Heat	Temp/Time
Bacillus subtilis	121°C/1 min	*Bacillus subtilis*	121°C/120 min
B. stearothermophilis	121°C/12 min	*B. stearothermophilis*	140°C/5 min
Clostridium botulinum	120°C/10 min	*Clostridium botulinum*	120°C/120 min
C. tetani	105°C/10 min	*C. tetani*	100°C/60 min

In general, reliable killing of the most heat-resistant species of sporeformers requires 121°C for 20 minutes in moist heat.

Vegetative cells also vary in their sensitivity to heat, though not to the same extent as spores. Among bacteria, the death conditions for moist heat range from 50°C for 3 minutes (*Neisseria gonorrhoeae*) to 60°C for 60 minutes (*Staphylococcus aureus*). It is worth noting that vegetative cells of sporeformers are just as susceptible as vegetative cells of non-sporeformers and that pathogens are neither more nor less susceptible than nonpathogens. Other microbes, including fungi, protozoa, and worms, are rather similar in their sensitivity to heat. Viruses can be surprisingly resistant to heat, with a tolerance range extending from 55°C for 2 to 5 minutes (adenoviruses) to 60°C for 600 minutes (hepatitis A virus). For practical purposes, all non-heat-resistant forms of bacteria, yeasts, molds, protozoa, worms, and viruses are destroyed by exposure to 80°C for 20 minutes.

Practical Concerns in the Use of Heat: Thermal Death Measurements

Adequate sterilization requires that both temperature and length of exposure be considered. As a general rule, higher temperatures allow shorter exposure times, and lower temperatures require longer exposure times. A combination of these two variables constitutes the **thermal death time,** or **TDT,** defined as the shortest length of time required to kill all test microbes at a specified temperature. The TDT has been experimentally determined for the microbial species that are common or important contaminants in various heat-treated materials. Another way to compare the susceptibility of microbes to heat is the **thermal death point (TDP),** defined as the lowest temperature required to kill all microbes in a sample in 10 minutes.

Many perishable substances are processed with moist heat. Some of these products are intended to remain on the shelf at room temperature for several months or even years. The chosen heat treatment must render the product free of agents of spoilage or disease. At the same time, the quality of the product and the speed and cost of processing must be considered. For example, in the commercial preparation of canned green beans, one of the cannery's greatest concerns is to prevent growth of the agent of botulism. From several possible TDTs (that is, combinations of time and temperature) for *Clostridium botulinum* spores, the cannery must choose one that kills all spores but does not turn the beans to mush. These many considerations produce an optimal TDT for a given processing method. Commercial canneries heat low-acid foods like green beans at 121°C for 30 minutes, a treatment that renders the

foods sterile. Because of such strict controls in canneries, cases of botulism from commercially canned foods are rare.

Common Methods of Moist Heat Control

The four ways moist heat is employed to control microbes are (1) steam under pressure, (2) nonpressurized steam, (3) boiling water, and (4) pasteurization.

Sterilization with Steam Under Pressure At sea level, normal atmospheric pressure is 15 pounds per square inch (psi), or 1 atmosphere. At this pressure, water will boil (change from a liquid to a gas) at 100°C, and the resultant steam will not go above 100°C, regardless of the method of heating. But boiling temperature is not hot enough to reliably kill all targeted microbes. The only way to raise the temperature of steam is to expose it to increased pressure. At higher pressures the temperature at which water boils and the temperature of steam both rise. For example, at a pressure of 20 psi (5 psi above normal), the temperature of steam is 109°C. As the temperature is increased to 10 psi above normal, the steam's temperature rises to 115°C, and at 15 psi above normal (a total of 2 atmospheres), it will be 121°C. It is not the pressure by itself that is killing microbes but the increased temperature it produces.

Such pressure-temperature combinations can be achieved only with a special device that can subject pure steam to pressures greater than 1 atmosphere. Health and commercial industries use an **autoclave** for this purpose, and a comparable home appliance is the pressure cooker. Autoclaves have a fundamentally similar plan: an enclosed metal chamber with an airtight door on one end and racks to hold materials **(figure 11.7).** Its construction includes a network of valves, pressure and temperature gauges, and ducts for regulating and measuring pressure and conducting the steam into the chamber. Sterilization is achieved when the steam condenses against the objects in the chamber and gradually raises their temperature.

Experience has shown that the most efficient pressure-temperature combination for achieving sterilization is 15 psi, which yields 121°C. It is possible to use higher pressure to reach higher temperatures (for instance, increasing the pressure to 30 psi raises the temperature to 132°C), but doing so will not significantly reduce the exposure time and can harm the items being sterilized.

The duration of the process is adjusted according to the bulkiness of the items in the load (thick bundles of material or large flasks of liquid) and how full the chamber is. The range of holding times varies from 10 minutes for light loads to 40 minutes for heavy or bulky ones; the average time is 20 minutes.

Nonpressurized Steam Selected substances that cannot withstand the high temperature of the autoclave can be subjected to **intermittent sterilization,** also called **tyndallization.**[2] This technique requires a chamber to hold the materials and a reservoir for boiling water. Items in the chamber are exposed to free-flowing steam for

30 to 60 minutes. This temperature is not sufficient to reliably kill spores, so a single exposure will not suffice. On the assumption that surviving spores will germinate into less resistant vegetative cells, the items are incubated at appropriate temperatures for 23 to

(a)

(b)

Figure 11.7 Sterilization using steam under pressure. (a) An autoclave for sterilizing media and cultures. **(b)** Cutaway section, showing autoclave components and the flow of steam through the autoclave. (a) © Kathy Park Talaro

2. Named for the British physicist John Tyndall, who did early experiments with sterilizing procedures.

24 hours, and then again subjected to steam treatment. This cycle is repeated for 3 days in a row. Because the temperature never gets above 100°C, highly resistant spores that do not germinate could survive even after 3 days of this treatment.

Boiling Water: Disinfection A simple boiling water bath or chamber can quickly decontaminate items in the clinic and home. Because a single processing at 100°C will not kill all resistant cells, this method can be relied on only for disinfection and not for sterilization. Exposing materials to boiling water for 30 minutes will kill most non-spore-forming pathogens, including resistant species such as the tubercle bacillus and staphylococci. Probably the greatest disadvantage with this method is that the items can be easily recontaminated when removed from the water.

Pasteurization: Disinfection of Beverages Fresh beverages such as milk, fruit juices, beer, and wine are easily contaminated during collection and processing. Because microbes have the potential to spoil these foods or cause illness, heat is frequently used to reduce the microbial load and destroy pathogens. **Pasteurization** is a technique of applying heat to consumable liquids to kill potential agents of infection and spoilage, while at the same time retaining the liquid's flavor and food value.

A widely used pasteurization technique is the **flash method,** that exposes the liquid to heat exchangers at 71.6°C for 15 seconds. This first method does not appreciably change flavor and nutrient content, and it is effective against certain resistant pathogens such as *Coxiella* and *Mycobacterium.* Although these treatments inactivate most viruses and destroy the vegetative stages of 97% to 99% of bacteria and fungi, they do not kill endospores or **thermoduric** microbes (mostly nonpathogenic lactobacilli, micrococci, and yeasts). Flash-pasteurized milk is not sterile after regular pasteurization. In fact, it can contain 20,000 microbes per milliliter or more, which explains why even an unopened carton of milk will eventually spoil. Many commercial dairies now produce *sterile milk* with an increased storage life of several months. This ultrapasteurized milk is processed using ultrahigh temperature (UHT) —134°C for 2 to 5 seconds.

One important aim in pasteurization is to prevent the transmission of milk-borne diseases from infected cows or milk handlers. The primary targets of pasteurization are non-spore-forming pathogens: *Salmonella* species (a common cause of food infection), *Campylobacter jejuni* (acute intestinal infection), *Listeria monocytogenes* (listeriosis), *Brucella* species (undulant fever), *Coxiella burnetii* (Q fever), *Mycobacterium bovis, M. tuberculosis,* and several enteric viruses.

Dry Heat: Hot Air and Incineration

Dry heat is not as versatile or as widely used as moist heat, but it has several important sterilization applications. The temperatures and times employed in dry heat vary according to the particular method, but in general they are greater than with moist heat. **Incineration** in a flame or electric heating coil is perhaps the most rigorous of all heat treatments. The flame of a Bunsen burner reaches 1,870°C at its hottest point, and furnaces/incinerators operate at temperatures of 800°C to 6,500°C. Direct exposure to such intense heat ignites and reduces microbes and other substances to ashes and gas.

Incineration of microbial samples on inoculating loops and needles using a Bunsen burner is a very common practice in the microbiology laboratory. This method is fast and effective, but it is also limited to metals and heat-resistant glass materials. Tabletop infrared incinerators (see table 11.4), an alternative to Bunsen burners, are safer and prevent splatter of the inoculum.

The hot-air oven provides another means of dry-heat sterilization. The so-called *dry oven* is usually electric (occasionally gas) and has coils that radiate heat within an enclosed compartment. Heated, circulated air transfers its heat to the materials in the oven. Depending on the type of oven and the material being decontaminated, a cycle takes 12 minutes to 4 hours to complete and involves temperatures of 150°C to 180°C.

Table 11.4 presents an illustrated summary of heating methods for microbial control.

The Effects of Cold and Desiccation

The principal benefit of cold treatment is to slow the growth of cultures and microbes in food and other perishable materials during processing and storage. *It must be emphasized that cold merely retards the activities of most microbes.* Although it is true that some sensitive microbes are killed by cold temperatures, most are not adversely affected by gradual cooling, long-term refrigeration, or deep-freezing. In fact, freezing temperatures, ranging from −70°C to −135°C, provide an environment that can preserve cultures of bacteria, viruses, and fungi for long periods. Recall from chapter 7 that some psychrotrophs grow very slowly even at freezing temperatures and can continue to secrete toxic products. Pathogens able to survive several months in the refrigerator are *Staphylococcus aureus; Clostridium* species (sporeformers); *Listeria*; and several types of yeasts, molds, and viruses. Outbreaks of *Salmonella* food infection traced backed to refrigerated foods such as ice cream and eggs are testimony to the inability of cold temperatures to reliably kill pathogens.

Most vegetative cells directly exposed to normal room air gradually become dehydrated, or **desiccated.*** Delicate pathogens such as *Streptococcus pneumoniae,* the spirochete of syphilis, and *Neisseria gonorrhoeae* can die after a few hours of air-drying, but many others are not killed and some are even preserved. Staphylococci and streptococci in dried secretions and the tubercle bacillus surrounded by sputum can remain viable in air and dust for lengthy periods. Many viruses (especially nonenveloped) and fungal spores can also withstand long periods of desiccation. Endospores of *Bacillus* and *Clostridium* are viable for millions of years under extremely arid conditions. Desiccation can be a valuable way to preserve foods because it greatly reduces the amount of water available to support microbial growth.

It is interesting to note that a combination of freezing and drying— **lyophilization***—is a common method of preserving microorganisms and other cells in a viable state for many years. Pure

* *desiccate* (des'-ih-kayt) To dry at normal environmental temperatures.
* *lyophilization* (ly-off″-il-ih-za'-shun) Gr. *lyein,* to dissolve, and *philein,* to love.

TABLE 11.4	Applications Using Heat for Sterilization and Disinfection	
Form of Heat	Equipment Temperature/Time	Medical and Commercial Uses and Limitations
Moist heat: Employs steam, boiling, or hot water; mode of action coagulates proteins		
Steam under pressure	Autoclave, pressure cooker; 121°C, pressure of 15 psi for 10 to 40 minutes, depending on load size; kills endospores.	Sterilization of heat-resistant materials made of glass, cloth, rubber, metal; media; a few plastics; canned food, spices. Decontamination of cultures. Not appropriate for moisture-repellent oil or powder.
Unpressurized steam (tyndallization)	Arnold's Sterilizer—100°C for 30 minutes on 3 days; items are incubated between treatments to allow endospores to germinate.	Limited uses in sterilizing some media and foods that cannot be autoclaved; can be used to disinfect but not sterilize some medical supplies
Pasteurization	Pasteurizer; flash method works at 71.6°C for 15 seconds; ultrahigh temperature (UHT) uses 134°C for a few seconds.	Disinfection of milk and dairy products to destroy milk-borne pathogens; UHT can sterilize for longer shelf life. Fruit juices, beer, and wine may be pasteurized to destroy contaminants and preserve the products.
Boiling water	Water bath or pan heated to 100°C; items are placed in water for 30 minutes to kill vegetative pathogens.	Limited to disinfection and sanitization of heat-resistant household objects such as eating utensils, clothing, sick room supplies, baby supplies, bedding, and water in emergencies; problem with recontamination
Dry heat: Employs dried hot air, flame, or heat coil; works by dehydration or combustion		
Dry oven	Electric or gas chamber heated to 150°C to 180°C for 2 to 4 hours	Materials that can withstand hot temperatures and dehydration; glassware, metals, powders, oils. Poor choice for liquids, rubber, plastics; time-consuming.
Incineration	Bunsen burner flame up to 1,800°C; small infrared incinerators to 800°C	Furnaces, incinerators up to 6,500°C For sterilizing tips of inoculation instruments; limited to high-temperature metals. Permanent disposal of waste material from laboratories; burning reduces bulky packages and garbage to ashes. For permanent disposal only.

A small flash pasteurizer used by dairies for calves' milk. © Goodnature Products

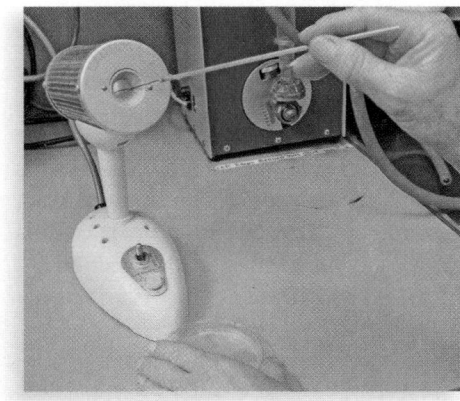

Dry heat incineration. Infrared incinerator with shield to prevent spattering of microbial samples during flaming. © Kathy Park Talaro

cultures are frozen instantaneously and exposed to a vacuum that rapidly removes the water. This method avoids the formation of ice crystals that would damage the cells. Although not all cells survive this process, enough of them do to permit future reconstitution of that culture.

As a general rule, chilling, freezing, and desiccation should not be construed as methods of disinfection or sterilization, because their antimicrobial effects are erratic and uncertain and one cannot be sure that pathogens subjected to them have been killed.

 Check Your Progress SECTION 11.2

9. What happens to microbes that encounter temperatures below their minimum and above their maximum growth temperatures?

10. Summarize the nature, mode of action, and effectiveness of moist and dry heat on spores and vegetative cells.

11. Explain the concepts of TDT and TDP, using examples.

12. What are the minimum TDTs for vegetative cells and endospores?

13. How can the temperature of steam be raised above 100°C? Which instrument is based on this technique and how is it used?

14. What are the primary purposes of pasteurization, and what is ultrapasteurization?

15. What are several microbial targets of pasteurization?

16. Explain why desiccation and cold are not reliable methods of disinfection but how they can still be effective in microbial control.

11.3 Physical Methods of Control: Radiation and Filtration

 Expected Learning Outcomes

14. Define *radiation* and assess the use of irradiation as a method of microbial control.

15. Differentiate between ionizing and nonionizing radiation in their modes of action.

16. Summarize the primary uses of ionizing and nonionizing radiation in sterilization and disinfection.

17. Describe the concept behind sterilization by filtration, including examples of applications.

Radiation as a Microbial Control Agent

Another form of energy that can serve as an antimicrobial agent is radiation. **Radiation** is defined as energy emitted from atomic activities and dispersed at high velocity through matter or space. It is characterized by a range of wavelengths known as the electromagnetic spectrum **(figure 11.8).** Although radiation exists in many energy states, we will consider only those types suitable for microbial control: gamma rays, X rays, and ultraviolet radiation.

Modes of Action of Ionizing Versus Nonionizing Radiation

The actual physical effects of radiation on microbes can be understood by visualizing the process of **irradiation,** or bombardment with radiation, at the cellular level **(figure 11.9).** When a cell is bombarded by certain waves or particles, its molecules absorb some of the available energy, leading to one of two consequences: (1) If the radiation ejects orbital electrons from an atom, it causes ions to form; this type of radiation is termed **ionizing radiation.** One of the most sensitive targets for ionizing radiation is DNA, which will undergo mutations on a broad scale. Secondary lethal effects appear to be chemical changes in organelles and the production of

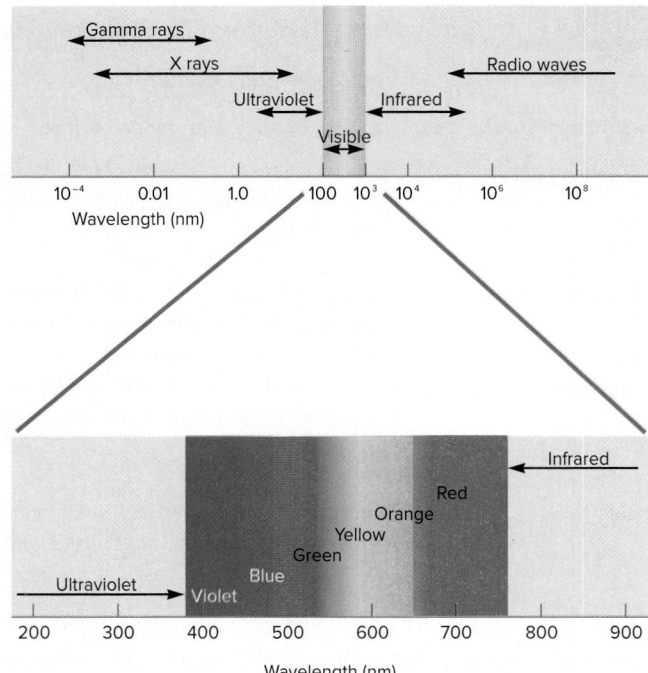

Figure 11.8 The electromagnetic spectrum. Waves range from the shortest gamma rays to the longest radio waves. Inset shows the spectrum of visible light between ultraviolet (UV) and infrared. Only ionizing radiation, UV, and infrared have applications in microbial control.

toxic substances. Gamma rays, X rays, and high-speed electrons are all ionizing in their effects. (2) **Nonionizing radiation,** best exemplified by UV, excites atoms by raising them to a higher energy state, but it does not ionize them. This atomic excitation, in turn, leads to the formation of abnormal bonds within molecules such as DNA and is thus a source of mutations (see figure 11.11).

Ionizing Radiation: Gamma Rays, X Rays, and Cathode Rays

Over the past several years, ionizing radiation has become safer and more economical to use, and its applications have expanded. It is a highly effective alternative for sterilizing materials that are sensitive to heat or chemicals. Because it sterilizes in the absence of heat, irradiation is a type of **cold** (or low-temperature) **sterilization.**[3] Devices that emit ionizing rays include gamma-ray machines containing radioactive cobalt, X-ray machines similar to those used in medical diagnosis, and cathode-ray machines that operate like the vacuum tube in a television set. Items are placed in these machines and irradiated for a short time with a carefully chosen dosage. The dosage of radiation is measured in grays (which has replaced the older term, *rads*). Depending on the application, exposure ranges from 5 to 50 kilograys (kGy; a kilogray is equal to 1,000 grays). Although all ionizing radiations can penetrate liquids and most solid materials, gamma rays are most penetrating, X rays are intermediate, and cathode rays least penetrating.

3. To clarify the use of the word *cold,* in this context, it only means the absence of heat. Beer manufacturers have sometimes used this terminology as well. When they say that their product is "cold-filtered," they mean that it has been freed of contaminants via filtration, without heating.

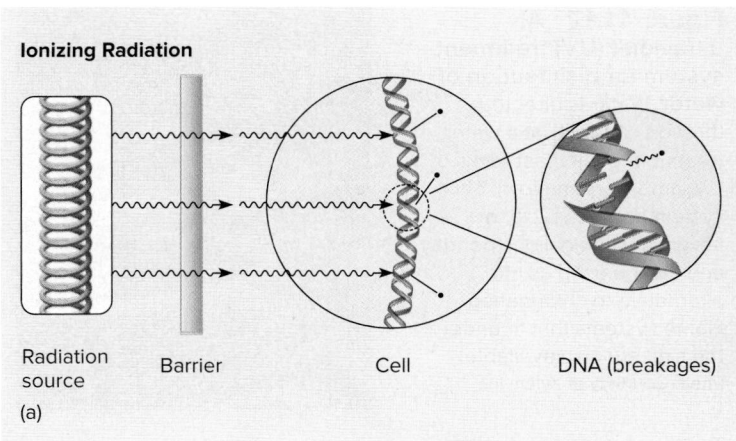

Ionizing Radiation

Radiation source | Barrier | Cell | DNA (breakages)
(a)

Nonionizing Radiation

(b)

DNA (abnormal bonds) See figure 11.10

UV does not penetrate No effect on cell

Radiation source | Barrier
(c)

Figure 11.9 Cellular effects of irradiation. (a) Ionizing radiation can penetrate a solid barrier, bombard a cell, enter it, and dislodge electrons from molecules. Breakage of DNA creates massive mutations. **(b)** Nonionizing radiation enters a cell, strikes molecules, and excites them. The effect on DNA is mutation by formation of abnormal bonds. **(c)** A solid barrier cannot be readily penetrated by nonionizing radiation.

Irradiation as a Method of Microbial Control

Foods have been subject to irradiation in limited circumstances for more than 50 years. From flour to pork and ground beef, to fruits and vegetables, radiation is used to kill bacterial pathogens, insects, and worms, and even to inhibit the sprouting of white potatoes. As soon as radiation is involved, however, consumers become concerned that ionizing radiation could make food less nutritious, unpalatable, or even unsafe. But irradiated food has been extensively studied, and each of these concerns has been addressed.

Raspberries exposed to radiation, 5 days later | Raspberries not treated with radiation, 5 days later

Figure 11.10 Sterilization with ionizing radiation. Many foods can be effectively sterilized by utilizing the penetrating power of ionizing radiation. The "radura" symbol (meaning treated with radiation; see the green image on the package on the left) applied to fresh foods signifies that the food has been irradiated and is required to be displayed on all food so treated. (photos): © Kathy Park Talaro

Figure 11.10 demonstrates how irradiation of fruit can greatly reduce spoilage and improve shelf life.

Irradiation may lead to a small decrease in the amount of thiamine (vitamin B_1) in food, but this change is small enough to be inconsequential. The irradiation process does produce short-lived free-radical oxidants, which disappear almost immediately (this same type of chemical intermediate is produced through cooking as well). Certain foods do not irradiate well and are not good candidates for this type of treatment. The white of eggs becomes milky and alfalfa seeds do not germinate properly. It should be emphasized that food is not made radioactive by the irradiation process, and many studies, in both animals and humans, have concluded that there are no ill effects from eating irradiated food. In fact, NASA relies on irradiated meat for its astronauts.

One of the potential benefits of irradiated foods has to do with infection control. It has been estimated that irradiation of 50% of the meat and poultry in the United States would result in 900,000 fewer cases of infection, 8,500 fewer hospitalizations, and 350 fewer deaths each year. Radiation is currently approved in the United States for the reduction of bacterial pathogens such as *E. coli* and *Salmonella* in beef and chicken, reduction of *Trichinella* worms in pork, and the reduction of pathogens and pests on fruits and vegetables. An additional benefit of irradiation is that microbes responsible for food spoilage are killed along with pathogens, leading to an increased shelf life. In any event, no irradiated food can be sold to consumers without clear labeling that this method has been used (figure 11.10).

Sterilizing medical products with ionizing radiation is a rapidly expanding field. Drugs, vaccines, medical instruments (especially plastics), and other delicate materials can be irradiated without harming them. Its main advantages include speed, high

Quick Search

Access YouTube to view the video "Irradiation in Food Processing."

penetrating power (it can sterilize materials through outer packages and wrappings), and the absence of heat. Its main disadvantages are potential dangers from factory exposure to radiation and possible damage to some materials. See table 11.5 for a summary of the applications of ionizing radiation.

Nonionizing Radiation: Ultraviolet Rays

Ultraviolet (UV) radiation ranges in wavelength from approximately 100 nm to 400 nm. It is most lethal from 240 nm to 280 nm (with a peak at 260 nm). In everyday practice, the source of UV rays is the germicidal lamp, which generates radiation at 254 nm. Owing to its lower energy state, UV radiation is not as penetrating as ionizing radiation. Because UV radiation passes readily through air, slightly through liquids, and only poorly through solids, the object to be disinfected must be directly exposed to it for full effect.

As UV radiation passes through a cell, it is initially absorbed by DNA. Specific molecular damage occurs on the pyrimidine bases (thymine and cytosine), which form abnormal linkages with each other called **pyrimidine dimers (figure 11.11).** These bonds occur between adjacent bases on the same DNA strand and interfere with normal DNA replication and transcription. The results are inhibition of growth and cellular death. In addition to altering DNA directly, UV radiation also disrupts cells by generating toxic photochemical products called free radicals. These highly reactive molecules interfere with essential cell processes by binding to DNA, RNA, and proteins. Ultraviolet rays are a powerful tool for destroying fungal cells and spores, bacterial vegetative cells, protozoa, and viruses. Bacterial spores are about 10 times more resistant to radiation than are vegetative cells, but they can be killed by increasing the time of exposure.

Figure 11.11 Formation of pyrimidine dimers by the action of ultraviolet (UV) radiation. This shows what occurs when two adjacent thymine bases on one strand of DNA are induced by UV rays to bond laterally with each other. The result is a thymine dimer shown in greater detail. Dimers can also occur between adjacent cytosines and thymine and cytosine bases. If they are not repaired, dimers can prevent that segment of DNA from being correctly replicated or transcribed. High levels of dimerization are lethal to cells.

Figure 11.12 An ultraviolet (UV) treatment system for disinfection of water. Water is directed through channels at a water treatment plant, past racks of UV lamps (green glow). This system has a capacity of several million gallons per day and can be used as an alternative to chlorination. Home systems that fit under the sink are also available. Photo Courtesy of Xylem Inc.

Ultraviolet radiation is usually aimed more at disinfection than at sterilization. Germicidal lamps can cut down on the concentration of airborne microbes by as much as 99%. Ultraviolet disinfection of air has proved effective in reducing postoperative infections, preventing the transmission of infections by respiratory droplets, and curtailing the growth of microbes in food-processing plants and slaughterhouses.

Ultraviolet irradiation of liquids requires special equipment to spread the liquid into a thin, flowing film that is exposed directly to a lamp. This method can be used to treat water **(figure 11.12)** and to purify other liquids (milk and fruit juices) as an alternative to heat. The surfaces of solid, nonporous materials such as walls and floors, as well as meat, nuts, tissues for grafting, and drugs, can be successfully disinfected with UV. Several types of small UV disinfection appliances have been marketed for personal items.

One major disadvantage of UV is its poor powers of penetration through solid materials such as glass, metal, cloth, plastic, and even paper. Another drawback to UV is the damaging effect of overexposure on human tissues, including sunburn, retinal damage, cancer, and skin wrinkling. These harmful results occur only if a person is directly exposed to the UV rays, but it necessitates placement of barriers to cut down on this possibility. **Table 11.5** summarizes UV applications.

Filtration—A Physical Removal Process

Filtration is an effective method to remove microbes from air and liquids. In practice, a fluid is strained through a filter with openings large enough for the fluid to pass through but too small for microorganisms to pass through **(figure 11.13a).**

Most modern microbiological filters are thin membranes of cellulose acetate, polycarbonate, and a variety of plastic materials (Teflon, nylon) whose pore size can be carefully controlled and standardized. Ordinary substances such as charcoal, diatomaceous earth, or unglazed porcelain are also used in some applications. Viewed microscopically, most filters are perforated by very precise, uniform pores **(figure 11.13b).** The pore diameters vary from coarse (8 μm) to ultrafine (0.02 μm), permitting selection of the minimum particle size to be trapped. Those with even smaller pore diameters permit true sterilization by removing viruses, and some will even remove large proteins. A sterile liquid filtrate is typically

TABLE 11.5	Applications Using Radiation for Sterilization and Disinfection	
Form of Radiation	**Equipment/Mode of Action**	**Medical and Commercial Uses and Limitations**
Ionizing radiation Sporicidal and penetrating	Cobalt radiation gives off gamma rays; others are X rays, cathode rays. Rays fragment DNA and create ionized bases; highly mutagenic.	Cold sterilization of packaged products such as medical supplies and instruments that cannot be heat sterilized Sterilization of foods, including fruits, lettuce and spinach, grains, spices, meats, and packaged meals Sterilization of drugs and vaccines Rapid and penetrating, prevents infection and spoilage
Nonionizing ultraviolet (UV) radiation Sporicidal and nonpenetrating	Germicidal lamps or generators set to deliver UV rays from 240 to 280 nm in wavelength; action creates abnormal bonds between adjacent thymines and cytosines and produces deletion mutations on DNA.	Mostly disinfection of air, some liquids, a few solids; reduces airborne pathogens in medical and dental offices, hospital rooms, and operating rooms Commercial use for clean rooms, food processing areas Works on liquids such as water, milk, drugs, vaccines if exposed to UV in thin layers Solid materials such as room surfaces, foods, tissue grafts, drugs, and instruments Lack of penetration through solids limits applications; direct exposure can be dangerous to eyes and skin.

Pallets of apples are being irradiated with a cobalt 60 machine that emits gamma rays. *Source:* Pacific Northwest National Laboratory

Tabletop chamber generates ultraviolet rays to disinfect a toothbrush. Courtesy of Violight

(b)

Figure 11.13 Membrane filtration. (a) Vacuum assembly for achieving filtration of liquids through suction. Inset shows filter as seen in cross section, with tiny passageways (pores) too small for the microbial cells to enter but large enough for liquid to pass through. **(b)** Scanning electron micrograph of *Enterococcus* trapped on the surface of a membrane filter (10,400×). Average cells measure about 0.5 µm to 1 µm in diameter, and average pore size is about 0.3 µm. (b) *Source:* Janice Haney Carr/CDC

produced by suctioning the liquid through a sterile filter into a presterilized container. These filters are also used to separate mixtures of microorganisms and to enumerate bacteria in water analysis.

Applications of Filtration Sterilization

Sterilization by means of filtration is used to prepare liquids that cannot withstand heat, including serum and other blood products, vaccines, drugs, IV fluids, enzymes, and media. Filtration has been employed as an alternative method for sterilizing milk and beer without altering their flavor. It is also an important step in water purification. Its use extends to filtering out particulate impurities (crystals, fibers) that can cause severe reactions in the body. It has the disadvantage of not removing soluble molecules (toxins) that can cause disease.

Filtration is a very effective means of removing airborne contaminants that are a common source of infection and spoilage. High-efficiency particulate air (HEPA) filters are widely used to provide a flow of sterile air to hospital rooms and sterile rooms.

Check Your Progress SECTION 11.3

17. Define *radiation* and *irradiation*.

18. Differentiate between the modes of action of ionizing radiation and ultraviolet radiation.

19. What are some advantages and disadvantages of ionizing radiation as a method of control?

20. Describe some applications of ionizing and ultraviolet radiation.

21. What are the advantages and disadvantages of using filtration as a method of control?

11.4 Chemical Agents in Microbial Control

 Expected Learning Outcomes

18. Define the terms that are used to describe chemical control of microbes.

19. Explain the desirable features of antimicrobial chemicals and what factors influence their effectiveness.

20. Describe the types of halogens, their modes of action, and their clinical and commercial applications.

21. Identify the characteristics of phenolic compounds and alcohols, explain how they act, and evaluate their primary uses.

22. Describe the characteristics of oxidizing agents and aldehydes, and summarize their main applications.

23. Differentiate among the sterilizing gases and aerosols, and describe how they are used.

24. Explain how detergents, soaps, and heavy metals can be employed in clinical, commercial, and household microbial control.

Chemical control of microbes probably emerged as a serious science in the early 1800s, when physicians used chloride of lime and iodine solutions to treat wounds and to wash their hands before surgery. At the present time, approximately 10,000 different antimicrobial chemical agents are manufactured; probably 1,000 of them are used routinely in the health care and home environments. A genuine need exists to avoid infection and spoilage, but the abundance of products available to "kill germs," "disinfect," "antisepticize," "clean and sanitize," "deodorize," "fight plaque," and "purify the air" indicates a preoccupation with eliminating microbes from the environment that, at times, seems excessive (**11.1 Making Connections**).

Antimicrobial chemicals occur in the liquid, gaseous, or even solid state. They serve as **disinfectants, antiseptics, sterilants** (chemicals that sterilize), degermers, or preservatives (chemicals that inhibit the deterioration of substances). In most cases, solid or gaseous antimicrobial chemicals are dissolved in water, alcohol, or a mixture of the two to produce a liquid solution. Solutions containing pure water as the solvent are termed **aqueous,** whereas those dissolved in pure alcohol or water-alcohol mixtures are termed **tinctures.**

Choosing a Microbicidal Chemical

The choice and appropriate use of antimicrobial chemical agents are of constant concern in medicine and dentistry. Although actual clinical practices of chemical decontamination vary widely, some desirable qualities in a germicide have been identified, including:

- rapid action in low concentrations,
- solubility in water or alcohol and long-term stability,
- broad-spectrum microbicidal action without being toxic to human and animal tissues,
- penetration of inanimate surfaces to sustain a cumulative or persistent action,
- resistance to becoming inactivated by organic matter,
- noncorrosive or nonstaining properties,
- sanitizing and deodorizing properties,
- affordability and ready availability

As of yet, no chemical can completely fulfill all of those requirements, but chlorhexidine and hydrogen peroxide approach this ideal. At the same time, we should question the rather overinflated claims made about certain commercial agents such as mouthwashes and disinfectant air sprays.

Germicides are evaluated in terms of their effectiveness in destroying microbes in medical and dental settings. The three levels of chemical decontamination procedures are *high, intermediate,* and *low* (**table 11.6**).

- High-level germicides kill endospores and, if properly used, are **sterilants**. Materials that necessitate high-level control are medical devices—for example, catheters, endoscopes, and implants—that are not heat-sterilizable and are intended to enter body tissues during medical procedures.
- Intermediate-level germicides kill fungal (but not bacterial) spores, resistant pathogens such as the tubercle bacillus, and viruses. They are used to disinfect items (respiratory equipment, thermometers) that come into intimate contact with the mucous membranes but are noninvasive.
- Low levels of disinfection eliminate only vegetative bacteria, vegetative fungal cells, and some viruses. They are used to clean materials such as electrodes, straps, and fixtures that touch the skin surfaces but not the mucous membranes.

Factors That Affect the Germicidal Activities of Chemical Agents

Several concepts of microbial control introduced in section 11.1 apply to germicides and other chemical agents. Many of these substances target the same cellular structures and functions previously discussed: cell walls and membranes, proteins, and nucleic acids. The effectiveness of the agents depends on several factors, including (1) the numbers and kinds of microbes that are present, (2) the kinds of materials that are being treated, (3) the time of exposure required, and (4) the strength and mode of action of the agent. Here, we focus on the latter two because of their importance to chemical agents.

Length of Exposure Most compounds require adequate contact time for the chemical to penetrate and to act on the microbes present. The composition of the material being treated can greatly affect this process. Smooth, solid objects are more rapidly disinfected

TABLE 11.6	Qualities of Chemical Agents Used in Health Care			
Agent	Target Microbes	Level of Germicidal Activity	Toxicity	Comments
Chlorine	Sporicidal (slowly)	Intermediate	Gas is highly toxic; solution irritates skin	Inactivated by organics; unstable in sunlight
Iodine	Sporicidal (slowly)	Intermediate	Can irritate tissue; toxic if ingested	Iodophors* are milder forms
Phenolics	Some bacteria, viruses, fungi	Low to intermediate	Can be absorbed by skin; can cause CNS damage	Poor solubility; expensive
Chlorhexidine*	Most bacteria, some viruses, fungi	Low to intermediate	Low toxicity	Fast-acting, mild, has residual effects
Alcohols	Most bacteria, viruses, fungi	Intermediate	Toxic if ingested; a mild irritant; dries skin	Flammable, fast-acting
Hydrogen peroxide,* stabilized	Sporicidal	High (sterilant)	Toxic to eyes; toxic if ingested	Improved stability; works well in organic matter
Quaternary ammonium compounds	Some bactericidal, virucidal, fungicidal activity	Low to intermediate	Irritating to mucous membranes; poisonous if taken internally	Weak solutions can support microbial growth; easily inactivated
Soaps	Certain very sensitive species	Very low	Nontoxic; few if any toxic effects	Used for removing soil, oils, debris
Mercurials	Weakly microbistatic	Low	Highly toxic if ingested, inhaled, absorbed	Easily inactivated
Silver nitrate	Bactericidal	Low	Toxic, irritating	Discolors skin
Glutaraldehyde*	Sporicidal	High (sterilant)	Can irritate skin; toxic if absorbed	Not inactivated by organic matter; unstable
Ortho-phthaldehyde*	Sporicidal, with time	Intermediate to high	Mild, no odor	less toxic than glutaraldehyde, highly stable, stains skin
Ethylene oxide gas*	Sporicidal	High (sterilant)	Very dangerous to eyes, lungs; carcinogenic	Explosive in pure state; good penetration; materials must be aerated
Dyes	Weakly bactericidal, fungicidal	Low	Low toxicity	Stains materials, skin

*These chemicals approach the ideal by having many of the following characteristics: broad spectrum, low toxicity, fast action, penetrating abilities, residual effects, stability, potency in organic matter, and solubility.

than are those with pores or pockets that can trap soil. An item contaminated with common biological matter such as serum, blood, saliva, pus, fecal material, or urine presents an additional barrier to effective disinfection. Large amounts of this organic material can hinder penetration and, in some cases, can form bonds that reduce a chemical's activity. Adequate cleaning of instruments and other reusable materials will ensure that the germicide or sterilant will better accomplish the job for which it was chosen.

Concentration of the Chemical Agent A chemical's strength or concentration is expressed in various ways, depending upon convention and the method of preparation. The content of many chemical agents can be formulated in more than one way. For dilutions, a small volume of the liquid chemical (solute) is diluted in a larger volume of solvent to achieve a certain ratio. For example, a common laboratory phenolic disinfectant such as Lysol is usually diluted 1:200; that is, one part of chemical (solute) has been added to 200 parts of water (solvent) by volume. Solutions such as chlorine that

are effective in very diluted concentrations are expressed in parts per million (ppm). In percentage solutions, the solute is added to solvent by weight or volume to achieve a certain percentage in the solution. Alcohol, for instance, is used in percentages ranging from 50% to 95%. In general, solutions of low dilution or high percentage have more of the active chemical (are more concentrated) and tend to be more germicidal, but expense and potential toxicity can necessitate using the minimum strength that is effective.

Standardized procedures for testing the effectiveness of germicides are summarized in Appendix B.

Categories of Chemical Agents

Several general groups of chemical compounds are widely used for antimicrobial purposes in medicine and commerce (table 11.6). Prominent agents include halogens, phenolic compounds, alcohols, oxidizers, aldehydes, gases, detergents, and heavy metals. These groups are surveyed in the following section from the standpoint of

11.1 MAKING CONNECTIONS

Pathogen Paranoia: "The Only Good Microbe Is a Dead Microbe"

The sensational publicity over influenza outbreaks, MRSA infections, *E. coli* and *Salmonella* foodborne illnesses, noroviruses, and many others has dramatically increased public awareness of microorganisms. This glut of coverage has also had certain benefits, such as greater emphasis on hand washing, food handling, and personal hygiene. But sometimes a little knowledge can be dangerous. The trend also seems to have escalated into an obsessive fear of "germs" lurking around every corner and a fixation on eliminating as many microbes as possible from the environment and the human body.

As might be expected, commercial industries have found a way to capitalize on these fears. A widespread array of commonplace household materials now contain chemicals that provide antibacterial or germicidal "protection." First it was hand soaps and dishwashing detergents, and eventually the list grew to include shampoos, laundry aids, hand lotions, deodorants, sponges, kitty litter, cutting boards, garbage bags, toys, toothpaste, and even building products.

Chemical agents that are frequently added to these products are phenolics called triclosan and triclocarban. These substances do indeed kill most pathogenic bacteria, but they do not reliably destroy viruses or fungi, and they have been culprits in skin rashes due to hypersensitivity. Some published reports have also linked them to some forms of hormone and muscle disorders. Others point to a possible accumulative effect they may have that could interfere with development in young children. Because of concerns for these possible toxic side effects, in 2016 the FDA directed industries to phase out all nonmedical uses by 2017.

Medical experts are concerned that the widespread overuse of antibacterial chemicals could favor the survival and growth of resistant strains of bacteria. One study reported that pathogens such as *Mycobacterium tuberculosis* and *Clostridium* are naturally resistant to triclosan and that E. coli and *Staphylococcus aureus* have developed decreased sensitivity to it. One fear has been that overuse could select for "superbugs" that can survive all attempts at disinfection.

Another outcome of excess contact with environmental germicides is to reduce our natural contact with microbes that is required to maintain the normal microbial residents of the body and stimulate immunities. Constant use of these agents could also shift the balance in our normal resident microbiota by killing off beneficial microbes that we now know are essential to our health.

An unfortunate side effect of negative views of microbes is how they foster the feeling that all microbes are harmful. We must not forget that most human beings manage to remain healthy despite the fact that they live in continual intimate contact with microorganisms. We really do not have to be preoccupied with microbes every minute or feel overly concerned that the things we touch, drink, or eat are sterile, as long as they are somewhat clean and free of pathogens. For most of us, resistance to infection is well maintained by our numerous host defenses. Infectious disease specialists urge a happy medium approach. Instead of filling the home with questionable germicidal products, they encourage cleaning with traditional soaps and detergents, reserving more potent products to reduce the spread of infection in clinical settings.

Can you name other undesirable consequences of mass usage of antimicrobial chemicals in addition to those mentioned here? *Answer available on Connect.*

each agent's specific forms and modes of action. Applications and methods of delivery are described in tables.

The Halogen Antimicrobial Chemicals

The **halogens*** are fluorine, bromine, chlorine, and iodine, a group of nonmetallic elements with similar chemical properties. Although they can exist in either the ionic (halide) or nonionic state, most halogens exert their antimicrobial effect primarily in the nonionic state, not the halide state (chloride, iodide, for example). Because fluorine and bromine are difficult and dangerous to handle and no more effective than chlorine and iodine, only the latter two are used routinely in germicidal preparations. These elements are highly effective components of disinfectants and antiseptics because they are microbicidal and not just microbistatic, and they are sporicidal with longer exposure. For these reasons, halogens are the active ingredients in nearly one-third of all antimicrobial chemicals currently marketed.

Chlorine and Its Compounds Chlorine has been used for disinfection and antisepsis for approximately 200 years. The major forms used in microbial control are liquid and gaseous chlorine (Cl_2), hypochlorites (O^-Cl), and chloramines (NH_2Cl). In solution, these compounds combine with water and release hypochlorous acid (HClO):

$$Cl_2 + H_2O \rightarrow HClO + HCl \text{ (hydrochloric acid)}$$
$$Ca(OCl)_2 + 2H_2O \rightarrow Ca(OH)_2 + 2HClO$$

Hypochlorous acid oxidizes the sulfhydryl (S—H) groups on the amino acid cysteine and breaks disulfide (S—S) bridges on numerous enzymes. The resulting denaturation of the enzymes is permanent and suspends metabolic reactions. It also damages the structure of DNA, RNA, and fatty acids. Death of almost all microorganisms usually occurs within 30 minutes, although endospores may require several hours. Chlorine compounds are less effective and relatively unstable if exposed to light, alkaline pH, and excess organic matter.

Iodine and Its Compounds Iodine (I_2) is a pungent blue-black element that forms a brown-colored solution when dissolved in

* *halogens* (hay'-loh-jenz) Gr. *halos,* salt, and *gennan,* to produce.

TABLE 11.7	Applications of Halogen Compounds	
Form of Chlorine	**Primary Applications**	**How Delivered/Concerns**
Chlorine Gas (Cl₂)	Large-scale disinfection of drinking water, sewage, and wastewater.	Chlorination to a concentration of 0.6 to 1.0 parts of chlorine per million parts of water destroys most vegetative pathogens.*
Hypochlorites (HClO) Bleach	Used extensively in sanitization and disinfection of food equipment, treatment of swimming pools, spas, drinking water, and fresh foods; for wound antisepsis and routine medical and household disinfection, deodorizing, and stain removal.	Common household bleach is a 5% solution of sodium hypochlorite; dilutions of 1:10-1:1,000 are highly effective germicides.
Chloramines (Dichloramine, Halazone)	An alternative to pure chlorine in treating water supplies; also as sanitizers and disinfectants; for treating wounds and skin surfaces.	Because standard gas chlorination of water is now believed to produce unsafe levels of trihalomethanes, some water districts are required to use chloramine treatment of water supplies.
Form of Iodine	**Primary Applications**	**How Delivered/Concerns**
Iodophors	Most common iodine for skin and mucous membranes; antiseptic prep for surgery and injections; for surgical handscrubs; to disinfect equipment and surfaces; possibly for burns; and may be an alternative preventive for eye infections in newborns.	A complex of iodine and a neutral protein polymer provides slow release and reduced toxicity or irritation of tissues; less prone to staining. Common products are povidone-iodine (PVI; Betadine), which contain 2% to 10% available iodine.
Elemental Iodine: Aqueous or Tinctures	Topical antiseptic prior to surgery; sometimes for burned or injured skin. Medium-level disinfection for plastic instruments, thermometers; tablet form available for disinfection of contaminated water.	Weak solutions of 1% to 3% in water or in alcohol tinctures. Aqueous solutions or tinctures of 5% to 10%; somewhat limited by their toxicity and tendency to stain.

Cantaloupes are disinfected with chlorine treated water to kill *Salmonella* carried on their surfaces.

Patient has been treated with a povidone-iodine solution in preparation for surgery.

*Some cyst-forming protozoans such as *Giardia* and *Cryptosporidium* can survive this level of chlorination. (left): *Source:* Paul Pierlott, USDA-ERRC/VGT; (right): Used with permission from 3M

water. It rapidly penetrates the cells of microorganisms, where it disturbs a variety of metabolic functions by interfering with the disulfide bonds of proteins. This mode of action is similar to that seen in chlorine, although iodine is less susceptible to inactivation by organic material. All classes of microorganisms can be killed if proper exposure times and iodine concentrations are used.

See **table 11.7** for information on the main uses and forms of halogens.

Phenol and Its Derivatives

Phenol (carbolic acid) is an acrid, poisonous compound derived from the distillation of coal tar. First adopted by Joseph Lister in 1867 as a surgical germicide, phenol was the dominant antimicrobial chemical until other phenolics with fewer toxic and irritating effects were developed. Solutions of phenol are now used only in certain limited cases, but it remains one standard against which other phenolic disinfectants are rated. The *phenol coefficient* quantitatively compares a chemical's antimicrobic properties to those of phenol. Substances chemically related to phenol are often referred to as phenolics.

Phenolics consist of one or more aromatic carbon rings with added functional groups **(figure 11.14)**. Among the more common phenolics are alkylated phenols (cresols), chlorinated phenols (see triclosan, discussed in 11.1 Making Connections), and bisphenols. In high concentrations, phenolics are cellular poisons, rapidly disrupting cell walls and membranes and precipitating proteins; in lower concentrations they inactivate certain critical enzyme systems. The phenolics will destroy vegetative bacteria (including the tuberculosis bacillus), fungi, and many viruses (not hepatitis B), but they are not reliably sporicidal. Their sustained activity in the

Phenol
(basic aromatic
ring structure)

o-cresol

p-cresol

Chlorophene
(a chlorinated phenol)

Hexachlorophene
(a bisphenol)

Figure 11.14 **Examples of phenolic structures.** All contain a basic aromatic ring, but they differ in the types of additional side groups, such as Cl and CH_3.

presence of organic matter and their detergent actions contribute to their usefulness. Unfortunately, the toxicity of many of the phenolics makes them a questionable choice as antiseptics.

Chlorhexidine

The compound chlorhexidine (Hibiclens, Hibitane) is a complex organic base containing chlorine and two phenolic rings. Its mode of action targets cell membranes by lowering surface tension and causing denaturation of proteins. At moderate to high concentrations it is bactericidal for both gram-positive and gram-negative bacteria but inactive against spores. Its effects on viruses and fungi vary. It possesses distinct advantages over many other antiseptics because of its mildness, low toxicity, and rapid action, and it is not absorbed into deeper tissues to any extent. It is the antiseptic cleanser of choice to control *Staphylococcus* MRSA and *Acinetobacter* outbreaks in hospitals.

Applications for phenolic compounds and chlorhexidine are given in **table 11.8**.

Alcohols as Antimicrobial Agents

Alcohols are colorless hydrocarbons with one or more —OH functional groups. Of several alcohols available, only ethyl and isopropyl are suitable for microbial control. Methyl alcohol is not particularly microbicidal, and more complex alcohols are either poorly soluble in water or too toxic for routine use. Alcohols are employed alone in aqueous solutions or as solvents for tinctures (with iodine, for example).

Alcohol's mechanism of action depends in part upon its concentration. Concentrations of 50% and higher dissolve membrane lipids, disrupt cell surface tension, and compromise membrane integrity. Alcohol that has entered the protoplasm denatures proteins through coagulation but only in alcohol-water solutions of 50% to 95%. Alcohol is the exception to the rule that higher concentrations of an antimicrobial chemical have greater microbicidal activity. Because water is needed for proteins to coagulate, alcohol shows a greater microbicidal activity at 70% concentration (that is, 30% water) than at 100% (0% water). Absolute alcohol (100%) dissolves cell membranes and inhibits cell growth but is generally not a protein coagulant.

TABLE 11.8	Applications of Phenolics and Chlorhexidine	
Form of Phenolic	**Primary Applications**	**How Delivered/Concerns**
Creolin	Common household version of phenol; for low or intermediate levels of disinfection in rooms, floors, and veterinary clinics.	1% to 3% emulsions that are combined with soap; can be too toxic for antisepsis; tend to be absorbed by membranes into the blood.
Bisphenols	Widely employed commercially, clinically, and in the home.	Orthophenyl phenol is a major ingredient in fungicides. This same phenolic is also found in some disinfectants used in the hospital and laboratory.
Hexachlorophene	Formerly a widespread additive of cleansing soaps (pHisoHex) for hospital and home use; may be used to control outbreaks of skin infections.	When hexachlorophene was found to be absorbed through the skin and a cause of neurological damage, it was no longer available without a prescription.
Triclosan (Dichlorophenoxyphenol)	Widely used antibacterial compound added to soaps, cosmetics, and many other household products.	It acts as both disinfectant and antiseptic and is broad-spectrum in its effects. New uses include surgical preps, handwashing, and bathing patients.
Form of Chlorhexidine	**Primary Applications**	**How Delivered/Concerns**
Chlorhexidine (Hibiclens, Hibitane)	Alcoholic or aqueous solutions are now commonly used for hand scrubbing, preparing skin sites for surgical incisions and injections, and patient cleansing. Solutions also serve as an obstetric and neonatal wash, a wound degermer, a mucous membrane irrigant, and a preservative for eye solutions.	Complex base containing chlorine and two phenolic rings; mode of action targets cell membranes and protein structure. At moderate to high concentrations, destroys both gram-positive and gram-negative bacteria but not spores. It may be fungicidal and virucidal. It is milder and less toxic than phenolics and is not absorbed into the skin.

Although useful in intermediate- to low-level germicidal applications, alcohol does not destroy bacterial spores at room temperature. Alcohol can, however, destroy resistant vegetative forms, including tuberculosis bacteria and fungal spores, provided the time of exposure is adequate. Alcohol tends to inactivate enveloped viruses more readily than nonenveloped viruses because of the surfactant effect on the envelope.

Hydrogen Peroxide and Related Germicides

Hydrogen peroxide (H_2O_2) is a colorless, caustic liquid that decomposes in the presence of light, metals, or catalase into water and oxygen gas. Early formulations were unstable and inhibited by organic matter, but manufacturing methods now permit synthesis of H_2O_2 so stable that even dilute solutions retain activity through several years of storage.

The germicidal effects of hydrogen peroxide are due to the direct and indirect actions of oxygen. Oxygen forms hydroxyl free radicals (—OH), which, like the superoxide radical (see chapter 7), are highly toxic and reactive to cells. Although most microbial cells produce catalase to inactivate the metabolic hydrogen peroxide, catalase cannot neutralize the amount of hydrogen peroxide penetrating these cells during disinfection and antisepsis. Hydrogen peroxide is bactericidal, virucidal, fungicidal, and, in higher concentrations, sporicidal. The need for heat-free rapid sterilization of small medical devices has led to the development of chambers that circulate combinations of peroxide and peracetic acid. A vaporized form of hydrogen peroxide is effective as an environmental sterilant used in decontaminating large spaces and objects (see figure 11.4).

Coverage on the applications of alcohols and peroxides is provided in **table 11.9.**

TABLE 11.9	Applications of Alcohols and Peroxides (and Other Oxidizing Agents)	
Form of Alcohol*	**Primary Applications**	**How Delivered/Concerns**
Ethyl Alcohol (Ethanol, Grain Alcohol)	Skin degerming agent and antiseptic; surfactant action removes skin oil, soil, and some microbes sheltered in deeper skin layers; occasionally used to disinfect electrodes and face masks.	Solutions of 70% to 95% are germicidal, inexpensive, nonirritating; limited by its rate of evaporation. Items need to be first cleaned and then soaked in alcohol for 15 to 20 minutes.
Isopropyl Alcohol (Rubbing Alcohol)	Some disinfection of objects, surfaces; limited skin cleansing.	More microbicidal and less expensive than ethanol, but these benefits must be weighed against its toxicity; inhaling its vapors can adversely affect the nervous system.
Form of Oxidizing Agent	**Primary Applications**	**How Delivered/Concerns**
Hydrogen Peroxide (H_2O_2)	Versatile use as an antiseptic, including skin and wound cleansing, bedsore care, and mouthwashing; disinfectant for soft contact lenses, surgical implants, plastics, bedding, and room interiors.	3% hydrogen peroxide—most common form—is useful in treating infections by anaerobic bacteria because of the lethal effects of the oxygen released.
Sterilizing H_2O_2	Vaporized H_2O_2 is a major sterilant; hydrogen peroxide plasma sterilizers used for industrial parts or medical items; for isolators, clean rooms, and space vehicles.	Hydrogen peroxide (35%) penetrates into delicate medical instruments, kills the most resistant microbes, and does not corrode small parts; vapors are sporicidal.
Peracetic Acid	Used to sterilize rooms, space shuttle; a chemical sterilant for medical devices.	Oxidizing agent—actions similar to those of hydrogen peroxide.
Ozone (O_3)	A gas used to disinfect air, water, industrial air conditioners, and cooling towers.	Difficult to handle in gaseous form

A nurse takes advantage of an alcohol gel hand cleansing station. These are very common in hospitals and clinics to encourage hand antisepsis.

Disinfection of contact lenses with peroxide-based cleaner. Note the bubbles produced, indicating the release of oxygen gas.

*Both forms of alcohol are in alcohol gels used for universal precautions and hand degerming in hospitals. (left) © Photo by Peter Morenus, University of Connecticut; (right) © Kathy Park Talaro

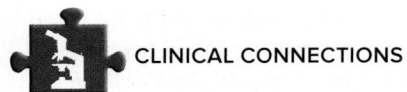

CLINICAL CONNECTIONS

When "Clean" is Not Enough

A number of clinical procedures rely on reusable instruments such as endoscopes and dental handpieces. Because these devices can become heavily contaminated by tissues and fluids, they need to undergo decontamination between patients to prevent transmission of infections such as hepatitis, tuberculosis, and HIV. These expensive diagnostic tools (depending on the tool, they can cost $10,000 to $80,000 each) have created a medical dilemma. Their intricate parts may trap infectious agents where they cannot be easily removed, and the tools are delicate and difficult to decontaminate. Traditional methods are either too damaging to the instruments (heat) or unable to sterilize them in a timely fashion between patients (radiation).

From 2013 to 2015 eight clusters of patients were infected and 20 died due to complications from contaminated endoscopes called duodenoscopes that are threaded through the mouth into the intestine. Hospitals were immediately forced to reevaluate how these devices were being cleaned or "reprocessed" between patients. In many cases the methods were found to be inadequate because they failed to eliminate a category of pathogens called CRE (carbapenem-resistant enterobacteriaceae). These are normal residents of the gastointestinal tract of many people, but they can cause deadly infections in compromised patients, and newer strains of them carry multiple-drug-resistant genes (see 12.3 Making Connections).

Ordinarily the entire scope was scrubbed, soaked in detergent, and rinsed in alcohol, which is considered a high level of disinfection. It was now apparent that this treatment was not rigorous enough, because the CRE bacteria had formed biofilms that were not being removed completely, and small amounts of them remained in the microscopic crevices of the tiny tools located on the tip of an endoscope **(figure 11.15).** About 700,000 endoscopic procedures are

Figure 11.15 Types of endoscope attachments. A close-up view of different endoscope tips shows a camera (top), biopsy tool (middle), and suction device (bottom). Others are equipped with lights and air or water outlets to flush or blow away debris. It is evident how readily microbes could become hidden or entrenched in these devices. *Source:* Linda Bartlett/National Cancer Institute

performed every year in the United States, making their safety a very high priority. Two governmental agencies (the CDC and FDA) intervened by requiring that this type of endoscope be sterilized through multistep techniques to remove debris, followed by extensive exposure to sterilants. Their effectiveness would have to be checked after every procedure by sampling and culturing. Chemicals that are currently recommended for this level of reprocessing are glutaraldehyde, peracetic acid/peroxide combination, and ethylene oxide gas.

Using lab techniques you learned in chapter 3, tell how would you conduct tests to determine whether sterilization of the endoscopes had been achieved. *Answer available on Connect.*

Aldehyde Sterilants and Disinfectants

Organic substances bearing a —CHO functional group (a strong reducing group) on the terminal carbon are called aldehydes. The two aldehydes used most often in microbial control are *glutaraldehyde* and *ortho-phthaldehyde.*

Glutaraldehyde is a yellow liquid with a mild odor. The molecule's two aldehyde groups favor the formation of polymers. The mechanism of activity involves cross-linking protein molecules on the cell surface. In this process, amino acids are alkylated, meaning that a hydrogen atom on an amino acid is replaced by the glutaraldehyde molecule itself. This irreversibly disrupts the activity of enzymes and other proteins within cells. Glutaraldehyde is rapid and broad-spectrum, and is officially labeled as a sterilant and high-level disinfectant. It is one of the few chemicals that can be used to sterilize delicate reusable medical and dental instruments (see **Clinical Connections, "When 'Clean' Is Not Enough",** and figure 11.15). It kills spores in 3 hours and fungi and vegetative bacteria (even *Mycobacterium* and *Pseudomonas*) in a few minutes. Viruses, including the most resistant forms, appear to be inactivated after relatively short exposure. A special device known as a

chemiclave (similar to an autoclave) vaporizes the solution under pressure to sterilize delicate instruments without heat. Glutaraldehyde retains its potency even in the presence of organic matter, is noncorrosive, does not damage plastics, and is less toxic or irritating than formaldehyde. Its principal disadvantage is that it is somewhat unstable, especially with increased pH and temperature.

Another aldehyde, ortho-phthalaldehyde (OPA), has been approved by the EPA as a high-level disinfectant. OPA is a pale blue liquid with a barely detectable odor. It has a mechanism of action similar to glutaraldehyde, is stable and nonirritating, and for most uses is much faster acting than glutaraldehyde. It is effective against vegetative bacteria, including *Mycobacterium* and *Pseudomonas,* fungi, and viruses. Chief among its disadvantages are an inability to reliably destroy spores and a tendency to stain proteins, including those in human skin.

Formaldehyde is a sharp, irritating gas that readily dissolves in water to form an aqueous solution called **formalin.** Pure formalin is a 37% solution of formaldehyde gas dissolved in water. The chemical is microbicidal through its attachment to nucleic acids and functional groups of amino acids. Formalin is an intermediate- to

high-level disinfectant, but its extreme toxicity (it is classified as a carcinogen) and irritating effects on the skin and mucous membranes greatly limit its clinical usefulness.

Gaseous Sterilants and Disinfectants

Processing inanimate substances with chemical vapors, gases, and aerosols provides a versatile alternative to heat or liquid chemicals. Currently the vapors and aerosols having the broadest applications are ethylene oxide (ETO), propylene oxide, and chlorine dioxide.

Ethylene oxide (ETO) and its relative propylene oxide (PO) are colorless gases at room temperature. The structure of ETO is shown below:

$$CH_2 - CH_2$$
$$\diagdown O \diagup$$

ETO is very explosive in air, a feature that can be eliminated by combining it with a high percentage of carbon dioxide or fluorocarbon. Like the aldehydes, ETO is a very strong alkylating agent, and it reacts vigorously with functional groups of DNA and proteins. Through these actions, it blocks both DNA replication and enzymatic actions. Ethylene oxide is one of a very few gases generally accepted for chemical sterilization because, when employed according to strict procedures, it is a sporicide. A specially designed ETO sterilizer is equipped with a chamber, gas ports, and controls for temperature, pressure, and humidity **(figure 11.16)**. Ethylene oxide is rather penetrating but relatively slow-acting, requiring from 90 minutes to 3 hours. Some items absorb ETO residues and must be aerated with sterile air for several hours after exposure to remove as much residual gas as possible. For all of its effectiveness, ETO has some unfortunate features. Its explosiveness makes it dangerous to handle; it

Figure 11.16 **Sterilization using an ethylene oxide sterilizer.** The machine is equipped with gas canisters containing ethylene oxide (ETO) and carbon dioxide, a chamber to hold items, and mechanisms that evacuate gas and introduce air.

can damage the lungs, eyes, and mucous membranes if contacted directly; and it is rated as a carcinogen by the government. Both ETO and propylene oxide are used mainly as disinfectant-sterilants for some foods and plastic items.

Chlorine dioxide is another gas used for sterilization. Despite the name, chlorine dioxide works in a completely different way from the chlorine compounds discussed earlier in the chapter. It is a strong alkylating agent, which disrupts proteins and is effective against vegetative bacteria, fungi, viruses, and endospores. Chlorine dioxide has mainly been used for the treatment of drinking water, wastewater, food-processing equipment, and medical waste, but it also has numerous applications in decontamination of rooms, buildings, and large spaces or objects.

Betapropiolactone (BPL) is a substance somewhat similar to ETO in applications. It is rapidly microbicidal when used as an aerosol or liquid to disinfect whole rooms and instruments, to sterilize bone and arterial grafts, and to inactivate viruses in vaccines.

A summary of applications for aldehydes and sterilizing gases is presented in **table 11.10**.

Chemicals with Surface Action: Detergents and Soaps

Detergents are polar molecules that act as surfactants. Most anionic detergents have limited microbicidal power. This includes most soaps. Much more effective are positively charged (cationic) detergents, particularly the **quaternary ammonium compounds** (usually shortened to *quats*). The activity of cationic detergents arises from the attachment of four organic radicals (the origin of the term quaternary) to a central, positively charged nitrogen atom. At least one of these radicals is a long-chain hydrocarbon surfactant that disrupts the cell membranes of sensitive microbes (see **table 11.11**). After a few minutes of contact, the cells burst and die.

The effects of detergents are varied. When used at high enough concentrations, quaternary ammonium compounds are effective against some gram-positive bacteria, viruses, fungi, and algae. In low concentrations, they exhibit only microbistatic effects. Drawbacks to the quats include their ineffectiveness against the tuberculosis bacterium, hepatitis virus, *Pseudomonas*, and spores at any concentration. Furthermore, their activity is greatly reduced in the presence of organic matter, and they require alkaline conditions to function. As a result of these limitations, quats are rated only for low to moderate-level disinfection in the clinical setting.

Soaps are alkaline compounds made by combining the fatty acids in oils with sodium or potassium salts. In usual practice, soaps alone are weak microbicides, and they destroy only highly sensitive forms such as the agents of gonorrhea, meningitis, and syphilis. The common hospital pathogen *Pseudomonas* is so resistant to soap that various species grow abundantly in soap dishes. Soaps function primarily as cleansing agents and sanitizers in industry, the home, and the medical setting. Combining germicides such as chlorhexidine or iodophor with soaps produces highly active antiseptics, degermers, and disinfectants. Vigorously brushing the hands with germicidal soap over a 15-second period can remove dirt, oil, and surface contaminants as well as some superficial

TABLE 11.10	Applications of Aldehydes and Sterilant Gases	
Form of Aldehyde	Primary Applications	How Delivered/Concerns
Glutaraldehyde	A sterilant for materials usually damaged by heat. Examples include respiratory therapy equipment, hemostats, fiberoptic endoscopes, kidney dialysis equipment, and dental instruments; an alternate vaccine preservative, and sanitizer for poultry carcasses.	Solutions diluted to 2% are relatively mild but require 2 to 4 hours' soaking. Usual commercial products are Cidex, Sporocidin. Instrument sterilization involves precleaning along with autoclaving to inactivate hepatitis B and other blood-borne viruses.
Ortho-phthalaldehyde (OPA)	High-level disinfectant similar in uses and effects to glutaraldehyde.	Faster acting and more stable than glutaraldehyde but less sporicidal.
Form of Gas	Primary Applications	How Delivered/Concerns
Ethylene Oxide (ETO)	An official sterilant for heat-sensitive plastics and delicate instruments in hospitals and industries—prepackaged medical supplies and disposable Petri dishes; used extensively to disinfect foods, spices, dried fruits, and drugs.	Carboxide and cryoxide are commercial products; gas is explosive and must be used with a stabilizer; ETO is rather toxic to humans; penetrating but slow, requiring 1 to 3 hours of exposure inside a special chamber.
Propylene Oxide	Sterilization of foods such as nuts, powders, starches, and spices.	Physical properties and mode of action similar to ETO but safer because it breaks down into a relatively harmless substance.
Chlorine Dioxide	Air and surface sterilant; for treatment of water, food-processing equipment, and medical waste; decontamination of whole rooms and space probes.	Similar in action to ETO but does not require a chamber; thus, effective for large spaces or objects.

Technician treats surgical instruments with glutaraldehyde prior to heat sterilization.
(left): Courtesy of AirClean Systems

Examples of prepackaged medical supplies sterilized in an ethylene oxide sterilizer
(right): Courtesy of Printpack, Inc.

TABLE 11.11	Applications of Detergents and Soaps	
Form of Detergents or Soaps	Primary Applications	How Delivered/Concerns
Quaternary Ammonium Compounds (Quats) Benzalkonium chloride	Mixed with cleaning agents to disinfect and sanitize floors, equipment surfaces, and restrooms; for sanitizing restaurant utensils, food-processing equipment, and clothing; common preservatives in ophthalmic solutions and cosmetics.	Names include benzalkonium chloride (Zephiran), the active ingredient in Lysol spray, and cetylpyridinium chloride (Ceepryn); in dilutions ranging from 1:100 to 1:1,000; the level of disinfection is too low for disinfecting medical instruments. Quats are active ingredients in many household disinfectant sprays.
Soaps Proper hand washing with soap is one of the most important control methods for health care and personal hygiene. © OJO Images/SuperStock RF	Cleansing agents and sanitizers in industry and the home; preparing instruments for heat sterilization; degerming patients' skin, routine hand washing by medical and dental personnel, surgical and preoperative hand scrubbing.	Alkaline salts of fatty acids; weak germicides with superior sudsing and wetting properties; can remove large amounts of surface soil, grease, and other debris; antimicrobial chemicals added to make germicidal soaps with greater disinfection power.

Quick Search
Go to CDC TV and watch the video "Put Your Hands Together" about the importance of hand washing.

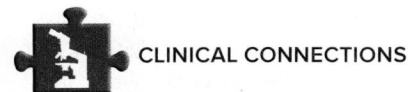
CLINICAL CONNECTIONS

The Quest for Sterile Skin

About 140 years ago, before sterile gloves were a routine part of medical procedures, the hands remained bare during surgery. Even then, medical practitioners understood the risk for infections and attempted to "sterilize" the hands of surgeons and their assistants. Several stringent (and probably very painful) techniques involving strong chemical germicides and vigorous scrubbing were practiced. Here are a few examples.

In Schatz's method, the hands and forearms were first cleansed by brisk scrubbing with liquid soap for 3 to 5 minutes, then soaked in a saturated solution of permanganate at a temperature of 110°F until they turned a deep mahogany brown. Next, the limbs were immersed in saturated oxalic acid until the skin became decolorized. Then, as if this were not enough, the hands and arms were rinsed with sterile limewater and washed in warm bichloride of mercury for 1 minute.

Park's method was even more like torture. First the surfaces of the hands and arms were rubbed completely with a mixture of cornmeal and green soap to remove loose dirt and superficial skin. Next a paste of water and mustard flour was applied to the skin until it began to sting. This potion was rinsed off in sterile water, and the hands and arms were then soaked in hot bichloride of mercury for a few minutes, during which time the solution was rubbed into the skin.

Another method once earnestly suggested for getting rid of microorganisms was to expose the hands to a hot-air cabinet to "sweat the germs" out of skin glands. Pasteur himself advocated a quick flaming of the hands to maintain asepsis.

The old dream of sterilizing the skin was finally reduced to some basic realities: The microbes entrenched in the epidermis and skin glands cannot be completely eradicated even with the most intense efforts, and the skin cannot be sterilized without also seriously damaging it. Because this is true for both medical personnel and their patients, the chance always exists that infectious agents can be introduced

Heavy growth of microbial colonies on a plate of blood agar prepared by having a classroom of 30 students touch its surface. The result was a mixed population of bacteria and fungi, many of which could be pathogens (clear zones of hemolysis). © Kathy Park Talaro

during invasive medical procedures. Safer surgery had to wait until 1890, when rubber gloves first became available to place a sterile barrier around the hands. Of course, this did not mean that skin cleansing and antiseptic procedures were abandoned or downplayed. A thorough scrubbing of the skin with antiseptic soap is still needed to remove the most dangerous source of infections—the superficial contaminants constantly picked up from whatever we touch.

What would be some of the more serious problems associated with sterilizing or trying to sterilize the skin? *Answer available on Connect.*

resident microbes, but it is unable to remove those deeply entrenched in the skin (see **Clinical Connections, "The Quest for Sterile Skin,"** and **figure 11.17**).

See table 11.11 for coverage on the main uses of detergents and soaps.

Heavy Metal Compounds

Various forms of the metallic elements mercury, silver, gold, copper, arsenic, and zinc have been applied in microbial control over several centuries. These are often referred to as heavy metals because of their relatively high atomic weights. However, from this list, only preparations containing mercury and silver still have any significance as germicides. Although some metals (zinc, iron) are needed

Figure 11.17 Graph showing some effects of hand scrubbing. Comparison of scrubbing over several days with a nongermicidal soap versus a germicidal soap. Germicidal soap has persistent effects on skin over time, reducing and keeping the microbial count low. Soap alone does not cause a sustained reduction in the microbial count, it even raises it. This effect is presumably due to the uncovering of skin layers with higher levels of normal residents.

Silver amalgam Gold foil

Figure 11.18 Demonstration of the oligodynamic action of heavy metals. A pour plate inoculated with saliva has small fragments of heavy metals pressed lightly into it. During incubation, clear zones indicating growth inhibition developed around both fragments. The slightly larger zone surrounding the amalgam (used in tooth fillings) probably reflects the combined effect of the silver and mercury it contains. © Kathy Park Talaro

in small concentrations as cofactors on enzymes, the higher molecular weight metals (mercury, silver, gold) can be very toxic, even in minute quantities (parts per million). This property of having antimicrobial effects in exceedingly small amounts is called an **oligodynamic*** **action (figure 11.18).** Heavy metal germicides contain either an inorganic or an organic metallic salt, and they come in the form of aqueous solutions, tinctures, ointments, or soaps.

Mercury, silver, and most other metals exert microbicidal effects by binding onto functional groups of proteins and inactivating them, rapidly bringing metabolism to a standstill (see figure 11.6c). This mode of action can destroy many types of microbes, including vegetative bacteria, fungal cells and spores, algae, protozoa, and viruses (but not endospores). A newer use of metals is their incorporation into medical supplies and hospital facilities. Silver is now added to catheters, IV lines, and prostheses to prevent biofilm formations and infections. Studies have demonstrated that copper has superior killing powers over chrome and aluminum when added to

* *oligodynamic* (ol″-ih-goh-dy-nam′-ik) Gr. *oligos,* little, and *dynamis,* power.

metal parts of lavatory fixtures, doors, and doorknobs. It destroys the most problematic drug-resistant pathogens such as MRSA, as well as *Clostridium difficile* vegetative cells, influenza virus, and gram-negative opportunists.

The main drawbacks to heavy metals are (1) metals (especially mercury) are very toxic to humans if ingested, inhaled, or absorbed through the skin, even in small quantities, for the same reasons that they are toxic to microbial cells; (2) they commonly cause allergic reactions; (3) large quantities of biological fluids and wastes neutralize their actions; and (4) microbes can develop resistance to metals.

Table 11.12 provides an overview of applications for heavy metals.

Dyes as Antimicrobial Agents

Dyes are important in staining techniques and as selective and differential agents in media; they are also a primary source of certain drugs used in chemotherapy. Their antimicrobial effects are apparently due to the way they insert into nucleic acids and cause mutations. There is also some evidence they can interfere with cell wall synthesis. Because aniline dyes such as crystal violet and malachite green are most active against gram-positive species of bacteria and some fungi, they are incorporated into solutions and ointments to treat skin infections (ringworm, for example). The yellow acridine dyes, acriflavine and proflavine, are sometimes utilized for antisepsis and wound treatment in medical and veterinary clinics. For the most part, applications for dyes will continue to be limited because dyes stain and have a narrow spectrum of activity.

Acids and Alkalis

Conditions of very low or high pH can destroy or inhibit microbial cells, but they are limited in applications due to their corrosive, caustic, and hazardous nature. Alkaline solutions are good for cleaning and some types of disinfection because they inactivate proteins. Ammonium hydroxide is a major component of detergents and cleansers, and a strong sodium hydroxide solution is one of the few substances that reliably destroys prions. Organic acids are widely used in food preservation because they prevent spore germination and bacterial and fungal growth and because they are generally regarded as safe to eat. Acetic acid (in the form of vinegar) is a pickling agent that inhibits bacterial growth; propionic acid is commonly

TABLE 11.12	Applications of Heavy Metals	
Form of Metal	**Primary Applications**	**How Delivered/Concerns**
Organic Mercury	Thimerosal (Merthiolate) and nitromersol (Metaphen) are weak antiseptics and infection preventives; can be preservatives in cosmetics and ophthalmic solutions.	Tinctures (0.001% to 0.2%) are fairly effective but a poor choice for broken skin because they are toxic and can delay healing. Mercurochrome is now considered one of the poorest of antiseptics.
Silver Sulfadiazine Ointment	Added to dressings, it effectively prevents infection in second- and third-degree burn patients.	A yellow ointment containing silver salts and a sulfa drug.
Silver Nitrate (AgNO₃)	Primarily as a topical germicide for mouth ulcers and occasionally root canals.	1% to 2% solutions have been used as antiseptics, but they discolor skin and other tissues.
Metallic Silver	Incorporated into catheters to prevent urinary tract infections in the hospital; added to paint, plastics, and steel as a way to control microbial survival on items such as toilet seats, stethoscopes, and even walls and floors in homes.	
Colloidal Silver	Mild germicidal ointments or rinses for the mouth, nose, eyes, and vagina.	

TABLE 11.13	Active Ingredients of Various Commercial Antimicrobial Products	
Product	**Specific Chemical Agent**	**Antimicrobial Category**
Lysol and Clorox Sanitizing Wipes	Dimethyl benzyl ammonium chloride	Detergent (quat)
Tilex and Lysol Mildew Remover	Sodium hypochlorites	Halogen
Ajax and Dial Antibacterial Hand Soap	Triclosan	Phenolic
Lysol Disinfecting Spray	Alkyl dimethyl benzyl ammonium/ethanol	Detergent (quat)/alcohol
ReNu Contact Lens Solution	Polyaminopropyl biguanide	Chlorhexidine
Wet Ones Antibacterial Moist Towelettes	Benzethonium chloride	Detergent (quat)
Noxzema Triple Clean	Triclosan	Phenolic
Scope Mouthwash	Ethanol	Alcohol
Purell Instant Hand Sanitizer	Ethanol	Alcohol
Pine-Sol	Pine oil and surfactant	Aldehydes, alcohols
Allergan Eye Drops	Sodium chlorite	Halogen

incorporated into breads and cakes to retard molds; lactic acid is added to sauerkraut and olives to prevent growth of anaerobic bacteria (especially the clostridia); and benzoic and sorbic acids are added to beverages, syrups, and margarine to inhibit yeasts.

For a look at the antimicrobial chemicals found in some common household products, see **table 11.13.**

 ### *Check Your Progress* SECTION 11.4

22. Describe situations that require high-level disinfection and those for which medium- or low-level disinfection is adequate.
23. What is the difference between a tincture and an aqueous solution?
24. Define *sterilant*, and name the principal sporicidal chemical agents.
25. Summarize the chief forms and uses of chlorine compounds.
26. What are the superior characteristics of iodophors over free iodine solutions?
27. What are the modes of action of alcohols and phenolics?
28. Why is hydrogen peroxide solution so effective against anaerobes?
29. Give the uses and disadvantages of the heavy metal chemical agents, glutaraldehyde, and the sterilizing gases.
30. What does it mean to say that a chemical has an oligodynamic action, and which compounds have this effect?

 ### CASE STUDY Part 2

The preliminary investigations linked nine diagnosed cases and 115 possible cases of HCV to the facility. It was the largest outbreak of health-care-associated hepatitis ever reported. Not only were 63,000 clinic patients tested for HCV, all of them were directed to have additional tests for hepatitis B virus (HBV) and human immunodeficiency virus (HIV).

In the meantime, the big question was: What went wrong at the endoscopy center? Teams of epidemiologists reviewed the records on the days the patients had procedures. Essentially, their investigation identified four potential problem areas and possible modes of transmission: staff to patient; contaminated endoscopes, which have to be reused and sterilized be-

tween patients; techniques for inserting the IV ports; and the method of preparing syringes for sedation drugs.

First they tested all of the health care workers at the clinic for HCV, none of whom registered a positive result. The endoscopes were found to have been scrubbed with an enzymatic cleaner, soaked in glutaraldehyde for 30 minutes, and flushed with alcohol. Other equipment was either new and sterile or disinfected in the clinic. The IV lines were inserted in a standard fashion with sterile needles and syringes. Protocols for preparing drugs for sedation were not standardized, however, and varied among the RN-anesthetists who did the procedures.

A report was made about an RN-anesthetist who was observed taking out a second dose of a drug from a large, multidose vial using the same syringe that had been used to take the first dose, although he did replace the needle. It was common practice to store these partly used vials and reuse them on other patients.

In 2010 a judge awarded the claimants in a lawsuit the huge sum of $500 million to be paid by the drug manufacturers and distributor for negligence in packaging of the drug. Several patients were awarded an additional $500 million in a civil lawsuit against the health maintenance organization that hired the endoscopy clinic. In 2013 the owner and two assistants were tried and convicted on a number of charges, including criminal neglect and failure to report unsafe medical procedures. The owner was found guilty of second-degree murder, because two of the patients died from the effects of HCV infection, and was sentenced to 18 years in prison. He has since been found guilty of medical fraud, overbilling, and other federal charges.

- *Why was it necessary to also test the patients for HBV and HIV?*

- *Why is it a mistake to use the same syringe, even with a new, sterile needle?*

To conclude this Case Study, go to Connect.

Additional coverage of these topics can be found in chapter 13.

A summary of this case can be found at http://www.cdc.gov/mmwr/preview/mmwrhtml/mm5719a2.htm.

Chapter Summary with Key Terms

11.1 Controlling Microorganisms

A. **Decontamination** procedures involve the destruction or removal of contaminants. Contaminants are defined as microbes present at a given place and time that are undesirable or unwanted.

B. **Sterilization** is a process that destroys or removes all microbes, including viruses.
 1. A **bactericide** is a chemical that destroys vegetative stages of bacteria. A **sporicide** can destroy bacterial endospores.
 2. **Bacteriostatic** agents inhibit or prevent the growth of bacteria on tissues or on other objects in the environment.
 3. A **germicide** or **microbicide** is a chemical that will kill any pathogenic microorganisms.

C. **Disinfection** refers to a physical or chemical process that destroys vegetative pathogens but not bacterial endospores on objects. An **antiseptic** is a chemical applied to living tissues to render them free of pathogens.

D. **Sanitization** is forceful cleaning that removes contaminants and makes objects hygienic and safe to use. **Degermation** is the same process applied to the body to remove potential pathogens.

E. Several factors influence the rate at which antimicrobial agents work. These factors are:
 1. Exposure time to the agent
 2. Numbers of microbes present
 3. Relative resistance of microbes (for example, endospores versus vegetative forms)
 4. Activity of the agent (microbicidal versus microbistatic)

F. How antimicrobial agents work: their **modes of action.** Agents affect cell wall synthesis, membrane permeability, and protein and nucleic acid synthesis and function.

11.2 Physical Methods of Control: Heat

A. **Moist heat** denatures proteins and DNA while destroying membranes.
 1. *Sterilization:* **Autoclaves** utilize steam under pressure (121°C) to sterilize heat-resistant materials, whereas intermittent sterilization uses steam at 100°C to sterilize more delicate items.
 2. *Disinfection:* **Pasteurization** subjects liquids to temperatures below 100°C and is used to lower the microbial load in liquids. Boiling water can be used to destroy vegetative pathogens in the home.

B. **Dry heat**, using higher temperatures than moist heat, can also be used to sterilize.
 1. **Incineration** can be carried out using a Bunsen burner or incinerator. Temperatures range between 800°C and 6,500°C.
 2. *Dry ovens* operate at temperatures of 150°C to 180°C.

C. Cold temperatures (refrigeration and freezing) are microbistatic and are commonly used to preserve food, media, and cultures.

D. Drying and **desiccation** lead to (often temporary) metabolic inhibition by reducing water in the cell.

11.3 Physical Methods of Control: Radiation and Filtration

A. *Radiation:* Energy in the form of **radiation** can be used for **cold sterilization,** which works by killing microbes without heat.
 1. **Ionizing radiation,** such as gamma rays and X rays, has deep penetrating power and works by causing breaks in the DNA of target organisms.

 2. **Nonionizing radiation** uses **ultraviolet** waves with very little penetrating power and works by creating dimers between adjacent pyrimidines, which interferes with replication.

B. *Using Filtration to Remove Microbes:* Filtration involves the physical removal of microbes by passing a gas or liquid through a fine filter, and can be used to sterilize air as well as heat-sensitive liquids.

11.4 Chemical Agents in Microbial Control

A. Chemicals are divided into **disinfectants, antiseptics, sterilants, sanitizers,** and **degermers** based on their level of effectiveness and the surfaces to which they are applied.

B. Antimicrobial chemicals are found as solids, gases, and liquids. Liquids can be either **aqueous** (water based) or **tinctures** (alcohol based).

C. **Halogens** are chemicals based on elements from group VII of the periodic table.
 1. Chlorine gas, hypochlorites, and chloramines all work by disrupting disulfide bonds and, given adequate time, are sporicidal.
 2. Iodine compounds include free iodine (I_2) and **iodophors** Iodine has a mode of action similar to chlorine and can also be sporicidal.

D. **Phenolics** work by disrupting cell membranes and precipitating proteins. They are bactericidal, fungicidal, and virucidal but not sporicidal. Bisphenols are mild forms for disinfection and antisepsis.

E. **Chlorhexidine** (Hibiclens, Hibitane) is a surfactant and protein denaturant with broad microbicidal properties, although it is not sporicidal. Solutions of chlorhexidine are used as skin degerming agents for preoperative scrubs, skin cleaning, and burns.

F. Ethyl and isopropyl **alcohol,** in concentrations of 50% to 90%, are useful for microbial control. Alcohols act as **surfactants,** dissolving membrane lipids and coagulating proteins of vegetative bacterial cells and fungi. They are not sporicidal.

G. **Hydrogen peroxide** produces highly reactive hydroxyl free radicals that damage protein and DNA while also decomposing to O_2 gas, which is toxic to anaerobes. Strong solutions of H_2O_2 are sporicidal.

H. Aldehydes such as **glutaraldehyde** kill microbes by alkylating protein and DNA molecules and can be used as sterilants.

I. Gases and aerosols such as **ethylene oxide (ETO),** propylene oxide, and chlorine dioxide are strong alkylating agents, all of which are sporicidal.

J. **Detergents** and **soaps**
 1. Cationic detergents known as **quaternary ammonium compounds (quats)** act as surfactants that alter the membrane permeability of some bacteria and fungi. They are not sporicidal.
 2. Soaps function by removing grease and soil that contain microbes.

K. **Heavy metals:** Compounds containing silver or mercury kill vegetative cells (but not spores) despite being highly diluted (**oligodynamic action**) by inactivating proteins.

L. **Dyes, acids,** and **alkalis** can also inhibit or destroy microbes but are limited in use.

 Level I. Knowledge and Comprehension

These questions require a working knowledge of the concepts in the chapter and the ability to recall and understand the information you have studied.

Multiple-Choice Questions

Select the correct answer from the answers provided. For questions with blanks, choose the combination of answers that most accurately completes the statement.

1. A microbicidal agent has what effect?
 a. sterilizes
 b. inhibits microorganisms
 c. is toxic to human cells
 d. destroys microorganisms

2. Microbial control methods that kill _____ are able to sterilize.
 a. viruses c. endospores
 b. the tubercle bacillus d. cysts

3. Any process that destroys the non-spore-forming contaminants on inanimate objects is
 a. antisepsis c. sterilization
 b. disinfection d. degermation

4. Sanitization is a process by which
 a. the microbial load on objects is reduced
 b. objects are made sterile with chemicals
 c. utensils are scrubbed
 d. skin is debrided

5. An example of an agent that lowers the surface tension of cells is
 a. phenol c. alcohol
 b. chlorine d. formalin

6. High temperatures _____ and low temperatures _____.
 a. sterilize, disinfect
 b. kill cells, inhibit cell growth
 c. denature proteins, burst cells
 d. speed up metabolism, slow down metabolism

7. The primary action of _____ heat is to _____.
 a. dry, destroy cell walls
 b. moist, kill vegetative cells
 c. dry, dissolve lipids
 d. moist, denature proteins

8. The standard temperature-pressure combination for an autoclave is
 a. 100°C and 4 psi c. 131°C and 9 psi
 b. 121°C and 15 psi d. 115°C and 3 psi

9. Microbe(s) that is/are the target(s) of pasteurization include(s)
 a. *Clostridium botulinum* c. *Salmonella* species
 b. *Mycobacterium* species d. both b and c

10. Ionizing radiation like _____ removes _____ from atoms.
 a. UV, protons c. gamma rays, electrons
 b. X-rays, the charge d. cathode rays, ions

11. The primary mode of action of nonionizing radiation is to
 a. produce superoxide ions
 b. make pyrimidine dimers
 c. denature proteins
 d. break disulfide bonds

12. The most versatile method of sterilizing heat-sensitive liquids is
 a. UV radiation c. beta propiolactone
 b. exposure to ozone d. filtration

13. _____ is the iodine antiseptic of choice for wound treatment.
 a. 8% tincture c. 5% Iodophor
 b. 5% aqueous d. 7.5% potassium iodide solution

14. A chemical with sporicidal properties is
 a. phenol c. quaternary ammonium compound
 b. alcohol d. glutaraldehyde

15. Silver sulfadiazine is used
 a. in antisepsis of burns c. to treat genital gonorrhea
 b. as a mouthwash d. to disinfect water

16. Detergents are
 a. high-level germicides
 b. surfactants
 c. excellent antiseptics
 d. used in disinfecting surgical instruments

17. Which of the following is an approved sterilant?
 a. chlorhexidine c. ethylene oxide
 b. betadyne d. ethyl alcohol

Case Study Review

1. How would one best describe the state of being a *carrier* of HCV?
 a. It is on the skin and transmitted through direct contact.
 b. The person shows symptoms and is jaundiced.
 c. The person has a chronic infection of the liver and harbors the virus.
 d. The person has had contact with blood in a medical setting.

2. The main issue in the case of clinic-acquired HCV infections was:
 a. use of unclean equipment
 b. a high number of carriers
 c. contamination of drugs
 d. health workers who were IV drug abusers

3. Outline the chain of events involved in the transmission of HCV from an infected patient to several other patients through the actions of a medical worker.

 Writing Challenge

For each question, compose a one- or two-paragraph answer that includes the factual information needed to completely address the question. Check Your Progress questions can also be used for writing-challenge exercises.

1. What is wrong with this statement: "Prior to vaccination, the patient's skin was sterilized with alcohol"? What would be the more correct wording?

2. Would it be proper to say that a patient has been disinfected? Why or why not?

3. Why are antimicrobials inhibited in the presence of organic matter?

4. Think of three situations in which the same microbe would be considered a serious contaminant in one case and completely harmless in another.

5. Explain what features of endospores make them so resistant to microbial control, and what sterilizing techniques could destroy them.

6. Explain some of the problems involved in sterilizing delicate and heat-sensitive instruments such as endoscopes, syringes, needles, catheters, and other instruments.

7. The shelf life and keeping qualities of fruit and other perishable foods are greatly enhanced through irradiation, saving industry and consumers billions of dollars. How would you personally feel about eating irradiated fruit? How about spices that had been sterilized with ETO?

 Concept Mapping

On Connect you can find an Introduction to Concept Mapping that provides guidance for working with concept maps and a sample of a concept map for this chapter.

 Level II. Application, Analysis, Evaluation, and Synthesis

These problems go beyond just restating facts and require higher levels of understanding and an ability to interpret, problem solve, transfer knowledge to new situations, create models, and predict outcomes.

 Critical Thinking

Critical thinking is the ability to reason and solve problems using facts and concepts. These questions can be approached from a number of angles, and in most cases, they do not have a single correct answer.

1. For each item on the following list, propose a reasonable method of sterilization. You cannot use the same method more than three times for the whole list. The method must sterilize, not just disinfect; and the method must not destroy the item or render it useless unless there is no other choice. After considering a workable method, think of a method that would not work. Note: Where an object containing something is given, you must sterilize everything (for example, both the jar and the petroleum jelly [Vaseline] in it). Some examples of sterilizing methods are autoclave, ethylene oxide gas, dry oven, and ionizing radiation.

 room air

 serum

 a pot of soil

 plastic Petri dishes

 heat-sensitive drugs

 cloth dressings

 leather shoes from a thrift shop

 a cheese sandwich

 a flask of nutrient agar

 an entire room (walls, floor, etc.)

 rubber gloves

 disposable syringes

 carcasses of cows with "mad cow" disease

 inside of a refrigerator

 wine

 a jar of Vaseline

 fruit in plastic bags

 milk

 orchid seeds

 metal instruments

 mail contaminated with anthrax spores

2. Using pasteurization techniques as a model, compare the TDTs and explain the relationships between temperature and length of exposure.

3. It may seem contradictory that lyophilization can be used to preserve foods yet also be used to keep cultures of microbes viable. Explain how these applications can be reconciled.

4. A supermarket/drugstore assignment: Look at the labels of 10 different products used to control microbes, and make a list of their active ingredients, their suggested uses, and information on toxicity and precautions.

5. Devise an experiment that will differentiate between bactericidal and bacteriostatic effects.

6. There is quite a bit of concern that chlorine used as a water purification chemical presents serious dangers. What alternative methods covered by this chapter could be used to purify water supplies and keep them safe from contamination?

7. Was the source patient in the case study most likely acutely or chronically infected? Explain.

8. Explain why surfactant chemicals are more effective against gram-negative bacteria than against gram-positive bacteria.

1. From **chapter 2, figure 2.20.** Study this illustration of a cell membrane. In what ways could alcohol damage this membrane? How would that harm the cell?

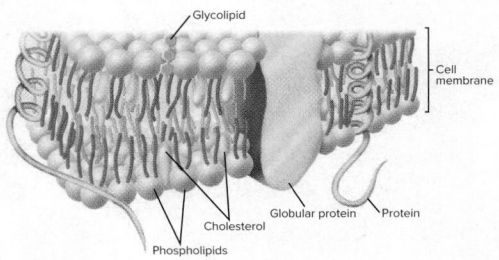

2. Explain what is happening with this graph that charts the reactions of a mesophilic bacterium to temperature. Blue shows the survival curve and brown the growth curve under various temperatures.

STUDY TIP: Quickly create a customized practice quiz for this chapter by going to your Connect SmartBook "My Reports" section. Don't have SmartBook and want to learn more? Go to **whysmartbook.com.**

12 Drugs, Microbes, Host—The Elements of Chemotherapy

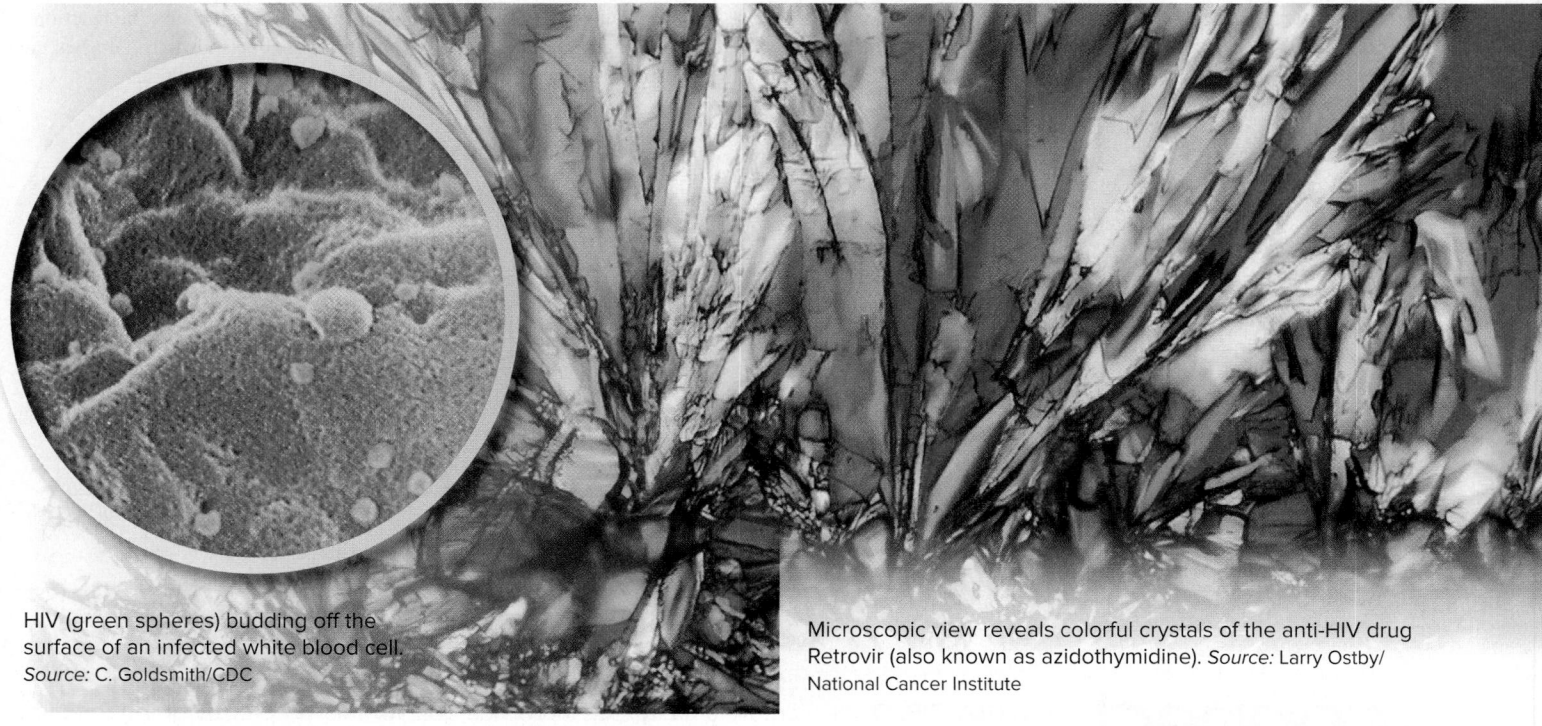

HIV (green spheres) budding off the surface of an infected white blood cell. *Source:* C. Goldsmith/CDC

Microscopic view reveals colorful crystals of the anti-HIV drug Retrovir (also known as azidothymidine). *Source:* Larry Ostby/ National Cancer Institute

CASE STUDY Part 1 An Ounce of Prevention and a Pound of Cure

Joan Mitchell, a 48-year-old registered nurse, had safely removed needles from intravenous **catheters** hundreds of times, yet all of those years of care and experience seemed to evaporate in an instant. The needle accidently jabbed with some force into her arm. She immediately reported it and began the **standard precautions** for a **needlestick.** She washed the site of the laceration with water and **Betadine,** as indicated by hospital protocol.

> "The needle accidently jabbed with some force into her arm."

Needlesticks are a regular occurrence in hospitals and clinics, and usually the greatest concern is the possibility of **infectious agents** being carried by the offending needle. Unfortunately for Nurse Mitchell, the patient had been suffering from **acquired immune deficiency syndrome** (AIDS) caused by **human immunodeficiency virus** (HIV). Within 2 hours it was necessary for her to begin a regimen of **anti-HIV drugs** that are routinely given for **postexposure prophylaxis.** The drug chosen was a combination of lamivudine and zidovudine (Combivir), a pill containing two different **reverse transcriptase inhibitors.** She was tested right away for HIV to establish a baseline for the possibility of a prior infection, and the result was negative.

Despite treatment, the AIDS patient died unexpectedly a few days later of respiratory failure apparently from *Pneumocystis* **pneumonia.** Unfortunately his medical history did not indicate other infections he may have had, and this lack of information turned the next several weeks into a nightmare for the nurse. She was supposed to take the Combivir for 6 weeks, but it caused **severe side effects,** including nausea, headache, muscle pain, weakness, abdominal pain, and rash. She discontinued taking it after 2 weeks with the hope that it had succeeded in preventing the HIV from infecting her.

During the next month, the site of Nurse Mitchell's injury became **abscessed** and painful, and she suffered from a **fever** up to 102°F. Because the wound was obviously infected, her doctor **biopsied** it and had routine **bacterial** and **fungal cultures** made, but they came back negative. The course of **dicloxacillin** she was taking did not seem to be working. A sign that the infection was spreading came in the form of an **enlarged lymph node** in her armpit.

- *What is meant by prophylactic drug treatment, and what are its purposes?*
- *Why is it necessary for two drugs to be given in the Combivir treatment?*

To continue the Case Study, go to Case Study Part 2 at the end of the chapter.

SCOPING OUT THE CHAPTER

Modern antimicrobial drugs have revolutionized medical treatment. Literally billions of lives have been saved since they were first introduced 80 years ago. Having an effective drug to treat most infectious diseases is now expected, whether bacterial, viral, fungal, protozoan, or helminth. Though we may take this availability for granted, there are numerous factors that complicate their use. In this chapter we will explore "the good, the bad, and the ugly" elements of antimicrobial drug therapy.

Drug Discovery Antimicrobial drugs come from many sources. Most of them, called antibiotics, are produced by certain bacteria or fungi, others are synthesized through chemical reactions alone, and some are made by combining the two methods.

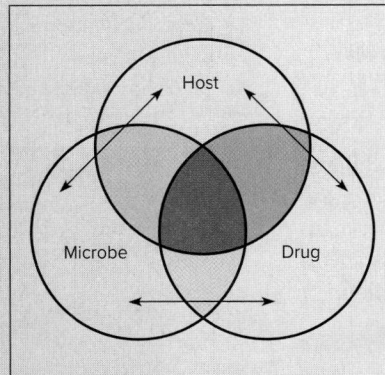

Do No Harm: Selective Toxicity Antibiotic literally means "against life", but the actual intention of these drugs is to target only microbial life. This is a very important guiding principle—that drugs do not harm humans while they are getting rid of the infectious agent.

How Antimicrobial Drugs Stop Infections Drugs are chemicals that can interfere with some specific microbial structure or function such as the cell wall, cell membrane, proteins, DNA, RNA, ribosomes, or metabolic pathways. While in contact with the drug, the microbes are either destroyed or severely inhibited and can no longer grow.

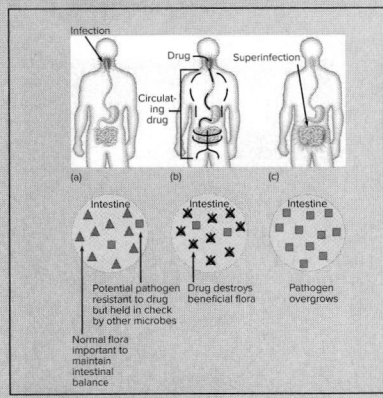

Toxicity and Other Side Effects The promise of antimicrobial therapy is spoiled by the numerous possibilities of adverse side effects. Drugs can harm body tissues and organs, disrupt the normal microbiota that help keep a balance in the body's organs, and induce allergies and hypersensitivities.

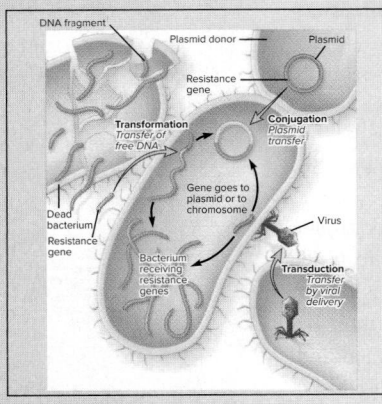

The Global Race Against Drug Resistance Microbes are very adept at rapidly altering their physiology and genetics to adapt to drugs, making them less effective. They may develop enzyme systems that dismantle the drugs, block the drug's entrance, expel the drugs, or use an alternate pathway that bypasses the drug's effects.

Choosing an Appropriate Drug Selection of an effective drug is guided by several factors. Among the most important considerations are the nature of the infectious agent, knowing which drugs it is sensitive to, the possible side effects of the drug, and the medical condition of the patient.

12.1 Principles of Antimicrobial Therapy

 Expected Learning Outcomes

1. State the goals of antimicrobial therapy.

2. Discuss the origins of antimicrobial drugs.

3. Differentiate between antibiotics, synthetic drugs, and semisynthetic drugs.

4. Describe some of the special strategies of antimicrobial therapy.

5. Explain what is meant by a drug's spectrum and how narrow- and broad-spectrum drugs differ.

6. Summarize the five modes of action of antimicrobial drugs, with specific examples of their effects.

More than a century ago in the United States, one out of three children was expected to die of an infectious disease before the age of 5. Early death or severe lifelong debilitation from scarlet fever, diphtheria, tuberculosis, meningitis, and many other bacterial diseases was a fearsome yet undeniable fact of life to most of the world's population.

The introduction of modern drugs to control infections in the 1930s was a medical revolution that has added significantly to the life span and health of humans. It is no wonder that for many years, antibiotics in particular were regarded as miracle drugs. In later discussions, we evaluate this misconception in light of the shortcomings of drug therapy. Although antimicrobial drugs have greatly reduced the incidence of certain infections, they have definitely not eradicated infectious disease and probably never will. In fact, in some parts of the world, mortality rates from infectious diseases are as high as before the arrival of antimicrobial drugs. Nevertheless, humans have been taking medicines to try to control diseases for thousands of years (**12.1 Making Connections**).

> **Quick Search**
> Look for "Antibiotics, Seven Wonders of the Microbe World" on the Internet.

The goal of **antimicrobial chemotherapy**[1] is deceptively simple: administer to an infected person a drug that destroys the infectious agent without harming the host's cells. In actuality this goal is rather difficult to achieve, because many often-contradictory factors must be taken into account. The ideal drug should be easily administered and able to reach the infectious agent anywhere in the body. It should be able to kill the infectious agent rather than merely inhibit its growth, and remain active in the body as long as needed, yet be safely and easily broken down and excreted. In short, the perfect drug does not exist, but by balancing drug characteristics against one another, a satisfactory compromise can usually be achieved (**table 12.1**).

The Origins of Antimicrobial Drugs

Modern antimicrobial drugs are highly varied in their origins. The type of drugs traditionally termed **antibiotics** are substances produced by the natural metabolic processes of certain microorganisms that can inhibit or destroy other microorganisms (**figure 12.1a**).

1. The word *chemotherapy,* meaning the use of chemicals in therapy, is often associated with the treatment of cancer, but its official meaning also includes antimicrobial treatment.

TABLE 12.1	Characteristics of the Ideal Antimicrobial Drug

- Selectively toxic to the microbe but nontoxic to host cells
- Microbicidal rather than microbistatic
- Remains potent long enough to act and is not broken down or excreted prematurely
- Is not subject to the development of antimicrobial resistance
- Complements or assists the activities of the host's defenses
- Remains active even when diluted in body fluids and tissues
- Readily delivered to the site of infection
- Reasonably priced
- Does not disrupt the host's health by causing allergies or predisposing the host to other infections

(a)

(b)

Figure 12.1 **(a) Mining for antibiotic-producing microbes.** One strategy is to seed a plate with a sample (here, soil) and then spray its surface with a fine mist of a bacteria and incubate it. Zones of inhibition (clear areas with no growth) surrounding several colonies indicate colonies that produce antibiotics. **(b) Two colonies with antibiotic secretions on their surfaces.** Left is *Penicillium chrysogenum,* with yellow droplets of penicillin; right is *Streptomyces coelicolor,* with blue droplets of actinorhodin, an experimental antibiotic. This release of the antibiotics is an important factor in the isolation and production of the drugs. (a, b left) © Kathy Park Talaro; (b right) Photo courtesy of Dr. Ramón Santamaría/IMB. CSIC/USAL. Salamanca, Spain

12.1 MAKING CONNECTIONS

From Witchcraft to Wonder Drugs

Early human cultures relied on various types of primitive medications concocted from plant, animal, and mineral products that had been found, usually through trial and error or accident, to have some curative effect upon ailments. In one ancient Chinese folk remedy, a fermented soybean curd was applied to skin infections. The Greeks used wine and plant resins (myrrh and frankincense), rotting wood, and various mineral salts to treat diseases. Some folk medicines were effective, but others either had no effect or were harmful. It is interesting that the Greek word *pharmakeutikos* originally meant the practice of witchcraft. These ancient remedies were handed down from generation to generation, but it was not until the Middle Ages that a specific disease was first treated with a specific chemical. Dosing syphilitic patients with inorganic arsenic and mercury compounds may have proved the ancient axiom: *Graviora quaedum sunt remedia periculus* ("Some remedies are worse than the disease").

In the late 1800s Paul Ehrlich was the first scientist to formulate basic theoretical concepts in chemotherapy. Ehrlich had observed that certain dyes affixed themselves to specific microorganisms and not to animal tissues. This observation led to the profound idea that if a drug was properly selective in its actions, it would zero in on and destroy a microbial target and leave human cells unaffected. His first discovery was an arsenic-based drug that was very toxic to the spirochete of syphilis but, unfortunately, was toxic to humans as well. After 606 attempts he arrived at a compound he called *salvarsan*. This drug had some therapeutic merit and was used for a few years, but it eventually had to be discontinued because it was still not selective enough in its toxicity. Ehrlich's work had laid important foundations for many of the developments to come.

Another pathfinder in early drug research was Gerhard Domagk, whose discoveries in the 1930s launched a breakthrough in therapy that marked the true beginning of broad-scale usage of antimicrobial drugs. Domagk showed that the red dye prontosil was chemically changed by the body into a compound with specific activity against bacteria. This substance was sulfonamide—the first *sulfa* drug. Once the structure of this drug was determined, it became possible to synthesize it on a wide scale and to develop scores of other sulfonamide drugs. Although these drugs had immediate applications in therapy (and still do), still another fortunate discovery ushered in the "age of antibiotics."

The discovery of antibiotics dramatically demonstrates how developments in science and medicine often occur through a combination of accident, persistence, collaboration, and vision. In 1928 in the London laboratory of Sir Alexander Fleming, a plate of *Staphylococcus aureus* became contaminated with the mold *Penicillium (notatum) chrysogenum* (see figure 12.1). Observing these plates, Fleming noted that the colonies of *Staphylococcus* were evidently being destroyed by some

A pioneer in the discovery of antibiotics. A Scottish physician, Sir Alexander Fleming, accidentally discovered penicillin in a culture plate contaminated with mold. From this mold he extracted the antibiotic, setting the scene for the development of the drug 10 years later. © Davies/Getty Images

activity of the nearby mold colonies. Struck by this curious phenomenon, he extracted from the fungus a compound he called penicillin and showed that it was responsible for the inhibitory effects.

A decade after his discovery, English chemists Howard Florey and Ernst Chain worked out methods for industrial production of penicillin to help in the war effort. Clinical trials conducted in 1941 ultimately proved penicillin's effectiveness, and by 1945 the drug had been used to save hundreds of thousands of lives. This event paved the way for an era of drug research and development that soon made penicillin only one of a large assortment of antimicrobial drugs. But in time, because of extreme overuse and misunderstanding of its capabilities, penicillin also became the model for one of the most serious drug problems—namely, drug resistance.

Once the first significant antibiotic was developed, the world immediately witnessed a scientific scramble to find more. Hundreds of investigators began searching for antibiotic-producing bacteria and fungi by screening of samples from soil, dust, muddy lake sediments, rivers, oceans, plant surfaces, compost heaps, sewage, skin, and even the hair and skin of animals. Finding a new antimicrobial substance is only a first step. The complete pathway of drug development from discovery to therapy can take more than a dozen years and cost several billion dollars.

Name some potential "therapies" you could employ to prevent and treat infections if you were a castaway on a desert island. *Answer available on Connect.*

Nature is a prolific producer of antimicrobial drugs. Many thousands of antibiotics have been discovered (although, surprisingly, only a relatively small number have actually been used clinically). Antibiotics are made primarily by aerobic spore-forming bacteria and fungi. The antibiotics appear to have some survival value for these microbes, because the genes for antibiotic production are preserved in evolution. Some experts theorize that microorganisms synthesize antibiotics to inhibit or destroy nearby competitors or

predators (antagonism); others propose that antibiotics are connected to the formation of spores.

The greatest numbers of antibiotics are derived from bacteria in the genera *Streptomyces* and *Bacillus* and molds in the genera *Penicillium* and *Cephalosporium* (**figure 12.1b** and **table 12.2**). Of these, species of *Streptomyces* are the most fertile producers. Nearly 9,000 different bioactive compounds have been isolated from them, and they are currently the source of one-third of all prescribed antibiotics.

TABLE 12.2	Selected Microbial Sources of Antibiotics	
Producer Genus	**Type of Microbe**	**Drugs Formed**
Penicillium	Mold	Penicillins Griseofulvin
Cephalosporium	Mold	Cephalosporins
Micromonospora	Bacteria	Gentamicin
Bacillus	Bacteria	Bacitracin Polymyxin B
Chromobacterium	Bacteria	Aztreonam
Streptomyces	Filamentous bacteria	Streptomycin Erythromycin Tetracycline Vancomycin Chloramphenicol Amphotericin B

TABLE 12.3	Terminology of Chemotherapy
Chemotherapeutic drug	Any chemical used in the treatment, relief, or prophylaxis of a disease.
Prophylaxis*	Use of a drug to prevent potential for infection in a person at risk.
Antimicrobial chemotherapy*	The use of chemotherapeutic drugs to control infection
Antimicrobials	All-inclusive term for any antimicrobial drug, regardless of its origin
Antibiotics*	Substances produced by the natural metabolic processes of some microorganisms that can inhibit or destroy other microorganisms
Semisynthetic drugs	Drugs that are chemically modified in the laboratory after being isolated from natural sources
Synthetic drugs	Antimicrobial compounds synthesized entirely in the laboratory through chemical reactions
Narrow spectrum (limited spectrum)	Antimicrobials effective against a limited array of microbial types; for example, a drug effective mainly on gram-positive bacteria
Broad spectrum (extended spectrum)	Antimicrobials effective against a wide variety of microbial types; for example, a drug effective against both gram-positive and gram-negative bacteria

* *prophylaxis* (proh″-fih-lak′-sis) Gr. *prophylassein,* to keep guard before. A process that prevents infection or disease in a person at risk.
* *chemotherapy* (kee″-moh-ther′-uh-pee) Gr. *chemieia,* chemistry, and *therapeia,* service to the sick. Use of drugs to treat disease.
* *antibiotic* (an-tee′-by-aw″-tik) Gr. *anti,* against, and *bios,* life.

The pharmaceutical industry has harnessed the power of these microbes by growing them in vast quantities and harvesting their products to be sold as drugs. Researchers have facilitated the work of nature by selecting mutant species that yield more abundant or effective products, by varying the growth medium, or by altering the procedures for large-scale industrial production.

A few classes of antimicrobial drugs, termed **synthetic,** are not considered antibiotics because they are synthesized entirely through chemical processes and not secreted by microorganisms. Synthetic drugs include sulfa drugs, quinolones (ciprofloacin), quinines, azoles and most antiviral drugs.

Probably the most common approach to drug production is the **semisynthetic** method, combining both natural and synthetic methods. Think of it as improving on nature by using chemical reactions to combine the basic antibiotic made by microorganisms with various preselected functional groups that tailor the action of the drug. Drugs produced in this way are designed to have new and improved functions over the original antibiotic. We can see examples of this technology in the design of the semisynthetic penicillins (see figure 12.7), cephalosporins and tetracyclines. The potential for using bioengineering techniques to design drugs seems almost limitless, and, indeed, several drugs have already been produced by manipulating the genes of antibiotic producers.

Antimicrobial drugs can be organized by their origin, range of effectiveness, and whether they are naturally produced or chemically synthesized. A few of the more important terms you will encounter are found in **table 12.3.**

Common Drug Strategies

You will encounter the terms **prophylaxis, combined therapy,** and drug **synergy** in several areas of this chapter. Each idea has its own special niche in drug therapy. With prophylaxis, drugs are administered to *prevent* an infection from happening. This is brought into play in cases of known or possible exposure. For example, certain dental patients are often pretreated with antibiotics prior to instrument procedures that could leak saliva into the bloodstream.

Combined therapy, also known as combination therapy, is a protocol for taking at leasto two drugs at a time to treat an infection. This is the usual method of treatment for tuberculosis, HIV/AIDS, and gonorrhea. Its main rationale is to prevent the survival of drug-resistant microbes, but it is also used for treating mixed infections by microbes that differ in the correct drug of choice.

Another advantage comes from the synergistic effect that two drugs working together can often provide. Synergy can produce an improved result over either drug used alone, and this enhanced effectiveness means that smaller quantities of the drugs can be used. Figure 12.18*b* shows a test result for this effect.

Interactions Between Drugs and Microbes

The goal of antimicrobial drugs is either to disrupt the cell processes or structures of bacteria, fungi, and protozoa or to inhibit the virus multiplication cycle. Most of the drugs used in chemotherapy interfere with the function of enzymes required to synthesize or assemble macromolecules, but a few destroy structures already formed in the cell. Above all, drugs should be **selectively toxic,** which means they can kill or inhibit the growth of microbes without simultaneously damaging host tissues. This concept of selective toxicity is central to chemotherapy. The best drugs are those that act

TABLE 12.4	General Actions of Drugs on Microbial Groups				
Microbe Type	**Drug Groups/ Examples**	**Effects of Drugs**	**Microbe Type**	**Drug Groups/ Examples**	**Effects of Drugs**
Bacteria	Penicillins	Cell wall damage and lysis	**Fungi**	Amphotericin B	Loss of membrane permeability
	Cephalosporins	Cell wall damage and lysis		Azoles	Loss of membrane permeability
	Bacitracin	Cell wall damage and lysis		Flucytosine	Inhibits DNA and RNA synthesis
	Aminoglycosides	Inhibit ribosomal protein synthesis	**Protozoa**	Quinines	Buildup of toxic wastes in the parasite's cells
	Macrolides				
	Erythromycin	Inhibits ribosomal protein synthesis		Metronidazole	Buildup of toxic free radicals
	Vancomycin	Cell wall damage and lysis	**Helminths**	Bendazoles	Inhibit glucose metabolism
	Tetracyclines	Inhibit protein synthesis on ribosomes		Diethylcarbamide	Kills larval forms
				Piperazine	Paralyzes muscular system
	Chloramphenicol	Inhibits protein synthesis on ribosomes		Niclosamide	Loosens worm hold
				Ivermectin	Inhibits neuromuscular system
	Fluoroquinolones	Block replication of DNA	**Viruses**	Oseltamivir	Prevents viral budding
	Rifampin	Stops mRNA synthesis		Cyclovirs	Stop viral replication
	Sulfa drugs	Inhibit folic acid metabolism		Azidothymidine	Blocks formation of DNA from RNA strand
	Trimethoprim	Inhibits folic acid metabolism			

specifically on microbial structures or functions not found in vertebrate cells. Examples of drugs with excellent selective toxicity are those that block the synthesis of the cell wall in bacteria (penicillins). They have low toxicity and few direct effects on human cells, because human cells lack a cell wall and are thus unaffected by this action of the antibiotic. Among the most toxic to human cells are drugs that act upon a structure common to both the infectious agent and the host cell, such as the cell membrane (for example, amphotericin B used to treat fungal infections). As the characteristics of the pathogen become more similar to those of the host cell, selective toxicity becomes more difficult to achieve and undesirable side effects are more likely to occur. We will cover the subject of toxicity in section 12.5.

Mechanisms of Drug Action

The primary basis of chemotherapy is to target an infectious agent with a chemical that irreversibly damages or inhibits it. This knowledge guides drug development as well. Most classes of antimicrobial drugs are active against some unique feature of microbial structure or metabolism. Observe **table 12.4** and take note of the diversity of cell types and possible effects of drug action. Overall, there are five major components that are useful drug targets in an actively dividing cell. These are the cell wall, the cell membrane, the genetic material, protein synthesis, and metabolic pathways. We cover the effects of drugs on these cell components based on **figure 12.2** and this outline:

1. inhibition of cell wall synthesis,
2. breakdown of the cell membrane structure or function,
3. interference with functions of DNA and RNA,
4. inhibition of protein synthesis, and
5. blockage of key metabolic pathways.

These targets are not completely discrete, and some have overlapping effects. For example, if DNA transcription is inhibited, protein

synthesis will be as well. Viruses are a special case and have very specialized targets for drugs (see table 12.6).

The Spectrum of an Antimicrobial Drug One result of drugs acting upon a particular microbial structure or function is that each drug has a given range of activity. This has been termed the drug's **spectrum.** Traditionally, **narrow-spectrum** drugs are effective over a limited range of cell types. This is usually because they target a specific component that is found only in certain bacteria. For instance, bacitracin blocks the elongation of the peptidoglycan in gram-positive bacteria but has no effect on most gram-negative bacteria because of the outer membrane on their cell wall that serves as a barrier to the drug. Polymyxin breaks down the outer membrane on gram-negative cells but does not target gram-positive cells because they lack this structure.

Drugs that are effective against a wider range of microbes are termed medium or broad spectrum, depending on the particular drug. For example, the **medium-spectrum** drug ampicillin is effective on both gram-positive and gram-negative bacteria, but not a wide range of them. **Broad-spectrum** drugs, such as tetracycline, have the greatest range of activity. They work on most gram-negative and gram-positive bacteria, rickettsias, mycoplasmas, and spirochetes. The broad-spectrum drugs usually exert their effects on common cell components such as ribosomes, which are found in all cells and mitochondria. Consult **table 12.5** to see an array of antibacterial drugs and their spectra.

1. Antimicrobial Drugs That Affect the Bacterial Cell Wall
The cell walls of most bacteria contain a rigid girdle of peptidoglycan, which protects the cell against rupture from hypotonic environments. Active cells must constantly synthesize new peptidoglycan and transport it to its proper place in the cell envelope. Drugs such as penicillins and cephalosporins react with one or more of the

Figure 12.2 Major targets of drugs acting on bacterial cells. Each mode of action correlates with the numbered section where it is discussed in the text.

TABLE 12.5	Spectrum of Effectiveness for Antibacterial Drugs					
Spectrum of Action	**Gram-Positive Bacteria**	**Gram-Negative Bacteria**	**Rickettsias**	**Chlamydias**	**Mycoplasmas**	**Examples of Drugs with This Spectrum**
Broad	+	+	+	+	+	Tetracyclines Quinolones Chloramphenicol Doripenem
Intermediate	+	+	−	V*	V*	Ampicillin Cephalexin (Keflex) Sulfa/trimethoprim Macrolides (erythromycin)
Narrow Group 1	+	−*	−	−	−	Penicillin G, V Vancomycin Bacitracin Rifampin Isoniazid (for TB)
Narrow Group 2	−	+	−	−	−	Polymyxin Streptomycin Cefaclor Aztreonam

*Some drugs in the list are effective on these microbes.

(a) Normal untreated cells. Left and center views are *Staphylococcus* cells; right view is the molecular structure of peptidoglycan, with chains of NAM (*N* muramic acid) and NAG (*N* glucosamine) linked together by cross bridges (green).

Exposure to penicillin

Weak points lacking peptidoglycan

Fragment of peptidoglycan

Membrane bulges out as water diffuses into cell.

Membrane breaks.

Cell lyses.

10,000X

10,000X

(b) Effects of treatment with penicillin. Left side shows the breakdown of a *Staphylococcus* cell exposed to penicillin. Center view shows actual cells undergoing lysis and cell death. Right indicates how penicillin blocks the formation of the cross bridges and dismantles the framework that holds the cell wall together.

Figure 12.3 **Effects of penicillin on gram-positive cell walls.**
(Top) *Source:* Janice Carr/CDC; (Bottom) © CNRI/Science Source

enzymes required to complete this process, causing the cell to develop weak points at growth sites and to become osmotically fragile **(figure 12.3).** Antibiotics that produce this effect are considered bactericidal, because the weakened cell usually undergoes **lysis.** However, cells that are old, inactive, or dormant and are not synthesizing peptidoglycan may resist these effects. (One exception is a class of antibiotics abbreviated as "-penems.")

Penicillins and cephalosporins bind and block peptidases that cross-link the glycan molecules, thereby interrupting the completion of the cell wall **(figure 12.3b).** Penicillins that do not penetrate the outer membrane are less effective against gram-negative bacteria. Many broad-spectrum penicillins (ampicillin) and cephalosporins (ceftriaxone) can pass into the cell walls of gram-negative species. Cycloserine inhibits the formation of the basic peptidoglycan subunits, and vancomycin hinders the elongation of the peptidoglycan.

2. Antimicrobial Drugs That Disrupt Cell Membrane Function
A cell with a damaged membrane invariably dies from disruptions in metabolism or complete breakdown and lysis. It does not even

have to be actively dividing to be destroyed. The antibiotic classes that damage cell membranes usually have specificity for particular microbial groups, based on differences in the types of lipids in their cell membranes.

Polymyxins interact with membrane phospholipids and cause leakage of proteins and nitrogen bases, particularly in gram-negative bacteria **(figure 12.4).** The polyene antifungal antibiotics (amphotericin B and nystatin) form complexes with the sterols on fungal membranes, which causes abnormal openings and seepage of small ions. Unfortunately this selectivity is not exact, and the similarities in membranes of microbial and animal cells means that most of these antibiotics have greater toxicity to humans cells.

3. Antimicrobial Drugs That Affect Nucleic Acid Synthesis
As you learned in chapter 9, the pathway that generates DNA and RNA involves an extensive series of enzyme-catalyzed reactions. Like any complicated process, it is subject to breakdown at many different points along the way, and inhibition at any point in the sequence can block subsequent events. Antimicrobial drugs

Figure 12.4 **Effects of drugs on membranes.** Polymyxin binds to the outer and cell membranes of gram-negative bacteria and creates abnormal openings that cause leakage and lysis.

interfere with nucleic acid synthesis by blocking synthesis of nucleotides, inhibiting replication, or stopping transcription. Because functioning DNA and RNA are required for proper translation as well, the effects on protein metabolism can be far-reaching.

Several antimicrobials inhibit DNA synthesis. The broad-spectrum quinolones inhibit DNA-unwinding enzymes or helicases, thereby stopping DNA replication and repair. Certain antiviral drugs are **analogs*** of purines and pyrimidines that will insert in the viral nucleic acid in place of a normal base. When this abnormal base stops synthesis, viral replication will be blocked (see table 12.6). The antifungal drug flucytosine is an analog of uracil that creates dysfunctional RNA. By inhibiting an enzyme that acts in purine synthesis, it also stops DNA replication. This double assault shuts down metabolism and causes cell death.

4. Antimicrobial Drugs That Block Protein Synthesis Most drugs that inhibit protein synthesis react with the ribosome-mRNA complex. Although human cells also have ribosomes, the ribosomes of eukaryotes are different in size and structure from those of prokaryotes, so these antimicrobials usually have a selective action against bacteria. One potential therapeutic consequence of drugs that bind to the prokaryotic ribosome is the damage they can do to eukaryotic mitochondria, which contain a prokaryotic type of ribosome.

Two possible targets of ribosomal inhibition are the 30S subunit and the 50S subunit **(figure 12.5)**. Aminoglycosides (streptomycin, gentamicin, for example) insert on sites on the 30S subunit and cause the misreading of the mRNA, leading to abnormal proteins. Tetracyclines block the attachment of tRNA on the A acceptor site

Quick Search
Search for the video "Antibiotic and Protein Interactions" on YouTube to see an animation of the effects of antibiotics on ribosomes.

and effectively stop further synthesis of that protein. Other antibiotics attach to sites on the 50S subunit in a way that prevents the formation of peptide bonds and blocks protein synthesis (chloramphenicol). A few inhibit translocation of the subunit during translation (erythromycin). The oxazolidinones are a newer drug class that affects the assembly of the completed ribosome by blocking the attachment site for the 50S and 30S subunits.

Figure 12.5 **Sites of inhibition on the prokaryotic ribosome and major antibiotics that act on these sites.** All have the general effect of blocking protein synthesis. Blockage actions are indicated by X.

* *analog* (an′-uh-log) A compound whose configuration closely resembles another compound required for cellular reactions.

5. Antimicrobial Drugs That Affect Metabolic Pathways Some drugs act by mimicking the normal substrate of an enzyme through a process called **competitive inhibition.** These drugs, termed **metabolic analogs,** are structurally similar to the natural substrate and compete with it for the active site on the enzyme. When the drug is supplied to a target cell in high concentrations, it will occupy the active site and displace the normal substrate. But the metabolic analog drug is a "dead end" and cannot function as required. As the enzyme is no longer able to produce a needed product, cellular metabolism slows or stops. Antibacterial drugs called sulfonamides and trimethoprim provide an example of this mode of action. Some antiviral drugs (example: retrovir) discussed in section 12.3 also function as competitive inhibitors.

Sulfonamides and trimethoprim interfere with folate metabolism by blocking enzymes required for the synthesis of tetrahydrofolate, which is needed by bacterial cells for the synthesis of folic acid and the eventual production of DNA and RNA and amino acids. Trimethoprim and sulfonamides are often given simultaneously to achieve a synergistic effect so that a lower dosage of each drug can be used. **Figure 12.6** illustrates sulfonamides' competition with PABA for the active site of the enzyme that synthesizes the folic acid precursor. The selective toxicity of these compounds is explained by the fact that mammals derive folic acid from their diet and so do not possess this enzyme system. This makes it possible to inhibit bacterial and protozoan parasites, which must synthesize folic acid, while leaving the human host unaffected.

✓ *Check Your Progress* SECTION 12.1

1. Differentiate between antibiotics and synthetic drugs.
2. Differentiate between narrow-spectrum and broad-spectrum antibiotics.
3. Can you determine why some drugs have narrower spectra than others? (Hint: Look at their mode of action.)
4. What is the major source of antibiotics and what appears to be the natural function of antibiotics in microbes that produce them?
5. What drug characteristics are most important to consider in therapy?
6. Why is it better for a drug to be microbicidal than microbistatic?
7. Explain the major modes of action of antimicrobial drugs, and give an example of each.
8. Explain the concept of competitive inhibition, and describe how a metabolic analog molecule inhibits metabolism.
9. What would be the extended effects of drugs that block transcription?

(a) The metabolic pathway needed to synthesize tetrahydrofolic acid (THFA) contains two enzymes that are chemotherapeutic targets. Under normal circumstances, PABA acts as a substrate for the enzyme pteridine synthetase, the product of which is needed for the eventual production of folic acid.

Structural differences

Sulfanilamide PABA

PABA (substrate)

Sulfa drug (inhibitor)

Sulfa drug (inhibitor)

Higher levels of sulfa drug more likely to bind to enzyme

Enzyme

(b) Left: Formula of a sulfanilamide molecule compared alongside one of PABA. Note that, despite a similarity in overall shape, sulfa drugs cannot be used to make folic acid. Right: The sulfa drug molecules can insert into the active site in the enzyme that would ordinarily fit the PABA. This will work only when the sulfa molecules are more prevalent and can outcompete with the PABA and keep the enzyme sites filled. Since little or no PABA can bind, the synthesis of folic acid will be inhibited.

Figure 12.6 Competitive inhibition as a mode of action. This example shows how sulfa drugs block a metabolic pathway used by bacteria to synthesize folic acid.

12.2 Survey of Major Antibacterial Drug Groups

 Expected Learning Outcomes

7. Describe the classes of antibacterial drugs that act on the cell wall and membrane, and list some of their applications.

8. Describe the classes of antibacterial drugs that act on DNA or RNA, and characterize their uses.

9. Indicate which types of antibacterial drugs block protein synthesis, and explain how they are used.

10. Describe some new classes of antibacterial drugs and their applications.

Scores of antimicrobial drugs are marketed in the United States. Although the medical and pharmaceutical literature contains a wide array of names for antimicrobials, most of them are variants of a small number of drug families. About 280 different antimicrobial drugs are currently classified in about 24 drug families. Drug reference books may give the impression that there are 10 times that many, because various drug companies assign different trade names to the very same generic drug. Ampicillin, for instance, is listed under 50 d̶ ̶ ̶ ̶ ̶ ̶ ̶ ̶ ̶ ̶ ̶ ̶ling
bacterial ̶ ̶ ̶ ̶ ̶ ̶ ̶ ̶ ̶ ̶ ̶ ̶ an-
tifungal, ̶ ̶ ̶ ̶ ̶ ̶ ̶ ̶ ̶ ̶ ̶ ̶ w.

Antib̶ ̶ ̶ ̶ ̶ ̶ ̶ ̶ ll

The be̶ ̶ ̶ ̶ ̶ ̶bon,
1-nitro̶ ̶ ̶ ̶ ̶ ̶ction
is to int̶ ̶ ̶ ̶ ̶ ̶wall,
leading ̶ ̶ ̶ ̶ ̶ ̶ ̶
death. ̶ ̶ ̶ ̶ ̶ ̶
antimi̶ ̶ ̶ ̶ ̶ ̶
beta-la̶ ̶ ̶ ̶ ̶ ̶
icillins̶ ̶ ̶ ̶ ̶ ̶
being ̶ ̶ ̶ ̶ ̶ ̶
represe̶ ̶ ̶ ̶ ̶ ̶

Handwritten note (overlaid):
Penicillin
1.) 3 parts → beta lactam ring, thiazolidine ring, side chain
2.) Pen G → 1st; narrow spectrum; DRPO; tx gram (+) strepto-cocci; tx gram (−) meningococci
Cephalosporin
less allergies
Given parenterally
Tx gram (−), (+)

Penicillin and its Relatives

The **penicillin** family of antibiotics, named for the parent compound, is a large, diverse group of compounds, most of which end in the suffix *-cillin*. Although penicillins could be completely synthesized in the laboratory from simple raw materials, it is more practical and economical to obtain natural penicillin through microbial fermentation. The major source of the drug is a unique strain of *Penicillium chrysogenum* that has been manipulated experimentally to synthesize 1,000 times more penicillin than the original strain. The natural product can then be used either in unmodified form or to make semisynthetic derivatives. All penicillins consist of three parts: a beta-lactam ring, a thiazolidine ring, and a variable side chain that dictates its microbicidal activity **(figure 12.7).**

Penicillin G was the first antibiotic and is the parent compound for all "-cillin" drugs. It is narrow spectrum and cannot be given by mouth. It is often considered the drug of choice for infections by sensitive gram-positive bacteria (streptococci) and some gram-negative bacteria (meningococci). A close relative, *penicillin V*, has similar uses but has been modified to be stable in the stomach acid and so can be taken orally. The narrow-spectrum penicillins *dicloxacillin* and *nafcillin* are often given in place of penicillin G or V. So many pathogens have become resistant to *methicillin* that it has largely been replaced by newer varieties.

Other semisynthetic penicillins such as *ampicillin* and *amoxicillin* have been chemically altered with side chains that help them move across the outer membrane of gram-negative cell walls. This increases their spectrum and makes them useful in treating many types of gram-negative infections. *Ticarcillin* and *piperacillin* have such an extended spectrum that they can be substituted for combinations of antibiotics.

Many bacteria produce enzymes that are capable of severing the beta-lactam ring of some penicillins, thereby destroying the mode of action. The enzymes are referred to as **penicillinases** or **beta-lactamases,** and they confer resistance on the bacteria that possess them. One way to bypass this problem was to develop penicillinase-resistant penicillins such as nafcillin and cloxacillin for treating infections caused by penicillinase-producing bacteria. Another strategy for reducing drug resistance has been to pair a beta-lactamase inhibitor with the penicillin to increase the longevity of the antibiotic in the presence of resistant bacteria. For example, clavulanic acid is often added to broad-spectrum penicillins to augment their effectiveness. *Clavamox* is a combination of amoxicillin and clavulanate marketed under the trade name Augmentin. *Zosyn*, a similar combination of the beta-lactamase inhibitor tazobactam and piperacillin, is used for a wide variety of systemic infections. These combined medications show an extended spectrum because of the synergy of the two drugs working together (see figure 12.18*b*).

All of the "-cillin" drugs are relatively mild and well tolerated because of their specific mode of action on cell walls (which humans lack). The primary problems in therapy include resistant strains of pathogens and allergy (which is altogether different from toxicity). Showing an allergy to any of the penicillin drugs will require avoidance of all of them, regardless of the type.

The Cephalosporin Group of Drugs

The **cephalosporin** antibiotics currently account for one-third of all antibiotics administered. Cephalosporins are similar to penicillins; they have a beta-lactam structure that can be synthetically altered **(figure 12.8)** and have a similar mode of action. The generic names of these compounds are often recognized by the presence of the root *cef, ceph,* or *kef* in their names.

The cephalosporins are versatile. They are relatively broad spectrum, resistant to most penicillinases, and cause fewer allergic reactions than penicillins. Although some cephalosporins are given orally, many are poorly absorbed from the intestine and must be administered **parenterally,*** by injection into a muscle or a vein.

Four generations of cephalosporins exist, with each group being more effective against gram-negative organisms than the generation

* *beta-lactam* (bey′-tuh-lak′-tam) The chemical name of a molecular group found in a number of drug families.

* *parenterally* (par-ehn′-tur-ah-lee) Gr. *para,* beyond, and *enteron,* intestine. A route of drug administration other than the gastrointestinal tract.

Methicillin First penicillin developed to resist effects of penicillinase.

Penicillin V withstands stomach acid and can be taken by mouth.

Ampicillin First penicillin to provide broad spectrum coverage for bacterial infections.

Penicillin G

Aminopenicillanic acid

Ticarcillin Extended spectrum; effective against gram-negative rods.

Nafcillin Narrow spectrum alternate to methicillin that resists penicillinase.

Cloxacillin Broad spectrum and resistant to penicillinase.

Figure 12.7 **Chemical structure and functions of semisynthetic penicillins.** All penicillins start with a central molecule called a nucleus that forms the basic framework (blue box). The nucleus is an inactive penicillin derivative—aminopenicillanic acid—which has a bonding site on the nitrogen on the beta lactam ring for addition of a side chain or R group. This "fine-tunes" the end product to create finished molecules with a specialized desired action. Here we see penicillins that have resistance to penicillinase (methicillin, nafcillin), a broader-activity spectrum (ticarcillin, ampicillin), and acid resistance (penicillin V).

before it. In addition, succeeding generations typically have improved dosing schedules and fewer side effects. First-generation cephalosporins (cephalothin and cefazolin) are most effective against gram-positive cocci and a few gram-negative bacteria. Second-generation forms (*cefaclor* and *cefonicid*) are more effective than the first-generation forms in treating infections by gram-negative bacteria such as *Enterobacter, Proteus,* and *Haemophilus*. Third-generation cephalosporins, such as *cephalexin* (Keflex) and *ceftriaxone* (Rocephin), are broad spectrum with especially well-developed activity against enteric bacteria that produce beta-lactamases. Fourth-generation drugs, such as *cefepime*, have the widest range of antimicrobial properties. They are effective with both gram-negative and gram-positive bacterial infections and are rapidly microbicidal.

Some recent additions to this drug family are two combination drugs that pair a broad spectrum cephalosporin with a beta-lactamase inhibitor. *Zerbaxa* is a combination of *ceftolozene* with *tazobactam* for treating resistant urinary tract and abdominal infections. *Avycaz,* which combines *ceftizadime* and *avibactam*, is being limited to the most life-threatening cases of kidney and abdominal infections.

Carbapenems and Monobactams Other beta-lactam drugs are the carbapenems or "penems" and aztreonam. *Imipenem* and *ertapenem* are broad-spectrum antibiotics with a mode of action similar to penicillin but with greater resistance to beta-lactamases. They are active in very low concentrations and can be taken by mouth with few side effects except for allergies. Starting in 2001 some carbapenem-resistant strains of gram-negative enteric bacteria began to emerge in hospital outbreaks. These pathogens, termed CRE, have continued to increase in numbers and are currently causing concern in some clinical settings (see 12.3 Making Connections).

Aztreonam, a monobactam, is a narrow-spectrum antibiotic for treating pneumonia, septicemia, and urinary tract infections by gram-negative aerobic bacilli. Aztreonam is especially useful when treating persons who are allergic to penicillin, because it is

R Group 1	Basic Nucleus	R Group 2	
	7 6 N1 2 / 5 4 3 COOH (S or O)		Cephalothin (first generation*)
	2nd		Cefotiam (second generation)
	3rd		Moxalactam (third generation)
	4th		Cefepime (fourth generation)

*New improved versions of drugs are referred to as new "generations."

Figure 12.8 The structure of cephalosporins. Like penicillin, their main nucleus consists of a beta-lactam ring (red) and a second ring (yellow). However, unlike penicillins, they have two sites for placement of R groups (at positions 3 and 7). This makes possible several generations of molecules with greater versatility in function and complexity in structure.

Isoniazid (INH) works by interfering with the synthesis of mycolic acid, a necessary component of the cell wall of acid-fast organisms. It is used to treat infections with *Mycobacterium tuberculosis* but is effective only against growing cells. It is often prescribed along with rifampin (see the section "Drugs That Act on DNA or RNA" below) and other drugs to treat tuberculosis. Oral doses are indicated for both active tuberculosis and prophylaxis in cases of a positive TB test. Ethambutol, a closely related compound, is another antituberculosis drug added to treatment regimens.

Antibiotics That Damage Bacterial Cell Membranes

Bacillus polymyxa is the source of the **polymyxins,** narrow-spectrum peptide antibiotics with a unique fatty acid component that contributes to their detergent activity (see figure 12.4). Only two polymyxins—B and E (also known as colistin)—have any routine applications, and even these are limited by their toxicity to the kidney. Either drug is indicated to treat drug-resistant infections caused by gram-negative rods such as *Pseudomonas aeruginosa* and *Acinetobacter baumannii*. It is also one of the topical agents in ointments to prevent skin infections, such as Neosporin.

chemically different and does not cross-react with antibodies made to penicillin.

Miscellaneous Cell Wall Inhibitors (Non-Beta-Lactam) *Vancomycin* is a narrow-spectrum antibiotic most effective in treating staphylococcal infections in cases of penicillin and methicillin resistance or in patients with an allergy to penicillins. It has also been chosen to treat *Clostridium* infections in children and endocarditis (infection of the lining of the heart) caused by *Enterococcus faecalis*. Because it is very toxic and hard to administer, vancomycin is usually restricted to the most serious, life-threatening conditions. Two relatives of vancomycin—Dalavancin and Orbactive—were recently introduced as alternative choices for skin and structure infections, especially those caused by MRSA.

Bacitracin is a narrow-spectrum peptide antibiotic produced by a strain of *Bacillus subtilis*. Its primary effect is to block the elongation of the peptidoglycan in gram-positive bacteria. Since it was first isolated, its greatest claim to fame has been as a major ingredient in a common drugstore antibiotic ointment (Neosporin) for combating superficial skin infections by streptococci and staphylococci. For this purpose, bacitracin is usually combined with neomycin (an aminoglycoside) and polymyxin.

Drugs That Act on DNA or RNA

Much excitement was generated by a class of synthetic drugs chem-

[handwritten note:] *Floxacin (Cipro = UTI, STD, GI, Osteomyelitis, respiratory infections* *Sparfloxacin / levo → Pneumonia, bronchitis, sinusitis*

These drugs topoisomer- the bacterial get in gram- ram-positive enzymes in at fluoroqui- tion to being ly absorbed

through four t two dozen olones, *nor-* any cases of astrointesti- d soft tissue *in* and *levo-* pneumonia, bronchitis, and sinusitis. Now that concerns have arisen regarding the overuse of quinolone drugs, the CDC is recommending careful

monitoring of their use to prevent the emergence of ciprofloxacin-resistant bacteria (see 12.3 Making Connections). Side effects that can limit the use of quinolones include damage to tendons and brain disturbances.

Another drug that disrupts nucleic acids is *rifampin*. The primary action of this drug is to block the action of RNA polymerase, thereby preventing transcription. It is somewhat limited in spectrum because the molecule cannot pass through the cell envelope of many gram-negative bacilli. It is mainly used to treat infections by selected gram-positive rods and cocci and a few gram-negative bacteria. Rifampin figures most prominently in treating mycobacterial infections, especially tuberculosis and leprosy, but it is usually given in combination with other drugs to prevent development of resistance. Rifampin is also recommended for prophylaxis in *Neisseria meningitidis* carriers and their contacts, and it is occasionally used to treat *Legionella, Brucella,* and *Staphylococcus* infections.

Drugs That Interfere with Protein Synthesis

The Aminoglycoside Drugs

Antibiotics composed of one or more amino sugars and an aminocyclitol (6-carbon) ring are referred to as **aminoglycosides (figure 12.9).** These complex compounds are exclusively the products of various species of soil **actinomycetes** in the genera *Streptomyces* and *Micromonospora*.

Subgroups and Uses of Aminoglycosides The aminoglycosides have a relatively broad antimicrobial spectrum because they inhibit protein synthesis by binding to one of the ribosomal subunits. They are especially useful in treating infections caused by aerobic gram-negative rods and certain gram-positive bacteria. *Streptomycin* is one of the oldest drugs but has gradually been replaced by

Figure 12.9 The structure of an aminoglycoside: Streptomycin. Colored portions of the molecule are found in all numbers of this drug class.

(a) **Tetracyclines**

(b) **Chloramphenicol**

(c) **Erythromycin**

Figure 12.10 Structures of three antibiotics that act on prokaryotic ribosomes. (a) Tetracyclines. These are named for their regular group of four rings. The several types vary in structure and activity by substitution at the four R groups. **(b)** Chloramphenicol. **(c)** Erythromycin, an example of a macrolide drug. Its central feature is a large lactone ring to which two hexose sugars are attached.

newer forms with less mammalian toxicity. It is still the antibiotic of choice for treating bubonic plague and tularemia, and is considered an effective antituberculosis agent. *Gentamicin* is less toxic and is widely administered for infections caused by gram-negative rods (*Escherichia, Pseudomonas, Salmonella,* and *Shigella*). Two relatively new aminoglycosides, tobramycin and amikacin, are also used for gram-negative infections; tobramycin is especially useful for treating *Pseudomonas* infections in cystic fibrosis patients.

Tetracycline Antibiotics

The first antibiotic in this class was *aureomycin*. It was used to synthesize terramycin, tetracycline, and several semisynthetic derivatives, commonly known as the **tetracyclines (figure 12.10a).** Their action of binding to ribosomes and blocking protein synthesis accounts for the broad-spectrum effects in the group.

Subgroups and Uses of Tetracyclines The scope of microorganisms inhibited by tetracyclines is very broad. It includes gram-positive and gram-negative rods and cocci, aerobic and anaerobic bacteria,

mycoplasmas, rickettsias, and spirochetes. Tetracycline compounds such as *doxycycline* and *minocycline* are administered orally to treat several sexually transmitted diseases, Rocky Mountain spotted fever, Lyme disease, typhus, *Mycoplasma* pneumonia, cholera, leptospirosis, acne, and even some protozoan infections. A newer derivative of minocycline—*tigecycline*—was developed to treat serious hospital skin and soft tissue infections in cases of drug resistance (*Acinetobacter* and MRSA). Although generic tetracycline is low in cost and easy to administer, its use can be limited by its side effects. In addition to gastrointestinal disruption due to changes in the normal microbiota and staining of the teeth, ingestion during pregnancy can interfere with fetal bone development (see table 12.9).

Chloramphenicol

Chloramphenicol is a potent broad-spectrum antibiotic with a unique nitrobenzene structure **(figure 12.10b).** Its primary effect on cells is to block peptide bond formation and protein synthesis. It is one type of antibiotic that is no longer derived from the natural source but is entirely synthesized through chemical processes. Although this drug is as broad spectrum as the tetracyclines, it is so toxic to human cells that its uses are restricted. A small number of people undergoing long-term therapy with this drug incur irreversible damage to the bone marrow that usually results in a fatal form of aplastic anemia.[2] Its administration is now limited to typhoid fever, brain abscesses, and rickettsial and chlamydial infections for which an alternative therapy is not available. Chloramphenicol should never be given in large doses repeatedly over a long time period, and the patient's blood must be monitored during therapy.

Macrolides and Related Antibiotics

Erythromycin is a representative of antibiotics termed macrolides.[3] Its structure consists of a large lactone ring with sugars attached **(figure 12.10c).** This drug has a moderate spectrum and fairly low toxicity. Its mode of action is to block protein synthesis by attaching to the 50S subunit of the ribosome. It is administered orally for *Mycoplasma* pneumonia, legionellosis, *Chlamydia* infections, pertussis, and diphtheria, and as a prophylactic drug prior to intestinal surgery. It also offers a useful substitute for dealing with penicillin-resistant streptococci and gonococci and for treating syphilis and acne. Newer semisynthetic macrolides include *clarithromycin* and *azithromycin* (Zithromax). Both drugs are useful for middle ear, respiratory, and skin infections and have been approved for *Mycobacterium* (MAC) infections in AIDS patients. Clarithromycin has additional applications in controlling infections and gastric ulcers caused by *Helicobacter pylori*. Azithromycin is one of the most prescribed antibiotics in the world. It is used to treat respiratory, gastrointestinal, and sexually transmitted infections. One serious side effect it causes in some people is heart arrhythmias, which limits its use in sensitive people.

Clindamycin is a broad-spectrum antibiotic derived from lincomycin. The tendency of clindamycin to cause adverse reactions in the gastrointestinal tract limits its applications to (1) serious infections in the large intestine and abdomen due to anaerobic bacteria

(*Bacteroides* and *Clostridium*), (2) infections with penicillin-resistant staphylococci, and (3) acne medications applied to the skin.

Drugs That Block Metabolic Pathways

Several of the drugs that interfere with metabolism are synthetic. These antimicrobials as a group do not originate from bacterial or fungal fermentations. Some were synthesized from aniline dyes, and others were originally isolated from plants. Although they have been largely replaced by antibiotics, a few of them remain essential to chemotherapy.

The most important drugs in this class are the **sulfonamides, or sulfa drugs.** Their name is derived from sulfanilamide, an early form of the drug **(figure 12.11).** These drugs function as metabolic analogs to block the synthesis of folic acid by bacteria (see figure 12.6). Dozens of these drugs have been developed, most of them narrow spectrum. Because of its solubility, sulfisoxazole is an effective choice for treating shigellosis, urinary tract infections, and certain protozoan infections. Silver sulfadiazine ointment and solution are prescribed for treatment of burns and eye infections. Another drug, *trimethoprim,* inhibits a second enzymatic step in the synthesis of folic acid. Because of this, trimethoprim is often given in combination with sulfamethoxazole to take advantage of the synergistic effect of the two drugs (Septra, Bactrim). This combination is one of the primary treatments for *Pneumocystis jiroveci* pneumonia (PCP), urinary tract infections, and otitis media.

Sulfones are compounds chemically related to the sulfonamides but lacking their broad-spectrum effects. They still remain important as key drugs in treating Hansen's disease (leprosy). The most active form is dapsone, usually given in combination with rifampin and clofazimine (an antibacterial dye) over long periods.

Newly Developed Classes of Antibiotics

Although most new antibiotics are formulated from the drug classes that already exist, a few classes of drugs with novel actions against bacteria have been added to drug regimens. Medical experts are urging physicians to prescribe these medications only when no other drugs are available to slow the development of drug resistance.

Figure 12.11 The structures of some sulfonamides. Boxed side chains attach at the —ONH group of the nucleus. **(a)** Sulfacetamide, **(b)** sulfadiazine, and **(c)** sulfisoxazole.

2. A failure of the blood-producing tissue that results in very low levels of blood cells.

3. Macrolide antibiotics get their name from the type of chemical ring structure (macrolide ring) they all possess.

Antiviral Chemotherapeutic Agents

The chemotherapeutic treatment of viral infections presents unique challenges. Throughout our discussion of infections caused by bacteria, fungi, protozoa, and helminths, we've emphasized differences in structure and metabolism that guide drug choice. With viruses, we are dealing with an infectious agent that relies upon the host cell for the vast majority of its metabolic functions. Disrupting viral function may require that we disrupt the metabolism of the host cell to a much greater extent than is desirable. Put another way, selective toxicity with regard to viral infection can be difficult to achieve because a single metabolic system is responsible for the well-being of both virus and host. A few viral diseases, such as measles, mumps, and hepatitis, are routinely prevented by effective vaccinations. Unfortunately, vaccines are unavailable for many serious viral diseases, creating a real need for antiviral medications.

Over the past few years several antiviral drugs have been developed that target specific points in the infectious cycle of certain types of viruses. But due to marked differences among viruses, antiviral compounds tend to be limited in their spectrum.

Most antiviral drugs are designed to block a step in completion of the virus cycle. Major modes of action include

1. barring penetration of the virus into the host cell;
2. blocking replication, transcription, and/or translation of viral genetic material; and
3. preventing the normal maturation of viral particles.

Although antiviral drugs protect uninfected cells by keeping viruses from being synthesized and released, most are unable to eliminate extracellular viruses or those in a latent or persistent state within the host's DNA. A summary covering the actions of several antiviral drugs appears in **table 12.6.**

Drugs for Treating Influenza

Zanamivir (*Relenza*) and oseltamivir (*Tamiflu*) are broader-spectrum antiinfluenza medications that block neuraminidase actions in both influenza A and influenza B viruses. Without the functions of neuraminidase, the virus cannot be budded off and released from the host cell, so cells cannot produce active viruses. Both types of drugs may be used prophylactically and must be given early in an infection to be most effective (table 12.6).

Amantadine and its relative, rimantadine, have been essentially discontinued as flu treatments because the current viral strains have developed almost complete resistance to them. They function by blocking envelope receptors that are essential for the uncoating and release of RNA from the influenza A virus.

Quick Search
See an animation of Relenzas's action by searching for "Relenza 2008WM" on YouTube.

Antiherpes Drugs

Many antiviral agents mimic the structure of nucleotides (i.e., act as *analogs*) and compete for sites on replicating DNA. Once one of these nucleotide analogs is incorporated into the growing chain of DNA, replication of that strand stops, interrupting the viral life cycle. **Acyclovir** (Zovirax) and the related compounds *valacyclovir*

(Valtrex), *famciclovir* (Famvir), and *penciclovir* (Denavir) all work in this manner and (with small exceptions from drug to drug) can be used either orally or topically to treat common herpesvirus infections such as oral and genital herpes, chickenpox, and shingles. A related drug, *ganciclovir* (Cytovene), is used parenterally to treat cytomegalovirus infection in patients with compromised immune systems.

An interesting aspect of some of these antiviral agents (specifically valacyclovir and famciclovir) is that they are activated by an enzyme encoded by the virus itself, making the drug active only in virally infected cells. The enzyme thymidine kinase is used by the virus to process nucleosides before incorporating them into viral RNA or DNA. When the inactive drug enters a virally infected cell, the virus's own thymidine kinase converts it to a working antiviral agent. In cells without viruses, the drug is never activated and normal DNA replication is allowed to continue.

Antihepatitis C Drugs

The chapter 11 Case Study gave a dramatic example of the danger of hepatitis C (HCV) infection. This infection has been difficult to manage because the usual drugs are either nonspecific (ribavirin) or have severe side effects (interferon). Pharmaceutical companies have now developed highly effective new drugs called *Sufosbuvir* and *Harvoni* that improve the cure rate and have fewer side effects and a shorter course of treatment. They both work to inhibit the RNA polymerase that HCV uses to replicate its genome. Sofobuvir is a nucleotide analog that acts much like the cyclovir drugs to block further synthesis of an RNA strand. The primary drawback is its cost, which can amount to nearly $100,000 per treatment series, but many medical experts feel that the 94–97% cure rate is worth the cost.

Drugs for Treating HIV Infection and AIDS

The global epidemic of HIV-related disease has been a powerful stimulus for drug discovery. Most of the newer drugs being introduced are designed to treat HIV infection and AIDS. There is now a considerable array of drugs and drug combinations to manage HIV, which can complicate treatment. All of the antiretrovial drugs target a specific part of the multiplication cycle of HIV. There are currently drugs that can inhibit or block every strategic step in the virus cycle (table 12.6).

One of the most important targets is the **reverse transcriptase (RT)** enzyme that HIV brings into the host cell. Recall from chapter 9 that HIV is a retrovirus with an RNA genome. Upon entering a host, the RNA is used as a template by reverse transcriptase to produce a DNA copy of this RNA. This activity offers two types of targets for anti-HIV drugs. The first type of anti-RT drugs are *nucleoside or nucleotide reverse transcriptase analogs* (NRTIs) that directly stop DNA synthesis through competitive inhibition. *Zidovudine* was the first NRTI and the first drug developed to treat HIV and AIDS. It is an analog of thymine that becomes incorporated into the DNA strand of HIV but terminates synthesis because it does not offer the correct sites for binding additional nucleotides. Other examples are stavudine and lamivudine.

Table 12.6 Actions of Selected Antiviral Drugs*

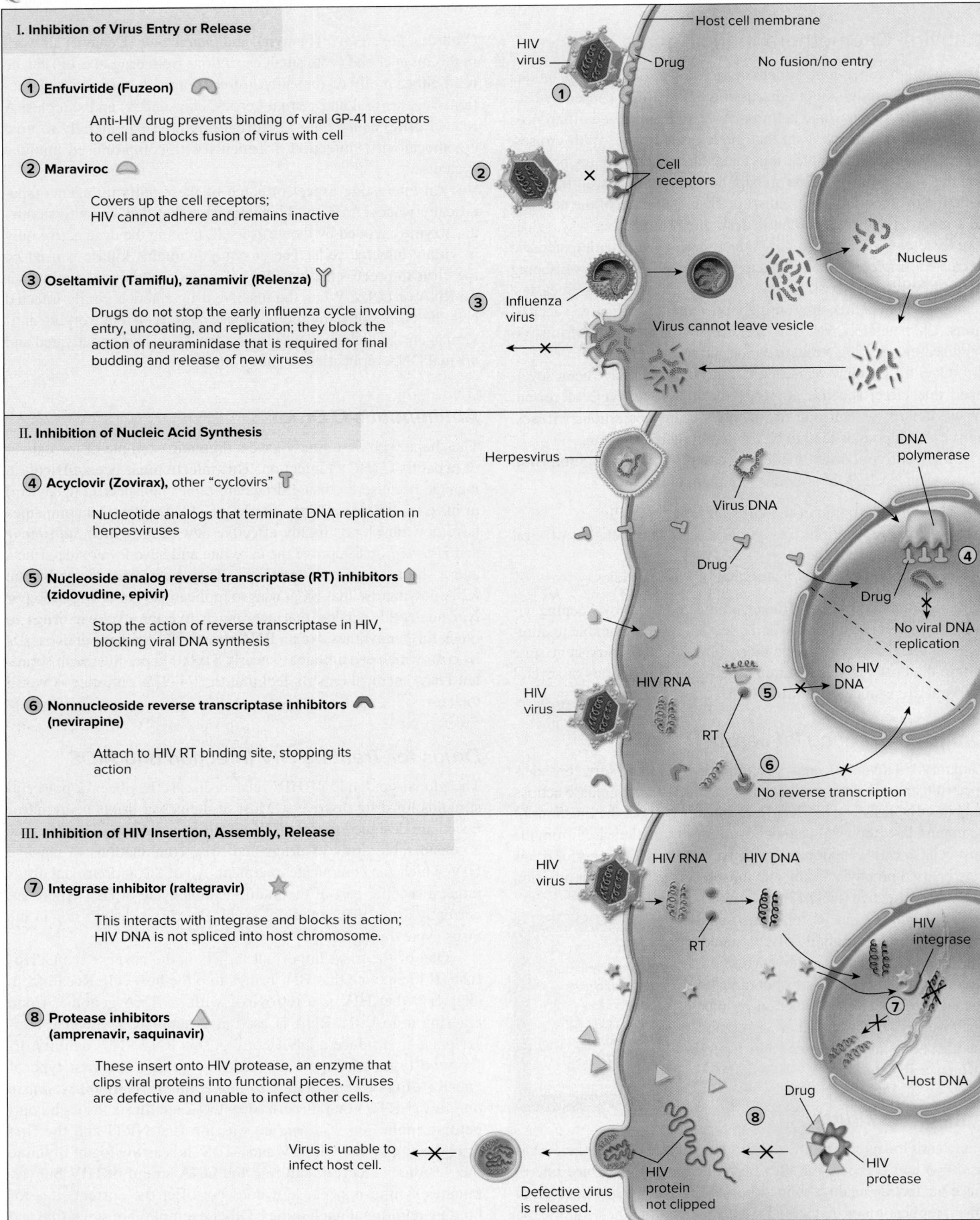

I. Inhibition of Virus Entry or Release

(1) Enfuvirtide (Fuzeon)

Anti-HIV drug prevents binding of viral GP-41 receptors to cell and blocks fusion of virus with cell

(2) Maraviroc

Covers up the cell receptors; HIV cannot adhere and remains inactive

(3) Oseltamivir (Tamiflu), zanamivir (Relenza)

Drugs do not stop the early influenza cycle involving entry, uncoating, and replication; they block the action of neuraminidase that is required for final budding and release of new viruses

Image labels: HIV virus · Host cell membrane · Drug · No fusion/no entry · Cell receptors · Nucleus · Influenza virus · Virus cannot leave vesicle

II. Inhibition of Nucleic Acid Synthesis

(4) Acyclovir (Zovirax), other "cyclovirs"

Nucleotide analogs that terminate DNA replication in herpesviruses

(5) Nucleoside analog reverse transcriptase (RT) inhibitors (zidovudine, epivir)

Stop the action of reverse transcriptase in HIV, blocking viral DNA synthesis

(6) Nonnucleoside reverse transcriptase inhibitors (nevirapine)

Attach to HIV RT binding site, stopping its action

Image labels: Herpesvirus · Virus DNA · DNA polymerase · Drug · No viral DNA replication · HIV virus · HIV RNA · No HIV DNA · RT · No reverse transcription

III. Inhibition of HIV Insertion, Assembly, Release

(7) Integrase inhibitor (raltegravir)

This interacts with integrase and blocks its action; HIV DNA is not spliced into host chromosome.

(8) Protease inhibitors (amprenavir, saquinavir)

These insert onto HIV protease, an enzyme that clips viral proteins into functional pieces. Viruses are defective and unable to infect other cells.

Image labels: HIV virus · HIV RNA · HIV DNA · RT · HIV integrase · Host DNA · Drug · HIV protease · HIV protein not clipped · Defective virus is released. · Virus is unable to infect host cell.

*Details of viral cycles are omitted for ease in observing drug effects.

TABLE 12.7	Overview of Anti-HIV Drugs, Effects, and Examples	
Category of Anti-HIV Drug	**Effects of Drug on Virus**	**Examples of Drugs**
Nucleotide and nucleoside reverse transcriptase inhibitors (NRTIs)	Drugs are chemical analogs that insert into the HIV DNA and prevent further addition of new bases. This terminates DNA synthesis and stops viral replication. This is a type of competitive inhibition.	Retrovir (azidothymidine [AZT], zidovudine), lamivudine (Epivir), abacavir, didanosine, stavudine, tenofovir, emtricitabine (Emtriva) Truvada, a combination of tenofovir and emtricitabine used for pre-exposure prophylaxis in high-risk persons.
Nonnucleoside reverse transcriptase inhibitors (NNRTIs)	Drugs interact with reverse transcriptase and block its active site; this is a type of noncompetitive inhibition that causes a shape change in the enzyme.	Nevirapine, efavirenz, delavirdine, rilpivirine
Protease inhibitors	HIV protease that completes viral assembly is blocked, leading to incomplete and dysfunctional viruses.	Amprenavir, atazanavir, darunavir, indinavir, saquinavir
Entry or fusion blockers	Maraviroc covers an entry receptor on the human cell (CCR5) and keeps the virus from binding; enfuvirtide (Fuzeon) stops the viral envelope from fusing with the cell membrane.	Maraviroc, enfuvirtide
Integrase inhibitors	Drugs interact with HIV integrase, which is required for the viral DNA to be inserted (integrated) into the human chromosome; this stops the action of the viral DNA.	Raltegravir, elvitegravir

The second type of anti-RT drugs are *nonnucleoside reverse transcriptase inhibitors* (NNRTIs) that attach to a site on the reverse transcriptase enzyme itself and alter its shape—a type of noncompetitive inhibition (see table 12.7). NNRTIs accomplish a similar goal in preventing reverse transcription of the HIV RNA, but they bind to the enzyme itself and prevent its reactive site from synthesizing DNA. They are usually given together or in combination with NRTIs.

Another enzyme essential to the virus's release is HIV protease, which is involved in cutting the proteins that are an essential part of the virus structure (capsids, enzymes). The class of drugs called *protease inhibitors* has the effect of blocking this enzyme and causing the assembly of faulty viruses that are not infective.

Newer classes of anti-HIV drugs are two that block entry or fusion of the virus at the beginning of its invasion and one that prevents viral integration into the host genome. The drug *maraviroc* keeps HIV out of cells by preventing the viruses from adhering to host cells, and without this step, they cannot invade. Enfuvirtide (*Fuzeon*) prevents infection when it stops the envelope of the virus from fusing with the host cell after it has attached. Another important enzyme targeted for HIV therapy is *HIV integrase*. This enzyme is essential for the final step in HIV DNA processing. Integrase inhibitors prevent this viral DNA from being incorporated or integrated into a host cell's chromosome. Because this step must occur for full viral function, the cycle is terminated by this action.

The strategy for HIV and AIDS treatment currently involves antiretroviral therapy (ART) that uses combinations of drugs to reduce the development of drug resistance. Combinations are available with two, three, or four drugs together in one pill. Most have at least one NRTI and one NNRTI, and some have protease inhibitors and an integrase inhibitor. They are used to treat ongoing HIV infection and AIDS, and for preexposure* and postexposure prophylaxis. In 2015 a new combination drug, Truvada, was approved as a prophylactic medication to prevent infection in people whose lifestyle increases their possible exposure to HIV. **Table 12.7** summarizes HIV drug types, their modes of action, and examples.

Interferons A sensible alternative to artificial antiviral drugs has been the human-based substance **interferon (IFN).** Interferon is a glycoprotein produced primarily by fibroblasts and leukocytes in response to various immune stimuli. It has numerous biological activities, including antiviral and anticancer properties. Studies have shown that it is a versatile part of animal host defenses, having a major role in natural immunities. Its mechanism is discussed in chapter 14.

The first investigations of interferon's antiviral activity were limited by the extremely minute quantities that could be extracted from human blood. Several types of interferon (Betaseron, Roferon) are currently produced by the recombinant DNA technology techniques outlined in chapter 10. Extensive clinical trials have tested interferon's effectiveness in viral infections and cancer. Some of the known therapeutic benefits of interferon include

1. reducing the time of healing and some of the complications in certain infections (mainly of herpesviruses);

*Pre-exposure prohylaxis (PrEP)—treatment given to uninfected persons to prevent infection, such as a daily dose of anti-HIV medication given to a person who is HIV negative — is now considered a viable way to prevent infection in people whose lifestyles may increase the risk of infection.

12.2 MAKING CONNECTIONS

Alternative Antimicrobial Therapies

"The enemy of my enemy is my friend."

This old saying might well describe a novel way to kill bacteria using viruses. This technique, called phage therapy, was first attempted over a hundred years ago after the discovery of bacteriophages. The concept behind it is rather simple: When you expose bacterial cells to their specific viruses, the viruses will invade and destroy them. If you apply the viruses in large quantity to treat a human or animal infection, they could lyse the bacteria and cure the infection. Phage therapy went out of favor for many years except in Russia, but it is currently making a comeback worldwide. An important push for renewed interest is the search for solutions to the high levels of drug resistance among bacteria. Bacteria do not become resistant to viruses as they do to drugs. Other significant advantages are that viruses are very narrow spectrum—they can target just one species of bacteria—and they would not be as toxic to humans as many drugs are.

For now, phage therapy in the United States is restricted to removing contaminants from meats, reducing the incidence of *Salmonella* and *Escherichia coli* in young farm animals, treating skin infections, and sanitizing food. In some clinics it is in experimental trials for treatment of skin ulcers, wounds, and urinary tract infections. It will require extensive testing before it becomes a first-line therapy.

Other novel approaches to controlling infections include the use of **probiotics**** and **prebiotics.** Probiotics are preparations of live microorganisms that are fed to animals and humans to modify the resident intestinal bacteria. These microorganisms can replace microbes lost during antimicrobial therapy or simply augment the normal microbiota that is already there. Culture-enriched dairy products are the latest commercial source of probiotics. Several new types of yogurt now claim specific health benefits can be gained from eating high numbers of special strains of bacteria. According to some published research findings, the ingestion of *Bifidobacterium lactis regularis* can help to maintain regular bowel function, and *Lactobacillus casei immunitas* may boost the immune function of the gut. Purposely eating such high numbers of bacterial cells—estimated at from 1 billion to 10 billion per

© Kathy Park Talaro

serving—is quite a departure from the usual view of microbes as disease agents. Bon appetit!

A probiotic approach is frequently used in female patients with recurring urinary tract infections. There is evidence that vaginal inserts or perineal swabs of *Lactobacillus* can restore a healthy acidic environment to the region and displace disease-causing microorganisms. Probiotics also seem to inhibit the development of food allergies.

Prebiotics are nutrients that encourage the growth of beneficial microbes in the intestine. For instance, fruit sugars such as fructans are thought to encourage the growth of *Bifidobacterium* in the large intestine and to discourage the growth of potential pathogens. Clearly, the use of these agents is a different type of antimicrobial strategy than we are used to, but they can provide an alternative to traditional antibiotics that have become less effective.

New attention is being paid to the curative properties of honey. Several studies have found that honey appears to limit infection, but even more, it can speed the healing of wounds and burns through its beneficial effects on the immune system. When applied to wounds, honey helps remove wound debris and promotes the growth of new tissue. A type of wound dressing marketed as Medihoney (antimicrobial active Manuka honey) has been approved by the FDA. Other home remedies that show pharmacologic potential are pomegranate, cranberry, and grape juices and bitter orange leaves.

Describe several benefits of using nondrug regimens and foods to treat or prevent infections. *Answer available on Connect.*

* *probiotic* (proh′-by-ah-tik′) Gr. *pro*, for, and *bios*, life.

2. preventing or reducing some symptoms of cold and papillomaviruses (warts);
3. slowing the progress of certain cancers, including bone cancer and cervical cancer, and certain leukemias and lymphomas; and
4. treating a rare cancer called hairy-cell leukemia, hepatitis C (a viral liver infection), genital warts, and Kaposi's sarcoma in AIDS patients.

Medical science is giving close scrutiny to alternate methods of preventing or treating infections beyond the usual antibiotics and synthesized drugs (**12.2 Making Connections**).

 Check Your Progress SECTION 12.3

12. Construct a chart to outline the applications of major groups of antifungal drugs, antiparasitic drugs, and antiviral drugs.
13. Explain why there are fewer antifungal, antiparasitic, and antiviral drugs than antibacterial drugs.
14. List the targets of antiviral drugs and their modes of action. What is the strategy behind taking several different types of drugs to treat HIV?
15. Summarize the origins and biological actions of interferon-based drugs.

12.4 Interactions Between Microbes and Drugs: The Acquisition of Drug Resistance

 Expected Learning Outcomes

15. Describe the development of microbial drug resistance, and explain how it is acquired.

16. Explain the major mechanisms by which microbes become resistant to drugs.

17. Review some of the ways to reduce levels of drug resistance.

18. Discuss how drug resistance spreads through a population of microbes and the long-term effects of this.

One unfortunate outcome of the use of antimicrobials is the development of microbial **drug resistance,** an adaptive response in which microorganisms begin to tolerate an amount of drug that would previously have been inhibitory. The development of mechanisms for resisting antimicrobial drugs is a result of the genetic versatility and adaptability of microbial populations. The property of drug resistance can be an intrinsic characteristic as well as acquired. Intrinsic drug resistance would be found in bacteria that must be resistant to any antibiotic that they produce or bacteria that have natural barriers that block the drug (the outer membrane of gram-negative bacteria, for example). This type of resistance is not the main problem with regard to antimicrobial chemotherapy. Of much greater importance is the acquisition of resistance by a microbe that was previously sensitive to the drug. In our context, the term *drug resistance* will refer to this type of acquired resistance.

How Does Drug Resistance Develop?

Contrary to popular belief, antibiotic resistance is not a recent phenomenon. Resistance to penicillin developed in some bacteria as early as 1940, 3 years before the drug was even approved for public use. The scope of the problem became apparent in the 1980s and 1990s when scientists and physicians observed treatment failures on a large scale.

Microbes become newly resistant to a drug after one of the following events occurs: (1) spontaneous mutations in critical chromosomal genes, or (2) acquisition of entirely new genes or sets of genes via transfer from another species. Chromosomal drug resistance usually results from spontaneous random mutations in bacterial populations. The chance that such a mutation will be advantageous is minimal, and the chance that it will confer resistance to a specific drug is lower still. Nevertheless, given the huge numbers of microorganisms in any population and the constant rate of mutation, such mutations do occur. The end result varies from slight changes in microbial sensitivity, which can be overcome by larger doses of the drug, to complete loss of sensitivity.

Resistance through intermicrobial transfer originates from chromosomal genes and plasmids called resistance (R) factors being transferred through the processes we considered in chapter 9: conjugation, transformation, or transduction **(figure 12.13).** Studies have shown that plasmids encoded with drug resistance are naturally present in microorganisms before they have been exposed

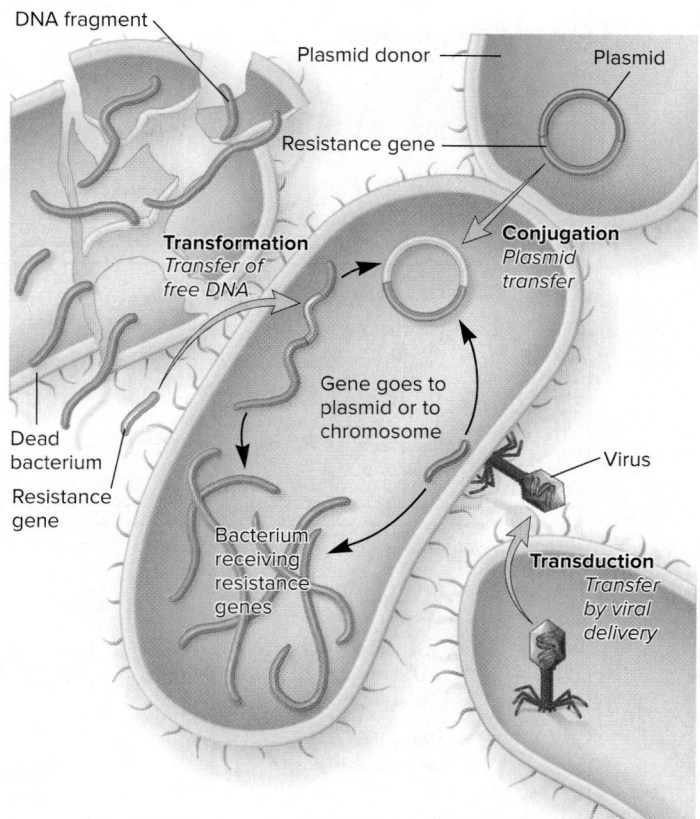

Figure 12.13 Transfer of drug resistance. Bacteria transmit R factors, often containing numerous genes for drug resistance by means of conjugation, transduction, and transformation. The widespread occurrence of such exchanges accounts for the rapid spread of drug resistance.

to the drug. Such traits are "lying in wait" for an opportunity to be expressed and to confer adaptability on the species. Many bacteria also maintain transposable drug-resistance sequences (transposons) that are duplicated and inserted from one plasmid to another or from a plasmid to the chromosome. Chromosomal genes and plasmids containing codes for drug resistance are faithfully replicated and inherited by all subsequent progeny. This sharing of resistance genes accounts for the rapid proliferation of drug-resistant species. A growing body of evidence points to the ease and frequency of gene transfers in biofilms between totally unrelated bacteria living in the body's microbiota and the environment.

Specific Mechanisms of Drug Resistance

Inside a bacterial cell, the net effect of these genetic transfers is one of the following, which actually causes the bacterium to be resistant (note that the numbers that follow correspond to the numbers seen in **figure 12.14**):

Quick Search
Explore this topic by going to "Animation of Antibiotic Resistance" on YouTube.

1. The development of alternate enzymes that inactivate the drug (occurs only when new genes are acquired).
2. Permeability or uptake of drug into bacterium is decreased or eliminated.

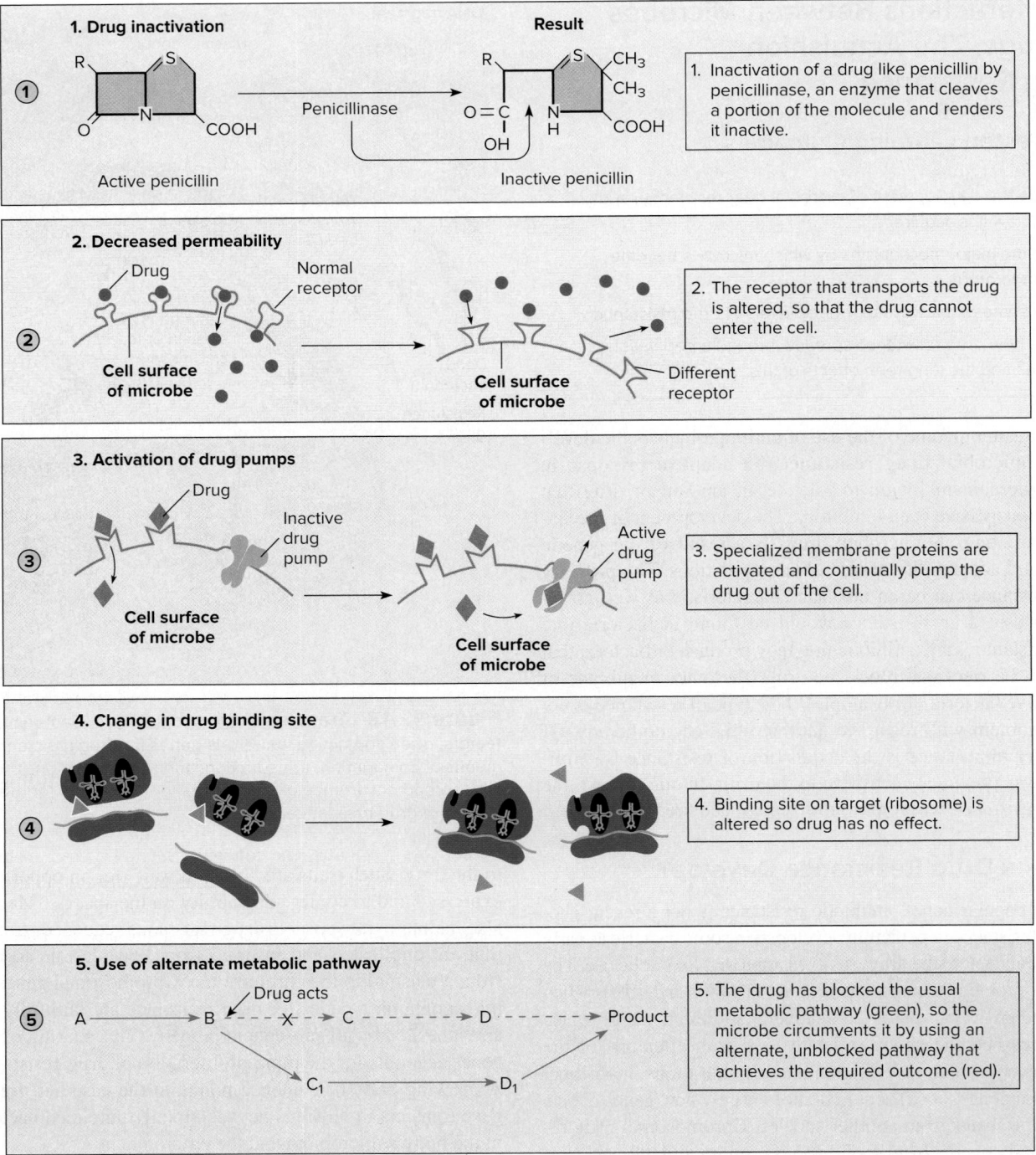

Figure 12.14 Examples of mechanisms of acquired drug resistance.

3. The microbe engages special drug transport pumps that remove the drug.
4. Binding sites for drug are decreased in number or affinity (can occur via mutation or acquisition of new genes).
5. An affected metabolic pathway is shut down or an alternate pathway is used (occurs due to mutation of original enzyme[s]).

Some bacteria can become resistant indirectly by lapsing into dormancy or, in the case of penicillin, by converting to a cell-wall-deficient form (L form) that penicillin cannot affect. A number of infections, including endocarditis, urinary infections, otitis media, osteomyelitis, and pneumonias, are more resistant to drugs because they develop as biofilms (see Clinical Connections).

Drug Inactivation Mechanisms

Microbes inactivate drugs by producing enzymes that permanently alter drug structure. One example, bacterial exoenzymes called

CLINICAL CONNECTIONS

Biofilms and Drug Resistance

In chapters 4 and 7 we described the development of microbial biofilms in various habitats. Recall that in these complex communities, certain "pioneer" microbes anchor themselves to a substrate by secreting a sticky matrix. This initial colonization stimulates a network of other microbes to migrate into the film.

It is estimated that around 60% of infections involve biofilms containing single or multiple species of microbes. Often the infections occur on natural tissues—for example, bacterial infections of the heart valves, middle ear, and teeth. The increased use of indwelling medical devices has created yet another habitat in which biofilms may become tenaciously attached. Catheters, artificial valves, and prosthetic devices can harbor persistent biofilm colonizations that are very difficult to treat and remove.

Microbes in infectious biofilms may be hundreds of times more drug resistant than the same free, unattached microbes. For one thing, microbes are protected by the impenetrable nature of the extracellular matrix—drugs may be blocked from entry. We also know that microbes communicate and regulate resistance mechanisms such as drug pumps, and the contact within the biofilm facilitates the transfer of drug-resistant plasmids and genes. Numerous studies have discovered the presence of persisters—microbes that remain embedded in the infection long after the drug has been discontinued, only to erupt at a later time in a new infection. These factors alone or in combination allow the persistence of the biofilm and a long-term source of infection for the patient.

With knowledge of biofilms, drug companies are researching newer drug strategies that can prevent or disrupt their formation. One concept already in use is to incorporate antimicrobic chemicals into implants, catheters, and other metal or plastic devices to inhibit microbial growth.

Would broad-spectrum or narrow-spectrum drugs be more effective in treating infections involving biofilms? Explain your reasoning. *Answer available on Connect.*

beta-lactamases, hydrolyze the beta-lactam ring structure of some penicillins and cephalosporins, rendering the drugs inactive. Two beta-lactamases—penicillinase and cephalosporinase—disrupt the structure of certain penicillin or cephalosporin molecules so their activity is lost **(figure 12.14, part 1).** So many strains of *Staphylococcus aureus* produce penicillinase that regular penicillin is rarely a possible therapeutic choice. Now that some strains of *Neisseria gonorrhoeae,* called PPNG,[4] have also acquired penicillinase genes, alternative drugs are required to treat gonorrhea. A large number of other gram-negative species are inherently resistant to some of the penicillins and cephalosporins because of naturally occurring beta-lactamases.

4. Penicillinase-producing *Neisseria gonorrhoeae.*

Decreased Drug Permeability or Increased Drug Elimination

The resistance of some bacteria can be due to a mechanism that prevents the drug from entering the cell and acting on its target. For example, the outer membrane of the cell wall of certain gram-negative bacteria is a natural blockade for some of the penicillin drugs. Aminoglycoside resistance is known to develop through changes in drug permeability caused by point mutations in proteins of the transport system or outer membrane **(figure 12.14, parts 2 and 3).**

Many bacteria possess multidrug-resistant (MDR) pumps that actively transport drugs and other chemicals out of cells. These pumps are proteins encoded by plasmids or chromosomes. They are stationed in the cell membrane and expel molecules by a proton-motive force similar to ATP synthesis **(figure 12.14, part 3).** They confer drug resistance on many gram-positive pathogens (*Staphylococcus, Streptococcus*) and gram-negative pathogens (*Pseudomonas, E. coli*). Because MDRs lack selectivity, one type of pump can expel a broad array of antimicrobial drugs, detergents, and other toxic substances.

Change of Drug Receptors

As most drugs act on a specific target such as protein, RNA, DNA, or membrane structure, microbes can circumvent drugs by altering the nature of this target. Bacteria can become resistant to aminoglycosides when point mutations in ribosomal proteins arise **(figure 12.14, part 4).** Erythromycin and clindamycin resistance is associated with an alteration on the 50S ribosomal binding site. Penicillin resistance in *Streptococcus pneumoniae* and methicillin resistance in *Staphylococcus aureus* are related to an alteration in the binding proteins in the cell wall. Several species of enterococci have acquired resistance to vancomycin through a similar alteration of cell wall proteins. Fungi can become resistant by decreasing their synthesis of ergosterol, the principal receptor for certain antifungal drugs.

Changes in Metabolic Patterns

The action of antimetabolites can be circumvented if a microbe develops an alternative metabolic pathway or enzyme **(figure 12.14, part 5).** Sulfonamide and trimethoprim resistance develops when microbes deviate from the usual patterns of folic acid synthesis. Fungi can acquire resistance to flucytosine by completely shutting off certain metabolic activities.

Natural Selection and Drug Resistance

So far we have been considering drug resistance at the cellular and molecular levels, but its full impact is felt only if this resistance occurs throughout the cell population. Let us examine how this might happen and its long-term therapeutic consequences.

Any large population of microbes is likely to contain a few individual cells that are already drug resistant because of prior mutations or transfer of plasmids **(figure 12.15a).** As long as the

Are We Losing the Drug Resistance Race?

"The rise of antibiotic resistance is a global health crisis. Antimicrobial resistance is on the rise in every region of the world. We are losing first line antibiotics. This makes a broad range of common infections much more difficult to treat, and replacement treatments are most costly, more toxic, and need much longer durations of treatment"

Margaret Chan, MD, WHO Director General.

It is a fact of nature that, when a large number of microbes are exposed to a variety of drugs, there will always be some genetically favored strains or types that survive in the presence of those drugs. One could even say that microbes are merely practicing what they do best—adapt and survive. Our very use of drugs is a catalyst that speeds their evolution toward drug resistance.

Approximately 2 million hospital-acquired infections causing 90,000 fatalities occur every year. About 70% of these infections are caused by bacteria resistant to at least one drug. For many years, concerned observers reported the gradual development of drug resistance in a number of pathogens, but during the past decade the scope of the problem has escalated. It is now a common event to discover microbes that have become resistant to relatively new drugs in a very short time. The clinical setting is a prolific source of drug-resistant strains of bacteria. This environment continually exposes large numbers of pathogens to a wide variety of drugs. The hospital also maintains patients with weakened defenses, making them highly susceptible to infections.

We can readily see the pattern by examining the case of *Staphylococcus aureus* and penicillin. In the 1950s, hospital strains began to show resistance to this drug, and because of indiscriminate use of penicillin, these strains became nearly 100% resistant to penicillin within 25 years. In response, drugmakers created methicillin as an alternative to penicillin, but within a few years MRSA (methicillin-resistant *S. aureus*) strains began to appear that ultimately came to tolerate nearly all antibiotics. To complicate this problem, strains of MRSA found their way from the clinic into the community. Fortunately MRSA was universally treatable with the drug vancomycin—at least for a while. Then there arose reports of VISA strains that were less sensitive, followed by reports of VRSA strains with complete resistance to this drug, all within a few years. It is this creeping pattern of drug resistance to newer replacement drugs that has health officials most concerned.

One of the more urgent alerts has arisen over carbapenem-resistant Enterobacteriaceae (CRE). This description refers to resistance among members of this family of gram-negative rods to the carbapenem drugs (imipenem, ertapenem), which had traditionally been used in cases of drug resistance to other beta-lactam drugs. Many new resistant strains of *Klebsiella pneumoniae, Enterobacter cloacae,* and *Escherichia coli* have emerged as significant hospital-acquired pathogens (see figure). In the hospital setting, enteric bacteria cause infections of the bloodstream and urinary tract. The fatality rate for compromised or debilitated patients can be as high as 40%. Similar to *Acinetobacter,* which was featured in the chapter 9 Case Study, the genes for carbapenem resistance are readily and rapidly transmitted by means of several mechanisms. Since 2001 the number of cases has risen by three to five times.

Drugs in Animal Feeds

Another practice that has contributed significantly to growing drug resistance is the addition of antimicrobials to livestock feed, with the idea of decreasing infections and thereby improving animal health and size. Eighty percent of all antibiotics used in the United States are added to animal food. Enteric bacteria such as *Salmonella, E. coli,* and enterococci that live as normal intestinal residents of these animals readily share resistance plasmids and are constantly selected and amplified by exposure to drugs. These pathogens subsequently "jump" to humans and cause drug-resistant infections, oftentimes at epidemic proportions. In a deadly outbreak of *Salmonella* infection in Denmark, the pathogen was found to be resistant to seven different antimicrobial drugs. The opportunistic pathogen called VRE (vancomycin-resistant enterococcus) has been traced to the use of a vancomycin-like drug in cattle feed. It is now one of the most tenacious of hospital-acquired infections, for which there are few drug choices.

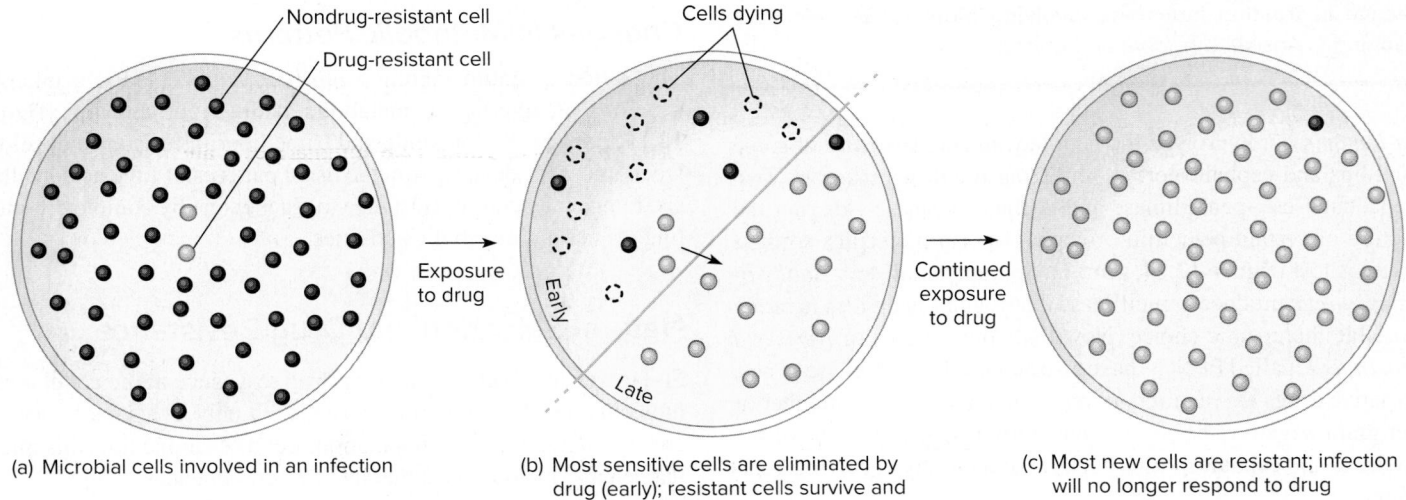

(a) Microbial cells involved in an infection

(b) Most sensitive cells are eliminated by drug (early); resistant cells survive and begin to grow (late)

(c) Most new cells are resistant; infection will no longer respond to drug

Figure 12.15 **A model of natural selection for drug resistance. (a)** Generally, any population of microbes will harbor a few cells carrying a prior mutation that confers drug resistance. **(b)** Environmental pressure (here, exposure to the drug) selects for survival of these cells. **(c)** In time, the resistant cells become the dominant cells.

Worldwide Drug Resistance

The drug dilemma has become a widespread problem, affecting all countries and socioeconomic groups. In general, the majority of infectious diseases, whether bacterial, fungal, protozoan, or viral, are showing increases in drug resistance. In parts of India, the main drugs used to treat cholera (furazolidone, ampicillin) have gone from being highly effective to essentially useless in 10 years. In Southeast Asia, 98% of gonococcus infections are multidrug resistant. Many strains of the malaria parasite have become resistant to the mainstay drug chloroquine, which means that it must be combined with an alternate drug or not given at all. The influenza drugs amantadine and rimantadine had to be discontinued because the virus had become almost completely resistant to these medications. In some parts of Africa, strains of the TB bacillus have developed such extreme resistance that there are no effective treatments available. To add to the problem, global travel and globalization of food products mean that drug-resistant pathogens can be spread around the world in a short time.

In the United States alone, the extra cost for treating drug-resistant infections is around $10 billion per year. In many developing countries, drugs are mishandled by overuse and underuse, either of which can contribute to drug resistance. Many countries that do not regulate the sale of prescription drugs make them readily available to purchase over the counter. For example, the antituberculosis drug INH (isoniazid) is sometimes used as a "lung vitamin" to improve health, and antibiotics are often taken in the wrong dose for undiagnosed conditions. These countries serve as breeding grounds for drug resistance that can eventually be carried to other countries.

"Bad Bugs—No Drugs"

The development of new antimicrobial drugs is considered an urgent goal. Medical experts are seeing an increase in the incidence of the "bad bugs–no drugs" dilemma. A case in point is the alarming trend of pathogens known collectively as ESKAPE. This acronym comes from the first letters of their generic names, but it also tells us that these six pathogens are very successful at escaping the effects of drugs and are getting better at it all the time. The bacteria—*Enterococcus, Staphylococcus aureus, Klebsiella,*

Acinetobacter, Pseudomonas, and *Enterobacter*—make up the majority of hospital-acquired infections, and are also the most drug-resistant. It is not unusual to isolate extremely drug-resistant strains or "superbugs" for which there are only two, one, and sometimes no drugs available. To avert a public health crisis, a number of governments are actively working with pharmaceutical companies on drug discovery and development, sometimes called the "new bugs–new drugs" approach. There may be renewed hope in gaining an advantage in this race with the microbes. Between 2013 and 2015, six new antibiotics and other drugs were approved by the FDA through an accelerated program that fast-tracks new drugs.

To see an excellent documentary addressing this crisis, go to the PBS Frontline website and search for "Hunting the Nightmare Bacteria."

What are some of the reasons microbes develop drug resistance so rapidly? *Answer available on Connect.*

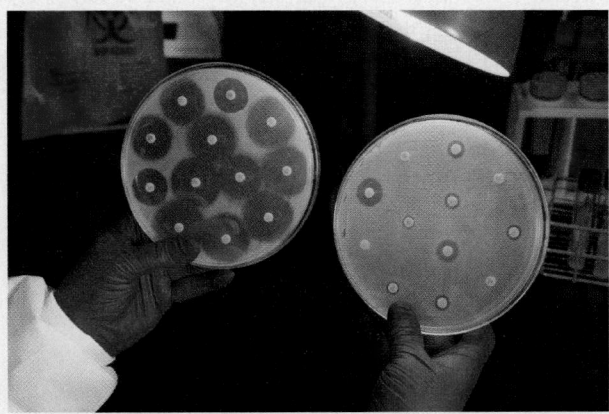

Plates comparing sensitivity test results for two bacterial cultures exposed to 12 different antibiotic discs. The plate on the left shows large, uniform zones of inhibition, indicating complete sensitivity of the culture to all 12 drugs. On the right is a sample of a CRE pathogen (here, *Enterobacter*) that has slight sensitivity to two of the drugs but is resistant to all of the rest. *Source:* James Gathany/CDC

drug is not present in the habitat, the numbers of these resistant forms will remain low because they have no particular growth advantage. But if the population is subsequently exposed to this drug **(figure 12.15*b*),** sensitive individuals are inhibited or destroyed, and resistant forms survive and proliferate. As the population continues to grow, offspring of these resistant microbes will inherit this drug resistance. In time, the replacement population will have more of the drug-resistant forms and can eventually become completely resistant **(figure 12.15*c*).** In ecological and evolutionary terms, the environmental factor (in this case, the drug) has put selection pressure on the population, allowing the better adapted microbe (the drug-resistant one) to survive and grow. The end result is a population that has evolved to a condition of drug resistance.

Natural selection for drug-resistant forms is apparently a common phenomenon. It takes place most frequently in various natural habitats, medical environments, and it also can occur within the

bodies of humans and animals during drug therapy. See **12.3 Making Connections** to find out some of the current implications of drug resistance. **Table 12.8** summarizes a number of actions that are being considered to slow its development.

Check Your Progress SECTION 12.4

16. Explain the phenomenon of drug resistance from the standpoint of microbial genetics (include a description of R factors).

17. Describe the five basic mechanisms of drug resistance.

18. Outline several situations or events that increase the development of drug resistant microbes.

19. Name several ways drug resistance can be prevented by medical personnel and patients (see table 12.8).

TABLE 12.8	Strategies to Limit Drug Resistance of Microorganisms

Drug Usage

- Physicians have the responsibility for making an accurate diagnosis and prescribing the correct drug therapy.
- Compliance with physicians' guidelines: The patient should take the correct dosage, by the best route, for the appropriate period. This diminishes the selection for mutants that can resist low drug levels and ensures elimination of the pathogen.
- Combined therapy with two or more drugs together increases the chances that at least one of the drugs will be effective and that a strain resistant to one drug will be eliminated by the other.

Drug Research

- Research focuses on developing shorter-term, higher-dose antimicrobials that are more effective, less expensive, and have fewer side effects.
- Pharmaceutical companies continue to seek new antimicrobial drugs with structures that are not readily inactivated by microbial enzymes or drugs with modes of action that are not readily circumvented.

Long-Term Strategies

- Proposals to reduce the abuse of antibiotics range from educational programs for medical workers to requiring justification for prescribing certain types of antibiotics.
- Especially valuable antimicrobials may be restricted in their use to only one or two types of infections.
- The addition of antimicrobials to animal feeds must be curtailed worldwide.
- Government programs that make effective therapy available to low-income populations should be increased.
- Vaccines should be used whenever possible to provide alternative protection.

12.5 Interactions Between Drugs and Hosts

 Expected Learning Outcomes

19. Summarize the adverse effects of drugs on the human body.
20. Describe some of the toxic effects of drugs on organs and tissues.
21. Discuss how drugs can cause damaging immune responses.
22. Explain the ways in which drugs alter the normal microbiota and the results.

Until now this chapter has focused on the interactions between antimicrobial drugs and the microorganisms they target. During an infection, the microbe is living in or on a host, which means that the drug is administered to the host, even though its target is the microbe. Because of this, the effect of the drug on the host must always be considered.

Although selective antimicrobial toxicity is the ideal constantly being sought, chemotherapy by its very nature involves contact with foreign chemicals that can harm human tissues. In fact, estimates indicate that at least 5% of all persons taking an antimicrobial drug experience some type of serious adverse reaction to it. Antibiotics cause one out of five emergency visits due to adverse drug reactions. The major **side effects** of drugs fall into one of three categories: direct damage to tissues through toxicity, allergic reactions, and disruption in the balance of normal microbial flora. The damage incurred by antimicrobial drugs can be short term and reversible or permanent, and it ranges in severity from cosmetic to lethal effects.

Toxicity to Organs

Drugs can adversely affect the following organs: the liver (hepatotoxic), kidneys (nephrotoxic), gastrointestinal tract, cardiovascular system and blood-forming tissue (hemotoxic), nervous system (neurotoxic), respiratory tract, skin, bones, and teeth.

Because the liver is responsible for metabolizing and detoxifying foreign chemicals in the blood, it can be damaged by a drug or its metabolic products. Injury to liver cells can result in enzymatic abnormalities, fatty liver deposits, hepatitis, and liver failure. The kidney is involved in excreting drugs and their metabolites. Some drugs irritate the nephron tubules, creating changes that interfere with their filtration abilities. Drugs such as sulfonamides can crystallize in the kidney and form stones that obstruct the flow of urine.

The most common complaint associated with oral antimicrobial therapy is diarrhea, which can progress to severe intestinal irritation or colitis. Although some drugs directly irritate the intestinal lining, the usual gastrointestinal complaints are caused by disruption of the intestinal microbiota (shown in figure 12.17).

Many drugs given for parasitic infections are toxic to the heart, causing irregular heartbeats and even cardiac arrest in extreme cases. Chloramphenicol can severely depress blood-forming cells in the bone marrow, resulting in either a reversible or a permanent (fatal) anemia. Some drugs hemolyze the red blood cells, others reduce white blood cell counts, and still others damage platelets or interfere with their formation, thereby inhibiting blood clotting.

Quick Search
Find the video "Important Side Effects and Toxicities" on YouTube.

Certain antimicrobials act directly on the brain and cause seizures. Others, such as aminoglycosides, damage nerves (very commonly, the eighth cranial nerve), leading to dizziness, deafness, or motor and sensory disturbances. When drugs block the transmission of impulses to the diaphragm, respiratory failure can result.

The skin is a frequent target of drug-induced side effects. The skin response can be a symptom of drug allergy or a direct toxic effect. Some drugs interact with sunlight to cause photodermatitis,

Figure 12.16
Drug-induced side effect. An adverse effect of tetracycline given to young children is the permanent brownish discoloration of tooth enamel continuing into adulthood. © Biophoto Associates/ Science Source

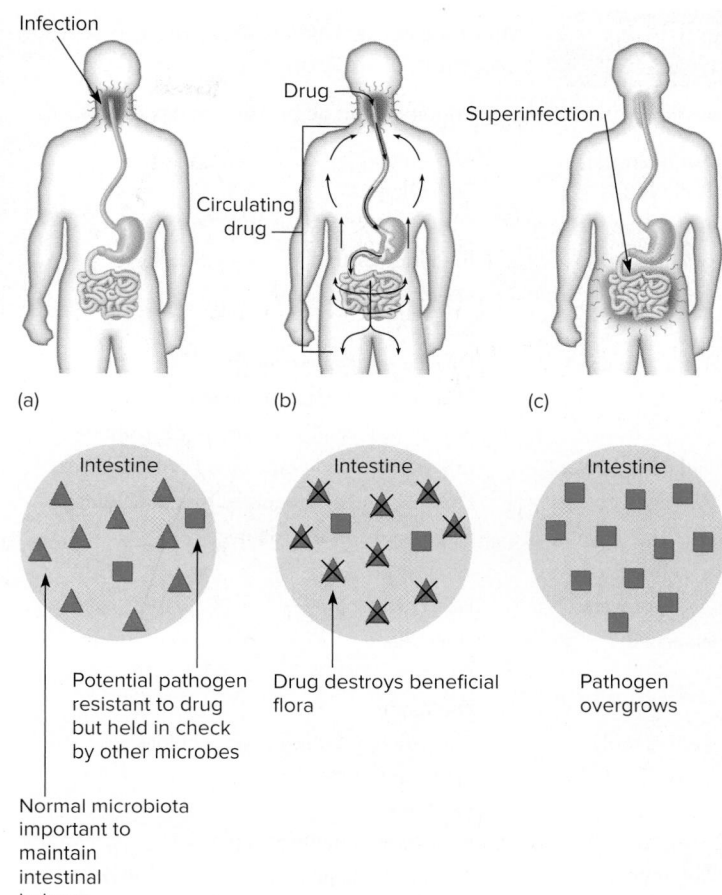

Figure 12.17 The role of antimicrobials in disrupting microbiota and causing superinfections. (a) A primary infection in the throat is treated with an oral antibiotic. **(b)** The drug is carried to the intestine and is absorbed into the circulation. **(c)** The primary infection is cured, but drug-resistant pathogens have survived and create an intestinal superinfection.

a skin inflammation. Tetracyclines are contraindicated (not advisable) for children from birth to 8 years of age because they bind to the enamel of the teeth, creating a permanent gray to brown discoloration **(figure 12.16).** Pregnant women should avoid tetracyclines because they cross the placenta and can be deposited in the developing fetal bones and teeth.

Allergic Responses to Drugs

One of the most frequent drug reactions is heightened sensitivity, or **allergy.** This reaction occurs because the drug acts as an antigen (a foreign material capable of stimulating the immune system) and stimulates an allergic response. This response can be provoked by the intact drug molecule or by substances that develop from the body's metabolic alteration of the drug. In the case of penicillin, for instance, it is not the penicillin molecule itself that causes the allergic response but a product, *benzylpenicilloyl.* Allergic reactions have been reported for every major type of antimicrobial drug, but the penicillins account for the greatest number of antimicrobial allergies, followed by the sulfonamides.

People who are allergic to a drug become sensitized to it during the first contact, usually without symptoms. Once the immune system is sensitized, a second exposure to the drug can lead to a reaction such as a skin rash (hives); respiratory inflammation; and, rarely, anaphylaxis—an acute, overwhelming allergic response that develops rapidly and can be fatal. This topic is discussed in greater detail in chapter 16.

Suppression and Alteration of the Microbiota by Antimicrobials

Most normal, healthy body surfaces, such as the skin, large intestine, outer openings of the urogenital tract, and oral cavity, provide numerous habitats for a virtual "garden" of microorganisms. These normal colonists or residents, called the microbiota, consist mostly of harmless or beneficial bacteria, but a small number of them can potentially be pathogens. We defer a more detailed discussion of this topic to chapter 13 and later chapters. Here we focus on the general effects of drugs on this population.

If a broad-spectrum antimicrobial drug is administered to treat infection, it will destroy microbes regardless of their roles in maintaining a balance, affecting not only the targeted infectious agent but also many others in sites far removed from the original infection **(figure 12.17).** When this therapy destroys beneficial resident species, resistant microbes that were once in small numbers begin

to overgrow and cause disease. This complication is called a **superinfection.**

Some common examples demonstrate how a disturbance in the microbiota leads to replacement pathogens and superinfection. A broad-spectrum cephalosporin used to treat a urinary tract infection by *E. coli* will cure the infection, but it will also destroy the lactobacilli in the vagina that normally maintain a protective acidic environment there. The drug has no effect, however, on *Candida albicans,* a yeast that also resides in normal vaginas. Released from the inhibitory environment provided by lactobacilli, the yeasts proliferate and cause symptoms. *Candida* can cause similar superinfections of the oropharynx (thrush) and the large intestine.

Oral therapy with tetracyclines, clindamycin, and broad-spectrum penicillins and cephalosporins is associated with a serious and potentially fatal condition known as *antibiotic-associated colitis* (pseudomembranous colitis). This condition is due to the overgrowth in the bowel of *Clostridium difficile,* an endospore-forming bacterium that is resistant to the antibiotic. It invades the intestinal lining and releases toxins that induce diarrhea, fever, and abdominal pain.

Refer to **table 12.9** for a general summary of major drug groups and their side effects.

TABLE 12.9	Major Adverse Toxic Reactions to Common Drug Groups
Antimicrobial Drug	**Primary Damage or Abnormality Produced**
Antibacterials	
Penicillin G	Skin allergy
Ampicillin	Diarrhea and enterocolitis
Cephalosporins	Inhibition of prothrombin synthesis
	Decreased circulation
	Nephritis
Tetracyclines	Diarrhea and enterocolitis
	Discoloration of tooth enamel
	Reactions to sunlight (photosensitization)
Chloramphenicol	Injury to red and white blood cell precursors
Aminoglycosides (streptomycin, gentamicin)	Diarrhea and enterocolitis; loss of hearing, dizziness, kidney damage
Azithromycin	Heart arrhythmias
Isoniazid	Hepatitis
	Seizures
	Dermatitis
Sulfonamides	Formation of kidney crystals; blockage of urine flow
	Hemolysis
	Reduction in number of platelets
Polymyxin	Damage to membranes of kidney tubules
	Weakened muscular responses
Quinolones (floxacins)	Headache, dizziness, GI distress, tendinitis
Rifampin	Damage to hepatic cells
	Dermatitis
Antifungals	
Amphotericin B	Disruption of kidney filtration
Flucytosine	Decreased number of white blood cells
Antiprotozoan Drugs	
Metronidazole	Nausea, vomiting
Chloroquine	Vomiting
	Headache
	Itching
Antihelminthics	
Niclosamide	Nausea, abdominal pain
Pyrantel	Irritation
	Headache, dizziness
Antivirals	
Acyclovir	Seizures, confusion
	Skin rash
Amantadine	Nervousness, light-headedness
	Nausea
AZT	Immunosuppression, anemia
Protease inhibitors	Hepatic damage
	Abnormal deposition of fats
	Diabetes

 Check Your Progress SECTION 12.5

20. Why do drugs that act on bacterial and fungal membranes generally have high toxicity?
21. Describe the series of events that gives rise to a superinfection, using an example.
22. What characteristic of antibiotics and other drugs causes hypersensitivity and allergic reactions?
23. Explain why it is better to use a narrow-spectrum drug instead of a broad-spectrum one, if possible.

12.6 The Process of Selecting an Antimicrobial Drug

 Expected Learning Outcomes

23. Outline some important factors to consider in drug therapy.
24. Describe the principles behind sensitivity testing for infectious agents.
25. Explain the uses of the minimum inhibitory concentration and the therapeutic index in drug selection.
26. Discuss medical considerations in final drug selection.
27. Summarize the problems in managing antimicrobial drugs.

Before actual antimicrobial therapy can begin, consideration must be given to at least three factors:

1. the nature of the microorganism causing the infection,
2. the degree of the microorganism's susceptibility (also called sensitivity) to various drugs, and
3. the overall medical condition of the patient.

Identifying the Agent

Identification of infectious agents from body specimens should be attempted as soon as possible. It is especially important that such specimens be taken before any antimicrobial drug is given, to prevent loss of the infectious agent. Direct examination of body fluids, sputum, or stool can provide a rapid detection method for bacteria or fungi. Newer methods of molecular diagnosis have greatly improved the speed of identification in many cases. A doctor often begins the therapy on the basis of such immediate findings. The choice of drug will be based on experience with drugs that are known to be effective against the microbe. For instance, if a sore throat appears to be caused by *Streptococcus pyogenes,* the physician may prescribe penicillin, because this species seems to be almost universally sensitive to it so far. If the infectious agent is not or cannot be isolated, epidemiological statistics may be required to predict the most likely agent in a given infection. For example, *Streptococcus pneumoniae* accounts for the majority of cases of meningitis in children, followed by *Neisseria meningitidis* and *Haemophilus influenzae.*

Kirby-Bauer Disc Diffusion Test

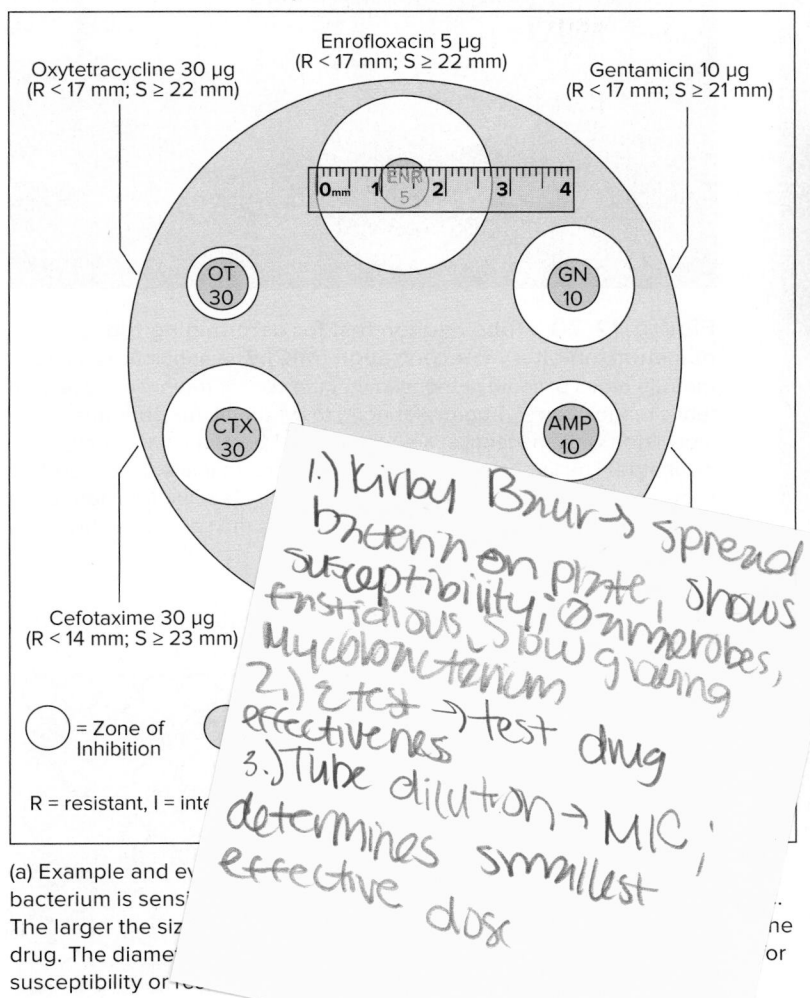

(a) Example and e~~xample~~
bacterium is sens~~itive~~
The larger the siz~~e~~
drug. The diame~~ter~~
susceptibility or r~~es~~

Figure 12.18 Techniques for prep~~aration~~ ~~of~~ **disc diffusion tests.** To see photographs of completed Kirby-Bauer tests, refer to the figure in 12.3 Making Connections. ~~(b~~) ~~valin,~~ MD, FRCP, Unité des Agents Antibactériens, Institut Pasteur, Paris, France;
(c) © Kathy Park Talaro

(handwritten notes overlaid:)
1.) Kirby Baur → spread baterin on plate, shows susceptibility; ① anaerobes, fastidious, slow growing Mycobacterium
2.) Etest → test drug effectivenes
3.) Tube dilution → MIC; determines smallest effective dose

(b) Result for gram-negative *Klebsiella pneumoniae,* with zones of inhibition indicating sensitivity to cefotaxime (CTX) and ampicillin and clavulanic acid (AMC). Observe the expanded zone between these two drugs that indicates their enhanced action, or synergy, when combined.

(c) Result for gram-positive *Staphylococcus aureus* using gentamicin (GM, left) and aztreonam (ATM, right). Although the bacteria are sensitive to both drugs, the GM has a few colonies growing in the zone of inhibition, indicating some cells that have developed resistance to that concentration of the drug.

Testing for the Drug Susceptibility of Microorganisms

Testing is essential in those groups of bacteria commonly showing resistance, primarily *Staphylococcus* species, *Neisseria gonorrhoeae, Streptococcus pneumoniae, Enterococcus faecalis,* and the aerobic gram-negative enteric bacilli. However, not all infectious agents require antimicrobial sensitivity testing. When certain groups, such as group A streptococci and most anaerobic bacteria, are known to be uniformly susceptible to penicillins, testing may not be necessary unless the patient is allergic to penicillin. Testing methods are available for fungi (see figure 12.21), protozoa, and viruses, although testing might not be done as often for these groups.

Selection of a proper antimicrobial agent begins by demonstrating the in vitro activity of several drugs against the infectious agent by means of standardized methods. In general, these tests involve exposing a pure culture of the bacterium to several different drugs and observing the effects of the drugs on growth.

The *Kirby-Bauer* technique is an agar diffusion test that provides useful data on antimicrobial susceptibility. In this test, the surface of a plate of special medium is spread with the test bacterium, and small discs containing a premeasured amount of antimicrobial are dispensed onto the bacterial lawn. After incubation, the zone of inhibition surrounding the discs is measured and compared with a standard for each drug **(figure 12.18).** The profile of antimicrobial sensitivity, or *antibiogram,* provides data for drug selection. The Kirby-Bauer procedure is less effective for bacteria that are anaerobic, highly fastidious, or slow-growing (*Mycobacterium*).

An alternative system that provides a quantitative rating of drug effectiveness is the *Etest*® **(figure 12.19).** In addition to providing a precise numerical rating (discussed in the next paragraph), this method allows testing for a wide variety of drugs and microbial types, including anaerobes, mycobacteria, and fungi.

More sensitive and quantitative results can also be obtained with tube dilution tests. First the antimicrobial is diluted serially in tubes of broth, and then each tube is inoculated with a small uniform sample of pure culture, incubated, and examined for growth (turbidity). The smallest concentration of drug in the series that visibly inhibits growth is called the **minimum inhibitory concentration,** or **MIC.** The MIC is useful in determining the smallest effective dosage of a drug and in providing a comparative index

Figure 12.19 Etest®: A quantitative method for determining sensitivity. This test uses a plastic strip that contains a premeasured gradient of a drug and is labeled with a scale that indicates varying concentrations of the drug in μg. The concentration ranges from higher at the top of the strip to lower toward the bottom. When the strips are placed on a plate swabbed with the test microbe and incubated, an oval area called the ellipse of inhibition develops adjacent to the strip. The lowest point on the scale where the ellipse begins to form corresponds with the minimum inhibitory concentration (IP= imepenem and TZ= ceftazidime). Etest® and the Etest gradient strip are registered trademarks of AB BIODISK. © Copyright AB BIODISK 2008. Re-printed with permission of AB BIODISK

Figure 12.20 Tube dilution test for determining the minimum inhibitory concentration (MIC). The antibiotic is diluted through tubes of liquid nutrient so that the concentration in successive tubes ranges from 6.4 μg/ml (far right) to 0.1 μg/ml. All tubes are inoculated with an identical amount of a test bacterium and then incubated. The first tube on the left is a control that lacks the drug and shows cloudiness indicative of normal growth. Reading from left to right, the MIC is the dilution of the first tube in the series that has no growth (is not cloudy). © Kathy Park Talaro

against other antimicrobials (**figures 12.20** and **12.21,** and **table 12.10**). In many clinical laboratories, these antimicrobial testing procedures are performed in automated machines that can test dozens of drugs and microbes simultaneously.

The MIC and the Therapeutic Index

The results of antimicrobial sensitivity tests guide the physician's choice of a suitable drug. If therapy has already commenced, it is imperative to determine if the tests bear out the use of that particular drug. Once therapy has begun, it is important to observe the

Figure 12.21 Microbroth dilution in a multiwell plate adapted for a pathogenic yeast. Drugs being tested are amphotericin B (AmB), flucytosine (5FC), fluconazole (Fluc), azole drugs (itraconazole [itra], ketoconazole [Keto], Voriconazole [vori], and caspofungin [caspo]). Pink indicates growth, blue indicates no growth. Reading across numbers 2–12 are increasing concentrations of the drugs in μg/ml. Row 1 has no drug. The first dilution with no growth is crossed out, indicating the MIC for that drug. Results are given in the box on the left. Image courtesy of David Ellis

TABLE 12.10	Comparative MICs (μg/ml) for Common Drugs and Pathogens				
Bacterium	Penicillin G	Ampicillin	Sulfamethoxazole	Tetracycline	Cefaclor
Staphylococcus aureus	4.0	0.05	3.0	0.3	4.0
Enterococcus faecalis	3.6	1.6	100.0	0.3	60.0
Neisseria gonorrhoeae	0.5	0.5	5.0	0.8	0.2
Escherichia coli	100.0	12.0	3.0	1–4.0	3.0
Pseudomonas aeruginosa	>500.0	>200.0	>100.0	>100.0	4.0
Salmonella species	12.0	6.0	10.0	1.0	0.8
Clostridium perfringens	0.16	NA	NA	3.0	12.0

NA = not available

patient's clinical response, because the in vitro activity of the drug is not always correlated with its in vivo effect. When antimicrobial treatment fails, the failure is usually due to

1. the inability of the drug to diffuse into that body compartment (the brain, joints, skin);
2. a few resistant cells in the culture that did not appear in the sensitivity test; or
3. an infection caused by more than one pathogen (mixed), some of which are resistant to the drug.

If therapy does fail, a different drug, combined therapy, or a different method of administration must be considered.

Many factors influence the choice of an antimicrobial drug besides microbial sensitivity to it. The nature and spectrum of the drug, its potential adverse effects, and the condition of the patient can be critically important. When several antimicrobial drugs are available for treating an infection, final drug selection advances to a new series of considerations. In general, it is better to choose the narrowest-spectrum drug of those that are effective if the causative agent is known. This decreases the potential for superinfections and other adverse reactions.

Because drug toxicity is of concern, it is best to choose the drug with high selective toxicity for the infectious agent and low human toxicity. The **therapeutic index (TI)** is defined as the ratio of the dose of the drug that is toxic to humans as compared to its minimum effective (therapeutic) dose. The closer these two figures are (the smaller the ratio), the greater is the potential for toxic drug reactions. For example, a drug that has a therapeutic index of

$$\frac{10\,\mu g/ml: \text{toxic dose}}{9\,\mu g/ml\,(\text{MIC})} \quad \boxed{TI = 1.1}$$

is a riskier choice than one with a therapeutic index of

$$\frac{10\,\mu g/ml}{1\,\mu g/ml} \quad \boxed{TI = 10}$$

Drug companies recommend dosages that will inhibit the microbes but not adversely affect patient cells. When a series of drugs being considered for therapy have similar MICs, the drug with the highest therapeutic index usually has the widest margin of safety.

Patient Factors in Choosing an Antimicrobial Drug

The physician must also take a careful history of the patient to discover any preexisting medical conditions that will influence the activity of the drug or the response of the patient. A history of allergy to a certain class of drugs should preclude the administration of that drug and any drugs related to it. Underlying liver or kidney disease will ordinarily necessitate the modification of drug therapy, because these organs play such an important part in metabolizing or excreting the drug. Infants, the elderly, and pregnant women require special precautions. For example, age can diminish gastrointestinal absorption and organ function, and most antimicrobial drugs cross the placenta and could affect fetal development.

The intake of other drugs must be carefully scrutinized, because incompatibilities can result in increased toxicity or failure of one or more of the drugs. For example, the combination of aminoglycosides and cephalosporins increases nephrotoxic effects; antacids reduce the absorption of isoniazid; and the interaction of tetracycline or rifampin with oral contraceptives can abolish the contraceptive's effect. Some drugs (penicillin with certain aminoglycosides, or amphotericin B with flucytosine) act synergistically, so that reduced doses of each can be used. Other concerns in choosing drugs include any genetic or metabolic abnormalities in the patient, the site of infection, the route of administration, and the cost of the drug.

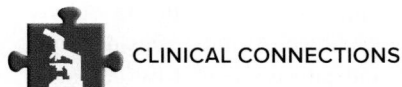 **CLINICAL CONNECTIONS**

An Antimicrobial Drug Dilemma

We began this chapter with a view of the exciting strides made in antimicrobial therapy, but we must be realistic about its downside. There is now a worldwide problem in the management of antimicrobial drugs, which rank second only to nervous system drugs in overall usage. The remarkable progress in treating many infectious diseases has spawned a view of antimicrobials as a "cure-all" for infections as diverse as the common cold and acne. And although it is true that nothing is as dramatic as curing an infectious disease with the correct antimicrobial drug, in many instances drugs have no effect and can even be harmful. The depth of this problem can perhaps be appreciated better with a few facts:

1. Roughly 200 million prescriptions for antimicrobials are written in the United States every year. One study found that 75% of antimicrobial prescriptions are for pharyngeal, sinus, lung, and upper respiratory infections. A fairly high percentage of these are viral in origin, and antibacterial drugs will have little or no effect.
2. Drugs are often prescribed without benefit of culture or susceptibility testing, even when such testing is clearly warranted.
3. Therapy is often applied in a "shotgun" approach for minor infections, which involves administering a broad-spectrum drug that destroys our protective resident microbes instead of a more specific narrow-spectrum one. This practice can lead to superinfections and other adverse reactions.
4. More expensive newer drugs are chosen when a less costly older one would be just as effective.
5. Tons of excess antimicrobial drugs produced in this country are exported to other countries, where controls are not as strict. Nearly 200 different antibiotics are sold over the counter in Latin American and Asian countries. Self-medication without understanding the correct medical indication is largely ineffectual, and unhealthy, but, worse yet, it is partly responsible for emergence of drug-resistant bacteria that subsequently cause epidemics.

The medical community recognizes the motivation for immediate therapy to protect a sick patient and for defensive medicine to provide the very best care possible. But in the final analysis, every allied health professional should be critically aware of not only the admirable and utilitarian nature of antimicrobials but also their limitations.

What efforts can people make in their everyday lives to help to solve these problems? *Answer available on Connect.*

Even when all the information is in, the final choice of a drug is not always straightforward. Consider the case of an elderly alcoholic patient with pneumonia caused by *Klebsiella* and complicated by diminished liver and kidney function. All drugs must be given parenterally because of prior damage to the gastrointestinal lining and poor absorption. Drug tests show that the infectious agent is sensitive to fourth-generation cephalosporins, gentamicin, imipenem, and ticarcillin. The patient's history shows previous allergy to the penicillins, so these would be ruled out. Cephalosporins are associated with serious bleeding in elderly patients, so this may not be a good choice. Aminoglycosides such as gentamicin are nephrotoxic and poorly cleared by damaged kidneys. Imipenem is probably the best choice because of its broad spectrum and low toxicity.

 Check Your Progress SECTION 12.6

24. How are dilution susceptibility tests and disc diffusion tests used to determine microbial drug sensitivity?
25. Briefly describe the Kirby-Bauer test and its purpose.
26. If a technician examined a Kirby-Bauer assay and found individual colonies of the plated microbe growing within the zone of inhibition for a drug, what should her conclusion be?
27. Explain the relationship between the minimum inhibitory concentration and the therapeutic index. What is optimum for each?
28. Discuss the general factors to consider in selecting an appropriate antimicrobial drug.

 CASE STUDY Part 2

After a few weeks, Nurse Mitchell was given a routine skin test for tuberculosis, and rather surprisingly, developed a raised red patch that indicated TB positive. A new specimen taken from the wound verified *Mycobacterium tuberculosis*. Unknown to the medical staff was the fact that the AIDS patient's lungs were infected with both *Pneumocystis* **and** tuberculosis. This now explained his sudden death, because tuberculosis is a deadly risk factor for AIDS patients, and he had not received any therapy for TB. The blood in the needle must have contained a high level of tubercle bacilli that were inoculated into the nurse's arm when the needle penetrated the skin.

Diagnosis of the tuberculosis skin infection called for an entirely different drug regimen. The dicloxacillin given early on was ineffective because this relative of penicillin does not have a mode of action that works on *M. tuberculosis*. Major problems with drug treatment for tuberculosis are that (1) the pathogen can already have or can develop drug resistance,

and (2) the TB pathogen takes longer to eliminate than most other pathogenic bacteria, even with effective drugs.

This meant that it was necessary to test for drug sensitivity, to choose drugs that would prevent resistance from happening, and to administer the drugs for several months. When tests indicated that the culture was sensitive to isoniazid, rifampin, and pyrazinamide, the nurse was started on this three-drug combined pill for 2 months, followed by 4 additional months on isoniazid and rifampin alone. By the end of this treatment, the infection was cured and Nurse Mitchell was relieved of her worst fears when a final HIV test came out negative.

■ *Suggest some reasons the cause of Nurse Mitchell's infection was not immediately evident.*

■ *What are the purposes of doing drug sensitivity tests for TB?*

To conclude this Case Study, go to Connect.

For more information on infections from needle sticks, go to http://emedicine.medscape.com/article/784812-overview.

 Chapter Summary with Key Terms

12.1 Principles of Antimicrobial Therapy
 A. **Chemotherapeutic drugs** are used to control microorganisms in the body. Depending on their source, these drugs are described as **antibiotics,** either **semisynthetic** or **synthetic.**
 B. Based on their mode and spectrum of action, they are described as **broad spectrum** or **narrow spectrum** and microbistatic or microbicidal.
 C. Strategic approaches to the use of chemotherapeutics include
 1. **Prophylaxis,** where drugs are administered to *prevent* infection in susceptible people.
 2. **Combined therapy,** where two or more drugs are given simultaneously, either to prevent the emergence of resistant species or achieve **synergism.**
 D. *Interactions Between Drugs and Microbes*
 1. The ideal antimicrobial is **selectively toxic,** highly potent, stable, and soluble in the body's tissues and fluids.
 2. It does not disrupt the immune system or microflora of the host and is exempt from drug resistance.

12.2 Survey of Major Antimicrobial Drug Groups
 A. *Drugs That Act on the Cell Envelope*
 1. **Penicillins** are beta-lactam-based drugs originally isolated from the mold *Penicillium chrysogenum.* The natural form is penicillin G, although various semisynthetic forms, such as ampicillin and methicillin, vary in their spectrum and applications.
 a. Penicillin is bactericidal, blocking completion of the cell wall, which causes eventual rupture of the cell.

b. Major problems encountered in penicillin therapy include allergic reactions and bacterial resistance to the drug through **beta-lactamase.**
2. **Cephalosporins** include both natural and semisynthetic drugs initially isolated from the mold *Cephalosporium*. Cephalosporins inhibit peptidoglycan synthesis (like penicillin) but have a much broader spectrum.
3. Bacitracin and **polymyxin** are narrow-spectrum antibiotics.
 a. Bacitracin prevents cell wall synthesis in gram-positive organisms and is used in antibacterial skin ointments.
 b. Polymyxin interferes with the cell membrane. It is added to skin ointments and can be used to treat *Pseudomonas* infections.
4. Vancomycin interferes with the early stages of cell wall synthesis and is used for life-threatening, methicillin-resistant staphylococcal infection.
5. Isoniazid (INH) blocks the synthesis of cell wall components in *Mycobacteria*. It is used in the treatment of tuberculosis.
B. *Drugs That Affect Nucleic Acid Synthesis*
 1. Fluoroquinolones (ciprofloxacin) represent a newer class of broad-spectrum synthetic drugs that have proven useful in a variety of infections.
 2. Rifampin interferes with RNA polymerase (thereby affecting transcription) and is primarily used for tuberculosis and leprosy infections.
C. *Drugs That Inhibit Protein Synthesis*
 1. **Aminoglycosides** include several narrow-spectrum drugs isolated from unique bacteria found in the genera *Streptomyces*. Examples include streptomycin, gentamicin, tobramycin, and amikacin.
 2. **Tetracyclines** and chloramphenicol are very broad-spectrum drugs isolated from *Streptomyces*. Both interfere with translation but their use is limited by adverse effects.
 3. Erythromycin and clindamycin both affect protein synthesis. Erythromycin is a broad-spectrum alternative for use with penicillin-resistant bacteria, while clindamycin is used primarily for intestinal infection by anaerobes.
D. *Antimetabolite Drugs*
 1. **Sulfonamides** are synthetic drugs that act as **metabolic analogs, competitively inhibiting** enzymes needed for nucleic acid synthesis.
 2. Trimethoprim is often used in combination with sulfa drugs.
 3. Dapsone is a narrow-spectrum drug used (often in combination with rifampin) to treat leprosy.
E. *Newly Developed Antibiotics* are useful when bacteria have developed resistance to more traditional drugs.
 1. Fosfomycin inhibits cell wall synthesis and is effective against gram-negative organisms.
 2. Synercid inhibits translation and is most effective against gram-positive cocci.
 3. Ketek is a new ketolide antibiotic with broad-spectrum effects.
 4. Oxazolidinones are another new class of synthetic drug that work by inhibiting the start of protein synthesis. They are useful for treatment of staphylococci that are resistant to other drugs (MRSA and VRE).

12.3 Drugs to Treat Fungal, Parasitic, and Viral Infections
A. *Antifungal Drugs*
 1. Amphotericin B and nystatin disrupt fungal membranes by detergent action.
 2. **Azoles** are synthetic drugs that interfere with membrane synthesis. They include ketoconazole, fluconazole, and miconazole.
 3. Flucytosine inhibits DNA synthesis. Due to fungal resistance, flucytosine must usually be used in conjunction with amphotericin.
B. *Drugs for Protozoan Infections*
 1. Quinine or the related compounds chloraquine, primaquine, or mefloquine are used to treat infections by the malarial parasite *Plasmodium*.
 2. Other antiprotozoan drugs include metronidazole, suramin, melarsopral, and nitrifurimox.
C. *Drugs for Helminth Infections* include mebendazole, thiabendazole, praziquantel, pyrantel, piperizine, and niclosamide.
D. *Drugs for Viral Infections* inhibit viral penetration, multiplication, or assembly. Because viral and host metabolism are so closely related, toxicity is a potential adverse reaction.
 1. **Acyclovir,** valacyclovir, famciclovir, and ribavirin act as nucleoside analogs, inhibiting viral DNA replication, especially in herpesviruses.
 2. Tamiflu acts to stop uncoating of the influenza A virus.
 3. Two varieties of reverse transcriptase inhibitors, protease inhibitors, integrase inhibitors, and entry/fusion inhibitors are the main categories of anti-HIV drugs, usually given in combinations of two or three.
 4. **Interferon** is a naturally occurring protein with both antiviral and anticancer properties.

12.4 Interactions Between Microbes and Drugs: The Acquisition of Drug Resistance
A. Microbes can lose their sensitivity to a drug through the acquisition of **resistance (R) factors.** Drug resistance takes the form of drug inactivation, decreased permeability to drug or increased elimination of drug from the cell, change in drug receptors, or change of metabolic patterns.
B. New approaches to antimicrobial therapy include the use of **probiotics** and prebiotics.

12.5 Interactions Between Drugs and Hosts
Side effects of chemotherapy include organ toxicity, allergic responses, and alteration of microflora.

12.6 The Process of Selecting an Antimicrobial Drug
A. Rapid identification of the infectious agent is important.
B. The pathogenic microbe should be tested for its susceptibility to different antimicrobial agents. This involves using standardized methods such as the *Kirby-Bauer* and **minimum inhibitory concentration (MIC)** techniques. The **therapeutic index** factors in the toxicity of a drug as compared with its MIC so as to guide proper drug selection.
C. Medical conditions of the patient such as age, allergy, disease, and pregnancy are also important guides.
D. The inappropriate use of drugs on a worldwide basis has led to numerous medical and economic problems.

Level I. Knowledge and Comprehension

These questions require a working knowledge of the concepts in the chapter and the ability to recall and understand the information you have studied.

? Multiple-Choice Questions

Select the correct answer from the answers provided. For questions with blanks, choose the combination of answers that most accurately completes the statement.

1. A compound synthesized by bacteria or fungi that destroys or inhibits the growth of other microbes is a(n)
 a. synthetic drug
 b. antibiotic
 c. antimicrobial drug
 d. competitive inhibitor

2. Which statement is *not* an aim in the use of drugs for antimicrobial chemotherapy? The drug should
 a. have selective toxicity
 b. be active even in high dilutions
 c. be broken down and excreted rapidly
 d. be microbicidal

3. Microbial resistance to drugs is acquired through
 a. conjugation
 b. transformation
 c. transduction
 d. all of these

4. R factors are _____ that contain a code for _____.
 a. genes, replication
 b. plasmids, drug resistance
 c. transposons, interferon
 d. plasmids, conjugation

5. When a patient's immune system becomes reactive to a drug, this is an example of
 a. superinfection
 b. drug resistance
 c. allergy
 d. toxicity

6. An antibiotic that disrupts the normal flora can cause
 a. the teeth to turn brown
 b. aplastic anemia
 c. a superinfection
 d. hepatotoxicity

7. Most antihelminthic drugs function by
 a. weakening the worms so they can be flushed out by the intestine
 b. inhibiting worm metabolism
 c. blocking the absorption of nutrients
 d. inhibiting egg production
 e. all of these

8. Select a drug or drugs that can prevent a viral nucleic acid from being replicated.
 a. retrovir
 b. acyclovir
 c. saquinavir
 d. both a and b

9. Which of the following effects do antiviral drugs *not* have?
 a. destroying extracellular viruses
 b. stopping virus synthesis
 c. inhibiting virus maturation
 d. blocking virus receptors

10. Which of the following modes of action would be most selectively toxic?
 a. interrupting ribosomal function
 b. dissolving the cell membrane
 c. preventing cell wall synthesis
 d. inhibiting DNA replication

11. The MIC is the _____ of a drug that is required to inhibit growth of a microbe.
 a. largest concentration
 b. standard dose
 c. smallest concentration
 d. lowest dilution

12. An antimicrobial drug with a _____ therapeutic index is a better choice than one with a _____ therapeutic index.
 a. low, high
 b. high, low

13. **Matching.** Select the mode of action for each drug in the left column; all but one choice is used.
 - *d* sulfonamides
 - *a* zidovudine
 - *g* penicillin
 - *b* tetracycline
 - *f* erythromycin
 - *h* quinolone
 - *c* relenza
 - *e* polymyxin

 a. reverse transcriptase inhibitor
 b. blocks the attachment of tRNA on the ribosome
 c. interferes with viral budding and release
 d. interferes with synthesis of folic acid
 e. breaks down cell membrane integrity
 f. prevents the ribosome from translocating
 g. blocks synthesis of peptidoglycan
 h. inhibits DNA gyrase

📖 Case Study Review

1. What was the purpose of giving postexposure prophylaxis to the nurse?
 a. to treat AIDS
 b. to prevent pneumocystis pneumonia
 c. to prevent HIV replication in cells
 d. to stimulate an immune response

2. What is a probable reason that tuberculosis therapy must continue for several months?
 a. to prevent drug resistance
 b. to keep the infection from spreading
 c. the TB bacillus grows very slowly
 d. the drugs are slow acting

3. Summarize the major problems with drug therapy as demonstrated in this case.

Writing Challenge

For each question, compose a one- or two-paragraph answer that includes the factual information needed to completely address the question. Check Your Progress questions can also be used for writing-challenge exercises.

1. Using the diagram as a guide, briefly explain how the three factors in drug therapy interact.

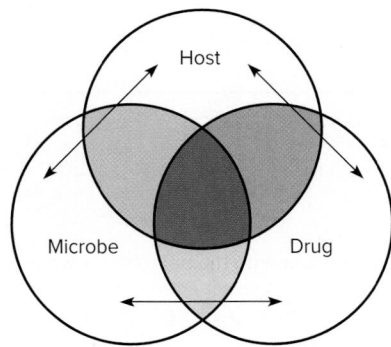

Host

Microbe Drug

2. Observe table 12.4 with regard to type of microbe and structures or functions that are affected by drugs. Write a paragraph or two that explains the different ways in which drugs can affect microbes and how this affects selectivity of drugs and their spectrum.

3. Drugs are often given to patients before going into surgery, to dental patients with heart disease, or to healthy family members exposed to contagious infections.
 a. What word would you use to describe this use of drugs?
 b. What is the purpose of this form of treatment?
 c. Explain some potential undesired effects of this form of therapy.
 d. Define probiotics and list some ways they are used.

4. Write an essay covering some of the main concerns in antimicrobial drug therapy, including resistance, allergies, superinfections, and other adverse effects.

5. a. Explain the basis for combined therapy.
 b. What are some reasons it could be helpful to use combined therapy in treating HIV infection?

6. Explain the kinds of tests that would differentiate between a broad or narrow spectrum antibiotic.

7. Summarize the primary reasons that we find ourselves in a "drug dilemma" with regard to antimicrobial drugs.

Concept Mapping

On Connect you can find an Introduction to Concept Mapping that provides guidance for working with concept maps and a sample of a concept map for this chapter.

Level II. Application, Analysis, Evaluation, and Synthesis

These problems go beyond just restating facts and require higher levels of understanding and an ability to interpret, problem solve, transfer knowledge to new situations, create models, and predict outcomes.

Critical Thinking

Critical thinking is the ability to reason and solve problems using facts and concepts. These questions can be approached from a number of angles, and in most cases, they do not have a single correct answer.

1. In 2015 the WHO surveyed 10,000 people on the subject of drug resistance. Seventy-five percent of them said that antibiotic resistance is due to the body becoming resistant to the drug. Explain how you would respond to someone having this misconception.

2. Explain a simple test one could do to determine if drug resistance is developing in a culture.

3. a. Your pregnant neighbor has been prescribed a daily dose of oral tetracycline for acne. Do you think this therapy is advisable for her? Why or why not?
 b. A woman has been prescribed a broad-spectrum oral cephalosporin for a strep throat. What are some possible consequences in addition to cure of the infected throat?

 c. A man has a severe case of sinusitis that is negative for bacterial pathogens. A physician prescribes an oral antibacterial drug for treatment. What is right or wrong with this therapy?

4. You have been directed to take a sample from a growth-free portion of the zone of inhibition in the Kirby-Bauer test and inoculate it onto a plate of nonselective medium.
 a. What does it mean if growth occurs on the new plate?
 b. What if there is no growth?

5. From the results shown in figure 12.21, determine which drugs could be used to treat the yeast infection. Taking into account their adverse side effects, which drug would probably be the best choice?

6. Explain why drugs that interfere with the prokaryotic ribosome can have harmful side effects on a human patient.

7. In cases in which it is not possible to culture or drug-test an infectious agent (such as middle ear infection), how would the appropriate drug be chosen?

8. Reviewing drug characteristics, choose an antimicrobial for each of the following situations (explain your choice):
 a. for an adult patient suffering from *Mycoplasma* pneumonia
 b. for a child with bacterial meningitis (drug must enter into cerebrospinal fluid)

c. for a patient who has chlamydiosis and an allergy to azithromycin
 d. for PPNG drug-resistant gonorrhea

9. a. Using table 12.10 as a reference, find and explain the differences between the results for *Staphylococcus* and *Pseudomonas*. Are any of these drugs broad spectrum?
 b. Explain the difference in the MIC of *E. coli* for penicillin versus ampicillin.
 c. Refer to figure 12.18a, take measurements, and interpret the results.

Visual Challenge

For the following figures a–e, research the chapter and book to find an appropriate drug to treat an infection with the microbe shown, and explain what the drug's effects on the microbe will be.

(a) Figure 18.1 (b) Figure 6.8b (c) Figure 5.15c

Envelope Capsid DNA core

Nucleus Bud scars

(d) Figure 5.26c (e) Figure 5.28

Female Anus Male Eggs

(a, b): © Eye of Science/Science Source

13 Microbe-Human Interactions:
Infection, Disease, and Epidemiology

A view of hundreds of Ebola viruses budding off the surface of an infected cell. The filamentous and wavy nature of the virus particles is a typical characteristic (25,000x). *Source:* National Institute of Allergy and Infectious Diseases (NIAID)/CDC

Medical personnel in Africa preparing to assist in containing the epidemic that devastated parts of West Africa in 2014. Note that the hazmat suits cover the workers completely so that no body surfaces are exposed to infected patients or their fluids. *Source:* Lindsey Horton/CDC

CASE STUDY Part 1 Fatal Filaments from Faraway Africa

In late December 2013 an 18-month-old child in a remote village in Guinea, West Africa, came down with an **acute** illness marked by high **fever,** severe **diarrhea,** and **vomiting.** He died two days later without being diagnosed. Within 2 weeks several family members and other villagers developed a similar illness that also led to fatalities. New **cases** began to appear in health care workers who had treated some of the sick patients. Local medical officers began to suspect a cholera **outbreak** based on the condition of the patients. This mystery **disease** escaped more serious notice until a member of the dead child's family traveled to a larger city, became ill, and died in the hospital. Local authorities were concerned but made no attempt to formally report it. When they suspected the unidentified disease might be **Lassa fever,** an outreach center for the **World Health Organization** (WHO) tested a blood sample and identified a **filovirus.** It took until March 2014 for them to finally clarify the diagnosis as **Ebola virus disease;** by then there had been 49 cases and 29 deaths.

Unfortunately the **virulence** of this virus, its small **infectious dose,** and the unpredictability of the **incubation period** were working against the patients and their caretakers. Very little attempt had been made to **isolate** the patients, and many people in both the clinics and the communities had been exposed. What had started as a local outbreak was becoming primed to explode into an Ebola **epidemic.** And that is precisely what happened. Within a month, cases appeared in the small African nations of Liberia and Sierra Leone, and by early summer of 2014 the disease had spread to several other countries and even beyond the continent of Africa. With thousands of cases and hundreds of deaths, a public health emergency

> "What had started as a local outbreak was becoming primed to explode into an Ebola epidemic."

was finally declared by officials in August 2014. Even with great attention being focused on stopping the spread of Ebola, there were not enough safeguards to prevent it from finding other exit routes.

In late September 2014 a Liberian citizen, Thomas Duncan, traveled to Texas to visit family. Although he arrived in a healthy state, within a few days he began to have nonspecific **symptoms** of fever, abdominal pain, and nasal drainage and sought treatment at a hospital. He was given antibiotics and sent home, but he was not tested for infection or questioned about his recent **travel history.** Soon after, he returned to the hospital with more severe symptoms of vomiting, diarrhea, muscle pain, and weakness. Once medical personnel learned he had flown in from Africa, he was placed in **strict isolation** and tested for Ebola virus. When the result came back as positive, the hospital immediately initiated protocols as required by the **Centers for Disease Control and Prevention** (CDC) to use personal protective gear (masks, gowns, gloves, biosafety suits) and **universal precautions,** but workers were not fully trained in techniques specific to Ebola. After a week of constant care, Mr. Duncan died. At some time during his stay, caretakers were exposed directly to his **infectious fluids.**

- *Explain why it took so long for the African outbreak to be recognized and acted upon.*

- *Give some reasons the protective gear did not prevent infection in the American nurses.*

To continue the Case Study, go to Case Study Part 2 at the end of the chapter.

SCOPING OUT THE CHAPTER

The lives of microbes and humans intersect in numerous ways, but none is more profound than the intimate contact with the 100 trillion microbes that occupy our bodies. These residents influence many aspects of our development and body functions. And to balance out this partnership, we nourish them and provide them with abundant habitats. Yet with all of its benefits, there is a downside to this close association, because sometimes we encounter pathogens that invade us and cause damage and disease. The goals of this chapter are to further explore the entire range of the microbe/human relationship that can provide both powerful benefits and do serious harm.

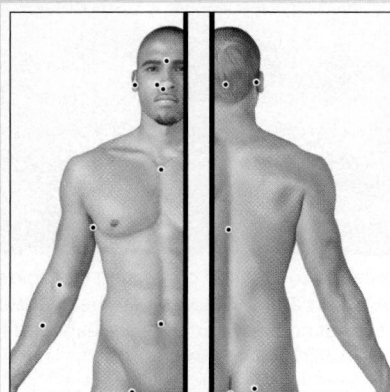

The Microbial Menagerie Our bodies are home to a complex community of tiny creatures that live and feed and reproduce there. All major groups are represented in the microbiota, from bacteria and viruses to fungi, protozoa, and archaea. This evolving coexistence has become an essential contributor to our nutrition, immunities, and general health.

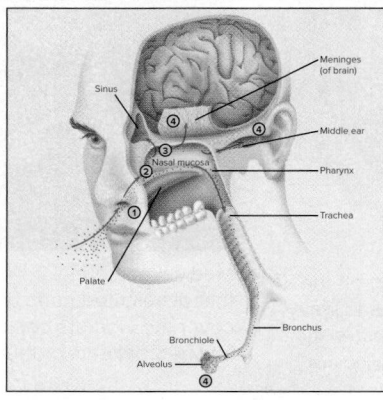

Infection Portals and Pathways Through the chain of infection, pathogens ultimately gain access to sterile tissues and grow there. Many exposed areas of the body provide an entrance that leads the way to vulnerable targets within the body. This figure depicts how the pneumococcus first enters the nasal passages and spreads from there to the lungs, ear, or brain.

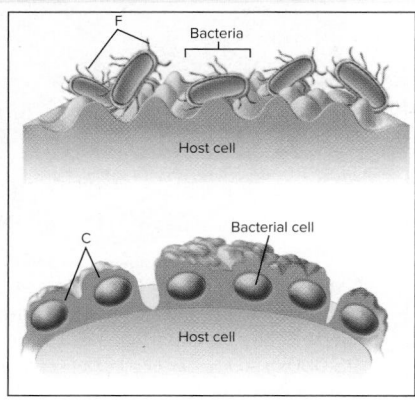

Tricks Invaders Play on Their Hosts Microbes come armed with a variety of special appendages such as capsules, hooks, or fibers for clinging to host cells and holding them in place. Once pathogens are successfully established in a tissue, they can begin the process of invasion.

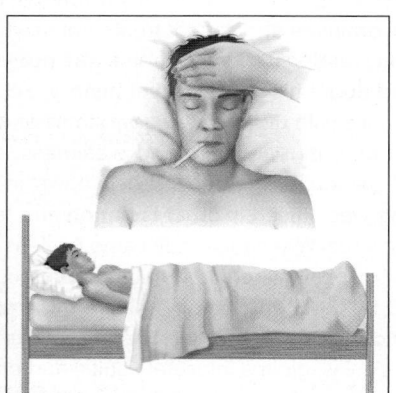

Toxins and Enzymes and Damage, Oh My! Once they reach a target site and begin to multiply, pathogens release factors that aid in their invasion or avoidance of host defenses. Many of these substances cause the damage that leads to signs and symptoms of disease.

How Microbe... Hosts In a late... infection, infect... from the body t... as body fluids... provides patho... new host and i... for the microbe's survival and spread.

...ens and ...can be ...t perspectives. ...their origins, ...o is infected, ...te of infection, ...ocation of outbreaks, and how they are recorded and reported.

[handwritten note:] Normal resident microbiota = indigenous; colonized the body short period of time = transients

13.1 We Are Not Alone

 Expected Learning Outcomes

1. Describe some of the major interactions between humans and the microbes that share our habitats.
2. Identify and define the terms associated with infectious diseases.
3. Discuss the characteristics of the normal microbiota and the types of functions they serve.
4. Briefly relate the sources and conditions that influence the development of microbiota in the major body systems.
5. Identify which bodily sites remain free of living organisms, and explain why this is necessary.
6. Explain the germ-free condition and how it is important to microbiologists.

The human body exists in a state of dynamic equilibrium with microorganisms. The healthy individual is able to maintain a balanced coexistence with them. But on occasion the balance is disrupted, allowing microorganisms to cause an infection or disease. In this chapter we explore each component of the human–microbe relationship, beginning with the nature and function of normal microbial residents, moving to the stages of infection and disease, and closing with a study of epidemiology and the patterns of disease in populations. These topics will set the scene for chapters 14 and 15, which deal with host defenses and the protections we use to fight off the agents of infectious disease.

Many of the interactions between the human body and microorganisms involve the development of biofilms (see 4.1 Making Connections). These associations result from the binding of microbial cells to our cells and to those of other microorganisms to form complex partnerships. As we discuss topics in this chapter, keep in mind that colonization of the body involves a constant "give and take" between us and them. The most significant of these interactions can be summarized as follows:

- Microbes provide a protective and stabilization effect on body surfaces by establishing themselves as normal residents.
- Microbes are involved in maturation of host defenses and development of the immune system.
- Microbes can invade and grow in sterile tissues, where they cause disease by damaging tissues and organs.

Contact, Colonization, Infection, Disease

In chapter 7 we first considered several of the basic interrelationships between humans and microorganisms. Most of the microbial inhabitants benefit from the nutrients and protective habitat the body provides. Our microbial partnerships run the gamut from mutualism to commensalism to parasitism and can have beneficial, neutral, or harmful effects.

Microbes that engage in mutual or commensal associations with humans belong to the **normal resident microbiota.** Other terms commonly used to refer to these residents are **indigenous***

indigenous (in-dih′-juh-nus) Belonging or native to.

microflora, normal flora, and commensals. Even though these microbes have colonized the body, they are more or less restricted to the outer surfaces without penetrating into sterile tissues or fluids. When a microbe has penetrated the host defenses, invaded sterile tissues, and multiplied, the result is an **infection,** and the microbe is considered an infectious agent or **pathogen.** If the infection causes damage or disruption to tissues and organs, it is now considered an **infectious disease.** Note that the term *disease* is defined as any deviation from health. There are hundreds of diseases caused by such factors as infections, diet, genetics, and aging. In this chapter, however, we discuss only diseases arising from the disruption of a tissue or organ caused by microbes or their products.

Figure 13.1 provides a diagrammatic view of the potential phases in infection and disease. Because of numerous factors relating to host resistance and degree of pathogenicity, not all contacts lead to infection and not all infections lead to disease. In fact, contact without infection and infection without disease are the rule. Before we consider further details of infection and disease, let us examine what happens when microbes colonize the body, thereby establishing a long-term, usually beneficial, relationship.

Resident Microbiota: The Human as a Habitat

The human body offers a seemingly endless variety of microenvironments and niches, with variations in temperature, pH, nutrients, moisture, and oxygen tension from one area to another. With such a wide range of habitats, it should not be surprising that the body supports an abundance of microbes. In fact, it is so favorable that our bodies hold 3 to 10 times more microbes than our own cells. The importance of this relationship has recently become the focus of a comprehensive initiative (**13.1 Secret World of Microbes**).

As shown in **table 13.1,** most areas of the body in contact with the outside environment harbor resident microorganisms. Mucosal surfaces provide a particularly attractive surface for attachment. Most other body sites, including internal organs, tissues, and the fluids they contain, are generally microbe-free (**table 13.2**). In fact, the presence of microbes in these areas is usually indicative of infection.

The vast majority of microbes that come in contact with the body are removed or destroyed by the host's defenses long before they are able to colonize a particular area. Those organisms that occupy the body for only short periods are known as **transients.** The remaining microbes that do become established more permanently are considered **residents.** One of their most important adaptations is to avoid the attention of the body's defenses. The resident organisms have coevolved along with their human hosts toward a complex relationship in which the effects of normal microflora are generally not harmful to the host, and vice versa.

Although generally stable, the microbiota fluctuates with general health, age, variations in diet, hygiene, hormones, and drug therapy. In many cases the microbiota actually benefits the human host by preventing the overgrowth of harmful microorganisms.

Quick Search Seek out the video "The Human Microbiome and What We Do to It" on YouTube.

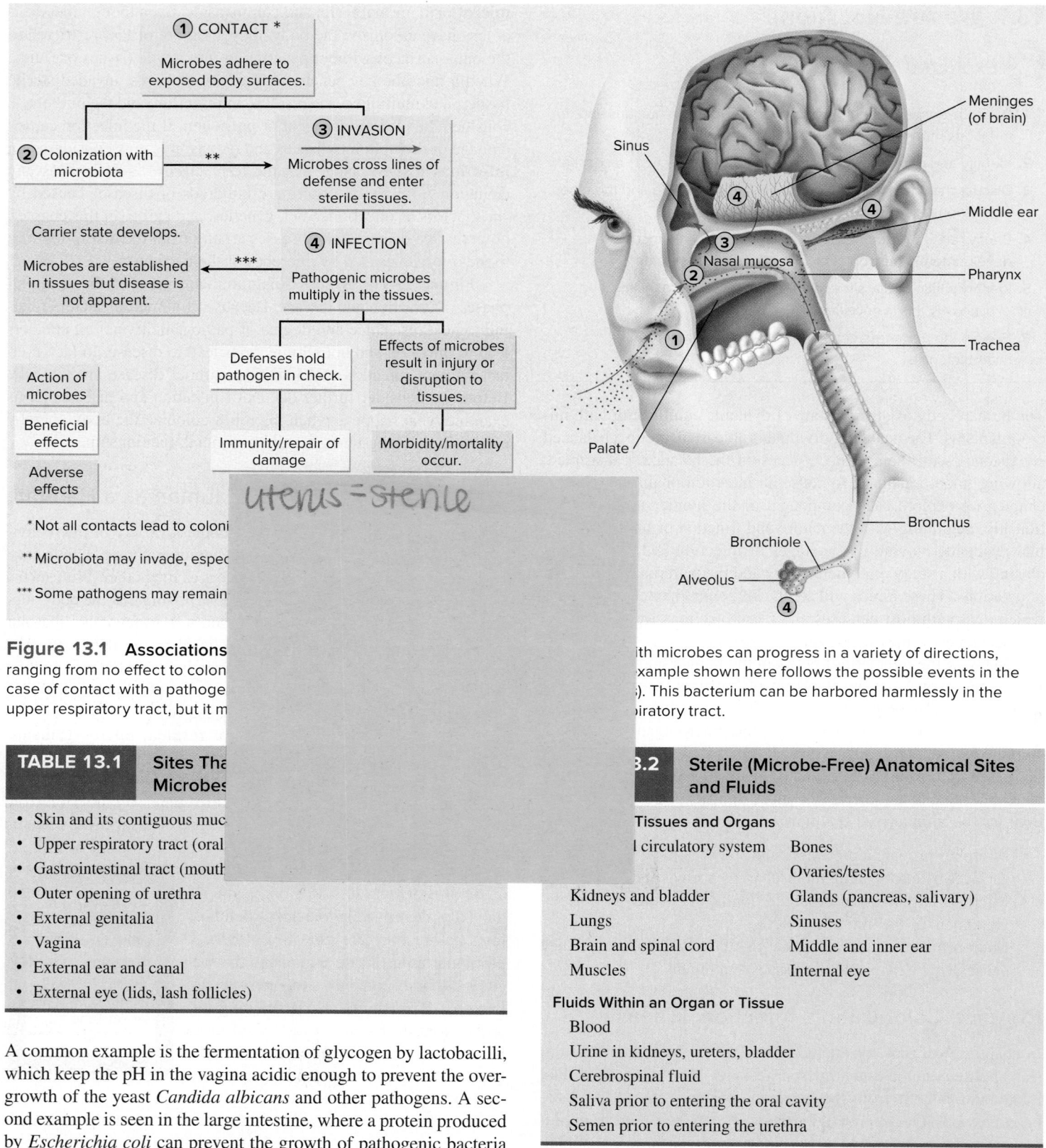

① CONTACT *
Microbes adhere to exposed body surfaces.

② Colonization with microbiota

③ INVASION
Microbes cross lines of defense and enter sterile tissues.

④ INFECTION
Pathogenic microbes multiply in the tissues.

Carrier state develops.
Microbes are established in tissues but disease is not apparent.

Action of microbes
Beneficial effects
Adverse effects

Defenses hold pathogen in check.

Effects of microbes result in injury or disruption to tissues.

Immunity/repair of damage

Morbidity/mortality occur.

* Not all contacts lead to coloni...

** Microbiota may invade, espec...

*** Some pathogens may remain...

utenis = sterile

Sinus
Meninges (of brain)
Middle ear
Nasal mucosa
Pharynx
Trachea
Palate
Bronchus
Bronchiole
Alveolus

Figure 13.1 Associations ... ith microbes can progress in a variety of directions, ranging from no effect to colon... xample shown here follows the possible events in the case of contact with a pathoge... s). This bacterium can be harbored harmlessly in the upper respiratory tract, but it m... piratory tract.

TABLE 13.1	**Sites Tha... Microbes**

- Skin and its contiguous muc...
- Upper respiratory tract (oral...
- Gastrointestinal tract (mouth...
- Outer opening of urethra
- External genitalia
- Vagina
- External ear and canal
- External eye (lids, lash follicles)

3.2	**Sterile (Microbe-Free) Anatomical Sites and Fluids**

Tissues and Organs

l circulatory system	Bones
	Ovaries/testes
Kidneys and bladder	Glands (pancreas, salivary)
Lungs	Sinuses
Brain and spinal cord	Middle and inner ear
Muscles	Internal eye

Fluids Within an Organ or Tissue
Blood
Urine in kidneys, ureters, bladder
Cerebrospinal fluid
Saliva prior to entering the oral cavity
Semen prior to entering the urethra

A common example is the fermentation of glycogen by lactobacilli, which keep the pH in the vagina acidic enough to prevent the overgrowth of the yeast *Candida albicans* and other pathogens. A second example is seen in the large intestine, where a protein produced by *Escherichia coli* can prevent the growth of pathogenic bacteria such as *Salmonella* and *Shigella*.

The generally antagonistic effect that "good" microbes have against intruder microorganisms is called **microbial antagonism.** Microbiota that exists in an established biofilm is unlikely to be displaced by incoming microbes. This antagonistic protection may simply be a result of a limited number of attachment sites in

the host site, all of which are stably occupied by normal flora. Antagonism may also result from the chemical or physiological environment created by the resident microbiota, which is hostile to other microbes.

In experiments performed with mice, scientists discovered that a species of the intestinal bacterium *Bacteroides* controlled the host's production of a defense compound that suppressed other microbes in the area. This is an extreme case of microbial antagonism, but the general phenomenon is of great importance to human health.

Generally the normal residents will have beneficial or benign effects only if the host is in good health with a fully functioning immune system, and if the microbiota remain in their natural microhabitats within the body. Hosts with compromised immune systems could very easily be infected by these residents (see table 13.4). We see this outcome when AIDS patients develop recurring bouts of pneumonia from *Streptococcus pneumoniae,* often carried as normal residents in the nasopharynx. Other microbiota-associated infections can occur when residents are introduced to a site that was previously sterile, as when *E. coli* from the large intestine gets into the bladder, resulting in a urinary tract infection.

Initial Colonization of the Fetus and Newborn

It has long been the consensus that the uterus and its contents are normally sterile during embryonic and fetal development and remain essentially germ-free until just before birth. A recent study now suggests that this is not the case. Samples taken from healthy placentas and amniotic sacs yielded a small collection of bacteria similar to those found in the oral cavity. So it now appears that our microbiomes begin to be established even before birth. As intriguing as these findings are, we do not yet understand the functions of the placental microbiota and what effects it may have on the fetus, whether negative or positive. Answering these questions will no doubt give rise to numerous research studies on the subject.

The early resident microbiota of fetuses is soon joined by a massive influx of additional microbes during birth and its aftermath. From the point of the amniotic sac breaking, microbes from the mother's vagina enter the womb. Even more extensive exposure occurs during the birth process itself, when the baby unavoidably comes into intimate contact with the birth canal **(figure 13.2).** Within 8 to 12 hours after delivery, the newborn typically has been colonized by bacteria such as streptococci, staphylococci, and lactobacilli, acquired primarily from its mother. The skin, gastrointestinal tract, and portions of the respiratory and genitourinary tracts all continue to be colonized as contact continues with family members, health care personnel, the environment, and food.

The nature of the microbiota initially colonizing the large intestine is greatly influenced by whether the baby receives breast milk or formula. Milk and breast tissues contain their own microbiomes that seed the baby's digestive tract with beneficial bacteria. A surprising part of neonatal research has revealed that mother's milk contains short carbohydrate chains (oligosaccharides) that cannot be digested by the infant. Only certain species of *Bifidobacterium* (*infantis*) produce enzymes to break them down. This ensures that the milk selects for the growth of these beneficial bacteria, and the baby's GI tract will become populated almost exclusively by them. It is likely that these bacteria provide

Figure 13.2 The origins of microbiota in newborns. Once a newborn baby emerges, it goes through a second phase of colonization from many sources. This includes exposure to the birth canal, breast milk or other foods, family members, medical personnel, and the environment. Within a short time the infant has developed its initial microbiota in major areas of the skin, oral cavity, and intestine.

protection against colonization by potentially harmful species and invasion by pathogens. In contrast, bottle-fed infants (receiving milk or a milk-based formula) tend to acquire a mixed population of coliforms, lactobacilli, enteric streptococci, and staphylococci. This combination is similar to an adult's microbiota and does not offer the same protective effect.

Milestones that contribute to further development of the microbiota are weaning, eruption of teeth, and introduction of the first solid food. Although exposure to microbes is unavoidable and even necessary for the maturation of the infant's microbiome, contact with pathogens is dangerous. The immune defenses of neonates have not yet fully developed, which makes them extremely susceptible to infection (see figure 13.9 and the discussion of STORCH in that section).

Indigenous Microbiota of Specific Regions

Although we tend to speak of the microbiota as a single unit, it contains a complex mixture of thousands of species, differing somewhat in quality and quantity from one individual to another. Studies have shown that most people harbor certain specially adapted bacteria, fungi, and protozoa and, occasionally, viruses and

13.1 Secret World of Microbes
The Human Microbiome—A Superorganism

Microbiologists have recognized for many years that the microbes residing on the human body exert profound effects on our health and disease. This is to be expected, because they outnumber our cells many times over and occupy about 2% to 5% of our body's mass. Knowledge of this intimate coexistence inspired the concept of a superorganism—the merger of humans and their microbiota into a single, functioning, interactive unit that shares nutrients, metabolism, and genetic information and stays in constant communication. Up until recently we had an incomplete understanding of the scientific details of our interactions with our microbiota and many of their functions. A complete picture has been hampered by having to rely on knowledge of cultured microbes, and many species that cannot be cultured have not been explored to any extent.

This all began to change when the National Institutes of Health launched a massive program to study the superorganism from a different perspective. The project, called the **Human Microbiome Project (HMP),** uses advanced technology in metagenomics to analyze the genomes of as many of the microbial residents as possible. Researchers took samples from 300 healthy volunteers, primarily from such locations as the oral cavity, skin, nose, digestive tract, and vagina. By extracting and sequencing the DNA, it was possible to identify a number of microbes that we already knew were part of the microbiota and to tabulate many others that were entirely new and unknown. After 5 years of analysis, the researchers estimated that the human body is home to at least 10,000 different species of microbes. Even more astonishing was the isolation of 8 million distinct microbial genes, which is 360 times more than our own genome holds. This means that, genetically, we are more microbe than we are human! The collective total of genetic material from all microbiota has been termed the human **microbiome.**

Preliminary results of the study provide some interesting insights. Representatives of all microbial groups were isolated, including bacteria (the largest group), archaea (the smallest group), fungi, protozoa, and viruses. Most of the microbes have a mutualistic relationship and do not cause disease, although a few pathogens, such as *Staphylococcus aureus* and *Helicobacter pylori,* were among the isolates. Although many bacteria were common to all volunteers, the precise content of microbiota is apparently unique to each individual, begins developing even before birth, and varies with age and sex. Because of these studies, we now know that many unculturable bacteria will not grow on artificial media because they have lost genes required for certain metabolic processes that are now derived from their hosts.

Approximately 500 species of bacteria and yeasts were associated with the skin, with notable differences depending on the location sampled.

For example, the skin of the hands and body creases had much greater microbial diversity than areas such as the ear and navel. Most bacterial isolates are members of four phyla: Actinobacteria, Firmicutes, Proteobacteria, and Bacteroidetes (discussed in chapter 4). See the figure for the sites sampled and the proportion of bacterial groups found in each site. One finding suggests that the skin harbors a large number and diversity of viruses. Most of the viruses appear to be bacteriophages, probably those that accompany bacteria also residing on the skin, but a number of human and other eukaryotic viruses were also discovered. Obviously this aspect of the microbiota has far greater complexity than was previously thought.

The intestinal microbiota is the largest and most varied, containing 500 to 1,000 different species of bacteria, fungi, archaea, and protozoa. The dominant phyla in this habitat are the Firmicutes and Bacteroidetes. Most of them occupy the colon, where they influence a number of physiological activities. The most common member is a species of *Bacteroides* that helps digest complex carbohydrates and contributes to the development of the protective mucous layer. There is evidence that these bacteria also signal intestinal cells to regulate development of the intestine in newborns. Other types of bacteria apparently communicate with cells of the immune system and assist in its development. Some newer research findings appear to suggest that microbes affect even brain development, biochemistry, and behavior.

The HMP has reemphasized the idea that a balanced microbiota is protective, and that a disturbed microbiota increases the susceptibility to a number of diseases. A well-established example can be seen with the colitis caused by *Clostridium difficile,* which arises when the normal intestinal residents have been eliminated by antibiotic therapy. Several studies are investigating the impact of an altered microbiota on other intestinal diseases such as irritable bowel syndrome, Crohn's disease, and colorectal cancer. Some findings indicate that composition of the microbiome can influence fat cell metabolism and can be a factor in obesity in certain individuals. There are also new data to show that the skin microbiota in diseases such as psoriasis, eczema, and acne differs significantly from that found in normal subjects. As hundreds of researchers continue to unravel the secrets of the microbiome, they are already on a quest for new strategies and therapies that may prevent these and many other diseases linked to our microbial passengers.

What effects on the development of our microbiome might be expected from excessive cleanliness in our habits and habitats? *Answer available on Connect.*

arthropods. **Table 13.3** provides an overview of the primary locations and types of normal residents.

Viruses as Normal Microbiota

The role of viruses as normal residents is complicated by their usual reputation as infectious agents. In addition, they are harbored inside cells, which makes them very different from most resident microbes.

But we now know that they can be harmless and even beneficial inhabitants of the body. Discoveries from the human genome sequencing project have shown that about 8% to 10% of the genetic material consists of endogenous retroviruses (ERVs). It has been postulated that these ERVs started out as ancient infections, but through coevolution the virus established a mutualistic existence with the human genetic material. Over time the viruses may have become an important factor in development and gene expression of their hosts.

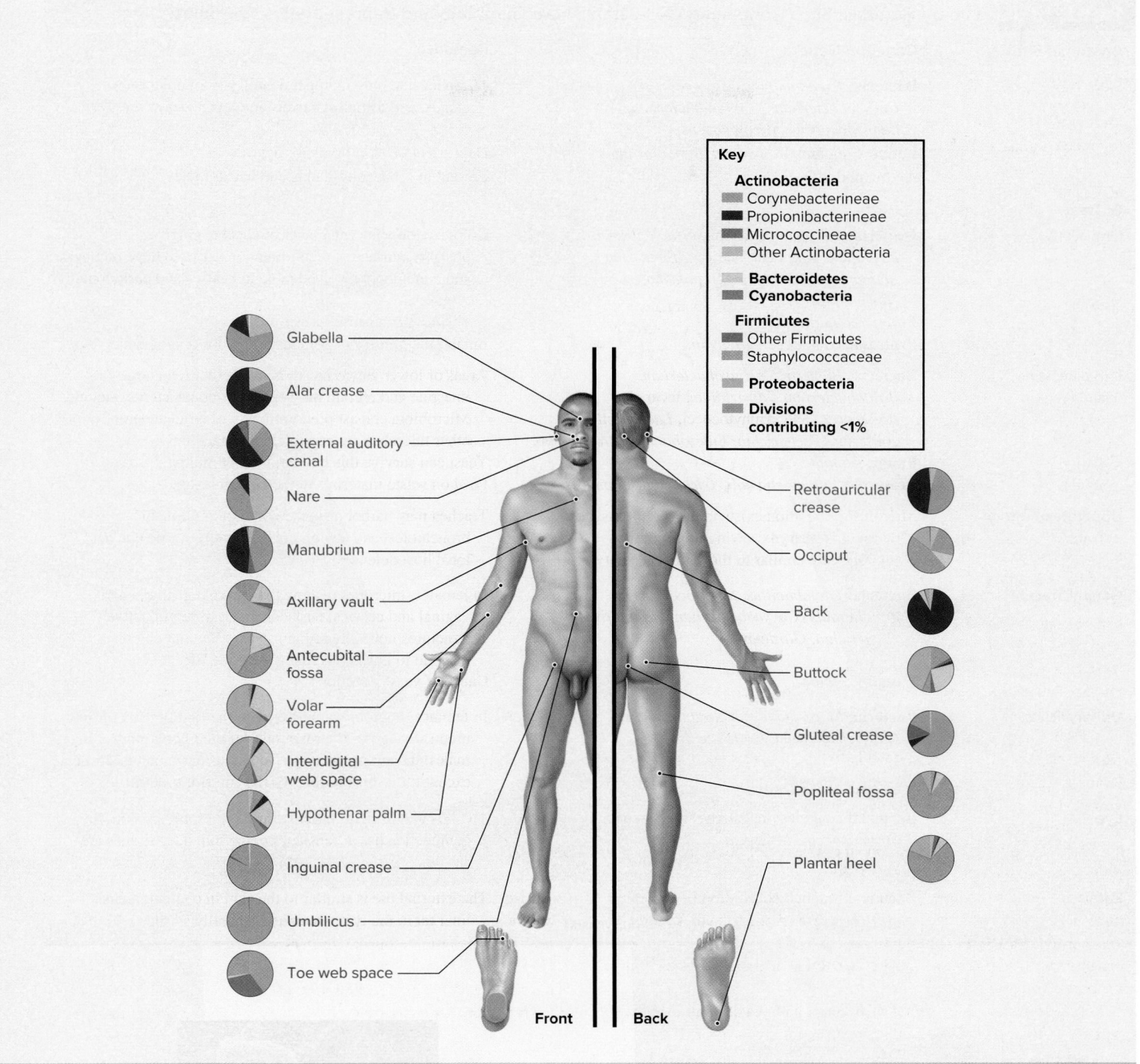

Colonizers of the Human Skin

The skin is the largest and most accessible of all organs. Its major layers are the epidermis, an outer layer of dead cells continually being sloughed off and replaced, and the dermis, which lies atop the subcutaneous layer of tissue **(figure 13.3a).** Depending on its location, skin also contains hair follicles and several types of glands; and the outermost surface is covered with a protective, waxy cuticle that can help

microbes adhere. The normal microbiota reside only in or on the dead cell layers, and except for areas around follicles and glands, it does not extend into the dermis or subcutaneous levels. The nature of the population varies according to site. Oily, moist skin supports a more prolific flora than dry skin. Humidity, occupational exposure, and clothing also influence its character. Transition zones where the skin joins with the mucous membranes of the nose, mouth, and external genitalia harbor a particularly rich set of microbial colonists.

TABLE 13.3 Life on Humans: Sites Containing Well-Established Microbiota and Representative Examples

Anatomic Sites	Common Genera	Remarks
Skin	**Bacteria:** *Pseudomonas, Micrococcus, Corynebacterium, Propionibacterium, Staphylococcus, Streptococcus*	Microbes live only in upper dead layers of epidermis, glands, and follicles; dermis and layers below are sterile.
	Fungi: *Candida, Malassezia, Rhodotorula*	Dependent on skin lipids for growth
	Arthropods: *Demodex* mite	Present in sebaceous glands and hair follicles
GI Tract		
Oral cavity	**Bacteria:** *Streptococcus, Neisseria, Veillonella, Fusobacterium, Lactobacillus, Bacteroides, Actinomyces, Eikenella, T̶____ma, Haemophilus*	Colonize the epidermal layer of cheeks, gingiva, pharynx; surface of teeth; found in saliva in huge numbers; some members involved in dental caries and periodontal diseases.
	Fungi: *Candida* spp.	*C. albicans* can cause thrush.
	Protozoa: *Entamoeba gi____*	Inhabit the gingiva of persons with poor oral hygiene.
Large intestine and rectum	**Bacteria:** *Bacteroides,* ____ *Bifidobacterium, Cl____* streptococci and sta____ coliforms (*Escher____*)	____ gastrointestinal tract other than large ____ have sparse or nonexistent residents. ____dominantly of strict anaerobes; ____otolerant or facultative.
	Fungi: *Candida*	____ habitat, but not molds.
	Protozoa: *Entamo____*	____ls in the large intestine.
Upper Respiratory Tract	Microbial popula____ throat, and ph____ microbes ar____	____ a sparse population; bronchi, ____alveoli are essentially sterile due to ____ses.
Genital Tract	**Bacteria:** *La____ diphthero____ Escheric____*	____obes occupy the external genitalia and ____ervical surfaces; internal reproductive ____ormally remain sterile. Vaginal colonists ____hormonal changes during life.
	Fungi: *Candida____*	____st infections.
Urinary Tract	**Bacteria:** *Staphylococcus, Str____ Corynebacterium, Lactobacillus*	____, microbiota exist only in the first portion of the ____ mucosa; the remainder of the tract is sterile. In ____, the entire reproductive and urinary tract is sterile ____pt for a short portion of the anterior urethra.
Eye	**Bacteria:** coagulase-negative Staphylococci, *Streptococcus, Neisseria*	The lids and follicles harbor similar microbes as skin; the conjunctiva has a transient population; deep tissues are sterile.
Ear	**Bacteria:** Staphylococci, diphtheroids	____ content; areas
	Fungi: *Aspergillus, Penicillium, Candid____*	____y sterile.

[Handwritten note overlaying table:]
Normal Skin Residents:
1.) Staphylococcus
2.) Corynebacterium
3.) Propionibacterium
4.) Yeasts
Mycobacterium = Smegma 3 mm

[Handwritten note:]
Lactobacilli & H. pylori can make it passed stomach acid

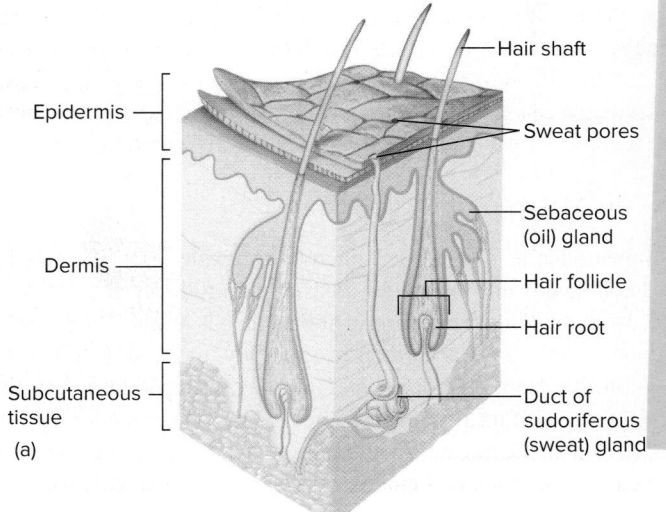

Hair shaft — Sweat pores — Sebaceous (oil) gland — Hair follicle — Hair root — Duct of sudoriferous (sweat) gland

Epidermis
Dermis
Subcutaneous tissue

(a)

a) The epidermis, ____ed), provides rich and ____d subcutaneous layers) are free of microbes. **(b)** A highly magnified view (7,500×) of skin layers reveals rod- and coccus-shaped bacteria adhering to skin cells (blue). (b): Janice Carr/CDC

Ordinarily there are two cutaneous populations. The **transients** cling to the skin surface but do not ordinarily grow there. They are acquired by routine contact and vary markedly from person to person and over time. The transients include any microbe a person has picked up, often species that do not ordinarily live on the body, and are greatly influenced by the hygiene of the individual.

The **resident populations** live and multiply in deeper layers of the epidermis and in glands and follicles (**figure 13.3b**). The composition of the resident microbiota is more stable, predictable, and less influenced by hygiene than are the transients. The normal skin residents consist primarily of bacteria (notably *Staphylococcus, Corynebacterium,* and *Propionibacterium*) and yeasts. Moist skin folds, especially between the toes, tend to harbor fungi, whereas lipophilic mycobacteria and staphylococci are prominent in sebaceous[1] secretions of the axilla, external genitalia, and external ear canal. One species, *Mycobacterium smegmatis,* lives in the waxy secretion, or *smegma,* on the external genitalia of men and women.

Microbial Residents of the Gastrointestinal Tract

The gastrointestinal (GI) tract receives, moves, digests, and absorbs food and removes waste. It encompasses the oral cavity, esophagus, stomach, small intestine, large intestine, rectum, and anus. Stating that the GI tract harbors residents may seem to contradict our earlier statement that internal organs are generally sterile, but it is not an exception to the rule. How can this be true? In reality, the GI tract is a long, hollow tube (with numerous pockets and curves), bounded by the mucous membranes of the oral cavity on one extreme and those of the anus on the other. Because the innermost surface of this tube is exposed to the environment, it is topographically outside the body, so to speak.

The shifting conditions of pH and oxygen tension and differences in the microscopic anatomy of the GI tract are reflected by the variations in or distribution of the microbiota (**figure 13.4**). Some microbes remain attached to the mucous epithelium or its associated structures, and others dwell in the lumen.* Although the abundance of nutrients invites microbial growth, the only areas that harbor appreciable permanent microbes are the oral cavity, large intestine, and rectum. The esophagus contains an extremely light load of microbiota, primarily bacteria swallowed with saliva. The stomach acid inhibits most microbes, although small numbers of lactobacilli and *Helicobacter pylori* (associated with stomach ulcers) can become established there. The small intestine has a sparse population of lactobacilli and streptococci except for its terminal segment, which has flora more similar to that of the adjacent large intestine.

Microbiota of the Mouth

The oral cavity has unique residents that are some of the most diverse and abundant of the body. Microhabitats, including the cheek epithelium, gingiva, tongue, floor of the mouth, and tooth enamel, provide numerous adaptive niches. By one estimate, there are more than 600 species residing there. The most common residents are aerobic *Streptococcus* species—*S. sanguis, S. salivarius, S. mitis*—

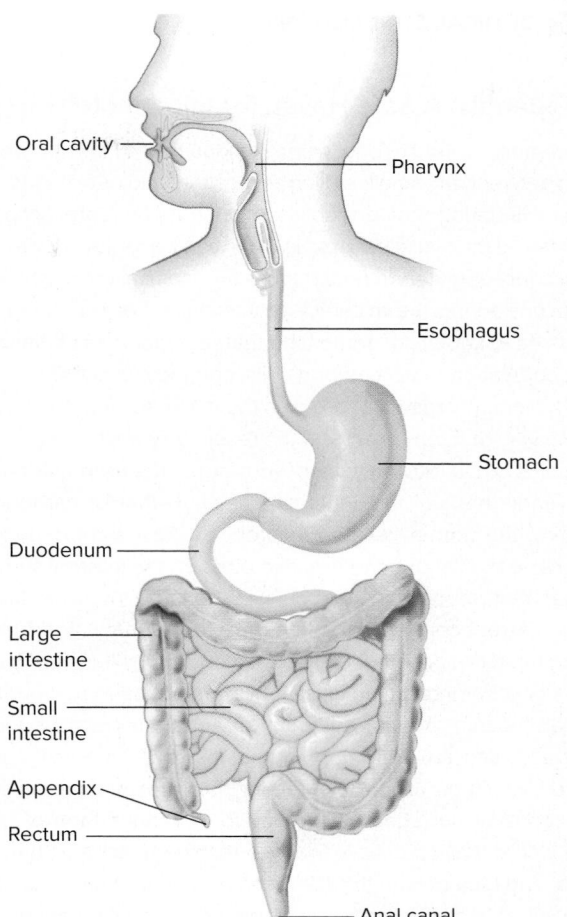

Figure 13.4 **Distribution of microbes in the GI tract.** Areas of the gastrointestinal tract that shelter major communities of resident microbes are highlighted in color. Noncolored areas do not harbor residents in significant numbers.

that colonize the smooth superficial epithelial surfaces. Two species, *S. mutans* and *S. sanguis,* make a major contribution to dental caries by forming sticky dextran slime layers in the presence of simple sugars. The adherence of dextrans to the tooth surface establishes the basis for a biofilm that attracts other bacteria.

Eruption of the teeth establishes an anaerobic habitat in the gingival crevice that favors the colonization by anaerobic bacteria that can be involved in dental caries and periodontal[2] infections. The instant that saliva is secreted from ducts into the oral cavity, it becomes laden with resident and transient bacteria. Saliva normally has a high bacterial count (up to 5×10^9 cells per milliliter), a fact that tends to make mouthwashes rather ineffective and a human bite quite dangerous.

Microbiota of the Large Intestine

Microbes inhabiting the intestinal tract have complex and profound interactions with the host. The large intestine (cecum and colon) and the rectum harbor a huge population of microbes (10^8–10^{11} per gram of feces). So abundant and prolific are these microbes that they constitute 30% or more of the fecal volume. Even an individual on a long-term fast passes feces consisting primarily of bacteria.

1. From *sebum,* a lipid material secreted by glands in the hair follicles.

* *lumen* (loo'-men) The space within a tubular structure.

2. Situated or occurring around the tooth.

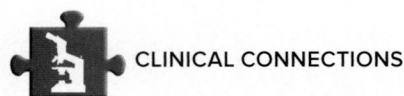

CLINICAL CONNECTIONS

The Appendix: A Safe Haven for Microbiota?

The appendix, a 4-inch-long wormlike pouch that sprouts from the junction between the small and large intestine, had been written off for years as a lingering remnant of tissue without any real functions. Evidence offered for its uselessness is that it can be removed without real medical consequences. In recent times the medical establishment has been reconsidering the accuracy of this claim. For one thing, immunologists have known for some time that the appendix contains a significant population of white blood cells, and they consider it a modest form of lymphatic organ that protects the intestines from pathogens in the same way that tonsils protect the respiratory tract.

Studies at Duke University have produced strong evidence that the appendix has an additional role to play—that of harboring and supporting the normal resident microbiota. Researchers uncovered an extensive biofilm of beneficial microbes on the internal surfaces of the organ that periodically reinoculates the large intestine and maintains the correct composition necessary for normal colon function—enabling rapid replacement of the colon's own biofilms after a case of diarrhea or other intestinal disruption. These experts reason that humans can function normally without an appendix because we don't need it as much today, in light of our improved medical care and sanitation, which greatly reduce the incidence of intestinal infections.

An unusual but effective therapy for a serious form of diarrhea caused by *Clostridium difficile* involves the transplantation of fecal microbiota. The idea behind this has been around for hundreds of years, but it involves implanting a fecal sample from a healthy donor directly into the infected colon or swallowing a pill containing normal microbiota. This technique is often more successful than taking antibiotics alone for this disease. It likely restores the normal microbial residents of the colon and the appendix, which inhibits the growth of the pathogen.

There are thousands of microbial species in a "normal" intestine. What could be some uninte̶̶̶̶̶̶ ̶̶̶̶̶̶̶̶̶̶̶̶̶̶̶̶̶̶̶̶̶̶̶̶̶̶̶̶̶ planting one person's intes̶̶̶̶̶̶ ̶̶̶̶̶̶̶̶̶̶̶̶̶̶̶̶̶̶̶̶ colon? *Answer available o̶*

The appendix has been ̶̶̶̶̶̶̶̶̶̶̶̶ but recent research is verif̶̶̶̶̶̶̶̶ of the normal microbiota (̶̶̶̶̶̶̶ the intestinal environmen̶̶̶̶̶̶ (*Bacteroides, Bifidobacteri̶̶*̶̶ **Coliforms*** such as *E. coli,* ̶̶̶̶ ent in smaller numbers. Ma̶̶̶̶ the feces, generating vitami̶̶̶̶ thiamine) and acids (acetic̶̶̶̶̶̶ value to the host. Occasiona̶̶̶̶̶̶ enzymes that convert disacc̶̶̶̶̶̶ steroid metabolism.

Intestinal bacteria contr̶̶̶̶̶̶̶̶̶̶ **skatole,*** amines, and gases (̶C̶O̶2̶,̶ ̶H̶2̶,̶ ̶C̶H̶4̶,̶ and H₂S). Intestinal gas

[handwritten note: Staph = nasal entrance / Neisseria = mucous / Haemophilus = throat / Nonhemolytic streptococci / urethra]

is known in polite circles as *flatus,* and the expulsion of it as *flatulence.* Some of the gas arises through the action of bacteria on dietary carbohydrate residues from vegetables such as cabbage, corn, and beans. The bacteria produce an average of 8.5 liters of gas daily, but only a small amount is ejected in flatus. Combustible gases occasionally form an explosive mixture in the presence of oxygen that has reportedly ignited during intestinal surgery and ruptured the colon!

Late in childhood many people lose the ability to secrete the enzyme lactase, a condition known as lactase deficiency or intolerance. Difficulties arise when an individual with this condition ingests milk or other lactose-containing dairy products. Lactose is a disaccharide that cannot be absorbed, and when lactose lies undigested, it increases the osmotic pressure in the gut, resulting in cramps, diarrhea, and intestinal distress. To complicate matters, the lactose can be digested by intestinal bacteria, which release gas and intestinal irritants. The recommended treatment for this deficiency is avoiding these foods or eating lactose-free substitutes.

Inhabitants of the Respiratory Tract

The first microorganisms to colonize the upper respiratory tract (nasal passages and pharynx) are predominantly oral streptococci. *Staphylococcus aureus* preferentially resides in the nasal entrance, nasal vestibule, and anterior nasopharynx, and *Neisseria* species take up residence in the mucous membranes of the nasopharynx behind the soft palate **(figure 13.5)**. Lower in the tract are assorted streptococci and species of *Haemophilus* that colonize the tonsils and lower pharynx. Conditions even lower in the respiratory tree (bronchi and lungs) are unfavorable habitats for permanent residents.

Microbiota of the Genitourinary Tract

̶̶̶̶̶̶̶̶̶s of the genitourinary tract that harbor resident microbes ̶̶̶̶̶̶̶ina and outer opening of the urethra in females and the

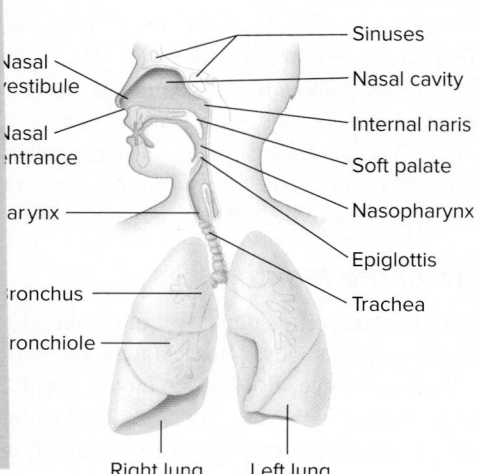

Figure 13.5 Colonized regions of the respiratory tract. The moist mucous blanket of the nasopharynx has well-entrenched resident microbes (indicated by color). Some colonization occurs in the upper trachea, but lower regions (bronchi, bronchioles, and lungs) lack resident microbes.

* *coliform* (koh′-lih-form) L. *colum,* a sieve. Gram-negative, facultatively anaerobic, and lactose-fermenting microbes.
* *skatole* (skat′-ohl) Gr. *skatos,* dung. One chemical that gives feces their characteristic stench.

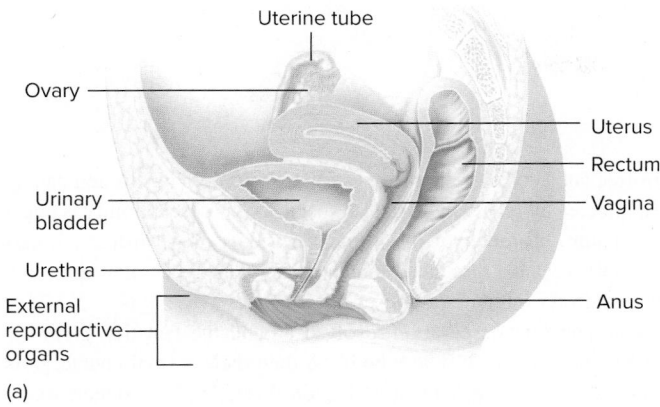

Uterine tube

Ovary

Urinary
bladder

Urethra

External
reproductive
organs

Uterus

Rectum

Vagina

Anus

(a)

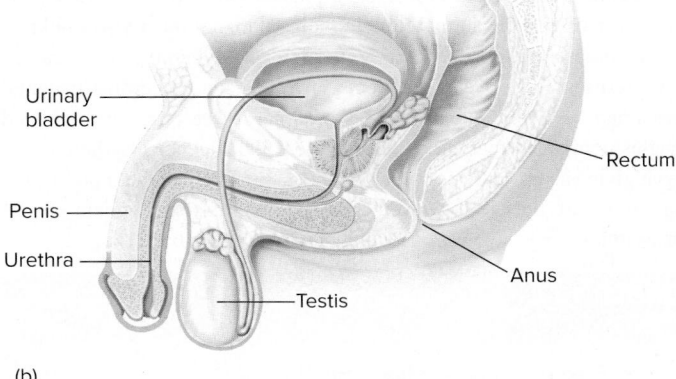

Urinary
bladder

Penis

Urethra

Rectum

Anus

Testis

(b)

Figure 13.6 Microbiota of the reproductive tract. (a) Female and **(b)** male genitourinary residents (location indicated by color).

anterior urethra in males **(figure 13.6).** The internal reproductive organs are kept sterile through physical barriers such as the cervical plug and other host defenses. The kidney, ureter, bladder, and upper urethra are presumably kept sterile by urine flow and regular bladder emptying. The shortness of the urethra in women (about 3.5 cm long), frequently leads to urinary tract infections. The principal residents of the urethra are nonhemolytic streptococci, staphylococci, corynebacteria, and occasionally, coliforms.

The vagina presents a notable example of how changes in physiology can greatly influence the composition of the normal microbiota. An important factor influencing these changes in women is the hormone estrogen. Estrogen normally stimulates the vaginal mucosa to secrete glycogen, which certain bacteria (primarily *Lactobacillus* species) ferment, thus lowering the pH to about 4.5. Before puberty a girl produces little estrogen and little glycogen and has a vaginal pH of about 7. These conditions favor the establishment of diphtheroids,[3] staphylococci, streptococci, and some coliforms. As hormone levels rise at puberty, the vagina begins to deposit glycogen, and the microbiota shift to the acid-producing lactobacilli. It is thought that the acidic pH of the vagina during this time prevents the establishment and invasion of microbes with potential to harm a developing fetus. The estrogen-glycogen effect continues, with minor disruption, throughout the childbearing years until menopause, when the microbiota return to

a mixed population similar to that of prepuberty. These transitions are not abrupt but occur over several months to years.

Maintenance of the Normal Microbiota

There is no question that the normal residents are essential to the health of humans and other animals. When living in balance with their host, the flora create an environment that may prevent infections and can enhance certain host defenses. In general, the microbes replace themselves naturally on a regular basis to maintain the types and numbers in their zones. However, because the exact content of the microbiota is not fixed, a number of changes can disrupt this balance. Use of broad-spectrum antibiotics, changes in diet, and underlying disease all have the potential to alter the makeup of the microbiota and tilt the system toward disease. A growing trend in therapy is the use of live cultures of known microbes in the form of **probiotics** (discussed in chapter 12). This essentially involves introducing pure cultures of known microbes into the body through ingestion or inoculation. The microbes chosen for this process are known to be beneficial and are considered nonpathogenic.

For a look into laboratory studies that address the effects of microbiota, see **13.1 Making Connections.**

 Check Your Progress SECTION 13.1

1. Describe the significant relationships that humans have with microbes.
2. Explain what is meant by *microbiota* and *microbiome* and summarize their importance to humans.
3. Differentiate between contamination, colonization, infection, and disease, and explain some possible outcomes in each.
4. How are infectious diseases different from other diseases?
5. Outline the general body areas that are sterile and those regions that harbor normal resident microbiota.
6. Differentiate between transient and resident microbes.
7. Explain the factors that cause variations in the microbiota of the newborn intestine and the vaginal tract.

13.2 Major Factors in the Development of an Infection

 Expected Learning Outcomes

7. Review the main stages in the development of an infection.
8. Categorize the different types and degrees of pathogens and differentiate pathogenicity from virulence.
9. Describe the differences among the portals of entry, and give examples of pathogens that invade by these means.
10. Explain what is meant by the infectious dose, using examples.
11. Describe the process of adhesion and various mechanisms by which microbes use it to gain entry.
12. Identify and discuss invasive factors and virulence factors.
13. Compare and contrast the major characteristics of exotoxins and endotoxins.

3. Any nonpathogenic species of *Corynebacterium.*

13.1 MAKING CONNECTIONS

A Germ-Free Existence

For years questions lingered about how essential the microbiota are to normal life and what functions they might serve. The need for animal models to further investigate these questions led eventually to development of laboratory strains of *germ-free,* or **axenic,*** mammals and birds. After the young mammals are taken from the mother aseptically by cesarean section, they are immediately transferred to a sterile isolator or incubator. The newborns must be fed by hand through gloved ports in the isolator until they can eat on their own, and all materials entering their chamber must be sterile. Mice, rats, rabbits, guinea pigs, monkeys, and dogs are some of the mammals raised in the germ-free state.

Experiments with germ-free animals are of two basic varieties: (1) general studies on how the lack of normal microbial residents influences the nutrition, metabolism, and anatomy of the animal; and (2) **gnotobiotic*** studies, in which the germ-free subject is inoculated either with a single type of microbe to determine its individual effect or with several known microbes to determine interrelationships. Results are validated by comparing the germ-free group with a conventional, normal control group. **Table 13.A** summarizes some major conclusions arising from studies with germ-free animals.

A dramatic characteristic of germ-free animals is that they live longer and have fewer diseases than normal controls, as long as they remain in a sterile environment. From this standpoint, it seems that the microbiota are not absolutely necessary for survival and may even be the source of infectious agents. But it is also clear that axenic life is highly impractical for humans and will probably never be an option. We will have to rely on germ-free animals for valuable insights into the powerful influence microbiota exert on development, biochemistry, and behavior.

Additional studies have revealed that microbiota contribute significantly to the development of the immune system. When germ-free animals are placed in contact with normal control animals, they gradually develop resident microbes similar to those of the controls. However,

* *axenic* (aye-zeen′-ik) Gr. *a,* not and *xeno,* strange or foreign
* *gnotobiotic* (noh″-toh-by-aw′-tik) Gr. *gnotos,* known and *biosis,* life

Laboratory technician handles germ-free mouse being reared in a sterile isolator. Most gnotobiotic studies are based on specially created strains of mice. © LookatSciences/Phototake

germ-free subjects are less tolerant of these microorganisms and can die when infected by relatively harmless species. This susceptibility is due to the immature character of the immune systems of germ-free animals. These animals have a reduced number of certain types of white blood cells and slower immune responses.

One puzzling observation revolved around the fact that germ-free mice had as much as 42% less body fat than their normal counterparts. Later experiments revealed that indigenous bacteria acted to increase the body fat of mice in two ways. First, certain bacteria in the gut helped to digest complex carbohydrates, making extra calories available to the mice. Second, these same bacteria secreted a substance that, by acting on mouse hormones, caused the animal to store more fat than it otherwise would.

Gnotobiotic experiments have clarified the dynamics of several infectious diseases. For years the precise involvement of microbes in dental caries had been ambiguous. Studies with germ-free rats, hamsters, and beagles confirmed that caries development is influenced by heredity, a diet high in sugars, and poor oral hygiene. Even when all these predisposing factors are present, however, germ-free animals still remain free of caries unless they have been inoculated with specific bacteria.

TABLE 13.A	A Sampling of Effects of the Germ-Free State
Germ-Free Animals Display	**Significance**
Enlargement of the cecum; other degenerative diseases of the intestinal tract of rats, rabbits, chickens	Microbes are needed for normal intestinal development.
Vitamin deficiency in some animals	Microbes can be a significant nutritional source of vitamins.
Underdevelopment of immune system in most animals	Microbes are needed to stimulate development of certain host defenses.
Absence of dental caries and periodontal disease in dogs, rats, hamsters	Microbes are key players in caries formation and gum disease.
Heightened susceptibility to enteric pathogens (*Shigella, Salmonella, Vibrio cholerae*) and to fungal infections	Normal bacterial residents are antagonistic against pathogens.
Lessened susceptibility to amebic dysentery	Normal resident microbes facilitate the completion of the life cycle of the amoeba in the gut.
Less body fat	Normal microbiota help to break down indigestible carbohydrates and increase fat storage in the body.
Increased anxiety and nervousness	Mice are much more sensitive to stressful conditions and produce increased levels of stress hormones.

Why is a cesarean section necessary for developing axenic mammals?
Answer available on Connect.

Figure 13.7 **A flow diagram that relates the events in entry, establishment, and exit of infectious agents.**

In this section we trace the course or "chain" of infection, a sequence of events that is summarized in **figure 13.7.** Pathogens are parasitic microbes whose relationship with a host results in infection and disease. The type and severity of infection depend on both the **pathogenicity** of the organism and the condition of the host. Pathogenicity, you will recall, is a broad concept that describes an organism's potential to cause infection or disease and is used to divide pathogenic microbes into one of two general groups.

True pathogens (primary pathogens) are capable of causing disease in healthy persons with normal immune defenses. They are generally associated with a specific, recognizable disease, which may vary in severity from mild (colds) to severe (malaria) to fatal (rabies). Other examples of true pathogens are the influenza virus, plague bacillus, and meningococcus.

Opportunistic pathogens cause disease when the host's defenses are compromised[4] or when they become established in a part of the body that is not natural to them. Opportunists are not considered pathogenic to a normal healthy person and, unlike primary pathogens, do not generally possess well-developed virulence properties. Examples of opportunistic pathogens include *Pseudomonas* species and *Candida albicans.* Factors that greatly predispose a person to infections, both primary and opportunistic, are shown in **table 13.4.**

As with many concepts, the idea of a pathogen being true or opportunistic is a convenience to simplify discussion. In reality, pathogenicity extends along a range from low to high. To recognize this, the CDC has adopted a system of biosafety categories for

pathogens based on their degree of pathogenicity and the relative danger in handling them. This system assigns microbes to one of four levels or classes. Microbes not known to cause disease in humans are assigned to class 1, and highly contagious viruses that pose an extreme risk to humans are classified as class 4. Microbes of intermediate virulence are assigned to class 2 or 3. Details of this system are presented in more detail in Appendix B.

The relative severity of the disease caused by a particular microorganism depends on its **virulence.**[*] Although the terms *pathogenicity* and *virulence* are often used interchangeably, virulence is the accurate term for describing the degree of pathogenicity. The virulence of a microbe is determined by its ability to establish itself in the host and cause damage. There is much involved in both of these steps. To invade, microbes must enter the host, attach firmly to host tissues, and survive the host defenses. To cause damage, microbes must grow in tissues, produce toxins, or induce a host response that is more injurious than protective. Any characteristic or structure of the microbe that contributes to the infection or disease state is called a **virulence factor.** Virulence can be due to a single factor or to multiple factors. In some microbes, the causes of virulence are clearly established, but in others, they are not as well understood. In the following section we examine the effects of virulence factors and their roles in the progress of an infection.

Becoming Established: Phase 1—Portals of Entry

To initiate an infection, a microbe enters the tissues of the body by a characteristic route, the **portal of entry (figure 13.8),** usually some sort of cutaneous or membranous boundary. The source of the infectious agent can be **exogenous,** originating from a source outside the body such as the environment or another person or animal, or **endogenous,** already existing on or in the body from microbiota or a latent infection.

For the most part, the portals of entry are the same anatomical regions that also support resident microbes: the skin, gastrointestinal tract, respiratory tract, and urogenital tract. The majority of pathogens have adapted to a specific portal of entry, one that provides a habitat for further growth and spread. This adaptation can be so restrictive that if certain pathogens enter the "wrong" portal,

TABLE 13.4	Factors That Weaken Host Defenses and Increase Susceptibility to Infection*

- Old age and extreme youth (infancy, prematurity)
- Genetic defects in immunity and acquired defects in immunity
- Surgery and organ transplants
- Organic disease: cancer, liver malfunction, diabetes
- Chemotherapy/immunosuppressive drugs
- Physical and mental stress
- Other infections

*These conditions compromise defense barriers or immune responses.

4. People with weakened immunity are often termed *immunocompromised.*

* *virulence* (veer-yoo-lents) L. *virulentia,* virus, poison.

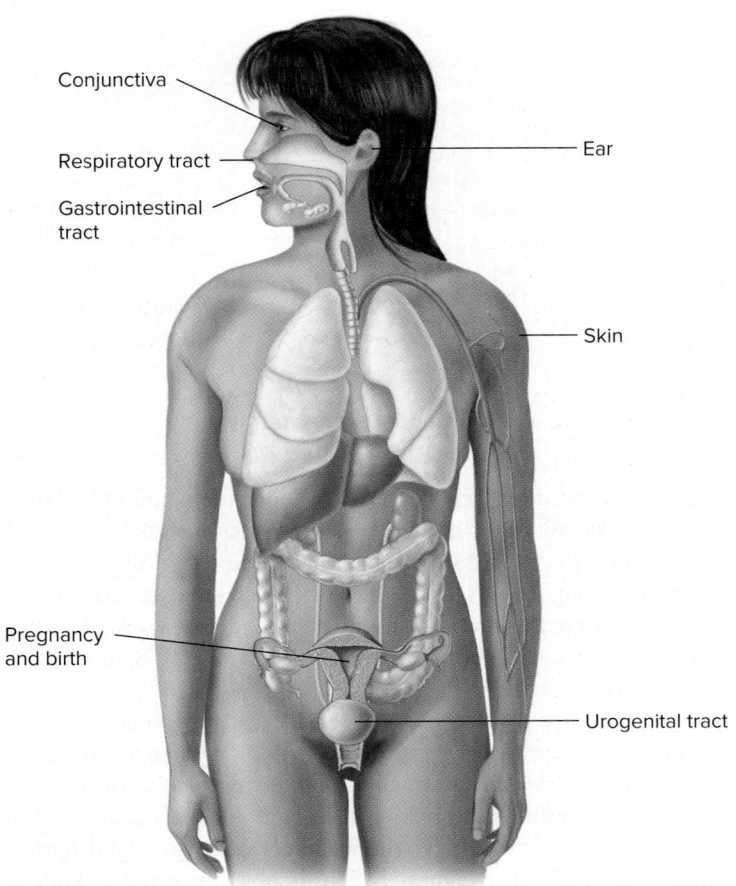

Conjunctiva

Respiratory tract

Gastrointestinal
tract

Ear

Skin

Pregnancy
and birth

Urogenital tract

Figure 13.8 Portals of entry. Infectious microbes first enter the body through characteristic routes. Most microbes have adapted to a specific means of entry. Only one of these (the uterus) is an internal portal of entry as it may allow microbes to access a developing fetus.

they will not be infectious. For instance, inoculation of the nasal mucosa with the influenza virus invariably gives rise to the flu, but if this virus contacts only the skin, no infection will result. Likewise, contact with athlete's foot fungi in small cracks in the toe webs can induce an infection, but inhaling the spores from this same fungus will not infect a healthy individual.

Occasionally an infectious agent can enter by more than one portal. For instance, *Mycobacterium tuberculosis* enters through both the respiratory and gastrointestinal tracts, and pathogens in the genera *Streptococcus* and *Staphylococcus* have adapted to invasion through several portals of entry such as the skin, urogenital tract, and respiratory tract. Understand that the same pathogen entering different portals of entry will likely lead to different infections or diseases. Infection of a skin follicle by *Staphylococcus aureus* will produce an abscess, but its invasion of the respiratory tract causes a serious form of pneumonia.

Infectious Agents That Enter the Skin and Associated Mucous Membranes

The skin is a very common portal of entry. The actual sites of entry are usually nicks, abrasions, and punctures (many of which are tiny and inapparent) rather than smooth, unbroken skin. *Staphylococcus aureus* (the cause of boils), *Streptococcus pyogenes* (an agent of

impetigo), the fungal dermatophytes, and agents of gangrene and tetanus gain access through damaged skin. The cold sore virus (herpes simplex) enters through the mucous membranes near the lips.

Some infectious agents create their own passageways into the skin using digestive enzymes. For example, certain helminth worms burrow through the skin directly to gain access to the tissues. Other infectious agents enter through bites. The bites of insects, ticks, and other animals offer an avenue to a variety of viruses, rickettsias, and protozoa. An artificial means for breaching the skin barrier is contaminated hypodermic needles by intravenous drug abusers. Users who inject drugs are predisposed to a disturbing list of well-known diseases: hepatitis, HIV/AIDS, tetanus, tuberculosis, osteomyelitis, and malaria. A resurgence of some of these infections is directly traceable to drug use. Contaminated needles often contain bacteria from the skin or environment that induce heart disease (endocarditis), lung abscesses, and chronic infections at the injection site.

Although the conjunctiva, the outer protective covering of the eye, is a relatively good barrier to infection, bacteria such as *Chlamydia trachomatis* (trachoma), and *Neisseria gonorrhoeae* readily attach to this membrane. Many childhood viruses (measles, mumps, rubella) can gain access through the eye. As was described in this chapter's Case Study, the ebolavirus readily invades the eye surface and other membranes through direct contact with infectious fluids.

The Gastrointestinal Tract Portal of Entry

The gastrointestinal tract is the portal of entry for pathogens contained in food, drink, and other ingested substances. These pathogens are adapted to survive digestive enzymes and abrupt pH changes. Most enteric pathogens possess specialized mechanisms for entering and localizing in the mucosa of the small or large intestine. The best-known enteric agents of disease are gram-negative rods in the genera *Salmonella, Shigella, and Vibrio* and in certain strains of *Escherichia coli*. Viruses that enter through the gut are poliovirus, hepatitis A virus, echovirus, and rotavirus. Important enteric protozoans are *Entamoeba histolytica* (amebiasis) and *Giardia lamblia* (giardiasis). Although the anus is not a typical portal of entry, it becomes one in people who practice anal sex.

The Respiratory Portal of Entry

The oral and nasal cavities are also the gateways to the respiratory tract, the portal of entry for the greatest number of pathogens. Because there is a continuous mucous membrane surface covering the upper respiratory tract, the sinuses, and the auditory tubes, microbes are often transferred from one site to another. The extent to which an agent is carried into the respiratory tree is based primarily on its size. In general, small cells and particles are inhaled more deeply than larger ones.

Infectious agents with this portal of entry include the bacteria of streptococcal sore throat, meningitis, diphtheria, and whooping cough and the viruses of influenza, measles, mumps, rubella, chickenpox, and the common cold. Pathogens that are inhaled into the lower regions of the respiratory tract (bronchioles and lungs) can cause pneumonia, an inflammatory condition of the lung. Bacteria (*Streptococcus pneumoniae, Klebsiella, Mycoplasma*) and fungi (*Cryptococcus* and *Pneumocystis*) are a few of the agents involved in pneumonias. Other agents causing unique recognizable lung diseases are *Mycobacterium tuberculosis* and fungal pathogens such as *Histoplasma*.

Urogenital Portals of Entry

The urogenital tract is the portal of entry for pathogens that are contracted by sexual means (intercourse or intimate direct contact). These conditions are termed **sexually transmitted infections (STIs)** or **sexually transmitted diseases (STDs).** The term *sexually transmitted infections* may be preferred in some cases because it recognizes the fact that many of the sexually related conditions are asymptomatic and do not lead to symptoms of disease, even though they are still transmissible. STIs and STDs account for an estimated 4% of infections worldwide, with approximately 20 million new cases each year, caused by just eight bacteria and viruses. **Table 13.5** provides the most recent available statistics (2014) and estimates of the most common sexually related infections and diseases.

The microbes of STI/STDs enter the skin or mucosa of the penis, external genitalia, vagina, cervix, and urethra. Some can penetrate an unbroken surface; others require a cut or abrasion. Once-predominant STDs such as syphilis and gonorrhea have been surpassed by a growing list led by genital warts (HPV), chlamydia, and trichomoniasis. Evolving sexual practices have increased the incidence of STI/STDs that were once uncommon, and diseases that were not originally considered sexually acquired are now so classified.[5] Other common sexually transmitted agents include HIV, herpes simplex virus, *Candida albicans* (a yeast), and hepatitis B virus.

Quick Search

See the videos "STD Symptoms" on the Internet and on YouTube.

Not all urogenital infections are transmitted sexually. Some of these infections are caused by displaced organisms, as when normal flora from the gastrointestinal tract cause urinary tract infections, or by opportunistic overgrowth of normal microbiota ("yeast infections").

5. Amebic dysentery, scabies, salmonellosis, and *Strongyloides* worms are examples.

TABLE 13.5	Estimated Case Numbers of Most Common Sexually Transmitted Infections/Diseases in the United States*	
STI/STD	Estimated Number of New Cases per Year (Incidence)	Estimated Number of Cumulative Cases over Extended Time Frame (Prevalence)
Human papillomavirus	14,000,000	79.200,000
Trichomoniasis	2,230,000	3,710,000
Chlamydiosis**	1,360,000	1,570,000
Herpes simplex	510,000	24,100,000
Gonorrhea**	355,000	800,000
Hepatitis B	40,000	420,000
New HIV infections	35,000	900,000
Syphilis**	21,000	117,000

*Data from CDC, 2015, based on reported statistics and clinical data; Prevalence data predicts the cases that are undiagnosed or not treated or cured.

** These are the only STDs that are officially reported by health authorities to the CDC.

Pathogens That Infect During Pregnancy and Birth

The placenta is an exchange organ formed by maternal and fetal tissues that separate the blood of the developing fetus from that of the mother yet permit diffusion of dissolved nutrients and gases to the fetus. The placenta is ordinarily an effective barrier against pathogens that have entered the maternal circulation. However, a few microbes, such as the syphilis spirochete, can cross the placenta, enter the umbilical vein, and spread by the fetal circulation into the fetal tissues **(figure 13.9).**

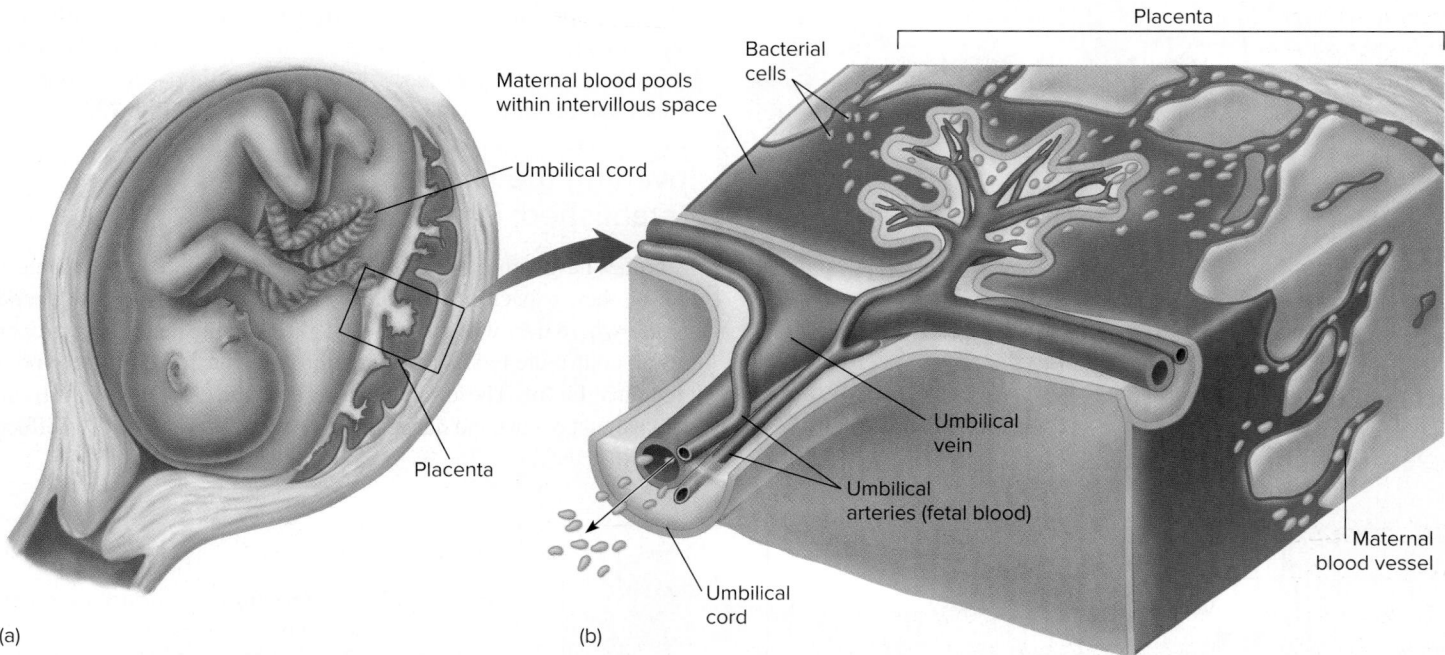

Figure 13.9 Transplacental infection of the fetus. (a) Fetus in the womb. **(b)** In a closer view, microbes are shown penetrating the maternal blood vessels and entering the blood pool of the placenta. They subsequently travel into the fetal circulation by way of the umbilical vein.

TABLE 13.6	Estimated Infectious Doses of Selected Pathogens*	
Agent of	Infectious Dose Estimate	Primary Route of Infection
Measles	1 virus	Respiratory
Shigellosis	1–10 cells	Ingestion
Norovirus gastroenteritis	18 viruses or less	Ingestion
Tuberculosis	10–50 cells	Various
Smallpox	10–100 viruses	Respiratory
Brucellosis	10–100 cells	Various
Viral encephalitis	10–100 viruses	Mosquito bite
Plague	100–500 cells	Flea bite
Gonorrhea	1,000 cells	Sexual contact
Anthrax	8,000–50,000 spores	Respiratory, cutaneous
Salmonellosis	10,000 cells	Ingestion
Cholera	100,000,000 cells	Ingestion

* Several of these agents are considered potential bioterror pathogens.

Not all neonatal infections occur through the placenta. Some, such as herpes simplex and chlamydiosis, develop perinatally when the infant is contaminated by the birth canal. The most prominent infections of fetus and neonate are grouped together in a unified cluster known by the acronym **STORCH** that medical personnel are on the alert to detect. STORCH stands for **s**yphilis, **t**oxoplasmosis, **o**ther diseases (hepatitis B, AIDS, and chlamydia), **r**ubella, **c**ytomegalovirus, and **h**erpes simplex virus. The most serious complications of STORCH infections are spontaneous abortion, congenital abnormalities, brain damage, prematurity, and stillbirths.

The Requirement for an Infectious Dose

Another factor crucial to the course of an infection is the quantity of microbial cells or particles that enters the portal of entry. For most agents, infection will proceed only if a minimum number, called the **infectious dose (ID),** is present. This number has been determined experimentally for many microbes. It is theorized that microbes with lower IDs tend to act locally and immediately, whereas those with higher IDs produce substances that must travel through the body to have their effect. **Table 13.6** presents examples of dose sizes for a number of pathogens, ranging from small to large infectious doses. Certain microbes (cholera, salmonellosis) have a significantly higher ID because they are inhibited or destroyed by acid as they pass through the stomach on their way into the lower GI tract. If the number of microbes entering a portal is below an infectious dose, infection and disease will generally not result. But if the quantity is far in excess of the ID, the onset of disease can be extremely rapid. Even weakly pathogenic species can be rendered more virulent with a large inoculum. Much of our knowledge of modes of infection and infectious dose has come from human and animal experimentation **(13.2 Making Connections).**

Attaching to the Host: Phase 2

Once a pathogen has penetrated the portal of entry in adequate numbers, its next must establish a stable association with the host's cells through **adhesion.** Such a firm attachment is almost always a prerequisite for invasion, because the body has so many mechanisms for flushing microbes and foreign materials from its tissues. Adhesion is dependent on binding between specific molecules on both the host and the pathogen, so a particular pathogen is limited to only those cells (and organisms) to which it can bind. Once attached, the pathogen is poised advantageously to invade the body compartments. Bacterial, fungal, and protozoal pathogens attach most often by appendages and surface structures such as fimbriae, pili, or flagella, and adhesive slimes or capsules; viruses attach by means of specialized receptors **(figure 13.10).** Larger pathogens like parasitic worms are mechanically fastened to the portal of entry by suckers, hooks, and barbs. Adhesion methods of various microbes and the diseases they cause are shown in **table 13.7.**

Invading the Host and Becoming Established: Phase 3

Microbes have numerous properties that improve their invasiveness, that is, their capacity to evade host defenses and enter into deeper tissues where they will grow and become established. These properties all contribute to virulence and thus are considered some type of **virulence factor.** These same factors determine the degree of tissue damage that occurs and hence the severity of the disease. The effects

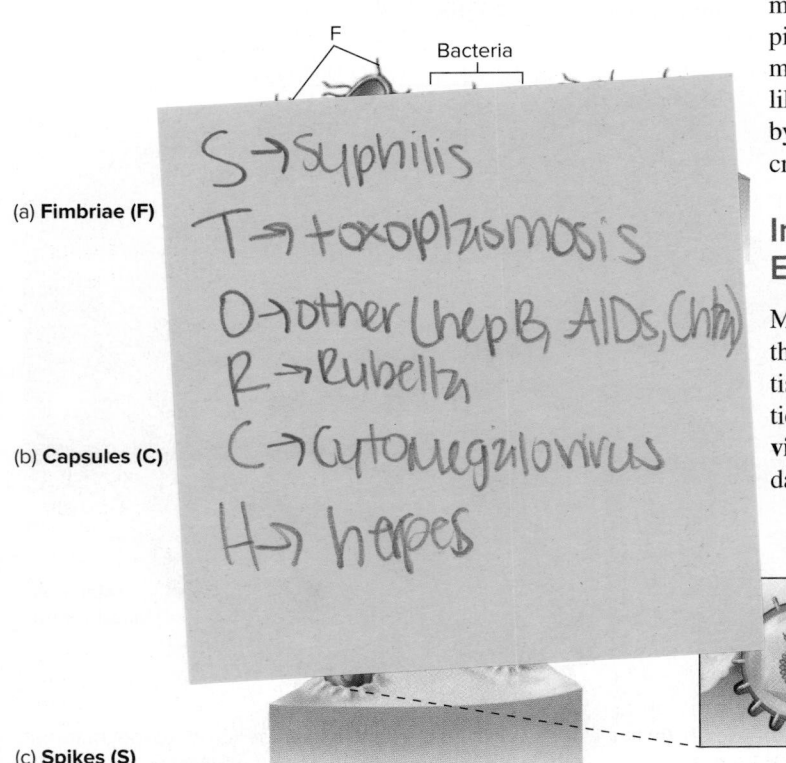

(a) **Fimbriae (F)**

(b) **Capsules (C)**

(c) **Spikes (S)**

F

Bacteria

Host cell

S → syphilis
T → toxoplasmosis
O → other (hep B, AIDs, Chla)
R → Rubella
C → cytomegalovirus
H → herpes

Figure 13.10 Mechanisms of adhesion by pathogens. (a) Fimbriae, minute bristlelike appendages. **(b)** Adherent extracellular capsules made of slime or other sticky substances. **(c)** Viral envelope spikes. See table 13.7 for specific examples.

TABLE 13.7	Adhesion Properties of Microbes	
Microbe	**Disease**	**Adhesion Mechanism**
Neisseria gonorrhoeae	Gonorrhea	Type IV pili attach to genital epithelium.
Escherichia coli	Diarrhea	Well-developed fimbriae adhere to intestinal cells.
Shigella and *Salmonella*	Gastroenteritis	Fimbriae attach to intestinal epithelium.
Vibrio cholerae	Cholera	Glycocalyx anchors to intestinal epithelium.
Treponema	Syphilis	Tapered hook embeds in host cell.
Mycoplasma	Pneumonia	Specialized tip fuses to lung epithelium.
Pseudomonas aeruginosa	Lung infections	Fimbriae and slime layer stick to host cells.
Streptococcus mutans	Dental caries	Dextran slime layer glues cocci to tooth surface.
Influenza virus	Influenza	Viral spikes dock to respiratory receptors.
Poliovirus	Polio	Capsid attaches to receptors on target cells.
HIV	AIDS	Viral spikes adhere to white blood cell receptors.
Giardia lamblia	Giardiasis	Suction disc on underside attaches to intestinal surface.
Trypanosoma	Trypanosomiasis	Flagellum assists in penetration of host cell.

of a pathogen's virulence factors on tissues vary greatly. Cold viruses, for example, invade and multiply but cause relatively little damage to their host. At the other end of the spectrum, pathogens such as the tetanus bacillus or rabies virus can severely damage or even kill their host.

Most of the common virulence factors fall into one or more of three categories: antiphagocytic effects, exoenzymes, and toxins. **Figure 13.11** illustrates the workings of some of these factors.

Antiphagocytic Factors

Microbes are likely to encounter resistance from host defenses when first entering the portal of entry. The

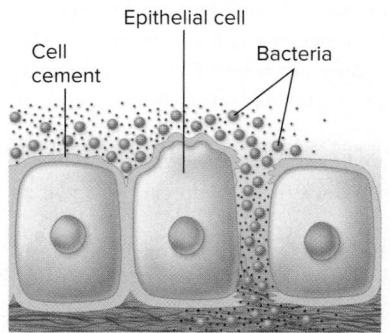

(a) **Exoenzymes**
Bacteria produce extracellular enzymes that dissolve barriers and penetrate through or between cells to invade underlying tissues.

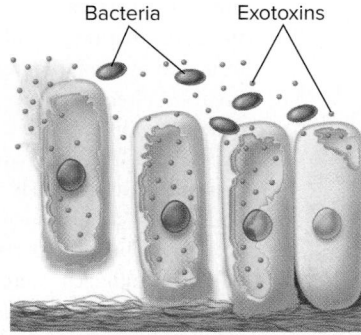

(b) **Toxins**
Toxins (primarily exotoxins) secreted by bacteria damage target cells, which die and begin to slough off.

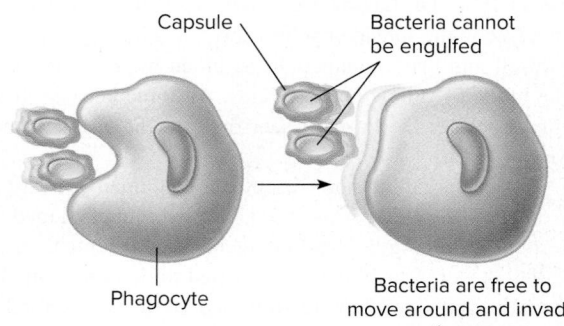

(c) **Blocked phagocytic response**
Bacteria have a property that enables them to escape phagocytosis and continue to grow and cause further infections.

(d) **Invasion factors**
Salmonella has a unique system to invade an intestinal cell. After adhering to the cell's microvilli with its fimbriae, it injects proteins into the cell with a probe-like structure. These proteins disrupt the actin filaments and cause the cell membrane to form a series of ruffles that surround the bacterium and pull it inside. It becomes shielded within a vacuole and begins to grow. From here, it can gain entrance to nearby tissues.

Figure 13.11 Pathological effects of virulence factors on host cells.

13.2 MAKING CONNECTIONS

Human Guinea Pigs

These days, human beings are not used as subjects for determining the effects of infectious disease, but in earlier times they were. The tradition of human experimentation dates well back into the eighteenth century, and the subject frequently was the experimenter himself. In some studies, mycologists inoculated their own skin and even that of family members with scrapings from fungal lesions to demonstrate that the disease was transmissible. In the early days of parasitology, it was not uncommon for a brave researcher to swallow worm eggs in order to study the course of his infection and the life cycle of the worm.

In a sort of reverse test, one of Robert Koch's German colleagues, named Max von Petenkofer, believed so strongly that cholera was *not* caused by a bacterium that he and his assistant swallowed cultures of *Vibrio cholerae.* Fortunately for them, they acquired only a mild form of the disease. Many self-experimenters have not been so fortunate.

One of the most famous cases is that of Jesse Lazear, a Cuban physician who worked with Walter Reed on the etiology of yellow fever in 1900. Dr. Lazear was convinced that mosquitoes were directly involved in the spread of yellow fever; and by way of proof, he allowed himself and two volunteers to be bitten by mosquitoes infected with the blood of yellow fever patients. All three became ill as a result of this exposure; Dr. Lazear's sacrifice was the ultimate one: he died of yellow fever.

A remarkable, more recent use of human guinea pigs occurred from 1954 to 1973, when the U.S. government staged 153 separate tests using biological warfare agents. The scientists in charge of so-called "Operation Whitecoat" wanted to determine the infectivity of certain agents and how readily they could be acquired through airborne or other means. A group of about 2,000 courageous volunteers agreed to be exposed to some highly virulent pathogens, including the agents of Q fever, tularemia, and leishmaniasis. None of the deadliest pathogens (anthrax, plague, smallpox) was tested on humans. The test subjects all shared some important characteristics: they were youngj male draftees, and they belonged to the Seventh-Day Adventist religion.

What made them uniquely suited to this highly secret study was their status as conscientious objectors—opposed to service as combat

A specially constructed chamber with vents for exposing test subjects to pathogens *Source:* US Army

soldiers on religious grounds. Many of them were assigned to other military positions of a medical or clerical type. One of the expectations was that they would consent to be subjects for experiments. The tests, run by the biological warfare program at Fort Detrick, Maryland, involved direct exposure to airborne pathogens released inside special chambers, or in some cases breathing aerosols of pure cultures released into the air. Frequently monkeys and actual guinea pigs were exposed simultaneously as a way to track the same infectious agent in animals for comparison.

A number of important discoveries were made. The infectious doses of several pathogens were determined by these studies, as were the effects of particle size. Other tests helped determine the effectiveness of vaccines. It is quite extraordinary that none of the men experienced long-term diseases as a result of these studies, given the high risks involved. It is very unlikely that this sort of testing would be sanctioned in the present era, but there is no doubt that the sacrifices of these men greatly advanced our understanding of the risks of biological warfare.

What are some reasons epidemiologists might have needed to use human subjects to test for the effects of pathogens? *Answer available on Connect.*

initial response comes from white blood cells called **phagocytes.** These cells can ordinarily engulf pathogens and destroy them by means of enzymes and antimicrobial chemicals (see chapter 14).

Antiphagocytic factors are a type of virulence factor used by some pathogens to avoid phagocytes or to circumvent some part of the phagocytic process **(figure 13.11c).** The most aggressive strategy involves bacteria that kill phagocytes outright. Species of both *Streptococcus* and *Staphylococcus* produce **leukocidins,** substances that are toxic to white blood cells. Some microorganisms secrete an extracellular surface layer (slime or capsule) that makes it physically difficult for the phagocyte to engulf them. *Streptococcus pneumoniae, Salmonella typhi, Neisseria meningitidis,* and *Cryptococcus neoformans* are notable examples. Some bacteria are well adapted to survival inside phagocytes after ingestion. For instance, pathogenic species of *Legionella* and *Mycobacterium* are readily engulfed but

are capable of avoiding further destruction. This survival inside phagocytes has special significance because it provides a place for the microbes to hide, grow, and be spread throughout the body.

Getting Inside Host Cells Intracellular pathogens in particular must gain entrance to a host cell to complete their life cycles. Rickettsias, chlamydias, and viruses have evolved specific virulence mechanisms for entering a cell through some variation of engulfment or fusion. Some bacterial pathogens invade their target cells by using a secretion system that inserts specialized virulence proteins directly into them **(figure 13.11d).** In the case of *Salmonella* and

> **Quick Search**
> Search on "*Salmonella* video HHMI Biointeractive" to view a short animation of this process.

E. coli, the virulence proteins disrupt the actin cytoskeleton of an intestinal cell and force the cell surface to form ruffles that pull the pathogens into the cell interior. Once inside, they multiply within vacuoles and use the cell as a passageway to break out into the underlying layers of tissues. Other pathogens (*Listeria* and *Shigella*) collect actin filaments and use them as motility organelles to move about in the cells.

Extracellular Enzymes

Many pathogenic bacteria, fungi, protozoa, and worms secrete **exoenzymes** that disrupt the structure of tissues. Other enzymes dissolve the host's defense barriers and promote the spread of microbes to deeper regions of the body. A few examples include:

1. mucinase, which digests the protective coating on mucous membranes and is a factor in amoebic dysentery;
2. keratinase, which digests the principal component of skin and hair and is secreted by fungi that cause ringworm;
3. collagenase, which digests the protein fibers of connective tissue and is an invasive factor for *Clostridium* species and some parasitic worms; and
4. hyaluronidase, which digests hyaluronic acid, the ground substance that cements animal cells together. This enzyme is an important virulence factor in staphylococci, clostridia, streptococci, and pneumococci (**figure 13.11***a*).

Some enzymes react with components of the blood. Coagulase, an enzyme produced by pathogenic staphylococci, causes clotting of blood or plasma. By contrast, the bacterial kinases (streptokinase, staphylokinase) activates a pathway that breaks down fibrin clots, which expedites the invasion of damaged tissues by the pathogen. In fact, a drug containing streptokinase (Streptase) is administered to patients with life-threatening thrombi and emboli.[6]

Bacterial Toxins: Poisonous Products

A **toxin** is a specific chemical product of microbes, plants, and some animals that has poisonous effects on other organisms. **Toxigenicity,** the power to produce toxins, is a genetically controlled characteristic of many bacterial species and is responsible for the adverse effects of a variety of diseases generally called **toxinoses.** Toxinoses in which the toxin is spread by the blood from the site of infection are called **toxemias** (tetanus and diphtheria, for example), whereas those caused by ingestion of toxins are **intoxications** (botulism). A toxin is named according to its specific target of action: neurotoxins act on the nervous system; enterotoxins act on the intestine; hemotoxins damage red blood cells; and nephrotoxins target the kidneys.

A more traditional scheme classifies bacterial toxins according to their origins (**figure 13.12**). A toxin molecule secreted by a living bacterial cell into the infected tissues is an **exotoxin.** A toxin that is not secreted but is released only after the cell is damaged or lysed is an **endotoxin.** Other important differences between the two groups are summarized in **table 13.8.**

(a) Target organs are damaged; heart, muscles, blood cells, intestinal tract show dysfunctions.

(b) General physiological effects—fever, malaise, aches, shock

Figure 13.12 The origins and effects of circulating exotoxins and endotoxin. (a) Exotoxins, given off by live cells, have highly specific targets and physiological effects. **(b)** Endotoxin, given off when the cell wall of gram-negative bacteria disintegrates, has more generalized physiological effects.

TABLE 13.8	Differential Characteristics of Bacterial Exotoxins and Endotoxin	
Characteristic	**Exotoxins**	**Endotoxin**
Toxicity	Toxic in small amounts	Toxic in higher quantities
Effects on the Body	Specific to a cell type (blood, liver, nerve)	Systemic and less specific: fever, inflammation, weakness, shock
Chemical Composition	Small proteins	Lipopolysaccharide of cell wall
Heat Denaturation at 60°C	Unstable	Stable
Toxoid Formation	Can be converted to toxoid*	Cannot be converted to toxoid
Immune Response	Stimulate antitoxins**	Does not stimulate antitoxins
Fever Stimulation	Usually not	Yes
Manner of Release	Secreted from live cell	Released from cell wall during lysis
Typical Sources	A few gram-positive and gram-negative bacteria	All gram-negative bacteria
Examples of Diseases	Tetanus, diphtheria, cholera, anthrax	Meningitis, endotoxic shock, salmonellosis

*A toxoid is an inactivated toxin used in vaccines.

**An antitoxin is an antibody that reacts specifically with a toxin.

6. These conditions are intravascular blood clots that can cause circulatory obstructions.

Exotoxins are proteins with a strong specificity for a target cell and extremely powerful, sometimes deadly, effects. They generally affect cells by damaging the cell membrane and initiating lysis or by disrupting intracellular function (**figure 13.11***b*). **Hemolysins**** are a class of bacterial exotoxins that disrupt the cell membrane of red blood cells (and some other cells, too). This damage causes the red blood cells to hemolyze—to burst and release hemoglobin pigment. Hemolysins that increase pathogenicity include the streptolysins of *Streptococcus pyogenes* and the alpha (α) and beta (β) toxins of *Staphylococcus aureus*. When colonies of bacteria growing on blood agar produce hemolysin, distinct zones appear around the colony. The pattern of hemolysis is one characteristic used to identify bacteria and determine their degree of pathogenicity.

The exotoxins of diphtheria, tetanus, and botulism, among others, attach to a particular target cell, become internalized, and interrupt an essential cell pathway. This type of exotoxin is specified as an **A-B toxin** (**figure 13.13**). so named because it consist of two parts, an A (active) component and a B (binding) component. The B component binds to a specific receptor molecule on the surface of the host cell, then the A component is moved across the host cell membrane. The A components of most A-B toxins interfere with the production of a normal cell protein. Loss of that protein's function then damages and disrupts a specific target cell or tissue. The consequences of cell disruption depend upon the target.

Examples include a toxin of *Clostridium tetani* that blocks the action of certain spinal neurons; the toxin of *Clostridium botulinum* that interferes with transmission of nerve-muscle stimuli; pertussis toxin that inactivates the respiratory cilia; and cholera toxin that provokes profuse salt and water loss from intestinal cells. Review the Case Study in chapter 8, which deals with some charactersistics of botulinum toxin.

Unlike exotoxins, of which there are many specific examples, there is just a single type of endotoxin. It is chemically a substance called lipopolysaccharide (LPS), which is part of the outer membrane of gram-negative cell envelopes. When gram-negative bacteria cause infections, some of the bacteria eventually lyse and release these LPS molecules into the infection site or into the circulation. Endotoxin differs from exotoxins in having a variety of systemic effects on tissues and organs. Depending upon the amounts present, endotoxin can cause fever, inflammation, hemorrhage, and diarrhea. Blood infection by gram-negative bacteria such as *Salmonella, Shigella, Neisseria meningitidis,* and *Escherichia coli* is particularly dangerous. It can lead to fatal endotoxic shock caused by large amounts of endotoxins being released into the blood stream.

Effects of Infection on Organs and Body Systems

In addition to the adverse effects of enzymes, toxins, and other factors, multiplication by a pathogen frequently weakens host tissues. Pathogens can obstruct tubular structures such as blood vessels, lymphatic channels, fallopian tubes, and bile ducts. Accumulated damage can lead to cell and tissue death, a condition called **necrosis.**

Figure 13.13 **The actions of A-B toxin of *Corynebacterium diphtheriae.*** The B part attaches to a specific receptor on the cell membrane, and the toxin is engulfed into a vacuole. The A and B parts separate, and the A portion is released into the cytoplasm, where it blocks the synthesis of a protein required for normal cell function. The diphtherotoxin is highly destructive of cells, especially those of the heart and nervous system.

Although viruses do not produce toxins or destructive enzymes, they destroy cells by multiplying in and lysing them. Many of the cytopathic effects of viral infection arise from the impaired metabolism and death of cells (see chapter 6).

The direct effects of virulence factors such as enzymes and toxins can cause significant harm to cells, tissues, and organs. It is important to note that many microbial diseases are actually the result of indirect damage, or the host's excessive or inappropriate response to a microorganism. This means that pathogenicity is not a trait inherent in microorganisms alone but is really a consequence of the interplay between microbe and host. The capsule of *Streptococcus pneumoniae* is a good example. It prevents the bacterium from being cleared from the lungs by phagocytic cells, leading to a powerful inflammatory response by white blood cells and a continuous influx of fluids into the lung spaces—a condition we know as pneumonia.

**hemolysins* (hee-mahl′-uh-sinz) Gt. *hemo*, blood, and *lysin,* to split.

Of course, it is easier to study and characterize the microbes that cause direct damage through toxins or enzymes. For this reason, these true pathogens were the first to be fully understood as the science of microbiology progressed. But in the last 15 to 20 years, microbiologists have come to appreciate exactly how important the relationship between microbe and host is, and this has greatly expanded our understanding of infectious diseases.

 Check Your Progress SECTION 13.2

8. Explain several ways in which true pathogens differ from opportunistic pathogens.

9. Distinguish between pathogenicity and virulence.

10. Explain the concept of portal of entry, and list the major portals of entry with examples of associated infections.

11. Differentiate between exogenous and endogenous infections.

12. What factors possibly affect the size of the infectious dose?

13. Describe several components of pathogens that are involved in microbial adhesion.

14. Define virulence factors and give examples of them in gram-positive and gram-negative bacteria, viruses, and parasites.

15. Discuss the effects of several virulence factors that help microbes invade hosts and evade host defenses.

16. Define *toxigenicity* and summarize the main features of exotoxins and endotoxins.

17. Which body cells or tissues are affected by hemolysins, leukocidins, hyaluronidase, kinases, tetanus toxin, pertussis toxin, and enterotoxin?

13.3 The Outcomes of Infection and Disease

 Expected Learning Outcomes

14. Describe the clinical stages of infection.

15. Use key terms to describe different patterns of infection.

16. Use correct terminology to explain the manifestations of infections and inflammation.

17. Discuss the major portals of exit and how they influence the end stages of infection and disease.

The Stages of Clinical Infections

As the body of the host responds to the invasive and toxigenic activities of a pathogen, it passes through four distinct phases of infection and disease: the incubation period, the prodromal stage, the period of invasion, and the convalescent period (**figure 13.14**).

The **incubation period** is the time from initial contact with the infectious agent (at the portal of entry) to the appearance of the first symptoms. During the incubation period, the agent is multiplying at the portal of entry but has not yet caused enough damage to elicit symptoms. Although this period is relatively well defined and predictable for each microorganism, it does vary according to host

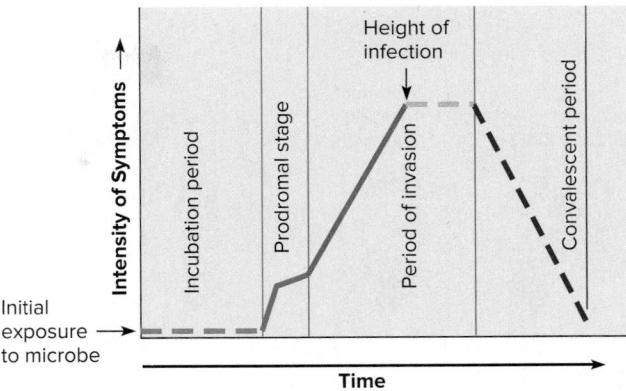

Figure 13.14 Stages in the course of infection and disease. Dashed lines represent periods with a variable length.

resistance, degree of virulence, and distance between the target organ and the portal of entry (the farther apart, the longer the incubation period). Overall, an incubation period can range from several hours in pneumonic plague to several years in leprosy. The majority of infections, however, have incubation periods ranging between 2 and 30 days.

The earliest notable symptoms of infection appear as a vague feeling of discomfort, such as head and muscle aches, fatigue, upset stomach, and general malaise. This short period (1–2 days) is known as the **prodromal stage.** The infectious agent next enters a **period of invasion,** during which it multiplies at high levels, exhibits its greatest toxicity, and becomes well established in its target tissues and organs. This period is often marked by fever and other prominent and more specific signs and symptoms, which can include cough, skin lesions, diarrhea, loss of muscle control, swelling, jaundice, discharge of exudates, or severe pain, depending on the pathogen. The length of this period is extremely variable.

As the patient begins to respond to the infection, the symptoms decline—sometimes dramatically, other times slowly. During the recovery that follows, called the **convalescent period,** the patient's strength and health gradually return as the immune response begins to clear the infectious agent and restore normal function to damaged tissues. In the event that the patient does not recover and dies, the infection is considered *terminal.*

The transmissibility of the microbe during these four stages must be considered on an individual basis. A few agents are released mostly during incubation (measles, for example); many are released during the invasive period (*Shigella*); and others can be transmitted during all of these periods (hepatitis B virus, norovirus).

Patterns of Infection

Patterns of infection are many and varied. In the simplest situation, a **localized infection,** the microbe enters the body and remains confined to a specific tissue (**figure 13.15a**). Examples of localized infections are boils, fungal skin infections, and warts.

Many infectious agents do not remain localized but spread from the initial site of entry to other tissues. In fact, spreading is necessary for pathogens such as rabies and hepatitis A virus, whose target tissue may be some distance from the site of entry. The rabies

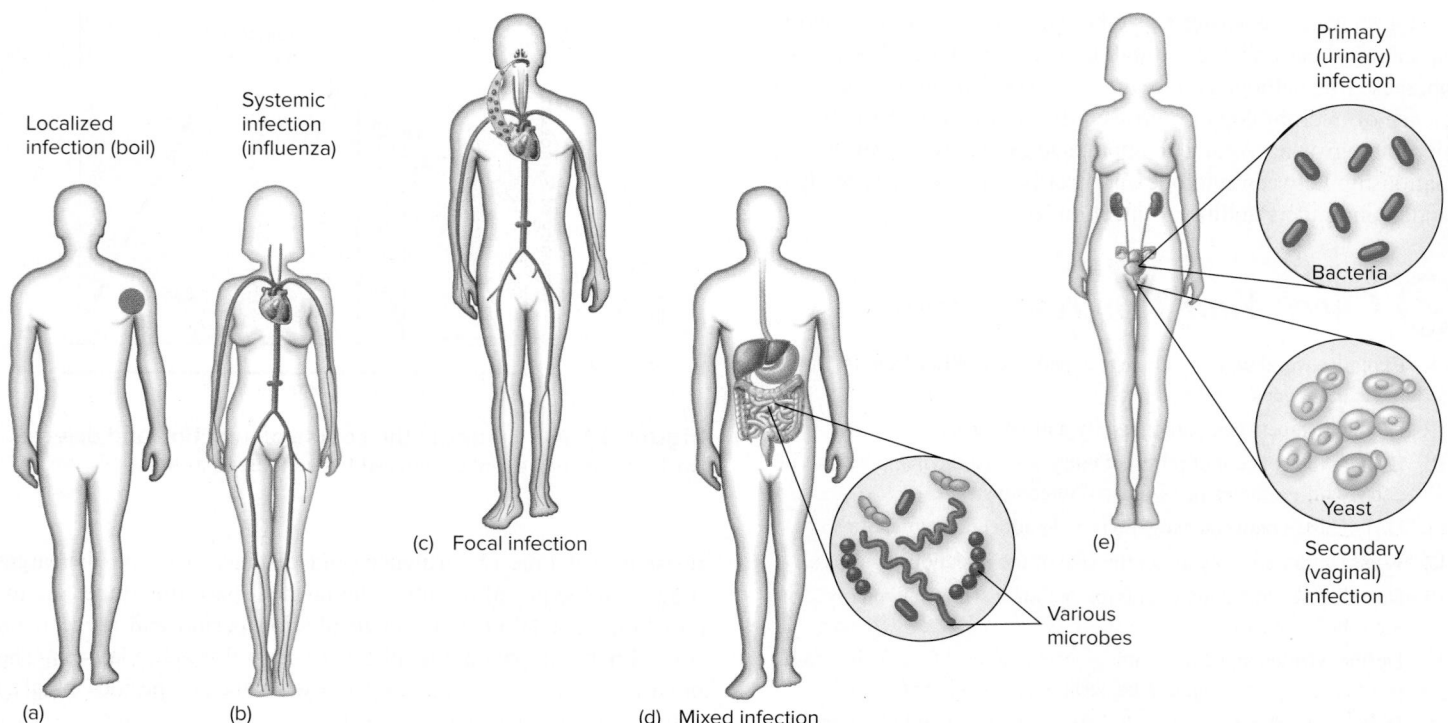

Localized
infection (boil)

Systemic
infection
(influenza)

(c) Focal infection

(a)

(b)

(d) Mixed infection

Various
microbes

Primary
(urinary)
infection

Bacteria

Yeast

(e)

Secondary
(vaginal)
infection

Figure 13.15 **The occurrence of infections with regard to location and sequence. (a)** A localized infection, in which the pathogen is restricted to one specific site. **(b)** Systemic infection, in which the pathogen spreads through circulation to many sites. **(c)** A focal infection occurs initially as a local infection, but circumstances cause the microbe to be carried to other sites systemically. **(d)** A mixed infection, in which the same site is infected with several microbes at the same time. **(e)** In a primary-secondary infection, an initial infection is complicated by a second one in the same or a different location and caused by a different microbe.

virus travels from a bite wound along nerve tracts to its target in the brain, and the hepatitis A virus moves from the intestine to the liver via the circulatory system. When an infection spreads to several sites and tissue fluids, usually in the bloodstream, it is called a **systemic infection (figure 13.15b).** Examples of systemic infections are viral diseases (measles, rubella, chickenpox, and AIDS); bacterial diseases (brucellosis, anthrax, typhoid fever, and syphilis); and fungal diseases (histoplasmosis and cryptococcosis). Infectious agents can also travel to their targets by means of nerves (as in rabies) or cerebrospinal fluid (as in meningitis).

A **focal infection** is said to exist when the infectious agent breaks loose from a local infection and is seeded or *disseminated* into other tissues **(figure 13.15c).** This pattern is exhibited by tuberculosis or by streptococcal pharyngitis, which gives rise to scarlet fever. In the condition called **toxemia,**[7] the infection itself remains localized at the portal of entry, but the toxins produced by the pathogens are carried by the blood to the actual target tissue. In this way, the cells initially infected by the bacterial cells can be different from the targets of their toxin. Examples include diphtheria and some staphylococcal diseases.

An infection is not always caused by a single microbe. In a **mixed infection,** several agents establish themselves simultaneously at the infection site **(figure 13.15d).** In some mixed infections, the microbes cooperate in breaking down a tissue. In other cases, one microbe creates an environment that enables another

microbe to invade. Gas gangrene, wound infections, dental caries, and human bite infections tend to be mixed. These are sometimes called **polymicrobial** diseases and may be the result of biofilm formation at the site of infection.

Some diseases are described according to the sequence of occurrence. When an initial, or **primary,** infection is complicated by another infection caused by a different microbe, this subsequent infection is termed a **secondary infection (figure 13.15e).** This pattern is seen in a child whose chickenpox (primary infection) becomes infected with *Staphylococcus aureus* (secondary infection). Another significant combination is seen in patients who develop secondary pneumonia as a complication of influenza. The secondary infection need not be in the same site as the primary infection, and it usually indicates weakened host defenses.

Infections that come on rapidly, with severe but short-lived effects, are called **acute infections.** Infections that progress and persist over a long period of time are **chronic infections.** A Guide to the Terminology of Infection and Disease illustrates other common words used to describe infectious diseases.

A Guide to the Terminology of Infection and Disease

Words in medicine have great power and economy. A single technical term can often replace a whole phrase or sentence, thereby saving time and space in charting. The beginning student may feel overwhelmed by what seems like a mountain of new words. However, having a grasp of a few root words and a fair amount of

7. Not to be confused with toxemia of pregnancy, which is a metabolic disturbance and not an infection.

anatomy can help you learn many of these words and even deduce the meaning of unfamiliar ones. Some examples of medical shorthand follow.

The suffix -*itis* means an inflammation and, when affixed to the end of an anatomical term, indicates an inflammatory condition in that location. Thus, meningitis is an inflammation of the meninges surrounding the brain; encephalitis is an inflammation of the brain itself; hepatitis involves the liver; gastroenteritis, the intestine; and otitis media, the middle ear. Although not all inflammatory conditions are caused by infections, many infectious diseases inflame their target organs.

The suffix -*emia* is derived from the Greek word *haeima,* meaning blood. When added to a word, it means "associated with the blood." Thus, septicemia means sepsis (infection) of the blood; bacteremia, bacteria in the blood; viremia, viruses in the blood; and fungemia, fungi in the blood. It is also applicable to specific conditions such as toxemia, gonococcemia, and spirochetemia.

The suffix -*osis* means "a disease or morbid process." It is frequently added to the names of pathogens to indicate the disease they cause: for example, listeriosis, histoplasmosis, toxoplasmosis, shigellosis, salmonellosis, and borreliosis. A variation of this suffix is -*iasis,* as in trichomoniasis and candidiasis.

The suffix -*oma* comes from the Greek word *onkomas* (swelling) and means tumor. Although the root is often used to describe cancers (lymphoma, melanoma), it is also applied in some infectious diseases that cause masses or swellings (tuberculoma, leproma).

Signs and Symptoms: Warning Signals of Disease

When an infection causes pathologic changes leading to disease, it is often accompanied by a variety of signs and symptoms. A **sign** is any objective evidence of disease as noted by an observer; a **symptom** is the subjective evidence of disease as sensed by the patient. Signs tend to be more precise than symptoms and are often measured. Both can be the result of the same underlying cause. For example, an infection of the brain may present with the sign of bacteria in the spinal fluid and the symptom of headache. Or a streptococcal infection may produce a sore throat (symptom) and inflamed pharynx (sign). Disease indicators that can be sensed and observed can qualify as either a sign or a symptom depending upon how they are reported. When a disease can be identified or described by a defined collection of signs and symptoms, it is termed a **syndrome.** Signs and symptoms with considerable importance in diagnosing infectious diseases are shown in **table 13.9.**

Signs and Symptoms of Inflammation

The earliest manifestations of disease generally result from the activation of the defense process called **inflammation.*** The inflammatory response includes cells and chemicals that are activated when there has been damage to tissues by a microbe or its products. This subject is discussed in greater detail in chapter 14, but as noted earlier, many symptoms and signs of infection are caused by the

inflammation (in-flam-aye'-shun) L. *inflammation,* to set on fire.

TABLE 13.9	Common Signs and Symptoms of Infectious Diseases
Signs*	**Symptoms**
Fever	Chills
Septicemia	Pain, irritation
Microbes in tissue fluids that should be sterile	Nausea
Abnormal chest sounds	Malaise, fatigue
Skin eruptions	Chest tightness
Leukocytosis	Itching
Leukopenia	Headache
Swollen lymph nodes	Weakness
Abscesses	Abdominal cramps
Tachycardia (increased heart rate)	Anorexia (lack of appetite)
Antibodies in serum	Sore throat

*The technical terms for signs are defined in the body of the text

mobilization of this system. Inflammation is always a likely cause of symptoms such as redness, pain, soreness, and swelling. A general term for a noticeable sign of infection or disease is **lesion.*** Some lesions typical of inflammation include **edema,*** the accumulation of fluid in an afflicted tissue; **granulomas** and **abscesses,** walled-off collections of inflammatory cells and microbes in the tissues; and **lymphadenitis,** swollen lymph nodes.

Rashes and other skin eruptions are some of the most common lesions in many diseases, and because they tend to mimic each other, it can be difficult to differentiate among diseases on this basis alone. Skin lesions can be restricted to the epidermis and its glands and follicles, or they can extend into the dermis and subcutaneous regions. The lesions of some infections (boils, chickenpox, athlete's foot, warts) undergo characteristic changes in appearance and location during the course of disease and thus fit more than one category.

Signs of Infection in the Blood

Changes in the number of circulating white blood cells, as determined by blood cell counts, are considered to be a significant sign of possible infection. **Leukocytosis*** is an increase in the level of white blood cells, whereas **leukopenia*** is a decrease. Other signs of infection revolve around the occurrence of a microbe or its products in the blood. The clinical term for blood infection, **septicemia,** refers to a general state in which microorganisms are multiplying in the blood and are present in large numbers. When small numbers of bacteria or viruses are found in the blood, the correct terminology is **bacteremia** or **viremia,** which means that these microbes are present in the blood but are not necessarily multiplying.

During infection, a normal host will invariably show signs of an immune response in the form of antibodies in the serum or some type of sensitivity to the microbe. This fact is the basis for

lesion lee'-zhun L. *laesio,* to hurt.
edema (uh-dee'-muh) Gr. *oidema,* swelling.
leukocytosis (loo″-koh'-sy-toh'-sis) From *leukocyte,* a white blood cell, and the suffix -*osis.*
leukopenia (loo″-koh'-pee'-nee-uh) From *leukocyte* and *penia,* a loss or lack of.

several serological tests used in diagnosing infectious diseases such as HIV infection or syphilis. Such specific immune reactions indicate the body's attempt to develop specific immunities against pathogens. We concentrate on this role of the host defenses in chapters 14 and 15.

Infections That Go Unnoticed

It is rather common for an infection to produce no noticeable or typical symptoms, even though the microbe is active in the host tissue. In these cases the person does not show overt indications of infection. Infections of this nature are known as **asymptomatic, subclinical,** or *inapparent* because the patient experiences no symptoms usually associated with disease and does not seek medical attention. However, it is important to note that most infections are attended by some sort of detectable *sign*. In section 13.4 we further address the significance of subclinical infections in the transmission of infectious agents.

The Portal of Exit: Vacating the Host

Earlier we introduced the idea that a parasite is considered *unsuccessful* if it does not have a provision for leaving its host and moving to other susceptible hosts. With few exceptions, pathogens depart by a specific avenue called the **portal of exit (figure 13.16).**

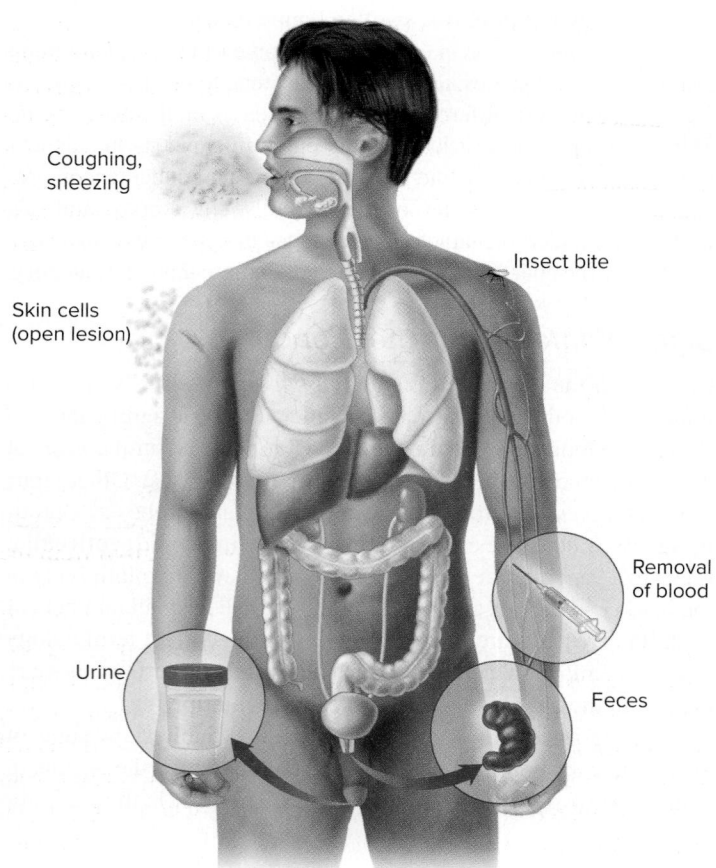

Coughing, sneezing

Skin cells (open lesion)

Insect bite

Removal of blood

Urine

Feces

Figure 13.16 Major portals of exit of infectious diseases. Pathogens are released from the body through secretion, excretion, discharge, or shedding.

In most cases the pathogen is shed or released from the body through secretion, excretion, discharge, or sloughed tissue. The high number of pathogens in these materials increases both virulence and the likelihood that the pathogen will reach other hosts. One of the most dramatic examples of this is ebolavirus (Case Study), which is at its most transmissible late in the disease and is released in massive numbers in every bodily fluid. In many cases the portal of exit is the same as the portal of entry, but a few pathogens use a different route. As we will see in section 13.4, the portal of exit concerns epidemiologists because it greatly influences the dissemination of infection in a population.

Respiratory and Salivary Portals

Mucus, sputum, nasal drainage, and other moist secretions are the fluids of escape for the pathogens that infect the lower or upper respiratory tract. The most forceful means of releasing these secretions are coughing and sneezing (see figure 13.19), although they can also be released during talking and laughing. Tiny particles of liquid released into the air form aerosols or droplets that can spread the infectious agent to other people. The agents of tuberculosis, influenza, measles, and chickenpox most often leave the host through airborne droplets. Droplets of saliva are also the exit route for several viruses, including those of mumps, rabies, and infectious mononucleosis.

Epithelial Cells

The outer layers of the skin and scalp are constantly being shed into the environment. A large proportion of household dust is actually composed of skin cells. A single person can shed several billion skin cells a day, and some persons, called shedders, disseminate massive numbers of bacteria into their immediate surroundings. Skin lesions and their exudates can serve as portals of exit in warts, fungal infections, boils, herpes simplex, smallpox, and syphilis.

Fecal Exit

Feces are a very common portal of exit. Some intestinal pathogens grow in the intestinal mucosa and create an inflammation that increases the motility of the bowel. This increased motility speeds up peristalsis, resulting in diarrhea. The more-fluid stool provides a rapid exit for the pathogen, as we see with dysentery, cholera, salmonellosis, and noroviruses. A number of helminth worms release cysts and eggs through the feces. Feces containing pathogens are a public health problem when allowed to contaminate drinking water or when used to fertilize crops.

Urogenital Tract

A number of agents involved in sexually transmitted infections leave the host in vaginal discharge or semen. This is also the source of neonatal infections such as herpes simplex, *Chlamydia,* and *Candida albicans,* which infect the infant as it passes through the birth canal. Less commonly, certain pathogens that infect the kidney are discharged in the urine; examples are the agents of leptospirosis, typhoid fever, tuberculosis, and schistosomiasis.

Removal of Blood or Bleeding

Although the blood does not have a direct route to the outside, it can serve as a portal of exit when it is removed or released through a vascular puncture made by natural or artificial means. Blood-feeding animals such as ticks, fleas, and mosquitoes are common transmitters of pathogens. Human immunodeficiency and hepatitis viruses are transmitted by shared needles or through small abrasions in a mucous membrane during sexual intercourse.

Because many microbes can be transferred through blood donation, the monitoring of both donors and donated blood has to be scrupulous. In the mid-1980s, several thousand people suffering from hemophilia were infected with HIV as a result of receiving blood-clotting factors contaminated with the virus. The current risk for transfusion-transmitted infections is about one case in 10 million transfusions. One factor that increases risk is the presence of emerging disease agents in the blood supply that have not yet been detected. This has already happened with West Nile virus and Chagas disease.

The Persistence of Microbes and Pathologic Conditions

The apparent recovery of the host does not always mean that the microbe has been completely removed or destroyed by the host defenses. After the initial symptoms in certain chronic infectious diseases, the infectious agent retreats into a state called chronic **persistence** or **latency.** In some cases of latency, the microbe can periodically become active and produce recurrent disease. The viral agents of herpes simplex, herpes zoster, hepatitis B, AIDS, and Epstein-Barr can persist in the host for long periods. The agents of syphilis, typhoid fever, tuberculosis, and malaria also enter into latent stages. The person harboring a persistent infectious agent may or may not shed it during the latent stage. If it is shed, such persons are chronic carriers who serve as sources of infection for the rest of the population.

Some diseases leave **sequelae*** in the form of long-term or permanent damage to tissues or organs. For example, meningitis can result in deafness, a strep throat can lead to rheumatic heart disease, Lyme disease can cause arthritis, and hepatitis B can initiate cirrhosis and liver cancer.

 Check Your Progress SECTION 13.3

18. Explain what is happening during each stage of infection. Compare and contrast: systemic, local, and focal infections; primary versus secondary infections; infection versus intoxication.
19. Discuss the distinguishing features of signs and symptoms, using examples.
20. Name some examples of infections and their portals of exit.
21. Using terminology from this section's "Guide to Terminology," define *urethritis, endotoxemia, chlamydiosis,* and *lymphoma*.

13.4 Epidemiology: The Study of Disease in Populations

 Expected Learning Outcomes

18. Define *epidemiology,* and summarize the major goals of its studies.
19. Relate the significant factors involved in the transmission of infectious diseases.
20. Explain the concept of carriers and describe several types.
21. Use examples to distinguish between the types of vectors, and discuss the importance of zoonoses.
22. Differentiate between communicable and noncommunicable infectious diseases.
23. Characterize the patterns of transmission for communicable diseases.

So far our discussion has revolved primarily around the impact of an infectious disease in individuals. We now turn our attention to the effects of diseases on the community—the realm of **epidemiology.*** By definition, this term involves the study of the frequency and distribution of disease and other health-related factors in defined human populations. It involves many disciplines—not only microbiology but also anatomy, physiology, immunology, medicine, psychology, sociology, ecology, and statistics—and it considers many diseases other than infectious ones, including heart disease, cancer, drug addiction, and mental illness.

The epidemiologist is a medical sleuth who collects clues on the causative agent, pathology, and sources and modes of transmission and tracks the numbers and distribution of cases of disease in the community. In fulfilling these demands, the epidemiologist looks for the "chain of infection" involving the who, when, where, how, why, and what that surrounds the spread of diseases. The outcome of these studies helps public health departments develop prevention and treatment programs and establish a basis for predictions.

Origins and Transmission Patterns of Infectious Microbes

Reservoirs: Where Pathogens Persist

One of the important concerns of epidemiology is discovering where pathogens originate and how they are transmitted. As we saw in the Case Study, these aspects are intensely scrutinized from the very first signs of an outbreak.

For an infectious agent to continue to exist and be spread, it must have a permanent place to reside. The **reservoir** is the primary habitat in the natural world from which a pathogen originates. Often it is a human or animal carrier, although soil, water, and plants are also reservoirs. The reservoir can be distinguished from the infection **source,** which is the individual or object from which an infection is actually acquired. An infectious agent that does not survive well outside of its host and is usually transmitted directly tends to have the same reservoir and source, whereas one that can remain viable outside of the host can have a different

* *sequelae* (su-kwee′-lee) L. *sequi,* to follow.

* *epidemiology* (ep″-ih-dee-mee-alh′-uh-gee) Gr. *epidemios,* prevalent.

Figure 13.17 Involvement of carriers in transmission of infectious agents. (a) An asymptomatic carrier is infected without symptoms. **(b)** Incubation, convalescent, and chronic carriers can transmit the infection either before or after the period of symptoms. **(c)** A passive carrier is contaminated but not infected and spreads the pathogen through physical contact.

reservoir and source. For example, gonorrhea has the same reservoir and source (the human body), whereas hepatitis A usually has a different reservoir (a human carrier) and source (contaminated food).

Living Reservoirs

Many pathogens continue to exist and spread because they are harbored by members of a host population. A person or animal with an obvious or *overt* symptomatic infection is clearly a source of infection, as opposed to a **carrier** who is, by definition, an individual who *inconspicuously* shelters a pathogen and spreads it to others unknowingly. Although human carriers are occasionally detected through routine screening (blood tests, cultures) and other epidemiological devices, they are unfortunately very difficult to discover and control. As long as a pathogenic reservoir is maintained, the disease will continue to exist in that population or location, and the potential for epidemics will be a constant threat. The duration of the carrier state can be short or long term, and the carrier may or may not have experienced disease due to the microbe.

Several situations can produce the carrier state. **Asymptomatic** (apparently healthy) carriers are infected, but as previously indicated, they show no outward symptoms **(figure 13.17a).** A few asymptomatic infections (such as gonorrhea and genital warts) can carry out their entire course without overt manifestations. **Figure 13.17*b*** demonstrates three types of carriers who show no indications of infection at the time they transmit the pathogen. *Incubation carriers* spread the infectious agent during the incubation period before symptoms have appeared. For example, patients with HIV infection can harbor and spread the virus for months and years before their infection is evident. Recuperating patients without symptoms are considered *convalescent carriers* when they continue to shed viable microbes and convey the infection to others after symptoms have subsided. Norovirus, the most common cause of gastroenteritis, can be transmitted for weeks after an infection is resolved.

An individual who continues to shelter the infectious agent for a long period after recovery is a *chronic carrier.* Patients who have recovered from tuberculosis, hepatitis, and herpes infections frequently carry the agent chronically. About one in 20 victims of typhoid fever harbors *Salmonella typhi* in the gallbladder for several

years and sometimes for life. The most infamous of these was "Typhoid Mary," a cook who spread the infection to hundreds of victims in the early 1900s.

The **passive carrier** state is of great concern during patient care (see the discussion of health-care-associated infections in section 13.5). Medical and dental personnel who must constantly handle materials that are heavily contaminated with body fluids and blood are at risk of picking up pathogens mechanically and accidentally transferring them to other patients **(figure 13.17c).** Proper hand washing, handling of contaminated materials, and aseptic techniques greatly reduce this likelihood.

Animals as Reservoirs and Sources Up to now we have lumped animals with humans in discussing living reservoirs or carriers, but animals deserve special consideration as vectors of infections. The word **vector** is used by epidemiologists to indicate a live animal that transmits an infectious agent from one host to another. (The term is sometimes misused to include any object that spreads disease.) The majority of vectors are arthropods such as fleas, mosquitoes, flies, and ticks, although larger animals can also spread infection; for example, mammals (rabies), birds (ornithosis), or lower vertebrates (salmonellosis).

By tradition, vectors are placed into one of two categories, depending upon the animal's relationship with the microbe. A **biological vector** actively participates in a pathogen's life cycle, serving as a site in which it can multiply or complete its development. A biological vector communicates the infectious agent to the human host by biting, aerosol formation, or touch. In the case of biting vectors, the animal can inject infected saliva into the blood (mosquito), defecate around the bite wound (flea), or regurgitate blood into the wound (tsetse fly).

Mosquito

Mechanical vectors are not necessary to the life cycle of an infectious agent and merely transport it without being infected. The external body parts of these animals become contaminated when they come into physical contact with a source of pathogens. The agent is subsequently transferred to humans indirectly by an intermediate such as food or occasionally by direct contact (as in certain eye infections). Houseflies are notorious mechanical

Housefly

vectors. They feed on decaying garbage and feces, and while they are feeding, their feet and mouthparts easily become contaminated. They also regurgitate juices onto food to soften and digest it. Flies spread more than 20 bacterial, viral, protozoan, and worm infections. Other nonbiting flies transmit tropical ulcers, yaws, and trachoma. Cockroaches, which have unsavory habits similar to those of flies, play a role in the mechanical transmission of fecal pathogens, as well as contributing to allergy attacks in asthmatic children.

Many vectors and animal reservoirs spread their own infections to humans. An infection indigenous to animals but naturally transmissible to humans is a **zoonosis.*** In these types of infections, the human is essentially a dead-end host and does not contribute to the natural persistence of the microbe. Some zoonotic infections

* *zoonosis* (zoh″-uh-noh′-sis) Gr. *zoion,* animal, and *nosos,* disease.

TABLE 13.10	Common Zoonotic Infections	
Disease/Agent	**Primary Animal Reservoirs**	**Mode of Infection**
Viruses		
Rabies	All mammals	Bite; aerosol
Yellow fever	Wild birds, mammals, mosquitoes	Mosquito bite
Viral fevers	Wild mammals	Various mosquitoes
Hantavirus	Rodents	Airborne wastes
Influenza	Chickens, swine	Airborne secretions
West Nile virus	Wild birds, mosquitoes	Mosquito bite
Bacteria		
Rocky Mountain spotted fever	Dogs, ticks	Tick bite
Psittacosis	Birds	Airborne secretions
Leptospirosis	Domestic animals	Animal wastes (urine)
Anthrax	Domestic animals	Airborne spores, cutaneous contact
Brucellosis	Cattle, sheep, pigs	Air- and food-borne
Plague	Rodents, fleas	Flea bite; aerosol
Salmonellosis	Variety of mammals, birds, and rodents	Food-borne
Tularemia	Rodents, birds, arthropods	Tick bite, airborne
Miscellaneous		
Ringworm	Domestic mammals	Cutaneous contact
Toxoplasmosis	Cats, rodents, birds	Food-borne, contact
Trypanosomiasis	Domestic and wild mammals	Tsetse fly; reduviid bug
Trichinosis	Swine, bears	Ingestion of cysts
Tapeworm	Cattle, swine, fish	Ingestion of eggs
Scabies	Domestic animals	Cutaneous contact

(rabies, for instance) can have multihost involvement, and others (such as plague) can have very complex cycles in the wild. Zoonotic spread of disease is promoted by close associations of humans with animals, and people in animal-oriented or outdoor professions are at greatest risk. At least 150 zoonoses exist worldwide; the most common ones are listed in **table 13.10.** Zoonoses make up a full 70% of all new emerging diseases worldwide. It is worth noting that zoonotic infections are impossible to completely eradicate without also eradicating the animal reservoirs. Attempts have been made to eradicate mosquitoes and certain rodents. During influenza epidemics, millions of chickens and thousands of pigs are slaughtered in the attempt to control its spread.

One technique that can provide an early warning signal for the occurrence of certain mosquito-borne zoonoses has been the use of *sentinel animals.* These are usually domestic animals (most often chickens or horses) that can serve as hosts for diseases such as West Nile fever, various viral encephalitides, and malaria. Sentinel animals are placed at various sites throughout the community, and their blood is monitored periodically for antibodies to the infectious

agents that would indicate a recent infection by means of a mosquito bite. The presence of infected animals provides useful data on the potential for human exposure, and it also helps establish the epidemiological pattern of the zoonosis, including where it may have spread.

Coccidioides & Blastomyces are spread by soil/dust

habitat in the ...ind their way ...saprobic and ...some are op- ...human hosts ...s, acquisition ...d epidemio-

...otozoa, hel- ...ental stages ...ens that live ...*tridium* that ...Pathogenic ...e spread by ...hookworm ...in the soil. Natural bodies of water carry fewer nutrients than soil does but still support pathogenic species such as *Legionella, Cryptosporidium,* and *Giardia.*

The Acquisition and Transmission of Infectious Agents

Infectious diseases can be categorized on the basis of how they are acquired. A disease is **communicable** when an infected host can transmit the infectious agent to another host and establish infection in that new host. Although this terminology is standard, one must realize that it is not the disease that is communicated but the microbe. Also be aware that the word *infectious* is sometimes used interchangeably with the word *communicable,* but this is not precise usage. The transmission of the agent can be direct or indirect, and the ease with which the disease is transmitted varies considerably from one agent to another. If the agent is highly communicable, especially through direct contact, the correct term to use is **contagious.** Influenza and measles move readily from host to host and thus are contagious, whereas leprosy is only weakly communicable. Because they can be spread through the population, communicable diseases are our main focus in the following sections.

In contrast, a **noncommunicable** infectious disease does *not* arise through transmission of the infectious agent from host to host. The infection and disease are acquired through some other, special circumstance. Noncommunicable infections occur primarily when a compromised person is invaded by his or her own microflora (as with certain pneumonias, for example) or when an individual has accidental contact with a facultative parasite that exists in a nonliving reservoir such as soil. Some examples are certain mycoses, acquired through inhalation of fungal spores, and tetanus, in which *Clostridium tetani* spores from a soiled object enter a cut or wound. Persons thus infected do not become a source of infection.

Patterns of Transmission in Communicable Diseases

The routes or patterns of disease transmission are many and varied. The spread of diseases is by direct or indirect contact with animate or inanimate objects and can be horizontal or vertical. The term *horizontal* means the disease is spread through a population from one infected individual to another; *vertical* signifies transmission from parent to offspring via the ovum, sperm, placenta, or milk. Although exact transmission patterns can vary, a general system of organization divides microorganisms into two major groups, as shown in **figure 13.18:** those transmitted by direct routes or those transmitted by indirect routes, which involve some sort of vehicle to transport them.

Modes of Direct Transmission For microbes to be directly transferred, some type of contact must occur between the skin or mucous membranes of the infected person and those of the infectee. It may help to think of this route as the portal of exit meeting the portal of entry without the involvement of an intermediate object or substance. Included in this category are fine droplets sprayed directly upon a person during sneezing or coughing (as distinguished from droplet nuclei that are transmitted some distance by air). Most sexually transmitted diseases are spread directly. In addition, infections that result from kissing or bites by biological vectors are direct. Most obligate parasites are far too sensitive to survive for long outside the host and can be transmitted only through direct contact. Diseases transmitted vertically from mother to baby fit in this contact category as well.

Routes of Indirect Transmission For microbes to be indirectly transmitted, the infectious agent must pass from an infected host to an intermediate conveyor and from there to another host. This form of communication is especially pronounced when the infected individuals contaminate inanimate objects, food, or air through their activities. The transmitter of the infectious agent can be either openly infected or a carrier.

Indirect Spread by Vehicles: Contaminated Materials The term **vehicle** specifies any inanimate material commonly used by humans that can transmit infectious agents. A **common vehicle** or **source** is a single material that serves as the source of infection for many individuals. Some specific types of vehicles are food, water, various biological products (such as blood, serum, and tissue), and fomites.

A **fomite** is an inanimate object that harbors and transmits pathogens. The list of possible fomites is as long as your imagination allows. Probably highest on the list would be objects commonly in contact with the public, such as doorknobs, telephones, push buttons, and faucet handles that are readily contaminated by touching. Shared bed linens, handkerchiefs, toilet seats, toys, eating utensils, clothing, personal articles, and syringes are other examples. Although paper money is impregnated with a disinfectant to inhibit microbes, pathogens are still isolated from bills as well as coins.

Outbreaks of food poisoning often result from the role of food as a common vehicle. The source of the agent can be soil, the handler, or a mechanical vector. Because milk provides a rich

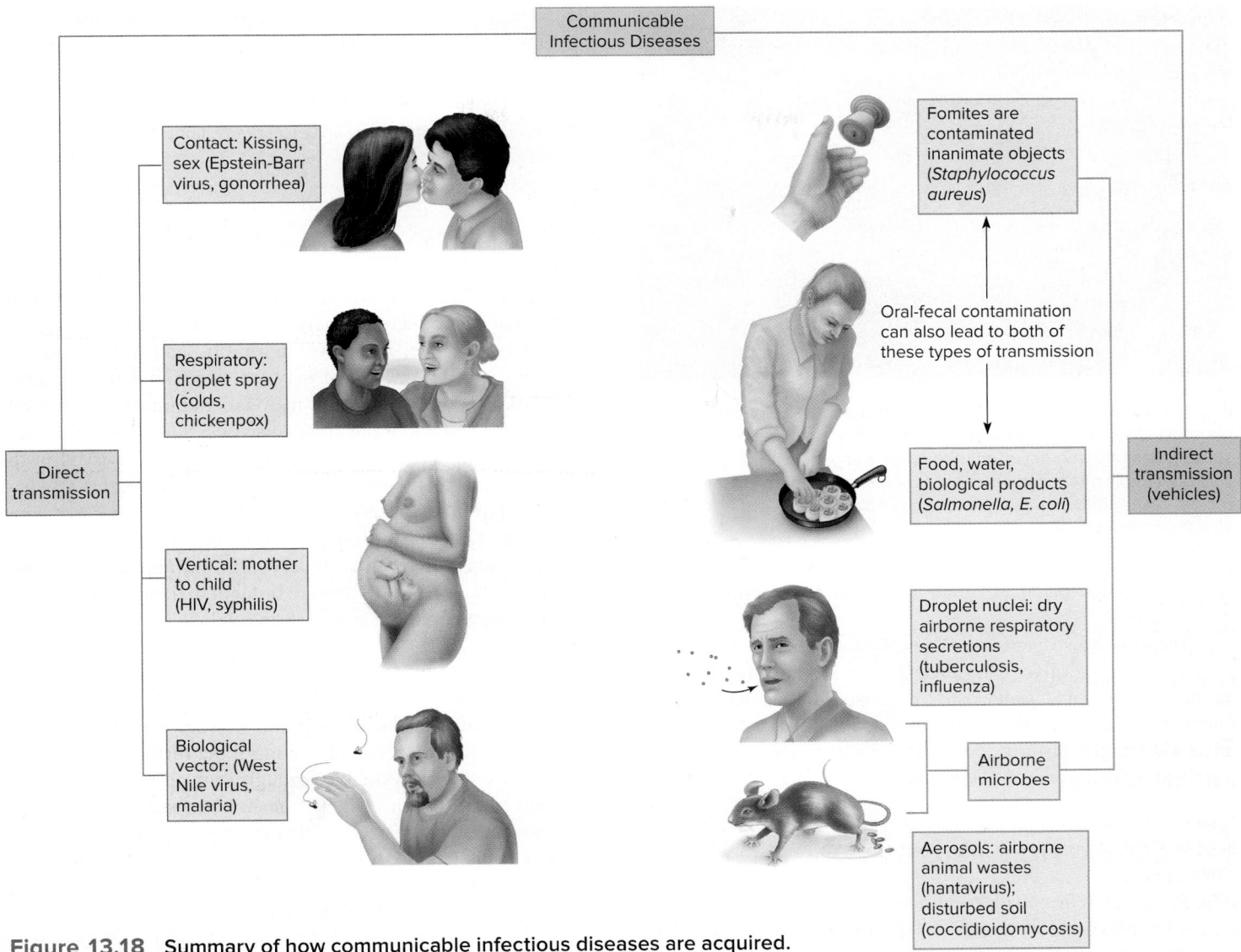

Figure 13.18 Summary of how communicable infectious diseases are acquired.

growth medium for microbes, it is a significant means of transmitting pathogens from diseased animals, infected milk handlers, and environmental sources of contamination. The agents of brucellosis, Q fever, salmonellosis, and listeriosis are transmitted by contaminated milk. Water that has been contaminated by excretions can carry *Salmonella, Vibrio* (cholera), viruses (hepatitis A, polio), *Leptospira*, and pathogenic protozoans (*Giardia, Cryptosporidium*).

The **oral-fecal route** is a special category of indirect transmission, which specifies that a vehicle first becomes contaminated through contact with fecal material and is then transported to someone's mouth. A common scenario would involve a fecal carrier contaminating food during handling and an unsuspecting person ingesting it. Hepatitis A, amebic dysentery, shigellosis, and typhoid fever are often transmitted this way. Oral-fecal transmission can

Quick Search
Learn about the effects of climate at "Infectious Diseases—Changing Planet" on YouTube.

also involve contaminated materials such as toys, diapers, and other inanimate objects. An example would be the norovirus, which is readily picked up through contact with public facilities touched by an infected person. Handwashing anyone?

Indirect Spread by Vehicles: Air as a Vehicle Unlike soil and water, outdoor air cannot provide nutritional support for most microbial growth and is less likely to transmit airborne pathogens. On the other hand, indoor air (especially in a closed space) can serve as an important medium for the suspension and dispersal of certain respiratory pathogens via droplet nuclei and aerosols.

Droplet nuclei are dried microscopic residues created when microscopic pellets of mucus and saliva are ejected from the mouth and nose. They are generated forcefully in an unstifled sneeze or cough (**figure 13.19**) or mildly during other vocalizations. Although the larger beads of moisture settle rapidly, smaller particles evaporate and remain suspended for longer periods and can be drawn deeply into the lungs. Droplet nuclei are implicated in the

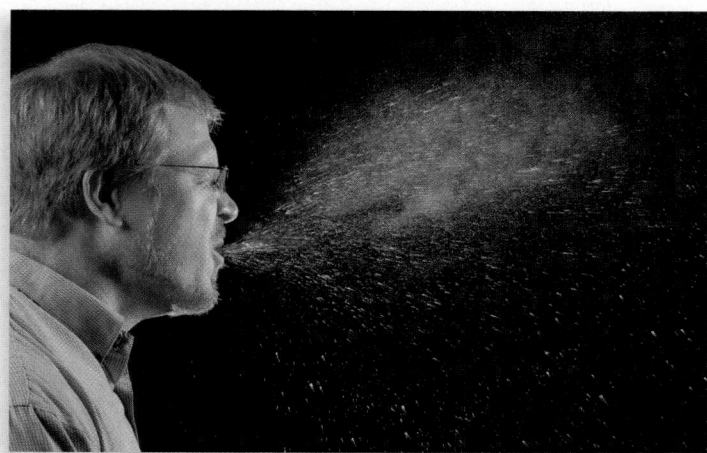

Figure 13.19 The explosiveness of a sneeze. Special photography dramatically captures droplet formation in an unstifled sneeze. Even the merest attempt to cover a sneeze with one's hand will reduce this effect considerably. When such droplets dry and remain suspended in air, they become droplet nuclei.
Source: James Gathany/CDC

spread of hardier pathogens such as the tubercle bacillus and the influenza virus.

Aerosols are suspensions of fine dust or moisture particles in the air that contain live pathogens. Q fever is spread by dust from animal quarters, and psittacosis by aerosols from infected birds. Outbreaks of coccidioidomycosis (a lung infection) often occur in areas where soil is disturbed by wind or digging, releasing clouds of dust bearing the spores of *Coccidioides*.

 Check Your Progress SECTION 13.4

22. Explain the differences between communicable and noncommunicable infectious diseases, and between (direct) contact and indirect modes of transmission.
23. Distinguish between reservoirs and sources, using examples.
24. State the primary differences between mechanical and biological vectors.
25. Define *zoonosis* and describe the difficulties in controlling zoonotic infections.
26. Explain what it means to be a carrier of infectious disease, and describe four ways in which humans can be carriers.
27. What is epidemiologically and medically important about carriers in the population?

13.5 The Work of Epidemiologists: Investigation and Surveillance

Expected Learning Outcomes

24. Explain the primary methods of tracking infections and diseases in a population.
25. Differentiate among the patterns of disease outbreaks according to frequency, number, and location.

26. Summarize the steps in Koch's postulates, and explain their importance to microbiologists.
27. Discuss important aspects of health-care-associated infections and their impact on patients in clinical settings.
28. Explain what is meant by universal precautions, and discuss how they are implemented.

Epidemiologists are concerned with all of the factors covered earlier in this chapter: virulence, portals of entry and exit, and the course of disease. But they are also interested in **surveillance;** that is, collecting, analyzing, and reporting data on the rates of occurrence, mortality, morbidity, and transmission of infections. A well-developed network of individuals and agencies at the local, district, state, national, and international levels keeps constant track of infectious diseases. Surveillance involves monitoring data for a large number of diseases observed in the medical setting and reported to public health authorities. By law, certain **reportable diseases,** also called notifiable diseases, must be reported to authorities; others are reported on a voluntary basis. Physicians and hospitals must contact the appropriate agency for all reportable diseases that are brought to their attention. Case reporting can focus on a single individual or collectively on group data.

Local public health agencies first receive the case data and determine how the data will be handled. In most cases, health officers investigate the history and movements of patients to trace their prior contacts and to control the further spread of the infection as soon as possible through drug therapy, immunization, and education. In sexually transmitted diseases, patients are asked to name their partners so that these persons can be notified, examined, and treated. It is very important to maintain the confidentiality of the persons in these reports. The principal government agency responsible for keeping track of infectious diseases nationwide is the Centers for Disease Control and Prevention (CDC) in Atlanta, Georgia, which is home to the National Notifiable Infectious Conditions Group (NNICG). The CDC publishes a weekly notice of diseases (the *Morbidity and Mortality Weekly Report*) that provides weekly and cumulative summaries of the case rates and deaths for about 50 notifiable and 65 reportable diseases, highlights important and unusual diseases, and presents data concerning disease occurrence in the major regions of the United States. This notice is available to anyone at http://www.cdc.gov/mmwr/. Ultimately the CDC shares its statistics on diseases with the World Health Organization (WHO) for worldwide tabulation and control.

Epidemiological Statistics: Frequency of Cases

The **prevalence** of a disease is the accumulated total of existing cases with respect to the entire population. It is usually reported as the percentage of the population having a particular disease and provides an indicator of how commonly a disease occurs in a population. Disease **incidence** measures the number of new cases over a certain time period, as compared with the general healthy population. This statistic,

also called the case rate or morbidity rate, indicates both the rate and the risk of infection at the time of reporting. The equations used to figure these rates follow:

$$\text{Prevalence} = \frac{\text{Total number of cases in population}}{\text{Total number of persons in population}} \times 100 = \%$$

$$\text{Incidence} = \frac{\text{Number of new cases}}{\text{Total number of susceptible persons}} \quad \begin{array}{l}\text{(Usually reported}\\ \text{per 100,000 persons)}\end{array}$$

As an example, let us use a classroom of 50 students exposed to a new strain of influenza. Before exposure, the prevalence and incidence in this population are both zero (0/50). If in one week, 5 out of the 50 people contract the disease, the prevalence is 5/50 = 10%, and the incidence is 1 in 10 (10 is used instead of 100,000 due to smaller population). If within a month 25 more people acquire the flu, the prevalence becomes 30/50 or 60%, and the incidence (the number of new infections) is 25/50, or 5 in 10. Using reported cases of AIDS as a real-life example, the most recent statistics show a cumulative total of cases (prevalence) between 1981 and 2014 of 1,350,000 cases, or 0.42% of the population. The incidence of 44,608 cases for just the year 2014 averages out to 16.6 cases per 100,000 people.

The changes in incidence and prevalence are usually followed over a seasonal, yearly, and long-term basis and are helpful in predicting trends **(figure 13.20).** Statistics of concern to the epidemiologist are the rates of disease with regard to sex, race, or geographic region. Also of importance is the **mortality rate,** which measures the total number of deaths in a population due to a certain disease. Over the past century, the overall death rate from infectious diseases has dropped, although the **morbidity*** rate, or the number of persons afflicted with infectious diseases, has remained relatively high.

Monitoring statistics also makes it possible to define how often a disease appears in a given population. An infectious disease that exhibits a relatively steady and predictable frequency over a long time period and is tied to a particular geographic locale is **endemic (figure 13.21a).** Often the reason for endemicity is the presence of a reservoir restricted to that location. For example, Lyme disease is endemic to certain areas of the United States where the tick vector is found. New cases are expected to appear in these areas every year. For a disease to be considered **sporadic,** occasional cases are reported at irregular intervals in widely dispersed locations **(figure 13.21b).** Tetanus and typhoid fever are reported sporadically in the United States (fewer than 500 cases a year in a random distribution).

When statistics indicate that the number of new cases of an endemic or sporadic disease is increasing beyond what is expected for that population, this trend is known as an **epidemic (figure 13.21c).** The time period is not defined. It can range from hours in food poisoning to years in STDs, such as gonorrhea or chlamydiosis.

* *morbidity* (mor-bih´-dih-tee) L. *morbidis,* sick. A condition of being diseased.

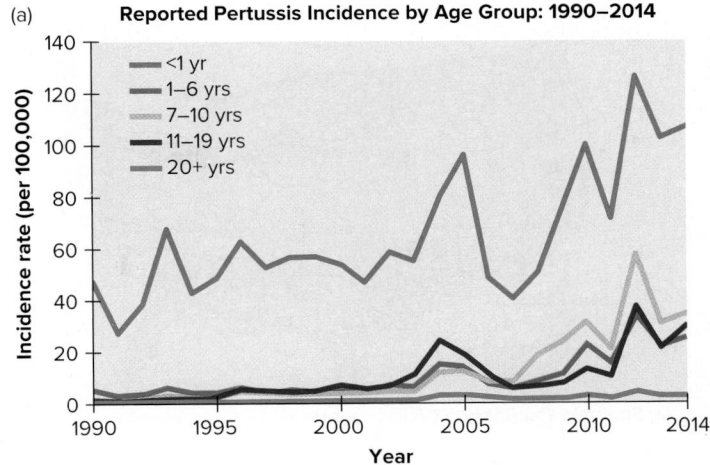

(a) **Reported Pertussis Incidence by Age Group: 1990–2014**

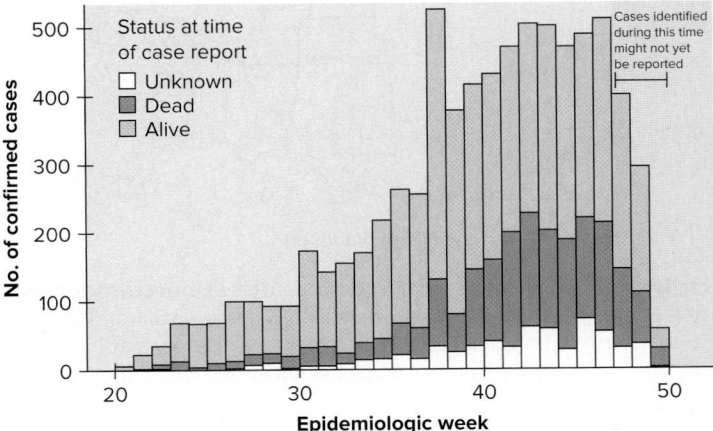

(b) **Number of Confirmed Cases of Ebola Virus Disease, by Epidemiologic Week and Status at Time of Case Report—Sierra Leone, May–December 2014**

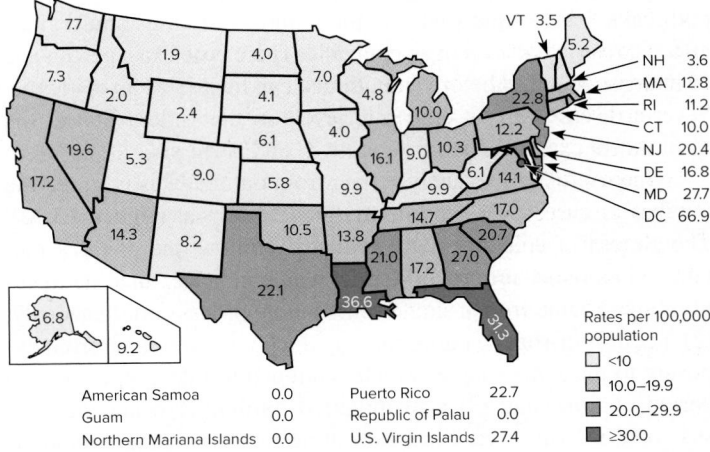

(c) **Rates of Diagnoses of HIV Infection Among Adults and Adolescents, 2014–United States and 6 Dependent Areas**
N = 44,609 Total Rate = 16.6

American Samoa	0.0	
Guam	0.0	
Northern Mariana Islands	0.0	
Puerto Rico	22.7	
Republic of Palau	0.0	
U.S. Virgin Islands	27.4	

Rates per 100,000 population
- <10
- 10.0–19.9
- 20.0–29.9
- ≥30.0

Note: Data include persons with a diagnosis of HIV infection regardless of stage of disease at diagnosis. All displayed data have been statistically adjusted to account for reporting delays, but not for incomplete reporting.

Figure 13.20 **Samples of epidemiological data as analyzed by the CDC. (a)** A graph that shows the reported incidence of pertussis according to age group over a 24-year period. **(b)** A histogram that tracks the Ebola epidemic in Africa over 8 weeks in 2014, comparing diagnosed versus unknown cases and mortality numbers. **(c)** A map that tracks new cases of HIV infection in the U.S. and territories for a single year. All data are from the CDC.

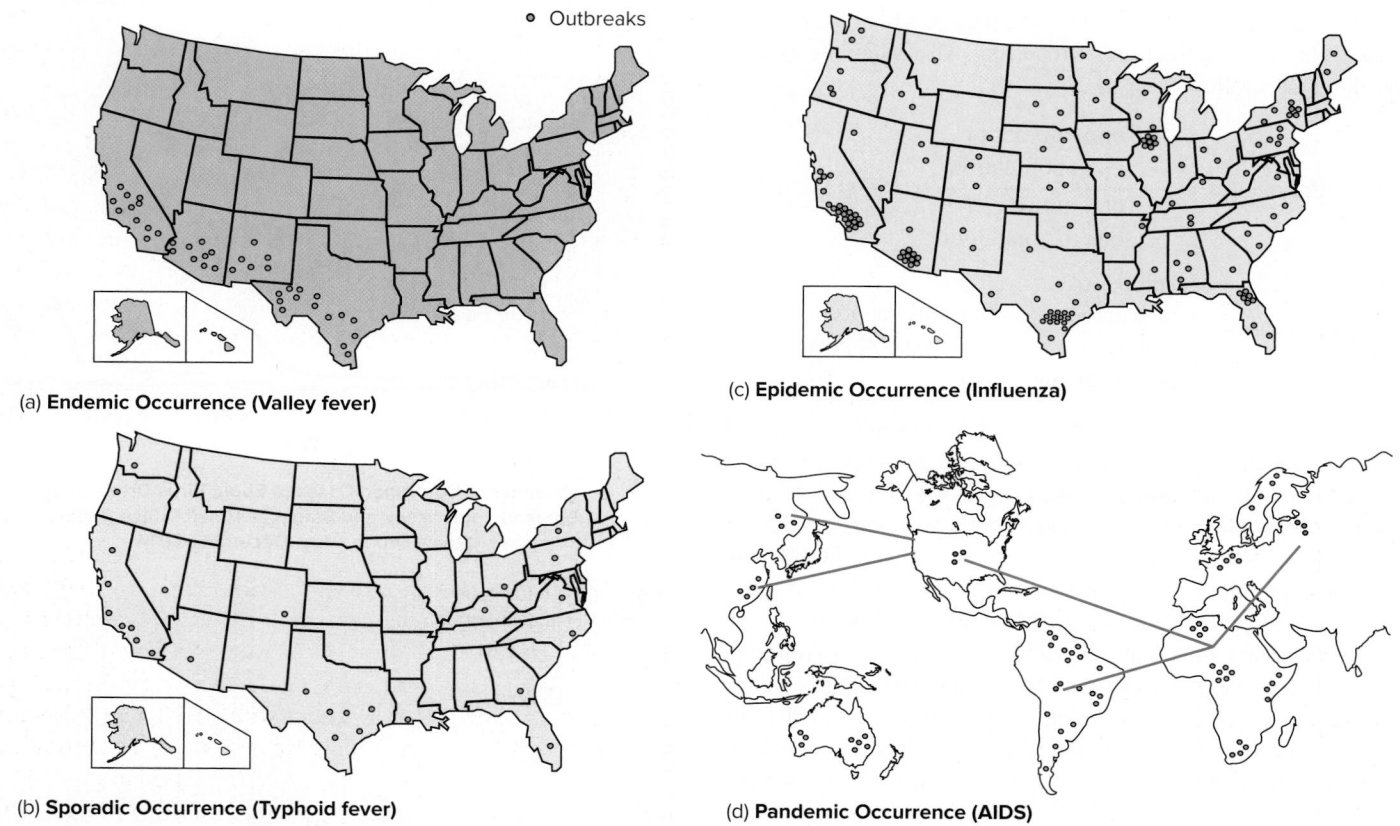

○ Outbreaks

(a) **Endemic Occurrence (Valley fever)**

(c) **Epidemic Occurrence (Influenza)**

(b) **Sporadic Occurrence (Typhoid fever)**

(d) **Pandemic Occurrence (AIDS)**

Figure 13.21 **Patterns of infectious disease occurrence. (a)** In endemic occurrence, cases are concentrated in one region at a relatively stable rate. **(b)** In sporadic occurrence, a few cases occur randomly over a wide area. **(c)** An epidemic is an increased number of cases that often appear in geographic clusters. The clusters may be local, as in the case of a restaurant-related food-borne epidemic, or nationwide, as is the case with influenza. **(d)** A pandemic results when an epidemic ranges across continents and borders.

The term **outbreak** is often used interchangeably with *epidemic,* but an outbreak is usually on a smaller scale in a more limited area. In this chapter's Case Study, the Ebola cases in Africa started as outbreaks, but as time went on, the numbers of cases and regions affected rose to the level of an epidemic. There is no exact percentage of increase required before an outbreak can qualify as an epidemic. Several diseases are at epidemic levels in the United States, the most recent examples being hepatitis C and pertussis.

One useful way to analyze data from outbreaks is to graph the number of cases over time in an epicurve format **(figure 13.22).** The pattern of epidemics will ordinarily follow one of three patterns: (1) a **point-source epidemic,** which indicates that the infectious agent came from a single source more or less simultaneously; (2) a **common-source epidemic,** in which all cases came from exposure to the same source, which continued to infect others over a period of time; and (3) a **propagated epidemic,** which shows a sustained increase over time, indicating that it is being communicated from person to person.

The spread of an epidemic across continents or international borders is a **pandemic,** as exemplified by AIDS and influenza **(figure 13.21d).** Review the Case Study in chapter 6 for an example of pandemic spread. One important epidemiological truism might be called the "iceberg effect," which refers to the fact that only a small portion of an iceberg is visible above the surface of the ocean,

with a much more massive part lying unseen below the surface. Regardless of case reporting and public health screening, a large number of cases of infection in the community go undiagnosed and unreported. In the instance of salmonellosis, approximately 40,000 cases are reported each year. Epidemiologists estimate that the actual number of infections is more likely somewhere between 1 million and 2 million. The iceberg effect can be even more lopsided for sexually transmitted diseases or for infections that are not reported to agencies. For a list of reportable diseases in the United States, refer to the Appendix on Connect.

Quick Search

To learn about the worst epidemics in recorded history, watch "10 Worst Epidemics" on the Discovery Channel.

Investigative Strategies of the Epidemiologist

Initial evidence of a new disease or an epidemic in the community is fragmentary. A few sporadic cases are seen by health care providers and are eventually reported to authorities. It can take several reports before any alarm is registered. Epidemiologists and public health departments play the part of medical detectives who must piece together odds and ends of data from similar cases and attempt to find any connections they may have. This often involves

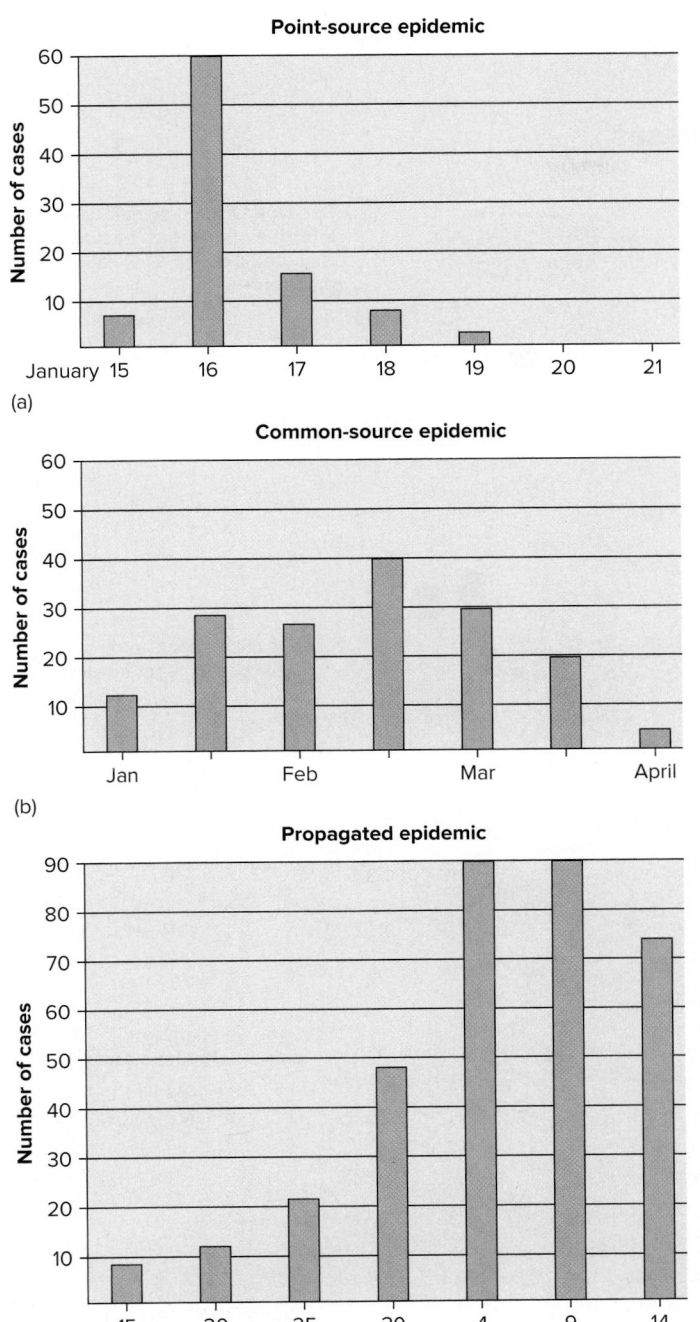

Figure 13.22 Epicurve graphs of the three main patterns typically shown by epidemics.

will help fine-tune the investigation. An example of an index case was the first case of hepatitis C discovered in the Las Vegas outbreak (see the Case Study in chapter 11).

All factors possibly impinging on the disease are scrutinized. Investigators search for *clusters* of cases indicating transmission between persons or a public (common) source or vehicle. They also look at possible contact with animals, contaminated food, water, and public facilities and at human interrelationships or changes in community structure. Out of this maze of case information, the investigators hope to recognize a pattern that indicates the source of infection so that they can quickly move to control it.

Hospital Epidemiology and Health-Care-Associated Infections

Infectious diseases that are acquired or develop during a stay in a medical facility are known as **health-care-associated infections** (HAIs) or **nosocomial* infections.** This concept seems incongruous at first thought, because a hospital or clinic is regarded as a place to get treatment for a medical problem, not a place to acquire one. Yet it is not uncommon for a surgical incision to become infected or a patient with a catheter to develop a urinary tract infection after being treated in a hospital or clinic. The rate of hospital-acquired infections can be as low as 0.1% or as high as 20% of all admitted patients, depending on the clinical setting or hospital. The average rate is about 4% (1 out of 25 patients). In light of the number of admissions, this adds up to 1 million to 2 million cases a year, which result in around 72,000 deaths. Nosocomial infections cost time and money as well as suffering. By one estimate, they amount to 8 million additional days of hospitalization a year and an increased cost of $5 billion to $10 billion.

So many factors unique to the hospital environment create conditions that favor HAIs that a certain number of them are virtually unavoidable. After all, the hospital both attracts and creates compromised patients, and it serves as a collection point for opportunistic pathogens. Some patients become infected when surgical procedures or lowered defenses permit resident microbes to invade their bodies. Other patients acquire infections directly or indirectly from contaminated catheters or ventilators and other medical equipment, other patients, medical personnel, visitors, air, and water.

The health care process itself increases the likelihood that infectious agents will be transferred from one patient to another. Treatments using reusable instruments such as respirators and endoscopes constitute a possible source of infectious agents. Indwelling devices such as catheters, prosthetic heart valves, grafts, and tracheostomy tubes form a ready portal of entry and habitat for infectious agents. Because such high numbers of the hospital population receive antimicrobial drugs during their stay, drug-resistant microbes are selected for at a much greater rate compared to outside the hospital.

working backward to reconstruct factors that could play some part in the outbreak or epidemic. A completely new disease requires even greater preliminary investigation, because the infectious agent must be isolated and linked directly to the disease **(13.3 Making Connections).** An early piece of information that drives an investigation is the **index case,** defined as the first case to bring the infection or disease to the attention of medical authorities. It may not even be the first case, but it is the first one reported. Usually other cases will follow, and the ensuing statistics

* *nosocomial* (nohz″-oh-koh′-mee-al) Gr. *nosos,* disease, and *komeion,* to take care of. Also known as health care-acquired infections.

13.3 MAKING CONNECTIONS

Koch's Postulates: Solving the Puzzle of New Diseases

An essential aim in the study of infection and disease is determining the precise **etiologic,** or causative, **agent.** In our modern technological age, we take it for granted that a certain infection is caused by a certain microbe, but such has not always been the case. More than 130 years ago, Robert Koch realized that in order to support the germ theory of disease, he would have to develop a standard for determining causation that would stand the test of scientific scrutiny. Out of his experimental observations on the transmission of anthrax in cows came a series of proofs, called **Koch's postulates,** that established the principal criteria for etiologic studies. A summary of these postulates is as follows:

1. find evidence of a particular microbe in every case of a disease,
2. isolate that microbe from an infected subject and cultivate it artificially in the laboratory,
3. inoculate a susceptible healthy animal with the laboratory isolate and observe the same resultant disease in that animal, and
4. re-isolate the agent from the diseased animal.

Valid application of Koch's postulates involves several critical details. Each culture must be pure, observed microscopically, and identified by means of characteristic tests; the first and second isolate must be identical; and the pathologic effects, signs, and symptoms of the disease in the first and second subject must be the same. Once established, these postulates were rapidly put to the test. Within a short time they had helped determine the causative agents of tuberculosis, diphtheria, and plague. Today most infectious diseases have been directly linked to a known agent.

Many infectious agents such as viruses and obligately parasitic bacteria can be difficult to culture or cannot be cultured. For these pathogens, the postulates have been modified to take this into account. It is now considered significant to repeatedly observe the infectious agent in tissue samples taken from people with the same disease. A good case in point is Whipple disease, a chronic intestinal infection that for many years could not be tied to a known causative agent. Finally medical researchers had sufficient evidence of a bacterium being consistently present in infected tissues to give an identity to the mystery pathogen, which they named *Tropheryma whipplei*. Modern sequencing techniques to analyze unculturable microbes and their genes have added a new type of evidence for fulfilling the postulates.

Koch's original postulates—or a modified version—continue to play an essential role in modern epidemiology. Nearly every year, new diseases arise or old ones spread out of their usual range. Notable examples include dengue fever, influenza, SARS, and West Nile fever. Viral diseases require some modifications to the postulates. These include that the virus must be cultivated in cell culture, it must be filterable through bacteriological filters, and it must cause a detectable specific immune response.

Explain how we can be so certain that HIV causes AIDS even though there is no effective animal model for it. *Answer available on Connect.*

① Specimen from patient ill with infection of unknown etiology is carried through an isolation procedure.

② A pure culture of the suspected agent is made.

Full microscopic and biological characterization

③ Inoculation of test animal

Observe animal for disease characteristics

Specimen taken

④ Pure culture and identification procedures

The microbe in the initial and second isolations and the disease in the patient and experimental animal must match for the postulates to be satisfied.

The most common HAIs involve surgical incisions and the respiratory tract, GI tract, skin, urinary tract, and blood (sepsis) **(figure 13.23)**. Gram-negative intestinal flora (*Escherichia coli, Klebsiella, Pseudomonas*) are cultured in more than half of patients with HAIs. Gram-positive bacteria (staphylococci and streptococci) and yeasts (*Candida*) make up most of the remainder. True pathogens such as the tubercle bacillus, *Salmonella,* hepatitis B, and influenza virus can be transmitted in the clinical setting as well.

The potential seriousness and impact of health-care-associated infections have required hospitals to develop committees that

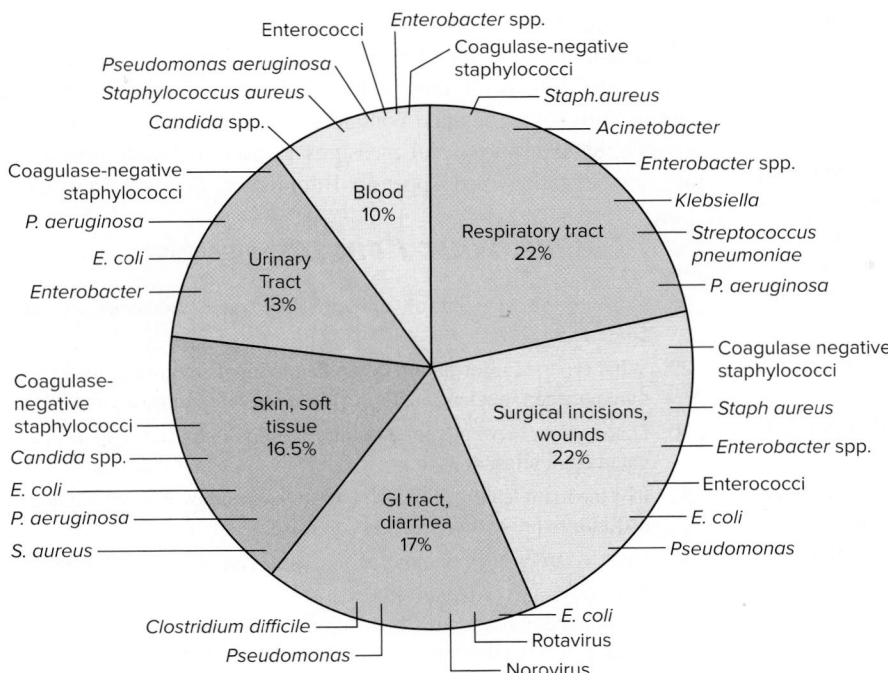

Figure 13.23 Prevalence of health-care-associated infections (HAIs). Relative frequency by body site. These are the most common isolates out of the nearly 500 pathogens that are known to cause HAIs. *Source:* Data from 2011 are compiled by the National Health Safety Network, a unit of the CDC.

hand washing, disinfection, and sanitization, as well as patient isolation. **Table 13.11** summarizes guidelines for the major types of isolation. The goal of these procedures is to limit the spread of infectious agents from person to person. The most rigorous level of precautions is associated with *surgical asepsis,* which involves all of the strategies listed previously plus ensuring that all surgical procedures are conducted under sterile conditions. This includes sterilization of surgical instruments, dressings, sponges and the like, as well as clothing personnel in sterile garments and scrupulously disinfecting the room surfaces and air.

Hospitals generally employ an *infection control officer* who not only implements proper practices and procedures throughout the hospital but also is charged with tracking potential outbreaks, identifying breaches in asepsis, and training other health care workers in aseptic technique. Among those most in need of this training are nurses and other caregivers whose work, by its very nature, exposes them to needlesticks, infectious secretions, blood, and physical contact with the patient. The same practices that interrupt the routes of infection in the patient can also protect the health care worker. It is for this reason that most hospitals have adopted universal precautions that recognize that all secretions from all persons in the clinical setting are potentially infectious and that transmission can occur in either direction.

monitor infectious outbreaks and develop guidelines for infection control and aseptic procedures. *Medical asepsis* includes practices that lower the microbial load in patients, caregivers, and the hospital environment. These practices include proper

TABLE 13.11	**Levels of Isolation Used in Clinical Settings**	
Type of Isolation*	**Protective Measures******	**To Prevent Spread of**
Enteric Precautions	Gowns and gloves must be worn by all persons having direct contact with patient; masks are not required; special precautions are taken for disposing of feces and urine.	Diarrheal diseases; *Shigella, Salmonella, E. coli, and Clostridium difficile;* cholera; hepatitis A; rotavirus; and giardiasis
Respiratory Precautions	Private room with closed door is necessary; gowns and gloves are not required; masks are usually indicated; items contaminated with secretions must be disinfected.	Tuberculosis, measles, mumps, meningitis, pertussis, rubella, chickenpox
Drainage and Secretion Precautions	Gowns and gloves are required; masks are not needed; contaminated instruments and dressings require special precautions.	Staphylococcal and streptococcal infections; gas gangrene; herpes zoster; burn infections
Strict Isolation	Private room with closed door is required; gowns, masks, and gloves must be worn; contaminated items must be wrapped and decontaminated.	Mostly highly virulent or contagious microbes; includes tuberculosis, some types of pneumonia, extensive skin and burn infections, herpes simplex and zoster
Reverse Isolation (Also Called Protective Isolation)	Same guidelines as for strict isolation are required; room may be ventilated by laminar airflow filtered through a high-efficiency particulate (HEPA) filter that removes airborne pathogens; infected persons must be barred.	Used to protect patients extremely immunocompromised by cancer therapy, surgery, genetic defects, burns, prematurity, or AIDS and therefore vulnerable to opportunistic pathogens

*Precautions are based upon the primary portal of entry and communicability of the pathogen.

**In all cases, visitors to the patient's room must report to the nurses' station before entering the room; all visitors and personnel must wash their hands upon entering and leaving the room.

Universal Blood and Body Fluid Precautions

Medical and dental settings require stringent measures to prevent the spread of infectious agents from patient to patient, from patient to worker, and from worker to patient. But even with precautions, the rate of such infections can be rather high. An established body of evidence indicates that more than one-third of health-care-acquired infections could be prevented by consistent and rigorous infection control methods.

Previously control guidelines were disease-specific, and clearly identified infections were managed with particular restrictions and techniques. With this arrangement, personnel tended to handle materials labeled *infectious* with much greater care than those that were not so labeled. The AIDS epidemic spurred a reexamination of that policy. Because of the potential for patients with undiagnosed infections, the CDC laid down more-stringent guidelines for handling patients and body substances. These guidelines have been termed **universal precautions (UPs),** because they are based on the assumption that all patient specimens could harbor infectious agents and so must be treated with the same degree of care. They also include body substance isolation (BSI) techniques to be used in known cases of infection.

It is worth mentioning that these precautions are designed to protect all individuals in the clinical setting—patients, workers, and the public alike. In general, they include techniques designed to prevent contact with pathogens and contamination and, if prevention is not possible, to take purposeful measures to decontaminate potentially infectious materials. See Appendix B for a more detailed list of UPs.

Check Your Progress SECTION 13.5

28. Outline the science of epidemiology and the work of an epidemiologist.
29. What types of information do epidemiologists require to differentiate between incidence and prevalence of infectious diseases?
30. Distinguish between point-source, common-source, and propagated epidemics, and explain what is meant by an index case.
31. List the main features of Koch's postulates and explain why it is so difficult to prove them for some diseases.
32. Health-care-associated infections can arise from what general sources?
33. Outline the major factors involved in health-care-associated infections and describe the levels of precaution that are followed to prevent HAIs.

CASE STUDY Part 2

Despite the attempts to prevent transmission of the virus, two people acquired Ebola infections from contact with Mr. Duncan. Both were nurses who had provided round-the-clock care during the time he was hospitalized. First, Nina Pham began to register a low-grade fever 4 days after his death. She was immediately tested for Ebola virus and moved into an isolation unit. Two days later Amber Vinson reported a fever and was also isolated. Both women tested positive for the virus. They were continuously monitored and provided intensive medical support at other hospitals, but they received no antiviral medicines because specific drugs were not yet available. Within 2 weeks both nurses recovered, were found to be free of the virus, and were released from isolation.

Even though these cases were the only ones that originated in the United States, they still fostered a wave of fear in the populace that this epidemic could evolve into a pandemic. In addition to the anguish of the two nurses and their families, repercussions were felt in the lives of many other people. In this situation, a carrier in the prodromal phase can actively transmit the virus, so it was necessary for public health officers to implement tracing, monitoring, and quarantining of all contacts. This included all hospital workers and their families, crews and passengers on Mr. Duncan's airline flight, and Mr. Duncan's family and contacts, which eventually totaled over 200 people. Several other American health care workers came down with Ebola, but all of them were infected while caring for patients in West Africa, and all but one of them survived.

The story of Ebola virus outbreaks is not a new one. The first cases, originally termed Ebola hemorrhagic fever, occurred near the Ebola River of Zaire (now the Central African

Republic) in 1976. Since then, there have been 24 additional outbreaks, but the one in 2014–2015 was by far the most severe. It affected 28,638 people and killed 11,315 in Africa, with the mortality rate varying from 25% to 90% (average is 50%). The epidemic was declared over in January 2016, following a period with no new case reports.

Virologists who have been studying the origins of the Ebola virus for 4 decades have long considered it a zoonosis transmitted by biological vectors. It has been isolated from primates such as great apes and monkeys, but it causes a deadly infection in these animals, which means they are probably not reservoir hosts. The current consensus is that species of fruit bats and insectivorous bats are the most likely reservoirs. We know that the source of most human infections is other humans and animals, especially forest animals that are butchered for meat. In humans it is transmitted only by direct contact with patient fluids such as blood, feces, vomit, saliva, urine, breast milk, and semen. It is likely that infections occur when these body fluids contaminate portals of entry such as exposed skin, cuts, and mucous membranes of the mouth, nose, and eyes. There is little evidence that infection occurs through airborne droplet nuclei, coughing, sneezing, food (except wild meat), fomites, or mosquitoes.

- *Explain why this epidemic of Ebola virus disease was not considered a pandemic.*
- *Provide some explanations why Ebola virus would be transmitted by fresh droplets but not droplet nuclei.*

To conclude this Case Study, go to Connect.

For additional coverage of this topic go to https://en .wikipedia.org/wiki/Ebolavirusdisease.

Chapter Summary with Key Terms

13.1 We Are Not Alone

A. The human body is in contact with microbes that are constantly present in the environment. Microbes are classified as either part of the normal residents of the body or as **pathogens,** depending on whether or not their colonization of the body results in infection.

B. *Resident Microbiota: The Human as Habitat*

1. **Normal resident microbiota** consist of a huge and varied population of microbes that reside permanently on all surfaces of the body exposed to the environment.

2. Colonization begins prior to birth and continues throughout life. Variations in normal residents occur in response to changes in an individual's age, diet, hygiene, and health.

3. Normal residents often provide some benefit to the host but are also capable of causing disease, especially when the immune system of the host is compromised or when they invade a normally sterile area of the body.

4. An important benefit provided by microbiota is **microbial antagonism.**

C. *Indigenous Microbiota of Specific Regions* Resident microbes include bacteria, fungi, and protozoa. They occupy the skin, mouth, gastrointestinal tract, large intestine, respiratory tract, and genitourinary tract.

13.2 Major Factors in the Development of an Infection

A. **Pathogenicity** is the ability of microorganisms to cause infection and disease; **virulence** refers to the degree to which a microbe can invade and damage host tissues.

1. **True pathogens** are able to cause disease in a normal healthy host with intact immune defenses.

2. **Opportunistic pathogens** can cause disease only in persons whose host defenses are compromised by *predisposing conditions.*

B. *Becoming Established: Phase 1—Portals of Entry*

1. Microbes typically enter the body through a specific **portal of entry.** Portals of entry are generally the same as those areas of the body that harbor microflora.

2. The minimum number of microbes needed to produce an infection is termed the **infectious dose,** and these pathogens may be classified as **exogenous** or **endogenous,** based on their source.

C. *Attaching to the Host: Phase Two* Microbes attach to the host cell, a process known as **adhesion,** by means of fimbriae, capsules, or receptors.

D. *Invading the Host and Becoming Established: Phase Three* Microbes that persist in the human host have developed mechanisms of disabling early defensive strategies.

E. *How Virulence Factors Contribute to Tissue Damage and Disease* **Virulence factors** are structures or properties of a microbe that lead to pathologic effects on the host. Some virulence factors are those that enable entry or adhesion, but many virulence factors lead directly to damage. These fall in three categories: enzymes, toxins, and the inhibition or destruction of phagocytes.

1. **Exoenzymes** digest epithelial tissues and permit **invasion** of pathogens.

2. **Toxigenicity** refers to a microbe's capacity to produce **toxins** at the site of multiplication. Toxins are divided

into **endotoxins** and **exotoxins,** and diseases caused by toxins are classified as **intoxications, toxemias,** or **toxinoses.**

13.3 The Outcomes of Infection and Disease

A. **Infectious diseases** follow a typical pattern consisting of an **incubation period, prodromal stage, period of invasion,** and **convalescent period.** Each stage is marked by symptoms of a specific intensity.

B. Infections are classified by

1. whether they remain localized at the site of inoculation **(localized infection, systemic infection, focal infection);**

2. the number of microbes involved in an infection, and the order in which they infect the body **(mixed infection, primary infection, secondary infection);** and

3. the persistence of the infection **(acute infection, chronic infection,** subacute infection).

C. **Signs** of infection refer to *objective* evidence of infection, and **symptoms** refer to *subjective* evidence. A **syndrome** is a disease that manifests as a predictable complex of symptoms.

D. After the initial infection, pathogens may remain in the body in a **latent** state and may later cause recurrent infections. Long-term damages to host cells are referred to as **sequelae.**

E. Microbes are released from the body through a specific **portal of exit** that allows them access to another host.

13.4 Epidemiology: The Study of Disease in Populations *Origins and Transmission Patterns of Infectious Microbes*

A. The **reservoir** of a microbe refers to its natural habitat; a pathogen's **source** refers to the immediate origin of an infectious agent.

B. **Carriers** are individuals who inconspicuously shelter a pathogen and spread it to others. Carriers are divided into **asymptomatic carriers,** incubation carriers, convalescent carriers, chronic carriers, and **passive carriers.**

C. **Vectors** refer to animals that transmit pathogens.

1. **Biological vectors** are animals that are involved in the life cycle of the pathogen they transmit; **mechanical vectors** transmit pathogens without being involved in the life cycle.

2. **Zoonoses** are diseases that can be transmitted to humans but are normally found in animal populations.

D. Agents of disease are either **communicable** or **noncommunicable,** with highly communicable diseases referred to as **contagious.**

1. Direct transmission of a disease occurs when a portal of exit meets a portal of entrance.

2. Indirect transmission occurs when an intermediary such as a **vehicle, fomite,** or **droplet nuclei** connects portals of entrance and exit.

13.5 The Work of Epidemiologists: Investigation and Surveillance

A. Epidemiologists are involved in the surveillance of **reportable** and notifiable diseases in populations.

B. Infectious diseases are tabulated with regard to their **prevalence** and **incidence.**

C. Populations are analyzed with regard to **morbidity, mortality,** age, sex, and geographic region.

D. Disease patterns are described as being **endemic, sporadic, epidemic,** or **pandemic** based on the frequency of their occurrence in a population.

E. Epidemiologists look for **index cases** and changes in the occurrence of diseases such as clusters of new cases. They compile evidence and **trace back** to the original source or reservoir.

F. **Health-care-acquired infections (HAIs)** are diseases acquired in a hospital or other clinical setting. Special vigilance is required to reduce their occurrence. Most HAIs involve the respiratory tract, surgical incisions, the GI tract, urinary tract, skin, and blood, Infectious agents may originate from the microbiota, medical devices, patients, health care workers, and the environment,

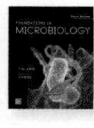

Level I. Knowledge and Comprehension

These questions require a working knowledge of the concepts in the chapter and the ability to recall and understand the information you have studied.

Multiple-Choice Questions

Select the correct answer from the answers provided. For questions with blanks, choose the combination of answers that most accurately completes the statement.

1. The best descriptive term for the resident microbes is
 a. commensals c. pathogens
 b. parasites d. contaminants

2. Resident microbiota is commonly found in the
 a. liver c. salivary glands
 b. kidney d. urethra

3. Normal resident microbes are absent from the
 a. pharynx c. intestine
 b. lungs d. hair follicles

4. Virulence factors include
 a. toxins c. capsules
 b. exoenzymes d. all of these

5. The specific action of hemolysins is to
 a. damage white blood cells
 b. cause fever
 c. damage red blood cells
 d. cause leukocytosis

6. The _____ is the time that lapses between encounter with a pathogen and the first symptoms.
 a. prodrome c. period of convalescence
 b. period of invasion d. period of incubation

7. A short period early in a disease that manifests with general malaise and achiness is the
 a. period of incubation c. sequela
 b. prodrome d. period of invasion

8. The presence of a few bacteria in the blood is termed
 a. septicemia c. bacteremia
 b. toxemia d. a secondary infection

9. A(n) _____ infection is acquired through exposure in a hospital setting.
 a. subclinical c. HAI (nosocomial)
 b. focal d. secondary

10. A/an _____ is a passive animal transporter of pathogens.
 a. zoonosis c. mechanical vector
 b. biological vector d. asymptomatic carrier

11. An example of a noncommunicable infection is
 a. measles c. tuberculosis
 b. leprosy d. tetanus

12. A general term that refers to an increased white blood cell count is
 a. leukopenia c. leukocytosis
 b. inflammation d. leukemia

13. The occurrence of Lyme disease mainly in areas where certain species of ticks live would define it as a(n) _____ disease.
 a. epidemic c. endemic
 b. pandemic d. sporadic

14. A positive antibody test for HIV would be a _____ of infection.
 a. sign c. syndrome
 b. symptom d. sequela

15. Which of the following would *not* be a portal of entry?
 a. the meninges c. skin
 b. the placenta d. small intestine

16. Which of the following is *not* a condition of Koch's postulates?
 a. isolate the causative agent of a disease
 b. cultivate the microbe in a lab
 c. inoculate a test animal to observe the disease
 d. test the effects of a pathogen on humans

Case Study Review

1. Which of these would be the index case for the Ebola epidemic?
 a. Nina Pham c. Thomas Duncan
 b. the baby in Guinea d. a dead monkey

2. Ebola virus is a(n)___virus classified in the___family.
 a. enveloped RNA, rhabdovirus
 b. naked DNA, adenovirus

c. enveloped DNA, herpesvirus.
d. enveloped RNA, filovirus.

3. Explain what features of the virulence, infectious dose, and incubation period of Ebola virus increase its transmission from person to person.

Writing Challenge

For each question, compose a one- or two-paragraph answer that includes the factual information needed to completely address the question. Check Your Progress questions can also be used for writing-challenge exercises.

1. What are the important clinical implications of blood or cerebrospinal fluid that tests positive for microbes?

2. What is the expected result if a compromised person is exposed to a true pathogen?

3. Explain how endotoxin gets into the bloodstream of a patient with endotoxic shock.

4. Give a brief description of how the A-B type of toxin works.

5. For each portal of entry, give a vehicle that carries the pathogen and the course it must travel to invade the tissues.

6. What are some important considerations about the portal of exit?

7. What does it mean for salmonellosis to have an iceberg effect?

8. a. Outline the five types of clinical isolation.
 b. Describe the work of an infection control officer.

9. Complete the following table.

	Exotoxins	Endotoxins
Chemical makeup		
General source		
Degree of toxicity		
Effects on cells		
Symptoms in disease		
Examples		

Concept Mapping

On Connect you can find an Introduction to Concept Mapping that provides guidance for working with concept maps and a sample of a concept map for this chapter.

Level II. Application, Analysis, Evaluation, and Synthesis

These problems go beyond just restating facts and require higher levels of understanding and an ability to interpret, problem solve, transfer knowledge to new situations, create models, and predict outcomes.

Critical Thinking

Critical thinking is the ability to reason and solve problems using facts and concepts. These questions can be approached from a number of angles, and in most cases, they do not have a single correct answer.

1. Discuss the relationship between the vaginal residents and the colonization of the newborn.

2. Can you think of some medical consequences of this relationship?

3. How could the microbiome cause some infections to be more severe and other infections to be less severe?

4. Each of the nine patient specimens listed below has produced a positive culture when inoculated and grown on appropriate media. Indicate whether this result is indicative of infection and explain why or why not.

Urine from urethra	Liver biopsy	Blood
Lung biopsy	Throat	Urine from bladder
Saliva	Cerebrospinal fluid	Semen

5. Use the following formula to explain the relationships among the several factors and what happens when they change:

$$\frac{\text{Infection}}{\text{(infectious disease)}} = \frac{\text{No. of organisms} \times \text{Virulence}}{\text{Host resistance}}$$

6. Assume that you have been given the job of developing a colony of germ-free chickens.
 a. What will be the main steps in this process?
 b. What possible experiments can you do with these animals?

7. Describe each of the following infections using correct technical terminology. (Descriptions may fit more than one category.) Use terms such as *primary, secondary, nosocomial, STD, mixed, latent, toxemia, chronic, zoonotic, asymptomatic, local, systemic, -itis, -emia.*

 Caused by needlestick in dental office
 Pneumocystis pneumonia in AIDS patient
 Bubonic plague from rat flea bite
 Diphtheria
 Undiagnosed chlamydiosis
 Acute necrotizing gingivitis
 Syphilis of long duration
 Large numbers of gram-negative rods in the blood
 A boil on the back of the neck
 An inflammation of the meninges

8. a. Suggest several reasons why respiratory, surgical, and gastrointestinal infections are the most common health-care-associated infections.
 b. Name several measures that health care providers must exercise at all times to prevent or reduce these types of infections.

9. Summarize the epidemiological findings in the three graphs in figure 13.20.

10. Looking at figure 13.20b. Which pattern of epicurve does it show?

Visual Challenge

1. Using figure 13.11 (reproduced below) as a guide, explain the processes occurring between Salmonella (pink) and the surface of a host cell (yellow) on both the left and the right side of the photo.

Source: National Institute of Allergy and Infectious Diseases (NIAID)/CDC

2. Observe the following maps (a)–(c) of three diseases. Determine which show endemic, sporadic, or epidemic pattern, and explain what features have helped you to make this decision.
(a-c) *Source:* Centers for Disease Control

Chlamydia. Incidence* Among Women — United States and U.S. Territories, 2012

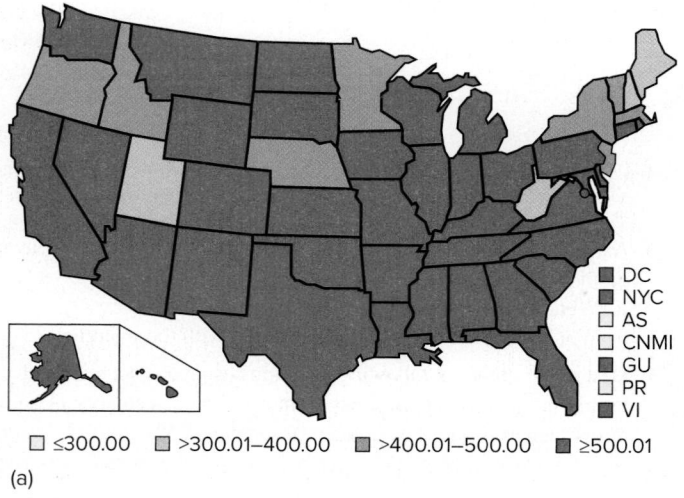

| ☐ ≤300.00 | ☐ >300.01–400.00 | ☐ >400.01–500.00 | ■ ≥500.01 |

(a)

Lyme Disease. Incidence* of Reported Confirmed Cases, by County—United States, 2013

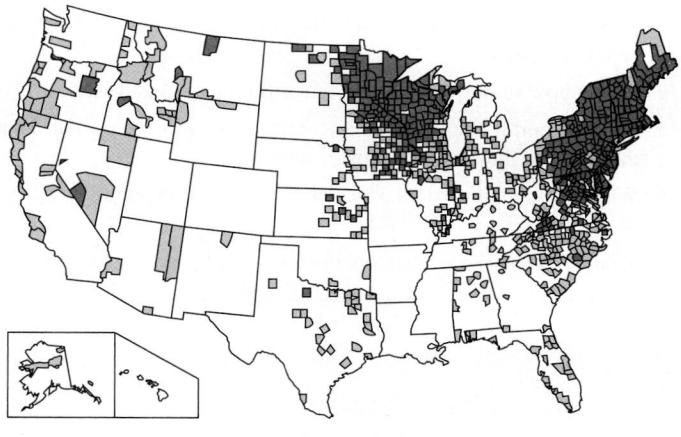

| ☐ 1.01–10.00 | ☐ 10.01–10.00 | ■ ≥100.01 |

* Per 100.000 population.

(b)

Dengue Fever and Dengue Hemorrhagic Fever. Number* of Reported Cases, by Location of Residence — United States and U.S. Territories, 2013

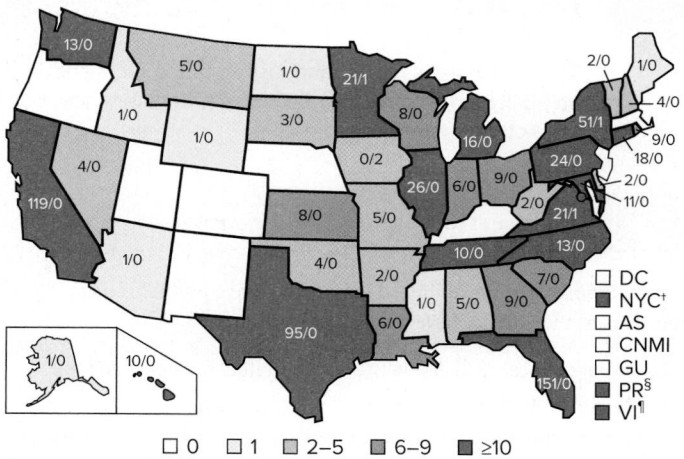

| ☐ 0 | ☐ 1 | ☐ 2–5 | ■ 6–9 | ■ ≥10 |

* Number of Dengue fever cases/number of Dengue Hemorrhagic fever cases. Data from the Division of Vector-Borne Diseases, National Center for Emerging and Zoonotic Infectious Diseases (ArboNET Surveillance).

[†] New York City reported cases 131/1.

[§] Puerto Rico locally acquired cases 9,557/153.

[¶] Virgin Islands reported cases 169/5.

(c)

3. Delineate what would have to occur for each of the diseases in question 2 to have a pandemic distribution.

14 An Introduction to Host Defenses and Innate Immunities

Lymphocytes

Macrophages

The conidial head of *Aspergillus fumigatus,* a one of the few molds that can invade and grow in human tissues and organs. *Source:* CDC

Section through a granuloma reveals a central compact mass of macrophages surrounded by a collar of lymphocytes. Courtesy of Sanjay Mukhopadhyay

CASE STUDY Part 1 When Phagocytes Fail

The 5-year-old boy's medical history included several episodes of **bacterial pneumonia,** starting at 7 months of age, that cleared up with antibiotic therapy. At some point his parents began to notice a troubling new set of symptoms and signs. He complained of **constant pain** in his **wrist** and **jawbone,** and the **tissues** overlying these bones were **swollen** and **inflamed.** They took him to the hospital, where he was admitted for lab tests. The results revealed no abnormalities, and the standard tests for infectious agents came out negative. When an X ray detected small spots in the bones of his wrist and jaw, a biopsy was performed that uncovered **granuloma-like lesions and signs of osteomyelitis.** Because of his history of recurrent infections, he was checked for **immune function,** but his B- and T-cell activities were normal. Based on the assumption that he had an infection of **unknown etiology,** he was placed on a **broad range of therapy** consisting of antibacterial, antifungal, and antiviral medications and went home as soon as his condition improved.

After 8 months, his parents took him back to the hospital. He was in a state of severe distress and was readmitted. This time, he had coughing spells, had difficulty breathing, and complained of chest and abdominal pain. He was immediately given antibiotics and **scanned** by computerized tomography (CT), which revealed a **mass in the liver** that extended into the ribcage. Physicians drew a sample

of fluid from the liver and sent it for testing. From the liver sample, technicians isolated a pure culture with greenish-gray, woolly colonies typical of the mold ***Aspergillus fumigatus.*** Microscopic preparations yielded branched, septate hyphae with **brushlike conidial heads** typical of this fungus.

It now became clear that the child had a serious **systemic fungal infection** that was likely responsible for his earlier episodes as well. This finding required a different regimen using a powerful **intravenous antifungal drug,** amphotericin B, backed up with interferon and rifampin. He responded well to the medications and was able to return home.

After considering the nature of these new infections, his doctors began to search for an unknown **loss of host defenses** that could account for the child's recurring **opportunistic** infections. They were aware, for instance, that systemic **aspergillosis** is one of the most common infections in children and adults with **defective phagocytes.**

> "Systemic aspergillosis is one of the most common infections in children and adults with defective phagocytes."

- *What physiological activities account for the inflammation and swelling?*

- *What is osteomyelitis, and what are some of its causes?*

To continue the Case Study, go to Case Study Part 2 at the end of the chapter.

SCOPING OUT THE CHAPTER

The human body is equipped with a deluxe set of defenses that are constantly brought into play with each challenge from foreign materials or pathogens. These defenses do their work behind the scenes, from the largest organs down to the smallest cluster of cells.

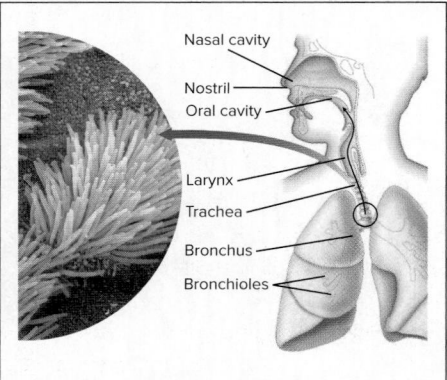

An Obstacle Course for Microbes Much of the body is protected by barriers that stop or deflect microbes near the body surfaces before they get too far. One example of these defenses is cilia in the trachea that brush foreign matter out of the respiratory system.

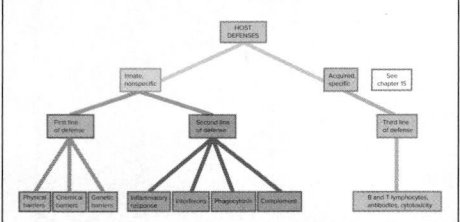

Some defenses are non-specific and present at birth and others are specific and must be acquired over time, but all are essential for complete immune protection. This chapter will survey the anatomy and physiology of the inborn and non-specific defenses that set up immediate zones of protection.

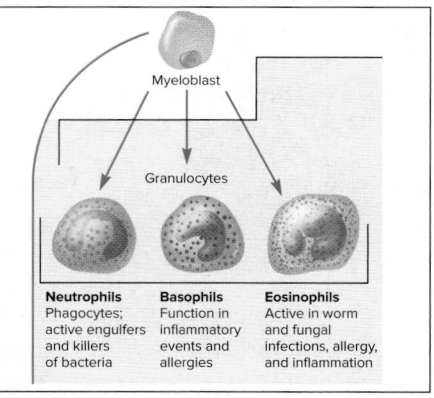

The Origins of Immunity: Blood cells Tissues in the bone marrow produce several different types of white blood cells that become the basis of our defense system. Some of them are phagocytes, some function during inflammation, and some react against specific infectious agents.

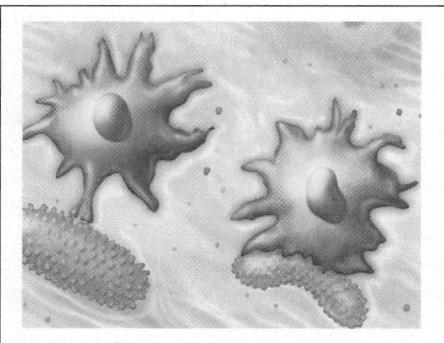

Wandering Search Parties Most blood cells are highly active and motile, with an ability to seek out and recognize abnormal or harmful cells and substances. They are also guides that recruit other cells to fight off infections.

A Flood of Protection: When Inflammation Strikes An injury to tissues and cells triggers a spontaneous series of reactions at the site of the damage. This event will alter the size and permeability of nearby blood vessels, stimulate migration of white blood cells, and cause an outpouring of chemical substances, all designed to contain pathogens and stimulate immunities.

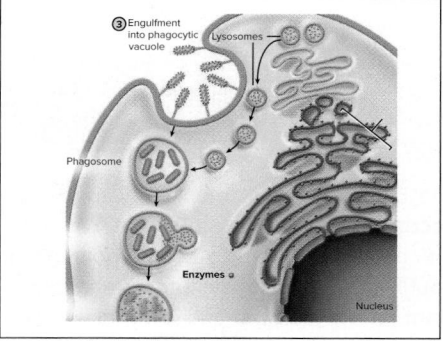

Phagocytes: Powerful Predators A special breed of white blood cells is stationed throughout the body to actively pursue foreign intruders, swallow them up, and kill and dismantle them.

(Obstacle Course for Microbes): Source: Courtesy Charles P. Daghlian/http://www.dartmouth.edu/~emlab/gallery

14.1 Overview of Host Defense Mechanisms

Expected Learning Outcomes

1. Summarize the characteristics of basic host defenses.
2. Differentiate between the three lines of defense, describing examples of each.
3. Explain the nature of the different types of innate, nonspecific defenses.

The survival of the host depends upon an elaborate network of defenses that keep harmful microbes and other foreign materials from entering the body and causing damage. Should they penetrate, additional host defenses are immediately summoned to prevent them from becoming established in tissues. Defenses involve barriers, cells, and chemicals, and they range from nonspecific to specific and from inborn or innate to acquired. This chapter introduces the main lines of defense intrinsic to all humans. Topics included in this survey are the anatomical and physiological systems that detect, recognize, and destroy foreign substances and the major activities that account for an individual's nonspecific resistance to infection and disease.

In chapter 13 we explored the human-microbe relationship, with emphasis on the role of microorganisms in disease. In this chapter, we examine the other side of the relationship—that of the host defending itself against microorganisms. As previously stated, in light of the unrelenting microbial contact with and colonization of humans, it is something of a miracle that we are not constantly infected and diseased. Our ability to overcome this continual assault on our bodies is the result of a fascinating, highly complex system.

To protect the body against pathogens, the immune system relies on an overlapping network of **host defenses** that operate on several levels. These include **innate, natural defenses** present at birth that provide nonspecific resistance to infection and **adaptive immunities** that are specific and must be acquired. These levels of protection are sometimes divided into categories called the *first, second,* and *third lines of defense* (**table 14.1**).

The *first line of defense* includes any barrier that blocks invasion at the portal of entry. This mostly nonspecific line of defense limits access to the internal tissues of the body. However, it is not considered a true immune response because it does not involve recognition of a specific foreign substance and is very general in its actions.

The *second line of defense* is a more internalized system of protective cells and fluids, which includes inflammation and phagocytosis. It acts rapidly at both the local and systemic levels once the first line of defense has been circumvented, and in most reactions it is also nonspecific.

The highly specific *third line of defense* is adaptive and acquired on an individual basis as each foreign substance is encountered by white blood cells called lymphocytes. Responses directed toward each different microbe produce unique protective substances and memory cells that can come into play if that microbe is encountered again. Details of how this third line of defense provides long-term immunity are discussed in chapter 15, whereas this chapter focuses on the first and second lines of defense.

The various levels of defense do not operate in a completely separate fashion; most defenses overlap and are even redundant in some of their effects. This cooperative reaction targets invading microbes on several different fronts simultaneously, making their survival unlikely. To present the interactive nature of host defenses, we start with basic concepts that will build a foundation for understanding the more complex reactions of the immune system to come later.

Barriers at the Portal of Entry: An Inborn First Line of Defense

A number of defenses are a normal part of the body's anatomy and physiology. These inborn, nonspecific defenses can be divided into physical, chemical, and genetic barriers that impede the entry of not only microbes but any foreign agent, whether living or not (**figure 14.1**).

Physical or Anatomical Barriers at the Body's Surface

The skin and mucous membranes of the respiratory and digestive tracts have several built-in defenses. The outermost layer (stratum corneum) of the skin is composed of epithelial cells that have become compacted, cemented together, and impregnated with an insoluble protein, keratin. The result is a thick, tough layer that is highly impervious and waterproof. Few pathogens can penetrate this unbroken barrier, especially in regions such as the soles of the feet or the palms of the hands, where the stratum corneum is much thicker than on other parts of the body. Other cutaneous barriers include hair follicles and skin glands. The hair shaft is periodically extruded, and the follicle cells are **desquamated,** * or shed. The flushing effect of sweat glands also helps remove microbes.

The mucocutaneous membranes of the digestive, urinary, and respiratory tracts and of the eye are moist and permeable. Despite the normal wear and tear upon these epithelia, damaged cells are rapidly replaced. The mucous coat on the free surface of some membranes impedes the entry and attachment of bacteria. Blinking and tear production (lacrimation) flush the eye's surface with tears and rid it of irritants. The constant flow of saliva helps carry microbes into the harsh conditions of the stomach. Vomiting and defecation also evacuate noxious substances or microorganisms from the body.

The respiratory tract is constantly guarded from infection by elaborate and highly effective adaptations. Nasal hair traps larger particles. Rhinitis, the inflammation of the nasal mucosa often seen with allergy and colds, creates a copious flow of mucus and fluids that helps to flush out the nasal passageways. In the respiratory tree (primarily the trachea and bronchi), a ciliated epithelium (called the

* *desquamated* (des′-kwuh-mayt) L. *desquamo,* to scale off. The casting off of epidermal scales.

TABLE 14.1	General Features of Host Defenses			
Line of Defense	Innate/ Acquired	Specific or Nonspecific	Development of Immunologic Memory	Examples
First	Innate	Nonspecific	No	Physical barriers: skin, tears, coughing, sneezing Chemical barriers: low pH, lysozyme, digestive enzymes Genetic barriers: resistance inherent in genetic makeup of host (pathogen cannot invade)
Second	Innate	Mostly nonspecific	No	Phagocytosis, inflammation, fever, interferon, complement
Third	Acquired	Specific	Yes	T lymphocytes, B lymphocytes, antibodies

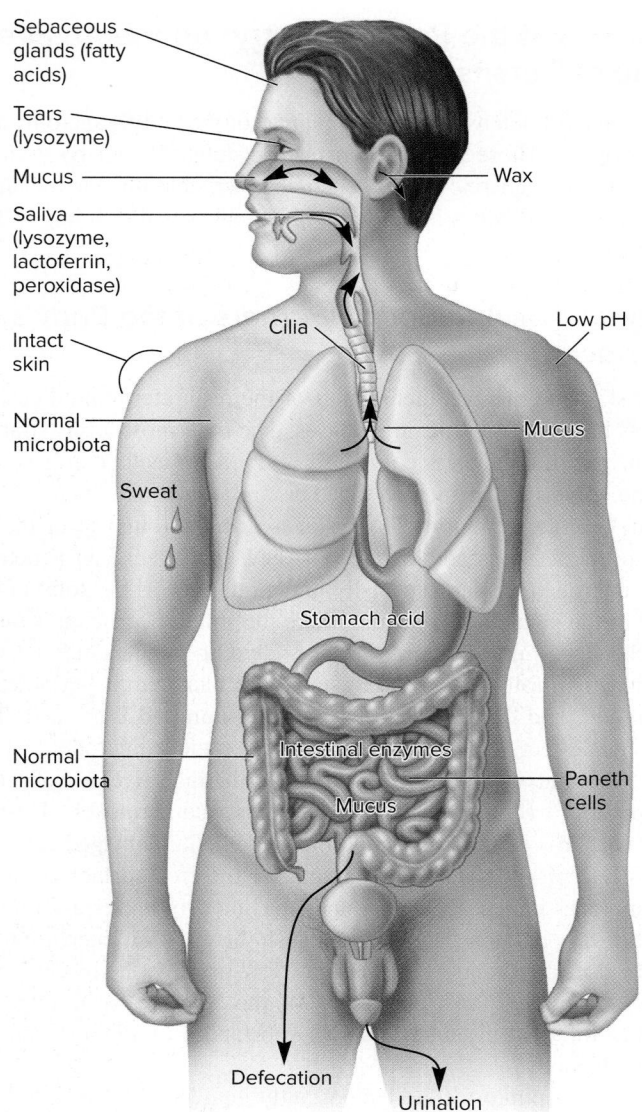

Figure 14.1 **Defenses.** Some nonspecific defense barriers that help prevent entry of microorganisms into the host's tissues.

body regions. Some bacterial residents of the large intestine secrete bacteriocins or antibiotics, which can inhibit or kill other bacteria.

Nonspecific Chemical Defenses

The skin and mucous membranes offer a variety of chemical defenses. Sebaceous secretions exert an antimicrobial effect, and specialized glands such as the meibomian glands of the eyelids lubricate the conjunctiva with an antimicrobial secretion. Other defenses in tears and saliva include **lysozyme** and **defensins.** Lysozyme is an enzyme that hydrolyzes the peptidoglycan in the cell wall of bacteria. Defensins are peptides produced by various cells and tissues that damage cell membranes and lyse bacteria and fungi. Skin cells produce a defensin called dermicidin that helps eliminate bacteria, and paneth cells in the intestine secrete defensins that can destroy or inhibit several types of infectious agents. The high concentrations of lactic acid and electrolytes in sweat and the skin's acidic pH and fatty acid content are also inhibitory to many microbes.

Internally, the hydrochloric acid content of the stomach renders protection against many pathogens that are swallowed. The intestine's digestive juices and bile are also potentially destructive to microbes. Even semen contains an antimicrobial chemical that inhibits bacteria, and the vagina has a protective acidic pH maintained by lactobacilli that are part of the microbiota.

Genetic Resistance to Infection

In certain cases, the genetic makeup of an individual is different enough to ensure protection from some pathogens. One explanation for this phenomenon is that some pathogens have such great specificity for one host species that they are incapable of infecting other species. Another way of putting it is, "Humans can't acquire distemper from cats, and cats can't get mumps from humans." This specificity is particularly true of viruses, which can invade only by attaching to a specific host receptor. But this defense does not hold true for zoonotic infectious agents that attack a broad spectrum of animals. Genetic differences in susceptibility can also exist within members of one

ciliary escalator) conveys foreign particles entrapped in mucus toward the pharynx to be removed **(figure 14.2).** Irritation of the nasal passage reflexively initiates a sneeze, which expels a large volume of air at high velocity. Similarly, the acute sensitivity of the bronchi, trachea, and larynx to foreign matter triggers coughing, which ejects irritants.

The genitourinary tract derives partial protection from the continuous trickle of urine through the ureters and from periodic bladder emptying that flushes the urethra.

The composition of microbiota and its protective effect were discussed in chapter 13. Even though the normal microbial residents do not constitute a physical or anatomical barrier, their presence can block the access of pathogens to epithelial surfaces and can create an unfavorable environment for pathogens by competing for limited nutrients or by altering the chemical make up of various

Figure 14.2 **The ciliary defense of the respiratory tree. (a)** The trachea and bronchi are structured to entrap and propel particles upward toward the pharynx. **(b)** Detail of trachea (5,000X) features tufts of cilia on the surface of epithelial cells that sweep away particles and microbes. (b): *Source:* Courtesy Charles P. Daghlian. http://www.dartmouth.edu/~emlab/gallery

species. Humans carrying a gene or genes for sickle-cell anemia are resistant to malaria. Genetic differences also exist in susceptibility to tuberculosis, leprosy, and certain systemic fungal infections.

The vital contribution of barriers is clearly demonstrated in people who have lost them or never had them. Patients with severe skin damage due to burns are extremely susceptible to infections; those with blockages in the salivary glands, tear ducts, intestine, and urinary tract are also at greater risk for infection. But as important as it is, the first line of defense alone cannot provide adequate protection. Because many pathogens find a way to circumvent the barriers by using their virulence factors (discussed in chapter 13), a whole new set of defenses—inflammation, phagocytosis, specific immune responses—is ready to react to them.

 Check Your Progress SECTION 14.1

1. Explain the functions of the three lines of defense and indicate which are the most essential to survival.
2. What is the difference between nonspecific host defenses and immune responses?
3. Differentiate innate defenses and acquired immunity. What are the general characteristics of each? Discuss the advantages of having multiple defenses (innate and acquired) against the same microbe.
4. List four innate defensive responses present in the body, and explain how each type functions to combat infection.

14.2 Structure and Function of the Organs of Defense and Immunity

 Expected Learning Outcomes

4. Describe the primary functions and organs of the immune system.
5. Describe several features of the recognition system of host defenses.
6. Characterize pattern recognition receptors and pathogen-associated molecular patterns.
7. Describe the microscopic anatomy of body compartments and how they interconnect.
8. Outline the anatomy and functions of the reticuloendothelial system.
9. Outline the origin and composition of the blood, and characterize the types of blood cells and their functions.
10. Explain the behavior of white blood cells with regard to the blood circulation and tissues.
11. Describe the characteristics and functions of the lymphatic system.

Immunology encompasses the study of all features of the body's second and third lines of defense. Although this chapter is concerned primarily with infectious microbes, be aware that immunology is central to the study of fields as diverse as cancer and allergy. In chapters 15, 16, and 17 you will see that many of the most powerful applications of immunology have been developed for use in health care, laboratory, or commercial settings.

The primary functions of a healthy immune system can be summarized as

1. surveillance of organs, tissues, and other compartments;
2. recognition and differentiation of normal components of the body and foreign material such as pathogens; and
3. attack against and destruction of foreign invaders.

Because infectious agents could potentially enter through any number of portals, the cells of the immune system constantly move about the body, surveying the tissues for potential pathogens (see figure 14.3). This process is carried out primarily by **white blood cells (WBCs)** or **leukocytes,** which have an innate capacity to recognize and differentiate any foreign material in the body, such as invading bacterial cells. Such foreign matter is often referred to as **nonself.** Normal cells of the body, called **self,** are monitored and recognized and are usually not attacked by the immune system. The ability to differentiate nonself and thus spare self from attack is central to the effectiveness of the immune system. In chapter 16 we will see how autoimmune diseases can arise when this system goes awry. Foreign cells must be recognized as a potential threat and dealt with appropriately, but normal, healthy cells should not come under attack by the immune defenses. Exceptions to this would be abnormal or damaged self such as cancer cells.

How Do White Blood Cells Carry Out Recognition and Surveillance?

White blood cells are motile, migratory cells that move about the tissues and body compartments in search of any foreign matter that may have gotten through the first line of defense. Even though each type of blood cell has a particular role to play in defense, most of them have some capacity to locate and interact with foreign matter **(figure 14.3).** They display special molecules on their membranes known as **pattern recognition receptors (PRRs),** which are like "feelers" for sensing pathogens. There are several different types of PRRs involved in this recognition, including kinases, lectins, and mannose-type molecules. The best-understood PRRs are the **toll-like receptors (TLRs),** which reside within the membranes of early responders such as phagocytes.

Although PRRs and TLRs are not specific to a particular microbe, they can immediately recognize and interact with molecules on the surface of many pathogens called **pathogen-associated molecular patterns,** or **PAMPs.** PAMPs are molecules shared by many microorganisms that act like "red flags" that signal white blood cells involved in innate immunity. The detection of PAMPs on pathogens by PRRs and TLRs provides an early alert of invasion, and triggers reactions that help to control pathogens before they can invade further. Examples of PAMPs include peptidoglycan, lipoteichoic acid, and lipopolysaccharide from bacterial cell walls, double-stranded RNA found in some viruses, zymosan from fungal cell walls, and bacterial flagellin. We will return to this topic again in section 14.4 on phagocytes.

Because these components are shared among many pathogens and not specific to a single cell type, only a small number of TLRs are needed to recognize a very wide variety of microbes. It is important to emphasize that this type of recognition is not specific or

Surveillance of body compartments by white blood cells

Pattern recognition receptor (PRRs)

Self molecules

Pathogen-associated molecular patterns (PAMPs)

(a) Recognition of self by WBCs

(b) Recognition of nonself by WBCs

(c) Destruction of nonself by WBCs

Figure 14.3 **To search, to recognize, and to eliminate are the functions of the immune system.** White blood cells (WBCs) as seen in this view are equipped with a highly developed sense of "touch." **(a, b)** As they sort through the tissues, they "feel" the surrounding environment by means of their pattern recognition receptors (PRRs), which detect what is foreign or nonself. **(c)** When nonself molecules called PAMPs are detected, WBCs mount a reaction to eliminate the microbes that carry them. PAMPs are pathogen-associated molecular patterns that many pathogens display. Self molecules do not trigger these reactions

selective for a single type of microbe, making it a type of nonspecific, innate immunity that should not be confused with the specific acquired immunity from the third line of defense.

Compartments and Connections of the Immune System

It will soon become clear, in this and subsequent chapters, that unlike many body systems, the immune "system" does not exist in a single, well-defined organ. Instead, it encompasses a large, complex, and diffuse network of cells and fluids that penetrate into every organ and tissue. It is this very arrangement that promotes the surveillance and recognition processes that help screen the body for microbes and harmful substances.

The body is partitioned into several fluid-filled spaces called the intracellular, extracellular, lymphatic, cerebrospinal, and circulatory compartments. Although these compartments are physically separated, they have numerous connections. Their structure and position permit extensive interchange and communication

(figure 14.4). The most extensive body compartments that participate in immune function are

1. the **reticuloendothelial system (RES),*****
2. the spaces surrounding tissue cells that contain **extracellular fluid (ECF),**
3. the **bloodstream,** and
4. the **lymphatic system.**

For effective immune responsiveness, the activities in one fluid compartment must be conveyed to other compartments. Let us see how this occurs by viewing tissue at the microscopic level (figure 14.4a). At this level, clusters of tissue cells are in direct contact with the reticuloendothelial system (RES) and the extracellular fluid (ECF), which bathes all cells and is an important site of exchange for nutrients and gases. Other compartments (vessels) that penetrate at this level are blood and lymphatic capillaries. This close association allows cells and chemicals that originate in the RES and ECF to seep or migrate into the blood and lymphatics; any products of a lymphatic reaction can be transmitted directly into the

White blood cells

Lymphatic capillaries

Tissue cells

Blood capillary

ECF

ECF

ECF

Reticular fibers

(a)

Blood

ECF

RE system

Lymphatics

ECF
(extracellular fluid)

(b)

Figure 14.4 Connections between the body compartments. (a) The meeting of the major fluid compartments at the microscopic level. The close proximity of these compartments at the cellular level allows for continuous exchange of cells, fluids, and molecules (arrows). **(b)** Schematic view of how main fluid compartments form a continuous cyclic system of exchange. Positioning of the compartments favors rapid circulation from one site to another, either through direct (→) or indirect (⋯→) connections.

blood through the connection between these two systems; and certain cells and chemicals originating in the blood can move through the vessel walls into the extracellular spaces and migrate into the lymphatic system (figure 14.4b).

The flow of events among these systems depends on where an infectious agent or foreign substance first intrudes. A typical progression might begin in the extracellular spaces and RES, move to the lymphatic circulation, and ultimately end up in the bloodstream. Regardless of which compartment is first exposed, an immune reaction in any one of them will eventually be communicated to the others at the microscopic level. An obvious benefit of such an integrated system is that no cell of the body, even the most isolated, is far removed from competent protection. Let us take a closer look at each of these compartments separately.

Immune Functions of the Reticuloendothelial System

The tissues of the body are permeated by a network of connective tissue fibers, or a reticulum, that originates in the cellular basal lamina, interconnects nearby cells, and meshes with the massive connective tissue network surrounding all organs. This network, or reticuloendothelial system **(figure 14.5),** is intrinsic to immune function because it provides a passageway within and between tissues and organs. It also is heavily endowed with white blood cells of the **mononuclear phagocyte system.** This system is waiting to attack passing foreign intruders as they arrive in the skin, lungs, liver, lymph nodes, spleen, and bone marrow. The RES serves an additional role in carrying extracellular fluid throughout its network.

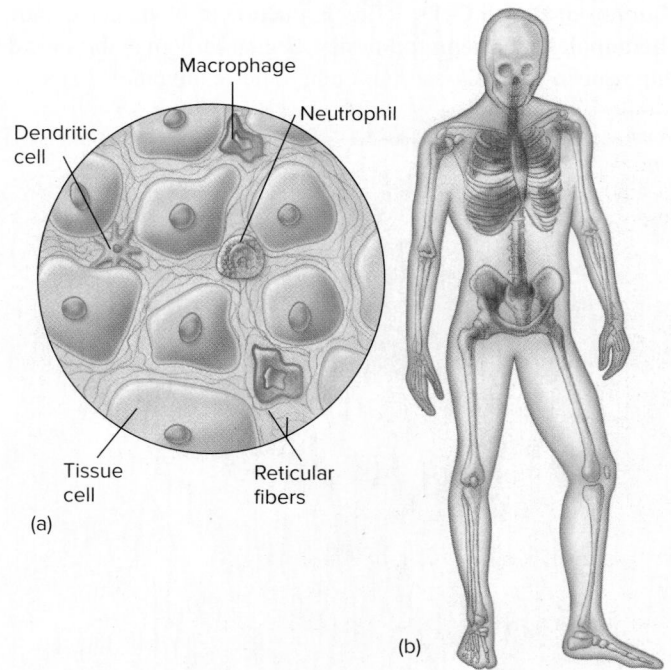

Dendritic cell

Macrophage

Neutrophil

Tissue cell

Reticular fibers

(a)

(b)

Figure 14.5 The reticuloendothelial system occurs as a continuous connective tissue framework throughout the body. (a) This system begins at the microscopic level with a fibrous support network (reticular fibers) enmeshing each cell. This web connects one cell to another within a tissue or organ and provides a niche for phagocytic white blood cells, which can crawl within and between tissues. **(b)** The degrees of shading in the body indicate variations in phagocyte concentration (darker = greater).

Origin, Composition, and Functions of the Blood

The circulatory system proper includes the heart, arteries, veins, and capillaries that circulate blood, and the lymphatic system, which includes lymphatic vessels and lymphatic organs (lymph nodes) that circulate lymph. As you will see, these two circulations parallel, interconnect with, and complement one another.

The substance that courses through the arteries, veins, and capillaries is **whole blood,** a liquid connective tissue consisting of **blood cells** (formed elements) suspended in **plasma.** One can see these two components with the naked eye when a tube of unclotted blood is allowed to sit or is spun in a centrifuge. The cells' density causes them to settle into an opaque layer at the bottom of the tube, leaving the plasma, a clear, yellowish fluid, on top **(figure 14.6). Serum** is essentially the same as plasma, except it is the clear fluid from clotted blood so it lacks the clotting proteins that plasma contains. Serum is often used in immune testing and therapy (chapter 15).

Fundamental Characteristics of Plasma Plasma contains hundreds of different chemicals produced by the liver, white blood cells, endocrine glands, and nervous system and absorbed from the digestive tract. The main component of this fluid is water (92%), and the remainder consists of proteins such as albumin and globulins (including antibodies); other immunochemicals; fibrinogen and other clotting factors; hormones; nutrients; ions and electrolytes; dissolved gases (O_2 and CO_2); and waste products (urea). These substances support the normal physiological functions of nutrition, development, protection, homeostasis, and immunity.

A Survey of Blood Cells The production of blood cells, known as **hemopoiesis,** or **hematopoiesis,*** begins early in embryonic development in the yolk sac (an embryonic membrane). Later it is

* *hematopoiesis* (hem″-mat-o-poy-ee′-sis) Gr. *haima,* blood, and *poiesis,* a making.

Figure 14.6 The macroscopic composition of whole blood.
(a) When uncoagulated blood sits in a tube, it stratifies into a clear layer of plasma, a thin layer of off-white material called the buffy coat (which contains the white blood cells), and a layer of red blood cells in the bottom. **(b)** Serum is the clear fluid that separates from clotted blood. Clotting causes the red and white cells to clump in the bottom of the tube in a single mass. *Source:* National Cancer Institute; Created by; Bruce Wetzel and Harry Schaefer (Photographers)

taken over by the liver and lymphatic organs, and it is finally assumed entirely and permanently by the red bone marrow. Although much of a newborn's red marrow is devoted to hematopoietic function, the active marrow sites gradually recede, and by the age of 4 years, only the ribs, sternum, pelvic girdle, flat bones of the skull and spinal column, and proximal portions of the humerus and femur are devoted to blood cell production.

The relatively short life of blood cells demands a rapid turnover that is continuous throughout a human life span. The primary precursor of new blood cells is a pool of undifferentiated cells called pluripotential **stem cells**[1] maintained in the marrow. During development, these stem cells proliferate and *differentiate*—meaning that immature or unspecialized cells develop the specialized form and function of mature cells. The primary lines of cells that arise from this process produce red blood cells (RBCs), white blood cells (WBCs, or leukocytes), and platelets (thrombocytes). The white blood cell lines are programmed to develop into several secondary lines of cells during the final process of differentiation **(figure 14.7).** These committed lines of WBCs are largely responsible for immune function.

The leukocytes* traditionally are evaluated by their reactions with a hematologic stain that contains a mixture of dyes and can differentiate cells by color and morphology. When this stain is used on blood smears and evaluated using the light microscope, the leukocytes appear either with or without noticeable colored granules in the cytoplasm; on that basis they are divided into two groups: **granulocytes** and **agranulocytes.** Greater magnification reveals that even the agranulocytes have tiny granules in their cytoplasm, so some hematologists also use the appearance of the nucleus to distinguish them. Granulocytes have a lobed nucleus, and agranulocytes have an unlobed, rounded nucleus. You can see both of these characteristics in circulating leukocytes shown in figure 14.7.

Granulocytes The types of granular leukocytes present in the bloodstream are neutrophils, eosinophils, and basophils. All three are known for prominent cytoplasmic granules that stain with some combination of acidic dye (eosin) or basic dye (methylene blue). Although these granules are useful diagnostically, they also function in numerous physiological events. Refer to figure 14.7 to view the cell types described.

Neutrophils* are distinguished from other leukocytes by their conspicuous lobed nuclei and their fine, pale lavender granules. In cells newly released from the bone marrow,

Neutrophil
© Harold J. Benson

1. Pluripotential stem cells can develop into several different types of blood cells; unipotential cells have already committed to a specific line of development.

* *leukocyte* (loo′-koh-syte) Gr. *leukos,* white, and *kytos,* cell. The whiteness of unstained WBCs is best seen in the white layer, or buffy coat, of sedimented blood.
* *neutrophil* (noo′-troh-fil) L. *neuter,* neither, and Gr. *philos,* to love. The granules are neutral and do not react markedly with either acidic or basic dyes. In clinical reports, they are often called "polys" or PMNs for short.

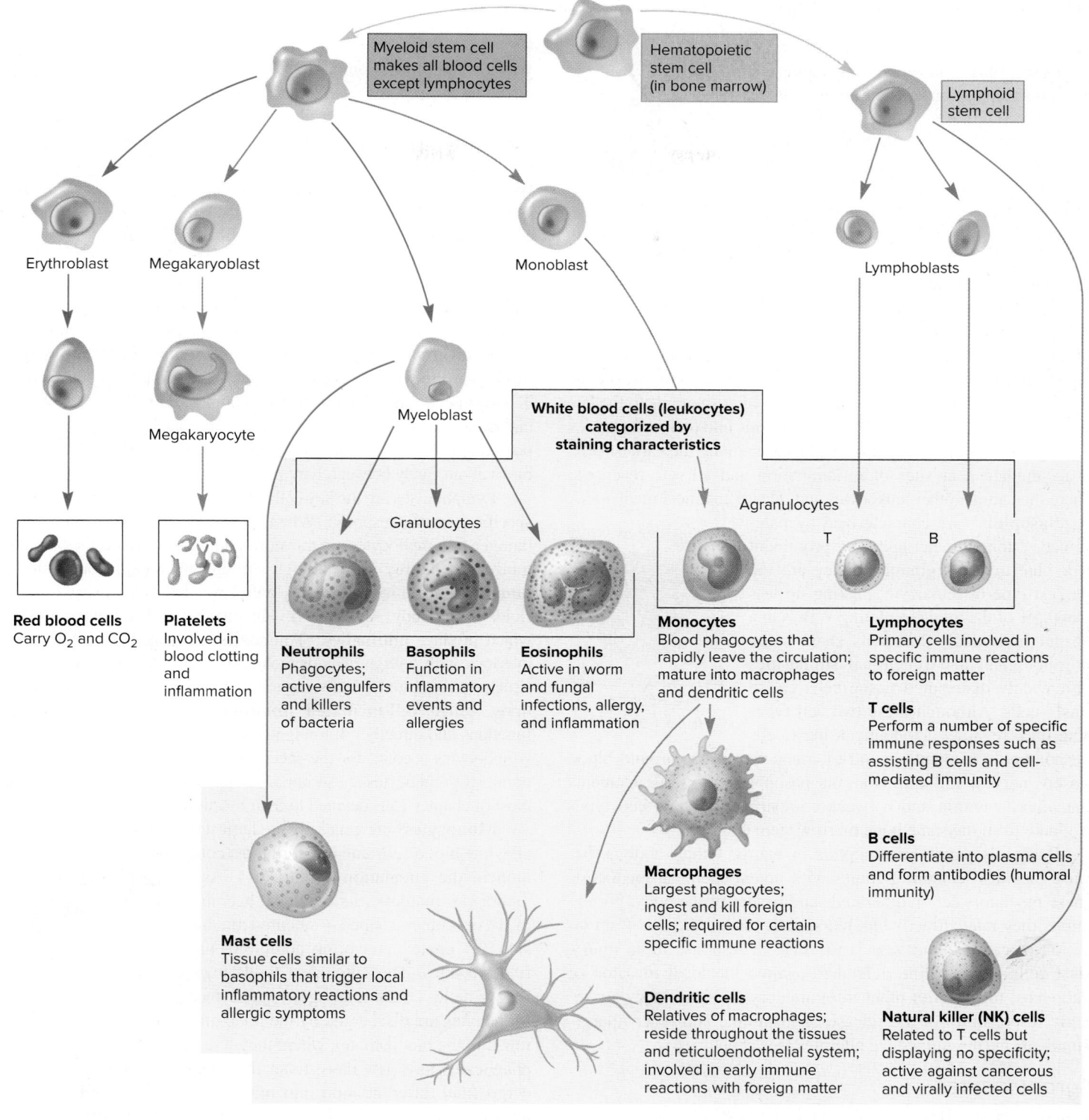

Figure 14.7 Simplified diagram of blood cell and platelet development. The details of several intermediate steps have been omitted. Undifferentiated stem cells in the red marrow give rise to several different cell lines that become increasingly specialized until mature cells are released into circulation. The yellow area indicates mature leukocytes (white blood cells). Photomicrographs of major circulating leukocytes are featured with the descriptions below.

the nuclei are horseshoe-shaped, but as they age, they form multiple lobes (up to five). These cells, also called polymorphonuclear neutrophils (PMNs), make up 55% to 90% of the circulating leukocytes—about 25 billion cells in the circulation at any given moment. The main work of the neutrophils is in phagocytosis. Their high numbers in both the blood and the tissues suggest that there is a constant challenge from resident microbiota and environmental sources. Most of the cytoplasmic granules carry digestive enzymes and other chemicals that degrade the phagocytosed materials (see the discussion of phagocytosis in section 14.4). The average neutrophil lives only about 2 days, spending much of this time in the tissues and only about 4 to 10 hours in circulation.

Labels within the figure:

- Myeloid stem cell makes all blood cells except lymphocytes
- Hematopoietic stem cell (in bone marrow)
- Lymphoid stem cell
- Erythroblast
- Megakaryoblast
- Monoblast
- Lymphoblasts
- Megakaryocyte
- Myeloblast
- White blood cells (leukocytes) categorized by staining characteristics
- Granulocytes
- Agranulocytes
- T
- B

Red blood cells Carry O_2 and CO_2

Platelets Involved in blood clotting and inflammation

Neutrophils Phagocytes; active engulfers and killers of bacteria

Basophils Function in inflammatory events and allergies

Eosinophils Active in worm and fungal infections, allergy, and inflammation

Monocytes Blood phagocytes that rapidly leave the circulation; mature into macrophages and dendritic cells

Lymphocytes Primary cells involved in specific immune reactions to foreign matter

T cells Perform a number of specific immune responses such as assisting B cells and cell-mediated immunity

B cells Differentiate into plasma cells and form antibodies (humoral immunity)

Mast cells Tissue cells similar to basophils that trigger local inflammatory reactions and allergic symptoms

Macrophages Largest phagocytes; ingest and kill foreign cells; required for certain specific immune reactions

Dendritic cells Relatives of macrophages; reside throughout the tissues and reticuloendothelial system; involved in early immune reactions with foreign matter

Natural killer (NK) cells Related to T cells but displaying no specificity; active against cancerous and virally infected cells

Eosinophils* are readily distinguished in a stain preparation by their larger, orange to red (eosinophilic) granules and bilobed nucleus. They are much more numerous in the bone marrow and the spleen than in the circulation, contributing only 1% to 3% of the total WBC count. Their granules contain peroxidase, lysozyme, and other digestive enzymes, as well as toxic proteins and inflammatory chemicals.

Eosinophil
© Harold J. Benson

The main protective action of eosinophils is their ability to attack and destroy large eukaryotic pathogens. They are also involved in inflammation and allergic reactions. Among their most important targets are the larval forms of worm parasites that cause ascariasis, filariasis, and schistosomiasis. The binding of eosinophils to the larval surface and the release of toxic compounds into their cells causes disintegration of the larvae. Eosinophils are among the earliest cells to accumulate near sites of inflammation and allergic reactions, where they attract other leukocytes and release chemical mediators.

Basophils* are characterized by pale-stained, 2-lobed nuclei and very prominent dark blue to black granules. They are the scarcest type of leukocyte, making up less than 0.5% of the total circulating WBCs in a normal individual. Basophils share some morphological and functional similarities with widely distributed tissue cells called **mast* cells.** Although these two cell types

Basophil
© Harold J. Benson

were once regarded as identical, mast cells are nonmotile elements bound to connective tissue around blood vessels, nerves, and epithelia, and basophils are motile elements that migrate within and between compartments. Both cell types originate from the same bone marrow stem cell line.

Basophils parallel eosinophils in many of their actions, because they also contain granules with potent chemical mediators. These mediators act on other cells and tissues of the body. For example, they may attract white blood cells toward the site of an infection or cause blood vessels to dilate in response to an injury. Mast cells are first-line defenders against the local invasion of pathogens; they recruit other inflammatory cells; and they are directly responsible for the release of histamine and other allergic stimulants during immediate allergies (see chapter 16).

Agranulocytes

Agranular leukocytes have globular, nonlobed nuclei and lack prominent cytoplasmic granules when viewed with the light microscope. The two general types are monocytes and lymphocytes. Refer to figure 14.7 to review their origins.

Although lymphocytes are the cornerstone of the third line of defense, which is the subject of chapter 15, their origin and morphology are described here so their relationship to the other blood components is clear. **Lymphocytes** are the second most common

WBC in the blood, comprising 20% to 35% of the total circulating leukocytes. The fact that their overall number throughout the body is among the highest of all cells indicates how important they are to immunity. One estimate suggests that about one-tenth of all adult body cells are lymphocytes, exceeded only by erythrocytes and fibroblasts.

Lymphocytes
© Harold J. Benson

In a stained blood smear, most lymphocytes are small, spherical cells with a uniformly dark, rounded nucleus surrounded by a thin fringe of clear cytoplasm, although in tissues they can become much larger and can even mimic monocytes in appearance. Lymphocytes exist as two functional types—the **B lymphocytes** (**B cells,** for short), and the thymus-derived or **T lymphocytes** (**T cells,** for short). In humans, B cells mature in special bone marrow sites. T cells mature in the thymus gland. Both populations of cells are transported by the bloodstream and lymph and move about freely between lymphoid organs and connective tissue.

Lymphocytes are the key cells of the third line of defense and the specific immune response. When stimulated by foreign substances (antigens), lymphocytes are transformed into activated cells that neutralize and destroy those foreign substances. B cells contribute to **humoral immunity,** defined as protective molecules carried in the fluids of the body. Activated B cells form specialized **plasma cells,** which produce **antibodies.*** Antibodies are protein molecules that interlock with foreign cells or molecules and participate in their destruction. Activated T cells engage in a spectrum of immune functions, called **cell-mediated immunity,** that modulate immune functions and directly kill foreign cells. The action of both classes of lymphocytes account for the specificity and memory typical of immunity. Lymphocytes are so important to the defense of the body that most of chapter 15 is devoted to their reactions.

Monocytes* are generally the largest of all white blood cells and the third most common in the circulation (3% to 7%). As a monocyte matures, its nucleus becomes oval- or kidney-shaped—indented on one side, off center, and often contorted with fine wrinkles. The cytoplasm holds many fine vacuoles containing digestive enzymes. Monocytes are discharged by the bone marrow into the bloodstream, where they live as

Monocyte
© Harold J. Benson

phagocytes for a few days. Later they leave the circulation to undergo final differentiation into **macrophages.*** Macrophages are key players in immunity, in general, responsible for

1. many types of specific and nonspecific phagocytic and killing functions (they assume the job of cellular housekeepers, "mopping up the messes" created by infection and inflammation);
2. processing foreign molecules and presenting them to lymphocytes; and
3. secreting biologically active compounds that assist, mediate, attract, and inhibit immune cells and reactions.

* *eosinophil* (ee″-oh-sin′-oh-fil) Gr. eos, dawn, rosy, and *philos,* to love. Eosin is a red, acidic dye attracted to the granules.
* *basophil* (bay′-soh-fil) Gr. *basis,* foundation. The granules attract to basic dyes.
* *mast* From Ger. *mast,* food. Early cytologists thought these cells were filled with food vacuoles.

* *antibody* (an′-tih-bahd″-ee) Gr. *anti,* against, and O. E., *bodig,* body.
* *monocyte* (mon′-oh-syte) From *mono,* one, and *cytos,* cell.
* *macrophage* (mak′-roh-fayj) Gr. *macro,* large, and *phagein,* to eat. They are the "large eaters" of the tissues.

TABLE 14.2	Characteristics of Leukocytes			
Cell Type	Prevalence in Circulation	Primary Function	Features	Appearance*
Neutrophils	55%–90% of white blood cells	General phagocytosis	Life span of 2 days, with only 4–10 hours spent in the circulation	Multilobed nuclei; small purple granules containing digestive enzymes
Eosinophils	1%–3% of white blood cells	Destruction of parasitic worms; mediators of allergy	Found in much higher numbers in the spleen and bone marrow	Bilobed nucleus with large orange granules containing toxic proteins, inflammatory mediators, and digestive enzymes
Basophils	0.5% of white blood cells	Active in allergy, inflammation, parasitic infections	Cytoplasmic granules contain histamines, prostaglandins, and other chemical mediators of the allergic response.	Pale-staining, constricted nuclei with dark blue to black granules
Monocytes	3%–7% of white blood cells	Phagocytosis, followed by final differentiation into macrophages and dendritic cells	Monocytes also secrete several chemicals that moderate the functions of the immune system.	Largest WBC; nuclei large, ovoid, and often indented—no cytoplasmic granules visible using a light microscope
Lymphocytes	20%–35% of white blood cells	Specific (acquired) immunity	Two types of lymphocytes exist. T cells are responsible for cell-mediated immunity, whereas B cells are responsible for humoral immunity.	Small spherical cells with uniformly staining dark, round nuclei

*See examples in figure 14.7.

We touch upon these functions in several ensuing sections.

Another product of the monocyte cell line is **dendritic cells,** named for their long, thin cell processes. Immature dendritic cells move from the blood to the reticuloendothelial system and lymphatic tissues, where they trap pathogens. Ingestion of bacteria and viruses stimulates dendritic cells to migrate to lymph nodes and the spleen to participate in reactions with lymphocytes.

Some of the major characteristics of leukocytes are summarized in **table 14.2.**

Erythrocyte and Platelet Lines These elements stay in the circulatory system proper and are not generally considered part of the immune system. Their development is also shown in figure 14.7.

Erythrocytes are produced by stem cells that go through several stages of differentiation and lose their nucleus just prior to entering the circulation. The resultant red blood cells are simple, biconcave sacs of hemoglobin that transport oxygen and carbon dioxide to and from the tissues. These are the most numerous of circulating blood cells, appearing in stains as small pink circles. Red blood cells do not ordinarily have immune functions, though they can be the target of immune reactions.

Platelets are formed elements in circulating blood that are *not* whole cells. They are formed by the disintegration of a large, multinucleate cell, the megakaryocyte. In stains, platelets are blue-gray with fine red granules and are readily distinguished from cells by their small size. Platelets function primarily in hemostasis (plugging broken blood vessels to stop bleeding) and in releasing chemicals that act in blood clotting and inflammation.

Components and Functions of the Lymphatic System

The lymphatic part of the circulatory system is a compartmentalized network of vessels, cells, and specialized accessory organs **(figure 14.8).** It begins in the farthest reaches of the tissues as tiny capillaries that transport a special fluid (lymph) through an increasingly larger system of vessels and filters (lymph nodes). It eventually connects to major vessels that drain back into the regular circulatory system. Some major functions of the **lymphatic** or **lymphoid system** relating to immune defenses are:

1. to provide an auxiliary route for the return of extracellular fluid to the circulatory system proper;
2. to act as a "drain-off" system for the inflammatory response; and
3. to render surveillance, recognition, and protection against foreign materials through a system of lymphocytes, phagocytes, and antibodies.

Lymphatic Fluid **Lymph** is a plasmalike liquid carried by the lymphatic circulation. It is formed when certain blood components move out of the blood vessels into the extracellular spaces and diffuse or migrate into the lymphatic capillaries. Because of this, the composition of lymph parallels that of plasma in many ways. It is made up of water, dissolved salts, and 2% to 5% protein (especially antibodies and albumin). Like blood, it also transports numerous white blood cells (especially lymphocytes) and miscellaneous materials such as fats, cellular debris, and infectious agents that have gained access to the tissue spaces. Unlike blood, red blood cells are not normally found in lymph.

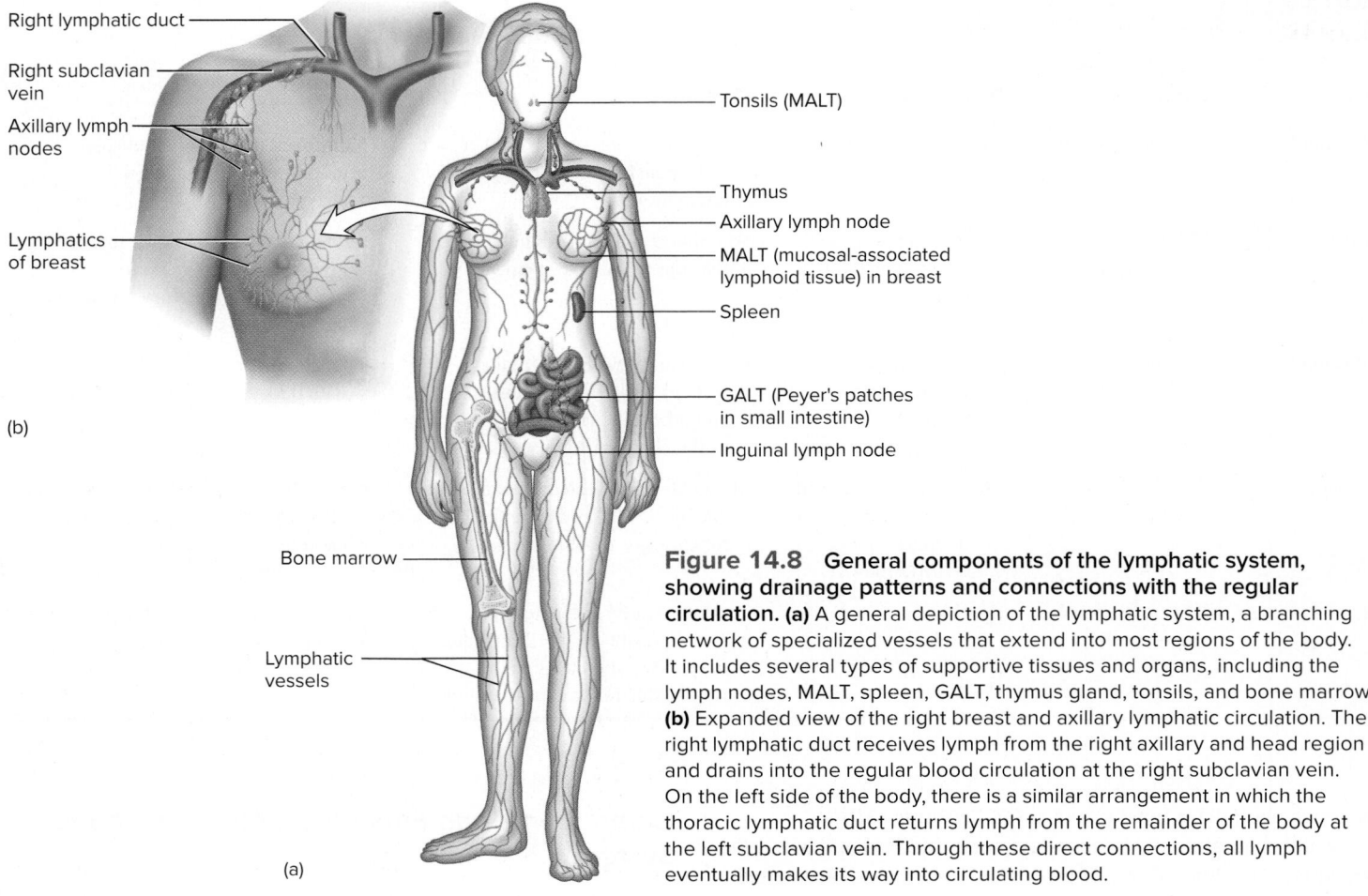

Right lymphatic duct

Right subclavian vein

Axillary lymph nodes

Lymphatics of breast

(b)

Tonsils (MALT)

Thymus
Axillary lymph node
MALT (mucosal-associated lymphoid tissue) in breast
Spleen

GALT (Peyer's patches in small intestine)
Inguinal lymph node

Bone marrow

Lymphatic vessels

(a)

Figure 14.8 **General components of the lymphatic system, showing drainage patterns and connections with the regular circulation. (a)** *A general depiction of the lymphatic system, a branching network of specialized vessels that extend into most regions of the body. It includes several types of supportive tissues and organs, including the lymph nodes, MALT, spleen, GALT, thymus gland, tonsils, and bone marrow.* **(b)** *Expanded view of the right breast and axillary lymphatic circulation. The right lymphatic duct receives lymph from the right axillary and head region and drains into the regular blood circulation at the right subclavian vein. On the left side of the body, there is a similar arrangement in which the thoracic lymphatic duct returns lymph from the remainder of the body at the left subclavian vein. Through these direct connections, all lymph eventually makes its way into circulating blood.*

Lymphatic Vessels The system of vessels that transport lymph is constructed along the lines of blood vessels (**figure 14.9**). The tiniest vessels, **lymphatic capillaries,** accompany the blood capillaries. The lymphatic capillaries permeate all parts of the body except the central nervous system and certain organs such as bone, placenta, and thymus. Their thin walls have a single layer of epithelial cells similar to blood capillaries, with loose junctions that allow free entry of extracellular fluid that collects from the circulatory system (see figure 14.4). Lymphatic vessels are found in particularly high numbers in the hands and feet and around the areolae of the breast.

Two important differences between the bloodstream and the lymphatic system should be mentioned (**figure 14.10**). First, given that one of the main functions of the lymphatic system is returning lymph to the circulation, lymph flows only in one direction, moving from the extremities toward the heart. Eventually it will be returned to the bloodstream through the thoracic duct and the right lymphatic duct near the base of the neck (see figure 14.8). The second difference concerns how lymph travels through the vessels of the lymphatic system. Whereas blood is transported through the body by means of a dedicated pump (the heart), lymph is moved only through the contraction of skeletal muscles that surround the lymphatic ducts. This dependence on muscle movement helps to explain the swelling of the hands and feet that sometimes occurs during the night (when muscles are inactive) yet goes away soon after waking.

The parasitic infection filariasis presents a dramatic example of what happens when the lymphatic drainage is blocked by infectious agents. Filarial worms from the infection become caught in the lymph nodes and channels of the extremities and plug them up. This prevents lymph from flowing into the accessory ducts and back to the circulation. The buildup of lymph massively distorts limbs and other body parts such as the scrotum and breasts.

Quick Search
To witness the devastation of this disease, type "filarial elephantiasis" into a search engine and click on images.

Lymphoid Organs Lymphatic organs and tissues with immune functions can be classified as primary and secondary, as summarized here:

Primary organs
 Thymus gland
 Bone marrow

Secondary organs and tissues
 Lymph nodes
 Spleen
 MALT—mucosal-associated lymphoid tissue
 SALT—skin-associated lymphoid tissue
 GALT—gut-associated lymphoid tissue (Peyer's patches)

(a) The finest level of lymphatic circulation begins with blind capillaries (green) that pick up foreign matter from the surrounding tissues and transport it in lymph away from the extremities via a system of small ducts.

(b) The ducts carry lymph into a circuit of larger ducts that ultimately flow into clusters of filtering organs, the lymph nodes.

(c) A section through a lymph node reveals the afferent ducts draining lymph into sinuses that house several types of white blood cells. Here, foreign material is filtered out and processed by lymphocytes, macrophages, and dendritic cells.

(d) and (e) Lymph continues to trickle from the lymph nodes via efferent ducts into a system of larger drainage vessels, which ultimately connect with large veins near the heart. In this way, cells and products of immunity continually enter the regular circulation.

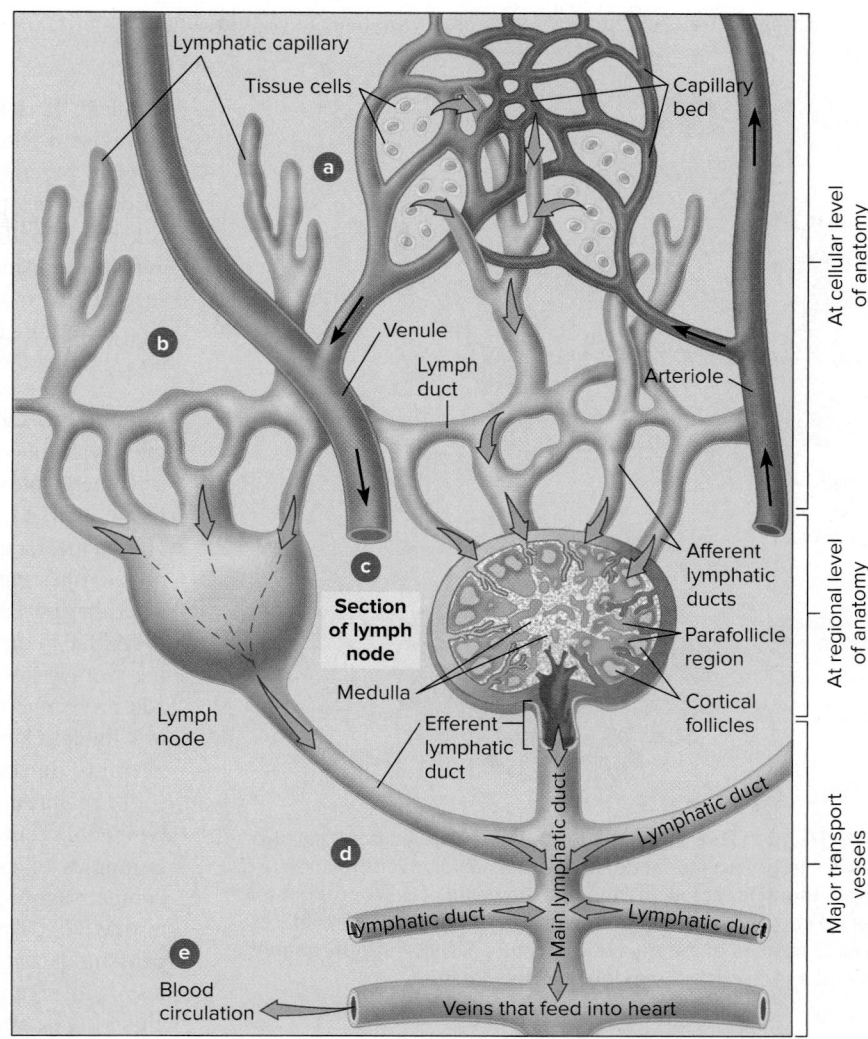

Figure 14.9 Scheme of circulation in the lymphatic vessels and lymph nodes.

The **primary lymphoid organs** include the thymus and bone marrow, because they are a source of immune cells and a "nursery" for developing white blood cells. They are the sites of origin and maturation of lymphocytes—white blood cells with specific immune function. They subsequently release these cells to populate the secondary lymphatic sites.

Secondary lymphoid organs, such as the spleen and lymph nodes, are circulatory-based locations where encounters with microbes and immune responses often take place. Associated lymphoid tissues are collections of cells widely dispersed throughout body tissues such as the skin and mucous membranes, ready to react with any locally entering infectious agents.

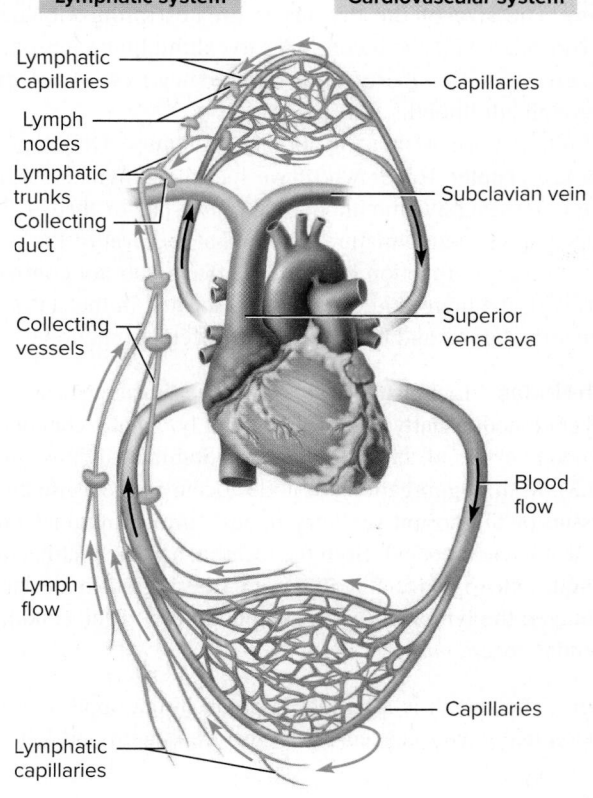

Figure 14.10 Comparison of the circulation patterns of the regular circulatory and lymphatic systems. The movement of lymphatic flow is in one direction (green) from lymphatic capillaries to collecting vessels and ducts to large lymphatic trunks to subclavian veins to the heart. The flow of blood, on the other hand, is cyclic, with blood continuously flowing through arteries to capillaries to veins to the heart and back around. With this combined system the lymphatics can collect excess tissue fluid and return it to the bloodstream. The two systems can also participate together in surveillance of the tissues for foreign invaders. Note: The lymphatic flow (green) is shown on only one side to keep the comparison uncluttered.

Figure 14.11 The thymus gland. Immediately after birth, the thymus is a large organ that nearly fills the region over the midline of the upper thoracic region. In the adult, however, it is proportionately smaller (to compare, see figure 14.8*a*). Section shows the main anatomical regions of the thymus. Immature T cells enter through the cortex and migrate into the medulla as they mature.

The Thymus Gland The **thymus** originates in the embryo as two lobes in the pharyngeal region that fuse into a triangular structure. It is located in the thoracic cavity near the tip of the sternum. The size of the thymus is greatest proportionately at birth **(figure 14.11),** and it continues to exhibit high rates of activity and growth until puberty, after which it begins to shrink gradually through adulthood.

Children born without a complete thymus (DiGeorge syndrome, see chapter 16) or who have had their thymus surgically removed are severely immunodeficient and fail to thrive. Adults have developed enough mature T cells that removal of the thymus or reduction in its function has milder effects. Do not confuse the thymus with the thyroid gland, which is located in the cervical region near the larynx and has an entirely different function.

Lymph Nodes Lymph nodes are small, encapsulated, bean-shaped organs stationed, usually in clusters, along lymphatic channels and large blood vessels of the thoracic and abdominal cavities (see figure 14.8). Major aggregations of nodes occur in the loose connective tissue of the armpit (axillary nodes), groin (inguinal nodes), and neck (cervical nodes). Both the location and the architecture of these nodes clearly specialize them for filtering out materials that have entered the lymph and providing appropriate cells and niches for immune reactions.

Spleen The spleen is a lymphoid organ in the upper portion of the abdominal cavity nestled below the diaphragm and left of the stomach. It is somewhat similar to a lymph node except that it serves as a filter for blood instead of lymph. While the spleen's primary function is to remove worn-out red blood cells from circulation, its most important immunologic function centers on the filtering of pathogens from the blood and their subsequent phagocytosis by resident macrophages. Adults whose spleens have been surgically removed can live a relatively normal life, but children who have failed to develop or lost their spleen are severely immunocompromised.

Associated Lymphoid Tissue Embedded throughout systems lined with mucous membranes are discrete bundles of lymphocytes and other white blood cells termed the **mucosal-associated lymphoid tissue** or **MALT.** The positioning of this widespread system provides a local, rapid mechanism for responding to the constant influx of microbes entering via the gastrointestinal, respiratory, urinary, and other portals of entry. The pharynx houses a prominent source of MALT in the form of tonsils. The breasts of pregnant and lactating women also become temporary sites of lymphoid tissues that add protective antibodies to breast milk (see colostrum, described in Clinical Connections, "The Immunological Benefits of Breast-Feeding," in chapter 15).

The intestinal tract houses the best-developed collection of lymphoid tissue, called **gut-associated lymphoid tissue,** or **GALT.** Examples of GALT include the appendix, and **Peyer's patches,** compact aggregations of lymphocytes in the ileum of the small intestine. GALT provides immune functions against intestinal pathogens and is a significant source of some types of antibodies. Other, less well organized collections of secondary lymphoid tissue include *skin-associated lymphoid tissue* and *bronchial-associated lymphoid tissue*.

 Check Your Progress SECTION 14.2

 5 How is surveillance of the tissues carried out, and what is responsible for it?

 6. Describe the processes through which the immune system distinguishes self from foreign cells and particles.

 7. What does the term *nonself* mean in reference to the immune system?

 8. What are the main components of the reticuloendothelial system, and why is it also called the mononuclear phagocyte system?

 9. How does the RES communicate with tissues and the vascular system?

 10. Explain the functions of stem cells, and summarize the stages of hematopoiesis.

 11. Differentiate between granulocytes and agranulocytes.

 12. Describe the main types of granulocytes and agranulocytes, their functions, and their incidence in the circulation.

 13. Describe the principal function of the two lymphocyte types, and differentiate between humoral and cell-mediated immunity.

 14. What is lymph, and how is it formed?

 15. Outline the functions of the lymphatic system and describe the actions of associated lymphatic tissues such as MALT and GALT.

14.3 Second-Line Defenses: Inflammation

 Expected Learning Outcomes

12. Describe the main events in the inflammatory reaction, and explain what is occurring in each.

13. Summarize the functions of cytokines, chemokines, and other inflammatory mediators.

14. Describe the mechanism behind fever, and explain its beneficial and harmful effects.

Now that we have introduced the principal anatomical and physiological framework of the immune system, we address some mechanisms that play important roles in host defenses: (1) inflammation, (2) phagocytosis, (3) interferon, and (4) complement. Because of the generalized nature of these defenses, they are primarily nonspecific in their effects, but they also support and interact with the specific immune responses described in chapter 15.

The Inflammatory Response: A Complex Concert of Reactions to Injury

At its most general level, **inflammation** is a reaction to any traumatic event in the tissues that attempts to restore homeostasis: It is a normal and necessary process that helps to clear away invading microbes and cellular debris left by immune reactions. Most people manifest inflammation in some way every day. It appears in the flare of a cut, the blistering of a burn, the painful lesion of an infection, and the symptoms of allergy, and at times it can even go unnoticed.

Inflammation is readily identifiable by a classic series of signs and symptoms characterized succinctly by four Latin terms: *rubor, calor, tumor,* and *dolor.* Rubor (redness) is caused by increased circulation and vasodilation in the injured tissues; calor (warmth) is the heat given off by the increased flow of blood; tumor (swelling) is caused by increased extracelluar fluid accumulating in the tissues; and dolor (pain) is caused by the stimulation of nerve endings from the pressure of swelling or by chemical mediators (**figure 14.12**). A fifth symptom, *loss of function,* is sometimes added for a complete picture of effects that can accompany inflammation. Although these manifestations are often unpleasant, they serve an important warning that injury has taken place and will set in motion responses that can save the body from further injury.

Factors that can elicit an inflammatory response include trauma from infection (the primary emphasis here), tissue injury or death, and specific immune reactions. Although the details of

> **Quick Search**
> To learn more about this process, search for videos on "Acute Inflammation."

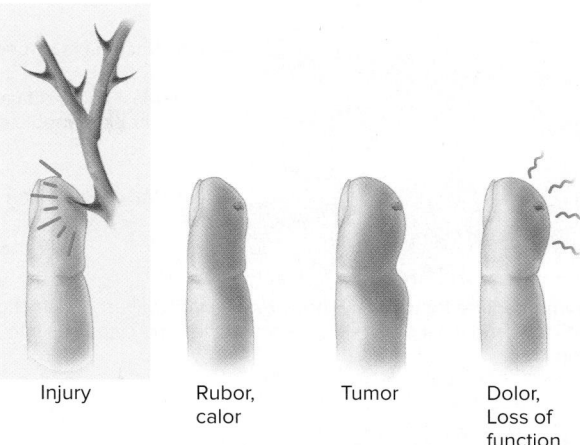

Injury | Rubor, calor | Tumor | Dolor, Loss of function

Figure 14.12 The response to injury. This classic series summarizes the reactions of the tissues to an assault. Each of the events is an indicator of one of the mechanisms of inflammation described in this section.

inflammation are very complex, its chief functions can be summarized as follows:

1. to mobilize and attract immune cells and chemicals to the site of the injury,
2. to set in motion mechanisms to repair tissue damage and localize and clear away harmful substances, and
3. to destroy and block microbes from further invasion (see figure 14.13).

The inflammatory response is a powerful defensive reaction, a means for the body to maintain stability and restore itself after an injury. But when it is chronic, it has the potential to actually *cause* tissue injury, destruction, and disease (see Clinical Connections, "The Downside of Inflammation").

The Stages of Inflammation

The inflammatory response is a dynamic sequence of events that can be acute (lasting from a few minutes or hours) to chronic (lasting for days, weeks, or years). Once the initial injury has occurred, a chain reaction takes place at the site of damaged tissue, summoning beneficial cells and fluids into the injured area. To envision this process in action, we start with an injury and observe the flow of major events at the microscopic level (**figure 14.13**).

Vascular Changes: Early Inflammatory Events

Following an injury, some of the earliest changes occur in the vasculature (arterioles, capillaries, venules) in the vicinity of the damaged tissue. These changes are controlled by nervous stimulation, **chemical mediators,** and **cytokines*** released by blood cells, tissue cells, and platelets in the injured area. Some mediators are *vasoactive;* that is, they affect the endothelial cells and smooth muscle cells of blood

* *cytokine* (sy′-toh-kyne) Gr. *cytos,* cell, and *kemein,* to move. A protein or polypeptide produced by WBCs that regulates host defenses.

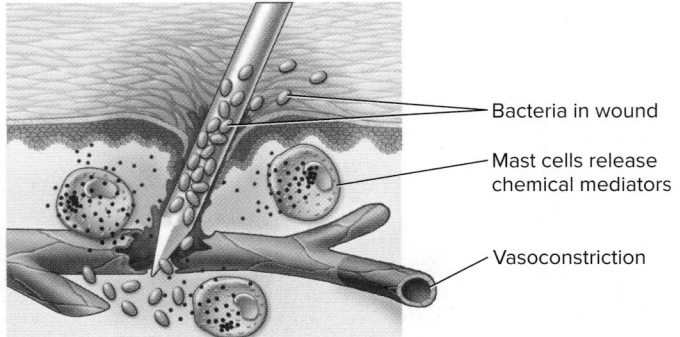

(a) **Injury/Immediate Reactions:** Blood vessels narrow (vasoconstriction); blood clots; and mast cells release chemokines and cytokines into injured area.

Bacteria in wound

Mast cells release chemical mediators

Vasoconstriction

Clot

Bacteria

Neutrophil

Seepage of plasma and migration of WBC out of blood vessels

Vasodilation

(b) **Vascular Reactions:** Nearby blood vessels dilate; increased blood flow; increased vascular permeability; increased leakage of fluid forms exudate.

Scab

Neutrophils

Pus

Fibrous exudate

(c) **Edema and Pus Formation:** Collection of fluid; edema/swelling; infiltration by neutrophils and formation of pus.

Scar

Lymphocytes

Fibroblast

Macrophage

(d) **Resolution/Scar Formation:** Macrophages, lymphocytes, and fibroblasts migrate in; initiate immune response and repair of injury; scar and loss of normal tissue.

Rubor/calor of inflammation

Edema (tumor) and dolor

Repaired or damaged tissue

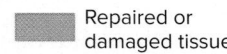 **Process Figure 14.13** **The major events in inflammation.**

vessels. Others are **chemotactic factors,** also called **chemokines,** that affect white blood cells. Inflammatory mediators cause fever, stimulate lymphocytes, prevent infection spread, and cause allergic symptoms (**figure 14.14** and **14.1 Making Connections**). Although the constriction of arterioles is stimulated first, it lasts for only a few seconds or minutes. Once a clot has formed to prevent blood loss, it is followed in quick succession by the opposite reaction, **vasodilation.** The overall effect of vasodilation is to increase the flow of blood into the area, which facilitates the influx of immune components and also causes redness and warmth (rubor and calor).

An Early Chemical Indicator of Inflammation

C-reactive protein* (CRP) is a pattern recognition receptor that is an early indicator of inflammation. It is a large protein synthesized by the liver that circulates in the blood. CRP is released in response to cytokines produced by macrophages during infections and other pathologic conditions. It attaches to PAMPs and other receptors on dead or injured body cells and infectious agents, an action that increases phagocytosis and promotes the complement system (see next section). In this way it serves as another source of localized, immediate, innate protection. Because it can be detected within the first few hours of the onset of inflammation, it is an effective marker for inflammation and can be measured with blood tests. Elevated levels of CRP can also be found in cardiovascular disease, cancer, tissue injury, and necrosis, and are often used to diagnose risks for these conditions.

Postcapillary venules, small veins that drain blood from the capillary beds, are important in several aspects of inflammation. When they constrict, this causes a slowing of circulation and a pooling of extracellular fluid at the site. Some vasoactive substances cause the endothelial cells in the walls of these venules to separate and form gaps through which blood-borne components exude into the extracellular spaces. Capillaries also become leaky under the influence of cytokines. The plasma that escapes from small vessels into the tissues is called the *exudate.* Accumulation of this fluid gives rise to local swelling and hardness known as **edema.** The edematous exudate contains varying amounts of plasma proteins, such as globulins, albumin, the clotting protein fibrinogen, blood cells, and cellular debris. Depending upon its content, the exudate may be clear (called *serous*), or it may contain red blood cells or **pus.** Pus is composed mainly of white blood cells, microbes, and the debris generated by phagocytosis. In some types of edema, the fibrinogen is converted to fibrin threads that enmesh the injury site (see figure 14.13c). Within an hour multitudes of neutrophils, responding chemotactically to special signaling molecules, converge on the injured site.

Unique Dynamic Characteristics of White Blood Cells In order for WBCs to leave the blood vessels and enter the tissues, they adhere to the inner walls of the smaller blood vessels such as capillaries and venules. From this position, they are poised to migrate out of the blood into the tissue spaces by a process called **diapedesis.***

Diapedesis, also known as transmigration, is aided by several related characteristics of WBCs. For example, they are actively motile and readily change shape. Their migration is assisted by the nature of the endothelial cells lining venules. Venules contain

* Not to be confused with the complement system.

* *diapedesis* (dye″-ah-puh-dee′-sis) Gr. *dia,* through, and *pedan,* to leap.

Figure 14.14 Major chemical mediators of the inflammatory response and their effects. See 14.1 Making Connections for more information on the effects of mediators.

complex adhesive receptors that develop increased "stickiness" under the influence of inflammatory mediators. This causes WBCs to adhere or marginate at the endothelial cells. From this location, they can readily crawl into the extracellular spaces (**figure 14.15**).

Another factor in the migratory habits of these WBCs is **chemotaxis,*** defined as the tendency of cells to migrate in response to a specific chemical stimulus released at a site of injury or infection. Through this means, cells swarm from many compartments to the site of infection and remain there to perform general functions such as phagocytosis, repair, and specific immune reactions. These basic properties are absolutely essential for the sort of intercommunication and deployment of cells required for most immune reactions.

The Benefits of Edema and Chemotaxis Both the formation of edematous exudate and the infiltration of neutrophils are physiologically beneficial activities. The influx of fluid dilutes toxic substances, and the fibrin clot can effectively trap microbes and prevent their further spread. The neutrophils that aggregate in the inflamed site are immediately involved in phagocytosing and destroying microbes, dead tissues, and particulate matter (by mecha-

nisms discussed in section 14.4 on phagocytosis). In some types of inflammation, accumulated phagocytes contribute to pus. Certain bacteria (streptococci, staphylococci, gonococci, and meningococci) are especially powerful attractants for neutrophils and are thus termed **pyogenic,** or pus-forming, bacteria.

Late Reactions of Inflammation Sometimes a mild inflammation can be resolved by edema and phagocytosis. Inflammatory reactions that last more than a few days attract a collection of monocytes, lymphocytes, and macrophages to the reaction site. Clearance of pus, cellular debris, dead neutrophils, and damaged tissue is performed by macrophages, the only cells that can engulf and dispose of such large masses. At the same time, specific immune reactions of acquired immunity are brought into play. B lymphocytes react with foreign molecules and cells by producing specific antimicrobial proteins (antibodies), and T lymphocytes kill intruders directly. Late in the process, the tissue undergoes various levels of repair and may be replaced by connective tissue in the form of a scar (see figure 14.13*d*). If the inflammation cannot be relieved or resolved in this way, it can become chronic and create long-term pathologic conditions discussed in **Clinical Connections,** "The Downside of Inflammation."

* *chemotaxis* (kee-moh-tak′-sis) NL. *chemo*, chemical, and *taxis*, arrangement.

14.1 MAKING CONNECTIONS

The Dynamics of Inflammatory Mediators

Just as the nervous system is coordinated by a complex communications network, so too is the immune system. Hundreds of small, active molecules are constantly being secreted to regulate, stimulate, suppress, and otherwise control the many aspects of cell development, inflammation, and immunity. These substances are the products of several types of cells, including monocytes, macrophages, lymphocytes, fibroblasts, mast cells, platelets, and endothelial cells of blood vessels. Their effects may be local or systemic, short term or long lasting, nonspecific or specific, protective or pathologic.

The field of cytokines has become so increasingly complex that we can include here only an overview of the major groups of important cytokines and other mediators. The major functional types can be categorized into

1. cytokines that mediate nonspecific immune reactions such as inflammation and phagocytosis,
2. cytokines that regulate the growth and activation of lymphocytes,
3. cytokines that activate immune reactions during inflammation,
4. hematopoiesis factors for white blood cells,
5. vasoactive mediators, and
6. miscellaneous inflammatory mediators.

Nonspecific Mediators of Inflammation and Immunity

- *Tumor necrosis factor* (*TNF*), a substance from macrophages, lymphocytes, and other cells. It increases chemotaxis and phagocytosis and stimulates other cells to secrete inflammatory cytokines. It also triggers fever (is an endogenous pyrogen) and stimulates blood coagulation. Unfortunately it is has the potential to cause damage to tissues when produced in excess.
- *Interferon alpha and interferon beta*, produced by leukocytes and fibroblasts, inhibit virus replication and cell division and increase the action of certain lymphocytes that kill other cells.
- *Interleukin* is a term that refers to a group of small peptides originally isolated from leukocytes. There are currently more than 50 known interleukins. We now know that other cells besides leukocytes can synthesize them and that they have a variety of biological activities.
 - *Interleukin-1* (*IL-1*), a product of macrophages and epithelial cells. It has many of the same biological activities as TNF, such as inducing fever and activating certain white blood cells.
 - *Interleukin-6*, secreted by macrophages, lymphocytes, and fibroblasts. Its primary effects are to stimulate the growth of B cells and to increase the synthesis of liver proteins such as C-reactive protein.
- *Various chemokines.* By definition, chemokines are cytokines that stimulate the movement and migration of white blood cells (chemotactic factors). Included among these are complement $C5_a$, interleukin-8, and platelet-activating factor.

Cytokines That Regulate Lymphocyte Growth and Activation

- *Interleukin-2*, the primary growth factor from T cells. Interestingly, it acts on the same cells that secrete it. It stimulates cell division and secretion of other cytokines. In B cells, it is a growth factor and stimulus for antibody synthesis.
- *Interleukin-4*, a stimulus for the development of B cells and production of antibodies that mediate allergies. It inhibits macrophage actions.
- *Granulocyte colony-stimulating factor* (*G-CSF*), produced by T cells, macrophages, and neutrophils. It stimulates the activation and differentiation of neutrophils.
- *Macrophage colony-stimulating factor* (*M-CSF*), produced by a variety of cells. M-CSF promotes the growth and development of stem cells that differentiate into macrophages.
- *Interferon gamma*, a T-cell-derived mediator whose primary function is to activate macrophages. It also promotes the differentiation of T and B cells, activates neutrophils, and stimulates diapedesis.

Vasoactive Mediators

- *Histamine*, a vasoactive mediator produced by mast cells and basophils. It causes vasodilation, increased vascular permeability, and mucus production. It functions primarily in inflammation and allergy.
- *Serotonin*, a mediator produced by platelets and intestinal cells. It causes smooth muscle contraction, inhibits gastric secretion, and acts as a neurotransmitter.
- *Bradykinin*, a vasoactive amine from the blood or tissues. It stimulates smooth muscle contraction and increases vascular permeability, mucus production, and pain. It is particularly active in allergic reactions.

Miscellaneous Inflammatory Mediators

- *Prostaglandins*, produced by most body cells, are complex chemical mediators that can have opposing effects (for example, dilation or constriction of blood vessels) and are powerful stimulants of inflammation and pain.
- *Leukotrienes* stimulate the contraction of smooth muscle and enhance vascular permeability. They are implicated in the more severe manifestations of immediate allergies (constriction of airways).
- *Platelet-activating factor*, a substance released from basophils, causes the aggregation of platelets and the release of other chemical mediators during immediate allergic reactions.

Which inflammatory mediators would be involved in rubor? in tumor? in dolor? in calor? *Answers available on Connect.*

Fever: An Adjunct to Inflammation

An important systemic component of inflammation is fever, defined as an abnormally elevated body temperature. Although fever is a nearly universal symptom of infection, it is also associated with certain allergies, cancers, and other organic illnesses. Fevers whose causes are undiagnosed are called fevers of unknown origin, or FUOs.

The body temperature is normally maintained by a control center in the hypothalamus region of the brain. This thermostat regulates the body's heat production and heat loss and sets the core temperature at around 37°C (98.6°F), with slight fluctuations (1°F) during a daily cycle. Fever is initiated when a circulating substance

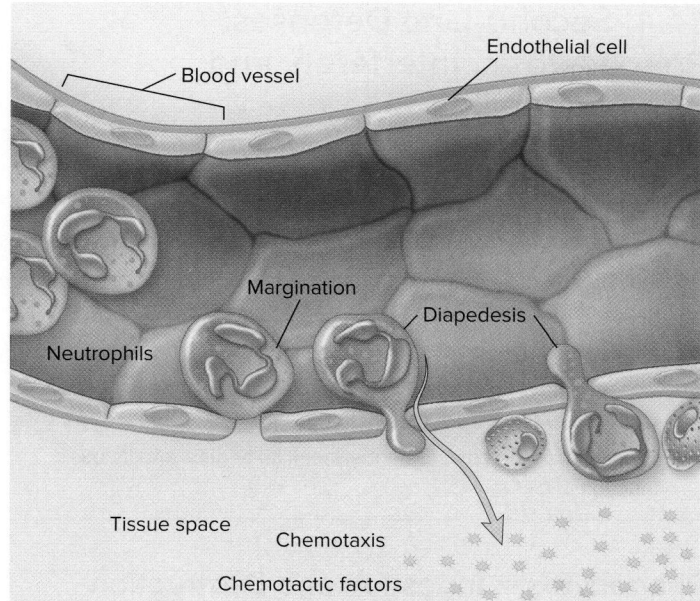

Figure 14.15 Diapedesis and chemotaxis of leukocytes.
Cross section of a venule depicts white blood cells squeezing themselves between spaces in the vessel wall through diapedesis. This process also indicates how the pool of leukocytes adheres to receptors in the endothelial wall. From this site, they are poised to migrate out of the vessel into the tissue space and respond to chemotactic factors released as part of the generalized inflammatory process.

called **pyrogen*** resets the hypothalamic thermostat to a higher setting. This change signals the musculature to increase heat production and peripheral arterioles to decrease heat loss through vasoconstriction. Fevers range in severity from low-grade (37.7°C to 38.3°C, or 100°F to 101°F) to moderate (38.8°C to 39.4°C, or 102°F to 103°F) to high (40.0°C to 41.1°C, or 104°F to 106°F).

Pyrogens are described as *exogenous* (coming from outside the body) or *endogenous* (originating internally). Exogenous pyrogens are products of infectious agents such as viruses, bacteria, protozoans, and fungi. One well-characterized exogenous pyrogen is endotoxin, the lipopolysaccharide found in the cell walls of gram-negative bacteria. Blood, blood products, vaccines, or injectable solutions can also contain exogenous pyrogens. Endogenous pyrogens are liberated by monocytes, neutrophils, and macrophages during the process of phagocytosis and appear to be a natural part of the immune response. Two potent pyrogens released by macrophages are interleukin-1 (IL-1) and tumor necrosis factor (TNF).

Fever is usually accompanied by chills. What would cause a febrile (feverish) person to periodically feel cold and tremble uncontrollably? The explanation lies in how the brain reacts to pyrogen and the natural physiological interaction between the the hypothalamus and the temperature of the blood. For example, if the thermostat has been reset (by pyrogen) at 102°F but the blood temperature is 99°F, the muscles are stimulated by the brain to contract involuntarily (shivering) as a means of producing more heat. In addition, the vessels in the skin constrict, creating a sensation of cold, and the piloerector muscles in the skin cause goose bumps to form.

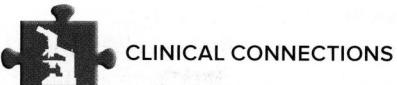

CLINICAL CONNECTIONS

The Downside of Inflammation

Not every aspect of inflammation is protective or results in the proficient resolution of tissue damage. As one looks over a list of diseases, it is rather striking how many of them are due in part or even completely to an overreactive or dysfunctional inflammatory response.

Some -*itis* reactions mentioned in chapter 13 are a case in point. Inflammatory exudates that build up in the brain in African trypanosomiasis, cryptococcosis, and other brain infections can be so injurious to the nervous system that impairment is permanent. Inflammatory reactions frequently lead to abscess, a swollen mass of neutrophils and dead, liquefied tissue that can harbor live pathogens in the center. Abscesses are a prominent feature of staphylococcal, amebic, and enteric infections.

Other pathologic manifestations of chronic diseases—for example, the tubercles of tuberculosis, the lesions of late syphilis, the disfiguring nodules of leprosy, and the cutaneous ulcers of leishmaniasis—are due to an aberrant tissue response called *granuloma formation* (see Case Study figure). Granulomas develop in response not only to microbes but also to inanimate foreign bodies that are difficult to break down. This condition is initiated when neutrophils and macrophages ineffectively and incompletely phagocytose the pathogens or materials involved in an inflammatory reaction. This failure to manage the debris of inflammation is sometimes referred to as "frustrated phagocytosis." Macrophages respond by storing ingested materials in vacuoles and becoming inactive. Over a given time period, macrophages fuse into giant, inactive multinucleate cells called foreign body giant cells. These sites are further infiltrated with lymphocytes. The resultant collections make the tissue appear granular—hence the name. A granuloma can exist in the tissue for months, years, or even a lifetime. This chapter's Case Study provides a striking example of massive proliferation of failed phagocytes.

One common cytokine, tumor necrosis factor (TNF), is a potent inflammatory mediator. It regulates a wide variety of early immune activities such as phagocytosis, and it is a powerful **pyrogen.** Unfortunately the overproduction of TNF is also involved in a number of diseases, such as septic shock, psoriasis, asthma, and autoimmunity. Several drugs (adalimumab [Humira] and etanercept [Enbrel], for example) have been developed as therapy to counteract the damaging effects of TNF.

Medical science is rapidly searching for additional applications based on new information about inflammatory mediators. One highly promising area appears to be the use of chemokine inhibitors that could reduce chemotaxis and the massive, destructive influx of leukocytes. Such therapy could ultimately be used to treat certain cancers, hardening of arteries, and Alzheimer disease.

Explain how inflammatory mediators stimulate tissue injury and disease. *Answer available on Connect.*

* *pyrogen* (py′-roh-jen) Gr. *pyr*, fire, and *gennan*, produce. Same origin as *funeral pyre* and *pyromaniac.*

Benefits and Treatment of Fever Medical experts have abundant evidence that fever serves an important biological function. Work with tissue cultures showed that increased temperatures stimulate the activities of T cells and increase the effectiveness of interferon. Artificially infected rabbits and pigs allowed to remain febrile survive at a higher rate than those given suppressant drugs. Fever appears to enhance phagocytosis of staphylococci by neutrophils in many mammalian species. The appearance of fever during infections strongly suggests that it is normal and beneficial reaction to invading microbes. Some other effects of fever include:

- It inhibits multiplication of temperature-sensitive microorganisms, such as the poliovirus, cold viruses, herpes zoster virus, systemic and subcutaneous fungal pathogens, *Mycobacterium* species, and the syphilis spirochete.
- It impedes the nutrition of bacteria by reducing the availability of iron. Research has demonstrated that during fever, the macrophages stop releasing their iron stores, which could retard several enzymatic reactions needed for pathogen growth.
- It increases metabolism and stimulates immune reactions and naturally protective physiological processes. It speeds up hematopoiesis, phagocytosis, and specific immune reactions.

Understanding that fever is possibly more beneficial than harmful, deciding whether to suppress it is an important medical question. Some advocates feel that a slight to moderate fever (99–101°F) in an otherwise healthy person should be allowed to run its course in light of its potential as an immune stimulant and its minimal side effects. All medical experts do agree that high and prolonged fevers or fevers in children or patients with cardiovascular disease, seizures, and respiratory ailments are risky and must be treated immediately with fever-suppressant drugs. The classic therapy for fever is an antipyretic drug such as aspirin or acetaminophen (Tylenol) that lowers the setting of the hypothalamic center and restores normal temperature. Any physical technique that increases heat loss (tepid baths, for example) can also help reduce the core temperature.

✅ *Check Your Progress* SECTION 14.3

16. Describe the major events in the inflammatory response, including the stimuli, physiological reactions, and signs and symptoms.

17. Of rubor, calor, dolor, and tumor, which are signs and which are symptoms?

18. What actions of the inflammatory and immune defenses account for swollen lymph nodes and leukocytosis?

19. What causes pus, and what does it indicate?

20. In what ways can edema be beneficial and in what ways is it harmful?

21. Describe the physiological events leading to fever and review the ways in which fever is naturally beneficial.

22. Explain the processes of diapedesis and chemotaxis, and show how they interrelate.

23. Briefly account for the origins and actions of the major types of inflammatory mediators and cytokines.

14.4 Second-Line Defenses: Phagocytosis, Interferon, and Complement

 Expected Learning Outcomes

15. Define *phagocytosis* and identify which cells serve this function.

16. Indicate the major stages of phagocytosis, and describe what is happening at each stage.

17. Explain how phagocytes kill pathogens.

18. Describe the origin of interferon, and explain its role in controlling viruses and other immune functions.

19. Characterize the complement system, its origins, and its basic functions.

Phagocytosis: Ingestion and Destruction by White Blood Cells

By any standard, a phagocyte represents an impressive piece of living machinery, meandering through the tissues to seek, capture, and destroy a target. The following is a summary of the major activities of phagocytes:

1. to survey the tissue compartments and target microbes, particulate matter (dust, carbon particles, antigen-antibody complexes), and injured or dead cells;
2. to ingest and eliminate these materials; and
3. to extract immunogenic information (antigens) from foreign matter.

Major Categories of Phagocytes

It is generally accepted that all cells have some capacity to engulf materials, but professional phagocytes do it for a living. The main types of phagocytes are neutrophils, monocytes, macrophages, and dendritic cells.

Neutrophils and Eosinophils As previously stated, neutrophils are general-purpose phagocytes that react early in the inflammatory response to bacteria and other foreign materials and to damaged tissue. A common sign of bacterial infection is a high neutrophil count in the blood (neutrophilia), and neutrophils are also a primary component of pus. Eosinophils are attracted to sites of parasitic infections and antigen–antibody reactions, though they play only a minor phagocytic role.

Extended Functions of Neutrophils Neutrophils are truly one of the "workhorses" of innate immunity. In addition to their phagocytic functions, they also have a distinct system for capturing pathogens called *neutrophil extracellular traps,* or NETS. Neutrophils are programmed to die after they have gone through their regular engulfment and killing of bacteria and other pathogens. But their death does not end their protective actions. Dying neutrophils throw out a fibrous matrix composed of DNA, enzymes, histones, and other cell contents that continue to stop invading microbes even after the neutrophils are completely lysed. The NET works at

several levels to trap and immobilize bacteria and fungi, degrade their virulence factors, kill them with microbicidal chemicals, and ultimately, prevent them from spreading.

Macrophages: Dynamic Scavengers After emigrating out of the bloodstream into the tissues, monocytes are transformed by various inflammatory mediators into macrophages. This process is marked by an increase in size and by enhanced development of lysosomes and other organelles **(figure 14.16).** At one time macrophages were classified as either fixed (adherent to tissue) or wandering, but this terminology can be misleading. All macrophages retain the capacity to move about. Whether they reside in a specific organ or wander depends upon their stage of development and the immune stimuli they receive. Specialized macrophages called *histiocytes* migrate to a certain tissue and remain there during their life span. Examples are alveolar (lung) macrophages; Kupffer cells in the liver; Langerhans cells in the skin; and macrophages in the spleen, lymph nodes, bone marrow, kidney, bone, and brain. Other macrophages do not reside permanently in a particular tissue but instead drift nomadically throughout the reticuloendothelial system. Dendritic cells—another product of the monocyte line—are concentrated in skin and the linings of mucous membranes, where they process foreign substances and prepare them for reactions with selected lymphocytes (see chapter 15).

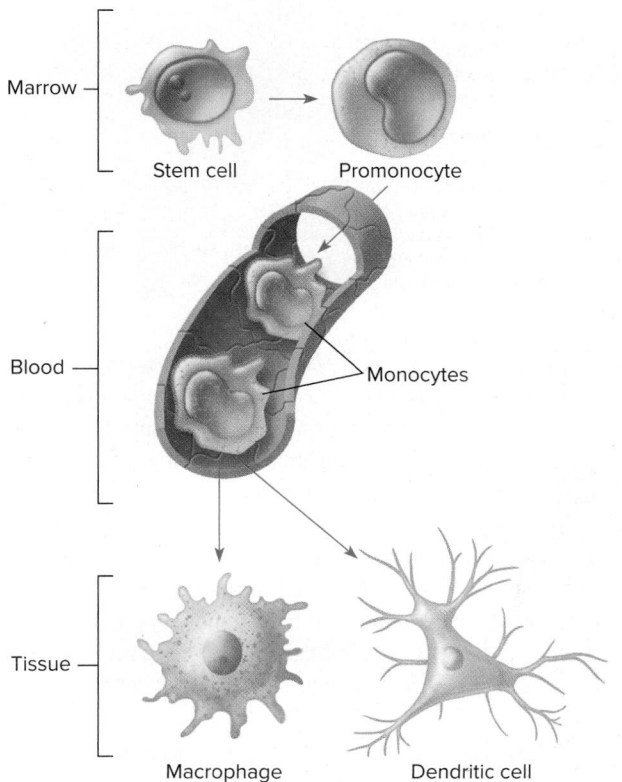

Figure 14.16 The developmental stages of monocytes and macrophages. The cells progress through maturational stages in the bone marrow and peripheral blood. Once in the tissues, macrophage and dendritic cells can remain nomadic or take up residence in a specific organ.

Marrow — Stem cell → Promonocyte

Blood — Monocytes

Tissue — Macrophage Dendritic cell

Mechanisms of Phagocytic Recognition, Engulfment, and Killing

The term **phagocytosis** literally means "eating cell process." But phagocytosis is more than just the physical act of engulfment, because phagocytes also actively attack and dismantle foreign cells using a wide array of antimicrobial substances. Phagocytosis can occur as an isolated event performed by a lone phagocytic cell responding to a minor irritant in its area or as part of the orchestrated events of inflammation described in section 14.3. The events in phagocytosis include chemotaxis, ingestion, phagolysosome formation, destruction, and elimination **(figure 14.17).**

> **Quick Search**
>
> Observe a video of active phagocytes by accessing "Bacterial Phagocytosis by Neutrophils" on YouTube.

Chemotaxis, Binding, and Ingestion Phagocytes migrate into a region of inflammation with a deliberate sense of direction, attracted by a gradient of stimulant products from the pathogen and host tissue at the site of injury.

Once a phagocyte encounters the pathogen, it uses its toll-like receptors (TLRs, discussed in section 14.2) to make contact with the pathogen (figure 14.17). Recall that TLRs are receptors that recognize and bind the pathogen-associated molecular pattern (PAMP) receptors of various microbes. There are about 10 different TLRs in the membranes of phagocytes. The exposed end of a receptor hooks on to a PAMP and immediately dimerizes or joins with a second TLR to encase the molecule **(figure 14.18).** This relays a signal into the nucleus that stimulates the intracellular phagocytic processes and the release of chemical mediators.

Phagolysosome Formation On the scene of an inflammatory reaction, phagocytes often trap cells or debris against the fibrous network of connective tissue or the wall of blood and lymphatic vessels. Once the phagocyte has "caught" its prey, it extends pseudopods that enclose the cells or particles in a pocket and internalize them in a vacuole called a **phagosome.**

In a short time, **lysosomes** migrate to the scene of the phagosome and fuse with it to form a **phagolysosome.** Other granules containing antimicrobial chemicals are released into the phagolysosome, forming a potent brew designed to poison and then dismantle the ingested material (see figure 14.17). The destructiveness of phagocytosis is evident by the death of bacteria within 30 minutes after contacting this battery of antimicrobial substances.

Destruction and Elimination Systems Destructive chemicals await the microbes in the phagolysosome. The oxygen-dependent system known as the **respiratory burst,** or oxidative burst, elaborates products of oxygen metabolism called **reactive oxygen intermediates (ROIs).** Myeloperoxidase, an enzyme found in granulocytes, forms halogen ions (OCl^-) that are strong oxidizing agents. Other products of oxygen metabolism such as hydrogen peroxide, the superoxide anion (O_2^-), activated or singlet oxygen ($^1O^-$), and the hydroxyl free radical (HO) separately and together have formidable killing power. This series of reactive oxygen products delivers a "knockout" punch necessary to kill aerobic pathogens such as fungi and many bacteria (see this chapter's

① Chemotaxis by phagocyte

Bacterial cells

PAMPs

② Adhesion of bacteria

Receptor on host cell

③ Engulfment into phagocytic vacuole

Lysosomes

Golgi apparatus

Rough endoplasmic reticulum

④ Phagosome

⑤ Phagolysosome formation

⑥ Killing and destruction of bacterial cells

Nucleus

Enzymes
Lysozyme
DNase
RNase
Proteases
Peroxidase

Reactive oxygen products
Superoxide ($O_2^{-\bullet}$)
Hydrogen peroxide (H_2O_2)
Singlet oxygen (1O_2)
Hydroxyl ion (OH^-)
Hypochlorite ion ($HClO^-$)

⑦ Release of residual debris

Process Figure 14.17 The sequential events in phagocytosis. **(1)** Phagocyte is attracted to bacteria. **(2)** Close-up view of process showing bacteria adhering to phagocyte receptors by their PAMPs. **(3)** Vacuole is formed around bacteria during engulfment. **(4)** Phagosome digestive vacuole results. **(5)** Lysosomes fuse with phagosome, forming a phagolysosome. **(6)** Enzymes and toxic oxygen products kill and digest bacteria. **(7)** Undigested particles are released. Inset: Scanning electron micrograph of a neutrophil (yellow) actively engaged in devouring anthrax bacteria (orange) (10,000x). © Dr. Brinkmann/Max Planck Institute for Infection Biology

Case Study). Other chemicals that come into play are lactic acid, lysozyme, and *nitric oxide* (NO), a powerful mediator that kills bacteria and inhibits viral replication. Cationic proteins that injure bacterial cell membranes and a number of **hydrolytic** enzymes complete the job. The bits of undigestible debris are released from the macrophage by exocytosis. As we shall see in chapter 15, macrophages and dendritic cells combine phagocytosis with further processing of microbial antigens required for specific immune responses with lymphocytes.

Interferon: Antiviral Cytokines and Immune Stimulants

Interferon (IFN) was described in chapter 12 as a small protein produced naturally by certain white blood and tissue cells. It is used in therapy against certain viral infections and cancer and can be used as an immune enhancer, as we saw in this chapter's Case Study. Although the interferon system was originally thought to be directed exclusively against viruses, it is now known to be involved also in defenses against other microbes and in immune regulation and intercommunication. Three major types are *interferon alpha,* a product of

lymphocytes and macrophages; *interferon beta,* a product of fibroblasts and epithelial cells; and *interferon gamma,* a product of T cells. See 14.1 Making Connections for a summary of interferon actions.

All three classes of interferon are produced in response to viruses, RNA, immune products, and a variety of foreign molecules. Their biological activities are extensive. In all cases, they bind to cell surfaces and induce changes in genetic expression, but the exact results vary. In addition to antiviral effects, discussed next, all three IFNs can inhibit the expression of cancer genes and have tumor suppressor effects. Interferon alpha and IFN beta stimulate phagocytes, and IFN gamma is an immune regulator of macrophages and T and B cells.

Characteristics of Antiviral Interferon

When a virus binds to the receptors on a host cell, a signal is sent to the nucleus that directs the cell to synthesize interferon. After transcription and translation of the interferon gene, newly synthesized interferon molecules are rapidly secreted by the cell into the extracellular space, where they bind to other host cells. The binding of interferon to a second cell induces the production of another class of

Toll-like receptor Foreign molecule (PAMP)

Nucleus

Macrophage

Cytokines
Interleukins
Inflammatory mediators

Figure 14.18 Phagocyte detection and signaling with toll-like receptors. Toll-like receptors (TLRs) span the membrane of phagocytes and other cells of the immune system. When a molecule such as a PAMP on a particular pathogen is recognized by this receptor, the TLRs merge and bind the foreign molecule. This induces production of chemicals and triggers engulfment.

proteins that inhibit viral multiplication by preventing the translation of viral proteins **(figure 14.19)**. Interferon is not virus-specific, so its synthesis in response to one type of virus will also protect against other types. Because this protein is an inhibitor of viruses, it has been a valuable treatment for a number of viral infections.

Other Roles of Interferon

Interferons are also important immune regulatory cytokines that activate or instruct the development of white blood cells. For example, interferon alpha produced by T lymphocytes activates a

subset of cells called natural killer (NK) cells. In addition, one type of interferon beta plays a role in the maturation of B and T lymphocytes and in inflammation. Interferon gamma inhibits cancer cells, stimulates B lymphocytes, activates macrophages, and enhances the effectiveness of phagocytosis.

Complement: A Versatile Backup System

Among its many overlapping functions, the immune system has another complex and multiple-duty system called **complement** that is brought into play at several levels. The complement system, named for its property of "complementing" immune reactions, consists of at least 30 blood proteins released by liver cells, lymphocytes, and monocytes. Complement factors are a form of pattern recognition receptor (PRR) that works in concert with inflammation and phagocytosis to destroy a wide variety of bacteria, viruses, and parasites. This system is generally nonspecific and innate, but it can also play a role in acquired immune responses. The three primary defensive features of the complement system are (1) the **membrane attack complex (MAC),** which kills pathogens directly, (2) the coating of pathogens with molecules that make them more attractive to phagocytes **(opsonization),** and (3) the recruitment of inflammatory cells and triggering of cytokine release. Having introduced the second and third topics previously, we will focus here on the main mechanisms of the membrane attack complex.

The membrane attack function of complement is a series of reactions like that of blood clotting, in which the first substance in a chemical series activates the next substance, which activates the next, and so on, until a desired end product is reached. Three different versions of the complement pathways exist **(figure 14.20).** Their main distinguishing features are how they are activated, major participating factors, and specificity. The end stages of all three converge at the same point and yield a similar end result, that is, destruction of a microbe or infected cell. Because the focus of damage is on the cell membrane, this portion of the complement defense is not as effective on microbes whose outer walls block access to the membrane, but is is effective on cells and viruses with exposed membranes.

The *classical pathway* is the most specific, activated mainly by the presence of antibody bound to microorganisms. It is this pathway that links complement to reactions of acquired immunity. The other two pathways are nonspecific and triggered by common foreign molecules on the surfaces of microbes. In the *lectin pathway,* a host serum protein or lectin binds a sugar called mannan present in the walls of bacteria and other microbes. The *alternative pathway* is initiated by complement proteins that bind to certain surface molecules of microbes (PAMPS),

Quick Search
For a dynamic look at the complement system, look up "Animation: Activation of Complement" online.

Virus
infection

Viral
nucleic acid

Assembly
of viruses

Virus
release

Degrades virus
nucleic acid

Blocks virus
replication

IFN
gene

Synthesis
of IFN

Attachment of IFN
to special receptor

Synthesis of antiviral proteins

Signals
activation of genes

Infected
cell

Nearby
cell

Figure 14.19 The antiviral activity of interferon. When a cell is infected by a virus, its nucleus is triggered to transcribe and translate the interferon (IFN) gene. Interferon diffuses out of the cell and binds to IFN receptors on nearby uninfected cells, where it induces production of proteins that eliminate viral genes and block viral replication. Note that the initial cell is not protected by its own IFN and that IFN does not prevent recipient cells from being invading by viruses, but the viruses are inactivated before they can start replicating.

ation

Classical Pathway	MB-Lectin Pathway	Alternative Pathway
Complement-fixing antibodies have rapid, specific effects	Mannose-binding lectin (MBL) binds mannose on pathogen surfaces. Nonspecific for bacteria and viruses	Molecules on surfaces of bacteria, viruses, and parasites. Nonspecific
C1q, C1r, C1s, Factors C4, C2, **C3**	MBL, MASP-1, MASP-2, Factors C4, C2, **C3**	Factors B, D and, **C3**

C3 convertase converts the C3 molecule to an activator of the cascade — $C3_b$.

C3 convertase enzyme

Gram-negative bacterial cell

C5

C3

$C5_b$ $C3_b$

$C3_a$ $C5_a$

(b) Cascade and Amplification. C5 factor is acted on by $C3_b$, which converts it to $C5_b$. $C5_b$ becomes bound to the membrane and serves as the starting molecule for the chain of events that assemble the complex in (c) and (d).

C8 C9

C6 $C5_b$

C7

Two products of the cascade reaction — $C3_a$ and $C5_a$ — have additional inflammatory functions. Both molecules stimulate mast cell degranulation, enhance chemotaxis of white blood cells, and act as inflammatory mediators.

(c) Polymerization. $C5_b$ is a reactive site for the final assembly of an attack complex. In series, C6, C7, and C8 aggregate with $C5_b$ and become integrated into the membrane. They form a substrate upon which the final component, C9, can bind. Up to 15 of these C9 units ring the central core of the final membrane attack complex (MAC).

Completed MAC

Lysis

(d) Membrane Attack. Insertion of MACs produces hundreds of tiny holes in the cell membrane. This can cause lysis and death of eukaryotic cells and many gram-negative bacteria.

Lysis of bacterial cell

Process Figure 14.20 Overview of the complement pathways leading to membrane attack. (a) All three pathways have different triggers and starting points, but they all converge at the same place, C3 convertase. This enzyme begins the series of reactions **(b,c,d)** that give rise to the membrane attack complex characteristic of the complement "machinery." The final result is the formation of tiny openings in the cell membrane and the destruction of the target pathogen.

although it may also be triggered spontaneously. Note that, because the complement numbers (C1 to C9) are based on the order of their discovery, some factors are not activated in numerical order.

Stages in the Complement Cascade

In general, the complement pathways go through four stages: initiation, amplification and cascade, polymerization, and membrane attack. The starting reaction requires some type of initiator molecule such as antibodies, lectins, or microbial surface receptors **(figure 14.20*a*)**, depending on the pathway. The presence of this initiator on the pathogen's membrane propels the chain of actions involving complement chemicals C1 through C4, with each step serving a dual purpose. We are omitting the fine details of the amplification part of the complement system, but the end result is a molecule, **C3,** that is a key factor in subsequent polymerization of the remaining complement factors in the membrane of the target cell. C3 is acted on by a convertase enzyme that splits it into factors $C3_a$ and **$C3_b$.** $C3_b$ forms the recognition site and anchor for the final series of reactions to come **(figure 14.20*b*).** $C3_b$ by itself is also an important factor in opsonization and thus facilitates phagocytosis. More information on opsonization can be found in chapter 15.

$C3_b$ moves the cascade reaction forward by splitting factor C5 into $C5_a$ and $C5_b$. One of these molecules—**$C5_b$**—is key to finishing up polymerization of the remaining complement factors described next. You will notice that the two cleavage products that do not continue in the cascade reaction—$C3_a$ and $C5_a$—still have a number of immune functions (figure 14.20—right side of *b*). Both of them are chemotactic factors that recruit white blood cells and trigger the release of inflammatory mediators such as histamine.

The final events in the complement series involve factors $C5_b$, C6, C7, and C8, which join together and give rise to an anchoring complex within the membrane **(figure 14.20*c*).** This is followed by the insertion of several **C9** molecules in the pattern of a ring-shaped formation that is the membrane attack complex (MAC) **(figure 14.20*d*).** The MAC is the primary destructive force of the complement system. It effectively perforates and lyses the membranes of gram-negative bacteria, some parasitic protozoans, enveloped viruses, and virally infected host cells. Except for the classical pathway, these reactions are nonspecific and can be active against a wide spectrum of microbes.

An Outline of Major Host Defenses

At this point, it should be apparent that the body's defenses often work in both parallel and redundant ways to fight pathogens. The result is a constant onslaught of defenses functioning at many levels. This can be illustrated by **figure 14.21,** organized by the three lines of defense that started out our discussion and that lead us naturally into chapter 15.

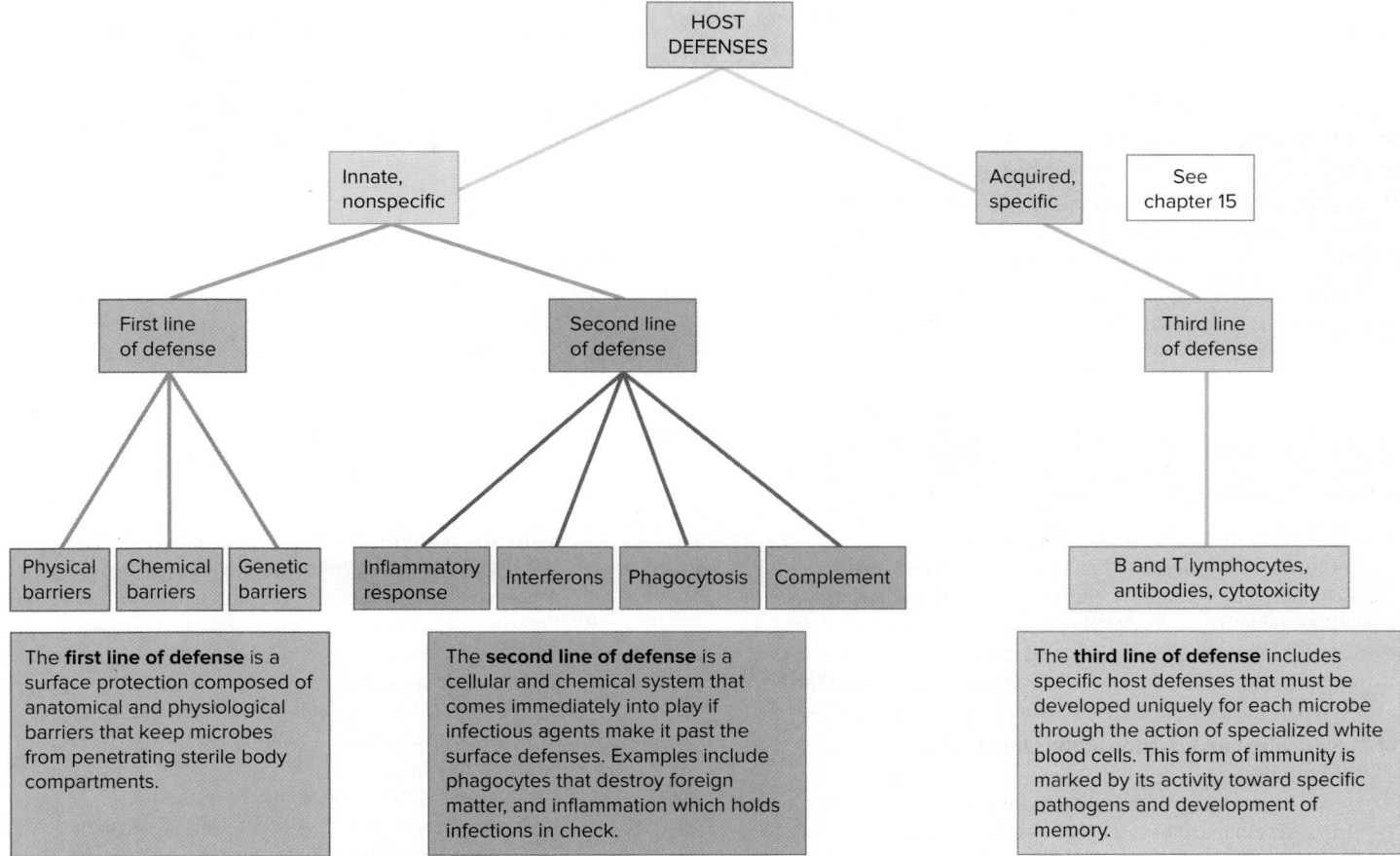

Figure 14.21 **Flowchart summarizing the major components of the host defenses.** Defenses are classified into one of two general categories: (1) innate and nonspecific or (2) acquired and specific. These can be further subdivided into the first, second, and third lines of defense, each being characterized by a different level and type of protection. The third line of defense, the most varied, is responsible for specific immunity. It is covered in greater detail in chapter 15.

 Check Your Progress SECTION 14.4

24. Describe the events that give rise to macrophages.

25. What are the types of macrophages, and what are their principal functions?

26. Outline the major phases of phagocytosis.

27. Briefly describe the three major types of interferon, their sources, and their biological effects.

28. Describe the mechanism by which interferon acts as an antiviral compound.

29. Describe the general actions of the complement system.

30. What are some factors that trigger the complement cascade?

31. Using figure 14.21 as a guide, give examples for each category shaded brown, purple, and gold.

 CASE STUDY Part 2

The boy's physicians examined his white blood cells using the nitroblue tetrazolium test, which monitors the effectiveness of phagocyte activity, and the neutrophil cytochrome assay, which can measure activities associated with the metabolism of oxygen. Both tests indicated a low rate of phagocyte responsiveness, which is highly diagnostic of a genetic disease known as chronic granulomatous disease (CGD).

About one in 200,000 children in the United States are diagnosed yearly with chronic granulomatous disease (CGD); of these, around 90% are males. CGD is somewhat complex in its inheritance, but there are essentially two forms. The X-linked type occurs as a defective gene on the X chromosome and is more severe. Over half of all cases are X-linked. The autosomal defect is located on a different chromosome and tends to be milder. In both types, the genetic codes carry errors that prevent phagocytes from producing certain key metabolic enzymes.

The neutrophils and macrophages of CGD patients can carry out the usual phagocytic activities such as amoeboid motion, chemotaxis, engulfment, and digestion of microbes in phagocytic vesicles, but they are unable to actively kill certain types of aerobic pathogens. They lack the enzymes to synthesize the battery of reactive oxygen intermediates (ROIs)—superoxide ion, hydrogen peroxide, and hydroxyl radical—that are essential to kill these microbes. As a result, the ineffective phagocytes harbor live microbes that can break away and continue their infectious path. In time these phagocytes shut down, die, and build up in the infected tissue in masses (granulomas).

This case is a striking example of how important every element of immunity can be. Although the components do overlap in some of their effects, each one carries out some essential function that cannot be totally duplicated. CGD individuals have a mostly intact immune response, but they lack a complete phagocyte defense that comes into play early in the infection process. Phagocytes (neutrophils, monocytes, and macrophages) are most active against bacteria and eukaryotic pathogens, but they also are important participants in many other immune reactions.

■ *What elements of host defenses and immunity would remain intact in a child with CGD?*

■ *What is a granuloma, and what factors in CGD cause it?*

To conclude this Case Study, go to Connect.

For additional information on this disease, go to http://emedicine.medscape.com/article/1116022-print. Other immune disorders are covered in chapter 16.

 Chapter Summary with Key Terms

14.1 Overview of Host Defense Mechanisms

The *first line of defense* is an inborn, nonspecific system composed of anatomical, chemical, and genetic barriers that block microbes at the portal of entry. The *second line of defense* is also inborn and nonspecific and includes protective cells and fluids in tissues, and the *third line of defense* is acquired and specific and is dependent on the function of T and B cells.

14.2 Structure and Function of the Organs of Defense and Immunity

A. **Immunology** is the study of **immunity,** which refers to the development of resistance to infectious agents by white blood cells (WBCs). WBCs conduct surveillance of the body, recognizing and differentiating **nonself** (foreign) cells by means of molecules or **pattern recognition receptors (PRRs)** in their membranes. PRRs recognize a number of indicators carried on foreign cells called **pathogen-associated molecular patterns (PAMPs).** Under normal conditions, **self** cells are left alone but nonself cells are destroyed by the concerted efforts of several types of immune reactions.

B. For purposes of immunologic study, the body is divided into three compartments: the blood, lymphatics, and the **reticuloendothelial system.** A fourth component, the extracellular fluid, surrounds the first three and allows constant communication between all areas of the body.

C. **Circulatory System: Blood and Lymphatics**

1. **Whole blood** consists of **blood cells** (formed by **hematopoiesis** in the bone marrow) dispersed in **plasma. Stem cells** in the bone marrow differentiate to produce white blood cells **(leukocytes),** red blood cells **(erythrocytes),** and megakaryocytes, which give rise to **platelets.**

2. Leukocytes are the primary mediators of immune function and can be divided into **granulocytes** (**neutrophils, eosinophils,** and **basophils**) and **agranulocytes** (**monocytes** and **lymphocytes**) based on their appearance. Monocytes later differentiate to become **macrophages.** Lymphocytes are divided into **B cells,** which produce **antibodies** as part of **humoral immunity,** and **T cells,** which participate in **cell-mediated immunity.** Leukocytes display both **chemotaxis** and **diapedesis** in response to chemical mediators of the immune system.

D. *Lymphatic System*
 1. The **lymphatic system** parallels the circulatory system and transports lymph while also playing host to cells of the immune system. Lymphoid organs and tissues include **lymph nodes,** the spleen, and the **thymus,** as well as areas of less well-organized immune tissues such as **GALT** and **MALT.**

14.3 Second-Line Defenses: Inflammation
 A. The inflammatory response is a complex reaction to tissue injury marked by *redness, heat, swelling,* and *pain.*
 1. Blood vessels narrow and then dilate in response to **chemical mediators, cytokines,** and **chemokines.**
 2. **Edema** swells tissues, helping prevent the spread of infection.

 3. WBCs, microbes, debris, and fluid collect to form **pus.**
 4. **Pyrogens** may induce fever.

14.4 Second-Line Defenses: Phagocytosis, Interferon, and Complement
 A. Phagocytosis: Partner to Inflammation
 1. Macrophages and neutrophils engage in **phagocytosis,** engulfing microbes in a phagosome. Uniting the phagosome with a **lysosome** results in destruction of the phagosome contents.
 2. Phagocytes use a type of PRRs called **toll-like receptors (TLRs)** to recognize and adhere to foreign markers on microbes such as PAMPs.
 B. **Interferon (IFN)** is a family of proteins produced by leukocytes and fibroblasts that inhibit the reproduction of viruses by degrading viral RNA or blocking the synthesis of viral proteins.
 C. **Complement** is an innate defense system that plays a role in the destruction of bacteria, viruses, and parasites. It causes reactions on the surfaces of cells which result in the formation of a **membrane attack complex** (MAC) that kills microbial cells by creating holes in their membranes. It also plays roles in inflammation and phagocytosis.

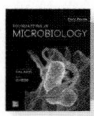

Level I. Knowledge and Comprehension

These questions require a working knowledge of the concepts in the chapter and the ability to recall and understand the information you have studied.

 Multiple-Choice Questions

Select the correct answer from the answers provided. For questions with blanks, choose the combination of answers that most accurately completes the statement.

1. An example/examples of a nonspecific chemical barrier to infection is/are
 a. unbroken skin
 b. lysozyme in saliva
 c. cilia in respiratory tract
 d. all of these

2. Which nonspecific host defense is associated with the trachea?
 a. lacrimation
 b. ciliary lining
 c. desquamation
 d. lactic acid

3. Which of the following blood cells function primarily as phagocytes?
 a. eosinophils
 b. basophils
 c. lymphocytes
 d. neutrophils

4. Which of the following is *not* a lymphoid tissue?
 a. spleen
 b. thyroid gland
 c. lymph node
 d. GALT

5. What is included in GALT?
 a. thymus
 b. Peyer's patches
 c. tonsils
 d. breast lymph nodes

6. A signaling molecule from microbes recognized by phagocytes is
 a. pyrogen
 b. PAMP
 c. complement
 d. lectin

7. Monocytes are _____ leukocytes that develop into _____.
 a. granular, phagocytes
 b. agranular, mast cells
 c. agranular, macrophages
 d. granular, T cells

8. Which of the following inflammatory signs specifies pain?
 a. tumor
 b. calor
 c. dolor
 d. rubor

9. Toll-like receptors are proteins on _____.
 a. phagocytes that recognize foreign molecules
 b. viruses that stimulate immune reactions
 c. skin cells that provide barriers to infection
 d. lymphocytes that damage parasitic worms

10. An example of an inflammatory mediator that stimulates vasodilation is
 a. histamine
 b. collagen
 c. complement $C5_a$
 d. interferon

11. _____ is an example of an inflammatory mediator that stimulates chemotaxis.
 a. Endotoxin
 b. Serotonin
 c. A fibrin clot
 d. Interleukin-2

12. An example of an exogenous pyrogen is
 a. interleukin-1 c. interferon
 b. complement d. endotoxin

13. _____ interferon, produced by T lymphocytes, activates cells called _____ and is involved in destroying viruses.
 a. Gamma, fibroblasts c. Alpha, natural killer cells
 b. Beta, lymphocytes d. Beta, fibroblasts

14. In what process is tumor necrosis factor (TNF) *not* involved?
 a. chemotaxis of phagocytes
 b. fever

c. the inflammatory response
d. the classic complement pathway

15. Which of the following substances is *not* produced by phagocytes to destroy engulfed microorganisms?
 a. hydroxyl radical c. hydrogen peroxide
 b. superoxide anion d. bradykinin

16. Which of the following is the end product of the complement system?
 a. properdin c. membrane attack complex
 b. cascade reaction d. complement factor C9

Case Study Review

1. What are signs of chronic granulomatous disease?
 a. unresolved inflammation
 b. opportunistic fungal infections
 c. low levels of white blood cells
 d. recurrent infections
 e. a, b, and c
 f. a, b, and d

2. What feature of CGD phagocytes is defective?
 a. the ability to engulf pathogens
 b. production of toxic forms of oxygen
 c. mechanisms to eliminate viruses
 d. chemotaxis

3. Explain why children with CGD would be more prone to infections from fungi and bacteria than from viruses.

Writing Challenge

For each question, compose a one- or two-paragraph answer that includes the factual information needed to completely address the question. Check Your Progress questions can also be used for writing-challenge exercises.

1. Use the lines on the figure to the right to locate and describe the eight components of the first line of defense.

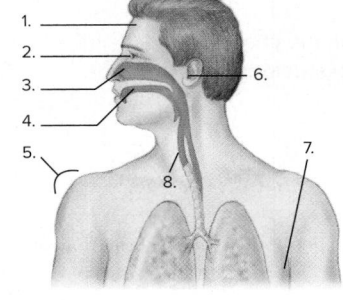

2. What effects do these defenses have on microbes? Describe the initial steps taken by the primary responders to foreign matter.

3. Explain what PRRs and TLRs are and the nature of their interaction with PAMPs.

4. Make a simple drawing of the functions of toll-like receptors.

5. Describe the mechanisms by which leukocytes migrate from the bloodstream into an area of infection. What chemical factors are they reacting to?

6. Prepare a simplified outline of the cell lines of hematopoiesis.

7. Differentiate between the functions of primary and secondary lymphoid organs

8. In what ways is a phagocyte a tiny container of disinfectants?

9. Macrophages perform the final job of removing tissue debris and other products of infection. Indicate some of the possible effects when these scavengers cannot successfully complete the work of phagocytosis.

10. Briefly outline what leads to the result shown in the figure on the right. Give the name and actions of this structure.

Concept Mapping

On Connect you can find an Introduction to Concept Mapping that provides guidance for working with concept maps and a sample of a concept map for this chapter.

Level II. Application, Analysis, Evaluation, and Synthesis

These problems go beyond just restating facts and require higher levels of understanding and an ability to interpret, problem solve, transfer knowledge to new situations, create models, and predict outcomes.

Critical Thinking

Critical thinking is the ability to reason and solve problems using facts and concepts. These questions can be approached from a number of angles, and in most cases, they do not have a single correct answer.

1. Suggest some reasons that there is so much redundancy of action and there are so many interacting aspects of immune responses.

2. a. What are some possible elements missing in children born without a functioning lymphocyte system?
 b. What is the most important component extracted in bone marrow transplants?

3. A patient's chart shows an increase in eosinophil levels.
 a. What does this cause you to suspect?
 b. What does it mean if the basophil levels are very high?
 c. What if the neutrophil levels are very high?

4. How can adults continue to function relatively normally after surgery to remove the thymus, tonsils, spleen, or lymph nodes?

5. An obsolete treatment for syphilis involved inducing fever by deliberately infecting patients with the agent of relapsing fever. An experimental AIDS treatment involved infecting patients with malaria to induce high fevers. Can you provide some possible explanations behind these peculiar forms of treatment?

6. Patients with a history of tuberculosis often show scars and other lesions in the lungs and experience recurrent infection. Account for these effects on the basis of the inflammatory response.

7. *Shigella, Mycobacterium,* and numerous other pathogens have developed mechanisms that prevent them from being killed by phagocytes.
 a. Suggest two or three factors that help them avoid destruction by the powerful antiseptics in macrophages.
 b. Suggest the potential implications that these infected macrophages can have on the development of disease.

8. Account for the several inflammatory symptoms that occur in the injection site when one has been vaccinated against influenza and tetanus.

9. What is the specific target of fever-suppressing drugs? Knowing that fever is potentially both harmful and beneficial, what are some possible guidelines for deciding whether to suppress it or not?

10. Discuss the multiple, overlapping functions of the immune system. Which tissues and organs are involved in the immune response, and how does each function?

Visual Challenge

1. Each numbered figure 1–4 represents a cell that provides an immune defense. Each lettered figure A–D shows a pathogenic microbe. Name the type of defense and match it with its primary target microbe and effects.

(2) © Harold J. Benson **(3)** © Harold J. Benson **(4)** © Harold J. Benson

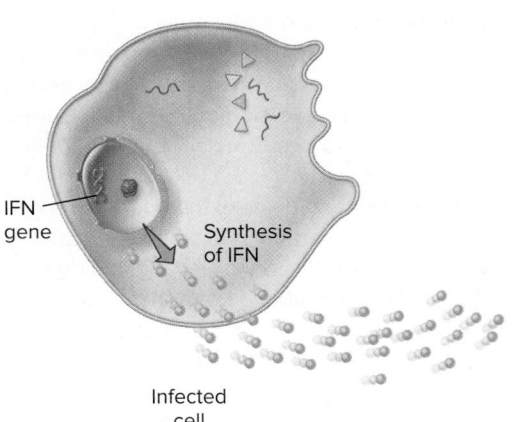

IFN gene — Synthesis of IFN

Infected cell

(1) Figure 14.19

A Figure 18.8
© Kathy Park Talaro

B Figure 23.21
Source: Oregon State Public Health Laboratory/ CDC-DPDx

C Table 6.2

D Case Study, Chapter 14
Source: CDC

15 Adaptive, Specific Immunity and Immunization

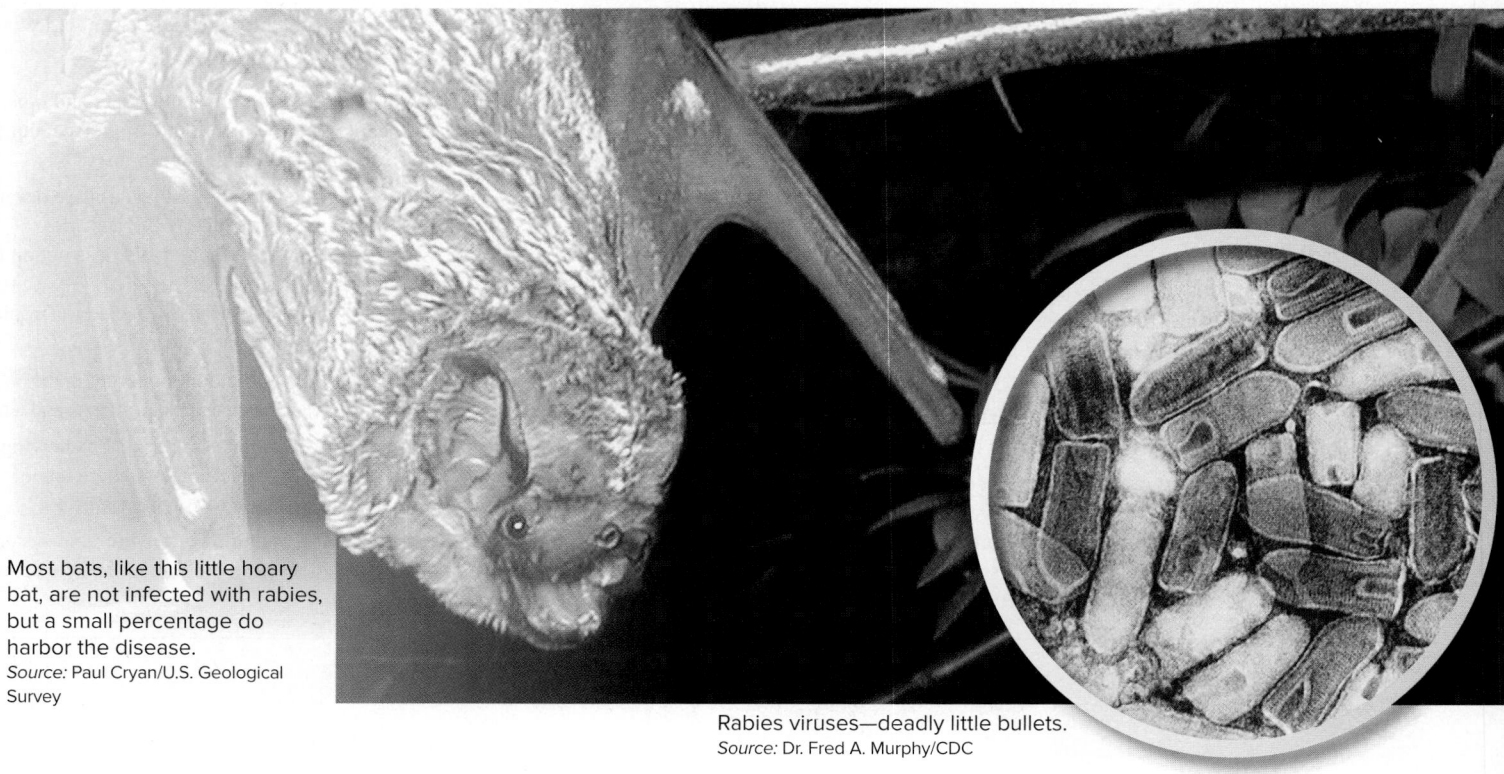

Most bats, like this little hoary bat, are not infected with rabies, but a small percentage do harbor the disease.
Source: Paul Cryan/U.S. Geological Survey

Rabies viruses—deadly little bullets.
Source: Dr. Fred A. Murphy/CDC

CASE STUDY Part 1 A Tale of Two Bat Bites

Case #1. In the fall of 2000, a 9-year-old Canadian boy awoke with a **fever and shooting pain** affecting his left arm. Over the next several days, he became **restless,** had difficulty sleeping, and developed tremors in his arm and hand. While being hospitalized for observation, he developed **increased salivation, hydrophobia,** uncontrollable lower body **contractions,** and visual **hallucinations.** His respirations became so labored that he required a breathing tube. Medical personnel were forced to consider the dreadful diagnosis of **rabies.** The only definitive diagnostic result showed **antibodies** against the rabies virus in his blood. A week after he first developed symptoms, his medical team decided to intervene with **postexposure prophylaxis** (PEP) therapy. This involved administering both **human rabies immune globulin** (HRIG) and five spaced-out doses of **rabies vaccine** in an effort to stop the infection from spreading.

Digging deeper to discover how the child may have become infected, the physicians learned that about a month before he showed symptoms, the father had captured and released a live bat trapped in the boy's bedroom. A day or so later, the child complained to his mother about a small **laceration** on his upper arm. It was cleaned and bandaged, but no further action was taken at the time.

Case #2. In October 2008, a 15-year-old Brazilian youth was **attacked by a bat,** which bit him in several places. The bat escaped

> "He developed increased salivation, hydrophobia, uncontrollable lower body contractions, and visual hallucinations."

and was not available for testing. Clinical workers washed his wounds and applied peroxide and iodine **antiseptics** to them. Even though he seemed fine, he was taken to a hospital for further evaluation. Four days after the injuries, concerned physicians began treating him with a version of PEP that included five doses of rabies vaccine but not rabies immune globulin.

Over the next 2 weeks the young man displayed early symptoms of rabies, including weakness, headache, fever, and a **prickling sensation around the bites.** Tests done on his blood and spinal fluid about this same time indicated antibodies specific to the rabies virus but no **virus antigen.** The medical staff had heard about a case of a young woman in Wisconsin who had survived rabies by being placed in a **medically induced coma** and given **antiviral drugs** for several weeks, and they decided to try this therapy on him. About 3 weeks from the time of the bite, it appeared that his condition had stabilized, and they removed him from sedation.

- *What are the basic immunological principles behind PEP with rabies immune globulin and rabies vaccine?*

To continue the Case Study, go to Case Study Part 2 at the end of the chapter.

SCOPING OUT THE CHAPTER

We recognize the first symptoms of an infection coming on—the scratchy throat, swollen lymph nodes, runny nose, earache. But what is not so evident is the entire army of immune cells working behind the scenes to contain it before it gets too far along. Among the most important soldiers in fighting infections are lymphocytes in partnership with macrophages. Like a special strike force, they can track down infectious agents, dismantle them, and call into service a set of weapons for immobilizing and killing them. And after that, they leave a long term back up force of trained cells ready to attack at the first sign of another infection. This is the most critical of all of our defenses, and counts as the major contributor to our immunities, whether to cold viruses or strep bacteria or the tetanus vaccine.

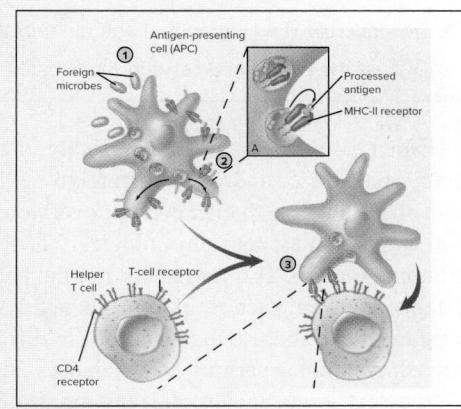

Antigens: Red Flags of Foreign Invaders
Infectious agents carry around molecules that are unique to them and can provoke the immune cells to go into attack mode. These antigens can be complex whole microbes or parts of microbes such as toxins.

Lymphocytes: Skilled Troops of Immunity
During development, millions of lymphocytes in the bone marrow each become programmed to go after a specific infectious agent and are primed for future battles with these microbes. Eventually they are deployed into lymphatic tissues throughout the body to be ready for attack.

Preparing for Battle Captured microbes are taken in by macrophages, where they are imprisoned and reduced to their most antigenic parts. The macrophages then display the antigens and make them available to lymphocytes. This brings about an activating process which prepares the T and B lymphocytes for their extended roles.

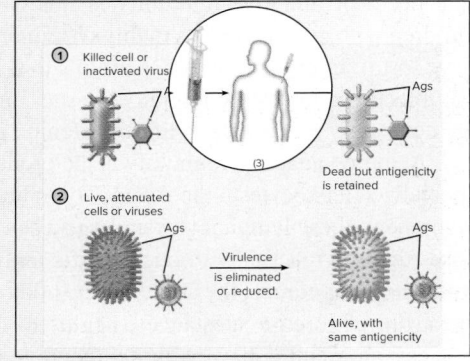

T cells: The Helpers and Killers Once they have been properly stimulated, T cells go down one of two different pathways, leading either to helpers that assist other cells get ready to fight infections or to killers that are involved in cell-to-cell combat. These cells bring smart bombs that can take out foreign cells with great precision.

B cells and Antibodies: A Protein Defense
Stimulated B cells go into active growth mode, making many clones of themselves that will produce their key defense molecules—antibodies. These weapons are released into circulation to surround and hold microbes hostage so that they can be dealt with and destroyed.

Vaccines and Other Therapies to Mimic Nature One significant advantage of having an organized and well-equipped strike team is that it can be trained for future conflicts. If an antigen is carefully selected and processed, it can be used to induce immunities before an infection has even occurred.

(bottom-left, inset): Source: Susan Arnold, Dr. Raowf Guirguis/National Cancer Institute

15.1 Specific Immunities: The Adaptive Line of Defense

 Expected Learning Outcomes

1. Summarize the general features of adaptive, acquired immunity.
2. Define *immunocompetence, antigens, specificity,* and *memory* as they relate to the immune system.
3. Outline the overall phases in a specific immune response.
4. Explain the functions and types of immune receptors, including where and how they develop.
5. Describe the major events in the origin of diversity of the immune system and features of the clonal selection theory.
6. Describe the development of antigen receptors on lymphocytes
7. Discuss the events in B-cell and T-cell maturation and how they differ.

In chapter 14 we described the capacity of the immune system to survey, recognize, and react to foreign cells and molecules. Our main focus in that chapter was to overview innate host defenses such as anatomical and physiological barriers, phagocytosis, and inflammation. We also provided essential background for the main systems involved, such as blood cells and lymphatic tissues. Refer to figure 14.21 and use the flowchart to refresh yourself on the relationships among various host defenses. An essential system of defense that is not innate but must be acquired will be the main subject of this chapter.

Normal humans have an extremely specific and powerful system for resisting infectious agents. This system is called **adaptive** or **acquired immunity,** sometimes known as the third line of defense. It is responsible for the long-term protection we develop through infections (measles or mumps) or vaccinations. The absolute need for adaptive immunity is impressively documented in children with genetic defects in this system or in AIDS patients who have lost it. Even with heroic measures to isolate the patient, combat infections, or restore lymphoid tissue, immunodeficient people are constantly vulnerable to life-threatening infections.

Acquired adaptive immunity is the product of a dual system of specialized leukocytes—the B and T lymphocytes. During fetal development, these lymphocytes undergo a selective process that prepares them to react only to one specific antigen. During this time, **immunocompetence,** the ability of the body to interact with a wide spectrum of foreign substances, begins to develop. Through this process, an infant is born with the theoretical *potential* to produce an immune response to hundreds of millions of different foreign molecules or antigens. But the completion of this immunocompetence takes many years, extending into late puberty.

Antigens figure very prominently in specific immunity. They are defined as any molecules that can stimulate a response by T and B cells. They consist of protein, polysaccharide, and other compounds from cells and viruses. Environmental chemicals can also be antigens, as we shall see in section 16.2 dealing with allergy. In fact, any exposed or released substance is potentially an antigen, even those from our own cells. For reasons we discuss later, our own antigens do not usually evoke a response from our own immune systems, but they may do so in other persons.

In chapter 14 we discussed pathogen-associated molecular patterns (PAMPs) that stimulate responses by phagocytic cells during an innate defense response. PAMPs are molecules shared by many types of microbes that stimulate a nonspecific response. In contrast, antigens are molecules unique to a microbe that stimulate specific immune responses. The two kinds of molecules do share two characteristics: (1) they are "parts" of foreign cells (microbes), and (2) they provoke a reaction by the white blood cells of the host.

Two features that make adaptive immunity very different from innate immunity are **specificity** and **memory.** Unlike mechanisms such as anatomical barriers or phagocytosis, acquired immunity is selective. For example, the antibodies produced against the chickenpox virus during an infection will protect against that virus but not against the measles virus. Demonstrating memory means that lymphocytes have been programmed to "recall" their first engagement with an invader and respond rapidly to that same invader during subsequent exposures, which is a critical feature of immunity. These topics will be covered in greater depth in several later sections.

An Overview of Specific Immune Responses

Immune responses are highly complex and regulated, and they represent one of the most elegant and coordinated networks of cells and chemicals in the body. To present this system in an organized and clear manner, we have found it helpful to divide it into separate but related sections, each detailing some event in development of the immune response. The sections of coverage are as follows:

I. Development and differentiation of the immune system (section 15.1)
II. Lymphocyte maturation, and the nature of antigens (section 15.2)
III. Immune reactions to antigens and the activities of T cells (section 15.3)
IV. Immune activities of B Cells and the production and actions of antibodies (section 15.3)

As we explore the information contained in these sections, we will be following a "map" of sorts. **Figure 15.1** lays out the flow of the main stages and will serve to coordinate the presentation. It also includes figure numbers that enlarge on the topics in that section.

Development of the Immune Response System

Before we examine lymphocyte development and function in greater detail, we must initially review concepts such as the unique structure of molecules (especially proteins), the characteristics of cell surfaces (membranes and envelopes), the ways that genes are expressed, immune recognition, and identification of self and nonself. Ultimately the shape and function of protein receptors and markers protruding from the surfaces of certain white blood cells are the result of genetic expression, and these molecules are responsible for specific immune recognition and, thus, immune reactions.

Markers on Cell Surfaces Involved in Recognition of Self and Nonself

Chapter 14 touched on the fundamental idea that cell surface receptors confer specificity and identity. A given cell can express several

I. Development of the Lymphocyte System. Lymphocytes arise from the same stem cells but differentiate into two distinct cell types early on. T cells mature in the thymus gland and B cells mature in specialized bone marrow sites. Once released, matured cells settle in lymphoid organs and serve as a constant attack force for infectious agents (see figures 15.3 and 15.4a).

II. Contact with Antigens and III. Presentation by Antigen-presenting Cells (APCs). Foreign cells bear molecules (antigens) that are recognized and engulfed by APCs such as dendritic cells. For most responses, T helper cells first receive the processed antigen from the APC and go on to activate B and other T cells (see figures 15.4b and 15.9).

IVA. Activation of T Cells and IVB. T-cell Responses. An activated T cell forms memory cells and differentiates into helper cells or cytotoxic cells. T-cell immunity is termed cell mediated because the whole T cell acts directly to destroy microbes, rather than by secreting molecules into the body fluids. (see figures 15.11, 15.12, and table 15.2).

VA. B-Cell Responses. A B cell activated by T helper cells undergoes a spurt of cell division, producing memory cells that provide a rapid recall of the antigen, and plasma cells that secrete proteins called antibodies. (see figure 15.13).

VB. Humoral Immunity*. Antibodies circulate in fluids (blood, ECF, and lymph) providing humoral immunity. The antibodies react specifically with the antigen and mark it for an enhanced response. (see figures 15.14 and 15.15, table 15.3).

* A traditional term for such body fluids is the "humors".

Vertical labels (left margin): Antigen-Independent Development | Antigen-Dependent Responses

Figure labels: **T-Cell Line** | **B-Cell Line** | Thymus | Special bone marrow sites | Lymphocyte stem cell maturation | Migration to and establishment of T and B cells in lymphoid organs such as lymph nodes, spleen, and MALT | Lymph node | Lymph node | **II Antigen contact** | **III Antigen presentation** | Processed antigen | Antigen | Dendritic cell displays antigen and presents it to T helper cell | Dendritic cell | B-cell receptor | T helper cell | Most B cells require activation by T helper cells | B cell | **IVA Antigen presentation to naive T cell** | **VA Activated B cell** | Activated T cell | Memory B cells | Cytokines | Memory T cells | Plasma cells secrete antibodies | **IVB** Develop receptors that differentiate them into → Helper T cells or Cytotoxic T cells | **VB** Antibodies | Blood vessel | **Cell-Mediated Immunity** | **Humoral Immunity**

Process Figure 15.1 Overview of the origins and events of adaptive immune responses. This summary lays out the plan of sections 15.2, 15.3, and 15.4.

different receptors, each type playing a distinct and significant role in detection, recognition, and cell communication. Major functions of immune receptors are as follows:

- to perceive and attach to nonself or foreign antigens,
- to promote the recognition of self antigens,
- to receive and transmit chemical messages among other cells of the system, and
- to aid in cellular development.

Because of their importance in the immune response, we concentrate here on the major receptors of lymphocytes and macrophages.

Major Histocompatibility Complex One set of genes that codes for human cell receptors is the **major histocompatibility complex (MHC).** This gene complex gives rise to a series of glycoproteins (called MHC molecules) found on all cells except red blood cells. The MHC is also known as the **human leukocyte antigen (HLA)** system. This receptor complex plays a vital role in recognition of self by the immune system and in rejection of transplanted tissues.

The functions of the MHC groups have been identified. **Class I MHC** genes code for markers that display unique characteristics of self and allow for the recognition of self molecules and the regulation of immune reactions. Coverage in section 15.4 will outline how some T cells require interaction with class I MHC before they can react with foreign cells.

The system is rather complicated in its details, but in general, each person inherits a particular combination of class I MHC (HLA) genes in a relatively predictable fashion. Although millions of different combinations and variations of these genes are possible among humans, the closer two people are related, the greater the probability their MHC profiles will be similar. Individual differences in the exact inheritance of MHC genes, however, make it highly unlikely that even closely related persons will express an identical MHC profile. This fact introduces an important recurring theme: Although humans are genetically the same species, the cells of each individual express molecules that are foreign (antigenic) to other humans. This is where the term *histo-* (tissue) *compatibility* (acceptance) originated. This fact necessitates testing for MHC and other antigens when blood is transfused and organs are transplanted (see blood types and graft rejection, chapter 16).

Class II MHC genes code for immune regulatory receptors. The cells that express class MHC II are grouped together by the term **antigen-presenting cells (APCs).** This includes macrophages, dendritic cells, and B cells. These are the only "professional" antigen presenters, in that they possess both class I and II receptors necessary for interactions with T cells. See **figure 15.2** for depictions of the two MHC classes.

Lymphocyte Receptors and Specificity to Antigen The part lymphocytes play in immune surveillance and recognition is very much a function of their receptors. B-cell receptors bind free antigens; T-cell receptors bind processed antigens together with the MHC molecules on the cells that present antigens to them. Because antigens are molecules, their chemical structures can vary over a very wide range, potentially exhibiting billions of uniquely different structures and shapes. The many sources of antigens include microorganisms as well as an awesome array of chemical

Class I MHC molecule found on all nucleated human cells

Class II MHC found on some types of white blood cells

Figure 15.2 Molecules of the human major histocompatibility complex.

compounds in the environment. One of the most fascinating questions in immunology is: How can the lymphocyte receptors be varied to react with such a large number of different antigens? After all, it is generally accepted that there will have to be a different lymphocyte receptor for each unique antigen. Some additional questions naturally follow: How can a cell accommodate enough genetic information to respond to possibly billions of different antigens? When, where, and how does the capacity to distinguish self from foreign tissue arise? To answer these questions, we must first introduce a central theory of immunity.

The Origin of Diversity and Specificity in the Immune Response

The Clonal Selection Theory and Lymphocyte Development
Research findings have shown that lymphocytes use slightly more than 500 genes to produce the tremendous repertoire of specific receptors they must display for antigens. The most widely accepted explanation for how this diversity is generated is called the **clonal selection theory.** According to this theory, early undifferentiated lymphocytes in the embryo and fetus undergo a continuous series of divisions and genetic changes that generate hundreds of millions of different cell types, each carrying a particular receptor specificity.

The mechanism, generally true for both B and T cells, can be summarized as follows: Certain stem cell lines in the bone marrow will develop into specialized white blood cells such as granulocytes, monocytes, or lymphocytes. The lymphocytic line of stem cells differentiates into either T cells or B cells. Cells destined to become B cells stay in the bone marrow; T cells "home" to the thymus. Mature B and T cells then migrate to secondary lymphoid tissues **(figure 15.3).** These secondary lymphoid tissues will constantly be resupplied with B and T cells through this series of activities in the primary lymphoid tissues.

By the time T and B cells reach the lymphoid tissues, each one is already equipped to respond to a single unique antigen. This

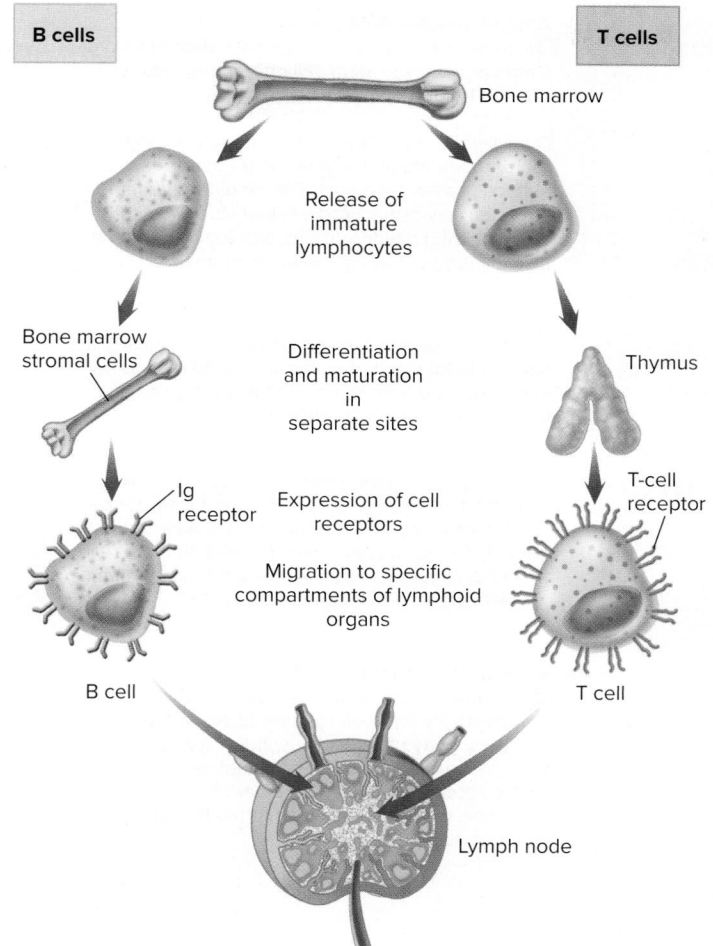

B cells

T cells

Bone marrow

Release of immature lymphocytes

Bone marrow stromal cells

Differentiation and maturation in separate sites

Thymus

Ig receptor

Expression of cell receptors

T-cell receptor

Migration to specific compartments of lymphoid organs

B cell

T cell

Lymph node

Figure 15.3 Major stages in the development of B and T cells.

amazing diversity is generated by extensive rearrangements of the gene segments that code for the antigen receptors on the T and B cells **(figure 15.4a).** These extensive genetic recombinations give rise to a huge assortment of lymphocytes.[1] Each genetically unique line of lymphocytes arising from this process is termed a **clone.** Keep in mind that the genetic code is expressed as a protein receptor of unique configuration on the surface of the lymphocyte, something like a "signpost" announcing its specificity and reactivity for an antigen.

This *proliferative* stage of lymphocyte development does not require the actual presence of foreign antigens. But it does require another important action—namely, the removal or "silencing" of clones of lymphocytes that can react against *self* MHC antigens. The presence of such "forbidden clones" could cause severe damage if the immune system mistakenly identifies self molecules as foreign and mounts a response against the host's own tissues. Thus, part of the clonal selection theory says that some clones are eliminated during development through clonal *deletion.* The removal of such potentially harmful clones is one basis of **immune tolerance**

or tolerance to self. Some diseases (such as autoimmunity) are thought to be caused by a loss of immune tolerance that can lead to immune reactions directed at self (covered in chapter 16).

The second stage of development—**clonal selection and expansion**—does require stimulation by antigens such as those that come from microbes. When an antigen enters the immune surveillance system, it encounters specific lymphocytes ready to recognize it. Such contact stimulates that clone to undergo mitotic divisions and expands it into a larger population of lymphocytes, all bearing the same specificity. This increases the capacity of the immune response for that antigen.

Two important generalities one can derive from the clonal selection theory are (1) that lymphocyte specificity is preprogrammed, existing in the genetic makeup before an antigen has ever entered the tissues; and (2) that each genetically distinct lymphocyte expresses only a single specificity and can react to only one type of antigen. Other important features of the lymphocyte response system are detailed in later sections.

The Specific B-Cell Receptor: An Immunoglobulin Molecule

In the case of B lymphocytes, the receptor genes that undergo the recombination process are those governing **immunoglobulin*** **(Ig)** synthesis. Immunoglobulins are large glycoprotein molecules that serve as the specific receptors of B cells and as antibodies. The basic immunoglobulin molecule is a composite of four polypeptide chains: a pair of identical heavy (H) chains and a pair of identical light (L) chains **(figure 15.5a).** Each light chain binds to a heavy chain, and the two heavy chains bind to each other with disulfide bonds, creating a symmetrical, Y-shaped arrangement.

The ends of the forks formed by the light and heavy chains contain pockets called the **antigen binding sites.** These sites can be highly variable in shape to fit a wide range of antigens. This extreme versatility is due to **variable (V) regions,** where amino acid composition is highly varied from one clone of B lymphocytes to another **(figure 15.5b).** The remainder of the light chains and heavy chains consist of **constant (C) regions** whose amino acid content does not vary greatly from one antibody to another. Figure 15.14 provides further details of antibody structure.

Development of B-Lymphocyte Receptors During Maturation

The genes that code for immunoglobulins lie on three different chromosomes. An undifferentiated lymphocyte has about 150 different genes that code for the variable region of light chains and about 250 genes for the **variable(V)** and **diversity (D) regions** of the heavy chains. The **constant (C)** regions and the **joining (J) regions** that link segments of the final molecule are represented by only a small number of genes. A result of the extensive genetic recombination that occurs during development is that only the selected V and D gene segments are active in the mature cell, and all the other V and D genes have been deleted (figure 15.5b). This is how the specificity of lymphocytes is maintained.

One can envision this process by comparing it to a molecular "cut and paste." The gene segments lie in an established sequence along a chromosome. A complex enzyme system randomly selects

1. Estimates of the theoretical number of possible variations that may be created vary from 10^{14} to 10^{18} different specificities—clearly enough to react with all potential antigens that might be encountered.

* *immunoglobulin* (im″-yoo-noh-glahb′-yoo-lin)

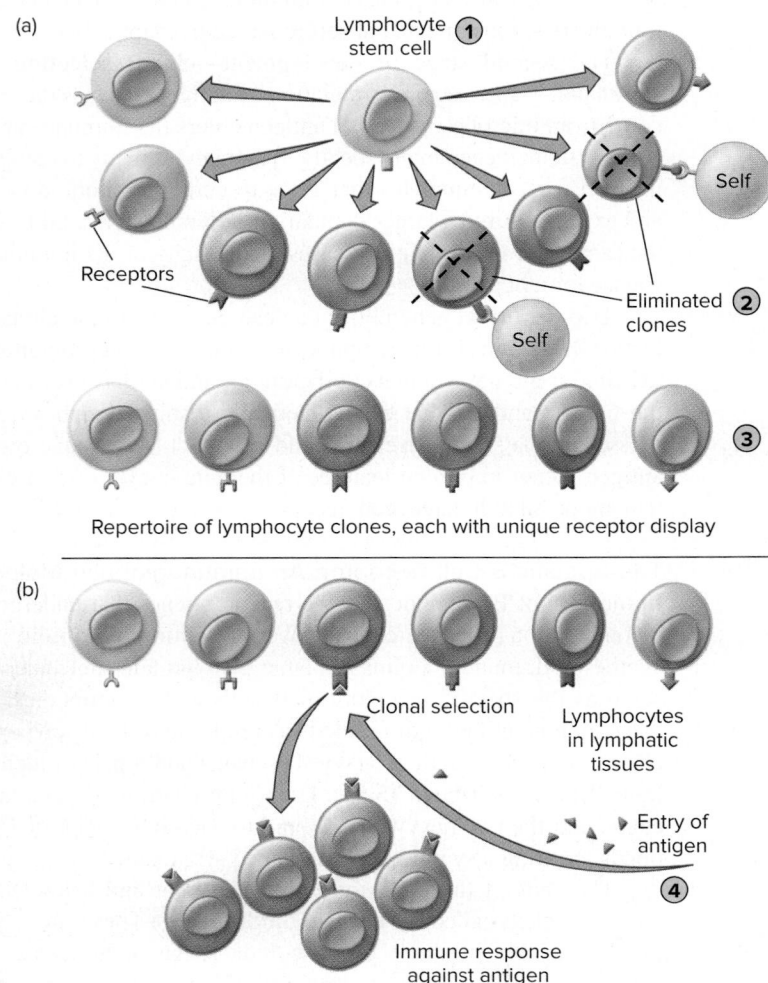

(a)

Lymphocyte ① stem cell

Receptors

Self

Eliminated ② clones

Self

③

Repertoire of lymphocyte clones, each with unique receptor display

(b)

Clonal selection

Lymphocytes in lymphatic tissues

Entry of antigen

④

Immune response against antigen

Figure 15.4 Overview of the clonal selection theory of lymphocyte development and diversity. Receptors are stylized to show diversity in shape that clonal selection generates. See figure 15.5 for the details of receptor development in B cells.

(a) **Antigen-Independent Period**
① During development of early lymphocytes from stem cells, a given stem cell undergoes rapid cell division to form numerous progeny.

During early cell differentiation, there are random rearrangements of the genes that code for cell surface protein receptors. The result is a large array of genetically distinct cells, called clones, each clone bearing a different receptor to react with only a single type of foreign molecule or antigen.

② At this time, any lymphocyte clones that carry a specificity for self molecules and could be harmful are eliminated from the pool of diversity. This is a way to achieve immune tolerance.

③ The specificity for a single antigen molecule is programmed into the lymphocyte and is set for the life of a given clone. The end result is an enormous pool of mature but naive lymphocytes that are ready to further differentiate under the influence of their "home" organs and immune stimuli.

(b) **Antigen-Dependent Period**
④ Lymphocytes migrate home to the lymphatic organs where they are situated to encounter antigens. Entry of a specific antigen selects only the lymphocyte clone or clones that carry matching surface receptors. This will trigger an immune response, which varies according to the type of lymphocyte involved.

and cuts out particular blocks of DNA and splices them together. All remaining unused gene segments are permanently removed from the genome of this cell, leaving only the selected segments, which will code for a specific polypeptide receptor. A summary of the steps in the process is as follows:

- For a heavy chain, a variable region gene segment and diversity region gene segment are selected from among the hundreds available and spliced to one joining region gene and one constant region gene.
- For a light chain, one variable, one joining, and one constant gene segment are spliced together.
- After transcription and translation of each gene complex into a polypeptide, a heavy chain combines with a light chain to form half an immunoglobulin; two of these combine to form a completed protein (figure 15.5b).

Once synthesized, the immunoglobulin product is transported to the cell membrane and inserted there to act as a receptor that expresses the specificity of that cell and to react with an antigen, as

shown in figure 5.5. It is notable that for each lymphocyte, the genes that were selected for the variable region, and thus for its specificity, will be locked in for the rest of the life of that lymphocyte and its progeny. But the constant region may be altered to provide different functional properties (see table 15.3). We discuss the different classes of Ig molecules further in section 15.4.

T-Cell Receptors for Antigen The T-cell receptor for antigen belongs to the same protein family as the B-cell receptor. It is similar to B cells in being formed by genetic modification, having variable and constant regions, being inserted into the membrane, and having an antigen binding site formed from two parallel polypeptide chains **(figure 15.6).** Unlike the immunoglobulins, the T-cell receptor is relatively small. It is equivalent to one fork of the B-cell receptor and is not secreted. Additional receptors of T cells are described in section 15.3.

With a working knowledge of some factors in the development of immune specificity, we can now continue our coverage of the immune response as outlined in figure 15.1.

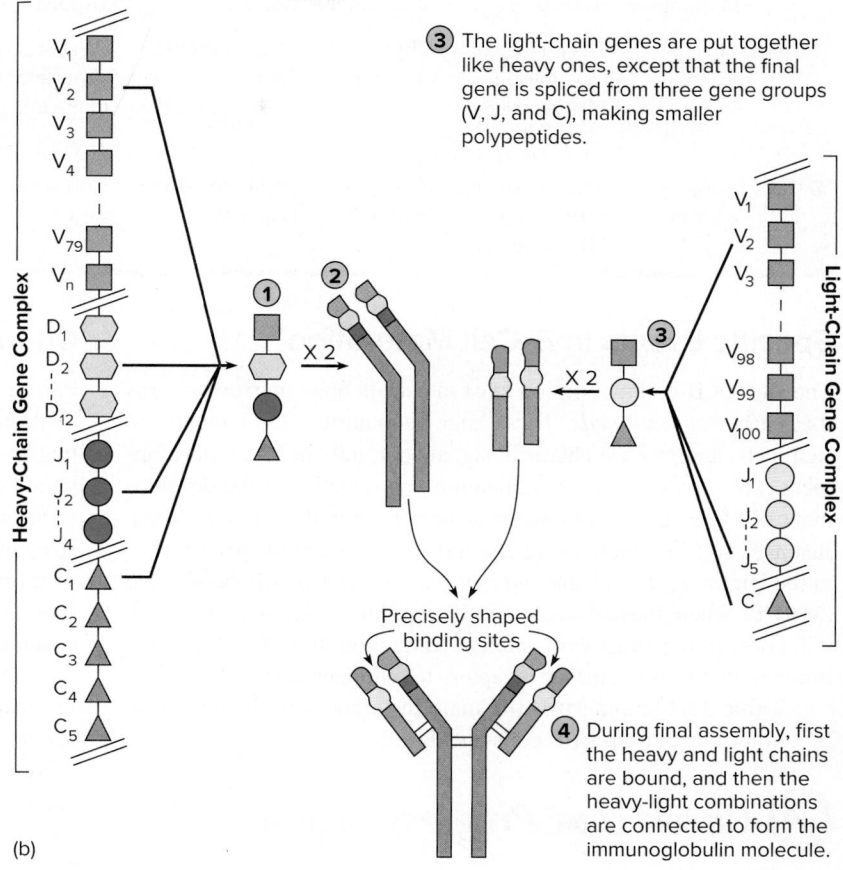

(1) The heavy-chain genes are composed of four separate segments (V, D, J, and C) that are (2) transcribed and translated to form the heavy polypeptide chains.

(3) The light-chain genes are put together like heavy ones, except that the final gene is spliced from three gene groups (V, J, and C), making smaller polypeptides.

Precisely shaped binding sites

(4) During final assembly, first the heavy and light chains are bound, and then the heavy-light combinations are connected to form the immunoglobulin molecule.

Figure 15.5 The basic structure and genetics of immunoglobulins. (a) Simple model of an immunoglobulin molecule. The main components are four polypeptide chains—two identical light chains and two identical heavy chains bound by disulfide bonds as shown. Each chain consists of a variable region (V) and a constant region (C). The variable regions of light and heavy chains form a binding site for antigen. **(b)** The final gene that codes for a heavy or light chain is assembled by splicing blocks of genetic material from several regions (1, 2, 3). These genes are transcribed and translated into the polypeptides that join to form the final molecule (4).

(b)

Figure 15.6 Structure of the T-cell receptor (TCR) for antigen and CD receptors. The structure of the TCR is similar to that of an immunoglobulin. It consists of two polypeptides that mimic the structure of one "arm" of an immunoglobulin. The TCR has variable (V) regions that can show high levels of diversity for antigens, and constant (C) regions that do not greatly vary. Another class of T-cell receptors called CD receptors, function in cell signaling. The CD4 and CD8 receptors will be discussed in section 15.3.

Specific Events in T-Cell Maturation

The further development and maturation of T cells are directed by the thymus gland and its hormones. The complexity of T-cell function, discussed in section 15.3, is partly due to different classes of T-cell surface molecules termed **clusters of differentiation** or **CD receptors** added during maturation. CD molecules serve multiple roles as cell receptors and may be involved in cell adhesion and communication in a variety of cells. They are denoted with a

number—CD1, CD2, and so on. Our focus will be primarily on two CD groups found on T cells: CD4 receptors on T helper cells and CD8 receptors on T cytotoxic cells. These CD receptors interact with a particular MHC marker on other white blood cells during immune reactions. As was shown in figure 15.3, mature T cells migrate to specific sites in lymphoid organs and constantly circulate from there. It is estimated that 25×10^9 T cells pass between the lymphatic and general circulation per day.

TABLE 15.1	Contrasting Properties of T Cells and B Cells							
	Site of Maturation	Immune Surface Markers	Circulation in Blood	Receptors for Antigen	Distribution in Lymphatic Organs	Require Antigen Presented with MHC	Product of Antigenic Stimulation	General Functions
T Cells	Thymus gland	T-cell receptor, CD molecules, MHC I receptors	High numbers	T-cell receptor (TCR)	Paracortical sites (interior to the follicles)	Yes	Helper and cytotoxic T cells and memory cells	Regulate immune functions, kill foreign and infected cells, synthesize cytokines
B Cells	Bone marrow	Immunoglobulin MHC I and MHC II receptors	Low numbers	Immunoglobulins D and M	Cortex (in follicles)	No	Plasma cells and memory cells	Produce antibodies to target, inactivate, and neutralize, antigens

Specific Events in B-Cell Maturation

The sites of B-cell maturation exist in certain bone marrow regions that harbor *stromal cells.* These huge cells nurture the lymphocyte stem cells and provide chemical signals that initiate B-cell development. The result of gene modification and selection is the development of hundreds of millions of distinct B cells (see figure 15.4a). Just as T cells do, these naive B lymphocytes travel to specific sites in the lymph nodes, spleen, and mucosal-associated lymphoid tissue (MALT), where they adhere to specific binding molecules. Here they will come into contact with antigens throughout life. B cells display immunoglobulins as surface receptors for antigens (see table 15.3).

Table 15.1 summarizes the main differences in the structure, functions, and actions of B cells and T cells.

 Check Your Progress SECTION 15.1

1. Discuss what is meant by immunocompetence, immune specificity and memory, and adaptive immunity.
2. What function do receptors play in specific immune responses and how can they be made to vary so widely?
3. Describe the major histocompatibility complex, and explain how it participates in immune reactions.
4. Explain the clonal selection theory of receptor specificity and diversity in lymphocytes.
5. Why must the body develop tolerance to self?
6. Trace the origin and development of T lymphocytes and of B lymphocytes.
7. What is happening during lymphocyte maturation?
8. Describe three ways in which B cells and T cells are similar and at least five major ways in which they are different.

15.2 The Nature of Antigens and Antigenicity

 Expected Learning Outcomes

8. Explain the characteristics of antigens, the property of antigenicity, and epitopes.
9. Discuss the main categories of antigens, based on function.

Characteristics of Antigens and Immunogens

Earlier we encountered the concept of an **antigen,**[2] which can now be more fully defined as any molecule or fragment of a molecule that has the *potential* to trigger a specific immune response by lymphocytes. This property is known as **antigenicity.** For a substance to be antigenic, it must meet several criteria related to foreignness, size, shape, and accessibility that are discussed next. A related set of more specific terms are **immunogen** and *immunogenicity.* An immunogen is a type of antigen that actually *does* induce a specific immune response when introduced into the body.

The difference between these terms is subtle and addresses the fact that some antigens may not fit all of the actual conditions to stimulate an immune reaction, whereas immunogens do. One example involves the small **polysaccharides** from the capsule of the bacterium *Haemophilus influenzae,* one cause of meningitis. These molecules are foreign and antigenic, but they are too small to trigger a lymphocyte response: They are antigenic but not immunogenic. To produce a vaccine from this type of antigen, it must be altered so that the immune system can more readily recognize and react to it (see the discussion of haptens later in this section). In most of the succeeding discussions on immune responsiveness, we will use the more general term *antigen* because it encompasses all possibilities of eliciting immune reactions.

One important characteristic of an antigen is that the immune cells react to it as nonself or as foreign, meaning that it is not a "normal" constituent of the body. Whole microbes or their parts, cells, or substances that arise from other humans, animals, plants, and various molecules all possess this quality of foreignness and thus are potentially antigenic to the immune system of an individual (**figure 15.7**). Molecules of complex composition such as proteins and protein-containing compounds prove to be more immunogenic than repetitious polymers composed of a single type of unit. Most materials that serve as antigens fall into these chemical categories (with examples):

- proteins and large polypeptides (enzymes, cell surface receptors, hormones, exotoxins);
- molecules that contain two or more different macromolecules bound together. Examples of these include:
 - lipoproteins (combinations of lipids and proteins) and lipopolysaccharides (combinations of lipids and carbohydrates). These substances originate from the cell membranes or cell walls of microbes.

2. *Antigen* (an'-tih-jen) means, literally, an **anti**body **gen**erator.

Figure 15.7 **Characteristics of antigens. (a)** Whole cells of bacteria, fungi, and viruses make good immunogens. So do plant parts and animal cells. **(b)** Complex molecules with several epitopes make good immunogens. **(c)** Poor immunogens include small, simple molecules not attached to a carrier molecule (1, 2) and larger molecules with repetitive chains of monomers that have reduced complexity (3, 4).

- nucleoproteins (DNA complexed to proteins); unbound DNA is too regular and repetitive in structure to be antigenic;
- polysaccharides (certain bacterial capsules) and glycolipids (mycolic acid of *M. tuberculosis*).

Effects of Molecular Shape and Size on Antigenicity

To initiate an immune response, a substance must also be large enough to "catch the attention" of the surveillance cells. Molecules with a molecular weight (MW) of less than 1,000 are seldom complete antigens, and those between 1,000 MW and 10,000 MW are weakly so. Complex macromolecules of at least 100,000 MW are the most immunogenic, a category also dominated by large proteins. Note that large size alone is not sufficient for antigenicity. Glycogen, a polymer of glucose with a highly repetitious structure,

has an MW over 100,000 and is not normally antigenic, whereas insulin, a protein with an MW of 6,000, can be antigenic.

A lymphocyte's capacity to discriminate differences in molecular shape is so fine that it recognizes and responds to only a portion of the antigen molecule. This molecular fragment, called the **epitope** or antigenic determinant, provides the primary signal that the molecule is foreign (figure 15.7b). The shape of this determinant fits like a key to the receptor "lock" of the lymphocyte, which then recognizes and responds to it. Certain amino acids accessible at the surface of proteins or carbohydrate side chains protruding from a protein receptor are

typical examples. Many foreign cells and molecules are very complex antigenically, with numerous determinants, each of which will elicit a separate and different lymphocyte response. Examples of these multiple, or *mosaic,* antigens include bacterial cell wall, membrane, flagellar, capsular, and toxin antigens; and viruses, which express various surface and core antigens.

Small foreign molecules that consist only of a determinant group and are too small by themselves to elicit an immune response are termed **haptens.** However, if such an incomplete antigen is linked to a larger carrier molecule, the combination develops immunogenicity (**figure 15.8**). The carrier group contributes to the size of the complex and enhances the orientation of the antigen, while the hapten serves as the epitope. Haptens include such molecules as vaccine antigens; drugs; metals; and household, industrial, and environmental chemicals. Many haptens develop immunogenicity in the body by combining with large carrier molecules such as serum proteins (see section 16.2).

Functional Categories of Antigens

Because each human being is genetically and biochemically unique (except for identical twins), the proteins and other molecules of one person can be antigenic to another. **Alloantigens** are cell surface markers and molecules that occur in some members of the same species but not in others. Alloantigens are the basis for an individual's blood group and major histocompatibility profile, and they are responsible for incompatibilities that can occur in blood transfusion or organ grafting.

Some bacterial proteins, called **superantigens,** are potent stimuli for T cells. Notable examples of these types of antigens are toxic shock toxin and enterotoxin of *Staphylococcus*. Reactions with superantigens may cause the massive release of cytokines, leading to cell death. The role of these compounds is covered in section 15.3.

Antigens that evoke allergic reactions, called **allergens,** will be characterized in detail in chapter 16.

On occasion, even a normal part of self can take on the character of an antigen. During lymphocyte differentiation, immune tolerance to self tissue occurs, but a few anatomical sites can contain sequestered (hidden) molecules that escape

Figure 15.8 The hapten-carrier phenomenon. (a) Haptens are antigens that are too small to be discovered by an animal's immune system and do not elicit a response. **(b)** When a hapten binds to a large molecule, the hapten serves as an epitope and stimulates a response and an antibody that is specific for it.

this assessment. Such molecules, called **autoantigens,** can occur in tissues (of the eye and thyroid gland, for example) that are walled off early in embryonic development before the surveillance system is in complete working order. Because tolerance to these substances has not yet been established, they can subsequently be mistaken as foreign; this mechanism appears to account for some types of autoimmune diseases such as rheumatoid arthritis (chapter 16).

With a working knowledge of some factors in the development of immune specificity, we can now continue our coverage of the immune response as outlined in figure 15.1.

 Check Your Progress SECTION 15.2

9. What are antigens, immunogens, and epitopes, and what role do they play in immune reactions?
10. How do foreignness, size, and complexity contribute to antigenicity?
11. Compare five unique types of antigens, and explain their functions or actions.

15.3 Immune Reactions to Antigens and the Activities of T Cells

 Expected Learning Outcomes

10. Describe the cooperative interactions between antigen-presenting cells, T cells, and B cells.
11. Discuss the actions of interleukins in the early reactions of recognition and activation.
12. Outline the main activities of cell-mediated immunities.
13. Analyze the relationships between the types of T cells and their receptors, and how they are activated.
14. Describe the primary functions of helper T cells and cytotoxic T cells.
15. Explain the concepts of natural killer cells and superantigens.

The basis for most immune responses is the encounter between antigens and white blood cells. Microbes and other foreign substances enter most often through the respiratory or gastrointestinal mucosa and less frequently through other mucous membranes or the skin. Antigens introduced intravenously become localized in

the liver, spleen, bone marrow, kidney, and lung. If introduced by some other route, antigens are carried in lymphatic fluid and concentrated by the lymph nodes. The lymph nodes and spleen are important in concentrating antigens in areas where they will come in contact with antigen-presenting cells (APCs) and lymphocytes. APCs and lymphocytes can subsequently circulate into fluid compartments to seek out the antigens for which they are specific.

The Role of Antigen Processing and Presentation

In most immune reactions, the antigen is in a "raw" state and must be further acted upon by **antigen-presenting cells (APCs)** before it is ready for contact with T cells. Three different cells can serve as APCs: **macrophages, dendritic* cells,** and B cells, although dendritic cells are the most common APC in the first contact with an antigen. Antigen-presenting cells modify the antigen so that it will be more immunogenic and recognizable. After processing is complete, the antigen is moved to the surface of the APC and bound to an MHC class II receptor to make it readily accessible to T cells during presentation **(figure 15.9).**

Quick Search
To learn more about this topic, go to "Our Immune System" on YouTube.

Before a T cell can respond to APC-bound antigens, certain conditions must be met. *T-cell-dependent antigens,* usually protein-based, require recognition steps between the APC, antigen, and lymphocytes. The first cells on the scene to assist in activating B cells and other T cells are a special class of **T helper cells (T_H).** The T-cell receptor (TCR) of this class of T cell will bind simultaneously with the class II MHC receptor on the APC and with the antigen (figure 15.9). A second interaction involves the binding of the T-cell CD4 receptor to the MHC of the APC. Once this identification step has occurred, a cytokine, **interleukin-1 (IL-1),** produced by the APC, activates this T helper cell. The T_H cell, in turn, produces a different cytokine, **interleukin-2 (IL-2),** which stimulates a general increase in activity of committed B and T cells. The manner in which B and T cells subsequently become activated by the APC–T helper cell complex and their individual responses to antigen are addressed below and in section 15.4.

* *dendritic* (den′-drih-tik) Gr. *dendron,* branching like a tree.

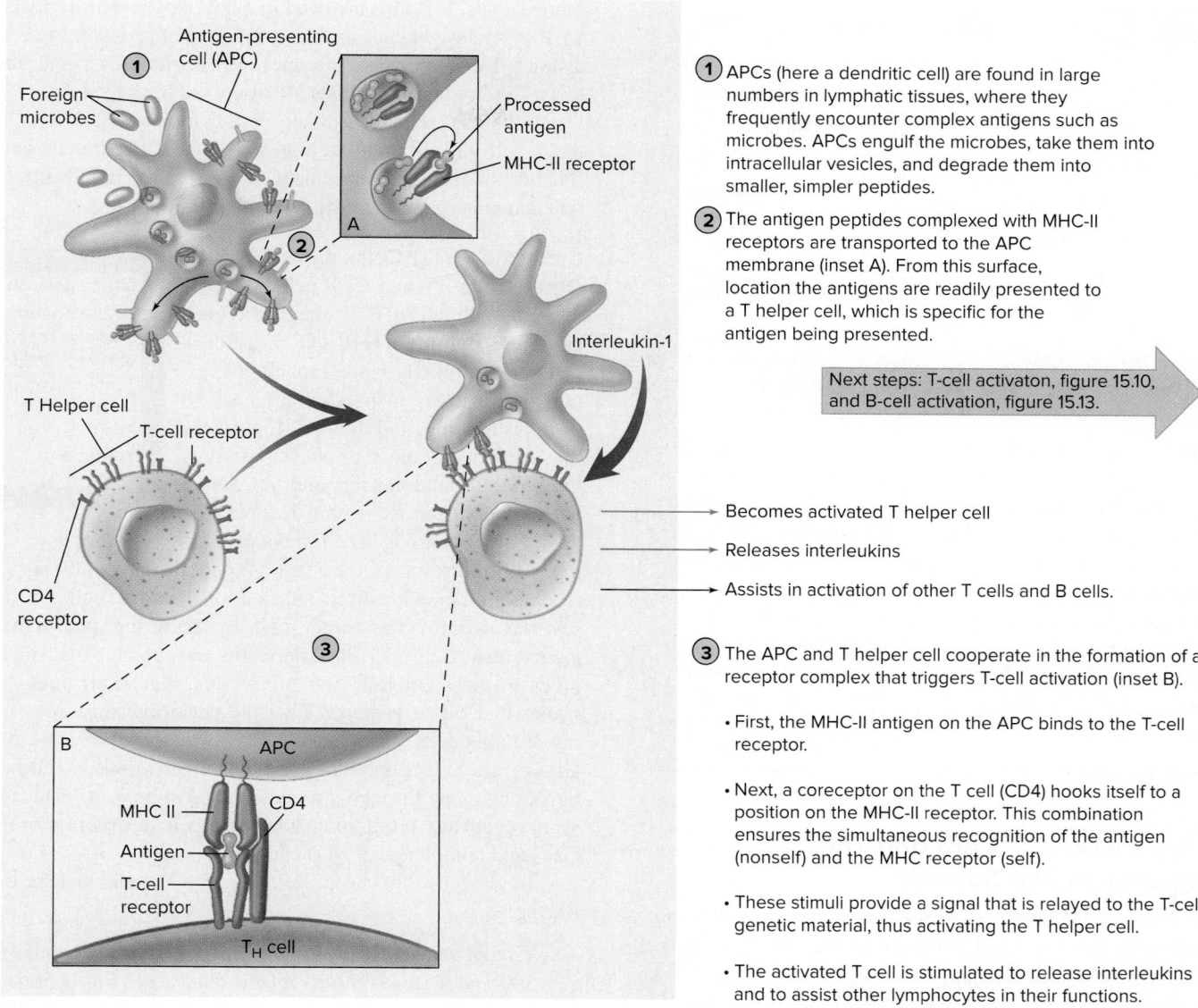

① APCs (here a dendritic cell) are found in large numbers in lymphatic tissues, where they frequently encounter complex antigens such as microbes. APCs engulf the microbes, take them into intracellular vesicles, and degrade them into smaller, simpler peptides.

② The antigen peptides complexed with MHC-II receptors are transported to the APC membrane (inset A). From this surface, location the antigens are readily presented to a T helper cell, which is specific for the antigen being presented.

Next steps: T-cell activaton, figure 15.10, and B-cell activation, figure 15.13.

→ Becomes activated T helper cell

→ Releases interleukins

→ Assists in activation of other T cells and B cells.

③ The APC and T helper cell cooperate in the formation of a receptor complex that triggers T-cell activation (inset B).

- First, the MHC-II antigen on the APC binds to the T-cell receptor.

- Next, a coreceptor on the T cell (CD4) hooks itself to a position on the MHC-II receptor. This combination ensures the simultaneous recognition of the antigen (nonself) and the MHC receptor (self).

- These stimuli provide a signal that is relayed to the T-cell genetic material, thus activating the T helper cell.

- The activated T cell is stimulated to release interleukins and to assist other lymphocytes in their functions.

Process Figure 15.9 Interactions between antigen-presenting cells (APCs) and T helper (CD4) cells required for T-cell activation. For T cells to recognize foreign antigens, the antigen must be processed and presented to them by a professional APC such as a dendritic cell. See figure 15.10 and figure 15.13 for next phases.

A few antigens can trigger a response from B lymphocytes without the cooperation of APCs or T helper cells. These *T-cell-independent antigens* are usually simple molecules such as carbohydrates with many repeating and invariable determinant groups. Examples include lipopolysaccharide from the cell wall of *Escherichia coli,* polysaccharide from the capsule of *Streptococcus pneumoniae,* and molecules from rabies and Epstein-Barr virus. Because so few antigens are of this type, most B-cell reactions require assistance from T helper cells.

T-Cell Responses and Cell-Mediated Immunity (CMI)

The responses of T cells, referred to as **cell-mediated immunities (CMIs)**, are among the most complex and diverse in the immune system. They involve several subsets of T cells that differ in their types of CD receptors and the precise ways they react against foreign antigens and cells (**table 15.2**). T cells are restricted, meaning that before they can be activated, they must have the antigen offered by an MHC complex on an APC (figure 15.9), to ensure recognition of self. All produce cytokines that, working together, carry out a spectrum of biological effects and immune functions. T cells differ notably from B cells in function. We will see that B cells combat foreign antigens by secreting molecules into the circulation, but in the case of T cells, the whole cell reacts directly in contact with target cells. T cells also stimulate other T cells, B cells, and phagocytes.

A T cell is initially sensitized by the binding of antigen/MHC to its receptors. This event activates the T cell, which transforms it in preparation for mitotic divisions, and targeted by interleukin signals. it divides to form one of the subsets of effector cells and memory cells that can interact with the antigen upon subsequent contact. Memory T cells are some of the longest-lived blood cells known (decades, rather than weeks or months for other lymphocytes).

TABLE 15.2	Characteristics of Subsets of T Cells	
Types	**Primary Receptors on T Cell**	**Functions/Important Features**
T helper cell 1 (T_H1)	CD4	Activates other CD4 and CD8 cells; secretes IL-2, tumor necrosis factor, and interferon gamma; responsible for delayed hypersensitivity; interacts with MHC-II receptors
T helper cell 2 (T_H2)	CD4	Drives B-cell proliferation; secretes IL-4, IL-5, IL-6, IL-10; can dampen T_H1 activity
T regulatory cell (T_{reg})	CD4, CD25	Involved in development of immune tolerance; suppression of pathological immune responses, inflammation, autoimmunity
T cytotoxic cell (T_C)	CD8	Destroys a target foreign cell by lysis; important in destruction of cancer cells, virus-infected cells; graft rejection; requires MHC I for function May have some regulatory functions.

The Activation of T Cells and Their Differentiation into Subsets

Mature T cells in lymphoid organs are primed to react with antigens that have been processed and presented to them by dendritic cells and macrophages. They recognize an antigen only when it is presented in association with a particular MHC carrier, but they differ in which carrier comes into play **(figure 15.10).** T cells with CD4 receptors recognize endocytosed peptides presented on MHC II, and T cells with CD8 receptors recognize peptides presented on MHC I.

T Helper (T_H) Cells: Activators of Specific Immune Responses
T helper (CD4) cells play a central role in regulating immune reactions to antigens, including those of B cells and other T cells. They are also involved in activating macrophages and increasing phagocytosis. They do this directly by receptor contact and indirectly by releasing cytokines such as interleukin-2, which stimulates the primary growth and activation of B and T cells, and interleukin-4, -5, and -6, which stimulate development of B cells. T helper cells are the most prevalent type of T cell in the blood and lymphoid organs, making up about 65% of this population. The severe depression of the CD4 class of T cells by HIV is a major factor in the immunopathology of AIDS.

When T helper cells are activated by antigen/MHC II, they differentiate into either **T helper 1** (T_H1) cells, or **T helper 2** (T_H2) cells, depending on what type of cytokines the antigen-presenting cells secrete. It is thought that if the dendritic cell (APC) secretes IL-1 or IL-12, the T cell will become a T_H1 cell that can activate

more T cells. It is also involved in delayed hypersensitivity (figure 15.10a). Delayed hypersensitivity is a type of response to allergens, distinct from immediate allergies such as hay fever and anaphylaxis. Both of these reactions are discussed in chapter 16.

If the APC secretes another set of cytokines (IL-4, IL-5, IL-6), the T cell will differentiate into a T_H2 cell. These cells have the function of secreting substances that influence B-cell differentiation and consequently affect the antibody response.

Cytotoxic T (T_C) Cells: Cells That Kill Other Cells

Quick Search
To see a short animation of this process, look for "Cell-Mediated Immune Response" on YouTube.

When CD8 cells are activated by antigen/MHC I, they differentiate into **T cytotoxic cells (T_C or killer T cells). Cytotoxicity** is the capacity of certain T cells to kill a specific target cell. It is a fascinating and powerful property that accounts for much of our immunity to foreign cells and cancer; and yet, under some circumstances, it can lead to disease. For a cytotoxic T cell to become activated, it must recognize a foreign peptide carried by an MHC-I receptor and mount a direct attack upon the target cell. The T_C cell severely injures the target cell by secreting **perforins**[3] and **granzymes** (figure 15.10b). Perforins are proteins that can punch holes in the membranes of target cells, and granzymes are enzymes that digest proteins. First the perforins cause ions to leak out of target cells and create a passageway for granzymes to enter. Granzymes induce the loss of selective permeability followed by target cell death through a process called *apoptosis.** The apoptosis is genetically programmed and results in destruction of the nucleus and complete cell lysis and death.

Target cells that can be destroyed by T_C cells include the following:

- Virally infected cells (figure 15.10b). Cytotoxic cells recognize and react against other cells that carry virus peptide MHC combinations expressed on their surface. Cytotoxic defenses are an essential protection against viral infections.
- Cancer cells. T cells constantly survey the tissues and immediately attack any abnormal cells they encounter **(figure 15.11).** The importance of this function is clearly demonstrated in the susceptibility of T-cell-deficient people to cancer (chapter 16).
- Cells from other animals and humans. Cytotoxic CMI is the most important factor in graft rejection. In this instance, the T_C cells attack the foreign tissues that have been implanted into a recipient's body.

Other Types of Killer Cells **Natural killer (NK)** cells are a type of lymphocyte related to T cells that are part of cytotoxic CMI. They circulate through the spleen, blood, and lungs, and are probably the first killer cells to attack cancer cells and virus-infected cells. They destroy these cells by similar mechanisms as cytotoxic T cells **(figure 15.12).** They are considered part of cell-mediated immunity because their activities are moderated by cytokines such as interleukin-12 and interferon.

3. *Perforin* is taken from the term perforate, to make holes in.
* *apoptosis* (ah-poh-toh′-sis) Gr., *apo*, away from, and *ptosis*, a falling or dropping.

MHC II CD4

Antigen TCR

APC

IL-2, IFN-γ

IL-4

(a) Reaction with CD4 cell

CD4 T cell

T_M Memory CD4 T cell

T_H2 Produces IL-4, IL-5, IL-6, and IL-10 (B-cell development and activation factors)

Activated B cell

Figure 15.9, part B

T_H1 Production of tumor necrosis factor and interferon gamma

Stimulate macrophages (also delayed hypersensitivity)

Ag enters with APC and activates CD4 or CD8 cell

T_H1

APC

MHC I

Antigen

CD8 receptor

T-cell receptor

(b) Reaction with CD8 cell

CD8 T cell

IL-2

T_M Memory CD8 T cell

Perforins Granzymes

TCR MHC I

Ag

CD8

T_C

Activated T_C cell recognizes infected self

Self cell infected with virus

Damaged and lysing cell

Process Figure 15.10 Overall scheme of T-cell activation and differentiation into different types of T cells. Antigen-presenting cells (APCs) present antigenic peptides to T cells bearing either CD4 (a) or CD8 (b) markers. **(a)** CD4 cells bind antigen/MHC-II complexes on APCs and, depending on the type of cytokine released by the APC, become either T_H1 or T_H2 cells. T_H1 cells synthesize IL-2 that activates CD8 cells and causes macrophages to destroy ingested microbes or to become more active. T_H2 cells secrete cytokines that enhance B-cell activation. **(b)** Upon binding antigen/MHC-I complex, CD8 cells become cytotoxic T (T_C) cells, which bind infected host cells displaying antigen/MHC-I complexes and release perforins and granzymes. This leads to the apoptosis (programmed cell death) of the infected cell.

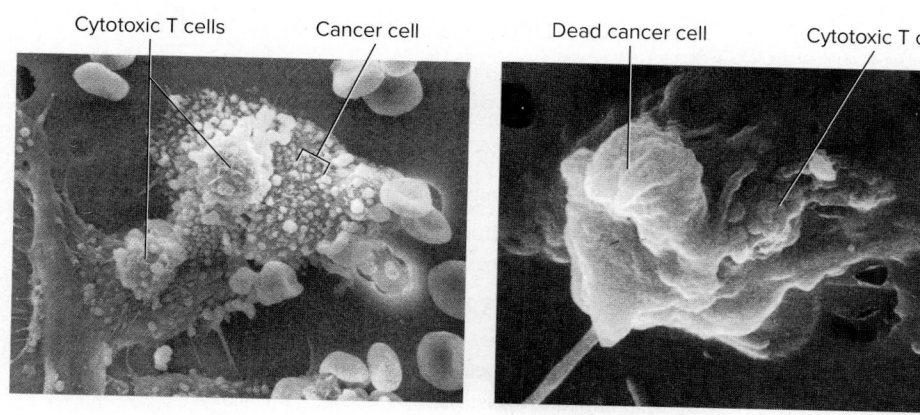

Cytotoxic T cells Cancer cell Dead cancer cell Cytotoxic T cell

Figure 15.11 Attack of the cytotoxic T cells. (Left) T cells recognize and bind to a cancer cell, which they perforate with many small holes. **(Right)** The cancer cell collapses and dies by apoptosis and the T cell remains alive and active. (left, right): *Source:* Susan Arnold, Dr. Raowf Guirguis/National Cancer Institute

1. NK cell releases perforins, which polymerize and form a hole in the foreign cell membrane.

2. Granzymes from NK cell enter perforin hole and degrade foreign cell proteins.

3. Foreign cell dies by apoptosis.

4. Macrophage engulfs and digests dying cell.

Figure 15.12 The action of natural killer (NK) cell.

Immune Regulation by T Cells

Most physiological processes operate under a system of checks and balances as a way to maintain normal stability of functions. This is also true of the immune system. Abnormal or out-of-control actions that could disrupt the beneficial effects of immunity are partly controlled by a subset of T_H cells called **regulatory T cells** or T_{regs} that express CD4 and CD25 receptors (see table 15.2). These cells were once called "suppressor T cells" for their modulating effects on immune responses of other T and B cells. Although the exact mechanisms are yet to be totally clarified, evidence is growing that they are responsible for maintaining immune tolerance and preventing the immune system from overreacting in a harmful or pathological way. They play important roles in moderating inflammation, allergy, autoimmunity, antibody production, cancer, and transplant rejection.

One well-defined mechanism of immune regulation involves the action of *programmed cell death protein-1* (PD-1), a receptor expressed by CD8 T cells. Its function is to "down regulate" or negatively regulate the immune system by acting as a checkpoint that can either 1) allow CD8 cells to become activated and attack their targets or 2) shut down the activation of CD8 cells to block an immune reaction. Most of the time this checkpoint "brake" on T cells maintains balanced immune function. It appears to be an important factor in developing tolerance for self antigens and preventing adverse events in pregnancy, transplant rejection, and autoimmune reactions. Once the PD-1 receptor has been engaged, it blocks further actions by the CD8 cell and causes it to undergo apoptosis and death, which is where the term PD-1 originated.

Malfunctions in the PD-1 system are a factor in the development of some types of cancer. Many cancer cells evade destruction by producing substances that inactivate the PD-1 receptor and put the brakes on the cytotoxic T-cell response that would ordinarily destroy those cancer cells. It is by this means that cancers become metastatic and spread. Two new successful anticancer drugs (Opdivo and Keytruda) are based on monoclonal antibodies that prevent cancer cells from accessing the PD-1 receptor, thus allowing activation of CD8 cells and destruction of the cancer cells. To see a short video explaining this go to https://www.youtube.com/watch?v=v9NBUeU3PG0.

T Cells and Superantigens

Most of the T-cell responses we have covered so far are primarily beneficial and protective. But there is one type of reaction that often has drastic consequences and may lead to serious diseases. Such is the case in exposure of T cells to **superantigens.** These antigens, found primarily in bacteria and viruses, are actually a form of virulence factor. Examples include enterotoxin given off by pathogenic staphylococci, certain toxins of group A streptococci, and proteins of Epstein-Barr virus. They can provoke overwhelming immune responses by large numbers of T cells regardless of specificity. Superantigen molecules are structured so that they can span both MHC-II receptors and some antigen receptors (TCR) on T cells. This event can "trick" large numbers of T cells into releasing massive amounts of cytokines such as tumor necrosis factor and interleukins-1 and -6. The overwhelming influx of such potent mediators leads to blood vessel damage, toxic shock, and multiorgan failure. This is thought to be the underlying mechanism of toxic shock syndrome and other serious infections.

✓ Check Your Progress SECTION 15.3

12. Describe the actions of an antigen-presenting cell.
13. Explain the difference between a T-cell-dependent and a T-cell-independent response.
14. Trace the immune response system, beginning with the entry of a T-cell-dependent antigen, antigen processing, presentation, the cooperative response among the macrophage and lymphocytes, and the reactions of activated B cells.
15. What are the actions of interleukins-1, -2, -4, and -12?
16. Why are the immunities involving T cells called cell-mediated?
17. Explain how the different types of T cells are activated and their responses.
18. What are the targets of cytotoxic cells and how do they destroy them?
19. Discuss how superantigens are different from other antigens and how they contribute to pathology.

15.4 Immune Activities of B Cells

💡 Expected Learning Outcomes

16. List and analyze the stages in activation of B lymphocytes, clonal expansion, and antibody formation.
17. Describe the structure and basic functions of the five major types of immunoglobulins.
18. Explain the actions of antibodies in protective immune reactions.
19. Analyze the primary and secondary responses to antigens and immunogens, including the importance of the anamnestic response and boosters.

Events in B-Cell Responses

Activation of B Lymphocytes: Clonal Selection, Expansion, and Antibody Production

The immunologic activation of most B cells requires a series of events as outlined in the numbered steps that follow and illustrated in **figure 15.13.**

1. **Clonal selection and binding of antigen.** In this case, a pre-committed B cell of a particular clonal specificity picks up the antigen on its Ig receptors and processes it into small peptide determinants. The antigen is then bound to the MHC-II receptors on the B cell. The MHC/Ag complex on the B cell is bound by T-cell receptors.

2. **Induction by chemical mediators.** The B cell receives developmental signals from macrophages and T cells (interleukin-2 and interleukin-6) and other growth factors, especially IL-4 and IL-5.

Quick Search
Watch "B Cells & Antibody" on YouTube to see this process animated.

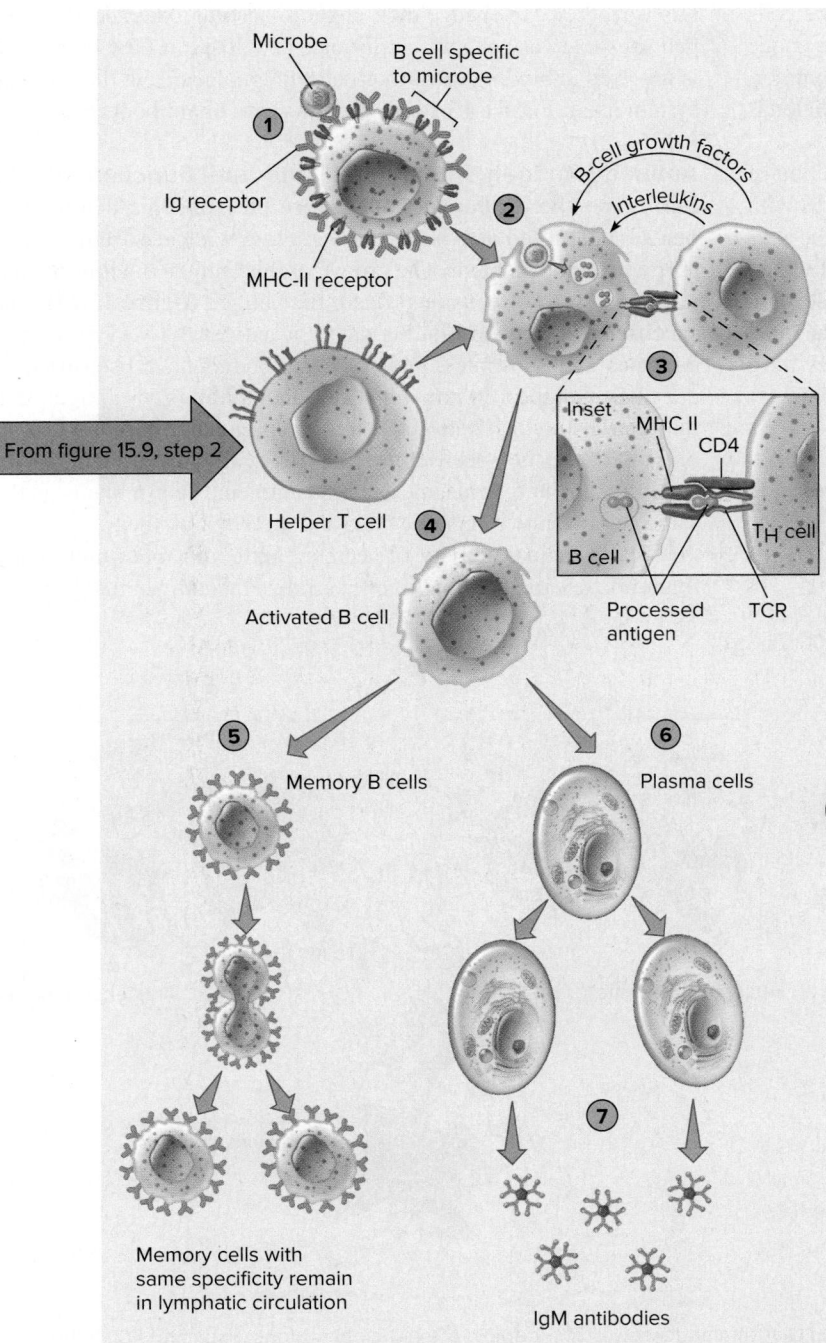

1 Clonal Selection and Antigen Binding
B cells can independently recognize microbes (example here is a virus) and their foreign antigens, and can bind them with their Ig receptors. This is how the initial selection of the antigen-specific B-cell clone occurs.

2 Antigen Processing and Presentation
Once the microbe is attached, the B-cell endocytoses it, processes it into smaller protein units, and displays these on the MHC-II complex (similar to other APCs). This event readies the antigen for presentation to a specific T_H cell.

3 B-Cell/T_H-Cell Cooperation and Recognition
For most B cells to become functional, they must interact with a T helper cell that bears receptors for antigen from the same microbe. This T cell may have been activated by an APC (see figure 15.9, part A). The two cells engage in linked recognition, in which the MHC-II receptor bearing antigen on the B cell binds to both the T-cell antigen receptor and the CD4 molecule on the T cell (inset).

4 B-Cell Activation
The T cell gives off additional signals in the form of interleukins and B-cell growth factors. The linked receptors and the chemical stimuli serve to activate the B cell. Such activation signals an increase in cell metabolism, leading to cell enlargement, proliferation, and differentiation.

5–6 Clonal Expansion/Memory Cells
The activated B cell undergoes numerous mitotic divisions, which expand the clone of cells bearing this specificity and produce memory cells and plasma cells. The memory cells are persistent, long-term cells that can react with the same antigen on future exposures.

7 Plasma Cells/Antibody Synthesis
The plasma cells are short-lived, active secretory cells that synthesize and release antibodies. These antibodies (here IgM) have the same specificity as the Ig receptor and circulate in the fluid compartments of the body, where they react with the same antigens and microbes shown in step 1.

Process Figure 15.13 Events in B-cell activation and antibody synthesis. These steps are tied to the numbered text. Only key receptors for these reactions are shown in the inset.

3. The combination of these stimuli on the membrane receptors causes a signal to be transmitted internally to the B-cell nucleus.

4. These events trigger B-cell activation. An activated B cell—referred to as a *lymphoblast*—enlarges and increases its synthesis of DNA and protein in preparation for entering the cell cycle and mitosis.

5–6. **Clonal expansion.** The stimulated B cell multiplies through successive mitotic divisions and produces a large population of genetically identical daughter cells. Some cells that stop short of becoming fully differentiated are **memory cells,** which remain for long periods to react with that same antigen at a later time. This reaction also expands the clone size, so that subsequent exposure to that antigen provides more cells with that specificity. This expansion of the clone size is one factor in the increased memory response. The more numerous progeny are large, specialized, terminally differentiated B cells called **plasma cells.**

7. **Antibody production and secretion.** The primary action of plasma cells is to secrete copious amounts of antibodies with the same specificity as the original receptor into the surrounding tissues **(figure 15.13, #7).** Although an individual plasma cell can produce around 2,000 antibodies per second, production does not continue indefinitely. The plasma cells do not survive for long, and they deteriorate after they have synthesized antibodies.

Products of B Lymphocytes: Antibody Structure and Functions

Earlier we saw that a basic **immunoglobulin (Ig)**[4] molecule

4. *Ig* is a shorthand term. Immunoglobulins are proteins found in the globulin fraction of the serum and have immune function as antibodies.

contains four polypeptide chains connected by disulfide bonds. Because antibodies are a type of immunoglobulin, they all feature this same basic structure. Let us review it using an IgG molecule as a model. Two functionally distinct segments called *fragments* can be differentiated. The two "arms" that bind antigen are termed **antigen binding fragments (Fabs),** and the rest of the molecule is the **crystallizable fragment (Fc),** so called because it can be crystallized in pure form. The amino-terminal end of each antigen binding fragment (consisting of the variable regions of the heavy and light chains) folds into a groove that will accommodate one epitope of the antigen. The presence of a special hinge region at the site of attachment between the Fabs and the Fc allows swiveling of the Fabs. In this way, they can change their angle to accommodate nearby antigen sites that vary slightly in distance and position. The Fc portion is involved in binding to various cells and molecules of the immune system itself. **Figure 15.14** shows two views of antibody structure.

Antibody-Antigen Interactions and the Function of the Fab The site on the antibody where the antigenic determinant binds is composed of *hypervariable regions* with an extremely variable amino acid content. The groove of this antigen binding site has a specific three-dimensional fit for the antigen **(figure 15.15).** The specificity of antigen binding sites for antigens is very similar to enzymes and substrates, and some antibodies have been used as enzymes. So specific are some immunoglobulins for antigen that they can distinguish between a single functional group of a few atoms. Because the specificity of the two Fab sites is identical for each antigen, an Ig molecule can bind antigenic determinants on the same cell or on two separate cells and thereby link them.

The principal activity of secreted antibodies is to unite with, immobilize, call attention to, or neutralize the antigen that initiated

(a)

(b)

Figure 15.14 Working models of antibody structure. (a) Diagrammatic view of IgG depicts the principal regions (Fabs and Fc) of the molecule. Note that the Fabs can swivel at the hinge region and provide flexibility in position. **(b)** Three-dimensional model of immunoglobulin shows the tertiary and quaternary structure achieved by intrachain and interchain bonds and the positioning of the light- and heavy-chain components. (b): *Source:* Dr. Timothy Vickers

Hypervariable region of Ab that binds Ag

(a) Good fit (b) No fit (c) Poor fit

Figure 15.15 Antigen-antibody binding. The union of antibody (Ab) and antigen (Ag) is characterized by a certain degree of fit and is supported by a multitude of weak linkages, especially hydrogen bonds and electrostatic attraction. A better fit (i.e., antigen in **(a)** vs antigen in **(c)**) provides greater lymphocyte stimulation during the activation stage.

the antibody's formation **(figure 15.16).** Antibodies called opsonins stimulate **opsonization,*** a process in which microorganisms or other particles are coated with specific antibodies so that they will be more readily recognized by phagocytes, which dispose of them. Opsonization has been likened to putting handles on a slippery object to provide a better grip.

Antibodies can aggregate or **agglutinate** cells by cross-linking them into large clumps. Agglutination renders microbes immobile and enhances their phagocytosis. **Precipitation** is a similar reaction

that occurs with small free antigen molecules. Both processes provide a basis for certain immune tests discussed in chapter 17. The interaction of an antibody with complement can result in the specific rupturing of cells and some viruses. This is seen in complement fixation, which refers to the classical pathway of complement activation (see figures 14.20 and 17.13).

In **neutralization** reactions, antibodies fill the surface receptors on a virus or the active site on a bacterial protein, which prevents them from attaching to their target cells. **Antitoxins** are special types of antibodies that neutralize bacterial exotoxins. It should be noted that not all antibodies are protective; some neither benefit nor harm, and a few, such as autoantibodies and those originating from allergic reactions, can actually cause diseases.

* *opsonization* (awp″-son-uh-zay′-shun) L., *opsonium,* food or provisions, and *izare,* to become.

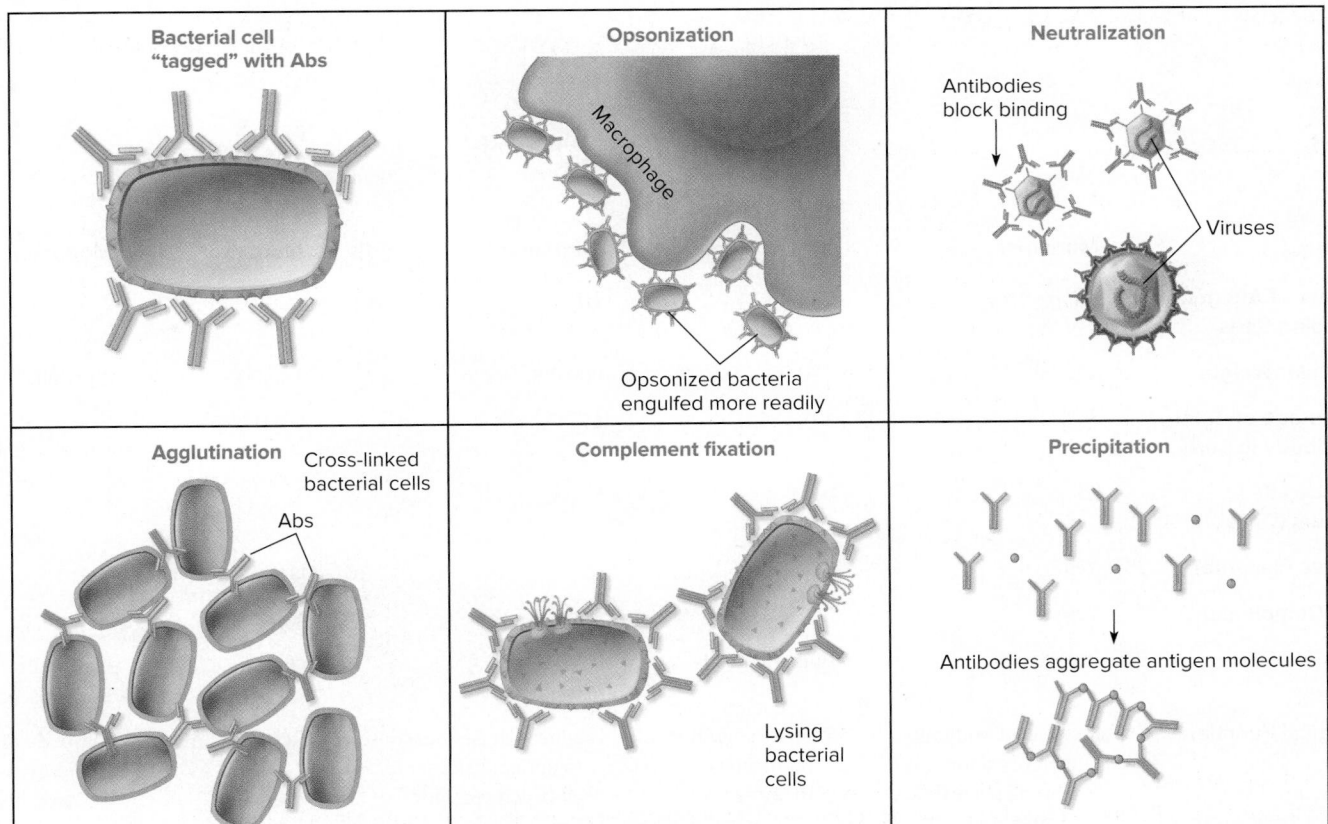

Bacterial cell "tagged" with Abs

Opsonization
Macrophage
Opsonized bacteria engulfed more readily

Neutralization
Antibodies block binding
Viruses

Agglutination Cross-linked bacterial cells
Abs

Complement fixation
Lysing bacterial cells

Precipitation
Antibodies aggregate antigen molecules

Figure 15.16 Summary of antibody functions. Complement fixation, agglutination, and precipitation are covered further in chapter 17.

Functions of the Crystallizable Fragment Although the Fabs bind antigen, the Fc has a different binding function. In most classes of immunoglobulin, the end of the Fc contains an effector that can bind to the membrane of cells, such as macrophages, neutrophils, eosinophils, mast cells, basophils, and lymphocytes. The effect of an antibody's Fc binding to a cell depends upon that cell's role. In the case of opsonization, the attachment of antibody to foreign cells and viruses exposes the Fcs to phagocytes.

Certain antibodies have regions on the Fc portion for fixing complement, and in some immune reactions, the binding of the Fc causes the release of cytokines. For example, the Fc end of the antibody of allergy (IgE) binds to basophils and mast cells, which causes the release of allergic mediators such as histamine. The size and amino acid composition of the Fc also determine an antibody's permeability, its distribution in the body, and its class.

Accessory Molecules on Immunoglobulins All antibodies contain molecules other than the basic polypeptides. Varying amounts of carbohydrates are affixed to the constant regions in most instances (see figure 15.14*b* and **table 15.3**). Two additional accessory molecules are the *J chain*, which helps keep the monomers of

IgA and IgM together, and the *secretory component,* which helps move IgA across mucous membranes. These proteins occur only in certain immunoglobulin classes.

The Classes of Immunoglobulins Immunoglobulins exist as structural and functional classes called *isotypes* (compared and contrasted in table 15.3). The differences in these classes are due primarily to variations in the Fc. The classes are differentiated with shorthand names—Ig, followed by a letter: IgG, IgA, IgM, IgD, IgE.[5]

The structure of **IgG** has already been presented. It is a monomer [6] produced by a plasma cell late in a primary response and by memory cells responding the second time to a given antigenic stimulus. It is by far the most prevalent antibody circulating throughout the blood, lymph, and extracellular fluids. It has numerous functions: It neutralizes toxins, opsonizes, and fixes complement, and it is the only antibody that crosses the placenta and provides protection to the fetus.

5. The letters correspond to the Greek letters gamma, alpha, mu, delta, and epsilon, which also refer to the distinct structures of their constant regions.

6. *Monomer* means "one unit" or "one part." Accordingly, *dimer* means "two units," *pentamer* means "five units," and *polymer* means "many units."

TABLE 15.3	**Characteristics of the Immunoglobulin (Ig) Classes**				
	IgG	IgA (dimer)	IgM	IgD	IgE
	Monomer	Dimer, Monomer	Pentamer	Monomer	Monomer
Number of Antigen Binding Sites	2	2 or 4	10	2	2
Molecular Weight	150,000	170,000–385,000	900,000	180,000	200,000
Percentage of Total Antibody in Serum	80%	13%	6%	0.001%	0.002%
Average Life in Serum (Days)	23	6	5	3	2.5
Crosses Placenta?	Yes	No	No	No	No
Fixes Complement?	Yes	No	Yes	No	No
Binds To	Phagocytes	Epithelial cells	NA	NA	Mast cells and basophils
Biological Function	Long-term immunity; memory antibodies; neutralizes toxins, viruses	Secretory antibody; on mucous membranes	Produced at first response to antigen; can serve as B-cell receptor	Receptor on B cells for antigen recognition	Antibody of allergy; worm infections

C = carbohydrate. J = J chain.

The two forms of **IgA** are (1) a **monomer** that circulates in small amounts in the blood, and (2) a **dimer** that is a significant component of the mucous and serous secretions of the salivary glands, intestine, nasal membrane, breast, lung, and genitourinary tract. The dimer, called secretory IgA, is formed in a plasma cell by two monomers held together by a J chain. To facilitate the transport of IgA across membranes, a secretory piece is later added by the epithelial cells of the mucosa. IgA coats the surface of these membranes and appears free in saliva, tears, colostrum, and mucus. It confers essential local immunity against enteric, respiratory, and genitourinary pathogens by excluding foreign intruders on mucosal membranes. Its contribution in protecting newborns who derive it passively from nursing is the focus of Clinical Connections.

IgM is a huge molecule composed of five monomers (making it a pentamer) attached by the Fc portions to a central J chain. With its 10 binding sites, this molecule has tremendous capacity for binding antigen. IgM is the first class synthesized following the host's first encounter with antigen. Its ability to agglutinate and fix *complement* make it an important antibody in many immune reactions. It circulates mainly in the blood and is far too large to cross the placental barrier.

IgD is a monomer found in minuscule amounts in the serum, and it does not fix complement, opsonize, or cross the placenta. Its main function is to serve as a receptor for antigen on B cells, usually along with IgM, and it is the triggering molecule for B-cell activation.

IgE is also a less common blood component unless one is allergic or has a parasitic worm infection. Its Fc region interacts with receptors on mast cells and basophils. Its biological significance is to stimulate an inflammatory response through the release of potent physiological substances by the basophils and mast cells. Because inflammation would enlist blood cells such as eosinophils and lymphocytes to the site of infection, it would certainly be one defense against parasites. Unfortunately, IgE has another, more insidious effect—that of mediating anaphylaxis, asthma, and certain other allergies (see chapter 16).

Evidence of Antibodies in Serum

Regardless of the site where antibodies are first secreted, a large quantity eventually ends up in the blood by way of the body's communicating networks. If one subjects a sample of **antiserum** (serum containing specific antibodies) to electrophoresis, the major groups of proteins migrate in a pattern consistent with their mobility and size. The albumins show up in one band, and the globulins in four bands called alpha-1 (α_1), alpha-2 (α_2), beta (β), and gamma (γ) globulins. Most of the globulins represent antibodies, which explains how the term *immunoglobulin* was derived. **Gamma globulin** is composed primarily of IgG, whereas β and α_2 globulins are a mixture of IgG, IgA, and IgM.

Monitoring Antibody Production over Time: Primary and Secondary Responses to Antigens

We can learn a great deal about how the immune system reacts to an antigen by studying the levels of antibodies in serum over time (**figure 15.17**). This level is expressed quantitatively as the **titer**,[7] or

7. *Titer* (ty'-tur) Fr. *titre*, standard. One method for determining titer is shown in figure 17.9.

Primary response. A latent period with no measurable antibody occurs early on. The first antibody to appear is IgM, followed later by IgG arising from activation of the first memory cells. Within weeks, the titer tapers back to low levels.

Secondary response. A latent period is lacking because other memory lymphocytes from the earlier response are immediately ready to react. A rapid rise in antibody titer, mainly of IgG, is sustained for several weeks. A smaller amount of IgM is also produced by naive B cells.

Figure 15.17 **A graphic view of the primary and secondary immune responses to antigenic stimuli.** Monitoring the antibody titer or concentration throughout the initial (primary) response and the second (secondary) response produces a distinct pattern. Note that the anamnestic response yields an amount of antibody that is nearly 1,000 times that of the primary response (the scale is logarithmic). This in effect "boosts" this defense to even more effective levels.

concentration, of antibodies. Upon the first exposure to an antigen or immunogen, the system undergoes a **primary response.** The earliest part of this response, the *latent period,* is marked by a lack of antibodies for that antigen, but much activity is occurring. During this time, the antigen is being concentrated in lymphoid tissue and is being processed by the correct clones of B lymphocytes. As plasma cells synthesize antibodies, the serum titer increases to a certain plateau and then tapers off to a low level over a few weeks or months.

When the class of antibodies produced during this response is tested, an important characteristic of the response is uncovered. It turns out that, early in the primary response, most of the antibodies are the IgM type, which is the first class to be secreted by plasma cells. Later, the class of the antibodies (but not their specificity) is switched to IgG or some other class (IgA or IgE).

When the immune system is exposed again to the same antigen or immunogen within weeks, months, or even years, a **secondary response** occurs. The rate of antibody synthesis, the peak titer, and the length of antibody persistence are greatly increased over the primary response. The rapidity and amplification seen in this response are attributable to the memory B cells that were formed during the primary response. Because of its association with recall, the secondary response is also called the **anamnestic* response.** The recall effect shortens the latent or lag period and yields a faster, stronger, and longer-lasting antibody response. The main factors in creating the anamnestic effect are that the memory B cells do not have to go through the early steps in activation, and they do not require as many signals to form plasma cells.

The advantage of this response is evident: It provides a quick and potent strike against subsequent exposures to infectious agents. This memory effect is the fundamental basis for **vaccine boosters,** which we discuss in section 15.6.

Monoclonal Antibodies: Useful Products from Cancer Cells

The value of antibodies as tools for locating or identifying antigens is well established. For many years, antiserum extracted from human or animal blood was the main source of antibodies for tests and therapy, but most antiserum has a basic problem. It contains **polyclonal antibodies,** meaning that it is a mixture of different antibodies because it reflects dozens of immune reactions from a wide variety of B-cell clones. This characteristic is to be expected, because several immune reactions may be occurring simultaneously; and even a single species of microbe can induce several different types of antibodies. Certain applications in immunology require a pure preparation of **monoclonal antibodies (MABs)** that originate from a single clone and have a single specificity for antigen. The development of human MABs and other novel approaches using genetic engineering are being widely pursued by pharmaceutical companies **(15.1 Making Connections).**

* *anamnestic* (an-am-ness′-tik) Gr. *anamnesis,* a recalling.

Check Your Progress SECTION 15.4

20. How are B cells activated, what events are involved in this process, and what happens after B cells are activated?
21. What are the functions of plasma cells, clonal expansion, and memory cells?
22. Describe the structure of immunoglobulin and the functions of the Fab and Fc portions.
23. Describe four or five ways that antibodies function in immunity
24. Describe the attachment of antibodies to antigens. What eventually happens to the antigens?
25. Contrast the primary and secondary response to antigens.
26. Explain the type, order of appearance, and amount of immunoglobulin in each response in figure 15.16 and what causes the differences.
27. What causes the latent period and the anamnestic response?

15.5 A Classification Scheme for Specific, Acquired Immunities

Expected Learning Outcomes

20. Outline the major categories of adaptive immunity.
21. Differentiate between natural and artificial immunities and between active and passive immunities.
22. Expand on the four combinations of the defining categories, using examples.

Humans develop their adaptive immunity by several general routes, as summarized in **figure 15.18.** These categories are based on the source of the immunity and whether the protection is produced by one's own immune system or whether it is received from another person. This classification system is a useful medical guide that can rapidly identify and explain the sources of immunity.

Defining Categories by Mode of Acquisition

Active immunity occurs when an individual receives an immune stimulus (microbe) that activates specific lymphocytes, causing an immune response such as production of antibodies. Active immunity is marked by several characteristics: (1) it is an essential attribute of an immunocompetent individual; (2) it creates a memory that renders the person ready for quick action upon reexposure to that same microbe; (3) it requires several days to develop; and (4) it can last for a relatively long time, sometimes for life. Active immunity can be stimulated by natural or artificial means.

Passive immunity occurs when an individual receives immune substances (primarily antibodies) that were produced actively by the immune system of another human or animal donor. Recipients are protected for a time even though they have not had actual exposure to the microbe. It is characterized by (1) lack of memory for the original antigen, (2) no production of new antibodies against that disease, (3) immediate onset of protection, and

15.1 MAKING CONNECTIONS

Of Mice and MABs?

The technology for producing monoclonal antibodies began with the discovery that tumors isolated from multiple myelomas in mice consist of identical plasma cells. These monoclonal plasma cells secrete a strikingly pure form of antibodies with a single specificity and continue to divide indefinitely. Immunologists recognized the potential in these plasma cells and devised a **hybridoma** approach to creating **m**onoclonal **a**nti**b**odies (**MABs.**) The basic idea behind this approach is to hybridize or fuse a **myeloma*** cell with a normal plasma cell from a mouse spleen to create an immortal cell that secretes a supply of functional antibodies with a single specificity. Most new technologies for making the hybridomas involve transgenic mice that have been engineered with a complete human immune system. This ensures that the antibodies made from the hybridoma contain no mouse immunoglobulin and will be less likely to cause adverse reactions to the mouse proteins.

This technology is being used for a wide variety of biomedical applications. Monoclonal antibodies have provided immunologists with excellent standardized tools for studying the immune system and for expanding disease diagnosis and treatment. Many of the successful applications thus far use MABs for diagnostic testing and research.

Monoclonals can be complexed with plant or bacterial toxins to form immunotoxins that attach to a target cell and poison it. The most exciting prospect of this therapy is that it can target a specified cancer cell and not normal cells. This type of monoclonal is currently employed in treatment of breast and colon cancers. Drugs such as Rituxan and Herceptin bind to cancer cells and trigger their death.

Pharmaceutical companies have created over 300 MABs as potential therapeutic agents. As of this date, about 70 are in use in some form of treatment, mainly for cancer and diseases caused by immune disorders such as allergies and autoimmunity. See **table 15.A** for examples of these drugs.

* *myeloma* (my-uh-loh′-muh) Gr. *myelos,* marrow, and *oma,* tumor. A malignancy of the bone marrow.

TABLE 15.A	Selected Monoclonal Antibody–Based Drugs	
Drug Names (Chemical and Trade)		**Used in Therapy for**
Cancer Drugs		
Trastuzumab	Herceptin	Breast cancer
Rituximab	Rituxan	B-cell disorders (lymphomas, leukemias)
Alemtuzumab	Campath	Chronic T-cell leukemia and lymphoma
Pembrolizumab	Keytruda	Metastatic melanoma and lung cancer
Other Applications		
Omalizumab	Xolair	Asthma
Vedolizumab	Entyvio	Ulcerative colitis
Infliximab	Remicade	Crohn's disease
Palivizumab	Synagis	Respiratory syncytial virus (RSV)
Adalimumab	Humira	Rheumatoid arthritis, psoriasis

Note that the chemical names are coded to end in *mab*—short for monoclonal antibody.

What would be the major drawback to using monoclonal antibodies that are made using a mouse that is not genetically engineered to have a human immune system? *Answer available on Connect.*

Summary of the technique for producing monoclonal antibodies by hybridizing myeloma tumor cells with normal plasma cells. Newer MABs take advantage of mice that have been genetically engineered to develop a human immune system.

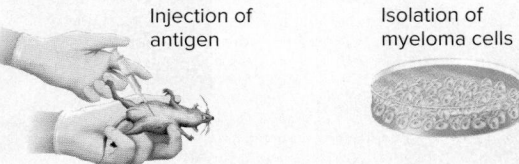

Injection of antigen

Isolation of myeloma cells

(a) A mouse is inoculated with an antigen having the desired specificity, and activated plasma cells are isolated from its spleen. Myeloma cells are cultured from a different mouse that has this cancer.

Mouse spleen cells producing antibody

Myeloma cells in culture

Fusion into hybridoma

(b) The two cell populations are mixed with polyethylene glycol, which causes some cells in the mixture to fuse and form hybridomas.

Selection of MAB-producing clones

(c) Surviving cells are cultured and separated into individual wells. Tests are performed on each hybridoma to determine the specificity of the antibody (Ab) it secretes.

Maintenance of hybridoma

Product

Hybridoma cells injected into mice for culture

Purified monoclonal antibodies extracted

(d) The hybridoma is maintained in a susceptible mouse for future use. A hybridoma with the desired specificity is grown in cell culture. Antibody product is isolated and purified.

Acquired Immunity

Natural Immunity
acquired through the normal life experiences
not induced through medical means

Artificial Immunity
produced purposefully through
medical procedures (also called immunization)

Active Immunity
results when a person
develops his own immune
response to a microbe
through infection.

Passive Immunity
results when a person
receives preformed
immunity through the
placenta or nursing.

Active Immunity (same as vaccination)
results when a person
develops his own
immune response to a
prepared microbial antigen.

Passive Immunity
results when a person is
given selected immune
substances made by
another individual.

Figure 15.18 **Categories of acquired immunities.** Natural immunities, which occur during the normal course of life, are either active (acquired from an infection and then recovering) or passive (antibodies donated by the mother to her child). Artificial immunities are acquired through medical practices and can be active (vaccinations with antigen, to stimulate an immune response) or passive (immune therapy with a serum containing antibodies). (left, left center): © EyeWire Collection/Getty Images RF; (center right): *Source:* Amanda Mills/CDC; (right): © Arthur Tilley/Getty Images RF

(4) short-term effectiveness because antibodies have a limited period of function. Ultimately, the recipient's body disposes of them. Passive immunity can also be natural or artificial in origin.

Natural immunity encompasses immunity that is acquired during any normal biological experiences of an individual but not through medical intervention.

Artificial immunity is protection from infection obtained through medical procedures. This type of immunity is induced by immunization with vaccines and immune serum.

Putting together all combinations of these categories, we arrive at four possible examples of origins of immunity.

1. Natural Active Immunity: Getting an Infection

After recovering from infectious disease, a person may be actively resistant to reinfection for a period that varies according to the disease. In the case of childhood viral infections such as measles, mumps, and rubella, this natural active stimulus provides nearly lifelong immunity. Other diseases result in a less extended immunity of a few months to years (such as pneumococcal pneumonia and shigellosis), and reinfection is possible. Even a subclinical infection can stimulate natural active immunity. This probably accounts for the fact that some people are immune to an infectious agent without ever having been noticeably infected with or vaccinated for it.

2. Natural Passive Immunity: Mother to Child

Natural, passively acquired immunity occurs as a result of the prenatal and postnatal, mother-child relationship. During fetal life, IgG

antibodies circulating in the maternal bloodstream are small enough to pass or be actively transported across the placenta. Antibodies against tetanus, diphtheria, pertussis, and several viruses regularly cross the placenta. This natural mechanism provides an infant with a mixture of many maternal antibodies that can protect it for the first few critical months outside the womb, while its own immune system is gradually developing active immunity. Depending upon the microbe, passive protection lasts anywhere from a few months to a year, but it eventually wanes. Most childhood vaccinations are timed so that there is no lapse in protection against common childhood infections.

Another source of natural passive immunity comes to the baby by way of mother's milk (**Clinical Connections**). Although the human infant acquires most natural passive immunity in utero and some through nursing, the milk-borne antibodies provide IgA antibodies that react against microbes entering the intestine. This unique protection is not available from transplacental antibodies.

Artificial Immunity: Immunization

Immunization is any clinical process that produces immunity in a subject. Because it is often used to give advance protection against infection, it is also called *immunoprophylaxis*. The details of this form of immunity are presented in section 15.6.

3. Artificial Active Immunization: Vaccination

The term *vaccination* originated from the Latin word *vacca* (cow), because the cowpox virus was used in the first preparation for active immunization against smallpox (see 15.2 Making Connections). Vaccination exposes a person to a specially prepared

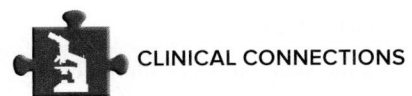
CLINICAL CONNECTIONS

The Immunological Benefits of Breast-Feeding

An advertising slogan from the past claims that cow's milk is "nature's most nearly perfect food." One could go a step further and assert that human milk is nature's *perfect* food for young humans. Clearly, it is loaded with essential nutrients, not to mention being available on demand from a readily portable, hygienic container that does not require refrigeration or warming. But there is another and perhaps even greater benefit. During lactation, the breast becomes a site for the proliferation of lymphocytes that produce IgA antibodies that protect the mucosal surfaces from local invasion by microbes.

International symbol for breast feeding
Courtesy http://www.mothering.com/articles/the-international-breastfeeding-symbol/ Artist Matt Daigle

The very earliest secretion of the breast, a thin, yellow milk called **colostrum,** is very high in these antibodies. They form a protective coating in the gastrointestinal tract of a nursing infant that guards against infection by a number of enteric pathogens (*Escherichia coli, Salmonella,* poliovirus, rotavirus). Protection at this level is especially critical because an infant's own antibodies and natural intestinal barriers are not yet developed. As with immunity in utero, the necessary antibodies will be donated only if the mother herself has active immunity to the microbe through a prior infection or vaccination.

A critical benefit of breast-feeding that has not been given adequate credit is the connection between nursing, the breast microbiome, and the neonatal microbiome. For many years milk was thought to be sterile unless the mother's breasts were infected. Now we know from studies on the human microbiome that breast milk and tissues harbor a diverse bacterial community that is distinct from that of the skin or the mouth. This discovery acknowledges that, in addition to its nutritional and immunological support, breast milk is an important early source of microbes that populate the newborn intestine and other body sites. A related function of the breast microbiome could be to supply bacterial stimuli for the continuing development of the baby's immune system.

For a number of years, the ready availability of artificial formulas and the changing lifestyles of women have reduced the incidence of breast-feeding. According to UNICEF, only about 38% of mothers worldwide breast-feed their babies for 6 months or more. Where adequate hygiene and medical care prevail, bottle-fed infants get through the critical period with few problems because the foods given them are relatively free of pathogens and they have received protection against some childhood infections in utero. Mothers in developing countries with untreated water supplies or poor medical services are strongly discouraged from using prepared formulas, because they can actually inoculate the baby's intestine with pathogens from the formula. Millions of neonates suffer from severe and life-threatening diarrhea that could have been prevented by sustained breast-feeding. The use of formula has been so damaging to infant health that the World Health Organization issued an *International Code on the Marketing of Breast Milk Substitutes,* discouraging the use of formula in the developing world.

Explain the reasons donated antibodies (either placental or in breast milk) are only a temporary protection. *Answer available on Connect.*

microbial (antigenic) stimulus, which then triggers the immune system to produce antibodies and memory cells to protect the person upon future exposure to that microbe. As with natural active immunity, the degree and length of protection vary. Commercial vaccines are currently available for many diseases (see table 15.5).

4. Artificial Passive Immunization: Immunotherapy

In immunotherapy, a patient at risk for acquiring a particular infection is administered a preparation that contains specific antibodies against that infectious agent. Pooled human serum from donor blood (gamma globulin) and immune serum globulins containing high quantities of antibodies are the usual sources. Immune serum globulins are used to protect people who have been exposed to certain diseases such as hepatitis A, rabies, and tetanus.

 Check Your Progress SECTION 15.5

28. Contrast active and passive immunity in terms of how each is acquired, how long it lasts, whether memory is triggered, how soon it becomes effective, and what immune cells and substances are involved.

29. Name at least two major ways in which natural and artificial immunities are different.

30. Explain why the passive transfer of T-cell immunities would be highly unlikely.

15.6 Immunization: Providing Immune Protection Through Therapy

 Expected Learning Outcomes

23. Explain the purposes of immunotherapy and immunization.

24. Describe the sources and uses of artificial passive immunization and artificial active immunization or vaccination.

25. Discuss which factors are involved in vaccine development and new strategies for developing vaccines.

26. Identify the major categories of vaccine antigens, citing examples.

27. Describe the medical guidelines for vaccination, its side effects, and how it relates to herd immunity.

Methods that actively or passively immunize people are widely used in disease prevention and treatment. The use of these terms is sometimes imprecise; thus, it should be stressed that active immunization, in which a person is administered some form of antigen, is synonymous with vaccination. Passive immunization, in which a person is given antibodies, is a type of **immunotherapy.**

Artificial Passive Immunization

The first attempts at passive immunization involved the transfusion of horse serum containing antitoxins to prevent tetanus and to treat patients exposed to diphtheria. Since then, antisera from animals have been largely replaced with products of human origin that function with various degrees of specificity. **Immune serum globulin (ISG),** sometimes called *gamma globulin,* contains immunoglobulin extracted from the pooled blood of at least 1,000 human donors. The method of processing ISG concentrates the antibodies to increase potency and eliminates potential pathogens (such as the hepatitis B and HIV viruses). It is a treatment of choice in preventing measles and hepatitis A and in replacing antibodies in immunodeficient patients. Most forms of ISG are injected intramuscularly to minimize adverse reactions, and the protection they provide lasts 2 to 3 months.

A serum product called **specific immune globulin (SIG)** is derived from a more defined group of donors. SIG preparations are made with serum from patients who are in a hyperimmune state after being vaccinated or infected by pertussis, tetanus, chickenpox, and hepatitis B. This chapter's Case Study examines how a specific antirabies immune globulin (HRIG) can be used in cases of rabies exposure. These globulins are preferable to ISG because they contain higher titers of specific antibodies obtained from a smaller pool of patients. Although useful for prophylaxis in persons who may be exposed to infectious agents, these sera can be limited in availability.

When a human immune globulin is not available, antisera and antitoxins of animal origin can be effective. Sera produced in horses are available for diphtheria, botulism, rabies, and spider and snake bites. Unfortunately, horse antigens often stimulate allergies such as serum sickness or anaphylaxis (chapter 16). With the development of monoclonal therapies, many of these older sources will be replaced by pure antibody drugs that have been tailored to just one specific pathogen, Although all of these passive immunities only last a relatively short time, they act immediately and can protect patients for whom no other useful medication or vaccine exists, such as immunocompromised patients who cannot be vaccinated.

Artificial Active Immunity: Vaccination

Active immunity can be conferred artificially by **vaccination**—exposing a person to material that is immunogenic but not pathogenic. The discovery of vaccination was one of the farthest-reaching and most important developments in medical science **(15.2 Making Connections).** The basic principle behind vaccination is to stimulate a primary response that primes the immune system for future exposure to a pathogen. If this pathogen enters the body, the secondary immune response will be immediate, powerful, and sustained. Second or additional doses of vaccines can also stimulate the anamnestic response.

Vaccines have profoundly reduced the prevalence and impact of many infectious diseases that were once common and often deadly. In this section, we survey the principles of vaccine preparation and important considerations surrounding vaccine indication and safety. Vaccines are also given specific consideration in later chapters on infectious diseases and organ systems.

Principles of Vaccine Preparation

A vaccine must be considered from the standpoints of immunogen selection, effectiveness, ease in administration, safety, and cost. In natural immunity, an infectious agent stimulates a relatively long-term protective response. In artificial active immunity, the objective is to obtain this same response with a modified version of the microbe or its components. Most vaccine preparations are based on one of the following antigen preparations **(figure 15.19):**

1. killed whole cells or inactivated viruses;
2. live, attenuated bacterial cells or viruses;
3. antigenic molecules derived from bacterial cells or viruses; or
4. genetically engineered microbes or microbial antigens.

The main qualities of an effective vaccine are listed in **table 15.4.**

Large, complex antigens such as whole cells or viruses are very effective immunogens. Depending on the vaccine, these are either killed or attenuated. Killed **or inactivated vaccines** are prepared by cultivating the desired strain or strains of a bacterium or virus and treating them with formalin, radiation, heat, or some other agent that kills the microbe but does not change its antigenic structure. One type of vaccine for the bacterial disease cholera is of this type. The IPV polio vaccine and some forms of influenza vaccine contain inactivated viruses. Because the microbe does not multiply, killed vaccines often require a larger dose and more boosters to be effective.

A number of vaccines are prepared from **attenuated** microbes. Attenuation is any process that substantially lessens or negates the virulence of viruses or bacteria. It is usually achieved by modifying the growth conditions or manipulating microbial genes in a way that eliminates virulence factors. Attenuation methods include long-term cultivation, selection of mutant strains that grow at colder temperatures (cold mutants), passage of the microbe through unnatural hosts or tissue culture, and removal of virulence genes. The vaccine for tuberculosis (BCG) was obtained after 13 years of subculturing the agent of bovine tuberculosis. Vaccines for measles, mumps, polio (oral vaccine), and rubella contain active, non-virulent viruses. The primary advantages of these preparations are that (1) viable microorganisms can multiply and produce infection (but not disease) like the natural organism, (2) they confer long-lasting protection, and (3) they usually require fewer doses and boosters than other types of vaccines.

Disadvantages of using live microbes in vaccines are that they require special storage facilities, can be transmitted to other people, and can conceivably mutate back to a virulent strain.

If the exact antigenic determinants that stimulate immunity are known, these substances can be isolated from microorganisms and serve as the basis for a vaccine **(figure 15.19b).** Vaccines made from bacterial cell parts are called **acellular** or **subcellular vaccines.** If isolated from viruses, they are called **subunit vaccines.** The antigen used in these vaccines may be extracted from cultures of the microbes, produced by genetic engineering, or synthesized chemically.

Examples of extracted antigens currently in use are the capsules of the pneumococcus and meningococcus, the protein surface antigen of anthrax, and the surface proteins of hepatitis B virus. A

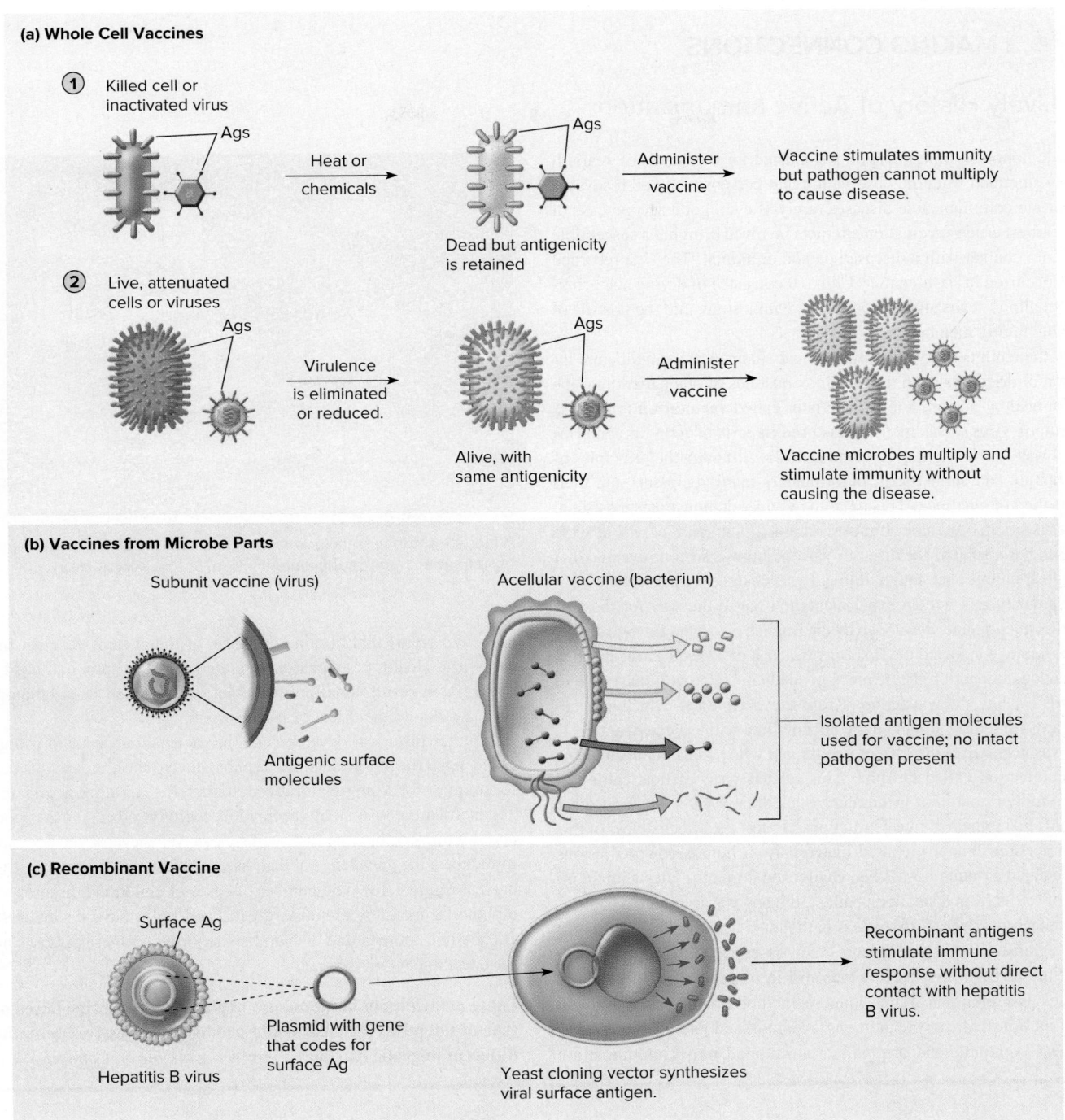

(a) Whole Cell Vaccines

① Killed cell or inactivated virus

Ags

Heat or chemicals →

Ags

Dead but antigenicity is retained

Administer vaccine →

Vaccine stimulates immunity but pathogen cannot multiply to cause disease.

② Live, attenuated cells or viruses

Ags

Virulence is eliminated or reduced. →

Ags

Alive, with same antigenicity

Administer vaccine →

Vaccine microbes multiply and stimulate immunity without causing the disease.

(b) Vaccines from Microbe Parts

Subunit vaccine (virus)

Acellular vaccine (bacterium)

Antigenic surface molecules

Isolated antigen molecules used for vaccine; no intact pathogen present

(c) Recombinant Vaccine

Surface Ag

Hepatitis B virus

Plasmid with gene that codes for surface Ag

Yeast cloning vector synthesizes viral surface antigen.

Recombinant antigens stimulate immune response without direct contact with hepatitis B virus.

Process Figure 15.19 Strategies in vaccine design. (a) Whole cells or viruses, killed or attenuated. **(b)** Acellular or subunit vaccines are made by disrupting the virus or cell to release antigenic molecules or components that can be isolated and purified. **(c)** Recombinant vaccines are made by isolating a gene for the antigen from the pathogen (here a hepatitis virus) and splicing it into a plasmid. Insertion of the recombinant plasmid into a cloning host (yeast) results in the production of large amounts of viral surface antigen to use in vaccine preparation.

special type of vaccine is the **toxoid,** which consists of a purified fragment of bacterial exotoxin that has been inactivated. By eliciting the production of antitoxins that can neutralize the natural toxin, toxoid vaccines provide protection against toxinoses such as diphtheria and tetanus.

A survey of major licensed vaccines and their indications is presented in **table 15.5.**

Development of New Vaccines

Despite considerable successes, dozens of bacterial, viral, protozoan, and fungal diseases still remain without a functional vaccine. At the present time, no reliable vaccines are available for malaria, HIV/AIDS, various diarrheal diseases (*E. coli, Shigella*), a number of respiratory diseases, and worm infections, which affect over

15.2 MAKING CONNECTIONS

The Lively History of Active Immunization

The basic notion of immunization has existed for thousands of years. It probably stemmed from the observation that persons who had recovered from certain communicable diseases rarely, if ever, got a second case of it. The earliest crude vaccination attempts involved bringing a susceptible person into contact with a diseased person or animal. The first recorded attempt occurred in sixth-century China. It consisted of drying and grinding up smallpox scabs and blowing them with a straw into the nostrils of vulnerable family members.

By the tenth century, this practice had changed to the deliberate inoculation of dried pus from the smallpox pustules of one patient into the arm of a healthy person, a technique later called **variolation** (variola is the smallpox virus). This method was used in parts of Asia for centuries before it was first brought to England in 1721. Although the principles of the technique had some merit, unfortunately many recipients and their contacts died of smallpox. This outcome vividly demonstrates a cardinal rule for a workable vaccine: It must contain an antigen that will provide protection but not cause the disease. Variolation was so controversial that any English practitioner caught doing it was charged with a felony.

Eventually this human experimentation paved the way for the first really effective vaccine, developed by the English physician Edward Jenner in 1796. Jenner conducted the first scientific test of a vaccine, one that had a tremendous impact on the advance of medicine. His work gave rise to the words *vaccine* and *vaccination* (from L., *vacca,* cow), which now apply to any immunity obtained by inoculation with selected antigens. Jenner was inspired by the case of a milkmaid who had been infected by a pustular infection called cowpox. This related virus afflicts cattle but causes a milder condition in humans. She explained that she and other milkmaids had remained free of smallpox. To test the effectiveness of this cross protection, Jenner prepared material from human cowpox lesions and inoculated a young boy. When challenged 2 months later with an injection of crusts from a smallpox patient, the boy was immune to it.

Jenner's discovery—that a less pathogenic agent could confer protection against a more pathogenic one—is especially remarkable in view of the fact that microscopy was still in its infancy and the nature of viruses was unknown. When Jenner's method proved successful and word of its significance spread, it was eventually adopted in many other countries. Eventually the original virus mutated into a unique strain

A modern setup for smallpox vaccination; note the two-pronged needle used to scratch the vaccine into the skin. *Source:* James Gathany/CDC

(*vaccinia* virus) that became the basis of the current vaccine. In 1979 the World Health Organization declared that smallpox had been eradicated. As a result, smallpox vaccination had been discontinued until recently, due to the threat of bioterrorism.

Other historical developments in vaccination included using heat-killed bacteria in vaccines for typhoid fever, cholera, and plague, and techniques for using neutralized toxins for diphtheria and tetanus. Throughout the history of vaccination, there have been vocal opponents and minimizers, but one fact seems clear: Whenever a vaccine has been introduced, the prevalence of that disease has declined dramatically (look back at figure 1.10). The number of cases of childhood diseases such as diphtheria, measles, mumps, rubella, and polio have declined 98% to 100% when compared to the numbers reported for these diseases prior to the development of vaccines.

Using principles of immunology, explain how a vaccine based on one type of microbe (vaccinia) could produce an effective immunity to a different microbe (variola). *Answer available on Connect.*

TABLE 15.4	Checklist of Requirements for an Effective Vaccine

- It should have a low level of adverse side effects or toxicity and not cause serious harm.
- It should protect against exposure to natural, wild forms of pathogen.
- It should stimulate both antibody (B-cell) response and cell-mediated (T-cell) response.
- It should have long-term, lasting effects (produce memory cells).
- It should work with minimal doses or boosters.
- It should be inexpensive, have a relatively long shelf life, and be easy to administer.

200 million people per year worldwide. Of all of the challenges facing vaccine specialists, probably the most difficult has been choosing a vaccine antigen that is safe and that properly stimulates immunity. Currently much attention is being focused on newer strategies for vaccine preparation that employ antigen synthesis, recombinant DNA, and gene cloning technology.

When the exact composition of an antigenic determinant is known, it is sometimes possible to artificially synthesize it. This ability permits preservation of antigenicity while greatly increasing antigen purity and concentration. A malaria vaccine currently being tested in areas of South America and Africa is composed of three synthetic peptides from the parasite. One candidate has been carried through 4 trial phases and is currently awaiting approval.

TABLE 15.5 Vaccines Currently Licensed in the United States

Disease/Vaccine Preparation	Trade Names	Recommended Usage/Comments
Contain live, attenuated bacteria		
Tuberculosis	BCG	For high-risk occupations only; protection variable
Typhoid	Vivotif	Mainly for travelers to endemic areas
Acellular vaccines (capsular polysaccharides or proteins)		
Anthrax	BioThrax	For protection in military recruits, occupationally exposed
Meningitis (meningococcal)**	Menactra, Menomune, Trumenba*	For protection in high-risk infants, military recruits, college students; For ages 10-25 years
Meningitis (*Haemophilus influenzae*)**	PedvaxHIB, ActHIB	For infants and children; may be administered with DTaP
Pneumococcal pneumonia	Prevnar, Pneumovax	Important for people at high risk: the young, elderly, and immunocompromised; moderate protection
Toxoids (formaldehyde-inactivated bacterial exotoxins)		
Diphtheria**	Dt	Used for boosters in teens and adults (uppercase letter indicates a higher dose of this antigen)
Tetanus**	Td	Used for boosters in teens and adults
Contain inactivated whole viruses		
Hepatitis A	Havrix, Vaqta	Protection for travelers, institutionalized people
Influenza	FluLaval, Flucelvax	For high-risk populations; requires constant updating for new strains; immunity not durable
Japanese encephalitis	JE-Vax	For those residing in endemic areas, lab workers
Poliomyelitis**	IPOL (inactivated polio)	Routine childhood vaccine; preferred over live vaccine
Rabies	Imovax, RabAvert	For victims of animal bites or otherwise exposed; effective
Contain live, attenuated viruses		
Adenovirus	Nasal	For immunizing military recruits
Chickenpox (varicella)**	Varivax	Routine childhood vaccine; immunity can diminish over time
Influenza	FluMist	No longer widely used
Measles (rubeola)**	Attenuvax	Routine childhood vaccine; very effective
Mumps (parotitis)**	Mumpsvax	Routine childhood vaccine; very effective
Rotavirus (RotaTeq)	RotaTeq, Rotarix	Live oral vaccine for infants at risk for rotavirus infection
Rubella**	Meruvax	Routine childhood vaccine; very effective
Smallpox and monkeypox	Dryvax	Since 2003 offered on voluntary basis for health care workers, some military personnel
Yellow fever	YF-Vax	Travelers, military personnel in endemic areas
Subunit viral vaccines		
Hepatitis B*, **	Engerix, Recombivax	Recommended for all children, starting at birth; also for health care workers and others at risk
Human papillomavirus*	Gardasil, Cervarix	To prevent infection in girls and boys and various cancers associated with HPV that may arise in adulthood.
Combined vaccines contain antigens for two or more diseases to cut down on the number of injections		
Diphtheria, tetanus, pertussis, *H. influenzae,* type b	DTaP, Tripedia, Infanrix	Routine infant and child vaccination
Diphtheria, tetanus, pertussis,	TriHIBit	Same as above
Diphtheria, tetanus, pertussis, hepatitis B, polio	Pediarix	Routine childhood vaccination
Hepatitis B and A	Twinrix	Combination of Havrix and Recombivax given at birth, one at 6 months of age
Measles, mumps, rubella	MMR II	Recommended childhood vaccination—all attenuated viruses
Measles, mumps, rubella, chickenpox	ProQuad	Recommended childhood vaccination—all attenuated viruses

*These vaccines are made by recombinant methods.
**Usually given in combination with other vaccines.
Source of Data: Centers for Disease Control and Prevention

Several biotechnology companies are using plants to mass produce vaccine antigens. Tests are underway to grow tomatoes, potatoes, and bananas that synthesize proteins from cholera, hepatitis, papillomavirus, and *E. coli* pathogens. This strategy aims to economically harvest vaccine antigens and make vaccines for populations that otherwise would not have access to them.

Genetically Engineered Vaccines

Some of the genetic engineering concepts introduced in chapter 10 offer novel approaches to vaccine development. These methods are particularly effective in designing vaccines for obligate parasites that are difficult or expensive to culture, such as the syphilis spirochete or the malaria parasite. Recombinant DNA technology provides a means of isolating the genes that encode various microbial antigens, inserting them into plasmid vectors, and cloning them in appropriate hosts. The outcome of recombination can be as varied as desired. For instance, the cloning host can be stimulated to synthesize and secrete a protein product (antigen), which is then harvested and purified (**figure 15.19*c***). Vaccines for hepatitis and human papillomavirus are currently being prepared in this way. Antigens from the agents of syphilis, *Schistosoma,* and influenza have been similarly isolated and cloned and are currently being considered as potential vaccine material.

Another technique using genetic recombination has been nicknamed the *Trojan horse* vaccine. The term derives from an ancient legend in which the Greeks sneaked soldiers into the fortress of their Trojan enemies by hiding them inside a large, mobile wooden horse. In the microbial equivalent, genetic material from a selected infectious agent is inserted into a live carrier microbe that is nonpathogenic. In theory, the recombinant microbe will multiply and express the foreign genes, and the vaccine recipient will be immunized against the microbial antigens. Vaccinia, the virus originally used to vaccinate for smallpox, and adenoviruses have proved practical agents for this technique. Vaccinia has been used as the carrier in experimental vaccines for AIDS, herpes simplex 2, leprosy, and tuberculosis.

DNA vaccines are another promising approach to immunization. The technique in these formulations is very similar to gene therapy as described in chapter 10, except in this case, microbial (not human) DNA is inserted into a plasmid vector and inoculated into a recipient (**figure 15.20**). The expectation is that the human cells will take up some of the plasmids and express the microbial DNA in the form of proteins. Because these proteins are foreign, they will be recognized during immune surveillance and cause B and T cells to be sensitized and form memory cells. Note: Even though the DNA is from a microbe, it will not cause an immune response because the structure of the DNA molecule by itself is not highly immunogenic.

Experiments with animals have shown that these vaccines are very safe and that only a small amount of the foreign antigen need be expressed to produce effective immunity. Another advantage to this method is that any number of potential microbial proteins can be expressed, making the antigenic stimulus more complex and improving the likelihood that it will stimulate both antibody and cell-mediated immunity. At the present time, vaccines for Lyme disease, hepatitis C, herpes simplex, influenza, tuberculosis, papillomavirus, malaria, and SARS are undergoing trials.

Routes of Administration and Side Effects of Vaccines

Most vaccines are injected by subcutaneous, intramuscular, or intradermal routes. Oral vaccines are available for only a few diseases (table 15.5), but they have some distinct advantages. An oral dose of a vaccine can stimulate protection (IgA) on the mucous membrane of the portal of entry. Oral vaccines are also easier to give, more readily accepted, and well tolerated. Other methods that show promise are an intranasal vaccine delivered into the nose by aerosol or drops (FluMist for influenza, for example) and skin patches.

Some vaccines are made more effective by adding an **adjuvant.*** An adjuvant is any compound that enhances immunogenicity and prolongs antigen retention at the injection site. The adjuvant precipitates the antigen and holds it in the tissues so that it will be

* *adjuvant* (ad′-joo-vunt) L., *adjuvans,* a thing that helps or aids.

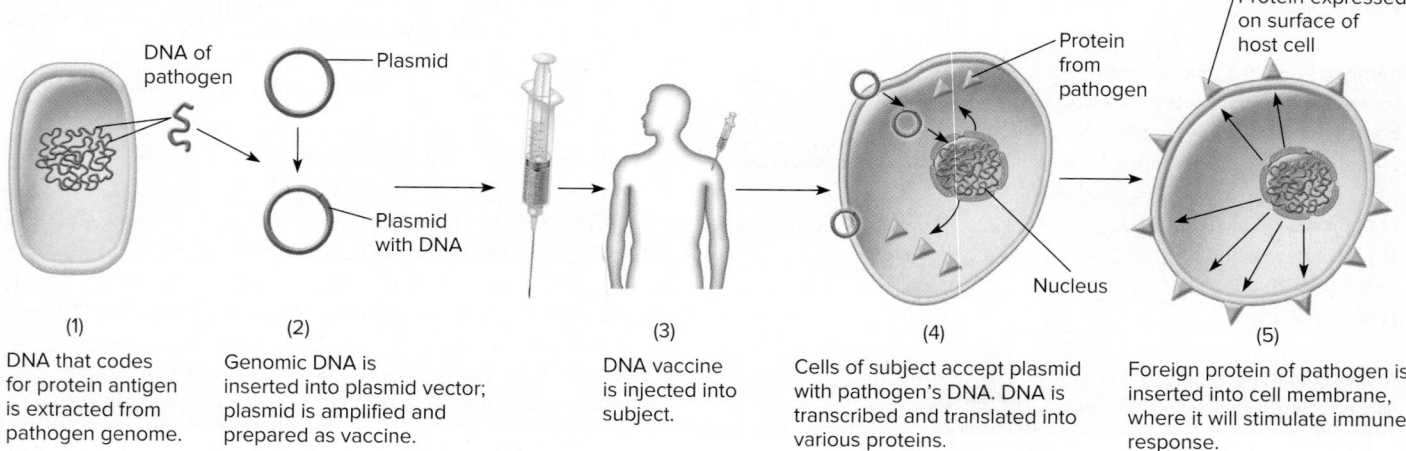

(1)	(2)	(3)	(4)	(5)
DNA that codes for protein antigen is extracted from pathogen genome.	Genomic DNA is inserted into plasmid vector; plasmid is amplified and prepared as vaccine.	DNA vaccine is injected into subject.	Cells of subject accept plasmid with pathogen's DNA. DNA is transcribed and translated into various proteins.	Foreign protein of pathogen is inserted into cell membrane, where it will stimulate immune response.

Figure 15.20 DNA vaccine preparation. DNA vaccines contain all or part of the pathogen's DNA, which is used to "infect" a recipient's cells. Processing of the DNA leads to production of an antigen protein that can stimulate a specific response against that pathogen.

released gradually. Its gradual release presumably facilitates contact with antigen-presenting cells and lymphocytes. Common adjuvants are alum (aluminum hydroxide salts), Freund's complete adjuvant (emulsion of mineral oil, water, and extracts of mycobacteria), and beeswax.

Vaccines must go through many years of trials in experimental animals and human volunteers before they are licensed for general use. Even after they have been approved, like all therapeutic products, they are not without complications. The most common of these are local reactions at the injection site, fever, allergies, and other adverse reactions. Relatively rare reactions (about one case out of 300,000 vaccinations) are panencephalitis (from measles vaccine), back-mutation to a virulent strain (from polio vaccine), disease due to contamination with dangerous viruses or chemicals, and neurological effects of unknown cause (from pertussis and swine flu vaccines). Some patients experience allergic reactions to the medium (eggs or tissue culture) rather than to vaccine antigens.

When known or suspected adverse effects have been detected, vaccines are altered or withdrawn. For example, the whole-cell pertussis vaccine was replaced by the acellular capsule (aP) form when it was associated with adverse neurological effects. The Lyme disease vaccine was withdrawn when recipients developed arthritis and other side effects. Polio vaccine was switched from live, oral to inactivated when too many cases of paralytic disease occurred from back-mutated vaccine stocks. Vaccine companies have also phased out certain preservatives, such as thimerosal, that could cause allergies and other medical problems in some patients.

Professionals involved in giving vaccinations must understand their inherent risks but also realize that the risks from the infectious disease almost always outweigh the chance of an adverse vaccine reaction. The greatest caution must be exercised in giving live vaccines to immunocompromised or pregnant patients, the latter because of possible risk to the fetus.

To Vaccinate: Why, Whom, and When?

Vaccination confers long-lasting, sometimes lifetime, protection in the individual, but an equally important effect is to protect public health. Vaccination is a proven method of establishing **herd immunity** in the population. According to this concept, individuals immune to a communicable infectious disease will not harbor it, thus reducing the occurrence of that pathogen. With a larger number of immune individuals in a population (herd), it will be less likely that an unimmunized member of the population will encounter the agent and become infected. In effect, collective immunity through mass immunization confers indirect protection on the nonimmune members (such as children). The most effective herd immunity outcome occurs when the healthiest members are vaccinated. Because most vaccine programs target the youngest and oldest (and thus weaker) people instead, herd immunity in many populations is not complete. Herd immunity maintained through immunization is considered a valuable factor in preventing epidemics.

> **Quick Search**
> "Get the Picture: Childhood Immunizations" on the CDCTV website emphasizes the importance of vaccinations.

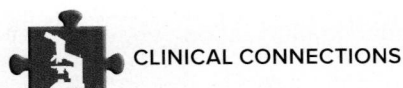

CLINICAL CONNECTIONS

Has Vaccination Become a Victim of Its Own Success?

Vaccination has always been considered one of humanity's triumphs, one that has made an enormous contribution to health and well-being. Most people are no longer at risk for such life-threatening diseases as tetanus, diphtheria, polio, and others. But lately, vaccination's good reputation is being tarnished. It began with a poorly designed 1998 British study that hinted at a link between autism and the MMR (mumps, measles, and rubella) vaccine. The publicity from this report spurred a controversy that led some parents to stop vaccinating their children against the usual childhood diseases. The study's results have since been thoroughly refuted, but the vaccine phobia persists. More than a dozen separate epidemiologic studies have discounted any connection between the MMR vaccine and autism, and additional studies show that the preservative thimerosal is also not correlated with autism (MMR is a live attenuated vaccine and would not contain this substance anyway). Recently the CDC reported findings of the latest in-depth study that found no correlation between vaccine type or vaccine number and the development of autism. In fact, the study identified a significant number of unvaccinated children who were diagnosed with some form of autism.

Many decisions not to vaccinate are based on misconceptions and emotion rather than science, yet consequences are being felt in some populations. Recent outbreaks of measles, mumps, meningitis, and pertussis have been tied to inadequate vaccinations. The reality is that these diseases are still around, and they pose a greater risk for young children than the possibility of a vaccine reaction. In fact, 1% of babies develop a severe brain condition from measles virus, and mumps, meningitis, and pertussis all have their own serious complications. Any reduction in vaccinations will lead to less herd immunity and create unknown pockets where children lack protection. Now more than ever before, health care workers have a responsibility to educate and counsel patients about the importance of routine immunizations for both individuals and the community.

Explain how vaccination has been responsible for the worldwide eradication of smallpox and the near elimination of polio and many other infectious diseases. *Answer available on Connect.*

Table 15.6 contains the recommended childhood and adolescent immunization schedule, categorized by age group and by medical condition. An additional table found on the CDC website at http://www.cdc.gov/mmwr/preview/mmwrhtml/su6201a3.htm lists the current recommended schedule for adult immunizations. Some vaccines are combination vaccines—mixtures of antigens from several pathogens as a way to decrease the number of injections given at one time (see the last listing on table 15.5). It is still common for several vaccines to be given simultaneously, as occurs in military recruits who receive as many as 15 injections within a few minutes

...mmended Immunization Schedule for Persons Aged 0 through 18 Years–2016*

...unization schedule for persons aged 0 through 18 years – **United States, 2016.**
...IND OR START LATE, SEE THE CATCH-UP SCHEDULE [FIGURE 2]).
...ed with the footnotes that follow. For those who fall behind or start late, provide catch-up vaccination at the earliest opportunity as indicated by the green bars in Figure 1.
...ween doses, see the catch-up schedule (Figure 2). School entry and adolescent vaccine age groups are shaded.

	Birth	1 mo	2 mos	4 mos	6 mos	9 mos	12 mos	15 mos	18 mos	19–23 mos	2-3 yrs	4-6 yrs	7-10 yrs	11-12 yrs	13–15 yrs	16–18 yrs
	1st dose	←-- 2nd dose --→			←------------------- 3rd dose -------------------→											
Rotavirus[2] (RV) RV1 (2-dose series); RV5 (3-dose series)			1st dose	2nd dose	See footnote 2											
Diphtheria, tetanus, & acellular pertussis[3] (DTaP: <7 yrs)			1st dose	2nd dose	3rd dose		←----- 4th dose -----→					5th dose				
Haemophilus influenzae type b[4] (Hib)			1st dose	2nd dose	See footnote 4		←-- 3rd or 4th dose, See footnote 4									
Pneumococcal conjugate[5] (PCV13)			1st dose	2nd dose	3rd dose		←----- 4th dose -----→									
Inactivated poliovirus[6] (IPV: <18 yrs)			1st dose	2nd dose	←------------------- 3rd dose -------------------→							4th dose				
Influenza[7] (IIV; LAIV)					Annual vaccination (IIV only) 1 or 2 doses					Annual vaccination (LAIV or IIV) 1 or 2 doses		Annual vaccination (LAIV or IIV) 1 dose only				
Measles, mumps, rubella[8] (MMR)					See footnote 8		←----- 1st dose -----→					2nd dose				
Varicella[9] (VAR)							←----- 1st dose -----→					2nd dose				
Hepatitis A[10] (HepA)							←----- 2-dose series, See footnote 10 -----→									
Meningococcal[11] (Hib-MenCY ≥6 weeks; MenACWY-D ≥9 mos; MenACWY-CRM ≥2 mos)					See footnote 11									1st dose		Booster
Tetanus, diphtheria, & acellular pertussis[12] (Tdap: ≥7 yrs)														(Tdap)		
Human papillomavirus[13] (2vHPV: females only; 4vHPV, 9vHPV: males and females)														(3-dose series)		
Meningococcal B[11]															See footnote 11	
Pneumococcal polysaccharide[5] (PPSV23)														See footnote 5		

Range of recommended ages for all children	Range of recommended ages for catch-up immunization	Range of recommended ages for certain high-risk groups	Range of recommended ages for non-high-risk groups that may receive vaccine, subject to individual clinical decision making	No recommendation

This schedule includes recommendations in effect as of January 1, 2016. Any dose not administered at the recommended age should be administered at a subsequent visit, when indicated and feasible. The use of a combination vaccine generally is preferred over separate injections of its equivalent component vaccines. Vaccination providers should consult the relevant Advisory Committee on Immunization Practices (ACIP) statement for detailed recommendations, available online at http://www.cdc.gov/vaccines/hcp/acip-recs/index.html. Clinically significant adverse events that follow vaccination should be reported to the Vaccine Adverse Event Reporting System (VAERS) online (http://www.vaers.hhs.gov) or by telephone (800-822-7967). Suspected cases of vaccine-preventable diseases should be reported to the state or local health department. Additional information, including precautions and contraindications for vaccination, is available from CDC online (http://www.cdc.gov/vaccines/recs/vac-admin/contraindications.htm) or by telephone (800-CDC-INFO [800-232-4636]).
This schedule is approved by the Advisory Committee on Immunization Practices (http://www.cdc.gov/vaccines/acip), the American Academy of Pediatrics (http://www.aap.org), the American Academy of Family Physicians (http://www.aafp.org), and the American College of Obstetricians and Gynecologists (http://www.acog.org).
NOTE: The above recommendations must be read along with the footnotes of this schedule.

Source of Data: Centers for Disease Control and Prevention

and children who receive boosters for DTaP and polio at the same time they receive the MMR vaccine. Experts doubt that immune interference (inhibition of one immune response by another) is a significant problem in these instances, and the mixed vaccines are carefully balanced to prevent this eventuality. The main problem with simultaneous administration is that side effects can be amplified.

Check Your Progress SECTION 15.6

31. Describe the preparation of killed vaccines; live, attenuated vaccines; subunit vaccines; recombinant vaccines; and DNA vaccines. Explain the advantages and disadvantages of
32. Explain the purposes of boosters, adjuvants, Trojan horse vaccines, and toxoids.
33. Describe the concept of herd immunity, and explain how vaccination contributes to its development in a community.
34. List some possible adverse side effects of vaccination.

CASE STUDY Part 2

Case #1. Despite intensive medical intervention on behalf of the Canadian boy, his condition deteriorated and he died about 6 weeks after the initial bat bite. By this time the laboratory had isolated rabies virus from the skin on the back of his neck and from his saliva and tears. Genetic testing of the virus revealed that it was a variant carried by silver-haired bats. Due to the boy's contact with his family and medical personnel, it was important that they also be given a course of PEP to prevent them from contracting the disease. This meant that around 60 people were injected with HRIG and rabies vaccine. Fortunately, none of his contacts became infected.

Case #2. The Brazilian youth was one of those rare cases in which medical intervention was successful. When he first came out of the coma, he could open his eyes and

respond, and he continued to improve from that point on. By November he was well enough to return home. He was the first recorded case in Brazil to survive rabies. Tests done from skin samples detected rabies virus RNA, which was found to be a variant of the virus carried by vampire bats. One finding that may have made a difference in his case was the early presence of neutralizing antibodies that continued to increase during his hospital stay.

Rabies is a type of encephalitis in which the virus migrates by nerve cell processes from the portal of entry, enters the brain, multiplies, and causes severe inflammation and damage to brain cells. It has the distinction of being the most deadly of all infectious diseases, being considered 100% fatal once symptoms appear. Worldwide, only 10 people have ever been known to survive the disease, and most of these survivors received some treatment, as in the cases presented

here. The disease is rather rare in humans but not in certain wild mammals. About two to three human cases are reported per year in the United States, but 3,000 to 4,000 animal cases are reported. The most prevalent animal carriers are raccoons, skunks, coyotes, foxes, and bats, but the virus can infect all mammals. Infection is most likely to occur following an animal bite or scratch, although exposure to saliva and other infectious fluids has been implicated.

■ *Provide some likely reasons the first boy died and the second did not.*

■ *Explain the strategy behind giving several injections of rabies vaccine.*

To conclude this Case Study, go to Connect.

To read more about rabies immunization, go to http://www.cdc.gov/rabies/exposure/index.html.

Chapter Summary with Key Terms

15.1 Specific Immunities: The Adaptive Line of Defense
Development of Lymphocyte Specificity/Receptors Acquired immunity involves the reactions of B and T lymphocytes to foreign molecules, or **antigens.** Before they can react, each lymphocyte must undergo differentiation into its final functional type by developing protein receptors for antigen, the specificity of which is randomly generated and unique for each lymphocyte.
 A. Development of the Immune Response System
 1. Genetic recombination and mutation during embryonic and fetal development produce billions of different lymphocytes, each bearing a different receptor. The process by which diversity is generated is **clonal selection.**
 2. **Immune tolerance** or tolerance to self occurs during this time.
 3. The receptors on B cells are **immunoglobulin (Ig)** molecules, and receptors on T cells are smaller unrelated glycoprotein molecules called **CD** markers.
 4. T-cell receptors bind antigenic determinants on **MHC** molecules, which in humans are called **human leukocyte antigens (HLA).**

15.2 The Nature of Antigens and Antigenicity
An antigen (Ag) is any substance that stimulates an immune response.
 A. Requirements for **antigenicity** include foreignness (recognition as nonself), large size, and complexity of cell or molecule.
 B. Foreign molecules less than 1,000 MW (**haptens**) are not antigenic unless attached to a larger carrier molecule.
 C. The epitope is the small molecular group of the foreign substance that is recognized by lymphocytes. Cells, viruses, and large molecules can have numerous antigenic determinants.
 D. Other antigens are mosaic alloantigens, superantigens, allergens, and autoantigens.

15.3 Immune Reactions to Antigens and the Activities of T Cells
 A. T-cell-dependent antigens must be processed by large phagocytes such as dendritic cells or macrophages called **antigen-presenting cells (APCs).**

 B. The presentation of a single antigen involves a direct collaboration between an APC, a T helper (T_H), and an antigen-specific B or T cell.
 C. Cytokines involved are **interleukin-1** from the APC, which activates the T_H cells, and **interleukin-2** produced by the T_H cell, which activates B and other T cells.
 D. **T helper cells (T_H or CD4)** are the dominant T cells.
 1. They regulate immune reactions to antigens and require interaction with MHC-II receptors on host cells.
 2. They activate macrophages, B cells, and other T cells by means of secreted interleukins.
 a. T_H1 cells function as T-cell activators and stimulate delayed hypersensitivity.
 b. T_H2 cells modulate B-cell functions and may suppress certain undesirable T cells' actions.
 E. **Cytotoxic T cells (T_C or CD8)** are responsive to MHC-I receptors on host cells and are sometimes called killer cells.
 1. They recognize and react to foreign cells, cancer cells, and graft tissues and secrete chemicals that directly damage and kill these cells.
 F. **Natural killer cells (NK)** are T-cell relatives responsible for early killing of infected cells and cancer cells in response to various cytokines.

15.4 Immune Activities of B Cells
 1. Once B cells process the antigen, interact with T_H cells, and are stimulated by B-cell growth and differentiation factors, they enter the cell cycle in preparation for mitosis and clonal expansion.
 2. Divisions give rise to **plasma cells** that secrete antibodies and **memory cells** that can react to that same antigen later.
 a. *Nature of antibodies* (immunoglobulins): A single immunoglobulin molecule (monomer) is a large Y-shaped protein molecule consisting of four polypeptide chains. It contains two identical regions with ends that form the **antigen binding fragments (Fabs)** that bind with a unique specificity to an antigen. The **crystallizable fragment (Fc)** determines the location and the function of the antibody molecule.

b. *Antigen-antibody (Ag-Ab) reactions* include **opsonization, neutralization, agglutination,** and complement fixation.

c. The five **antibody** or **immunoglobulin classes,** which differ in size and function, are **IgG, IgA** (secretory Ab), **IgM, IgD,** and **IgE.**

3. *Antibodies in serum* (**antiserum**)

a. The first introduction of an Ag to the immune system produces a **primary response,** with a gradual increase in Ab titer.

b. The second contact with the same Ag produces a **secondary,** or **anamnestic, response** due to memory cells produced during initial response.

c. **Monoclonal antibodies** (MABs) are pure forms of immunoglobulins produced in hybridoma cells. They may be tailored to high specificities and have numerous applications in medicine and research.

15.5 A Classification Scheme for Specific, Acquired Immunities
Immunities acquired through B and T lymphocytes can be classified by a simple system.

A. **Active immunity** results when a person is challenged with antigen that stimulates production of antibodies. It creates memory, takes time, and is lasting.

B. In **passive immunity,** preformed antibodies are donated to an individual. It does not create memory, acts immediately, and is short term.

C. **Natural immunity** is acquired as part of normal life experiences.

D. **Artificial immunity** is acquired through medical procedures such as immunization.

15.6 Immunization: Providing Immune Protection Through Therapy

A. Passive **immunotherapy** includes administering immune serum globulin and specific immune globulins pooled from donated serum to prevent infection and disease in those at risk; antisera and antitoxins from animals are occasionally used.

B. Active immunization is synonymous with **vaccination** and provides an antigenic stimulus that does not cause disease but can produce long-lasting, protective immunity. **Vaccines** are made with:

1. **Killed** whole cells or **inactivated viruses** that do not reproduce but are antigenic

2. Live, **attenuated** cells or viruses that are able to reproduce but have lost virulence

3. **Acellular** or **subunit** components of microbes such as surface antigens or neutralized toxins (**toxoids**)

4. Genetic engineering techniques, including cloning of antigens, recombinant attenuated microbes, and DNA-based vaccines

5. Boosters (additional doses) are often required.

C. Vaccination increases **herd immunity,** protection provided by mass immunity in a population.

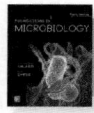

Level I. Knowledge and Comprehension

These questions require a working knowledge of the concepts in the chapter and the ability to recall and understand the information you have studied.

? Multiple-Choice Questions

Select the correct answer from the answers provided. For questions with blanks, choose the combination of answers that most accurately completes the statement.

1. Which of these characteristics is not a major factor in adaptive or acquired immunity?
 a. specificity
 b. chemotaxis
 c. recognition
 d. memory

2. The primary B-cell receptor is
 a. IgD b. IgA c. IgE d. IgG

3. In humans, B cells mature in the _____ and T cells mature in the _____.
 a. GALT, liver
 b. bursa, thymus
 c. bone marrow, thymus
 d. lymph nodes, spleen

4. Small, simple molecules are _____ antigens.
 a. poor b. never c. effective d. mosaic

5. Which type of cell actually secretes antibodies?
 a. T cell b. macrophage c. plasma cell d. monocyte

6. CD4 cells are _____ cells and CD8 cells are _____ cells.
 a. killer, suppressor
 b. helper, cytotoxic
 c. cytotoxic, helper
 d. B, T

7. Helper T cells receive antigen from _____, and cytotoxic T cells receive antigen from _____.
 a. macrophages, B cells
 b. class II MHC, class I MHC
 c. viruses, bacteria
 d. class I MHC, class II MHC

8. The cross-linkage of antigenic cells by antibodies is known as
 a. opsonization
 b. a cross-reaction
 c. agglutination
 d. complement fixation

9. The greatest concentration of antibodies is found in the _____ fraction of the serum.
 a. gamma globulin
 b. albumin
 c. beta globulin
 d. alpha globulin

10. T _____ cells assist in the functions of certain B cells and other T cells.
 a. sensitized
 b. cytotoxic
 c. helper
 d. natural killer

11. Cytotoxic T cells are important in controlling
 a. virus infections
 b. allergy
 c. autoimmunity
 d. all of these

12. Vaccination is synonymous with _____ immunity.
 a. natural active
 b. artificial passive
 c. artificial active
 d. natural passive

13. Which of the following can serve as antigen-presenting cells (APCs)?
 a. T cells
 b. B cells
 c. macrophages
 d. dendritic cells
 e. b, c, and d

14. A living microbe with reduced virulence that is used for vaccination is considered
 a. a toxoid
 b. attenuated
 c. denatured
 d. an adjuvant

15. A vaccine that contains parts of viruses is called
 a. acellular
 b. recombinant
 c. subunit
 d. attenuated

16. Widespread immunity that protects the population from the spread of disease is called
 a. seropositivity
 b. cross-reactivity
 c. epidemic prophylaxis
 d. herd immunity

17. DNA vaccines contain _____ DNA that stimulates cells to make _____ antigens.
 a. human, RNA
 b. microbial, protein
 c. human, protein
 d. microbial, polysaccharide

18. What is the purpose of an adjuvant?
 a. to kill the microbe
 b. to stop allergic reactions
 c. to improve the contact between the antigen and lymphocytes
 d. to make the antigen more soluble in the tissues

19. **Multiple Matching.** Place all possible matches in the spaces at the left of each Ig.
 _____ IgG
 _____ IgA
 _____ IgD
 _____ IgE
 _____ IgM

 a. found in mucous secretions
 b. a monomer
 c. a dimer
 d. has greatest number of Fabs
 e. first antibody made during the primary response
 f. first antibody made during the secondary response
 g. crosses the placenta
 h. fixes complement
 i. involved in allergic reactions
 j. primarily a surface receptor on B cells

20. **Multiple Matching.** An outline summarizing the general host defenses covered in chapters 14 and 15 was presented in figure 14.21. You may want to use this resource to review major aspects of immunity and to guide you in answering this question. In the blanks on the left, place the letters of all of the host defenses and immune responses in the right column that can fit the description.
 ____ vaccination for tetanus
 ____ lysozyme in tears
 ____ immunization with horse serum
 ____ in utero transfer of antibodies
 ____ booster injection for diphtheria
 ____ colostrum
 ____ interferon
 ____ action of neutrophils
 ____ injection of gamma globulin
 ____ recovery from a case of mumps
 ____ edema
 ____ humans having protection from canine distemper virus
 ____ stomach acid
 ____ cilia in trachea

 a. active
 b. passive
 c. natural
 d. artificial
 e. acquired
 f. innate, inborn
 g. chemical barrier
 h. mechanical barrier
 i. genetic barrier
 j. specific
 k. nonspecific
 l. inflammatory response
 m. second line of defense

Case Study Review

1. What does the rabies vaccine contain that rabies immunoglobulin does *not?*
 a. antibodies
 b. inactive virus
 c. live virus
 d. animal serum

2. What types of cells could be involved in specific immune reactions against the rabies virus?
 a. T cells
 b. B cells
 c. macrophages
 d. dendritic cells
 e. a, b, and d
 f. all of these

3. Describe the specific components of acquired immunity that would be necessary to control rabies virus.

Writing Challenge

For each question, compose a one- or two-paragraph answer that includes the factual information needed to completely address the question. Check Your Progress questions can also be used for writing-challenge exercises.

1. Using words and arrows, complete a flow outline of an immune response, beginning with entrance of antigen; include processing, cell interaction, involvement of cytokines, and the end results for B and T cells.

2. a. Explain why most immune reactions result in a polyclonal collection of antibodies.
 b. How do monoclonal antibodies differ from this?
 c. Describe several applications of monoclonals in medicine.

3. Give an immunological explanation as to why people can get influenza every year even though they have been vaccinated.

4. How can the anamnestic response be explained in immunologic terms?

5. Differentiate between humoral and cell-mediated immunity.

6. Combine information on the functions of different classes of Ig to explain the exact mechanisms of natural passive immunity (both transplacental and colostrum-induced).

Concept Mapping

On Connect you can find an Introduction to Concept Mapping that provides guidance for working with concept maps and a sample of a concept map for this chapter.

Level II. Application, Analysis, Evaluation, and Synthesis

These problems go beyond just restating facts and require higher levels of understanding and an ability to interpret, problem solve, transfer knowledge to new situations, create models, and predict outcomes.

Critical Thinking

Critical thinking is the ability to reason and solve problems using facts and concepts. These questions can be approached from a number of angles, and in most cases, they do not have a single correct answer.

1. Cells contain built-in suicide genes to self-destruct by apoptosis under certain conditions. Can you explain why development of the immune system might depend in part on this sort of adaptation?

2. a. Describe what could be summarized as the three Rs of immunity: recognize, react, remember.

 b. Why would it be necessary for the T cells to bind both antigen and self (MHC) receptors?

3. Double-stranded DNA is a large, complex molecule, but it is not generally immunogenic unless it is associated with proteins or carbohydrates. Can you think why this might be so? (Hint: How universal is DNA?)

4. Explain how it is possible for people to give a false-positive reaction in blood tests for syphilis, AIDS, and infectious mononucleosis.

5. Describe the relationship between an antitoxin, a toxin, and a toxoid.

6. It is said that a vaccine does not prevent infection; rather, it primes the immune system to undergo an immediate response to prevent an infection from spreading. Explain what is meant by this statement, and outline what is happening at the cellular/molecular level from the time of vaccination until subsequent contact with the infectious agent actually occurs.

7. A study by the CDC reported that less than 80% of children in some areas of the United States are being adequately vaccinated. Several million children are at high risk for infections.

 a. Name some factors that account for this trend.

 b. Explain how the reduction in vaccination impacts herd immunity.

8. a. From tables 15.6 and http://www.cdc.gov/mmwr/preview/mmwrhtml/su6201a3.htm, determine the vaccines you have been given and those for which you will require periodic boosters.

 b. Suggest vaccines you may need in the future.

9. When traders and missionaries first went to the Hawaiian Islands, the natives there experienced severe disease and high mortality rates from smallpox, measles, and certain STDs. Explain what immunologic factors are involved in the devastating effects of new diseases in previously unexposed populations.

Visual Challenge

1. Using figure 15.13, label the lines and arrows, and explain what is happening at numbers 1, 2, and 3.

2. Examine figure 6.6c and determine which components would act as epitopes or antigens.

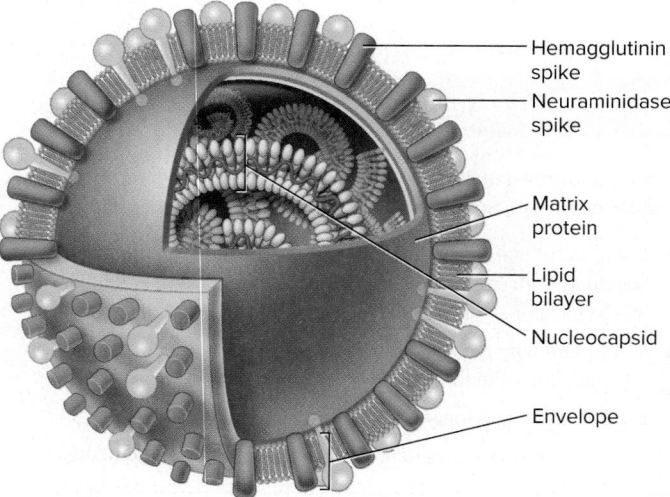

Hemagglutinin spike

Neuraminidase spike

Matrix protein

Lipid bilayer

Nucleocapsid

Envelope

Antibodies (IgM) agglutinate RBCs due to incorrect blood transfusion.

Packs of washed red blood cells ready for transfusions.
Source: U.S. Department of Health & Human Services/National Institutes of Health

CASE STUDY Part 1 A Deadly Transfusion Reaction

In late 2007 a 67-year-old male patient at a medical center in Florida underwent a minor procedure that required a **blood transfusion.** In anticipation of this surgery, the patient's blood had been **typed** by a lab technician a few days earlier. The man was transfused with one pint of **type A$^+$** blood during the surgery. Within a few hours he had a sudden onset of fever, chills, and difficulty in breathing. When a vital sign check noted **tachycardia,*** **hypotension,**** and **urine** with a **pinkish tinge,** an emergency team was called in to evaluate his condition and administer treatment. A blood sample taken at this time showed unusual levels of **hemoglobin** in the plasma and **clumping** of **red blood cells** (RBCs). Despite being surrounded by state-of the-art medical technology, nothing could be done to save his life. The patient died just a few hours after his operation from heart failure, shock, and **disseminated intravascular coagulation** (DIC).*** Attending physicians concluded that he had experienced an **acute hemolytic transfusion reaction** (AHTR).

The usual cause of a transfusion reaction is the presence of RBCs of the wrong blood type in the patient's body. A pathological reaction occurs when these **donated RBCs** carry **antigens** on their surface that react with IgM and IgG antibodies that already exist in the plasma of the recipient. When the antibodies combine with the antigens on the RBCs, the complement system in the blood is activated, which leads to the formation of **membrane attack complexes** and **hemolysis** of the donated red blood cells. Because such a reaction can occur even if only 5 to 20 ml of blood is transfused, the 500-ml transfusion was more than enough to be fatal.

What seemed most puzzling to investigators in this case was that the records showed that the **donated blood** (type A$^+$) had been matched to the patient's by **cross-matching** just minutes before the transfusion. This procedure involves mixing a drop of the patient's blood with a drop of the blood to be transfused and observing the mixture macroscopically and microscopically for evidence of **agglutination.** Technicians had reported that the cross-match was successful and showed no evidence of a reaction.

> "Despite being surrounded by state-of the-art medical technology, nothing could be done to save his life."

- *What is responsible for the ABO and Rh blood types, and why does transfusion of an incompatible blood type lead to such severe complications?*

- *What causes the abnormal signs indicative of a transfusion reaction?*

To continue the Case Study, go to Case Study Part 2 at the end of the chapter.

* Tachycardia is an increased heart rate.

** Hypotension is decreased blood pressure.

*** DIC is a clumping of RBCs in blood vessels throughout the body.

SCOPING OUT THE CHAPTER

Up to now, we have focused primarily on the ways our powerful and intricate immune systems defend us against infection, but by their very nature, these systems also carry tremendous potential to cause injury and disease. In most instances an immune dysfunction presents itself in commonplace but miserable symptoms of hay fever or skin rashes. But abnormal immune function is also the basis of debilitating or life-threatening diseases such as asthma, anaphylactic shock, rheumatoid arthritis, and cancer. In this chapter we get to examine the flip side of this double-edged "sword" of immunity.

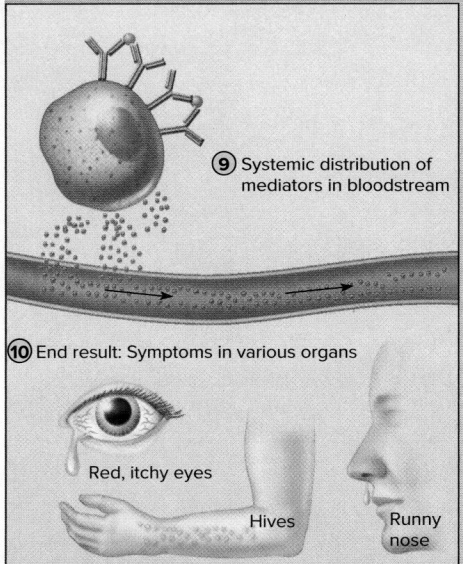

⑨ Systemic distribution of mediators in bloodstream

⑩ End result: Symptoms in various organs

Red, itchy eyes

Hives

Runny nose

B cells and antibodies are involved in immediate allergies such as hay fever and asthma.

Allergies: The Immune System Running Rampant Allergies arise when the immune system has been primed to overreact to commonplace substances such as pollen grains, foods, cat hair, and poisonous plants. These reactions use the same systems in place for protective immunity—the B and T lymphocytes. But the end results damage the respiratory system, skin, intestinal tract, and other organs. Some of these reactions are rapid and others take time, but in both cases, they cause medical problems, often for life.

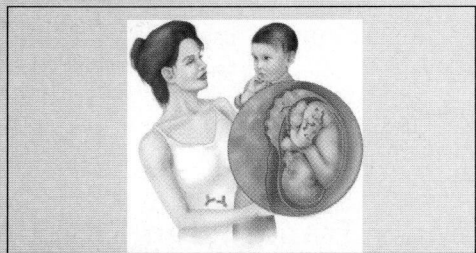

Blood Types and Complications Humans inherit markers on their red blood cells that determine their blood types. Because one person's blood markers can be foreign to another person, it is necessary to check for incompatibilities during transfusions and between pregnant women and their fetuses.

T cells are responsible for delayed allergies such as poison ivy sensitivity.

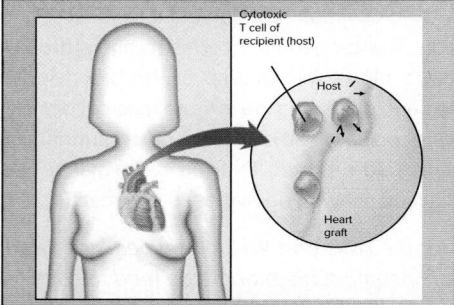

Cytotoxic T cell of recipient (host)

Host

Heart graft

Immune Interference in Transplants Immune-based diseases may also arise from medical interventions such as transplants and grafts of organs or tissues that are often incompatible with the recipient. T cells inhabiting the person or the graft can become sensitized and cause life-threatening rejection of the tissue or organ.

Autoimmunity: A Case of Mistaken Identity A devastating and unexpected immune dysfunction happens when a person's tissues and organs are targeted by his own immune system. Diseases such as rheumatoid arthritis, lupus, and multiple sclerosis are the consequences of "self attacking self."

Weakened Immune Function The immune system is the silent partner in human health, but when it is missing or compromised, the body is subjected to out-of-control infections. Immune deficiencies are either inborn and genetic, caused by lack of development in critical blood cells types, or acquired after birth through infections and cancers.

16.1 The Immune Response: A Two-Sided Coin

 Expected Learning Outcomes

1. Summarize the main categories of immunopathology and their medical consequences.
2. Discuss the factors involved in allergies and hypersensitivities.

Our previous discussions of the immune response centered primarily around its numerous beneficial effects. The precisely coordinated system that recognizes and destroys invading microbes is clearly protective, but it also presents another side—a side that promotes rather than prevents disease. In this chapter, we survey **immunopathology,** the study of disease states associated with overreactivity or underreactivity of the immune response **(figure 16.1).** In the cases of **allergy** and **autoimmunity,** the tissues are innocent bystanders to immune attacks that do not distinguish between one's own cells and foreign antigens. In **grafts** and **transfusions,** a recipient reacts to the foreign tissues and cells of another individual. In **immunodeficiency** diseases, immune function is incompletely developed, suppressed, or destroyed. **Cancer** falls into a special category, because it is both a cause and an effect of immune dysfunction. As we shall see, one fascinating by-product of studies of immune disorders has been our increased understanding of the basic workings of the immune system.

Overreactions to Antigens: Allergy/Hypersensitivity

The term **allergy** means a condition of altered reactivity or exaggerated immune responsiveness that is manifested by inflammatory symptoms. Although it is sometimes used interchangeably with **hypersensitivity,** some experts use the term *allergy* in reference to immediate reactions (hay fever, asthma) and use *hypersensitivity* in reference to other forms of overreactions to antigens (contact dermatitis). Allergic and hypersensitive individuals are acutely sensitive to repeated contact with antigens called **allergens** that do not noticeably affect other "normal" individuals. Although the general effects of these immune pathologies are usually detrimental, we must emphasize that they involve the very same types of immune reactions as protective immunities. These include humoral and cell-mediated actions, the inflammatory response, phagocytosis, and complement. Such an association means that all humans have the potential to develop allergy or hypersensitivity under particular circumstances.

Originally allergies and hypersensitivities were defined as either immediate or delayed, depending upon the time lapse between contact with the allergen and onset of symptoms. Subsequently they were esedifferentiated as humoral versus cell-mediated. But as information accumulated on these types of dysfunctional immune responses, it became evident that, although useful, these schemes oversimplified what is really a very complex spectrum of reactions.

Figure 16.1 Overview of diseases of the immune system (immunopathologies). Just as the system of T cells and B cells provides necessary protection against infection and disease, the same system can cause serious and debilitating conditions by overreacting or underreacting to immune stimuli.

TABLE 16.1	Classification of Immunopathologies Caused by Overreactions to Antigens		
Type		**Systems and Mechanisms Involved**	**Examples**
I	Immediate allergies	IgE-mediated; involves mast cells, basophils, and allergic mediators	Anaphylaxis, allergies such as hay fever, asthma
II	Antibody-mediated incompatibilities	IgG, IgM antibodies act upon cells with complement and cause cell lysis; includes some autoimmune diseases	Blood group incompatibility, pernicious anemia; myasthenia gravis
III	Immune complex diseases	Antibody-mediated inflammation; circulating IgG complexes deposited in basement membranes of target organs; includes some autoimmune diseases	Systemic lupus erythematosus; rheumatoid arthritis; serum sickness; rheumatic fever
IV	T-cell-mediated hypersensitivities	Delayed hypersensitivity and cytotoxic reactions in tissues	Infection reactions; contact dermatitis; graft rejection; some types of autoimmunity

The most widely accepted classification, first introduced by immunologists P. Gell and R. Coombs, includes four major categories of immune system pathologies involving overreactions to antigens: type I ("common" allergy and anaphylaxis), type II (IgG- and IgM-mediated cell damage), type III (immune complex diseases), and type IV (delayed hypersensitivity) **(table 16.1).** In general, types I, II, and III involve a B-cell–immunoglobulin response, and type IV involves a T-cell response (figure 16.1). The antigens that elicit these reactions can be exogenous, originating from outside the body (microbes, pollen grains, and foreign cells and proteins), or endogenous, arising from self tissue (autoimmunities).

One of the reasons allergies and hypersensitivities are easily mistaken for infections is that both involve damage to the tissues and thus trigger the inflammatory response, as described in chapter 14. Many symptoms and signs of inflammation (redness, heat, skin eruptions, edema, and granuloma) are prominent features of allergies and are caused by some of the same chemical mediators.

 Check Your Progress SECTION 16.1

1. Explain what is meant by immunopathology and give some examples of this condition.
2. Define *allergy/hypersensitivity* and explain what accounts for the reactions that occur in these conditions.
3. What is involved in the four categories of B-cell and T-cell-mediated immunopathologies outlined by Gell and Combs?
4. What does it mean for a reaction to be immediate or delayed?

16.2 Allergic Reactions: Atopy and Anaphylaxis

 Expected Learning Outcomes

3. Describe general characteristics of allergic reactions.
4. Outline the major allergen categories, giving examples.
5. Explain the mechanism of immediate allergies, including the concepts of IgE, mast cells, sensitization, and provocation.

6. Summarize the physiological effects of allergies that cause major symptoms.
7. Define the types of atopic allergies and describe their major features.
8. Explain the process of anaphylaxis and its outcome.
9. Relate the primary methods of diagnosing, treating, and preventing allergies.

Type 1 allergies share a similar physiological mechanism, are immediate in onset, and are associated with exposure to specific antigens. However, it is convenient to recognize two levels of severity: **Atopy*** is any chronic allergy, such as hay fever or eczema, whose effects are localized to a region and usually not life-threatening; and **anaphylaxis*** is a systemic, sometimes fatal reaction that involves airway obstruction and circulatory collapse. In the following sections we consider the allergens, routes of inoculation, mechanisms of disease, and specific syndromes involved in type I allergies.

Modes of Contact with Allergens

Allergies exert profound medical and economic impact. Allergists (physicians who specialize in treating allergies) estimate that about 10% to 30% of the population is prone to atopic allergy. It is generally acknowledged that self-treatment with over-the-counter medicines accounts for significant underreporting of cases. The 40 million people afflicted with hay fever (15% to 20% of the population) together spend about half a billion dollars annually for medical treatment. The monetary loss due to employee debilitation and absenteeism is immeasurable. The majority of allergies are relatively mild, but certain forms, such as asthma and anaphylaxis, may require hospitalization and can cause death.

The predisposition for atopic allergies has a strong familial association. Be aware that what is hereditary here is a generalized *susceptibility,* not an allergy to a specific substance. For example, a parent who is allergic to ragweed pollen can have a child who is allergic to cat hair. The prospect of a child's developing an atopic allergy is at least 25% if one parent is atopic, increasing up to 50% if grandparents

* *atopy* (at′-oh-pee) Gr. *atop,* out of place.
* *anaphylaxis* (an″-uh-fih-lax′-us) Gr. *ana,* excessive, and *phylaxis,* protection.

CLINICAL CONNECTIONS

Are Some Allergies a Consequence of Clean Living?

Allergists have been observing an increased incidence of allergies such as asthma, allergic rhinitis, and other atopic allergies that is not tied to genetic factors or better diagnosis. Most of these increases have occurred in more affluent regions of the United States and Europe. Although this phenomenon cannot easily be explained, many medical experts think it is connected to something called the **hygiene or biodiversity hypothesis.** This explanation basically says that ultraclean lifestyles and living environments prevent the kind of contact with microbes that is required for the normal development of the immune system. The effects of modern housekeeping, lifestyles, and antimicrobial drugs, mean that children are not exposed as much to noninfectious environmental microbes and viral and parasitic infections. Some support for this exists in the fact that levels of allergies are lower in developing countries and rural areas, where children have greater exposure to soil, infectious agents, and household pets.

Although the factors involved appear to be complex, it is thought that contact with certain types of microbes initiates protective responses and shapes the future direction of immune system development. The immune systems of children who lack sufficient microbial contact become programmed to react to allergens such as household chemicals, animal and plant molecules, and even food, which sets the scene for long-term allergies. Because people are unlikely to return to less clean lifestyles, researchers are on the trail of finding out what kinds of stimuli can inoculate people against allergies. It may turn out that "a little dirt can't hurt" and may even be beneficial.

Explain how the hygiene hypothesis relates to the overuse of household disinfectants and antibiotics as discussed in chapters 11 and 12. *Answer available on Connect.*

	Common Allergens, Classified by Portal of Entry		
Inhalants	**Ingestants**	**Injectants**	**Contactants**
Pollen	Food (milk, peanuts,	Hymenopteran	Topical
Dust	wheat, shellfish,	venom (bee,	drugs
Mold	soybeans,	wasp)	Cosmetics
spores	nuts, eggs,	Insect bites	Heavy metals
Dander	fruits)	Drugs	Detergents
Animal hair	Food additives	Vaccines	Formalin
Insect parts	Oral drugs	Serum	Rubber
Formalin	(aspirin,	Enzymes	Solvents
Aerosols	penicillin)	Hormones	Dyes

(Table 16.2)

industrial and household products, cosmetics, food, and drugs are commonly of this type. **Table 16.2** lists a number of common allergenic substances and their portal of entry.

Allergens typically enter through the epithelia of the respiratory tract, gastrointestinal tract, and skin. The mucosal surfaces of the gut and respiratory system present a thin, moist surface that is normally quite penetrable. The dry, tough keratin coating of skin is less permeable, but access still occurs through tiny breaks, glands, and hair follicles. It is worth noting that the organ of allergic expression may or may not be the same as the portal of entry.

Airborne environmental allergens such as pollen, house dust, dander (shed skin scales), or fungal spores are termed *inhalants.* Each geographic region harbors a particular combination of airborne substances; this combination varies with the season and humidity **(figure 16.2a).** Pollen, the most common offender, is given off seasonally by the reproductive structures of pines and flowering plants (weeds, trees, and grasses). Unlike pollen, mold spores are released throughout the year and are especially profuse in moist areas of the home and garden. Airborne animal hair and dander (skin flakes), feathers, and the saliva of dogs and cats are common sources of allergens.

House dust is a complex mixture of organic debris, soil, human skin cells, and a wide array of bacteria and fungi. The component of house dust that appears to account for most dust allergies is not the debris or microbes but the decomposed skeletons of tiny mites that commonly live in this dust **(figure 16.2b).** Some people are allergic to their work, in the sense that they are exposed to allergens on the job. Examples include florists, woodworkers, farmers, drug processors, and plastics manufacturers whose work can aggravate inhalant and contact allergies.

Allergens that enter by mouth, called *ingestants,* often cause food allergies. *Injectant* allergies are an important adverse side effect of injected drugs or other substances used in diagnosing, treating, or preventing disease. A natural source of injectants is venom from stings by hymenopterans, a family of insects that includes honey-bees and wasps. *Contactants* are allergens that enter through the skin. Many contact allergies are of the type IV, delayed variety discussed in section 16.5.

Mechanisms of Allergy: Sensitization and Provocation

What causes some people to sneeze and wheeze every time they step out into the spring air, while others suffer no ill effects? We can answer this question by examining what occurs in the tissues of the allergic individual that does not occur in the normal person. What

or siblings are also afflicted. The actual basis for atopy appears to be the inheritance of genes that favor allergic antibody (IgE) production, increased reactivity of mast cells, and increased susceptibility of target tissue to allergic mediators. Allergic persons often exhibit a combination of syndromes, such as hay fever, eczema, and asthma.

Other factors that affect the onset of allergy are age, infection, and geographic locale. New allergies tend to crop up throughout an allergic person's life, especially as new exposures occur after moving or changing lifestyle. In some persons, atopic allergies last for a lifetime; others "outgrow" them, and still others suddenly develop them later in life. Some features of allergic pathology are not yet completely explained.

The Nature of Allergens and Their Portals of Entry

As with other antigens, allergens have certain immunogenic characteristics. Not unexpectedly, proteins are more allergenic than carbohydrates, fats, or nucleic acids. Some allergens are haptens, nonproteinaceous substances with a molecular weight of less than 1,000 that can form complexes with carrier molecules in the body (shown in figure 15.9). Organic and inorganic chemicals found in

National Allergy Bureau
Pollen and Mold Report*

Location: Flower Mound, TX Date: Sept 23, 2015
Counting Station: Allergy Medical Group of the North Area

Trees	Moderate concentration	Total count: 15–89 / m³
Weeds	High concentration	Total count: 499 / m³
Grass	Moderate concentration	Total count: 15–19 / m³
Mold	Low concentration	Total count: <6499 / m³

(a) *Count standards vary with the type of pollen and mold spore.

(b)

(c)

Figure 16.2 **Monitoring airborne allergens. (a)** The air in heavily vegetated places with a mild climate is especially laden with allergens such as pollen and mold spores. These counts vary seasonally. **(b)** An SEM of the dust mite *Dermatophagoides* (2,000x), which feeds primarily on human skin cells in house dust, and is found in abundance in bedding and carpets. Airborne mite feces and particles from their exoskeletons are an important source of allergies. **(c)** Scanning electron micrograph of a pollen grain from a morning glory flower (5,000x). Millions of these are released from a single flower. (b): Courtesy Sandia National Laboratories; (c): *Source:* Louisa Howard/Dartmouth College

is not as easy to answer is the question of why allergy is so common, affecting over 40% of the world's population. What useful function, if any, can it possibly serve? See 16.1 Making Connections for some possible explanations.

In general, allergies develop in stages (**figure 16.3**). The initial encounter with an allergen provides a **sensitizing dose** that primes the immune system for a subsequent encounter with that allergen but generally elicits no signs or symptoms. The memory cells and immunoglobulin produced during this encounter are then ready to react with a subsequent **provocative dose** of the same allergen. It is this dose that precipitates the first signs and symptoms that an allergy exists. Despite numerous anecdotal reports of people showing an allergy upon first contact with an allergen, it is generally believed that these individuals unknowingly had contact at some previous time. Fetal exposure to allergens from the mother's bloodstream is one possibility, and foods can be a prime source of "hidden" allergens such as penicillin.

The Physiology of IgE-Mediated Allergies

During primary contact and sensitization, the allergen penetrates the portal of entry (**figure 16.3, steps 1–6**). When large particles such as pollen grains, hair, and spores encounter a moist membrane, they are first captured by dendritic cells, which process the allergen and present it to T helper cells in the lymph nodes. Here, the T cells interact with clones of B cells, which they activate to respond to the specific allergen. This causes proliferation of plasma cells, following a sequence similar to the typical immune response to an antigen. These plasma cells produce **immunoglobulin E** (IgE), the antibody of allergy. IgE is different from other immunoglobulins in having an Fc region with great affinity for mast cells and basophils. The binding of IgE to these

cells in the tissues sets the scene for the reactions that occur upon repeated exposure to the same allergen (**figure 16.3, steps 7–10**).

The Role of Mast Cells and Basophils

The most important characteristics of mast cells and basophils relating to their roles in allergy are these:

1. Their ubiquitous location in tissues. Mast cells are located in the connective tissue of virtually all organs, but particularly high concentrations exist in the lungs, skin, gastrointestinal tract, and genitourinary tract. Basophils circulate in the blood but migrate readily into tissues.

2. Their capacity to bind IgE during sensitization (figure 16.3). Each cell carries 30,000 to 100,000 cell receptors that bind 10,000 to 40,000 IgE antibodies.

SEM of mast cell with ruffled membrane and receptors for IgE
Source: Oliver, Janet M., and Wilson, Bridget. The Cell: An Image Library, CIL:222. Available at http://www.cellimagelibrary.org

3. Their cytoplasmic granules (secretory vesicles), which contain physiologically active cytokines (histamine, serotonin—introduced in chapter 14).

4. Their tendency to **degranulate** (figures 16.3 and 16.4) or release the contents of the granules into the tissues when triggered by a specific allergen through the IgE bound to them.

Let us now see what occurs when sensitized cells are challenged with allergen a second time.

(a) **Sensitization/IgE Production**

① Allergens enter through mucous membrane and are picked up by a dendritic cell.

② Dendritic cell processes allergen and presents it to a T helper cell in lymph node.

③ B cell activated by T$_H$ cell reacts to allergen.

④ B cell divides to form numerous plasma cells.

⑤ Synthesis of IgE specific to allergen

Fc fragments

IgE-primed mast cells remain in tissues for extended periods.

⑥ IgEs bind with their Fc fragments to mast cell receptors.

Granules with inflammatory mediators

Mast cell in tissue

(b) **Subsequent Exposure to Allergen and Provocation**

⑦ Allergen is encountered again and goes directly to primed mast cells.

⑧ Allergen attaches to IgE on mast cells and triggers degranulation and release of allergic mediators.

stimulates

⑨ Systemic distribution of mediators in bloodstream

⑩ End result: Symptoms in various organs

Red, itchy eyes

Hives

Runny nose

Process Figure 16.3 A schematic view of cellular reactions during the allergic response. (a) Sensitization (initial contact with sensitizing dose), includes stages 1–6. **(b)** Provocation (later contacts with provocative dose), stages 7–10.

The Provocativel Contact with Allergen

After sensitization, the IgE-primed mast cells can remain in the tissues for years. Even after long periods without contact, a person can retain the capacity to react immediately upon reexposure. The next time allergen molecules contact these sensitized cells, they bind across adjacent receptors and stimulate degranulation. As chemical mediators are released, they diffuse into the tissues and bloodstream. Cytokines give rise to numerous local and systemic reactions, many of which appear quite rapidly **(figure 16.3, step 10)**. Thus, the symptoms of allergy are caused, not by the direct action of allergen on tissues, but by the physiological effects of mast cell mediators on target organs.

Cytokines, Target Organs, and Allergic Symptoms

Numerous substances involved in mediating allergy (and inflammation) have been identified. The principal chemical mediators produced by mast cells and basophils are histamine, serotonin, leukotriene, platelet-activating factor, prostaglandins, and bradykinin

(figure 16.4). These chemicals, acting alone or in combination, account for the tremendous scope of allergic symptoms. Targets of these mediators include the skin, upper respiratory tract, gastrointestinal tract, and conjunctiva. The general responses of these organs include rashes, itching, redness, rhinitis, sneezing, diarrhea, and shedding of tears. Systemic targets include smooth muscle, mucous glands, and nervous tissue. Because smooth muscle is responsible for regulating the size of blood vessels and respiratory passageways, changes in its activity can profoundly alter blood flow, blood pressure, and respiration. Pain, anxiety, agitation, and lethargy are also attributable to the effects of mediators on the nervous system.

Histamine* is the most profuse and fastest-acting allergic mediator. It is a potent stimulator of smooth muscle, glands, and eosinophils. Histamine's actions on smooth muscle vary with location. It *constricts* the smooth muscle layers of the bronchioles and intestine, thereby causing labored breathing and increased intestinal motility. In contrast, histamine *relaxes* vascular smooth muscle and

* *histamine* (his′-tah-meen) Gr. *histio,* tissue, and amine.

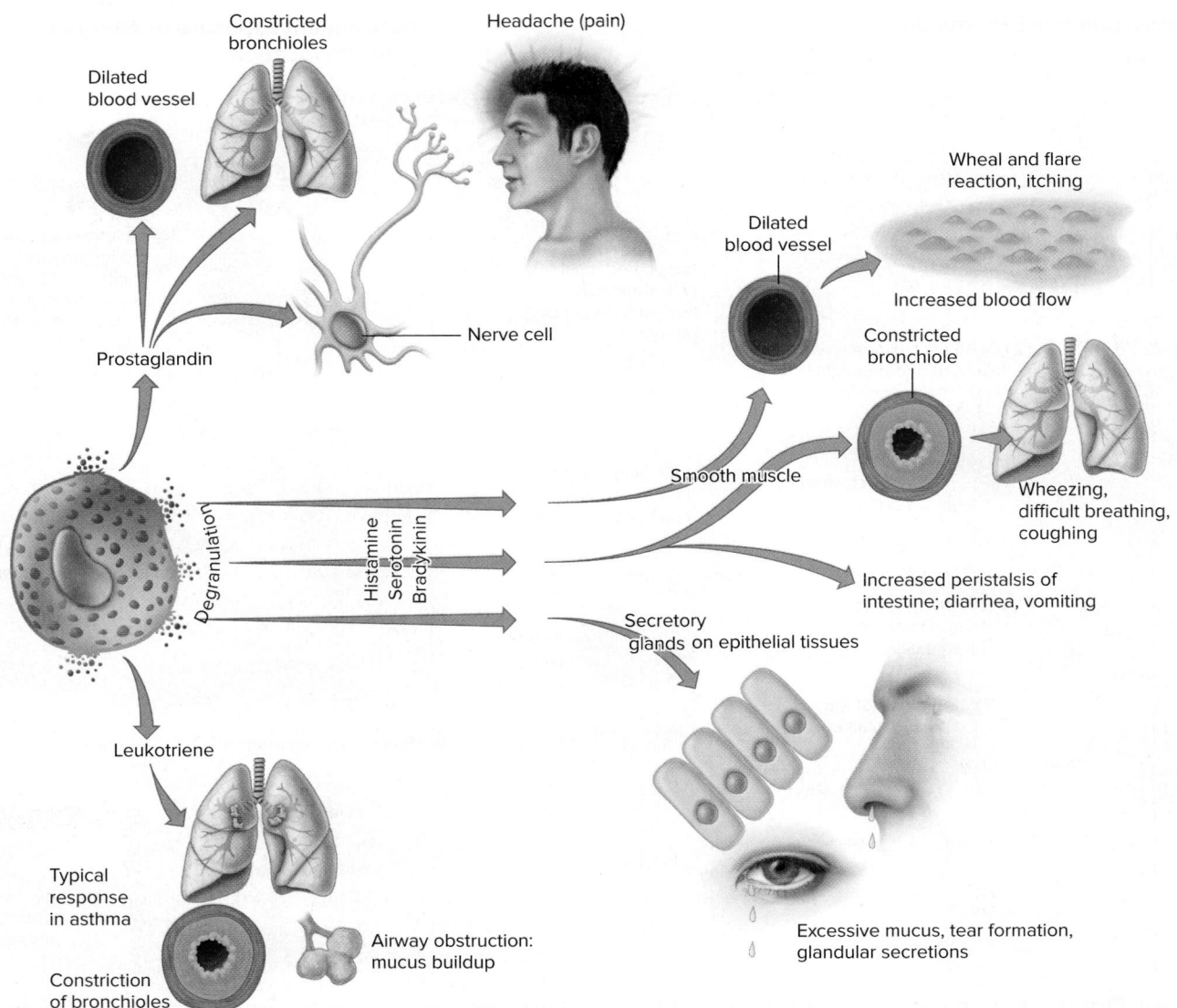

Figure 16.4 **The spectrum of reactions to inflammatory cytokines released by mast cells and the common symptoms they elicit in target tissues and organs.** Note the extensive overlapping effects.

dilates arterioles and venules. It is responsible for the **wheal*** **and flare reaction** in the skin (see figure 16.6), pruritis (itching), and headache. More severe reactions such as anaphylaxis can be accompanied by edema and vascular dilation, which lead to hypotension, tachycardia, circulatory failure, and, frequently, shock. Salivary, lacrimal, mucous, and gastric glands are also histamine targets.

Although the role of **serotonin*** in human allergy is uncertain, its effects appear to complement those of histamine. In experimental animals, serotonin increases vascular permeability, capillary dilation, smooth muscle contraction, intestinal peristalsis, and respiratory rate, but it diminishes central nervous system activity.

A type of **leukotriene,*** known as the "slow-reacting substance of anaphylaxis," induces gradual contraction of smooth muscle.

This leukotriene is responsible for the prolonged bronchospasm, vascular permeability, and mucous secretion of the asthmatic individual. Other leukotrienes stimulate the activities of polymorphonuclear leukocytes.

Platelet-activating factor is a lipid released by basophils, neutrophils, monocytes, and macrophages. The physiological response to this factor is similar to that of histamine, including increased vascular permeability, pulmonary smooth muscle contraction, pulmonary edema, hypotension, and a wheal and flare response in the skin.

Prostaglandins* are a group of powerful inflammatory agents. Normally these substances regulate smooth muscle contraction. For example, they stimulate uterine contractions during delivery. In allergic reactions, they are responsible for vasodilation, increased vascular permeability, increased sensitivity to pain, and bronchoconstriction. Certain anti-inflammatory drugs work by preventing the actions of prostaglandins.

* *wheal* (weel) A smooth, slightly elevated, temporary welt (wheal) that is surrounded by a flushed patch of skin (flare).
* *serotonin* (ser″-oh-toh′-nin) L. *serum,* fluid, and *tonin,* stretching.
* *leukotriene* (loo″-koh-try′-een) Gr. *leukos,* white blood cell, and *triene,* a chemical suffix.

* *prostaglandin* (pross″-tah-glan′-din) From prostate gland. The substance was originally isolated from semen.

16.1 MAKING CONNECTIONS

The Origins of Allergy

Why would humans and other mammals evolve an allergic response that is capable of doing so much harm and even causing death? It is unlikely that this limb of immunity exists merely to make people miserable; it must have a role in protection and survival. What are the underlying biological functions of IgE, mast cells, and the array of potent cytokines? Analysis has revealed that, although allergic persons have high levels of IgE, trace quantities are present even in the sera of nonallergic individuals, just as mast cells and inflammatory chemicals are also part of normal human physiology. In chapter 14 you learned that inflammatory mediators serve valuable functions, such as increasing blood flow and vascular permeability to summon essential immune components to an injured site. They are also responsible for increased mucus secretion, gastric motility, sneezing, and coughing, which help expel noxious agents. The difference is that, in allergic persons, the quantity and quality of these reactions are excessive and uncontrolled.

SEM of cockroach chitin exoskeleton and its complex allergenic surface.
Source: Janice Carr/CDC

Some surprising insights have emerged from studies at the University of California. It seems that a significant source of many allergies is the polysaccharide *chitin*. This ubiquitous compound is found in insect exoskeletons, fungal cell walls, and parasitic worms. The digestion of chitin by enzymes present in the human body causes the release of smaller antigen molecules. These set off powerful immune and inflammatory reactions. There is even a specific pathway in lymphocytes and other white blood cells for responding to chitin and its antigens. It is theorized that this system originally functioned in the attack against worms, fungi, and other pathogens containing chitin, but modern hygiene and therapy have greatly reduced these infections. The remnant of this formerly protective mechanism now actually causes "out of control" reactions such as asthma and other respiratory allergies. This appears to be one basis of allergies to cockroaches and dust mites and the increase in childhood asthma.

What are some potential benefits of allergy? *Answer available on Connect.*

Bradykinin* is related to a group of plasma and tissue peptides known as kinins that participate in blood clotting and chemotaxis. In allergy, it causes prolonged contraction of the bronchioles, dilation of peripheral arterioles, increased capillary permeability, and increased mucus secretion.

Specific Diseases Associated with IgE- and Mast-Cell-Mediated Allergy

The mechanisms just described are basic to hay fever, allergic asthma, food allergy, drug allergy, eczema, and anaphylaxis. In this section we cover the main characteristics of these conditions, followed by methods of detection and treatment.

Atopic Diseases

Hay fever is a generic term for **allergic rhinitis,*** a seasonal reaction to inhaled plant pollen or molds or a chronic, year-round reaction to a wide spectrum of airborne allergens or inhalants (see table 16.2). The targets are typically respiratory membranes, and the symptoms include nasal congestion; sneezing; coughing; profuse mucus secretion; itchy, red, and teary eyes; and mild bronchoconstriction.

Allergic asthma* is a respiratory disease characterized by episodes of impaired breathing due to severe bronchoconstriction. The airways of asthmatic people are exquisitely responsive to minute amounts of inhalant allergens, food, or other stimuli, such as

infectious agents. The symptoms of asthma range from occasional, annoying bouts of difficult breathing to fatal suffocation. Labored breathing, shortness of breath, wheezing, cough, and ventilatory **rales*** are present to one degree or another. The respiratory tract of an asthmatic person is chronically inflamed and severely overreactive to allergy chemicals, especially leukotrienes and serotonin from pulmonary mast cells. Other pathologic components are thick mucous plugs in the air sacs and lung damage that can result in long-term respiratory compromise. An imbalance in the nervous control of the respiratory smooth muscles is apparently involved in asthma, and the episodes are influenced by the psychological state of the person, which strongly supports a neurological connection.

The number of asthma sufferers in the United States is estimated at more than 20 million, with nearly one-third of them children. For reasons that are not completely understood, asthma and deaths from it have been increasing over the last 30 years, even though effective agents to control it are more available now than they have ever been before. Surprisingly, asthma is more common in developed regions of the world (the United States and parts of Europe) than in less developed countries in Asia and Africa, It has been suggested that modern insulated buildings have created indoor conditions that harbor higher concentrations of airborne allergens and ozone. This suggests a pronounced influence of environmental factors on the development of allergies and asthma **(16.1 Making Connections).**

Quick Search

For an overview, watch the video "Allergies—Hypersensitivity Type I" on YouTube.

* *bradykinin* (brad″-ee-kye′-nin) Gr. *bradys,* slow, and *kinein,* to move.
* *rhinitis* (rye-nye′-tis) Gr. *rhis,* nose, and *itis,* inflammation.
* *asthma* (az′-muh) The Greek word for gasping.

* *rales* (rails) Abnormal breathing sounds.

Atopic dermatitis is an intensely itchy inflammatory condition of the skin, also known as **eczema.*** Sensitization occurs through ingestion, inhalation, and, occasionally, skin contact with allergens. It usually begins in infancy with reddened, vesicular, weeping, encrusted skin lesions (**figure 16.5a**). It then progresses in childhood and adulthood to a dry, scaly, thickened skin condition. Lesions can occur on the face, scalp, neck, and inner surfaces of the limbs and trunk. The itchy, painful lesions cause considerable discomfort, and they are often predisposed to secondary bacterial infections. An anonymous writer once aptly described eczema as "the itch that rashes" or "one scratch is too many but one thousand is not enough."

Food Allergy

The ordinary diet contains a vast variety of compounds that are potentially allergenic. Although food allergies have an intestinal mode of entry, they can affect the skin and respiratory tract as well. Gastrointestinal symptoms include vomiting, diarrhea, and abdominal pain. In severe cases, nutrients are poorly absorbed, leading to growth retardation and failure to thrive in young children. Other manifestations of food allergies include rashes, hives (**figure 16.5b**), rhinitis, asthma, and occasionally anaphylaxis. Classic food hypersensitivity involves IgE and degranulation of mast cells, but not all reactions involve this mechanism. The most common food allergens come from peanuts, fish, cow's milk, eggs, shellfish, and soybeans.[1] A recent investigation of babies who were considered at risk for peanut allergy concluded that eating a small amount of peanuts early in life greatly reduced their developing sensitivity to peanuts. Allergists are now thinking that in some cases, certain foods should be introduced earlier to prevent sensitization of children to that food.

> **Quick Search**
>
> Learn some fascinating facts from the video "Animation Describing Anaphylaxis" on YouTube.

* *eczema* (eks´-uh-mah; also ek-zeem´-uh) Gr. *ekzeo*, to boil over.

1. Do not confuse food allergy with food intolerance. Many people are lactose intolerant, for example, due to a deficiency in the enzyme that degrades the sugar in milk.

Drug Allergy

Modern chemotherapy has been responsible for many medical advances. Unfortunately it has also been hampered by the fact that drugs are foreign compounds capable of stimulating allergic reactions. In fact, allergy to drugs is one of the most common side effects of treatment (present in 5% to 10% of hospitalized patients). Depending upon the allergen, route of entry, and individual sensitivities, virtually any tissue of the body can be affected, and reactions range from mild atopy to fatal anaphylaxis. Compounds implicated most often are antibiotics (penicillin is number one in prevalence), synthetic antimicrobials (sulfa drugs), aspirin, opiates, and contrast dye used in X rays. The actual allergen is not the intact drug itself but a hapten given off when the liver processes the drug. Some forms of penicillin sensitivity are due to the presence of small amounts of the drug in meat, milk, and other foods and to exposure to *Penicillium* mold in the environment.

Anaphylaxis: A Powerful Systemic Reaction to Allergens

The term **anaphylaxis** was first used to denote a reaction of animals injected with a foreign protein. Although the animals showed no response during the first contact, upon reinoculation with the same protein at a later time, they exhibited acute symptoms—itching, sneezing, difficult breathing, prostration, and convulsions—and many died in a few minutes. Two clinical types of anaphylaxis are seen in humans. **Cutaneous anaphylaxis** is the wheal and flare inflammatory reaction to the local injection of allergen. **Systemic anaphylaxis,** also called **anaphylactic shock,** is characterized by sudden respiratory and circulatory disruption that can be fatal in a few minutes. In humans, the allergen and route of entry are

Source: Rob Flynn/USDA

Figure 16.5 Skin manifestations in atopic allergies. (a) Atopic dermatitis, or eczema. Vesicular, encrusted lesions are typical in afflicted infants. This condition is prevalent enough to account for 1% of pediatric care. **(b)** Skin rash caused by an allergy to strawberries. The slightly raised red lesions, termed urticaria or hives, can sometimes merge to form a solid red rash.
(a): © DermNet New Zealand Trust; (b): © Kathy Park Talaro

(a)

(b)

variable, though bee stings and injections of antibiotics or serum are implicated most often. Honeybee and other hymenopteran stings are one of the most frequent causes of anaphylactic shock. Bee venom often creates a sensitivity that can last for decades after exposure.

The underlying physiological events in systemic anaphylaxis parallel those of atopy, but the concentration of chemical mediators and the strength of the response are greatly amplified. The immune system of a sensitized person exposed to a provocative dose of allergen responds with a sudden, massive release of chemicals into the tissues and blood, which act rapidly on the target organs. The throat swells, respiration is compromised, blood pressure falls, and the heart may stop. Anaphylactic persons have been known to die within 15 minutes from complete airway blockage. Peanut allergies, the leading cause of death related to food allergy, are notorious for their rapid onset and tiny allergen dose. Fatal cases have occurred from merely kissing a person who had eaten peanuts or sharing a drinking glass that had been in contact with peanuts.

Diagnosis of Allergy

Because allergy mimics infection and other conditions, it is important to determine if a person is actually allergic. If possible or necessary, it is also helpful to identify the specific allergen or allergens. Allergy diagnosis involves several levels of tests, including nonspecific, specific, in vitro, and in vivo methods.

A test that can distinguish whether a patient has experienced an allergic attack measures elevated blood levels of tryptase, an enzyme released by mast cells. The enzyme level increases during an allergic response. Several types of specific in vitro tests can determine significant signs of allergy from a patient's blood sample. A differential blood cell count can indicate the levels of basophils and eosinophils—a higher level of these indicates allergy. The leukocyte histamine-release test measures the amount of histamine released from the patient's basophils when exposed to a specific allergen. Serological tests that use radioimmune assays to reveal the quantity and type of IgE are also clinically helpful (see chapter 17).

Skin Testing

Skin testing can precisely detect atopic or anaphylactic sensitivities. With this technique, a patient's skin is injected, scratched, or pricked with a small amount of a pure allergen extract. Drug companies can supply a large battery of standard allergen extracts containing common airborne allergens (plant and mold pollen) and more unusual allergens (mule dander, theater dust, bird feathers). Unfortunately, skin tests for food allergies using food extracts are unreliable in most cases.

In patients with numerous allergies, the allergist maps the skin on the inner aspect of the forearms or back and injects the allergens intradermally according to this predetermined pattern (**figure 16.6a**). Approximately 20 minutes after antigenic challenge, each site is appraised for a wheal response indicative of histamine release. The diameter of the wheal is measured and rated on a scale of 0 (no reaction) to 4+ (greater than 15 mm). **Figure 16.6b** shows skin test results for a person with extreme inhalant allergies. A newer rapid test provides an alternate system for testing that has greater accuracy than skin testing and does not require injection of allergen.

> **Quick Search**
> Find out more about this process by searching the Internet for videos, such as "Allergy Skin Tests—How They're Performed" on YouTube.

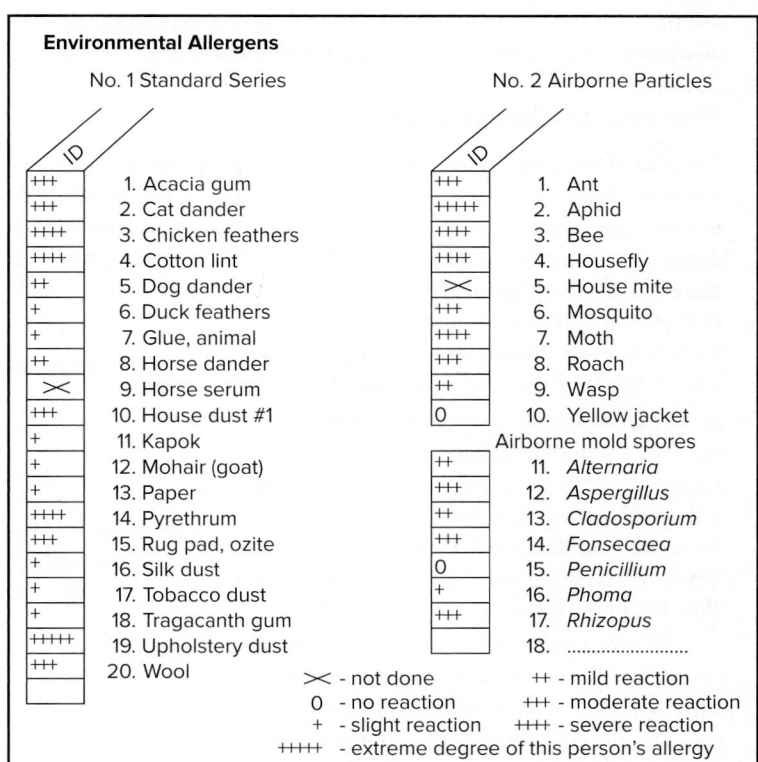

Figure 16.6 A method for conducting a skin test for type I allergies. The back (or forearm) is mapped and injected with a selection of allergen extracts. The allergist must be very aware of potential anaphylaxis attacks triggered by these injections. **(a)** View of skin prick test reactions for arthropod allergens, ranging from severe (spider) to 0 (caterpillar web). **(b)** A skin test record for some common environmental allergens in a patient with severe multiple allergies, some showing even more than 4+s.
(a): *Source:* Dr. Frank Perlman, M.A. Parson/CDC

Corticosteroids keep the plasma cell from synthesizing IgE and inhibit T cells.

Cromolyn acts on the surface of mast cell; no degranulation

Antihistamines, aspirin, epinephrine, theophylline counteract the effects of cytokines on smooth muscle and membrane targets.

Allergen

Plasma cell

Avoidance of allergen

IgE

Monoclonal drugs that inactivate IgE

Figure 16.7 **Strategies for circumventing allergic attacks.** These range from avoidance of the allergen, inhibiting immune reactions, inhibiting IgE release, stopping degranulation, to preventing the effects of cytokines.

Treatment and Prevention of Allergy

In general, the methods of treating and preventing allergy involve (1) avoiding the allergen, although this may be very difficult in many instances; (2) taking drugs that block the action of lymphocytes, mast cells, or chemical mediators; and (3) undergoing allergen desensitization therapy.

It is not possible to completely prevent initial sensitization, because there is no way to tell in advance if a person will develop an allergy to a particular substance. There has been a trend to prevent food allergies in babies by not feeding them solid foods for the first several months. Unfortunately babies could still be exposed through breast milk that contains allergens ingested by the mother. Rigorous cleaning and air conditioning can reduce contact with airborne allergens, but it is not feasible to isolate a person from all allergens, which is the reason drugs are so important in control.

Therapy to Counteract Allergies

The aim of antiallergy medication is to block the progress of the allergic response somewhere along the route between IgE production and the appearance of symptoms **(figure 16.7).** Oral anti-inflammatory drugs such as corticosteroids inhibit the activity of lymphocytes and thereby reduce the production of IgE, but they also have dangerous side effects and should not be taken for prolonged periods. Some drugs block the degranulation of mast cells and reduce the levels of inflammatory cytokines. The most effective of these are diethylcarbamazine and cromolyn. Asthma and rhinitis sufferers can find relief with a drug that blocks synthesis of leukotriene (montelukast [Singulair]) and a monoclonal antibody that inactivates IgE (omalizumab [Xolair]).

Widely used medications for preventing symptoms of atopic allergy are **antihistamines,** the active ingredients in most over-the-counter allergy-control drugs. Antihistamines interfere with histamine activity by binding to histamine receptors on target organs. Although many of them have major side effects such as

drowsiness, many of the newer antihistamines lack this side effect because they do not cross the blood-brain barrier. Other drugs that relieve inflammatory symptoms are aspirin and acetaminophen, which reduce pain by interfering with prostaglandin, and theophylline, a bronchodilator that reverses spasms in the respiratory smooth muscles. Persons who suffer from anaphylactic attacks are urged to carry at all times injectable epinephrine (adrenaline) and an identification tag indicating their sensitivity. An aerosol inhaler containing epinephrine can also provide rapid relief. Epinephrine reverses constriction of the airways and slows the release of allergic mediators.

Approximately 70% of allergic patients benefit from controlled injections of specific allergens as determined by skin tests. This technique is a form of allergen immunotherapy called **desensitization,** or **hyposensitization,** that prevents reactions between allergen, IgE, and mast cells. The allergen preparations contain pure, preserved suspensions of plant antigens, venoms, dust mites, dander, and molds. A newer experimental method of hyposensitization involves sublingual therapy, in which a tiny dose of premeasured allergen is placed under the tongue to be absorbed. It is being tested for peanut and other food allergies and even rhinitis and dermatitis with some success.

The immunologic basis of this treatment is open to differences in interpretation. One theory suggests that injected allergens stimulate the formation of high levels of allergen-specific IgG **(figure 16.8)**

B Cell / Plasma Cell

IgG "blocking antibodies"

Mast Cell with previous IgE

Allergen

No allergen to bind with mast cell

No degranulation No allergic symptoms

IgE

IgG binds allergens

Figure 16.8 **The blocking antibody theory for allergic desensitization.** An injection of allergen causes IgG antibodies to be formed instead of IgE; these blocking antibodies crosslink and effectively remove the allergen before it can react with the IgE in the mast cell.

instead of IgE. It has been proposed that these IgG **blocking antibodies** remove allergen from the system before it can bind to IgE, thus preventing the degranulation of mast cells. It is also possible that allergen delivered in this fashion combines with the IgE itself and takes it from circulation before it can react with the mast cells.

 Check Your Progress SECTION 16.2

5. Give some examples of allergens and explain their routes of access to the body.
6. Describe several factors that influence types and severity of allergic responses.
7. Describe the events in sensitization and provocation.
8. How are atopic allergies similar to anaphylaxis, and how are they different?
9. Trace the course of a pollen grain through sensitization and provocation in allergies. Include in the discussion the role of mast cells, basophils, IgE, and allergic mediators.
10. Relate the roles of major mediators to the signs and symptoms of allergy.
11. Outline the target organs and symptoms of the principal atopic diseases and their diagnosis and treatment.
12. Explain how hyposensitization is achieved and suggest a credible mechanisms by which it might work.

16.3 Type II Hypersensitivities: Reactions That Lyse Foreign Cells

 Expected Learning Outcomes

10. Explain the requirements for developing a type II hypersensitivity.

11. Define what is meant by blood groups, explain how they are inherited and expressed, and relate the primary medical concerns of blood transfusions.

12. Discuss how blood is typed and notated.

13. Explain the special concerns about the Rh factor, the causes of hemolytic disease of the newborn, and how to prevent it.

The diseases termed type II hypersensitivities are a complex group of syndromes that involve complement-assisted destruction (lysis) of cells by antibodies (IgG and IgM) directed against those cells' surface antigens. This category includes transfusion reactions and some types of autoimmunities (discussed in section 16.6). The cells targeted for destruction are often red blood cells, but other cells can be involved.

Chapters 14 and 15 described the functions of unique surface markers on cell membranes. Ordinarily these molecules play essential roles in transport, recognition, and development, but they become medically important when the tissues of one person come in intimate contact with the body of another person. Blood transfusions and organ donations introduce **alloantigens** (molecules that differ in the same species) on donor cells that are recognized by lymphocytes of the recipient. These reactions are not really immune dysfunctions in the way that allergy and autoimmunity are. The immune system is in fact working normally, but it is not equipped to distinguish between the beneficial foreign cells of a transplanted tissue and the harmful cells of a microbe.

The Basis of Human ABO Antigens and Blood Types

The existence of human blood types was first demonstrated by an Austrian pathologist, Karl Landsteiner, in 1904. While studying incompatibilities in blood transfusions, he found that the serum of one person could clump the red blood cells of another. Landsteiner identified four distinct types, subsequently called the **ABO blood groups.**

Like the **MHC/HLA** antigens on white blood cells described in chapter 15, the ABO antigen markers on red blood cells are genetically determined and composed of glycoproteins. These ABO antigens are inherited as two (one from each parent) of three alternative **alleles:*** A, B, or O. A and B alleles are dominant over O and codominant with one another. As **table 16.3** indicates, this mode of inheritance gives rise to four blood types (phenotypes), depending on the particular combination of genes (genotypes). Thus, a person with an *AA* or *AO* genotype has type A blood; genotype *BB* or *BO* gives type B; genotype *AB* produces type AB; and genotype *OO* produces type O. Some important points about the blood types are:

1. they are named for the dominant antigen(s);
2. the RBCs of type O persons lack the A and B antigens, but they do have other types of antigens; and
3. tissues other than RBCs carry A and B antigens which can complicate organ transplantation.

* *allele* (ah-leel´) Gr. *allelon*, of one another. An alternate form of a gene for a given trait.

TABLE 16.3	Characteristics of ABO Blood Groups					
				Incidence of Type in United States		
Genotype	Blood Type	Antigen Present on Erythrocyte Membranes	Antibody in Plasma	Among Whites (%)	Among Asians (%)	Among Those of African and Caribbean Descent (%)
AA, AO	A	A	Anti-b	41	28	27
BB, BO	B	B	Anti-a	10	27	20
AB	AB	A and B	Neither anti-a nor anti-b	4	5	7
OO	O	Neither A nor B	Anti-a and anti-b	45	40	46

Figure 16.9 **The genetic/molecular basis for the A and B antigens (receptors) on red blood cells.** In general, persons with blood types A, B, and AB inherit a gene or genes for enzymes that add a terminal sugar to the basic RBC receptor. Type O persons do not have these enzymes and lack a terminal sugar.

A diagram of the AB antigens and blood types is shown in **figure 16.9.** The A and B genes each code for an enzyme that adds a terminal carbohydrate to RBC surface molecules during maturation. RBCs of type A contain an enzyme that adds *N*-acetylgalactosamine to the molecule; RBCs of type B have an enzyme that adds D-galactose; RBCs of type AB contain both enzymes that add both carbohydrates; and RBCs of type O lack the genes and enzymes to add a terminal molecule and are not antigenic.

Antibodies Against A and B Antigens

Although an individual does not normally produce antibodies in response to his or her own RBC antigens, the serum can contain antibodies that react with blood of another antigenic type, even though contact with this other blood type has *never* occurred. These preformed antibodies account for the immediate and intense quality of transfusion reactions. As a rule, type A blood contains antibodies (anti-b) that react against the B antigens on type B and AB red blood cells. Type B blood contains antibodies (anti-a) that react with A antigen on type A and AB red blood cells. Type O blood contains antibodies against both A and B antigens. Type AB blood does not contain antibodies against either A or B antigens[2] (see table 16.3). Where do these anti-a and anti-b antibodies originate? Immunologists have evidence that they develop in early infancy through exposure to certain antigens that are widely distributed in nature. These antigens are surface molecules on bacteria and plant cells that mimic the structure of A and B antigens and stimulate the production of the antibodies that cause transfusion reactions. Once the immune system has been primed to form these antibodies, they will continue to circulate in the blood throughout life.

Antiserum A Antiserum B

Agglutinated blood Unagglutinated blood

Type A Type B

Type AB Type O

Figure 16.10 **Interpretation of blood typing.** In this test, a drop of blood is mixed with antiserum A, containing antibodies to type A antigen (blue), and antiserum B, containing antibodies to type B antigen (yellow). If that particular antigen is not present, the red blood cells in that droplet do not agglutinate, and form a uniform suspension. If that antigen is present, agglutination occurs and the RBCs form visible clumps (small red particles). The patterns and interpretations for the four blood types are shown. © Kathy Park Talaro

Clinical Concerns in Transfusions

The presence of ABO antigens and a, b antibodies underlies several clinical concerns in giving blood transfusions. First, the individual blood types of donor and recipient must be determined. By use of a standard technique, drops of blood are mixed with **antisera** that contain antibodies against the A and B antigens and are then observed for the evidence of agglutination **(figure 16.10).**

Knowing the blood types involved makes it possible to determine which transfusions are safe to do. The general rule of compatibility is that the RBC antigens of the donor must not be agglutinated by antibodies in the recipient's blood **(figure 16.11).** The ideal

2. Why would this be true? The answer lies in the first sentence of the paragraph.

Type A Donor **Type B Recipient**

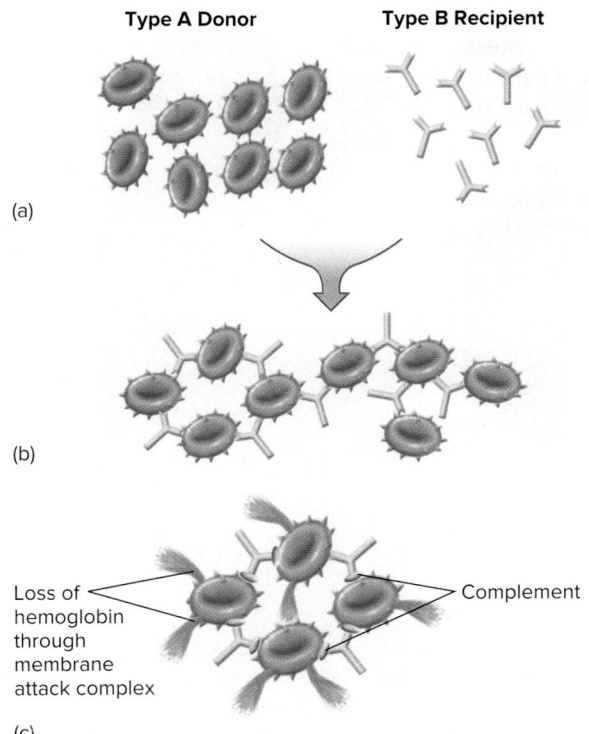

(a)

(b)

Loss of
hemoglobin
through
membrane
attack complex

Complement

(c)

Figure 16.11 Microscopic view of a transfusion reaction.
(a) Incompatible blood types. The red blood cells of the type A donor
contain antigen A, whereas the serum of the type B recipient contains
anti-a antibodies that can agglutinate donor cells. **(b)** Agglutination
complexes can block the circulation in vital organs. **(c)** Activation of
the complement by antibody on the RBCs causes hemolysis and
anemia. This sort of incorrect transfusion is very rare because of the
great care taken by blood banks to ensure a correct match.

practice is to transfuse blood that is a perfect match (A to A, B to B).
But even in this event, blood samples must be cross-matched before
the transfusion because other blood group incompatibilities can ex-
ist. This test involves mixing the blood of the donor with the serum
of the recipient to check for agglutination.

Under certain circumstances (medical emergencies, the battle-
field) the concept of universal transfusions can be used. To appreci-
ate how this works, we must apply the rule stated in the previous
paragraph. Type O blood lacks A and B antigens and will not be
agglutinated by other blood types, so it could theoretically be used
in any transfusion. We call a person with this blood type a **univer-
sal donor.** Because type AB blood lacks agglutinating antibodies,
an individual with this blood could conceivably receive any type of
blood. Type AB persons are consequently called **universal recipients.**
Although both types of transfusions involve antibody incompatibil-
ities, these are of less concern because of the dilution of the donor's
blood in the circulatory system of the recipient. Additional RBC
markers that can be significant in transfusions are the Rh, MN, and
Kell antigens, but they do not create the same transfusion risks as
ABO types.

Transfusion of the wrong blood type causes differing degrees
of adverse reactions. The severest reaction is massive hemolysis
when the donated red blood cells react with recipient antibody and
trigger the complement cascade (figure 16.11). The resultant

destruction of red cells leads to systemic shock and kidney failure
brought on by the blockage of glomeruli (blood-filtering apparatus)
by cell debris. Death is a common outcome. Other reactions caused
by RBC destruction are fever, anemia, and jaundice. A transfusion
reaction is managed by immediately halting the transfusion, admin-
istering drugs to remove hemoglobin from the blood, and beginning
another transfusion with red blood cells of the correct type. See this
chapter's Case Study for a description of a transfusion reaction.

The Rh Factor and Its Clinical Importance

Another RBC antigen of major clinical concern is the **Rh factor** (or
D antigen). This factor was first discovered in experiments explor-
ing the genetic relationships among animals. Rabbits inoculated
with the RBCs of rhesus monkeys produced an antibody that also
reacted with human RBCs. Further tests showed that this monkey
antigen (termed Rh for rhesus) was present in about 85% of humans
and absent in the other 15%. The details of Rh inheritance are more
complicated than those of ABO; but in simplest terms, a person's
Rh type results from a combination of two possible alleles—a dom-
inant one that codes for the factor and a recessive one that does not.
A person inheriting at least one Rh gene will be Rh^+; only those
persons inheriting no Rh genes are Rh^-. The "+" or "−" appearing
after a blood type refers to the Rh status of the person, as in O^+ or
AB^-. However, unlike the ABO antigens, exposure to environmen-
tal antigens does not sensitize Rh^- persons to the Rh factor. The
only ways one can develop antibodies against this factor are through
placental sensitization or incorrect transfusion.

Hemolytic Disease of the Newborn and Rh Incompatibility

The potential for placental sensitization occurs when a mother is
Rh^- and her unborn child is Rh^+. The obvious intimacy between
mother and fetus makes it possible for fetal RBCs to leak into the
mother's circulation during childbirth, especially when the detach-
ment of the placenta creates avenues for fetal blood to enter the
maternal circulation. The mother's immune system detects the for-
eign Rh factors on the fetal RBCs and is sensitized to them by pro-
ducing antibodies and memory B cells. The first Rh^+ child is
usually not affected because the process begins so late in pregnancy
that the child is born before maternal sensitization is completed.
However, the mother's immune system has been strongly primed
for a second contact with this factor in a subsequent pregnancy
(figure 16.12*a*).

If a mother has been sensitized and becomes pregnant again
with an Rh^+ fetus, fetal blood cells escape into the maternal circula-
tion late in pregnancy and elicit a memory response. The fetus is at
risk when the maternal anti-Rh antibodies cross the placenta into
the fetal circulation, where they affix to fetal RBCs and cause com-
plement-mediated lysis. The outcome is a potentially fatal **hemo-
lytic disease of the newborn (HDN)** called *erythroblastosis
fetalis.** This term is derived from the presence of immature nucle-
ated RBCs called erythroblasts in the blood. They are released into
the infant's circulation to compensate for the massive destruction of

* *erythroblastosis fetalis* (eh-rith″-roh-blas-toh′-sis fee-tal′-is).

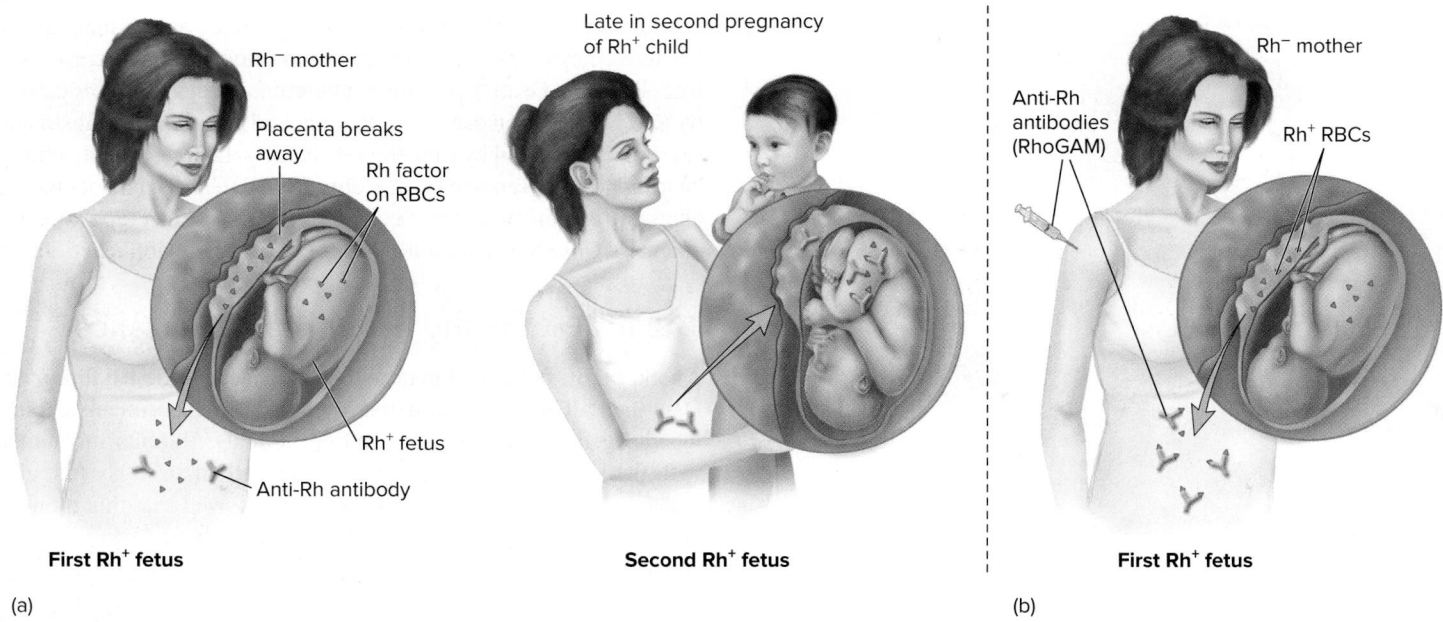

Figure 16.12 Rh factor incompatibility can result in RBC lysis and hemolytic disease of the newborn. (a) A naturally occurring blood cell incompatibility results when a Rh$^+$ fetus develops within a Rh$^-$ mother. Initial sensitization of the maternal immune system occurs when fetal blood crosses through the placenta near the time of birth. In most cases, the fetus develops normally and is born without illness. A subsequent pregnancy with an Rh$^+$ fetus results in a severe hemolysis of fetal RBCs. **(b)** Control of incompatibility: Anti-Rh antibody (RhoGAM) can be administered to Rh$^-$ mothers during pregnancy to help bind, inactivate, and remove any Rh factor that may be transferred from the fetus. In some cases, RhoGAM is administered before sensitization occurs.

RBCs stimulated by maternal antibodies. Additional symptoms are severe anemia, jaundice, and enlarged spleen and liver.

Maternal-fetal incompatibilities are also possible in the ABO blood group, but adverse reactions occur less frequently than with Rh sensitization because the antibodies to these blood group antigens are IgM rather than IgG and are unable to cross the placenta in large numbers. The maternal-fetal relationship itself is a fascinating instance of foreign tissue not being rejected, despite the extensive potential for contact. It is thought that the placenta forms a barrier to keep the fetus isolated in its own antigen-free environment. The placenta is surrounded by a dense, many-layered envelope that prevents the passage of maternal cells; and it actively absorbs, removes, and inactivates circulating antigens.

Preventing Hemolytic Disease of the Newborn

Once sensitization of the mother to Rh factor has occurred, all other Rh$^+$ fetuses will be at risk for hemolytic disease of the newborn. Prevention requires a careful family history of an Rh$^-$ pregnant woman. It can predict the likelihood that she is already sensitized or is carrying an Rh$^+$ fetus. It must take into account other children she has had, their Rh types, and the Rh status of the father. If the father is also Rh$^-$, the child will be Rh$^-$ and free of risk, but if the father is Rh$^+$, the probability that the child will be Rh$^+$ is 50% or 100%, depending on the exact genetic makeup of the father. If there is any possibility that the fetus is Rh$^+$, the mother must be passively immunized with antiserum containing antibodies against the Rh factor (*Rh$_0$ [D] immune globulin,* or RhoGAM[3]). This antiserum, injected

3. RhoGAM: Immunoglobulin fraction of human anti-Rh serum, prepared from pooled human sera.

at 28 to 32 weeks and again immediately after delivery, keeps the maternal immune system from recognizing fetal RBCs that have escaped into the mother's circulation. This prevents the sensitization of her immune system to Rh factor (**figure 16.12b**). Anti-Rh antibody must be given with each pregnancy that involves an Rh$^+$ fetus. It is ineffective if the mother has already been sensitized by a prior Rh$^+$ fetus or an incorrect blood transfusion, which can be determined by a serological test.

As in ABO blood types, the Rh factor should be matched for a transfusion, although it is acceptable to transfuse Rh$^-$ blood if the recipient's Rh type is not known.

Check Your Progress SECTION 16.3

13. What is the mechanism of type II hypersensitivity?
14. Explain why the tissues of some people are antigenic to others, and why this would not be a problem except for transfusions and transplants.
15. What is the molecular basis of the ABO and Rh blood groups?
16. Where do we derive our natural hypersensitivities to the A or B antigens that we do not possess?
17. How does a person become sensitized to Rh factor?
18. Describe the sequence of events in an Rh incompatibility between mother and fetus.
19. Explain the rules of transfusion. Illustrate what will happen if type A blood is accidentally transfused into a type B person.

16.4 Type III Hypersensitivities: Immune Complex Reactions

 Expected Learning Outcomes

14. Describe the background features of immune complex reactions.

15. Differentiate between the major types of immune complex diseases, and discuss their physiological effects.

Type III hypersensitivity involves the reaction of soluble antigen with antibody and the deposition of the resulting complexes in basement membranes of epithelial tissue. It is similar to type II because it involves the production of IgG and IgM antibodies after repeated exposure to antigens and the activation of complement. Type III differs from type II because its antigens are not attached to the surfaces of cells. The interaction of these antigens with antibodies produces free-floating complexes that can be deposited in the tissues, causing an **immune complex reaction** or disease. This category includes therapy-related disorders (serum sickness and the Arthus reaction) and a number of autoimmune diseases (such as glomerulonephritis and lupus erythematosus).

Mechanisms of Immune Complex Diseases

After initial exposure to a profuse amount of antigen, the immune system produces large quantities of antibodies that circulate in the fluid compartments. When this same antigen enters the system a second time, it reacts with the antibodies to form antigen-antibody complexes **(figure 16.13).** These complexes summon the usual inflammatory components such as complement and neutrophils, which would ordinarily eliminate antibody-antigen (Ab-Ag) complexes as part of the normal immune response. In an immune complex disease, however, these complexes are so abundant that they deposit in the **basement membranes**[4] of epithelial tissues and become inaccessible. In response to these events, neutrophils release lysosomal granules that digest tissues and cause destructive inflammatory reactions at these sites. The symptoms of type III hypersensitivities are due in great measure to this pathology.

Types of Immune Complex Disease

During the early tests of immunotherapy using animals, hypersensitivity reactions to serum and vaccines were common. In addition to anaphylaxis, two syndromes, the **Arthus reaction**[5] and **serum sickness,** were identified. These syndromes are associated with certain types of passive immunization (especially with animal serum).

Serum sickness and the Arthus reaction are like anaphylaxis in requiring sensitization and preformed antibodies. Characteristics that set them apart from anaphylaxis are (1) they depend upon IgG, IgM, or IgA (precipitating antibodies) rather than IgE; (2) they require large doses of antigen (not the small dose required for anaphylaxis); and (3) their symptoms are delayed by a few hours to days. The Arthus reaction and serum sickness differ from each other in some important ways. The Arthus reaction is a *localized* dermal injury due to inflamed blood vessels in the vicinity of any injected antigen. Serum sickness is a *systemic* injury initiated by antigen-antibody complexes that circulate in the blood and settle into membranes at various sites.

The Arthus Reaction

The Arthus reaction is usually an acute response to a second injection of vaccines (boosters) or drugs at the same site as the first injection. In a few hours, the area becomes red, hot to the touch, swollen, and very painful. These symptoms are mainly due to the destruction of tissues in and around the blood vessels and the release of histamine from mast cells and basophils. Although the reaction is usually self-limiting and rapidly cleared, intravascular blood clotting can occasionally cause necrosis and loss of tissue.

Phases:

Antibody combines with excess soluble antigen, forming large quantities of Ab-Ag complexes.

Circulating immune complexes become lodged in the basement membranes of epithelia in blood vessels, kidney, skin and other sites.

Complement factors trigger release of histamine and other inflammatory mediators.

Neutrophils migrate to sites of Ab-Ag complexes and release enzymes and chemokines that severely damage the target tissues and organs.

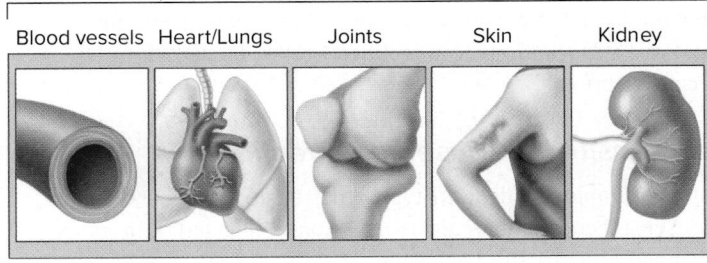

Major organs where immune complexes are deposited

Figure 16.13 Pathology of immune complex disease. Deposition of circulating Ab-Ag complexes in tissues results in major organ damage.

4. **Basement membranes** are basal partitions of epithelia that normally filter out circulating antigen-antibody complexes.

5. Named after Maurice Arthus, the physiologist who first identified this localized inflammatory response.

Serum Sickness

Serum sickness was named for a condition that appeared in soldiers after repeated injections of horse serum to treat tetanus. It can also be caused by injections of animal hormones and drugs. The immune complexes enter the circulation; are carried throughout the body; and are eventually deposited in blood vessels of the kidney, heart, skin, and joints (figure 16.13). The condition can become chronic, causing symptoms such as enlarged lymph nodes, rashes, painful joints, swelling, fever, and renal dysfunction.

 ### *Check Your Progress* SECTION 16.4

20. Contrast type II and type III hypersensitivities with respect to type of antigen, antibody, and manifestations of disease.

21. Explain what occurs in immune complex diseases and give examples.

16.5 Immunopathologies Involving T Cells

 ### Expected Learning Outcomes

16. Define what is meant by *type IV hypersensitivities.*

17. Describe the origins of allergies to infectious agents and contact dermatitis.

18. Outline the primary mechanisms in contact dermatitis.

19. Discuss the involvement of T cells in organ transplantation and the different ways that grafted tissues may be rejected.

20. Describe the categories of grafts and how rejection is prevented.

Type IV Delayed-Type Hypersensitivity

The adverse immune responses we have covered so far are explained primarily by B-cell involvement and antibodies. A notable difference exists in type IV hypersensitivity, which involves primarily the T-cell branch of the immune system. Type IV immune dysfunction has traditionally been known as delayed hypersensitivity because the symptoms arise one to several days following the second contact with an antigen. In general, type IV diseases result when T cells respond to antigens displayed on self tissues or transplanted foreign cells. Examples of type IV hypersensitivity include delayed allergic reactions to infectious agents, contact dermatitis, and graft rejection.

Hypersensitivity Caused by Infections

A classic example of a delayed-type hypersensitivity occurs when a person sensitized by tuberculosis infection is injected with an extract (tuberculin) of the bacterium *Mycobacterium tuberculosis.* The so-called tuberculin reaction is an acute skin inflammation at the injection site appearing within 24 to 48 hours. So useful and diagnostic is this technique for detecting present or prior tuberculosis

that it is the chosen screening device. Other infections that use similar skin testing are leprosy, syphilis, histoplasmosis, toxoplasmosis, and candidiasis. This form of hypersensitivity arises from a specific class of T cells (T_H1) that receive the processed allergens from dendritic cells. Activated T_H cells release cytokines that recruit inflammatory cells such as macrophages, neutrophils, and eosinophils. The buildup of fluid and cells at the site gives rise to a red papule characteristic of a positive tuberculin test (see https://commons.wikimedia.org/wiki/File:Mantoux_test.jpg). Other type IV reactions that occur in chronic infection (tertiary syphilis, for example), cause granuloma formation that can lead to severe organ damage.

Contact Dermatitis

The most common delayed allergic reaction, **contact dermatitis,** is caused by exposure to resins in poison ivy or poison oak **(16.2 Making Connections),** to simple haptens in household and personal articles (jewelry, cosmetics, elasticized undergarments), and to certain drugs. Like immediate atopic dermatitis, the reaction to these allergens usually requires a sensitizing and a provocative dose. The allergen first penetrates the outer skin layers, is processed by Langerhans cells (skin dendritic cells), and is presented to T cells. When subsequent exposures attract CD8 lymphocytes and macrophages to this area, these cells give off enzymes and inflammatory cytokines that severely damage the epidermis in the immediate vicinity **(figure 16.14a).** This response accounts for the intensely itchy papules and blisters that are the early symptoms **(figure 16.14b).** As healing progresses, the epidermis is replaced by a thick, horny layer. Depending upon the dose and the sensitivity of the individual, the time from initial contact to healing can take up to 2 weeks. Many allergists conduct a variation on skin testing called a patch test to detect delayed-type hypersensitivities to contact allergens **(figure 16.15).**

T Cells in Relation to Organ Transplantation

Transplantation or grafting of organs and tissues is a common medical procedure. Although it is life-giving, this technique is plagued by the natural tendency of lymphocytes to seek out foreign antigens and mount a campaign to destroy them through **graft rejection.** The bulk of the damage that occurs in graft rejections can be attributed to expression of cytotoxic T cells and other killer cells described in chapter 15. This section covers the mechanisms involved in graft rejection, tests for transplant compatibility, reactions against grafts, prevention of graft rejection, and types of grafts.

The Genetic and Biochemical Basis for Graft Rejection: T-Cell-Mediated Recognition of Foreign MHC Receptors

In chapter 15 we discussed the role of **major histocompatibility** (MHC or HLA) genes and surface markers in immune function. In general, the genes and markers in MHC classes I and II are extremely important in recognizing self and in regulating the immune response. These molecules also set the events of graft rejection in motion. The MHC genes of humans are inherited from among a

(a)

(b)

1. Lipid-soluble catechols are absorbed by the skin.

2. Dendritic cells close to the epithelium pick up the allergen, process it, and display it on MHC receptors.

3. Previously sensitized T_H1 ($CD4^+$) cells recognize the presented allergen.

4. Sensitized T_H1 cells are activated and secrete cytokines (IFN, TNF).

5. These cytokines attract macrophages and cytotoxic T cells to the site.

6. Macrophages release mediators that stimulate a strong, local inflammatory reaction. Cytotoxic T cells directly kill cells and damage the skin. Fluid-filled blisters result.

Process Figure 16.14 Contact dermatitis. (a) Stages in the development of of contact dermatitis. **(b)** A case of contact dermatitis on the leg caused by poison ivy. Note the zones of inflammation, crusted lesions, and clusters of blisters where the allergen was initially deposited. (b): © Kathy Park Talaro

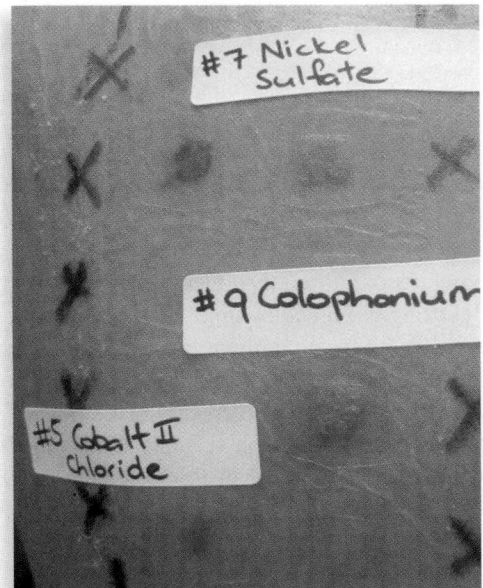

Figure 16.15 Patch testing for contact dermatitis. Most tests are done with household chemicals, cosmetics, plant materials, and metals. Small allergen-impregnated patches are applied to the skin and held in place for 48 hours. Most reactions shown here are negative, but three have varying degrees of positive reactions. The strongest reaction is for colophonium (below #9), a plant resin used in many products; nickel (below #7) produces a moderate positive; and colbalt (below #5) produces a weak reaction. © DermNet New Zealand Trust

large pool of genes, so the cells of each person can exhibit variability in the pattern of cell surface molecules. The pattern is identical in different cells of the same person and can be similar in related siblings and parents, but the more distant the relationship, the less likely that the MHC genes and markers will be similar. When donor tissue (a graft) displays surface molecules of a different MHC type than the recipient, the T cells of the recipient (also called the host) will recognize its foreignness and mount an immune attack against the donated tissue (**figure 16.16***a*).

Host Rejection of Graft When certain T cells of a host recognize foreign class I MHC markers on the surface of cells in grafted tissues or organs such as the heart, kidney, or liver, they release interleukin-2 as part of a general immune mobilization. The effect is to expand the helper and cytotoxic T cells specific to the antigens displayed by the donated cells. The cytotoxic cells bind to the grafted tissue and secrete lymphokines that begin the rejection process within 2 weeks of transplantation. Late in this process, antibodies formed against the graft contribute to immune damage. A final blow is the destruction of the vascular supply, promoting death of the grafted tissue or organ.

Graft Rejection of Host Some types of immunodeficiencies prevent the host from rejecting a graft. But this failure may not protect the host from serious damage, because graft incompatibility is a two-way phenomenon. Some grafted tissues (especially bone

16.2. MAKING CONNECTIONS

Pretty, Pesky, Poisonous Plants

As a cause of allergic contact dermatitis (affecting about 10 million people a year), nothing can compare with a single family of plants belonging to the genus *Toxicodendron*. At least one of these plants—poison ivy, poison oak, or poison sumac—flourishes in the forests, woodlands, or along the trails of most regions of America. The allergen in these plants, an oil called urushiol, has such extreme potency that a pinhead-sized amount could spur symptoms in 500 people, and it is so long-lasting that botanists must be careful when handling 100-year-old plant specimens. Although degrees of sensitivity vary among individuals, it is estimated that 85% of all Americans are potentially hypersensitive to this compound. Some people are so acutely sensitive that even the most minuscule contact, such as handling pets or clothes that have touched the plant or breathing vaporized urushiol, can trigger an attack.

Humans first become sensitized by contact during childhood. Individuals at great risk (firefighters, hikers) are advised to determine their degree of sensitivity using a skin test, so that they can be adequately cautious and prepared. Some odd remedies include skin potions containing bleach, buttermilk, ammonia, hair spray, and meat tenderizer. Commercial products are available for blocking or washing away the urushiol. Allergy researchers have found that oral vaccines containing a form of urushiol could desensitize experimental animals. An effective method using poison ivy desensitization injection is currently available to people with extreme sensitivity.

Learning to identify these common plants can prevent exposure and sensitivity. One old saying that might help warns, "Leaves of three, let it be; berries white, run with fright." Search YouTube for videos that help you differentiate between these three poisonous plants.

Explain why a single, first-time exposure to poison oak can sometimes result in a dermatitis reaction. *Answer available on Connect.*

Poison oak Poison sumac Poison ivy

(left, middle): *Source:* Robert H. Mohlenbrock @ USDA-NRCS PLANTS Database/USDA SCS. 1991. Southern wetland flora: Field office guide to plant species. South National Technical Center, Fort Worth; (right): *Source:* Jennifer Anderson. United States, IA, Scott Co., Davenport, Nahant Marsh. 2001

marrow) contain an indigenous population called passenger lymphocytes. This makes it quite possible for the graft to reject the host, causing **graft versus host disease (GVHD) (figure 16.16*b*).** Because any host tissue bearing MHC markers foreign to the graft can be attacked, the effects of GVHD are widely systemic and toxic. A papular, peeling skin rash is the most common symptom. Other organs affected are the liver, intestine, muscles, and mucous membranes. Previously, GVHD occurred in approximately 30% of bone marrow transplants within 100 to 300 days of the graft. This percentage is declining with the development of better screening and selection of tissues.

Classes of Grafts Grafts are generally classified according to the genetic relationship between the donor and the recipient. Tissue transplanted from one site to another on an individual's body is known as an **autograft.** Typical examples are skin replacement in burn repair and the use of a vein to fashion a coronary artery bypass. In an **isograft,** tissue from an identical twin is used. Because isografts do not contain foreign antigens, they are not rejected, but this type of grafting has obvious limitations. **Allografts** also called **allogeneic grafts,** the most common type of grafts, are exchanges between genetically different individuals belonging to the same species (two humans). A close genetic correlation is sought for most allograft transplants (see next section). A **xenograft** is a tissue exchange between individuals of different species. Until rejection can be better controlled, most xenografts are experimental or for temporary therapy only.

Avoiding and Controlling Graft Incompatibility

Graft rejection can be averted or lessened by directly comparing the tissue of the recipient with that of potential donors. Several tissue matching procedures are used. In the *mixed lymphocyte reaction (MLR),* lymphocytes of the two individuals are mixed and incubated. If an incompatibility exists, some of the cells will become activated and proliferate. *Tissue typing* is similar to blood typing,

(a) (b)

Figure 16.16 **Potential reactions in transplantation. (a)** The host's immune system (primarily cytotoxic T cells) encounters the cells of the donated organ (heart) and mediates the rejection of the organ. **(b)** Grafted tissue (bone marrow) contains endogenous T cells that recognize the host's tissues as foreign and mount an attack on many tissues and organs. The recipient will develop symptoms of graft versus host disease.

except that specific antisera are used to disclose the HLA antigens on the surface of lymphocytes. In most grafts (one exception is bone marrow transplants), the ABO blood type must also be matched. Although a small amount of incompatibility is tolerable in certain grafts (liver, heart, kidney), a closer match is more likely to be successful, so the closest match possible is sought.

Practical Examples of Transplantation

Today transplantation is a recognized medical procedure whose benefit is reflected in the fact that more than 100,000 transplants take place each year in the United States. It has been performed on every major organ, including parts of the brain. The most frequent transplant operations involve skin, liver, heart, kidney, coronary artery, cornea, and blood stem cells. The sources of organs and tissues are live donors (kidney, skin, bone marrow, liver), cadavers (heart, kidney, cornea), and fetal tissues. In the past decade, we have witnessed some unusual types of grafts. For instance, the fetal pancreas has been implanted as a potential treatment for diabetes and fetal brain tissues for Parkinson's disease.

Allogeneic hematopoietic stem cell transplantation is one of the most common medical procedures for patients suffering from immune deficiencies, aplastic anemia, leukemia and other cancers, and radiation damage. The three primary sources of stem cells used for this procedure are bone marrow, peripheral blood, and umbilical cord blood. These transplants are attempted in only the most severe cases because of a relatively high mortality rate associated with them. They are also costly, ranging from $100,000 to $250,000 for the entire process. For bone marrow transplants, a compatible donor with a closely matched MHC profile is sought. This person is sedated, and a bone marrow sample is aspirated by inserting a special needle into an accessible marrow cavity. The most favorable sites are the crest and spine of the ilium (major bone of the pelvis). During this procedure, between 500 and 800 ml of bone marrow are extracted with a special syringe. In a few weeks, the donor's marrow

will naturally replace itself. Implanting the harvested bone marrow is rather convenient, because it is not necessary to place it directly into the marrow cavities of the recipient. Instead, it is dripped intravenously into the circulation, and the new marrow cells automatically settle in the appropriate bone marrow regions.

To prepare recipients for the infusion procedure, they must be treated with chemotherapy and whole-body irradiation, a procedure designed to destroy their own blood stem cells and thus prevent rejection of the new stem cells. Within 2 weeks to a month after infusion, the grafted cells become established in the host. Because donor lymphoid cells can still cause GVHD, antirejection drugs may be necessary. A surprising consequence of bone marrow transplantation is that a recipient's blood type may change to the blood type of the donor.

Advances in stem cell technology have made it possible to harvest stem cells directly from the blood of donors by using *leukapheresis** to separate them and flow cytometry to identify and collect them (see discussion in chapter 7). Another potential source is the umbilical cord blood from a newborn infant. A decided advantage to cord blood is that its stem cells have yet to develop the donor's receptors and are less likely to cause transplant rejections.

* The process of selectively removing and collecting WBCs from the circulating blood, then returning the rest of the blood components back to the circulation.

 Check Your Progress SECTION 16.5

22. Explain the molecular/cellular basis for a host rejecting the graft and a graft rejecting the host.

23. Compare the four types of grafts.

24. What does it mean to say that tissues from two different individuals show a close match?

25. Describe the procedures involved in a bone marrow transplant and stem cell isolation.

16.6 Autoimmune Diseases: An Attack on Self

Expected Learning Outcomes

21. Outline the pathology of autoimmune diseases.

22. Explain the origins of autoimmunity and describe which persons are most often targeted by it.

23. Review the major types of autoimmune diseases and the adverse medical effects they have.

The immune diseases we have covered so far are all caused by foreign antigens. In the case of autoimmunity, individuals actually develop hypersensitivity to their own cells. This pathologic process accounts for **autoimmune diseases,** in which **autoantibodies,** T cells, and in some cases, both, mount an abnormal attack against antigens that originate from one's own cells. The scope of autoimmune diseases is extremely varied. In general, they are either *systemic,* involving several major organs, or *organ-specific,* involving only one organ or tissue. They usually fall into categories II or III hypersensitivity, depending upon how the autoantibodies bring about injury. Some major autoimmune diseases, their targets, and basic pathology are presented in **table 16.4.** (For a reminder of the four Gell and Combs categories, refer to table 16.1.)

Genetic and Gender Correlation in Autoimmune Disease

In most cases the precipitating cause of autoimmune disease remains obscure, but we do know that susceptibility is determined by genetics and influenced by gender and environmental factors. Cases cluster in families, and even unaffected members tend to develop autoantibodies for that disease. More direct evidence comes from studies of the major histocompatibility gene complexes. Particular genes in the class I and class II major histocompatibility complexes coincide with certain autoimmune diseases. For example, autoimmune joint diseases such as rheumatoid arthritis and ankylosing spondylitis are more common in persons with the B-27 human leukocyte antigen (HLA) type; systemic lupus erythematosus, Graves' disease, and myasthenia gravis are associated with the B-8 HLA. Why autoimmune diseases (except ankylosing spondylitis) afflict more females than males also remains a mystery. Females are more susceptible during childbearing years than before puberty or after menopause, suggesting possible disruption of the immune system by hormones and pregnancy.

The Origins of Autoimmune Disease

Because otherwise healthy individuals show low levels of autoantibodies, it is suspected that there is a normal function for them. A moderate, regulated amount of autoimmunity is probably required to dispose of old cells and cellular debris. Disease apparently arises when this regulatory or recognition apparatus goes awry. Although there is not yet a complete explanation for autoimmune diseases, several theories have been proposed.

The **sequestered antigen theory** explains that during embryonic growth, some tissues are immunologically privileged; that is, they are sequestered behind anatomical barriers and cannot be surveyed by the immune system. Examples of these sites are regions of the central nervous system, which are shielded by the meninges and blood-brain barrier; the lens of the eye, which is enclosed by a thick sheath; and antigens in the thyroid and testes, which are sequestered behind an epithelial barrier. In time, the antigen becomes exposed

TABLE 16.4	Selected Autoimmune Diseases		
Disease	Target	Type of Hypersensitivity	Characteristics
Systemic lupus erythematosus (SLE)	Systemic	II and III	Inflammation of many organs; antibodies against red and white blood cells, platelets, clotting factors, nucleus DNA
Rheumatoid arthritis and ankylosing spondylitis	Systemic	III and IV	Vasculitis; frequent target is joint lining; antibodies against other antibodies (rheumatoid factor)
Scleroderma	Systemic	II	Excess collagen deposition in organs; antibodies formed against many intracellular organelles
Hashimoto's thyroiditis	Thyroid	II	Destruction of the thyroid follicles
Graves' disease	Thyroid	II	Antibodies against thyroid-stimulating hormone receptors
Pernicious anemia	Stomach lining	II	Antibodies against receptors prevent transport of vitamin B_{12}
Myasthenia gravis	Muscle	II	Antibodies against the acetylcholine receptors on the nerve-muscle junction alter function
Type I diabetes	Pancreas	II	Antibodies stimulate destruction of insulin-secreting cells
Multiple sclerosis	Myelin	II and IV	T cells and antibodies sensitized to myelin sheath destroy neurons
Goodpasture syndrome (glomerulonephritis)	Kidney	II	Antibodies to basement membrane of the glomerulus damage kidneys
Rheumatic fever	Heart	II	Antibodies to group A streptococci cross-react with heart valve tissue

by means of infection, trauma, or deterioration and is perceived by the immune system as a foreign substance.

According to the clonal selection theory, the immune system of a fetus develops tolerance to self by eradicating or silencing all self-reacting lymphocyte clones, called **forbidden clones,** while retaining only those clones that react to foreign antigens. Some of these forbidden clones may survive; and because they have not been subjected to this tolerance process, they can attack tissues carrying self molecules, which they mistake as antigens.

The **theory of immune deficiency** proposes (1) that mutations in the receptor genes of some lymphocytes render them reactive to self and (2) that T-cell regulatory cells that normally maintain tolerance to self have become dysfunctional. These would both set the scene for autoimmunity. Abnormal expression of MHC II markers on cells that don't normally express them has been found to cause other immune reactions against self. In a related phenomenon, T-cell activation has been known to mistakenly "turn on" B cells that can react with self antigens. This phenomenon is called the bystander effect.

Some autoimmune diseases appear to be caused by **molecular mimicry,** in which microbial antigens bear molecular determinants similar to normal human cells. An infection with this type of microbe could stimulate antibodies that can cross-react with tissues. This is one purported explanation for the pathology of rheumatic fever. Another probable example of mimicry leading to autoimmune disease is the skin condition psoriasis. Although the etiology of this condition is complex and involves the inheritance of certain types of MHC alleles, infection with group A streptococci also plays a role. Scientists report that T cells primed to react with streptococcal surface proteins also react with keratin cells in the skin, causing them to proliferate. For this reason, psoriasis patients often report flare-ups after a streptococcal throat infection.

Other autoimmune disorders such as type I diabetes and multiple sclerosis are likely triggered by *viral infection*. Viruses can noticeably alter cell receptors, thereby causing immune cells to attack the tissues bearing viral receptors.

Some researchers believe that many, if not most, autoimmune diseases will someday be found to have an underlying microbial cause or influence. The basis for this could be molecular mimicry, viral alteration of host antigens, or the presence of as yet undetectable microbes in the sites affected by the autoimmunity.

Examples of Autoimmune Disease

Systemic Autoimmunities

One of the most severe chronic autoimmune diseases is **systemic lupus erythematosus*** (**SLE**), or *lupus*. This name originated from the characteristic butterfly-shaped rash positioned across the nose and cheeks **(figure 16.17a)**. Apparently ancient physicians thought the rash resembled a wolf bite (*lupus* is Latin for "wolf"). Although the manifestations of the disease vary considerably, all patients produce autoantibodies against a great variety of organs and tissues. Antibodies to intracellular materials such as the nucleoprotein of the nucleus and mitochondria are an additional diagnostic feature. In addition to a facial rash, lesions also occur on the skin of the

(a) (b)

Figure 16.17 Common autoimmune diseases. (a) Systemic lupus erythematosus. One symptom is a prominent rash extending across the nose, cheeks and forehead. Papules and blotches can also occur on the chest and limbs. **(b)** Rheumatoid arthritis commonly targets the synovial membranes of joints. Chronic inflammation thickens this membrane, erodes the articular cartilage, and fuses the joint. These effects severely limit motion and can swell and distort the joints as evidenced by the second and third panel. (a): © DermNet New Zealand Trust; (b): © Mediscan/Alamy Stock Photo

limbs and body. Other affected organs are the kidneys, bone marrow, nervous system, joints, muscles, heart, and GI tract.

In SLE, autoantibody-autoantigen complexes appear to be deposited in the basement membranes of various organs. Kidney failure, blood abnormalities, lung inflammation, myocarditis, and skin lesions are the predominant symptoms. The disease often cycles between periods of flare-ups and remission. One form of chronic lupus (called discoid) is influenced by exposure to UV radiation and primarily afflicts the skin. The etiology of lupus is still not completely understood. It is not known how such a generalized loss of self-tolerance arises, though viral infection or loss of T-cell suppressor function is suspected. The fact that women of childbearing years account for 90% of cases indicates that immune suppression during pregnancy may be involved. The diagnosis of SLE can usually be made with a combination of symptoms and blood tests. Antibodies against nuclear DNA and RNA and various tissues (detected by indirect fluorescent antibody or radioimmune assay techniques) are common, and a positive test for the lupus factor (an antinuclear factor) is indicative of the disease. Treatment aims at controlling flare-ups with steroids, methotrexate, and the monoclonal antibody belimumab (Benlysta).

Rheumatoid arthritis (RA),* another systemic autoimmune disease, incurs progressive, debilitating damage to the joints. In some patients, the lung, eye, skin, and nervous systems are also involved. In the joint form of the disease, autoantibodies form immune complexes that bind to the synovial membrane of the joints and activate phagocytes and stimulate release of cytokines. Chronic inflammation leads to scar tissue and joint destruction. The joints in the hands and feet are affected first, followed by the knee and hip joints **(figure 16.17b)**. The precipitating cause in rheumatoid arthritis is not known, though infectious agents such as Epstein-Barr virus have been suspected. The most common feature of the disease is the presence of an IgM antibody, called rheumatoid factor (RF), directed

* *systemic lupus erythematosus* (sis-tem′-ik loo′-pis air″-uh-theem-uh-toh′-sis) L. *lupus*, wolf, and *erythema*, redness.

* *rheumatoid arthritis* (roo′-muh-toyd ar-thry′-tis) Gr. *rheuma*, a moist discharge, and *arthron*, joint.

against other antibodies. This does not cause the disease but is used mainly in diagnosis. More drugs have been developed to treat rheumatoid arthritis than other autoimmune diseases. They include methotrexate, TNF inhibitors (etanercept [Enbrel], adalimumab [Humira], infliximab [Remicade]), rituximab (Retuxan), abatacept (Orencia), and tocilizumab (Actrema). All of these inhibit some key immune cell or substance responsible for the pathology of RA.

Autoimmunities of the Endocrine Glands

On occasion, the thyroid gland is the target of autoimmunity. The underlying cause of **Graves' disease** is the attachment of autoantibodies to receptors on the follicle cells that secrete the hormone thyroxin. The abnormal stimulation of these cells causes the overproduction of this hormone and the symptoms of hyperthyroidism. In **Hashimoto's thyroiditis,** both autoantibodies and T cells are reactive to the thyroid gland, but in this instance, they reduce the levels of thyroxin by destroying follicle cells and by inactivating the hormone. As a result of these reactions, the patient suffers from hypothyroidism.

The pancreas and its hormone, insulin, are other autoimmune targets. Type I **diabetes mellitus** is caused by a dysfunction in insulin production by beta cells in the pancreas or its utilization by cells. This form is associated with autoantibodies and sensitized T cells that damage the beta cells. A complex inflammatory reaction leading to lysis of these cells greatly reduces the amount of insulin secreted. For several years it has been proposed that type I diabetes is linked to infection by the coxsackievirus, a common cause of colds and enteric infections, possibly due to molecular mimicry.

Neuromuscular Autoimmunities

Myasthenia gravis* is named for the pronounced muscle weakness that is its principal symptom. Although the disease afflicts all skeletal muscle, the first effects are usually felt in the muscles of the eyes and throat. In some cases it can progress to complete loss of muscle function and death. Its cause is unknown, but it appears most often in females less than 40 years old and in males more than 60 years old, and it is often associated with having other autoimmune diseases. The classic syndrome is caused by autoantibodies binding to the receptors for acetylcholine, a chemical required to transmit a nerve impulse across the synaptic junction to a muscle. The immune attack so severely damages the muscle cell membrane that transmission is blocked and paralysis ensues. Treatment usually includes immunosuppressive drugs and therapy to remove the autoantibodies from the circulation. Experimental therapy using immunotoxins to destroy lymphocytes that produce autoantibodies shows some promise.

Multiple sclerosis (MS) is a paralyzing neuromuscular disease associated with lesions in the insulating myelin sheath that surrounds neurons in the white matter of the central nervous system. The underlying pathology involves damage to the sheath by both T cells and autoantibodies; this damage severely compromises the capacity of neurons to send impulses. The principal motor and sensory symptoms are muscular weakness and tremors, difficulties in speech and vision, and some degree of paralysis. Most MS patients first experience symptoms as young adults, and they tend to have remissions (periods of relief) alternating with recurrences of disease throughout their lives.

* *Myasthenia gravis* (my-uss"-theen'-ee-uh grah'vis.) L *myo,* muscle and *aesthene,* weakness.

Despite years of research, no definitive cause for MS has been pinpointed. It appears to arise from a combination of genetics, environmental exposure, and possibly viral infections. It is immune-mediated, but the antigen that triggers the attack on the myelin sheath has not yet been identified. A number of studies have examined the involvement of viral infections of the brain such as measles, herpes, and rubella. The only one that has a verifiable connection to MS is Epstein-Barr virus, but isolating the virus alone is not enough to show causation.

The disease can be treated with oral medications that suppress immune functions of lymphocytes, including cortisone, beta interferon, and monoclonal antibody drugs (natalizumab or Tysabri). An unfortunate side effect of many of these therapies is that they compromise normal immune function and increase the risks for infections and cancer. There has been some interest in techniques to restore the normal immune system through autologous (self) stem cell transplants taken from the marrow of the MS patient, which would bypass some of these complications.

 Check Your Progress SECTION 16.6

26. Explain the pathologic process in autoimmunity.
27. Draw diagrams that explain five possible mechanisms for the development of autoimmunity.
28. Describe four major types of autoimmunity, comparing target organs and symptoms.

16.7 Immunodeficiency Diseases and Cancer: Compromised Immune Responses

 Expected Learning Outcomes

24. Outline the categories of immunodeficiency diseases.

25. Describe the origins and effects of primary immune diseases caused by B-cell, T-cell, or combined defects.

26. Relate examples of secondary immunodeficiencies.

27. Describe the characteristics of cancer, and explain the mechanism of oncogenesis.

28. Explain how immune function relates to the development of cancer.

It is a marvel that development and function of the immune system proceed as normally as they do. On occasion, however, an error occurs and a person is born with or develops weakened immune responses. In many cases, these "experiments of nature" have provided penetrating insights into the exact functions of certain cells, tissues, and organs because of the specific signs and symptoms shown by the immunodeficient individuals. The pronounced consequences of **immunodeficiencies** are recurrent, overwhelming infections, often with opportunistic microbes, and adverse effects on overall health. Dysfunctional immunities can be grouped into two general categories: **primary diseases,** present at birth (congenital) and usually stemming from genetic errors, and **secondary diseases,** acquired after birth and caused by natural or artificial agents **(table 16.5).**

TABLE 16.5	General Categories of Immunodeficiency Diseases with Selected Examples

Primary Immune Deficiencies (Genetic)

B-Cell Defects (Low Levels of B Cells and Antibodies)

Agammaglobulinemia (X-linked, non-sex-linked)
Hypogammaglobulinemia
Selective immunoglobulin deficiencies

T-Cell Defects (Lack of All Classes of T Cells)

Thymic aplasia (DiGeorge syndrome)
Chronic mucocutaneous candidiasis

Combined B-Cell and T-Cell Defects (Severe combined immunodeficiency disease (SCID))

Adenosine deaminase (ADA) deficiency
X-SCID due to an interleukin defect
Wiskott-Aldrich syndrome
Ataxia-telangiectasia

Phagocyte Defects

Chédiak-Higashi syndrome
Chronic granulomatous disease of children
 (discussed in the chapter 16 Case Study)
Lack of surface adhesion molecules

Complement Defects

Lacking one of C components
Hereditary angioedema
Associated with rheumatoid diseases

Secondary Immune Deficiencies (Acquired)

From Natural Causes

Infection: AIDS, leprosy, tuberculosis, measles
Other disease: cancer, diabetes
Nutrition deficiencies
Stress
Pregnancy
Aging

From Immunosuppressive Agents

Irradiation
Severe burns
Steroids (cortisones)
Drugs to treat graft rejection and cancer
Removal of spleen

Primary Immunodeficiency Diseases

Deficiencies affect both specific immunities such as antibody production and less-specific ones such as phagocytosis, as featured in the chapter case study. Check out **figure 16.18** to survey the places in the normal sequential development of lymphocytes where defects can occur and the possible consequences. In many cases the deficiency is due to an inherited abnormality, though the exact nature of the abnormality is not known for a number of diseases. Because the development of B cells and T cells departs at some point, an individual can lack one or both cell lines. It must be emphasized, however, that some deficiencies affect other cell functions. For example, a T-cell deficiency can affect B-cell function because of the role of

CD4 T helper cells. In some deficiencies, the lymphocytes in question are completely absent or is present at very low levels, whereas in others, lymphocytes are present but do not function normally.

Clinical Deficiencies in B-Cell Development or Expression

Genetic deficiencies in B cells usually appear as an abnormality in immunoglobulin expression. In some instances, only certain immunoglobulin classes are absent; in others, the levels of all types of immunoglobulins (Ig) are reduced. A significant number of B-cell deficiencies are X-linked (also called sex-linked) recessive traits, meaning that the gene occurs on the X chromosome and the disease appears primarily in male children.

The term **agammaglobulinemia** literally means the absence of gamma globulin, the fraction of serum that contains immunoglobulins (Ig). Because it is very rare for Ig to be completely absent, some physicians prefer the term **hypogammaglobulinemia.*** T-cell function in these patients is usually normal. The symptoms of recurrent, serious bacterial infections usually appear about 6 months after birth. The bacteria most often implicated are pyogenic cocci, *Pseudomonas,* and *Haemophilus influenzae;* and the most common infection sites are the lungs, sinuses, meninges, and blood. Many Ig-deficient patients can have recurrent infections with viruses and protozoa, as well. Patients often manifest a wasting syndrome and may have a reduced life span, but modern therapy has improved their prognosis. About the only treatments available for this condition are passive immunotherapy with immune serum globulin and continuous antibiotic therapy.

The lack of a particular class of immunoglobulin is a relatively common condition. Although genetically controlled, its underlying mechanisms are not yet clear. IgA deficiency is the most prevalent, occurring in about 1 person in 600. Such persons have normal quantities of B cells and other immunoglobulins, but they are unable to produce IgA. Consequently, they lack protection against local microbial invasion of the mucous membranes and suffer recurrent respiratory and gastrointestinal infections. The usual treatment using Ig replacement does not work, because conventional preparations are high in IgG, not IgA.

Clinical Deficiencies in T-Cell Development or Expression

Due to T cells' critical role in immune defenses, a genetic defect in T cells results in a broad spectrum of disease, including severe opportunistic infections, wasting, and cancer. In fact, a dysfunctional T-cell line is usually more devastating than a defective B-cell line because T helper cells are required to assist in most specific immune reactions. The deficiency can occur anywhere along the developmental spectrum, from thymus to mature, circulating T cells.

Abnormal Development of the Thymus The most severe of the T-cell deficiencies involve a thymus gland that is underdeveloped (hypoplastic) or congenitally absent (aplastic). The disease associated with thymic dysfunction, **DiGeorge syndrome,** is caused by failure in the embryonic development of the third and

* *hypogammaglobulinemia* (hy′-poh-gem-ah-glob-yoo-lin-ee′-mee-ah) Gr. *hypo,* denoting a lowered level of Ig.

Figure 16.18 The stages of development and the functions of B cells and T cells, whose failure causes severe immunodeficiencies. Dotted lines represent the phases in development where breakdown can occur.

fourth pharyngeal pouches, and in most cases is associated with a deletion mutation in chromosome 22. The adverse effects on thymic function range from reduced immune activity to complete lack of cell-mediated immunity, especially of the CD4 or helper class of T cells. This makes makes children highly susceptible to persistent infections by fungi, protozoa, and viruses. Common, usually benign childhood infections such as chickenpox, measles, or mumps can be fatal if not diagnosed and treated right away. Even vaccinations using attenuated microbes pose a danger. Part of the severity of Di-George syndrome is the lack of antibodies resulting from deficiencies in helper T cells that are needed to activate B cells. Other symptoms of this syndrome are heart defects, reduced growth, wasting of the body, unusual facial characteristics **(figure 16.19)**, and an increased incidence of lymphatic cancer. The only therapy that could restore their immune system is a thymus transplant. Other conditions are alleviated by surgery and drugs to treat the recurrent infections.

Severe Combined Immunodeficiencies: Dysfunction in B and T Cells

Severe combined immunodeficiencies (SCIDs) are the most dire and potentially lethal of the immunodeficiency diseases because they involve a loss of both T and B lymphocyte functions. Some SCIDs are due to the complete absence of the lymphocyte stem cell in the marrow; others are attributable to the dysfunction of B cells and T cells later in development. Infants with SCIDs usually manifest the T-cell deficiencies within days after birth by developing

Figure 16.19 A composite showing facial appearance of a child with DiGeorge syndrome, sometimes referred to as velocardiofacial syndrome. Altered development of the facial regions of the embryo leads to some degree of anatomical abnormalities. Examples of the most common features include low-set, deformed earlobes; prominent ears; wide-set eyes; folds in the eyelids; wide nose bridge and small nostrils; small jaw and chin; a small, bow-shaped mouth; and the absence of a philtrum (the vertical furrow between the nose and upper lip). Not every child manifests all of these together.

candidiasis, sepsis, pneumonia, or systemic viral infections. This debilitating condition appears to have several forms. In the two most common forms, Swiss-type agammaglobulinemia and thymic alymphoplasia, the numbers of all types of lymphocytes are extremely low, the blood antibody content is greatly diminished, and the thymus and cell-mediated immunity are poorly developed. Both diseases are due to a genetic defect in the development of the lymphoid cell line.

A rare form of SCID is **adenosine deaminase (ADA) deficiency,** which is caused by an autosomal recessive defect in the metabolism of adenosine. In this case, lymphocytes develop, but a metabolic product builds up to abnormal levels and selectively destroys the cells. Infants with ADA deficiency are subject to recurrent infections and severe wasting typical of severe deficiencies. Some SCID cases are due to a developmental defect in receptors for B and T cells. One of these, termed **X-SCID,** is a deficiency in interleukin receptors that occurs in about 1 in 50,000 to 100,000 births, Because it is X-linked, it occurs most often in male infants and was responsible for the disease of David, the child in the "plastic bubble" **(figure 16.20).** As heart-wrenching as David's life was, he also provided a means to study and understand the disease in a way that had never been possible. Later studies found that he had inherited a defective gene for essential molecules in several interleukin receptors. This prevented both T and B cells from receiving signals that control growth, development, and responsiveness. Both cytotoxic and humoral immunities were shut down, leaving the body vulnerable to infections and cancers.

Because of their profound lack of specific adaptive immunities, children with SCID require the most rigorous kinds of aseptic techniques to protect them from opportunistic infections. Nowadays a better option for long-term survival is an immediate bone marrow or stem cell transplant right after the disease is diagnosed, rather than placing them into sterile bubbles. Because they lack functioning lymphocytes, the requirement for a bone marrow match is not as important with SCID children. Although transplanting bone marrow has been about 50% successful in curing the disease, it is complicated by graft versus host disease. This outcome can be corrected by selectively removing the reactive T cells from the donated marrow. Some infants have benefited from fetal liver or stem cell grafts, and some ADA-deficient patients can remain healthy through periodic transfusions of blood containing large amounts of the normal enzyme. A more lasting treatment for all types of SCID has emerged from gene therapy. So far medical geneticists have cured two dozen children with SCID by inserting normal genes to replace the defective genes (see figure 10.15).

Secondary Immunodeficiency Diseases

Secondary acquired deficiencies in B cells and T cells are caused by one of four general agents: (1) infection, (2) organic disease, (3) chemotherapy, or (4) radiation.

The most recognized infection-induced immunodeficiency is **acquired immunodeficiency syndrome** or **AIDS.** This syndrome is caused when several types of immune cells, including T helper cells, monocytes, macrophages, and antigen-presenting cells, are infected by the human immunodeficiency virus (HIV). It is generally thought that

> **Quick Search**
>
> See a slide show summarizing David's story at "Bubble Boy 40 Years Later" from *CBS News.*

Figure 16.20 An answer to the mystery of David. David Vetter, the best-known SCID child, lived all but the last 2 weeks of his life in a sterile environment to isolate him from the microorganisms that could have quickly ended his life. When David was 12, his doctors attempted a bone marrow transplant using his sister's bone marrow, but the marrow harbored Epstein-Barr virus. Because he lacked any form of protective immunities against this oncogenic virus, a cancer spread rapidly through his body. © Baylor College of Medicine, Public Affairs

the depletion of T helper cells and functional impairment of immune responses ultimately account for the cancers and opportunistic protozoan, fungal, and viral infections associated with this disease. (See chapter 25 for an extensive discussion of HIV and AIDS.) Other infections that can deplete immunities are tuberculosis, leprosy, and malaria.

Cancers that target the bone marrow or lymphoid organs can be responsible for extreme malfunction of both humoral and cellular immunity. In leukemia, the massive numbers of cancer cells compete for space and literally displace the normal cells of the bone marrow and blood. Plasma cell tumors produce large amounts of nonfunctional antibodies, and thymus gland tumors cause severe T-cell deficiencies.

The Role of the Immune System in Cancer

The term **cancer** comes from the Latin word for crab and presumably refers to the appendage-like projections that a spreading tumor develops. A cancer is defined as the new growth and spread of abnormal cells. The disease is also known by the synonym **neoplasm,*** or by more specific terms that usually end in the suffix *-oma.* **Oncology*** is the field of medicine that specializes in cancer.

An abnormal growth or tumor is generally characterized as benign or malignant. A **benign tumor** is a self-contained mass within an organ that does not spread into adjacent tissues. The mass is usually slow-growing, rounded, and not greatly different from its tissue of origin. Benign tumors do not ordinarily cause death unless they grow into critical spaces such as the heart valves or brain ventricles. The feature that most distinguishes a **malignant tumor** (cancer) is uncontrolled growth of abnormal cells within normal tissue. As a general rule, cancer originates in cells such as skin and bone marrow cells that have retained

* *neoplasm* (nee'-oh-plazm) Gr. *neo.,* new, and *plasm,* formation.
* *oncology* (awn-kaw'-luh-gee″) Gr. *onkos,* cancer.

the capacity to divide, whereas mature cells that have lost this power (some neurons, for instance) do not commonly become cancerous.

Research findings link cancers to genetic damage or inherited genetic predispositions that alter the function of certain genes called oncogenes found normally in all cells. These genetic alterations disrupt the normal cell division cycle and *transform* a normal cell into a cancer cell. Cancer development involves many complex interactions between genes, their products, and external signals received by the cell. There is no single mechanism. Consider the fact that at least eight different genetic defects lead to colon cancer and nearly that many are implicated in breast cancers.

It is increasingly evident that susceptibility to certain cancers is inherited. At least 30 neoplastic syndromes that run in families have been described. A particularly stunning discovery is that some cancers in animals are actually due to endogenous viral genomes passed from parent to offspring in the gametes. Many cancers reflect an interplay of both hereditary and environmental factors. An example is lung cancer, which is associated mainly with tobacco smoking. But not all smokers develop lung cancer, and not all lung cancer victims smoke tobacco.

What is the role of the immune system in controlling cancer? To understand the connection, we need to review the actions of certain lymphocytes discussed in chapter 15. A concept that readily explains the detection and elimination of cancer cells is **immune surveillance.** Many experts postulate that cells with cancer-causing potential arise constantly in the body but that the immune system ordinarily discovers and destroys these cells, thus keeping cancer in check. Experiments with mammals have amply demonstrated that components of cell-mediated immunity interact with tumors and their antigens. The primary types of cells that operate in surveillance and destruction of tumor cells are cytotoxic T cells, natural killer (NK) cells, and macrophages. It appears that these cells recognize abnormal or foreign surface markers on the tumor cells and destroy them by mechanisms shown in figure 15.16. Antibodies help destroy tumors by interacting with macrophages and natural killer cells. This involvement of the immune system in destroying cancerous cells has inspired a number of specific immunotherapies.

How do we account for the commonness of cancer in light of this powerful range of defenses against tumors? To an extent, the answer is simply that, as with infection, the immune system can and does fail. In some cases, the cancer may not be immunogenic enough; it may display self markers and not be targeted by the surveillance system. In other cases, the tumor antigens may have mutated to escape detection. As we saw earlier, patients with immunodeficiencies such as AIDS and SCID are more susceptible to various cancers because they lack essential T-cell or cytotoxic functions. It may turn out that most cancers are associated with some sort of immunodeficiency, even a slight or transient one. See the discussion of immune regulation by T cells in section 15.3.

An ironic outcome of lifesaving medical procedures is the possible suppression of a patient's immune system, which could leave them susceptible to cancer. Although radiation and anticancer drugs are the first line of therapy for many types of cancer, both agents are extremely damaging to the bone marrow and other body cells. Some immunosuppressive drugs that prevent graft rejection by T cells can likewise suppress responses against neoplastic cells.

Check Your Progress SECTION 16.7

29. In general, how do primary immunodeficiencies and acquired immunodeficiencies differ in their basic causes?
30. What kinds of symptoms accompany B-cell defects, T-cell defects, and combined defects, and why is this the case?
31. Give examples of specific diseases that involve each type of defect.
32. Define *cancer,* and differentiate between a benign tumor and a malignant tumor, using examples.
33. Describe the relationship between cancer and the immune system.

CASE STUDY Part 2

Hospital investigators reviewing the events in the case were faced with explaining the fatal blood transfusion to the patient's family. What could have gone wrong? Going by the first typing of the patient's blood and the cross-match with the transfused blood, nothing unusual would have been expected. But there was no question that he suffered an extreme transfusion reaction. When the patient's blood was retyped after death, it turned out that he was actually O$^+$, the most common blood type among humans. One possible explanation for the adverse outcome is that he reacted to a blood group other than ABO. But his blood had the same Rh factor as the transfused blood and was compatible for Rh factor. And in cases of the minor RBC antigens such as MN and Kell, the reactions are usually quite mild and never result in mass hemolysis and death. As the investigation continued, it became evident that the cause of this tragedy was not a mistake in the technology of the blood typing or some unusual characteristic of the blood, but rather it was placed squarely on human error in record keeping.

The patient in this case was caught in a worst-case scenario. First, he had the misfortune of being mistaken for his hospital roommate, and second, he was given type A blood, one of the most serious incompatibilities for type O. The blood vial was labeled with his correct name but was used to take blood from the man in the adjacent bed. This same blood was also used in the cross-match, so it did not react with the transfusion blood. A safer system used by many hospitals to prevent this sort of error is to have a microchip embedded in the patient's armband that reads out his blood type. It is compared to the blood being transfused and sounds an alarm if the bloods are incompatible.

■ *Why do people carry antibodies against other blood types even when they have never been sensitized by prior contact with that blood?*

■ *If the transfusion reaction had not progressed so rapidly, what blood could have been safely given to the patient and why?*

To conclude this Case Study, go to Connect.

To read more about the subject of transfusion reactions, go to http://emedicine.medscape.com/article/780074-overview.

 Chapter Summary with Key Terms

16.1 The Immune Response: A Two-Sided Coin

The study of disease states involving malfunctions of the immune system is called immunopathology.

A. **Allergy** and hypersensitivity are terms for an exaggerated, misdirected expression of certain immune responses. "Allergy" is used mainly for immediate type reactions and "hypersensitivity" for other forms of overreactions to allergens.

B. Autoimmunity involves abnormal responses to self antigens. A deficiency or loss in immune function is called **immunodeficiency.**

C. Immediate-onset allergies involve contact with **allergens,** antigens that affect certain people; susceptibility is inherited.

D. Allergens enter through four portals: Inhalants are breathed in (pollen, dust); ingestants are swallowed (food, drugs); injectants are inoculated (drugs, bee stings); contactants react on skin surface (cosmetics, glue).

16.2 Allergic Reactions: Atopy and Anaphylaxis

A. On first contact with allergen, specific B cells react with allergen and form a special antibody class called IgE, which affixes by its Fc receptor to mast cells and basophils.

1. This **sensitizing dose** primes the allergic response system.

2. Upon subsequent exposure with a **provocative dose,** the same allergen binds to the IgE–mast cell complex.

3. This causes **degranulation,** release of intracellular granules containing mediators (**histamine, serotonin, leukotriene, prostaglandin),** with physiological effects such as vasodilation and bronchoconstriction.

4. Conditions that are common types of atopy are **eczema,** food allergies, and **asthma;** severe systemic reaction is **anaphylaxis.**

5. Other symptoms are rash, itching, redness, increased mucous discharge, pain, swelling, and difficulty in breathing.

B. Diagnosis of allergy can be made by a histamine release test on basophils; serological assays for IgE; and skin testing, which injects allergen into the skin and mirrors the degree of reaction.

C. Control of allergy involves drugs to interfere with the action of histamine, inflammation, and release of cytokines from mast cells. **Desensitization** or hyposensitzation therapy involves the administration of purified allergens.

16.3 Type II Hypersensitivities: Reactions That Lyse Foreign Cells

A. Type II reactions involve the interaction of antibodies, foreign cells, and complement, leading to lysis of the foreign cells. In transfusion reactions, humans may become sensitized to special antigens on the surface of the red blood cells of other humans.

B. The **ABO blood groups** are genetically controlled: Type A blood has A antigens on the RBCs; type B has B antigens; type AB has both A and B antigens; and type O has neither antigen. People produce antibodies against A or B antigens if they lack these antigens. Antibodies can react with antigens if the wrong blood type is transfused.

C. **Rh factor** is another RBC antigen that becomes a problem if an Rh⁻ mother is sensitized by an Rh⁺ fetus. A second fetus can receive antibodies she has made against the factor and develop hemolytic disease of the newborn. Prevention involves therapy with Rh immune globulin.

16.4 Type III Hypersensitivities: Immune Complex Reactions

Exposure to a large quantity of soluble foreign antigens (serum, drugs) stimulates antibodies that produce small, soluble Ab-Ag complexes. These **immune complexes** are trapped in various organs and tissues, which incites a damaging inflammatory response.

16.5 Immunopathologies Involving T Cells

A. *Type IV Delayed Hypersensitivity*

1. A delayed response to antigen involving the activation of and damage by T cells.

2. Skin response to allergens, including infectious agents. Example is tuberculin reaction.

3. Contact dermatitis is caused by exposure to plants (ivy, oak) and simple environmental molecules (metals, cosmetics).

4. Cytotoxic T cells acting on allergen elicit a skin reaction.

B. **Graft Rejection:** Reaction of cytotoxic T cells directed against foreign cells of a grafted tissue; involves recognition of foreign human leukocyte antigens (HLA) by T cells and rejection of tissue.

1. Host may reject graft; graft may reject host.

2. Types of grafts include: **autograft,** from one part of body to another; **isograft,** between identical twins; **allograft,** between two members of same species; **xenograft,** between two different species.

3. All major organs may be successfully transplanted.

4. Allografts require tissue match (HLAs must correspond); rejection is controlled with drugs.

16.6 Autoimmune Diseases—An Attack on Self

A. In certain type II and type III hypersensitivities, the immune system has lost tolerance to self molecules (**autoantigens)** and forms **autoantibodies** and sensitized T cells against them. Disruption of function can be systemic or organ-specific.

B. **Autoimmune diseases** are genetically determined and more common in females.

C. Examples include **systemic lupus erythematosus, rheumatoid arthritis, diabetes mellitus,** and **multiple sclerosis.**

16.7 Immunodeficiency Diseases and Cancer: Compromised Immune Responses

I. Immunodeficiency Diseases

Components of the immune response system are absent. Deficiencies involve B and T cells, phagocytes, and complement.

A. Primary immunodeficiency is genetically based, congenital; defect in inheritance leads to lack of B-cell activity, T-cell activity, or both.

B. B-cell defect is called **agammaglobulinemia;** patient lacks antibodies; serious recurrent bacterial infections result. In Ig deficiency, one of the classes of antibodies is missing or deficient.

C. In T-cell defects, the thymus is missing or abnormal. In **DiGeorge syndrome,** the thymus fails to develop; afflicted children experience recurrent infections with eukaryotic pathogens and viruses; immune response is generally underdeveloped.

D. In **severe combined immunodeficiency (SCID),** both limbs of the lymphocyte system are missing or defective; no adaptive immune response exists; fatal without replacement of bone marrow or other therapies.

E. Secondary (acquired) immunodeficiency is due to damage after birth (infections, drugs, radiation). **HIV/AIDS** is the most common of these; T helper cells are main target; deficiency manifests in numerous opportunistic infections and cancers.

II. Cancer: The Role of the Immune System

A. **Cancer** is characterized by overgrowth of abnormal tissue, also known as a **neoplasm.** It can arise when normal **immune surveillance** by lymphocytes malfunctions and fails to detect and destroy tumors and cancers, allowing them to grow and spread.

1. **Tumors** can be **benign** (a nonspreading local mass of tissue) or **malignant** (a cancer) that spreads (metastasizes) from the tissue of origin to other sites.
2. Cancers occur in nearly every cell type (except mature, nondividing cells).

B. Cancer cells appear to share a common basic mechanism involving a genetic alteration that **transforms** a normally dividing cell into one that grows out of control.

Level I. Knowledge and Comprehension

These questions require a working knowledge of the concepts in the chapter and the ability to recall and understand the information you have studied.

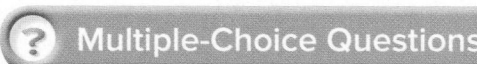

Multiple-Choice Questions

Select the correct answer from the answers provided. For questions with blanks, choose the combination of answers that most accurately completes the statement.

1. Pollen is which type of allergen?
 a. contactant c. injectant
 b. ingestant d. inhalant

2. B cells are responsible for which conditions?
 a. asthma c. tuberculin reactions
 b. anaphylaxis d. both a and b

3. Which hypersensitivities are T-cell mediated?
 a. type I c. type III
 b. type II d. type IV

4. The contact with allergen that results in symptoms is called the
 a. sensitizing dose c. provocative dose
 b. degranulation dose d. desensitizing dose

5. Production of IgE and degranulation of mast cells are involved in
 a. contact dermatitis c. Arthus reaction
 b. anaphylaxis d. both a and b

6. The direct, immediate cause of allergic symptoms is the action of
 a. the allergen directly on smooth muscle
 b. the allergen on B lymphocytes
 c. allergic mediators released from mast cells and basophils
 d. IgE on smooth muscle

7. Theoretically, type _____ blood can be donated to all persons because it lacks _____.
 a. AB, antibodies c. AB, antigens
 b. O, antigens d. O, antibodies

8. An example of a type III immune complex disease is
 a. serum sickness c. graft rejection
 b. contact dermatitis d. atopy

9. Type II hypersensitivities are due to
 a. IgE reacting with mast cells
 b. activation of cytotoxic T cells
 c. IgG-allergen complexes that clog epithelial tissues
 d. complement-induced lysis of cells in the presence of antibodies

10. Production of autoantibodies may be due to
 a. emergence of forbidden clones of B cells
 b. production of antibodies against sequestered tissues
 c. infection-induced change in receptors
 d. All of these are possible.

11. Rheumatoid arthritis is an _____ that affects the _____.
 a. immunodeficiency disease, muscles
 b. autoimmune disease, nerves
 c. allergy, cartilage
 d. autoimmune disease, joints

12. A positive tuberculin skin test is an example of
 a. a delayed-type allergy c. acute contact dermatitis
 b. autoimmunity d. eczema

13. Contact dermatitis can be caused by
 a. pollen grains
 b. chemicals absorbed by the skin
 c. microbes
 d. proteins found in foods

14. Which disease would be most similar to AIDS in its pathology?
 a. X-linked agammaglobulinemia
 b. SCID
 c. ADA deficiency
 d. DiGeorge syndrome

15. Which of these immunopathologies could be treated with a stem cell transplant?
 a. arthus reaction d. a and b only
 b. SCID e. b and c only
 c. anaphylaxis

16. How is the immune system involved in development of cancer?
 a. failure in immune surveillance
 b. mutation in cytotoxic T cells
 c. autoantibody formation
 d. overreaction to environmental chemicals

 Case Study Review

1. Which blood type antibodies would the patient normally have carried in his plasma? (Explain your answer.)
 a. anti-a alone
 b. anti-b alone
 c. anti-Rh alone
 d. anti-a and -b
 e. anti-a, -b, and -Rh
 f. neither anti-a nor anti-b

2. If the patient had a blood type different from O$^+$, which one would *not* have resulted in his death? (Explain your answer.)
 a. type A$^-$
 b. type B$^+$
 c. type AB$^-$
 d. type O$^-$

3. Given that it is the donated RBCs and not the patient's normal RBCs that are hemolyzed, how can a hemolytic transfusion reaction still be fatal?

 Writing Challenge

For each question, compose a one- or two-paragraph answer that includes the factual information needed to completely address the question. Check Your Progress questions can also be used for writing-challenge exercises.

1. Compare and contrast atopic allergy and type IV (delayed) hypersensitivity as to mechanism, symptoms, eliciting factors, and allergens.
2. Go through the sequence of events that gives rise to an anaphylactic response to peanuts.
3. How does a baby inherit Rh$^+$ blood from an Rh$^-$ mother?
4. Why is a hemolytic transfusion reaction considered a type of hypersensitivity?
5. Describe three circumstances that might cause antibodies to develop against self tissues.
6. Explain how people with autoimmunity could develop antibodies against intracellular components (nucleus, mitochondria, and DNA).

 Concept Mapping

On Connect you can find an Introduction to Concept Mapping that provides guidance for working with concept maps and a sample of a concept map for this chapter.

 Level II. Application, Analysis, Evaluation, and Synthesis

These problems go beyond just restating facts and require higher levels of understanding and an ability to interpret, problem solve, transfer knowledge to new situations, create models, and predict outcomes.

 Critical Thinking

Critical thinking is the ability to reason and solve problems using facts and concepts. These questions can be approached from a number of angles, and in most cases, they do not have a single correct answer.

1. Suggest some possible physiological benefits of allergy.
2. A 3-week-old neonate develops severe eczema after being given penicillin therapy for the first time. Explain what probably has happened.
3. Why would a person be allergic to strawberries when he eats them but show a negative skin test to them?
4. a. Where in the course of type I allergies do antihistamine drugs, cortisone, and desensitization work? Exactly what do they do?
 b. Compare the sites of action of montelukast (Singulair) and omalizumab (Xolair).

5. Explain how cleaner living and medical advances could contribute to allergies.

6. Why would it be necessary for an Rh⁻ woman who has had an abortion, miscarriage, or an ectopic pregnancy to be immunized against the Rh factor?

7. How can a person prevent becoming allergic to poison oak? What is the basis for infectious allergy?

8. Why are primary immunodeficiencies considered "experiments of nature"?

9. a. Explain why babies with agammaglobulinemia do not develop opportunistic infections until about 6 months after birth.

 b. Explain why people with B-cell deficiencies can benefit from artificial passive immunotherapy. Explain whether vaccination would work for them.

 c. Explain why T-cell deficiencies usually cause more severe effects than B-cell deficiencies.

10. In what ways can cancer be both a cause and a symptom of immunodeficiency?

 Visual Challenge

1. Looking at **figure 15.8,** reproduced here, explain how this phenomenon pertains to allergies.

2. Observe the blood typing results shown here and label the blood types, including Rh. Which types are the most and least common?

Like tiny green sausages, these vibrios (13,000x) cause the majority of food infections that originate from shellfish. *Source:* Janice Haney Carr/CDC

Seafood: A hidden source of pathogens.
© PhotoBliss/Alamy Stock Photo RF

 CASE STUDY Part 1 A Hospital Outbreak of Seafood Sickness

Personnel working in a Taiwanese hospital during a spring heat wave were alerted to a possible outbreak of **health-care-associated infections.** The first evidence came in a report from the nursing staff about nine patients in two adjacent wards with severe **diarrhea, fever, nausea,** and **vomiting** several hours after eating a seafood meal. An **infection control** team was immediately brought in to monitor and record all hospital-wide cases of **gastroenteritis.** Stool and food samples were sent to their clinical laboratory and the local public health department for testing. Over the next 48 hours the staff logged 16 additional cases spread among five wards in two separate buildings.

Specimens were screened for **common enteric pathogens,** including *Salmonella* species, *Shigella* species, *Staphylococcus aureus, Vibrio* species., *Campylobacter jejuni, Bacillus cereus,* and *Clostridium* species. Plating for initial isolation was performed on **blood agar, MacConkey agar,** and other selective and differential media. **Gram stains** of the stool samples were performed early on, but

> "A hanging drop slide viewed with dark-field microscopy revealed curved and comma-shaped rods with darting motility."

they were inconclusive, yielding mostly gram-negative rods mixed with some gram-positive rods and cocci. After 2 days, colonies grew in the presence of air on blood agar and on **selective media** containing **bile.** A second Gram stain of pure cultures indicated that the isolates were **gram-negative rods,** and a **hanging drop slide** viewed with **dark-field microscopy** revealed **curved** and comma-shaped rods with **darting motility. Case Study table 1** summarizes the profiles of the potential pathogens in this case. They can be compared with results from actual patient and food specimens as a means of narrowing the initial identification.

- *Explain the reason for using bile as a selective agent and the rationales for the preliminary biochemical tests listed in table 1.*

- *Given the data so far, identify the possible infectious agents that best fit the pathogen profile.*

To continue the Case Study, go to Case Study Part 2 at the end of the chapter.

CASE STUDY TABLE 1					
TEST	*Salmonella*	*Shigella*	*Vibrio*	*Campylobacter*	Patient and Food Isolate Results
Motility	+	–	+	+	+
Shape	Straight rods	Straight rods	Curved rods	Curved, spiral rods	Curved rods
Oxidase	–	–	+	+	+
Growth on 6% salt medium	–	–	+	+	+
Oxygen requirements	Facultative	Facultative	Aerobic	Microaerophilic	Aerobic
Fermentation of glucose	+	+	+	–	+

SCOPING OUT THE CHAPTER

Any time an infection rears its ugly head, the occasion always begins with a mystery. What exactly is the cause of this illness? Is that red patch on my foot a fungus or a wart? Is this rash measles or chickenpox? Is my sore throat caused by strep or a cold? Fortunately, modern medicine has reached a point in diagnosis that most infections don't remain a mystery for very long. Specimens are taken, cultures are made and inspected, symptoms assessed, and high powered diagnostic tests ordered—all usually within a few hours after seeking medical attention. This chapter focuses on what goes on behind the scenes of a clinical microbiology laboratory as it processes samples, identifies pathogens, and helps support medical treatment.

Directly Examining Microbes: Gathering Background Data The earliest instrument of the microbiology laboratory was the microscope, and even now, a magnified specimen or culture is a powerful tool for revealing unique characteristics of the cells or viruses discovered there. Observing microbes grown on a variety of specialized media can be an indispensable way to develop morphological and biochemical profiles of a suspected infectious agent.

Checking out DNA and RNA Another essential method for detecting and identifying pathogens is to extract their nucleic acids, develop a genetic profile of their DNA or RNA, and compare it with profiles of known organisms or viruses. The uniqueness of a genetic profile makes this one of the most specific methods available, although it does require additional technology and expertise.

Tests Based on Antibodies and Antigens

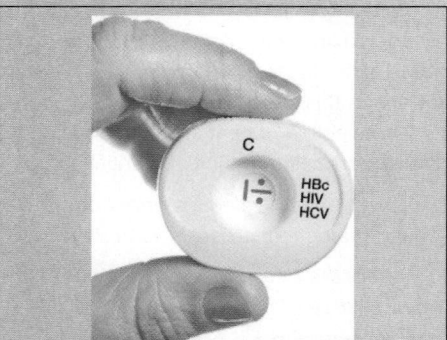

Labs On the Go: Rapid Testing Some of the newest diagnostic and identification techniques are based on compact, quick tests that can be done in a medical office, at home, or in the field. These tests are generally designed to detect the possible presence of pathogens from specimens and provide immediate feedback on the possibility of an infection that can read out as a simple color change.

(1) Cells infected with herpesvirus
(2) Cells infected with herpesvirus # 6

Identifying Microbes With Antibodies Another form of immune testing is the use of specially-made antibodies to interact with unknown pathogens to form a visible reaction. This technique supports the identification of viral and bacterial pathogens isolated from a variety of samples, including the body, food, and water.

Looking for Signs of Infection in the Blood Most infections leave behind a tell-tale sign--antibodies that have been formed in response to the pathogen. When detected in a blood or serum sample, they can provide the immune history of an individual and a possible diagnosis. Tests of this type rely on the specificity that antibodies have for their targets, so that a positive reaction can verify that a person has had some contact with a pathogen such as HIV.

(top-left): Courtesy of bioMerioux, Inc.; (top-right): Source: Zaidenstein R, Sadik C, Lerner L, Valinsky L, Kopelowitz J, Yishai R, et al. Clinical characteristics and molecular subtyping of Vibrio vulnificus illnesses, Israel. *Emerg Infect Dis* [10.3201/eid1412.080499]. 2008 Dec [2013 July].; (bottom-left): Courtesy of MedMira Inc.; (bottom-middle): Source: Zaki Salahuddin. Laboratory of Tumor Cell Biology/National Cancer Institute

17.1 An Overview of Clinical Microbiology

 Expected Learning Outcomes

1. Describe what is involved in the main categories of laboratory identification of infectious agents.
2. Explain several techniques in specimen collection and the concerns in acquiring good cultures.
3. Summarize the main procedures in isolation, cultivation, and identification of pathogens.

In earlier chapters (3, 4, 5, 6, and 10) we introduced a variety of laboratory procedures used in cultivating and isolating microorganisms from samples. These techniques are an important part of studying and characterizing microbes, whether it be for research purposes or diagnosing infectious diseases. For many professionals and even beginning microbiology students, the most pressing concern is *how to identify unknown microbes* in specimens or **cultures.** The methods microbiologists use to identify bacteria to the level of genus and species fall into three main categories: **phenotypic,** which includes a consideration of morphology (microscopic and macroscopic) as well as physiology or biochemistry; **immunologic,** which entails analysis of the blood (serology) and other fluids; and **genotypic** (or genetic) techniques.

Data from a cross section of such tests can produce a unique profile of each isolate. Increasingly, genetic means of identification are being used as a sole resource for identifying bacteria. As universally used databases of genetic profiles of microbes become more complete, these types of analyses provide a more accurate and speedy way of identifying microbes than was possible even a decade ago. There are still many organisms, however, that must be identified through traditional means. Biochemical tests and rapid testing systems continue to be a mainstay of the clinical laboratory. Serology is a highly reliable procedure for some diseases, and is still favored as a way to screen large numbers of people rapidly and economically. In the next sections we emphasize cultural, serological, and genetic methods of analysis. Additional information on the methods for identifying bacteria can be found in Appendix C. The specialized techniques required for identification and diagnosis of viruses and viral diseases are covered in section 17.6.

Phenotypic Methods

Microscopic Morphology

Traits that can factor in identification of bacteria are combinations of cell shape and size, Gram stain reaction, acid-fast reaction, and special structures, including endospores, granules, and capsules. Electron microscope studies can pinpoint additional structural features (such as the cell wall, flagella, pili, and fimbriae). Microscopic analysis is also used to evaluate specimen contents prior to culturing. This can often narrow the causative agent as bacterial, fungal, or protozoan.

Macroscopic Morphology

Cultural traits that can be assessed with the naked eye are also useful in diagnosis. These include the appearance of growth, including texture, size, shape, pigment of colonies, speed of growth, and reactions to special types of selective and differential media.

Physiological/Biochemical Characteristics

Physiological and biochemical characteristics have been the traditional mainstay of bacterial identification. Enzymes and other biochemical properties of bacteria are fairly reliable and stable expressions of the chemical identity of each species. Dozens of diagnostic tests exist for determining the presence of specific enzymes and for assessing nutritional and metabolic activities. Examples include tests for fermentation of sugars; capacity to digest or metabolize complex polymers such as proteins and polysaccharides; production of gas; presence of enzymes such as catalase, oxidase, and decarboxylases; and sensitivity to antimicrobic drugs. Special rapid identification test systems that record the major biochemical reactions of a culture have streamlined data collection and analysis.

Genotypic Methods

Examining the genetic material itself has revolutionized the identification and classification of bacteria. There are many advantages of genotypic methods over phenotypic methods, when they are available. The primary advantage is that actually culturing the microorganisms is not always necessary. In recent decades, scientists have come to realize that there are many more microorganisms that we can't grow in the lab compared with those that we can (3.1 Secret World of Microbes). Another advantage is that genotypic methods are increasingly automated, and results are obtained very quickly, often with more precision than with phenotypic methods.

Immunologic Methods

Bacteria and other microbes have surface and other molecules called antigens that are recognized by the immune system. One immune response to antigens is the production of antibodies that are designed to bind tightly to the antigens. The nature of the antibody response can be determined from a blood (or other tissue) sample. The presence of specific antibodies to a suspected pathogen is highly suggestive of infection. This is often easier than testing for the microbe itself, especially in the case of viral infections. Most HIV testing entails examination of a person's blood for presence of antibody to the virus. Laboratory kits based on this technique are available for immediate identification of a number of pathogens. These methods are covered in greater detail in section 17.4.

On the Track of the Infectious Agent: Specimen Collection

Regardless of the method of diagnosis, specimen collection is the common point that guides the health care decisions of every member of a clinical team. Indeed, the success of identification and

treatment depends on how specimens are collected, handled, and stored. Specimens can be taken by a clinical laboratory scientist or medical technologist, nurse, physician, or even by the patient himself. However, it is imperative that general aseptic procedures be used, including sterile sample containers and other tools to prevent contamination from the environment or the patient. **Figure 17.1a** delineates the most common sampling sites and procedures.

In sites that normally contain resident microbiota, care should be taken to sample only the infected site and not surrounding areas. One common type of collection system is the sterile swab transport tube **(figure 17.1b)**. This allows for sampling, transporting, and maintaining the specimen in its original condition. When using a swab to take a sample from throat and nasopharyngeal regions, the swab should not touch the tongue, cheeks, or saliva. Saliva is an especially undesirable contaminant because it contains millions of bacteria per milliliter, most of which are normal microbiota. Sputum, the mucus secretion that coats the lower respiratory surfaces, especially the lungs, is discharged by coughing or taken by catheterization to avoid contamination with saliva. Depending on the nature of the lesion, skin can be swabbed or scraped with a scalpel to expose deeper layers. The mucous lining of the vagina, cervix, or urethra can be sampled with a swab or applicator stick.

If sterile urine is required, it is usually removed aseptically from the bladder with a thin tube called a catheter. Another method, called a "clean catch," is taken by washing the external urethra and collecting urine midstream during urination. The latter method inevitably incorporates a few normal microbiota into the sample, but these can usually be differentiated from pathogens in an actual infection. Sometimes diagnostic techniques require first-voided "dirty catch" urine.

Sterile materials such as blood, cerebrospinal fluid, and tissue fluids must be taken by sterile needle aspiration. Antisepsis of the puncture site is extremely important in these cases. Additional sources of specimens are the vagina, eye, ear canal, nasal cavity (all by swab), and diseased tissue that has been surgically removed (biopsied).

After proper collection, the specimen is promptly transported to a lab and stored appropriately (usually refrigerated) if it must be held for a time. Nonsterile samples in particular, such as urine, feces, and sputum, are especially prone to deterioration at room temperature. The overgrowth of normal microbiota in these samples could interfere with isolation of the pathogens. It could also alter the numbers and proportions of cells, making analysis more difficult. Many transport devices contain nonnutritive maintenance media (so the microbes do not grow), a buffering system, and an aerobic or sometimes anaerobic environment as necessary for maintaining microbes that require oxygen or are damaged by it.

A summary of specimen collection guidelines can be found in **table 17.1.**

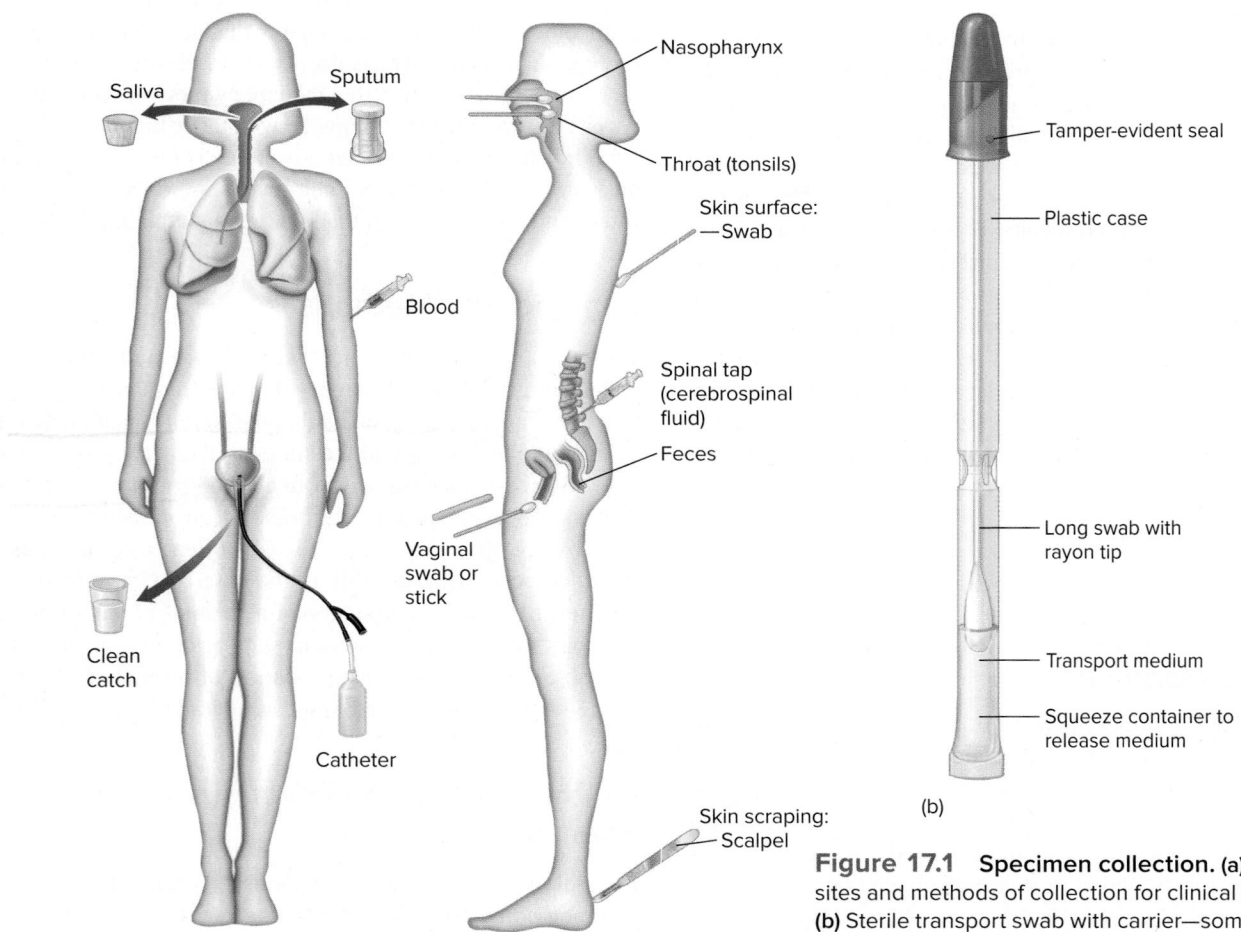

Figure 17.1 Specimen collection. (a) Sampling sites and methods of collection for clinical laboratories. **(b)** Sterile transport swab with carrier—sometimes called a culturette. This example is a type of aerobic transport swab.

TABLE 17.1	General Guidelines for Specimen Collection*	
Type of Specimen	**Collection Method**	**Comments**
Abscess of skin or membrane or decubitus ulcer	Debride surface with sterile wipe; suction fluid with sterile needle and syringe.	Aspiration of tissue samples preferred to swab; transfer to anaerobic transport system.
Anaerobic cultures	Collect from deeper tissue below the surface, using sterile syringe.	Sample must not be exposed to air; transport immediately in anaerobic system to lab.
Blood	Prep area of skin with iodine; use Vacutainer or Bactec blood collection tubes.	Inoculate blood culture bottle immediately.
Bone marrow	Prepare site for surgical incision; extract sample with special needle.	Most samples taken from sternum or ilium; placed in blood bottle for culture.
Cerebrospinal fluid (CSF)	Remove CSF aseptically by lumbar puncture.	Place CSF into sterile tubes; do not refrigerate; transport immediately to lab for processing.
Feces	Take a small specimen into sterile container, cover and transport.	Refrigerate samples held over 1 hour; culture is mainly to rule out enteric pathogens; special kits available to detect cysts and trophozoites of protozoans and ova and larvae of intestinal worms.
Genital/urinary tract	Sterile swab of cervical mucus; swab urethral membrane or insert into lumen and twist.	Plate directly onto selective culture media with high CO_2 atmosphere or collect with anaerobic transport swab system; process immediately.
Respiratory tract, lower	Have patient cough to loosen phlegm and expectorate sputum into sterile cup for transport.	Patient should brush teeth and rinse mouth with water prior to any sampling. If coughing does not work, induce sputum by having patient inhale sterile saline from a nebulizer.
Respiratory tract, upper	Cleanse and debride oral cavity membranes; vigorously swab area of lesion; throat cultures are taken by swabbing the posterior pharynx, tonsils, and inflamed sites.	Tissue aspirates of infected tissue preferred; throat samples are processed for strep throat; culturing for other throat pathogens (except viruses) is done when specifically requested; care must be taken to avoid normal tissue and saliva.
Urine (two methods)	Sample by clean midstream catch into a sterile container.	Clean external genitalia, separate labia (female) or pull back foreskin (male) and urinate for a short time, followed by collection of 100- to 200-cc sample (midstream); samples not immediately processed must be refrigerated.
	Sample by sterile catheter inserted into bladder; urine is drained into sterile container.	Clean and rinse the urethral opening; only a specimen catheter should be used to sample for bladder infection; patients with indwelling catheters will always have microbes in their bladders.

*For a detailed account of collecting and managing specimens, log on to http://www.dshs.state.tx.us/LAB/bac_guidelines.shtm#guidelines.

Special Considerations in Specimen Analysis

The routes taken in specimen analysis come under the heading of (1) direct tests of microscopic, immunologic, or genetic characteristics that provide immediate clues as to the identity of the microbes in the sample; and (2) cultivation, isolation, and identification of pathogens using general and specific tests that may require several days for results **(figure 17.2).**

Most test results fall into two categories: presumptive data, which place the isolated microbe (isolate) in a preliminary category such as a genus; and more specific, confirmatory data, which provide more definitive evidence for a species. Some tests are more important for some groups of bacteria than for others. The time required for analysis may take a few minutes in a streptococcal sore throat to several days in tuberculosis.

Some diseases are diagnosed without the need to identify microbes from specimens. Serological tests on a patient's serum can detect signs of an antibody response. One method that clarifies whether a positive test indicates current or prior infection is to take two samples several days apart to see if the antibody titer is rising. Skin testing can pinpoint a delayed allergic reaction to a microorganism and can be used to screen the general population for exposure to infectious agents such as rubella or tuberculosis.

Microbiology is experiencing an era of tremendous change in availability of testing choices. The methods used by clinical microbiology labs are governed by cost, availability of specialized materials and equipment, and established practices. **Clinical Connections** introduces some of the newer technologies that are no doubt going to be the wave of the future.

Quick Search

For a glimpse of the brave new world of the microbiology lab, search the Internet for "Work Cell Automation Concept."

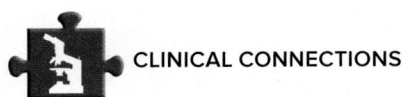

CLINICAL CONNECTIONS

The Changing Scene of Clinical Microbiology

Although it is true that most testing to identify pathogens must be carried out in a laboratory setting, medical researchers are continuing to look for ways to support the diagnosis of infections right at a patient's bedside or even at home. Such tests, called **point of care tests** (POCTs), have the benefits of speeding up treatment and giving patients greater charge of their own medical choices. These tests are usually performed in the presence of the patient and provide immediate feedback, similar to taking blood pressure or monitoring blood sugar, except that POC tests aim to detect infections or infectious agents.

POC tests range from nonspecific, over-the-counter kits that can help determine whether an infection is present (vaginal pH tests and ear checks that detect fluid in the middle ear, for example) to **rapid diagnostic tests** (RDTs) that identify specific microbes or diseases in a few minutes. Rapid *Staphylococcus* and *Streptococcus* kits used in clinics for several years to detect possible skin and throat infections were among the earliest examples. But these have been joined by dozens of other RDTs. There is now some type of RDT available for most major infectious agents, including Ebola virus and parasitic protozoa. Recent examples include an oral test for HIV; small inoculation modules for hepatitis C and B; and a panel for malaria **(figures *a, b, c*)**. Depending on the particular microbe and site of infection, these tests can be performed directly from saliva, vaginal fluids, blood, urine, and food, and do not require isolation of the infectious agent or a great deal of technical skill to read. The accuracy of RDTs is not yet as high as that of traditional cultural testing, but they are good early screening tools for possible infections.

An even more technologically advanced form of microbial identification (but not an RDT) is MALDI, which stands for matrix-assisted laser desorption/ionization. MALDI uses a mass spectrometer to ionize the organic molecules in cells with a UV laser and analyze their content. In practice, a sample of microbe (like an isolated colony) is mixed on a small metal plate **(figure *d*)** with a matrix chemical and dried. The plate is placed in a spectrometer and bombarded by a UV laser until the sample has been completely ionized. Irradiation emits a specific spectrum of ions that is unique to each microbe, which can be graphically analyzed and displayed. When the pattern is compared with a computer data bank that compiles spectra of known microbes, an immediate identification is possible. It is limited only by the numbers of species that have been programmed into the data bank. This process can be completely automated and, except for the initial equipment costs, will greatly reduce the time and cost of testing. Although it has a relatively high level of accuracy, it is still not as accurate as traditional laboratory tests, but this is likely to change with increased usage.

Looking back at genome analysis in chapter 10, you will no doubt recall that the DNA or RNA profile of an organism is a powerful tool of identification and the most specific of all tests available. In this chapter, we cover the use of hybridization probes (FISH) and PCR. As more and more tests are developed, genome testing is gradually becoming the method of choice for many diseases. Companies are developing miniature "labs on a chip" or microarray biochips that can identify several pathogens at once. They consist of small platforms (like slides) that have been loaded with probes made to hybridize with specific genes of pathogens. When a sample of unknown nucleic acid is inserted onto the biochip and allowed to hybridize, any probes that have reacted with the nucleic acid will report by fluorescing. Fluorescence detectors create an image of the pattern that can identify the probable pathogen(s). So far, these have had the most extensive use in detecting some forms of enteric pathogens in food and water.

Do you expect that sometime in the future, the traditional microbiology labs, with media, Petri dishes, biochemical tests, microscopes, and the like, will no longer be needed? Base your answer on your own experiences and observations. *Answer available on Connect.*

(a) An oral home test rapidly tests for HIV antibodies in saliva.
© Kathy Park Talaro

(b) Rapid diagnostic test uses a finger-prick blood drop to measure antibodies to hepatitis B and C viruses and HIV. *Courtesy of MedMira Inc.*

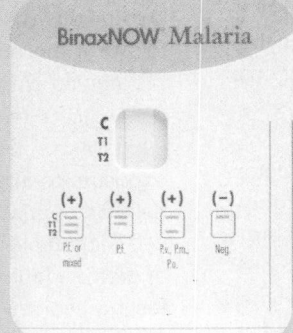

(c) A rapid diagnostic panel that identifies antigens from different species of *Plasmodium*, the agent of malaria, using a small drop of whole blood. *Courtesy of Alere, Inc.*

(d) A metal plate with tiny dots for placement of samples to be analyzed by MALDI technology.
© MedSci/Alamy Stock Photo

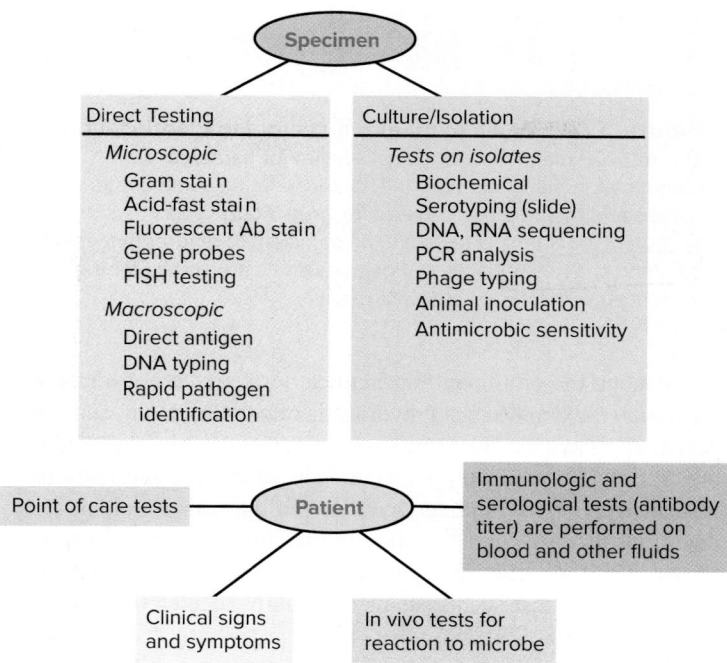

Figure 17.2 A scheme for specimen isolation and identification.

 Check Your Progress SECTION 17.1

1. Summarize the major techniques in identifying and diagnosing pathogens and infections.
2. Describe the general principles in specimen collection.
3. Explain why it is important to prevent microbes from growing in specimens, and why speed is so important in the clinical laboratory.
4. Summarize the kinds of tests that are used to identify microbes and detect signs of infection.

17.2 Phenotypic Methods

 Expected Learning Outcomes

4. Describe some direct methods of testing a specimen.
5. Summarize the aims in selection of culture techniques and media.
6. Explain the basis of biochemical and other methods of testing.
7. Describe how flowcharts and comparison tables are used in identification of bacterial pathogens.

Immediate Direct Examination of Specimen

Direct microscopic observation of a fresh or stained specimen is one of the most rapid methods of determining presumptive and sometimes confirmatory characteristics. Stains most often employed for bacteria are the Gram stain (see 3.2 Making Connections) and the acid-fast stain (see figure 3.8). For many species, these ordinary stains can be useful, but they do not work with certain organisms.

For these types of microbes, direct immunofluorescent antibody (DIFA) tests can highlight the presence of the microbe by means of antibodies labelled with fluorescent dyes **(figure 17.3)**. DIFA tests are particularly useful for bacteria, such as the syphilis and Lyme disease spirochetes that are not readily cultivated in the laboratory.

Another way that specimens can be analyzed is through *direct antigen testing*. This technique is similar to direct fluorescent antibody testing in that known antibodies are used to identify antigens on the surface of bacterial isolates. But in direct antigen testing, the reactions can be seen with the naked eye (see figure 17.8*b*). Quick test kits that greatly speed clinical diagnosis are available for *Staphylococcus aureus, Streptococcus* species, *Neisseria gonorrhoeae, Haemophilus influenzae, Neisseria meningitidis,* and *Salmonella.* However, when the microbe is very sparse in the specimen, direct testing is less effective, and identification will require more sensitive methods.

(a)

(b)

Figure 17.3 Direct immunofluorescent antibody testing.
(a) A setup that positively identifies the syphilis spirochete compared to a negative result. **(b)** Photomicrograph results for *Borrelia burgdorferi,* the spirochete of Lyme disease. Fluid from a lesion observed with a UV microscope reveals numerous fluorescing spirochetes. The two colors indicate different surface antigens reacting with antibodies specific to *B. burgdorferi.* (b): *Source:* National Institute of Allergy and Infectious Diseases (NIAID)/CDC

Cultivation of Specimen

Isolation Media

Such a wide variety of media exist for microbial isolation that a certain amount of preselection must occur, depending on the nature of the specimen. When the sample contains such small numbers of the pathogen that it could be easily lost or overgrown by competing microbes, the specimen is often cultured in enrichment media that can amplify the pathogen. In this chapter's Case Study, for instance, 1% to 2% salt is added to favor the growth of any *Vibrio* that may be present. In specimens such as feces, mucus, or urine that may have high bacterial counts and a diversity of species, selective media are chosen to make the target microbes more prominent by inhibiting microbes that are part of the normal microbiota. These media can be moderately inhibitory, such as those that grow only gram-positive or gram-negative bacteria, or highly inhibitory, being specific to a single group or species. One example is a medium used to screen clinical specimens that will grow only *Listeria monocytogenes,* a serious foodborne pathogen. This agar medium contains six selective agents, three of which are antibiotics. This type of media has the benefit of selecting the pathogen from the sample in a single step.

In most cases, specimens are also inoculated into differential media that define such characteristics as reactions in blood (blood agar) and fermentation patterns (mannitol salt and MacConkey agar). A patient's blood is usually cultured in a special bottle of broth that can be periodically sampled for growth. Numerous other examples of isolation, differential, and biochemical media were presented in chapter 3. So that subsequent steps in identification will be as accurate as possible, all work must be done from isolated colonies or pure cultures, because working with a mixed or contaminated culture gives misleading and inaccurate results. From such isolates, clinical microbiologists obtain information about a pathogen's microscopic morphology and staining reactions, cultural appearance, motility, oxygen requirements, and biochemical characteristics.

Biochemical Testing

The physiological reactions of bacteria to nutrients and other substrates provide useful indirect evidence of the types of enzyme systems present in a particular species. Many of these tests are based on the following scheme:

The microbe is cultured in a medium with a special substrate and then tested for a particular end product. The presence of the end product indicates that the enzyme is expressed in that species; its absence means it lacks the enzyme for utilizing the substrate in that particular way. These types of reactions are particularly meaningful in bacteria, which are haploid and generally express their genes for utilizing a given nutrient.

Figure 17.4 Rapid miniature tests for lab identification. The API 20E manual biochemical system for bacterial identification. Each cupule is inoculated with the same bacterial culture and the strip is incubated for a few hours. This cross section of biochemical tests can be sufficient to differentiate many common species of bacteria. Positive results are shown for a 20-test strip. Courtesy of bioMerioux, Inc.

Among the prominent biochemical tests are carbohydrate fermentation (acid and/or gas); hydrolysis of gelatin, starch, and other polymers; enzyme actions such as catalase, oxidase, and coagulase; and various by-products of metabolism. Many are presently performed with rapid, miniaturized systems that can simultaneously determine up to 23 characteristics in small individual cups or spaces **(figure 17.4).** An important plus, given the complexity of biochemical profiles, is that such systems are readily adapted to computerized analysis.

Common schemes for identifying bacteria are somewhat artificial but convenient. They are based on easily recognizable characteristics such as motility, oxygen requirements, Gram stain reactions, shape, spore formation, and various biochemical reactions. Schemes can be set up as flowcharts **(figure 17.5)** or keys that trace a route of identification by offering pairs of opposing characteristics (positive versus negative, for example) from which to select. Eventually, an endpoint is reached, and the name of a genus or species that fits that particular combination of characteristics appears. Diagnostic tables structured like this chapter's Case Study that provide more complete information are preferred by many laboratories because variations from the general characteristics used on the flowchart can be misleading. Both systems are used in this text.

Miscellaneous Tests

In many situations, morphological and biochemical tests are not definitive or relevant and need to be supplemented by other methods.

> **Quick Search**
> Access "Bacterial Infections and Involved Species" on the Internet for a useful overview of major pathogens and the body systems they infect.

Bacteria host viruses called bacteriophages that are very species- and strain-specific. Such selection by a virus for its host is useful in typing some bacteria, primarily *Staphylococcus* and *Salmonella.* The technique of *phage typing* involves inoculating a lawn of cells onto a Petri dish, mapping it off into blocks, and applying a different phage to each block. Cleared areas corresponding to lysed cells indicate sensitivity to that phage. Phage typing is chiefly used for tracing strains of bacteria in epidemics.

Animals may be needed to cultivate bacteria such as *Mycobacterium leprae* and *Treponema pallidum,* whereas avian embryos and cell cultures are used to grow rickettsias, chlamydias, and viruses. Animal inoculation is also occasionally used to test bacterial or fungal virulence.

Figure 17.5 flowchart (a) Cocci

```
                           Cocci
              ┌──────────────┴──────────────┐
           Gram (+)                       Gram (−)
     ┌────────┴────────┐          ┌──────────┴──────────┐
Catalase (+),    Catalase (−),   Aerobic,          Anaerobic,
irregular        pairs, chain    oxidase (+),      oxidase (−),
clusters,        arrangement     catalase (+)      catalase (−)
tetrads
                 Streptococcus    Neisseria         Veillonella
                 Enterococcus     Branhamella
                                  Moraxella
   ┌──────┴──────┐
Strictly      Facultative
aerobic       anaerobic

Micrococcus   Staphylococcus
              Planococcus
(a)
```

Figure 17.5 flowchart (b) Bacilli

```
                                    Bacilli
              ┌────────────────────────┴────────────────────────┐
           Gram (+)                                           Gram (−)
     ┌────────┴────────┐            ┌──────────────────┬──────────────────┐
 Sporeformer      Non-sporeformer  Aerobic,      Facultative        Aerobic, oxidase (+),
                                   oxidase (+),   anaerobic,         curviform shape
                                   not curved     oxidase (−),
                                                  ferment glucose
              ┌────────┴────────┐
          Acid-fast       Non-acid-fast
                       ┌──────┴──────┐
                    Regular      Pleomorphic

Bacillus     Mycobacterium  Lactobacillus  Corynebacterium    Pseudomonas
Clostridium  Nocardia       Listeria       Propionibacterium  Alcaligenes
(b)
```

Motile: Escherichia, Enterobacter, Citrobacter, Proteus, Salmonella, Erwinia
Nonmotile: Shigella, Klebsiella
Ferment glucose: Vibrio
Do not utilize glucose: Campylobacter

Figure 17.5 **Flowchart key that separates genera of gram-positive and gram-negative bacteria often isolated from specimens.** **(a)** Cocci and **(b)** bacilli of clinical interest.

Antimicrobial sensitivity tests are not only important in determining the drugs to be used in treatment (see figure 12.18), but the patterns of sensitivity can also be used in presumptive identification of some species of *Streptococcus, Pseudomonas,* and *Clostridium.* Antimicrobials are also used as selective agents in many media.

Determining Clinical Significance of Cultures

Questions that can be difficult but necessary to answer in this era of debilitated patients and opportunists are: Is an isolate clinically important, and how can one know whether it is a contaminant or just part of the normal microbiota? The number of microbes in a sample is one useful criterion. For example, a few colonies of *Escherichia coli* in a urine sample can simply indicate normal microbiota, whereas several hundred can mean active infection. In contrast, the presence of a single colony of a true pathogen such as *Mycobacterium tuberculosis* in sputum or an opportunist in sterile sites such as cerebrospinal fluid or blood is highly suggestive of its role in disease. Furthermore, the repeated isolation of a relatively pure culture of any microorganism can mean it is an agent of disease, though care must be taken in this diagnosis. Another challenge

facing the clinical laboratory is that of differentiating a pathogen from species in the normal microbiota that are similar in morphology from their more virulent relatives. Medical microbiologists are able to apply their expertise, supported by the tools of laboratory science, to solve even the most difficult identifications.

17.3 Genotypic Methods

 Expected Learning Outcome

8. Explain the different variations on genetic testing and how they can be used in identification and diagnosis.

DNA Analysis Using Genetic Probes

The exact order and arrangement of the DNA code is unique to each organism. With a technique called *hybridization,* it is possible to identify a bacterial species by analyzing segments of its DNA. This requires small fragments of single-stranded DNA (or RNA) called

probes that are known to be complementary to the specific sequences of DNA from a particular microbe. The test is conducted by extracting unknown test DNA from cells in specimens or cultures and binding it to special blotter paper. After several different probes have been added to the blotter, it is observed for visible signals that the probes have become fixed (hybridized) to the test DNA. The binding of probes onto several areas of the test DNA indicates close correspondence and makes positive identification possible. This forms the basis for certain DNA analysis techniques discussed in chapter 10.

A form of DNA hybridization with numerous applications for rapid identification of unknown bacteria is **fluorescent in situ hybridization,** or **FISH** (see figure 10.6). An uncultured specimen such as blood, a skin lesion, or a throat swab can be smeared on a slide and reacted with a nucleic acid probe that carries a peptide and a fluorescent dye marker. The probe can detect a specific segment of ribosomal RNA. If the unknown cells are a match to the probe, they will retain the dye and fluoresce under a UV microscope **(figure 17.6).** This has decreased the time for identification of septic blood cultures to 90 minutes instead of 24 hours.

Another genetic-based technique often brought in to analyze outbreaks or epidemics is **pulse-field gel electrophoresis** (PFGE). This rapid method of DNA typing makes a restriction fragment fingerprint of the suspected pathogens on a single gel that identifies and compares the isolates alongside each other. It is especially useful to trace a causative agent to the same source or reservoir. An example of its application can be found at http://wwwnc.cdc.gov/eid/article/8/5/01-0373-f1.

Roles of the Polymerase Chain Reaction and Ribosomal RNA in Identification

The polymerase chain reaction (PCR) is one of the most valuable high technology tools for DNA and RNA analysis. A review of chapter 10 (see figure 10.8) will remind you how it can rapidly synthesize hundreds of thousands and even millions of copies of a particular segment of DNA. It is both highly sensitive and specific, amplifying minute quantities from unknown samples that would otherwise be lost. It is a widely-used technique of forensics and molecular biology, but it can also be applied to the identification of microbes. With this method, laboratories can often completely dispense with traditional culturing techniques and go directly to diagnosis in a few hours.

PCR-based tests are used routinely for detecting HIV, Lyme disease, the human papillomavirus, tuberculosis, hepatitis, and a number of other bacterial and viral infections **(Figure 17.7).** In many laboratories, the PCR or a similar rapid test is the method of choice for identifying gonorrhea and chlamydiosis. A version of this technique called real-time PCR can simultaneously pull DNA from a sample, register the amount that is amplified, and identify it.

One particularly versatile aspect to PCR is that it can be combined with other technologies for additional applications. For example, if the sample being analyzed contains mainly RNA (from an RNA virus or ribosome), the initial introduction of reverse transcriptase (RT) will convert the RNA in the sample to DNA, and this DNA can then be amplified by PCR. This technique—RT-PCR—is valuable for verifying RNA-based viruses such as rabies virus (see figure 17.17*f*). DNA of PCR origin can be analyzed with the other tools of technology: it can be probed, hybridized, profiled, and sequenced.

One of the most viable indicators of evolutionary relatedness and affiliation is comparison of the nucleotide sequence of ribosomal RNA, a major component of ribosomes. Ribosomes have the same function (protein synthesis) in all cells, and their nucleic acid content tends to remain more or less stable over long periods. Thus, any major differences in the sequence, or "signature," of the rRNA is likely to indicate some distance in ancestry. This technique is powerful at two levels: It is effective for differentiating general group differences (it was used to separate the three superkingdoms of life discussed in chapter 1), and it can be fine-tuned to identify species.

Fluorescent molecule

PNA for *S. aureus*

Rinse

Gram stain

Result = staphylococci
Question = *S. aureus?*

In situ hybridization

Result = (+) for *S. aureus*

Positive blood culture tube

Figure 17.6 **Peptide-nucleic acid (PNA) FISH test for *Staphylococcus aureus* performed on a blood culture.** Once staphylococci are established by microscopic examination, a fluorescent peptide-nucleic acid combination is hybridized to a sample on a slide. UV illumination reveals a positive reaction of fluorescing cocci in clusters. (right) Image provided by AdvanDx, Inc.

Figure 17.7 **PCR testing for identification of pathogens.** PCR gels (B, C) identify isolates of *Vibrio vulnificus* from patients with unusual symptoms. Band patterns appearing in lanes 1-4 are specific for the cytotoxin genes of *V. vulnificus.* M = molecular size standard. See figure 10.5 to refresh your memory on how these DNA profiles are made. *Source:* Zaidenstein R, Sadik C, Lerner L, Valinsky L, Kopelowitz J, Yishai R, et al. Clinical characteristics and molecular subtyping of Vibrio vulnificus illnesses, Israel. *Emerg Infect Dis* [10.3201/eid1412.080499]. 2008 Dec [2013 July].

 Check Your Progress SECTIONS 17.2 AND 17.3

5. Describe what is involved in direct specimen testing, presumptive and confirmatory tests, cultivating and isolating the pathogen, biochemical testing, and gene probes.

6. Summarize the important aspects for determining if a clinical isolate is involved in infection.

7. Describe the applications of PCR in identification techniques.

17.4 Immunologic Methods

 Expected Learning Outcomes

9. Describe the background aims of immunologic testing.

10. Identify how antigen-antibody reactions are detected and quantified.

11. Differentiate between agglutination and precipitation, and describe how they are used in diagnosis.

12. Explain the basic methods behind the Western blot and complement fixation tests.

13. Interpret the outcome of direct and indirect immunofluorescent antibody testing.

The antibodies formed during an immune reaction are important in combating infection, but they hold additional practical value. Characteristics of antibodies such as their quantity or specificity can reveal the history of a patient's contact with microorganisms or other antigens. This is the underlying basis of serological testing. **Serology** is the branch of immunology that traditionally deals with in vitro

diagnostic testing of serum. Serological testing is based on the familiar concept that antibodies have extreme specificity for antigens, so when a particular **antigen** is exposed to its specific **antibody,** it will bind tightly to it. Being able to visualize this interaction provides a powerful tool for detecting, identifying, and quantifying antibodies—or for that matter, antigens. The scheme works both ways, depending on the situation. One can detect or identify an unknown antibody using a known antigen, or one can use an antibody of known specificity to help detect or identify an unknown antigen **(figure 17.8).** Modern serological testing has grown into a field that tests more than just serum. Urine, cerebrospinal fluid, whole tissues, and saliva can also be used to determine the immunologic status of patients. These and other immune tests are helpful in confirming a suspected diagnosis or in screening a certain population for disease.

General Features of Immune Testing

The strategies of immunologic tests are diverse, and they underline some of the brilliant and imaginative ways that antibodies and antigens can be used as tools. We summarize them under the headings of agglutination, precipitation, complement-fixation, fluorescent antibody tests, and immunoassay tests. First, we overview the general characteristics of immune testing and then look at each type specifically.

The most effective serological tests have a high degree of specificity and sensitivity **(figure 17.9).** Specificity is the property of a test to focus upon only a certain antibody or antigen and not to react with unrelated or distantly related ones. **Sensitivity** means that the test can detect even very small amounts of antibodies or antigens that are the targets of the test. A test with high sensitivity will diagnose even patients who are weakly positive. Modern systems using monoclonal antibodies have greatly improved specificity, and those using radioactivity, enzymes, and electronics have improved sensitivity.

Visualizing Antigen-Antibody Interactions

The primary basis of most tests is the binding of an antibody (Ab) to a specific molecular site on an antigen (Ag). Because this reaction cannot be readily seen without an electron microscope, tests involve some type of endpoint reaction visible to the naked eye or with regular magnification that tells whether the result is positive or negative. In the case of large antigens such as cells, Ab binds to Ag and creates large clumps or aggregates that are visible macroscopically or microscopically **(figure 17.10a).** Smaller Ab-Ag complexes that do not result in readily observable changes will require special indicators in order to be visualized. Endpoints are often revealed by dyes or fluorescent reagents that can tag molecules of interest. Similarly, radioactive isotopes incorporated into antigens or antibodies constitute sensitive tracers that are detectable with photographic film.

An antigen-antibody reaction can be used to read a **titer,** or the quantity of antibodies in serum. Titer is determined by serially diluting a sample in tubes or in a multiple-welled microtiter plate and mixing it with antigen **(figure 17.10b).** It is determined by observing the highest dilution of serum that produces a visible reaction with an antigen. The titer is the denominator of this dilution. The more a sample can be diluted and yet still react with antigen, the greater is the concentration of antibodies in that sample and the higher is its titer. Possible interpretations of test results are discussed in **Clinical Connections, "When Positive Is Negative."**

(a1) In serological diagnosis of disease, a blood sample is tested for the presence of antibody using an antigen of known specificity. A positive reaction is usually evident as some visible sign, such as color change or clumping, that indicates a specific interaction between antibody and antigen. (The reaction at the molecular level is rarely observed.)

(b1) An unknown microbe is mixed with serum containing antibodies of known specificity, a procedure known as serotyping. Macroscopic or microscopic clumping indicates a correct match between antibody and antigen and permits identification of the microbe.

(a2) Test of unknown serum on card with five different known antigens and a positive control. Clumps indicate precipitated antigens and a possible infection in the patient.

(b2) Unknown test colonies from streak plate mixed with antiserum against *Neisseria meningitidis;* agglutination on left is positive for that pathogen.

Figure 17.8 Basic principles of serological testing using antibodies and antigens. (a) Serological diagnosis **(b)** Direct antigen testing (a2) Reproduced by kind permission of Thermo Fisher Scientific Inc.; (b2) *Source:* Ortika BD, Habib H, Dunne EM, Porter BD, Satzke C: Production of latex agglutination reagents for pneumococcal serotyping. BMC Res Notes 2013, 6:49.

Figure 17.9 Specificity and sensitivity in immune testing. (a) This test shows specificity in which an antibody (Ab) attaches with great exactness with only one type of antigen (Ag). **(b)** Sensitivity is demonstrated by the fact that Ab can detect antigens even when the antigen is greatly diluted.

(a)

(b)

Agglutination*

Epitope

Whole cell

+

Antigen

Antibody

Microscopic appearance of clumps

Precipitation*

Cell-free molecule in solution

Epitope

Antigen

+

Antibody

Microscopic appearance of precipitate

(a) Agglutination involves clumping of whole cells; precipitation is the formation of antigen-antibody complexes in cell-free solution. Both reactions can be observed by noticeable clumps or precipitates in test tubes (see (b) and figure 17.11).

The Tube Agglutination Test

Unagglutinated cells

Reaction +++ ++ + + − −

Agglutinated cells

Dilution 1/20 1/40 1/80 1/160 1/320 1/640 Control (no serum)

(b) The tube agglutination test. A sample of patient's serum is serially diluted with saline. The dilution is made in a way that halves the number of antibodies in each subsequent tube. An equal amount of the antigen (here, blue bacterial cells) is added to each tube. The control tube has antigen, but no serum. After incubation and centrifugation, each tube is examined for agglutination clumps as compared with the control, which will contain unagglutinated cells and no clumps. The titer is equivalent to the denominator of the dilution of the last tube in the series that shows agglutination.

*Although IgG is shown as the Ab, IgM is also involved in these reactions.

Figure 17.10 Antigen-Antibody Testing. (a) Cellular/molecular view of agglutination and precipitation reactions that produce visible antigen-antibody complexes. **(b)** An agglutination test for determining the antibody titer of serum.

Agglutination and Precipitation Reactions

The essential differences between agglutination and precipitation are in size, solubility, and location of the antigen. In agglutination, the antigens are whole cells such as red blood cells or bacteria with determinant groups on the surface. In precipitation, the antigen is a soluble molecule. In both instances, when Ag and Ab are optimally combined so that neither is in excess, one antigen is interlinked by several antibodies to form an insoluble, three-dimensional aggregate so large that it cannot remain suspended and it settles out.

Agglutination Tests

Agglutination is discernible because the antibodies cross-link the antigens to form visible clumps. Agglutination tests are performed routinely by blood banks to determine ABO and Rh (Rhesus) blood types in preparation for transfusions. In this type of test, antisera containing antibodies against the blood group antigens on red blood cells are mixed with a small sample of blood and read for the presence or absence of clumping (see figure 16.10). The Widal test is an example of a tube agglutination test for diagnosing salmonelloses and undulant fever. In addition to detecting specific antibody, it also gives the serum titer.

Numerous variations of agglutination testing exist. The rapid plasma reagin (RPR) test is one of several tests commonly used to test for antibodies to syphilis. The cold agglutinin test, named for antibodies that react only at lower temperatures (4°C to 20°C), was developed to diagnose *Mycoplasma* pneumonia. The *Weil-Felix reaction* is an agglutination test sometimes used in diagnosing rickettsial infections.

In some tests, agglutinating antigens are affixed to the surface of an inert particle. In **latex agglutination** tests, the inert particles are tiny latex beads. Kits based on agglutination are available for assaying pregnancy hormone in the urine; for identifying *Candida* yeasts, staphylococci, streptococci, and meningococci (see figure 17.8*b2*); and for diagnosing rheumatoid arthritis.

Precipitation Tests

In **precipitation** reactions, the soluble antigen is precipitated (made insoluble) by an antibody. This reaction is observable in a test tube in which an antigen sample has been carefully laid over an antiserum solution. At the point of contact, a cloudy or opaque zone forms. One example of this technique is the **VDRL** (Venereal Disease Research Laboratory) test that also detects antibodies to syphilis. Although it is a good screening test, it contains a heterophilic antigen (cardiolipin) that may give rise to false positive results.

Although precipitation is a useful detection tool, the precipitates are so easily disrupted in liquid media that most precipitation reactions are carried out in agar gels. These substrates are sufficiently soft to allow the reactants (Ab and Ag) to freely diffuse, yet firm enough to hold the Ab-Ag precipitate in place. One technique with applications in microbial identification and diagnosis of disease is the double diffusion (Ouchterlony) method **(figure 17.11).** It is called double diffusion because it involves diffusion of both antigens and antibodies. The test is performed by punching a pattern of small wells into an agar medium and filling them with test antigens and antibodies. A band forming between two wells

I.

Side view

I. In one method of setting up a double diffusion test, wells are punctured in soft agar, and antibodies (Ab) and antigens (Ag) are added in a pattern. As the contents of the wells diffuse toward each other, a number of reactions can result, depending on whether antibodies meet and precipitate antigens.

II.

Precipitation bands

II. Example of test pattern and results. Antigen (Ag) is placed in the center well and antibody (Ab) samples are placed in outer wells. The control contains known Abs to the test Ag. Note bands that form where Ab/Ag meet. The other wells (1, 2) contain unknown test sera. One is positive and the other is negative. Double bands indicate more than one antigen and antibody that can react.

III.

III. Actual test results for detecting infection with the fungal pathogen *Histoplasma*. Numbers 1 and 4 are controls and 2, 3, 5, and 6 are patient test sera. Can you determine which patients have the infection and which do not?

Figure 17.11 Precipitation reactions. The example shown here is a double diffusion (Ouchterlony) test in a semisolid matrix. (III): © National Institute Slide Bank/The Wellcome Centre for Medical Sciences

indicates that antibodies from one well have met and reacted with antigens from the other well. Variations on this technique provide a means of identifying unknown antibodies or antigens.

The Western Blot for Detecting Proteins

The **Western blot test** involves the **electrophoretic** separation of proteins, followed by an immunoassay to detect these proteins. This test is a counterpart of the Southern blot test for identifying DNA, described in chapter 10. It is a highly specific and sensitive method to identify or verify a particular protein (antibody or antigen) in a sample **(figure 17.12)**. First, the test material is electrophoresed in a gel to separate out particular bands. The gel is then transferred to a special blotter that binds the reactants in place. The blot is developed by incubating it with a solution of antigen or antibody that has been labeled with radioactive, fluorescent, or luminescent labels. Sites of specific binding will appear as a pattern of bands that can be compared with known positive and negative samples. This is usually the second test used to verify the status of people who are antibody-positive for HIV in the ELISA test (described in section 17.5), because it tests more types of antibodies and is less subject to

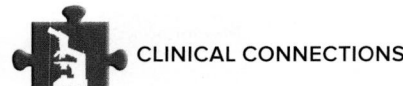

CLINICAL CONNECTIONS

When Positive Is Negative: How to Interpret Serological Test Results

What does it mean if a patient's serum gives a positive reaction—is **seropositive**—in a serological test? In most situations it indicates that antibodies specific for a particular microbe have been detected in the serum sample. But one must be cautious in proceeding to the next level of interpretation. The mere presence of antibodies does not necessarily indicate that the patient has a disease but only that he or she has possibly had contact with a microbe or its antigens through infection or vaccination. In screening tests for determining a patient's history (rubella, for instance), finding a measurable titer of antibodies can be significant, because it shows that the person has some protection against that infectious agent.

When the test is being used to diagnose ongoing infection or disease, however, it is necessary to show a rising titer of antibodies over time. The figure indicates how such a test can be used to diagnose patients who have nonspecific symptoms that could fit several diseases. Lyme disease, for instance, can be mistaken for arthritis or viral infections. In the first group, note that the antibody titer against *Borrelia burgdorferi,* the causative agent, increased steadily over a 6-week period. A control group that shared similar symptoms did not exhibit a rise in titer for antibodies to this microbe. Clinicians call samples collected early and late in an infection *acute* and *convalescent* sera.

Another important consideration in testing is the occasional appearance of **biological false positives.** These results show a positive reaction even though, in reality, the patients are not or have not been infected by the microbe. False positives, such as those in syphilis and HIV testing, arise when antibodies or other substances present in the serum cross-react with the test reagents, producing a positive result. Such false results may require retesting by a method that greatly minimizes cross-reactions.

Define a *false negative reaction.* What might cause such a reaction? *Answer available on Connect.*

misinterpretation and false positives than are other antibody tests. The technique has significant applications for detecting microbes and their antigens in specimens.

Successive tests on an HIV+ patient over 30 days reveals an increase in band intensity over time. This is due to continued formation of anti-HIV antibodies.

Figure 17.12 **The Western blot procedure.** Examples shown here test for antibodies to several HIV antigens. The test strips are prepared by electrophoresing the major HIV surface and core antigens and then blotting them onto special filters. The test strips are incubated with a serum sample and developed with radioactive or colorimetric labels. Sites where HIV antigens have bound antibodies show up as bands. A positive control strip (SRC) containing antibodies for all HIV antigens serves as a comparison.

Interpretation of bands to report a positive test

- Labels correspond with glycoproteins (gp) or proteins (p) that are part of HIV-1 antigen structure.
- The test is considered positive if bands occur at at least two locations: gp160 or gp120 and p31 or p24.
- The test is considered negative if no bands are present for any HIV antigen.
- The test is considered indeterminate if bands are present but not at the primary locations. This result may require retesting at a later date.

© Genelabs Diagnostics Pte Ltd.

Complement Fixation

In chapter 14 we introduced the functions of complement as a defense system for destroying foreign cells (see figure 14.20). An antibody that requires the actions of complement to complete the lysis of its antigenic target cell is termed a **lysin** or cytolysin. When lysins act in conjunction with the intrinsic complement system on red blood cells,

the cells hemolyze (lyse and release their hemoglobin). This lysin-mediated hemolysis is the basis of a type of test called **complement fixation,** or **CF (figure 17.13).** Complement fixation tests are used to diagnose some viral, rickettsial and fungal infections, but are often replaced by more sensitive and accurate ELISA and genetic tests.

Complement fixation testing uses four components—antibody, antigen, complement, and sensitized sheep red blood cells—and it is conducted in two stages.

In the first stage, the test antigen is allowed to react with the test antibody (at least one must be of known identity) in the absence of complement. If the antibody and antigen are specific for each other, they form complexes. To this mixture, purified complement proteins from guinea pig blood are added. If the antibody and antigen have complexed during the previous step, they attach, or fix, the complement to them, thus preventing it from participating in further reactions.

The extent of any complement fixation is determined in the second stage by means of sheep red blood cells (RBCs) with surface lysin molecules. The sheep RBCs serve as an indicator complex that can also fix complement. Contents of the stage 1 tube are mixed with the stage 2 tube and observed with the naked eye for hemolysis.

- If hemolysis *does not* occur, it means that the complement was used up by the first stage Ab-Ag complex and that the unknown antigen or antibody was indeed present. This result is considered positive.
- If hemolysis *does* occur, it means that unfixed complement from tube 1 reacted with the RBC complex instead, thereby causing lysis of the sheep RBCs. This result is negative for the antigen or antibody that was the target of the test.

A relative of the CF test, the antistreptolysin O (ASO) titer test, measures the levels of antibody against the streptolysin toxin, an important hemolysin of group A streptococci. A serum sample is exposed to known suspensions of streptolysin and then allowed to incubate with RBCs. Lack of hemolysis indicates antistreptolysin antibodies in the patient's serum that have neutralized the streptolysin and prevented hemolysis. This can be a useful procedure to verify scarlet fever, rheumatic fever, and other related streptococcal syndromes.

Miscellaneous Serological Tests

A test that relies on changes in cellular activity as seen microscopically is the *Treponema pallidum* **immobilization (TPI)** test for syphilis. The impairment or loss of motility of the *Treponema* spirochete in the presence of test serum and complement indicates that the serum contains anti-*Treponema pallidum* antibodies. In **toxin neutralization** tests, a test serum is incubated with the microbe that produces the toxin. If the serum inhibits the growth of the microbe, one can conclude that antitoxins to inactivate the toxin are present.

Serotyping is an antigen-antibody technique for identifying, classifying, and subgrouping certain bacteria into categories called serotypes, using antisera for cell antigens such as the capsule, flagellum, and cell wall. It is widely used in typing *Salmonella* species and strains and is the basis for identifying the numerous serotypes of streptococci (see figure 17.8, part *b*1). The Quellung test, which identifies serotypes of the pneumococcus, involves a precipitation reaction in which antibodies react with the capsular polysaccharide. Although the reaction makes the capsule seem to swell, it is actually creating a zone of Ab-Ag complexes on the cell's surface.

Stage 1			Reaction System	Stage 2		
Positive patient's serum			Sheep red blood cells with lysins on surface	Complement fixes to antibodies; RBCs do not lyse.		

Figure 17.13 **Complement fixation test.** In this example, two serum samples are being tested for antibodies to a certain infectious agent. In reading this test, one observes the cloudiness of the tube. If it is cloudy, the RBCs are not hemolyzed and the test is positive. If it is clear and pink, the RBCs are hemolyzed and the test is negative.

In Vivo Testing

Probably the first immunologic tests were performed not in a test tube but on the body itself. A classic example of one such technique is the **tuberculin test,** which uses a small amount of purified protein derivative (PPD) from *Mycobacterium tuberculosis* injected into the skin. The appearance of a red, raised, thickened lesion in 48 to 72 hours can indicate previous exposure to tuberculosis. In practice, in vivo tests employ principles similar to serological tests, except in this case, an antigen or an antibody is introduced into a patient to elicit some sort of visible reaction. Like the tuberculin test, some of these diagnostic skin tests are useful for evaluating infections due to fungi (coccidioidin and histoplasmin tests, for example) or allergens. Testing for allergic reactions was covered in chapter 16.

Fluorescent Antibody and Immunofluorescent Testing

The property of certain dyes to emit visible light in response to ultraviolet radiation was discussed in chapter 3. This property of fluorescence has found numerous applications in diagnostic immunology. The fundamental tool in immunofluorescent testing is a **fluorescent antibody**—a monoclonal antibody labeled by a fluorescent dye (fluorochrome).

The two ways that fluorescent antibodies (FABs) can be used are shown in **figure 17.14.** In **direct testing,** an unknown test specimen or antigen is fixed to a slide and exposed to a fluorescent antibody solution of known composition. If the antibodies are complementary to antigens in the material, they will bind to it. After the slide is rinsed to remove unattached antibodies, it is observed with the fluorescent microscope. Fluorescing cells or specks indicate the presence of Ab-Ag complexes and a positive result. These tests are valuable for identifying and locating antigens on the surfaces of cells or in tissues and in identifying the disease agents of syphilis, gonorrhea, chlamydiosis, whooping cough, Legionnaires' disease, plague, trichomoniasis, meningitis, and listeriosis.

In **indirect testing** methods, the fluorescent antibodies are antibodies made to react with the Fc region of another antibody (remember that antibodies can be antigenic). In this scheme, an antigen of known identity (a bacterial culture, for example) is combined with a test serum of unknown antibody content. The fluorescent antibody solution that can react with the unknown antibody is applied and rinsed off to visualize whether the serum contains antibodies that have affixed to the antigen. A positive test shows fluorescing aggregates or cells, indicating that the fluorescent antibodies have combined with the unlabeled antibodies. In a negative test, no fluorescent complexes will appear. This technique is frequently used to diagnose syphilis and various viral infections.

Visible fluorescence

Fluorescent microscopy

Unknown antigen (usually cell or tissue)

+

Antibody labeled with fluorescent dye

Figure 17.14 Immunofluorescent testing. Direct Testing: Unidentified antigen (Ag) is directly tagged with fluorescent Ab. See figure 17.3 to view an example of actual test results.

(b)

Known Ag

Ab1 in serum

No Ab in serum

Ab2 fluorescently labeled; specific for Ab1

Ab2 attaches to Ab1—visible fluorescence

Positive

Ab2 cannot attach—no fluorescence

Negative

(a)

Figure 17.15 Immunofluorescent testing. Indirect Testing: (a) Ag of known identity is used to assay unknown Ab; a positive reaction occurs when the second Ab (with fluorescent dye) affixes to the first Ab. **(b)** Indirect immunofluorescent stain of cells infected with two different viruses. Cells with green fluorescing nuclei contain cytomegalovirus; cells with yellow fluorescing nuclei contain adenovirus. (b) CHEMICON® International, Inc.

 Check Your Progress SECTION 17.4

8. What is the basis of serology and serological testing?
9. Differentiate between specificity and sensitivity.
10. Describe several general ways that Ab-Ag reactions are detected.
11. What do *seropositivity* and *seronegativity* mean?
12. What is a false positive test result and what are some possible causes?
13. Explain how agglutination and precipitation reactions are alike and how they are different.
14. Give examples of several tests that employ agglutination and precipitation reactions.
15. What is meant by complement fixation? What are cytolysins?
16. Describe the principles behind direct and indirect fluorescent antibody tests.

17.5 Immunoassays: Tests with High Sensitivity

Expected Learning Outcome

14. Describe the concepts behind the main types of immunoassays, and discuss their uses in diagnosis.

15. Explain how in vivo testing differs from in vitro testing.

The elegant tools of the microbiologist and immunologist are being used increasingly in athletics, criminology, government, and business to test for trace amounts of substances such as hormones, metabolites, and drugs. But traditional techniques in serology are not refined enough to detect a few molecules of these chemicals. Extremely sensitive alternative methods that permit rapid and accurate measurement of trace antigens or antibodies are called **immunoassays.** Examples of the technology for detecting an antigen or antibody in minute quantities include radioactive isotope labels, enzyme labels, and sensitive electronic sensors. Many of these tests are based on specifically formulated monoclonal antibodies.

Radioimmunoassay (RIA)

Antibodies or antigens labeled with a radioactive isotope can be used to pinpoint minute amounts of a corresponding antigen or antibody. Very complex in practice, these assays compare the amount of radioactivity present in a sample before and after incubation with a known, labeled antigen or antibody. The labeled substance competes with its natural, nonlabeled partner for a reaction site. Large amounts of a bound radioactive component indicate that the unknown test substance was not present. The amount of radioactivity is measured with an isotope counter or a photographic emulsion (autoradiograph). Radioimmunoassay has been employed to measure the levels of insulin and other hormones and to diagnose allergies, chiefly by the radioimmunosorbent test (RIST) for measurement of IgE in allergic patients and the radioallergosorbent test (RAST) to standardize allergenic extracts.

Enzyme-Linked Immunosorbent Assay (ELISA)

The **ELISA,** also known as enzyme immunoassay (EIA), contains an enzyme-antibody complex that can be used as a color tracer for antigen-antibody reactions. The enzymes used most often are horseradish peroxidase and alkaline phosphatase, both of which release a dye (chromogen) when exposed to their substrate. This technique also relies on a solid support such as a plastic microtiter plate that can *adsorb* (attract on its surface) the reactants **(figure 17.16).**

The **indirect ELISA** can capture antibodies in a serum sample. As with other indirect tests, the final positive reaction is achieved by means of an antibody-antibody reaction. The indicator antibody is complexed to an enzyme that produces a color change with positive serum samples **(figure 17.16a, b).** The starting reactant is a known antigen that is adsorbed to the surface of a

(a) **Indirect ELISA,** comparing a positive vs. negative reaction. This is the basis for HIV screening tests.

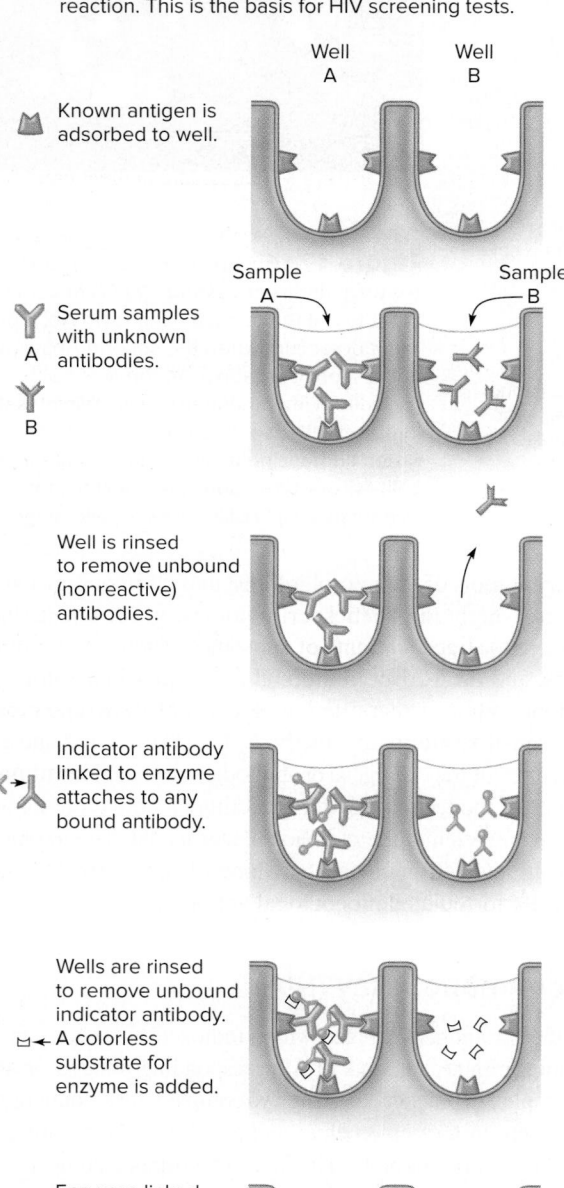

Known antigen is adsorbed to well.

Serum samples with unknown antibodies.

Well is rinsed to remove unbound (nonreactive) antibodies.

Indicator antibody linked to enzyme attaches to any bound antibody.

Wells are rinsed to remove unbound indicator antibody. A colorless substrate for enzyme is added.

Enzymes linked to indicator Ab hydrolyze the substrate, which releases a dye. Wells that develop color are positive for the antibody; colorless wells are negative.

(b) **Microtiter ELISA Plate with 96 Tests for HIV Antibodies.** Colored wells indicate a positive reaction.

(c) **Capture or Antibody Sandwich ELISA Method.** Note that an antigen is trapped between two antibodies. This test is used to detect the measles virus.

Antibody specific to test antigen is adsorbed to well.

Test antigen is added; if complementary, antigen binds to antibody.

Enzyme

Enzyme-linked antibody specific for test antigen then binds to different antigenic site forming a sandwich.

Enzyme's substrate (□) is added, and reaction produces a visible color change (●).

Figure 17.16 Methods of ELISA testing. © Hank Morgan/Science Source

well. To this, the unknown test serum is added. After rinsing, an enzyme-Ab reagent that can react with the unknown test antibody is placed in the well. The substrate for the enzyme is then added, and the wells are scanned for color changes. Color development indicates that all the components reacted and that the antibody was present in the patient's serum. This is the common screening test for the antibodies to HIV, various rickettsial species, hepatitis A and C, the cholera vibrio, and *Helicobacter,* a cause of gastric ulcers. Because false positives can occur, a verification test may be necessary (such as Western blot for HIV).

In **capture ELISA** (or sandwich) tests, a known antibody is adsorbed to the bottom of a well and incubated with a solution containing unknown antigen **(figure 17.16c).** After excess unbound components have been rinsed off, an enzyme-antibody indicator that can react with the antigen is added. If antigen is present, it will attract the indicator-antibody and hold it in place. Next, the substrate for the enzyme is placed in the wells and incubated. Enzymes affixed to the antigen will hydrolyze the substrate and release a colored dye. Thus, any color developing in the wells is a positive result. Lack of color means that the antigen was not present and that the subsequent rinsing removed the enzyme-antibody complex. The capture technique is used to detect antibodies to hantavirus, rubella virus, and *Toxoplasma.*

17.6 Viruses as a Special Diagnostic Case

Expected Learning Outcome

16. Summarize the protocols in identifying the causative agent in a viral infection.

CASE STUDY Part 2

This case provides some insight into the protocol behind a clinical identification process. Preliminary laboratory investigations are designed to narrow the scope of possible infectious agents using as few tests as possible. The selective media and the second Gram stains informed the technologists relatively early on that the gram-positive *Bacillus cereus, Staphylococcus aureus,* and *Clostridium* were not involved in the outbreak, even though two of them are common hospital pathogens. Being able to eliminate them saved the lab from doing unnecessary specialized testing.

The microbiologists at the medical center were able to identify the pathogen as belonging to the genus *Vibrio* based on the fact that *Campylobacter,* the only other genus that would have fit, is known to exhibit spiral-shaped cells and is microaerophilic and nonfermentative in its physiology. The next step was to perform additional tests that could differentiate among the three major species in genus *Vibrio,* all of which are associated with intestinal infections. The panel of standard tests performed on the patient and food isolates is provided in **Case Study table 2,** with expected results from the three pathogens for comparison.

All of the methods discussed so far—phenotypic, genotypic, and immunologic—are applicable to many different types of microorganisms. Viruses can present special difficulties because they are not cells and they are more labor intensive to culture in the laboratory. **Figure 17.17** presents an overview of various techniques available to diagnose viral infections. These range from observing symptoms, to direct microscopic examination, to cultivation, and to serological and genetic analysis.

Quick Search

Access "Viral Infections and Involved Species" on the Internet for a useful summary of major pathogens and the body systems they infect.

Check Your Progress SECTIONS 17.5 AND 17.6

17. Explain the differences between direct and indirect procedures in immunoassay tests, giving examples.

18. How does a positive reaction in an ELISA test appear? How many wells are positive in figure 17.16*b*?

19. Provide some reasons in vivo tests would not be as reliable as immunoassay tests.

20. Explain how a clinical technologist would have determined that the causative agent in this chapter's Case Study was a virus instead of *Vibrio* bacteria.

21. Observing figure 17.17, indicate whether each method (a) through (f) is phenotypic, genotypic, or immunologic.

■ *Which species is responsible for the outbreak of gastroenteritis, and what additional confirmatory tests can verify this identification?*

■ *Why is it so important to identify the infectious agent to the level of species?*

To conclude this Case Study, go to Connect.

For more information on *Vibrio* detection, go to http://www.fda.gov/Food/FoodScienceResearch/LaboratoryMethods/ucm070830.htm.

CASE STUDY TABLE 2

TEST	*Vibrio cholerae*	*Vibrio parahaemolyticus*	*Vibrio vulnificus*	Patient Isolate	Food Isolate
Sucrose*	+	–	–	–	–
Lactose*	–	–	+	–	–
Arabinose*	–	+	–	+	+
ONPG**	+	–	+	–	–
Growth in 0% salt	+	–	–	–	–
Growth in 6% salt	–	+	+	+	+

* These are sugars for testing fermentation.
** ONPG tests for slow lactose fermentation.

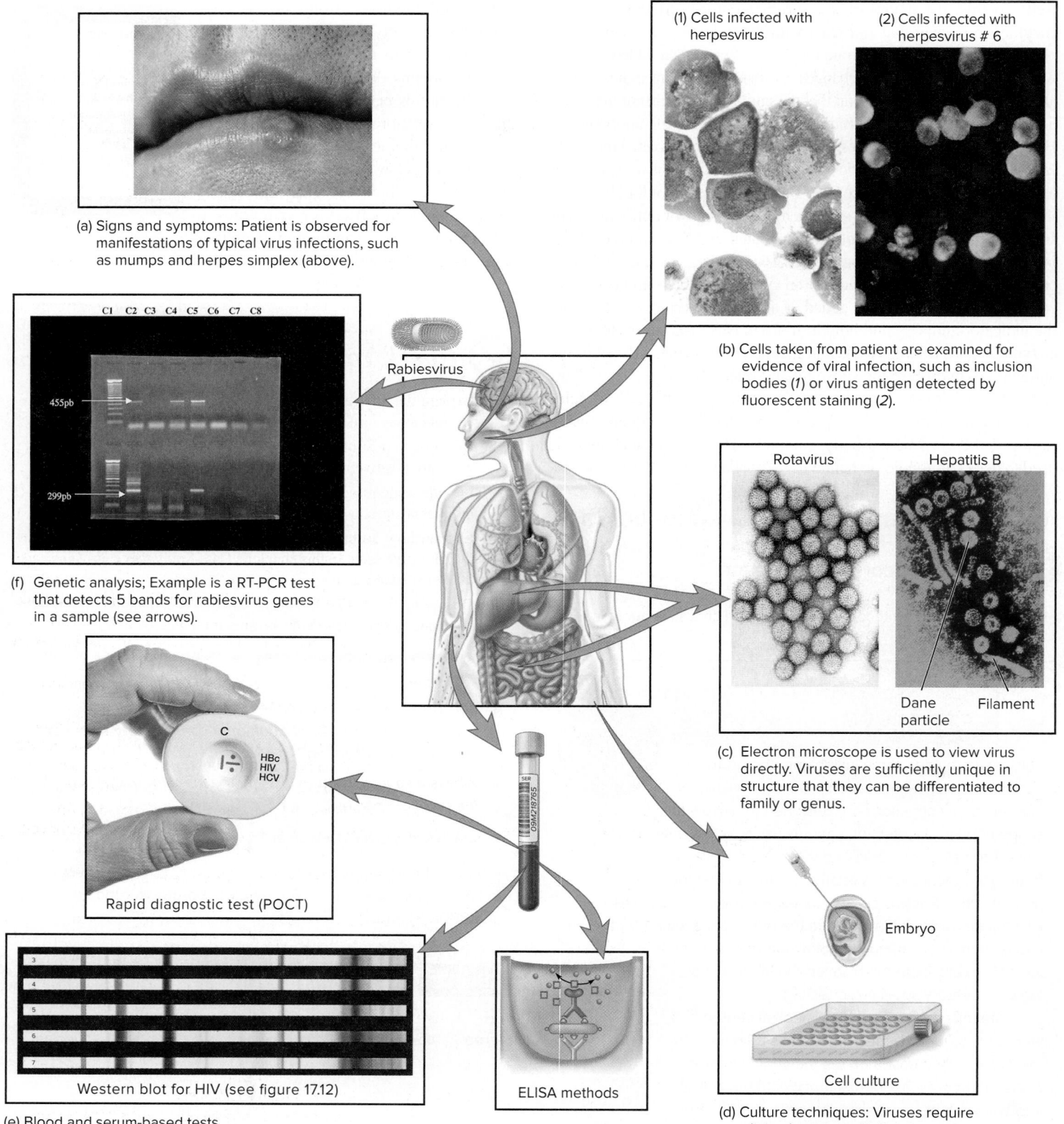

(a) Signs and symptoms: Patient is observed for manifestations of typical virus infections, such as mumps and herpes simplex (above).

(1) Cells infected with herpesvirus

(2) Cells infected with herpesvirus # 6

(b) Cells taken from patient are examined for evidence of viral infection, such as inclusion bodies (1) or virus antigen detected by fluorescent staining (2).

Rabiesvirus

C1 C2 C3 C4 C5 C6 C7 C8

455pb

299pb

(f) Genetic analysis; Example is a RT-PCR test that detects 5 bands for rabiesvirus genes in a sample (see arrows).

Rotavirus Hepatitis B

Dane Filament
particle

(c) Electron microscope is used to view virus directly. Viruses are sufficiently unique in structure that they can be differentiated to family or genus.

C
HBc
HIV
HCV

Rapid diagnostic test (POCT)

Embryo

Cell culture

(d) Culture techniques: Viruses require a living host to multiply.

3
4
5
6
7

Western blot for HIV (see figure 17.12)

ELISA methods

(e) Blood and serum-based tests

Figure 17.17 Summary of methods used to diagnose viral infections. (a) *Source:* Dr. Hermann/CDC; (b1–2) *Source:* Zaki Salahuddin. Laboratory of Tumor Cell Biology/National Cancer Institute; (c, left) *Source:* CDC; (c, right) *Source:* Bryon Skinner/CDC; (e) © Genelabs Diagnostics Pte Ltd.; (rapid diagnostic test) Courtesy of MedMira Inc.; (f) *Source:* 2008 Araújo et al. licensee BioMed Central Ltd.

 Chapter Summary with Key Terms

17.1 An Overview of Clinical Microbiology
A. **Phenotypic** methods are those that assess microscopic morphology, macroscopic morphology, physiological and biochemical characteristics, and chemical composition.
B. **Genotypic** methods examine the genetic composition of a microorganism.
C. **Immunologic** methods exploit the host's antibody reaction to microbial antigens for purposes of diagnosis.
D. Diagnosis begins with accurate **specimen collection,** which means the sampling of body sites or fluids that are suspected to contain the infectious agent. Strategies vary depending on the source.
E. Laboratory methods yield either presumptive data or confirmatory data. Sometimes pathogens are diagnosed based solely on signs and symptoms in the patient. The route of diagnosis depends on the particular microbe and the technologies available to the lab.

17.2 Phenotypic Methods
A. The most obvious phenotypic characteristics of a microbe are gleaned from its **microscopic** and **macroscopic** appearance. Many infections can be diagnosed by microscopic examination of fresh or stained microorganisms from specimens. Direct antigen or antibody testing can also be performed on uncultured specimens.
B. Cultivation of specimens allows the examination of **cultural** characteristics such as colony morphology, growth characteristics, or both, which can be useful in identifying microbes.
C. **Biochemical** tests determine whether microbes possess particular enzymes or metabolic pathways that can serve to identify them.
D. Of utmost importance is the determination of whether an identified microbe is actually causing an infection or is simply a bystander, or member of the normal **microbiota.**

17.3 Genotypic Methods
A. Genetic methods of identification are being used increasingly in diagnostic microbiology.
B. **Probes** can detect specific genetic patterns, and advanced technology can characterize unique DNA sequences.
C. An increasingly common procedure is the **polymerase chain reaction (PCR)** to amplify DNA present in a sample and apply labeled hybridization probes specific to known genetic signatures of microbes.
D. RNA in ribosomes and other cultures or specimens (RNA viruses) can be converted to DNA and treated with methods of DNA analysis.

17.4 Immunologic Methods
A. **Serology** is a science that tests a patient's serum to detect signs of infection, for example, finding specific antibodies for a particular microbe.
B. The basis of serological tests is that **antibodies** (Abs) bind to their specific **antigens** (Ags) in vitro. Known Ags can be used to react with Abs in an unknown serum sample. The reverse is also true; known Abs can be used to detect and type unidentified Ags.

C. These Ab-Ag reactions are visible in the form of obvious clumps and precipitates, color changes, or the release of radioactivity. Test results are read as positive or negative.
D. Desirable properties of tests are high **specificity** and **sensitivity.**
E. Types of Tests
1. In **agglutination** tests, antibody cross-links whole-cell antigens, forming complexes that settle out and form visible clumps in the test chamber.
 a. Examples are tests for blood type, some bacterial diseases, and viral diseases.
 b. These tests can determine the amount of antibody in serum, or **titer.**
2. Double diffusion **precipitation** tests involve the diffusion of Ags and Abs in a soft agar gel, forming zones of Ab/Ab complexes where they meet and react.
3. In **immunoelectrophoresis,** migration of serum proteins in gel is combined with precipitation by antibodies.
4. The **Western blot** test separates antigen into bands. After the gel is affixed to a blotter, it is reacted with a test specimen and developed by radioactivity or with dyes.
5. A **complement fixation** test detects **lysins**—antibodies that fix complement and can lyse target cells. It involves first mixing test Ag and Ab with complement and then with sensitized sheep RBCs. If the complement is fixed by the Ab-Ag, the RBCs remain intact, and the test is positive. If RBCs are hemolyzed, specific antibodies are lacking.
6. In direct assays, known marked Ab is used to detect unknown Ag (microbe).
 a. In indirect testing, known Ag reacts with unknown Ab, and the reaction is made visible by a second Ab that can affix to and identify the unknown Ab.
 b. **Immunofluorescent** testing uses **fluorescent antibodies** (FABs tagged with fluorescent dye) either directly or indirectly to visualize cells or cell aggregates that have reacted with the FABs.

17.5 Immunoassays: Tests of Great Sensitivity
Immunoassays are highly sensitive tests for Ag and Ab.
A. In *radioimmunoassay,* Ags or Abs are labeled with radioactive isotopes and traced.
B. The **enzyme-linked immunosorbent assay (ELISA)** can detect unknown Ag or Ab by direct or indirect means. A positive result is visualized when a colored product is released by an enzyme-substrate reaction.
C. With in vivo testing, Ags are introduced into the body directly to determine a patient's immunologic condition.

17.6 Viruses as a Special Diagnostic Case
A. Viral diseases are diagnosed through patient observation supported by microscopic, cultural, serological, and genetic tests.

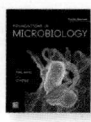

Level I. Knowledge and Comprehension

These questions require a working knowledge of the concepts in the chapter
and the ability to recall and understand the information you have studied.

Multiple-Choice Questions

Select the correct answer from the answers provided. For questions with blanks, choose the combination of answers that most accurately completes the
statement.

1. **Multiple Matching.** Match each of the following culture results to
 the most likely interpretation, and explain your choice.
 a. probable infection
 b. normal microbiota
 c. environmental
 contaminant

 _____ 1. Isolation of two colonies of *E. coli*
 on a plate streaked from the urine
 sample
 _____ 2. Fifty colonies of *Streptococcus
 pneumoniae* on culture of sputum
 _____ 3. A mixture of 80 colonies of various
 streptococci on a culture from a
 throat swab
 _____ 4. Colonies of black bread mold on
 selective media used to isolate
 bacteria from stool
 _____ 5. Blood culture bottle with heavy
 growth

2. Which of the following methods is most sensitive for identifying
 different strains of a microbe?
 a. microscopic examination
 b. hemolysis on blood agar
 c. DNA analysis
 d. agglutination test

3. In agglutination reactions, the antigen is a _____; in precipitation
 reactions, it is a _____.
 a. soluble molecule, whole cell
 b. whole cell, soluble molecule
 c. bacterium, virus
 d. protein, carbohydrate

4. Which reaction requires complement?
 a. hemagglutination
 b. precipitation
 c. hemolysis
 d. toxin neutralization

5. A patient with a _____ titer of antibodies to an infectious agent
 generally has greater protection than a patient with a _____ titer.
 a. high, low
 b. low, high
 c. negative, positive
 d. old, new

6. Direct immunofluorescent tests use a labeled antibody to identify
 a. an unknown microbe
 b. an unknown antibody
 c. fixed complement
 d. agglutinated antigens

7. The Western blot test can be used to identify
 a. unknown antibodies
 b. unknown antigens
 c. specific DNA
 d. both a and b

8. An example of an in vivo serological test is
 a. indirect immunofluorescence
 b. radioimmunoassay
 c. tuberculin test
 d. complement fixation

9. Which of the following specimens must be removed by sterile
 techniques?
 a. feces
 b. urine
 c. upper respiratory tract
 d. blood

Case Study Review

1. On what basis were both *Shigella* and *Salmonella* eliminated as
 possible causes of the outbreak pathogen?
 a. Gram reaction
 b. fermentation of glucose
 c. oxidase results
 d. motility

2. Which *Vibrio* test results does *V. parahaemolyticus* most resemble?
 a. *V. cholerae*, salt tolerance
 b. *V. vulnificus*, salt tolerance
 c. *V. cholerae*, sucrose fermentation
 d. *V. vulnificus*, ONPG

3. What other forms of testing from this chapter could differentiate the
 Vibrio pathogens?

Writing Challenge

For each question, compose a one- or two-paragraph answer that includes the factual information needed to completely address the question. Check Your
Progress questions can also be used for writing-challenge exercises.

1. Why do we interpret positive hemolysis in the complement fixation
 test to mean negative for the test substance?

2. What information can the antibody titer tell us about the immune
 status of a person?

3. Briefly describe the principles and give an example of the use of a
 specific test using immunoelectrophoresis, Western blot, complement
 fixation, fluorescent testing (direct and indirect), and immunoassays
 (direct and indirect ELISA).

4. Differentiate between the serological tests used to identify isolated
 cultures of pathogens and those used to diagnose disease from
 patients' serum as shown in figure 17.8.

5. How would the staining and cultural tests have differed if
 Staphylococcus or *Clostridium* had been the cause of the outbreak of
 gastroenteritis described in this chapter's Case Study?

6. Explain what must be done to identify RNA viruses using genetic
 techniques.

Concept Mapping

On Connect you can find an Introduction to Concept Mapping that provides guidance for working with concept maps and a sample of a concept map for this chapter.

Level II. Application, Analysis, Evaluation, and Synthesis

These problems go beyond just restating facts and require higher levels of understanding and an ability to interpret, problem solve, transfer knowledge to new situations, create models, and predict outcomes.

Critical Thinking

Critical thinking is the ability to reason and solve problems using facts and concepts. These questions can be approached from a number of angles, and in most cases, they do not have a single correct answer.

1. How would you explain to a biology class that in the next decade, some diseases currently thought to be noninfectious will probably be found to be caused by microbes?

2. In what way could the extreme sensitivity of the PCR method be a problem when working with clinical specimens?

3. Why do some tests for antibody in serum (such as for HIV and syphilis) require backup verification with additional tests at a later date?

4. a. Look at figure 17.10*b*. What is the titer as shown?

 b. If the titer had been 40, would a different interpretation be made as to the immune status of the patient?

 c. Which patients are positive in figure 17.11, part III?

 d. What would it mean if a test 2 weeks later revealed a titer of 280?

 e. What would it mean if no agglutination had occurred in any tube?

5. Observe figure 17.15 and make note of the several steps in the indirect ELISA test. What four essential events are necessary to develop a positive reaction (besides having antibody A)? Hint: What would happen without rinses?

6. Distinguish between point of care tests that can be used to diagnose infection and those that can be used to identify pathogens, using examples.

Visual Challenge

1. From **chapter 3, fig 3.17*a*** (reproduced on the right) shows triple sugar iron agar. What biochemical characteristics does this figure illustrate? How could this medium be used to begin the identification of the isolated bacteria in this chapter's Case Study?

2. Starting with figure 17.5 and using data from Case Study tables 1 and 2, devise a flowchart that outlines the steps in identifying all of the *Vibrio* species.

1 2 3 4 5

© Kathy Park Talaro

STUDY TIP: Quickly create a customized practice quiz for this chapter by going to your Connect SmartBook "My Reports" section. Don't have SmartBook and want to learn more? Go to **whysmartbook.com.**

18 The Gram-Positive and Gram-Negative Cocci of Medical Importance

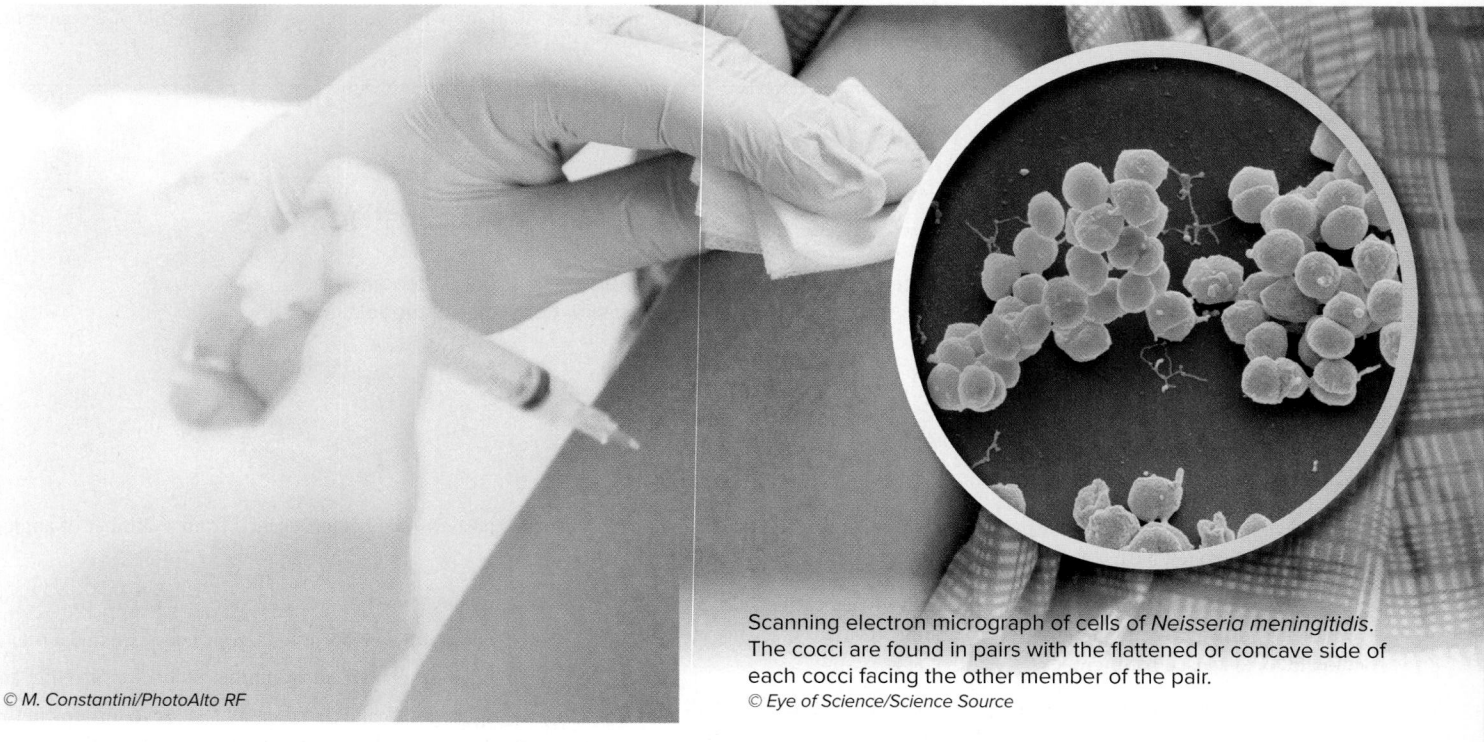

Scanning electron micrograph of cells of *Neisseria meningitidis*. The cocci are found in pairs with the flattened or concave side of each cocci facing the other member of the pair.
© *Eye of Science/Science Source*

© *M. Constantini/PhotoAlto RF*

 CASE STUDY Part 1 A Rapidly Advancing Pathogen Deserves an Equally Rapid Response

Some **signs** and **symptoms** are more likely than others to get the attention of medical professionals. Patients suffering chest pain or sudden paralysis receive immediate attention because heart attacks and strokes are medical emergencies that require prompt treatment if there is to be any hope of a successful outcome. Similarly, **headache** and **fever,** accompanied by a **stiff neck**—the hallmarks of **meningitis**—will never fail to rouse medical personnel into immediate action.

> "It's only been eight hours, and already two students have died."

On the morning of March 10, 2010, the Oklahoma State Department of Health (OSDH) was notified of a 7-year-old boy admitted to the hospital with meningitis, an **inflammation** of the tissue covering the **brain** and **spinal cord.** A swift, definitive diagnosis was necessary for a number of reasons: Meningitis may result from infection with bacteria, viruses, fungi, or parasites, each of which requires different treatment, and if, as suspected, the meningitis was bacterial in nature, proper medical intervention must begin immediately, because even when treated, the disease has a **mortality rate** approaching

15%. Furthermore, bacterial meningitis is **contagious,** and the appearance of a single case could be a harbinger of more cases to come. A **spinal tap** was ordered, and laboratory culture and analysis of **cerebrospinal fluid** from the patient revealed the presence of ***Neisseria meningitidis,*** a **gram-negative diplococcus** that is the second most common cause of the disease. The patient was admitted to the intensive care unit and successfully treated, while four household members of the patient were **prophylactically** treated with the antibiotics **rifampin** or **ceftriaxone.** Unfortunately, this was not the end of the case.

■ *What are the signs and symptoms of meningococcal meningitis?*

■ *How is* N. meningitidis *commonly spread?*

To continue the Case Study, go to Case Study Part 2 at the end of the chapter.

Pathogen Profiles and Systems Profiles

Scoping Out the Chapter does not appear in chapters 18–25 but will return in chapter 26. Beginning with this chapter we are introducing two features called *Pathogen Profiles* and *Systems Profiles*. Pathogen Profiles are short boxes describing the characteristics of major pathogens in the chapter. Each profile also includes a figure indicating the organs and systems most commonly infected by the pathogen. Systems Profiles categorize the major pathogens found in each chapter by the organ system(s) each pathogen generally affects, the name of the infection or disease it causes, and certain signs and symptoms associated with it.

Systems Profile 18.1	Pathogenic Gram-Positive and Gram-Negative Cocci					
Organism	Skin/Skeletal	Nervous/ Muscle	Cardiovascular/ Lymphatic/ Systemic	Gastrointestinal	Respiratory	Urogenital
Staphylococcus aureus	1. Boils, carbuncles 2. Impetigo 3. Scalded skin syndrome 4. Osteomyelitis		1. Endocarditis 2. Toxic shock syndrome	Food intoxication	Pneumonia	
Coagulase-negative staphylococci	Surgical infections		Endocarditis			Urinary tract infections
Streptococcus pyogenes (group A)	1. Pyoderma (Impetigo) 2. Erysipelas 3. Necrotizing fasciitis		1. Scarlet fever 2. Rheumatic fever		1. Pharyngitis (strep throat) 2. Sinusitis	Glomerulo-nephritis
Streptococcus agalactiae (group B)	Wound infections	Neonatal meningitis	1. Neonatal sepsis 2. Endocarditis		Neonatal pneumonia	
Enterococci	Wound and surgical infections					Healthcare-associated urinary tract infections
Viridans streptococci			Subacute endocarditis	Dental caries		
Streptococcus pneumoniae		Pneumococcal meningitis (adult)			1. Pneumonia 2. Otitis media 3. Sinusitis	
Neisseria gonorrhoeae	Gonococcal opthalmia					Gonorrhea
Neisseria meningitidis		Meningococcal meningitis				

Gram-positive and gram-negative cocci are among the most significant infectious agents of humans as well as prevalent members of the normal microbiota of the skin, oral cavity, and intestine. Because infections with these bacteria tend to stimulate pus formation, they are often referred to collectively as the **pyogenic cocci.** The most common infectious species in this group belong to four genera: *Staphylococcus, Streptococcus, Enterococcus,* and *Neisseria.* See **Systems Profile 18.1** for an outline of organs and systems targeted by these pathogens.

18.1 General Characteristics of the Staphylococci

 Expected Learning Outcomes

1. Name the five most important human pathogens in the genus *Staphylococcus.*

2. Describe the ability of *S. aureus* to withstand extremes of salt, pH, temperature, and desiccation, and how these abilities contribute to pathogenicity.

3. Name the enzymes and toxins found in *S. aureus,* and explain their role in increasing the virulence of the species.

4. Summarize the cutaneous, systemic, and toxigenic diseases caused by *S. aureus*, along with how the body defends against infection.

5. Differentiate between staphylococcal species in the laboratory through the examination of physiological and biochemical traits.

6. Explain strategies used to prevent and treat staphylococcal infections, and how infection with drug-resistant strains of *Staphylococcus* affects these strategies.

The genus *Staphylococcus* is a common inhabitant of the skin and mucous membranes, and it accounts for a considerable proportion of human infections (often called "staph" infections). Its spherical cells are arranged primarily in irregular clusters and occasionally in short chains and pairs **(figure 18.1).** Although typically gram-positive, older cultures and clinical specimens sometimes do not stain true. As a group, the staphylococci lack spores and flagella and may be encapsulated.

Currently 47 species have been placed in the genus *Staphylococcus,* but the most important human pathogens are:

1. *Staphylococcus aureus;*[1]
2. *Staphylococcus epidermidis, Staphylococcus capitis, Staphylococcus hominis;* and
3. *Staphylococcus saprophyticus.*

1. From L. *aurum,* gold. Named for the tendency of some strains to produce a golden-yellow pigment.

Figure 18.1 **Views of *Staphylococcus aureus* shape and arrangement.** This genus owes its name to the grapelike clusters it commonly displays. (*Left*) A Gram stain shows gram-positive cocci in irregular clusters (1,000×). (*Right*) A colorized scanning electron micrograph of methicillin resistant *Staphylococcus aureus* cells being engulfed by a neutrophil (7,500×). (left): © Eye of Science/Science Source; (right): *Source:* National Institute of Allergy and Infectious Diseases (NIAID)/CDC

Of these, *S. aureus* is considered the most serious pathogen, although the other species have become increasingly associated with opportunistic infections and can no longer be regarded as harmless commensals. As a group, the staphylococci are implicated, annually in the United States, in nearly 80,000 deaths and a not insignificant percentage of healthcare-associated infections. The following sections focus on the significant characteristics of each species separately.

Growth and Physiological Characteristics of *Staphylococcus aureus*

Staphylococcus aureus produces round, opaque colonies (**figure 18.2**) and grows at an optimum temperature of 37°C, though it can grow at any temperature between 10°C and 46°C. The species is a facultative anaerobe whose growth is enhanced in the presence of O_2 and CO_2. Its nutrient requirements can be satisfied by routine laboratory media, and most strains are metabolically versatile; that is, they can digest proteins and lipids, and ferment a variety of sugars. This species is considered one of the most resistant of all non-spore-forming pathogens, with well-developed capacities to withstand high salt (7.5%–10%), extremes in pH, and high temperatures (up to 60°C for 60 minutes). It also remains viable after months of air drying and resists the effects of many disinfectants and antibiotics. These properties contribute to the reputation of *S. aureus* as a troublesome hospital pathogen.

Virulence Factors of Staphylococcus aureus

Staphylococcus aureus exhibits a great variety of virulence factors, with no single toxin or enzyme completely responsible for the virulence seen in the species (**table 18.1**).

Pathogenic *S. aureus* typically produces **coagulase,** an enzyme that clots plasma (see figure 18.6*b*). The precise importance of coagulase to the disease process remains uncertain. It may be that coagulase causes fibrin to be deposited around staphylococcal cells. Fibrin can stop the action of host defenses such as phagocytosis, or

Figure 18.2 **Blood agar plate growing *Staphylococcus aureus*.** Some strains show two zones of hemolysis. The relatively clearer inner zone is caused by α-toxin, whereas the outer zone is fuzzy and appears only if the plate has been refrigerated. This outer zone is the β, or "hot-cold," hemolysin that shows up when the plate is refrigerated. © Kathy Park Talaro

it may promote staphylococcal adherence to tissues. Because 97% of all human isolates of *S. aureus* produce this enzyme, its presence is considered the most diagnostic species characteristic.

An enzyme that appears to promote invasion is hyaluronidase, or the "spreading factor," which digests the intercellular "glue" (hyaluronic acid) that binds connective tissue in host tissues. Other enzymes associated with *S. aureus* include staphylokinase, which

TABLE 18.1	Major Virulence Factors of *Staphylococcus aureus*	
Name	**Enzyme/Toxin**	**Effect**
Coagulase	Enzyme	Coagulates blood plasma
Hyaluronidase	Enzyme	Digests connective tissue of the host
Staphylokinase	Enzyme	Digests blood clots
Lipase	Enzyme	Digests oils, allowing bacteria to more easily colonize the skin
Penicillinase	Enzyme	Inactivates penicillin, rendering the bacterium resistant
Hemolysins (α, β, γ, δ)	Toxin	Lyse red blood cells
Leukocidin	Toxin	Lyses neutrophils and macrophages
Enterotoxins	Toxin	Induce nausea, vomiting, and diarrhea
Exfoliative toxins (A, B)	Toxin	Cause desquamation of the skin
Toxic shock syndrome toxin	Toxin	Induces fever, vomiting, rash, organ damage

digests blood clots; a nuclease that digests DNA (DNase); and lipases that help bacteria colonize oily skin surfaces. Enzymes that inactivate penicillin (penicillinase) or other drugs are produced by a majority of strains, and many isolates show multiple resistance.

The toxic products of this species include blood cell toxins (hemolysins and leukocidins), intestinal toxins, and epithelial toxins. **Hemolysins** lyse red blood cells, an effect that can be seen in the laboratory (figure 18.2). The most important, in terms of biological effect, is alpha-toxin (α-toxin), which lyses red blood cells while also causing damage to leukocytes, renal tissue, and both skeletal and heart muscle. Other hemolysins are designated by the Greek letters β, δ, and γ. (Note that the Greek letters used to describe the hemolysins do not correspond to the Greek letters describing patterns of hemolysis in chapter 13. For example, alpha-toxin produces beta-hemolysis.)

Other staphylococcal exotoxins include **leukocidin,** which damages cell membranes of neutrophils and macrophages, causing them to lyse. This toxin probably helps incapacitate the phagocytic line of defense. Some strains produce exotoxins called **enterotoxins** that act upon the gastrointestinal tract of humans. A few strains produce an **exfoliative*** **toxin** that separates the epidermal layer from the dermis and causes the skin to peel away. This toxin is responsible for staphylococcal scalded skin syndrome, in which the skin looks burned (see figure 18.5*b*). The most recent toxin brought to light is **toxic shock syndrome toxin (TSST).** The presence of this toxin in victims of toxic shock syndrome indicates its probable role in the development of this dangerous condition. The contributions of these toxins and enzymes to disease are discussed in the section on pathology.

Epidemiology and Pathogenesis of S. aureus

It is surprising that a bacterium with such great potential for virulence as *Staphylococcus aureus* is a common, intimate human associate. The microbe is present in most environments frequented by humans and is readily isolated from fomites. Colonization of some infants begins within hours after birth and continues throughout life. Anywhere from 20% to 60% of healthy adults may carry *S. aureus*, and in these instances the pathogen tends to be harbored intermittently rather than chronically. Carriage occurs mostly in the anterior nares (nostrils) and to a lesser extent in the skin, nasopharynx, and intestine. Usually this colonization is not associated with symptoms, nor does it ordinarily lead to disease in carriers or their contacts. Circumstances that predispose an individual to infection include poor hygiene and nutrition, tissue injury, preexisting primary infections, diabetes mellitus, and immunodeficiency states. *Staphylococcus aureus* is the third most common cause of infections in the newborn nursery and surgical wards. The "hospital strains" can readily spread in an epidemic pattern within and outside the hospital.

A serious concern has arisen from the increase in community infections by strains of *S. aureus* called MRSA (**m**ethicillin-**r**esistant *S. aureus*). Several outbreaks have been reported in prison inmates, athletes, and schoolchildren. The infections are spread by contact with skin lesions and have proved to be very difficult to treat and control.

The Scope of Staphylococcal Disease

Depending on the degree of invasion or toxin production by *S. aureus*, disease ranges from localized to systemic. A local staphylococcal infection often presents as an inflamed, fibrous lesion enclosing a core of pus called an **abscess (figure 18.3*a*).** Toxigenic disease—disease that is due to the presence of toxins rather than the bacterium itself—can present as a toxemia if the toxins are produced in the body, or as food intoxication if *S. aureus* toxins present in food are ingested.

Localized Cutaneous Infections

Staphylococcus aureus usually invades the skin through wounds, follicles, or skin glands. The most common infection is a mild, superficial inflammation of hair follicles termed **folliculitis (figure 18.3*b*)** or glands (hidradenitis). Although these lesions are usually resolved with no complications, they can lead to infections of subcutaneous tissues. A **furuncle*** (boil) results when the inflammation of a single hair follicle or sebaceous gland progresses into a

exfoliative (eks-foh′-lee-ay″-tiv) L. *exfoliatio,* falling off in layers.

furuncle (fur′-unkl) L. *furunculus,* little thief.

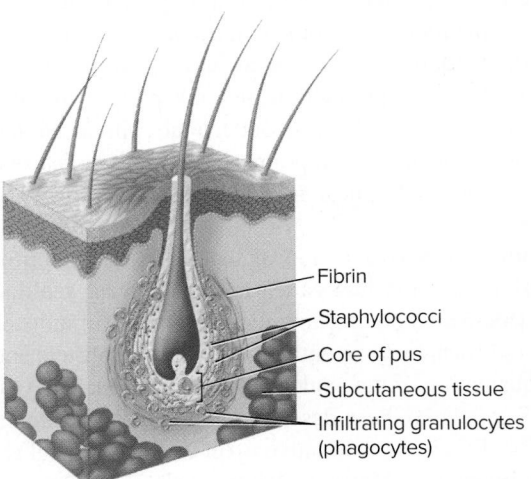

Fibrin
Staphylococci
Core of pus
Subcutaneous tissue
Infiltrating granulocytes (phagocytes)

(a) Sectional view of a boil or furuncle, a single pustule that develops in a hair follicle or gland and is the classic lesion of the species. The inflamed infection site becomes abscessed when masses of phagocytes, bacteria, and fluid are walled off by fibrin.

(b) Appearance of folliculitis caused by *S. aureus*. Note the clusters of inflamed papules and pustules.

(c) An abscess on the knee caused by methicillin-resistant *Staphylococcus aureus* (MRSA).

Figure 18.3 **Cutaneous lesions of *Staphylococcus aureus*.** Fundamentally, all are skin abscesses that vary in size, depth, and degree of tissue involvement. (b) © DermNet New Zealand Trust; (c): © Gregory Moran

large, red, and extremely tender abscess or pustule **(figure 18.3c).** Furuncles often occur in clusters (furunculosis) on parts of the body such as the buttocks, armpit, and back of the neck, where skin rubs against other skin or clothing. A **carbuncle*** is a deeper lesion created by aggregation of a cluster of furuncles into one large mass (sometimes as big as a baseball). It is usually found in areas of thick, tough skin such as the back of the neck. Carbuncles are extremely painful and can even be fatal in elderly patients when they give rise to systemic disease. One staphylococcal skin infection not confined to follicles and skin glands is the *bullous* type of **impetigo.*** It is characterized by bubblelike epidermal swellings that can break and peel away like a localized form of scalded skin syndrome (see figure 18.5a). This form of impetigo is most common in newborns.

Miscellaneous Systemic Infections

Most systemic staphylococcal infections have a focal pattern spreading from a local cutaneous infection to other sites via the bloodstream. In **osteomyelitis (figure 18.4)** the pathogen is established in the highly vascular metaphyses of a variety of bones (often the femur, tibia, ankle, or wrist). Abscess formation in the affected area results in an elevated, tender lump leading to necrosis or breakage of the bony tissue. Symptoms of osteomyelitis include fever, chills, pain, and muscle spasm. This form of osteomyelitis is seen most frequently in growing children, adolescents, and intravenous drug abusers. Another type, secondary osteomyelitis, develops after traumatic injury (compound fracture) or surgery in cancer and diabetes patients.

Systemic staphylococcal infections affect a variety of organs. Because these bacteria inhabit the nasopharynx, they can be aspirated into the lungs and cause a form of pneumonia involving multiple lung abscesses and symptoms of fever, chest pain, and bloody sputum. Although staphylococci account for only a small proportion of such pneumonia cases, the fatality rate is 50%. Most cases occur in infants and children suffering from cystic fibrosis and measles; this form of pneumonia is also one of the most serious complications of influenza in elderly patients.

Staphylococcal bacteremia causes a high mortality rate among hospitalized patients with chronic disease. Its primary origin is bacteria that have escaped from an infection site or from colonized medical devices (catheters, shunts). Circulating bacteria transported to the kidneys, liver, and spleen often form abscesses and give off toxins into the circulation. One consequence of staphylococcal bacteremia is a fatal form of endocarditis associated with the colonization of the heart's lining, cardiac abnormalities, and rapid destruction of the valves. Infection of the joints can produce a deforming arthritis (pyoarthritis). A severe form of meningitis (accounting for less than 10% of cases) occurs when *S. aureus* invades the brain.

Toxigenic Staphylococcal Disease

Diseases associated with toxin production by *S. aureus* are food intoxication, scalded skin syndrome, and toxic shock syndrome

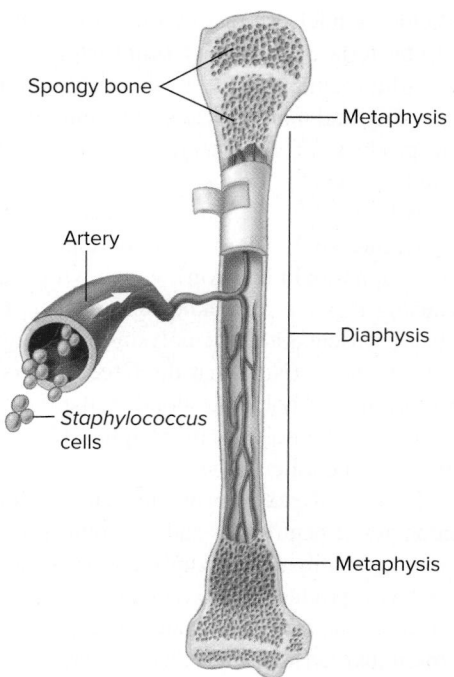

Figure 18.4 **Staphylococcal osteomyelitis in a long bone.** In the most common form, the bacteria spread in the circulation from some other infection site, enter the artery, and lodge in the small vessels in bony pockets of the marrow. Growth of the cells causes inflammation and damage that manifest as swelling and necrosis.

(18.1 Making Connections). Enterotoxins produced by certain strains are responsible for a common type of food poisoning in the United States. This illness is associated with foods such as custards, sauces, cream pastries, processed meats, potato salad, or dressings that have been contaminated during preparation and then left unrefrigerated for a few hours. Because of the high salt tolerance of *S. aureus,* even foods containing salt as a preservative are not exempt. The toxins produced by the multiplying bacteria do not noticeably alter the food's taste or smell. Enterotoxins are heatstable (inactivation requires 100°C for at least 30 minutes), so heating the food after toxin production may not prevent disease. The ingested toxin acts upon the gastrointestinal epithelium and stimulates nerves, with acute symptoms of cramping, nausea, vomiting, and diarrhea that appear in 2 to 6 hours. Recovery is rapid, usually within 24 hours.

Children with *S. aureus* infection of the umbilical stump or eyes are susceptible to a toxemia called **staphylococcal scalded skin syndrome (SSSS).** Upon reaching the skin, the toxin induces a painful, bright red flush over the entire body that first blisters and then causes desquamation of the epidermis **(figure 18.5).** The vast majority of SSSS cases have been described in infants and children under the age of 4. This same exfoliative toxin causes the local reaction in bullous impetigo, which can afflict people of all ages.

Host Defenses Against *S. aureus*

Humans have a well-developed resistance to staphylococcal infections. Studies performed on human volunteers established that an injection of several hundred thousand staphylococcal cells into

* *carbuncle* (car'-bunkl) L. *carbunculus,* little coal.
* *anaphylaxis* (an''-uh-fih-lax'-us) Gr. *ana,* excessive, and *phylaxis,* protection.

18.1 MAKING CONNECTIONS

Tampons and TSS

Toxic shock syndrome (TSS) was first identified as a discrete clinical entity in 1978 in young women using vaginal tampons, even though the disease probably existed before that time. Two discoveries helped to clarify the link between TSS and tampon use. The first was that ultra absorbent brands of tampons bind magnesium ions that are normally present in the blood; the second was that in the absence of these ions, *Staphylococcus aureus* produces much greater levels of exotoxins, including the toxin responsible for toxic shock syndrome. These findings also helped to explain why TSS was most likely to occur when menstrual flow was low. During periods of high flow, even the most absorbent tampons cannot absorb all the magnesium present; during low-flow days, magnesium may be completely depleted, leading to an increase in toxin production.

Toxic shock syndrome toxin is a type of superantigen that initiates a massive immune response by T cells when it enters the bloodstream. What results is an outpouring of cytokines that can wreak havoc on several systems. Symptoms are fever, rash, low blood pressure, shock, vomiting, and diarrhea. Potentially fatal complications occur in the renal, hepatic, and nervous systems.

A small percentage of TSS cases are seen in women using diaphragms or sponge-type contraceptives, as well as in children, men, non-menstruating women, and patients recovering from nasal surgery. The evidence implicating ultra absorbent tampons as a major factor in the disease was so compelling that those tampons were taken off the market, and tampon packages now contain a warning that the least-absorbent tampon that is effective should be used. From the highest yearly count of 880 TSS cases in 1981, the number has gradually decreased to a current average of about 100.

Provide an explanation for how cases of TSS have developed in men, children, and surgical patients. *Answer available on Connect.*

(a) (b)

(c)

Figure 18.5 **Effects of staphylococcal toxins on skin.**
(a) Bullous impetigo, a localized staphylococcal infection with filled and collapsed bullae. **(b)** Staphylococcal scalded skin syndrome results when exfoliative toxin produced in local infections causes blistering and peeling away of the outer layer of skin. **(c)** Photomicrograph of a segment of skin affected with SSSS. The point of epidermal shedding, or desquamation, is between the epidermis and the dermis. The lesion will heal because the level of separation is so superficial.
(a): © Kathy Park Talaro; (b): © National Institute Slide Bank/The Wellcome Centre for Medical Sciences; (c): © Rejendra Singh

Labels on image (c): Epidermis; Site of separation; Intact dermis

unbroken skin was not sufficient to cause abscess formation. Yet when sutures containing only a few hundred cells were sewn into the skin, a classic lesion rapidly formed, indicating that the presence of a foreign body to provide a surface for biofilm production dramatically lowered the infectious dose. Specific antibodies are produced against most of the staphylococcal antigens, though none of these antibodies appears to be effective (except for the antitoxin to SSSS toxin). The most powerful defense lies in the phagocytic response by neutrophils and macrophages. When aided by the opsonic action of complement, phagocytosis effectively disposes of staphylococci that have gained access to tissues. Inflammation and cell-mediated immunity, which stimulate abscess formation, also help contain staphylococci and prevent their further spread.

Other Important Staphylococci

Because they lack the coagulase present in most *S. aureus,* other *Staphylococcus* species are called the **coagulase-negative staphylococci.** Several species are included in this group, some of human origin and others of nonhuman, mammalian origin. Although this group was once considered clinically nonsignificant, its importance has greatly increased over the past 20 years. Coagulase-negative staphylococci currently account for a large proportion of healthcare-associated and opportunistic infections in immunocompromised patients.

The normal habitats of ***Staphylococcus epidermidis*** are the skin and mucous membranes. *Staphylococcus hominis* lives in skin areas rich in apocrine glands; and *S. capitis* populates the scalp, face, and external ear. From these sites, the bacteria are positioned to enter breaks in the protective skin barrier. Infections usually occur after surgical procedures, such as insertion of shunts, catheters, and various prosthetic devices that require incisions through the skin. It appears that any foreign object can support biofilm formation by these bacteria, which develop thick adherent capsules. It is particularly alarming that these infections are increasing because of

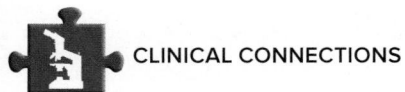

CLINICAL CONNECTIONS

Biofilms in Infections

Many species of gram-positive cocci produce slimes or capsules that contribute to biofilm formation in the body. These thick tenacious films increase adherence in infections and can lead to a variety of complications. For example, staphylococci lay down a slimy matrix that binds them to surfaces and allows them to colonize catheters, artificial valves, and other indwelling medical devices (see the chapter 4 Case Study). Certain species of oral streptococci are active initiators of plaque on teeth and are ultimately responsible for many types of oral diseases. The streptococci that cause endocarditis produce thick vegetations on damaged areas of the heart valves and heart lining. Such biofilms can have several adverse effects. Acting as an impenetrable shield, they hide the infectious agent and make it inaccessible to the host defenses. They can also block the effects of antimicrobic drugs by slowing passage into the embedded microbes. In some infections the biofilm holds the infectious agent in the body, where it becomes a continued source of pathogens. The subsequent release of bacteria into the circulation may give rise to sepsis and focal infections.

How could drugs that affect biofilm formation be used in the treatment of infection with biofilm-forming bacteria? *Answer available on Connect.*

(a) **Catalase test**

Coagulase-positive

Coagulase-positive
Staphylococcus aureus

Coagulase negative
Other *Staphylococcus* species

Streptococcus species

(b) **Coagulase test**

Figure 18.6 Tests for differentiating the genus *Staphylococcus* from *Streptococcus* and for identifying *Staphylococcus aureus*. (a) Staphylococci produce catalase to break down the hydrogen peroxide generated during oxidative metabolism. Catalase is lacking in streptococci. In the *plate catalase test,* adding a 3% solution of hydrogen peroxide to a bacterium growing on a plate will cause the sample to bubble vigorously if catalase is present and to remain undisturbed if catalase is lacking. (b) Staphylococcal coagulase is an enzyme that reacts with factors in plasma to initiate clot formation. In the coagulase test, a tube of plasma is inoculated with the bacterium. If it remains liquid, the test is negative. If the plasma develops a lump or becomes completely clotted, the test is positive. © Barry Chess

technological advances in maintaining patients. Although *S. epidermidis* is not as invasive or toxic as *S. aureus,* it can also cause endocarditis, bacteremia, and urinary tract infections.

Less is known about the distribution and epidemiology of *S. saprophyticus.* It is an infrequent resident of the skin, lower intestinal tract, and vagina, but its primary importance is in urinary tract infections. For unknown reasons, urinary *S. saprophyticus* infection is found almost exclusively in sexually active young women and is the second most common cause of urinary infections in this group.

Identification of *Staphylococcus* Isolates in Clinical Samples

Staphylococci are frequently isolated from pus, tissue exudates, sputum, urine, and blood. Primary isolation is achieved by inoculation on sheep or rabbit blood agar; or in heavily contaminated specimens, selective media such as mannitol salt agar are used. In addition to culturing, the specimen can be Gram stained and observed for irregular clusters of gram-positive cocci.

Because differentiating among gram-positive cocci is not possible using colonial and morphological characteristics alone, other presumptive tests are required (see figure 17.5). Traditionally, *Staphylococcus* has been differentiated from *Streptococcus* by the production of catalase, which is produced in the former but not the latter **(figure 18.6a),** and from *Micrococcus* through biochemical and cultural differences **(figure 18.7).** *S. aureus* may be distinguished from other species of *Staphylococcus* as it is the only

species within the genus to produce coagulase (figure 18.6b). Newer identification methods rely on antibody-coated latex beads or the polymerase chain reaction (PCR) to provide a far more rapid identification, reducing the time required from 2 or 3 days to just a few hours. The PCR method has the added advantage of identifying in a single step those *S. aureus* isolates that are methicillin resistant (MRSA).

Quick Search
To see how quickly antibody-coated latex beads can be used to identify *S. aureus,* search the Internet for "Staphaurex test for Staph aureus."

Differentiating the common coagulase-negative staphylococci from one another relies primarily upon the novobiocin resistance of *S. saprophyticus,* mannose fermentation in *S. epidermidis,* the lack of urease production in *S. capitis,* and lack of anaerobic growth in *S. hominis.* Commercial systems that allow rapid identification of staphylococcal species based on their biochemical and physiological characteristics (figure 18.7) have largely replaced the need to perform a great number of individual tests **(table 18.2).**

Figure 18.7 Miniaturized test system used in further identification of *Staphylococcus* isolates. Microbiologists can separate 23 species of gram-positive cocci with this system, including 20 clinically important species of *Staphylococcus*. Each cupule contains a sugar or other biochemical substrate important to the differentiation of gram-positive cocci, along with a chemical indicator. The combination of positive and negative results obtained allows for the identification of an unknown species. Courtesy of bioMerioux, Inc.

Clinical Concerns in Staphylococcal Infections

The staphylococci are notorious for their acquisition of resistance to new drugs, and they continue to defy attempts at medical control. Because resistance to common drugs is likely, antimicrobic susceptibility tests are essential in selecting a correct therapeutic agent.

Ninety-five percent of strains of *S. aureus* have acquired genes for penicillinase, which makes them resistant to the traditional drugs penicillin and ampicillin. Equally problematic are strains of **MRSA** that carry multiple resistance to a wide range of antimicrobics, including methicillin, gentamicin, cephalosporins, tetracycline, erythromycin, and even quinolones. The CDC estimates about 82,000 new cases per year, with 11,500 deaths caused by this "superbug." A few strains have acquired resistance to all major drug groups except vancomycin. Most of these isolates are found in hospitals (HA-MRSA), where they account for 80% of staphylococcal infections. Now that there have been reports of vancomycin resistance in *S. aureus,* infectious disease specialists are keeping close track of this trend.

The latest cause for concern is a form of resistant *S. aureus* acquired outside of the hospital during community contact (CA-MRSA). This strain causes the usual staph-type skin infections, but it has also been implicated in fatal systemic diseases. It is contracted most often from locations where people congregate and may share fomites, such as in gyms, schools, and prisons. Although its numbers are increasing, its overall incidence is still far below that of hospital MRSA. CA-MRSA is also sensitive to a wider array of antibiotics, so more treatment options are available. The CDC is currently being urged to include this strain on the notifiable diseases list to monitor its epidemiology.

Treatment of Staphylococcal Infections

Selecting the correct drug to treat staphylococcal infections is highly dependent on culture and sensitivity testing of the staphylococcal isolates. The less resistant forms of *S. aureus* can generally be treated with cephalexin, sulfa drugs, tetracyclines, or clindamycin. If the strain is MRSA, there are still a few drugs that will generally work, including vancomycin, ceftaroline, linezolid, and daptomycin. It may be necessary to administer two of these in combination to lessen the development of further drug resistance. For cases of VISA and VRSA, quinupristin/dalfopristin is the last drug of choice.

TABLE 18.2	Separation of Clinically Important Species of *Staphylococcus*

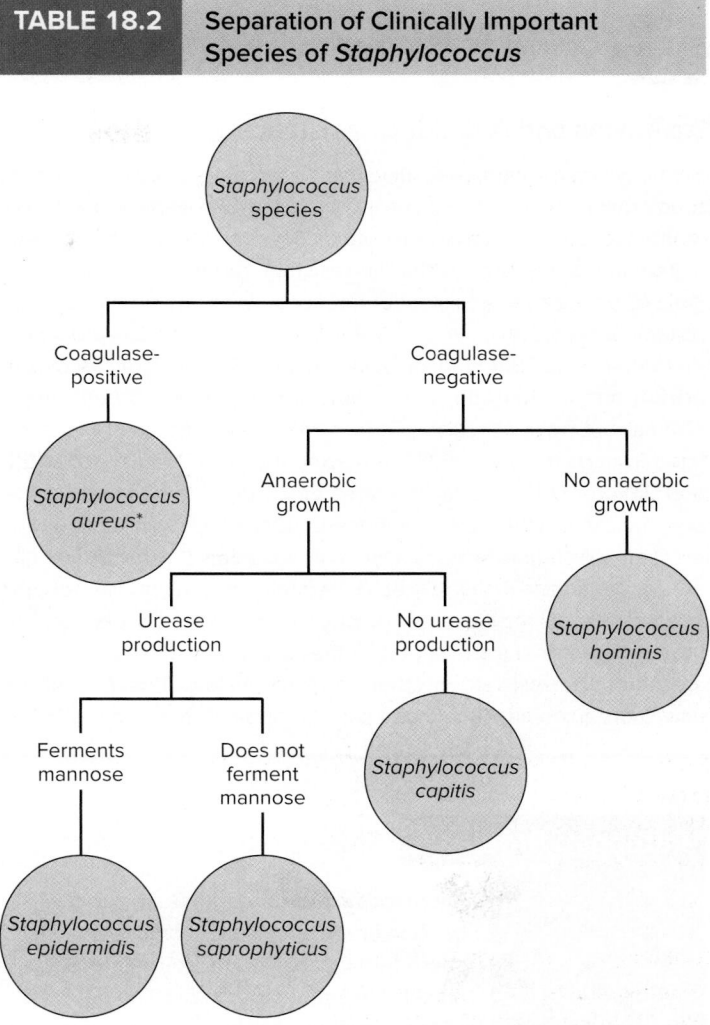

*A few strains of *S. aureus* are coagulase-negative.

Clinical experience has shown that abscesses will not respond to therapy unless they are surgically perforated and cleared of pus and foreign bodies. Severe systemic conditions such as endocarditis, septicemia, osteomyelitis, pneumonia, and toxemia respond slowly and require intensive, lengthy therapy by oral or injected drugs or some combination of the two.

Prevention of Staphylococcal Infections

Because humans are the principal reservoir of the pathogenic staphylococci, there will probably always be carriers and infections. It is difficult to block the colonization of the human body, but one can minimize the opportunity for healthcare-associated infections by careful hygiene and adequate cleansing of surgical incisions and burns. Controlling the increasing incidence of healthcare associated outbreaks, especially among people at greatest risk (surgical patients and neonates), is a real challenge for hospital staff. Patients or hospital personnel who are asymptomatic carriers or have frank infections are the most common source. The burden of effective deterrence again falls to meticulous hand washing, proper disposal of infectious dressings and discharges, isolation of people with open lesions, and attention to indwelling catheters and needles.

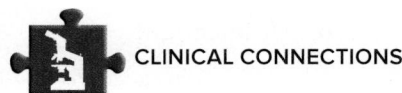

CLINICAL CONNECTIONS

Acronyms and Antibiotic Resistance

So many strains of gram-positive cocci have developed resistance to drugs that specialists have resorted to using acronyms to keep easier track of them. The five most significant strains in healthcare-associated infections are methicillin-resistant *Staphylococcus aureus* (MRSA), vancomycin-resistant enterococci (VRE), methicillin-resistant *Staphylococcus epidermidis* (MRSE), drug-resistant *Streptococcus pneumoniae* (DRSP), vancomycin-resistant *Staphylococcus aureus* (VRSA), and vancomycin intermediate *Staphylococcus aureus* (VISA). The current rates of resistance (based on how often a strain is isolated from patients) are 63% for MRSA, 30% for VRE, 90% for MRSE, and 22% for DRSP. So far the rates of VRSA have been very low—only 14 cases have been diagnosed. VISA, which shows slightly increased resistance to vancomycin, is considered a harbinger of a potential increase in the rate of VRSA infections. The number of VISA infections is on the increase, numbering well over 50 per year. Although these strains are named for their primary drug resistance, they are often also resistant to other drugs (cephalosporins, macrolides, and fluoroquinolones). In many cases it appears that the resistance has been acquired through transfer of resistance plasmids. As a result, the prevalence of these strains is increasing and will continue to be a serious challenge for health care workers.

In response to this trend, drug companies have developed several new antimicrobic agents as alternate treatments to target these resistant strains. Quinupristin/dalfopristin (Synercid) is one such alternative, a combination antibiotic recommended for treatment of some types of VRE and staphylococcal infections. Moving beyond combinations of naturally occurring compounds, a number of semisynthetic drugs have been developed to combat bacterial resistance. Daptomycin (Cubicin) is often the first choice for treating skin infections and bacteremia caused by any of several species of *Staphylococcus, Streptococcus,* and *Enterococcus,* whereas telavancin (Vibativ) targets skin infections caused by the same three genera of bacteria. Two other semisynthetics, linezolid (Zyvox) and ceftaroline (Teflaro), are used to treat skin and lung infections caused by MRSA or DRSP.

Explain why VISA is of concern to clinicians and why VRSA could be a more serious problem than MRSA. *Answer available on Connect.*

Pathogen Profile #1 *Staphylococcus aureus*

Source: Janice Carr/CDC

Microscopic Morphology Gram-positive cocci in irregular clusters; nonmotile; nonspore-forming. May form biofilm infections on catheters and other indwelling devices.

Identified by Growth on high-salt (7.5% NaCl or more) media, Gram reaction, and arrangement. Fermentation of sugars distinguishes *Staphylococcus* from *Micrococcus;* catalase production distinguishes *Staphylococcus* from *Streptococcus.* Coagulase production distinguishes *S. aureus* from other species of *Staphylococcus.* Commercially available rapid identification tests rely on antibody-coated latex beads that bind specifically with *S. aureus.*

Habitat Carried by 20% to 60% of healthy persons in the nostrils, skin, nasopharynx, and intestine. Very resistant to harsh environmental conditions and routinely present on fomites.

Virulence Factors *S. aureus* possesses enzymes that destroy host tissue (hyaluronidase), digest blood clots (staphylokinase), colonize oily skin (lipase), and resist the effects of penicillin (penicillinase). Toxins (leukocidins) destroy neutrophils and macrophages, lyse red blood cells (hemolysins), and cause damage throughout the body (enterotoxins, exfoliative toxins, toxic shock syndrome toxin).

Primary Infections/Disease Local cutaneous infections include folliculitis, furuncles, and carbuncles, as well as bullous impetigo. Systemic infections include osteomyelitis, pneumonia, and endocarditis. Diseases due to *S. aureus* toxins include food intoxication, scalded skin syndrome, and toxic shock syndrome.

Control and Treatment Control of healthcare-associated infection relies on careful hygiene, adequate cleansing of surgical incisions and burns; isolation of persons with open lesions; and barring of *S. aureus* carriers from sensitive areas such as operating rooms and nurseries. Special concern is paid to the strains known as MRSA, which have high levels of drug resistance. Community-acquired infections are controlled through disinfection of shared environments and equipment. Treatment involves intensive chemotherapy, often with multiple antimicrobics. The problem of MRSA strains requires antimicrobial susceptibility testing to select a correct chemotherapeutic agent. Many cutaneous lesions require perforation and drainage prior to antimicrobic therapy.

When possible, other preventive measures must be instituted. For example, hospital workers who carry *S. aureus* nasally may be barred from nurseries, operating rooms, and delivery rooms. Carriers can be treated for several months with a combination of antimicrobic drugs such as bactroban and dicloxacillin to reduce the source of this pathogen in clinical settings. A vaccine aimed at providing protection against *S. aureus* is currently in clinical trials.

 Check Your Progress SECTION 18.1

1. List the most clinically significant pathogens in the genus *Staphylococcus,* and identify the species that is defined by production of coagulase.

2. Discuss why food intoxication in salty foods, such as ham, is often due to the growth of *S. aureus.*

3. Explain how unbroken skin, the phagocytic action of neutrophils and macrophages, inflammation, and cell-mediated immunity all help to protect against staphylococcal infection.

4. List the strategies commonly used to prevent and treat staphylococcal infections.

5. Explain the meanings of the acronyms *MRSA, VRE, MRSE, DRSP, VRSA,* and *VISA.*

18.2 General Characteristics of the Streptococci and Related Genera

 Expected Learning Outcomes

7. Name the most important human pathogens in the genus *Streptococcus.*

8. Summarize the virulence factors of *S. pyogenes,* as well as the diseases and long-term complications associated with *S. pyogenes* infection.

9. Recall the epidemiology and pathogenesis of the streptococcal species most often associated with human disease and how these groups are separated in the laboratory.

10. Explain strategies used to prevent and treat streptococcal infections.

11. Relate the pathogenesis and epidemiology of *S. pneumoniae.*

The genus ***Streptococcus**** includes a large and varied group of bacteria. Some are commensals or pathogens in humans and animals; others are free-living in the environment. Members of this group are known for the arrangement of cocci in long, beadlike chains, especially when grown in a liquid culture. The length of these chains varies, and it is common to find them in pairs **(figure 18.8).** The general shape of the cells is spherical, but they can also appear ovoid or rodlike, especially in actively dividing young cultures.

Streptococci are non-spore-forming and nonmotile (except for an occasional flagellated strain), and they can form capsules and slime layers. They are facultative anaerobes that ferment a variety of sugars, usually with the production of lactic acid (homofermentative). Streptococci do not form catalase, but they do have a peroxidase system for inactivating hydrogen peroxide, which allows their survival in the presence of oxygen. Most pathogenic forms are fastidious in nutrition and require enriched media for cultivation. Colonies are usually small, nonpigmented, and glistening. Most members of the genus are quite sensitive to drying, heat, disinfectants, and drugs. A few members, such as the pneumococci and enterococci, have developed significant drug resistance.

Species of *Streptococcus* were first classified using a system devised by Rebecca Lancefield in the 1930s. This system divided

* *Streptococcus* (strĕp′tə-kŏk′əs) Gr. *streptos,* winding, twisted. The chain arrangement is the result of division in only one plane.

Figure 18.8 One of the truly great sights in microbiology: A freshly isolated *Streptococcus* with its long, intertwining chains. These occur only in liquid media and are especially well developed if nutrients are in limited supply (1,000×). © Kathy Park Talaro

streptococcal isolates into 17 alphabetically named Lancefield groups (group A, group B, etc.) based on cell wall carbohydrates displayed by each species. The system is somewhat incomplete—for example, some streptococcal species possess no cell wall carbohydrates and need additional methods of identification. One technique commonly employed in the clinical laboratory uses the reaction in blood agar of a given culture to begin the identification of streptococcal isolates **(figure 18.9*a–c*). Table 18.3** summarizes the streptococcal groups based on their habitat, hemolysis, and pathogenicity.

Many species of *Streptococcus* exist as microbiota of the oral cavity, whereas others live as commensals of domestic animals, including pets, and may colonize or infect humans. Human disease is most often associated with ***S. pyogenes, S. agalactiae, viridans streptococci, S. pneumoniae,*** and ***Enterococcus faecalis*** (formally known as *Streptococcus faecalis*).

Beta-Hemolytic Streptococci: *Streptococcus pyogenes*

By far the most serious streptococcal pathogen of humans is *Streptococcus pyogenes,* a beta-hemolytic species that is the main representative of group A. It is a relatively strict pathogen, inhabiting the throat, nasopharynx, and occasionally the skin of humans. The involvement of this species in severe disease is partly due to the substantial array of surface antigens, toxins, and enzymes it can generate.

Cell Surface Antigens and Virulence Factors

Streptococci display numerous surface antigens **(figure 18.10).**

- C-carbohydrates, specialized polysaccharides found on the surface of the cell wall, are the basis for Lancefield groups. Their apparent contribution to pathogenesis is to protect the bacterium from being dissolved by the lysozyme defense of the host.

- Lipoteichoic acid bound to fimbriae contributes to the adherence of *S. pyogenes* to epithelial cells in the skin or pharynx.

- Another type-specific molecule is the M-protein, of which about 80 different subtypes exist. This substance makes up

Streptococcus pyogenes with zones of β-hemolysis

Streptococcus pneumoniae displaying α-hemolysis

(a) (b)

Figure 18.9 **Hemolysis patterns on blood agar may be used to separate streptococci into major subgroups. (a)** Throat cultures with colonies of *Streptococcus pyogenes* showing β-hemolysis. Note clear zones with completely lysed red blood cells. **(b)** Colonies of *S. pneumoniae* demonstrate the greenish zones, with partial lysis of RBCs, typical of α-hemolysis. **(c)** Differentiation of streptococcal species based on hemolysis and drug resistance. *Some group D species are beta- or non-hemolytic. (a and b): © Kathy Park Talaro

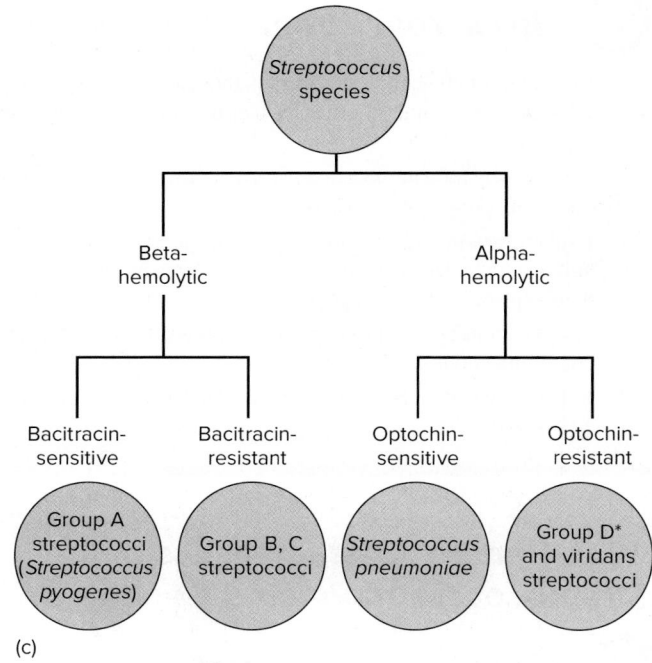

(c)

TABLE 18.3	Major Species of *Streptococcus* and Related Genera			
Species	**Lancefield Group**	**Hemolysis Type**	**Habitat**	**Pathogenicity to Humans**
S. pyogenes	**A**	Beta (β)	**Human throat**	**Skin, throat infections, scarlet fever**
S. agalactiae	**B**	β	**Human vagina, cow udder**	**Neonatal, wound infections**
S. equi	C	β	Various mammals	Rare, in abscesses
S. dysgalactiae	C	β	Cattle	Rare
Enterococcus faecalis	**D**	α, β, N	**Human, animal intestine**	**Endocarditis, UTI***
E. faecium, E. durans	D	Alpha (α)	Human, animal intestine	Similar to *E. faecalis*
S. bovis	D	N	Cattle	Subacute endocarditis, bacteremia
S. anginosus	F, G, L	β	Humans, dogs	Endocarditis, URT** infections
S. sanguinis	**H**	α	**Human oral cavity**	**Endocarditis**
S. salivarius	K	N	Human saliva	Endocarditis
Lactococcus lactis	N	V	Dairy products	Very rare
S. mutans	**NI***	N	**Human oral cavity**	**Dental caries**
S. uberis, S. acidominimus	NI	V	Domestic mammals	Rare
S. pneumoniae	**NI**	α	**Human respiratory tract**	**Bacterial pneumonia**

Note: Species in bold type are the most significant sources of human infection and disease. N = none; V = varies.
*Urinary tract infection. **Upper respiratory tract. ***No group C-carbohydrate identified.

spiky surface projections that contribute to virulence by resisting phagocytosis and improving adherence.

- A capsule made of hyaluronic acid (HA) is formed by most *S. pyogenes* strains. The HA is chemically indistinguishable from HA found in human tissues, so it does not provoke an immune response from the host.
- C5a protease catalyzes the cleavage of the C5a protein of the complement system. Experiments suggest that this not only hinders the complement-associated aspects of the immune system but also interferes with the chemotactic response of neutrophils.

Major Extracellular Toxins Group A streptococci owe some of their virulence to the effects of hemolysins called **streptolysins.** The two types are streptolysin O (SLO) and streptolysin S (SLS).[2] (Both types cause beta-hemolysis of sheep blood agar.) Both hemolysins rapidly injure many cells and tissues, including leukocytes, liver, and heart muscle.

2. In SLO, O stands for oxygen because the substance is inactivated by oxygen. SLO is produced by most strains of *S. pyogenes*. In SLS, S stands for serum because the substance has an affinity for serum proteins. SLS is oxygen-stable.

(a) (b)

Figure 18.10 **Cutaway view of group A *Streptococcus*. (a)** The outermost fringe consists of fimbriae composed in part of M-protein antigens. Other layers making up the cell envelope are the capsule, protein antigens, and C-carbohydrate antigens. **(b)** Detail of M-protein and its position in the cell envelope.

A key toxin in the development of scarlet fever (discussed in a later section of this chapter) is **erythrogenic*** (also known as the pyrogenic) toxin. This toxin is responsible for the bright red rash typical of this disease, and it also induces fever by acting upon the temperature regulatory center. Only lysogenic strains of *S. pyogenes* that contain genes from a temperate bacteriophage can synthesize this toxin.

Some of the streptococcal toxins (erythrogenic and streptolysin O) contribute to increased tissue injury by acting as **superantigens** (see chapter 15). These toxins increase the stimulation of certain T lymphocytes. Overactive T cells proliferate and produce a cascade of immune responses that release *tumor necrosis factor* and other cytokines, which lead to vascular injury. This is the likely mechanism behind the severe pathology of streptococcal toxic shock syndrome and necrotizing fasciitis.

Major Extracellular Enzymes Several enzymes for digesting macromolecules are virulence factors of the group A streptococci. Streptokinase, similar to staphylokinase, activates a pathway leading to the digestion of fibrin clots and may play a role in invasion. Hyaluronidase breaks down the binding substance in connective tissue and promotes spreading of the pathogen into the tissues, while streptodornase (DNase) liquefies purulent discharges by hydrolyzing DNA.

Epidemiology and Pathogenesis of Streptococcus pyogenes

Streptococcus pyogenes has traditionally been linked with a diverse spectrum of disease. Before the era of antibiotics, it accounted for a major portion of serious human infections and deaths from such diseases as rheumatic fever and puerperal sepsis. Although its importance has diminished over the past several years, recent epidemics have reminded us of its potential for sudden and serious illness **(18.2 Making Connections).**

Humans are the only significant reservoir for *S. pyogenes*. About 5% to 15% of the population are subclinical carriers of virulent strains. Infection is generally transmitted through direct contact, droplets, and occasionally fomites. The bacteria gain a foothold at periods of lowered host resistance, primarily on the skin or in the pharynx. The incidence and types of infections are altered by climate, season, and living conditions. Skin infections occur more frequently during the warm temperatures of summer and fall, and pharyngeal infections increase in the winter months. Children 5 to 15 years of age are the predominant group affected by both types of infections. In addition to local cutaneous and throat infections, *S. pyogenes* can give rise to a variety of systemic infections and progressive sequelae if not properly treated.

Skin Infections When virulent streptococci invade a nick in the skin or the mucous membranes of the throat, an inflammatory primary lesion is produced. Pyogenic infections appearing after local invasion of the skin are pyoderma or erysipelas; those developing in the throat are pharyngitis or tonsillitis.

Streptococcal **impetigo,** or **pyoderma,** is marked by burning, itching papules that break and form a highly contagious yellow crust **(figure 18.11*a*).** Impetigo often occurs in epidemics among schoolchildren, and it is also associated with insect bites, poor hygiene, and crowded living conditions. A slightly more invasive form of skin infection is **erysipelas.*** The pathogen usually enters through a small wound or incision on the face or extremities and eventually spreads to the dermis and subcutaneous tissues. Early symptoms are edema and redness of the skin near the portal of entry, fever, and chills. The lesion begins to spread outward, producing a slightly elevated edge

that is noticeably red, hot, and often vesicular **(figure 18.11*b*).** Depending on the depth of the lesion and how the infection progresses, cutaneous lesions can remain superficial or produce long-term systemic complications. Some severe cases involving large areas of skin have been fatal.

Most people associate streptococci with the condition called *strep throat* or, more technically, **streptococcal pharyngitis.** Most humans will acquire this particular infection at some time during

(a)

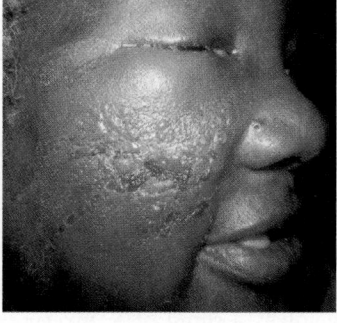

Figure 18.11 **Streptococcal skin infections. (a)** Impetigo lesions on the face. **(b)** Erysipelas of the face. Tender, red, puffy lesions have a sharp border that can burst and release fluid. Also note the release of purulent discharge from the eye.
(a): © Dr. P. Marazzi/Science Source; (b): *Source:* Dr. Thomas F. Sellers, Emory University/CDC

(b)

* *erythrogenic* (eh-rith″-roh-jen′-ik) Gr. *erythros,* red, and *gennan,* to produce. The term *pyrogenic* refers to its fever-causing effects.

* *erysipelas* (er″-ih-sip′-eh-las) Gr. *erythros,* red, and *pella,* skin.

18.2 MAKING CONNECTIONS

Invasive Group A Streptococci and "Flesh-Eating" Syndrome

Streptococcal infections are "occupational diseases of childhood" that usually follow a routine and uncomplicated course. The greatest cause for concern are those few occasions when such infections erupt into far more serious ailments. One dramatic example is **necrotizing fasciitis,** * a complication of *S. pyogenes* infection that has received heavy publicity as the "flesh-eating disease." It should be emphasized that cases of this disease are rather rare, but its potential for harm is high. It can begin with an innocuous cut in the skin and spread rapidly into nearby tissue, causing severe disfigurement and even death.

There is really no mystery to the pathogenesis of necrotizing fasciitis. It begins very much like impetigo and other skin infections: Streptococci on the skin are readily introduced into small abrasions or cuts, where they begin to grow rapidly. These strains of group A streptococci release special enzymes and toxins that greatly increase their invasiveness and virulence. Some of the toxins acting as superantigens can trigger harmful immune responses.

> **Quick Search**
>
> To see just how much damage *S. pyogenes* can do, search for images of "necrotizing fasciitis."

The enzymes digest the connective tissue in skin, and their toxins poison the epidermal and dermal tissues. As the flesh dies, it separates and sloughs off, forming a pathway for the bacteria to spread into deeper tissues such as muscle. More dangerous infections involve a mixed infection with anaerobic bacteria, systemic spread of toxins to other organs, or both. It is true that some patients have lost parts of their limbs and faces and others have required amputation, but early diagnosis and treatment can prevent these complications. Fortunately, even virulent strains of *Streptococcus pyogenes* are not usually drug resistant.

Explain what is meant by the terms necrotizing and fasciitis. Does the disease really eat flesh? *Answer available on Connect.*

Streptococci

Connective tissue

Damage to connective tissue, muscle

The phases of *Streptococcus pyogenes* induced necrotizing fasciitis.

Necrotizing fasciitis.
© DermNet New Zealand Trust

* *necrotizing fasciitis* (nee'-kroh-ty"-zing fass"-ee-eye'-tis) Gr. *nekrosis,* deadness, and L. *fascia,* the connective tissue sheath around muscles and other organs.

their lifetime. The organism multiplies in the tonsils or pharyngeal mucous membranes, causing redness, edema, enlargement, and extreme tenderness, which make swallowing difficult and painful **(figure 18.12).** These symptoms can be accompanied by fever, headache, nausea, and abdominal pain. Other signs include a **purulent*** exudate over the tonsils; swollen lymph nodes; and occasionally, white, pus-filled abscesses on the tonsils.

Systemic Infections Streptococcal throat infection can lead to **scarlet fever** (scarlatina) when it involves a strain of *S. pyogenes*

carrying a prophage that codes for erythrogenic toxin. Systemic spread of this toxin results in high fever and a bright red, diffuse rash over the face, trunk, inner arms and legs, and even the tongue. Within 10 days the rash and fever usually disappear, often accompanied by sloughing of the epidermis. Many cases of pharyngitis and even scarlet fever are mild and uncomplicated, but on occasion they lead to severe sequelae. The dissemination of streptococci into the lymphatics and the blood can give rise to septicemia, an infrequent complication limited to people who are already weakened by underlying disease. Another relatively rare infection is *S. pyogenes* pneumonia, which accounts for less than 5% of bacterial pneumonias and is usually secondary to influenza or some other lung

* *purulent* (puh'-roo-lent) L. *purulentus,* inflammation. In reference to pus.

Figure 18.12 The appearance of the throat in pharyngitis and tonsillitis. The pharynx and tonsils become bright red and suppurative. The white areas are pus-containing nodules. In severe cases, where the tonsils meet, they are said to be "kissing." © Scott Camazine/PhotoTake

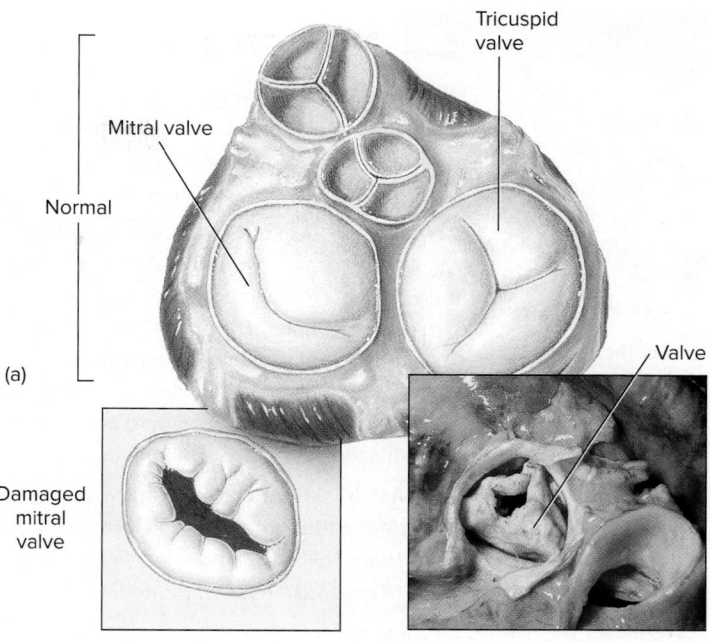

Figure 18.13 The cardiac complications of rheumatic fever. Pathologic processes of group A streptococcal infection can extend to the heart. In this example, it is believed that cross-reactions between host-produced antibodies to streptococci and heart proteins have a gradual destructive effect on the valves (especially the mitral valve). Scarring and deformation change the capacity of the valves to close and shunt the blood properly. **(a)** Normal valves, viewed from above. **(b)** Left inset depicts scar tissue on a damaged mitral valve; right inset is a photograph of the valve revealing the narrowing typical of rheumatic scarring. (b): *Source:* Dr. Edwin P. Ewing, Jr./CDC

infection. One of the most severe and invasive infections is *streptococcal toxic shock syndrome.* It is the result of a profound bacteremia and deep tissue infection and rapidly progresses to multiple organ failure. Even with treatment, around 30% of patients die.

Long-Term Complications of Group A Infections

Even when an initial group A streptococcal infection appears to be resolved, occasional long-term sequelae can appear within a few weeks. The two most important ones are (1) **rheumatic* fever (RF),** a delayed inflammatory condition of the joints, heart, and subcutaneous tissues; and (2) **acute glomerulonephritis (AGN),** a disease of the kidney glomerulus and tubular epithelia. In both cases the body's own immune system is responsible for the pathology of the condition. In RF, damage to cardiac tissue occurs as part of a type II hypersensitivity reaction. This reaction comes about when antibodies formed against the streptococcal cell wall and membranes cross-react with molecules in the host that resemble the bacterial antigens. In AGN, damage can be traced to a type III hypersensitivity, occurring when immune complexes deposited in the epithelia inflame and damage the filtering apparatus (see chapter 16).

Rheumatic Fever Rheumatic fever usually follows an overt or subclinical case of streptococcal pharyngitis in children. Its major clinical features are carditis, an abnormal electrocardiogram, painful arthritis, chorea, nodules under the skin, and fever.[3] The course of the syndrome extends from 3 to 6 months, usually without lasting

damage. In patients with severe carditis, however, extensive damage to the heart valves and muscle can occur **(figure 18.13).** Although the degree of permanent damage does not usually reveal itself until middle age, it is often extensive enough to require the replacement of damaged valves. Heart disease due to RF is on the wane because of medical advances in diagnosing and treating streptococcal infections, but it is still responsible for thousands of deaths each year in the United States and many times that number in the rest of the world.

Acute Glomerulonephritis In AGN the kidney cells are so damaged that they cannot adequately filter blood. The first symptoms are nephritis (appearing as swelling in the hands and feet and low urine output); increased blood pressure; and, occasionally, heart failure. Urine samples are abnormal, with high levels of red and white blood cells and protein. Many cases of AGN clear up spontaneously, but some become chronic and lead to kidney failure.

Two antibodies can give long-term protection against group A infections. One is the type-specific antibody produced in response to the M-protein, though it will not prevent infections with a different strain (which explains why a person can have recurring cases of streptococcal pharyngitis). The other is a neutralizing antitoxin against the erythrogenic toxin that prevents the fever and rash of scarlet fever. Serological tests for antistreptococcal antibodies are used primarily to detect continuing or recent streptococcal infection and to judge a patient's susceptibility to rheumatic fever and glomerulonephritis.

* *rheumatic* (roo-mat'-ik) Gr. *rheuma,* flux. Involving inflammation of joints, muscles, and connective tissues.

3. Carditis is inflammation of heart tissues; chorea is a nervous disorder characterized by involuntary, jerky movements.

Pathogen Profile #2 *Streptococcus pyogenes*

Modified image reprinted with permission from Medscape Reference (http://emedicine.medscape.com/), 2013, available at: http://emedicine.medscape.com/article/228936-overview.

Microscopic Morphology Gram-positive cocci arranged in chains and pairs; very rarely motile; non-spore-forming.

Identified by Results of a catalase test are used to distinguish *Streptococcus* (negative) from *Staphylococcus* (positive). Beta-hemolysis and sensitivity to bacitracin are hallmarks of *S. pyogenes*. Rapid methods of identification use monoclonal antibodies to detect the C-carbohydrate found on the cell surface of *S. pyogenes*. Such tests provide accurate identification in as little as 10 minutes.

Habitat A fairly strict parasite, *S. pyogenes* is found in the throat, nasopharynx, and occasionally the skin of humans. From 5% to 15% of persons are asymptomatic carriers.

Virulence Factors *S. pyogenes* possesses several cell surface antigens that serve as virulence factors. C-carbohydrate helps prevent the bacterium from being dissolved by the lysozyme of the host; fimbriae on the outer surface of the cell enhance adherence of the bacterium; M-protein helps the cell resist phagocytosis while also improving adherence; and C5a protease catalyzes the cleavage of the C5a protein of the complement system, inhibiting the actions of complement. Most strains of *S. pyogenes* are covered with a capsule composed of hyaluronic acid (HA) identical to the HA found in host cells, preventing an immune response by the host. Two different hemolysins, streptolysin O (SLO) and streptolysin S (SLS), cause damage to leukocytes, and liver and heart muscle, whereas erythrogenic toxin produces fever and the bright red rash characteristic of *S. pyogenes* disease. Invasion of the body is aided by several enzymes that digest fibrin clots (streptokinase), connective tissue (hyaluronidase), or DNA (streptodornase).

Primary Infections/Disease Local cutaneous infections include pyoderma (impetigo) or the more invasive erysipelas. Infection of the tonsils or pharyngeal mucous membranes can lead to streptococcal pharyngitis (strep throat), which, if left untreated, may lead to scarlet fever. Rarer infections include streptococcal toxic shock syndrome, *S. pyogenes* pneumonia, and necrotizing fasciitis. Long-term complications of *S. pyogenes* infections include rheumatic fever and acute glomerulonephritis.

Control and Treatment Control of *S. pyogenes* infection involves limiting contact between carriers of the bacterium and immunocompromised potential hosts. Patients should be isolated, and care must be taken when handling infectious secretions. As the bacterium shows little drug resistance, treatment is generally a simple course of penicillin.

The possibility of one of these serious or long-term complications is the reason that severe sore throats should be taken seriously. A simple throat swab can distinguish between group A streptococci and other causes so that antibiotics, if called for, can be administered immediately.

Group B: *Streptococcus agalactiae*

Several other species of beta-hemolytic *Streptococcus* in groups B, C, and D live among the normal microbiota of humans and other mammals and can be isolated in clinical specimens from diseased human tissue. The group B streptococci (GBS), represented by the species *S. agalactiae,* demonstrate clearly how the distribution of a pathogen can change in a relatively short time. The species has been associated with cattle, in which it is a frequent cause of bovine mastitis.[4] It has also become a resident in the human vagina, pharynx, and large intestine. Since its colonization of humans, there has been a dramatic increase in serious infections in newborns and compromised people. The CDC estimates there are approximately 20,000 cases per year.

Streptococcus agalactiae is primarily implicated in neonatal meningitis, wound and skin infections, and endocarditis. Elderly people suffering from diabetes and vascular disease are particularly susceptible to wound infections. Because of its location in the vagina, GBS can be transferred to the infant during delivery, sometimes with dire consequences. An early-onset infection develops a few days after birth and is accompanied by sepsis, pneumonia, and high mortality.

This bacterial pathogen is the most prevalent cause of neonatal pneumonia, sepsis, and meningitis in the United States and Europe. Approximately 2,200 babies a year acquire infection in the United States alone. A later complication arises in 2 to 6 weeks, with symptoms of meningitis—fever, vomiting, and seizures. About 20% of children have long-term neurological damage. Because most cases occur in the hospital, personnel must be aware of the risk of passively transmitting this pathogen, especially in the neonatal and surgical units. Pregnant women should be screened for colonization in the third trimester and immunized with globulin and treated with a course of antibiotics if infection is found.

Group D Enterococci and Groups C and G Streptococci

Enterococcus faecalis, E. faecium, and *E. durans* are collectively referred to as "enterococci" because they are normal colonists of the human large intestine. Two other members of group D, *Streptococcus bovis* and *S. equinus,* are nonenterococci that colonize other animals and occasionally humans. Infections caused by *E. faecalis* arise most often in elderly patients undergoing surgery and affect the urinary tract, wounds, blood, the endocardium, the appendix, and other intestinal structures. Enterococci are emerging as serious opportunists in the health care setting, primarily because of the rising incidence of multidrug-resistant strains, especially vancomycin-resistant enterococci (VRE).

Groups C and G are common microbiota of domestic animals but are frequently isolated from the human upper respiratory tract.

4. Inflammation of the mammary glands.

Positive reaction Negative reaction

Figure 18.14 A rapid, direct test kit for diagnosis of group A infections. With this method, a patient's throat swab is mixed with a drop of latex beads and monoclonal antibodies. (Left) In a positive reaction, the C-carbohydrate on group A streptococci causes visible clumping. (Right) A smooth, milky reaction is negative. (left, right): © Diagnostic Products Corporation/Siemens Healthcare

Occasionally they imitate group A strep in causing pharyngitis and glomerulonephritis. More often they cause bacteremia and disseminated deep-seated infections in severely compromised patients.

Laboratory Identification Techniques

As mentioned previously, the failure to recognize group A streptococcal infections (even very mild ones) can have devastating effects. Rapid cultivation and diagnostic techniques that will ensure proper treatment and prevention measures are essential. Several companies have developed rapid diagnostic test kits to be used in clinics or offices to detect group A streptococci from pharyngeal swab samples. These tests, based on monoclonal antibodies that react with the C-carbohydrates of group A, are highly specific and sensitive **(figure 18.14)**.

Today's large clinical labs rarely use traditional microbiological techniques to identify streptococcal species. Instead, monoclonal antibodies supplied in commercially available test kits can be enriched and then used to rapidly identify most samples. The speed and accuracy of these tests can rarely be matched by methods that depend on culturing samples. In some cases, however—including small or isolated labs, and equivocal results when using serological tests—cultural techniques are still relied upon, and in fact they remain the gold standard for identification of any isolate. **Table 18.4** summarizes the characteristics that are used to separate the various beta-hemolytic streptococci, and **figure 18.15** illustrates some of these tests (see also figure 18.14).

Treatment and Prevention of Group A, B, and D Streptococcal Infections

Antimicrobic therapy is aimed at curing infection and preventing complications. Most strains of *S. pyogenes* continue to be very sensitive to penicillin or one of its derivatives, although more penicillin resistant strains are being encountered. In cases of pharyngitis, small children usually receive an intramuscular injection of 600,000 units of benzathine penicillin to achieve the necessary circulating levels for a period of 10 days; older children receive 900,000 units, and adults receive 1.2 million units. An alternative therapy often used for impetigo is oral cephalexin or dicloxacillin.

The only certain way to arrest rheumatic fever or acute glomerulonephritis is to treat the preceding infection, because once these two pathologic states have developed, there are no specific treatments.

TABLE 18.4 Scheme for Differentiating Beta-Hemolytic Streptococci

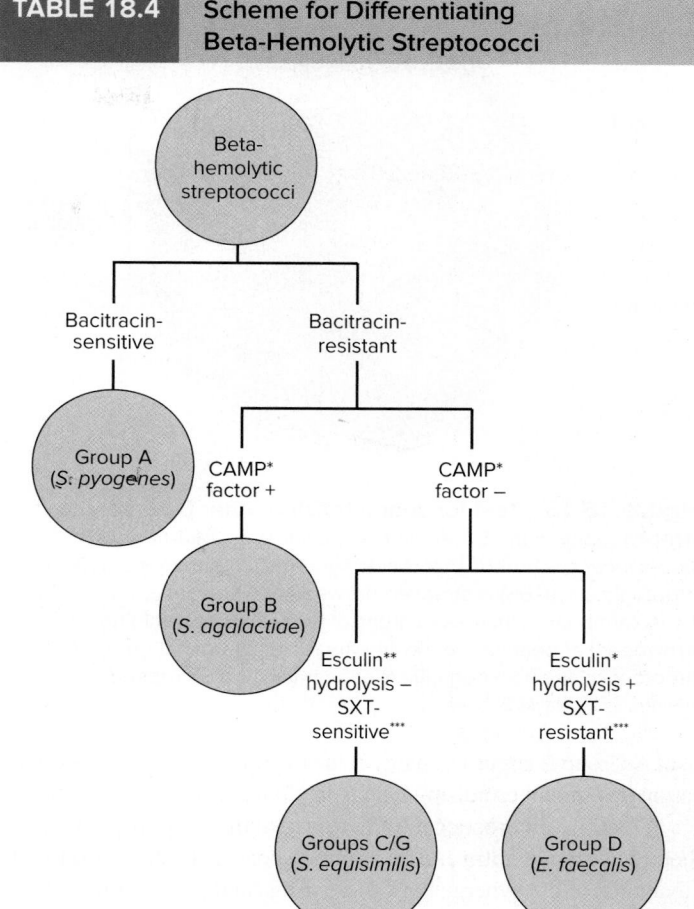

*Name is derived from the first letters of the names of its discoverers. CAMP is a diffusible substance of group B, which lyses sheep red blood cells in the presence of staphylococcal hemolysin.

**A sugar that can be split into glucose and esculetin. Group D streptococci can accomplish this in the presence of 40% bile.

***Sulfamethoxazole and trimethoprim. The test is performed (like bacitracin) with discs containing this combination drug.

Some physicians recommend that people with a history of rheumatic fever or recurring streptococcal pharyngitis receive continuous, long-term penicillin prophylaxis. Mass prophylaxis has also been indicated for young people exposed to epidemics in boarding schools, military camps, and other institutions and for known carriers. Tonsillectomy (removal of the tonsils) has been used extensively in the past but with questionable effectiveness, because these bacteria can infect other parts of the throat. In hospitals, carriers of *S. pyogenes* should not be allowed to work with surgical, obstetric, or immunocompromised patients. Patients with group A infections must be isolated, and high-level precautions must be practiced in handling infectious secretions.

The antibiotic treatment of choice for group B streptococcal infection is penicillin G; alternatives are vancomycin and gentamicin. Some physicians advocate routine penicillin prophylaxis in colonized mothers and infants, but others fear that this practice can increase the rate of allergies and infections by resistant strains. Artificial passive immunization with human immunoglobulins is currently being considered for treatment and prevention in mothers and infants at high

SXT

Bacitracin

(+)
CAMP test

Figure 18.15 **Test for characterizing other β-hemolytic streptococci.** In the CAMP test, a special strain of β-hemolytic *Staphylococcus aureus* that reacts synergistically with group B streptococcus (GBS) is streaked between two smears of the sample. If the isolate is GBS, increased areas of hemolysis (resembling an arrowhead) appear where the smears meet. (Note that group B streptococci are also negative for bacitracin and SXT tests.)
Courtesy of Dr. David Schlaes

Figure 18.16 **Effects of streptococcal colonization.** Bacterial vegetations can be seen along the upper edge of a surgically opened mitral valve. The vegetations (whitish patches) constantly release bacteria into the circulation. Image reprinted with permission from Allen Patrick Burke, MD, University of Maryland School of Medicine, published by Medscape Reference (http://emedicine.medscape.com/), 2013, available at: http://emedicine.medscape.com/article/1954887-overview.

risk. A Group B streptococcal vaccine to protect women from colonization has shown promising results in clinical trials.

Treating enterococcal infection usually requires combined therapy with penicillin and an aminoglycoside (gentamicin) to take advantage of the synergism of these drugs and to overcome antimicrobic resistance.

Alpha-Hemolytic Streptococci: The Viridans Group

The **viridans*** category includes streptococci of human origin not entirely groupable by Lancefield serology. They are the most numerous and widespread residents of the oral cavity (gingiva, cheeks, tongue, saliva) and are also found in the nasopharynx, genital tract, and skin. These species can cause serious systemic infections, although most of them are opportunists and lack a full set of toxins and enzymes common in more virulent streptococci. The one characteristic shared by major species (*Streptococcus mitis, S. mutans, S. anginosus, S. salivarius, S. sanguinis*) is alpha-hemolysis. Other characteristics used to distinguish them are beyond the scope of this textbook.

Because viridans streptococci are not highly invasive, their entrance into tissues usually occurs through dental or surgical instrumentation and manipulation. These organisms are constant inhabitants of the gums and teeth, and even chewing hard candy or brushing the teeth can provide a portal of entry for them. Dental procedures can lead to bacteremia, meningitis, abdominal infection, and tooth abscesses. But the most important complication of all viridans streptococcal infections is **subacute endocarditis.** In this condition, blood-borne bacteria settle on areas of the heart lining or valves that have previously been injured by rheumatic fever, valve surgery, or the like. Colonization of these surfaces leads to a

thick biofilm called a *vegetation* (**figure 18.16**). As the disease progresses, vegetations increase in size, occasionally releasing masses of bacteria into the circulation. These masses, or emboli, can travel to the lungs and brain, blocking circulation and damaging those organs. Because of its gradual and somewhat concealed course, endocarditis is considered subacute rather than acute or chronic. Symptoms and signs range from fever and heart murmur to weight loss and anemia.

Endocarditis is diagnosed almost exclusively by blood culture, and repeated blood samples positive for bacteremia are highly suggestive of it. The goal in treatment is to completely destroy the microbes in the vegetations, which usually can be accomplished by long-term therapy with penicillin G. Because persons with preexisting heart conditions are at high risk for this disease, they sometimes receive prophylactic antibiotics prior to dental and surgical procedures.

Another very common dental disease involving viridans streptococci is dental caries. In the presence of sugar, *S. oralis* and *S. gordonii* (the *S. mutans* group) produce slime layers made of glucose polymers that adhere tightly to tooth surfaces. These sticky polysaccharides are the basis for plaque, the adhesive white material on teeth that becomes coinfected with other bacteria and fosters dental disease.

Streptococcus pneumoniae: The Pneumococcus

High on the list of significant human pathogens is ***Streptococcus pneumoniae,*** a unique species that was formerly called *Diplococcus* until its genetic similarity to the streptococci was demonstrated. *S. pneumoniae* is also referred to as the **pneumococcus.** This bacterium is involved in the majority of **bacterial pneumonias,** which primarily affect immunocompromised patients. It also is a major cause of **meningitis** and **otitis media** in young children. Gram stains of sputum specimens from pneumonia patients reveal small, lancet-shaped cells arranged in pairs and short chains (see Pathogen Profile #3). Because it is fastidious, *S. pneumoniae* must be cultured on blood agar, on which it produces smooth or mucoid colonies and alpha-hemolysis (see figure 18.9*b*). Growth is improved by the presence of 5% to 10% CO_2, and cultures die in the presence of oxygen because they lack catalase and peroxidases.

All pathogenic strains form rather large capsules, this being their major virulence factor. These colonies appear smooth on solid media. Rough strains lack a capsule and are nonvirulent. Capsules

* *viridans* (vih´-rih-denz) L. *viridis,* green. This term comes from the green color that develops around colonies that produce alpha-hemolysis.

help the streptococci escape phagocytosis, which is the major host defense in pyogenic infections. They contain a polysaccharide antigen called the specific soluble substance (SSS) that varies chemically among the pneumococcal types and stimulates antibodies of varying specificity. So far, 90 different capsular types (specified by numerals 1, 2, 3, . . .) have been identified.

Epidemiology and Pathology of the Pneumococcus

From 5% to 50% of all people carry *S. pneumoniae* as part of the normal microbiota in the nasopharynx. Although infection is often acquired endogenously from one's own microbiota, it occasionally occurs after direct contact with respiratory secretions or droplets from carriers. *Streptococcus pneumoniae* is very delicate and does not survive long out of its habitat. Factors that predispose patients to pneumonia are old age, the season (rate of infection is highest in the winter), other lung diseases or viral infections that weaken defenses, and living in close proximity to infected people.

The Pathology of *Streptococcus pneumoniae* Healthy people commonly inhale microorganisms into the respiratory tract without serious consequences because of the host defenses present there. Pneumonia is likely to occur when mucus containing a load of bacterial cells is aspirated from the pharynx into the lungs of susceptible individuals who have lowered defenses. Passing into the bronchioles and alveoli, the pneumococci multiply and induce an overwhelming inflammatory response. This is marked by exudation of fluids into the lungs. In a form of pneumococcal pneumonia termed **lobar pneumonia,** this fluid accumulates in the alveoli along with red and white blood cells. As the infection and inflammation spread rapidly through the lung, the patient can actually "drown" in his own secretions. If this mixture of exudate, cells, and bacteria solidifies in the air spaces, a condition known as **consolidation*** occurs **(figure 18.17)**. In infants and the elderly, the areas of infection are usually spottier and centered more in the bronchi than in the alveoli (bronchial pneumonia).

Symptoms of pneumococcal pneumonia are chills, shaking, rapid breathing, and fever. The patient can experience severe pain in the chest wall, cyanosis (due to compromised oxygen intake), a cough that produces rusty-colored (bloody) sputum, and abnormal breathing sounds. Systemic complications of pneumonia are pleuritis and endocarditis, but pneumococcal bacteremia and meningitis are the greatest danger to the patient.

It is common for this agent to gain access to the chamber of the middle ear by way of the eustachian tube and cause a middle ear infection called **otitis media (figure 18.18)**. This occurs readily in children under 2 years of age because of their relatively short eustachian tubes. Otitis media is the third most common childhood disease in the United States, and the pneumococcus accounts for the majority of cases. The severe inflammation in the small space of the middle ear induces acutely painful earaches and sometimes even temporary deafness **(figure 18.19)**.

> **Quick Search**
> To see how a tympanostomy tube is surgically inserted into the eardrum, search the Internet for "Ventilation Tubes for Ear Infections."

* *consolidation* L. *consolidatio,* to make firm.

(a)

(b)

Figure 18.19 Pneumococcal otitis media. (a) The acute inflammation and internal pressure of a middle ear infection are evident in the redness and bulging of the tympanum as seen during otoscopic examination. **(b)** For children with chronic infections a tympanostomy tube may be surgically inserted into the tympanum to equalize the pressure of the middle ear with that of the environment, minimizing pain and pressure. (a, b) © Clinica Claros/ Phototake

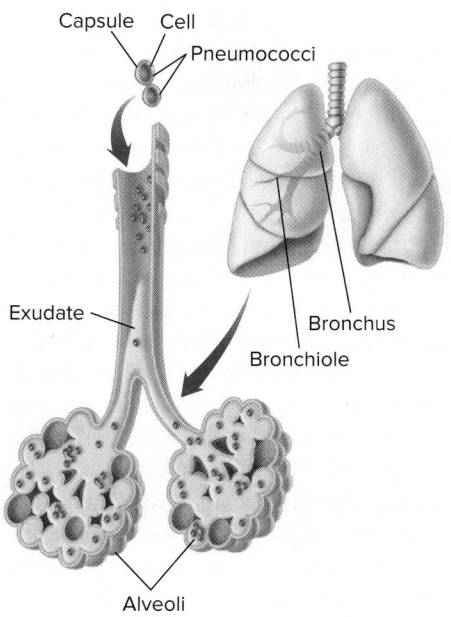

Capsule Cell
Pneumococci
Exudate
Bronchus
Bronchiole
Alveoli

Figure 18.17 The course of bacterial pneumonia. As the pneumococcus traces a pathway down the respiratory tree, it provokes intense inflammation and exudate formation. The blocking of the bronchioles and alveoli by consolidation of inflammatory cells and products is shown.

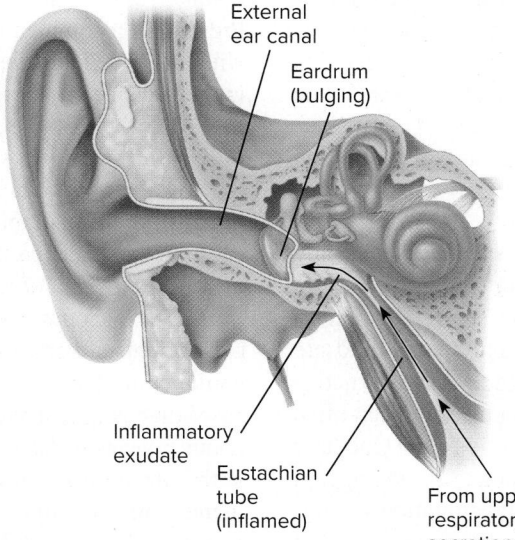

External ear canal
Eardrum (bulging)
Inflammatory exudate
Eustachian tube (inflamed)
From upper respiratory secretions

Figure 18.18 Cutaway view of ear anatomy indicating the route of infection and its primary targets. The direct connection between the pharynx and middle ear via the eustachian tube makes this possible.

Pathogen Profile #3 *Streptococcus pneumoniae* (pneumococcus)

Microscopic Morphology Gram-positive, lancet (arrowhead) shaped cells arranged in pairs and short chains.

Identified by Gram stain is often enough to provide presumptive identification. Alpha-hemolysis distinguishes pneumococcus from beta-hemolytic streptococci, and its sensitivity to optochin distinguishes it from group D and viridans streptococci. Identification can be confirmed by testing for bile solubility and fermentation of inulin (pneumococcus produces a positive reaction in both cases). A serological test called the Quellung reaction uses antibodies to detect the bacterial capsule and provides a rapid, definitive test for the organism.

Source: Dr. Mike Miller/ CDC

Habitat Between 5% and 10% of adults and 20% to 40% of children carry *S. pneumoniae* as part of their nasopharyngeal microbiota. Because the organism is quite delicate, it rarely survives outside the host.

Virulence Factors The major virulence factor is a thick capsule which helps streptococcus escape phagocytosis. Nonencapsulated strains are avirulent. Over 90 different capsular serotypes exist.

Primary Infections/Disease *Streptococcus pneumoniae* is the most common cause of community-acquired pneumonia, particularly among the elderly and those suffering from weakened host defenses (asthma, viral lung diseases, or smoking). Pneumococcus is a major causative agent of meningitis in all age groups and is the leading cause of adult meningitis. The bacterium can also travel to the middle ear—primarily in young children with short eustachian tubes—leading to otitis media, where swelling of the middle ear leads to painful earaches.

Control and Treatment *S. pneumoniae* can be prevented in most cases through vaccination. Pneumovax contains capsular antigens from 23 of the most commonly encountered serotypes and is recommended for adults over 65 and those from 19 to 65 with underlying health issues. Prevnar 13 provides protection against 13 different serotypes and a four-dose series of the vaccine is recommended for children between 2 and 59 months of age and adults older than 65 years. Treatment of pneumococcal infection relies on penicillin in the 85% of cases where the organism is susceptible. In cases where penicillin is ineffective, or contraindicated, cephalosporins, sulfonamides, quinolones, or telithromycin are commonly used.

In young children, *S. pneumoniae* is a common agent of upper respiratory tract infections that can spread to the meninges and cause meningitis. The pneumococcus accounts for 10% to 50% of reported cases of meningitis depending on the age group. It is currently the most common cause of this disease in adults.

Healthy people have high natural resistance to the pneumococcus. The effects of mucus and ciliary responses of the respiratory tract help capture and flush out transient organisms. Respiratory phagocytes are also essential in the eradication process, but this system works only if the capsule is coated by opsonins in the presence of complement. After recovery, the individual will be immune to future infections by that particular pneumococcal type. This immunity is the basis for a successful vaccine (see subsequent section on prevention).

Laboratory Cultivation and Diagnosis A specimen is usually necessary before diagnosing pneumococcal infections because a number of other microbes can cause pneumonia, sepsis, and meningitis. Blood cultures, sputum, pleural fluid, and spinal fluid are common specimens. A Gram stained specimen is very instructive (see Pathogen Profile #3), and presumptive identification can often be made by this method alone. Also highly definitive is the Quellung reaction, a serological test in which sputum is mixed with anticapsular antisera and observed microscopically. Precipitation of antibodies on the surface of the capsule gives the appearance of swelling. Alpha-hemolysis helps differentiate the pneumococcus from the beta-hemolytic streptococci. Diagnosis can then be confirmed by testing for the agent's sensitivity to optochin, a positive bile solubility, and positive inulin fermentation tests.

Treatment and Prevention of Pneumococcal Infections The treatment of choice for pneumococcal infections has traditionally been a large daily dose of penicillin G or penicillin V. In recent times, however, clinics have reported increased cases of drug-resistant strains. In fact, an average of 15% of isolates, termed DRSP, have shown multidrug resistance. Drug sensitivity testing is therefore essential. Alternative drugs are cephalosporins, sulfonamides, quinolones, and telithromycin. The Centers for Disease Control and Prevention recommends daily drug prophylaxis to protect children with sickle-cell anemia against recurrent pneumococcal infections. Untreated cases in these children have a mortality rate up to 30%.

Along with more widespread antibiotic resistance, the need to prevent recurrences has shifted the focus of treatment toward the development of active immunity. Fortunately, *S. pneumoniae* is one of the few streptococcal species for which an effective vaccination exists. There are two vaccines. An older vaccine (Pneumovax) consists of capsular antigens from 23 of the most frequently encountered serotypes. It is indicated for older adults and patients at particularly high risk, including those with sickle-cell anemia, lack of a spleen, congestive heart failure, lung disease, diabetes, and kidney disease. This vaccine is effective for about 5 years in 60% to 70% of those vaccinated. Immunization with a pneumococcal **conjugate*** vaccine (Prevnar 13) is recommended for all children younger than five years and adults older than 65 years as an effective preventative therapy for otitis media and meningitis.

* *conjugate* vaccines consist of the capsule polysaccharide complexed to a protein carrier to make it more immunogenic in young children.

 Check Your Progress SECTION 18.2

6. List the clinically significant pathogens in the genus *Streptococcus,* and summarize the virulence factors seen in the group.

7. Describe the immediate effects and long-term complications of *S. pyogenes* infection.

8. List the drugs commonly used to treat streptococcal infection, and explain why combined drug therapy is sometimes required.

9. Differentiate between pneumococcal disease typically seen in the very young and that seen in the elderly.

18.3 The Family Neisseriaceae: Gram-Negative Cocci

Expected Learning Outcomes

12. Name the most important human pathogens in the genus *Neisseria,* along with the body systems affected by each.

13. Compare the differences in pathologies of *N. gonorrhoeae* infection in men, women, and children.

14. Summarize the pathogenesis and epidemiology of *N. meningitidis.*

15. Recall the medically important genera, besides *Neisseria,* that contain gram-negative cocci and coccobacilli.

Members of the Family Neisseriaceae are residents of the mucous membranes of warm-blooded animals. Most species are relatively innocuous commensals, but two are primary human pathogens with far-reaching medical impact. The genera contained in this group are *Neisseria, Branhamella,* and *Moraxella.* Of these, *Neisseria* has the greatest clinical significance.

One distinguishing feature of the *Neisseriae** is their cellular morphology. Rather than being perfectly spherical, the cells are bean-shaped and paired, with their flat sides touching **(figure 18.20).** None develops flagella or spores, but capsules can be found on the pathogens. The cells are gram-negative, possessing an outer membrane with lipopolysaccharides in the cell wall and, in many cases, fimbriae.

* *Neisseriae** (ny-serr'-ee-uh) After the German physician Albert Neisser who first observed the agent of gonorrhea in 1879.

Figure 18.20
Transmission electron micrograph of *Neisseria* (52,000×). This cross section shows how the cocci are flattened on their adjacent sides. Note the fine fimbriae that arise from their surfaces. © Charles C. Brinton, Jr., John A. Tainer, Michael E. Piques and Lisa Craig

Most *Neisseria* are strict parasites that do not survive long outside the host, particularly where hostile conditions of drying, cold, acidity, or light prevail. *Neisseria* species are aerobic or microaerophilic and have an oxidative form of metabolism. They produce catalase, enzymes for fermenting various carbohydrates, and the enzyme cytochrome oxidase that can be used in identification. The pathogenic species, *N. gonorrhoeae* and *N. meningitidis,* require complex enriched media and grow best in an atmosphere containing additional CO_2. We shall concentrate on other features of the pathogenic *Neisseria* in the following sections.

Neisseria gonorrhoeae: The Gonococcus

Gonorrhea* has been known as a sexually transmitted disease since ancient times. Its name originated with the Greek physician Claudius Galen, who thought that it was caused by an excess flow of semen. For a fairly long period in history, gonorrhea was confused with syphilis. Later, microbiologists went on to cultivate *N. gonorrhoeae,* also known as the **gonococcus,** and to prove conclusively that it alone was the etiologic agent of gonorrhea.

Factors Contributing to Gonococcal Pathogenicity

The virulence of the gonococcus is due chiefly to the presence of fimbriae and other surface molecules that promote mutual attachment of cocci to each other and invasion and infection of epithelial tissue. In addition to their role in adherence, the fimbriae also seem to slow phagocytosis by macrophages and neutrophils. Another contributing factor in pathogenicity is a protease that cleaves the secretory antibody (IgA) on mucosal surfaces and stops its protective effects.

Epidemiology and Pathology of Gonorrhea

Gonorrhea is a strictly human infection that occurs worldwide and ranks among the top five sexually transmitted diseases (STDs). Although 300,000 cases are reported in the United States each year, it is estimated that the actual incidence is much higher—in the millions if one counts asymptomatic infections. Most cases occur in young adults (18–24 years old) with multiple sex partners. Figures on the prevalence of sexually transmitted diseases, including gonorrhea, syphilis, and chlamydia often show a fluctuating pattern corresponding to periods of societal turmoil or changing sexual attitudes. The two large increases in the prevalence of gonorrhea since 1940 **(figure 18.21)** provide an interesting historical perspective. The first occurs during World War II, when casual sexual encounters involving servicemen spiked. The second took place during the sexual revolution of the 1960s, when reliance on oral contraceptives rather than condoms to prevent pregnancy led to an epidemic in gonorrhea cases. The relationship of gonorrhea with infertility, especially in women, is discussed in **18.3 Making Connections.**

Experiments done with male volunteers revealed that an infectious dose can range from 100 to 1,000 colony-forming units. *Neisseria gonorrhoeae* does not survive more than 1 or 2 hours on fomites and is most infectious when transferred through direct contact. Except for neonatal infections, the gonococcus is acquired when the pathogen is introduced into a genital or extragenital

* *gonorrhea* (gä-n-'rē⁻-) Gr. *gonos,* seed, and *rhein,* to flow.

Gonorrhea—Rates of Reported Cases by Year, United States, 1941–2014

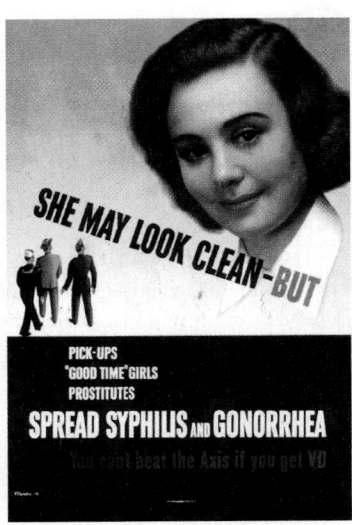

Figure 18.21 The rate of sexually transmitted disease is related to societal changes. (a) The rate of gonorrhea infection shows sharp increases during World War II and during the sexual revolution of the 1960s. **(b)** A government poster from the 1940s meant to warn soldiers of the risk of contracting a sexually transmitted disease. **(b):** *Source:* U.S. National Library of Medicine

(rectum, eye, or throat) portal of entry. After attaching to the epithelial surface by fimbriae, the bacteria invade the underlying connective tissue. In 2 to 6 days, this process results in an inflammatory reaction that may or may not produce noticeable symptoms. The infection is asymptomatic in approximately 10% of males and 50% of females. It is this phenomenon that is most important in the

persistence and spread of the pathogen. The following sections survey the several manifestations of gonorrhea.

Genital Gonorrhea in the Male Infection of the urethra causes urethritis, with painful urination and a purulent discharge in most cases **(figure 18.22b).** Infection tends to be limited to the distal urogenital tract, but it can occasionally spread from the urethra to the prostate gland and epididymis **(figure 18.22a).** Scar tissue formed in the spermatic ducts during healing of an invasive infection can render the individual infertile. This outcome is becoming increasingly rare with improved diagnosis and treatment regimens.

Figure 18.22 Gonorrheal damage to the male reproductive tract. (a) Frontal view of the male reproductive tract. (Left side) The normal, uninfected state. The sperm-carrying ducts are continuous from the testis to the urethral opening. (Right side) The route of ascending gonorrhea complications. Infection begins at the tip of the urethra, ascends the urethra through the penis, and passes into the vas deferens. Occasionally, it can even enter the epididymides and testes. (Inset) Damage to the ducts carrying sperm can create scar tissue and blockage, which reduce sperm passage and can lead to sterility. **(b)** A milky or yellow discharge from the urethra is a common sign of gonorrheal infection.
(b): *Source:* Renelle Woodfall/CDC

![Puzzle piece icon] **18.3 MAKING CONNECTIONS**

Pelvic Inflammatory Disease and Infertility

The National Center for Health Statistics estimates that more than 6 million women in the United States have impaired fertility. There are many different reasons for infertility, but the leading cause is pelvic inflammatory disease, or PID. PID is caused by infection of the upper reproductive structures in women, namely the uterus, fallopian tubes, and ovaries. These organs lack normal microbiota. When bacteria from the vagina are spread higher in the tract, they initiate an infection that may or may not be noticeable to the patient. Inflammation caused by the infection can be acute, resulting in pain, abnormal vaginal discharge, fever, and nausea, or it can be chronic, with less noticeable symptoms. In acute cases, women usually seek care; in some ways these can be considered the lucky ones. If the inflammation is curbed at an early stage by using antibiotics to kill the bacteria, chances are better that the long-term sequelae of PID can be avoided.

The most notable long-term consequence is tubal infertility. Inflammatory repair processes, especially in the fallopian tubes, can lead to the deposition of scar tissue that narrows the lumen in the tubes, in some cases closing them off completely. If the lumen is only narrowed, fertilization could still occur, but a fertilized egg would be unable to travel through the tube and implant in the uterine wall. In some cases fertilized eggs implant in the tube walls or even leave the fallopian tubes and implant elsewhere in the abdominal cavity. Both of these situations are known as ectopic pregnancies. Women with a history of PID have a 7 to 10 times greater chance of experiencing an ectopic pregnancy than other women. Ectopic pregnancy is a life-threatening situation. An embryo growing in the tube usually causes the tube to rupture in about 12 weeks, and an embryo in the abdominal cavity can cause the same complication as a tumor. Surgical intervention is usually required in either case to save the woman's life.

Chlamydia infection is the leading cause of PID, followed closely by *N. gonorrhoeae* infection. But other bacteria, perhaps also including normal microbiota of the reproductive tract, can also cause PID if they are traumatically introduced into the uterus. Intercourse, tampon usage, the use of an intrauterine contraceptive device, and even douching can encourage the transmission of bacteria into the upper genital tract. Women who suspect for any reason that they may have a reproductive tract infection should always seek diagnosis and treatment as early as possible.

Describe the possible events leading to invasion of infectious agents into the sterile tissues of the reproductive tract. *Answer available on Connect*

Genitourinary Gonorrhea in the Female The proximity of the genital and urinary tract openings increases the likelihood that both systems can be infected during sexual intercourse. A mucopurulent or bloody vaginal discharge occurs in about half the cases, along with painful urination if the urethra is affected. Major complications occur when the infection ascends from the vagina and cervix to higher reproductive structures such as the uterus and fallopian tubes **(figure 18.23)**. One disease resulting from this progression is

Figure 18.23 Ascending gonorrhea in women. (Left) Normal state. (Right) In ascending gonorrhea, the gonococcus is carried from the cervical opening up through the uterus and into the fallopian tubes. On rare occasions, it can escape into the peritoneum and invade the ovaries, or cause peritonitis. Pelvic inflammatory disease (PID) is a serious complication that can lead to scarring in the fallopian tubes, ectopic pregnancies, and mixed anaerobic infections.

salpingitis,* also known as **pelvic inflammatory disease (PID)**, a condition characterized by fever, abdominal pain, and tenderness. It is not unusual for the microbe to become involved in mixed infections with anaerobic bacteria. The buildup of scar tissue from these infections can block the fallopian tubes, causing sterility and ectopic pregnancies. (Note: PID can also be caused by *Chlamydia* bacteria and, less frequently, by nonsexually transmitted bacteria; see 18.3 Making Connections.)

Extragenital Gonococcal Infections in Adults Extragenital transmission and carriage of the gonococcus can occur under some circumstances. Anal intercourse can lead to proctitis, and oral sex can result in pharyngitis and gingivitis. Careless personal hygiene can account for self-inoculation of the eyes and a serious form of conjunctivitis. In a small number of cases, the gonococcus enters the bloodstream and is disseminated to the joints and skin. Involvement of the wrist and ankle can lead to chronic arthritis and a painful, sporadic, papular rash on the limbs. Rare complications of gonococcal bacteremia are meningitis and endocarditis.

Gonococcal Infections in Children Infants born to gonococcus carriers are in danger of being infected as they pass through the birth canal. Because of the potential harm to the fetus, physicians usually screen pregnant mothers for its presence. Gonococcal eye infections are very serious and often manifest sequelae such as keratitis, ophthalmia neonatorum, and even blindness **(figure 18.24)**. A universal precaution to prevent those complications is the instillation of antibiotics, silver nitrate, or other antiseptics into the conjunctival sac of newborn babies. The pathogen may also infect the pharynx and respiratory tract of neonates. Finding gonorrhea in children other than babies is strong evidence of sexual abuse by infected adults, and it mandates child welfare consultation along with thorough bacteriologic analysis.

Clinical Diagnosis and Control of Gonococcal Infections

Diagnosing gonorrhea is relatively straightforward. The presence of gram-negative diplococci in neutrophils from urethral, vaginal, cervical,

* *salpingitis* (sal″-pin-jy′-tis) Gr. *salpinx*, tube, and *itis*, inflammation. An inflammation of the fallopian tubes.

Figure 18.24 **Gonococcal ophthalmia neonatorum in an infant infected during birth.** The infection is marked by intense inflammation and edema; if allowed to progress, it causes damage that can lead to blindness. Fortunately, this infection is completely preventable and treatable. *Source:* CDC

Gonococci

Neutrophil

Figure 18.25 **Gram stain of urethral pus from a patient with gonorrhea (1,000x).** Note the intracellular (phagocytosed) gram-negative diplococci in polymorphonuclear leukocytes (neutrophils). *Source:* Bill Schwartz/CDC

or eye exudates is especially diagnostic because gonococci tend to be engulfed and remain viable within phagocytes **(figure 18.25)**. Thus, a simple Gram stain can provide at least presumptive evidence of gonorrhea. Other tests used to identify *Neisseria gonorrhoeae* and differentiate it from related species are discussed in a later section.

Approximately 350,000 new cases of gonorrhea are reported each year in the United States and *N. gonorrhoeae* has progressively developed resistance to each of the antimicrobials used for treatment of gonorrhea. Currently the only treatment recommended by the CDC is combination therapy with ceftriaxone and azithromycin. Beyond treating just the gonococcal infection, combination therapy also typically resolves concurrent STDs such as chlamydiosis, a common complication in many gonorrhea infections.

Although gonococcal infections stimulate local production of antibodies and activate the complement system, these responses do not produce lasting immunity, and some people experience recurrent infections.

Gonorrhea is a reportable infectious disease, which means that any physician diagnosing it must forward the information to a public health department. The follow-up to this finding involves tracing sexual partners to offer prophylactic antibiotic therapy. There is a pressing need to seek out and treat asymptomatic carriers and their sexual contacts, but complete control of this group is nearly impossible. Other control measures include education programs that emphasize the effects of all STDs and promote safer sexual practices such as the use of condoms.

Neisseria meningitidis: The Meningococcus

Another serious human pathogen is *Neisseria meningitidis*, a bacterium known commonly as the **meningococcus** and usually associated with epidemic **cerebrospinal meningitis.** Important factors in meningococcal invasiveness are a polysaccharide capsule, adhesive fimbriae, and IgA protease. Although 12 different strains of capsular antigens exist, serotypes A, B, and C are responsible for most cases of infection. Another virulence factor with potent pathologic effects is the lipopolysaccharide (**endotoxin**) released from the cell wall when the microbe lyses.

Pathogen Profile #4 *Neisseria gonorrhoeae* (gonococcus)

Source: U.S. Centers for Disease Control and Prevention - Medical Illustrator/CDC

Microscopic Morphology Gram-negative, flattened cocci, arranged as diplococci; very rarely motile; non-spore-forming.

Identified by Visualization of gram-negative diplococci in neutrophils. Because gonococcus tends to remain alive after being engulfed by neutrophils, their appearance in urethral, vaginal, cervical, or eye exudates is presumptive evidence of infection. Biochemical and DNA testing can both be used to confirm identification.

Habitat A strictly human infection, *N. gonorrhoeae* is an obligate parasite and may be found in the mucosal cells of the genitourinary tract, eye, rectum, and throat.

Virulence Factors Fimbriae and other cell surface molecules serve as the primary virulence factor by promoting attachment of the cocci both to themselves and to the surface of mucosal cells. Gonococcus also produces a protease that cleaves IgA, preventing it from functioning.

Primary Infections/Disease Genital gonorrhea can lead to urethritis and painful urination in males while in females, infection can cause pelvic inflammatory disease and ectopic pregnancy. The buildup of scar tissue within the spermatic ducts of men or the fallopian tubes of women can cause sterility. Infants born to gonococcus carriers can be infected as they pass through the birth canal, often infecting the eyes and potentially causing blindness.

Control and Treatment Because of widespread resistance to a growing list of antimicrobials, the only recommended treatment for *N. gonorrhoeae* infection is dual therapy with ceftriaxone and azithromycin. A large proportion of gonorrhea infections are complicated by infection with another sexually transmitted disease (most commonly chlamydia), and multi-drug treatment has the added benefit of generally resolving these infections as well. Control of future infections depends on the emphasis of safe sexual practices such as the use of condoms.

Epidemiology and Pathogenesis of Meningococcal Disease

The diseases of *N. meningitidis* generally have a sporadic or epidemic incidence in late winter or early spring. The continuing reservoir of infection is humans who harbor the pathogen in the nasopharynx. Because this bacterium does not survive long in the environment, meningococci are usually acquired through close contact with secretions or droplets. The carriage state, which can last from a few days to several months, exists in 3% to 30% of the adult population and can exceed 50% in institutional settings. The scene is set for transmission when carriers live in close quarters with nonimmune individuals, as might be expected in college dormitories, day care facilities, and military barracks. The highest-risk groups are young children (6–36 months old) and older children and young adults (10–20 years old). *N. meningitidis* is the second most common cause of meningitis in adults, after *Streptococcus pneumoniae*.

Upon reaching its portal of entry in the nasopharynx, the meningococcus attaches there with fimbriae. In many people, this can result in simple asymptomatic colonization. In more vulnerable individuals, however, the meningococci are engulfed by epithelial cells of the mucosa and penetrate into the nearby blood vessels, along the way damaging the epithelium and causing pharyngitis. If the bacteria remain confined to the bloodstream, the result is **meningococcemia,** marked initially by cough, headache, and sore throat, and later by malaise, nausea, vomiting, and stiffness in the joints. The pathogen sheds endotoxin into the systemic circulation, stimulating certain classes of white blood cells to release cytokines. Damage to the blood vessels caused by a rush of cytokines triggers hemorrhage and coagulation, leading to vascular damage, and a lack of oxygen perfusion of the tissues, especially in the limbs.

Crops of lesions called **petechiae* (figure 18.26)** develop on the trunk and appendages in about half of the cases. In a small number of cases, meningococcemia progresses to shock, necrosis of tissues in extremities, and loss of limbs. Generalized intravascular clotting, cardiac failure, and death can occur within a few hours.

In other cases, instead of remaining within the blood vessels, the bacteria rapidly cross the blood-brain barrier, permeate the meninges, and multiply within the cerebrospinal fluid, resulting in **meningitis.** Symptoms of meningitis include fever, sore throat, headache, stiff neck, convulsions, vomiting, and photophobia **(figure 18.27)**. These symptoms may be missing in infants and the elderly. The permanent effects of central nervous system infection are cognitive impairment, learning disabilities, and deafness in about 10% to 20% of patients.

* *petechiae* (pee-tee′-kee-ee) Small, nonraised, purple blotches caused by hemorrhage into the skin. Larger spots are called *ecchymoses* (ek″-ih-moh′-seez).

Figure 18.26 One clinical sign of meningococcemia. The release of endotoxins during meningococcemia may cause hemorrhage, coagulation, and vascular damage, leading to necrosis of tissue, especially in the extremities. This infant displays necrosis of both legs due to meningococcemia. *Source:* Mr. Gust/CDC

Pathogen Profile #5 *Neisseria meningitidis* (meningococcus)

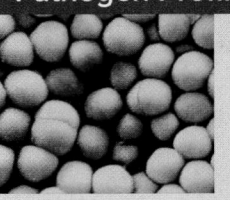

© CNRI/Science Source

Microscopic Morphology Gram-negative, flattened cocci, found growing as diplococci. Very rarely motile, non-spore-forming.

Identified by Presence of gram-negative diplococci in blood, cerebrospinal fluid, or within the cytoplasm of neutrophils. A positive oxidase test, the ability to ferment maltose, and an inability to ferment sucrose or lactose helps to distinguish *N. meningitidis* from normal microbiota of the oropharynx. Rapid identification methods rely on the detection of *N. meningitidis* DNA or antigens specific to the pathogen within the blood or cerebrospinal fluid.

Habitat Humans are the sole reservoir for *Neisseria meningitidis*, with 3% to 30% of the adult population harboring the pathogen in their nasopharynx at any given time. When carriers live with nonimmune individuals in close quarters—college dormitories, military barracks, day care centers—transmission is common. The pathogen does not survive for long outside the body.

Virulence Factors Virulence factors of *N. meningitidis* include an antiphagocytic polysaccharide capsule and fimbriae that facilitate attachment of the bacterium to cells of the nasopharynx. The pathogen also releases large quantities of a potent endotoxin into the systemic circulation.

Primary Infections/Disease There are two overlapping conditions that together constitute meningococcal disease. Meningococcemia is marked by the appearance of petechiae or ecchymoses on the skin or mucous membrane. These lesions are the result of endotoxin-induced subcutaneous hemorrhage. Meningococcemia may progress to include generalized intravascular clotting, tissue necrosis in the extremities, cardiac failure, and death. Meningitis refers to inflammation of the meninges— the membranes surrounding the brain and spinal cord—and most commonly affects young children between 6 and 36 months, older children, and young adults between 10 and 20 years old. Even when treated, meningitis has a mortality rate of up to 15%. The permanent effects of cerebrospinal meningitis include cognitive impairment and deafness in 10% to 20% of patients.

Control and Treatment Vaccination provides protection against *N. meningitidis* infection and is recommended for all persons between 11 and 18 years of age as well as those in high-risk populations, including students, military recruits, and people exposed during outbreaks. Available vaccines include: MCV4 (Menactra) for persons between 2 and 55 years of age, MPSV4 (Menomune) for those older than 55, and Tremenba specifically for ages 10–25.

Cerebrospinal fluid

Meninges

Nasal cavity

Initial infection site

Palate

Clinical Diagnosis of Meningococcal Disease

Suspicion of meningococcal infection constitutes a medical emergency, and differential diagnosis must be done with great haste and accuracy because complications can arise rapidly and with lethal consequences. Cerebrospinal fluid and blood samples are stained and observed directly for the typical gram-negative diplococci. Cultivation may be necessary to differentiate the bacterium from other species (see subsequent section on laboratory diagnosis). Specific rapid tests are also available for detecting the capsular polysaccharide or the cells directly from specimens without culturing.

Immunity, Treatment, and Prevention of Meningococcal Infection

The infection rate in most populations is rather low, indicating that most people have a natural immunity to the meningococcus. This probably develops during the early years of life as one is exposed to the meningococcus and its close relatives. Resistance is due to the the production of antibodies against the capsular polysaccharides in groups A and C and against membrane antigens in group B.

Figure 18.27 Dissemination of the meningococcus from a nasopharyngeal infection. Bacteria spread to the roof of the nasal cavity, which borders a highly vascular area at the base of the brain. From this location, they can enter the blood and escape into the cerebrospinal fluid. Infection of the meninges leads to meningitis and an inflammatory purulent exudate over the brain surface.

Because even treated meningococcal disease has a mortality rate of up to 15%, it is vital that therapy begin as soon as possible with one or more drugs. It can even be given while tests for the causative agent are under way. The treatment of choice for meningococcal meningitis is a third-generation cephalosporin such as ceftriaxone that is broad spectrum and crosses the blood-brain barrier. Patients may also require treatment for shock and intravascular clotting.

When family members, medical personnel, or children in day care have come in close contact with infected people, preventive therapy with rifampin or ceftriaxone may be warranted. Vaccination is recommended for all persons between 11 and 18 years as well as those in high-risk populations, including students, military recruits, and people exposed during outbreaks. Vaccines that are currently available include: MCV4 (Menactra), a meningococcal conjugate vaccine recommended for persons between 2 and 55 years of age; MPSV4 (Menomune), recommended for those older than 5; and two recombinant DNA vaccines (Trumenba and Bexsero), specific for serotype B and recommended for children and young adults between the ages of 10 and 25.

Differentiating Pathogenic from Nonpathogenic *Neisseria*

It is usually necessary to differentiate true pathogens from normal *Neisseria* species that also live in the human body and can be present in infectious fluids. Immediately after collection, specimens are streaked on modified Thayer-Martin (MTM) medium or chocolate agar and incubated in a high CO_2 atmosphere. Presumptive identification of the genus is obtained by a Gram stain and oxidase testing on isolated colonies. Further testing may be necessary to differentiate the two pathogenic *Neisseria* from one another, from other oxidase-positive bacteria, and from normal microbiota of the oropharynx and genitourinary tract that can be confused with the pathogens. Sugar oxidation, growth patterns, nitrate reduction, and pigment production are useful differentiation tests at this point (**table 18.5**). Several rapid method identification kits have been developed for this purpose. Many laboratories now rely on a rapid PCR test that can detect gonococcal DNA directly from a urine sample.

TABLE 18.5 Scheme for Differentiating Gram-Negative Cocci and Coccobacilli

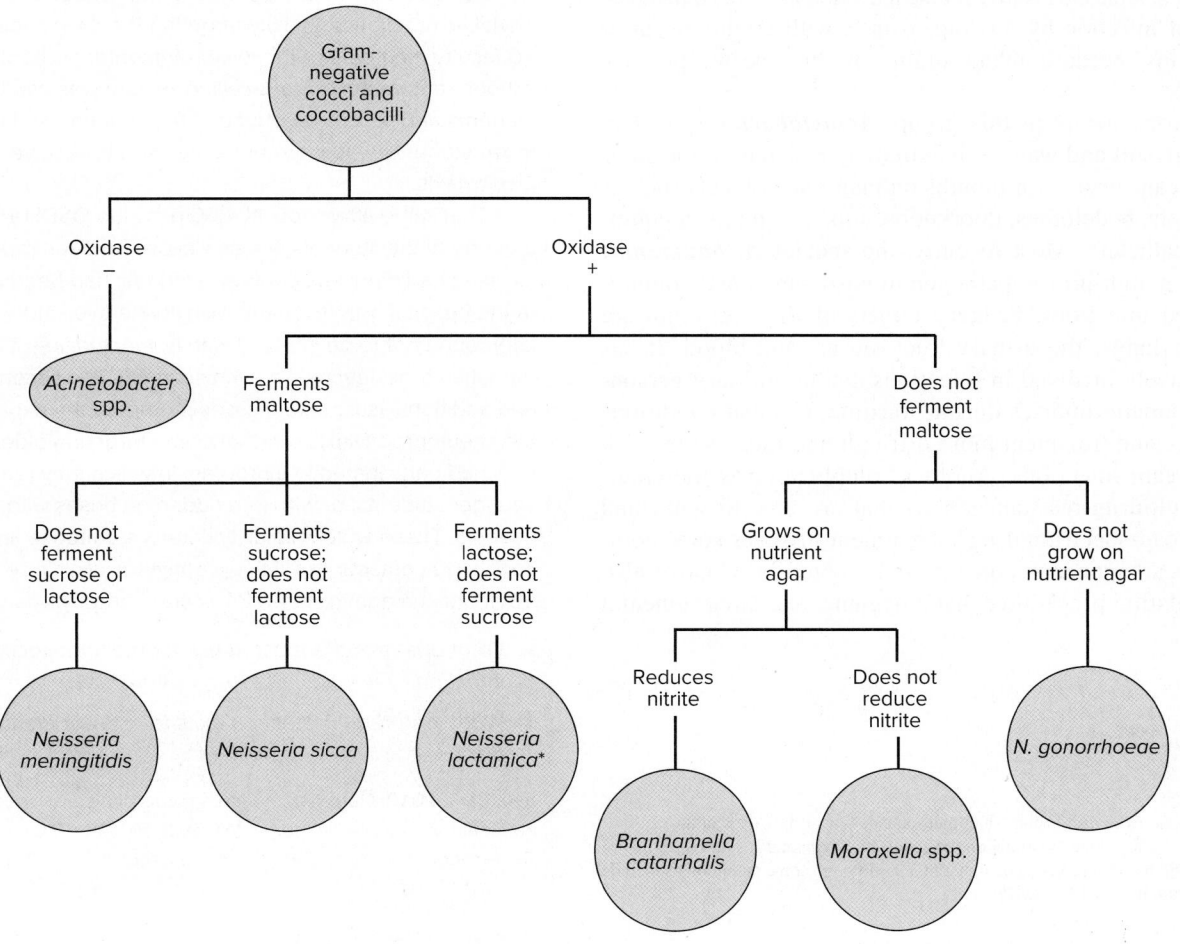

*A weak pathogen, found in the nasopharynx of children and sometimes mistaken for *N. meningitidis*.

Other Genera of Gram-Negative Cocci and Coccobacilli

The genera *Moraxella and Acinetobacter* are included in the same family as *Neisseria* because of morphological and biochemical similarities. Most species are either relatively harmless commensals of humans and other mammals or are saprobes living in soil and water. In the past few years, however, two species in particular have emerged as significant opportunists in hosts with deficient immune functions. The first species, **Moraxella (Branhamella) catarrhalis,*** is found in the normal human nasopharynx and can cause purulent disease. It is associated with several clinical syndromes such as meningitis, endocarditis, otitis media, bronchopulmonary infections, and neonatal conjunctivitis. Adult patients with leukemia, alcoholism, malignancy, diabetes, or rheumatoid disease are the most susceptible to it.

Moraxella species often appear as short, plump rods as well as cocci, and some exhibit a twitching motility. These bacteria are widely distributed on the mucous membranes of domestic mammals and humans and are generally regarded as weakly pathogenic or nonpathogenic. Some species are rarely implicated in ear infections and conjunctivitis in humans.

Because *Moraxella* is morphologically similar to the gonococcus and the meningococcus, biochemical methods are needed to discriminate it. Key characteristics are the complete lack of carbohydrate fermentation and positive nitrate reduction in *M. B. catarrhalis.* Treatment of infection by this organism is with erythromycin or cephalosporins, because many strains of this species produce penicillinase.

The second genus in this group, **Acinetobacter** is widely distributed in soil and water and extremely resistant to the environment. It can survive for months on inanimate objects such as faucets, toilets, bedclothes, doorknobs, sinks, respiratory equipment, and catheters. Most recently, the species **A. baumannii** has become a high-profile pathogen in nosocomial and community-acquired infections. Primary targets of the bacterium are wounds, the lungs, the urinary tract, burns, and blood. It has been extensively involved in infections among military personnel with traumatic injuries during wartime. Combat conditions make control and treatment more difficult and increase the risk for bloodstream infections. Another complication is the emergence of multidrug-resistant strains that are able to withstand the usual broad-spectrum drugs. Treatment involves some combination of carbapenems, colistin, and ampicillin. Control also includes isolation procedures, hand hygiene, and environmental disinfection.

* *Branhamella catarrhalis* (bran″-hah-mel′-ah cah-tahr-al′-is) After Sarah Branham, a bacteriologist working on the genus *Neisseria,* and L. *catarrhus,* to flow. We will use this genus name in her honor, although some taxonomists include this bacterium in genus **Moraxella.**

Check Your Progress SECTION 18.3

10. Describe the epidemiology and pathology of *N. gonorrhoeae*, and compare them to those of *N. meningitidis.*

11. List the short-term and long-term consequences of *N. gonorrhoeae* infection and compare them to those of *N. meningitidis.*

12. Recall the epidemiology and pathology of the medically important organisms in the genera *Moraxella and Acinetobacter.*

CASE STUDY Part 2

The following morning, over the course of 2 hours, OSDH received reports of three more cases of suspected meningococcal disease, including one fatality. All four patients attended the same lower elementary school (prekindergarten through second grade), which was housed in a single building on a large campus. The campus included three other buildings providing classroom space for a total of 1,850 pre-K through twelfth-grade students. The occurrence of the additional cases within such a short time frame prompted state and local authorities to initiate measures to control the outbreak.

At noon a clinic was opened at the school to provide prophylactic antibiotics, initially targeting the 443 students and 50 faculty members of the lower elementary school. The school's automated parent-calling system was used to keep students and parents apprised of the situation and to notify those students who were being advised to receive chemoprophylaxis.

That same afternoon, at 4:00 p.m., the OSDH was notified that one of the students whose illness had been reported that morning had died, and two new students had been admitted to the hospital with fever and rash (these two students were later found not to be infected with *N. meningitidis*, but this information took several days to ascertain). The second death and additional (suspected) cases prompted the expansion of the chemoprophylaxis clinic, this time targeting older students who had participated in a program in which they read to younger students, or who had ridden on buses with any of the patients. These wider eligibility criteria resulted in an additional 400 contacts. In all, the prophylaxis clinic operated for 16 hours and administered 846 chemoprophylaxis doses.

■ *What age group is most at risk for meningococcal infection?*

■ *What long-term protection exits to prevent meningococcal disease?*

See: CDC. 2012. Outbreak of meningococcal disease associated with an elementary school—Oklahoma, March 2010. *MMWR* 61:217–21.
To conclude this Case Study, go to Connect.

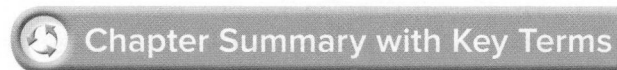

Gram-positive and gram-negative cocci are frequently called **pyogenic,** due to their stimulation of pus formation.

18.1 General Characteristics of the Staphylococci

Nonmotile, gram-positive cocci arranged in irregular clusters; facultative anaerobes; fermentative; salt-tolerant, catalase-positive.

A. *Staphylococcus aureus* produces a number of virulence factors. Enzymes include **coagulase** (a confirmatory characteristic), hyaluronidase, staphylokinase, nuclease, and penicillinase. Toxins are **beta-hemolysins, leukocidin, enterotoxin, exfoliative toxin,** and toxic shock syndrome toxin. Microbe carried in nasopharynx and skin.

1. *Skin Infections:* Local **abscess** occurs at site of invasion of hair follicle, gland; manifestations are **folliculitis, furuncle, carbuncle,** and bullous **impetigo.**

2. *Toxic Disease:* Food intoxication due to enterotoxin; **staphylococcal scalded skin syndrome (SSSS),** a skin condition that causes desquamation; **toxic shock syndrome,** toxemia in women due to infection of vagina, associated with using tampons.

3. *Other:* **Osteomyelitis,** bacteremia leading to endocarditis, and pneumonia.

B. *Principal Coagulase-Negative Staphylococci:* **S. epidermidis,** a normal resident of skin and follicles; an opportunist and one of the most common causes of nosocomial infections, chiefly in surgical patients with indwelling medical devices or implants. **S. saprophyticus** is a urinary pathogen.

C. *Treatment: S. aureus* has multiple resistance to antibiotics, especially penicillin, ampicillin, and methicillin **(MRSA).** Abscesses require debridement and removal of pus. Staphylococci have extreme resistance to harsh environmental conditions.

D. Staphylococcal infections are frequently complicated by their tendency to form biofilms on medical devices.

18.2 General Characteristics of the Streptococci and Related Genera

A large, varied group of gram-positive bacteria, containing the genera *Streptococcus* and *Enterococcus.* Cocci are in chains of various lengths; nonmotile, non-spore-forming; often encapsulated; catalase-negative. Classified into Lancefield groups (A–R) according to the type of serological reactions of the cell wall carbohydrate; also characterized by type of hemolysis. The most important sources of human disease are beta-hemolytic **S. pyogenes** (group A), **S. agalactiae** (group B), *Enterococcus faecalis* (group D), and alpha-hemolytic **S. pneumoniae** and the **viridans streptococci.**

A. *Beta-Hemolytic Streptococci:* **S. pyogenes** is the most serious pathogen of family; produces several virulence factors, including C-carbohydrates, M-protein (fimbriae), streptokinase, hyaluronidase, DNase, **streptolysins** (SLO, SLS), **erythrogenic toxin.**

1. Skin infections include pyoderma (strep **impetigo**) and **erysipelas** (deeper, spreading skin infection).

2. Systemic conditions include *strep throat* or **pharyngitis;** may lead to toxemia, called **scarlet fever**—also causes pneumonia.

3. Sequelae caused by immune response to streptococcal toxins include **rheumatic fever** and **glomerulonephritis.**

B. Other *Beta-Hemolytic:* **Streptococcus agalactiae** (group B), a pathogen increasingly found in the human vagina; causes neonatal, wound, and skin infections, particularly in debilitated persons. *Enterococcus faecalis* (sometimes beta-hemolytic) and other enteric group D species are normal flora of intestine; cause opportunistic urinary, wound, and surgical infections.

C. *Treatment:* Group A and group B are treated primarily with some type of penicillin (G is very effective); sensitivity testing may be necessary for enterococci; no vaccines available.

D. *Alpha-Hemolytic Streptococci:* The **viridans streptococci** *S. mitis, S. salivarius, S. mutans,* and *S. sanguinis* constitute oral flora in saliva and can form biofilms on tissues. Several species are implicated in dental caries. Principal non-oral infections are **subacute endocarditis,** mass colonization of heart valves following dental procedures.

E. *S. pneumoniae,* the **pneumococcus,** is a heavily encapsulated, lancet-shaped diplococcus; capsule is an important virulence factor—90+ types; reservoir is nasopharynx of normal healthy carriers.

1. The pneumococcus is the most common cause of bacterial pneumonia; attacks patients with weakened respiratory defenses; symptoms include fever/chills, cyanosis, cough.

2. **Otitis media,** inflammation of middle ear, is common in children.

3. It is the most common cause of meningitis in adults.

4. Vaccination available.

18.3 The Family Neisseriaceae: Gram-Negative Cocci

Primary pathogens in the genus *Neisseria* are common residents of mucous membranes. *Neisseria* species are bean-shaped diplococci that have capsules and fimbriae; they are oxidase-positive, non-spore-forming, nonmotile; pathogens are fastidious and do not survive long in the environment.

A. *Neisseria gonorrhoeae:* The **gonococcus,** cause of **gonorrhea,** tends to be located intracellularly in pus cells; among the most common STDs; may be transmitted from mother to newborn; asymptomatic carriage is common in both sexes.

1. Symptoms of gonorrhea in males are urethritis and discharge; infection of deeper reproductive structures may cause scarring and infertility. Symptoms of gonorrhea in females include vaginitis and urethritis. Ascending infection may lead to **salpingitis (PID),** mixed anaerobic infection of abdomen; common cause of sterility and ectopic tubal pregnancies due to scarred fallopian tubes. Extragenital infections may be anal, pharyngeal, conjunctivitis, septicemia, arthritis.

2. Infections in newborns cause eye inflammation, occasionally infection of deeper tissues and blindness; can be prevented by prophylaxis immediately after birth.

3. Preferred treatment is a combination of ceftriaxone and azithromycin due to increasing drug-resistant strains; no vaccine exists.

B. *Neisseria meningitidis:* The **meningococcus** is a prevalent cause of **meningitis.** Infection begins when bacteria enter bloodstream, pass into the cranial circulation, multiply in meninges and brain; very rapid onset; initial disease symptoms are neurological; **endotoxin** released by pathogen causes hemorrhage and shock; can be fatal; treated with third-generation cephalosporins; vaccines exist for groups A and C.

C. *Other Genera of Gram-Negative Cocci and Coccobacilli*
 1. *Other Gram-Negative Cocci and Coccobacilli:* Genus *Moraxella* are short rods that colonize mammalian mucous membranes. *Moraxella* (*Branhamella*) *catarrhalis,* common member of throat flora; is an opportunist in cancer, diabetes, alcoholism.
 2. *Acinetobacter baumannii,* nosocomial and community-acquired infection of wounds, lungs, urinary tract, and blood. Many drug-resistant strains.

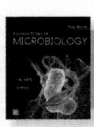

Level I. Knowledge and Comprehension

These questions require a working knowledge of the concepts in the chapter and the ability to recall and understand the information you have studied.

Multiple-Choice Questions

Select the correct answer from the answers provided. For questions with blanks, choose the combination of answers that most accurately completes the statement.

1. Which of the following is/are a pyogenic coccus/cocci?
 a. *Streptococcus*
 b. *Staphylococcus*
 c. *Neisseria*
 d. all of these

2. The coagulase test is used primarily to differentiate *Staphylococcus aureus* from
 a. other staphylococci
 b. streptococci
 c. micrococci
 d. enterococci

3. The symptoms in scarlet fever are due to
 a. streptolysin
 b. coagulase
 c. erythrogenic toxin
 d. alpha-toxin

4. The most severe streptococcal diseases are caused by
 a. group B streptococci
 b. group A streptococci
 c. pneumococci
 d. enterococci

5. Rheumatic fever damages the _____, and glomerulonephritis damages the _____.
 a. skin, heart
 b. joints, bone marrow
 c. heart valves, kidney
 d. brain, kidney

6. _____ hemolysis is the partial lysis of red blood cells due to bacterial hemolysins.
 a. Gamma-
 b. Alpha-
 c. Beta-
 d. Delta-

7. Viridans streptococci commonly cause
 a. pneumonia
 b. meningitis
 c. subacute endocarditis
 d. otitis media

8. Which of the following strains of pathogens has the lowest incidence?
 a. VRE
 b. VRSA
 c. MRSA
 d. MRSE

9. Otitis media is a/an _____ infection often caused by _____ .
 a. bone, *Staphylococcus aureus*
 b. eye, *N. gonorrhoeae*
 c. middle ear, *Streptococcus pneumoniae*
 d. brain, *N. meningitidis*

10. Which genus of bacteria has pathogens that can cause blindness?
 a. *Streptococcus*
 b. *Staphylococcus*
 c. *Neisseria*
 d. *Moraxella*

11. An important test for identifying *Neisseria* is
 a. production of oxidase
 b. production of catalase
 c. sugar fermentation
 d. beta-hemolysis

12. A complication of genital gonorrhea in both men and women is
 a. infertility
 b. pelvic inflammatory disease
 c. arthritis
 d. blindness

13. What virulence factor is associated with the development of skin hemorrhages in meningitis?
 a. skin invasion by *N. meningitidis*
 b. blood clots
 c. erysipelas
 d. endotoxins in the blood

14. Which infectious agent of those covered in the chapter would most likely be acquired from a contaminated doorknob?
 a. *Staphylococcus aureus*
 b. *Streptococcus pyogenes*
 c. *Neisseria meningitidis*
 d. *Streptococcus pneumoniae*

15. Which of the following effects is/are caused by formation of an infectious biofilm?
 a. disseminated infection
 b. failure of drug therapy
 c. colonization of tissues
 d. a and b
 e. b and c
 f. all of these

16. **Single Matching.** Match each description in the right-hand column with the one word in the left-hand column that fits it. (See index and glossary.)
 _____ furuncle
 _____ osteomyelitis
 _____ coagulase
 _____ erythrogenic toxin
 _____ rheumatic fever
 _____ beta-hemolysis
 _____ consolidation
 _____ viridans streptococci
 _____ erysipelas
 _____ endocarditis
 _____ streptolysin
 _____ streptokinase

 a. complete red blood cell lysis
 b. substance involved in heart valve damage
 c. enzyme that dissolves clots
 d. enzyme of pathogenic *S. aureus*
 e. cutaneous infection of group A streps
 f. solidification of lung alveoli
 g. unique pathologic feature of *N. gonorrhoeae*
 h. a boil
 i. cause of tooth abscesses
 j. focal infection of long bones
 k. heart colonization by oral streps
 l. cause of scarlet fever symptoms
 m. long-term sequelae of strep throat

 Case Study Review

1. The first step in the clinical diagnosis of bacterial meningitis is a Gram stain of blood or cerebrospinal fluid. What other bacterial species could be mistaken for *N. meningitidis* at this stage of the identification process?
 a. *Moraxella catarrhalis*
 b. *Staphylococcus epidermidis*
 c. *Streptococcus agalactiae*
 d. *Enterococcus faecalis*

2. Vaccination can provide protection against *N. meningitidis*, as eventually happened in this case. A vaccination also exists for what other common cause of meningococcal disease?
 a. *Staphylococcus aureus*
 b. *Streptococcus pyogenes*
 c. *Streptococcus pneumoniae*
 d. *Acinetobacter baumannii*

3. In this case, the first step taken to treat students potentially exposed to meningitis was to administer antibiotics, while a meningococcal vaccination clinic was organized for the area a week after the outbreak. What were the benefits of fighting the outbreak with both antibiotics and vaccination?

 Writing Challenge

For each question, compose a one- or two-paragraph answer that includes the factual information needed to completely address the question. Check Your Progress questions can also be used for writing-challenge exercises.

1. Explain the actions of each of the following that make them virulence factors:
 a. hemolysins
 b. leukocidin
 c. kinases
 d. hyaluronidase
 e. coagulase

2. What conditions favor staph food poisoning?

3. Describe what is meant by HA-MRSA and CA-MRSA and why they are important.

4. a. Describe the major group A streptococcal infections.
 b. Why is a "strep throat" a cause for concern?

5. Discuss the apparent pathology at work in rheumatic fever and acute glomerulonephritis.

6. What are the recommendations for pneumococcal vaccination?

7. a. Compare and contrast the characteristics of male and female genital gonorrhea.
 b. Describe at least two extragenital complications of gonorrhea.

 Concept Mapping

On Connect you can find an Introduction to Concept Mapping that provides guidance for working with concept maps along with concept mapping activities for this chapter.

 Level II. Application, Analysis, Evaluation, and Synthesis

These problems go beyond just restating facts and require higher levels of understanding and an ability to interpret, problem solve, transfer knowledge to new situations, create models, and predict outcomes.

 Critical Thinking

Critical thinking is the ability to reason and solve problems using facts and concepts. These questions can be approached from a number of angles, and in most cases, they do not have a single correct answer.

1. You have been handed the problem of diagnosing gonorrhea from a single test. Which one will you choose and why?

2. You have been called upon to prevent outbreaks of SSSS in the nursery of a hospital where this strain of *S. aureus* has been isolated.
 a. What will be the main concerns?
 b. What procedures will address these concerns?

3. You have been given the assignment of obtaining some type of culture that could help identify the cause of otitis media in a small child.
 a. What area might you sample?
 b. Why is this disease so common in children?

4. a. Name the three species of pyogenic cocci most commonly implicated in neonatal disease.
 b. Explain how infants might acquire such diseases.

5. Explain why it is so important to take a detailed cardiovascular and infection history of dental patients.

6. Case Study. An elderly man with influenza acquires a case of pneumonia. Gram-positive cocci isolated from his sputum give beta-hemolysis on blood agar; the infection is very difficult to treat. Later, it is shown that the man shared a room with a patient with a bone infection. Isolates from both infections were the same.
 a. What is the probable species?
 b. Justify your answer.
 c. What treatment will be necessary?
 d. What special precautions should be taken by persons attending to the patient? Why?

Visual Challenge

1. From **chapter 14, figure 14.2a.** How does this aspect of the body's defenses help protect against *Streptococcus pneumoniae* infection?

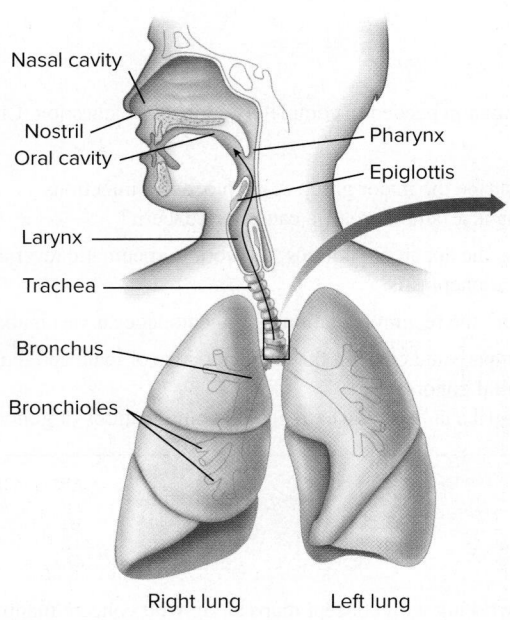

Nasal cavity
Nostril
Oral cavity
Pharynx
Epiglottis
Larynx
Trachea
Bronchus
Bronchioles
Right lung Left lung

2. From **chapter 17, figure 17.8b1.** Explain how this technique can be used to diagnose infections with group A streptococci.

Isolated colony, identity *unknown*
Antibodies of *known* identity
ANTISERUM
Macroscopic reaction
Molecular reaction
Ag on microbe
Ab

STUDY TIP: Quickly create a customized practice quiz for this chapter by going to your Connect SmartBook "My Reports" section. Don't have SmartBook and want to learn more? Go to **whysmartbook.com.**

Electron micrograph of a single *Listeria monocytogenes* cell.
(left): *Source:* Dr. Balasubr Swaminathan & Pegggy Hayes, photo by Elizabeth White/CDC; (right): © BananaStock/PunchStock RF

 CASE STUDY Part 1 Three Patients

Dr. Mike Hauser was doing everything he could to fight back against **myeloma,** a cancer that attacks plasma cells, the blood cells responsible for **antibody production.** The 68-year-old podiatrist was dealing as best he could with the anemia and lowered immune response that are part of both the disease and its treatment, and he seemed to be regaining his strength. After breakfast with his family, he went back to planning an upcoming bowhunting trip, a welcome respite from talk of cancer and chemotherapy. Several hundred miles from Dr. Hauser's Colorado home, Paul Schwarz was also eating breakfast, the 92-year-old crediting his long life and good health, at least in part, to a diet rich in fruit. At roughly the same time, Michelle Wakley, 6 months pregnant, was enjoying a breakfast of fresh **cantaloupe** with her family in Indianapolis. Michelle and her husband were looking forward to their second child. Within days, Mike Hauser and Paul Schwarz would be dead, and Michelle Wakley would be admitted to the hospital in premature labor.

All three patients eventually sought medical attention for what at first seemed to be the flu, a not uncommon diagnosis in September. Doctors, however, knew that severe headache and flulike symptoms in pregnant women, the elderly, and those who are **immune compromised**

> "Every day when I woke up my head hurt. My legs were killing me. … Kind of like when you get the flu, your body aches. It was painful!… and I had the chills. I should've gone to the hospital but knowing … you get flulike symptoms when you're pregnant, I didn't go."
>
> **Michelle Wakley**

were ominous signs, suggesting infection with *Listeria monocytogenes,* a diagnosis confirmed by the presence of the bacteria in the blood and cerebrospinal fluid of each patient. In most people **listeriosis** is a mild illness, which is rarely treated, but in susceptible hosts the disease follows a well-established pattern, as bacteria spread through the body, causing **septicemia, meningitis,** and **stillbirth** or premature delivery in the case of late-term pregnancy. Although treatable with antibiotics, in these cases the disease had progressed too far; Mike Hauser and Paul Schwarz died, and Michelle Wakley's daughter Kendall was born 3 months premature.

And they were only the beginning. Because listeria is a **reportable disease,** the cases were forwarded to the CDC for potential epidemiological investigation. On September 7, after receiving reports of at least 55 cases of listeriosis, the CDC began coordinating a multistate investigation into the source of the outbreak.

- *What is the difference between food infection and food intoxication? Into which category does listeriosis fall?*

- *What foods are commonly implicated in cases of listeriosis?*

To continue the Case Study, go to Case Study Part 2 at the end of the chapter.

19.1 Medically Important Gram-Positive Bacilli

 Expected Learning Outcome

1. Describe how cellular characteristics are used to categorize gram-positive bacilli into major groups.

The gram-positive bacilli can be subdivided into three general groups based on the presence or absence of endospores and the characteristic of acid-fastness. Further levels of separation correspond to oxygen requirements and cell morphology. Organization of the most important gram-positive pathogens can be seen in **table 19.1** and **Systems Profile 19.1**.

 Check Your Progress SECTION 19.1

1. Explain why *Bacillus, Clostridium,* and *Mycobacterium* are found in separate groups despite all being gram-positive bacilli.

19.2 Gram-Positive Spore-Forming Bacilli

 Expected Learning Outcomes

2. Recall the general characteristics of the genus *Bacillus*.
3. Distinguish between cutaneous and pulmonary anthrax, and describe how anthrax is controlled.
4. State the general characteristics of the genus *Clostridium,* and distinguish between *Bacillus* and *Clostridium*.
5. Recall the organisms responsible for myonecrosis and what is required for infection and disease to progress.

6. Clarify why *Clostridium difficile*–associated disease is associated with the use of antibiotics.
7. Compare the course of botulism infection and disease with that of tetanus.
8. Contrast common botulism with infant botulism, and relate prevention and treatment strategies for both.
9. Compare food intoxication caused by *Bacillus cereus* with that caused by *Clostridium perfringens*.

Most endospore-forming bacteria are gram-positive, motile, rod-shaped forms in the genera *Bacillus* and *Clostridium*. An endospore is a dense survival unit that develops in a vegetative cell in response to nutrient deprivation (**figure 19.1;** see figure 4.22). The extreme resistance to heat, drying, radiation, and chemicals accounts for the survival, longevity, and ecological niche of sporeformers, and it is also a significant factor in their pathogenicity.

General Characteristics of the Genus *Bacillus*

The genus *Bacillus* includes a large assembly of mostly saprobic bacteria widely distributed in the earth's habitats. *Bacillus* species are aerobic and catalase-positive, and, though they have varied nutritional requirements, none is fastidious. The group is noted for its versatility in degrading complex macromolecules, and it is also a common source of antibiotics. Because the primary habitat of many species is the soil, spores are continuously dispersed by means of dust into water and onto the bodies of plants and animals. Despite their ubiquity, the two species with primary medical importance are *B. anthracis,* the cause of anthrax, and *B. cereus,* the cause of one type of food poisoning.

Bacillus anthracis *and Anthrax*

Bacillus anthracis is among the largest of all bacterial pathogens, composed of block-shaped, angular nonmotile rods 3 to 5 μm long

TABLE 19.1 Scheme for Differentiating Gram-Positive Bacilli

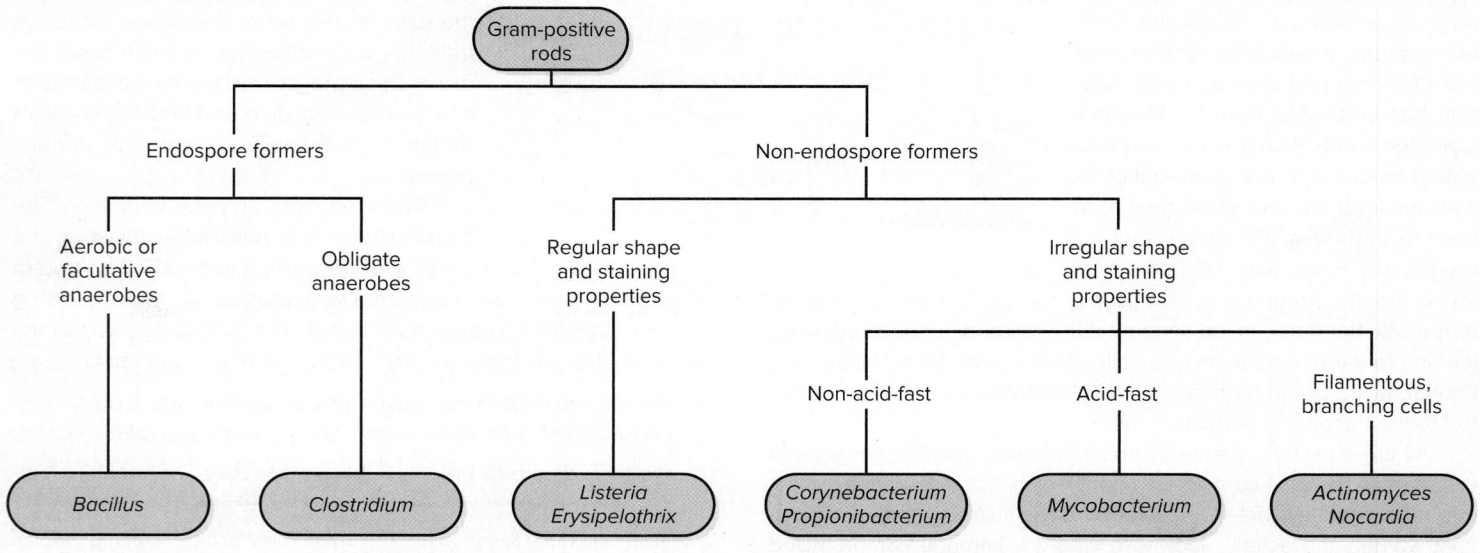

Systems Profile 19.1	Pathogenic Gram-Positive Bacilli				
Organism	**Skin/Skeletal**	**Nervous/Muscle**	**Cardiovascular/ Lymphatic/Systemic**	**Gastrointestinal**	**Respiratory**
Bacillus anthracis	Cutaneous anthrax				Pulmonary anthrax
Bacillus cereus				Food poisoning	
Clostridium perfringens	Gas gangrene (myonecrosis)	Gas gangrene (myonecrosis)		Food poisoning (mild)	
Clostridium tetani		Tetanus			
Clostridium difficile				C. difficile–associated disease	
Clostridium botulinum	Wound botulism	Botulism (wound, infant, and food-borne)			
Listeria monocytogenes		Meningitis (in immunocompromised hosts)	Septicemia	Diarrhea	Sore throat
Erysipelothrix rhusiopathiae	Erysipelas				
Corynebacterium diphtheriae	Cutaneous diphtheria				Diphtheria
Propionibacterium acne	Acne				
Mycobacterium tuberculosis		Recurrent tuberculosis	Extrapulmonary tuberculosis		Primary tuberculosis
Mycobacterium leprae	Leprosy	Leprosy (tuberculoid)	Leprosy (lepromatous)		

Figure 19.1 **Examples of endospore-forming pathogens. (a)** The morphological appearance of *Bacillus anthracis,* showing centrally placed endospores and a streptobacillus arrangement (600×). **(b)** *Clostridium botulinum* (1,000×). Its typical tennis racquet morphology is created by terminal spores that swell the sporangium. (a): *Source:* CDC; (b): *Source:* Dr. George Lombard/CDC

and 1 to 1.2 μm wide and central spores that develop under all growth conditions except in the living body of the host **(figure 19.1a).** Its virulence factors include a polypeptide capsule and exotoxins that in varying combinations produce edema and cell death. For centuries, **anthrax*** has been known as a zoonotic disease of herbivorous livestock (sheep, cattle, goats). It has an important place in the history of medical microbiology because it was Robert Koch's

model for developing his postulates in 1877; and later Louis Pasteur used the disease to prove the usefulness of vaccination.

The anthrax bacillus is a facultative anaerobe that undergoes its cycle of vegetative growth and sporulation in the soil. Animals become infected while grazing on grass contaminated with spores. When the pathogen is returned to the soil in animal excrement or carcasses, it can sporulate and become a long-term reservoir of infection for the animal population. The majority of anthrax cases are reported in livestock from Africa, Asia, and the Middle East. Most recent human cases in the United States have occurred in textile

* *anthrax* (an's-thraks) Gr. *anthrax,* carbuncle.

Pathogen Profile #1 *Bacillus anthracis*

Source: CDC

Microscopic Morphology Large gram-positive bacilli, often present in long chains. Cells are block shaped with endospores developing in the center of the cell.

Identified by Gram reaction and endospore formation. *Bacillus* is aerobic, whereas *Clostridium* is a strict anaerobe. Physiological and biochemical tests are used to differentiate *B. anthracis* from other members of the genus.

Habitat Both vegetative growth and endospore formation occur in soil, with spores easily spreading to water, plants, and the bodies of animals. The majority of human infection occurs in workers handling animal hair, hides, or products made from them.

Virulence Factors Cells are covered in a thick capsule that inhibits phagocytosis and produce exotoxins that cause the death of macrophages. Lysosomes within the dying macrophages release their contents, leading to capillary thrombosis, cardiovascular shock, and rapid death.

Primary Infections/Disease Cutaneous anthrax is the most common and least dangerous form of infection and occurs when *B. anthracis* spores enter broken skin. Growth of the bacterium is noted as a papule that becomes necrotic and gives rise to a painless black eschar. Without antibiotic treatment, mortality can be as high as 20%. Pulmonary anthrax (woolsorter's disease) may occur when a relatively small number (8,000–50,000) of spores are inhaled and germinate in the lungs. Antibiotic treatment must commence before a critical concentration of toxin accumulates; otherwise, mortality may exceed 99%.

Control and Treatment Anthrax is rare in the United States, typically less than 1 case per year, primarily due to vaccination of animals such as cattle and regulations surrounding the importation of animal hides. A vaccine also exists for military personnel and those in high-risk occupations.

workers handling imported animal hair or hide or products made from them. In 2008 a drum maker in London died of inhalational anthrax, contracted while scraping goat hides to be used as drumheads. The hides had been imported from Africa, where anthrax is endemic. Remarkably similar incidents occurred in New York in 2006 and Scotland in 2005. Because of effective control procedures, the number of cases in the United States is extremely low (fewer than 10 a year).

The circumstances of human infection depend upon the portal of entry. The most common and least dangerous of all forms is **cutaneous anthrax,** caused by spores entering the skin through small cuts and abrasions. Germination and growth of the pathogen in the skin are marked by the production of a papule that becomes increasingly necrotic and later ruptures to form a painless, black **eschar*** **(figure 19.2).**

A far more destructive infection is **pulmonary anthrax** (woolsorter's disease) associated with the inhalation of airborne spores, either from animal products or from contaminated soil. The infectious dose is relatively small—8,000 to 50,000 spores. Upon reaching the lungs, the spores are phagocytosed and transported to lymph nodes, where they germinate and secrete exotoxins that enter the circulatory system. The toxins attach to membranes of macrophages and gain entry into these cells through engulfment. The toxins are highly lethal, causing massive macrophage death and release of chemical mediators. The pathologic effects of these events are wide-ranging: capillary thrombosis, cardiovascular shock, and rapid death in over 99% of untreated cases. The gastrointestinal form of infection is a rare (1 U.S. case in the last 70 years) but deadly condition that follows a course similar to pulmonary anthrax but is acquired through the ingestion of contaminated meat.

The terrorist attacks of 2001 focused a great deal of attention on the threat of bioterrorism, but it was hardly the first time the

Figure 19.2 Cutaneous anthrax. In early stages, the tissue around the site of invasion is inflamed and edematous, later developing into a thick, necrotic lesion called the eschar. It usually separates and heals spontaneously. *Source:* CDC

virulent nature of anthrax had been considered as a weapon of war **(19.1 Making Connections).**

Methods of Anthrax Control Active cases of anthrax are treated with clindamycin, doxycycline, or ciprofloxacin. Because drug therapy targeting the bacteria itself does nothing to lessen the effects of toxemia, people can still die from it. Antibiotics are generally given along with a second drug, Raxibacumab, which uses monoclonal antibodies to bind to one of the toxins secreted by *B. anthracis,* preventing it from entering cells and dramatically

* *eschar* (ess's-kar) Gr. *eschara,* scab.

19.1 MAKING CONNECTIONS

A Bacillus That Could Kill Us

Mechanisms of biological warfare are frequently based on the pathogenic potential of microorganisms. Because spore-forming bacteria such as *Bacillus* are so hardy, they have held a particular attraction to scientists as research subjects and, in one case, as the basis of a deadly bomb.

During World War II, the mounting outrage over German bombings of London forced the British and American governments to collaborate on a powerful means of striking back. Given the knowledge that pneumonic anthrax was rapidly fatal in most cases, a highly secret council designed a bomb containing the spores of *Bacillus anthracis*. In the original plan, the detonation of these deadly projectiles over six German cities would have caused a shower of spores to be inhaled by humans and animals. Some estimates predicted that this could have wiped out the population over hundreds of square miles. In 1943 the British government conducted a bomb test on Gruinard, a small island near Scotland, using sheep as test subjects. Late in the war, a plant in Indiana actually began to manufacture the bombs, but by this time the war had started to wind down. An unfortunate long-term result was the creation of a dangerous source of the infectious agent on "anthrax island." This island remained uninhabitable until its decontamination in 1986.

During the cold war, the U.S. Army's Special Operations Division conducted a chilling experiment using *Bacillus subtilis*. The plan was to simulate a possible method of germ warfare by releasing spore aerosols into the air of a crowded Washington, DC, airport and bus terminal. Agents carried suitcases to spray the air, while hundreds of unsuspecting people inhaled "large and acceptably uniform" doses, presumably with no ill effects.

This study and several others like it were based on the disturbing assumption that *B. subtilis* is a harmless air contaminant. It should be clear even to beginning students that many species of bacteria are opportunists and are capable of causing disease, especially if present in large numbers. In fact, *B. subtilis* can cause lung and blood infections in people with weakened immunity. The tests were able to determine that if smallpox virus had been used instead of *B. subtilis,* 300 people in 93 cities would have become infected. Similar exposures at terminals in several other large cities would have spread the disease rapidly to thousands of others. It was also shown that this contamination procedure could be accomplished without detection.

The potential threat from biological warfare is not only a concern of the past. Recent reports indicate that rogue states and terrorist organizations may have access to germ agents, including anthrax spores and botulinum toxin (another product of a gram-positive sporeformer). The fear that these agents could be incorporated into bombs or missiles to be

Anthrax island—once a forbidden zone
© Stan Pritchard/Alamy Stock Photo

spread over long distances has spurred international monitoring for potential sources of deadly agents. All U.S. military personnel are vaccinated to protect them against the use of biowarfare agents. The anthrax attacks of late 2001 and the continuing tensions abroad underscore our need to be educated on the potential dangers.

In late 2001, only one week after the terrorist attack on the World Trade Center, anthrax-tainted letters were mailed to several media outlets and two senators, resulting in a total of 22 cases of anthrax and five deaths. The media raised loud alarms about the these anthrax attacks, but in perspective more than 10 times this many people are killed annually by lightning strikes. So in reality, this attack, while certainly tragic, was not nearly the national threat that was feared. Epidemiologists were quick to trace the route of the envelopes implicated in the attack and to prescribe prophylactic antibiotics (ciprofloxacin or doxycycline) to more than 32,000 people who may have come in contact with the letters. If the attack had consisted of multiple pathogens, been initiated in more sites, involved multiple modes of transmission, or employed antibiotic-resistant agents, the effects could have been considerably worse. Such a drastic bioterrorism attack would have tested the capacity of our health care system, our ability to manufacture large quantities of drugs, even our food distribution networks. All of our safety nets would have been stretched, perhaps to the point of breaking.

How many years elapsed before anthrax bacteria were eliminated from the contaminated island? Explain why it could persist there for so long. *Answer available on Connect.*

decreasing the severity of the disease. A vaccine containing live spores and a toxoid prepared from a special strain of *B. anthracis* are used to protect livestock in areas of the world where anthrax is endemic. A vaccine, Biothrax, based on a purified toxoid, is recommended for people in high-risk occupations and military personnel. Effective vaccination requires five inoculations given over 18 months, with yearly boosters. The potential side effects reported for this vaccine have spurred research and development of alternative vaccines. Animals that have died from anthrax must be burned or chemically decontaminated before burial to prevent establishing the microbe in the soil; and imported items containing animal hides, hair, and bone should be gas-sterilized.

Other Bacillus *Species Involved in Human Disease*

Bacillus cereus is a common airborne and dust-borne contaminant that multiplies very readily in cooked foods such as rice, potato, and meat dishes. The spores survive short periods of cooking and reheating; when the food is stored at room temperature, the spores germinate and release enterotoxins. Ingestion of toxin-containing food causes nausea, vomiting, abdominal cramps, and diarrhea. There is no specific treatment, and the symptoms usually disappear within 24 hours.

For many years, most common airborne *Bacillus* species were dismissed as harmless contaminants with weak to nonexistent pathogenicity. However, infections by these species are increasingly

TABLE 19.2	Differentiation of Important Spore-Forming Species			
Species	**Oxygen Requirements**	**Motility**	**Disease in Humans**	**Treatment**
Bacillus anthracis	Aerobe	–	Cutaneous anthrax Pulmonary anthrax Gastrointestinal anthrax	Antibiotics Vaccines for high risk Antibiotics
Bacillus cereus	Facultative anaerobe	+	Food poisoning	None; disease self-limiting
Clostridium perfringens	Strict anaerobe	–	Gas gangrene Food poisoning (mild)	Debridement; antibiotics; oxygen therapy None; disease self-limiting
Clostridium difficile	Strict anaerobe	+/–	Antibiotic-associated colitis	Withdrawal of antibiotics; administration of probiotics; fecal microbiota transplant
Clostridium tetani	Strict anaerobe	+	Tetanus	Vaccination; passive immunization
Clostridium botulinum	Strict anaerobe	+/–	Botulism	Antitoxin

reported in immunosuppressed and intubated patients and in drug addicts who do not use sterile needles and syringes. Two important contributing factors are the abundance of spores in the environment and how ineffective the usual methods of disinfection and antisepsis are when it comes to spores.

The Genus *Clostridium*

Another genus of gram-positive, spore-forming rods that is widely distributed in nature is ***Clostridium.**** It is differentiated from *Bacillus* on the basis of being anaerobic and catalase-negative (table 19.2). The large genus (over 120 species) resides in a variety of habitats. Saprobic members are frequently found in soil, sewage, vegetation, and organic debris; and commensals inhabit the bodies of humans and other animals. Infections caused by pathogenic species are not normally communicable but occur when spores are introduced into injured skin.

Clostridia produce oval or spherical spores that often swell the vegetative cell (see figure 19.1*b*). Spores are produced only under anaerobic conditions. Their nutrient requirements are complex, and they can decompose a variety of substrates. They can also synthesize organic acids, alcohols, and other solvents through fermentation. This capacity makes some clostridial species essential tools of the biotechnology industry (see chapter 27). Other extracellular products, primarily **exotoxins,** play an important role in various clostridial diseases such as botulism and tetanus.

The Role of Clostridia in Infection and Disease

Clostridial disease can be divided into (1) wound and tissue infections, including myonecrosis, antibiotic-associated colitis, and tetanus; and (2) food intoxication of the perfringens and botulism varieties. Most of these diseases are caused by the release of potent exotoxins that act on specific cellular targets. In fact, the exotoxins responsible for these diseases are among the most poisonous substances on earth, being millions of times more toxic than strychnine or arsenic.

Gas Gangrene/Myonecrosis

The majority of clostridial soft tissue and wound infections are caused by ***Clostridium perfringens,*** *C. novyi,* and *C. septicum.* The spores of these species can be found in soil, on human skin, and in the human intestine and vagina. The disease they cause has the common name **gas gangrene*** in reference to the gas produced by the bacteria growing in the tissue. It is technically termed anaerobic cellulitis or **myonecrosis.*** The conditions that predispose a person to gangrene are surgical incisions, compound fractures, diabetic ulcers, septic abortions, puncture and gunshot wounds, and crushing injuries contaminated by spores from the body or the environment.

Because clostridia are not highly invasive, infection requires damaged or dead tissue that supplies growth factors and an anaerobic environment. The low oxygen tension results from an interrupted blood supply and the growth of aerobic contaminants that deplete oxygen. Due to this interaction, gas gangrene is considered a type of mixed infection. These conditions stimulate spore germination, rapid vegetative growth in the dead tissue, and release of exotoxins. *Clostridium perfringens* produces several physiologically active toxins; the most potent one, *alpha-toxin* (lecithinase), causes red blood cell rupture, edema, and tissue destruction (**figure 19.3**). Additional virulence factors that enhance tissue destruction are collagenase, hyaluronidase, and DNase. The gas formed in tissues, due to fermentation of muscle carbohydrates, can also destroy muscle structure.

Extent and Symptoms of Infection Two forms of gas gangrene have been identified. In anaerobic cellulitis, the bacteria spread within damaged necrotic muscle tissue, producing toxins and gas, but the infection remains localized and does not spread into healthy tissue. The pathology of true myonecrosis is more destructive and mimics some aspects of necrotizing fasciitis. Toxins produced in large muscles, such as the thigh, shoulder, and buttocks, diffuse into nearby healthy tissue and cause local necrosis there. This damaged tissue then supports continued clostridial growth, toxin formation, and gas production. The disease can progress through an entire

* *Clostridium* (klaws-trid′s-ee-um) Gr. *closter,* spindle.

* *gangrene* (gang′s-green) Gr. *gangraina,* an eating sore. A necrotic condition associated with the release of gases.
* *myonecrosis* (my″-oh-neh-kro′ssis) Gr. *myo,* muscle, and *necros,* dead.

Figure 19.3 Growth of *Clostridium perfringens* (plump rods), causing gas formation and separation of the fibers. Histological cross section of skeletal muscle fibers displaying clostridial myonecrosis. Courtesy of Franciso Uzal

Figure 19.4 The clinical appearance of myonecrosis. This patient developed myonecrosis after injecting cocaine. Note the severe degree of involvement, with blackening, general tissue destruction and bubbles on the skin caused by gas formation in underlying tissue. *Clostridium septicum* was isolated from the wound. Image reprinted with permission from Medscape Reference (http://emedicine.medscape.com/), 2013, available at: http://emedicine.medscape.com/article/217943-overview.

limb or body area, destroying tissues as it goes **(figure 19.4)**. Initial symptoms of pain, edema, and a bloody exudate in the lesion are followed by fever, tachycardia, and blackened necrotic tissue filled with bubbles of gas. Gangrenous infections of the uterus due to septic abortions and clostridial septicemia are particularly serious complications. If treatment is not initiated early, the disease is invariably fatal.

Treatment and Prevention of Gangrene One of the most effective ways to prevent clostridial wound infections is immediate and rigorous cleansing and surgical repair of deep wounds, pressure sores (bedsores), compound fractures, and infected incisions. **Debridement*** of diseased tissue eliminates the conditions that promote the spread of gangrenous infection. This is most difficult in the intestine or body cavity, where only limited amounts of tissue can be removed. Surgery is accompanied by antibiotic therapy to

Figure 19.5 Hyperbaric treatment for wound infections and diabetic ulcers. This therapy involves holding patients in a chamber that exposes them to higher levels of oxygen to inhibit anaerobic infection and to promote healing. © Charlie Neuman/San Diego Union-Tribune/Zuma Press/Corbis

control the infection. The preferred treatment is clindamycin, supplemented with penicillin for the 5% of clostridial species that are clindamycin-resistant. Hyperbaric oxygen therapy, in which the affected part is exposed to an increased oxygen tension in a pressurized chamber, can also lessen the severity of infection **(figure 19.5)**. The increased oxygen content of the tissues blocks further bacterial multiplication and toxin production, while simultaneously promoting healing. Extensive myonecrosis of a limb may call for surgical removal, or amputation. Because there are so many different antigen subtypes in this group, active immunization is not possible.

Tetanus, or Lockjaw

Tetanus* is a neuromuscular disease whose alternate name, lockjaw, refers to an early effect of the disease on the jaw muscle. The etiologic agent, ***Clostridium tetani,*** is a common resident of cultivated soil and the gastrointestinal tracts of animals. Spores usually enter the body through accidental puncture wounds, burns, the umbilicus, frostbite, and crushed body parts.

The incidence of tetanus is low in North America. Most cases occur among geriatric patients and intravenous drug abusers. The incidence of neonatal tetanus—predominantly the result of an infected umbilicus or circumcision—is higher in cultures that apply dung or mud to these sites to arrest bleeding or as a customary ritual. The disease accounts for several hundred thousand infant deaths a year in developing countries.

The Course of Infection and Disease The risk for tetanus occurs when spores of *Clostridium tetani* are forced into injured tissue. But the mere presence of spores in a wound is not sufficient to initiate infection because the bacterium is unable to invade damaged tissues readily. It is also a strict anaerobe, and the spores cannot become established unless tissues at the site of the wound are necrotic and poorly supplied with blood, conditions that favor germination.

As the vegetative cells multiply, various metabolic products are released into the infection site. Of these, the most serious is **tetanospasmin,** a potent neurotoxin that accounts for the major symptoms

* *debridement* (dih-breed′-ment) Surgical removal of dead or damaged tissue.

* *tetanus* (tet′s-ah-nus) Gr. *tetanos,* to stretch.

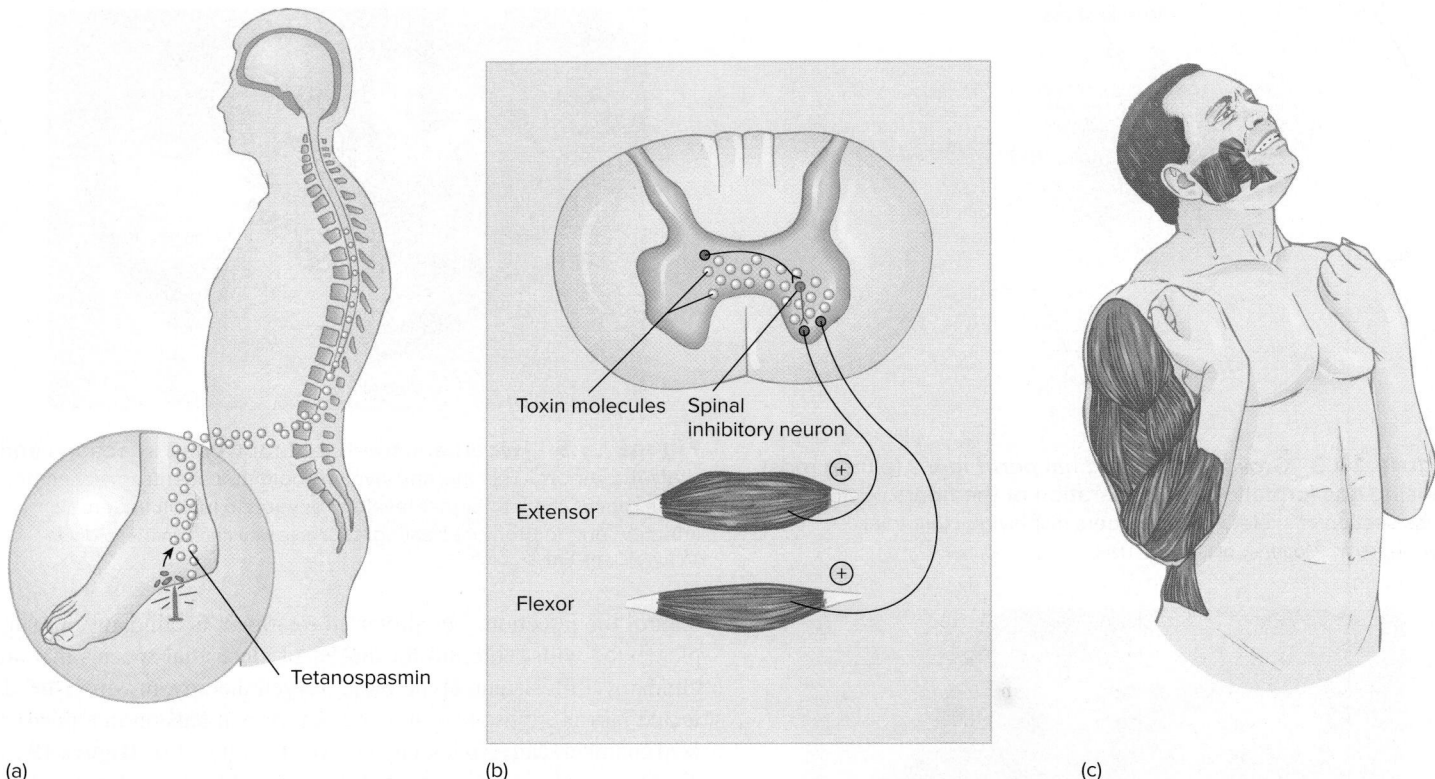

(a) (b) (c)

Figure 19.6 **The events in tetanus.** **(a)** After traumatic injury, bacilli infecting the local tissues secrete tetanospasmin, which is absorbed by the peripheral axons and is carried to the target neurons in the spinal column (not to scale). **(b)** In the spinal cord, the toxin attaches to the junctions of regulatory neurons that inhibit inappropriate contraction. Released from inhibition, the muscles, even opposing members of a muscle group, receive constant stimuli and contract uncontrollably. **(c)** Muscles contract spasmodically, without regard to regulatory mechanisms or conscious control. Note the clenched jaw typical of *risus sardonicus*.

of tetanus. The toxin spreads to nearby motor nerve endings in the injured tissue, binds to them, and travels by axons to the ventral horns of the spinal cord **(figure 19.6).** In the spinal column, the toxin binds to specific target sites on the spinal neurons that are responsible for inhibiting skeletal muscle contraction. The toxin inhibits the release of neurotransmitter, and only a small amount is required to initiate the symptoms. The incubation period varies from 4 to 10 days, and shorter incubation periods signify a more serious condition.

Tetanospasmin alters the usual regulatory mechanisms for muscle contraction. As a result, the muscles are released from normal inhibition and begin to contract uncontrollably. Powerful muscle groups are most affected, and the first symptoms are clenching of the jaw, followed in succession by extreme arching of the back, flexion of the arms, and extension of the legs **(figure 19.7).** Lockjaw confers the bizarre appearance of *risus sardonicus* (sarcastic grin) that looks eerily as though the person is smiling (see figure 19.6*c*). These contractions are intermittent and extremely painful and may be forceful enough to break bones, especially the vertebrae. Death is most often due to paralysis of the respiratory muscles and respiratory failure. The fatality rate, ranging from 10% to 70%, is highest in cases involving delayed medical attention, a short incubation time, or head wounds. Full recovery requires a few weeks, and other than transient stiffness, no permanent damage to the muscles usually remains.

Treatment and Prevention of Tetanus Tetanus treatment is aimed at reducing the level of toxemia and providing supportive care for the patient. A patient with a clinical appearance suggestive of

Figure 19.7 **Neonatal tetanus.** Baby with neonatal tetanus, showing spastic paralysis of the paravertebral muscles, which locks the back into a rigid, arched position. Also note the abnormal flexion of the arms and legs. Courtesy of Dr. T.F. Sellers

tetanus should immediately receive antitoxin therapy with human tetanus immune globulin (TIG). Although the antitoxin inactivates circulating toxin, it will not counteract the effects of toxin already bound to neurons. Other methods include thoroughly cleansing and removing the afflicted tissue, controlling infection with metronidazole or penicillin, and administering muscle relaxants. The patient may require a respirator, and a tracheostomy[1] is sometimes performed to prevent complications such as aspiration pneumonia or lung collapse.

1. The surgical formation of an air passage by perforation of the trachea.

Pathogen Profile #2 *Clostridium perfringens, Clostridium tetani*

Clostridium perfringens

Clostridium tetani

Source: CDC

Microscopic Morphology Gram-positive bacilli, present singly or in pairs. Endospores distend the cell, altering its shape. In *C. tetani,* spores are located terminally, whereas in *C. perfringens,* spores are central to terminal in location.

Identified by Gram reaction and endospore formation. *Clostridium* is differentiated from *Bacillus* as the former is typically a strict anaerobe and the latter is not. A variety of physiological, biochemical, and cultural tests are used to distinguish individual species. Many laboratories employ ELISA, PCR, or toxicity testing as a means of differentiation.

Habitat Endospores of both species of bacteria may be found in soil. In addition, *C. perfringens* may be found in the human intestine and vagina, whereas *C. tetani* is found in the gastrointestinal tract of animals.

Virulence Factors Clostridial species produce some of the most potent exotoxins ever encountered, and these are responsible for most of each species' pathological effects. *C. perfringens* produces several enzymes that cause tissue destruction, including collagenase, hyaluronidase, and DNase. The species' most potent toxin, alpha-toxin

(lecithinase), causes red blood cell rupture, edema, and tissue damage. The primary virulence factor in *C. tetani* is tetanospasmin, a neurotoxin that accounts for the major symptoms of tetanus.

Primary Infections/Disease *C. tetani* is responsible for tetanus. The primary disease associated with *C. perfringens* is variously known as myonecrosis, gas gangrene, or anaerobic cellulitis. The bacterium is also responsible for clostridial gastroenteritis, a mild and widespread form of food poisoning.

Control and Treatment Gas gangrene is prevented through thorough cleansing of deep wounds and surgical debridement of diseased tissue to eliminate anaerobic environments. Vaccination with tetanus toxoid is the primary means of preventing tetanus. Human tetanus immune globulin (TIG) can alleviate symptoms of the disease by binding to toxin that is already in the body but has yet to bind to neurons, while the infection itself is treated with metronidazole or penicillin. Supportive care may include the use of assisted ventilation and muscle relaxants.

Tetanus is one of the world's most preventable diseases, chiefly because of effective vaccines (DTaP, Td) containing tetanus toxoid. During World War II, only 12 cases of tetanus occurred among 2,750,000 wounded soldiers who had been previously vaccinated. The recommended vaccination series for babies consists of three injections given 2 months apart, followed by booster doses about 1 and 4 years later. Childhood immunization provides protection for 10 years. Additional protection against neonatal tetanus may be achieved by vaccinating women prior to pregnancy so they will have antibodies to pass on to a fetus. Toxoid should also be given to injured persons who have never been immunized, have not completed the series, or whose last booster was received more than 10 years previously. The vaccine can be given simultaneously with passive TIG immunization to provide immediate and long-term protection.

Clostridium difficile *Infection*

A clostridial disease most commonly called **C-diff infection,** or **CDI,** is the second most common intestinal disease after salmonellosis in industrialized countries. It is caused by *Clostridium difficile,* a normal resident of the intestine that is usually present in low numbers. Most instances of this infection are traced to therapy with broad-spectrum drugs such as clindamycin, ceftriaxone, or ciprofloxacin (which is the origin for the older name of the disease, *antibiotic-associated* colitis). It is a major cause of diarrhea in hospitals.

Over the last several years, *Clostridium difficile* infections have become both more common and more severe. In 2013 the CDC estimated 250,000 infections and 14,000 deaths from *C. difficile,* statistics that surpass even those of MRSA infections. Many cases today involve a strain that produces about 20 times the usual amount of exotoxins. This strain is far more likely to be found

outside the hospital and is not always associated with antimicrobic therapy. The use of gastric acid inhibitors such as famotidine (Pepcid) and omeprazole (Prilosec) increase prevalence of the disease, as less gastric acidity may improve survival of the pathogen in the stomach and intestine.

Although *C. difficile* is relatively noninvasive, it can superinfect the colon when the normal microbiota have been disrupted. It produces enterotoxins that trigger necrosis in the wall of the intestine. The predominant symptom is diarrhea commencing late in therapy or even after therapy has stopped. More severe cases exhibit abdominal cramps, fever, and leukocytosis. The colon is inflamed and gradually sloughs off loose, membrane-like patches called *pseudomembranes* consisting of fibrin and cells **(figure 19.8).** If the condition is not arrested, cecal perforation and death can result.

(a)

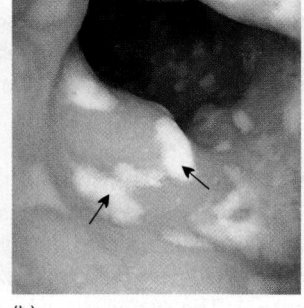
(b)

Figure 19.8 **Antibiotic-associated colitis as imaged by a sigmoidoscope during colonoscopy. (a)** View of a normal colon. **(b)** Arrows indicate white patches or pseudomembranes that are caused by inflammation and desquamation of epithelial cells that line the colon.
(a): © Gastrolab/Science Source; (b): © Dr. Paul L. Beck, University of Calgary

Pathogen Profile #3 *Clostridium difficile*

Source: CDC

Microscopic Morphology Gram-positive bacilli, present singly or in short chains. Endospores are subterminal and distend the cell, altering its shape.

Identified by Gram reaction and endospore formation. *Clostridium* is differentiated from *Bacillus* as the former is typically a strict anaerobe and the latter is not. ELISA is often used to detect toxins of *C. difficile* in fecal samples.

Habitat Found in small numbers as part of the normal microbiota of the intestine.

Virulence Factors Enterotoxins that cause epithelial necrosis of the colon.

Primary Infections/Disease *Clostridium difficile* infection (CDI) refers to disease caused by the overgrowth of *C. difficile*. Symptoms may range from diarrhea to inflammation of the colon, cecal perforation, and, rarely, death. Although *C. difficile* is ordinarily present in low numbers, treatment with broad-spectrum antibiotics may disrupt the normal microbiota of the colon, leading to a *C. difficile* superinfection.

Control and Treatment Mild cases generally respond to withdrawal of the antibiotic. Severe cases are treated with oral vancomycin or metronidazole, along with probiotics or fecal microbiota transplants to restore the normal microbiota.

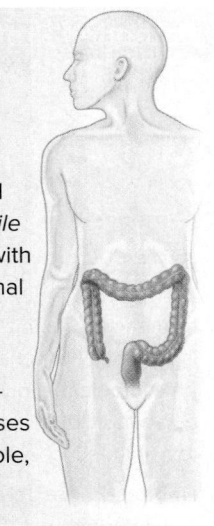

Mild, uncomplicated cases respond to withdrawal of antibiotics and replacement therapy for lost fluids and electrolytes. More severe infections are treated with oral vancomycin or metronidazole and replacement cultures to restore normal microbiota. Fecal microbiota transplantation, a procedure in which feces from a healthy donor are transferred via enema, swallowed capsules, or colonoscopy directly to the colon of patients suffering *C. difficile* infection, has gained prominence—though not FDA approval—in recent years. Anecdotal evidence suggests that the technique produces a rapid cure in 80% to 90% of patients, as the transplanted microbiota quickly establishes itself throughout the colon. Because of the potential risks and ethical questions involved in purposefully inoculating a patient with another person's microbiota, federal agencies have been hesitant to oversee the process. In 2016 the FDA began to increase oversight of fecal microbiota transplants, tightening the regulations governing the procedure.

In the clinical setting, stringent precautions are necessary to prevent the spread of the agent from infected persons, who shed large numbers of spores in their feces, to other patients who may be on antimicrobic therapy. Immediate diagnosis is obtained by a rapid ELISA that detects toxins in fecal samples. Work is progressing on a vaccine containing *C. difficile* toxoid for use in high-risk populations.

Clostridial Food Poisoning

Two *Clostridium* species are involved in food poisoning. *Clostridium botulinum* produces a severe intoxication, usually from home-canned food. *Clostridium perfringens,* type A, accounts for a mild intestinal illness that is one of the most common forms of food poisoning worldwide.

Epidemiology of Botulinum Food Poisoning *Clostridium botulinum* is a spore-forming anaerobe that commonly inhabits soil and water and, occasionally, the intestinal tract of animals. It is distributed worldwide but occurs most often in the Northern Hemisphere. The species has eight distinctly different types (designated A, B, C$_\alpha$ and C$_\beta$, D, E, F, and G), which vary in distribution among animals, regions of the world, and type of exotoxin produced. Human disease is usually associated with types A, B, E, and F, and animal disease with types A, B, C, D, and E.

Botulism* is an intoxication usually associated with eating improperly canned or poorly preserved foods, though it can occur as a result of infection. Until recent times it was relatively prevalent and commonly fatal, but modern techniques of food preservation and medical treatment have reduced both its incidence and its fatality rate. However, botulism is a common cause of death in livestock that have grazed on contaminated food and in aquatic birds that have eaten decayed vegetation.

There is a high correlation between cultural dietary preferences and food-borne botulism. In the United States, the disease is often associated with low-acid vegetables (green beans, corn) and, occasionally, meats, fish, and dairy products. Most botulism outbreaks occur in home-processed foods, including canned vegetables, smoked meats, and cheese spreads. The demand for prepackaged convenience foods such as vacuum-packed cooked vegetables and meats has created a new risk, but commercially canned foods are only rarely a source of botulism.

One of the most recent outbreaks of botulism traced to a commercial source occurred in 2007 among people who had consumed canned chili sauce. The cans were traced to a food plant that apparently had defects in its sterilization procedures. Due to the potential for additional cases, 90 other food products processed by that plant were recalled and millions of cans removed from stores. Most foods contained peppers, potatoes, or meats. This outbreak was one of the first incidents of commercial botulism in more than a decade, and it emphasized the reality that preventing botulism must always remain a priority for canneries. The constant presence of *C. botulinum* spores in the soil and on produce means that there is zero tolerance for errors in quality control.

Pathogenesis of Botulism The factors in food processing that lead to botulism depend upon several circumstances. Spores are present on the vegetables or meat at the time of gathering and are difficult to remove by washing alone. When contaminated food is bottled and steamed in a pressure cooker that does not reach reliable pressure and temperature, some spores survive (botulinum spores are highly heat-resistant). At the same time, the pressure is sufficient

* *botulism* (boch's-oo-lizm) L. *botulis,* sausage. The disease was originally linked to spoiled sausage.

to evacuate the air and create anaerobic conditions ideal for spore germination and vegetative growth. One of the products of metabolism is **botulinum toxin,** the most potent microbial toxin known.

Bacterial growth may not be evident in the appearance of the bottle or can or in the food's taste or texture, and only minute amounts of toxin may be present. Swallowed toxin enters the small intestine and is absorbed into the lymphatics and circulation. From there, it travels to its principal site of action, the neuromuscular junctions of skeletal muscles **(figure 19.9).**

Botulinum toxin prevents the release of the neurotransmitter acetylcholine, that initiates the signal for muscle contraction. The usual time before onset of symptoms is 12 to 72 hours, depending on the size of the dose. Neuromuscular symptoms first affect the muscles of the head and include double vision, difficulty in swallowing,

> **Quick Search**
> To read about an unusual case of botulism involving prison-brewed alcohol, review the chapter 8 Case Study, or search the Internet for "Utah botulism."

and dizziness, but there is no sensory or mental lapse. Although nausea and vomiting can occur at an early stage, they are not common. Later symptoms are descending muscular paralysis and respiratory failure. In the past, death resulted from stoppage of respiration, but mechanical respirators have reduced the fatality rate to about 10%. A fascinating medical use for targeting unwanted muscle contractions with the toxin is discussed in **19.2 Making Connections.**

Infant and Wound Botulism Food botulism is a result of ingesting botulism toxin already present in food. By contrast, in infant and wound botulism, *C. botulinum* causes infections when endospores germinate in the body and produce the toxin in vivo.

Infant botulism was first described in the late 1970s in children between the ages of 2 weeks and 6 months who had ingested spores. It is currently the most common type of botulism in the United States, with approximately 80 to 100 cases reported annually. The exact food source is not always known, although raw honey and homemade baby food have been implicated in some cases, while others have been attributed to ingesting small amounts of dust or dirt. Apparently the immature state of the neonatal intestine and resident microbiota allows the spores to gain a foothold, germinate, and produce neurotoxin. As in adults, babies exhibit flaccid paralysis, usually with a weak sucking response, generalized loss of tone ("floppy baby syndrome"), and respiratory complications. Although adults can also ingest botulinum spores in contaminated vegetables and other foods, the adult intestinal tract normally inhibits this sort of infection.

In wound botulism, the spores enter a wound or puncture much as in tetanus, but the symptoms are similar to those of food-borne botulism. Increased cases of this form of botulism are being reported in injecting drug users. The rate of infection is highest in people who inject black tar heroin into the skin.

Treatment and Prevention of Botulism Differentiating botulism from other neuromuscular conditions usually involves testing food samples, intestinal contents, and feces for evidence of toxin or *C. botulinum.* Treatment involves administration of botulinum antitoxin, supplied by the CDC, and which must be administered early for greatest effectiveness. The antitoxin does not reverse the paralysis but does stop its progression and can dramatically shorten the course of the disease. Patients are also managed with intensive respiratory and cardiac support care. Infectious botulism is treated with penicillin to control the microbe's growth and toxin production.

Botulism will always be a potential threat for people who consume home-preserved foods. Preventing it depends on educating the public about the proper methods of preserving and handling canned foods. Pressure cookers should be tested for accuracy in sterilizing, and home canners should be aware of the types of food and conditions likely to cause botulism (see food microbiology in chapter 27). Other effective preventives include addition of preservatives such as sodium nitrite, salt, or vinegar (which lowers pH). Bulging cans or bottles that look or smell spoiled should be discarded, and all home-bottled foods should be boiled for 10 minutes before eating, because the toxin is heat-sensitive and is rapidly inactivated at 100°C.

Clostridial Gastroenteritis *Clostridium perfringens* spores contaminate many kinds of food, but those most frequently involved in disease are animal flesh (meat, fish) and vegetables (beans) that

Figure 19.9 The pathophysiological effects of botulinum toxin. (a) The relationship between the motor neuron and the muscle at the neuromuscular junction. **(b)** In the normal state, acetylcholine released at the synapse crosses to the muscle and creates an impulse that stimulates muscle contraction. **(c)** In botulism, the toxin enters the motor end plate and attaches to the presynaptic membrane, where it blocks release of the neurotransmitter, prevents impulse transmission, and keeps the muscle from contracting. This causes flaccid paralysis.

Motor neuron end plate

Muscle cell membrane

Synapse

(a)

Vacuole

Acetylcholine

(b)

Presynaptic membrane

Botulinum toxin

(c)

19.2 MAKING CONNECTIONS

Botox: No Wrinkles. No Headaches. No Worries?

In 2015, people paid more than $2 billion to have one of the most potent toxins on earth injected into their face. The toxin, of course, is botulinum toxin type A, or **Botox,** a medical form of botulism toxin. The story of how these injections came to be the most common cosmetic procedure in the United States is a welcome respite from talk of flesh-eating bacteria and germ warfare.

It has long been known that death from *Clostridium botulinum* intoxication is due to paralysis of the respiratory muscles. In fact, it was even known that botulinum toxin causes death by interfering with the release of acetylcholine, a neurotransmitter that causes the contraction of skeletal muscles. The trick was finding a safe application for this remarkable action.

In 1989 Botox was first approved to treat cross-eyes and uncontrollable blinking, two conditions resulting from the inappropriate contracting of muscles around the eye. Success soon led to Botox treatment for a variety of disorders that cause painful contraction of neck and shoulder muscles. A much wider use of Botox was "off-label"—doctors found that injecting facial muscles with the toxin relaxed these muscles and reduced wrinkling of the overlying skin. The "lunch hour facelift" became so common that it won official FDA approval in 2002. In a surprise twist, patients undergoing Botox treatment for wrinkles reported fewer headaches, especially migraines. After clinical trials found this result to be widespread and reproducible, Botox injections were approved for treating chronic migraines and are now a routine part of therapy.

Is Botox a miracle drug? Not so fast. In the rush to embrace the admittedly dramatic results of Botox injections, most patients have paid scant

Before

After

Effects of Botox. By preventing the transmission of nerve impulses to muscles, botulism toxin reduces the appearance of wrinkles, as seen here in the lower photograph. Excessive amounts of the toxin or poor positioning of the injections can lead to an inability to raise the eyebrows or eyelids. (both): Courtesy of M. Kayser, MD, F.A.C.S.

attention to potential problems that can arise from the use of a potent paralytic agent. The dominant side effect of Botox treatment is excessive paralysis of facial muscles resulting from poorly targeted injections. Other results include drooping eyelids, facial paralysis, slurred speech, and drooling. Even if the treatment works perfectly, patients are generally unable to move their eyebrows; some become unable to frown or squint. Although most experts agree that Botox is relatively safe, some rare cases of severe reactions and deaths have been reported. Reacting to a spate of adverse events, the FDA in 2009 required all packages of Botox to display a "black box warning"—one of the strongest safety actions the agency can take—informing doctors and consumers of dangers associated with the toxin. This action came as a result of more than 200 incidents in which the toxin spread from the initial injection site and led to symptoms of botulism poisoning. In most of these cases the toxin was being used for unapproved therapies, such as controllimg spastic limbs in children suffering from cerebral palsy.

Botox effects are not permanent. As the toxin wears off, the wrinkles (or headaches, as the case may be) return, and every 4 to 6 months the treatment must be repeated. This last fact has been welcomed by plastic surgeons because a permanent solution to wrinkles would rule out any repeat clientele. Malpractice lawsuits are also reduced, because by the time a patient has consulted a lawyer, any consequences of a botched treatment have usually worn off.

Explain why the effects of Botox are short term. *Answer available on Connect.*

have not been cooked thoroughly enough to destroy the spores. When these foods are cooled, spores germinate and the germinated cells multiply, especially if the food is left unrefrigerated. When the food is eaten without adequate reheating, live *C. perfringens* cells enter the small intestine and release enterotoxin. The toxin, acting upon epithelial cells, initiates acute abdominal pain, diarrhea, and nausea in 8 to 16 hours. Recovery is rapid, and deaths are extremely rare. *Clostridium perfringens* also causes enterocolitis similar to that caused by *C. difficile*. This infectious type of diarrhea is acquired from contaminated food, or it may be transmissible by inanimate objects.

Differential Diagnosis of Clostridial Species

Although clostridia are common isolates, their clinical significance is not always immediately evident. Diagnosis frequently depends on the microbial load, the persistence of the isolate on resampling, and the condition of the patient. Laboratory differentiation relies on testing morphological and cultural characteristics, exoenzymes,

carbohydrate fermentation, reaction in milk, and toxin production and pathogenicity. Some laboratories use a sophisticated method of gas chromatography that analyzes the chemical differences among species. Other valuable procedures are direct ELISA testing of isolates, toxicity testing in mice or guinea pigs, serotyping with antitoxin neutralization tests, and PCR analysis of samples.

 Check Your Progress SECTION 19.2

2. Why are *Bacillus* infections commonly associated with exposure to soil and dust?

3. Compare the three common forms of anthrax, and identify people at increased risk for contracting the disease.

4. What are the common elements of puncture wounds, crush injuries, and compound fractures that encourage the growth of *Clostridium perfringens?* Explain the rationale for debridement and oxygen therapy for such injuries.

5. What is the relationship between the normal microbiota of the large intestine, antibiotic treatment, and the development of *Clostridium difficile*–associated disease?

6. Botulism is commonly associated with ingestion of specific types of food prepared in a particular manner. What is it about both the food and preparation that encourages infection?

7. Compare the effect of tetanospasmin (tetanus toxin) with the effect of botulinum on muscles.

8. In what way is the ingested agent responsible for adult botulism different from that responsible for infant botulism? Why?

19.3 Gram-Positive Regular Non-Spore-Forming Bacilli

 Expected Learning Outcomes

10. Relate the severity of listeriosis with the immunocompetency of its host.

11. Explain why people in certain occupations are at increased risk of becoming infected with *Erysipelothrix rhusiopathiae*.

The non-spore-forming gram-positive bacilli are a mixed group of genera subdivided on the basis of morphology and staining characteristics. One loose aggregate of seven genera is characterized as **regular** because they stain uniformly and do not assume pleomorphic shapes. Regular genera include *Lactobacillus, Listeria, Erysipelothrix, Kurthia, Caryophanon, Brochothrix,* and *Renibacterium.*

The predominant pathogens in this group are *Listeria monocytogenes* and *Erysipelothrix rhusiopathiae. Lactobacillus* is widely distributed in the environment and is a common resident of the intestinal tract and vagina of humans. Some species are also important in processing dairy products. Only rarely are they pathogens.

An Emerging Food-Borne Pathogen: *Listeria monocytogenes*

*Listeria monocytogenes** ranges in morphology from coccobacilli to long filaments in palisades formation. Cells show tumbling motility, with one to four flagella, and do not produce capsules or spores. *Listeria* is not fastidious and is resistant to cold, heat, salt, pH extremes, and bile. The bacterium is an intracellular pathogen during part of its life cycle. It displays a unique multiplication cycle in which it replicates within the cytoplasm of a host cell after inducing its own phagocytosis. By moving directly from cell to cell, *Listeria* is able to avoid the humoral immune system, enhancing its own virulence **(figure 19.10).**

* *Listeria* (lis-ter′s-ee-ah) For Joseph Lister, the English surgeon who pioneered antiseptic surgery; (mah″-noh-sy-toj′s-uh-neez) For its effect on monocytes.

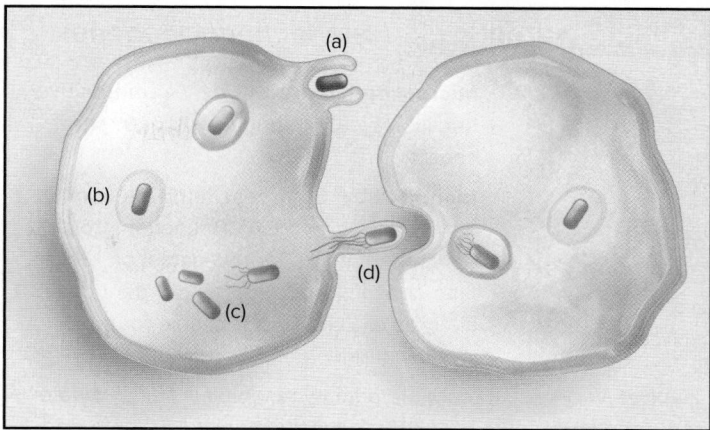

Figure 19.10 Multiplication cycle of *Listeria monocytogenes*. (a) Beginning in the intestinal mucosa, *Listeria monocytogenes* invades cells by inducing its own phagocytosis. **(b)** Once in the host cell, the bacterium lyses the phagosome in which it is enclosed and multiplies rapidly within the cytoplasm. **(c)** Mature bacteria develop long proteinaceous tails derived from the actin of the host cell. These propel them from one host cell to the next and **(d)** facilitate the spread of the pathogen to other cells.

Epidemiology and Pathology of Listeriosis

The distribution of *L. monocytogenes* is so broad that its reservoir has been difficult to determine. It has been isolated all over the world from water, soil, plant materials, and the intestines of healthy mammals (including humans), birds, fish, and invertebrates. Apparently, the primary reservoir is soil and water, whereas animals, plants, and food are secondary sources of infection. Although most cases of **listeriosis** are associated with ingesting contaminated dairy products, poultry, and meat, a highly unusual outbreak in 2011—responsible for 33 deaths and one miscarriage—was linked to contaminated cantaloupes (see this chapter's Case Study). Recent epidemics have spurred an in-depth investigation into the prevalence of *L. monocytogenes* from these sources. The pathogen has been isolated in 12% of ground beef and in 15% of chicken carcasses. It was also present in 6% of luncheon meats, hot dogs, and cheeses. Aged cheeses made from raw milk are of special concern because *Listeria* readily survives through long storage and can grow during refrigeration. Listeriosis is the third most prevalent cause of food-borne fatalities, after salmonellosis and toxoplasmosis.

A predisposing factor in listeriosis seems to be the weakened condition of host defenses in the intestinal mucosa, because studies have shown that immunocompetent individuals are rather resistant to infection. Listeriosis in normal adults is often a mild or subclinical infection with nonspecific symptoms of fever, diarrhea, and sore throat. However, listeriosis in immunocompromised patients, fetuses, and neonates usually affects the brain and meninges and results in septicemia. The death rate is around 20% due to the pathogen's virulence. Pregnant women are especially susceptible to infection, which is transmitted to the infant prenatally when the microbe crosses the placenta or perinatally through the birth canal. Intrauterine infections are systemic and usually result in miscarriage and fetal death. Neonatal infections that localize in the meninges cause extensive damage to the nervous system if not treated at an early stage.

Pathogen Profile #4 *Listeria monocytogenes*

Source: CDC

Microscopic Morphology Gram-positive bacilli, typically found singly or in short chains.

Identified by Gram reaction, type of motility, hemolysis, and CAMP reaction. Cold enrichment helps improve isolation but is time-consuming. Assays dependent on DNA analysis or ELISA are available for rapid identification.

Habitat Very widespread; the primary reservoir is soil and water, and secondary sources of infection include animals, plants, and food. Most cases are associated with contaminated dairy products, poultry, and meat.

Virulence Factors *L. monocytogenes* replicates within the cytoplasm of a host cell after inducing its own phagocytosis. By moving directly from cell to cell, the bacterium avoids the humoral immune system, enhancing its own virulence.

Primary Infections/Disease Listeriosis is often mild or subclinical in immunocompetent adults but may affect the brain and meninges of immunodeficient patients. Pregnant women are highly susceptible to infection and may pass the bacterium to the fetus across the placenta or during birth, commonly resulting in fetal death or severe damage to the nervous system.

Control and Treatment Control of *L. monocytogenes* centers on pasteurization and adequate cooking of foods that may be contaminated with animal feces. Recommended drugs include ampicillin/gentamicin and trimethoprim/sulfamethoxazole.

Diagnosis and Control of Listeriosis

Diagnosing listeriosis can be delayed by difficulties in isolating the pathogen. Isolation can be improved by using a procedure called cold enrichment, in which the specimen is held at 4°C and periodically plated onto media, but this procedure can take 4 weeks. *Listeria monocytogenes* can be differentiated from the nonpathogenic *Listeria* and from other bacteria to which it bears a superficial resemblance by comparing traits such as cell shape, arrangement, motility, production of catalase, and positive CAMP reaction. Rapid diagnostic kits using ELISA, immunofluorescence, and DNA analysis are now available for direct testing of foods.

Antibiotic therapy should be started as soon as listeriosis is suspected. Ampicillin and gentamicin are the first choices, followed by trimethoprim/sulfamethoxazole. Prevention can be improved by adequate pasteurization temperatures and by cooking foods that could be contaminated with animal feces. The U.S. Department of Agriculture implemented regulations designed to prevent the incorporation of *Listeria* into processed meat and poultry products. Under these rules, companies that use less stringent methods of microbial control will be subject to more frequent government testing, while those employing more effective strategies will be subject to fewer inspections.

Erysipelothrix rhusiopathiae: A Zoonotic Pathogen

Epidemiology, Pathogenesis, and Control

*Erysipelothrix rhusiopathiae** is a gram-positive rod widely distributed in animals and the environment. Its primary reservoir appears to be the tonsils of healthy pigs. It is also a normal microbiota of other vertebrates and is commonly isolated from sheep, chickens, and fish. It can persist for long periods in sewage, seawater, soils, and foods. The pathogen causes epidemics of swine erysipelas and sporadic infections in other domestic and wild animals.

Figure 19.11 Erysipeloid on the hand of an animal handler.
© DermNet New Zealand Trust

Humans at greatest risk for infection are those who handle animals, carcasses, and meats, such as slaughterhouse workers, butchers, veterinarians, farmers, and fishermen.

The common portal of entry in human infections is a scratch or abrasion on the hand or arm. The microbe multiplies at the invasion site to produce a disease known as **erysipeloid,** characterized by inflamed, dark red lesions that burn and itch **(figure 19.11).** Although the lesions usually heal without complications, rare cases of septicemia and endocarditis do arise. Inflamed red sores on the hands of people in high-risk occupations suggest erysipeloid, but the lesions must be cultured for confirmatory diagnosis. The condition is treated with penicillin or erythromycin. Swine erysipelas can be prevented by vaccinating pigs, but the vaccine does not protect humans. Animal handlers can lower their risk by wearing protective gloves.

 Check Your Progress SECTION 19.3

9. Compare the effects of listeriosis in healthy adults, pregnant women, fetuses, and neonates.
10. Why do erysipeloids commonly appear on the hands, as opposed to other body parts?

* *Erysipelothrix* (er″-ih-sip′s-eh-loh-thriks) Gr. *erythros*, red, *pella*, skin, and *thrix*, filament. *Rhusiopathiae* (ruz″-ee-oh-path′s-ee-eye) Gr. *rhusios*, red, and *pathos*, disease.

19.4 Gram-Positive Irregular Non-Spore-Forming Bacilli

 Expected Learning Outcomes

12. Relate the stages of diphtherial disease to infection and metabolism of its infectious agent.

13. Outline methods for the detection, prevention, and treatment of diphtheria.

The **irregular** non-spore-forming bacilli tend to be pleomorphic and to stain unevenly. Of the 20 genera in this category, *Corynebacterium, Mycobacterium,* and *Nocardia* have the greatest clinical significance. These three genera are grouped together because of similar morphological, genetic, and biochemical traits. They produce catalase and possess mycolic acids and a unique type of peptidoglycan in the cell wall. The following sections discuss the primary diseases associated with *Corynebacterium,* a similar genus called *Propionibacterium, Mycobacterium, Actinomyces,* and *Nocardia.*

Corynebacterium diphtheriae

Although several species of **Corynebacterium*** are found on the human body, most human disease is associated with **C. diphtheriae.** In general morphology, this bacterium is a straight or somewhat curved rod that tapers at the ends, but it has many pleomorphic variants, including club, filamentous, and swollen shapes. Older cells are filled with metachromatic (polyphosphate) granules and can occur in side-by-side palisades arrangement **(figure 19.12).**

Epidemiology of Diphtheria

For hundreds of years, **diphtheria*** was a significant cause of morbidity and mortality, but for more than five decades, both the number of

** Corynebacterium* (kor-eye″-nee-bakter′s-ee-um) Gr. *koryne,* club, and *bakterion,* little rod.
** diphtheriae* (dif-thee′-ree-ah) Gr. *diphthera,* membrane.

Palisades arrangement Pleomorphism

Figure 19.12 *Corynebacterium diphtheriae.* A photomicrograph of *Corynebacterium diphtheriae,* showing pleomorphism and palisades arrangement (1200×). *Source:* CDC

cases and the fatality rate have steadily declined throughout the world. The current rate for the entire United States is 0.01 cases per million population with the last confirmed case seen in 2003. Recent outbreaks of the cutaneous form of the disease have occurred among Native American populations and the homeless. Because many populations harbor a reservoir of healthy carriers, the potential for diphtheria is constantly present. Most cases occur in nonimmunized children from 1 to 10 years of age living in crowded, unsanitary situations.

Pathology of Diphtheria

Exposure to the diphtheria bacillus usually results from close contact with the droplets from human carriers or active infections and occasionally with fomites or contaminated milk. The clinical disease proceeds in two stages: (1) local infection by *Corynebacterium diphtheriae* and (2) toxin production and toxemia. The most common location of primary infection is in the upper respiratory tract (tonsils, pharynx, larynx, and trachea). *Cutaneous diphtheria* usually starts as a secondary infection manifesting as deep, erosive ulcers that are slow to heal **(figure 19.13a).** The bacterium becomes established by means

(a)

(b)

Figure 19.13 **Clinical aspects of diphtheria. (a)** Cutaneous diphtheria appears as a deep erosive ulcer that is slow to heal. **(b)** Respiratory diphtheria includes gross inflammation of the pharynx and tonsils marked by grayish patches (a pseudomembrane) and swelling over the entire area. (a): *Source:* CDC: (b): © Mediscan/Alamy Stock Photo

of virulence factors that assist in its attachment and growth. The cells are not ordinarily invasive and usually remain localized at the portal of entry. This form of the disease is on the rise in the United States.

Diphtherotoxin and Toxemia Although infection is necessary for disease, the cardinal determinant of pathogenicity is the production of **diphtherotoxin.** The ability of some *C. diphtheriae* strains to produce toxin is the result of infection with a lysogenic bacteriophage that carries the genes for the toxin. This cytotoxin consists of two polypeptide fragments. Fragment B binds to and is endocytosed by mammalian target cells in the heart and nervous system. Fragment A interacts metabolically with factors in the cytoplasm and arrests protein synthesis (see figure 13.13).

The toxin affects the body on two levels. The local infection produces an inflammatory reaction, sore throat, nausea, vomiting, enlarged cervical lymph nodes, severe swelling in the neck, and fever. One life-threatening complication is the **pseudomembrane,** a greenish-gray film that develops in the pharynx from the solidification of fluid expressed during inflammation **(figure 19.13b).** The pseudomembrane is so leathery and tenacious that attempts to pull it away result in bleeding; and if it forms in the airways, it can cause asphyxiation.

The most dangerous systemic complication is **toxemia,** which occurs when the toxin is absorbed from the throat and carried by the blood to certain target organs, primarily the heart and nerves. The action of the toxin on the heart causes myocarditis and abnormal EKG patterns. Cranial and peripheral nerve involvement can cause muscle weakness and paralysis. Although toxic effects are usually reversible, patients with inadequate treatment often die from asphyxiation, respiratory complications, or heart damage.

Diagnostic Methods for Corynebacterium

Diphtheria has such great potential for harm that often the physician must make a presumptive diagnosis and begin treatment before the bacteriologic analysis is complete. A gray membrane and swelling in the throat are somewhat indicative of diphtheria, although several diseases present a similar appearance. Epidemiological factors such as living conditions, travel history, and immunologic history (a positive Schick test) can also aid in initial diagnosis.

A simple stain of *C. diphtheriae* isolates with alkaline methylene blue reveals pleomorphic granulated cells. Other methods include the Elek test, an assay that relies on antibodies to detect toxins, as well as amplification of specific DNA sequences specific to *C. diphtheriae*. It is important to differentiate *C. diphtheriae* from "diphtheroids"—similar species often present in clinical materials that are not primary pathogens. *Corynebacterium xerosis** normally lives in the eye, skin, and mucous membranes and is an occasional opportunist in eye and postoperative infections. *Corynebacterium pseudodiphtheriticum,** a normal inhabitant of the human nasopharynx, can colonize natural and artificial heart valves.

Treatment and Prevention of Diphtheria

The adverse effects of toxemia are treated with diphtheria antitoxin (DAT) derived from horses. Prior to injection, the patient must be

tested for allergy to horse serum and be desensitized, if necessary. The infection is treated with antibiotics from the penicillin or erythromycin family. Bed rest, heart medication, and tracheostomy or bronchoscopy to remove the pseudomembrane may be indicated. Diphtheria can be easily prevented by a series of vaccinations with toxoid, usually given as part of a mixed vaccine against tetanus and pertussis (DTaP). Five vaccines are recommended for children under 7 years of age beginning at age 2 months, with boosters at 4 months, 6 months, 15 to 18 months, and 4 to 6 years. Children from ages 7 to 18 and adults who are not immune to the toxin can be immunized with two doses of tetanus, diphtheria, and pertussis (Tdap) vaccine.

The Genus *Propionibacterium*

*Propionibacterium** resembles *Corynebacterium* in morphology and arrangement, but it differs by being aerotolerant or anaerobic and nontoxigenic. The most prominent species is *P. acnes,** a common resident of the **pilosebaceous** glands of human skin and occasionally the upper respiratory tract. The primary importance of this bacterium is its relationship with the familiar acne vulgaris lesions of adolescence. Acne is a complex disease influenced by genetic and hormonal factors as well as by the structure of the epidermis, but it is also an infection. *Propionibacterium* is occasionally involved in infections of the eye and artificial joints.

 Check Your Progress SECTION 19.4

11. What is the difference between *C. diphtheriae* infection and toxemia?

12. List and describe the means by which *C. diphtheriae* infection is prevented, diagnosed, and treated.

19.5 Mycobacteria: Acid-Fast Bacilli

 Expected Learning Outcomes

14. Compare the clinical manifestations of tuberculosis at different stages of the disease.

15. Understand how tuberculosis is detected and treated.

16. Contrast the two forms of leprosy.

17. Relate the effect of *M. leprae* to the physical manifestation of the disease.

18. Explain the significance of nontuberculous mycobacterial diseases.

The genus *Mycobacterium** is distinguished by its complex layered structure containing high-molecular-weight mycolic acids and waxes. The high lipid content of the cell wall imparts the characteristic of **acid-fastness** and is responsible for the resistance of the group to

xerosis (zee-roh's-sis) Gr. *xerosis,* parched skin.
pseudodiphtheriticum (soo''-doh-dif-ther-it's-ih-kum) Gr. *pseudes,* false, and *diphtheriticus,* of diphtheria.

Propionibacterium (pro''-pee-on''-ee-bak-tee's-ree-um) Gr. *pro,* before, *prion,* fat, and *backterion,* little rod. Named for its ability to produce propionic acid.
acnes (ak's-neez) Gr. *akme,* a point. (The original translators incorrectly changed an *m* to an *n*.)
pilosebaceous (py''-loh-see-bay's-shus) L. *pilus,* hair, and *sebaceous,* tallow.
Mycobacterium (my''-koh-bak-tee's-ree-um) Gr. *myces,* fungus, and *bakterion,* little rod.

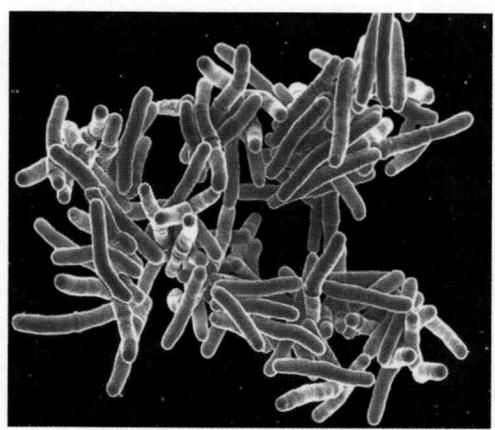

Figure 19.14 The microscopic morphology of mycobacteria.
In a scanning electron micrograph of *Mycobacterium tuberculosis* the
irregular morphology and filimentous nature of the species can be
seen. *Source:* National Institute of Allergy and Infectious Diseases (NIAID)/CDC

(a)

Cords WBCs

(b)

Figure 19.15 Diagnosing tuberculosis. (a) The cultural
appearance of *Mycobacterium tuberculosis*. Colonies with a typical
granular, waxy pattern of growth. **(b)** Cord formation in infected tissue.
Strands are elongate, massed filaments of TB bacilli. Note that they
have stained red, denoting acid-fastness. (a): *Source:* Dr. George Kubica/
CDC; (b): From ASM News, Vol. 59(12), Dec. 1993, courtesy of Pascal Meylan

drying, acids, and various germicides. The cells of mycobacteria are
long, slender, straight, or curved rods with a slight tendency to be fila-
mentous or branching **(figure 19.14).** Although they usually contain
granules and vacuoles, they do not form capsules, flagella, or spores.

Most mycobacteria are strict aerobes that grow well on simple
nutrients and media. Compared with other bacteria, the growth rate
is generally slow, with generation times ranging from 2 hours to
several days. Some members of the genus exhibit colonies contain-
ing yellow, orange, or pink carotenoid pigments that require light
for development; others are nonpigmented. Many of the 50 myco-
bacterial species are saprobes living free in soil and water, and
several are highly significant human pathogens. Worldwide, hun-
dreds of millions of people are afflicted with tuberculosis and lep-
rosy. Certain opportunistic species loosely grouped into a category
called NTM (**n**ontuberculous **m**ycobacteria) have become an in-
creasing problem in immunosuppressed patients.

Mycobacterium tuberculosis: The Tubercle Bacillus

The **tubercle*** bacillus is a long, thin rod that grows in sinuous
masses or strands called cords. Unlike many pathogens, it produces
no exotoxins or enzymes that contribute to infectiousness. Most
strains contain complex waxes and a cord factor **(figure 19.15)** that
contribute to virulence by preventing the mycobacteria from being
destroyed by the lysosomes of macrophages. Their survival contrib-
utes to further invasion and persistence as intracellular parasites.

Epidemiology and Transmission of Tuberculosis

Mummies from the Stone Age, ancient Egypt, and Peru provide
unmistakable evidence that tuberculosis (TB) is an ancient human
disease. In fact, it was such a prevalent cause of death that it was
called "Captain of the Men of Death" and "White Plague."

Tuberculosis is an extraordinarily complex disease exerting
profound effects on the health and economy of much of the world's
population. People in developing countries are often infected as

infants and harbor the microbe for many years until the disease is
reactivated in young adulthood. Estimates indicate that possibly
one-third of the world's population and 15 million people in the
United States carry the TB bacillus.

Cases in the United States show a strong correlation with the
age, sex, and recent immigration history of the patient. The highest
case rates occur in people over 65 years of age, as well as in new
immigrant populations from certain areas of Southeast Asia, Latin
America, and Africa and in AIDS patients. For recent trends in the
epidemiology of tuberculosis, see **19.3 Making Connections.**

The agent of tuberculosis is transmitted almost exclusively by
fine droplets of respiratory mucus suspended in the air. The tuber-
cle bacillus is very resistant and can survive for 8 months in fine
aerosol particles. Although larger particles become trapped in the
mucus and expelled, tinier ones can be inhaled into the bronchioles
and alveoli. This effect is especially pronounced among people
sharing closed, small rooms with limited access to sunlight and
fresh air. Factors that significantly affect a person's susceptibility to

* *tubercle* (too's-ber-kul) L. *tuberculum,* a swelling or knob.

19.3 MAKING CONNECTIONS

The Threat from Tuberculosis

After 80 years of decline, a tiny wax-coated bacillus that hides in human lungs has re-emerged as a serious threat to the world's health. The hope for eradicating tuberculosis is becoming an unlikely prospect in parts of Africa, Southeast Asia, and Latin America. A new infection occurs somewhere in the world every second, and at least 2 billion people have been exposed to it. With 1.4 million deaths per year, TB is second only to AIDS as a cause of death.

The United States has fared somewhat better. After a sharp increase in reported tuberculosis cases due in no small measure to the spread of HIV, reported cases declined steadily. There were 9,421 reported cases of TB in the United States in 2014—the last year for which complete statistics are available—a 1.5% reduction compared to 2013. In spite of these encouraging numbers, complacency must be avoided like the proverbial plague, as many of the underlying conditions that led to the reemergence of tuberculosis still exist.

One especially troubling trend both worldwide and in the United States has been the increasing proportion of *Mycobacterium tuberculosis* infections resistant to the first-line drugs isoniazid and rifampin. These multidrug-resistant tuberculosis (MDR-TB) cases required treatment with second-line drugs that were less effective, more toxic, and costlier. Continuing this trend, we are now seeing a rise of extensively drug-resistant TB (XDR-TB) infections, which are defined as MDR TB plus resistance to any fluoroquinolone and to any of the three second-line injectable drugs (i.e., amikacin, kanamycin, capreomycin). Data from the World Health Organization indicate that 9.7% of patients with MDR-TB have XDR-TB.

Major factors contributing to new TB cases in the United States come from immigration, drug addiction, the HIV epidemic, reduced government support of TB programs, and homelessness. In 2014, 66% of reported cases

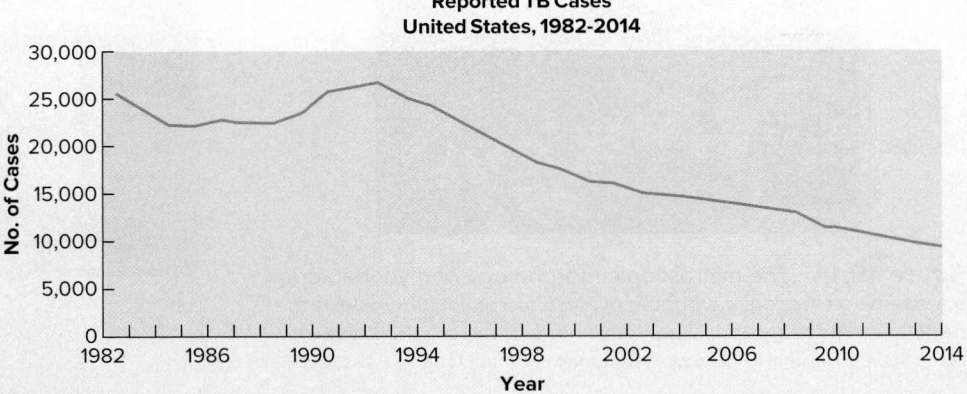

Reported TB Cases United States, 1982-2014

Source: CDC

occurred in foreign-born residents. These populations are less likely to have access to good medical care and to comply with the necessarily long course of therapy. Fully one-fifth of cases occur in AIDS patients. The mortality rate in this group is very high: Up to 80% die within 16 weeks of diagnosis.

The lack of well-funded TB control programs has contributed to the epidemic by discontinuing low-cost treatments, limiting surveillance, and not following up on treatment regimens. Experts at the CDC believe that patients with active TB who do not follow their treatment regimen are a breeding ground for many of the resistant strains of TB seen today. Implementation of a program of directly observed therapy (DOT), wherein health care workers visit each patient to observe the actual ingestion of each day's medication, has greatly contributed to the decrease in case rates, because it ensures that drugs are taken according to the correct schedule. At-home treatment also saves money, because hospitalization for TB patients can cost thousands of dollars even for a few days of care.

Suggest reasons TB is more prevalent among the homeless, drug addicts, and AIDS patients. *Answer available on Connect.*

tuberculosis are inadequate nutrition, debilitation of the immune system, poor access to medical care, lung damage, and genetics.

The Course of Infection and Disease

A clear-cut distinction can be made between infection with the tubercle bacillus and the disease it causes. In general, humans are rather easily infected with the bacillus but are resistant to the disease. Estimates project that only about 5% to 10% of infected people actually develop a clinical case of tuberculosis. Untreated tuberculosis progresses slowly and is capable of lasting a lifetime, with periods of health alternating with episodes of morbidity. The majority (85%) of TB cases are contained in the lungs, even though disseminated tubercle bacilli can give rise to tuberculosis in any organ of the body. The major clinical manifestations are primary tuberculosis, latent (reactivation) tuberculosis, and disseminated (extrapulmonary) tuberculosis, covered in the following sections and **figure 19.16a.**

Initial Infection and Primary Tuberculosis The minimum infectious dose for lung infection is around 10 cells. As part of the body's

immune response the bacilli are phagocytosed by alveolar macrophages where they multiply intracellularly. This period of hidden infection is asymptomatic or accompanied by mild fever, but some cells escape from the lungs into the blood and lymphatics. After 3 to 4 weeks, the immune system mounts a complex, cell-mediated assault against the bacilli. The large influx of mononuclear cells into the lungs plays a part in the formation of specific infection sites called **tubercles.** Tubercles are granulomas with a central core containing TB bacilli and enlarged macrophages, and an outer wall made of fibroblasts, lymphocytes, and neutrophils **(figure 19.16b).** Although this response further checks spread of infection and helps prevent the disease, it also carries a potential for lung damage. Frequently, the centers of tubercles break down into necrotic, **caseous***

Quick Search
To view an animation of *M. tuberculosis* infection, search YouTube for "CDC tuberculosis (TB) Transmission and Pathogenesis."

* *caseous* (kay's-see-us) L. *caseus,* cheese. The material formed resembles cheese or curd.

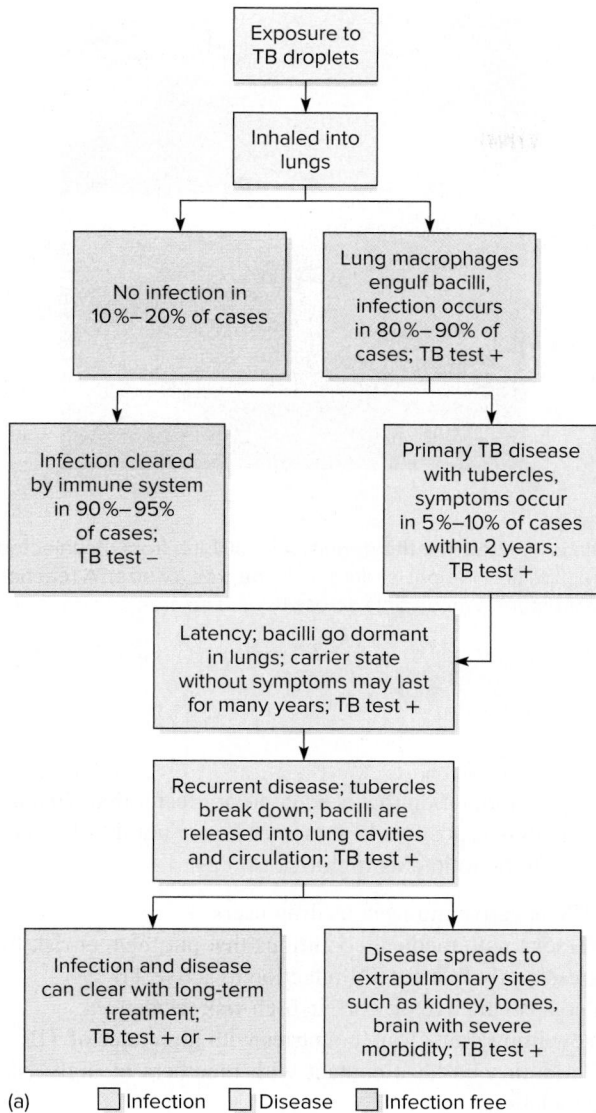

(a) ☐ Infection ☐ Disease ☐ Infection free

lesions that gradually heal by calcification when lung tissue is replaced by calcium deposits. The response of T cells to *M. tuberculosis* proteins also causes a cell-mediated immune response evident in the **tuberculin reaction** used in diagnosis and epidemiology **(figure 19.17).**

Latent and Recurrent Tuberculosis Although the majority of TB patients recover more or less completely from the primary infection or disease, live bacilli can remain latent and become reactivated weeks, months, or years later, especially in people with weakened immunity. In reactivated tuberculosis, tubercles filled with masses of bacilli expand and drain into the bronchial tubes and upper respiratory tract. Gradually, the patient experiences more severe symptoms, including violent coughing, greenish or bloody sputum, low-grade fever, anorexia, weight loss, extreme fatigue, night sweats, and chest pain. It is the gradual wasting of the body that accounts for an older name for tuberculosis—*consumption*. Untreated reactivated disease has nearly a 60% mortality rate.

Extrapulmonary Tuberculosis During the course of reactivated TB, the bacilli disseminate rapidly to sites other than the lungs. Organs most commonly involved in **extrapulmonary TB** are the regional lymph nodes, kidneys, long bones, genital tract, brain, and meninges. Because of the debilitation of the patient and the high load of tubercle bacilli, these complications are usually grave.

Renal tuberculosis results in necrosis and scarring of the renal medulla and the pelvis, ureters, and bladder. Genital TB often damages the reproductive organs in both sexes. Tuberculosis of the bone and joints is a common complication. The spine is a frequent site of infection, though the hip, knee, wrist, and elbow can also be involved. Degenerative changes can collapse the vertebrae, resulting in abnormal curvature of the thoracic or lumbar regions. Neurological damage stemming from compression on nerves can cause extensive paralysis and sensory loss.

Tubercular meningitis is the result of an active brain lesion seeding bacilli into the meninges. Over a period of several weeks, the infection of the cranial compartment can create mental deterioration, permanent retardation, blindness, and deafness. Untreated tubercular meningitis is invariably fatal, and even treated cases can have a 30% to 50% mortality rate.

Clinical Methods of Detecting Tuberculosis

Clinical diagnosis of tuberculosis traditionally includes some combination of these techniques:

1. tuberculin or immunologic testing,
2. radiography (X rays) of the chest,
3. direct identification of acid-fast bacilli (AFB) in sputum or some other specimen, and
4. cultural isolation and identification.

(b)

Figure 19.16 Features of tuberculosis. (a) The staging of tuberculosis from initial contact to infection, latency, and disease. Disease is the result of chronic infection and does not occur with every infected person. With proper treatment, all stages may be controlled and cured. Infection and disease may induce a permanent positive reaction to the tuberculin test. **(b)** Section of a tubercle shows how the granuloma infiltrate has set up a dense collar of fibroblasts, lymphocytes, and epithelioid cells. The core of this tubercle is a caseous (cheesy) material containing the bacilli.

Figure 19.17 **Skin testing for tuberculosis. (a)** The Mantoux test. Tuberculin is injected into the dermis. A small bleb from the injected fluid develops but will be absorbed in a short time. **(b)** After 48 to 72 hours, the skin reaction (induration) is rated by its degree (or size). A reaction of less than 5 mm is negative in all cases. **(c)** Category 3 positive Mantoux test. Measurement is made only across the raised area and does not include the reddened margin.

Final diagnosis of overt or latent TB cannot be made on a single test alone but requires an overall medical evaluation.

Tuberculin Sensitivity and Testing Because infection with the TB bacillus can lead to delayed hypersensitivity to tuberculoproteins, testing for hypersensitivity has been an important way to screen populations for tuberculosis infection. The tuberculin test, called the **Mantoux test,** involves local injection of tuberculin, a purified protein derivative (PPD), taken from culture fluids of *M. tuberculosis*. The injection is done intradermally into the forearm to produce an immediate small bleb. After 48 and 72 hours, the site is observed for a red wheal called an **induration,** which is measured and classified according to size (figure 19.17).

The current practices for interpreting tuberculin tests are focused on selected groups known to have higher risk for tuberculosis. It is no longer a routine screening method among populations of children or adults who are not within the target groups. The reasoning behind this change is to allow more focused screening and to reduce expensive and unnecessary follow-up tests and treatments. Guidelines for test groups and methods of interpreting tests are listed in the following summary.

Category 1. Induration (skin reaction) that is equal to or greater than *5 mm* is classified as positive in persons who

- Have had contact with actively infected TB patients
- Are HIV positive
- Have past history of tuberculosis as determined through chest X rays
- Are organ transplant recipients
- Are immunosuppressed for other reasons

Category 2. Induration that is equal to or greater than *10 mm* is classified as positive in persons who are not in category 1 but who fit the following high-risk groups:

- HIV-negative intravenous drug users
- Persons with medical conditions that put them at risk for progressing from latent TB infection to active TB
- Persons who live or work in high-risk residences
- New immigrants from countries with high rates of TB
- Children who have contact with members of high-risk adult populations
- Mycobacteriology laboratory personnel

Category 3. Induration that is equal to or greater than *15 mm* is classified as positive in persons who do not meet criteria in categories 1 or 2.

A positive reaction in a person from one of the categories is fairly reliable evidence of recent infection or a latent infection, but a diagnosis of TB should not be made based on this result alone. Vaccination with the tuberculosis vaccine (BCG) can also cause a TB positive result, so clinicians must weigh a patient's history, especially among individuals who have immigrated from countries where the vaccine is routinely given. Another cause of a false positive reaction is the presence of an infection with a closely related species of *Mycobacterium*.

A negative skin test usually indicates that ongoing TB infection is not present. In some cases, it may be a false negative, meaning that the person is infected but is not yet reactive. One cause of a false negative test may be that it is administered too early in the infection, requiring retesting at a later time. Subgroups with severely compromised immune systems, such as those with AIDS, advanced age, and chronic disease, may be unable to mount a reaction even though they are infected. Skin testing may not be a reliable diagnostic indicator in these populations.

The In Vitro TB Test

The CDC has accepted two new tests for aiding in TB diagnosis and ruling out most false positives. Both the QuantiFERON-TB Gold test and the T-SPOT TB test are indirect methods of detecting infection based on a blood test. A fresh sample of blood is incubated with synthetic *M. tuberculosis* proteins for about a day. If the blood contains lymphocytes with a prior sensitivity to the tuberculoproteins, the cells will release interferon gamma. The amount of this substance can be measured by machine. The advantage to this test is that it is rapid, it is easy to read, it does not sensitize the patient to TB (which tuberculin can do), and it is not affected by prior BCG vaccination. This will make it particularly useful in ruling out TB in immigrants who have been vaccinated.

Radiography and Tuberculosis Chest X rays may be suggestive of—but are never diagnostic of—TB. They may, however, be used to rule out the possibility of pulmonary TB in a person who has a positive reaction to the tuberculin skin test and no symptoms of disease. X-ray films reveal abnormal radiopaque patches whose appearance and location can be very indicative. Primary tubercular infection presents the appearance of fine areas of infiltration and enlarged lymph nodes in the lower and central areas of the lungs. Secondary tuberculosis films show more extensive infiltration in the upper lungs and bronchi and marked tubercles **(figure 19.18)**. Scars from older infections often show up on X rays and can furnish a basis for comparison with which to identify newly active disease.

Acid-Fast Staining Acid-fast staining of sputum or other specimens may be used to detect *Mycobacterium,* with several variations of the technique currently in use. The Ziehl-Neelsen stain produces bright red acid-fast bacilli (AFB) against a blue background. Fluorescence staining shows luminescent yellow-green bacilli against a dark background **(figure 19.19)**. The fluorescent acid-fast stain is becoming the method of choice because it is easier to read and provides a more striking contrast.

Laboratory Cultivation and Identification *Mycobacterium tuberculosis* infection is most accurately diagnosed by isolating and

Figure 19.19 A fluorescent acid-fast stain of *Mycobacterium tuberculosis* from sputum. Using this technique, a cluster of mycobacterial cells fluoresces bright yellow under ultraviolet light. Smears are evaluated in terms of the number of AFB seen per field. This quantity is then applied to a scale ranging from 0 to 4+, 0 being no AFB observed and 4+ being more than 9 AFB per field. *Source:* CDC

identifying the causative agent in pure culture. Because of the specialized expertise and technology required, this is not done by most clinical laboratories as a general rule. In the United States, the handling of specimens and cultures suspected of containing the pathogen is strictly regulated by federal laws.

Diagnosis that differentiates between *M. tuberculosis* and other mycobacteria must be accomplished as rapidly as possible so that appropriate treatment and isolation precautions can be instituted. Cultures are incubated under varying temperature and lighting conditions to clarify thermal and pigmentation characteristics and are then observed for signs of growth. Several newer cultivation schemes have shortened the time to several days instead of the 6 to 8 weeks once necessary. Other identification techniques use DNA probes to detect specific genetic markers and can confirm positive specimens early in the infection. Rapid diagnosis is particularly important for public health and treatment considerations.

Management and Prevention of Tuberculosis

Treatment of TB involves administering drugs for a sufficient period of time to kill the bacilli in the lungs, organs, and macrophages, usually 6 to 9 months in uncomplicated cases (but as long as 24 months in more complex infections). Because many strains of *M. tuberculosis* are resistant to at least one of the drugs commonly used for treatment, the most commonly followed anti-TB chemotherapeutic regimen begins with an initial phase of four drugs—isoniazid (INH), rifampin (RIF), ethambutol (EMB), and pyrazinamide (PZA)—taken daily for eight weeks. After this, a continuation phase involves daily doses of INH and RIF for 18 weeks. At this point treatment is generally deemed completed, although modifications are made in some circumstances (HIV infection, drug resistance, pregnancy, or the treatment of children for example). A one-pill regimen called *Rifater* combines isoniazid, rifampin, and pyrazinamide, and may be supplemented with ethambutol if needed. The simplicity of the one pill regimen increases the rate of patient compliance, a common problem with anti-TB therapy. In 2012 the first new drug in 40 years—bedaquiline (Sirturo)—was introduced primarily for cases of multi- and extensively-drug-resistant

Tubercles

Figure 19.18 X ray of a secondary tubercular infection shows dozens of tubercles and cavities throughout the lung.
© BioPhoto Associates/Science Source

Pathogen Profile #5 *Mycobacterium tuberculosis*

Source: Janice Carr/CDC

Microscopic Morphology Acid-fast rods, often present as long, enmeshed filaments of cells.

Identified by Acid-fast reaction, morphology, biochemical testing, and DNA tests. Infection with *M. tuberculosis* is usually inferred through some combination of tuberculin and serological testing, X rays, and laboratory tests.

Habitat Humans are the primary habitat, with over 15 million people in the United States and 2 billion worldwide carrying the pathogen.

Virulence Factors The acid-fast cell wall of *M. tuberculosis* provides it with protection against lysozyme and many antibacterial agents. Resistance to drying and a very low infectious dose (10 cells) facilitate infection, while resistance to multiple drugs makes antibiotic therapy challenging. Cord factor is responsible for the long strands of cells often seen upon examination and is correlated with increased virulence.

Primary Infections/Disease Primary tuberculosis manifests as tubercles within the lungs and may be marked only by mild fever. In recurrent tuberculosis, bacilli within the tubercles drain into the bronchi and upper respiratory tract. Symptoms gradually become worse, with the patient experiencing violent coughing, green or bloody sputum, low-grade fever, anorexia, weight loss, fatigue, night sweats, and chest pain. Untreated reactivated disease has a 60% mortality rate. Extrapulmonary tuberculosis occurs when the bacterium disseminates to sites other than the lungs, most commonly lymph nodes, kidneys, long bones, genital tract, brain, and meninges. Prognosis is exceedingly poor at this point.

Control and Treatment Control of *M. tuberculosis* relies on early detection of those infected or at risk of becoming infected. Treatment consists of multidrug therapy lasting from 6 to 24 months. Although the BCG vaccine is available, because of the low incidence of TB in the United States and the vaccine's relative ineffectiveness, it is only recommended for certain health professionals and military personnel.

tuberculosis. Clinicians expect it to significantly reduce the spread of these dangerous strains of TB, but severe side effects limit its use to those instances where no other drug combination is effective. When evaluating the effectiveness of chemotherapy, the presence of a negative culture or a gradual decrease in the number of AFB on a smear indicates success. However, a cure will not occur if the patient does not comply with drug protocols, which accounts for many relapses.

Although it is essential to identify and treat people with active TB, it is equally important to seek out and treat those in the early stages of infection or at high risk of becoming infected. Treatment groups are divided into tuberculin-positive "converters," who appear to have a latent infection, and tuberculin-negative people in high-risk groups such as the contacts of tubercular patients. The standard treatment is a daily dose of INH for 9 months or RIF for 4 months. An equally effective, and far simpler, treatment for patients who fit a specific criteria is a single dose of INH and rifapentine (RPT) once weekly for 12 weeks. This treatment is recommended for patients older than 12 years who are not receiving antiretroviral treatment for HIV, not infected with a resistant strain of *M. tuberculosis*, and not pregnant or planning to become pregnant during the course of treatment. In the hospital, the use of UV lamps in air-conditioning systems and negative pressure rooms to isolate TB patients can help control the spread of infection.

A vaccine based on the attenuated "bacille Calmette-Guérin" (BCG) strain of *M. bovis* is often given to children in countries that have high rates of tuberculosis. Studies have shown that the success rate of vaccinations is around 80% in children and less than that (20% to 50%) in adults. The length of protection varies from 5 to 15 years. Because the United States does not have as high an incidence as other countries, BCG vaccination is not generally recommended except among certain health professionals and military personnel who may be exposed to TB carriers. Vaccine trials using an attenuated strain of *M. tuberculosis* are currently under way.

Mycobacterium leprae: The Leprosy Bacillus

Mycobacterium leprae, the cause of leprosy, was first detected by a Norwegian physician named Gerhard Hansen, and it is sometimes called the Hansen bacillus in his honor. The general morphology and staining characteristics of the leprosy bacillus are similar to those of other mycobacteria, but it is exceptional in two ways: (1) It is a strict parasite that has not been grown in artificial media or human tissue cultures, and (2) it is the slowest growing of all the species. *Mycobacterium leprae* multiplies within host cells in large packets called globi at an optimum temperature of 30°C. *Mycobacterium lepromatosis*, a species first described in 2008, is thought to be responsible for a few cases of a unique form of leprosy endemic to Mexico.

Leprosy* is a chronic, progressive disease of the skin and nerves known for its extensive medical and cultural ramifications. From ancient times, leprosy victims were stigmatized because of their severe disfigurement and the belief that leprosy was a divine curse. As if the torture of the disease were not enough, leprosy patients once suffered terrible brutalities, including imprisonment under the most gruesome conditions. The modern view of leprosy is more enlightened. We know that it is not readily communicated and that it should not be accompanied by social banishment. Because of the unfortunate connotations associated with the term *leper* (a person who is shunned or ostracized), more acceptable terms such as *leprosy patient, Hansen's disease patient,* or *leprotic* are preferred.

Epidemiology and Transmission of Leprosy

The incidence of leprosy has been declining due to a worldwide control effort. World Health Organization (WHO) estimates about

* *leprosy* (lep′-roh-see) Gr. *lepros*, scaly or rough. Translated from a Hebrew word that refers to uncleanliness.

175,000 cases, mostly in endemic areas of Asia, Africa, Central and South America, and the Pacific islands. Leprosy is not restricted to warm climates, however, because it has been reported in Siberia, Korea, and northern China. The disease is also reported in a few limited locales in the United States, including parts of Hawaii, Texas, Louisiana, Florida, and California. The total number of new cases reported nationwide is from 50 to 100 per year, many of which are associated with recent immigrants.

The mechanism of transmission among humans is yet to be fully verified. Theories propose that the bacillus is directly inoculated into the skin through contact with a leprosy patient, that mechanical vectors are involved, or that inhalation of droplet nuclei is a factor.

Although the human body was long considered the sole host and reservoir of the leprosy bacillus, it is now clear that armadillos also harbor *M. leprae* and may develop a granulomatous disease similar to leprosy. Although the risk of contracting leprosy from armadillos is exceedingly low, zoonotic infection is thought to be responsible for the 30 to 40 new cases of leprosy seen each year in U.S.-born citizens who have never traveled to regions where the disease is prevalent.

Because the leprosy bacillus is not highly virulent, most people who come into contact with it do not develop clinical disease. As with tuberculosis, it appears that health and living conditions influence susceptibility and the course of the disease. An apparent predisposing factor is some defect in the regulation of T cells. Mounting evidence also indicates that some forms of leprosy are associated with a specific genetic marker. Long-term household contact with leprotics, poor nutrition, and crowded conditions increase the risks of infection. Many people become infected as children and harbor the microbe through adulthood.

The Course of Infection and Disease

Once *M. leprae* has entered a portal of entry, macrophages successfully destroy the bacilli, and there are no initial manifestations of infection. But in a small percentage of cases, a weakened or slow macrophage and T-cell response leads to intracellular survival of the pathogen. The usual incubation period varies from 2 to 5 years, with extremes of 3 months to 40 years. The earliest signs of leprosy appear on the skin of the trunk and extremities as small, spotty lesions colored differently from the surrounding skin **(figure 19.20)**. In untreated cases, the bacilli grow slowly in the skin macrophages and Schwann cells of peripheral nerves, and the disease progresses to one of several outcomes.

A useful rating system for leprosy was developed by D. S. Ridley and W. H. Jopling. At the two extremes are tuberculoid leprosy (TL) and lepromatous leprosy (LL) **(table 19.3)**; and in between are borderline tuberculoid (BT), borderline (BB), and borderline lepromatous (BL) leprosy. Patients can have more than one form of leprosy simultaneously, and one type can progress to another. Currently, epidemiologists at the CDC and World Health Organization are transitioning to a system in which leprosy is classified as **paucibacillary** (few bacteria found in skin scrapings, otherwise classified as tuberculoid or borderline tuberculoid leprosy) or **multibacillary** (many bacteria found in skin scrapings, corresponding to borderline, borderline lepromatous, or lepromatous leprosy).

Tuberculoid (Paucibacillary) leprosy, the most superficial form, is characterized by asymmetrical, shallow skin lesions

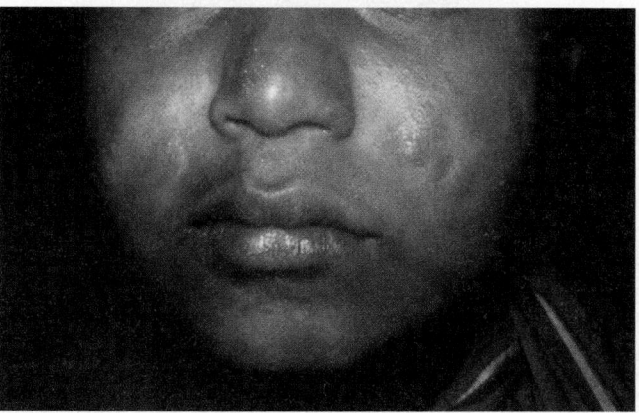

Figure 19.20 Leprosy. (a) Tuberculoid (paucibacillary) Infection often appears as hypopigmented patches, especially in dark skinned persons. **(b)** Borderline (multibacillary) leprosy is the most common form of the disease. Tuberculoid-like lesion are present in greater numbers, and peripheral nerves in the area are affected, causing weakness and a loss of sensation. (a, b): Source: CDC

TABLE 19.3	The Two Major Clinical Forms of Leprosy
Tuberculoid (Paucibacillary) Leprosy	**Lepromatous (Multibacillary) Leprosy**
Few bacilli in lesions	Many bacilli in lesions
Shallow skin lesions in many areas	Numerous deeper lesions concentrated in cooler areas of body
Loss of pain sensation in lesions	Sensory loss more generalized; occurs late in disease
No skin nodules	Gross skin nodules
Occasional mutilation of extremities	Mutilation of extremities common
Reactive to lepromin*	Not reactive to lepromin
Lymph nodes not infiltrated by bacilli	Lymph nodes massively infiltrated by bacilli
Well-developed cell-mediated (T-cell) response	Poorly developed T-cell response

*Lepromin is an extract of the leprosy bacillus injected intradermally, like tuberculin, to detect delayed allergy to leprosy.

Figure 19.21 **A clinical picture of lepromatous (multibacillary) leprosy.** Infection of the nose, lips, chin, and brows produces moderate facial deformation, typical of lepromas. *Source:* CDC

Figure 19.22 **Deformation of the hands caused by borderline leprosy.** The clawing and wasting are chiefly due to nerve damage that interferes with musculoskeletal activity. Individuals in a later phase of the disease can lose their fingers. © Medicshots/ Alamy Stock Photo

containing very few bacilli (figure 19.20). Microscopically, the lesions appear as thin granulomas and enlarged dermal nerves. Damage to these nerves usually results in local loss of pain reception and feeling. This form has fewer complications and is more easily treated than other types of leprosy.

Lepromatous (Multibacillary) leprosy is responsible for the disfigurations commonly associated with the disease. It is marked by chronicity and severe complications due to widespread dissemination of the bacteria. Leprosy bacilli grow primarily in macrophages in cooler regions of the body, including the nose, ears, eyebrows, chin, and testes. As growth proceeds, the face of the afflicted person develops folds and granulomatous thickenings, called lepromas, which are caused by massive intracellular overgrowth of *M. leprae* (**figure 19.21**). Advanced LL causes a loss of sensitivity that predisposes the patient to trauma and mutilation, secondary infections, blindness, and kidney or respiratory failure. Diffuse lepromatous leprosy, a unique form of the disease endemic to Mexico, is thought to be caused by *M. lepromatosis,* a species first described in 2008.

The condition of borderline (multibacillary) leprosy patients can progress in either direction along the scale, depending upon their treatment and immunologic competence. The most severe effect of intermediate forms of leprosy is early damage to nerves that control the muscles of the hands and feet. The subsequent wasting of the muscles and loss of control produce drop foot and claw hands (**figure 19.22**). Sensory nerve damage can lead to trauma and loss of fingers and toes.

Diagnosing Leprosy

Leprosy is diagnosed by a combination of symptomatology, microscopic examination of lesions, and patient history. A simple yet effective field test in endemic populations is the feather test. A skin area that has lost sensation and does not itch may be an early symptom. Numbness in the hands and feet, loss of heat and cold sensitivity, muscle weakness, thickened earlobes, and chronic stuffy nose are additional evidence. Laboratory diagnosis relies upon the detection of acid-fast bacilli in smears of skin lesions, nasal discharges, and tissue samples. Knowledge of a patient's prior contacts with leprotics also supports diagnosis. Laboratory isolation of the leprosy bacillus is difficult and not ordinarily attempted.

Treatment and Prevention of Leprosy

Leprosy can be controlled by drugs, but therapy is most effective when started before permanent damage to nerves and other tissues has occurred. Because of an increase in resistant strains, multidrug therapy is necessary. Tuberculoid (paucibacillary) leprosy can be managed with rifampin and dapsone for 12 months. Lepromatous (multibacillary) leprosy requires a combination of rifampin, dapsone, and clofazimine until the number of AFB in skin lesions has been substantially reduced (requiring up to 2 years).

Preventing leprosy requires constant surveillance of high-risk populations to discover early cases, chemoprophylaxis of healthy persons in close contact with leprotics, and isolation of leprosy patients. No vaccine for leprosy is currently available, but it is thought that the bacille Calmette-Guérin (BCG) vaccine that is used in many countries to prevent tuberculosis also provides some protection against infection with the leprosy bacillus.

Infections by Nontuberculous Mycobacteria (NTM)

For many years, most mycobacteria were thought to have low pathogenicity for humans. Saprobic and commensal species are isolated so frequently in soil, drinking water, swimming pools, dust, air, raw milk, and even the human body that both contact and asymptomatic infection appear to be widespread. However, the recent rise in opportunistic and healthcare-associated mycobacterial infections has demonstrated that many species are far from harmless.

Pathogen Profile #6 *Mycobacterium leprae*

Source: CDC

Microscopic Morphology Acid- fast rods.

Identified by Laboratory isolation of *M. leprae* is difficult and so is rarely attempted. Instead, leprosy (Hansen's disease) is diagnosed through a combination of microscopic examination of lesions, patient history, and symptomatology, especially the loss of sensation on the skin.

Habitat Humans and armadillos are the sole reservoir for *M. leprae*.

Virulence Factors *M. leprae* is not highly virulent. Predisposing factors for infection include defects in T-cell regulation, poor overall health, inadequate nutrition, and long-term household contact with leprotics.

Primary Infections/Disease Most people exposed to *M. leprae* do not go on to develop the disease. For those who do, an incubation period of several years is followed by the earliest signs of leprosy, small spotty lesions appearing on the trunk or extremities.

Untreated disease may progress to tuberculoid (paucibacillary) leprosy—the milder form, marked by loss of sensation—or lepromatous (multibacillary) leprosy, the more severe form of the disease. Lepromatous leprosy is commonly associated with disfigurement. Some forms of leprosy may initially present as borderline (multibacillary) and may progress toward the more or less severe subtypes, depending on treatment and immunologic competence.

Control and Treatment Control of *M. leprae* relies on early detection of those infected, chemoprophylaxis of healthy persons in close contact with leprotics, and isolation of leprosy patients. Because of an increase in resistant strains, treatment involves multidrug therapy and is most effective when started before permanent damage to nerves and other tissues has occurred.

Disseminated Mycobacterial Infection in AIDS Bacilli of the *Mycobacterium avium* complex (MAC) frequently cause secondary infections in AIDS patients with low T-cell counts. These common soil bacteria usually enter through the respiratory tract, multiply, and rapidly disseminate. In the absence of an effective immune counterattack, the bacilli flood the body systems, especially the blood, bone marrow, bronchi, intestine, kidney, and liver. Recommended treatment is combined drug therapy using clarithromycin and ethambutol, occasionally supplemented with rifabutin. This therapy usually has to be given for several months or even years.

Nontuberculous Lung Disease Pulmonary infections caused by commensal mycobacteria have symptoms like a milder form of tuberculosis, but they are not communicable. *Mycobacterium kansasii* infection is endemic to urban areas in the midwestern and southwestern United States and parts of England. It occurs most often in adult White males who already have emphysema or bronchitis. *Mycobacterium fortuitum* complex causes postsurgical skin and soft-tissue infection and pulmonary complications in immunosuppressed patients.

A form of TB caused by *M. bovis* was once quite common but is now extremely rare due to efforts to control the disease in cattle. Occasional clusters of cases have appeared, all probably from imported cheese made with unpasteurized milk.

Miscellaneous Mycobacterial Infections An infection by *M. marinum* has been labeled fish tank granuloma because of the strong relationship between the disease and work around aquariums. Interestingly, the same disease was referred to as swimming

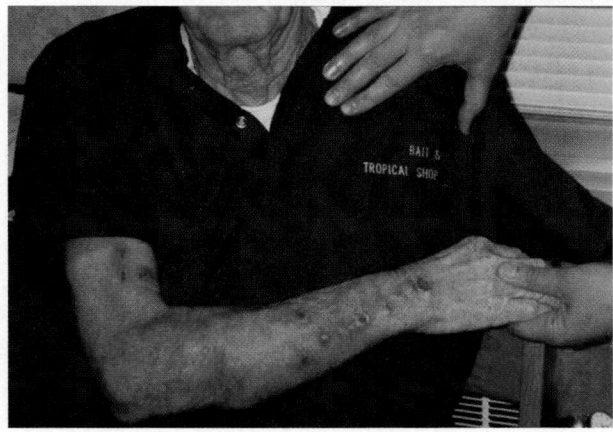

Figure 19.23 Chronic fish tank granulomas on the arm of a fish store employee. Image reprinted with permission from Medscape Drugs & Diseases (http://emedicine.medscape.com/), 2016, available at: http://emedicine.medscape.com/article/223363-overview.

pool granuloma for many years but consistent chlorination of pools has led to its eradication in pools. The disease begins when *M. marinum*, found in both fresh and salt water, enters a cut or scrape. A nodule begins to form, which then enlarges, ulcerates, and drains **(figure 19.23)**. The granuloma can clear up spontaneously, but it can also persist and require long-term treatment.

Mycobacterium scrofulaceum causes an infection of the cervical lymph nodes in children living in the Great Lakes region, Canada, and Japan. Infection is apparently associated with ingestion of food or milk and generally affects the oral cavity and cervical lymph nodes. In most cases the infection is without complications,

but certain children develop *scrofula,* in which the affected lymph nodes ulcerate and drain.

Research on a chronic intestinal syndrome called Crohn's disease has uncovered a strong association between the disease and a form of *Mycobacterium paratuberculosis.* When technicians used a PCR technique to analyze the DNA of colon specimens, it was found that 65% of Crohn's disease patients tested positive for *M. paratuberculosis.* Further studies have shown that this bacterium is found in cow's milk, which may well be the source of infection.

 Check Your Progress SECTION 19.5

13. Outline the differences in sites of infection and pathology among primary, reactivated, and extrapulmonary tuberculosis.

14. Compile a list of the advantages, disadvantages, and information gained through TB skin tests, immunological blood tests, chest X rays, acid-fast staining, and cultural isolation and identification.

15. Explain how and why antibacterial treatment for tuberculosis differs from that for other bacterial pathogens.

16. List several differences between lepromatous and tuberculoid leprosy.

17. Explain the association between *M. leprae* infection and deformation of the fingers and toes.

18. List the diseases and at-risk populations associated with infection by *Mycobacterium avium, Mycobacterium kansasii,* and *Mycobacterium marinum.*

19.6 Actinomycetes: Filamentous Bacilli

 Expected Learning Outcomes

19. Identify and describe the pathogenic actinomycetes.

20. Describe the types of infections attributable to bacteria in the genus *Nocardia.*

A group of bacilli closely related to *Mycobacterium* are the pathogenic **actinomycetes.*** These are nonmotile filamentous rods that may be acid-fast. Certain members produce a mycelium-like growth and spores reminiscent of fungi. They also produce chronic granulomatous diseases. The main genera of actinomycetes involved in human disease are *Actinomyces* and *Nocardia.*

Actinomycosis

Actinomycosis is an endogenous infection of the cervicofacial, thoracic, or abdominal regions by species of *Actinomyces* living normally in the human oral cavity, tonsils, and intestine. The cervicofacial form of disease can be a common complication of tooth extraction, poor oral hygiene, and rampant dental caries. The lungs, abdomen, and uterus are also sites of infection.

Figure 19.24 Clinical signs of actinomycosis. Early periodontal lesions work their way to the surface in cervicofacial disease. The lesions eventually break out and drain. Note the dimpling around the lesion, indicating loss of underlying bone to the infection. *Source:* CDC

Cervicofacial disease is caused by **A. israelii,** which enters a damaged area of the oral mucous membrane and begins to multiply there. Diagnostic signs are swollen, tender nodules in the neck or jaw that give off a discharge containing macroscopic (1–2 mm) sulfur granules **(figure 19.24).** In most cases, infection remains localized, but bone invasion and systemic spread can occur in people with poor health. Thoracic actinomycosis is a necrotizing lung disorder that can project through the chest wall and ribs. Abdominal actinomycosis is a complication of burst appendixes, gunshot wounds, ulcers, or intestinal damage. Uterine actinomycosis has been increasingly reported in women using intrauterine contraceptive devices. Infections are treated with surgical drainage and drug therapy with penicillin, amoxicillin, erythromycin, or doxycycline.

Compelling evidence indicates that oral actinomyces play a strategic role in the development of plaque and dental caries. Studies of the oral environment show that *A. viscosus* and certain oral streptococci are the first microbial colonists of the tooth surface. Both groups have specific tooth-binding powers and can adhere to each other and to other species of bacteria (see chapter 21).

Nocardiosis

*Nocardia** is a genus of bacilli widely distributed in the soil. Most species are not infectious, but **N. brasiliensis** is a primary pulmonary pathogen, and *N. asteroides* and *N. caviae* are opportunists. Nocardioses fall into the categories of pulmonary, cutaneous, or subcutaneous infection. Most cases in the United States are reported

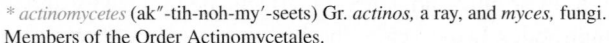

* *actinomycetes* (ak″-tih-noh-my′-seets) Gr. *actinos,* a ray, and *myces,* fungi. Members of the Order Actinomycetales.

* *Nocardia* (noh-kar′-dee-ah) After Edmund Nocard, the French veterinarian who first described the genus.

in patients with deficient immunity, but a few occur in normal individuals.

Pulmonary nocardiosis is a form of bacterial pneumonia with pathology and symptoms similar to tuberculosis. The lung develops abscesses and nodules and can consolidate. Often, the lesions extend to the pleura and chest wall and disseminate to the brain, kidneys, and skin **(figure 19.25).** *Nocardia brasiliensis* is also one of the chief etiologic agents in mycetoma, discussed in chapter 22.

 Check Your Progress SECTION 19.6

19. Compare the types of infections caused by the primary members of the actinomycetes.

Figure 19.25 **Nocardiosis.** This case of pulmonary disease has extended from the shoulders and trunk to the skin of the upper arm. *Source:* CDC

 CASE STUDY Part 2

The investigation initiated in response to the illnesses was one of thousands, big and small, done every year to track down the source of an outbreak. The first step in such an inquiry is decidedly low-tech, with investigators quizzing ill persons as to their actions over the weeks leading up to the outbreak, searching for behaviors that could have put them at risk for infection; with *Listeria*, the questioning centered on food and drink. Investigators found that 93% of ill persons reported consuming fresh cantaloupe in the month before falling ill. This was unusual, as *Listeria* is generally associated with unpasteurized milk, soft cheeses, and processed meats, although outbreaks in 2009 and 2010 had been linked to sprouts and celery. Investigators from the Colorado Department of Public Health and Environment collected all available cantaloupes from local markets and from patients' homes. Several patients specifically reported eating "Rocky Ford" cantaloupes, which are grown only in the Rocky Ford region of Colorado. These were traced back to Jensen Farms and had been shipped from July 24 through September 10.

A second front in the investigation centered on molecular typing of the *Listeria* bacterium, generating a DNA profile—or fingerprint—of the strain responsible for infection. This phase of the effort was carried out by PulseNet, a national network of laboratories that use a DNA profiling technique called pulsed-field gel electrophoresis (PFGE) to distinguish strains at the molecular level. PulseNet analysis found five strains of *Listeria monocytogenes* associated with the outbreak. A team headed by FDA inspectors began an investigation of Jensen Farms, a fixture in southeastern Colorado, where four generations had grown cantaloupe, hay, and alfalfa. The current owners, Ryan and Eric Jensen, had made a number of changes to a packing shed just prior to the 2011 harvest. Investigators discovered several problems that may have contributed to the outbreak. Among them were the following:

- A refrigeration drain line allowed water to pool directly under packing facility equipment, and testing of the water revealed the presence of one of the outbreak strains of *Listeria monocytogenes*.

- The floor of the packing facility was not easily cleanable, creating a refuge for bacteria.
- Jensen Farms had recently installed packing equipment that had previously been used in the harvesting of potatoes. This created a potential problem because potatoes, unlike cantaloupe, are generally cooked prior to being consumed. The equipment had visible dirt and corrosion, and was also designed in a manner that prevented adequate cleaning.
- Removal of an initial antimicrobial wash meant that a single contaminated melon could spread bacteria to other packing equipment and potentially cross-contaminate every melon that passed through the facility.

Jensen Farms immediately halted sales of cantaloupes and a nationwide recall was instituted, but this was of no help to those already affected. Nationwide, 147 persons reported illness, 143 of these were hospitalized, and 33 deaths were attributed to the outbreak, making it the third deadliest food-related outbreak in U.S. history. Mike Hauser and Paul Schwarz, along with Michelle and Kendall Wakely, fairly represent those hit hardest by the outbreak. The median age of those infected was 78, and those who died had a median age of 81. Seven cases were diagnosed in newborns or pregnant women, and one miscarriage was reported. An onslaught of medical claims and civil suits related to the outbreak led Jensen Farms to file for bankruptcy in May 2012; Eric and Ryan Jensen pleaded guilty to misdemeanor charges in October 2013. Each received five years probation and six months home detention. Each also was ordered to pay $150,000 in restitution and perform 100 hours of community service.

- *Explain the importance of PFGE subtyping of Listeria in this case.*

- *Milk and meat are kept cool during processing in order to slow the growth of E. coli. Would cooling be an effective method of controlling Listeria ?*

To learn more about this case, go to http://www.cdc.gov and search "jensen farms cantaloupe."

To conclude this Case Study, go to Connect.

Chapter Summary with Key Terms

19.1 Medically Important Gram-Positive Bacilli
The gram-positive bacilli include about a dozen genera of medically significant bacteria. They are differentiated on the basis of endospores, acid-fastness, and cell morphology.

19.2 Gram-Positive Spore-Forming Bacilli
Endospore-forming rods include species in the genera *Bacillus* and *Clostridium.*
 A. *Bacillus* is an aerobic genus, the most prominent member of which, *Bacillus anthracis,* is the causative agent of **anthrax,** a zoonosis that exists in **cutaneous, pulmonary,** and **gastrointestinal** forms.
 B. *Clostridium* is an anaerobic genus whose members exert their effects primarily through the release of powerful **exotoxins,** each of which attack specific tissues.
 1. *Clostridium perfringens* causes **gas gangrene,** a soft tissue and muscle infection. Ingestion of cells may cause clostridial food poisoning.
 2. *C. difficile* causes antibiotic-associated colitis, a gastrointestinal infection acquired during drug therapy.
 3. *C. tetani* causes **tetanus** (lockjaw) when the toxin **tetanospasmin** affects cells in the central nervous system.
 4. *C. botulinum* is the causative agent of **botulism,** a food intoxication associated with home-canned foods.
 5. Infant and wound botulism are unusual types of infectious botulism contracted like tetanus.

19.3 Gram-Positive Regular Non-Spore-Forming Bacilli
Non-spore-forming rods are divided among those that have **regular** and irregular shapes and staining properties. Straight, nonpleomorphic rods stain evenly and include the genera *Listeria* and *Erysipelothrix.*
 A. *Listeria monocytogenes* causes **listeriosis,** a food infection associated with contaminated meat and dairy products.
 B. *Erysipelothrix rhusiopathiae* is a zoonosis normally carried by swine. It is the cause of **erysipeloid** among people who handle animals or animal products.

19.4 Gram-Positive Irregular Non-Spore-Forming Bacilli
Irregular non-spore-forming rods include *Corynebacterium* and *Propionibacterium.*
 A. *Corynebacterium diphtheriae* is a pleomorphic aerobic rod with metachromatic granules and a palisades arrangement. It is the causative agent of both respiratory and cutaneous **diphtheria.**
 B. *Propionibacterium acnes* is an anaerobic rod associated with acne vulgaris lesions.

19.5 Mycobacteria: Acid-Fast Bacilli
Acid-fast bacilli contain large amounts of mycolic acid and waxes; they are strict aerobes that are widely distributed in the environment.
 A. *Mycobacterium tuberculosis* causes tuberculosis (TB). The disease is divided into primary pulmonary infection, involving the formation of **tubercles** in the lungs; reactivated TB, which occurs in people with latent or reactivated disease; and disseminated **extrapulmonary TB,** due to the TB bacilli being carried throughout the body.
 B. *Mycobacterium leprae* causes **leprosy,** a chronic disease that begins in the skin but quickly moves to the nerves. **Tuberculoid (paucibacillary) leprosy** is a superficial infection, whereas **lepromatous (multibacillary) leprosy** causes more severe disfigurement.
 C. Nontuberculous mycobacteria (NTM) are common environmental species that may be involved in disease. Examples include *M. avium* complex and *M. marinum.*

19.6 Actinomycetes: Filamentous Bacilli
The genera *Actinomyces* and *Nocardia* are filamentous bacteria related to mycobacteria that may cause chronic infection of the skin and soft tissues.
 A. *Actinomyces* species such as *A. israelii* are responsible for diseases of the oral cavity and intestines.
 B. *Nocardia brasiliensis* causes pulmonary disease similar to TB as well as cutaneous nodules.

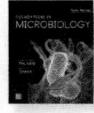

Level I. Knowledge and Comprehension

These questions require a working knowledge of the concepts in the chapter and the ability to recall and understand the information you have studied.

Multiple-Choice Questions

Select the correct answer from the answers provided. For questions with blanks, choose the combination of answers that most accurately completes the statement.

1. What is/are the usual habitat(s) of endospore-forming bacteria that are agents of disease?
 a. intestines of animals c. water
 b. soil d. foods

2. Most *Bacillus* species are
 a. true pathogens c. nonpathogens
 b. opportunistic pathogens d. commensals

3. Many clostridial diseases require a/an _____ environment for their development.
 a. living tissue c. aerobic
 b. anaerobic d. low pH

4. *Clostridium perfringens* causes
 a. myonecrosis c. antibiotic-induced colitis
 b. food poisoning d. both a and b

5. The action of tetanus exotoxin is on the
 a. neuromuscular junction
 b. sensory neurons
 c. spinal interneurons
 d. cerebral cortex

6. The action of botulinum toxin is on the
 a. spinal nerves c. neuromuscular junction
 b. cerebellum d. smooth muscle

7. An infection peculiar to swine causes _____ when transmitted to humans.
 a. anthrax c. tuberculosis
 b. diphtheria d. erysipeloid

8. TB is spread by
 a. contaminated fomites c. respiratory droplets
 b. food d. vectors

9. The form of leprosy associated with severe disfigurement of the face is
 a. paucibacillary c. facio-cranial
 b. multibacillary d. papular

10. Soil mycobacteria can be the cause of
 a. tuberculosis c. fish tank granuloma
 b. leprosy d. erysipeloid

11. Caseous lesions containing inflammatory white blood cells are
 a. lepromas c. eschars
 b. pseudomembranes d. tubercles

12. Which infectious agent is an obligate parasite?
 a. *Mycobacterium tuberculosis*
 b. *Corynebacterium diphtheriae*
 c. *Mycobacterium leprae*
 d. *Clostridium difficile*

13. Which infection(s) would be categorized as a zoonosis?
 a. anthrax c. gas gangrene
 b. diphtheria d. both a and b

14. Actinomycetes are _____ that cause _____.
 a. filamentous rods, chronic granuloma infections
 b. pleomorphic bacteria, erysipeloid
 c. sporeformers, tooth decay
 d. hyphae, pseudomembranes

15. **Multiple Matching.** Choose all descriptions that fit.
 _____ *Bacillus* 1. is acid-fast
 _____ *Clostridium* 2. cells irregular, pleomorphic
 _____ *Mycobacterium* 3. can form metachromatic granules
 _____ *Corynebacterium* 4. regular-shaped rods
 _____ *Listeria* 5. forms endospores
 6. is psychrophilic
 7. is primarily anaerobic
 8. is aerobic
 9. is associated with dairy products
 10. cells can be elongate filaments
 11. shows palisades arrangement

16. **Matching.** Match the disease with the principal portal of entry.
 _____ 1. anthrax a. skin
 _____ 2. botulism b. gastrointestinal tract
 _____ 3. gas gangrene c. traumatized tissue
 _____ 4. antibiotic colitis d. respiratory tract
 _____ 5. tetanus e. urogenital tract
 _____ 6. diphtheria f. placenta
 _____ 7. listeriosis g. none of these
 _____ 8. tuberculosis
 _____ 9. leprosy

Case Study Review

1. Which of the conditions below would make someone more susceptible to infection with *Listeria monocytogenes*?
 a. being pregnant
 b. being elderly
 c. undergoing chemotherapy
 d. all of these would make a person more susceptible

2. During this outbreak, some people sickened with listeriosis tested positive for *Listeria monocytogenes* in their bloodstream, but a DNA profile of the bacterium showed that it was not the outbreak strain. What likely conclusion can be drawn from this information?
 a. It is a false positive.
 b. The person is not part of the outbreak.
 c. *Listeria monocytogenes* mutates frequently, making DNA profiling unreliable.
 d. The person has already begun to recover from the disease.

3. No *Listeria monocytogenes* was discovered in the field where the cantaloupes were grown, and the original source of the bacterium in this case was never found. The FDA report on the farm made mention of a truck that was used to carry rotted or broken cantaloupe to a nearby ranch where they were fed to cattle. What could be the importance of this fact, and what steps would you take to investigate it further?

Writing Challenge

For each question, compose a one- or two-paragraph answer that includes the factual information needed to completely address the question. Check Your Progress questions can also be used for writing-challenge exercises.

1. a. What is the role of spores in infections?
 b. Describe the general distribution of spore-forming bacteria.

2. a. What characteristics of *Clostridium* contribute to its pathogenicity?
 b. Compare the toxigenicity of tetanospasmin and botulinum toxin.
 c. What predisposes a patient to clostridial infection?

3. a. Outline the epidemiology of the major wound infections and food intoxications of *Clostridium*.
 b. What is the origin of the gas in gas gangrene?

4. a. What causes the jaw to "lock" in lockjaw?
 b. How would it be possible for patients with no noticeable infection to present with symptoms of tetanus?

5. How are botulism and tetanus alike and different?

6. a. Why is listeriosis a serious problem even with refrigerated foods?
 b. Which groups are most at risk for serious complications?

7. Why is erysipeloid an occupation-associated infection?

8. a. What are the distinctive morphological traits of *Corynebacterium?*
 b. How can the pseudomembrane be life threatening?
 c. What is the ultimate origin of diphtherotoxin?

9. a. Outline the unique characteristics of *Mycobacterium.*
 b. What is the epidemiology of TB?
 c. Differentiate between TB infection and TB disease.

d. What are tubercles?
e. Describe the applications of BCG vaccine.

10. a. What characteristics make *M. leprae* different from other mycobacteria?
 b. Differentiate between tuberculoid and lepromatous leprosy.

11. a. What is the importance of NTM?
 b. Describe the effects of *Mycobacterium avium* complex in AIDS patients.

12. a. Describe the bacteria in the actinomycete group, and explain what characteristics make them similar to fungi.
 b. Briefly describe two of the common diseases caused by this group.

 Concept Mapping

On Connect you can find an Introduction to Concept Mapping that provides guidance for working with concept maps along with concept mapping activities for this chapter.

 Level II. Application, Analysis, Evaluation, and Synthesis

These problems go beyond just restating facts and require higher levels of understanding and an ability to interpret, problem solve, transfer knowledge to new situations, create models, and predict outcomes.

 Critical Thinking

Critical thinking is the ability to reason and solve problems using facts and concepts. These questions can be approached from a number of angles, and in most cases, they do not have a single correct answer.

1. a. What is the main clinical strategy in preventing gas gangrene?
 b. Why does it work?

2. a. Why is it unlikely that diseases such as tetanus and botulism will ever be completely eradicated?
 b. Name some bacterial diseases in this chapter that could be completely eradicated and explain how.

3. Why is the cause of death similar in tetanus and botulism?

4. a. Why does botulinum toxin not affect the senses?
 b. Why does botulism not commonly cause intestinal symptoms?

5. Account for the fact that boiling does not destroy botulism spores but does inactivate botulinum toxin.

6. Adequate cooking is the usual way to prevent food poisoning. Why doesn't it work for *Clostridium perfringens* and *Bacillus* food poisoning?

7. a. Why do patients who survive tetanus and botulism often have no sequelae?
 b. How has modern medicine improved the survival rate for these two diseases?

8. What would be the likely consequence of diphtheria infection alone without toxemia?

9. How can one tell that acne involves an infection?

10. Do you think the spittoons of the last century were effective in controlling tuberculosis? Why or why not?

11. a. Provide an explanation for the statement that TB and leprosy are "family diseases."
 b. What, if anything, can be done about multidrug-resistant tuberculosis?
 c. Explain an important rationale for *not* administering BCG vaccine in the United States to the general public.

12. Carefully compare figures 19.11 and 19.23.
 a. Describe the main differences observable in the lesions of these two conditions.
 b. Explain how you would go about diagnosing them differentially.

13. Which diseases disucssed in this chapter have no real portal of exit?

14. Eighty-six people at a St. Patrick's Day celebration developed diarrhea, cramps, and vomiting after eating a traditional dinner of corned beef, cabbage, potatoes, and ice cream. Symptoms appeared within 10 hours of eating, on average. Of those afflicted, 85 had eaten corned beef and one had not. The corned beef had been cooked in an oven the day before the event, stored in the refrigerator, and sliced and placed under heat lamps 1½ hours before serving.
 a. Use this information to diagnose the food poisoning agent.
 b. What sorts of lab tests could one do to verify it as this bacterium?
 c. What would you have done to prevent this incident?

15. An outbreak of gastrointestinal illness was reported in a day care center. Of 67 people, 14 came down with symptoms of nausea, cramps, and diarrhea within 2 to 3 hours of eating. Food items included fried rice, peas, and apple rings. Symptoms occurred in 14 of the 48 who had eaten the fried rice and in none of those who had

not eaten it. The rice had been cooked the night before and refrigerated. The next morning, it had been fried with leftover chicken and held without refrigeration until lunchtime.

a. On the basis of symptoms and food types, what do you expect the etiological agent to be?

b. What would you expect to culture from the food?
c. To what other disease is this one similar?
d. How can it be differentiated from the agent and disease in question 14?

 Visual Challenge

1. From **chapter 7, figure 7.11.** Which of the thioglycollate tubes in the figure would be indicative of *Bacillus* and which would be characteristic of *Clostridium?* Explain why this is true.

© Terese M. Barta, Ph.D.

2. From **chapter 3, figure 3.8.** What type of bacterial infection could be diagnosed using the stain shown in the figure? What does this result indicate?

© Steven R. Spilatro, Department of Biology, Marietta College, Marietta, OH

STUDY TIP: Quickly create a customized practice quiz for this chapter by going to your Connect SmartBook "My Reports" section. Don't have SmartBook and want to learn more? Go to **whysmartbook.com.**

The Gram-Negative Bacilli of Medical Importance

2011

2012

No. of Cases

Bordetella pertussis, the gram-negative bacterium responsible for whooping cough, as viewed through a scanning electron microscope at 5,000x.
© Dennis Kunkel Microscopy, Inc./Phototake

Number of Confirmed and Probable Pertussis Cases Reported, by Week of Onset—Washington, January 1, 2011—June 16, 2012.

CASE STUDY Part 1 The More Things Change, the More They Stay the Same

On April 3, 2012, in response to an increase in the number of reported cases of pertussis, the Washington State Secretary of Health declared an epidemic. Within 11 weeks the number of **reported** cases in the state had soared to 2,520 for the year, an increase of 1,300% compared to the same period in 2011 and the highest number of cases since 1942. Pertussis, commonly known as *whooping cough,* looks at first like nothing more sinister than a common cold—runny nose, coughing, and congestion—but soon worsens to include uncontrollable coughing followed by a deep inhalation of air whistling through the congested larynx, causing the "whooping" sound for which the disease is known.

Whooping cough results from infection with ***Bordetella pertussis,*** a gram-negative bacillus that spreads rapidly from person to person. Historically, infants and young children are hardest hit by the disease, with infants having the highest incidence rate and accounting for the

> "It's the largest epidemic I've ever seen. They don't have time to call and say who's positive and who's negative."
>
> **Becky Neff, RN**

majority of deaths. Due primarily to an aggressive vaccination campaign in the United States, pertussis cases dropped from an average of 175,000 a year in the 1940s to 1,010 cases by 1976. While the group most affected by the 2012 Washington State outbreak were children under 1 year of age, high rates of disease were seen in 10-, 13-, and 14-year-olds, despite the fact that most had completed the entire five-dose vaccination series recommended by the ACIP (Advisory Committee for Immunization Practices).

- *How is* Bordetella pertussis *spread?*
- *What conclusion can be drawn from the fact that many adolescents who had completed the immunization series were still hard hit by pertussis?*

To continue the Case Study, go to Case Study Part 2 at the end of the chapter.

The gram-negative bacilli comprise a large group of non-spore-forming bacteria adapted to a wide range of habitats and niches. Members of this group are classified in Phylum Proteobacteria, a large assembly of 500 different genera that are highly diverse in metabolism and pathogenicity. Because the group is so large and complex and many of its members are not medically important, we have included mainly those that are human pathogens.

A large number of representative genera are inhabitants of the large intestine (enteric); some are zoonotic; some are adapted to the human respiratory tract; and still others live in soil and water. It is useful to distinguish between true pathogens (*Salmonella, Yersinia pestis, Bordetella,* and *Brucella,* for example), which are infectious to the general population, and the opportunists (*Pseudomonas* and coliforms), which may be resident microbiota or saprobes that cause infection in people with weakened host defenses (**Systems Profile 20.1**).

20.1 Aerobic Gram-Negative Nonenteric Bacilli

 Expected Learning Outcomes

1. Explain the effect of the virulence factor common to all gram-negative bacteria.
2. Identify those people most at risk of developing a *Pseudomonas* infection.

The aerobic gram-negative bacteria of medical importance are a loose assortment of genera including the soil bacteria *Pseudomonas* and *Burkholderia;* the zoonotic pathogens *Brucella* and *Francisella;* and *Bordetella* and *Legionella,* which are mainly human pathogens. The gram-negative rods can be organized in several ways. To simplify coverage here, they are grouped into three major categories based on oxygen requirements and fermentation of lactose (**table 20.1**).

One notable component of all gram-negative rods is a lipopolysaccharide in the outer membrane of the cell wall that acts as an **endotoxin** (see chapter 13). The release of endotoxin in the blood from gram-negative septicemia can have severe and far-reaching pathophysiological effects (**20.1 Making Connections**).

Pseudomonas: The Pseudomonads

The pseudomonads are members of the Family Pseudomonadaceae, a large group of free-living bacteria that live primarily in soil, seawater, and fresh water. They also colonize plants and animals and are frequent contaminants in homes and clinical settings. These small, gram-negative rods have a single polar flagellum (**figure 20.1**), produce oxidase and catalase, and do not ferment carbohydrates. Although they ordinarily obtain energy aerobically through oxidative metabolism, some species can grow anaerobically if provided with a salt such as nitrate. Many species produce green, brown, red, or yellow pigments that diffuse into the medium and change its color.

*Pseudomonas** species are highly versatile. They can adapt to a wide range of habitats and manage to extract needed energy from minuscule amounts of dissolved nutrients. Most members can grow in a medium containing minerals and a simple organic compound. This adaptability accounts for their constant presence in the environment and their capacity to thrive on hosts. They degrade

* *Pseudomonas* (soo″-doh-moh′-nas) Gr. *pseudes,* false, and *monas,* a unit.

TABLE 20.1 Gram-Negative Pathogens

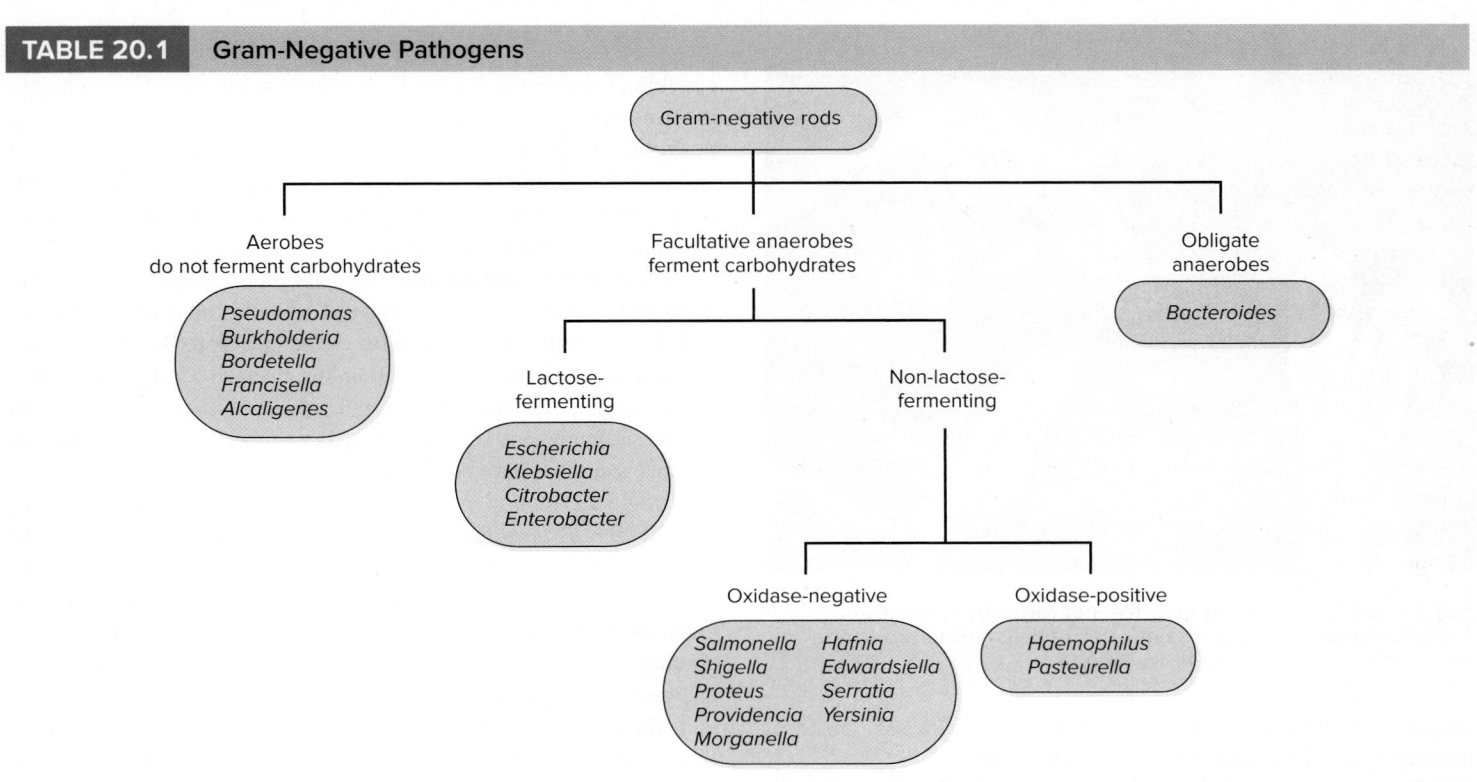

Systems Profile 20.1	Pathogenic Gram-Negative Bacilli					
Organism	Skin/Skeletal	Nervous/ Muscle	Cardiovascular/ Lymphatic/ Systemic	Gastrointestinal	Respiratory	Urogenital
Pseudomonas aeruginosa	Complication of burns, skin rashes	Corneal ulcers, meningitis	Endocarditis		Complication of cystic fibrosis	Urinary tract infections
Francisella tularensis			Tularemia			
Bordetella pertussis					Whooping cough	
Legionella spp.					1. Pneumonia 2. Pontiac fever	
Escherichia coli				Acute diarrhea		Urinary tract infections
Escherichia coli O157:H7				Acute diarrhea Colitis		Hemolytic uremic syndrome
Klebsiella pneumoniae	Wound infections	Meningitis			Pneumonia	Urinary tract infections
Proteus spp.	Wound infections				Pneumonia	Urinary tract infections
Salmonella typhi			Septicemia	Typhoid fever		
Salmonella spp.				Acute diarrhea		
Shigella spp.		Neuron damage		Acute diarrhea and bacillary dysentery		
Yersinia pestis			Bubonic plague		Pneumonic plague	
Haemophilus influenzae	Conjunctivitis	Meningitis				

Figure 20.1 Scanning electron micrograph of several *Pseudomonas aeruginosa* cells (large rods) attached to the skin surface. *Source:* Janice Haney Carr/CDC

numerous extracellular substances by means of protease, amylase, pectinase, cellulase, and numerous other enzymes.

Pseudomonads have extensive impact on ecology, agriculture, and commerce. They are important decomposers and bioremediators,

capable of degrading hundreds of natural substrates. These characteristics make them useful in cleaning up oil spills and clearing pesticides (see chapter 27). Various species have roles as food spoilage agents and plant pathogens. Some species produce antibiotics termed pseudomycins that are being tested for their effectiveness in treating fungal infections.

Pseudomonas aeruginosa is a common inhabitant of soil and water and an intestinal resident in about 10% of healthy people. On occasion it can be isolated from saliva or even the skin. Because the species is resistant to soaps, dyes, certain disinfectants, drugs, and drying, it is a frequent contaminant of ventilators, intravenous solutions, and anesthesia equipment. Even disinfected instruments, utensils, bathroom fixtures, and mops have been incriminated in hospital outbreaks.

In a pattern similar to that of the enteric bacteria discussed in section 20.3, *P. aeruginosa* is a typical opportunist. It is unlikely to cross healthy, intact anatomical barriers; thus, its infectiousness results from invasive medical procedures or weakened host defenses. Once in the tissues, *P. aeruginosa* expresses virulence factors including exotoxins, a phagocytosis-resistant slime layer, and various enzymes that degrade host tissues. It also causes endotoxic shock.

The most common health-care associated *Pseudomonas* infections occur in compromised hosts with severe burns, neoplastic

* *aeruginosa* (uh-roo″-jih-noh′-suh) L. *aeruginosa,* full of blue-green copper rust.

20.1 MAKING CONNECTIONS

Gram-Negative Sepsis and Endotoxic Shock

Blood infection is one of the most serious consequences with gram-negative bacteria, even if the species involved is not a virulent pathogen. Nationwide, over 100,000 patients a year die from a condition called **septic shock.** Shock is a pathologic state of low blood pressure accompanied by a reduced amount of blood circulating to vital organs, particularly the brain, heart, lungs, and kidneys. Its major symptoms and signs are nausea, tachycardia, cold and clammy skin, and weak pulse. Damage to the organs can elicit respiratory failure, coma, heart failure, and death in a few hours. Although the endotoxins of all gram-negative bacteria can cause shock, the most common clinical cases are due to gram-negative rods.

bacteria to release profuse amounts of LPS into the bloodstream. Some of the cytokines are pyrogenic and induce fever, some activate the complement cascade, and others promote intravascular blood coagulation. The end results are circulation failure, tissue damage, hypotension, and shock.

Ironically, traditional treatments with antibiotics can compound the problem of septic effects because these drugs can increase the lysis of bacteria and the release of endotoxins. Drug companies are actively developing new drugs that will block the effects of endotoxin. One clinical

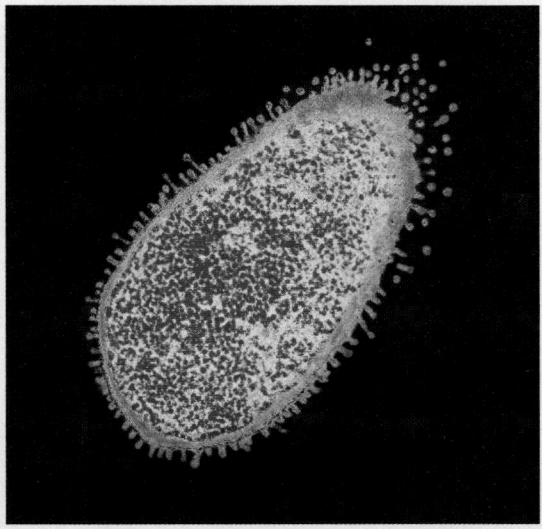

LPS
Polysaccharide
Lipid A
Peptidoglycan

Gram-negative bacilli

Section of cell wall and outer membrane with lipopolysaccharide (LPS) layer

Composition of one LPS unit. Lipid A is the most responsible for pathogenesis.

Section of gram-negative cell wall showing the location and structure of lipopolysaccharide.

Micrograph of gram-negative rod in the process of lysis. Small blue spheres are molecules of lipopolysaccharide being shed. © Dr. Kari Lounatmaa/Science Source

The adverse factor in gram-negative sepsis is the presence of an endotoxin called **lipopolysaccharide (LPS)** in the outer membrane of the gram-negative cell wall. The component that accounts for most of the adverse effects is *lipid A* embedded in the external layer of the membrane. This lipid is active only after it is liberated by bacteria that are growing or lysed by host defenses and other factors. In general, lipid A is a potent immune stimulator that triggers the secretion of interleukins, tumor necrosis factor, and other cytokines by macrophages. In a local infection with only small amounts of endotoxin, the effects of the cytokines are protective. But massive sepsis attracts large numbers of macrophages that cause the dying

trial is testing an endotoxin antagonist that would work on all forms of lipid A. Other possible treatments contain monoclonal antibodies specific for endotoxin. Although these drugs are not yet available, they show some promise in controlling the more severe effects of endotoxemia.

Describe a scenario that could lead to shock, beginning with a hospitalized diabetic patient, and that ends with endotoxemia. *Answer available on Connect.*

disease, and premature birth. Cystic fibrosis patients in particular suffer from a lung condition in which mucus collects and creates a protective biofilm tenaciously colonized by *Pseudomonas aeruginsa. Pseudomonas*-related septicemia can disseminate into such conditions as endocarditis, meningitis, and bronchopneumonia. Even healthy people can be subject to outbreaks of skin rashes, urinary tract infections, and external ear infections picked up in community hot tubs and swimming pools. Sponges and washcloths are common sources of these bacteria and can lead to local infections when they are used to clean skin. **(figure 20.2).** Wearers of contact lenses are also vulnerable to corneal ulcers from contaminated storage and disinfection solutions.

Unusual characteristics of *P. aeruginosa* infections are a grapelike odor and the noticeable color that appears in tissue

Figure 20.2 *Pseudomonas* **skin infection.** The subcutaneous blue color is due to the production of a blue-green pigment that diffuses out from the infection. © DermNet New Zealand Trust

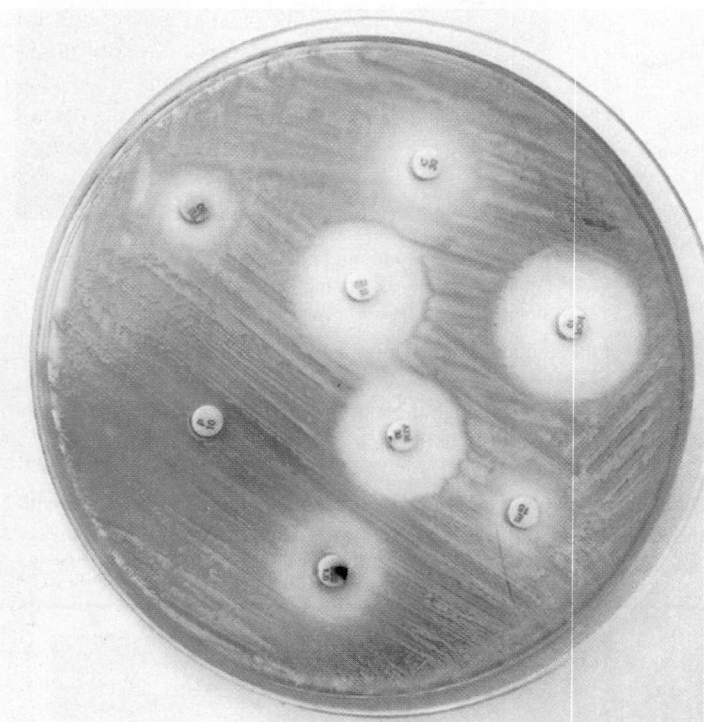

Pathogen Profile #1 *Pseudomonas aeruginosa*

Source: Janice Carr/CDC

Microscopic Morphology Small, gram-negative rods with a single polar flagellum.

Identified by Gram reaction and cellular morphology. *P. aeruginosa* is aerobic and does not ferment lactose, traits that distinguish it from members of the Enterobacteriaceae. Cultures typically produce a greenish pigment that fluoresces under ultraviolet light, as well as a distinctive odor.

Habitat *P. aeruginosa* is commonly found in soil and water and is an intestinal resident of about 10% of healthy people. The species is a frequent colonist of plants and animals and a routine contaminant of the home and clinical setting.

Virulence Factors Virulence factors include pili that aid in attachment of the bacterium to host cells, a phagocytosis-resistant slime layer that contributes to the formation of biofilms, enzymes that degrade host tissues, and exotoxins that damage or destroy neutrophils and lymphocytes. The LPS layer of the outer membranes also causes endotoxic shock.

Primary Infections/Disease *P. aeruginosa* is a prime example of an opportunistic pathogen, with the most serious infections seen in patients with severe burns, neoplastic disease, or cystic fibrosis, in many cases leading to endocarditis, bronchopneumonia, or meningitis. Otherwise-healthy persons may acquire skin rashes, urinary tract infections, or outer ear infections, often connected to the use of community hot tubs and swimming pools.

Control and Treatment The metabolic versatility of *Pseudomonas* makes controlling the bacterium difficult; soap, mops, and sponges are all commonly implicated in outbreaks. Effective treatment makes susceptibility testing an absolute necessity. Effective drugs include third- and fourth-generation cephalosporins, aminoglycosides, and polymixins.

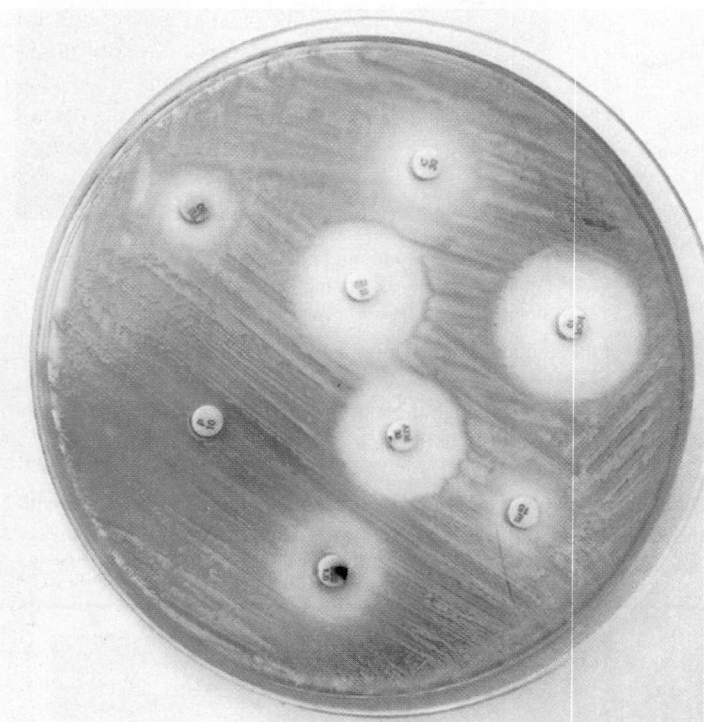

Figure 20.3 Characteristics of *Pseudomonas aeruginosa*. *Pseudomonas* produces a blue to green pigment that diffuses into the medium. *Pseudomonas* also commonly displays drug resistance, as demonstrated by the weak or incomplete zones of inhibition surrounding several of the antibiotic discs. © Kathy Park Talaro

exudates ("blue pus"). The color is due to the bacterium's production of a blue-green or greenish-yellow pigment (pyocyanin) that can fluoresce **(figure 20.3)**. Lab identification relies on a battery of tests similar to those given for the Enterobacteriaceae (see figure 20.9). The notorious multidrug resistance of *P. aeruginosa* makes testing for the drug sensitivity of most isolates

imperative. Drugs which have had some success in controlling *Pseudomonas* infection include third- and fourth-generation cephalosporins, aminoglycosides such as amikacin, and polymixins like colistin.

 Check Your Progress SECTION 20.1

1. Briefly describe the human infections caused by *Pseudomonas*.
2. How can antibiotic treatment of a gram-negative infection lead to septic shock?

20.2 Related Gram-Negative Aerobic Rods

 Expected Learning Outcomes

3. Name the genera of bacteria that are physiologically similar to *Pseudomonas* and also account for a variety of opportunistic infections.
4. Outline the pathology and epidemiology of *Brucella*.
5. Explain the epidemiology of *Francisella tularensis* and the pathology of infection with the bacterium.
6. Recall the epidemiology of pertussis.
7. Describe the stages of whooping cough, and identify those persons most likely to experience the disease.
8. Identify the bacterium responsible for legionellosis and the environmental conditions that promote its growth.

One group of proteobacteria is represented by the genera *Burkholderia* and *Stenotrophomonas*. These genera share similarities with *Pseudomonas* in many of their physiological and pathogenic characteristics. Both are obligate aerobes that can utilize a variety of organic substrates but do not ferment sugars. They can be readily

Dense infection with nodules

(a)

(b)

Figure 20.4 Lung Infection due to *Burkholderia pseudomallei.* **(a)** X ray indicates cavities in the middle lung zone where infection is concentrated. **(b)** Gram stain of sputum reveals typical gram-negative rods with bipolar staining (1,500×). (a) *Source:* Dr. Thomas Hooten/CDC; (b) *Source:* Andrew J. Brent, "Misdiagnosing Melioidosis," *Emerging Infectious Diseases*, 13(2), February 2007, Fig. A. Centers for Disease Control and Prevention

cultured on blood agar, MacConkey agar, and other common lab media. They occupy a wide variety of habitats in soil, water, and related environmental locations, where they have established niches in nutrient recycling, but are also known as opportunists that readily invade a compromised human host.

Burkholderia cepacia is a common isolate from moist environments; it appears to play an important ecological role in controlling fungal disease on plants. It is also active in biodegradation of a wide variety of toxic substances, which would ordinarily make it a useful addition to bioremediation programs. Unfortunately it also has a potential for opportunistic infections in the respiratory tract, the urinary tract, and occasionally the skin. Pneumonia caused by *B. cepacia* holds heightened risks for cystic fibrosis patients, whose lung abnormalities favor the entrenchment of pathogens. Because of its tendency to mutate rapidly and resist antimicrobic drugs, finding a successful treatment can be a challenge. Coinfection with *Pseudomonas* may lead to severe respiratory compromise and death in 80% of patients.

Burkholderia pseudomallei is a relative of *B. cepacia* that resides in the soil and water of tropical climates. The disease it causes, melioidosis, is prevalent in many areas of Southeast Asia and parts of Africa, India, and the Middle East. It is generally acquired through a penetrating injury by a soiled object or inhalation from an environmental reservoir rather than from an infected person. Manifestations of wound infection are localized skin nodules along with symptoms of fever and muscle aches. Lung infection gives rise to bronchitis and pneumonia **(figure 20.4)**, which are attended by cough and chest pain. Chronic infection may progress to septicemia, endotoxic shock, and abscesses in organs such as the liver and brain. In persons with compromised immune systems, the infection can cause severe disseminated disease that is occasionally fatal.

Stenotrophomonas maltophilia. The natural habitat of this bacterium appears to be fresh water and soil associated with plant roots. It is also a member of the fecal microbiota in humans and may be isolated from domestic animals. It appears in the clinical setting primarily as a contaminant of disinfectants, dialysis equipment, respiratory equipment, water dispensers, and catheters. One factor that favors colonization of the environment appears to be its formation of biofilms, which protect it against drugs and disinfectants.

It is a prominent clinical isolate most often found in respiratory, blood, spinal fluid, and ophthalmic specimens. Higher rates of infection occur in cancer patients and people with indwelling invasive devices. Although this species is one of the most drug-resistant of bacteria, therapy with a combination of trimethoprim and sulfamethoxazole is usually effective.

Brucella and Brucellosis

Undulant fever, Malta fever, and Bang disease[1] are synonyms for **brucellosis,** a zoonosis transmitted to humans from infected animals or contaminated animal products harboring ***Brucella.**** Two species of these tiny, gram-negative coccobacilli are *B. abortus* (from cattle) and *B. suis* (from pigs). Animal brucellosis results from an infection of the placenta and fetus that can cause abortion, as typified by Bang disease in cattle. Humans infected with either of these two agents experience a severe febrile illness but not abortion.

Brucellosis occurs worldwide, with concentrations in Europe, Africa, India, and Latin America. It is associated predominantly with occupational contact in slaughterhouses, livestock handling, and the veterinary trade. Infection takes place through contact with blood, urine, placentas, and consumption of raw milk and cheese. Human-to-human transmission is rare. Brucellosis is also a common disease of wild herds of bison and elk. Cattle sharing grazing land with these wild herds often suffer severe outbreaks of Bang disease.

In humans, *Brucella* enters through damaged skin or mucous membranes of the digestive tract, conjunctiva, and respiratory tract.

One of *Brucella*'s most effective virulence factors lies in its capacity to survive and grow inside of macrophages. These infected macrophages transport the pathogen into the bloodstream, creating focal lesions in the liver, spleen, bone marrow, and kidney. The cardinal manifestation of human brucellosis is a fluctuating pattern of fever, which is the origin of the common name, undulant fever. It is also accompanied by chills, profuse sweating, headache, muscle pain and weakness, and weight loss. Fatalities are not common, though the syndrome can last for a few weeks to a year, even with treatment. Such chronic infections can be linked to the hidden bacteria carried by macrophages.

The patient's history can be very helpful in diagnosis, as are serological tests of the patient's blood **(figure 20.5)** and blood culture of the pathogen. Newer genetic tests are being developed but are not yet widely used. A combination of doxycycline and streptomycin taken for 6 weeks is usually effective in controlling infection. Prevention is effectively achieved by testing and elimination of infected animals, quarantine of imported animals, and pasteurization of milk. Though several types of animal vaccines are available, those developed for humans are ineffective or unsafe. The status of this pathogen as a potential germ warfare agent makes the need for a reliable vaccine even more urgent.

Quick Search
To learn more about agents of bioterrorism, search the CDC website for "bioterrorism."

1 After B. L. Bang, a Danish physician.
* *Brucella* (broo-sel′-uh) After David Bruce, who isolated the bacterium.

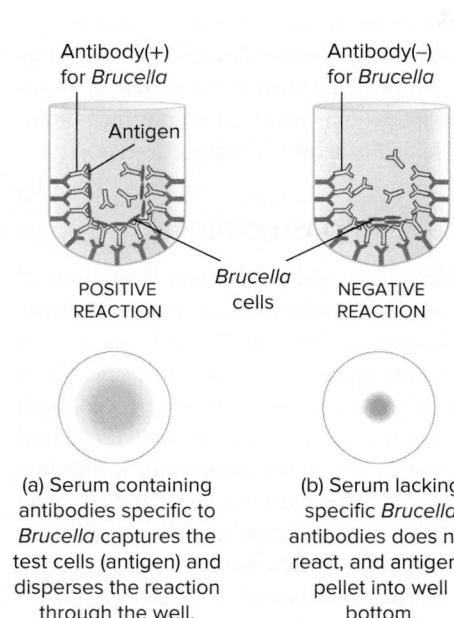

(a) Serum containing antibodies specific to *Brucella* captures the test cells (antigen) and disperses the reaction through the well.

(b) Serum lacking specific *Brucella* antibodies does not react, and antigens pellet into well bottom.

(c) The numbered lanes are individual serum samples. The serum titer can be determined by diluting it through wells A to H, then looking for the last well in a series with a positive reaction. The titer is the dilution of that well.

Figure 20.5 **Agglutination titer test for brucellosis and its interpretation.** See the Visual Challenge at the end of this chapter for an enlargement of part c. (c) Picture courtesy of Vircell

Francisella tularensis and Tularemia

*Francisella tularensis** is the causative agent of **tularemia,** a zoonotic disease of assorted mammals endemic to the Northern Hemisphere. Its characteristics are similar enough to *Yersinia pestis,* the cause of plague (discussed in section 20.5) that the two species were once included in the same genus, *Pasteurella.* Because tularemia has been associated with outbreaks of disease in wild rabbits, it is sometimes called *rabbit fever.*

Tularemia is distributed throughout numerous animal reservoirs and vectors in northern Europe, Asia, and North America but not in the tropics. This disease is noteworthy for its complex epidemiology and spectrum of symptoms. Although rabbits and rodents (muskrats and ground squirrels) are the chief reservoirs, other wild animals (skunks, beavers, foxes, opossums) and some domestic animals are implicated as well. Of the more than 50 bloodsucking arthropod vectors known to carry *F. tularensis,* about half also bite humans. Ticks are most often involved, followed by biting flies, mites, and mosquitoes.

Tularemia is strikingly varied in its portals of entry and disease manifestations. In a majority of cases, infection results when the skin or eye is inoculated through contact with infected animals, animal products, contaminated water, and dust. Vector bites are also a frequent source of infection, but the disease is not communicated from human to human. With an estimated infective dose of between 10 and 50 organisms, *F. tularensis* is often considered one of the most infectious of all bacteria. Along with plague and several other virulent pathogens, it is currently listed as a pathogen of concern on the lists of bioterrorism agents.

Francisella is similar to *Brucella* in being an intracellular pathogen that survives and spreads in macrophages to multiple sites. After an incubation period ranging from a few days to 3 weeks, acute symptoms of headache, backache, fever, chills, and malaise appear. Clinical manifestations include ulcerative skin lesions, swollen lymph glands, conjunctival inflammation, sore throat, intestinal disruption, and pulmonary involvement. The death rate in systemic and pulmonic forms is 10%, but proper treatment with gentamicin or streptomycin reduces it to almost zero. Because the intracellular persistence of *F. tularensis* can lead to relapses, antimicrobial therapy must not be discontinued prematurely. Protection is afforded by protective gloves, masks, and eyewear for laboratory workers and other occupationally exposed personnel. No vaccine is currently available.

Bordetella pertussis and Relatives

*Bordetella pertussis** is a minute, encapsulated coccobacillus responsible for **pertussis,** or **whooping cough,** a communicable childhood affliction that causes an acute respiratory syndrome. Contrary to the common impression that the disease is mild and self-limiting, it often causes severe, life-threatening complications in babies. *Bordetella parapertussis* is a closely related species that causes a milder form of the infection.

The primary route of infection is direct contact with inhaled droplets or aerosols given off during the coughing stage of infection. About half of pertussis cases occur in children between birth and 4 years of age.

The primary virulence factors of *B. pertussis* are (1) adhesion molecules that specifically recognize and bind to ciliated respiratory epithelial cells and (2) toxins that destroy and dislodge ciliated cells (a primary host defense). The loss of the ciliary mechanism leads to buildup of mucus and blockage of the airways. The initial phase of pertussis is the *catarrhal stage,* marked by nasal drainage and congestion, sneezing, and occasional coughing. As symptoms deteriorate into the *paroxysmal stage,* the child experiences recurrent, persistent coughing. Fits of several abrupt, hacking coughs are followed by a deep inhalation that pulls air through the congested larynx and produces a "whoop" **(figure 20.6).** Many of the complications of pertussis are due to compromised respiration. Standard therapy and prophylaxis requires a weeklong course of azithromycin or clarithromycin.

An acellular vaccine (aP) is available for all routine immunizations, starting at 6 weeks of age. This vaccine contains toxoid and other antigens and is well tolerated when given with DT vaccine (DTaP). The vaccine is administered in a five-dose series between

> **Quick Search**
> To better understand why pertussis is known as whooping cough, point your browser to "Sounds of Whooping Cough."

* *Francisella tularensis* (fran-sih-sel′-uh too-luh-ren′-sis) After Edward Francis, one of its discoverers, and Tulare County, California, where the agent was first identified.

* *Bordetella pertussis* (bor-duh-tel′-uh pur-tus′-is) After Jules Bordet, a discoverer of the agent, and L. *per,* severe, and *tussis,* cough.

Pathogen Profile #2 *Bordetella pertussis*

Source: Janice Carr/CDC

Microscopic Morphology Small gram-negative coccobacillus typically found singly or in pairs. It is encapsulated and nonmotile.

Identified by Gram reaction, cellular morphology, and motility. The bacterium is aerobic and is distinguished from similar species using biochemical tests. *B. pertussis* is oxidase positive but urease, nitrate reductase, and citrate negative. The polymerase chain reaction is used in many laboratories to detect *B. pertussis* DNA.

Habitat Humans, and perhaps some higher primates, are the only known reservoirs.

Virulence Factors The primary virulence factors are fimbriaelike adhesion molecules that allow *B. pertussis* to recognize and bind to ciliated respiratory epithelial cells and exotoxins that destroy these host cells after the bacterium has bound.

Primary Infections/Disease *B. pertussis* is the etiological agent of whooping cough, which occurs in two stages. The catarrhal stage

occurs as mucus builds up in the airways and is marked by nasal drainage, congestion, sneezing, and occasional coughing. The second, or paroxysmal, stage manifests as episodes of persistent coughing followed by deep inhalations that produce a characteristic "whoop" as air is pulled through the congested larynx. Complications of pertussis are generally due to compromised respiration. The disease is responsible for only a few deaths a year in the United States but as many as 300,000 worldwide.

Control and Treatment Even with a robust vaccination program, pertussis has reached epidemic levels in many parts of the United States, primarily due to the fact that the childhood vaccine series does not provide long-term protection. Teenagers and adults often contract a mild form of the disease that can nevertheless be passed on to infants who are not yet vaccinated and have a much greater risk of contracting serious disease. The vaccine consists of a five-dose series with an additional booster given to adolescents and adults. Standard therapy for pertussis is a 1-week course of azithromycin or clarithromycin.

Figure 20.6 Young child suffering the coughing stages of pertussis (whooping cough). The majority of young children with this condition will require hospitalization for monitoring. *Source:* CDC

ages 2 months and 6 years. Due to changes in disease epidemiology **(figure 20.7),** it is clear that more booster doses are necessary. In response, the Centers for Disease Control and Prevention (CDC) now recommends an additional booster containing diphtheria and tetanus antigens, along with an acellular pertussis component, at 11 or 12 years of age. Further isolation and barrier precautions are implemented in child and health care settings to prevent infected workers from coming in contact with children or infants, who represent the highest-risk group for serious disease.

In 2013 a strain of pertussis was identified that was missing the protein pertactin, a protein that helps the bacteria attach to the

lining of the airways. Pertactin is one of several components of all pertussis vaccines. The CDC is currently conducting studies in the United States to determine whether an increase in pertactin deficient *B. pertussis* strains is one of the factors contributing to the increase in the number of reported pertussis cases. Current evidence suggests however that pertussis vaccines continue to prevent disease caused by both pertactin-positive and pertactin-negative pertussis strains since other components of the vaccines provide protection.

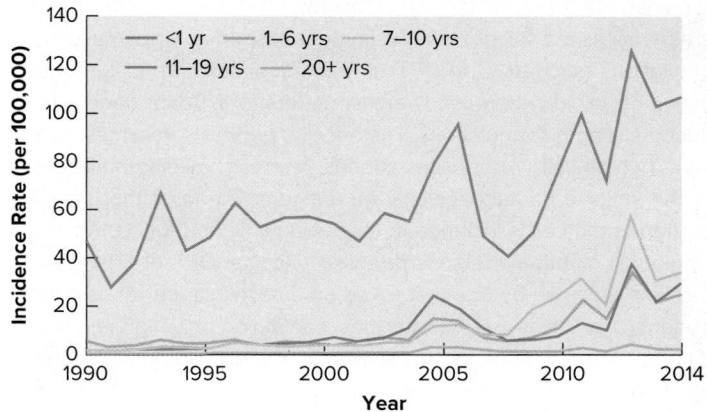

Figure 20.7 Incidence of pertussis (whooping cough) in the United States. Despite the availability of a vaccine, the number of pertussis cases continue to rise. Children under one year of age have not received the full vaccination series and are at greatest risk of infection. The CDC now recommends an additional booster at age 11 or 12.

Pathogen Profile #3 *Legionella pneumophila*

Source: Janice Carr/CDC

Microscopic Morphology Small gram-negative rods. *L. pneumophila* is motile and does not produce a capsule.

Identified by *Legionella* is fastidious and difficult to isolate, with disease often diagnosed by symptomatology and patient history. Laboratory identification of the pathogen usually relies on some combination of DNA probes, fluorescent antibody staining of urinary and respiratory specimens, or cultivation on charcoal yeast extract agar, a medium selective for *Legionella*.

Habitat *Legionella* is widely distributed in freshwater habitats, where amoebas in the water serve as the primary reservoir. Human infections are commonly traced to tap water, spas, ponds, air conditioners, and misting systems.

Virulence Factors Most strains contain pili that allow the bacterium to bind to respiratory epithelial cells and have the ability to actively grow within macrophages.

Primary Infections/Disease Two distinct clinical forms of legionellosis exist. Shared symptoms include fever, cough, diarrhea, and abdominal pain. While the less severe form, Pontiac fever, is self-limiting, Legionnaires' disease—the more severe manifestation—progresses to pneumonia and has a mortality rate of 3% to 30%.

Control and Treatment Control is difficult due to the widespread nature of the pathogen, but cleaning and chlorination of artificial aquatic habitats provides some benefits. Levofloxacin is the primary treatment for Legionnaires' disease. Pontiac fever does not require antimicrobic treatment.

The genus *Alcaligenes** is another group of nonfermentative, oxidase-positive, motile bacteria in the same family as *Bordetella.* Members of *Alcaligenes* live primarily in soil and water habitats, and they may also become established as normal microbiota. The most common clinical species, *A. faecalis,* is isolated from feces, sputum, and urine. It is occasionally associated with opportunistic infections such as pneumonia, septicemia, and meningitis.

Minding Your *T*'s and *D*'s

The vaccines used to provide protection against pertussis, diphtheria, and tetanus provide an alphabet soup of acronyms—DPT, DTwP, DTaP, Tdap, DT and Td—that can confuse even trained medical personnel, potentially placing patients in danger. In order to lessen this confusion, we offer this brief primer.

Diphtheria and tetanus vaccines elicit the production of protective antibodies through the use of **toxoids,** inactivated exotoxins that are harmless but still provoke an immune response. Uppercase abbreviations (such as *D* in DPT or *T* in Tdap) refer to full-strength doses of toxoid, whereas lowercase letters (*d* in Tdap) denote reduced-strength formulations. Historically, pertussis vaccines contained whole-cell (w) pertussis antigens, whereas newer formulations of the vaccine contain acellular (a) antigens, meaning that only a portion of the cell is included in the vaccine. During the 1990s the whole-cell diphtheria/tetanus/pertussis vaccine (DPT or DTwP) series was replaced by one that relied on an acellular pertussis component (DTaP), whereas DT vaccines, containing only high levels of diphtheria and tetanus antitoxins, were given to children who could not tolerate the pertussis antigen. Current recommendations require administration of a Tdap booster—Boostrix or Adacel—during early adolescence. Tetanus and diphtheria boosters (Td) are recommended every 10 years, or after potential exposure to tetanus.

Legionella and Legionellosis

*Legionella** is a bacillus unrelated to other strictly aerobic gram-negative bacteria. Although the bacterium was originally described in the late 1940s, it was not clearly associated with human disease until 1976. The incident that brought it to the attention of medical microbiologists was a sudden outbreak of pneumonia that afflicted 200 American Legion members attending a convention and killed 29 of them. After 6 months of painstaking analysis, epidemiologists isolated the pathogen and traced its source to contaminated air-conditioning vents in the Legionnaires' hotel. The news media latched onto the name **Legionnaires' disease,** and the experts named their new isolate *Legionella.*

Legionellas are weakly gram-negative motile rods that range in morphology from cocci to filaments. They have fastidious nutrient requirements and can be cultivated only in special media (**figure 20.8***a*) and cell cultures. Several species or subtypes have been characterized, but *L. pneumophila* (lung-loving) is the one most frequently isolated in infections.

Legionella appears to be widely distributed in aqueous habitats as diverse as tap water, cooling towers, spas, ponds, and other fresh waters. Its survival and persistence in these habitats is now known to be related to *Legionella*'s close associations with free-living amoebas (**figure 20.8***b*). The bacteria can be released during aerosol formation and are often carried for long distances. Cases have been traced to supermarket vegetable sprayers and the fallout from the Mount Saint Helens volcano.

Studies since 1976 have determined that Legionnaires' disease is not a new disease and that earlier outbreaks occurred sporadically in conjunction with accidental exposure to contaminated environments. The disease is now known to be worldwide and prevalent in males over 50 years of age. Healthcare-associated infections occur most often in elderly patients hospitalized with diabetes, malignant

** Alcaligenes* (alk′-uh-lij′-uh-neez) ME. *alkaly,* salt, and L. *generatus* to produce.

** Legionella* (lee″-jun-ell′-uh).

Legionella bacteria

Amoeba cell

(a)

(b)

Figure 20.8 Appearance of *Legionella pneumophila*. (a) A mixed culture of *Legionella* and other isolated bacteria on charcoal yeast extract medium. Arrows point to *Legionella* colonies, which are differentiated by their blue periphery and pale center. **(b)** *Legionella* living intracellularly in the amoeba *Vermamoeba vermiformis* (4,000×). Amoebas inhabiting natural waters are the reservoir for this pathogen and a means for it to survive in rather hostile environments. The pathogenesis of *Legionella* in humans is likewise dependent on its uptake by and survival in phagocytes. (both): *Source:* CDC

disease, transplants, alcoholism, and lung disease. Patients with the disease cannot communicate it to other persons.

Two major clinical forms of the disease are *Legionnaires' disease* and *Pontiac fever*. The symptoms of both are a rising fever (up to 41°C), cough, diarrhea, and abdominal pain. Legionnaires' disease is the more severe of the two, progressing to lung consolidation and impaired respiration and organ function. It has a fatality rate of 3% to 30%. In contrast, Pontiac fever does not lead to pneumonia and rarely causes death.

Legionellosis is diagnosed by symptomatology and the patient's history. Laboratory analyses include fluorescent antibody staining of specimens, cultivation on charcoal yeast extract (CYE) agar, and DNA probes. The disease is treated with levofloxacin or azithromycin. *Legionella*'s wide dispersal in aquatic habitats makes control difficult, though chlorination and regular cleaning of artificial habitats are of some benefit.

 Check Your Progress SECTION 20.2

3. List the four genera of bacteria that cause opportunistic infections and are physiologically similar to *Pseudomonas*.

4. Compare the genera *Brucella* and *Francisella*. Include the pathology for each disease, reservoirs, portal of entry, and treatment.

5. What is the pathologic effect of whooping cough?

6. Who is most likely to contract whooping cough, and how would the disease be recognized?

7. What is unusual about the reservoir of *Legionella?* What is the epidemiological pattern of the disease?

20.3 Identification and Differential Characteristics of Family Enterobacteriaceae

 Expected Learning Outcomes

9. Recall the medically important members of the Enterobacteriaceae and the characteristics they have in common.

10. Name the biochemical tests commonly used to differentiate common opportunistic enterics.

11. Explain the importance of the three major surface antigens seen in gram-negative enteric rods.

One large family of gram-negative bacteria that exhibits a considerable degree of relatedness is the Enterobacteriaceae. Although many members of this group inhabit soil, water, and decaying matter, they are also common occupants of the large bowel of humans and animals. All members are small (the average is 1 μm by 2–3 μm) non-spore-forming rods. The bacteria in this group grow best in the presence of air, but they are facultative and can ferment carbohydrates by an anaerobic pathway. This group is probably the most common one isolated in clinical specimens—both as normal microbiota and as agents of disease.

Enteric pathogens are the most frequent cause of diarrheal illnesses **(20.2 Making Connections),** which account for an annual mortality rate of 3 million people and an estimated 4 billion infections worldwide. Counting the total number of cases that receive medical attention, along with estimates of unreported, subclinical cases, enteric illness is probably responsible for more morbidity than any other disease. Prominent enteric pathogens include *Salmonella, Shigella,* and strains of *Escherichia,* many of which are harbored by human carriers.

Gram-negative enterics (along with *Pseudomonas* and *Acinetobacter*) account for more than 30% of all isolates in healthcare associated infections (see figure 13.22). Their widespread involvement in hospital-acquired infections can be attributed to their constant presence in the hospital environment and their capacity to resist drying, disinfectants, and drugs. Enterics are commonly isolated from soap dishes, sinks, and invasive devices such as tracheal cannulas and indwelling catheters. The most important enteric opportunists are *E. coli, Klebsiella, Enterobacter, Serratia, Proteus,* and *Citrobacter.* Although these pathogens live in the intestine, the diseases they cause usually involve other body systems such as the lungs and the urinary tract.

The genera in this family share several key characteristics:

- glucose fermentation;
- nitrate reduction;
- oxidase-negative;
- in most, presence of flagella (except in *Shigella* and *Klebsiella*); and
- regular, straight rods.

A battery of biochemical tests is commonly used to identify the principal genera and species. Although the details of this process

20.2 MAKING CONNECTIONS

Diarrheal Disease

Diarrhea is an acute syndrome of the intestinal tract in which the volume, fluid content, and frequency of bowel movements increase. It is usually a symptom of **gastroenteritis** (inflammation of the lining of the stomach and intestine) and can be accompanied by severe abdominal pain. Diarrhea is generated by several pathologic states—most commonly infection, intestinal disorders, and food poisoning. Although the human large intestine ordinarily harbors a huge microbial population, most agents of diarrhea are not members of this normal gut microbiota but are acquired through contaminated food or water.

Infectious diarrhea occurs by one of two basic mechanisms: (1) The secretion of enterotoxins, or (2) Invasion of the intestinal epithelium by the microbe. In the toxigenic type of disease, bacteria release **enterotoxins** that bind surface receptors of the small intestine. These toxins disrupt the physiology of epithelial cells and cause increased secretion of electrolytes and water loss, a condition called secretory diarrhea. The organism itself does not invade the tissues. Secretory diarrhea is characterized by its large volume, and there is little blood in the stool. This is the mechanism of cholera (see chapter 21) and some types of *Escherichia coli* and *Shigella* infectious damage.

In a more invasive diarrheal disease, the microbe invades the tissues of the small or large intestine and damages cells that line the mucosa and underlying tissue layers. Signs and symptoms include pain in the rectum, blood in the stool, and ulceration of the inner lining. *Salmonella,* different strains of *Shigella* and *E. coli, Campylobacter,* and *Entamoeba histolytica* are responsible for this type of intestinal disease. Regardless of the cause, the loss of fluid that accompanies diarrhea results in severe dehydration and sometimes death. It is the second leading cause of death in children under five years old, and is responsible for killing approximately 760, 000 children every year. Infants are especially vulnerable to diarrheal illness because of their smaller fluid reserves and undeveloped immunity.

See 21.2 Making Connections to review a successful support treatment that offsets some of the complications of diarrhea.

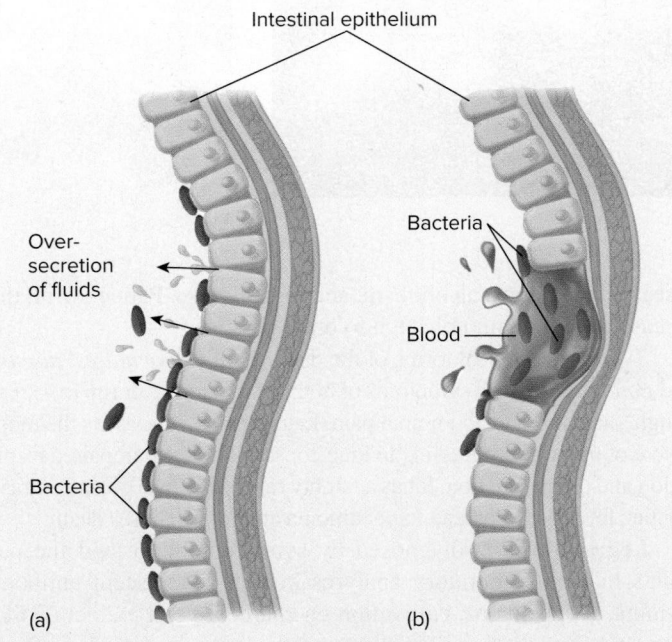

Mechanisms of infectious diarrhea. **(a)** In a toxigenic infection, the microbe remains on the surface of epithelial cells and secretes toxin into the cells. This results in disruption of cell metabolism and loss of electrolytes and water. **(b)** In an invasive infection, the microbe destroys epithelial cells and forms ulcerations, resulting in loss of the intestinal lining and bleeding.

Explain how toxigenic disease with no intestinal damage can be as severe as invasive disease. *Answer available on Connect.*

are beyond the scope of this text, a summary reviews the major steps **(figure 20.9).**

Because fecal specimens contain such a high number of normal residents, they can be inoculated first into enrichment media (selenite or GN—gram-negative—broth) that inhibit the normal microbiota and favor the growth of pathogens. Enrichment and nonfecal specimens are streaked onto selective, differential media such as MacConkey, EMB, or Hektoen enteric agar and incubated for 24 to 48 hours. Colonies developing on these media will provide the initial separation into lactose-fermenters and non-lactose-fermenters (**figure 20.10;** see figure 3.17*c*).

Isolated colonies are subcultured on triple-sugar iron (TSI) agar slants for further analysis. As is evident in figure 20.9, additional tests can differentiate isolates to the level of genus **(table 20.2).** One series of reactions, the **IMViC** (indole, methyl red, Voges-Proskauer, *i* for pronunciation, and citrate), is a traditional panel that can be used to differentiate among several genera **(table 20.3).** In general, whether an isolate is clinically significant depends upon the specimen's origin. For example, normal microbiota isolated from stool specimens do not usually signify infection in that site, but the same organisms isolated from extraintestinal sites in sputum, blood, urine, and spinal fluid are considered likely agents of opportunistic infections.

The Family Enterobacteriaceae is traditionally divided into two subcategories. The **coliforms** include *Escherichia coli* and other gram-negative normal enteric microbiota that ferment lactose rapidly (within 48 hours). **Noncoliforms** are generally non-lactose-fermenting or slow lactose-fermenting bacteria that are either normal microbiota or regular pathogens. Although minor exceptions occur, this distinction has considerable practical merit in laboratory, clinical, and epidemiological terms. The following outline provides a framework for organizing the discussion of the Enterobacteriaceae.

Coliforms in Normal Microbiota: Rapid lactose-fermenting enteric bacteria that are normal residents and opportunistic (some serotypes of *E. coli* are true pathogens). Examples include:

Escherichia coli	*Klebsiella*
Enterobacter	*Hafnia*
Serratia	*Citrobacter*

Colonies from selective media inoculated into triple-sugar iron (TSI)

Rapid lactose fermentation on (TSI)

Figure 20.9 **Outline of biochemical traits for separating selected enteric genera.** Flowchart is based on the usual or dominant characteristics of examples used here.

Figure 20.10 **Isolation media for enterics, showing differentiating reactions. (a)** Levine's eosin methylene blue (EMB) agar. **(b)** Hektoen enteric agar. (See table 20.2.) (both): © Kathy Park Talaro

Noncoliforms in Normal Microbiota: Lactose-negative bacteria that are opportunistic, normal gut residents:

Proteus *Morganella*
Providencia *Edwardsiella*

True Pathogenic Enterics

Salmonella typhi, S. enterica
Shigella dysenteriae, S. flexneri, S. boydii, S. sonnei
Yersinia enterocolitica, Y. pseudotuberculosis

True Pathogenic Nonenteric

Yersinia pestis

Antigenic Structures and Virulence Factors

The gram-negative enterics have complex surface antigens that are important in pathogenicity and are also the basis of immune responses **(figure 20.11).** By convention, they are designated **H,** the **flagellar antigen; K,** the **capsule** and/or **fimbrial antigen;** and **O,** the **somatic** or **cell wall antigen.**[2] Not all species carry the H and K antigens, but all have O, the lipopolysaccharide implicated in endotoxic shock. Most species of gram-negative enterics exhibit a variety of

2. H comes from the German word *hauch* (breath) in reference to the spreading pattern of motile bacteria grown on moist solid media. K is derived from the German term *kapsel.* O is for *ohne-hauch,* or nonmotile.

TABLE 20.2	Protocol in Isolating and Identifying Selected Enteric Genera*

- **MacConkey agar** contains bile salts and crystal violet to inhibit gram-positive bacteria. It contains lactose and neutral red dye to indicate pH changes and differentiate the rapid lactose-fermenting bacteria from non-lactose-fermenters or slow fermenters. Lactose (+) colonies are red to pink from the accumulation of acid acting on the neutral red indicator (see figure 3.17c).

- **Levine's EMB** agar also contains bile salts, plus eosin and methylene blue dyes that cause lactose-fermenters to develop a dark nucleus and sometimes a metallic sheen over the surface. Non-lactose-fermenters are pale lavender and nonnucleated (**figure 20.10a**).

- **Hektoen enteric agar** also contains bile salts and is especially good for isolating pathogens. The bromthymol blue and acid fuchsin indicators differentiate the lactose-fermenters (salmon pink to orange) from the non-lactose-fermenters (green, blue-green) and also detect the production of H_2S gas, which turns the center of a colony black (**figure 20.10b**).

- **Utilization of lactose**, a disaccharide, requires two enzymes: a permease to bring lactose into the cell and β-galactosidase to split it into the monosaccharides glucose and galactose. Rapid fermenters have both enzymes. Slow fermenters have only galactosidase. **ONPG** is chemical shorthand for a test that detects these slow lactose-fermenters.

- **TSI** is a nonselective medium that indicates some combination of fermentation reactions (primarily lactose and glucose and possibly sucrose) by means of a phenol red indicator dye. It also reveals gas and **hydrogen sulfide (H_2S) production** (see figure 3.17a). Hydrogen sulfide gas is a product of the reduction of sulfur and is indicated by reacting with iron salts to form a black precipitate of ferric sulfide.

- The **indole test** indicates the capacity of an isolate to cleave a compound called indole off the amino acid tryptophan. If cleaving occurs, Kovac's reagent reacts with indole to form a bright red ring at the surface of the tube.

- The **methyl red (MR)** test indicates a form of glucose fermentation in which large amounts of mixed acids accumulate in the medium. The pH is lowered to around 4.2, so that methyl red dye remains red when added to the tube.

- The **Voges-Proskauer (VP)** test determines whether the product of glucose fermentation is a neutral metabolite called acetyl-methylcarbinol (acetoin). This substance reacts with Barritt's reagent to form a rosy-red tinge in the medium.

- **Citrate media** contain citrate as the only usable carbon source and a pH indicator, bromthymol blue. If an isolate can utilize this carbon source, it produces alkaline by-products, resulting in a conversion of the medium's color from green (neutral) to blue (alkaline).

- **Lysine** and **ornithine decarboxylase (LDC, ODC)** detect the presence of enzymes that remove carbon dioxide from amino acids. The end products are alkaline amines that raise the pH.

- **Urea media** contain 2% urea, a nitrogenous waste product of mammals, and phenol red pH indicator. The enzyme urease hydrolyses urea into ammonium, which raises the pH of the medium and turns the indicator bright pink.

- **Phenylalanine (PA)** deaminase, an enzyme produced primarily by the *Proteus* group, removes an amino group from the amino acid phenylalanine to produce phenylpyruvic acid. This reaction is visualized with ferric chloride, which turns olive green when it reacts with the phenylpyruvic acid.

- **Motility** is tested by a hanging drop slide or motility test medium (see figure 3.15b).

*The metabolic basis for some tests is discussed in chapter 8.

TABLE 20.3	The IMViC Tests for Differentiating Common Opportunistic Enterics			
Genus	Indole	Methyl Red	Voges-Proskauer	Citrate
Escherichia	+	+	−	−
Citrobacter	+	+	−	+
Klebsiella/ Enterobacter	+/−	−	V*	+
Serratia	−	V*	+	+
Proteus	+	−	−	+
Providencia	+	+	−	+
Pseudomonas	−	−	−	V*

*V = Species variable.

subspecies or serotypes caused by slight variations in the chemical structure of the HKO antigens. These are usually identified by serotyping, a technique in which specific antibodies are mixed with a culture to measure the degree of agglutination or precipitation (see chapter 17). Not only is serotyping helpful in classifying some species, but it also helps pinpoint the source of outbreaks by a particular serotype, making it an invaluable epidemiological tool.

Figure 20.11 Antigenic structures in gram-negative enteric rods. Variations in the composition of these antigens provide the basis for the serological types found in most genera.

Capsule (K antigen, or V_i in *Salmonella*)

Somatic (O antigen, or cell wall antigen)

Flagellar (H antigen)

The pathogenesis of enterics is tied to their production of endotoxins, exotoxins, capsules, fimbriae, and molecules that favor attachment to host cells. These factors function in invasion and destruction of tissues. Although most enterics are commensals, a significant number have a potential to rapidly alter their pathogenicity. As we saw in chapters 9 and 12, gram-negative bacteria freely transfer chromosomal or plasmid-borne genes that can mediate drug resistance, toxigenicity, and other adaptive traits. When this transference occurs in mixed populations of fecal residents (for example, in the intestines of carriers), it is possible for nonpathogenic or less pathogenic residents to acquire greater virulence (see the discussion of *E. coli* in section 20.4). The most common virulence genes transferred contain the codes for toxins, capsules, hemolysins, and fimbriae that promote invasion of the intestinal or urinary epithelia.

 Check Your Progress SECTION 20.3

8. Name the key characteristics shared by the medically important members of the Family Enterobacteriaceae.
9. Explain what is meant by IMViC.
10. With regard to gram-negative enterics, what do the letters O, H, and K stand for?
11. Which bacteria in the Family Enterobacteriaceae are true enteric pathogens and which ones are opportunists?
12. Briefly describe the methods used to isolate and identify enterics.
13. What is the basis of serological tests, and what is their main use for enterics?

20.4 Coliform Organisms and Diseases

 Expected Learning Outcomes

12. Differentiate among the major enteric pathologies seen among various strains of *E. coli*.

13. Explain the role of *E. coli* in infantile and traveler's diarrhea.

14. State the significance of *E. coli* O157:H7.

15. Explain the importance of carbepenem-resistant Enterobacteriaceae infections and identify the organisms responsible for the majority of these infections.

Escherichia coli: The Most Prevalent Enteric Bacillus

Escherichia coli* is the best-known coliform, largely because of its use as a subject for laboratory studies. Although it is called the colon bacillus and sometimes regarded as the predominant species in the intestine of humans, *E. coli* is actually outnumbered 9 to 1 by strictly anaerobic bacteria of the gut (*Bacteroides* and *Bifidobacterium*) that are not readily cultured. Its prevalence in clinical specimens and infections is due to its being the most common aerobic and nonfastidious bacterium in the gut. Even though *E. coli* is usually a commensal, it is far from harmless. Many of the 150 strains are not highly infectious, but some have developed greater virulence

through plasmid transfer, and others are opportunists. Pathogenic *E. coli* strains are grouped into **pathotypes**—strains of a single bacterial species that possess a common set of virulence factors. Six *E. coli* pathotypes are associated with diarrhea.

- **Enterohemorrhagic** *E. coli* (**EHEC**) causes hemorrhagic colitis and may lead to hemolytic uremic syndrome (HUS), which can cause severe damage to the kidney. Because much of its virulence is derived from the shiga toxin aquired from *Shigella,* it is also known **Shiga toxin producing *E. coli*,** or **STEC.**
- **Enterotoxigenic** *E. coli* (**ETEC**) causes a severe diarrheal illness brought on by two exotoxins, termed heat-labile toxin (LT) and heat-stable toxin (ST), that stimulate heightened secretion and fluid loss. In this way, it mimics the pathogenesis of cholera (see figure 21.11). This serotype of *E. coli* also has fimbriae that provide adhesion to the small intestine.
- **Enteroinvasive** *E. coli* (**EIEC**) causes an inflammatory disease similar to **dysentery*** from *Shigella* that involves invasion and ulceration of the mucosa of the large intestine.
- **Enteropathogenic** strains of *E. coli* (**EPEC**) are linked to a wasting form of infantile diarrhea whose pathogenesis is not well understood.
- **Enteroaggregative** *E. coli* (**EAEC**) cells aggregate and colonize the intestinal mucosa; enterotoxins they release damage the host cells and induce inflammation, leading to diarrhea. Pediatric diarrhea is commonly caused by EAEC.
- **Diffusely adherent** *E. coli* (**DAEC**) is thought to be a minor cause of pediatric diarrhea.

Clinical Diseases of E. coli

Most clinical diseases of *E. coli* are transmitted exclusively among humans. Pathogenic strains of *E. coli* are frequent agents of infantile diarrhea, the greatest single cause of mortality among babies. Worldwide, 15% of childhood deaths are due to diarrhea. The rate of infection is higher in crowded tropical regions where sanitary facilities are poor, water supplies are contaminated, and adults carry pathogenic strains to which they have developed immunity.

The immature, nonimmune neonatal intestine has no protection against these pathogens entering in unsanitary food or water. The practice of preparing formula from dried powder mixed with contaminated water can be tantamount to inoculating the infant with a pathogen. Often the babies are already malnourished, and further loss of body fluids and electrolytes is rapidly fatal. Mothers in developing countries especially are being urged to nurse their infants so that the children get better and safer nutrition as well as gastrointestinal immunity (see Clinical Connections, "The Immunological Benefits of Breast-Feeding," in section 15.5).

Despite the popular belief that "Montezuma's revenge," "Delhi belly," and other travel-associated gastrointestinal diseases are caused by more exotic pathogens, up to 70% of traveler's diarrhea cases are due to an enterotoxigenic strain of *E. coli.* Travelers pick up more virulent strains while eating or drinking contaminated foods. Within 5 to 7 days, victims are stricken with profuse, watery diarrhea, low-grade fever, nausea, and vomiting. Some travelers develop more chronic enteroinvasive dysentery that resembles *Shigella* or *Salmonella* infection.

* *Escherichia* (ess-shur-eek′-ee-uh) After the German physician Theodor Escherich.

* *disentery* (dis′-en-ter″-ee) Gr. *dys,* bad, and *enteria,* bowels.

Oral antimicrobics may be effective in the early phases of infection. Over-the-counter preparations that slow gut motility (kaolin or *loraperamide* [Imodium]) afford symptomatic relief, but they can also serve to retain the pathogen longer. They are probably not as effective as bismuth salicylate mixture (Pepto-Bismol), which counteracts the enterotoxin and also provides an antimicrobic effect.

The Role of Escherichia coli *in Food Infections*

The most highly publicized strain of *Escherichia coli* is an enterohemorrhagic strain famous for its involvement in infections traced to fast-food hamburgers that first appeared in 1982. Since that time, it has been commonly referred to by its antigen profile: **O157:H7,** which stands for somatic (O) type 157 and flagellar (H) type 7. This bacterial strain owes much of its virulence to a cell wall receptor that can fuse with the host cell membrane, creating a direct port into which the bacteria secrete toxins and other proteins into the host cell. This strain is also referred to as STEC—for shiga toxin producing *E. coli*—because it has acquired the gene for shiga toxin from *Shigella*. The toxin enters the host cell, binds ribosomes, and disrupts protein synthesis, leading to death and shedding of intestinal cells (see 20.2 Making Connections).

The pathogen has a reservoir in the intestine of cattle, who are unaffected by the bacterium. From there, it can enter the food chain in contaminated beef, water, and fresh vegetables. Commercially supplied hamburger is practically a "recipe" for transmission, because of the policy in meat processing plants to mix meats from hundreds of sources. A single animal carrier can potentially contaminate thousands of pounds of beef. The infectious dose is a mere 100 cells—even a small bit of uncooked meat could contain this amount. It has also become a contaminant in vegetables, fruit drinks, and groundwater into which animal waste has seeped. Approximately 2,000 to 4,000 cases per year are currently reported to the CDC; the actual number of cases is likely to be many times higher.

This strain of *E. coli* causes a range of symptoms and signs, from gastroenteritis and dysentery to fever and acute abdominal pain, similar to shigellosis. In about 10% of cases, a much more severe form of disease called hemolytic uremic syndrome (HUS) appears. This condition is caused by the effects of the shiga toxin on the blood and kidney, leading to hemolysis and kidney damage and failure. Children, the elderly, and immunocompromised people have the highest risk for this complication. Antibiotic treatments often are not effective.

The emergence of this pathogen has prompted the food and safety inspection section of the U.S. Department of Agriculture (USDA) to increase its monitoring of fresh meats and sausages. The primary way to rapidly identify it is with selective media in the lab or using serological tests that detect the somatic (O) and flagellar (H) antigens **(figure 20.12)**. Currently, testing for O157:H7 is performed randomly, and it is not required by slaughterhouses and meat suppliers, but it has been successful in identifying pathogens in meats and has resulted in the recall of millions of pounds of hamburger. Largely because of this pathogen, the USDA now requires that all fresh meat and poultry products bear small warning labels to remind the consumer of safer handling instructions (see 20.3 Making Connections).

Miscellaneous Infections

Escherichia coli often invades sites other than the intestine. For instance, it causes 50% to 80% of urinary tract infections (UTI) in

Figure 20.12 **Rapid identification of *E. coli* (EHEC).** Rapid serotyping of *E. coli* is accomplished using latex beads coated with antibodies that recognize and bind to specific somatic or cellular antigens. Here the red clumps in the left sample indicate that the O111 bacteria have bound with the O111 antibody-coated latex beads. The sample on the right tested negative. *E. coli* O111 is, like *E. coli* O157:H7, a shiga toxin-producing strain of *E. coli*. © Peggy Greb/USDA/ Science Source

healthy people. Urinary tract infections usually result when the urethra is invaded by its own endogenous bacterial colonists. The infection is more common in women because their relatively short urethras promote ascending infection to the bladder (cystitis) and occasionally the kidneys. Patients with bladder catheters are also at risk for *E. coli* urethritis. Other extraintestinal infections in which *E. coli* is involved are neonatal meningitis, pneumonia, septicemia, and wound infections. These often complicate surgery, endoscopy, tracheostomy, catheterization, renal dialysis, and immunosuppressant therapy.

E. coli *and the Coliform Count*

Because of its prominence as a normal intestinal bacterium in most humans, *E. coli* is currently one of the indicator bacteria to monitor fecal contamination in water, food, and dairy products. According to this rationale, if *E. coli* is present in a water sample, fecal pathogens such as *Salmonella,* viruses, or even pathogenic protozoa may also be present. Coliforms such as *E. coli* are used because they are present in larger numbers, can survive in the environment, and are easier and faster to detect than true pathogens. If a certain number of coliforms are detected in a sample, the water is judged unsafe to drink. This topic is covered in more detail in chapter 26.

Other Coliforms

Other coliforms of clinical importance mostly as opportunists are *Klebsiella, Enterobacter, Serratia,* and *Citrobacter* (see figure 20.9). These bacteria cause infection primarily when the natural host defenses fail, often in conjunction with medical procedures. Various members of this group can infect nearly any organ, but the most serious complications involve septicemia and endotoxemia. Because of their widespread occurrence, their drug resistance, and the poor medical state of the patients they infect, complete control will probably never be possible.

In addition to inhabiting the intestines of humans and animals, ***Klebsiella****** species are also found in the respiratory tracts of normal individuals. This colonization leads to the chronic lung infections with which the species is associated. Infection by this organism is promoted by the large capsule, which prevents phagocytosis **(figure 20.13)**. Some strains also produce toxins. The most important species in this group is ***Klebsiella pneumoniae,*** a frequent

* *Klebsiella* (kleb-see-el′-uh) After Theodor Klebs, a German bacteriologist.

Pathogen Profile #4 *Escherichia coli*

Source: CDC

Microscopic Morphology Small, motile, gram-negative rods.

Identified by Gram reaction, morphology, and physiological and biochemical tests. *E. coli* is a facultative anaerobe that ferments lactose. It produces positive results for both the methyl red and indole tests and negative reactions for the Voges-Proskauer and citrate tests, results that distinguish it from other common enteric genera.

Habitat *E. coli* is a common inhabitant of the digestive tract of most warm-blooded animals.

Virulence Factors All strains of *E. coli* possess a capsule and LPS endotoxin. Individual serotypes express a variety of other virulence factors, including fimbriae, which allow attachment to host cells; heat-labile (LT) and heat-stable (ST) toxins that stimulate increased secretion and water loss (causing diarrhea); and shiga toxin, which can lead to hemorrhagic colitis.

Primary Infections/Disease *E. coli* is responsible for up to 70% of traveler's diarrhea, with some strains causing more severe enteroinvasive disease with ulceration of the mucosa of the large intestine.

Enteropathogenic strains of the bacterium are a leading cause of infantile diarrhea in developing countries. *E. coli* O157:H7 causes hemorrhagic colitis and, in about 10% of cases, progresses to hemolytic uremic syndrome (HUS), which can cause kidney failure. Up to 80% of urinary tract infections can be traced to *E. coli* originating from the body's own residents.

Control and Treatment Control of *E. coli* rests on preventing the entry of the bacterium into the food chain and water supply. Proper disinfection of water, isolation of potable water and sewage, care in harvesting fruits and vegetables, pasteurization of milk and fruit juices, along with a aseptic slaughtering of meat animals all contribute to this goal.

Treatment of *E. coli* infections in otherwise healthy people may be contraindicated because it often increases the speed with which endotoxin is released into the system. Preparations that slow gut motility (diphenoxylate/atropine [Lomotil]) relieve symptoms but allow the pathogen to remain in the gut for a longer period, increasing the risk of HUS. Bismuth salicylate (Pepto-Bismol), which both counteracts toxins and provides an antimicrobic effect, is the best treatment in most cases.

Figure 20.13 The mucoid colonies of *Klebsiella pneumoniae* are evidence of the bacterium's heavy capsule. *Source:* CDC

Figure 20.14 **The two faces of *Serratia marcescens*.** (*Left*) A culture incubated at 20-30°C produces colonies with a distinctive red pigment (prodigiosin). (*Right*) A culture incubated at 38°C (body temperature) grows but stops producing this pigment and has off-white colonies (10×). Cultures from infections will not usually show the distinctive red pigment because the incubation temperature is too high. (both) © Kathy Park Talaro

secondary invader and cause of nosocomial pneumonia, meningitis, bacteremia, wound infections, and UTIs.

Enterobacter and a closely related genus, *Hafnia,* commonly inhabit soil, sewage, and dairy products. Urinary tract infections account for the majority of clinical diseases associated with *Enterobacter.* The bacterium is also isolated from surgical wounds, spinal fluid, sputum, and blood. Although sometimes regarded as a minor pathogen, this organism is lethal when introduced into the bloodstream. In one outbreak of *Enterobacter* septicemia associated with contaminated intravenous fluids, 150 people were infected and 9 died.

Citrobacter, a regular inhabitant of soil, water, and the human colon, is an occasional opportunist in urinary tract infections and bacteremia in debilitated persons.

Serratia occurs naturally in soil and water as well as in the intestine. A common species, *Serratia marcescens,** produces an intense red pigment when grown at temperatures between 20°C and 30°C, and produces little or no pigment above 37°C **(figure 20.14).** This production of pigment has been used in the past in a misguided attempt to track movements of air currents in hospitals and over

* *Serratia* (sur-at′-ee-uh) In memory of the great physicist S. Serrati and *marcescens* (mar-sess′-uns) for "fading away."

cities and even to demonstrate transient bacteremia after dental extraction. Unfortunately this species is far from benign and is fully capable of invading a compromised host. *Serratia* pneumonia, particularly prevalent in alcoholics, is transmitted by contaminated respiratory-care equipment and infusions. The organism is often implicated in burn and wound infections, as well as in fatal septicemia and meningitis in immunosuppressed patients.

Carbepenem-Resistant Enterobacteriaceae

Of urgent concern is a group of bacteria referred to as **carbepenem-resistant Enterobacteriaceae,** or **CRE.** These bacteria, mostly species of *Klebsiella, Enterobacter,* and certain strains of *E. coli*—always troublesome opportunists in the hospital setting—have taken on added importance as they have become resistant to many classes of antibiotics, particularly carbepenems, which are frequently the antibiotic of last resort for these infections. The resistant bacteria have acquired genes for one of three different enzymes, known as KPC (*Klebsiella pneumoniae* carbapenemase), NDM (New Delhi Metallo-beta-lactamase) or VIM (Verona Integron-Mediated Metallo-β-lactamase), any one of which provides resistance to carbapenem antibiotics. KPC is the most common of the three enzymes, especially in the United States, and was first identified in 2001.

An estimated 140,000 healthcare-associated Enterobacteriaceae infections occur in the United States each year; with about 9,300 of these caused by CRE. Each year, approximately 600 deaths result from infections caused by carbapenem-resistant *Klebsiella spp.* and carbapenem-resistant *E. coli.* Patients whose care requires indwelling devices like ventilators, urinary or intravenous catheters, or endoscopes are at particular risk. Duodenoscopes—a specialized type of endoscope inserted through the mouth and down to the intestine to study the bile ducts, pancreatic duct and gallbladder—have been implicated in the transfer of CRE from patient to patient. Read more about the difficulty in sterilizing reusable medical devices in chapter 11.

> **Quick Search**
>
> To learn more about the CRE outbreak at the UCLA medical center, search the Internet for "A Killer on the Loose at UCLA."

✓ *Check Your Progress* SECTION 20.4

14. Explain the difference between enterohemorrhagic, enterotoxigenic, enteroinvasive, enteropathogenic, enteroaggregative, and diffusely adherent strains of *E. coli.*

15. Describe the kinds of infections for which *E. coli* is primarily responsible.

16. Justify treating *E. coli* O157:H7 differently from other strains of the species.

17. Relate the pathology of hemolytic uremic syndrome.

18. Explain the concern that CRE infection causes in the healthcare community.

20.5 Noncoliform Enterics

 Expected Learning Outcomes

16. Differentiate between true noncoliform enteric pathogens and opportunistic genera.

17. Distinguish the pathologies of typhoidal and nontyphoidal salmonellosis.

18. Identify the possible sources of *Shigella* infection.

19. Recall the etiologic agent of plague, and create a simple diagram of the infection cycle.

20. Recall the diseases attributable to infection with species in the genus *Haemophilus,* and identify which is most severe.

Opportunists: *Proteus* and Its Relatives

The three closely related genera ***Proteus,*** **Morganella,**** and **Providencia**** are saprobes in soil, manure, sewage, and polluted water and commensals of humans and other animals. Despite widespread distribution, they are ordinarily harmless to a healthy individual. *Proteus* species swarm on the surface of moist agar in an unmistakable concentric pattern **(figure 20.15).** They are commonly involved in urinary tract infections, wound infections, pneumonia, septicemia, and occasionally infant diarrhea. *Proteus* urinary infections appear to stimulate renal stones and damage due to the urease they produce and the rise in urine pH this causes. *Morganella morganii* and *Providencia* species are involved in similar infections. One opportunist, *P. stuartii,* is a frequent invader of burns. Because of widespread antibiotic resistance, therapy is guided by susceptibility testing.

True Enteric Pathogens: *Salmonella* and *Shigella*

The salmonellae and shigellae are distinguished from the coliforms and the *Proteus* group by having well-developed virulence factors,

* *Proteus* (pro'-tee-us) After Proteus, a Greek sea god who could assume many forms.
* *Morganella* (mor-gan-ell'-uh) after T. H. Morgan.
* *Providencia* (prah-vih-den'-see-uh) after Providence, RI.

Figure 20.15 The wavelike swarming pattern of *Proteus vulgaris* is typical of this genus. Growth on agar creates concentric rings rather than discrete colonies. This makes isolation of colonies in mixtures containing *Proteus* difficult. © K. Talaro

being primary pathogens, and not being normal residents of humans. The illnesses they cause—called **salmonelloses** and **shigelloses,** respectively—show some gastrointestinal involvement and diarrhea but often affect other systems as well.

Typhoid Fever and Other Salmonelloses

The naming of bacteria in the genus *Salmonella** is more complicated than that seen with other organisms. The genus contains two species, *Salmonella enterica*, and *Salmonella bongori* (a species found almost exclusively in cold-blooded animals like lizards). *Salmonella enterica* is divided into six subspecies; *enterica*, *salamae*, *arizonae*, *diarizonae*, *houtenae*, and *indica*. Additionally, each isolate is identified as belonging to one of over 2,500 serotypes, based on its expression of O and H antigens. The serotype typically reflects the source of the isolate, such as Typhimurium (found in rodents) or Newport (the city of isolation). Therefore, the formal designation of a Salmonella serotype is its genus-species-subspecies name, followed by "serotype" and the serotype name. To make matters even more confusing, the formal name is often abbreviated to include only the genus and serotype. Examples include:

- *Salmonella enterica* subspecies *enterica* serotype Typhimurium
- *Salmonella* serotype Newport
- *Salmonella* ser. Enteritidis

* *Salmonella* (sal-moh-nel′-uh) After Daniel Salmon, an American pathologist.

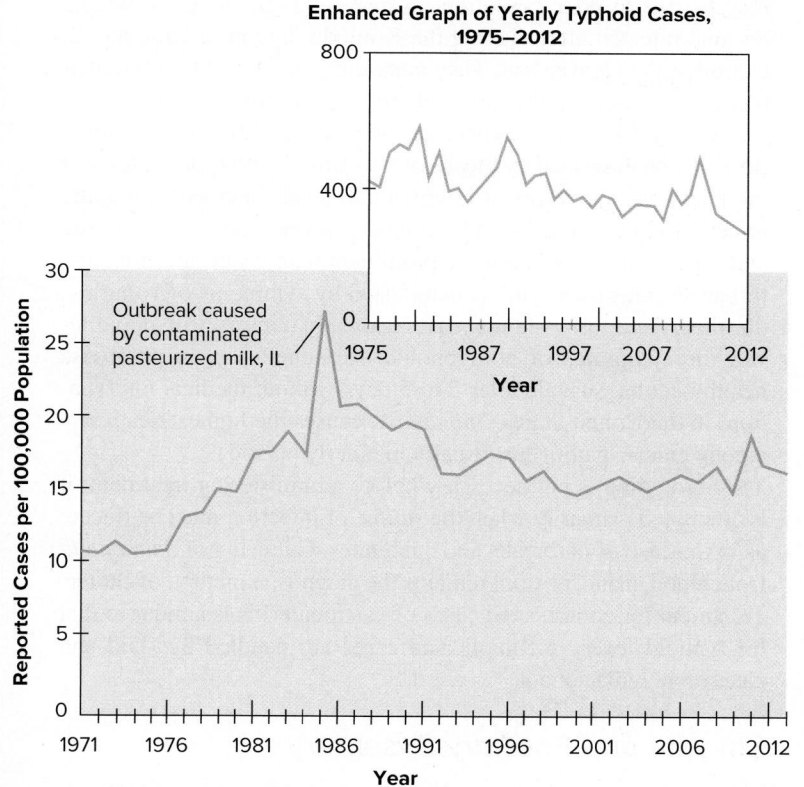

Enhanced Graph of Yearly Typhoid Cases, 1975–2012

Figure 20.16 **Prevalence of salmonelloses from 1971 to 2013.** The vast majority of cases are nontyphoidal gastroenteritis and food poisoning. Cases of typhoid fever are depicted with an inset. This disease is sporadic and is often brought back to the United States by travelers.

Commonly the word "serotype" is also eliminated from the full name (*Salmonella* Typhimurium), but this naming convention is discouraged. Occasionally the serotype is written as a species (*Salmonella typhimurium*), a practice that is simply incorrect because serotypes have no taxonomic standing.

Salmonellae are flagellated, grow readily on most laboratory media, and survive outside the host in environments such as fresh water and freezing temperatures. These pathogens are resistant to chemicals such as bile and dyes (the basis for isolation on selective media) and do not lose virulence after long-term artificial cultivation. The complexity of identifying *Salmonella* serotypes usually requires specialists at the CDC.

Typhoid fever is so named because it bears a superficial resemblance to typhus, a rickettsial disease, even though the two diseases are otherwise very different. Typhoid fever is caused by *Salmonella* serotype Typhi. *Salmonella* serotype Paratyphi causes paratyphoid, a very similar disease; both bacteria are considered typhoidal *Salmonella*. In the United States, the incidence of typhoid fever has remained at a steady rate since the early 1970s, appearing sporadically **(figure 20.16).** Of the 200 to 400 hundred cases reported annually, roughly half are imported from endemic regions. In other parts of the world, typhoid fever is still a serious health problem responsible for 200,000 deaths each year and millions of cases.

The typhoid bacillus usually enters the alimentary canal along with water or food contaminated by feces, although it is occasionally spread by close personal contact. Because humans are the exclusive hosts for *Salmonella* serotype Typhi, asymptomatic carriers are important in perpetuating and spreading typhoid. Even 6 weeks after convalescence, about half of recovered patients still shed the bacillus. A small number of people chronically carry the bacilli for longer periods in the gallbladder; from this site the bacilli are constantly released into the intestine and feces.

The number of bacilli that must be swallowed to initiate infection is between 1,000 and 10,000. Within the small intestine, the bacilli adhere to the mucosa and then initiate an invasive infection that leads eventually to septicemia. Symptoms are fever and abdominal pain, along with alternating diarrhea and constipation. The typhoid bacillus infiltrates the mesenteric lymph nodes and the phagocytes of the liver and spleen. In some people the small intestine develops areas of ulceration that are vulnerable to hemorrhage, perforation, and peritonitis **(figure 20.17).** Its presence in the general circulation may lead to nodules or abscesses in the liver or urinary tract. Prompt treatment greatly reduces the possibility of death.

A preliminary diagnosis of typhoid fever can be based upon the patient's history and presenting symptoms, supported by a rising antibody titer (see chapter 17). However, definitive diagnosis requires isolation of the typhoid bacillus. Either ciprofloxacin or ceftriaxone is the drug of choice, though some strains are resistant to these. Antimicrobial drugs are usually effective in treating the chronic carrier, but surgical removal of the gallbladder may be necessary in individuals with chronic gallbladder inflammation. Two vaccines are available: One is a live attenuated oral vaccine, and the other is based on the capsular polysaccharide. Both provide temporary protection for travelers and military personnel.

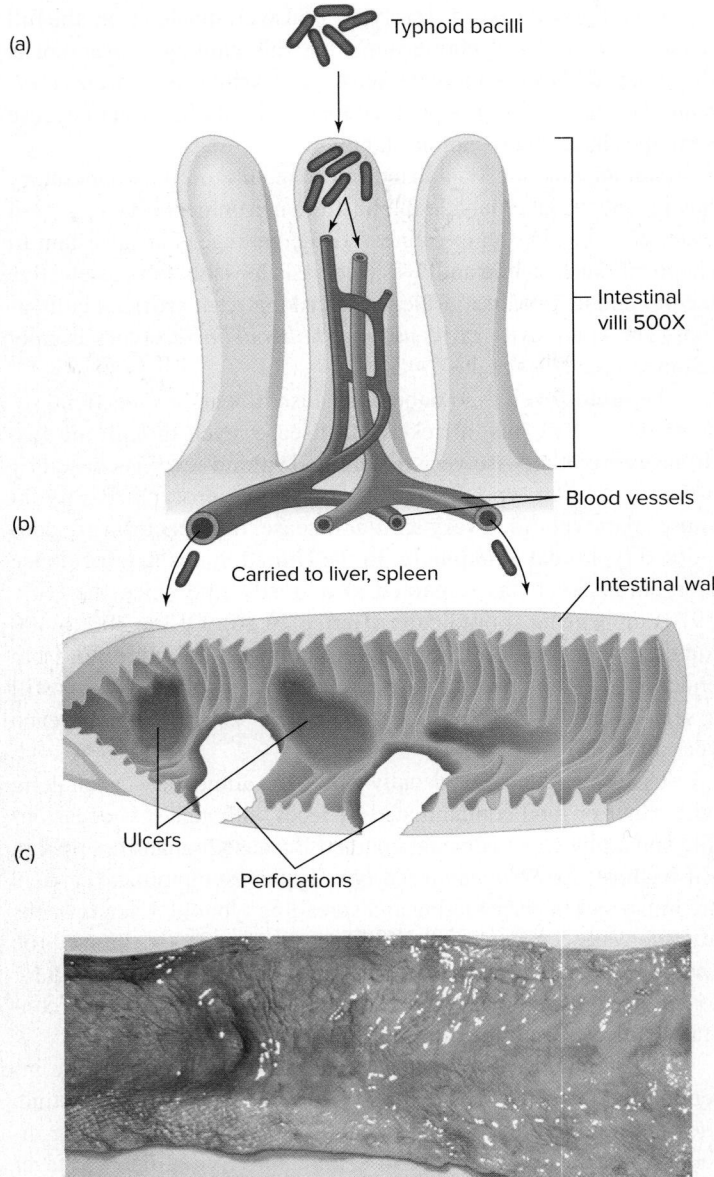

(a)

Typhoid bacilli

Intestinal
villi 500X

(b)

Blood vessels

Carried to liver, spleen

Intestinal wall

(c)

Ulcers

Perforations

(d)

Figure 20.17 **The phases of typhoid fever. (a)** Ingested cells invade the small intestinal lining. **(b)** From this site, they enter the bloodstream, causing septicemia and endotoxemia. **(c)** Infection of the lymphatic tissue of the small intestine can produce varying degrees of ulceration and perforation of the intestinal wall. **(d)** Image of an infected jejunum; note large areas of ulceration and a deep perforation. (d): *Source:* CDC

Animal Salmonelloses

Salmonelloses other than typhoid fever are designated as food poisoning or gastroenteritis and are usually less severe than typhoid fever. The nontyphoidal members all belong to the same species (*Salmonella enterica*) and one of the six subspecies (*enterica, salamae, arizonae, diarizonae, houtenae,* and *indica*), but have different serotypes. Of the 2,500 different serotypes, half of all infections are caused by the serotypes Enteritidis, Typhimurium, Newport, and Javiana.

Nontyphoidal salmonelloses are more prevalent than typhoid fever and are currently holding steady at 40,000 to 50,000 cases a year.

The CDC estimates that the true incidence is actually much higher, around 1.5 million cases per year (see figure 20.16). Unlike typhoid, all strains are zoonotic in origin, but humans may become carriers under certain circumstances. *Salmonella* are normal intestinal microbiota in cattle, poultry, rodents, and reptiles. Animal products such as meat and milk can be readily contaminated during slaughter, collection, and processing. There are inherent risks in eating poorly cooked beef or unpasteurized fresh or dried milk, ice cream, and cheese. A particular concern is the contamination of foods by rodent feces. Several outbreaks of infection have been traced to unclean food storage or to food-processing plants infested with rats and mice.

It is estimated that one out of every three chickens is contaminated with *Salmonella,* and other poultry such as ducks and turkeys are also affected. Eggs are a particular problem because the bacteria may actually enter the egg while the shell is being formed in the chicken. The poultry industry has begun using a probiotic containing a mixture of normal microbiota sprayed on newly hatched chicks to colonize them. This inoculation appears to curb the acquisition of *Salmonella* in a high percentage of treated animals. In general, one should always assume that poultry and poultry products harbor *Salmonella* and handle them accordingly, with clean techniques and adequate cooking (**20.3 Making Connections**). Drug resistance in the salmonellas is on the rise, some of which can be traced to the practice of adding antibiotics to animal feeds.

Most cases are traceable to a common food source such as milk or eggs. Outbreaks have occurred in association with milk, homemade ice cream made with raw eggs, and Caesar salad. Some cases may be due to poor sanitation. In one outbreak, about 60 people became infected after visiting the Komodo dragon (a large lizard) exhibit at the Denver zoo. They apparently neglected to wash their hands after handling the rails and fence of the dragon's cage.

The symptoms, virulence, and prevalence of *Salmonella* infections can be described by means of a pyramid, with typhoid fever at the peak, gastroenteritis at intermediate levels, and asymptomatic infection at the base. In typhoid fever, elevated body temperature and septicemia are much more prominent than gastrointestinal disturbances. Gastroenteritis is manifested by symptoms of vomiting, diarrhea, fluid loss, and mucosal lesions. In contrast to typhoid fever, the symptoms of nontyphoidal salmonelloses, in otherwise healthy adults, subside after 2 to 5 days without medical intervention. In the United States, *Salmonella* causes the highest death rate among enteric pathogens, usually in elderly persons.

Serotyping is not necessary before administering treatment; it is attempted primarily when the source of infection must be traced, as in the case of outbreaks and epidemics. Culturing of *Salmonella* from blood, urine, or stool remains the diagnostic method of choice. Treatment for complicated cases of gastroenteritis is similar to that for typhoid fever; uncomplicated cases are handled by fluid and electrolyte replacement.

Shigella *and Bacillary Dysentery*

*Shigella** causes a common but often incapacitating dysentery called shigellosis, which is marked by fever, nausea, painful abdominal cramps, and frequent defecation of watery stool filled

* *Shigella* (shih-gel'-uh) After K. Shiga, a Japanese physician.

20.3 MAKING CONNECTIONS

Avoiding Gastrointestinal Infections

Most food infections or other enteric illnesses caused by *Salmonella, Shigella, Vibrio, Escherichia coli,* and other enteric pathogens are transmitted by a combination of the Five *F*'s—feces, food, fingers, flies, and fomites—as well as by water. Individuals can protect themselves from infection by taking precautions when preparing food and traveling; communities can protect their members by following certain public health measures.

In the Kitchen

A little knowledge and good common sense go a long way. Food-related illnesses can be prevented by

1. Use of good sanitation methods while preparing food. This includes liberal hand washing, disinfecting surfaces, sanitizing utensils, and preventing contamination by mechanical vectors such as flies and cockroaches.
2. Use of temperature. Sufficient cooking of meats, eggs, and seafood is essential. Bringing meat up to 155°F (68°C) for 10 minutes would kill most food-borne pathogens. Refrigeration and freezing prevent multiplication of bacteria in fresh food that is not to be cooked (but remember, it does not kill them).
3. Sell-by dates are guidelines for supermarkets to remove unsold fresh meat off their shelves. They also give consumers information about how fresh the products are. Storage recommendations issued by the USDA are that chicken and fish be cooked or frozen no later than 2 days after its sell-by date. You should cook or freeze beef, pork, and lamb no more than 5 days after their sell-by dates.*
4. Do not taste uncooked batter made with raw eggs or ground meats before cooking. Do not consume rare or pink hamburger.

*For more information, consult the food and safety inspection service of the USDA on the Internet.

Supermarket label emphasizes precautions in handling, storing, and cooking meat and poultry products, as a way to avoid food-borne disease. *Source:* United States Department of Agriculture. www.fsis.usda.gov

While Traveling

1. Combat the Five *F*'s with the Five *B*'s—bread, bananas, beer, bottled beverages, and boiled water. In other words, food and drink that are packaged or heated.
2. Do not brush your teeth or wash fruit or vegetables in tap water.
3. Pack antidiarrheal medicines and water disinfection tablets.

Community Measures

1. Water purification: most enteric pathogens are destroyed by ordinary chlorination
2. Pasteurization of milk
3. Detection and treatment of carriers
4. Restriction of carriers from food handling
5. Constant vigilance during floods and other disasters that result in sewage spillage and contamination of food and drink
6. Implementation of PulseNet database for following outbreaks of enteric pathogens

What are some preventive measures for safely consuming fruits and vegetables that won't be heated or cooked? *Answer available on Connect.*

with mucus and blood. The etiologic agents (*Shigella dysenteriae, S. sonnei, S. flexneri,* and *S. boydii*) are primarily human pathogens, though they can infect apes. All produce a similar disease that can vary in intensity. They are nonmotile, nonencapsulated, and not fastidious. The shigellae resemble some types of pathogenic *E. coli* so closely that they are placed in the same subgroup.

Although *S. dysenteriae* causes the severest form of dysentery, it is uncommon in the United States and occurs primarily in the Eastern Hemisphere. In the past decade, the prevalent agents in the United States have been *S. sonnei* and *S. flexneri,* which cause approximately 20,000 reported cases each year, half of them in children between 1 and 10 years of age. Shigellosis is acquired primarily by eating fecally contaminated foods. A quick guide to remembering the sources is "feces transferred to food, fingers, flies, and sometimes fomites (Five *F*'s)." It is also acquired through direct person-to-person contact, largely because of the small infectious dose required (100–200 cells). The disease is mostly associated

with lax sanitation, malnutrition, and crowding, and is spread epidemically in day care centers, prisons, mental institutions, nursing homes, and military camps. As in other enteric infections, *Shigella* can establish a chronic carrier condition in some people that lasts several months.

Shigellosis is different from salmonellosis in that *Shigella* invades the villus cells of the large intestine, rather than the small intestine. In addition, it is not as invasive as *Salmonella* and does not perforate the intestine or invade the blood. It enters the intestinal mucosa by means of lymphoid cells in Peyer's patches. Once in the mucosa, *Shigella* instigates an inflammatory response that causes extensive tissue destruction. The most important virulence factors of Shigella are toxins. The release of endotoxin causes fever, and enterotoxin damages the mucosa and villi. Local areas of erosion give rise to bleeding and heavy mucus secretion **(figure 20.18).** *Shigella dysenteriae* produces a heat-labile exotoxin (shiga toxin) that has a number of effects, including injury to nerve

Pathogen Profile #5 *Salmonella typhi, Salmonella enterica*

Source: Janice Haney Carr/CDC

Microscopic Morphology Small, motile, gram-negative rods.

Identified by Gram reaction, morphology, and biochemical and serological tests. Because of the complexity of the genus, with thousands of different serotypes, specialized antibody-based tests are generally required to achieve full identification, and this process is not often attempted by most laboratories.

Habitat Humans are the exclusive hosts for *Salmonella enterica* serotypes Typhi and Paratyphi, whereas other serotypes of *Salmonella* are normal intestinal residents of cattle, poultry, rodents, and reptiles.

Virulence Factors In addition to LPS endotoxin, *Salmonella* species produce a capsule (referred to as the V$_i$ antigen) that increases the infectivity of the bacterium. The entire genus is robust and survives well outside the host environment. Antibiotic resistance is common in many strains.

Primary Infections/Disease *Salmonella enterica* serotype *Typhi* is the causative agent of typhoid fever, a rare (400 U.S. cases per year) yet serious illness that progresses from fever and diarrhea to septicemia and peritonitis. Untreated, the disease has a mortality

rate of about 20%. Two of the three strains of *Salmonella enterica* serotype Paratyphi cause paratyphoid, a disease very similar to typhoid fever. The remainder of the approximately 2,500 serotypes of *Salmonella enterica* may cause *Salmonella* food poisoning and gastroenteritis (nontyphoidal salmonelloses), with over 40,000 reported cases per year in the United States (the number of unreported U.S. cases may be as high as 1.5 million annually). The disease is marked by fever, vomiting, diarrhea, fluid loss, and mucosal lesions.

Control and Treatment Because humans are the sole host of the Typhi and Paratyphi serotypes, asymptomatic carriage must be controlled, either through antibiotic treatment—ciprofloxacin or ceftriaxone—or removal of the gallbladder, where the bacterium often resides. Because other species of *Salmonella* exist as part of the microbiota of a wide variety of animals, elimination of the bacterium is effectively impossible, and control focuses on preventing ingestion of live bacteria. Adequate sanitation during food preparation, proper hand washing, and refraining from consuming unwashed fruits or vegetables, inadequately cooked meat and eggs, and unpasteurized milk all contribute to this goal. In otherwise healthy people, nontyphoidal salmonellosis is self-limiting, and symptoms subside after 2 to 5 days.

cells and damage to the intestine. Shiga toxin also causes damage to the kidneys by the same mechanism seen in *E. coli* O157:H7–related hemolytic uremic syndrome.

Diagnosis is complicated by the coexistence of several alternative candidates for bloody diarrhea such as *E. coli* and the protozoans *Entamoeba histolytica* and *Giardia lamblia*. Isolation and identification follow the usual protocols for enterics. Infection is

treated by fluid replacement and oral drugs such as ciprofloxacin and sulfatrimethoprim (SxT) unless drug resistance is detected, in which case azithromycin or cephalosporins are prescribed. Prevention follows the same steps as for salmonellosis (see 20.3 Making Connections). There is no vaccine yet available.

Fresh Foods and Infection

Fresh produce is responsible for half of all food-borne infections. The most common sources in order of occurrence are lettuce, sprouts, juice, melon, tomatoes, and spinach. One major reason for this association is that produce is consumed raw and thus not subject to heating that would reduce the microbial load. The predominant agents of infection are *E. coli, Salmonella, Listeria,* and *Shigella.* If produce is grown in contaminated soil and water and not properly cleaned or stored, it may harbor and even grow pathogens. The consumption of such products has increased tremendously in the past 10 years, and this has been accompanied by thousands of outbreaks all over the world. In a well-studied case, several outbreaks of diarrheal disease were reported in restaurant patrons who had consumed parsley. Tracing back the source of infection, investigators found a strain of *Shigella* in salads and appetizers. The parsley, grown in Baja, Mexico, had been stored and shipped in ice made from contaminated, unchlorinated water. This pathogen is notorious for having a low infectious dose, and this one source ultimately led to dozens of infections nationwide.

Figure 20.18 The appearance of the large intestinal mucosa in Shigella (bacillary) dysentery. (a) A rendering depicts patches of blood and mucus, erosion of the lining, and absence of perforation. **(b)** A section through the large intestine reveals points of hemorrhage that release bloody discharges. *Source:* CDC

Other outbreaks of food infection that received significant publicity are shown in the following table.

Pathogen	Number of Cases	Food Involved	Probable Source of Pathogen
E. coli O157:H7	907	Cucumbers	Soil contaminated by livestock
Listeria	146	Cantaloupe	Contaminated packing equipment
Salmonella	561	Alfalfa sprouts	Field contamination?
Shigella	406	Bean dip	Produce contaminated by human carrier

Due to the globalization of food, these products can originate from anywhere in the world, from local farms to countries thousands of miles away. A national information-sharing system—PulseNet—has been set up to investigate and tabulate these outbreaks, using a powerful, specific tool called pulsed-field gel electrophoresis. Additional information on this subject can be accessed at "CDC Foodborne Outbreaks."

Figure 20.19 A Gram stained preparation of *Yersinia pestis* in the blood of an infected mouse. The cells exhibit a distinctive bipolar morphology in which the ends of the cell stain darker—making it look like a "safety pin" (1,000×). *Source:* CDC

The Enteric Yersinia *Pathogens*

The genus ***Yersinia**** contains three species, all of which cause zoonotic infections called *yersinoses*. *Yersinia enterocolitica* and *Y. pseudotuberculosis* are intestinal inhabitants of wild and domestic animals that cause enteric infections in humans, and *Y. pestis* is the nonenteric agent of bubonic plague.

Yersinia enterocolitica has been isolated from healthy and sick farm animals, pets, wild animals, and fish, as well as fruits, vegetables, and drinking water. During the incubation period of about 4 to 10 days, the bacteria invade the small intestinal mucosa, and some cells enter the lymphatics and are harbored intracellularly in phagocytes. Inflammation of the ileum and mesenteric lymph nodes gives rise to severe abdominal pain that mimics appendicitis. Although it shares many characteristics with *Y. enterocolitica*, *Yersinia pseudotuberculosis* infections tend to be less serious and center upon lymph node inflammation rather than mucosal involvement.

Yersinia pestis and Plague

The word **plague**[4] conjures up visions of death and morbidity unlike any other infectious disease. Although pandemics of plague have probably occurred since antiquity, the first one that was reliably chronicled killed an estimated 100 million people in the sixth century. The last great pandemic occurred in the late 1800s and was transmitted around the world, primarily by rats transported in ships. The disease was brought to the United States through the port of San Francisco around 1906. Eventually infected rats mingled with native populations of rodents and gradually spread the populations throughout the West and Midwest. The cause of this dread disease is a tiny, harmless-looking gram-negative rod called ***Yersinia pestis,*** with unusual bipolar staining and capsules **(figure 20.19).**

Virulence Factors

Strains of the plague bacillus are just as virulent today as in the Middle Ages. Some of the virulence factors that coincide with its high morbidity and mortality are capsular and envelope proteins, which protect against phagocytosis and foster intracellular growth. The bacillus also produces coagulase, which clots blood and is involved in clogging the esophagus in fleas and obstructing blood vessels in humans. Other factors that contribute to pathogenicity are endotoxin and a highly potent murine toxin.

The Complex Epidemiology and Life Cycle of Plague

The plague bacillus exists naturally in many animal hosts throughout the world, even though the incidence of disease has been reduced in most areas. Plague still exists endemically in large areas of Africa, South America, the Mideast, Asia, and the former Soviet Union, where it sometimes erupts into epidemics. In the United States, sporadic cases (usually fewer than 10) occur as a result of contact with wild and domestic animals. No cases of human-to-human transmission have been recorded since 1924. Persons most at risk for developing plague are veterinarians and people living and working near woodlands and forests. Cases are restricted to locations west of the Mississippi River in the United States.

The epidemiology of plague is among the most complex of all diseases. It involves several different types of vertebrate hosts and flea vectors, and its exact cycle varies from one region to another. A general scheme of the cycle is presented in **figure 20.20.** Humans can develop plague through contact with wild animals **(sylvatic plague),** domestic or semidomestic animals **(urban plague),** or infected humans.

The Animal Reservoirs The plague bacillus occurs in 200 different species of mammals. The primary long-term *endemic reservoirs* are various rodents such as mice and voles that harbor the

* *Yersinia* (yur-sin´-ee-uh) After Alexandre Yersin, a French bacteriologist.
4. From the Latin *plaga,* meaning to strike, infest, or afflict with disease, calamity, or some other evil.

Pathogen Profile #6 *Shigella sonnei, S. dysenteriae, S. flexneri, S. boydii*

Microscopic Morphology Small, nonmotile, gram-negative rods.

Identified by Gram reaction, morphology, and physiological and biochemical tests. Motility can be used to distinguish *Shigella* (nonmotile) from *E. coli* (motile).

Habitat *Shigella* is primarily a pathogen of humans, whose sources include the Five *F*'s (feces, food, fingers, flies, and fomites).

Source: CDC

Virulence Factors The primary virulence factor of the genus is heat-labile shiga toxin, which causes injury to nerve cells, the intestine, and kidneys.

Primary Infections/Disease Shigellosis results when *Shigella* (usually *S. sonnei*) invades the mucosa of the large intestine, with the initial damage being caused by an inflammatory response. As the disease progresses, endotoxins promote fever, while enterotoxins damage the intestinal mucosa and villi, leading to bleeding and heavy mucus secretion.

Control and Treatment As in control of *Salmonella*, control of *Shigella* aims to prevent entry of the pathogen into the food chain. Adequate sanitation during food preparation, proper hand washing, and refraining from consuming unwashed fruits or vegetables, inadequately cooked meat and eggs, and unpasteurized milk all contribute to this goal. Treatment involves the administration of oral antimicrobics such as ciprofloxacin or sulfatri-methoprim (SxT), whereas resistant strains may be treated with azithromycin or cephalosporins.

organism but do not develop the disease. These **hosts** spread the disease to other mammals called *amplifying hosts* that become infected with the bacillus and experience massive die-offs during epidemics. These hosts, including rats, ground squirrels, chipmunks, and rabbits, are the usual sources of human plague. The particular mammal that is most important in this process depends on the area of the world. Other mammals (camels, sheep, coyotes, deer, dogs, and cats) can also be involved in the transmission cycle.

Flea Vectors The principal agents in the transmission of the plague bacillus from reservoir hosts to amplifying hosts to humans are fleas. These tiny, bloodsucking insects (see figure 23.26) have a special relationship with the pathogen. After a flea ingests a blood meal from an infected animal, the bacilli multiply in its gut. In fleas that effectively transmit the bacillus, the esophagus becomes blocked by coagulated blood. Being unable to feed properly, the ravenous flea jumps from animal to animal in a futile attempt to get nourishment. During this process, regurgitated infectious material is inoculated into the bite wound.

Normally the flea transmits the pathogen within that population, but many fleas are not host-specific and will attempt to feed on other species, even humans. Fleas of rodents such as rats and squirrels are most often vectors in human plague, though occasionally the human flea is involved. Humans can also be infected by handling infected animals, animal skins, or meat and by inhaling droplets.

Pathology of Plague

The number of bacilli required to initiate a plague infection is small—perhaps 3 to 50 cells. The manifestations of infection lead

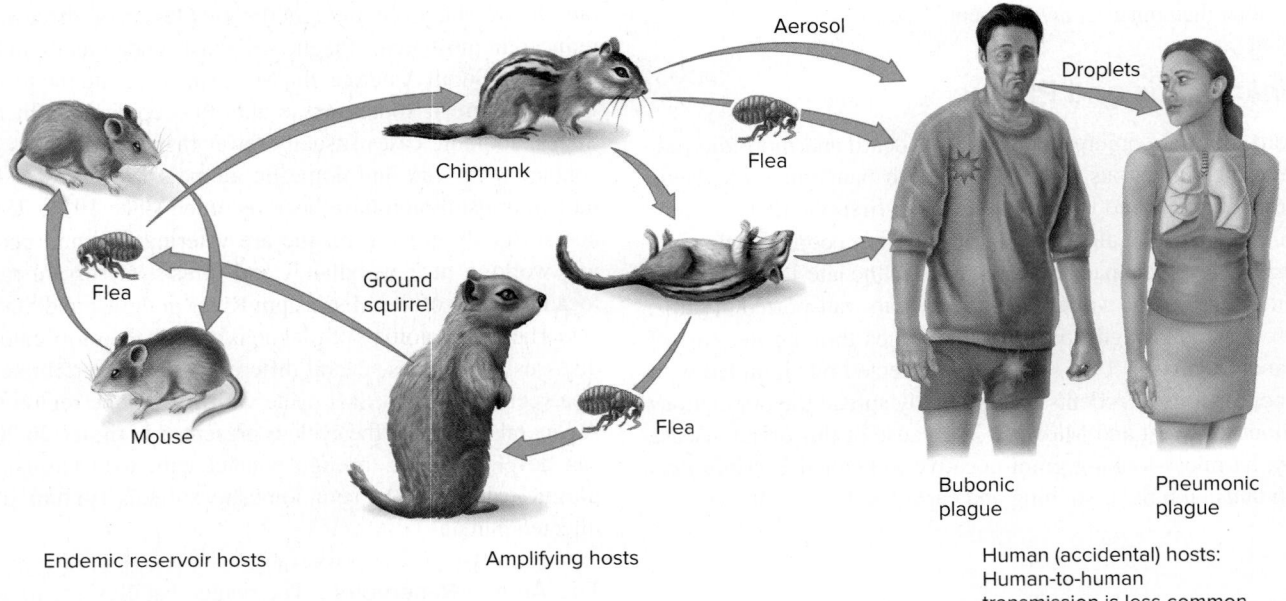

Figure 20.20 The infection cycle of *Yersinia pestis,* simplified for clarity.

Pathogen Profile #7 *Yersinia pestis*

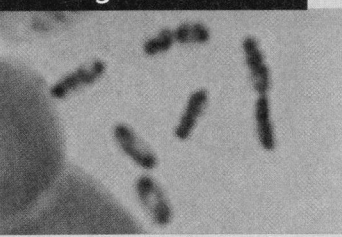

Source: CDC

Microscopic Morphology Small, gram-negative, encapsulated rod. The cells often stain darker at the end, giving them a "safety pin" appearance.

Identified by Gram reaction, morphology, and biochemical and serological tests. Along with symptoms, knowledge of patient's travel to endemic regions often can aid diagnosis.

Habitat *Y. pestis* is found in over 200 small rodents, which do not develop disease but serve as the primary endemic reservoir of the bacterium. Fleas are the primary arthropod vectors, spreading the bacterium from the primary hosts to amplifying hosts such as rats, squirrels, and rabbits.

Virulence Factors Capsular and envelope proteins that allow *Y. pestis* to survive and multiply within phagocytic cells. The bacterium also produces coagulase, which coagulates blood and causes the obstruction of blood vessels. *Y. pestis* has a small infective dose (3–50 cells), and produces both exotoxins and endotoxin.

Primary Infections/Disease Humans contract plague through the bite of an infected flea, handling infected animals, skins,

meat, or inhaling droplets. Symptoms of bubonic plague include fever, chills, headache, exhaustion, and the development of enlarged lymph nodes called bubos. Septicemic plague occurs as massive growth of bacteria in the blood, and release of virulence factors leads to disseminated intravascular coagulation, subcutaneous hemorrhage, and necrosis. The name "black death" is a reference to the darkening of the skin at this point. In pneumonic plague, infection is localized to the lungs and is contagious through sputum and aerosols.

Control and Treatment Plague is held at bay through quarantines during periods of epidemic disease and by reducing the population of endemic hosts and fleas through the use of traps, poisons, and insecticides. Treatment of plague must commence quickly, as death can occur in as little as 2 days after the appearance of symptoms. Streptomycin, gentamicin, doxycycline, and ciprofloxacin are all effective, and treated disease has a 90% to 95% survival rate.

to bubonic, septicemic, or pneumonic plague. In **bubonic plague,** the plague bacillus multiplies in the flea bite, enters the lymph, and is filtered by the local lymph nodes. Infection causes necrosis and swelling of the node, called a **bubo,*** typically in the groin (inguinal) and less often in the armpit (axilla) or neck (cervical). The incubation period lasts 2 to 8 days, ending abruptly with the onset of fever, chills, headache, nausea, weakness, and tenderness of the bubo.

Cases of bubonic plague often progress to massive bacterial growth in the blood termed **septicemic plague.** The release of virulence factors causes disseminated intravascular coagulation, subcutaneous hemorrhage, and purpura that may degenerate into necrosis and gangrene. Because of the visible darkening of the skin, the plague has often been called the "black death." In **pneumonic plague,** infection is localized to the lungs and is highly contagious through sputum and aerosols. Without proper treatment, it is invariably fatal.

Diagnosis, Treatment, and Prevention Because death can ensue as quickly as 2 to 4 days after the appearance of symptoms, prompt diagnosis and treatment of plague are imperative. The patient's history, including recent travel to endemic regions, symptoms, and laboratory findings from bubo aspirates, helps establish a diagnosis. Streptomycin is the antibiotic of choice. Treated plague has a 90% to 95% survival rate.

The menace of plague is proclaimed by its status as one of the internationally quarantinable diseases (the others are cholera and yellow fever). In addition to quarantine during epidemics, plague is

controlled by trapping and poisoning rodents near urban and suburban communities and by dusting rodent burrows with insecticide to kill fleas. These methods, however, cannot begin to control the reservoir hosts, so the potential for plague will always be present in endemic areas, especially as humans encroach into rodent habitats.

Oxidase-Positive Nonenteric Pathogens in Family Pasteurellaceae

Pasteurella multocida

Pasteurella is a zoonotic genus that occurs as normal microbiota in animals and is mainly of concern to the veterinarian. Of the six recognized species, ***Pasteurella multocida**** is responsible for the broadest spectrum of opportunistic infections. For example, poultry and wild fowl are susceptible to cholera-like outbreaks, and cattle are especially prone to epidemic outbreaks of hemorrhagic septicemia or pneumonia known as "shipping fever." The species has also adapted to the nasopharynx of the household cat and is normal microbiota in the tonsils of dogs. Because many hosts are domesticated animals with relatively close human contact, zoonotic infections are an inevitable and serious complication.

Animal bites or scratches, usually from cats and dogs, cause a local abscess that can spread to the joints, bones, and lymph nodes. Patients with weakened immune function due to liver cirrhosis or rheumatoid arthritis are at great risk for septicemic complications

* *bubo* (byoo'-boh) G. *boubon*, the groin.

* *Pasteurella* (pas″-teh-rel′-uh mul-toh-see′-duh) From Louis Pasteur, plus L. *multi,* many, and *cidere,* to kill.

involving the central nervous system and heart. Patients with chronic bronchitis, emphysema, pneumonia, or other respiratory diseases are vulnerable to pulmonary failure. Contrary to the antibiotic resistance of many gram-negative rods, *P. multocida* and other related species are susceptible to amoxicillin, and doxycycline is an effective alternative.

Haemophilus: The Blood-Loving Bacilli

Haemophilus[5] cells are small (1 × 2 µm) gram-negative encapsulated rods **(figure 20.21).** The members of this group tend to be fastidious and sensitive to drying, temperature extremes, and disinfectants. Even though their name means blood-loving, none of these organisms can grow on blood agar alone without special techniques. *Haemophilus* members require certain factors released by lysed red blood cells for energy metabolism. These factors are made available to them in media such as chocolate agar (a form of cooked blood agar; see chapter 3) and Fildes medium. Some *Haemophilus* species are normal colonists of the upper respiratory tract or vagina, and others (primarily *H. parainfluenzae* and *H. ducreyi*) are virulent species responsible for childhood meningitis and chancroid.

Haemophilus influenzae was originally named after it was isolated from patients with "flu" about 100 years ago. For over 40 years it was erroneously considered the causative agent until the real agent, the influenza virus, was discovered. The potential of this species to act as a pathogen remained in question until it was clearly shown to be an agent of acute **bacterial meningitis** in humans, caused primarily by the b serotype. The case rates in children have declined over the past years, likely because of an increased emphasis on vaccination **(figure 20.22).** Other prominent causes of meningitis are **N. meningitidis** and **S. pneumoniae.** Together these three agents comprise 85% of all cases of bacterial meningitis. The effect of vaccines and better diagnosis and treatments have reduced the numbers of these infections in recent times.

In contrast to *Neisseria* meningitis, *Haemophilus* meningitis is not associated with epidemics in the general population but tends to

5 Also spelled *Hemophilus*.

Figure 20.21 *Haemophilus influenzae* **as seen with fluorescent microscopy.** During the influenza epidemic of 1918, *H. influenzae* was found in the sputum of many influenza patients and was mistakenly thought to be the cause of the disease. *Source:* CDC

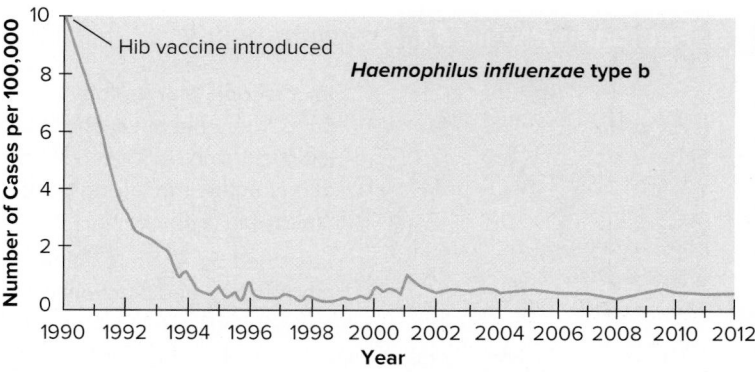

Figure 20.22 **Meningitis in the United States.** Graph of case rates for *Haemophilus influenzae*. Note the decrease in childhood meningitis seen by 1991, the point at which most children under 5 years of age were being immunized against *Haemophilus influenzae*.

occur sporadically in day care and family settings. It is transmitted by close contact with nose and throat discharges. Healthy adult carriers are the usual reservoirs of the bacillus.

Haemophilus meningitis is very similar to meningococcal meningitis (see chapter 18), with symptoms of fever, vomiting, stiff neck, and neurological impairment. Untreated cases have a fatality rate of nearly 90%; but even with prompt diagnosis and aggressive treatment, 20% of children sustain residual disability. Other important diseases caused by *H. influenzae* are an intense form of epiglottitis common in older children and young adults that may require immediate intubation or tracheostomy to relieve airway obstruction. This species is also an agent of otitis media, sinusitis, pneumonia, and bronchitis.

Haemophilus meningitis is usually treated with ceftriaxone as well as dexamethasone. Outbreaks of disease in families and day care centers may necessitate rifampin prophylaxis for all contacts. Routine vaccination with a subunit vaccine (Hib) based on type b polysaccharide is recommended for all children, beginning at age 2 months, with three follow-up boosters. It is frequently given in combination with DTaP.

Haemophilus influenzae subtype *aegyptius* is the primary agent of acute communicable **conjunctivitis,** sometimes called pinkeye. The subconjunctival hemorrhage that accompanies infection imparts a bright pink tinge to the sclera **(figure 20.23).** The disease occurs primarily in children, is distributed worldwide, and is spread through contaminated fingers and shared personal items

Figure 20.23 Acute conjunctivitis, or pinkeye, caused by *H. influenzae* subtype *aegyptius*. © Biophoto Associates/Science Source

as well as mechanically by gnats and flies. It is treated with antibiotic eyedrops.

Haemophilus ducreyi is the agent of **chancroid** (soft chancre), a sexually transmitted disease prevalent in the tropics and subtropics that afflicts mostly males. It is transmitted by direct contact with infected lesions and is favored by sexual promiscuity and unclean personal habits. After an incubation period lasting 2 to 14 days, lesions develop in the genital or perianal area. First to appear is an inflammatory macule that evolves into a painful, necrotic ulcer similar to those found in lymphogranuloma venereum and syphilis (see chapter 21). Often the regional lymph nodes develop into bubolike swellings that burst open. Treatment with azithromycin or ceftriaxone is usually effective.

Haemophilus parainfluenzae and ***H. aphrophilus,*** members of normal oral and nasopharyngeal microbiota, are involved in infective endocarditis in adults who have underlying congenital or rheumatic heart disease. Such infections typically stem from routine dental procedures, periodontal disease, or some other oral injury.

 Check Your Progress SECTION 20.5

19. Name three genera of noncoliform enteric bacteria that are opportunistic pathogens.

20. Compare typhoidal and nontyphoidal salmonelloses with regard to the number of cases normally seen, common sources of infection, and pathology of each disease.

21. Make a comparison chart for *Shigella* and *Salmonella* based on differences in morphology, epidemiology, and pathogenesis.

22. What are the Five *F*'s and how do they relate to enteric pathogens?

23. Who is most at risk for contracting plague and why?

24. Which body systems are commonly infected by *Haemophilus* species?

25. Describe the epidemiology and pathology of *Haemophilus influenzae* meningitis.

 CASE STUDY **Part 2**

The credit for lowering pertussis rates so dramatically from 1940 to 1976 went to the DTwP vaccine, which consisted of antigens against diphtheria, tetanus (the *D* and *T*), and a whole-cell pertussis component. The whole-cell designation meant that the vaccine contained whole (though very much dead) pertussis cells, which, when injected into the body, instigated the production of a powerful and long-lasting immune response. Unfortunately, the whole-cell pertussis vaccine was also linked to a large number of

local reactions (swelling, pain, and redness at the injection site) along with a smaller number of more serious neurological complications. In response to this, in 1992 the fourth and fifth doses of the pertussis vaccine were changed to contain an acellular pertussis component (DTaP). The acellular version of the vaccine contained toxoid along with a number of other antigens from *B. pertussis*, but no whole cells. In 1997 the vaccine recommendation was again changed, so that all five doses contained the acellular pertussis component. The end result of this shift in vaccine strategy was that in 2012, children 14 years old and younger most likely received acellular vaccines for the complete series, whereas those 16 to 19 years of age received whole-cell vaccines for the first three doses and acellular vaccines for the last two. Adolescents who were 15 years old in 2012 may have fallen into either category, depending on when they received their vaccinations.

Incidence of Pertussis Among Persons Aged ≤ 19 Years, by Patient Age and Vaccines Received – National Notifiable Diseases Surveillance System, United States, January 1, 2011–June 14, 2012

According to a 2010 study, the DTaP vaccine works well for 2 years, after which protection begins to wane. Five years after vaccination with the acellular vaccine, less than 30% of the vaccine effectiveness remains. This information moved the Advisory Committee on Immunization Practices to recommend an additional booster for persons age 11 through 18 who have completed the full childhood pertussis series.

■ *What type of pertussis vaccines were received by persons who were older than 20 in 2012?*

■ *What correlation exists between the type of vaccine received and the likelihood of illness?*

See: CDC. 2012. Pertussis epidemic—Washington, 2012. *MMWR* 61:517–22.

To conclude this Case Study, go to Connect.

3. *Francisella tularensis* has which portal(s) of entry?
 a. tick bite c. respiratory
 b. intestinal d. all of these

4. A classic symptom of pertussis is
 a. labored breathing c. convulsions
 b. paroxysmal coughing d. headache

5. The severe symptoms of pertussis are due to what effect(s)?
 a. irritation of the glottis by the microbe
 b. pneumonia
 c. the destruction of the respiratory epithelium
 d. blocked airways
 e. both c and d are involved

6. *Escherichia coli* displays which antigens?
 a. capsular c. flagellar
 b. somatic d. all of these

7. Which of the following is *not* an opportunistic enteric bacterium?
 a. *Serratia* b. *Klebsiella* c. *Proteus* d. *Shigella*

8. Which of the following is a major difference between *Salmonella* and *Shigella* infections?
 a. mode of transmission
 b. likelihood of septicemia
 c. the portal of entry
 d. presence/absence of fever and diarrhea

9. A complication/complications of typhoid fever is/are
 a. neurological damage c. liver abscesses
 b. intestinal perforation d. b and c

10. *Shigella* is transmitted by
 a. food c. feces
 b. flies d. all of these

11. The bubo of bubonic plague is a/an
 a. ulcer where the flea bite occurred
 b. granuloma in the skin
 c. enlarged lymph node
 d. infected sebaceous gland

12. *Haemophilus influenzae* is _____ and requires special _____ for growth.
 a. motile, temperatures c. intracellular, sample swabs
 b. encapsulated, minerals d. fastidious, blood factors

13. Which of the following is *not* associated with *H. influenzae* infections?
 a. fever b. flu c. stiff neck d. headache

14. Which of the following are primarily zoonoses?
 a. tularemia d. brucellosis
 b. salmonellosis e. pasteurellosis
 c. shigellosis f. bubonic plague

15. **Single Matching.** Match the infectious agent with the disease it is most associated with.
 _____ *Francisella tularensis* a. dysentery
 _____ *Yersinia pestis* b. local abscess
 _____ *Escherichia coli* O157:H7 c. chancroid
 _____ *Shigella species* d. enteric fever
 _____ *Salmonella enterica* e. whooping cough
 _____ *Salmonella typhi* f. meningitis
 _____ *Pseudomonas aeruginosa* g. typhoid fever
 _____ *Bordetella pertussis* h. hemolytic uremic syndrome
 _____ *Legionella pneumophila* i. bubonic plague
 _____ *H. influenzae aegyptius* j. Pontiac fever
 _____ *Haemophilus influenzae* k. folliculitis
 _____ *Haemophilus ducreyi* l. rabbit fever
 _____ *Pasteurella multocida* m. pinkeye

Case Study Review

1. One vaccine administered to protect against whooping cough also contains antigens meant to provide protection against
 a. toxoplasmosis d. *Pseudomonas*
 b. tetanus e. *Proteus*
 c. dysentery

2. Which vaccine utilizes intact *Bordetella pertussis* cells to initiate an immune response?
 a. DTaP d. Td
 b. DT e. Tdap
 c. DTwP

3. When treating teenagers and adults afflicted with pertussis (who often show only mild disease symptoms), health care professionals will generally ask what type of work the patient does and who else lives in the household. What exactly is the health care worker trying to determine, and why is this important?

Writing Challenge

For each question, compose a one- or two-paragraph answer that includes the factual information needed to completely address the question. Check Your Progress questions can also be used for writing-challenge exercises.

1. Write a short essay that summarizes the circumstances in which gram-negative bacteria are involved in opportunistic infections.

2. What are unique features in the epidemiology of *E. coli* O157:H7 (STEC)?

3. Explain several practices an individual can use to avoid enteric infection and disease at home and when traveling.

4. a. List the bacteria from this chapter for which general, routine vaccines are given.
 b. For which special groups of bacteria are there vaccines?
 c. For which bacteria is there no vaccines?

5. Briefly outline the zoonotic infections in this chapter, and describe how they are spread to humans.

 Concept Mapping

On Connect you can find an Introduction to Concept Mapping that provides guidance for working with concept maps along with concept mapping activities for this chapter.

 Level II. Application, Analysis, Evaluation, and Synthesis

These problems go beyond just restating facts and require higher levels of understanding and an ability to interpret, problem solve, transfer knowledge to new situations, create models, and predict outcomes.

 Critical Thinking

Critical thinking is the ability to reason and solve problems using facts and concepts. These questions can be approached from a number of angles, and in most cases, they do not have a single correct answer.

1. What is the logic behind testing for *E. coli* to detect fecal contamination of water?

2. Identify the genera with the following characteristics from figure 20.9:
 a. Lactose (−), phenylalanine and urease (−), citrate (+), ONPG (−)
 b. Lactose (+), motility (−), VP (−), indole (+)
 c. Lactose (+), motility (+), indole (−), H_2S (+)
 d. Lactose (−), phenylalanine and urease (+), H_2S (−), citrate (+)

3. Given that so many infections are caused by gram-negative opportunists, what would you predict will happen as the number of compromised patients increases, and on what do you base these predictions?

4. An infectious dose of several million cells in enteric infections seems like a lot.
 a. In terms of the size and abundance of microbes, could you see a cluster containing that many cells with the naked eye?
 b. Refer to chapter 7 on microbial growth cycles. About how long would it take an average bacterial species to reach the infectious dose of a million cells, starting from a single cell?
 c. Why do enteric diseases require a relatively higher infectious dose than nonenteric diseases?
 d. Explain why antimicrobic therapy used on gram-negative bacterial infections could actually lead to illness rather than cure it.

5. Students in our classes sometimes ask how it is possible for a single enteric carrier to infect 1,000 people at a buffet or for a box turtle to expose someone to food infection. We always suggest that they use their imagination. Provide a detailed course of events that would lead to these types of outbreaks.

6. Explain the reasons for an increase in numbers of pertussis cases. (Think vaccine.)

7. Compare and contrast the pathology, diagnosis, and treatment of meningococcal meningitis and *Haemophilus influenzae* meningitis.

8. Several persons working in an exercise gym acquired an acute disease characterized by fever, cough, pneumonia, and headache. Treatment with erythromycin cleared it up. The source was never found, but an environmental focus was suspected.
 a. What do you think might have caused the disease?
 b. People in a different gym got skin lesions after sitting in a redwood hot tub. Which pathogen could have caused that?

9. A 3-year-old severely ill child was admitted to a hospital with symptoms of diarrhea, fever, and malaise. Laboratory testing showed abnormal renal and liver values and anemia. She had no history of previous illness, and her food history was a recent meal of teriyaki beef consumed at a local restaurant.
 a. What was the probable pathogen?
 b. What was the likely source?
 c. What is the pathologic effect of the pathogen?

10. a. Name six bacteria dicussed in this chapter or in chapter 19 that could be used in biological warfare.
 b. What are some possible ways these bacteria could be used in warfare?
 c. What is your personal opinion of using microorganisms as weapons?

11. "There is no circumstance [in which] you can cook out 230 million bacteria. I'm not willing to take the risk that one pathogen isn't going to survive"—Ron Schnitzer (microbiologist). Comment on this quote. What do you think this microbiologist is referring to?

1. Use **figure 20.5 a,b** as a reference guideline for interpretation. Lanes numbered 1 through 6 are different patient sera and A through H are increasing dilutions of these sera. Identify the samples that show *Brucella* infection and give their titers. Which positive sample has the fewest antibodies? What conclusion can be made if a second test done on this same patient a week later indicates no change? What is the conclusion if all lane wells are positive for this patient?

Picture courtesy of Vircell

STUDY TIP: Quickly create a customized practice quiz for this chapter by going to your Connect SmartBook "My Reports" section. Don't have SmartBook and want to learn more? Go to **whysmartbook.com.**

Miscellaneous Bacterial Agents of Disease

Electron micrograph of Coxiella burnetii cells.
Source: National Institute of Allergy and Infectious Disease

Source: Dr. Edwin P. Ewing, Jr./CDC

CASE STUDY Part 1 A different kind of tourism

Often a good patient history is as important as any laboratory test. During the fall of 2014 five patients with similar complaints were seen by doctors in the New York area. All five presented with fever, fatigue, chills, and headache, a combination of **signs** and **symptoms** so general that they provide little diagnostic information. Three of the five reported preexisting medical conditions: one patient with atrial fibrillation and kidney stones, one with Parkinson disease and osteoarthritis, and one with multiple sclerosis. Questioning revealed that all five patients had recently traveled to Germany to receive injections of processed cells from sheep fetuses, a treatment known as **live cell therapy,** with at least two of the patients traveling to Germany twice a year in each of the previous five years to receive the treatments. Despite a complete lack of clinical evidence since the procedure was first introduced in the 1930s, the procedure is commonly advertised as having anti-aging effects and as a treatment for a variety of ailments, including those

> "Live Cell Therapy injections have long been a celebrity secret in preserving a youthful appearance and prolonging vital health."
>
> —From an Internet advertisement promoting live cell therapy

displayed by the three patients reporting preexisting conditions.

Live cell therapy is considered a form of **xenotransplantation,** as living cells are transplanted across species boundaries, and is associated with serious risks, including **anaphylaxis** and exposure of the recipient to a number of **zoonotic** infections. The procedure has been linked in the past to **vasculitis, encephalitis, clostridial** infections, **paresis** (weakening of the limbs), and death.

- *Why did the patients in this case travel to Germany to obtain treatment?*

- *What groups are normally at greatest risk for contracting zoonotic infections?*

To continue the Case Study, go to Case Study Part 2 at the end of the chapter.

Systems Profile 21.1 Miscellaneous Bacterial Pathogens

Organism	Skin/Skeletal	Nervous/Muscle	Cardiovascular/ Lymphatic/ Systemic	Gastrointestinal	Respiratory	Urogenital
Treponema pallidum pallidum	Skin rash (secondary syphilis)	Tertiary syphilis	Secondary syphilis Tertiary syphilis			Chancre (primary syphilis)
Leptospira interrogans		Leptospirosis	Leptospirosis			Leptospirosis
Borrelia burgdorferi	Lyme disease	Lyme disease	Lyme disease			
Vibrio cholerae				Secretory diarrhea		
Vibrio parahaemolyticus				Gastroenteritis		
Campylobacter jejuni				Gastroenteritis		
Helicobacter pylori				Gastritis, peptic and duodenal ulcers		
Rickettsia prowazekii	Epidemic typhus	Epidemic typhus				
Rickettsia rickettsii	Rocky Mountain spotted fever	Rocky Mountain spotted fever	Rocky Mountain spotted fever			
Ehrlichia spp.			Ehrlichiosis			
Coxiella burnetii			Q fever			
Bartonella henselae	Local papules		Cat scratch disease			
Chlamydia trachomatis	1. Neonatal conjunctivitis 2. Trachoma					Chlamydiosis, lymphogranuloma venereum
Mycoplasma pneumoniae					Atypical pneumonia	

A number of bacterial pathogens do not fit the usual categories of gram-positive or gram-negative rods or cocci. This group includes spirochetes and curviform bacteria, obligate intracellular parasites such as rickettsias and chlamydias, and mycoplasmas. This chapter covers not only those agents but also the mixed bacterial infections that are responsible for the diseases of the oral cavity **(Systems Profile 21.1).**

21.1 The Spirochetes

 Expected Learning Outcomes

1. Differentiate among the different stages of syphilis.
2. Explain the different ways in which syphilis infections may be diagnosed.
3. List the nonsyphilitic treponematoses.
4. Justify the strategies used to prevent leptospirosis.
5. Identify the vectors and behaviors associated with *Borrelia* infection.

Spirochetes typically display a helical form and flagellar mode of locomotion that appear especially striking in live, unstained preparations using the dark-field or phase-contrast microscope.

Other traits include a typical gram-negative cell wall and a well-developed periplasmic space that encloses the flagella (called endoflagella or periplasmic flagella) **(figure 21.1a)**. Although internal flagella are constrained somewhat like limbs in a sleeping bag, their flexing propels the cell by rotation and even crawling motions. The spirochetes are classified in the Phylum Spirochaetes, which contains three families and 13 genera. The majority of spirochetes are free-living saprobes or commensals of animals and are not primary pathogens. Major pathogens are found in three genera: *Treponema, Leptospira,* and *Borrelia* **(figures 21.1b** and 21.6).

Treponemes: Members of the Genus *Treponema*

Treponemes are thin, somewhat regular, coiled bacteria that live in the oral cavity, intestinal tract, and perigenital regions of humans and animals. The pathogens are strict parasites with complex growth requirements that necessitate cultivating them in live cells. Diseases caused by *Treponema* are called treponematoses. The

Quick Search
Point your browser toward YouTube and search for "Spirochetes" to see their fascinating mode of motility.

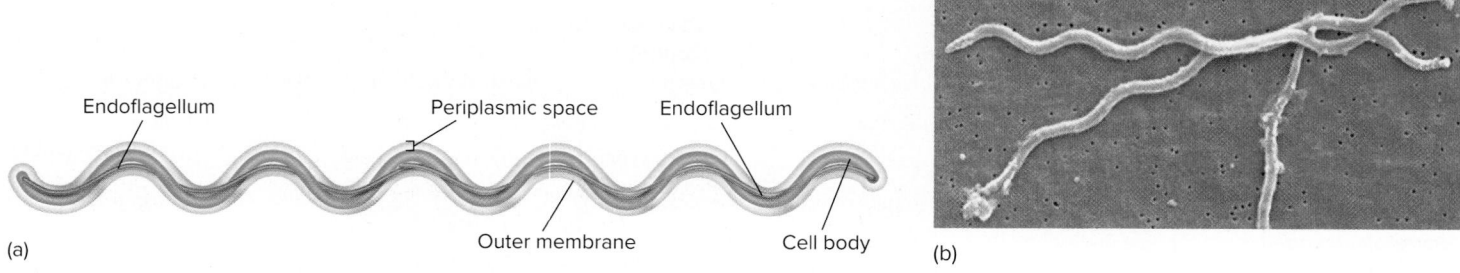

Endoflagellum Periplasmic space Endoflagellum

Outer membrane Cell body

(a) (b)

Figure 21.1 **Typical spirochete. (a)** Representation of general spirochete morphology with endoflagella inserted in the opposite poles, lying beneath the outer membrane and within the periplasmic space. **(b)** Members vary in the number of coils. For example, Borrelias have 3 to 10 loose, irregular coils (1,500×). (b): *Source:* Janice Haney Carr/CDC

subspecies *Treponema pallidum pallidum** is responsible for venereal and congenital syphilis **(Pathogen Profile #1)**; the subspecies *T. p. endemicum* causes nonvenereal endemic syphilis, or bejel; and *T. p. pertenue* causes yaws. *Treponema carateum* is the cause of pinta. Infection begins in the skin, progresses to other tissues in gradual stages, and is often marked by periods of healing interspersed with relapses. The major portion of this discussion centers on syphilis, and any mention of *T. pallidum* refers to the subspecies *T. p. pallidum*. Other treponemes of importance are involved in infections of the gingiva.

Treponema pallidum: *The Spirochete of Syphilis*

The origin of *syphilis** is an obscure yet intriguing topic of speculation. The disease was first recognized at the close of the fifteenth century in Europe, a period coinciding with the return of Columbus from the West Indies, which led some medical scholars to conclude that syphilis was introduced to Europe by early explorers. DNA analyses indicate that treponemal disease is ancient and widespread. Even wild primates carry strains of *Treponema* that are related to the human forms. Whatever its origins, once it became sexually transmitted, the pathogen was ultimately carried worldwide.

Epidemiology and Virulence Factors of Syphilis

Although infection can be induced in laboratory animals, the human is evidently the sole natural host and source of *T. pallidum*. It is an extremely fastidious and sensitive bacterium that cannot survive for long outside the host, being rapidly destroyed by heat, drying, disinfectants, and other adverse conditions. It survives a few minutes to hours when protected by body secretions and about 36 hours in stored blood. Research with human subjects has demonstrated that the risk of infection from an infected sexual partner is 12% to 30%. The bacterium may also be passed from an infected mother to the fetus in utero. Syphilitic infection through blood transfusion or exposure to fomites is rare.

* *Treponema pallidum* (trep″-oh-nee′-mah pal′-ih-dum) Gr. *trepo,* turn, and nema, thread; L. *pallidum,* pale. The spirochete does not stain with the usual bacteriologic methods.

* The term *syphilis* first appeared in a poem entitled "Syphilis sive Morbus Gallicus" by Fracastorius (1530) about a mythical shepherd whose name eventually became synonymous with the disease from which he suffered.

Syphilis, like other sexually transmitted diseases (STDs), has experienced periodic increases during times of social disruption. After experiencing a decline during the 1990s, case reports are once again increasing (see figure 18.22). In fact, the case rate has more than tripled since 2000. Major contributors to this trend are gay and bisexual men, who now account for 83% of new infections. Other increases are associated with prostitution and intravenous drug abuse. Because many cases go unreported, the actual incidence is likely to be several times higher than these reports show. Syphilis continues to be a serious problem worldwide, especially in Africa and Asia. Persons with syphilis often suffer concurrent infection with other STDs such as chlamydia, herpes simplex, gonorrhea, and AIDS.

A Dark Event in Human Experimentation

One of the most disturbing events in the study of syphilis occurred in the United States. Beginning in 1932, the U.S. government conducted a study called the "Tuskegee Study of Untreated Syphilis in the Negro Male," which eventually involved 399 indigent African-American men living in the South. Infected men were recruited into the study, which sought to document the natural progression of the disease. These men were never told that they had syphilis and were never treated for it, even after penicillin was shown to be an effective cure. The study ended in 1972 after it became public. In 1997 the federal government issued a public apology for having permitted the study to proceed for so long and began paying millions of dollars in compensation to the victims and their heirs.

Pathogenesis and Host Response

Brought into direct contact with mucous membranes or abraded skin, *T. pallidum* binds avidly by its hooked tip to the epithelium. The number of cells required for infection using human volunteers was established at 57 organisms. At the binding site, the spirochete multiplies and penetrates the capillaries within a short time. Once in the general circulation, the pathogen grows in virtually any tissue.

Specific factors that account for the virulence of the syphilis spirochete appear to be its outer membrane proteins. It produces no toxins and does not appear to kill cells directly, but rather causes damage by invoking an inflammatory response in the host. Studies have shown that phagocytes are active against it and several types of antitreponemal antibodies are formed, but cell-mediated immune

Pathogen Profile #1 *Treponema pallidum (T. pallidum pallidum)*

Source: CDC

Microscopic Morphology Thin, gram-negative spirochetes.

Identified by The slender nature of *T. pallidum* often makes it invisible in a Gram stain. Wet mounts of fluid taken from suspected syphilitic lesions may be examined with dark-field microscopy for motile spirochetes. Immunofluorescent staining and DNA probes may also be used to detect the pathogen. Infection may be determined using serological testing to detect antibodies to *T. pallidum*.

Habitat Humans are the only host of *T. pallidum*.

Virulence Factors The hooked tip of the spirochete is responsible for initial binding of the pathogen to epithelial cells. Proteins within the outer membrane of *T. pallidum* induce an inflammatory response in the host, eventually leading to tissue damage.

Primary Infections/Disease Infection with *T. pallidum* causes syphilis, a sexually transmitted disease. The disease has three distinct stages, separated by periods of latency. Primary syphilis occurs within 3 months of infection and is marked by the formation of a chancre at the site of infection. Secondary syphilis appears a few weeks to several months later and is marked by fever, headache, sore throat, lymphadenopathy, and a red or brown rash that occurs on all skin surfaces. Like chancres, the lesions spontaneously heal after a few weeks. Complications of secondary syphilis may occur in the bones, hair follicles, joints, eyes, liver, and brain and may remain for years. Tertiary syphilis occurs in about 30% of untreated syphilis cases—often many years later—and is marked by the formation of swollen tumors called gummas, along with neural and cardiovascular complications. Congenital syphilis occurs when *T. pallidum* passes from the circulation of a pregnant woman to the placenta, infecting the fetus and causing a wide variety of symptoms.

Control and Treatment Penicillin G is the treatment of choice for syphilis. Control of the disease relies on detection and treatment of sexual contacts of syphilitic patients. The use of latex condoms reduces but does not eliminate the chance of infection because *T. pallidum* may infect sites that are not normally protected by a condom.

responses are unable to contain it. The primary lesion occurs when the spirochetes invade the spaces around arteries and stimulate an inflammatory response. Organs are damaged when granulomas form at these sites and block circulation.

Clinical Manifestations Untreated syphilis is marked by distinct clinical stages designated as primary, secondary, and tertiary syphilis. These phases exhibit multiple signs and symptoms, which cause it to "imitate" many other diseases (**table 21.1**). It also has latent periods of varying duration during which the disease is quiescent. The spirochete appears in the lesions and blood during the primary and secondary stages and, thus, is communicable at these times. Syphilis is largely noncommunicable during the tertiary stage, though it can be transmitted during early latency.

Primary Syphilis The earliest indication of syphilis infection is the appearance of a hard **chancre*** at the site of inoculation, after an incubation period that varies from 9 days to 3 months (**figure 21.2**). The chancre begins as a small, red, hard bump that enlarges and breaks down, leaving a shallow crater with firm margins. The base of the chancre beneath the encrusted surface swarms with spirochetes. Most chancres appear on the internal and external genitalia, but about 20% occur on the lips, nipples, or fingers or around the anus. Because genital lesions tend to be painless, they may escape notice in some cases. Lymph nodes draining the affected region become enlarged and firm, but systemic symptoms are usually

* *chancre* (shang′-ker) Fr. for canker; from L. *cancer,* crab. An injurious sore.

Figure 21.2 Primary syphilis lesion: chancre. The chancre is a dense, off-colored patch that ulcerates. Common locations for chancres include the genitals and mouth. *Source:* Dr. Gavin Hart, Dr. N. J. Fiumara/CDC

absent. The chancre heals spontaneously without scarring in 3 to 6 weeks, but this healing is deceptive, because the spirochete is entering a period of tremendous systemic activity.

Secondary Syphilis About 3 weeks to 6 months (average is 6 weeks) after the chancre heals, the secondary stage appears. By then many systems of the body have been invaded, and the signs and symptoms are more profuse and intense. Initially fever, headache, and sore throat occur, followed by lymphadenopathy and a peculiar red or brown rash that breaks out on all skin surfaces, including the palms and the soles (**figure 21.3**). Like the chancre, the

TABLE 21.1	Syphilis: Stages, Symptoms, Diagnosis, and Control			
Stage	**Average Duration**	**Clinical Setting**	**Diagnosis**	**Treatment**
Incubation	3 weeks	No lesion; treponemes adhere and penetrate the epithelium; after multiplying, they disseminate.	Asymptomatic phase	Not applicable
Primary	2–6 weeks	Initial appearance of chancre at inoculation site; intense treponemal activity in body; chancre later disappears.	Dark-field microscopy; VDRL, FTA-ABS, MHA-TP testing	Benzathine penicillin G, 2×10^6 units; doxycycline
Primary latency	2–8 weeks	Healed chancre; little scarring; treponemes in blood; few, if any, symptoms.	Serological tests (+)	As above
Secondary	2–6 weeks after chancre leaves	Skin, mucous membrane lesions; hair loss; patient highly infectious; fever, lymphadenopathy; symptoms can persist for months.	Dark-field testing of lesions; serological tests (+)	As above
Latency	6 months to 8 or more years	Treponemes quiescent unless relapse occurs; lesions can reappear.	Seropositive blood test	As above
Tertiary	Variable, up to 20 years	Neural, cardiovascular symptoms; gummas develop in organs; seropositivity.	Treponeme may be demonstrated by DNA analysis of tissue.	As above

lesions contain viable spirochetes and disappear spontaneously in a few weeks. The major complications, occurring in the bones, hair follicles, joints, liver, eyes, and brain, can linger for months and years.

Latency and Tertiary Syphilis After resolution of secondary syphilis, about 30% of infections enter a highly varied latent period that can last for 20 years or longer. Latency is divisible into early and late phases, and though antitreponemal antibodies are readily detected, the spirochete itself is not. The final stage of disease—late, or tertiary, syphilis—is quite rare today because of widespread use of antibiotics to treat other infections. By the time a patient reaches this phase, numerous pathologic complications occur in susceptible tissues and organs. Cardiovascular syphilis is a late complication of the disease. Infection of the small blood vessels that supply blood to the heart and aorta causes constriction and blockage. In time, compromised circulation to these organs can give rise to heart failure and aortic aneurysms.

In one form of tertiary syphilis, painful swollen syphilitic tumors called **gummas*** develop in tissues such as the liver, skin, bone, and cartilage **(figure 21.4)**. Gummas are usually benign and only occasionally lead to death, but they can impair function. Neurosyphilis can involve any part of the nervous system, but it shows particular affinity for the blood vessels in the brain, cranial nerves, and dorsal roots of the spinal cord. Destruction of parts of the spinal cord can lead to muscle wasting and loss of activity and coordination. Other manifestations include severe headaches, convulsions, mental derangement, atrophy of the optic nerve, blindness, and the Argyll-Robertson pupil reaction. The latter condition being small pupils that do not react to light (figure 21.4). This is perhaps the most common sign still seen today and is caused by adhesions along the inner edge of the iris that fix the pupil's position into an irregular-shaped circle.

Congenital Syphilis *Treponema pallidum* can pass from a pregnant woman's circulation into the placenta and can be carried throughout the fetal tissues. An infection leading to congenital syphilis can occur in any of the three trimesters, though it is most common in the second and third. The pathogen inhibits fetal growth and disrupts critical periods of development with varied consequences, ranging from mild to the extremes of spontaneous miscarriage or stillbirth. Early congenital syphilis encompasses the period

Figure 21.3 Symptom of secondary syphilis. The skin rash in secondary syphilis can form on most areas of the body. The rash does not hurt or itch and can persist for months. *Source:* CDC

* *gumma* (goo′-mah) L. *gummi,* gum. A soft tumorous mass containing granuloma tissue.

(a)

Pupil with an
(b) irregular shape

Figure 21.4 The pathology of late, or tertiary, syphilis. (a) An ulcerating syphilis tumor, or gumma, appears on the nose of this patient. Gummas can also be internal. **(b)** The Argyll-Robertson pupil constricts into an irregular-shaped opening, indicating damage to the nerves that control the iris. The iris itself may have prominent areas of discoloration. (a): *Source:* J. Pledger/CDC; (b): © K. Talaro

from birth to 2 years of age and is usually first detected 3 to 8 weeks after birth. Infected infants often demonstrate such signs as nasal discharge, skin eruptions and loss, bone deformation, and nervous system abnormalities. The late form gives rise to an unusual assortment of stigmata in the bones, eyes, inner ear, and joints, and causes the formation of Hutchinson's teeth **(figure 21.5).** The number of congenital syphilis cases is closely tied to the incidence in adults; because it is sometimes not diagnosed, some children sustain life-long disfiguring disease.

Figure 21.5 Congenital syphilis. A common characteristic of late congenital syphilis is notched, barrel-shaped teeth (Hutchinson's teeth). *Source:* CDC

Figure 21.6 A bright-field view of *Treponema pallidum*, highlighted by a special stain. *Treponema* spirochetes generally display 6 to 14 coils. © Melba Photo Agency/Alamy Stock Photo RF

Clinical and Laboratory Diagnosis The pattern of syphilis imposes many complications on diagnosis. Not only do the stages mimic other diseases, but their appearance can be so separated in time as to seem unrelated. The chancre and secondary lesions must be differentiated from various bacterial, fungal, and parasitic infections, tumors, and even allergic reactions. Overlapping symptoms of concurrent, sexually transmitted infections such as gonorrhea or chlamydiosis can further complicate diagnosis. The clinician must weigh presenting symptoms, patient history, and microscopic and serological tests in rendering a definitive diagnosis.

Although spirochetes like *Treponema* do not stain well with traditional dyes, certain special techniques using silver can make them more visible with a bright-field microscope **(figure 21.6).** A faster and less involved technique for diagnosing primary, early congenital, and, to a lesser extent, secondary syphilis is dark-field microscopy of a suspected lesion (see Pathogen Profile #1). The lesions are gently squeezed or scraped to extract clear serous fluid. A wet mount prepared from the exudate is then observed for the characteristic size, shape, and motility of *T. pallidum*. A single negative test is insufficient to exclude syphilis, so follow-up tests are recommended. Another microscopic test for discerning the spirochete directly in samples is direct immunofluorescent staining with monoclonal antibodies (see figure 17.3). Patient samples can also be tested with a DNA probe specific to various spirochete gene sequences.

Testing Blood for Syphilis Serological tests are based upon detection of antibody formed in response to *T. pallidum* infection (see table 21.1). Several of the tests (rapid plasma reagin [RPR], VDRL, Kolmer) are variations on the original test developed by Wasserman using cardiolipin, a natural constituent of many cells, as the antigen. Although anticardiolipin antibodies are not specific for syphilis, the test is an effective way to screen the population for people who may be infected. This includes high-risk groups such as homosexual men, male and female prostitutes, people with other STDs, and pregnant women.

With positive results, it is important to detect an elevated antibody titer indicative of active infection and to rule out residual antibodies from a prior cured infection. Because the most common screening tests (RPR and VDRL) are based on reactions to a substance found normally in human tissue, biological false positives can occur, especially in patients with other infections or

immunopathologies. A more specific test is needed for those who are suspected of having a false-positive result.

Typical of these specific tests is the *T. pallidum* microhemagglutination assay (MHA-TP), which employs red blood cells that have been coated with treponemal antigen. Agglutination of the cells by serum indicates antitreponemal antibodies and infection. Another standard test is an indirect immunofluorescent method called the FTA-ABS (**f**luorescent **t**reponemal **a**ntibody **abs**orbance) test. The test serum is first absorbed with treponemal cells and reacted with antihuman globulin antibody labeled with fluorescent dyes. If antibodies to the treponeme are present, the fluorescence on the outside of these cells is highly visible with a fluorescent microscope. Another variant, the *T. pallidum* immobilization (TPI) test, mixes live syphilis spirochetes with test serum to assay the spirochete's loss of motility. These tests are highly sensitive and specific and can rule out false-positive results. Children with suspected cases of congenital syphilis may be verified by a Western blot test.

Treatment and Prevention Penicillin G retains its status as a wonder drug in the treatment of all stages and forms of syphilis. It is given parenterally in large doses with benzathine or procaine to maintain a blood level lethal to the treponeme for at least 7 days (see table 21.1). Alternative drugs (tetracycline and doxycycline) are less effective and indicated only if penicillin allergy has been documented. It is important that all patients be monitored for compliance or possible treatment failure.

The core of an effective prevention program depends upon detection and treatment of the sexual contacts of syphilitic patients. Public health departments and physicians are charged with the task of questioning patients and tracing their contacts. All individuals identified as being at risk, even if they show no signs of infection, are given immediate prophylactic penicillin in a single, long-acting dose. Protective immunity does arise in humans and in experimentally infected rabbits, which raises the prospect of an effective immunization program in the future. The cloning of treponemal surface antigens using recombinant DNA technology will support development of vaccines and new diagnostic testing methods.

Nonsyphilitic Treponematoses

The other treponematoses are ancient diseases that closely resemble syphilis in their effects, though they are rarely transmitted sexually or congenitally. These infections, known as bejel, yaws, and pinta, are endemic to certain tropical and subtropical regions of the world. The treponemes that cause these infections are nearly indistinguishable from those of syphilis in morphology and behavior. The diseases are slow developing and progressive and involve primary, secondary, and tertiary stages. They begin with local invasion by the treponeme into the skin or mucous membranes and its subsequent spread to subcutaneous tissues, bones, and joints. Drug therapy with penicillin, erythromycin, or tetracycline remains the treatment of choice for these treponematoses.

Bejel Bejel is also known as endemic syphilis and nonvenereal childhood syphilis. The pathogen, the subspecies *T. pallidum endemicum,* is harbored by a small reservoir of nomadic and seminomadic people in arid areas of the Middle East and North Africa. It is a chronic, inflammatory childhood disease transmitted by direct contact or shared

(a) (b)

Figure 21.7 Endemic treponematoses. (a) Skin and membrane nodules in a young boy with endemic syphilis (bejel). **(b)** The clinical appearance of yaws. The bowing of the lower leg is seen in some children with congenital syphilis, and in patients with yaws.
(a, b): *Source:* CDC

household utensils and other fomites and is facilitated by minor abrasions or cracks in the skin or a mucous membrane. Often the infection begins as small, moist patches in the oral cavity (**figure 21.7a**) and spreads to the skin folds of the body and to the palms.

Yaws Yaws is a West Indian name for a chronic disease known by the regional names bouba, frambesia tropica, and patek. It is endemic to warm, humid, tropical regions of Africa, Asia, and South America. The microbe, subspecies *T. pallidum pertenue,* is readily spread by direct contact with skin lesions or fomites. Crowded living conditions and poor community or personal hygiene are contributing factors. The earliest sign is a large, abscessed papule called the "mother yaw," usually on the legs or lower trunk. After the initial lesion has healed, a secondary crop of moist nodular masses develop in the skin, periosteum, and bones but do not penetrate to the viscera (**figure 21.7b**). Yaws can be prevented by improved hygiene, shielding minor skin injuries from contamination, and surveillance for new cases.

Pinta The names *mal del pinto* and *carate* are regional synonyms for pinta, a chronic skin infection caused by *T. p. carateum.** Transmission evidently requires several years of close personal contact accompanied by poor hygiene and inadequate health facilities. Even though pinta is not currently widespread, the disease is still found in isolated populations inhabiting the tropical forest and valley regions of Latin America. Infection begins in the skin with a dry, scaly papule reminiscent of psoriasis or leprosy. In time, pigmented secondary macules and blanched tertiary lesions appear. Pinta is not life-threatening, but it often creates scars on the afflicted area.

Leptospira and Leptospirosis

Leptospires are typical spirochetes marked by 12 to 18 tight, regular, individual coils with a bend or hook at one or both ends. There are only two species in the genus: ***Leptospira interrogans,*** which causes

* *carateum* (kar-uh′-tee-um) From *carate,* the South American name for pinta.
* *Leptospira interrogans* (lep″-toh-spy′-rah in-terr′-oh-ganz) Gr. *leptos,* slender or delicate, and *speira,* a coil; *interrogans* because it has a single hook that makes it look like a question mark.

leptospirosis in humans and animals, and *L. biflexa,* a harmless, free-living saprobe. The two species are serologically, genetically, and physiologically distinct. *Leptospira interrogans* demonstrates nearly 200 serotypes distributed among various animal groups, which accounts for the extreme variations in leptospirosis among humans.

Epidemiology and Transmission of Leptospirosis

Leptospirosis is a zoonosis associated with wild animals such as rodents, skunks, raccoons, foxes, and some domesticated animals, particularly horses, dogs, cattle, and pigs. Although these reservoirs are distributed throughout the world, the disease is concentrated mainly in the tropics. Leptospires shed in the urine of an infected animal can survive for several months in neutral or alkaline soil or water. Infection occurs almost entirely through contact of skin abrasions or mucous membranes with animal urine or some environmental source containing urine. It is not associated with animal bites, inhalation, or human contact. In the United States, about 100 cases of leptospirosis are reported annually, with about half of the cases occurring in Hawaii. Persons most commonly affected include older children and young adults exposed to water polluted with animal urine and soldiers involved in jungle training.

Pathology of Leptospirosis and Host Response

Leptospirosis proceeds in two phases, and its principal targets are the kidneys, liver, brain, and eyes. During the early, or leptospiremic, phase, the pathogen appears in the blood and cerebrospinal fluid. Symptoms are sudden high fever, chills, headache, muscle aches, conjunctivitis, and vomiting. During the second, or immune, phase, the blood infection is cleared by natural defenses. This period is marked by milder fever, headache due to leptospiral meningitis, and *Weil's syndrome,* a cluster of symptoms characterized by kidney invasion, hepatic disease, jaundice, anemia, and neurological disturbances. Long-term disability and even death can result from injury to the kidneys and liver, but they occur primarily with virulent strains and in elderly patients.

Diagnosis, Treatment, and Prevention

A history of environmental exposure, along with presenting symptoms, can support initial diagnosis of leptospirosis, but definitive diagnosis relies on dark-field microscopy of specimens, *Leptospira* culture, and serological tests. Because leptospiral infection stimulates a strong humoral response, it is possible to test the patient's serum for its antibody titer. A fast, specific, and effective test called the macroscopic slide agglutination test is most often employed for routine screening. Live or formalinized *L. interrogans* is mixed with the patient's serum and observed for agglutination or lysis with a dark-field microscope.

Early treatment, prior to the fifth day of illness, with penicillin or doxycycline rapidly reduces symptoms and shortens the course of disease, but delayed therapy is less effective. A vaccine made from killed cells is available for pets, but not for humans. Protection is less than 100% because the vaccine confers protection only to a few specific strains. The best controls are to wear protective footwear and clothing and to avoid swimming or wading in natural waters used by animals.

Borrelia: Arthropod-Borne Spirochetes

Members of the genus *Borrelia** are larger than other pathogenic spirochetes, ranging from 0.2 to 0.5 μm in width and from 10 to 20 μm in length, and they contain 3 to 10 irregularly spaced and loose coils (see figure 21.1*b*) with many (30–40) periplasmic flagella. The nutritional requirements of *Borrelia* are so complex that the bacterium can be grown artificially only on specially formulated media.

Human infections with *Borrelia,* termed borrelioses, are all transmitted by some type of arthropod vector, usually ticks or lice. The two most important human diseases are relapsing fever and Lyme disease.

Epidemiology of Relapsing Fever

Borrelia hermsii, the cause of tick-borne relapsing fever, is carried by soft ticks of the genus *Ornithodoros.* The mammalian reservoirs of this zoonosis are squirrels, chipmunks, and other wild rodents, and the human is generally an accidental host. The spirochetes mature and persist in the salivary glands and intestines of the tick. As a result, both the bite itself and attempts to scratch it can initiate infection. Tick-borne relapsing fever occurs sporadically in the United States, usually in campers, backpackers, and forestry personnel who frequent the higher elevations of western states. The incidence of infection is higher in endemic areas of the tropics, especially where rodents have easy access to dwellings.

Epidemics of louse-borne relapsing fever occur whenever famine, war, or natural disasters are coupled with poor hygiene, crowding, and inadequate medical attention. Such conditions favor the survival and spread of the louse vector *Pediculus humanus,* which harbors the spirochete *B. recurrentis* in its body cavity. A host is infected when lice are smashed and scratched into a wound or the skin. Louse-borne fever is most common in parts of sub-Saharan Africa, China, and Afghanistan.

Pathogenesis and the Nature of Relapses The pathologic manifestations are similar in tick- and louse-borne relapsing fever. After a 2- to 15-day incubation period, patients experience the abrupt onset of high fever, shaking chills, headache, and fatigue. Later features of the disease include nausea, vomiting, muscle aches, and abdominal pain. Extensive damage to the liver, spleen, heart, kidneys, and cranial nerves occurs in many cases. Half of the patients hemorrhage profusely into organs, and some develop a rash on the shoulders, trunk, and legs. Untreated cases are often lengthy and debilitating and may have a 40% mortality rate.

As the name *relapsing fever* indicates, the fever follows a fluctuating course that is explained by changes in the spirochete and the attempts of the immune system to control it **(figure 21.8).** *Borrelia* have adopted a remarkable strategy for evading the immune system and avoiding destruction. They change surface antigens during growth, so that, in time, the antibodies formed against the earlier antigens lose effectiveness. Cells in the new antigen phase survive, multiply, and cause a second wave of symptoms. Over time the immune system forms new antibodies, but it is soon challenged with yet another antigenic phase. A single strain has been known to generate 24 distinct serological types. Eventually cumulative immunity against the variety of antigens develops, and complete recovery can occur.

* *Borrelia* (boh-ree'-lee-ah) Named after Amédé Borrel, a French bacteriologist

21.1 MAKING CONNECTIONS

The Disease Named for a Town

In the 1970s an enigmatic cluster of arthritis cases appeared in the town and surrounding suburbs of Old Lyme, Connecticut. This phenomenon caught the attention of nonprofessionals and professionals alike, whose persistence and detective work ultimately disclosed the unusual nature and epidemiology of Lyme disease. The process of discovery began in the home of Polly Murray, who, along with her family, was beset for years by recurrent bouts of stiff neck, swollen joints, malaise, and fatigue that seemed vaguely to follow a rash from tick bites. When Mrs. Murray's son was diagnosed as having juvenile rheumatoid arthritis, she became skeptical. Conducting her own literature research, she began to discover inconsistencies. Rheumatoid arthritis was described as a rare, noninfectious disease, yet over an 8-year period she found that 30 of her neighbors had experienced similar illnesses. Eventually this cluster of cases and several others were reported to state health authorities.

The reports caught the attention of Dr. Allen Steere, a rheumatologist with a Centers for Disease Control and Prevention (CDC) background. He was able to forge the vital link between the case histories, the disease symptoms, and the presence of unique spirochetes in ticks preserved by some of the patients. These same spirochetes had been previously characterized in 1981 by Dr. Willy Burgdorfer, though he did not realize their importance at the time.

In the years since Lyme disease was formally characterized, retrospective studies showed that this is not really a new disease and was first reported in Europe at the turn of the twentieth century. Recent PCR analysis of tick museum specimens from 50 years ago documents the presence of *Borrelia burgdorferi*. It is now thought that Lyme disease has been present in North America for centuries.

Explain how an infection such as Lyme disease mimics other noninfectious diseases. *Answer available on Connect.*

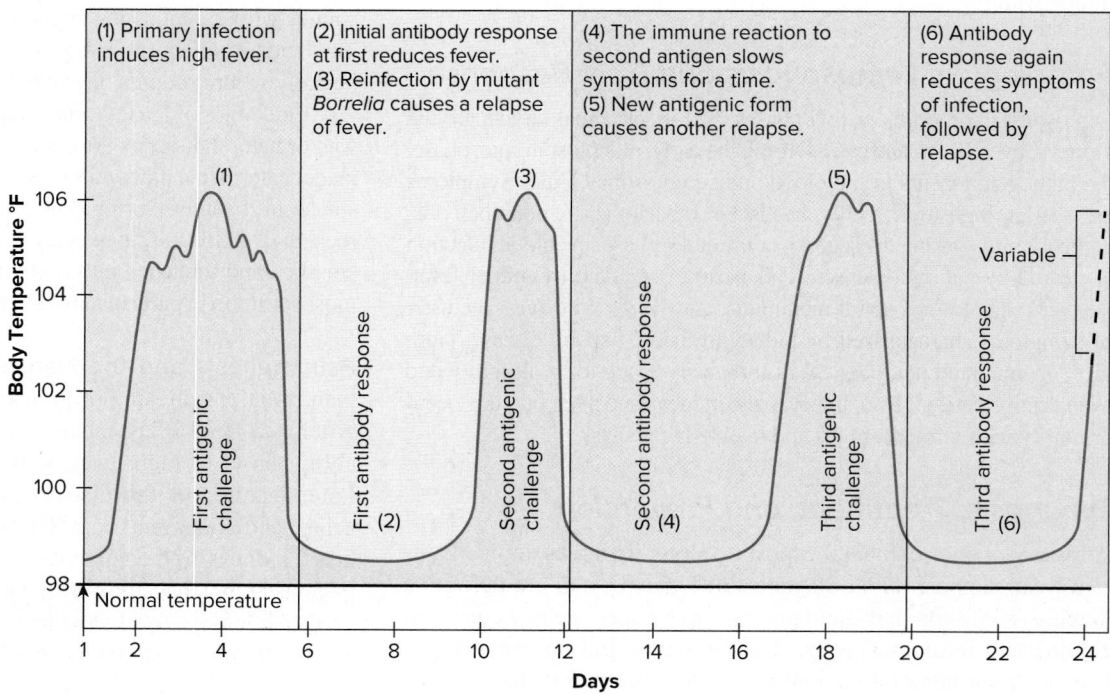

Figure 21.8 The pattern in relapsing fever, based on symptoms (fever) over time. The antigen phase changes can continue for several more days.

Diagnosis, Treatment, and Prevention A patient's history of exposure, clinical symptoms, and the presence of *Borrelia* in stained blood smears are very definitive evidence of borreliosis. Except for pregnant women and young children, doxycycline or tetracycline is the treatment of choice, with erythromycin serving as an alternative. Because vaccines are not available, prevention of relapsing fever is dependent upon controlling rodents and avoiding tick bites. Louseborne relapsing fever can be controlled with improved hygiene.

Borrelia burgdorferi *and Lyme Disease*

Lyme disease, the most prominent borreliosis in the United States (see **21.1 Making Connections**), is caused by *Borrelia*

*burgdorferi** and is transmitted primarily by hard ticks of the genus *Ixodes*. In the northeastern part of the United States, *Ixodes scapularis* (the black-legged deer tick) passes through a complex 2-year cycle that involves two principal hosts **(figure 21.9)**. The larva or nymph stage feeds on the white-footed mouse, where it picks up the infectious agent. The nymph is relatively nonspecific and will try to feed on nearly any type of vertebrate; thus, it is the form most likely to bite humans. The adult tick reproductive phase of the cycle is completed on deer. In California, the transmission cycle involves *Ixodes pacificus* and the dusky-footed

* *burgdorferi* (berg-dor′-fer-eye) Named for its discoverer, Dr. Willy Burgdorfer.

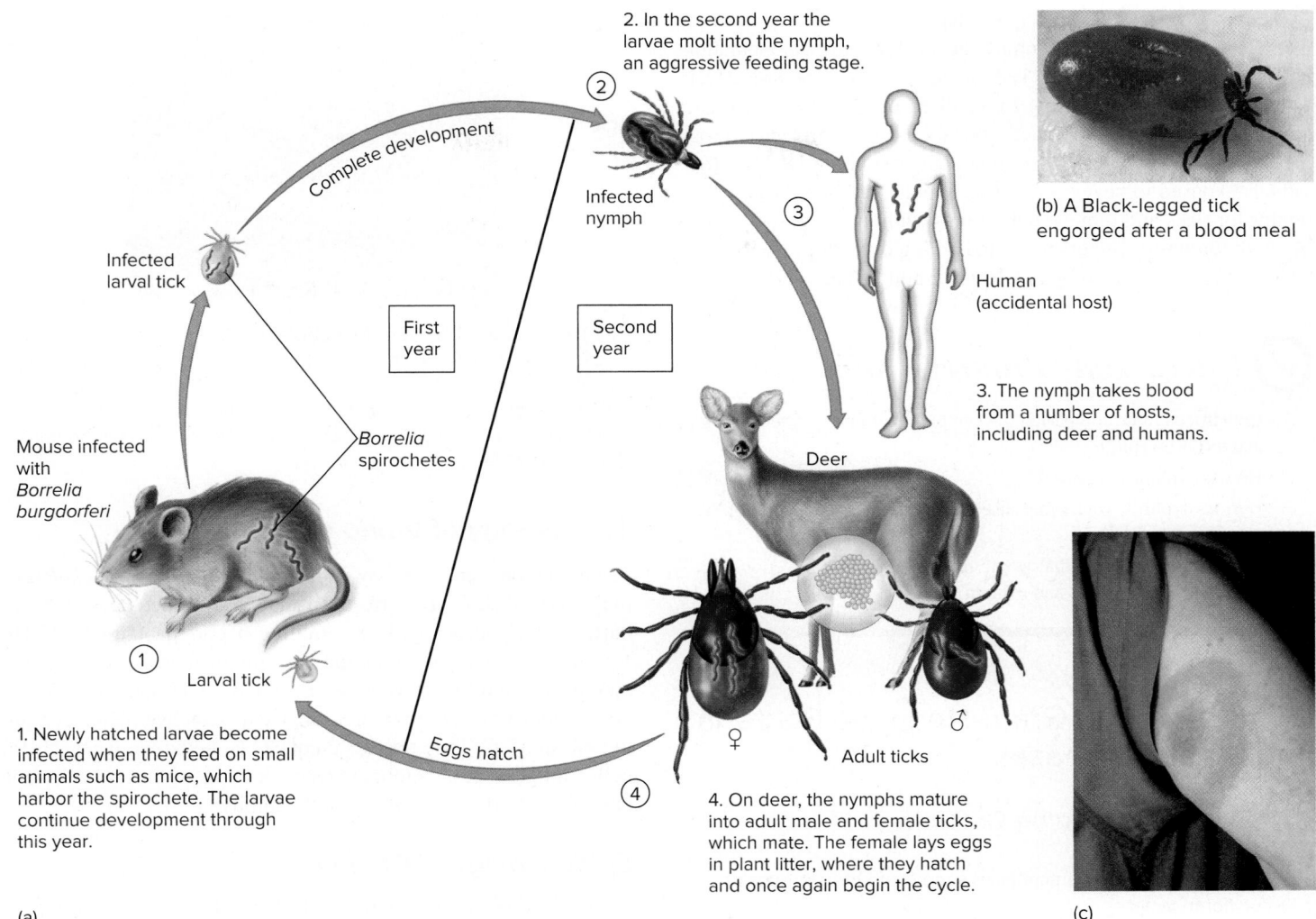

Figure 21.9 The cycle of Lyme disease in the northeastern United States. (a) The disease is tied intimately into the life cycle of a tick vector, which generally is completed over a 2-year period. The exact hosts and species of tick vary from region to region but still display this basic pattern. **(b)** Black-legged ticks, *Ixodes scapularis,* transmit Lyme disease to humans and animals during feeding. **(c)** About 70% to 80% of Lyme disease cases present with erythema migrans, a red, often bull's-eye-shaped rash surrounding the initial infection site. (b): *Source:* Dr. Gary Alpert - Urban Pests - Integrated Pest Management (IPM)/CDC; (c): *Source:* CDC

woodrat reservoir. In 2016 a new species of bacteria, provisionally named *Borrelia mayonii,* was discovered in the blood of six people with suspected Lyme disease, all living in the upper midwestern United States. The illness caused by *B. mayonii* seems similar to that caused by *B. burgdorferi,* but it may be less likely to cause the bull's-eye rash characteristic of the disease.

The incidence of Lyme disease is showing a gradual upward trend from about 10,000 cases per year in 1991 to around 24,000 in 2014. This increase may be partly due to improved diagnosis, but it also reflects changes in the numbers of hosts and vectors. The greatest concentrations of Lyme disease are in areas having large populations of mice and deer. Most of the cases have occurred in New York, Pennsylvania, Connecticut, New Jersey, Rhode Island, and Maryland, though the number in the Midwest and West is growing. Highest-risk groups include hikers, backpackers, and people living in newly developed communities near woodlands and forests. Peak seasons for infection are the summer and early fall.

Lyme disease is nonfatal but often evolves into a slowly progressive syndrome that mimics neuromuscular and rheumatoid conditions. An early symptom in 70% to 80% of cases is a rash at the site of

a larval tick bite. The lesion, called *erythema migrans,* can appear something like a bull's-eye, with a raised erythematous ring that gradually spreads outward and a pale central region (see **figure 21.9c**). In other cases it may be a series of red papules or a spotty rash. Other early symptoms are fever, headache, stiff neck, and dizziness.

If not treated or if treated too late, the disease can advance to the second stage, during which cardiac and neurological symptoms (facial palsy) develop. After several weeks or months, a crippling polyarthritis can attack joints, especially in the European strain of the agent. Some people acquire chronic neurological complications that are severely disabling.

Diagnosis of Lyme disease can be difficult because of the range of symptoms it presents. Most suggestive are the ring-shaped lesions, isolation of spirochetes from the patient, and early serological testing with an ELISA method that tracks a rising antibody titer (see chapter 17). Tests for spirochetal DNA in specimens is especially helpful for late-stage diagnosis. Early treatment with doxycycline or amoxicillin is effective, and other antibiotics such as ceftriaxone and azithromycin are used in late Lyme disease therapy.

Because dogs can also acquire the disease, a vaccine has been marketed to protect them. A human vaccine for high-risk populations has been discontinued due to lack of sales over fear of side effects. The best means of preventing Lyme disease is to avoid the vector. Anyone involved in outdoor activities should wear protective clothing, boots, leggings, and insect repellent containing **DEET.**[*] Individuals exposed to heavy infestation should routinely inspect their bodies for ticks and remove ticks gently without crushing, preferably with forceps or fingers protected with gloves, because it is possible to become infected by tick feces or body fluids.

 Check Your Progress SECTION 21.1

1. Construct a table comparing the symptoms of primary, secondary, and tertiary syphilis.
2. How is syphilis diagnosed?
3. Besides syphilis, what other diseases are caused by bacteria in the genus *Treponema?*
4. List several behaviors that would put one at risk for infection with *Leptospira* or *Borrelia* and explain why.

21.2 Curviform Gram-Negative Bacteria and Enteric Diseases

 Expected Learning Outcomes

6. Relate the physical characteristics seen in the three groups of curviform bacteria.
7. Understand the pathogenesis of cholera.
8. Name the usual source of infection for each genus of commonly acquired curviform bacteria.
9. Explain the adaptations present in *Helicobacter* that help the pathogen survive in the stomach.

Three groups of curviform bacteria are represented by the following families, genera, and characteristics:

Vibrionaceae *Vibrio**[*]	Campylobacteraceae *Campylobacter**[*]	Helicobacteraceae *Helicobacter**[*]
Comma-shaped rods with one flagellum	Short spirals or curved rods with one flagellum	Tight spirals and curved rods with several polar flagella

Many of the pathogens in these groups share adaptations to survival in the often inhospitable environment of the intestine. They readily move within the mucous coating and avoid being swept away by intestinal motility. Species of *Helicobacter* are uniquely adapted to survival in the stomach with its high acid content.

Figure 21.10 The agent of cholera. Vibrio cholerae, showing its characteristic curved shape. Visible on some cells is a single polar flagellum.Members of the genus *Vibrio* (here *Vibrio vulnificus*) possess a characteristic curved shape and are sometimes referred to as "comma bacteria" (10,000×). *Source:* Louisa Howard/Dartmouth Electron Microscope Facility

The Biology of *Vibrio cholerae*

The most prominent species of pathogenic vibrios is **Vibrio cholerae,**[*] the cause of cholera. A freshly isolated specimen reveals quick, darting cells occasionally resembling a comma **(figure 21.10).** *Vibrio* shares many cultural and physiological characteristics with members of the Enterobacteriaceae, a closely related family. They are fermentative and grow on ordinary or selective media containing bile at 37°C. They possess unique O (somatic) antigens, H (flagella) antigens, and membrane receptor antigens that provide some basis for classifying members of the family.

Epidemiology of Cholera

Epidemic cholera, or Asiatic cholera, has been a devastating disease for centuries. Although the human intestinal tract was once thought to be the primary reservoir, it is now known that the pathogen is free-living in certain endemic regions.

The pattern of cholera transmission and the onset of epidemics are greatly influenced by the season of the year and the climate. Cold, acidic, dry environments inhibit the migration and survival of *Vibrio,* whereas warm, monsoon, alkaline, and saline conditions favor them. The disease has persisted in a pandemic pattern since 1961, when the *El Tor* biotype began to prevail worldwide. This strain survives longer in the environment, infects a higher number of people, and is more likely to be chronically carried than any other strain. One of the worst epidemics in recent history started shortly after the 2010 earthquake in Haiti and continues to the present time. In the five years since the earthquake, the WHO has reported 750,000 cases and 9,000 deaths, most of which were probably preventable. The overwhelming disruption to vital services and sanitation greatly complicated control measures. It ranks among the top seven causes of morbidity and mortality, affecting several million people in endemic regions of Asia and Africa.

In nonendemic areas such as the United States, the microbe is spread by water and food contaminated by asymptomatic carriers, but it is relatively uncommon. Sporadic outbreaks occasionally

[*] The active ingredient in OFF! and Cutter repellents.
[*] *Vibrio* (vib′-ree-oh) L. *vibrare,* to shake.
[*] *Campylobacter* (kam″-pih-loh-bak′-ter) Gr. *campylo,* curved, and *bacter,* rod.
[*] *Helicobacter* (hee″-lih-koh-bak′-ter) Gr. *helicos,* a coil.

[*] *cholerae* (kol′-ur-ee) Gr. *chole,* bile. The bacterium was once named *V. comma* for its comma-shaped morphology.

occur along the Gulf of Mexico, and the cholera vibrio is sometimes isolated from shellfish in this region.

Pathogenesis of Cholera: Toxigenic Diarrhea

After being ingested with food or water, *V. cholerae* must pass through the acid environment of the stomach. To compensate, the size of the infectious dose is quite high (10^8 cells). The dose may be reduced by certain types of food that can more readily shelter the pathogen. At the mucosa of the duodenum and jejunum, the vibrios penetrate the mucous barrier and come to rest near the surface of the epithelial cells. The cholera vibrio is strictly an epipathogen that does not enter cells or deeper tissues. The virulence is due entirely to an enterotoxin called *cholera toxin* (CT) that disrupts the normal physiology of intestinal cells. When this toxin binds to specific intestinal receptors, a secondary signaling system is activated. Under the influence of this system, the cells shed large amounts of electrolytes into the intestine, an event that is accompanied by profuse water loss **(figure 21.11)**. Most cases of cholera are mild or self-limited, but in children and weakened individuals, the disease can strike rapidly and violently.

(a) Normal

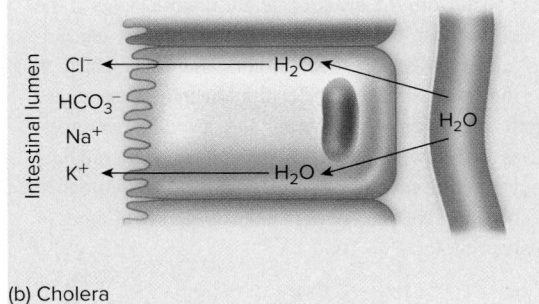

(b) Cholera

① The *Vibrio cholerae* cell comes to rest in the protective mucous coating near the cell surface and secretes the cholera toxin.

② The toxin has an affinity for specialized receptors on the glycocalyx and binds there.

③ The active portion of the toxin is released, is transported through the membrane, and enters the cytoplasm.

④ It becomes a signal in a system that converts inactive adenyl cyclase into an active state.

⑤ This enzyme converts ATP into a molecule called cyclic AMP (cAMP). The cAMP is used by the cell to control a major membrane pump for negative ions.

⑥ The result is that the membrane begins to actively pump Cl^- and HCO_3^- into the intestinal lumen. One additional effect of the toxin is that it overrides the usual controls for the adenyl cyclase/cAMP system so that the cell continues to pump out these ions for an extended time.

⑦ Positive ions (Na^+ and K^+) follow the anions and are also lost into the intestinal fluid, along with large amounts of water, causing secretory diarrhea and dehydration.

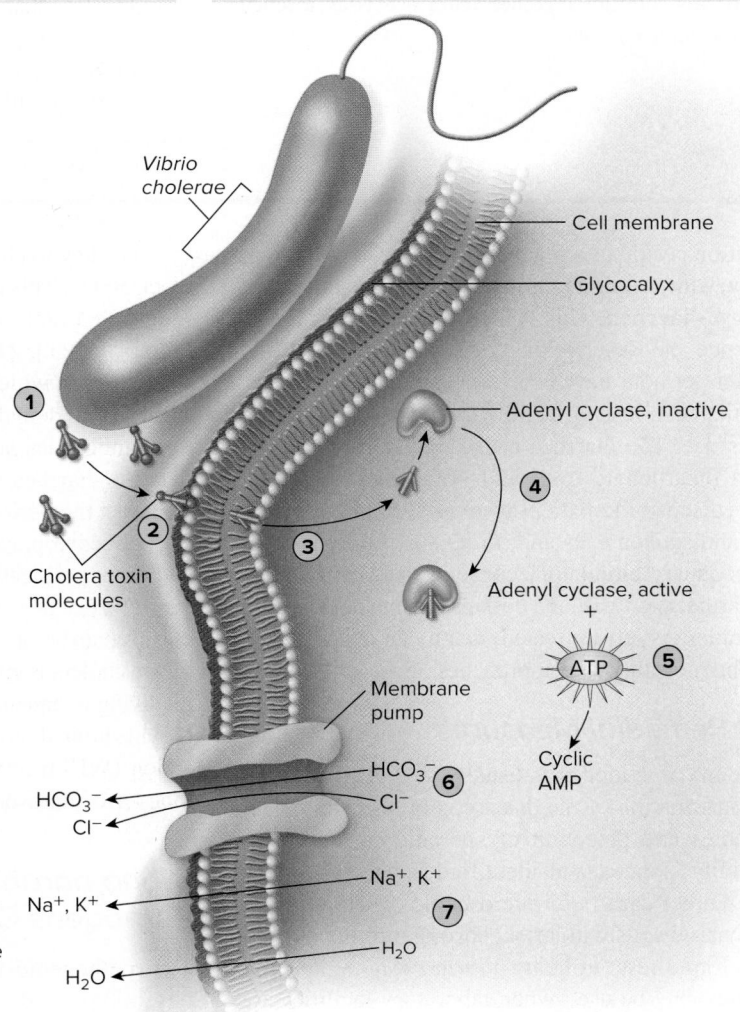

(c)

Process Figure 21.11 Alterations in intestinal function caused by cholera toxin. (a) Normal actions of intestinal absorption. **(b)** General actions of cholera toxin on electrolytes and water. **(c)** Magnified view of intestinal cell reacting to cholera toxin.

21.2 MAKING CONNECTIONS

Oral Rehydration Therapy

Drinking a simple sugar-salt mixture is often a miracle cure for cholera. This solution, developed by the World Health Organization and termed oral rehydration therapy (ORT), consists of a mixture of sodium chloride, sodium bicarbonate, potassium chloride, and glucose or sucrose dissolved in water **(table 21.A).** When administered early in amounts ranging from 100 to 400 ml/hour, the solution can restore patients in 4 hours, literally bringing them back from the brink of death. It works even if the individual has diarrhea, because the particular combination of ingredients is well tolerated and rapidly absorbed. Infants and small children who once would have died now survive so often that the mortality rate for treated cases of cholera is near zero. This therapy has several advantages, especially for countries with few resources. It does not require medical facilities, high-technology equipment, or complex medication protocols. It is also inexpensive, noninvasive, fast-acting, and useful for a number of diarrheal diseases besides cholera.

How could one put this solution together from everyday kitchen supplies? *Answer available on Connect.*

TABLE 21.A	Recommendations for Oral Rehydration Therapy (ORT)	
Standard Composition of Oral Rehydration Solution (ORS)	**Grams per Liter (g/l)**	**Concentration in Millimoles**
NaCl	3.5 g/l	90 (Na) 80 (Cl)
$NaHCO_3$	2.5 g/l	30 (HCO_3)
KCl	1.5 g/l	20 (K)
Glucose*	20.0 g/l	111

Guidelines for Preparation and Treatment
- Components are dissolved in 1 liter of water that has been boiled to disinfect it. It is thoroughly mixed and tasted for saltiness. It should be only slightly salty. This product is similar to a commercial product called Pedialyte or Oralyte.
- Moderately dehydrated children and adults should receive frequent small amounts of ORS at the rate of 4 to 8 ounces per hour over 6 hours. Plain water can be administered in between ORS administrations.
- To check the patient for the return of normal hydration, test the skin turgor (pinch it to see if a fold remains), and observe whether the eyes are still sunken.

*Sucrose or rice powder may be substituted if glucose is not available.

After an incubation period of a few hours to a few days, symptoms begin abruptly with vomiting, followed by copious watery feces called **secretory diarrhea.** This voided fluid contains flecks of tissue debris, hence the description "rice-water stool." Fluid losses of nearly 1 liter per hour have been reported in severe cases, and an untreated patient can lose up to 50% of body weight during the course of the disease. The diarrhea causes reduced blood volume, acidosis from bicarbonate loss, and potassium depletion. These conditions predispose the patient to muscle cramps, severe thirst, flaccid skin, and sunken eyes and, in young children, coma and convulsions. Secondary circulatory consequences can include hypotension, tachycardia, cyanosis, and collapse from shock within 18 to 24 hours. If cholera is left untreated, death can occur in less than 48 hours, and the mortality rate approaches 55%.

Diagnosis and Remedial Measures

During epidemics, clinical evidence is usually sufficient to diagnose cholera. But confirmation of the disease is often required for epidemiological studies and detection of sporadic cases. *Vibrio cholerae* can be readily isolated and identified in the laboratory from stool samples. Direct dark-field microscopic observation reveals characteristic curved cells with brisk, darting motility as confirmatory evidence. Rapid tests utilizing *Vibrio cholerae* specific monoclonal-antibodies may be used when laboratory facilities are unavailable—such as in remote locations or after natural disasters—but the tests lack the sensitivity and specificity needed to confirm the presence of the pathogen. Difficult or elusive cases can be traced by detecting a rising antitoxin titer in the serum.

The key to cholera therapy is prompt replacement of water and electrolytes. This can be accomplished by various rehydration techniques that replace the lost fluid and electrolytes (oral rehydration therapy [ORT]; **21.2 Making Connections**).

Cases in which the patient is unconscious or has complications from severe dehydration require intravenous replenishment as well. Oral antibiotics such as doxycycline and azithromycin can terminate the diarrhea in 48 hours, and they also promote recovery and diminish the period of vibrio excretion.

Effective prevention is contingent upon proper sewage disposal and water purification. Detecting and treating carriers with mild or asymptomatic cholera are serious goals but ones that are frequently difficult to attain because of inadequate medical provisions in those countries where cholera is endemic. Vaccines are available for travelers and people living in endemic regions. Both are whole-cell killed vaccines that provide limited protection for about 2 years. The World Health Organization (WHO) recommends immunization in endemic areas, while long-term improvements to water quality and sanitation are undertaken.

Vibrio parahaemolyticus and *Vibrio vulnificus*: Pathogens Carried by Seafood

Two pathogenic relatives of *V. cholerae* share its morphology, physiology, and ecological adaptation. *Vibrio parahaemolyticus* and *V. vulnificus* are both salt-tolerant inhabitants of coastal waters and associate with marine invertebrates. In temperate zones, vibrios survive over the winter by settling into the ocean sediment; when resuspended by upwelling during the warmer seasons, they

become incorporated into the food web, eventually growing on fish, shellfish, and other edible seafood.

Features of V. parahaemolyticus *Gastroenteritis*

Vibrio parahaemolyticus infection, an acute form of **gastroenteritis,** was first described in Japan more than 30 years ago. The great majority of cases appear in individuals who have eaten raw, partially cooked, or poorly stored seafood. The vehicles most often implicated are squid, mackerel, sardines, crabs, tuna, shrimp, oysters, and clams. Outbreaks tend to be concentrated along coastal regions during the summer and early fall. The incubation period of nearly 24 hours is followed by explosive, watery diarrhea accompanied by nausea, vomiting, abdominal cramps, and sometimes fever. *Vibrio* toxins cause symptoms that last about 72 hours but can persist for 10 days.

Food infection caused by *V. vulnificus* is similar in symptoms to that caused by *V. parahaemolyticus*. It is most often associated with ingesting raw oysters and can have a much more severe outcome in critically ill patients with diabetes or liver disease. It is the leading cause of death from food-borne illness in some areas.

Treatment of severe gastroenteritis can require fluid and electrolyte replacement and occasionally antibiotics. Control measures aim to keep the bacterial count in all seafood below the infective dose by continuous refrigeration during transport and storage, sufficient cooking temperatures, and prompt serving. Consumers of raw oysters (or other shellfish) should understand their possible risk. In some regions, the U.S. Department of Agriculture mandates that health warnings be posted about this danger in markets and restaurants.

Review the chapter 17 Case Study for information on the identification of *Vibrio* species.

Diseases of the *Campylobacter* Vibrios

Campylobacters are curved or spiral bacilli, often appearing in S-shaped or gull-winged pairs **(figure 21.12)**. Their polar flagellum imparts active spinning motility. Their physiological profile includes being microaerophilic, oxidase-positive, and nonfermenting. These bacteria are common residents of the intestinal tract, genitourinary tract, and oral cavity of birds and mammals. Species of *Campylobacter* most significant in medical and veterinary practice are *C. jejuni* and *C. fetus.*

Campylobacter jejuni *Enteritis*

*Campylobacter jejuni** has emerged as a pathogen of such imposing proportions that it is considered one of the most important causes of bacterial gastroenteritis worldwide **(Pathogen Profile #2).** In the United States, approximately 1.3 million cases occur every year. Epidemiological and pathologic studies have shown that this species is a primary pathogen transmitted through contaminated beverages and food, especially water, milk, meat, and chicken.

Ingested *C. jejuni* cells travel to the mucosa at the last segment of the small intestine (ileum) near its junction with the colon. They adhere, burrow through the mucus, and are taken in by intestinal cells. The pathogen disrupts the cytoskeleton, damages the epithelium, and penetrates the intestinal wall. After an incubation period of 1 to 7 days, acute symptoms of headache, fever, abdominal pain, and bloody or watery diarrhea develop. The mechanisms of pathology involve a heat-labile enterotoxin called CJT that stimulates secretory diarrhea like cholera. An occasional sequela of this infection is the development of a neurological disease called Guillain-Barré syndrome.

Diagnosis of *C. jejuni* enteritis is usually made with fecal samples and occasionally blood samples. This species is microaerophilic and thermophilic, and isolation requires special culturing media. It can be grown on CCD medium, under conditions of reduced oxygen, and at 42°C, typical avain body temperature. More rapid presumptive diagnosis can be obtained from examination of feces with a dark-field microscope, which accentuates the characteristic curved rods and darting motility.

Resolution of infection occurs in most instances with rehydration and electrolyte balance therapy. In more severely affected patients, it may be necessary to administer erythromycin or azithromycin. Because vaccines are yet to be developed, prevention depends upon rigid sanitary control of water and milk supplies and care in food preparation.

Traditionally of interest to the veterinarian, *C. fetus* (subspecies *venerealis*) causes a sexually transmitted disease of sheep, cattle, and goats. Its role as an agent of abortion in these animals has considerable economic impact on the livestock industry. About 40 years ago, the significance of *C. fetus* as a human pathogen was first uncovered, though its exact mode of transmission in humans is yet to be clarified. This bacterium appears to be an opportunistic pathogen that attacks debilitated persons or women late in pregnancy. Diseases to which *C. fetus* has been linked are meningitis, pneumonia, arthritis, septicemic infection in the newborn, and occasionally, sexually transmitted proctitis in adults.

Helicobacter pylori: Gastric Pathogen

Although the human stomach is too harsh for most bacteria, it serves as a primary habitat for an unusual spiral bacterium, *Helicobacter pylori.* Like *Campylobacter,* it is microaerophilic and oxidative, but it differs in having multiple sheathed polar flagella. Not only does it thrive in the

Figure 21.12 Scanning micrograph of *Campylobacter jejuni.* The S form is seen here, but the pathogen also appears as a shorter, comma-shaped bacillus and in a longer, spiral form. *Source:* Emerging Infectious Diseases http://wwwnc.cdc.gov/eid/

* *jejuni* (jee-joo′-nye) L. *jejunum.* The small section of intestine between the duodenum and the ileum.

Pathogen Profile #2 *Campylobacter jejuni*

Source: CDC

Microscopic Morphology Small, curved, gram-negative rods.

Identified by Dark-field examination of stool samples, revealing curved cells with a darting motility provides presumptive identification. Positive identification requires culturing the pathogen on specialized media under conditions of low oxygen concentration.

Habitat *C. jejuni* is a common resident of the intestinal tract, genitourinary tract, and oral cavity of birds and mammals.

Virulence Factors The primary virulence factor in *C. jejuni* is CJT, a heat-labile enterotoxin that stimulates secretory diarrhea.

Primary Infections/Disease *C. jejuni* is one of the most important causes of gastroenteritis, with over 1.3 million cases of the disease occurring every year in the United States. Common symptoms include fever, diarrhea, and abdominal pain within 2 to 5 days after exposure to the organism. Illness typically lasts about 1 week and is self-limiting, although it is estimated that *C. jejuni* is responsible for about 75 deaths annually in the United States, mostly among people with compromised immune systems.

Control and Treatment Prevention depends on proper sanitary control of water and milk supplies, along with care in food harvesting and preparation. Rehydration therapy is usually adequate to resolve infection, whereas antibiotics—erythromycin or azithromycin—may be necessary in more severe cases.

acidic environment, but evidence has clearly linked it to a variety of gastrointestinal ailments. It is known to cause an inflammation of the stomach lining called gastritis and is implicated in 90% of stomach and duodenal ulcers. It is also an apparent cofactor in the development of a common type of stomach cancer called adenocarcinoma.

This novel pathogen was first detected by J. Robin Warren in 1979 in stomach biopsies from ulcer patients. He and an assistant, Barry J. Marshall, isolated the microbe in culture and even tested its effects by swallowing a good-sized inoculum. Both developed a short-term case of gastritis. For their extraordinary discovery of this microbe and its connection to gastric diseases, the two scientists were awarded the Nobel Prize in Physiology or Medicine in 2005.

Continued research has revealed that the bacterium is present in a large proportion of people. It occurs in the stomachs of 25% of healthy middle-aged adults and in more than 50% of adults over 60 years of age. *Helicobacter pylori* is probably transmitted from person to person by the oral-oral or oral-fecal route. It can also be spread by houseflies acting as mechanical vectors. It seems to be acquired early in life and carried asymptomatically until its activities begin to damage the digestive mucosa. Because other animals are also susceptible to *H. pylori* and even develop chronic gastritis, it has been proposed that the disease is a zoonosis transmitted from an animal reservoir. Several other species of this genus have been isolated from cats, dogs, and other mammals.

Other studies have helped to explain how infection with the pathogen progresses, eventually leading to the formation of a peptic ulcer. First, it bores through the outermost mucus that lines the epithelial tissue. Then it attaches to specific binding sites on the cells and entrenches itself **(figure 21.13)**. It turns out that one receptor specific for *Helicobacter* is the same receptor as type O blood, which accounts for the finding that people with this blood type have a 1.5 to 2 times higher rate of ulcers than people in other blood groups. Another protective adaptation is the formation of urease, an enzyme that converts urea into ammonium and bicarbonate, both alkaline compounds that can neutralize stomach acid. As the immune system recognizes and attacks the pathogen, certain white blood cells damage the epithelium to some degree, leading to chronic active gastritis. In some people these lesions lead to deeper erosions and ulcers and eventually can lay the groundwork for cancer to develop.

Helicobacter pylori is isolated primarily from biopsy specimens. Serological detection of *H. pylori* antigens in the stool has become a favored diagnostic test as it is noninvasive and has a high degree of sensitivity.

Understanding the underlying pathology has useful applications in treatment. Gastritis and ulcers have traditionally been treated with drugs (cimetidine [Tagamet], or ranitidine [Zantac]) that suppress symptoms by slowing the secretion of acid in the stomach. They must be taken continuously for indefinite periods, and relapses are common. Current recommended therapy is 2 to 4 weeks of clarithromycin to eliminate the bacterial infection, along with stomach acid inhibitors. This regimen can cure the infection and eliminate symptoms.

Stomach mucosa *Helicobacter* cells

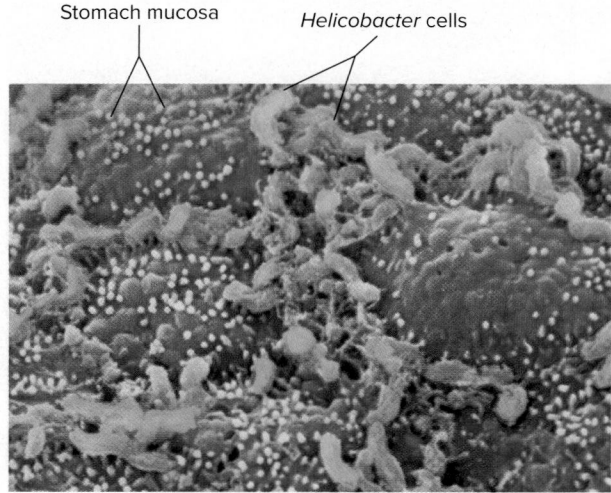

Figure 21.13 The causative agent of stomach ulcers.
Colorized scanning electron microscope view of *Helicobacter pylori* nestled into the stomach lining. Characteristics include loose, shallow coils and multiple flagella (2,500×). © CNRI/Science Source

 Check Your Progress SECTION 21.2

5. How could one differentiate the genera *Vibrio, Campylobacter,* and *Helicobacter?*

6. Relate the successful use of oral rehydration therapy to the pathogenesis of cholera.

7. Which two species of curviform bacteria are associated with consumption of seafood?

8. Briefly describe the nature of food infection in species of *Vibrio* and the diseases of *Campylobacter.*

9. What diseases involve *Helicobacter pylori* infection?

21.3 Medically Important Bacteria of Unique Morphology and Biology

Expected Learning Outcomes

10. Recall the characteristics seen in the *Rickettsia* that make them unique.

11. Understand the epidemiology and pathology of Rocky Mountain spotted fever.

12. Name the human pathogens within the genera *Ehrlichia* and *Anaplasma* and the diseases they cause.

13. Recognize behaviors or activities that would increase the risk of contracting Q fever.

14. Recall the body systems commonly affected by species in the genus *Chlamydia.*

15. Recall the names and characteristics of diseases attributable to infection with *Chlamydophila* species.

Pathogenic bacteria that exhibit atypical morphology, physiology, and behavior include: (1) rickettsias and (2) chlamydias, which are obligately parasitic gram-negative coccobacilli, and (3) mycoplasmas, which are highly pleomorphic bacteria that lack cell walls (see figure 21.24). The three groups are not closely related but are included together because of similar morphology and pathogenicity.

Order Rickettsiales

The Order Rickettsiales contains about two dozen species of pathogens, mostly in the genus **Rickettsia.*** Other members include **Ehrlichia,*** **Anaplasma,** and **Orientia**. These organisms are known commonly as **rickettsias,** or rickettsiae, and the diseases they cause are called rickettsioses.

The rickettsias are all obligate to their host cells and require live cells for cultivation; they also spend part of their life cycle in the bodies of arthropods, which serve as vectors. Rickettsioses are among the most important emerging diseases. Six of the 14 recognized diseases have been identified in the last 20 years.

(a) (b)

Figure 21.14 The order *Rickettsiales.* **(a)** Transmission electron micrograph of a mouse cell that had been infected with the rickettsial organism *Orientia tsutsugamushi,* the causative agent of "scrub" typhus, also known as "river," or "flood," typhus. **(b)** Transmission electron micrograph of an *Orientia tsutsugamushi* cell budding from the surface of a mouse cell that had been experimentally infected. (a, b): *Source:* CDC

Morphological and Physiological Distinctions of Rickettsias

Rickettsias possess a gram-negative cell wall, binary fission, metabolic pathways for synthesis and growth, and both DNA and RNA. They are among the smallest cells, ranging from 0.3 to 0.6 µm wide and from 0.8 to 2.0 µm long. They are nonmotile pleomorphic rods or coccobacilli **(figure 21.14).**

The precise nutritional requirements of the rickettsias have been difficult to demonstrate because of a close association with host cell metabolism. Their obligate parasitism originates from an inability to metabolize AMP, an important precursor to ADP and ATP, which they must obtain from the host. Rickettsias are generally sensitive to environmental exposure, although *R. typhi* can survive several years in dried flea droppings.

Distribution and Ecology of Rickettsial Diseases

The Role of Arthropod Vectors The rickettsial life cycle depends upon a complex exchange between blood-sucking arthropod[1] hosts and vertebrate hosts (see section 23.8). Eight tick genera, two fleas, and one louse are involved in the spread of rickettsias to humans. Humans accidently enter the zoonotic life cycles through occupational contact with the animals except in the cases of louse-borne typhus and trench fever. Most vectors apparently harbor rickettsias with no ill effect, but others, like the human body louse, die from typhus infection and do not continuously harbor the pathogen. In certain vectors (Rocky Mountain spotted fever ticks), rickettsias are transferred to offspring by the eggs of an infected female. Such continuous inheritance of the microbe through multiple generations of ticks creates a long-standing reservoir.

These arthropods feed on the blood or tissue fluids of their mammalian hosts, but not all of them transmit the rickettsial pathogen through direct inoculation with saliva. Ticks directly inoculate the skin lesion from their mouths, but fleas and lice harbor the infectious agents in their intestinal tracts. During their stay on the host's body, these latter insects defecate or are smashed, thereby releasing the rickettsias onto the skin or into a wound. Ironically, scratching the bite helps the pathogen invade deeper tissues.

* *Rickettsia* (rik′-ett′-see-ah) After Howard Ricketts, an American bacteriologist who worked extensively with this group.
* *Ehrlichia* (ur-lik′-ee-ah) After Paul Ehrlich, a German immunologist.

1. Ticks are in the Class Arachnida, and lice and fleas are in the Class Insecta, both in the Phylum Arthropoda.

General Factors in Rickettsial Pathology and Isolation

A common target in rickettsial infections is the endothelial lining of the small blood vessels. The bacteria recognize, enter, and multiply within endothelial cells, causing necrosis of the vascular lining. Among the immediate pathologic consequences are vasculitis, perivascular infiltration by inflammatory cells, vascular leakage, and thrombosis. These pathologic effects are manifested by skin rash, edema, hypotension, and gangrene. Intravascular clotting in the brain accounts for the stuporous mental changes and other neurological symptoms that sometimes occur.

Isolation of most rickettsias from clinical specimens requires a suitable live medium and specialized laboratory facilities, including controlled access and safety cabinets. The usual choices for routine growth and maintenance are the yolk sacs of embryonated chicken eggs, chick embryo cell cultures, and, to a lesser extent, mice and guinea pigs.

Specific Rickettsioses

Rickettsioses can be differentiated on the basis of their clinical features and epidemiology as (table 21.2):

1. the typhus group,
2. the spotted fever group,
3. scrub typhus, and
4. ehrlichiosis and anaplasmosis.

Rickettsia prowazekii and Epidemic Typhus

Epidemic, or louse-borne, **typhus*** has been a constant accompaniment to war, poverty, and famine. The extensive investigations of

* *typhus* (ty'-fus) Gr. *typhos,* smoky or hazy, underlining the mental deterioration seen in this disease. The term *typhus* is commonly confused with typhoid fever, an unrelated enteric illness caused by *Salmonella typhi.*

Dr. Howard Ricketts and Stanislas von Prowazek in the early 1900s led to the discovery of the vector and the rickettsial agent but not without mortal peril; both men died of the very disease they investigated. *Rickettsia prowazekii* was named in honor of their pioneering efforts.

Epidemiology of Epidemic Typhus Humans are the sole hosts of human body lice and the only reservoirs of *R. prowazekii.* The louse spreads infection by defecating into its bite wound or other breaks in the skin. Infection of the eye or respiratory tract can take place by direct contact or inhalation of dust containing dried louse feces, but this is a rarer mode of transmission.

The transfer of lice is increased by crowding, infrequently changing clothing, and sharing clothing. The overall incidence of epidemic typhus in the United States is very low, with no epidemics since 1922. Although no longer common in regions of the world with improved standards of living, epidemic typhus presently persists in regions of Africa, Central America, and South America.

Disease Manifestations and Immune Response in Typhus After entering the circulation, rickettsias pass through an intracellular incubation period of 10 to 14 days. The first clinical manifestations are sustained high fever, chills, frontal headache, and muscular pain. Within 7 days, a generalized rash appears, initially on the trunk, and then spreads to the extremities. Personality changes, oliguria (low urine output), hypotension, and gangrene complicate the more severe cases. Mortality is lowest in children and as high as 40% to 60% in patients over 50 years of age.

Recovery usually confers resistance to typhus, but in some cases, the rickettsias are not completely eradicated by the immune response and enter into latency. After several years, a milder recurring form of the disease, known as *Brill-Zinsser disease,* can appear. This disease is seen most often in people who have immigrated from endemic areas and is of concern mainly because these immigrants provide a continuous reservoir of the pathogen.

TABLE 21.2 Characteristics of Major Rickettsias Involved in Human Disease

Disease Group	Species	Disease	Vector	Primary Reservoir	Mode of Transmission to Humans	Where Found
Typhus	*Rickettsia prowazekii*	Epidemic typhus	Body louse	Humans	Louse feces rubbed into bite; inhalation	Worldwide
	R. typhi (mooseri)	Murine typhus	Flea	Rodents	Flea feces rubbed into skin; inhalation	Worldwide
Spotted Fever	*R. rickettsii*	Rocky Mountain spotted fever	Tick	Small mammals	Tick bite; aerosols	North and South America
	R. akari	Rickettsialpox	Mite	Mice	Mite bite	Worldwide
Scrub Typhus	*Orientia tsutsugamushi*	—	Immature mite	Rodents	Bite	Asia, Australia, Pacific Islands
Human Ehrlichiosis	*Ehrlichia chaffeensis*	Human monocytic ehrlichiosis	Tick	—	Tick bite	Similar to Rocky Mountain spotted fever
Human Anaplasmosis	*Anaplasma phagocytophilum*	Human granulocytic anaplasmosis	Tick	Deer, rodents	Tick bite	Unknown

Treatment and Prevention of Typhus Standard chemotherapy for typhus is doxycycline or chloramphenicol. Despite antibiotic therapy, however, the prognosis can be poor in patients with advanced circulatory or renal complications. Eradication of epidemic typhus is theoretically possible by exterminating the vector. Widespread dusting of human living quarters with insecticides has provided some environmental control, and individual treatment with an antilouse shampoo or ointment is also effective.

Epidemiology and Clinical Features of Endemic Typhus

The agent for endemic typhus is *Rickettsia typhi,* which shares many characteristics with *R. prowazekii* except that *R. typhi* has pronounced virulence. Synonyms for this rickettsiosis are endemic typhus, murine (mouse) typhus, and flea-borne typhus. The disease is endemic to certain parts of Central and South America and the Southeast, Gulf Coast, and Southwest regions of the United States. The native rodents and opossums in these areas are a reservoir for *R. typhi* and transmit it to fleas feeding on blood. Humans become infected through flea bites and occasionally inhalation of dried feces from infected rats. In the United States, most reported cases arise sporadically among workers in rat-infested industrial sites.

The clinical manifestations of endemic typhus include fever, headache, muscle aches, and malaise. After 5 days, a skin rash, transient in milder cases, begins on the trunk and radiates toward the extremities. Symptoms dissipate in about 2 weeks. Doxycycline and chloramphenicol are effective therapeutic agents, and various pesticides are available for vector and rodent control.

Rocky Mountain Spotted Fever: Epidemiology and Pathology

The rickettsial disease with greatest impact on people living in North America is Rocky Mountain spotted fever (RMSF), named for the place it was first seen—the Rocky Mountains of Montana and Idaho **(Pathogen Profile #3).** Ricketts identified the etiologic agent *Rickettsia rickettsii* in smears from infected animals and patients and later discovered that it was transmitted by ticks. Despite its geographic name, this disease occurs infrequently in the western United States. The majority of cases are concentrated in the Southeast and eastern seaboard regions **(figure 21.15).** It also occurs in Canada and Central and South America. Infections occur most frequently in the spring and summer, when the tick vector is most active. The yearly rate of RMSF averages six to seven cases per million population, with fluctuations coinciding with weather and tick infestations. The incidence has been generally increasing since 2000.

The principal reservoirs and vectors of *R. rickettsii* are hard ticks such as the wood tick (*Dermacentor andersoni),* the American dog tick (*D. variabilis,* among others), and the Lone Star tick

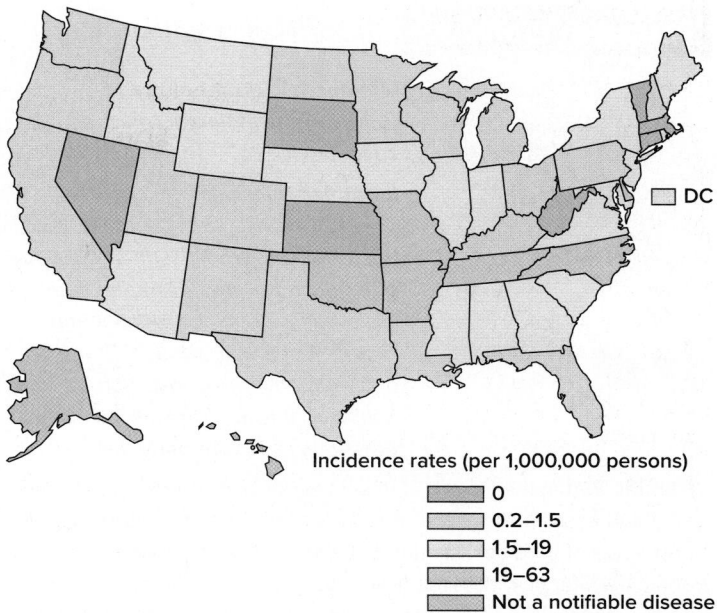

Incidence rates (per 1,000,000 persons)

- 0
- 0.2–1.5
- 1.5–19
- 19–63
- Not a notifiable disease

Figure 21.15 Trends in infection for Rocky Mountain spotted fever. A summary map showing the distribution of Rocky Mountain spotted fever cases. The disease is not reportable in Alaska or Hawaii. *Source of data:* CDC

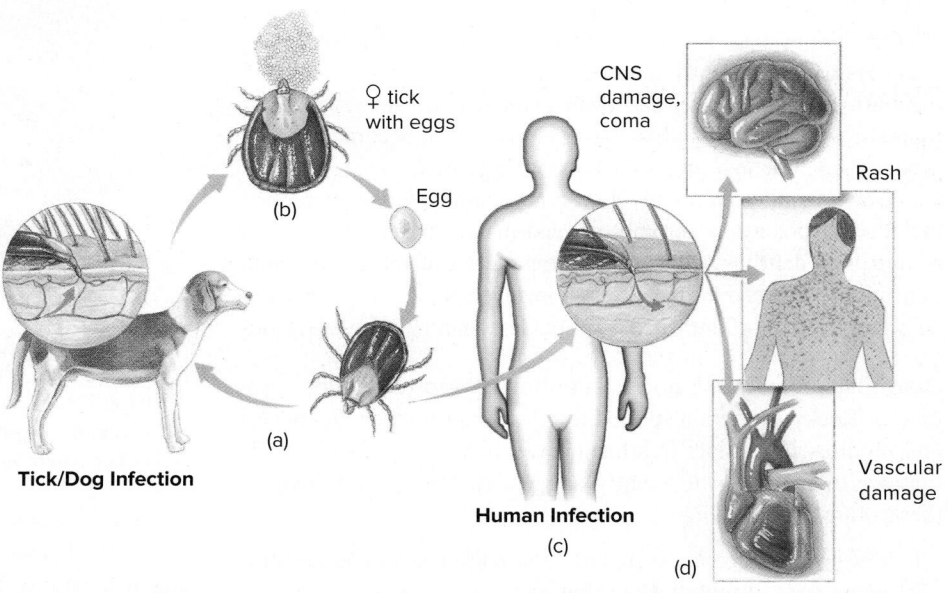

Figure 21.16 The transmission cycle in Rocky Mountain spotted fever. Dog ticks and wood ticks (*Dermacentor*) are the principal vectors. (a) Ticks are infected from a mammalian reservoir during a blood meal. **(b)** Transovarial passage of *Rickettsia rickettsii* to tick eggs serves as a continual source of infection within the tick population. Infected eggs produce infected adults. **(c)** A tick attaches to a human, embeds its head in the skin, feeds, and sheds rickettsias into the bite. **(d)** Systemic involvement includes severe headache, fever, rash, coma, and vascular damage such as blood clots and hemorrhage.

(*Amblyomma americanum*). The dog tick is probably most responsible for transmission to humans because it is the major vector in the southeastern United States **(figure 21.16).**

Pathogenesis and Clinical Manifestations of Spotted Fever
After 2 to 4 days incubation, the first symptoms are sustained fever,

Pathogen Profile #3 *Rickettsia rickettsii*

Source: CDC

Microscopic Morphology *Very small, gram-negative pleomorphic rods or coccobacilli.*

Identified by *Cultivation of R. rickettsii is difficult, and detection of bacterial cells from tissue samples is often accomplished using fluorescent antibodies. Definitive identification may be achieved using PCR and a presumptive diagnosis can be achieved through an ELISA test that detects a rising antibody titer.*

Habitat Within the United States, *R. rickettsii* is found in the Southeast and eastern seaboard regions, where it is transmitted by several species of hard tick, including the wood tick, American dog tick, and Lone Star tick. With the tick serving as a biological vector, the bacterium is spread between mammalian reservoirs and humans.

Virulence Factors Proteins residing on the outer membrane of *R. rickettsii* are responsible for binding to the endothelial cells of

the host. The bacterium is able to induce its own phagocytosis by cells of the body and multiply within the cytoplasm, eventually leading to disease.

Primary Infections/Disease Symptoms of Rocky Mountain spotted fever (RMSF) are first noticed 2 to 4 days after infection and include fever, chills, headache, and muscle pain. Shortly thereafter, a red spotted rash appears and worsens, occasionally becoming necrotic in some areas. Progression of the disease includes cardiovascular disruption and involvement of the central nervous system. Fatality results in 10% of untreated cases but less than 1% of treated cases.

Control and Treatment Early doxycycline therapy is the treatment of choice for RMSF. Prevention of the disease is synonymous with prevention of tick bites when involved in outdoor activities, and the use of protective clothing, boots, and insect repellent containing DEET is recommended.

chills, headache, and muscular pain. The distinctive spotted rash appears within 1 to 3 days after the prodromium **(figure 21.17).** Early lesions are slightly mottled like measles, but later ones are macular, maculopapular, and even petechial. In the severest untreated cases, the enlarged lesions merge and can become necrotic, predisposing to gangrene of the toes or fingertips.

Grave manifestations of disease are cardiovascular disruption, including hypotension, thrombosis, and hemorrhage. Conditions of restlessness, delirium, convulsions, tremor, and coma are alarming indications of central nervous system involvement. Fatalities occur in an average of 10% of untreated cases but less than 1% of treated cases.

Diagnosis, Treatment, and Prevention of Spotted Fever

Any case of Rocky Mountain spotted fever is a cause for great concern and requires immediate treatment, even before laboratory confirmation. Indications sufficiently suggestive to start antimicrobic therapy are the following:

1. a cluster of symptoms, including sudden fever, headache, and rash;
2. recent contact with ticks or dogs; and
3. possible occupational or recreational exposure in the spring or summer.

Early diagnosis can be made by staining rickettsias directly in a tissue biopsy using fluorescent antibodies. Isolating rickettsias from the patient's blood or tissues is desirable, but it is expensive and requires specially qualified personnel and laboratory facilities. Specimens taken from the rash lesions are suitable for PCR assay, which is very specific and sensitive and can circumvent the need for culture. Because antibodies appear relatively soon after infection, a change in serum titer detected through an ELISA test can confirm a presumptive diagnosis.

The drug of choice for suspected and known cases is doxycycline administered for 1 week. Chloramphenicol is an alternate choice used for pregnant patients or if an allergic reaction to tetracycline-class

Figure 21.17 The spotted appearance of the rash in RMSF. This case occurred in a child several days after the onset of fever. It may occur on most areas of the body. *Source: CDC*

antibiotics is noted. Preventive measures follow the pattern for Lyme disease and other tick-borne disease: wearing protective clothing, using insect sprays, and fastidiously removing ticks.

Emerging Rickettsioses

Other genera that share similarities with *Rickettsia* are **Ehrlichia** and **Anaplasma,** two closely related obligate parasites spread by ticks. Although these rickettsial pathogens have been known as pathogens of dogs, horses, and other mammals for some time, infections caused by them have emerged more recently in humans. *Ehrlichia chaffeensis* is the cause of human monocytic ehrlichiosis (HME), and *Anaplasma phagocytophilum* is the cause of human granulocytic anaplasmosis (HGA). The diseases are reported in many regions of the United States and Europe and appear to be on the increase at the present time. Both human pathogens cause an acute flulike disease that ranges from mild to severe and can even

be fatal. In both pathogens, white blood cells are the primary targets of infections.

Human monocytic ehrlichiosis is associated with contact with the Lone Star tick. The rickettsia enters the tick bite and is phagocytosed by monocytes and macrophages, which can lead to cell death and leukopenia. It is carried to many organs, leaving widespread inflammation in its path. The main symptoms are fever, muscle pains, headache, and a rash. Approximately 850 cases per year are reported to the CDC.

Anaplasma phagocytophilum is the cause of human granulocytic anaplasmosis (HGA). The primary reservoirs and vectors for the pathogen are very similar to those of *Borrelia burgdorferi*. It is carried by white-footed mice and deer, and its vectors are ticks in the genus *Ixodes*. The primary targets of infection are neutrophils and other granulocytes. The pathogen disrupts neutrophil function and causes diminished immunities. The symptoms are similar to ehrlichiosis but can also involve the respiratory and gastrointestinal tracts, kidney, and liver. Approximately 1,000 cases are reported each year in the United States, but epidemiologists suspect the true number of cases is much higher.

Most patients recover rapidly with no lasting effects, but around 5% of older chronically ill patients die from disseminated infection. Rapid diagnosis is enabled by PCR tests and indirect fluorescent antibody tests. It can be critical to differentiate or detect coinfection with the Lyme disease *Borrelia,* which is carried by the same tick. Doxycycline will clear up most infections within 7 to 10 days.

Coxiella and *Bartonella:* Other Vector-Borne Pathogens

Q fever[2] was first described in Queensland, Australia. Its origin was mysterious for a time, until Harold Cox working in Montana and Frank Burnet in Australia discovered the agent later named *Coxiella burnetii.** This bacterium is similar to rickettsias in being an intracellular parasite, but it is much more resistant because it produces an unusual type of spore **(figure 21.18).** It is apparently harbored by a wide assortment of vertebrates and arthropods, especially ticks, which play an essential role in transmission between wild and domestic animals. Humans acquire infection largely by means of environmental contamination and airborne spread. Sources of infectious material include urine, feces, milk, and airborne particles from infected animals. The primary portals of entry are the lungs, skin, conjunctiva, and gastrointestinal tract.

Coxiella burnetii has been isolated from most regions of the world. In the United States, it occurs sporadically, primarily in the western states, and it is believed that most cases probably go undetected. People at highest risk are farm workers, meat cutters, veterinarians, laboratory technicians, and consumers of raw milk. The clinical manifestations typical of *Coxiella* infection are abrupt onset of fever, chills, head and muscle ache, and, occasionally, a rash. Disease is sometimes complicated by pneumonitis, hepatitis, and endocarditis. Mild or subclinical cases resolve spontaneously, and more severe cases respond to doxycycline therapy. A vaccine is available in many parts of the world and is used on military personnel in the United States. People working with livestock should

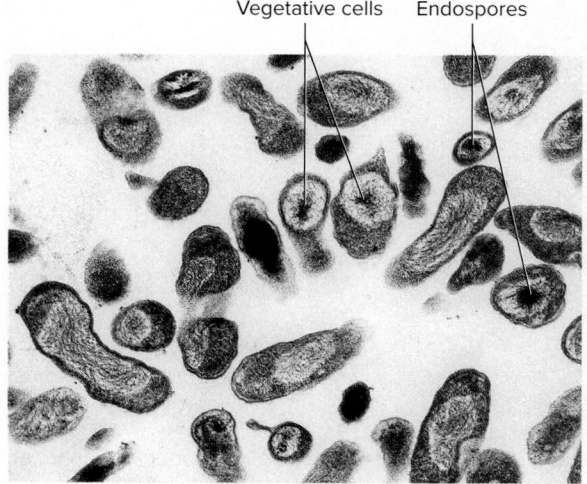

Vegetative cells Endospores

Figure 21.18 The agent of Q fever. The vegetative cells of *Coxiella burnetii* produce unique endospores (visible in most of the cells above) that are released when the cell disintegrates. Free spores survive outside the host and are important in transmission (100,000×). *Source:* National Institute of Allergy and Infectious Diseases (NIAID); Rocky Mountain Laboratories, NIH/CDC

avoid contact with excrement and secretions and should observe decontamination precautions.

The Family Bartonellaceae contains the genus ***Bartonella,**** which now includes species formerly known as *Rochalimaea*. These small, gram-negative rods are fastidious but not obligate intracellular parasites, readily cultured on blood agar. *Bartonella* species are currently considered a group of emerging pathogens. One disease with a long history is **trench fever,** once a common condition of soldiers in battle. The causative agent, ***Bartonella quintana,*** is carried by lice. Most cases occur in endemic regions of Europe, Africa, and Asia. Highly variable symptoms can include a 5- to 6-day fever (hence, 5-day, or quintana, fever); leg pains, especially in the tibial region (shinbone fever); headache; chills; and muscle aches. A macular rash can also occur. The microbe can persist in the blood long after convalescence and is responsible for later relapses.

Bartonella henselae is the most common agent of **cat-scratch disease (CSD),** an infection connected with being clawed or bitten by a cat. The pathogen can be isolated in over 40% of cats, especially kittens. There are approximately 25,000 cases per year in the United States, 80% of them in children 2 to 14 years old. Symptoms start after 1 to 2 weeks, with a cluster of small papules at the site of inoculation. In a few weeks, the lymph nodes along the lymphatic drainage swell and can become pus-filled **(figure 21.19).** Most infections remain localized and resolve in a few weeks, but drugs such as doxycycline and erythromycin can be effective therapies. The disease can be prevented by thorough degerming of a cat bite or scratch.

Bartonella is recognized as an important emerging pathogen in AIDS patients. It is the cause of bacillary angiomatosis, a severe cutaneous and systemic infection. The cutaneous lesions arise as reddish nodules or crusts that can be mistaken for Kaposi's sarcoma. Systems most affected are the liver and spleen, and symptoms are fever, weight loss, and night sweats. Treatment is similar to that for CSD.

2. For "query," meaning to question, or of unknown origin.
 * *Coxiella burnetii* (kox′-ee-el′-uh bur′-net-ee′-eye)

* *Bartonella* (barr″-tun-el′-ah) After A. L. Barton, a Peruvian physician who first described the genus.

Figure 21.19 Cat-scratch disease. Primary nodules appear at the site of the scratch or bite after about 21 days. Note the scab formation, swelling, and inflammation around the bite infection. © K. Talaro

Other Obligate Parasitic Bacteria: The Chlamydiaceae

Even though they are not close relatives of rickettsias, members of Family Chlamydiaceae are also obligate intracellular parasites that depend on certain metabolic constituents of host cells for growth and maintenance. They show further resemblance to the rickettsias with their small size and pleomorphic morphology, but they are markedly different in several aspects of their life cycle. The species of greatest medical significance is **Chlamydia trachomatis,*** a very common pathogen involved in sexually transmitted, neonatal, and ocular disease (trachoma). A related genus, **Chlamydophila,** includes **C. pneumoniae,** the cause of one type of atypical pneumonia; and **C. psittaci,*** a zoonosis of birds and mammals that causes ornithosis in humans.

The Biology of Chlamydia and Related Forms

Chlamydias alternate between two distinct stages: (1) a metabolically inactive, infectious form called the **elementary body** that is released by the infected host cell; and (2) a noninfectious, actively dividing form called the **reticulate body** that grows within host cell vacuoles. The reticulate body completes the cycle by forming new elementary bodies (**figure 21.20**). Elementary bodies are shielded by a rigid, impervious envelope that ensures survival outside the eukaryotic host cell. Reticulate bodies are energy parasites, lacking enzyme systems for catabolizing glucose and synthesizing ATP. They do possess ribosomes and pathways for synthesizing proteins, DNA, and RNA.

* *Chlamydia trachomatis* (klah-mid′-ee-ah trah-koh′-mah-tis) Gr. *chlamys,* a cloak, and *trachoma,* roughness.
* *psittaci* (sih-tah′-see) Gr. *ipsittacus,* a parrot.

Diseases of Chlamydia trachomatis

The reservoir of pathogenic strains of *Chlamydia trachomatis* is the human body. The microbe shows astoundingly broad distribution within the population, often being carried with no symptoms. Elementary bodies are transmitted through contact with secretions. Although infection occurs in all age groups, disease is most severe in infants and children. The two human strains are the *trachoma* strain, which attacks the squamous or columnar cells of mucous membranes in the eyes, genitourinary tract, and lungs; and the *lymphogranuloma venereum* (LGV) strain, which invades the lymphatic tissues of the genitalia.

Chlamydial Diseases of the Eye The two forms of chlamydial eye disease, **ocular trachoma** and **inclusion conjunctivitis,** differ in their patterns of transmission and ecology. Ocular trachoma, an infection of the epithelial cells of the eye, is an ancient disease and a major cause of blindness in certain parts of the world. Although a few cases occur yearly in the United States, several million cases occur endemically in parts of Africa and Asia. Transmission is favored by contaminated fingers, fomites, flies, and a hot, dry climate.

The first signs of infection are a mild conjunctival exudate and slight inflammation of the conjunctiva. These are followed by marked infiltration of lymphocytes and macrophages into the

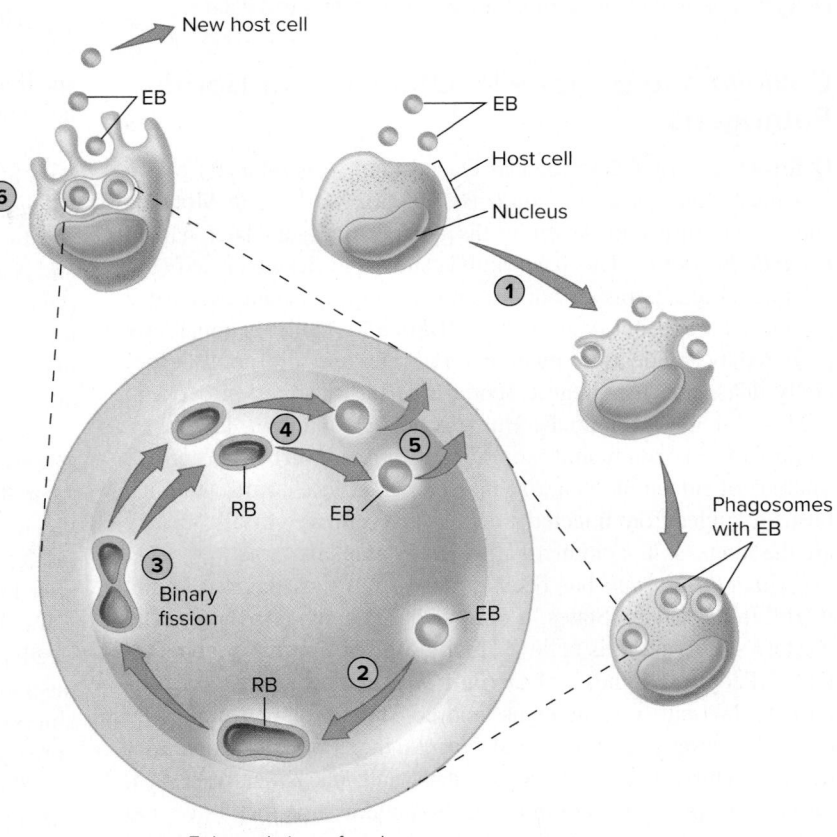

Enlarged view of cycle in phagosome

Process Figure 21.20 The life cycle of Chlamydia. (1) The infectious stage, or elementary body (EB), is taken into phagocytic vesicles by the host cell. **(2)** In the phagosome, each elementary body develops into a reticulate body (RB). **(3)** Reticulate bodies multiply by regular binary fission. **(4–5)** Mature RBs become reorganized into EBs. **(6)** Completed EBs are released from the host cell.

Pathogen Profile #4 *Chlamydia trachomatis*

Reticulocyte bodies in the vacuole of an infected cervical cell.
Source: Dr. Lance Liotta Laboratory/ National Cancer Institute

Microscopic Morphology Very small, gram-negative pleomorphic rods.

Identified by Definitive identification involves cultivation in chicken, mice, or tissue cell lines but is rarely done due to time and cost constraints. Immunofluorescent staining of specimens and PCR are commonly employed.

C. trachomatis resides within the human body, where it is often carried without symptoms. It invades primarily epithelial and lymphatic cells.

Virulence Factors The cell wall of *Chlamydia* is able to inhibit phagolysosome fusion, allowing reticulate bodies to survive and multiply within phagosomes. Elementary bodies are encased in an impervious envelope that ensures survival outside of the host cell.

Primary Infections/Disease *Chlamydia trachomatis* is the causative agent of chlamydiosis, nongonococcal urethritis, pelvic inflammatory disease, and lymphogranuloma venereum. It also causes the eye infections ocular trachoma and inclusion conjunctivitis.

Control and Treatment Condoms provide limited protection as the pathogen can invade multiple cell types. Newborns are treated prophylactically with erythromycin or doxycycline; the treatment of urogenital infections relies on doxycycline or azithromycin.

infected area. As these cells build up, they impart a pebbled (rough) appearance to the inner aspect of the upper eyelid (**figure 21.21a**). In time, a vascular pseudomembrane of exudate and inflammatory leukocytes forms over the cornea, a condition called *pannus* that lasts a few weeks. Chronic and secondary infections can lead to corneal damage and impaired vision. Early treatment of this disease with azithromycin is highly effective and prevents all of the complications. It is a tragedy that in this day of preventive medicine, millions of children worldwide will develop blindness for lack of a few dollars' worth of antibiotics.

Inclusion conjunctivitis is usually acquired through contact with secretions of an infected genitourinary tract. Infantile conjunctivitis develops 5 to 12 days after a baby has passed through the birth canal of its infected mother and is the most prevalent form of conjunctivitis in the United States (100,000 cases per year). The initial signs are conjunctival irritation, a profuse adherent exudate,

redness, and swelling (**figure 21.21b**). Although the disease usually heals spontaneously, trachoma-like scarring occurs often enough to warrant routine eye prophylaxis of all newborns (as for gonococcal infection) with antibiotics such as erythromycin and doxycycline.

Sexually Transmitted Chlamydial Diseases It has been estimated that *C. trachomatis* is carried in the reproductive tract of up to 10% of the population, with even higher rates among promiscuous individuals. About 70% of infected women harbor it asymptomatically on the cervix, while 10% of infected males show no signs or symptoms. The potential for this disease to cause long-term reproductive damage has initiated its listing as a reportable disease since 1995. Statistics now show that chlamydiosis is the most prevalent bacterial STD. The reported incidence for 2015 was over 1.4 million cases, but its true incidence probably exceeds that level by 10 times. The occurrence of this infection in young, sexually

(a)

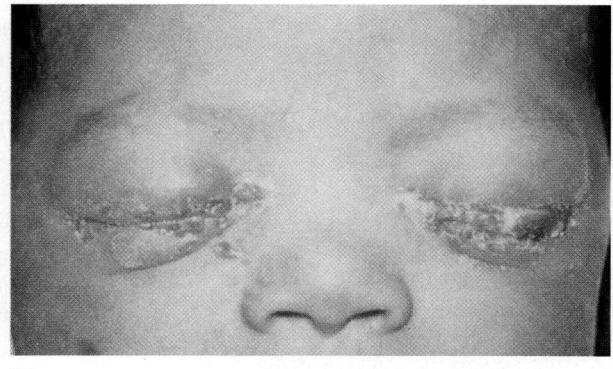

(b)

Figure 21.21 **The pathology of primary ocular chlamydial infection. (a)** Ocular trachoma, an early, pebblelike inflammation of the conjunctiva and inner lid. (Note: The eyelid has been retracted to make the lesion more visible.) **(b)** Inclusion conjunctivitis (in this case due to gonococcal infection) in a newborn. Within 5 to 6 days, an abundant, watery exudate collects around the conjunctival sac. This is currently the most common cause of ophthalmia neonatorum. (a): © Western Ophthalmic Hospital/Science Source; (b): *Source:* J. Pledger/CDC

active teenagers is increasing by 8% to 10% per year. Medically and socioeconomically, its clinical significance now eclipses gonorrhea, herpes simplex 2, and syphilis.

A syndrome among males with chlamydial infections is an inflammation of the urethra called nongonococcal urethritis (NGU). This diagnosis is derived from the symptoms that mimic gonorrhea yet do not involve gonococci. Women with symptomatic chlamydial infection have cervicitis accompanied by a white drainage, endometritis, and salpingitis (pelvic inflammatory disease—PID). As is often the rule with sexually transmitted diseases, chlamydia frequently appears in mixed infections with the gonococcus and other genitourinary pathogens, thereby greatly complicating treatment.

When a particularly virulent strain of *Chlamydia* chronically infects the genitourinary tract, the result is a severe, often disfiguring disease called **lymphogranuloma venereum**.[3] The disease is endemic to regions of South America, Africa, and Asia but occasionally occurs in other parts of the world. Its incidence in the United States is about 500 cases per year. Chlamydias enter through tiny nicks or breaks in the perigenital skin or mucous membranes and form a small, painless vesicular lesion that often escapes notice. Other acute symptoms are headache, fever, and muscle aches. As the lymph nodes near the lesion begin to fill with granuloma cells, they enlarge and become firm and tender **(figure 21.22)**. These nodes, or buboes, can cause long-term lymphatic obstruction that leads to chronic, deforming edema of the genitalia and anus.

Identification, Treatment, and Prevention of Chlamydiosis

Because chlamydias reside intracellularly, specimen sampling requires enough force to dislodge some of the cells from the mucosal surface. Genital samples are taken with a swab inserted a few centimeters into the urethra or cervix, rotated, and removed. Although the most reliable diagnosis comes from culture in chicken embryos, mice, or cell lines, this procedure is too costly and time-consuming to be used routinely in STD clinics; however, it is an essential part of diagnosing neonatal infections. The most sensitive and specific tests currently available are a direct assay of specimens using immunofluorescence and a PCR-based probe. Methods useful in diagnosing inclusion conjunctivitis are Giemsa or iodine stains **(figure 21.23)**, but they are not recommended for

3. Also called tropical bubo or lymphogranuloma inguinale.

Figure 21.22 **The clinical appearance of advanced lymphogranuloma venereum in a man.** A chronic local inflammation blocks the lymph channels, causing swelling and distortion of the tissues near the scrotum. *Source:* CDC

Figure 21.23 **Cells in tissue culture infected with** *Chlamydia trachomatis*. The dark-staining inclusion bodies (arrows) indicate a phagosome containing *Chlamydia* in various stages of development. These can be diagnostic of chlamydiosis. *Source:* CDC

urogenital specimens because of low sensitivity and the possibility of obtaining false-negative results in asymptomatic patients.

Urogenital chlamydial infections are most effectively treated with drugs that act intracellularly, such as doxycycline and azithromycin. Because of the high carrier rate and the difficulty in detection, prevention of chlamydial infections is a public health priority. As a general rule, sexual partners of infected people should be treated with drug therapy to prevent infection, and sexually active people can achieve some protection with a condom.

Chlamydophila Species

Chlamydophila is a genus of obligate pathogens; it contains members that used to be included with *Chlamydia*. One of these is *Chlamydophila pneumoniae*, a strict human pathogen with characteristics that distinguish it from other species in the genus. It has been linked to a type of respiratory illness that includes pharyngitis, bronchitis, and pneumonitis. It is usually a mild illness in young adults, though it can cause a severe reaction in asthmatic patients that is responsible for increased rates of death in this group. Some evidence exists to implicate it in heart disease.

Chlamydophila psittaci

The term *psittacosis* was adopted to describe a pneumonia-like illness contracted by people working with imported parrots and other psittacine birds in the last century. As outbreaks of this disease appeared in areas of the world having no parrots, it became evident that other birds could carry and transmit the microbe to humans and other animals. The term **ornithosis**[*] is sometimes used to refer to the same infection.

Psittacosis is a worldwide zoonosis that is carried in a latent state in wild and domesticated birds but becomes active under stressful conditions such as overcrowding. In the United States, poultry have been subject to extensive epidemics that killed as many as 30% of flocks. Infection is communicated to other birds, mammals, and humans by contaminated feces and other discharges that become airborne and are inhaled. Sporadic cases in the United States occur mainly among poultry and pigeon handlers.

The symptoms of human psittacosis mimic those of influenza and pneumococcal pneumonia. Early manifestations are fever, chills,

* *ornithosis* (or″-nih-thoh′-sis) Gr. *ornis*, bird. More than 90 species of birds harbor *C. psittaci*.

frontal headache, and muscle aches, and later ones are coughing and lung consolidation. Unchecked, infection can lead to systemic complications involving the meninges, brain, heart, or liver. Although most patients respond well to doxycycline or erythromycin therapy, recovery is often slow and fraught with relapses. Control of the disease is usually attempted by quarantining imported birds and by taking precautions in handling birds, feathers, and droppings.

 Check Your Progress SECTION 21.3

10. What do rickettsias and chlamydias derive from the host?
11. Provide an example of a species, vector, and disease for each of the four major groups of rickettsioses.
12. Where is Rocky Mountain spotted fever commonly found? What symptoms and activities would justify treatment for the disease prior to laboratory confirmation?
13. Why do ranchers have an increased risk of becoming infected with *Coxiella burnetii*? Who else would have a similar level of risk?
14. How are chlamydias transmitted?
15. Describe the major complications of eye infections and STDs.
16. Name three diseases caused by species within the genus *Chlamydia*.
17. How do the respiratory infections attributable to the two species of *Chlamydophila* differ?

21.4 Mollicutes and Other Cell-Wall-Deficient Bacteria

 Expected Learning Outcomes

16. Describe the characteristics and diseases of mycoplasmas.
17. Describe how L forms may arise.

Bacteria in the Class Mollicutes, also called the **mycoplasmas,** are among the smallest self-replicating microorganisms. All of them naturally lack a cell wall **(figure 21.24a),** and except for one genus, all species are parasites of animals and plants. The two most clinically important genera are *Mycoplasma* and *Ureaplasma*. Disease of the respiratory tract has been primarily associated with *Mycoplasma pneumoniae; M. hominis* and *Ureaplasma urealyticum* are implicated in urogenital tract infections.

Biological Characteristics of the Mycoplasmas

Without a rigid cell wall to delimit their shape, mycoplasmas are exceedingly pleomorphic. The small (0.3–0.8 μm), flexible cells assume a spectrum of shapes, ranging from cocci and filaments to doughnuts, clubs, and helices (see figure 4.16). Mycoplasmas are not strict parasites, and they can grow in cell-free media, generate metabolic energy, and synthesize proteins with their own enzymes. However, most are fastidious and require complex media containing sterols, fatty acids, and preformed purines and pyrimidines. Mycoplasmas are sometimes referred to as membrane parasites because they acquire certain necessary lipids from host cell membranes **(figure 21.24b).** Infections are chronic and difficult to eliminate, because mycoplasmas bind tenaciously to receptor sites

(a)

(b)

Figure 21.24 The morphology of mycoplasmas. (a) A color-enhanced scanning electron micrograph of *Mycoplasma pneumoniae* (10,000×). Note pleomorphic shapes and elongate attachment tips. The cells use these to anchor themselves to host cells. **(b)** Figure depicts how *M. pneumoniae* becomes a membrane parasite that adheres tightly and fuses with the host cell surface. This fusion makes destruction and removal of the pathogen very difficult. (a): © VEM/Science Source

on cells of the respiratory and urogenital tracts and they are not easily removed by usual defense mechanisms.

Mycoplasma pneumoniae *and Atypical Pneumonia*

Mycoplasma pneumoniae is a human parasite that is the most common agent of primary atypical pneumonia (PAP).[4] This syndrome is atypical in that its symptoms do not resemble those of pneumococcal pneumonia. Mycoplasmal pneumonia is transmitted by aerosol droplets among people confined in close living quarters, especially families, students, and the military. Community resistance to this pneumonia is high; only 3% to 10% of those exposed become infected, and fatalities are rare.

Mycoplasma pneumoniae binds to specific receptors of the epithelial cells of the respiratory tract. Gradual spread of the bacteria over the next 2 to 3 weeks disrupts the cilia and damages the respiratory epithelium. The first symptoms—fever, malaise, sore

4. Primary atypical pneumonia can also be caused by rickettsias, chlamydias, respiratory syncytial viruses, and adenoviruses.

throat, and headache—are not suggestive of pneumonia. A cough is not a prominent early symptom, and when it does appear, it is mostly unproductive. As the disease progresses, nasal symptoms, chest pain, and earache can develop. The lack of acute illness in most patients has given rise to the nickname "walking pneumonia."

Diagnosis Because a culture can take 2 or 3 weeks, early diagnosis of mycoplasma pneumonia is difficult and relies chiefly on close clinical observation to rule out other bacterial or viral agents. Stains of sputum appear devoid of bacterial cells, leukocyte counts are within normal limits, and X-ray findings are nonspecific. Serological tests based on complement fixation, immunofluorescence, and indirect hemagglutination are useful later in the disease.

Doxycycline and azithromycin inhibit mycoplasmal growth and help to rapidly diminish symptoms, but they do not stop the shedding of viable mycoplasmas. Patients frequently experience relapses if treatment is not continued for 14 to 21 days. Preventive measures include controlling contamination of fomites, avoiding contact with droplet nuclei, and reducing aerosol dispersion.

Other Mycoplasmas

Mycoplasma genitalium and *Ureaplasma urealyticum* are regarded as weak reproductive pathogens. They are frequently encountered in samples from the urethra, vagina, and cervix of newborns and adults. These species initially colonize an infant at birth and subsequently diminish through early and late childhood. A second period of colonization and persistence is initiated by the onset of sexual intercourse.

Evidence linking genital mycoplasmas to human disease is substantial and growing every year. *Ureaplasma urealyticum* is implicated in some types of nonspecific or nongonococcal urethritis and prostatitis. There is increasing evidence that this mycoplasma plays a role in opportunistic infections of the fetus and fetal membranes. It appears to cause some cases of miscarriage, stillbirth, premature birth, and respiratory infections of newborns.

Mycoplasma genitalium is an agent of STDs. It is very common in the genital tract of both women and men, and it often accompanies other STDs. In women, it is one cause of PID, cervicitis, urethritis, and vaginitis. Studies have also shown it can be carried into the uterus by spermatozoa. In males, *M. genitalium* can cause a form of NGU, prostatitis, and epididymitis.

Bacteria That Have Lost Their Cell Walls

Exposure of typical walled bacteria to certain drugs (penicillin) or enzymes (lysozyme) can result in wall-deficient bacteria called L forms or L-phase variants. L forms are induced or occur spontaneously in numerous species and can even become stable and reproduce themselves, but they are not naturally related to mycoplasmas.

L Forms and Disease

The role of certain L forms in human and animal disease is a distinct possibility, but proving etiology has been complicated because infection is difficult to verify. One theory proposes that antimicrobic therapy with cell-wall-active agents induces certain infectious agents to become L forms. In this wall-free state, they resist further treatment with these drugs and remain latent until the therapy ends,

at which time they reacquire walls and resume their pathogenic behavior. Infections involving L-phase variants of group A streptococci, *Proteus,* and *Corynebacterium* have occasionally been reported. In a number of chronic pyelonephritis and endocarditis cases, cell-wall-deficient bacteria have been the only isolates. Research on people with a chronic intestinal syndrome called Crohn's disease has uncovered a strong association with a wall-deficient form of *Mycobacterium avium paratuberculosis* (MAP), a relative of the TB bacillus. When technicians used a PCR technique to analyze the DNA of colon specimens, it was found that 65% of Crohn's patients tested positive for *M. a. paratuberculosis.* It is likely to be at least a cofactor for developing this disease.

 Check Your Progress SECTION 21.4

18. Explain the unique features of mycoplasmas.
19. What are the pathologic effects of *Mycoplasma pneumoniae?*
20. Which body systems are affected by mycoplasma infection?
21. Explain why penicillin therapy may result in the formation of L forms but tetracycline therapy would not.

21.5 Bacteria in Dental Disease

 Expected Learning Outcomes

18. Discuss the formation of plaque on tooth surfaces.
19. Relate the formation of plaque to periodontitis.

The relationship between humans and their oral microbiota is a complex, dynamic microecosystem. The mouth contains a diversity of surfaces for colonization, including the tongue, teeth, gingiva, palate, and cheeks; and it provides numerous aerobic, anaerobic, and microaerophilic microhabitats for the estimated 600 different oral species with which humans coexist. The habitat of the oral cavity is warm, moist, and greatly enriched by the periodic infusion of food. In most humans this association remains in balance with little adverse effect, but in people with poor oral hygiene, it teeters constantly on the brink of disease.

The Structure of Teeth and Associated Tissues

Dental diseases can affect nearly any part of the oral cavity, but most of them involve the dentition (teeth) and the surrounding supportive structures, collectively known as the **periodontium*** (gingiva, ligaments, membrane, bone) **(figure 21.25).** A tooth is composed of a flared crown that protrudes above the gum and a root that is inserted into a bony socket. The outer surface of the crown is protected by a dense coating of enamel, an extremely hard, noncellular material composed of tightly packed rods of hydroxyapatite crystals $[(Ca_{10}(OH)_2(PO_4)_6]$. The root is surrounded by a layer of cementum, which is anchored by ligaments to the periodontal membrane that

* *periodontium* (per″-ee-oh-don′-shee-um) Gr. *peri,* around, and *odous,* tooth. Gums, bones, and cementum.

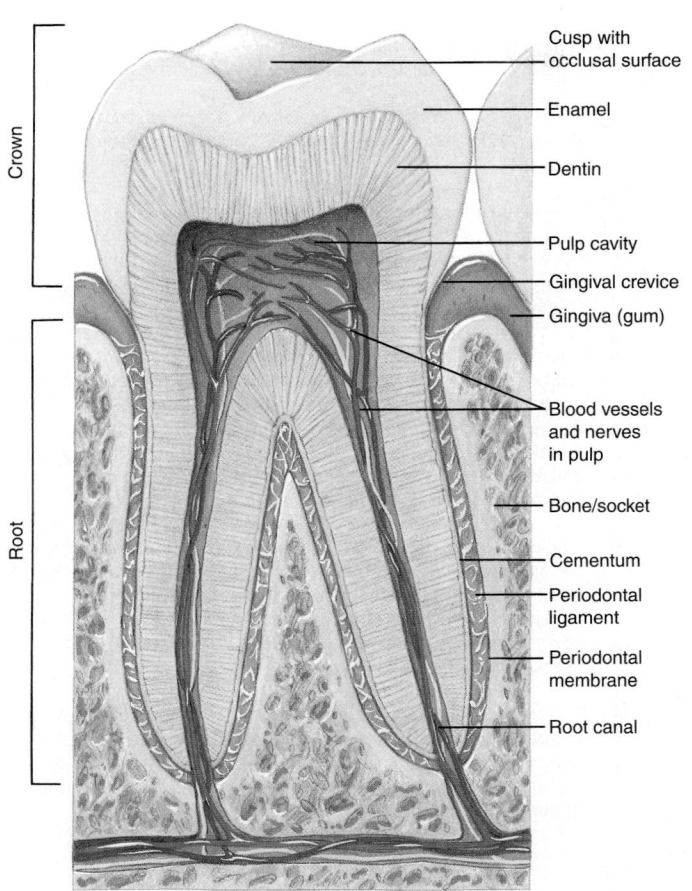

Figure 21.25 The anatomy of a tooth.

Tooth Surface
(enamel, root)

↓

Pellicle Formation

↓

Plaque Formation

↙ ↘

Acidification and Cavitation Calculus (Tartar) Formation

↓ ↓

Dental Caries
Damage to enamel Gingivitis/Periodontal Disease

↓ ↓

Tooth Destruction
(exposure of dentin, pulp, root) Bone Resorption

↘ ↙

Tooth
Loss

Figure 21.26 Summary of the events leading to dental caries, periodontal disease, and bone and tooth loss.

lines the socket. The major portion of the tooth inside the crown and root is composed of a highly regular calcified material called dentin, and the core contains a pulp cavity that supplies the living tissues with blood vessels and nerves. The root canal is the portion of the pulp that extends into the roots. The space surrounding the teeth is protected by the gingiva (gum), a soft tissue composed of connective tissue covered by a mucous membrane. The primary sites for dental infections are grooves in the enamel, especially the cusps, and the crevice, or sulcus, formed where the gingiva meets the tooth.

Dental pathology generally affects both hard and soft tissues **(figure 21.26).** Although both categories of disease are initiated when microbes adhere to the tooth surface and produce dental plaque, their outcomes vary. In the case of dental caries, the gradual breakdown of the enamel leads to invasive disease of the tooth itself. In soft-tissue disease, calcified plaque damages the delicate gingival tissues and predisposes them to bacterial invasion of the periodontium. These conditions are responsible for the loss of teeth, though dental caries are usually implicated in children and periodontal infections in adults.

Hard-Tissue Disease: Dental Caries

Dental caries* is the most common human disease. It is a complex **biofilm infection** of the dentition that gradually destroys the enamel

and often lays the groundwork for the destruction of deeper tissues. It occurs most often on tooth surfaces that are less accessible and harder to clean and on those that provide pockets or crevices where bacteria can cling. Caries commonly develop on enamel pits and fissures, especially those of the occlusal (grinding) surfaces, though they can also occur on the smoother crown surfaces and subgingivally on the roots.

Several views have been suggested to explain how dental caries originate. For some time experts knew that sugar, microbes, and acid were involved in tooth decay. It was studies with germ-free animals that finally showed that no single factor can account for caries—all three are necessary. Caries development occurs in many phases and requires multiple interactions involving the anatomy, physiology, diet, and microbiota of the host.

Plaque and Dental Caries Formation

A freshly cleaned tooth is a perfect landscape for colonization by microbes **(figure 21.27).** Within a few moments it develops a thin, mucous coating called the acquired pellicle, which is made up of adhesive salivary proteins. This structure presents a potential substrate upon which certain bacteria first gain a foothold. The process of colonization of the tooth follows a classic pattern of biofilm formation. The most prominent pioneering colonists belong to the genus *Streptococcus*. These gram-positive cocci have adhesive receptors such as lectins and slime layers that allow them to cling to the tooth surfaces and to each other, forming a foundation for the mature biofilm aggregate known as **plaque.*** Fed by a diet high in

caries (kar′-eez) L. rottenness.

plaque (plak) Fr. a patch.

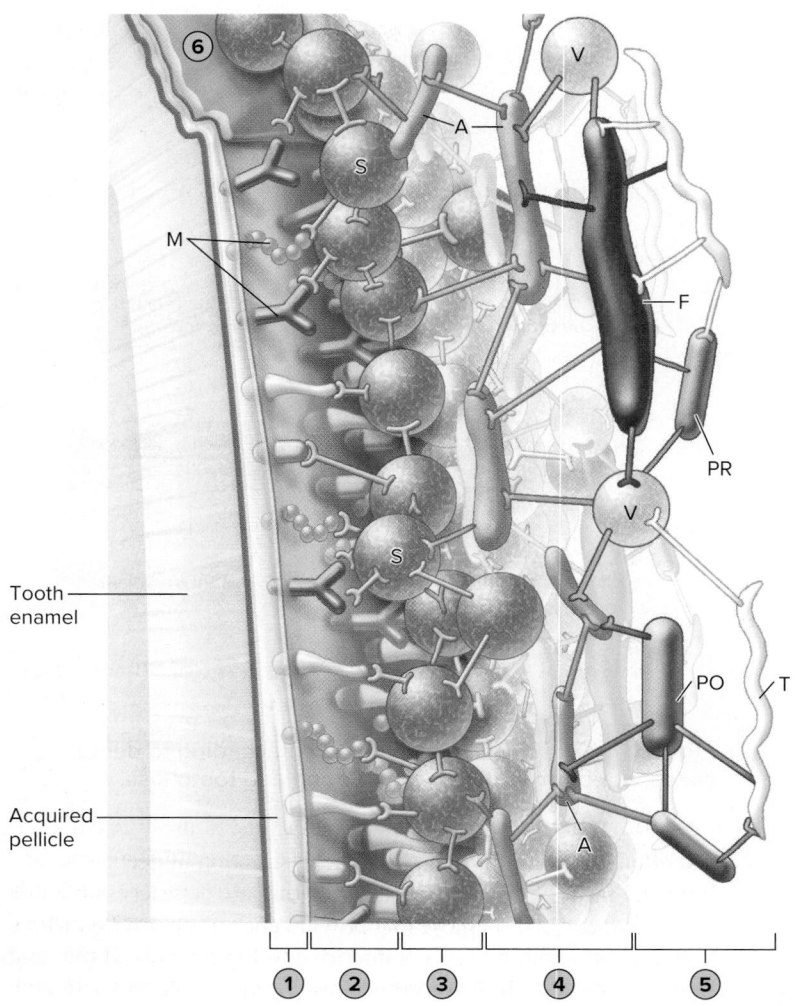

1. A freshly cleaned tooth surface immediately develops a thin layer of salivary glycoproteins (the acquired pellicle).

2. Fibers of proteins, antibodies, salivary enzymes, bacterial debris, and other salivary molecules adhere to the pellicle (M).

3. The earliest tooth colonizers are the *Streptococcus mutans* group(s) (*S. oralis* and *S. gordonii*). These bacteria have specific receptors that adhere to the outer pellicle molecules. The streptococci likewise bind to each other, forming the initial base of plaque.

4. The next phase involves cell-cell signaling and coaggregation with additional colonists. The most common bacteria to add to the biofilm during this phase are filamentous rods in the genus *Actinomyces* (A). Other species of *Streptococcus* (the *mutans* group) use dietary carbohydrates to secrete glucans that add bulk to the matrix and serve as a source of sugars.

5. Once this initial framework has been laid down, it enters a second phase of aggregation which creates the final dense mat of plaque. Bacteria that colonize at this point are frequently anaerobes such as *Fusobacterium* (F), *Porphyromonas* (PO), *Prevotella* (PR), *Veillonella* (V), and *Treponema* (T).

6. See upper left. Initial damage to the enamel occurs when streptococci near the enamel surface ferment the sugars in plaque to lactic, acetic, and other acids. When these acids are trapped against the tooth surface and etch through it, a dental caries has developed.

Process Figure 21.27 Stages in the formation of plaque biofilm and dental caries. The process involves specific binding to enamel by a multilayered biofilm of oral microbes that interact, recognize, and aggregate.

sucrose, glucose, and certain complex carbohydrates, the *Streptococcus mutans* group secretes sticky polymers of glucose called fructans and glucans that form the majority of the plaque biofilm. As these primary invaders continue to build up on the tooth surface, they aggregate with thin, branching *Actinomyces* cells. As the plaque biofilm thickens, it creates an oxygen-free microenvironment that favors colonization by anaerobic bacteria. Among these secondary invaders are species of *Fusobacterium, Porphyromonas,* and *Treponema.* A microscopic view of plaque reveals a rich and varied network of bacteria and their products along with epithelial cells and fluids **(figure 21.28).**

If mature plaque is not removed from sites that readily trap food, it can develop into a dental caries (see figure 21.27). The role of plaque in caries development is related directly to the fermentation of dietary carbohydrates by streptococci. In the denser regions

(a)

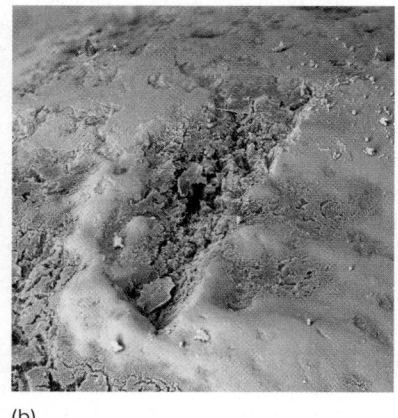

(b)

Figure 21.28 The microscopic appearance of plaque. (a) Scanning electron micrograph of plaque displaying the development of a rich biofilm (1,800×). **(b)** Scanning electron micrograph of a dental cavity filled with bacterial plaque. (a): © Steve Gschmeissner/SPL/Getty Images RF; (b): © Eye of Science/Science Source

of plaque, the acid can accumulate in direct contact with the enamel surface and lower the pH to below 5, which is acidic enough to begin to dissolve (decalcify) the calcium phosphate of the enamel in that spot. This initial lesion can remain localized in the enamel (first-degree caries) and can be repaired with various inert materials (fillings). Once the deterioration has reached the level of the dentin (second-degree caries), tooth destruction speeds up, and the tooth can be rapidly destroyed. Exposure of the pulp (third-degree caries) is attended by severe tenderness and toothache, and the chance of saving the tooth is diminished.

Soft-Tissue and Periodontal Disease

Gingival and periodontal disease are so common that 97% to 100% of the population has some manifestation of it by age 45. The most common predisposing condition occurs when the plaque becomes mineralized (calcified) with calcium and phosphate crystals. This process produces a hard, porous substance called **calculus** above and below the gingiva that can induce varying degrees of periodontal damage (see **figure 21.28***b*). Other factors that contribute to gingival and periodontal disease are diabetes, smoking, immune deficiency, and stress.

Calculus accumulating in the gingival sulcus causes abrasions to the delicate gingival membrane, and the chronic trauma causes a pronounced inflammatory reaction. The damaged tissues become a portal of entry for a variety of bacterial residents. These include genera such as *Actinobacillus, Porphyromonas, Bacteroides, Fusobacterium, Prevotella,* and *Treponema* spirochetes. The anaerobic bacteria in these infections outnumber aerobes by 100 to 1. In response to the mixed infection, the damaged area becomes infiltrated by white blood cells, which cause further inflammation and tissue damage **(figure 21.29***a*). The initial signs of **gingivitis** are swelling, loss of normal contour, patches of redness, and increased bleeding of the gingiva. Spaces or pockets of varying depth also develop between the tooth and the gingiva.

If this condition persists, a more serious disease called **periodontitis** results. This is the natural extension of the disease into the periodontal membrane and cementum. The deeper involvement results in chronic inflammation, loss of the ligament, and deeper pockets in the sulcus. It can cause bone resorption severe enough to loosen the tooth in its socket. If the condition is allowed to progress, the tooth can be lost **(figure 21.29***b*).

Chronic periodontal infections may lead to necrotizing ulcerative gingivitis (NUG), formerly called trench mouth or Vincent disease. This disease is a synergistic infection involving *Treponema vincentii, Bacteroides forsythus,* and fusobacteria **(figure 21.30)**. These pathogens together produce several invasive factors that cause rapid advancement into the periodontal tissues. The condition is associated with severe pain, bleeding, pseudomembrane formation, and necrosis. NUG usually results from poor oral hygiene, altered host defenses, or prior gum disease, and is not considered communicable. It does respond well to broad-spectrum antibiotics.

Factors in Dental Disease

Nutrition and eating patterns are closely tied to oral diseases. People whose diet is high in refined sugar (sucrose, glucose, and fructose) tend to have more caries, especially if these foods are eaten constantly throughout the day without brushing the teeth. The practice of putting a baby down to nap with a bottle of fruit juice or formula can lead to rampant dental caries ("nursing bottle caries"). In addition to diet, numerous anatomical, physiological, and hereditary factors influence oral diseases. The structure of the tooth enamel can be influenced by genetics and by environmental factors such as fluoride, which strengthens the enamel bonds. Inhibitory factors in saliva such as antibodies and lysozyme can also help prevent dental disease by inhibiting bacterial attachment and growth.

The best control of dental diseases comes from preventive dentistry, including regular brushing and flossing to remove plaque, because stopping plaque buildup automatically reduces caries and calculus production. Mouthwashes are relatively ineffective in

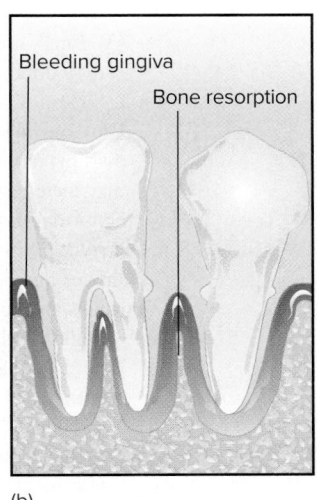

(a) (b)

Figure 21.29 Stages in soft-tissue infection, gingivitis, and periodontitis. (a) Calculus buildup and gingivitis. **(b)** Late-stage periodontitis, with tissue destruction, deep pocket formation, loosening of teeth, and bone loss.

Figure 21.30 SEM of a sample of exudate from a gingival pocket (7,040×). Note the wide variations in sizes and shapes of bacteria which create a mixed infection that invades periodontal tissues. © Scimat/Science Source

controlling plaque formation because of the high bacterial content of saliva and the relatively short acting time of mouthwashes.

Once calculus has formed on teeth, it cannot be removed by brushing but can be dislodged only by special mechanical procedures (scaling) in the dental office. One of the most exciting prospects is the possibility of a vaccine to protect against the primary colonization of the teeth. Some success in inhibiting plaque formation has been achieved in experimental animals with vaccines raised against the whole cells of cariogenic *Streptococcus* and the fimbriae of species of *Actinomyces.*

 Check Your Progress SECTION 21.5

22. What is the relationship between dietary carbohydrates, plaque, and dental caries?
23. Explain the stepwise progression of plaque deposition to periodontal disease.
24. Describe areas of the teeth and gums that support growth of microbes and infections.
25. Explain the colonization of teeth and the development of a biofilm.

 CASE STUDY Part 2

In June 2014 a resident of Canada sought medical care for fever, pain, and erythema at the site of a recent injection. A few weeks earlier she too had traveled to Germany, where she had undergone injections of fetal sheep cells two days before the New York patients were treated. In July she was diagnosed with Q fever and the Public Health Agency of Canada notified German authorities in September 2014. At the time of notification, public health authorities in Germany were investigating an outbreak of human Q fever associated with inhalation exposure to a sheep flock. The flock was used for the production of fetal sheep cell injections by the German physician who treated the Canadian woman and the five New York patients.

In September, patients who had received live cell therapy at the German clinic between January and July 2014 were informed of their potential exposure to *Coxiella burnetii,* the organism responsible for Q fever. Immunoflorescence testing revealed *C. burnetii*-specific antibodies in all six patients, with especially high IgM titers, suggesting acute infection. The

patients were successfully treated with doxycycline, although three patients reported lingering symptoms of the disease (fatigue, chills, sweats, and difficulty sleeping) up to 10 months after exposure. Public health officials in New York notified the CDC and posted a report on the CDC's Epidemic Information Exchange to solicit additional cases, but none were found.

Xenotransplantation practices like live cell therapy are prohibited in both the United States and Canada, but patients are of course free to travel to other countries where the procedure is less tightly regulated. In Germany, for instance, live cell therapy is permitted if the cells are prepared by a doctor for use only in his or her own patients. According to an assessment by the World Health Organization, between 1994 and 2009 xenotransplantation procedures were identified in 12 countries, of which nine had no clear national regulations on the process.

■ *What other diseases in this chapter are considered zoonotic in origin?*

See: CDC. 2015. Q Fever Outbreak Among Travelers Who Received Live Cell Therapy—United States and Canada, 2014. *MMWR* 64:1071–1073.

To conclude this Case Study, go to Connect.

 Chapter Summary with Key Terms

21.1 The Spirochetes
 A. Spirochetes are helical, flexible bacteria that move by periplasmic flagella. Several *Treponema* species are obligate parasitic spirochetes with 8 to 12 regular spirals; best observed under dark-field microscope; cause treponematoses.
 1. Direct observation of treponeme in tissues and blood tests is important in diagnosis.
 2. Cannot be cultivated in artificial media; treatment by large doses of penicillin or doxycycline.
 B. **Treponemes: Members of the Genus *Treponema***
 1. *Syphilis: **Treponema pallidum*** causes complex progressive disease in adults and children.
 a. Sexually transmitted syphilis is acquired through close contact with a lesion; untreated sexual disease occurs in stages over long periods.
 (1) At site of entrance, multiplying treponemes produce a primary lesion, a hard ulcer or **chancre,** which disappears as the microbe becomes systemic.
 (2) Secondary syphilis is marked by skin rash, fever, damage to mucous membranes. Both primary and secondary syphilis are communicable. Latent period establishes pathogen in tissues.

 (3) Final, noncommunicable **tertiary stage** is marked by tumors called **gummas** and life-threatening cardiovascular and neurological effects.
 b. **Congenital syphilis** is acquired transplacentally; disrupts embryonic and fetal development; survivors may have respiratory, skin, bone, teeth, eye, and joint abnormalities if not treated.
 2. *Nonsyphilitic treponematosis:* Slow progressive cutaneous and bone diseases endemic to specific regions of tropics and subtropics; usually transmitted under unhygienic conditions.
 a. Bejel is a deforming childhood infection of the mouth, nasal cavity, body, and hands.
 b. Yaws occurs from invasion of skin cut, causing a primary ulcer that seeds a second crop of lesions.
 c. Pinta is superficial skin lesion that depigments and scars the skin.
 C. *Leptospira interrogans* causes leptospirosis, a worldwide zoonosis acquired through contact with urine of wild and domestic animal reservoirs. Spirochete enters cut, multiplies in blood and spinal fluid. Long-term infections may affect the kidneys and liver.

D. *Borrelia* species are loose, irregular spirochetes that cause borreliosis; infections are vector-borne (mostly by ticks). Borrelias cause recurrent fever and other symptoms because the spirochetes repeatedly change antigenically and force the immune system to keep adapting.
1. Both *B. hermsii* and *B. recurrentis* cause relapsing fever.
2. *Borrelia burgdorferi.*
 a. Causes **Lyme disease,** a syndrome that occurs endemically in several regions of the United States.
 b. A zoonosis carried by mice and spread by a hard tick (*Ixodes*) that lives on deer and mice.
 c. Tick bite leads to fever and a prominent ring-shaped rash; if left untreated, may cause cardiac, neurological, and arthritic symptoms.
 d. Can be controlled by antibiotics and by avoiding tick contact.

21.2 Curviform Gram-Negative Bacteria and Enteric Diseases
A. Vibrios are short spirals or sausage-shaped cells with polar flagella.
 1. *Vibrio cholerae* causes **epidemic cholera,** a human disease that is distributed worldwide in natural waters.
 a. Organism is ingested in contaminated food and water and infects the surface of the small intestine, is not invasive.
 b. Severity of cholera due to potent cholera toxin that causes electrolyte and water loss through secretory diarrhea.
 c. Resulting dehydration leads to muscle, circulatory, and neurological symptoms.
 d. Treatment is with oral rehydration (electrolyte and fluid replacement); vaccine available.
 2. *Vibrio parahaemolyticus* causes food infection associated with seawater and seafood; symptoms similar to mild cholera.
B. *Campylobacter* species: *C. jejuni* is a common cause of severe **gastroenteritis** worldwide; acquired through water contaminated by animal feces and meats, eggs, and other food products; enteritis is due to enterotoxin; *C. fetus* causes diseases in pregnant women and fatal septicemia in neonates.
C. *Helicobacter pylori* is a helical cell adapted to the stomach lining that may invade it and cause gastric diseases such as ulcers and gastritis. Treatment includes antibiotics and antacid drugs.

21.3 Medically Important Bacteria of Unique Morphology and Biology
A. **Rickettsias** are tiny, gram-negative rods or cocci that are metabolic and intracellular parasites; diseases treatable with doxycycline and chloramphenicol. *Rickettsia* causes rickettsioses; most are zoonoses spread by arthropod vectors.
 1. *Epidemic typhus:* Caused by *Rickettsia prowazekii;* carried by lice; disease starts with high fever, chills, headache; rash occurs; Brill-Zinsser disease is a chronic, recurrent form.
 2. *Endemic (murine) typhus:* Zoonosis caused by *R. typhi,* harbored by mice and rats; occurs sporadically in areas of high flea infestation; symptoms like epidemic form but milder.
 3. *Rocky Mountain spotted fever:* Etiologic agent is *R. rickettsii;* zoonosis carried by dog and wood ticks (*Dermacentor*); most cases on eastern seaboard; infection causes distinct spotted, migratory rash; effects may include heart damage, CNS damage; disease controlled by drug therapy and avoidance of ticks.
 4. *Ehrlichia* and *Anaplasma* are two species of rickettsias that infect humans. These tick-borne bacteria cause human monocytic ehrlichiosis and human granulocytic anaplasmosis. Both cause acute febrile infections.
B. *Coxiella* and *Bartonella:* **Other Vector-Borne Pathogens**
 1. *Q fever:* Caused by *Coxiella burnetii;* agent has a resistant spore form that can survive out of host; a zoonosis of domestic animals; transmitted by air, dust, unpasteurized milk, ticks; usually inhaled, causing pneumonitis, fever, hepatitis.
 2. *Bartonella:* A related genus is the cause of trench fever, spread by lice; and cat-scratch disease, a lymphatic infection associated with a clawing or biting injury by cats.
C. *Chlamydia* is another genus of tiny, gram-negative coccobacilli that are obligate intracellular parasites of animals.
 1. Organisms in the genus *Chlamydia* pass through a transmission phase involving a hardy elementary body and an intracellular reticulate body that has pathologic effects.
 a. *Chlamydia trachomatis:* A strict human pathogen that causes eye diseases and STDs. Ocular trachoma is a severe infection that deforms the eyelid and cornea and may cause blindness. **Conjunctivitis** occurs in babies following contact with birth canal; prevented by ocular prophylaxis after birth.
 b. *C. trachomatis* causes very common bacterial STDs: nongonococcal urethritis in males, and cervicitis, salpingitis (PID), infertility, and scarring in females; also lymphogranuloma venereum, a disfiguring disease of the external genitalia and pelvic lymphatics.
 c. Obligate pathogens in the related genus *Chlamydophila* are involved in pneumonia and **psittacosis,** a zoonosis associated with birds.

21.4 Mollicutes and Other Cell-Wall-Deficient Bacteria
A. **Mycoplasmas** naturally lack cell walls and are thus highly pleomorphic; not obligate parasites but require special lipids from host membranes; diseases treated with doxycycline and azithromycin.
 1. *Mycoplasma pneumoniae:* Causes primary atypical pneumonia; pathogen slowly spreads over interior respiratory surfaces, causing fever, chest pain, and sore throat.
 2. *L forms:* Bacteria that normally have cell walls but have transiently lost them through drug therapy. They may be involved in certain chronic diseases.

21.5 Bacteria in Dental Disease
A. The oral cavity contains hundreds of microbial species that participate in interactions between themselves, the human host, and the nutritional role of the teeth and mouth.
 1. **Hard-Tissue (Tooth) Disease Dental caries** is a slow, progressive breakdown of irregular areas of tooth enamel; it begins with colonization of teeth by slime-forming species of *Streptococcus, Actinomyces,* and numerous other oral bacteria. This process forms a layer of thick, adherent material called **plaque.** Acid formed by agents in plaque dissolves the inorganic salts in enamel and may lead to caries and tooth damage.
B. **Disease of Tooth Support Structures** Infection occurs in the **periodontium** (gingiva and surrounding tissues); it starts when plaque forms on root of tooth and is mineralized to a hard concretion called **calculus;** this irritates tender gingiva; inflammatory reaction and swelling create **gingivitis** and pockets between tooth and gingiva invaded by bacteria (spirochetes and anaerobic bacilli); tooth socket may be involved **(periodontitis),** and the tooth may be lost.

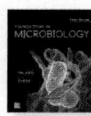

Level I. Knowledge and Comprehension

These questions require a working knowledge of the concepts in the chapter and the ability to recall and understand the information you have studied.

Multiple-Choice Questions

Select the correct answer from the answers provided. For questions with blanks, choose the combination of answers that most accurately completes the statement.

1. *Treponema pallidum* is cultured in/on
 a. blood agar
 b. animal tissues
 c. serum broth
 d. eggs

2. A gumma is
 a. the primary lesion of syphilis
 b. a syphilitic tumor
 c. the result of congenital syphilis
 d. a damaged aorta

3. The treatment of choice for syphilis is
 a. chloramphenicol
 b. antiserum
 c. penicillin
 d. sulfa drugs

4. Which of the treponematoses is/are *not* (an) STD(s)?
 a. yaws
 b. pinta
 c. syphilis
 d. both a and b

5. Lyme disease is caused by _____ and spread by _____ .
 a. *Borrelia recurrentis*, lice
 b. *Borrelia hermsii*, ticks
 c. *Borrelia burgdorferi*, fleas
 d. *Borrelia burgdorferi*, ticks

6. Which of the following conditions may occur in untreated Lyme disease?
 a. arthritis
 b. rash
 c. heart disorder
 d. a and b
 e. all of these

7. Relapsing fever is spread by
 a. lice
 b. ticks
 c. animal urine
 d. a and b

8. The primary habitat of *Vibrio cholerae* is
 a. intestine of humans
 b. intestine of animals
 c. natural waters
 d. exoskeletons of crustaceans

9. The most effective therapy for cholera is
 a. oral doxycycline
 b. oral rehydration therapy
 c. antiserum injection
 d. oral vaccine

10. Rickettsias and chlamydias are similar in being
 a. free of a cell wall
 b. the cause of eye infections
 c. carried by arthropod vectors
 d. obligate intracellular bacteria

11. Which of the following is *not* an arthropod vector of rickettsioses?
 a. mosquito
 b. louse
 c. tick
 d. flea

12. Chlamydiosis caused by *C. trachomatis* attacks which structure(s)?
 a. eye
 b. urethra
 c. fallopian tubes
 d. all of these

13. What stage(s) of *Chlamydia* is/are infectious?
 a. reticulate body
 b. elementary body
 c. vegetative cell
 d. both a and b

14. Psitticosis is a _____ infection associated with _____ .
 a. rickettsial, parrots
 b. chlamydial, mice
 c. chlamydial, birds
 d. rickettsial, flies

15. Mycoplasmas attack the _____ of host cells.
 a. nucleus
 b. cell walls
 c. ribosomes
 d. cell membranes

16. The earliest process at the basis of most dental disease is
 a. acquired pellicle
 b. acid release
 c. enamel destruction
 d. plaque accumulation

17. Dental caries are directly due to
 a. microbial acid etching away tooth structures
 b. buildup of calculus
 c. death of tooth by root infection
 d. the acquired pellicle

18. Necrotizing ulcerative gingivitis is a _____ infection.
 a. contagious
 b. mixed
 c. spirochete
 d. systemic

19. **Single Matching.** Match each disease in the left column with its vector or vectors in the right column.
 _____ leptospirosis
 _____ Lyme disease
 _____ murine typhus
 _____ psitticosis
 _____ relapsing fever
 _____ lymphogranuloma venereum
 _____ cat-scratch disease
 _____ epidemic typhus
 _____ Rocky Mountain spotted fever
 _____ Q fever
 _____ cholera
 _____ anaplasmosis

 a. wild animals
 b. flea
 c. tick
 d. birds
 e. louse
 f. domestic animals
 g. none of these

20. **Single Matching.** Match each disease in the left column with its portal of entry in the right column.
 _____ Q fever
 _____ psitticosis
 _____ dental caries
 _____ NUG
 _____ mycoplasma
 _____ syphilis
 _____ leptospirosis
 _____ lymphogranuloma venereum
 _____ cholera
 _____ Lyme disease
 _____ trachoma
 _____ *Campylobacter* infection
 _____ gastric ulcers

 a. skin
 b. mucous membrane
 c. respiratory tract
 d. urogenital tract
 e. eye
 f. oral cavity
 g. gastrointestinal tract

 Case Study Review

1. The patients in this case were probably treated with
 a. fluoroquinolones
 b. specific immunoglobulin
 c. vaccination
 d. doxycycline

2. What other risk factors could have alerted physicians to the possibility of Q fever in this case?
 a. Recent hospitalization
 b. consumption of raw milk
 c. Working in a veterinarians office
 d. advanced age
 e. b and c

3. What characteristics of *Coxiella* are important factors in its transmission to humans?

 Writing Challenge

For each question, compose a one- or two-paragraph answer that includes the factual information needed to completely address the question. Check Your Progress questions can also be used for writing-challenge exercises.

1. Describe the conditions leading to congenital syphilis and the long-term effects of the disease.

2. a. Trace the route of the infectious agent from a tick bite to infection in relapsing fever.
 b. Explain the events in infection that give rise to relapses.

3. a. In what ways are dental diseases mixed infections?
 b. Discuss the major factors in the development of dental caries and periodontal infections.

4. a. Which diseases in this chapter are zoonoses?
 b. Name them and the major vector involved.

 Concept Mapping

On Connect you can find an Introduction to Concept Mapping that provides guidance for working with concept maps along with concept mapping activities for this chapter.

 Level II. Application, Analysis, Evaluation, and Synthesis

These problems go beyond just restating facts and require higher levels of understanding and an ability to interpret, problem solve, transfer knowledge to new situations, create models, and predict outcomes.

 Critical Thinking

Critical thinking is the ability to reason and solve problems using facts and concepts. These questions can be approached from a number of angles, and in most cases, they do not have a single correct answer.

1. a. Why does syphilis have such profound effects on the human body?
 b. Why is long-term immunity to syphilis so difficult to achieve?

2. a. In view of the fact that cholera causes the secretion of electrolytes into the intestine, explain what causes the loss of water.
 b. Explain why ORT is so effective in restoring victims of cholera.

3. What would be the best type of vaccine for cholera?

4. a. Explain the general relationships of the vector, the reservoir, and the agent of infection.
 b. Can you think of an explanation for Lyme disease having such a low incidence in the southern United States? (Hint: In this region, the larval stages of the tick feed on lizards, not on mice.)

5. Humans are accidental hosts in many vector-borne diseases. What does this indicate about the relationship between the vector and the microbial agent?

6. Name four bacterial diseases for which the dark-field microscope is an effective diagnostic tool.

7. Describe the conditions that lead to disease by L forms.

8. Which two infectious agents covered in this chapter would be the most resistant to the environment and why are they so resistant?

9. a. In what way is the oral cavity an ecological system?
 b. What can cause an imbalance in the oral ecology?
 c. Explain how it is possible for the gingiva and tooth surface to provide an anaerobic habitat.
 d. What are some logical ways to prevent dental disease besides removing plaque?

10. Have Koch's postulates been carried out for Lyme disease? Explain.

11. A journalist returning from a trip experienced severe fever, vomiting, chills, and muscle aches, followed by symptoms of meningitis and kidney failure. Early tests were negative for septicemia; throat cultures were negative; and penicillin was an effective treatment. Doctors believed the patient's work in the jungles of South America was a possible clue to his disease. What do you think might have been the cause?

12. A man went for a hike in the mountains of New York State and later developed fever and a rash. What two totally different diseases might he have contracted, and what could have been the circumstances of infection?

Visual Challenge

1. Identification of a unique skin rash can often be the first step in diagnosing a disease. What infectious agents are indicated by the rashes below?

Source: Susan Lindsley/CDC

Source: James Gathany/CDC

STUDY TIP: Quickly create a customized practice quiz for this chapter by going to your Connect SmartBook "My Reports" section. Don't have SmartBook and want to learn more? Go to **whysmartbook.com**.

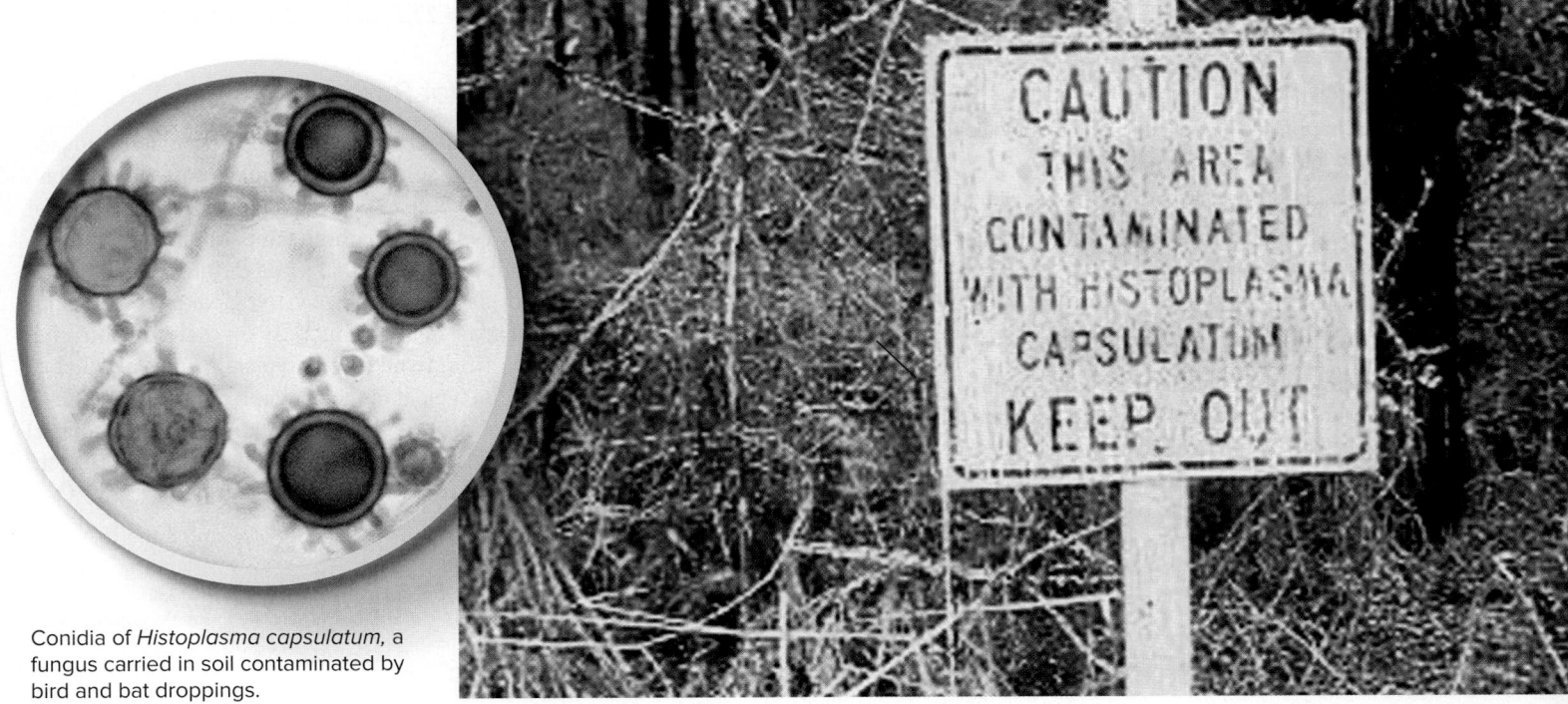

Conidia of *Histoplasma capsulatum,* a fungus carried in soil contaminated by bird and bat droppings.
Source: CDC

© Tom Volk

CASE STUDY Part 1 Sun, Fun, and Fungi

Histoplasmosis Outbreak Among Day-Camp Attendees—Nebraska, June 2012

Mention microbial infections and most people think of hospitals, not summer camp. But being outdoors in the sunshine and fresh air is not without risk. In June 2012 a camp counselor in Omaha, Nebraska, was diagnosed with **histoplasmosis,** a fungal infection with symptoms similar to pneumonia: fever, headache, and respiratory distress. Although most people suffering from histoplasmosis will recover without medical treatment, the very young, the very old, and people with chronic lung conditions are at greater risk for developing severe disease, which can on rare occasions prove fatal. In areas of the eastern and central United States, where *Histoplasma capsulatum* is **endemic,** dirt contaminated with bird and bat **guano** often contains **spores** of the fungus. Disturbing such contaminated dirt is a known risk factor for developing the disease.

> "The counselors spent the week raking leaves, cleaning picnic tables, digging fire pits, and moving firewood."

Because the sick camp counselor was one of 32 that worked at the camp, and the camp served 797 children, the Douglas County Department of Health (DCDH) and the Nebraska Department of Health and Human Services (NDHHS) began an investigation to determine the extent and source of the outbreak, so as to prevent further infections. Initial testing revealed 17 counselors (53%) whose blood serum or urine contained **antigens** from *Histoplasma capsulatum*; these were deemed confirmed cases of the disease. Of 153 responses to a questionnaire sent to camp attendees, 17 campers (11%) had confirmed or suspected histoplasmosis (suspected cases were those in which illness was present but laboratory results were inconclusive or unavailable).

- *What activities would be associated with an increased risk of* Histoplasma capsulatum *infection?*
- *What other fungal pathogens are found in the United States?*

To continue the Case Study, go to Case Study Part 2 at the end of the chapter.

22.1 Fungi as Infectious Agents

 Expected Learning Outcomes

1. Distinguish between the characteristics indicative of true versus opportunistic fungal pathogens.

2. Relate opportunistic fungal infections to conditions that predispose patients to infection.

3. Explain the geographical limits associated with fungal pathogens.

4. List the most important antifungal defenses found in humans.

5. Understand why direct examination of patient samples is often part of the diagnostic process.

Molds and yeasts are so widely distributed in air, dust, and fomites, and even in our normal microbiota, that humans are incessantly exposed to them. The fact that the planet's surface is totally dusted with spores led noted California mycologist W. B. Cooke to christen it "our moldy earth." Fortunately, because of the relative resistance of humans and the comparatively nonpathogenic nature of fungi, most exposures do not lead to overt infection. Of the estimated 100,000 fungal species, only about 300 have been linked to disease in animals, though among plants, fungi are the most common and destructive of

all pathogens. Human mycotic disease, or **mycosis,*** is associated with true fungal pathogens that exhibit some degree of virulence or with opportunistic pathogens that take advantage of defective host resistance (**Systems Profile 22.1, tables 22.1** and **22.2**).

Primary/True Fungal Pathogens

A **true** or **primary** fungal **pathogen** is a species that can invade and grow in a healthy, non-compromised animal host. This behavior is contrary to the metabolism and adaptation of fungi, most of which are inhibited by the relatively high temperature and low oxygen tensions of a warm-blooded animal's body. But a small number of fungi have the morphological and physiological adaptations required to survive and grow in this habitat. By far their most striking adaptation is a switch from hyphal cells typical of the mycelial or mold phase to yeast cells typical of the parasitic phase (figure 22.1). This biphasic characteristic of the life cycle is termed **thermal dimorphism*** because it is initiated by changing temperature. In general, these organisms grow as molds at 30°C and as yeasts at 37°C.

* *mycosis* (my-koh′-sis) pl. mycoses; Gr. *mykos,* fungi, and *osis,* a disease process.
* *dimorphism* (dy-mor′-fizm) Gr. *dimorphos,* having two forms; the property of existing in two distinct cell types.

TABLE 22.1	Representative Fungal Pathogens, Degree of Pathogenicity, and Habitat	
Microbe	**Disease/Infection***	**Primary Habitat and Distribution**
I. Primary True Pathogens		
Histoplasma capsulatum	Histoplasmosis	Soils high in bird guano; Ohio and Mississippi valleys of United States; Mexico; Central and South America; Africa
Blastomyces dermatitidis	Blastomycosis	Presumably soils, but isolation has been difficult; southern Canada; Midwest, Southeast, Appalachia in United States; along drainage of major rivers
Coccidioides immitis	Coccidioidomycosis	Highly restricted to alkaline desert soils in southwestern United States (California, Arizona, Texas, and New Mexico)
Paracoccidioides brasiliensis	Paracoccidioidomycosis	Soils of rain forests in South America (Brazil, Colombia, Venezuela)
II. Pathogens with Intermediate Virulence		
Sporothrix schenckii	Sporotrichosis	In soil and decaying plant matter; widely distributed
Genera of dermatophytes (*Microsporum, Trichophyton, Epidermophyton*)	Dermatophytosis (various ringworms or tineas)	Human skin, animal hair, soil throughout the world
III. Secondary Opportunistic Pathogens		
Candida albicans	Candidiasis	Normal flora of human mouth, throat, intestine, vagina; also normal in other mammals, birds; ubiquitous
Aspergillus spp.	Aspergillosis	Soil, decaying vegetation, grains; common airborne contaminants; extremely pervasive in environment
Cryptococcus neoformans	Cryptococcosis	Pigeon roosts and other nesting sites (buildings, barns, trees); worldwide distribution
Pneumocystis jirovecii	*Pneumocystis* pneumonia (PCP)	Upper respiratory tract of humans, animals
Genera in Mucorales (*Rhizopus, Absidia, Mucor*)	Mucormycosis	Soil, dust; very widespread in human habitation

*Specific mycotic infections are usually named by adding *-mycosis, -iasis,* or *-osis* to the generic name of the pathogen.

Systems Profile 22.1 — Fungal Pathogens

Organism	Skin/Skeletal	Nervous/ Muscle	Cardiovascular/ Lymphatic/ Systemic	Gastrointestinal	Respiratory	Urogenital
Histoplasma capsulatum					Histoplasmosis (Ohio Valley fever)	
Coccidioides immitis	Coccidioidomycosis	Meningitis	Fungemia		Coccidioidomycosis (valley fever)	
Blastomyces dermatitidis	Cutaneous blastomycosis				Pulmonary blastomycosis (North American blastomycosis)	
Paracoccidioides brasiliensis	Paracoccidioidomycosis				Paracoccidioidomycosis (South American blastomycosis)	
Sporothrix schenckii	Cutaneous sporotrichosis				Pulmonary sporotrichosis	
Madurella spp.	Mycetoma					
Dermatophytes (*Trichophyton* spp., *Microsporum* spp., *Epidermophyton* spp.)	Ringworm					
Candida albicans	Thrush			Esophageal and anal infections		Vulvovaginal candidiasis
Cryptococcus neoformans		Meningitis			Pulmonary cryptococcosis	
Pneumocystis jirovecii					*Pneumocystis* pneumonia	
Aspergillus spp.	Ocular infections				Aspergillosis	

TABLE 22.2 — Comparison of True and Opportunistic Fungal Infections

Characteristic of Fungus/Disease	True Pathogenic Infections	Opportunistic Infections
Degree of Virulence	Well-developed	Limited
Condition of Host	Resistance high or low	Resistance low
Primary Portal of Entry	Respiratory	Respiratory mucocutaneous
Nature of Infection	Usually primary pulmonary and systemic; often asymptomatic	Varies from superficial skin to pulmonary and systemic; usually symptomatic
Nature of Immunity	Well-developed, specific	Weak, short-lived
Infecting Form	Primarily conidial	Conidial or mycelial
Thermal Dimorphism	Highly characteristic	Absent
Habitat of Fungus	Soil	Varies from soil to flora of humans and animals
Geographic Location	Restricted to endemic regions	Distributed worldwide

Emerging Fungal Pathogens

An opportunistic fungal pathogen is different from a true pathogen in several ways (table 22.2). An opportunist has weak to nonexistent invasiveness or virulence, and the host's defenses must be impaired to some degree for the microbe to gain a foothold. Although some species have both mycelial and yeast stages in their life cycles, they usually do not show thermal dimorphism. Opportunistic mycoses vary in their manifestations from superficial and benign colonizations to spreading systemic disease that is rapidly fatal.

Opportunistic mycotic infections are an emerging medical concern. By one estimate they account for around 10% of all healthcare associated infections. This presents a dilemma for hospitals—having to maintain patients with conditions that inhibit their ability to fight off infections, yet also keeping them from getting infections. In the past these types of patients would have died from their cancer or diabetes, but more often now they are being kept alive by medical technology. Like any other environment, hospitals and clinics are a common habitat for fungi, even when stringent disinfection procedures are followed. Environmental sampling for molds and yeasts shows that they are abundant in tap and bottled water, sinks, showers, and the air, as well as on the walls and floors of patients' rooms. Some opportunists also come from the normal microbiota. Such being the case, exposure is constant and hard to

(1) When fungal spores from the environment gain entrance to a warm-blooded animal, they germinate into yeasts and remain in this phase in the host.

(2) Yeast cells leaving the animal host return to the environment and revert to the sporulating hyphal state. These conversions can be demonstrated on artificial media in the laboratory.

Spores

Hyphae

Hyphal colonies

Natural Habitat
Saprobic (free-living)
Mycelial phase
Temperature (<30°C)
Reproduction through sporulation

(1) Increased temperature Reduced O₂ Suboptimal nutrients

Spore

Bud

Yeast

Animal Habitat
Parasitic
Yeast phase
Temperature (35°C–40°C)
Reproduction through budding or endospores

Yeast colonies

(2) Decreased temperature Optimal O₂ content Improved nutrients

Figure 22.1 **The general changes associated with thermal dimorphism.** Dimorphism is a phenomenon primarily seen in true pathogens and is triggered by growth temperature. Left is the free-living hyphal phase and right is the parasitic yeast phase. Insets reveal the cultural appearance of these phases. (left): *Source:* James Gathany/CDC; (right): © Kathy Park Talaro

TABLE 22.3	Common Opportunistic Fungi and Conditions That Predispose Patients to Them
Pathogen	**Associated with**
Candida	Antibiotic therapy, catheters, diabetes, corticosteroids,* immunosuppression**
Aspergillus	Leukemia, corticosteroids, tuberculosis, immunosuppression, IV drug abuse
Cryptococcus	Diabetes, tuberculosis, cancer, corticosteroids, immunosuppression
Zygomycota species	Diabetes, cancer, corticosteroids, IV therapy, third-degree burns

*Anti-inflammatory drugs are often given for chronic lung diseases and to transplant patients to prevent rejection.

**Includes AIDS and genetic conditions that compromise host defenses.

control. **Table 22.3** presents a list of the most common fungal opportunists and medical conditions that may predispose patients to these infections.

Some fungal pathogens exist in a category between true pathogens and opportunists. These species are not inherently invasive but can grow when inoculated into the skin wounds or abrasions of healthy people. Examples are *Sporothrix,* the agent of a subcutaneous infection, and the **dermatophytes,*** which cause ringworm and

athlete's foot. Adhering to host cells and growing readily at body temperature are factors that appear to enhance their pathogenicity.

Epidemiology of the Mycoses

Most fungal pathogens do not require a host to complete their life cycles, and the infections they cause are not communicable. Notable exceptions are some dermatophyte and *Candida* species that naturally inhabit the human body and are transmissible. For the remainder of the pathogens, exposure and infection occur when a human happens upon fungal spores in the environment (usually air, dust, or soil). True fungal pathogens, unlike opportunists, are distributed in a predictable pattern that coincides with the pathogen's adaptation to the specific climate, soil, or other factors of a relatively restricted geographic region **(figure 22.2).**

The prevalence of fungal infections has traditionally been difficult to measure because few of the reportable infections monitored by the CDC are caused by fungi. Dermatophytoses are probably the most prevalent. It is thought that at least 90% of all humans will acquire ringworm or athlete's foot at least once during their lifetimes. Estimates provided by routine skin testing indicate that millions of people have experienced true mycoses, though most cases probably go undiagnosed or misdiagnosed.

Epidemics of mycotic infections can occur after mass exposure to a common source (see Case Study). Coccidioidomycosis is a

* *dermatophyte* (der-mah'-toh-fyte″) Gr. *dermos,* skin, and *phyte,* plant. Fungi were once included in the plant kingdom.

Figure 22.2 Distribution of the four true fungal pathogens. The exposure of humans to a given pathogen is greatly influenced by where they live or travel. *Histoplasma* and *Blastomyces* occupy somewhat the same areas of North America, and their incidence is correlated with the course of the Mississippi River and its tributaries. A unique form of *Histoplasma* occurs in Africa and is the only true fungal pathogen on a continent other than the Americas.

particular hazard of construction workers and people living in the paths of windstorms. Transmissible fungal infections such as dermatophytoses are readily passed by means of shared personal articles, public facilities, swimming pools, gymnasiums, and contact with infected animals. Candidiasis can be communicated during sexual contact and from mother to child at birth.

Pathogenesis of the Fungi

Mycoses involve complex interactions among the portal of entry, the nature of the infectious dose, the virulence of the fungus, and host resistance. Fungi enter the body mainly via respiratory, mucous, and cutaneous routes. In general, the agents of primary mycoses have a respiratory portal (spores inhaled from the air); subcutaneous agents enter through inoculated skin (trauma); and cutaneous and superficial agents enter through contamination of the skin surface. Spores, hyphal elements, and yeasts can all be infectious, but spores are most often involved because of their durability and abundance.

Thermal dimorphism greatly increases virulence by enabling fungi to tolerate the relatively high temperatures and low oxygen tensions of the body. Fungi in the yeast form can be more invasive than hyphal forms because they grow more rapidly and spread through tissues and blood, whereas those producing hyphae tend to localize along the course of blood vessels and lymphatics.

Specific factors contributing to fungal virulence are the subject of much research. A number of studies have isolated cell wall antigens in pathogens such as *Blastomyces* that impart virulence by enhancing adhesion to host cells and suppressing phagocytosis and inflammation. Fungi may also produce capsules, hydrolytic enzymes, toxins, inflammatory stimulants, and allergens, all of which generate strong host responses.

The human body is extremely resistant to establishment of fungi. Among its numerous antifungal defenses are the normal integrity of the skin, mucous membranes, and respiratory cilia, but the most important defenses are cell-mediated immunity, phagocytosis, and the inflammatory reaction. Long-term protective immunity can develop for some of the true pathogens, but for the rest, reinfection is a distinct possibility.

Diagnosis of Mycotic Infections

Satisfactory diagnosis of fungal infections depends mainly upon isolating and identifying the pathogen in the laboratory. Accurate and speedy diagnosis is especially critical to the immunocompromised patient, who must have prompt antifungal chemotherapy. For example, a patient with systemic *Candida* infection can die if the infection is not detected and treated within 5 to 7 days. Because therapy can vary among the pathogenic fungi, identification to the species level is often necessary.

A suitable specimen can be obtained from sputum, skin scrapings, skin biopsies, cerebrospinal fluid, blood, tissue exudates, urine, or vaginal samples, as determined by the patient's symptoms. Routine laboratory procedures include isolation, microscopic and macroscopic examination, histological stains, serology, animal inoculation, and genetic testing **(figure 22.3).**

Because culture of the sample can require several days, the immediate direct examination of fresh samples is recommended. Wet mounts can be prepared by mixing a small portion of the sample on a slide with saline, water, or potassium hydroxide to clear the specimen of background debris. The relatively large size and unique appearance of many fungi help them stand out. Large, spherical

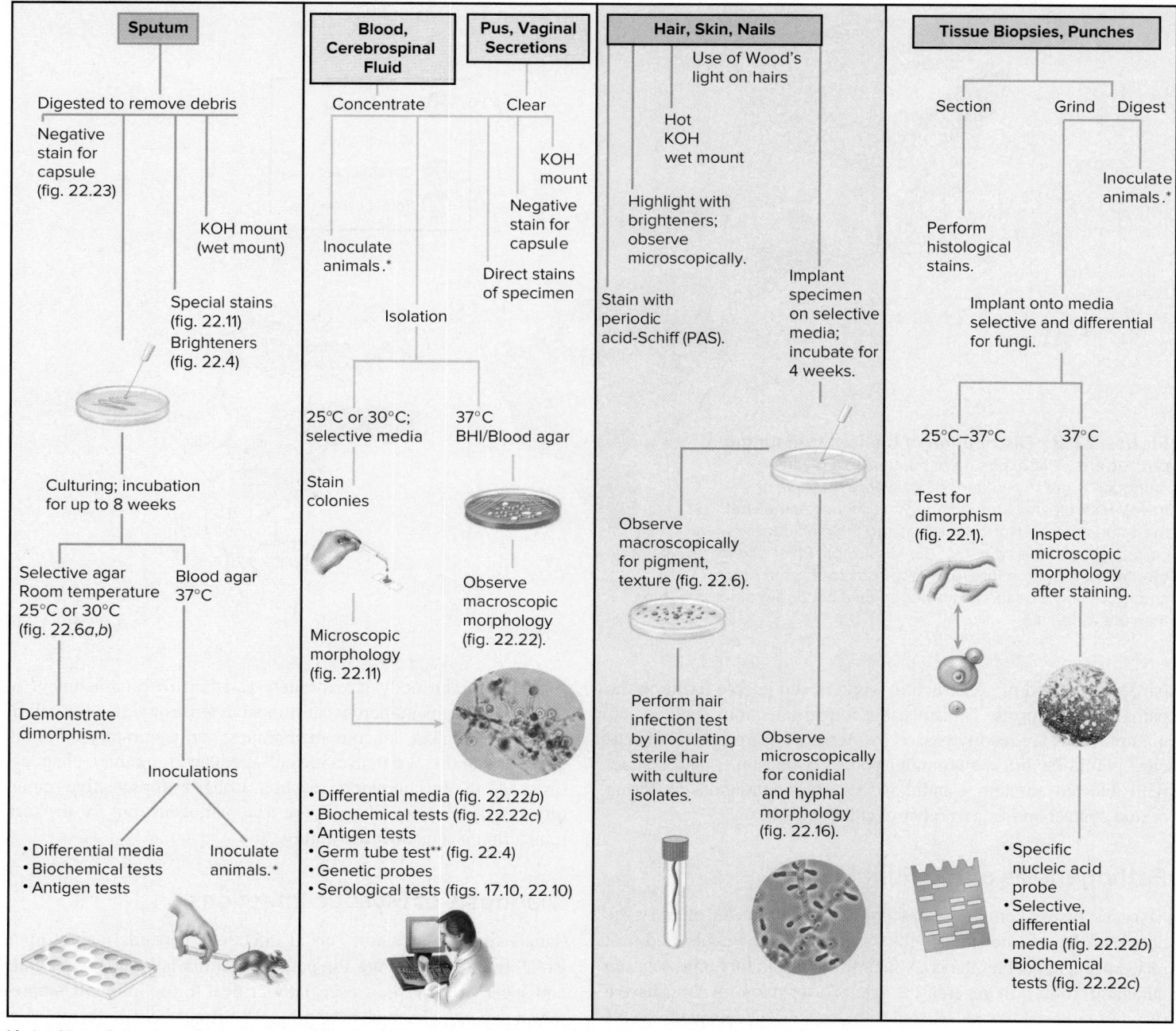

*Animal inoculation is performed only to help diagnose systemic mycoses when other methods are unavailable or indeterminant.
**Some yeasts, when incubated in serum for 2–4 hours, sprout tiny hyphal tubes called germ tubes.

Figure 22.3 **Methods of processing specimens and identifying fungal isolates.** References indicate figures showing these techniques. (left, middle): *Source:* CDC; (right): *Source:* Dr. Libero Ajello/CDC

budding cells are evidence of yeasts, whereas elongate, branching strands suggest hyphae. A variety of stains or brighteners provide striking images of fungi from specimens and cultures (**figure 22.4**).

The fungal pathogen can be isolated with various solid media such as Sabouraud's dextrose agar, Mycosel agar, inhibitory mold medium, and brain-heart infusion agar. These media can be enriched with blood or made selective with chloramphenicol and gentamicin to inhibit bacteria or cyclohexamide to slow the growth of undesirable fungal contaminants. Cultures are incubated at room temperature, 30°C, and 37°C to observe dimorphic conversion. This may require several days to weeks. Gross colonial morphology, such as

the color and texture of the colony's surface and underside, can be very distinctive. Initial identification of the pathogen can be followed by additional confirmatory tests.

Antigen tests that capture fungus-specific surface molecules can be useful in many cases. Several PCR-based detection systems have been developed for fungal pathogens that will continue to speed diagnosis. Other examples of tests are given during discussions of specific diseases later in the chapter.

In vitro tests to detect antifungal antibodies in serum (see figure 22.10) can be a useful diagnostic tool in some infections. In vivo skin testing for delayed hypersensitivity to fungal antigens is

(a)

(b)

Figure 22.4 **Special tissue stains called brighteners can amplify the presence of fungal elements in specimens.** They bind tightly to the carbohydrates of the fungal surface. **(a)** A fluorescent view of *Aspergillus fumigatus* taken from a thoracic specimen (500×). **(b)** *Candida albicans* highlighted by brightener. (a, b): *Source:* CDC

mostly used to trace epidemiological patterns. Its limitations are that it may not verify ongoing infection and considerable cross-reactivity exists among *Histoplasma, Blastomyces,* and *Coccidioides.* A negative test in a healthy person, however, can rule out infection by these fungi.

With increasing travel and globalization, it is not unusual for visitors to endemic regions to acquire infections that emerge after returning home. Here, knowledge of patients' history and versatility of the medical personnel become vital for treatment and survival.

Control of Mycotic Infections

Treatment for fungal infections is based primarily on antifungal drugs, covered in chapter 12 and on a disease-by-disease basis in this chapter. A summary is provided in **table 22.4.** Immunization is not usually effective against fungal infections, but work is proceeding on vaccines for coccidioidomycosis and histoplasmosis. Prevention measures are limited to masks and protective clothing to reduce contact with spores and, in some cases, surgical removal of damaged tissues.

TABLE 22.4	Characteristics of Antifungal Drugs
Drug Target/Drug	**Comments**
Plasma Membrane	
Polyenes	Disrupt the plasma membrane and cause leakage of the cytoplasm
Amphotericin B	Amphotericin B is highly toxic but most effective for treating life-threatening infections.
Nystatin	Nystatin is too toxic for systemic use, but it can be used topically.
Azoles	Interfere with ergosterol synthesis, leading to defective cell membranes; used to treat a variety of systemic and localized fungal infections
Imidazoles	Examples are ketoconazole (Nizoral), miconazole, and clotrimazole.
Triazoles	Examples are fluconazole and itraconazole (Sporanox).
Allylamines	Inhibit an enzyme in ergosterol synthesis; administered topically to treat dermatophyte infections
Naftifine, terbinafine (Lamisil)	Terbinafine can be taken orally.
Cell Wall	
Echinocandins	Block the synthesis of a polysaccharide component of the cell wall in many pathogenic fungi
Caspofungin, anidulafungin micafungin	
Cell Division	
Griseofulvin	Used to treat skin and nail infections; concentrates in the dead keratinized layers of the skin; active only against fungi that invade keratinized cells
Nucleic Acid Synthesis	
Flucytosine	Used for systemic yeast infections; inhibits an enzyme required for nucleic acid synthesis; not effective against most molds

 Check Your Progress SECTION 22.1

1. Outline the differences between true and opportunistic fungal pathogens with regard to the habitat of the fungus, its geographic location, nature of infection, and immunity developed.
2. What medical conditions commonly predispose patients to mycotic infection?
3. Name the three genera containing opportunistic fungal pathogens.
4. Name the locations where each of the four true pathogenic fungal genera would be found in the United States.
5. Why are humans typically resistant to fungal infection?
6. Why is direct examination of patient samples particularly useful when diagnosing a mycotic infection?
7. Give an overview of the major steps in fungal identification.

22.2 Organization of Fungal Diseases

 Expected Learning Outcomes

6. Summarize the aspects of systemic mycotic infection common to all fungal pathogens.

7. Recall the habitat, geographic distribution, and diseases associated with *Histoplasma capsulatum, Coccidioides immitis, Blastomyces dermatitidis,* and *Paracoccidioides brasiliensis.*

8. Summarize the pathology of each of the four fungal pathogens.

9. Understand the means by which infections with *Histoplasma capsulatum, Coccidioides immitis, Blastomyces dermatitidis,* and *Paracoccidioides brasiliensis* are diagnosed in the laboratory.

Fungal infections can be presented in several schemes, depending on the purpose. Traditional methods are based on taxonomic group, location of infection, and type of pathogen. Chapter 5 presents a taxonomic breakdown of the fungi. Most pathogens belong to the Phylum Ascomycota; a few important species are scattered through Phylums Basidiomycota (*Cryptococcus*) and Zygomycota (*Mucor*). This chapter presents the mycoses according to the type and level of infection they cause and their degree of pathogenicity. The order and location of coverage are:

1. true pathogens: **systemic, subcutaneous, cutaneous,** and **superficial mycoses (figure 22.5);** and
2. **opportunistic mycoses** (starting in section 22.6).

Systemic Infections by True Pathogens

The infections of primary fungal pathogens can all be described by the same general model:

- They are restricted to certain endemic regions of the world.
- Infection occurs when soil or other matter containing the fungal conidia is disturbed and the spores are inhaled into the lower respiratory tract.
- The spores germinate in the lungs into yeasts or yeastlike cells and produce an asymptomatic or mild **primary pulmonary infection (PPI)** that parallels tuberculosis.
- In a small number of hosts, this infection becomes systemic and creates severe, chronic lesions.
- In a few cases spores are inoculated into the skin, where they form localized granulomatous lesions.
- All diseases result in immunity that can be long term and that manifests clinically as an allergic reaction to fungal antigens.

Histoplasmosis: Ohio Valley Fever

The most common true pathogen is ***Histoplasma capsulatum,**** the cause of histoplasmosis. This disease has probably afflicted humans since antiquity, though it was not described until 1905 by Dr. Samuel Darling. Through the years, it has been known by various synonyms: Darling's disease, Ohio Valley fever, and reticuloendotheliosis.

* *Histoplasma capsulatum* (his″-toh-plaz′-mah kap″-soo-lay′-tum) Gr. *hist,* tissue, and *plasm,* shape; L. *capsula,* small box.

The skin and its attendant structures provide many potential sites for invasion, including the scalp, smooth skin, hair, and mucous membranes. Differing depths of involvement are: superficial, consisting of extremely shallow epidermal colonizations; cutaneous, involving the stratum corneum and occasionally the upper dermis; and subcutaneous, occurring after a puncture wound has introduced the fungus deeper into the subcutaneous tissues. In systemic (deep) mycoses, the fungus disseminates from the lungs or other sites into the circulation. Fungemia leads to infection of the brain, kidneys, and other organs.

Figure 22.5 Levels of invasion by fungal pathogens. Some species can invade more than one level.

Certain aspects of its current distribution and epidemiology suggest that it has been an important disease as long as humans have practiced agriculture.

Biology and Epidemiology of *Histoplasma capsulatum*
Histoplasma capsulatum is typically dimorphic. Growth on media below 35°C is characterized by a white or brown, hairlike mycelium; and growth at 37°C produces a creamy white, textured colony **(figure 22.6).**

(a) (b)

Figure 22.6 Macroscopic appearance of *Histoplasma capsulatum.* **(a)** Grown at 25°C, *H. capsulatum* produces a white mycelium. **(b)** When incubated at 37°C, the fungus grows as a creamy yeast colony. (a, b): *Source:* CDC

Pathogen Profile #1 *Histoplasma capsulatum*

Histoplasma "fish-eye" yeasts infecting a macrophage
Source: Dr. J.T. McClellan, Lexington, KY/CDC

Microscopic Morphology Typically dimorphic, *H. capsulatum* produces a white or brown hairlike mycelium when grown below 35°C, whereas growth above 37°C results in creamy, white textured yeast colonies.

Identified by Tiny intracellular yeasts in macrophages or free yeasts in samples of sputum or cerebrospinal fluid. Because infection with *H. capsulatum* is relatively common, especially in endemic areas, simple detection of antibodies is not indicative of disease. Complement fixation and immunodiffusion can detect a rising antibody titer, consistent with ongoing infection.

Habitat *H. capsulatum* is endemically distributed on all continents except Australia, with its highest rate of incidence occurring in the eastern and central regions of the United States (the Ohio Valley). It is found in moist soils high in nitrogen content and is associated with areas rich in bird or bat guano.

Virulence Factors The principal virulence factor of all primary pathogenic fungi is the ability to grow dimorphically, being able to flourish both as a saprobic mycelium and as a pathogenic yeast.

Primary Infections/Disease Histoplasmosis (Ohio Valley fever) presents in many ways. Initial infection is usually asymptomatic and may lead to mild symptoms including aches and coughing, while more severe cases may progress to encompass fever, night sweats, and weight loss. Patients with defective cell-mediated immunity (AIDS patients, among others) can develop severe multiorgan disease, which may progress to lethality.

Control and Treatment Because *H. capsulatum* is so widespread in endemic areas, absolute avoidance is impractical. In these cases, respiratory protection will reduce the number of inhaled fungal spores when performing high-risk activities (exploring caves, cleaning bat and bird roosts, and so on). Treatment is not necessary in most cases, but amphotericin B and itraconazole are the antifungals of choice.

Histoplasma capsulatum is endemically distributed on all continents except Australia. Its highest rates of incidence occur in the eastern and central regions of the United States (the Ohio Valley). This fungus appears to grow most abundantly in moist soils high in nitrogen content, especially those supplemented by bird and bat **guano.***

A useful tool for determining the distribution of *H. capsulatum* is to inject a fungal extract called *histoplasmin* into the skin and monitor for allergic reactions. Application of this test has verified the extremely widespread distribution of the fungus. In high-prevalence areas such as southern Ohio, Illinois, Missouri, Kentucky, Tennessee, Michigan, Georgia, and Arkansas, 80% to 90% of the population show signs of prior infection by the age of 20. Histoplasmosis incidence in the United States is estimated at about 500,000 cases per year, with several thousand of them requiring hospitalization and a small number resulting in death.

The spores of the fungus are probably dispersed by the wind and, to a lesser extent, animals. The most striking outbreaks of histoplasmosis occur when concentrations of spores have been dislodged by humans working in parks, bird roosting areas, and old buildings. People of both sexes and all ages incur infection, but adult males experience the majority of cases, probably due to higher levels of occupational exposure.

Infection and Pathogenesis of *Histoplasma* Histoplasmosis presents a formidable array of manifestations. It can be benign or severe, acute or chronic, and it can show pulmonary, systemic, or cutaneous lesions. Inhaling a small dose of microconidia (only 5 spores are required) into the deep recesses of the lung establishes a primary pulmonary infection that is usually asymptomatic. Its primary location of growth is in the cytoplasm of phagocytes such as

macrophages. Within these cells, it flourishes and is carried to other sites **(figure 22.7)**. Some people experience mild symptoms such as aches, pains, and coughing, but a few develop more severe symptoms, including fever, night sweats, and weight loss. Primary cutaneous histoplasmosis, in which the agent enters via the skin, is very rare.

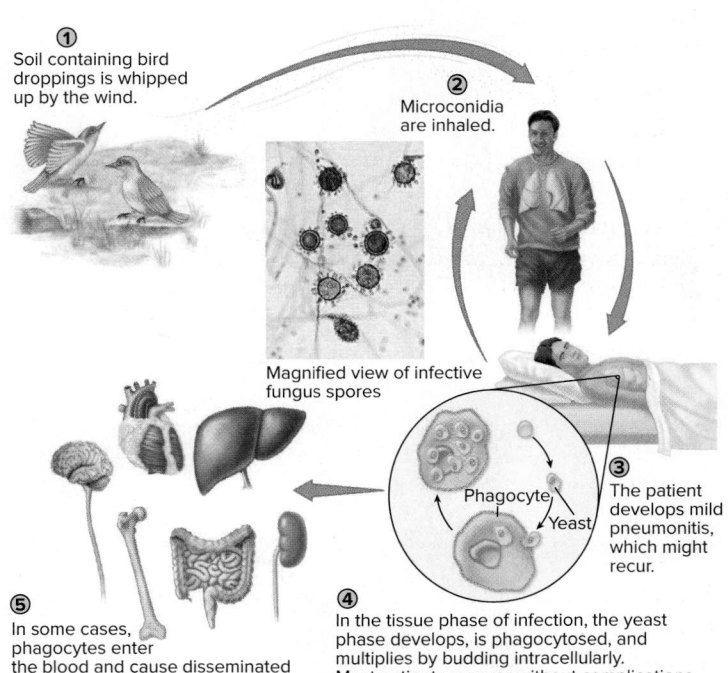

① Soil containing bird droppings is whipped up by the wind.

② Microconidia are inhaled.

Magnified view of infective fungus spores

③ The patient develops mild pneumonitis, which might recur.

Phagocyte / Yeast

④ In the tissue phase of infection, the yeast phase develops, is phagocytosed, and multiplies by budding intracellularly. Most patients recover without complications.

⑤ In some cases, phagocytes enter the blood and cause disseminated disease in a number of organs.

Process Figure 22.7 Events in *Histoplasma* infection and histoplasmosis. Infection begins with inhaled spores (microconidia) that give rise to the more invasive yeast phase. *Source:* CDC

* *guano* (gwan'-oh) Sp. *huanu,* dung. An accumulation of animal manure.

The most serious systemic forms of histoplasmosis occur in patients with defective cell-mediated immunity such as AIDS patients. In children, this can lead to liver and spleen enlargement, anemia, circulatory collapse, and death. Adults with systemic disease can acquire lesions in the brain, intestine, adrenal gland, heart, liver, spleen, lymph nodes, bone marrow, and skin. Persistent colonization of patients with emphysema and bronchitis causes *chronic pulmonary histoplasmosis,* a complication that has signs and symptoms similar to those of tuberculosis.

Diagnosis and Control of Histoplasmosis Discovering *Histoplasma* in clinical specimens is a substantial diagnostic indicator. Usually it appears as spherical, "fish-eye" yeasts intracellularly in macrophages and occasionally as free yeasts in samples of sputum and cerebrospinal fluid. Isolating the agent and demonstrating dimorphism help to confirm infection, but this step can require up to 12 weeks. Complement fixation and immunodiffusion serological tests can support a diagnosis by showing a rising antibody titer. Because a positive histoplasmin test does not indicate a new infection, it is not useful in diagnosis.

Undetected or mild cases of histoplasmosis resolve without medical management, but chronic or disseminated disease calls for systemic chemotherapy. The principal drug, amphotericin B, which disrupts the fungus membranes, is administered in daily intravenous doses for a few days to a few weeks. Under some circumstances, itraconazole is the drug of choice. Surgery to remove affected masses in the lungs or other organs can also be effective.

Coccidioidomycosis: Valley Fever

Coccidioides immitis* is the etiologic agent of coccidioidomycosis. Although the fungus has probably lived in soil for millions of years, human encounters with it are relatively recent and coincide with the encroachment of humans into its habitat. This unique and fascinating fungus demonstrates the greatest virulence of all mycotic pathogens.

Biology and Epidemiology of Coccidioides The morphology of *C. immitis* is very distinctive. At 25°C, it forms a moist, white to brown colony with abundant, branching, septate hyphae. These hyphae fragment into thick-walled, blocklike **arthroconidia** (arthrospores) at maturity (inset photo, **figure 22.8**). On special media incubated at 37°C to 40°C, an arthrospore germinates into the parasitic phase, a spherical cell called a **spherule.** This structure swells into a giant sporangium that cleaves internally to form numerous endospores that look like bacterial endospores but lack their resistance traits **(figure 22.8, step 3).**

* *Coccidioides immitis* (kok-sid″-ee-oy′-deez ih′-mih-tis) From *coccidia,* a sporozoan, and L. *immitis,* fierce. The original discoverers thought the microbe looked like a protozoan.

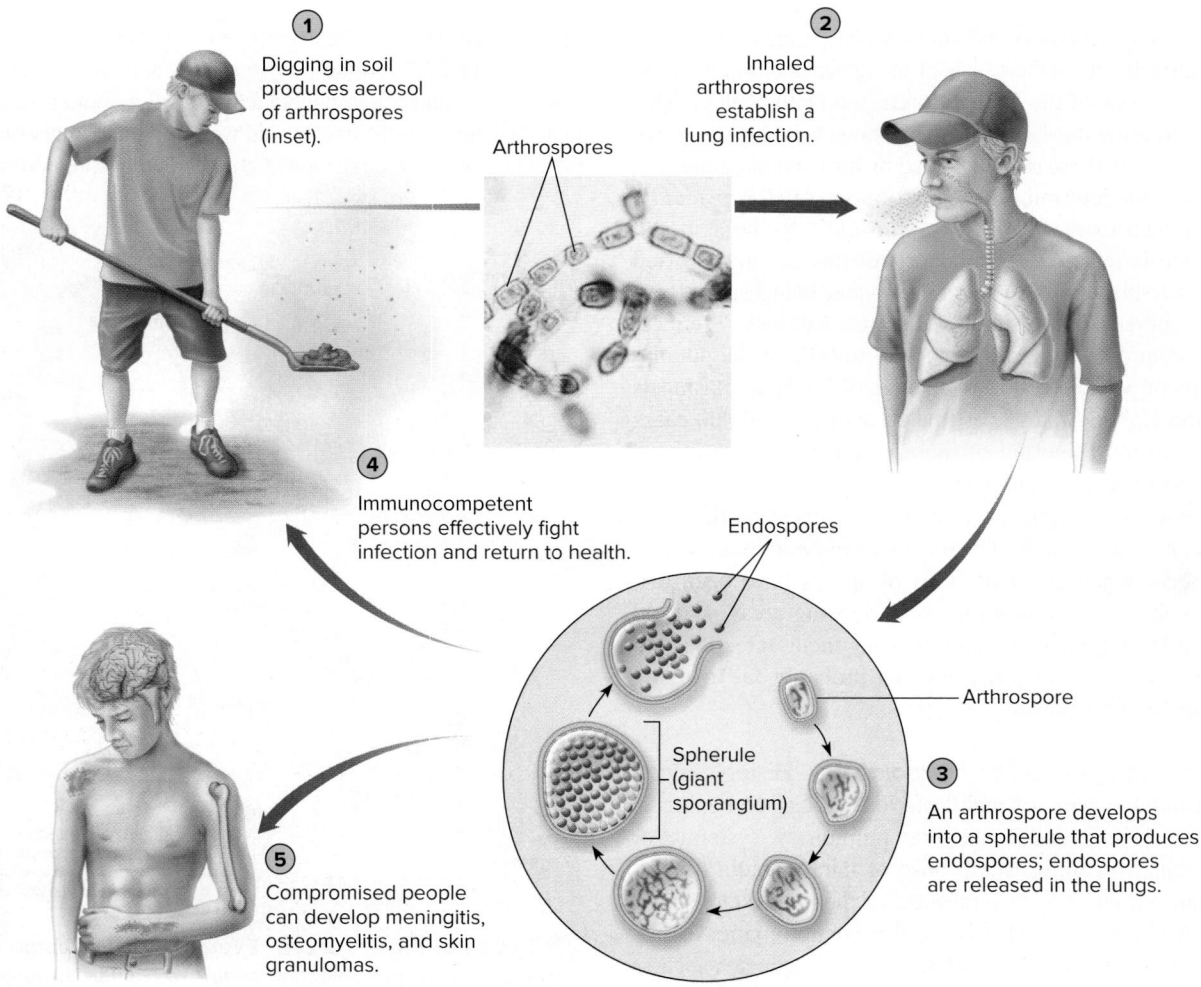

① Digging in soil produces aerosol of arthrospores (inset).

Arthrospores

② Inhaled arthrospores establish a lung infection.

④ Immunocompetent persons effectively fight infection and return to health.

Endospores

Arthrospore

Spherule (giant sporangium)

③ An arthrospore develops into a spherule that produces endospores; endospores are released in the lungs.

⑤ Compromised people can develop meningitis, osteomyelitis, and skin granulomas.

Process Figure 22.8 **Events in *Coccidioides* infection and coccidioidomycosis.** *Source:* CDC

Pathogen Profile #2 | *Coccidioides immitis*

Spherule stage of *C. immitis* from the lung
© Science Source

Microscopic Morphology At 25°C, forms moist, white to brown colonies with branching septate hyphae that fragment into thick-walled arthroconidia. At 37°C to 40°C, the arthrospores germinate into large spherical cells called spherules.

Identified by Appearance of spherules in the sputum, spinal fluid, and biopsies. Identification is enhanced by the cultivation of mycelia growth and induction of spherules. Immunodiffusion and latex agglutination tests on serum samples aid in early diagnosis.

Habitat *C. immitis* is endemic to many areas and occasionally found in many more where it may have been carried by wind or animals. A warm, dry environment with an abundance of salt and carbon favors settlement of the fungus. Highest incidence of infection with *C. immitis* occurs in the southwestern United States—particularly the San Joaquin Valley in California— as well as parts of Arizona, Mexico, and Central and South America.

Virulence Factors Like other pathogenic fungi, *C. immitis* has the ability to grow dimorphically, as both a saprobic mycelium and a pathogenic spherule.

Primary Infections/Disease Coccidioidomycosis is inapparent in about 60% of those infected with the pathogen, while in the remaining 40%, infection manifests with coldlike symptoms and malaise. In most cases, complete recovery and lifelong immunity result. Some people seem genetically predisposed to develop more serious disease, with the development of nodular growths (fungomas) and cavity formation in the lungs that compromises respiration. In about 7% of cases, endospores disseminate into the major organs of the body, leading to more severe disease. Being of Asian, Hispanic, or African descent, immunocompromised, or pregnant correlates with an increased risk of developing severe disease.

Control and Treatment Minimizing contact with *C. immitis* spores—through oiling dirt roads, planting vegetation to reduce dust, and wearing respiratory protection when digging—is the best defense. Treatment is not necessary in most cases, but amphotericin B and itraconazole are the antifungals of choice.

Coccidioides immitis occurs endemically in various natural reservoirs and casually in areas where it has been carried by wind and animals. Conditions favoring its settlement in a given habitat include high carbon and salt content and a semiarid, relatively hot climate. The fungus has been isolated from soils, plants, and numerous vertebrates. The natural history of *C. immitis* follows a cyclic pattern— a period of dormancy in winter and spring, followed by growth in summer and fall. Growth and spread are greatly increased by cycles of drought and heavy rains, followed by windstorms.

Skin testing has disclosed that the highest incidence of coccidioidomycosis (8,000–10,000 reported cases per year) is in the southwestern United States. Especially concentrated reservoirs exist in Southern Arizona and the San Joaquin Valley of California, from which it derives its common name. Outbreaks are usually associated with farming activity, archeological digs, landslides, and dust storms. Valley fever also extends into areas of Mexico and South and Central America. Several outbreaks have been reported in travelers and workers in these areas.

Infection and Pathogenesis of Coccidioidomycosis

The arthrospores of *C. immitis* are lightweight and readily inhaled. Testing has found that only a single spore is required to initiate an infection. In the lung they convert to spherules, which swell, sporulate, burst, and release spores that continue the cycle. In 60% of patients this primary pulmonary infection is inapparent; in the other 40% it is accompanied by coldlike symptoms such as fever, chest pain, cough, headaches, and malaise. In uncomplicated cases, complete recovery and lifelong immunity are the rule.

All persons inhaling arthrospores probably develop some degree of infection, but certain groups have a genetic susceptibility

that gives rise to more serious disease. In about 5 out of 1,000 cases, the primary infection does not resolve and progresses with varied consequences, including infection of the skin, bones, and central nervous system. Chronic progressive pulmonary disease is manifested by nodular growths called **fungomas*** and cavity formation in the lungs, which compromise respiration. Dissemination of the endospores into major organs takes place in about 7% of cases. Increased risk for more severe, extrapulmonary disease occurs among certain ethnic groups, people with impaired immunity, pregnant women, and AIDS patients (**figure 22.9**).

Diagnosis and Control of Coccidioidomycosis

Diagnosis of coccidioidomycosis is straightforward when the highly distinctive spherules are found in sputum, spinal fluid, and biopsies. This

* *fungomas* (fun-joh′-mah) A fungus tumor or growth.

Figure 22.9 Disseminated coccidioidomycosis manifested by cutaneous lesions. (left, right): © Joshua Fierer, M.D

Pathogen Profile #3 *Blastomyces dermatitidis*

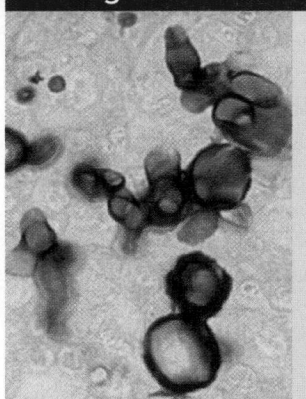

Yeast phase
Source: CDC

Microscopic Morphology Typically dimorphic, *B. dermatitidis* grows as a white to tan, septate mycelium when grown at lower temperature, whereas high temperature growth results in creamy, white, wrinkled yeast colonies.

Identified by Smears displaying large, ovoid, thick-walled yeast cells with oversize buds provide presumptive identification. Serological tests are not usually useful.

Habitat *B. dermatitidis* is found almost exclusively in the eastern United States, where it inhabits rich organic matter found in forest soil, decaying wood, and animal manure.

Virulence Factors Like other pathogenic fungi, *B. dermatitidis* has the ability to grow dimorphically, as both a saprobic mycelium and a pathogenic yeast. Additionally, a very small infectious dose (10 to 100 spores) is sufficient to initiate infection.

Primary Infections/Disease Infection with *B. dermatitidis*, known variously as Gilchrist disease, Chicago disease, and North American blastomycosis, usually results in some pulmonary symptoms, including cough, chest pain, hoarseness, and fever. Chronic disease may produce lung abscesses and tumorlike nodules that can be mistaken for cancer. Cutaneous disease is seen more commonly with *B. dermatitidis* than with other fungal pathogens and usually begins as a nodule on the face or hands. About 20% of infections progress to include the skin, bones, prostate, or (rarely) central nervous system. Death may occur in up to 5% of cases.

Control and Treatment Control, as with other fungal disease, relies on not inhaling fungal spores. Treatment is not necessary in most cases, but amphotericin B or azole drugs such as itraconazole or voriconazole are the antifungals of choice.

finding is further supported by isolation of typical mycelia and spores on Sabouraud's agar and induction of spherules.

All cultures must be grown in closed tubes or bottles and opened in a biological containment hood to prevent laboratory infections. Newer specific antigen tests have been effective tools to identify and differentiate *Coccidioides* from other fungi. Immunodiffusion (**figure 22.10**) and latex agglutination tests on serum samples are excellent screens for detecting early infection. Skin tests using an extract of the fungus (coccidioidin or spherulin) are of primary importance in epidemiological studies.

Figure 22.10 Immunodiffusion test for coccidioidomycosis. Small wells are punched into agar, and patients' sera are placed in wells number 2, 3, 5, 6. Wells 1 and 4 contain known positive antibodies. The center well contains a known fungal antigen such as coccidioidin. Lines of precipitation forming between the outer well and the inner well indicate a reaction between antibodies in the serum and antigen diffusing from the center well. Only one patient is positive for this test. © Sean Bauman, Ph.D./Immuno-Mycologics, Inc.

The majority of patients do not require treatment. However, in people with disseminated disease, amphotericin B is administered intravenously. One of the azole drugs can have some benefit, but it may require a longer term of therapy (up to 12 months). Minimizing contact with the fungus in its natural habitat has been of some value. For example, oiling roads and planting vegetation help reduce spore aerosols, and using dust masks while excavating soil prevents workers from inhaling spores.

Biology of Blastomyces dermatitidis: *North American Blastomycosis*

Blastomyces dermatitidis, * the cause of **blastomycosis,** is another fungal pathogen endemic to the United States. Alternative names for this disease are Gilchrist disease, Chicago disease, and North American blastomycosis. The dimorphic morphology of *Blastomyces* follows the pattern of other pathogens. Colonies of the hyphal phase are uniformly white to tan, with a thin, septate mycelium and simple ovoid conidia (**figure 22.11a**). Temperature-induced conversion results in a wrinkled, creamy-white colony composed of large, heavy-walled yeasts with buds nearly as large as the mother cell.

Studies indicate that *B. dermatitidis* inhabits rich organic matter found in forest soil, decaying wood, and animal manure. Its life cycle features dormancy during the warmer, dryer times of the year with growth and sporulation during the colder, wetter seasons. In general, blastomycosis occurs from southern Canada to southern Louisiana and from Minnesota to Georgia. Cases have also been

* *Blastomyces dermatitidis* (blas″-toh-my′-seez der″-mah-tit′-ih-dis) Gr. *blastos,* germ, *myces,* fungus, *dermato,* skin, and *itis,* inflammation.

(a) (b)

Figure 22.11 **The dimorphic nature of *Blastomyces dermatitidis*. (a)** Hyphal filaments give rise to conidia that resemble tiny lollipops. **(b)** Fluorescent antibody staining of lung tissue reveals the thick walls present in the yeast stage of *Blastomyces dermatitidis*. Arrow indicates a budding yeast cell. (a, b): *Source:* CDC

Figure 22.12 **Cutaneous blastomycosis.** Lesions are most common on the face, neck and extremities. Note the black vegetative plaques and purple discoloration around the lesion. Both are typically seen in cutaneous infection. Image reprinted with permission from Medscape Drugs & Diseases (http://emedicine.medscape.com/), 2016, available at: http://emedicine.medscape.com/article/296870-overview.

reported in Central America, South America, Africa, and the Middle East. Humans, dogs, cats, and horses are the chief targets of infection, which is usually acquired by inhaling conidia-laden dust.

Infection and Pathology in Blastomycosis The primary portal of entry of *B. dermatitidis* is the respiratory tract, though it also may enter through accidental inoculation. Inhaling only 10 to 100 conidia is enough to initiate infection. As conidia convert to yeasts and multiply, they encounter macrophages in the lung. A large proportion of primary pulmonary infections are probably symptomatic to some degree.

Mild disease is accompanied by cough, chest pain, hoarseness, and fever. More severe, chronic blastomycosis can progress from the lungs to the skin and numerous other organs. Lung abscesses and tumorlike nodules are often mistaken for cancer or tuberculosis. Compared with other fungal diseases, the chronic cutaneous form of blastomycosis is rather common. It frequently begins on the face or limbs as a subcutaneous nodule that erupts to the skin surface **(figure 22.12)**. Yeasts disseminating into the bone induce symptoms of arthritis and osteomyelitis. Involvement of the central nervous system can bring on headache, convulsions, coma, and mental confusion. Chronic systemic blastomycosis of the spleen, liver, and urogenital tract can last for weeks to years and eventually impair the host defenses.

Laboratory Diagnosis and Therapy Microscopic smears of specimens showing large, ovoid yeasts with broad-based buds can provide the most reliable diagnostic evidence of blastomycosis (see **figure 22.11*b***). Dimorphic cultures are highly desirable, although they can require several weeks to develop. Serological tests are generally not useful. Although disseminated infections were once nearly 100% fatal, modern drugs have greatly improved the prognosis. The drug of choice for disseminated and cutaneous disease is amphotericin B, and milder cases typically respond to one of the azole drugs.

Paracoccidioidomycosis

The remaining dimorphic fungal pathogen would be a front-runner in a competition for the most tongue-twisting scientific name. ***Paracoccidioides brasiliensis*** * causes paracoccidioidomycosis, also known as

* *Paracoccidioides brasiliensis* (pair″-ah-kok-sid″-ee-oy′-deez brah-sil″-ee-en′-sis) Named for its superficial resemblance to *Coccidioides* and its prevalence in Brazil.

paracoccidioidal granuloma and South American blastomycosis. Because of its relatively restricted distribution, this disease is the rarest of the primary mycoses. *Paracoccidioides* forms a small, nondescript colony with scanty, undistinctive spores at room temperature, but it develops an unusual yeast form at 37°C. The large mother cells sprout small, narrow-necked buds that radiate around the periphery.

Paracoccidioides has been isolated from the cool, humid soils of tropical and semitropical regions of South and Central America, particularly Brazil, Colombia, Venezuela, Argentina, and Paraguay. People most often afflicted with paracoccidioidomycosis are rural agricultural workers and plant harvesters. Factors that can influence the progress of infection are poor nutrition and impaired host resistance. Most infections occur in the lungs or the skin, and most of the time they are benign, self-limited events that go completely unnoticed. In a small minority of patients, the pathogen invades the lungs, skin and mucous membranes (especially of the head), and lymphatic organs.

Paracoccidioidomycosis is diagnosed by standard procedures previously described for the other mycoses in this chapter. Serological testing can be instrumental in diagnosing and monitoring the course of infection. Principal treatment drugs for disseminated disease, in order of choice, are ketoconazole, amphotericin B, and sulfa drugs.

 Check Your Progress SECTION 22.2

8. Differentiate between systemic, subcutaneous, cutaneous, and superficial infections.

9. Briefly outline the life cycles of *Histoplasma capsulatum*, *Coccidioides immitis*, and *Blastomyces dermatitidis*.

10. Name the diseases caused by each of the four primary fungal pathogens.

11. Identify the common habitats of *Histoplasma capsulatum*, *Coccidioides immitis*, *Blastomyces dermatitidis*, and *Paracoccidioides brasiliensis*.

12. Outline the course of a typical infection and disease with each of the four primary fungal pathogens.

13. Compare the laboratory results diagnostic of infection with each of the four primary fungal pathogens.

22.3 Subcutaneous Mycoses

 Expected Learning Outcomes

10. Describe the progression of a typical subcutaneous mycosis.

11. Summarize the major characteristics of *Sporothrix schenckii*.

12. Recall the characteristics of chromoblastomycosis and phaeohyphomycosis.

13. Describe the progression of events leading to mycetoma and chromoblastomycosis, as well as the prognosis of each disease.

When certain fungi are transferred from soil or plants directly into traumatized skin, they can invade the damaged site. Such infections are termed subcutaneous because they involve tissues within and just below the skin. Most species in this group are greatly inhibited by the higher temperatures of the blood and viscera, and only rarely do they disseminate. Nevertheless, these diseases are progressive and can destroy the skin and associated structures. Mycoses in this category are sporotrichosis, chromoblastomycosis, phaeohyphomycosis, and mycetoma.

The Natural History of Sporotrichosis: Rose-Gardener's Disease

The cause of **sporotrichosis**, ***Sporothrix schenckii,*** * is a very common saprobic fungus that exhibits both mycelial and yeast phases **(figure 22.13)**. *Sporothrix* resides in warm, temperate, and moist areas of the tropics and semitropics, where it decomposes plant matter in soil and humus. The incidence of sporotrichosis is highest in Africa, Australia, Mexico, and Latin America. A prick by a rose thorn is the classic origin of the cutaneous form of infection, which has given rise to its common name, rose-gardener's disease. Most episodes of human infection follow penetration by thorns, wood, sphagnum moss, bare roots, and other vegetation, although bites and scratches from cats have been found responsible for an increasing

* *Sporothrix schenckii* (spoh´-roh-thriks shenk´-ee-ee) Gr. *sporos,* seed, and *thrix* hair. Named for B.R. Schenck, who first isolated it.

(a) (b)

Figure 22.13 The microscopic morphology of *Sporothrix schenckii*. (a) Hyphal phase: Spores develop as floral clusters borne on conidiophores or as single spores pinched off the edge of a hypha (550×). **(b)** Tiny ovate budding yeasts typical of the tissue phase (800×). (a): *Source:* Dr. Libero Ajello/CDC; (b): *Source:* Dr. Lucille K. Georg/CDC

Figure 22.14 The clinical appearance of lymphocutaneous sporotrichosis. A primary sore is accompanied by a series of nodules and ulcers running along the lymphatic channels of the arm. *Source:* CDC

number of recent cases. Horticulturists, gardeners, farmers, and basket weavers incur the majority of cases. Overt disease is uncommon unless a person has inadequate immune defenses or is exposed to a large inoculum or a virulent strain. Other mammals—notably horses, dogs, cats, and mules—are also susceptible to sporotrichosis.

Pathology, Diagnosis, and Control of Sporotrichosis

In lymphocutaneous sporotrichosis, the fungus grows at the site of penetration and develops a small, hard, nontender nodule within a few days to months. This subsequently enlarges, becomes necrotic, breaks through to the skin surface, and drains **(figure 22.14)**. Infection often progresses along the regional lymphatic channels, leaving a chain of lesions at various stages of development. This condition can persist for several years, though it does not spread in the circulation. When *Sporothrix* conidia are drawn into the lungs, primary pulmonary sporotrichosis can result.

Discovery of the agent in tissue exudates, pus, and sputum is difficult because there are usually so few cells in the infection. Special stains can be used to highlight the tiny, cigar-shaped yeasts. Typical colony morphology and serological tests can confirm the diagnosis. The sporotrichin skin test can be used to determine prior infection. Itraconazole is the preferred treatment.

An older but effective drug for sporotrichosis is potassium iodide, given orally in milk or applied topically to open lesions. The fungus cannot withstand heat; thus, local applications of heat packs to lesions have helped resolve some infections. People with occupational exposure should cover their bare limbs as a preventive measure.

Chromoblastomycosis and Phaeohyphomycosis: Diseases of Pigmented Fungi

Chromoblastomycosis* is a progressive subcutaneous mycosis characterized by highly visible **verrucous*** lesions. The principal etiologic agents are a collection of widespread soil saprobes with

* *Chromoblastomycosis* (kroh˝-moh-blas˝-toh-my-koh´-sis) Gr. *chroma,* color, and *blasto,* germ. The fungal body is highly colored in vitro.
* *verrucous* (ver-oo´-kus) Tough, warty.

dark-pigmented mycelia and spores. *Fonsecaea pedrosoi, Phialophora verrucosa,* and *Cladophialophora carrionii* are among the most common causative species. **Phaeohyphomycosis,*** a closely related infection, differs primarily in the causative species and the appearance of the infectious agent in tissue. Chromoblastomycotic agents produce very large, thick, yeastlike bodies called sclerotic cells, whereas the agents of phaeohyphomycosis remain typically hyphal.

The fungi associated with these infections are of low inherent virulence, and none exhibits thermal dimorphism. Infection occurs when body surfaces, especially the legs and feet, are penetrated by soiled vegetation or inanimate objects. Chromoblastomycosis occurs throughout the world, with peak incidence in the American subtropics and tropics. Men with rural occupations who go barefoot are most vulnerable. After a very long (2- to 3-year) incubation period, a small, colored, warty plaque, ulcer, or papule appears. Because this lesion is generally not painful, many patients do not seek treatment and can go for years watching it progress to a more advanced state. Unfortunately, scratching the lesions can aggravate the spread of infection on the body and provoke secondary bacterial infections.

Chromoblastomycosis is frequently confused with cancer, syphilis, yaws, and blastomycosis. Diagnosis is based on the clinical appearance of the lesions, microscopic examination of biopsied lesions, and the results of culture. Therapy for the condition includes heat, drugs, surgical removal of early nodules, and amputation in advanced cases. Some success has been seen with amphotericin B, terbinafine, or combined therapy with itraconazole and flucytosine.

The phaeohyphomycoses are among the strangest oddities of medical science. The etiologic agents are soil fungi or plant pathogens with brown-pigmented mycelia. Victims of the disease are usually highly compromised patients who have become inoculated with these fungi, which are so widespread in household and medical environments that contact is inevitable. Genera of fungi commonly involved in these mycoses are *Alternaria, Aureobasidium, Curvularia, Dreschlera, Exophiala, Phialophora,* and *Wangiella.* Starting at the site of inoculation, the fungus grows into the skin, often creating crusty skin lesions and abcesses. As it invades deeper into the subcutaneous tissues, it often forms disfiguring cysts that have to be surgically removed. In patients with underlying conditions such as endocarditis, diabetes, and leukemia, the fungi can spread into the bones, brain, and lungs.

Mycetoma: A Complex Disfiguring Syndrome

Another disease elicited when soil microbes are accidently implanted into the skin is **mycetoma,** a mycosis usually of the foot or hand that looks superficially like a tumor. It is also called *madura foot* for the Indian region where it was first described. About half of all mycetomas are caused by fungi in the genera *Pseudallescheria* or *Madurella,* though filamentous bacteria called actinomycetes (*Nocardia,* discussed in chapter 19) are often implicated. Mycetomas are endemic to equatorial Africa, Mexico, Latin America, and the Mediterranean, and cases occur sporadically in the United States.

Infection begins when bare skin is pierced by a thorn, sliver, leaf, or other type of sharp plant debris. The infectious spores and hyphae grow in the damaged epidermis, creating dark granules containing mycelium (**figure 22.15b**). First to appear is a localized

Figure 22.15 Mycetoma (a) Macroscopic view of madura foot features ulceration, swelling, and scarring. **(b)** Microscopic view of biopsy with masses of dark, thick-walled hyphae that has replaced much of the normal tissue. (a): © Medicshots/Alamy Stock Photo; (b): *Source: Dr. Libero Ajello/CDC*

abscess in the subcutaneous tissues, which gradually swells and drains (**figure 22.15a**). Untreated cases that spread to the muscles and bones can cause pain and loss of function in the affected body part. Mycetoma has an insidious, lengthy course and is exceedingly difficult to treat.

 Check Your Progress SECTION 22.3

14. What weakness of host defenses typically contributes to subcutaneous mycoses?
15. Describe the disease, habitat, pathology, and diagnostic indicators of *Sporothrix schenckii.*
16. What are the risk factors associated with contracting chromoblastomycosis and phaeohyphomycosis?
17. List the similarities and differences encountered when comparing mycetoma and chromoblastomycosis.

22.4 Cutaneous Mycoses

 Expected Learning Outcomes

14. Name the three dermatophytic genera and the diseases they cause.
15. Understand the pathology, epidemiology, and treatment of ringworm.

Fungal infections strictly confined to the nonliving epidermal tissues (stratum corneum) and its derivatives (hair and nails) are termed **dermatophytoses.** Common terms used in reference to these diseases are **ringworm,** because they tend to develop in circular, scaly patches (see figure 22.17), and **tinea.*** About 39 species in the genera *Trichophyton, Microsporum,* and *Epidermophyton* are involved in various dermatophytoses. The causative agent of a given type of ringworm differs from person to person and from place to place and is not restricted to a particular genus or species (**table 22.5**).

* *Phaeohyphomycosis* (fy″-oh-hy″-foh-my-coh′-sis) Gr. *phaeo,* brown, and *hypho,* thread.

* *tinea* (tin′-ee-ah) L. a larva or worm. Early observers thought they were caused by worms.

TABLE 22.5	The Dermatophyte Genera and Diseases		
Genus	Name of Disease	Principal Targets	How Transmitted
Trichophyton	Ringworm of the scalp, body, beard, and nails Athlete's foot	Hair, skin, nails	Human to human, animal to human
Microsporum	Ringworm of scalp Ringworm of skin	Scalp hair Skin; not nails	Animal to human, soil to human, human to human
Epidermophyton	Ringworm of the groin and nails	Skin, nails; not hair	Strictly human to human

Characteristics of Dermatophytes

The dermatophytes are so closely related and morphologically similar that they can be difficult to differentiate. Various species exhibit unique macroconidia, microconidia, and unusual types of hyphae. In general, *Trichophyton* produces thin-walled, smooth macroconidia and numerous microconidia; *Microsporum* produces thick-walled, rough macroconidia and sparser microconidia; and *Epidermophyton* has ovoid, smooth, clustered macroconidia and no microconidia (**figure 22.16**).

Epidemiology and Pathology of Dermatophytoses

The natural reservoirs of dermatophytes are other humans, animals, and the soil (**22.1 Making Connections**). Important factors that promote infection are the hardiness of the dermatophyte spores (they can last for years on fomites); presence of abraded skin; and intimate contact. Most infections exhibit a long incubation period (months), followed by localized inflammation and allergic reactions to the fungal proteins. As a general rule, infections acquired from animals and soil cause more severe reactions than do infections acquired from other humans, and infections eliciting stronger immune reactions are resolved faster.

(a) (b) (c)

Figure 22.16 Examples of dermatophyte spores. (a) Regular, numerous microconidia of *Trichophyton* (750×). **(b)** Macroconidia of *Microsporum gypseum,* a cause of ringworm in cats, dogs, and humans (1,000×). **(c)** Smooth-surfaced macroconidia in clusters characteristic of *Epidermophyton* (1,000×). (a, b, & c): *Source:* CDC

Dermatophytic fungi exist throughout the world. Although some species are endemic, they tend to spread rapidly to nonendemic regions through travel. Dermatophytoses endure today as a serious health concern, not because they are life-threatening but because of the extreme discomfort, stress, pain, and unsightliness they cause.

An Atlas of Dermatophytoses

In this section, the dermatophytoses are organized according to the area of the body they affect, their mode of acquisition, and their pathologic appearance. Both the common English name and body site and its equivalent Latin name (tinea) are given.

Quick Search
To appreciate the true variety of dermatophytoses, point your browser to "Tinea" and search images.

Ringworm of the Scalp (Tinea Capitis)
This mycosis results from the fungal invasion of the scalp and the hair of the head, eyebrows, and eyelashes (**figure 22.17a**). Very common in children, tinea capitis is acquired from other children and adults or from domestic animals. Manifestations range from small, scaly patches (gray patch) to a severe inflammatory reaction (kerion) to destruction of the hair follicle and permanent hair loss.

Ringworm of the Beard (Tinea Barbae)
This tinea, also called "barber's itch," afflicts the chin and beard of adult males. Although

(a)

(b)

Figure 22.17 Ringworm lesions vary in appearance. (a) Tinea capitis with deep scalp involvement and complete hair loss in the affected region. **(b)** Tinea faciei, a dermatophytic infection of the face. (a, b): *Source:* CDC

22.1 MAKING CONNECTIONS

The Keratin Lovers

The dermatophytic fungi are especially well adapted to breaking down keratin, the primary protein of the epidermal tissues of vertebrates (skin, nails, hair, feathers, and horns). Their affinity for this compound gives them the name *keratophiles*. Examination of infected hairs indicates that these fungi attach to the hair surface and penetrate into its cortex. In time, they grow along the hair's length to the follicle, where they initiate a skin infection. A study of dermatophyte ecology reveals a gradual evolutionary trend from saprobic soil forms that digest keratin but do not parasitize animals to soil forms that occasionally parasitize animals to species that are dependent on live animals. Some species can infect a broad spectrum of animals, and others are specific to one particular animal species or region of the body.

One adaptive challenge faced by relatively new fungal parasites is that they are likely to cause severe reactions in the host's skin and to be attacked by the host defenses and eliminated. Thus, the more successful fungi equilibrate with the host by reducing their activity (growth rate, sporulation) and lessen the inflammatory response. Eventually, these dermatophytes become such "good parasites" that they colonize the host for life. A striking example is *Trichophyton rubrum*, a fungus that causes a form of athlete's foot. It has such a tenacious hold and is so hard to cure that its carriers have been called the "*T. rubrum* people."

This microscopic view of a human hair shows dermatophyte hyphae growing along the hair and penetrating into its center.
Courtesy of Toshio Kanbe

What makes it easier for dermatophytes to gain a foothold in the body as compared with other pathogens? *Answer available on Connect.*

once a common aftereffect of unhygienic barbering, it is now contracted mainly from animals. Infection of the non-bearded areas of the face is less common, and referred to as Tinea faciei (**figure 22.17b**).

Ringworm of the Body (Tinea Corporis) This extremely prevalent infection of humans can appear nearly anywhere on the body's glabrous (smooth and bare) skin. The principal sources are other humans, animals, and soil, and it is transmitted primarily by direct contact and fomites (clothing, bedding). The infection usually appears as one or more scaly reddish rings on the trunk, hip, arm, neck, or face. The ringed pattern is formed when the infection radiates from the original site of invasion into the surrounding skin. Depending on the causal species and the health and hygiene of the patient, lesions vary from mild and diffuse to florid and pustular.

Ringworm of the Groin (Tinea Cruris) Sometimes known as "jock itch," crural ringworm occurs mainly in males on the groin, perianal skin, scrotum, and, occasionally, the penis. The fungus thrives under conditions of moisture and humidity created by profuse sweating or tropical climates. It is transmitted primarily from human to human and is pervasive among athletes and persons living in close situations (ships, military quarters).

Ringworm of the Foot (Tinea Pedis) Tinea pedis is known by a variety of synonyms, including athlete's foot and jungle rot. The disease is clearly connected to wearing shoes, because it is uncommon in cultures where the people customarily go barefoot. (But as you have seen, bare feet have other fungal risks!) Conditions that encase the feet in a closed, warm, moist environment increase the possibility of infection. Tinea pedis is a known hazard in shared

facilities such as shower stalls, public floors, and locker rooms. Infections begin with small blisters between the toes that burst, crust over, and can spread to the rest of the foot and nails (**figure 22.18**).

Ringworm of the Hand (Tinea Manuum) Infection of the hand by dermatophytes is nearly always associated with concurrent infection of the foot. Lesions usually occur on the fingers and palms of one hand, and they vary from white and patchy to deep and fissured.

Ringworm of the Nail (Tinea Unguium) Fingernails and toenails, being masses of keratin, are often sites for persistent fungus colonization. The first symptoms are usually superficial white patches in the nail bed. A more invasive form causes thickening,

Figure 22.18 Ringworm of the extremities. *Trichophyton* infection spreading over the foot in a "moccasin" pattern. The chronicity of tinea pedis is attributed to the lack of fatty-acid-forming glands in the feet. *Source:* CDC

distortion, and darkening of the nail. Nail problems caused by dermatophytes are on the rise as more women wear artificial fingernails, which can provide a portal of entry into the nail bed.

Diagnosis of Ringworm

Dermatologists are most often called upon to diagnose dermatophytoses. Occasionally, the presenting symptoms are so dramatic and suggestive that no further testing is necessary, but in most cases, direct microscopic examination and culturing are needed. Diagnosis of tinea of the scalp (tinea capitis) caused by some species of *Microsporum* is aided by use of a long-wave ultraviolet lamp that causes infected hairs to fluoresce. Samples of hair, skin scrapings, and nail debris treated with heated KOH show a thin, branching fungal mycelium if infection is present. Culturing specimens on selective media and growth on sterile hair can be important diagnostic aids.

Treatment of the Dermatophytoses

Ringworm therapy is based on the knowledge that the dermatophyte is feeding on dead epidermal tissues. These regions undergo constant replacement from living cells deep in the epidermis, so if multiplication of the fungus can be blocked, the fungus will eventually be sloughed off along with the skin or nail. Unfortunately, this takes time. By far the most satisfactory choice for therapy is a topical antifungal agent. Ointments containing terbinafine, miconazole, or tolnaftate are applied regularly for several weeks. Some drugs work by speeding up loss of the outer skin layer. Gentle debridement of skin and ultraviolet light treatments can have some benefit.

 Check Your Progress SECTION 22.4

18. Distinguish among the diseases caused by members of the genera *Trichophyton, Microsporum,* and *Epidermophyton.*
19. List the factors involved in contracting ringworm, and outline the progression of a typical infection.
20. How does simply inhibiting the growth of a fungus eventually lead to curing a dermatophytic infection?

22.5 Superficial Mycoses

 Expected Learning Outcome

16. Explain the significance of superficial mycoses.

Agents of superficial mycoses involve the outer epidermal surface and are ordinarily innocuous infections with cosmetic rather than inflammatory effects. **Tinea versicolor**[1] is caused by the yeast *Malassezia furfur,* a normal inhabitant of human skin that feeds on the high oil content of the skin glands. Even though this yeast is very common (carried by nearly 100% of humans tested), in some people its growth elicits mild, chronic scaling and interferes with production of pigment by melanocytes. The trunk, face, and limbs take on a mottled appearance **(figure 22.19).** The disease is most pronounced in young people

Figure 22.19 Tinea versicolor. Mottled, discolored skin pigmentation is characteristic of superficial skin infection by *Malassezia furfur.* © Medical-on-Line/Alamy Stock Photo

Figure 22.20 Example of superficial mycosis. A hair shaft exhibiting black piedra due to infection with *Piedraia hortae.* Infection is usually limited to the scalp but may be found on hairs of the beard, moustache, and pubic region. *Source:* CDC

who are frequently exposed to the sun. Recommended therapy zinc pyrithione shampoo scrubbed onto affected area and rinsed off after a few minutes, daily for 3 to 4 weeks. Other skin conditions in which *M. furfur* is implicated are folliculitis, psoriasis, and seborrheic dermatitis. It is also occasionally associated with systemic infections and catheter-associated sepsis in compromised patients.

Piedras* are marked by a tenacious, colored concretion forming on the outside surface of hair shafts **(figure 22.20).** In **white piedra,** caused by *Trichosporon beigelii,* a white to yellow adherent mass develops on the shaft of scalp, pubic, or axillary hair. At times, the mass is invaded secondarily by brightly colored bacterial contaminants, with startling results. **Black piedra,** caused by *Piedraia hortae,* is characterized by dark-brown to black gritty nodules, mainly on scalp hairs. Neither piedra is common in the United States.

* *Piedra* (pee-ay′-drah) Sp. stone. Hard nodules of fungus formed on the hair.

 Check Your Progress SECTION 22.5

21. Briefly describe three superficial mycoses.

1. *Versicolor* refers to the color variations this mycosis produces in the skin. Also called pityriasis.

22.6 Opportunistic Mycoses

 Expected Learning Outcomes

17. Name the most common opportunistic fungal pathogens.

18. Explain the epidemiology of candidiasis.

19. Relate the options available for treatment of common fungal opportunists.

Figure 22.21 *Candida albicans* **infection of the mouth (oral thrush).** Courtesy of University of Adelaide, Australia

In section 22.1 we introduced the concept of opportunistic fungal infections and their predisposing factors (see table 22.3). The yeast *Candida* is the dominant opportunistic pathogen, and it accounts for the majority of invasive infections. The mold *Aspergillus* is the second in incidence, accounting for the most lung infections. Other fungi that are appearing more often in clinical isolates are *Crypto-coccus, Alternaria, Paecilomyces, Fusarium, Rhizopus,* and *Torulopsis*—all genera that were once considered "harmless" air contaminants.

Infections by *Candida:* Candidiasis

*Candida albicans,** an extremely widespread yeast, is the major cause of candidiasis (also called candidosis or moniliasis). Manifestations of infection run the gamut from short-lived, superficial skin irritations to overwhelming, fatal systemic diseases. Microscopically, *C. albicans* has budding cells of varying size that may form both elongate pseudohyphae and true hyphae (see figure 22.22*a*); macroscopically, it forms an off-white, pasty colony with a yeasty odor.

Epidemiology of Candidiasis

Candida albicans occurs as normal microbiota in the pharynx, genitalia, large intestine, or skin of 20% of humans. A healthy person with an unimpaired immune system can hold it in check, but the risk of invasion increases with infancy, old age, pregnancy, drug therapy, immunodeficiency, and trauma. Any situation that maintains the yeast in contact with moist skin provides an avenue for infection. Although candidiasis is usually endogenous and not contagious, it can be spread in nurseries or through surgery, childbirth, and sexual contact. *Candida albicans* and its close relatives account for around 70% of health-care-associated fungal infections.

Diseases of Candida albicans

Candida albicans causes local infections of the mouth, pharynx, vagina, skin, alimentary canal, and lungs, and it can also disseminate to internal organs. The mucous membranes most frequently involved are the oral cavity and vagina. **Thrush** is a white, adherent, patchy infection affecting the membranes of the oral cavity or throat, usually in newborn infants and elderly, debilitated patients **(figure 22.21). Vulvovaginal candidiasis (VC),** known more commonly as *yeast infection,* has widespread

occurrence in adult women, especially those who are taking oral antibiotics or those who are diabetic or pregnant, all conditions that can disrupt the normal vaginal microbiota. Candidal vaginitis also poses a risk for neonates, which can be infected during childbirth, and it can be transmitted to male partners during sexual intercourse. The chief symptoms of VC are a yellow to white discharge, inflammation, painful ulcerations, and itching. The most severe cases spread from the vagina and vulva to the perineum and thighs.

Candidal attack of keratinized structures such as skin and nails, called **onychomycosis,** is often brought on by predisposing occupational and anatomical factors. People whose occupations require their hands or feet to be constantly immersed in water are at risk for finger and nail invasion. **Intertriginous** infection occurs in moist areas of the body where skin rubs against skin, as beneath the breasts, in the armpit, and between folds of the groin. **Cutaneous candidiasis** can also complicate burns and produce a scaldlike rash on the skin of neonates (a form of diaper rash).

Of all the areas of the gastrointestinal tract, *Candida* most often infects the esophagus and the anus. Esophageal candidiasis, which afflicts 70% of AIDS patients, causes painful, bleeding ulcerations, nausea, and vomiting.

Candidal blood infection usually becomes systemic in patients chronically weakened by surgery, bone marrow transplants, advanced cancer, or intravenous drug addiction. In fact, blood infection with *C. albicans* results in more deaths than any other fungal pathogen. Principal targets of systemic infections are the urinary tract, endocardium, and brain. Patients with valvular disease of the heart are vulnerable to candidal endocarditis, usually caused by other species (*C. tropicalis* and *C. parapsilosis*). *Candida krusei* has reportedly caused terminal infections in bone marrow transplant patients and in recipients of anticancer therapy. Another problem is the production of biofilms on artificial joints, catheters, and heart valves. In most of these cases, the colonization is very drug resistant and may require removal of the device until it can be controlled.

* *Candida albicans* (kan´-dih-dah al´-bih-kanz) L. *candidus,* glowing white, and *albus,* white.

* *onychomycosis* (ahn″-ih-koh-my-koh´-sis) Gr. *onyx,* nail. Infection of the nails by *Candida* or *Aspergillus.*
* *Intertriginous* (in″-ter-trij´-ih-nus) L. *inter,* between, and *trigo,* to rub.

Pathogen Profile #4 *Candida albicans*

Microscopic Morphology Macroscopically, *C. albicans* produces off-white colonies, whereas microscopic examination reveals budding yeast cells, pseudohyphae, and true hyphae.

Identified by Visualization of budding yeast cells and pseudohyphae in primary specimens provides presumptive evidence of infection, whereas growth on differential media containing trypan blue can differentiate *Candida* from *Cryptococcus* species. Definitive identification may be achieved using biochemical testing or amplification of *Candida*-specific DNA sequences.

Source: CDC

Habitat 20% of humans carry *C. albicans* in their pharynx, genitalia, large intestine, or skin as part of their normal microbiota.

Virulence Factors *C. albicans* is a prototypical opportunist, displaying little virulence in those with an unimpaired immune system but quickly causing disease when host defenses are compromised by young age, pregnancy, drug therapy, immunodeficiency disease, or trauma.

Primary Infections/Disease Infection with *C. albicans* causes a number of local infections. Thrush is a white adherent infection affecting the membranes of the oral cavity or throat, most prevalent in newborns and elderly patients. Vulvovaginal candidiasis—yeast infection—occurs among adult women, especially those who are diabetic, pregnant, or taking oral antibiotics, all conditions that alter the normal vaginal microbiota. Onychomycosis occurs when *C. albicans* infects keratinized structures such as skin and nails. Intertriginous infection is seen where skin rubs against skin—commonly under the breasts, in armpits, and in folds of the groin. Cutaneous candidiasis may be a complication of burns or diaper rash. Systemic *Candida* infection is generally only seen in patients weakened by cancer, surgery, bone marrow transplants, or intravenous drug abuse and is a serious complication. *Candida* biofilm formation on artificial joints, catheters, and heart valves often requires removal of the device until the infection can be controlled.

Control and Treatment Because *C. albicans* is such a successful opportunist, candidiasis generally returns unless the underlying conditions predisposing the patient to infection are successfully treated. Treatment of cutaneous infection relies on topical antifungals (azoles and polyenes), whereas terbinafine is used to treat onychomycosis, and amphotericin B or fluconazole is effective against systemic infections. Vaginal candidiasis is sexually transmissible, requiring treatment of both partners to avoid reinfection.

Laboratory Techniques in the Diagnosis of Candidiasis

A presumptive diagnosis of *Candida* infection is made if budding yeast cells and pseudohyphae are found in specimens from localized infections **(figure 22.22a)**. Specimens are cultured on standard fungal media incubated at 30°C. Identification is complicated by the numerous species of *Candida* and other look-alike yeasts. Growth on a selective, differential medium containing trypan blue can easily differentiate *Candida* species from the yeast *Cryptococcus* **(figure 22.22b)**. Confirmatory evidence of *C. albicans* can also be obtained by the germ tube test, the presence of chlamydospores, and multiple-panel systems that test for biochemical characteristics **(figure 22.22c)**. A sensitive PCR technique has been developed for identifying this species directly from clinical specimens.

Treatment of Candidiasis

Because candidiasis is almost always opportunistic, it will return in many cases if the underlying disease is not also treated. Therapy for superficial mucocutaneous infection consists of topical antifungal agents (azoles and polyenes). Terbinafine has been approved for treating onychomycosis. Amphotericin B and fluconazole are usually effective in systemic infections. Recurrent bouts of vulvovaginitis are managed by topical azole ointments, available as over-the-counter drugs. Vaginal candidiasis is sexually transmissible; thus, it is very important to treat both sex partners to avoid reinfection.

Cryptococcus neoformans and Cryptococcosis*

Another widespread resident of human habitats is the fungus ***Cryptococcus neoformans.**** This yeast has a spherical to ovoid shape, with small, constricted buds and a large capsule that is important in its pathogenesis **(figure 22.23)**. Its role as an opportunist is supported by evidence that healthy humans have strong resistance to it and that frank infection occurs primarily in debilitated patients. *Cryptococcus gattii* is a newly identified species that has been seen primarily in the Northwestern United States (Oregon and Washington), Western Canada and Australia. It is very similar to *C. neoforman* in all respects except that it is more likely to cause infections in healthy hosts, and has a worse prognosis. Most **cryptococcoses** center around the respiratory, central nervous, and mucocutaneous systems.

* *cryptococcosis* (krip″-toh-kok-oh′-sis) Infections by *Cryptococcus.* Older names are torulosis and European blastomycosis.
* *Cryptococcus neoformans* (krip″-toh-kok′-us nee″-oh-for′-manz) Gr. *kryptos,* hidden, *kokkos,* berry, *neo,* new, and *forma,* shape.

Figure 22.22 **Detection of *Candida albicans*. (a)** Gram stain of *Candida albicans* in a vaginal smear reveals gram-positive budding yeasts, a pseudohypha, and hyphae (750×). In many cases of vaginal candidiasis, infection is detected during a routine Pap smear. **(b)** Colonies of *Candida albicans* appear pale blue on trypan medium, whereas *Cryptococcus* are dark blue. **(c)** Rapid yeast identification system using biochemical reactions that separate 40 different species of yeast. (a): Courtesy of Danny L. Wiedbrauk; (b): Courtesy of Glen S. Bulmer; (c): Photo courtesy of REMEL, Inc.

Epidemiology of C. neoformans

The primary ecological niche of *C. neoformans* is associated with birds. It is prevalent in urban areas where pigeons congregate, and it proliferates in the high nitrogen content of droppings that accumulate on pigeon roosts. Particles containing dried yeast cells are readily scattered into the air and dust. The species is sporadically

Figure 22.23 *Cryptococcus neoformans* **from infected spinal fluid stained negatively with India ink.** Halos around the large spherical yeast cells are thick capsules. Also note the buds forming on one cell. Encapsulation is a useful diagnostic sign for cryptococcosis, although the capsule is fragile and may not show up in some preparations (150×). Photo courtesy of Gordon Love, M.D. VA, North CA Healthcare System, Martinez, CA. G. Love, et al., "Large Cryptococcus neoformans Isolated from Brain Abscess," *JCM,* 22 (6): p. 1069, Reprinted by permission of American Society for Microbiology

isolated from dairy products, fruits, and the healthy human body (tonsils and skin), where it is considered a transient. The highest rates of cryptococcosis occur among patients with AIDS. It causes a severe form of meningitis that is frequently fatal. Other conditions that predispose to infection are steroid treatment, diabetes, and cancer. The disease exists throughout the world, with highest prevalence in the United States. Outbreaks have been reported in construction workers exposed to pigeon roosts, but it is not considered communicable among humans.

Pathogenesis of Cryptococcosis

The primary portal of entry for *C. neoformans* is respiratory, and most lung infections are subclinical and rapidly resolved. A few patients with pulmonary cryptococcosis develop fever, cough, and nodules in their lungs. The escape of *C. neoformans* into the blood is more common in those with weakened host defenses and causes severe complications, including infection of the brain and meninges. The tumorlike masses formed in these locations can cause headache, mental changes, coma, paralysis, eye disturbances, and seizures. In some cases, the infection disseminates into the skin, bones, and viscera **(figure 22.24).**

Diagnosis and Treatment of Cryptococcosis

The first step in diagnosis of cryptococcosis is negative staining of specimens to detect encapsulated budding yeast cells that do not occur as pseudohyphae. Isolated colonies can be used to perform screening tests that presumptively differentiate *C. neoformans* from the seven other cryptococcal species. Cryptococcal antigen can be

Figure 22.24 Cryptococcosis. An early case of cutaneous cryptococcosis. In the most serious cases, seen in patients with impaired immune function, the infection may progress to include the brain and meninges. *Source:* © Gordon Love, M.D. VA, North CA Healthcare System, Martinez, CA

Figure 22.25 *Pneumocystis jirovecii* cells in the lung tissue (500x; stained with methamine silver) of an AIDS patient. *Source:* CDC

detected in a specimen by means of serological tests, and DNA probes can make a positive genetic identification. Systemic cryptococcosis requires immediate combined therapy with amphotericin B and flucytosine for two weeks, followed by fluconazole over a period of weeks or months.

Pneumocystis jirovecii and *Pneumocystis* Pneumonia

Although *Pneumocystis jirovecii* was discovered in 1909, it remained relatively obscure until it was suddenly propelled into clinical prominence as the agent of ***Pneumocystis jiroveci*[2] pneumonia (PCP).** PCP is the most frequent opportunistic infection in AIDS patients, most of whom will develop one or more episodes during their lifetimes.

Because this unicellular microbe superficially has characteristics of both fungi and protozoa, its taxonomic status has been debated. The sequence of its ribosomal RNA indicates that it is very similar genetically to the yeast *Saccharomyces*. It differs from most other fungi because it lacks ergosterol, has a weak cell wall, and is an obligate parasite.

Unlike most of the human fungal pathogens, less is known about the life cycle or epidemiology of *Pneumocystis*. Evidently the parasite is a common, relatively harmless resident of the upper respiratory tract. Contact with the agent is so widespread that in some populations, a majority of people show serological evidence of infection. Until the AIDS epidemic, symptomatic infections by this organism were very rare, occurring only among elderly patients or premature infants who were severely debilitated or malnourished.

Pneumocystis is probably spread in droplet form between humans. In people with intact immune defenses, it is usually held in check by lung phagocytes and lymphocytes, but in those with deficient immune systems, it multiplies intracellularly and extracellularly. The massive numbers of parasites adhere tenaciously to the lung pneumocytes and cause an inflammatory condition. The lung epithelial cells slough off, and a foamy exudate builds up. Symptoms are nonspecific and include cough, fever, shallow respiration,

and cyanosis. Improved therapy for AIDS patients (see chapter 25) has reduced the incidence of PCP and its mortality rate.

Because human strains of the species have not yet been successfully cultured, PCP must be diagnosed by symptoms and direct examination of lung secretions and tissue **(figure 22.25).** A DNA amplification probe can provide rapid confirmation of infection, but is prone to false positives. Traditional serological tests for antigens and antibodies are not useful diagnostic tools for *Pneumocystis*.

Traditional antifungal drugs that inhibit the synthesis of ergosterol are ineffective against *Pneumocystis* pneumonia because it lacks this sterol. The primary treatment is sulfamethoxazole and trimethoprim given over a 21-day period. This therapy should be applied even if disease appears mild or is only suspected. Pentamidine in an aerosol form is also given as a prophylactic measure in patients with a low T-cell count. The airways of patients in the active stage of infection often must be cleared by suctioning.

Aspergillosis: Diseases of the Genus *Aspergillus*

Molds of the genus *Aspergillus** are possibly the most pervasive of all fungi. The estimated 600 species are widely dispersed in dust and air and can be isolated from vegetation, food, and compost matter. Of the eight species involved in human diseases, the thermophilic fungus *A. fumigatus* accounts for the most cases. Aspergillosis is almost always an opportunistic infection, lately posing a serious threat to AIDS, leukemia, and immunodeficient patients. It is also involved in allergies and toxicoses.

Aspergillus infections usually occur in the lungs. People breathing clouds of conidia from the air of granaries, barns, and silos are at greatest risk. In healthy people with extensive exposure, the spores simply germinate in the lungs and form *fungus balls*. Similar benign, noninvasive infections occur in colonization of the sinuses, ear canals, eyelids, and conjunctiva **(figure 22.26a),** whereas in the more invasive form of aspergillosis, the fungus produces a necrotic pneumonia and disseminates to the brain, heart, skin, and a wide range of other organs. Systemic aspergillosis in severely ill hospitalized patients has a poor prognosis.

2. Formerly known as *Pneumocystis carinii* or *Pneumocystis (carinii) jiroveci.*

* *Aspergillus* (as″-per-jil′-us) L. *aspergere,* to scatter.

(a)

(b)

Figure 22.27 *Absidia corymbifera,* **a common air contaminant. (a)** *A. corymbifera* may cause a serious case of mucormycosis in susceptible patients. Note the sporangium releasing large spores and the nonseptate hyphae (450×). **(b)** Cutaneous mucormycosis as a complication of tattooing. (a): Image courtesy of doctorfungus.org © 2010; (b): Image courtesy David Ellis, Adelaide Women's and Children's Hospital

(b)

Figure 22.26 **Appearance of aspergillosis (a)** This chest radiograph shows probable aspergillosis with an aspergilloma, or fungus ball in the upper lobe of the right lung (arrow). **(b)** Scanning electron micrograph of the conidial head of *Aspergillus,* (400×), These molds are not dimorphic and exist only in the hyphal stage. (a): *Source:* CDC; (b): *Source:* Janice Haney Carr/CDC

An obvious branching, septate mycelium with characteristic conidial heads in a specimen is presumptive evidence of aspergillosis **(figure 22.26b).** A culture of the specimen that produces abundant colonies and growth at 37°C will rule out its being only an air contaminant. Molecular methods include a PCR test that can identify *Aspergillus*-specific DNA and a serological test that detects fungal antigen. Noninvasive disease can be treated by surgical removal of the fungus tumor or by local drug therapy of the lung using voriconazole, Cresemba (a newly approved azole drug), or amphotericin. Combined therapy with two of these drugs is the most effective treatment for systemic disease.

Mucormycosis

Members of the Division Zygomycota are extremely abundant saprobic fungi found in soil, water, organic debris, and food. Their large, prolific sporangia release multitudes of lightweight spores that literally powder the living quarters of humans and usually do little harm besides spoiling foods and rotting fruits and vegetables. But an increasing number of critically ill patients are contracting a disease called **mucormycosis,** also known as **zygomycosis.** Common airborne molds in the genera *Rhizopus, Mucor,* and *Absidia* **(figure 22.27a)** are most often involved in these mycoses.

Underlying debilities that seem to increase the risk for mucormycosis are leukemia, malnutrition, and skin damage **(figure 22.27b).** Uncontrolled diabetes is a special risk because it reduces the pH in the blood and tissue fluids, which enhances invasion of the fungus. The fungus enters by way of the nose, lungs, skin, or oral cavity. In rhinocerebral mucormycosis, infection begins in the nasal passages or pharynx and then grows in the blood vessels and up through the palate into the eyes and brain. Invasion of the lungs occurs most often in leukemia patients, and gastrointestinal mucormycosis is a complication in children suffering from protein malnutrition. Unfortunately, the outlook for most of these patients is poor.

Diagnosis of mucormycosis rests on clinical grounds and biopsies or smears. Very large, thick-walled, nonseptate hyphae scattered through the tissues of a gravely ill diabetic or leukemic patient are very suggestive of zygomycosis. Attempts to isolate the molds from specimens are frequently met with failure, and confusion of

22.2 MAKING CONNECTIONS

Contaminated Steroids Lead to Fungal Meningitis

In late 2012 a rare outbreak of fungal meningitis struck across the United States, killing 64 people and sickening over 753. A compounding pharmacy in Massachusetts was quickly identified as the source of the fungus, which was allowed to contaminate several thousand doses of methylprednisolone acetate, a steroid used to treat pain and swelling of the joints. Inspection of the New England Compounding Center revealed a poorly controlled—many would say filthy—laboratory environment, where spores borne on the wind could easily enter the manufacturing process.

Regrettably, dirty manufacturing facilities and large-scale outbreaks of disease are more common than we'd care to admit. What turned out to be noteworthy—especially to microbiologists—was the identity of the contaminant itself, an obscure fungus called *Exserohilum rostratum.* A plant pathogen that favors warm, moist climates, *E. rostratum* typically infects corn, sorghum, grass, and other monocot plants and fights them to a draw; being too weak to cause the death of the plant, while too resilient to be completely eliminated. The tenacious fungus is rewarded in the end, however, being first in line to devour the remains of its botanical host once the plant dies. The fungi's rapid growth rate, faster than any other related species, serves it well in the cornfield, while also making it a prime candidate for colonization of a

Conidia of *Exserohilum rostratum*
Source: CDC

dirty laboratory—an unfortunate confluence of events that could easily have been avoided.

Given the practices of the New England Compounding Center, the events leading to the deaths of so many people were entirely predictable. Top executives of the company had, for several years, blatantly broken the law in search of ever-greater profits. Documents related to cleanliness in the lab were routinely falsified and employees were ordered to prioritize speed over aseptic production techniques. Autoclave times were even shortened to allow the processing of a greater quantity of medication, according to court documents. In the fall of 2014 the federal government, along with many states, filed 131 charges against 14 people associated with the company. Two executives, including the owner of the NECC, were charged with 25 counts of racketeering and second-degree murder; if convicted, they could face life in prison. During bankruptcy proceedings, a $200 million fund was established to help compensate those who became sick from tainted steroids and the families of those who died.

The New England Compounding Center was faulted for, among other things, turning off the air conditioning to their manufacturing plant every evening. Why is this detail important? *Answer available on Connect.*

the isolates with air contaminants is also a problem. Treatment, which is effective only if commenced early in the infection, consists of surgical removal of infected areas accompanied by high doses of amphotericin B, posaconazole, or Cresemba.

Miscellaneous Opportunists

It is increasingly clear that nearly any fungus can be implicated in infections when the host's immune defenses are severely compromised. Although it is still relatively unusual to become infected by common airborne fungi, these infections are currently on the rise.

Geotrichosis is a rare mycosis caused by *Geotrichum candidum* **(figure 22.28a),** a mold commonly found in soil, in dairy products, and on the human body. The mold is not highly aggressive and is primarily involved in secondary infection in the lungs of tubercular or highly immunosuppressed patients.

Species of *Fusarium* **(figure 22.28b),** a common soil inhabitant and plant pathogen, occasionally infect the eyes, toenails, and burned skin. Mechanical introduction of the fungus into the eye (for example, by contaminated contact lenses) can induce a severe mycotic ulcer. In 2006 an outbreak of keratitis was traced to *Fusarium* in contact lens cleaning solution. It can also infect the nail bed and cause a condition like tinea unguium. *Fusarium* colonization of patients with widespread burns has occasionally led to mortalities.

Unusual opportunistic infections have been reported for other commonplace fungi such as *Alternaria, Malassezia,* and *Exserohilum*

(a) (b)

Figure 22.28 **Common fungi that can cause uncommon infections. (a)** *Geotrichum* produces blocky arthrospores (400×). **(b)** *Fusarium* displays crescent-shaped conidia (800×). (a): *Source:* Dr. Lucille K. Georg/CDC; (b): *Source:* CDC

(22.2 Making Connections). The widespread genus *Penicillium* is not usually infectious, although one species, *P. marneffei,* is a true pathogen restricted to certain parts of the world (Southeast Asia) and is occasionally isolated from cancer and AIDS patients.

 Check Your Progress SECTION 22.6

22. What two genera account for the majority of opportunistic fungal infections?

23. What are the treatments for localized and systemic candidiasis?

24. List the conditions that can increase the risk of developing candidiasis.

25. Outline the diseases caused by *Candida albicans* and the area of the body where each commonly manifests.

26. Summarize the antifungal regimens recommended for treating *Cryptococcus, Pneumocystis, Aspergillus,* and Zygomycota infections.

27. Briefly outline the epidemiology and pathology of *Pneumocystis jirovecii.*

22.7 Fungal Allergies and Intoxications

 Expected Learning Outcome

20. Distinguish between fungal infections, fungal allergies, and fungal intoxications.

It should be amply evident by now that fungi are constantly present in the air we breathe. In most people, inhaled fungal spores are totally without effect; in a few, they instigate infection; and in still others, they stimulate hypersensitivity. So significant is the impact of **fungal allergies** that airborne spore counts are performed by public health departments as a measure of potential risk (see figure 16.2). The most important fungal allergies are seasonal asthma and other chronic respiratory conditions that occur after occupational exposure to airborne mold spores. These allergies are typically named for the type of worker most likely afflicted: farmer's lung, teapicker's lung, barkstripper's disease.

Another effect of contact with fungi or their products is **mycotoxicosis,** a disease caused by ingesting or inhaling fungal toxins. One of the most potent of these toxins is **aflatoxin,*** a product of *Aspergillus flavus* growing on grains, corn, and peanuts. Although the mold itself is relatively innocuous, the toxin is carcinogenic and hepatotoxic. It is especially lethal to turkeys, ducklings, pigs, and calves that have eaten contaminated feed. In humans, it appears to be a cofactor in liver cancer. The potential hazard aflatoxin poses for domesticated animals and humans necessitates constant monitoring of peanuts, grains, nuts, vegetable oils, animal feed, and even milk. A quick exposure of seeds to ultraviolet radiation is a useful screening test because they glow when contaminated by the toxin.

Toxic fungi can also be a source of illness sometimes termed "sick building syndrome." These toxinoses are due to inhalation of toxins (and spores) from buildings heavily contaminated by certain "black molds." The most notable of these is *Stachybotrys chartarum* (see 5.1 Secret World of Microbes), whose toxins have been known to cause severe hematologic and neurological damage, especially in animals. Recent publicity for toxic molds has caused widespread fungus phobias and spawned a large industry for environmental mold control. Although human illness with toxic fungi has been reported with increased frequency, it is not always possible to link an outbreak to a specific fungus or toxin. For example, sick building syndrome has causations other than molds, and not all molds growing in homes or buildings are highly toxic.

 Check Your Progress SECTION 22.7

28. Discuss the importance of fungal infections, fungal allergies, and mycotoxicoses.

* *aflatoxin* (af″-lah-toks′-in) Acronym for *Aspergillus flavus* toxin.

 CASE STUDY **Part 2**

Prior to the beginning of camp, the counselors participated in a camp cleanup, where the 12 campsites were prepared for the summer. Preparations included raking leaves, cleaning picnic tables, digging fire pits, and moving firewood. Counselors reported that they found bird and bat guano on picnic tables and the dirt floor of camp shelters at two of the campsites. Campers arrived a week later and took part in nature walks, outdoor games, wilderness training, and arts and crafts, but did not engage in high-risk activities like those performed by the counselors earlier in the summer. Analysis of data collected by the DCDH and NDHHS determined that campers assigned to campsites where guano was seen were more than twice as likely to become ill than those campers assigned to campsites more than

20 yards away. In fact, the data showed that the farther away a camper was from a site where guano was found, the less likely they were to become ill. Based on this report, the camp was relocated to a different park, and the city's parks and recreation division was given recommendations to prevent bat roosting, recognize potentially contaminated areas, and properly deal with such sites.

■ *Explain why simply having antibodies specific for H. capsulatum in the bloodstream is not evidence of current infection.*

■ *In what people would infection with H. capsulatum be especially serious?*

See: CDC. 2012. Histoplasmosis outbreak among day-camp attendees—Nebraska, June 2012. *MMWR* 61(37):747–48.

To conclude this Case Study, go to Connect.

 Chapter Summary with Key Terms

22.1 Fungi as Infectious Agents
A. Microscopic fungi (molds and yeasts) are widespread in nature. A few members cause fungal diseases or **mycoses.**
B. Infectious fungi occur in groups based upon the virulence of the pathogen and the level of involvement, whether **systemic, subcutaneous, cutaneous, or superficial.**
C. **True** or **primary pathogens** have virulence factors that allow them to invade and grow in a healthy host.
 1. They are also **thermally dimorphic,** occurring as hyphae in their natural habitat and converting to yeasts while growing as parasites at body temperature (37°C).
 2. Inhaled spores initiate **primary pulmonary infection (PPI),** which can spread to the skin and other systems.
 3. Diseases are not transmissible and cause a long-term allergic reaction to fungal proteins.
 4. True pathogens are endemic to certain geographic areas.
D. **Opportunistic fungal pathogens** are normal flora or common inhabitants of the environment that invade mainly patients whose host defenses are compromised.
 1. They are only weakly virulent and lack thermal dimorphism.
 2. Infections may be local, cutaneous, or systemic and are currently on the rise, due to more weakened patients.
E. Immunity to fungal infections consists primarily of nonspecific barriers, inflammation, and cell-mediated defenses. Antibodies and antigens may be used to detect disease in some cases.
F. Diagnosis and identification require
 1. microscopic examination of stained specimens,
 2. culturing of pathogen in selective and enriched media, and
 3. specific biochemical and in vitro serological tests.
 4. Skin testing for true and opportunistic pathogens can determine prior disease but is not useful in diagnosis.
G. Control of fungal infections involves drugs such as intravenous amphotericin B, flucytosine, azoles (fluconazole), and nystatin.

22.2 Organization of Fungal Diseases
A. Histoplasmosis (Ohio Valley fever): The agent is *Histoplasma capsulatum,* distributed worldwide but most prevalent in the soil of eastern and central regions of the United States.
 1. Inhaled conidia produce primary pulmonary infection that may progress to systemic involvement of a variety of organs and chronic lung disease.
B. Coccidioidomycosis (valley fever): *Coccidioides immitis* has blocklike **arthroconidia** in the free-living stage and **spherules** containing endospores in the lungs.
 1. The agent lives in alkaline soils in semiarid, hot climates and is endemic to the southwestern United States.
 2. Arthrospores inhaled from dust create spherules and nodules in the lungs.
C. **Blastomycosis** (Chicago disease): The agent is *Blastomyces dermatitidis,* a free-living species distributed in soil of a large section of the midwestern and southeastern United States.
 1. Inhaled conidia convert to yeasts and multiply in lungs.
 2. Symptoms include cough and fever; chronic cutaneous, bone, and nervous system complications may develop.

D. Paracoccidioidomycosis (South American blastomycosis): *Paracoccidioides brasiliensis* is distributed in various regions of Central and South America. Lung infection occurs through inhalation or inoculation of spores, and systemic disease is not common.

22.3 Subcutaneous Mycoses
A. **Sporotrichosis** (rose-gardener's disease) is caused by *Sporothrix schenckii,* a free-living fungus that accidently infects the appendages and lungs. The lymphocutaneous variety occurs when contaminated plant matter penetrates the skin and the pathogen forms a local nodule, then spreads to nearby lymph nodes.
B. **Chromoblastomycosis** is a deforming, deep infection of the tissues of the legs and feet by soil fungi that enter a traumatic injury.
C. **Mycetoma** is a progressive, tumorlike disease of the hand or foot due to chronic fungal infection; may lead to loss of the body part.

22.4 Cutaneous Mycoses
A. **Dermatophytoses** are known as **tinea, ringworm,** and athlete's foot. They are caused by fungi called **dermatophytes** in the genera *Trichophyton, Microsporum,* and *Epidermophyton.* These species are adapted to growing in keratinized epidermis (skin, hair, nails).
B. Infections are communicable among humans, animals, and soil.
 1. Infection is facilitated by moist, chafed skin.
 a. Ringworm of scalp (**tinea capitis**) affects scalp and hair-bearing regions of head; hair may be lost.
 b. Ringworm of body (**tinea corporis**) occurs as inflamed, red ring lesions anywhere on smooth skin.
 c. Ringworm of groin (**tinea cruris**) affects groin and scrotal regions.
 d. Ringworm of foot and hand (**tinea pedis** and **tinea manuum**) is spread by exposure to public surfaces; occurs between digits and on soles.
 e. Ringworm of nails (**tinea unguium**) is a persistent colonization of the nails of the hands and feet that distorts the nail bed.

22.5 Superficial Mycoses
Superficial mycoses affect the outermost epidermal structures. **Tinea versicolor** causes mild scaling, mottling of skin; **white piedra** is whitish or colored masses on the long hairs of the body; and **black piedra** causes dark, hard concretions on scalp hairs.

22.6 Opportunistic Mycoses
A. Candidiasis is caused by *Candida albicans* and other *Candida* species, common yeasts that normally reside in the mouth, vagina, intestine, and skin. Infection predominates in cases of lowered resistance and arises from normal microbiota or is transmissible through intimate contact.
 1. **Thrush** occurs as a thick, white, adherent growth on the mucous membranes of mouth and throat.
 2. **Vulvovaginal candidiasis (VC)** (*yeast infection*) is a painful inflammatory condition of the female genital region that causes ulceration and whitish discharge.
 3. **Cutaneous candidiasis** occurs in chronically moist areas of skin and in burn patients, and onychomycosis is candidal infection of the nails.

B. **Cryptococcosis** is caused by ***Cryptococcus neoformans*** and ***Cryptococcus gattii,*** widespread encapsulated yeasts that inhabit soils around pigeon roosts; common infection of AIDS, cancer, or diabetes patients.
 1. Infection of lungs leads to cough, fever, and lung nodules.
 2. Dissemination to meninges and brain can cause severe neurological disturbance and death.
C. ***Pneumocystis jirovecii*** is a small, unicellular fungus that causes PCP, the most prominent opportunistic infection in AIDS patients. This pneumonia forms secretions in the lungs that block breathing and can be rapidly fatal if not controlled with medication.
D. Aspergillosis is caused by ***Aspergillus,*** a very common airborne soil fungus. Inhalation of spores causes fungus balls in lungs; it also causes invasive disease in the eyes, heart, and brain.

E. **Mucormycosis** is caused by molds in genus *Rhizopus, Mucor,* and *Absidia.* These usually harmless air contaminants invade the membranes of the nose, eyes, and brain of people with acidosis (diabetes, malnutrition), with severe consequences.
F. Other significant opportunists are *Geotrichum* (lungs) and *Fusarium* (eyes and nails).

22.7 Fungal Allergies and Intoxications
A. Airborne fungal spores are common sources of atopic allergies.
B. Contact with fungal toxins leads to **mycotoxicoses.** Most cases are caused by eating poisonous or hallucinogenic mushrooms or food.
C. **Aflatoxins** are dangerous toxins produced by *Aspergillus flavus.* When ingested in contaminated food, they may cause cancer and liver damage.

Level I. Knowledge and Comprehension

These questions require a working knowledge of the concepts in the chapter and the ability to recall and understand the information you have studied.

Multiple-Choice Questions

Select the correct answer from the answers provided. For questions with blanks, choose the combination of answers that most accurately completes the statement.

1. The ability of a fungus to alternate between hyphal and yeast phases in response to temperature is a type of
 a. sporulation
 b. conversion
 c. dimorphism
 d. binary fission

2. Which phase of a fungal life cycle is best adapted to growing in a host's body?
 a. yeast
 b. hyphal
 c. conidial
 d. filamentous

3. Primary pathogenic fungi differ from opportunistic fungi in being
 a. more virulent
 b. less virulent
 c. more prevalent
 d. more transmissible

4. True pathogenic fungi
 a. are transmissible from person to person
 b. cause primary pulmonary infections
 c. are endemic to specific geographic areas
 d. both b and c

5. Example(s) of fungal infection(s) that is/are communicable is/are
 a. dermatomycosis
 b. candidiasis
 c. sporotrichosis
 d. both a and b

6. Histoplasmosis has the greatest endemic occurrence in which region?
 a. midwestern United States
 b. southwestern United States
 c. Central and South America
 d. North Africa

7. Coccidioidomycosis is endemic to which geographic region in the United States?
 a. Southwest
 b. Northeast
 c. Pacific Northwest
 d. Southeast

8. Skin testing with antigen is a useful epidemiological procedure for
 a. histoplasmosis
 b. candidiasis
 c. coccidioidomycosis
 d. a and c

9. A mycetoma is a
 a. fungal tumor of the lung
 b. collection of pus in the brain caused by *Cryptococcus*
 c. deeply invasive fungal infection of the foot or hand
 d. type of ringworm of the scalp

10. Dermatophytic fungi attack the _____ in the _____.
 a. epithelial cells, lungs
 b. melanin, stratum corneum
 c. keratin; skin, nails, hair
 d. bone, foot

11. Example(s) of dermatophyte genus/genera is/are
 a. *Epidermophyton*
 b. *Sporothrix*
 c. *Trichophyton*
 d. a and c
 e. all of these

12. *Candida albicans* is the cause of _____, an infection of the
_____.
 a. onychomycosis, root canal
 b. moniliasis, lungs
 c. thrush, mouth
 d. salpingitis, uterus

13. Cryptococcosis is associated with
 a. blowing winds c. pigeon droppings
 b. plant debris d. excavation of ancient dwellings

14. Which fungus does *not* commonly cause systemic infection?
 a. *Blastomyces* b. *Cryptococcus*
 b. *Histoplasma* c. *Malassezia*

15. Which of the following is the cause of mycotoxicosis?
 a. *Mucor* b. *Pneumocystis*
 b. *Aspergillus* c. *Cryptococcus*

16. **Single Matching.** Match the disease with its common name.
 _____b____ coccidioidomycosis a. tinea
 _____a____ histoplasmosis b. San Joaquin Valley fever
 _____d____ blastomycosis c. South American blastomycosis
 _____e____ sporotrichosis d. Chicago disease
 _____c____ paracoccidioidomycosis e. rose-gardener's disease
 _____a____ dermatophytosis f. yeast infection
 _____f____ candidiasis g. Ohio Valley fever

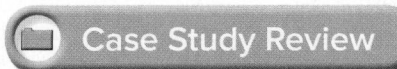

Case Study Review

1. The patients in this case were probably treated with
 a. penicillin c. specific immune globulin
 b. amphotericin B d. vaccination

2. The geographic location of infection was important in this case. If the students in the case had all visited Brazil, what fungal pathogens would have been suspect? (Choose all that apply.)
 a. *Histoplasma* c. *Coccidioides*
 a. *Blastomyces* d. *Paracoccidioides*

3. If you were tasked with preventing an outbreak similar to the one seen in the Case Study, what recommendations would you make to workers preparing to open a similar camp the following year?

Writing Challenge

For each question, compose a one- or two-paragraph answer that includes the factual information needed to completely address the question. Check Your Progress questions can also be used for writing-challenge exercises.

1. Describe the concept behind an immunodiffusion test. Explain how the histoplasmin test is done.

2. a. What are the dermatophytoses?
 b. What is meant by the term *keratophile?*
 c. What do the fungi actually feed upon?
 d. To what do the terms *ringworm* and *tinea* refer?

3. a. What is the difficulty in ridding the epidermal tissues of certain dermatophytes?
 b. What is necessary in treatment?

4. Describe three forms of fungal toxinoses and the agents involved in them.

Concept Mapping

On Connect you can find an Introduction to Concept Mapping that provides guidance for working with concept maps along with concept mapping activities for this chapter.

Level II. Application, Analysis, Evaluation, and Synthesis

These problems go beyond just restating facts and require higher levels of understanding and an ability to interpret, problem solve, transfer knowledge to new situations, create models, and predict outcomes.

Critical Thinking

Critical thinking is the ability to reason and solve problems using facts and concepts. These questions can be approached from a number of angles, and in most cases, they do not have a single correct answer.

1. Give reasons most humans are not constantly battling fungal infections, considering how prevalent fungi are in the environment.

2. a. Name several medical conditions that compromise the immunities of patients (see table 22.3).
 b. Is there any such thing as a harmless contaminant with these patients?
 c. Describe the consequences if a virulent fungal pathogen infects a severely compromised patient.

3. Using the general guidelines presented in chapter 17 and figure 22.3, explain how you can tell for sure if a fungal isolate is the actual causative agent or merely a contaminant.

4. a. Explain why some dermatophytes are considered good parasites.
 b. Why would infections from soil and animal species be more virulent than ones from fungi adapted to humans?
 c. Suggest a reason adult males have more cases of histoplasmosis.

5. What features of the lungs make them a favorable site for fungal infections?

6. a. Describe the difficulties in diagnosing endemic fungal infections that are acquired by travelers or nonnatives to endemic areas.
 b. Which forms of diagnostic testing will improve the likelihood of rapid detection of such infections?

7. A male researcher wanting to determine the pathogenesis of a certain fungus that often attacks women did the following experiment on himself. He taped a piece of gauze heavily inoculated with the fungus to the inside of his thigh and left it there for several days, periodically moistening it with sterile saline. After a time he experienced a severe skin reaction that resembled a burn and the skin peeled off.
 a. Suggest a species of fungus that can cause this effect.
 b. What does this experiment tell you about the factors in pathogenesis related to portal of entry and gender?

8. A worker involved in demolishing an older downtown building is the victim of a severe lung infection.
 a. Without any lab data, what two fungal infections do you suspect?
 b. What is your reasoning?
 c. How would you differentiate them?

Visual Challenge

1. Using figure 22.3 as a guide, proceed to go through the procedures to diagnose this infection.

2. What would you expect to see microscopically from a scraping of the white patch? Describe other diagnostic procedures for this infection.

Source: CDC

© Kathy Park Talaro

 connect® **STUDY TIP:** Quickly create a customized practice quiz for this chapter by going to your Connect SmartBook "My Reports" section. Don't have SmartBook and want to learn more? Go to **whysmartbook.com**.

23 The Parasites of Medical Importance

An electron micrograph of *Naegleria fowleri*. The eyes and mouth of its "face" are actually feeding structures. Courtesy of Dr. Francine Marciano-Cabral

WARNING
DO NOT ALLOW WATER TO ENTER YOUR NOSE
NAEGLERIA FOWLERI
AN AMOEBA COMMON TO THERMAL POOLS MAY ENTER CAUSING A RARE INFECTION AND DEATH

AVISO
NO DEJAR QUES SE META AGUA POR LA NARIZ
NAEGLERIA FOWLERI
ES UNA AMIBA QUE SE ENCUENTRA EN MANANTIALES CAL DOS Y PUEDE SER CAUSA DE UNA INFECC ON GRAVE O LA MUERTE SI SE ENTRA AGUA POR LA NARIZ

© birdandhike.com

 CASE STUDY Part 1 Keep Your Head Above Water

Like the winter flu season, the warm temperatures and sunny days of springtime bring with them an increase in a number of diseases: food poisoning from undercooked hamburgers, **Lyme disease** from tick bites, **hookworm** from walking barefoot in new grass. But when a 28-year-old man was declared **brain-dead** only 3 days after his symptoms first appeared, there was more than the usual cause for concern.

When first admitted to the hospital, the man was mentally altered and highly combative. His symptoms had begun the previous day with a sudden **headache, stiff neck,** and **backache.** He was **vomiting** intermittently and on the day of his admission developed a **fever.** Suspecting **meningitis,** doctors performed a **spinal tap,** revealing **cerebrospinal fluid** that was cloudy, indicative of infection. Despite antibiotic treatment, the man was declared brain-dead 2 days after admission. Four months later, a 51-year-old woman from northern Louisiana was admitted to the hospital under similar circumstances, her stay mirroring that of the younger man. She died after 5 days in the hospital.

The Centers for Disease Control and Prevention examined both patients and determined that they died of **primary amebic**

> "Despite antibiotic treatment, the man was declared brain-dead 2 days after his admission to the hospital."

meningoencephalitis (PAM), an exceedingly rare disease caused by infection of the brain with the protozoan **Naegleria fowleri.** Sometimes referred to as the **"brain-eating amoeba,"** for the widespread destruction of brain tissue it causes, *N. fowleri* is commonly found in warm lakes and rivers in the southern United States, especially during the summer months. Infection begins when the amoeba enters the nose while swimming, diving, or even simply ducking underwater. From there, it migrates along the **olfactory nerve** to the brain, initiating disease in which death is a virtual certainty. Why such a commonly encountered pathogen results in so few infections—32 cases since 2002—is not known.

- *Why doesn't* Naegleria fowleri *cause infection when it is swallowed?*

- *What precautions do you think have been recommended to prevent infection with* N. fowleri?

To continue the Case Study, go to Case Study Part 2 at the end of the chapter.

23.1 The Parasites of Humans

Expected Learning Outcome

1. Describe the changing epidemiology of parasitic disease.

Parasitology is traditionally the study of eukaryotic parasites. It encompasses protozoa (unicellular animal-like organisms) and helminths (worms), which are sometimes collectively called the macroparasites, as opposed to the microparasites (bacteria and viruses). This inherently fascinating and practical subject will be presented as a compact survey of the major clinically important protozoa, helminths, and some of the arthropod vectors. This chapter covers elements of their morphology, life cycles, epidemiology, pathogenic mechanisms, and methods of control. Arthropods, which are often included in the field of parasitology, are discussed as vectors for parasitic, bacterial, and viral infections.

One cannot underestimate the impact of parasites on the everyday lives of humans. The World Health Organization (WHO) estimates that parasites cause nearly 20% of all infectious diseases. We were first introduced to the parasitic diseases in chapter 5 and its Case Study. Many of these are considered a type of neglected tropical disease; such diseases afflict at least 20% of the people on earth. Although these diseases are less prevalent in industrialized countries, most humans will, at some point in their lives, have some form of parasite infection, in hidden or overt form. The distribution of parasitic diseases is influenced by many factors, including travel, immigration, and the increased number of immunocompromised patients. Because parasites can literally travel around the world, it is not uncommon to discover exotic parasitic diseases even in the United States.

The current increase in patients with severe immune suppression is changing the epidemiological pattern of parasitic diseases. AIDS patients in particular play host to a variety of opportunists that dominate their clinical picture. More and more, parasites are emerging as the agents of life-threatening diseases.

 Check Your Progress SECTION 23.1

1. How do travel, immigration, and AIDS all affect the epidemiology of parasitic infection?

23.2 Major Protozoan Pathogens

Expected Learning Outcomes

2. Recall the identifying attributes of the four commonly recognized groups of protozoa.

3. Identify the amoebas generally seen as human pathogens and the portals of entry for each.

4. Depict the life cycle seen in *Entamoeba histolytica* infection.

5. Contrast the pathogenesis, treatments, and common outcomes of *Entamoeba, Naegleria,* and *Acanthamoeba* infections.

6. Classify and describe the important characteristics of *Balantidium coli.*

The introduction to **protozoa** in chapter 5 described them as single-celled, animal-like microbes, most of which have some form of motility. The four commonly recognized informal groups

Systems Profile 23.1	Parasitic Protozoan Pathogens				
Organism	Skin/Skeletal	Nervous/Muscle	Cardiovascular/ Lymphatic/Systemic	Gastrointestinal	Urogenital
Entamoeba histolytica				Chronic diarrhea dysentery	
Naegleria fowleri		Meningoencephalitis			
Acanthamoeba		Meningoencephalitis			
Balantidium coli				Chronic diarrhea	
Trichomonas vaginalis					Trichomoniasis
Giardia intestinalis				Chronic diarrhea	
Trypanosoma brucei		African sleeping sickness			
Trypanosoma cruzi			Chagas disease		
Leishmania spp.	Cutaneous leishmaniasis		Systemic leishmaniasis	Intestinal damage	
Plasmodium falciparum, P. vivax, P. ovale, P. malariae, P. knowlesi			Malaria		
Toxoplasma gondii		Subacute encephalitis			
Sarcocystis spp.				Acute diarrhea	
Cryptosporidium				Gastroenteritis	
Cyclospora spp.				Acute diarrhea	

are sarcodinians (amoebas that move with pseudopods), ciliates that move with cilia, flagellates that use flagella, and apicomplexans that lack typical organelles for movement.

Of the estimated 100,000 species of protozoans, about 25 are important pathogens of humans **(table 23.1).** They collectively cause billions of infections worldwide.

Although the life cycles of pathogenic protozoa can vary a great deal, most of them propagate by simple asexual cell division of the active feeding cell, called a **trophozoite.** Many species undergo formation of a **cyst** that can survive for periods outside the host and can also function as an infective body (see figure 5.24). Others have a more complex life cycle that includes asexual and sexual phases (see discussion of apicomplexans in section 23.4) and a variety of host animals.

Drugs used to treat protozoan infections were introduced in chapter 12. There are currently about a dozen different drug types available. **Table 23.2** outlines a selection of drugs that target major protozoan parasites along with their general modes of action. Because these drugs target eukaryotic cells and they usually work by disrupting common cell processes, many of them have severe side effects and some of them are quite toxic to human cells.

Infective Amoebas

Entamoeba histolytica *and Amebiasis*

Amoebas are widely distributed in aqueous habitats and are frequent parasites of animals, but only a few of them have the necessary virulence to invade tissues and cause serious pathology.

The most significant pathogenic amoeba is ***Entamoeba histolytica.**** The relatively simple life cycle of this parasite alternates between a large trophozoite that is motile by means of pseudopods and a smaller, compact, nonmotile cyst (see figure 5.24). The trophozoite lacks most of the organelles of other eukaryotes, and it has a large single nucleus that contains a prominent nucleolus called a *karyosome.* Amoebas from fresh specimens are often packed with food vacuoles containing host cells and bacteria. The mature cyst is encased in a thin yet tough wall and contains four nuclei as well as distinctive bodies called *chromatoidals,* which are actually dense clusters of ribosomes.

Epidemiology of Amebiasis Humans are the primary hosts of *E. histolytica.* Infection is usually acquired by ingesting food or drink contaminated with cysts released by an asymptomatic carrier. The cyst form of the amoeba is thought to be carried in the intestine of one-tenth of the world's population, and it kills around 100,000 people a year. Its geographic distribution is partly due to local sewage disposal and fertilization practices. Occurrence is highest in tropical regions (Africa, Asia, and Latin America), where night soil[1] or untreated sewage is sometimes used to fertilize crops, and sanitation of water and food is not always adequate. Although the prevalence of the disease is substantially lower in the United States, as many as 10 million people could harbor the agent asymptomatically.

* *Entamoeba histolytica* (en″-tah-mee′-bah his″-toh-lit′-ih-kuh) Gr. *ento,* within, *histos,* tissue, and *lysis,* a dissolution.
1. Human excrement used as fertilizer.

TABLE 23.1	Major Pathogenic Protozoa, Infections, and Primary Sources	
Protozoan/Disease*	**Organ/System Primarily Affected****	**Reservoir/Source**
Amoeboid Protozoa		
Amebiasis: *Entamoeba histolytica*	Gastrointestinal tract	Human/water and food
Brain infection: *Naegleria, Acanthamoeba*	Nervous system	Free-living in water
Ciliated Protozoa		
Balantidiosis: *Balantidium coli*	Gastrointestinal tract	Zoonotic in pigs
Flagellated Protozoa		
Giardiasis: *Giardia intestinalis*	Gastrointestinal tract	Zoonotic/water and food
Trichomoniasis: *Trichomonas vaginalis*	Urogenital tract	Human
Hemoflagellates		
Trypanosomiasis: *Trypanosoma brucei, T. cruzi*	Nervous system	Zoonotic/vector-borne
Leishmaniasis: *Leishmania donovani, L. tropica, L. brasiliensis*	Skin	Zoonotic/vector-borne
Apicomplexan Protozoa		
Malaria: *Plasmodium vivax, P. falciparum, P. malariae, P. ovale, P. knowlesi*	Cardiovascular system	Human/vector-borne
Toxoplasmosis: *Toxoplasma gondii*	Nervous system	Zoonotic/vector-borne
Cryptosporidiosis: *Cryptosporidium*	Gastrointestinal tract	Free-living/water, food
Isosporosis: *Cystoisospora belli*	Gastrointestinal tract	Dogs, other mammals
Cyclosporiasis: *Cyclospora cayetanensis*	Gastrointestinal tract	Water/fresh produce

*Many flagellates are named in honor of one of their discoverers: for example, *T. cruzi,* O. Cruz; *Leishmania,* W. Leishman; *donovani,* C. Donovan; *Giardia,* A. Giard; and *brucei,* D. Bruce. Geographic location (*gambiense, rhodesiense, tropica, brasiliensis, mexicana*) is another basis for naming. Names are also derived from a certain characteristic of the organism: for example, *vaginalis,* vaginal habitat.

**Although signs and symptoms of infection may often be observed throughout the body, initial infection is often restricted to a single organ or system.

Pathogen Profile #1 *Entamoeba histolytica*

Source: CDC-DPDx

Microscopic Morphology The large trophozoite has a single nucleus with a prominent nucleolus called a karyosome and is often packed with food vacuoles. Mature cysts contain four nuclei, along with dense clusters of ribosomes called chromatoidals.

Identified by Visualization of trophozoites or cysts during microscopic examination of fecal smears; ELISA testing, specific to cells.

Habitat Humans are the primary host for *E. histolytica*. Occurrence is highest in tropical regions—Africa, Asia, and Latin America—where untreated sewage is used to fertilize crops. Prevalence is much lower in the United States but may still encompass 10 million asymptomatic carriers. Cysts are continuously shed in the feces of chronic carriers.

Virulence Factors Hardy cysts; lytic enzymes that promote invasiveness and induce tissue damage.

Primary Infections/Disease Amebiasis begins when ingested cysts undergo excystment in the small intestine. *E. histolytica* secretes enzymes that digest tissue, causing ulcerations of the intestinal mucosa, abdominal pain, fever, diarrhea, and weight loss. More severe infection may cause hemorrhage, appendicitis, and tumorlike growths called amoebomas. Extraintestinal infection may encompass the liver, pulmonary system, and, rarely, the spleen, adrenals, kidney, skin, and brain. Severe forms of the disease have about a 10% mortality rate.

Control and Treatment Control centers on adequate disinfection of water. Because cysts are resistant to levels of chlorine commonly used in water treatment, boiling or iodine treatment of water is required. Amebiasis may be treated with iodoquinol, metronidazole, or paromomycin.

TABLE 23.2 Guidelines for Antiprotozoan Drugs

Infectious Agent	Typical Infection	Drugs of Choice	Mechanism of Drug Action
Entamoeba histolytica	Amebiasis	Metronidazole*	Specific for anaerobes; inhibits DNA repair
		Iodoquinol	Acts locally in intestine; unknown mode of action
		Paromomycin	Inhibits protein syntheses
Giardia intestinalis	Giardiasis	Tinidazole	Selectively blocks parasite metabolism
		Nitazoxanide	Selectively blocks parasite metabolism
Plasmodium	Malaria	Chloroquine	Interferes with metabolism and use of hemoglobin
		Atovaquone-proguanil	Interferes with parasitic electron transport chain
		Mefloquine	Interferes with metabolism and use of hemoglobin
		Artemether	Redox reagent; produces reactive oxygen radicals
Toxoplasma gondii	Toxoplasmosis	Pyrimethamine Sulfadiazine	Together with sulfadiazine, inhibits enzymes in the pathway
Trichomonas vaginalis	Trichomoniasis	Metronidazole	Together with sulfadiazine, inhibits enzymes in the pathway for folate synthesis
Trypanosoma brucei	Sleeping sickness	Suramin	Inhibits NADH oxidation; blocks aerobic respiration
		Pentamidine	Blocks nucleic acid synthesis
		Eflornithine	Inhibits deamination of amino acids
		Melarsoprol	Arsenic-based, very toxic, inhibits many enzymes
Trypanosoma cruzi	Chagas disease	Benznidazole	Interferes with protein synthesis and microtubule formation
		Nifurtimox	Breaks up DNA and creates superoxide radicals

*Metronidazole is also called Flagyl. It is selectively toxic to anaerobic protozoa like *Entamoeba, Giardia,* and *Trichomonas,* which lack mitochondria.

Epidemics of amebiasis are infrequent but have been documented in prisons, hospitals, juvenile care institutions, and communities where water supplies were polluted. Amoebic infections can also be transmitted by some forms of sexual contact.

Life Cycle, Pathogenesis, and Control of *E. histolytica* Amebiasis begins when viable cysts are swallowed and arrive in the small intestine (**figure 23.1;** more detail is shown in figure 5.27), where the alkaline pH and digestive juices of this environment stimulate excystment. The cyst releases four trophozoites (**figure 23.1c**), which are swept into the **cecum*** and large intestine. There the

* *cecum* (see′-kum) L. *caecus,* blind. The pouchlike anterior portion of the large intestine near the appendix.

(a) (b) **Trophozoite** (c) **Mature Cyst**

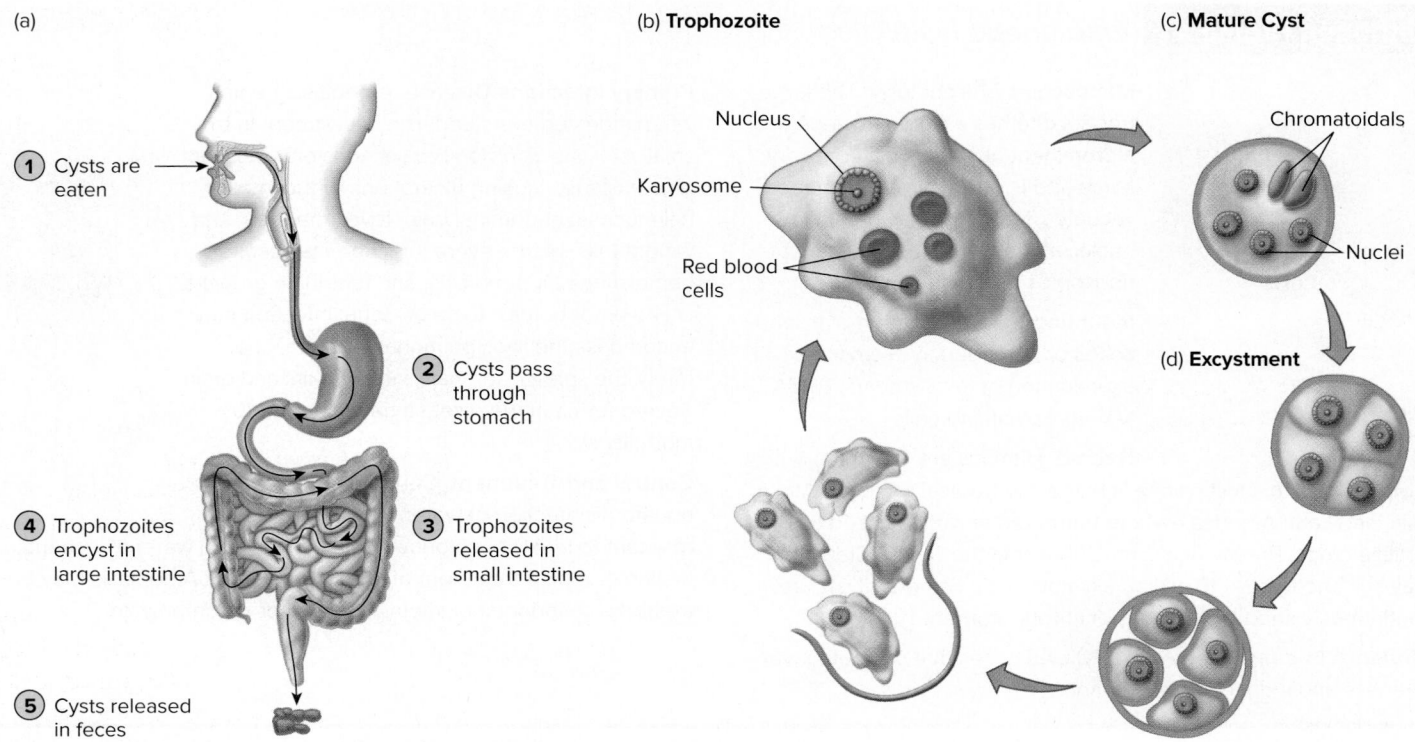

1. Cysts are eaten
2. Cysts pass through stomach
3. Trophozoites released in small intestine
4. Trophozoites encyst in large intestine
5. Cysts released in feces

Nucleus
Karyosome
Red blood cells
Chromatoidals
Nuclei
(d) **Excystment**

Process Figure 23.1 **Cellular forms of *Entamoeba histolytica*. (a)** Simplified view of route of infection, development, and release. **(b)** A trophozoite containing a single nucleus, a karyosome, and red blood cells. **(c)** A mature cyst with four nuclei and two blocky chromatoidals. **(d)** Stages in excystment. Divisions in the cyst create four separate cells, or metacysts, that differentiate into trophozoites and are released.

trophozoites attach by fine pseudopods and multiply, actively moving about and feeding. In about 90% of patients, infection is asymptomatic or very mild, and the trophozoites do not invade beyond the most superficial layer. The severity of the infection can vary with the strain of the parasite, inoculation size, diet, and host resistance.

As hinted by its species name, tissue damage is one of the formidable characteristics of untreated *E. histolytica* infection. Clinical amebiasis exists in intestinal and extraintestinal forms. The initial targets of intestinal amebiasis are the cecum, appendix, colon, and rectum. The amoeba secretes enzymes that dissolve tissues, and it actively penetrates deeper layers of the mucosa, leaving erosive ulcerations **(figure 23.2)**. This phase is marked by dysentery (bloody, mucus-filled stools), abdominal pain, fever, diarrhea,

and weight loss. The most life-threatening manifestations of intestinal infection are hemorrhage, perforation, appendicitis, and tumorlike growths called amoebomas.

Extraintestinal infection occurs when amoebas invade the viscera of the peritoneal cavity. The most common site of invasion is the liver. Here, abscesses that contain necrotic tissue and trophozoites cause amebic hepatitis. Another regular complication is pulmonary amebiasis. Less frequent targets of infection are the spleen, adrenals, kidney, skin, and brain. Severe forms of the disease result in about a 10% fatality rate.

Entamoeba is harbored by chronic carriers whose intestines favor the encystment stage of the life cycle. Cyst formation cannot occur in active dysentery because the feces are so rapidly flushed from the body, but after recuperation, cysts are continuously shed in feces.

Initial diagnosis involves microscopic examination of a fecal smear for the presences of trophozoites or cysts typical of *E. histolytica* **(figure 23.3)**. More definitive identification can be obtained with a stained smear or with a rapid ELISA test using specific monoclonal antibodies. It may be necessary to supplement laboratory findings with clinical data, including symptoms, X rays, and the patient's history, which might indicate a source of exposure.

Effective treatment for amebic dysentery usually involves the use of drugs such as metronidazole (Flagyl), tinidazole, or paromomycin, or nitazoxanide. Other drugs are given to relieve diarrhea and cramps, while lost fluid and electrolytes are replaced by oral or intravenous therapy. Infection with *E. histolytica* provokes antibody formation against several antigens, but long-term immunity is unlikely and reinfection can occur. Prevention depends upon concerted efforts similar to those for other enteric diseases (see 20.3 Making Connections).

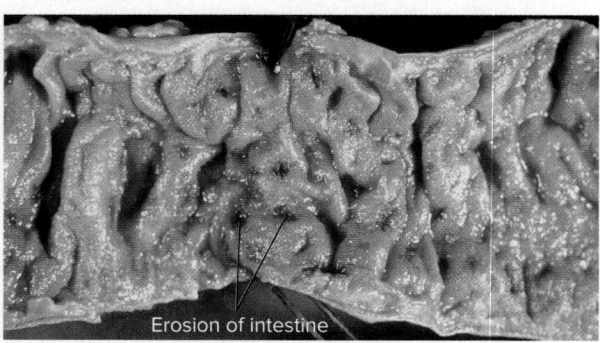

Erosion of intestine

Figure 23.2 Appearance of amebiasis and dysentery lesions of the large intestine. Red erosions are sites of amebic damage to the intestinal mucosa (pointers). *Source:* CDC

 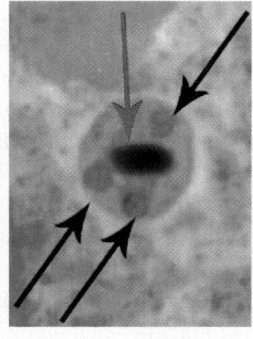

(a) (b)

Figure 23.3 **Microscopic appearance of *Entamoeba histolytica* in fresh specimen. (a)** A trophozoite (20 μm) is filled with its primary nutrient—red blood cells (black spheres). **(b)** A cyst (12 μm) with three nuclei (black arrows) and a single chromatoidal (red arrow). (a, b): *Source:* CDC-DPDx

Because regular chlorination of water supplies does not kill cysts, more rigorous methods such as boiling or iodine are required.

Amebic Infections of the Brain

Two common free-living protozoans that cause a rare and usually fatal infection of the brain are ***Naegleria fowleri**** and ***Acanthamoeba* spp.*** Both genera are accidental parasites that invade the body only under unusual circumstances. The trophozoite of *Naegleria* is a small, flask-shaped amoeba that moves by means of a single, broad pseudopod **(figure 23.4)**. It forms a rounded, thick-walled, uninucleate cyst that is resistant to temperature extremes and mild chlorination. *Acanthamoeba* has a large, amoeboid trophozoite with spiny pseudopods and a double-walled cyst.

Both *N. fowleri* and *Acanthamoeba* ordinarily inhabit standing fresh or brackish water, lakes, puddles, ponds, hot springs, moist soil, and even swimming pools and hot tubs. They are especially abundant in warm water with a high bacterial count.

Most cases of *Naegleria* infection reported worldwide occur in people who have been swimming in warm, natural bodies of fresh

* *Naegleria fowleri* (nay-glee′-ree-uh fow′-ler-eye) After F. Nagler and N. Fowler.
* *Acanthamoeba* (ah-kan″-thah-mee′-bah) Gr. *acanthos,* thorn.

(a) (b)

Figure 23.4 **Free-living infective amoebas (a)** *Naegleria fowleri* has thick, blunt pseudopods and is filled with food vacuoles. **(b)** *Acanthamoeba* has spinelike processes extending from its pseudopodia. (a, b): *Source:* CDC-DPDx

water. Infection can begin when amoebas are forced into human nasal passages as a result of swimming, diving, or other aquatic activities. Once the amoeba is inoculated into the favorable habitat of the nasal mucosa, it burrows in, multiplies, and subsequently migrates into the brain and surrounding structures. The result is primary acute meningoencephalitis, a rapid, massive destruction of brain and spinal tissue that causes hemorrhage and coma and invariably ends in death within a week or so.

Unfortunately, *Naegleria* meningoencephalitis advances so rapidly that treatment usually proves futile. Studies have indicated that early therapy with amphotericin B, sulfadiazine, or tetracycline in some combination can be of some benefit. Because of the wide distribution of the amoeba and its hardiness, no general means of control exists. Public swimming pools and baths must be adequately chlorinated and checked periodically for the amoeba.

Acanthamoeba differs from *Naegleria* in its portal of entry, invading broken skin, the conjunctiva, and occasionally the lungs and urogenital epithelia. Although it causes a meningoencephalitis somewhat similar to that of *Naegleria,* the course of infection is lengthier. At special risk for infection are people with traumatic eye injuries, contact lens wearers, and swimmers in contaminated water. Ocular infections can be avoided by attention to injured eyes and using sterile solutions to store and clean contact lenses.

An Intestinal Ciliate: *Balantidium coli*

Most ciliates are free-living in various aquatic habitats, where they assume a role in the food web as scavengers or as predators upon other microbes. Several species are pathogenic to animals, but only one, ***Balantidium**** *coli,* occasionally establishes infection in humans. This large ciliate (about the size of a period on this page) exists in both trophozoite and cyst states **(figure 23.5).** Its surface is covered by orderly, oblique rows of cilia; and a shallow depression, the cytostome (cell mouth), is at one end. These parasites will grow only under anaerobic conditions in the presence of the host's bacterial microbiota.

* *Balantidium* (bal″-an-tid′-ee-um) Gr., a small sac.

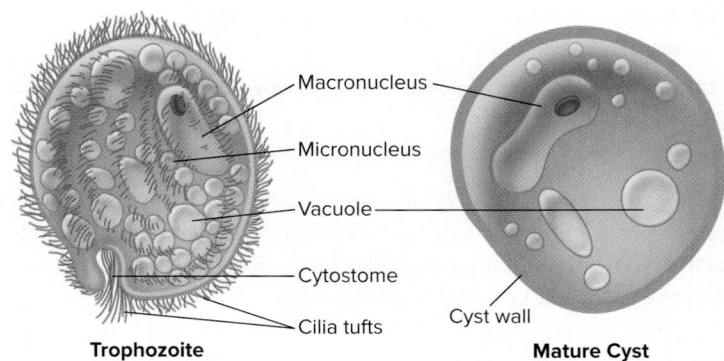

Trophozoite Mature Cyst

Figure 23.5 **Microscopic views of *Balantidium coli*.** The morphology of a trophozoite and a mature cyst.

<div style="text-align:right">

Quick Search

To view the deliberate movements of amoebas like *Naegleria fowleri,* search the Internet for "*Naegleria fowleri.*"

</div>

The large intestine of pigs and, to a lesser extent, sheep, cattle, horses, and primates constitutes the natural habitat of *B. coli,* which are shed as cysts in the feces of these animals. Although pigs are the most common source of *balantidiosis,* it can also be spread by contaminated water and from person to person in institutional settings. Healthy humans and animals are resistant to infection, but compromised hosts are vulnerable to attack.

Like cysts of *E. histolytica,* those of *B. coli* resist digestion in the stomach and small intestine and liberate trophozoites that immediately burrow into the epithelium with their cilia. The resultant erosion of the intestinal mucosa produces varying degrees of irritation and injury, leading to nausea, vomiting, diarrhea, dysentery, and abdominal colic. The protozoan rarely penetrates the intestine or enters the blood. The treatment of choice is oral tetracycline, but if this therapy fails, iodoquinol, nitrimidazine, or metronidazole can be given. Preventive measures are similar to those for controlling amebiasis, with additional precautions to prevent food and drinking water from being contaminated with pig manure.

 Check Your Progress SECTION 23.2

2. List the four groups into which protozoa are commonly placed, along with the characteristics used to do so.

3. Describe the basic pathology of amebiasis. How and where does it invade?

4. Provide several behaviors that would increase the chance of contracting *Entamoeba, Naegleri,* and *Acanthamoeba* infections.

5. Outline the course of typical *Entamoeba, Naegleri,* and *Acanthamoeba* infections.

6. Relate the life cycle, pathogenesis, and control of *Entamoeba histolytica* to that of *Balantidium coli.*

23.3 The Flagellates (Mastigophorans)

 Expected Learning Outcomes

7. Describe the important characteristics of *Trichomonas vaginalis* and the pathology of infection.

8. Understand the importance of cysts to the transmission of *Giardia.*

9. Describe the four developmental stages of hemoflagellates, and identify those stages that are infectious.

10. Compare the vectors involved in sleeping sickness, Chagas disease, and leishmaniasis.

11. Compare the pathologies of sleeping sickness, Chagas disease, and leishmaniasis.

Pathogenic flagellates play an extensive role in human diseases, ranging from milder or self-limited infections (trichomoniasis and giardiasis) to debilitating and life-threatening, vector-borne diseases (trypanosomiasis and leishmaniasis).

Trichomonads: *Trichomonas* Species

Trichomonads are small, pear-shaped protozoa with four anterior flagella and an undulating membrane, which together produce a

characteristic twitching motility in live preparations. They exist only in the trophozoite form and do not produce cysts. Three trichomonads intimately adapted to the human body are *Trichomonas vaginalis, T. tenax,* and *T. hominis* (**figure 23.6**).

The most important pathogen, ***Trichomonas** *vaginalis,*** causes a sexually transmitted disease called **trichomoniasis.** The reservoir for this protozoan is the human urogenital tract, and about 70% of infected people are asymptomatic carriers. Because *T. vaginalis* has no protective cysts, it is a relatively strict parasite that does not survive for long out of the host. Transmission is primarily through contact between genital membranes, although it can occasionally be transmitted through communal bathing and public facilities. The incidence of infection is highest among sexually active young women already infected with other sexually transmitted diseases such as gonorrhea or chlamydia. Approximately 7 million new cases of trichomoniasis occur every year in the United States, placing it near the top of the most common STDs.

Symptoms and signs of trichomoniasis in the female include a foul-smelling, green-to-yellow vaginal discharge; vulvitis; cervicitis; and urinary frequency and pain. Severe inflammation of the infection site causes edema, chafing, and itching. Males with overt clinical infections experience persistent or recurring urethritis, with a thin, milky discharge and occasionally prostate infection.

Infection may be preemptively diagnosed by demonstrating the active swimming trichomonads in a wet film of exudate, but

* *Trichomonas* (trik″-oh-moh′-nus) Gr. *thrix,* hair, and *monas,* unit.

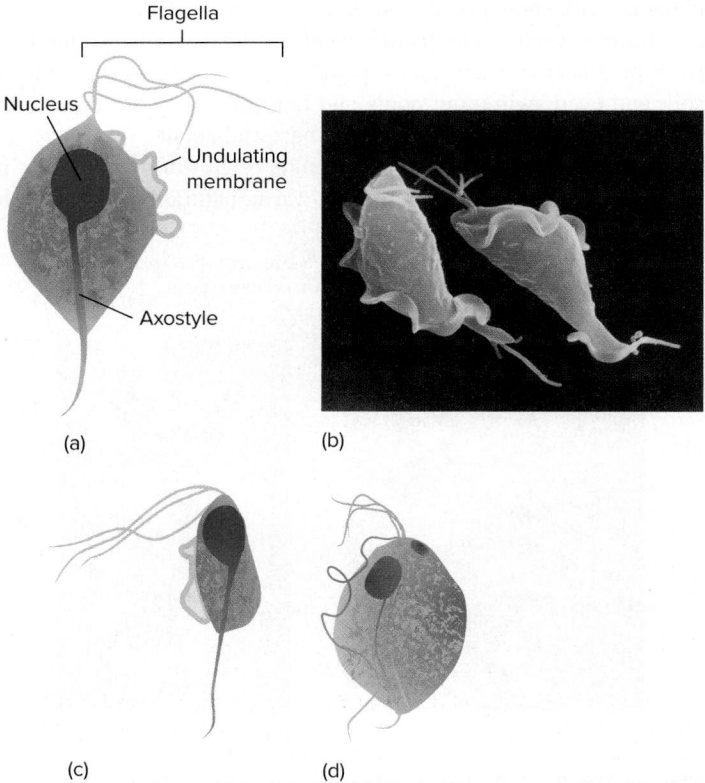

Figure 23.6 The trichomonads of humans. (a) *Trichomonas vaginalis,* a urogenital pathogen. **(b)** Micrograph at 500×. **(c)** *T. tenax,* a gingival form (infection is rare). **(d)** *T. hominis,* an intestinal species (infection is rare). (b): © David M. Phillips/Science Source

culture of the parasite is the gold standard. A number of commercial serological and DNA amplification techniques exist for detection of the parasite. The infection has been successfully treated with oral and vaginal metronidazole, and it is essential to treat both sex partners to prevent "ping-pong" reinfection.

Giardia intestinalis and Giardiasis

Giardia intestinalis* is a pathogenic, flagellated protozoan first observed by Antonie van Leeuwenhoek in his own feces. For 200 years, it was considered a harmless or weak intestinal pathogen, and only since the 1950s has its prominence as a cause of diarrhea been recognized. In fact, it is the most common flagellate isolated in clinical specimens. Observed straight on, the trophozoite has a unique, symmetrical heart shape with organelles positioned in such a way that it resembles a face **(figure 23.7*a,b*).** Four pairs of flagella emerge from the ventral surface, which is concave and acts

* *Giardia intestinalis* (jee′-ard-ee-uh in-tes′-Ti-n‾a′lis) Also called *G. lamblia.*

Trophozoite **Cyst**

(a)

(b) (c)

Figure 23.7 Identification of trophozoites and cysts of *Giardia intestinalis.* **(a)** The face of a trophozoite with eyes (paired nuclei). Side view shows the ventral depression it uses for suction. **(b)** Stained example of trophozoite from a specimen (1,000×). The nuclei are evident but flagella are less observable. **(c)** A DNA probe test uses FISH (fluorescent in situ hybridization) technology to differentiate *Giardia* species using colored dyes. *G. intestinalis* glows blue, whereas *G. muris* glows green. In both forms, the cyst wall antigens are shown in red. (b): © Donald Lehman, Dept. of Medical Laboratory Sciences, University of Delaware; (c): Courtesy Stanley Erlandsen

like a suction cup for attachment to a substrate. *Giardia* cysts are small, compact, and multinucleate (figure 23.7*a*).

Giardiasis (gee′-ar-dy′-uh-sis) has a complex epidemiological pattern. The protozoan has been isolated from the intestines of beavers, cattle, coyotes, cats, and human carriers, all of which may serve as reservoirs. Although both trophozoites and cysts escape in the stool, the cysts play a greater role in transmission. Unlike other pathogenic flagellates, *Giardia* cysts can survive for 2 months in the environment. Cysts are usually ingested with water and food or swallowed after close contact with infected people or contaminated objects. Infection can occur with a dose of only 10 to 100 cysts.

Reported cases cases of giardiasis for 2015 were 11,665, a slight decrease compared to the previous few years, and only a fraction of the actual number of cases. The best estimate for the incidence of giardiasis is 1.2 million cases per year. There are several possible modes of transmission. *Giardia* epidemics have been traced to water from fresh mountain streams as well as chlorinated municipal water supplies in several states. Infections can occur in hikers and campers who used what they thought was clean water from sources in remote mountain areas. Because wild mammals such as muskrats and beavers are intestinal carriers, they could account for cases associated with drinking water from these sources. Obviously, checking water for purity with the naked eye is unreliable, because the cysts are too small to be detected.

Cases of food-borne illness have been traced to carriers who contaminate food through unclean personal habits. Day care centers have reported outbreaks associated with diaper changing and contact with other types of fomites. *Giardia* is also a well-known cause of traveler's diarrhea, which sometimes does not appear until the person has returned home.

Ingested *Giardia* cysts enter the duodenum, germinate, and travel to the jejunum to feed and multiply. Feeding trophozoites damage the epithelial cells causing edema and infiltration by white blood cells, but these effects are reversible. Typical symptoms include diarrhea, abdominal pain, and flatulence, lasting a few weeks.

Diagnosis of giardiasis can be difficult because the organism is not shed constantly in the feces. New developments include a promising DNA assay for detection of the cysts in samples (figure 23.7*c*) and a rapid screening test that differentiates *Giardia* from another parasite, *Cryptosporidium,* in stool samples. The infection can be eradicated with tinidazole or metronidazole therapy.

Giardia has forced many water agencies have had to rethink their policies on water maintenance and testing. The agent is killed by boiling, ozone, and iodine, but, unfortunately, the amount of chlorine used in municipal water supplies does not destroy the cysts. People who must use water from remote sources should assume that it is contaminated and take measures to disinfect it.

Hemoflagellates: Vector-Borne Blood Parasites

Other parasitic flagellates are called *hemoflagellates* because of their propensity to live in the blood and tissues of the human host. Members of this group, which include species of ***Trypanosoma**** and ***Leishmania,*** have several distinctive characteristics in common. They are obligate parasites that cause life-threatening and

* *Trypanosoma* (try-pan′-oh-soh-mah) Gr. *trypanon,* borer, and *some,* body.

TABLE 23.3	Cellular and Infective Stages of the Hemoflagellates			
Genus/Species	Amastigote	Promastigote	Epimastigote	Trypomastigote
Leishmania	Intracellular in human macrophages	Found in sand fly gut; **infective to humans**	Does not occur	Does not occur
Trypanosoma brucei	Does not occur	Does not occur	Present in salivary gland of tsetse fly	In biting mouthparts of tsetse fly; **infective to humans**
Trypanosoma cruzi	Intracellular in human macrophages, liver, heart, spleen	Occurs	Present in gut of reduviid (kissing) bug	In feces of reduviid bug; **transferred to humans**

debilitating zoonoses; they are spread by blood-sucking insects that serve as intermediate hosts; and they are acquired in specific tropical regions. Hemoflagellates have complicated life cycles and undergo morphological changes as they pass between vector and human host. To keep track of these developmental stages, parasitologists have adopted the following set of terms:

amastigote (ay-mas′-tih-goht) Gr. *a-*, without, and *mastix,* whip. The form lacking a free flagellum.

promastigote (proh-mas′-tih-goht) Gr. *pro-*, before. The stage bearing a single, free, anterior flagellum.

epimastigote (ep″-ih-mas′-tih-goht) Gr. *epi-*, upon. The flagellate stage, which has a free anterior flagellum and an undulating membrane.

trypomastigote (trih″-poh-mas′-tih-goht) Gr. *trypanon*, auger. The large, fully formed stage characteristic of *Trypanosoma.*

These stages suggest an evolutionary transition from the simple, nonmotile amastigote form to the highly complex trypomastigote form. All four stages are present in *T. cruzi,* but *T. brucei* lacks the amastigote and promastigote forms, and *Leishmania* has only the amastigote and promastigote stages **(table 23.3)**. In general, the vector host serves as the site of differentiation for one or more of the stages, and the parasite is infectious to humans only in its fully developed stage.

Trypanosoma *Species and Trypanosomiasis*

Trypanosoma species are distinguished by their infective stage, the trypomastigote, an elongate, spindle-shaped cell with tapered ends that is capable of eel-like motility. Two types of trypanosomiasis are distinguishable on a geographic basis. *Trypanosoma brucei* is the agent of African sleeping sickness, and *T. cruzi* is the cause of Chagas disease, which is endemic to the Americas. Both exhibit a biphasic lifestyle alternating between a vertebrate and an insect host.

Trypanosoma brucei *and Sleeping Sickness*

Trypanosomiasis has greatly affected the living conditions of Africans since ancient times. Today, at least 50 million people are at risk, and 7,000 to 10,000 cases are reported yearly. It imposes an additional

hardship when it attacks domestic and wild mammals. Two subspecies of *Trypanosoma brucei* exist, each of which is responsible for a distinct form of the disease. The Gambian (West African) form is caused by the subspecies *T. b. gambiense,* and the Rhodesian (East African) form is caused by *T. b. rhodesiense* **(figure 23.8a)**. These geographically isolated forms of the disease are associated with different ecological niches of the principal tsetse fly vectors (genus *Glossina*). In the West African form, the fly inhabits the dense vegetation along rivers and forests typical of that region, whereas the East African form is adapted to savanna woodlands and lakefront thickets.

The cycle begins when a tsetse fly becomes infected after feeding on an infected reservoir host, such as a wild animal (antelope, lion, hyena), domestic animal (cow, goat), or human **(figure 23.8c)**. In the fly's gut, the trypanosome multiplies, migrates to the salivary glands, and develops into the infectious stage. When the fly bites its host, it releases trypomastigotes into the wound. At this site, the trypanosome multiplies and produces a sore called the primary chancre. From there, the pathogen moves into the lymphatics and the blood **(figure 23.8b)**.

Although an immune response occurs, it is counteracted by an unusual adaptation. The trypanosome has a genetic program that regulates constant change in its surface antigens over the course of the infection. This may give rise to at least 100 different antigens, which prevents a lasting immunity. The result is a long-term, uncontrolled infection. The Rhodesian form is acute and advances to brain involvement in 3 to 4 weeks and to death in a few months. The Gambian form, on the other hand, has a longer incubation period, is usually chronic, and may not affect the brain for several years.

Trypanosomiasis involves the lymphatics and areas surrounding blood vessels. Early symptoms include intermittent fever, enlarged spleen, swollen lymph nodes, and joint pain. In time the central nervous system is affected, the initial signs being personality and behavioral changes that progress to lassitude and sleep disturbances. The disease is commonly called sleeping sickness, but in fact, uncontrollable sleepiness occurs primarily in the day and is followed by sleeplessness at night. Signs of advancing neurological deterioration are muscular tremors, shuffling gait, slurred speech, seizures, and local paralysis. Death results from coma, secondary infections, or heart damage.

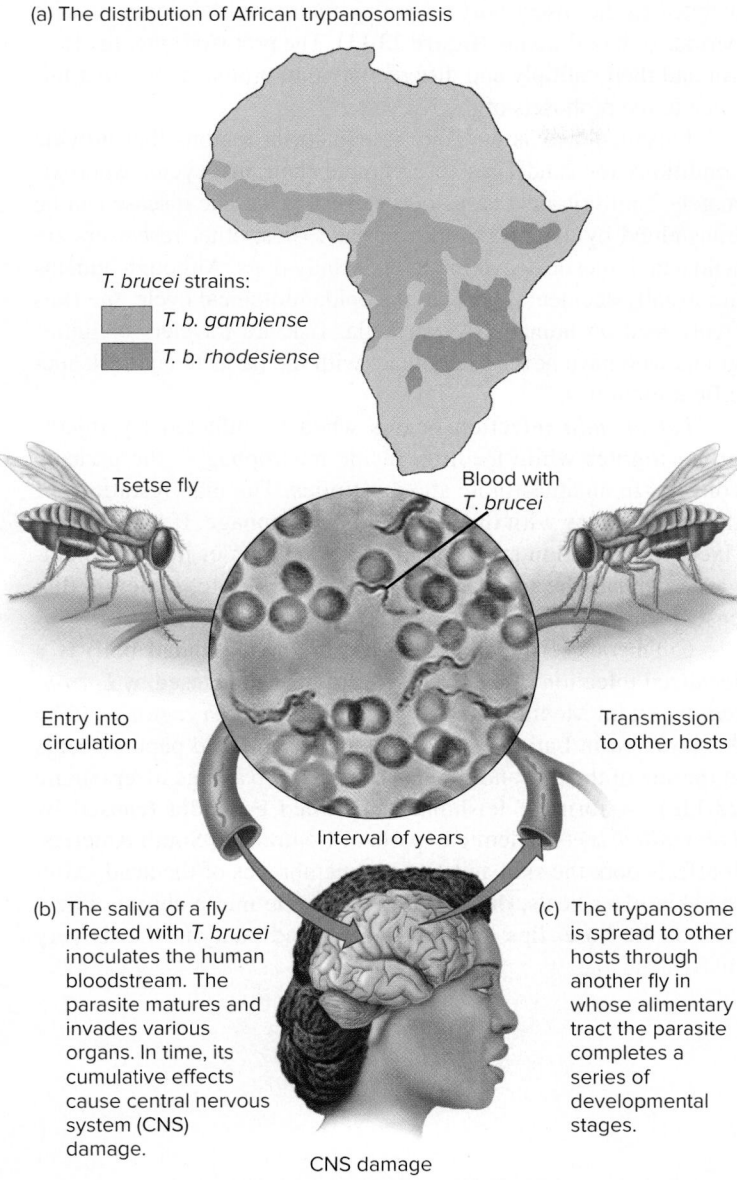

(a) The distribution of African trypanosomiasis

T. brucei strains:
- T. b. gambiense
- T. b. rhodesiense

Tsetse fly

Blood with
T. brucei

Entry into
circulation

Transmission
to other hosts

Interval of years

(b) The saliva of a fly infected with *T. brucei* inoculates the human bloodstream. The parasite matures and invades various organs. In time, its cumulative effects cause central nervous system (CNS) damage.

(c) The trypanosome is spread to other hosts through another fly in whose alimentary tract the parasite completes a series of developmental stages.

CNS damage

Figure 23.8 The generalized cycle between humans and the tsetse fly vector. Image by Piotr Rotkiewicz

Sleeping sickness may be suspected if the patient has been bitten by a distinctive fly while living or traveling in an endemic area. Trypanosomes are readily demonstrated in the blood, spinal fluid, or lymph nodes. An immunologic test—the card agglutination test for *Trypanosoma* (CATT)—is a rapid field assay that detects antibodies to the pathogen.

Chemotherapy is most successful if administered prior to nervous system involvement. Brain infection must be treated with the highly toxic drugs melarsoprol or suramin. A less toxic alternative is eflornathine. Several drug companies operating with the WHO have made these drugs available to all patients diagnosed with this disease.

Control of trypanosomiasis in western Africa, where humans are the main reservoir hosts, involves eliminating tsetse flies by applying insecticides, trapping flies, or destroying shelter and breeding sites. In eastern regions, control is less practical because large mammals are the hosts, and flies are less concentrated in specific sites.

Trypanosoma cruzi *and Chagas Disease*

Chagas disease accounts for 12 million to 19 million cases, thousands of deaths, and untold economic loss and human suffering in Latin America. The life cycle of its etiologic agent, *Trypanosoma cruzi* (see figure 5.26), parallels that of *T. brucei,* except that the insect hosts are "kissing,"[2] or **reduviid, bugs** (genus *Triatoma*) that harbor the trypanosome in the hindgut and discharge it in feces. These bugs live in numerous areas of Central and South America and Mexico and may extend to the southern parts of the United States **(figure 23.9).** They often share the living quarters of human and mammalian hosts. Infection occurs when the bug bites a person, often near the corners of the eyes or mouth, and immediately defecates. The human inoculates this wound by accidently rubbing the bug feces into it. Infection is also possible when bug feces fall onto the mucous membranes or conjunctiva, through contact with blood, and across the placenta. The course of the illness is somewhat

2. This nickname arises from the bug's tendency to bite the corner of the mouth.

(a)

Cycle in
wild animals

Reduviid bug

T. cruzi cell

Cycle in human habitats

Bite of bug

(c) Reduviid bug
(*Triatoma infestans*)

(b)

Figure 23.9 American trypanosomiasis (Chagas disease).
(a) Simplified view of disease transmission cycle. A reduviid bug infected with *Trypanosoma cruzi* carries the trypomastigote stage of the parasite in its feces. This is passed to humans, or other mammals, when the bug takes a blood meal. Trypomastigotes transform into amastigotes and multiply within infected tissues, causing symptoms of Chagas disease. When a reduviid bug bites an infected human, amastigotes pass from the human to the bug, completing the cycle.
(b) Map showing distribution of the vectors/disease (green).
(c) Reduviid bug (*Triatoma infestans*). (c): *Source:* Dr. Erwin Huebner, University of Manitoba, Winnipeg, Canada/National Human Genome Research Institute

similar to African trypanosomiasis, marked by a local lesion (Romaña's sign), fever, and swelling of the lymph nodes, spleen, and liver **(figure 23.10a).** Particular targets are the heart muscle and large intestine, both of which harbor masses of amastigotes. The infection of these organs causes enlargement and severe disruption in function **(figure 23.10b).** Chronically infected people account for 30% to 40% of heart disease in endemic regions.

Diagnosis of Chagas disease is made through microscopic examination of blood and serological testing. Although treatment is sometimes attempted with two drugs, nifurtimox and benzonidazole, they are not effective if the disease has progressed too far, and they have damaging side effects. It has been shown that the prevalence of Chagas disease can be reduced both by controlling the reduviid bugs that carry the trypanosome as well as by controlling the rodent population that often serves as hosts for the insect vector. Insecticides are of some benefit, but the bugs are very tough and capable of resisting doses that readily kill flies or mosquitoes. Mass collections of reduviid bugs and housing improvements to keep them out of human dwellings have also been effective. Of concern to the Centers for Disease Control and Prevention (CDC) are recent cases of Chagas disease in Texas, indicating a possible change in vector range **(23.1 Making Connections).** There is also greater possibility of the pathogen in donated blood, necessitating increased screening for it.

Leishmania *Species and Leishmaniasis*

Leishmaniasis is a zoonosis transmitted among various mammalian hosts by female **phlebotomine*** flies (sand flies) that require a blood meal to produce mature eggs. *Leishmania* species have

* *phlebotomine* (flee-bot'-oh-meen) Gr. *phlebo,* vein, and *tomos,* cutting. So called because they feed on blood.

adapted to the insect host by timing their own development with periods of blood-taking **(figure 23.11).** The protozoa enter the fly's gut and then multiply and differentiate into promastigotes that migrate to the proboscis of the fly.

Leishmaniasis is endemic to equatorial regions that provide conditions for sand flies to complete their life cycle. Approximately 2 million new cases occur every year. The disease can be transmitted by over 50 species of sand flies; other reservoirs are wild and domesticated animals, especially dogs. Although humans are usually accidental hosts in the epidemiological cycle, the flies freely feed on humans. At particular risk are travelers or immigrants who have never had contact with the parasite and lack specific immunities.

Leishmania infection begins when an infected fly injects promastigotes while feeding. Inside macrophages, the parasite converts to an amastigote and multiplies. The manifestations of the disease vary with the fate of the macrophage. If they remain fixed, the infection stays localized in the skin or mucous membranes, but if the infected macrophages migrate, systemic disease occurs.

Cutaneous leishmaniasis (oriental sore, Baghdad boil) is a localized infection of the capillaries of the skin caused by *L. tropica* in certain Mediterranean, African, and Indian regions and by *L. mexicana* in Latin America. A single small, red papule occurs at the site of the bite and spreads laterally into a large ulcer (figure 23.11c). A form of leishmaniasis called espundia (caused by *L. brasiliensis*) is endemic to parts of Central and South America. It affects both the skin and mucous membranes of the head. After the skin ulcer heals, the protozoa can settle into a chronic infection of the nose, lips, palate, gingiva, and pharynx that is very disfiguring.

(a)

Abnormal enlargement
(b)

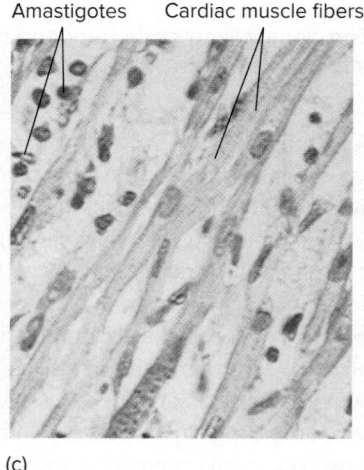
Amastigotes Cardiac muscle fibers
(c)

Figure 23.10 Conditions associated with Chagas disease. (a) A sign of acute infection is Romaña's sign—swelling of the eye or mouth at the entry point for the pathogen. **(b)** Heart pathology in chronic disease: Ventricles show abnormal swelling and enlargement. **(c)** Section of heart muscle with amastigotes filling the spaces between muscle fibers. (a): *Source:* CDC; (b, c): Courtesy of Emeritus Professor Wallace Peters: Formerly Honorary Director, Centre for Tropical Antiprotozoal Chemotherapy Nortwick Park Institute for Medical Research

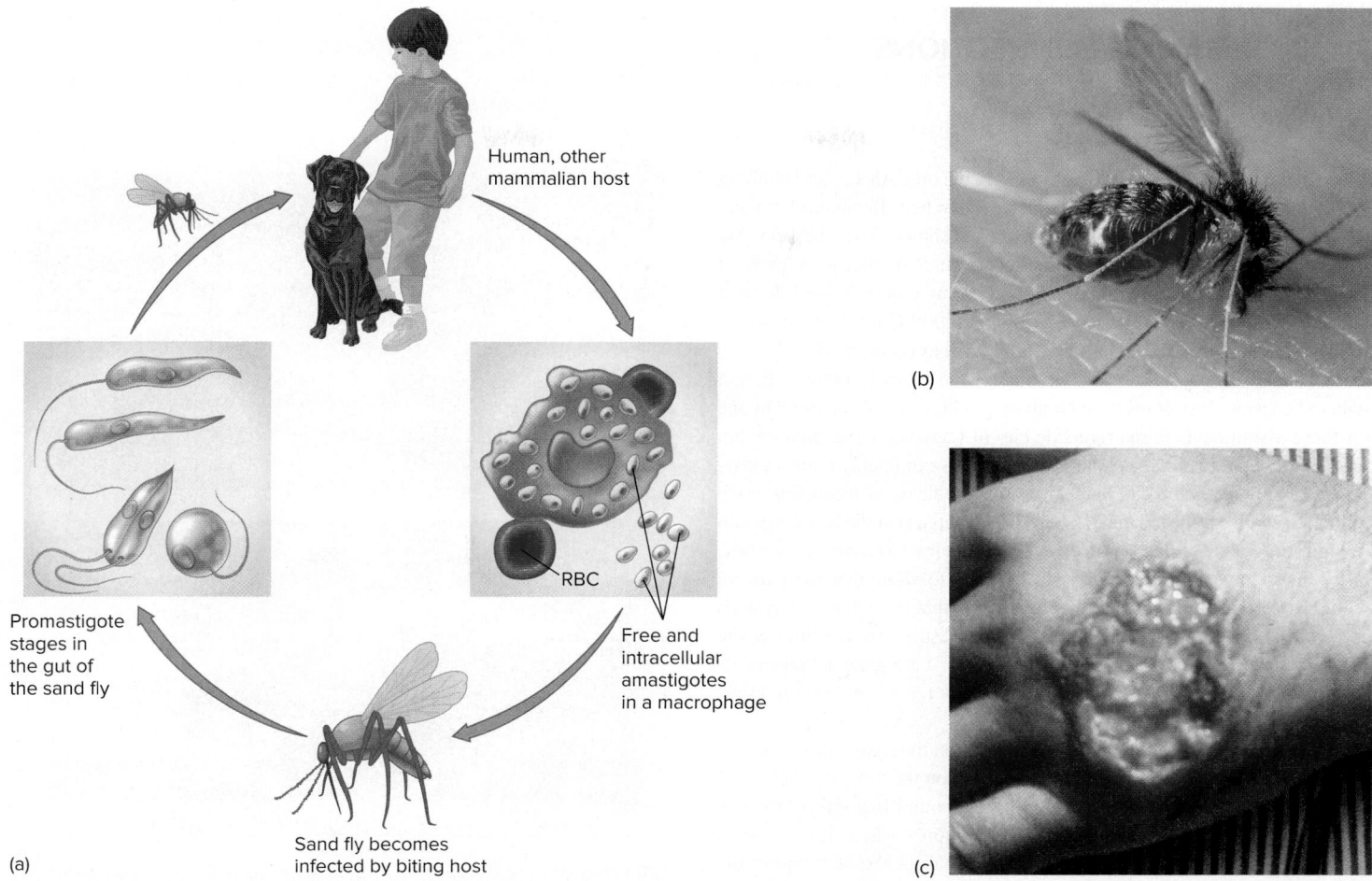

Figure 23.11 **The life cycle and pathology of *Leishmania species.* (a)** Humans and other mammals harbor the amastigote stage in the vascular organs. This stage is infectious to sand flies, which acquire it while feeding on infected blood. After development in the sand fly gut, promastigote forms are injected into another mammalian host, and a second round of development is initiated. **(b)** Image of feeding sand fly. **(c)** View of a sore or ulcer developing on the skin of the hand. (b) © www.raywilsonbirdphotography.co.uk; (c): *Source:* CDC

Systemic (visceral) leishmaniasis that occurs in parts of Africa, Latin America, and China appears gradually after 2 to 18 weeks. It presents with a high, intermittent fever; weight loss; and enlargement of internal organs, especially the spleen, liver, and lymph nodes. The most deadly form, *kala azar,* is 75% to 95% fatal in untreated cases. Death is due to the destruction of blood-forming tissues, anemia, secondary infections, and hemorrhage.

Diagnosis is complicated by the similarity of leishmaniasis to several granulomatous bacterial and fungal infections. The organism can usually be detected in the cutaneous lesion or in the bone marrow during visceral disease. Additional tests include cultivating the parasite, serological tests, and leishmanin skin testing. Chemotherapy with injected pentostam or glucantime is usually effective, though it has toxic side effects. Some disease prevention has been achieved by applying insecticides and eliminating sand fly habitats. Vaccines for both humans and dogs are currently under development.

Check Your Progress SECTION 23.3

7. Describe *Trichomonas vaginalis* with respect to transmission and life cycle.

8. Compare the relative importance of protozoan cysts to the epidemiology of trichomoniasis and giardiasis.

9. Compare the infective stages and means of vector transfer in the two types of trypanosomiasis and leishmaniasis.

10. Differentiate between amastigote, promastigote, epimastigote, and trypomastigote and indicate which stages are infectious to humans.

11. Why do two forms of sleeping sickness exist? How do they differ?

12. Outline the major organs affected by sleeping sickness, Chagas disease, and leishmaniasis.

13. What factor influences whether *Leishmania* infection manifests as a cutaneous or systemic disease?

23.1 MAKING CONNECTIONS

Who's on Your Money?

Check your wallet. George Washington is on the one, Alexander Hamilton on the ten, if you're lucky enough to have a hundred, Benjamin Franklin. Good choices all. Statesmen, presidents, inventors. The question you should be asking yourself at this point in your study of microbiology is, of course, "Why are there no microbiologists on my money?" Well, for that you'd have to take a look at the Brazillian 10,000 Cruzado note, which sports a likeness of none other than Carlos Chagas on the front.

Chagas was physician and microbiologist in the early 1900s in Brazil, where he spent most of his time combating outbreaks of malaria. On one of these assignments in the riverside city of Lassance he studied the behavior of the large bloodsucking reduviid bug—commonly called a kissing bug in deference to its habit of biting the faces of its victims each night in search of a blood meal. Chagas discovered that the bug carried in its intestines a flagellated protozoan that he named *Trypanosoma cruzi*, after his mentor Oswaldo Cruz. He went on to show that the parasite could be spread through the bite of a bug, and that those bitten displayed a specific collection of signs and symptoms, especially swelling of the heart muscle. His work was recognized as one of the great achievements in parasitology, and he was twice nominated for the Nobel Prize in Physiology or Medicine.

Even more interesting than Chagas' face on the right side of the bill, though, is the life cycle of *Trypanosoma cruzi* on the left side. The reduviiad bug is shown defecating into an open wound (caused by its own bite), releasing the parasite into the blood supply where it can encyst within the muscle of the heart. A second break in the skin shows the parasite reentering a reduviid bug, infecting a new vector and completing the life cycle. More information than you'd expect to find on your money. But there's more. Look at the back of your bills. The White House, the U.S. Treasury, the U.S. Capitol. Important places where important work took place. What's on the backside of the 10,000 Cruzado? Carlos Chagas' laboratory, of course.

100 years after the discovery of *Trypanosoma cruzi*, 8 million people in Mexico, Central America, and South America are infected with the parasite and 12,00 people a year will die of the disease.
Source: Barry Chess

What techniques could be employed to prevent the spread of Chagas disease? *Answer available on Connect.*

23.4 Apicomplexan Parasites

Expected Learning Outcomes

12. Recall the most important apicomplexan parasites.

13. Explain the endemic occurrence of malaria.

14. Diagram the life cycle of *Plasmodium*.

15. Describe the pathogenesis of malaria.

16. Explain the strategies used to combat malaria.

17. Describe the life cycle of *Toxoplasma gondii*.

18. Explain the occasional presence of *Cryptosporidium* in domestic water sources.

19. Relate *Sarcocystis, Cystoisospora, Cyclospora,* and *Babesia* to their specific route of infection.

Among the most unusual protozoa are the apicomplexans. These tiny parasites live in a variety of animal hosts. They are distinct from other protozoans in lacking well-developed locomotor organelles in the trophozoite state, though some forms do have a flexing, gliding form of locomotion, and others have flagellated sex cells. They are also notable for living as intracellular pathogens in host cells. Characteristics of the apicomplexan life cycle that add greatly to their complexity are alternations between sexual and asexual phases and between different animal hosts. Most members of the group also form specialized infective bodies that are transmitted by vectors, food, water, or other means, depending on the parasite. The most important human pathogens belong to *Plasmodium* (malaria), *Toxoplasma* (toxoplasmosis), and *Cryptosporidium* (cryptosporidiosis).

Plasmodium: The Agent of Malaria

Throughout human history, including prehistoric times, malaria has been one of the greatest afflictions, in the same ranks as bubonic plague, influenza, and tuberculosis. Even now, as the dominant protozoan disease, it threatens 40% of the world's population every year. The origin of the name is from the Italian words *mal,* bad, and

aria, air. The superstitions of the Middle Ages explained that evil spirits or mists and vapors arising from swamps caused malaria, because many victims came down with the disease after this sort of exposure. Of course, we now know that a swamp was mainly involved as a habitat for the mosquito vector.

The agent of malaria is an obligate intracellular sporozoan in the genus *Plasmodium,* which contains five species: *P. malariae, P. vivax, P. falciparum, P. ovale,* and *P. knowlesi.* The human and some primates are the primary vertebrate hosts for these species, which are geographically separate and show variations in the pattern and severity of disease. All forms are spread primarily by the female *Anopheles* mosquito and occasionally by shared needles, blood transfusions, and from mother to fetus. The detailed study of malaria is a large and engrossing field that is far too complicated for this text. We cover only the general aspects of its epidemiology, life cycle, and pathology.

Epidemiology and Events in Plasmodium *Infection*

Although malaria was once distributed throughout most of the world, the control of mosquitoes in temperate areas has successfully restricted it mostly to a belt extending around the equator **(figure 23.12)**. Despite this achievement, approximately 200 million new cases are still reported each year, about 90% of them in Africa. The most frequent victims are children and young adults, of whom at least 650,000 die annually. The reported cases in the United States, generally 1,000 to 2,000 new cases a year, are mainly in new immigrants.

Life Cycle Development of the malarial parasite is divided into two distinct phases: the asexual phase, carried out in the human, and the sexual phase, carried out in the mosquito **(figure 23.13)**. The asexual phase and infection begin when an infected female *Anopheles* mosquito injects saliva containing anticoagulant into the capillary in preparation for taking a blood meal. In the process, she inoculates the blood with spindle-shaped, asexual cells called **sporozoites.** The sporozoites circulate through the body and gravitate to the liver in a short time. Within liver cells, the sporozoites undergo asexual division called **schizogony,** which generates

* *sporozoites* (spor-oh-zoh-eyetz′) Gr. *sporo,* seed, and *zoon,* animal.
* *schizogony* (shizz′-aw-gon-ee). Gr. *schizo,* to divide, and *gone,* seed. Multiple division of the nucleus, followed by division of the cytoplasm.

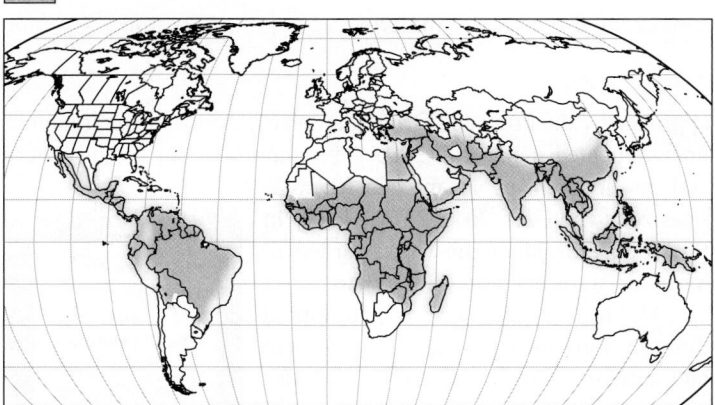

☐ Distribution of Malaria

Figure 23.12 The malaria belt. Pink zones outline the major regions that harbor malaria. The malaria belt corresponds to a band around the equator.

numerous daughter parasites, or **merozoites.** This phase of *exoerythrocytic development* lasts from 5 to 16 days, depending upon the species of parasite. Its end is marked by eruption of the liver cell, which releases from 2,000 to 40,000 mature merozoites into the circulation.

During the *erythrocytic phase,* merozoites invade red blood cells (RBCs) and convert in a short time to circular (ring) trophozoites. This stage feeds upon hemoglobin, grows, and undergoes multiple divisions to produce a cell called a *schizont,* which is filled with more merozoites. Bursting red cells liberate merozoites to infect more red cells. Eventually, certain merozoites differentiate into two types of specialized gametes called *macrogametocytes* (female) and *microgametocytes* (male). Because the human does not provide a suitable environment for the next phase of development, this is the end of the cycle in humans.

The sexual phase (sporogony) results when a mosquito draws infected RBCs into her stomach, where fertilization occurs. The resultant diploid cell (oocyst) implants into the stomach wall of the mosquito and forms sporozoites that lodge in the salivary glands. This event completes the sexual cycle and makes the sporozoites available for infecting the next human host.

Pathogenesis After a 10- to 16-day incubation period, symptoms of malaise, fatigue, vague aches, and nausea are followed by bouts of chills, fever, and sweating. These symptoms occur at 48- or 72-hour intervals, as a result of the synchronous rupturing of RBCs. Their interval, length, and regularity reflect the type of malaria. Patients with falciparum malaria, the most malignant type, often manifest persistent fever, cough, and weakness for weeks without relief. Complications of malaria are hemolytic anemia from lysed blood cells and organ enlargement and rupture due to cellular debris that accumulates in the spleen, liver, and kidneys. Falciparum malaria has the highest death rate in the acute phase, especially in children. Certain kinds of malaria are subject to relapses because some infected liver cells harbor dormant infective cells for up to 5 years.

Some populations are naturally resistant to *Plasmodium* infection. One dramatic example occurs in Africans who carry one copy of the gene that codes for sickle-cell hemoglobin. Although such genetic carriers are relatively healthy, they produce some abnormal RBCs that prevent the parasite from deriving sufficient nutrition to multiply. The continued existence of sickle-cell anemia in these geographic areas can be directly attributed to the increased survival of these carriers.

Diagnosis and Control of Malaria

Malaria can be diagnosed definitively by the discovery of a typical stage of *Plasmodium* in stained blood smears (see figure 23.13). Several rapid malaria tests have been developed to detect the parasite in early stages of the infection. Other indicators are the patient's residence or travel in endemic areas and symptoms such as recurring chills, fever, and sweating.

The world's health communities once hoped to decrease the impact of malaria and possibly even eradicate it. This goal has had many setbacks. In fact, malaria is on the rise in some regions. The reasons for this failure involve the sheer adaptive and survival capacity of both the parasite and the vector. Standard chemotherapy regimens have selected for drug-resistant strains of the malarial parasite, and mosquitoes have developed resistance to some common insecticides used to control them.

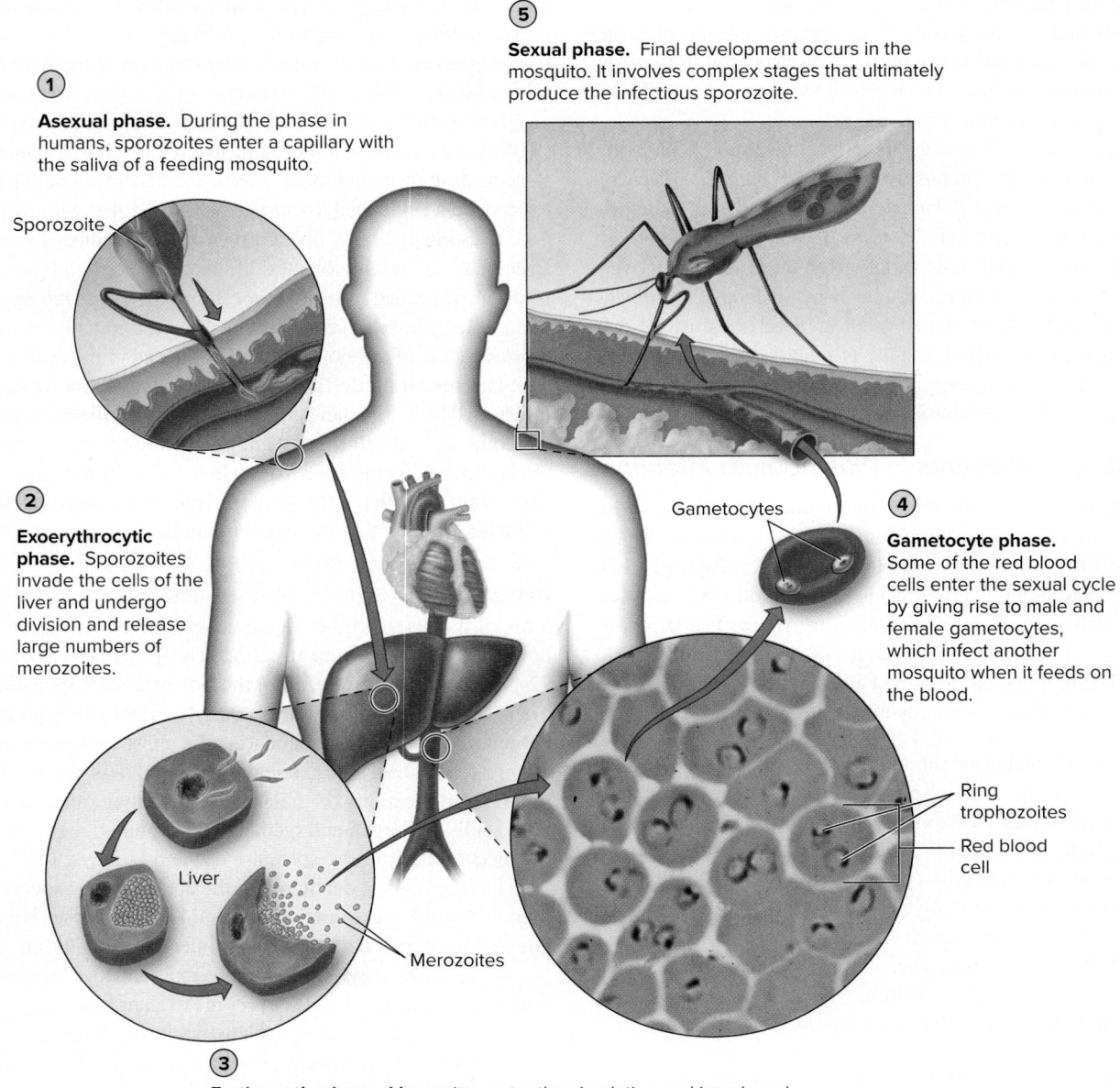

⑤

Sexual phase. Final development occurs in the mosquito. It involves complex stages that ultimately produce the infectious sporozoite.

①

Asexual phase. During the phase in humans, sporozoites enter a capillary with the saliva of a feeding mosquito.

Sporozoite

②

Exoerythrocytic phase. Sporozoites invade the cells of the liver and undergo division and release large numbers of merozoites.

Liver

Merozoites

Gametocytes

④

Gametocyte phase. Some of the red blood cells enter the sexual cycle by giving rise to male and female gametocytes, which infect another mosquito when it feeds on the blood.

Ring trophozoites

Red blood cell

③

Erythrocytic phase. Merozoites enter the circulation and invade red blood cells. Infection gives rise to a prominent phase—the ring trophozoite—in blood cells. Shizogony of this form produces additional merozoites that burst out and continue the infection cycle.

Process Figure 23.13 The life and transmission cycle of *Plasmodium*, the cause of malaria. Courtesy of Stephen B. Aley, PhD, Univ. of Texas at El Paso

Most current treatments for malaria are based on some form of quinine drug. All of the drugs in this class are toxic to the parasite by preventing it from processing hemoglobin in its food vacuoles. This results in an accumulation of waste products and cell death. Chloroquine is the drug of choice in areas of the world where resistance is not a problem. In areas of the world where resistant forms predominate, a course of malarone is used. Artemisinin-based combination therapy (ACT) is the latest salvo in the malaria wars. It combines two drugs with different metabolic targets in a single pill to allow for shorter—and therefore more likely to be completed—treatment cycles, while also slowing the development of drug resistant strains of *Plasmodium*. Coartem, an ACT drug combining lumefantrine and artemether (a derivative of artemisinin), was approved by the Food and Drug Administration for the treatment of uncomplicated malaria in 2009. The importance of the discovery and development of artemisinin was emphasized when the 2015 Nobel Prize in Physiology or Medicine was awarded to Youyou Tu for her work in the field.

Malaria prevention is attempted through long-term mosquito abatement and human prophylaxis. This includes elimination of water that could serve as a breeding site and broadcasting of insecticides to reduce populations of adult mosquitoes, especially near human dwellings.

Quick Search

What's the connection between malaria and a gin and tonic? Point your browser to "Herbs and Empires" and view the 3-minute video.

Pathogen Profile #2 *Plasmodium malariae, P. vivax, P. falciparum, P. ovale, P. knowlesi*

Sporozoites of
Plasmodium falciparum
in blood smear
Source: CDC-DPDx/Wilcox
A. Manual for the
Microscopical Diagnosis of
Malaria in Man, Washington:
U.S. Department of Health,
Education and Welfare, 1960

Microscopic Morphology Each stage of the *Plasmodium* life cycle has a unique morphology, with the circular ring trophozoite stage, seen within the erythrocytes of infected persons, being the most recognizable.

Identified by The presence of any stage of the *Plasmodium* life cycle within stained blood smears offers definitive diagnosis, as does serological testing. Preliminary diagnosis is often based on residence in or travel to endemic areas, along with such symptoms as recurring chills, fever, or sweating.

Habitat Humans are the primary vertebrate hosts of the five species of *Plasmodium* that cause malaria, whereas the female *Anopheles* mosquito serves as a biological vector in the spread of the disease. World regions that harbor malaria cluster near the equator, with 90% of new cases occurring in Africa.

Virulence Factors Multiple stages to the life cycle and multiple antigenic types enhance virulence of all species.

Primary Infections/Disease Symptoms of malaria appear after a 10- to 16-day incubation period and begin with malaise, fatigue, aches, and nausea, and then proceed to bouts of chills, fever, and sweating. These symptoms occur at 48- to 72-hour intervals and correspond to the synchronous rupture of red blood cells. *P. falciparum* produces the most severe symptoms, and patients suffering from this type of malaria may manifest uninterrupted fever, cough, and weakness for many weeks. Complications of malaria include hemolytic anemia from lysed red blood cells and organ enlargement and rupture due to cellular debris, which accumulates in the spleen, liver, and kidneys.

Control and Treatment Vector control is key to the prevention of malaria, with elimination of mosquito breeding sites and use of insecticide-laced bed nets both playing important roles. Prophylactic drugs include atovaquone/proguanil, chloroquine, mefloquine, and artemether/lumefantrine, among others, and these same drugs can be used to treat active disease.

Researchers are in the early stages of producing genetically altered mosquitoes that are immune to the malarial parasite. The immune insects can then be released in large numbers where they will mate with wild mosquitoes, spreading the resistance gene. The hope is to one day make mosquitoes resistant to the parasite without eradicating the insect (which would leave a void in the ecosystem that perhaps would be filled by something even worse). In the meantime, humans can reduce their risk of infection considerably by the following:

- using bed nets treated with insecticides, screens, and repellents **(figure 23.14);**
- remaining indoors at night; and
- taking weekly doses of prophylactic drugs such as atovaquone/proguanil (Malarone), doxycycline, or chloroquine.

People with a recent history of malaria must be excluded from blood donation. Even with massive efforts undertaken by the Roll Back Malaria initiative, the prevalence of malaria in endemic areas remains high.

Malaria and Its Elusive Vaccine

Several animal and human models have proved that effective vaccination for malaria is possible, but a practical vaccine has proved more difficult to develop. The problems arise from the four different species each having different sporozoite, merozoite, and gametocyte stages. In addition, each species can also have different antigenic types of sporozoites and merozoites. A sporozoite vaccine would prevent liver infection, whereas a merozoite vaccine would prevent infection of red blood cells. The best vaccines would provide long-term protection against all the phases and variants so that if one immunity were to fail, another could come into play.

The most advanced vaccine currently in development is RTS,S, targeted exclusively against *Plasmodium falciparum*. In late 2015 the vaccine received a favorable recommendation from the World Health Organization, based on its phase III clinical trials, which indicated a 30% to 40% reduction in malaria among those receiving the vaccine. Targeted at the exoerythocytic phase of the *P. faliparum*, the vaccine is designed to prevent the parasite from infecting the liver where it can

Figure 23.14 Insecticide treated bed nets (ITNs). *Anopheles* mosquitoes that spread malaria feed between dusk and dawn. Sleeping under a net treated with insecticide prevents mosquito bites during the night. If enough households are protected with netting, the number of mosquitoes, as well as their life span, is reduced, protecting even those members of the community not shielded by nets. © Andy Crump, TDR, World Health Organization/Science Source

mature, multiply, and re-enter the bloodstream. A larger pilot study—400,000 to 800,000 people in 3 to 5 locations— is currently being planned, so that any long-term effects of the drug can be studied.

Coccidian Parasites

The Order Coccidiorida of the Apicomplexa contains a number of members that are zoonotic in domestic mammals and birds and can cause severe disease in humans. These unusual protozoans exist in several stages, including dormant oocysts and pseudocysts.

Toxoplasma gondii *and Toxoplasmosis*

Toxoplasma gondii* is an intracellular apicomplexan parasite with an extensive global distribution, leading some experts to speculate that it infects the majority of the world's population at some time in their lives. In most cases toxoplasmosis is asymptomatic or mild, however disease can be severe and even terminal in fetuses and in immunodeficient patients, especially those with AIDS. *Toxoplasma gondii* is a highly successful parasite with so little host specificity that it can attack at least 200 species of birds and mammals. Studies have traced its primary reservoir and hosts to members of the feline family, both domestic and wild, and their rodent prey.

To follow the transmission of toxoplasmosis, we must look at the general stages of *Toxoplasma*'s life cycle in the cat **(figure 23.15a)**. The parasite undergoes a sexual phase in the intestine and is then released in feces, where it becomes an infective *oocyst* that survives in moist soil for several months. Ingested oocysts release an invasive asexual tissue phase called a tachyzoite that infects many different tissues and often causes disease in the cat. Eventually these forms enter a cyst state in tissues, called a pseudocyst. The parasite does not cycle in cats alone and is spread by oocysts to intermediate hosts, usually rodents and birds. The cycle returns to cats when they eat these infected prey animals.

Other vertebrates become a part of this transmission cycle **(figure 23.15b).** Herbivorous animals such as cattle and sheep ingest oocysts that persist in the soil of grazing areas and then develop pseudocysts in their muscles and other organs. Carnivores such as canines are infected by eating pseudocysts in the tissues of carrier animals.

Humans appear to be constantly exposed to the pathogen. The rate of prior infection, as detected through serological tests, can be as high as 90% in some populations. Many cases are caused by ingesting pseudocysts in raw or undercooked meat, which is customary fare in certain ethnic foods, such as raw spiced beef, raw pork, and ground lamb. The grooming habits of cats spread fecal oocysts on their body surfaces, and unhygienic handling of them presents an opportunity to ingest oocysts. Infection can also occur when tachyzoites cross the placenta to the fetus.

Most cases of toxoplasmosis are asymptomatic or marked by mild symptoms such as sore throat, lymph node enlargement, and low-grade fever. In patients whose immunity is suppressed by infection, cancer, or drugs, the outlook is grim. Chronic, persistent *Toxoplasma* infection can produce extensive brain lesions and can create fatal disruptions of the heart and lungs. A pregnant woman with toxoplasmosis has a 33% chance of transmitting the infection to her fetus. Congenital infection occurring in the first or second

* *Toxoplasma gondii* (toks″-oh-plaz′-mah gawn′-dee-eye) L. *toxicum,* poison, and Gr. *plasma,* molded and *gundi,* a small rodent.

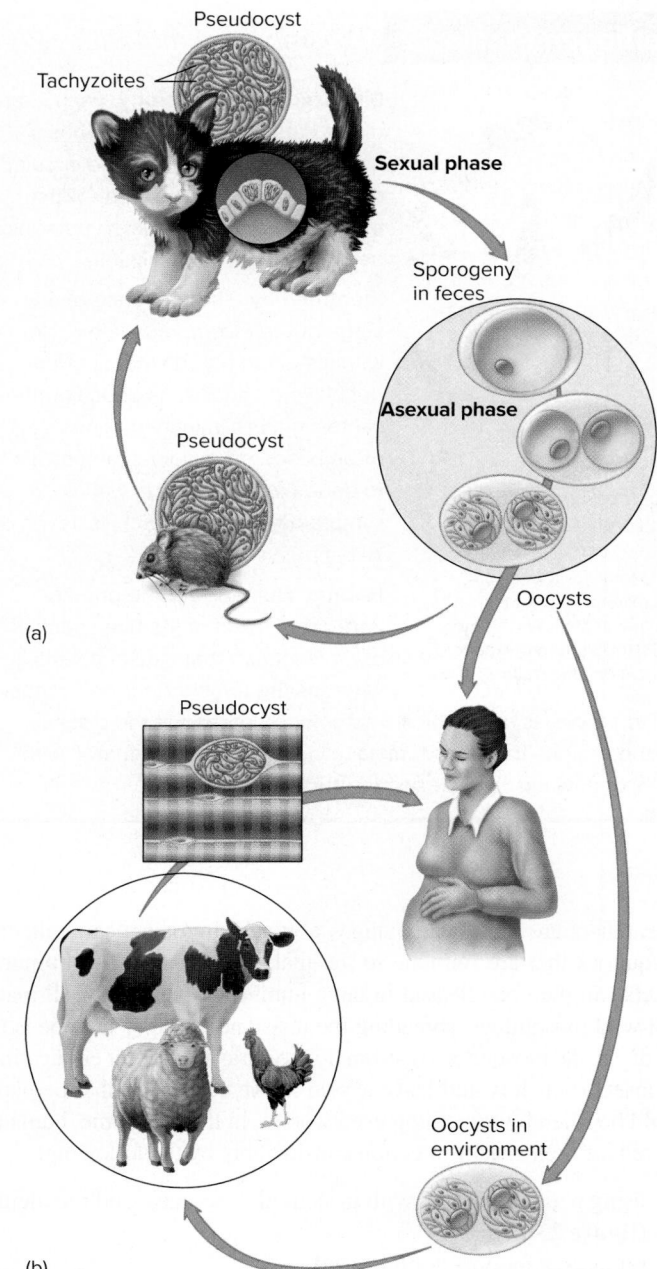

Figure 23.15 The life cycle and morphological forms of ***Toxoplasma gondii.* (a)** The cycle in cats and their prey. **(b)** The cycle in other animal hosts. The zoonosis has a large animal reservoir (domestic and wild) that becomes infected through contact with oocysts in the soil. Humans can be infected through contact with cats or ingestion of pseudocysts in animal flesh. Infection in pregnant women is a serious complication because of the potential damage to the fetus.

trimester is associated with stillbirth and severe abnormalities such as liver and spleen enlargement, liver failure, hydrocephalus, convulsions, and damage to the retina that can result in blindness.

Toxoplasmosis in AIDS and other immunocompromised patients can result from a primary acute or reactivated latent infection. The acute form is a disseminated, rapidly fatal disease with massive brain involvement that is sometimes seen in AIDS patients. Trophozoites and cysts clustered in intracerebral lesions cause seizures, altered mental state, and coma.

Toxoplasmosis can be differentiated by means of serological tests that detect antitoxoplasma antibodies, especially those for IgM, which appears early in infection. Disease can also be diagnosed by culture and histological analysis.

The most effective drugs are pyrimethamine and sulfadiazine alone or in combination. Because they do not destroy the cyst stage, they must be given for long periods to prevent recurrent infection.

In view of the fact that the oocysts are so widespread and resistant, hygiene is of paramount importance in controlling toxoplasmosis. Adequate cooking or freezing of meat below −20°C destroys both oocysts and tissue cysts. Oocysts can also be avoided by washing the hands after handling cats or soil that is possibly contaminated with cat feces. Pregnant women should be especially attentive to these rules and should avoid contact with cat excrement.

(a) (b)

23.16 Other apicomplexan parasites. (a) Light micrograph of gallbladder epithelium with numerous *Cryptosporidium* organisms (arrows) along the luminal surfaces of the epithelial cells. **(b)** Graphic of *Cystoisospora belli* oocysts to show internal sporocysts. (a): *Source:* CDC

Sarcocystis *and Sarcocystosis*

Although *Sarcocystis* species are parasites of cattle, swine, and sheep, some have an interesting transmission cycle that involves passage through humans as the final host. Domestic animals are intermediate hosts that pick up infective cysts while grazing on grass contaminated by human excrement. Humans are subsequently infected when they eat beef, pork, or lamb containing tissue cysts, but human-to-human transmission does not occur. Sarcocystosis is an uncommon infection mostly caused by ingesting uncooked meats. Initial symptoms of intestinal sarcocystosis are diarrhea, nausea, and abdominal pain, which peak in a few hours after ingestion but can persist for about 2 weeks. There is no specific treatment at the present time.

Cryptosporidium: *An Intestinal Pathogen*

Cryptosporidium* is a genus of apicomplexan parasites that infect a variety of mammals, birds, and reptiles, causing a disease called cryptosporidiosis. For many years, cryptosporidiosis was considered an intestinal ailment exclusive to calves, pigs, chickens, and other poultry, but it is clearly a zoonosis as well. The organism exhibits a life cycle similar to that of *Toxoplasma,* with a hardy intestinal oocyst and a tissue phase. In transmission pattern, it is very similar to *Giardia.*

Cryptosporidiosis exists worldwide, but its highest prevalence is in areas with unreliable water and food sanitation. The carrier state occurs in 3% to 30% of the population in some developing countries. The susceptibility of the general public to this pathogen has been amply demonstrated by frequent outbreaks. In the 1990s, 370,000 people developed *Cryptosporidium* gastroenteritis from tap water in Wisconsin. More recently, a similar outbreak sickened 2,800 people in Oregon. In both cases, contamination of water reservoirs by livestock wastes was thought to be responsible for the spread of the pathogen. Studies indicate that at least one-third of all fresh surface waters harbor this parasite. Recently the use of ozone or ultraviolet light to disinfect recreational water sources has become more widespread as both methods are effective against *Cryptosporidium.*

Infection is initiated when ingested oocysts release sporozoites that penetrate the intestinal cells and live intracellularly in them. The prominent symptoms in patients with clinical disease mimic other types of gastroenteritis, with headache, sweating, vomiting, severe abdominal cramps, and diarrhea. AIDS patients may experience chronic persistent cryptosporidial diarrhea, which is one criterion used to diagnose AIDS. The agent can be detected in fecal samples with indirect immunofluorescence and by acid-fast staining of biopsy tissues **(figure 23.16a).** Immunocompetent patients may be successfully treated with nitazoxanide, but immunodeficient patients tend to suffer relapses regardless of the treatment regimen.

Cystoisospora belli *and Coccidiosis*

Cystoisospora (formerly known as *Isospora*) is an intracellular intestinal parasite with a distinctive stage featuring oocysts containing infective sporocysts **(figure 23.16b).** Of several species that colonize mammals, apparently only ***Cystoisospora belli**** infects humans. The parasite is transmitted in fecally contaminated food or drink and through laboratory contact. Not much is known about the incidence of infection because it is usually asymptomatic or self-limited. The more prominent symptoms are malaise, nausea and vomiting, diarrhea, fatty stools, abdominal colic, and weight loss. In cases requiring treatment, trimethoprim-sulfamethoxazole is the drug of choice.

Cyclospora cayetanensis *and Cyclosporiasis*

Cyclospora cayetanensis is an emerging protozoan pathogen related to *Cystoisospora*. Since its first occurrence in 1979, hundreds of outbreaks have been reported in North America. Its mode of transmission is oral-fecal, and most cases have been associated with consumption of fresh produce and water, presumably contaminated with feces. Cyclosporiasis occurs worldwide; although primarily of human origin, it is not spread directly from person to person. Outbreaks have been traced to imported raspberries, fresh greens, and drinking water. The parasite has also been identified as a significant cause of diarrhea in travelers.

The disease is acquired when oocysts enter the small intestine and invade the mucosa. After an incubation period of about 1 week, symptoms of watery diarrhea, stomach cramps, bloating, fever, and

* *Cryptosporidium* (krip″-toh-spo-rid′-ee-um). Gr. *cryptos,* hidden, and *sporos,* seed.

* *Cystoisospora belli* (sis′-toh′-eye-saus′-poh-rah bel′-eye) Gr. *iso,* equal, and *sporos,* seed; L. *bellum,* war.

muscle aches appear. Patients with prolonged diarrheal illness experience anorexia and weight loss.

Diagnosis can be complicated by the lack of recognizable oocysts in the feces. Techniques that improve identification of the parasite are examination of fresh preparations under a fluorescent microscope and an acid-fast stain of a processed stool specimen. A PCR-based DNA test can also be used to identify *Cyclospora* and differentiate it from other coccidian parasites. Most cases of infection have been effectively controlled with a combination of trimethoprim and sulfamethoxazole for a week. Traditional antiprotozoan drugs are not effective. Some cases of disease may be prevented by cooking or freezing food to kill the oocysts.

Babesia Species and Babesiosis

Babesiosis, or piroplasmosis, is historically important for two firsts in microbiology. *Babesia** was the first protozoan found to be responsible for a disease—specifically, redwater fever.[3] This is an acute febrile hemolytic disease of cattle characterized by the destruction of erythrocytes and the presence of hemoglobin in the urine, hence its name. The disease was also the first one associated with an arthropod vector—a tick. Although the morphology and life cycle of *Babesia* are not completely understood, they probably mirror those of the plasmodia.

Human babesiosis is a zoonosis caused by a species of *Babesia* from wild rodents that has been accidently introduced by an *Ixodes* tick bite. This infection occurs throughout the world, and was once thought to be very rare or even eliminated from the United States. Finding increased case reports spurred the CDC to place it on the notifiable disease list in 2011. Of special concern has been its transmission through blood transfusions. It appears to be endemic to several areas of the Northeast and Midwest, with about 1,700 cases per year. It parallels malaria in its symptoms and effects, destroying red blood cells and inducing fever.

 Check Your Progress SECTION 23.4

14. List the major characteristics of the apicomplexan parasites.
15. Explain why malaria is a greater concern in some geographic areas than in others.
16. What is the name of the infective stage of the malaria parasite in humans and mosquitoes?
17. What stage of development of the malaria parasite is the ring form?
18. Outline some of the problems which make malaria difficult to control.
19. Identify the phases of the *Plasmodium* life cycle that take place in mosquitoes, blood cells, and liver cells.
20. Explain the means by which cats, rodents, birds, domestic animals, and people are infected with *Toxoplasma gondii*.
21. Why don't common methods of water treatment reliably exclude *Cryptosporidium* from domestic water supplies?
22. Compare the infectious routes and targets of *Cryptosporidium*, *Cystoisospora, Cystoiospora, Cyclospora,* and *Babesia*.

23.5 A Survey of Helminth Parasites

 Expected Learning Outcomes

20. Recall the three categories of parasitic helminths.
21. Distinguish between intermediate hosts and definitive hosts.
22. Describe the four basic helminth transmission cycles.
23. Describe the strategies used to diagnose and control helminth infection.

So far, our survey has been limited to single-celled infectious agents; however, **helminths,*** particularly the adults, are comparatively large, multicellular animals with specialized tissues and organs similar to those of the hosts they parasitize. They are amazingly diverse, ranging from barely visible roundworms (0.3 mm) to huge tapeworms 25 meters long and varying from oblong- and crescent-shaped to whiplike or elongate flat ribbons. Their behavior and their adaptations to humans are at once fascinating and grotesque. Imagine having long, squirming white worms suddenly expelled from your nose, feeling and seeing a worm creeping beneath your skin, or having a worm slither across the front of your eye, and you have some idea of the conspicuous nature of these parasites.

In the following section, we present major facets of helminth epidemiology, life cycles, pathology, and control, followed by an accounting of the major worm diseases. A short review of taxonomy from chapter 5 may be helpful. Helminths are invertebrate worms classified into two phylums. Members of Phylum Aschelminthes are commonly known as nematodes or roundworms. They have elongate, somewhat curved, cylindrical bodies. Members of Phylum Platyhelminthes, or flatworms, have thin, flattened bodies. Parasitic flatworms come in two varieties: (1) trematodes called flukes that have ovoid, compact bodies, and (2) cestodes called tapeworms with long, ribbonlike, segmented bodies. **Table 23.4** presents a selected list of medically significant helminths in the categories of nematodes, trematodes, and cestodes.

General Life and Transmission Cycles

The complete life cycle of helminths includes the fertilized egg (embryo), larval, and adult stages. In the majority of helminths, adults derive nutrients and reproduce sexually in a host's body. In nematodes, the sexes are separate and usually different in appearance; in trematodes, the sexes can be either separate or **hermaphroditic,** meaning that male and female sex organs are in the same worm; cestodes are generally hermaphroditic. For a parasite's continued survival as a species, it must complete the life cycle by transmitting an infective form, usually an egg or larva, to the body of another host, either of the same or a different species. By convention, the host in which larval development occurs is the **intermediate (secondary) host,** and adulthood and mating occur in the **definitive (final) host.** A transport host is an intermediate host that experiences no parasitic development but is an essential link in the completion of the cycle.

* *Babesia* (bab-ee′-see″-uh) Named for its discoverer, Victor Babes.
3. Also known as bovine babesiosis, Texas fever, or tick fever.

* Gr. *helmins,* worm.

TABLE 23.4	Major Helminths of Humans and Their Modes of Transmission		
Classification	**Common Name of Disease or Worm**	**Life Cycle Requirement**	**Spread to Humans by**
Nematodes (Roundworms)			
Intestinal Nematodes			Oral-fecal means
Infective in egg (embryo) stage			
Trichuris trichiura	Whipworm	Humans	Fecal pollution of soil with eggs
Ascaris lumbricoides	Ascariasis	Humans	Fecal pollution of soil with eggs
Enterobius vermicularis	Pinworm	Humans	Close contact and fomites
Infective in larval stage			
Necator americanus	New World hookworm	Humans	Fecal pollution of soil with eggs
Ancylostoma duodenale	Old World hookworm	Humans	Fecal pollution of soil with eggs
Strongyloides stercoralis	Threadworm	Humans; may live free	Fecal pollution of soil with eggs
Trichinella spiralis	Trichina worm	Pigs, wild mammals	Consumption of meat containing larvae
Tissue Nematodes			Burrowing of larva into tissue
Wuchereria bancrofti	Filariasis	Humans, mosquitoes	Mosquito bite
Onchocerca volvulus	River blindness	Humans, black flies	Fly bite
Loa loa	Eye worm	Humans, mangrove flies, or deer flies	Fly bite
Trematodes (Flukes)			
Schistosoma japonicum	Blood fluke	Humans and snails	Ingestion of fresh water containing larval stage
S. mansoni	Blood fluke	Humans and snails	Ingestion of fresh water containing larval stage
S. haematobium	Blood fluke	Humans and snails	Ingestion of fresh water containing larval stage
Opisthorchis sinensis	Chinese liver fluke	Humans, snails, fish	Consumption of fish
Fasciola hepatica	Sheep liver fluke	Herbivores (sheep, cattle)	Consumption of water and water plants
Paragonimus westermani	Oriental lung fluke	Humans, mammals, snails, freshwater crabs	Consumption of crabs
Cestodes (Tapeworms)			
Taenia saginata	Beef tapeworm	Humans, cattle	Consumption of undercooked or raw beef
T. solium	Pork tapeworm	Humans, swine	Consumption of undercooked or raw pork
Diphyllobothrium latum	Fish tapeworm	Humans, fish	Consumption of undercooked or raw fish
Hymenolepis nana	Dwarf tapeworm	Humans	Oral-fecal means; close contact

Although individual variations occur in life and transmission cycles, most can be described by one of four basic patterns (**figure 23.17**). In general, sources for human infection are contaminated food, soil, water, or infected animals; and routes of infection are by oral intake or penetration of unbroken skin. Humans are the definitive hosts for many of the parasites listed in table 23.4, and in about half, they are also the sole biological reservoir. In other cases, animals or insect vectors serve as reservoirs or are required to complete worm development. In the majority of helminth infections, the worms must leave their host to complete the entire life cycle.

General Epidemiology of Helminth Diseases

The helminths have a far-reaching impact on human health, economics, and culture. Worldwide health organizations estimate around 2.5 billion helminth infections at any one time worldwide. The statistics of the most common forms are as follows:

- ascariasis—1 billion,
- trichuriasis—700 million,
- hookworm—600 million,
- schistosomiasis—200 million, and
- filariasis—120 million.

Systems Profile 23.2	Parasitic Helminths					
Organism	**Skin/Skeletal**	**Nervous/ Muscle**	**Cardiovascular/ Lymphatic/Systemic**	**Gastrointestinal**	**Respiratory**	**Urogenital**
Ascaris lumbricoides				Intestinal distress		
Trichuris trichiura		Muscle pain		Diarrhea		
Enterobius vermicularis	Anal itching			Diarrhea		
Necator americanus	Dermatitis			Intestinal distress	Pneumonitis	
Ancylostoma duodenale				Intestinal distress		
Strongyloides stercoralis	Dermatitis			Intestinal distress	Pneumonitis	
Trichinella spp.	Myositis	Encephalitis		Intestinal distress		
Wuchereria bancrofti	Dermatitis		Elephantiasis			
Onchocerca volvulus (w/Wolbachia)	River blindness	Loss of optic nerve				
Schistosoma spp.			Blood infection	Intestinal Schistosomiasis	Cough	Urinary schistosomiasis
Opisthorchis sinensis				Liver and intestinal disease	Pleural pain, abscess	
Fasciola hepatica				Liver and intestinal disease		
Taenia saginata				Abdominal pain		
Tania solium		Cysticercosis	Cystercercosis	Intestinal distress		

Although individuals from all societies and regions play host to worms at some time in their lives, the highest case rates occur among children in rural areas of the tropics and subtropics. In most developed countries with temperate climates, the major parasitic infections are pinworm and trichinosis, though tropical parasitic worms are often encountered in immigrants and travelers.

The continuing cycle between parasites and humans is often the result of practices such as using human excrement for fertilizer, defecating in open soil, exposing bare feet to soil or standing water, and eating undercooked or raw meat and fish. The burden of helminths is borne largely by the developing world. There is a clear association between poverty and malnutrition and higher levels of disease and death. Heavy worm loads can retard the physical and mental development of children. It is ironic indeed that some cultures have accepted parasites as their burden for thousands of years. Review chapter 5's Case Study, which covers many of these neglected tropical diseases.

Pathology of Helminth Infestation

Worms exhibit numerous adaptations to the parasitic habit. They have specialized mouthparts for attaching to tissues and for feeding, enzymes with which they liquefy and penetrate tissues, and a cuticle or other covering to protect them from the host defenses. In addition, their organ systems are usually reduced to the essentials: getting food and processing it, moving, and reproducing. Helminths entering by mouth undergo a period of development in the intestine that is often followed by spread to other organs. Helminths that burrow directly into the skin or are inserted by an insect bite quickly penetrate the circulation and are carried throughout the body. In both cases, most worms do not remain in a single location, and they migrate by various mechanisms through major systems, inflicting injury as they go.

Adult helminths eventually take up final residence in the intestinal mucosa, blood vessels, lymphatics, subcutaneous tissue, skin, liver, lungs, muscles, brain, or even the eyes. Traumatic damage occurring in the course of migration and organ invasion is primarily due to tissue feeding, toxic secretions, and blockages within organs due to high worm load. Pathologic effects can include enlargement and swelling of organs, hemorrhage, weight loss, and anemia caused by the parasites' feeding on blood. Despite the popular misconception that intestinal worms cause weight loss by competing for food, such weight loss is more likely caused by malnutrition, malabsorption of food and vitamins due to intestinal damage, lack of appetite, vomiting, and diarrhea.

One common antihelminthic defense is an increase in granular leukocytes called *eosinophils* that have a specialized capacity to destroy worms (see chapter 14). Although antibodies and sensitized T cells are also formed against worms, their effects are rarely long term, and reinfection is always a distinct possibility. The inability of the immune system to completely eliminate worm infections appears to be due to the worms' relatively large size, migratory habits, and inaccessibility. The incompleteness of immunity has hampered the development of vaccines for parasitic diseases and has emphasized the use of chemotherapeutic agents instead.

Elements of Diagnosis and Control

Diagnosis of helminthic infestation may require several steps. A differential blood count showing eosinophilia and serological tests

Cycle A

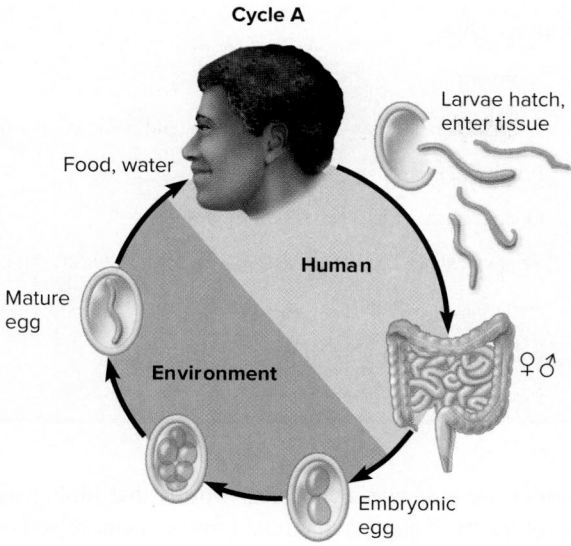

Larvae hatch, enter tissue

Food, water

Human

Mature egg

Environment

Embryonic egg

In cycle A, the worm develops in intestine; egg is released with feces into environment; eggs are ingested by new host and hatch in intestine (examples: *Ascaris, Trichuris*).

Cycle B

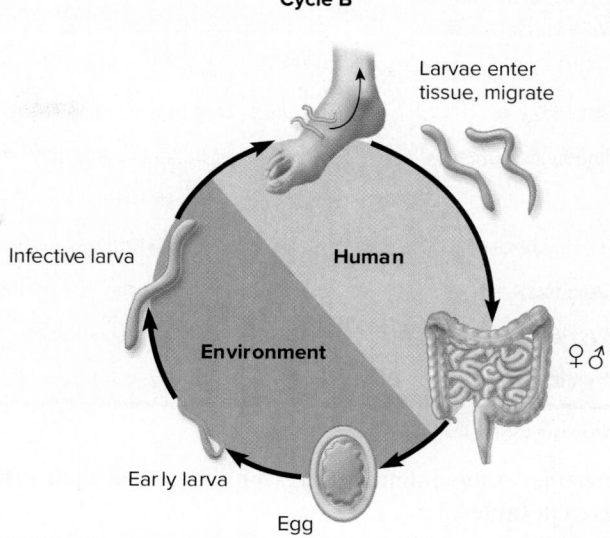

Larvae enter tissue, migrate

Infective larva

Human

Environment

Early larva

Egg

In cycle B, the worms mature in intestine; eggs are released with feces; larvae hatch and develop in environment; infection occurs through skin penetration by larvae (example: *hookworms*).

Cycle C

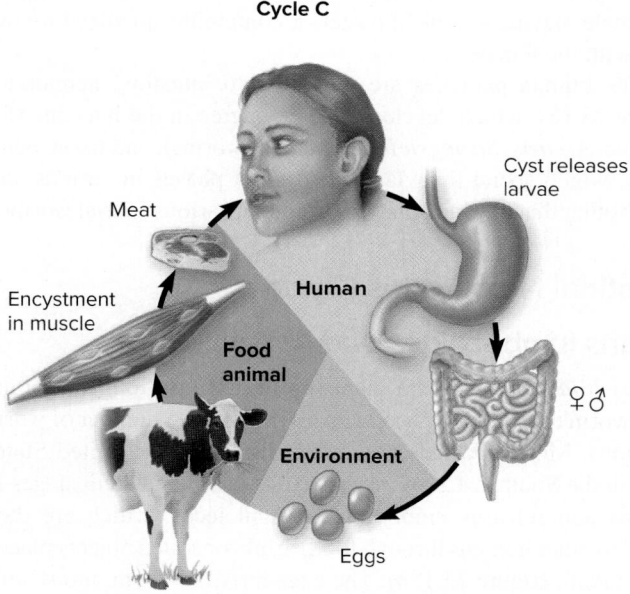

Cyst releases larvae

Meat

Encystment in muscle

Human

Food animal

Environment

Eggs

In cycle C, the adult matures in human intestine; eggs are released into environment; eggs are eaten by grazing animals (intermediate host); larval forms encyst in tissue; humans (definitive host) eating animal flesh are infected (example: *Taenia*).

Cycle D

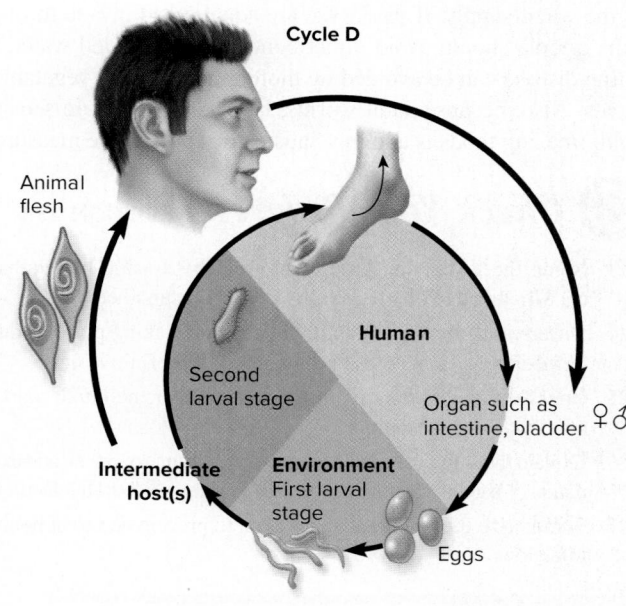

Animal flesh

Human

Second larval stage

Organ such as intestine, bladder

Intermediate host(s)

Environment First larval stage

Eggs

In cycle D, eggs are released from human; humans (definitive host) are infected through ingestion or direct penetration by larval phase (examples: *Opisthorchis* and *Schistosoma*).

Figure 23.17 **Four basic helminth life and transmission cycles.**

indicating sensitivity to helminths both provide indirect evidence of worm infestation. A history of travel to the tropics or immigration from those regions is also helpful, even if it occurred years ago, because some flukes and nematodes persist for decades. The most definitive evidence, however, is the discovery of eggs, larvae, or adult worms in stools, sputum, urine, blood, or tissue biopsies. The worms are sufficiently distinct in morphology that positive identification can be based on any stage, including eggs.

Although several useful antihelminthic medications exist, the cellular physiology of the eukaryotic parasites resembles that of humans, and drugs toxic to them can also be toxic to humans. Some antihelminthic drugs act to suppress a metabolic process that is more important to the worm than to the host. Others inhibit the worm's movement or prevent it from holding a position in an organ such as the intestine. Therapy is also based on a drug's greater toxicity to the helminths or acting locally in the

TABLE 23.5	Antihelminthic Therapeutic Agents and Indications for Their Use	
Drug	**Effect**	**Used for**
Albendazole	Blocks key step in worm metabolism	Filariasis, ascariasis, strongyloidiasis, and some adult worms
Diethylcarbamazine	Unknown but kills microfilariae	Filariasis, loiasis
Ivermectin	Blocks nerve transmission	Onchocerciasis, ascariasis, strongyloidiasis
Mebendazole*	Blocks key step in worm metabolism	Trichuriasis (whipworm), ascariasis, hookworm, trichinosis
Piperazine*	Paralyzes worm so it can be expelled in feces	Ascariasis, pinworm, hookworm
Praziquantel	Interferes with worm metabolism	Schistosomiasis, other flukes, tapeworm
Pyrantel	Paralyzes worm so it can be expelled in feces	Pinworm, hookworm, trichinosis

*No longer used in the United States.

intestine. Antihelminthic drugs of choice and their effects are given in **table 23.5.**

Preventive measures are aimed at minimizing human contact with the parasite or interrupting its life cycle. In areas where the worm is transmitted by fecally contaminated soil and water, disease rates are significantly reduced through proper sewage disposal, using sanitary latrines, avoiding human feces as fertilizer, and disinfection of the water supply. If the larvae are aquatic and invade through the skin, people should avoid direct contact with infested water. Food-borne disease can be avoided by thoroughly cooking vegetables and meats. Also, because adult worms, larvae, and eggs are sensitive to cold, freezing foods is a highly satisfactory preventive measure.

 Check Your Progress SECTION 23.5

23. Name the taxonomic groups into which parasitic helminths are classified and tell how they differ from one another.
24. Differentiate between helminthic activities taking place in the intermediate host and those occurring in the definitive host.
25. Contrast the important points of the four basic helminth transmission cycles with examples.
26. Explain the significance of eosinophilia, a history of travel, and examination of patient specimens when diagnosing a helminth infection.
27. Summarize the activities undertaken to prevent and treat helminth infections.

23.6 Nematode (Roundworm) Infestations

 Expected Learning Outcomes

24. Identify the transmission cycle of each of the most common intestinal nematodes.
25. Summarize the mode of infestation and pathology of *Necator, Strongyloides,* and *Trichinella.*
26. Identify risk factors for contracting trichinellosis, and list several preventative measures.
27. Differentiate between tissue nematodes and intestinal nematodes.
28. List the vectors associated with each of the common filarial worms.
29. Describe the signs and symptoms associated with the most common tissue nematodes.

Nematodes* are elongate, cylindrical worms that biologists consider one of the most abundant animal groups (a cup of soil or mud can easily contain millions of them). The vast majority are free-living soil and freshwater worms, but around 200 are parasitic, including 50 species that affect humans. Nematodes bear a smooth, protective outer covering, or cuticle, that is periodically shed as the worms grow. They are essentially headless, tapering to a fine point anteriorly. The sexes are distinct, and the male is often smaller than the female, having a hooked posterior containing spicules for coupling with the female.

The human parasites are divided into intestinal nematodes **(figure 23.18),** which develop to some degree in the intestine (for example, *Ascaris, Strongyloides,* and hookworms), and tissue nematodes, which spend their larval and adult phases in various soft tissues other than the intestine (for example, various filarial worms).

Intestinal Nematodes (Cycle A)

Ascaris lumbricoides *and Ascariasis*

*Ascaris lumbricoides** is a giant (up to 300 mm long) intestinal roundworm that probably accounts for the greatest number of worm infections. Most reported cases of ascariasis in the United States occur in the Southeast. *Ascaris* spends its larval and adult stages in humans and releases embryonic eggs in feces, which are then spread to other humans through food, drink, or soiled objects placed in the mouth (figure 23.18*b*). The eggs thrive in warm, moist soils but are sensitive to sunlight, high temperatures, and drying. After ingested eggs hatch in the human intestine, the larvae embark upon an odyssey in the tissues. First they penetrate the intestinal wall and enter its lymphatic and circulatory drainage. From there, they are swept into the heart and eventually arrive at the capillaries of the lungs. From this point, the larvae migrate up the respiratory tree to the glottis. Worms entering the throat are swallowed and returned to the small intestine, where they reach adulthood and reproduce, producing up to 200,000 fertilized eggs a day.

Even as adults, male and female worms are not attached to the intestine and retain some of their exploratory ways. They are known to invade the biliary channels of the liver and gallbladder, and on occasion the worms emerge from the mouth and nares. Severe

* *Nematodes* (nem′-ah-tohd) Gr. *nema,* thread.
* *Ascaris lumbricoides* (as′-kah-ris lum″-brih-koy′-dees) Gr. *askaris,* an intestinal worm; L. *lumbricus,* earthworm, and Gr. *eidos,* resemblance.

(a)

(b)

Figure 23.18 *Ascaris lumbricoides.* **(a)** A mass of *Ascaris lumbricoides* worms, which had been passed by a child in Kenya. **(b)** A microscopic view (128x) of a fertile *Ascaris lumbricoides* egg (left) and larva (right) in a fecal sample. (a): *Source:* James Gathany/CDC; (b: left, right): *Source:* CDC-DPDx

inflammatory reactions mark the migratory route, and allergic reactions such as bronchospasm, asthma, or skin rash can occur.

Trichuris trichiura *and Whipworm Infection*

The common name for *Trichuris** *trichiura*—whipworm—refers to its likeness to a miniature buggy whip. Although the worm is morphologically different from *Ascaris,* it has a similar transmission cycle, with humans as the sole host. Trichuriasis has its highest incidence in areas of the tropics and subtropics with widespread fecal contamination of the soil. Embryonic eggs deposited in the soil are not immediately infective and continue development for 3 to 6 weeks in this habitat. Ingested eggs hatch in the small intestine, where the larvae attach, penetrate the outer wall, and develop into adults. Mature adults move to the large intestine and gain a hold with their long, thin heads, while the thicker tail dangles free in the lumen. Following sexual maturation and fertilization, the females eventually lay 3,000 to 5,000 eggs daily into the bowel. The entire

cycle requires about 90 days, and untreated infestation can last up to 2 years.

Worms burrowing into the intestinal mucosa can pierce capillaries, cause localized hemorrhage, and provide a portal of entry for secondary bacterial infection. Heavier infestations can cause dysentery, loss of muscle tone, and rectal prolapse, which can prove fatal in children.

Enterobius vermicularis *and Pinworm Infection*

The pinworm or seatworm, ***Enterobius vermicularis,**** was first discussed in chapter 5 (see figure 5.29). Enterobiasis is the most common worm disease of children in temperate zones. Around 30% of children tested are found to be infected. The transmission of the pinworm follows a course similar to that of the other enteric parasites. Freshly deposited eggs have a sticky coating that causes them to lodge beneath the fingernails and to adhere to fomites. Upon drying, the eggs become airborne and settle in house dust. Worms are ingested from contaminated food and drink, and self-inoculation from one's own fingers is common.

Eggs hatching in the small intestine release larvae, which migrate to the large intestine, mature, and then mate. Itching is brought on when the mature female emerges from the anus and lays eggs. Although infestation is not fatal and most cases are asymptomatic, the afflicted child can suffer from disrupted sleep and sometimes nausea, abdominal discomfort, and diarrhea. A simple, rapid test can be done by pressing a piece of transparent adhesive tape against the anal skin and then applying it to a slide for microscopic evaluation.

Intestinal Nematodes (Cycle B)

The Hookworms

Two major agents of human hookworm infestation are ***Necator americanus,**** endemic to the New World, and ***Ancylostoma duodenale,**** endemic to the Old World, though the two species overlap in parts of Latin America. Otherwise, with respect to transmission, life cycle, and pathology, they are usually lumped together. Humans are sometimes invaded by cat or dog hookworm larvae, sustaining a dermatitis called creeping eruption. The *hook* refers to the adult's oral cutting plates by which it anchors to the intestinal villi and its curved anterior end **(figure 23.19).** Unlike other intestinal worms, hookworm larvae hatch outside the body and infect by penetrating the skin.

Hookworm infestation remains endemic to the tropics and subtropics and is absent in arid, cold regions. In recent times, its worldwide incidence has decreased. In the past, hookworm plagued parts of the southern United States, unfortunately sapping its victims' strength and stigmatizing them. The decline of the disease can be attributed to such simple changes as improved sanitation (indoor plumbing) and better standards of living (being able to afford shoes).

* *Trichuris* (trik-ur′-is) Gr. *thrix,* hair, and *oura,* tail.

* *Enterobius vermicularis* (en″-ter-oh′-bee-us ver-mik″-yoo-lah′-ris) Gr. *enteron,* intestine, and *bios,* life; L. *vermiculus,* a small worm.
* *Necator americanus* (nee-kay′-tor ah-mer″-ih-cah′-nus) L. *necator,* a murderer, and *americanus,* of New World origin.
* *Ancylostoma duodenale* (an″-kih-los′-toh-mah doo-oh-den-ah′-lee) Gr. *ankylos,* curved or hooked, *stoma,* mouth, and *duodenale,* the intestine.

23.2 MAKING CONNECTIONS

Neglected Parasitic Infections in the United States

Mention the word *parasites* and most people think of developing countries, but parasitic infections are not unknown in the United States, with 300,000 people infected with the parasite that causes Chagas disease, and 60,000,000 suffer from toxoplasmosis. Our collective mindset makes the situation even worse, as many people go undiagnosed and untreated because they don't understand that they are at risk and lack access to appropriate care. In some areas of the United States, neurocysticercosis is the single most common infectious cause of seizures, with some 2,000 people diagnosed every year. *Toxocara canis*, a roundworm that infects dogs, causes an estimated 70 cases of blindness in children annually. The actual figure is probably quite higher, but adequate surveillance of the disease is lacking. Compounding the problem are health care providers who are unfamiliar with parasitic infections and fail to diagnose or treat them properly.

The CDC has targeted five parasitic infections as public health priorities—Chagas disease, neurocysticercosis, toxocariasis, toxoplasmosis, and trichomoniasis—based on the number of people infected, the severity of the illness, and the ability to prevent and treat them. It is clear that a multifaceted approach is needed to combat a problem that manages to be both massive and invisible at the same time, and this is just what the CDC has undertaken. By increasing awareness among physicians and the public, improving diagnostic testing, and making available needed drugs, it is hoped that the toll of these diseases on individuals and the strain they place on our health care system can both be reduced.

How could U.S. Immigration and Customs officials be of use in the fight against parasitic infections? *Answer available on Connect.*

After being deposited into soil by defecation, hookworm eggs hatch into fine **filariform*** larvae that instinctively climb onto grass or other vegetation to improve the probability of contact with a host. Ordinarily the parasite enters sites on bare feet such as hair follicles, abrasions, or the soft skin between the toes, but cases have occurred in which mud splattered on the ankles of people wearing shoes. Infection has even been reported in people handling soiled laundry.

On contact, the hookworm larvae actively burrow into the skin. After several hours they reach the lymphatic or blood circulation

* *filariform* (fih-lar′-ih-form) L. *filum*, a thread.

and are immediately carried into the heart and lungs. The larvae proceed up the bronchi and trachea to the throat, where they are swallowed with sputum and arrive in the small intestine. There they anchor, feed on tissues, and mature.

Initial penetration with larvae often evokes a localized dermatitis called "ground itch," and movement of the larvae to the lungs can cause symptoms of pneumonia. The potential for injury is greatest with chronic infections, when heavy worm burdens cause vomiting and bloody diarrhea. Because of significant blood loss, iron-deficient anemia can develop, and infants are especially susceptible to hemorrhagic shock. Chronic fatigue, listlessness, apathy, and anemia worsen with chronic and repeated infections.

An experimental hookworm vaccine, now in phase II trials in Brazil, has generated a great deal of interest. Underwritten by a grant from the Gates Foundation, this could be the first vaccine available for a helminth disease.

Strongyloides stercoralis *and* Strongyloidiasis

The agent of strongyloidiasis, or threadworm infection, is *Strongyloides stercoralis*.* This nematode is exceptional because of its minute size and its capacity to complete its life cycle either within the human body or outside in moist soil. Its distribution and life cycle, similar to hookworms, afflicts an estimated 75 million to 100 million people worldwide. Infection occurs when soil larvae penetrate the skin. The worm then enters the circulation, is carried to the respiratory tract and swallowed, and then enters the small intestine to complete development. Although adult *S. stercoralis* lay eggs in the gut just as hookworms do, the eggs can hatch into

Figure 23.19 The hookworms. (a) Adult and egg stages of *Necator americanus* with size grids for comparison (grids are not to actual scale). **(b)** Cutting teeth on the mouths of (top) *N. americanus* and (bottom) *Ancylostoma braziliense*. (b: top, bottom): *Source:* CDC

* *Strongyloides stercoralis* (stron″-jih-loy′-deez ster″-kor-ah′-lis) Gr. *strongylos*, round, and *stercoral*, pertaining to feces.

Pathogen Profile #3 *Trichinella spiralis*

Trichinella worms encysted
in muscle
Source: National Human
Genome Research Institute

Microscopic Morphology Round larval worms are 1 to 2 mm in length and coiled cysts are about 1 mm in diameter.

Identified by Antibody detection. Clinical signs and symptoms vary, depending on the stage of infection, but include gastrointestinal distress, muscle pain, weakness, and eosinophilia. Visualization of encysted worms following muscle biopsy and microscopic examination offer confirmation of infection.

Habitat The entire *Trichinella* life cycle is spent within the body of a mammalian host, with pigs, bears, cats, dogs, and rats all commonly serving as reservoirs.

Primary Infections/Disease Trichinellosis is acquired by eating the meat of an infected animal, with most cases being traced to consumption of undercooked swine or bear meat. Initial symptoms include diarrhea, nausea, abdominal pain, fever, and sweating as the worm invades the intestinal mucosa, followed by muscle and joint pain, weakness, and shortness of breath as larvae encyst in the muscles. In most cases symptoms eventually subside, although involvement of the heart or brain, a rare occurrence, can lead to life-threatening complications.

Control and Treatment Adequate cooking, freezing, or smoking of pork and wild meats will kill larvae. Treatment with albendazole and steroids is useful during the initial stages of infection, but there is no cure once larvae have encysted in the muscles.

larvae in the colon and remain entirely in the host's body to complete the cycle. Eggs can likewise exit with feces and go through an environmental cycle. These numerous alternative life cycles greatly increase the chance of transmission and the likelihood for chronic infection.

The first hint of threadworm infection is usually a red, intensely itchy skin rash at the site of entry. Mild migratory activity in an otherwise normal person can escape notice, but heavy worm loads can cause symptoms of pneumonitis and eosinophilia. The nematode activities in the intestine produce bloody diarrhea, liver enlargement, bowel obstruction, and malabsorption. In immunocompromised patients, there is a risk of severe, disseminated infestation involving numerous organs. Hardest hit are AIDS patients, transplant patients on immunosuppressant drugs, and cancer patients receiving irradiation therapy, who can die if not treated promptly with ivermectin.

Trichinella spiralis *and Trichinellosis*

The life cycle of ***Trichinella spiralis**** is spent entirely within the body of a mammalian host, usually a carnivore or omnivore such as a pig, bear, cat, dog, or rat. In nature, the parasite is maintained in an encapsulated (encysted) larval form in the muscles of these animal reservoirs and is transmitted when other animals prey upon them. Humans acquire **trichinellosis** (also called trichinosis) by eating the meat of an infected animal (usually wild or domestic swine or bears), but humans are essentially dead-end hosts **(figure 23.20).**

Yearly estimates of trichinellosis average about 10,000 cases worldwide, with fewer cases occurring in regions of the world where religious or cultural traditions disapprove of the eating of pork. In the United States, pork is generally frozen or fully cooked—

both of which kill the *Trichinella* worm—prior to eating, and people have become more aware of the dangers of raw or undercooked dishes containing pork. As a result, only 10 to 20 cases are seen each year, and most of these are traced to the eating of wild game, particularly bear meat.

The cyst envelope is digested in the stomach and small intestine, which liberates the larvae. After burrowing into the intestinal mucosa, the larvae reach adulthood and mate. An unusual aspect of this worm's life cycle is that males die and are expelled in the feces, while females incubate eggs and shed live larvae. These penetrate the intestine and enter the lymphatic channels and blood. Final development occurs when the coiled larvae are encysted in the skeletal muscle (figure 23.20). At maturity, the cyst is about 1 mm long and can be observed by careful inspection of meat. Although larvae can deteriorate over time, they have also been known to survive for years.

The first symptoms of trichinellosis mimic influenza or viral fevers, with diarrhea, nausea, abdominal pains, fever, and sweating. The second phase, brought on by the mass migration of larvae and their entrance into muscle, produces intense muscle and joint pain, shortness of breath, and pronounced eosinophilia. The most serious life-threatening manifestations are heart and brain involvement. Although the symptoms eventually subside, a cure is not available once the larvae have encysted in muscles. The most effective preventive measures are adequately cooking, freezing, or smoking pork and wild meats.

Tissue Nematodes

As opposed to the intestinal worms previously described, tissue nematodes complete their life cycle in human blood, lymphatics, or skin. The most important members are **filarial worms.** Their most notable characteristics are having thin, filamentous bodies; being spread by biting arthropods; and causing chronic, deforming diseases.

* *Trichinella spiralis* (trik″-ih-nel′-ah spy-ral′-is) A little, spiral-shaped hair worm.

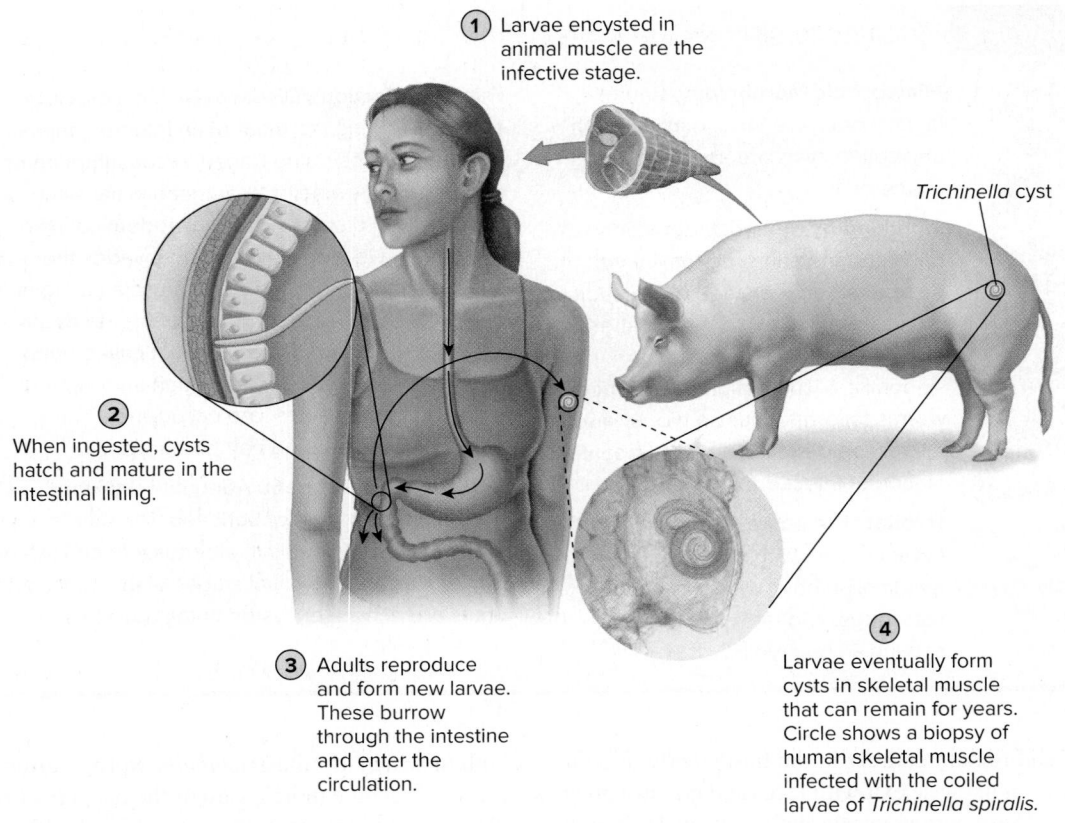

① Larvae encysted in animal muscle are the infective stage.

Trichinella cyst

② When ingested, cysts hatch and mature in the intestinal lining.

③ Adults reproduce and form new larvae. These burrow through the intestine and enter the circulation.

④ Larvae eventually form cysts in skeletal muscle that can remain for years. Circle shows a biopsy of human skeletal muscle infected with the coiled larvae of *Trichinella spiralis*.

Process Figure 23.20 The cycle of trichinellosis. © Pr. Bouree/Science Source

Filarial Nematodes and Filariasis

The filarial parasites are characterized by the production of tiny microfilarial larvae that circulate in the blood and lymphatics. Their anatomy is relatively simple, but they have a complex biphasic life cycle that alternates between humans and blood-sucking mosquito or fly vectors. The species responsible for most **filariases** are *Wuchereria bancrofti* and *Onchocerca volvulus*.

Wuchereria bancrofti and Bancroftian Filariasis *Wuchereria bancrofti** causes **bancroftian filariasis,** a common affliction in tropical regions where its vectors live (female *Culex, Anopheles,* and *Aedes* mosquitoes). The WHO estimates there are 120 million cases worldwide distributed in a pattern similar to malaria. The microfilariae are highly attuned to their vectors, such that they swarm to the skin capillary circulation to coincide with the times the mosquito feeds. The microfilariae complete development in the flight muscles of the mosquito, and the resultant larvae gather in its mouthparts. Infection begins when a mosquito bite deposits the infective larvae into the skin. These larvae penetrate the lymphatic vessels, where they develop into long-lived, reproducing adults. Mature female worms expel huge numbers of microfilariae directly into the circulation throughout their life cycle **(figure 23.21).**

The first symptoms of larval infestation in humans are inflammation and dermatitis in the skin near the mosquito bite, lymphadenitis, and testicular pain. The most conspicuous and bizarre sign of

chronic bancroftian filariasis is profound swelling of the scrotum or legs called **elephantiasis.*** This edema is caused by inflammation and blockage of the main lymphatic channels, which prevent return of lymph to the circulation and cause large amounts of fluid to

* *elephantiasis* (el″-eh-fan-ty′-ah-sis) Gr. *elephas,* elephant.

Figure 23.21 Bancroftian filariasis. Elephantiasis occurs when filarial worms block the main lymphatic channels, causing lymph to accumulate in the extremities. Inset: *Wuchereria bancrofti* microfilaria in a blood smear. *Source:* CDC; (inset): *Source:* CDC-DPDx/Oregon State Public Health Laboratory

* *Wuchereria bancrofti* (voo″-kur-ee′-ree-ah bang-krof′-tee) After Otto Wucherer and J. Bancroft.

accumulate in the extremities (figure 23.21). The condition can predispose to ulceration and infection. In extreme cases, the patient can become physically incapacitated and suffer from the stigma of having to carry massively swollen body parts. This condition occurs only in individuals in endemic regions after 1 to 5 decades of chronic infection. The microfilariae can be killed by oral medications (albendazole, ivermectin), but the lymphatic form of the disease is incurable.

Onchocerca volvulus and River Blindness

*Onchocerca volvulus** is a filarial worm transmitted by small biting vectors called black flies. These voracious flies can attack in such large numbers that it is not uncommon to be bitten several hundred times a day. Onchocerciasis is found in rural settlements along rivers bordered by overhanging vegetation, which are typical breeding sites of the black fly. Most cases of this disease occur in 30 endemic countries in Africa. Estimates from the WHO project approximately 25 million infections and nearly 1 million blind from its effect on the eyes.

The life and transmission cycles are similar to those of *Wuchereria,* although the larvae are deposited into the bite wound and develop into adults in the immediate subcutaneous tissues within 1 to 2 years after initial contact. Microfilariae produced by the adult female migrate via the blood to many locations but especially the eyes. It is the blood phase that infects other feeding black flies.

As the larvae and adult worms degenerate in the skin and the subcutaneous tissues, they provoke inflammatory reactions, rashes, and nodules in these regions that can persist for years. Invasion into the eye and its surrounding tissues gives rise to its most debilitating complication—river blindness. Recent experiments provide convincing evidence that the eye damage is actually due to a coinfection with a bacterium called *Wolbachia.* The bacterium is an endosymbiont that the larval *Onchocerca* require for their normal development. Over time, the bacterial antigens given off by the infection elicit strong inflammatory reactions that permanently damage the cornea, retina, and the optic nerve. This discovery means that it may now be possible to break the cycle of the disease by using antibiotics such as tetracyclines to destroy the bacteria in worms and the antiprotozoan drug *ivermectin* to kill the microfilarial larvae themselves. These treatments, along with an insecticide program to control the black flies, are the major strategies for containing the disease—one of the current aims of the WHO.

Loa loa: The African Eye Worm

Another nematode that makes disturbing migratory visits across the eye and under the skin is *Loa loa.** This worm is larger and more apparent than *O. volvulus.* The agent occurs in parts of western and Central Africa and is spread by the bites of dipteran flies. Its life cycle is similar to *Onchocerca* in that the larvae and adults remain more or less confined to the subcutaneous tissues. The first signs of disease are itchy, marble-sized lesions called *calabar swellings* at the site of penetration. These nodules are transient, and the worm lives for only a year or so, but the infestation is often recurrent. Because the nematode is temperature-sensitive, it is drawn to the body surface in response to warmer temperature and is frequently felt slithering beneath the skin or observed beneath the conjunctiva. One method of treatment is to carefully pull the worm from a small hole in the conjunctiva following application of local anesthesia. Some infections can be suppressed with high doses of diethylcarbamazine.

Dracunculus medinensis* and Dracontiasis

Dracontiasis, caused by the guinea worm, once caused untold misery in residents of India, the Middle East, and Central Africa. The parasite is carried by *Cyclops,* a microscopic arthropod commonly found in still water. When people ingest water from a contaminated community pond or open well, the larvae are freed from their *Cylops* host and penetrate into subcutaneous tissues of the intestine, where they form highly irritating lesions.

An intense effort by the World Health Organization in conjunction with the Carter Foundation began eradicating the disease by converting ponds to wells, filtering water through a fine nylon sieve, and treating the water with an insecticide. So successful was this strategy that dracontiasis cases have dropped from 3.5 million cases in 1986 to 24 in 2015 and will soon be eliminated from the earth.

 Check Your Progress SECTION 23.6

28. Name the intestinal nematodes in which infestation begins with the ingestion of helminth eggs.
29. How do adult *Ascaris* get into the intestine?
30. How do adult hookworms get into the intestine?
31. How do microfilariae get into the blood?
32. Outline the major similarities and differences among *Necator, Ancylostoma, Strongyloides,* and *Trichinella.*
33. How is trichinellosis contracted? How can it be prevented?
34. How does the life cycle of intestinal nematodes differ from that of tissue nematodes?
35. Construct a table providing the name of the infestation and the vector responsible for transmission of *Wuchereria bancrofti, Onchocerca volvulus, Loa loa,* and *Dracunculus medinensis.*
36. Which worms can be found in the eye? Which cause blindness?

23.7 Flatworms: The Trematodes and Cestodes

 Expected Learning Outcomes

30. Identify the intermediate and definitive hosts in trematode infestation.
31. Recall the stages of the *Schistosoma* life cycle.
32. Discuss the anatomical adaptation of tapeworms to their intestinal habitat.
33. Describe the major characteristics of *Taenia saginata* and *Taenia solium.*

* *Onchocerca volvulus* (ong″-koh-ser′-kah volv′-yoo-lus) Gr. *onkos,* hook, and *kerkos,* tail; L. *volvere,* to twist.
* *Loa loa* (loh′-ah loh′-ah).

* *Dracunculus medinensis* (drah-kung′-kyoo-lus meh-dih-nen′-sis) The little dragon of Medina. Also called guinea worm.

The fluke is named for the characteristic ovoid or flattened body found in most adult worms, especially liver flukes such as *Fasciola hepatica*. Blood flukes, also called schistosomes, have a more cylindrical body plan. They are also called **trematodes** (Gr. *trema*, a hole) because of the muscular sucker containing a mouth (hole) at the fluke's anterior end (see figure 5.28*b*). Flukes have digestive, excretory, neuromuscular, and reproductive systems but lack circulatory and respiratory systems. Except in the blood flukes, both male and female reproductive systems occur in the same individual and occupy a major portion of the body. The trematode life cycles are closely adapted to agricultural practices such as flood irrigating and using raw sewage as fertilizer. In human fluke cycles, animals such as snails or fish are usually the intermediate hosts, and humans are the definitive hosts.

Blood Flukes: Schistosomes (Cycle D)

Schistosomiasis has afflicted humans for thousands of years. The disease is caused by ***Schistosoma*** mansoni, S. japonicum,* or *S. haematobium.*[4] All three species are morphologically and geographically distinct but share similar life cycles, transmission methods, and general disease manifestations. The disease is endemic to 52 countries located in Africa, South America, the Middle East, and the Far East. Schistosomiasis is one of the most prominent helminth diseases, probably affecting over 200 million people at any one time. Recent increases in its occurrence in Africa have been attributed to new dams on the Nile River, which have provided additional habitat for snail hosts.

The schistosome parasite demonstrates intimate adaptations to both humans and certain species of freshwater snails required to complete its life cycle. The cycle begins when infected humans release eggs into irrigated fields or ponds either by deliberate fertilization with excreta or by defecating or urinating directly into the water. The egg hatches in the water and gives off an actively swimming ciliated larva called a **miracidium*** **(figure 23.22*a*),** which instinctively swims to a snail and burrows into a vulnerable site, shedding its ciliated covering in the process. In the body of the snail, the miracidium multiplies and transforms into a larger, fork-tailed swimming larva called a **cercaria*** **(figure 23.22*b*).** Cercariae are given off by the thousands into the water by infected snails.

Upon contact with a human, cercariae attach themselves to the skin by ventral suckers and penetrate hair follicles. They pass into small blood and lymphatic vessels and are carried to the liver. Here the schistosomes achieve sexual maturity, and the male and female worms remain permanently entwined to facilitate mating **(figure 23.22*c*).** In time, the pair lodges in small blood vessels. *Schistosoma mansoni* and *S. japonicum* end up in the mesenteric venules of the small intestine, and *S. haematobium* enters the venous plexus of the bladder. While attached to these intravascular sites, the worms feed upon blood, and the female lays eggs that are eventually voided in feces (*Schistosoma mansoni* and *S. japonicum*) or urine (*S. haematobium*).

(a) The miracidium phase, which infects the snail (300×).

(b) The cercaria phase, which is released by snails and burrows into the human host (5,000×).

(c) Schistosoma mansoni trematodes, with a long slender female worm cradled within a groove (right) on the ventral canal of the larger, thicker male to facilitate mating.

Figure 23.22 Stages in the life cycle of *Schistosoma*.
(a): © Sinclair Stammers/Science Source; (b): Courtesy of Harvey Blankespoor; (c: left, right): *Source:* CDC-DPDx

The first symptoms of infestation are itchiness in the vicinity of cercarial entry, followed by fever, chills, diarrhea, and cough. The most severe consequences, associated with chronic infections, are hepatomegaly and liver disease, splenomegaly, bladder obstruction, and blood in the urine. Occasionally eggs carried into the central nervous system and heart cause chronic granulomatous lesions.

In parts of Africa where this disease is endemic, epidemiologists are attempting a unique form of biological control. They have introduced into local lakes a species of small fish that feeds on snails, which greatly reduces the hosts for cercariae and the possibility for infection.

Liver and Lung Flukes (Cycle D)

Several trematode species that infest humans may be of zoonotic origin. The Chinese liver fluke, ***Opisthorchis** (**Clonorchis**) sinensis,**

* *Schistosoma* (skis″-toh-soh′-mah) Gr. *schisto*, split, and *soma*, body.

4. The species are named after P. Manson; Japan; and Gr. *haem*, blood, and *obe*, to like.

* *miracidium* (my″-rah-sid′-ee-um) Gr. *meirakidion*, little boy.

* *cercaria* (sir-kair′-ee-uh) Gr. *kerkos*, tail.

* *Opisthorchis sinensis* (oh″-piss-thor′-kis sy-nen′-sis) Gr. *opisthein*, behind, *orchis*, testis, and *sinos*, oriental.

completes its sexual development in mammals such as cats, dogs, and swine. Its intermediate development occurs in snail and fish hosts. Humans ingest cercariae in inadequately cooked or raw freshwater fish and crustaceans. Larvae hatch and crawl into the bile duct, where they mature and shed eggs into the intestinal tract. Feces containing eggs are passed into standing water that harbors the intermediate snail host. The cycle is complete when infected snails release cercariae that invade fish or invertebrates living in the same water.

The liver fluke *Fasciola hepatica** is a common parasite in sheep, cattle, goats, and other mammals and is occasionally transmitted to humans **(figure 23.23).** Periodic outbreaks in temperate regions of Europe and South America are associated with eating wild watercress. The life cycle is very complex, involving a mammal as the definitive host, fecal release of eggs into water, hatching of eggs into miracidia, invasion of freshwater snails (the intermediate host), development and release of cercariae, encystment of cercariae on a water plant, and ingestion of the cysts by a mammalian host eating the plant. Humans develop symptoms of vomiting, diarrhea, hepatomegaly, and bile obstruction only if they are chronically infected by a large number of flukes.

The adult lung fluke *Paragonimus westermani,** endemic to Asia, India, and South America, occupies the pulmonary tissues of various reservoir mammals, usually carnivores such as cats, dogs, foxes, wolves, and weasels, and has two intermediate hosts—snails and crustaceans. Humans accidently contract infection by eating

* *Fasciola hepatica* (fah-see′-oh-lah heh-pat′-ih-kah) Gr. *fasciola,* a band, and *hepatos,* liver.
* *Paragonimus westermani* (par″-ah-gon′-ih-mus wes-tur-man′-eye) Gr. *para,* to bear, and *gonimus,* productive; after Westerman, a researcher in flukes.

Figure 23.23 *Fasciola hepatica,* **the sheep liver fluke (2×).**
Flukes like this are a good example of hermaphroditism, having both sex organs in the same worm. © Sinclair Stammers/Science Source

undercooked crustaceans. After being released in the intestine, the worms migrate to the lungs, where they can cause cough, pleural pain, and abscess.

Cestode (Tapeworm) Infections (Cycle C)

The **cestodes,** or tapeworms, exhibit extreme adaptations, consisting of little more than a tiny holdfast connected to a chain of flattened sacs that contain the reproductive organs. Several anatomical features of the adult emphasize their adaptations to an intestinal existence (**figure 23.24*a,b*** and figure 5.28*a*). The small **scolex,** or head of the worm, has suckers or hooklets for clinging to the intestinal wall, but it has no mouth because nutrients are absorbed directly through the body of the worm. The importance of the scolex in anchoring the worm makes it a potential target for antihelminthic drugs. The head is attached by a neck to the **strobila,** a long ribbon composed of individual body segments called **proglottids,** which are generated by the neck. The contents of the proglottids are dominated by male and female reproductive organs. Each segment also contains a primitive neuromuscular system that permits feeble movements. The proglottids taper from the smaller, newly formed, and least mature segments nearest the neck to the larger, older ones containing ripe eggs at the distal end of the worm. The proglottids are either shed intact in feces after detaching from the strobila or break open in the intestine and expel free eggs into the feces.

All tapeworms for which humans are the definitive hosts require an intermediate animal host. The principal causes of human **taeniasis** are *Taenia saginata,* the beef tapeworm, and *Taenia solium,* the swine tapeworm. *Diphyllobothrium latum* is acquired through raw fish **(23.3 Making Connections).** The beef tapeworm follows a cycle C pattern, in which the larval worm is ingested in raw beef; and the pork tapeworm follows both a cycle C and a modified cycle A, in which the eggs are ingested and hatch in the body.

*Taenia saginata** is one of the largest tapeworms, composed of up to 2,000 proglottids and anchored by a scolex with suckers. Taeniasis caused by the beef tapeworm is distributed worldwide but is reported most often where cattle ingest proglottids or eggs from contaminated fields and water, and where raw or undercooked beef is eaten. The eggs hatch in the small intestine, and the released larvae migrate throughout the organs of the cow **(figure 23.24*c*).** Ultimately, they encyst in the muscles, becoming **cysticerci,** young tapeworms that are the infective stage for humans. When humans ingest a live cysticercus in beef, it is uncoated and firmly attaches to the intestine by the scolex, where it develops into an adult tapeworm. Humans are not known to acquire infection by ingesting the eggs. For such a large organism, it is remarkable how few symptoms a tapeworm causes. Patients may complain of vague abdominal pain and nausea and discover proglottids in their stool.

* *cestodes* (sess′-tohd) L. *cestus,* belt, and *ode,* like.
* *scolex* (skoh′-leks) Gr. scolos, worm.
* *strobila* (stroh-bih′-lah) Gr. *strobilos,* twisted.
* *proglottids* (proh-glot′-id) Gr. *pro,* before, and glotta, the *tongue.*
* *Taenia saginata* (tee′-nee-ah saj-ih-nah′-tah) L. *taenia,* a flat band, *sagi,* a pouch, and *natare,* to swim.

23.3 MAKING CONNECTIONS

Waiter, There's a Worm in My Fish

Various forms of raw fish or shellfish are part of the cuisine of many cultures. Popular examples are Japanese sashimi and sushi. Although these delicacies are generally safe, fish flesh often harbors various types of helminths. Humans are not usually natural hosts for these parasites, so the worms do not fully invade, but some of their effects can be quite distressing.

One of the common fish parasites transmitted this way is **Anisakis,** a nematode parasite of marine vertebrates, fish, and mammals. The symptoms of anisakiasis, usually appearing within a few hours after a person ingests the half-inch larvae, are due to the attempts of the larvae to burrow into the tissues. The first symptom is often a tingling sensation in the throat, followed by acute gastrointestinal pain, cramping, and vomiting, which mimic appendicitis. At times the worms are regurgitated or discovered during surgery, but in many cases they are probably expelled without symptoms. Another worm that can hide in raw or undercooked freshwater fish is the tapeworm *Diphyllobothrium latum,* common in the Great Lakes, Alaska, and Canada. Because humans serve as its definitive host, this worm can develop in the intestine and cause long-term symptoms.

Students always ask about the safety of sushi and seem uncertain about how to eat it and still avoid parasites. One good safeguard is to patronize a reputable sushi bar, because a competent sushi chef carefully examines the fish for the readily visible larvae. Another is to avoid raw salmon and halibut, which are more often involved than tuna and octopus. Fish can also be freed of live parasites by freezing for 5 days at −20°C.

One common perception is that fish from saltwater are safer to eat raw than those from fresh water. Based on the examples here, what do you think is true? *Answer available on Connect.*

(a) *Taenia solium* tapeworm scolex displaying four suckers and two rows of hooklets.

(b) Adult *Taenia saginata.* The arrow points to the scolex; the remainder of the tape, called the strobila, has a total length of 5 meters.

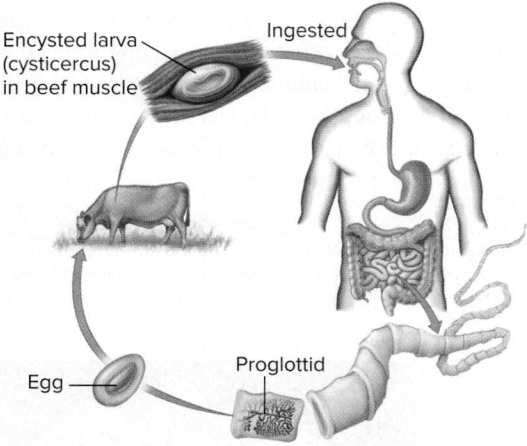

Encysted larva (cysticercus) in beef muscle

Ingested

Egg

Proglottid

(c) A generalized diagram of the life cycle of the beef tapeworm *T. saginata.*

Figure 23.24 Features of *Taenia* tapeworms. (a): *Source:* CDC; (b): © Eric Grave/Science Source

Taenia solium differs from *T. saginata* by being somewhat smaller, having a scolex with hooklets and suckers to attach to the intestine, and being infective to humans in both the cysticercus and egg stages. The disease is endemic to regions where pigs eat fecally contaminated food and humans consume raw or partially cooked pork (Southeast Asia, South America, Mexico, and eastern Europe). The cycle involving ingestion of cysticerci is nearly identical to that of the beef tapeworm (figure 23.24*c*).

A different form of the disease, called **cysticercosis,** occurs when humans ingest pork tapeworm eggs from food or water. Although humans are not the usual intermediate hosts, the eggs can still hatch in the intestine, releasing tapeworm larvae that migrate to all tissues. They finally settle down to form peculiar cysticerci, or bladder worms, each of which has a small capsule resembling a little bladder. They do the most harm when they lodge in the heart muscle, eye, or brain, especially if a large number of bladder worms occupy tissue space **(figure 23.25).** It is not uncommon for patients to exhibit seizures, psychiatric disturbances, and other signs of neurological impairment.

Figure 23.25 *Taenia solium* cysticerci within a section of porcine muscle. *Source:* Dr. Green/CDC

 Check Your Progress SECTION 23.7

37. Provide examples of intermediate and definitive hosts for *Schistosoma japonicum, Opisthorchis sinensis, Fasciola hepatica,* and *Paragonimus westermani*.

38. What are the stages in *Schistosoma* development? Where does each stage occur?

39. Justify the strategy of using the scolex of a tapeworm as a target for antihelminthic drugs.

40. Discuss the similarities and differences encountered when comparing the two major *Taenia* species.

23.8 The Arthropod Vectors of Infectious Disease

 Expected Learning Outcomes

34. Differentiate among the arthropod vectors of disease.

35. Describe the relationship between arthropod vectors and the parasites they carry.

The **arthropods** are a phylum of invertebrates that all share a body plan with an exoskeleton and jointed legs. Members include insects, arachnids, and crustaceans. Of the millions of species of arthropods known, a few thousand must feed on the blood or tissue fluids of a host sometime during their life cycle. In this adaptation, they are considered a type of **ectoparasite** because they remain on the outside of the host's body.

The types of ectoparasites discussed here not only feed on blood, they also transmit infectious microbes to the host. Animals with this behavior are considered *biological vectors* (see chapter 13). This means that they foster the survival of a pathogen (as do fleas with the plague bacillus) and may even provide an environment for the pathogen to complete its own life cycle (sand flies with *Leishmania*: see figure 23.11). In this way, they are alternate host organisms. Depending on the specificity of the vector and microbe, the diseases range from zoonoses that affect many different animals (trypanosomiasis) to those that attack a specific group (malaria). The types of arthropods that most often serve as vectors are insects: mosquitoes, fleas, lice, flies; and arachnids: ticks and mites. Although all vector groups are found worldwide, certain species may be found only in a particular geographic region such as the tropics. Other vectors may be endemic to a specific continent or country (Africa or South America, for example). *As a general rule, the location of the vector will dictate the origin and distribution of the disease.* **Table 23.6** provides a survey of vectors and the diseases they transmit.

Mosquitoes (Spanish for "little fly") are cosmopolitan insects in the Class Diptera. Approximately 3,500 species inhabit regions from the tropics to the Arctic. The adult can fly up to 300 miles and live for 2 months. All but a few species require blood and an aquatic habitat to complete their life cycle. The eggs are laid in standing water and go through embryonic, larval, and adult development there. An interesting note is that only the female mosquito actually sucks blood from the host—she needs blood protein to complete the development of her eggs. The male feeds primarily on plant juices. The female penetrates the skin with her proboscis to the level of a capillary, where she injects a small amount of anticoagulant so that the blood will flow freely into her digestive tract **(figure 23.26a)**. This is the route by which infections take place. Mosquitoes are

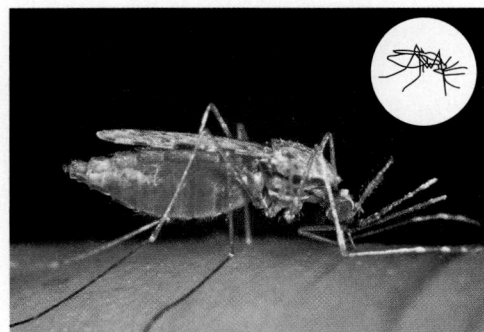
(a) Mosquito feeding on human

(b) Flea (note blood in gut).

(c) Louse.

(d) Hard tick, *Amblyomma*, A: shown actual size; B and C: feeding stages; D and E: stages of engorgement.

(e) Soft tick, *Ornithodoros*, ventral and dorsal views.

Figure 23.26 **An array of arthropod vectors shown enlarged and actual size (circled) for comparison. Note the mosquito, flea, and tick engorged with blood.** (a): *Source:* James Gathany/CDC; (b): *Source:* CDC; (c): *Source:* James Gathany/CDC; (d, e): *Source:* Dr. James Ochi/CDC-DPDx

TABLE 23.6	Common Arthropods That Are Biological Vectors of Disease		
Insect Vectors	**Diseases Transmitted**	**Type of Microbe**	**Distribution**
Mosquitoes			
Anopheles	Malaria	Apicomplexan protozoan	In a band around the equator♦
Anopheles/Culex	Filariasis	Nematode	Worldwide in tropics♦
Culex	Dengue fever	Arbovirus	Asia, Africa, Europe, Americas♣
Culex	West Nile fever	Arbovirus	Africa, Middle east, Americas♣
Aedes	Yellow fever	Arbovirus	Tropics, subtropics♣
Aedes	Zika virus disease	Arbovirus	Africa, Southeast Asia, Americas
Various	Encephalitides	Arboviruses	Worldwide♣
Fleas			
Xenopsylla	Bubonic plague	Gram-negative bacteria	Worldwide♦
Ctenocephalides	Murine typhus	*Rickettsia*	Worldwide☆
Body lice	Epidemic typhus	*Rickettsia*	Worldwide☆
Pediculus	Trench fever	*Bartonella* bacteria	Asia, Africa, Europe☆
	Relapsing fever	*Borrelia* spirochetes	Asia, Africa☆
Biting flies			
Tsetse	African trypanosomiasis	Flagellated protozoan	Central Africa♦
Sand fly	Leishmaniasis	Flagellated protozoan	Africa, India, Latin America♦
Black fly	Onchocerciasis	Nematode	Africa, Latin America♦
Biting bugs			
Triatoma	Chagas disease	Flagellated protozoan	Central and South America♦
Arachnid Vectors	**Diseases Transmitted**	**Type of Microbe**	**Distribution**
Hard ticks			
Dermacentor *Ixodes*	Rocky Mountain spotted fever	*Rickettsia*	North, Central, and South America☆
Amblyomma (several ticks can be vectors of the same disease)	Q fever	*Rickettsia*	Europe, Africa, Asia, N. America☆
	Human ehrlichiosis	*Rickettsia*	North America☆
	Lyme disease	*Borrelia* spirochetes	Worldwide☆
	Tularemia	Gram-negative bacteria	United States♦
	Babesiosis	Apicomplexan protozoan	Eastern/Midwestern United States♦
	Various arboviruses		Europe, Africa, Asia♣
Soft ticks			
Ornithodoros	Relapsing fever	Borrelia spirochete	Africa, Asia, Europe, Americas☆
Location of descriptions of diseases: ♦ Ch 23 ☆ Ch 21 ♣ Ch 25 ♠ Ch 20			

probably the dominant vector on earth with regard to numbers of infections. The CDC estimates that mosquito-borne diseases such as malaria, filariasis, Dengue fever, and other forms of arbovirus encephalitis affect 700 million people and that 3 million people, most of them children, die every year from these diseases.

Fleas are laterally flattened, wingless insects with well-developed jumping legs and prominent probosci for piercing the skin of warm-blooded animals **(figure 23.26b).** They are known for their extreme longevity and resistance, and many are notorious in their nonspecificity, passing with ease from wild or domesticated mammals to humans. In response to mechanical stimulation and warmth, fleas jump onto their targets and crawl about, feeding as they go. Infection takes place when a carrier flea regurgitates blood into the wound. The most common flea-borne diseases are bubonic plague (see figure 20.20) and murine typhus.

Lice are small, soft insects equipped with biting or sucking mouthparts **(figure 23.26c).** The lice of humans usually occupy head and body hair (*Pediculus humanus*) or pubic, chest, and axillary hair (*Phthirus pubis*). They feed by gently piercing the skin and sucking blood and tissue fluid. Infection develops when the louse or its feces are inadvertently crushed and rubbed into wounds, skin, eyes, or mucous membranes. Lice are involved in transmitting epidemic typhus and relapsing fever.

Quick Search
To observe an *Ixodes* tick in the empty and full state, search for "engorged tick" on Google images.

Distributed the world over are 810 species of **ticks,** all of them ectoparasites and about 100 of them carriers of infectious diseases **(figure 23.26*d,e*).** Ticks can be broadly characterized as hard bodied (ixodid), or soft bodied (argasid). Ticks exhibit the broadest host range of all vectors, able to feed on every vertebrate group except fish.

Hard (ixodid) ticks have tiny compact bodies suitable for hitchhiking along as their hosts wander through forest, savanna, or desert regions. Depending upon the species, ticks feed during larval, nymph, and adult metamorphic stages. The longevity of ticks is formidable; metamorphosis can extend for 2 years, and adults can survive for 4 years away from a host without feeding. The tiny, unengorged ticks cling to vegetation and attach to clothing of people. They crawl to skin surfaces and embed their mouthparts. As blood enters their gut, they expand significantly. Ixodid ticks are implicated in a number of bacterial, rickettsial, and viral diseases.

Soft (argasid) ticks with their flexible bodies seek sheltered habitats in caves, barns, burrows, or even birdcages. They parasitize wildlife during nesting seasons and domesticated animals throughout the year. Endemic relapsing fever is transmitted by this type of tick.

 Check Your Progress SECTION 23.8

41. Identify the types of arthropods that commonly serve as biological vectors of disease and summarize their characteristics.

42. Explain the differences between insects and arachnids and their role as vectors.

43. Explain several ways that vectors spread parasites to other hosts.

 CASE STUDY **Part 2**

Of concern to the CDC was the fact that neither patient could be linked to the warm-water springs, lakes, or rivers that are breeding grounds for the amoeba. As an investigation of both cases began, health officials worried that they could be dealing with a strain of *Naegleria fowleri* that had become more virulent or that had evolved a new means of entering the body.

One commonality investigators did discover was that both patients used neti pots filled with warm tap water to rinse their sinuses. A neti pot resembles a small teapot with a long spout, and is used by some people to rinse the nasal passages with a saline solution as a treatment for colds, allergies, or congestion. Investigators collected water samples from both homes and testing revealed the presence of *N. fowleri* in each case. These two incidents were the first time

PAM was associated with the presence of *N. fowleri* in household plumbing served by treated municipal water supplies, as well as the first time that infection with the amoeba was linked to use of a neti pot or similar nasal irrigation device.

- *N. fowleri grows at temperatures up to 46°C, but can survive for many hours at even higher temperatures. What biological structure could* N. fowleri *use to survive the hottest temperatures seen in household water systems?*

- *What additional precautions as to the avoidance of infection may have been issued based on the information seen in this part of the Case Study?*

Yoder, J. et. al. 2012. Primary amebic meningoencephalitis deaths associated with sinus irrigation using contaminated tap water. *Clin Infect Dis* 55:e79–e85.

To conclude this Case Study, go to Connect.

 Chapter Summary with Key Terms

23.1 The Parasites of Humans
A. Parasitology covers **protozoa** and **helminths** (worms), which live on the body of a host. Parasites are spread to humans by other humans, animal hosts, and vectors.
B. Parasitic diseases have existed since ancient times but are still a major public health problem today. International travel of people and materials has facilitated the spread of parasites and their vectors worldwide. The increase in immunosuppressed people, particularly those with AIDS, has caused an increase in formerly obscure protozoan infections.
 1. Some protozoans propagate only as **trophozoites** (active feeding stage found in host), whereas others alternate between a trophozoite and a **cyst** (dormant, resistant body).
 2. Some have complex life cycles with sexual and asexual phases carried out in more than one host.

23.2 Major Protozoan Pathogens
A. **Infective Amoebas**
 Entamoeba histolytica is the cause of amebiasis, or amebic dysentery, a worldwide human infection that affects approximately 500 million people in the tropics.
 1. The protozoan alternates between a trophozoite and a cyst. Cysts are spread through unsanitary water and food.
 2. Ingested cysts release trophozoites that invade the large intestine and cause ulceration and dysentery.
 3. Severe cases may occur when the pathogen penetrates into deeper layers.
 4. Amebic brain infections are caused by *Naegleria fowleri* and *Acanthamoeba* species, free-living inhabitants of natural waters. Primary acute meningoencephalitis is acquired through nasal contact with water or traumatic eye damage. Infiltration of the brain is usually fatal.

B. **Infectious Ciliates**

The only important ciliate pathogen is *Balantidium coli,* an occupant of the intestines of domestic animals such as pigs and cattle. It is acquired by humans when cyst-containing food or water is ingested. The trophozoite erodes the intestine and elicits intestinal symptoms.

23.3 The Flagellates (Mastigophorans)

A. *Trichomonas* is an animal parasite or commensal. The principal human pathogen is *T. vaginalis,* the cause of a very common STD known as **trichomoniasis** that infects the vagina, cervix, and urethra.

B. *Giardia intestinalis* is an intestinal flagellate that causes **giardiasis.** Its natural reservoir is animal intestines, and the source of infection is cyst-contaminated fresh water and food and causes severe diarrhea that may be chronic in immunodeficient patients.

C. **Hemoflagellates** are members of *Trypanosoma* that occur in blood during infection. The pathogens have complex life cycles, with various mastigote phases that develop in the insect and human hosts. Diseases are known as **trypanosomiases.**

1. *T. brucei* causes **African sleeping sickness,** spread by **tsetse flies** and harbored by reservoir mammalian hosts. The biting fly inoculates skin with the trypanosome, which multiplies in the blood, and damages the spleen, lymph nodes, and brain. Chronic disease symptoms are sleep disturbances, tremors, paralysis, and coma.

2. *T. cruzi* causes **Chagas disease,** a disease endemic to Latin America. Its cycle involves the **reduviid** (kissing) **bug** vector. Infection occurs when bug feces are inoculated into a cutaneous portal. Chronic inflammation occurs in the organs (especially heart and brain).

3. *Leishmania* contains geographically separate species that cause **leishmaniasis,** a zoonosis of wild animals that is transmitted by the bite of **phlebotomine** (sand) flies. Infected macrophages carry the pathogen into the skin and bloodstream, giving rise to fever, enlarged organs, and anemia. Kala azar is the most severe and fatal form.

23.4 Apicomplexan Parasites

Apicomplexans are tiny obligate intracellular parasites that have complex life cycles and lack motility organelles in their mature stage. Infectious forms include sporozoites, fecal cysts called oocysts, and tissue cysts. Most diseases are zoonotic and vector-borne.

A. *Plasmodium* is the cause of malaria; the human is the primary host for the asexual phase of parasite; the female *Anopheles* **mosquito** is the vector/host for the sexual phase.

1. **Malaria** is the most common parasitic disease, distributed primarily in a belt around the equator.

2. Infective forms for humans (**sporozoites**) enter blood with mosquito saliva, penetrate liver cells, multiply, and form hundreds of **merozoites.** These multiply in and lyse red blood cells.

3. Symptoms include episodes of chills, fever, and sweating; anemia; and organ enlargement.

4. *Toxoplasma gondii* lives naturally in cats that harbor oocysts in the GI tract.

a. **Toxoplasmosis** is acquired by ingesting raw or rare meats containing tissue cysts or by accidently ingesting oocysts from substances contaminated by cat feces.

b. In humans, infection is usually mild and flulike. Immunodepressed (AIDS) patients or fetuses (fetal toxoplasmosis) suffer brain and heart damage.

5. *Cryptosporidium* is a vertebrate pathogen that exists in both tissue and oocyst phases. Most cases of cryptosporidiosis are caused by handling animals with infections or by drinking contaminated water. Infection causes enteric symptoms.

B. **Coccidian Parasites**

1. *Cystoisospora* is an uncommon human intestinal parasite transmitted through fecal contamination that causes coccidiosis.

2. *Cyclospora* causes a diarrheal illness when its oocysts are ingested in fecally contaminated produce and water.

23.5 A Survey of Helminth Parasites

A. The parasitic **helminths,** or worms, are multicellular animals with specialized mouthparts and adaptations such as reduction of organs, protective cuticles, and complex life cycles.

B. Adult worms mate and produce fertile eggs that hatch into larvae that mature in several stages to adults. In some worms, the sexes are separate; others are **hermaphroditic.**

1. Adults live in a **definitive host,** whereas larval development occurs in the **intermediate host.**

2. Eggs and larvae may develop in the same host, the external environment, or an intermediate host.

3. Transmission can involve eggs, larvae, and insect vectors.

4. Helminth pathology arises from worms feeding on and migrating through tissues and accumulation of worms and worm products.

5. Useful antihelminthic drugs control worms by paralyzing their muscles and causing them to be shed or interfering with metabolism and killing them.

6. Other control measures are improved sanitation of food and water, protective clothing, cooking meat adequately, and controlling vectors.

23.6 Nematode (Roundworm) Infestations

Roundworms are filamentous with protective cuticles, circular muscles, a complete digestive tract, and separate sexes with well-developed reproductive organs.

A. *Ascaris lumbricoides* is indigenous to humans. Eggs ingested with foods hatch into larvae and burrow through the intestine into the circulation. From there, they travel to the lungs and pharynx and are swallowed. The adult worms complete the reproductive cycle in the intestine.

B. *Trichuris trichiura,* the whipworm, is a small parasite transmitted like *Ascaris* but restricted to the intestine throughout development.

C. *Enterobius vermicularis,* the pinworm, is a common childhood infection confined to the intestine. Eggs are picked up from surroundings and swallowed. After hatching, they develop into adults, and the females release eggs at the anus; self-inoculation is common.

D. Hookworms have characteristic curved ends and hooked mouths. The two major species, *Necator americanus* and *Ancylostoma duodenale,* share a similar life cycle, transmission, and pathogenesis. Humans shed eggs in feces, which hatch in soil and burrow into the skin of lower legs. Larvae travel from blood to lungs and are swallowed. Adult worms reproduce in the intestine and complete the cycle.

E. *Strongyloides stercoralis,* the threadworm, is a tiny nematode that completes its life cycle in humans or in moist soil. Larvae emerging from soil-borne eggs penetrate the skin and migrate to the lungs, are swallowed, and complete development in the intestine.

F. *Trichinella spiralis* causes **trichinellosis,** a zoonosis in which humans are a dead-end host. Disease is acquired from eating undercooked or raw pork or bear meat containing encysted larvae. Larvae migrate from the intestine to blood vessels, muscle, heart, and brain, where they enter dormancy.

G. **Filarial worms** are long, threadlike worms with tiny larvae (microfilariae) that circulate in blood and reside in various organs; are spread by various biting insects.

 1. *Wuchereria bancrofti* causes **bancroftian filariasis,** a tropical infestation spread by mosquitoes. The vector deposits the larvae, which move into lymphatics and develop. Chronic infection causes blockage of lymphatic circulation and **elephantiasis,** which swells the extremities.

 2. *Onchocerca volvulus* causes river blindness, an African disease spread by small, river-associated insects that feed on blood. Inoculated larvae migrate to skin to mature and form nodules in the eye that can destroy the cornea.

 3. *Loa loa,* the African eye worm, is spread by the bite of small flies. The worm migrates around under the skin and may enter the eye.

23.7 Flatworms: The Trematodes and Cestodes

Trematodes, or *flukes,* are flatworms with leaflike bodies bearing suckers; most are hermaphroditic.

A. Blood flukes and *Schistosoma* species. **Schistosomiasis** is a prevalent tropical disease. Adult flukes live in humans and release eggs into water. The early larva, or **miracidium,** develops in the freshwater snail into a second larva, or **cercaria.** This larva penetrates human skin and moves into the liver to mature; adults migrate to intestine or bladder and shed eggs, giving rise to chronic organ enlargement.

B. Zoonotic flukes include the Chinese liver fluke *Opisthorchis sinensis,* which cycles between mammals and snails and fish. Humans are infected when eating fish containing larvae, which invade the liver. *Fasciola hepatica* cycles between herbivores, snails, and aquatic plants. Humans are infected by eating raw aquatic plants; the fluke lodges in the liver.

C. **Cestodes or tapeworms:** These flatworms have long, very thin, ribbonlike bodies **(strobila)** composed of sacs

(proglottids) and a **scolex** that grips the host intestine. Each proglottid is an independent unit adapted to absorbing food and making and releasing eggs.

 1. *Taenia saginata* is the beef tapeworm for which humans are the definitive host. Animals are infected by grazing on land contaminated with human feces. Humans ingest raw beef in which the larval form **(cysticercus)** has encysted. The larva attaches to the small intestine and becomes an adult tapeworm.

 2. *T. solium* is the pork tapeworm that infects humans through cysticerci or by ingesting eggs in food or drink **(cysticercosis).** The larvae hatch and encyst in many organs and can damage them.

23.8 The Arthropod Vectors of Infectious Disease

Some invertebrates are ectoparasites that feed on blood and/or tissue fluids as part of their life cycles. Many of them serve as biological vectors by transmitting infectious agents during this process.

A. **Mosquitoes** are insects that require an aquatic habitat for development. Only females take blood from humans to complete egg production. They are transmitters of major diseases, with malaria the most prominent of these; others include filariasis (roundworms) and zoonoses such as arbovirus encephalitis.

B. **Fleas** are tiny, highly motile insects with flattened bodies that feed on warm-blooded animals. They are not specific to their hosts, so they mainly carry zoonotic diseases such as plague (bacteria) and murine typhus (rickettsias).

C. **Lice** are small soft insects that attach to head and body hair, feeding on blood and tissue fluids while releasing feces that contaminate the wound. Epidemic typhus (rickettsias) and relapsing fever (borrelias) are the most serious diseases they spread.

D. Other insect vectors include flies: for instance, the tsetse fly spreads African trypanosomiasis; the sand fly transmits leishmaniasis; the reduviid bug spreads Chagas disease.

E. **Ticks** are arachnids that cling to vegetation and hitch a ride on passing host bodies. Larvae, nymph, and adults feed by piercing the skin of the host, sucking blood, and swelling many times their size.

 1. *Hard* or *ixodid* ticks (*Dermacentor, Ixodes*) have small, compact, somewhat rigid bodies; they transmit several rickettsial, borrelial, and viral diseases.

 2. *Soft* or *argasid* ticks (*Ornithodoros*) have more flexible outer bodies; they transmit relapsing fever.

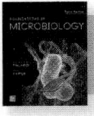

Level I. Knowledge and Comprehension

These questions require a working knowledge of the concepts in the chapter and the ability to recall and understand the information you have studied.

 Multiple-Choice Questions

Select the correct answer from the answers provided. For questions with blanks, choose the combination of answers that most accurately completes the statement.

1. All protozoan pathogens have a _____ phase.
 a. cyst c. trophozoite
 b. sexual d. blood

2. *Entamoeba histolytica* primarily invades the
 a. liver c. small intestine
 b. large intestine d. lungs

3. *Giardia* is a/an _____ that invades the _____.
 a. flagellate, large intestine c. ciliate, large intestine
 b. amoeba, small intestine d. flagellate, small intestine

4. Hemoflagellates are transmitted by
 a. mosquito bites c. bug feces
 b. insect vectors d. contaminated food

5. *Plasmodium* reproduces sexually in the _____ and asexually in the _____.
 a. liver, red blood cells c. human, mosquito
 b. mosquito, human d. red blood cell, liver

6. In the exoerythrocytic phase of infection, *Plasmodium* invades the
 a. blood cells c. salivary glands
 b. heart muscle d. liver

7. The oocyst of *Toxoplasma* is found in _____, and a pseudocyst is found in _____ .
 a. humans, cats c. feces, tissue
 b. cats, humans d. tissue, feces

8. A person can acquire toxoplasmosis from
 a. pseudocysts in raw meat c. cleaning out the cat litter box
 b. oocysts in air d. all of these

9. All adult helminths produce
 a. cysts and trophozoites c. fertilized eggs and larvae
 b. scolex and proglottids d. hooks and cuticles

10. The _____ host is where the larva develops, and the _____ host is where the adults produce fertile eggs.
 a. intermediate, definitive c. secondary, transport
 b. definitive, intermediate d. primary, secondary

11. Antihelminthic medications work by
 a. paralyzing the worm
 b. disrupting the worm's metabolism
 c. causing vomiting
 d. both a and b

12. Host defenses that are most active in worm infestations:
 a. phagocytes c. killer T cells
 b. antibodies d. eosinophils

13. Currently the most common nematode infestation worldwide is
 a. hookworm c. tapeworm
 b. ascariasis d. trichinosis

14. Hookworm diseases are spread by
 a. the feces of humans
 b. mosquito bites
 c. contaminated food
 d. microscopic invertebrates in drinking water

15. Trichinosis can only be spread from human to human by
 a. cannibalism c. raw pork
 b. flies d. contaminated water

16. The swelling of limbs typical of elephantiasis is due to
 a. allergic reaction to the filarial worm
 b. granuloma development due to inflammation by parasites
 c. lymphatic circulation being blocked by filarial worm
 d. heart and liver failure due to infection

17. Which of the following is *not* considered an insect vector?
 a. flea c. tick
 b. mosquito d. tsetse fly

18. The _____ mosquito feeds on blood, which is required for development of _____.
 a. male, larvae c. female, GI tract
 b. male, parasites d. female, eggs

19. **Single Matching.** Match the disease or condition with the causative agent, and indicate for each whether the agent is a protozoan (P) or a helminth (H).
 1. ____ ____ amebic dysentery a. *Plasmodium vivax*
 2. ____ ____ Chagas disease b. *Enterobius vermicularis*
 3. ____ ____ tapeworm c. *Onchocerca vovulus*
 4. ____ ____ hookworm d. *Toxoplasma gondii*
 5. ____ ____ African sleeping sickness e. *Trichuris trichiura*
 6. ____ ____ pinworm f. *Entamoeba histolytica*
 7. ____ ____ filariasis g. *Necator americanus*
 8. ____ ____ amebic meningoencephalitis h. *Trypanosoma brucei*
 9. ____ ____ malaria i. *Wuchereria bancrofti*
 10. ____ ____ toxoplasmosis j. *Trypanosoma cruzi*
 11. ____ ____ trichinellosis k. *Taenia saginata*
 12. ____ ____ whipworm l. *Naegleria fowleri*
 13. ____ ____ river blindness m. *Trichinella spiralis*

20. **Single Matching.** Match the disease with its primary mode of transmission or acquisition in humans.
 1. ____ amebic dysentery a. eating poorly cooked beef or pork
 2. ____ Chagas disease b. ingestion of cysts in water contaminated by wild animal feces
 3. ____ cyclosporiasis c. ingestion of food contaminated by human feces
 4. ____ toxoplasmosis
 5. ____ giardiasis d. bite from a reduviid bug
 6. ____ tapeworm e. bite from a black fly
 7. ____ filariasis f. water or food contaminated with human feces containing eggs
 8. ____ schistosomiasis
 9. ____ ascariasis g. contact with cats or ingesting rare or raw meat
 10. ____ river blindness
 h. bite from an *Anopheles* mosquito
 i. ingesting fecally contaminated water or produce
 j. freshwater snail vector releases infectious stage

Case Study Review

1. What cellular structures would *Naegleria fowleri* use for motility?
 a. cilia c. flagella
 b. pseudopods d. None; *N. fowleri* is nonmotile.

2. Infection with which organism could produce symptoms similar to those seen in primary amebic meningoencephalitis?
 a. *Balantidium coli* c. *Taenia solium*
 b. *Plasmodium falciparum* d. *Trichomonas vaginalis*

3. Provide at least two reasons that primary amebic meningoencephalitis is more common in the summer months.

Writing Challenge

For each question, compose a one- or two-paragraph answer that includes the factual information needed to completely address the question. Check Your Progress questions can also be used for writing-challenge exercises.

1. As a review, compare the four major groups of pathogenic protozoa according to overall cell structure, locomotion, infective state, and mode of transmission.

2. a. What are the primary functions of the trophozoite and the cyst in the life cycle?
 b. In which diseases is a cyst stage an important part of the infection cycle?

3. a. Describe the life cycle and host range of *Toxoplasma gondii*.
 b. How are humans infected?
 c. What are the most serious outcomes of infection?

4. a. In what ways are helminths different from other parasites, and why are there so many parasitic worm infections worldwide?
 b. What is the nature of human defenses against them?

5. In what ways is trichinellosis different from other worm infections?

6. a. Which parasites are zoonotic?
 b. Which are strictly human parasites?
 c. Which are borne by vectors?
 d. Which cause intestinal symptoms?
 e. Which are hermaphroditic?
 f. Which are associated with rare or raw meat?

Concept Mapping

On Connect you can find an Introduction to Concept Mapping that provides guidance for working with concept maps along with concept mapping activities for this chapter.

Level II. Application, Analysis, Evaluation, and Synthesis

These problems go beyond just restating facts and require higher levels of understanding and an ability to interpret, problem solve, transfer knowledge to new situations, create models, and predict outcomes.

Critical Thinking

Critical thinking is the ability to reason and solve problems using facts and concepts. These questions can be approached from a number of angles, and in most cases, they do not have a single correct answer.

1. Explain why a person with overt symptoms of intestinal *Entamoeba histolytica* infection is unlikely to transmit infection to others.

2. a. Explain why *Trichomonas vaginalis* is less likely to be transmitted by casual contact.
 b. What is meant by "ping-pong" infection, and why must both sex partners be treated for trichomoniasis?

3. Explain why only female mosquitoes are involved in malaria and elephantiasis.

4. Which parasitic diseases could conceivably be spread by contaminated blood and needles?

5. For which diseases can one *not* rely upon chlorination of water as a method of control? Explain why this is true.

6. a. If a person returns from traveling afflicted with trypanosomiasis or leishmaniasis, is he or she generally infective to others? Explain.
 b. Explain why there is no malaria above 6,000 feet in altitude.
 c. Account for all of the protective effects offered by insecticide-treated bednets.

7. Give some of the reasons that there are no really effective human vaccines so far for parasitic infections.

8. a. After comparing the maps shown in figures 23.8, 23.9, and 23.12, provide reasons to account for the patterns of distribution shown.
 b. Describe some factors that would affect the changes in the patterns.

9. a. What one simple act could in time eradicate ascariasis?
 b. Hookworm?
 c. Beef tapeworm?

10. To achieve a cure for tapeworm, why must the antihelminthic drug either kill the scolex or slacken its grip?

11. Students sometimes react with horror and distress when they discover that cats and dogs carry parasites to humans. Give an example of a disease for each of these animals that can be spread to humans, and explain how to avoid these diseases and still enjoy your pet.

12. a. Why is it necessary for most parasites to leave their host to complete the life cycle?
 b. What are some ways to prevent completion of the life cycle?
 c. What is the benefit to parasites of having numerous hosts?
 d. What are the disadvantages in having more than one host?

13. East Indian natives habitually chew on betel nuts as an alkaloidal stimulant and narcotic. A side effect is phlegm collection, which is eliminated by frequent spitting. The incidence of *Strongyloides*

infections is relatively low in this population. Can you account for this? (No, the betel nuts are not effective antihelminthics.)

14. In New York City, four Orthodox Jewish patients brought into an emergency clinic with seizures and other neurological symptoms tested positive for antibodies to the pork tapeworm, *Taenia solium.* This was embarrassing to them because they do not eat pork for religious reasons. It also became something of a medical mystery, because there were no immediate explanations for how they could have become infected. Later it was shown that the patients had recently employed housekeepers and cooks who were free of symptoms yet also seropositive for the worm infection. Use this case history to give a plausible explanation for the transmission of the infection.

 Visual Challenge

1. From **chapter 5, figure 5.29.** Correlate the pinworm cycle with the correct type of cycle in **figure 23.17.**

2. Observe figures **a, b, c,** and **d** and identify the parasite shown.

(a)
Source: CDC

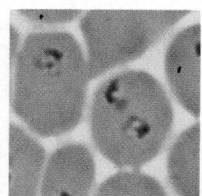

(c)
Courtesy of Stephen B. Aley, PhD, Univ. of Texas at El Paso

(b)
Source: CDC-DPDx

(d)
Source: CDC-DPDx

24 Introduction to Viruses That Infect Humans: The DNA Viruses

Varicella-zoster virus is responsible for both chickenpox (right) and shingles (center).
(left and right): *Source:* CDC; (center): © SPL/Science Source

CASE STUDY Part 1 Childhood Diseases

A skin rash can indicate anything from an annoyance (an allergic reaction to a new laundry detergent) to a global health emergency (smallpox), and the nature of the **rash** often provides the initial clues as to its cause. In this case, a 94-year-old woman presented to her physician with a widespread rash. She appeared healthy on initial examination and had no fever but had developed a **vesicular rash** over her face, limbs, and trunk over the last 48 hours. Reddened vesicular rashes—essentially small blisters or **lesions** filled with clear fluid—concentrated in the center of the body are a hallmark of **chickenpox.** The woman mentioned that she had never had chickenpox and hadn't had recent contact with anyone (grandchildren presumably) who may have had the disease. She did volunteer that her husband had been diagnosed with **opthalmic zoster (shingles)** several weeks previously. Based on the initial examination, the physician suspected chickenpox and prescribed **famicyclovir,** an antiviral medication.

The next day during a follow-up exam, many of the lesions had crusted over, typical of chickenpox. Blood tests revealed the presence of **varicella-zoster virus** (VZV) specific **IgM antibodies,** but no IgG; a second blood test 2 days showed the same result, but with an increased IgM titer. Two weeks later, a third blood test revealed the presence of **IgG antibodies** specific for VCV.

> "Over the course of several days, the lesions began to crust over."

Varicella-zoster virus, a member of the **herpesvirus** family, is responsible for varicella, more commonly known as chickenpox. The virus spreads easily to people who have never had the disease or received the chickenpox vaccine. Coughing and sneezing release the virus into the air, and the virus can also be contracted by inhaling viral particles released from chickenpox lesions. The illness typically lasts 5 to 10 days, after which the person is immune to a second bout of the disease. In general, having the disease when you are older results in more severe symptoms and a longer course of illness, with complicating symptoms ranging from **dehydration** to **encephalitis.**

Over the course of the next 2 weeks the patient seemed to recover without incident; all lesions were crusted over and the earliest lesions had begun to heal. Seventeen days after initially presenting to her doctor the patient developed paralysis on one side of her body. Imaging revealed an acute infarct, confirming that the patient had indeed had a stroke. She died one month later.

- Why is the husband's opthalmic zoster infection important in this case?

- What is the significance of the three serum antibody tests done on the woman?

To continue the Case Study, go to Case Study Part 2 at the end of the chapter.

24.1 Viruses in Human Infections and Diseases

 Expected Learning Outcomes

1. Recall the general characteristics of viruses.
2. Understand the progression and scope of viral infection.
3. Explain how viral structure is related to viral classification.

Viruses are the smallest parasites with the simplest biological structure. Essentially they are particles composed of DNA or RNA molecules surrounded by a protein coat. They are completely dependent on a host cell for their continued existence. Viruses have special adaptations for entering a cell and instructing the cell's genetic and molecular machinery to produce and release new viruses.

Animal viruses are divided into families based on the nature of the nucleic acid (DNA or RNA), the type of capsid, and whether or not an envelope is present. All DNA viruses that infect humans are double-stranded except for the parvoviruses, which contain single-stranded DNA. All RNA viruses that infect humans are single-stranded except for the double-stranded reoviruses. The genome of RNA viruses can be further described as segmented (consisting of more than one molecule) or nonsegmented (consisting of a single molecule). The envelope is derived from host cell membranes as the virus buds off the nuclear envelope, cell membrane, or Golgi membrane, and it contains spikes that interact with the host cell. See **tables 24.1** and 25.1 for a simple classification of DNA and RNA viruses.

TABLE 24.1	DNA Virus Groups That Infect Humans

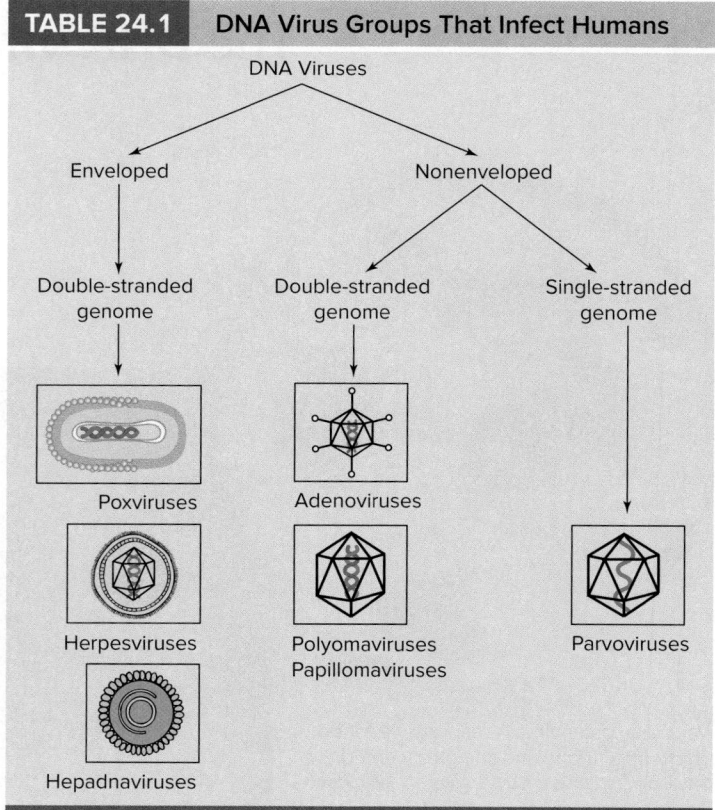

Systems Profile 24.1	Pathogenic DNA Viruses					
Virus	**Skin/Skeletal**	**Nervous/ Muscle**	**Cardiovascular/ Lymphatic/ Systemic**	**Gastrointestinal**	**Respiratory**	**Urogenital**
Variola Virus	Smallpox		Viremia			
Molluscum Contagiosum Virus	Molluscum contagiosum					Genital molluscum
Herpes Simplex Virus 1 and 2	1. Oral herpes 2. Whitlows 3. Keratitis	Encephalitis				Genital herpes
Varicella-Zoster Virus	Chickenpox	Shingles				
Cytomegalovirus (CMV)	Salivary glands		CMV infection			
Epstein-Barr Virus	Oral cavity, glands		Infectious mononucleosis			
Human Herpesvirus 6 and 7	Roseola					
Hepatitis B Virus	Jaundice		Hepatitis			
Adenovirus	Conjunctivitis				Colds	
Human Papillomavirus	Warts					Genital warts
JC Virus		Progressive multifocal leukoencephalopathy				
Parvovirus B19	Erythema infectiosum (fifth disease)		Red blood cell damage			

Important Medical Considerations in Viral Diseases

Target Cells

Because viruses have specific receptors for interacting with host cells, their infectiousness is limited to a particular host or cell type in most cases. Viruses infect most types of tissues, including the nervous system (polio and rabies), liver (hepatitis), respiratory tract (influenza, colds), intestine (polio), skin and mucous membranes (herpes), and immune system (AIDS). Many DNA viruses are assembled and budded off the nucleus, whereas most RNA viruses multiply in and are released from the cytoplasm. The presence of assembling and completed viruses in the cell gives rise to various telltale disruptions called cytopathic effects. The lytic infection of cells often destroys them, which accounts for the sometimes severe pathology and loss of function associated with viral infections.

Scope of Infections

Viral diseases range from very mild or asymptomatic infections such as colds to destructive and life-threatening diseases such as rabies and AIDS. Many viral infections occur primarily in the young and are readily transmitted through droplets. Although we tend to consider these infections as self-limited and inevitable, even measles, mumps, chickenpox, and rubella can manifest severe complications. Many viruses are strictly human in origin, but an increasing number (hantavirus, West Nile fever virus, and viral encephalitis) are zoonoses transmitted by vectors.

The course of viral diseases starts with invasion at the portal of entry by a few virions and a primary infection. In some cases the viruses replicate locally and disrupt the tissue, and in others the virus enters the blood or travels along neuron pathways to tissues far from the initial entry point. Common manifestations of virus infections include rashes, fever, muscle aches, respiratory involvement, and swollen glands. Some diseases of unknown etiology are thought to have a viral basis. Examples include type I diabetes and multiple sclerosis.

Protection from viral infections arises from the combined actions of interferon, neutralizing antibodies, and cytotoxic T cells. Because of the effective immune response to viruses, infections frequently result in lifelong immunities, and many viruses have been adaptable for vaccination.

Viral Persistence, Latency, and Oncogenicity

Many human viral infections such as colds and influenza have a rapid course and result in a lytic cycle of infection (**figure 24.1**). Such infections generally end with the virus being eliminated by the body within a relatively short time. But a few important viruses establish themselves in longer-term **persistent infections** that last for many years and sometimes for life. Two types of persistence are chronic viral infections and latent viral infections, or latency (**figure 24.1, steps 3 and 4**). In a **chronic infection** the virus is detectable in tissue samples and is still multiplying at a slow rate, but symptoms of infection are usually mild or absent. Examples of viruses with chronic effects are HIV and hepatitis B virus (HBV). **Latent infections** result when the virus enters a dormant phase inside host cells and becomes inactive after a lytic infection. It is not

Process Figure 24.1 Overview of possible effects of viral infections on host cells. (1) Infection is initiated by the entry of virus genetic material into a host cell. (2) In lytic infections, the virus is assembled and released with destruction of the host cell. In persistent infections, the virus does not destroy the cell. (3) In one instance—latency—the viral nucleic acid remains inactive and new viruses are not released, but the disease may recur. (4) In the case of chronic infections the virus continues to be active. (5) The genetic material of some viruses is oncogenic and can transform the host cells into cancer cells that grow in an uncontrolled manner.

multiplying, generally not detectable, and does not cause symptoms during latency. Because the virus is still present (usually as a nucleic acid), it can become reactivated and give rise to recurrent lytic infections. The herpesviruses are prominent examples of latent viruses.

The mechanisms by which viruses become persistent are probably multiple and vary with the type of virus. Some viruses integrate their genetic material into that of the host (HIV), while others remain as separate stable genetic elements called *episomes* that stay in the nucleus but do not integrate into a host chromosome (herpesviruses).

Some persistent viruses are **oncogenic** by virtue of their tendency to induce cancer. The most well-established human

oncogenic viruses and the cancers they are associated with are as follows:

DNA Viruses

Epstein-Barr virus	Burkitt lymphoma
Human herpesvirus 8	Kaposi sarcoma
Hepatitis B	Hepatic carcinoma
Certain papillomaviruses	Cervical and penile cancer

RNA Viruses

Retrovirus HTLV 1	Leukemia
Hepatitis C	Liver cancer

Tracing the involvement of viruses in human cancers has been stymied by the obvious inability to carry out human experiments. Even if a virus is discovered in a tumor, this does not immediately prove causation. When a correlation is found to be repeatable and specific, and is supported by a suitable animal model, the case for virus oncogenicity is more readily verified.

Viruses cause cancer through a variety of genetic effects, some of which were covered in chapter 16. Some may carry a viral onco-gene into the human DNA. Others may insert into a site on the ge-nome and activate a human oncogene. Some viruses do not insert into the genome but appear to regulate or promote the cell cycle in a way to stimulate uncontrolled cell growth. In all cases of virally induced cancers, a normal cell is transformed into a cancer cell and a tumor results.

Congenital and Perinatal Transmission

Several viruses can cross the placenta from an infected mother to the embryo or fetus. For example, about one in three children born to untreated HIV-positive mothers is infected. Infection of the fetus can cause developmental disturbances and permanent defects in the child that are present at birth (congenital). Viruses with known **teratogenic*** effects include rubella and cytomegalovirus. Another risk for infants involves infection around the time of birth, as occurs with hepatitis B and herpes simplex viruses.

Viral Diagnosis

Viral diseases are diagnosed by a variety of clinical and laboratory methods, including symptoms, isolation in cell or animal culture, and serological testing for antibodies. The use of antigen detection assays is on the increase, especially those based on monoclonal antibodies. Rapid nucleic acid probes directed against DNA or RNA are another option for several viruses. These have the advantage of being so sensitive that they could conceivably detect the viral nucleic acid in a single infected cell. Review figure 17.17 for a summary of identification techniques.

Overview of DNA Viruses

Classification of viruses is based on the structure of their capsid, the presence of an envelope, and the type of nucleic acid they contain.

* *teratogenic* (tur″-ah-toh-jen′-ik) Gr. *teratos,* monster. Any process that causes physical defects in the embryo or fetus.

Based on this scheme, animal viruses are divided into seven fami-lies of DNA-containing viruses and fourteen families of RNA-containing viruses. Family names are recognizable by the suffix -*viridae* (for example, Hepadnaviridae). Within a single family, many genera exist, each ending with the suffix -*virus* (for example, herpesvirus). A single genus may contain many viral "types" or strains, which may not differ significantly from one another.

Because we are concerned less with an exhaustive taxonomic classification of all viruses and more with those viruses of impor-tance to humans, we employ a more relaxed scheme of classification in which viral *groups* occupy the same hierarchical level as a family but contain only agents of human concern and are identified by the suffix -*virus.* Using this system of classification, DNA viruses that cause human disease are placed into six groups (see table 24.1).

 Check Your Progress SECTION 24.1

1. Discuss the structures found in a typical virus.
2. Differentiate among lytic infection, latent infection, chronic infec-tion, and transformation.
3. Relate the importance of structure to viral classification.
4. What accounts for the affinity of viruses for certain hosts or tissues?
5. What accounts for the symptoms of viral diseases?
6. In what ways can people carry viruses?

24.2 Enveloped DNA Viruses: Poxviruses

 Expected Learning Outcomes

4. Recall the structures seen in poxviruses that make them unique.
5. Understand why vaccination against smallpox has been curtailed.
6. Recall the poxviruses that can infect humans and the diseases they cause.

Classification and Structure of Poxviruses

Poxviruses produce eruptive skin pustules called pocks or **pox,** which leave small, depressed scars (pockmarks) upon healing. The poxviruses are distinctive because they are the largest and most complex of the animal viruses **(figure 24.2),** and they multiply in the cytoplasm in well-defined sites called factory areas, which ap-pear as inclusion bodies in infected cells.

Among the best-known poxviruses are variola, the cause of **smallpox,** and vaccinia, a closely related virus used in vaccina-tions. Other members are carried by domestic or wild animals and may occasionally be transmitted to humans. A common feature of all poxvirus infections is a specificity for the cytoplasm of epider-mal cells and subcutaneous connective tissues. In these sites, they produce the typical pox lesions, and some may later stimulate cell growth and tumor formation.

Smallpox: A Perspective

Largely through the World Health Organization's comprehensive global efforts, smallpox is now a disease of the past. It is one of the

Figure 24.2 The structure of poxviruses. Poxviruses are larger and more complex than other viruses. They have an envelope and an outer membrane with a rough texture of tubules. They have a central nucleus-like core that contains the DNA strand. They lack the usual nucleocapsid found in other viruses. (Right) Photomicrograph of smallpox virus (300,000×). *Source:* CDC

few infectious agents that now exists only in government laboratories. At one time, smallpox was ranked as one of the deadliest infectious diseases, leaving no civilization untouched. When introduced by explorers into naive populations (those never before exposed to the disease) such as Native Americans and Hawaiians, smallpox caused terrible losses and was instrumental in destroying those civilizations.

The twentieth century saw one of mankind's greatest achievements when, as a result of a worldwide health campaign that focused on immunization and isolation, the last case of smallpox was seen in 1977. In 1999 the World Health Organization recommended the destruction of all remaining smallpox stocks, which at this point existed only in laboratory freezers in the United States and the former Soviet Union. Despairing at the loss of such a novel subject that could be used for research, scientists in both countries balked at this request. Since that time, reports have come out that certain Russian labs have experimented with selecting strains of the virus to withstand delivery by bombs or missiles. There remains concern that the virus has been removed from Russia and sold to terrorist groups.[1]

Disease Manifestations Exposure to smallpox usually occurred through inhalation of droplets or skin crusts. Infection was associated with fever, malaise, and prostration and, later, a rash that began in the pharynx, spread to the face, and progressed to the extremities. Initially, the rash was macular, evolving in turn to papular, vesicular, and pustular before eventually crusting over, leaving nonpigmented sites pitted with scar tissue **(figure 24.3).** The two principal forms of smallpox were variola minor and variola major. Variola major was a highly virulent form that caused toxemia, shock, and intravascular coagulation. People who survived any form of smallpox nearly always developed lifelong immunity.

Smallpox Vaccination The smallpox vaccine uses a single drop of vaccinia virus punctured into the skin with a double-pronged needle. In a successful take, a large, scablike pustule with a reddened

Figure 24.3 Appearance of pocks in smallpox infections. Newer pox on face are elevated pustules filled with liquid, while older pox, on torso, have begun to crust over. *Source:* Dr. Michael Schwartz/CDC

periphery develops at the infection site, where it leaves a permanent scar **(figure 24.4).** The vaccine provides significant long-term protection against variola as well as other poxvirus infections. In 2009 a U.S. military service member was stricken with progressive vaccinia (PV), a rare but often fatal complication of the smallpox

Figure 24.4 Smallpox vaccination. A forked needle is dipped into the smallpox vaccine (which actually contains vaccinia virus) and used to puncture the skin repeatedly in a small area. The inset shows that in a successful take, a large, scablike crust will develop over 3 to 4 weeks, eventually falling off and leaving a permanent scar. (top): *Source:* James Gathany/CDC; (bottom): *Source:* CDC

1. If you would like to know more about this subject, read *Demon in the Freezer* by Richard Preston, published in 2002 by Random House.

vaccination. This was the first instance of PV seen since 1987 and came on the heels of a new policy under which service members and medical personnel were vaccinated against smallpox as a defense against a potential bioterrorist attack. Risks like this, along with the fact that smallpox no longer poses an immediate threat, are the reason that smallpox vaccination has been discontinued for most populations.

Other Poxvirus Diseases

A much more common poxvirus is the molluscipoxvirus. It causes a skin disease called **molluscum contagiosum.*** This disease is distributed throughout the world, with highest incidence occurring in certain Pacific islands. In endemic regions, it is primarily an infection of children and is transmitted by direct contact and fomites.

Skin lesions take the form of smooth, waxy nodules on the face, trunk, and limbs. The infection can also be spread by sexual intercourse and is most common in sexually active young people in the United States. The lesions of the STD form are small, smooth macules that occur on the genital area and thighs. They may be spread to other body sites through scratching and self-inoculation. AIDS patients suffer from an atypical version of the disease, which attacks the skin of the face and forms tumorlike growths.

Treatment requires destruction of the virus by freezing (cryotherapy), electric cautery, and chemical agents applied directly to the lesions.

Many mammalian groups host some sort of poxvirus infection. Cowpox, rabbitpox, monkeypox, mousepox, camelpox, and elephantpox are among the types that occur (chickenpox though is actually a herpesvirus infection, not a poxvirus, and it does not occur in chickens!) However, only monkeypox and cowpox appear to be capable of infecting humans symptomatically.

Historically, monkeypox has been seen primarily in Africa, occurring sporadically in humans having contact with monkeys, squirrels, or rats—traditional vectors of the disease. Since the discontinuation of smallpox vaccination, which also protected against monkeypox, more person-to-person transmission of monkeypox has been seen, an unintended consequence of the smallpox eradication effort.

Although cowpox is an eruptive cutaneous disease that can develop on cows' udders and teats, cows are not the actual reservoirs of this virus. Other mammals such as rodents, cats, and even zoo animals can carry it. Human infection is rare and usually confined to the hands, although the face and other cutaneous sites can be involved.

 Check Your Progress SECTION 24.2

7. What structural attributes differentiate smallpox from most other viruses?
8. Justify the selective use of the smallpox vaccine.
9. List the poxviruses that cause symptomatic human infection.

* *molluscum contagiosum* (mah-lusk′-uhm kahn″-taj-ee-oh′-sum) L. *molluscus*, soft. In reference to the soft, rounded, cutaneous papules.

24.3 Enveloped DNA Viruses: The Herpesviruses

 Expected Learning Outcomes

7. Characterize the herpesviruses that commonly infect humans.
8. Describe the range of pathologies associated with herpes simplex virus infection.
9. Summarize the means by which herpes simplex infection is diagnosed, treated, and controlled.
10. Understand the relationship between varicella and herpes zoster.
11. Understand the etiological agent, epidemiology, diagnosis, and control of varicella-zoster virus, cytomegalovirus, and Epstein-Barr virus.
12. Recall the complications associated with EBV infection.
13. Relate human herpesviruses 6, 7, and 8 to the diseases they cause.

The **herpesviruses*** were named for the tendency of some herpes infections to produce a rash that spreads (creeps). The term is used to designate a large family whose members include the following:

* herpes simplex 1 and 2 (HSV), the cause of fever blisters and genital infections;
* varicella-zoster virus (VZV), the cause of chickenpox and shingles;
* cytomegalovirus (CMV), which infects the salivary glands and other viscera;
* Epstein-Barr virus (EBV), associated with infection of the lymphoid tissue;
* human herpesviruses 6 and 7 (HHVs), which cause roseola; and
* herpesvirus 8 (KSHV), which is implicated in Kaposi sarcoma.

An alternate system for naming herpesviruses assigns them a number. Thus, 1 and 2 are HSV, 3 is VZV, 4 is EBV, 5 is CMV, 6 and 7 are HHVs, and 8 is KSHV. A prominent feature of the family is that all members show latency and cause recurrent infections. This behavior involves entry of viral DNA into the nucleus, where it becomes an extrachromosomal particle called an episome. Two of the viruses in this group also have a well-documented association with cancers.

Virtually everyone becomes infected with at least some of the herpesviruses, usually without adverse effect. The clinical complications of latency and recurrent infections become more severe with advancing age, cancer chemotherapy, or other conditions that compromise the immune defenses. The herpesviruses are among the most common and serious opportunists among AIDS patients. Nearly 95% of this group will experience recurrent bouts of skin, mucous membrane, intestinal, and eye disease from these viruses.

The herpesviruses are among the larger viruses, with a diameter ranging from about 150 nm to 200 nm **(figure 24.5).** They are enclosed within a loosely fitting envelope that contains glycoprotein spikes. Like other enveloped viruses, herpesviruses are prone to deactivation by organic solvents or detergents and are unstable outside the host's body. The icosahedral capsid houses a core of double-stranded DNA. Replication occurs primarily within the nucleus, and viral release is usually accompanied by cell lysis.

* *herpesviruses* (her′-peez) Gr. *herpein*, to creep.

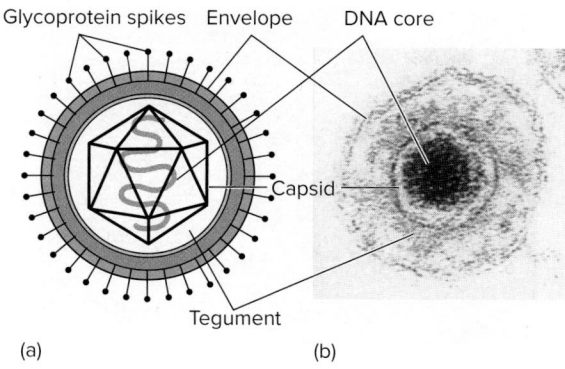

(a) (b)

Figure 24.5 **Herpesviruses. (a)** A schematic of the general structure of a herpesvirus alongside **(b)** a herpes simplex virus particle (30,000×). (b): *Source:* Laboratory of Tumor Cell Biology, Bernard Kramarsky photographer, National Cancer Institute

General Properties of Herpes Simplex Viruses

Humans are susceptible to two varieties of **herpes simplex** viruses: **HSV-1,** usually characterized by lesions on the oropharynx, and **HSV-2,** usually producing lesions on the genitalia. The viruses are very similar in the types of cells they infect, so they are not specific to these sites. **Table 24.2** provides a summary of major characteristics and differences. Even though other mammals can be experimentally infected, humans appear to be the only natural reservoir for herpes simplex virus.

Epidemiology of Herpes Simplex

Herpes simplex infection occurs globally in all seasons and among all age groups. Because these viruses are relatively sensitive to the

TABLE 24.2	Comparative Epidemiology and Pathology of Herpes Simplex, Types 1 and 2	
	HSV-1	HSV-2
Usual Etiologic Agent of*	Herpes labialis Ocular herpes Gingivostomatitis Pharyngitis	Herpes genitalis
Transmission	Close contact, usually of face	Sexual or intimate contact
Latency	Occurs in trigeminal ganglion	Occurs primarily in sacral ganglia
Skin Lesions	On face, mouth	On internal, external genitalia, thighs, buttocks
Complications		
Whitlows	Among personnel working on oral cavity	Among obstetric, gynecological personnel
Neonatal encephalitis	Causes up to 30% of cases**	Causes most cases through contact with birth canal

*The other herpes simplex type can be involved in this infection, though not as commonly.
**Due to mothers infected genitally by HSV-1 or contamination of the neonate by oral lesions.

environment, transmission usually involves direct exposure to secretions containing the virus. People with active lesions are the most significant source of infection, though studies indicate that genital herpes can be transmitted even when no lesions are present. Occasionally cases have been traced to contact with moist secretions on inanimate objects.

Primary infections tend to be age-specific: HSV-1 frequently occurs in infancy and early childhood, and by adulthood, most people exhibit some serological evidence of infection. On the other hand, primary infection with HSV-2 occurs most frequently between the ages of 14 and 29, a pattern that reflects the sexual route of transmission. In fact, genital herpes is one of the most common STDs in the United States with an estimated 780,000 new cases each year. Cases of HSV-1 genital infection and HSV-2 oral infections are probably due to autoinoculation with contaminated hands or oral sexual contact.

The Nature of Latency and Recurrent Attacks

In approximately 20% to 50% of primary infections, herpes simplex viruses multiply in the distal regions of sensory neurons and are carried to the ganglia. Type 1 HSV enters the trigeminal, or fifth cranial nerve, which has extensive innervations in the oral region **(figure 24.6).** Type 2 HSV usually becomes latent in the ganglion of the lumbosacral spinal nerve trunk. The exact details of this unusual mode of latency are not yet fully understood, but the virus remains inside the nucleus of the neuron in a nonproliferative stage for a variable time. Recurrent infection is triggered by various stimuli such as fever, UV radiation, stress, or mechanical injury, all of which reactivate the virus genes. Newly formed viruses migrate to the body surface and produce a local skin or membrane lesion, often in the same site as a previous infection. The number of such attacks can range from one to dozens each year.

The Spectrum of Herpes Infection and Disease

Herpes simplex infections, often called simply herpes, usually target the mucous membranes. The virus enters cracks or cuts in the

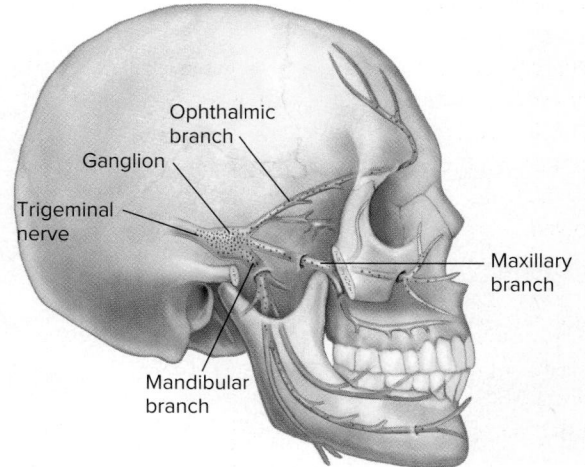

Figure 24.6 **The site of latency and routes of recurrence in herpes simplex, type 1.** Following primary infection, the virus invades the trigeminal, or fifth cranial, nerve and is harbored in its ganglion. Various provocative stimuli cause it to migrate back to the skin surface by the mandibular, maxillary, or ophthalmic branch.

membrane surface and then multiplies in basal and epithelial cells in the immediate vicinity. This results in inflammation, edema, cell lysis, and a characteristic thin-walled vesicle. The main diseases of herpes simplex viruses are facial herpes (oral, optic, and pharyngeal), genital herpes, neonatal herpes, and disseminated disease.

Type 1 Herpes Simplex in Children and Adults

Herpes labialis, otherwise known as fever blisters or cold sores, is the most common recurrent HSV-1 infection. Vesicles usually crop up on the mucocutaneous junction of the lips or on adjacent skin (**figure 24.7**). A few hours of tingling or itching precede the formation of one or more vesicles at the site. Pain is most acute in the earlier stage as vesicles ulcerate; lesions crust over in 2 or 3 days and heal completely in a week.

Although most herpetic infections remain localized, a condition known as herpetic **gingivostomatitis** can occur as the primary manifestation, especially in young children. Inflammation of the oral mucosa can involve the gums, tongue, soft palate, and lips, sometimes with ulceration and bleeding. A common complication in adolescents is pharyngitis, a syndrome marked by sore throat, fever, chills, swollen lymph nodes, and difficulty in swallowing.

Herpetic keratitis (also called ocular herpes) is an infective inflammation of the eye in which a latent virus travels into the ophthalmic rather than the mandibular branch of the trigeminal nerve. In some cases, it is introduced into the eye by contaminated fingers or contact lenses. Preliminary symptoms are a gritty feeling in the eye, conjunctivitis, sharp pain, and sensitivity to light. Some patients develop characteristic branched or opaque corneal lesions as well. In 25% to 50% of cases, keratitis is recurrent and chronic and can interfere with vision.

Type 2 Herpes Infections

Type 2 herpes infection usually coincides with sexual maturation and an increase in sexual encounters. **Genital herpes** (herpes genitalis) starts out with malaise, anorexia, fever, and bilateral swelling and tenderness in the groin. Then, intensely sensitive vesicles break

Figure 24.8 Genital herpes. Vesicles start out separate, but become confluent and ulcerate into red erosions so tender that any contact can be highly painful. *Source:* CDC

out in clusters on the genitalia, perineum, and buttocks (**figure 24.8**). The chief symptoms are urethritis, painful urination, cervicitis, and itching. After a period of days to weeks, the vesicles ulcerate and develop a light-gray exudate before healing. Bouts of recurrent genital herpes are usually less severe than the original infection and are triggered by menstruation, stress, and concurrent bacterial infection.

Herpes of the Newborn

Although HSV infections in healthy adults are annoying, unpleasant, and painful, only rarely are they life-threatening. However, in the neonate and fetus (**figure 24.9**), HSV infections are very destructive and can be fatal. Most cases occur when infants are contaminated by the mother's reproductive tract immediately before or during birth, but they have also been traced to hand transmission from the mother's lesions to the baby. Infections are evenly split between type 1 and type 2, and both types act similarly to one another. In infants whose disease is confined to the mouth, skin, or eyes, the mortality rate is 30%, but disease affecting the central nervous system has a 50% to 80% mortality rate.

Because of the danger of herpes to fetuses and newborns and also the increase in the number of cases of genital herpes, it is now

Figure 24.7 Type I herpes simplex. Recurrent lesions of herpes labialis (cold score, fever blister). Tender, itchy papules erupt on the perioral regions and progress to vesicles that burst, drain, and scab over. These sores and fluid are highly infectious and should not be touched. *Source:* CDC

Figure 24.9 Neonatal herpes simplex. Babies can be born with the lesions or develop them 1 to 2 weeks after birth. © Dr. M.A. Ansary/Science Source

standard procedure to screen pregnant women for the herpesvirus early in their prenatal care. Women with a history of recurrent infections must be monitored for any signs of viral shedding during the last period of pregnancy. Acyclovir may be prescribed to lessen viral shedding during the third trimester. In the absence of lesions or symptoms, vaginal birth is safe. Any evidence of an outbreak at the time of delivery will require a cesarean section.

Miscellaneous Herpes Infections

When either form of herpesvirus enters a break in the skin, a local infection can develop. This route of infection is most often associated with occupational exposure or severely damaged skin. A hazard for health care workers who handle patients or their secretions without hand protection is a disease called herpetic **whitlow.*** Workers in the fields of obstetrics, gynecology, dentistry, and respiratory therapy are probably at greatest risk for contracting it. Whitlows are deep-set, usually occur on one finger, and are extremely painful and itchy (**figure 24.10**). Afflicted personnel should not work with patients until the whitlow has healed, usually in 2 to 3 weeks.

Life-Threatening Complications

Although herpes simplex encephalitis is a rare complication of type 1 infection, it is probably the most common sporadic form of viral encephalitis in the United States. The infection disseminates along nerve pathways to the brain or spinal cord. The effects on the CNS begin with headache and stiff neck and can progress to mental disturbances and coma. The fatality rate in untreated cases is 70%. Patients with underlying immunodeficiency are more prone to severe, disseminated herpes infection than are immunocompetent patients. Of greatest concern are patients receiving organ grafts, cancer patients on immunosuppressive therapy, those with congenital immunodeficiencies, and AIDS patients.

Diagnosis, Treatment, and Control of Herpes Simplex

Small, painful vesicles on the mucous membranes of the mouth or genitalia, lymphadenopathy, and exudate are typical diagnostic

* *whitlow* (hwit′-loh) ME *white,* flaw. An abscess on the distal portion of a finger.

Figure 24.10 Herpetic whitlows. These painful, deep-set vesicles can become inflamed and necrotic and are difficult to treat. They occur most frequently in dental and medical personnel or people who carelessly touch lesions on themselves or others. © Kathy Park Talaro

symptoms of herpes simplex. Further diagnostic support is available by examining scrapings from the base of such lesions stained with Giemsa, Wright, or Papanicolaou (Pap) methods (**figure 24.11**). The presence of multinucleate cells, giant cells, and inclusion bodies inside the nucleus can help establish herpes infection. This method, however, will not distinguish among HSV-1, HSV-2, or other herpesviruses, which require more specific subtyping. Direct tests on specimens or cell cultures using fluorescent antibodies or DNA probes can differentiate among HSV-1, HSV-2, and closely related herpesviruses.

Laboratory culture and specific tests are essential for diagnosing severe, disseminated herpes infection. A specimen of tissue or fluid is introduced into a primary cell line such as monkey kidney or human embryonic kidney tissue cultures and is then observed for cytopathic effects within 24 to 48 hours.

Several agents are available for treatment. Acyclovir (Zovirax) is the most effective therapy developed to date that is nontoxic and highly specific to HSV. Famciclovir and valacyclovir are alternate drugs. Topical medications applied to genital and oral lesions cut the length of infection and reduce viral shedding. Systemic therapy is available for more serious complications such as herpes keratitis and disseminated herpes. Daily doses of oral valacyclovir (Valtrex) have been shown to be an effective suppressive therapy for genital herpes, preventing recurrent outbreaks for at least 6 months in more than half of those adhering to the regimen.

Although the barrier protection afforded by condoms can diminish the prospects of spreading genital herpes sexually, people with active infection should avoid any sexual contact. Mothers with cold sores should observe extreme care in handling their newborns; infants should never be kissed on the mouth. Medical and dental workers who deal directly with patients can reduce their risk of exposure by wearing gloves.

Giant cell with multiple nuclei

Normal cells

Inclusion

Figure 24.11 Diagnosis of herpesvirus infection from stained specimen. A direct Papanicolaou (Pap) smear of a cervical scraping shows enlarged (multinucleate giant) cells and intranuclear inclusions typical of herpes simplex, type 2. This technique is not specific for HSV, but most other herpesviruses do not infect the reproductive mucosa. © Dr. Walker/Science Source

Pathogen Profile #1 *Herpes simplex virus 1 and 2 (HSV-1, HSV-2)*

Source: CDC

Microscopic Morphology Enveloped virus with an icosahedral capsid; 150 to 200 nm in diameter.

Identified by Clinical presentation, examination of stained scrapings from the base of lesions, growth in tissue culture, antibody tests, PCR.

Habitat Humans are the only natural reservoir for herpes simplex virus.

Virulence Factors Latency with entry of the HSV episome into the nucleus of neurons.

Primary Infections/Disease HSV-1 is usually characterized by lesions on the oropharynx, whereas HSV-2 causes lesions primarily on the genitalia. Both types of HSV can cause neonatal herpes simplex, a disease of the mouth, skin, eyes, or nervous system, with a mortality rate approaching 80%. Herpetic whitlows are deep, painful vesicles on the skin caused when either type of HSV enters a cut. Whitlows are associated with health care workers who don't use gloves when touching infected patients or their secretions. Herpes simplex encephalitis is a rare but serious complication of HSV-1 infection, with mortality rates of up to 70%.

Control and Treatment The use of barrier protection—gloves for health care procedures, condoms for sexual activity—is the best strategy to prevent herpes infection. Treatment with acyclovir reduces both the length of infection and the intensity of viral shedding. Oral valacyclovir (Valtrex) is often used as a suppressive therapy, preventing recurrent outbreaks for as long as 6 months at a time.

The Biology of Varicella-Zoster Virus

Another herpesvirus causes both **varicella*** (commonly known as **chickenpox**) and a recurrent infection called **herpes zoster,*** or **shingles.*** Because the same virus causes two forms of disease, it is known by a composite name, **varicella-zoster virus (VZV).** At one time, the differences in clinical appearance and time separation between varicella and herpes zoster were sufficient to foster the notion that they were caused by different infectious agents. In time, it was noted that young children acquired chickenpox from family members afflicted with shingles, giving the first hint that the two diseases were related (see this chapter's Case Study). More recently doctors used Koch's postulates to follow the two diseases in a single patient. They demonstrated that the same virus caused both diseases and that zoster is a reactivation of a latent varicella virus **(figure 24.12).**

Epidemiological Patterns of VZV Infection

Humans are the only natural hosts for varicella-zoster virus. The virus is harbored in the respiratory tract, but it is communicable from both respiratory droplets and the fluid of active skin lesions. Infected persons are most infectious a day or two prior to the development of the rash. The dried scabs are not infectious because the virus is unstable and loses infectivity when exposed to the environment. Patients with shingles are a source of infection for nonimmune children, but the attack rate is significantly lower than with exposure to chickenpox. Only in rare instances will a child acquire chickenpox more than once, and even a subclinical case can result in long-term immunity. Immunity protects against reinfection with chickenpox but not against an attack of shingles.

Varicella (Chickenpox)

The respiratory epithelium serves as the chief portal of entry and the initial site of viral replication for varicella-zoster virus, which initially produces no symptoms. After an incubation period of 10 to 20 days, the first symptoms to appear are fever and an abundant rash that begins on the scalp, face, and trunk and radiates in sparse crops to the extremities. Skin lesions progress from small red macules to itchy blisters that encrust and drop off, usually healing completely but sometimes leaving a tiny scar (figure 24.12*a*). Lesions number from a few to hundreds and are more abundant in adolescents and adults than in young children.

Herpes Zoster (Shingles)

Most cases of varicella result in the entry of the varicella-zoster virus into ganglia of neurons that supply innervation to skin dermatomes, especially thoracic and trigeminal nerves (figure 24.12*b*). VZV becomes latent in these ganglia and may later reemerge with its characteristic asymmetrical distribution on the skin of the trunk or head or occasionally on the neck or limbs (figure 24.12*c*).

Shingles develops abruptly after reactivation by such stimuli as X-ray treatments, immunosuppressive and other drug therapy, surgery, or developing malignancy. It may be associated with declining immune function and is more common in older patients. About 1 million cases occur per year in the United States. The virus is believed to migrate down the ganglion to the skin, where multiplication resumes and produces crops of tender, persistent vesicles. Inflammation of the ganglia and the pathways of nerves causes pain and tenderness (radiculitis) that can last for several months. Involvement of cranial nerves can lead to eye inflammation and ocular and facial paralysis. A vaccine (Zostavax) has recently been approved as a prevention for shingles in patients over 60. It is indicated only for people who have already had a case of chickenpox.

* *varicella* (var″-ih-sel′-ah) The Latin diminutive for smallpox or variola.
* *zoster* (zahs′-tur) Gr. *zoster,* girdle.
* *shingles* (shing′-gulz) L. *cingulus,* cinch. (Both *zoster* and *shingles* refer to the beltlike encircling nature of the rash.)

Pathogen Profile #2 *Varicella-zoster virus (VZV)*

Source: CDC

Microscopic Morphology Enveloped virus with an icosahedral capsid; 150 to 200 nm in diameter.

Identified by Nearly always by clinical presentation alone.

Habitat Humans are the only natural reservoir for varicella-zoster virus.

Virulence Factors Latency, especially in ganglia.

Primary Infections/Disease Initial infection with VZV results in varicella (chickenpox), which is marked by numerous small, red lesions that eventually encrust and drop off. Reemergence of VZV after a long latency period—often many decades—results in zoster (shingles), a highly localized, painful rash that may last several months.

Control and Treatment The addition of a live, attenuated chickenpox vaccine to the recommended childhood vaccination schedule has dramatically lessened the occurrence of the disease. Although a small number of vaccinated children still acquire chickenpox, the course of the disease is far milder. A shingles vaccine (Zostavax) is recommended for persons over the age of 60. Treatment for both conditions generally focuses on the alleviation of discomfort and the prevention of secondary bacterial infections. In some cases, acyclovir or famciclovir may be administered to combat the virus directly.

Figure 24.12 **The relationship between varicella (chickenpox) and zoster (shingles) and the clinical appearance of each. (a)** First contact with the virus (usually in childhood) results in a macular, papular, vesicular rash distributed primarily on the face and trunk. **(b)** The virus becomes latent in the dorsal ganglia of nerves that supply dermatomes of mid-thoracic nerves and the cranial nerves that supply facial regions. **(c)** The clinical appearance of shingles. Early symptoms are acute pain in the nerve root and redness of the dermatome, followed by a vesicular-papular rash on the chest and back that is usually asymmetrical and does not cross the midline of the body. (a): *Source:* CDC; (c): © DermNet New Zealand Trust

Diagnosis, Treatment, and Control of VZV Infection

The cutaneous manifestations of varicella and shingles are sufficiently characteristic for ready clinical recognition. Supportive evidence for a diagnosis of varicella usually comes from recent close contact with a source or with an active case of VZV infection. A diagnosis of shingles is supported by the distribution pattern of skin lesions and confirmed from multinucleate giant cells in stained smears prepared from vesicle scrapings. However, unequivocal identification is best done with fluorescent antibody detection of viral antigen in skin lesions, DNA typing, or culture.

Uncomplicated varicella is self-limited and requires no therapy aside from alleviation of discomfort. Secondary bacterial infections that can cause dangerous complications (especially streptococcal and staphylococcal) are prevented by application of antimicrobic ointments. Drugs that can be beneficial for systemic disease are intravenous acyclovir or famciclovir. Varicella-zoster immune globulin (Varizig) is a passive immunotherapy containing antibodies against the varicella-zoster virus. It may be administered up to 96 hours after exposure to reduce symptoms of viral infection.

A live, attenuated vaccine (Varivax) is currently recommended for routine childhood vaccination against chickenpox. Vaccination is indicated for children between 2 months and 12 years of age. Immunity will be longer lasting if two doses are given. Some children can still acquire chickenpox, but it is a milder case. Older children and adults with risk factors may also be vaccine candidates. Because some parents regard the disease as mild and inevitable, they purposely expose their children to infected individuals as a means of "immunizing" them. This practice is not recommended because of the potential complications of VZV infection.

The Cytomegalovirus Group

Another group of herpesviruses, the **cytomegaloviruses*** **(CMVs),** are named for their tendency to produce giant cells with nuclear and cytoplasmic inclusions **(figure 24.13).** These viruses, also termed salivary gland virus and cytomegalic inclusion virus, are among the most ubiquitous pathogens of humans.

Epidemiology of CMV Disease

The testing of human populations has indicated that antibodies to CMV developed during a prior infection are very common. For instance, 50% of women of childbearing age had CMV antibodies, and nearly 10% of newborns reveal some evidence of infection as well, making CMV the most prevalent fetal infection. Cytomegalovirus is transmitted in saliva, respiratory mucus, milk, urine, semen, and cervical secretions. Transmission usually involves intimate exposure such as sexual contact, vaginal birth, transplacental infection, blood transfusion, and organ transplantation. As with other herpesviruses, CMV is commonly carried in a latent state in white blood cells such as monocytes.

Infection and Disease

Most healthy adults and children with primary CMV infection are asymptomatic. However, three groups that develop a more virulent form of disease are fetuses, newborns, and immunodeficient adults. The incidence of congenital CMV infection occurs in about 20% of pregnancies in which the mother has concurrent infection. Newborns with clinical disease exhibit enlarged liver and spleen, jaundice, capillary bleeding, microcephaly, and ocular inflammation. In some cases, damage can be so severe and extensive that death follows within a few days or weeks. Many babies who survive develop long-term neurological sequelae, including hearing and visual disturbances and even mental retardation. In the United States, about 5,000 babies a year are affected. Perinatal CMV infection, which occurs after exposure to the mother's vagina, is chiefly asymptomatic, although pneumonitis and a mononucleosis-like syndrome can develop during the first 3 months after birth.

* *cytomegaloviruses* (sy″-toh-may″-gah-loh-vy′-rus) Gr. *cyto,* cell, and *megale,* big, great.

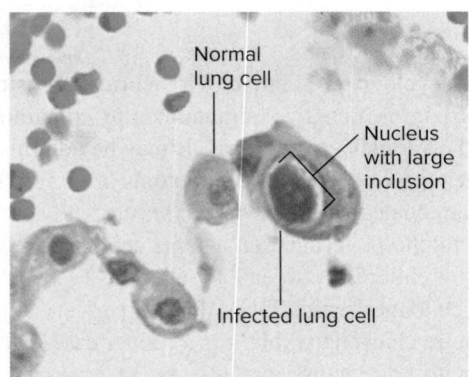

Figure 24.13 **Identifying cytomegaloviruses.** Infection of lung cells with cytomegalovirus shows the cellular enlargement and distortion of the nucleus by a large inclusion, called an "owl's eye" effect (1,500×). *Source:* Dr. Edwin P. Ewing, Jr./CDC

Cytomegalovirus mononucleosis* is a syndrome characterized by fever and lymphocytosis, somewhat similar to the disease caused by Epstein-Barr virus. Although it can occur in children, it is chiefly an illness of adults. Disseminated cytomegalovirus is a severe opportunistic infection of AIDS patients. It produces overwhelming systemic disease, with fever, severe diarrhea, hepatitis, pneumonia, and high mortality. Invasion of the retina can lead to blindness. Most patients undergoing kidney transplantation and half of those receiving bone marrow develop CMV pneumonitis, hepatitis, myocarditis, meningoencephalitis, hemolytic anemia, and thrombocytopenia.

Diagnosis, Treatment, and Prevention of CMV Infection

In neonates, CMV infection must be differentiated from toxoplasmosis, rubella, and herpes simplex. In adults, it must be distinguished from Epstein-Barr infection and hepatitis. The cytomegalovirus can be isolated from specimens taken from virtually all organs as well as from epithelial tissue. Cell enlargement and prominent inclusions in the cytoplasm and nucleus are suggestive of CMV. The virus can be cultured and tested with monoclonal antibody against the early nuclear antigen. Polymerase chain reaction assays, which detect viral DNA, have become the most commonly used clinical detection method.

Drug therapy is reserved for immunosuppressed patients. The main drugs are ganciclovir and valganciclovir. The development of a vaccine is hampered by the lack of an animal that can be infected with human cytomegalovirus.

Epstein-Barr Virus

Although the lymphatic disease known as **infectious mononucleosis** was described more than a century ago, its cause was finally discovered through a series of accidental events starting in 1958. First, Michael Burkitt working w ith African children detected an unusual malignant tumor (Burkitt lymphoma) that appeared to be infectious. Later, Michael Epstein and Yvonne Barr cultured a virus from tumors that showed typical herpesvirus morphology. Proof that the two diseases had a common cause was completed when a laboratory technician accidently acquired mononucleosis while working with the Burkitt lymphoma virus. The **Epstein-Barr virus (EBV)** shares morphological and antigenic features with other herpesviruses. In addition, it contains genes that, when expressed, transform certain lymphocytes into malignant cells.

Epidemiology of Epstein-Barr Virus

This virus has a firmly established niche in human lymphoid tissue and salivary glands. Salivary excretion occurs in about 20% of otherwise normal individuals. Direct oral contact and contamination with saliva are the principal modes of transmission, although transfer through blood transfusions and organ transplants is also possible.

* *mononucleosis* (mah″-noh-noo″-klee-oh′-sis) Gr. *mono,* one, *nucleo,* nucleus, and *sis,* state.

The nature of infection depends on the patient's age at first exposure, socioeconomic level, geographic region, and genetic predisposition. In less-developed regions of the world, infection usually occurs in early childhood. Such early and chronic infection means that children living in Africa have higher rates of Burkitt lymphoma. Likewise, residents of certain areas of Asia show high levels of nasopharyngeal carcinoma, another tumor linked to EBV.

In industrialized countries, exposure to EBV is usually postponed until adolescence and young adulthood. This delay is generally ascribed to improved sanitation and less frequent and intimate contact with other children. Because nearly 70% of college-age Americans have never had EBV infection, this population is vulnerable to infectious mononucleosis, often dubbed the "kissing disease" or "mono." However, by midlife, 90% to 95% of all people, regardless of geographic region, show serological evidence of infection. EBV is one of the herpesviruses that researchers suspect as a trigger for multiple sclerosis. A National Institutes of Health study found an association between EBV infection and the presence of T cells that attack the myelin sheath, triggering an autoimmune disease.

Diseases of EBV

The epithelium of the oropharynx is the portal of entry for EBV during the primary infection. From this site, the virus moves to the parotid gland, which is the major area of viral replication and latency. After extended dormancy in B cells, the virus is reactivated by some ill-defined mechanism. In some people, the entire course of infection and latency is asymptomatic.

The symptoms of infectious mononucleosis are sore throat, high fever, and cervical lymphadenopathy, which develop after a long incubation period (30–50 days). Many patients also have a gray-white exudate in the pharynx **(figure 24.14)**, a skin rash, and enlarged spleen and liver. A notable sign of mononucleosis is sudden leukocytosis, consisting initially of infected B cells and later T cells. The strong, cell-mediated immune response is decisive in controlling the infection and preventing complications. Some people can become chronically infected with EBV, but a link to debilitating illnesses like chronic fatigue syndrome is now thought to be unlikely.

Tumors and Other Complications Associated with EBV

Burkitt lymphoma is a B-cell malignancy that usually develops in the jaw and grossly swells the cheek **(figure 24.15)**. Most cases are reported in central African children 4 to 8 years old, and a small number are seen in North America. The prevalence in Africa may be associated with chronic coinfections with other diseases, such as malaria, that weaken the immune system and predispose the body to tumors. One theory to explain the neoplastic effect is that persistent EBV infection in young children transforms B lymphocytes into immortalized cells. During this time, chronic infections overwhelm the T-cell cytotoxic response needed to control the overgrowth of these B-cell clones, which develop into malignant tumors.

Unlike Burkitt lymphoma, **nasopharyngeal carcinoma** is a malignancy of epithelial cells that occurs in older Chinese and African men. Although much about this disease is still to be explained, high antibody titer to EBV is a consistent finding. It is not known how the virus invades epithelial cells (not its usual target), why the incidence of malignancy is so high in particular areas, or to what extent dietary carcinogens are involved.

In general, any person with an immune deficiency (especially of T cells) is highly susceptible to Epstein-Barr virus. Organ transplant patients carry a high risk because their immune systems are weakened by measures to control rejection. The oncogenic potential of the virus is another factor. It has been associated with some types of lymphatic tumors in kidney transplant patients and is commonly isolated from AIDS-related lymphomas. AIDS patients also develop oral *hairy leukoplakia,* white adherent plaques on the tongue due to invasion of the epithelium by EBV. Lesions can be reduced by oral acyclovir therapy.

Figure 24.14 Infectious mononucleosis. This view of the pharynx shows the swollen tonsils and throat tissues, a gray-white exudate, and an inflammation of the palate that can be of some use in diagnosis. Kane, Kay et al., *Color Atlas and Synopsis of Pediatric Dermatology, 2/e* © 2010 McGraw-Hill Education. Reproduced with permission of McGraw-Hill Education

Figure 24.15 EBV-associated tumor. Burkitt lymphoma, a malignancy of B cells commonly associated with chronic Epstein-Barr infection. *Source:* Robert S. Craig/CDC

(a) (b)

Figure 24.16 Histology of lymphocytes infected with Epstein-Barr virus (1,000×). (a) Blood smear of normal lymphocytes, with regular, round nuclei and small cytoplasmic bands. **(b)** Blood smear of patient with infectious mononucleosis. Note a large atypical lymphocyte with irregular nucleus and indented border (arrows). (a): © Harold J. Benson; (b): © Joaquin Carrillo-Farga/Science Source

Figure 24.17 Roseola infantum in 12-month-old baby. Disease is marked by confluent rash over face, neck, and trunk. This viral infection is the most common febrile disease in young babies. Courtesy of Stephen Dewhurst

Diagnosis, Treatment, and Prevention of EBV Infection

Because clinical symptoms in EBV infection are common to many other illnesses, laboratory diagnosis is necessary. A differential blood count that shows lymphocytosis, neutropenia, and large, atypical lymphocytes with lobulated nuclei and vacuolated cytoplasm is suggestive of EBV infection **(figure 24.16).** Serological assays to detect antibody against the capsid and DNA are helpful, as are direct viral antigen tests using probes and labeled antibodies.

The usual treatments for infectious mononucleosis are directed at symptomatic relief of fever and sore throat. In more severe cases antiviral agents like acyclovir can be given to block virus replication directly while corticosteroids can reduce airway obstruction. Burkitt lymphoma requires systemic chemotherapy with an anticancer drug such as cyclophosphamide or vincristine. These measures, along with surgical removal of the tumor, have considerably reduced the mortality rate.

Diseases of Herpesviruses 6, 7, and 8

Human herpesvirus 6 (HHV-6), also known as human B-lymphotropic virus, was originally isolated from infected lymphocytes (see figure 17.17*b*). It has characteristics similar to CMV, but it is genetically distinct from the other herpesviruses. The virus can enter and replicate in T lymphocytes, macrophages, and salivary gland tissues. It is probably transmitted by close contact with saliva and other secretions. It is among the most common herpesviruses, with up to 95% prevalence in tested human populations. It is the cause of **roseola** (also known as **roseola infantum**), an acute febrile disease in babies between 2 and 12 months of age **(figure 24.17).** It begins with fever that can spike as high as 105°F (40°C) and is followed a few days later by a faint maculopapular rash over the neck, trunk, and buttocks. The disease is usually self-limited, and the children recover spontaneously with support care.

Infected adults present with mononucleosis-like symptoms and may develop lymphadenopathy and hepatitis. Patients with renal or bone marrow transplants often acquire HHV-6 infections that suppress the function of the grafted tissues and are possibly a factor in rejection. Because HHV-6 is known for causing encephalitis and neurological disease, it is another herpesvirus that could play a role in multiple sclerosis (MS). Over 70% of MS patients have positive serological signs of HHV-6 infection, and the brain lesions of many of them have active virus infection. Although this is not conclusive proof of etiology, it is the first crucial step in connecting this disease to a virus. Much additional study will be necessary before specialists can verify any link between MS and herpesviruses. Experts consider HHV-6 infection an important emerging disease.

The fact that herpesviruses are persistent, can develop latency in the cell nucleus, and have demonstrated oncogenesis has spurred an active search for the roles of these viruses in cancers of unknown cause. A number of clinical studies have disclosed a relationship between HHV-6 and Hodgkin's lymphoma, oral carcinoma, and certain T-cell leukemias. Keep in mind that isolating a virus or its antigens from a tumor is not sufficient proof of causation and requires further evidence, but it is highly suggestive of some role in cancer. Another human herpesvirus, HHV-7, is very closely related to HHV-6, and it causes similar diseases in children and adults.

The latest herpesvirus was isolated from Kaposi sarcoma, a tumor of AIDS patients. This virus, termed *human herpesvirus 8* (*HHV-8*) or *Kaposi sarcoma–associated herpesvirus* (*KSHV*), is not found in other normal tissues or in patients without sarcoma. Its primary target cells are endothelial cells and B cells. The genetic mechanisms and pathology behind this cancer are an ongoing focus of study.

Pathogen Profile #3 *Epstein-Barr virus (EBV)*

Source: CDC

Microscopic Morphology Enveloped virus with an icosahedral capsid; 150 to 200 nm in diameter.

Identified by Differential blood count, which reveals increased lymphocytes and decreased neutrophils, along with large, atypical lymphocytes, is suggestive of infection. Serological assays can confirm the diagnosis.

Habitat Epstein-Barr virus is carried in the lymphoid tissue and salivary glands of more than 90% of the world's population.

Virulence Factors Latency, along with the ability to easily incorporate its DNA into that of its host.

Primary Infections/Disease Symptomatic infection with EBV leads to mononucleosis, a disease marked by sore throat, high fever, and cervical lymphadenopathy occurring after a long (30–50 day) incubation period. Some patients also display a gray-white exudate in the throat, skin rash, and an enlarged spleen and liver. Cancer is a potential complication of EBV infection, with Burkitt lymphoma, a B-cell cancer, and nasopharyngeal carcinoma both being linked to previous EBV infection.

Control and Treatment Treatment is directed at the alleviation of symptoms such as fever and sore throat. Acyclovir may be administered in some cases and corticosteroids may be used in the case of severe pharyngitis that obstructs the airway. Burkitt lymphoma is treated with anticancer drugs and surgery.

 Check Your Progress SECTION 24.3

10. What are the common characteristics of herpesviruses?
11. List the herpesviruses known to infect humans.
12. Describe the ways in which HSV-1 differs from HSV-2.
13. Summarize the signs, symptoms, and epidemiology associated with herpes simplex 1 and 2 infections, herpetic keratitis, newborn herpes, herpetic whitlows, and herpes simplex encephalitis.
14. List the measures undertaken to diagnose, treat, and prevent transmission of herpes simplex virus.
15. Explain the manner in which shingles is a complication of chickenpox.
16. Summarize the signs, symptoms, and epidemiology associated with varicella-zoster virus, cytomegalovirus, and Epstein-Barr virus.
17. What cancers are related to infection with Epstein-Barr virus?
18. What viruses are associated with roseola infantum and Kaposi sarcoma?

24.4 The Viral Agents of Hepatitis

 Expected Learning Outcomes

14. Contrast the major characteristics of the three most common agents of viral hepatitis.

15. Understand the epidemiology, diagnosis, pathology, and management of hepatitis B.

When certain viruses infect the liver, they cause **hepatitis,** an inflammatory disease marked by necrosis of hepatocytes and a mononuclear response that swells and disrupts the liver architecture. This pathologic change interferes with the liver's excretion of bile pigments such as bilirubin into the intestine. When bilirubin, a greenish-yellow pigment, accumulates in the blood and tissues, it causes **jaundice,** * a yellow tinge in the skin and eyes. Characteristics of the three principal viruses responsible for human hepatitis are summarized in **table 24.3.**

The major group of DNA viruses involved in hepatitis are the **hepadnaviruses.** * These enveloped viruses have never been grown in tissue culture and have an unusual genome containing both double- and single-stranded DNA. Hepadnaviruses show a decided tropism for the liver, where they persist and usually trigger liver cell carcinoma. One member of the group, **hepatitis B virus (HBV),** causes a common form of hepatitis in humans, and related viruses cause hepatitis in other vertebrates.

Other hepatitis viruses are RNA based and are covered in chapter 25. We include them here primarily to contrast basic characteristics of viral hepatitis agents. Incidence of the major forms is presented in table 24.3.

Hepatitis A virus (HAV) is a nonenveloped RNA enterovirus transmitted through contaminated food. In general, HAV disease is far milder, shorter term, and less virulent than the other forms. Another significant RNA virus is **hepatitis C virus (HCV),** a flavivirus that causes many cases of transfusion hepatitis. HCV is involved in a chronic liver infection that can go undiagnosed, later leading to severe liver damage and cancer. It is spread primarily by exposure to blood and blood products (see the chapter 11 Case Study). It is considered one of the most common unreported agents of hepatitis (an estimated 3 million cases nationally). Another agent is hepatitis E (HEV), a newly identified RNA virus that causes a disease similar to HAV. It is associated with poor hygiene and sanitation and is spread through fecally contaminated foods.

An unusual virus, hepatitis D, or delta agent, is a defective RNA virus that cannot produce infection unless a cell is also infected with

* *jaundice* (jon′-dis) Fr. *jaunisse,* yellow. Also called icterus.
* *hepadnaviruses* (hep″-ah-dee″-en-ay-vy′-rus) Gr. *hepatos,* liver, plus DNA and virus.

TABLE 24.3	Principal Morphological and Pathologic Features of the Major Hepatitis Viruses		
Property/Diseases	HAV*	HBV	HCV*
Biology			
Nucleic acid	RNA	DNA	RNA
Cell culture	+	−	−
Envelope	−	+	+
Epidemiology	Endemic and epidemic	Endemic	Endemic
Reservoir	Active infections	Chronic carrier	Chronic carrier
Transmission	Oral-fecal; water- or food-borne	Contact with blood or serum; sexual and intimate contact	Contact with blood, serum; intimate contact
Incubation Period	2–7 weeks	1–6 months	2–8 weeks
Symptoms	Fever, GI tract disorder	Fever, rash, arthritis	Similar to HBV
Jaundice	1 in 10	Common	Common
Onset/Duration	Acute, short	Gradual, chronic	Acute to chronic
Serious Complications	Uncommon	Chronic active hepatitis, cirrhosis, hepatic cancer	Chronic inflammation, cirrhosis, cancer
2013 Incidence (cases per 100,000 population)	0.6	1.0	0.7
Availability of Vaccine	+	+	−
Diagnostic Tests to Differentiate	+	+	+

*Details of these viruses are covered in chapter 25.

HBV. **Hepatitis D virus** has a circular RNA genome more closely related to viroids than other viruses. HDV infection can only occur simultaneously or subsequent to HBV infection because the delta agent lacks the genes for structural proteins and is reliant on hepatitis B virus for completing its multiplication cycle. When HBV infection is accompanied by the delta agent, the disease becomes more severe and is more likely to progress to permanent liver damage.

Hepatitis B Virus and Disease

Because a cell culture system for propagating hepatitis B virus is not yet available, by far the most significant source of information about its morphology and genetics has been obtained from viral fragments, intact virions, and antibodies, which are often abundant in patients' blood. One of these blood-borne pieces, the *Dane particle,* often seen in electron micrographs of infected blood, is considered the intact virion. Blood-borne viral components are shown in **figure 24.18.**

Epidemiology of Hepatitis B

An important factor in the transmission pattern of hepatitis B virus is that it is a chronic infection of the liver, which continuously seeds the blood with viruses. Electron microscopic studies have revealed up to 10^7 virions per milliliter of infected blood. Even a minute amount of blood or serum (10^{-6} to 10^{-7} ml) can transmit infection. The abundance of circulating virions is so high and the minimal dose so low that such simple practices as sharing a toothbrush or a razor can transmit infection. HBV has also been detected in semen and vaginal secretions, and it is currently believed to be transmitted

Dane particle Filamentous form Envelope

Figure 24.18 **Hepatitis B.** Blood from hepatitis B virus–infected patients presents an array of particles, including Dane particles (the intact virus), the envelope (or surface shell), and filaments that are polymers of the envelope antigen. *Source:* CDC

by these fluids in certain populations. Spread of the virus by means of close contact in families or institutions is also clearly involved.

The incidence and risk for hepatitis B are highest among people having unprotected sexual intercourse with multiple partners, drug addicts, and certain occupations, including people who conduct medical procedures involving blood or blood products such as serum. The incidence of HBV infection and carriage among drug addicts who routinely share needles is very high. Homosexual males constitute another high-risk group because such practices as

anal intercourse can traumatize membranes and permit transfer of virus into injured tissues. Heterosexual intercourse can also spread infection but is less likely to do so. Infection of a newborn during birth occurs readily and predisposes to development of the carrier state and increased risk of liver cancer in the child. The role of mosquitoes, which can harbor the virus for several days after biting infected persons, has been well documented in the tropics and in some parts of the United States.

> **Quick Search**
>
> Asians and Pacific Islanders are far more likely to be chronically infected with hepatitis B. To learn more, search the Internet for "The Link Between Hepatitis B and Liver Cancer: The Asian American Community" and search videos.

The tropism of HBV for the liver becomes a significant epidemiological complication in some people. Persistent or chronic carriers harbor the virus and continue to shed it for many years. Worldwide, 300 million people are estimated to carry the virus, but it is most prevalent in Africa, Asia, and the western Pacific and is least prevalent in North America and Europe.

Pathogenesis of Hepatitis B Virus

The hepatitis B virus enters the body through a break in the skin or mucous membrane or by injection into the bloodstream. Eventually, it reaches the liver cells (hepatocytes), where it multiplies and releases viruses into the blood during an incubation period of 4 to 24 weeks (7 weeks average). Surprisingly, the majority of those infected exhibit few overt symptoms and eventually develop an immunity to HBV. But some people experience malaise, fever, chills, anorexia, abdominal discomfort, and diarrhea. In people with more severe disease, the symptoms and the aftermath of hepatic damage vary widely. Fever, jaundice, rashes, and arthritis are common reactions **(figure 24.19)**, and a smaller number of patients develop glomerulonephritis and arterial inflammation. Most patients experience complete liver regeneration and restored function. However, a small percentage of patients develops chronic liver disease

in the form of necrosis or cirrhosis (permanent liver scarring and loss of tissue).

The association of HBV with **hepatocellular carcinoma**[2] is based on these observations:

1. certain hepatitis B antigens are found in malignant cells and are often detected as integrated components of the host genome;
2. persistent carriers of the virus are more likely to develop this cancer; and
3. people from areas of the world with a high incidence of hepatitis B (Africa and the Far East) are more frequently affected. This connection is probably a result of infection early in life and the long-term carrier state.

In general, people with chronic hepatitis are 200 times more likely to develop liver cancer, though the exact role of the virus is still the object of molecular analysis. One explanation is that HBV infection increases the rate of cell division in liver cells. As these cells accumulate mutations, oncogenes become activated and cause malignancy.

Diagnosis and Management of Hepatitis B Infection

Hepatitis symptoms are so similar that differential diagnosis on this basis alone is unlikely. A careful examination of possible risk factors could distinguish between hepatitis B and A. For example, HBV infection is associated with occupational exposure, drug abuse, and relatively long incubation period. Serological tests can detect either virus antigen or antibodies. Serological testing permits detection of the surface antigen very early in the infection. These same tests are essential for screening blood destined for transfusions, semen in sperm banks, and organs intended for transplant.

It is possible for mothers who are carriers of HBV to transmit the infection to the fetus transplacentally and highly likely that infection occurs during delivery. Unfortunately, infection with this virus early in life is a known risk factor for chronic liver disease. In light of the increasing incidence of HBV-positive pregnant women, routine testing of all pregnant women is now an important prenatal screen. It has been added to the standard procedures to detect infections that are especially harmful to fetuses and neonates, known collectively as STORCH (see chapter 13).

Mild cases of hepatitis B are managed by symptomatic treatment and supportive care. Chronic infection can be controlled with recombinant interferon, which stops virus multiplication and prevents liver damage in up to 60% of patients. Active disease may also be treated with nucleoside analogs such as entecavir, tenofovir, and lamivudine, which inhibit the DNA polymerase of the virus. Passive immunization with hepatitis B immune globulin (HBIG) gives significant protection to people who have been exposed to the virus through needle puncture, broken blood containers, or skin and mucosal contact with blood. Other groups for whom prophylaxis is highly recommended are the sexual contacts of actively infected people or carriers and neonates born to infected mothers.

The primary prevention for HBV infection is vaccination. The most widely used vaccines (Recombivax, Energix) contain the pure surface antigen cloned in yeasts. They are given in three doses over

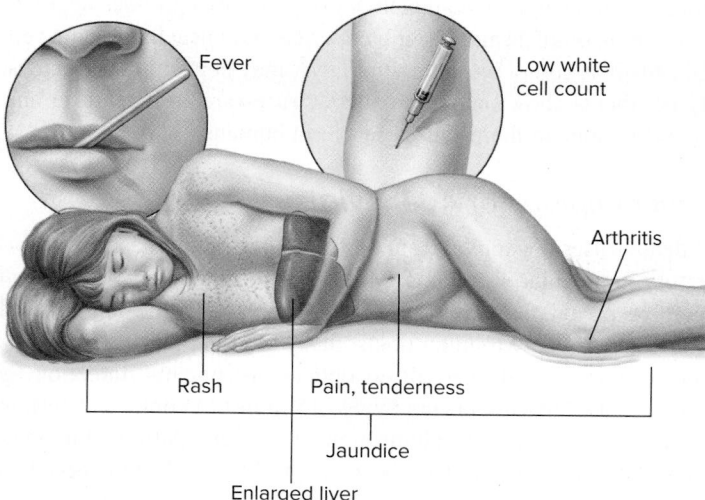

Figure 24.19 The clinical features of hepatitis B. Among the most prominent symptoms are fever, jaundice, and intestinal discomfort. The liver is also enlarged and tender.

2. HOC. Also called hepatoma, a primary malignant growth of hepatocytes.

Pathogen Profile #4 *Hepatitis B virus (HBV)*

Source: CDC

Microscopic Morphology Small, enveloped virus about 42 nm in diameter.

Identified by Serological tests (radioimmunoassay or ELISA) can detect either the virus itself or antibodies manufactured in response to infection.

Habitat Hepatitis B virus is chronically carried, and this population serves as the reservoir for infection. The virus remains infective for days in dried blood, months in serum at room temperature, and decades if frozen. Transfusions or injections of blood products, such as plasma and clotting factors, served as prime means of infection prior to routine testing of donated blood. Currently, kidney dialysis equipment, reusable needles (such as those used for acupuncture), tattooing and piercing equipment, and needleless vaccine guns have all been linked to infection.

Virulence Factors Persistence in liver cells and integration into the genome, leading to oncogenesis.

Primary Infections/Disease Most people infected with HBV show no symptoms and eventually develop immunity to the virus. Initial

symptomatic infection is marked by malaise, fever, and discomfort, with rashes, arthritis, and jaundice occurring in some cases. Most patients experience complete recovery, whereas a small number may develop chronic liver disease as a result of necrosis or cirrhosis. Long-term complications include a greatly increased risk of hepatocellular carcinoma (liver cancer).

Control and Treatment Mild HBV infections are treated by managing symptoms, whereas chronic infection may be treated with interferon, which stops virus multiplication. Passive immunization with hepatitis B immune globulin (HBIG) is administered to those who may have been exposed to the virus through needle punctures or other contact with blood. Extreme care when handling needles, broken blood vials, and splashed blood or serum is essential to reduce the risk of occupational exposure to the virus. Vaccination is available and is recommended for persons under 18 years of age and for adults with potential medical or occupational exposure to blood and blood products.

18 months, with occasional boosters. Immunization to protect against hepatitis B is part of the routine schedule of childhood vaccinations recommended by the Advisory Committee on Immunization Practices.

 Check Your Progress SECTION 24.4

19. Summarize the symptoms, complications, route of transmission, and type of nucleic acid found in hepatitis A, B, and C viruses. What is the effect of the delta agent?
20. How is hepatitis B virus commonly spread, diagnosed, and treated?
21. What groups are most at risk for developing hepatitis B?
22. What is the nature of the vaccines for hepatitis B, and who should receive them?

24.5 Nonenveloped DNA Viruses

 Expected Learning Outcomes

16. Recall the structure, epidemiology, and diseases caused by adenovirus.
17. Describe the epidemiology and pathology of papillomaviruses.
18. Explain how HPV infection can be prevented.
19. Understand the epidemiology and pathology of the human polyomaviruses.
20. Recall the characteristics of parvovirus that make it unique.
21. Identify the most important human parvovirus and the disease it causes.

The Adenoviruses

During a study seeking the cause of the common cold, a new virus was isolated from the **adenoids*** of young children. It turned out that this virus, termed **adenovirus,** was not the sole agent of colds but one of numerous others (described in chapter 25). It was also the first of about 80 strains of nonenveloped, double-stranded DNA viruses classified as adenoviruses, including 30 types associated with human infection **(figure 24.20).** Besides infecting lymphoid tissue, adenoviruses have a preference for the respiratory and intestinal epithelia and the conjunctiva. Adenoviruses produce aggregations of incomplete and assembled viruses that usually lyse the cell. In cells that do not lyse, the viral DNA may persist in the nucleus. Experiments show that some adenoviruses are oncogenic in animals but none of them causes cancer in humans.

Epidemiology of Adenoviruses

Adenoviruses are spread from person to person by means of respiratory and ocular secretions. Most cases of conjunctival infection involve preexisting eye damage and contact with contaminated sources such as swimming pools, dusty working places (shipyards and factories), and unsterilized optical instruments. Infection by adenoviruses usually occurs by age 15 in most people, and certain individuals become chronic respiratory carriers. Although the reason is obscure, military recruits are at greater risk for infection than

* *adenoids* (ad'-eh-noyd) Gr. *adeno,* gland, and *eidos,* form. A lymphoid tissue (tonsil) in the nasopharynx that sometimes becomes enlarged.

Figure 24.20 Distinctive adenovirus structures at high magnification. Colorized electron micrograph of adenovirus particles (220,000×). *Source:* Dr. G. William Gary, Jr/CDC

the general population, and respiratory epidemics are common in armed services installations. Civilians, even of the same age group and living in close quarters, seldom experience such outbreaks.

Respiratory, Ocular, and Miscellaneous Diseases of Adenoviruses

The patient infected with an adenovirus is typically feverish, with acute rhinitis, cough, and inflammation of the pharynx and enlarged cervical lymph nodes. Certain adenoviral strains produce an acute follicular lesion of the conjunctiva. Usually one eye is affected by a watery exudate, redness, and partial closure. A deeper and more serious complication is **keratoconjunctivitis,** an inflammation of the conjunctiva and cornea.

Acute hemorrhagic cystitis in children has been traced to an adenovirus. This self-limiting illness of 4 to 5 days duration is marked by hematuria (blood in the urine), frequent and painful urination with occasional episodes of fever, bedwetting, and suprapubic pain. Because the virus apparently infects the intestinal

epithelium and can be isolated from some diarrheic stools, adenovirus is also considered an agent of infantile gastroenteritis.

Severe cases of adenovirus infection, especially in immunocompromised hosts, can be treated with the antiviral drug cidofovir, but nephrotoxicity is a concern. An attenuated oral vaccine effective against two strains of the virus is routinely administered to military recruits, but no vaccine is available to the public.

Papilloma- and Polyomaviruses

Two families of small, nonenveloped double-stranded DNA viruses are **papillomaviruses*** and **polyomaviruses.*** They are similar in having circular DNA and causing persistent infections and tumors. They differ in their size and the makeup of their genomes.

Epidemiology and Pathology of the Human Papillomaviruses

A papilloma is a squamous epithelial growth commonly referred to as a **wart,** or **verruca,** caused by over 100 different strains of **human papillomavirus (HPV).** Some warts are specific for the mucous membranes; others invade the skin. The appearance and seriousness of the infection vary somewhat from one anatomical region to another. Painless, elevated, rough growths on the fingers and occasionally on other body parts are called **common, or seed, warts (figure 24.21a).** These commonly occur in children and young adults. **Plantar warts** are deeper, painful papillomas on the soles of the feet; flat warts are smooth, skin-colored lesions that develop on the face, trunk, elbows, and knees. A special form of verruca known as **genital warts** is a prevalent STD and is linked to some types of cancer. Warts are transmissible through direct contact with a wart or contaminated fomites, and they can also spread on the same person by autoinoculation. The incubation period varies from 2 weeks to more than a year. Papillomas are extremely common in all human populations.

Genital Warts: An Insidious Papilloma Genital warts are the latest concern among young, sexually active people. Although this

* *papillomaviruses* (pap″-il-oh′-mah) L. *papilla,* pimple, and Gr. *oma,* tumor. A nodular epithelial tumor.
* *polyomaviruses* (paw″-lee-oh′-mah) L. *poly,* many. A glandular tumor caused by a virus.

Figure 24.21 Human papillomas. (a) Common warts. Virus infection induces a neoplastic change in the skin and growths on the fingers, face, or trunk. One must avoid picking, scratching, or otherwise piercing the lesions to avoid spread and secondary infections. **(b)** Chronic genital warts (condylomata acuminata) that have extended into the labial, perineal, and perianal regions. (a): © DermNet New Zealand Trust; (b): *Source:* CDC

(a)

Condylomata acuminata

(b)

disease is not new, case reports have increased to such an extent that it is now regarded as the most common STD in the United States, with millions of new cases per year. There are probably in excess of 30 million carriers of 35 types of HPV associated with genital warts (most commonly types 6, 11, 16, 18, and 31). The virus invades the external and internal genital membranes, especially of the vagina and the head of the penis. Wart morphology ranges from tiny, flat, inconspicuous bumps to extensive, branching, cauliflower-like masses called **condylomata acuminata* (figure 24.21b)**. Infection is also common in the cervix, urethra, and anal skin. Most cases (90%) present with no or minimal symptoms, which can interfere with detection and diagnosis.

The most disturbing aspect of the sudden increase in HPV infection is its strong association with cancer of the reproductive tract, especially the cervix and penis. Ten HPV types increase the risk for developing reproductive cancer, but numbers 16 and 18 account for 70% of metastatic tumors. These viruses are latent in the genome of epithelial cells and contain two oncogenes that transform genital wart lesions into malignant tumors. In the United States, oncogenic HPVs cause over 17,000 cancers in women (mostly cervical) and 9,000 cancers (mostly oropharyngeal) in men each year. Infectious disease specialists are also reporting an increasing number of HPV-related oropharyngeal cancers, which is now the leading cause of mouth and throat cancers in both men and women.

Susceptibility to this infection and its potential for cancer can be greatly reduced by early detection and treatment of both the patient and the sexual partner. Such early diagnosis depends on thorough inspection of the genitals and, in women, a Papanicolaou (Pap) smear to screen for abnormal cervical cells **(figure 24.22)**.

Diagnosis, Treatment, and Prevention of HPV Infection

The warts caused by papillomaviruses are usually distinctive enough to permit reliable clinical diagnosis without much difficulty. However,

* *condylomata acuminata* (kahn″-dee-loh′-mah-tah ah-kyoo″-min-ah′-tah) Gr. *kondyloma*, knob, and L. *acuminatus*, sharp, pointed. Not to be confused with condyloma lata, a broad, flat lesion of syphilis.

a biopsy and histological examination can help clarify ambiguous cases. A sensitive DNA probe can detect HPV by means of a hybridization procedure. This technique has the advantage of also determining the type of HPV in the infection. The Pap smear procedure is a highly effective microscopic method for detecting cells that are infected or converting to cancer. Both provide valuable assessments of the patient's risk for developing cervical cancer.

Although most common warts regress over time, they cause sufficient discomfort and cosmetic concern to require treatment. Treatment strategies for all types of warts include physical removal of the affected skin or membrane by cauterization, freezing, or laser surgery. Additionally, chemicals may be directly applied to warts to speed their removal, with podophyllin typically applied to genital warts and salicylic acid to all others. Because treatments may not destroy the virus in all cells, warts can recur.

Three effective HPV vaccines—Cervarix, Gardasil, and Gardasil 9—have been approved by the FDA. All three are based on viral protein surface receptors made by recombinant DNA technology. Protection is aimed at the most prevalent strains that cause infection and cancer. Bivalent Cervarix contains antigens only from the two HPV types associated with cervical cancer (16 and 18). Quadrivalent Gardasil contains antigens from four types of HPV (6, 11, 16, 18) that are responsible for 70% of cervical cancers and 90% of genital warts. The newest vaccine, Gardasil 9 contains antigens from 5 additional types of HPV (31, 33, 45, 52, and 58) that, altogether, are responsible for 90% of cervical cancers and 90% of genital warts. Immunization is now recommended for both males and females at the age of 11 or 12, prior to becoming sexually active.

The Polyomaviruses

In the mid-1950s the search for a leukemia agent in mice led instead to the discovery of a new virus capable of inducing an assortment of tumors—hence its name, polyomavirus. Among the most important human polyomaviruses are the JC virus and BK virus.[3] Although many polyomaviruses are fully capable of transforming host cells and inducing tumors in experimental animals, none causes cancer in its natural host.

Epidemiology and Pathology On the basis of serological surveys, it appears that infections by JC virus and BK virus are commonplace throughout the world. Because the majority of infections are asymptomatic or mild, not a great deal is known about their mode of transmission, portal of entry, or target cells, but urine and respiratory secretions are suspected as possible sources of the viruses. The main complications of infection occur in patients who have cancers or are immunosuppressed. **Progressive multifocal leukoencephalopathy* (PML)** is an uncommon but generally fatal infection by JC virus that attacks accessory brain cells and gradually demyelinizes certain parts of the cerebrum. Infection with BK viruses, usually associated with renal transplants, causes complications in urinary function. By the time PML is diagnosed, the patient is usually immunocompromised with extensive brain damage. A prevention for BK infection is to treat all renal transplant patients with human leukocyte interferon.

Koilocyte Normal cervical cell

Figure 24.22 Pap smear of cervix in patient with HPV infection. The major sign is the koilocyte, an abnormal cell with a marked clear space around the enlarged abnormal nucleus. Normal cells are labeled for comparison. Detection of such abnormal cells is indicative of a precancerous condition. © Roger Dachez/Science Source

3. The initials signify the cancer patients from whom the viruses were first isolated.

* *leukoencephalopathy* (loo″-koh-en-sef″-uh-lop′-uh-thee) Gr. *leuko*, white, *cephalos*, brain, and *pathos*, disease.

Pathogen Profile #5 *Human papillomavirus (HPV)*

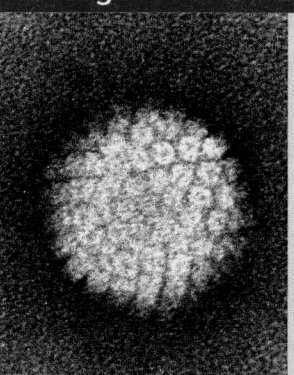

Source: Laboratory of Tumor Virus Biology/National Cancer Institute

Microscopic Morphology Nonenveloped virus about 55 nm in diameter.

Identified by Clinical appearance. Biospy and histological examination, DNA probes, and PCR can be used to confirm clinical diagnosis.

Habitat HPV is common in human carriers throughout the world.

Virulence Factors Ability to persist in epithelial cells of the genitalia; can insert in genome and cause oncogenesis.

Primary Infections/Disease Human papillomavirus is the causative agent of verruca, or warts. The various manifestations of HPV infection—common warts, plantar warts, and genital warts—are the result of different strains of the virus. Infection with some types of HPV both causes genital warts and increases the risk of developing cancer of the reproductive tract, especially of the cervix and penis;

types 16 and 18 account for 70% of metastatic tumors. Infections of the mouth and throat can give rise to oropharyngeal cancers.

Control and Treatment HPV is transmissible through direct contact with a wart or contaminated fomites; therefore, control rests with barrier protection such as gloves or condoms, as appropriate. Treatment involves removal of the wart using chemicals such as podophyllin or physical methods such as surgery, freezing, or cauterization. While these treatments remove the wart, virally infected cells almost always remain and may reappear. Three vaccines—Cervarix, Gardasil and Gardasil 9—effective against those strains of HPV most closely linked to cancer of the reproductive tract have been approved, and their administration is part of the recommended vaccination schedule for children eleven to eighteen.

Source: CDC

Nonenveloped Single-Stranded DNA Viruses: The Parvoviruses

The **parvoviruses*** **(PVs)** are unique among the human viruses in having single-stranded DNA molecules. They are also notable for their extremely small diameter (18–26 nm) and genome size. Parvoviruses are indigenous to and can cause disease in several mammalian groups: for example, distemper in cats, an enteric disease in adult dogs, and a potentially fatal cardiac infection in puppies.

The most important human parvovirus is B19, the cause of **erythema infectiosum (figure 24.23).** It is also known as fifth disease because it is the fifth rash of childhood (the four others being rubella, rubeola, scarlet fever, and scarletina). It is a common childhood infection spread primarily by droplets. Once it enters the circulation, it selectively attacks and lyses erythrocytes. Often the infection goes unnoticed, though the child may have a low-grade fever and a bright red rash on the cheeks. This same virus can be more dangerous in children with immunodeficiency or sickle-cell anemia because of its destruction of red blood stem cells. Adults can acquire the infection, and they often experience a rash along with joint swelling and pain. If a pregnant woman contracts infection and transmits it to the fetus, a severe and often fatal anemia can result.

One type of parvovirus, the adeno-associated virus (AAV), is another example of a defective virus that cannot replicate in the host cell without the function of a helper virus (in this case, an adenovirus). The possible impact of this coinfection on humans is still obscure.

Figure 24.23 **Typical "slapped face" rash of human parvovirus (B19).** Confluent red rash on the cheek gives this child a "slapped face" appearance. This is one symptom of erythema infectiosum, a common childhood ailment caused by human parvovirus (B19). © DermNet New Zealand Trust

 Check Your Progress SECTION 24.5

23. Describe the general structure of an adenovirus.
24. List the diseases caused by adenoviruses and explain how they are spread.
25. Compare the locations and other characteristics of the three types of warts; what types are most serious and why?
26. What is the importance of the HPV vaccines?
27. Describe the unique structure of parvoviruses.
28. Explain the importance of B19.

* *parvoviruses* (par″-voh-vy′-rus) L. *parvus,* small.

CASE STUDY Part 2

A common characteristic of herpesviruses is their ability to cause latent infections and then reemerge in the future, and VZV is no different in this regard. When a varicella infection has resolved, the virus typically retreats to dorsal root ganglia and remains there. Years later—most likely responding to a combination of decreased cellular immunity and unknown environmental cues—the virus may reactivate, resulting in a painful maculopapular rash called herpes zoster, and more commonly known as shingles. Those with active zoster lesions can spread varicella zoster virus to people who have never had chickenpox or been vaccinated. If this happens, the newly infected person may develop varicella (chickenpox), but not zoster (shingles). In this case the woman's infection was almost certainly attributable to her husband's zoster.

A rare complication of both varicella and zoster is delayed paralysis on one side of the body (hemiplegia). One can speculate that there was a connection between the illness and delayed paralysis in this case, but a much more likely explanation is that a 94-year-old woman succumbed to a stroke, without the involvement of a virus. In any case, no postmortem was ordered, so blood and tissue samples that might have supported such a connection were never collected.

Chickenpox is the epitome of a childhood disease. Prior to the introduction of a chickenpox vaccination in 1995, 4 million people would get chickenpox each year in the United States. Although most cases were mild and self-limiting, 10,600 people were hospitalized and 100 to 150 died each year. Despite its occasional severity, the disease was treated as a rite of childhood, with parents routinely holding "pox parties" where infected and uninfected children could mingle, ensuring that children got the disease when they were young (and in the case of families with several children, that all children had the disease simultaneously).

- *Many of the complications of chickenpox involve bacterial infections that are able to enter the body more easily because of breaks in the skin caused by the lesions. How would you describe this type of infection?*

- *Aside from complications of chickenpox, why else might a "pox party" be a bad idea?*

For a synopsis of this Case Study, search the Internet for "Chickenpox at Ninety four."

To conclude this Case Study, go to Connect.

Chapter Summary with Key Terms

24.1 Viruses in Human Infections and Diseases
 A. Viruses are minute intracellular parasites containing a genome of RNA or DNA. They appropriate the metabolic machinery of their host to ensure their reproduction.
 B. Viruses infect a variety of hosts, although the host range of a specific virus is usually quite limited. The effects of viral infection range from mild to fatal, again depending on the particular virus. Some viruses persist in the host in a **chronic** or **latent** state; others are potentially **oncogenic** or **teratogenic.**
 C. Identification and diagnosis of viral infection depends on culture, microscopy, genetic probes, and serology. Limited treatments are available for viral infection.
 D. DNA viruses are classified based on the presence of an envelope and whether they contain double- or single- stranded DNA.

24.2 Enveloped DNA Viruses: Poxviruses
 A. **Poxviruses** are large, complex viruses that produce skin lesions called **pox** (pocks).
 1. **Variola** is the agent of **smallpox,** the first disease to be eliminated from the world by vaccination.
 2. **Molluscum contagiosum** is caused by a molluscipoxvirus. It causes an STD with waxy skin nodules.
 3. Cowpox and monkeypox represent other human diseases caused by poxviruses.

24.3 Enveloped DNA Viruses: The Herpesviruses
 A. **Herpesviruses** are persistent latent viruses that cause recurrent infections and are implicated in neoplastic transformations.
 1. **Herpes simplex virus (HSV):** Both **HSV-1** and **HSV-2** create lesions of the mucous membranes, with type 1 affecting the lips, eyes, and oropharynx and type 2 usually affecting the genitals. Infections are treated with valacyclovir or related compounds.

2. **Varicella-zoster virus (VZV):** Responsible for both **varicella (chickenpox)** and **zoster (shingles).** Chickenpox is the familiar childhood infection; zoster occurs when virus in the spinal/cranial nerves is reactivated. Vaccines now offer protection against both aspects of the virus.

3. **Cytomegalovirus** is perhaps the most common herpesvirus. Healthy adults may be asymptomatic; more serious disease is seen in neonates, newborns, and immunocompromised patients. Drug therapy is reserved for immunocompromised patients.

4. **Epstein-Barr virus (EBV)** is a herpesvirus of the lymphoid and glandular tissue. Leads to **mononucleosis** and potentially **Burkitt lymphoma** or **nasopharyngeal carcinoma.** Disseminated EBV infection is a complication of AIDS and organ transplant patients.

5. **Herpesviruses 6** and **7** are the cause of **roseola infantum** in children; **herpesvirus 8** is the viral agent of Kaposi sarcoma, a tumor of AIDS patients.

24.4 The Viral Agents of Hepatitis
 A. **Hepadnaviruses:** One cause of viral hepatitis, an inflammatory disease of liver cells that may result from several different viruses.
 1. **Hepatitis B virus (HBV)** causes **hepatitis B.** Infection is via body fluids and is marked by fever and **jaundice.** Chronic carriage may lead to liver disease and liver cancer. Management is by vaccine, immune serum globulin, interferon, and antiviral chemotherapy.
 B. Other agents of viral hepatitis (HAV, HCV) are RNA viruses covered in the next chapter (25).

24.5 Nonenveloped DNA Viruses
 A. **Adenoviruses:** Common infectious agent of the lymphoid organs, respiratory tract, eyes. Diseases include the common

cold, **keratoconjunctivitis** (eye infections), and cystitis (urinary tract infections).
B. **Papilloma- and polyomaviruses.**
 1. **Human papillomaviruses (HPVs)** are the causative agent of **papillomas, verrucas, common warts, plantar warts,** and **genital warts,** the last of which may progress to **condylomata acuminata.** Chronic latent infection with HPV is associated with cervical and penile cancer. Treatment includes surgery and interferon. Vaccination is available for prevention.

 2. **Polyomaviruses** include JC virus, which causes **progressive multifocal leukoencephalopathy,** and BK virus, which causes complications in renal transplants.
C. **Nonenveloped Single-Stranded DNA Viruses: The Parvoviruses Parvoviruses:** Human parvovirus (B19) causes **erythema infectiosum,** a relatively mild infection common in children.

Level I. Knowledge and Comprehension

These questions require a working knowledge of the concepts in the chapter and the ability to recall and understand the information you have studied.

? Multiple-Choice Questions

Select the correct answer from the answers provided. For questions with blanks, choose the combination of answers that most accurately completes the statement.

1. Which of the following is *not* caused by a poxvirus?
 a. molluscum contagiosum c. cowpox
 b. smallpox d. chickenpox

2. In general, zoonotic viral diseases are _____ in humans.
 a. mild and asymptomatic c. severe and systemic
 b. self-limited d. not contagious

3. Which virus is used in smallpox vaccination?
 a. variola c. cowpox
 b. vaccinia d. varicella

4. Herpes simplex 1 usually causes _____, and herpes simplex 2 usually causes _____.
 a. cold sores, genital herpes c. canker sores, fever blisters
 b. fever blisters, cold sores d. shingles, stomatitis

5. Herpetic whitlows are _____ infections of the _____.
 a. pox, oral cavity c. EBV, skin
 b. CMV, lymph nodes d. herpes, fingers

6. Neonatal herpes simplex is usually acquired through
 a. contaminated mother's hands
 b. other babies in the newborn nursery
 c. crossing the placenta
 d. an infected birth canal

7. _____ is an effective treatment for herpes simplex lesions.
 a. Amantadine c. Acyclovir
 b. Interferon d. Cortisone

8. Which herpesvirus is most commonly associated with a dangerous fetal infection?
 a. herpes simplex c. EBV
 b. herpes zoster d. CMV

9. Varicella and zoster are caused by
 a. the same virus
 b. two different strains of VZV
 c. herpes simplex and herpes zoster
 d. CMV and VZV

10. A common sign of hepatitis is
 a. liver cancer c. anemia
 b. jaundice d. bloodshot eyes

11. A virus associated with chronic liver infection and cancer is
 a. hepatitis C c. hepatitis A
 b. hepatitis B d. both a and b

12. Benign epithelial growths on the skin of fingers are called
 a. polyomas c. whitlows
 b. verrucas d. pox

13. Parvoviruses are unique because they contain
 a. a double-stranded RNA genome
 b. reverse transcriptase
 c. a single-stranded DNA genome
 d. an envelope without spikes

14. Adenoviruses are the agents of
 a. hemorrhagic cystitis c. common cold
 b. keratoconjunctivitis d. all of these

15. Which of the following viruses causes chronic persistent infections?
 a. hepatitis B c. parvovirus
 b. smallpox d. varicella-zoster virus

16. Which of the following viruses can be oncogenic in humans?
 a. papillomavirus c. adenovirus
 b. Epstein-Barr virus d. both a and b

17. **Multiple Matching.** Match the virus with the disease or condition. (Some viruses may match more than one condition.)

 _____ herpes simplex 1 a. gingivostomatitis
 _____ variola b. condylomata acuminata
 _____ polyoma c. chickenpox
 _____ varicella zoster d. erythema infectiosum
 _____ herpes simplex 2 e. fever blisters
 _____ adenovirus f. mononucleosis
 _____ papilloma g. smallpox
 _____ Epstein-Barr h. genital herpes
 _____ cytomegalovirus i. leukoencephalopathy
 _____ parvovirus j. roseola
 _____ herpesvirus 6 k. cancer
 l. keratoconjunctivitis

 Case Study Review

1. The varicella-zoster virus that causes chickenpox is classified as a
_____.

 a. herpes virus
 b. pox virus
 c. papillomavirus
 d. parvovirus

2. For which viral infections are vaccines available to the general public?

 a. varicella-zoster
 b. human papillomavirus
 c. hepatitis C virus
 d. adenovirus
 e. both a and b

3. The woman in this case contracted chickenpox from her husband, yet the husband was suffering from shingles. Why did the woman not contract shingles?

 Writing Challenge

For each question, compose a one- or two-paragraph answer that includes the factual information needed to completely address the question. Check Your Progress questions can also be used for writing-challenge exercises.

1. Outline 10 characteristics of viruses that add significantly to their medical impact and complications.

2. Explain the involvement of viruses in cancer and congenital infections.

3. Briefly clarify the difference between chronic and latent viral infections.

4. Develop a table for the herpesviruses that summarizes virus name, tissues infected, names of diseases, and special characteristics.

5. a. Compare the two types of human herpes simplex virus according to the types of diseases they cause, body areas affected, and complications.
 b. Why is neonatal infection accompanied by such severe effects?
 c. How are herpes simplex infections diagnosed and treated?

6. Why does mononucleosis occur most commonly among college-age Americans?

7. Name the DNA viruses that are oncogenic, sexually transmitted, spread by blood and blood products, and spread by respiratory droplets, kissing, and other nonsexual intimate contact.

 Concept Mapping

On Connect you can find an Introduction to Concept Mapping that provides guidance for working with concept maps along with concept mapping activities for this chapter.

 Level II. Application, Analysis, Evaluation, and Synthesis

These problems go beyond just restating facts and require higher levels of understanding and an ability to interpret, problem solve, transfer knowledge to new situations, create models, and predict outcomes.

 Critical Thinking

Critical thinking is the ability to reason and solve problems using facts and concepts. These questions can be approached from a number of angles, and in most cases, they do not have a single correct answer.

1. Discuss the pros and cons of salvaging or destroying the smallpox virus stocks.

2. Describe several measures health care workers can take to avoid whitlows.

3. Explain why a baby whose mother has genital herpes is not entirely safe even with a cesarean birth.

4. a. What specific host defenses do immunodeficient, cancer, and AIDS patients lack that make them so susceptible to the viral diseases in this chapter?
 b. Given the ubiquity of most of these viruses, what can health care workers do to prevent transmission of infection to other patients and themselves?

5. a. Name two different defective viruses that require another virus for function.
 b. Explain how they might interrelate with the host virus.

6. Explain this statement: One acquires chickenpox from others, but one acquires shingles from oneself.

7. A man wants to divorce his wife because he believes her genital herpes could only have been acquired sexually. Her doctor says she is infected with type 1. Suggest a more likely source.

8. Weigh the following observation: High titers of antibodies for EBV are found in a leukemia patient; a chronically tired, ill businessman; a healthy military recruit; and an AIDS patient. Comment on the probable significance of antibodies to EBV in human serum.

9. Give some pros and cons of exposing children to chickenpox to infect them and provide subsequent immunity, as opposed to vaccination.

10. What type of virus is most likely to be transmitted by vaccination guns and acupuncture tools?

11. What two different DNA viral infections can be identified with a Pap smear? Explain what abnormalities one would see.

Visual Challenge

1. Using **figure 24.1** and **table 24.1** as references, match each virus (a–d) with its usual infection pattern (1–5). Explain the outcomes of the disease process for each virus.

 STUDY TIP: Quickly create a customized practice quiz for this chapter by going to your Connect SmartBook "My Reports" section. Don't have SmartBook and want to learn more? Go to **whysmartbook.com**.

CHAPTER **25** The RNA Viruses That Infect Humans

Source: Jim Goodson, M.P.H./CDC

Transmission electron micrograph of a single measles virus particle
Source: CDC

 CASE STUDY **Part 1** **This Can't Be From That**

In 2011, an 11 year-old boy first presented to his family physician with a rapidly progressing collection of disturbing **neurological** symptoms. Four months earlier the boy had been a straight-A student in the fifth grade, with a healthy interest in basketball and soccer, when he began to have trouble completing assignments in school. His coordination and attention suffered; at home he would drop utensils and fall asleep during meals. Over the next month his **symptoms intensified;** he was inattentive and **confused** and would sometimes fall asleep while walking. His **coordination deteriorated** and he began to experience involuntary hand and arm movements that became increasingly severe. He lost weight as his appetite decreased.

The boy was referred to a pediatric neurologist, who ordered a **lumbar puncture, magnetic resonance imaging** (MRI) of the brain, and an **electroencephalogram** (EEG). The lumbar puncture and MRI were normal. The EEG showed a number of abnormalities that unfortunately did little to explain the boys rapidly worsening neurological condition. His cognitive and motor skills continued to decline and included frequent falling, the asking of meaningless questions, and **increased aggression.** By the third month after the onset of symp-

"He no longer appeared to recognize his family members' faces or voices."

toms, the boy had deteriorated dramatically. His speech became difficult to understand, he experienced worsening **spasticity** and rapid decline in mental status; he became **incontinent** and was unable to eat or drink. He did not fix on or follow objects, and he no longer appeared to recognize his family members' faces or voices.

In 2012 the boy was admitted to the hospital. He was unresponsive to questions, could not follow commands, and continued to exhibit **spastic movements** of the arms and legs. An EEG displayed results indicative of **subacute sclerosing panencephalitis** (SSPE), a neurodegenerative disease with no treatment and no cure. After 14 days the boy was discharged from the hospital and entered home hospice care, where he died 43 months later, in 2015.

■ *Subacute sclerosing panencephalitis is a complication of what viral infection?*

■ *How long does it take for SSPE to appear after the initial infection?*

To continue the Case Study, go to Case Study Part 2 at the end of the chapter.

The RNA viruses that cause disease in humans are a remarkably diverse group of microbes. Viruses are assigned to one of 12 families on the basis of their envelope, their capsid, and the nature of their RNA genome (**table 25.1**). RNA viruses are etiologic agents in a number of serious and prevalent human diseases, including influenza, AIDS, hepatitis, and viral encephalitis. Many experts believe that these viruses and several others have the potential of becoming the most important disease threats of the future (**25.1 Making Connections**).

25.1 Enveloped Segmented Single-Stranded RNA Viruses

Expected Learning Outcomes

1. Describe the structure of influenza virus and its spikes.
2. Understand what is meant by antigenic shift and antigenic drift.
3. Recall the epidemiology and complications associated with influenza.
4. Describe the production of influenza vaccine.
5. Recall the most important bunyaviruses found in the United States, and explain how infection commonly occurs.

The Biology of Orthomyxoviruses: Influenza

Orthomyxoviruses are spherical particles with an average diameter of 80 nm to 120 nm. There are three distinct influenza virus types (A, B, C), but type A causes the most cases of infection and is the one we discuss in the greatest detail. The virus attaches to and multiplies in cells of the respiratory tract (**figure 25.1**). It is engulfed and uncoated, and the segments of its RNA genome enter the nucleus to be transcribed and replicated. Gene expression gives rise

to glycoprotein spikes and capsid proteins. Finished viruses are assembled and budded off the cell with an envelope. Key to the many aspects of virus behavior and pathology are the glycoprotein spikes—termed **hemagglutinin (H)** and **neuraminidase (N).** Virologists have characterized 17 different H subtypes and 9 N subtypes and have used these numbers to distinguish them. Influenza viruses of human origin are primarily of the H1, H2, and H3 and N1 and N2 subtypes. **Table 25.2** is a listing of significant influenza epidemics and the strains involved.

The name *hemagglutinin* is derived from its property of agglutinating (clumping) red blood cells, a property useful in identifying the virus. Hemagglutinin is the most important virulence factor because it is needed to bind to the respiratory mucosal receptors and induce viral entry. Neuraminidase is an enzyme that breaks down the respiratory mucus, keeps viruses from sticking together, and assists in budding from the host cell. Both types of spikes undergo frequent alterations due to genetic changes (**figure 25.2**).

The ssRNA genome of the influenza virus is known for its extreme variability. Research has shown that genetic changes are very frequent in the areas of the glycoprotein spikes recognized by the host immune response but very rare in the areas used for attachment to the host cell. In this way the virus can continue to attach to host cells while managing to decrease the effectiveness of the host response to its presence. This constant genetic change of the glycoproteins through mutation is called **antigenic drift**—the antigens gradually change their amino acid composition, resulting in decreased ability of host memory cells to recognize them.

A change with even more dramatic effects is known as **antigenic shift.** The genome of the virus consists of just 10 genes encoded on eight separate RNA strands. Antigenic shift is the substitution of one of those genes or strands with a gene or strand from an influenza virus originating from a different animal host. This is made possible when a strain of influenza A virus infects both humans and swine and a different strain of influenza A virus

TABLE 25.1 RNA Viruses with Examples of Diseases

Systems Profile 25.1	Pathogenic RNA Viruses					
Virus	Skin/Skeletal	Nervous/Muscle	Cardiovascular/ Lymphatic/Systemic	Gastrointestinal	Respiratory	Urogenital
Influenza Virus					Influenza	
Hantavirus					Hantavirus pulmonary syndrome	
Parainfluenza Virus					Croup	
Mumps Virus			Mumps			
Morbillivirus	Measles	Subacute sclerosing panencephalitis				
Respiratory Syncytial Virus (RSV)					RSV disease	
Rabies Virus		Rabies				
SARS-Associated Coronavirus, MERS-Associated Coronavirus					Severe respiratory distress (SARS, MERS)	
*Rubivirus**	Rubella					
Hepatitis C Virus	Jaundice			Hepatitis		
St. Louis Encephalitis Virus		Encephalitis				
Yellow Fever Virus		Yellow fever		Yellow fever		
Dengue Fever Virus		Dengue fever	Dengue hemorrhagic fever			
Human Immunodeficiency Virus 1 and 2			HIV infection and AIDS			
Poliovirus		Poliomyelitis		Enteric infection		
Hepatitis A Virus				Hepatitis		
Rhinovirus					Common cold	
Norovirus				Acute diarrhea		
Rotavirus				Acute diarrhea		
Ebola Virus			Ebola hemorrhagic fever			
Prions		Creutzfeldt-Jakob disease				

*Infection during pregnancy causes various birth defects in the developing fetus.

infects birds and swine. The genes of all of these viruses code for the same important influenza proteins (including H and N)—but the actual sequence of the genes is different in the different types of viruses. When a bird virus and a human virus infect a single swine host, it is not uncommon for both virus types to infect the same host cell. Then the viral packaging step (figure 25.1, step 6) can randomly produce a human influenza virus that contains seven human influenza virus RNA strands plus a single bird influenza virus RNA strand **(figure 25.3).** When that virus infects a human, the protein receptors from the bird virus will not be recognized by memory cells, so there will not be a rapid response. Antibodies from any prior influenza virus will not be protective. As a result, there are deficiencies in both individual and herd immunities and epidemics become a real possibility. Experts have traced the flu pandemics of 1918, 1957, 1968, 1977, and 2009 to virus strains that were hybrid recombinants from bird and pig viruses that were also infective to humans.

To characterize each virus strain, virologists have developed a special nomenclature that chronicles the influenza virus type, the presumed animal of origin, the location, and the year of origin. For example, A/duck/Ukraine/63 identifies the strain isolated from domestic or migratory fowl; likewise, A/seal/Mass/1/80 is the causative agent of seal influenza. From this convention come such expressions as "swine flu" and "Hong Kong flu" (see table 25.2).

Emerging Avian Influenza Viruses

Starting in 2003, strains of influenza A viruses that usually infect birds underwent an antigenic shift and began to infect humans. The first of these emerging bird flu viruses, *H5N1,* originated from poultry in Southeast Asia and caused a highly virulent disease with about a 60% fatality rate in humans. All of these cases apparently arose through bird-to-human transmission, and there was no

1 Virus (structure shown in a cutaway view) adsorbs to a respiratory epithelial cell by hemagglutinin spikes and fuses with the membrane.

Nucleocapsids

Hemagglutinin spike

Neuraminidase spike

Envelope forming

2 The virus is endocytosed into a vacuole and uncoated to release its 8 nucleocapsid segments into the cytoplasm.

4 (+) Sense RNA is used to synthesize glycoprotein spikes inserted into the host membrane.

6 Release of mature virus occurs when viral parts gather at the cell membrane and are budded off with an envelope containing spikes.

5 The (+) sense RNA strands are used to synthesize new (−) sense RNA strands. These are assembled into nucleocapsids and transported out of the nucleus to the cell membrane.

3 The nucleocapsids are transported into the nucleus. There the (−) sense RNA strand (black) is transcribed into a (+) sense strand (red) that will be translated into viral proteins that make up the capsid and spikes.

Process Figure 25.1 Stages in influenza virus multiplication cycle.

evidence of person-to-person transmission. In 2013 another bird flu virus, *H7N9,* appeared in Chinese patients. By the middle of 2016 at least 807 cases and 151 deaths had been attributed to the flu, almost all of then in China. The vast majority of the cases involved direct contact with birds, often at live poultry markets, indicating that while the strain moved quickly from bird to human, person to person transmission was far less efficient.

Epidemiology and Pathology of Influenza A

Influenza is an acute, highly contagious respiratory illness afflicting people of all ages and marked by seasonal regularity and pandemics at irregular intervals. One of its first names was an Italian derivative of *un influenza di freddo,* an "influence of mystery." This name has remained and is now contracted to the vernacular "flu." Periodic

TABLE 25.2	Human Influenza Virus Types and Variant Forms	
Type	**H/N Subtype**	**Strain/History**
A	H1N1	Spanish flu pandemic of 1918
		A/New Jersey/76* (swine flu outbreak)
		A/USSR/90/77
		A/Texas/36/91
		Pandemic H1N1/09
	H2N2	A/Singapore/57 (Asian flu pandemic)
		A/Japan/62
		A/Taiwan/64
	H3N2	A/Hong Kong/68 (Hong Kong flu pandemic)
		A/Johannesburg/33/94
B	None	B/Harbin/07/94
C	None	C/JHB/2/66

*The last number is the year the virus appeared.

Binding sites used to anchor virus to host cell receptors (low rate of mutation)

Site for antibody binding (high rate of mutation)

Viral envelope

Figure 25.2 **Schematic drawing of hemagglutinin (H) of influenza virus.** Blue boxes depict site used to attach virus to host cells; green circles depict sites for anti-influenza antibody binding.

Duck influenza virus

H RNA
N RNA

Human influenza virus

H RNA
N RNA

H
N

Human influenza virus
with duck H spike

Immune system has
no recall for duck
antigens.

Figure 25.3 Antigenic shift event. Where swine, birds (ducks here), and humans live close together, the swine can serve as a melting pot for creating "hybrid" influenza viruses that are not rapidly neutralized by the human immune system.

outbreaks of respiratory illness that were probably influenza have been noted throughout recorded history. Note that the term *flu* is often used as a shorthand for any short-term acute illness such as "stomach flu." Influenza does not affect the gastrointestinal tract in humans.

Influenza pandemics are accelerated by the contagiousness of the virus and widespread migrations or travel. They also occur in cycles, with episodes spaced about a decade apart. By compiling extensive serological profiles of populations and the influenza strains to which they are immune, the Centers for Disease Control and Prevention (CDC) have a scientific basis for predicting expected epidemic years and the probable strains involved. One advantage

of forecasting the expected viral subtypes is that the most vulnerable people can be vaccinated ahead of time with the appropriate virus strain(s).

Mode of Influenza Transmission

Inhalation of virus-laden aerosols and droplets constitutes the major route of influenza infection, although fomites can play a secondary role. The virus survives for 24 hours on nonliving surfaces. Transmission is greatly facilitated by crowding and poor ventilation in living quarters during the late fall and winter. In the tropics, infection occurs throughout the year, without a seasonal distribution. Occupational contact with chickens, other poultry, and swine constitutes a special high-risk category for disease. Rates of infection are highest among school-age children (30%) and lowest in adults (6%). The overall mortality rate for influenza A is about 0.1% of cases. Influenza or influenzal pneumonia rank among the top 10 causes of death in the United States, usually among elderly persons and small children.

Infection and Disease

The influenza virus binds primarily to ciliated cells of the respiratory mucosa. Infection causes the rapid destruction of these cells, releasing a load of viruses. Stripping the respiratory epithelium to the basal layer eliminates protective ciliary clearance and leads to severe inflammation and irritation. The illness is further aggravated by fever, headache, myalgia, pharyngeal pain, shortness of breath, and repetitive bouts of coughing. As the normal ciliated columnar epithelium is restored in a week or two, the symptoms usually abate.

Patients with emphysema or cardiopulmonary disease, along with very young, elderly, or pregnant patients, are more susceptible to serious complications. Most often, the weakened host defenses predispose patients to secondary bacterial infections, especially pneumonia caused by *Streptococcus pneumoniae* and *Staphylococcus aureus*. Although less common than bacterial pneumonia, primary influenza virus pneumonia is a dangerous complication that can cause rapid death—even of people in their prime. Other complications are bronchitis, meningitis, and neuritis.

Diagnosis, Treatment, and Prevention of Influenza

Except during epidemics, influenza must be differentiated from other acute respiratory diseases. Rapid serological tests can detect influenza antigens in a pharyngeal specimen in less than 30 minutes but have poor sensitivity, often producing false negative results. Assays which detect and amplify the RNA of the virus are the gold standard, and can be completed less than two hours.

The primary treatments for influenza include three prescription drugs: Zanamivir (Relenza), oseltamivir (Tamiflu), and peramivir (Rapivab). These drugs both target the function of neuraminidase as a way to block viral budding and release (see table 12.6). These drugs have replaced amantadine and rimantadine in routine therapy due to the viruses' high levels of resistance to these older drugs. They must be taken as early as possible after infection to be beneficial. If clinical signs point to influenza, it is recommended that treatment be started immediately, without waiting for laboratory confirmation. Primary effects of the drugs

25.1 MAKING CONNECTIONS

A Global Viral Threat

Over the last several decades, some of humankind's most deadly viral diseases have been vanquished. Two cases in point are polio, now restricted to a few small areas of the globe, and smallpox, which has been eliminated entirely. Yet these successes are being overshadowed by a rise in the incidence of new and emerging viruses. Since 1967 at least 20 new viruses have been identified during an outbreak or epidemic of human disease **(table 25.A)**. All but one of these viruses is an RNA virus, and most of them are zoonotic in origin, so in that sense they are new only to humans.

What accounts for the appearance of these viral diseases? One explanation is that RNA viruses mutate and evolve very rapidly. This creates opportunities for them to adapt in a short time to host species other than their natural ones. To quote Dr. Eddie Holmes at the Centers for Infectious Disease Dynamics:

> "What a virus wants to do is simply to self-replicate. And if it means jumping species boundaries to find that cell, that will happen. What viruses are interested in . . . is to reproduce quickly with as much progeny as possible."

Another major factor in the changing face of viruses on the world scene has to do with alterations in human ecology. We have much greater exposure to animals and animal habitats that were formerly isolated in remote areas of the world. When you combine this situation with movements to cities, increased population size, and rapid modes of travel, there is a "perfect storm" for transmission of any virus that has made the jump to humans. A particularly illustrative example of this was the West African Ebola outbreak of 2014–2016, which accounted for more than 28,000 cases and over 11,300 deaths (see the chapter 13 Case Study). The next largest outbreak on record took place in 2000–2001 in Uganda and was responsible for 224 deaths. Urbanization is clearly a driver of increasingly severe viral outbreaks.

Not only do we have to be vigilant about unidentified viruses, we also have to stay aware of the threat of emerging and reemerging viruses—those that have been around for a time but are increasing in numbers or range. High on the list of emerging viruses are the mosquito-borne arboviruses. West Nile fever virus, which had been endemic to Africa for years, has since become well-entrenched in North America while Dengue virus and Zika virus have been steadily extending their northward range in parts of Asia and the Americas. Virus watchers are also concerned about the continuing spread of Chikungunya arbovirus. In 2014, transmission of Chikungunya virus was identified in Florida, Puerto Rico, and the U.S Virgin Islands. It causes debilitating arthritis and encephalitis that can be highly damaging to children.

Fortunately the methods we use to track the emergence of new viruses have improved as well, allowing scientists to characterize novel agents in a fraction of the time it took even a few years ago. When severe acute respiratory syndrome (SARS) emerged in China in 2002, it took scientists 3 months to identify the pathogen as a new type of coronavirus and notify the World Health Organization. In the meantime the ensuing epidemic infected over 8,000 people and killed nearly 1,000. In late 2012 and early 2013, 17 people were hospitalized with respiratory distress brought on by infection with a virus thought to be similar to the SARS virus. A new technology known as *deep sequencing* allowed scientists to swiftly identify areas of the genome where the new virus differed from previously identified coronaviruses. The virus—now known as Middle East Respiratory Syndrome Coronavirus, or MERS-CoV— was found to be unique, and WHO was warned of the potential dangers just 3 weeks after the second patient began to show symptoms. Ultimately this new means of viral classification allows for a more rapid identification of the source of the virus, better treatment, and smaller outbreaks.

Why do viruses that have jumped hosts (such as from bats to humans) cause a much more severe infection? *Answer available on Connect.*

TABLE 25.A	Novel Virus Discoveries Since 1967		
Microbe	**Source**	**Infection**	**Year of Emergence**
Marburg virus	Bats	Hemorrhagic fever	1967
Lassa fever virus	Rodents	Lassa fever	1969
Monkeypox virus	Monkeys, rodents	Monkeypox	1970
Ebola virus	Bats	Hemorrhagic fever	1976
Human T-cell virus	Human?	Lymphoma, leukemia	1980
Human immuno-deficiency virus	Chimpanzees	AIDS	1981
Hepatitis C virus	Humans	Hepatitis, liver diseases	1989
Hantavirus	Mice	Respiratory distress syndrome	1993
Hendra virus	Bats	Encephalitis	1994
New variant CJD prion	Cattle	Creutzfeldt-Jakob disease	1996
Nipah virus	Bats	Encephalitis, pneumonia	1998
Human *metapneumovirus*	Humans	Bronchiolitis	2001
SARS virus	Palm civet	Severe acute respiratory syndrome	2002
H5N1 flu virus	Birds	Influenza	2003
Chikungunya virus	birds, small mammals	Chikungunya	2005
Astrovirus	Mink?	Encephalitis	2007
Lujo virus	Rodents	Hemorrhagic fever	2008
H1N1 flu virus	Swine, birds	Influenza	2009
Heartland virus	Ticks	Flulike disease	2012
MERS-CoV	Camels	Middle East respiratory syndrome	2012
H7N9 flu virus	Birds	Influenza	2013
Zika virus	Monkeys, rodents	Zika fever	2015

Pathogen Profile #1 *Influenza virus*

Source: CDC

Microscopic Morphology Enveloped, spherical virus, 80 to 120 nm in diameter, possessing a segmented genome of seven or eight strands of single-stranded RNA.

Identified by Serological tests to detect influenza antigens in pharyngeal samples, virus isolation in cell culture. Genetic analysis is the gold standard for detection and can be used to determine the origin, or specific strain, of a virus.

Habitat Influenza virus infects humans, swine, and birds.

Virulence Factors Hemagglutinin (H) and neuraminidase (N) spikes. Ability to undergo antigenic shift.

Primary Infections/Disease Influenza is a respiratory illness marked by seasonal regularity along with irregular pandemics. Infection of cells in the respiratory tract leads to the death of ciliated epithelial cells and causes inflammation and irritation of the respiratory epithelium. Other symptoms include fever, headache, myalgia, pharyngeal pain, coughing, and shortness of breath. As the epithelium is replaced by healthy cells, usually over a week or two, symptoms abate.

Control and Treatment Vaccine, which must be repeated for each year's dominant influenza strains, remains the best preventative measure. Three prescription drugs are currently recommended for the treatment of influenza—zanamivir (Relenza), oseltamivir (Tamiflu), and peramivir (Rapivab) —each of which interferes with viral entry into the cell. These drugs typically decrease both the duration of the disease and the severity of symptoms. Two older drugs, amantadine and rimantadine are not currently recommended because of high levels of resistance.

include reduction of symptoms, shortened disease duration, and slowed virus transmission.

Nonspecific treatments, mainly to control symptoms, include fluids, bed rest, and nonaspirin pain relievers and anti-inflammatory drugs. Aspirin is not indicated because it significantly increases the risk for **Reye's syndrome,** a rare disease that strikes the brain, liver, and kidney and is characterized by the fatty degeneration of those organs.

Influenza Vaccination Because of the changing nature of the antigens on the viral surface, a result of antigenic drift and shift, annual vaccination is considered the best way to avoid infection. Anyone over the age of 6 months can take the vaccine, and it is especially recommended for high-risk groups or those who have extensive contact with the public. New guidelines add young children as a target group for vaccination.

Preventing influenza infections and epidemics is one of the top priorities for public health officials. In that the standard inactivated whole-virus vaccine requires nearly a year to prepare, much advance planning takes place. Here is the general time line with the events involved:

1. In January, three or four virus variants are selected according to evidence that predicts the most prominent strains to circulate that year. For example, the 2016–2017 vaccines contained A/California/7/2009 (H1N1)pdm09-like virus; A/Hong Kong/4801/2014 (H3N2)-like virus; B/Brisbane/60/2008-like virus (B/Victoria lineage). Four component vaccines add an additional virus, B/Phuket/3073/2013-like virus (B/Yamagata lineage).
2. CDC delivers the correct virus seed stock to manufacturers.
3. From February to May, viruses are grown in millions of fertilized eggs from a special breed of chickens.
4. The viruses are harvested, processed, and inactivated by chemical treatment to create the vaccine.
5. In June/July, the vaccine is tested for purity, safety, and sterility.
6. In August/September, the vaccine is dispensed, prepared for injection, and shipped to medical centers and pharmacies.
7. In October/November, the vaccination programs begin.

Currently the majority of influenza vaccines are produced by this method, but two vaccines (Flublok, Flucelvax) not grown in eggs are available. Not using eggs to make the vaccine has clear benefits, including shorter lead times in production, faster alterations to accommodate new strains midseason, and less likelihood of allergies.

Several types of influenza vaccines exist. Traditional vaccination, approved for the majority of individuals 6 months or older, is an inactivated trivalent vaccine (containing antigens from 3 viral strains) injected intramuscularly. A newer alternative is an intradermal vaccine that injects the same three inactivated strains into the skin, using a needle one-tenth the size of the traditional vaccine. Currently approved only for adults between 18 and 64 years of age, intradermal vaccination is helpful for those who have a fear of needles. An additional benefit is that the vaccination requires 40% less antigen than a traditional vaccine, helping to guard against vaccine shortages by allowing more doses to be manufactured from a given amount of antigen.

A serious complication of influenza vaccines has been **Guillain-Barré syndrome.** It is a neurological condition that can arise in approximately 1 in 100,000 vaccine recipients. This syndrome appears to be an autoimmunity induced by viral proteins and marked by varying degrees of demyelination of the peripheral nervous system, leading to weakness and sensory loss. Most patients recover function, but the disease can also be debilitating and fatal.

Other Viruses with a Segmented Genome: Bunyaviruses and Arenaviruses

Between them the **bunyavirus*** and **arenavirus*** groups contain hundreds of viruses whose normal hosts are arthropods or rodents. Although none of the viruses is a natural parasite of humans, they can be transmitted zoonotically and periodically cause epidemics in human populations. Examples of important bunyavirus diseases are California encephalitis, spread by mosquitoes; Rift Valley fever, an African disease transmitted by sand flies; and Korean hemorrhagic fever, caused by a hantavirus and transmitted by rodents. Other bunyaviruses are discussed in section 25.4 along with other arboviruses because most of them are transmitted by insects and ticks.

The most important American bunyavirus is **Sin Nombre** (no name) **hantavirus.** The first cases came to light in 1993 in the Four Corners area of New Mexico, when a cluster of unusual cases was reported to the CDC. The victims were struck with high fever, lung edema, and pulmonary failure. Most of them died within a few days. Similar cases were reported sporadically in Arizona, Colorado, and other western states.

The disease was termed **hantavirus pulmonary syndrome (HPS).** Comprehensive studies of rodent populations have revealed that the virus is carried by deer and harvest mice. It is probably spread by dried animal wastes that have become airborne when rodent nests are disturbed. The disease occurs sporadically at a rate of 25 to 50 cases per year but with a high (33%) mortality rate.

During the summer of 2012, 10 confirmed cases of hantavirus infection—and three deaths—occurred among visitors to Yosemite National Park, with 9 of the 10 cases occurring in people staying in tent cabins in the same campground. Inspection revealed mice to be nesting within the walls of several cabins. Outbreaks like this, which exemplify the ease with which humans can intermingle with a large rodent reservoir, are a primary reason that hantavirus pulmonary syndrome is considered an emerging disease.

Four major arenavirus diseases are Lassa fever, found in parts of Africa; Argentine hemorrhagic fever, typified by severe weakening of the vascular bed, hemorrhage, and shock; Bolivian hemorrhagic fever; and lymphocytic choriomeningitis, a widely distributed infection of the brain and meninges. The viruses are closely associated with a rodent host and are continuously shed by the rodent throughout its lifetime. Transmission of the virus to humans is through aerosols and direct contact with the animal or its excreta. These diseases vary in symptomatology and severity from mild fever and malaise to complications such as hemorrhage, renal failure, cardiac damage, and shock syndrome. The lymphocytic choriomeningitis virus can cross the placenta and infect the fetus. Pathologic effects include hydrocephaly, blindness, deafness, and mental retardation. So dangerous are these viruses to laboratory workers that the highest level of containment procedures and sterile techniques must be used in handling them.

* *bunyavirus* (bun′-yah-vy″-rus) From Bunyamwera, an area of Africa where this type of virus was first isolated.
* *arenavirus* (ah-ree′-nah-vy″-rus) L. *arena,* sand. In reference to the appearance of virus particles in electron micrographs.

 Check Your Progress SECTION 25.1

1. Describe the structure and functions of the hemagglutinin and neuraminidase spikes in influenza virus.
2. What is unusual about the genome of influenza virus?
3. Explain how antigenic shift and drift differ and how they affect the immune status of populations.
4. Explain the course of infection and disease and the complications of influenza.
5. Explain generally how flu vaccines are prepared. For which groups is vaccination indicated?
6. Give examples of bunyaviruses and arenaviruses, and briefly describe the nature of the diseases they cause.
7. Describe the manner in which hantavirus is typically contracted.

25.2 Enveloped Nonsegmented Single-Stranded RNA Viruses

 Expected Learning Outcomes

6. Recall the enveloped viruses possessing a nonsegmented RNA genome.
7. Describe the epidemiology, pathology, diagnosis, treatment, and prevention of mumps and measles.
8. Identify those most at risk of contracting respiratory syncytial virus.
9. Name the common reservoirs of rabies.

Paramyxoviruses

The important human paramyxoviruses are *Paramyxovirus* (parainfluenza and mumps viruses), *Morbillivirus* (measles virus), and *Pneumovirus* (respiratory syncytial virus), all of which are readily transmitted through respiratory droplets. The envelope of a paramyxovirus possesses specialized glycoprotein spikes that initiate attachment to host cells. They also bear fusion (F) spikes that initiate the cell-to-cell fusions typical of these viruses (**figure 25.4**). A series of multiple cell fusions produces a **syncytium,*** or **multinucleate giant cell** with cytoplasmic inclusion bodies, a cytopathic effect that is useful in diagnosis.

Epidemiology and Pathology of Parainfluenza

One type of infection by *Paramyxovirus,* called **parainfluenza,** is as widespread as influenza but usually more benign. It is spread by droplets and respiratory secretions that are inhaled or inoculated into the mucous membranes by contaminated hands. Parainfluenzal respiratory disease is seen most frequently in children, most of whom have been infected by the age of 6. Newborns lacking passive antibodies are particularly susceptible, and they develop more severe symptoms. The usual effects of parainfluenza are minor upper respiratory symptoms (a cold), bronchitis, bronchopneumonia, and laryngotracheobronchitis (croup). **Croup*** manifests as labored

* *syncytium* (sin-sish′-yum) Gr. *syn,* together, and *kytos,* cell.
* *croup* (kroop) Scot. *kropan,* to cry aloud.

Figure 25.4 **The effects of paramyxoviruses.** When they infect a host cell, paramyxoviruses induce the cell membranes of adjacent cells to fuse into large multinucleate giant cells, or syncytia. This fusion allows direct passage of viruses from an infected cell to uninfected cells. Through this means, the virus evades antibodies.

and noisy breathing accompanied by a hoarse cough that is most common in infants and young children.

Diagnosis, Treatment, and Prevention of Parainfluenza

Presenting symptoms typical of a cold are often sufficient to presume respiratory infection of viral origin. Determining the actual viral agent, however, is difficult and usually unnecessary in older children and adults whose infection is usually self-limited and benign. Primary infection in infants can be severe enough to be life-threatening. So far no specific chemotherapy is available, but supportive treatment with immune serum globulin or ribavirin (Virazole), a nucleoside analog, can be of benefit.

Mumps: Epidemic Parotitis

Another infection caused by *Paramyxovirus* is **mumps** (Old English for lump or bump). So distinctive are the pathologic features of this disease that Hippocrates clearly characterized it several thousand years ago as a self-limited, mildly epidemic illness associated with painful swelling at the angle of the jaw **(figure 25.5).** Also called **epidemic parotitis,** this infection typically targets the parotid salivary glands, but it is not limited to this region. The mumps virus bears morphological and antigenic characteristics similar to the parainfluenza virus and has only a single serological type.

Epidemiology and Pathology of Mumps Humans are the exclusive natural hosts for the mumps virus. It is communicated

Figure 25.5 **The external appearance of swollen parotid glands in mumps (parotitis).** Usually both sides are affected, though parotitis affecting one side occasionally develops. *Source:* Patricia Smith; Barbara Rice/CDC

primarily through salivary and respiratory secretions. Infection occurs worldwide, with epidemic increases in the late winter and early spring in temperate climates. High rates of infection arise among crowded populations or communities with low levels of immunity. Most cases occur in children under the age of 15, and as many as 40% are subclinical. Because lasting immunity follows any form of mumps infection, no long-term carrier reservoir exists in the population. The incidence of mumps has been reduced in the United States to several hundred cases per year, although a 2009 outbreak was responsible for nearly 1,500 cases, and a similar outbreak in 2006 sickened over 6,500 people. These outbreaks as well as those of measles are associated with a lapse in vaccination of babies and young children.

After an average incubation period of 2 to 3 weeks, symptoms of fever, nasal discharge, muscle pain, and malaise develop. These may be followed by inflammation of the salivary glands (especially the parotids), producing the classic gopherlike swelling of the cheeks on one or both sides that can induce considerable discomfort (see figure 25.5). The virus can multiply in other organs, especially the testes, ovaries, thyroid gland, pancreas, meninges, heart, and kidney. Despite the invasion of multiple organs, the prognosis for most infections is complete, uncomplicated recovery, with permanent immunity.

Complications in Mumps In 20% to 30% of young adult males, mumps infection localizes in the epididymis and testis, usually on one side only. The resultant syndrome of orchitis and epididymitis may be rather painful, but no permanent damage usually occurs. The popular belief that mumps readily causes sterilization of adult males is still held, despite medical evidence to the contrary. Perhaps this notion has been reinforced by the tenderness that continues long after infection and by the partial atrophy of one testis that occurs in about half the cases. Permanent sterility due to mumps is very rare.

Diagnosis, Treatment, and Prevention of Mumps

Mumps can be tentatively diagnosed in a child with swollen parotid glands and a known exposure 2 or 3 weeks previously. Because swelling is not always present and the incubation period can range from 7 to 23 days, a practical diagnostic alternative is to test the serum by a direct fluorescent antibody or ELISA method.

The general pathology of mumps is mild enough that symptomatic treatment to relieve fever, dehydration, and pain is usually adequate. A live, attenuated mumps vaccine (part of the MMR vaccine) is given routinely at 12 to 15 months of age, followed by at least one booster. A separate single vaccine is available for adults who require protection. Although the antibody titer is lower than that produced by wild mumps virus, protection usually lasts at least a decade.

Measles: *Morbillivirus* Infection

Measles, an acute disease caused by *Morbillivirus,* is also known as **red measles*** and **rubeola.*** A somewhat similar disease—rubella or German measles—is caused by an unrelated togavirus (section 25.3). Some differentiating criteria for these two forms are summarized in **table 25.3.** Despite a tendency to think of red measles as a mild childhood illness, several recent outbreaks have reminded us that measles can kill babies and young children and that it is a frequent cause of death worldwide.

Epidemiology of Measles

Measles is one of the most contagious infectious diseases, transmitted principally by respiratory aerosols. There is no reservoir other than humans, and a person is infectious during the periods of incubation, prodromium, and the skin rash but not usually during convalescence. Epidemic spread is favored by crowding, low levels of herd immunity, malnutrition, and inadequate medical care. Several outbreaks of measles have been linked to the lack of immunization in children or the failure of a single dose of vaccine in many children. The importance of vaccinating children cannot be overemphasized, and the latest trends toward fear of vaccines are unfounded. Parents need to be reminded of the dangers of measles and its complications. In the United States, the incidence of measles is sporadic, usually less than 100 cases per year.

Infection, Disease, and Complications of Measles

The measles virus invades the mucosal lining of the respiratory tract during an incubation period of nearly 2 weeks. The initial

* *measles* (mee′-zlz) Dutch *maselen,* spotted.
* *rubeola* (roo-bee′-oh-lah) L. *ruber,* red.

Koplik's spots

(a)

(b)

Figure 25.6 Signs and symptoms of measles. (a) Koplik's spots, named for the physician who described them in the 1800s, are tiny white lesions with a red border that form on the inside of the mouth adjacent to the molars. **(b)** Appearance of measles rash. Individual lesions are flat or slightly raised bumps (maculopapular) distributed over most of the body. (a and b): *Source:* CDC

symptoms are sore throat, dry cough, headache, conjunctivitis, lymphadenitis, and fever. In a short time, unusual oral lesions called **Koplik's spots** appear as a prelude to the characteristic red maculopapular **exanthem*** that erupts on the head and then progresses to the trunk and extremities, until most of the body is covered **(figure 25.6).** The rash gradually coalesces into red patches that fade to brown.

A small number of children develop laryngitis, bronchopneumonia, encephalitis, and bacterial secondary infections such as otitis media and sinusitis. Children afflicted with leukemia or thymic deficiency are especially predisposed to pneumonia because of their lack of the natural T-cell defense. Undernourished children may have severe diarrhea and abdominal discomfort, which add to their debilitation.

* *exanthem* (eg-zan′-thum) Gr. *exanthema,* to bloom or flower. An eruption or rash of the skin.

TABLE 25.3	The Two Forms of Measles					
	Synonyms	**Etiology**	**Primary Patient**	**Complications**	**Skin Rash**	**Koplik's Spots**
Measles	Rubeola, red measles	Paramyxovirus: *Morbillivirus*	Child	SSPE,* pneumonia	Present	Present
German Measles	Rubella, 3-day measles	Togavirus: *Rubivirus*	Child/fetus	Congenital defects**	Present	Absent

*Subacute sclerosing panencephalitis.
**When transmitted in utero.

The most serious complication is **subacute sclerosing panencephalitis (SSPE),*** a progressive neurological degeneration of the cerebral cortex, white matter, and brain stem. Its incidence is approximately one case in a million measles infections, and it afflicts primarily male children and adolescents. The pathogenesis of SSPE appears to involve a defective virus, one that has lost its ability to form a capsid and be released from an infected cell. Instead, it spreads unchecked through the brain by cell fusion, gradually destroys neurons and accessory cells, and breaks down myelin. The disease is known for profound intellectual and neurological impairment. The course of the disease invariably leads to coma and death in a matter of months or years (see this chapter's Case Study).

Diagnosis, Treatment, and Prevention of Measles

The patient's age, a history of recent exposure to measles, and the season of the year are all useful epidemiological clues for diagnosis. Clinical characteristics such as dry cough, sore throat, conjunctivitis, lymphadenopathy, fever, and especially the appearance of Koplik's spots and a rash are presumptive of measles.

Treatment relies on reducing fever, suppressing cough, and replacing lost fluid. Complications require additional remedies to relieve neurological and respiratory symptoms and to sustain nutrient, electrolyte, and fluid levels. Therapy includes antibiotics for bacterial complications and doses of vitamin A for children.

> **Quick Search**
> How did fear of vaccines take a lead role in the measles outbreak of 2015? Search the Internet for "Vaccines: An unhealthy Skepticism" and watch the video.

Vaccination with attenuated viral vaccine administered by subcutaneous injection achieves immunity that persists for about 20 years. It must be stressed that because the virus is live, it can cause an atypical infection sometimes accompanied by a rash and fever. Measles immunization is recommended for all healthy children at the age of 12 to 15 months (MMR vaccine, with mumps and rubella), and a booster is given before a child enters school. A single antigen vaccine (Meruvax) is also available for older patients who require protection against measles alone. As a general rule, anyone who has had the measles is protected against reinfection.

Respiratory Syncytial Virus: RSV Infections

As its name indicates, **respiratory syncytial virus (RSV),** also called *Pneumovirus,* infects the respiratory tract and produces giant multinucleate cells. Outbreaks of droplet-spread RSV disease occur regularly throughout the world, with peak incidence in the winter and early spring. Children 6 months of age or younger are especially susceptible to serious disease of the respiratory tract. Approximately 5 in 1,000 newborns are affected, making RSV the most prevalent cause of respiratory infection in this age group. An estimated 100,000 children are hospitalized with RSV infection each year in the United States. The mortality rate is highest for children who have complications such as prematurity, congenital disease, and immunodeficiency.

The epithelia of the nose and eye are the principal portals of entry, and the nasopharynx is the main site of RSV replication. The first symptoms of primary infection are fever that lasts for 3 days, rhinitis, pharyngitis, and otitis. Infections of the bronchial tree and lung parenchyma give rise to symptoms of croup that include acute bouts of coughing, wheezing, **dyspnea,*** and abnormal breathing sounds (rales). Adults and older children with RSV infection can experience cough and nasal congestion but are frequently asymptomatic.

Diagnosis, Treatment, and Prevention of RSV Diagnosis of RSV infection is more critical in babies than in older children or adults. The afflicted child is conspicuously ill, with signs typical of pneumonia and bronchitis. The preferred diagnostic technique is reverse transcriptase-polymerase chain reaction (RT-PCR), which detects viral RNA.

Some newer treatments include Synagis, a monoclonal antibody that blocks viral attachment to cells, and RSV immunoglobulin obtained from people with high RSV antibody titers. Both can greatly reduce complications and the need for hospitalization. An antiviral drug, ribavirin, can also be administered as an inhaled aerosol. Supportive measures include drugs to reduce fever, providing ventilation, and treatment for secondary bacterial infection if present. An effective vaccine has not yet been developed.

Rhabdoviruses

The most conspicuous **rhabdovirus*** is the **rabies*** virus, genus *Lyssavirus.** The particles of this virus have a distinctive bullet-like appearance, round on one end and flat on the other. Additional features are a helical nucleocapsid and spikes that protrude through the envelope **(figure 25.7).** The family contains approximately

* *dyspnea* (dysp'-nee-ah) Gr. *dyspnoia,* difficulty in breathing.
* *rhabdovirus* (rab'-doh-vy"-rus) Gr. *rhabdos,* rod. In reference to its bullet, or bacillary, form.
* *rabies* (ray'-beez) L. *rabidus,* rage or fury.
* *lyssavirus* (lye'-suh-vy"-rus) Gr. *lyssa,* madness.

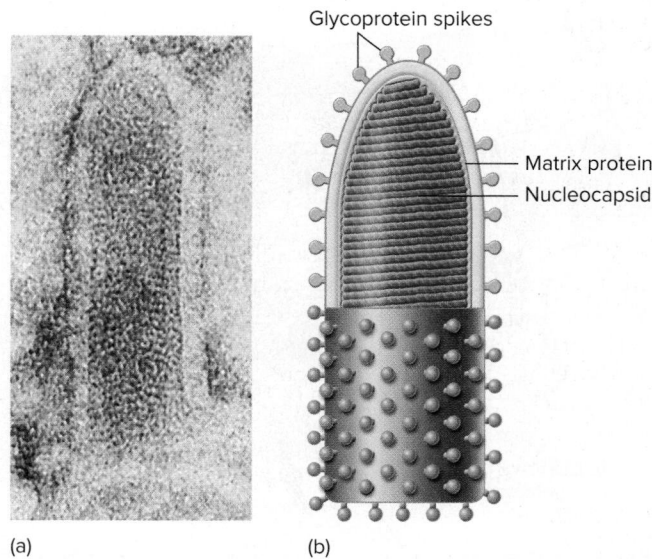

Figure 25.7 The structure of a rhabdovirus. (a) Electron-micrograph of virion shows internal serrations, which represent the tightly coiled nucleocapsid (36,700×). **(b)** A schematic model of the virus, showing its major features. (a): *Source:* CDC

60 different viruses, but only the rabies lyssavirus, and emerging Bas-Congo virus, infect humans.

Epidemiology of Rabies

Rabies is a slow, progressive zoonotic disease characterized by a fatal meningoencephalitis. It is distributed nearly worldwide, except for 34 countries that have remained rabies-free by practicing rigorous animal control. The primary reservoirs of the virus are wild mammals such as canines, skunks, raccoons, cats, and bats. These can spread the infection to domestic cats and dogs. Cats are so important in spreading the virus that most states require vaccination of pet cats as well as dogs. Both wild and domestic mammals can spread the disease to humans through bites, scratches, and inhalation of droplets. The annual worldwide total for human rabies is estimated at about 55,000 cases, but only a tiny number of these cases occur in the United States. Most U.S. cases of rabies occur in wild animals (about 6,000 cases per year), while dog rabies has declined.

The epidemiology of animal rabies in the United States varies with geographic region and vector. The most common wild animal reservoir hosts have changed from foxes to skunks to raccoons. Regional differences in the dominant reservoir also occur. Bats and skunks are the predominant carriers of rabies in much of the country while raccoons dominate in the East. Foxes are a reservoir of the virus in parts of Alaska, Arizona, and Texas **(figure 25.8)**. Review the Case Study in chapter 15 to refresh yourself on other aspects of rabies.

Infection and Disease

Infection with rabies virus typically begins when an infected animal's saliva enters a puncture site. Occasionally, the virus is inhaled or inoculated orally. The virus remains up to a week at the trauma site, where it multiplies. It then gradually enters nerve endings and advances toward the ganglia, spinal cord, and brain. Viral multiplication throughout the brain is eventually followed by

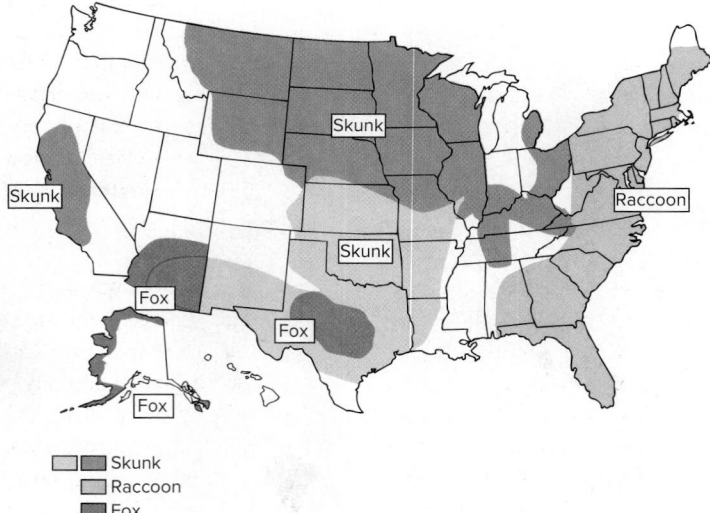

Figure 25.8 **Distribution of rabies in the United States.** Rabies is found in 10 distinct geographic areas. In each area, a particular animal is the reservoir as illustrated by different shades of color. Not shown is the occurrence of insectivorous bats that cause sporadic cases of rabies in wild animals throughout the country.

migration to such diverse sites as the eye, heart, skin, and oral cavity. The infection cycle is completed when the virus replicates in the salivary glands and is shed into the saliva. Clinical rabies proceeds through several distinct stages and ends in death except in rare cases.

Clinical Phases of Rabies

The average incubation period of rabies is 1 to 2 months or more, depending upon the wound site, its severity, and the inoculation dose. The incubation period is shorter in facial, scalp, or neck wounds because of greater proximity to the brain. The prodromal phase begins with fever, nausea, vomiting, headache, fatigue, and other nonspecific symptoms. Some patients continue to experience pain, burning, prickling, or tingling sensations at the wound site.

In the *furious* phase of rabies, the first acute signs of neurological involvement are agitation, disorientation, seizures, and twitching. Spasms in the neck and pharyngeal muscles lead to severe pain upon swallowing. As a result, attempts to swallow or even the sight of liquids brings on *hydrophobia* (fear of water). Throughout this phase, the patient is fully coherent and alert. When rabies enters the *dumb* phase, a patient is not hyperactive but paralyzed, disoriented, and stuporous. Ultimately, both forms, when untreated, progress to the coma phase, resulting in death from cardiac or respiratory arrest. In the past, humans were never known to survive a full-blown case of rabies, but recently 3 patients recovered after receiving intensive, long-term treatment.

Diagnosis and Management of Rabies

When symptoms appear after a rabid animal attack, the disease is readily diagnosed. But the diagnosis can be obscured when contact with an infected animal is not clearly defined or when symptoms are absent or delayed. Anxiety, agitation, and depression can pose as a psychoneurosis; muscle spasms resemble tetanus; and encephalitis with convulsions and paralysis mimics a number of other viral infections. Often the disease is diagnosed at autopsy. Criteria indicative of rabies are intracellular inclusions (Negri bodies) in nervous tissue, isolation of rabies virus from saliva or brain tissue, and demonstration of rabies virus antigens in specimens of the brain, serum, cerebrospinal fluid, or cornea using immunofluorescent methods.

Rabies Prevention and Control A bite from a wild or stray animal demands assessment of the animal, meticulous care of the wound, and a specific treatment regimen. A wild mammal, especially a skunk, raccoon, coyote, or bat that bites without provocation, is presumed to be rabid, and therapy is immediately commenced. If the animal is captured, brain samples and other tissue are examined for verification of rabies. Healthy domestic animals are observed closely for signs of disease and sometimes quarantined. Preventive therapy is initiated if any signs of rabies appear.

After an animal bite, the wound should be scrupulously washed with soap or detergent and water, followed by debridement and application of a povidine-iodine antiseptic, such as Betadine, which can decrease the risk of infection by 90%. Rabies is one of the few infectious diseases for which a combination of passive and active postexposure immunization is indicated and successful. Initially, the wound is infused with human rabies immune globulin (HRIG) to impede the spread of the virus, and globulin is also injected intramuscularly to

provide immediate systemic protection. A full course of vaccination is started simultaneously. The current vaccine of choice is the **human diploid cell vaccine (HDCV).** This potent inactivated vaccine is cultured in human embryonic fibroblasts. The routine in postexposure vaccination entails intramuscular or intradermal injection on the first, third, seventh, and fourteenth days. High-risk groups such as veterinarians, animal handlers, and laboratory personnel should receive three doses to protect against possible exposure.

Control measures such as vaccination of domestic animals, elimination of strays, and strict quarantine practices have helped reduce the virus reservoir. In recent years, the United States and other countries have utilized a live oral vaccine made with a vaccinia virus that carries the gene for the rabies virus surface antigen. The vaccine has been incorporated into bait (sometimes peanut butter sandwiches!) placed in the habitats of wild reservoir species such as skunks and raccoons.

 Check Your Progress SECTION 25.2

8. Describe the structural characteristics common to the paramyxoviruses, rhabdoviruses, and filoviruses.

9. Describe the steps in the production of multinucleate giant cells during a viral infection.

10. Which viruses produce multinucleate giant cells, and what impact does cell fusion have on the spread of infection?

11. Name two examples of *Paramyxovirus* and describe the diseases they cause.

12. How is mumps spread and how is it prevented?

13. Summarize the epidemiology, diagnosis, treatment, and prevention of measles.

14. Describe the epidemiological cycle in rabies.

15. Which animals in the United States are most frequently involved as carriers of rabies?

16. Describe the route of infection and the virus's pathologic effects.

17. Describe the indications for preexposure and postexposure rabies treatment and vaccination.

25.3 Other Enveloped RNA Viruses: Coronaviruses, Togaviruses, and Flaviviruses

 Expected Learning Outcomes

10. Recall the diseases caused by coronaviruses.

11. Understand the epidemiology, diagnosis, and prevention of rubella.

12. Describe the pathology of rubella.

13. Summarize the transmission and pathology of the hepatitis C virus.

Coronaviruses

Coronaviruses[1] are relatively large RNA viruses with distinctive, widely spaced spikes on their envelopes. These viruses are common

1. Named for the resemblance of the viral spikes to a crown.

in domesticated animals and are responsible for epidemic respiratory, enteric, and neurological diseases in pigs, dogs, cats, and poultry. Thus far, five types of human coronaviruses have been characterized. One of these is an etiologic agent of the common cold. The same virus is also thought to cause some forms of viral pneumonia and myocarditis. Another human coronavirus is very closely associated with the intestine and may have a role in some human enteric infections.

Severe Acute Respiratory Syndrome–Associated Coronavirus

Historically, most coronaviruses were considered mildly pathogenic, and the infections tended to be relatively innocuous in healthy persons. In 2002, reports of an acute respiratory illness started to filter in from Asia. The disease was given the name **SARS** for **severe acute respiratory syndrome.** By early 2003 the World Health Organization (WHO) issued a global health alert about the new illness. In a short time, scientists had sequenced the entire genome of the causative agent, revealing it to be a new type of coronavirus.

The epidemic was contained by mid-year, but in less than a year, it had sickened more than 8,000 people. About 9% of those died. Most cases were concentrated in China and Southeast Asia. Several dozen countries, from Australia and Canada to the United States, reported cases, many of which originated in people who had traveled to Asia or had contact with travelers. Close contact (direct or droplet) is involved in its transmission. Eventually, no further cases were reported and the virus had evidently disappeared. The most recent research has linked its origin from fruit bats in Southeast Asia.

Symptoms begin with a fever of above 38°C (100.4°F) and progress to body aches and an overall feeling of malaise. Early in the infection, there seems to be little virus in the patient and a low probability of transmission. Within a week, viral numbers surge and transmissibility is very high. After 3 weeks, viral levels decrease significantly and symptoms subside. Patients may or may not experience classical respiratory symptoms, but severe cases of the illness can result in respiratory distress and death.

Diagnosis of the disease relies first on exclusion of other likely agents, using a Gram stain (for bacterial pneumonia) and identification of influenza and RSV viruses. Acute and convalescent sera should be collected to document a rise in antibodies against the coronavirus. Specimens can be sent to reference labs where PCR will be performed to confirm the diagnosis. There is no specific treatment other than supportive care.

In late 2012 a novel coronavirus caused 130 infections and 58 deaths in patients in the United Kingdom and Arabian peninsula. Christened Middle East respiratory syndrome coronavirus (MERS-CoV), it was, like the virus that causes SARS, similar to coronaviruses found in bats in certain parts of the world. Since 2012, 27 countries have reported a total of 1,806 infections, and the virus has been responsible for 643 deaths. The largest outbreak of the virus outside of the Arabian Peninsula was in the Republic of Korea in 2015, where 185 confirmed cases and 36 deaths were reported. Only two cases of the disease have been seen in the United States, both occurring in healthcare providers who lived and worked in Saudi Arabia. The World Health Organization currently recommends testing for MERS-CoV in patients with unexplained respiratory

illness, especially for persons residing in or traveling from parts of the world known to be affected.

Rubivirus: The Agent of Rubella

Togaviruses are nonsegmented, single-stranded RNA viruses with a loose envelope.[2] There are several important members, including *Rubivirus,* the agent of rubella, and certain arboviruses.[3] **Rubella,** or German measles, was first recognized as a distinct clinical entity in the mid-eighteenth century and was considered a benign childhood disease until its teratogenic effects were discovered. It was first observed that cataracts often developed in neonates born to mothers who had contracted rubella in the first trimester. Epidemiological investigations over the next generation confirmed the link between rubella and many other congenital defects as well.

Epidemiology of Rubella

Rubella is an endemic disease with worldwide distribution. Infection is initiated primarily through contact with respiratory secretions. The virus is shed during the prodromal phase and up to a week after the rash appears. Because the virus is only moderately communicable, close living conditions are required for its spread. Although epidemics and pandemics of rubella once regularly occurred in 6- to 9-year cycles, the introduction of vaccination has essentially stopped this pattern in the United States. Most cases are reported among adolescents and young adults in military training camps, colleges, and summer camps. The greatest concern is that nonimmune women of childbearing age might be caught up in this cycle, raising the prospect of congenital rubella.

Infection and Disease

Two forms of rubella can be distinguished: Postnatal infection develops in children or adults, and congenital (prenatal) infection of the fetus is expressed in the newborn as various types of birth defects.

Postnatal Rubella During an incubation period of 2 to 3 weeks, the virus multiplies in the respiratory epithelium, infiltrates local lymphoid tissue, and enters the bloodstream. Early symptoms include malaise, mild fever, sore throat, and lymphadenopathy. The rash of pink macules and papules first appears on the face and progresses down the trunk and toward the extremities, advancing and resolving in about 3 days **(figure 25.9).** Adult rubella is often accompanied by joint inflammation and pain rather than a rash. Except for an occasional complication, postnatal rubella is generally mild and produces lasting immunity.

Congenital Rubella Because the rubella virus is teratogenic, its transmission to a fetus can result in a serious complication called **congenital rubella.** The mother is able to transmit the virus even if she is asymptomatic. Fetal injury varies according to the time of infection. It is generally accepted that infection in the first trimester is most likely to induce miscarriage or multiple permanent defects

Figure 25.9 One sign of postnatal rubella is a maculopapular rash similar to measles, but less intensely red. *Source:* CDC

in the newborn such as cardiac abnormalities, ocular lesions, deafness, and mental and physical retardation. Less drastic sequelae that usually resolve in time are anemia, hepatitis, pneumonia, carditis, and bone infection.

Diagnosis and Prevention of Rubella

Rubella mimics other diseases and is often asymptomatic, so it should not be diagnosed on clinical grounds alone. The confirmatory methods of choice are serological testing and virus isolation systems. IgM tests can determine recent infection, and a rising titer is a clear indicator of continuing rubella infection. The latex-agglutination card, a simple, miniaturized test, provides a rapid method for IgM assay.

Postnatal rubella is generally benign and requires only symptomatic treatment. Because no specific therapy for rubella is available, most control efforts are directed at maintaining herd immunity with attenuated rubella virus vaccine. This vaccine is usually given to children in the combined form (the MMR vaccination) at 12 to 15 months and a booster at 4 or 6 years of age. One complication regarding this method of protection is that the rubella vaccine does not always provide lasting immunity. Because congenital rubella is such a serious concern, testing of adult women is often recommended to confirm prior infection and protective antibodies. The current recommendation for nonpregnant, antibody-negative women is immediate immunization. Because the vaccine contains live virus and a teratogenic effect is possible, sexually active women must utilize contraception for 4 weeks after the vaccine is administered. Antibody-negative pregnant women should not be vaccinated and must be monitored for rubella infection.

Hepatitis C Virus

Hepatitis C is a type of hepatitis caused by an RNA virus called a **flavivirus.** Hepatitis C is sometimes referred to as the "silent epidemic" because more than 3 million Americans are infected with the virus, but it takes many years to cause noticeable symptoms. In the United States, epidemiologists estimate that at least 30,000 new infections occur every year. Liver failure from hepatitis C is one of the most common reasons for liver transplants in the U.S. See table 24.3 for a complete comparison of the viral hepatitides.

2. Reminiscent of a toga.

3. A group of viruses transmitted by arthropods. See section 25.4.

Transmission and Epidemiology

This virus is acquired in ways similar to hepatitis B (HBV). It is most commonly transmitted through blood contact such as in blood transfusions or needle sharing by injecting drug users. It may occasionally be acquired through transfer of other body fluids or across the placenta. Unlike HBV, sexual contact is not a major mode of transmission for HCV. See the Case Study in chapter 11 to revisit a dramatic HCV outbreak.

Pathogenesis and Virulence Factors

Hepatitis C virus is so adept at establishing chronic infections that researchers are studying the ways that it evades immunologic detection and destruction. The virus's core protein seems to play a role in the suppression of cell-mediated immunity as well as in the production of various cytokines.

> **Quick Search**
> To view a video of the hepatitis C life cycle, search the Internet for "The Life Cycle of the Hepatitis C Virus."

Before a test was available to test blood products for this virus, it seems to have been frequently transmitted through blood transfusions. Hemophiliacs who were treated with clotting factor prior to 1985 were infected at a high rate with HCV. Once blood began to be tested for HIV and screened for so-called "non-A non-B" hepatitis, the risk of contracting HCV from blood was greatly reduced. The current risk for transfusion-associated HCV is thought to be 1 in 2 million units transfused.

Because HCV was not recognized sooner, a relatively large percentage of the population is infected. Eighty percent of the 3.5 million carriers in the United States are suspected to be chronic carriers. It also has a very high prevalence in parts of South America and Central Africa and in China.

Signs and Symptoms

People have widely varying experiences with this infection. It shares characteristics of hepatitis B disease, but it is much more likely to become chronic. Without treatment, 75% to 85% will remain infected for life. (In contrast, only about 6% of persons who acquire hepatitis B after the age of 5 will be chronically infected.) While it is possible to have severe symptoms without permanent liver damage, it is more common to have chronic liver disease without overt symptoms. As the infection progresses, HCV disease may give rise to liver cancer. Infection is usually diagnosed with a blood test for antibodies to the virus.

Prevention, Treatment, and Cure

There is currently no vaccine for hepatitis C due to the extreme antigenic variation it shows. In 2014 the FDA approved a new combination of drugs that for the first time offers a cure for hepatitis C. Ledipasvir and sofosbuvir inhibit two proteins crucial to the replication of the hepatitis C RNA genome. Marketed as a single pill under the trade name Harvoni, a 12 week course of the drug results in a 93-99% cure rate, meaning that no hepatitis C virus is detectable in the blood three months after treatment is completed. For those patients for whom the drug doesn't work, or who cannot afford the $90,000 a 12-week course of therapy typically costs, other treatments are still available. A combination of pegylated interferon and ribivirin can reduce the viral load dramatically. Although not curative, damage to the liver can be prevented or reduced.

 Check Your Progress SECTION 25.3

18. Name the body systems typically infected by coronaviruses.
19. Describe how rubella is spread, diagnosed, and prevented.
20. Compare postnatal rubella with congenital rubella.
21. In what ways is rubella different from red measles?
22. What measures are used to prevent congenital rubella?
23. Outline a typical course of hepatitis C infection, its epidemiology, and its treatment.

25.4 Arboviruses: Viruses Spread by Arthropod Vectors

 Expected Learning Outcomes

14. Understand the relationship between arthropod vectors and the spread of arboviruses.

15. Describe the pathology of arboviral disease.

16. Recall the arboviruses and vectors responsible for encephalitides and hemorrhagic fevers.

Vertebrates are hosts to more than 400 viruses transmitted primarily by arthropods. To simplify presentation, these viruses are often lumped together in a loose grouping called the **arboviruses** (**ar**thropod-**bo**rne **viruses**). The major arboviruses pathogenic to humans belong to the togaviruses (*Alphavirus*), flaviviruses (*Flavivirus*), some bunyaviruses (*Bunyavirus* and *Phlebovirus*), and reoviruses (*Orbivirus*). The chief vectors are blood-sucking arthropods such as mosquitoes, ticks, flies, and gnats. Most types of illness caused by these viruses are mild, undifferentiated fevers, though some cause severe encephalitides (plural of *encephalitis*) and life-threatening hemorrhagic fevers. Although these viruses have been assigned taxonomic names, they are more often known by their common names, which are based on geographic location and primary clinical profile.

Epidemiology of Arbovirus Disease

Because arthropod vectors are found worldwide, so too are the arboviruses they carry. The vectors and viruses tend to be clustered in the tropics and subtropics, but many temperate zones report periodic epidemics **(figure 25.10)**. A given arbovirus type may have very restricted distribution, even to a single isolated region; some range over several continents; and others can spread along with their vectors.

The Influence of the Vector

All aspects of the arbovirus life cycle are closely tied to the ecology of the vectors. Factors that weigh most heavily are the arthropod's

Figure 25.10 **Worldwide distribution of *Aedes aegypti*.** The geographic distribution of arthropod borne viruses mirrors that of the insect vectors that carry them. Distribution of the *Aedes aegypti* mosquito, which serves as a vector for chikungunya virus, dengue virus and Zika virus is indicated on the map. Blue areas correspond to regions free of the mosquito while red areas equate to a 100% chance of encountering the vector. In general the colder northern and southern regions are free of the vector. *Source:* National Center for Biotechnology Information.

life span, the availability of food and breeding sites, and climatic influences such as temperature and humidity. Most arthropod vectors become infected by feeding on the blood of hosts. Infections show a peak incidence from late spring through early fall, when the arthropod is actively feeding and reproducing. Warm-blooded vertebrates also maintain the virus during the cold and dry seasons. Humans can serve as dead-end, accidental hosts, as in equine encephalitis and Colorado tick fever, or they can be a maintenance reservoir, as in dengue fever and yellow fever. Risk of arboviral infection is greatest in wilderness areas where encounters with arthropod vectors are frequent. Clearing remote forest habitats for human colonization greatly increases this form of contact.

Arboviral diseases have a great impact on humans. Although exact statistics are unavailable, there is a consensus that millions of people acquire infections each year and thousands of them die. The uncertain nature of vector and viral cycles frequently results in sudden, unexpected epidemics, sometimes with previously unreported viruses. Travelers and military personnel entering endemic areas are at special risk because, unlike the natives of that region, they have no immunity to the viruses.

General Characteristics of Arbovirus Infections

Our coverage of the arboviruses will be focused on these main areas:

1. the manifestations of disease in humans;
2. aspects of the most important North American viruses; and
3. methods of diagnosis, treatment, and control.

Febrile Illness and Encephalitis

One type of disease elicited by arboviruses is an acute, undifferentiated fever, often accompanied by rash. These infections, typified by dengue fever and Colorado tick fever, are usually mild and leave no long-term effects. Prominent symptoms are fever, prostration, headache, myalgia, orbital pain, muscle aches, and joint stiffness. About midway through the illness, a maculopapular or petechial rash can erupt over the trunk and limbs.

When the brain, meninges, and spinal cord are invaded, symptoms of viral encephalitis occur. The viruses cycle between wild animals (primarily birds) and mosquitoes or ticks, but humans are not usually reservoir hosts. The disease begins with a bite, which releases the virus into the tissues and nearby lymphatics. Prolonged viremia establishes the virus in the nervous system, where inflammation can cause swelling and damage to the brain, nerves, and meninges. Symptoms are extremely variable and can include coma, convulsions, paralysis, tremor, loss of coordination, memory deficits, changes in speech and personality, and heart disorders. In a small number of cases, there will be some degree of permanent brain damage. Young children and the elderly are most sensitive to injury by arboviruses.

West Nile virus is the most prevalent cause of arboviral disease in the United States. Mosquitoes generally become infected when they feed on birds infected with the virus. Infected mosquitoes can then bite and transmit the virus to humans. Because 80% of people infected show no symptoms, the 2,060 cases reported to the CDC in 2015 most likely represented only a small fraction of the actual number of cases. For the 1 in 5 people who do suffer symptoms, fever, headache, body aches, joint pains, vomiting, diarrhea, or rash can be expected. Recovery from this less severe type of West Nile virus disease is typically complete, but fatigue and weakness can last for weeks or months. Fewer than 1% of people who are infected will develop neurological illness such as encephalitis or meningitis, and people over 60 or with certain underlying medical conditions (cancer, diabetes, hypertension, kidney disease or organ transplant recipients) are at greatest risk for serious illness. Approximately 10% of the people who develop neurologic infection with West Nile virus will die.

Colorado tick fever (*CTF*) is the most common tick-borne viral fever in the United States. Restricted in its distribution to the Rocky Mountain states, it occurs sporadically during the spring and summer, corresponding to the time of greatest tick activity and human contacts. Two hundred to three hundred endemic cases are reported each year, and deaths are rare.

Eastern equine encephalitis (*EEE*) is endemic to an area along the eastern coast of North America and Canada. The usual pattern is sporadic cases, but occasional epidemics can occur in humans and horses. High periods of rainfall in the late summer increase the chance of an outbreak, and disease usually appears first in horses and caged birds. The case fatality rate is among the highest (35%) for all encephalitides.

LaCrosse encephalitis virus (LACV) is responsible for roughly 100 cases of viral encephalitis a year, primarily in the Midwest and across the Appalachian mountains. Severe disease—which is rare, most people show no symptoms of infection—occurs most often in children under 16 years of age.

St. Louis encephalitis (*SLE*) appears sporadically, primarily in the Eastern and Central states. Silent infection is very common, and the total number of cases is probably thousands of times greater than the 5 to 10 reported each year. The seasons of peak activity are spring and summer, depending on the region and species of mosquito.

Chikungunya virus has garnered more attention in the United States since infections were seen on Caribbean islands in 2013, the first time it had been detected in the Americas. In contrast to many arboviral diseases, the majority of people infected with chikungunya

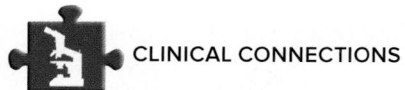

CLINICAL CONNECTIONS

Three Viruses, One Vector

Recently, increasing attention has been paid to three formerly obscure viral diseases that have become more prevalent, especially in the Americas. Dengue fever virus is actually a collection of four similar viruses that cause over 100 million cases of disease each year. the milder form of the disease is marked by high fever, severe headache, muscle, bone and joint pain, and mild bleeding from the nose or gums. Severe cases, termed dengue hemorrhagic fever, includes most of the above symptoms plus a period of excessive capillary permeability, in which the fluid component of the blood leaks from the blood vessels, potentially leading to failure of the circulatory system, shock and death. The symptomatology of chikungunya is similar to mild dengue, with headache, muscle and joint pain, lasting a week or so, although the joint pain may persist for months.

Zika virus, which had been seen previously in Africa, Southeast Asia, and some Pacific Islands, achieved notoriety when an outbreak in Brazil was thought to be linked to cases of Guillain-Barré syndrome and an increase in birth defects, especially microcephaly. After a months-long examination of affected infants, the CDC confirmed that the virus can cause abnormal brain development. They issued an alert advising pregnant women to postpone travel to any region where Zika virus was currently being transmitted. A more general alert advised everyone traveling to these regions to take extra precautions—to wear long-sleeved shirts and long pants, use mosquito repellent, and stay indoors as much as possible—to prevent mosquito bites. The virus was also found to be present in the semen of infected men and several cases were identified in which women were infected with the virus after having unprotected sex with an infected man. In response, the CDC issued guidelines meant to reduce sexual transmission of the virus, especially when the woman is pregnant. Although some cases result in the serious complications mentioned above, usually an infection with Zika virus causes only a mild illness similar to dengue fever or chikungunya.

Aedes mosquitoes serve as biological vectors for chikungunya, dengue, and Zika viruses. In this photo of an *Aedes aegypti* mosquito, both the fascicle that the mosquito uses to feed, and the insect's abdomen are filled with blood from a recent meal and appear red. *Source:* James Gathany/CDC

The common denominator of all three viruses is the *Aedes* mosquito, an aggressive, daytime feeder that is found throughout the Americas and is particularly adept at surviving in a hostile environment. Its eggs can remain viable for several months without water, meaning that even if all active mosquitoes in an area were killed, the population could completely recover in as little as 2 weeks after the addition of water to dried eggs (rainfall, for example). According to the World Health Organization, it may only be a matter of time before these viruses, like West Nile virus a decade before, become endemic to the United States.

Unlike the *Anopheles* mosquito that spreads malaria, the *Aedes* mosquito is most active during the day. What protective strategy used against the *Anopheles* mosquito is ineffective against *Aedes* mosquitoes? *Answer available on Connect.*

virus become symptomatic, suffering fever, headache, muscle and joint pain, nausea and vomiting, and occasionally rash, clinical features very similar to dengue fever. Most symptoms abate in 7 to 10 days and mortality is rare.

Infection with **Zika virus** produces symptoms very similar to infection with chikungunya or dengue viruses, and the three viruses are carried by the same biological vector (see Clinical Connections). Zika concerns were amplified when a 2015 outbreak in Brazil led to reports of viral infection being linked to the autoimmune disease Guillain-Barré syndrome, as well as birth defects (most notably microcephaly) in the children of women who were infected while pregnant. In 2016 the CDC recommended that pregnant women not travel to areas where Zika virus transmission was ongoing, a geographic distinction that included some parts of Florida. The Zika virus was also found to be transmissible through sexual intercourse, and abstinence or the use of a condom is recommended if the partner of a pregnant woman has recently

traveled to at-risk areas. Clinical trials of a Zika vaccine are currently underway.

Hemorrhagic Fevers

Certain arboviruses, principally the yellow fever and dengue fever viruses, cause capillary fragility and disrupt the blood clotting system. These effects lead to localized bleeding, fever, and shock, which explains the common name for the disease. Hemorrhagic fevers are caused by a variety of viruses, are carried by a variety of vectors, and are distributed globally. Reservoir animals are usually small mammals, although yellow fever and dengue fever can be harbored in the human population.

Until Zika virus took center stage in 2015, the best-known arboviral disease has historically been **yellow fever.** Although this disease was at one time widespread in its distribution, mosquito control measures have eliminated it in many countries, including

the United States. Two patterns of transmission occur in nature. One is an urban cycle between humans and the mosquito *Aedes aegypti,* which reproduces in standing water in cities. The other is a sylvan cycle, maintained between forest monkeys and mosquitoes. Most cases in the Western Hemisphere occur in the jungles of Brazil, Peru, and Colombia during the rainy season. Infection begins acutely with fever, headache, and muscle pain. In some patients the disease progresses to oral hemorrhage, nosebleed, vomiting, jaundice, and liver and kidney damage, with significant mortality rates.

Dengue* fever is caused by a flavivirus and is also carried by the *Aedes* mosquitoes. Although mild infection is the usual pattern, a form called dengue hemorrhagic shock syndrome can be lethal. This disease is also called "breakbone fever" because of the severe pain it induces in the muscles and joints. The illness is endemic to Southeast Asia and India, and several epidemics have occurred in South and Central America, the Caribbean, and Mexico. It is a serious risk factor for about 40% of the world's population. The WHO has estimated that more than 100 million cases occur every year, with about 500,000 hemorrhagic fever cases. It is thought that climate change, including heavy rains and warmer weather that favor the mosquito life cycle, is a contributing factor in several major epidemics across Asia.

Quick Search

To see how many cases of West Nile disease have been reported in your neighborhood, search the Internet for "USGS disease maps viewer" to get county-specific information on arboviral infections.

Concern over the possible spread of the disease in the United States led the CDC to add dengue fever to the notifiable diseases list and to survey mosquito populations in states along the southern border. Several states, primarily in the southeastern United States, harbor the *Aedes* mosquitoes that serve as vectors for dengue virus. Contact with infected human carriers could establish the virus in these mosquito populations. About 200 locally acquired cases per year are diagnosed in the United States, mostly in Puerto Rico, Texas, Hawaii, and Florida.

Diagnosis, Treatment, and Control of Arbovirus Infections

Except during epidemics, detecting arboviral infections can be difficult. The patient's history of travel or contact with vectors, along with serum analysis, is highly supportive of a diagnosis. In most cases specific diagnosis of arbovirus infection is based on detection of arboviral RNA in the blood or serological detection of antibodies against specific viral proteins. Treatment of the various arbovirus diseases relies completely on support measures to control fever, convulsions, dehydration, shock, and edema.

The most reliable arbovirus vaccine is a live, attenuated yellow fever vaccine that provides relatively long-lasting immunity. Vaccination is a requirement for travelers in the tropics and can be administered to entire populations during epidemics. A dengue fever vaccine was approved for use in Mexico, the Philippines and Brazil in late 2015. The live virus vaccine reduced the chance of developing dengue by 60% but is only approved for use in people between 9 and 45 years of age who live in dengue endemic areas.

** dengue* (den′-gha) A corruption of Sp. "dandy" fever. Other synonyms are dengue shock syndrome and stiffneck fever.

 Check Your Progress SECTION 25.4

24. Name several activities that increase the risk of acquiring an arbovirus infection.
25. Summarize the signs and symptoms of encephalitis and hemorrhagic fever.
26. Specify the arboviruses responsible for encephalitides and hemorrhagic fevers and their proper vectors.
27. How is the cycle of the viruses maintained in the wild?
28. Describe the symptoms of the encephalitis type of infection (EEE, SLE) and the hemorrhagic fevers (yellow fever).
29. Discuss the factors in the worldwide spread of arboviral diseases.

25.5 Retroviruses and Human Diseases

 Expected Learning Outcomes

17. Describe the main characteristics of retroviruses that make them unique.
18. Identify the causative agent of acquired immunodeficiency syndrome.
19. Describe the structural features of the human immunodeficiency virus and the host cell membrane that are required for infection to occur.
20. Discuss the routes of HIV transmission.
21. Understand the progression of HIV infection.
22. Describe how testing for HIV infection is accomplished.
23. Recall how AIDS is defined.
24. Understand why an HIV vaccine has not been developed.
25. Describe the mode of action of each class of retroviral agent used to treat infection with HIV.
26. Understand the purpose of using combination therapy to treat HIV infection.
27. Describe the diseases associated with HTLV-I.

HIV Infection and AIDS

The sudden emergence of AIDS in the early 1980s focused an enormous amount of public attention, research studies, and financial resources on the virus and its disease.

The first cases of AIDS were seen by physicians in Los Angeles, San Francisco, and New York City. They observed clusters of young male patients with one or more of a complex of symptoms: severe pneumonia caused by *Pneumocystis jirovecii* (ordinarily a harmless fungus), a rare vascular cancer called Kaposi sarcoma, sudden weight loss, swollen lymph nodes; and general loss of immune function. Eventually, virologists at the Pasteur Institute in France isolated a novel retrovirus, later named the **human immunodeficiency virus (HIV).** The cluster of symptoms resulting from the degradation of the immune system by HIV was termed **acquired immunodeficiency syndrome,** or **AIDS,** by the medical community.

Characteristics of Human Retroviruses

HIV is a **retrovirus,** in the genus *Lentivirus.* Most retroviruses have the potential to cause cancer, produce dire, often fatal diseases, and

alter the host's DNA in profound ways. They are named "retroviruses" because they reverse the usual order of transcription. They contain an enzyme called **reverse transcriptase (RT)** that catalyzes the replication of double-stranded DNA from single-stranded RNA. The association of retroviruses with their hosts can be so intimate that viral genes are permanently integrated into the host genome. Not only can this retroviral DNA be incorporated into the host genome as a provirus that can be passed on to progeny cells, but some retroviruses also transform cells and regulate certain host genes.

The most prominent human retroviruses are HIV and the T-cell lymphotropic viruses I and II (HTLV-I and HTLV-II). The two types of HIV are HIV-1, which is the dominant form in most of the world, and HIV-2, found mainly in parts of Africa. HTLV type I is associated with leukemia and lymphoma (discussed in a later section).

HIV and other retroviruses display structural features typical of enveloped RNA viruses **(figure 25.11)**. The outermost component is a lipid envelope with transmembrane glycoprotein spikes and knobs that mediate viral adsorption and fusion to the host cell. HIV can only infect host cells that present the required receptors, which is a combination receptor consisting of the CD4 marker plus a co-receptor. The virus uses these receptors to gain entrance to several types of leukocytes and tissue cells **(figure 25.11b)**.

Epidemiology of HIV Infection

One important question about HIV—*Where did it come from?*—seems to have been answered. Researchers have been comparing the genetics of HIV with various African monkey viruses called simian immunodeficiency viruses, or SIVs. The genetic sequences in these viruses led researchers to conclude that HIV is a hybrid virus that has evolved from two separate monkey SIVs. One of the SIVs has as its natural host the greater spot-nosed monkey, and the other infects red-capped mangabeys. Apparently, when these two monkey viruses infected chimpanzees, a third type of virus emerged that contained genetic sequences from both SIVs. This new type of SIV was probably transmitted to humans, possibly when they used the chimps for food. The crossover into humans probably occurred in the early part of the 1900s; the earliest record we have of human infection is a blood sample preserved from an African man who died in 1959.

HIV probably remained in small isolated villages, causing sporadic cases and mutating into more virulent strains that were readily transmitted from human to human. When this pattern was combined with changing social and sexual practices and increased immigration and travel, a pathway was opened up for rapid spread of the virus pandemically.

Modes of Transmission

HIV transmission occurs mainly through two forms of contact: sexual intercourse and transfer of blood or blood products. Babies

Figure 25.11 The general structure of HIV. (a) The envelope contains two types of glycoprotein (GP) spikes; the capsid contains two identical RNA strands and several molecules of reverse transcriptase, protease, and integrase. **(b)** The snug attachment of HIV glycoprotein molecules (GP-41 and GP-120) to their specific receptors on a human cell membrane. These receptors are CD4 and a co-receptor, which control docking with the host cell and fusion with its cell membrane. **(c)** Electron micrograph of HIV (yellow) infecting a T cell (blue). *Source:* Seth Pincus, Elizabeth Fischer and Austin Athman, National Institute of Allergy and Infectious Diseases, National Institutes of Health

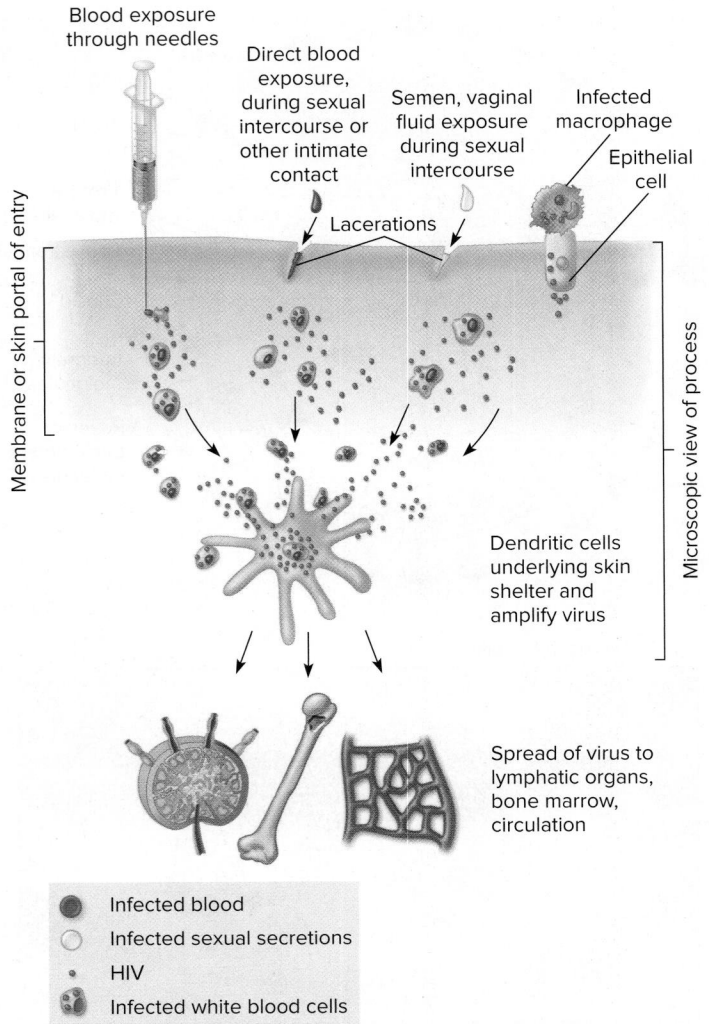

Figure 25.12 Primary sources and suggested routes of infection by HIV.

can also be infected before or during birth, as well as through breast feeding. The mode of transmission is similar to that of hepatitis B virus, except that the HIV does not survive for as long outside the host and is far more sensitive to heat and disinfectants. And HIV is not transmitted through saliva, as hepatitis B can be.

In general, HIV is spread only by direct and rather specific routes **(figure 25.12).** Because the blood of HIV-infected people harbors high levels of free virus in early and late stages of infection and high levels of infected leukocytes throughout infection, any form of intimate contact involving transfer of blood (trauma, needle sharing) can be a potential source of infection. Semen and vaginal secretions also harbor free virus and infected white blood cells, which accounts for another source in sexual transmission. The virus can be isolated from urine, tears, sweat, and saliva but in such small numbers (less than 1 virion per cubic millimeter) that these fluids are not considered sources of infection.

AIDS Morbidity

AIDS first became a notifiable disease at the national level in 1984, and it has continued in an epidemic pattern, although the number of people developing the disease in the United States has decreased since 1994. This situation is due to the advent of effective therapies that prevent the progression of HIV infection to full-blown AIDS. But in developing countries, which are hardest hit by the HIV epidemic, access to these lifesaving drugs has been limited. And even in the United States, despite treatment advances, HIV infection and AIDS are the eighth most common cause of death among people ages 25 to 34, although they have fallen out of the top 10 list for overall causes of death. Unfortunately, new HIV infections are remaining at a steady rate of 35,000 to 50,000 per year.

Figure 25.13 contrasts the number of new infections in the United States, broken down by state of residence, sex, and behavior.

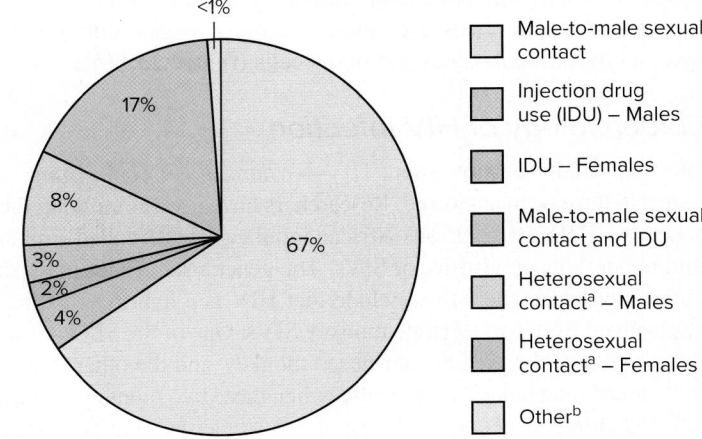

(a) **Diagnosis of HIV Infection Among Adults and Adolescents, by Transmission Category, 2014**

- Male-to-male sexual contact
- Injection drug use (IDU) – Males
- IDU – Females
- Male-to-male sexual contact and IDU
- Heterosexual contact[a] – Males
- Heterosexual contact[a] – Females
- Other[b]

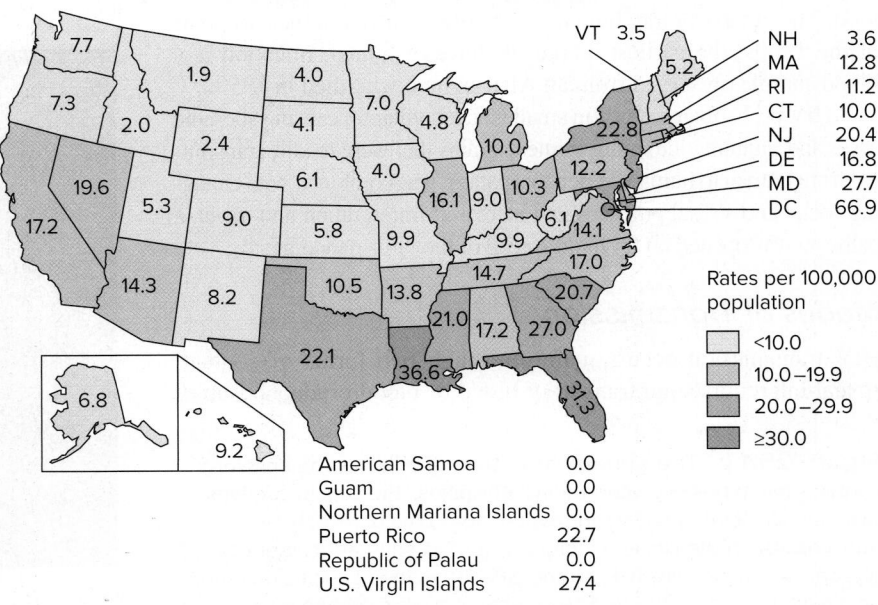

(b) **Rates of HIV Infection Among Adults and Adolescents, 2014**

NH	3.6
MA	12.8
RI	11.2
CT	10.0
NJ	20.4
DE	16.8
MD	27.7
DC	66.9

Rates per 100,000 population

- <10.0
- 10.0–19.9
- 20.0–29.9
- ≥30.0

American Samoa	0.0
Guam	0.0
Northern Mariana Islands	0.0
Puerto Rico	22.7
Republic of Palau	0.0
U.S. Virgin Islands	27.4

Figure 25.13 Patterns of HIV infections. (a) Percentage of new infections by mode of transmission. **(b)** Incidence of new infections by state. *Source of Data:* Centers For Disease Control

Men account for 81% of new infections. Sixty-seven percent of all new infections are acquired through male homosexual activity. Men having sex with other men (a group labeled MSM) have increased susceptibility because of the practice of anal sex, which is known to lacerate the rectal mucosa and can provide an entrance for viruses from semen into the blood. The receptive partner (whether male or female) is the more likely of the two to become infected. In addition, bisexual men are a significant factor in spreading the virus to women.

Especially in large metropolitan areas, many intravenous drug users (IDUs) are also HIV carriers. Infection from contaminated needles continues to grow rapidly, and it is another significant factor in the spread of HIV to the heterosexual population.

HIV infection and AIDS have been reported in every country. Parts of Africa and Asia have the highest case rates. In 2014 the WHO estimated that 37 million people were living with HIV or AIDS worldwide. Even with improved treatments, the disease continues to have devastating repercussions. Reports from the CDC estimate approximately 1.2 million people in the United States living with HIV/AIDS. A number of these people may be in the latent phase of the disease and not yet aware of their infection.

In most parts of the world heterosexual intercourse is the primary mode of transmission. In the industrialized world, the overall rate of heterosexual infection has increased dramatically in the past several years, especially in adolescent and young adult women. In the United States, about 25% of HIV infections arise from unprotected sexual intercourse with an infected partner of the opposite sex.

Now that donated blood is routinely tested for antibodies to HIV, transfusions are no longer considered a serious risk. Because there can be a lag period of a few weeks to several months before antibodies appear in an infected person, it is remotely possible to be infected through donated blood. Rarely, organ transplants can carry HIV, so they too are tested. Other blood products (serum, coagulation factors) were once implicated in HIV infection. Thousands of hemophiliacs died from AIDS in the 1980s and 1990s. It is now standard practice to heat-treat any therapeutic blood products to destroy all viruses.

A small percentage of HIV infections (around 1%) occur in people without apparent risk factors. This does not mean that some other unknown route of spread exists. Factors such as patient denial, unavailability of history, death, or uncooperativeness make it impossible to explain every case. We should note that not everyone is susceptible to HIV infection, and not everyone who becomes infected or is antibody-positive develops AIDS. In the case of natural resistance, HIV researchers have discovered people with a mutated coreceptor (primarily CCR5) that is used by the virus to fuse with its host cell. About 1% of the population has inherited a pair of these mutated genes, which when expressed, give rise to abnormal receptors that keep HIV from entering cells. This confers a greatly reduced susceptibiity to HIV infection. Researchers are focusing on this knowledge as a basis for developing new types of anti-HIV drugs. About 5% of people who are antibody-positive remain free of disease, indicating that some people can develop immunity or are less susceptible to infection.

Treatment of HIV-infected mothers with combination antiretroviral therapy (cART) has dramatically decreased the rate of maternal to infant transmission during pregnancy. Current estimates indicate that fewer than 200 HIV-infected infants are born each year in the United States. The cost of perinatal prevention strategies ($1,000 per pregnancy) and the scarcity of medical counseling in underserved areas have led to an increase in maternal transmission of HIV in developing countries, at the same time that the developed world has seen a marked decrease.

Medical and dental personnel are not considered a high-risk group. A relatively small number of medical and dental workers have acquired HIV or become antibody-positive as a result of clinical accidents. A health care worker involved in an accident in which gross inoculation with contaminated blood occurs (as in the case of a needlestick) has a less than 1 in 1,000 chance of becoming infected. Experts emphasize that transmission of HIV will not occur through casual contact or routine patient care procedures and that universal precautions for infection control (see chapter 13) provide protection for both worker and patient.

Some Harsh Statistics

A powerful way to drive home the staggering effect of HIV and AIDS is to examine some of its statistics in everyday terms (data generated by WHO and CDC up to 2016):

- Globally, 37 million people are infected with HIV or have AIDS; 2.6 million are children.
- In 2015, 1.1 million people died of AIDS.
- Each day, almost 5,800 people are newly infected.

Parts of Africa are harder hit than other regions. Consider the following:

- 70% of cases worldwide occur in sub-Saharan Africa.
- One in 20 adults is infected with HIV.
- Fourteen million African children are orphaned by AIDS.

The United States has its own disturbing set of statistics:

- Over 1.2 million people are currently infected with HIV.
- Nearly 45,000 new HIV infections are reported each year.
- African Americans represent 12% of the U.S. population but account for 44% of new HIV infections.
- Almost 7,000 people diagnosed with AIDS die each year.
- HIV infection is the eighth leading cause of death in people 25 to 34 years old, and the fifth leading cause of death for African American men ages 35 to 54.

Pathogenesis and Virulence Factors of HIV

As summarized in figure 25.12, HIV enters a mucous membrane or the skin and is phagocytosed by a dendritic cell. In the dendritic cell, the virus grows and is shed from the cell without killing it. New viruses are taken up and amplified by macrophages in the skin, lymph organs, bone marrow, and blood. One of the great ironies of HIV is that it infects and destroys many of the very cells needed to combat it, especially the helper (CD4 or T4) class of lymphocytes. It also infects monocytes, dendritic cells, and macrophages.

Once the virus is inside a target cell, its reverse transcriptase converts its RNA into DNA. Although initially many viruses produce a lytic infection, in some cells the DNA becomes inactive in the nucleus of the host cell and its DNA becomes integrated into host DNA **(figure 25.14).** This event accounts for the lengthy course of the disease. Because different host cells are in different stages of infection, some host cells are releasing new viruses and being lysed, and new T cells are constantly being infected.

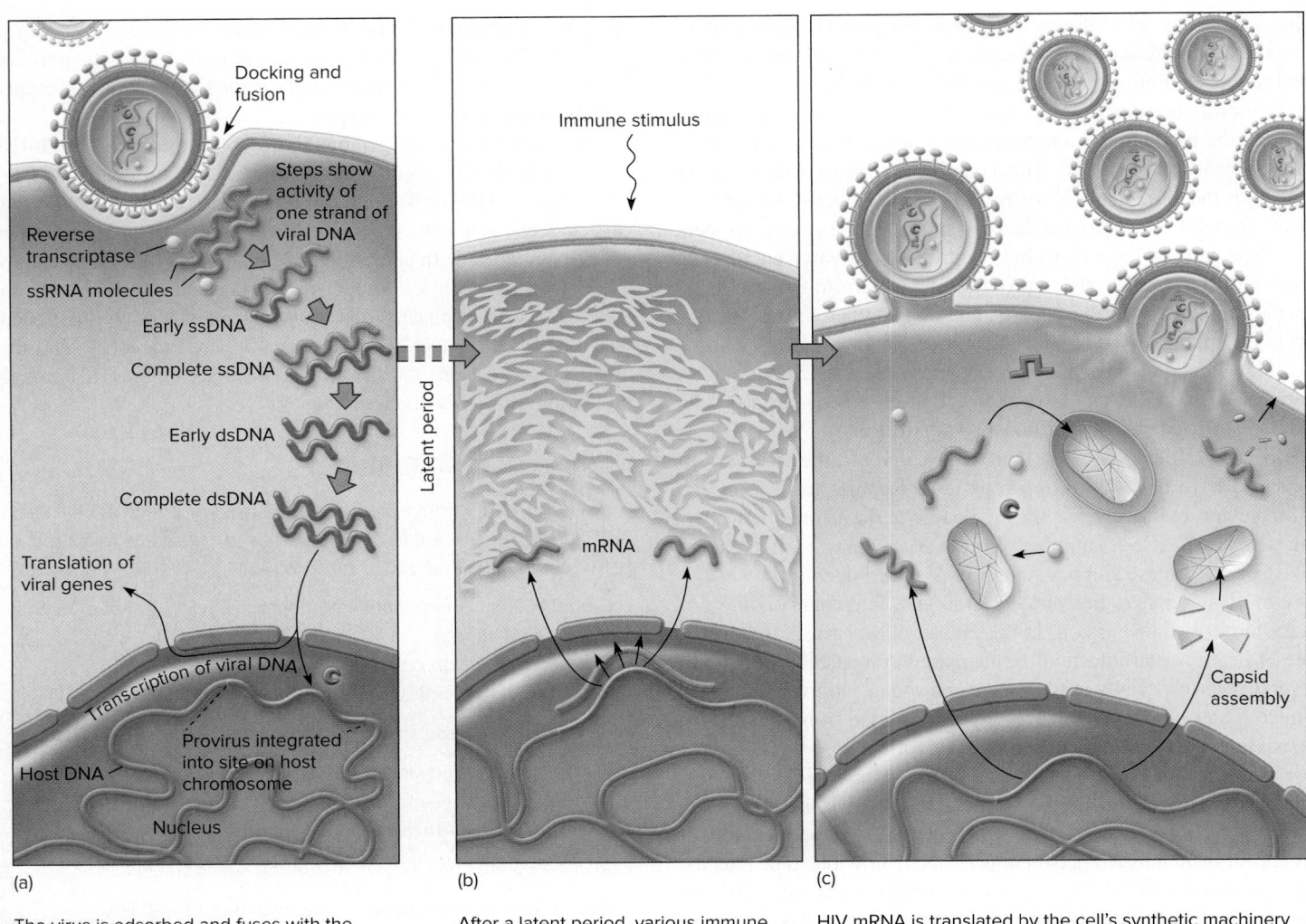

(a)

(b)

(c)

The virus is adsorbed and fuses with the cell. The twin RNAs are uncoated. Reverse transcriptase catalyzes the synthesis of a single complementary strand of DNA (ssDNA). This single strand serves as a template for synthesis of a double strand (ds) of DNA. In latency, dsDNA is inserted into the host chromosome as a provirus.

After a latent period, various immune activators stimulate the infected cell, causing reactivation of the provirus genes and production of viral mRNA.

HIV mRNA is translated by the cell's synthetic machinery into virus components (capsid, reverse transcriptase, spikes), and the viruses are assembled. Budding of mature viruses lyses the infected cell.

Figure 25.14 The general multiplication cycle of HIV.

The primary effects due directly to HIV infection are extreme leukopenia, with lowered levels of lymphocytes in particular. Both T cells and monocytes undergo extensive die-offs through programmed cell suicide (apoptosis). The CD4 memory clones and stem cells are among the prime targets. HIV also causes formation of giant syncytial T cells, which allow the spread of viruses directly from cell to cell, followed by mass destruction of the syncytia. The central nervous system is affected when infected macrophages cross the blood-brain barrier and spread HIV into brain cells. It appears that some of the viral envelope proteins have a direct toxic effect on the brain's glial cells and other cells. Research has shown that some peripheral nerves become demyelinated and the brain becomes inflamed.

The secondary effects of HIV infection are the opportunistic infections and malignancies associated with destruction of essential CD4 functions needed to control pathogens. These are summarized in **25.2 Making Connections.**

Stages, Signs, and Symptoms of HIV Infection and AIDS

The clinical spectrum of HIV infection ranges from acute early symptoms to the end stage symptoms of AIDS. To understand the progression, closely follow **figures 25.15** and **25.16.** Pathology in HIV infection is directly tied to two factors: (1) the level of viruses, and (2) the level of T cells in the blood. Figure 25.16 presents relative amounts of viruses and T cells over the course of the infection and disease. Note that the figure is showing the pattern of HIV infection in the absence of medical intervention or chemotherapy.

The initial infection is acute and often attended by vague, mononucleosis-like symptoms that soon disappear. This phase is marked by high levels of free virus in the blood (figure 25.16, phase I), followed by a rapid drop (figure 25.16, phase II). Observe that the antibody levels rise (figure 25.16, phases II and III) at the same

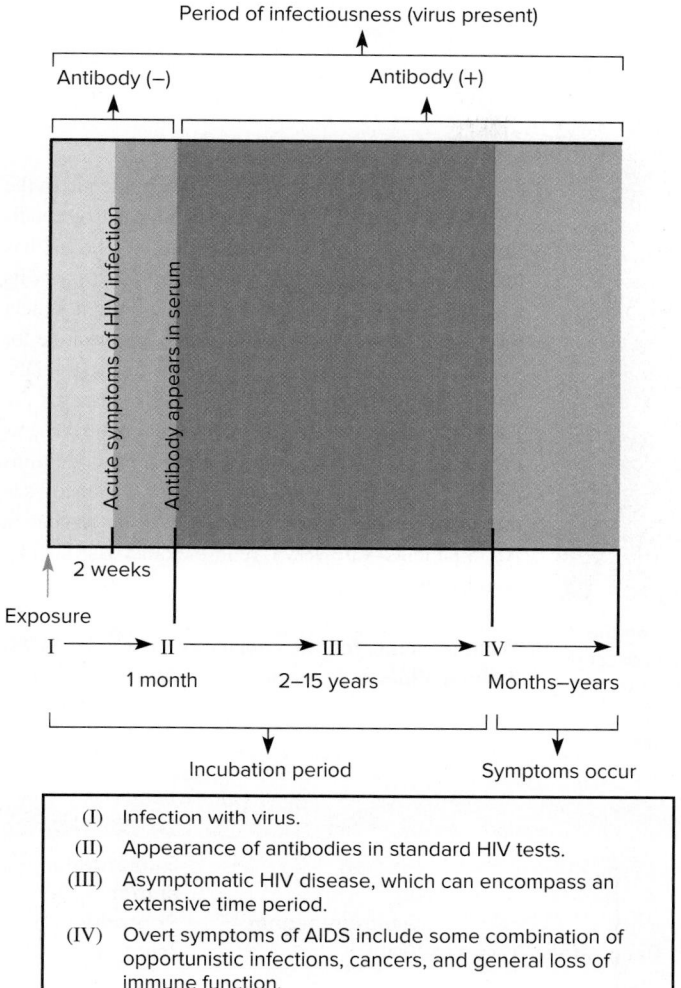

Figure 25.15 Time line in HIV infection, asymptomatic HIV disease, and AIDS.

In the figure, the following labels appear:

Period of infectiousness (virus present)

Antibody (−) Antibody (+)

Acute symptoms of HIV infection
Antibody appears in serum

2 weeks

Exposure

I → II → III → IV →

1 month 2–15 years Months–years

Incubation period Symptoms occur

(I) Infection with virus.
(II) Appearance of antibodies in standard HIV tests.
(III) Asymptomatic HIV disease, which can encompass an extensive time period.
(IV) Overt symptoms of AIDS include some combination of opportunistic infections, cancers, and general loss of immune function.

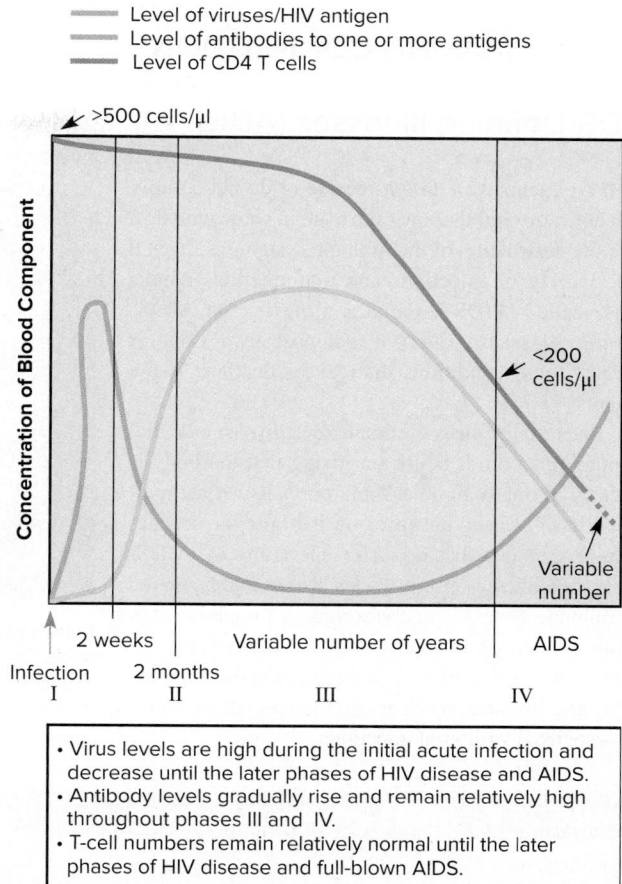

Level of viruses/HIV antigen
Level of antibodies to one or more antigens
Level of CD4 T cells

>500 cells/μl

Concentration of Blood Component

<200 cells/μl

Variable number

2 weeks Variable number of years AIDS

Infection 2 months

I II III IV

- Virus levels are high during the initial acute infection and decrease until the later phases of HIV disease and AIDS.
- Antibody levels gradually rise and remain relatively high throughout phases III and IV.
- T-cell numbers remain relatively normal until the later phases of HIV disease and full-blown AIDS.

Figure 25.16 Changes in the quantity of virus antigen, antibody, and T cells in circulation over time.

time that the virus load is dropping. The antibodies are responsible for neutralizing the free viruses in circulation during this stage. One feature of the ongoing HIV infection is a period of mostly asymptomatic disease (sometimes called latency) that varies in length from 2 to 15 years, with the average being about 10 years (figure 25.15, phase III). Another important occurrence during the mid-to-late asymptomatic periods is how the number of T cells in the blood steadily decreases (figure 25.16, phase III). This is critical to the progression of the disease as T helper cells are responsible for regulating B-cell antibody production and macrophage stimulation. As the number of T helper cells declines, so too does the efficiency of the macrophage response and the ability to produce antibodies. Once the CD4 cell levels fall below 200 cells/mm³ (μl) of blood, symptoms of AIDS appear (figure 25.16, phase IV).

Initial symptoms of AIDS may be fatigue, diarrhea, weight loss, and neurological changes, but most patients first notice this phase of infection because of one or more opportunistic infections or neoplasms. These **AIDS-defining illnesses** are detailed in 25.2 Making Connections. Other disease-related symptoms appear to accompany severe immune deregulation, hormone imbalances, and metabolic disturbances. Pronounced wasting of body mass is a consequence of weight loss, diarrhea, and poor nutrient absorption. Protracted fever, fatigue, sore throat, and night sweats are significant and debilitating. Both a rash and generalized lymphadenopathy in multiple lymph nodes are presenting symptoms in many AIDS patients.

Some of the most virulent complications are neurological. Lesions occur in the brain, meninges, spinal column, and peripheral nerves. Patients with nervous system involvement show some degree of withdrawal, persistent memory loss, spasticity, sensory loss, and progressive AIDS dementia.

Diagnosis of HIV Infection

A person is diagnosed as having HIV infection if he or she has tested positive on a series of tests for the human immunodeficiency virus. This diagnosis is not the same as having AIDS.

Most viral testing is based on detection of antibodies specific to the virus in serum or other fluids, which allows for the rapid, inexpensive screening of large numbers of samples. Testing usually proceeds at two levels. The initial screening tests include the older ELISA and newer latex agglutination and rapid antibody tests. The advantage of these tests is that results are available within minutes instead of the days to weeks previously required for the return of results from ELISA testing. Several kits have been approved by the Food and Drug Administration for home use. Most kits require the user to submit a small sample of blood or saliva, which is then sent to

25.2 MAKING CONNECTIONS

AIDS-Defining Illnesses (ADIs)

In HIV patients who do not receive or do not comply with antiretroviral therapy (and even in some who do), the slow destruction of the immune system results in a wide variety of infectious and noninfectious conditions, called AIDS-associated illnesses or AIDS-defining illnesses (ADIs). It is almost always one or more of these conditions that causes death in AIDS patients.

Because the virus eventually destroys an essential immune function, it is not surprising that the body is beset by normally harmless microorganisms, many of which have been living in or on the host for decades without causing disease. The spectrum of AIDS-associated illnesses also provides insight into how vital the immune system is in controlling or mitigating cancerous changes in our cells. AIDS patients are at increased risk for Burkitt lymphoma, Kaposi sarcoma (KS), and invasive cervical carcinomas, all of which are associated with viral infections.

Kaposi sarcoma lesions on the back. The flat red tumors occur in almost any tissue and frequently develop in clusters.
Source: National Cancer Institute

Since the beginning of the AIDS epidemic in the early 1980s, the CDC has maintained a list of conditions that are part of the case definition. The list has been modified periodically over almost 30 years of its existence. One of the ways that people currently meet the case definition for AIDS is if they are positive for the virus *and* experience one or more of these ADIs. The ADIs are listed in **table 25.B.** The diseases are listed according to the organ system where the presenting symptoms might be found. (Some of the conditions may be listed in more than one column.) You can see that many of them—or at least the way they occur in AIDS patients—are very rare in the otherwise healthy population.

What accounts for the enormous variety in AIDS-defining illnesses? *Answer available on Connect.*

TABLE 25.B	AIDS-Defining Illnesses				
Skin and/or Mucous Membranes (Includes Eyes)	Nervous System	Cardiovascular and Lymphatic System or Multiple Organ Systems	Respiratory Tract	Gastrointestinal Tract	Genitourinary and/or Reproductive Tract
Cytomegalovirus retinitis (with loss of vision)	Cryptococcosis, extrapulmonary	Coccidioidomycosis	Candidiasis	Candidiasis of esophagus, GI tract	Invasive cervical carcinoma
Herpes simplex chronic ulcers (>1 month duration)	HIV encephalopathy	Cytomegalovirus (other than liver, spleen, nodes)	Herpes simplex bronchitis	Herpes simplex chronic ulcers (>1 month duration)	Herpes simplex chronic ulcers (>1 month duration)
Kaposi sarcoma	Lymphoma, primarily in brain	Histoplasmosis, disseminated or extrapulmonary	*Mycobacterium avium* complex (MAC)	or esophagitis	
	Progressive multifocal leukoencephalopathy	Burkitt lymphoma	Tuberculosis (*M. tuberculosis*)	Isosporiasis, (diarrhea caused by *Cystoisospora*)	
	Toxoplasmosis of the brain	Immunoblastic lymphoma	*Pneumocystis jirovecii* pneumonia	chronic intestinal (>1 month duration)	
		Mycobacterium kansasii, disseminated	Pneumonia, recurrent in 12-month period	Cryptosporidiosis, chronic intestinal (>1 month duration)	
		Mycobacterium tuberculosis, disseminated			
		Salmonella septicemia, recurrent wasting syndrome			

a testing laboratory to be analyzed. The client can call an automated phone service, enter his or her anonymous testing code, and receive results over the phone. Over-the-counter tests that detect HIV antibodies in oral fluids and saliva can be performed and read in the privacy of home, similar to a pregnancy test (see Clinical Connections, "The Changing Scene of Clinical Microbiology," in section 17.1) Improvements in technology have increased their reliability, but positive results must still be confirmed with a more specific test.

Although the approved tests just described are largely accurate, around 1% of results are false positives, and they always require

follow-up with a more specific test called *Western blot* analysis (see figure 17.12). This test detects several different anti-HIV antibodies and can usually rule out false positive results.

Another inaccuracy can be false negative results that occur when testing is performed before the onset of detectable antibodies, a period called a window. To rule out this possibility, persons who test negative but are at some risk for exposure should be tested a second time 3 to 6 months later.

Blood and blood products are sometimes tested for HIV antigens (rather than for HIV antibodies) to determine whether a test is

possibly a false negative. For many AIDS patients, it is necessary to check viral loads in order to monitor the effectiveness of drugs.

In the United States, people are diagnosed with AIDS if they meet the following criteria: (1) they are positive for the virus, *and* (2) they fulfill one of these additional criteria:

- They have a CD4 (helper T-cell) count of fewer than 200 cells per microliter of blood.
- Their CD4 cells account for fewer than 14% of all lymphocytes.
- They experience one or more of a CDC-provided list of AIDS-defining illnesses (ADIs), as surveyed in 25.2 Making Connections.

Preventing HIV Infection

Avoidance of sexual contact with infected persons is a cornerstone of HIV prevention. Abstaining from sex is an obvious prevention method, although those who are sexually active can also take steps to decrease their risk. Epidemiologists cannot overemphasize the need to screen prospective sex partners and to follow a monogamous sexual lifestyle. The only sure way to avoid infection is for all sexually active people to consider every partner to be infected unless proven otherwise. Barrier protection (condoms) should be used when having sex with anyone whose HIV status is not known with certainty to be negative. Although avoiding intravenous drugs is an obvious deterrent, many drug addicts do not or will not choose this option. In such cases, risk can be decreased by not sharing syringes or needles or by cleaning needles with bleach and then rinsing before another use.

In 2012 the FDA for the first time approved a drug specifically for pre-exposure prophylaxis (PrEP). Marketed under the trade name Truvada and consisting of tenofovir and emtricatabine, the single daily pill has been shown to reduce the risk of HIV infection in people who are at high risk by up to 92%, provided it is taken every day.

From the very first years of the AIDS epidemic, the potential for creating a vaccine has been regarded as slim, because the virus presents many seemingly insurmountable barriers. Among them, HIV becomes latent in cells, and its cell surface antigens mutate rapidly. Although it does elicit immune responses, it is apparently not completely controlled by them. In view of the great need for a vaccine, however, none of those facts has stopped the medical community from moving ahead.

Currently, dozens of potential HIV vaccines are in clinical trials overseen by an international agency—the HIV Vaccine Trials Network. None of the trials has yet produced an effective vaccine, and even those that showed some promise have been fraught with problems. Even with the large-scale efforts, HIV specialists do not expect a viable vaccine to be approved and available for several more years.

Treating HIV Infection and AIDS

It must be forcefully stated: There is no established cure for HIV infection or AIDS. None of the therapies does more than slow the progress of the disease or diminish symptoms.

Clear-cut guidelines exist for treating people who test HIV-positive. These guidelines are updated regularly, and they differ depending on whether a person is completely asymptomatic or experiencing some of the manifestations of HIV infection, and also whether they have previously been treated with antiretroviral

agents. A person diagnosed with AIDS receives treatment for HIV infection as well as drugs to prevent or treat a variety of opportunistic infections and other ADIs such as wasting disease. These treatment regimens vary according to each patient's profile and needs.

Treatment strategies against HIV aim to disrupt some part of its multiplication cycle (see tables 12.6 and 12.7). The first effective anti-HIV drugs were described as nucleoside reverse transcriptase inhibitors (NRTIs) and consisted of synthetic nucleoside analogs that inhibit the reverse transcriptions process, including azidothymidine (AZT), didanosine (ddI), lamivudine (Epivir, 3TC), and stavudine (d4T). They interrupt the HIV multiplication cycle by mimicking the structure of actual nucleosides being added to viral DNA by reverse transcriptase. Because these drugs lack all of the correct binding sites for further DNA synthesis, viral replication and the viral cycle are terminated (**figure 25.17a**). Non-nucleoside reverse transcriptase inhibitors (NNRTIs) include etravirine and

Location of reaction

☐ External to cell

☐ Cytoplasm

(a) **Reverse transcriptase blockers.**
A prominent group of drugs (AZT, ddI, 3TC) act as nucleoside analogs to inhibit reverse transcriptase. They are inserted in place of the natural nucleotide by reverse transcriptase but block further action of the enzyme and synthesis of viral DNA.

(b) **Protease inhibitors that cause abnormal viruses to be released.**
Protease inhibitors plug into the active sites on HIV protease. This enzyme is necessary to cut long HIV protein strands and produce smaller protein units. During budding viruses incorporate this uncut nonfunctioning protein. The resultant viruses are unable to mount an infection.

Figure 25.17 Mechanisms of action of anti-HIV drugs (simplified to show basic drug effects).

Pathogen Profile #3 — *Human immunodeficiency virus 1 and 2 (HIV-1, HIV-2)*

HIV budding off cell

Late stage in budding

Completed HIV virion

Source: Dr. Matthew Gonda/
Laboratory of Cell and Molecular
Structure. National Cancer Institute

Microscopic Morphology Enveloped virus with an RNA genome. The envelope contains 72 glycoprotein spikes that mediate adsorption to the host cell.

Identified by Two-stage serological testing. The initial screening is completed using a sensitive ELISA or similar rapid antibody test that detects the presence of anti-HIV antibodies, while also producing some false positive results. All positive results are then rescreened using a very specific Western blot test, which relies on several different anti-HIV antibodies and rules out any false positives.

Habitat Human immunodeficiency virus resides within the blood, breast milk, semen, and vaginal secretions of HIV-infected persons.

Virulence Factors Spikes that mediate attachment to host cells, reverse transcriptase, rapid mutation, formation of syncytia.

Primary Infections/Disease HIV is the causative agent of acquired immunodeficiency syndrome (AIDS). Initial infection with HIV is marked by vague, mononucleosis-like symptoms that quickly dissipate. This is followed by an asymptomatic phase that may last from 2 to 15 years and coincides with a gradual depletion of CD4 T cells in the blood. When these cells fall below 200 cells per mm^3 of blood, the opportunistic diseases characteristic of AIDS begin to appear. These AIDS-defining illnesses (ADIs) not only signify the progression of HIV to full-blown AIDS but are also the cause of death for most AIDS patients.

Control and Treatment Control relies on preventing exposure to the HIV virus, which, for most people, involves safe sexual practices, from abstention, to monogamy, to barrier protection. Utilizing pre-exposure prophylactic therapy can greatly reduce the risk of infection. Those who choose to inject illegal drugs must use clean needles as contaminated needles are commonly implicated in infection. No vaccine exists for HIV. Treatment of HIV relies on highly active antiretroviral therapy (HAART), in which a combination of reverse transcriptase inhibitors, protease inhibitors, fusion inhibitors, and integrase inhibitors are used to interrupt the virus in different stages of its replication cycle. It is important to note that HIV infection is incurable, and treatment serves only to slow or prevent the progression of HIV infection to AIDS.

rilpivarine, bind to the reverse transcriptase enzyme and restructure it so that it cannot properly function. Another important class of drugs, the protease inhibitors **(figure 25.17*b*),** block the action of the HIV enzyme (protease) involved in the final assembly and maturation of the virus. Examples of these drugs include ritonavir and darunavir.

Integrase nuclear strand transfer inhibitors (INSTIs) inhibit the viral enzyme integrase, which is responsible for the integration of the viral DNA into the DNA of the infected host cell. First line drugs in this class include elvitegravir and dolutegravir. Finally, fusion inhibitors (enfuvirtide) and entry inhibitors (maraviroc) block the entry of HIV into potential host cells, typically by interacting with receptors like CCR5 on the host cell. Because fusion and entry inhibitors have severe side effects, including hepatotoxicity, they are generally used only when other drug regimens have failed to keep a patients viral load at low levels.

The most effective strategy in the battle against HIV has been the development of **highly active antiretroviral therapy (HAART).** In this treatment regimen, multiple drugs are used simultaneously to interrupt the HIV multiplication cycle at several different points, so that even if a virus managed to evade one drug due to a fortuitous mutation, viral progression would be halted by a second or third drug. Several forms of HAART exist, and generally rely on at least three drugs from two different categories. For example, an integrase plus two NRTIs or a protease inhibitor plus two NRTIs. This therapy has been shown to reduce viral load to undetectable levels and facilitate the improvement of immune function. It has also reduced the incidence of viral drug resistance, because the virus would have to undergo at least two separate mutations simultaneously to develop resistance. Patients who are HIV-positive but asymptomatic can avoid AIDS with this therapy as well. The primary drawbacks to anti-HIV drugs are high cost, toxic side effects, drug failure due to patient noncompliance, and an inability to completely eradicate the virus (see **Clinical Connections, "'Curing' HIV"**). An example of this type of therapy is Genvoya, which contains elvitegravir (an integrase inhibitor) along with emtricitabine and tenofovir alafenamide (both NRTIs). Cobicistat, which has no anti-HIV activity of its own, is used to enhance the effects of elvitegravir while reducing adverse side effects. Adhering to Genvoya's one-pill-once-daily treatment regimen is far easier than some older forms of HAART, which required patients to take up to 30 pills a day.

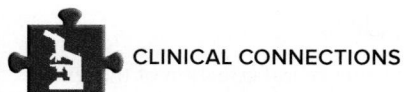

CLINICAL CONNECTIONS

"Curing" HIV

There is no cure for HIV infection. True, but one case calls for a short explanation.

The so-called "Berlin Patient" was infected with HIV and also suffered from acute myelogenous leukemia (AML), a cancer of bone marrow cells responsible for the production of certain white blood cells. Treatment of AML often involves treating the patient with a large dose of radiation that kills all of the bone marrow cells, both healthy and diseased, followed by a bone marrow transplant. In this case doctors in Berlin used bone marrow from a donor who possessed a rare mutation that made his cells highly resistant to HIV infection. More than 3 years after the transplant, and without benefit of antiretroviral drug treatment, the patient has HIV levels below the limit of detection. Unfortunately the utility of bone marrow transplants to cure HIV infection is limited, as the risk of the transplant is greater than the risk of a lifetime reliance on antiretroviral drugs.

Treatment for serious illnesses often includes a risk of severe side effects. What side effects could (possibly) be expected if the treatment discussed here were attempted on another patient? *Answer available on Connect.*

Human T-Cell Lymphotropic Viruses

Leukemia is the general name for at least four different malignant diseases of the white blood cell forming elements originating in the bone marrow. Leukemias are all acquired, rather than being inherited; some forms are acute and others are chronic. Leukemias have many causes, some of which are thought to be viral. The **human T-cell lymphotropic virus I (HTLV-I)** is associated with a form of leukemia called *adult T-cell leukemia.*

All leukemias first manifest with conditions such as easy bruising or bleeding, paleness, fatigue, and recurring minor infections. Underlying pathologies that account for these effects are anemia, platelet deficiency, and immune dysfunction brought about by the disturbed lymphocyte ratio and function. Some cases of adult T-cell leukemia first present with cutaneous T-cell lymphoma accompanied by dermatitis, with thickened, scaly, ulcerative, or tumorous skin lesions. Other complications are lymphadenopathy and dissemination of the tumors to the lung, spleen, and liver.

The possible mechanisms by which this retrovirus induces cancer are not entirely clear. One hypothesis is that the virus carries an oncogene that, when spliced into a host's chromosome and triggered by various carcinogens, can immortalize the cell and deregulate the cell division cycle. One of HTLV-I's genetic targets seems to be the gene and receptor for interleukin-2, a potent stimulator of T cells.

Adult T-cell leukemia was first described by physicians working with a cluster of patients in southern Japan. Later, a similar clinical disease was described in Caribbean immigrants. In time, it was shown that these two diseases were the same. Although more common in Japan, Europe, and the Caribbean, a small number of cases occurs in the United States. The disease is not highly transmissible;

studies among families show that repeated close or intimate contact is required. Because the virus is thought to be transferred in infected blood cells, blood transfusions and blood products are potential agents of transmission. Intravenous drug users could spread it through needle sharing. The retrovirus HTLV-II is a close relative of HTLV-I but is not as clearly associated with a distinct disease.

 Check Your Progress SECTION 25.5

30. What are retroviruses, and how are they different from other viruses?
31. Give examples of some human retroviruses and the diseases they cause.
32. Explain the function of reverse transcriptase.
33. Explain the importance of GP-41, GP-120, and CD4 to HIV infection.
34. Briefly explain the behaviors most likely to spread HIV.
35. Differentiate between HIV infection, HIV disease, and AIDS.
36. What are the primary target cells and pathologic effects of HIV over time?
37. Give a comprehensive definition of AIDS.
38. Describe the changes over time in virus antigen levels, antibody levels, and CD4 T cells in the blood of an HIV-infected individual.
39. List the major opportunistic bacterial, fungal, protozoan, viral, and other secondary diseases that are AIDS-defining conditions.
40. Explain how HIV affects the immune system and creates vulnerability to certain pathogens and other conditions.
41. Explain why two separate tests are generally used to diagnose HIV infection.
42. Compare the actions of NRTIs, NNRTIs, protease inhibitors, fusion inhibitors, and integrase inhibitors.
43. What is the rationale behind the use of HAART treatment?
44. Explain strategies for vaccine development; include in your answer the primary drawbacks to development of an effective vaccine.
45. Identify the virus responsible for adult T-cell leukemia.

25.6 Nonenveloped Single-Stranded and Double-Stranded RNA Viruses

 Expected Learning Outcomes

28. Identify the characteristics of poliovirus that enhance its transmission.
29. Describe the range of pathologies seen in poliovirus infection.
30. Recall how polio is diagnosed, treated, and prevented.
31. Understand the epidemiology, diagnosis, and prevention of infectious hepatitis.
32. Understand why rhinovirus infections are typically limited to the upper respiratory tract.
33. Describe the epidemiology and pathology of rhinovirus and calicivirus infections.
34. Describe the unique characteristic of reoviruses and recall the diseases they cause.

(a)

(b)

Figure 25.18 **Typical structure of a picornavirus. (a)** A poliovirus, a type of picornavirus that is one of the simplest and smallest viruses (30 nm). It consists of an icosahedral capsid shell around a tightly packed molecule of RNA. **(b)** A crystalline mass of stacked poliovirus particles in an infected host cell (300,000×).
(b): *Source:* Dr. Joseph J. Esposito; F. A. Murphy/CDC

Picornaviruses and Caliciviruses

As suggested by the prefix, picornaviruses are named for their small (pico) size and their RNA core **(figure 25.18).** Important representatives include *Enterovirus* and *Rhinovirus,* which are responsible for a broad spectrum of human neurological, enteric, and other illnesses **(table 25.4);** and *Cardiovirus,* which infects the brain and heart in humans and other mammals. The following discussion on human picornaviruses centers upon the poliovirus and other related enteroviruses, the hepatitis A virus, and the human rhinoviruses (HRVs).

TABLE 25.4	Selected Representatives of the Human Picornaviruses	
Genus	**Representative**	**Primary Diseases**
Enterovirus	Poliovirus	Poliomyelitis
	Coxsackievirus A	Focal necrosis, myositis
	Coxsackievirus B	Myocarditis of newborn
	Echovirus	Aseptic meningitis, enteritis, others
	Enterovirus 72	Hepatitis A
Rhinovirus	Rhinovirus	Common cold
Cardiovirus	Cardiovirus	Encephalomyocarditis
Aphthovirus	Aphthovirus	Foot-and-mouth disease (in cloven-foot animals)

Poliovirus and Poliomyelitis

Poliomyelitis* (polio) is an acute enteroviral infection of the spinal cord that can cause neuromuscular paralysis. Because it often affects small children, it is also called infantile paralysis. No civilization or culture has escaped the devastation of polio.

The poliovirus has a naked capsid (figure 25.18*a*) that confers chemical stability and resistance to acid, bile, and detergents. By this means, the virus survives the gastric environment and other harsh conditions, which contributes to its ease in transmission.

Epidemiology of Polio

Sporadic cases of polio can break out at any time of the year, but its incidence is more pronounced during the summer and fall. The virus is passed within the population through food, water, hands, objects contaminated with feces, and mechanical vectors. Due to worldwide vaccination programs, the number of cases was reduced from 350,000 in 1988 to 74 by 2015. There have been no cases of wild polio in the Americas since 1991.

After being ingested, polioviruses adsorb to receptors of mucosal cells in the oropharynx and intestine **(figure 25.19).** Here, they multiply in the mucosal epithelia and lymphoid tissue. Multiplication results in large numbers of viruses being shed into the throat and feces, and some of them leak into the blood.

* *poliomyelitis* (poh″-lee-oh-my″-eh-ly′-tis) Gr. *polios,* gray, *myelos,* medulla, and *itis,* inflammation.

Figure 25.19 **The stages of infection and pathogenesis of poliomyelitis. (a)** First, the virus is ingested and carried to the throat and intestinal mucosa. **(b)** The virus then multiplies in the tonsils. Small numbers of viruses escape to the regional lymph nodes and blood. **(c)** The viruses are further amplified and cross into certain nerve cells of the spinal column and central nervous system. **(d)** Last, the intestine actively sheds viruses.

Pathogen Profile #4 *Poliovirus*

Microscopic Morphology Small (30 nm), nonenveloped virus with a nonsegmented RNA genome.

Identified by Growing of virus in tissue culture, serological detection of antipoliovirus antibodies.

Habitat Humans are the reservoir; infection is transmitted through food, water, and close contact with infected people and their excretions. Afghanistan and Pakistan are the only countries in the world where polio remains endemic.

Source: Dr. Joseph J. Esposito; F. A. Murphy/CDC

Virulence Factors Naked capsid that provides resistance to acid, bile, and detergents, increasing ease of transmission.

Primary Infections/Disease Most people infected with polio show no symptoms, whereas about 5% will have minor symptoms including fever, fatigue, nausea, headache, stiffness in the neck and back, and pain in the limbs, which often resolves completely. Less than 1% of polio cases result in permanent paralysis of the limbs (usually the legs). Of those paralyzed, 5% to 10% die when the paralysis affects the respiratory muscles. Poliovirus multiplies in the gastrointestinal tract (which sheds viruses), and exerts its pathology within cells of the central nervous system, while signs and symptoms appear in the muscles.

Control and Treatment Two different polio vaccines exist, with the live, attenuated "Sabin" vaccine recommended in most of the world and the inactivated "Salk" vaccine recommended in the United States and parts of Europe. Treatment of the disease is limited to relieving pain and providing artificial ventilation if respiratory failure occurs.

Most infections are expressed in a short-term, mild viremia. Some persons develop mild nonspecific symptoms of fever, headache, nausea, sore throat, and myalgia. If the viremia persists, viruses can be carried to the central nervous system through its blood supply. The virus then spreads along specific pathways in the spinal cord and brain. Being **neurotropic,*** the virus infiltrates the motor neurons of the anterior horn of the spinal cord, though it can also attack spinal ganglia, cranial nerves, and motor nuclei. Nonparalytic disease involves the invasion but not the destruction of nervous tissue. It gives rise to muscle pain and spasm, meningeal inflammation, and vague hypersensitivity. The progression of infection and disease is shown in **figure 25.20.**

Paralytic Disease

Invasion of motor neurons causes various degrees of flaccid paralysis over a period of a few hours to several days. Depending on the level of damage to motor neurons, paralysis of the muscles of the legs, abdomen, back, intercostals, diaphragm, pectoral girdle, and bladder can result. In rare cases of **bulbar poliomyelitis,** the brain stem, medulla, or even cranial nerves are affected. This situation leads to loss of control of cardiorespiratory centers, requiring mechanical respirators **(figure 25.21a).** In time, the unused muscles begin to atrophy, growth is slowed, and severe deformities of the trunk and limbs develop. Common sites of deformities are the spine, shoulder, hips, knees, and feet **(figure 25.21b).** Because motor function, but not sensation, is compromised, the crippled limbs are often very painful.

In recent times, a condition called post-polio syndrome (PPS) has been diagnosed in long-term survivors of childhood infection. PPS manifests as a progressive muscle deterioration that develops in about 25% to 40% of patients several decades after their original polio attack.

Diagnosis, Treatment, and Prevention of Polio

Polio is mainly suspected when epidemics of neuromuscular disease occur in the summer in temperate climates. Poliovirus can usually be isolated by inoculating cell cultures with stool or throat washings in the early part of the disease. The stage of the patient's infection can also be demonstrated by testing serum samples for the type and amount of antibody.

Treatment of polio rests largely on alleviating pain and suffering. During the acute phase, muscle spasm, headache, and associated discomfort can be alleviated by pain-relieving drugs. Respiratory failure may require artificial ventilation maintenance. Prompt physical therapy to diminish crippling deformities and to retrain muscles is recommended after the acute febrile phase subsides.

* *neurotropic* (nu″-roh-troh′-pik) Having an affinity for the nervous system.

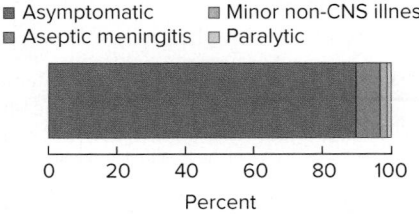

Figure 25.20 Diagrammatic representation of possible outcomes of poliovirus infection.

Figure 25.21 Images of polio. (a) An external respirator, or iron lung, was used to provide adequate ventilation for polio patients suffering from paralysis of the respiratory muscles. With the exception of the head, the entire body was enclosed in the vessel and modulation of the air pressure within the iron lung forced air in and out of the patient's lungs. Most patients spent a week to 10 days in the respirator, but in cases of permanent paralysis, a lifetime of confinement was required. **(b)** Deformed limbs are a common outcome of paralytic polio. (a and b): *Source: CDC*

The mainstay of prevention is vaccination as early in life as possible, usually in four doses starting at about 2 months of age. Adult candidates for immunization are travelers and members of the armed forces. The two forms of vaccine currently in use are inactivated poliovirus vaccine (IPV), known as the Salk[4] vaccine, and oral poliovirus vaccine (OPV), known as the Sabin[5] vaccine. Both are prepared from animal cell cultures and are trivalent (combinations of the three serotypes of the poliovirus). Both vaccines are effective, but one may be favored over the other under certain circumstances.

For many years, the Sabin vaccine was used in the United States because it is easily administered by mouth, but it is not free of medical complications. It contains an attenuated virus that can multiply in vaccinated people and be spread to others. In very rare instances—1 out of 2.4 million doses—the attenuated virus reverts to a neurovirulent strain that causes disease rather than protects against it. Numerous instances of paralytic polio have occurred among children with hypogammaglobulinemia who have been mistakenly vaccinated. There is a tiny risk (about 1 case in 4 million) that an unvaccinated family member will acquire infection and disease from a vaccinated child. For these reasons, public health officials have revised their recommendations and now favor IPV for all immunizations in the United States.

> **Quick Search**
> To read about a collision between polio and politics, search the Internet for "CIA Polio Ruse to Capture Osama Bin Laden."

The efforts of a WHO campaign, called National Vaccination Days, have significantly reduced the global incidence of polio. The last cases of polio infection are confined to a few pockets in Pakistan and Afghanistan. Unfortunately, the success of this campaign has been plagued by outbreaks in countries that barred the use of the oral polio vaccine because of political upheavals and concerns about its safety.

Nonpolio Enteroviruses

A few viruses related to poliovirus commonly cause short-term, usually mild infections. The most common of these are **coxsackieviruses*** A and B, **echoviruses,*** and nonpolio enteroviruses. They are similar to the poliovirus in many of their epidemiological and infectious characteristics. The incidence of infection is highest from late spring to early summer in temperate climates and is most frequent in infants and persons living under unhygienic circumstances.

Specific Types of Enterovirus Infection

Enteroviral infections are either subclinical or fall into the category of "undifferentiated febrile illness," characterized by fever, myalgia, and malaise. Symptoms are usually mild and last only a few days. The initial phase of infection is intestinal, after which viruses enter the lymph and blood and disseminate to other organs. The outcome of this infection largely depends on the organ affected. An overview of the more severe complications follows.

4. Named for Dr. Jonas Salk, who developed the vaccine in 1954.

5. Named for Dr. Albert Sabin, who had the idea to develop an oral attenuated vaccine in the 1960s.
 * *coxsackieviruses* (kok-sak'-ee-vy"-rus) Named for Coxsackie, New York, where the viruses were first isolated.
 * *echoviruses* (ek'-oh-vy"-rus) An acronym for **e**nteric **c**ytopathic **h**uman **o**rphan **virus.**

Figure 25.22 Acute viral hemorrhagic conjunctivitis. In the early phase of disease, the eye is severely inflamed, with the sclera and inner eyelids becoming bright red from subconjunctival hemorrhages. Later, edema of the lids can cause complete closure of the eye. © Kathy Park Talaro

Important Complications

Children are more prone than adults to lower respiratory tract illness, namely bronchitis, bronchiolitis, croup, and pneumonia. All ages, however, are susceptible to the "common cold syndrome" of enteroviruses. **Pleurodynia*** is an acute disease characterized by recurrent sharp, sudden intercostal or abdominal pain accompanied by fever and sore throat. Although nonpolio enteroviruses are less virulent than the polioviruses, rare cases of coxsackievirus and echovirus paralysis, aseptic meningitis, and encephalitis occur. Even in severe childhood cases involving seizures, ataxia, coma, and other central nervous system symptoms, recovery is usually complete.

Eruptive skin rashes (exanthems) that resemble the rubella rash are other manifestations of enterovirus infection. Coxsackievirus can cause a peculiar pattern of lesions on the hands, feet, and oral mucosa (hand-foot-mouth disease), along with fever, headache, and muscle pain. Acute hemorrhagic conjunctivitis is an abrupt inflammation associated with subconjunctival bleeding, serous discharge, painful swelling, and sensitivity to light **(figure 25.22).** Spread of the virus to the heart in infants can cause extensive damage to the myocardium, leading to heart failure and death in nearly half of the cases. Heart involvement in older children and adults is generally less serious, with symptoms of chest pain, altered heart rhythms, and pericardial inflammation.

Hepatitis A Virus and Infectious Hepatitis

One enterovirus that tends to affect the digestive tract alone is the **hepatitis A virus** (**HAV;** enterovirus 72), the cause of infectious, or short-term, hepatitis. Although this virus is not related to the hepatitis B or C viruses, it shares their tropism for liver cells. Otherwise, the two viruses are different in almost every respect (see table 24.3). The hepatitis A virus is a cubical picornavirus relatively

resistant to heat and acid but sensitive to formalin, chlorine, and ultraviolet radiation. There appears to be only one major serotype of this virus.

Epidemiology of Hepatitis A

Hepatitis A virus is spread through the oral-fecal route, but the details of transmission vary from one area to another. In general, the disease is associated with deficient personal hygiene and lack of public health measures. In countries with inadequate sewage control, most outbreaks are associated with fecally contaminated water and food. The United States has a yearly reported incidence of 2,000 to 5,000 cases. Most of these result from close institutional contact, unhygienic food handling, eating shellfish, sexual transmission, or travel to other countries. Occasionally, hepatitis A can be spread by blood or blood products. In developing countries, children account for most cases due to exposure early in life, whereas in North America and Europe, more cases appear in adults. Because the virus is not carried chronically, the principal reservoirs are asymptomatic, short-term carriers or people with clinical disease.

The Course of HAV Infection

Swallowed hepatitis A virus multiplies in the small intestine during an incubation period of 2 to 6 weeks. The virus is then shed in the feces, and in a few days, it enters the blood and is carried to the liver. Most infections are either subclinical or accompanied by vague, flu-like symptoms. In more overt cases, the presenting symptoms are loss of appetite, nausea, diarrhea, fever, pain and discomfort in the region of the liver, and darkened urine. Jaundice is present in only about 1 in 10 cases. Occasionally hepatitis A occurs as a fulminating disease and causes liver damage, but this manifestation is quite rare. Because the virus is not oncogenic and does not predispose to liver cancer, complete uncomplicated recovery results.

Diagnosis and Control of Hepatitis A

A patient's history, liver and blood tests, serological tests, and viral identification all play a role in diagnosing hepatitis A and differentiating it from the other forms of hepatitis. Certain liver enzymes are elevated, and leukopenia is common. Diagnosis is aided by detection of anti-HAV IgM antibodies produced early in the infection and tests to identify hepatitis A antigen or virus directly in stool samples.

There is no specific treatment for hepatitis A once the symptoms begin. Patients who receive immune serum globulin early in the disease usually experience milder symptoms than patients who do not receive it. Prevention of hepatitis A is based primarily on immunization. Two inactivated viral vaccines are approved for protection against hepatitis A, HAVRIX and VAQTA. Both vaccines are administered in two doses, between 6 and 18 months of age. TWINRIX is an inactivated vaccine that provides protection against both hepatitis A and B, but is currently approved only for adults. Pooled immune serum globulin is also recommended for travelers and armed forces personnel planning to enter endemic areas, contacts of known cases, and occupants of various institutions during epidemics. Control of this disease can be improved by sewage treatment, hygienic food handling and preparation, and adequate cooking of shellfish.

* *Pleurodynia* (plur″-oh-din′-ee-ah) Gr. *pleura,* rib, side, and *odyne,* pain.

Human Rhinovirus (HRV)

The **rhinoviruses*** are a large group of picornaviruses (more than 110 serotypes) associated with the common cold. Although the majority of characteristics are shared with other picornaviruses, two distinctions set rhinoviruses apart. First, they are sensitive to acidic environments such as that of the stomach, and second, their optimum temperature of multiplication is not normal body temperature but 33°C, the average temperature in the human nose.

Virologists have subjected rhinovirus (type 14) to a detailed structural analysis. The results of this study provided a striking three-dimensional view of its molecular surface and also explained why immunity to the rhinovirus has been so elusive. The capsid subunits are of two types: protuberances (knobs), which are antigenically diverse among the rhinoviruses, and indentations (pockets), which exist only in two forms **(figure 25.23).** Because the antigens on the surface are the only ones accessible to the immune system, a successful vaccine would have to contain hundreds of different antigens, so it is not practical. Unfortunately the recessed antigens are too inaccessible for immune actions. These factors make the development of a vaccine unlikely, but it could provide a basis for developing drugs to block the pocket receptors.

Epidemiology and Infection of Rhinoviruses

Rhinovirus infections occur in all areas and all age groups at all times of the year. Epidemics caused by a single type arise on occasion, but usually many strains circulate in the population at one time. As mutations increase and herd immunity to a given type is established, newer types predominate. People acquire infection from contaminated hands and fomites, and to a lesser extent from droplet nuclei. Children are highly susceptible to colds and often transmit the virus to the rest of a family. Although other animals have rhinoviruses, interspecies infections have never been reported. After an incubation period of 1 to 3 days, the patient can experience some combination of headache, chills, fatigue, sore throat, cough, a mild nasal drainage, and atypical pneumonia. Natural host defenses and nasal antibodies have a beneficial local effect on the infection, but immunity is short-lived.

Control of Rhinoviruses

The usual therapy is to force fluids and relieve symptoms with various cold remedies and cough syrups that contain nasal decongestants, antihistamines, and analgesics. The actual effectiveness of most of these remedies (of which there are hundreds) is rather questionable. Owing to the extreme diversity of rhinoviruses, prevention of colds through vaccination will probably never be a medical reality. Some protection is afforded by such simple measures as hand washing and care in handling nasal secretions.

Caliciviruses

Caliciviruses* are an ill-defined group of enteric viruses found in humans and mammals. The best-known human pathogen is the

* *rhinoviruses* (ry′-noh-vy″-rus) Gr. *rhinos,* nose.
* *caliciviruses* (kal′-ih-sih-vy″-rus) L. *calix,* the cup of a flower. These viruses have cup-shaped surface depressions.

Figure 25.23 Antigen structure of a rhinovirus. (a) The surface of a rhinovirus is composed of knobs and pockets. The antigens on the knobs are extremely variable in shape among the scores of viral strains, and those in the pockets do not vary among the strains. **(b)** The knobs are readily accessible to the immune system, and antibodies formed against them will neutralize the virus (circle). Unfortunately, being fully protected against rhinoviruses would require making antibodies that react with 100 different knob antigens. It is true that reactions against pocket antigens would be universal, but these molecules are too inaccessible for immune reactions.

norovirus also known as the **Norwalk agent,** named for an outbreak of gastroenteritis that occurred in Norwalk, Ohio, from which a new type of virus was isolated. The Norwalk virus is now believed to cause nearly 90% of all cases of viral gastroenteritis. It is transmitted oro-fecally in schools, restaurants, cruise ships, and nursing homes, and through contaminated water and shellfish. Infection occurs throughout the year among people of all ages. Onset is acute, accompanied by nausea, vomiting, cramps, diarrhea, and chills; recovery is rapid and complete.

Reoviruses: Segmented Double-Stranded RNA Viruses

Reoviruses have an unusual double-stranded RNA genome and both an inner and an outer capsid (see table 25.1). Two of the best-studied viruses of the group are *Rotavirus* and *Reovirus.* Named for

(a) (b)

Figure 25.24 Diagnosing gastroenteritis. (a) A sample of feces from a child with gastroenteritis as viewed by electron microscopy. Note the unique "spoked-wheel" morphology that can be used to identify *Rotavirus* (150,000×). **(b)** An infant suffering from chronic rotavirus gastroenteritis shows evidence of malnutrition, failure to thrive, and stunted growth. (a): *Source:* Bryon Skinner/CDC; (b): From Farrar and Lambert, *Pocket Guide for Nurses: Infectious Diseases,* 1984. Williams and Wilkins, Baltimore, MD. Reprinted by permission of Dr. William Edmund Farrar, Jr.

its wheel-shaped capsid, *Rotavirus** (**figure 25.24a**) is a significant cause of diarrhea in newborn humans, calves, and piglets. Because the viruses are transmitted via fecally contaminated food, water, and fomites, disease is prevalent in areas of the world where access to adequate sanitation is lacking. Globally, *Rotavirus* is the primary viral cause of mortality and morbidity resulting from diarrhea. It accounts for nearly 50% of all the cases, resulting in the deaths of over 500,000 children. The effects of infection vary with the nutritional state, general health, and living conditions of the infant. Babies from 6 to 24 months of age lacking maternal antibodies have the greatest risk for fatal disease. These children present with symptoms of watery diarrhea, fever, vomiting, dehydration, and shock. The intestinal mucosa can be damaged in a way that chronically compromises nutrition, and long-term or repeated infections can retard growth (**figure 25.24b**).

In the United States, rotavirus infection is relatively common, but its course is generally mild. Children are treated with oral replacement fluid and electrolytes. Two vaccines (RotaTeq, Rotarix) based on various versions of live, attenuated viruses are currently available. Initial results indicate that the vaccines reduced serious illness by 85% to 98% in the groups of infants tested. They are both easily administered oral vaccines but rather costly.

*Reovirus** is not considered a significant human pathogen. Adults who were voluntarily inoculated with the virus developed coldlike symptoms. The virus has been isolated from the feces of children with enteritis and from people suffering from an upper respiratory infection and rash, but most infections are asymptomatic. Although a link to biliary, meningeal, hepatic, and renal syndromes has been suggested, a causal relationship has never been proved.

 Check Your Progress SECTION 25.6

46. Describe the epidemiology and progress of polio infection and disease.
47. Why is poliovirus spread so effectively?
48. Compare paralytic and nonparalytic polio.
49. Explain why the Salk vaccine is currently recommended over the Sabin vaccine in the United States.
50. What characteristics of enteric viruses cause them to be readily transmissible?
51. Describe some common ways that hepatitis A is spread.
52. What are its primary effects on the body?
53. List several activities that reduce the incidence of infectious hepatitis.
54. What is the common cold, and which groups of viruses can be involved in its transmission?
55. Why is it so difficult to control?
56. Describe the most important calicivirus and the disease it causes.
57. Summarize the epidemiology of noroviruses.
58. What viruses possess a double-stranded RNA genome and what diseases do they cause?

25.7 Prions and Spongiform Encephalopathies

 Expected Learning Outcomes

35. Recall the structure of prions and the diseases they cause.
36. Explain the pathogenesis, transmission, and epidemiology of prion disease.
37. Understand how prion infection is diagnosed, treated, and prevented.

Devastating diseases such as subacute sclerosing panencephalitis and progressive multifocal leukoencephalopathy arising from persistent viral infections of the CNS have previously been described. They are characterized by a long incubation period, and they can progress over months or years to a state of severe neurological impairment. These types of chronic infections are caused by known viruses with conventional morphology. But there are other types of CNS infectious diseases for which no traditional microorganism has been isolated. The most serious of these are the diseases associated with **prions.**[6]

You may recall from chapter 6 that prions are proteinaceous infectious particles that appear to lack genetic material. Prions are incredibly hardy "pathogens." They are highly resistant to chemicals, radiation, and heat. They can withstand prolonged autoclaving (**table 25.5**). They are known to cause diseases called **transmissible spongiform encephalopathies** (**TSEs**), neurodegenerative

TABLE 25.5	Properties of the Agents of Spongiform Encephalopathies

Consist entirely of protein with no nucleic acid

Very resistant to chemicals, radiation, and heat (can withstand autoclaving)

Do not present virus morphology in electron microscopy of infected brain tissue

Not integrated into nucleic acid of infected host cells

Do not elicit inflammatory reaction or cytopathic effects in host

Do not elicit antibody formation in host

Responsible for vacuoles and abnormal fibers forming in brain of host

Transmitted only by close, direct contact with infected tissues and secretions

(a) (b)

Figure 25.25 The microscopic effects of spongiform encephalopathy. (a) Normal cerebral cortex section, showing neurons and glial cells. **(b)** Section of cortex in CJD patient shows numerous holes. The open spaces correspond to sites once occupied by brain cells that have been destroyed. This gives the tissue a spongy appearance. It is such massive loss of neurons and glial cells that gives rise to the deadly outcome in the disease (750×). (a): © M. Abbey/Science Source; (b): *Source:* Sherif Zaki; MD; PhD; Wun-Ju Shieh; MD; PhD; MPH/CDC

diseases with long incubation periods but rapid progressions once they begin. The human TSEs are **Creutzfeldt-Jakob disease (CJD),** kuru, Gerstmann-Strussler-Scheinker disease, and fatal familial insomnia. TSEs are also found in animals and include a disease called **scrapie** in sheep and goats, transmissible mink encephalopathy, and bovine spongiform encephalopathy (BSE). This last disease, commonly known as mad cow disease, has been a big concern in recent years due to its apparent link to a variant form of CJD human disease in Great Britain.

Pathogenesis and Effects of CJD

The pathogenesis of the prions involves an alteration in the structure of a normal host protein (called PrPC) found in mammalian brains. Abnormal PrPSc has acquired a significant change in its shape through mutation or some other means. The effect of this alteration is that the abnormal PrPSc itself becomes catalytic and able to spontaneously convert other normal human PrPC proteins into the abnormal form (see figure 6.21). This becomes a self-propagating chain reaction that creates a massive accumulation of altered PrPSc that results in nerve cell death, spongiform damage (that is, holes in the brain), and severe loss of brain function. Autopsies of the brain of CJD patients reveal spongiform lesions **(figure 25.25)** as well as tangled protein fibers (neurofibrillary tangles) and enlarged astroglial cells. These changes occur primarily in the gray matter of the CNS. The PrPSc are not apparently antigenic and do not stimulate an immune response.

Symptoms of CJD include altered behavior, dementia, memory loss, impaired senses, delirium, and premature senility. Uncontrollable muscle contractions continue until death, which usually occurs within 1 year of diagnosis.

Transmission and Epidemiology

The transmissible form of CJD is not a highly communicable disease in that ordinary contact with infected people is not known to transmit the prions. Direct or indirect contact with infected brain tissue or cerebrospinal fluid is one factor in its acquisition. The classic form of CJD is considered endemic and occurs at the rate of one case per million persons in the United States per year, mainly in the elderly. Using the term *transmissible agent* may be a bit

misleading, because some cases of CJD arise through genetic mutation of the PrPC gene rather than through contact with infectious material. In fact, up to 15% of CJD cases appear to develop after inheritance of a mutated gene.

In the late 1990s it became apparent that humans were contracting a variant form of CJD (vCJD) after ingesting meat from cattle that had been afflicted by bovine spongiform encephalopathy. Presumably, meat products had been contaminated with fluid or tissues infected with the prion, although the exact food responsible for the transmission has not been pinpointed. When experimenters purposely infect cattle with prions, they subsequently detect the agent in the retina, dorsal root ganglia, parts of the digestive tract, and the bone marrow of the animals. Even so, the risk of contracting vCJD from the ingestion of meat is extremely small, even in countries such as Great Britain, where a significant number of livestock have been found to have BSE. There the risk of infection is estimated to be 1 case per 10 billion meat servings. As of 2011 a total of 224 cases of vCJD had occurred worldwide. The median age at death of patients with vCJD is 28 years. In contrast, the median age at death of patients with classic CJD is 68 years.

Health care professionals should be aware of the possibility of CJD in patients, especially when surgical procedures are performed, as cases have been reported of transmission of CJD via contaminated surgical instruments. Due to the heat and chemical resistance of prions, normal disinfection and sterilization procedures are usually not sufficient to eliminate them from instruments and surfaces. The latest CDC guidelines for handling of CJD patients in a health care environment should be consulted. CJD has also been transmitted through corneal grafts and administration of contaminated human growth hormone. Currently, no cases have been documented of transmission through blood products, although scientists have shown that it is possible in laboratory experiments. Experiments suggest that vCJD seems to be more transmissible through blood than is classic CJD. For that reason, blood donation programs screen for possible exposure to BSE by asking about travel and residence history.

Culture and Diagnosis

It can be very difficult to diagnose CJD. Definitive diagnosis requires examination of biopsied brain or nervous tissue, and this procedure is usually considered too risky because of both the trauma induced in the patient and the undesirability of contaminating surgical instruments and operating rooms. Electroencephalograms and magnetic resonance imaging can provide important clues. A new test has been developed that can detect prion proteins in cerebrospinal fluid, but it is not yet widely used because it has a relatively high error rate.

Prevention and/or Treatment

Prevention of either type of CJD relies on avoiding any contact with infected tissues. No known treatment exists for CJD, and patients inevitably die. Medical intervention focuses on easing symptoms and making the patient as comfortable as possible.

 Check Your Progress SECTION 25.7

59. Describe the characteristics of the agents responsible for spongiform encephalopathies.

60. Explain how bovine spongiform encephalopathy can be transmitted to humans.

61. Make a flowchart to explain the mechanism of how prions can cause a progressive effect, even though they cannot divide by themselves.

62. Other than infection, how else may Creutzfeldt-Jakob disease be acquired?

63. What measures would be necessary to control infections with these agents?

 CASE STUDY **Part 2**

While in the hospital, samples of the boy's blood and cerebrospinal fluid were examined. His serum measles IgG level was markedly elevated, as was his cerebrospinal fluid measles IgG level; serum measles IgM was negative.

The boy had received a single dose of measles vaccine in the Philippines at 8 months of age, yet contracted the disease when he was a year old. Vaccination during early infancy has been associated with a less efficient production of antibodies when compared to children vaccinated later, likely as a result of interference by circulating maternal antibodies. Lack of adequate protection could also be due to the fact that

the boy received only a single dose of vaccine. The Centers for Disease Control recommends that all children get two doses of measles, mumps, rubella (MMR) vaccine to ensure adequate protection. The first dose is given at 12 to 15 months of age, with a second dose between 4 and 6 years of age.

- *What is the significance of the boy's IgG and IgM levels?*

- *What type of side effects are occasionally seen with the measles vaccine? Why?*

See; Liko J, Guzman-Cottrill JA, Cieslak PR. *Notes from the Field: Subacute Sclerosing Panencephalitis Death — Oregon, 2015.* MMWR Morb Mortal Wkly Rep 2016;65:10–11.

To conclude this Case Study, go to Connect.

 Chapter Summary with Key Terms

25.1 Enveloped Segmented Single-Stranded RNA Viruses
 A. **Orthomyxoviruses:** Influenza viruses have a segmented genome that encourages **antigenic shift, antigenic drift,** and glycoprotein spikes (**hemagglutinin** and **neuraminidase**), which are needed to bind to host cells.
 B. **Bunyaviruses:** Zoonotic viruses. *Hantavirus* is responsible for Korean hemorrhagic fever and *hantavirus pulmonary syndrome.*
 C. **Arenaviruses:** Lassa virus causes hemorrhagic fever and is spread by rodent secretions.

25.2 Enveloped Nonsegmented Single-Stranded RNA Viruses
 A. **Paramyxoviruses:** Nonsegmented genome. *Paramyxovirus* causes **parainfluenza** and **mumps.** *Morbillivirus* is responsible for **measles** (roseola). *Paramyxovirus* and *Morbillivirus* are prevented through vaccination (MMR). *Pneumovirus* (**respiratory syncytial virus, RSV**) is the causative agent of **croup** and is treated with aerosol ribavirin.

 B. **Rhabdoviruses:** *Lyssavirus* is the cause of **rabies,** a lethal zoonosis. Treatment requires both active and passive immunization.

25.3 Other Enveloped RNA Viruses: Coronaviruses, Togaviruses, and Flaviviruses

 A. **Coronaviruses** are one cause of the common cold. Other coronaviruses are responsible for **severe acute respiratory syndrome (SARS)** and **middle east respiratory syndrome (MERS).**

 B. **Togaviruses: Rubella** (German measles) is of minor concern to healthy adults but can be **teratogenic** (cause birth defects) in pregnant women. Control is through immunization (MMR).

 C. Flaviviruses: **Hepatitis C virus** is a major cause of hepatitis transmitted through blood contact. Chronic infection may cause liver damage and cancer.

25.4 Arboviruses: Viruses Spread by Arthropod Vectors

A. **Arboviruses:** Large group (400+) of viruses spread by arthropod vectors and harbored by birds and other animals.

B. Examples of viruses include *togaviruses, bunyaviruses,* and *reoviruses;* diseases include West Nile encephalitis, Chikungunya, and eastern equine encephalitis, **yellow fever, dengue fever,** and **Zika.**

25.5 Retroviruses and Human Diseases

A. **HIV Infection and AIDS: Retroviruses** are enveloped, single-stranded RNA viruses that encode the enzyme **reverse transcriptase,** which allows the viruses to convert their single-stranded RNA genome to a double-stranded DNA genome.

B. The most prominent retrovirus is **human immunodeficiency virus** (HIV types 1 and 2), the causative agent of **acquired immunodeficiency syndrome (AIDS).** AIDS is marked by the presence of antibodies to HIV along with some combination of opportunistic infections and a CD4 T-cell count of <200/μl of blood. In addition, other common symptoms include fever, weight loss, and night sweats.

 1. HIV is found in blood, semen, and vaginal secretions and can be transmitted by any activity that causes these fluids to be exchanged (sex, sharing needles, childbirth, etc.).

 2. HIV attacks the cells of the immune system, beginning with T lymphocytes and macrophages.

 3. The first signs of AIDS are the **AIDS-defining illnesses (ADIs),** opportunistic infections such as *Pneumocystis jirovecii* **pneumonia (PCP)** and cancers such as **Kaposi sarcoma.**

 4. Treatment of HIV relies on highly active antiretroviral therapy (HAART), in which a combination of reverse transcriptase inhibitors (AZT, ddI, 3TC), protease inhibitors (ritonavir, saquinavir), fusion inhibitors, and integrase inhibitors are used to interrupt the virus in different stages of its replication cycle. Drugs are also given to prevent/treat opportunistic infections.

C. **Human T-cell Lymphotropic Viruses** Other retroviruses include **HTLV-I** and **HTLV-II.** The former causes adult T-cell leukemia, whereas the latter may be involved in some forms of brain disease.

25.6 Nonenveloped Single-Stranded and Double-Stranded RNA Viruses

A. **Picornaviruses** are the smallest human viruses and include

 1. **Poliovirus,** the causative agent of polio.

 2. **Coxsackievirus,** which is responsible for respiratory infections, hand-foot-mouth disease, and conjunctivitis.

 3. **Hepatitis A virus,** which is responsible for the mildest form of hepatitis.

 4. **Rhinovirus,** the most prominent agent of the common cold.

B. **Caliciviruses,** the most common of which is **norovirus** or **Norwalk agent,** a common cause of viral gastroenteritis.

C. **Reoviruses: Segmented Double-Stranded RNA Viruses Reoviruses,** the most important of which is *Rotavirus,* a leading cause of severe infantile diarrhea worldwide.

25.7 Prions and Spongiform Encephalopathies

Prions are infectious agents consisting exclusively of protein. They are responsible for **transmissible spongiform encephalopathies** such as mad cow disease, **scrapie** in animals, and **Creutzfeldt-Jakob disease** in humans.

Level I. Knowledge and Comprehension

These questions require a working knowledge of the concepts in the chapter and the ability to recall and understand the information you have studied.

 Multiple-Choice Questions

Select the correct answer from the answers provided. For questions with blanks, choose the combination of answers that most accurately completes the statement.

1. Which receptors of the influenza virus are responsible for binding to the host cell?
 a. hemagglutinin c. type A
 b. neuraminidase d. capsid proteins

2. The primary site of attack in influenza is the
 a. small intestine c. skin
 b. respiratory epithelium d. meninges

3. Which of these happen(s) in the case of antigenic shift in influenza A?
 a. single mutations in hemagglutinin
 b. recombination of RNA segments between bird and human strains
 c. change from influenza A to influenza B
 d. both a and b

4. Infections with _____ virus cause the development of multinucleate giant cells.
 a. rabies c. pneumo
 b. influenza d. corona

5. For which disease is an exanthem (skin rash) *not* a symptom?
 a. measles c. coxsackievirus infection
 b. rubella d. parainfluenza

6. A common, highly diagnostic sign of measles is
 a. viremia c. sore throat
 b. red rash d. Koplik's spots

7. Viruses that cause serious diseases in infants are _____ and _____.
 a. mumps, calicivirus
 b. respiratory syncytial virus, rotavirus
 c. coxsackievirus, HTLV-II
 d. bunyavirus, cardiovirus

8. For which disease are active and passive immunization given simultaneously?
 a. influenza c. measles
 b. yellow fever d. rabies

9. What property of the retroviruses enables them to integrate into the host genome?
 a. the RNA they carry
 b. presence of glycoprotein receptors
 c. presence of reverse transcriptase
 d. a positive-sense genome

10. Which of the following conditions is *not* associated with HIV infection?
 a. *Pneumocystis jirovecii* pneumonia
 b. Kaposi sarcoma
 c. dementia
 d. adult T-cell leukemia

11. Polio and hepatitis A viruses are _____ viruses.
 a. arbo
 b. enteric
 c. cold
 d. syncytial

12. Rhinoviruses are the most common cause of
 a. conjunctivitis
 b. gastroenteritis
 c. hand-foot-mouth disease
 d. the common cold

13. Which of the following is *not* a characteristic of the agents of spongiform encephalopathies?
 a. highly resistant
 b. associated with tangled protein fibers in the brain
 c. are naked fragments of RNA
 d. cause chronic transmissible disease

14. **Multiple Matching.** Select all descriptions that fit each type of virus.
 _____ orthomyxovirus
 _____ rhabdovirus
 _____ paramyxovirus
 _____ reovirus
 _____ morbillivirus
 _____ poliovirus
 _____ retrovirus
 _____ hantavirus
 a. enveloped
 b. double-stranded RNA
 c. single-stranded RNA
 d. genome in segments
 e. icosahedral capsid
 f. helical nucleocapsid
 g. nonenveloped

15. **Multiple Matching.** Match the virus with its primary target organ or site of attack. Some viruses may attack more than one organ.
 _____ influenza
 _____ HIV
 _____ mumps virus
 _____ measles virus
 _____ hepatitis A virus
 _____ hantavirus
 _____ dengue virus
 _____ encephalitis virus
 _____ poliovirus
 _____ rabies virus
 _____ rubella virus
 _____ rotavirus
 a. brain
 b. parotid gland
 c. respiratory tract
 d. T lymphocytes
 e. kidney
 f. spinal cord
 g. liver
 h. intestinal tract
 i. heart
 j. eye

16. **Matching.** Match the virus with its primary mode of transmission. Some viruses may have more than one mode.
 _____ influenza
 _____ HIV
 _____ mumps virus
 _____ measles virus
 _____ hepatitis A virus
 _____ hantavirus
 _____ chikungunya virus
 _____ poliovirus
 _____ rabies virus
 _____ rubella virus
 _____ rotavirus
 a. respiratory droplets
 b. sexual transmission
 c. ingestion (oral-fecal)
 d. arthropod bite
 e. contact with mammals
 f. fomites

Case Study Review

1. During which periods of a measles infection is a person not usually contagious?
 a. incubation
 b. prodromium
 c. skin rash
 d. convalescence

2. Match each of the statements below to measles or German measles.
 a. Transmitted by a togavirus.
 b. Koplik spots are a sign.
 c. subacute sclerosing panencephalitis is a complication.
 d. Also known as rubella.
 e. May cause congenital defects.

3. Why would it be especially important for someone who worked in an infant nursery to be vaccinated against the measles?

Writing Challenge

For each question, compose a one- or two-paragraph answer that includes the factual information needed to completely address the question. Check Your Progress questions can also be used for writing-challenge exercises.

1. Discuss the similarities in the symptoms of influenza, parainfluenza, mumps, measles, rubella, RSV, and rhinoviruses.

2. a. What is a teratogenic virus?
 b. Which RNA virus has this potential?

3. Explain why patients with AIDS have deficiencies in antibody formation.

4. a. Discuss some ways that HIV has *not* been documented to spread.
 b. How is the brain affected by HIV infection?
 c. What is the most sensible interpretation for a seropositive or seronegative result in HIV testing?
 d. What is meant by the "window" regarding antibody presence in the blood?
 e. Are there any actual cures for AIDS? Explain your answer.

5. a. For which of the RNA viruses are vaccines available?
 b. For which of the RNA viruses are there specific drug treatments?

6. Explain why infections such as mumps, measles, polio, rubella, and RSV regularly infect children and not adults.

Concept Mapping

On Connect you can find an Introduction to Concept Mapping that provides guidance for working with concept maps along with concept mapping activities for this chapter.

Level II. Application, Analysis, Evaluation, and Synthesis

These problems go beyond just restating facts and require higher levels of understanding and an ability to interpret, problem solve, transfer knowledge to new situations, create models, and predict outcomes.

Critical Thinking

Critical thinking is the ability to reason and solve problems using facts and concepts. These questions can be approached from a number of angles, and in most cases, they do not have a single correct answer.

1. a. Explain the relationship between herd immunity and the development of influenzal pandemics.
 b. Why will herd immunity be lacking if antigenic shift or drift occurs?

2. a. Explain how the antibody content of a patient's sera could be used to predict which strain of influenza predominated during a previous epidemic.
 b. Explain the precise way that cross-species influenza infections give rise to new and different strains of virus.
 c. Relate this answer to the avian influenza virus and how it could cause a human pandemic.

3. When a newborn infant is positive for HIV antibodies in both ELISA and Western blot tests, what conclusions can you make about the baby's condition? What other information do you need to be sure that this is the right conclusion?

4. a. Explain what characteristics of virus and host make zoonotic diseases possible and difficult to control.

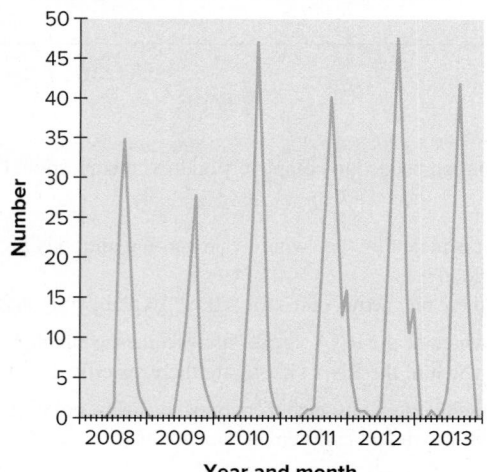

Encephalitis/Meningitis, Arboviral, California Serogroup. Number of reported cases, by month of onset—United States, 2008–2013

b. Make a diagram that details the progression of arboviruses from deep in the Amazon rain forest to Florida or from Central Africa to Seattle, Washington (as happened with West Nile fever virus).
 c. Observe the graph showing case reports for arboviral encephalitis from 2008–2013. Describe the general pattern that is evident and analyze the factors that would cause it.

5. a. Can you think of a plausible series of events that could explain how HIV came to be so widespread?
 b. Explain why there can be such a discrepancy between the total number of reported cases of AIDS and the projected number of cases estimated by health authorities.
 c. Using figure 25.15 and table 12.6 as references, outline the strategies of anti-HIV drugs and their effects as correlated with the virus cycle.

6. Explain the characteristics of HIV and its infection that make it a challenging target for the immune system and for drugs.

7. a. What precautions can a person take to prevent himself or herself from contracting HIV infection?
 b. How can a health care worker prevent possible infection?

8. If wild-type polio disappeared from the Western Hemisphere by 1991, how do you explain the 11 cases of polio reported in the United States in 2000?

9. Various over-the-counter cold remedies control or inhibit inflammation and depress the symptoms of colds. Is this beneficial or not? Support your answer.

10. Late in the spring, a young man from rural Idaho developed fever, loss of memory, difficulty in speech, convulsions, and tremor and lapsed into a coma. He tested negative for bacterial meningitis and had no known contact with wild animals. He survived but had long-term neurological disabilities. What are the possible diseases he might have had, and how would he have contracted them?

11. Biopsies from the liver and intestine of an otherwise asymptomatic 35-year-old male show masses of acid-fast bacilli throughout the tissue. Explain the pathology going on here and provide a preliminary diagnosis.

1. Refer to **figures 25.2** and **25.23,** and compare and contrast how both viruses use surface receptors to evade an immune reaction and neutralizing antibodies.

Binding sites used to anchor virus to host cell receptors (low rate of mutation)

Site for antibody binding (high rate of mutation)

Viral envelope

STUDY TIP: Quickly create a customized practice quiz for this chapter by going to your Connect SmartBook "My Reports" section. Don't have SmartBook and want to learn more? Go to **whysmartbook.com.**

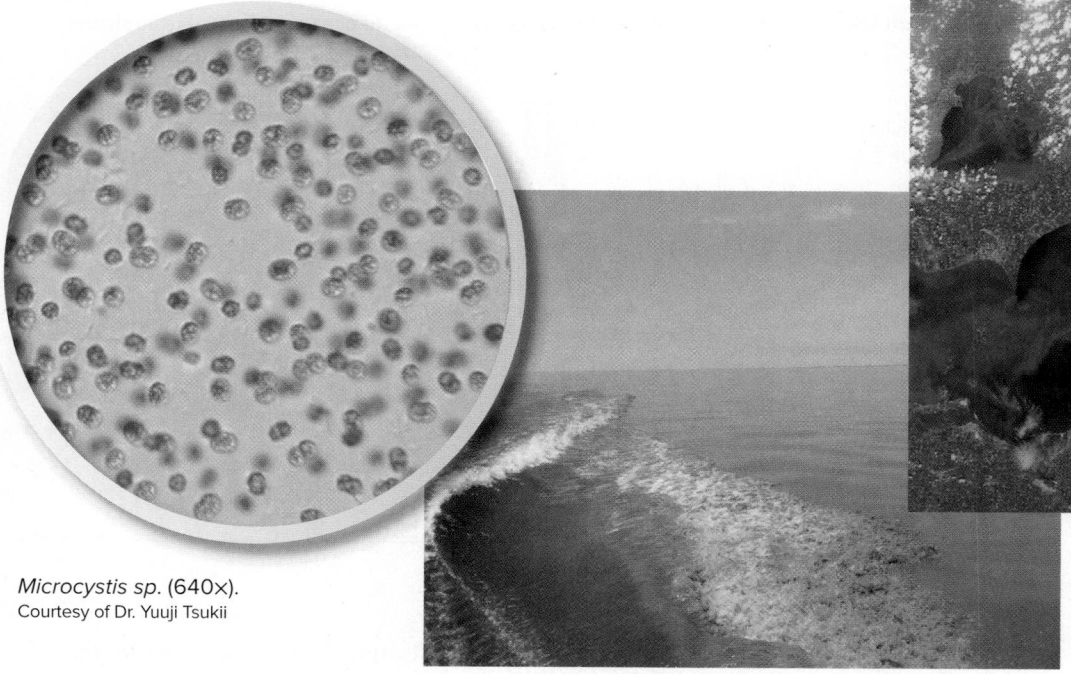

Microcystis sp. (640×).
Courtesy of Dr. Yuuji Tsukii

A harmful algae bloom in Lake Erie.
Source: NOAA Great Lakes CoastWatch

Satellite image of Lake Erie taken on August 22, 2015. Toledo, Ohio, sits at the extreme western (left) shore of the lake. *Source:* NOAA Great Lakes CoastWatch

CASE STUDY Part 1 May I Get a Glass of Water?

On August 2, 2014, a **DO NOT DRINK** order was issued for 500,000 people in and around Toledo, Ohio. The order stated that tap water should not be used for drinking, brushing teeth, or preparing infant formula, food, or ice. The cause of the order was an increased level of **microcystins,** a toxin produced by several species of the cyanobacterial genus *Microcystis*. Environmental conditions had allowed the cyanobacterium to grow at a prodigious rate producing a **harmful algae bloom** (HAB). The immediate effects of ingesting microcystins include nausea, vomiting, and dizziness, but liver **toxicity** is the most concerning issue, with both short-term liver failure and longer-term liver cancer reported. The danger is real. One hundred patients developed liver failure (and 50 died) when water contaminated with microcystins was used at a dialysis clinic in Brazil. The **Environmental Protection Agency** has deemed that water with greater than 1.6 **parts per billion** (ppb) microcystin is unsafe to drink, and because the toxin is **heat stable,** boiling—the usual advice for water contaminated by microbes—is not helpful. In fact, because boiling reduces the volume of the water, it simply concentrates the toxin.

Shortly after the water advisory was issued, the effects could be seen throughout the area. Stores quickly sold out of bottled water and shipments destined for other parts of the state were rerouted to Toledo. The Ohio National Guard trucked in water from distant areas

> "The Ohio National Guard was used to deliver pre-made infant formula."

and delivered **premixed infant formula** and military meals. Restaurants were urged to shut down, and the Ohio Department of Rehabilitation and Corrections converted its milk bottling plant to package safe water for the area. Toledo knew what to do when the city's water was deemed unsafe to drink, because it had been in the same situation before.

In the 1920s the city of Toledo established a water intake (called a crib) at Maumee Bay on the western end of Lake Erie. The lake's large size and the location of the **crib** near the water inlet for the lake were thought to offer an ideal source of clean water, distant from the industrial wastes and **raw sewage** that were routinely dumped into the rivers of Toledo. Unfortunately, over the next several decades many factors conspired to favor the growth of toxin-producing microcystis in Lake Erie.

- *What environmental factors favor the growth of microorganisms like algae and cyanobacteria?*

- *Boiling is commonly recommended when water is contaminated with E.coli. What does boiling do to make the water safe to drink?*

To continue the Case Study, go to Case Study Part 2 at the end of the chapter.

SCOPING OUT THE CHAPTER

Remove the microbes, and life on Earth stops. Plants don't grow, animals starve, lakes and oceans die. Even the nonliving aspects of the planet—the amount and type of nitrogen and phosphorous available to plants and animals, for instance—depends on the actions of microorganisms. Microbes are not always the heroes, though. Overgrowth of algae or bacteria can poison water supplies and cut off oxygen to other members of aquatic communities. In this chapter we discuss the impact of microorganisms on the world at large.

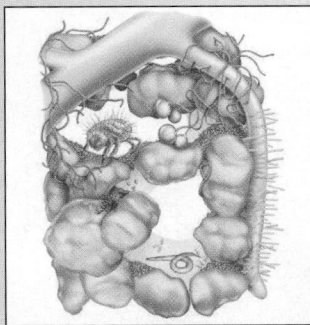

A single spoonful of soil contains a vibrant collection of microorganisms working together. Bacteria, fungi, protozoans, even animals like mites and nematode worms, exist in complex associations that decompose organic matter, supply nitrogen to plant roots, and enrich the surrounding soil, making it so much more than just a pile of dirt.

Boiling, acidic, sulfur filled pits— areas that would seem to be anathema to life—harbor complex microbial communities. The biogeochemical interactions found here are essential to the cycling of iron, sulfur, manganese, phosphorous, and other inorganic molecules through the environment.

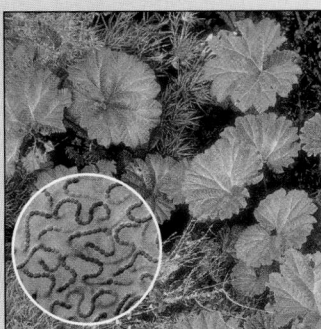

Mutualistic associations between green plants and cyanobacteria like *Nostoc* are responsible for nitrogen fixation, ensuring that atmospheric nitrogen gas is converted to a form usable by plants and animals.

In the developed world we take for granted the safety of our drinking water, but safe water takes hard work. Testing for the presence of coliforms like *E. coli* is just one step in the management of drinking water.

(top: right): © Robert Glusic/Corbis RF; (bottom: left): © Steven P. Lynch; (inset): © McGraw-Hill Education/Don Rubbelke, photographer; (bottom: right): © Harold J. Benson

26.1 Ecology: The Interconnecting Web of Life

 Expected Learning Outcomes

1. Define *microbial ecology* and describe what it entails.

2. Characterize the organization of the biosphere, from the highest to the lowest levels.

3. Differentiate between habitat and niche, using examples.

"When we try to pick out anything by itself, we find it hitched to everything else in the universe."

California naturalist John Muir

The study of microbes in their natural habitats is known as **microbial ecology;** the study of the practical uses of microbes in food processing, industrial production, and biotechnology is known as industrial or **applied microbiology** (chapter 27). The two areas actually overlap to a considerable degree—largely because most natural habitats have been altered by human activities. Human intervention in natural settings has changed the earth's warming and cooling cycles, increased wastes in soil, polluted water, and altered some of the basic relationships between microbial, plant, and animal life. Now that humans are also beginning to release new, genetically recombined microbes into the environment and to alter the genes of plants, animals, and even themselves, what does the future hold? Although this question has no clear answer, we know one thing for certain: Microbes—the most vast and powerful resource of all—will be silently working in nature.

In chapter 7 we first touched upon the widespread distribution of microorganisms and their adaptations to most habitats of the world, from extreme to temperate. Regardless of their exact location or type of adaptation, microorganisms necessarily interact with their environment in complex and extraordinary ways. Microbial ecology is focused on interactions between microbes and their environment and the effects of those interactions on the earth. Unlike studies that deal with the activities of a single organism or its individual characteristics in the laboratory, ecological studies are aimed at the interactions taking place between organisms and their environment at many levels at any given moment. Therefore, ecology is a broad-based science that merges many subsciences of biology as well as geology, physics, and chemistry.

Ecological studies deal with both the biotic and the abiotic components of an organism's environment. **Biotic*** factors are defined as any living or dead organisms[1] that occupy an organism's habitat. **Abiotic*** factors include nonliving components such as atmospheric gases, minerals, water, temperature, and light. A term used to denote the collective interactions occurring between the biotic and abiotic factors in a given location is **ecosystem.** An ecosystem can vary in size from a single root of a plant to a whole forest, and it serves as the basic unit of ecological function.

1. Biologists make a distinction between nonliving and dead. A nonliving object has never been alive, whereas a dead organism was once alive but no longer is.
* *biotic* (by-ah′-tik) Gr. *bios,* living.
* *abiotic* (ay″-by-ah′-tik) Gr. *a,* not, and *bios,* living.

The Organization of the Biosphere

The earth initially may seem like a random, chaotic place, but it is actually an incredibly organized, fine-tuned machine. Ecological relationships exist at several levels, ranging from the entire earth all the way down to a single organism **(figure 26.1).** The most all-encompassing of these levels, the **biosphere,** contains all physical locations on

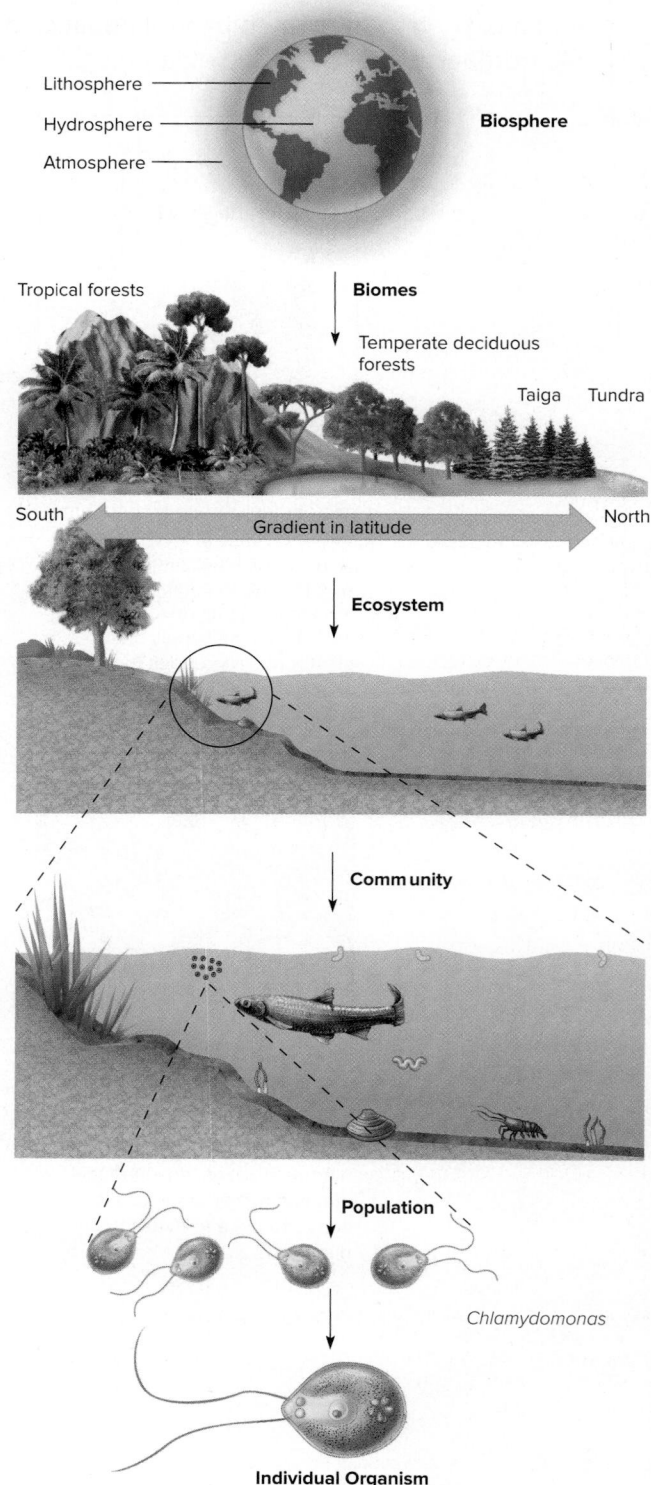

Figure 26.1 Levels of organization in an ecosystem, ranging from the biosphere to the individual organism.

earth that support life, including the thin envelope of life that sur-
rounds the earth's surface and extends some distance into the crust.
This global system comprises the **hydrosphere** (water), the **litho-
sphere** (3 to 6 miles into the earth's crust), and the **atmosphere** (a
few miles into the air). The biosphere maintains or creates the con-
ditions of temperature, light, gases, moisture, and minerals required
for life processes. The biosphere can be naturally subdivided into
terrestrial and aquatic realms. The terrestrial realm is usually dis-
tributed into particular climatic regions called **biomes,*** defined as
large assemblages of organisms that have adapted to a distinct ge-
ography and climate and are characterized by dominant life forms,
usually plants. Particular biomes include grassland, desert, moun-
tain, and tropical rain forest. The aquatic biosphere is generally di-
visible into freshwater and marine realms. We now know that the
earth's crust also supports a vast and diverse number of life forms,
estimated to be equal to or even greater than life as we know it in
aquatic and terrestrial realms.

Biomes and aquatic ecosystems are generally composed of
mixed assemblages of organisms that live together at the same
place and time and that usually exhibit well-defined nutritional or
behavioral interrelationships. These clustered associations are
called **communities.** Although most communities are identified by
their easily visualized dominant plants and animals, they also con-
tain a complex assortment of bacteria, fungi, algae, protozoa, and
even viruses. The basic unit of community structure is the **population,**
defined as a group of organisms of the same kind. For organisms
with sexual reproduction, this level is the species. Populations of
organisms can also be classified to a subspecies level if needed,
often as strains or colonies. The organizational unit of a population
is the individual organism, and each multicellular organism, in turn,
has its own levels of organization (organs, tissues, cells).

Ecosystems are generally balanced, with each organism exist-
ing in its particular habitat and niche. The **habitat** is the physical
location in the environment to which an organism has adapted. In
the case of microorganisms, the habitat is frequently a *microenvi-
ronment,* where particular qualities of oxygen, temperature, light,
and nutrient content are somewhat stable. The **niche** is the overall
role that a species (or population) serves in a community. This in-
cludes such activities as nutritional intake (what it eats), position in
the community structure (what is eating it), and rate of population
growth. A niche can be broad (such as scavengers that feed on
nearly any organic food source) or narrow (microbes that decom-
pose cellulose in forest litter or that fix nitrogen).

Note that microbes exist as communities in and on plants and
animals as well, including humans. Pure cultures are seldom found
anywhere in nature.

 Check Your Progress SECTION 26.1

1. Present in outline form the levels of organization in the biosphere.
 Define the terms *ecosystem, biome, community,* and *population.*
2. Compare the concepts of habitat and niche using *Chlamydomonas*
 (figure 26.1) as an example.

* *biomes* (by′ ohmz) Gr. *bios,* life and *ome,* group. An extensive community of
organisms whose makeup is determined by climatic region.

26.2 Energy and Nutritional Flow in Ecosystems

Expected Learning Outcomes

4. Describe the flow of energy through an ecosystem.
5. Analyze trophic structures and nutritional relationships of a
 community.
6. Outline several types of ecological associations with examples.

All living things must obtain nutrients and a usable form of energy
from their abiotic and biotic environments. The energy and nutri-
tional relationships in ecosystems can be described in a number of
convenient ways. An **energy pyramid** or **food chain** provides a
simple summary of the general trophic (feeding) levels, designated
as producers, consumers, and decomposers, and traces the flow and
quantity of available energy from one level to another **(figure 26.2).**
It is worth noting that microorganisms are the only life forms that
exist at all three major trophic levels. The nutritional roles of micro-
organisms in ecosystems are summarized in **table 26.1.**

Life would not be possible without **producers,** because they
provide the fundamental energy source that drives the trophic pyra-
mid. Producers are the only organisms in an ecosystem that can

Figure 26.2 A trophic and energy pyramid. The relative size
of the blocks indicates the number of individuals that exist at a given
trophic level. The arrow on the right indicates the amount of usable
energy from producers to top consumers. Both the number of
organisms and the amount of usable energy decrease with each
trophic level. Decomposers are an exception because they can feed
from all trophic levels (gray arrows). Blocks at the left indicate the
general nutritional types and levels that correspond with the pyramid.

TABLE 26.1 The Major Roles of Microorganisms in Ecosystems

Role	Description of Activity	Examples of Microorganisms Involved
Primary producers	Photosynthesis	Algae, bacteria, sulfur bacteria
	Chemosynthesis	Chemolithotrophic bacteria in thermal vents
Consumers	Predation	Free-living protozoa that feed on algae and bacteria; some fungi that prey upon nematodes
Decomposers	Degrading dead organisms and wastes	Soil saprobes (primarily bacteria and fungi) that degrade cellulose, lignin, and other complex macromolecules
	Mineralization of organic compounds	Soil bacteria that reduce organic compounds to inorganic compounds such as CO_2 and minerals
Cycling agents for biogeochemical cycles	Recycling compounds containing carbon, nitrogen, phosphorus, sulfur	Specialized bacteria that transform elements and keep them cycling from the biotic to the abiotic and back to the biotic phases of the biosphere
Parasites	Living and feeding on hosts	Viruses, bacteria, protozoa, fungi, and worms that play a role in population control

synthesize organic carbon compounds such as glucose by assimilating (fixing) inorganic carbon (CO_2) from the atmosphere. Being able to use CO_2 as a sole carbon source means that producers are **autotrophs.** Most of them are **photoautotrophs,** such as plants and cyanobacteria, that convert the sun's energy into chemical bond energy (covered in chapter 8). A smaller but not less important amount of CO_2 assimilation is brought about by bacteria called **chemoautotrophs.** These organisms derive energy from simple inorganic compounds such as ammonia, sulfides, and hydrogen by using redox reactions. In certain ecosystems (for thermal vents, see 7.2 Secret World of Microbes), chemoautotrophs are the only primary producers that support the energy pyramid.

Consumers feed on other living organisms and obtain energy from bonds present in the organic substrates they contain. The category includes animals, protozoa, fungi, and some bacteria. A pyramid usually has several levels of consumers, ranging from *primary consumers* (grazers or herbivores), which consume producers; to *secondary consumers* (carnivores), which feed on primary consumers; to *tertiary consumers,* which feed on secondary consumers; and up to *quaternary consumers* (usually the last level), which feed on tertiary consumers. **Figures 26.3** and **26.4** show specific organisms at these levels.

Decomposers, primarily microbes inhabiting soil and water, break down and absorb the organic matter of dead organisms, including plants, animals, and other microorganisms. Because of their biological function, decomposers are active at all levels of the food pyramid. Without this important nutritional class of saprobes, the biosphere would stagnate and die. The work of decomposers is to break organic matter down into inorganic minerals and gases that can be cycled back into the ecosystem, especially for the use of primary producers. This process, also termed **mineralization,** is so efficient that almost all biological compounds can be reduced by some type of decomposer. Numerous microorganisms decompose cellulose and lignin, polysaccharides from plant cell walls that account for the vast bulk of detritus in soil and water. Complex macromolecules from animal and plant bodies are also broken down by an assortment of bacteria and fungi. Surprisingly, decomposers can also break down most human-made

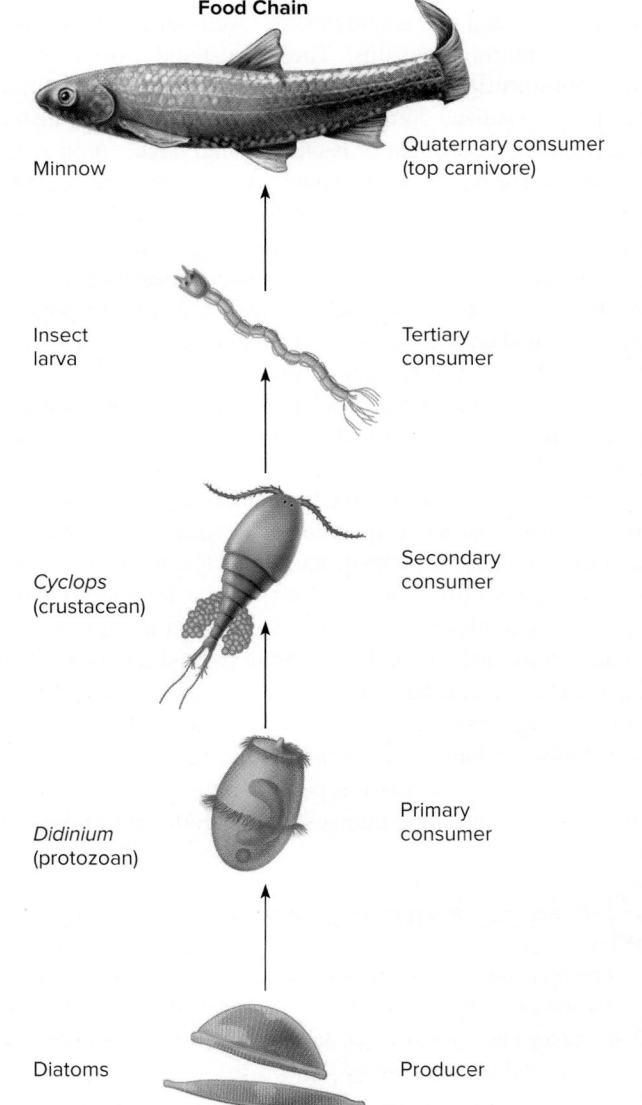

Figure 26.3 Food chain. A food chain is the simplest way to present specific feeding relationships among organisms, but it may not reflect the total nutritional interactions in a community (figure not to scale).

Food Web

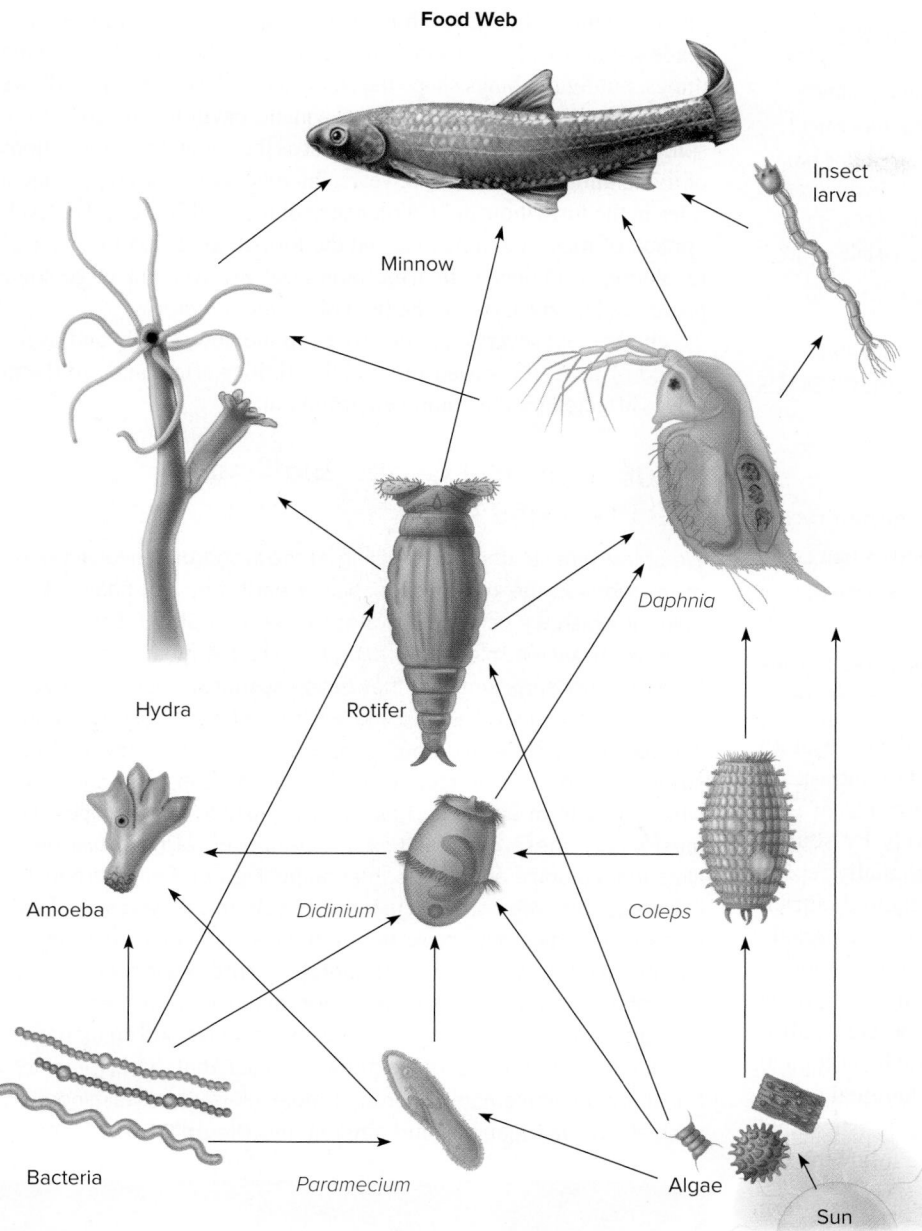

Figure 26.4 **Food web.** More complex trophic patterns are accurately depicted by a food web, which traces the multiple feeding options that exist for most organisms. Note: Arrows point toward the consumers. Compare this pattern of feeding with the chain in figure 26.3 (organisms not to scale).

energy is transferred to the next level, a large proportion (as high as 90%) of the energy will be lost in a form (primarily heat) that cannot be fed back into the system. Thus, the amount of energy available decreases at each successive trophic level. This energy loss also decreases the actual number of individuals that can be supported at each successive level.

The most basic image of a feeding pathway can be provided by a **food chain.** Although it is a somewhat simplistic way to describe feeding relationships, a food chain helps identify the types of organisms that are present at a given trophic level in a natural setting (see figure 26.3). Feeding relationships in communities are more accurately represented by a multichannel food chain, or a **food web** (see figure 26.4). A food web reflects the actual nutritional structure of a community. It can help to identify feeding patterns typical of herbivores (plant eaters), carnivores (flesh eaters), and omnivores (feed on both plants and flesh).

Ecological Interactions Between Organisms in a Community

Whenever complex mixtures of organisms associate, they develop various dynamic interrelationships based on nutrition and shared habitat. These relationships, some of which were described in earlier chapters, include mutualism, commensalism, parasitism, competition, synergism (cross-feeding), predation, and scavenging.

Mutually beneficial associations (**mutualism**), such as that of protozoans living in the termite intestine, are so well evolved that the two members require each other for survival. In contrast, **commensalism** is one-sided and independent (these terms describing relationships between organisms echo the terms we described in chapter 7). Although the action of one microbe favorably affects another, the first microbe receives no benefit. Many commensal unions involve *co-metabolism,* meaning that the waste products of the first microbe are useful nutrients for the second one, a process called **syntrophy. Parasitism** is an intimate relationship whereby a parasite derives its nutrients and habitat from a host that is usually harmed in the process. In **competition,** one microbe gives off antagonistic substances that inhibit or kill susceptible species sharing its habitat. A **predator** is a form of consumer that actively seeks out and ingests live prey (protozoa that prey on algae and bacteria). **Scavengers** are nutritional jacks-of-all-trades; they feed on a variety of food sources, ranging from live cells to dead cells and wastes.

compounds that are not naturally found on earth. This process, referred to as **bioremediation,** is discussed in 26.2 Making Connections. Often, bioremediation involves several types of microbes known as a *consortium* that collectively decompose and mineralize substances.

The pyramid in figure 26.2 illustrates several limitations of ecosystems with regard to energy. Unlike nutrients, which can be passed among trophic levels, recycled, and reused, energy does not cycle. Maintenance of complex interdependent trophic relationships such as those shown in figures 26.3 and 26.4 requires a constant input of energy at the producer level. As

 Check Your Progress SECTION 26.2

3. Compare autotrophs and heterotrophs; producers and consumers.
4. Where in the energy and trophic schemes do decomposers enter?
5. Using figures 26.3 and 26.4, point out specific examples of producers; primary, secondary, and tertiary consumers, herbivores; primary, secondary, and tertiary carnivores; and omnivores.
6. What is mineralization, and which organisms are responsible for it?

26.3 The Natural Recycling of Bioelements

 Expected Learning Outcomes

7. Summarize the main concepts pertaining to biogeochemical cycles.
8. Discuss the primary participants in and activities of the carbon cycle.
9. Describe the forms in which nitrogen is found and the contributions of microbes to the nitrogen cycle.
10. Indicate the main components of the sulfur and phosphorus cycles.

Environmental ecosystems are exposed to the sun, which constantly infuses them with a renewable source of energy. In contrast, the bioelements and nutrients that are essential components of cells and multicellular organisms are supplied exclusively by sources somewhere in the biosphere and are not being continually replenished from outside the earth. In fact, the lack of a required nutrient in the immediate habitat is one of the chief factors limiting organismic and population growth. It is for these reasons that there must be continuous and sustained **recycling** of elements and nutrients in the biosphere. Essential elements such as carbon, nitrogen, sulfur, phosphorus, oxygen, and iron are recharged through biological, geologic, and chemical mechanisms called **biogeochemical cycles.** Although these cycles vary in certain specific characteristics, they share several general qualities, as summarized in the following list:

- All elements ultimately originate from a nonliving, long-term reservoir in the atmosphere, the lithosphere, or the hydrosphere. They cycle in pure form (N_2) or as compounds (PO_4).
- Elements make the rounds between the abiotic environment and the biotic environment.
- Recycling maintains a necessary balance of nutrients in the biosphere so that they do not build up or become unavailable.
- Cycles are complex systems that rely upon the interplay of producers, consumers, and decomposers. Often the waste products of one organism become a source of energy or building material for another.
- All organisms participate directly in recycling, but only certain categories of microorganisms have the metabolic pathways for converting inorganic compounds from one nutritional form to another.

The English biologist James Lovelock has postulated a concept called the **Gaia* theory,** after the mythical Greek goddess of earth. Its primary idea proposes that the biosphere contains a diversity of

* *Gaia* (guy'-uh).

habitats and niches favorable to life because living things have made it that way. Not only does the earth shape the character of living things, but living things shape the character of the earth. After all, we know that the compositions of the aquatic environment, the atmosphere, and even the soil would not exist as they do without the actions of living things. For billions of years, microbes have played prominent roles in the formation and maintenance of the earth's crust, the development of rocks and minerals, and the formation of fossil fuels. This revolution in understanding the biological involvement in geologic processes has given rise to the field of *geomicrobiology.*

In the next several sections we examine how, jointly and over a period of time, the varied microbial activities affect and are themselves affected by the abiotic environment.

Biogeochemical Cycles, Biofilms, and the Living Earth

We started this textbook marveling at the wondrous involvement of microbes in the dynamics of planet earth and are finally at a point at which we can explore these roles in depth. Because the earth is an ancient place and microbes are the most ancient life forms, it is not surprising that they have essentially coevolved over several billion years. This is especially true of the lithosphere and hydrosphere, where microbes penetrate miles below the earth's surface and oceans. In fact, microbes dominate in most habitats, where they form stable and usually long-lasting communities. In most cases, these communities are composed of microbial biofilms that colonize and shape the composition of their microenvironments. So when we discuss the geochemical cycles, realize that most participants have adapted to a particular niche in a shared habitat, where they cooperate in processing elements through their various phases as a survival mechanism. Geomicrobiologists are actively studying these communities and so far have documented hundreds of bacterial species that participate in transforming bioelements of organic molecules as well as minerals such as iron, manganese, and other metals **(figure 26.5).**

Figure 26.5 A hot spring in Yellowstone National Park.
Extremes of temperature and pH, along with liberal quantities of phosphorus and sulfur, create a complex habitat that many microbes have evolved to exploit. Areas of intense geochemical activity are often home to complex microbial communities. © Robert Glusic/Corbis RF

Atmospheric Cycles

The Carbon Cycle

Carbon is the fundamental atom in all biomolecules and accounts for at least one-half of the dry weight of biomass, defined as the total matter produced by biological systems in a given ecosystem. The **carbon cycle** is more intimately associated with the energy transfers and trophic patterns in the biosphere than are other elements. Carbon exists predominantly in the mineral state and as an organic reservoir in the bodies of organisms. A smaller amount of carbon also exists in the gaseous state as carbon dioxide (CO_2), carbon monoxide (CO), and methane (CH_4). In general, carbon is recycled through ecosystems via carbon fixation, respiration, or fermentation of organic molecules, limestone decomposition, and methane production. A convenient starting point from which to trace the movement of carbon is with carbon dioxide gas, which occupies a central position in the cycle and represents a large common pool that diffuses into all parts of the ecosystem **(figure 26.6)**. As a general rule, the cycles of oxygen and hydrogen are closely allied to the carbon cycle.

The principal users of the atmospheric carbon dioxide pool are photosynthetic autotrophs (photoautotrophs) such as plants, algae, and bacteria. An estimated 165 billion tons of organic material per year are produced by terrestrial and aquatic photosynthesis. Although we don't yet know exactly how many autotrophs exist in the earth's crust, a small amount of CO_2 is used by these bacteria

Free CO_2 in atmosphere

Volcanos

Respiration

Combustion

Respiration

Respiration

Photosynthesis (plants, algae)

CO_3^{2-} (carbonate in sediments)

Organic carbon decomposed by microorganisms (fungi, bacteria)

Organic carbon taken in by consumers (animals, protozoa)

Figure 26.6 The carbon cycle. This cycle traces carbon from the CO_2 pool in the atmosphere to the primary producers (green), where it is fixed into protoplasm. Organic carbon compounds are taken in by consumers (blue) and decomposers (yellow) that produce CO_2 through respiration and return it to the atmosphere (pink). Combustion of fossil fuels and volcanic eruptions also add to the CO_2 pool (gray). Some of the CO_2 is carried into inorganic sediments by organisms that synthesize carbonate (CO_3) skeletons. In time, natural processes acting on exposed carbonate skeletons can liberate CO_2.

(chemoautotrophs) that derive their energy from bonds in inorganic chemicals. A review of figure 8.28 reveals that phototrophs use energy from the sun to fix CO_2 into organic compounds such as glucose that can be used in synthesis. Photosynthesis is also the primary means by which the atmospheric supply of O_2 is regenerated (see the case study for this chapter).

Just as photosynthesis removes CO_2 from the atmosphere, other modes of generating energy, such as respiration and fermentation, can be used to remove and return it. As you may recall from the discussion of aerobic respiration in chapter 8, in the presence of O_2, organic compounds such as glucose are degraded completely to CO_2, with the release of energy and the formation of H_2O. Carbon dioxide is also released by anaerobic respiration and by certain types of fermentation reactions.

A small but important phase of the carbon cycle involves certain limestone deposits composed primarily of calcium carbonate ($CaCO_3$). Limestone is produced when marine organisms such as mollusks, corals, protozoans, and algae form hardened shells by combining carbon dioxide and calcium ions from the surrounding water. When these organisms die, the durable skeletal components accumulate in marine deposits rich in limestone. As these immense deposits are gradually exposed by geologic upheavals or receding ocean levels, various decomposing agents liberate CO_2 and return it to the CO_2 pool of the water and atmosphere.

The complementary actions of photosynthesis and respiration, along with other natural CO_2-releasing processes such as limestone erosion and volcanic activity, have maintained a relatively stable atmospheric pool of carbon dioxide. Recent figures show that this balance is being disturbed as humans burn *fossil fuels* and other organic carbon sources. Fossil fuels, including coal, oil, and natural gas, were formed through millions of years of natural biological and geologic activities. Humans are so dependent upon this energy source that, over the past 55 years, the concentration of CO_2 in the atmosphere has steadily increased from 316 ppm to 400 ppm. Although this increase may seem slight and insignificant, most climatologists now feel it has begun to disrupt the delicate temperature balance of the biosphere (**26.1 Making Connections**).

Compared with carbon dioxide, methane gas (CH_4) plays a secondary part in the carbon cycle, though it can be a significant product in anaerobic ecosystems dominated by **methanogens** (methane producers), first discussed in chapter 7. In general, when methanogens reduce CO_2 by means of various oxidizable substrates, they give off CH_4. The practical applications of methanogens are covered in chapter 27 in a section on sewage treatment, and their contribution to the greenhouse effect is also discussed in 26.1 Making Connections.

The Nitrogen Cycle

Nitrogen (N_2) gas is the most abundant component of the atmosphere, accounting for nearly 79% of air volume. As we will see, this extensive reservoir in the air is largely unavailable to most organisms. Only about 0.03% of the earth's nitrogen is combined (or fixed) in some other form such as nitrates (NO_3), nitrites (NO_2), ammonium ion (NH_4^+), and organic nitrogen compounds (proteins, nucleic acids).

The **nitrogen cycle** is relatively more intricate than other cycles because it involves such a diversity of specialized microbes to maintain the flow of the cycle. In many ways, it is actually more of

26.1 MAKING CONNECTIONS

Greenhouse Gases, Fossil Fuels, Cows, Swamps, and Global Warming

The sun's radiant energy helps maintain the stability of the earth's temperature and climatic conditions. As radiation impinges on the earth's surface, much of it is absorbed, but a large amount of the infrared (heat) radiation bounces back into the upper levels of the atmosphere. For billions of years, the atmosphere has been insulated by a layer of gases (primarily CO_2; CH_4; water vapor; and nitrous oxide, N_2O) formed by natural processes such as respiration, mineralization, and methanogenesis, which are linked to biogeochemical cycles. This layer traps a certain amount of the reflected heat yet also allows some of it to escape into space. As long as the amounts of heat entering and leaving are balanced, the mean temperature of the earth will not rise or fall in an erratic or life-threatening way. Although this phenomenon, called the **greenhouse effect,** is popularly viewed in a negative light, it must be emphasized that its function for eons has been primarily to foster life.

The greenhouse effect has lately been a matter of concern because *greenhouse gases* appear to be increasing at a rate that could disrupt the temperature balance. In effect, a denser insulation layer will trap more heat energy and gradually heat the earth. The amount of CO_2 released collectively by respiration, anaerobic microbial activity, fuel combustion, and volcanic activity has increased more than 30% since the beginning of the industrial era. By far the greatest increase in CO_2 production results from human activities such as combustion of fossil fuels, burning forests to clear agricultural land, and manufacturing. Deforestation has the added

Methane production from swamps may have an even greater effect on global warming than CO_2.
Source: U.S. Fish & Wildlife Service

impact of removing large areas of photosynthesizing plants that would otherwise consume some of the CO_2.

Originally, experts on the greenhouse effect were concerned primarily about higher CO_2 levels, but it now appears that the other greenhouse gases combined may have as great a contribution as CO_2, and they, too, are increasing. One of these gases, methane (CH_4), is released by swamps and other natural sources, and it also comes from the gastrointestinal tracts of ruminant animals. Its levels have doubled over the past century. Methane traps 21 times more heat than does carbon dioxide. Other greenhouse gases such as nitrous oxide and sulfur dioxide (SO_2) are also increasing through automobile and industrial pollution.

There is little doubt in the scientific community that global warming is a reality. It has been documented that the mean temperature of the earth has increased by approximately 1.0°C since 1860. If the rate of increase continues, by 2050 a rise in the average temperature of 4°C to 5°C will escalate the melting of the polar ice caps and raise the levels of the ocean. Some experts predict extreme weather and climate change, including massive flooding of coastal regions, changes in rainfall patterns, expansion of deserts, and long-term climatic disruptions. Some of these effects have already been set in motion.

Throughout this text, we have emphasized the adaptability of microbes. Why is global warming considered such a serious problem even for them? *Answer available on Connect.*

a nitrogen "web" because of the array of adaptations that occur. Higher plants can utilize two inorganic forms of nitrogen, NO_3^- and NH_4^+; animals must receive nitrogen in organic form from plants or other animals; however, microorganisms can use all forms of nitrogen: NO_2^-, NO_3^-, NH_4^+, N_2, and organic nitrogen. The cycle includes four basic types of reactions: nitrogen fixation, ammonification, nitrification, and denitrification **(figure 26.7).**

Nitrogen Fixation The biosphere is most dependent upon the only process that can remove N_2 from the air and convert it to a form usable by living beings. This process, called **nitrogen fixation,** is the beginning step in the synthesis of virtually all nitrogenous compounds. Nitrogen fixation is brought about primarily by nitrogen-fixing bacteria in soil and water, though a small amount is formed through nonliving processes involving lightning. Nitrogen-fixing microbes have developed a unique enzyme system capable of breaking the triple bonds of the N_2 molecule and reducing the N atoms, an anaerobic process that requires the expenditure of considerable ATP. The primary product of nitrogen fixation is the

ammonium ion, NH_4^+. Nitrogen-fixing bacteria live free or in a symbiotic relationship with plants. Among the common free-living nitrogen fixers are the aerobic *Azotobacter* and *Azospirillum,* certain members of the anaerobic genus *Clostridium,* and the cyanobacteria *Anabaena* and *Nostoc.*

Root Nodules: Natural Fertilizer Factories A significant symbiotic association occurs between **rhizobia*** (bacteria in the genera such as *Rhizobium, Bradyrhizobium,* and *Azorhizobium*) and **legumes*** (plants such as soybeans, peas, alfalfa, and clover that characteristically produce seeds in pods). The infection of legume roots by these gram-negative, motile, rod-shaped bacteria causes the formation of special nitrogen-fixing organs called **root nodules (figure 26.8).** This interaction can be highly specific, requiring a unique rhizobial species or strain for each type of legume. Nodulation begins when rhizobia colonize specific sites on root hairs. From

* *rhizobia* (ry-zoh′-bee-uh) Gr. *rhiza,* root, and *bios,* to live.
* *legumes* (leg′-yoom) L. *legere,* to gather.

Figure 26.8 Nitrogen fixation through symbiosis. (a) Events leading to formation of root nodules. Cells of the bacterium *Rhizobium* attach to a legume root hair and cause it to curl. Invasion of the legume root proper by *Rhizobium* initiates the formation of an infection thread that spreads into numerous adjacent cells. The presence of bacteria in cells causes nodule formation. **(b)** Mature nodules within the roots of a legume. **(c)** *Gunnera* plants are involved in a mutualistic interaction with the cyanobacterial genus *Nostoc* (inset). *Nostoc* enters the base of the plant stem and initiates a symbiotic relationship, providing the plant with fixed nitrogen and receiving organic carbon in return. (b): Soil and Water Conservation Society (SWCS). 2000. *Soil Biology Primer. Rev. ed.* Ankeny, IA: Soil and Water Conservation Society; (c): © Steven P. Lynch; (c: inset): © McGraw-Hill Education/ Don Rubbelke, photographer

Process Figure 26.7 Simplified events in the nitrogen cycle. (1) Nitrogen fixation: Gaseous nitrogen (N_2) is acted on by nitrogen-fixing bacteria, which give off ammonia (NH_4^+). **(2) Nitrification:** Ammonia is converted to nitrite (NO_2^-) and nitrate (NO_3^-) by nitrifying bacteria. **(3) Organic Synthesis:** Plants, algae, and bacteria use nitrates to synthesize nitrogenous organic compounds (proteins, amino acids, nucleic acids). **(4) Assimilation:** Organic nitrogen compounds are used by animals and other consumers. **(5) Ammonification:** nitrogenous macromolecules from wastes and dead organisms are converted to NH_4^+ by ammonifying bacteria. NH_4^+ can be directly recycled into nitrates. **(6) Denitrification:** specialized bacteria reduce nitrates to nitrogen gas, which is returned to the atmosphere.

there, the bacteria invade deeper root cells and induce the cells to form tumorlike masses. The bacterium's enzyme system supplies a constant source of reduced nitrogen to the plant, and the plant furnishes nutrients and energy for the activities of the bacterium. The legume uses the NH_4^+ to aminate (add an amino group to) various carbohydrate intermediates and thereby synthesize amino acids and other nitrogenous compounds that are used in plant and animal synthesis.

Plant-bacteria associations have great practical importance in agriculture, because an available source of nitrogen is often a limiting factor in the growth of crops. The self-fertilizing nature of legumes makes them valuable food plants in areas with poor soils and in countries with limited resources. It has been shown that crop health and yields can be improved by inoculating legume seeds with pure cultures of rhizobia, because the soil is often deficient in the proper strain of bacteria for forming nodules.

Ammonification, Nitrification, and Denitrification In another part of the nitrogen cycle, nitrogen-containing organic matter is decomposed by various bacteria (*Clostridium, Proteus,* for example)

that live in the soil and water. Organic detritus consists of large amounts of protein and nucleic acids from dead organisms and nitrogenous animal wastes such as urea and uric acid. The decomposition of these substances splits off amino groups and produces NH_4^+. This process is thus known as **ammonification.** The ammonium released can be reused by certain plants or converted to other nitrogen compounds, as discussed next.

The oxidation of NH_4^+ to NO_2^- and NO_3^- is called **nitrification.** It is an essential conversion process for generating the most oxidized form of nitrogen (NO_3). This reaction occurs in two phases and involves two different kinds of chemoautotrophic bacteria in soil and water. In the first phase, certain gram-negative genera such as *Nitrosomonas, Nitrosospira,* and *Nitrosococcus* oxidize NH_3 to NO_2^- as a means of generating energy. Nitrite is rapidly acted upon by a second group of nitrifiers, including *Nitrobacter, Nitrosospira,* and *Nitrococcus,* which perform the *final* oxidation of NO_2^- to NO_3^-. Nitrates can be **assimilated** (incorporated into cell metabolism) through several routes by a variety of organisms (plants, fungi, and bacteria).

The nitrogen cycle is completed through **denitrification,** a series of reduction reactions that convert nitrate to atmospheric nitrogen. These reactions play a significant role in some forms of anaerobic respiration that use nitrate and other nitrogen-oxygen compounds as terminal electron acceptors. The first step, which involves the reduction of nitrate to nitrite, is so common that hundreds of different bacterial species can do it. Several species of *Bacillus, Pseudomonas, Spirillum, Rhodobacter,* and *Thiobacillus* can carry out denitrification to completion as follows:

$$NO_3^- \rightarrow NO_2^- \rightarrow NO \rightarrow N_2O \rightarrow N_2 \text{ (gas)}$$

The final product of this series returns nitrogen gas to its primary reservoir. Along the way, other gases—notably, nitrous oxide (N_2O)—are given off during incomplete denitrification. This is the main source of N_2O in the atmosphere and an important contributor to the greenhouse effect.

Sedimentary Cycles

The Sulfur Cycle

The **sulfur cycle** and the phosphorus cycle are similar in that both elements exist mostly in solid form and originate from natural sedimentary deposits in rocks, oceans, and lakes and not from the atmosphere. Sulfur exists in the elemental form (S) and as hydrogen sulfide gas (H_2S), sulfate (SO_4), and thiosulfate (S_2O_3). Most of the oxidations and reductions that convert one form of inorganic sulfur to another are accomplished by bacteria. Plants and many microorganisms can assimilate only SO_4, and animals must have an organic source. Sulfur is found in the amino acids cysteine and methionine, both of which are essential to protein function.

One of the most remarkable contributors to the cycling of sulfur in the biosphere are the thiobacilli. These gram-negative, motile rods flourish in mud, sewage, bogs, mining drainage, and brackish springs that can be inhospitable to organisms that require complex organic nutrients. But the metabolism of these specialized lithotrophic bacteria is adapted to extracting energy by oxidizing elemental sulfur, sulfides, and thiosulfate. One species, *Thiobacillus thiooxidans,* is so efficient at this process that it

secretes large amounts of sulfuric acid into its environment, as shown by the following equation:

$$Na_2S_2O_3 + H_2O + 2O_2 \rightarrow Na_2SO_4 + H_2SO_4 \text{ (sulfuric acid)}$$

The marvel of this bacterium is its ability to create and survive in the most acidic habitats on the earth. It also plays an essential part in the phosphorus cycle, and its relative, *T. ferrooxidans,* participates in the cycling of iron. Other bacteria that can oxidize sulfur to sulfates are the photosynthetic sulfur bacteria mentioned in chapter 8.

The sulfates formed from oxidation of sulfurous compounds are assimilated into biomass by a wide variety of organisms. The sulfur cycle reaches completion when inorganic and organic sulfur compounds are reduced. Bacteria in the genera *Desulfovibrio* and *Desulfuromonas* anaerobically reduce sulfates to hydrogen sulfide (H_2S) or metal sulfide as the final step in electron transport. Sites in ocean sediments and mud where these bacteria live usually emanate a strong, rotten-egg stench from H_2S and may be blackened by the iron they contain.

The Phosphorus Cycle

Phosphorus is an integral component of DNA, RNA, and ATP, and all life depends upon a constant supply of it. It cycles between the abiotic and biotic environments almost exclusively as inorganic **phosphate** (PO_4) rather than its elemental form **(figure 26.9).** The chief inorganic reservoir is phosphate rock, which contains the insoluble compound fluorapatite, $Ca_5(PO_4)_3F$. Before it can enter biological systems, this mineral must be *phosphatized*—converted into more soluble PO_4^{3-} by the action of acid. Phosphate is released naturally when the sulfuric acid produced by *Thiobacillus* dissolves phosphate rock. Soluble phosphate in the soil and water is the principal source for autotrophs, which fix it onto organic molecules and pass it on to heterotrophs in this form. Organic phosphate is returned to the pool of soluble phosphate by decomposers, and it is finally cycled back to the mineral reservoir by slow geologic processes such as sedimentation. Because the low phosphate content of many soils can limit productivity, phosphate is added to soil to increase agricultural yields. The excess runoff of fertilizer into the hydrosphere is often responsible for overgrowth of aquatic pests (see eutrophication in section 26.5 and this chapter's Case Study).

Other Forms of Cycling

The involvement of microbes in cycling elements and compounds can be escalated by the introduction of toxic substances into the environment. Such toxic elements as arsenic, chromium, lead, and mercury, as well as hundreds of thousands of synthetic chemicals introduced into the environment over the past hundred years, are readily caught up in cycles by microbial actions. Some of these chemicals will be converted into less harmful substances, but others, such as PCB and heavy metals, persist and flow along with nutrients into all levels of the biosphere. If such a pollutant accumulates in living tissue and is not excreted, it can be accumulated by living things through the natural trophic flow of the ecosystem. These processes form the basis of **bioaccumulation** and **biomagnification.** Microscopic producers such as bacteria and algae begin the accumulation process. Consumers biomagnify the toxic substances by feeding on contaminated organisms, thereby increasing the amounts at each trophic level.

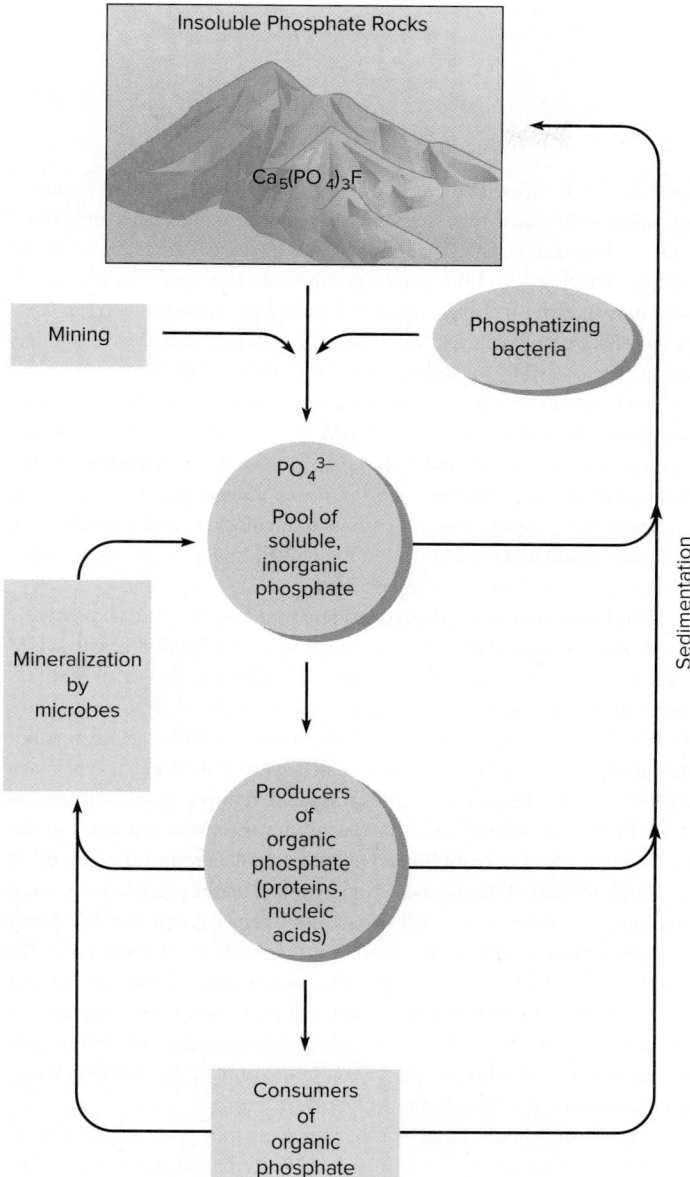

Figure 26.9 **The phosphorus cycle.** The pool of phosphate existing in sedimentary rocks is released into the ecosystem either naturally by erosion and microbial action or artificially by mining and the use of phosphate fertilizers. Soluble phosphate (PO_4^{3-}) is cycled through producers, consumers, and decomposers back into the soluble pool of phosphate, or it is returned to sediment in the aquatic biosphere.

One example of this is mercury compounds used in household antiseptics and disinfectants, agriculture, and industry. Mercury precipitates proteins by attaching to functional groups and is most toxic in the ethyl or methyl mercury form. Bacteria methylate mercury in sediments and contribute to its biomagnification through the food web. Recent studies have disclosed increased mercury content in fish taken from oceans and freshwater lakes in North America and even in canned tuna, adding to the risk in consumption of these products. This is another global problem that is being addressed by bioremediation (**26.2 Making Connections**).

 Check Your Progress SECTION 26.3

7. Outline the general characteristics of a biogeochemical cycle.
8. What are the major sources of carbon, nitrogen, phosphorus, and sulfur?
9. In what natural forms is carbon found? Name three ways carbon is returned to the atmosphere and a way it is fixed into organic compounds.
10. What carbon compound is the least available for the majority of life forms and why is this the case?
11. Describe nitrogen fixation, ammonification, nitrification, and denitrification and how microbes are involved.
12. What form of nitrogen is required by plants? By animals?
13. Summarize the main stages in the cycling of sulfur and phosphorus.
14. Explain the processes of bioaccumulation and biomagnification of toxic elements and the problems they cause.

26.4 Terrestrial Microbiology: The Composition of the Lithosphere

 Expected Learning Outcomes

11. Describe the structure of soil and how it varies in content.
12. Discuss the activities of microbes in the rhizosphere.
13. Explain how bioremediation relates to soil and water.

Descriptions such as "soiled" or "dirty" may suggest to some that soil is an undesirable, possibly harmful substance; or its appearance might suggest a somewhat homogeneous, inert substance. At the microscopic level, however, soil is a dynamic ecosystem that supports complex interactions between numerous geologic, chemical, and biological factors. This rich region, part of the lithosphere, teems with microbes, serves a dynamic role in biogeochemical cycles, and is an important repository for organic detritus and dead terrestrial organisms.

The abiotic portion of soil is a composite of mineral particles, water, and atmospheric gas. The development of soil begins when geologic sediments are mechanically disturbed and exposed to weather and microbial action.

Rock decomposition releases various-size particles ranging from rocks, pebbles, and sand grains to microscopic morsels that lie in a loose aggregate (**figure 26.10**). The porous structure of soil creates various-size pockets or spaces that provide numerous microhabitats. Some spaces trap moisture and form a liquid phase in which mineral ions and other nutrients are dissolved. Other spaces trap air that will provide gases to soil microbes, plants, and animals. Because both water and air compete for these pockets, the water content of soil is directly related to its oxygen content. Water-saturated soils contain less oxygen, and dry soils have more. Gas tensions in soil can also vary vertically. In general, the concentration of O_2 decreases and that of CO_2 increases with the depth of soil. Aerobic and facultative organisms tend to occupy looser, drier soils, whereas anaerobes would adapt to waterlogged, poorly aerated soils.

26.2 MAKING CONNECTIONS

Bioremediation: The Pollution Solution?

The soil and water of the earth have long been considered convenient repositories for solid and liquid wastes. Humans have been burying solid wastes for thousands of years, but the process has escalated in the past 50 years. Every year in the United States, about 250 million tons of pollutants, industrial wastes, and garbage are deposited into the natural environment. Often, this dumping is done with the mistaken idea that naturally occurring microbes will eventually biodegrade (break down) waste material.

Landfills currently serve as a final resting place for hundreds of castoffs from an affluent society, including yard wastes, paper, glass, plastics, wood, textiles, rubber, metal, paints, and solvents. This conglomeration is dumped into holes and is covered with soil. Although it is true that many substances are readily biodegradable, materials such as plastics and glass are not. Successful biodegradation also requires a compost containing specific types of microorganisms, adequate moisture, and oxygen. The environment surrounding buried trash provides none of these conditions. Large, dry, anaerobic masses of plant materials, paper, and other organic materials will not be successfully attacked by the aerobic microorganisms that dominate in biodegradation. As we continue to fill up hillsides with waste, the future of these landfills is a prime concern. One of the most serious of these concerns is that they will be a source of toxic compounds that seep into the ground and water.

Pollution of groundwater, the primary source of drinking water for 100 million people in the United States, is an increasing problem. Because of the extensive cycling of water through the hydrosphere and lithosphere, groundwater is often the final collection point for hazardous chemicals released into lakes, streams, oceans, and even garbage dumps. Many of these chemicals are pesticide residues from agriculture (dioxin, selenium, 2,4-D), industrial hydrocarbon wastes (PCBs), and hydrocarbon solvents (benzene, toluene). They are often hard to detect, and, if detected, are hard to remove. Recently, the technique of hydraulic fracturing (fracking), in which chemicals are injected into the Earth under high pressure to liberate trapped oil, has come under scrutiny as a potential source of groundwater pollution.

Quick Search

To see how one company is using algae in bioremediation efforts, search the Internet for "Algal Bioremediation for Waste Treatment."

For many years, polluted soil and water were simply sealed off or dredged and dumped in a different site, with no attempt to get rid of the pollutant. But now, with greater awareness of toxic wastes, many Americans are adopting an attitude known as NIMBY (Not in my backyard!), and environmentalists are troubled by the long-term effects of contaminating the earth.

In a search for solutions, waste management has turned to **bioremediation**—using microbes to break down or remove toxic wastes from water and soil. Some of these waste-eating microbes are soil and water residents with a natural capacity to decompose even artificial substances. Because the normal, unaided process occurs too slowly, most cleanups are accomplished by commercial bioremediation services that treat the contaminated soil with oxygen, nutrients, and water to increase the rate of microbial action. Through these actions, levels of pesticides such as 2,4-D can be reduced to 96% of their original levels, and solvents can be reduced from 1 million parts per billion (ppb) to 10 ppb or less.

Among the most important bioremedial microbes are species of *Pseudomonas* and *Bacillus* and various toxin-eating fungi. Although much recent work has focused on creating "superbugs" through genetic engineering, public resistance to releasing genetically modified organisms (or GMOs) in the environment is high. Thankfully, naturally occurring biodegraders are plentiful, and efforts to optimize their performance have had some success. A bacterium showing the most versatile capacity for bioremediation is *Geobacter*. This extremophile thrives on heavy metals and other toxicants. It has been used successfully to clean up petroleum spills and uranium waste.

So far about 35 recombinant microbes have been created for bioaugmentation (the release of selected GMOs to improve bioremediation). Species of *Rhodococcus* and *Burkholderia* have been engineered to decompose PCBs, and certain forms of *Pseudomonas* now contain genes for detoxifying heavy metals, carbon tetrachloride, and naphthalene. With over 3,000 toxic waste sites in the United States alone, the need for effective bioremediation is a top priority.

The passage of time has given us renewed insights into the problems of remediating huge contamination events such as oil spills. Twenty-eight years ago, the oil tanker *Exxon Valdez* ran aground on Prince William Sound in Alaska. Almost 41 million liters of crude oil were spilled into the beautiful, pristine wilderness of the sound. Initially, steam was applied under high pressure, a technique that seemed to work. The primary benefit of steam cleaning the shores was quick removal of large quantities of oil. But an adverse effect of the cleaning was to wipe out many of the bacteria that could have facilitated a more complete removal of the oil.

Another part of the cleanup strategy relied on the presence of natural communities of microorganisms that could help degrade the oil. Many microorganisms—even those inhabiting the rocks on the shores of the sound—can use oil as a source of carbon and energy. However, without added nutrients, bacterial metabolism and bioremediation often do not proceed very quickly. Understanding this, environmental microbiologists attempted to accelerate bioremediation of the oil spill by applying fertilizers containing nitrogen and phosphorus.

Today, the shores superficially appear as they were before the oil spill. Closer examination, unfortunately, reveals that oil still remains. The high-pressure cleaning forced it deeper into the rocky shores, inaccessible to microbial action. The tragedy of the *Exxon Valdez* serves as a forceful reminder of the impact of humans on nature and the limitations inherent in microbial cleanup of toxic substances.

What are some other solutions to cleaning up or getting rid of wastes and pollutants, besides microbial action? *Answer available on Connect.*

Workers steam-cleaning the site of the *Exxon Valdez* oil spill of 1989. Was bioremediation successful?
© Vanessa Vick/Science Source

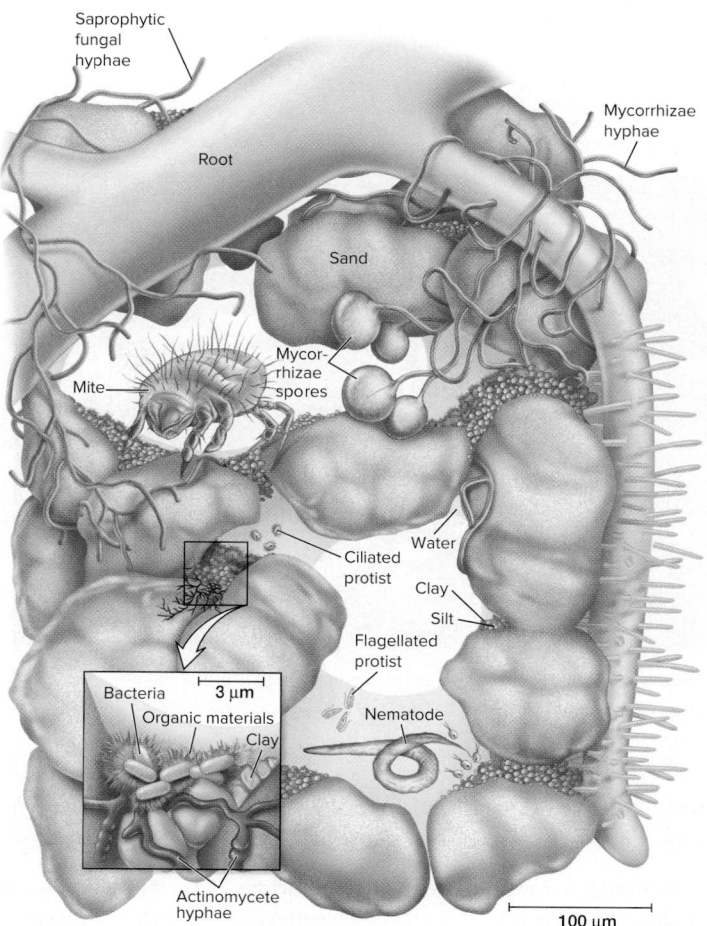

Figure 26.10 The soil habitat. A typical soil habitat contains a mixture of clay, silt, and sand along with soil organic matter. Roots and animals (e.g., nematodes and mites), as well as protozoa and bacteria, consume oxygen, which rapidly diffuses into the soil pores where the microbes live. Note that two types of fungi are present: mycorrhizal fungi, which derive their organic carbon from plant roots; and saprophytic fungi, which help degrade organic material.

Within the superstructure of the soil are varying amounts of **humus,*** the slowly decaying organic litter from plant and animal tissues. This soft, crumbly mixture holds water like a sponge. It is also an important habitat for microbes that decompose the complex litter and recycle nutrients.

The humus content varies with climate, temperature, moisture and mineral content, and microbial action. Warm, tropical soils have a high rate of humus production and microbial decomposition. Because nutrients in these soils are swiftly released and used up, they do not accumulate. Fertilized agricultural soils in temperate climates build up humus at a high rate and are rich in nutrients. The very low content of humus and moisture in desert soils greatly reduces its microbial biota, rate of decomposition, and nutrient content. Bogs are likewise nutrient-poor due to a slow rate of decomposition of the humus caused by high acid content and lack of oxygen. Humans can artificially increase the amount of humus by mixing plant refuse and animal wastes with soil and allowing natural decomposition to occur, a process called *composting.* Composting is a very active

* *humus* (hyoo'-mus) L., earth.

metabolic process that generates a great deal of heat. The temperature inside a well-maintained compost can reach 80°C to 100°C.

Living Activities in Soil

The rich culture medium of the soil supports a fantastic array of microorganisms (bacteria, fungi, algae, protozoa, and viruses). A gram of moist loam soil with high humus content can have a microbe count as high as 10 billion, each competing for its own niche and microhabitat. Some of the most distinctive biological interactions occur in the **rhizosphere,** the zone of soil surrounding the roots of plants, which contains associated bacteria, fungi, and protozoa (see figure 26.10). Plants interact with soil microbes in a truly synergistic fashion. Studies have shown that a rich microbial community grows in a biofilm around the root hairs and other exposed surfaces. Their presence stimulates the plant to exude growth factors such as carbon dioxide, sugars, amino acids, and vitamins. These nutrients are released into fluid spaces, where they can be readily captured by microbes. Bacteria and fungi likewise contribute to plant survival by releasing hormonelike growth factors and protective substances. They also convert minerals into forms usable by plants. We saw numerous examples in the nitrogen, sulfur, and phosphorus cycles.

We previously observed that plants can form close symbiotic associations with microbes to fix nitrogen. Other mutualistic partnerships between plant roots and microbial fungi are **mycorrhizae.*** These associations occur when various species of basidiomycetes, ascomycetes, or zygomycetes attach themselves to the roots of vascular plants **(figure 26.11)**. The plant feeds the fungus through photosynthesis, and the fungus sustains the relationship in several ways. By extending its mycelium into the rhizosphere, it helps anchor the plant and increases the surface area for capturing water from dry soils and minerals from poor soils. Plants with mycorrhizae can inhabit severe habitats more successfully than plants without them.

The topsoil, which extends a few inches to a few feet from the surface, supports a host of burrowing animals such as nematodes, termites, and earthworms. Many of these animals are decomposer-reducer organisms that break down organic nutrients through digestion and also mechanically reduce or fragment the size of particles so that they are more readily mineralized by microbes. Aerobic bacteria initiate the digestion of organic matter into carbon dioxide and water and generate minerals such as sulfate, phosphate, and nitrate, which can be further degraded by anaerobic bacteria. Fungal

Figure 26.11 Mycorrhizae, symbiotic associations between fungi and plant roots, favor the absorption of water and minerals from the soil. *Source:* USDA

26.1 Secret World of Microbes
Just How Old Is That Body?

When a body is found, crime scene investigators and medical examiners are often called upon to estimate the time of death, whether in real life or, more famously, on television. For short periods of time, body temperature, rigor mortis, and lividity (the manner in which blood settles after death) provide a good estimate. After a few days, however, bodies begin to decompose, helped along by saprophytic microorganisms. Until recently, little was known about the precise speed with which this occurred, and the rate at which decomposition was affected by the environment. Enter the Forensic Anthropology Center at the University of Tennessee, more commonly known as the "Body Farm." On a 2.5-acre wooded area surrounded by barbed wire, as many as 40 human bodies can be found in various states of decay. Whether shaded by a tree, exposed to the sun, clothed or not, even partially buried, each corpse is attended to by researchers and graduate students, who carefully record the rate and manner at which decomposition occurs. Beginning with the first corpse—labeled 1-81, for the month and year of his death—the actions of bacteria, fungi, and insects have been meticulously cataloged, and the decomposition process divided into four stages: fresh, bloat, decay, and dry. Research on the Body Farm has allowed forensic scientists to estimate the postmortem interval with greater accuracy than ever before. Beyond the obvious support of criminal cases, scientists have also begun to study the hundreds of different chemicals produced during decomposition, using some of the chemicals to train dogs to track human corpses, an activity carried out not only with regard to the search for murder victims but also after natural disasters. Since patient 1-81, the success of the Body Farm has spawned five additional outdoor

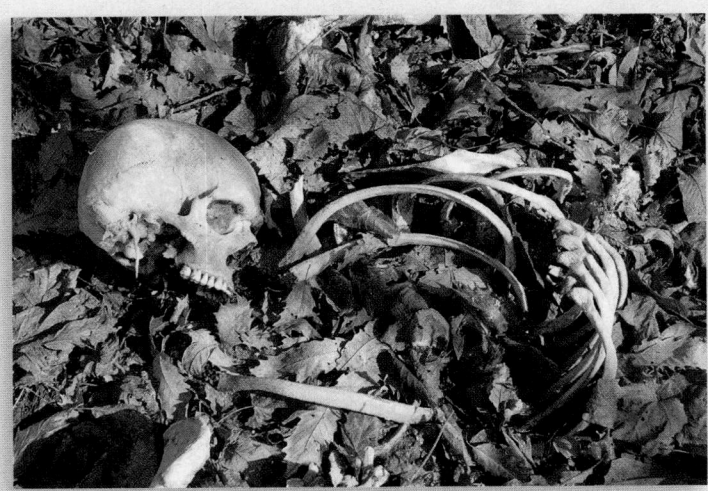

The Forensic Anthropology Center studies the manner in which environmental factors affect the actions of saprophytic microorganisms.
© University of Tennessee

laboratories across the United States. It seems that death may in fact be a growth industry.

How would you expect environmental factors like temperature and sunlight to affect a body's rate of decomposition? *Answer available on Connect.*

enzymes increase the efficiency of soil decomposition by hydrolyzing complex natural substances such as cellulose, keratin, lignin, chitin, and paraffin **(26.1 Secret World of Microbes).**

The soil is also a repository for agricultural, industrial, and domestic wastes such as insecticides, herbicides, fungicides, manufacturing wastes, and household chemicals. Applied microbiologists, using expertise from engineering, biotechnology, and ecology, work to explore the feasibility of harnessing indigenous soil microbes to break down undesirable hydrocarbons and pesticides (see 26.2 Making Connections).

26.5 The Microbiology of the Hydrosphere

Expected Learning Outcomes

14. Recall the main aspects of the hydrologic cycle.
15. Describe the structure of aquatic ecosystems.
16. Explain how aquatic environments vary in nutrient content.
17. Relate the principles involved in water monitoring.
18. Summarize the primary methods of assessing the microbiological quality of water.

The Hydrologic Cycle

Water occupies nearly three-fourths of the earth's surface. In the same manner as minerals, the earth's supply of water is continuously cycled between the hydrosphere, atmosphere, and lithosphere **(figure 26.12).** The **hydrologic cycle** begins when surface water (lakes, oceans, rivers) exposed to the sun and wind evaporates and enters the vapor phase of the atmosphere. Living beings contribute to this reservoir by various activities. Plants lose moisture through transpiration (evaporation through leaves), and all aerobic organisms give off water during respiration. Airborne moisture accumulates in the atmosphere, most conspicuously as clouds.

Water is returned to the earth through condensation or precipitation (rain, snow). The largest proportion of precipitation falls back into surface waters, where it circulates rapidly between running water and standing water. Only about 2% of water seeps into the earth or is bound in ice, but these are very important reservoirs. **Table 26.2** shows how water is distributed in the various surface compartments. Surface water collects in extensive subterranean pockets produced by the underlying layers of rock, gravel, and sand. This process forms a deep groundwater source called an **aquifer.** The water in aquifers circulates very slowly and is an important replenishing source for surface water. It can resurface through

Figure 26.12 The hydrologic cycle. The largest proportion of water cycles through evaporation, transpiration (together referred to as evapotranspiration), and precipitation between the hydrosphere and the atmosphere. Other reservoirs of water exist in the groundwater or deep storage aquifers in sedimentary rocks. Plants add to this cycle by releasing water through transpiration, and heterotrophs release it through respiration.

TABLE 26.2	Distribution of Water on Earth's Surface	
Water Source	**Water Volume, in Cubic Miles**	**Percentage of Total Water**
Oceans	317,000,000	97.24
Ice caps, glaciers	7,000,000	2.14
Groundwater	2,000,000	0.61
Freshwater lakes	30,000	0.009
Inland seas	25,000	0.008
Soil moisture	16,000	0.005
Atmosphere	3,100	0.001
Rivers	300	0.0001
Total		100

Source: U.S. Geological Survey.

springs, geysers, and hot vents, and it is also tapped as the primary supply for one-fourth of all water used by humans.

Although the total amount of water in the hydrologic cycle has not changed over millions of years, its distribution and quality have been greatly altered by human activities. Two serious problems have arisen with aquifers. First, as a result of increased well drilling, land development, and persistent local droughts, the aquifers in many areas have not been replenished as rapidly as they have been depleted. As these reserves are used up, humans will have to rely on other delivery systems such as pipelines, dams, and reservoirs, which can further disrupt the cycling of water. Second, because water picks up materials when falling through air or percolating through the ground, aquifers are also important collec-

tion points for pollutants. As we will see, the proper management of water resources is one of the greatest challenges of this century.

The Structure of Aquatic Ecosystems

Surface waters such as ponds, lakes, oceans, and rivers differ to a considerable extent in size, geographic location, and physical and chemical character. Although an aquatic ecosystem is composed primarily of liquid, it is predictably structured, and it contains significant gradients or local differences in composition. Factors that contribute to the development of zones in aquatic systems are sunlight, temperature, aeration, and dissolved nutrient content. These variations create numerous macro- and microenvironments for communities of organisms. An example of zonation can be seen in the schematic section of a freshwater lake in **figure 26.13.**

A lake is stratified vertically into three zones, or strata. The uppermost region, called the **photic zone,** extends from the surface

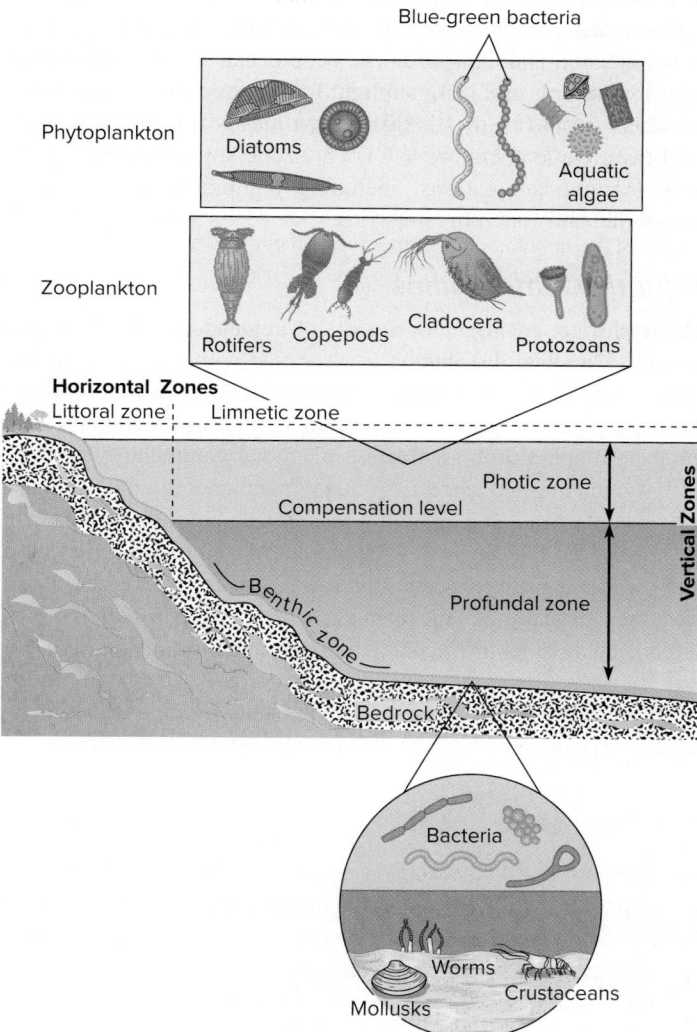

Figure 26.13 Zones of life in a freshwater lake. Depth changes in this ecosystem create significant zones that differ in light penetration, temperature, and community structure. Plankton suspended in upper zones float with the currents, whereas benthic organisms remain in or near the sediments.

to the lowest limit of sunlight penetration. Its lower boundary (the compensation depth) is the greatest depth at which photosynthesis can occur. Directly beneath the photic zone lies the **profundal* zone,** which extends from the edge of the photic zone to the lake sediment. The sediment itself, or **benthic* zone,** is composed of organic debris and mud, and it lies directly on the bedrock that forms the lake basin. The horizontal zonation includes the shoreline, or **littoral* zone,** an area of relatively shallow water. The open, deeper water beyond the littoral zone is the **limnetic* zone.**

Marine Environments

The ocean exhibits extreme variations in salinity, depth, temperature, hydrostatic pressure, and mixing. Even so, it supports a great abundance of bacteria and viruses, the extent of which has only been appreciated in very recent years. It contains a unique zone where the river meets the sea called an *estuary.* This region fluctuates in salinity, is very high in nutrients, and supports a specialized microbial community. It is often dominated by salt-tolerant species of *Pseudomonas* and *Vibrio.* Another important factor is the tidal and wave action that subjects the coastal habitat to alternate periods of submersion and exposure. The deep ocean, or **abyssal zone,** is poor in nutrients and lacks sunlight for photosynthesis, and its tremendous depth (up to 10,000 meters) makes it oxygen-poor and cold (average temperature 4°C). This zone supports communities with extreme adaptations, including halophilic, psychrophilic, barophilic, and anaerobic lifestyles.

Aquatic Communities

The freshwater environment is a site of tremendous microbiological activity. Microbial distribution is associated with sunlight, temperature, oxygen levels, and nutrient availability. The uppermost portion is the most productive self-sustaining region because it contains large numbers of **plankton,*** a floating microbial community that drifts with wave action and currents. A major member of this assemblage is the phytoplankton, containing a variety of photosynthetic algae and cyanobacteria. The phytoplankton provide nutrition for **zooplankton,** microscopic consumers such as protozoa and invertebrates that filter, feed, prey, or scavenge. The plankton support numerous other trophic levels such as larger invertebrates and fish. With its high nutrient content, the deeper region also supports an extensive variety and concentration of organisms, including aquatic plants, aerobic bacteria, and anaerobic bacteria actively involved in recycling organic detritus.

The Microbial Loop

Microbes are the only collection of organisms that participate at all levels of an aquatic ecosystem. One of their most important roles is to drive a central processing scheme ecologists call the **microbial loop.** This term denotes the dynamic interplay that keeps nutrients cycling around in a continuous loop, so that they can be made available to

the entire trophic structure. Microbes contribute to this system in numerous ways, most notably as producers, primary consumers, decomposers, and mineralizers. Their actions maintain a pool of nutrients called the *dissolved organic matter* (DOM) **(figure 26.14).**

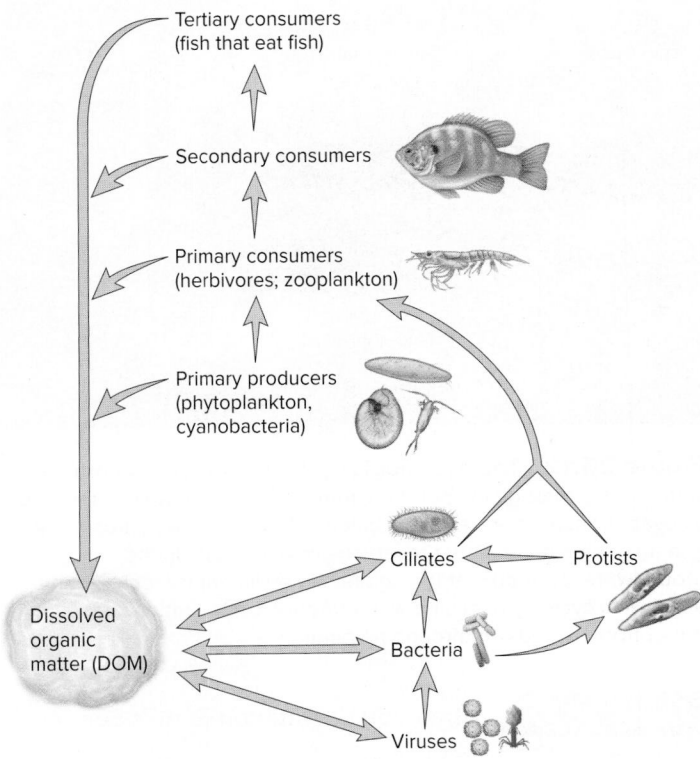

Figure 26.14 The microbial loop. This describes the complex nutrient cycle in aquatic habitats driven primarily by microorganisms. All organisms contribute to a pool of dissolved organic matter, but the protists, bacteria, and viruses keep it in dynamic flux through a constant process of removal and replenishment. The pattern of loop flow is shown in gray.

Larger bodies of standing water develop gradients in temperature or thermal stratification, especially during the summer **(figure 26.15).** The upper region, called the *epilimnion,* is warmest, and the deeper *hypolimnion* is cooler. Between these is a buffer zone, the **thermocline,** which ordinarily prevents the mixing of the two. Twice a year, during the warming cycle of spring and the cooling cycle of fall, temperature changes in the water column break down the thermocline and cause the water from the two strata to mix. Mixing disrupts the stratification and creates currents that bring nutrients up from the sediments, a process called *upwelling* (figure 26.15*b*).

Because oxygen is not very soluble in water and is rapidly used up by the plankton, its concentration exists in a gradient from highest in the epilimnion to lowest at the lower levels. In general, the amount of oxygen that can be dissolved is dependent on temperature, and warmer water holds less oxygen than cooler water. Of all the characteristics of water, the greatest range occurs in nutrient levels. Nutrient-deficient aquatic ecosystems are called **oligotrophic.*** Species that can make a living on such starvation rations are *Hyphomicrobium* and *Caulobacter.* These bacteria have special

* *profundal* (proh-fun´-dul) L. *pro,* before, and *fundus,* bottom.
* *benthic* (ben´-thik) Gr. *benthos,* depth of the sea.
* *littoral* (lit´-or-ul) L. *litus,* seashore.
* *limnetic* (lim-neh´-tik) Gr. *limne,* marsh.
* *plankton* (plang´-tun) Gr. *planktos,* wandering.

* *oligotrophic* (ahl″-ih-goh-trof´-ik) Gr. *oligo,* small, and *troph,* to feed.

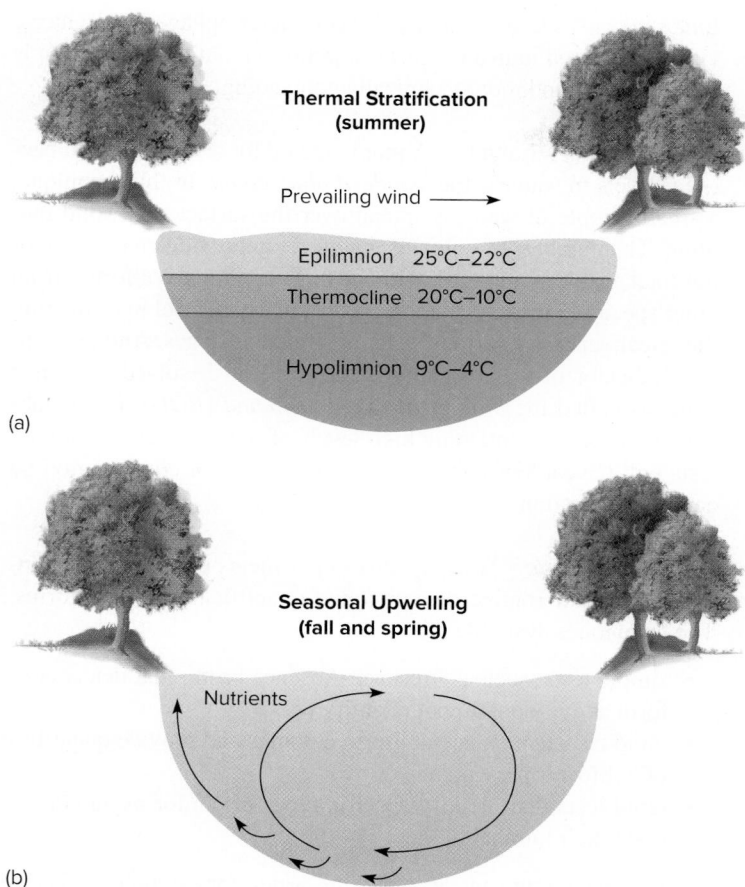

Thermal Stratification (summer)

Prevailing wind ⟶

Epilimnion	25°C–22°C
Thermocline	20°C–10°C
Hypolimnion	9°C–4°C

(a)

Seasonal Upwelling (fall and spring)

Nutrients

(b)

Figure 26.15 Profiles of a lake. (a) During summer, a lake becomes stabilized into three major temperature strata. **(b)** During fall and spring, cooling or heating of the water disrupts the temperature strata and causes upwelling of nutrients from the bottom sediments.

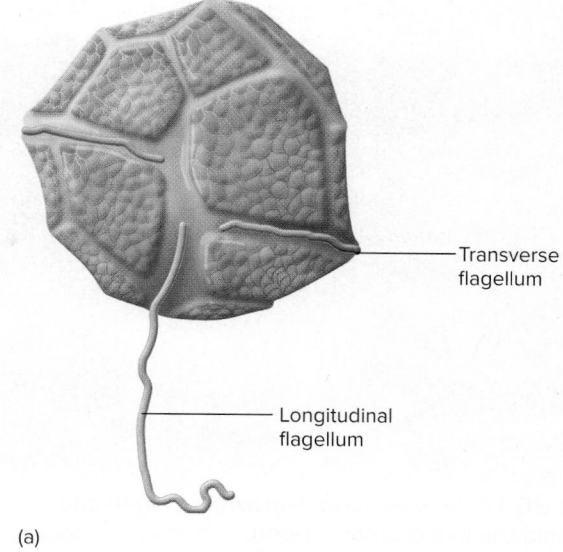

— Transverse flagellum

— Longitudinal flagellum

(a)

(b)

Figure 26.16 Red tides. (a) Single-celled red algae called dinoflagellates (*Karenia* shown here) bloom in high-nutrient, warm seawater and impart a noticeable red color to it, as shown in (b). **(b)** An aerial view of Texas coastline in the midst of a massive red tide, caused by a bloom of *K. brevis*. (b): *Source:* NOAA News Archive 120910, NOAA/Department of Commerce

stalks that capture even minuscule amounts of hydrocarbons present in oligotrophic habitats. Viruses are also well represented in aquatic habitats, with oceans and lakes containing as many as 10 million viral particles per milliter. Most of these viruses pose no danger to humans, but as parasites of bacteria, they appear to be a natural control mechanism for these populations.

At the other extreme are waters overburdened with organic matter and dissolved nutrients. Some nutrients are added naturally through seasonal upwelling and disasters (floods or typhoons). One outcome of upwelling in the ocean is the phenomenon known as *red tides* **(figure 26.16).** These periodic events are caused by dinoflagellates that grow rapidly in the enriched water and produce a potent muscle toxin. Shellfish that filter feed can concentrate these microbes and their toxins in their tissues. When people eat contaminated clams, mussels, and oysters, they are at risk for paralytic shellfish poisoning. People in coastal areas are cautioned not to eat shellfish during those months associated with red tides, which varies from region to region.

Another significant alteration of natural waters comes from effluents from sewage, agriculture, and industry that contain heavy loads of organic debris or nitrate and phosphate fertilizers. The addition of excess quantities of nutrients to aquatic ecosystems, termed **eutrophication,** * often wreaks havoc on the communities

involved. The sudden influx of abundant nutrients, along with warm temperatures, encourages a heavy surface growth of cyanobacteria and algae called a bloom **(figure 26.17).** This heavy mat of biomass effectively shuts off the oxygen supply to the lake below. The oxygen content below the surface is further depleted by aerobic heterotrophs that actively decompose the organic matter. The lack of oxygen greatly disturbs the ecological balance of the community. It causes massive die-offs of strict aerobes (fish, invertebrates), and only anaerobic or facultative microbes will survive.

Water Monitoring to Prevent Disease

Microbiology of Drinking Water Supplies We do not have to look far for overwhelming reminders of the importance of safe water. Worldwide epidemics of cholera have killed thousands of people, and an outbreak of *Cryptosporidium* in Wisconsin affecting over 400,000 people was traced to a contaminated municipal water supply. In a large segment of the world's population, the lack of sanitary water is responsible for billions of cases of diarrheal illness

* *eutrophication* (yoo″-troh-fih-kay′-shun) Gr. *eu,* good.

Figure 26.17 **Heavy surface growth of algae and cyanobacteria in a eutrophic pond.** © McGraw-Hill Education/Pat Watson, photographer

that kills nearly a million children each year. In the United States, more than 10 million people develop water-borne illness every year.

Good health is dependent upon a clean, **potable** (drinkable) water supply. This means the water must be free of pathogens; dissolved toxins; and disagreeable turbidity, odor, color, and taste. As we shall see, water of high quality does not come easily, and we must look to microbes as part of the problem and part of the solution.

Through ordinary exposure to air, soil, and effluents, surface waters usually acquire harmless, saprobic microorganisms. But along its course, water can also pick up pathogenic contaminants. Among the most prominent water-borne pathogens of recent times are the protozoans *Giardia* and *Cryptosporidium;* the bacteria *Campylobacter, Salmonella, Shigella, Vibrio,* and *Mycobacterium;* and hepatitis A and noroviruses. Some of these agents (especially encysted protozoans) can survive in natural waters for long periods without a human host, whereas others are present only transiently and are rapidly lost. The microbial content of drinking water must be continuously monitored to ensure that the water is free of infectious agents.

Attempting to survey water for specific pathogens can be very difficult and time-consuming, so most assays of water purity are more focused on detecting fecal contamination. High fecal levels can mean the water contains pathogens and is consequently unsafe to drink. Thus, wells, reservoirs, and other water sources can be analyzed for the presence of various **indicator bacteria.** These species are intestinal residents of birds and mammals, and they are readily identified using routine lab procedures.

Enteric bacteria most useful in the routine monitoring of microbial pollution are gram-negative rods called *coliforms* and enteric *streptococci,* which survive in natural waters but do not multiply there. Finding them in high numbers thus implicates recent or high levels of fecal contamination. Environmental Protection Agency standards for water sanitation are based primarily on the levels of coliforms, which are described as gram-negative, lactose-fermenting, gas-producing bacteria such as *Escherichia coli, Enterobacter,* and *Citrobacter.* Fecal contamination of marine waters that poses a risk for gastrointestinal disease is more readily correlated with gram-positive cocci, primarily in the genus

Enterococcus. Occasionally, coliform bacteriophages and enteric viruses are good indicators of fecal pollution, but their detection is more difficult and more technically demanding.

Water Quality Analysis A rapid method for testing the total bacterial levels in water is the *standard plate count.* In this technique, a small sample of water is spread over the surface of a solid medium. The numbers of colonies that develop provide an estimate of the total viable population without differentiating coliforms from other species. This information is particularly helpful in evaluating the effectiveness of various water purification stages. Another general indicator of water quality is the level of dissolved oxygen it contains called the *biological oxygen demand* (*BOD*). It is established that water containing high levels of organic matter and bacteria will have a lower oxygen content because of consumption by aerobic respiration.

Coliform Assays Water quality departments employ some standard assays for routine detection and quantification of coliforms. The techniques available are

- simple tests, such as Presence-Absence broth, that detect coliform activity but do not quantify it;
- rapid tests that isolate coliform colonies and provide quantities of coliforms present; and
- rapid tests that identify specific types of coliforms and determine their numbers.

In many circumstances (drinking water, for example), it is important to *differentiate* between facultative **coliforms** (*Enterobacter*) that are often found in other habitats (soil, water) *and* true **fecal coliforms** that live mainly in human and animal intestines. At one time, it was too difficult to differentiate *E. coli* from the closely related species of *Citrobacter, Klebsiella,* and *Enterobacter,* so laboratories instead simply reported whether a sample contained any of these isolates. All of these bacteria ferment lactose and are phenotypically similar. The terminology adopted was "coli*form-,*" meaning *E. coli*-like. In other words, such bacteria present in the sample were not necessarily *E. coli.*

Recently, microbiologists have noted serious problems with the use of coliform counts to indicate fecal contamination. The main issue is that the three other bacterial genera already mentioned, among others, are commonly found growing in nonfecal environments such as fresh water and plants that eventually become food. In other words, if the tests are not specifically for *E. coli,* you can't be sure that feces are present.

New Environmental Protection Agency Regulations change the way water quality testing is approached. These new rules require routine monitoring of public water systems, with the monitoring frequency correlated to the number of persons supplied by the system. For example, a public water system serving 2,500 people would be required to monitor the water supply twice a month, whereas one serving 2,500,000 would be required to monitor at least 420 times monthly. These samples would be tested for the presence of total coliforms, and a positive result would require several follow-up tests from the same area of the system. Only if a follow-up test is positive would specific testing for *E. coli* be required. Although the new rules alter the manner in which testing

will be accomplished, the tests themselves, as outlined in the following paragraphs, will remain essentially the same.

Standard Coliform Testing **Presence-Absence broth,** as its name implies, tests simply for the presence of coliforms in a water sample, whereas other selective and differential media can be used to identify *E. coli* specifically (**figure 26.18*a,b***). The **membrane filter** method is a widely used rapid method suitable for the field or lab to process and test larger quantities of water. This method is more suitable for dilute fluids, such as drinking water, that are relatively free of particulate matter, and it is less suitable for water containing heavy microbial growth or debris. This technique is related to the method described in chapter 11 for sterilizing fluids by filtering out microbial contaminants, except that in this system, the filter containing the trapped microbes is the desired end product. The steps in membrane filtration are diagrammed in **figure 26.18*c,d***. After filtration, the membrane filter is placed in a Petri dish containing selective and differential media. After incubation, both total and fecal coliform colonies can be counted and often presumptively identified by their distinctive characteristics on these media (**figure 26.18*e***).

Another more time-consuming but useful technique is the **most probable number (MPN)** procedure, which detects coliforms by a series of **presumptive, confirmatory,** and **completed** tests. The MPN procedure requires several dozen tubes of media to be inoculated, incubated, and evaluated over the course of 3 days. The results are then subjected to a mathematical analysis, which determines the statistically most likely number of coliforms (see Appendix B and table B.2). Its main drawback, beyond time and expense, is that it does not differentiate between total and fecal coliforms.

When water tests negative for coliforms, it is generally considered fit for human consumption. While the new regulations—more correctly known as the *Revised Total Coliform Rule*—do allow for a small number of total coliform-positive samples, there is no acceptable level for fecal coliforms, enterococci, viruses, or pathogenic protozoans in drinking water. Water that will not be consumed, but is used for fishing or swimming, is permitted to have counts up to about 126 coliforms per 100 ml.

(c) Membrane filter technique. The water sample is filtered through a sterile membrane filter assembly and collected in a flask.

(d) The filter is removed and placed in a small Petri dish containing a differential selective medium such as M-FD endo agar and incubated.

(a) Phenol red broth supplemented with lactose is often used as a preliminary indicator of the presence of coliforms. The medium contains lactose and a pH indicator, along with nutrients needed to support bacterial growth. The presence of coliforms leads to fermentation of the lactose, lowering the pH and causing the medium to turn from red (negative) to yellow (positive). A small inverted tube (called a Durham tube) is placed within the larger culture tube to trap any gas that is produced by the bacteria.

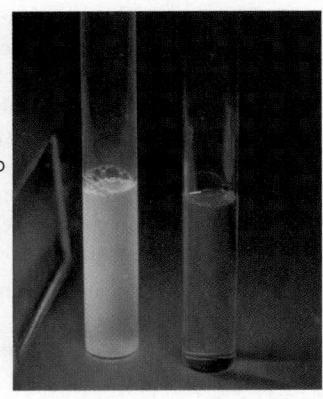

(b) EC medium with MUG is used to detect *Escherichia coli*. Selective agents inhibit gram-positive bacteria, while MUG (4-methylumbelliferyl-β-D-glucuronide) acts as a substrate that only *E. coli* can hydrolyze to produce a fluorescent product as seen in the tube on the left (a negative reaction is seen in the tube on the right). Typically, in a positive reaction, gas from fermentation is also observed (note the bubbles at the surface of the media in the tube on the left).

(e) On M-FD endo medium, colonies of *Escherichia coli* often yield a noticeable metallic sheen. The medium permits easy differentiation of various genera of coliforms, and the grid pattern can be used as a guide for rapidly counting the colonies.

Figure 26.18 **Rapid methods of water analysis for coliform contamination.** (a): © Harold J. Benson; (b): Courtesy of Biokar Diagnostics; (e): © Kathy Park Talaro

 Check Your Progress SECTIONS 26.4 AND 26.5

15. Describe the composition of the soil, the rhizosphere, and humus.
16. Outline the modes of cycling water through the lithosphere, hydrosphere, and atmosphere.
17. What are the roles of precipitation, condensation, respiration, transpiration, surface water, and aquifers?
18. What causes the formation of the epilimnion, hypolimnion, and thermocline?
19. What is upwelling and how is it important?
20. Why must water be subjected to microbiological analysis?
21. What are the characteristics of good indicator organisms, and why are they monitored rather than pathogens?
22. Give specific examples of indicator organisms and water-borne pathogens.
23. Describe two methods of water analysis.

 CASE STUDY Part 2

If you were designing an ecosystem to favor the growth of algae and cyanobacteria, you couldn't do much better than the area encompassing Lake Erie. The lake is surrounded by more cropland and pasture than any of the other Great Lakes and has far less undisturbed forests and wetlands. Consequently the Lake Erie watershed contains higher amounts of phosphorous (from fertilizer) and nitrogen (from animal waste) than any of the other Great Lakes. Lake Erie is also the shallowest of the Great Lakes, with a depth of just 210 feet, meaning it gets far more sunlight and is warmer than the other lakes. By comparison, Lake Superior is over 1,300 feet deep. Together these factors ensure that photosynthesis occurs exceptionally well, making Lake Erie the most productive of the Great Lakes. Throughout the latter half of the twentieth century, harmful algae blooms in Lake Erie had become increasingly common, and in 2011 the microcystin levels in Maumee Bay briefly reached 1,200 PPB, more than 750 times the level considered safe. Toledo has long been aware of its water problems and has been working on a number of solutions. Over 80% of the phosphorous that enters the lake does so during severe rainstorms, so rules have been established that prohibit fertilization when rain is in the forecast. Broadcast application of fertilizer has also been discouraged, as this leaves the fertilizer on top of the soil and encourages runoff. Climate change has increased the number of storm runoff events, and an increase in the severity of storms has in many cases driven cyanobacteria far below the surface of the lake and closer to subsurface water intakes. The city of Toledo has installed a number of buoys that transmit information about water temperature, wind, and cyanobacteria concentration in the area of the water intake, allowing additional purification measures to be implemented if needed.

■ *Those who study the Great Lakes describe Lake Erie as having "2% of the water but 50% of the fish." How might continued harmful algal blooms affect this description?*

To read more on this topic, search the Internet for "Ohio Sea Grant HABS."

To conclude this Case Study, go to Connect.

 Chapter Summary with Key Terms

26.1 Ecology: The Interconnecting Web of Life
This chapter emphasizes microbial activities that help maintain, sustain, and control the life support systems on the earth. This includes the natural roles of microorganisms in the environment and their contributions to the ecological balance, including soil, water, and mineral cycles.
 A. Microbial ecology deals with the interaction between the environment and microorganisms. The environment is composed of **biotic** (living or once-living) and **abiotic** (nonliving) components. The combination of organisms and the environment make up an **ecosystem.**
 B. **Ecosystem Organization**
 1. Living things inhabit only that area of the earth called the **biosphere,** which is made up of the **hydrosphere** (water), the **lithosphere** (soil), and the **atmosphere** (air).
 2. The biosphere consists of **biomes,** large-scale terrestrial systems adapted to distinct climatic and geographic environments, and aquatic ecosystems.
 3. Biomes contain **communities,** assemblages of coexisting organisms.
 4. Communities consist of **populations,** groups of like organisms of the same species.
 5. The space within which an organism lives is its **habitat;** its role in community dynamics is its **niche.**

26.2 Energy and Nutritional Flow in Ecosystems
 A. Organisms derive nutrients and energy from their habitat.
 B. Their collective trophic status relative to one another is summarized in a **food** or **energy pyramid.**
 C. At the beginning of the chain or pyramid are **producers**—organisms that synthesize large, complex organic compounds from small inorganic molecules.
 D. The levels above producer are occupied by **consumers,** organisms that feed upon other organisms.
 E. **Decomposers** are consumers that obtain nutrition from dead organisms and recycle and **mineralize** nutrients.
 F. **Bioremediation** is the process by which microbes, or communities of microbes, decompose chemicals that are harmful to the environment and its inhabitants.
 G. **Bioaccumulation** results when nondegradable substances are passed through trophic levels.

26.3 The Natural Recycling of Bioelements
 A. **Atmospheric Cycles**
 Key compounds in the **carbon cycle** include carbon dioxide, methane, and carbonates.
 1. Carbon is fixed when autotrophs (photosynthesizers) add carbon dioxide to organic carbon compounds.

2. The **nitrogen cycle** requires four processes and several types of microbes.
 a. In **nitrogen fixation,** atmospheric N_2 gas (the primary reservoir) is converted to NO_2^-, NO_3^-, or NH_4^+ salts.
 b. **Ammonification** is a stage in the degradation of nitrogenous organic compounds (proteins, nucleic acids) by bacteria to ammonium.
 c. Some bacteria **nitrify** NH_4^+ by converting it to NO_2^- and to NO_3^-.
 d. **Denitrification** is a multistep microbial conversion of NO_3^- to atmospheric nitrogen.
B. **Sedimentary Cycles**
 1. In the sulfur cycle, environmental sulfurous compounds are converted into useful substrates and returned to the inorganic reservoir through the action of microbes.
 2. The chief compound in the phosphorus cycle is phosphate (PO_4) found in certain mineral rocks. Microbial action on this reservoir makes it available to be incorporated into organic phosphate forms.
 3. Microorganisms often cycle and help accumulate heavy metals and other toxic pollutants that have been added to habitats by human activities.

26.4 Terrestrial Microbiology: The Composition of the Lithosphere
A. Soil is a dynamic, complex ecosystem that accommodates a vast array of microbes, animals, and plants coexisting among rich organic debris, water and air spaces, and minerals.

B. Numerous biological activities occur in soil, including decomposition of organic matter in **humus,** mineralization, and biogeochemical cycling.

26.5 The Microbiology of the Hydrosphere
A. The surface water, atmospheric moisture, and groundwater are linked through a **hydrologic cycle** that involves evaporation and precipitation. Living things contribute to the cycle through respiration and transpiration.
B. Aquatic ecosystems are supported by the **microbial loop,** a close community of microbes that maintain nutrient flow and cycling.
C. The diversity and distribution of water communities are related to sunlight, temperature, aeration, and dissolved nutrients. Phytoplankton and zooplankton drifting in the uppermost zone constitute a microbial community that supports the aquatic ecosystem.
D. Water Monitoring
 1. Providing **potable water** is central to prevention of water-borne disease.
 2. Water is constantly surveyed for **indicator bacteria** such as *E. coli,* **coliforms,** and enterococci, that may signal fecal contamination.
 3. Assays for possible water contamination include the standard plate count and membrane filter tests to identify and enumerate total and fecal coliforms.

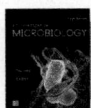

Level I. Knowledge and Comprehension

These questions require a working knowledge of the concepts in the chapter and the ability to recall and understand the information you have studied

Multiple-Choice Questions

Select the correct answer from the answers provided. For questions with blanks, choose the combination of answers that most accurately completes the statement.

1. Which of the following is *not* a major subdivision of the biosphere?
 a. hydrosphere
 b. lithosphere
 c. stratosphere
 d. atmosphere

2. A/an _____ is defined as a collection of populations sharing a given habitat.
 a. biosphere
 b. community
 c. biome
 d. ecosystem

3. The quantity of available nutrients _____ from the lower levels of the energy pyramid to the higher ones.
 a. increases
 b. decreases
 c. remains stable
 d. cycles

4. Which of the following is/are considered a greenhouse gas?
 a. CO_2
 b. CH_4
 c. N_2O
 d. all of these

5. Root nodules contain _____, which can _____.
 a. *Azotobacter,* fix N_2
 b. *Nitrosomonas,* nitrify NH_3^-
 c. rhizobia, fix N_2
 d. *Bacillus,* denitrify NO_3^-

6. Which element(s) has/have an inorganic reservoir that exists primarily in sedimentary deposits?
 a. nitrogen
 b. phosphorus
 c. sulfur
 d. both b and c

7. Which of the following bacteria would be the most accurate indicator of fecal contamination?
 a. *Enterobacter aerogenes*
 b. *Thiobacillus acidophilus*
 c. *Escherichia coli*
 d. *Staphylococcus aureus*

8. The floating assemblage of microbes and animals that drifts on or near the surface of large bodies of water is the _____ community.
 a. abyssal
 b. benthic
 c. littoral
 d. plankton

9. An oligotrophic ecosystem would be most likely to exist in a/an
 a. ocean
 b. high mountain lake
 c. tropical pond
 d. polluted river

10. Which of the following does not vary predictably with the depth of the aquatic environment?
 a. dissolved oxygen
 b. temperature
 c. penetration by sunlight
 d. salinity

 Case Study Review

1. The cyanobacteria in Lake Erie would best be described as _____.
 a. decomposers
 b. tertiary consumers
 c. primary producers
 d. heterotrophs
 e. secondary consumers

2. Increased average water temperature in Lake Erie as a result of climate change would most likely have what effect?
 a. decreased harmful algal blooms, as the higher temperatures would quickly kill most algae

 b. decreased fish populations as a result of eutrophication
 c. longer algal blooms due to longer periods of warm water
 d. b and c
 e. a, b, and c

3. During the summer of 2015 a harmful algal bloom took place off the coast of Central California. In the weeks and months that followed, officials banned the sale of clams, oysters, and crabs, and several sea lions in the area were found sick or dead. How do these events relate to one another?

 Writing Challenge

For each question, compose a one- or two-paragraph answer that includes the factual information needed to completely address the question. Check Your Progress questions can also be used for writing-challenge exercises.

1. a. What factors cause energy to decrease with each trophic level?
 b. How is it possible for energy to be lost and the ecosystem to still run efficiently?
 c. Are the nutrients on the earth a renewable resource? Why or why not?

2. Compare and contrast root nodules with mycorrhizae.

3. Because elemental poisons, such as heavy metals, cannot be further degraded even by microbes, what is a possible fate of these metals?

4. Draw a diagram that follows the effects of CO_2 on climate and the interactions between CO_2 levels and phytoplankton.

5. Explain the significant ways in which the ocean is involved in the carbon cycle.

6. Explain what is meant by the "microbial loop."

 Concept Mapping

On Connect you can find an Introduction to Concept Mapping that provides guidance for working with concept maps along with concept mapping activities for this chapter.

 Level II. Application, Analysis, Evaluation, and Synthesis

These problems go beyond just restating facts and require higher levels of understanding and an ability to interpret, problem solve, transfer knowledge to new situations, create models, and predict outcomes.

 Critical Thinking

Critical thinking is the ability to reason and solve problems using facts and concepts. These questions can be approached from a number of angles, and in most cases, they do not have a single correct answer.

1. Biologists can set up an ecosystem in a small, sealed aquarium that continues to function without maintenance for years. Describe the minimum biotic and abiotic components it must contain to remain balanced and stable.

2. Observe the carbon and nitrogen cycles and explain those places in the cycles where interactions in biofilms would be important factors.

3. a. Is the greenhouse effect harmful under ordinary circumstances?
 b. List the primary greenhouse gases and explain their effects on the earth's temperature.
 c. What occurrence has made them dangerous to the global ecosystem?
 d. What could each person do on a daily basis to cut down on the potential for disrupting the delicate balance of the earth's ecosystems?

4. Why are organisms in the abyssal zone of the ocean necessarily halophilic, psychrophilic, barophilic, and anaerobic?

5. a. What eventually happens to the nutrients that run off into the ocean with sewage and other effluents?

 b. Why can high mountain communities usually dispense with water treatment?

6. If we are to rely on microorganisms to biodegrade wastes in landfills, aquatic habitats, and soil, list some ways that this process could be made more efficient.

 Visual Challenge

1. From **chapter 3, figure 3.17c.** If this medium streaked with 1 loopful of well water showed the result below, would you report that coliforms were present in the well, and is the water potable? Explain your answer.

© Kathy Park Talaro

2. From **chapter 8, figure 8.27.** What process does this represent? How does it link to the biogeochemical cycles from this chapter?

STUDY TIP: Quickly create a customized practice quiz for this chapter by going to your Connect SmartBook "My Reports" section. Don't have SmartBook and want to learn more? Go to **whysmartbook.com.**

Escherichia coli O157:H7 (6,836×).
Source: Janice Haney Carr/CDC

Clostridium botulinum (1,000×).
Source: CDC

Staphylococcus aureus (20,000×).
Source: Janice Haney Carr/CDC

 CASE STUDY Part 1 **Something You Can Really Sink Your Teeth Into**

Near the end of October 2015, hospitals in Washington and Oregon treated over three dozen people suffering from **vomiting** and **bloody diarrhea,** leading to 14 hospitalizations. Stool samples revealed the presence of **shiga toxin-producing *E. coli*** (STEC), and further testing identified all of the ill people as being infected with a strain of the bacterium known as STEC O26. On November 4 the Washington State Department of Health and the Oregon Health Authority announced as part of their **investigation** into the **outbreak** that all of the ill persons had eaten at one of several locations of Chipotle Mexican Grill in the week prior to becoming ill. Chipotle voluntarily closed 43 locations in Washington and Oregon while the U.S. Food and Drug Administration began testing various **food** items for the presence of the **suspect bacteria.** Over the next month the outbreak grew to 53 people in nine states, and the CDC noted that more cases may appear in the near future because of the long lag time **(2 to 3 weeks)** between a person becoming ill and when the illness is reported.

On April 21, 2015, a hospital in Fairfield County, Ohio, reported a case of suspected **botulism** to the Ohio Department of Health. Botulism is a severe, potentially fatal, paralytic illness marked in its early stages by **blurred vision,** slurred speech, dizziness, and difficulty swallowing. Later, as the **botulinum neurotoxin** spreads throughout the

> "Everyone knows what potlucks are like. They know food poisoning can happen anywhere."

body, facial paralysis, an impaired gag reflex, and **musculoskeletal weakness** become evident. Eventually **paralysis of the respiratory muscles** leads to death. Within two hours of the first case being reported, four others arrived at the emergency room; the **first death** occurred that afternoon. In all, 25 patients were confirmed as having botulism, the worst outbreak in 40 years. All of them had eaten at the same church potluck meal **2 days** previously.

On July 30, 2012, the emergency department at a military hospital examined 13 persons complaining of nausea, vomiting, **abdominal pain**, or diarrhea. Nine others were later identified who had similar symptoms but did not seek medical attention. All were determined to be suffering from staphylococcal food poisoning, and staphylococcal **enterotoxin** was isolated from a single dish at a lunch party that all the patients attended. The average time between eating at the party and the onset of severe illness was **3 hours.**

■ *What is the difference between food infection and food intoxication?*

■ *Which of these cases involved food infection and which were intoxication?*

To continue the Case Study, go to Case Study Part 2 at the end of the chapter.

SCOPING OUT THE CHAPTER

We've reached the end. We've spent quite a while discussing what a microbe is; how it survives; how we can help it to grow, or kill it; and how it causes harm, or not. Not terribly different from humans thousands of years ago discovering new plants and animals for the first time. Which were dangerous? poisonous? delicious? Which could pull a plow, or supply us with milk or eggs? In this chapter we discuss the myriad ways in which we use microbes to our own ends. In many ways, microbes are the twenty-first century's version of a domesticated animal.

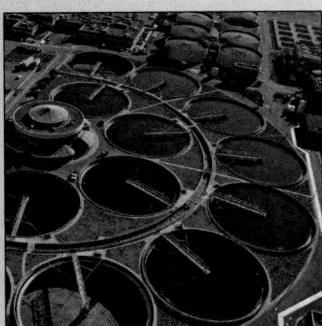

A modern water reclamation plant does exactly what would happen in nature, only faster. We allow bacteria, algae, and protozoans to do what comes naturally to them—degrade organic waste, turning toxic sewage into little more than nutrient rich water.

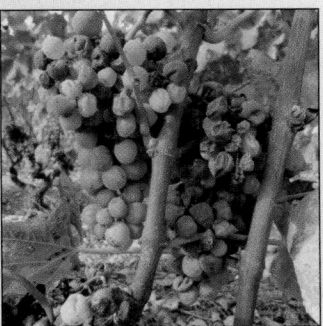

The biochemical pathways of microbial metabolism can be harnessed to create many of our foods—turning grapes into wine, milk into yogurt, causing bread to rise, and even adding the holes to your Swiss cheese.

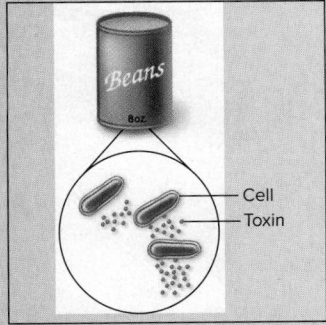

One person's delicacy is another person's garbage, such is the line between microbial enhancement and microbial spoilage. And while we may differ on whether kimchi or limburger is delectable or disgusting, food-borne illness is always bad, and its prevention is a top priority in any kitchen.

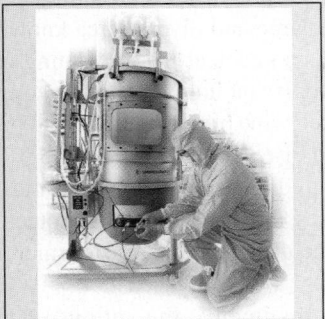

Using microbes to make wine or digest sewage is nice, but nothing they don't do already. With the advent of genetic engineering we now have the ability to use microbes to make proteins from other species. From vaccines to pharmaceuticals to commercial enzymes, we are now doing things in microbial cells that we used to do in brick-and-mortar factories.

(top: left): © Kent Knudson/PhotoLink/Getty Images RF; (top: right): © David Lytle; (bottom: right): © Photo courtesy of Sartorius Stedim Biotech

27.1 Applied Microbiology and Biotechnology

Expected Learning Outcomes

1. Describe the primary interests of applied and industrial microbiology.

2. Outline the steps in treatment of municipal water supplies.

3. Summarize the stages in sewage and wastewater treatment.

In this chapter we address the diverse ways microbes are harnessed for specific metabolic tasks that can provide practical benefits to humans. One area of interest, called **applied microbiology,** takes advantage of the microbes living in natural habitats to treat wastewater and bioremediate damaged environments. Because of this, it correlates directly with several concepts introduced in chapter 26 on environmental microbiology. Another area—**industrial microbiology**—explores the use of microbes in making a wide variety of food, medical, manufacturing, and agricultural products. Although this may be an "artificial" use of microbes, it applies many of the principles we learned about the metabolic and adaptive potentials of microorganisms covered in chapters 7, 8, and 9.

In chapter 26 we witnessed the profound involvement of microbes in the natural world. Although our daily encounters with them usually go unnoticed, human and microbial life are clearly intertwined on many levels. It is no wonder that long ago humans began to discover their potential for practical applications. These efforts eventually gave rise to the large and diverse area known as **biotechnology.** Biotechnology has an ancient history, dating back nearly 6,000 years to those first observant humans who discovered that grape juice left sitting produced wine or that bread dough properly infused with a starter would rise. Today, biotechnology has become a fertile ground for hundreds of applications in industry, medicine, agriculture, food sciences, and environmental protection, and it has even come to include the genetic alterations of microbes and other organisms.

Most biotechnological systems involve the actions of bacteria, yeasts, and molds that are able to synthesize a certain food, drug, organic acid, alcohol, or vitamin. Many such end products are obtained through industrial **fermentation,** a general term used here to refer to the mass, controlled culture of microbes to produce desired organic compounds. It also includes the use of microbes in sewage control, metal mining, and bioremediation. This term in biotechnology is not used in the same sense as metabolic fermentation, even though some processes (beer, wine production) do involve biochemical fermentation as described in chapter 8.

These subjects could easily be the inspiration for a whole book, let alone a single chapter. Our coverage of necessity only touches on some basic concepts important for rounding out your background in microbiology. The rest of section 27.1 is devoted to management of water and sewage; sections 27.2, 27.3, and 27.4 cover food microbiology; and section 27.5 concerns industrial biotechnology.

Microorganisms in Water and Wastewater Treatment

Management of Drinking Water

Most drinking water comes from rivers, aquifers, and springs. Only in remote, undeveloped, or high mountain areas is this water used in its natural form. In most cities, it must be treated before it is supplied to consumers. Water supplies such as deep wells that are relatively clean and free of contaminants require less treatment than those from surface sources laden with wastes. The stepwise process in water purification as carried out by most cities is shown in **figure 27.1.**

Treatment begins with the impoundment of water in a large reservoir such as a dam or catch basin that serves the dual purpose of storage and sedimentation. The access to reservoirs is controlled to avoid contamination by animals, wastes, and runoff water. In addition, overgrowth of cyanobacteria and algae that add undesirable qualities to the water is prevented by pretreatment with copper sulfate (0.3 ppm). Sedimentation to remove large particulate matter is also encouraged during this storage period.

Next the water is pumped to holding ponds or tanks, where it undergoes further settling, aeration, and filtration. Larger contaminants are filtered out, while smaller particles are aggregated together and allowed to settle. After several rounds of filtration/settling, the water may be softened to remove minerals and adjust

Process Figure 27.1 The major steps in water purification as carried out by a modern municipal treatment plant. (1) Water destined for the tap is pumped from lakes and rivers and screened **(2)** to remove large objects such as tree branches. **(3)** Chemicals are added to cause aggregation of small particles, which settle **(4)** in large tanks. **(5)** Calcium and magnesium—responsible for water hardness—are removed, the water is allowed to settle again **(6)**, and the pH is lowered to about 7.8 through the addition of carbon dioxide **(7)**. After filtering the water through gravel, sand, and anthracite **(8)**, it is disinfected with chlorine or chloramines. Chemicals, including fluoride and corrosion inhibitors, may be added at this point **(9)**, and the now potable water is pumped to large storage tanks **(10)**. **(11)** Potable water is delivered to individual households through municipal pipes.

its pH—water that is too acidic or basic is hard on plumbing and may even leach elements like lead from the pipes, as happened in Flint, Michigan, in 2015. In some communities fluoride or corrosion inhibitors are added at this point. Finally, the water is disinfected through the addition of chlorine or, more commonly, chloramines (see chapter 11). A few water plants in the United States utilize peroxide or ozone for final disinfection, but these methods are expensive and have no residual antimicrobial effects, an important point as water moves through the delivery system. The final quality varies, but most tap water has a slight odor or taste from disinfection.

> **Quick Search**
> To view a video on the complete process of water purification, search the Internet for "Water: The Best Deal Around."

Management of Wastewater and Sewage

In many parts of the world, the same water that serves as a source of drinking water is also used as a dump for solid and liquid wastes (**figure 27.2**). Continued pressure on the finite water resources may require reclaiming and recycling of contaminated water such as sewage. **Sewage** is the used wastewater draining out of homes and industries that contains a wide variety of chemicals, debris, and microorganisms. The dangers of typhoid, cholera, and dysentery linked to the unsanitary mixing of household water and sewage have been a threat for centuries. In current practice, some sewage is treated to reduce its microbial load before release, but a large quantity is still being emptied raw (untreated) into the aquatic environment primarily because heavily contaminated waters require far more stringent and costly methods of treatment than are currently available to most cities.

Sewage contains large amounts of solid wastes, dissolved organic matter, and toxic chemicals that pose a health risk. To remove all potential health hazards, treatment typically requires three phases: The primary stage separates out large matter; the secondary stage reduces remaining matter and can remove some toxic substances; and the tertiary stage completes the purification of the water (**figure 27.3**). Microbial activity is an integral part of the overall process. The systems for sewage treatment are massive engineering marvels.

In the **primary stage** of treatment, floating bulkier materials such as paper, plastic waste, and bottles are skimmed off. The remaining smaller, suspended particulates are allowed to settle. Sedimentation in settling tanks usually takes 2 to 10 hours and leaves a mixture rich in organic matter. This aqueous residue is carried into a **secondary stage** of active microbial decomposition, or biodegradation. In this stage, a diverse community of natural bioremediators (bacteria, algae, and protozoa) aerobically decomposes the remaining particles of wood, paper, fabrics, petroleum, and organic molecules inside a large digester tank (**figure 27.4**). This forms a suspension of material called *sludge* that tends to settle out and slow the process. Most processing plants improve the microbial decomposition of the sludge through a system of aeration, stirring, and recirculation, which forms **activated sludge.** This action mineralizes a large amount of organic matter into sulfates, nitrates, phosphates, carbon dioxide, and water. Certain volatile gases such as hydrogen sulfide, ammonia, nitrogen, and methane may also be released. Water from this process is siphoned off and carried to the **tertiary stage,** which involves further filtering and chlorinating prior to discharge. Such reclaimed sewage water is usually used to water golf courses and parks rather than for drinking, or it is gradually released into large bodies of water.

In some cases, the solid waste that remains after aerobic decomposition is harvested and reused. Its rich content of nitrogen, potassium, and phosphorus makes it a useful fertilizer. But if the waste contains large amounts of nondegradable or toxic substances, it must be disposed of properly. In many parts of the world, the sludge, which still contains significant amounts of simple but useful organic matter, is used as a secondary source of energy. Further digestion is carried out by microbes in sealed chambers called bioreactors, or **anaerobic digesters.** The digesters convert components of the sludge to swamp gas, primarily methane with small amounts of

Figure 27.2 **The many uses of water.** © Dr. Parvinder Sethi

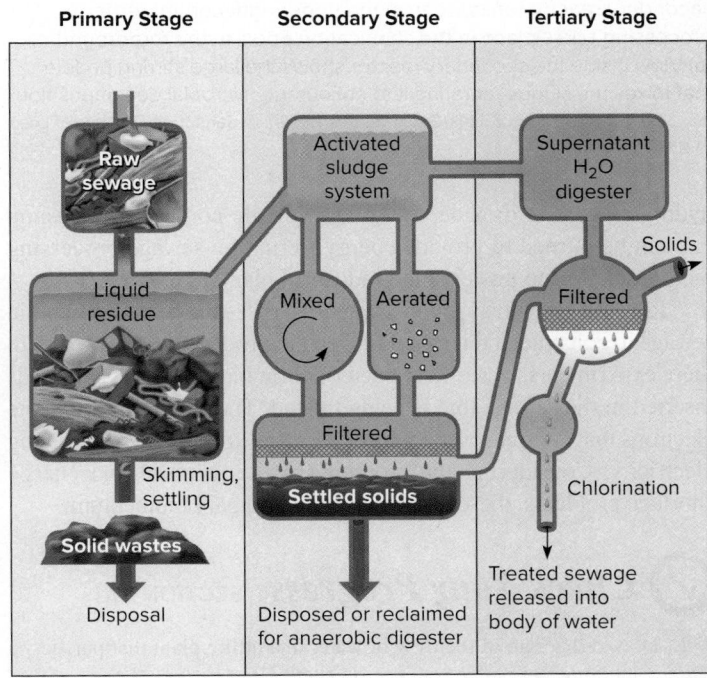

Figure 27.3 The primary, secondary, and tertiary stages in sewage treatment.

(a)

(b)

Figure 27.4 Treatment of sewage and wastewater.
(a) Wastewater treatment plants can process several millions gallons of raw sewage a day. In this plant, biodegradation occurs in the secondary reactor tanks, seen in the upper right corner; tertiary processing takes place in the clarification pools in the foreground. **(b)** View inside the secondary reactor shows the large stirring paddle that mixes the sludge, aerating it to encourage microbial decomposition. (a): © Kent Knudson/PhotoLink/Getty Images RF; (b): © Sanitation Districts of Los Angeles County

hydrogen, carbon dioxide, and other volatile compounds. Swamp gas can be burned to provide energy to run the sewage processing facility itself or to power small industrial plants.

Recently, scientists found a way to harness the bacteria found in sewage to construct a microbial fuel cell to produce usable energy. In these experiments, wastewater bacteria form biofilms on special rods inserted in the sewage that is being treated. These biofilms generate electrons that are transferred via copper wires to cathodes, producing electricity. Considering the mounting waste disposal and energy shortage problems, these technologies should gain momentum.

 Check Your Progress SECTION 27.1

1. Draw a diagram of the flow of water in a utility plant that purifies drinking water.
2. How is water decontaminated and made safe to drink?

3. Describe the three stages of sewage treatment.
4. What is activated sludge?
5. What does an anaerobic digester do?
6. How does treatment of drinking water and sewage vary in purpose and techniques?

27.2 The Microbiology of Food

 Expected Learning Outcome

4. Describe the major ways that microbes are involved with food.

All human food—from vegetables to caviar to cheese—comes from some other organism, and rarely is it obtained in a sterile, uncontaminated state. Food is but a brief stopover in the overall scheme of biogeochemical cycling. This means that microbes and humans are in direct competition for the nutrients in food, and we must be constantly aware that microbes' fast growth rates give them the winning edge. Somewhere along the route of procurement, processing, or preparation, food becomes contaminated with microbes from the soil, the bodies of plants and animals, water, air, food handlers, or utensils. The final effects depend upon the types and numbers of microbes and whether the food is cooked or preserved. In some cases, specific microbes can even be added to food to obtain a desired effect. The effects of microorganisms on food can be classified as beneficial, detrimental, or neutral to humans, as summarized by the following outline:

Beneficial effects
 Food is fermented or otherwise chemically changed by the addition of microbes or microbial products to alter or improve flavor, taste, or texture.
 Microbes can serve as food.
Detrimental effects
 Food poisoning or food-borne illness
 Food spoilage
 Growth of microbes makes food unfit for consumption; adds undesirable flavors, appearance, and smell; destroys food value
Neutral effects
 The presence or growth of microbes that do not cause disease or change the nature of the food

As long as food contains no harmful substances or organisms, its suitability for consumption is largely a matter of taste. But what tastes like rich flavor to some may seem like decay to others. The test of whether certain foods are edible is guided by culture, experience, and preference. The flavors, colors, textures, and aromas of many cultural delicacies are supplied by bacteria and fungi. Chocolate, pickled cabbage, fermented fish, and Limburger cheese are notable examples. If you examine the foods of most cultures, you will find some foods that derive their delicious flavor from microbes.

27.3 Microbial Fermentations in Food Products

 Expected Learning Outcomes

5. Characterize the basic science behind the use of microbes in food fermentations.

6. Describe the microbiological processes involved in bread making, alcoholic beverage production, and fermented plant products.

7. Outline the fermentation processes related to the preparation of dairy products.

8. Explain several ways that microbes may be used as foods.

In contrast to methods that destroy or keep out unwanted microbes, many culinary procedures deliberately add microorganisms and encourage them to grow. Common substances such as bread, cheese, beer, wine, yogurt, and pickles are the result of **food fermentations.** These reactions actively encourage biochemical activities that impart a particular taste, smell, or appearance to food. The microbe or microbes can occur naturally on the food substrate, as in sauerkraut, or they can be added as pure or mixed samples of known bacteria, molds, or yeasts called **starter cultures.** Many food fermentations are synergistic, with a series of microbes acting in concert to convert a starting substrate to the desired end product. Because large-scale production of fermented milk, cheese, bread, alcoholic brews, and vinegar depends upon inoculation with starter cultures, considerable effort is spent selecting, maintaining, and preparing these cultures and excluding contaminants that can spoil the fermentation. Most starting raw materials are of plant origin (grains, vegetables, beans) and, to a lesser extent, of animal origin (milk, meat).

Bread Making

Microorganisms accomplish three functions in bread making:

1. leavening the flour-based dough,
2. imparting flavor and odor, and
3. conditioning the dough to make it workable.

Leavening is achieved primarily through the release of gas to produce a porous and spongy product. Without leavening, bread dough remains dense, flat, and hard. Although various microbes and leavening agents can be used, the most common ones are various strains of the baker's yeast *Saccharomyces cerevisiae.* Other gas-forming microbes such as coliform bacteria, certain *Clostridium* species, heterofermentative lactic acid bacteria, and wild yeasts can be employed, depending on the type of bread desired.

Yeast metabolism requires a source of fermentable sugar such as maltose or glucose. Because the yeast respires aerobically in bread dough, the chief products of maltose fermentation are carbon dioxide and water rather than alcohol (the main product in beer and wine). Other contributions to bread texture come from kneading, which incorporates air into the dough, and from microbial enzymes, which break down flour proteins (gluten) and give the dough elasticity.

Besides carbon dioxide production, bread fermentation generates other volatile organic acids and alcohols that impart delicate flavors and aromas. These are especially well developed in home-baked bread, which is leavened more slowly than commercial bread. Yeasts and bacteria can also impart unique flavors, depending upon the culture mixture and baking techniques used. The pungent flavor of rye bread, for example, comes in part from starter cultures of lactic acid bacteria such as *Lactobacillus plantarum, L. brevis, L. bulgaricus, Leuconostoc mesenteroides,* and *Streptococcus thermophilus.* Sourdough bread gets its unique tang from *Lactobacillus sanfrancisco.*

Production of Beer and Other Alcoholic Beverages

The production of alcoholic beverages takes advantage of another useful property of yeasts. By fermenting carbohydrates in fruits or grains anaerobically, they produce ethyl alcohol, as shown by this equation:

$$C_6H_{12}O_6 \rightarrow 2C_2H_5OH + 2CO_2$$
(Yeast + Sugar = Ethanol + Carbon dioxide)

Depending upon the starting materials and the processing method, alcoholic beverages vary in alcohol content and flavor. The principal types of fermented beverages are beers, wines, and spirit liquors.

Brewing

The earliest evidence of beer brewing appears in ancient tablets by the Sumerians and Babylonians around 6000 BC. The starting ingredients for both ancient and present-day versions of beer, ale, stout, porter, and other variations are water, malt (barley grain), hops, and special strains of yeasts. The steps in brewing include malting, mashing, adding hops, fermenting, aging, and finishing.

For brewer's yeast to convert the carbohydrates in grain into ethyl alcohol, the barley must first be sprouted and softened to make its complex nutrients available to yeasts. This process, called **malting,*** releases amylases that convert starch to dextrins and maltose, and proteases that digest proteins. Other sugar and starch supplements added in some forms of beer are corn, rice, wheat, soybeans, potatoes, and sorghum. After the sprouts have been separated, the remaining malt grain is dried and stored in preparation for mashing.

The malt grain is soaked in warm water and ground up to prepare a **mash.** Sugar and starch supplements are then introduced to the mash mixture, which is heated to a temperature of about 65°C to 70°C. During this step, the starch is hydrolyzed by amylase and simple sugars are released. Heating this mixture to 75°C stops the activity of the enzymes. Solid particles are next removed by settling and filtering. **Wort,*** the clear fluid that comes off, is rich in dissolved carbohydrates. It is boiled for about 2.5 hours with **hops,** the dried scales of the female flower of *Humulus lupulus* (**figure 27.5),** to extract the bitter acids and resins that give aroma and flavor to the finished product.[1] Boiling also caramelizes the sugar and imparts a golden or brown color, destroys any bacterial contaminants that can destroy flavor, and concentrates the mixture. The filtered and cooled supernatant is then ready for the addition of yeasts and fermentation.

Primary fermentation begins when wort is inoculated with a species of *Saccharomyces* that has been specially developed for beer

* *malting* (mawlt′-ing) Gr. malz, to soften.

* *wort* (wurt) O.E. wyrt, a spice

1. This substance, called humulus, provides some rather interesting side effects besides flavor. It is a moderate sedative, a diuretic, and a mild antiseptic.

Figure 27.5 Female Flowers of the Japanese hop plant. This dried herb gives beer some of its flavor and aroma. *Source:* Carole Ritchie/USDA-NRCS PLANTS Database

Processing Step		Outcome
Grape pressing		Formation of must with fruit sugars
Heat sterilization		Elimination of contaminants
Yeast inoculation		Addition of desired organisms
Fermentation of must		Alcohol production from sugars
	Tank	
Storage in barrels to age		Development of final wine bouquet
	Barrel	
Filtration and collection		Removal of yeast and particles
Bottling		

making. Top yeasts such as *Saccharomyces cerevisiae* function at the surface and are used to produce the higher alcohol content of *ales*. Bottom yeasts such as *S. uvarum* (*carlsbergensis*) function deep in the fermentation vat and are used to make other beers. In both cases, the initial inoculum of yeast starter is aerated briefly to promote rapid growth and increase the load of yeast cells. Shortly thereafter, an insulating blanket of foam and carbon dioxide develops on the surface of the vat and promotes anaerobic conditions. During 8 to 14 days of fermentation, the wort sugar is converted chiefly to ethanol and carbon dioxide. The diversity of flavors in the finished product is partly due to the release of small amounts of glycerol, acetic acid, and esters. Fermentation is self-limited, and it essentially ceases when a concentration of 3% to 6% ethyl alcohol is reached.

Freshly fermented, or "green," beer is **lagered,*** meaning it is held for several weeks to months in vats near 0°C. During this maturation period, yeast, proteins, resin, and other materials settle, leaving behind a clear, mellow fluid. Lager beer is subjected to a final filtration step to remove any residual yeasts that could spoil it. Finally, it is carbonated with carbon dioxide collected during fermentation, and packaged in kegs, bottles, or cans.

Preparation of Wine and Liquors

Wine is traditionally considered any alcoholic beverage arising from the fermentation of grape juice, but practically any fruit can be rendered into wine. The essential starting point is the preparation of **must,** the juice given off by crushed fruit that is used as a substrate for fermentation. In general, grape wines are either white or red. The color comes from the skins of the grapes, so white wine is prepared either from white-skinned grapes or from red-skinned grapes that have had the skin removed. Red wine comes from the red- or purple-skinned varieties. Major steps in making wine include must preparation (crushing), fermentation, pasteurization, storage, and aging **(figure 27.6).**

The right amount of sugar is essential for proper fermentation. The must should contain 12% to 25% glucose or fructose, so the art

(a)

* *lagered* (law'-gurd) Gr. lager, to store or age.

Figure 27.6 Wine making. (a) General steps in wine making. **(b)** Wine fermentation vats in a large commercial winery. (b): © James Hardy/PhotoAlto RF

(b)

27.1 Secret World of Microbes
The Noble Rot

Harvesting grapes for wine making requires the balancing of several factors: sugar, acid, and tannin levels within the grapes, time left in the growing season, even the expected onset of inclement weather such as heat or hail, which can damage the crop. In eighteenth-century Germany, grape producers traditionally began their harvest only when the estate owner gave his approval. During the 1775 growing season, in the Rheingau region, the messenger delivering the order to harvest the grapes was delayed for several weeks, time enough for a fungus to infest much of the vineyard. Cutting their losses, the grapes were culled and given to local farmers who used them to create an unexpectedly enjoyable sweet wine. The fungus, *Botrytis cinerea*, came to be valued throughout Europe for the taste it imparted to the wine, while the dry decay and partial dehydration of the grapes it produced was called the "Noble Rot." Look back at figure 1.5*b*, to view this interesting fungus. Today, botrytized wines can be found all over the world; if *Botrytis* is not naturally present in the environment it may be imported and the spores sprayed over the grapes. The growing fungus slowly dehydrates the grapes and concentrates the solids responsible for the final flavor of the wine. Some of the finest botrytized wines owe their flavor to the fact the grapes were picked one at a time, when each grape had exactly the correct amount of Noble Rot.

In what way are botrytized wines and sourdough bread similar?
Answer available on Connect.

Semillon grapes in southwest France, displaying Noble Rot.
© David Lytle

of wine making begins in the vineyard. Grapes are harvested when their sugar content reaches 15% to 25%, depending on the type of wine to be made. Grapes from the field carry a mixed biofilm on their surface called the *bloom* that can serve as a source of wild yeasts **(27.1 Secret World of Microbes).** Some winemakers allow these natural yeasts to dominate, but many wineries inoculate the must with a special strain of *Saccharomyces cerevisiae,* variety *ellipsoideus.* To discourage yeast and bacterial spoilage agents, winemakers usually treat grapes with sulfur dioxide or potassium metabisulfite. The inoculated must is thoroughly aerated and mixed to promote rapid aerobic growth of yeasts, but when the desired level of yeast growth is achieved, anaerobic alcoholic fermentation is begun.

The temperature of the vat during fermentation must be carefully controlled to facilitate alcohol production. The length of fermentation varies from 3 to 5 days in red wines and from 7 to 14 days in white wines. The initial fermentation yields ethanol concentrations reaching 7% to 15% by volume, depending upon the type of yeast, the source of the juice, and ambient conditions. The fermented juice (raw wine) is decanted and transferred to large vats to settle and clarify. Before the final aging process, it is flash-pasteurized to kill microorganisms and filtered to remove any remaining yeasts and sediments. Wine is aged in wooden casks for varying time periods (months to years), after which it is bottled and stored for further aging. During aging, nonmicrobial changes produce aromas and flavors (the bouquet) characteristic of a particular wine.

The fermentation processes discussed thus far can only achieve a maximum alcoholic content of 17%, because concentrations above this level inhibit the metabolism of the yeast. The fermentation product must be distilled to obtain higher concentrations such as those found in liquors. During distillation, heating the liquor separates the more volatile alcohol from the less volatile aqueous phase. The alcohol is then condensed and collected. The alcohol content of distilled liquors is rated by *proof,* a measurement that is usually two times the alcohol content. Thus, 80 proof vodka contains 40% ethyl alcohol.

Distilled liquors originate through a process similar to wine making, although the starting substrates can be extremely diverse. In addition to distillation, liquors can be subjected to special treatments such as aging to provide unique flavor or color. Vodka, a colorless liquor, is usually prepared from fermented potatoes, and rum is distilled from fermented sugarcane. Assorted whiskeys are derived from fermented grain mashes; rye whiskey is produced from rye mash, and bourbon from corn mash. Brandy is distilled grape, peach, or apricot wine.

Other Fermented Plant Products

Fermentation provides an effective way of preserving vegetables, as well as enhancing flavor with lactic acid and salt. During pickling fermentations, vegetables are immersed in an anaerobic salty solution (brine) to extract sugar and nutrient-laden juices. The salt also disperses bacterial clumps, and its high osmotic pressure inhibits proteolytic bacteria and sporeformers that can spoil the product.

Sauerkraut is the fermentation product of cabbage. Cabbage is washed, wilted, shredded, salted, and packed tightly into a fermentation vat. Weights cover the cabbage mass and squeeze out its

juices. The fermentation is achieved by natural cabbage microbiota or by an added culture. The initial agent of fermentation is *Leuconostoc mesenteroides,* which grows rapidly in the brine and produces lactic acid. It is followed by *Lactobacillus plantarum,* which continues to raise the acid content to as high as 2% (pH 3.5) by the end of fermentation. The high acid content restricts the growth of spoilage microbes.

Fermented cucumber pickles come chiefly in salt and dill varieties. Salt pickles are prepared by washing immature cucumbers, placing them in barrels of brine, and allowing them to ferment for 6 to 9 weeks. The brine can be inoculated with *Pediococcus cerevisiae* and *Lactobacillus plantarum* to avoid unfavorable qualities caused by natural microbiota and to achieve a more consistent product. Fermented dill pickles are prepared in a somewhat more elaborate fashion, with the addition of dill herb, spices, garlic, onion, and vinegar.

Natural vinegar is produced when the ethyl alcohol in fermented plant juice is oxidized to acetic acid, which is responsible for the pungent odor and sour taste. Although a reasonable facsimile of vinegar could be made by mixing about 4% acetic acid and a dash of sugar in water, this preparation would lack the traces of various esters, alcohol, glycerin, and volatile oils that give natural vinegar its pleasant character. Vinegar is actually produced in two stages. The first stage is similar to wine or beer making, in which a plant juice is fermented to alcohol by *Saccharomyces.* The second stage involves aerobic metabolism carried out by acetic acid bacteria in the genera *Acetobacter* and *Gluconobacter.* These bacteria oxidize the ethanol in a two-step process, as shown here:

$$2C_2H_5OH + \tfrac{1}{2}O_2 \rightarrow CH_3CHO + H_2O$$
Ethanol　　　　　　　　Acetaldehyde

$$CH_3CHO + \tfrac{1}{2}O_2 \rightarrow CH_3COOH$$
Acetaldehyde　　　　Acetic acid

The abundance of oxygen necessary in commercial vinegar making is furnished by exposing inoculated raw material to air by arranging it in thin layers in open trays, allowing it to trickle over loosely packed beechwood twigs and shavings, or aerating it in a large vat. Different types of vinegar are derived from substrates such as apple cider (cider vinegar), malted grains (malt vinegar), and grape juice (wine vinegar).

Microbes in Milk and Dairy Products

Milk has a highly nutritious composition. It contains an abundance of water and is rich in minerals, protein (chiefly casein), butterfat, sugar (especially lactose), and vitamins. It starts its journey in the udder of a mammal as a sterile substance, but as it passes out of the teat, it is inoculated by the animal's normal biota. Other microbes can be introduced by milking utensils. Because milk is a nearly perfect culture medium, it is highly susceptible to

microbial growth. When raw milk is left at room temperature, a series of bacteria ferment the lactose, produce acid, and alter the milk's content and texture (**figure 27.7a**). This progression can occur naturally, or it can be induced, as in the production of cheese and yogurt.

In the initial stages of milk fermentation, lactose is rapidly attacked by *Streptococcus lactis* and *Lactobacillus* species (**figure 27.7b**). The resultant lactic acid accumulation and lowered pH cause the milk proteins to coagulate into a solid mass called the **curd.** Curdling also causes the separation of a watery liquid called **whey** on the surface. Curd can be produced by microbial action or by an enzyme, **rennin** (casein coagulase), which is isolated from the stomach of unweaned calves.

Cheese Production

Since 5000 BC, various forms of cheese have been produced by spontaneous fermentation of cow, goat, or sheep milk. Present-day, large-scale cheese production is carefully controlled and uses only freeze-dried samples of pure cultures. These are first inoculated into a small quantity of pasteurized milk to form an active starter

(a)

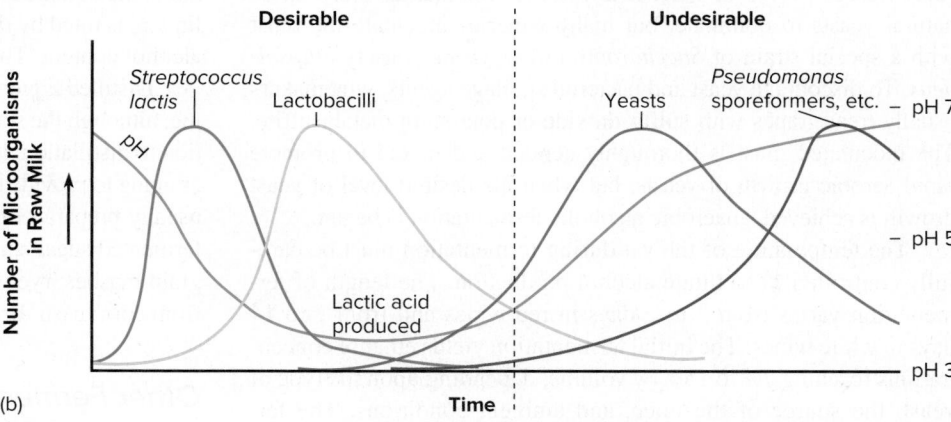

(b)

Figure 27.7　Microbes at work in milk products. (a) Litmus milk is used to indicate pH and consistency changes in milk resulting from microbial action. The first tube is an uninoculated control. The second tube has a white, decolorized zone indicative of litmus reduction. The third tube shows acid production (pink), and its proteins have formed a loose curd. In the fourth tube, digestion of milk proteins has caused complete clarification or peptonization of the milk. The fifth tube has a solid curd overlaid by a clear fluid, the whey. **(b)** Chart depicting changes in the numbers and types of microorganisms and the pH of raw milk as it incubates. Early acid production by streptococci and lactobacilli favors desirable effects such as curds and whey needed for flavor and texture. Late growth spoilage agents turn milk alkaline and cause undesirable changes such as protein digestion. (a): © Kathy Park Talaro

culture. This amplified culture is subsequently inoculated into a large vat of milk, where rapid curd development takes place. Such rapid growth is desired because it promotes the overgrowth of the desired inoculum and prevents the activities of undesirable contaminants. Rennin is usually added to increase the rate of curd formation.

After its separation from whey, the curd is rendered to produce one of the 20 major types of soft, semisoft, or hard cheese (**figure 27.8**). The composition of cheese is varied by adjusting water, fat, acid, and salt content. Cottage and cream cheese are examples of the soft, more perishable variety. After light salting and the optional addition of cream, they are ready for consumption without further processing. Other cheeses acquire their character from "ripening," a complex curing process involving bacterial, mold, and enzyme reactions that develop the final flavor, aroma, and other features characteristic of particular cheeses.

The distinctive traits of soft cheeses such as Limburger, Camembert, and Liederkranz are acquired by ripening with a reddish-brown mucoid coating of yeasts, micrococci, and molds. The microbial enzymes permeate the curd and ferment lipids, proteins, carbohydrates, and other substrates. This process leaves assorted acids and other by-products that give the finished cheese powerful aromas and delicate flavors. Semisoft varieties of cheese such as Roquefort, bleu, or Gorgonzola are infused and aged with a strain of *Penicillium roqueforti* mold. Hard cheeses such as Swiss, cheddar, and Parmesan develop a sharper flavor by aging with selected bacteria. The pockets in Swiss cheese come from entrapped carbon dioxide formed by *Propionibacterium,* which is also responsible for its bittersweet taste.

Other Fermented Milk Products

Yogurt is formed by the fermentation of milk by *Lactobacillus bulgaricus* and *Streptococcus thermophilus.* These organisms produce organic acids and other flavor components and can grow in such numbers that a gram of yogurt regularly contains 100 million bacteria. Live cultures of *Lactobacillus acidophilus* are an important additive to acidophilus milk, which is said to benefit digestion and to help maintain the normal biota of the intestine. Fermented milks

Figure 27.8 Cheese making. The curd-cutting stage in the making of cheddar cheese. © Joe Munroe/Science Source

such as kefir, koumiss, and buttermilk are a basic food source in many cultures.

Microorganisms as Food

At first, the thought of eating bacteria, molds, algae, and yeasts may seem odd or even unappetizing. We do eat their macroscopic relatives such as mushrooms, truffles, and seaweed, but we are used to thinking of the microscopic forms as agents of decay and disease or, at most, as food flavorings. The consumption of microorganisms is not a new concept. In Germany during World War II, it became necessary to supplement the diets of undernourished citizens by adding yeasts and molds to foods.

At present, most countries are able to produce enough food for their inhabitants, but in the future, countries with exploding human populations and dwindling arable land may need to consider microbes as a significant source of protein, fat, and vitamins. Several countries already commercially mass-produce food yeasts; bacteria; and, in a few cases, algae. Although eating microbes has yet to win total public acceptance, their use as feed supplements for livestock is increasing. A technology that shows some promise in increasing world food productivity is single-cell protein (SCP). This material is produced from waste materials such as molasses from sugar refining, petroleum by-products, and agricultural wastes. Quorn, a mycoprotein product made from the fungus *Fusarium venenatum,* is sold worldwide as a high protein alternative to meat.

Health food stores carry bottles of dark green pellets or powder that are a culture of a spiral-shaped cyanobacterium called *Spirulina.* This microbe is harvested from the surface of lakes and ponds, where it grows in great mats. In some parts of Africa and Mexico, *Spirulina* has become a viable alternative to green plants as a primary nutrient source. It can be eaten in its natural form or added to other foods and beverages.

27.4 Microbial Involvement in Food-Borne Diseases

 Expected Learning Outcomes

9. Explain the importance of microbes as agents of food-borne illnesses.

10. Classify the types of food poisoning, providing examples of each.

11. Identify the major methods of food preservation.

12. Describe methods for preventing food-borne diseases.

Diseases caused by ingesting food are usually referred to as **food poisoning,** and although this term is often used synonymously with microbial food-borne illness, not all food poisoning is caused by microbes or their products. Several illnesses are caused by poisonous plant and animal tissues or by ingesting food contaminated by pesticides or other poisonous substances. **Table 27.1** summarizes the major types of bacterial food-borne disease.

TABLE 27.1	Major Forms of Bacterial Food Poisoning		
Disease	Microbe	Foods Involved	Comments/with Chapter References
Food Intoxications: Caused by Ingestion of Foods Containing Preformed Toxins			
Staphylococcal enteritis	*Staphylococcus aureus*	Custards, cream-filled pastries, ham, dressings	Very common; symptoms come on rapidly; usually nonfatal; chapter 18
Botulism	*Clostridium botulinum*	Home-canned or poorly preserved low-acid foods	Rarely seen in commercially prepared foods; can be fatal; chapter 19
Bacillus cereus enteritis	Bacillus cereus	Reheated rice, potatoes, puddings, custards	Mimics staphylococcal enteritis; usually self-limited; chapter 19
Food Infections: Caused by Ingestion of Live Bacterial Pathogens That Invade the Intestine			
Campylobacteriosis	*Campylobacter jejuni*	Raw milk, raw chicken, shellfish and meats	Very common; animals are carriers of other species; chapter 21
Salmonellosis	*Salmonella typhimurium* and *S. enteritidis*	Poultry, eggs, dairy products, meats	Very common; can be severe and life-threatening; chapter 20
Perfringens enterotoxemia	*Clostridium perfringens*	Inadequately cooked meats	Vegetative cells produce toxin within the intestine; chapter 19
Shigellosis	Various *Shigella* species	Unsanitary cooked food; fish, shrimp, potatoes, salads	Carriers and flies contaminate food; the cause of bacillary dysentery; chapter 20
Vibrio enteritis	*Vibrio parahaemolyticus* *Vibrio vulnificus*	Raw or poorly cooked seafoods	Microbe lives naturally on marine animals; chapter 21
Listeriosis	*Listeria monocytogenes*	Poorly pasteurized milk, cheeses	Most severe in fetuses, newborns, and the immunodeficient; chapter 19
Escherichia enteritis	*Escherichia coli*	Contaminated raw vegetables, cheese	Various strains can produce infantile and traveler's diarrhea; chapter 20
	E. coli O157:H7	Raw or rare beef, vegetables, water	The cause of hemolytic uremic syndrome; chapter 20

Food poisoning of microbial origin can be divided into two general categories **(figure 27.9). Food intoxication** results from the ingestion of exotoxin secreted by bacterial cells growing in food. The absorbed toxin disrupts a particular target such as the intestine (if an enterotoxin) or the nervous system (if a neurotoxin). The symptoms of intoxication vary from bouts of vomiting and diarrhea (staphylococcal intoxication) to severely disrupted muscle function (botulism). In contrast, **food infection** is associated with the ingestion of live, intact microbial cells that target the intestine. In some cases, they infect the surface of the intestine, and in others, they invade the intestine and other body structures. Most food infections manifest some degree of diarrhea and abdominal distress (see 20.2 Making Connections). It is important to realize that disease symptoms in food infections such as *Clostridium perfringens* enterotoxemia can be initiated by toxins released by microbes growing in the infected tissue rather than in the food.

Reports of food poisoning are escalating in the United States and worldwide. Outbreaks attributed to common pathogens (*Salmonella, Escherichia coli, Vibrio,* hepatitis A, and various protozoa) have doubled in the past 25 years. The Centers for Disease Control and Prevention (CDC) estimate that nearly one-sixth of the population suffers each year from some form of food poisoning. A major factor in this changing pattern is the mass production and distribution of processed food such as raw vegetables, fruits, and meats. Improper handling can lead to gross contamination of these products with soil or animal wastes.

Estimated Incidence of Food-Borne Illness in the United States*

Illnesses	48,000,000 cases
Hospitalizations	128,000 cases
Deaths	3,000 cases

*One-third of all reported cases result from eating restaurant food.

Many reported food poisoning outbreaks occur where contaminated food has been served to large groups of people, but most cases probably occur in the home and are not reported. On occasion, even commercially prepared foods can be a source of infections.

Data collected by food microbiologists indicate that the most common bacterial food-borne pathogens are *Campylobacter, Salmonella, Shigella,* and *Staphylococcus aureus.* The dominant protozoa causing food infections are *Giardia, Cryptosporidium,* and *Toxoplasma.* The top viruses are noroviruses and hepatitis A viruses. These microbes and their diseases were previously discussed in chapters 18, 19, 20, and 21 (see table 27.1 along with chapters 23 and 25).

Food Intoxication **Food Infection**

Cell
Toxin

Cells

Ingestion

Improper cooking technique

Bacteria
multiply
in food

Ingestion

Target organs

Intestinal
infection

(a) (b)

Care in Harvesting, Preparation

Reduction of Microbial Load

Heat

Canning | Pasteurization | Cooking

Radiation

Filtration

Prevention of Growth

Maintenance
temperature

Hot | Cold
freezing

Preservative
additives

SALT | Sugar | Nitrogen salts
Gas | Vinegar

Figure 27.10 The primary methods to prevent food poisoning and food spoilage.

Figure 27.9 Food-borne illnesses of microbial origin.
(a) Food intoxication. The toxin is released by microbes growing in the food. After the toxin is ingested, it acts upon its target tissue and causes symptoms. (b) Food infection. The infectious agent comes from food or is introduced into it through poor food processing and storage. After the cells are ingested, they invade the intestine and cause symptoms of gastroenteritis.

Prevention Measures for Food Poisoning and Spoilage

The current approach used to regulate the food industry is called Hazard Analysis & Critical Control Points, or HACCP, and it is adapted from procedures crafted for the space program in the 1970s. It involves principles that are more systematic and scientific than previous random-sampling quality procedures. The program focuses on the identification, evaluation, control, and prevention of hazards at all stages of the food production process. Recent data show that some food-borne illnesses have begun to decline, due in part to implementation of HACCP procedures.

It will never be possible to avoid all types of food-borne illness because of the ubiquity of microbes in air, water, food, and the human body. But most types of food poisoning require the growth of microbes in the food. In the case of food infections, an infectious

dose (sufficient cells to initiate infection) must be present, and in food intoxication, enough cells to produce the toxin must be present. Thus, food poisoning or spoilage can be prevented by proper food handling, preparation, and storage. The methods shown in **figure 27.10** are aimed at preventing the incorporation of microbes

into food, removing or destroying microbes in food, and keeping microbes from multiplying.

Preventing the Incorporation of Microbes into Food

Most agricultural products such as fruits, vegetables, grains, meats, eggs, and milk are naturally exposed to microbes. Vigorous washing reduces the levels of contaminants in fruits and vegetables, whereas meat, eggs, and milk must be taken from their animal source as aseptically as possible. Aseptic techniques are also essential in the kitchen. Contamination of foods by fingers can be easily remedied by hand washing and proper hygiene, and contamination by flies or other insects can be stopped by covering foods or eliminating pests from the kitchen. Care and common sense also apply in managing utensils. It is important to avoid cross-contaminating food by not using the same cutting board for meat and vegetables without disinfecting it between uses.

Preventing the Survival or Multiplication of Microbes in Food

Because it is not possible to eliminate all microbes from certain types of food by clean techniques alone, a more efficient approach is to preserve the food by physical or chemical methods. Hygienically preserving foods is especially important for large commercial companies that process and sell bulk foods and must ensure that products are free from harmful contaminants. Regulations and standards for food processing are administered by two federal agencies: the Food and Drug Administration (FDA) and the U.S. Department of Agriculture (USDA).

Temperature and Food Preservation

Heat is a common way to destroy microbial contaminants or to reduce the load of microorganisms. Commercial canneries preserve food in hermetically sealed containers that have been exposed to high temperatures over a specified time period. The temperature used depends upon the type of food, and it can range from 60°C to 121°C, with exposure times ranging from 20 minutes to 115 minutes. The food is usually processed at a thermal death time (TDT; see chapter 11) that will destroy the main spoilage organisms and pathogens but will not alter the nutrient value or flavor of the food. For example, tomato juice must be heated to between 121°C and 132°C for 20 minutes to ensure destruction of the spoilage agent *Bacillus coagulans*. Most canning methods are rigorous enough to sterilize the food completely, but some only render the food "commercially sterile," which means it contains live bacteria that are unable to grow under normal conditions of storage.

Another use of heat is **pasteurization,** usually defined as the application of heat below 100°C to destroy nonresistant bacteria and yeasts in liquids such as milk, wine, and fruit juices. The heat is applied in the form of steam, hot water, or even electrical current. The most prevalent technology is the *high-temperature short-time* (*HTST*), or flash method, using a pasteurizer with extensive networks of tubes that expose the liquid to 72°C for

15 seconds. An alternative method, ultrahigh-temperature (UHT) pasteurization, steams the product until it reaches a temperature of 134°C for at least 2 seconds. Although milk processed this way is not actually sterile, it is often marketed as sterile, with a shelf life of up to 3 months. Older methods involve large bulk tanks that hold the fluid at a lower temperature for a longer time, usually 62.3°C for 30 minutes.

Cooking temperatures used to boil, roast, or fry foods can render them free or relatively free of living microbes if carried out for sufficient time to destroy any potential pathogens. A quick warming of chicken or an egg is inadequate to kill microbes such as *Salmonella*. In fact, any meat is a potential source of infectious agents and should be adequately cooked. Because most meat-associated food poisoning is caused by nonsporulating bacteria, heating the center of meat to at least 80°C and holding it there for 30 minutes is usually sufficient to kill pathogens. Roasting or frying food at temperatures of at least 200°C or boiling it will achieve a satisfactory degree of disinfection.

Any perishable raw or cooked food that could serve as a growth medium must be stored to prevent the multiplication of bacteria that have survived during processing or handling. Because most foodborne bacteria and molds that are agents of spoilage or infection can multiply at room temperature, manipulation of the holding temperature is a useful preservation method **(figure 27.11).** A good general directive is to store foods at temperatures below 4°C or above 60°C.

Regular refrigeration reduces the growth rate of most mesophilic bacteria by 10 times, although some psychrotrophic microbes can continue to grow at a rate that causes spoilage. This factor limits the shelf life of milk, because even at 7°C, a population could go from a few cells to a billion in 10 days. Pathogens such as *Listeria monocytogenes* and *Salmonella* can also continue to grow in refrigerated foods. Freezing is a longer-term method for cold preservation. Foods can be either slow-frozen for 3 to 72 hours at −15°C to −23°C or rapidly frozen for 30 minutes at −17°C to −34°C. Because freezing cannot be counted upon to kill microbes, rancid, spoiled, or infectious foods will still be unfit to eat after freezing and defrosting. *Salmonella* is known to survive several months in frozen chicken and ice cream, and *Vibrio parahaemolyticus* can survive in frozen shellfish. For this reason, frozen foods should be defrosted rapidly and immediately cooked or reheated. However, even this practice will not prevent staphylococcal intoxication if the toxin is already present in the food before it is heated.

Foods such as soups, stews, gravies, meats, and vegetables that are generally eaten hot should not be maintained at warm or room temperatures, especially in settings such as cafeterias, banquets, and picnics. The use of a hot plate, chafing dish, or hot water bath will maintain foods above 60°C, well above the incubation temperature of food-poisoning agents.

As a final note about methods to prevent food poisoning, remember the simple axiom: "When in doubt, throw it out."

Irradiation

Ultraviolet (nonionizing) lamps are commonly used to destroy microbes on the surfaces of foods or utensils, but they do not

Figure 27.11 **Temperatures favoring and inhibiting the growth of microbes in food.** Most microbial agents of disease or spoilage grow in the temperature range of 15°C to 40°C. To ensure safety, foods should not be allowed to remain at temperatures between 4°C and 60°C for more than two hours. The red area represents temperatures at which bacteria grow most rapidly while yellow areas allow for microbial growth at a reduced rate; green areas correspond to greatly decreased rates of microbial growth. Preventing unwanted growth in foods in long-term storage is best achieved by refrigeration (4°C) or freezing (0°C or lower). Preventing microbial growth in foods intended to be consumed warm in a few minutes or hours requires maintaining the foods above 60°C.

penetrate far enough to sterilize bulky foods or food in packages. Food preparation areas are often equipped with UV radiation devices that are used to destroy spores on the surfaces of cheese, breads, and cakes and to disinfect packaging machines and storage areas.

Food itself is usually sterilized by gamma rays, electron beams, or X rays, because these ionizing rays can penetrate denser materials. It must also be emphasized that this method does not cause the targets of irradiation to become radioactive.

Although public opinion polls routinely find resistance to eating irradiated food, there is little scientific evidence to support the belief that food irradiation is harmful. Small changes in flavor, odor, texture, and the level of some vitamins have been noted, but these changes are generally so minor that it is difficult to distinguish between irradiated and nonirradiated foods. More than three decades of research have shown that irradiation is a safe and effective means of greatly reducing the risk of food-borne illness, lengthening shelf life, and even delaying ripening of some fruits and vegetables (slowing the sprouting of potatoes, for example). The safety of irradiated food is endorsed by the Food and Drug Administration (FDA), the Centers for Disease Control and Prevention

(CDC), the U.S. Department of Agriculture (USDA), and the World Health Organization (WHO). Although irradiation is approved for treatment of beef, pork, poultry, fish, eggs, spices, grain, and some fruits and vegetables, less than 10% of these products are sterilized this way.

Other Forms of Preservation

The addition of chemical preservatives to many foods can prevent the growth of microorganisms that could cause spoilage or disease. Preservatives include natural chemicals such as salt (NaCl) or table sugar and artificial substances such as ethylene oxide. The main classes of preservatives are organic acids, nitrogen salts, sulfur compounds, oxides, salt, and sugar.

Organic acids, including lactic, benzoic, and propionic acids, are among the most widely used preservatives. They are added to baked goods, cheeses, pickles, carbonated beverages, jams, jellies, and dried fruits to reduce spoilage from molds and some bacteria. Nitrites and nitrates are used primarily to maintain the red color of cured meats (hams, bacon, and sausage). By inhibiting the germination of *Clostridium botulinum* spores, they also prevent botulism intoxication, but their effects against other microorganisms are limited. Sulfite prevents the growth of undesirable molds in dried fruits, juices, and wines and retards discoloration in various foodstuffs. Ethylene and propylene oxide gases disinfect various dried foodstuffs. Their use is restricted to fruit, cereals, spices, nuts, and cocoa.

The high osmotic pressure contributed by hypertonic levels of salt plasmolyzes bacteria and fungi and removes moisture from food, thereby inhibiting microbial growth. Salt is commonly added to brines, pickled foods, meats, and fish. However, it does not retard the growth of pathogenic halophiles such as *Staphylococcus aureus,* which grows readily even in 7.5% salt solutions. The high sugar concentrations of candies, jellies, and canned fruits also exert an osmotic preservative effect. Other chemical additives that function in preservation are alcohols and antibiotics. Alcohol is added to flavoring extracts, and antibiotics are approved for treating the carcasses of chickens, fish, and shrimp.

Food can also be preserved by **desiccation,** a process that removes moisture needed by microbes for growth by exposing the food to dry, warm air. Solar drying was traditionally used for fruits and vegetables, but modern commercial dehydration is carried out in rapid-evaporation mechanical devices. Drying is not a reliable microbicidal method, however. Numerous resistant microbes such as micrococci, coliforms, staphylococci, salmonellae, and fungi survive in dried milk and eggs, which can subsequently serve as agents of spoilage and infections.

In 2006 the FDA approved the spraying of bacteriophages onto ready-to-eat meat products. The bacteriophages are specific for *Listeria* and will kill this pathogen in cold cuts and poultry that are usually not cooked before consumption. Efforts are currently underway to use *Salmonella*—specific bacteriophage—in the same manner, but it has not yet been approved for use.

 Check Your Progress SECTIONS 27.2–27.4

7. Which microbes are used as starter cultures in bread, beer, wine, cheeses, and sauerkraut?

8. Define *malting, must, wort,* and *leavening.*

9. Outline the steps in beer making.

10. List the steps in wine making.

11. What are curds and whey, and what causes them?

12. Explain the meaning of fermentation from the standpoint of biotechnology.

13. Describe five types of fermentations.

14. What are the main criteria regarding the safety of foods?

15. Differentiate between food infection and food intoxication.

16. When are microbes in food harmless?

17. What is the rule of temperature with regard to food safety?

27.5 General Concepts in Industrial Microbiology

 Expected Learning Outcomes

13. Recall the major concepts in industrial microbiology.

14. Describe industrial fermentations and the role of the fermentor.

15. Outline the steps in industrial mass production of organic substances.

16. Describe several types of products made with fermentation technology and how they are used.

"Never underestimate the power of the microbe."
—Jackson W. Foster,
pioneering microbiologist in the
development of antibiotics

Virtually any large-scale commercial enterprise that enlists microorganisms to manufacture consumable materials is part of the realm of industrial microbiology. Here, the term pertains primarily to bulk production of organic compounds such as antibiotics, hormones, vitamins, acids, solvents **(table 27.2),** and enzymes **(table 27.3).** Many of the processing steps involve fermentations similar to those described in food technology, but industrial processes usually occur on a much larger scale, produce a specific compound, and involve numerous complex stages. The aim of industrial microbiology is to produce chemicals that can be purified and packaged for sale or for use in other commercial processes. Thousands of tons of organic chemicals worth several billion dollars are produced by this industry every year. To create just one of these products, an industry must determine which microbes, starting compounds, and growth conditions work best. The research and development involved require an investment of 10 to 15 years and billions of dollars.

The microbes used by fermentation industries are mutant strains of fungi or bacteria that selectively synthesize large amounts

TABLE 27.2	Industrial Products of Microorganisms		
Chemical	**Microbial Source**	**Substrate**	**Applications**
Pharmaceuticals			
Cephalosporins	*Cephalosporium*	Glucose	Antibacterial antibiotics, broad spectrum
Penicillins	*Penicillium chrysogenum*	Lactose	Antibacterial antibiotics, broad and narrow spectrum
Vitamin B_{12}	*Pseudomonas*	Molasses	Dietary supplement
Steroids (hydrocortisone)	*Rhizopus, Cunninghamella*	Deoxycholic acid, stigmasterol	Treatment of inflammation, allergy; hormone replacement therapy
Food Additives and Amino Acids			
Citric acid	*Aspergillus, Candida*	Molasses	Acidifier in soft drinks; used to set jam; candy additive; fish preservative; retards discoloration of crabmeat; delays browning of sliced peaches
Xanthan	*Xanthomonas*	Glucose medium	Food stabilizer; not digested by humans
Acetic acid	*Acetobacter*	Any ethylene source, ethanol	Food acidifier; used in industrial processes
Miscellaneous			
Ethanol	*Saccharomyces*	Beet, cane, grains, wood, wastes	Additive to gasoline (gasohol)
Acetone	*Clostridium*	Molasses, starch	Solvent for lacquers, resins, rubber, fat, oil
Glycerol	Yeast	By-product of alcohol	Explosive (nitroglycerine)
Dextran	*Klebsiella, Acetobacter, Leuconostoc*	Glucose, molasses, sucrose	Polymer of glucose used as adsorbents, blood expanders, and in burn treatment; a plasma extender; used to stabilize ice cream and in sugary syrup and candies

TABLE 27.3	Industrial Enzymes and Their Uses	
Enzyme	**Source**	**Application**
Amylase	*Aspergillus, Bacillus, Rhizopus*	Flour supplement, desizing textiles, mash preparation, syrup manufacture, digestive aid, precooked foods, spot remover in dry cleaning
Cellulase	*Aspergillus, Trichoderma*	Denim finishing ("stone-washing"), digestive aid, increase digestibility of animal feed, degradation of wood or wood by-products
Hyaluronidase	Various bacteria	Medical use in wound cleansing, preventing surgical adhesions
Keratinase	*Streptomyces*	Hair removal from hides in leather preparation
Pectinase	*Aspergillus, Sclerotina*	Clarifies wine, vinegar, syrups, and fruit juices by degrading pectin, a gelatinous substance; used in concentrating coffee
Proteases	*Aspergillus, Bacillus, Streptomyces*	To clear and flavor rice wines, process animal feed, remove gelatin from photographic film, recover silver, tenderize meat, unravel silkworm cocoon, remove spots
Rennet	*Mucor*	To curdle milk in cheese making
Streptokinase	*Streptococcus*	Medical use in clot digestion, as a blood thinner

of various metabolic intermediates, or **metabolites.** Two basic kinds of metabolic products are harvested by industrial processes: (1) *Primary metabolites* are produced during the major metabolic pathways and are essential to the microbe's function. (2) *Secondary metabolites* are by-products of metabolism that may not be critical to the microbe's function **(figure 27.12).** In general, primary products are compounds such as amino acids and organic acids synthesized during the logarithmic phase of microbial growth; and secondary products are compounds such as vitamins, antibiotics, and steroids synthesized during the stationary phase (see chapter 7).

Most strains of industrial microorganisms have been chosen for their high production of a particular primary or secondary metabolite. Certain mutated strains of yeasts and bacteria can produce 20,000 times more metabolite than a wild strain of that same microbe.

Industrial microbiologists have several tricks to increase the amount of the chosen end product. First, they can manipulate the growth environment to increase the synthesis of a metabolite. For instance, adding lactose instead of glucose as the fermentation substrate increases the production of penicillin by *Penicillium.* Another strategy is to select microbial strains that genetically lack a feedback system to regulate the formation of end products, thus encouraging mass accumulation of this product. Many syntheses occur in sequential fashion, wherein the waste products of one organism become the building blocks of the next. During these *biotransformations,* the substrate undergoes a series of slight modifications, each of which gives off a different by-product **(figure 27.13).** The production of an antibiotic such as tetracycline requires several microorganisms and 72 separate metabolic steps.

From Microbial Factories to Industrial Factories

Industrial fermentations begin with microbial cells acting as living factories. When exposed to optimum conditions, they multiply in massive numbers and synthesize large volumes of a desired product. Producing appropriate levels of growth and fermentation requires cultivation of the microbes in a carefully controlled environment. This process is basically similar to culturing bacteria in a test tube of nutrient broth. It requires a sterile medium containing appropriate nutrients, protection from contamination, provisions for introduction of sterile air or total exclusion of air, and a suitable temperature and pH.

Many commercial fermentation processes have been worked out on a small scale in a lab and then *scaled up* to a large commercial venture. An essential component for scaling up is a **fermentor**

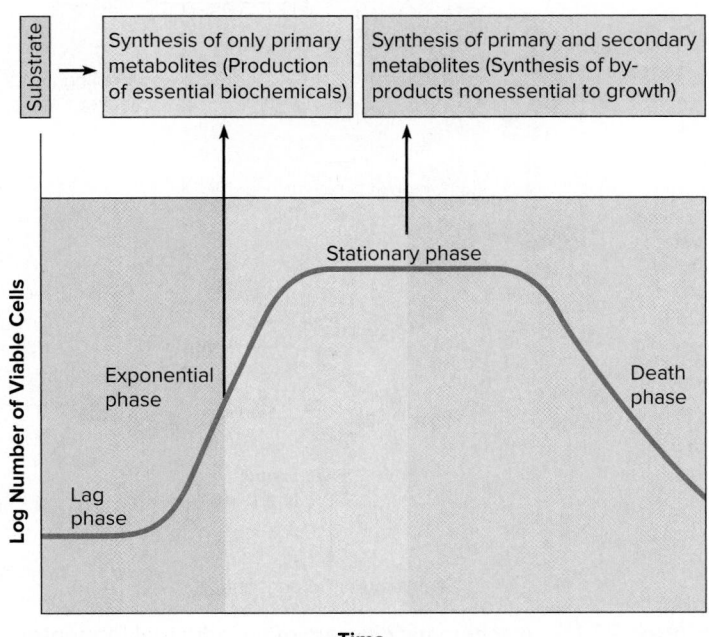

Figure 27.12 The origins of primary and secondary microbial metabolites harvested by industrial processes.

Figure 27.13 **An example of biotransformation by microorganisms in the industrial production of steroid hormones.** Desired hormones such as cortisone and progesterone can require several steps and microbes, and rarely can a single microbe carry out all the required synthetic steps. **(a)** Beginning with starting compound Δ^4-pregnene-3,20-dione (blue), **(b)** 11-dehydrocortisone (orange) is synthesized in three stages by two species of mold. **(c)** Cortisone (yellow) is synthesized in two stages by a pair of molds. **(d)** 15α-hydroxyprogesterone (green) is synthesized in a single step by a pair of molds.

or **bioreactor (figure 27.14),** a device in which mass cultures are grown, reactions take place, and product develops. Some fermentors are large tubes, flasks, or vats, but most industrial types are metal cylinders with built-in mechanisms for stirring, cooling, monitoring, and harvesting product **(figure 27.15).** Fermentors are made of materials that can withstand pressure and are rustproof, nontoxic, and leakproof. They range in holding capacity from small, 5-gallon systems used in research labs to larger, 5,000- to 100,000-gallon vessels and, in some industries, to tanks of 250 million to 500 million gallons.

For optimum yield, a fermentor must duplicate the actions occurring in a tiny volume (a test tube) on a massive scale. Most microbes performing fermentations have an aerobic metabolism, and the large volumes make it difficult to provide adequate oxygen. Fermentors have a built-in device called a *sparger* that aerates the medium to promote aerobic growth. Paddles (*impellers*) located in the central part of the fermentor increase the contact between the

Figure 27.14 **A small-scale bioreactor used in vaccine production.** Such elaborate systems require the highest levels of sterility and clean techniques. © Photo courtesy of Sartorius Stedim Biotech

Figure 27.15 **A schematic diagram of an industrial fermentor for mass culture of microorganisms.** Such instruments are equipped to add nutrients and cultures; to remove product under sterile or aseptic conditions; and to aerate, stir, and cool the mixture automatically.

microbe and the nutrients by vigorously stirring the fermentation mixture. Their action also maintains its uniformity.

Substance Production

The general steps in mass production of organic substances in a fermentor are illustrated in **figure 27.16.** These can be summarized as follows:

1. introduction of microbes and sterile media into the reaction chamber;
2. fermentation;
3. **downstream processing** (recovery, purification, and packaging of product); and
4. removal of waste.

All phases of production must be carried out aseptically and monitored (usually by computer) for rate of flow and quality of product. The starting raw substrates include crude plant residues, molasses, sugars, fish and meat meals, and whey. Additional chemicals can be added to control pH or to increase the yield. In *batch fermentation,*

the substrate is added to the system all at once and taken through a limited run until product is harvested. In *continuous feed systems,* nutrients are continuously fed into the reactor and the product is siphoned off throughout the run.

Ports in the fermentor allow the raw product and waste materials to be recovered from the reactor chamber when fermentation is complete. The raw product is recovered by settling, precipitation, centrifugation, filtration, or cell lysis. Some products come from this process ready to package, whereas others require further purification, extraction, concentration, or drying. The end product is usually in a powder, cake, granular, or liquid form that is placed in sterilized containers. The waste products can be siphoned off to be used in other processes or discarded, and the residual microbes and nutrients from the fermentation chamber can be recycled back into the system or removed for the next run.

Fermentation technology for large-scale cultivation of microbes and production of microbial products is versatile. Table 27.2 itemizes some of the major pharmaceutical substances, food additives, and solvents produced by microorganisms. Some newer technologies employ extremophiles and their enzymes to run the processes at high or low temperatures or in high-salt conditions. Hyperthermophiles have been adapted for high-temperature detergent and enzyme production. Psychrophiles are used for cold processing of reagents for molecular biology and medical tests. Halophiles are effective for processing of salted foods and dietary supplements.

Pharmaceutical Products

Health care products derived from microbial biosynthesis are antibiotics, hormones, vitamins, and vaccines. The first mass-produced antimicrobic was penicillin, which came from *Penicillium chrysogenum,* a mold first isolated from a cantaloupe in Wisconsin. The current strain of this species has gone through 70 years of selective mutation and screening to increase its yield. (The original wild *P. chrysogenum* synthesized 60 mg/ml of medium, and the latest isolate yields 85,000 mg/ml.) The semisynthetic penicillin derivatives are produced by introducing the assorted side-chain precursors to the fermentation vessel during the most appropriate phase of growth. Currently, a semisynthetic form of the antimalaria drug artemisinin is being produced by bioengineered *Saccharomyces cervisiae.*

Several steroid hormones used in therapy are produced industrially. Corticosteroids of the adrenal cortex, cortisone and cortisol (hydrocortisone), are invaluable for treating inflammatory and allergic disorders, and female hormones such as progesterone or estrogens are the active ingredients in birth control pills. For years, the production of these hormones was tedious and expensive because it involved purifying them from slaughterhouse animal glands or chemical syntheses. In time, it was shown that, through biotransformation, various molds could convert a precursor compound called diosgenin into cortisone. By the same means, stigmasterol from soybean oil could be transformed into progesterone.

Some vaccines are also adaptable to mass production through fermentation. Vaccines for *Bordetella pertussis, Salmonella typhi, Vibrio cholerae,* and *Mycobacterium tuberculosis* are produced in large batch cultures. *Corynebacterium diphtheriae* and *Clostridium tetani* are propagated for the synthesis of their toxins, from which toxoids for the DT vaccines are prepared.

Figure 27.16 **The general layout of a fermentation plant for industrial production of drugs, enzymes, fuels, vitamins, and amino acids.**

27.1 MAKING CONNECTIONS

The Bioterrorism-Biotechnology Connection

Our increasing use of genetic technology to manipulate organisms, combined with an extensive knowledge of the genomes of many pathogens, brings to mind a question. Could terrorists use these same techniques and information to create more effective weapons of bioterrorism? The answer is troubling but is fortunately tempered by some good news as well.

Startlingly effective pathogens have been created by accident in the laboratory. In one incident, what started as a contraceptive vaccine to help control Australia's mouse population resulted in a recombinant mousepox virus that, by inactivating part of the immune system, quickly killed infected mice, even those that had been previously vaccinated against the virus. As mousepox is related to smallpox, it wasn't a great leap to imagine creating a smallpox virus that was even more virulent than wild-type smallpox.

Another easily imagined scenario is the incorporation of antibiotic-resistance genes into highly virulent bacterial pathogens. The anthrax scare of late 2001 was tempered somewhat by the fact that anthrax is easily treated with the antibiotic ciprofloxacin. If the anthrax had been a strain engineered to carry a gene for ciprofloxacin resistance, its power both to kill and to induce fear would have been much greater. This is not idle speculation; when several scientists from Biopreparat, a network of laboratories in Russia and Kazakhstan, defected to the West, they claimed they had created a form of *Yersinia pestis,* the bacterium responsible for plague, that was resistant to 16 different antibiotics.

Several companies have developed techniques to engineer a wide variety of bacteria with traits such as production of new antibiotics. A California company used this technology to create a strain of *E. coli* that was 32,000 times **less** sensitive to the antibiotic cefotaxime. After this persuasive demonstration of its new technology, the company destroyed the resultant strain at the request of the American Society for Microbiology, which was concerned about the potential of the resistant strain falling into the wrong hands.

But all is not lost. Pathogens engineered in the laboratory would probably struggle to survive in the outside world. It is extremely difficult to engineer a pathogen that can retain its newly enhanced (or acquired) virulence yet remain transmissible. Any new trait would soon be lost unless it conferred some selective advantage on its bearer. The questions then become: How quickly will the pathogen lose its virulence, and how many people will be infected before that happens? Clearly, the rapid discovery of pathogens is a key aspect of keeping people safe.

Beginning in 2003 the BioWatch system was implemented to provide just such rapid detection. Relying on air samplers in more than 30 cities to collect air in public places like train stations and airports, some samplers even sat atop public buildings. As air traveled through filters in the sampler, microbes would be trapped. Once a day, the filters were collected and screened for the presence of plague, tularemia, smallpox, and other pathogens. Unfortunately BioWatch was prone to false alarms, with over 50 recorded since the system was first put into service. Based on testing carried out secretly at the Pacific Northwest National Laboratory

A BioWatch air sampler in the Washington, D.C. subway Courtesy of Lawrence Livermore National Laboratory

in Washington State, the assays used to detect pathogens as part of BioWatch were found to be unsuitable for the task and the search began for more specific methods.

In 2010 a highly advanced technology that screens for pathogens was introduced by the Lawrence Livermore National Laboratory. This device, called the Microbial Detection Array, is loaded with nearly 400,000 known DNA probes that interact with more than 2,000 viruses and about 900 bacteria. Its plan is similar to the microarrays described in chapter 10 for identifying genes.

In practice, DNA is extracted from a sample of interest and labeled with a fluorescent dye. The processed sample is loaded onto the array chip, which is about the size of a glass slide. After hybridization, the chip is scanned and analyzed by computers that detect any of the probes that are adhering to the sample DNA and provide a readout of that sample's identity. This versatile technology has numerous benefits and applications. Most results will be available in about 24 hours and can be focused on detecting just those pathogens with the highest bioterrorism potential. To illustrate the detection array's accuracy and sensitivity, when its makers used it to screen a new rotavirus vaccine, they were able to detect a pig virus that had contaminated the vaccine. Right now, its potentials for use in biodefense, law enforcement, diagnosis, and industry seem almost unlimited.

What are some of the major ethical and medical concerns surrounding the development of harmful technologies? *Answer available on Connect.*

Miscellaneous Products

An interesting use of the powers of certain microbes has been the development and industrial production of natural *biopesticides* using *Bacillus thuringiensis*. During sporulation, these bacteria produce intracellular crystals that can be toxic to certain insects. When the insect ingests this endotoxin, its digestive tract breaks down and it dies, but the material is relatively nontoxic to other organisms. Commercial dusts are available to suppress caterpillars, moths, and worms on various agricultural crops and trees. A strain of this bacterium is also being considered to control the mosquito vector of malaria (chapter 23) and the black fly vector of onchocerciasis (river blindness; chapter 23).

Enzymes are critical to chemical manufacturing, the agriculture and food industries, textile and paper processing, and even laundry and dry cleaning. The advantage of enzymes is that they are very specific in their activity and are readily produced and released by microbes. Mass quantities of proteases, amylases, lipases, oxidases, and cellulases are produced by fermentation technology (see table 27.3). The wave of the future appears to be custom designing enzymes to perform a specific task by altering their amino acid content. Other compounds of interest that can be mass-produced by microorganisms are amino acids, organic acids, solvents, and natural flavor compounds to be used in air fresheners and foods.

In exploring the unlimited possibilities for applications of this technology, biotechnology-based industries have even ventured into developing products related to bioterrorism (**27.1 Making Connections**).

 ## *Check Your Progress* SECTION 27.5

18. Describe the aims of industrial microbiology.
19. Differentiate between primary and secondary metabolites.
20. What is meant by biotransformation?
21. Describe a fermentor and explain how it is scaled up for industrial use.
22. Explain what is meant by *downstream processing*.
23. List some specific examples of products derived from industrial processes.

 ## CASE STUDY Part 2

The outbreak at the military lunch revolved around a dish called perlo, which contained chicken, sausage, and rice. The day before the lunch, chicken was cooked in a pot of boiling water; the meat was removed from the bones, by hand, and combined with cooked sausages, onions, seasonings, and rice. The pot of cooked perlo was placed in an unheated oven overnight. The following day the perlo, which was still warm, was transferred to a slow cooker and reheated before being served. Most likely, *Staphylococcus aureus* was present on the hands of one of the persons who removed the cooked chicken from the bones (gloves were not worn during this process). *S. aureus* growth and toxin production is optimal at 34°C to 41°C (93°F to 106°F), and storing the dish in an unheated oven overnight provided

a perfect environment for both growth and toxin production. Staphylococcal enterotoxins are resistant to heat, and though the *S. aureus* bacteria may have been killed when the perlo was reheated, the toxins were still fully active. The speed with which staphylococcal intoxication occurs (as early as 2 hours after lunch, in this case) is always a clue as to the offending organism. Further, the halotolerance of *Staphylococcus* makes it a prime suspect when a dish is highly salty (containing ham or sausage, for instance). Fortunately food intoxication due to *S. aureus* generally passes quickly and has no lingering effects.

The botulism case was traced to homemade potato salad. Every ill person reported having eaten some at the potluck, and leftover potato salad tested positive for the presence of botulinum toxin. The person who prepared the potato salad used home-canned potatoes that had been prepared in a boiling water canner, which does not kill endospores of *Clostridium botulinum*. Acidic foods, like tomatoes or pickles, with a pH below 4.6 inhibit germination of *C. botulinum* endospores, but potatoes are a low-acid food and don't possess this inhibitory property. Further, the home canning process creates an anaerobic environment, and if endospores of *C. botulinum* (an obligate anaerobe) are present, they will germinate, grow, and produce toxins during storage of the food. Finally, in contrast to staphylococcal enterotoxin, botulinum neurotoxin is heat labile and boiling the potatoes for 10 minutes prior to making the potato salad would have inactivated the toxins. Fortunately no other patients died, but 29 were hospitalized.

The Chipotle outbreak may never be solved. The slower pace of an *E. coli* infection compared to the intoxications seen in the other two cases make epidemiological investigation far more difficult. In the week to 10 days it took for most patients to present to a physician, restaurant trash had been thrown away, supplies of meat and vegetables had been exhausted and reordered several times, and countertops, refrigerators, and work surfaces had been cleaned many times over. The fact that several different Chipotle locations were involved in the outbreak indicated that the source of infection was probably a food ingredient delivered to many restaurants from a single supplier and not a sick Chipotle employee or other isolated source. By the end of 2015 the outbreak seemed to have run its course, and it was thought that whatever had caused the problem was no longer in the restaurant's supply chain. On February 1, 2016, the CDC declared the outbreak to be over.

- *Why would potatoes be a common vehicle for botulism?*
- *Food safety guidelines recommend that large batches of food be transferred to shallow containers prior to refrigeration. Why is this?*

To learn more about these three cases, search the Internet for "Large Outbreak of Botulism Associated with a Church Potluck Meal," "Outbreak of Staphylococcal Food Poisoning from a Military Unit Lunch Party," or "Multistate Outbreaks of Shiga Toxin-Producing Escherichia coli O26 Infections Linked to Chipotle Mexican Grill Restaurants."

To conclude this Case Study, go to Connect.

Chapter Summary with Key Terms

27.1 Applied Microbiology and Biotechnology
Biotechnology is the practical application of microbiology in the manufacture of food, industrial chemicals, drugs, and other products. Many of these processes use mass, controlled microbial **fermentations** and bioengineered microorganisms.
A. Microorganisms in Water and Wastewater Treatment
1. Drinking water is rendered safe by a purification process that involves storage, sedimentation, settling, aeration, filtration, and disinfection.
2. Sewage or used wastewater can be processed to remove solid matter, dangerous chemicals, and microorganisms.
3. Microbes biodegrade the waste material or sludge. Solid wastes are further processed in **anaerobic digesters.**

27.2 The Microbiology of Food
A. Microbes and humans compete for the rich nutrients in food. Many microbes are present on food as harmless contaminants; some are used to create flavors and nutrients; others may produce unfavorable reactions.
1. The beneficial effects of microorganisms on food are based mainly on their ability to ferment or chemically change it to alter flavor, taste, or texture. Microbes can also serve as food themselves.
2. The detrimental effects of microorganisms are their ability to cause food-borne illness and to spoil food, making it unsuitable for consumption.
3. Microbes in or on food may result in no consequences at all. In that case, their effect on food is considered neutral.

27.3 Microbial Fermentations in Food Products from Plants
A. Microbes can impart desirable aroma, flavor, or texture to foods. Bread, alcoholic beverages, some vegetables, and some dairy products are infused with pure microbial strains to yield the necessary fermentation products.
1. Baker's yeast, *Saccharomyces cerevisiae,* is used to **leaven** bread dough by giving off CO_2.
2. Beer making involves the following steps: Barley is sprouted (**malted**) to generate digestive enzymes, transformed to **mash,** and heated with **hops** to produce a **wort.** Wort is fermented by yeast and **lagered** in large tanks before it is carbonated and packaged.

3. Wine is started by fermentation of **must** (fruit juices). Whiskey, vodka, brandy, and other alcoholic beverages are distilled to increase their alcohol content.
4. Vegetable products, including sauerkraut, pickles, and soybean derivatives, can be pickled in the presence of salt or sugar.
5. Vinegar is produced by fermenting plant juices first to alcohol and oxidizing the alcohol to acetaldehyde and acetic acid.
6. Most dairy products are produced by microbes acting on nutrients in milk. In cheese production, milk proteins are coagulated to a solid **curd** that separates from the watery **whey.**
7. Mass-cultured microbes such as yeasts, molds, and bacteria can serve as food. Proteins derived from microbes are currently added to animal feeds.

27.4 Microbial Involvement in Food-Borne Disease
A. A number of bacteria, protozoa, and viruses cause spoilage and food poisoning.
B. Prevention Measures
1. Microbial growth that leads to spoilage and food poisoning can be avoided by high temperature and pressure treatment (canning) and **pasteurization** for disinfecting milk and other heat-sensitive beverages.
2. Refrigeration and freezing inhibit microbial growth.
3. Irradiation sterilizes or disinfects foods for longer-term storage.
4. Alternative preservation methods include additives (salt, nitrites), treatment with ethylene oxide gas, and drying.

27.5 General Concepts in Industrial Microbiology
A. Industrial microbiology involves the large-scale commercial production of organic compounds such as antibiotics, vitamins, amino acids, enzymes, and hormones using specific microbes in carefully controlled fermentation settings.
B. Microbes are chosen for their production of a desired **metabolite;** several different species can be used to biotransform raw materials in a stepwise series of metabolic reactions.
C. Fermentations are conducted in massive culture devices called **fermentors** that have special mechanisms for adding nutrients, stirring, oxygenating, altering pH, cooling, monitoring, and harvesting product.

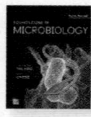
Level I. Knowledge and Comprehension

These questions require a working knowledge of the concepts in the chapter and the ability to recall and understand the information you have studied.

Multiple-Choice Questions

Select the correct answer from the answers provided. For questions with blanks, choose the combination of answers that most accurately completes the statement.

1. Which of the following is *not* a stage in purification of drinking water?
a. filtration
b. chlorination
c. activated sludge
d. sedimentation

2. Milk is usually pasteurized by
a. the high-temperature short-time method
b. ultrapasteurization
c. batch method
d. electrical currents

3. Substances given off by yeasts during fermentation are
 a. alcohol c. organic acids
 b. carbon dioxide d. all of these

4. Which of the following is added to facilitate milk curdling during cheese making?
 a. lactic acid b. salt c. *Lactobacillus* d. rennin

5. Secondary metabolites of microbes are formed during the _____ phase of growth.
 a. exponential b. stationary c. lag d. death

6. The dried, presprouted grain that is soaked to activate enzymes for beer is
 a. hops b. malt c. wort d. mash

7. A food maintenance temperature that generally will prevent food poisoning is
 a. below 4°C c. room temperature
 b. above 60°C d. both a and b

8. Most cases of food intoxication are caused by
 a. *Salmonella enteritidis* c. *Bacillus cereus*
 b. *Staphylococcus aureus* d. *Giardia lamblia*

9. The large tanks used in industrial production of antibiotics are termed
 a. digesters b. fermentors c. spargers d. biotransformers

10. Which of the following microbes produces a natural biopesticide?
 a. *Penicillium chrysogenum* c. *Bacillus thuringiensis*
 b. *Saccharomyces cerevisiae* d. *Lactobacillus bulgaricus*

 ## Case Study Review

1. Chopped garlic often contains *C. botulinum* endospores and is commonly sold submerged in olive oil, which creates an anaerobic environment. What is required to be added to garlic sold in this manner?
 a. sodium chloride
 b. citric acid
 c. water
 d. A warning label describing the symptoms of botulism.

2. During the Chipotle outbreak, some people were sickened with *E. coli*, but a DNA profile of the bacterium in their feces showed that it was not the outbreak strain. What conclusion can be drawn from this information?
 a. The person was not sickened by *E. coli*.
 b. The person is not part of the outbreak.
 c. *Escherichia coli* mutates frequently, making DNA profiling unreliable.
 d. The person has already begun to recover from the disease.

3. If the bacterium in question during the perlo outbreak had been *E. coli* instead of *S. aureus*, how might the outbreak have changed?

 ## Writing Challenge

For each question, compose a one- or two-paragraph answer that includes the factual information needed to completely address the question. Check Your Progress questions can also be used for writing-challenge exercises.

1. Every year, supposedly safe municipal water supplies cause outbreaks of enteric illness.
 a. How in the course of water analysis and treatment might these pathogens be missed?
 b. Why is there less tolerance for a fecal coliform in drinking or recreational water than for other bacteria?

2. Summarize the basic processes in food fermentations, using examples from beer, wine, and dairy products.

3. Outline the major methods used to prevent foodborne illness and spoilage. Explain the adage "When in doubt, throw it out" from a microbiological perspective.

4. Explain the differences between primary and secondary metabolites as they relate to industrial microbiology.

 ## Concept Mapping

On Connect you can find an Introduction to Concept Mapping that provides guidance for working with concept maps along with concept mapping activities for this chapter.

Level II. Application, Analysis, Evaluation, and Synthesis

These problems go beyond just restating facts and require higher levels of understanding and an ability to interpret, problem solve, transfer knowledge to new situations, create models, and predict outcomes.

Critical Thinking

Critical thinking is the ability to reason and solve problems using facts and concepts. These questions can be approached from a number of angles, and in most cases, they do not have a single correct answer.

1. Describe three important potential applications of bacteria that can make methane.

2. a. Describe four food-preparation and food-maintenance practices in your own kitchen that could expose people to food poisoning, and explain how to prevent them.
 b. What is a good general rule to follow concerning food of questionable quality?

3. a. What is the purpose of boiling the wort in beer preparation?
 b. What are hops used for?
 c. If fermentation of sugars to produce alcohol in wine is anaerobic, why do winemakers make sure that the early phase of yeast growth is aerobic?

4. Predict the differences in the outcome of raw milk that has been incubated for 48 hours versus pasteurized milk that has been incubated for the same length of time.

5. Explain the ways that biotransformation by synergistic microorganisms is harnessed in industrial microbiology.
 a. Outline the steps in the production of an antibiotic, starting with a newly discovered mold.
 b. Review chapter 10 and locate examples of recombinant DNA technology being applied to industrial processes.

6. Log on to http://www.cdc.gov/pulsenet/ to learn about the PulseNet system and explain how it would have been applied in the three outbreaks described in this chapter's Case Study.

Visual Challenge

1. From **chapter 7, figure 7.6.** What method of food preservation exploits the principles that are illustrated here?

Hypotonic Solution

Hypertonic Solution

2. From **chapter 8, figure 8.23.** Correlate the stages in this figure with corresponding steps in wine making, starting with grapes (figure 27.6). Where do champagne bubbles come from? What causes wine to go sour?

STUDY TIP: Quickly create a customized practice quiz for this chapter by going to your Connect SmartBook "My Reports" section. Don't have SmartBook and want to learn more? Go to **whysmartbook.com.**

Appendix A

Detailed Steps in the Glycolysis Pathway

Steps in the Glycolysis Pathway The first portion of glycolysis involves activation of glucose, which is followed by oxidation reactions of the glucose fragments, the synthesis of ATP, and the formation of pyruvic acid. *Although each step of metabolism is catalyzed by a specific enzyme, we will not mention it for most reactions.* The following outline lists the principal steps of glycolysis. To see an overview of glycolysis, refer to figure 8.17.

1. **Glucose** is phosphorylated by means of an **ATP** acting with the enzyme hexokinase. The product is glucose-6-phosphate. (Numbers in chemical names refer to the position of the phosphate on the carbon skeleton.) This is a way of "priming" the system and keeping the glucose inside the cell.

2. Glucose-6-phosphate is converted to its isomer, fructose-6-phosphate, by phosphoglucoisomerase.

3. Another ATP is spent in phosphorylating the first carbon of fructose-6-phosphate, which yields fructose-1,6-diphosphate.

Up to this point, no energy has been released, no oxidation-reduction has occurred, and, in fact, 2 ATPs have been used. In addition, the molecules remain in the 6-carbon state.

4. Now doubly activated, fructose-1,6-diphosphate is split into two 3-carbon fragments: glyceraldehyde-3-phosphate (G-3-P) and dihydroxyacetone phosphate (DHAP). These molecules are isomers, and DHAP is enzymatically converted to G-3-P, which is the more reactive form for subsequent reactions.

The effect of the splitting of fructose diphosphate is to double every subsequent reaction, because where there was once a single molecule, there are now two to be fed into the remaining pathways.

5. Each molecule of glyceraldehyde-3-phosphate becomes involved in the single oxidation-reduction reaction of glycolysis, a reaction that sets the scene for ATP synthesis. Two reactions occur simultaneously and are catalyzed by the same enzyme, called glyceraldehyde-3-phosphate dehydrogenase. The coenzyme **NAD$^+$** picks up electrons and a proton from G-3-P, forming **NADH.** This step is accompanied by the addition of an inorganic phosphate (PO$_4^{3-}$) to form an unstable bond on the third carbon of the G-3-P substrate. The product of these reactions is diphosphoglyceric acid.

In aerobic organisms, the NADH formed during step 5 will undergo further reactions in the electron transport system, where the final electron acceptor will be oxygen, and an additional 3 ATPs will be generated per NADH. In organisms that ferment glucose anaerobically, the NADH will be oxidized back to NAD$^+$, and the hydrogen acceptor will be an organic compound.

6. One of the high-energy phosphates of diphosphoglyceric acid is donated to ADP via substrate-level phosphorylation, resulting in a molecule of ATP. The product of this reaction is 3-phosphoglyceric acid.

7,8. During this phase, a substrate for the synthesis of a second ATP is made in two substeps. First, 3-phosphoglyceric acid is converted to 2-phosphoglyceric acid through the shift of a phosphate from the third to the second carbon. Then, the removal of a water molecule from 2-phosphoglyceric acid produces phosphoenolpyruvic acid, which is charged to release a high-energy phosphate.

9. In the final reaction of glycolysis, phosphoenolpyruvic acid gives up its high-energy phosphate to form a second ATP via substrate-level phosphorylation. This reaction, catalyzed by pyruvate kinase, also produces pyruvic acid (pyruvate), a compound with many roles in metabolism.

The 2 ATPs formed during steps 6 and 9 are examples of substrate-level phosphorylation, in that the high-energy phosphate is transferred directly from a substrate to ADP. These reactions are catalyzed by kinases. Because both molecules that arose in step 4 undergo these reactions, an overall total of **4** ATPs is generated in the partial oxidation of a glucose to 2 pyruvic acids. However, 2 ATPs were expended for steps 1 and 3, so the net number of ATPs available to the cell from these reactions is **2**.

TABLE A.1	Twenty Amino Acids and Their Abbrevations*

Nonpolar

Glycine (Gly) Alanine (Ala) Valine (Val) Leucine (Leu) Isoleucine (Ile)

Tryptophan (Trp) Proline (Pro) Cysteine (Cys) Methionine (Met) Phenylalanine (Phe)

Polar

Asparagine (Asn) Glutamine (Gln) Tyrosine (Tyr) Serine (Ser) Threonine (Thr)

Electrically Charged

ACIDIC

Aspartic Acid (Asp) Glutamic Acid (Glu)

BASIC

Arginine (Arg) Lysine (Lys) Histidine (His)

*The basic skeleton is in yellow; R groups are in purple, blue, or green, depending on the nature of their composition.

Appendix B

Tests and Guidelines

Methods for Testing Sterilization and Germicidal Processes

Most procedures used for microbial control in the clinical setting are either physical methods (heat, radiation) or chemical methods (disinfectants, antiseptics). These procedures are so crucial to the well-being of patients and staff that their effectiveness must be monitored in a consistent and standardized manner. Particular concerns include the time required for the process, the concentration or intensity of the antimicrobic agent being used, and the nature of the materials being treated. The effectiveness of an agent is frequently established through controlled microbiological analysis. In these tests, a biological indicator (a highly resistant microbe) is exposed to the agent, and its viability is checked. If this known test microbe is destroyed by the treatment, it is assumed that the less resistant microbes have also been destroyed. Growth of the test organism indicates that the sterilization protocol has failed. The general categories of testing include:

Heat The usual form of heat used in microbial control is steam under pressure in an autoclave. Autoclaving materials at a high temperature (121°C) for a sufficient time (15–40 minutes) destroys most bacterial endospores. To monitor the quality of any given industrial or clinical autoclave run, technicians insert a dried test strip of *Bacillus stearothermophilus,* a spore-forming bacterium with extreme heat resistance. After autoclaving, the ampule is incubated at 56°C (the optimal temperature for this species) and checked for growth.

Radiation The effectiveness of ionizing radiation in sterilization is determined by placing special strips of dried spores of *Bacillus sphaericus* (a common soil bacterium with extreme resistance to radiation) into a packet of irradiated material.

Filtration The performance of membrane filters used to sterilize liquids may be monitored by adding *Pseudomonas diminuta* to the liquid. Because this species is a tiny bacterium that can escape through the filter if the pore size is not small enough, it is a good indicator of filtrate sterility.

Gas sterilization Ethylene oxide gas, one of the few chemical sterilizing agents, is used on a variety of heat-sensitive medical and laboratory supplies. The most reliable indicator of a successful sterilizing cycle is the sporeformer *Bacillus subtilis,* variety *niger.*

Germicidal assays U.S. hospitals and clinics commonly employ more than 250 products to control microbes in the environment, on inanimate objects, and on patients. Controlled testing of these products' effectiveness is usually performed not in the clinic itself but by the chemical or pharmaceutical manufacturer. A number of standardized in vitro tests are currently available to assess the effects of germicides. Most of them use a non-sporeforming pathogen as the biological indicator.

The Phenol Coefficient (PC)

The older disinfectant phenol has been the traditional standard by which other disinfectants are measured. In the PC test, a water-soluble phenol-based (phenolic) disinfectant (Amphyl, Lysol) is tested for its bactericidal effectiveness as compared with phenol. These disinfectants are serially diluted in a series of *Salmonella choleraesuis, Staphylococcus aureus,* or *Pseudomonas aeruginosa* cultures. The tubes are left for 5, 10, or 15 minutes and then subcultured to test for viability. The PC is a ratio derived by comparing the following data:

$$PC = \frac{\text{Greatest dilution of the phenolic that kills test bacteria in 10 min but not in 5 min}}{\text{Greatest dilution of phenol giving same result}}$$

The general interpretation of this ratio is that chemicals with lower phenol coefficients have greater effectiveness. The primary disadvantage of the PC test is that, because it is restricted to phenolics, it is inappropriate for the vast majority of clinical disinfectants.

Use Dilution Test

An alternative test with broader applications is the use dilution test. This test is performed by drying the test culture (one of those species used in the PC test) onto the surface of small stainless steel cylindrical carriers. These carriers are then exposed to varying concentrations of disinfectant for 10 minutes, removed, rinsed, and placed in tubes of broth. After incubation, the tubes are inspected for growth. The smallest concentration of disinfectant that kills the test organism on 10 carrier pieces is the correct dilution for use.

Filter Paper Disc Method

A quick measure of the inhibitory effects of various disinfectants and antiseptics can be achieved by the filter paper disc method. First, a small (½-inch) sterile piece of filter paper is dipped into a disinfectant of known concentration. This is placed on an agar medium seeded with a test organism (*S. aureus* or *P. aeruginosa*), and the plate is incubated. A zone devoid of growth (zone of inhibition) around the disc indicates the capacity of the agent to inhibit growth **(figure B.1).** Like the antimicrobic sensitivity test (see figure 12.18), this test measures the minimum inhibitory concentration of the chemical. In general, chemicals with wide zones of inhibition are effective even in high dilutions.

Guidelines for Universal Precautions in the Clinical Setting

The significance of these procedures for infection control is introduced in chapter 13.

1. Barrier precautions, including masks and gloves, should be taken to prevent contact of skin and mucous membranes with patients' blood or other body fluids. Because gloves can

FIGURE B.1 A filter paper disc containing hydrogen peroxide produces a broad clear zone of inhibition in a culture of *Staphylococcus aureus.*
Source: © Kathy Park Talaro

develop small invisible tears, double gloving further decreases the risk. For protection during surgery, venipuncture, or emergency procedures, gowns, aprons, and other body coverings should be worn. Dental workers should wear eyewear and face shields to protect against splattered blood and saliva.

2. More than 10% of health care personnel are pierced each year by sharp (and usually contaminated) instruments. These accidents carry risks not only for AIDS but also for hepatitis B, hepatitis C, and other diseases. Preventing inoculation infection requires vigilant observance of proper techniques.

 • All disposable needles, scalpels, or sharp devices from invasive procedures must immediately be placed in puncture-proof containers for sterilization and final discard.

 • Under no circumstances should a worker attempt to recap a syringe, remove a needle from a syringe, or leave unprotected used syringes where they pose a risk to others.

 • Reusable needles or other sharp devices must be heat-sterilized in a puncture-proof holder before they are handled.

 • If a needlestick or other injury occurs, immediate attention to the wound, such as thorough degermation and application of strong antiseptics, can prevent infection.

3. Dental handpieces should be sterilized between patients, but if this is not possible, they should be thoroughly disinfected with a high-level disinfectant (peroxide, hypochlorite). Blood and saliva should be removed completely from all contaminated dental instruments and intraoral devices prior to sterilization.

4. Hands and other skin surfaces that have been accidentally contaminated with blood or other fluids should be scrubbed immediately with a germicidal soap. Hands should likewise be washed after removing rubber gloves, masks, or other barrier devices.

5. Because saliva can be a source of some types of infections, barriers should be used in all mouth-to-mouth resuscitations.

6. Health care workers with active, draining lesions of the skin or mucous membrane must refrain from handling patients or equipment that will come into contact with other patients. Pregnant health care workers risk infecting their fetuses and must pay special attention to these guidelines. Personnel should be protected by vaccination whenever possible.

Isolation procedures for known or suspected infections should still be instituted on a case-by-case basis. See table 13.11.

An additional consideration with regard to infectious agents is covered by the biosafety levels of microbes being handled in the laboratory, ranging from 1 (least level of pathogenicity) up to 4 (highest level of pathogenicity). This topic is also discussed in chapter 13. An overview of the containment levels and concerns are presented in **table B.1.**

TABLE B.1	Primary Biosafety Levels and Agents of Disease	
Personnel handling infectious agents in the laboratory must be protected from possible infection through special risk management or containment procedures.		
Biosafety Level	**Facilities and Practices**	**Risk of Infection and Class of Pathogens**
1	Standard, open bench, no special facilities needed; typical of most microbiology teaching labs; access may be restricted.	Low infection hazard; class 1 microbes not generally considered pathogens and will not invade the bodies of healthy persons; *Micrococcus luteus, Bacillus megaterium, Lactobacillus, Saccharomyces.*
2	At least Level 1 facilities and practices; plus personnel must be trained in handling pathogens; lab coats and gloves required; safety cabinets may be needed; biohazard signs posted; access restricted.	Agents with moderate potential to infect; class 2 pathogens can cause disease in healthy people but can be contained with proper facilities; pathogens that belong to class 2 include *Staphylococcus aureus, Escherichia coli, Salmonella* spp.; pathogenic helminths; hepatitis A, B, and rabies viruses; *Cryptococcus* and *Blastomyces;* HIV.
3	Minimum of Level 2 facilities and practices; manipulation performed in safety cabinets with special containment features; all personnel require protective clothing; no unsterilized materials can leave the lab; personnel are monitored and vaccinated.	Agents can cause severe or lethal disease especially when inhaled; spp. 3 microbes include *Mycobacterium tuberculosis, Francisella tularensis, Yersinia pestis, Brucella* spp., *Coxiella burnetii, Coccidioides immitis,* yellow fever, and western equine encephalitis.
4	Minimum of Level 3 facilities and practices; facilities have highest levels of controlled access; clothing changes and showers required for all people entering and leaving; materials must be autoclaved or gas-sterilized prior to entering and leaving lab.	Agents being handled are highly virulent microbes that pose extreme risk for morbidity and mortality when inhaled in droplet or aerosol form; class 4 microbes include flaviviruses, arenaviruses (Lassa fever virus), and filoviruses (Ebola and Marburg viruses).

TABLE B.2 Most Probable Number Chart for Evaluating Coliform Content of Water

The Number of Tubes in Series, with Positive Growth at Each Dilution							
10 ml	1 ml	0.1 ml	MPN* per 100 ml of Water	10 ml	1 ml	0.1 ml	MPN* per 100 ml of Water
0	1	0	0.18	5	0	1	3.1
1	0	0	0.20	5	1	0	3.3
1	0	0	0.40	5	1	1	4.6
2	0	0	0.45	5	2	0	4.9
2	0	1	0.68	5	2	1	7.0
2	2	0	0.93	5	2	2	9.5
3	0	0	0.78	5	3	0	7.9
3	0	1	1.1	5	3	1	11.0
3	1	0	1.1	5	3	2	14.0
3	2	0	1.4	5	4	0	13.0
4	0	0	1.3	5	4	1	17.0
4	0	1	1.7	5	4	2	22.0
4	1	0	1.7	5	4	3	28.0
4	1	1	2.1	5	5	0	24.0
4	2	0	2.2	5	5	1	35.0
4	2	1	2.6	5	5	2	54.0
4	3	0	2.7	5	5	3	92.0
5	0	0	2.3	5	5	4	160.0

*Most probable number of cells per sample of 100 ml.

FIGURE B.2 The most probable number (MPN) procedure for determining the coliform content of a water sample. In the presumptive test, each set of tubes of broth is inoculated with a water sample reduced by a factor of 10. After incubation, the sets of tubules are examined and rated for gas production (for instance, 0 means no tubes with gas, 1 means one tube with gas, 2 means two tubes with gas). Applying this result to Table C.2 will indicate the probable number of cells present in 100 ml of the water sample. Confirmation of coliforms can be achieved by confirmatory tests on additional media, and complete identification can be made through selective and different media and staining (Chapter 20).

Tests for Water Analysis

Figure B.2 and table B.2 provide standards used in analyzing water samples using the most probable number (MPN) method, as described in chapter 26. The figure indicates the steps in the laboratory procedure, and the table provides estimated counts for coliforms, based on the number of tubes with growth.

Appendix C

General Classification Techniques and Taxonomy of Bacteria

| TABLE C.1 | Summary of Techniques Used in Bacterial Identification and Classification, Chapter 4, Section 4.6 |

- **Microscopic Morphology** Traits that can be valuable aids to identification are combinations of cell shape and size, Gram stain reaction, acid-fast reaction, and special structures, including endospores, granules, and capsules. Electron microscope studies can pinpoint additional structural features (such as the cell wall, flagella, pili, and fimbriae).
- **Macroscopic Morphology** Appearance of colonies, including texture, size, shape, pigment, speed of growth and patterns of growth in broth and gelatin media.
- **Physiological/Biochemical Characteristics** Enzymes and other biochemical properties of bacteria are fairly reliable and stable expressions of the "chemical identity" of each species. Diagnostic tests determine the presence of specific enzymes and assess nutritional and metabolic activities. Examples include fermentation of sugars; digestion of complex polymers such as proteins and polysaccharides; presence of catalase, oxidase, and decarboxylases; sensitivity to antimicrobic drugs.
- **Chemical Analysis** Analyzing the types of specific structural substances that the bacterium contains, such as the chemical composition of peptides in the cell wall and lipids in membranes.
- **Serological Analysis** Bacteria display molecules called antigens that are recognized by the immune system. One immune response to antigens is to produce antibodies that are designed to bind tightly to the antigens. This response is so specific that antibodies can be used as a means of identifying bacteria in specimens and cultures.
- **Genetic and Molecular Analysis** Examining the genetic material itself has revolutionized the identification and classification of bacteria.

G + C Base Composition The overall percentage of guanine and cytosine (the G + C content as compared with A + T content) in DNA is a general indicator of relatedness because it is a trait that does not change rapidly. Bacteria with a significant difference in G + C percentage are less likely to be genetically related. See the revised classification scheme that is partly based on this percentage (table C.2).

DNA Analysis Using Genetic Probes It is possible to identify a bacterial species by analyzing its genetic material. This uses small fragments of DNA (or RNA) called probes that are complementary to the specific sequences of DNA from a particular microbe. Unknown test DNA from specimens or cultures is bound to a special surface. After several different probes have been added, the plate is observed for visible signs that the probes have become fixed to the test DNA. The binding of probes onto areas of the test DNA indicates close correspondence and makes identification possible.

Nucleic Acid Sequencing and rRNA Analysis One of the most valuable indicators of evolutionary relatedness is the sequence of nitrogen bases in ribosomal RNA, a major component of ribosomes. Ribosomes have the same function (protein synthesis) in all cells, and they tend to remain stable in their nucleic acid content over long periods. Thus, any major differences in the sequence, or "signature," of the rRNA is likely to indicate some distance in ancestry.

This technique is effective for differentiating general group differences (it was used to separate the three superkingdoms of life discussed in chapter 1), and it can identify species. Elements of these and other identification methods are presented in more detail in chapters 10, 16, and 17.

| TABLE C.2 | Organization of *Bergey's Manual of Systematic Bacteriology*, Second Edition |

Taxonomic Rank	Representative Genera
Volume 1. The Archaea and the Deeply Branching and Phototrophic Bacteria	
Domain Archaea	
Phylum Crenarchaeota	*Thermoproteus, Pyrodictium, Sulfolobus*
Phylum Euryarchaeota	
Class I. Methanobacteria	*Methanobacterium*
Class II. Methanococci	*Methanococcus*
Class III. Halobacteria	*Halobacterium, Halococcus*
Class IV. Thermoplasmata	*Thermoplasma, Picrophilus,*
Class V. Thermococci	*Ferroplasma*
Class VI. Archaeoglobi	*Thermococcus, Pyrococcus*
Class VII. Methanopyri	*Archaeoglobus*
	Methanopyrus

TABLE C.2 (continued)	
Taxonomic Rank	**Representative Genera**
Volume 1. The Archaea and the Deeply Branching and Phototrophic Bacteria (*continued*)	
Domain Archaea	
Phylum Aquificae	*Aquifex, Hydrogenobacter*
Phylum Thermotogae	*Thermotoga, Geotoga*
Phylum Thermodesulfobacteria	*Thermodesulfobacterium*
Phylum "Deinococcus-Thermus"	*Deinococcus, Thermus*
Phylum Chrysiogenetes	*Chrysogenes*
Phylum Chloroflexi	*Chloroflexus, Herpetosiphon*
Phylum Thermomicrobia	*Thermomicrobium*
Phylum Nitrospira	*Nitrospira*
Phylum Deferribacteres	*Geovibrio*
Phylum Cyanobacteria	*Prochloron, Synechococcus,*
Phylum Chlorobi	*Pleurocapsa, Oscillatoria, Anabaena, Nostoc, Stigonema Chlorobium, Pelodictyon*
Volume 2. Domain Bacteria: The Proteobacteria	
Phylum Proteobacteria	
Class I. Alphaproteobacteria	*Rhodospirillum, Rickettsia, Bartonella, Caulobacter, Rhizobium, Brucella, Nitrobacter, Methylobacterium, Beijerinckia, Hyphomicrobium*
Class II. Betaproteobacteria	*Neisseria, Burkholderia, Alcaligenes, Comamonas, Nitrosomonas, Methylophilus, Thiobacillus*
Class III. Gammaproteobacteria	*Chromatium, Leucothrix, Legionella, Pseudomonas, Moraxella, Acinetobacter, Azotobacter, Vibrio, Escherichia, Klebsiella, Proteus, Salmonella, Shigella, Yersinia, Haemophilus*
Class IV. Deltaproteobacteria	*Desulfovibrio, Bdellovibrio, Myxococcus, Polyangium*
Class V. Epsilonproteobacteria	*Campylobacter, Helicobacter*
Volume 3. Domain Bacteria: The Low G + C Gram-Positive Bacteria	
Phylum Firmicutes	
Class I. Clostridia	*Clostridium, Peptostreptococcus, Eubacterium, Desulfotomaculum, Heliobacterium, Veillonella*
Class II. Mollicutes	*Mycoplasma, Ureaplasma, Spiroplasma, Acholeplasma Bacillus, Caryophanon, Paenibacillus,*
Class III. Bacilli	*Thermoactinomyces, Lactobacillus, Streptococcus, Enterococcus, Listeria, Leuconostoc, Staphylococcus*
Volume 4. Domain Bacteria: The High G + C Gram-Positive Bacteria	
Phylum Actinobacteria	
Class Actinobacteria	*Actinomyces, Micrococcus, Arthrobacter, Corynebacterium, Mycobacterium, Nocardia, Actinoplanes, Propionibacterium, Streptomyces, Thermomonospora, Frankia, Actinomadura, Bifidobacterium*
Volume 5. Domain Bacteria: The Planctomycetes, Spirochaetes, Fibrobacteres, Bacteriodetes, and Fusobacteria	
Phylum Planctomycetes	*Planctomyces, Gemmata*
Phylum Chlamydiae	*Chlamydia, Chlamydiophila*
Phylum Spirochaetes	*Spirochaeta, Borrelia, Treponema, Leptospira*
Phylum Fibrobacteres	*Fibrobacter*
Phylum Acidobacteria	*Acidobacterium*
Phylum Bacteriodetes	*Bacteroides, Porphyromonas, Prevotella, Flavobacterium, Sphingobacterium, Flexibacter, Cytophaga*
Phylum Fusobacteria	*Fusobacterium, Streptobacillus*
Phylum Verrucomicrobia	*Verrucomicrobium*
Phylum Dictyoglomus	*Dictyoglomus*

Answers to Multiple-Choice, Matching, and Case Study Questions

Chapter 1
1. d
2. c
3. d
4. c
5. d
6. a
7. c
8. b
9. d
10. c
11. c
12. b
13. d
14. order top to bottom: 3, 7, 4, 2, 8, 5, 6, 1
15. c
16. d, not cellular and not alive

Case Study 1 Questions
1. b
2. b
3. Answer appears in Connect

Chapter 2
1. c
2. c
3. b
4. c
5. a
6. d
7. c
8. b
9. c
10. b
11. b
12. d
13. b
14. c
15. c
16. c
17. a
18. d
19. d
20. b

Case Study 2 Questions
1. b
2. b
3. Answer appears in Connect

Chapter 3
1. c
2. b
3. c
4. d

5. b
6. d
7. b
8. b
9. c
10. c
11. a
12. b
13. c
14. d
15. b
16. c
17. abf, df, abf, ef, def, af, bf, bf

Case Study 3 Questions
1. d
2. b
3. Answer appears in Connect

Chapter 4
1. a
2. d
3. d
4. a
5. c
6. a
7. c
8. b
9. d
10. b
11. b
12. b
13. d
14. c
15. c
16. c
17. c
18. d

Case Study 4 Questions
1. c
2. b
3. Answer appears in Connect

Chapter 5
1. b
2. d
3. d
4. a
5. b
6. c
7. c
8. b
9. b
10. c
11. a
12. d

13. b
14. c
15. d
16. b
17. b
18. e
19. c
20. h
21. g
22. j
23. i
24. d
25. a
26. f

Case Study 5 Questions
1. f
2. c
3. Answer appears in Connect

Chapter 6
1. c
2. d
3. d
4. b
5. d
6. a
7. a.
8. d
9. b
10. b
11. c
12. d
13. a
14. d
15. d

Case Study 6 Questions
1. b
2. d
3. Answer appears in Connect

Chapter 7
1. c
2. a
3. a
4. c
5. c
6. b
7. a
8. a
9. b
10. b
11. c
12. d
13. c
14. c

15. b
16. c

Case Study 7 Questions
1. b
2. e
3. Answer appears in Connect

Chapter 8
1. b
2. a
3. c
4. d
5. d
6. b
7. b
8. b
9. b
10. a
11. c
12. a
13. d
14. b
15. c
16. d
17. b
18. c
19. c
20. c
21. c
22. c, a, abc, a, b, a, c, c

Case Study 8 Questions
1. c
2. b
3. Answer appears in Connect

Chapter 9
1. b
2. e
3. b
4. b
5. c
6. c
7. c
8. a
9. b
10. a
11. e
12. b
13. d
14. b
15. c
16. b
17. d
18. d, f, b, g, e, a, i, c/e/h

Case Study 9 Questions
1. b
2. e
3. Answer appears in Connect

Chapter 10
1. c
2. c
3. d
4. a
5. b
6. c
7. c
8. e
9. d
10. h, c, f, a, g, b, e, d

Case Study 10 Questions
1. b
2. c
3. Answer appears in Connect

Chapter 11
1. d
2. c
3. b
4. a
5. c
6. b
7. d
8. b
9. d
10. c
11. b
12. d
13. c
14. d
15. a
16. b
17. c

Case Study 11 Questions
1. c
2. c
3. Answer appears in Connect

Chapter 12
1. b
2. c
3. d
4. b
5. c
6. c
7. b
8. d

9. a
10. c
11. c
12. b
13. d, a, g, b, f, h, c, e

Case Study 12 Questions
1. c
2. c
3. Answer appears in Connect

Chapter 13
1. a
2. d
3. b
4. d
5. c
6. d
7. b
8. c
9. c
10. c
11. d
12. c
13. c
14. a
15. a
16. d

Case Study 13 Questions
1. b
2. d
3. Answer appears in Connect

Chapter 14
1. b
2. b
3. d
4. b
5. b
6. b
7. c
8. c
9. a
10. a
11. c
12. d
13. c
14. d
15. d
16. c

Case Study 14 Questions
1. f
2. b
3. Answer appears in Connect

Chapter 15
1. b
2. a
3. c
4. a
5. c
6. b
7. b
8. c
9. a
10. c
11. a
12. c
13. e
14. b
15. c
16. d
17. b
18. c
19. IgG bfgh
 IgA abc
 IgD bj
 IgE bi
 IgM deh
20. adej, fgk, bej, bcej, adej, bcej, fm, fklm, bdej, acej, fklm, cfi, fgk, fhk

Case Study 15 Questions
1. b
2. f
3. Answer appears in Connect

Chapter 16
1. d
2. d
3. d
4. c
5. b
6. c
7. b
8. a
9. d
10. d
11. d
12. a
13. b
14. d
15. b
16. a

Case Study 16 Questions
1. d
2. d
3. Answer appears in Connect

Chapter 17

1. b, a, b, c, a
2. c
3. b
4. c
5. a
6. a
7. d
8. c
9. d

Case Study 17 Questions

1. c
2. b
3. Answer appears in Connect

Chapter 18

1. d
2. a
3. c
4. b
5. c
6. b
7. c
8. b
9. c
10. c
11. a
12. a
13. d
14. a
15. f
16. h, j, d, l, m, a, f, i, e, k, b, c

Case Study 18 Questions

1. a
2. c
3. Answer appears in Connect

Chapter 19

1. b
2. c
3. b
4. d
5. c
6. c
7. d
8. c
9. b
10. c
11. d
12. c
13. a
14. a
15. 4.5.8/ 4.5.7/ 1.2.8.10/ 2.3.8.11/ 4.5.8.9
16. a/d, b, c, b c, d, b, d, a

Case Study 19 Questions

1. d
2. b
3. Answer appears in Connect

Chapter 20

1. b
2. b
3. d
4. b
5. e
6. d
7. d
8. b
9. d
10. d
11. c
12. d
13. b
14. all but c
15. l, i, h, a, d, g, k, e, j, m, f, c, b

Case Study 20 Questions

1. b
2. c
3. Answer appears in Connect

Chapter 21

1. b
2. b
3. c
4. d
5. d
6. e
7. d
8. c
9. b
10. d
11. a
12. d
13. b
14. c
15. d
16. d
17. a
18. b
19. a, c, b, d, c/e, g, f, e, c, f, g, c
20. g, c, f, f, c, d, a, d, g, a, e, g, g

Case Study 21 Questions

1. d
2. e
3. Answer appears in Connect

Chapter 22

1. c
2. a
3. a
4. d
5. d
6. a
7. a
8. d
9. c
10. c
11. d
12. c
13. c
14. d
15. b
16. b, g, d, e, c, a, f

Case Study 22 Questions

1. b
2. a, d
3. Answer appears in Connect

Chapter 23

1. c
2. b
3. d
4. b
5. b
6. d
7. c
8. d
9. c
10. a
11. d
12. d
13. b
14. a
15. a
16. c
17. c
18. d
19. 1. f, P
 2. j, P
 3. k, H
 4. g, H
 5. h, P
 6. b, H
 7. i, H
 8. l, P
 9. a, P
 10. d, P
 11. m, H
 12. e, H
 13. c, H
20. c, d, i, g, b, a, h, j, f, e

Case Study 23 Questions

1. b
2. c
3. Answer appears in Connect

Chapter 24

1. d
2. c
3. b
4. a
5. d
6. d
7. c
8. d
9. a
10. b
11. d
12. b
13. c
14. d
15. a
16. d
17. e/a, g, i, c, h, l, b/k, f/k, f, d, j

Case Study 24 Questions

1. a
2. e
3. Answer appears in Connect

Chapter 25

1. a
2. b
3. b
4. c
5. d
6. d
7. b
8. d
9. c
10. d
11. b
12. d
13. c
14. a, c, d, f; a, c, f; a, c, f; b, d, e, g; a, c, f; c, e, g; a, c, e; a, c, d, f
15. c, ad, b, ac, hg, c, af, hf, a, cij, h
16.
 a. Influenza, mumps, rubella, measles
 b. HIV
 c. Hepatitis A virus, polio, rotavirus
 d. western equine encephalitis virus
 e. rabies, hantavirus
 f. rotavirus, measles, polio, mumps, rubella, influenza

Case Study 25 Questions

1. d
2. a-G, b-M, c-M, d-G, e-G
3. Answer appears in Connect

Chapter 26

1. c
2. b
3. b
4. d
5. c
6. d
7. c
8. d
9. b
10. d

Case Study 26 Questions

1. c
2. d
3. Answer appears in Connect

Chapter 27

1. c
2. a
3. d
4. d
5. b
6. b
7. d
8. b
9. b
10. c

Case Study 27 Questions

1. b
2. b
3. Answer appears in Connect

Glossary

abiogenesis The belief in spontaneous generation as a source of life.

abiotic Nonliving factors such as soil, water, temperature, and light that are studied when looking at an ecosystem.

ABO blood group system Developed by Karl Landsteiner in 1904; the identification of different blood groups based on differing isoantigen markers characteristic of each blood type.

abscess An inflamed, fibrous mass enclosing a core of pus.

abyssal zone The deepest region of the ocean; a sunless, high-pressure, cold, anaerobic habitat.

acid-fast A term referring to the property of mycobacteria to retain carbol fuchsin even in the presence of acid alcohol. The staining procedure is used to diagnose tuberculosis.

acidic A solution with a pH value below 7 on the pH scale.

acidic fermentation An anaerobic degradation of pyruvic acid that results in organic acid production.

acquired immunodeficiency syndrome See **AIDS**

acquired immunodeficiency syndrome/AIDS The complex signs and symptoms of immune system destruction seen during the late phase of human immunodeficiency virus (HIV) infection.

acquired or adaptive immunity Refers to specific immunity developed by lymphocytes responding to antigens.

actin Fine protein fibers within cells that contribute to structure and support.

actinomycetes A group of filamentous, funguslike bacteria.

active immunity Immunity acquired through direct stimulation of the immune system by antigen.

active site The specific region on an apoenzyme that binds substrate. The site for reaction catalysis.

active transport Nutrient transport method that requires carrier proteins in the membranes of the living cells and the expenditure of energy.

acute Characterized by rapid onset and short duration.

acyclovir A synthetic purine analog that blocks DNA synthesis in certain viruses, particularly the herpes simplex viruses.

adapt/adaptation The act of successfully adjusting to a new environment, often made possible by random genetic changes that provide an advantage under changed environmental conditions.

adenine (A) A purine nitrogen base found in DNA and RNA that pairs with the pyrimidine thymine.

adenosine deaminase (ADA) deficiency An immunodeficiency disorder and one type of SCIDS that is caused by an inborn error in the metabolism of adenine. The accumulation of adenine destroys both B and T lymphocytes.

adenosine triphosphate (ATP) A nucleotide that is the primary source of energy for cells.

adenovirus Nonenveloped DNA virus that causes colds; transmissible via respiratory and ocular secretions.

adhesion The process by which microbes gain a stable foothold at the portal of entry; often involves a specific interaction between the molecules on the microbial surface and the receptors on the host cell.

adjuvant In immunology, a chemical vehicle that enhances antigenicity, presumably by prolonging antigen retention at the injection site.

adsorption A process of adhering one molecule onto the surface of another molecule.

aerobe A microorganism that lives and grows in the presence of free gaseous oxygen (O_2).

aerobic respiration Respiration in which the final electron acceptor in the electron transport chain is oxygen (O_2).

aerosols Airborne suspensions of fine dust or moisture particles that contain live pathogens.

aflatoxin From *Aspergillus flavus* toxin, a mycotoxin that typically poisons moldy animal feed and can cause liver cancer in humans and other animals.

agammaglobulinemia The absence of functioning antibodies in serum. A related term, hypogammaglobulinemia, means there is a reduced level of antibodies, rather than a complete absence.

agar A polysaccharide found in seaweed and commonly used to prepare solid culture media.

agglutination The aggregation caused by antibodies when they cross-link cells or large particles into clumps that settle.

agranulocyte One form of leukocyte (white blood cell), having globular, nonlobed nuclei and lacking prominent cytoplasmic granules.

AIDS-defining illness A group of serious and life-threatening diseases that occur primarily in HIV-positive people. When a person acquires one of these illnesses, he or she is diagnosed with the advanced stage of HIV infection known as AIDS.

alcoholic fermentation An anaerobic degradation of pyruvic acid that results in alcohol production.

algae Photosynthetic, plantlike organisms that generally lack the complex structure of plants; they may be single-celled or multicellular and inhabit diverse habitats such as marine and freshwater environments, glaciers, and hot springs.

allergen A substance that provokes an allergic response.

allergy The altered, usually exaggerated, immune response to an allergen. sometimes used interchangeably with hypersensitivity.

alloantigen Antigens that vary in exact composition among the members the same species, which is what causes incompatibilities in blood types.

allograft Relatively compatible tissue exchange between nonidentical members of the same species. Also called homograft.

allosteric Pertaining to the altered activity of an enzyme due to the binding of a molecule to a region other than the enzyme's active site.

amastigote The rounded or ovoid nonflagellated form of the *Leishmania* parasite.

Ames test A method for detecting mutagenic and potentially carcinogenic agents based upon the genetic alteration of nutritionally defective bacteria.

amination The addition of an amine ($—NH_2$) group to a molecule.

amino acids The building blocks of protein. Amino acids exist in 20 naturally occurring forms that impart different characteristics to the various proteins they compose.

aminoglycoside A complex group of drugs derived from soil actinomycetes that impair ribosome function and have antibiotic potential. Example: streptomycin.

ammonification Phase of the nitrogen cycle in which ammonia is released from decomposing organic material.

amphibolism Pertaining to the metabolic pathways that serve multiple functions in the breakdown, synthesis, and conversion of metabolites.

amphipathic Relating to a compound that has contrasting characteristics, such as hydrophilic-hydrophobic or acid-base.

amphitrichous Having a single flagellum or a tuft of flagella at opposite poles of a microbial cell.

anabolism The energy-consuming process of incorporating nutrients into protoplasm through biosynthesis.

anaerobe A microorganism that grows best, or exclusively, in the absence of oxygen, and which does not use oxygen in its metabolism.

anaerobic digesters Closed chambers used in a microbial process that converts organic sludge from waste treatment plants into useful fuels such as methane and hydrogen gases. Also called bioreactors.

anaerobic respiration Respiration in which the final electron acceptor in the electron transport chain is an inorganic molecule containing sulfate, nitrate, nitrite, carbonate and other salts, rather than molecular oxygen gas.

analog In chemistry, a compound that closely resembles another in structure.

anamnestic response In immunology, an augmented response or memory related to a prior stimulation of the immune system by antigen. It boosts the levels of antibodies.

anaphylaxis An exaggerated systemic allergic reaction to antigen that leads to severe respiratory and cardiac complications.

anion A negatively charged ion.

anoxygenic Any reaction that does not produce oxygen; usually in reference to the type of photosynthesis occurring in anaerobic photosynthetic bacteria.

anthrax A zoonotic disease of herbivorous livestock. The anthrax bacillus is a facultative parasite and can infect humans in a number of ways. In its most virulent form, it can be fatal.

antibiotic A chemical substance produced by one microorganism that can inhibit or kill another microbe, even in minute amounts.

antibody A large protein molecule evoked in response to an antigen that interacts specifically with that antigen.

anticodon The triplet sequence in transfer RNA that is complementary to the triplet sequence of messenger RNA (the codon).

antigen Any cell, particle, or chemical that has properties that allow it to stimulate a specific immune response by B cells or T cells. See the related term **immunogen.**

antigen binding site Specific regions at the ends of the antibody molecule that recognize specific antigens. These sites have numerous shapes to fit a wide variety of antigens.

antigen-presenting cell (APC) A macrophage or dendritic cell that ingests and degrades an antigen and subsequently places the antigenic determinant molecules on its surface for recognition by CD4 T lymphocytes.

antigenic drift Minor antigenic changes in the influenza virus due to mutations in the spikes' genes.

antigenic shift Major changes in the influenza virus due to recombination of viral strains from two different host species.

antigenicity The property of a substance to stimulate a specific immune response such as antibody formation.

antihistamine A drug that counters the action of histamine and is useful in allergy treatment.

antisepsis/antiseptic Use of a growth-inhibiting agent (antiseptic) on tissues to prevent infection.

antiserum Antibody-rich serum derived from people who have recovered from specific infections such as hepatitis; sometimes taken from the blood of animals deliberately immunized against an infectious or toxic antigen. Used in passive immune therapy.

antitoxin Globulin fraction of serum that neutralizes a specific toxin. Also refers to the specific antitoxin antibody itself.

apoenzyme The protein part of an enzyme, as opposed to the nonprotein or inorganic cofactors.

appendages Accessory structures that sprout from the surface of bacteria. They can be divided into two major groups: those that provide motility and those that enable adhesion.

aquifer A subterranean water-bearing stratum of permeable rock, sand, or gravel.

arboviruses Arthropod-borne viruses including togaviruses, reoviruses, flaviviruses, and bunyaviruses. These viruses generally cause mild, undifferentiated fevers and occasionally cases of severe encephalitis and hemorrhagic fever.

archaea Prokaryotic single-celled organisms of primitive origin that have unusual anatomy, physiology, and genetics, and live in harsh habitats; when capitalized (**Archaea**), the term refers to one of the three domains of living organisms as proposed by Woese.

arthropod A phylum of invertebrate animals with jointed exoskeletons. Includes insects such as mosquitoes and arachnids such as ticks that may be vectors of infectious diseases.

Arthus reaction An immune complex phenomenon that develops after repeat injection. This localized inflammation results from aggregates of antigen and antibody that bind, complement, and attract neutrophils.

artificial immunity Specific protection acquired through medical means.

ascospore A spore formed within a saclike cell (ascus) of Ascomycota following nuclear fusion and meiosis.

ascus Special fungal sac in which haploid spores are created.

asepsis A condition free of viable pathogenic microorganisms.

aseptic technique Methods of handling microbial cultures, patient specimens, and other sources of microbes in a way that prevents infection of the handler and others who may be exposed.

asthma An allergic respiratory disease marked by extreme sensitivity to allergens.

asymptomatic An infection that produces no noticeable symptoms even though the microbe is active in the host tissue.

asymptomatic carrier A person with an inapparent infection who shows no symptoms of being infected yet is able to pass the disease agent on to others.

atmosphere That part of the biosphere that includes the gaseous envelope up to 14 miles above the earth's surface. It contains gases such as carbon dioxide, nitrogen, and oxygen.

atom The smallest particle of an element to retain all the properties of that element.

atomic number (AN) A measurement that reflects the number of protons in an atom of a particular element.

atomic weight The average of the mass numbers of all the isotopic forms for a particular element.

atopy Allergic reaction classified as type I, with a strong familial relationship; caused by allergens such as pollen, insect venom, food, and dander; involves IgE antibody; includes symptoms of hay fever, asthma, and skin rash.

ATP synthase A unique enzyme located in the mitochondrial cristae, bacterial cell membrane, and chloroplast grana that harnesses the flux of hydrogen ions to the synthesis of ATP.

attenuate To reduce the virulence of a pathogenic bacterium or virus by passing it through a nonnative host or by long-term subculture.

autoantibody An antibody formed by a person against his own self molecules, called autoantigens, that are part of his cells.

autoantigen Molecules that are inherently part of self but are perceived by the immune system as foreign.

autoclave A sterilization chamber that allows the use of steam under pressure to sterilize materials. The most common temperature/pressure combination for an autoclave is 121°C and 15 psi.

autograft Tissue or organ surgically transplanted to another site on the same subject.

autoimmune disease/autoimmunity Literally, an immune reaction that targets self. This pathologic condition arises from the production of antibodies against autoantigens, for example in rheumatoid arthritis and multiple sclerosis.

autotroph A microorganism that requires only inorganic nutrients and whose sole source of carbon is carbon dioxide.

axenic A sterile state such as a pure culture. An axenic animal is born and raised in a germ-free environment. See **gnotobiotic.**

axial filament A type of flagellum (called an endoflagellum) that lies in the periplasmic space of spirochetes and is responsible for locomotion. Also called periplasmic flagellum.

azoles Five-ringed compounds used in antifungal therapy. They are usually recognized by their -*azole* suffix, e.g., ketoconazole, clotrimazole.

bacillus Bacterial cell shape that is basically cylindrical (longer than it is wide); plural is bacilli.

back-mutation A mutation that counteracts an earlier mutation, resulting in the restoration of the original DNA sequence.

bacteremia The presence of viable bacteria in circulating blood.

bacteria (plural of *bacterium*) Category of prokaryotes with peptidoglycan in their cell walls and a single, circular chromosome. This group of small cells is widely distributed in the earth's habitats.

bacterial chromosome A circular body in bacteria that contains the primary genetic material. Also called **nucleoid.**

bacterial meningitis Inflammation of the meninges as a result of bacterial infection.

bactericide An agent that kills bacterial vegetative cells.

bacteriophage A virus that specifically infects bacteria; phage means "to eat."

bacteriostatic Any process or agent that inhibits bacterial growth.

barophile A microorganism that thrives under high (usually hydrostatic) pressure.

basement membrane A thin layer (1–6 μm) of protein and polysaccharide found at the base of epithelial tissues.

basic A solution with a pH value above 7 on the pH scale.

basidiospore A sexual spore that arises from a basidium. Found in basidiomycota fungi.

basidium A reproductive cell created when the swollen terminal cell of a hypha develops filaments (sterigmata) that form spores.

basophil A motile polymorphonuclear leukocyte that binds IgE. The basophilic cytoplasmic granules contain mediators of anaphylaxis and atopy.

benign tumor A self-contained growth within an organ that does not spread into adjacent tissue.

benthic zone The sedimentary bottom region of a pond, lake, or ocean.

beta oxidation The degradation of long-chain fatty acids. Two-carbon fragments are formed as a result of enzymatic attack directed against the second or beta carbon of the hydrocarbon chain. Aided by coenzyme A, the fragments enter the tricarboxylic acid (TCA) cycle and are processed for ATP synthesis.

beta-lactamase An enzyme secreted by certain bacteria that cleaves the beta-lactam ring of penicillin and cephalosporin and thus provides for resistance against the antibiotic. See **penicillinase.**

binary fission The formation of two new cells of approximately equal size as the result of parent cell division.

binomial system Scientific method of assigning names to organisms that employs two names to identify every organism—genus name plus species name.

biochemistry The study of organic compounds produced by (or that are components of) living things. The four main categories of biochemicals are carbohydrates, lipids, proteins, and nucleic acid.

bioenergetics The study of the production and use of energy by cells.

bioethics The study of biological issues and how they relate to human conduct and moral judgment.

biofilm A complex aggregate of interacting microbial cells that adhere to each other and to surfaces by means of a polysaccharide matrix. Biofilms permit communication among participants, which facilitates their survival and adaptation.

biogenesis Belief that living things can only arise from others of the same kind.

biogeochemical cycle A process by which matter is converted from organic to inorganic form and returned to various nonliving reservoirs on earth (air, rocks, and water) where it becomes available for reuse by living things. Elements such as carbon, nitrogen, and phosphorus are constantly cycled in this manner.

bioinformatics A field that uses computer technology to manage and analyze large sets of biological data, including genome and protein sequences.

biological vector An animal that not only transports an infectious agent but plays a role in the life cycle of the pathogen, serving as a site in which it can multiply or complete its life cycle. It is usually an alternate host to the pathogen.

biomes Particular climate regions in a terrestrial realm.

bioremediation The use of microbes to reduce or degrade pollutants, industrial wastes, and household garbage.

biosphere Habitable regions comprising the aquatic (hydrospheric), soil-rock (lithospheric), and air (atmospheric) environments.

biotechnology The use of organisms (microbes, plants, animals) or their products in the commercial or industrial realm.

biotic Living factors such as parasites, food substrates, or other living or once-living organisms that are studied when looking at an ecosystem.

blocking antibody The IgG class of immunoglobulins that competes with IgE antibody for allergens, thus blocking the degranulation of basophils and mast cells.

blood cells Cellular components of the blood consisting of red blood cells, primarily responsible for the transport of oxygen and carbon dioxide, and white blood cells, primarily responsible for host defense and immune reactions.

bradykinin An active polypeptide that is a potent vasodilator released from IgE-coated mast cells during anaphylaxis.

broad spectrum A term to denote drugs that affect many different types of bacteria, both gram-positive and gram-negative as well as atypical bacteria such as rickettsias and chlamydias.

brucellosis A zoonosis transmitted to humans from infected animals or animal products; causes a fluctuating pattern of severe fever in humans as well as muscle pain, weakness, headache, weight loss, and profuse sweating. Also called **undulant fever.**

bubo The swelling of one or more lymph nodes due to inflammation.

bubonic plague The form of plague in which bacterial growth is primarily restricted to the lymph and is characterized by the appearance of a swollen lymph node referred to as a bubo.

budding See **exocytosis.**

bulbar poliomyelitis Complication of polio infection in which the brain stem, medulla, or cranial nerves are affected. Leads to loss of respiratory control and paralysis of the trunk and limbs.

calculus Dental deposit formed when plaque becomes mineralized with calcium and phosphate crystals. Also called **tartar.**

Calvin cycle The recurrent photosynthetic pathway characterized by CO_2 fixation and glucose synthesis. Also called the dark reactions.

cancer Any malignant neoplasm that invades surrounding tissue and can metastasize to other locations. A carcinoma is derived from epithelial tissue, and a sarcoma arises from proliferating mesodermal cells of connective tissue.

capsid The protein covering of a virus' nucleic acid core. Capsids exhibit symmetry due to the regular arrangement of subunits called capsomers. See **icosahedron.**

capsomer A subunit of the virus capsid shaped as a triangle or disc.

capsule In bacteria, the loose, gel-like covering or slime made chiefly of simple polysaccharides. This layer is protective and can be associated with virulence.

carbohydrate A compound containing primarily carbon, hydrogen, and oxygen in a 1:2:1 ratio.

carbon cycle That pathway taken by carbon from its abiotic source to its use by producers to form organic compounds (biotic), followed by the breakdown of biotic compounds and their release to a nonliving reservoir in the environment (mostly carbon dioxide in the atmosphere).

carbon fixation Reactions in photosynthesis that incorporate inorganic carbon dioxide into organic compounds such as sugars. This occurs during the Calvin cycle and uses energy generated by the light reactions. This process is responsible for the vast majority of productivity on earth.

carbuncle A deep staphylococcal abscess joining several neighboring hair follicles.

carotenoid Yellow, orange, or red pigments that are accessories that trap light for photosynthesis.

carrier In epidemiology, a person who asymptomatically harbors infectious agents and inconspicuously spreads them to others. In chemistry, a chemical that can accept an atom, chemical radical, or electron from one compound and pass it to another chemical.

caseous lesion Necrotic area of lung tubercle superficially resembling cheese. Typical of tuberculosis.

catabolism The chemical breakdown of complex compounds into simpler units to be used in cell metabolism.

catalyst A substance that alters the rate of a reaction without being consumed or permanently changed by it. In cells, enzymes are catalysts.

catalytic site The niche in an enzyme where the substrate is converted to the product (also active site).

cation A positively charged ion.

cecum The intestinal pocket that forms the first segment of the large intestine. Also called the appendix.

cell An individual membrane-bound living entity; the smallest living unit capable of an independent existence.

cell-mediated immunity The type of immune responses brought about by T cells, such as cytotoxic, regulatory, and helper effects. An activated T cell interacts directly with antigen-bearing cells in order to bring about its end result.

cellulose A long, fibrous polymer composed of β-glucose; one of the most common substances on earth.

cephalosporins A group of antibiotics isolated from the fungus *Cephalosporium* that block the synthesis of the cell wall and can be used for a variety of infections.

cercaria The free-swimming larva of the schistosome trematode that emerges from the snail host and can penetrate human skin, causing schistosomiasis.

cestode The common name for tapeworms that parasitize humans and domestic animals.

chancre The primary sore of syphilis that forms at the site of penetration by *Treponema pallidum*. It begins as a hard, dull red, painless papule that erodes from the center.

chancroid A lesion that resembles a chancre but is soft and is caused by *Haemophilus ducreyi*.

chemical bond A link formed between molecules when two or more atoms share, donate, or accept electrons.

chemical mediators Small molecules that are released during inflammation and specific immune reactions that allow communication between the cells of the immune system and facilitate surveillance, recognition, and attack.

chemiosmotic theory An explanation for ATP formation that is based on the formation of a proton (H^+) gradient across a membrane during electron transport. Movement of the protons back across an ATP synthase causes the formation of ATP.

chemoautotrophs An organism that relies upon inorganic chemicals for its energy and carbon dioxide for its carbon. Also called a chemolithotroph.

chemokines Chemical mediators (cytokines) that stimulate the movement and migration of white blood cells.

chemostat A growth chamber with an outflow that is equal to the continuous inflow of nutrient media. This steady-state growth device is used to study such events as cell division, mutation rates, and enzyme regulation.

chemotaxis The tendency of cells to move in response to a chemical gradient (toward an attractant or to avoid adverse stimuli). In inflammation, it refers to the movement of blood cells in reaction to chemical signals.

chemotherapy The use of chemical substances or drugs to treat or prevent disease.

chemotroph An organism that oxidizes compounds to feed on nutrients.

chickenpox A papulo-pustular rash caused by infection by varicella-zoster virus.

chimera A product formed by the fusion of two different organisms.

chitin A polysaccharide similar to cellulose in chemical structure. This polymer makes up the horny substance of the exoskeletons of arthropods and certain fungi.

chlamydias Tiny gram negative bacteria that are obligate parasites inside the vacuoles of host cells; members cause a type of STD and pneumonia.

chlorophyll A group of mostly green pigments that are used by photosynthetic eukaryotic organisms and cyanobacteria to trap light energy to use in making chemical bonds.

chloroplast An organelle containing chlorophyll that is found in photosynthetic eukaryotes.

cholesterol Best-known member of a group of lipids called steroids. Cholesterol is commonly found in cell membranes and animal hormones.

chromatin The genetic material of the nucleus. Chromatin is made up of nucleic acid and stains readily with certain dyes.

chromophore The chemical radical of a dye that is responsible for its color and reactivity.

chromosome The tightly coiled bodies within cells, composed of DNA, that are the primary locations for genetic material.

chronic infection An infection that persists over a long duration, with symptoms being mild or absent.

cilium (plural: *cilia*) Eukaryotic structure similar to flagella that propels a protozoan through the environment.

class In the levels of classification, the division of organisms that follows phylum.

clonal selection theory A conceptual explanation for the development of lymphocyte specificity and variety during maturation of the immune system.

clone A colony of cells (or group of organisms) derived from a single cell (or single organism) by asexual reproduction. All units share identical characteristics. Also used as a verb to refer to the process of producing a genetically identical population of cells or genes.

cloning host An organism such as a bacterium or a yeast that receives and replicates a foreign piece of DNA inserted through genetic engineering techiques.

cluster of differentiation (CD) Receptors on mature T cells that serve in cell communication; react with MHC receptors for immune recognition by antigen processing cells.

coagulase A plasma-clotting enzyme secreted by *Staphylococcus aureus*. It contributes to virulence and is involved in forming a fibrin wall that surrounds staphylococcal lesions.

coagulase-negative staphylococci Opportunistic species that are usually part of the normal flora of the skin and mucous membranes. They cause infection when host defenses are low and lack many virulence factors such as coagulase.

coccobacillus An elongated coccus; a short, thick, oval-shaped bacterial rod.

coccus A spherical-shaped bacterial cell; plural is cocci

codon A specific sequence of three nucleotides in DNA or mRNA that constitutes the genetic code for a particular amino acid.

coenzyme A complex organic molecule, several of which are derived from vitamins (e.g., nicotinamide or niacin, riboflavin). A coenzyme operates in conjunction with an enzyme. Coenzymes serve as transient carriers of specific atoms or functional groups during metabolic reactions.

cofactor An enzyme accessory. It can be organic, such as coenzymes, or inorganic, such as Fe^{2+}, Mn^{2+}, Zn^{2+}, or other metallic ions.

cold sterilization The use of nonheating methods such as radiation or filtration to sterilize materials.

coliform A collective term that includes normal enteric bacteria that are gram-negative and lactose-fermenting.

colony A macroscopic cluster of cells appearing on a solid medium, each arising from the multiplication of a single cell.

colostrum The clear yellow early product of breast milk that is very high in secretory antibodies. Provides passive intestinal protection.

commensalism An unequal relationship described as one member (A) deriving benefit from another member (B), without harming or benefiting (B).

common (seed) warts Painless, elevated rough growths on the fingers or occasionally other parts of the body.

communicable infection Capable of being transmitted from one individual to another.

community The interacting mixture of populations in a given habitat.

competitive inhibition The blockage of microbial growth through a metabolic analog drug that inserts on the active site of an essential metabolic enzyme and prevents further action of that enzyme.

complement/complement fixation In immunology, protein components that act in a defined sequence (cascade) when set in motion either by an antigen-antibody complex or by factors from the lectin or alternative pathways. When complement adheres to a certain antibody or receptor, it is said to be fixed.

complementary DNA (cDNA) DNA created by using reverse transcriptase to synthesize DNA from RNA templates.

complex (nonsynthetic) medium Culture medium containing at least one component that is chemically undefined.

compounds Molecules that are a combination of two or more different elements.

concentration The expression of the amount of a solute dissolved in a certain amount of solvent. It may be defined by weight, volume, or percentage.

condylomata acuminata Extensive, branched masses of genital warts caused by infection with human papillomavirus.

congenital rubella Transmission of the rubella virus to a fetus *in utero*. Injury to the fetus is generally much more serious than it is to the mother.

conidia Asexual fungal spores shed as free units from the tips of fertile hyphae.

conjugation In bacteria, the contact between donor and recipient cells associated with the transfer of genetic material such as plasmids. Can involve special (sex) pili. Also a form of sexual recombination in ciliated protozoans.

conjunctivitis Sometimes called "pinkeye." A *Haemophilus* infection of the subconjunctiva that is common among children, easily transmitted, and treated with antibiotic eyedrops.

constitutive enzyme An enzyme present in bacterial cells in constant amounts, regardless of the presence of substrate. Enzymes of the central catabolic pathways are typical examples.

consumer An organism that feeds on producers or other consumers. It gets all nutrients and energy from other organisms (also called **heterotroph**). May exist at several levels, such as primary (feeds on producers), and secondary (feeds on primary consumers).

contagious Very readily communicable; transmissible by direct contact with infected people and their fresh secretions or excretions.

contaminant An impurity; any undesirable material or organism. A culture into which unknown microbes have been introduced is contaminated.

convalescence Recovery; the period between the end of a disease and the complete restoration of health in a patient.

corepressor A molecule that combines with an inactive repressor to form an active repressor. The resultant combined molecule attaches to the operator gene site and inhibits the activity of structural genes in line with the operator.

covalent bond A chemical bond formed by the sharing of electrons between two atoms.

Creutzfeld-Jakob disease (CJD) A spongiform encephalopathy caused by infection with a prion. The disease is marked by dementia, impaired senses, and uncontrollable muscle contractions.

crista The infolded inner membrane of a mitochondrion that is the site of the respiratory chain and oxidative phosphorylation.

culture The visible accumulation of microorganisms in or on a nutrient medium. Also, the propagation of microorganisms with various media.

curd The coagulated milk protein used in cheese making.

cutaneous Second level of skin, including the stratum corneum and occasionally the upper dermis.

cutaneous anthrax The mildest form of anthrax, caused by the entrance of bacterial spores into small nicks or openings in the skin, their germination, and the formation of a dark necrotic lesion called an eschar.

cutaneous candidiasis A mycosis occurring in the skin and membranes of compromised patients and neonates as a result of infection by *Candida albicans*.

cyanobacteria Widespread and ecologically important photosynthetic bacteria; evidence indicates their role in the evolution of chloroplasts in eukaryotic cells.

cyst The resistant, dormant, but infectious form of protozoans. Can be important in spread of infectious agents such as *Entamoeba histolytica* and *Giardia lamblia*.

cysteine Sulfur-containing amino acid that is essential to the structure and specificity of proteins.

cytochrome A group of heme protein compounds whose chief role is in electron and/or hydrogen transport occurring in the last phase of aerobic respiration.

cytokine A chemical substance produced by white blood cells and tissue cells that regulates development, inflammation, and immunity.

cytopathic effect The degenerative changes in cells associated with virus infection. Example: the formation of multinucleate giant cells (Negri bodies), the prominent cytoplasmic inclusions of nerve cells infected by rabies virus.

cytoplasm Dense fluid encased by the cell membrane; the site of many of the cell's biochemical and synthetic activities.

cytosine (C) One of the nitrogen bases found in DNA and RNA, with a pyrimidine form.

cytoskeleton A fine intracellular network of fibers, filaments, and other strands that function in support and shape of cells.

cytotoxic/cytotoxicity Having the capacity to destroy specific cells. The cytotoxic class of T cells attacks cancer cells, virus-infected cells, and eukaryotic pathogens.

death phase A dramatic downturn in numbers occurs due to lack of nutrition, adverse environmental factors, and accumulation of wastes. A greater number of cells are dying in this phase.

debridement Trimming away devitalized tissue and foreign matter from a wound.

decomposer A consumer that feeds on organic matter from the bodies of dead organisms. These microorganisms feed from all levels of the food pyramid and are responsible for recycling elements (also called saprobes).

decomposition The breakdown of dead matter and wastes into simple compounds that can be directed back into the natural cycle of living things.

decontamination The removal or neutralization of an infectious, poisonous, or injurious agent from a site.

definitive host The organism in which a parasite develops into its adult or sexually mature stage. Also called the **final host.**

degeneracy The property of the genetic code that allows an amino acid to be specified by several different codons.

degerm/degermation To physically remove surface oils, debris, and soil from skin and wounds to reduce the microbial load.

degranulation The release of cytoplasmic granules, as when cytokines are secreted from mast cell granules.

dehydration synthesis During the formation of a carbohydrate bond, the step in which one carbon molecule gives up its OH group and the other loses the H from its OH group, thereby producing a water molecule. This process is common to all polymerization reactions.

denature/denaturation The loss of normal characteristics (shape, configuration) resulting from some molecular alteration. Used in reference to loss of normal activity by proteins when their 3-D structure has been altered by heat or chemicals.

dendritic cell A large, antigen-processing cell characterized by long, branchlike extensions of the cell membrane.

denitrification The end of the nitrogen cycle when nitrogen compounds are returned to the reservoir in the air.

dental caries A mixed infection of the tooth surface that gradually destroys the enamel and may lead to destruction of the deeper tissue.

deoxyribonucleic acid (DNA) The nucleic acid often referred to as the "double helix." DNA carries the master plan for an organism's heredity.

deoxyribose A 5-carbon sugar that is an important component of DNA.

dermatophytoses Superficial mycoses such as athlete's foot and ringworm associated with certain fungi with an affinity for the skin, hair, and nails. These fungi are called **dermatophytes.**

desensitization/hyposensitization A therapeutic exposure to known allergens designed to build tolerance and eventually prevent allergic reaction. Hyposensitization is gaining in use because it emphasizes that the allergy is reduced not completely removed.

desiccate/desiccated To dry thoroughly. To preserve by drying.

diabetes mellitus A disease involving compromise in insulin function. In one form, the pancreatic cells that produce insulin are destroyed by autoantibodies, and in another, the pancreas does not produce sufficient insulin.

diapedesis The migration of intact blood cells between endothelial cells of a blood vessel such as a venule.

differential medium A single substrate that discriminates between groups of microorganisms on the basis of differences in their appearance due to different chemical reactions.

differential stain A technique that utilizes two dyes to distinguish between different microbial groups or cell parts by color reaction.

diffusion The passive dispersal of molecules, ions, or microscopic particles propelled down a concentration gradient by spontaneous random motion to achieve a uniform distribution.

DiGeorge syndrome A genetic defect that causes the loss of thymus function; results in abnormally low or absent T cells and other developmental abnormalities.

dimer A molecule that is composed of two monomers bound together.

dimorphic In mycology, the tendency of some pathogens to alter their growth form from mold to yeast in response to rising temperature.

diphtheria Infection by *Corynebacterium diphtheriae*. It is transmitted by human carriers or contaminated milk, and the primary infection is in the upper respiratory tract. Several forms of this infection may be fatal if untreated. Vaccination on the recommended schedule can prevent infection.

diphtherotoxin Exotoxin responsible for the effects seen in diphtheria. The toxin affects the upper respiratory system, peripheral nervous system, and the heart.

diplococcus Spherical or oval-shaped bacteria, typically found in pairs.

direct, or total, cell count 1. Counting total numbers of individual cells being viewed with magnification. 2. Counting isolated colonies of organisms growing on a plate of media as a way to determine population size.

disaccharide A sugar containing two monosaccharides. Example: sucrose (fructose + glucose).

disinfection/disinfectant The destruction of pathogenic nonsporulating microbes or their toxins, usually on inanimate surfaces. A disinfectant is a chemical used for this purpose.

division In the levels of classification, an alternate term for *phylum*.

DNA polymerase Enzyme responsible for the replication of DNA. Several versions of the enzyme exist, each completing a unique portion of the replication process.

DNA profiling, DNA fingerprinting A method of identifying patterns of specific genetic markers unique to an individual for the purpose of identification. Used in forensics, medicine, anthropology, and microbial identification

DNA sequencing A process of determining the exact order of nucleotides in a segment of DNA, ranging from small pieces to whole chromosomes or genomes. The newest technologies are rapid machine-based methods that use lasers, cameras, or microscopes to read fluorescent nucleotides on a strand of DNA .

DNA vaccine A newer vaccine preparation based on inserting DNA from pathogens into host cells to encourage them to express the foreign protein and stimulate immunity.

domain In the levels of classification, the broadest general category to which an organism is assigned. Members of a domain share only one or a few general characteristics.

droplet nuclei The dried residue of fine droplets produced by mucus and saliva sprayed while sneezing and coughing. Droplet nuclei are less than 5 μm in diameter (large enough to bear a single bacterium and small enough to remain airborne for a long time) and can be carried by air currents.

drug resistance An adaptive response in which microorganisms begin to tolerate an amount of drug that had previously been inhibitory.

dry heat Air with a low moisture content that has been heated from 160 to several thousand degrees Celsius.

dysentery Infectious diarrhea characterized by blood and mucus in the feces.

dyspnea Difficulty in breathing.

ecosystem A collection of organisms together with its surrounding physical and chemical factors existing in a given place and time.

eczema An acute or chronic allergy of the skin associated with itching and burning sensations. Typically, red, edematous, vesicular lesions erupt, leaving the skin scaly and sometimes hyperpigmented.

edema The accumulation of excess fluid in cells, tissues, or serous cavities. Also called swelling.

electrolyte Any compound that ionizes in solution and conducts current in an electrical field.

electron A negatively charged subatomic particle that is distributed around the nucleus in an atom.

electrophoresis The separation of molecules by size and charge through exposure to an electrical current.

element A substance comprising only one kind of atom that cannot be degraded into two or more substances without losing its chemical characteristics.

ELISA Abbreviation for enzyme-linked immunosorbent assay, a very sensitive serological test used to detect antibodies in diseases such as AIDS. Two types are **indirect ELISA** and **capture ELISA.**

emerging disease Newly identified diseases that are becoming more prominent.

endemic disease A native disease that exists continuously in a geographic region. This pattern may reflect a vector or environmental source.

endergonic reaction A chemical reaction that occurs with the absorption and storage of surrounding energy. Antonym: **exergonic.**

endocytosis The process whereby solid and liquid materials are taken into the cell through membrane invagination and engulfment into a vesicle.

endoenzyme An intracellular enzyme that functions primarily within the cell compartment, as opposed to enzymes that are secreted.

endogenous Originating or produced within an organism or one of its parts.

endoplasmic reticulum (ER) An intracellular network of flattened sacs or tubules with or without ribosomes on their surfaces.

endospore A small, dormant, resistant derivative of a bacterial cell that germinates under favorable growth conditions into a vegetative cell. The bacterial genera *Bacillus* and *Clostridium* are typical sporeformers.

endosymbiosis The evolutionary process through which prokaryotic cells came together in a mutually beneficial association that gave rise to eukaryotic cells during billions of years of coevolution.

endotoxin A bacterial intracellular toxin that is not ordinarily released (as is exotoxin). Endotoxin is composed of a phospholipid-polysaccharide complex that is an integral part of gram-negative bacterial cell walls. Endotoxins can cause severe shock and fever.

energy of activation The minimum energy input necessary for reactants to form products in a chemical reaction.

energy pyramid An ecological model that shows the energy flow among the organisms in a community. It is structured like the food pyramid, but shows how energy is reduced from one trophic level to another.

enriched medium A nutrient medium supplemented with blood, serum, or some growth factor to promote the multiplication of fastidious microorganisms.

enterohemorrhagic *E. coli* (EHEC) A strain of *E. coli,* type 0157: H7, that secretes toxins that provoke secretory diarrhea and intestinal hemorrhage.

enteroinvasive Predisposed to invade the intestinal tissues.

enteropathogenic Pathogenic to the alimentary canal.

enterotoxigenic *E. coli* (ETEC) Strains of *E. coli* that produce toxins that attack the GI tract; effects are similar to cholera.

enterotoxin A bacterial toxin that specifically targets intestinal mucous membrane cells. Enterotoxigenic strains of *Escherichia coli* and *Staphylococcus aureus* are typical sources.

enveloped virus A virus whose nucleocapsid is enclosed by a membrane derived in part from the host cell. It usually contains exposed glycoprotein spikes specific for the virus.

enzyme A protein biocatalyst that facilitates metabolic reactions.

enzyme induction One of the controls on enzyme synthesis. This occurs when enzymes are synthesized only when suitable substrates are present.

enzyme repression The inhibition of enzyme synthesis by the end product of a catabolic pathway.

eosinophil A leukocyte whose cytoplasmic granules readily stain with red eosin dye. Eosinophils function primarily in the destruction of large parasites and in allergy and inflammation.

epidemic A sudden and simultaneous increase in the number of cases of a certain disease in a community.

epidemic parotitis (mumps) Disease caused by infection with *Paramyxovirus* and marked by swelling in and around the parotid salivary glands.

epidemiology The study of the factors affecting the prevalence and spread of disease within a community.

epimastigote The trypanosomal form found in the tsetse fly or reduviid bug vector. Its flagellum originates near the nucleus, extends along an undulating membrane, and emerges from the anterior end.

epitope The portion of a foreign cell or virus that is the precise stimulus for an immune response. Also called the **antigenic determinant.**

Epstein-Barr virus (EBV) Herpesvirus linked to infectious mononucleosis, Burkitt lymphoma, and nasopharyngeal carcinoma.

erysipelas An acute, sharply defined inflammatory disease specifically caused by hemolytic *Streptococcus*. The eruption is limited to the skin but can be complicated by serious systemic symptoms.

erysipeloid An inflammation resembling erysipelas but caused by *Erysipelothrix*, a gram-positive rod. The self-limited cellulitis that appears at the site of an infected wound, usually the hand, comes from handling contaminated fish or meat.

erythrocytes (red blood cells) Blood cells involved in the transport of oxygen and carbon dioxide.

erythrogenic toxin An exotoxin produced by lysogenized group A strains of β-hemolytic streptococci that is responsible for the severe fever and rash of scarlet fever in the nonimmune individual. Also called a pyrogenic toxin.

eschar A dark, sloughing scab that is the lesion of anthrax and certain rickettsioses.

essential nutrient Any ingredient such as a certain amino acid, fatty acid, vitamin, or mineral that cannot be formed by an organism and must be supplied in the diet. A growth factor.

ester bond A covalent bond formed by reacting carboxylic acid with an OH group:

$$(R - \overset{\overset{\displaystyle O}{\displaystyle \|}}{C} - O - R')$$

Olive and corn oils, lard, and butter are examples of triacylglycerols—esters formed between glycerol and three fatty acids.

ethylene oxide A potent, highly water-soluble gas invaluable for gaseous sterilization of heat-sensitive objects such as plastics, surgical and diagnostic appliances, and spices. Potential hazards are related to its carcinogenic, residual, and explosive nature. Ethylene oxide is rendered nonexplosive by mixing with 90% CO_2 or fluorocarbon.

etiologic agent The microbial cause of disease; the pathogen.

Eukarya One of the three domains (sometimes called superkingdoms) of living organisms, as proposed by Woese; contains all eukaryotic organisms.

eukaryotic cell A cell that differs from a prokaryotic cell chiefly by having a nuclear membrane (a well-defined nucleus), membrane-bounded subcellular organelles, and mitotic cell division.

eutrophication The process whereby dissolved nutrients resulting from natural seasonal enrichment or industrial pollution of water cause overgrowth of algae and cyanobacteria to the detriment of fish and other large aquatic inhabitants.

evolution Scientific principle that states that living things change gradually through hundreds of millions of years, and these changes are expressed in structural and functional adaptations in each organism. Evolution presumes that those traits that favor survival are preserved and passed on to following generations, and those traits that do not favor survival are lost.

exanthem An eruption or rash of the skin.

exergonic A chemical reaction associated with the release of energy to the surroundings. Antonym: endergonic.

exfoliative toxin A poisonous substance that causes superficial cells of an epithelium to detach and be shed. Example: staphylococcal exfoliatin. Also called an epidermolytic toxin.

exocytosis The process that releases enveloped viruses from the membrane of the host's cytoplasm.

exoenzyme An extracellular enzyme chiefly used to hydrolyze nutrient macromolecules that cannot readily enter the cell. This enzyme is secreted into the environment, where it may function in saprobic decomposition of organic debris or support the invasion of tissues by pathogens.

exogenous Originating from outside the body.

exon A section of eukaryotic DNA coding for a corresponding portion of mRNA that is translated into proteins. Intervening stretches of DNA that are not expressed are called introns. During transcription, exons are separated from introns and are spliced together into a continuous mRNA transcript.

exotoxin A toxin (usually protein) that is secreted and acts upon a specific cellular target. Examples: botulin, tetanospasmin, diphtheria toxin, and erythrogenic toxin.

exponential Pertaining to the use of exponents, numbers that are typically written as superscripts to indicate how many times a factor is to be multiplied. Exponents are used in scientific notation to render large, cumbersome numbers into small, workable quantities.

extrapulmonary tuberculosis A condition in which tuberculosis bacilli have spread to organs other than the lungs.

extremophiles Organisms capable of living in harsh environments, such as extreme heat or cold.

facilitated diffusion The passive movement of a substance across a plasma membrane from an area of higher concentration to an area of lower concentration utilizing specialized carrier proteins.

facultative Pertaining to the capacity of microbes to adapt or adjust to variations; not obligate. Example: The presence of oxygen is not required for a facultative anaerobe to grow, even though it can use oxygen if it is present.

family In the levels of classification, a mid-level division of organisms that groups more closely related organisms than previous levels. An order is divided into families.

fastidious Requiring strict, narrow nutritional or environmental conditions for growth. Said of bacteria.

fecal coliforms Any species of gram-negative lactose-positive bacteria (primarily *Escherichia coli*) that live primarily in the intestinal tract and not the environment. Finding evidence of these bacteria in a water or food sample is substantial evidence of fecal contamination and potential for infection (See **coliform**).

fermentation The extraction of energy through anaerobic degradation of substrates into simpler, reduced metabolites. In large industrial processes, fermentation can mean any use of microbial metabolism to manufacture organic chemicals or other products.

fermentor A large tank used in industrial microbiology to grow mass quantities of microbes that can synthesize desired products. These devices are equipped with means to stir, monitor, and harvest products such as drugs, enzymes, and proteins in very large quantities.

fertility (F) factor Donor plasmid that allows synthesis of a pilus in bacterial conjugation. Presence of the factor is indicated by F^+, and lack of the factor is indicated by F^-.

filament A helical structure composed of proteins that is part of bacterial flagella.

fimbria, fimbriae Short, numerous surface appendages on some bacteria that provide adhesion but not locomotion.

final electron acceptor A molecule that receives the electrons (is reduced)at the end of a pathway where electrons are being transported. For example, in aerobic organisms,the final acceptor in the electron transport chain is O_2, whereas in organisms with anaerobic respiration, it is some other compound such as SO_2^{2-}, or NO_3^-.

flagellum (Plural: *flagella*) A long appendage used to propel an organism through a fluid environment. Flagella of bacteria and eukaryotes are similar in general function but differ significantly in structure.

fluid mosaic model A conceptualization of the molecular architecture of cellular membranes as a bilipid layer containing proteins. Membrane proteins are embedded to some degree in this bilayer, where they float freely about.

fluorescence/fluorescent antibodies The property possessed by certain minerals and dyes to emit visible light when excited by ultraviolet radiation. One application is to combine a fluorescent dye with a specific antibody to detect the presence of an unknown antigen.

focal infection Occurs when an infectious agent breaks loose from a localized infection and is carried by the circulation to other tissues.

folliculitis An inflammatory reaction involving the formation of papules or pustules in clusters of hair follicles.

fomite Virtually any inanimate object an infected individual has contact with that can serve as a vehicle for the spread of disease.

food chain A simple straight-line feeding sequence among organisms in a community.

food fermentations Addition to and growth of known cultures of microorganisms in foods to produce desirable flavors, smells, or textures. Includes cheeses, breads, alcoholic beverages, and pickles.

food infection A form of food-borne illness associated with ingesting living pathogenic microbes that invade the intestine. The damage caused by microbes growing in the body cause the main symptoms, which are usually gastroenteritis (example is salmonellosis).

food intoxication A form of food-borne illness that is the result of ingesting microbial toxins given off by bacteria growing in the food. The toxins cause the main symptoms (example is botulism).

food poisoning Any form of illness acquired from ingesting foods; sources include microbes, chemicals, plants, and animals (also food-borne disease).

food web A complex network that traces all feeding interactions among organisms in a community (see **food chain).** This is considered to be a more accurate picture of food relationships in a community than a food chain.

formalin A 37% aqueous solution of formaldehyde gas; a potent chemical fixative and microbicide.

frameshift mutation An insertion or deletion mutation that changes the codon reading frame from the point of the mutation to the final codon. Almost always leads to a nonfunctional protein.

fructose One of the carbohydrates commonly referred to as sugars. Fructose is commonly fruit sugars.

functional group In chemistry, a particular molecular combination that reacts in predictable ways and confers particular properties on a compound. Examples: —COOH, —OH, —CHO.

fungicide A chemical that can kill fungal spores, hyphae, and yeast.

fungus/fungi Heterotrophic unicellular or multicellular eukaryotic organism that may take the form of a larger macroscopic organism, as in the case of mushrooms, or a smaller microscopic organism, as in the case of yeasts and molds.

furuncle A boil; a localized pyogenic infection arising from a hair follicle.

Gaia theory The concept that biotic and abiotic factors sustain suitable conditions for one another simply by their interactions. Named after the mythical Greek goddess of earth.

GALT Abbreviation for **gut-associated** lymphoid tissue. Includes **Peyer's patches.**

gamma globulin The fraction of plasma proteins high in immunoglobulins (antibodies). Preparations from pooled human plasma containing normal antibodies make useful passive immunizing agents against pertussis, polio, measles, and several other diseases.

gas gangrene Disease caused by a clostridial infection of soft tissue or wound. The name refers to the gas produced by the bacteria growing in the tissue. Unless treated early, it is fatal. Also called **myonecrosis.**

gastroenteritis Inflammation of the lining of the stomach and intestine. May be caused by infection, intestinal disorders, or food poisoning.

gene A site on a chromosome that provides information for a certain cell function. A specific segment of DNA that contains the necessary code to make a protein or RNA molecule.

gene probe Short strands of single-stranded nucleic acid that hybridize specifically with complementary stretches of nucleotides on test samples and thereby serve as a tagging and identification device.

gene therapy The introduction of normal functional genes into people with genetic diseases such as sickle-cell anemia and cystic fibrosis. This is usually accomplished by a virus vector.

generation time Time required for a complete fission cycle—from parent cell to two new daughter cells. Also called **doubling time.**

genetic engineering/bioengineering A field involving deliberate alterations (recombinations) of the genomes of microbes, plants, and animals through special technological processes.

genetics The science that studies inheritance of biological characteristics.

genital warts A prevalent STD linked to some forms of cancer of the reproductive organs. Caused by infection with human papillomavirus.

genome The complete set of chromosomes and genes in an organism.

genotype The genetic makeup of an organism as inherited from parents. The genotype is ultimately responsible for an organism's phenotype, or expressed characteristics.

genus In the levels of classification, the second most specific level. A family is divided into several genera.

germ theory of disease A theory first originating in the 1800s that proposed that microorganisms can be the cause of diseases. The concept is so well established in the present time that it is considered a fact.

germicide An agent lethal to non-endospore-forming pathogens.

giardiasis Infection by the *Giardia* flagellate. The most common mode of transmission is contaminated food and water. Symptoms include diarrhea, abdominal pain, and flatulence.

gingivitis Inflammation of the gum tissue in contact with the roots of the teeth.

glomerulonephritis An inflammatory autoimmune condition in response to streptococcal toxins.

gluconeogenesis The formation of glucose (or glycogen) from noncarbohydrate sources such as protein or fat. Also called glyconeogenesis.

glucose One of the carbohydrates commonly referred to as sugars. Glucose is characterized by its 6-carbon structure.

glutaraldehyde A yellow acidic liquid used in antimicrobial control. Because glutaraldehyde kills spores, it is considered a sterilant.

glycan A polysaccharide.

glycerol A 3-carbon alcohol, with three OH groups that serve as binding sites.

glycocalyx A filamentous network of carbohydrate-rich molecules that coats cells.

glycogen A glucose polymer stored by cells.

glycolysis The energy-yielding breakdown of glucose to pyruvic or lactic acid (occurs in fermentation). It is often called anaerobic glycolysis because no molecular oxygen is consumed in the processes.

glycosidic bond A bond that joins monosaccharides to form disaccharides and polymers.

gnotobiotic Literally means known (gnoto) and life (bios). Referring to experiments performed on germ-free animals that have been inoculated with known species of microbes.

Golgi apparatus An organelle of eukaryotes that participates in packaging and secretion of molecules. Also known as **Golgi complex.**

gonococcus Common name for *Neisseria gonorrhoeae,* the agent of gonorrhea.

graft versus host disease (GVHD) A condition associated with a bone marrow transplant in which T cells in the transplanted tissue mount an immune response against the recipient's (host) normal tissues.

graft/graft rejection Live tissues or organs taken from a donor and transplanted into a recipient to replace damaged or missing tissues such as skin, bone, blood vessels, heart, bone marrow. When the donor displays antigens that are recognized and trigger an immune attack, the grafted tissue or organ may be lost.

Gram stain A differential stain for bacteria useful in identification and taxonomy. Gram-positive organisms appear purple from crystal violet-mordant retention, whereas gram-negative organisms appear red after loss of crystal violet and absorbance of the safranin counterstain.

grana Discrete stacks of chlorophyll-containing thylakoids within chloroplasts.

granulocyte A mature leukocyte that contains noticeable granules in a Wright stain. Examples: neutrophils, eosinophils, and basophils.

granuloma A solid mass or nodule of inflammatory tissue containing modified macrophages and lymphocytes. Usually a chronic pathologic process of diseases such as tuberculosis or syphilis.

granzymes Cytotoxic enzymes produced by cytotoxic T cells and natural killer cells that participate in cell death.

Graves' disease A malfunction of the thyroid gland in which autoantibodies directed at thyroid cells stimulate an overproduction of thyroid hormone (hyperthyroidism).

greenhouse effect The capacity to retain solar energy by a blanket of atmospheric gases that redirects heat waves back toward the earth.

group translocation A form of active transport in which the substance being transported is altered during transfer across a plasma membrane.

growth curve A graphical representation of the change in population size over time. This graph has four periods known as lag phase, exponential or log phase, stationary phase, and death phase.

growth factor An organic compound such as a vitamin or amino acid that must be provided in the diet to facilitate growth. An essential nutrient.

guanine (G) The purine nitrogen base found in DNA and RNA that pairs with cytosine.

Guillain-Barré syndrome A neurological complication of influenza vaccination. Approximately 1 in 100,000 recipients of the vaccine will develop this autoimmune disorder, which is marked by varying degrees of weakness and sensory loss.

gumma A nodular, infectious granuloma characteristic of tertiary syphilis.

gyrase An enzyme that functions in packaging of DNA in bacterial chromosomes by preventing excess distortion of the molecule as it is coiling.

HAART Short for highly active antiretroviral therapy, the drug cocktail that is used as a treatment for HIV infection and AIDS.

habitat The environment to which an organism is adapted.

halogens A group of related chemicals with antimicrobial applications. The halogens most often used in disinfectants and antiseptics are chlorine and iodine.

halophile A microbe that needs a high concentration of salt for growth or one that tolerates high salt concentrations.

hapten An incomplete or partial antigen. Although it constitutes the determinative group and can bind antigen, hapten cannot stimulate a full immune response without being carried by a larger protein molecule.

Hashimoto's thyroiditis An autoimmune disease of the thyroid gland that damages the thyroid follicle cells and results in decreased production of thyroid hormone (hypothyroidism).

hay fever A form of atopic allergy marked by seasonal acute inflammation of the conjunctiva and mucous membranes of the respiratory passages. Symptoms are itching, coughing, and runny nose.

health care-associated/nosocomial infection An infection not present upon admission to a hospital, clinic, or care home, but incurred while being treated there.

helical Having a spiral or coiled shape. Said of certain virus capsids and bacteria.

helminth A term that designates parasitic worms such as roundworms and flatworms.

hemagglutinin (H) A molecule that causes red blood cells to clump or agglutinate. A major surface receptor on the influenza virus needed for entry.

hemolysin Any biological agent that is capable of destroying red blood cells and causing the release of hemoglobin. Many bacterial pathogens produce exotoxins that act as hemolysins.

hemolytic disease of the newborn Incompatible Rh factor between mother and fetus causes maternal antibodies to enter the fetus and trigger complement-mediated destruction of fetal RBCs (hemolysis).

hemolytic uremic syndrome (HUS) A kidney disease caused by infection with *E.coli* 0157:H7.

hemopoiesis The process by which the various types of blood cells are formed by stem cells in the bone marrow. Also called hematopoiesis.

hepadnavirus Enveloped DNA viruses with a predisposition to affect the liver. Hepatitis B is the most serious form.

hepatitis Inflammation and necrosis of the liver, often the result of viral infection.

hepatitis A virus (HAV) Enterovirus spread by contaminated food responsible for short-term (infectious) hepatitis.

hepatitis B virus (HBV) Hepadnavirus that is the causative agent of serum hepatitis.

hepatitis C virus (HCV) A blood-borne RNA virus that is the third most common cause of hepatitis.

hepatocellular carcinoma A liver cancer associated with infection with hepatitis B virus.

herd immunity A collective acquired immunity in a population that reduces the incidence of a pathogen and makes it less likely that nonimmune individuals will contract and spread infection. One aim of vaccination is to induce herd immunity.

heredity Genetic inheritance.

hermaphroditic Containing the sex organs for both male and female in one individual.

herpes zoster A recurrent infection caused by latent chickenpox virus. Its manifestation on the skin tends to correspond to dermatomes and to occur in patches that "girdle" the trunk. Also called **shingles.**

herpetic keratitis Corneal or conjunctival inflammation due to herpesvirus type 1.

heterotroph An organism that relies upon organic compounds for its carbon and energy needs.

hexose A 6-carbon sugar such as glucose and fructose.

hierarchies Levels of power. Arrangement in order of rank.

histamine A cytokine released when mast cells and basophils release their granules. An important mediator of allergy, its effects include smooth muscle contraction, increased vascular permeability, and increased mucus secretion.

histone Proteins associated with eukaryotic DNA. These simple proteins serve as a scaffolding to compact and condense DNA into chromosomes, while also enhancing or preventing the expression of certain genes.

holoenzyme An enzyme complete with its apoenzyme and cofactors.

hops The ripe, dried fruits of the hop vine (*Humulus lupulus*) that is added to beer wort for flavoring.

host range The limitation imposed by the characteristics of the host cell on the type of virus that can successfully invade it.

human diploid cell vaccine (HDCV) A vaccine made using cell culture that is currently the vaccine of choice for preventing infection by rabies virus.

human herpesvirus-6 (HHV-6) The herpesvirus that causes roseola and may possibly be linked to chronic neurological diseases.

human immunodeficiency virus (HIV) A retrovirus that causes acquired immunodeficiency syndrome (AIDS).

human papillomavirus (HPV) A group of DNA viruses whose members are responsible for common, plantar, and genital warts.

humoral immunity Protective molecules (mainly antibodies) carried in the fluids of the body.

hybridization A process that matches complementary strands of nucleic acid (DNA-DNA, RNA-DNA, RNA-RNA). Used for locating specific sites or identifying exact sequences of nucleic acids.

hybridoma An artificial cell line that produces monoclonal antibodies. It is formed by fusing (hybridizing) a normal antibody-producing cell with a cancer cell, and it can produce pure antibody indefinitely.

hydrogen bond A weak chemical bond formed by the attraction of forces between molecules or atoms—in this case, hydrogen and either oxygen or nitrogen. In this type of bond, electrons are not shared, lost, or gained.

hydrologic cycle The continual circulation of water between hydrosphere, atmosphere, and lithosphere.

hydrolysis A process in which water is used to break bonds in molecules. Usually occurs in conjunction with an enzyme.

hydrophilic The property of attracting water. Molecules that attract water to their surface are called hydrophilic.

hydrophobic The property of repelling water. Molecules that repel water are called hydrophobic.

hydrosphere That part of the biosphere that encompasses water-containing environments such as oceans, lakes, rivers.

hypertonic A solution that, when compared to a reference solution, has a higher concentration of solute and less water; Placed opposite a hypotonic solution, it will cause water to to diffuse more rapidly from the hypotonic solution, thus creating a greater osmotic pressure.

hyphae The tubular threads that make up filamentous fungi (molds). This web of branched and intertwining fibers is called a mycelium.

hypogammaglobulinemia An inborn disease in which the gamma globulin (antibody) fraction of serum is greatly reduced. The condition is associated with a high susceptibility to pyogenic infections.

hypothesis A tentative explanation of what has been observed or measured.

hypotonic A solution that, when compared with a reference solution, is less concentrated, i.e. has more water and less solute. When a membrane is placed between this solution and a hypertonic solution, water will move from the hypotonic side to the hypertonic side at a higher rate.

icosahedron A regular geometric figure having 20 surfaces that meet to form 12 corners. Some virions have capsids that resemble icosahedral crystals.

immune complex reaction Type III hypersensitivity of the immune system. It is characterized by the reaction of soluble antigen with antibody, and the deposition of the resulting complexes in basement membranes of epithelial tissue.

immune surveillance The continual function of macrophages, cytotoxic T cells, and natural killer cells in identifying and destroying cancer cells within the body.

immune tolerance The ability of the immune system to remain nonreactive to normal cell molecules.

immunity An acquired resistance to an infectious agent due to prior contact with that agent.

immunoassays Extremely sensitive tests that permit rapid and accurate measurement of trace antigen or antibody.

immunocompetence The ability of the body to recognize and react with multiple foreign substances.

immunodeficiency Immune function is incompletely developed, suppressed, or destroyed.

immunogen Any antigen molecule that, after being processed by the immune system, will stimulate protective T and B cell responses.

immunoglobulin (IG) The chemical class of proteins to which antibodies belong.

immunology The study of internal body defenses that protect against infection. This includes phagocytosis, inflammation, and acquired immunities.

immunopathology A disease state initiated by incorrect or harmful immune responses. Some immune diseases are caused by overreactions (allergies) and others are due to the lack of a functional immune response (immunodeficiencies).

immunotherapy Preventing or treating infectious diseases by administering substances that produce artificial immunity. May be active or passive.

impetigo Contagious skin infection marked by pustules and vesicles; caused by *S. aureus* and *S. pyogenes*. Also known as **pyoderma**.

IMViC Abbreviation for four identification tests: **i**ndole production, **m**ethyl red test, **V**oges-Proskauer test (**i** inserted to concoct a wordlike sound), and **c**itrate as a sole source of carbon. This test was originally developed to distinguish between *Enterobacter aerogenes* (associated with soil) and *Escherichia coli* (a fecal coliform).

in vitro Literally means "in glass," signifying a process or reaction occurring in an artificial environment, as in a test tube or culture medium.

in vivo Literally means "in a living being," signifying a process or reaction occurring in a living thing.

incidence In epidemiology, the number of new cases of a disease occurring during a time period.

incineration Destruction of microbes by subjecting them to extremes of dry heat. Microbes are reduced to ashes and gas by this process.

inclusion A relatively inert body in the cytoplasm such as storage granules, glycogen, fat, or some other aggregated metabolic product. Also called **inclusion body**.

incubate To isolate a sample culture in a temperature-controlled environment to encourage growth.

incubation period The period of time from the initial contact with an infectious agent to the appearance of the first symptoms.

indicator bacteria In water analysis, any easily cultured bacteria that may be found in the intestine and can be used as an index of fecal contamination. The category includes coliforms and enterococci. Discovery of these bacteria in a sample means that pathogens may also be present.

induced mutation Any alteration in DNA that occurs as a consequence of exposure to chemical or physical mutagens.

inducible operon A system of gene regulation in which a gene is ordinarily repressed unless the substrate of the structural proteins is present and induces it to become active; otherwise the operon is blocked and the RNA polymerase cannot transcribe.

induction Process by which an individual accumulates data or facts and then formulates a general hypothesis that accounts for those facts.

infection The entry, establishment, and multiplication of pathogenic organisms within a host.

infectious disease The state of damage or toxicity in the body caused by an infectious agent.

infectious dose The estimated number of microbial cells or units required to establish an infection.

inflammation A natural, nonspecific response to tissue injury that protects the host from further damage. It stimulates immune reactivity and blocks the spread of an infectious agent.

inoculation The implantation of microorganisms into or upon culture media.

inorganic chemicals Molecules that are composed of elements other than the basic framework of carbon and hydrogen, and which are usually much simpler chemically than organic molecules.

interferon (IFN) Naturally occurring polypeptide produced by fibroblasts and lymphocytes that can block viral replication and regulate a variety of immune reactions.

interleukin Molecules (interleukin-1, or IL-1) secreted by macrophages that stimulate lymphocyte function. In response to IL-1 T cells release yet another interleukin (IL-2), which amplifies T-cell response by activating additional T cells. T helper cells stimulated by IL-2 stimulate B-cell proliferation and promote antibody production.

intermediate host The organism in which the larval development of a parasite occurs.

intoxication Poisoning that results from the ingestion of a toxin and the subsequent effects of that toxin as it spread into body tissues.

intron A contracted version of "intragenic regions." These are segments on split genes of eukaryotes that do not code for polypeptide. They serve other functions. See **exon**.

iodophor A combination of iodine and an organic carrier that is a moderate-level disinfectant and antiseptic.

ionic bond A chemical bond in which electrons are transferred and not shared between atoms.

ionization The aqueous dissociation of an electrolyte into ions.

ionizing radiation Radiant energy consisting of short-wave electromagnetic rays (X ray) or high-speed electrons that cause dislodgement of electrons on target molecules, which creates create ions.

irradiation The application of radiant energy for diagnosis, therapy, disinfection, or sterilization.

irregular (in shape) Refers to bacteria that stain unevenly or display cell to cell variation in size and/or shape (pleomorphic) within a single species.

isograft Transplanted tissue from one monozygotic twin to the other; transplants between highly inbred animals that are genetically identical.

isolation The separation of microbial cells by serial dilution or mechanical dispersion on solid media to create discrete colonies.

isotonic Solutions having the same osmotic pressure, such that, when separated by a semipermeable membrane, show no net movement of solvent in either direction.

isotope A version of an element that is virtually identical in all chemical properties to another version except that their atoms have slightly different atomic masses.

jaundice The yellowish pigmentation of skin, mucous membranes, sclera, deeper tissues, and excretions due to abnormal deposition of bile pigments. Jaundice is associated with liver infection, as with hepatitis B virus and leptospirosis.

Kaposi's sarcoma A malignant or benign neoplasm that appears as multiple hemorrhagic sites on the skin, lymph nodes, and viscera and apparently involves the metastasis of abnormal blood vessel cells. It is a clinical feature of AIDS.

keratoconjunctivitis Inflammation of the conjunctiva and cornea.

killed or inactivated vaccine A whole cell or intact virus preparation in which the microbes are dead or preserved and cannot multiply but are still capable of conferring immunity.

kingdom In the levels of classification, the second division from more general to more specific. Each domain is divided into kingdoms.

Koch's postulates A procedure to establish the specific cause of disease. In all cases of infection: (1) The agent must be found; (2) inoculations of a pure culture must reproduce the same disease in animals; (3) the agent must again be present in the experimental animal; and (4) a pure culture must again be obtained.

Koplik's spots Tiny red blisters with central white specks on the mucosal lining of the cheeks. Symptomatic of measles.

Krebs cycle Metabolic cycle that is linked to glycolysis and receives acetyl groups generated when pyruvic acid is decarboxylated; it provides $NADH_1$, $FADH_2$, and ATP; also known as TCA cycle and citric acid cycle.

L form L-phase variants; wall-less forms of some bacteria that are induced by drugs or chemicals. These forms can be involved in infections.

labile In chemistry, molecules or compounds that are chemically unstable in the presence of environmental changes.

lactose One of the carbohydrates commonly referred to as sugars. Lactose is commonly found in milk.

lactose (*lac*) operon Control system that manages the regulation of lactose metabolism. It is composed of three DNA segments, including a regulator, a control locus, and a structural gene.

lag phase The early phase of population growth during which little sign of growth occurs. It is this period that readies the cells for the rapid expansion to come.

lager The maturation process of beer, which is allowed to take place in large vats at a reduced temperature.

lagging strand The newly forming DNA strand that is discontinuously replicated in short segments (Okazaki fragments) because the template cannot be continuously read by the DNA polymerase.

latency The state of being inactive and not multiplying. Example: a latent virus or latent infection.

leading strand The newly forming DNA strand that is replicated in a continuous fashion without fragments because it is oriented in the correct 3′ to 5′ direction for the DNA polymerase.

leaven To lighten food material by entrapping gas generated within it. Example: the rising of bread from the CO_2 produced by yeast or baking powder.

legionellosis/ legionnaire's disease Pulmonary infection by *Legionella* bacteria. Gram-negative rods survive inside amoebas in aquatic habitats. Some conditions may be fatal.

lepromatous leprosy Severe, disfiguring leprosy characterized by widespread dissemination of the leprosy bacillus in deeper lesions.

leprosy A chronic progressive disease of the skin and nerves caused by a slow growing, strictly parasitic bacterium, Mycobacterium leprae. Also known as **Hansen disease.**

lesion A wound, injury, or some other pathologic change in tissues.

leukocidin A heat-labile substance formed by some pyogenic cocci that impairs and sometimes lyses leukocytes.

leukocytes White blood cells. The primary infection-fighting blood cells.

leukocytosis An abnormally large number of leukocytes in the blood, which can be indicative of acute infection.

leukopenia A lower-than-normal leukocyte count in the blood, which can be indicative of blood infection or disease.

leukotriene An unsaturated fatty acid derivative of arachidonic acid. Leukotriene functions in chemotactic activity, smooth muscle contractility, mucus secretion, and capillary permeability.

ligase An enzyme required to join nucleotides together to complete the final attachment of the ends of two fragments of DNA; this enzyme also works in DNA technology to attach sticky ends of DNA pieces during splicing.

light-dependent reactions The series of reactions in photosynthesis that are driven by the light energy (photons) absorbed by chlorophyll. They involve splitting of water into hydrogens and oxygen, transport of electrons by NADP, and ATP synthesis.

light-independent reactions The series of reactions in photosynthesis that can proceed with or without light. It is a cyclic system that uses ATP from the light reactions to incorporate or fix carbon dioxide into organic compounds, leading to the production of glucose and other carbohydrates (also called the Calvin cycle).

limnetic zone The deep-water region beyond the shoreline.

lipid A term used to describe a variety of substances that are not soluble in polar solvents such as water but will dissolve in nonpolar solvents such as benzene and chloroform. Lipids include triglycerides, phospholipids, steroids, and waxes.

lipopolysaccharide A molecular complex of lipid and carbohydrate found in the bacterial cell wall. The lipopolysaccharide (LPS) of gram-negative bacteria is an endotoxin with generalized pathologic effects such as fever.

listeriosis Infection with *Listeria monocytogenes,* usually due to eating contaminated dairy products, poultry, or meat. It is usually mild in healthy adults but can produce severe symptoms in neonates and immunocompromised adults.

lithosphere That part of the biosphere that encompasses the earth's crust, including rocks and minerals.

littoral zone The shallow region along a shoreline.

lobar pneumonia Infection involving whole segments (lobes) of the lungs, which may lead to consolidation and plugging of the alveoli and extreme difficulty in breathing.

localized infection An infection in which a microbe enters a specific tissue, infects it, and remains confined there.

locus (plural, **loci**) The specific site on a DNA molecule where a particular gene or sequence of nucleotides is located. In diploid organisms, the locus is occupied by two allele copies, one on each of the chromosome pairs.

logarithmic or log phase Maximum rate of cell division during which growth is geometric in its rate of increase. Also called **exponential growth phase.**

lophotrichous Describing bacteria having a tuft of flagella at one or both poles.

lumen The cavity within a tubular organ.

lymphadenitis Inflammation of one or more lymph nodes. Also called lymphadenopathy.

lymphatic or lymphoid system A system of vessels and organs that serve as sites for development of immune cells and immune reactions. It includes the spleen, thymus, lymph nodes, and GALT.

lymphocyte The second most common form of white blood cells.

lyophilization Freeze-drying; the separation of a dissolved solid from the solvent by freezing the solution and evacuating the solvent under vacuum. A means of preserving the viability of cultures.

lysin A complement-fixing antibody that destroys specific targeted cells. Examples: hemolysin and bacteriolysin.

lysis The physical rupture or deterioration of a cell.

lysogeny The indefinite persistence of bacteriophage DNA in a host without bringing about the production of virions. A lysogenic cell can revert to a lytic cycle, the process that ends in lysis.

lysosome A cytoplasmic organelle containing lysozyme and other hydrolytic enzymes.

lysozyme An enzyme that attacks the bonds on bacterial peptidoglycan. It is a natural defense found in tears and saliva.

macromolecules Large, molecular compounds assembled from smaller subunits, most notably biochemicals.

macronutrient A chemical substance required in large quantities (phosphate, for example).

macrophage A white blood cell derived from a monocyte that leaves the circulation and enters tissues. These cells are important in nonspecific phagocytosis and in regulating, stimulating, and cleaning up after immune responses.

macroscopic Visible to the naked eye.

major histocompatibility complex (MHC) See **HLA.**

major histocompatibility complex/human leukocyte antigen Major histocompatibility complex (MHC) and human leukocyte antigen (HLA). These alternate terms refer to a closely linked cluster of genes that program cell surface glycoprotein markers that control immune interactions between cells and are involved in immune recognition and some hypersensitivies, transfusion reactions, and graft rejections.

malignant tumor Cancerous growth characterized by uncontrolled growth or abnormal cells within normal tissue.

malt The grain, usually barley, that is sprouted to obtain digestive enzymes and dried for making beer.

MALT (Mucosal-Associated Lymphoid Tissue) Active lymphocytes distributed within the mucosal surfaces of respiratory and other systems.

maltose One of the carbohydrates referred to as sugars. A fermentable sugar formed from starch.

Mantoux test An intradermal screening test for tuberculin hypersensitivity. A red, firm patch of skin at the injection site greater than 10 mm in diameter after 48 hours is a positive result that indicates current or prior exposure to the TB bacillus.

marker Any trait or factor of a cell, virus, or molecule that makes it distinct and recognizable. Example: a genetic marker.

mash In making beer, the malt grain is steeped in warm water, ground up, and fortified with carbohydrates to form mash.

mass number (MN) Measurement that reflects the number of protons and neutrons in an atom of a particular element.

mast cell A nonmotile connective tissue cell implanted along capillaries, especially in the lungs, skin, gastrointestinal tract, and genitourinary tract. Like a basophil, its granules store mediators of allergy.

matrix The dense ground substance between the cristae of a mitochondrion that serves as a site for metabolic reactions.

matter All tangible materials that occupy space and have mass.

maximum temperature The highest temperature at which an organism will grow.

mechanical vector An animal that transports an infectious agent but is not infected by it, such as houseflies whose feet become contaminated with feces.

medium (plural, **media**) A nutrient used to grow organisms outside of their natural habitats.

membrane, cell A molecular "skin" surrounding the cytoplasm of cells that functions in protection, transport, communication, and secretion.

memory cell A long-lived progeny of a sensitized lymphocyte that remains in circulation and is genetically programmed to react rapidly the next time it encounters its antigen.

meningitis An inflammation of the membranes (meninges) that surround and protect the brain. It is often caused by bacteria such as *Neisseria meningitidis* (the meningococcus) and *Haemophilus influenzae*.

merozoite The motile, infective stage of an apicomplexan parasite that comes from a liver or red blood cell undergoing multiple fission.

mesophile Microorganisms that grow at intermediate temperatures—usually between 20 and 40 degrees C.

messenger RNA (MRNA) A single-stranded transcript that is a copy of the DNA template that corresponds to a gene; it generally carries the genetic code for a protein.

metabolic analog A chemical, usually a drug, that mimics the natural substrate of an enzyme and vies for its active site. This reaction, called competitive inhibition, can block the metabolic pathway that requires that enzyme.

metabolism A general term designating the totality of chemical processes occurring in a cell.

metabolites Small organic molecules that are intermediates in the stepwise biosynthesis or breakdown of macromolecules.

metagenomics The analysis of genomes recovered from mixed environmental samples that provides information on the diversity of microbe types without isolating or culturing them.

methanogens Microbes that produce methane gas.

MIC Abbreviation for minimum inhibitory concentration. The lowest concentration of antibiotic needed to inhibit bacterial growth in a test system.

microaerophile An aerobic bacterium that requires oxygen at a concentration less than that in the atmosphere.

microbe See **microorganism.**

microbial loop A process occurring in natural aquatic habitats that maintains the cycling of nutrients from producers to consumers and back to producers. It is powered primarily by mineralizing and decomposing bacteria and protozoa.

microbiology A specialized area of biology that deals with living things ordinarily too small to be seen without magnification, including bacteria, archaea, fungi, protozoa, and viruses.

microbiota The native types of bacteria, fungi, and viruses that normally reside on the body. An older term is normal flora.

microfilaments Cellular cytoskeletal element formed by thin protein strands that attach to cell membrane and form a network though the cytoplasm. Responsible for movement of cytoplasm.

micronutrient A chemical substance required in small quantities (trace metals, for example).

microorganism A living thing ordinarily too small to be seen without magnification; an organism of microscopic size.

microscopic Invisible to the naked eye.

microscopy Science that studies structure, magnification, lenses, and techniques related to use of a microscope.

microtubules Long tubular structures made of protein that function in the structure, shape, and movement of eukaryotic cells.

mineralization The process by which decomposers (bacteria and fungi) convert organic debris into inorganic and elemental form. It is part of the recycling process.

minimum inhibitory concentration (MIC) The smallest concentration of drug needed to visibly control microbial growth.

minimum temperature The lowest temperature at which an organism will grow.

miracidium The ciliated first-stage larva of a trematode. This form is infective for a corresponding intermediate host snail.

missense mutation A mutation in which a change in the DNA sequence results in a different amino acid being incorporated into a protein, with varying results.

mitochondrion A double-membrane organelle of eukaryotes that is the main site for aerobic respiration. It contains its own chromosome and ribosomes.

mitosis Somatic cell division that preserves the somatic chromosome number.

mixed acid fermentation An anaerobic degradation of pyruvic acid that results in more than one organic acid being produced (acetic acid, lactic acid, succinic acid, etc.).

mixed culture A container growing two or more different, known species of microbes.

mixed infection Occurs when several different pathogens interact simultaneously to produce an infection. Also called a synergistic infection.

moist heat Heat combined with some form of water, including hot water, boiling water, or steam, generally ranging from 60°C to 135°C.

molds The filamentous fungi. composed of elongate hyphae.

molecule A distinct chemical substance that results from the combination of two or more atoms.

molluscum contagiosum Poxvirus-caused disease that manifests itself by the appearance of small lesions on the face, trunk, and limbs. Often associated with sexual transmission.

monoclonal antibody An antibody produced by a clone of lymphocytes that respond to a particular antigenic determinant and generate identical antibodies only to that determinant. See **hybridoma.**

monocyte A large mononuclear leukocyte normally found in the lymph nodes, spleen, bone marrow, and loose connective tissue. This type of cell makes up 3% to 7% of circulating leukocytes.

monomer A simple molecule that can be linked by chemical bonds to form larger molecules.

mononuclear phagocyte system A collection of monocytes and macrophages scattered throughout the extracellular spaces that function to engulf and degrade foreign cells and substances.

monosaccharide A simple sugar such as glucose that is a basic building block for more complex carbohydrates.

monotrichous Describing a microorganism that bears a single flagellum.

morbidity A diseased condition.

morphology The study of organismic structure.

mortality rate Total number of deaths in a population attributable to a particular disease.

most probable number (MNP) Coliform test used to detect the concentration of contaminants in water supplies.

motility Self-propulsion.

MRSA/methicillin-resistant *Staphylococcus aureus* This refers to a strain of *S. aureus* that resists not only methicillin but a wide variety of nonpenicillin drugs such as cephalosporins, gentamycin, tetracycline, and erythromycin. These "superbugs" are common in the hospital and are increasingly found in community settings.

Multibacillary Containing many bacilli.

multinucleate giant cell A cytopathic effect that arises from the fusion of the cytoplasmic membranes of several infected cells. Characteristic of infection with paramyxoviruses.

mumps See **epidemic parotitis.**

must Juices expressed from crushed fruits that are used in fermentation for wine.

mutagen Any agent that induces a change in the composition of DNA. Examples: certain chemical substances, ultraviolet radiation, radioactive elements.

mutant strain A subspecies of microorganism that has undergone a mutation, causing expression of a trait that differs from other members of that species.

mutation A permanent inheritable alteration in the DNA sequence or content of a cell.

mutualism Organisms living in a close, mutually beneficial relationship.

mycelium The filamentous mass that makes up a mold. Composed of hyphae.

mycetoma A chronic fungal infection usually afflicting the feet, typified by swelling and multiple draining lesions. Example: maduromycosis or Madura foot.

mycoplasma Species of mycoplasma are among the smallest self-replicating microorganisms. Mycoplasma naturally lack a cell wall. Most species are parasites of animals and plants.

mycorrhizae Various species of fungi adapted in an intimate, mutualistic relationship to plant roots.

mycosis Any disease caused by a fungus.

mycotoxicosis Illness resulting from eating poisonous fungi.

myonecrosis Another name for **gas gangrene.**

nanobes Cell-like particles, found in sediments and other geologic deposits, that some scientists speculate are the smallest bacteria. Short for **nanobacteria.**

narrow spectrum Denotes drugs that are selective and limited in their effects. For example, they inhibit either gram-negative or gram-positive bacteria but not both.

nasopharyngeal carcinoma A malignancy of epithelial cells that occurs in older Chinese and African men and is associated with exposure to Epstein-Barr virus.

natural immunity Specific protection from disease acquired through normal, nonmedical life processes.

necrosis A pathologic process in which cells and tissues die and disintegrate.

necrotizing faciitis Severe invasive disease of skin, muscle, and connective tissue, usually caused by group A streptococci.

negative feedback Enzyme regulation of metabolism by the end product of a multienzyme system that blocks the action of a "pacemaker" enzyme at or near the beginning of the pathway.

negative stain A staining technique that renders the background opaque or colored and leaves the object unstained so that it is outlined as a colorless area.

nematode A common name for helminths called roundworms.

neoplasm A synonym for tumor.

neuraminidase A glycoprotein found in the envelope of influenza virus that facilitates release of new viruses from the host cell.

neurotropic Having an affinity for the nervous system. Most likely to affect the spinal cord.

neutralization In immunology, a type of antibody-antigen reaction which results in the blockage of a virus receptor or toxic protein by an antibody. In chemistry, the process of combining an acid and a base until they reach a balanced proportion, with a pH value close to 7.

neutron An electrically neutral particle in the nuclei of all atoms except hydrogen.

neutrophil A mature granulocyte present in peripheral circulation, exhibiting a multilobular nucleus and numerous cytoplasmic granules that retain a neutral stain. The neutrophil is an active phagocytic cell in bacterial infection.

niche In ecology, an organism's biological role in or contribution to its community.

Nicotinamide adenine dinucleotide/NAD/NADH Abbreviations for the oxidized/reduced forms of nicotinamide adenine dinucleotide, an electron carrier. Nicotinamide is another name for the vitamin niacin.

nitrification Phase of the nitrogen cycle in which ammonium is oxidized.

nitrogen cycle The pathway followed by the element nitrogen as it circulates from inorganic sources in the nonliving environment to living things and back to the nonliving environment. The longtime reservoir is nitrogen gas in the atmosphere.

nitrogen fixation A process occurring in certain bacteria in which atmospheric N_2 gas is converted to a form (NH_4) usable by plants.

nitrogenous base A ringed compound of which pyrimidines and purines are types.

nomenclature A set system for scientifically naming organisms, enzymes, anatomical structures, etc.

noncoliforms Lactose-negative enteric bacteria in normal flora.

noncommunicable An infectious disease that is not established through transmission of an infectious agent via another infected host.

nonionizing radiation Method of microbial control, best exemplified by ultraviolet light, that causes the formation of abnormal bonds within the DNA of microbes, increasing the rate of mutation. The primary limitation of nonionizing radiation is its inability to penetrate beyond the surface of an object.

nonpolar A term used to describe an electrically neutral molecule formed by covalent bonds between atoms that have the same or similar electronegativity.

nonself Cells or molecules recognized by the immune system as containing foreign markers. Nonself can stimulate the immune system to react and eliminate the foreign substances.

nonsense codon A triplet of mRNA bases that does not specify an amino acid but signals the end of a polypeptide chain. Also called **stop codon.**

nonsense mutation A mutation that changes an amino acid-based codon into a stop codon, leading to premature termination of a protein during translation.

Norwalk agent One of a group of Caliciviruses, Norwalk agent causes gastrointestinal distress and is commonly transmitted in schools, camps, cruise ships, and nursing homes.

nucleocapsid In viruses, the close physical combination of the nucleic acid with its protective covering.

nucleoid The undifferentiated space or region in the cytoplasm that contains the bacterial genetic material. It is naked and not enclosed in a membrane.

nucleolus A granular mass containing RNA that is contained within the nucleus of a eukaryotic cell.

nucleosome Structure in the packaging of eukaryotic DNA formed when the DNA strands wrap around histone proteins, forming an arrangement that looks like beads on a chain.

nucleotide The basic structural unit of DNA and RNA; each nucleotide consists of a phosphate, a sugar (ribose in RNA, deoxyribose in DNA), and a nitrogenous base such as adenine, guanine, cytosine, thymine (DNA only), or uracil (RNA only).

nucleus 1) In chemistry, the central core of an atom, composed of protons and neutrons. 2) In cell structure, a prominent spherical organelle that is encased by a double membrane or envelope; it is the site where the majority of a cell's DNA is housed.

numerical aperture (NA) In microscopy, a measure of the light passing from the object and into the objective to maximize optical clarity and resolution.

nutrient Any chemical substance that is acquired from the environment and used for metabolism and growth. Macronutrients are required in large amounts, and micronutrients in small amounts.

nutrition The acquisition of chemical substances by a cell or organism for use as an energy source or as building blocks of cellular structures.

obligate/strict Without alternative; restricted to a particular way of life. Example: An obligate parasite survives and grows only in a host; an obligate aerobe must have oxygen to grow; an obligate anaerobe is destroyed by oxygen.

Okazaki fragment In replication of DNA, a segment formed on the lagging strand where biosynthesis is conducted in a discontinuous manner as required by the DNA polymerase orientation.

oligodynamic action A chemical having antimicrobial activity in minuscule amounts. Example: Certain heavy metals are effective in a few parts per billion.

oligotrophic Nutrient-deficient ecosystem.

oncogene A naturally occurring type of gene that when activated can transform a normal cell into a cancer cell.

oncology The study of neoplasms, their cause, disease characteristics, and treatment.

onychomycosis A fungal infection of the nail and nailbed. Also called tinea unguium.

operator In an operon sequence, the DNA segment where transcription of structural genes is initiated.

operon A genetic operational unit that regulates metabolism by controlling mRNA production. In sequence, the unit consists of a regulatory gene, inducer or repressor control sites, and structural genes.

opportunistic infection A situation where ordinarily nonpathogenic or weakly pathogenic microbes cause disease in an immunologically compromised host.

opportunistic pathogen A microbe that infects a host when the body's defense system is vulnerable, causing an **opportunistic infection.**

opsonization The process of stimulating phagocytosis by affixing molecules (opsonins such as antibodies and complement) to the surfaces of foreign cells or particles.

optimum temperature The temperature at which a species shows the most rapid growth rate.

orbitals The pathways of electrons as they rotate around the nucleus of an atom.

order In the levels of classification, the division of organisms that follows class. Increasing similarity may be noticed among organisms assigned to the same order.

organelle A small component of eukaryotic cells that is bounded by a membrane and specialized in function.

organic chemicals Molecules that are structured with a basic framework of the elements carbon and hydrogen; usually these are larger than inorganic molecules and contain additional elements as well.

ornithosis Worldwide zoonosis carried in the latent state in wild and domesticated birds; caused by *Chlamydia psittacosis.*

osmophile A microorganism that thrives in a medium having high osmotic pressure.

osmosis The diffusion of water across a selectively permeable membrane in the direction towards lower water concentration and higher solute concentration.

osteomyelitis A focal infection of the internal structures of long bones, leading to pain and inflammation. Often caused by *Staphylococcus aureus.*

otitis media Acute infection of the middle ear, usually by the pneumococcus, that causes severe pain and drainage.

outer membrane A structure made of lipids in the outer layer of the cell wall of gram-negative bacteria.

oxidation In chemical reactions, the loss of electrons by one reactant. An oxidizing agent is a molecule that causes electron loss or removes electrons from a substrate.

oxidation-reduction Redox reactions, in which paired sets of molecules participate in electron transfers.

oxidative phosphorylation The synthesis of ATP using energy released during the electron transport phase of respiration.

oxygenic Any reaction that gives off oxygen; usually in reference to the result of photosynthesis in eukaryotes and cyanobacteria.

palindrome In the DNA code, a short segment of paired nitrogen bases that have the same sequence when read in opposite directions. Palindromes have significance as transposable elements, regulatory protein targets, and in DNA splicing. In ordinary language, a palindrome is a word, verse, number, or sentence that reads the same forward or backward.

palisades The characteristic bacterial arrangement resembling a row of fence posts created by cells snapping together.

pandemic An epidemic that has spread across a wide geographic area, including more than one continent or international border, and can extend to a worldwide occurrence.

papilloma Benign, squamous epithelial growth commonly referred to as a wart.

parainfluenza A respiratory disease (croup) caused by infection with Paramyxovirus.

parasitism/parasite A close interaction in which one organism (the parasite) lives on or within another organism (the host), from which it obtains nutrients and receives protection. The parasite produces some degree of harm to the host.

parenteral Administering a substance into a body compartment other than through the gastrointestinal tract, such as via intravenous, subcutaneous, intramuscular, or intramedullary injection.

passive carrier Persons who mechanically transfer a pathogen without ever being infected by it, such as health care workers who don't wash their hands adequately between patients.

passive immunity Specific resistance that is acquired indirectly by donation of preformed immune substances (antibodies) produced in the body of another individual.

passive transport Nutrient transport method that follows basic physical laws and does not require direct energy input from the cell.

pasteurization Heat treatment of perishable fluids such as milk, fruit juices, or wine to destroy heat-sensitive vegetative cells, followed by rapid chilling to inhibit growth of survivors and germination of spores. It prevents infection and spoilage.

pathogen Any agent—usually a virus, bacterium, fungus, protozoan, or helminth—that infects body tissues and causes disease.

pathogen-associated molecular pattern (PAMP) Molecules on the surfaces of pathogenic microbes that are recognized by phagocytes and trigger immune responses against the microbes.

pathogenicity The capacity of microbes to cause disease.

Paucibacillary Containing just a few bacilli.

pelvic inflammatory disease (PID) An infection of the uterus and fallopian tubes that has ascended from the lower reproductive tract. Caused by gonococci and chlamydias.

penicillinase An enzyme that hydrolyzes penicillin; found in penicillin-resistant strains of bacteria.

penicillins A large group of naturally occurring and synthetic antibiotics produced by *Penicillium* mold and active against the cell wall of bacteria.

pentose A monosaccharide with five carbon atoms per molecule. Examples: arabinose, ribose, and xylose.

peptide Molecule composed of short chains of amino acids, such as a dipeptide (two amino acids), a tripeptide (three), and a tetrapeptide (four).

peptide bond The covalent union between two amino acids that forms between the amine group of one and the carboxyl group of the other. The basic bond of proteins.

peptidoglycan A network of polysaccharide chains cross-linked by short peptides that forms the rigid part of bacterial cell walls. Gram-negative bacteria have a smaller amount of this rigid structure than do gram-positive bacteria.

perforin A protein secreted by cytotoxic and natural killer cells that attacks the cell membrane of infectious agents and forms pores.

period of invasion The period during a clinical infection when the infectious agent multiplies at high levels, exhibits its greatest toxicity, and becomes well established in the target tissues.

periplasmic space An open area between the cell wall and cell membrane in the cell envelopes of bacteria. Gram negative bacteria have a a more extensive space than do gram positive bacteria.

peritrichous In bacterial morphology, having flagella distributed over the entire cell.

pertussis Infection by *Bordetella pertussis*. A highly communicable disease that causes acute respiratory syndrome. Pertussis can be life-threatening in infants, but vaccination on the recommended schedule can prevent infection. Also called **whooping cough.**

petechiae Minute hemorrhagic spots in the skin that range from pinpoint- to pinhead-size.

Peyer's patches Oblong lymphoid aggregates of the gut located chiefly in the wall of the terminal and small intestine. Along with the tonsils and appendix, Peyer's patches make up the **gut-associated lymphoid tissue** that responds to local invasion by infectious agents.

pH The symbol for the negative logarithm of the H ion concentration; p (power) or $[H^+]_{10}$. A system for rating acidity and alkalinity.

phagocytosis A type of endocytosis in which the cell membrane actively engulfs large particles or cells into vesicles. A **phagocyte** is a cell specialized for doing this.

phagolysosome A body formed in a phagocyte, consisting of a union between a vesicle containing the ingested particle (the phagosome) and a vacuole of hydrolytic enzymes (the lysosome).

phenetic Based on phenotype, or expression of traits.

phenotype The observable characteristics of an organism produced by the expression of its genetic potential (genotype). Includes any morphological and physiological traits.

phlebotomine Pertains to a genus of very small midges or blood-sucking (phlebotomous) sand flies and to diseases associated with those vectors such as kala-azar, Oroya fever, and cutaneous leishmaniasis.

phosphate An acidic salt containing phosphorus and oxygen that is an essential inorganic component of DNA, RNA, and ATP.

phospholipid A class of lipids that compose a major structural component of cell membranes.

phosphorylation Process in which inorganic phosphate is added to a compound.

photic zone The aquatic stratum from the surface to the limits of solar light penetration.

photoautotroph An organism that utilizes light for its energy and carbon dioxide chiefly for its carbon needs.

photolysis Literally, splitting water with light. In photosynthesis, this step frees electrons and gives off O_2.

photon A subatomic particle released by electromagnetic sources such as radiant energy (sunlight). Photons are the ultimate source of energy for photosynthesis.

photophosphorylation The process of electron transport during photosynthesis that uses captured light energy to synthesize ATP from ADP and phosphate.

photosynthesis A process occurring in plants, algae, and some bacteria that traps the sun's energy and converts it to ATP in the cell. This energy is used to fix CO_2 into organic compounds.

phototroph A microbe that uses sunlight for energy and CO_2 as a carbon source.

phylogenetic A classification system based on evolutionary relationships. Also called phyletic.

phylum In the levels of classification, the third level of classification from general to more specific. Each kingdom is divided into numerous phyla. Sometimes referred to as **division.**

physiology The study of the function of an organism.

pili (singular, **pilus**) Small, stiff filamentous appendages in gram-negative bacteria that function in DNA exchange during bacterial conjugation.

pinocytosis The engulfment, or endocytosis, of liquids by extensions of the cell membrane.

plague Zoonotic disease caused by infection with *Yersinia pestis*. The pathogen is spread by flea vectors and harbored by various rodents.

plankton/planktonic Minute animals and protozoa (zooplankton) or algae (phytoplankton) that float and drift in the limnetic zone of bodies of water.

plantar warts Deep, painful warts on the soles of the feet as a result of infection by human papillomavirus.

plaque In virus propagation methods, the clear zone of lysed cells in tissue culture or chick embryo membrane that corresponds to the area containing viruses. In dental application, the filamentous mass of microbes that adheres tenaciously to the tooth and predisposes to caries, calculus, or inflammation.

plasma The carrier fluid element of blood.

plasma cell A progeny of an activated B cell that actively produces and secretes antibodies against a specific microbe.

plasmids Extrachromosomal genetic units characterized by several features. A plasmid is a double-stranded DNA that is smaller than and replicates independently of the cell chromosome; it bears genes that are not essential for cell growth; it can bear genes that code for adaptive traits; and it is transmissible to other bacteria.

platelets Formed elements in the blood that develop when megakaryocytes disintegrate. Platelets are involved in hemostasis and blood clotting.

pleomorphism Normal variability of cell shapes in a single species; bacteria having diverse shapes are called pleomorphic.

pneumococcus Common name for *Streptococcus pneumoniae*, the major cause of bacterial pneumonia.

pneumonia An inflammation of the lung leading to blockage of the respiratory tract that compromises breathing.

pneumonic plague The acute, frequently fatal form of pneumonia caused by *Yersinia pestis*.

polar Term to describe a molecule with an asymmetrical distribution of charges. Such a molecule has a negative pole and a positive pole.

Polar flagellum Description of flagella that are attached at one or both ends of the cell.

poliomyelitis An acute enteroviral infection of the spinal cord that can cause neuromuscular paralysis.

polyclonal antibodies In reference to a collection of antibodies with mixed specificities that arose from more than one clone of B cells.

polymer A macromolecule made up of a chain of repeating units. Examples: starch, protein, and DNA.

polymerase An enzyme that produces polymers through catalyzing bond formation between building blocks (polymerization).

polymerase chain reaction (PCR) A technique that amplifies segments of DNA for testing. Using denaturation, primers, and heat-resistant DNA polymerase, the number can be increased several million-fold.

polymyxin A mixture of antibiotic polypeptides from *Bacillus polymyxa* that are particularly effective against gram-negative bacteria.

polypeptide A relatively large chain of amino acids linked by peptide bonds.

polyribosomal complex An assembly line for mass production of proteins composed of a chain of ribosomes involved in mRNA transcription.

polysaccharide A carbohydrate that can be hydrolyzed into a number of monosaccharides. Examples: cellulose, starch, and glycogen.

population A group of organisms of the same species living simultaneously in the same habitat. A group of different populations living together constitutes the community level.

porin Transmembrane proteins of the outer membrane of gram-negative cells that permit transport of small molecules into the periplasmic space but bar the penetration of larger molecules.

portal of entry Characteristic route of entry for an infectious agent; typically a cutaneous or membranous route.

portal of exit Characteristic route through which a pathogen departs from the host organism.

positive stain Technique in which dye affixes to a specimen and imparts color to it. It takes advantage of the ready binding of bacterial cells to dyes.

potable Describing water that is relatively clear, odor-free, and safe to drink.

pox The thick, elevated pustular eruptions of various viral infections. Also spelled *pocks*.

poxvirus A group of large complex DNA viruses that create pox skin lesions. Agents of smallpox and molluscum contagiosum.

prevalence The total cumulative number of cases of a disease in a certain area and time period.

primary infection An initial infection in a previously healthy individual that is later complicated by an additional (secondary) infection.

primary pulmonary infection (PPI) Disease that results from the inhalation of fungal spores and their germination in the lungs. This may serve as a focus of infection that spreads throughout the body.

primary response The first response of the immune system when exposed to an antigen.

primary structure Initial protein organization described by type, number, and order of amino acids in the chain. The primary structure varies extensively from protein to protein.

primers Short segments of RNA that serve as binding sites for DNA polymerase to start adding nucleotides to the new strand of DNA.

prion A concocted word to denote "proteinaceous infectious agent"; a cytopathic protein associated with the slow-virus spongiform encephalopathies of humans and animals.

probes Small fragments of single-stranded DNA (or RNA) that are known to be complementary to a specific sequence of DNA being studied.

probiotic A microbe or mixture of microbes that may be used to balance the normal microbiota and improve gastrointestinal health.

prodromium A short period of nonspecific symptoms at the end of the period of incubation that is usually the earliest indication of an infection.

producer An organism that synthesizes complex organic compounds from simple inorganic molecules. Examples would be photosynthetic microbes and plants. These organisms are solely responsible for originating food pyramids and are the basis for life on earth (also called autotroph).

proglottid The egg-generating segment of a tapeworm that contains both male and female organs.

progressive multifocal leukoencephalopathy (PML) An uncommon, fatal complication of infection with JC virus (polyomavirus).

prokaryotic cell Small cell lacking special structures such as a nucleus and organelles. All prokaryotes are microorganisms.

promastigote A morphological variation of the trypanosome parasite responsible for leishmaniasis.

prophage A lysogenized bacteriophage; a phage that is latently incorporated into the host chromosome instead of undergoing viral replication and lysis.

prophylactic/prophylaxis Any device, method, or substance that prevents a disease from developing.

prostaglandin A hormonelike substance that regulates many body functions. Prostaglandin comes from a family of organic acids containing 5-carbon rings that are essential to the human diet.

protein Predominant organic molecule in cells, formed by long chains of amino acids.

proton An elementary particle that carries a positive charge. It is identical to the nucleus of the hydrogen atom.

protoplast A bacterial cell whose cell wall is completely lacking and that is vulnerable to osmotic lysis.

protozoa A group of single-celled, eukaryotic organisms that feed on other cells and usually have a locomotor organelle.

pseudohypha A chain of easily separated, spherical to sausage-shaped yeast cells partitioned by constrictions rather than by septa.

pseudomembrane A tenacious, noncellular mucous exudate containing cellular debris that tightly blankets the mucosal surface in infections such as diphtheria and pseudomembranous enterocolitis.

pseudopods Appendages responsible for motility of protozoa such as amoebas. Also called "false feet."

psychrophile A microorganism that thrives at low temperatures (0°C–20°C), with a temperature optimum of 0°C–15°C.

pulmonary Occurring in the lungs. Examples include pulmonary anthrax and pulmonary nocardiosis.

pulmonary anthrax The more severe form of anthrax, resulting from inhalation of spores and resulting in a wide range of pathological effects, including death.

pure culture A container growing a single species of microbe whose identity is known.

purine A nitrogen base that is an important encoding component of DNA and RNA. It is composed of a basic two-ring structure and is larger than a pyrimidine. The two most common purines are adenine and guanine.

pus The viscous, opaque, usually yellowish matter formed by an inflammatory infection. It consists of serum exudate, tissue debris, leukocytes, and microorganisms. *Purulent* refers to a puslike condition.

pyogenic Pus-forming; especially pertaining to the pyogenic cocci: pneumococci, streptococci, staphylococci, and neisseriae.

pyrimidine A nitrogen base that helps form the genetic code on DNA and RNA. It is composed of a single-ring structure and is smaller than a purine. Thymine, cytosine, and uracil are the most important pyrimidines.

pyrimidine dimer The union of two adjacent pyrimidines on the same DNA strand, brought about by exposure to ultraviolet light. It is a form of mutation.

pyrogen A substance that causes a rise in body temperature. It can come from infectious agents (exogenous) or from polymorphonuclear leukocytes and macrophages (endogenous pyrogens).

Q fever A disease first described in Queensland, Australia, initially dubbed Q for "query" to denote a fever of unknown origin. Q fever is now known to be caused by a rickettsial infection.

quaternary ammonium compounds See **quats.**

quaternary structure The most complex protein structure, characterized by the formation of large, multiunit proteins by more than one of the polypeptides. This structure is typical of antibodies and some enzymes that act in cell synthesis.

quats/quaternary ammonium compounds A family of surfactants called **quaternary ammonium compounds,** termed **quats** for short. These detergents are weakly microbicidal and are used as disinfectants, sanitizers, and preservatives.

quorum sensing A phenomenon occurring among microbes in a biofilm in which the members signal each other and coordinate their functions.

rabies The only rhabdovirus that infects humans. Zoonotic disease characterized by fatal meningoencephalitis.

radiation Electromagnetic waves or rays given off from an energy source. Includes gamma rays, X rays, ultraviolet, visible light, and others.

reactants Molecules entering or starting a chemical reaction.

reactive oxygen intermediate (ROI) Any charged oxygen radical (e.g. superoxide ions, peroxide) given off during oxygen metabolism that can be toxic to microbial and other cells.

real image An image formed at the focal plane of a convex lens. In the compound light microscope, it is the image created by the objective lens.

recombinant DNA technology A technology, associated with genetic engineering, that deliberately modifies the genetic structure of an organism to create novel products, microbes, animals, plants, and viruses.

recombination A type of genetic transfer in which DNA from one organism is donated to another. The resultant cell is termed *recombinant.*

recycling A process that converts unusable organic matter from dead organisms back into their essential inorganic elements and returns them to their nonliving reservoirs to make them available again for living organisms. This is a common term that means the same as *mineralization* and *decomposition.*

redox Denoting an oxidation-reduction reaction.

reduction In chemistry, the gain of electrons. A reducing agent is one that causes the addition of electrons to a substrate.

reemerging disease Previously identified disease that is increasing in occurrence.

refraction In optics, the bending of light as it passes from one medium to another with a different index of refraction.

regular (in shape) Refers to bacteria with uniform staining properties and that are consistent in shape from cell to cell within a particular species.

regulator DNA segment that codes for a protein capable of repressing an operon.

rennin The enzyme casein coagulase, which is used to produce curd in the processing of milk and cheese.

reovirus Respiratory enteric orphan virus. Virus with a double-stranded RNA genome and both an inner capsid and an outer capsid. Not a significant human pathogen.

replication In DNA synthesis, the semiconservative mechanisms that ensure precise duplication of the parent DNA strands.

replication fork The Y-shaped point on a replicating DNA molecule where the DNA polymerase is synthesizing new strands of DNA.

reportable disease Any disease that needs regular and frequent monitoring of individual cases so that it can be managed and prevented. These are diseases that must be reported to health authorities by law.

repressible operon A system of gene regulation that is normally active but can be blocked when a build-up of the operon's product occurs; this causes transcription by the RNA polymerase to be stopped, i.e. *repressed.*

repressor The protein product of a repressor gene that combines with the operator and arrests the transcription and translation of structural genes.

reservoir In epidemiology, the natural host or habitat of a pathogen that is its long-term origin.

resident microbiota The deeper, more stable microflora that inhabit the skin and exposed mucous membranes, as opposed to the superficial, variable, transient population.

resistance (R) factor Plasmids, typically shared among bacteria by conjugation, that provide resistance to the effects of antibiotics.

resolving power The capacity of a microscope lens system to accurately distinguish between two separate entities that lie close to each other. Also called resolution.

respiratory chain In cellular respiration, a series of electron-carrying molecules that transfers energy-rich electrons and protons to molecular oxygen. In transit, energy is extracted and conserved in the form of ATP.

respiratory syncytial virus (RSV) An RNA virus that infects the respiratory tract. RSV is the most prevalent cause of respiratory infection in newborns.

restriction endonuclease An enzyme present naturally in bacteria that cleaves specific locations (palindromes) on DNA. It is an important means of inactivating viral genomes, and it is also used in genetic engineering to splice genes.

reticuloendothelial system (RES) A network of fibers and phagocytic cells (macrophages) that permeates the tissues of all organs. Examples: Kupffer cells in liver sinusoids, alveolar phagocytes in the lung, and microglia in nervous tissue. Also known as the mononuclear phagocyte system.

retrovirus A group of RNA viruses (including HIV) that have the mechanisms for converting their genome into a double strand of DNA that can be inserted on a host's chromosome.

reverse transcriptase (RT) The enzyme possessed by retroviruses that carries out the reversion of RNA to DNA—a form of reverse transcription.

Reye's syndrome A sudden, usually fatal neurological condition that occurs in children after a viral infection. Autopsy shows cerebral edema and marked fatty change in the liver and renal tubules.

Rh factor An isoantigen that can trigger hemolytic disease in newborns due to incompatibility between maternal and infant blood factors.

rhabdovirus Family of bullet-shaped viruses that includes rabies.

rheumatic fever Chronic group A *Streptococcus*-associated syndrome that damages the heart valves.

rhinoviruses Picornaviruses associated with the common cold. Transmission is through human-to-human contact, and symptoms typically are short-lived. Most effective control is effective hand washing and care in handling nasal secretions.

rhizobia Bacteria that live in plant roots and supply supplemental nitrogen that boosts plant growth.

rhizosphere The zone of soil, complete with microbial inhabitants, in the immediate vicinity of plant roots.

ribonucleic acid (RNA) The nucleic acid responsible for carrying out the hereditary program transmitted by an organism's DNA.

ribose A 5-carbon monosaccharide found in RNA.

ribosome A bilobed macromolecular complex that coordinates the codons of mRNA with tRNA anticodons and, in so doing, constitutes the protein assembly site. It is composed of ribosomal RNA (rRNA) and protein.

ribozyme An RNA-containing catalyst that serves to cut and splice RNA during gene processing, protein synthesis, and other RNA-based functions.

rickettsias Medically important family of tiny, intracellular parasitic bacteria, commonly carried by ticks, lice, and fleas. Cause of Rocky Mountain spotted fever and typhus.

Rifampin An antibiotic used primarily in the treatment of mycobacterial infection.

ringworm A superficial mycosis caused by various dermatophytic fungi. This common name is actually a misnomer.

RNA polymerase The enzyme that converts the genetic code provided by DNA into a complementary RNA molecule.

root nodules Small growths on the roots of legume plants that arise from a symbiotic association between the plant tissues and bacteria (Rhizobia). This association allows fixation of nitrogen gas from the air into a usable nitrogen source for the plant.

roseola Disease of infancy, usually self-limiting, caused by infection with human herpesvirus-6.

Rotavirus Virus with a double-stranded RNA genome and both an inner capsid and an outer capsid. Transmitted by fecal contamination, it is common in areas with poor sanitation. *Rotavirus* typically causes diarrheal disease and can be fatal.

rough endoplasmic reticulum (RER) Microscopic series of tunnels that originates in the outer membrane of the nuclear envelope and is used in transport and storage. Large numbers of ribosomes, partly attached to the membrane, give the rough appearance.

rubella Commonly known as German measles. Rubella is caused by *Rubivirus*, a member of the togavirus family. Postnatal rubella is generally a mild condition; congenital rubella poses a risk of birth defects and results when virus passes from infected mother to fetus.

rubeola (red measles) Acute disease caused by infection with *Morbillivirus*.

saccharide Scientific term for sugar. Refers to a simple carbohydrate with a sweet taste.

salmonelloses Illnesses caused by infection of the noncoliform *Salmonella* pathogens. The cause of typhoid fever, *Salmonella* food poisoning, and gastroenteritis.

salpingitis Inflammation of the fallopian tubes.

sanitize/sanitization To clean inanimate objects using soap and degerming agents so that they are safe and free of high levels of microorganisms.

saprobe A microbe that decomposes organic remains from dead organisms. Also known as a saprophyte or saprotroph.

sarcina A cubical packet of 8, 16, or more cells; the cellular arrangement of the genus *Sarcina* in the family Micrococcaceae.

saturation The complete occupation of the active site of a carrier protein or enzyme by the substrate.

scarlet fever Serious group A *Streptococcus* infection that causes redness, shedding of the skin, and high fever.

schistosomiasis Infection by blood fluke, often as a result of contact with contaminated water in rivers and streams. Symptoms include fever, chills, diarrhea, and cough. Infection may be chronic.

schizogony A process of multiple fission whereby first the nucleus divides several times, and subsequently the cytoplasm is subdivided for each new nucleus during cell division.

scientific method Principles and procedures for the systematic pursuit of knowledge, involving the recognition and formulation of a problem, the collection of data through observation and experimentation, and the formulation and testing of a hypothesis.

scolex The anterior end of a tapeworm characterized by hooks and/or suckers for attachment to the host.

secondary infection An infection that complicates a preexisting primary infection.

secondary response The rapid rise in antibody titer following a repeat exposure to an antigen that has been recognized from a previous exposure. This response is brought about by memory cells left behind after the primary exposure.

secondary structure Protein structure that occurs when the functional groups on the outer surface of the molecule interact by forming hydrogen bonds. These bonds cause the amino acid chain either to twist, forming a helix, or to pleat into an accordion pattern called a β-pleated sheet.

selectively toxic Property of an antimicrobic agent to be highly toxic against its target microbe while being far less toxic to other cells, particularly those of the host organism.

self Natural markers of the body that are recognized by the immune system and generally do not stimulate an immune response.

semiconservative replication In DNA replication, during the synthesis of new DNA strands (the daughter DNA), the parent strand template DNA is retained in the final molecule.

semisolid media Nutrient media with a firmness midway between that of a broth (a liquid medium) and an ordinary solid medium; motility media.

semisynthetic Antibiotics that have been isolated as natural products of bacteria, fungi, or other living sources, and are then chemically modified in the industrial setting to achieve a desired action.

sensitizing dose The initial effective exposure to an antigen or an allergen that stimulates an immune response but not symptoms.

sepsis The state of putrefaction; the presence of pathogenic organisms or their toxins in tissue or blood.

septic shock Blood infection resulting in a pathological state of low blood pressure accompanied by a reduced amount of blood circulating to vital organs. Endotoxins of all gram-negative bacteria can cause shock, but most clinical cases are due to gram-negative enteric rods.

septicemia Systemic infection associated with microorganisms multiplying in circulating blood.

septum A partition or cellular cross wall, as in certain fungal hyphae.

sequela A morbid complication that follows a disease.

serology The branch of immunology that deals with *in vitro* diagnostic testing of serum.

seropositive Showing the presence of specific antibodies in a serological test. Indicates ongoing or recent infection.

serotonin A vasoconstrictor that inhibits gastric secretion and stimulates smooth muscle.

serotyping The subdivision of a species or subspecies into an immunologic type, based upon antigenic characteristics.

serum The clear fluid expressed from clotted blood that contains dissolved nutrients, antibodies, and hormones but not cells or clotting factors.

serum sickness A type of immune complex disease in which immune complexes enter circulation, are carried throughout the body, and are deposited in the blood vessels of the kidney, heart, skin, and joints. The condition may become chronic.

severe acute respiratory syndrome (SARS) A severe respiratory disease caused by infection with a newly described Coronavirus.

severe combined immunodeficiencies A collection of syndromes occurring in newborns caused by a genetic defect that knocks out both B-and T-cell types of immunity. There are several versions of this disease, termed **SCIDS** for short.

sewage Liquid runoff or wastewater from communities.

sexually transmitted disease (STD) An infection resulting from pathogens that enter the body via sexual intercourse or intimate, direct contact.

shigellosis An incapacitating dysentery caused by infection with *Shigella* bacteria.

shingles A skin condition occurring on the trunk or head that is caused by the reactivation of a latent varicella-zoster virus. Also known as **herpes zoster.**

side effects Unintended, usually harmful consequences of chemotherapy, including damage to organs, allergy, and disruption of normal microbiota.

sign Any abnormality uncovered upon physical diagnosis that indicates the presence of disease. A sign is an objective assessment of disease, as opposed to a symptom, which is the subjective assessment perceived by the patient.

silent mutation A mutation that, because of the degeneracy of the genetic code, results in a nucleotide change in both the DNA and mRNA but not the resultant amino acid and thus, not the protein.

simple stain Type of positive staining technique that uses a single dye to add color to cells so that they are easier to see. This technique tends to color all cells the same color.

smallpox Disease caused by infection with variola virus. Smallpox has been eradicated worldwide, and variola virus exists today only in a few government laboratories.

smooth endoplasmic reticulum (SER) A microscopic series of tunnels lacking ribosomes that functions in the nutrient processing function of a cell.

solution A mixture of one or more substances (solutes) that cannot be separated by filtration or ordinary settling.

solvent A dissolving medium.

somatic (O or cell wall) antigen One of the three major antigens commonly used to differentiate gram-negative enteric bacteria.

source The person, location, or object that supplies the pathogen in an infection. This is different from the reservoir, which is where the pathogen lives.

Southern blot A technique that separates fragments of DNA using electrophoresis and identifies them by hybridization.

species In the levels of classification, the most specific level of organization.

specificity Limited to a single, precise characteristic or action.

spheroplast A gram-negative cell whose peptidoglycan, when digested by lysozyme, remains intact but is osmotically vulnerable.

spike A receptor on the surface of certain enveloped viruses that facilitates specific attachment to the host cell.

spirillum A type of bacterial cell with a rigid spiral shape and external flagella; plural is spirilla.

spirochete A coiled, spiral-shaped bacterium that has endoflagella and flexes as it moves.

spontaneous generation Early belief that living things arose from vital forces present in nonliving, or decomposing, matter.

spontaneous mutation A mutation in DNA caused by random mistakes in replication and not known to be influenced by any mutagenic agent. These mutations give rise to an organism's natural, or background, rate of mutation.

sporadic Description of a disease that exhibits new cases at irregular intervals in unpredictable geographic locales.

sporangium A fungal cell in which asexual spores are formed by multiple cell cleavage. This division results in sporangiospores.

spore A differentiated, specialized cell form that can be used for dissemination, for survival in times of adverse conditions, and/or for reproduction. Spores are usually unicellular and may develop into gametes or vegetative organisms.

sporicide A chemical agent capable of destroying bacterial endospores.

sporotrichosis A subcutaneous mycosis caused by *Sporothrix schenckii;* a common cause is the prick of a rose thorn.

sporozoite One of many minute elongated bodies generated by multiple division of the oocyst. It is the infectious form of the malarial parasite that is harbored in the salivary gland of the mosquito and inoculated into the victim during feeding.

sporulation The process of spore formation.

staphylococci (singular, **staphylcoccus**) Coccus-shaped bacteria arranged in irregular clusters due to their division in several different planes.

start codon The nucleotide codon AUG that codes for the first amino acid in protein sequences.

starter culture Pure cultures of bacteria, molds, or yeasts inoculated into substrates for bulk processing, as in the preparation of fermented foods, beverages, and pharmaceuticals.

stationary growth phase Survival mode in which cells either stop growing or grow very slowly and the population number evens out.

stem cells Undifferentiated cells found in the bone marrow and other organs that have the capacity to develop or differentiate into the lines of blood cells (pluripotential).

sterile Completely free of all microbial life forms, including spores and viruses.

sterilization Any process that completely removes or destroys all viable microorganisms, including viruses, from an object or habitat. Material so treated is sterile.

STORCH Acronym for common infections of the fetus and neonate. STORCH stands for Syphilis, Toxoplasmosis, Other diseases (hepatitis B, AIDS and chlamydiosis), Rubella, Cytomegalovirus, and Herpes simplex virus.

strain In microbiology, a set of descendants cloned from a common ancestor that retain the original characteristics. Any deviation from the original is a different strain.

streptococci (singular, **streptococcus**) Coccus-shaped bacteria arranged in chains due to their division in one plane.

streptolysin A hemolysin produced by streptococci.

strict, or obligate, anaerobe An organism that does not use oxygen gas in metabolism and cannot survive in oxygen's presence.

stroma The matrix of the chloroplast that is the site of the dark reactions.

subacute endocarditis An infection of the heart lining due to invasion of the blood by oral streptococci.

subacute sclerosing panencephalitis (SSPE) A complication of measles infection in which progressive neurological degeneration of the cerebral cortex invariably leads to coma and death.

subcellular vaccine A vaccine against isolated microbial antigens rather than against the entire organism.

subclinical A period during an infection when no noticeable manifestations occur.

subculture To make a second-generation culture from a well-established colony of organisms.

subcutaneous The deepest level of the skin structure.

substrate The specific molecule upon which an enzyme acts.

subunit vaccine A vaccine preparation that contains only antigenic fragments such as surface receptors from the microbe. Usually in reference to virus vaccines.

sucrose One of the carbohydrates commonly referred to as sugars. Common table or cane sugar.

sulfonamide Antimicrobial drugs that interfere with the essential metabolic process of bacteria and some fungi. Also known as **sulfa drugs.**

superantigens Microbial toxins and proteins that can activate T cells nonspecifically and often lead to an overpowering reaction such as toxic shock.

superficial mycosis A fungal infection located in hair, nails, and the epidermis of the skin.

superinfection An infection occurring during antimicrobic therapy that is caused by an overgrowth of drug-resistant microorganisms.

surfactant An agent that reduces surface tension and forms a water-soluble interface. Derived from **surf**ace-**act**ive agent. Examples: detergents, soaps, wetting agents, and dispersing agents.

sylvatic Denotes the natural presence of disease among wild animal populations. Examples: sylvatic (sylvan) plague, and rabies.

symbiosis/symbionts An intimate association between two or more individuals that are termed symbionts; is sometimes used incorrectly as a synonym for mutualism.

symptom The subjective evidence of infection and disease as perceived by the patient.

syncytium A multinucleated protoplasmic mass formed by consolidation of individual cells.

syndrome The collection of signs and symptoms that, taken together, paint a portrait of the disease.

synergism The coordinated or correlated action by two or more drugs or microbes that results in a heightened response or greater activity.

syphilis A sexually transmitted bacterial disease caused by the spirochete *Treponema pallidum*.

systemic infection *Systemic* means occurring throughout the body; said of infections that invade many compartments and organs via the circulation.

T helper cell/helper T cell A class of thymus-stimulated lymphocytes that facilitate various immune activities such as assisting the activation of B cells and macrophages. Also called T_H or **helper T cell.**

T-even phages A group of related bacterial viruses (T-2, T-4, and T-6) that infect *Escherichia coli* bacteria.

taxa Taxonomic categories.

taxonomy The formal system for organizing, classifying, and naming living things.

temperate phage A bacteriophage that enters into a less virulent state by becoming incorporated into the host genome as a prophage instead of in the vegetative or lytic form that eventually destroys the cell.

template The strand in a double-stranded DNA molecule that is used as a model to synthesize a complementary strand of DNA or RNA during replication or transcription.

teratogenic Causing abnormal fetal development.

tertiary structure Protein structure that results from additional bonds forming between functional groups in a secondary structure, creating a three-dimensional mass.

tetanospasmin The neurotoxin of *Clostridium tetani,* the agent of tetanus. Its chief action is directed upon the inhibitory synapses of the anterior horn motor neurons.

tetanus A neuromuscular disease caused by infection with *Clostridium tetani.* Usual portals of entry include puncture wounds, burns, umbilical stumps, frostbite sites, and crushed body parts. Vaccination repeated at the recommended times can prevent infection. Also called lockjaw.

tetracyclines A group of broad-spectrum antibiotics with a complex 4-ring structure.

tetrads Groups of four.

theory A collection of statements, propositions, or concepts that explains or accounts for a natural event.

therapeutic index (TI) The ratio of the toxic dose to the effective therapeutic dose that is used to assess the safety and reliability of the drug.

thermal death point The lowest temperature that achieves sterilization in a given quantity of broth culture upon a 10-minute exposure. Examples: 55°C for *Escherichia coli*, 60°C for *Mycobacterium tuberculosis*, and 120°C for spores.

thermal death time The least time required to kill all cells of a culture at a specified temperature.

thermal dimorphism A property of certain fungal pathogens of converting from hyphal to yeast form in response to increased temperature.

thermocline A temperature buffer zone in a large body of water that separates the warmer water (the epilimnion) from the colder water (the hypolimnion).

thermoduric Resistant to the harmful effects of high temperature.

thermophile A microorganism that thrives at a temperature of 50°C or higher.

thrush *Candida albicans* infection of the oral cavity.

thylakoid Vesicles of a chloroplast formed by elaborate folding of the inner membrane to form "discs." Solar energy trapped in the thylakoids is used in photosynthesis.

thymine (T) One of the nitrogen bases found in DNA but not in RNA. Thymine is a pyrimidine that pairs with adenine.

thymus Butterfly-shaped organ near the tip of the sternum that is the site of T-cell maturation.

tincture A medicinal substance dissolved in an alcoholic solvent.

tinea Ringworm; a fungal infection of the hair, skin, or nails.

tinea versicolor A condition of the skin appearing as mottled and discolored pigmentation as a result of infection by the yeast *Malassezia furfur.*

titer In immunochemistry, a measure of antibody levels in a patient's serum or specimen, as determined by agglutination methods.

toll-like receptors (TLRs) Small surface receptors on phagocytes that are specialized to bind and react to pathogens.

topoisomerases Enzymes that can add or remove DNA twists and thus regulate the degree of supercoiling.

toxemia An abnormality associated with certain infectious diseases. Toxemia is caused by toxins or other noxious substances released by microorganisms circulating in the blood.

toxic shock syndrome Toxemia caused by *Staphylococcus aureus* and associated with use of tampons.

toxigenicity The tendency for a pathogen to produce toxins. It is an important factor in bacterial virulence.

toxin A specific chemical product of microbes, plants, and some animals that is poisonous to other organisms.

toxinosis Disease whose adverse effects are primarily due to the production and release of toxins.

toxoid A toxin that has been rendered nontoxic but is still capable of eliciting the formation of protective antitoxin antibodies; used in vaccines.

trace elements Micronutrients (zinc, nickel, and manganese) that occur in small amounts and are involved in enzyme function and maintenance of protein structure.

transcription mRNA synthesis; the process by which a strand of RNA is produced against a DNA template.

transduction The transfer of genetic material from one bacterium to another by means of a bacteriophage vector.

transfection Introduction of genetic material into an animal or plant cell using a viral or bacterial vector.

transfer RNA (tRNA) A transcript of DNA that specializes in converting RNA language into protein language.

transformation In microbial genetics, the transfer of genetic material contained in "naked" DNA fragments from a donor cell to a competent recipient cell.

transfusion Infusion of whole blood, red blood cells, or platelets directly into a patient's circulation.

transgenic organisms/ genetically modified organisms (GMOs) Cells or organisms into which foreign DNA has been introduced. Includes recombinant plants, animals, and microbes created through genetic engineering.

transients In normal microbiota, the assortment of superficial microbes whose numbers and types vary depending upon recent exposure. The deeper-lying residents constitute a more stable population.

translation Also termed protein synthesis; the process of decoding a messenger RNA code into a polypeptide.

transmissible spongiform encephalopathies (TSEs) Progressive brain infections caused by unusual pathogens called prions, leading to loss of brain structure and function.

transposon A DNA segment with an insertion sequence at each end, enabling it to migrate to another plasmid, to the bacterial chromosome, or to a bacteriophage.

trematode A fluke or flatworm parasite of vertebrates.

trichomoniasis Sexually transmitted disease caused by infection by the trichomonads, a group of protozoa. Symptoms include urinary pain and frequency and foul-smelling vaginal discharge in females or recurring urethritis with a thin milky discharge in males.

triglyceride A type of lipid composed of a glycerol molecule bound to three fatty acids.

triplet A grouping of three nucleotides that serves as the genetic code for protein synthesis. Synonymous with codon.

trophozoite A vegetative protozoan (feeding form) as opposed to a resting (cyst) form.

true pathogen A microbe capable of causing infection and disease in healthy persons with normal immune defenses.

trypomastigote The infective morphological stage transmitted by the tsetse fly or the reduviid bug in African trypanosomiasis and Chagas disease.

tubercle In tuberculosis, the granulomatous well-defined lung lesion that can serve as a focus for latent infection.

tuberculin reaction A diagnostic test in which an intradermal injection of a purified protein (tuberculin) extract from *M. tuberculosis* elicits an immune response, seen as a small red bump, in those persons previously exposed to tuberculosis.

tuberculoid leprosy A superficial form of leprosy characterized by asymmetrical, shallow skin lesions containing few bacterial cells.

tularemia Infection by *Francisella tularensis.* A zoonotic disease of mammals common to the northern hemisphere. Occasionally called rabbit fever. Portal of entry and symptoms are varied.

turbid Cloudy appearance of nutrient solution in a test tube due to growth of microbe population.

tyndallization Fractional (intermittent) sterilization designed to destroy spores indirectly. A preparation is exposed to flowing steam for an hour, followed by incubation to permit spore germination. The resultant new vegetative cells are destroyed by repeated steaming and incubation.

typhoid fever A form of salmonelloses. It is highly contagious. Primary symptoms include fever, diarrhea, and abdominal pain. Typhoid fever can be fatal if untreated.

typhus Rickettsia infection characterized by high fever, chills, frontal headache, muscular pain, and a generalized rash within 7 days of infection. In more severe cases, a personality change, low urine output, hypotension, and gangrene can cause complications. Mortality is high in older adults.

ubiquitous, ubiquity The state of being everywhere at the same time; omnipresent. Said of microbes which are known to live in the vast majority of the habitats on earth.

ultraviolet (UV) radiation Radiation with an effective wavelength from 240 nm to 260 nm. UV radiation induces mutations readily but has very poor penetrating power.

undulant fever See **brucellosis.**

universal donor In blood grouping and transfusion, blood whose erythrocytes bear no A nor B antigens and so may be given to any person in an emergency.

universal precautions (UPs) Centers for Disease Control and Prevention guidelines for health care workers regarding the prevention of disease transmission when handling patients and body substances.

uracil (U) The only nitrogen base existing in RNA but not in DNA. It is a pyrimidine and the mate of adenine during transcription.

urban plague Plague passed to humans through contact with domestic animals or other humans.

vaccination The process of inoculation with a selected microbial antigen (vaccine) in order to stimulate an immunity to that microbe.

vaccine Originally used in reference to inoculation with the cowpox or vaccinia virus to protect against smallpox. In general, the term now pertains to injection of whole microbes (killed or attenuated), toxoids, or parts of microbes as a prevention or cure for disease.

vacuoles In the cell, membrane-bounded sacs containing fluids or solid particles to be digested, excreted, or stored.

valence The combining power of an atom based upon the number of electrons it can either take on or give up.

variable (v) region The antigen-binding fragment of an immunoglobulin molecule, consisting of a combination of heavy and light chains whose molecular conformation is specific for the antigen.

varicella See **chickenpox.**

varicella-zoster virus Herpesvirus responsible for the diseases chickenpox and shingles.

variolation A hazardous, outmoded process of deliberately introducing smallpox material scraped from a victim into the nonimmune subject in the hope of inducing resistance.

VDRL A flocculation test that detects syphilis antibodies. An important screening test. The abbreviation stands for **V**enereal **D**isease **R**esearch **L**aboratory.

vector An animal that transmits infectious agents from one host to another, often a biting or piercing

arthropod such as the tick, mosquito, or fly, but it can also include birds and mammals. Vectors convey infectious agents mechanically by simple contact or biologically, with the parasite developing in the vector.

vector₂ A genetic element such as a plasmid or a bacteriophage used to introduce genetic material into a cloning host as a part of recombinant DNA technology.

vehicle An inanimate material (solid object, liquid, or air) that serves as a transmission agent for pathogens.

verruca A flesh-colored wart. This self-limited tumor arises from accumulations of growing epithelial cells. Example: papilloma warts.

vesicle A blister characterized by a thin-skinned, elevated, superficial pocket inflated with serum.

vibrio A curved, rod-shaped bacterial cell.

viremia The presence of viruses in the bloodstream.

virion An elementary virus particle in its complete morphological and thus infectious form. A virion consists of the nucleic acid core surrounded by a capsid, which can be enclosed in an envelope.

viroid An infectious agent that, unlike a virion, lacks a capsid and consists of a closed circular RNA molecule. Although known viroids are all plant pathogens, it is conceivable that animal versions exist.

virtual image In optics, an image formed by diverging light rays; in the compound light microscope, the second, magnified visual impression formed by the ocular from the real image formed by the objective.

virucide A chemical agent that inactivates viruses, especially on living tissue.

virulence In infection, the relative capacity of a pathogen to invade and harm host cells.

virulence factor A product of microbes such as an enzyme or toxin that increases the microbe's invasiveness or pathogenicity.

virus Microscopic, acellular agent composed of nucleic acid surrounded by a protein coat.

vitamins Components of coenzymes critical to nutrition and the metabolic function of coenzyme complexes.

vulvovaginal candidiasis (VC) Vaginal STD caused by *Candida albicans* and often associated with disruption of the normal vaginal flora.

wart An epidermal tumor caused by papillomaviruses. Also called a **verruca.**

Western blot test A procedure for separating and identifying antigen or antibody mixtures by two-dimensional electrophoresis in polyacrylamide gel, followed by immune labeling.

wheal/flare A welt; marked, slightly red, usually itchy bumps on the skin that change in size and shape as they spread. It is surrounded by a red patch (the flare). The reaction is triggered by cutaneous contact or intradermal injection of allergens in sensitive individuals.

whey The residual fluid from milk coagulation that separates from the solidified curd.

white piedra A fungus disease of hair, especially of the scalp, face, and genitals, caused by *Trichosporon beigelii*. The infection is associated with soft, mucilaginous, white-to-light-brown nodules that form within and on the hair shafts.

whitlow A deep inflammation of the finger or toe, especially near the tip or around the nail. Whitlow is a painful herpes simplex virus infection that can last several weeks and is most common among health

care personnel who come in contact with the virus in patients.

whole blood A liquid connective tissue consisting of blood cells suspended in plasma.

whooping cough See **pertussis.**

Widal test An agglutination test for diagnosing typhoid fever and determining the antibody titer.

wild type The natural, nonmutated form of a genetic trait.

wort The clear fluid derived from soaked mash that is fermented for beer.

xenograft The transfer of a tissue or an organ from an animal of one species to a recipient of another species.

yaws A tropical disease caused by *Treponema pertenue* that produces granulomatous ulcers on the extremities and occasionally on bone but does not produce central nervous system or cardiovascular complications.

yeasts Single-celled, budding fungi.

yellow fever Best-known arboviral disease. Yellow fever is transmitted by mosquitoes. Its symptoms include fever, headache, and muscle pain that can proceed to oral hemorrhage, nosebleeds, vomiting, jaundice, and liver and kidney damage.

zoonosis An infectious disease indigenous to animals that humans can acquire through direct or indirect contact with infected animals.

zooplankton The collection of nonphotosynthetic microorganisms (protozoa, tiny animals) that float in the upper regions of aquatic habitat and together with phytoplankton comprise the plankton.

zygospore A thick-walled sexual spore produced by the zygomycete fungi. It develops from the union of two hyphae, each bearing nuclei of opposite mating types.

Index

Note: In this index entries followed by a lower-case f refer to figures, entries followed by a lower-case t refer to tables, and entries printed in **boldface** type indicate definitions or introductory discussions.